W9-AFC-610

THE PICTURE OF THE TAOIST GENII PRINTED ON THE COVER
of this book is part of a painted temple scroll, recent but traditional, given to
Mr Brian Harland in Szechuan province (1946). Concerning these four divinities,
of respectable rank in the Taoist bureaucracy, the following particulars have been
handed down. The title of the first of the four signifies 'Heavenly Prince', that
of the other three 'Mysterious Commander'.

At the top, on the left, is Liu *Thien Chün*, Comptroller-General of Crops and
Weather. Before his deification (so it was said) he was a rain-making magician
and weather forecaster named Liu Chün, born in the Chin dynasty about +340.
Among his attributes may be seen the sun and moon, and a measuring-rod or
carpenter's square. The two great luminaries imply the making of the calendar, so
important for a primarily agricultural society, the efforts, ever renewed, to reconcile
celestial periodicities. The carpenter's square is no ordinary tool, but the gnomon
for measuring the lengths of the sun's solstitial shadows. The Comptroller-General
also carries a bell because in ancient and medieval times there was thought to be
a close connection between calendrical calculations and the arithmetical acoustics
of bells and pitch-pipes.

At the top, on the right, is Wên *Yuan Shuai*, Intendant of the Spiritual Officials
of the Sacred Mountain, Thai Shan. He was taken to be an incarnation of one of
the Hour-Presidents (*Chia Shen*), i.e. tutelary deities of the twelve cyclical characters
(see Vol. 4, pt. 2, p. 440). During his earthly pilgrimage his name was Huan Tzu-
Yü and he was a scholar and astronomer in the Later Han (b. +142). He is seen
holding an armillary ring.

Below, on the left, is Kou *Yuan Shuai*, Assistant Secretary of State in the Ministry
of Thunder. He is therefore a late emanation of a very ancient god, Lei Kung.
Before he became deified he was Hsin Hsing, a poor woodcutter, but no doubt an
incarnation of the spirit of the constellation Kou-Chhen (the Angular Arranger),
part of the group of stars which we know as Ursa Minor. He is equipped with
hammer and chisel.

Below, on the right, is Pi *Yuan Shuai*, Commander of the Lightning, with his
flashing sword, a deity with distinct alchemical and cosmological interests. According
to tradition, in his early life he was a countryman whose name was Thien Hua.
Together with the colleague on his right, he controlled the Spirits of the Five
Directions.

Such is the legendary folklore of common men canonised by popular acclamation.
An interesting scroll, of no great artistic merit, destined to decorate a temple wall,
to be looked upon by humble people, it symbolises something which this book has
to say. Chinese art and literature have been so profuse, Chinese mythological
imagery so fertile, that the West has often missed other aspects, perhaps more
important, of Chinese civilisation. Here the graduated scale of Liu Chün, at first
sight unexpected in this setting, reminds us of the ever-present theme of quantitative measurement in Chinese culture; there were rain-gauges already in the Sung
(+12th century) and sliding calipers in the Han (+1st). The armillary ring of
Huan Tzu-Yü bears witness that Naburiannu and Hipparchus, al-Naqqāsh and
Tycho, had worthy counterparts in China. The tools of Hsin Hsing symbolise that
great empirical tradition which informed the work of Chinese artisans and technicians
all through the ages.

SCIENCE AND CIVILISATION IN CHINA

In all the world there be no better workmen for buildings than the inhabitants of China.

GALEOTE PEREIRA
c. 1577

The Chinese have their Contrivances for everything.

DOMINGO DE NAVARRETE
1676

C'est le pays le plus peuplé et le mieux cultivé qu'il y ait au monde; il est arrosé de plusieurs grandes rivières, et coupé d'une infinité de canaux que l'on y fait pour faciliter le commerce. Le plus remarquable est celui que l'on nomme le *canal royal*, qui traverse toute la Chine.

DENIS DIDEROT
1752

As regards the people who protect and manage the dykes and channels of the nine rivers and the four lakes, they are the same in all ages; they did not learn their business from Yü the Great, they learnt it from the waters.

Shen Tzu
c. 4th century

Those who know how to manage ships learnt from boats and not from Wo the Shipman. Those who can think learnt from themselves, and not from the Sages.

Kuan Yin Tzu
8th century

中國科學技術史

李約瑟 著

SCIENCE AND CIVILISATION IN CHINA

BY

JOSEPH NEEDHAM, F.R.S.

MASTER OF GONVILLE AND CAIUS COLLEGE, CAMBRIDGE
FOREIGN MEMBER OF ACADEMIA SINICA

With the collaboration of

WANG LING, PH.D.

PROFESSORIAL FELLOW IN THE INSTITUTE OF ADVANCED STUDIES
AUSTRALIAN NATIONAL UNIVERSITY, CANBERRA

and

LU GWEI-DJEN, PH.D.

FELLOW OF LUCY CAVENDISH COLLEGE, CAMBRIDGE

VOLUME 4

PHYSICS AND PHYSICAL TECHNOLOGY

PART III: CIVIL ENGINEERING AND NAUTICS

CAMBRIDGE
AT THE UNIVERSITY PRESS
1971

Published by the Syndics of the Cambridge University Press
Bentley House, 200 Euston Road, London N.W.1
American Branch: 32 East 57th Street, New York, N.Y.10022

Library of Congress Catalogue Card Number: 54-4723

Standard Book Number: 521 07060 0

Printed in Great Britain
at the University Printing House, Cambridge
(Brooke Crutchley, University Printer)

To the memory of

CHI CHHAO-TING

Historian of China's water ways and works

a friend beside the Chialing River
economic and financial leader in a resurgent land

and of

HERBERT CHATLEY

Once Professor of Engineering at Thang-shan College
and
Chief Engineer of the Huang-po Conservancy

an 'Old China Hand' who loved the Chinese people
historian of the engineers of Cathay and Manzi

this volume is
dedicated

*The Syndics of the Cambridge University Press
desire to acknowledge with gratitude certain financial aid
towards the production of this book, afforded by the
Bollingen Foundation*

CONTENTS

LIST OF ILLUSTRATIONS

LIST OF TABLES

NOTE

On the system of romanisation of Chinese characters adopted in this work, see Vol. 1, pp. 23 ff.

LIST OF ABBREVIATIONS

The following abbreviations are used in the text and footnotes. For abbreviations used for journals and similar publications in the bibliographies, see pp. 701 ff.

B   Bretschneider, E., *Botanicon Sinicum*.

BCFA   Britain–China Friendship Association.

CCL   *Chê Chiang Lu* (Biographies of [Chinese] Engineers, Architects, Technologists and Master-Craftsmen). 1 to 6 See Chu Chhi-Chhien & Liang Chhi-Hsiung (*1* to *6*); 7 See Chu Chhi-Chhien, Liang Chhi-Hsiung & Liu Ju-Lin (*1*); 8, 9 See Chu Chhi-Chhien & Liu Tun-Chên (*1, 2*).

CFC   Chao Ju-Kua, *Chu Fan Chih* (Records of Foreign Peoples), +1225.

CKHW   *Chung-Kuo Hsin Wên* (Chinese People's Republic News Bulletin, in Chinese).

CLPT   Thang Shen-Wei *et al.* (ed.), *Chêng Lei Pên Tshao* (Reorganised Pharmacopoeia), ed. of +1249.

CPCRA   Chinese People's Association for Cultural Relations with Foreign Countries.

CSHK   Yen Kho-Chün (ed.), *Chhüan Shang-ku San-Tai Chhin Han San-Kuo Liu Chhao Wên* (Complete Collection of Prose Literature (including fragments) from Remote Antiquity through the Chhin and Han Dynasties, the Three Kingdoms, and the Six Dynasties), 1836.

CTS   Liu Hsü, *Chiu Thang Shu* (Old History of the Thang Dynasty), +945.

EB   *Encyclopaedia Britannica*.

ERE   *Encyclopaedia of Religion and Ethics* (ed. Hastings).

G   Giles, H. A., *Chinese Biographical Dictionary*.

HCCC   Yen Chieh (ed.), *Huang Chhing Ching Chieh* (monographs by Chhing scholars on classical subjects), 1829, contd. 1860.

HTCKM Shang Lu (ed.), *Hsü Thung Chien Kang Mu*, = *Thung Chien Kang Mu Hsü Pien* (Continuation of the Short View of the Comprehensive Mirror (of History, for Aid in Government)), +1476, pr. after +1500.

K   Karlgren, *Grammata Serica* (dictionary giving the ancient forms and phonetic values of Chinese characters).

KCCY   Chhen Yuan-Lung, *Ko Chih Ching Yuan* (Mirror of Scientific and Technological Origins), an encyclopaedia of +1735.

KHTT   Chang Yü-Shu (ed.), *Khang-Hsi Tzu Tien* (Imperial Dictionary of the Khang-Hsi reign-period), +1716.

LCCCC   Li Chao-Hsiang, *Lung-Chiang Chhuan Chhang Chih* (Record of the Shipyards on the Dragon River, at Nanking), +1553.

MCPT   Shen Kua, *Mêng Chhi Pi Than* (Dream Pool Essays), +1089.

N   Nanjio, B., *A Catalogue of the Chinese Translations of the Buddhist Tripiṭaka*, with index by Ross (3).

NCCS   Hsü Kuang-Chhi, *Nung Chêng Chhüan Shu* (Complete Treatise on Agriculture), +1639.

NCNA   *New China News Agency Bulletin.*

PWYF   Chang Yü-Shu (ed.), *Phei Wên Yün Fu* (encyclopaedia), +1711.

R   Read, Bernard E. *et al.*, Indexes, translations and précis of certain chapters of the *Pên Tshao Kang Mu* of Li Shih-Chen. If the reference is to a plant, see Read (1); if to a mammal, see Read (2); if to a bird, see Read (3); if to a reptile, see Read (4 or 5); if to a mollusc, see Read (5); if to a fish, see Read (6); if to an insect, see Read (7).

SCTS   *Chhin-Ting Shu Ching Thu Shuo* (imperial illustrated edition of the *Historical Classic*), 1905.

SKCS   *Ssu Khu Chhüan Shu* (Complete Library of the Four Categories), +1782; here the reference is to the *tshung-shu* collection printed as a selection from one of the seven imperially commissioned MSS.

SKCS/TMTY   *Ssu Khu Chhüan Shu Tsung Mu Thi Yao* (Analytical Catalogue of the *Complete Library of the Four Categories*), +1782; the great bibliographical catalogue of the imperial MS. collection ordered by the Chhien-Lung emperor in +1772.

STTH   Wang Chhi, *San Tshai Thu Hui* (Universal Encyclopaedia), +1609.

T   Tunhuang Archaeological Research Institute numbers of the Chhien-fo-tung cave-temples. If an identification is given according to the system of Hsieh Chih-Liu in his *Tunhuang I Shu Hsü Lu* (Shanghai, 1955) the Institute number and the Pelliot number are also given, but if a single number is given it is the Institute number. A valuable concordance table of the three systems is given in Hsieh's book, and a still more complete one in Chhen Tsu-Lung (1).

TCKM   Chu Hsi *et al.* (ed.), *Thung Chien Kang Mu* ((Short View of the) *Comprehensive Mirror* (*of History*), *for Aid in Government*), classified into Headings and Subheadings); the *Tzu Chih Thung Chien* condensed, a general history of China, +1189; with later continuations.

TCTC Ssuma Kuang, *Tzu Chih Thung Chien* (Comprehensive Mirror (of History) for Aid in Government), +1084.

TH Wieger, L. (1), *Textes Historiques*.

TKKW Sung Ying-Hsing, *Thien Kung Khai Wu* (The Exploitation of the Works of Nature), +1637.

TPYC Li Chhüan, *Thai Pai Yin Ching* (Manual of the White and Gloomy Planet (of War, Venus)), treatise on military and naval affairs, +759.

TPYL Li Fang (ed.), *Thai-Phing Yü Lan* (the Thai-Phing reign-period (Sung) Imperial Encyclopaedia), +983.

TSCC Chhen Mêng-Lei *et al.* (ed.), *Thu Shu Chi Chhêng* (the Imperial Encyclopaedia of +1726). Index by Giles, L. (2).

TSFY Ku Tsu-Yu, *Tu Shih Fang Yü Chi Yao* (The Historian's Geographical Companion) begun before +1666, finished before +1692; but not printed till the end of the eighteenth century (1796 to 1821).

TTN Tu Yu, *Thung Tien* (Comprehensive Institutes), a reservoir of source material on political and social history, *c.* +812.

TW Takakusu, J. & Watanabe, K. *Tables du Taishō Issaikyō* (*nouvelle édition* (*Japonaise*) *du Canon bouddhique chinoise*). Index-catalogue of the Tripiṭaka.

WCTY/CC Tsêng Kung-Liang (ed.), *Wu Ching Tsung Yao* (*Chhien Chi*), military encyclopaedia, first section, +1044.

WHTK Ma Tuan-Lin, *Wên Hsien Thung Khao* (Comprehensive Study of (the History of) Civilisation), +1319.

WPC Mao Yuan-I, *Wu Pei Chih* (Treatise on Armament Technology), +1628.

YHSF Ma Kuo-Han (ed.), *Yü Han Shan Fang Chi I Shu* (Jade-Box Mountain Studio Collection of (reconstituted and sometimes fragmentary) Lost Books), 1853.

YTFS Li Chieh, *Ying Tsao Fa Shih* (Treatise on Architectural Methods), +1097, pr. +1103, repr. +1145.

ACKNOWLEDGEMENTS

LIST OF THOSE WHO HAVE KINDLY READ THROUGH SECTIONS IN DRAFT

The following list, which applies only to this volume, brings up to date those printed in Vol. 1, pp. 15 ff., Vol. 2, p. xxiii, Vol. 3, pp. xxxix ff., Vol. 4, pt. 1, p. xxi, and Vol. 4, pt. 2, p. xli.

Mr R. C. Anderson (Greenwich)	Nautics (Shipping).
Prof. Guy Beaujouan (Montrouge)	Nautics (Navigation and Voyages).
Dr Asit K. Biswas (Ottawa)	Hydraulics.
The late Mr Andrew Boyd (London)	Building Technology, Bridges.
The late Dr Herbert Chatley (Bath)	Hydraulics.
Mr Wells Coates (London)	Nautics (Sails).
Mr F. R. Cowell (Kemsing)	Perspective.
Mr Basil Davidson (London)	Nautics (Voyages).
Dr R. D. Davies (Cambridge)	Bridges.
Prof. Edwin Doran (College, Texas)	Nautics.
Prof. V. Elisséeff (Paris)	Both sections.
Prof. James M. Fitch (New York)	Building Technology.
Mr Klaus Flessel (Tübingen)	Hydraulics.
Mrs Else Glahn (Copenhagen)	Building Technology.
Mr Philip Grierson (Cambridge)	Nautics (Voyages).
Dr Trevor Hodge (Cambridge)	Building Technology.
Prof. Huang Jen-Yü (New York)	Hydraulics.
Mr Bryan J. Hudson (Hongkong)	Hydraulics.
Mr John Hunter (Thaxted)	Building Technology.
Mr David H. Kelley (Lubbock, Texas)	Nautics (pre-Columbian contact).
Mr James Kirkman (Mombasa)	Nautics (Voyages).
Mr N. E. Lee (Victoria, N.S.W.)	Nautics.
Mr Alfred Lieber (Jerusalem)	Nautics (Voyages).
Prof. Lo Jung-Pang (Davis, California)	Roads, Walls, Bridges, Nautics.
Dr Ian McPherson (Cambridge)	Nautics (Voyages).
Sir Leslie Martin (Cambridge)	Building Technology.
Mr J. V. Mills (La Tour de Pcilz, Vaud)	Nautics (Navigation and Voyages).
Mr J. S. Morrison (Cambridge)	Nautics (Shipping).
Lt.-Cdr George Naish (Greenwich)	Nautics (Shipping).
Dr Anthony Pearson (Cambridge)	Hydraulics.
Prof. Luciano Petech (Rome)	Both sections.
The late Dr Victor Purcell (Cambridge)	Nautics (Voyages).
Mr Francisco Quintanilha (Cambridge)	Nautics (Voyages).
Mr Nathan Silver (Cambridge)	Building Technology.

Prof. A. W. Skempton (London)	Roads, Walls, Building, Bridges, Hydraulics.
Mr E. G. Sterland (Bristol)	Roads, Walls, Bridges.
Miss Barbara E. Ward (Bideford)	Nautics (Shipping).
Lt.-Cdr D. W. Waters (Greenwich)	Nautics.
Prof. Gene Weltfish (New York)	Nautics (pre-Columbian contact).
Mr Clough Williams-Ellis (Penrhyndeudraeth)	Building Technology.
The late Mr G. R. G. Worcester (Windlesham)	Nautics (Shipping).

AUTHOR'S NOTE

PURSUING our exploration of the almost limitless caverns of Chinese scientific history, so much of which has never yet come to the knowledge and recognition of the rest of the world, we now approach the glittering veins of physics and physical technology; a subject which forms a single whole, constituting Volume Four, though delivered to the reader in three separate volumes. First come the physical sciences themselves (Vol. 4, pt. 1), and then their diverse applications in all the many branches of mechanical engineering (Vol. 4, pt. 2), civil and hydraulic engineering, and nautical technology (Vol. 4, pt. 3).

With the opening chapter we find ourselves at a focal point in the present study, for mechanics and dynamics were the first of all the conquests of modern science. Mechanics was the starting-point because the direct physical experience of man in his immediate environment is predominantly mechanical, and the application of mathematics to mechanical magnitudes was relatively simple. But ancient and medieval China belonged to a world in which the mathematisation of hypotheses had not yet brought modern science to birth, and what the scientific minds of pre-Renaissance China neglected might prove almost as revealing as that which aroused their interest and investigation. Three branches of physics were well developed among them, optics (Section 26g), acoustics (26h), and magnetism (26i); mechanics was weakly studied and formulated, dynamics almost absent. We have attempted to offer some explanation for this pattern but without any great conviction, and better understanding of the imbalance must await further research. The contrast with Europe, at least, where there was a different sort of one-sidedness, is striking enough, for in Byzantine and late medieval times mechanics and dynamics were relatively advanced while magnetic phenomena were almost unknown.

In optics the Chinese of the Middle Ages kept empirically more or less abreast of the Arabs, though greatly hampered in theory by the lack of that Greek deductive geometry of which the latter were the inheritors. On the other hand they never entertained that peculiar Hellenistic aberration according to which vision involved rays radiating from, not into, the eye. In acoustics the Chinese proceeded along their own lines because of the particular and characteristic features of their ancient music, and here they produced a body of doctrine deeply interesting but not readily comparable with those of other civilisations. Inventors of the bell, and of a great variety of percussion instruments not known in the West, they were especially concerned with timbre both in theory and practice; developing their unique theories of melodic composition within the framework of a twelve-note gamut rather than an eight-note scale. At the end of the +16th century Chinese mathematical acoustics succeeded in solving the problem of equal temperament just a few decades before its solution was reached in the West (Section 26h, 10). Lastly, Chinese investigation of magnetic phenomena and their practical application constituted a veritable epic. Men were arguing in China about the cause of the declination of the magnetic needle, and using it at sea, before Westerners even knew of its directive property.

Readers pressed for time will doubtless welcome once more a few suggestions. In the chapters which we now present it is possible to perceive certain outstanding traditions of Chinese physical thought and practice. Just as Chinese mathematics was indelibly algebraic rather than geometrical, so Chinese physics was wedded to a prototypic wave-theory and perennially averse to atoms, always envisaging an almost Stoic continuum; this may be seen in Section 26b and followed through in relation to tension and fracture (c, 3) and to sound vibrations (h, 9). Another constant Chinese tendency was to think in pneumatic terms, faithfully developing the implications of the ancient concept of *chhi* (= *pneuma*, *prāṇa*).[a] Naturally this shows itself most prominently in the field of acoustics (Section 26h, 3, 7, etc.), but it was also connected with some brilliant successes in the field of technology such as the inventions of the double-acting piston-bellows and the rotary winnowing-fan (Section 27b, 8), together with the water-powered metallurgical blowing-engine (27h, 3, 4, direct ancestor of the steam engine itself). It was also responsible for some extraordinary insights and predictions in aeronautical pre-history (27m, 4). Traditions equally strong and diametrically opposite to those of Europe also make their appearance in the purely technical field. Thus the Chinese had a deep predilection for mounting wheels and machinery of all kinds horizontally instead of vertically whenever possible; as may be followed in Section 27 (h, k, l, m).

Beyond this point, guidance to the reader is not very practicable since so many different preoccupations are involved. If he is interested in the history of land transport he will turn to the discussion of vehicles and harness (Section 27e, f), if he delights, like Leviathan, in the deep waters, a whole chapter (29) will speak to him of Chinese ships and their builders. The navigator will turn from the compass itself (26i, 5) to its fuller context in the haven-finding art (29f); the civil engineer, attracted by a survey of those grand water-works which outdid the 'pyramides of Aegypt', will find it in Section 28f. The folklorist and the ethnographer will appreciate that 'dark side' of history where we surmise that the compass-needle, most ancient of all those pointer-readings that make up modern science, began as a 'chess-man' thrown on to a diviner's board (26i, 8). The sociologist too will already find much of interest, for besides discussing the place of artisans and engineers in feudal-bureaucratic society (27a, 1, 2, 3), we have ventured to raise certain problems of labour-saving invention, man-power, slave status and the like, especially with regard to animal harness (27f, 2), massive stone buildings (28d, 1), oared propulsion (29g, 2), and water-powered milling and textile machinery (27h).

Many are the ways in which these volumes link up with those which have gone before. We shall leave the reader's perspicacity to trace how the *philosophia perennis* of China manifested itself in the discoveries and inventions here reported. We may point out, however, that mathematics, metrology and astronomy find numerous echoes; in the origins of the metric system (Section 26c, 6), the development of lenses (g, 5), and

[a] Cf. the wise footnote of Sarton (1), vol. 3, p. 905.

the estimation of pitch-pipe volumes (h, 8)—or the rise of astronomical clocks (Section 27j), the varying conceptions of perspective (28d, 5) and the planning of hydraulic works (f, 9). Similarly, much in the present volumes points forward to chapters still to come. All uses of metal in medieval Chinese engineering imply what we have yet to say on metallurgical achievements; in the meantime reference may be made to the separate monograph *The Development of Iron and Steel Technology in China*, published as a Newcomen Lecture[a] in 1958. In all mentions of mining and the salt industry it is understood that these subjects will be fully dealt with at a later stage. All water-raising techniques remind us of their basic agricultural purpose, the raising of crops.

As for the discoveries and inventions which have left permanent mark on human affairs, it would be impossible even to summarise here the Chinese contributions. Perhaps the newest and most surprising revelation (so unexpected even to ourselves that we have to withdraw a relevant statement in Vol. 1) is that of the six hidden centuries of mechanical clockwork which preceded the clocks of +14th-century Europe. Section 27j is a fresh though condensed treatment of this subject, incorporating still further new and strange material not available when the separate monograph *Heavenly Clockwork* was written in 1957 with our friend Prof. Derek J. de Solla Price, now of Yale University.[b] It still seems startling that the key invention of the escapement should have been made in a pre-industrial agrarian civilisation among a people proverbially supposed by bustling nineteenth-century Westerners to take no account of time.[c] But there are many other equally important Chinese gifts to the world: the development of the magnetic compass (Section 26i, 4, 6), the invention of the first cybernetic machine (27e, 5), both forms of efficient equine harness (27f, 1), the canal lock-gate (28f, 9, v)[d] and the iron-chain suspension bridge (28e, 4). The first true crank (Section 27b, 4), the stern-post rudder (Section 29h), the man-lifting kite (Section 27m)—we cannot enumerate them all.

In these circumstances it seems hardly believable that writers on technology have run up and down to find reasons why China contributed nothing to the sciences, pure or applied. At the beginning of a recent popular *florilegium* of passages on the history of technology one comes across a citation from the +8th-century Taoist book *Kuan Yin Tzu*,[1] given as an example of 'oriental rejection of this world and of worldly activity'. It had been culled from an interesting essay on religion and the idea of progress, well known in the 'thirties and still stimulating, the author of which, led astray by the old rendering of Fr Wieger, had written: 'It is obvious that such beliefs can afford no basis for social activity and no incentive to material progress.' He was, of course, concerned to contrast the Christian acceptance of the material world with 'oriental' otherworldliness, in which the Taoists were supposed to participate. Yet in almost every one of the inventions and discoveries we here describe the Taoists and Mohists were intimately involved (cf. e.g. Sections 26c, g, h, i, 27a, c, h, j, 28e, 29f, h). As it happened, we had ourselves studied the same *Kuan Yin Tzu* passage and given parts of it in

[a] Needham (32), cf. (31).
[b] Needham, Wang & Price (1), cf. Needham (38).
[c] Cf. Needham (55, 56).
[d] Cf. Needham (57).

[1] 關尹子

d

translation at an earlier stage;[a] from this it can be seen that Wieger's version[b] was no more than a grievously distorted paraphrase. Far from being an obscurantist document, denying the existence of laws of Nature (a concept totally unheard-of by the original writer)[c] and confusing reality with dream, the text is a poem in praise of the immanent Tao, the Order of Nature from which space and time proceed, the eternal pattern according to which matter disperses and reassembles in forms ever new; full of Taoist relativism, mystical but in no way anti-scientific or anti-technological, on the contrary prophesying of the quasi-magical quasi-rational command over Nature which he who truly knows and understands the Tao will achieve. Thus upon close examination, an argument purporting to demonstrate the philosophical impotence of 'oriental thought' turns out to be nothing but a figment of occidental imagination.

Another method is to admit that China did something but to find a satisfying reason for saying nothing about it. Thus a recent compendious history of science published in Paris maintains that the sciences of ancient and medieval China and India were so closely bound to their peculiar cultures that they cannot be understood without them. The sciences of the ancient Greek world, however, were truly sciences as such, free of all subordination to their cultural matrix and fit subjects with which to begin a story of human endeavour in all its abstract purity. It would be much more honest to say that while the social background of Hellenistic science and technology can be taken for granted because it is quite familiar to us from our schooldays onwards, we do not yet know much about the social background of Chinese and Indian science, and that we ought to make efforts to get acquainted with it. In fact, of course, no ancient or medieval science and technology can be separated from its ethnic stamp,[d] and though that of the post-Renaissance period is truly universal, it is no better understandable historically without a knowledge of the milieu in which it came to birth.

Finally, many will be desirous of looking into questions of intercultural contacts, transmissions and influences. Here we may only mention examples still puzzling of inventions which occur almost simultaneously at both ends of the Old World, e.g. rotary milling (Section 27d, 2) and the water-mill (h, 2). Parallels between China and ancient Alexandria often arise (for instance in Section 27b) and the powerful influence of Chinese technology on pre-Renaissance Europe appears again and again (26c, h, i; 27b, d, e, f, g, j, m; 28e, f; 29j). Important inventions in hydraulic engineering travelled westwards; and despite the supposed conservatism of sailors, there was hardly any Western century in the past twenty which did not see the adoption of some nautical technique from the East.

In a brilliant *ponencia* at the Ninth International Congress of the History of Sciences at Barcelona in 1959 Professor Willy Hartner raised the difficult question of how far anyone can ever anticipate anyone else. What does it mean to be a precursor or a predecessor? For those who are interested in intercultural transmissions this is a vital point. In European history the problem has assumed acute form since the school of Duhem acclaimed Nicholas d'Oresme and other medieval scholars as the precursors of

[a] Vol. 2, pp. 449 and 444.
[c] Cf. Section 18 in Vol. 2 above.

[b] Originally (4), p. 548.
[d] Cf. Vol. 3, p. 448.

Copernicus, Bruno, Francis Bacon, Galileo, Fermat and Hegel. Here the difficulty is that every mind is necessarily the denizen of the organic intellectual medium of its own time, and propositions which may look very much alike cannot have had quite the same meaning when considered by minds at very different periods. Discoveries and inventions are no doubt organically connected with the milieu in which they arose. Similarities may be purely fortuitous. Yet to affirm the true originality of Galileo and his contemporaries is not necessarily to deny the existence of precursors, so long as that term is not taken to mean absolute priority or anticipation; and in the same way there were many Chinese precursors or predecessors who adumbrated scientific principles later acknowledged—one thinks immediately of Huttonian geology (Vol. 3, p. 604), the comet tail law (p. 432) or the declination of the magnetic needle (Section 26*i*). So much for science more or less pure; in applied science we need hesitate less. For example, the gaining of power from the flow and descent of water by a wheel can only have been first successfully executed once. Within a limited lapse of time thereafter the invention may have occurred once or twice independently elsewhere, but such a thing is not invented over and over again. All subsequent successes must therefore derive from one or other of these events. In all these cases, whether of science pure or science applied, it remains the task of the historian to elucidate if possible how much genetic connection there was between the precursor and the great figures which followed him. Did they know certain actual written texts? Did they work by hearsay? Did they first conceive their ideas alone and find them unexpectedly confirmed? As Hartner says, the variations range from the certain to the impossible.[a] Often hearsay seems to have been followed by a new and different solution (cf. Section 27*j*, *l*). In our work here presented to the reader he will find that we are very often quite unable to establish a genetic connection (for example between the suspension of Ting Huan and that of Jerome Cardan, in Section 27*d*, 4; or between the rotary ballista of Ma Chün and that of Leonardo, in Sections 27*a*, 2 and 30*i*, 4), but in general we tend to assume that when the spread of intervening centuries is large and the solution closely similar, the burden of proof must lie on those who desire to maintain independence of thought or invention. On the other hand the genetic connection can sometimes be established with a high degree of probability (for example, in the matters of equal temperament, Section 26*h*, 10, sailing-carriages, Section 27*e*, 3, and the kite, the parachute and the helicopter, Section 27*m*). Elsewhere one is left with strong suspicions, as with regard to the water-wheel escapement clock (Section 27*j*, 6).[b]

Although every attempt has been made to take into account the most recent research in the fields here covered, we regret that it has generally not been possible to mention work appearing after May 1968.

We have not printed a contents-table of the entire project since the beginning of

[a] Many a surprise is still in store for us. After the discovery by Al-Ṭaṭāwī in 1924 that Ibn al-Nafīs (+1210 to +1288) had clearly described the pulmonary circulation (cf. Meyerhof (1, 2); Haddad & Khairallah), it was long considered extremely unlikely that any hint of this could have reached the Renaissance discoverer of the same phenomenon, Miguel Servetus (cf. Temkin, 2). But now O'Malley (1) has found a Latin translation of some of the writings of Ibn al-Nafīs published in +1547.

[b] On the criterion of genetic connection in the history of science and technology in general, see Needham (45).

Vol. 1, and it has now been felt desirable to revise it in prospectus form.[a] So much work has now been done in preparation for the later volumes that it is possible to give their outline subheadings with much greater precision than could be done seven years ago. More important, perhaps, is the division into volumes. Here we have sought to retain unaltered the original numbering of the successive Sections, as for the needs of cross-referencing we must. Vol. 4, as originally planned, included physics, all branches of engineering, military and textile technology and the arts of paper and printing. As will be seen, we now entitle Vol. 4 *Physics and Physical Technology*, Vol. 5 *Chemistry and Chemical Technology* and Vol. 6 *Biology and Biological Technology*. This is a logical division, and Vol. 4 concludes very reasonably with Nautics (29) for in ancient and medieval times the techniques of shipping were almost entirely physical. Similarly Vol. 5 starts with Martial Technology (30), for in this field and in those times the opposite was the case; the chemical factor was essential. We found not only that we must embody iron and steel metallurgy therein (hence the slight but significant change of title), but also that without the epic of gunpowder, the fundamental discovery of the first known explosive and its development through five pre-occidental centuries, the history of Chinese military technique could not be written. With Textiles (31) and the other arts (32) the same argument was found to apply, for so many of the processes (retting, fulling, dyeing, ink-making) allied them to chemistry rather than to physics. Of course we could not always consistently adhere to this principle; for instance, no discussion of lenses was possible without some knowledge of glass technology, and this had therefore to be introduced at an early stage in the present volume (26*g*, 5, ii). For the rest, it is altogether natural that Mining (36), Salt-winning (37) and Ceramic Technology (35) should find their place in Vol. 5. The only assymetry is that while in Vols. 4 and 6 the fundamental sciences are dealt with at the beginning of the first part, in Vol. 5 the basic science, chemistry, with its precursor, alchemy, is discussed in the second part. This probably matters the less because in ready response to the critics who found Vol. 3 too heavy and bulky for comfortable meditative evening reading, the University Press has decided to produce the present volume in three physically separate parts, each being as usual independent and complete in itself.

In Vol. 1, pp. 18 ff., we gave details of the plan of the work (conventions, bibliographies, indexes, etc.) to which we have since closely adhered, and we promised that in the last volume a list would be given of the editions of the Chinese books used. It now seems undesirable to wait so long, and thus for the convenience of readers with knowledge of the Chinese language, we append to the present volume an interim list of these editions down to the point now reached. We are grateful to Miss Léonie Callaghan of Canberra for carrying out the bulk of the work involved. Two further related points may be mentioned here. First, in this and recent volumes some page references to Chinese texts are placed in round brackets without *recto* and *verso* letters; this directs to the modern edition as against the old one. Secondly, the reader is reminded that where in these volumes two statements disagree, that in the latest of the series is to be preferred. *Experientia docet*.

[a] An extract of this contents-table relevant to the present volume will be found on pp. 929–31 below.

China to Europeans has been like the moon, always showing the same face—a myriad peasant-farmers, a scattering of artists and recluses, an urban minority of scholars, mandarins and shopkeepers. Thus do civilisations acquire 'stereotypes' of one another. Now, raised upon the wings of the space-ship of linguistic resource and riding the rocket of technical understanding (to use an Arabic trope), we intend to see what is on the other side of the disc, and to meet the artisans and engineers, the ship-wrights and the metallurgists of China's three-thousand-year-old culture.

In our note at the beginning of Vol. 3 we took occasion to say something of the principles of translation of old scientific texts and of the technical terms contained in them.[a] Since this is the first volume largely devoted to the applied sciences we are moved to insert a few reflections here on the present position of the history of technology, a discipline which has suffered even more perhaps than the history of science itself from that dreadful dichotomy between those who know and those who write, the doers and the recorders. If men of scientific training, with all their handicaps, have contributed far more than professional historians to the history of science and medicine (as is demonstrably true), technologists as a whole have been even less well equipped with the tools and skills of historical scholarship, the languages, the criticism of sources, and the use of documentary evidence. Yet nothing can be more futile than the work of a historian who does not really understand the crafts and techniques with which he is dealing, and for any literary scholar it is hard to acquire that familiarity with things and materials, that sense of possibilities and probabilities, that understanding of Nature's ways, in fact, which comes (in greater or lesser measure) to everyone who has worked with his hands whether at the laboratory bench or in the factory workshop. I always remember once studying some medieval Chinese texts on 'light-penetration mirrors' (*thou kuang chien*[1]), that is to say, bronze mirrors which have the property of reflecting from their polished surfaces the designs executed in relief on their backs. A non-scientific friend was really persuaded that the Sung artisans had found out some way of rendering metal transparent to light-rays, but I knew that there must be some other explanation and it was duly found (cf. Section 26g, 3). The great humanists of the past were very well aware of their limitations in these matters, and sought always, so far as possible, to gain acquaintance with what my friend and teacher, Gustav Haloun, used half-wistfully, half-ironically to call the *realia*. In a passage we have already quoted (Vol. 1, p. 7), another outstanding sinologist, Friedrich Hirth, urged that the Western translator of Chinese texts must not only translate, he must identify, he must not only know the language but he must also be a collector of the objects talked about in that language. The conviction was sound, but if porcelain or cloisonné could (at any rate in those days) be collected and contemplated with relative ease, how much more difficult is it to acquire an understanding of machinery, of tanning or of pyrotechnics, if one has never handled a lathe, fitted a gear-wheel or set up a distillation.

What is true of living humanists in the West is also true of some of the Chinese

---

[a] Cf. Needham (34).

[1] 透光鑑

scholars of long ago whose writings are often our only means of access to the techniques of past ages. The artisans and technicians knew very well what they were doing,
but they were liable to be illiterate, or at least inarticulate (cf. the long and illuminating
text which we have translated in Section 27a, 2). The bureaucratic scholars, on the
other hand, were highly articulate but too often despised the rude mechanicals whose
activities, for one reason or another, they wrote about from time to time. Thus even the
authors whose words are now so precious were often more concerned with their
literary style than with the details of the machines and processes which they mentioned. This superior attitude was also not unknown among the artists, back-room
experts (like the mathematicians) of the officials' *yamens*, so that often they were more
interested in making a charming picture than in showing the precise details of
machinery when they were asked to limn it, and now sometimes it is only by comparing one drawing with another that we can reach certainty about the technical content.
At the same time there were many great scholar-officials throughout Chinese history
from Chang Hêng in the Han to Shen Kua in the Sung and Tai Chen in the Chhing
who combined a perfect expertise in classical literature with complete mastery of the
sciences of their day and the applications of these in artisanal practice.

For all these reasons our knowledge of the development of technology is still in a
lamentably backward state, vital though it is for economic history, that broad meadow
of flourishing speculation. In a recent letter, Professor Lynn White, who has done as
much as anyone else in the field, wrote memorable words with which we fully agree:
'The whole history of technology is so rudimentary that all one can do is to work very
hard, and be happy when one's errors are corrected.'[a] On every hand pitfalls abound.
On a single page of a recent most authoritative and admirable collective treatise, one of
our best historians of technology can first suppose Heron's toy windmill to be an
Arabic interpolation, though the *Pneumatica* never passed to us through that language,
and a moment later assert that Chinese travellers in +400 saw wind-driven prayer-
wheels in Central Asia, a story based on a mistranslation now just 125 years old. The
same authoritative treatise says that Celtic wagons of the −1st century had hubs
equipped with roller-bearings, and we ourselves at first accepted this opinion. We
learnt in time, however, that examination of the actual remains preserved at Copenhagen makes this highly improbable, and that reference to the original paper in Danish
clinches the matter—the pieces of wood which came out from the hub-spaces when
disinterred were flat strips and not rollers at all. We have often been saved from other
such mistakes only by the skin of our teeth, so to say, and it is not in a spirit of criticism, but rather to demonstrate the difficulties of the work, that we draw attention to
them.

Certain safeguards one can always try to obtain. There is no substitute for actually
seeing for oneself in the great museums of the world, and the great archaeological sites;

[a] Truly all the conclusions in the present volume must be regarded as provisional. Faithfully though
we seek to apply the comparative method, our final assessment is often only a bridge built upon wide-
spaced and insecure piers looming out of the mists. A single new and crucial fact can sometimes change
the whole complexion of what seemed a fairly stable pattern. Our successors will doubtless see more
clearly—but how it all happened Allah knoweth best.

there is no substitute for personal intercourse with the practising technicians themselves. To be sure the scholarly standard of any particular work must necessarily
depend upon the ground which is covered. Only the specialist using intensive methods
—a Rosen elucidating the tangled roots of ophthalmic lenses or a Drachmann exploring Roman oil-presses—can afford the time to go into a matter *au fond* and bring truth
wholly out of the well. We have tried to do this only in very few fields, such as that of
medieval Chinese clockwork, because our aim is essentially extensive and pioneering.
There is no escape, much must be taken on trust. If we are deficient in our knowledge
of the objects of occidental archaeology, it is because we have laboured to study *in situ*
those of the Chinese culture-area, our primary responsibility. If we had been able at
that time to visit the museum in Copenhagen where the Dejbjerg wagons are kept we
might have been more wary of accepting current statements about them, but—ὁ βίος
βραχύς ἡ δὲ τέχνη μακρή, the craft is long but life is short. On the other side of the scale a
deep debt of gratitude is owing to the President and Council of Academia Sinica for
generous facilities which enabled me in 1958, together with Dr Lu Gwei-Djen, to visit
or revisit many of the great museums and archaeological sites in China.

But not with archaeologists only must one converse. One must follow the example of
little Dr Harvey (of Caius College). In the seventeenth century John Aubrey tells us of
a conversation he had with a sow-gelder, a countryman of little learning but much
practical experience and wisdom. He told him that he had met Dr William Harvey,
who had conversed with him above two or three hours, and 'if he had been', the man
remarked, 'as stiffe as some of our starcht and formall Doctors, he had known no more
than they'.[a] A Kansu carter threw light upon the harness not only of our own time but
indirectly of the Han and Thang, Szechuanese iron-workers were well able to help our
understanding of how Chhiwu Huai-Wên in +545 made co-fusion steel, and a
Peking kite-maker could reveal with his simple materials those secrets of the cambered
wing and the airscrew which lie at the heart of modern aeronautical science. Nor may
the technicians of one's own civilisation be neglected, for a traditional Surrey wheelwright can explain how wheels were 'dished' by the artisans of the State of Chhi two
thousand years and more ago. A friend in the zinc industry disclosed to us that the
familiar hotel cutlery found today all over the world is made essentially of the medieval
Chinese alloy *paktong*,[1] a nautical scholar from Greenwich demonstrated the significance of the Chinese lead in fore-and-aft sailing, and it took a professional hydraulic
engineer to appreciate at their true value the Han measurements of the silt-content of
river-waters. As Confucius put it, *San jen hsing, pi yu wo shih*,[2] 'Where there are three
men walking together, one or other of them will certainly be able to teach me something'.[b]

The demonstrable continuity and universality of science and technology prompts a
final observation. Some time ago a not wholly unfriendly critic of our previous volumes
wrote, in effect: this book is fundamentally unsound for the following reasons. The

---

[a] See Keynes (2), pp. 422, 436, 437.
[b] *Lun Yü*, VII, xxi.

[1] 白銅   [2] 三 人 行 必 有 我 師

authors believe (1) that human social evolution has brought about a gradual increase in man's knowledge of Nature and control of the external world, (2) that this science is an ultimate value and with its applications forms today a unity into which the comparable contributions of different civilisations (not isolated from each other as incompatible and mutually incomprehensible organisms) all have flowed and flow as rivers to the sea, (3) that along with this progressive process human society is moving towards forms of ever greater unity, complexity and organisation. We recognised these invalidating theses as indeed our own, and if we had a door like that of Wittenberg long ago we would not hesitate to nail them to it. No critic has subjected our beliefs to a more acute analysis, yet it reminded us of nothing so much as that letter which Matteo Ricci wrote home in +1595 to describe the various absurd ideas which the Chinese entertained about cosmological questions:[a] (One), he said, they do not believe in solid crystalline celestial spheres, (Item) they say that the heavens are empty, (Item) they have five elements instead of the four so universally recognised as consonant with truth and reason, etc. But we have made our point.

A decade of fruitful collaboration came to an end when early in 1957 Dr Wang Ling[1] (Wang Ching-Ning[2]) departed from Cambridge to the Australian National University at Canberra, where he is now Professorial Fellow in sinology at the Institute of Advanced Studies. Neither of us will ever forget the early years of the project when our organisation was finding its feet, and a thousand problems had to be solved (with equipment much less adequate than now) as we went along. In the present volume Dr Wang's co-operation was particularly valuable and fruitful in Sections $28f$ and $29h$. The essential continuity of day-to-day collaboration with Chinese scholars was however happily preserved at his departure by the arrival of a still older friend, Dr Lu Gwei-Djen,[3] late in 1956. Among other posts, Dr Lu had been Research Associate at the Henry Lester Medical Institute, Shanghai, Professor of Nutritional Science at Ginling College, Nanking, and later in charge of the Field Co-operation Offices Service in the Department of Natural Sciences at UNESCO headquarters in Paris. With a basis of wide experience in nutritional biochemistry and clinical research, she is now engaged in pioneer work for the biological and medical part of our plan (Vol. 6). Probably no single subject in our programme presents more difficulties than that of the history of the Chinese medical sciences. The wealth of the literature, the systematisation of the concepts (so different from those of the West), the use of ordinary and philosophical words in special senses so as to constitute a subtle and precise technical terminology, and the strangeness of certain important branches of therapy—all demand great efforts if the result is to give, as has not yet been given, a true picture of Chinese medicine. It is fortunate that time permits our soundings to chart the true bottom. Concurrently Dr Lu has participated in the extensive revision of the present volume for press, work which has taken her into the very different waters of hydraulic technology and shipping.

[a] Cf. Vol. 3, p. 438.

[1] 王鈴   [2] 王靜寧   [3] 魯桂珍

A year later (early in 1958) we were joined by Dr Ho Ping-Yü,[1] then Reader in Physics at the University of Malaya, Singapore. Primarily an astro-physicist by training, and the translator of the astronomical chapters of the *Chin Shu*, he was happily willing to broaden his experience in the history of science by devoting himself to the study of alchemy and early chemistry, helping thus to lay the foundations for the relevant volume (Vol. 5). Such work had been initiated some years earlier by yet another friend, Dr Tshao Thien-Chhin,[2] when a Research Fellow of Caius College, before his return to the Biochemical Institute of Academia Sinica at Shanghai. Dr Tshao had been one of my wartime companions, and while in Cambridge made a most valuable study of the alchemical books in the *Tao Tsang*.[a] Dr Ho Ping-Yü was able to extend this work with great success in many directions. Although Dr Ho is now Professor of Chinese at the University of Malaya, Kuala Lumpur, he has been able to rejoin us in Cambridge for a period for the further preparation of the volume on chemistry and chemical technology.

It is good to record that already a number of important sub-sections of both these volumes (5 and 6) have been written. The publication of some of them in draft form facilitates criticism and aid by specialists in the different fields.

Lastly, an occidental collaborator appears with us on the title-page of the first part of this volume, Mr Kenneth Robinson, one who combines most unusually sinological and musical knowledge. Professionally he is an educationalist, and with a Malayan background in teachers' training, frequented as Director of Education in Sarawak the villages and long-houses of the Dayaks and other peoples, whose remarkable orchestras seemed to him to evoke the music of the Chou and Han. We were fortunate indeed that he was willing to undertake the drafting of the Section on the recondite but fascinating subject of physical acoustics, indispensable because it was one of the major interests of the scientific minds of the Chinese Middle Ages. He is thus the only participator in this enterprise so far who has contributed direct authorship as well as research activity. Another European colleague, Mr John Combridge, of the Engineering Department of the General Post Office, has greatly added to our understanding of medieval Chinese clockwork, especially by experiments with working models.

Once again it is a pleasure to offer public gratitude to those who have helped us in many different ways. First, our advisers in linguistic and cultural fields unfamiliar to us, notably Prof. D. M. Dunlop for Arabic, Dr Shackleton Bailey for Sanskrit, Dr Charles Sheldon for Japanese, and Prof. G. Ledyard for Korean. Secondly, those who have given us special assistance and counsel, Mr E. G. Sterland in mechanical engineering, Prof. Lo Jung-Pang in the history of transport, Prof. A. W. Skempton and the late Dr Herbert Chatley in hydraulic engineering, Mr J. V. Mills in navigation and Cdr. George Naish and Cdr. D. W. Waters in nautics. Thirdly, all those whose names will be found in the adjoining list of readers and kind critics of Sections in draft or proof form. But only Dr Dorothy Needham, F.R.S., has weighed every word in these volumes and our debt to her is incalculable.

[a] Cf. Vol. 1, p. 12.

[1] 何丙郁   [2] 曹天欽

Once again we renew our warmest thanks to Mrs Margaret Anderson for her indispensable and meticulous help with press work, and to Mr Charles Curwen and Mr Ian McMaster for acting as our agents-general with regard to the ever-increasing flood of current Chinese literature on the history and archaeology of science and technics. A particularly generous service was given to the project recently by Mr Walter Sheringham, who carried out an expert valuation of our working library on an honorary basis. Miss Muriel Moyle has continued to provide her very detailed indexes, the excellence of which has been saluted by many reviewers. As the enterprise continues, the burden of typing and secretarial work seems to grow beyond expectation, and we have had many occasions to recognise that a good copyist is like the spouse in Holy Writ, precious beyond rubies. Thus we most gratefully acknowledge the help of the late Mrs Betty May, Miss Margaret Webb, Miss Jennie Plant, Mrs Evelyn Beebe, Miss June Lewis, Mr Frank Brand, Mrs W. M. Mitchell, Miss Frances Boughton, Mrs Gillian Rickaysen and Mrs Anne Scott McKenzie.

The part played by publisher and printer in a work such as this, considered in terms either of finance or technical skill, is no less vital than the research, the organisation and the writing itself. Few authors could have more appreciation of their colleagues executive and executant than we for the Syndics and the Staff of the Cambridge University Press. Among the latter formerly was our friend Frank Kendon, for many years Assistant Secretary, whose death occurred after the appearance of Volume 3. Known in many circles as a poet and literary scholar of high achievement, he was capable of divining the poetry implicit in some of the books which passed through the Press, and the form which his understanding took was the bestowal of infinite pains to achieve the external dress best adapted to the content. I shall always remember how when *Science and Civilisation in China* was crystallising in this way, he 'lived with' trial volumes made up in different styles and colours for some weeks before arriving at a decision most agreeable to the author and his collaborators—and what was perhaps more important, equally so to thousands of readers all over the world. Infinite debt we owe also to our friend Mr Peter Burbidge, now Production Manager of the Cambridge University Press, who has watched over the complicated travail of the successive volumes as Editor, with warm appreciation and enthusiasm, from the first beginnings of the project. No trouble was too much on its behalf.

To the Master and Fellows of the Hall of the Annunciation, commonly called Gonville and Caius College, a family of immediate colleagues, I can offer only inadequate words. I do not know where else conditions so perfect for carrying out an enterprise such as this could be found, a peaceful workshop in the topographical centre of the University and all its libraries, between the President's apple-tree and the Porta Honoris. Here our study of Chinese shipping, the results of which are reported in this volume, has had a special appropriateness, for these rooms knew as their previous occupant my own tutor, Sir William Bate Hardy, F.R.S. Hardy as a discoverer and organiser, one of the founders of modern cytology, biophysics and food preservation technology, has now long himself become part of the history of science—but he was also a master-mariner of fame in his day. He would have delighted to know that the breath of the sea still penetrates to the court which Dr Caius expressly left open to receive it. So also would that much earlier 'mathematical practitioner', Edward Wright, Fellow from 1587 to 1596, *vir morum simplicitate et candore omnibus gratus*, the author of 'Certaine Errors in Navigation' and one of the greatest of Elizabethan boffins. Meanwhile, the daily appreciation and encouragement of every one in the Society helps us to surmount all the difficulties of the task. Nor could I omit meed of thanks to the Head of the Department of Biochemistry and its Staff for the indulgent understanding which they showed to a colleague seconded, as it were, to another universe.

The above paragraph was written before I was elevated to the responsibility of being Master myself, but the work continues nothing changed, and my gratitude to my colleagues ever increases.

The financing of the research work for our project has always been difficult and still presents serious problems. We are nevertheless deeply indebted to the Wellcome Trust, whose exceptionally generous support has relieved us of all anxiety concerning the biological and medical volume. We cannot forbear from offering our deepest gratitude for this to its Scientific Consultant, formerly long its Chairman, the late Sir Henry Dale, O.M., F.R.S. An ample benefaction by the Bollingen Foundation, elsewhere acknowledged, has assured the adequate illustration of the successive volumes. To Dato Lee Kong-Chian of Singapore we are beholden for a splendid contribution towards the expenses of research for the chemical volume, and Prof. Ho's work towards this has been made possible by sabbatical leave from the University of Malaya. Here we wish to pay a tribute to the memory of a great physician and servant of his country, Wu Lien-Tê, of Emmanuel College, already Major in the Chinese Army Medical Corps before the fall of the Chhing dynasty, founder long ago of the Manchurian Plague Prevention Service and pioneer organiser of public health work in China. During the last year of his life Dr Wu exerted himself to help in securing funds for our work, and his benevolence in this will always be warmly remembered. Certain kind well-wishers of our enterprise grouped themselves together some time ago in a committee of 'Friends of the Project' with a view to securing further necessary financial support, and to our friend the late Victor Purcell, C.M.G., who kindly accepted the honorary secretaryship of this committee, our best thanks are recorded. We are grateful indeed to Prof. E. Pulleyblank, Sir Eric Ashby, F.R.S. and Dr E. H. Carr, F.B.A., who have continued the oversight of this operation. At various periods during the studies which see the light in these volumes we have also received financial help from the Universities' China Committee and from the Managers of the Ocean Steamship Company acting as Trustees of funds bequeathed by members of the Holt family, more recently from the American Philosophical Society; for all this we record most grateful thanks.

28. CIVIL ENGINEERING

(a) INTRODUCTION

No ancient country in the world did more in civil engineering, both as to scale and skill, than China, yet very little has been done towards making known the history of it. Perhaps this is the less surprising when we reflect that competence in civil, preferably hydraulic, engineering is rarely combined with sinological knowledge and a good understanding of Chinese historical literature, nor much more often with the opportunity of travelling over the country to study the vestiges of the great engineering works of former times. However, a beginning has been made, and in the present Section we shall try to sketch some of their most important features, beginning with roads and walls, going on to bridges, and then devoting the major part of our space to the great public works of hydraulic engineering in which the Chinese excelled.

There seems to be no general history of Chinese civil engineering (*thu mu kung chhêng*[1]) in the vernacular, and even well-balanced accounts of the development of the science in the West, essential for comparative purposes, are not too easy to find.[a] Chinese literature does of course embody a wealth of notable books on water conservancy and control, but few which treat of the history of the techniques concerned in a modern manner, the authors preferring to discuss the geographical and economic aspects of the great works. Again, there was almost no coherent treatment of bridge-building until the Society for the History of Chinese Architecture took up the matter during the past thirty years, and published a number of important studies in its journal. All helpful sources will be mentioned in detailed reference as we proceed.

(b) ROADS

'Good roads, canals and navigable rivers', wrote Adam Smith in +1776, 'by diminishing the expense of carriage, put the remote parts of a country more nearly upon a level with those in the neighbourhood of the town. They are upon that account the greatest of all improvements.'[b] His opinion is none the less just if we recognise, as is unavoidable, that the greatest highway systems of the ancient and medieval worlds were planned and constructed with strategic intent. Of the engineering techniques, as well as the geographical pattern, of the famous Roman roads, much is known, since besides the numerous remains which it has been possible to excavate, there are detailed literary descriptions of theory and practice.[c] We know how the largest stones were laid

---

[a] For antiquity see Merckel (1) and, better, Leger (1); for the Renaissance there is Parsons (2). The only modern general survey is that of Straub (1), unfortunately too short. Pannell's illustrated history (1), confined to occidental material, appeared too late to be of assistance to us. So also the collected articles of Merdinger (1).

[b] (1), p. 62. For a history of road engineering in general see Schreiber (1).

[c] The classical description of the Roman road system, and still the most complete, is that of Bergier (1), but it has long been out-dated by the results of modern archaeology, as in Leger (1), especially p. 157.

[1] 土木工程

at the bottom to form the *statumen*, rubble and chips of different sizes laid above for the *ruderatio*, then the nucleus of sand or gravel, or of broken pottery and bricks cemented with lime,[a] the whole being topped with flat stone slabs to form the *summa crusta* or *summum dorsum*. Kerbs were often provided. The body of the Roman road thus occupied an excavated trough as deep as five or six feet, some three times the depth required for a modern road. Sometimes the lower layers extended widely beyond the breadth of the actual road itself, with a ditch on each side, sometimes the road was accompanied by a drainage canal of substantial size, sometimes it ran on embankments or through cuttings, and elsewhere it might have retaining-walls along the sides of steep slopes. Such were the *viae munitae*, but besides these fully paved roads the Romans also used graded earth tracks (*viae terrenae*) and gravelled-surface side-roads (*viae glareatae*).

It has often been observed that roads in the Roman style resembled to some extent a series of walls lying horizontally. The methods of their engineers were for long the object of great admiration on the part of archaeologists, but, as des Noëttes (1, 9) pointed out,[b] they were in truth primitive and ill-suited to their purpose. Allowing nothing for expansion and contraction due to temperature, frost fissures and unequal drainage, they depended on thickness and rigidity,[c] while the more successful modern methods, culminating in the compacted chips of McAdam,[d] and all their asphalted and other developments, depend on thinness and elasticity. These appear to be medieval in origin,[e] but Chinese roads of similar light and elastic type long preceded them, as we shall see.

Although we can trace the origin of highways back to prehistoric tracks, bronze-age ridgeways and the like,[f] impressive and complex systems do not develop until the rise of strong and centralised government. Hence the Persian Royal Road of the early — 5th century from Susa (the capital in the mountains north of Basra) to Sardis (the most westerly city in the Iranian empire, near the port of Ephesus in Asia Minor), a distance of some 1,400 miles; and another road which led eastwards, about as far, to Sogdiana.[g] Hence also the remarkable road-system, at least as large, and covering much more difficult terrain in the Andes, built by the Inca State and its predecessors.[h]

and pl. III; Gregory (1); Forbes (6), (11), pp. 126 ff., (22); Merckel (1); Birk (1). Among specific papers those of Birk (2) and Hertwig (1) may be mentioned. There are recent maps of parts of the system in van der Heyden & Scullard (1), charts 53, 60, cf. figs. 289, 290, 291, 293, 294, 443; Bengtson & Milojčić (1), charts 30, 31.

[a] Cf. Vol. 4, pt. 2, p. 219.

[b] Since supported by Forbes (6) and others.

[c] When a paved road breaks to pieces, and the slabs become up-ended in all directions, it becomes much worse than no road at all. A Chinese example, near the Salween River, is shown in Gregory (1), fig. 12, a striking photograph. The upkeep cost of paved roads is also of course relatively high.

[d] See Gregory (1), pp. 220 ff. McAdam's achievement was to show that the foundation of the road may be the subsoil and need not be stone, if the sole above it consists of pieces of stone of the right kind and size, with a self-cementing carpet above. For modern practice see, e.g., Spielmann & Elford (1).

[e] On medieval European land communications see Lopez (2); Forbes (22).

[f] On the prehistoric north–south trade routes of amber across Europe cf. de Navarro (1); Gregory (1), pp. 28 ff.

[g] Cf. Calder (1); Forbes (11), pp. 130 ff. See chart 23 in van der Heyden & Scullard (1), and charts 11*b*, 12*c*, 17 in Bengtson & Milojčić (1).

[h] Cf. von Hagen (2, 3); Saville (1).

Comparable road-building work, perhaps not so extensive, was also carried on in Maurya India, judging from indications in the *Arthaśāstra*.[a]

(1) NATURE AND EXPANSION OF THE NETWORK

Gazing down upon the Old World during the few centuries before and after the turn of the era, some demiurge might have seen, as in a slow-motion film, the appearance and radiation of two dendritic systems of highway communication springing from two different centres, one near the western coast about the middle of the Italic peninsula, the other near the great bend of the Yellow River where it swings round the Shansi mountains to flow eastwards to the Yellow Sea. The vision would have resembled somewhat the radiation of blood-vessels from the body of the foetal bird to make their ramifying way all over its yolk-store of food—and the bio-sociological analogy is not invalid altogether, for the tax-goods coming in would pass the legions on their outward ways. Could the Romans have ever succeeded in conquering the Parthians and Persians the two road systems might have met, perhaps, anastomosing somewhere west of Sinkiang, but this was not to be. The octopus-like arms expanded independently, each in a world of its own, their builders troubled only occasionally by the vaguest rumours of another system too far away to matter.

There is a curious parallel between the Roman and the Chinese systems in that both, after the +3rd century, fell into a long period of decay, but while Europe became parcelled out into feudal kingdoms and domains with poor communications except by sea, the role of the Chinese highways passed over to an immense system of navigable rivers and artificial canals, leaving only the mountain roads to continue their age-old function. As the chief sources of knowledge about the ancient Chinese network and its growth the dynastic histories take first place, together with the abundant remains of the historical geographers, a numerous tribe among the Chinese literati.[b] And always, as would be expected in a feudal-bureaucratic society, the central government concerned itself with the construction and maintenance of the principal routes of communication.

In China [wrote Adam Smith], and in several other governments of Asia, the executive power charges itself both with the reparation of the high-roads and with the maintenance of the navigable canals. In the instructions which are given to the governor of each province, these objects, it is said, are constantly recommended to him, and the judgment which the court forms of his conduct is very much regulated by the attention which he appears to have paid to this part of his instructions. This branch of public policy, accordingly, is said to be very much attended to in all those countries, but particularly in China, where the high-roads,

---

[a] Shamasastry tr., pp. 46, 48, 194, 334, and below, p. 5. According to Strabo, xv, 1, xi, a road 10,000 stadia long ran from the north-west frontier to the capital. With the stadion at 0·11 mile, this means some 1,100 miles. Cf. Anon. (82).

[b] Valuable modern monographs on the history of communications are not wanting: e.g. Lao Kan (2); Pai Shou-I (1) and Lo Jung-Pang (6). The history of ancient travel has been discussed by Chiang Shao-Yuan (1, 1), and the post-station system, at which we shall take a more particular look below, by Lou Tsu-I in various writings, especially (1). Unfortunately no one has studied ancient Chinese road engineering specifically from the technological point of view.

and still more the navigable canals, it is pretended, exceed very much everything of the same kind which is known in Europe.[a]

These features may be illustrated by some of the oldest records of road-building in the Chinese culture-area which have come down to us. A verse in the *Shih Ching* (Book of Odes) expresses admiration of the roads in the neighbourhood of the Chou capital:[b]

> The roads of Chou are (smooth) as a whetstone,
> Straight as an arrow('s flight);
> Ways where the lords and officials pass,
> Ways where the common people look on.

As this folk-song is considered rather ancient, perhaps of the −9th century, in the Western Chou period, it may refer either to the Wei Valley, Kuan-chung[1] ('within the passes') as it was later called, or to the eastern capital and domains of the High King near the later site of Lo-yang.[c] The route between Chhang-an (1) and Lo-yang (6) must certainly be one of the most ancient tracks in all China.[d] When we come to the *Chou Li* (Record of Institutions (lit. Rites) of the Chou Dynasty), that −2nd-century compilation of the ideal structure of the feudal-bureaucratic State, we have much more detailed information on the technical terms for roads, though it seems to incorporate two distinct traditions, probably from different earlier feudal States. In the entry for the Ssu Hsien[2] (Director of Communications) we read:[e]

He studies the maps of the nine provinces in order to obtain a perfect knowledge of the mountains, forests, lakes, rivers and marshes, and to understand the (natural) routes of communication.

[Comm. When mountains and forests present obstacles, he cuts through them. When rivers and lakes offer impediment, he bridges them.]

He lays out the five kinds of canal and the five kinds of road, planting trees and hedges along them for defence. All (special points, passes and junctions) have guard-posts, and he knows the ways and roads that lead to them.

[Comm. The five kinds of canal (*kou*[3]) are *sui*[4] (ditches), *kou*[3] (conduits), *hsüeh*[5] (or *hsü*,[5] small canals), *kuei*[6] (or *kuai*,[6] medium canals), and *chhuan*[7] (great canals). The five kinds of road (*thu*[8]) are *ching*[9] (paths or ways), *chen*[10] (larger, paved, ways), *thu*[8] (one-width roads), *tao*[11] (two-width roads), and *lu*[12] (three-width roads).]

If there is alarm in the empire he fortifies the roads and difficult points, halts wanderers, and guards the positions with his men, letting past the barriers only those with the imperial seal.

---

[a] (1), p. 305. He goes on to minimise the civil engineering works of China, regarding the accounts of them as the exaggerations of 'weak and wondering travellers' or 'stupid and lying missionaries', but his reaction from Chinoiserie led him into error. We shall quote later on from some of the missionaries, who were no more than truthful (pp. 22, 33, 135, 142, 205, 208, 211, 363, 379).

[b] Mao, no. 203, tr. auct. adjuv. Legge (8), vol. 2, p. 353; Waley (1), p. 318; Karlgren (14), p. 154. The poem was quoted by Mencius (*Mêng Tzu*, v (2), vii, 8, tr. Legge (3), p. 267). Another folk-song also refers to the roads of Chou, but whether as 'winding and tedious' or 'flat and even' is not clear from the archaic language used; cf. Mao, no. 162, tr. Legge (8), vol. 2, p. 247; Waley (1), p. 151; Karlgren (14), p. 105.   [c] Cf. Yetts (17).

[d] Here the numbers in brackets refer to the Map in Fig. 711 and its accompanying tables, where the characters for all the place-names will also be found.

[e] Ch. 7, p. 26a (ch. 30), tr. auct. adjuv. Biot (1), vol. 2, pp. 198 ff.

¹ 關中   ² 司險   ³ 溝   ⁴ 遂   ⁵ 洫   ⁶ 澮   ⁷ 川
⁸ 涂   ⁹ 徑   ¹⁰ 畛   ¹¹ 道   ¹² 路

The systematisation of the capacities of roads and canals, doubtless largely schematic, appears in the passages devoted to the Sui Jen[1] (Grand Extensioner, or Minister of Agriculture).[a]

This is how he organises the countryside. Between each farm there is a ditch (*sui*[2]) with a path (*ching*[3]) along it. Past every ten farms there runs a conduit (*kou*[4]) with a way (*chen*[5]) alongside. Past every hundred farms there runs a small canal (*hsüeh*[6]) with a one-width road (*thu*[7]) accompanying it. Past every thousand farms there runs a medium-sized canal (*kuei*[8]) with a two-width road (*tao*[9]) along its bank. Past every ten thousand farms there runs a large canal (*chhuan*[10]) with a three-width road (*lu*[11]) at its side. Such are the communications in the imperial domains.

[Comm. The five grades of roads are all to connect the country and the capital for carriages and pedestrians. (Apart from men) paths will take only horses and oxen, the wider (paved) ways will take large hand-carts, a one-width road will take a single chariot, a two-width road will take two abreast, and a three-width road will take three abreast. One may make the country roads the same width as the ring-roads of cities.]

Now we know what was meant by a 'two-width road'.[b] But another text in the same book has more spacious ideas. Under the heading of Chiang Jen[12] (Master-Builders) we find[c] that in the capital the main streets (*ching thu*[13]) are to carry nine chariots abreast, the ring-roads (*huan thu*[14]) are to carry seven, and the country roads (presumably imperial highways, *yeh thu*[15]) are to carry five (Fig. 712). Furthermore, capitals of feudal princes are to have their main streets of the seven-width grade, their ring-roads five-width, and their approach roads three-width. Other cities must not exceed the five-width grade for their broadest streets, with all their other roads at the three-width level. Perhaps there is no discrepancy if the grandeur of the Chou (or Han) capital is at issue only in this second text.

During the Warring States period there was much road-building activity both for military and commercial purposes but the details are still unclear. The State of Chhin, however, as we shall see, had been particularly busy, and the works achieved may well have been a great factor in its success. As soon as the whole empire was for the first time united under Chhin Shih Huang Ti in −221, he embarked upon his celebrated policy of metrological standardisation and fixed, among other things, the gauge of chariot-wheels.[d] In −220 he made a tour of inspection in Kansu and Shensi, after which

[a] Ch. 4, pp. 24b, 25a (ch. 15), tr. auct. adjuv. Biot (1), vol. 1, p. 341. Note the decimal progression, with regard to Vol. 3, pp. 82 ff., 89.

[b] An interesting Indian parallel of a few centuries later occurs in the *Arthaśāstra* (Shamasastry tr. p. 53). Roads to military stations are to be 48 ft. wide, royal chariot roads in the countryside 24 ft., elephant forest roads 12 ft., ordinary chariot roads 7½ ft., cattle ways 6 ft. and paths for men 3 ft. Inca data from Peru range from 75 ft. (processional) through 45 ft. and 24 ft. ('regulation') to 15 ft. (quite frequent). Widths of 12 ft. and 6 ft. only occur in the communications of the subsidiary cultures (von Hagen, 3).

[c] Ch. 12, pp. 17b, 18a, 20a (ch. 43), tr. Biot (1), vol. 2, pp. 564 ff.

[d] *Shih Chi*, ch. 6, pp. 12a, 13b, tr. Chavannes (1), vol. 2, pp. 130, 135. The text says that the Chhin double-pace (*pu*) was fixed at 6 ft., and that the chariot gauge was uniformised throughout the empire. The gauge has always been taken as one double-pace, and indeed the *Chou Li* (*Khao Kung Chi*) says that 'the distance between the wheels is 6 ft.' (ch. 12, p. 24a, tr. Biot (1), vol. 2, p. 580). But what this means

[1] 遂人　　[2] 遂　　[3] 徑　　[4] 溝　　[5] 畛　　[6] 洫
[7] 涂　　[8] 澮　　[9] 道　　[10] 川　　[11] 路　　[12] 匠人
[13] 經涂　　[14] 環涂　　[15] 野涂

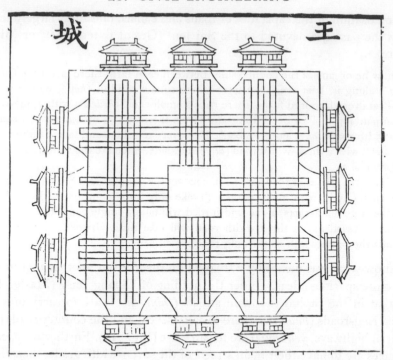

Fig. 712. Diagram of an idealised imperial or princely city, with its thoroughfares, showing the traditional decumane-cardinal plan. From *San Li Thu*, ch. 4, p. 2*b* (cf. pp. 73, 80ff.).

he ordered the construction of a vast set of arterial post-roads, 'speed-ways' (*chhih tao*[1]) or 'straight-ways' (*chih tao*[2]) radiating from the capital at Chhang-an, near modern Sian, especially to the north, north-east, east and south-east.[a]

in modern terms is a little uncertain. If Wu Chhêng-Lo's figure for the Chhin and Chhien Han foot is taken (27·65 cm.) it would be equivalent to 5·44 ft.; if the foot of Chhin State, followed by Hsin and Hou Han, is taken (23·1 cm.), it would be 4·54 ft. The *Chou Li* (*Khao Kung Chi*) says elsewhere that 'the widths of roads are measured with the chariot gauge as the unit' (ch. 12, p. 17*a*, tr. Biot (1), vol. 2, p. 562), and later commentators say, assuming a Chou date for the *Chou Li*, that this was 8 ft. But while the Chou double-pace had 8 ft., not 6 ft., the Chou foot (19·9 cm.) was so short that it gives a gauge very little different from the former of our two figures, 5·22 ft.

We can gain more assurance from archaeological excavations of vehicles (cf. Vol. 4, pt. 2, pp. 77 ff., 246 ff.). Chariot gauges in the Shang are as wide as 7·07 ft., in the early Chou they average about 6·55 ft., in late Chou 5·71 ft., and in the Warring States period 5·41 ft., ranging down to 4·59 ft. There was thus a continuing reduction in gauge. Recent excavations of the city gates of the Han capital at Chhang-an show four traffic lanes within a width of 19·7 ft., i.e. 4·92 ft. for each one. There is thus fair agreement between textual and archaeological evidence; leading us incidentally to the interesting conclusion that Chhin and Han chariot gauges were quite close to the standard railway gauge of modern times, 4·71 ft. (4 ft. 8½ in.). It follows further that the three-width roads of the *Chou Li* would have been of the order of 15 ft. broad, and the nine-width roads 45 to 50 ft. As for Western parallels, there seems to have been an opposite tendency, gauge gradually widening from the 3·77 ft. of early Roman vehicles to the characteristic Romano-British rut distances which vary from 4·50 to 4·83 ft. (Lee, 1)—again approximating to 'standard' gauge. I am much indebted to Dr Lo Jung-Pang for discussion in correspondence helping to establish this note. Cf. Vol. 2, pp. 210, 214, 553, Vol. 4, pt. 2, pp. 250, 253.

[a] *Shih Chi*, ch. 6, p. 14*b*, cf. Chavannes (1), vol. 2, p. 139 and *TH*, p. 211. It must be remembered that the capital was very much in the west of the united empire. Communications to the south and south-west presented great natural difficulties, to which we shall return (pp. 15, 19 below).

[1] 馳道   [2] 直道

Though contemporary descriptions are not available, it is worth giving one from only a few years afterwards. About −178 Chia Shan,[1] one of the emperor Wên Ti's counsellors, presented him with a hortatory essay, entitled *Chih Yen*[2] (Words to the Point), in which he analysed the causes of good government and civil confusion, particularly criticising Chhin Shih Huang Ti. After decrying the luxury of the palaces built at Hsien-yang, he continued:[a]

He also ordered the building of the post-roads all over the Empire, east to the uttermost bounds of Chhi and Yen, south to the extremities of Wu and Chhu, around the lakes and rivers, and along the coasts of the sea; so that all was made accessible. These highways were 50 paces wide, and a tree was planted every 30 ft. along them. The road was made very thick and firm at the edge, and tamped with metal rammers (*chin chhui*[3]). The planting of the green pine-trees[b] was what gave beauty to the roads. Yet all this was done (only) so that (Chhin Shih Huang Ti's) successors (on the throne) should not have to take circuitous routes.

Later commentators were a little puzzled by the statement about the structure of the roads (*hou chu chhi wai*[4]), some thinking that they were lined by walls on each side, like raised causeways (*yung tao*[5]),[c] others that the tamping referred simply to the consolidation of the edges, especially where there was an embankment.[d] That little trace of these roads remained in later ages presumably implies that they were less massively built than the Roman roads.[e] Yet if they consisted chiefly of rubble and gravel tamped down in the manner of pisé walls (see on, p. 38), they were more elastic and therefore much more modern in conception. Such 'water-bound macadam' was in fact the traditional material of Chinese highways in all periods.[f]

As for the width, it is generally agreed that the '50 paces' of the *Chhien Han Shu* was a scribal error for 50 ft.[g] so that the imperial highways would have been approximately nine-width roads equivalent to the broadest described in the *Chou Li*. They were thus rather larger than most of the Roman roads.[h] The inner lanes of these nine-lane thoroughfares were apparently reserved for the equipages of the emperor himself and authorised members of the ruling house; messengers, officials and merchants using the outer ones.[i]

[a] *Chhien Han Shu*, ch. 51, p. 2*a*, tr. auct. adjuv. Lo Jung-Pang (6). See also Chhü Shou-Yo (1).
[b] Cf. Chhen Jung (1), pp. 21, 25.
[c] Fu Chhien (+2nd century). [d] Yen Shih-Ku (+7th century).
[e] Interesting remains have however been recorded. The Chhin imperial highway uncovered near Ling-ling (18) in Hunan was as wide and flat as a dry river-bed (*Hu-nan Thung Chih*, ch. 33, p. 9*b*).
[f] Cf. Tan Pei-Ying (1). Here everything depends upon the choice of rock for the chips and the cementing material.
[g] I.e. 32·7 of our feet if the Chou foot is taken, or 45·8 and 37·8 ft. if the Chhien Han and Hou Han standards are taken respectively. Archaeological finds indicate a breadth around 40 ft. The Chinese pace was always a double pace; 6 ft. until the end of the Sui, 5 ft. from the beginning of the Thang onwards.
[h] Estimates for these vary; Leger (1) considered that they were rarely wider than 25 ft., but there is some evidence for occasional broader stretches.
[i] An incident of −47, told in *Chhien Han Shu*, ch. 10, p. 1*b* (cf. Dubs (2), vol. 2, p. 374), confirms something of this description, and other demonstrative passages may be found, but it is rather difficult to believe, with some, that the central imperial lanes were so sacred that the common people could never cross them except by over-bridges. Such an arrangement all over the empire would have been far too expensive. We suspect that the taboo petered out some distance away from the capital. But daring to travel on the central lane was a high misdemeanour, and in one case a marquis was executed for doing so.

[1] 賈山   [2] 至言   [3] 金椎   [4] 厚築其外   [5] 甬道

Let us look at the course of the imperial highways shown on the map (Fig. 711).[a] A more easterly centre, San-chhuan (6) in the neighbourhood of modern Loyang, was chosen as the hub of the system, the road from Chhang-an (1) negotiating the Han-ku Pass (12) much as the Lung-hai railway does today,[b] and traversing the smaller centres of Hua-yin (57) and Hung-nung (58)[c] in the Yellow River valley. Thence the Eastern road went straight to Lin-tzu (8) in Shantung, capital of the former State of Chhi, following the Chi[1] River and passing places not now exactly identifiable. Branching from San-chhuan, the North-eastern road went up diagonally through Hopei to Chi (7), near modern Peking, capital of the former State of Yen, probably passing through Han-tan (62) and Chung-shan (63), well-known cities of the Warring States period. This road followed for a long way the old course of the Yellow River (cf. pp. 240 ff. below) after crossing it not far from Loyang, and water transport probably proved useful for the conveyance of the road metal. The longest road was that which struck off south-east towards the mouth of the Yangtze. Skirting first the northern edge of the vast valley of the Huai[2] River, it came to Chhen-liu (9) and then to Phei (10) after which it turned south, crossed the Yangtze in the vicinity of modern Nanking [54][d] and made its way past the Chiangsu lakes to Kuei-chi (11) at or near Suchow, the capital of the former State of Wu. Nearly as long as the South-eastern road was the Southern one. This did not go through San-chhuan but over the Hsiung-erh Shan[3] by the Wu Pass (13) direct from Chhang-an to Wan (14),[e] after which it went south-wards, crossing the Han[4] River near Hsiang-yang (71) and reaching Nan-chün (15),[f] i.e. Ying, the capital of the former State of Chhu. Chhin Shih Huang Ti's opening of the south did not stop there, however, for the road crossed the Yangtze somewhere near the debouchment of the Tung-thing Lake[5] after following the river's windings, and so came to Chhang-sha (16). It then proceeded up the Hsiang[6] River valley past Hêng-yang (17) to its terminus at Ling-ling (18). Although it was now going eastwards again this was no mistake, for as we shall see later on (pp. 299 ff.) the upper waters of the Hsiang were made to connect in Chhin times by a remarkable canal with the upper waters of the West River (Hsi Chiang[7]) of Kuangtung, thus permitting the transport of arms and supplies for the conquest of the Cantonese State of Nan Yüeh.[g] Such was the way in which the order was carried out to link the ancient countries of Chhi and Yen, Wu and Chhu, with the capital of the Chhin empire.

It remains to speak of the Great North Road, the only one for which we have any details concerning its construction. In −212 Mêng Thien,[8] one of the First Emperor's

---

[a] For further explanations the reader may like to turn to the geographical introduction, with its maps, in Vol. 1, pp. 55 ff. In what follows, numbers in round brackets refer to the place-names given in the accompanying table, while those in square brackets refer to Table 4 and the map in Fig. 35 of Vol. 1.

[b] Cf. Anon. (57).   [c] Cf. Vol. 2, p. 367.

[d] Ferries are of course implied; the Yangtze was not bridged anywhere below the gorges until our own time.

[e] This was the famous metallurgical centre of the Warring States and Han periods; cf. Vol. 5, Sect. 30d.

[f] Later Chiang-ling; cf. Vol. 2, p. 367.   [g] Cf. p. 441 below.

[1] 濟   [2] 淮河   [3] 熊耳山   [4] 漢水   [5] 洞庭湖   [6] 湘江
[7] 西江   [8] 蒙恬

Table 60. *Place-names for the map of road communications in ancient
China* (Fig. 711) *and for the map of civil engineering works* (Fig. 859)

Notes: (1) Roads west of An-hsi graded off for the most part into caravan tracks without sharp distinction.

(2) During the course of Chinese history a single place often bore as many as half-a-dozen different names. In the table the most ancient name comes first, generally speaking, the modern name last.

(3) The locations of the older and newer cities of different dynastic periods are often not quite identical; they may be a few miles apart. Here for convenience they are treated as synonymous.

(4) Numbers in square brackets are a cross-reference to Table 4 and the map in Fig. 35 of Vol. 1.

1 Chhang-an [9] 長安	12 Han-ku Kuan 函谷關 (Pass)
= Hsi-an (Sian) 西安	13 Wu Kuan 武關 (Pass)
Cf. Vol. 1, pp. 58, 103, 124, 181	14 Nan-yang 南陽
2 Hsien-yang 咸陽	= Wan 宛
= Wei-chhêng 渭城	Cf. Vol. 5, Sects. 30, 36
Cf. Vol. 1, p. 100	15 Nan 南
Vol. 4, pt. 2, p. 130	= Nan-chün 南郡
3 Kan-chhüan Shan 甘泉山	= Ying 郢
= Kan-chhüan Kung 甘泉宮	(former capital of Chhu State)
(the temple founded by Shao Ong)	= Chiang-ling 江陵
= Shun-hua 淳化	= Lin-chiang 臨江
= Yün-yang 雲陽	= Ching-chou 荊州
Cf. Vol. 1, p. 108	Cf. Vol. 2, pp. 191, 197, 198
Vol. 4, pt. 1, pp. 122, 315	16 Chhang-sha [56] 長沙
Vol. 5, Sect. 33	17 Hêng-yang 衡陽
4 Kan-chhüan 甘泉	= Hêng-shan [53] 衡山
5 Chiu-yuan 九原	18 Ling-ling 零陵
= Wu-yuan 五原	19 Kuei-lin [61] 桂林
6 San-chhuan 三川	20 Hsiang 象
= Lo-yang [8] 洛陽	= Tshang-wu 蒼梧
7 Yen 燕 (former capital of Yen State)	= Wu-chou 梧州
= Chi 薊 (near modern I-hsien 易縣)	21 Wu-kung 武功
= Pei-ching (Peking) [50] 北京	22 Fu-fêng 扶風
Cf. below, pp. 75 ff.	Cf. Vol. 4, pt. 2, p. 39
Vol. 1, pp. 139 ff.	23 Old Pao-chi 寶雞
8 Lin-tzu 臨菑	= Chhen-tshang 陳倉
= Chhi 齊	24 Thien-shui 天水
(former capital of Chhi State)	25 Lung-hsi 隴西
9 Chhen-liu 陳留	26 Ting-hsi 定西
= Ta-liang 大梁	27 Chin-chhêng 金城
= Pien-ching 汴京	= Kao-lan 皋蘭
= Khai-fêng 開封	= Lan-chou (Lanchow) [7] 蘭州
10 Phei 沛	28 Yung-têng 永登
= Phei-hsien 沛縣	29 Wu-shao Ling 烏鞘嶺 (Pass)
(north of Hsüchow 徐州)	30 Wu-wei 武威
11 Wu 吳 (former capital of Wu State)	= Liang-chou (Liangchow) 涼州
= Kuei-chi 會稽	= Sera Metropolis (by mistake for
= Su-chou (Suchow) 蘇州	Chhang-an)

Table 60 (*continued*)

31 Yung-chhang 永昌 (in Kansu)
    Cf. Vol. 1, p. 237
32 Shan-tan 山丹
    Cf. Vol. 4, pt. 2, p. 402
33 Chang-yeh 張掖
    = Kan-chou (Kanchow) 甘州
34 Kao-thai 高臺
35 Chiu-chhüan 酒泉
    = Su-chou (Suchow) 肅州
36 Chia-yü Kuan 嘉峪關
    (Western Gate of the Great Wall)
    Cf. Vol. 1, Fig. 14
37 Yü-mên 玉門
38 An-hsi 安西
    = Kua-chou 瓜州
39 Tun-huang [45] 敦煌
    = Sha-chou 沙州
40 Yü-mên Kuan 玉門關 (Jade Gate)
41 I-wu 伊吾
    = Hami 哈密
    = Qomul
42 Lou-lan 樓蘭
43 Yü-ni 嶺匿
    = Shan-shan 鄯善
    = Erh-chhiang 婼羌
    = Issedon Serica
    = Charkliq
44 Chhieh-mo 且末
    = Cherchen
45 Kao-chhang 高昌
    = Shan-shan 鄯善
    = Karakhoja (Qarākhoja)
    = Turfan
46 Chiao-ho 交河
    = Piala
    = Yarkhoto
47 Yen-chhi 焉耆
    = Karashahr (Qarāshahr)
48 Wei-li 尉犁
    = Kalgaman
    = Kurla
49 Pei-ti 北地
    = Ning-hsien 寧縣
50 An-ting 安定
    = Phing-liang 平涼

51 Yung 雍
    (former capital of Chhin State)
    = Fêng-hsiang 鳳翔
52 Hsiao Kuan 蕭關 (Hsiao Pass)
53 Hui-chung Kung 回中宮
    = Ku-yuan 固原
54 Li-yang 攊 (or 櫟) 陽
55 Shang-chün 上郡
56 Yü-lin [11] 榆林
57 Hua-yin 華陰
    near Thung-kuan [5] 潼關
58 Hung-nung 弘 (or 宏) 農
    = Kuo-lüeh 虢略
    Cf. Vol. 1, p. 94
       Vol. 2, p. 367
59 Ho-tung 河東
60 Chin-yang 晉陽
    = Thai-yuan 太原
61 Tai 代
    = Tai-chün 代郡
62 Han-tan 邯鄲
63 Chung-shan 中山
    Cf. Vol. 5, Sect. 30
64 Lang-ya 琅邪
    = Lang-yeh 琅琊
    (the kuan-thai 觀臺, observation
    terrace)
    Cf. Chavannes (1), vol. 2, p. 144
65 Lu-chiang 廬江
66 Chiu-chiang 九江
67 Chhing-chiang 清江
68 Gan(-hsien) 贛 (縣)
69 Chhü-chiang (Kukong) 曲江
    = Shao-kuan 韶關
    = Shao-chou 韶州
70 Nan-hai 南海
    = Kuang-chou (Canton) [28] 廣州
71 Hsiang-yang 襄陽
    Cf. Vol. 5, Sects. 30, 34
72 Mien-hsien 沔縣
73 Pao-chhêng 褒城
74 Han-chung [18] 漢中
    = Nan-chêng 南鄭
75 Fêng-hsien 鳳縣
    = Shuang-shih-phu 雙石舖

Table 60 (*continued*)

76 Liu-pa 留壩	99 Chü-yen 居延
77 Mei-hsien 郿縣	= Edsin (or Etsin) Gol
78 Chou-chih 盩厔	100 Yung-chhang 永昌 (in Yunnan)
79 Ning-chhiang 寧羌 (or 強)	= Pao-shan 保山
80 Chao-hua 昭化	101 Shan-hai Kuan 山海關
81 Chien-mên Kuan 劍門關 (Sword-gate Pass)	(Eastern Gate of the Great Wall)
82 Mien-yang 綿陽	102 Yeh-lang 夜郎
83 Shu [59] 蜀	= Thung-tzu 桐梓
(former capital of Shu State)	103 Lang-chou 郎州
= Chhêng-tu 成都	= Tsun-i 遵義
84 Pa [58] 巴	104 I-chou 益州
(former capital of Pa State)	= Chhêng-chiang 澂江
= Chung-chhing (Chungking) 重慶	105 Chü-mi 且彌
85 Pa-yü Kuan 巴峪關 (Pass)	= Chen-hsi 鎮西
86 Phing-chhêng 平城	= Pa-li-khun 巴理坤
= Ta-thung (Tatung) 大同	= Barköl
87 Fei-hu Khou 飛狐口	106 I-chih 移支
(Flying-fox Pass)	= Ti-hua 廸化
88 Tzu-kuei 秭歸	= Urumchi
89 Pho-tao 僰道	107 Jung-yang 榮陽
= I-pin 宜賓	= Chêng-chou (Chêngchow) 鄭州
90 Tien 滇	108 Ku-pei-khou 古北口
(capital of tributary State of Tien)	109 Nan-khou 南口
= Tien-chhih 滇池	= Chü-yung Kuan 居庸關 (Gate)
= Khun-ming (Kunming) [25] 昆明	110 Chang-chia-khou 張家口
91 Yeh-yü 葉 (or 楪) 榆	= Wan-chhüan 萬全
= Ta-li [62] 大理	= Kalgan
92 Tsang-kho 牂 (or 牱) 牁 (or 牱 or 柯)	111 Tzu-ching Kuan 紫荊關 (Pass)
93 Yüeh-sui (越嶲 or 巂)	112 Phing-hsing Kuan 平型關 (Pass)
= Chhiung-tu 邛都	113 Yen-mên 雁門
94 Ho-phu 合浦	– Yu-yü 右玉
= Lei-chou 雷州	114 Ning-hsia 寧夏
= Hai-khang 海康	= Yin-chhuan 銀川
Cf. p. 669 below	115 Hsi-ning 西寧
95 Chiao-chou 交州	116 Yang Kuan 陽關 (Gate)
= Hanoi	117 Kao-chhüeh 高闕
= Kattigara	118 Shen-yang 瀋陽
Cf. Vol. 1, pp. 178, 183	= Mukden
96 Han-kuang 洽光 (or 洸)	119 Lin-thao 臨洮
= Han-khuang 合洭	= Min-chou 岷州
97 Ying-tê (Yingtak) 英德	120 Su-chou (Suchow) 宿州
= Chên-yang 湞陽	121 Shê-hsien 葉縣
98 Old Kuei-yang 桂陽	122 Shou-chou 壽州
= Yuan-ling 沅陵	123 Ta-ming 大名
	124 Lin-chhing 臨清

Table 60 (*continued*)

Place-name Index

Table 60 (*continued*)

Pien-ching, 9	Ta-thung, 86	Wu-kung, 21
	Tai, 61	Wu-shao-ling, 29
Qarākhoja, 45	Tai-chün, 61	Wu-wei, 30
Qarāshahr, 47	Thai-yuan, 60	Wu-yuan, 5
	Thien-shui, 24	
San-chhuan, 6	Thung-kuan, 57	Yang-kuan, 116
Sera Metropolis, 30, 1	Thung-tzu, 102	Yarkhoto, 46
Sha-chou, 39	Ti-hua, 106	Yeh-lang, 102
Shan-hai Kuan, 101	Tien, 90	Yeh-yü, 91
Shan-shan, 43, 45	Tien-chhih, 90	Yen, 7
Shan-tan, 32	Ting-hsi, 26	Yen-chhi, 47
Shang-chün, 55	Tsang-kho, 92	Yen-mên, 113
Shao-chou, 69	Tshang-wu, 20	Yin-chhuan, 114
Shao-kuan, 69	Tsun-i, 103	Ying, 15
Shen-yang, 118	Tun-huang, 39	Ying-tê, 97
Shê-hsien, 121	Turfan, 45	Yingtak, 97
Shou-hsien, 122	Tzu-ching-kuan, 111	Yü-lin, 56
Shu, 83	Tzu-kuei, 88	Yü-mên, 37
Shuang-shih-phu, 75		Yü-mên Kuan, 40
Shun-hua, 3	Urumchi, 106	Yü-ni, 43
Sian, 1		Yu-yü, 113
Su-chou (Chiangsu), 11	Wan, 14	Yuan-ling, 98
Su-chou (Kansu), 35	Wan-chhüan, 110	Yüeh-sui, 93
Su-chou (Anluri), 120	Wei-chhêng, 2	Yün-yang, 3
	Wei-li, 48	Yung, 51
Ta-li, 91	Wu, 11	Yung-chhang (Kansu), 31
Ta-liang, 9	Wu-chou, 20	Yung-chhang (Yunnan), 100
Ta-ming, 123	Wu Kuan, 13	Yung-têng, 28

Table 61. *Highway names on the map of road communications in ancient China* (Fig. 711)

A	Old Silk Road		
B	Lien-yün Tao	連雲道	(Linked-Cloud Road)
C	Pao-yeh Tao	褒斜道	(Pao-Yeh Road)
	= Pei-chan Lu	北棧路	(Northern Trestle and Gallery Road)
	= The New Road		
D	Thang-lo Tao	儻駱道	(Lo Valley Road)
	= Lo-ku Tao	駱谷道	
E	Tzu-wu Tao	子午道	(North–South Road)
F	Chhen-tshang Tao	陳倉道	(Chhen's Granary Road)
	= The Old Road		
G	Chin-niu Tao	金牛道	(Road of the Golden Oxen)
	= Shih-niu Tao	石牛道	(Road of the Stone Oxen)
H	Mi-tshang Tao	米倉道	(Rice Granary Road)
J	Fei-hu Tao	飛狐道	(Flying-Fox Road)
K	Wu-chhih Tao	五尺道	(Five-Foot Way)
L	Thien-Shan Pei Lu	天山北路	(Road north of the Thien-Shan)
M	Thien-Shan Nan Lu	天山南路	(Road south of the Thien-Shan)
N	Nan-Shan Pei Lu	南山北路	(Road north of the Nan-Shan)
P	Liu-chung Lu	柳中路	(Road through the Willows)
R	Ling-shan Tao	靈山道	(Magic Mountains Road)

most important generals, whose name will always be connected with the Great Wall (see on, p. 52), was ordered to build a road from Hsien-yang (2), the imperial capital just across the Wei[1] River from Chhang-an, up through the Kan-chhüan districts (3, 4) and out through the Wall, striding across the Ordos Desert to the northernmost line of the Yellow River. Here, somewhere not far west of the modern steel city of Pao-thou,[2] there was a fortified outpost called Chiu-yuan (5), probably a vantage-point for observing the affairs of the Huns and other nomadic peoples,[a] and doubtless also a trading centre for their products.[b] The texts expressly say that this road was carried in a straight line through the mountains by cuttings and across the valleys by embankments (*chhien shan yin ku chih thung chih*[3]).[c]

It is interesting to assess the total length of these imperial highways which fanned out from the metropolitan region of Kuan-chung in the −3rd century. The figures are approximately as shown in Table 62.[d]

[a] In this context the question has been asked (not at all an irrelevant one for transport problems) at what time the snowshoe and the ski were known in Chinese culture. From time to time popular journals reproduce drawings from Chinese and Korean encyclopaedias which show hunters on skis (*mu ma*,[4] 'wooden horses'); ascribing to them of course the usual legendary datings (e.g. Mathys, 1). One of these was kindly brought to our attention by my colleague Dr F. P. Bowden in 1955. The texts, first studied by Laufer (39), take a little digging out, but they indicate conclusively (cf. Needham, 49) that the knowledge the Chinese had of skis and snowshoes came from their contact with the northern Turkic peoples early in the +7th century. No earlier references have been found. In +629 the Pa-yeh-ku[5] (Bayirku, a branch of the Thieh-lê[6] (Tölös) tribe, cf. Vol. 3, p. 612, Vol. 4, pt. 1, p. 49 and Beer, Needham *et al.*), who 'riding on pieces of wood, hunt deer over the ice', first brought tribute (*Thung Tien*, ch. 199 (p. 1081.1), *Hsin Thang Shu*, ch. 217B, p. 7b; *Thang Hui Yao*, ch. 98 (p. 1754), etc.). By +640 the Liu-kuei[7] tribe of Turkic nomads north of Lake Baikal sent representatives to court; 'as their country is so soon covered with frost and snow they use wooden boards 6 in. wide and 7 ft. long with which to glide over the piled-up ice and go hunting the deer and other animals' (*Thung Tien*, ch. 200 (p. 1084.2), *Hsin Thang Shu*, ch. 220, p. 12b). The Pa-hsi-mi[8] tribe, known from +649, also used these *mu ma*[4] for hunting (*Thung Tien*, ch. 200 (p. 1084.2), *Hsin Thang Shu*, ch. 217B, p. 9a). About the same time there were the Chieh-ku[9] (or Chieh-ka-ssu,[10] i.e. Kirghiz) people beyond the Uighurs, 'who ride on wooden horses (*mu ma*[4]) over (snowy) hill and dale like flying' (*Thung Tien*, ch. 200 (pp. 1084.3, 1085.1), *Hsin Thang Shu*, ch. 217B, p. 10b, *Thang Hui Yao*, ch. 100 (p. 1784)). To their east one came to three tribes of 'Wooden-Horse Turks' (Mu-ma Thu-chüeh[11]). These people 'commonly go very fast over the ice by binding boards of bent wood on their feet, and then treading with a propulsive motion of the base of the spine they suddenly shoot forward a hundred paces with great rapidity' (*Hsin Thang Shu*, ch. 217B, p. 11b). Finally the same technique is recorded for the Shih-wei,[12] a branch of the Chhi-tan people (*Thung Tien*, ch. 200 (p. 1083.3), *Thang Hui Yao*, ch. 96 (p. 1721)).

The Chinese evidence thus supports the accepted archaeological view that the snowshoe originated in neolithic N.E. Siberia as early as the −3rd millennium, and spread westwards to Scandinavia, south to Korea and Japan, and eastwards to N. America, giving rise to skis only in the Old World however; cf. Davidson (1, 2), Dresbeck (1). A Korean picture of wooden snowshoes, likely predecessors of skis, is given in Buschan *et al.* (1), vol. 2, pt. 1, p. 656, fig. 1. The oldest reference for Japan appears to be +912, when the troops of Fujiwara no Toshihito were equipped with snowshoes. But the climatic environment of traditional Chinese culture never induced the adoption of either skis or snowshoes.

Accurate European knowledge of skis (see Luther, 1) came some four centuries later than in China. But Luther urges, following +18th-century suggestions, that ancient peoples using skis and snowshoes were the material basis of the legends of hoofed men (hippopeds and bovipeds), both in East and West. Such beings certainly abound in the *Shan Hai Ching* (chs. 2, 3, 18, etc.) which, like ancient Western texts, often places them in Hyperborean regions. Luther's euhemeristic explanation is not unplausible, but there are others also (cf. Vol. 3, pp. 504 ff.).   [b] See Yü Ying-Shih (1).

[c] *Shih Chi*, ch. 6, p. 24a (cf. Chavannes (1), vol. 2, p. 174); ch. 88, p. 2a (cf. Bodde (15), p. 55). See also *TH*, p. 218.   [d] Based on Lo Jung-Pang (6).

[1] 渭河   [2] 包頭   [3] 塹山堙谷直通之   [4] 木馬   [5] 拔野古
[6] 鐵勒   [7] 流鬼   [8] 拔悉彌   [9] 結骨   [10] 黠戛斯
[11] 木馬突厥   [12] 室韋

Table 62. *Lengths of Chhin imperial highways*

	Chhin and Chhien Han *li*
Hsien-yang (Chhang-an) to Lo-yang	950
Lo-yang north-eastwards to Chi (Yen)	2,000
Lo-yang eastwards to Lin-tzu (Chhi)	1,800
Lo-yang south-eastwards to Kuei-chi (Wu)	3,800
Hsien-yang southwards to Ling-ling (Chhu)	3,300
Hsien-yang northwards to Chiu-yuan	1,910
Total	13,760

Taking the Chhin and early Han *li* as equivalent to 0·309 mile, this amounts to some 4,250 miles—not very unlike the distance of 3,740 miles which Gibbon estimated as the length of first-rate Roman road running continuously from the Antonine Wall in Scotland to Jerusalem.[a] Further comparisons will arise as we go on.

There is evidence that the imperial highways system was only part of the road-building activity of the Chhin State and dynasty. The city of Yung (51), its former capital, was certainly connected with Hsien-yang and Chhang-an by a road running north of the Wei River through Fu-fêng (22) and Wu-kung (21) on higher ground than the present line of the Lung-hai railway. By −209, when the dynasty fell, some kind of road was in existence connecting Hêng-yang (17) with Nan-hai (70, Canton) via Chhü-chiang (69, Kukong) and Ying-tê (97, Yingtak) just as the Yüeh-han railway does today. This must have penetrated the gorges of the Nan Ling[1] [35] mountains. In later centuries it became a very important north–south route between Canton and Peking, though for long the bulk of the traffic went round by the canal near Ling-ling. But the really heroic test of the ancient road-builders was the challenge of the Chhin-ling Shan[2] [3], that great range of mountains shutting off Chhang-an and the Wei valley from the south-west. This we shall discuss in a moment, since the works of Chhin and Han cannot easily be separated; here it is only needful to recall that the State of Chhin made conquest of the Szechuanese States of Shu and Pa as early as −316, and the work of its unknown engineers must have been a primary factor in the achievement.[b] By the end of the century the Chhin rulers were pushing on their colonisation across the barbarian uplands of Yunnan. Ssuma Chhien says, laconically:[c] 'Under the Chhin, Chhang An[3] planned and constructed the Five-Foot Way (Wu-chhih Tao[4]).' Starting across the Yangtze from Pa (84, modern Chungking), already connected by road with Shu (83), it ran down through the hills of Kweichow via Yeh-lang (102) and Lang-chou (103) to reach the Kunming Lakes[d] and the Erh-hai[5] Lake

[a] Gregory (1), p. 59, citing the *Decline and Fall*, vol. 1, p. 81.

[b] The date of the first military road through the Chhin-ling Shan parallels curiously the construction of the Appian Way in −312.

[c] Ch. 116, p. 2*a*.

[d] Tien Chhih[6] or Khun-ming Hu,[7] and Fu-hsien Hu.[8]

[1] 南嶺 [2] 秦嶺山 [3] 常頞 [4] 五尺道 [5] 洱海 [6] 滇池

[7] 昆明湖 [8] 撫仙湖

at Tien (90), I-chou (104) and Yeh-yü (91) respectively. This was a traverse very similar to that of the present-day road between Kunming and Chungking, continuous with the Burma Road, which took so much traffic during the Second World War. The commentators add that the Five-Foot Way (K) got its name[a] from the abundance of stretches where it could not be more than five feet wide, since it often skirted precipices and included many hanging galleries (*chan tao*[1]), i.e. wooden balconies jutting out from perpendicular cliffs and carrying the road along.[b] Here for the first time we encounter a method of construction about which there will soon be much to say.

If the Chhin capital had had five major imperial roads radiating from it, the same metropolis under the Former Han emperors, rulers of a still more closely knit subcontinental nation, had as many as seven. We may consider them best not by boxing the compass, but in ascending order of magnitude and engineering interest. To the north-west there was a further opening-up of the Kansu and Shensi hill-country north of the Wei valley, with new roads to An-ting (50) and Hui-chung Kung (53), the latter completed in −108. To the south-east the distance to Chhang-sha and points south was shortened by a road direct from the capital to Hsiang-yang (71) passing south of the Liu-ling[2] and Hsin-khai Ling[3] mountains. Northward the capital was newly connected with Chiu-yuan (5, now Wu-yuan) by a road through Shang-chün (55) which, though more roundabout, probably traversed easier and more inhabited country than the Great North Road built by Mêng Thien. More important than any of these was the use of the long Fên[4] River valley in Shansi to throw a road north-eastwards through Chin-yang (60, modern Thai-yuan) to Yen (7, Peking), considerably shorter than the former one, though still necessitating a Yangtze crossing near Thung-kuan (57). From Yen communications were pushed through to Korea about −129, on the occasion of the conquest of that country by Phêng Wu,[5] the General Wade of the Chhien Han,[c] whose road must have passed through the Wall in the neighbourhood of Shan-hai Kuan (101).

In the eastern region the Han road-builders were particularly active, laying down a network of highways all over the North China Plain. Han-tan (62) was connected eastwards to Lintzu (8), south-eastwards to Phei (10), and southwards to Chhen-liu (9), whence a road continued to Nan-yang (14) joined by one from Lo-yang (6). Another new road struck off south from Phei to Chiu-chiang (66), after which it descended to Chhing-chiang (67) and linked west with Chhang-sha (16) just as the Nan-hsün railway does today. Phei was also connected with the coast at Lang-ya (64), a place

---

[a] Road names, with their key letters, will be found assembled in Table 61.
[b] From the amounts of timber recorded as having been used in this (and other ancient Chinese roads), it must have been in part a 'corduroy road'. Three-inch diameter saplings, laid down like sleepers and tightly wired together, make an excellent short-life road which will float in liquid mud, but of course it is expensive in timber. Here is another factor in ancient China's deforestation (cf. Vol. 5, Sect. 30d). Corduroy roads were a prominent feature of medieval Russian technology (cf. Mongait (1), pp. 299, 300, 306, fig. 1b).
[c] *Shih Chi*, ch. 30, p. 3b (cf. Chavannes (1), vol. 3, p. 549; Watson (1), vol. 2, p. 82); *Chhien Han Shu*, ch. 24B, p. 6a (cf. Swann (1), p. 243). The name of the road-building general may have been Phêng Wu-Chia[6] (the ancient texts vary).

[1] 棧道   [2] 流嶺   [3] 新開嶺   [4] 汾河   [5] 彭吳   [6] 彭吳賈

famous from the time of Chhin Shih Huang Ti. Further extensions were made by the Hou Han engineers, as we shall see, but in ancient times no penetration was made of the Fukien amphitheatre, where the country was too difficult and the people still barbarous.

Out towards the far west the case was very different. The desert oases formed less barrier to intercourse than the forested mountains, and the moment has come to examine the Chinese end of the Old Silk Road (A), an incurably romantic subject. Here we may not repeat what has already been said about east–west communications at an earlier stage,[a] but we must see how the Chinese highway network joined on to the caravan trails around the Taklamakan Desert in Sinkiang (Chinese Turkestan). Leaving behind the complex of ancient roads in the Wei valley at Pao-chi (23) the road to the West passes through the gorges of the upper waters by Thien-shui (24)[b] and then lifts itself up over the Hua-chia Ling [1] passes to descend to Chin-chhêng (27, modern Lanchow) on the Yellow River beyond the great bends, for two millennia the hub of routes in this western region. Then over the Wu-shao Ling Pass (29) to Wu-wei (30), anciently known, by mistake for Chhang-an, as Sera Metropolis. After Wu-wei comes Yung-chhang (31), site of the settlement about −30 of captured Roman legionaries,[c] and then after the crossing of many streams and alluvial fans coming down from the lofty Nan Shan [2] [13] (Chhi-lien Shan [3]) [14] on the left to be lost in the Gobi Desert on the right, Chiu-chhüan (35, modern Suchow) is reached. All this while, the road has been protected upon its right-hand side by the extension of the Great Wall also constructed in the Han period, and at the same time a road was made north-eastwards to the lost city of Chü-yen (99) by the Edsin Gol, certainly another listening-post for tribal intentions. Now the road leaves the guardianship of the Wall and strikes off across wild desert country to An-hsi (38, anciently Kua-chou, the Melon City), a fortified post marking the first fork of the Old Silk Road. All this 'Kansu Corridor' was settled after the victory over the Huns in −121.

Thenceforward we have to think of the circumnavigation of the Taklamakan Desert in the Tarim Basin by two main routes. Southerly, there was the Road North of the Southern Mountains (Nan-Shan Pei Lu [4]), i.e. the enormous Tibetan Khun-lun Shan [5] [10], passing first (N) through Tun-huang (39, anciently Sha-chou, the City of the Sands)[d] and then forking again, one way keeping to the foot-hills to reach Kashgar through Yü-ni (43, Charkliq) and Chhieh-mo (44, Cherchen), the other passing by way

---

[a] Vol. 1, pp. 181 ff., and Fig. 32. I cannot forbear from saying once again that no part of China is more interesting and beautiful than her 'Wild West', where Chinese culture meets that of the nomadic peoples, and where every place-name in the sandy wastes, with the snowy ghosts of the Nan Shan or the Thien Shan shining in the distance, bears witness to the heroic deeds of the Chinese soldiers and administrators of ancient times, manning the outposts of a truly great civilisation.

[b] Just to the south is Mai-chi Shan,[6] the site of later Buddhist cave-temples.

[c] Cf. Vol. 1, p. 237.

[d] Just to the south is Chhien-fo-tung [7] (Mo-kao-khu [8]), the most famous site of later Buddhist cave-temples, so often referred to in this book.

| [1] 華家嶺 | [2] 南山 | [3] 祁連山 | [4] 南山北路 | [5] 崑崙山 |
| [6] 麥積山 | [7] 千佛洞 | [8] 莫高窟 | | |

of the lost city of Lou-lan (42) beside the doomed lake Lop Nor.[a] West of Tun-huang and Yü-mên Kuan (40) the routes were caravan trails rather than made roads, but up to those points the roads were fully equipped with forts, beacons and post-stations in −107, during the campaign against the Lou-Lan State.[b] By −101 the line of forts and beacons was continued to Lou-lan itself, and perhaps also to Yü-ni. Between −77 and −59 this was the only loop of the Old Silk Road open to traffic.

North of the Taklamakan Desert there was the Road South of the Celestial Mountains (Thien-Shan Nan Lu[1]), running along their foothills (M) towards Kashgar. It was first reached in −101 through Lou-lan to Wei-li (48, Kurla), but closed by the hostile Lou-lan people, who had moved northward, during the period just mentioned. This was the route which the armies of Kan Yen-Shou[2] and Chhen Thang[3] employed in −36 in their expedition against the Hun Khan (the Shanyu) in Sogdiana. The Thien-Shan Nan Lu was reached in quite another way towards the end of the Former Han period by a road (P) which forked north-westwards at An-hsi (38) and struck boldly across the Gobi to I-wu (41, north-east of modern Hami, Qomul).[c] It then passed west to Kao-chhang (45, Turfan), rejoining the older Lou-lan route at Wei-li (48). This road, later known as the Liu-chung Lu[4] (Road through the Willows), was not complete until about +5, and not in real use till +73, when the I-wu region was occupied by the Chinese.[d] After this, it superseded altogether the old roads through Lou-Lan, where Lop Nor was drying up, and the city itself soon to be buried in the sands until the coming of modern archaeologists. Still later an even more northerly route made its appearance, the Road North of the Celestial Mountains (Thien-Shan Pei Lu[5]), which penetrated their eastern end (L) by taking off from Hami (41) and running to the lake at Chü-mi (105, Barköl), thence westwards to Ti-hua (106, Urumchi) and so to Kazakhstan. Eventually, especially in times of general imperial peace, the bulk of the trade and migration moved this way, down to the days indeed when war supplies for China rolled from the Turk-Sib railway along the southern edge of the Tsungarian plain. But there is ground for thinking that this northern route was the ancient way in which rhubarb travelled to Rome, passing through the hands of tribes north of the Caspian Sea,[e] so it reasonably demands mention here in the context of the peripheral social blood-vessels of the empires of the Han.

[a] Stein (10) has given a graphic account of his explorations of ancient roads in this region. On the Han Great Wall which protected Tunhuang, see p. 49 below.

[b] All this region had been prospected in the travels of Chang Chhien, already described in Vol. 1, p. 173, between −138 and −126. Dr Lu Gwei-Djen and I had the pleasure of studying the Han beacons and remains of forts on the Suchow–Tunhuang road once again in 1958. Cf. Fig. 721.

[c] These names are evocative. Certain inscriptions in this region are unforgettable. The central drum-tower at Chiu-chhüan (Suchow) carries the words 'The sound that both Chinese and Barbarians hear' (*Shêng chen Hua I*[6]). On the north arch: 'Straight to the desert' (*Pei thung sha-mo*[7]). On the south arch: 'Face to face with the Chhi-lien Shan' (*Nan wang Chhi-lien*[8]). On the eastern arch: 'Straight to the holy mountains of China' (*Tung thung hua yo*[9]). And on the western arch: 'Towards the outpost of I-wu' (*Hsi chien I-wu*[10]). So I saw it first in 1943 with Mr R. Alley, Mr Sun Kuang-Chün, Mr Wang Wan-Shêng and other friends.   [d] The initial conquest had actually been in −58.

[e] Cf. Hudson (1), p. 94; Laufer (1), p. 548; Yule (2), vol. 1, pp. 290, 292, vol. 2, p. 247. Cf. also Vol. 1, p. 183 and Fig. 32.

[1] 天山南路   [2] 甘延壽   [3] 陳湯   [4] 柳申路   [5] 天山北路
[6] 聲震華夷   [7] 北通沙漠   [8] 南望祈連   [9] 東通華嶽   [10] 西建伊吾

That lines of forts were established along the great roads in the neighbourhood of the northern and north-western *limes* is most certain, and the same system was probably used in Yunnan and other parts of the wild south-west. But besides these there were also fortified causeways (*yung tao*[1]) running for relatively short distances in and around Kuan-chung, where the defence of a line of communication seemed particularly important; perhaps they were something like the Long Walls of the Piraeus. One such work connected the Ao Granary, a vital strongpoint, with the Yellow River quays near Jung-yang (107).[a]

The greatest engineering work of the Chhien Han road-builders, however, was the consolidation of the passes through the towering Chhin-ling Shan [3], those routes which had been pioneered by the Chhin people as early as the −4th century.[b] To explain this we must return upon our steps and take up the story again from that date. The problem was to find routes across the great mountain divide between Shensi (Kuan-chung) in the north, where the Chhin capital was located, and the Szechuanese basin, where the half-barbarian States of Shu and Pa, occupying rich agricultural land in a key economic area,[c] were inviting annexation by the aggressive Chhin State. A deep intervening valley, that of the Han[2] River, permitted the surveyors to take their breath before finding their way down or across the head waters of the Chialing[3] River and into Szechuan. First then there was the range which brings Thai-Pai Shan[4] with its snows in sight from every vantage-point on the northern slopes of the Wei valley, and afterwards there were the less grim ranges of Mi-tshang Shan[5] and Ta-pa Shan.[6] Even today, the modern motor-road which connects Szechuan with Shensi passes through remarkable scenery, but in ancient times it must have been even wilder and more difficult country. This region saw the most spectacular work of the engineers of Chhin and Han.

The focal way-stations were three small cities on the upper reaches of the Han River, Mien-hsien (72), Pao-chhêng (73) and Han-chung (74). All the roads which went through them were in existence when the First Emperor ascended the throne in −221, but they needed, and received, great improvement under the Han. The oldest road (F) started from Chhen-tshang (old Pao-chi, 23),[d] crossed the Wei River, ran up a small tributary through the Ta-san[7] Pass and came to the little town of Fêng-hsien (75).[e] Then it turned westwards to the upper waters of the Chialing before doubling back into the

---

[a] Cf. p. 270. Another such fortified road did good service in +139, when the Chhiang tribes invaded the region around Fu-fêng (22).

[b] Lin Chao (1); Wiens (1, 2); Hisamura Yukari (1) and Lo Jung-Pang (6) have made detailed studies of this road-system. A debt, here gratefully acknowledged, is owing to Dr Lo for giving us the privilege of using his monograph before publication.

[c] Cf. Vol. 1, pp. 114 ff. and *passim*.

[d] Hence its name, the Chhen-tshang Tao[8] (Chhen's Granary Road), but it was also known as the Old Road (Ku Tao[9]).

[e] When this Section was first drafted (1953) the Chhêngtu–Paochi railway was in full construction along a closely similar route, to join at Paochi with the Lung-hai extension past Lanchow over the Old Silk Road. In 1958 I much enjoyed a journey along the latter line. Cf. Lan Tien (1); Chang Ching-Chih (1); Wang Yu-Chi (1); Anon. (54, 55, 56).

---

[1] 甬道 [2] 漢水 [3] 嘉陵江 [4] 太白山 [5] 米倉山
[6] 大巴山 [7] 大散關 [8] 陳倉道 [9] 故道

2-2

mountains to find its way to Mien-hsien. The first Chhin improvement (B) was to shorten the western detour by striking through the mountains directly from Fêng-hsien by way of Liu-pa (76) to Pao-chhêng; this was known as the Lien-yün Tao[1] (Linked-Cloud Road). The passes on this route did not exceed 7,000 ft., but no less than one third of its 430 *li* was composed of wooden trestles built over the beds of roaring torrents, or actually shelved into the cliff higher up by means of timber baulk brackets driven into holes in the rock face.[a] An occidental parallel for this *chan tao*[2] system[b] was the mountain road-building of Tiberius (r. +14 to +37) and Trajan (r. +98 to +117),[c] but it was rather limited in comparison, and both these emperors were, as we should look at it, Hou Han people. The southern portion of the Chhen-tshang Road now fell out of general use. It is interesting that still to this day the modern motor road follows approximately the route of the Lien-yün Tao,[d] which has been the principal line of communication through the mountains since the Sung dynasty.[e]

The second Chhin improvement (C) chose a still more easterly traverse. Starting from the town of Mei-hsien (77) on a road running south of the Wei River from Chhang-an, not north, it ran up the steep valley of the Yeh[3] tributary through the Yeh-yü[4] Pass and came out on the high mountain country surrounding Thai-Pai Shan to the west. Here it found the southward-pointing valley of the Pao[5] or Thai-pai[6] River, where a place still bears the name of Thung-chhê-pa[7] (Passing-place for Carts), and dropped down to Pao-chhêng from near Liu-pa. It thus short-circuited the northern portion of the Lien-yün Road just as that had cut out the southern portion of the Chhen-tshang Road.[f] First constructed with innumerable trestle and gallery sections in the −4th century, it was extensively enlarged and repaired about −260, in −120 and +66.[g] Geography gave it the name of the Pao-yeh Tao,[8] but because of its engineering works it was also known as the Pei-chan Lu[9] (the Northern Trestle Road). Early in the +3rd century Chuko Liang described the massive pillars and beams which upheld the road

---

[a] A photograph of such a 'shelf-road' under construction in the Po-Yigrong gorge on the Tibetan border forms the frontispiece of Kingdon Ward (13).

[b] The word *chan* originally meant brackets for a bed (Kelling (1), p. 100). A garbled account of the Chinese 'cientao' was given by Sir Wm. Chambers (2) in +1773.

[c] Cf. Merckel (1), p. 252. A *chan tao* is to be seen in the left background of Pieter Breughel the Elder's painting of the Conversion of St Paul (+1567), now in the Kunst-historisches Museum at Vienna.

[d] But the railway runs more to the west, adopting the Chialing R. valley.

[e] The Mongolian armies used it in attacking Szechuan during the conquest of the Sung. Its upkeep in Chhing times is described in Wiens (2), p. 146.

[f] Traffic had to use the Chhen-tshang Road again, however, for some time after −206, when the Pao-yeh Road trestles were destroyed in military operations by Chang Liang at the order of Liu Pang; cf. Vol. 1, p. 102. The story is told in *Shih Chi*, ch. 55, p. 5b (cf. Watson (1), Vol. 1, p. 139); *Chhien Han Shu*, ch. 40, p. 4b and elsewhere.

[g] It was still kept up in the Thang period (+8th to +10th centuries) when it formed the most important north–south line of communications. Under the Northern Wei in +507 a road had been built between Paochi (23) and a point south of Mei (77) on the Pao-Yeh Tao but it never became important. It was called the Stonegate Gallery (Shih-mên-ko Tao[10]).

---

[1] 連雲道　　　[2] 棧道　　　[3] 斜　　　[4] 斜峪關　　　[5] 褒水
[6] 太白河　　　[7] 同車壩　　　[8] 褒斜道　　　[9] 北棧路　　　[10] 石門閣道

through the ravines which it traversed. About what happened in −120 we know a good deal. Ssuma Chhien says:[a]

> After this, someone presented a memorial to the emperor proposing that the Pao-yeh Road should be adapted for the transport of grain by boat. The matter was referred to the Censor-in-chief Chang Thang,[1] who enquired into the matter and reported as follows: 'In order to reach Shu (Szechuan) (traffic) goes through the Marches of the Old Road (Ku-tao[2])[b] where there are many rocky descents and long detours; if now we pierce a (better) road between the Pao and the Yeh the gradients will be much less difficult and the distance will be shorter by 400 *li*. Since the Pao river flows into the Mien river,[c] and since the Yeh descends into the Wei, we should be able to use them for grain transport by boat. The grain would come up from Nan-yang along the Han, the Mien and the Pao; then from the point where the Pao becomes too shallow the grain will be transported in carts to the headwaters of the Yeh about 100 or more *li*, after which other boats will take it down the Wei (to the capital). Thus the grain of Han-chung will become available, and that from east of the mountains can come in unlimited quantities, much more easily than it does past Ti-chu[3].[d] Lastly, the abundance of wood and bamboo, both small and large, in the Pao and Yeh valleys, useful for construction, is comparable to that in Pa and Shu itself.' The emperor approved the project.
>
> Chang Thang's son, Chang Ang,[4] was (accordingly) appointed Administrator of Han-chung, and he recruited several tens of thousands of men who re-constructed the Pao-yeh Road over a total distance of more than 500 *li*. The road was in effect convenient and shorter (than the others), but the rivers proved to be too violent and too much encumbered with rocks to allow of the grain transport by boats which had been envisaged.[e]

This passage is of much interest as it illustrates the close coordination between land and water transport which characterised Chinese planning for more than two millennia.

Between the periods of rule of Tiberius and Trajan, the Pao-yeh Road, incorporating the southern portion of the older Lien-yün Road, was repaired again after some three centuries of continuous use, and what that involved may be estimated by the figures which have come down to us—766,800 man-days of work, carried out by 2,690 convict workers.[f] Much of the information about these roads in Han times comes from still existing stele inscriptions, which Lao Kan (2) discusses.[g] Dating, for example, from +57 and +63, they mention surveyors[h] as well as *corvée* labourers. Besides the miles of trestles and galleries, there were 623 small bridge-spans and five large ones,

---

[a] *Shih Chi*, ch. 29, p. 5*a*, *b*, tr. auct. adjuv. Chavannes (1), vol. 3, pp. 529 ff.; Watson (1), vol. 2, pp. 74 ff.; parallel passage in *Chhien Han Shu*, ch. 29, pp. 4*b*, 5*a*.

[b] Another name for the region traversed by the northern portion of the old Chhen-tshang Road, hence the appellation.

[c] Another name for the upper waters of the Han near Mien-hsien.

[d] A rocky barrier and rapids, harmful to traffic, in the middle of the Yellow River, at the southern border of Shansi. Also called San-mên Hsia[5] (the Three-Gate Gorge), it is the site of the magnificent dam now under construction; cf. Têng Tsê-Hui (1); Li Fu-Tu (1). See pp. 274 ff. below.

[e] The scheme was abandoned by −116, but the consolidation of the road was very much worth while. Cf. Lo Jung-Pang (6), pp. 52, 56.

[f] Wang Chhang (1), ch. 5, pp. 3*b* ff.; *Pao-chhêng Hsien Chih*, ch. 8, p. 2; Yang Lien-Shêng (11), p. 5.

[g] Cf. Huang Shêng-Chang (1).

[h] Many of these were apparently 'government slaves', cf. Vol. 4, pt. 2, pp. 35 ff.

[1] 張湯   [2] 故道   [3] 砥柱   [4] 張卬   [5] 三門陝

while 64 post-station rest-houses were provided along a stretch of 258 *li* (about 86 miles). The names of the original designers of the Chhen-tshang, Lien-yün and Pao-yeh road systems are lost in the mists of the past, but we know that Fan Sui [1] was in charge of the repairs about −260, and the steles commemorate the names of Wang Hung,[2] Hsün Mao,[3] Chang Yü[4] and Han Tshen,[5] all civil engineers of the Hou Han. They deserved well of their nation.

Besides the roads already described, there were two more easterly ways over the Chhin-ling Shan. One of these (D) started from Chou-chih (78), a town nearer to Chhang-an on the road south of the Wei River than Mei-hsien. Following up the Liu-yeh[6] River, a tributary, it came out in high mountain country east, not west, of Thai-Pai Shan, and found its way to Han-chung over passes between 7,000 and 9,000 ft. high, through the Lo[7] Valley. Hence arose its name, the Thang-lo Tao.[8] In later ages it was little used, except perhaps in the Thang period.

A still more easterly route was taken by the Tzu-wu Tao[9] (North–South Road), which went straight up into the hills south of the capital,[a] and passing over a high plateau made for the valley of the Tzu-wu[10] River, whence Han-chung was reached on either bank of the Han River. This route (E) was pioneered in the time of Wang Mang (+5), but it gave little relief from the difficulties of the others, for its highest passes were as severe as those of the Thang-lo Road and its lowest stretches passed through swampy ground. Moreover it covered troublesome terrain for the whole of its 660 *li* (c. 220 miles). It therefore fell into general disuse after +126, though it had proved helpful after +107, when the Pao-yeh trestle bridges and galleries had been destroyed in a revolt of the Chhiang tribes. And as late as the Thang (+8th century) it was a regular courier route; indeed the Post Office maps of modern times still mark it so.

Not only in the Chhin-ling Shan but all over China, *chan tao* trestle-work and galleries[b] were a prime feature of road engineering through the centuries,[c] so mountainous is the country. In the +17th century Lecomte wrote:[d]

The Civil Government of the Chinese does not only preside over the Towns, but extends also over the Highways, which they make handsome and easily passable....

The Road from Si-ngan-fou (Sian) to Hamtchoum (Hanchung) is one of the strangest pieces of work in the world. They say, for I myself have never yet seen it, that upon the side of some Mountains which are perpendicular and have no shelving they have fixed large beams into them, upon the which beams they have made a sort of Balcony without rails, which reaches thro' several Mountains in that fashion; those who are not used to these sort of

---

[a] The village of Tzu-wu-chen[11] still bears witness to its alignment.

[b] Illustrations are not easily available; Gregory (1), fig. 13, gives a very poor example. We offer in Fig. 713 (pl.) the excellent one from Anon. (26), fig. 283. Galleries in this traditional style give access to the cave-temples in the perpendicular cliff of Mai-chi Shan. Cf. Vol. 2, Fig. 40.

[c] In subsequent ages they were constantly under repair, attended to by men such as Chia San-Tê[12] in the +6th century, and Kao Ti[13] or Chia Han-Fu[14] in the +17th.

[d] (1), p. 304. Cf. Martini, *Atlas Sinensis*, p. 48 (+1655).

[1] 范睢	[2] 王弘	[3] 荀茂	[4] 張宇	[5] 韓岑
[6] 留業水	[7] 駱谷	[8] 儻駱道	[9] 子午道	[10] 子午河
[11] 子午鎮	[12] 賈三德	[13] 高第	[14] 賈漢復	

Galeries, travel over them in a great deal of pain, afraid of some ill accident or other. But the People of the place are very hazardous; they have Mules used to these sort of Roads, which travel with as little fear or concern over these steep and hideous precipices as they would do in the best or plainest Heath.

And a thousand years earlier, the great poet Li Pai, who passed over the Chhinling Mountain roads more than once, wrote a famous poem on 'The Road to Szechuan'.

> . . .To the West, starting from the Great White Mountain,[a] it was said
> There was a bird-track that cut across to the mountains to Szechuan;
> But the earth of the hills crumbled[b] and heroes perished.[c]
> So afterwards they made sky-ladders and hanging-bridges—
> Above, high beacons of rock that turn back the chariot of the sun,
> Below, whirling eddies that meet the clashing torrent and turn it away;
> . . .It would be easier to climb to Heaven than walk the Szechuan Road.[d]

The construction of cliff-galleries was also naturally adopted in Tibet.[e] We hear of 'suspended roads' (*hsüan lu*[1]) in the travels of the monk Hui-Chhao to and from India around +726 on his pious missions.[f] The Chinese also used half-tunnelling, i.e. excavation of half the cross-section of the road into the cliff-face with a rock overhang (*khung tao*[2] or *hsüeh ching*[3]), just as the Romans did in the time of Tiberius and Trajan. Sometimes the rock will support an almost full cross-section; astonishing examples of great length are seen in the paths of the trackers or haulers of boats through some of the gorges of the Yangtze (Fig. 880, pl.). A number of these galleries exist in and around the San-mên Gorge, where the rocks bear many ancient inscriptions; all this has been the subject of a valuable recent monograph.[g] But the modern motor-roads have also made much use of the technique, as can be seen, for example, on the Szechuan–Shensi highway in the Chhinling Mountains just north of Kuang-yuan[4] near Chao-hua (80).[h] It is probable that the balcony roads of the Chinese were sometimes suspended from chains since, as we shall see, the invention of the iron-chain suspension bridge occurred so early among them (cf. p. 193 below); if so, it would have anticipated the use of the device in Europe, which is attested[i] for the first opening of the Gotthard Pass through the Alps about +1236. This was surely well within the powers of the Taoist patron saint of hanging-gallery builders, Lu Phi Kung[5].[j]

It would be hard to over-estimate the importance of the Chhinling passes in

a Thai-Pai Shan.

b Anyone who has travelled much in the mountainous western regions of China will have made personal acquaintance with landslides (*khai shan*[6]) and what they can do. Cf. p. 33.

c This refers to five envoys sent by an ancient king of Shu to fetch five daughters of the king of Chhin; they are supposed to have perished on the return journey.

d Tr. Waley (13), pp. 38 ff.; another translation in Alley (3), p. 48.

e Mason (1), p. 576; Edgar (2).   f Cf. Fuchs (4).

g Anon. (33); we shall return to it in connection with the rock-cut canal at this place; pp. 277 ff. below.

h I knew this place well during the second world war. There are Buddhist cave-shrines in the cliffs here, recently described by Shih Yen (1).

i Cf. Straub (1), p. 54; Imberdis (1).

j *Lieh Hsien Chuan*, ch. 48, tr. Kaltenmark (2), p. 151.

1 懸路   2 孔道   3 穴徑   4 廣元   5 鹿皮公   6 開山

Chinese history.[a] Twice at least did these transmontane roads become the paths of flight of emperors seeking the sanctuary of mountain-battlemented Szechuan. But much more important was the assimilation of Pa and Shu by Chhin in the −4th century, since the vast natural resources of Szechuan were placed at the disposal of the rulers of the Wei Valley, and this, together with the great irrigation projects which they also completed, must have been a cardinal factor in the first political unification of all China. And later the Old Silk Road was fed by the Linked-Cloud Trestle Ways, for much of the silk exported to Rome was Szechuanese in origin.

We are now in a position to take a look at the far-winding roads in the Land of the Four Rivers and the Land South of the Clouds to which the reduction of the Chhin-ling was, as it were, the gate. Restored by a pause in the upper Han or Mien valley, the principal route to Shu (Chhêng-tu, 83) directed itself south-westwards from Chao-hua (80), having quickly reached the head waters of the Chia-ling[1] River by the Yang-phing[2] Pass or taken a slightly easier route through the Chhi-phan[3] Pass amidst magnificent scenery from the town of Ning-chhiang (79). After negotiating the impressive Chien-mên (Sword-Gate) Pass (81) the way was clear along the edge of the great basin to the capital. This road (G) was known from Chhin times onwards as the Chin-niu[4] or Shih-niu Tao[5] (Road of the Golden, or the Stone, Oxen).[b]

It had an alternative, probably rather later in date, which instead of leaving from Mien-hsien (72) took off more directly southward from Han-chung (74) over the Mi-tshang Shan[6] range (hence its name the Mi-tshang Tao, or Rice-Granary Road), and crossed the Chialing River much lower down to enter the Szechuan basin from the east rather than the north. This (H) was much less used, as it involved daunting defiles such as the Pa-yü[7] Pass.

Both these roads had sections built on trestles or galleries like those through the Chhin-ling, so it was said that the trestle-road from Shu to Kuan-chung extended for a thousand *li*.[c]

In the −2nd century much wider horizons began to open. The emperor Han Wu Ti began to meditate the annexation of the Cantonese State of Nan Yüeh and the other smaller States of the south from Fukien to Annam. His ambassador at Canton in −135, Thang Mêng,[8] noticed that the produce of Szechuan found its way down the West River, and argued that in that case naval and military forces could come too. But to do

---

[a] Cf. Liu Sung-Nien's painting, Shu Tao,[9] c. +1190 (Liu Hai-Su (1), vol. 2, pl. 39); Fessler (1), pl. on p. 68.

[b] The motor-road still follows for many miles the course of this road, which can constantly be seen, though the railway north of Mien-yang (82) now takes a trace at higher altitude. The Chin-niu Tao had large flat paving-stones, like so many old roads in Western China, but these were (in 1943) mostly broken and up-ended, making the surface very rough going. Between Tzu-thung[10] and Sword-Gate Pass (81) it was bordered on each side by huge cedars, which the local people said that Chuko Liang himself (or Chang Fei[11]) had planted (cf. p. 20 above).

[c] We know hardly anything about the builders of these roads, but the name of Chang Wên[12] has come down to us as having been occupied with strategic roads south of the Chhin-ling Shan in −202. Cf. Chu Chhi-Chhien & Liang Chhi-Hsiung (5).

[1] 嘉陵江	[2] 陽平關	[3] 七盤關	[4] 金牛道	[5] 石牛道
[6] 米倉山	[7] 巴峪關	[8] 唐蒙	[9] 蜀道	[10] 梓潼
[11] 張飛	[12] 章文			

this, a road would have to be made to the Tsang-kho River (92), a tributary of the northern branch of the Phan Chiang,[1] which falls into the West River.[a] Accordingly in −130 the road was undertaken. The *Chhien Han Shu* says:[b]

> Thang Mêng and (the Szechuanese) Ssuma Hsiang-Ju[2] opened up for the first time (the territory of) the south-western I[3] peoples. Cutting through the mountains (*tso shan*[4]), they constructed a great highway more than a thousand *li* in length, in order to extend (the territory of) Pa and Shu (Szechuan). In this (undertaking) the population of Pa and Shu became wearied. (At the same time also) Phêng Wu[5] opened (communications to the north-east) right through to the Wei-mo[6] and Chao-hsien[7] tribes (Korea) where (soon after) there was established the province of Tshang-hai[8]....[c]

Those who constructed the roads numbered several tens of thousands of (conscripted) men, for whom (supplies for) meals and provisions (for daily food) were carried on the backs of porters for a thousand *li*. Of the supplies sent, on an average, one picul (weight) out of more than ten *chung* (capacity) reached (its destination).

The new southern road was thus intended to transport men and supplies south-eastwards from Pho-tao (89, I-pin) in Szechuan through Yeh-lang (102) in Kweichow to Kuangsi and Kuangtung, but it had the effect of opening up Yunnan also. It was long and hard in the building, and the opposition of local princes at first prevented its use, but less than two decades later, boats of more than one deck, or their parts, were being carried along this road to help the forces fighting the southerners.[d] The second was the road already mentioned (p. 16) which opened up the Korean provinces, among which Lo-lang was later one. The information that only about 2% of the supplies reached their destination is interesting, but may not mean that all the loss was due to corruption *en route*, for the carriers and road guards had themselves to be supplied. Meanwhile the imperial roads of the Chhin dynasty were still being kept up, as we know from several statements in the same chapter of the *Chhien Han Shu*,[e] where repairs and maintenance (*shan tao*[9]) are mentioned. And the organisation of the south-west went steadily on, for a road was pushed up into the tangled mountain country [32, 33] of the upper Yangtze to Yüeh-sui (93), near the long north–south valley road through Hsi-chhang[10] which formed the rear-most lateral line of communications during the second world war. About the same time, too, the Burma road was extended to Yung-chhang (Pao-shan, 100), between the Salween and the Mekong Rivers.

It only remains to sketch a few of the major developments of the Later Han period. In +27 an important strategic road on the northern frontier was built by Tu Mao,[11] a military engineer who also constructed many fortifications, beacon stations and

---

[a] The story is told in *Shih Chi*, ch. 116, pp. 2b ff. (tr. Watson (1), vol. 2, p. 291 ff.) and abridged in Cordier (1), vol. 1, p. 235.

[b] *Chhien Han Shu*, ch. 24B, pp. 6a, b, tr. Swann (1), pp. 242, 246, mod. auct. Parallel passage in *Shih Chi*, ch. 30, p. 3b (tr. Chavannes (1), vol. 3, p. 549; Watson (1), vol. 2, p. 82). Cf. Hervouet (1).

[c] In −128.   [d] See further on p. 441 below.

[e] Ch. 24B, p. 16b, tr. Swann (1), pp. 305, 308.

[1] 盤江	[2] 司馬相如	[3] 夷	[4] 鑿山	[5] 彭吳	[6] 穢貉
[7] 朝鮮	[8] 滄海	[9] 繕道	[10] 西昌	[11] 杜茂	

transport depots. Starting from the old city of Tai (61) on the Shansi road to Peking, it led up through the Fei-hu[1] Pass, from which it gained its name (the Flying-Fox Road), over the uplands north of Wu-thai Shan[2] (so famous later for its mountain abbeys) to reach Phing-chhêng (86) near Ta-thung, a focal point for the guard of the Great Wall.[a] This road (J) was more than 300 *li* in length over difficult country. Other +1st-century Hou Han activities in the north we have already discussed—the opening of new routes in Sinkiang and the easternmost crossing of the Chhin-ling Shan (pp. 18, 22 above).

This was the region also that saw the activities of another engineer of note, Yü Hsü[3] (*fl.* +110 to +136). When stationed at I-chou,[4] somewhere between Lüeh-yang[5] and Chao-hua (80) on the upper Chia-ling River over from the Han Valley, he found that

> the roads carrying traffic (in that district) were very difficult and dangerous. Neither boats nor carts could get through (the defiles), so donkeys and pack-horses were used, which cost five times the value of all that was transported. Therefore Yü Hsü himself led out his officers and clerks on tours of inspection through the river-gorges from Chhü[6] down to Hsia-pien,[7] and he caused the rocks to be fired and the wood to be cut down for several tens of *li*, so as to open the road for boats to pass.[b]

He thus constructed what was probably a trackers' path, able also to take some wheeled traffic, along a river made navigable.[c] A commentator explains:[d]

> To the east of Hsia-pien for more than 30 *li* there is a gorge, where the rocks formed great barriers in the way of the spring freshets. This led to floods in spring and summer, spoiling the harvest and damaging the villages. So Yü Hsü made his men set (great) fires to the rocks and then lead water on to them, so that they split in pieces and could be removed with crowbars. Afterwards there was no more worry about flooding.

This is far from being the only mention of the fire-setting method in Han texts.[e] The resolution of the old highway engineers, ill-equipped as they were, can be guessed from the description[f] of Li Hsi[8] directing the burning and splitting of rocks about +160.

But perhaps the most extensive Hou Han road-building took place in the south. In +35 a road was thrown along the mountains north of the Yangtze from Pa (84, Chungking) to Tzu-kuei (88) in order to by-pass the dangerous Wu Shan[9] gorges [21]. A very much larger network took the form of a cross in Chiangsi, Hunan, Kuangtung

---

[a] *Hou Han Shu*, ch. 52, p. 6*b*, and further in Lao Kan (2).

[b] *Hou Han Shu*, ch. 88, p. 4*b*, tr. auct. Chhü = Hsing-chou,[10] and Hsia-pien = Chhêng-hsien.[11]

[c] A similar work was undertaken in the Lungmên gorge near Loyang in +844 by a monk, Tao-Yü;[12] see Yang Lien-Shêng (11), pp. 15 ff. These rapids were not till then navigable.

[d] Hsiao Chhang, in *Hsü Hou Han Shu*, tr. auct. Cf. Lo Jung-Pang (6), p. 60.

[e] And similar techniques were practised in quite late times. The use of vinegar as well as water for the explosive force of steam is described in engineering instructions of +1739 issued especially for this region and preserved in the *Chao-hua Hsien Chih*[13] (tr. Wiens (2), p. 147). Cf. p. 278 below, and Sandström (1), p. 29.

[f] Chu Chhi-Chhien & Liang Chhi-Hsiung (5).

[1] 飛狐關   [2] 五台山   [3] 虞詡   [4] 益州   [5] 略陽   [6] 沮
[7] 下辯   [8] 李翕   [9] 巫山   [10] 興州   [11] 成縣   [12] 道遇
[13] 昭化縣志

and Kuangsi. The N.W.–S.E. limb, starting from Yuan-ling (old Kuei-yang, 98) which was in good communication with the Tung-thing Lake by water, came down to Ling-ling (18) and so to Han-kuang (96), after which it joined the old road to Nan-hai (70, Canton) at Chên-yang (97, Yingtak). The N.E.–S.W. limb, starting from Chhing-chiang (67) on the Former Han road from Wu, struck south up the Gan River to Gan-hsien (68) and thence to Chhü-chiang (69, Kukong), much as a good motor highway does now, joining there the old road to Nan-hai, but continuing also to Han-kuang. All this system was complete by +31. Subsequent developments were more military in conception. The second limb was continued all the way to Chiao-chou (95), i.e. Indo-Chinese Hanoi, the Kattigara of ancient times, by +83, passing Tshang-wu (20), the water-transport way-station for traffic coming down from the 'Magic Canal'[a] near Kwei-lin (19). A further mixed land-and-water route was opened in +41, when the general Ma Yuan,[1] campaigning against an Annamese rebellion,[b] built a road 1,000 li long from Ho-phu (94)[c] in the Lei-chou[2] peninsula to Chiao-chou. It is note-worthy that this southern road network of the +1st century contrasts with the earlier road-building in Kansu, Szechuan, Yunnan and Kweichow, in that the sacrifices and suffering of the people seem to have been much less than in those regions of wilder nature, and the benefits of the new communications more quickly felt by the inhabi-tants.

We are now in a position to make a very rough assessment of the extent of the road-building in the Chhin and Han periods. The approximate distances measured with a cartographical perambulator on a map such as that shown in Fig. 711 may be listed as shown in Table 63. They amount to just under 65,000 li.

Converted into miles by the appropriate factors[d] this means a total mileage of some 19,500, but with such a method the error of under-estimation of circuitousness must be considerable, and we shall not be wide of the mark if we accept a final estimate of between 20,000 and 25,000 miles for the main roads in existence by the end of the Han period. Estimates of the greatest extent of the Roman roads vary considerably, but the Peutinger Tables and the Antonine Itinerary seem to indicate something of the order of 48,500 miles, with about 2,400 in Britain and some 9,000 in Italy.

It thus becomes tempting to try to compare the road-building of the Chinese and Roman empires in terms of mileage per 1,000 sq. miles of territory. But while we may reasonably take the drainage areas of the Huang Ho, the Yangtze and the Hsi Chiang as the oikoumene which Chinese civilisation was organising, and accept a figure of 1,532,000 sq. miles for these three basins as an estimate of the empire, it is much more difficult to obtain a satisfactory figure for the extent of that of Rome.[e] At its greatest

[a] See p. 299 below.   [b] See p. 442 below.
[c] See p. 669 below.
[d] For Chhin and Chhien Han the li was 0·309 mile, for Hou Han 0·258 mile (cf. Wu Chhêng-Lo, 1). The total neglects possible decay of some earlier stretches while later roads were built.
[e] Little help is forthcoming from standard reference books, and I am most grateful to my colleague Mr Guy Griffith, F.B.A., for his kindness in working out an estimate which we could use in the above computations.

[1] 馬援   [2] 雷州

Table 63. *The extent of road-building in the Chhin and Han periods*

		Chhin or Han *li*
Chhin imperial roads (value from Table 62)		13,760
Chhin non-imperial roads		
Chhin-ling Shan routes	3,100	
Chin-niu Tao and Mi-tshang Tao	2,000	
Shu to Pa	750	
Wu-chhih Tao	2,950	
Hêng-yang to Nan-hai	1,150	9,950
Chhien Han roads		
Old Silk Road (Chhang-an to An-hsi including Edsin Gol)	5,200	
Nan-Shan Pei Lu (to Lou-lan and Yü-ni only)	4,200	
New Chiu-yuan road	2,600	
Short-haul Shensi roads	1,000	
Yen road via Shansi	2,800	
Korea road	2,550	
North China Plain system	7,300	
Hsiang-yang short cut	1,300	
Szechuan–Kweichow road	2,600	
Burma road extension	350	29,900
Hou Han roads		
Tzu-wu Tao	700	
Liu-chung Road (to I-wu only)	1,000	
Fei-hu Tao	400	
Wu Shan by-pass	1,400	
Hunan–Chiangsi–Kuangsi–Kuangtung system, including re-building of the old road to Canton)	7,750	11,250
		64,860

extent, towards the end of the reign of Trajan (+117) it seems to have covered well nigh two million square miles, but this was a maximum not long sustained, and when his successor Hadrian abandoned Armenia and Mesopotamia to the Parthians, the area of Roman rule fell to approximately 1,763,000 sq. miles. It is interesting that Gibbon long ago ventured a figure of 'above 1,600,000' sq. miles for the Hadrianic period.[a] Table 64 shows then that the two road systems are quite comparable.

It may well be, however, that the Roman relative figures are too high, for one must remember that on the outskirts of the empire, and especially throughout North Africa

a (1), vol. 1, p. 44. He was in fact citing—and with evident mistrust—an earlier eighteenth-century writer, but the estimate is very close to one arrived at in modern research; for Beloch (1), p. 507, gives an area equivalent to 1,310,000 sq. miles for the end of Augustus' reign (+14), to which 305,000 sq. miles would have to be added for the provinces permanently annexed thereafter, bringing the area to a Hadrianic level of 1,615,000 sq. miles in all. Mr Griffith believes that his figure may well be some 10% too high; it depends on how much one includes of such regions as the Egyptian and North African desert. Nevertheless we retain it for comparison, since if one is to take the maximum figure for the land-mass incorporated into the Chinese empire one must also take the maximum figures for the Roman dominion; it is equally difficult to be sure just how far the Chinese writ ran in the sub-tropical Yunnanese forests or the moorlands of the Korean border.

and much of the Near East, the Roman paved *viae militares* probably tailed off into caravan tracks like those of Sinkiang and other 'Western regions', which we took care to omit from our estimate of the Chinese highway mileages. Such factors are extremely difficult to estimate, but their operation might bring down the Roman relative figure to some 20 miles per thousand sq. miles or even less.[a] A full study of the problem would involve comparative estimates of population in the two empires. While this would take us too far here, it is at least quite certain that a high proportion of the area of the Roman empire was very thinly populated as compared with the same regions at the present day;[b] and the necessity of linking far-flung administrative and defensive centres across deserted country would naturally have tended to stretch the mileage of the interconnecting roads. To some extent this must have applied in China too. If we knew more about the population in particular provinces we could make a better guess about the solidity of particular stretches of road.

Table 64. *Road-building in the Chinese and Roman empires*

	Area in sq. miles	Road mileage	Road mileage per 1,000 sq. miles
Roman Empire			
Trajanic (before +117)	1,963,000	48,500	24·7
Hadrianic (after +117)	1,763,000	48,500	27·5
Gibbon's estimate	1,600,000	48,500	30·3
Chinese Empire			
Hou Han (*c.* +190)	1,532,000	22,000	14·35

Broadly speaking, however, it may be said that down to the +3rd century the Chinese network attained from 55 to 75 % of the size of the Roman. In this difference geo-political circumstances may have been important. The lesser relative mileage of the Chinese network must surely have had something to do with the greater navigability of the inland rivers and the greater use of artificial water-ways in China, as compared with Europe. One must conversely reflect that the Roman road system was, in a way, an exo-skeleton, for the heart of the empire was a vast (and potentially stormy) body of water, the Mediterranean Sea, and while this certainly facilitated maritime transport the Romans seem to have felt the necessity of laying down roads of great length all round the basin. On the other hand the Chinese highway system radiated out from a particular centre, Chhang-an, over a very large continuous land mass, constituting, as it were, the endo-skeleton of the empire. In any case, we have every reason to admire the planning and the construction of the ancient Chinese roads, the firmness of which, to borrow agreeably from Gibbon's page, has not entirely yielded to the effort of fifteen centuries.

[a] Similarly the Chinese figure would rise to 16·9 if one adopted the estimate of area of the Khang-Hsi atlas (see Vol. 3, p. 585), just under 1·3 million sq. miles (cf. Barrow (1), p. 575). And see p. 35 below.
[b] Though the figures of Beloch are now universally considered too low.

There is space for only a very brief treatment of the subsequent development of the Chinese highway system. Its rapid rise during five formative centuries has now been sketched, and broadly speaking it declined in importance relative to the water-ways during the successive ten. The pattern of government enterprise nevertheless persisted. In the Han, the planning, review and construction of all major roads was the charge of the Grand Secretary, or Minister of Works (Yü-Shih Ta-Fu [1]), but the building of post-stations, forts, rest-houses and bridges was under the control of the Director of the Imperial Architectural and Engineering Department (Chiang-Tso Ta-Chiang [2]). Since the roads, like all other great public works, were manned by conscripted *corvée* labour, assisted by convict or 'State slave' labour,[a] as well as that of military troops, the personnel and man-power problems were at first the responsibility of the Ssu-Li Chiao-Wei,[3] but this office became gradually confined to criminal investigation and police affairs alone, so that the Minister of Works (Ssu-Khung [4]) and the Superintendent-General of Convict Labour (Thu Ssu-Khung [5]) took over. We have glanced already *en passant* at some of the logistics problems involved, but it would be interesting to have some estimates of the costs of construction. To this matter Lo Jung-Pang (6) has addressed himself,[b] with the result that a figure of some £55,000 per mile emerges for the Szechuan–Kweichow road of −130 and no less than £109,000 per mile for the re-construction of the Pao-yeh road through the Chhin-ling mountains about +65. These estimates are curiously similar to some which have been made for the Roman roads. The 36-ft. Via Appia, if built today, would, it is thought, cost some £105,000, and smaller 12-ft. roads about £41,000 per mile.[c]

That a slow and general decline of the highways took place after the Han is almost certain. The last mention of the use of government relay carriages occurs in the year +186, and for official travel they were replaced by the post-horse system, carts being used only for the transport of goods and for the movement of families of the better sort. At the same time the Later Han suffered from a shortage of horses, Lo-yang having but one stable instead of the six (said to have housed ten thousand animals) which the capital of Chhang-an had had two centuries earlier.[d] Wagons drawn by oxen, mules or donkeys[e] came into more widespread use, and on the feeder roads and pathways the South Chinese custom of riding in litters borne on men's shoulders prevailed more widely.[f] Indeed, a connection with the invention of the wheelbarrow in the

[a] Cf. Vol. 4, pt. 2, pp. 23 ff.

[b] With all due reservations, as he himself says, for such estimates necessarily involve a number of assumptions about prices, monetary values, purchasing power, etc.

[c] Rose (1, 2).

[d] Perhaps because of its intimate, though so often hostile, contacts with the Huns and other nomadic peoples, the Former Han Chinese were very rich in horses. No less than 300,000 were maintained in 36 pasturages on the northern borders. Chariots of private persons were numbered in thousands. When the mother of Lou Hu,[6] an eminent physician, died about +4, her funeral was honoured by a train of between two and three thousand carriages.

[e] Neither mules (*lo* [7]) nor asses (*lü* [8]) were native to China, but were soon introduced from Central and Western Asia when the Old Silk Road began to function.

[f] As late as the second world war I myself made many quite long journeys in the western parts of China in this way, and in all the mountainous provinces it must still persist in some measure.

[1] 御史大夫   [2] 將作大匠   [3] 司隸校尉   [4] 司空   [5] 徒司空
[6] 樓護   [7] 騾   [8] 驢

Later Han is surmised, not implausibly, by some.[a] Then there was a general tendency to replace civil by military control as the centuries went by, and though the use of *corvée* and convict labour continued, the highway management as well as the post-station system came under the Ministry of War (Ping Pu[1]) at least from Sui and Thang onwards.

Though the network had decayed, it should not be thought that the highways were despised during the period of disunion of the Nan Pei Chhao (c. +380 to +580). There were notable road-builders in the +5th century, such as An Nan,[2] originally a tribes-man from Liao-tung, who built a number of roads for the Northern Wei dynasty which included cuttings and embankments like those of Mêng Thien of old. So far did this house appreciate its highways that we can find in the settlement of the official hierarchy of +493 the first mention of a Regius Professor of Geographical Communications (Fang I Po-Shih[3]), alongside those for Astronomy and Medicine in the Imperial University.[b] The Sui were also active in road-building. Apart from a number of new highways in the central provinces, Yang Ti constructed about +607 a military road (*yü tao*[4]) some 3,000 *li* long and 100 ft. wide running parallel with the Great Wall, which he also extensively repaired, from Yü-lin (56) to Peking (7).[c] The road network of the Thang was very thoroughly organised,[d] but if we may judge from the number of land-transport post-stations (1,297) each 30 *li* apart, giving a total of some 38,900 *li*, and take for computation the long Thang *li* equivalent to 0·348 mile,[e] the total mileage was not more than 13,550, a distinct reduction from the late Han figure. It is interest-ing to compare a map of the chief Thang routes with that in Fig. 711. Apart from a new route into Yunnan south from Yüeh-sui (93), that province, together with Kweichow and Kuangsi, had almost reverted to nature, but on the other hand the upper and lower Yangtze Valley regions were connected by a road by-passing the gorges completely, and the Chiangsi-Hunan-Kuangtung network was well kept up. We also see, for the first time, a highway running down from Wu (11) in Chekiang to the great ports of the Fukien coast which were to be such famous centres of inter-national trade in the Sung. North of the 32nd parallel there was little essential change— the Old Silk Road, the Chhin-ling passes, the Road of the Golden Oxen, the northern highways to the Wall and to Peking (7), the ways to Nan-yang (14) and to Shantung, all radiating from the eternal city, Chhang-an, remained very much as they had been in the youth of the Chinese empire.

[a] Lo Jung-Pang (6). Details of the invention and its spread have already been given in Vol. 4, pt. 2, pp. 258 ff.

[b] *Wei Shu*, ch. 113, p. 21 b.

[c] For the road, see *TCTC*, ch. 180 (p. 5631); *TCKM*, ch. 36, p. 129 b. For the Wall, see *TCTC*, ch. 180 (p. 5632), ch. 181 (pp. 5638, 5641); *TCKM*, ch. 36, p. 130 a, ch. 37, pp. 2 b, 4 b; Cordier (1), vol. 1, p. 395; *TH*, p. 1278.

[d] Abundant details are given by Lou Tsu-I (1). Cf. Schafer (13), pp. 13 ff., 16 ff., (16), pp. 20 ff.

[e] If we take the short Thang *li* equivalent to 0·274 mile, the result is reduced to 10,650 miles. One is always doubtful which to use; the long was for general purposes, the short for astronomy, medicine and imperial paraphernalia (cf. Beer, Needham *et al.* (1) and Wu Chhêng-Lo's book (1), so often quoted).

[1] 兵部　　　[2] 安難　　　[3] 方驛博士　　　[4] 御道

The remarks of Lecomte and other early Western observers, accepted rather un-willingly by Adam Smith, on the responsibility of the central government for road-building and maintenance as well as all other forms of public communication in feudal-bureaucratic China, were quite correct, and applied from the time of the first unification onwards as we have seen; but in many periods at any rate the government was interested primarily, and sometimes even exclusively, in those roads and water-ways which were significant for tax-grain transportation and the conveyance of official messages. The upkeep of a multitude of local roads and paved pathways devolved, therefore, upon the people themselves, acting in their co-operative capacity under village elders and small-town worthies.[a] In this context religious associations, such as the Taoist Yellow Turbans about +180, later so politically important,[b] or the Buddhist fraternities afterwards, played a significant part. Making good (*thien sai tao lu*[1]) was nothing less than a pious duty (*i chü*[2], *shan chü*[3]).[c] Thus in the course of time, quite apart from the ancient and medieval imperial highways, China's landscape became shot through with millions of miles of well-paved paths, suitable chiefly for pedestrians, porters with carrying-poles,[d] pushers of wheelbarrows,[e] and men carry-ing litters. Rough unpaved cart-tracks predominated only in the eastern plains. Those who, like the author, have followed these paved ways past woods and rice-fields for many a mile cannot think of them without intense nostalgia.[f]

Quite a number of names of the engineers who were responsible for the surveying and making of these ways have come down to us—one might mention Lu Min[4] in the Thang, Li Yü-Chhing[5] in the Sung, and Hu Shao-Chi[6] in the Chhing.[g] One of the most remarkable was a woman Taoist, known to us only as Chia Ku Shan Shou[7] (the Valley-Loving Mountain Immortal), who directed the building of a mountain road in Fukien in +1315. There was a long tradition of such privately initiated roads going back to the Han or even earlier, and their total mileage far outstripped that of the government main roads as the ages passed.

Foreigners were usually much impressed by China's land communication system. The Japanese monk Ennin, who travelled widely in the Thang between +838 and +847, had many things to complain of, but never depreciated the roads, with their milestones, sign-posts, watch-towers, ferries and bridges.[h] Though nineteenth-

---

[a] See in this connection an exposé in Needham (43).

[b] Cf. *TH*, p. 773.

[c] A striking European parallel is provided in the hospitallers called Fratres Pontifices, founded in +1189, and the builders of the great bridge at Avignon (*ERE*, vol. 2, p. 856). Such religious engineers echo the Mohists (cf. Vol. 2, p. 165). Books of Taoist–Buddhist piety like the *Kung Kuo Ko*[8] always emphasised the spiritual value of road- and bridge-building (cf. Yang Lien-Shêng (11), pp. 13 ff.).

[d] Cf. Vol. 4, pt. 1, p. 21.   [e] Cf. Vol. 4, pt. 2, pp. 258 ff.

[f] Cf. Needham (44). We illustrate by pictures of the old roads at Ku-shan near Fuchow in Fukien (Fig. 714, pl.), at Ko-lo-shan near Chungking in Szechuan (Fig. 715, pl.), and at Pa-ta-chhu near Peking (Fig. 716, pl.). Fig. 717 (pl.) shows the road that climbs into the hills from the Yangtze at Li-chuang in southern Szechuan. For a more spectacular example, a portion of the pilgrims' way ascend-ing Thai Shan in Shantung, see Mullikin (1), p. 711.

[g] Their biographies will be found among those collected by Chu Chhi-Chhien *et al.* (1–9).

[h] Cf. Reischauer (2), pp. 175 ff., (3), pp. 142 ff.

[1] 填塞道路   [2] 義舉   [3] 善舉   [4] 路旻   [5] 李虞卿
[6] 胡紹箕   [7] 夾谷山壽   [8] 功過格

century travellers grumbled about the Chinese highways,[a] which fell into a decline under the Chhing while those of Europe steadily improved, they inspired the admiration of Westerners in the +17th century. After the gloomy accounts of merchants and missionaries it is rather surprising to read Lecomte's words of +1696:

One can't imagine what care they (the Chinese) take to make the common Roads convenient for passage. They are fourscore foot broad or very near it; the Soil of them is light, and soon dry when it has left off raining.[b] In some Provinces there are on the right and left hand Causeways for the foot Passengers, which are on both sides supported by long rows of Trees, and ofttimes terrassed with a wall of eight or ten foot high on each side, to keep Passengers out of the fields. Nevertheless these Walls have breaks, where Roads cross one the other, and they all terminate at some great Town.[c]

And he goes on to speak of triumphal arches,[d] milestones, and 'little Towers...on which they set up the Emperors Standard' near lodges for soldiers or country militia— in a word, the post-stations and guard-houses to which we must turn in a moment.[e]

In the twentieth century a vast system of modern roads suitable for motor traffic has been constructed in China.[f] On the Old Silk Road today lorries and land-rovers speed back and forth. The network has spread to Lhasa and far afield in Tibet[g] and Sinkiang, generating a subsidiary net in the formerly untrodden Tsaidam Basin.[h] Such roads, too, have foreshadowed the extensive railway system which today unites the Indo-Chinese with the Korean border, stretching before long from Kuang-chou to Kazakhstan.[i] Only those who have had the opportunity of seeing for themselves what

[a] Generally speaking (cf. Monnier, 1), but there were exceptions. Thus John Barrow, travelling with the Macartney embassy in +1793—not at all an indulgent observer—was well impressed with two mountain roads. He tells us (1), p. 530, that between the upper waters of the Chhien-thang River at Chiang-shan and those of the Hsin-chiang at Yü-shan there was 'a very fine causeway, judiciously led through the defiles of the mountains'. Further south, through the Nan-ling range between Gan-hsien (68) and Nan-hsiung, Barrow admired (p. 543) 'a very well-paved road, carried in a zig-zag manner over the very highest point, where a pass was cut to a considerable depth through a granite rock; a work that had evidently not been accomplished with any moderate degree of labour or expence'. Cf. Macartney himself, in Cranmer-Byng (2), p. 193. One must always remember comparative standards. Smiles (1), vol. 1, pp. 162 ff., on the state of the roads around Manchester about +1730, is well worth reading.

[b] This remark is interesting in connection with our estimate of the light and elastic nature of the Chhin and Han roads (p. 7 above). Even today, however, the prolonged rainy season of the monsoon climate plays havoc with the modern motor roads, as I know from many personal experiences—often not without their compensations. The landslides on the Burma Road, so well described by Tan Pei-Ying (1), are paralleled in all the mountainous provinces; and most of China is mountainous.

[c] (1), p. 302. He goes on to complain, it is true, of the dust-storms of North China. De Navarrete (+1661) was equally enthusiastic (Cummins (1), pp. 182 ff.).

[d] See further, pp. 142 ff. below.

[e] This may be very suitably illustrated by a cut (Fig. 718, pl.), from G. L. Staunton (1), pl. 17. For a corresponding description see Barrow (1), p. 408.

[f] A background can be obtained from the accounts of Alley (1); Anon. (36) and Phan Kuang-Chhiung (1); a more recent brief treatment in Chang Po-Chun (1). A detailed map showing the position during the second world war is the four-sheet 1:2,000,000 U.S. Govt. Asia Transportation Map. The building of the Burma Road at this time has been described in a very readable book by Tan Pei-Ying (1), and the Ledo Road by Anders (1). Topological methods are now being applied for the rationalisation of transport problems on the roads of contemporary China; cf. Hua Lo-Kêng (1).

[g] Cf. Beba (1); Chang Po-Chun (2).   [h] Cf. Ku Lei (1).

[i] An account of China's railways after the second world war is given in Anon. (58), and, apart from special articles already mentioned, Ting Wan-Chêng (1) has written on the new trans-Mongolian railway to Ulan Bator. Railway alignments, as we have seen, frequently follow those of the most ancient roads.

a single trunk road or a single line of railway can mean in terms of enlightenment and benefit to a province of many million people theretofore almost medieval can truly appreciate the work of the highway and railway engineers of modern China.[a] For more than two millennia their fathers before them had served the same civilising mission.

(2) THE POST-STATION SYSTEM

Once the work of the road engineers was completed, it remained to weld it into a great social institution by establishing and deploying a whole army of messengers, coach-men and station-masters along the lines of communication throughout the empire's length and breadth. Such a system arises naturally as soon as developing society attains an imperial level of organisation. We find it, perhaps for the first time, in the great road already mentioned (p. 2) which integrated Achaemenid Persia, for according to good authority this had 111 relay stations and the whole length could be travelled by troops in 90 days.[b] 'Nothing mortal', said Herodotus, 'travels so fast as the Persian messengers.'[c] From their organisation, already in full service at the beginning of the − 5th century, the plan descended to Ptolemaic Egypt and reached its occidental apogee from the time of Augustus (− 31) onwards.[d] The *cursus publicus* of the Roman Empire ran everywhere from furthest Britain to uppermost Egypt and from Finisterre to Basra, with its stables (*mutationes*) every ten miles or so apart, and its rest-houses (*mansiones*) about every thirty miles. Station-masters (*mancipes*) provided horses or carriages (*rhedae*) for the couriers (*tabellarii*) or for people on official business carry-ing warrants (*diplomata*) authorising them to use the public transport. And the Chinese knew of this system of the Romans, for Yü Huan, writing about Roman Syria (Ta-Chhin) *c.* +264, remarked,

They hoist flags, beat drums, use small carriages with white canopies, and have post-offices, mounted couriers, cantonal offices and post-stations just as we do in China...Every ten *li* there is a cantonal office and every thirty *li* there is a post-station.[e]

So the world of Gallienus, seen from afar off, appeared to be reasonably civil and well organised.

The Chinese had indeed possessed a governmental 'neuro-vascular' receptor and effector system, from time immemorial, so to say. Shang records of the − 14th century, comparable with those of Babylonia and Ancient Egypt, speak of systematic reports from frontier regions,[f] a fact which gives colour to the *Shuo Wên*'s definition of the

[a] Mention may here be justly made of the first modern Chinese railway engineer, Chan Thien-Yu[1] (1860–1919), who built the Ching-pao railway (Peking to Pao-thou via Kalgan). Biography by Hsü Chi-Hêng (1).

[b] Herodotus, v, 52 ff.   [c] viii, 98.

[d] It is always with some surprise that one finds it necessary to regard the Roman Empire as a rela-tively late, a Hou Han, institution. The best monograph on the Roman post-station system is probably still that of Hudemann (1), but more briefly see Friedländer (1).

[e] *Wei Lüeh*, cit. *San Kuo Chih*, ch. 30, pp. 32*b*, 33*a*; tr. auct. adjuv. Hirth (1), p. 70.

[f] Cf. Kuo Mo-Jo (2).

[1] 詹天佑

term *yu*[1] as 'a station on the border for the transmission of despatches'.[a] In the fragmentation period of Chou feudalism each potentate had his own system of tracks, but in the heyday of the Chou High Kings there had been full centralisation; as we may see from the saying of Confucius, reported by his greatest exponent, that 'the radiation of virtue is faster than the transmission of (imperial) orders by stages and couriers (*chih yu*[2])'.[b] This remark would have been made, it is curious to note, at a time exactly contemporary with the functioning of the Persian Royal Road, *c.* −495. By the time of the compilation of the *Chou Li* in the −2nd century the system is taking the very definite shape which it conserved for two millennia. Referring to the Almoners-General (I Jen[3]), the text says:[c]

> In principle, along all the roads of the Empire and the (feudal) States there is a rest-house (*lu*[4]) every ten *li* where food and drink may be had. Every thirty *li* there is an overnight rest-house (*su*[5]) with lodgings (*lu shih*[6]) and a (government) grain-store. Every fifty *li* there is a market (*shih*[7]) and a station (*hou kuan*[8]) with an abundant stock of supplies.

> [Comm. The *lu* was like our *yeh hou thu*[9] with stables, and the *su* was like our *thing*.[10] The *hou kuan* included a watch-tower. Between every two *shih* there were three *lu* and one *su*.]

Thus the Han commentator introduces the terminology which proved through the centuries most persistent. Broadly speaking the main roads were equipped from Han to Sung times with a post-office (*yu*[1]) every five *li*, a cantonal office (*thing*[10]) every ten *li*, and a post-station (*chih*[11]) every thirty *li*.[d] These short distances were undoubtedly chosen so that flag and drum signals, or the fire and smoke of beacons, could readily give and receive information. The postal clerks (*chhêng yu li*[12]) kept records of the despatches which they transmitted—in Han times these *hsi*[13] were written on foot-long wooden strips contained in bamboo tubes closed with a spring lock contrivance—and the cantonal officers (*thing chang*[14]) policed the road and its neighbouring districts with their guards. At the post-stations (*chiu chih*,[15] *chuan shê*[16]) there were stables and couriers in readiness for the relay service (*i*,[17] *jih*[18])[e] under the authority of a station-master (*chuan li*,[19] *chih chang*[20]). Strategic points in the network had veritable sorting offices (*yu thou*,[21] *yu chü*[22]), centres where mail was collected and distributed. By −77 the post-station system extended as far west as Lou-lan (42) and Pao-shan (100).[f] By +89

---

[a] Ch. 5, p. 80*b*.

[b] *Mêng Tzu*, II (1), i, 12; tr. Legge (3), p. 60.

[c] Ch. 4, pp. 5*b*, 6*a* (ch. 13); tr. Biot (1), vol. 1, p. 288; Lo Jung-Pang (6), p. 101.

[d] Sometimes two or more offices were present at one place (hence *yu-thing*, *yu-chih*).

[e] These words were of rather wide meaning, applicable to the riders as well as the horses, and to the relay system itself; usable also as verbs meaning to transmit, or to carry, government despatches.

[f] According to the sources collated by Lo Jung-Pang (6), there were 29,635 cantonal offices under the Chhien Han, and 12,445 in Hou Han times. If these were all about 10 *li* apart on roads, the total mileage would have been 296,350 *li* in the first case, and 124,450 *li* in the second; i.e. some 7·45 times the mileage presented in Table 63 above for the Chhien Han, and 1·93 times that for the Hou Han. This would raise the road mileage for the Chinese empire in Table 64 to 42,500 and the relative figure of miles per 1,000 sq. miles to 27·7, i.e. entirely equivalent to the Roman empire estimates. But the discrepancy is so

[1] 郵	[2] 置郵	[3] 遺人	[4] 廬	[5] 宿	[6] 路室
[7] 市	[8] 候館	[9] 野候徒	[10] 亭	[11] 置	[12] 丞郵吏
[13] 檄	[14] 亭長	[15] 廄置	[16] 傳舍	[17] 驛	[18] 馹
[19] 傳吏	[20] 置長	[21] 郵頭	[22] 郵窠		

tropical fruits were sent with the aid of the relay service from Nan-hai (70) to the capital. The post-station rest-houses had washing and sleeping accommodation for officials and authorised travellers, restaurants—and also cells for prisoners moving under guard. Often the way-stations had some private inns or hostelries (*kho shê*,[1] *ni lü*[2]) for those who lacked the right to use government facilities. Our footsteps in these volumes have often led us to the rest-houses of ancient and medieval China without our noticing it; for Liu Pang, the founder of the Han dynasty, was at the start of his career a *thing chang*[3] at Ssu-shui[4] near Phei (10),[a] a piquant story of the −2nd century had Ssuma Hsiang-Ju in just such a setting,[b] and was it not at the Chhen-chhiao[5] rest-house that Chao Khuang-Yin, the founder of the Sung dynasty,[c] was invested with the imperial yellow?

Though the post system reached a particularly high development in the Thang period, with a force of some 21,500 officers strung out along the roads and managed by 100 high officials in the capital, the terminology did not change.[d] In the Sung, however, the post-station service has become the 'hot-foot relay' (*chi chiao ti*[6]) and the stations *ma ti*[7] or *ma phu*[8].[e] With the growing militarisation previously mentioned, the couriers appear in the Yuan dynasty as *phu ping*,[9] serving in *chan-chhih*[10] (Mong. *jamči*) under controllers (*tho-tho-ho-sun*,[11] another Mongolian transliteration).[f] Though corruption was prevalent at all levels, the +14th-century service actually attempted to work to a timetable. After this there was little change until the arrival of the telegraph and modern road-building in the nineteenth century.

We may glance at a few further points of interest concerning ancient times, however. The speeds attained by the couriers have been thoroughly studied by Lo Jung-Pang (6), who concludes that a fair average was some 120 miles in 24 hr.,[g] though

large that we are reluctant to adopt the cantonal offices criterion as the best guide to total road mileage. It is surely probable that many of these offices, after all somewhat like police or gendarme stations, were located far off the beaten track of the great highways, and, at the least, a great many of them would have been on secondary roads not counted in the estimates of Table 63. If this is so, we can hardly assume, as Lo Jung-Pang (6) does, that the number of post-offices was double that of the cantonal offices, since the former were certainly located along the main lines of communication. At the same time, the high figures for the numbers of local offices and stations may be a warning that the Chinese estimates in Tables 63 and 64 are distinctly too low. The greatest difficulty, perhaps, is to distinguish between main and secondary roads in both empires.

    [a] Cf. Vol. 1, p. 102.   [b] Cf. Vol. 4, pt. 2, p. 234.   [c] Cf. Vol. 1, p. 131.
    [d] On the Thang system see especially Lou Tsu-I (1). The Japanese modelled theirs on it. The most famous Japanese road was of course the Tōkaidō,[12] running along the coast from Osaka to Tokyo; its 53 stages were the subject of a series of wood-block prints by Hiroshige (cf. Satchell, 1). Its route is now followed by one of the fastest and most modern of railways.
    [e] Shen Kua has an interesting account of the service in his time (*MCPT*, ch. 11, para. 15; cf. Hu Tao-Ching (1), vol. 1, pp. 416 ff.). See also Koiwai Hiromitsu (1).
    [f] The Yuan system is perhaps the best known of all, for chs. 19, 416 to 19, 426 inclusive of the *Yung-Lo Ta Tien* have survived, and have been reproduced in facsimile by the Tōyō Bunko, to which is also owing the publication of the relevant monograph of Haneda Tōru (1). There is besides an important study by Olbricht (1), and a valuable paper by Phan Nien-Tzhu (1). Although the word *chan* was used to transliterate a Mongolian word, it means a stopping-place, and survives in modern Chinese as the term for a railway station.
    [g] This speed, about 5 miles per hr., is much the same as that estimated as average for the Roman post-couriers (Ramsay, 1).

| [1] 客舍 | [2] 逆旅 | [3] 亭長 | [4] 泗水 | [5] 陳橋 | [6] 急脚遞 |
| [7] 馬遞 | [8] 馬舖 | [9] 舖兵 | [10] 站赤 | [11] 脫脫禾孫 | [12] 東海道 |

rather more could be attained by men who galloped 12 hr. from dawn to dusk at an average speed of 11 miles per hr. The literature contains a number of exceptional records for the Liao and Chhing periods ranging up to 260 miles a day, explicable only by conditions which permitted continuous travel through the night. Beacon signalling[a] alone gave faster results. In −74 the news of the death of the emperor Chao Ti was transmitted by this means at a speed of some 27 miles per hr.

In the Chhien Han period, when the roads were at their best, an elaborate service of government relay chariots, carts and carriages was also in use.[b] Besides the holders of noble titular ranks such as Kung Chhêng[1] (Riders in Public Transport) and Ssu-chhê Shu-chang[2] (Quadriga Elders),[c] the roads were full of officials in chariots carrying imperial commands, proceeding on tours of inspection, assuming or retiring from posts, or supervising the government monopolies and other enterprises, including the roads and canals themselves. The bureaucracy was always sensitive about the misuse of transport, and one of the crimes with which the general Yang Phu[d] was charged in −111 was that of asking for a relay carriage to go to the frontier, but actually using it to go home instead. Public transport was correspondingly a mark of imperial favour, as we noticed at a very early stage[e] in connection with the scientific-technological congress of +5 when government coaches were provided for the doctors. About eight years earlier, when an assembly of literati was convoked by the emperor Ai Ti, the eminent scholar Kung Shêng,[3] who had arrived first, suggested that government vehicles should be sent for them, for such transport had already been granted to physicians and sorcerers (i wu[4]). The emperor said: 'Did you then, Sir, come by private vehicle?', and when Kung replied that he had, official carriages were ordered for all the others.[f] About the speeds attained in the relay carriage system we know little that is certain, but to judge by an urgent journey of the Prince of Chhang-i in −74, it could rise as high as 9 miles per hour. Horses were changed every thirty li at the post-stations.

In the Han, carriages and mounts were under the control of the Thai Phu[5] (Grand Keeper of Equipages), but another bureau, the Regulations Department (Fa Tshao[6]), looked after the post-system, and yet another, the Commandant's Department (Wei Tshao[7]), was responsible for the public vehicles. Each province or commandery had its Superintendent of Posts (Tu Yu[8]) and was divided into a number of postal districts. Each provincial government had also its civilian Tao-chhiao Yuan-li[9] (Secretarial Aides for Highways and Bridges), its Chin Yuan[10] (Fords Officers), and its military Kuan Tu-Wei[11] (Commandants of Passes and Barriers). These last had to check the credentials of travellers and goods, collect internal customs, prevent smuggling, and

---

[a] Cf. Vol. 5, Sect. 30*j*.
[b] Of the great wealth and variety of Han vehicles we have seen something in Sect. 27*e*, *f* (Vol. 4, pt. 2); it can be studied further in albums such as those of Anon. (22) and Liu Chih-Yuan (1).
[c] Cf. Loewe (2).   [d] See further, p. 441 below.
[e] Vol. 1, p. 110; cf. Vol. 6, in Sect. 38.   [f] *Chhien Han Shu*, ch. 72, p. 19*a*.

[1] 公乘   [2] 駟車庶長   [3] 龔勝   [4] 醫巫   [5] 太僕
[6] 法曹   [7] 尉曹   [8] 督郵   [9] 道橋掾吏   [10] 津掾
[11] 關都尉

maintain general security. This reminds us that one of the more unexpected functionaries of the Ministry of Works was the Prefect of Credentials and Tokens (Fu Chieh Ling¹). All passports (*chuan*²) had to bear the seal of this Minister.ᵃ The *corvée* and taxation system always applied for the needs of the post-service, local people having to pay special taxes (*ma khou chhien*³) for the horses, and to supply a tithe of food for the travellers. The service was thus exposed to trouble from two sides. On the one hand there was the perennial resentment of the Chinese population, which disliked passports, *corvée* and direct taxation, and which therefore showed in general a passive hostility to the road authorities and all their works. But on the other hand there was the armed resistance of the tribal peoples, who saw very clearly that the road network was a major factor in the expansion of the Chinese empire, and therefore cut the roads and destroyed the post-stations whenever they could. This was what the Chhiang did in Kansu in −41, and their example was well followed by the south-western tribes in +93. But in spite of everything, the Chinese highway network, and the post-station system which was inseparable from it, constituted a cardinal factor in the advance of East Asian civilisation. Those who are aware of the facts which we have here rehearsed know what to make of the commonly received opinion that 'the Chinese have no great record as road builders'.ᵇ

(c) WALLS, AND THE WALL

There can be no doubt that the most ancient form of walling in China, both for houses and un-roofed enclosures, was that of *terre pisé* or tamped earth (*thien ni*⁴).ᶜ Removable elongated boxes or forms (*pan*⁵ or *kan*⁶) without tops or bottoms are used, dry earth being rammed within them at successively higher levels as the wall rises (Figs. 719, 720, pl.).

This shuttering resembles that used today for confining concrete while setting.ᵈ So

---

ᵃ Cf. Vol. 1, pp. 97, 104. Though passports were abolished in −168 they were reintroduced in −153, and remained generally in force, with respites of freedom in the more peaceful times, for many centuries. They caused much popular resentment, and not only among the educated scholars.

ᵇ Leading article in *The Times*, 27 June 1964.

ᶜ Full descriptions by Hommel (1), pp. 293 ff., and Boehling (1). The process was sometimes found surprising by neighbouring nations. For instance, an accurate account of the making of *terre pisé* walls was given in +1307 by Rashīd al-Dīn al-Hamadānī in his *Jāmiʿ al-Tawārīkh* (tr. Klaproth (3), p. 345).

ᵈ The thought is worth a moment's pause. There was something notable (though probably not noticed) in the adoption of this age-old device by François Lebrun (disciple of the great Louis Vicat) when in 1835 he brought concrete above ground for the first time since the Romans had used it for their wall-fillings, and began to build architectural structures of concrete alone (Skempton, 6). As has been said before (Vol. 4, pt. 2, p. 219, in connection with the use of crushed pottery tiles for hydraulic cement), no adequate study has yet been made of the history of Chinese practices concerning mortar, cement and concrete (cf. Lea (1); Davey (1), pp. 97 ff.); fundamental though the subject is for all aspects of civil engineering. Another interesting reflection is that the introduction of the iron reinforcement of concrete was a fresh illustration of the principle used so long in that natural material the bamboo (cf. Vol. 4, pt. 2, p. 61), the principle of two-phase components, now iron for tension and concrete for compression, as W. B. Wilkinson in 1854 was the first to recognise (Skempton, 6). Finally, we shall see later on in this volume (p. 178) how Chinese engineers of the +7th century anticipated in their stone segmental arch bridges the design brought to such exquisite perfection in reinforced concrete by Robert Maillart during the first decades of the present century.

¹ 符節令　　　² 傳　　　³ 馬口錢　　　⁴ 填泥　　　⁵ 版　　　⁶ 榦

basic was the compaction process that the word which expresses it, *chu*,[1] came in course of time to form one half of the standard term for architecture in general (*chien chu*[2]).[a] The platform foundations of ancient buildings were also of rammed earth (*ta hang*,[3] *hang thu*[4]). In Chinese practice it was customary to use rubble stone without

Fig. 719. Tamped earth (*terre pisé*) walling under construction; a drawing from the *Erh Ya*, ch. 2, p. 6*b*. The caption says that the boards of the shuttering are also called *yeh*.

binding material as the foundation of walls, and to spread a layer of thin bamboo stems between each pisé block so as to hasten thorough drying-out. The ubiquity of *terre pisé* walling in Chinese culture certainly has something to do with two features which we shall later find characteristic of Chinese architecture, first that walls were not in general weight-bearing, and secondly that buildings were provided with generously overhanging eaves (pp. 65, 103).

In modern times European travellers have seen so much more *terre pisé* in China than in their own continent that they have thought of it as a Chinese invention. But the compression of earth in climbing shuttering, raised lift upon lift till the wall has reached its full height, was well known to Pliny, who wrote:[b]

Are there not, moreover, in Africa and Spain walls made of earth that are called 'framed walls', because they are made in a frame of two boards, one on each side, and so are packed or rammed rather than constructed; and do they not last for ages, undamaged by rain, wind and fire, and stronger than any cement? Spain still sees the watchtowers of Hannibal,[c] and his forts of earth walling placed on the mountain ridges.

[a] One can see in *chu* the bamboo rammer, the wooden shuttering, the worker (represented by a mouth) and the work done (cf. K 1019).

[b] *Hist. Nat.* xxxv, xlviii, tr. auct. adjuv. Rackham (2), vol. 9, p. 385 and J. St Loe Strachey.

[c] He was there from −221 to −219 preparing for the war against Rome, a great event which few think of as contemporaneous with the more successful conquest of another empire by Chhin Shih Huang Ti.

[1] 築   [2] 建築   [3] 打硪   [4] 夯土

Indeed in rural England and France these techniques have never died out.[a] In many English counties boundary walls with roofs of tile or thatch, looking extremely Chinese, can be found, and the old Devonshire saying about earth walling—'Give it a good hat and a good pair of boots and it will last for ever'—could have come straight out of the *Lu Pan Ching*.[b]

In late Chou times bricks were mostly of adobe, i.e. sun-dried mud (*ni phi*,[1] *thu phei*[2]), such as may frequently be seen in China still,[c] but by the Han period baked bricks (*chuan*[3]) were becoming general (Figs. 721, 722, pls.). Plaster (*wu*[4]) was then already used, and often covered with paintings.[d] Many different brick sizes have been current through the ages, and now different shapes are found in different regions, e.g. in the north-east 12 in. × 9 in. × 6 in.; in the south-west 6 in. × 5 in. × 1 in., and this latter tile-like form is seen also in the Great Wall (15 in. × 7½ in. × 1½ in.).[e] Besides the bonds with which we are familiar in the West, the Chinese have long used a 'box bond', in which stretchers are placed vertically between layers of horizontal ones, the interior being filled with earth and rubble. Fig. 723 (pl.), a family temple in Szechuan, shows this. Or two horizontal layers may intervene (Fig. 724). Often bricks of two different thicknesses, but similar in other dimensions, are used. Then there is the Chinese cross bond, in which the stretchers are placed in groups of three.[f]

Besides the ordinary baked bricks[g] of the Warring States and Han periods, the Chinese craftsmen were the first to master the art of moulding and firing large hollow blocks of terracotta brick (*khung hsin chuan*[5]) ornamented with very intricate scenes and patterns (Fig. 725, pl.).[h] These were used mostly for tomb walling, as in the Warring States burials at the Erh-li Kang site near Chêngchow.[i] The stamped bricks of Han tombs are well known, and have afforded many glimpses of life at that time,

---

[a] See Davey (1), pp. 20 ff., but especially the monograph by Williams-Ellis, Eastwick-Field & Eastwick-Field (1). *Terre pisé* is characteristic of the Lyonnais and Catalonia, but similar methods using a wetter mix without shuttering live on in Devonshire, South Wales and Wiltshire as what is known as 'cob' and 'chalk mud'. There is a considerable eighteenth- and early nineteenth-century literature on rammed earth walling. Like Persians in China, Scots were sometimes surprised at what they saw in rural England, as we see from the diary of Miss Lindsay of Glasgow in 1802.

[b] From this geographical distribution one would seem to be in presence of a technique which had spread both east and west from ancient Mesopotamia, but its presence there seems uncertain. Lloyd (1), pp. 456, 460, speaks first of 'solid walls of rammed clay, pisé' from −4000 onwards, but then of pisé 'prepared with a rectangular wooden mould, open at top and bottom'. That however is adobe, not pisé. Since reading this I have been informed by Prof. Seton Lloyd, through the kind intermediation of Miss Munn-Rankin, that he knows of no Mesopotamian example of a wall made by ramming earth between frames. Sun-dried adobe bricks must therefore have been meant, and not pisé at all. Thus it is strange that both China and Europe should have had the technique.

[c] Adobe survives as 'clay-lump' buildings in East Anglia to this day.

[d] Kelling & Schindler (1); Fischer (2).   [e] Mirams (1); Spencer (1).

[f] See further Hommel (1), pp. 279 ff. and Arnaiz (1); but especially Liu Chih-Phing (1).

[g] Burnt or baked bricks had been used extensively in Mesopotamia and the Indus Valley, and also, though not commonly, in ancient Egypt (Davey (1), pp. 22 ff., 64 ff., 76 ff.; Petrie (2); Capart (1); Lloyd (1); Briggs (2), p. 41). They seem to have been an innovation at Rome in the time of Vitruvius (late −1st century), who worked mostly with sun-dried bricks. Thus they spread, like so many other things, both east and west. Cf. the quotation from Fang I-Chih in Vol. 4, pt. 2, p. 219.

[h] Cf. Davey (1), p. 89 and pl. xxxix.

[i] Anon. (23), pp. 50 ff. and pl. 20, fig. 1. Dimensions average here 3½ ft. × 1 ft. × 6 in. Openings are both round and rectangular.

[1] 泥坯   [2] 土坯   [3] 磚   [4] 圬   [5] 空心磚

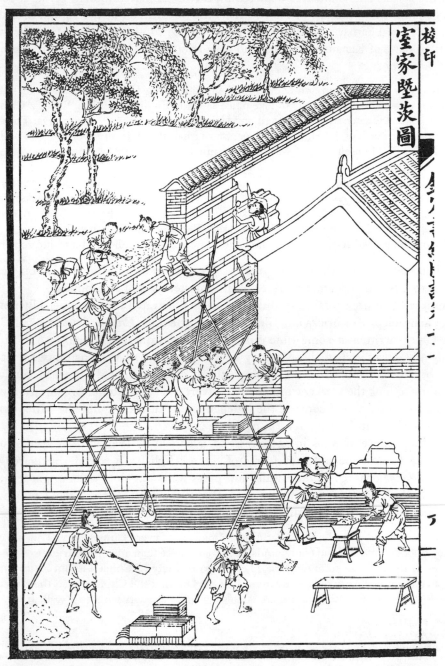

Fig. 724. A late Chhing representation of bricklayers at work. Box bond with the horizontal stretchers in double layers and a coping of five layers of headers. Plasterers are also seen. From *SCTS*, ch. 31, Tzu Tshai (Medhurst (1), p. 240; Karlgren (12), p. 48), whence the caption: 'Preparing the home of a family for roofing.'

from the scenes and characters depicted on them.[a] It may be worth giving a curious passage concerning the use of baked bricks for fortification at the capital of a short-lived peripheral barbarian dynasty, the Hsia[1] (+407 to +431), whose Hunnish rulers governed parts of Kansu and Shensi and were overthrown by the Thopa Wei people.[b] The *Chin Shu* says:[c]

In this year (+412) the reign name was changed to Fêng-Hsiang. (The emperor) Holien Pho-Pho[2] chose Chhihkan A-Li[3] to take office as Chief Engineer (Chiang Tso Ta Chiang[4]), and to mobilise 100,000 workers of the I and Hsia peoples north of the mountains. They were to go to a place north of the Shuofang River and south of the Heishui River, there to build the capital. Pho-Pho himself said, 'As I have just united the world and attained the empire over ten thousand regions, I shall name it Thung-Wan.'[5]

(Chhihkan) A-Li was extremely skilled and clever, but also cruel and violent. He caused the workers to bake bricks (*chêng thu*[6]) to make the city wall. (He used to test the bricks) and if a hammer blow would make a depression as much as an inch deep, he would have the worker (responsible) killed and buried inside the wall.[d] Pho-Pho thought that this showed much loyalty on the part of A-Li, and gave him the responsibility of all buildings and repairs.

We shall meet again with this disagreeable Hun technician in the Section on weapon metallurgy, for he was an armourer and ironmaster as well as a builder.

Chinese literature contains, or did contain, an illustrated treatise especially devoted to the manufacture of bricks and tiles. This was the *Tsao Chuan Thu Shuo*[7] of Chang Wên-Chih,[8] written at some time during the second quarter of the +16th century. Chang was an official of the Ministry of Works who had responsibility for certain State brickfields, and wrote his book partly in order to remedy the bad conditions prevailing among the workers and partly to avert the confusion of organisation which had led to the suicide of some of the independent contractors. But the Ming dynasty decayed before much of his reform could be achieved.

The importance of walls in ancient times is shown by the fact that several words were used to describe different forms of them. High walls round courtyards were called *chhiang*[9] or *yung*,[10] house walls and party walls *pi*,[11] low walls in gardens, etc., *yuan*.[12] Sirén (5) has given a graphic description of the walls of traditional China, which everyone who knows the country will recognise as true.

Walls, walls, and yet again walls, form the framework of every Chinese city. They surround it, they divide it into lots and compounds, they mark more than any other structures the basic features of the Chinese communities. There is no real city in China without a surrounding wall, a condition which indeed is expressed by the fact that the Chinese used the same word *chhêng*[13] for a city and a city wall; there is no such thing as a city without a wall. It would be

---

[a] Cf. pp. 112, 150, 187.   [b] Cf. Eberhard (9), pp. 149 ff.

[c] Ch. 130, p. 3b, tr. auct.

[d] This is the same story as that which is the basis of the Mêng Chiang Nü ballad (cf. p. 53) where it relates to the Great Wall. Here it may be brought in because it was the standard thing to say about such activities. Cf. Eberhard (20) however. The laying of bricks in a half-dried condition was old Mesopotamian practice.

[1] 夏   [2] 赫連勃勃   [3] 叱干阿利   [4] 將作大匠   [5] 統萬

[6] 蒸土   [7] 造甎圖說   [8] 張問之   [9] 牆   [10] 墉

[11] 壁   [12] 垣   [13] 城

just as inconceivable as a house without a roof. These walls belong not only to the provincial capitals or other large cities, but to every community, even small towns and villages. There is hardly a village of any age or size in northern China which has not at least a mud wall, or remnants of a wall, around its huts and stables. No matter how poor and inconspicuous the place, however miserable the mud houses, however useless the ruined temples, however dirty and ditch-like the sunken roads, the walls are still there, and as a rule kept in better condition than any other constructions. Many a city in north-western China which has been partly demolished by war and famine and fire, where no house is left standing and where no human being lives, still retains its crenellated walls with their gates and watchtowers. Those bare brick walls, sometimes rising over a moat, or again simply from the open level ground where the view of the distance is unblocked by buildings, often tell more of the ancient greatness of the city than the houses or temples. Even when such city walls are not of a very early date (hardly any walls now standing are older than the Ming)[a] they are nevertheless ancient-looking, with their battered[b] brickwork and broken battlements. Repairs or rebuilding have done little to refashion them or change their proportions.

Sirén was writing several decades ago, and in fact the exigencies of modern life have now led to the disappearance of some parts of city and village walls,[c] but it will be a long time before travellers cease to marvel at their size and ubiquity.

Even the ancient Asian nomads surrounded their camps with earthen ramparts,[d] and there can be no doubt that the essentially settled agrarian culture of China erected walls round its earliest cities. As Creel (2) has remarked, the character for capital (*ching*[1]) has as its oldest graph (K 755) a picture of a guard-house over a city gate. And city walls we find as early as the −15th century. Those of the Shang city of this period, Ao,[2] just north of modern Chêng-chou, some 65 ft. wide at the base, enclosed an area about 2,100 yards square.[e] Similar excavations and studies have been made of Chou feudal capitals such as that of the State of Chao at Han-tan (62) in Hopei, founded in −386, where the main rectangular enclosure has sides some 1,530 yards in length, with walls originally as high as 50 ft. on a base again 65 ft. wide.[f] All such walls consisted essentially of successive layers of tamped earth (*terre pisé*) averaging from three to four inches thick. When Chinese city-walls are cut through today for modern transport improvements, these layers can still be seen.[g] Whether or not the walls of Shang and Chou cities were always faced with sun-dried (adobe) bricks we cannot be sure.

[a] This applies, of course, only to the facings; the cores of many walls are very old. Also it is not true of the Great Wall.

[b] I.e. backward sloping.

[c] In north-west China in 1958 I found that the removal of city-walls was much welcomed as symbolic of the modernisation and pacification of a countryside which had known rebellion, ethnic strife and banditry for twenty centuries.

[d] Cf. Uchida (2).

[e] Chêng Tê-Khun (9), vol. 2, pp. 17, 27 ff., 39 ff.; Watson (2), pp. 61 ff. Anyang was not founded till after −1400.

[f] Watson (2), pp. 121 ff.; cf. Vol. 1, p. 94. See also Chêng Tê-Khun (9), vol. 3, pp. 18 ff.

[g] Li Chi (2) has reckoned that of 163 city-walls built before −722 no less than ten were still in use in 1928. Of 585 constructed between −722 and −207 as many as 74 still remained.

[1] 京　　[2] 隞

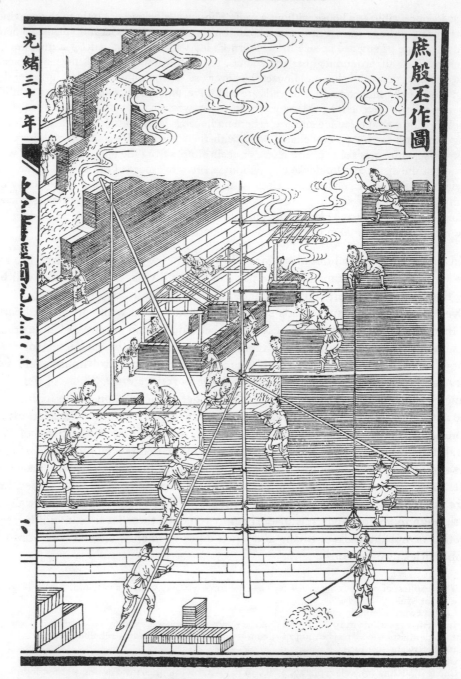

Fig. 726. A late Chhing representation of the construction of a city-wall. On a solid foundation of large bricks or dressed stone the rampart walls of tamped earth contained by headers and stretchers of smaller thin bricks are rising into the clouds. From *SCTS*, ch. 32, Shao Kao (Medhurst (1), p. 242; Karlgren (12), p. 48). The walls are those of the capital city of the Chou dynasty, and the caption quotes the passage which records that the work was done by the artisans and people of the defeated kingdom of the Shang.

By the end of the − 3rd century the art of fortification had made so much progress,[a] and the unification of the empire had afforded such abundance of men and materials, that the walls of the Former Han capital at Chhang-an, still traceable some five miles north-west of modern Sian, were on an altogether greater scale.[b] Along a circuit of some 16 miles arose a rampart wall some 50 ft. high and 40 ft. broad at the top, still devoid of enfilading bastions, but it did not stand alone; it was backed by a terreplein more than 200 ft. wide raised about 20 ft. above the surrounding land level, and protected by a moat 150 ft. wide with a depth of about 15 ft. The terreplein was once covered with buildings, perhaps garrison dwellings. Its inner face, as well as the berm outside the wall and the scarp and counterscarp of the moat, can still be made out. From the outer edge of the counterscarp to the foot of the inner face of the terreplein the total width of the works around the whole perimeter must have been of the order of 480 ft.[c] All this was an effective reply to the poliorcetic arts which had developed during the Warring States period, such as fire-setting in mine tunnels to cause the collapse of walls, or the diversion of rivers to wash away their foundations. The traditional builder of these fortifications of Chhang-an about − 200 was Yangchhêng Yen.[1]

The cores of Chinese city-walls were always of earth or rubble (hence perhaps the 'earth' radical in the word for city-wall or city, *chhêng*[2]),[d] but in later centuries they were usually provided with outer, and often also inner, facings made of large grey burnt bricks laid in lime mortar. Occasionally, where stone was plentiful, as in Szechuan, they were revetted with dressed stone blocks in regular courses of equal height.[e] A traditional Chinese representation of the foundation and revetment of a city wall is shown in Fig. 726, where we see also the in-fill of rubble.[f] Fig. 727 (pl.) shows part of the walls of modern Sian, which run three miles from east to west,[g] and which I remember seeming interminably long when I saw them for the first time as I arrived by train at dawn from Pao-chi.[h]

[a] Cf. Vol. 2, p. 165.

[b] See for example the interesting studies of Bishop (6) and Oshima (1). Map in Anon. (*43*), p. 80. The walls of the Thang city of Chhang-an, 26 miles in circumference, to the south of the modern city, are currently under investigation by Chinese archaeologists; cf. *CKHW*, 28 Dec. 1961, and derivative notices.

[c] The foundations of the great Wei Yang Kung[3] palace are also still to be seen at the west side of the Han city. They consist of a series of five superimposed rectangular terraces like the terreplein on a base of 450 × 145 yards with its long axis north and south, so arranged as to lead up a processional way to a small rectangular terrace at the northern end still over 50 ft. above the surrounding land level. I have vivid memories of a visit paid to this romantic but desolate site with Dr Dorothy Needham and Dr Tshao Thien-Chhin in 1945 when the rain beat down upon the Han paving-tiles still abundantly scattered about among the trees.

[d] Like the Roman *opus incertum* and *opus reticulatum* (cf. Vitruvius, II, viii, 1, and Blake, 1), covered with stucco.

[e] Like the Roman *opus isodomum* (cf. Vitruvius, II, viii, 5).

[f] In many dynasties there was the equivalent of the Royal Engineers, the *chuang chhêng ping*[4] of the army, regular units of *chün chiang*,[5] artisans skilled in fortification as well as other branches of military technology. See further Yang Lien-Shêng (11), pp. 30 ff.

[g] This is taken from the inward side and may be complemented by a splendid air photograph (probably von Castel's) of the succession of external bastions, reproduced in Gutkind (1), pl. 9.

[h] Cf. p. 17 above.

[1] 陽成延   [2] 城   [3] 未央宮   [4] 壯城兵   [5] 軍匠

Chinese city-walls are never complete without their watch-towers and gate-towers, usually single structures in two or three storeys, with the up-turning roof-corners so characteristic of Chinese buildings, and set directly over the opening in the city-wall. The latter is sometimes protected by a curving curtain-wall which makes the gate a double one. The drum-tower in the centre of the city follows a similar pattern, for it is often a two- or three-storey pavilion set upon a solid rectangular fort tunnelled spaciously in both directions at right angles. An alternative type of city entrance, probably rather older, flanked the gateway by two such pavilion-bearing towers. But whatever may be the plan, walls and bastions are invariably battered, i.e. they slope markedly inwards to the top; contrasting thus with the perpendicular walls so often seen in medieval Western castles. To illustrate the style of Chinese defensive architecture,[a] Fig. 728 (pl.) shows a Thang-period fresco from Tunhuang, and Figs. 729 and 730 (pls.) give different views of the Ming gate-fortress at Chia-yü-kuan (36), where the Old Silk Road passes out through the Great Wall. A typical wall of a small Szechuanese city is seen in Fig. 731 (pl.).[b]

The massive Chinese wall did not always rise straight out of the ground or the water-filled moat. It frequently had a supporting platform or plinth, just as if it were any other building (see on, p. 91). But since many of the plinth designs had re-entrant mouldings, the placing of the heavy wall above them gave a remarkable effect of elegance and lightness. As Mirams has pointed out,[c] the battered line of the superstructure, instead of starting from the line of the inset die of the base, as is usually the practice in the West, starts from the face of the corona. Hence the spreading static effect of Western classical designs is altogether avoided.

The Great Wall (the Ten-Thousand Li City-Wall—Wan Li Chhang Chhêng,[1] as it is called in Chinese)[d] has of course been in the reader's mind during the previous paragraphs. It notably stirred the imagination of eighteenth-century Europeans. In +1778 Dr Johnson

talked with an uncommon animation of travelling into distant countries; that the mind was enlarged by it, and that an acquisition of dignity of character was derived from it. He expressed a particular enthusiasm with respect to visiting the Wall of China. I catched it for the moment, and said I really believed I should go and see the Wall of China had I not children, of whom it was my duty to take care. 'Sir (said he), by doing so, you would do what would be of importance in raising your children to eminence. There would be a lustre reflected upon them from your spirit and curiosity. They would be at all times regarded as the children of a man who had gone to view the Wall of China. I am serious, Sir.'[e]

Though Boswell never saw the Shansi mountains or the Yellow River, his friend's confidence was not misplaced. Stretching from Chinese Turkestan to the Pacific in a

---

[a] We must not anticipate here the discussions of Sect. 30 h below.

[b] Cf. Schmitthenner (1), p. 272.

[c] (1), p. 48.

[d] A literary appellation, less known in the West, was Tzu Sai,[2] the Purple Barrier, so called, says the *Ku Chin Chu*, ch. 2, because the earth of the Shansi hills looked purple, or, in another tradition, because the grass was purple at Yen-mên (113).

[e] Boswell (1), vol. 3, p. 292.

[1] 萬里長城　　　[2] 紫塞

line of well over two thousand miles (nearly a tenth of the earth's circumference), the Wall has been considered the only work of man which could be picked out by Martian astronomers. To visualise its equivalent in Europe one must think of a continuous structure reaching from London to Leningrad or from Paris to Bucarest. The Roman empire had its *limes* or frontier fortifications covering stretches between rivers and other natural borders, but they never attained anything like this length. The longest was the Limes Germaniae et Rhaetiae[a] which connected the Rhine and the Danube across South Germany, but it did not exceed 350 miles and its defences were confined to earthworks and timber forts.[b] As for the walls across the British isthmus, the longest was but a fifth of this,[c] while the Limes Syriae, though extending for some 625 miles, had no continuous line of defence at all.[d]

There is no lack of travellers' descriptions of the Great Wall,[e] but studies based on modern historical scholarship are few and far between, whether in Chinese or Western languages.[f] Figs. 732 and 733 (pls.) give good impressions of it winding over the mountains of Hopei and Shensi, and Fig. 734 (pl.) shows it in its lesser glory along the western marches in Kansu, where it has lost its stone facings (if indeed it ever had them in those parts), and runs along as a ridge of compacted loess with numerous breaks. In some places it is almost overwhelmed by the desert sands (Fig. 735, pl.). We know of no accurate count of the number of wall-towers and isolated watch-towers in the Wall's neighbourhood, but those who have walked long distances along the line estimate that some 20,000 of the former and 10,000 of the latter are still standing, and that at the time of its maximum strength the Wall embodied at least 5,000 more of each of the two types.[g]

It will be worth while to examine briefly the component parts of the Great Wall as shown on the map in Fig. 711. Its present main line is, broadly speaking, that of the Ming period, but different parts of it are built along alignments of earlier walls dating from very different periods, as we shall see. Let us begin from the eastern end, Shan-hai-kuan (101),[h] where one of the portals on the road to Manchuria and Korea bears the

---

[a] Begun by Vespasian in +74 and furnished with an earthen rampart by Domitian in +83. After the mass crossing by the Franks in +260 the frontier was withdrawn to Rhine and Danube.

[b] Some stone watch-towers were added by Hadrian in +122, and Caracalla built a short stretch of stone wall about +215.

[c] The Vallum Antonini Pii of +81 (evacuated +184) extended only 37 miles as an earthwork. The Vallum Hadriani of +126 (evacuated +383) was faced with stone but did not exceed 73 miles. For a sketch of present remains, cf. Mothersole (1).

[d] It was established by Trajan about +100 against the Arsacid Parthians, but had only guard-posts every 12 miles and forts every 28 miles, with connecting roads but no rampart.

[e] Cf. de Navarrete (+1665), in Cummins (1), vol. 2, p. 219.

[f] Indeed we have not been fortunate enough to find one. The work of Clapp (1), with its excellent chart, is perhaps the best, but it is strictly geographical. Beyond this there is Geil (3), well illustrated but the most anecdotal and chaotic of all his books; Hayes (1), more businesslike but small, and more recently the *romancé* history of Lum (1) based solely on translated sources. Silverberg (1) we have not seen. Academia Sinica should produce a standard work. The monograph of Wang Kuo-Liang (1), though brief, is excellent on the historical side, but we had access to it too late to make the best use of it. Still more tantalising was my experience with a much newer study, that of Shou Phêng-Fei (1) printed in 1961. This I purchased at Sian in 1964, only to find, before I could enjoy it, that it had been withdrawn from sale and could not be delivered. The reason for this remains a mystery. I saw only that it has good maps.

[g] Cf. Geil (3), p. 327.

[h] Numbers in brackets refer to Table 60. Those in square brackets refer to Table 4 and Fig. 35 in Vol. 1.

inscription *Thien-Hsia ti I Kuan*[1] (The World's First Gate). Ascending steeply on to the mountains overlooking the Hopei plain, the Wall snakes along ridge after ridge in a section broken by only one important gate, that of Ku-pei-khou (108), through which passed the road to Jehol, so important for its imperial summer residence during the Chhing. Then almost due north of Peking (7) at a point known as the Eastern Bifurcation, it divides into two, the Outer or Northern Wall running along the border of Shansi province, the Inner descending to a point some 125 miles further south among the mountains and then rejoining the main defence line some 30 miles east of the Yellow River. Within this lozenge are the important cities of Chang-chia-khou (110, Kalgan) and Ta-thung (86), both giving access to Inner Mongolia, and both reached through a famous check-point in the Inner Wall, the Chü-yung-kuan[2] gate[a] (Fig. 736, pl.) at the Nan-khou pass (109). Here came through the old road to Urga; and the new railway to Ulan-Bator, the modern reincarnation of the Mongolian capital, takes exactly the same route, issuing from the Outer Wall just north of Ta-thung and striking off across the Gobi desert. Further south along the Inner Wall is the site of the old fortress of Tzu-ching-kuan (111) beside which no doubt passed the road from Tai (61) already described (p. 26). Similarly the great road from the south through Shansi pierced the Inner Wall at a more westerly point, Phing-hsing-kuan (112). Besides the two cities already mentioned, the lozenge contained the important Han fortress of Yen-mên (113).

The Outer or Northern Wall is undoubtedly based on an alignment dating from the +5th and +6th centuries. After Northern Wei tentatives the first construction was undertaken by the Eastern Wei in +543,[b] but it was carried on far more energetically by the Northern Chhi between +552 and +563,[c] especially in a great effort of 3,000 *li* in +556[d] which cost heavily in men and material and almost bankrupted the State.[e] The Inner Wall line is of uncertain time,[f] and it has a very curious branch wall running south for some 230 miles along the edge of the Shansi uplands overlooking the Hopei plain. This is certainly somewhere near the site of the −5th-century wall built by the feudal State of Chungshan,[g] but the line is more probably of much later date, perhaps traced by the Shansi people of Western Yen about +390 against the Later Yen State of Hopei and Shantung which soon afterwards annexed them;[h] or alternatively the remains of the frontier between two transient States in the Wu Tai period (+907 to +960).[i]

---

[a] As this gate led to Lamaist regions, it was reconstructed in +1343 with three *stūpas* over the arch and an abundance of Buddhist carvings, together with inscriptions in six languages, Lantsha (Nepalese Sanskrit), Tibetan, 'Phags-pa Mongolian, Uighur, Hsi-Hsia and Chinese. We owe to Murata & Fujieda (*1*, 1) an elaborate monograph on this 'pagoda-bridge' (*kuo chieh tha*[3]), as it was called. Nan-khou is today a favourite view-point for visitors from Peking (cf. Chin Shou-Shen (1), Schulthess (1), pl. 64).

[b] *TCKM*, ch. 32, p. 59*a*.　　　　　　　[c] *TCKM*, ch. 33, p. 91*b*, ch. 34, pp. 63*b*, 64*a*.
[d] *Pei Chhi Shu*, ch. 4, p. 26*a*; *TCKM*, ch. 34, p. 20*a*.　　　[e] *TCKM*, ch. 34, p. 36*a*.
[f] Wang Kuo-Liang (1) considers it also of Northern Chhi date.
[g] Cf. Vol. 1, Fig. 12.　　　　　　　　　[h] Cf. Herrmann (1), map 29 (iii).
[i] Cf. Herrmann (1), map 41 (i, iv). Wang (1) ascribes it also to the Northern Chhi but its purpose then is not obvious; cf. Herrmann (1), map 33.

[1] 天下第一關　　　　[2] 居庸關　　　　[3] 過街塔

Once across the Yellow River, the Wall runs south-west across the Ordos Desert, choked and overblown with sand, rising north-west again to meet the river higher up in the vicinity of Ninghsia (114). The first part of this traverse is very ancient for it was essentially[a] the line of a frontier wall built in −353 by the State of Wei; it passes the outpost city of Yülin (56) and the vanished gates of the Chhin and Han imperial roads which led to the north of the great bend of the Yellow River (pp. 14, 16).[b] In the neighbourhood of Lanchow (27) the layout again becomes complex. From Ninghsia the remains of an inner wall follow for a long way the right bank of the River, and from Lanchow north-westwards a wall in better preservation protects closely the Old Silk Road, but besides these an outer wall strikes off across desert country to rejoin the latter in the neighbourhood of Liangchow (30). The dates of these modifications are not clearly known. North-west of Liangchow the line continues as a loess wall with periodical stone towers, taking in the Pai-thing Ho[1] Valley which ends in lakes in the Gobi outside the Wall but rejecting the Jo-shui[2] which leads out to Edsin Gol,[c] until it curls inward to the fortress of Chia-yü-kuan (36)[d] and ends a few miles further in towards the Chhi-lien Shan. But although this was the 'Last Gate of the World' in Ming times and for many centuries earlier, it was not so in the Han (−2nd century). A further wall of loess, now completely missing where at right angles to the prevailing winds, but otherwise standing 12 ft. high at minimum, and dotted with many 30-ft. towers, continued along the Su-lo Ho[3] past An-hsi (38) to surround Tun-huang (39) in a protective embrace.[e] The roads to the West (cf. pp. 17, 18) passed out through Yu-mên-kuan (40)[f] and Yang-kuan (116) respectively.

A very mysterious extension of the Great Wall exists in the form of the 'Chhing-hai Loop' in the neighbourhood of Lanchow.[g] Originating from the western junction of the Lanchow Loop with the Outer Wall, it passes south-westerly in an arc enclosing Hsi-ning (115) and Kumbum, short of Lake Kokonor itself, crosses the Yellow River and returns in a curve to the neighbourhood of Lanchow giving off at least one major spur southward on the way. This seems to be a +4th-century alignment, for the country it encloses was contested between the Chhien Liang and Hou Chao States (+314 to +376), and became independent as Hsi Chhin (Western Chhin) between +385 and

---

[a] But not exactly, for the ancient wall ran more directly north–south, nearer the east limb of the Yellow River's great detour.

[b] This stretch, known as the North Shensi Frontier Wall, was the scene of the chief reconstruction work under the Sui. In +585 a length of 700 *li* with ten fortresses connecting the eastern passage of the Yellow River with Ninghsia was completed (*TCKM*, ch. 36, p. 9 *a*, *b*; cf. *TH*, p. 1250). In +607 and +608 the line between Yü-lin and the Ta-thung region was put in order (*TCKM*, ch. 36, p. 131 *a*, ch. 37, pp. 2 *b*, 4 *b*; cf. *TH*, p. 1278), thus linking the work of the Han and the Northern Chhi. The Sui Wall may have been rather more east–west, however.

[c] Except for a short spur near Chin-tha,[4] built doubtless to protect the road to Chü-yen (99), cf. p. 17 above.

[d] Cf. Vol. 1, Fig. 14.

[e] This Han part of the Great Wall was first surveyed by Sir Aurel Stein; see his classical studies (1, 2, 3, 4 and especially 9). We showed it already in Vol. 1, Fig. 16.

[f] This gate-fortress was already in existence before −111 (Hsia Nai (2), pp. 76, 168). On the exact position of the two see Hibino Takeo (1).

[g] This was first surveyed by Geil (3).

[1] 白亭河   [2] 弱水   [3] 疏勒河   [4] 金塔

+431.[a] At this time the Thu-yü-hun[1] tribes were a great menace in the Tibetan hills to the west, and it was doubtless partly against them that this loop was constructed.[b]

Thus, in sum, the Wall has had periods of importance and periods of decay. After the Chhin and Han (−3rd to +3rd centuries) there was little maintenance. The Wei, Chhi and Sui accomplished major reconstruction (+6th and +7th centuries) but the seven centuries of Thang and Sung passed without upkeep or any fresh building. In the Yuan and Chhing the Wall lost all significance and this is why its present state is essentially Ming.[c]

The sight that paternal preoccupations withheld from James Boswell became in due time the lot of Captain Parish of the Royal Artillery, Lord Macartney's military attaché, who in +1793 gazed upon the Wall at the gate of Ku-pei-khou when the ambassador's company was on its way to attend the Chhien-Lung emperor in Jehol.[d] Parish measured and surveyed as much as he could of the walls and towers in that neighbourhood,[e] and his estimate of the Wall's magnitude formed the basis of a famous statement by his colleague, Macartney's private secretary, John Barrow. The Great Wall, wrote Barrow,[f]

is so enormous, that admitting, what I believe has never been denied, its length to be fifteen hundred miles,[g] and the dimensions throughout pretty much the same as where it was crossed by the British Embassy,[h] the materials of all the dwelling-houses of England and Scotland, supposing them to amount to 1,800,000, and to average on the whole 2,000 cubic feet of masonry or brick-work, are barely equivalent to the bulk or solid contents of the Great Wall of China. Nor are the projecting massy towers of stone and brick included in this calculation. These alone, supposing them to continue throughout at bow-shot distance, were calculated to contain as much masonry and brickwork as all London....

The approximate dimensions of wall and towers in the eastern stretches along the Hopei and Shansi borders may be seen in the accompanying diagram (Fig. 737), based upon Parish and the data of many travellers subsequent to him.[i] The granite blocks of the stone foundations are often as large as 14 ft. × 3 ft. or 4 ft., those of the stone facings for the rubble core some 5 ft. × 2 ft. × 1½ ft.; if the facing (always about 5 ft. thick) is of brick-work, it contains seven or eight thicknesses (cf. p. 45 above).[j] There are eight to twelve towers per mile, at distances ranging from 100 to 200 yards. To compare the constructional methods of past ages with those which would be employed

---

[a] Cf. Herrmann (1), map 29 (i, iii).

[b] Cf. *Thung Tien*, ch. 190 (pp. 1021.1 ff.). Wang Kuo-Liang (1) seems to think that its northern part was a construction of the Sui time, and Lo Chê-Wên (2) that its southern part was Chhin.

[c] Our knowledge of the Wall suffers from the defect of insufficiently close collaboration between archaeologists and historians. A Stein or a Clapp actually visits and measures existing remains of the Wall for his charts, and a Wang Kuo-Liang ransacks the historical literature, but the two columns never seem to meet.

[d] For a brief biography of Parish, see Cranmer-Byng (2), p. 313. Macartney (1), pp. 110 ff., gives the ambassador's own observations.

[e] His report was printed in G. L. Staunton (1), vol. 2, pp. 186 ff.

[f] (1), p. 334. Barrow himself had remained behind in Peking with Dr Dinwiddie (cf. Vol. 4, pt. 2, p. 475).   [g] Here he under-estimated by some 25%, as we shall see.

[h] This was of course an over-estimate.

[i] Geil (3) and other references already cited. Most recently, Gower (1), p. 142.

[j] A common size is 15 in. × 7½ in. × 3½ in. Analysis in Brazier (1).

[1] 吐谷渾

for a similar work today would lead us far into conjecture; one can only say that an immense amount of man-handling of the blocks on slides must have been used, with suitable tackle for laying in place.[a]

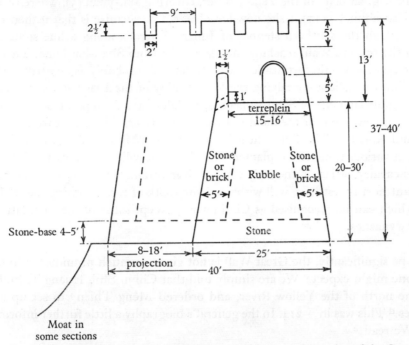

Fig. 737. Approximate dimensions of wall and towers in the eastern stretches of the Great Wall (data of Parish and others).

Reinforcements of wood, and even of iron, appear to have been used occasionally in certain sections of the wall. The *Kuei-Hsin Tsa Chih* (*Hsü Chi*) of about +1298 tells us that people living along the Great Wall used to find within it, in places where it broke down after heavy rain, lengths (*kan*[1]) of an extremely hard wood which had been used by the builders centuries before.[b] These were good for spear-shafts. In view of the term used (cf. p. 38 above) these may well have been boards which had been part of the shuttering, but it is not excluded that they were baulks intended for reinforcement only, since Stein (10) found that the walls of Han forts in the Tarim Basin were generally formed of brushwood fascines and wild poplar trunks alternating with layers of tamped clay. Perhaps the same wood served both purposes. It could moreover have been piling.[c] 'If you ram down piles of "thousand-year wood"', wrote Fang I-Chih in his +17th-century technical encyclopaedia, 'much energy is saved; pine and cypress wood can last for centuries without decay.'[d]

[a] The stretches were probably built upwards to the maximum elevations from each side so that the terreplein could serve for bringing up more material. On cranes see Vol. 4, pt. 2, pp. 95 ff. A comparative estimate of cost (cf. p. 30 above) would also be interesting. Cf. p. 409 below.

[b] Ch. 1, p. 41 *a*, *b*. [c] Cf. Vol. 4, pt. 2, p. 52.

[d] *Wu Li Hsiao Shih*, ch. 8, p. 38 *b*.

[1] 榦

There is every evidence that the first Great Wall of Chhin Shih Huang Ti took quite a different course from the present main line. How far north-west of Lanchow it started we know not, but it certainly passed Ninghsia (114) and then kept all the way north of the great bend of the Yellow River, covering Wu-yuan (5), where there was a gate at Kao-chhüeh (117), a site now long lost.[a] The surmise is that it then ran east-wards through the southern steppes of Inner Mongolia along a line some distance north of the present Wall, reaching the sea not far from Shan-hai-kuan. There is also evidence that the Chhin fortifications were extended, probably as an earthwork, along a great elliptical salient containing the lower valley of the Liao[1] River in Manchuria and having the later city of Shen-yang (118, Mukden) as a focal point, so as to reach the sea again somewhere near the mouth of the Yalu[2] River, the present frontier between China and Korea.[b] This alignment is known as the 'Willow Pale' (Liu Thiao Pien[3]) because at various times it was planted with thick-set willow-trees as a defence against nomadic cavalry, as last in the early Chhing. But although the Han wall and towers in the distant north-west have well withstood the tooth of time, nothing has so far been found which can be recognised as Chhin work, except some of the foundations used by later dynasties.

Perhaps significantly, the Great Wall is not given as much prominence in the *Shih Chi* as one might expect.[c] We are simply told that Chhin Shih Huang Ti built a wall along the north of the Yellow River, and ordered Mêng Thien[4] to set up a line of fortresses.[d] This was in −214. In the general's biography a little further information is given. We read:[e]

After Chhin had unified the world (in −221), Mêng Thien was sent to command a host of 300,000 men to drive out the Jung and Ti (barbarians) along the northern (marches). He took from them the territory to the south of the (Yellow) River,[f] and built a Great Wall, con-structing its defiles and passes in accordance with the configurations of the terrain. It started at Lin-thao[5,g] and extended to Liao-tung,[h] encompassing a distance of more than 10,000 *li*.[i] After crossing the (Yellow) River, it wound northwards, touching Mount Yang[6].[j]

[a] Some authorities, such as Wang Kuo-Liang (1), place Kao-chhüeh south of the Yellow River.
[b] This is no doubt why the *Thung Tien* says, ch. 186 (p. 992.1), that the Chhin Great Wall started from what afterwards became the Lo-Lang province of the Han. Hence Wang Kuo-Liang (1) shows it running some way south of modern Pyong-yang in Korea.
[c] The social and political aspects of this great engineering work are mentioned elsewhere (Sect. 6a in Vol. 1 above, and Sect. 47 below).
[d] Ch. 6, p. 22a (Chavannes (1), vol. 2, p. 168).
[e] Ch. 88, p. 1b, tr. Bodde (15), p. 54.
[f] I.e. the Ordos region within the great bend.
[g] Modern Min-chou,[7] in Kansu (119), a short distance south of Lanchow.
[h] The Manchurian peninsula of the same name today.
[i] With the Chhin and Chhien Han *li* at 2·01 to the kilometre, this would mean, if taken literally, some 3,080 miles. The present line of the Wall, and the Pale, from Lanchow to the mouth of the Yalu River, would be about 1,700 miles, but the Chhin Wall took an immense detour north of the great bend of the Yellow River.
[j] A hill in the low range to the north of the Yellow River at the great bend, not far west of Wu-yuan (5). The range is now called the Yin Shan [34].

[1] 遼河   [2] 鴨綠江   [3] 柳條邊   [4] 蒙恬   [5] 臨洮   [6] 陽山
[7] 岷州

But of the immense organisation which the task must have involved, of the supply trains, of the surveying and planning, no word has come down to posterity.

A work planned on so vast a scale seems to have aroused superstitious fears of undue interference with the given pattern of Nature.[a] During the course of the work Chhin Shih Huang Ti went on his tour to the coast of Chekiang and Shantung, where he died, and in −209 Mêng Thien fell a victim to the intrigues of the eunuch Chao Kao[1] and the first emperor's feeble successors. The *Shih Chi* reports[b] lengthy speeches on the occasion when Mêng Thien received the envoy bringing the order that he should commit suicide. Finally he said:

Indeed I have a crime for which I merit death. Beginning at Lin-thao, and extending to Liao-tung, I made ramparts and ditches over more than 10,000 *li*, and in that distance it is impossible that I did not cut through the veins of the earth (*pu nêng wu chüeh ti mo tsai*[2]). This is my crime.[c]

He then swallowed poison and so killed himself.

Now although there are good reasons for thinking that the story is a literary invention which Ssuma Chhien felt that he had to reproduce from some earlier source, it has considerable interest as one of the oldest references to geomantic doctrine (*fêng-shui*[3]).[d] In his conclusion to the chapter, the great historian did his best to discredit it, saying that Mêng Thien's death was the result of his lack of humanitarian regard for the people, and nothing whatever to do with the veins of the earth.[e] But there is intrinsic interest in the story, and it may well have some connection with ancient engineering controversies about the proper methods[f] of dealing with watercourses, at which we shall have to glance later on.[g]

The building of the Great Wall can only be viewed in its correct historical perspective by realising that what was done in the Chhin was not so much the construction of a continuous work entirely *de novo*, but the extension and linking of a number of walls which had been built previously by the various Warring States. Their purpose was to break the shock tactics of the nomadic horse-archers, or of cavalry belonging to feudal States which had adopted such tactics. This was first perhaps brought out by a Ming scholar, Tung Yüeh,[4] in his *Chhi Kuo Khao*[5] (Investigation of the Seven States),[h] and has been set forth by a number of modern authors.[i]

We have already sketched the pre-Chhin walls in Vol. 1, Fig. 12, based on a chart

---

[a] The Wall, as is well known, has remained for centuries a focal point of Chinese folksong and legend. The story that one of the workers (at least) was buried in it gave rise to the famous *Ballad of Mêng Chiang Nü*[6] (tr. Wymsatt, and Needham & Liao) on which see Ku Chieh-Kang (1); Hsiao Yü (1); especially Lu Kung (1). The burying of objects in walls was only one department of builders' magic in China, on which Eberhard (20) has written, drawing from the *Lu Pan Ching*[7] (cf. Vol. 4, pt. 2, p. 44).

[b] Ch. 88, pp. 4*b* ff.   [c] Tr. Bodde (15), p. 61.

[d] See Vol. 2, pp. 359 ff., Vol. 4, pt. 1, pp. 239 ff. Han mentions of the selection of suitable sites for buildings and fortifications often use words which imply divination, e.g. 'chan hsiang ti shih'[8] in *Hou Han Shu*, ch. 88, p. 4*b*. The fear of 'interfering with the earth (*tung thu*[9])' embarrassed mining and engineering not infrequently in later centuries (cf. Yang Lien-Shêng (11), pp. 68 ff.).

[e] P. 6*a*.   [f] I.e. the most natural, hence least impious, methods.

[g] Sect. 28*f*, see pp. 249 ff. below.   [h] Cf. especially ch. 3, pp. 8*a*, 18*a*.

[i] E.g. Hsü Chü-Chhing (1); Wang Kuo-Liang (1); Mu Shou-Chhi (1); Lattimore (1, 2); Puini (1).

[1] 趙高   [2] 不能無絕地脈哉   [3] 風水   [4] 董說   [5] 七國考
[6] 孟姜女   [7] 魯班經   [8] 占相地勢   [9] 動土

constructed by Lattimore.[a] They began in the late −4th century with a number of works of defence against the steppe nomads. About −300 the Chhin State built one wall from the Thao[1] River in Kansu[b] north-eastwards to somewhere near the point in northern Shensi where the present Great Wall turns northward again, so that it joined an earlier Wei wall of −353 which ran further north-eastwards along the edge of the Ordos desert towards the descending loop of the Yellow River; this line was later followed more or less (as we have seen) by the Sui Great Wall. The State of Chao, under Wu Ling,[2] also about −300, built another running from Kao-chhüeh (117) eastwards north of the Yellow River to somewhere between modern Kalgan and Peking.[c] The State of Yen, about −290, built a third fortification from near the eastern end of the Chao wall to the lower valley of the Liao River in Manchuria; this was the precursor of the Willow Pale. Both these were very near, if not exactly on, the line taken by the Great Wall more than half a century later. But the walls were not only to keep out the barbarians or to prevent marginal Chinese from joining them, for some were built along the boundaries of individual feudal States. In −353 Wei built the north–south one in Shensi to protect its western territories from Chhin,[d] and later, after this had proved useless, another one further back across the Yellow River valley somewhere near Loyang. Similarly Chhi, in northern Shantung, had long before (c. −450) built a wall running east and west north of Lu State, to protect itself from the growing force of Chhu in the south. At the beginning of the −3rd century, Chhu also set up a wall somewhere fairly high above Nan-yang (14) protecting the northern side of the Han Valley, as a defence against Chhin. 'It was as if', says Lattimore, 'the "cellular" units of walled cities, each with its adherent population, were grouped in agglomerations of cells, each with a wall which identified it as a major unit compounded of minor units.' One cannot suppose that the resources of these minor 'metazoan' social entities permitted of much more than a continuous dyke and ditch to aid defence; probably these early walls were something like the series of Romano-British 'Devil's Dykes' between the ancient fen and forest with which we are familiar in East Anglia. It is very unlikely that they were faced with brick or stone as were large sections of the later Great Wall.

The best estimate of the Great Wall's length puts it at 3,930 miles if all its branch walls are counted, and 2,150 miles if the main line alone is taken.[e] The distances divide (going westwards) as shown in Table 65. These figures are really remarkable when the relatively primitive state of transportation in the Chhin and Han is taken into consideration, as Western historians of technology, for example Horwitz (6), have agreed.[f]

---

[a] (1), pp. 336, 389, 403, 414, 430, 434. One of the chief sources of our information is *Shih Chi*, ch. 110, on the Huns. For other maps of the walls see Herrmann (1), map 16; Thung Shih-Hêng (1), p. 3.

[b] Starting from about the same point as the Chhin Great Wall did later.

[c] This long projecting western tongue was certainly intended to take in the fertile lands between the Yellow River and the Gobi desert which have been irrigable since ancient times. Cf. p. 272.

[d] The name of its civil engineer has come down to us—Lung Ku.[3]

[e] Details of the branches and loops are best sought in the chart of Clapp (1).

[f] A present-day Chinese estimate officially accepted is given by Chin Shou-Shen (1) as the equivalent of 3,720 miles.

[1] 洮河   [2] 武靈   [3] 龍賈

Table 65. *Lengths of the divisions of the Great Wall*

		Miles
Main Line:	Shanhaikuan to the Eastern Bifurcation	300
	Eastern Bifurcation to the Yellow River (Outer Wall)	500
	North Shensi Frontier Wall	350
	Ninghsia to Liangchow (direct)	250
	Kansu, along the Old Silk Road to Chiayükuan	450
	Chiayükuan to Yümên-kuan and Yang-kuan	300
		2,150
Loops:	Manchurian Extension (Willow Pale)	400
	Eastern Bifurcation to the Yellow River (Inner Wall)	400
	South Branch Wall (Shansi border)	230
	Lanchow Loop	250
	Chhinghai Loop	400
	Miscellaneous offshoots	100
		1,780
	Total	3,930

As to the effectiveness of the Great Wall in keeping out the troops of nomadic horse-men, it was probably considerable.[a] Any breaking-down of the wall, or building ramps up to it, would allow time for the arrival of Chinese reinforcements. At an earlier stage in this book,[b] evidence was considered which suggests that the impenetrability of the Great Wall was a factor in the initiation of a series of shocks in tribal relations which transmitted themselves like a chain reaction to cause disturbances and invasions at the nomad-settlement frontier in Western Europe. Chinese engineering skill, and the genius of that people at that time for the organisation of mass labour projects, might therefore be said to have outplayed the protective capacities of the Roman Empire. For while the Romans were quite capable of building, in Hadrian's time, a wall and line of forts across the narrow neck of northern England, they never attempted to build what would have been the true counterpart to Chhin Shih Huang Ti's wall, namely one reaching from the mouth of the Rhine to the mouth of the Danube.[c] For half a millen-nium the Great Wall fulfilled its purpose, and only after the end of the +3rd century, when imperial Rome had been very largely 'barbarised', did the contracting power of the Chinese central State permit in its turn the establishment of numerous Turkic and Hunnish principalities north of the Yangtze and south of the Wall.[d]

[a] Contrary to the opinion of Gibbon (1), vol. 2, p. 82.
[b] Vol. 1, p. 185 above.
[c] It is interesting that a suggestion for just such a wall was proposed by the Anonymus of about +370 (*De Rebus Bellicis*, xx, tr. Thompson & Flower (1), pp. 72, 122).
[d] I confess to some hesitation in reproducing this argument. It is true that before +300 large numbers of Germanic people had been absorbed into the Roman Empire as individuals, but the mass break-through came just after the time of Alaric, who had sacked the city of Rome in +410. Neverthe-less, it may well be true that in the half-millennium from −200 to +300 the Chinese defences acted as a kind of sea-wall which turned nomadic expansion westwards upon a course which could not be diverted.

The most ironical thing about the Wall was that medieval Europeans (and Arabs) were under the impression that their own forefathers had built it. The biblical characters Gog and Magog,[a] according to many versions of the Alexander-Romance,[b] had been driven eastwards by Alexander the Great, and confined, with twenty-two nations of evil men, behind an iron wall which divine assistance had helped him to build.[c] In the last days they would break through his gate and overrun the world (Fig. 738).[d] As de Goeje (1) maintained long ago,[e] there can hardly be any doubt that this legendary engineering work (which constantly appears in Western medieval maps) was an echo of the real Great Wall itself. The story is mentioned in the Koran,[f] in connection with 'Him of the Horns' (Dhū al Qarnain), i.e. Alexander the Great,[g] so that it must have been contained in Syriac sources. Some Arab travellers were said to have seen the Wall of Gog and Magog, notably Sallām al-Tarjamān (the Interpreter),[h] who gave an account of it afterwards to the geographer Ibn Khurdādhbih (d. c. +912), and read to him his report to the Caliph.[i] Sallām had seen an 'iron gate' in the Wall, and others averred that it was made of alternate courses, red and black, of bronze and iron or lead. Efforts have been made to identify this iron gate (after all, a rather common geographical name in all parts of the world) with passes in the Urals, but Togan (2) suggests more plausibly that it was the Talka pass[j] in the Thien-Shan range. In any case both Franks and Saracens knew of a Great Wall, vague though their knowledge might be, and throughout the middle ages they ascribed its origin to the Macedonian world-conqueror.[k] But by the time that His Excellency Ysbrants Ides,[l] Ambassador from His Czarish Majesty to the Emperor of China, rode with his cavalcade through the Nan-khou Gate in October 1693, Europeans well knew who deserved the credit.

[a] Hostile to Israel; see Ezekiel xxxviii. 1–6 and the Apocalypse xx. 7, 8. Cf. Josephus, *Archaeologia*, 1, 123. We have already had occasion to mention them in Vol. 1, p. 168.

[b] Cf. Vol. 4, pt. 2, p. 572; and p. 674 below.

[c] See, e.g., Lloyd Brown (1), p. 99; Cary (1), pp. 18, 130 ff., 295 ff.; Baltrušaitis (1), pp. 184 ff.; and the special monograph of A. R. Anderson (1). This fear of the peoples of Eastern Asia is very reminiscent of the converse fear which they had of Westerners—as witness the numerous devils with blue eyes and red hair depicted on the Tunhuang frescoes, a feature which much impressed me during my first residence at Chhien-fo-tung.

[d] From the *Revelationes* of Pseudo-Methodius, ed. Michael Furter at Basel in +1499. Cf. Sackur (1).

[e] And Marquart (1), p. 86. C. E. Wilson (1) demurs, but his arguments are weak.

[f] *Holy Qu'rān*, Surah XVIII, verses 82 ff.; R. Bell edn. p. 281.

[g] Confusion with Moses.

[h] He was perhaps a Jew of Khazaria (see Vol. 3, p. 681), and certainly interpreter to one of the Abbasid Caliphs, in all likelihood al-Wāthiq (+842 to +847; Beckford's Vathek), the predecessor of al-Mutawakkil. Sallām was sent by his master on a mission of investigation about +843 because of a rumour that the wall of Alexander had been breached.

[i] The somewhat fabulous passage in Ibn Khurdādhbih's *Kitāb al-Masālik w'al-Mamālik* of +846 has been translated by Barbier de Meynard (2), pp. 190 ff.

[j] This pass connects the town of Kuldja on the Ili River with more easterly points in Sinkiang, and with China, by traversing the mountains to the north of the valley.

[k] We are much indebted to Prof. D. M. Dunlop for some of the information contained in this paragraph, given to us before the appearance of his discussion of the subject in his book on the Khazars, (1), pp. 190 ff.

[l] His own account, p. 60.

¶ Quomodo Gog et Magog exeūtes de caspys mōtibus obtinent terrā Israhel

¶ In nouiſſimis vero temporibus ſecunduʒ Ezechie-
lis prophetiam que dicit. In nouiſſimo die conſuma-
tiōis mundi exiet Gog et Magog in terram Israel: ꝗ
ſunt gentes et reges quos recluſit Alexander magnus
in finibus aquilonis et in finibus ſeptentrionis Gog
et Magog Moſech et Thubal et Anog et Ageg et

Fig. 738. Gog and Magog breaking through Alexander's gate in the last days to overrun the world, a page from the *Revelationes* of Pseudo-Methodius (Furter ed.), one of the versions of the medieval Alexander-Romance corpus. After Cary (1). The caption at the top says: 'How Gog and Magog issuing from the Caspian Mountains shall obtain the land of Israel.' Below: 'Truly in the last days, the time of consummation of the history of the world, according to the prophet Ezekiel, Gog and Magog shall come forth into the land of Israel, for these are the peoples and the kings that Alexander the Great shut up in the extremest northern and eastern regions; Gog and Magog, Meschech and Thubal, Anog and Ageg and...' Cf. Ezekiel xxxviii and xxxix, where of course there is nothing about the Wall.

(d) BUILDING TECHNOLOGY

(1) INTRODUCTION

Though architecture (*chien chu*[1]) is a subject which lies so near to the fine arts that it hardly comes within the scope of the present book, it has a technological basis which we could not dare to omit. This problem will face us again, as in ceramics, where the fundamental discoveries, such as that of porcelain, will concern us deeply, but not the styles and details of the artistic triumphs of the potters. In all such instances the reader must have recourse to the abundant literature which already exists on the aesthetic aspects of Chinese civilisation; it lies off our course, and we can only glance at it.

To the first European visitors in the sixteenth and seventeenth centuries, Chinese buildings must have seemed very strange. They attracted more detailed interest and study in the Chinoiserie period of the eighteenth century, as witness Sir William Chambers's *Designs of Chinese Buildings*, etc., of +1757, in which, however, the principles of the building construction were only half-understood.[a] In the following century there appeared a meritorious account by an army doctor, John Lamprey, who wrote in 1866; and later one of the less inspired contributions of Edkins (15), the sinologist-missionary to whom in general we owe much.

Recent studies, however, have partially made up for former neglect. There are the impressive folios of Sirén (1) and Boerschmann (1, 8) on Chinese architecture in general, largely devoted, of course, to the more important works, temple halls, palace buildings, and the like. For temple architecture, the early work of Combaz (4) on the imperial temples in Peking was extended by Boerschmann (2) and Prip-Møller (1) to Buddhist and Taoist temples in many parts of the country. The imperial palaces have been described and photographed in luxurious style by Combaz (3); Sirén (3); Ogawa (1); and Okuyama, Ito, Tsuchiya & Ogawa (1). Sirén (4) devoted a special work to the walls and gates of the city of Peking, while studies of its plan are due to Chu Chhi-Chhien & Yeh Kung-Chao (1) and Rasmussen (1) among others. A different kind of architectural complex is formed by the imperial tombs, with their huge embattlemented tumuli, and their sacrificial halls among the hills, preceded by long avenues of stone figures; the description of these was tackled first by Combaz (2) and afterwards by Bouillard & Vaudescal (1); Grantham (1) and other writers. For the pagodas scattered throughout the country Boerschmann's account (4, 7) is the standard one, while Combaz (5) traced their origin and evolution from the Indian *stūpa*.

After some time the reader begins to feel, however, that he is suffering from a surfeit of beautiful photographs and too much archaeology and comparative religion, and would prefer more precise information regarding the functional basis of the building construction. At the same time he also feels that he would like to give less study to the highest flights of Chinese architecture, and more to the regular, common-

---

[a] Lord Macartney's acute and appreciative comments on Chinese architecture (in Cranmer-Byng (2), p. 272) are well worth reading.

[1] 建築

place, but regionally diverse and often very attractive, dwellings of the townsman and the farmer. Thirdly, tired of looking at so many examples of the exquisite curving roofs, he would like some clearer ideas about the history of the remarkable style which produced them. These needs have evoked some response in the West, if not as yet a wholly satisfactory one.

The first study which provided good drawings of the timber construction, roof supports, etc., of a Chinese Buddhist temple was that of Hildebrand (1), some sixty years ago, but it gave no Chinese technical terms. This gap was filled more or less well by the monograph of Kelling (1), who had the advantage of being able to study the great Sung compendium of architectural practice, the *Ying Tsao Fa Shih* (see on, p. 84), lithographically reproduced in 1920. Defects this monograph may have, but there is nothing else of the kind available,[a] and among its merits is the fact that it devotes particular attention to domestic architecture. So do the excellent papers of Spencer (1) and Skinner (2), more recent in date.[b] On the historical side, there is little save the monograph of Bulling (2) (originally a thesis), but Kelling & Schindler investigated ancient building technique in China with philological methods, and in Japanese there is a noteworthy treatise by Ito Seizo (1).

The best all-round book in a Western language on Chinese architecture has for many years been that of Mirams (1); its illustrations, though few, were chosen with great care, and its text, though brief, is clear and helpful.[c] It has been supplemented,

[a] More or less parallel are the works of Baltzer (1, 2) and Yoshida (1) on Japanese architecture. The biggest contribution in this field is the six-volume treatise of Amanuma Shun-ichi (1), and the most complete bibliography that of Murata Jirō (1).

[b] For some of the finest examples of Chinese private houses and gardens the reader is referred to the album of photographs by Juan Mien-Chhu (Henry Inn), edited by Li Shao-Chhang with contributions by several other scholars. Stress is well laid upon the inseparability of the house and garden in traditional China.

[c] The excellent conspectus of Boyd (2) and the albums of Hsü Ching-Chih (1) and Speiser (1) unfortunately appeared too late to be of assistance to us in writing this Section. For comparative purposes it would naturally be desirable to have at hand, when studying Chinese architecture, some general account of the building styles of all the principal cultures. Unfortunately nothing seems to exist which quite meets this need. Moreover, the Chinese sections of books which purport to describe the architecture of all peoples are in general very bad, and to this the well-known treatise of Fletcher (1) is no exception. Least open to such criticism is the *History* of Choisy (1) with its beautiful and clear drawings, but this is now half a century old. It is better, therefore, to have recourse to specific works. Comparisons with India may be assisted by the work of Fergusson (1), unfortunately now somewhat outdated, and conspicuously lacking in understanding as regards Chinese architecture. Though a pioneer in the European study of building in Asia, he could write of the Chinese, 'there is not perhaps a more industrious or, till the late wars, happier, people on the face of the globe, but they are singularly deficient in every element of greatness, whether political or artistic'. On India Fabri (1) and Wu No-Sun (1) are now available. For the building of ancient Egypt, we may select Choisy (2); Capart (1); and Flinders Petrie (2). Ancient Greek achievements may be followed in Perrot & Chipiez (1), vol. 7, or in Dinsmoor (1). For Roman construction there is Choisy (3); Blake (1), and of course Neuburger (1). Byzantine and Romanesque buildings have been analysed by Choisy (4) and Jackson (1). On Muslim ones Briggs (1) and Cresswell (2, 3) assist. On Western European medieval Gothic architecture there is an abundance of books familiar to all, such as that of Jackson (2). But for ancient and medieval building materials and technology as such, more important in the present context, aids to study are rarer. We now have the wide but thorough survey of Davey (1). The reader should also know of the valuable monographs of Salzman (1); Innocent (1); and Briggs (2, 3), dealing with all aspects of the subject. A special contribution on medieval building carpentry has been made by Moles (1). The paper of Murphy (1), though very appreciative of past Chinese architectural achievements, is mostly concerned with the perpetuation of some of their distinctive stylistic elements in modern steel and concrete buildings. He himself was a pioneer in this.

but not superseded, by the more extensive account of Sickman & Soper (1).[a] From the photographs of Boerschmann (3) an excellent idea can be gained, by those who have never been in China, of the way in which Chinese builders sited their constructions so as to blend into the scenery and topography in the most intimate way.

For those who understand Chinese, of course, a vaster field of literature lies open, albeit less well documented pictorially perhaps than the Western productions. Useful summaries have been written by the doyen of Chinese architectural historians, Liang Ssu-Chhêng (3, 1). An enormous mass of information is contained in the Chinese *Journal of the Society for Research in Chinese Architecture*, recourse to which is indispensable for anyone wishing to penetrate beyond the surface of the subject.[b] The Society has also published monographs and portfolios of plates, such as the contribution of Liang Ssu-Chhêng & Liu Chih-Phing (1) on some of the most essential features of Chinese building construction.[c] Twenty-five years later the second author produced an invaluable systematic treatise on Chinese architecture and its development (Liu Chih-Phing, 1).[d] And in recent years there has been a spate of valuable publications on domestic architecture,[e] architectural detail,[f] and the like, which we may mention again as need arises. Lastly, a bookcase of modern Chinese architectural works would not be complete without some of the excellent albums of photographs now available (e.g. Anon. 37). And the study of the history of building technology has newly arisen in neighbouring countries of the culture-area.[g]

(2) THE SPIRIT OF CHINESE ARCHITECTURE

In no other field of expression have the Chinese so faithfully incarnated their great principles that Man cannot be thought of apart from Nature, and that man is not to be divided from social man. Not only in the great constructions of temples and palaces,

[a] Note may be taken of the parallel work on Japanese architecture, Paine & Soper (1). We also now have Sadler (1); Alex (1); With (1); Kirby (1) and Engel (1). On Japanese temples see the splendid monographs of Tange & Kawazoe (1) and Soper (2).

[b] The Society was always greatly concerned with the restoration and protection of ancient buildings as national cultural monuments, and during the past dozen years immense efforts have been successfully made in this field—as I myself have seen in 1952 and 1958. For an example of the discussions which go on one may look at an article by Liang Ssu-Chhêng (11). They have at times been rather violent as traditionalism conflicted with practical demands, and veneration for antiquity with rejection of the heritage. Glimpses of them in English will be found in a recent book (somewhat eccentric not only in its shape) by Hsü Ching-Chih (1), an architect notable as co-designer of the Nanking Museum (1935), a vast building in Liao style, but constructed, down to the last bracket-arm, in reinforced concrete. He recognises that this was not the solution of the problem for modern Chinese architecture, which has tended to place traditional roofs, like elaborate hats, on ferro-concrete boxes; and honestly declares that the problem is not yet solved. As an introduction to Chinese building technology Hsü's book has few of the merits of others just mentioned, but it is particularly useful for anyone interested in the dreadful modern constructions of Kuomintang China and Thaiwan, buildings which have ignored the cultural heritage more completely than the work of any of the marxist architects of People's China.

[c] See Hummel (21).

[d] We greatly regret that this was not at our disposal when the present Section was written. A similar work by Yao Chhêng-Tsu, Chang Chih-Kang & Liu Tun-Chên (1) was also unavailable to us. It contains a valuable glossary of technical terms.

[e] Liu Tun-Chên (4); Chang Chung-I et al. (1).

[f] Anon. (16, 38, 39).

[g] See, for example, the Korean historical vocabulary of architecture issued in 1955 (Anon. 40).

but also in the domestic buildings scattered as farmsteads or collected in villages and towns, there was embodied throughout the ages a feeling for cosmic pattern and the symbolism of the directions, the seasons, winds and constellations.[a] Preferring moreover to construct all buildings for the living in the relatively impermanent media of wood and tile, bamboo and plaster, the Chinese made the use of horizontal spaces the keynote of their architecture and planning;[b] and though some buildings might be of one or two storeys above a ground floor, their height was strictly subordinated to the large-scale horizontal perspectives of which they might form a part. Significantly, the pagoda, the typical Chinese form of the vertical spire, retained to the end the apartness of its foreign origin, and continues to grace isolated hills outside cities, removed from the characteristic architectonic wholes, though still related, as nothing could fail to be in China, with the landscape quality of all the planning. Chinese architecture was always with, not against, Nature; it did not spring suddenly out of the ground as if aiming to pierce the sky like European 'Gothic' building,[c] nor did it force trees and plants into Renaissance straight lines, lozenges and triangles. It took every possible advantage of the natural beauty of the site, of woods and hills; and this is true not only of such marvels of art as the Summer Palace (I Ho Yuan[1]) near Peking (Fig. 739, pl.)[d] but also of the ordinary Szechuanese farmhouse surrounding its threshing-floor and backed by clumps of bamboos, at the head of its valley of terraced rice-fields (Fig. 740, pl.).

The fundamental conception of Chinese architecture arranges one or more courtyards (*thing yuan*[2]) to compose, sometimes in a very complex way, a general walled

[a] Cf. the discussion in Sections 13 *d* (Vol. 2) and 26 *h* (Vol. 4, pt. 1) above. We shall touch upon these matters again, pp. 73, 76, 112, 121, 143 below.

[b] I remember once sitting long with a Chinese friend on a bench in the great square surrounding one of the finest of the cathedrals of Europe, Chartres, and meditating together, in the spirit of Henry Adams, on the implications of the extraordinary contrast between Chinese horizontal lines and European vertical soaring. Of course the European tradition also contains horizontal elements, as in the vistas of baroque palace gardens. And it has sometimes been said that the medieval cathedrals, like the skyscrapers of modern Western culture, were necessitated by the cramped conditions of the fortified city or the general centripetal tendency of urban life. But such explanations cannot be the whole story. Chinese cities, though invariably fortified, were never very constricted, and left plenty of room for noble buildings both secular and religious. One cannot get rid of the suspicion that the soaring Gothic arch and spire was the manifestation of a theology which laid its greatest emphasis on transcendence—but for all three of the *san chiao* (Confucianism, Taoism and Buddhism), the divine was primarily immanent in the world. Why then should a building seek to take flight above and beyond the world?

[c] Even the theme of pyramid, *ziggurat* and *teocalli*, beloved of ancient civilisations, was softened in China into the rounded breasts of tumuli or artificial hills; or tapered to the blest elegance of pagoda towers (cf. p. 137); or flattened down to the concentric terraces of cosmic altars (cf. Vol. 3, p. 257); or soberly transformed into a scientific instrument (cf. Vol. 3, p. 296).

[d] The famous Lei family of Peking engineers and architects, founded by Lei Li[3] (*fl.* +1520 to +1565), was long connected with this. The beginning of it was the Garden of Clear Ripples (Chhing I Yuan[4]) established at Wan-shou Shan[5] west of Peking by the Chhien-Lung emperor in +1751 and +1761 (cf. Malone (2), pp. 111 ff.). Lei Shêng-Chhêng[6] (+1729 to +1792) must have been the main architect in these years and was succeeded by his son Lei Chia-Hsi[7] (+1764 to 1825). Then when the Dowager Empress Tzhu-Hsi[8] constructed the present Garden of Peaceful Longevity (I Ho Yuan[1]) between 1886 and 1891, she could still call upon a member of the Lei family, Lei Thing-Chhang,[9] a descendant of Lei Li in the tenth generation. For details of this remarkable family see Chu Chhi-Chhien & Liang Chhi-Hsiung (3), pp. 107 ff., (4), pp. 84 ff. Cf. pp. 75, 81 below.

[1] 頤和園　　　[2] 庭院　　　[3] 雷禮　　　[4] 清漪園　　　[5] 萬壽山
[6] 雷聲澂　　　[7] 雷家璽　　　[8] 慈禧　　　[9] 雷廷昌

'compound' (*ssu ho yuan*[1])[a] (cf. Fig. 741). The main longitudinal axis is always (or ideally) north–south, but the chief buildings (*chêng thing*,[2] *chêng fang*[3]), or halls (*tien*[4]) are always placed transversely to it. They thus come one behind another, with the main entrance of each always in the centre of the long south-facing side.[b] These rectangular buildings may or may not connect, by means of open galleries (*lang*[5]) variously planned, with rows of smaller buildings (*phei thing*,[6] *hsiang fang*[7]) flanking the courtyards on both sides. On this system, enlargement is never carried out by adding to height, but by continual duplication of existing units, and growth in breadth or preferably depth. The entrance to a large composition is likely to be through a gate (*thai*[8]) which may resemble a city-gate in having a barrel-vaulted way below and a pavilion above (Fig. 742, pl.), or simply cap parallel tunnels with a heavy roof.[c] The halls may occasionally be of two storeys (in this case they are called *ko*[9]), and small open pavilions (*thing*[10]) or pavilions of more than two storeys (*lou*[11]) may be used, often away from the main axis, for the diversification of the general plan (Fig. 743, pl.). In domestic buildings, the courtyard system, as Liang Ssu-Chhêng (*3*) well points out, has the great merit that yards and gardens are made a part of the building, and not something additional and separate. The spaces between the low houses supply abundant air,[d] and all the windows look out internally on to plants and trees in the garden courtyards.[e]

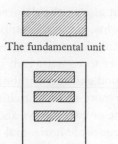

The fundamental unit

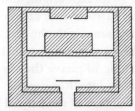

Three halls in one courtyard

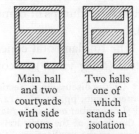

Main hall, connecting and side galleries, side rooms

Main hall and two courtyards with side rooms

Two halls one of which stands in isolation

Fig. 741. Typical ground-plans of Chinese buildings.

[a] Of course there are many exceptions to such a broad generalisation. In southern cities deep plots with narrow frontages are filled by the buildings, which have loggia arcades over the pavement in front of them (rather like Berne or Annecy). Courtyard and compound planning has been greatly modified during the past century, even in rural conditions. On the great variety of ground-plans see further, pp. 121 and 133 below.

[b] But the entry building or porch (*chhien thing*[12]) faces north, hence it is called the 'reversed room' (*tao tso*[13]).

[c] Cf. Mirams (*1*), opp. pp. 70, 71, and Fig. 754 (pl.). The entrance may also take the form of a hall with a large central gate (Mirams (*1*), opp. p. 69), or it may be of the triumphal *phai-lou*[14] type, whether roofed or bare (see on, p. 142).

[d] Dr John Caius, our second founder, who directed that one side of his Chapel Court should never be built up, in order to allow the passage of healthful breezes (+1573), would have appreciated Chinese architecture greatly if he could have known of it. Sometimes we suspect that he did know something of it, presumably through Portuguese sources, for he laid out his new foundation with a succession of symbolical gates named, like *phai-lou*, successively, the Gates of Humility, Virtue, and Honour.

[e] It might thus be said that while within the family there was probably less privacy in China than in Europe, there was much more for the family *vis-à-vis* the external world.

[1] 四合院   [2] 正廳   [3] 正房   [4] 殿   [5] 廊   [6] 配廳
[7] 廂房   [8] 臺   [9] 閣   [10] 亭   [11] 樓   [12] 前廳
[13] 倒座   [14] 牌樓

Thus man is not isolated from Nature.[a] As old Magalhaens wrote three centuries ago:[b]

There are two things to be observ'd; the first, that all the Cities and all the Palaces...are so built, that the Gates and Principal Apartments look towards the South; the second, that whereas we build our Lodgings one Story above another, the Chinese build upon the same Level one within another—so that we possess the Air and they the Earth.

The long rear wall is almost always unbroken by doors or windows, and forms, as it were, the ultimate statement of the plan, though not its climax, since the largest hall will be placed somewhere north of the central point, and there will be a diminuendo of constructions behind (i.e. north of) it. For a concrete illustration in the Chinese style of draughtsmanship of how the system works out, we may glance at a plan of a Confucian temple (Fig. 744) taken from the *Shêng Hsien Tao Thung Thu Tsan*[1] (Comments on Pictures of the Saints and Sages, Transmitters of the Tao).[c] Here may be seen the triumphal gateways (*phai-lou*; see on, p. 142), the succession of halls and avenues on the central axis, the arrangement of two subsidiary parallel axes, the enclosing walls, and divers pavilions for special uses.[d] Even domestic architecture, moreover, has an informally liturgical character, which reflects the ancient prescriptions of the books of social ceremonial, such as the *Li Chi*. Fig. 745, well illustrating this, shows a bird's-eye view of a traditional home in Peking.[e]

Another characteristic of the Chinese house, seen as soon as it rises above the level of importance of the simple thatched cottage, is that it is based upon a platform. This was presumably utilitarian from earliest times, to raise the living quarters and the passages between them above the mire of farmyard and caravanserai. But with the passage of time this facility developed into one of the most majestic elements of the full style, joining with others such as the great emphasis placed on roofs, and the advantage invariably taken of sloping ground.[f] In very important buildings, the terrace platforms may exceed 6 ft. in height, and may be constructed of white marble, access to them

[a] One of my most dominant impressions for some time after first returning to Europe from China was the sense of loss of intimate contact with the weather. The wooden lattice windows covered with paper (often torn), the thin plaster walls, the open verandahs outside every room, the sound of rain-water dripping in the courtyards and small patios, warmth made individual by fur-lined gowns and charcoal fires—everything gave a consciousness of Nature's moods, of rain, snow, wind and sunshine, from which one is utterly isolated in European housing.

[b] (1), p. 271.

[c] A small Confucian tract put together by a local official, Huang Thung-Fan,[2] in +1629.

[d] A particularly good perspective plan of this kind is seen on an inscribed stone stele of a temple at Jung-ho in Shansi (Wang Shih-Jen, 1).

[e] Rasmussen (1), p. 6; Liu Tun-Chên (4), pls. 90-92.

[f] In some large compositions the main entrance is at the lowest point, so that the visitor wanders through the series of courtyards and halls ever ascending. This trait has been mentioned already (Vol. 2, p. 164) in connection with a Taoist temple near Kunming, and it is found often also in Confucian temples such as that at Chhêng-kung, Yunnan, illustrated in Fig. 37 (Vol. 2). My memories of these beautiful places are heightened by the fact that when I first began to explore them in 1942 no one in China had much time to spare for conducting foreigners around. Thus with solitary steps I penetrated into an enchanted world of exquisite ancient buildings, able in silence to receive their full message of Chinese cultural values incarnated in wood and stone.

[1] 聖賢道統圖贊　　　[2] 黃同樊

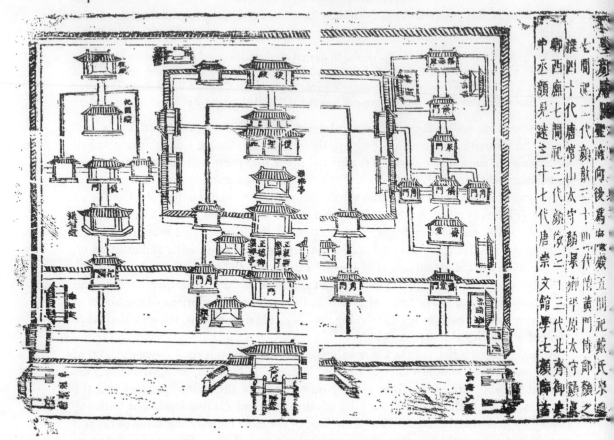

Fig. 744. Plan of a Confucian temple taken from a roughly printed popular edition of the *Shêng Hsien Tao Thung Thu Tsan*. It is that of Yen Hui, one of the 'four associates' of Confucius, and his favourite pupil (cf. Watters (2), pp. 2 ff.), at Chhüfou east of the temple of Confucius himself. Symmetry dominates the placing of the subsidiary fasting pavilions, and the sacrificial shrines of the sage's distinguished descendants. The entrance is through the *phai-lou* and gateway at the bottom. The text on the right mentions some of the men whose lesser votive altars are within the compound—e.g. on the east side, Yen Chih-Thui of the Sui (34th generation), the well-known author of the *Yen shih Chia Hsün*, and two Thang governors (40th generation); on the west side Yen Shih-Ku (37th generation), the famous Thang historical scholar.

being gained by two central stairways flanking a central inclined 'spirit-path' carved in high relief, besides which there will also be staircases at the sides of the courtyard. 'The wooden pillars', says Sirén (1), 'rise above the supporting terraces, which often reach considerable heights, like tall trees on mounds and hillocks. The lines of the far-projecting curving roofs suggest the long wavering branches of the cryptomerias, and if there are any walls, they almost disappear in the play of light and shade produced by the broad eaves, the open galleries,[a] the lattice-work of the windows, and the balus-

---

[a] For the numerous open passages between the buildings within a single compound, the language contains a special word, *khun*.[1,2,3] Though bone and bronze forms of it seem to be lacking, it is strongly evocative of the ground-plan of an ancient architect. See further on plans below, p. 143.

1 壼   2 㙥   3 㾛

trades.' This point is important, for the walls in Chinese buildings are indeed of secondary significance; it is the terraces and overhanging roofs which are decisive for all outer aspect. As we shall shortly see more fully, the walls of Chinese buildings are always curtain walls; they take no share as bearing walls in the support of the structure.[a] In contrast to Greek structures, Sirén goes on, the gables have no function but to terminate the long hall; they lack every architectural emphasis and in many cases are not even intended to be seen. This follows from the choice of the long side as the main façade and the transverse position on the axis of the composition.

The main features of Chinese building may be summarised then[b] as follows: (a) emphasis on the roof, and its construction in sweeping curves, (b) formal grouping of buildings round rectangular courts, and marked attention to axis, (c) frankness of construction, the supporting pillars of the massive roof timbering being clearly visible, even when partly engaged in walls, (d) a lavish use of colour, not only in roof tiles, but on painted columns, lintels and beams, richly bracketed cornices, and broad expanses of plastered walls.[c] Here the third item of this observer particularly arrests our attention. Indeed, as another modern Western architect has said, the quality of greatest interest in traditional Chinese building is perhaps that it was functionally and structurally direct and honest.[d] The structural elements are distinct and explicit, all decoration being based on them. Clarity and rationality appears in plan, section and elevation, and in the high degree of harmony between the three. 'Chinese buildings, for all the sophisticated aesthetic that controls every part, have a look of being built by a master-craftsman or architect-engineer, as indeed they were.'

And all was under the aegis of the element Wood. Though we shall ponder upon this again,[e] we cannot forget it here. In spite of early knowledge of arch and vault, masonry and brickwork were always confined to terraces, defensive works, walls, tombs and pagodas. No Chinese house could be a proper dwelling for the living, or a proper place of worship for the gods, unless it were built in wood and roofed with tile. Immense consequences followed. The timber frame and the screen wall provided large span, compact supports, maximal unobstructed space, standardisation of planning, and flexibility of use. The timber structure blossomed into the elaborate roof which became the main feature of the building, and considerable height and monumentality, when occasion demanded, was not beyond its powers. As we shall see in more detail later, the most typical Chinese building was a rectangular hall on a terraced base, marked out by wooden columns joined together in a complex trabeate system. The heavy and overhanging roof was supported not by means of triangular trusses but by a network of beams of diminishing length, separated by struts and superimposed one over the other across the rectangle. Since the purlins were supported directly by this trabeation, any desired curve could be given to the roof. All round the building the

[a] This was well stated by Lamprey a century ago. He also appreciated fully that in Chinese planning, extent of area covered was much more important than height of structure.

[b] In the words of Murphy (1).

[c] See here especially Anon. (38); Anon. (16).

[d] Boyd (1), in a lapidary introduction used also in the following paragraphs.

[e] Cf. pp. 90, 167 below.

5

Fig. 745. Sketch of a traditional Chinese home in Peking (after Rasmussen).

beams were cantilevered outwards to form the generous overhang of the eaves, and as time went on there arose an extremely elaborate development of cantilever brackets piled up in tiers, enlarging thus the overhang to reach a maximum in Thang and Sung times, and embodying the most ingenious carpentry. The basic ground-plan was capable of expansion in all directions—lengthwise by unit-repetition, crosswise by verandah bays or 'transepts', and upwards by the addition of storeys in numerous forms, e.g. low verandahs with a high internal hall, or galleries in several storeys surrounding a lofty interior space. And marvellous polychrome ornament was added by the decorators.[a]

The words 'standardisation of planning' may have caught the reader's eye in the

[a] Who can forget, having once seen, the corner pavilions high on the grey walls of the Forbidden City at Peking (Fig. 746, pl.), their roof-tiles of imperial yellow surmounting their fretted walls and doors of a faded murrey colour?

preceding paragraph, and perhaps they awakened a curiosity which deserves satisfaction. Modern architecture has indeed been influenced more by Chinese (and Japanese) conceptions than is usually supposed. One basic Chinese characteristic has always been the addition at will of repeating units keyed to the size and scale of human beings— pillar-intervals or bays (*chien*[1]) in buildings,[a] and spaces in open-air courts. Such 'modules' occur in the theory and practice of modern architects such as le Corbusier (1),[b] some of whom (e.g. Frank Lloyd Wright) themselves worked in Japan, as Murphy did in China.[c] Le Corbusier's 'modulor' is a series of predetermined lengths intended for use as building measures, and generated by the application of the *sectio aurea* (0·618) to the height of a man taken as 6 ft.[d] It thus derives from Pythagoras through Fibonacci and Dürer.[e] But the harmonious assembly of units each fixed to the human scale is even more deeply Chinese, because it was universally, not occasionally, practised in that civilisation, a working norm rather than an aesthetic theory. Unit-repetition flexibly available for a variety of different purposes is now acclimatising itself in the West on other grounds also, having proved its value for example in the architecture of modern scientific laboratories. In Chinese tradition faithfulness to the human scale doubtless had some connection with the natural limitations of timber construction without geometrical trusses, but contemporary builders all over the world, though now far more capable than medieval Europeans of erecting structures entirely out of scale with individual men, are coming to appreciate more and more the sober humanism of the Chinese style, which was certainly not conditioned by the materials alone. Its multiplication of relatively small spaces horizontally is in many ways more satisfying than the attempt to bridge ever larger and loftier spaces, only to dwarf their human inhabitants.

In this connection it is significant that the traditional Chinese architects and builders were extremely conscious of standard dimensions and right proportionality. Soper remarked in passing that the Sung manual of about +1100, the *Ying Tsao Fa Shih*, which we shall study in detail presently (p. 84), uses a particular proportion as a modulor, namely the end elevation (*kuang*,[2] *kao tu*[3]) of the horizontal corbel bracket arm (the *hua kung*,[4] cf. pp. 93, 95).[f] Actually, this had been the discovery of Liang Ssu-Chhêng (9) when he was studying the Tu-Lo Ssu[5] temple at Chi-hsien, a Liao

---

[a] In the Thang and other dynasties there were sumptuary laws which regulated by this measure the sizes of the buildings in which commoners and officials of different ranks were entitled to dwell.

[b] This was first brought to our attention by Dr Otto van der Sprenkel.

[c] Cf. p. 76 below. The definition of Japanese room sizes by the number of standard rectangles of *tatami* matting which they are to contain is a system that has become very widely known in the West. Cf. Maraini (1), p. 337. It was originally Chinese (see p. 89 below).

[d] Cf. Entwistle (1).

[e] Cf. d'Arcy Thompson (2), pp. 511, 643, 649.

[f] Sickman & Soper (1), p. 252. In fact he said the depth, but this is rather confusing. Presented with a series of timber beams of rectangular cross-sections (as in the immediately following quotation) we tend to think of the long side as the breadth and the short side as the height, because that is how one would lay them down on the ground. But in building technology, for obvious strength-of-materials reasons, the beam is fixed with its long side vertically. Hence we should speak of this as the height, and of the short side as the thickness. Engineers and architects always call the vertical dimension of a beam in place the depth.

[1] 間   [2] 廣   [3] 高度   [4] 華栱   [5] 獨樂寺

dynasty (late +10th-century) structure.[a] This module was called a *tshai*,[1] an area measured in Chhing times as 2×1 *tou-khou*,[2] a *tou-khou* being a relative dimension from 6 in. down to 1 in. in all of eleven different sizes (6, $5\frac{1}{2}$, 5, $4\frac{1}{2}$, etc.). The height of one corbel arm (*kung*[3]), plus that of the wooden blocks (*tou*[4]) carried on each end, was 2 *tou-khou*, and the thickness of the corbel arm was 1 *tou-khou*; making 1 *tshai*.[1] The corbel arm alone had the height of 1·4 *tou-khou*, and this was called a *tan-tshai*[5].[b] All other measurements of the building were derived as multiples of these.

In Sung times the module system was a little different. One *tshai*[1] measured 10×15 *fên*[6], where *fên*[c] was a relative dimension ranging from 0·6 in. down to 0·1 in. The height of corbel arm plus block (10×21 *fên*) was then called *tsu-tshai*,[7] the corbel arm rise alone (10×15 *fên*) was the *tan-tshai*,[5] and the block height alone (10×6 *fên*) was the *chhi*.[8]

Although the term *tshai*[1] had this special technical significance, its use was certainly related to the fact that the timber beams and baulks came in a variety of predetermined sizes, also called *tshai*.[1] 'The height and thickness of the corbel brackets must always correspond with (those of) the standard timbers.'[d] The *Ying Tsao Fa Shih* says:[e]

As for the roof-structures of houses and halls, everything depends on the standard size of the materials chosen (*chieh i tshai wei tsu*[9]).[f] There are eight standard sizes (of the height and thickness of timber baulks), and these are used in accordance with the size of the building. [Measurements follow, ranging from 9×6 in. to $4·5 \times 3$ in. Notes specify the type of buildings, partly in terms of numbers of rooms, for which each grade of cross-section is suitable.]

When the measure called *chhi*,[8] six standard parts (*fên*) in height and ten in thickness, is placed on top of the *tan-tshai*,[5] the whole is called the *tsu-tshai*[7].[g]

The height (*kuang*[10]) of each standard beam is divided into 15 standard parts, and the thickness (*hou*[11]) (always) corresponds to 10 of these.

Thus the height and depth of the roof, the length, curvature and trueness of the members, and the ratios of column and post heights (*chü chê*,[12] lit. raising and cutting) (in the structural cross-section adopted for any particular ground plan), together with the right use of square and compass, plumbline and ink-box—all proportion and rule depends on the system of standard timber dimensions and the standard divisions of these.

This system was undoubtedly inherited from Thang practice, probably not new even then. Thus the entire Chinese building was designed in terms of standard modules, and modulors of variable absolute size, none ever out of scale with man himself. Right proportion was thus safeguarded, and relational harmony preserved, whatever the magnitude of the structure.[h]

---

[a] Cf. p. 131 below. Hsü Ching-Chih (1), pp. 215 ff., has some interesting things to say about the standard Chinese modules, but he does not adequately explain them.

[b] Lit. 'single, or unit, timber'. See *YTFS*, ch. 30, pp. 11*a*, 19*a*, *b* (vol. 3, pp. 183, 198).

[c] This technical term is not to be confused with the usual name for tenths of an inch; the commentary says it was pronounced with a different tone.

[d] *YTFS*, ch. 4, p. 4*a* (vol. 1, p. 78).

[e] Ch. 4, pp. 1*a* ff. (vol. 1, pp. 73 ff.), tr. auct. with Dr Else Glahn.

[f] Cf. the parallel statement in ch. 1, p. 8*b* (emended in vol. 1, p. 15).

[g] Lit. 'full timber'.

[h] Cf. the quotation on p. 82 below. Later on, we shall come upon the use of basic units of a relative character in anatomy, physiology and medicine also (Sects. 43, 44). For instance, the biological inch for

[1] 材   [2] 斗口   [3] 栱   [4] 枓   [5] 單材   [6] 分   [7] 足材

[8] 栔   [9] 皆以材爲祖   [10] 廣   [11] 厚   [12] 舉折

To unravel the moulding of the various building patterns of China by the concrete needs of the different parts of the social order at various stages would constitute a special task in itself, and an important one. By Western historians of Chinese architecture it has been strangely neglected, and only now are the Chinese themselves beginning the task, as in the monograph of Liu Tun-Chên (4). For example, the large joint-family system, in which the sons did not quit the ancestral compound on marriage, must have evoked powerfully the differentiation of a multiplicity of halls and courtyards within the single enclosure. So also the tendency of great families in the Han to establish what were almost factories in their large dwellings[a] must have worked in the same direction. After the Thang the wealthy family was not so often a centre of production of commodities, but then the effects of artisanal family production on urban buildings could certainly be studied in Chhang-an and Hangchow[b] as well as the great cities of subsequent times. Rural building must always have centred round the farm as an agricultural production-unit, and domestic architecture must have waxed and waned in accordance with the prosperity of countrysides and the particular social strata (poor peasants, rich peasants, scholar-gentry, etc.) in question. As for the formative influence of the bureaucratic-feudal State on Chinese public architecture, it was evidently capable of large-scale planning from the start,[c] and its splendid works were always essentially secular in spirit. Immanentist, ethical, hierarchical, liturgical, axial, symmetrical—these were the qualities of Confucian architectonics. Taoist influence was on the side of immanence too, of course, but it tended to softer, less severe, formulations, finding architectural expression in beautifully sited buildings and romantic ensembles, developing the garden and the artificial landscape. Buddhism went along with the Taoists in these matters, adding however the pagoda derived from the Indian *stūpa*, and the *phai-lou*, the triple or five-span gateway derived from the Indian *toraṇa*, familiar from Sānchī.[d] Typically, the enceinte wall of the compound turned into long cloisters facing inwards, the gates held their place on the main south–north axis, the pagoda, at first central, was duplicated symmetrically, or pushed to one side or northward, and finally exiled to the outer grounds; north of its primitive mid-

---

any particular human being of whatever age depended upon the length of a particular bone in the hand. This sophisticated application of a relative measure threw at once into relief disorders of normal proportionality, and facilitated the location of desired points in the anatomical pattern. How characteristic these ideas and practices were, whether in building or biology, of a culture saturated with the organic view of the universe!

[a] Cf. Vol. 4, pt. 2, p. 26.

[b] Here a special case was constituted by the 'great towers of stone' which Marco Polo saw, solid and moated. In fact these buildings were fireproof warehouses (*tha fang*[1]), caravanserai with more than a thousand rooms rented to merchants for their goods according to need on monthly leases. The descriptions in *Tu Chhêng Chi Shêng*, ch. 12 (p. 100) and *Mêng Liang Lu*, ch. 19 (p. 299) have been translated and discussed in Moule (11), (15), pp. 24 ff. and Gernet (2), p. 38. For an eye-witness account of the Canton towers in 1903 see R. D. Thomas (1), p. 3. Cf. p. 90 below.

[c] Cf. what was said in Vol. 4, pt. 1, p. 53, about the organisation of the meridian arc survey early in the +8th century.

[d] On these see pp. 142 ff. below.

[1] 塌房

line position came the worship-hall, north again the lecture-hall, and still further north the dormitories and living-quarters.[a]

As Soper (2) has emphasised, in his luminous discussion of Chinese and Japanese temple planning, there was never any disjunction or dividing line between the secular and the sacred in East Asian architecture. Temple buildings were repeatedly compared with those of palaces, palace buildings were often converted into temples, and temples once again secularised for use as schools, hospitals, or government offices. The very word *ssu*,[1] later so denominative of Buddhist temples,[b] had meant a government bureau in the Han. Must we not see in this ambiguity an outward and visible sign of the fundamentally organic and integrated quality of Chinese thought and feeling?[c]

For the temple itself, the sacred grove (often extensive woodland) was generally sufficient protection, but lay folk sometimes felt the need of more, and thus the element of security also had its effects on Chinese building. No one who has travelled in China's north-west can forget the fortified villages (*pao*[2]) surrounded with their crumbling walls of loess, hardly distinguishable at a distance from the circumambient hills.[d] Elsewhere, as in the far south-east, migrations of people in times of disturbance led to the construction of veritable fortified apartment-houses, some fairly conventional in their rectangular planning but others with great originality making use of vast cylindrical edifices with inward-opening dwellings of many storeys (cf. p. 134 below). So far the social significance of the infinite variety of China's building patterns has been studied insufficiently whether in east or west. Though the historian of science and technology cannot do the job, he can at any rate voice the need for it.

Whatever the forces which moulded the Chinese building trade, its achievement was truly extraordinary. It mirrored in hard structural materials the outstanding genius of this people for combining the rational and the romantic. Harmonising intellect and emotion, it wedded the science of the erection and disposition of edifices to the art of

---

[a] See Soper (2), pp. 23 ff., on the 'Kudara (i.e. Pakche) Plan', so called because the Japanese obtained it about +590 from Korean country. It was broadly that of older Chinese temples such as Ho-Tung Ssu[3] founded at Chingchow in +380 and Yung-Ning Ssu[4] at Loyang (+516). Much about this type of foundation can be gained from the writings of the monk Tao-Hsüan[5] (+596 to +667), who described an imaginary temple in India in his *Chung Thien-Chu Shê-Wei Kuo Chih-Yuan Ssu Thu Ching*[6] (Illustrated Description of the Jetavana Monastery in Srāvastī in Central India), and real ones in China in his *Lü Hsiang Kan Thung Chuan*[7] (Miscellaneous Temple Traditions according with the Vinaya Regulations) and his *Kuan-Chung Chhuang-Li Chieh-Than Thu Ching*[8] (Illustrated Treatise on the Method of setting up Ordination Altars used in Kuan-Chung). He himself used earlier records, especially Ling-Yü's[9] *Shêng Chi Chi*[10] (Records of Holy Places), c. +585. The oldest Buddhist temple architecture in China can be dimly visualised from what Chai Jung[11] erected between +189 and +193 (*Hou Han Shu*, ch. 103, p. 13b, *San Kuo Chih*, ch. 49 (*Wu*, ch. 4), p. 2b). This certainly had a court enclosed by cloister galleries, with apparently a combined pagoda and worship-hall at the centre. Cf. Soper (2), p. 39.
[b] Cf. Vol. 2, p. 56.   [c] Cf. Vol. 2, pp. 154, 302 ff., 498, etc.
[d] Here another special case would be constituted by the lofty towers of stone found attached to domestic buildings in the Tibetan marches among the mountains of Sikang (cf. Stein, 3, 4). These served the same purpose as that of the round towers of old Ireland, protection for the inhabitants in case of raid and rapine.

[1] 寺   [2] 堡   [3] 河東寺   [4] 永寧寺   [5] 道宣
[6] 中天竺舍衛國祇洹寺圖經   [7] 律相感通傳   [8] 關中創立戒壇圖經
[9] 靈裕   [10] 聖迹記   [11] 笮融

landscape design in such a way that Nature remained dominant, free from subjection to an imposed architectural pattern, and rather uniting with the works of Man in a larger synthesis.

(3) THE PLANNING OF TOWNS AND CITIES

If the individual family dwelling, the temple, or the palace was so elaborately and attentively planned, set out, indeed, as a highly integrated organic pattern,[a] it would naturally be expected that urban planning would also show a considerable degree of organisation. The question is not however quite so simple, for in China there was a rather marked difference between the spontaneously growing village or rural settlement and the town planned from above. It would not be desirable here to anticipate what will have to be said later on about the Chinese town in its social and economic context,[b] but a few words on the subject cannot be avoided.

As Gutkind has pointed out, in the only review we have of Chinese town and village plans,[c] the villages tended to grow up, as elsewhere in the world, along roads, paths and other lines of communication, according to a sort of 'ribbon development'.[d] They arose at cross-road points, or where three ways met. At all times they possessed a great deal of unofficial self-government, and formed integrated community groups, dominated by one or more homogeneous clans. The sense of community was very real, especially when the village grew up far away from a line of communication at some point most convenient for the working of an agricultural area surrounded by upland or forest.

The Chinese town, on the other hand, was not a spontaneous accumulation of population, nor of capital or facilities of production, nor was it only or essentially a market-centre; it was above all a political nucleus, a node in the administrative network, and the seat of the bureaucrat who had replaced the ancient feudal lord. Originally, before the −1st millennium, the proto-feudal chieftains appropriated the centres of assembly where the people exchanged commodities and came together for the seasonal festivals. There was therefore no distinction throughout Chinese history between the feudal castle and the town; the town *was* the castle, and was built so that it could serve as the protection and refuge, as well as the administrative centre, of the surrounding countryside. Towns and cities in China were not the creation of burghers, and never achieved any degree of autonomy with regard to the State. They existed for the sake of the country and not vice versa; they were planned as rational fortified patterns imposed from above upon carefully chosen portions of the earth's surface. Hence they did not necessarily grow, indeed they shrank as often as they expanded, while the exoskeleton of walls remained in being to be refilled with flesh perhaps during a later dynasty. Their population was merely a sum-total of individuals, each of whom was closely linked with the village from which the family had originated, and where its ancestral clan temple still stood. While the European city or borough

[a] Cf. all that has been said concerning the universal tendencies of Chinese philosophy towards the organic as opposed to the mechanical (esp. Vol. 2).
[b] Sect. 47 below.   [c] Cf. also Haverfield (1); Miyazaki (2, 1).
[d] A typical plan is given in Gutkind (1), fig. 27, due originally to J. L. Buck.

developed from within outwards, centred on its agora, forum, cathedral, market-place, and halls of municipality and guilds, the external fortifications were of the essence of the Chinese city (the same word *chhêng*[1] still means both),[a] and the key points were the central drum-tower (*ku lou*[2]) and the offices (*ya mên*[3]) of the governors, civil and military.[b]

It is probable that all Chinese cities were laid out, since the Chou period, in a rectangular manner closely resembling that of the Roman *castra*.[c] There was the great east–west street corresponding to the *via principalis*, and the great north–south one cutting it at right angles like the *via decumana*.[d] Some indeed have thought[e] that certain ancient forms of characters betray an older circular type of wall. This is argued from the word *i*[4] (rad. no. 163) (K683) which means a *hsien*[5] city,[f] and appears on bones and bronzes as a ring-wall with a human figure kneeling beside it. So also *kuo*,[6] later meaning a suburb, shows (K774) an outer wall, apparently circular, with two

K683   K774   K1184   K1006   K76   K1197e

well-drawn gate-towers. This came to be written in later times with the addition of the abbreviated form of the radical *i*[7] just mentioned, generally used in combinations. A number of existing derivatives of this radical have meanings connected with urban settlement, for example, the moat, *yung*,[8] in which (K1184) it is combined with water. But these arguments assume that in the difficult media of bone and bronze the Shang and Chou scribes were really capable of distinguishing clearly between the round and the square, which may perhaps be doubted. It is true, however, that the word *ying*,[9] to build (K843*f*), which originally meant the demarcation of an encampment (hence the fires at the top of the character), is related to *kung*,[10] a palace, which in its ancient form (K1006) shows distinctly two square rooms under a roof. The word *lü*,[11] written like it

---

[a] It is composed of the phonetic 'completion' combined with the radical 'earth'.

[b] Dr Trevor Hodge points out a certain parallel in the Greek acropolis of Mycenean times, a fortified hill that held the king's palace and the main temples but served also as a stronghold for all the peasants of the surrounding districts in time of war. The agora was something quite different, connected with democratic life, and outside the acropolis.

[c] So also apparently the Etruscan 'terramare'; cf. Piganiol (1); Säflund (1); Barocelli (1). And some of the Maurya cities in ancient India; cf. F. W. Thomas (2), p. 476. Occasionally medieval Western foundations followed the Roman system, e.g. the 'bastides' such as Aigues-mortes, established by St Louis (+1226 to +1270).

[d] We have not come upon specific technical town-planning terms for these, but there is a close parallel with the terms for the paths, balks or headlands running between fields—*chhien*[12] generally north–south, and *mo*[13] generally east–west. These will come into some prominence at a later stage (pp. 258, 261, 267 below).

[e] E.g. Herrmann (12).

[f] This term, which has denoted the lowest rank of city possessing a magistrate since Chhin times, may be connected with a cognate word *hsüan*,[14] meaning 'to hang', e.g. the plumb-line, for buildings, and metaphorically for justice. It was also the custom in ancient China to hang up boards bearing written laws outside the government offices in towns (cf. *Chou Li*, ch. 1, p. 16*b* (ch. 2); tr. Biot (1), p. 34). The District Commissioner administering the law on the verandah of the only building for many miles around is thus the image of the Chinese magistrates of the −2nd millennium.

[1] 城   [2] 鼓樓   [3] 衙門   [4] 邑   [5] 縣   [6] 郭   [7] 阝
[8] 邕   [9] 營   [10] 宮   [11] 呂   [12] 阡   [13] 陌   [14] 懸

today, was however originally a graph of two round pitch-pipes seen from above
(K 76). On the other hand, one of the ancient forms of *pang*,[1] a State, now written with
the city radical, has the typical ground-plan of a square city, with what might be
branching ways leading from its gate (K 1197*e*).[a]

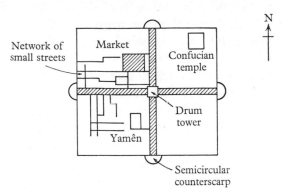

Fig. 747. Typical city plan.

In any case, the most typical form of a Chinese city would be as seen in the adjoining
sketch.[b] The palace of the feudal lord, and afterwards the yamen of the civil official,
were usually in the 'front' or southern part, with the market-places situated more to
the north. There universally followed from the main crosswise plan a rectangular grid
system of streets,[c] dividing the city into blocks, each of which was called a *li*[2] or *fang*.[3]
Often these also were separated by walls and gates, constituting areas under the
authority of a subsidiary of the city magistrate. Later developments led in Sui and
Thang to a concentric series of walled enclosures, the palace innermost (*kung chhêng*[4]),
then the administrative offices (*huang chhêng*[5]), and finally the people's city with its

[a] The late Prof. E. A. Gutkind and I discussed in correspondence the meaning of the fact that
Chinese city-plans were almost always square or rectangular rather than circular, rounded or irregular.
We agree in recognising a strong cosmological element in the tradition, connected no doubt with the
ancient and widespread idea that the heavens were round while the earth was square (cf. Vol. 3, pp. 211 ff.,
220, 498; Vol. 4, pt. 1, pp. 262 ff.; Forke (6), pp. 52 ff.; Granet (5), pp. 90 ff., 154, 345 and *passim*). The
unsophisticated early Chou cosmology surely visualised the heavens as round because the starry sky
seems to the observer like a hollow spherical dome rotating continuously above him in a circular manner.
The parallel idea that the earth was square surely arose from the simplest way of dividing the azimuth,
into the four cardinal points. Hence, then, not only the theory of the well-field system and its symbolic
ideograph (cf. p. 256 below), but also the rectangular zones of civilisation (p. 76 below)—and the
perennial tradition that the proper layout for a city must be more or less square. On cosmological
parallels in the genesis of cities in European and other civilisations see the monograph of Rykwert (1).

[b] Many Chinese drawings and paintings give life to this pattern; see, e.g., *SCTS*, ch. 16, p. 38*a*. One
of the best historical studies of the planning of a Chinese city is that of Ho Ping-Ti (3) for Loyang in the
+6th century. He draws attention to the great size of some medieval Chinese cities, as shown by the
following comparative figures in square miles: Roman and medieval London, 0·52; Loyang (+300), 3·9;
Byzantium (+447), 4·63; Rome (+300), 5·28; Baghdad (+800), 11·6 (walled part, 1·75 only); Chhang-
an (+750), 30·0, 'the greatest walled city ever built by men'; Peking (+1410), c. 24·0.

[c] Cf. Neuburger (1), pp. 270 ff., for many similar examples in European antiquity. Corresponding
with the handicraft character of Chinese industry, there was no distinction between shop and workshop,
and the trades tended to congregate in specialised streets, as one may see at Chhêngtu today.

[1] 邦　　　[2] 里　　　[3] 坊　　　[4] 宮城　　　[5] 皁城

external walls (*tu chhêng*[1]).[a] The word *tu* (capital) was written in some of its earliest forms with the city radical already mentioned, combined with mouths and the footsteps of people walking in the streets (K45g'). Chhang-an had no less than nine main streets in each direction.[b]

Still at the present day, the principal walls of cities in China form a square or a rectangle (cf. Fig. 751, pl.), though there are many exceptions. The very long walls of Nanking follow the local topographical configuration, and some large cities such as Fuchow have very irregular outlines. Occasionally there occurred a circle or ellipse, as in the case of Sung Shanghai.[c] Nearly all cities still have wide empty spaces left within the walls, available for kitchen-gardens and even farms. Sometimes ribbon-development occurring along an important road outside the city-gates became later incorporated within the walls, forming an elongated protrusion, as we see in the Tungkuan[2] district at Lanchow. Another Kansu city, Thienshui, came thus to consist of five walled towns joined together in a row.[d]

K45g'

In Chinese cities, a rather large population was sometimes packed into a confined area. In general the builders did not resort to multiple storeys, party walls were constantly used to separate dwellings of different families, and even the wealthy had rather restricted space, but every courtyard, no matter how small, became something of a garden by the use of plants and small trees in pots, no 'grounds' being provided. This meant that population density could reach high levels. And yet a sense of seclusion was preserved. In Peking, figures for residential areas reached 55,000 per square mile, and for working areas 85,000. But the city maintained a garden character owing to the abundance of trees, which paradoxically were more numerous within the walls than outside, and to this day Peking seen from some such vantage-point resembles a forest, with only the roofs of the most important buildings visible above the tree tops (Fig. 748, pl.).[e]

The Chinese garden is a goodly subject of enquiry and instruction in itself.[f] Though its motive throughout the ages was of course primarily aesthetic, our own preoccupations will bring us back to it later on in connection with botanical and zoological collec-

[a] Cf. Peking in Figs. 751 (pl.), 752.
[b] Cf. p. 5 and Fig. 712 above. For a general air view of the modern city see Gutkind (1), pl. 62.
[c] Plan in Gutkind (1), fig. 41, after Oberhummer. There is also the round citadel (Thuan Chhêng[3]), of Sung or Yuan date, near the Northern Lake at Peking; cf. p. 79.
[d] Many small sketch-plans of Chinese cities will be found in Herrmann (1).
[e] This statement was substantially true as late as 1952, but during the past decade the exigencies of status as the capital of a great modern nation, and perhaps a half-conscious desire to show that China can build steel-frame structures as high as anyone else, has led to the erection of some buildings of skyscraper type which run the risk of spoiling Peking's perfection. One hopes that this tendency may be held in check, for besides running counter to the great canons of Chinese architecture it is not at all imposed by constriction of building land. It is fair to say, however, that many of the new buildings of Peking, though very monumental and massive, still harmonise well enough with the old horizontal emphasis. The problem and its background has been considered by Skinner (1).
[f] See the great work of Sirén (8), besides some shorter writings (11). Wilson (1) is also valuable, and the fine album of Juan Mien-Chhu & Li Shao-Chhang (1). The Chinese garden has perhaps been somewhat overshadowed in the European mind by those of Japan (cf. Harada, 1), but unjustly. Cf. Sugimura Yuzo (1). On miniature gardens, which began in China, and their symbolism, see Stein (2).

[1] 都城   [2] 東關   [3] 圜城

tions.[a] From the formal arrangement likely in Han palaces and imperial residences there developed in south China during the Liu Chhao period a new naturalistic style certainly connected with the growth of Taoism.[b] This reached its peak in the Thang, to be followed by a revival of formalism expressing itself in the incorporation of exotic plants and rocks amidst the pools and buildings.[c] By the early Ming the characteristic styles of Chinese garden-planning were fully developed, and we have accounts both contemporary[d] and modern[e] of their abundant beauties (Figs. 749, 750, pls.). Many details are preserved of the life and work of the great architect-gardeners such as Chang Jan[1] and Chang Lien.[2] Among the most important books on gardening design and horticulture is the *Yuan Yeh*[3] of Chi Chhêng.[4] These three men all lived in the +17th century. From that time onwards Chinese garden technique exerted profound effects on Europe, the romantic style displacing the geometrical.[f] As a canon of taste 'sharawadgi' came to stay,[g] and 'Capability Brown' was a Taoist without knowing it.[h]

The city of Peking that we see now is mainly a creation of the Yuan, Ming and Chhing dynasties.[i] Its planning brought to highest perfection everything of which the Chinese city was capable (Fig. 751, pl.).[j] Centred upon the 'Purple Forbidden City' (Tzu

---

[a] Sects. 38, 39 below.

[b] Murakami Yoshimi (*1*). It is not easy to find traces in China of the layout of Han gardens, but one formal plan of the +5th century still exists below the great rock of Sigiriya in Ceylon, crowned with its palace-temple.

[c] Murakami Yoshimi (*2*). Cf. Schafer (*13*) on Thang exoticism.

[d] In +1533 the scholar and painter Wên Chêng-Ming[5] produced his *Cho Chêng Yuan Thu*[6] (Pictures of the Garden of an Unsuccessful Official), the famous garden started some time before by Wang Huai-Yü.[7] This work has been reproduced and translated by Kerby & Chung Mo-Tsung (*1*).

[e] For example, the famous Yen Shêng Kung Fu[8] garden at Chhü-fou in Shantung has recently been described by Yang Hung-Hsün & Wang Shih-Jen (*1*).

[f] The story has been told by Lovejoy (*3*). Further studies of the influence on Europe are due to Bald (*1*); Chhen Shou-Yi (*2*); Baltrušaitis (*2*) and Laske (*1*). We shall return to it in Vol. 6 (Sect. 38).

[g] See Vol. 2, p. 361.

[h] Lancelot Brown (+1715 to +1783), protagonist of naturalism in landscape-garden design, as at Kew and Blenheim.

[i] See the historical account of Chu Chhi-Chhien & Yeh Kung-Chao (*1*), and general descriptions by Bretschneider (*5*); Favier (*1*); Fabre (*1*) and others. The names of many of the architects who contributed to it are known. Their line might be said to begin with Khung Yen-Chou[9] who rebuilt the capital with great magnificence for the Chin (Tartar) dynasty after the fall of the Liao about +1120. On him, see Fan Chhêng-Ta's *Lan Phei Lu*[10] (Grasping the Reins) of +1170 (p. 7a). Under the Mongols foreign experts were much in evidence, such as the Arab architect Yeh-Hei-Tich-Erh[11] and the Nepalese metal-founder and decorator A-Ni-Ko,[12] but as far as we know they built in pure Chinese style and must have had Chinese advisers and staff. The outstanding architect of the early Ming period was Chhen Kuei[13] (*fl.* +1406), and for the later Ming work there were Chhin Liang[14] and the first of the famous family of Peking architects, Lei Li.[15]

[j] Of the literature on city planning a brief account will be given shortly (p. 87). It is a commonplace to say that the Chinese capital was copied by all the peoples of the surrounding culture-area for their centres of government, notably Japan (cf. Sansom (*1*), p. 108, (*2*), vol. 1, p. 82), but it is literally true. Nara,[16] laid out in +710, derived from Thang Chhang-an, but its imperial palace of Heijō[17] was deeply influenced by the Sui capital of Ta-hsing[18] near Peking, set out at the end of the +6th century by Yüwên Khai (cf. Vol. 4, pt. 2 *sub voce*). The question has been studied by Iida Sugashi (*2*). Some of the Thang palaces at Chhang-an are currently under excavation; Lanciotti (*3*); Anon. (*63*). In another direction we have now a useful album of the traditional architecture of Inner Mongolia (Anon. *41*).

---

[1] 張然	[2] 張連	[3] 園冶	[4] 計成	[5] 文徵明
[6] 拙政園圖	[7] 王槐雨	[8] 衍聖公府	[9] 孔彥舟	[10] 攬轡錄
[11] 也黑迭兒	[12] 阿尼哥	[13] 陳珪	[14] 秦梁	[15] 雷禮
[16] 奈良	[17] 平城	[18] 大興		

Chin Chhêng [1]),[a] with extremely broad roads transecting the capital in both directions, and with the superb axis leading from the centre to the cosmic temples near the southernmost gate,[b] it has excited profound admiration in modern architects and writers such as Murphy, Gothein, and Rasmussen.[c] The first speaks of this axis as the greatest in the world today. Southward it runs from the Bell Tower through the central pavilion on the Coal Hill, through the main transverse buildings of the Forbidden City with their magnificent yellow roofs, and through the towering wall-gate of Chhien Mên,[2] to end five miles away, at the South Gate, Yung-Ting Mên,[3] between the Altar of Agriculture and the Altar of Heaven (Fig. 752). The formal grouping of buildings, says Murphy, is marked not by rigid symmetry, but by that nice feeling for balance which is characteristic of all Chinese art. By avoiding exact duplication on each side of an axis, sufficient variation to avoid monotony could be introduced. This is seen, for instance, in the handling of the beautiful artificial lakes with which Peking is so well provided.[d] A stream was led in at the north-west corner of the present city, and then expanded into a series of lakes (Pei Hai,[4] Chung Hai,[5] Nan Hai[6]) forming, as it were, a parallel but sinuous western axis subsidiary to the main one which runs north–south through the centre of the Forbidden City, indeed through the imperial throne itself. All three lakes lie within the confines, formerly walled, of the Imperial City (Chiu Huang Chhêng[7]), the next largest of the concentric rectangles which constitute the whole capital.[e] The moat of the Forbidden City is supplied by a branch canal from the northern lake, and the moat in turn feeds the Stream of Aureate Water which is crossed (at no. 9 in Fig. 752) by the five marble bridges (Fig. 753, pl.) between the Meridian Gate (no. 8) and the Gate of Supreme Harmony (no. 10). Similarly the southern lake provides the ceremonial stream under the marble bridges which give access from the south to the grand entrance at the Thien An Mên.[8] The water finally finds its way out at the south-east corner of the city. To give an idea of the grandeur of the view which lies before the visitor who passes northwards through the Gate of

[a] See Sirén (3), and on the symbolism Ayscough (2). Air photographs in Mirams (1), opp. pp. 35, 38, 39.

[b] Cf. Hu Chia (1). Wu No-Sun (1) likens it, with diagrams, to a *ziggurat* in one plane.

[c] As it did in one of the greatest of its adopted sons, Antoine Gaubil, S.J. (cf. Vol. 3, *sub voce*), whose *Description de la Ville de Pékin* (11) appeared in +1758 in the *Philosophical Transactions of the Royal Society*.

[d] While thinking of the genius of the city plan, there is another aspect which should not be forgotten, namely the drainage system. About the middle of the +16th century the 24 square miles within the city wall were provided with an underground network of brick sewers no less than 195 miles in total length (personal communication from Dr Liang Ssu-Chhêng). Elizabethan London had nothing comparable to show. This main drainage subsequently became silted up, but in 1951 was restored to full use.

[e] The reader will be reminded of the ancient schematism which depicted the radiation of Chinese culture in the form of a series of concentric rectangles; cf. Fig. 204 in Vol. 3, p. 502. It may not be irrelevant here to pay a tribute to the acute psychology of the Kung-chhan-tang in the Liberation period. The Forbidden City could have been made into a kind of Kremlin, but instead it was thrown open to the Chinese people as their greatest national museum, while the locus of government was settled just outside in the south-west quarter of the Imperial City. This was a symbolic act of restraint in the highest Taoist tradition. Cf. Thang Lan (1).

[1] 紫禁城   [2] 前門   [3] 永定門   [4] 北海   [5] 中海   [6] 南海
[7] 舊皇城   [8] 天安門

Heavenly Peace, we may quote the excellent description of Murphy (cf. Fig. 754, pl.):

At the middle of the southern wall of the Forbidden City is the finest architectural unit in the country, the great Wu Mên[1] (Meridian Gate), a central building some two hundred feet long, on a balustraded terrace, flanked by a pair of square, sixty-foot pavilions. The four-hundred-foot composition is raised on a wall base fifty feet high, plastered in dark red, and pierced by five arched tunnel entrances. Projecting three hundred feet south are two flanking wings of the wall base; at the outer ends of these, a second pair of pavilions repeat those of the main group. The effect is one of overpowering majesty and breath-taking beauty.[a]

To sum up, then, we find (in the words of a contemporary English architect)[b] a series of demarcated spaces, each opening into the next but screened and stopped from it each time by walls, gateways, buildings overhead, and the tension heightened at chosen points, as some climax is approached, by such incidents as the bow-shaped stream with its marble balustrades and five parallel marble bridges. Between the constituent parts there is remarkable balance and interdependence. The contrast with the Renaissance palace is striking, for there the open vista, as at Versailles, is concentrated upon a single central building, the palace as something detached from the town. The Chinese conception was much grander and more complex, for in one composition there were hundreds of buildings, and the palace itself was only part of the larger organism of the whole city with its walls and avenues. Although so strongly axial, there was no single dominating centre or climax, but rather a series of architectural experiences. Hence the bathos of an anti-climax had no place in such designs. Even the Thai Ho Tien[2] is not the climax, for the composition flows on northwards past it and behind it. The Chinese conception also shows more subtlety and variety; it invites a diffusion of interest. The whole length of an axis is not revealed at once, but rather a succession of vistas none of which is overpowering in scale. Sometimes, in the approach to the final objective, the visitor is brought back to ponder a stretch of the approach which has just been negotiated—as in the case of the imperial Ming tomb of Thai Tsu at Nanking.[c] Thus the Chinese form of the great architectural ensemble, which attained its highest level already in the early +15th century with the Temple and Altar of Heaven in Peking (Fig. 755, pl.),[d] combined a meditative humility attuned to Nature with a poetic grandeur to form organic patterns unsurpassed by any other culture.

[a] (1), pp. 365 ff.
[b] Mr Andrew Boyd, in a private communication (cf. Boyd, 1) on which the first part of this paragraph is based. The second part draws upon the impressions of Mr Francis Skinner.
[c] The base of the main building is penetrated by a staircase in a barrel-vaulted tunnel, which brings the visitor to a terrace at the far side against the mountain, whence he returns round either end of the building up ramps or stairs to reach the top terrace which faces back along the main approach axis. Finally the main hall is entered through one of three arched openings. Similar arrangements exist, for example, at the tomb of the Yung-Lo emperor (Chhêng Tsu) north of Peking. Cf. p. 144 below.
[d] Cf. Mirams (1), opp. p. 40.

[1] 午門   [2] 太和殿

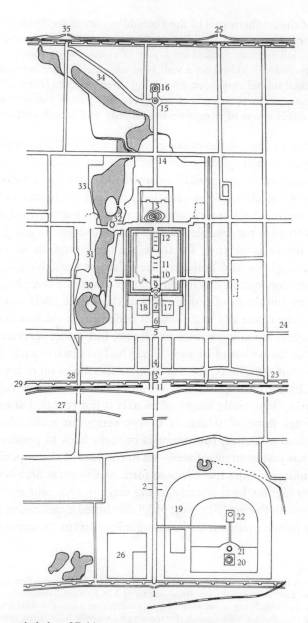

Fig. 752. Plan of the capital city of Peking to show its axial pattern; north at the top (after Hu Chia, 1).

Explanation of Fig. 752

1 South Gate, Gate of Everlasting Stability (Yung-Ting Mên) of the outer city (formerly called the 'Chinese' city)
2 Bridge of Heaven (Thien Chhiao) market, traditional centre of folk entertainments
3 Front Gate (Chhien Mên) of the inner city (formerly called the 'Tartar', i.e. Manchu, city)
4 China's Gate (Chung-Hua Mên)
5 Square of the Gate of Heavenly Peace, to which five marble bridges give access
6 Gate of Heavenly Peace (Thien-An Mên), entrance of the Imperial City (Chiu Huang Chhêng)
7 Gate of Correct Deportment (Tuan Mên)
8 Meridian Gate (Wu Mên), i.e. Noon Gate or South Gate, entrance of the Purple Forbidden City (Tzu Chin Chhêng)
9 The five marble bridges over the Stream of Aureate Water (Chin Shui Ho)
10 Gate of Supreme Harmony (Thai-Ho Mên); to the left the Gate of the Floribundant West, or Western Floriate Gate (Hsi-Hua Mên); to the right the Gate of the Floribundant East, or Eastern Floriate Gate (Tung-Hua Mên)
11 Hall of Supreme Harmony (Thai Ho Tien), the throne-hall, backed by the Hall of Central Harmony (Chung Ho Tien) and the Hall of the Preservation of Harmony (Pao Ho Tien)
12 Complex of inner palace apartments (Ku Kung), having the Hall of Reverence for Peace (Chhin An Tien) at their northern end, behind which the Gate of the Warlike Spirit (of the North, Shen-Wu Mên) terminates the enclosure. Now the Imperial Palace Museum
13 'Coal Hill' (Ching Shan) gardens and pavilions
14 Site of the Gate of Earthly Peace (Ti-An Mên)
15 Drum Tower (Ku Lou)
16 Bell Tower (Chung Lou)
17 Imperial Ancestral Temple (Thai Miao), now People's Palace of Culture (Lao-Tung Jen-Min Wên-Hua Kung)
18 Temple of the Altar of the Land and Grain (Shê Chi Than), now Sun Yat-Sen Park (Chung-Shan Kung-Yuan)
19 Temenos of the Altar and Temple of Heaven (Thien Than)
20 Altar of Heaven, the Orbed Concentric Platforms (Huan Chhiu Thai); cf. Vol. 3, p. 257
21 Hall of the Infinite Canopy of Heaven (Huang Chhiung Yü)
22 Hall of Prayer for the Year (Chhi Nien Tien), the blue-tiled circular edifice built in +1420 and rebuilt in +1530 and +1751, commonly called the Temple of Heaven
23 Gate of Veneration of Letters (Chhung-Wên Mên)
24 Road leading to the Imperial Observatory (Kuan Hsiang Thai) on the eastern wall of the inner city; cf. Vol. 3, p. 451 and *passim*
25 Northeast Gate, Gate of Peaceful Stability (An-Ting Mên) of the inner city
26 Temenos of the Altar and Temple of Agriculture (Hsien Nung Than)
27 Glazed Tile Factory Street (Liu-Li Chhang), famous centre of bookshops, art and antique shops
28 Gate of Peace and Harmony (Ho-Phing Mên)
29 Gate of the Proclamation of Military Might (Hsüan-Wu Mên), like gates 3, 23 and 28, between the inner and the outer cities
30 Southern Lake (Nan Hai)
31 Central Lake (Chung Hai)
32 The Round Fort (Thuan Chhêng), perhaps the throne-castle of Khubilai Khan; behind it, on the island to the north, the prominent White Stupa (Pai Tha) built in +1651
33 North Lake (Pei Hai)
34 Back Lake (Hou Hai) with the Front Lake (Chhien-Hai) to the south of it
35 Northwest Gate, Gate of Victorious Virtue (Tê-Shêng Mên) of the inner city

N.B. The plan covers only a north south strip sufficiently wide to include the two inner concentric rectangles of the capital city. The Purple Forbidden City (Tzu Chin Chhêng) is seen at the centre surrounded by its moat. Outside it and enclosing the string of lakes, south of the Gate of Earthly Peace (Ti-An Mên, no. 14) and north of the great east–west road in front of the Gate of Heavenly Peace (Thien-An Mên, no. 6) is the Imperial City (Chiu Huang Chhêng). It is substantially square save for the indentation of its south-western corner. Of the people's cities only parts of three east–west boundary walls can be seen, enclosing the so-called 'Tartar' (Manchu) city above the middle one, and the 'Chinese' city below it.

(4) BUILDING SCIENCE IN CHINESE LITERATURE

Presumably owing to the fact that architectural employment was not considered a very suitable occupation for a Confucian scholar, Chinese literature is relatively poor in writings on the subject.[a] However, the earliest dictionary, the *Erh Ya*, dating from Chou and early Han times, has a special chapter[b] devoted to matters connected with building (Shih Kung;[1] Explanations Concerning Palaces and Halls). In this we find a good many technical terms which retained their meaning afterwards with little or no change.[c] Later encyclopaedias often have similar sections. In Chhing times a number of scholars made useful studies to elucidate the meaning of ancient architectural words and expressions.[d]

The chief literary tradition which involved actual architectural plans was that of the *San Li Thu*[2] (Illustrations of the Three Rituals). Two books of this name were composed in the Later Han period (+2nd century), one by Chêng Hsüan,[3] the famous commentator, the other by Juan Shen,[4] his contemporary. At some subsequent time they were probably combined under the editorship of Liang Chêng,[5] and an important series of illustrations was added about +600 by Hsiahou Fu-Lang,[6] to whose authorship the book was ascribed in the Sui bibliography. Then in or around +770 a further recension was made by Chang I.[7] All this work was afterwards lost, but not before some of it had been used by Nieh Chhung-I[8] in the definitive text of +956, made under the leadership of Tou Yen,[9] who provided a preface.[e] A little more than a century later, Nieh's work was referred to by Shen Kua, who had doubts whether the pictures then existing could be considered authoritative.[f] The text of Nieh has come down to us, having been re-edited for the last time in +1676 by the Manchu prince Nalan Chhêng-Tê.[10]

The reason why this material was preserved even as well as it was lay in the desire of scholars to interpret the rubrics in the ancient liturgical and ceremonial texts.[g] Many books of the Sung and later simply illustrated these as their authors saw fit, without any traditional basis. The plans of architectural interest are found in ch. 4 of the *San Li Thu*, and include the Ming Thang[11] (cosmic palace-temple),[h] Kung Chhin

---

[a] Reviews by Demiéville (4), probably the best account, and Yetts (10).

[b] Ch. 5.   [c] See on, pp. 92ff.

[d] Notably Chiao Hsün[12] in his *Chhün Ching Kung Shih Thu*[13] (Illustrated Treatise on the Plans, Technical Terms, and Uses of the Houses, Palaces, Temples, and other buildings described in the various Classics). This was late in the +18th century, and will be found in *HCCC*, *Hsü pien*, chs. 359, 360. A similar work was the *Shih Kung Hsiao Chi*[14] (Brief Record of Buildings and Palace Halls), by Chhêng Chêng-Chün[15] (*HCCC*, *Hsü pien*, ch. 535).

[e] The ancient fragments contained in this have been assembled by Ma Kuo-Han, *YHSF*, ch. 28, pp. 44a ff.

[f] *MCPT*, ch. 19, para. 1.

[g] A modern interpretation of this kind of material will be found in Kelling & Schindler (1), p. 102, taken from Couvreur (3).

[h] Cf. Granet (5), pp. 178 ff.; Soothill (5). See also Vol. 3, p. 189 above.

---

[1] 釋宮	[2] 三禮圖	[3] 鄭玄	[4] 阮諶	[5] 梁正
[6] 夏侯伏郎	[7] 張鎰	[8] 聶崇義	[9] 竇儼	[10] 納蘭成德
[11] 明堂	[12] 焦循	[13] 羣經宮室圖	[14] 釋宮小記	[15] 程徵君

Chih[1] (public halls and domestic apartments), and the Wang Chhêng[2] (princely city)[a] (Fig. 712). At various times efforts were continued to throw further light on the buildings which must have been visualised by the writers of the Rituals, e.g. the *I Li Shih Kung*[3] (Explanations concerning the Buildings referred to in the Personal Conduct Ritual) by Li Jo-Kuei[4] about +1193; and the early +18th-century *Kung Shih Khao*[5] (Study of Halls and Buildings) by Jen Chhi-Yün.[6] The planning of imperial ancestral temples was a related study.[b]

All this, however, was on the purely scholarly level, concerned with general layout rather than with the techniques of construction, and somewhat remote from practical architects and building workers.[c] The names of many of the latter have come down to us, however, and it may well be that there was contact between the two streams of tradition. In the +8th century Chang I may have known such a man as Khang Kung-Su.[7] In the +10th Nieh Chhung-I had great architects among his contemporaries, Mêng Tê-Yü[8] the palace-builder,[d] Li Huai-I[9] the ironmaster and designer of gates and imperial residences, Hu I[10] and Kuo Chung-Shu,[11] painters and architectural draughtsmen. In the Chhing, Nalan the Manchu must certainly have known the Lei family, including men such as Lei Fa-Ta,[12] and his son Lei Chin-Yü[13] who built the first part of the Yuan Ming Yuan[14] summer palace.

The practical architects[e] also had a literary tradition. From early times there must have been manuals of procedure for all the varied trades carried on in, or under the auspices of, the imperial workshops (*shang fang*[15]).[f] The emperor Yuan of the Liang dynasty tells us that in the time of Tshao Tshao (+3rd century; founder of the Wei State in the San Kuo period) there were elaborate rules for the erection of all kinds of buildings.[g] But not even the titles of such manuals have been preserved in the official lists, and that of even the most important of them would not have survived if it had not been mentioned by certain Sung writers. This was the *Mu Ching*[16] (Timberwork Manual), authorship of which was attributed to a famous builder, Yü Hao[17,18] who flourished in the decades +965 to +995, i.e. at the beginning of the Sung. He constructed the Khai-Pao Tha[19] (pagoda) at Khaifêng, which was universally regarded as

[a] There are also 'maps' of the schematic fiefs of the feudal lords (Chiu Fu,[20] cf. Vol. 3, p. 502 above), the Ching Thien[21] system (see pp. 256ff. below), and the theoretical irrigation arrangement (Kou Hsü,[22] discussed on pp. 4, 254).

[b] See, e.g., the *Miao Chih Thu Khao*,[23] written about +1685 by the famous scholar Wan Ssu-Thung.[24]

[c] The reader will remember the traditions associated with the name of the patron saint of artisans, Kungshu Phan (Lu Pan), described at some length in Vol. 4, pt. 2, pp. 43 ff. His effigy in Fig. 756, pl., suitably opens this story.

[d] Cf. *Chien Chieh Lu*[25] by Ho Kuang-Yuan[26] (almost his contemporary), ch. 1, pp. 1a ff.

[e] It is interesting that etymologically 'architect' (chief of the workmen) is identical with its corresponding term *ta chiang*,[27] just as 'archimandrite' (director of the enclosure) parallels *fang chang*,[28] the Buddhist name for an abbot.

[f] Cf. Vol. 4, pt. 2, pp. 18 ff. above.   [g] *Chin Lou Tzu*, ch. 1, p. 12b.

[1] 宮寢制	[2] 王城	[3] 儀禮釋宮	[4] 李如圭	[5] 宮室考
[6] 任啓運	[7] 康誓素	[8] 孟德預	[9] 李懷義	[10] 胡翼
[11] 郭忠恕	[12] 雷發達	[13] 雷金玉	[14] 圓明園	[15] 尙方
[16] 木經	[17] 喩皓	[18] 預浩	[19] 開寶塔	[20] 九服
[21] 井田	[22] 溝洫	[23] 廟制圖考	[24] 萬斯同	[25] 鑒誡錄
[26] 何光遠	[27] 大匠	[28] 方長		

a marvel of art, but struck by lightning and destroyed about + 1040.[a] The fact that his book was not recorded in the official bibliography is significant as showing that building technique was regarded as too 'mechanical' for inclusion among scholarly works.[b] There was probably also a social barrier, for Yü Hao was a Master-Carpenter (Tu Liao Chiang[1]), while the man who built upon his work to produce the greatest architectural book in Chinese history was a 'white-collar' Assistant in the Directorate of Buildings and Construction (Chiang Tso Chien Chhêng[2]).

Shen Kua, the great Sung scholar of scientific and technical interests, so well known to us, wrote about Yü Hao in a passage which is worth quoting in full.[c] He said:

Methods of building construction are described in the *Timberwork Manual*, which, some say, was written by Yü Hao.

(According to that book), buildings have three basic units of proportion (*fên*[3]);[d] what is above the cross-beams follows the Upperwork Unit, what is above the ground floor follows the Middlework Unit, and everything below that (platforms, foundations, paving, etc.) follows the Lowerwork Unit.

The length of the cross-beams will naturally govern the lengths of the uppermost cross-beams as well as the rafters, etc. Thus for a (main) cross-beam of 8 ft. length, an uppermost cross-beam of $3\frac{1}{2}$ ft. length will be needed. (The proportions are maintained) in larger and smaller halls. This (2·28) is the Upperwork Unit.

Similarly, the dimensions of the foundations must match the dimensions of the columns to be used, as also those of the (side-) rafters, etc. For example, a column 11 ft. high will need a platform $4\frac{1}{2}$ ft. high. So also for all the other components, corbelled brackets (*kung*[4]), projecting rafters (*tshui*[5]), other rafters (*chüeh*[6]), all have their fixed proportions. All these follow the Middlework Unit (2·44).

Now below of ramps (and steps) there are three kinds, steep, easy-going and intermediate. In palaces these gradients are based upon a unit derived from the imperial litters. Steep ramps (*chün tao*[7]) are ramps for ascending which the leading and trailing bearers have to extend their arms fully down and up respectively (ratio 3·35).[e] Easy-going ramps (*man tao*[8])

---

[a] *Kuei Thien Lu*, ch. 1, pp. 1 a ff.; *Mo Chi*[9] (Things Silently Recorded) by Wang Chih[10] (+11th century), p. 40b; *MCPT*, ch. 18, para. 15.

[b] It is paradoxical (and perhaps significant) that from the brush of Li Ao[11] (c. +775 to +844), one of the philosophical precursors of Neo-Confucianism (cf. Vol. 2, pp. 452, 494), there has come down to us a small tractate with a very similar title, the *Wu Mu Ching*.[12] It is listed among various books on technical subjects about +1135 in *Yü Chien*, ch. 7, p. 4a. But in fact it only concerns the game of dicing, and its title could be translated 'Manual of the Five (Throws of the) Wooden (Dice)'. Nevertheless, Yü Hao was greatly admired by some of the outstanding scholars of the early Sung, such as Ouyang Hsiu.

[c] *MCPT*, ch. 18, para. 2; tr. auct. Cf. Hu Tao-Ching (1), vol. 2, p. 570, (2), p. 177.

[d] Here equivalent to *fên*.[13] Cf. p. 68 above.

[e] It will have been obvious that the first two units of proportion are derived by simple division. The three now given for the ramps represent in the same way the relation between the heights above ground of the two ends of the emperor's litter during its ascent. The proportions of the human body are taken from standard figures given in the *Huang Ti Nei Ching Thai Su* of +678 (ch. 13, p. 87.2) and the *I Tsung Chin Chien* of +1742 (ch. 71, p. 14a) as follows: trunk and legs 6·2 ft., upper arm 1·7 ft., lower arm 1·65 ft. Such data came from long before the Thang and changed little or not at all afterwards. To determine the absolute gradients it would be necessary to know the lengths of the standard litters, but we have not gone into this. It is interesting to find here yet again a modulor unit based upon the proportions of the human body (cf. pp. 68, 69).

---

[1] 都料匠   [2] 將作監丞   [3] 分   [4] 栱   [5] 榱
[6] 桷   [7] 峻道   [8] 慢道   [9] 默記   [10] 王銍
[11] 李翱   [12] 五木經   [13] 份

Fig. 757. 'Building the New City in the East Country'; a late Chhing representation of the events recounted in *SCTS*, ch. 33, Lo Kao (Medhurst (1), pp. 248 ff., Karlgren (12), pp. 51 ff.), i.e. the foundation of the eastern capital of the Chou kingdom at Lo-i (Wang-chhêng) a couple of miles from modern Loyang. Here it illustrates the technical terms 'upperwork' and 'lowerwork' (see text). The Duke of Chou (Chou Kung) is seen in the centre encouraging the craftsmen.

are those for which the leaders use elbow length and the trailers shoulder height (ratio 1·38); intermediate ones (*phing tao*[1]) are negotiated by the leaders with downstretched arms and trailers at shoulder height (ratio 2·18). These are the Lowerwork Units.[a]

The book (of Yü Hao) had three chapters. But builders (*thu mu chih kung*[2]) in recent years have become much more precise and skilful (*yen shan*[3]) than formerly. Thus for some time past the old *Timberwork Manual* has fallen out of use. But (unfortunately) there is hardly anybody capable of writing a new one. To do that would be a masterpiece in itself!

This passage would have been written about +1080. Within twenty years the man capable of doing the job which Shen Kua saw was necessary had arisen and completed it. This was Li Chieh,[4] and the title of his book was *Ying Tsao Fa Shih*[5] (Treatise on Architectural Methods).[b]

The date of Li Chieh's birth is not sure, but he was already a subordinate official in the Bureau of Imperial Sacrifices when Shen Kua was about to produce his *Mêng Chhi Pi Than*. Moving to the Directorate of Buildings and Construction in +1092, he must have shown immediate and outstanding promise as an architect, for his revision of the old treatises was commissioned in +1097, completed by +1100, and printed three years later. He was a distinguished practising builder as well as a writer, for he erected administrative offices, palace apartments, gates and gate-towers, and the ancestral temple of the Sung dynasty, as well as Buddhist temples. Li Chieh says in his preface that he studied long and minutely the practices and orally transmitted rules of the master-carpenters and other responsible artisans.[c]

It is of much interest that Li Chieh never quite succeeded in fusing the scholarly and the technical traditions. His method was to quote many ancient and medieval texts, with great reverence, in the earlier chapters, then to describe the practice of his time, and finally to enunciate rules (*thiao*[6]), which are always based on practice and have little or nothing to do with the texts. Introductory sections (including a Khan Hsiang,[7] Critical Revision)[d] discuss meanings of old terms, and are followed by the Rules and Regulations (Chih Tu[8]) which form the main body of the book. These deal systematically with one department after another, comprising:

[a] Completion of a job of upperwork and lowerwork is seen in Fig. 757.

[b] What Shen Kua says does not quite fit in with another tradition mentioned by Yetts (8) according to which Li Chieh's book was a definitive revision of a previous work with the same name which had been commissioned about +1070 and finished in +1091. We have already considered his biography in some detail in Vol. 4, pt. 2, p. 37. Cf. Chhen Chung-Chhih (1).

[c] A MS. copy of Li Chieh's book was reproduced photolithographically in 1919 (reduced) and 1920 (facsimile), but this took no account of variations in other MSS. and printed fragments, so that the whole was reissued with many changes, and in colour, in 1925. On the earlier editions see Demiéville (4); on the later one Hsieh Kuo-Chên (1) and Yetts (8, 9). The close relation of Li Chieh's work to that of Yü Hao and similar predecessors can be seen in the fact that the official Sung bibliography does not know his work by the name which it afterwards bore, but as *Hsin Chi Mu Shu*[9] (New Collection of Timberwork Treatises). Of course this may have been the title, or the intended title, of the earlier manual which had been commissioned about +1070. The stock of copies of Li Chieh's book was destroyed in the sack of Khaifêng in +1126, but a few specimens got through to the south, and it was reprinted from new blocks in +1145.

[d] Followed by Tsung Shih,[10] General Glossary, and Tsung Li,[11] General Instructions.

[1] 平道	[2] 土木之工	[3] 嚴善	[4] 李誡	[5] 營造法式
[6] 條	[7] 看詳	[8] 制度	[9] 新集木書	[10] 總釋
[11] 總例				

Hao chai[1]	Moats and fortifications	Chu tso[8]	Bamboo work
Shih tso[2]	Stonework	Wa tso[9]	Tiling
Ta mu tso[3]	Greater woodwork	Ni tso[10]	Wall building
Hsiao mu tso[4]	Lesser woodwork	Tshai hua tso[11]	Painting and decorating
Tiao tso[5]	Wood-carving	Chuan tso[12]	Brickwork
Hsüan tso[6]	Turning and drilling	Yao tso[13]	Glazed tile making
Chü tso[7]	Sawing		

The last chapters deal with Job accounting (Kung hsien[14]); Materials (Liao li[15]), including some interesting paint compositions; and the Classification of crafts (Chu tso têng ti[16]). From the first the book had contained excellent illustrations (see Figs. 759, 762, 763, 773, 774), and the *editio princeps* of 1925 used elaborate colour-printing to reproduce designs in which the tints had formerly been indicated only by name. It also added to chs. 30 and 31 appendices of diagrams prepared by an old master builder, Ho Hsin-Kêng,[17] who for many years had been in charge of imperial and public works in Peking, to show how the technical terms of the trade had changed since the time of Li Chieh.[a] The attention given in the *Ying Tsao Fa Shih* to the basic construction and the shaping of the woodwork is striking since this is missing from European building manuals until the end of the eighteenth century.[b]

The only other surviving architectural document of the Sung is an illustrated MS. on the buildings and fittings of some Buddhist abbeys in south-east China, written by a Japanese monk, Gikai,[18] in +1259, and entitled *Wu Shan Shih Chha Thu*.[19] But its technological detail is not very impressive. A more artisanal tradition is represented by a Ming book, *Ying Tsao Chêng Shih*[20] (Right Standards of Building Construction), evidently closely related to the *Lu Pan Ching* discussed in Vol. 4, pt. 2, pp. 44 ff.

No work by another individual ever took the place of the *Ying Tsao Fa Shih*, but subsequent dynasties issued more or less official technical compilations. In the Yuan there was a *Yuan Nei Fu Kung Tien Chih Tso*[21] (Regulations for (Construction) Work on Palaces and Public Buildings, authorised by the Imperial Directorate of Architecture),[c] but it is lost. Similar material existed in the Ming, and early in the +18th century there was a *Chhin-Ting Kung Pu Kung Chhêng Tso Fa*[22] (Official Manual of

[a] His explanatory notes are printed in red. The willingness of Li Chieh himself to learn from experienced artisans and craftsmen is worth emphasising a little further. It had been a long tradition in Taoism, as witness the story of Pien the Wheelwright (Vol. 2, p. 122 above), and Liu Tsung-Yuan's old gardener (Vol. 2, p. 577 above). In the Thang, Han Yü,[23] though so Confucian a scholar, had written a famous essay on what he had learnt from the mason Wang Chhêng-Fu[24] (*Ku Wên Hsi I*, ch. 12, pp. 29 a ff. tr. Margouliés (1), p. 195, (3), p. 178). Cf. p. 315 and Fig. 905 (pl.).

[b] Briggs (2), p. 141. It is interesting, however, to compare a book such as that of Mathurin Jousse, *Le Théâtre de l'Art de Carpentier* (+1627), with the Chinese work of five hundred years earlier. The real comparison of course is with Villard de Honnecourt (+1237), and again the Chinese advantage is extraordinary. More will be said below (p. 107) on the position of the *Ying Tsao Fa Shih* in the history of technical drawing.

[c] Cf. des Rotours (1), pp. 253, 461.

[1] 壕寨	[2] 石作	[3] 大木作	[4] 小木作	[5] 彫作
[6] 旋作	[7] 鋸作	[8] 竹作	[9] 瓦作	[10] 泥作
[11] 彩畫作	[12] 塼作	[13] 窰作	[14] 功限	[15] 料例
[16] 諸作等第	[17] 賀新賽	[18] 義介	[19] 五山十刹圖	
[20] 營造正式	[21] 元內府宮殿制作	[22] 欽定工部工程做法		
[23] 韓愈	[24] 王承福			

Constructional Engineering drawn up by the Ministry of Works upon Imperial Order).[a] A bulky MS. without title, but containing regulations for building and furnishing imperial palaces, which (from internal evidence) covers the dates +1727 to +1750, has been described by Hummel (20) and Malone (1). It deals largely with the Yuan Ming Yuan summer palace (started in +1709 and destroyed by foreign troops in 1860), but has more to do with the economics than the technique of building.[b]

A quite different tradition of literature which is relevant here is that of the descriptions which have come down to us, in prose or poetry, of cities, palaces and temples. From the Later Han period onwards rhapsodical odes on the successive capitals became a distinct literary genre, and so it came about that when the great collection called the *Wên Hsüan* (General Anthology of Prose and Verse) was brought together by a prince of the Liang in +530, the section entitled Ching Tu[1] took pride of place in the opening chapters.[c] A pair of odes on the western and the eastern capitals, Chhang-an (Sian) and Loyang (*Hsi Tu Fu*[2] and *Tung Tu Fu*[3]), was written by the famous historian Pan Ku[4] about +87;[d] and there followed soon afterwards (in +107) a second pair by the eminent astronomer and mathematician Chang Hêng[5] (the *Hsi Ching Fu*[6] and *Tung Ching Fu*[7]).[e] Since the two capitals symbolised the ethos of the Former and the Later Han respectively these odes have attracted the close attention of modern historians of thought and culture (notably Hughes, 9) interested in their differentiation; but the poetical phraseology is notoriously difficult, the technical terms obscure, and the descriptions of buildings and layouts naturally somewhat vague. Still, a careful study by a historian of Chinese building technology would be well worth while. Chang Hêng also left a third ode, the *Nan Tu Fu*,[8] on his birthplace, the ancient city of Wan[9] (Nanyang[10]), long a centre of the iron and steel industry;[f] this was written in +110 during a period of temporary retirement. In the following centuries the tradition continued and about +270 Tso Ssu[11] produced a set of poetical descriptions of the capitals of the Three Kingdoms, Shu, Wu and Wei.[g] Panegyrics of Han palaces also exist, for instance Wang Wên-Khao's[12] *Lu Ling-Kuang Tien Fu*,[13] written about +140, an

[a] In +1726 and +1734. Liang Ssu-Chhêng (9), p. 9, cites it as *Kung Chhêng Tso Fa Tsê Li*.[14] Choisy (1) and Boerschmann (1) mention a MS. *Kung Chhêng Tso Fa*,[15] not further specified, with which they worked.

[b] There must be a good deal of such material extant, for in the summer of 1952 I came across a similar MS. on a stall in the Tung An Shih Chhang market in Peking; I mentioned it to Chinese specialist friends, but do not know whether it was acquired for some library.

[c] By this time all the odes had been illustrated by eminent painters in scroll-paintings, but none have survived. Sullivan (3) has assembled what is known of them.

[d] *Wên Hsüan*, ch. 1; tr. Margouliés (2); Hughes (9).

[e] *Wên Hsüan*, chs. 2, 3. The former was translated into German long ago by von Zach (2), cit. in Bulling (2), p. 51, and englished by Gutkind (1), p. 318. Both appear in translation in von Zach (6) and Hughes (9).

[f] Cf. Sect. 30*d* in Vol. 5 below.

[g] *Shu Tu Fu*,[16] *Wu Tu Fu*[17] and *Wei Tu Fu*,[18] in *Wên Hsüan*, chs. 4, 5, 6; tr. von Zach (6). The three capitals were Chhêngtu, Suchow and Yeh (Hsiangchow) respectively.

[1] 京都	[2] 西都賦	[3] 東都賦	[4] 班固	[5] 張衡
[6] 西京賦	[7] 東京賦	[8] 南都賦	[9] 宛	[10] 南陽
[11] 左思	[12] 王文考	[13] 魯靈光殿賦	[14] 工程做法則例	
[15] 工程作法	[16] 蜀都賦	[17] 吳都賦	[18] 魏都賦	

elaborate account, but more of the ornamentation than the building.[a] The *Ying Tsao Fa Shih* quotes judiciously from these sources in its opening chapters. All in all they constitute a valuable quarry for architectural history not yet fully exploited.

The *San Fu Huang Thu*[1] (Description of the Three Metropolitan Cities), attributed to Miao Chhang-Yen,[2] may date from this time, but more probably from the +3rd century; it had originally drawings and charts, and is considered a fairly reliable source for details about the Later Han capital, Chhang-an.[b] Then in the +6th century came the famous *Lo-yang Chhieh-Lan Chi*[3] (Description of the Buddhist Temples and Monasteries at Loyang).[c] In the Sung the city was still so beautiful that it merited another monograph, the *Lo-yang Ming Yuan Chi*[4] (Record of the Celebrated Gardens of Loyang), written by Li Ko-Fei[5] about +1080. But the time was soon coming when the victories of barbarians would act as a powerful stimulus for the reminiscent description of cities and their buildings as they had been before the storms burst. Thus it was that the *Tung Ching Mêng Hua Lu*[6] (Dreams of the Glories of the Eastern Capital), dealing with Khaifêng, was written just twenty years after its fall. The pattern imposed by the Chin Tartars was repeated a century or two later by the Mongols, giving us at least four books about the Sung capital at Hangchow, among which we need mention here only the *Tu Chhêng Chi Shêng*[7] (The Wonder of the Capital) of +1235, and the *Mêng Liang Lu*[8] (The Past seems a Dream) of +1275.[d] Finally, we have descriptions of the palaces and public buildings of both Yuan and Ming. Hsiao Hsün's[9] book on the former, *Ku Kung I Lu*,[10] has already more than once been mentioned,[e] and the same category might include the *Yuan Shu Tsa Chi*[11] written about Peking by Shen Pang[12] in +1593, and now newly republished from a lone copy preserved in Japan.[f] Not long afterwards came the *Ming Kung Shih*[13] of Liu Jo-Yü,[14] written about +1620 and dealing with Peking also.[g] It would certainly be possible to derive from these graphic accounts of the appearance and life of the great cities matter of interest both for architecture and town planning, but an exposition of it would require specific researches not yet complete, so that here we can do no more than call attention to the field.

A third great class of literature gives further opportunities of reconstructing the architecture and layouts of medieval Chinese cities and public buildings, namely the indefatigable researches of archaeologists and local antiquarians.[h] Nearly every city

[a] *Wên Hsüan*, ch. 11; tr. v. Zach (3) completely, and Waley (11) partially. We have come across this already in Vol. 4, pt. 2, p. 131.

[b] There is also a *San Fu Chiu Shih*[15] (Stories of the Three Cities), written certainly much later.

[c] Cf. the reconstruction of the city plan from its descriptions by Lao Kan (3).

[d] Parts of both these have been translated and discussed by Moule (5, 15) and Gernet (2). Marco Polo's descriptions of this and other cities will not be forgotten, and are easily referred to.

[e] E.g. Vol. 4, pt. 2, pp. 133, 508.

[f] Yuan-phing[16] was the old name of Lu-kou-chhiao (cf. p. 183 below).

[g] Cf. Hirth (17).

[h] It may be well to remind the reader of what was said in Vol. 2, pp. 390 ff., 393 ff., about the strength of critical humanistic studies in Chinese culture.

[1] 三輔皇圖　　　[2] 苗昌言　　　[3] 洛陽伽藍記　　　[4] 洛陽名園記
[5] 李格非　　　[6] 東京夢華錄　　　[7] 都城紀勝　　　[8] 夢粱錄
[9] 蕭洵　　　[10] 故宮遺錄　　　[11] 宛署雜記　　　[12] 沈榜
[13] 明宮史　　　[14] 劉若愚　　　[15] 三輔舊事　　　[16] 宛平

has its 'local gazetteer' (*fang chih*[1]), a work often in many successive recensions on the history and topography of the place, always including traditions of buildings and building plans.[a] But most attention, of course, was given to the capitals. The line could begin with Wei Shu's[2] *Liang Ching Hsin Chi*[3] (New Records of the Two Capitals), written during the first half of the +8th century.[b] Though only one chapter of this work has survived until now, it laid the foundation for the subsequent work of Sung Min-Chhiu[4] about +1075 entitled *Chhang-an Chih*[5] (History of the City of Eternal Peace). About the same time Liu Ching-Yang[6] and Lü Ta-Fang[7] were commissioned to make a historical map of the city, and this they did on a scale of 2 in. to the mile, identifying and marking the sites of ancient buildings.[c] This *Chhang-an Thu Chi*[8] served in turn as the basis for more elaborate plans produced about +1330 by Li Hao-Wên,[9] an imperial tutor, in his *Chhang-an Chih Thu*[10] (Maps to illustrate the History of the City of Eternal Peace).[d] Chinese traditional archaeology laboured on continuously into the modern period, as may be seen by the work of Hsü Sung[11] in 1810 with the title *Thang Liang Ching Chhêng Fang Khao*[12] (Studies on the Districts in the Two Thang Capitals).[e]

Recent studies have shown, however, that any reconstruction of ancient and medieval buildings in the light of textual evidence only is extremely difficult. For interpreting the texts correctly iconographic evidence, whenever it can be obtained, is indispensable. Fortunately we are not entirely without it. From the Warring States period, from the Han and Chin dynasties, there are carved vessels, moulded bricks and tomb models, all making a considerable contribution; presently (pp. 126 ff.) we shall illustrate various points by their means. For the Thang period, however, we are much better off, for we have the iconographic wealth of the fresco paintings of the Tunhuang cave-temples, many abounding in architectural detail (Figs. 356, 728, 758, pls.). For a good while past Chinese historians of architecture have been guided by these (cf. the excellent paper of Liang Ssu-Chhêng, 6), and in a series of valuable papers Bulling (2, 9) has pioneered by attempting reconstructions of the architectural plans of the buildings depicted in the frescoes. These may be divided into two groups, actual cities and temples of the period, and scenes in Buddhist paradises. For example, cave no. 61 is filled with a panorama of Wu-thai Shan,[13] mountain home of numerous abbeys from the Liu Chhao period onwards,[f] and one at least of those represented has been identified with +9th-century buildings still extant today[g] (cf. p. 130 and Fig. 789 (pl.) below). From one or other of the frescoes it has been possible already to reconstruct a

---

[a] See the more extended discussion in Vol. 3, pp. 517 ff.
[b] We have met with Wei Shu already in Vol. 4, pt. 2, p. 471.
[c] Cf. *Yün Lu Man Chhao*, ch. 2, p. 11a, and for further context, Vol. 3, p. 547.
[d] Most of what survives of these maps and plans has been collected in Anon. (9).
[e] On Hsü Sung, cf. Vol. 3, p. 525. Besides the systematic scholarly works here mentioned, one must not forget the remarks on buildings in the desultory notes of Chinese writers of all ages—for example, in the *Chhang Wu Chih*[14] (Notes on Life's Staples) written by Wên Chen-Hêng[15] about +1595.
[f] On this panorama see a special paper by Hibino Takeo (2).
[g] Cf. Liang Ssu-Chhêng (4).

[1] 方志	[2] 韋述	[3] 兩京新記	[4] 宋敏求	[5] 長安志
[6] 劉景陽	[7] 呂大防	[8] 長安圖記	[9] 李好文	[10] 長安志圖
[11] 徐松	[12] 唐兩京城坊考		[13] 五台山	[14] 長物志
[15] 文震亨				

dozen interesting designs, often including a peripheral cloister studded with small pavilion towers. As for the paradises, the settings where Buddhas and Bodhisattvas are assembled to witness sacred dances, acrobatics, the births of new souls from lotuses, and so on, all to the accompaniment of orchestral music,[a] invariably comprise halls, pavilions, galleries, forecourts, pools, platforms and bridges in elaborate ensembles. Here again everything can be translated into accurate ground-plan and precise elevation.[b] We reproduce in Fig. 758 (pl.) a picture of the Western paradise of Amida, painted about +700, from cave no. 172.

A further source of information about the position of architectural science in the different periods would be the titles and powers of officials concerned. The *Chou Li*, in its account of the Artisan-Carpenters and Engineers (*chiang jen*[1]), says that they have to construct capital cities (*ying kuo*[2]) and buildings (*ying shih*[3]).[c] It includes some schematic information about the plan of a standard city, with its three gates on each side, nine transverse main streets,[d] central palace, northern market, and so on. It also indicates that buildings were measured in 9-ft. mat-lengths (*chiu chhih chih yen*[4]),[e] and street breadths in 6-ft. chariot-gauges.[f] The Sung *Shih Wu Chi Yuan* tells us[g] that in Chhin and Han there were officials called Chiang Tso[5] (Directors of Workmen)[h] but no special bureau. A Directorate of Buildings appeared in the Northern Chhi dynasty (+550 to +557) as the Chiang Tso Ssu,[6] and in the Sui and Thang[i] it became the Chiang Tso Chien.[7] Before the Sui, architectural officials were appointed only when there was need of palace or government buildings, at other times the posts were left vacant. Gradually changes of meaning occurred, and apparently in the Yuan period, the Chiang Tso Yuan[8] was no longer a building department, but rather a part of the imperial workshops connected with the working of precious metals, and the weaving of valuable textiles. Here again we can see how useful a monograph on the general history of the imperial workshops would be.[j]

---

[a] A picture of one of these orchestras has already been given in Vol. 4, pt. 1, Figs. 313, 314.

[b] The work thus begun needs systematic extension. Only in this way will it be possible to assess some of Bulling's theses, for example that the accent on axiality is missing from pre-Thang and early Thang temple plans. There may be something in this, but we are still more reluctant to accept her view that the crowds in the paradise pictures represent living people vested to play the parts of Buddhas and Bodhisattvas at the periodical temple festivals.

[c] Ch. 12, pp. 15b ff. (ch. 43) (*Khao Kung Chi*); tr. Biot (1), vol. 2, pp. 555 ff.

[d] Cf. above, pp. 6, 73.

[e] Here is another origin of that module or standard increment system which continues to delight modern Western architects (e.g. J. M. Richards (1), in 1962) who come in contact with it in Japanese domestic building practice. They are astonished that it so long preceded industrial production of standard components. Cf. p. 67 above.

[f] Cf. above, pp. 5 ff.

[g] Ch. 6, p. 42b.

[h] Also Chiang Tso Ta Chiang[9] and Ta Chiang Chhing[10] later.

[i] Des Rotours (1), pp. 476 ff.   [j] Cf. Vol. 4, pt. 2, pp. 18 ff.

[1] 匠人   [2] 營國   [3] 營室   [4] 九尺之筵   [5] 將作
[6] 將作寺   [7] 將作監   [8] 將作院   [9] 將作大匠   [10] 大匠卿

(5) PRINCIPLES OF CONSTRUCTION

In what has gone before we have been able to catch a glimpse of some of the essential principles of Chinese building, the use of walls as pure screens and partitions which carry no weight; the dependence on firm geometrical grids of pillars and beams, allowing great emphasis to be placed on the roof; the employment of raised platform foundations which balance the roof aesthetically; and the fact that buildings are nearly always meant to be approached from their sides rather than their ends. The moment has come to penetrate further into the matter. We shall find that certain obvious questions are capable of illuminating answers, while to others there is still no satisfactory solution.

Of this second class is the problem of basic materials. Why was it that the Chinese throughout their history systematically built in wood and tile, bamboo and plaster, never making use of the stone which in other civilisations such as Greece, India and Egypt left such durable monuments behind? I have often felt that if a full answer to this question could be obtained, it would throw light on many wider aspects of cultural difference.[a] It certainly cannot be said that China had no stone suitable for great buildings analogous to those of Europe and Western Asia, but it was used only for tomb-construction, steles and monuments (in which typical woodwork details were frequently imitated),[b] and for pavements of roads, courts and paths.[c] Perhaps further knowledge of social and economic conditions might throw light on the matter, for the forms of slavery known in China in different ages seem never to have paralleled those occidental usages which could despatch thousands of human beings at a time to hard labour in the quarries.[d] In Chinese civilisation there is absolutely no parallel to those great sculptured friezes of Assyria or Egypt[e] which depict the harnessing of large numbers of workers in the transportation of enormous monoliths for carving or building. It might indeed seem that no rule could have been more absolute than that of a Chhin Shih Huang Ti, the builder of the first Great Wall, and of course there is no doubt that in ancient and medieval China very great labour forces could be mobilised by *corvée*,[f] but what matters is the state of society in which the characteristic forms of

---

[a] For instance, there seems certainly to be some relation between aspiring monumentality in stone and the influences of mystical religion. But the Chinese mood was essentially secular, loving life and Nature. Hence the gods had to conform, to sit and be worshipped in buildings identical with the halls of families and palaces, or not to be worshipped at all.

[b] Cf. the frank imitation in stone of bundles of reeds so common in the Maya architecture of Yucatan.

[c] Exceptionally also for forts and strongholds. A particularly interesting exception (noticed already on p. 69 above) was constituted by the fireproof warehouses (*tha-fang*[1]) of the city of Hangchow, described in +13th-century books. These caravanserais were stone towers like castle keeps surrounded by waterways. See the discussion of Moule (11), as also (5, 15).

[d] The whole question of slavery in ancient China will receive consideration in the concluding Sections of this book. Meanwhile see Vol. 4, pt. 2, pp. 23 ff.

[e] Cf. the famous fresco at al-Bersheh showing the mode of transporting a colossus from the quarries, figured by Wilkinson (1) as the frontispiece to his second volume. Or Klebs (2), fig. 40, p. 61. Cf. too Vol. 4, pt. 2, pp. 74, 92.

[f] Cf. Vol. 4, pt. 2, pp. 22, 330 and 21, 30 38 above.

[1] 塌坊

Chinese architecture were originally determined, and it may well be that some connection exists between the timberwork style and the absence of mass slavery.[a] On quite another plane, the ancient symbolic-correlation philosophy[b] was perhaps also involved, for if stone was regarded as belonging to the element earth it would have been proper for use only upon and under the ground, while wood was an element in itself, occupying a middle position between earth and the fiery *chhi* of the heavens, hence the only fitting substance with which to build. Such a philosophy might have been but the expression of that genial sobriety and sensible dislike of extravagance so characteristic of Chinese culture. Why try to dominate posterity? 'One can create something, it is true,' said Chi Wu-Fou, China's greatest writer on gardens (+1634), 'which will last for a thousand years, but no one can tell who will be living after a hundred. Let it suffice to create a spot for pleasure and ease, which envelops the dwelling with harmonious stillness.'[c] Lastly, we must not lose sight of the fact that nearly the whole of China was, at one time or another, subject to earthquakes, so that experience may have shown the flexibility and elasticity of wood to be preferable to the unyielding but collapsible weight of stone. But all these notions are speculative, and the question remains.[d]

Much more fruitful is that other question which so many foreigners in China must have asked themselves: what can be the nature and origin of the curving roof, that most characteristic and beautiful feature of Chinese buildings? The idea that it derives from an ancient desire to imitate the catenary curves of tents and mat-sheds has been a popular cliché of tourists for centuries.[e] No one, however, has found any authority for this, either literary or archaeological. Besides, it is the wrong kind of answer. What we really need to know first is how (i.e. by what constructional method) the Chinese roof succeeds in getting its curve. This can be appreciated by examining Figs. 759 and 760. The former gives a schematic cross-section of a hall, based upon the drawings in the *Ying Tsao Fa Shih*.[f]

The great supporting columns rest upon plinths which form part of the platform foundation, and they are fixed together by massive tie-beams at various heights above the floor. Above them rise in tiers the main cross-beams which carry the roof; these are upheld by many queen-posts arranged at suitable positions, with usually only one

[a] Cf. Sect. 30 in Vol. 5 below for another correlation between a profound cultural trait and the circumstances in which Chinese civilisation crystallised. An alternative way of putting the matter would of course be to relate it to the specific agricultural methods of China, so extravagant in man-power and time, and so dangerous for rulers if interfered with. Of course heavy building in stone does not always imply the use of slave labour; the Greek temples were built almost entirely by freemen.

[b] Cf. Vol. 2, pp. 261 ff, above.   [c] *Yuan Yeh*, ch. 1, p. 6a.

[d] I am indebted to Miss Elizabeth Vellacott for suggestions in this paragraph.

[e] Lamprey (1); von Fries (1); and Edkins (15) all long ago took the trouble to contest this. Cf. Arlington & Lewisohn (1), p. 325. It would be at least equally plausible to derive the curve from brush-writing. Let anyone look at the uppermost strokes of the calligraphers' *tshang*[1] (granary) or *san*[2] (umbrella), and ponder on the aesthetic standard thus set.

[f] For all concrete detail the reader is referred to the books mentioned in the introduction, p. 59 above. We can be concerned here only with the abstract skeleton. In the paragraph which follows reference to Fig. 759 should be made for the Chinese technical terms. Cf. p. 99.

[1] 倉   [2] 傘

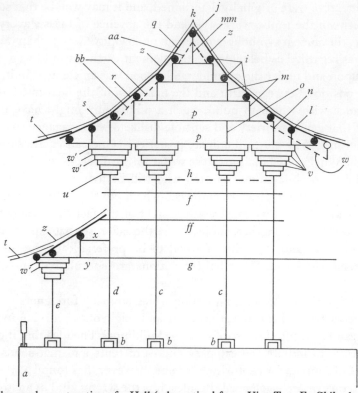

Fig. 759. Timberwork construction of a Hall (schematised from *Ying Tsao Fa Shih*, ch. 30, pp. 8*a* ff.).

Key:

a	platform *thai*,[1] *chieh chi*[2] (with balustrade *kou lan*[3])
b	stone plinths for the wooden columns[a] *chu chhu*[4]
c	principal columns[b] *chu*,[5] *ying chu*,[6] *chin chu*[7]
d	external principal column supporting eaves[b] *lao yen chu*,[8] *yen chu*[9]
e	short external principal column supporting aisle eaves[b] *hsiao yen chu*[10]
f	main tie-beams[c] *ta ê fang*[11] ('forehead beam')
ff	other tie-beams *fang*[12]
g	lowest main tie-beam *khua khung fang*[13] ('bestriding emptiness beam')
h	flat boards for coffer ceiling *phing pan fang*[14]
i	purlins *hêng*,[15] *lin*,[16] *lin thiao tzu*[17]
j	ridge-pole (uppermost purlin) *chi hêng*;[18] *tung*,[19] *fu*[20] (can be purlins in general)
k	roof ridge *chêng chi*[21]
l	purlin supported from eave column *chêng hsin hêng*[22]
m	upper outer queen-posts (or blocks) *tho tun*[23] ('main beam stands')
mm	king-post *chu ju chu*[24] (*cho*,[25,26] poetic word also applied to all vertical posts in the roof-timbering), *shan chu*[27]
n	lowest outer queen-post *kua chu*,[28] *thung chu*[29]

[a] Note the absence of a 'ground-sill' or tie-beam along the ground as in medieval European frame buildings. Yet the ground-sill is very prominent in China, not absenting itself from main doors and great gates. But it is concerned only with the woodwork planking of the curtain-walls.

[b] All these are connected together along the length of the building by tie-beams.

[c] Medieval 'bressumers'. Strictly, they should run directly under the lowest corbel brackets.

[1] 臺	[2] 階基	[3] 鈎闌	[4] 柱礎	[5] 柱
[6] 楹柱	[7] 金柱	[8] 老檐柱	[9] 檐柱	[10] 小檐柱
[11] 大額枋	[12] 枋	[13] 跨空枋	[14] 平板枋	[15] 桁
[16] 檁	[17] 檁條子	[18] 脊桁	[19] 棟	[20] 桴
[21] 正脊	[22] 正心桁	[23] 柁墩	[24] 株儒柱	[25] 梲
[26] 棳	[27] 山柱	[28] 瓜柱	[29] 童柱	

central king-post at the top under the ridge-pole.[a] Longitudinal beams or purlins are placed at any desired points towards the ends of these cross-beams, and to these purlins the rafters are fixed. A closely similar arrangement is used for aisles, galleries or verandahs, built in a lean-to manner. The fundamental unit is thus a 'two-dimensional' framework capable of endless variation and adaptability;[b] this is known as the Method of Frame Construction (*Ku chia chieh kou fa*[1]). Based on the platform of rammed earth (*chu thu wei thai*[2]) it goes back in essence, as archaeological discoveries have shown, to the Shang dynasty (−2nd millennium).[c] In ancient times only the cross-beams (*chia liang*[3]) were used, but as time went on it was found that this placed

[a] An assembly of two cross-beams, being decorated before put into position, is seen in *SCTS*, ch. 31, p. 7 *a*.
[b] Various cross-sections noted by Spencer (1) are shown in Fig. 760 *j–n*.
[c] Liang Ssu-Chhêng (*3*). See further on p. 122 below.

[1] 骨架結構法   [2] 築土爲臺   [3] 架梁

Continuation of Key for Fig. 759

o	inner queen-posts *chhen kho mu*[1] ('under-clothes')	*w*	cantilever principal rafters (see on, p. 95) *ang*[13]
p	main cross-beams[a] *chia liang*[2] (no. 1, no. 2, etc.), also *liang*[3]	*w'*	false cantilever principal rafters (see on, p. 97) *ang*[13]
q	upper rafters *nao chhuan*[4] ('brain rafters'); all rafters may be called *chhuan*,[5] *chüeh*[6]	*x*	extension beam supporting purlin of aisle roof *tan pu liang*,[14] *ju fu*[15]
r	middle rafters *hua chia chhuan*[7] (i.e. those corresponding to the layers of cross-beams)	*y*	main aisle extension beam *thao chien pao thou liang*,[16] *ju fu*[15]
s	lower, or eave, rafters *yen chhuan*[8] ('eave-rafters')	*z*	tile surface alignment, with boarding (*wang pan*[17]) beneath
t	cantilever eave rafters[b] *fei yen chhuan*,[9] *tshui*[10] ('flying rafters')	*aa*	inverted V-brace supporting the king-post (only in Sung and pre-Sung buildings) *chha shou*[18] (derived from the ancient *jen tzu kung*[19])
u	lowest corbel bracket unit *thou chhiao tou kung*[11] ('head of the tail-feathers')	*bb*	side braces connecting either cross-beam with cross-beam or cross-beam with purlin (only in Sung and pre-Sung buildings) *chha shou*,[18] *tho chiao*[20]
v	superimposed corbel bracket units (both parallel with, and at right angles to, the beams)[c] *tou kung*[12]		

[a] Medieval 'somers'. The uppermost one was called the 'wind-beam'.
[b] Analogous to the 'firrings' of medieval European wood construction.
[c] The transverse corbel brackets (or bracket arms) are termed *hua kung*.[21] The height of the standard *hua kung* selected is taken in the *Ying Tsao Fa Shih* as the module for the whole building (cf. pp. 67 ff. above). The longitudinal corbel brackets, which it is not possible to show in the diagram, have several names. The *kua tzu*[22] is a longitudinal bracket arm midway between the wall and the eaves purlin; the *man kung*[23] is a longer one which is fixed above it so as to extend the support. The *ling kung*[24] is a longitudinal bracket arm supporting the eaves purlin. The longitudinal tie-beams, also not seen in the diagram, are all termed *fang*,[25] with qualifying adjectives.

[1] 襯科木	[2] 架梁	[3] 梁	[4] 腦椽	[5] 椽
[6] 桷	[7] 花架椽	[8] 檐椽	[9] 飛檐椽	[10] 槜
[11] 頭翹科栱	[12] 科栱	[13] 昂	[14] 單步梁	[15] 乳栿
[16] 桃尖抱頭梁	[17] 望板	[18] 叉手	[19] 入字栱	[20] 托脚
[21] 華栱	[22] 瓜子	[23] 慢栱	[24] 令栱	[25] 枋

N.B. The sketch is purely schematic and takes no account of the sizes and strengths of the various component timbers. The proportions of the building as shown in the diagram are rather inelegant, but have been chosen for convenience of demonstration on the page of a book. At the same time it is true that Chinese monumental buildings always tended to be much larger in length than in depth, and the cross-section here depicted is not unlike that of the famous hall of the Hōryūji temple in Japan, built in +670.

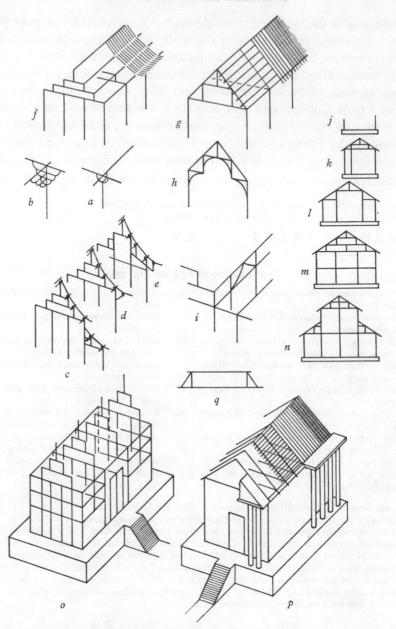

Fig. 760. Diagrams to elucidate Chinese and Western building construction (explanations in text).

a   The principle of corbel brackets (*tou kung*)
b   A more complex example
c   Typical arrangement of roof and subsidiary aisle roof in a Chinese building. No principal rafters;
    longitudinal purlins supported on the transverse frameworks in any desired profile
d   Eave rafters and flying rafters on bracket-supported purlins
e   Eave purlins supported from the ends of cantilever principal rafters or lever arms (*ang*)
f   Characteristic structure of a Chinese building
g   Characteristic structure of a European building
h   Corbelled half-arches of an English hammer-beam roof
i   Inverted V-braces of early medieval Chinese buildings (*jen tzu kung*)
j, k, l, m, n   Typical Chinese transverse frames of columns (*chu*), tie-beams (*fang*) and cross-beams (*liang*)
o, p   Comparison of the fundamental Chinese building design with that of Greek and Gothic building
    (see pp. 62, 65, 102). Normally the Greek gable covered the perimetral colonnade.
    'Forked hand' struts forming a trapezoidal truss (see p. 101).

excessive tension at the junction between columns and beams, failures tending to occur there. The improvement was therefore introduced of inserting a number of corbel brackets (*tou kung*[1]) between the top of the column and the cross-beam. The *tou* was so called because it was a block of wood resembling a capacity-measure (*tou*[2]) in shape, and the *kung* or bow-piece was the double elbow-shaped arm supporting one of these on each side. Corbel brackets successively longer were then piled on top of one another at the capital of the column, so as to form what were essentially corbelled arches of wood supporting the cross-beams. The *tou kung* branched forth not in one direction only but in both, i.e. as well parallel as transverse to the long axis of the building, thus supporting both longitudinal and crosswise beams (see Figs. 760 *a*, *b* and 761, pl.). The original name for them was *lu*[3] or *luan*,[4] words which afterwards came to be applied occasionally to the king-post.[a] The *Ying Tsao Fa Shih* has many illustrations of them (Fig. 762), showing how they were fitted with tenons and mortises (*shun mao*[5]), pegs and dowels (*kan*,[6] *chhüan*[7]) (Fig. 763). Sometimes they are hidden by panelling.

The typical arrangement of roof and subsidiary roof for an aisle or verandah is shown in Fig. 760 (*c*), but during the Thang and before, it was usual to support the eaves with their rafters and flying rafters by means of purlins on brackets (Fig. 760*d*). There then developed another system, that of placing eave purlins upon 'cantilever principal rafters' or 'lever arms' (*ang*), Fig. 764 (pl.), fixed to the interior framework and piercing the bracket-arm clusters in a direction approximately parallel with the slope of the roof above (Figs. 760*e* and 765).[b] This system died out during the Yuan period,[c] but

---

[a] It may be noted that in the *Chou Li*, ch. 11, p. 23 *a*, *b*, the word *luan* is applied to the sharply curving ends of bells of oval cross-section—we do not know which usage was the older. Cf. Vol. 4, pt. 1, p. 196.

[b] Most extant temple buildings of the +7th century onwards have them, both in China and Japan (cf. pp. 100, 109, 130). For other drawings see Mirams (1), opp. p. 58.

[c] This is true only of China; in Japan the *ang* has lived on down to our own time. On a visit to both countries in 1964 I was able to compare and photograph successively temple roof construction at close quarters. In China I was impressed by the prominence of the cantilever principal rafters, first in the +9th-century woodwork of Fo-Kuang Ssu[8] (cf. p. 130); then in half a dozen temple halls dating from between +1030 and +1180 (Shan-Hua Ssu[9] and the two Hua-Yen Ssu[10] at Ta-thung, Chin Tzhu[11] near Thai-yuan, all in Shansi, the Hsüan Miao Kuan[12] at Suchow (Chiangsu), minutely described by Liu Tun-Chên (5), and Chieh-Thai Ssu[13] near Peking). No trace of *ang* is left, of course, in any of the monumental Ming or Chhing buildings in Peking or elsewhere. Then in Japan I admired the expected *ang* in the +7th-century Hōryūji[14] buildings near Nara, as also those in the exquisitely beautiful Byōdōin[15] beside its little lake at Uji (+11th century). But they could still be seen in full employment in the gatehouse of the Engakuji[16] at Kamakura, not earlier than +14th or +15th century, and before long I found them in double rows in the main halls of the Shinyōdō[17] and Hyakumamben Chionji[18] temples at Kyoto, rebuilt in the +18th or even the 19th century. Finally, by the kindness of the Rev. Yamasaki Teruo, I was able to inspect the gallery and interior of the gatehouse of the Kōmyōji[19] temple at Kamakura, rebuilt in the early 19th century, and to see how the upper of the two rows of *ang* still retained its original structural function while the lower was 'false' and purely decorative. A consistent difference between the *ang* of the two cultures is that in China they end prismatically while in Japan they are cut off at right angles and the flat surface painted in contrasting colour. I owe much gratitude to my friend Dr Nakayama Shigeru for his kindness in helping our studies in Japan.

---

[1] 枓栱   [2] 斗   [3] 櫨   [4] 欒   [5] 榫卯
[6] 杆   [7] 栓   [8] 佛光寺   [9] 善化寺   [10] 華嚴寺
[11] 晉祠   [12] 玄妙觀   [13] 戒台寺   [14] 法隆寺   [15] 平等院
[16] 圓覺寺   [17] 眞如堂   [18] 百萬遍知恩寺   [19] 光明寺

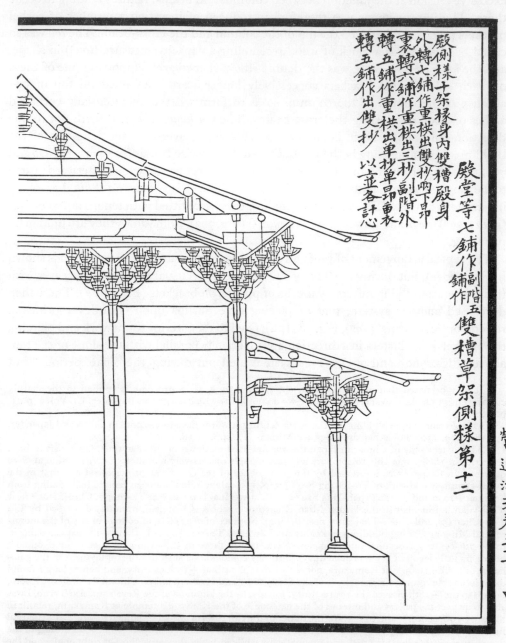

殿側樣十架椽身內雙槽殿身
外轉七鋪作重栱出雙抄兩下昂
裏轉六鋪作重栱出三抄副階外
轉五鋪作重栱出單抄單昂重栱
轉五鋪作出雙抄以並各計心

殿堂等七鋪作副階五鋪作雙槽草架側樣第十二

營造法式卷三十一　四

Fig. 762. Diagram of three corbel bracket assemblies, from a cross-section of a hall (*Ying Tsao Fa Shih*, ch. 31, p. 4*a*). The curving profile of purlins and the consequent curve of the roof line is well seen. Lever arms (*ang*) are present (see pp. 95, 99).

left permanent traces in the form of 'birds'-beak' ends on the corbel brackets (as indicated in Fig. 759w'); these may be called 'false cantilever principal rafters'.[a] In most regions it was customary to raise up the four corners of the roof above the main roof edge line (cf. p. 128 and Fig. 766, pl.), and in the south the roof ridge itself was sometimes made to rise towards each end of the building in a graceful curve (*chhui chi*,[1] *shou chi*[2]).[b] Main roofs may be either hipped or gabled, but in important buildings the gables seldom reach down to the eaves; they are cut at half their height or less, the slant of the roof continuing below and around them.[c] This combination, as one of the pundits has justly remarked, is most imposing and harmonious.[d]

Whatever the details of the roof-supporting system might be, it is important to bear in mind that the skeleton of the building as a whole stood up of itself, needing neither base, walls nor roof. The two-dimensional trabeate frames with their cross-beams and tie-beams were connected longitudinally by tie-beams below as well as other tie-beams[e] and purlins above, so that a three-dimensional continuum almost like a cubical crystal lattice came into being. As we shall see (p. 103), this was the veritable ancestor (collateral if not direct) of the steel-frame lattice of modern building technology. Far different was it from the solid casing walls of ancient and medieval Europe.[f]

We are now in a position to understand another radical difference between Chinese and occidental architecture. The curved roof and all that that implies was impossible in the West because the West was wedded to straight and rigid sloping principal rafters, that is to say, a transverse slanting element; while in China, on the contrary, the most important element was the longitudinal purlin, groups of which could be assembled according to any profile desired, by adjusting the framework itself.[g] Some of the European principal rafters, of course, actually formed sides of the triangular roof-trusses, which, though they have their internal king-posts and queen-posts, present

[a] Cf. Sirén (5); Mirams (1), opp. p. 59; Sickman & Soper (1), p. 265.

[b] Cf. Mirams (1), opp. pp. 56, 74.

[c] The technical terms for the various types of roofs are given in Anon. (*37*), p. 6, simplified in Hsü Ching-Chih (1), pp. 38, 220. The roof of a building with plain gables is called 'hard-edged' (*ying shan*[3]), one where the purlins project to give 'eaves' 5 to 8 rafters deep is called 'hanging' (*hsüan shan*[4]). A hipped roof covers a 'verandah hall' (*wu tien*[5]), and a hipped gable roof, as here described, is called 'curtailed' (*hsieh shan*[6]). The pyramidal roof of a square pavilion is called *tshuan chien*.[7] All these have one tier of eaves (*tan yen*[8]), but often a second tier is added below a short clerestory; this is called *chhung yen*.[9]

[d] Sirén (1), vol. 4, p. 22.

[e] Named *phing thuan*,[10] and the highest under the ridge-pole *chi thuan*[11] (cf. Liu Chih-Phing (1), fig. 219).

[f] To some extent the European medieval 'half-timber' building (cf. Briggs (2), pp. 131 ff.; Davey (1), pp. 40 ff.) was a structure inherently stable, even when the upper storeys overhung, as they often did. But the individual timbers employed were always so small relative to the whole building that there was no possibility of any wide openings, and the walls were still weight-bearing, with the disadvantage of being half composed of inflammable material.

[g] This is well seen in the roof assembly for a small gate-house which I photographed at Chhien-fo-tung in 1958 (Fig. 767, pl.). The single cross-beam carries a ridge-pole and four purlins by means of king- and queen-posts of different lengths. Fig. 768 (pl.) shows the skeleton of a larger construction, the framework of a new commune headquarters, at Ho-thang near Lo-phing in north-eastern Chiangsi. The columns, tie-beams, cross-beams, and posts for the purlins are clearly seen (1964).

[1] 垂脊　　　[2] 獸脊　　　[3] 硬山　　　[4] 懸山　　　[5] 廡殿　　　[6] 歇山
[7] 攢尖　　　[8] 單檐　　　[9] 重檐　　　[10] 平槫　　　[11] 脊槫

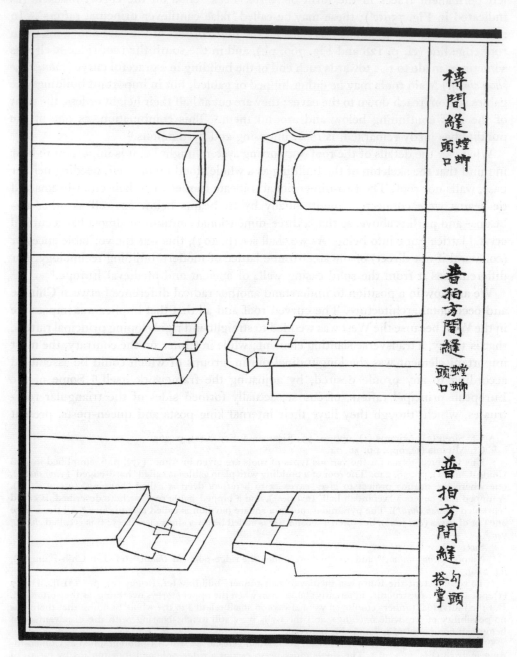

Fig. 763. Tenon and mortice work; forms of jointing in tie-beams or cross-beams (*Ying Tsao Fa Shih*, ch. 30, p. 17*b*). Allowance must be made for distortion of the isometrics in copying.

only a superficial similarity with the Chinese transverse frame (Fig. 760*f*, *g*).[a] The corbelled half-arches of English hammer-beam roofs (Fig. 760*h*) were to some extent analogous with the *tou-kung* construction of China, but in no way liberated the West from its dependence on the sloping principal rafter. So long as that dominated, the purlins could never be in any relation other than that of a straight line, and therefore the common rafters which they carried were also necessarily straight, and often single. In China, on the other hand, the common rafters, though individually straight, were multiple, ending at every third or every fourth purlin if not more often, and thus permitting the tiles to descend in a smooth curve.[b] Curiously, when the European roof did take on a curve, it was convex, not concave, as in the Mansard roof, where the queen-posts of the trusses are placed very much towards the sides of the building.[c] It will be seen that the Chinese roof never had any principal rafters at all. The most obvious analogues were the cantilever 'principals' already referred to (Fig. 760*e*), but these, though lying indeed under the purlins, served only as a support for the purlins of the eaves, which could take their due place in the curve of the cross-section. The cantilevers never dominated the cross-section and indeed died out after a time, leaving only the downward-pointing beaks of some of the corbel brackets as a kind of vestigial organ. In the Ming and Chhing periods even the brackets themselves tended to become more and more ornamental, their functions being less necessary because of a greater use of horizontal joists projecting from the pillars and a much increased size and strength of the main beams.[d] Originally only in clusters over the columns, they ended by forming a kind of decorative frieze all along the outer beams under the eaves.

What was the origin of the 'cantilever principal rafters' or 'lever arms' (*ang*), constituting as they did a seemingly intrusive triangular element amidst the stalwart rectitude (Fig. 760*j–n*) of the characteristic Chinese transverse frames[e]? To answer this one has to go back into history.[f] In anicent times Chinese building technique had made considerable use of double slanting joists meeting at a point like the European roof-truss (cf. Fig. 760*i*). But these inverted V-braces (*jen tzu kung*[1] or *chha shou*[2]) had at first little constructional importance; they were used mainly as an ornamental device between the longitudinal tie-beams, diversifying the appearance of the building as seen from the front. They must have been in use as early as the +1st century,

[a] See here the excellent monograph of Hodge (1) on Greek roofs, and also particular papers, e.g. (2). Before about −460 the Greeks sometimes dispensed with principal rafters over certain parts of their temples (as in the Megaron of Demeter at Gaggera), laying purlins directly from transverse wall to wall, but they never departed from the straight line laid down by the triangular roof-trusses over the cella. Sometimes also the tiles were laid directly on the purlins.

[b] There were various ratio methods of plotting this, e.g. the *chü chê*[3] system in the Sung and the *chü chia*[4] system in the Chhing.

[c] Convex roofs for a Yang culture and concave ones for a Yin one? But see also p. 249 below.

[d] Sirén (1), vol. 4, p. 72; (5); Sickman & Soper (1), p. 284; Chang Fo-Kuei (1).

[e] The frames are technically known as *liang chia*[5] (beam frame-works); this term is not to be confused with its inverse, used for the cross-beams (lit. frame-beams), cf. p. 93 above. An example of the interchangeability of adjectival nouns in Chinese technical terminology.

[f] There is argument about the first appearance of this member. The word certainly appears in the *Ching-Fu Tien Fu* (mid +3rd century), but some commentators gave it a different meaning. Li Chieh in the *YTFS* accepted it here in the normal sense, however (cf. Soper (2), pp. 99 ff.).

[1] 人字栱   [2] 叉手   [3] 舉折   [4] 舉架   [5] 梁架

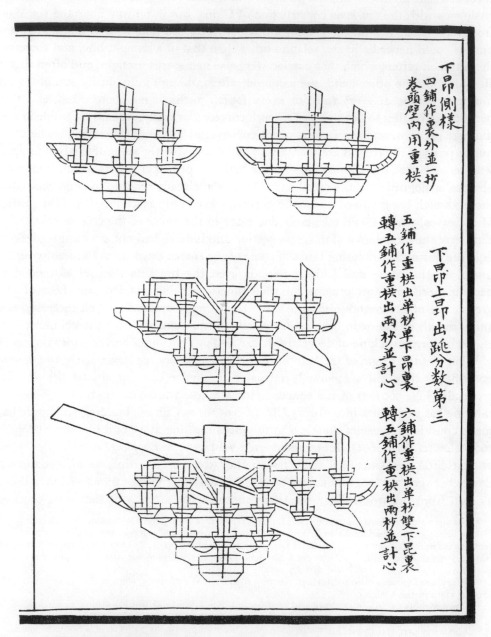

下昂側樣
四鋪作裏外並一抄
卷頭壁內用重栱

下昂上昂出跳分數第三

五鋪作重栱出單抄單下昂裏
轉五鋪作重栱出兩抄並計心

六鋪作重栱出單抄雙下昂裏
轉五鋪作重栱出兩抄並計心

Fig. 765. Bracket-arm clusters containing *ang* cantilevers; drawings from the *Ying Tsao Fa Shih* (ch. 30, p. 6*a*).

because the *Shih Ming* dictionary defines them under the name *hsieh chu*[1] (slanting struts).[a] They appear not infrequently in Northern Wei buildings as depicted in the Yün-kang cave-temples (dating between +460 and +535),[b] and in a famous design of about +652 (or +700) on a lintel of the Ta Yen Tha[2] (Great Wild Goose Pagoda) at Sian.[c] Other examples of similar dates occur in Korean tombs and cave-temples as well as in China.[d] There was one employment, however, in which these inverted V-braces made a significant contribution to structural stability, for sometimes they were used between the uppermost cross-beam and the ridge-pole instead of a king-post. One can see this still today in the +7th-century cloister galleries at the Hōryūji temple in Japan,[e] and (as part of a much more complex roof construction) in the +9th-century Fo-Kuang Ssu hall on Wu-thai Shan (Fig. 769, pl.).[f] The curve of the roof was of course quite unaffected by the subjacent triangle. This system is extremely rare, if not unique, in Chinese buildings still extant, for by the end of the Thang period the king-post had universally superseded it. Nevertheless this did not mean that the *chha shou* ('forked hand' struts) disappeared entirely; they continued to exist for some time as strengthening elements for the king-post.[g] Not only that, but they could be repeated at each end of the lower cross-beams, connecting sometimes beam with beam, sometimes beam with purlin,[h] and forming thus a true kind of trapezoidal truss (Fig. 760q);[i] but because of their multiple character never in any way dictating the profile of the roof curve, which continued to depend solely on the layout of beams and purlins.[j] One can

[a] Cf. Kelling (1), p. 125. The *Lu Ling-Kuang Tien Fu* calls them *chih chhêng*[3] (forked branch-like props).

[b] Grousset (1), vol. 1, p. 321; Bulling (2), fig. 73/75, 73/76; Sickman & Soper (1), pl. 160; Forman & Forman (1), pl. 33. So also in the Northern Chou (late +6th-century) frescoes at Mai-chi Shan, cave no. 4 (personal observations, 1958).

[c] Bulling (2), p. 24 and fig. 77/78; Sickman & Soper (1), p. 244, fig. 17; Liu Chih-Phing (*1*), fig. 315. Cf. p. 139 below.

[d] Cf. Bulling (2), pl. C, figs. 8, 9, 11; pl. E, figs. 1, 2; also fig. 134/156. For Korea see Liu Chih-Phing (*1*), fig. 315; Bulling (2), fig. 133/136/154.

[e] Sickman & Soper (1), p. 234, fig. 15.

[f] Liang Ssu-Chhêng (*4*, 2); Liu Chih-Phing (*1*), fig. 220; Sickman & Soper (1), p. 248, fig. 20.

[g] One can see this well in Liu Chih-Phing's elevations (*1*), figs. 197, 198, 199, 299, 502. In Japan they continued in their original function down to the end of the Tokugawa period, as at Nikko.

[h] Again, in Liu Chih-Phing's elevations (*1*), fig. 197 shows them all the way down, figs. 299 and 502 show them at three out of four possible levels, and figs. 198 and 199 show them at the uppermost level under the ridge-pole only. Here the illustrations in the 1925 edition of the *Ying Tsao Fa Shih* are instructive. *Chha shou* at each level are to be seen in ch. 30 (p. 180), ch. 30 app. (p. 216) and ch. 31 (pp. 5–15 incl.). Elevations with none at all occur in ch. 30 app. (p. 217) and ch. 31 app. (pp. 31–52 incl.). This clearly shows that they had disappeared by the Chhing period and probably long before the Ming. They were, in fact, an archaic feature which did not survive into the fully developed phase of Chinese building technology.

[i] The point is noteworthy since it has sometimes been said that no form of truss occurred in traditional Chinese technology (cf. pp. 141, 145 below). It was too simplified to say, as Choisy did (1), vol. 1, p. 185: 'Nous assurons la stabilité de nos charpentes par des triangles indéformables; les Chinois recourent à des rectangles rigides.' But on the whole precision joincry substituted for diagonal bracing.

[j] This remains strictly true even in certain very exceptional cases where the builders used what might be called a hyper-*ang*, running straight up under the purlins almost from eaves to ridge-pole. Such structures are found in the late +12th-century Great South Gate of the Tōdaiji[4] temple at Nara in Japan and in the Kuang-Shêng Ssu[5] temple built in +1319 at Chao-chhêng in Shansi (see Lin Hui-Yin & Liang Ssu-Chhêng (1), pp. 41 ff.). Cf. Sickman & Soper (1), pp. 267, 282; Soper (2), pp. 155 ff., 286.

[1] 斜柱 [2] 大雁塔 [3] 枝牚 [4] 東大寺 [5] 廣勝寺

now see how easily the idea of the *ang* or 'cantilever principal rafters' could have arisen; they were nothing but an extension of the principle of the *chha shou* or 'forked hand' struts to solve the problem of the widely overhanging eaves.[a]

It may well be that the curving roof is ultimately referable to the fact that from the most ancient times the 'front' of a Chinese building was the longer side, instead of (as in Egypt, Greece, and medieval Europe) the shorter one (cf. Fig. 760*o* and *p*). It was therefore natural enough to use complicated transverse framework partitions,[b] since they would not spoil the perspective. They would be seen, if at all, end on. On the other hand, if the perspective had to be through the longitudinal axis of a building, roofing was necessary, whether of beams or trusses, which would leave that axis free. The desire for a vista found its ultimate expression in the Gothic vault. The natural tendency was to have rafters springing straight from the top of the side-wall to the ridge-pole, following the geometrical pattern set by the end-wall or gable. The importance of end-walls in Western architecture is of course obvious; one has only to think of the pediment of the Greek temple and the west front of the Gothic cathedral. Even when European buildings were meant to be approached towards their longer sides, a pediment and columns were felt to be necessary as in the Palladian villa,[c] or dormer windows in Dutch and Danish town houses, recalling pediments.[d] On the other hand, the Chinese building, with no longitudinal vista, could fashion its transverse networks of wooden beams, and their upper contours, at will.[e]

Whatever we may now think about the 'tent-theory', it is clear that the upturned roof-edge in China had the practical effect of admitting the maximum amount of slanting winter sunlight and the minimum amount of downpouring summer sunlight. It also reduced the height of the roof while keeping a steep pitch for the upper part and a wide span at the eaves; and thus it reduced the lateral wind-pressure. This property must have been very important in reducing movements about the bases of the columns which simply rested on their stone plinths and were not generally taken into the ground. Another practical effect of the curving concave roof may have been the shooting of snow and rain-water well off the eaves into the courtyard away from the edge of the platform.[f] But of course it must always have given the greatest aesthetic satisfaction, and Lamprey was probably not far wrong in his suggestion that it was the

[a] The process can be traced in the sketches of Liu Chih-Phing (*1*), figs. 315, 316. The relation between the *chha shou* and the *ang* can be appreciated from diagrams such as those in *Ying Tsao Fa Shih*, ch. 30 app. (pp. 216, 218, 220, 221). Cf. Liang Ssu-Chhêng (*9*), pl. 4.

[b] Attention may be drawn here to a striking parallel between Chinese house construction and Chinese naval architecture. Just as the house or hall depends upon a series of transverse frameworks supporting the roof, so the junk or ship depends upon a series of transverse bulkheads, there being neither keel, stempost nor sternpost. This will be explained in detail below, pp. 390 ff.

[c] Cf. Rasmussen (*1*), p. 73.

[d] Cf. Rasmussen (*1*), p. 124.

[e] Cf. Sirén (*1*), vol. 4, p. 19: 'The marked development of the roof on Chinese halls is due to the fact that the entrance is not placed at one of the gable ends but in the middle of the south façade, the importance of which is brought out, not only by the gallery or open portico of free-standing pillars, but also by the wide sweep of the roof.'

[f] One of the problems of medieval European builders was how to prevent the drip of water from the roofs, which injured the foundations (Briggs (*2*), p. 210). Hence the proliferation of parapet-gutters, spouts, gargoyles, etc.

imprint of a particular taste and genius upon a structure intrinsically capable of responding.[a]

While examining the foregoing diagrams it has been repeatedly seen that no part of the weight of the roof or structural beams was taken in China by the walls. They are but membranes clothing a well-shaped structure of bone. Complete freedom was thus attained for the placing of doors and windows, and for their construction in delicate woodwork and lattice.[b] A building could be remodelled without any danger of collapse, and all openings in walls could be as large as desired; indeed in the hot climate of the south, one whole side of the hall could be, and often was, left open. Liang Ssu-Chhêng (3) and others have drawn attention to the fact that this style of building followed exactly the same principles as those of modern steel and concrete construction; the skeletal framework is the essential thing, and the walls with their openings simply clothe it. Medieval European buildings, with their buttresses and flying buttresses,[c] showed a tendency towards the same solution. It remains to be demonstrated, however, that Chinese building methods had any direct influence on the European escape from bondage to weight-bearing walls. This was perhaps more probably a slow intrinsic development, occurring first because of the resolve to build ever higher with ever greater daylight illumination of the interior (as in King's College Chapel and other buildings of the Perpendicular style); and then in a later phase because of the availability of metallic building materials with which fireproof construction could be attained.[d]

The use of iron columns and beams goes much further back historically than is generally thought; we shall see some unexpected examples in Chinese bridges (p. 151).[e] Later on (in Sect. 30d), we shall read of a hall built at Canton about +950 on twelve pillars of cast iron each 12 ft. long.[f] Iron tie-bars were used fairly frequently to secure late medieval and Renaissance vaulting.[g] Cast-iron beams were employed as furnace lintels in +17th-century England, and the strengthening of monumental buildings by inserted wrought-iron armature systems ('reinforced masonry') was carried out extensively in France from +1667 onwards.[g] Enormous wrought-iron roof-trusses were installed in certain French buildings just before the Revolution.[h] But the real break-

---

[a] How persistent the transverse frames and curving array of purlins are in modern Chinese building can be seen from Fig. 770 (pl.), a photograph taken at Lanchow in 1943. Although these workers' houses then under construction have a western look at first sight, the walls are not weight-bearing and the essential skeleton is in fully traditional style.

[b] *Ling chhuang*.[1] On the innumerable designs for lattice see Dye (1), and Vol. 3, pp. 95, 112. Its interstitial spaces were, and still are, covered with thin paper. Glass was never used until late times. In this, Europe led the way from the +1st century onwards. On the ornamentation of all Chinese woodwork components see Anon. (16) and Anon. (38).

[c] The cantilever principal rafter (*ang*) has been rightly compared with the flying buttress. The dramatic, upthrusting lines secured by the full exposure of the *ang* preferred by the architects of the Sung evoke 'something like the visual excitement given by Middle Gothic flying buttresses, and are, in the same way, as structurally alive as they seem' (Sickman & Soper (1), p. 262).

[d] On the general story see Gloag & Bridgwater (1); Watterson (1); Giedion (2), pp. 101 ff.

[e] Cf. Hamilton (4).

[f] Cf. *Yang-chhêng Ku Chhao*, ch. 8, p. 41a. On pagodas of cast iron see p. 141 below.

[g] Skempton (6).   [h] See Bannister (1); cf. Skempton (6).

[1] 櫺窗

through was made by Charles Bage (+1752 to 1822) with a five-storey flax mill completed at Shrewsbury in +1797 and still standing in good repair today.[a] Cast-iron beams supported by cast-iron columns were joined by segmental arches of brick, thus forming the first multi-storey iron-frame building. Here lateral stability was still provided by massive external walls, and it took forty years before portal bracing could be achieved so that a three-dimensional iron lattice would stand up of itself—as the comparable wooden lattices of China had done for so long. This was the period of the Crystal Palace and the Sheerness Boat Store.[b] Cast iron, wrought iron and steel were all used in the first buildings of 'skyscraper' type, such as the ten-storey Home Insurance Building put up in Chicago in 1884, but the complete steel skeleton followed very soon afterwards.[c] One effect of these developments was to make it possible to replace the walls of buildings almost wholly by transparent sheets of glass. The rebuilt cathedral at Coventry shows today how dramatic the result can be. But probably very few of the architects and engineers who participated in this great movement realised that a definitive escape from weight-bearing walls had already been accomplished by their Chinese predecessors during the previous two or three thousand years.

(i) *Drawings, models and calculations*

Any history of Chinese building technology, however brief, should devote some attention to the records which exist of the preparatory work carried out by the builders and architects of old. Here we can give only a few examples from the surviving literary and graphic material.

The +5th-century *Shih Shuo Hsin Yü*,[1] speaking of a complex of buildings erected by the emperor Wên of the Wei State (+220 to +226), says:[d]

The Ling Yün Thai[2] had towers and temples most elaborately and ingeniously built. All the pieces of timber were first weighed, so that there was a perfect balance (between the sides of the buildings). This was why the high buildings showed no sign of leaning over or collapsing, though some of the storeys were quite lofty, and often shook and vibrated in strong winds. The emperor Ming (+227 to +239), on mounting some of the towers, was alarmed at (what he thought was) a dangerous situation, so he caused one of them to be supported by an additional large column. (Some time after this was done) the tower collapsed and was destroyed. People talking about this result said that it was due to getting the weight unbalanced (*chhing chung li phien ku yeh*[3]).

---

[a] For descriptions of this and its historical context, see Johnson & Skempton (1); Skempton & Johnson (1); Skempton (2); Hamilton (3). During the years just preceding Bage's Shrewsbury mill, several similar buildings had been erected by William Strutt (+1756 to 1830) but their beams were still of timber.

[b] See Skempton (1).

[c] It was built by W. le B. Jenney; see Hamilton (5); Skempton (3). A number of ancillary inventions may well have been limiting factors in the development of steel and concrete buildings: e.g. lifts from 1820 onwards, but their safety devices only from 1852 (Otis); electric light, central generators and underground cables (Edison, 1879); steam heating through pipes (Tredgold, 1836), and so on.

[d] Ch. 3A, p. 32b, tr. auct.

[1] 世說新語　　　[2] 陵雲臺　　　[3] 輕重力偏故也

The commentary of Liang date adds further details about these buildings from the *Lo-yang Kung Tien Pu*[1] (Notebook on the Halls and Palaces of Loyang), since lost. The story gives a glimpse of the preparatory experiments of the early builders, spoilt, not for the first or last time in history, by the interference of highly placed laymen.[a]

Pilot projects were also used to evaluate costs. About +1197 the Neo-Confucian philosopher, Huang Kan,[2] found himself charged with rebuilding the walls of An-chhing, a city of which he was governor; so he began by constructing a trial length which enabled a fair estimate to be made of the expense in man-power and materials.[b] After that he pushed on the work with all speed, and it was successfully completed before the Jurchen Chin armies could attack the city. Similar methods must have been used in hydraulic engineering works (cf. p. 333 below).

In the Liu Chhao period drawings and models also appear. About +491 (according to the *Nan Shih*):[c]

Tshui Yuan-Tsu[3] said to the emperor that his nephew Tshui Shao-Yu[4] was coming to the capital, and that he was exceptionally skilled in coloured architectural drawings (*pan chhui*[5]). Tshui suggested that the emperor should order him to make a model (*mu*[6]) of (new?) palace buildings, and retain him. But the emperor did not feel able to follow this suggestion, and so, after making painted diagrams of the palaces, Tshui Shao-Yü went home.

One would give a good deal to have these late +5th-century drawings and models now, so that we could see how much progress had been made beyond the simple designs seen in the tomb-models of the Warring States and Han periods (cf. Figs. 783, 784, 786, pls.). Miniaturisation was evidently coming into use. Models are often heard of in the Sung. For example:[d]

Certain workers made for Sun Chhêng-Yu[7] (a +10th-century general) a small model of Li Shan[8] (mountain), complete with streams, bridges, houses, pavilions and paths, made of a kind of cake mixed with camphor. Subsequently a model in wax was made.

This would have been one of those landscaping designs, such as must have been prepared for the summer palaces which we can visit today. The same source[e] tells also of a maker of architectural models (*yang*[9]) who prepared one of some kind of grotto-pavilions for Kuo Tshung-I[10] about the same time.

It was just at this period that miniature buildings were greatly employed as a sort of interior decoration.[f] When the Japanese monk Jōjin[11] visited the capital Khaifêng in +1072 he noted that the great hall of the chief abbey of the Chhan school there had a

---

[a] Many stories are found in Chinese literature about alarm caused by towers and pagodas swaying in the wind. Presently we shall see what advice the master-builder Yü Hao gave in a case of this kind (p. 141 below). As for the difficulties encountered by technical experts, the reader will remember the amusing passage about the house built for Kaoyang Ying (Vol. 2, p. 72 above).

[b] *Sung Shih*, ch. 430, p. 3a; cf. Yang Lien-Shêng (11), p. 81.

[c] Ch. 47, p. 6a, tr. auct.

[d] *Chhing I Lu*,[12] ch. 2, p. 23b; tr. auct.   [e] Ch. 2, p. 4a.

[f] Cf. Sickman & Soper (1), pp. 257, 278 ff.

[1] 洛陽宮殿簿   [2] 黃幹   [3] 崔元祖   [4] 崔少游   [5] 班倕
[6] 模   [7] 孫承祐   [8] 驪山   [9] 樣   [10] 郭從義
[11] 成尋   [12] 清異錄

ceiling that was 'all set out with (model) treasure-halls'. Libraries especially were treated in this way, the bookcases being crowned with whole temples *in piccolo*. We can gain an excellent idea of what Jōjin saw because there still exists at Ta-thung in the north a splendid library at the Lower Hua-Yen Ssu[1] temple built under the Liao in +1038. Round the ends and the back wall of a building 85 ft. long run two double-storey roofed wall-cupboards for the sūtras, diversified with higher pavilions and connected together over a central doorway with a magnificent model pavilion on a flying bridge of cantilever type (Fig. 771, pl.).[a] Beautifully modelled buildings have also often been used as reliquaries.

In the +17th century Chiang Shao-Shu[2] wrote an interesting note on men who had specialised in drawings and paintings of architecture.[b] Though these were not plans, they doubtless helped clients and builders alike. The style was called 'sharp-edge painting' (*chieh hua*[3]), distinguishing it from that of the vaguer forms of misty landscapes.[c]

The painting of palaces and houses is a very difficult thing. It requires great precision to be satisfactory. General Li[d] was widely acknowledged as a great expert in this, but people do not know that there have been many others. In the Thang there was Yin Chi-Chao;[4] Hu I[5] and Wei Hsien[6] worked in the Wu Tai period. There was no one to equal them. Later on, Kuo Shu-Hsien[7] had an outstanding personality; though fully acquainted with the use of the compass, carpenter's square, water-level and plumb-line, he was in no wise embarrassed by these instruments. I once saw his *Pi Shu Kung Thu*[8] (Drawing of the Summer Palace); there were thousands of *tshui*[9] (eave-rafters) and *chüeh*[10] (rafters), yet none was missing from the drawing. Another expert, Wang Ku-Yün,[11] flourished in the late defeated (Ming) dynasty; he also was successful, but not as good as Kuo. The masterpieces of the latter were the *Hsien Shan Lou Ko Thu*[12] (Towers and Halls of the Mountain of the Immortals), and *Tuan Yang Ching Tu Thu*[13] (Sailing Races at the Fifth Month Festival). The compositions of these paintings are very elegant, and every stroke is refined, fitting its place exactly....

The commentary adds that artists who wished to succeed in this speciality usually became skilled also in building calculations (*mu ching suan fa*[14]).[e] The paintings of Kuo and Wang were apparently devoted entirely to architectural constructions and details, without any extraneous figures or landscape. The importance attributed in China to drawings of buildings may be seen by the fact that this class of picture was one of the ten divisions of the catalogue of all the drawings and paintings in the collection of the

[a] See Sickman & Soper (1), p. 279, pls. 180, 181 *a*; Forman & Forman (1), pl. 39. I had the great pleasure of visiting this temple in 1964. Cf. Vol. 4, pt. 2, pp. 547 ff., on the rotating sūtra-repositories.
[b] *Yün Shih Chai Pi Than*[15] (Jottings from the Sounding-Stone Studio), ch. 2, p. 11 *a*, tr. auct.
[c] See Waley (19), pp. 184 ff.
[d] This must refer either to Li Ssu-Hsün[16] (+651 to +720) or to his son Li Chao-Tao,[17] both generals and both famous painters.
[e] Examples of such calculations, dealing with labour, materials, and time, accompanied by diagrams, will be found in the *Shu Shu Chiu Chang* (Mathematical Treatise in Nine Sections), by Chhin Chiu-Shao (+1247), esp. chs. 13 and 14.

[1] 華嚴寺 [2] 姜紹書 [3] 界畫 [4] 尹繼昭 [5] 胡翼
[6] 衛賢 [7] 郭恕先 [8] 避暑宮圖 [9] 檼 [10] 桷
[11] 王孤雲 [12] 仙山樓閣圖 [13] 端陽競渡圖 [14] 木經算法
[15] 韻石齋筆談 [16] 李思訓 [17] 李昭道

emperor Hui Tsung, *c.* +1120 (*Hsüan-Ho Hua Phu*[1]).[a] What kind of perspective was used in Chinese architectural paintings will engage us presently.

Of Chinese city-maps and plans we have already spoken in various connections,[b] but this is perhaps the place to reproduce the most famous, that of Suchow (Fig. 772, pl.), carved on stone in +1229 and still preserved there.[c] Also of the Sung, but rather earlier (+10th or +11th century), are the plans of the Thang palaces still to be seen engraved on stone in the Pei Lin Museum at Sian. When the sites of the Hsing-Chhing Kung[2] and Ta-Ming Kung[3] palaces in that city were being excavated in 1934, stone steles with inscribed plans (*thu shih*[4]) were found, with the positions and elevations of buildings on them carefully drawn to scale with ruled lines.[d] As these certainly derive from Thang originals they are quite comparable with the oldest European architectural plan, that of the Abbey of St Gall, made with red ink on parchment about +820 and still treasured in the Stiftsbibliothek at St Gallen in Switzerland.[e] Still more venerable is the plan of the Tōdaiji temple at Nara in Japan, drawn on hempen cloth and dated +756.[f] The dating of architectural drawings and models has much comparative interest, because Salzman (1) and other historians have drawn attention to their singular absence in early times in Europe. Few if any architectural drawings have reached us from before the time of Villard de Honnecourt (*c.* +1240),[g] and Salzman finds no model before +1390.

This is why the *Ying Tsao Fa Shih* of +1103 is such a landmark. The excellence of its constructional drawings raises an issue of some importance. Nothing so far mentioned in this discussion constitutes (or could have constituted) what we would now call 'working drawings'. But the shapes of the component parts of the frameworks are so clearly delineated by Li Chieh's drawing-office clerks (cf. Figs. 773, 774) that we can at last almost speak of working drawings in a modern sense—perhaps for the first time in any civilisation.[h] Engineers of our own time are often inclined to wonder why

[a] This was the period of activity of another celebrated architectural painter, Chao Po-Chü,[5] on whom Tronsdale (1) has written. He translates an amusing passage from the *Thu Hua Chien Wên Chih*[6] (Observations on Drawing and Painting) written by Kuo Jo-Hsü[7] soon after +1074, which describes the daunting technological expertise necessary for success in this field. From +1180 we have the Japanese scroll-painting Shigisan Engi Emaki,[8] the architectural drawing and perspective in which has been discussed at length by Armbruster (1).

[b] Pp. 63, 80, 87, 88 above, and Vol. 3, p. 547.

[c] See Chavannes (8); Moule (15), pp. 51 ff.; Chhien Yung (1); Liu Tun-Chên (5), and Vol. 3, pp. 278, 551. It was made by Lü Yen,[9] Chang Yün-Chhêng[10] and Chang Yün-Ti.[11] Moule also demonstrates the extraordinary accuracy of the city-maps of Hangchow, which date from at least +1274 (p. 12). I had the pleasure of studying the original stele of the Suchow map at the old Confucian College there (now a Middle School), with Dr Dorothy Needham and Dr Lu Gwei-Djen in 1964.

[d] Viewed and photographed with Dr Lu Gwei-Djen in 1958. Cf. Wei Chü-Hsien (1), p. 211. On Thang Chhang-an Schafer (14) is to be read.

[e] Viewed with Dr Lu Gwei-Djen in 1956. Cf. Reinhardt (1). I except the Madeba mosaic of Hadrianic Jerusalem (cf. Perowne, 1), which is a perspective view rather than a plan.

[f] Ishida & Wada (1), pl. 161. [g] Cf. elsewhere, p. 296 and Vol. 4, pt. 2, pp. 229, 404.

[h] The illustrations were probably much better originally than they are now, for some of the details cannot be correct, and it is reasonable to attribute this to the work of the Chhing copyist, who was not 'in the trade'. The awaited studies of Dr Else Glahn will, we are sure, throw more light on this.

[1] 宣和畫譜　[2] 興慶宮　[3] 大明宮　[4] 圖石　[5] 趙伯駒
[6] 圖畫見聞誌　[7] 郭若虛　[8] 信貴山緣起繪卷　[9] 呂挺
[10] 張允成　[11] 張允廸

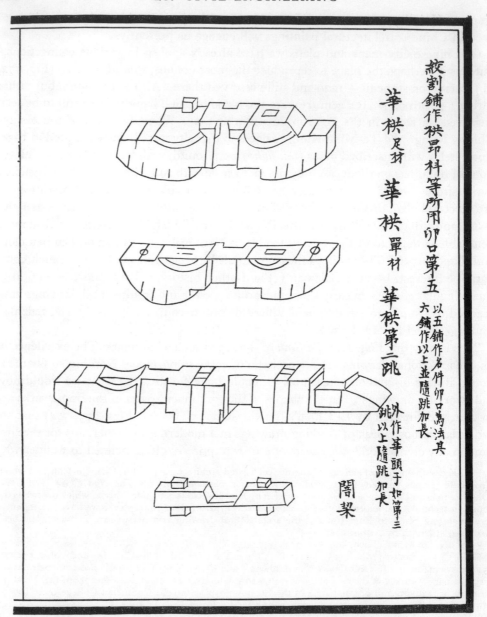

Fig. 773. Working drawings in the *Ying Tsao Fa Shih*; three transverse corbel brackets (*hua kung*).
Ch. 30, p. 11*a*.

the technical drawings of ancient and medieval times were so extremely bad.[a] What remains from the Hellenistic world is so distorted as to need much interpretation,[b] and the machine drawing of the Arabs is notoriously obscure. The builders of the medieval

---

[a] Sociological reasons probably had an important part to play in this. We have touched already (p. 82 above, and Vol. 4, pt. 2, pp. 1, 2, 37) upon certain relevant factors in the Chinese situation.

[b] Consider, for example, Drachmann (7, 9).

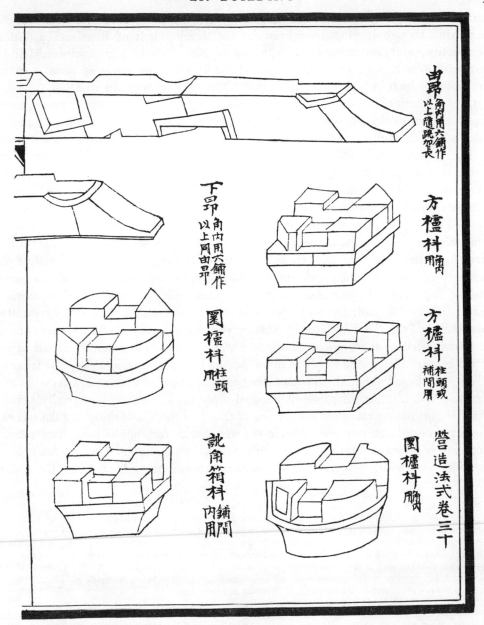

Fig. 774. Working drawings in the *Ying Tsao Fa Shih*; five examples of bracket-arm bases (*tou*) and two cantilever arms (*ang*). Ch. 30, p. 16*a*.

European cathedrals were no better draughtsmen, and the +15th-century Germans, even Leonardo himself, produced little that was clearer than sketches, brilliant though these sometimes were.[a] If then there was no Western parallel to the *Ying Tsao Fa Shih*

[a] There are two standard histories of technical drawing for the West, that of Feldhaus (24) and that of Neduloha (1); the former concentrating on instruments and materials more than the latter, but both mainly occupied with post-Renaissance developments. See further Heymann (1); Booker (1).

we must face the fact that Euclidean geometry (which Europe had and the Chinese did not), vital though it must have been for the development of Renaissance optical perspective,[a] and fundamental too for the rise of modern experimental science,[b] had no power to give Europe precedence over China in the appearance of good working drawings, at least in building construction.[c] The very reverse was true.[d]

Of the computational work necessary before any building was even started many traces remain in the work of Li Chieh. But before leaving the question of building calculations, attention must be drawn to a point which was already apparent at an earlier stage,[e] namely the derivation of perhaps the majority of Chinese technical terms in solid geometry from the preoccupations of builders. Examples of this are numerous and interesting.[f] The parallelepiped with two square surfaces was called *fang pao thao*,[1] *pao* meaning (as still today) a fortified village, and *thao* a rampart of tamped earth. This figure is of course just that of each block of *terre pisé* within its shuttering. By extension, the word for cylinder was derived (*yuan pao thao*[2]). The term for a parallelepiped with no square surface (*tshang*[3]) was taken directly from 'granary'. Sometimes the appellations seem to have originated from tools, as in the case of the pyramid (*fang chui*[4]) and cone (*yuan chui*[5]), *chui* by itself meaning a carpenter's boring tool, awl or gimlet. The lower corners of a pyramid[g] were called *yang ma*,[6] and this the *Ying Tsao Fa Shih* uses[h] as a technical term for the eave corners or horns; perhaps the original significance was phallic.[i] It is certainly as old as the +3rd century.[j] Then the word 'pavilion' occurs in the terms *fang thing*,[7] the frustum of a square-based pyramid, and *yuan thing*,[8] the frustum of a cone. The frustum of a pyramid with a rectangular base of unequal sides was known as *chhu*[9] (really *chhu*[10]) *thung*,[11] *chhu* having reference to the straw of thatched roofs, and *thung* to a flat top like the shaven head of a boy. Similarly, in the wedge with rectangular base, *chhu mêng*,[12] both words refer to roofing, and the latter means the ridge-pole. The curious expression for a wedge with a trapezoid base, *yen chhu*,[13] comes undoubtedly from the under-

---

[a] Cf. pp. 113 ff. immediately below.

[b] Cf. Needham (45) and the accompanying contributions.

[c] Drawings of machines and their parts in China did not match up to the standard set by Li Chieh, but even Europeans could not produce good drawings of this kind until the +16th and +17th centuries.

[d] We shall find a striking parallel to this in Sects. 38 and 45. Accurate illustration of plant forms in China preceded the work of the +16th-century European pharmaceutical botanists by some four hundred years. But that is more understandable since geometry was less necessary for it. The content of this paragraph emerged in conversations with my colleagues of the Newcomen Society, Mr Hugh Clausen and Mr Rex Wailes, with a background of Brixham shipwrights.

[e] Cf. the mathematical Section, Vol. 3, p. 97 above.

[f] Some of these interpretations were recognised by the Sung mathematical commentator, Li Chi,[14] in his *Chiu Chang Suan Shu Yin I*[15] (Explanations of Meanings and Sounds of Words occurring in the Nine Chapters of Mathematical Art).

[g] And hence sometimes, by extension, the pyramid as a whole.

[h] E.g. ch. 5, p. 6a.

[i] Cf. 'male' and 'female' in modern electrical technology.

[j] One finds it in the *Ching-Fu Tien Fu*[16] of Ho Yen[17] (*CHSK*, San Kuo sect., ch. 39, p. 5b).

---

[1] 方堡壔	[2] 圓堡壔	[3] 倉	[4] 方錐	[5] 圓錐	[6] 陽馬
[7] 方亭	[8] 圓亭	[9] 芻	[10] 蒭	[11] 童	[12] 芻甍
[13] 羨除	[14] 李藉	[15] 九章算術音義		[16] 景福殿賦	
[17] 何晏					

ground passages made to give access to tombs, or in the recesses of imperial palaces.[a] *Chhien tu*,[1] a prism, combines words for moat and a low wall. The wall motif is also obvious in the terms for the frustum of a wedge, *chhêng*,[2] *yuan*,[3] and again *chhien*.[1]

This derivation of geometrical terms from building operations illuminates once again the practical and empirical genius of the Chinese people.

(ii) *Perspective*

Architectural draughtsmanship raises the question of the Chinese attitude towards perspective.[b] Post-Renaissance European drawing has studiously followed the rules of convergent perspective, based upon the science of optics, according to which lines and planes on each side of the observer's position, though in fact parallel, appear to meet at a vanishing-point upon the horizon. It has commonly been held that this is the only possible kind of perspective, and that it was not known to the Chinese nor used by them until its introduction by the scientifically minded Jesuits early in the +17th century. The second of these statements is undoubtedly true, but the first, if we take the word perspective in its broad sense, is certainly at fault; for the Chinese were necessitated to introduce a sense of distance into their pictures, and did so successfully by a number of conventions which were not those of Europe.

The introduction of convergent or optical perspective into China has been considered in detail by Pelliot (27, 28), Hirth (9), Laufer (28) and others. There is no question that Louis Buglio, S.J. (Li Lei-Ssu;[4] +1606 to +1682), made the Western methods known by giving to the emperor three pictures in which the rules were perfectly followed. Gradually there appeared a mixture of styles: European missionary painters[c] began to paint Chinese subjects in semi-European manner or themselves learnt the Chinese ways; Chinese artists began to paint in European style. The efforts of the Westerners were at first modified by the fact that to the Chinese the Western perspective seemed wrong;[d] Fig. 775 (pl.) shows a painting which suggests that the drawing was modified to suit the Chinese taste in this matter.[e] Among the earliest Chinese to draw in the Western manner were Tung Chhi-Chhang[5] (+1555 to +1636) and Wu Li[6] (+1632 to +1718), but convergent perspective was not fully incorporated in the traditional Chinese style until the famous set of drawings for the *Kêng Chih Thu*[f] by Chiao Ping-Chên[7] in +1696. In +1629 already Francisco Sambiasi (Pi Fang-Chi[8])

---

[a] As in the A Fang Kung, the palace of Chhin Shih Huang Ti (*Shih Chi*, ch. 6, p. 25 a).

[b] It would have been possible to treat of this subject in the Section on physics, but its relations with practical draughtsmanship are so close that this seemed a better place.

[c] The most famous are Jean-Denis Attiret (Wang Chih-Chhêng;[9] +1702 to +1768) and Joseph Castiglione (Lang Shih-Ning;[10] +1688 to +1766). For a new biographical study of the latter see Ichida Mikinosuke (2).

[d] Cf. Gombrich (1), p. 227.　　　　　　　　　　　[e] Laufer (28).

[f] See the full discussion, in Vol. 4, pt. 2, pp. 166 ff., and here especially Hirth (9).

| 1 塾堵 | 2 城 | 3 垣 | 4 利類思 | 5 董其昌 |
| 6 吳歷 | 7 焦秉貞 | 8 畢方濟 | 9 王致誠 | 10 耶世寧 |

had issued a small tractate on the laws of perspective.[a] Later on (+1729) came the *Shih Hsüeh*[1] (Science of Seeing), written by Nien Hsi-Yao[2].[b]

Yet it is clear that from the Han period onward, Chinese draughtsmen had had a great sense of distance in pictures. They had been conscious of the problem of projection, how to represent three-dimensional space upon a plane surface. The terms *kao yuan*[3] (high distance), *shen yuan*[4] (deep distance), and *phing yuan*[5] (level distance), not lightly to be equated with background, middle distance, and foreground, had been in use among Chinese painters for centuries (March, 3). One of the canons of Hsieh Ho,[6] the great theoretician of painting in the +5th century,[c] had been 'the right distribution of space' (*ching ying wei chih*[7]), which must have meant perspective of some kind (Hirth, 12). Certain Sung painters such as Li Chhêng[8] have remained notable for their handling of distances, and in the Yuan, one of the great mistakes of beginners was said to be 'not distinguishing between the near and the far' (*yuan chin pu fên*[9]).

In general it may be said that in Chinese drawing distance had always been represented by height, so that one object standing behind another had been drawn above it, and not necessarily smaller.[d] This has the result of giving to many Chinese pictures the character of bird's-eye views. Everything is seen as if from a height.[e] The style is already present in the oldest Chinese landscape pictures still existing (−1st century).[f] Might this not be one reason why the word *thu*[10] has always retained an ambiguity,[g] being applied equally to maps and charts and to drawings and paintings? A curious consequence of this Chinese style has been pointed out by Wells.[h] In the European scene, the spectator feels that he has the scene thoroughly under control. 'He looks into it, it is all *before* him, and even if it is seen from a great height, it is seen, so to speak, from the top of a solid cliff. With the Chinese style the ground surface starts from the

---

[a] Curiously entitled *Shui Hua Erh Ta*[11] (Replies to Questions on Sleep and on Painting).

[b] A scholarly painter, the pupil of the Jesuit artist Joseph Castiglione (Lang Shih-Ning); but better known as one of the Directors of Potteries at Ching-tê-chen, hence the term Nien Yao[12] for certain excellent wares produced under his authority.

[c] Cf. Waley (18). His book, the *Ku Hua Phin Lu*, was continued c. +550 by the *Hsü Hua Phin Lu*. See Wang Po-Min (1).

[d] Rasmussen (1), p. 30, compared this system with the top-to-bottom arrangement of Chinese written or printed characters. He clearly appreciated the bottom-to-top succession of objects receding in space in Chinese pictures, but it would be impossible to sustain his statement that they give no impression of a third dimension. He is also wrong in supposing that the Chinese did not make the details of the far distant parts of their landscapes smaller and the colour lighter. Wang Wei[13] in the Thang distinctly stated these principles (Elisséev, 1).

[e] The so-called 'perspective cavalière', sometimes spoken of as 'zenith perspective'.

[f] Bulling (11) discusses the two hollow-tile door-panels of a tomb excavated near Chêngchow dating c. −60 (Fig. 777, pl.), and the painted ceiling vault of a Shansi tomb of c. +10. In the former, continuous depth recession is achieved by the zigzag lines of road and garden wall, the picture being built up by the repeated use of standard stamps on the wet clay. In the painting, clouds lift to disclose an L-shaped farmhouse below and mountains far away. For her sources see Wang Yü-Kang (1) and Yang Mo-Kung & Hsieh Hsi-Kung (1). For the oldest Chinese wall-paintings so far discovered, c. −50, see Li Ching-Hua *et al.* (1) and Kuo Mo-Jo (7).

[g] For an example, see p. 536 in the geography Section (Vol. 3).

[h] (1), p. 35.

---

[1] 視學	[2] 年希堯	[3] 高遠	[4] 深遠	[5] 平遠
[6] 謝赫	[7] 經營位置	[8] 李成	[9] 遠近不分	[10] 圖
[11] 睡畫二答	[12] 年窰	[13] 王維		

distance and slips past under the spectator's feet to a goal infinitely beyond, i.e. below and perhaps behind him. In some cases this produces a feeling of uncertainty, of falling into the scene.' Sometimes, too, as Bachhofer (1) has pointed out in his elaborate studies of perspective in Chinese art, there seem to be a series of plane ground-surfaces, each with its own vanishing-point; for example, in the terraces which form the setting for the assemblies of Buddhas and Boddhisattvas in the Tunhuang frescoes.[a] But this is rather unusual.

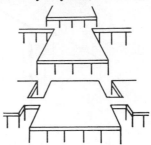

Fig. 776. A form of perspective seen in the Chhien-fo-tung frescoes—a series of superimposed vanishing-points. Cf. Fig. 758 (pl.).

Let it be accepted, then, that on the whole there is no true vanishing-point in Chinese drawing, and no exact rules of foreshortening. The horizon boundary was not felt to be important; the spectator was not compelled to participate in the drawing by his very physical position.[b] How then was it possible for the Chinese to delineate, as they did, the 'sharp-edge' quality of buildings? What they employed was 'parallel perspective',[c] i.e. a system in which lines which were parallel in fact remained so in the drawing. Many of our illustrations show how this works out;[d] many more could easily be added.[e] The convention, reduced to its simplest elements, can be appreciated by comparing it with the same drawing in convergent or optical perspective: Fig. 778. March (1) has noted the paradox that while the European system depended tacitly on the non-Euclidean postulate that parallel lines meet at infinity, the Chinese, who had little or no Euclidean geometry, remained faithful to the postulate that parallel lines never meet at all, even in pictures. It is clear, too, that the Chinese convention could never show more than three surfaces of the interior of a parallelepiped, while the post-Renaissance convention could show five. The three surfaces are often referred to in Chinese writings on painting, for example, by Ta Chung-Kuang[1] in his *Hua Chhüan*[2] (The Painting Basket), a Chhing book, where he speaks of *shih khan san mien*[3] (seeing three sides of rocks).[f] That the Chinese never attempted to solve the problem of showing five surfaces is simply another result of the fact that they lacked geometrical optics.[g] There remains a further paradox, namely that the projection which they did adopt is closely similar to some of those which architects and engineers use today for

[a] Cf. p. 89 above and Fig. 758 (pl.), and Vol. 1, Fig. 23.
[b] There is a good summary in Jenyns (1), p. 130.
[c] Sometimes called, not very happily, 'linear perspective' (March, 2).
[d] E.g. Figs. 724, 728 (pl.), 860.
[e] See, for instance, *Chieh Tzu Yuan Hua Chuan*, pt. 1 (p. 286), rep. Combaz (6), p. 113.
[f] P. 5a. Elsewhere (p. 7a) he refers to the representation of distance, saying that 'if in the painting of the trees there is no distinction between "outer" and "inner" (*shu wu piao li*[4]), the artist has not understood the art of "hiding and exposing" (*yin chien chih fang*[5])'.
[g] Cf. Vol. 3, pp. 91 ff., Vol. 4, pt. 1, pp. 78 ff., 86, 98, etc. When a writer such as Jacot (1) says that 'the superiority of the art of Graeco-Roman culture in contrast to the Oriental is due to long years of scientific study', he is not, as he supposes, condemning the Chinese convention, but simply stating that the European convention, and more particularly the post-Renaissance European convention, was profoundly influenced by deductive geometry while the Chinese was not.

¹ 笪重光　　² 畫筌　　³ 石看三面　　⁴ 樹無表裏　　⁵ 隱見之方

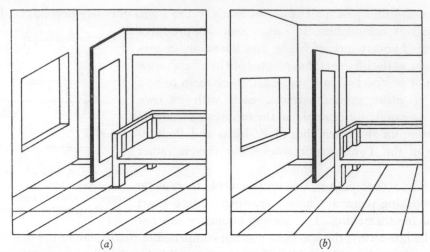

(a)   (b)

Fig. 778. Diagrams to show the contrast between (a) Chinese 'parallel perspective' or axonometric drawing, and (b) convergent or optical perspective. Redrawn from March (1).

mechanical or structural 'working drawings'.[a] It is therefore unlikely that the absence of convergent perspective was a limiting factor at any time in China during the eotechnic phases of mechanical invention.

Parallel perspective can be found already in the drawing of the scenes carved in relief in the stone tomb-shrines of the Han period (Chu Wei, Wu Liang, etc.).[b] Diagonal lines strike off from the front line of the picture, with figures or buildings along them. This is also to be seen in many of the Han reliefs from Szechuan published by Rudolph & Wên. The convention continued in the works of the famous +4th-century painter,[c] Ku Khai-Chih,[1] but with subtle modifications. Thereafter it was never relinquished. It has been justly pointed out[d] that in early Chinese drawing there is no single point of view at which the spectator is supposed to stand, and from which his glance radiates from right to left. In all cases he has to be imagined as standing point blank in front of that part of the surface on which the object is presented. One comes back again, therefore, to the impression that Chinese conventions of draughtsmanship involved a 'minimum of subjectivity'.[e] There was a 'multiple station-point',

[a] Representations in parallel or non-diminishing perspective have been found extremely useful in modern technology, because, though planes therein do not diminish where they are supposedly receding, the drawings provide an adequate pictorial view of an object, and the distances in three principal directions can be scaled. Such drawings are called in general 'axonometric' projections, and they are either 'isometric', 'dimetric' or 'trimetric' depending on whether one, two or three different scales are used for height, width and depth. As long as two setting-out lines for principal directions are 90° apart (the others can be any division of 270°) one side of the rectilinear object depicted can be projected directly in plan or elevation, which makes the representation advantageous, and quick to draw. Also not only distances but angles and curves are 'true' on that plane. Such axonometric drawings, sometimes also called oblique parallel projections, were the type used in the traditional Chinese system, the front elevation being always chosen as the 'true' plane.

[b] Cf. Vol. 3, Fig. 125, Vol. 4, pt. 1, Fig. 301, etc. See further Fischer (2); Fairbank (1); Wells (1, 2); March (1, 2) and others.

[c] His biography has been translated by Chhen Shih-Hsiang (2).

[d] Wells (1), p. 18.   [e] Wells (1), p. 21.

[1] 顧愷之

or rather a 'hovering or dynamic view-region', not a viewpoint at all.[a] One cannot help correlating, therefore, the domination of the individual eye in convergent optical perspective, with the emphasis on the individual so characteristic of the European Renaissance. Although it is unsatisfactory to have to treat of the Chinese draughtsmanship of all ages within the limits of the very short exposition possible here, it might perhaps be said that the hovering view-region, the parallel perspective, and the representation of scene depth by height, were all indications of an attitude towards Nature at once humbler and more social than that of Western man.[b] Chinese parallel perspective represents distance, yes, but it is not founded upon the idea that *I* personally, the Spectator, am more distant from one part than from another.

It is fortunate that we have a striking defence of the diffuse view-region principle written in the Sung period, and paradoxically by the statesman-scientist Shen Kua, whom we so often quote. As already mentioned, Li Chhêng (*d. c.* +985) made some experiments towards a kind of optical perspective, as did Chang Tsê-Tuan later, which is one of the reasons why his Chhing-Ming Shang Ho Thu (painted *c.* +1120, cf. pp. 165, 359, 463, 648) makes such an immediate appeal to the modern eye. It was about +1080 that Shen Kua wrote:[c]

> Then there was Li Chhêng, who when he depicted pavilions and lodges amidst mountains, storeyed buildings, pagodas and the like, always used to paint the eaves as seen from below. His idea was that 'one should look upwards from underneath, just as a man standing on level ground and looking up at the eaves of a pagoda can see its rafters and its cantilever eave rafters'. This is all wrong. In general the proper way of painting a landscape is to see the small from the viewpoint of the large (*i ta kuan hsiao*[1]), just as one looks at artificial mountains in gardens (as one walks about). If one applies (Li's method) to the painting of real mountains, looking up at them from below, one can only see one profile at a time, and not the wealth of their multitudinous slopes and profiles, to say nothing of all that is going on in the valleys and gorges, and in the lanes and courtyards with their dwellings and houses. If we stand to the east of a mountain its western parts would be on the vanishing boundary of far-off distance, and vice versa. Surely this could not be called a successful painting? Mr Li did not understand the principle of 'seeing the small from the viewpoint of the large'. He was certainly marvellous at diminishing accurately heights and distances, but should one attach such importance to the angles and corners of buildings?

[a] March (2); Wells (2), p. 219. Sullivan (1), p. 144, speaks of 'shifting perspective'.

[b] Western man, we write, because the beginnings of convergent or optical perspective can be traced back to the Greek — 5th century, with Agatharchos of Samos and his scene-painting (Sarton (1), vol. 1, p. 95; Frankfort (3); Schäfer (1), chs. 4, 5, esp. pp. 54 ff.). The beginnings of the deliberate use of optical perspective in western Europe are indeed seen in the works of Ambrogio Lorenzetti of Siena (*fl.* +1344), but this was because (Sarton (1), vol. 3, p. 111) he had mastered not only Euclidean geometry but the optical discoveries of the Arabs such as al-Haitham. Or someone else had mastered them and explained them to him. Convergent perspective developed slowly; pictures by Paul of Limbourg, for example (+1416), do not show it. The work of Albrecht Dürer (+1471 to +1528) in this connection is well known (cf. Rasmussen (1), p. 30). It is important to note that the earlier medieval European works seem not to have followed any system; they did not, for instance, develop parallel perspective and then give it up. On the history of perspective drawing in the West see Poudra (1); Wolff (2).

[c] *MCPT*, ch. 17, para. 6, cf. Hu Tao-Ching (1), vol. 1, pp. 546 ff.; tr. auct. adjuv. Schwarz in Tsung Pai-Hua (1), p. 27 copied by Sullivan (1), p. 143, abridged (2), p. 195.

[1] 以大觀小

Thus the small viewpoint of the stationary individual eye was condemned in favour of the large scanning view-area from which the artist, embodying in himself the insights of a whole troupe of observers, could attempt to convey the scene in its totality. The mountains and their detail were but small, what was large was the painter's mind and vision. Such was the orthodox attitude in all Chinese aesthetics.

A further point of great interest is that the Chinese developed an 'informative' as well as a 'representational' kind of perspective drawing. Two paintings of Ku Khai-Chih, the Bedroom Scene[a] and the Scene of the Emperor's Palanquin,[b] have been analysed from this point of view by Wells (1). At first sight the drawing seems to be very erratic; certain lines which ought to converge actually diverge, so that it is possible to speak paradoxically of an 'inverted or divergent perspective'. Upon investigation it appears that what was done was to swing round the fronts or sides of the bed and the litter so as to make them appear nearer, and thus give to the spectator certain items of information which otherwise he would not have had. Similar examples of divergent perspective can be traced in less marked form on Han reliefs.[c] It has been suggested that the motivation was similar to that of the child when it draws a human being with arms and fingers wide outstretched, though persons are rarely seen in this position (Wells, 2).[d] But in view of the creative discoveries of modern graphic and plastic art, the wise will not be hasty to condemn these liberties which ancient and medieval Chinese artists took with bald reality. Besides this, we also find examples of marked divergent perspective drawing in certain geometrical frieze motifs[e] in the Tunhuang cave-temple frescoes from the Northern Wei to the Sung periods.[f] Lastly, it is interesting that both parallel and divergent perspective radiated from China to many other parts of Asia, especially to Tibet and the countries of the South Seas such as Java and Siam.[g]

[a] Reproduced e.g. by Binyon (1), pl. 1; Waley (19), pl. v.

[b] Waley (19), pl. iv.

[c] And in very many later pictures strikingly, for instance the silk banner painting from Tunhuang (the Paradise of Bhaishajyaguru, the Buddha of Healing), c. +9th century, figured in Waley (19), pl. xx. Cf. Fig. 302 in Vol. 4, pt. 1; here, Fig. 757.

[d] 'Drawing a picture more true than one's visual impressions', says Frankfort (3), pp. 36 ff., expounding the views of Schäfer (1), pp. 332 ff., on the absence of optical perspective in ancient Egyptian art. It may well be, as he says, that the Greek invention of this way of drawing was connected with a sharp philosophical distinction between the world of appearance and the world of the mind—the 'physio-plastic' *versus* the 'ideoplastic'.

[e] In what I call the 'block frieze' the painters were accustomed to underline their compositions with a row of blocks of different colours and designs (possibly intended to represent the ends of beams), shown salient and re-entrant as if nearer or further from the observer. The drawing gives an impression of uncertainty, as when the eye is presented with a pattern of blocks so illuminated as to leave one in doubt whether they are steps below or a ceiling above. In one case (cave no. 303, Sui; Anon. (10), pl. 24) the blocks carry balustrades as if they were fancifully part of a city-wall, but the drawing is such that the eye oscillates violently between right-pointing beam-ends seen from below and left-pointing beam-ends seen from above. But if the picture is turned upside down the ambiguity vanishes. I paid special attention to these friezes on my visit of 1958. They illustrate well the fact that (as we shall shortly see) all optical illusions are connected with the perception of depth, but how much of the effect produced was intentional it is hard to say.

[f] Caves nos. 288, 431, 435 (Northern Wei); 303, 390, 402, 404, 417, 420, 424 (Sui); 307, 328 (Sung). Some of these friezes can be seen in Anon. (10), pls. 11, 12, 13, 24; Chhang Shu-Hung (3), figs. 33, 34, 41, 54, 207. This list is of course incomplete.

[g] See Auboyer (1).

Beyond this stage of analysis sinology and comparative art history have not penetrated. It is possible, however, that experimental psychology may be able to illuminate the problem from quite a different angle, for current investigations are revealing what may be physiological differences in distance perception between different peoples. Appreciation of depth and perspective may not be quite the same for all human beings. Although we cannot hope to do justice to these studies it would not be in accord with the spirit of a book on the history of science to ignore them.[a]

If a square white card of 6 in. side is placed at a distance of 8 ft. from an observer, and a number of smaller cards are presented to him alongside it at a distance of only 4 ft., he may be asked to choose which of them appears to him to be of exactly the same size as the more distant one. It has been found that practically no subject ever chooses the 3-in. card, though in fact this is the one which produces a retinal image of exactly the same size as the further card. A card of intermediate size is always selected. In other words no one (of whatever culture) chooses the card which would be correct according to the laws of optical perspective,[b] but a larger one instead. 'Mathematical' perspective is thus generally felt to be too drastic. This tendency is known as 'phenomenal regression to the "real" object', an effect defined by one of its first discoverers, Thouless (1), as 'the general tendency in perception for phenomenal (or apparent) characters to be intermediate between the characters indicated by retinal stimulation and the "real" characters of the perceived object'; and it holds good not only for size, as in the example just given, but also for shape.[c] Alternatively stated, in the perception of objects, every subject sees (i.e. immediately experiences) not the sensory characters indicated by peripheral stimulation (such as the image on the retina), but a compromise between these and the 'real' characters of the physical object itself, provided he has adequate perceptual cues as to what these 'real' characters are.[d]

[a] For helpful discussions on the subject of the following paragraphs I am indebted to Mr F. J. Pedler, Dr Richard Gregory and Miss Charmian Shopland.

[b] As when one looks through a lensless pinhole camera. Distant objects seem curiously small and near ones uncommonly large.   [c] Though not always for colour or brightness.

[d] A universal accompaniment of 'phenomenal regression' is 'size constancy', the fact that if we look at a small object such as a wristwatch or a seal-stone within a considerable range of distances, it always appears to be the same size, though the images which it forms on the retina are much smaller when it is further away than when it is near. Cerebral-mental compensation is in fact going on; perceptual enlargement of the more distant image. This was one of the effects which led Bishop Berkeley to write his *New Theory of Vision* (+1709); cf. Wolf (2), pp. 668 ff.

Worth noting too is the fact that features which we generally associate with depth-perception form a common property of all optical illusions (cf. Gregory, 1). The well-known Ponzo figure and the Muller-Lyer lines are examples of this. Consider also Baltrušaitis (3); Escher (1); Gombrich (1); Campbell (1).

In the same way we may judge as different the size of two apparent objects which can be proved to give exactly the same-sized retinal image, magnifying too much the thing which we think is the further away. This is the case with the apparent gross enlargement of the sun or moon seen near the horizon—a phenomenon discussed in ancient China (as we have seen in the amusing story of Confucius and Hsiang Tho, Vol. 3, pp. 225 ff.) and recognised already then to be subjective, by Chang Hêng as well as Ptolemy. The moon low on the horizon looks larger than when it rides high because it seems to be further away—again the process of distance enlargement by cerebral-mental compensation is at work. On this illusion see Dember & Uibe (1); Kaufman & Rock (1).

It is now thought that the distance and constancy systems may be two parallel and semi-independent mechanisms of sensory adjustment, capable of giving conflicting information. Of course objects sensed as being very distant look smaller than near objects, but size constancy works over a surprisingly wide range of distance. The changeover zone will vary greatly according to the conditions.

A convenient measure of this tendency is given by what is called the Phenomenal Regression Index

$$\frac{\log p - \log s}{\log r - \log s},$$

where p stands for a numerical measure of the phenomenal or apparent character (here the dimension of the card chosen), s for a measure of the stimulating character presented (here the 3-in. card), and r for a measure of the 'real' character of the object (here the 6-in. card). On this scale a zero value would indicate no influence of the 'real' character on the perception of perspective, i.e. that the cerebral-mental interpretation of vision was strictly following optical principles; while a value of unity would indicate an over-riding influence of the 'real' character, i.e. that appreciation of perspective was quite absent. Thouless himself first showed (2) that for European (British) subjects there were wide differences in the index (with an extreme range from 0·2 to nearly unity), but that statistically significant correlations could be made with age, sex, intelligence and artistic training.[a] He then went on to show (3) a significant difference between representative groups of Indian and British students, the index for the former being 0·76 while that for the latter was 0·61. Subsequently the question attracted much attention in Africa, where it had been found that drawings in approximately optical perspective were not understood by Africans.[b] Beveridge (1) found higher index values for West African draughtsmen than for Scottish students. Bush & Culwick (1) then reported values from East Africa as follows: Haia (Tanganyikan) Africans 0·82, Arabs of partly African descent 0·58, and Europeans 0·54. They could not establish any correlation between the index and the level of education.

Now it is clear that for anyone having a Phenomenal Regression Index higher than zero, a distant object, by comparison with a nearer one, will appear larger than it theoretically should; and the higher the index the greater the effect. The possible implications of this for the understanding of Asian art were immediately appreciated by Thouless himself (3), who illustrated an example of 'parallel perspective' from a +16th-century Mogul painting. In his view the striking differences in drawing technique between European and Asian artists and architectural draughtsmen were not so much due to the influence of geometry and optical science on the one hand, or to philosophical preferences on the other, as to demonstrable 'built-in' differences in perception between the peoples, analogous to the data studied by physical anthropology. With a high regression index it would be natural that an object producing a convergent retinal image should be perceived as parallel-sided. This striking conclusion still suffers from one experimental deficiency and invites one basic doubt. No quantitative figures, so far as we know, are yet available for any people of the Chinese culture-area; and even if it were proved that their index is uniformly high, this difference from Europeans would have to carry a heavy burden if all the traditions of distance-representation and 'parallel perspective' in Chinese drawing were to be imputed to it.[c]

[a] His tests were not confined to the simple perspective experiment here described.
[b] Cf. Pedler (1). Prof. J. M. Fitch (priv. comm.) has noted similar difficulties.
[c] The difference might be the effect, rather than the cause, of the differing traditions of representation in art. Here the African results are perhaps a test case, since there was no background of that kind.

Nevertheless it might be plausible to say that the Western development of optical perspective was aided by the low regression index of Europeans no less than by the life-work of Euclid.[a]

Thouless (3) had two points to add. He regarded Asian artists as strongly 'eidetic', excelling, that is, in vivid and precise imagination, held 'in the mind's eye' as they wielded their brushes in front of their scrolls. It is true that Chinese painters rarely painted from life, but in quiet recollection. This would have led naturally to a magnification of size as the eidetic image was brought nearer than the 'real' object by a person with a high regressive index. So far could this go, says Thouless, that even the 'divergent perspective' described above could become quite comprehensible. For in fact precisely this effect is produced when we look through a telescope at parallel lines receding from us, as on a railway or in a long corridor. How delighted Ku Khai-Chih would have been with such a demonstration!

(6) NOTES ON THE HISTORICAL DEVELOPMENT OF BUILDING

(i) *Words and traditions*

As would naturally be expected, the technique of building goes so far back in history that it is worth while to look at what has become embedded in the structure of the ideographic language itself.[b] There are three main radicals which have to do with dwellings, *yen*[1] (no. 53) which must originally have depicted a lean-to shelter against a cliff, *hsüeh*[2] (no. 116) which was originally a drawing of a cave- or pit-dwelling in rock or loess, and *mien*[3] (no. 40) which is frankly a roof. From these three origins derive the

Rad. no. 53   Rad. no. 116   Rad. no. 40   K413 later forms   K725

greater number of characters representing houses or parts of houses, though the technical terms for constructional parts are nearly all obtained, as we have seen, from the wood radical. From *yen*, the cliff-shelter, come such words as *thing*[4] (courtyard), *hsü*[5] (a side-house), *yu*[6] (the space under the eaves), *wu*[7] (gallery or verandah), and *khu*[8] (carriage-shed, treasury, arsenal). These are but a few examples. *Hsüeh*, not so prolific, gave *chhuang*[9] (window) and *tou*[10] (drain). Many familiar words, however, arose from *mien*, such as *kung*[11] (palace hall, already analysed, p. 72 above, K1006), *shih*[12] (a private house, K413), *chia*[13] (the family), *thang*[14] (reception-hall),[c] *chhin*[15] (bedroom)

---

[a] The Chinese convention of course continues in contemporary Chinese drawing in traditional style, and sophisticated taste can learn to appreciate it just as the Chinese themselves have become accustomed both to their own and to the Western styles. Europeans first came to know of Chinese painting in the +17th century. Joachim von Sandrart seems to have been the first to write of it (+1675) but his account was garbled to a degree (Sullivan, 4).

[b] For fuller discussion than can be given here see Kelling & Schindler (1).

[c] This derives from the word *shang*,[16] high and admirable, which has (the lord's) mouth under a roof; one can even see the ridge-pole (K725 *b*, *c*, *h*).

[1] 广   [2] 穴   [3] 屵   [4] 庭   [5] 序   [6] 序   [7] 廡
[8] 庫   [9] 窗   [10] 竇   [11] 宮   [12] 室   [13] 家   [14] 室
[15] 寢   [16] 尙

and *lao*[1] (stables). *Kung-shih* formed an antithetical couplet; the former perhaps meant originally the communal men's house, while *shih* meant the individual house-holder's house, and the words always retained this sense of communal and individual apartments. The derivatives of the *mien* radical radiated in due course into fields much wider than that of building. For instance, the word family (*chia*[2]) shows clearly (K 32) that all homesteads were originally farms, for the roof has a pig underneath it. The word for peace (*an*[3]) shows a roof (K 146*b*) with a woman underneath it. The word for cold (*han*[4]) shows (K 143*b*) a roof with a man, a mat, and firewood underneath it. *Hsiu*,[5] the resting-place, that technical term so important in Chinese astronomy (cf. Vol. 3, pp. 231 ff. above), also shows (K 1029) a house with a man and a mat inside it.

K 32   K 146*b*   K 143*b*   K 1029   K 1003   K 1160   K 661*h*

Tsung,[6] ancestral, a word with such far-reaching import in Chinese culture, represents (K 1003) the symbol for a sign or omen set up within a house; it refers therefore essentially to the ancestral shrine or temple. The usual term for temple however, *miao*,[7] comes from the lean-to shelter radical, but we do not yet know the significance of the objects which its earliest forms portray inside it (K 1160). They seem closely related to the character for 'early morning' and the court ceremony which took place at that time, so that the reference may be to early morning worship. *Chhin*,[8] the sleeping-room, seems to have a brush underneath a roof, probably because at first sleeping-rooms were also store-rooms[a] (K 661*h*).

Ancient tradition among the Chinese as to their earliest dwellings was rather precise. In a famous passage, the *Li Chi* says:[b]

Formerly the ancient kings had no halls or houses (*kung shih*[9]). In winter they lived in caves (*ying khu*[10]) which they had excavated, while in summer they lived in nests (*tsêng chhao*[11]) which they had framed.

There can be no doubt that these winter-dwellings were really holes in the ground, for shallow circular pits some 3 to 4 ft. deep and 9 to 15 ft. in diameter have been found by archaeologists investigating the black-pottery Lung-shan culture.[c] Some of the dugouts (*hsüeh*[12]) are much larger than this,[d] and all were covered by thatched roofs.

---

[a] Cf. a derivation of the word 'storey'.
[b] Ch. 9 (Li Yün), p. 50*b*, tr. Legge (7), vol. 1, p. 369.
[c] See Eberhard (24); Anon. (43), p. 15. Chêng Tê-Khun (9), vol. 1 and suppl. Cf. Vol. 1, p. 83.
[d] When conditions permitted, as in the loess country, it was easy to excavate roomy dwellings, such as those still used in the north-west, some of which I have myself often visited. Cf. Fig. 7 in Vol. 1. Fuller & Clapp (1) have given a description of these cave-dwellings; see also Creel (2), p. 56; Franck (1); and especially Lung Fei-Liao (1).

[1] 牢   [2] 家   [3] 安   [4] 寒   [5] 宿   [6] 宗   [7] 廟
[8] 寢   [9] 宮室   [10] 營窟   [11] 橧巢   [12] 穴

Frequently the dugouts used as storage-pits are beehive-shaped (6 ft. deep, 6 ft. diameter at the bottom but only 2 to 3 ft. at the top), and it is to be suspected that this age-old shape was perpetuated above ground for the poorest of the common people as late as the Thang, since one can see many low beehive-shaped huts of reeds or thatch painted in lifelike manner in the Tunhuang frescoes.[a]

Fig. 779. Beehive huts from Thang frescoes at Chhien-fo-tung (Tun-huang), cave no. 236.

The fullest excavation of a late neolithic village site is that which has been carried out at Pan-pho-tshun[1] in Shensi near Sian.[b] So impressive is the extent and detail of this, bespeaking a population much denser than that of Europe at the same time (*c.* −2500) and more like that of Egypt or Mesopotamia, that much of it has been roofed over and made into a national museum. The house floors, mostly circular in outline (about 16 ft. in diameter) but also oblong, are sunk 3 ft. or more below the surface of the ground (Fig. 780, pl.), surrounded by a low wall some $1\frac{1}{2}$ ft. in height, and have a hearth in the centre. Poles on each side of this supported the upper part of the roof, which certainly had a central hole, and a further row of poles outside the peripheral wall sustained the rim of the thatched or plastered covering. Beside the dwelling-pits, which have been studied in many other provinces, e.g. Yunnan, these villages have also many storage-pits.

The anthropological and symbolic-cosmographical aspects of these ancient dwellings have been surveyed in a *tour d'horizon* by Stein (3, 4). The upper Yang part with its orifice allowed smoke to escape, and light and rain to enter, by the same route as the human occupants took to climb in and out; the lower Yin part fostered the hearth and the *impluvium*.[c] One of the five *lares* of later times, Chung-liu,[2] took his name from the 'central drip', and the term came to stand for the tank in the *patio*. So also the stylised cupolas of subsequent ceilings were called *thien ching*[3] or *thien chhuang*,[4] and while *chhuang* itself meant a window, as we have seen, *chhuang*[5] came to mean a vent or a flue.[d]

[a] For example, cave-temples 108, 150, 197, 199.
[b] See Anon. (*25*); Chêng Tê-Khun (*9*), vol. 1, pp. 59, 75 ff., 132 ff.; Watson (*2*), p. 39; Liu Tun-Chên (*4*), pls. 1–5; Anon. (*43*), pls. 1, 2; Hsia Nai (*5*). It is a Middle Yang-shao site, cf. Vol. 1, pp. 81 ff. I was able to visit it with great profit in 1964.
[c] Stein was able to find many parallels for the ancient Chinese dwellings in those still existing among tribal peoples such as the Koryaks, the people of Kamchatka and the Alaskan Eskimos. Pursuing the concept of the house into the realms of macrocosmic–microcosmic thought, he pointed out connections between, e.g. the central pillar or ladder and the central cosmic mountain (Mt Meru, Khunlun Shan, cf. Vol. 3, pp. 565 ff.), or the upward and downward paths of the human souls at death (cf. Vol. 2, p. 490), etc., further to follow which would take us too far astray. We must be content to refer only to parallel explorers of the history of ideas in ancient building, Lethaby (1), and Hentze (6). Cf. p. 73 above.
[d] It is natural to ponder the symbolic significance of the great cruciform royal tombs of the Shang period, vast amplifications in a way of the semi-subterranean dwellings in which at one time all living men and women had passed the winter, and linking on by their shape to the designs of the Ming Thang or cosmic temple of later imperial authority, a cross of some kind in plan. With their ceilings 20 ft. underground and their ramps up to 65 ft. in length they are very impressive structures (Fig. 781, pl.). For a review of what is known about them see Chêng Tê-Khun (*9*), vol. 2, pp. 60 ff.; Watson (*2*), pp. 69 ff.

[1] 半坡村   [2] 中霤   [3] 天井   [4] 天窗   [5] 囱

In spite of the supposed seasonal alternation in types of dwelling, it may be that they derive from separate contributions to Chinese cultural evolution, the semi-subterranean caves or pits Tungusic, and the 'nests' Thai.[a] In any case, the 'nests' have of course left no permanent traces of themselves, but it is probable that they were rough shelters or houses built on piles, taking advantage perhaps of jungle trees as supports. If this was so, one can see how they could have given rise to the great tradition of columnar trabeate woodwork in Chinese architecture. It may be, too, that the raised field-watchers' or harvesters' huts (lu[1]), which are so common a feature of the western

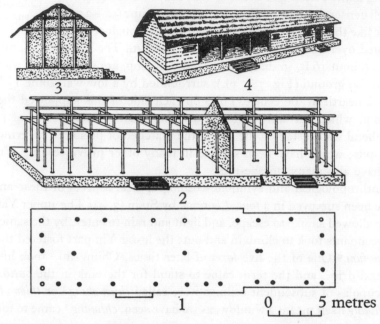

Fig. 782. Foundation of a Shang ceremonial building (A4) at Hsiao-thun in the area of the capital of Anyang; c. −1250 (Chêng Tê-Khun (9), vol. 2, fig. 11, after Shih Chang-Ju).
1   The foundation of tamped earth (terre pisé) with steps still in position on one side, and all the pillar bases in their original positions; length 80 ft., width 26 ft., height of platform 3 ft.
2   Reconstruction of the skeletal timber structure.
3   Probable appearance from one end.
4   Probable general appearance, suggesting a votive temple for ancestors.

[a] So Eberhard (2, 3). The contrast here between the round and the square invites thought on the outcome of this basic dichotomy in later Chinese building—a point raised with us in correspondence by the late Dr H. E. Stapleton of Jersey. Perhaps, in the beginning, the circular plan was more natural for the pit, and the rectangular one for the hut raised on poles. But though the latter certainly dominated throughout Chinese history, the former was never forgotten. In connection with the Ming Thang some very remarkable monuments and buildings of a circular character were erected, as by the empress Wu Tsê Hou in the Thang dynasty (cf. Sect. 30 d), and this tradition can still today be seen in the Altar and Temple of Heaven at Peking. Furthermore we shall shortly see how the need of fortifying the multiple dwelling of a clan community led in some parts of China to extraordinary circular castellate structures (p. 134). Finally, there were the round tents of the nomads, the yurts of the Mongols, and these also have crystallised into actual round dwelling-houses in modern Chinese Mongolia (cf. Liu Tun-Chên (4), pls. 36–41; Anon. (41), pl. 107). Cf. p. 73 above.

[1] 廬

China landscape at the right season of the year, have come down almost unchanged from this remote antiquity.[a] At all events, there can be no doubt that the type of building which was later on to become so universal in China, namely the raising of wooden pillars upon stone plinths above a platform of tamped earth (*hang thu*[1]) foundations, with *terre pisé* walls afterwards thrown around the structure, had already reached a highly developed state in the Shang period (−2nd millennium). This is proved by the excavations at Anyang.[b] Fig. 782 illustrates the plan and reconstruction of one of these long buildings with thirty column bases, most of which still retained their stones.[c] True to type, the main entrance seems to have been in the middle of one of the 80-ft. long sides, but the chief axis of the building was north–south instead of east–west, as became customary later.[d]

The folk-songs in the *Shih Ching* (Book of Odes), which may go back at least as far as the −8th century, have several passages of architectural interest, and though not perhaps very informative, they are worth quoting. For example:[e]

> Of old, Tan-Fu the duke[f]
> Made kiln-like dwelling-pits with roofs,[g]
> As yet (the people) had no houses.[h]
>
> Of old, Tan-Fu the duke
> At coming of day galloped his horses,
> Going west along the river bank
> Till he came to the foot of Mount Chhi,
> Where with the lady Chiang
> He came to look for a home.
> The plain of Chou was very fertile,
> Its celery and sow-thistle sweet as rice-cakes,
> 'Here we will make a start; here take counsel,

[a] Besides, one must not forget that many of the tribal peoples formerly on the fringes of Chinese society, such as the Miao, have lived in pile-dwellings until the present day (cf. Liu Tun-Chên (4), pls. 43, 48).

[b] See Li Chi (1) and later Academia Sinica publications, summarised and digested by Eberhard (24); Chêng Tê-Khun (9), vol. 2, pp. 20 ff., 27 ff., 41, 44, 50 ff., 247; Watson (2), pp. 62 ff. Cf. Liu Tun-Chên (4), pls. 6–8. Of course large numbers of the smaller structures at Anyang and other Shang sites were pit-dwellings of the *hsüeh* type, often oval as well as round or square.

[c] Chêng Tê-Khun (9), vol. 2, p. 52; Creel (2), p. 59.

[d] Shang platform foundations as large as 80 ft. × 26 ft., as here, are quite common. The largest so far found measures 280 ft. × 48 ft. approximately. It is interesting that there are no signs of courtyard house plans in the Shang cities.

[e] The Mien-mien song (Mao no. 237); tr. Legge (8), III, i, 3 (p. 437); Karlgren (14), p. 189; Waley (1), p. 248.

[f] Tan-Fu was one of the semi-legendary chiefs of the proto-Chou people, dated traditionally in the −14th and −13th centuries.

[g] On this see Stein (3). What we now know of the pottery kilns of the neolithic and the Shang periods fully justifies this interpretation of a phrase which has given some trouble to commentators. Cf. Anon. (43), p. 16.

[h] Waley (1) remarks that this was not because they were not able to build, but because the ancient shelters were used until the new city could be planned and started. We believe, on the other hand, that the common people continued to live in semi-subterranean dwellings till long after the end of the Shang period, at any rate in some places.

[1] 夯 土

Here notch our tortoise'.[a]
It says 'Stop', it says, 'Halt,
Build houses here'.

So he halted, so he stopped.
And left and right
He drew the boundaries of big plots and little,
He opened up the ground, he counted the acres
From west to east;
Everywhere he took the task in hand.

Then he summoned his Master of Works (Ssu Khung[1])[b]
Then he summoned his Master of Lands (Ssu Thu[2])
And charged them with the building of houses.
Dead straight was the plumbline,
The planks were lashed to hold (the earth);
They made the Hall of Ancestors, very venerable.
They tilted in the earth with a rattling,
They pounded it with a dull thud,
They beat the walls with a loud clang,
They pared and chiselled them with a faint *phing, phing*;
Three hundred rod-lengths[c] all rose up,
The drummers could not hold out.[d]

They raised the outer gate;
The outer gate soared high.
They raised the inner gate;
The inner gate was very strong.
They raised the great earth-mound[e]
Whence excursions of war might start...

Here the chief technical interest is the ramming of earth for the foundations and the walls of a temple. Another of these songs describes the erection of a feudal palace.[f]

(The Lord) resembles and succeeds his forebears,
He builds a house of a hundred cubits[c]
To the west and south are its doors,
There will he live and dwell, there laugh and talk.

They bind (the shuttering frames) one over the other,
They pound (the earth in them, it sounds) *thak, thak*;
This will keep out the wind and rain, the birds and the rats,
Here will the Lord be eaves-covered...[g]

[a] Scapulimancy, cf. Vol. 2, pp. 347 ff. above.   [b] Note how old these titles are.
[c] Lit. 100 *tu*[3] measures. The *tu* was 50 ft. (*wu chang*), a measure especially used for walling, and the word also meant a wall. Cf. p. 111 above.
[d] The workers worked so enthusiastically that they outdid those who were setting the rhythm.
[e] Altar of Earth.
[f] The Ssu Kan song (Mao no. 189); tr. Legge (8), II, iv, 5 (p. 303); Waley (1), p. 282; Karlgren (14), p. 130.
[g] Cf. *Mêng Tzu*, VII (2), xxxiv, 2, who says he dislikes eaves (*tshui*[4]) luxuriously broad.

[1] 司空   [2] 司徒   [3] 堵   [4] 檼

Level is the courtyard, straight are the columns (*ying*[1])[a]
Pleasant and cheerful the halls of reception
Ample the living-rooms, where the Lord can be at peace...

Again we have the moulds for the tamped earth, and a reference to the eaves.

Among other classical references to building[b] and its traditions we may notice one in the *I Ching* and another in the *Mo Tzu* book. The Great Appendix in the former (perhaps of Warring States time, certainly Early Han) remarks:[c]

In primitive times people dwelt in caves and lived in forests. Later sages made the change to buildings (*kung shih*[2]). At the top was a ridge-pole (*tung*[3]), and sloping down from it there was the roof (*yü*[4]) to keep off wind and rain.

And it goes on to name the Kua (no. 34) from which the sages might be supposed to have derived their inspiration, in the course of its sketch of social development. Mo Ti, as usual, attacks what he considers undue luxury and elaboration:[d]

Master Mo said: 'Before the art of building halls and houses (*kung shih*[2]) was known, the people lingered among the hillsides and lived in caves or pit-dwellings (*hsüeh*[5]) where it was damp and injurious to health. Thereafter the sage-kings made halls and houses. The guiding principles for buildings were these, that (the house should be built) high enough to avoid damp and moisture, (that the walls should be) thick enough to keep out the wind and the cold, and (the roof) strong enough to stand snow, frost, rain and dew; lastly, that the partition walls in the palaces should be high enough to observe the proprieties of (separate accommodation for) the sexes. These things are sufficient, and any expense of money or labour which does not bring additional utility should not be permitted.

On the whole, it will be seen that these references do not give us much of the ancient technicalities. But a glance at the *Erh Ya*, the most ancient dictionary (already mentioned), in its special section on buildings (Shih Kung[6]), will bring to light at once some twenty or so of the technical terms with which we have now become familiar (e.g. *fu*[7] and *cho*[8]); these it duly explains. There are also, of course, others which are rarer or which long ago became obsolete.[e] Nevertheless, it remains clear from this that a substantial part of the technical vocabulary of building construction as we now know it was used by the architects of the Chhin and Early Han (−3rd century).[f]

---

[a] We have already come across (Vol. 4, pt. 2, p. 189 above) a possible reference to the stone bases of columns, not very much subsequent to this in date, in the *Kuo Yü* in connection with millstones.

[b] Li Chieh in the beginning of the +12th century opened his great work with a selection of such passages.

[c] Pt. 2, ch. 2, tr. Wilhelm (2), Eng. ed., vol. 1, p. 359. Cf. Vol. 2, p. 307.

[d] Ch. 6, p. 13a, tr. Mei Yi-Pao (1), p. 22, mod. auct.

[e] Such as *erh*[9] which meant one of the vertical posts in the roof timbering.

[f] We cannot here discuss the many subsequent repositories of technical terms, such as the *Shu Hsü Chih Nan* (Literary South-Pointer) of +1126 (ch. 16, pp. 1a ff.), or Ming books such as the *Piao I Lu* (Notices of Strange Things), ch. 4. All the encyclopaedias have relevant sections which could be drawn upon. Another interesting study would be the attributions of the various architectural processes or objects to the legendary inventors (cf. Vol. 1, pp. 51 ff. above), as in the Sung *Shih Wu Chi Yuan*, ch. 8.

---

[1] 楹   [2] 宮室   [3] 棟   [4] 宇   [5] 穴   [6] 釋宮   [7] 桴
[8] 㮰   [9] 梲

(ii) *Periods and styles*

We may accept, therefore, that the use of tamped earth platforms, halls with many wooden pillars standing on stone bases, and suitable simple roofs, was widespread from at least the −13th century onwards. It must have been towards the end of the Chou period that the invention of the corbel brackets (*tou kung*) took place, for they are characteristic of all Han buildings.[a] Of course none of these structures have themselves survived, but fortunately the Han people imitated their woodwork in pottery models[b] as well as in the stone of tomb-chambers and funeral steles. The latter were not mere slabs with inscriptions, but rather resembled models of towers some 12 ft. high, crowned with elaborate roofs the timberwork of which was faithfully reproduced in stone. A number of these were discovered in rural Szechuan by Segalen, de Voisins & Lartigue; for example that set up to commemorate Chao Chia-Phing at Chhü-hsien near Pao-ning (± 1st century).[c] It is interesting in that it shows both forms of *tou kung*, a straight type and a type with an S-shaped curve. Another fine example is the stele of Kao I,[1] an official who died in +209 and whose tomb is at Ya-an in Sikang province on the borders of the Tibetan mountains.[d] Besides these, there are the funeral chambers. Massive *tou kung* are carved in the stone tomb-shrine of Chu Wei,[2] one of Wang Mang's generals, who died about +45.[e] So also, in the two-storeyed hall which appears in the bas-reliefs of the Hsiao-thang Shan[3] tombs[f] (Fig. 783, pl.), the capitals are drawn in such a way as to indicate that the columns were topped by a series of successively longer *tou kung* (c. +125). Sometimes we see them only end-on, as in the Szechuanese moulded brick of an entrance-gate flanked by towers (*chhüeh*[4]).[g] And

---

a We can just make them out on what is perhaps our oldest Chinese architectural picture, the bronze bowl from Huihsien shown already in Vol. 4, pt. 1, Fig. 299 (Liu Tun-Chên (4), pls. 10, 11). Cf. Fig. 300. This is of the −4th century, but already the design of a two-storeyed hall, with roofed galleries and side pavilions on the lower level, is clearly visible. Pottery moulds of the late Chou period (see Anon. (43), pl. 51) show what seem to be personified *tou kung*. On late Chou and Han building in general see conveniently Watson (2), pp. 122 ff.; but the best discussion of the Han style in architecture is still no doubt Pao Ting *et al.* (1). Cf. too Chhen Ming-Ta (1).

b Cf. Anon. (42).

c Pl. 30; reproduced in Bulling (2), fig. 6/4. We illustrate (Fig. 785, pl., from Sickman & Soper (1), pl. 156A) the tomb of Fêng Huan, dating from +121.

d Segalen *et al.*, pls. 47, 48; reproduced in Bulling (2), fig. 5/3. Cf. Mirams (1), opp. p. 19. The stele of Shen Fu-Chün, which we have encountered before in connection with the invention of the wheelbarrow, also shows corbel brackets (Vol. 4, pt. 2, Fig. 508).

e See W. Fairbank (1). This tomb, at Chin-hsiang in Shantung, has figures much more naturally portrayed, and set in parallel perspective, than the later tomb-shrines of the +2nd century. She suggests that this was because in the former the artists were representing in stone the paintings on plaster which we know the Han people had (cf. Fischer (2); Watson (1), pls. 100–5), while in the latter they were copying the stamped bricks and tiles which decorated, with rather mechanical repetition, Han buildings and tombs. Chu Wei's tomb was buried for five centuries, but Shen Kua knew it and wrote about it in +1085 (*MCPT*, ch. 19, para. 8).

f *Chin Shih So*, Shih sect. ch. 1 (pp. 89–92); cf. Chavannes (9), fig. 45; Bulling (2), figs. 67/67, 67/68; Kelling & Schindler (1), p. 110. The chief occupant was Kuo Chü,[5] an official famous for his filial piety.

g Liu Chih-Yuan (1), fig. 63; Anon. (22), figs. 80, 81; Rudolph & Wên (1), pl. 86; Hsü Ching-Chih (1), pl. 19. Other good representations of entrance gateways are to be seen in Liu Chih-Yuan (1), fig. 65; Anon. (22), figs. 16, 17.

¹ 高頤   ² 朱鮪   ³ 孝堂山   ⁴ 闕   ⁵ 郭巨

again they are visible at the top of the tower in that wonderful representation of a
country manor of the Han period given us on another brick (Fig. 784, pl.),[a] with its
various yards separated by roofed *terre pisé* walls. On yet another, where we get a
closer view of the gentry performing their ritual salutations, the significance of the
hang-thu platform is apparent.[b]

The Hsiao-thang Shan picture introduces, however, a new point, namely, that in the
Han period it is impossible to find any curvature in the roofs. Not until long after its
close, perhaps not till the +6th century, does the discovery appear to have been made
that the transverse frames permitted any desired variation from the straight line.[c] This
stiffness is shown by many other bas-reliefs depicting buildings, such as the famous
examples at the tomb of Wu Liang [1] (*c.* +147).[d] It is shown also by the existing stone
shrines themselves in which these carvings are contained,[e] and by numerous tomb-
models in pottery.[f] Of these a great number have in recent years been excavated in and
near Canton. To illustrate a building of a rather different type from any so far mentioned,[g] we reproduce a remarkable model of a fortified manor-house self-dated at
+76 (Fig. 786, pl.).[h] When the pieces are taken apart, animals in their byres, and
numerous farming and domestic activities can be seen.[i] Interesting reconstructions of
Han palace buildings are now being attempted.[j]

A characteristic Han feature was the use of caryatides as columns. This appears
fairly frequently in the *Chin Shih So* rubbings from Wu Liang Tzhu and other sites,
but there is little trace of it in later times, save that columns of important buildings
were (and still are) sometimes carved in high relief. Dragons, for instance, twine
around them, as in the Confucian temple at Chhang-ting (Fukien) and elsewhere.[k]
We illustrate the columns of the main hall (Shêng-Mu Tien) at Chin Tzhu in Shansi
(Fig. 787, pl.).

During the period between the Han and the Sui, there was a gradual development in
complexity, the *tou kung* becoming more and more elaborate, and the roofs showing a

---

[a] Liu Chih-Yuan (*1*), fig. 64; Anon. (*22*), figs. 18, 19; Liu Tun-Chên (*4*), pl. 18, cf. pl. 12; Anon. (*43*),
pl. 85 B.

[b] Anon. (*22*), figs. 24, 25.

[c] The further development of turning up the eaves at the corners to add lightness and elegance did
not come about till much later, the Wu Tai and early Sung periods (+10th century), and it is associated
with the name of Yü Hao himself (p. 81). Cf. Sickman & Soper (*1*), p. 256.

[d] *Chin Shih So*, Shih sect. ch. 3 (pp. 66, 67); Jung Kêng (*1*), pp. 46 b–48 b; Chavannes (*9*); Bulling (*2*),
figs. 12/69/70, 12/69/71, 69/71 A; Kelling & Schindler (*1*), p. 110.

[e] Fairbank (*1*). The main approach led, as one would expect, to the long south side.

[f] Cf. Laufer (*3*), pp. 40, 42; Maspero (*17*); Bulling (*2*), fig. 105/109; Kelling & Schindler (*1*), p. 90.

[g] But mem. p. 70 above.

[h] Anon. (*42*), fig. 40, pp. 31 ff.; (*43*), pl. 89.

[i] Criss-cross lines shown on representations of walls and friezes of buildings between columns in Han
tomb-reliefs and models sometimes look as if complex trusses were intended, but this is almost certainly
an illusion. They must surely be ornamentation. There were, however, the inverted V-braces, on which
see p. 101 above.

[j] See Wang Shih-Jen (*2*). A *pi yung* (College, and Hall for the Veneration of Elders) has been
excavated in the Western suburb of Sian (old Chhang-an); see Thang Chin-Yü (*1*).

[k] Marco Polo was impressed by some of these, cf. ch. 75 (Moule & Pelliot ed.).

[1] 武梁

more and more concave tendency.[a] By the earliest years of the Thang period this was fully established, for we find it not only in the oldest wooden buildings of China still extant[b] but also in the still older structures conserved in Japan.[c] The problem of its origin and spread has prompted a question about it different from any of those we have already considered (pp. 91, 97, 102), and some scholars are inclined to regard it as an example of the influence of South-east Asian roof forms.[d] There is certainly little difficulty in finding examples of all three types of roof curve (the concave slope, the curving ridge and the uplifted eave corners) in the traditional architecture of Indonesia, Indo-China, Siam, Cambodia and the Philippines, but archaeological proof has not yet been provided that they were there before the Liu Chhao and Sui periods. Further research will be required to eliminate the converse possibility that all these forms were inspired by Chinese influence radiating along the trade-routes in medieval times.

The chief new development in the Liu Chhao period was that of pagoda construction. The pagoda (*tha*[1]), as it eventually flourished, was a combination of the ancient Han towers of several storeys (known from pottery models, e.g. Fig. 788, pl.)[e] and the *stūpa* forms from India, which ultimately imposed upon the structure its various plastically curving silhouettes. We shall say more about it presently; here we would rather dwell for a moment on the ancient Chinese inclination to build upwards.[f] Chou

[a] The process can be followed to some extent in the Tunhuang frescoes. For example, in Northern Wei (late +4th century onwards) cave no. 257 (Anon. (*10*), pls. 4, 5). In Western Wei (+6th-century) cave no. 285 (Anon. (*10*), pl. 18 and Yeh Chhien-Yü (*1*), fig. 5). In Sui (+581 to +618) cave no. 290 (Anon. (*10*), pls. 25–31). At these periods the curvature, though slight, is already perceptible or even quite apparent. Then one finds it fully developed in the architectural paintings of the late +6th-century painter Chan Tzu-Chhien[2] (cf. Waley (*19*), pp. 109, 134, 140) preserved in the Imperial Palace Museum at Peking (see Liu Tun-Chên (*4*), pls. 22, 23).

[b] The temples of Wu-thai Shan[3] in Shansi, Nan-Chhan Ssu[4] (+8th century) and Fo-Kuang Ssu[5] (+9th century); see immediately below. The temple hall depicted on the stone lintel of the Ta-Yen Tha pagoda at Sian (+652 to +701) also shows the curving roof slope strikingly; cf. Sickman & Soper (*1*), fig. 17.

[c] The Golden Hall (Kondō[6]) and the five-storeyed pagoda (Gojū-no-tō[7]) at the Hōryūji[8] temple near Nara, buildings of *c*. +712 copied from predecessors of *c*. +623; as also the portable model Tamamushi[9] shrine preserved there, which cannot be later than the mid +7th century. Cf. p. 95.

[d] Notably Sullivan (*1*), pp. 119, 137, (*2*), p. 228. The great Sung architect Yü Hao (cf. p. 81) has, I think, been wrongly enlisted in this cause. According to the *Hou Shan Than Tshung*[10] (Hou-Shan Table Talk) by Chhen Shih-Tao[11] (*c*. +1090), when Yü Hao came north to Khaifêng from Hangchow he was deeply impressed by the 'Tower Gate' of the Hsiang-Kuo Ssu[12] temple, a Thang structure. After closely examining it for a long time, he said, 'They were certainly capable enough (in those days); the only thing is, they didn't understand how to curve up their eaves (toward the corners).' The last three words in brackets, inserted by the translator, Soper (*1*), are surely right—what Yü Hao brought northwards was the uplifting of the eave corners, not the curving slope of the roof profile, which had long been there. The incident would have occurred about +970.

[e] Maspero (*17*); Anon. (*37*), pl. 5; Anon. (*43*), pl. 88; Sickman & Soper (*1*), pl. 156B; Bulling (*2*), figs. 31/22, 22/23, 32/24; Ao Chhêng Lung *et al*. (*1*), pls. 66–9, examples from a tomb of +182 at Wangtu, with sectional elevation drawings.

[f] This tendency there was, but it does not contradict what was said earlier about the dominance of the horizontal line in Chinese architectural planning, first because great towers were rare, and secondly because they were almost always kept somewhat at a distance from the plan of the layout as a whole. This continued to be true for the Buddhist pagodas.

[1] 塔	[2] 展子虔	[3] 五臺山	[4] 南禪寺	[5] 佛光寺
[6] 金堂	[7] 五重塔	[8] 法隆寺	[9] 玉蟲厨子	[10] 後山談叢
[11] 陳師道	[12] 相國寺			

and Han tower-building has been somewhat overshadowed by the dominance of Buddhist 'spires' in later ages, but if the pagoda owed much of its ancestry to the *stūpa*, the high raised platform (*thai*[1]) and the towers to which it gave rise were not without a family connection with the *ziggurat*.[a] The *thai* of the Chou period was an elevated platform or terrace of pounded earth faced with brick or possibly stone and carrying sometimes a building on top.[b] In the *Tso Chuan* there are many references to the *thai* of the feudal States as places used for diplomatic audiences, interviews between rulers, feasting, imprisonment, last-ditch stands against enemies, look-outs for the regular watch, and last but not least astronomical and meteorological observations.[c] The term Ling Thai[2] or Numinous Tower, as we know from Section 20 on astronomy,[d] came to be the standard appellation for an observatory. A very ancient reference to this occurs in the *Shih Ching* (Book of Odes) which celebrates the willingness of the people to build one for Wên Wang.[e] Flattened and extended, but equally numinous, the Altar of Heaven (Thien Than[3]) was, and is, a *thai*.[f] Hughes has drawn attention to the emphasis on very lofty towers in some of the Han odes, notably Pan Ku's *Hsi Tu Fu*,[g] where the dizzy height of the Ching-Kan Lou,[4] built by Han Wu Ti under Taoist influence to ensure communication with the aerial spirits, is assessed at some 375 ft. Probably this was a wooden tower based on a high stone-faced platform.[h] Although nothing of this kind survived, the impressive guard-houses (*chhüeh*[5]) over the gates of great Chinese cities (cf. Fig. 742, pl.) remained until our own time to testify to the awe-inspiring nature of the formula.[i] One curious feature of the Liu Chhao period is a number of representations of pavilions standing on a base of tall piles, and looking rather like early nineteenth-century designs for iron lighthouses.[j] These seem to be found first in Buddhist frescoes of the Northern Wei dynasty, and strangely recall the ancient pile-dwelling traditions previously referred to, but they left no permanent mark on Chinese building after the Thang. By the end of the Liu Chhao both curving roofs and pagoda towers had completed their basic development.

---

a Cf. Sickman & Soper (1), pp. 210, 219 ff.; Hughes (9), pp. 115 ff. The Mesopotamian echo has been particularly remarked upon by Christian (1), his comments arising from discussion of the Ta-Yen Tha pagoda at Sian, on which see p. 139 below. Cf. Ling Shun-Shêng (2, 3) and p. 543 below.

b Cf. *Tao Tê Ching*, ch. 64: 'A tree big as a man's embrace begins as a tiny shoot; a *thai* nine storeys high is only a heap of earth to start with.'

c Excavation of such a *thai* has recently been made at Chhêngtu (Yang Yu-Jun, 1). A three-tiered platform with retaining walls, some 300 ft. square, it probably dates from the early years of the State of Shu, c. −660.

d Vol. 3, pp. 189, 207, 284 and *passim*.

e Mao no. 242; tr. Legge (8), III, i, 8, (p. 456); Karlgren (14), p. 197; Waley (1), p. 259.

f See the recent exhaustive history of Ishibashi Ushio (1).

g *Wên Hsüan*, ch. 1, pp. 8 b, 9 a, tr. Hughes (9), pp. 32 ff. So also the *Kan Chhüan Fu* of Yang Hsiung, tr. von Zach (6).

h The names of some of the Han builders of such towers have come down to us, e.g. Thao Chhien.[6]

i Cf. Vol. 1, Fig. 14, and Lin Yü-Thang (7), pl. 19. Like so many other survivals of the archaic, this architectural theme also lived on, and had perhaps its greatest development, in Japan, where the typical feudal castle (many examples of which still exist) was essentially an elaborately roofed cluster of buildings on a lofty stone base, the latter characterised by a concave profile. Cf. Paine & Soper (1); Murasawa Fumio (1); Kirby (1); Guillan (1).

j Bulling (2), figs. 28/19, 28/19A, 24/16, 24/16A, 26/17, 26/18, 27/18A, etc.; Anon. (37), pl. 35.

¹ 臺   ² 靈臺   ³ 天壇   ⁴ 井幹樓   ⁵ 闕   ⁶ 陶謙

The Thang period seems to have been a time of architectural experimentation.[a] It is from the Thang that we possess the oldest existing Chinese wooden building.[b] One of the halls of the Fo-Kuang Ssu[1] (Buddha's Aureole Temple) near Tou-tshun-chen in the Wu-thai Shan[2] foothills (Shansi) constitutes the next oldest; it was discovered and thoroughly studied by Liang Ssu-Chhêng during the Second World War (2, 3, 4).[c] It stands on a high platform adjusted to the slope of the mountain (Fig. 789, pl.), and its massive columns, beams, *ang* and *tou kung* have withstood all ravages of time since +857.[d] Probably also from before the end of the same century is the main hall of the Confucian temple at Chêng-ting[3] in Hopei.[e] The only other wood-work from this century which has remained in China is perhaps that of some of the galleries outside the cave-temples at Tun-huang.[f] Magnificent reconstructions of

[a] See Liang Ssu-Chhêng (6).

[b] This is the main hall of the Nan-Chhan Ssu[4] temple at Li-chia-chuang up the Pai-ta-hsing village valley in the foothills of Wu-thai Shan. A surviving inscription dates the last restoration at +782. Little has been published about it since its discovery in 1953, but it is illustrated by Liu Chih-Phing (1), figs. 211, 212. Rather small and simple in construction, it demonstrates its antiquity by the use of inverted V-braces or forked-hand struts (cf. p. 100) on each side of the king-posts and as supports of queen-posts, and by the fact that no cantilever rafters (*ang*, cf. p. 95) trouble the symmetry of the bracket-arm clusters. In 1964 I had the good fortune to be able to visit and study both this temple and the neighbouring Fo-Kuang Ssu;[1] warmest thanks are due to my friends Mr Hou Tshun-Hsi and Dr Shen Chhêng-Shu of Thaiyuan for organising our expedition. Nan-Chhan Ssu, however, is not the oldest wooden building of the Chinese culture-area. Temples and pagodas dating from the previous century have been preserved in Japan, successive restorations faithfully copying the original structures. We have just mentioned (p. 128) Hōryūji[5] near Nara, where the Kondō and the five-storeyed pagoda date from perhaps as early as +680, certainly not later than +714. The three-storeyed pagoda at the Hōkkiji[6] dates from +685, and that at the Yakushiji[7] from *c.* +698 or +720. All of them have *ang*, like Fo-Kuang Ssu. Community of design imposes the inclusion of these in any history of Chinese building technique, all the more as we know that the architects of some of the oldest Japanese Buddhist temples were Chinese monks.

Of this we shall see a classical example immediately below in Chien-Chen and Ju-Pao towards the end of the +8th century. But builders and architects had gone to Pakche in Korea from the Liang dynasty in south China in +541, and on to Japan in +552. A temple-builder (*tsao ssu kung*[8]) arrived there in +577, and expert bronze-founders and tilers (*lu-phan po-shih*,[9] *wa po-shih*[10]) in +588. Six centuries later Shunjōbō-Chōgen[11] got the assistance of two distinguished Chinese engineers and builders, Chhen Ho-Chhing[12] and his brother Chhen Chu-Fo,[13] as well as a Japanese bronze-founder Kusakabe no Koresuke,[14] in their epic rebuilding of the Tōdaiji[15] at Nara from +1168 to +1195. Engineering particulars of this can be found in the *Tōdaiji Zōritsu Kuyōki*[16] of +1452. One cannot attempt to separate the two traditions.

[c] See also Ecke (4); Liu Chih-Phing (1), figs. 205–10, 219, 220; Sirén (1), vol. 4, p. 68, pl. 115; Anon. (37), p. 8, pls. 37–40; Anon. (26), p. 460; Sickman & Soper (1), pp. 246 ff. An excellent wooden model is in the Imperial Palace Museum at Peking. It is interesting that the oldest wooden building in England dates from just about the same time—the church at Greensted in Essex. But it is a small matter in comparison with the Fo-Kuang Ssu, for it consists of little more than a stockade of split logs, the roof-beams of which perished long ago. On Wu-thai Shan cf. E. S. Fischer (1).

[d] There is a fairly close parallel to Fo-Kuang Ssu still extant in Japan, the Kondō at Tōshōdaiji[17] on the western outskirts of Nara. Its date must be between +760 (the last year of the famous Chinese abbot Chien-Chen[18]) and +815 (the date of the death of his architectural disciple Ju-Pao[19]). On Chien-Chen (Kanshin) see now Chou I-Liang (2); Andō Kōsei (1).

[e] See Liang Ssu-Chhêng (8); Sickman & Soper (1), p. 254.

[f] Personal observations in 1943 and 1958. The remains of the balconies outside caves nos. 224 and 305 seemed to be Thang in date (Fig. 790, pl.); those of caves nos. 202 and 232 seemed to be Sung. In

[1] 佛光寺	[2] 五臺山	[3] 正定	[4] 南禪寺	[5] 法隆寺
[6] 法起寺	[7] 藥師寺	[8] 造寺工	[9] 鑪盤博士	[10] 瓦博士
[11] 俊乘坊重源	[12] 陳和卿	[13] 陳鑄佛	[14] 草部是助	[15] 東大寺
[16] 東大寺造立供養記	[17] 唐招提寺	[18] 鑑眞	[19] 如寶	

the great halls of the palaces at the Thang capital, Chhang-an, are now being made.[a]

Next oldest to the extant Thang buildings is the great hall of the Chen-Kuo Ssu [1] (Protection-of-the-Nation Temple) at Phing-yao [2] in Shansi, built in +963 under the short-lived Eastern Han dynasty;[b] after which follows the more famous Tu-Lo Ssu [3] (Joy-in-Solitude Temple) at Chi-hsien [4] in Hopei.[c] The main gate and the Kuan-Yin Ko [5] hall of this temple were built under the Liao rule in +984. This latter is a very large work, a three-storeyed building containing a statue of Kuan-Yin (the Goddess of Mercy) more than 60 ft. high (Fig. 791, pl.).[d] The architect arranged a space (ching [6]) in the centre so that the statue could penetrate through all three floor-levels. More than a dozen different kinds of *tou kung* were used to suit the different positions. A few other buildings also date from the early Sung;[e] rather less than a century later came the great octagonal wooden tower of nine storeys at Ying-hsien [7] in Shansi (+1056).[f] Its total height is just under 200 ft. In this masterpiece nearly sixty different types of *tou kung* can be found (Fig. 792, pl.).

Thus from the +10th century only half a dozen buildings have lasted.[g] There are, however, at least fifty dating from before the +14th century which have also been carefully investigated.[h] Buildings of the Ming, the +15th century, are comparatively

some cases the original inscriptions with dates can still easily be read; I did not copy them, but afterwards found a correspondence between Pelliot and the Chinese architectural historians on the subject, published by Liang Ssu-Chhêng (7). Thus the wooden gallery of cave no. 431 was built and dedicated by the Exarch of Tunhuang, Tshao Yen-Lu,[8] in +980, and that of nos. 443 and 444 was the work of his predecessor Tshao Yen-Kung [9] four years before. Photographs of the former before and after restoration are given by Chhang Shu-Hung (1), figs. 12, 13.

[a] On the basis of textual and iconographic evidence as well as excavation of the sites; see Kuo I-Fu (1); Liu Chih-Phing & Fu Hsi-Nien (1); Hsü Ming (1).

[b] See Anon. (37), pl. 43.   [c] Elaborate description in Liang Ssu-Chhêng (9).

[d] Anon. (37), p. 8, pls. 46–8; Anon. (26), p. 492; Bulling (2), p. 25, fig. 82/83; Liu Chih-Phing (1), fig. 197; Sickman & Soper (1), pp. 275 ff., pl. 177. A beautiful model of this temple is in the Imperial Palace Museum in Peking.

[e] Description in Sickman & Soper (1), pp. 261 ff.

[f] At the Fo-Kung Ssu [10] (Buddha-Palace Temple). Description in Sickman & Soper (1), p. 274, pl. 176A. Cf. Liu Chih-Phing (1), figs. 126–8; Boerschmann (4); Anon. (37), pls. 54, 55; Anon. (26), p. 491.

[g] A very few examples of domestic architecture have been placed in this period. By the kindness of Mr Rewi Alley and Mr Courtney Archer I am able to illustrate the Chhen-chia Lou [11] at Shantan [12] in Kansu, a courtyard house in the shape of a double L with walls and framework very markedly battered (Fig. 793, pl.). In spite of age and rotting timbers it successfully withstood the earthquake of 1954 which almost wrecked the rest of the city, perhaps because of this battering. Its date however may well be as late as the +13th century.

[h] A useful list of extant (and lost) Sung, J/Chin and Liao buildings between +970 and +1252 is given by Hung Huan-Chhun (1), p. 41. The two latter periods have been the subject of a special study by Sekino & Takeshima (1). A newly studied temple hall dating probably from the early years of the +11th century has recently been described by Chhi Ying-Thao (1), and two J/Chin halls (+12th century) by Tu Hsien-Chou (1) and Li Chu-Chün et al. (1). In 1964 I was able to visit and photograph in detail seven temple halls of the Sung, Liao and J/Chin periods. I particularly recall two Taoist temples: the Hall of the Holy Mother (Shêng-Mu Tien [13]) at Chin Tzhu [14] near Thaiyuan, built c. +1030, with its alternating true and false cantilever rafters (see Lin Hui-Yin & Liang Ssu-Chhêng (1); Sickman & Soper (1), p. 263); and the Temple of the Mystery of Mysteries (Hsüan-Miao Kuan [15]) at Suchow in Chiangsu, built c. +1179 (see Liu Tun-Chên, 5).

[1] 鎮國寺	[2] 平遙	[3] 獨樂寺	[4] 薊縣	[5] 觀音閣
[6] 井	[7] 應縣	[8] 曹延祿	[9] 曹延恭	[10] 佛宮寺
[11] 陳家樓	[12] 山丹	[13] 聖母殿	[14] 晉祠	[15] 玄妙觀

common. Of course certain sites have a continuous history going back to Han times. At Chhêngtu in Szechuan there is a famous hall known as the Ta-Chhêng Tien.[1] About +1190 Fei Kun[2] was convinced that the building then standing dated from the end of the Han. In his *Liang Chhi Man Chih*[3] (Bridge Pool Essays) he wrote:[a]

The Hall of Great Accomplishment in Chhêngtu was built in the Chhu-Phing reign-period (+190 to +193). It is indeed great and grand. The month and year of its construction are still recorded in an inscription on the eastern side, written in *li* style by the Han people. So the hall is still standing magnificently after a thousand years. It may be compared with the Ling-Kuang palace in Lu (Shantung).[b] In the *ping-chhen* year of the Shao-Hsing reign-period (+1136) the emperor Kao Tsung, at the request of the president of the provincial academy, Fan Chung-Shu,[4] wrote with his own hand four characters 'Ta Chhêng chih Tien',[5] for this building. Later on, when Hu Shih-Chiang[6] came as Special Commissioner to Szechuan and Shensi, he visited the hall, and surveyed the beams and pillars. He decided to replace some parts which had rotted, and he added several thousand tiles, but he did not dare to undertake any modification of the ancient structure.

There is nothing impossible in this long survival, as is evident from the oldest building which we now have, but of course we do not know how skilled Fei Kun was at recognising a true Han building when he saw one. The present structure seems to be of much later date, probably not even containing any parts from the time of Fei Kun himself.

We have tended naturally to concentrate attention on the oldest and the most splendid of Chinese buildings, partly because their essential structure and the history of its development can best be brought out in this way, but it would be inexcusable to say nothing of the domestic architecture of the country, fulfilling its homely functions in a thousand beautiful forms over thirty-five degrees of latitude and sixty of longitude. Something indeed of the nature of a debt is owing from one who for years has derived intense pleasure from the buildings of one or another home, farm, inn or temple where he has stayed, in Chinese villages and towns of nearly a score of provinces. One publication is now available, that of Liu Tun-Chên (4), which can convey a measure of this delight to any reader, for it deals solely with the variations of domestic architecture in the different regions of the country.[c] We are privileged, however, to illustrate some of the styles from another source, so far unpublished, leaves from the sketch-book of a visiting English architect in China in 1954 (Figs. 794, 795, pls.).[d] Here one can see the flat roofs of mud and wheat-straw used in the north from Kansu to Hopei combined with verandah and lattice windows,[e] the stepped and shaped gables of Hunan, Chiangsi and Kweichow, horned gables of shrines to the tutelary field-gods in Hupei, and the Cantonese farmhouse with its ridge terminals, central ornament, and recessed bay entrance surmounted by decorative carving. Convex barrel roofs somewhat like

---

[a] Ch. 6, p. 1 a, tr. auct.
[b] Cf. p. 86 above.
[c] Cf. also Spencer (1); Arnaiz (1); Boyd (2); Penn (1) and others.
[d] Cf. Skinner (2).
[e] Cf. Liu Tun-Chên (4), pls. 50, 52, 93, and diagrams of the flat roof construction in fig. 2.

[1] 大成殿   [2] 費袞   [3] 梁溪漫志   [4] 范仲殳   [5] 大成之殿
[6] 胡世將

those of railway carriages occur in Liaoning,[a] and in Kansu province the principle is extended to veritable barrel-vaults of adobe brick that look like rows of Nissen huts, probably a valuable adaptation to a notoriously active earthquake region.[b] The barrel-vault is also found in the cave-dwellings excavated in the loess hillsides of northern Shensi,[c] and in the stone dwellings which the people build nearby in the image of them.[d] Particularly attractive effects are produced when a village is composed of courtyard houses in repeating units, as in the sketch made in Honan.[e] Alternate blank and recessed bays, as in the Hunan drawing, also provide a charming neighbourhood ensemble. In the south-western provinces of Szechuan and Yunnan the urban picture takes on an almost Spanish-Mexican character, in the sense that long expanses of blank white or grey wall capped only with coloured tiles alternate with highly ornamented and brightly coloured entrance gates and porticos.[f] Rural manor-houses in Szechuan stand out against the groves of bamboos with their half-timbering and white plastered walls (Fig. 796, pl.).[g] Anhui, on the other hand, can boast a series of large courtyard farm-houses with exquisite interior carvings on beams and balconies.[h] For the wealthier farmers and the scholar-gentry the courtyard system lasted all through the ages; Fig. 797 (pl.) shows two interesting sets of farmstead buildings, tomb-furniture of Ming date.[i]

The most extraordinary types of Chinese rural dwellings are those of the Hakka[1] people in Fukien province,[j] known only to few, even among Chinese, until their recent study by Liu Tun-Chên.[k] The need for security among an originally somewhat hostile indigenous population led to the development of fortified clan community 'apartment houses'. Sometimes these adopt the normal rectangular ground-plan, the place of the highest and most northerly temple hall being taken by a massive block (*thang*[2]) of three or four storeys, while long wings (*hêng*[3]), of height declining in stages, occupy the east

---

[a] Liu Tun-Chên (4), pl. 53. A tendency to rounded convexity, softening the sharp edge of the roof ridge, is also found in Ming and Chhing ornamental buildings; brought about by using two uppermost queen-posts of equal height instead of a king-post, it is called the rolled mat-shed style (*chüan phêng*[4]). Cf. Liu Chih-Phing (1), figs. 301, 302.

[b] Personal observations along the Lung-hai Railway in 1958; cf. Liu Chih-Phing (1), fig. 37.

[c] Cf. Fig. 7 in Vol 1, and Rudofsky (1), figs. 15–18.

[d] Though we should distinguish between a cave and a house, in China both the caves and the cave-shaped dwellings are called *tung*,[5] a fact which leads to some confusion among interpreters and foreign guests visiting the homes of the revolutionary leaders in and around Yenan. An interesting life-story of a peasant builder of both kinds of dwellings has been recorded by Myrdal & Kessle (1), pp. 12 ff. A *fang*[6] must thus be defined as a structure with column-beam lattice and roof.

[e] Cf. Liu Tun-Chên (4), pl. 76.   [f] Worcester (14), p. 123.

[g] Anon. (37), pl. 143; Liu Tun-Chên (4), pls. 112, 113; cf. Dye (2). The example from Nan-chhi is closely similar to that which I knew well at Lichuang near by, the war-time headquarters of the Institute of History and Philology of Academia Sinica. This was where Prof. Wang Ling and I first met, in 1943.

[h] This is the neighbourhood of Huichow,[7] which has merited a special architectural monograph, that of Chang Chung-I *et al.* (1). Cf. Liu Tun-Chên (4), pls. 30, 31, 74, 75, 104, 131.

[i] White (2); Bulling (3).

[j] Hakka people are descendants of inhabitants of Honan and other more northerly provinces who migrated to the south in times of disturbance, and settled there. The dates of the migrations are rather obscure, being variously put in the +4th, +9th, +12th and +13th centuries.

[k] (4), pp. 44, 47 ff., figs. 8–10, and pls. 105 (see also 125), 114, 115, 116, 117 (see also 126), 118, 119, and 120.

[1] 客家   [2] 堂   [3] 橫   [4] 捲棚   [5] 洞   [6] 房   [7] 徽州

and west sides, with an assembly-hall in the centre. But elsewhere the plan is circular (Fig. 798, pl.), three or four storeys of inward-facing apartments with balconies for individual families forming the periphery and looking down on a central circular courtyard around which are set guest-rooms, washing-places and yards for pigs and poultry. An assembly-hall and the ancestral chapel are arranged diametrically opposite the main entrance, while lavatories, milling and pounding sheds, also brick-built, occupy lateral positions outside the main perimeter.

It remains to add a few words about other aspects of building in the different periods. The earliest type of roof-material[a] was no doubt the fully grown hollow bamboo stem split in half longitudinally, convenient lengths being then laid in rows with the concavities alternately facing outwards and inwards.[b] This very corrugated arrangement was afterwards carried out in half-burnt grey tile (wa^1), which weathers to very attractive colours.[c] Most of the Han tomb models suggest this roofing.[d] It reached the climax of its capabilities when the tiles were made in earthenware and covered with a bright ceramic glaze; such are the orange-yellow roofs of Peking's imperial palaces, the green roofs of temples, and the deep blue roofs of the Temple of Heaven and its ancillary buildings.[e] Slates are used where locally available, and in the north and north-west there are (as we have seen) many houses with flat roofs made of a thatch of branches and reeds surfaced over with beaten mud. The Han stone tomb-shrines were roofed with slabs of stone, but the use of stone in housing for any purpose other than the plinths of columns is now found only in the Tibetan culture-area, and in a narrow zone along the Thai-hang Shan (between Shansi and Hopei).

Anciently, the flooring was nothing but the packed earth of the foundation, and earthen floors continue in use in many parts of rural China to this day. However, lime cement floors are favoured in the south, while floors of brick or stone are commoner in the north; both of these go back many centuries. In large or important buildings, whether public or private, the floors have often for many centuries been made of broad wooden planks.

Numerous travel accounts of north China have familiarised Western people with the widespread use of a simple form of central heating in domestic houses there; this is the *khang*,[2] a raised built-in divan along one side of the room, made of sun-dried bricks or often simply of tamped earth, under which a fire of any available fuel is stoked up from outside. All the family sleep on this.[f] What is not so generally known is that the

---

[a] See Eastwood (1) for a comparative study, and more generally Briggs (2) and Davey (1).

[b] Bamboo and reed matting must also have been used from high antiquity, and still frequently forms the covering for sheds and boats. We recall Vol. 2, p. 488.

[c] The tiled roofs were the first strong impression of beauty which I received on the day when I first arrived in China (the autumn of 1942), while the plane was descending over the city of Kunming.

[d] The eaves tiles ended in discoidal 'stoppers' (*wa tang*[3]) decorated with all kinds of ornamental designs.

[e] These tiles and their uses have been given a special monograph by Boerschmann (5). Cf. also Yetts (6), but now especially Liu Chih-Phing (1), pp. 132 ff. and figs. 504, 525. The glazed and coloured tiles of China inspired a rhapsodical description from the pen of a +17th-century Jesuit—Gabriel de Magalhaens (1), p. 352, Eng. ed., p. 326.

[f] The writer has had his own experiences with *khang*; though hard, they can be very comfortable in icy weather, but the straw in the mud bricks is liable to catch fire and set the whole structure smouldering, as happened in his presence on one occasion.

¹ 瓦   ² 炕   ³ 瓦當

device was common in Han times, for tomb-model houses show it (Laufer, 3). This raises the question of its possible relation to the hypocaust heating so elaborately developed by the Romans.[a] The first mention of that appears to be relatively late (+ 1st century),[b] though Vitruvius describes the essentials of it in connection with the heating of baths.[c] At present there is no way of deciding whether either civilisation influenced the other, or whether the invention was approximately contemporary and independent.

At any rate the indigenous central heating did not fail to make an impression on the Jesuits. Gabriel de Magalhaens, writing about + 1660, thus described it:[d]

This Coal[e] is brought from certain Mountains two Leagues distant from the City (Peking), and it is a wonderful thing that the Mine has never fail'd, notwithstanding that for above these four Thousand Years not only this City so large and populous, but also the greatest part of the Province, has consum'd such an incredible quantity, there not being any one Family, tho' never so poor, which has not a Stove heated with this Coal that lasts and preserves a Heat much more violent than Charcoal. These Stoves are made of Brick like a Bed or Couch three or four Hands Breadth high, and broader or narrower according to the number of the Family; Here they lie and sleep upon Matts or Carpets; and in the day time sit together, without which it would be impossible to endure the great Cold of the Climate. On the side of the Stove there is a little Oven wherein they put the Coal, of which the Flame, the Smoak and Heat spread themselves to all the sides of the Stove, through Pipes made on purpose, and have a passage forth through a little opening, and the Mouth of the Oven, in the which they bake their Victuals, heat their Wine, and prepare their Cha or The; for that they always drink their Drink hot.... The Cooks of the Grandees and Mandarins, as also the Tradesmen that deal in Fire, as Smiths, Bakers, Dyers, and the like, both Summer and Winter make use of this Coal; the Heat and Smoak of which are so violent, that several Persons have been smother'd therewith; and sometimes it happens that the Stove takes Fire, and that all that are asleep upon it are burnt to Death....

Every invention has its inconveniences.

All buildings have to be dulcified with furniture. In the field of cabinet-making China developed along unique and characteristic lines,[f] which had a powerful influence on Europe in the eighteenth century.[g] In the present book we can do no more than mention the carpentry techniques required, referring the reader to a number of useful books and papers which exist on the subject.[h] Archaeological evidence also has been studied.[i] It should not be supposed, however, that the Chinese awaited the activities

---

[a] See Vetter (1) for a comparative study, and Garrison (2). Could the Finnish *sauna* have been an intermediate common source? The question is raised by Dr E. Röttig in correspondence.

[b] Statius (b. +45), *Silvae*, i, 5, 59; see Neuburger (1), pp. 258 ff.

[c] v, 10.

[d] (1), p. 10. Cf. Trigault (tr. Gallagher, p. 311) who also mentioned coal, and in the following century van Braam-Houckgeest (1), Fr. ed. vol. 1, p. 266, Eng. ed. vol. 2, p. 65.

[e] On coal, see on, Sects. 30d, 36d.

[f] 'No wooden pins, unless absolutely necessary; no glue, where it may be avoided; no turning wheresoever—these are the three fundamental rules of the Chinese cabinet-maker' (Ecke, 6).

[g] Cf. Reichwein (1).

[h] Roche (1); Dupont (1); Kates (1); Stone (1); Ecke (2, 6); Cescinsky (1); Yetts (11); Lo Wu-Yi (1).

[i] For example, the Northern Chou (late + 6th-century) frescoes at Mai-chi Shan depict furniture, and actual specimens of Sung date have been recovered from tombs.

of Western scholars before discussing furniture in print; already in about +1090 Huang Po-Ssu[1] wrote of the special tables made in Peking (*Yen Chi Thu*[2]), and the line was continued by Ko Shan[3] with his *Thieh Chi Phu*[4] of +1617, and in our own time Chu Chhi-Chhien (3). These are only a few titles taken at random from a literature which the specialists must explore more fully. As for the various kinds of furniture in ancient times, and their special names, information will be found in Maspero (17) and Kelling & Schindler (1). It is thought that the primary unit of furniture was the low platform or dais (*tha*,[5] later *khang*[6]), which, in various shapes and sizes, served for kneeling or sleeping on, or as a table or arm-rest. Possibly the radicals *chhiang*[7] (frame, bed) and *phien*[8] (slip, strip) were originally pictographs of this piece of furniture. Examples from as early as the −4th century, from princely tombs of the Chhu State, together with lacquered tables, stools, beds and the like, have been recovered in recent years.[a]

Here it is worth pausing for a moment to consider the peregrinations of the chair. Many reflective travellers in modern times must have found it odd that while the chair was so universal a feature of all Chinese civilisation, echoing its ubiquity in ancient Egypt, Greece and Rome,[b] throughout the intervening length and breadth of Asia (and in Japan) people squatted, knelt or reclined, with or without cushions, on the floor. It seems that no comparative history of the chair as yet exists, but still the conclusion of Laufer[c] nearly half a century ago holds good, namely that the chair was not known or used in China before the Early Han, and did not come into general acceptance before the end of the Later Han. Datings are now rather later; habitual use of the camp-stool or folding chair cannot be established before the early +3rd century, and the wooden frame chair did not find wide employment till the end of the Thang (+9th century). There may have been a double evolution, the wooden dais indigenously evolving into a seat with back and arms,[d] and the folding chair with a seat of cloth or leather arriving from somewhere in Central Asia. In the time of the emperor Ling Ti (+168 to +187) this was known as the *hu chhuang*[9] (barbarian bed), and the name of a man (Ching Shih[10]), who was the first to produce it on a large scale, has come down to us.[e] The correspondence of dates has induced Ecke (2) and others to assume that the introduction of the frame chair (*i*[11]) occurred at the same time, and by the same intermediaries, as that of Buddhism, but for such a link there is little plausibility, and it would be more reasonable to suspect the journeys of Chang Chhien or some of his

---

[a] See Anon. (24); Anon. (43), pls. 67, 71B. Dr Lu Gwei-Djen and I had the privilege of examining these before publication when we visited the Honan Provincial Archaeological Institute at Chêngchow in 1958.

[b] Cf. Richter (1) on ancient Western furniture.

[c] (3), p. 235.

[d] A Han bronze showing this exists, see Stone (1), p. 4. This would have been the development which led to the transfer of the word *i*,[11] which had originally meant the *Catalpa* tree, to mean chair, as it does today. Cf. Fujita Toyohachi (1).

[e] Cf. *San Tshai Thu Hui*, Chhi section, ch. 12, p. 14a.

---

¹ 黃伯思   ² 燕几圖   ³ 戈汕   ⁴ 蝶几譜   ⁵ 榻   ⁶ 匟

⁷ 爿   ⁸ 片   ⁹ 胡床   ¹⁰ 景師   ¹¹ 椅

successors. Certain at least it is that folding stools are represented in Gandhāra art,[a] and we can safely believe that Bactria was the region which transmitted the thrones of Ra and Zeus to every Chinese farmstead. Byzantine influence may have played a part later. The question really is what similarity there could have been between Chinese and Mediterranean life which ensured the steady spread of the upright sitting custom throughout the Chinese culture-area.[b]

(7) PAGODAS, TRIUMPHAL GATEWAYS AND IMPERIAL TOMBS

The pagoda is a great feature of the Chinese landscape. One chooses the word landscape advisedly, since (as we have already noted) the half-foreign origin of the structure from Indian Buddhism generally prevented it from arising within the city walls to compete with the drum-towers and gate-towers of cosmic-imperial authority. Its ancestor, the *stūpa* or *dagoba*, an artificial hemispherical mound, had also had cosmic or microcosmic significance, since it was a model of the whole world or at least the central sacred mountain;[c] and it contained Buddhist relics at its heart,[d] whence the superimposed parasols of honour from which perhaps the storeys of the pagoda ultimately derive.[e] Its situation in or near Buddhist abbeys at some distance from the town gradually brought about a syncretistic connection with Taoist *fêng-shui* geomancy,[f] until the time came when no *hsien* city was complete without a pagoda (*tha*[1]) to harmonise the telluric influences by standing firm near by, on the most suitable isolated hill. Everyone who has lived in China has favourite pagodas of his own, and as in private duty bound, I would like to recall the beautiful towers which overlook the junction of the rivers south of the city of Mienyang in Szechuan,[g] and the eastern gate of Lanchow in Kansu. Sometimes they are found in groups of three (*chhün tha*[2]), for example the great trinity near Tali in Yunnan, or the smaller group near Chia-hsing (Kashing) on the Grand Canal. These free-standing spires were never campaniles, though bell-casting was so ancient in China—but that does not mean that they were not often ornamented with innumerable small bells (*thieh ma*;[3] somewhat like Swiss cow-bells), which still hang from the eaves and make music when moved by the wind.[h] Pagodas have as many as a dozen storeys (*chi*[4]) with or without external galleries (*wai*

---

[a] Hence, no doubt, the chairs in the Tunhuang cave-temple frescoes, e.g. no. 285 (dated +538), and nos. 196, 200, 202, all Thang.

[b] We are indebted to Dr C. P. Fitzgerald for reminding us of the interest of this example of cultural diffusion. His own book on the subject (10) appeared too late to guide us in this survey.

[c] Cf. Vol. 3, pp. 565 ff., Vol. 4, pt. 2, pp. 529 ff.

[d] This is well seen at Dedigama in Ceylon, where the excavated contents of the relic chamber (*garbha*) of the Sutighara *cetiya* (or *dagoba*) built by Parākrama Bāhu I (+1153 to +1186) are preserved in an adjacent museum. I had the pleasure of visiting this in 1958.

[e] On the development of the pagoda from the *stūpa* or *dagoba* see, e.g., Combaz (5); Bulling (2). It is interesting to recognise here the implicit conflict between Indian and Sinic cosmism.

[f] Cf. Vol. 2, pp. 359 ff. and Vol. 4, pt. 1, pp. 239 ff. above.

[g] Cf. Needham & Needham (1), p. 250.

[h] The sound of these aeolian bells on the façades of the Tunhuang cave-temples in the night, surrounded by the quiet of the desert, is a memory never to be forgotten by those who have heard it. Cf. Fig. 799 (pl.).

[1] 塔   [2] 羣塔   [3] 鐵馬   [4] 級

lang[1]) and are sometimes square, sometimes polygonal, rarely circular; they may be of wood, more often of brick, rarely of stone.[a] They became essentially superimposed chapels and were never intended as dwellings, even for monks. A particular type, the Thien-Ning[2] style, so called from a famous monastery near Peking,[b] has a more or less unbroken tower from ground level to about a third of its height, the galleries and storeys being repeated only above that level.

Somewhat of this kind is the oldest extant pagoda in China, that at Sung-Yo Ssu[3] on the sacred mountain of Sung Shan in Honan. It is of brick (*chuan*[4]) and was constructed in +523 (Fig. 800, pl.) under the Northern Wei.[c] There are fifteen storeys and the structure is twelve-sided. The arrangement exemplifies the statement in Thao Chhien's[5] biography in the *Hou Han Shu*, that for pagodas (*fou-thu*[6,7]),[d] one builds below a double (Chinese) tower (*chung lou*[8]) and above it one piles (Indian) shrines (*chin phan*[9]).[e] But the beautiful spire on Sung Shan represents already a very sophisticated stage of development, and we must assume that the earlier towers were much simpler. The evidence is that they were square in plan, with successive repetitions of the basic storey unit, each with its own roof, the whole diminishing regularly with height, and being crowned with mast and discs. In the +4th century there seem usually to have been only three storeys, but the Northern Wei emperor, Thopa Hung,[10] erected a famous one of seven at Ta-thung in +467. This was probably something like the archaic six-storeyed Satmahal-prāsādaya still to be seen at Polonnaruwa in Ceylon (Fig. 801, pl.). Rock-cut pagodas of similar pattern, though miniature in size, also exist in the Yün-kang cave-temples (*c.* +500).[f] A very magnificent tower of nine storeys was built of wood at Loyang in +513, but it was destroyed by fire in +534.[g] Many of the brick pagodas reproduced in their own medium as ornamentation the *tou kung* brackets which would have been rendering service in a wooden building,[h] for example the tower at Hsing-Chiao Ssu south of Sian, built in +669 and repaired in +838.[i]

[a] Classical works on pagodas are those of Boerschmann (4) and Tokiwa & Sekino (1). Cf. W. C. Milne (1), a paper now a century old, but still interesting. Add Alley (5).

[b] Thien-Ning Ssu[2] near Peking, with its +11th- or early +12th-century pagoda, built under the Liao dynasty (Sickman & Soper (1), pp. 272 ff., pl. 173A; Anon. (37), pl. 63).

[c] Sirén (1), pl. 105; Sickman & Soper (1), pl. 158A and p. 230; Anon. (26), p. 449; Anon. (37), pls. 16, 17; Bulling (2), fig. 40/32.

[d] Such was the old term; according to Chu Chün-Shêng's commentary on the *Shuo Wên* (*Shuo Wên Thung Hsün Ting Shêng*) the word *tha* is first found in an inscription of +536. Presumably *fou-thu* transliterated *buddha*.

[e] Ch. 103, p. 13*b*.

[f] Anon. (37), pl. 15; Sickman & Soper (1), pl. 157A.

[g] *Lo-yang Chhieh-Lan Chi*, ch. 1 (pp. 10 ff.), tr. Sickman & Soper (1), p. 229; the dates are due to Sirén (5).

[h] The writer of the *Ju Shu Chi* in +1170 noticed this and remarked on it with admiration (ch. 4, pp. 1*b*, 4*a*).

[i] Bulling (2), figs. 46/39, 46/40; Sickman & Soper (1), pl. 162B and p. 242. This was built as a memorial to Hsüan-Chuang. Another tower which shows this well is the Iron-Coloured Pagoda at Khaifêng built of glazed brick in +1041 (see Lung Fei-Liao (2); Sickman & Soper (1), pl. 166A; Anon. (37), pls. 52, 53). An example of a different kind is the Wu-Liang Tien temple, with its brick barrel vaults, at Suchow (see Sirén (1), vol. 4, p. 37).

[1] 外廊   [2] 天寧寺   [3] 嵩嶽寺   [4] 磚   [5] 陶謙   [6] 浮屠
[7] 浮圖   [8] 重樓   [9] 金盤   [10] 拓跋弘

From these early Thang centuries many beautiful monuments remain, such as the Ta-Yen Tha[1] (Great Pagoda of the Wild Geese) at the Tzhu-Ên Ssu[2] (Loving-Kindness Temple) at Sian. This was Hsüan-Chuang's headquarters, and it was built in +652 (he had returned from India in +645), then repaired in +704. A square, rather squat, brick structure, of seven storeys, it is still to be seen south of the present city of Sian (Fig. 802, pl.);[a] as also is the Hsiao-Yen Tha,[3] slenderer and taller, with thirteen storeys (+708).[b] The most natural development from the square plan was of course the octagonal one, and the majority of later pagodas are of this design. As it would take us too far to expatiate on their styles and beauties, we shall conclude with a reminder only that the south had also its great tradition of tower-building, as is seen for example by the twin pagodas of Zayton (Chhüan-chou) in Fukien, executed in stone with brick cores (+1150 to +1250), elaborate corbel brackets imitative of woodwork in the so-called 'Indian style', and markedly upward-curving eave-corners.[c]

Entirely true to its principle of building from repeatable single units or modules (cf. p. 67), Chinese culture contains many small square one-storey buildings which represent, so to say, the pagoda's base or storey in isolation, the single cell apart from the body as a whole.[d] As we shall shortly see more fully in connection with bridges (p. 167), the arch, in the shape of the barrel vault, was known and used in China probably in the Chou and certainly in the Han. But the true arch with keystone was not greatly employed in the building of pagodas; the commoner construction was the corbelled vault. These can be studied particularly well in some of the small one-storey shrine buildings, of which the most famous example is the Ssu-Mên Tha[4] at Shen-Thung Ssu[5] in Shantung,[e] dating from +544. This has a corbelled vault sustained by a central pillar. Many structures of this kind are known, with or without central pillars, but always in use as shrines.[f] On the desert across from Chhien-fo-tung one comes upon just such a square one-room chapel, sheltering the image of a Thang abbot, who gazes for ever into the west towards the cliff face with its caves. Somewhat analogous is a small building on a forested hill behind (i.e. to the south of) the great Taoist abbey of Lou-Kuan Thai[6] near Chou-chih south-west of Sian.[g] It is known as the 'Alchemy Tower' (Lien-Tan Lou[7]), and consists of a single brickwork chamber entered by one

---

[a] Bulling (2), fig. 44/37; Sickman & Soper (1), pl. 162A and p. 241; Anon. (37), pl. 29; Forman & Forman (1), p. 229.

[b] Bulling (2), fig. 42/35; Anon. (37), pl. 31. These square tapering structures occur over the whole of the Chinese culture-area, e.g. in Korea, where a notable specimen in moulded brick exists at the Punhoang[8] temple, built in the Silla Kingdom in +634 under Queen Sŏndŏk Yŏwang[9] (cf. Vol. 3, p. 297). Originally in nine storeys it now has but three. Cf. Bulling (2), fig. 56/46 and Li Chhêng-Fan (1).

[c] Their description afforded matter for a classical monograph by Ecke & Demiéville (1). Cf. also Mirams (1), pp. 82 ff.; Sickman & Soper (1), pp. 260, 267; Anon. (37), pl. 66. There is now also a monograph on the pagodas of Chiangsu (Anon. 47).

[d] Ecke (7) has a special paper devoted to these.

[e] Boerschmann (4), p. 366; Sirén (5); Anon. (37), pl. 20; Forman & Forman (1), pp. 93, 95; Liu Chih-Phing (1), figs. 137, 138; Sickman & Soper (1), pl. 157B.

[f] Cf. Liu Chih-Phing (1), figs. 139–41; Sickman & Soper (1), p. 240.

[g] Together with Dr Dorothy Needham, Dr Tshao Thien-Chhin and Dr Chhiu Chhiung-Yün, I had the pleasure of inspecting and sketching it in 1945.

[1] 大雁塔   [2] 慈恩寺   [3] 小雁塔   [4] 四門塔   [5] 神通寺
[6] 樓觀台   [7] 煉丹樓   [8] 芬皇寺   [9] 善德女王

door and roofed by a corbelled vault, the layers of bricks rising from the angles to form a series of squinches. An octagon is formed where they meet, then a circular space, and finally a square.[a] The brick was a burnt red brick unlike anything now commonly seen in the neighbourhood, and was arranged in billet mouldings outside

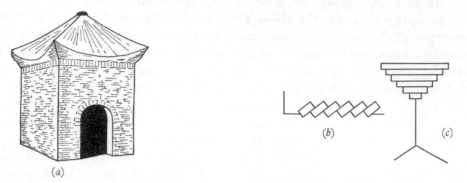

(a)   *(b)*   *(c)*

Fig. 803. The 'Alchemy Tower' (Lien-Tan Lou) at the Taoist abbey of Lou-Kuan Thai near Chou-chih south-west of Sian, Shensi.

   a General view
   b Exterior brick billet moulding (in plan)
   c Interior corbel vaulting of squinch type supporting an octagon, then a circle and finally a small
      square.

Date probably Wu Tai or early Sung.

strongly reminiscent of those of the +7th-century Hsing-Chiao Ssu pagoda. This little pavilion, so interesting for its proto-scientific associations, may also be of the Thang period, but as its eaves curl up at the corners it should perhaps rather be dated Wu Tai or early Sung. We do not doubt that in those days it was the scene of alchemical experiments.

It is not to be thought that the ancient Indian *stūpa* was completely absorbed into the Chinese pagoda and wrapt up aloft into the seventh heaven. In a thousand different shapes, it continued on the ground, to be used throughout the length and breadth of China, primarily for tombs and pious aedicules. In this it continued one of its earliest functions. Thus the desert in the neighbourhood of Chhien-fo-tung is strewn with exquisite shapes (Figs. 804, 805, pls.) commemorating monks of the Sung or Yuan, the coloured plasterwork scoured and polished with the force of a thousand autumn sandstorms. By contrast both of place and profile other stone lotuses sleep like magic mushrooms in the sunshine of a Shantung glade (Fig. 806, pl.). Beside the Old Silk Road we see other forms, and among the barren hills of the San-wei Shan a stalwart Thang tomb of considerable size shows that many structures can be almost intermediate between the *stūpa* and the pagoda.

As for the technical principles of pagoda building, depending as it did upon frameworks of wood and the bonding of bricks or masonry, they were really only the exten-

---

   [a] This kind of geometrical coffering is very old in Chinese architecture, for one may see it done in stone in the ceilings of the I-Nan tombs in Shantung, *c.* +193 (Tsêng Chao-Yü *et al.* (*1*), pl. 18, fig. 4; Watson (*1*), pl. 114).

sion of the techniques of all building to a particular specialised field. It may be guessed, however, that the simple truss forms represented by the sloping struts (*hsieh chu, chha shou,* cf. p. 100 above) proved especially useful in high wooden towers. There is a significant story about Yü Hao[1] in this connection. As the reader will remember from p. 81, Yü Hao was the Master-Builder who constructed the Khai-Pao Pagoda in Khaifêng[a] in +989, as well as other famous buildings, and the author of the *Mu Ching* (Timberwork Manual). In the *Mêng Chhi Pi Than* Shen Kua has an entertaining story about the advice he gave to another artisan-architect some ten years later.[b]

When Mr Chhien (Wei-Yen[2]) was Governor of the two Chekiang provinces, he authorised the building of a wooden pagoda at the Fan-Thien Ssu[3] (Brahma-Heaven Temple) in Hangchow with a design of twice three storeys. While it was under construction General Chhien went up to the top and was worried because it swayed a little. But the Master-Builder explained that as the tiles had not yet been put on, the upper part was still rather light, hence the effect. So then they put on all the tiles, but the sway continued as before. Being at a loss what to do, he privately sent his wife to see the wife of Yü Hao with a present of golden hairpins, and enquire about the cause of the motion. (Yü) Hao laughed and said: 'That's easy, just fit in struts (*pan*[4]) to settle the work, fixed with (iron) nails, and it will not move any more.' The Master-Builder followed his advice, and the tower stood quite firm. This is because the nailed struts filled in and bound together (all the members) up and down so that the six planes (above and below, front and back, left and right) were mutually linked like the cage of the thorax. Although people might walk on the struts, the six planes grasped and supported each other, so naturally there could be no more motion. Everybody acknowledged the expertise thus shown.

Surely we have to deal here with slanting struts inserted in an otherwise purely rectangular reticulum—diagonal wind bracing.

A remarkable department of pagoda building was that which made them of cast iron, or more often of bronze.[c] These masterpieces have aroused the astonishment and admiration of foreign travellers in China from Ennin[d] in the +9th century to Bernard in the 19th.[e] The oldest existing iron pagoda,[f] at Yü-Chhüan Ssu[5] (Jade Springs Temple) at Tangyang[6] in Hupei, which dates from +1061, is of a very considerable

---

[a] This tower, one of the greatest glories of the city, was a slender octagonal eleven-storey structure in wood, some 360 ft. in height, and probably with curved-up eave-corners at each level, then a relatively new thing. It was burnt down in +1037 but soon rebuilt, though with only nine storeys, as we know from the enthusiastic description of the Japanese monk Jōjin, who climbed it in +1072. How long it lasted after that we do not know. Cf. Sickman & Soper (1), pp. 256, 274.

[b] *MCPT*, ch. 18, para. 15, tr. auct. Cf. Hu Tao-Ching (1), vol. 2, p. 613, (2), p. 186.

[c] Many details in Boerschmann (4), pp. 336 ff.; cf. Needham (32), pls. 34, 35. They were not of course all of one piece, but were constructed of interlocking plates of metal.

[d] This Japanese monk, so often quoted by us, visited Wu-thai Shan in +840 and mentions eight iron pagodas or rather *stūpas* on the terraces there, all set up by the empress Wu Tsê-Thien about +695. Another such monument, seen by him near Laichow in Shantung, had been erected as a votive offering by a naval commander, Wang Hsing-Tsê,[7] in +665; it was 10 ft. high and had seven storeys. See Reischauer (2), pp. 190, 237, 240, 243, 245, (3), p. 205.

[e] W. D. Bernard (1), vol. 2, p. 431.

[f] Unless, that is, the cast-iron pagoda at Ningpo, dating from +961, is still standing.

[1] 喻皓　　[2] 錢惟演　　[3] 梵天寺　　[4] 板　　[5] 玉泉寺
[6] 當陽　　[7] 王行則

size, being 70 ft. high and having thirteen storeys.[a] Its weight is some 53 tons.[b] Another smaller one (of nine storeys) is at Kan-Lu Ssu[1] (Sweet Dew Temple) at Chen-chiang[2] in Chiangsu.[c] Local tradition dates it from the time of Li Tê-Yü,[3] the geographer and minister of State (+787 to +849),[d] who founded the temple, but more probably it was the work of Phei Chhü[4] (+1078 to +1086).[e] In other cases a masonry core may be clothed with cast-iron plates, as at Pei-tu-tshun north-west of Sian, where a Ming (+15th-century) pagoda of this kind rises to a height of 74 ft. in nine storeys.[f] Smaller ones wholly in bronze are quite numerous.[g]

Another gift from Indian to Chinese architecture was the triumphal gateway, or *phai-lou*,[5] a free-standing gate of wood or stone, with superimposed beam lintels, erected for commemorative or triumphal purposes on an approach to a tomb, temple or palace,[h] or even across any road or village path. Its name implies that it was to bear aloft a notice, often an epigram. The traveller on the stone pathways of Szechuan comes upon relatively simple ones from time to time, proclaiming the name of a virtuous widow or a popular magistrate. Greater occasions call for three, five or seven arches in a row (Fig. 807, pl.). Lecomte, in the seventeenth century, wrote:[i]

> The town (Ningpo) is still full of Monuments called by the Chinese Paifam (*phai-fang*[6]) or Pailou, and by us Triumphal Arches, which are very frequent in China.[j]
>
> They consist in three great Arches abreast, built with long Marble Stones. That in the middle is much higher than the other two. The four Columns which support them are sometimes round, but oftner square, made of one only Stone placed on an irregular Basis. In some this Basis is not to be seen, whether they never had any, or that thro' Age it was sunk into the Ground. They have no Capitals, but the Trunk is fastned into the Architrave, if we will give

---

[a] This temple was founded in the +6th century by Prince Kuang of Chin[7] for the celebrated monk Chih-I,[8] founder of the Thien Thai school of Buddhism (cf. Vol. 2, p. 407). The names of 115 monks and 57 novices, spiritual descendants of this divine, are cast in the metal panels of the pagoda.

[b] Photograph in Boerschmann (4), reproduced in Needham (32), fig. 34.

[c] This, known to foreigners as Gutzlaff's Pagoda, was the one that impressed Bernard. It had been rebuilt in +1583, after a failure of the core, by two monks, Hsing-Chhêng[9] and Kung-Chhi.[10] During the Opium Wars the British thought of taking it away, but refrained; however, it fell again in 1868 and now only two storeys are left.

[d] Cf. Vol. 3, p. 544; Reischauer (3), pp. 212 ff.

[e] Yet another, at Chhung-Hsüeh Ssu[11] (Reverence-for-Learning Temple) at Chi-ning in Shantung, has nine storeys. This dates from +1105, when its construction was either financed or directed by a woman, Mrs Chhang,[12] the wife of Hsü Yung-An.[13] Two additional storeys were added about +1582, raising the height to 74 ft.; at which it still stands. This was the occasion when Wang Tzu,[14] a local official, wrote an account of the repairs and embellishment, still contained in the local gazetteer, and comparable with older commemorative writings of technological interest which we discuss elsewhere (pp. 173, 203).

[f] Photograph in Boerschmann (4), reproduced in Needham (32), pl. 35.

[g] Certain places also have chapels built entirely of cast-bronze components. One of these (often visited by the writer) is at a Taoist temple near Kunming in Yunnan, another in the Peking summer Palace (I Ho Yuan); see the beautiful photograph in J. Thomson (1); cf. Geil (2), p. 152.

[h] The prototype is always considered to be the stone gateway (*toraṇa*), four of which surround the −1st-century *dagoba* or *tope* at Sānchī, in India, facing the quarters of the world. Cf. Combaz (1); Bruhl & Lévi (1), pl. 11; Kramrisch (1), pls. 10, 22, 24.

[i] (1), p. 88.   [j] They were in fact never arches in the strict sense.

[1] 甘露寺	[2] 鎮江	[3] 李德裕	[4] 裴璩	[5] 牌樓
[6] 牌坊	[7] 晉廣王	[8] 智顗	[9] 性成	[10] 功琪
[11] 崇學寺	[12] 常氏	[13] 徐永安	[14] 王梓	

that Name to some Figures over the Pillars. The Frize is better distinguished, but too high in proportion to the rest; they adorn it with Inscriptions, Figures and Embossed Sculptures of a wonderful beauty, with Knots wrought loose one within another, with Flowers curiously carved, and Birds flying as it were from the Stone, which in my Mind are Master-pieces.

I myself had no less respect for these dignified structures when I first came upon them, though in the city of Kunming they are of painted wood. The roofs which they generally carry are constructed with *tou kung* and cross-beams in exactly the same way as those of houses, and the whole set of gates, if of wood, is often stayed by sloping struts on each side (Fig. 808, pl.). An excellent monograph has been devoted to the *phai-lou* by Chu Chhi-Chhien (2).[a]

Early in this section reference was made to the imperial tombs as one of the great forms of Chinese architectural achievement. If we close it on the same subject, this is not because of any particular regard for the imperial system as such, but because the whole pattern which they constitute is perhaps the greatest example of the co-option of wide tracts of landscape as part of an architectonic whole.[b] The Chinese of today appreciate the achievement more than ever, conscious as they are that the tomb-temples are at least as much a monument to the architects and building workers who made and designed them, as they are to the emperors whose lot it was to order them and to be buried in them. From the Han to the Sui dynasties there remain nothing but the tumuli (not yet excavated), and from the Thang and later there are only the tumuli and some incomplete lines of battered statues (Fig. 809, pl.).[c] At a number of places, such as Shenyang (Mukden) in Manchuria, there are tomb-temples of the early Chhing emperors which are still in perfect preservation (Fig. 742, pl.). But the greatest masterpiece is certainly that complex of Ming tombs in the mountains north of Peking known as the 'Thirteen Tombs' (Shih-San Ling[1]).[d] These are disposed over a wide valley in the hills, each battlemented tumulus being located usually on the slope of a salient spur between two side-valleys, and fronted by a complex of halls and temples in a vast compound peopled with trees. The pilgrim advances along the road from the capital, meeting first with a splendid *phai-lou* of five arches (+1540), and then a solid triumphal gate-tower with three barrel-vaulted tunnels (Ta-Hung Mên[2]). From thence he can descry in the distance misty mountains surrounding the mouth of a wide valley. Next he reaches a colossal pavilion (Pei Thing[3]) open to the weather on each side through four great arches, and containing the largest inscribed stele in China, poised on

a Cf. also Volpert (1).

b The western environs of Hangchow, covering an area some four or five miles square, also provide a superb example of this. The West Lake region, with its causeways, islands, temples and pagodas, is surrounded on three sides by hills or mountains.

c See Combaz (2); Chhen Chung-Chhih (1).

d Descriptions by Bouillard (1); Bouillard & Vaudescal (1); Grantham (1). Cf. Favier (1), p. 310; Fabre (1), pp. 225 ff.; Arlington & Lewisohn (1), pp. 317 ff. At the time of the restoration of the Chhang Ling,[4] the tomb of the Yung-Lo emperor, in 1935, an excellent architectural account of it with many scale drawings was published in Peking (Anon. 3). Of Ming emperors there were sixteen, but the first is buried at Nanking, the resting-place of the second is not known, and the seventh was not interred in the valley of the tombs as his reign was considered a regency.

1 十三陵　　　 2 大紅門　　　 3 碑享　　　 4 長陵

a stone tortoise (+ 1420). This building is guarded by four ceremonial columns carrying stylised clouds (Thien Chu Hua Piao [1]). Then, as the processional path through the
fields of grain curves slowly to the right, a long range of stone statues on each side is
encountered, camels and elephants, horses and mythological animals, civil and military
officials (Fig. 810, pl.); this ends in another *phai-lou* (the Lung-Fêng Mên [2]). The
traveller now crosses two bridges (or rather, he used to do so, for they are today
partially washed away)[a] and begins to be able to make out the majestic roofs of the
temples, backed by their tumuli, on each side of the valley reaching back as far as the
eye can see. Along a serpentine way paved with great flat stones he reaches at last the
tomb of the Yung-Lo emperor (+ 1424), the greatest of the family, with its encircling
wall and gatehouses (the Chhang Ling). To the right, in the first courtyard, as he
enters, he passes a pavilion housing the steles which record the duty enjoined upon the
local city magistrate by the first Chhing emperor to maintain in perpetuity these
monuments of a conquered dynasty (Figs. 811, 812, pls.). Before him he has the main
ancestral hall (Ling-Ên Tien [3]), today empty, but still sustained by its twenty-four
giant cedar columns, each 12 ft. in circumference and 60 ft. high.[b] Penetrating through
this, he gains still further courts, and passing the altar in the sacred grove comes at last
to the Spirit Tower (Ling Thai [4]) carrying a great pavilion (Ming Lou [5]) which shelters
another stele (Fig. 813, pl.). It will take him a good half-hour to make the round of the
walls (Pao Chhêng [6]) which enclose the tumulus. And from the tower itself he will
enjoy a magnificent view of the whole valley, and meditate upon the sublime sense of
organic plan which conceived the whole pattern of landscape and buildings,[c] as well as
the manifestation of the genius of a people in the skill of its architects and builders.

---

[a] I leave this description just as I wrote it in 1952, after an unforgettable visit in the company of
Mr Rewi Alley. The great tomb-temples were then in poor repair again after the years of the Second
World War, and at the Chhang Ling we had to push through tangled thickets and long grass among
the halls and altars. Though in such conditions there was romantic beauty, the situation completely
changed in the ensuing decade, and when I came there again eight years later with Prof. Yeh Chhi-Sun
and Dr Lu Gwei-Djen as well as Mr Alley, all buildings, courts and gardens were magnificently
restored and kept up. One could take tea outside the main gate, and buses brought out the people of
Peking to enjoy their possession just as those of Granada are accustomed to enjoy the Alhambra and the
Generalife. The Ming Tombs have also changed in two other perhaps more fundamental ways (cf.
Needham, 46). Systematic excavation has been begun, and the first tomb to be opened, that of the Wan-
Li emperor (the Ting Ling [7]), has yielded a mass of extraordinary treasures; cf. Anon. (*12*); Hsia Nai (4).
We were privileged to view this before it was opened to the public. In the three large halls built some
60 ft. below the surface of the mound behind the range of temple halls on the surface, one finds the
throne of the emperor, his empress and his principal consort, together with the dais on which their
coffins rested. Particularly striking are the doors of white jade about 12 ft. high, kept in place by colossal
bronze lintels about the thickness of a man's body. Secondly the valley floor, where we had to dig out
our vehicle from the mud in 1952, is now covered with many feet of water, for in 1958 a dam was built
across its mouth, largely by the voluntary labour of the people of Peking (Fig. 814, pl.). Although this is
primarily for power and the irrigation of the dusty North China plain, the reservoir has much increased
the beauty of the site as a whole, since it mirrors in its waters ten or twelve of the tomb-temples at the
foot of their mountain background. On the epic of its construction see Fig. 877 (pl.) and Yen Yao-Ching
et al. (*1*); Tan Ai-Ching (*1*); Chao Yung-Shen (*1*).

[b] Perhaps these were a product of the voyages of the admiral Chêng Ho (cf. pp. 487 ff.).

[c] All honour to the geomancers Master Wang Hsien [8] of Shantung, and Master Liao Chhiung-Ching [9]
of Chiangsi, who appear to have directed the builders in their auspicious work.

| [1] 天柱華表 | [2] 龍鳳門 | [3] 陵恩殿 | [4] 靈臺 | [5] 明樓 |
| [6] 寶城 | [7] 定陵 | [8] 王顯 | [9] 廖瓊靜 | |

PLATE CCLXXXII

Fig. 713. A trestle gallery road (*chan tao*) through the Chhin-ling Mountains (Anon. *26*).
See pp. 20 ff.

PLATE CCLXXXIII

Fig. 714. The old road at Ku-shan near Fuchow in Fukien.

Fig. 715. A typical rural path, at Ko-lo-shan near Chungking in Szechuan (orig. photo., 1943).

PLATE CCLXXXIV

Fig. 716. Old road at Pa-ta-chhu near Peking (orig. photo., 1958).

Fig. 717. The way into the hills from the Yangtze at Li-chuang in southern Szechuan (orig. photo., 1943).

PLATE CCLXXXV

Fig. 718. A military guard at a post-station in +1793 (from Staunton, 1). See pp. 33 ff.

PLATE CCLXXXVI

Fig. 720. Tamped earth (*terre pisé*) walling under construction; poles used as shuttering, near Sian (orig. photo., 1964).

Fig. 721. Brickwork facing and tamped earth core of a Han watch-tower on the ancient *limes*; at Thien-shui-ching (Sweetwater Well) beside the desert road between Anhsi and Chhien-fo-tung, in the far north-west of Kansu (orig. photo., 1958).

Fig. 722. Contemporary adobe brick walling for a building under construction in the oasis at Chhien-fo-tung (orig. photo., 1958). Upright layer bonding on a foundation of baked brick and stones.

PLATE CCLXXXVII

Fig. 723. Box bond brickwork in the walls of the old Lo family temple beside the
Chungking–Chhêngtu road, Szechuan (orig. photo., 1943).

Fig. 725. Hollow stamped and fired brick from a Han tomb-chamber (British Museum, after Davey, 1).
Two human figures stand, *inter alia*, underneath porticoes with two-tiered roofs.

PLATE CCLXXXVIII

Fig. 727. The walls of Sian in 1938 (Bishop, 6). In the foreground, one of the ramps giving access to the ramparts. As in many other cities, shrinkage of population had given place for much agricultural land within the walls.

PLATE CCLXXXIX

Fig. 728. The style of Chinese defensive architecture as seen in a Thang fresco painting (cave no. 268 at Chhien-fo-tung; T 217, P 70). Its general position in the cave can be seen in Vincent (1), pl. 29. One notices the battered, i.e. inward sloping, wall surfaces, and the galleried barracks pavilions surmounting corner towers and gate-towers. The dresses and uniforms of the period (c. +660) are also noteworthy. There has been some doubt about the subject of the picture. Waley (19), p. 124, pl. 17, thought it represented the fight between the armies of Kusinagara and Magadha for the possession of the Buddha's relics; but later, noticing that one side have lances only and the other side only spears, he proposed (in Gray & Vincent (1), pp. 13, 58, col. pl. 44) that it shows the young Śākyas engaging in military exercises outside the city of Kapilavastu, superintended by the Buddha when still the young prince Śākyamuni. Photo: Lo Chi-Mei, 1943.

PLATE CCXC

Fig. 729. Part of the ramparts of the Ming fort at Chia-yü-kuan, guarding the western end of the
Great Wall (orig. photo., 1943). Cf. Fig. 14 in Vol. 1.

Fig. 730. The fort at Chia-yü-kuan; an inclined ramp leading to one of the barracks pavilions on
the walls (orig. photo., 1943).

PLATE CCXCI

Fig. 731. A typical wall of a small city; part of the fortifications of Chiating (Loshan) in western Szechuan (orig. photo., 1943).

Fig. 732. The Great Wall near the Nan-khou pass, north of Peking (photograph of 1962, Chin Shou-Shen, 1). Note one of the slotted access staircases in the right-hand lower corner. Cf. Fig. 13 in Vol. 1. See pp. 46 ff.

PLATE CCXCIII

Fig. 733. The Great Wall further west, near Lien-hua-chhih, in north-western Shensi, on the border of the old province of Ninghsia. Range after range the mountains descend to the Ordos desert, and along the ridge of one of them can be seen the wall marked prominently by its 'mile-castles' (photograph of 1909, Geil, 3).

Fig. 734. The Great Wall protecting the Old Silk Road in the Kansu panhandle (cf. Vol. 1, p. 59); now only a ridge of compacted loess with many breaks (orig. photo., 1943).

PLATE CCXCIV

Fig. 735. Towers of the 'First Frontier Wall' (the Great Wall) along the boundary between northern Shensi and Inner Mongolia, seen from near Chhang-lo-pao village in the neighbourhood of the city of Yülin (photograph of 1920, Clapp, 1). Cf. Fig. 15 (in Vol. 1) and Fig. 721. The Wall has here been almost buried by the sands of the Ordos desert.

Fig. 736. A check-point in the Inner Wall, the Buddhist gate of Chü-yung-kuan in the Nan-khou pass (orig. photo., 1964). See p. 48.

PLATE CCXCV

Fig. 739. The Summer Palace (I Ho Yuan) at Peking (Anon. (*37*), pl. 123). In its present form it dates from 1888, but much of the palace-temple-park complex formed part of the 'Garden of Clear Ripples' (Chhing I Yuan) built by the Chhien-Lung emperor in +1751 and extended in +1761.

PLATE CCXCVI

Fig. 740. A typical farmhouse; near Shaoshan in Hunan (orig. photo., 1964).

PLATE CCXCVII

Fig. 742. Entrance to the tomb-temple of the first Chhing emperor at Shenyang (Mukden) in Liaoning (orig. photo., 1952).

Fig. 743. Central octagonal pavilion, sited on the main axis, of the 'Blue Goat Temple' (Chhing Yang Kung) in Chhêngtu, Szechuan (orig. photo., 1943). It has been suggested that this Taoist symbol derives from the Paschal Lamb of the Nestorians (cf. Vol. 2, p. 160).

PLATE CCXCVIII

Fig. 746. Corner tower on the wall of the Imperial Palace in Peking (roof-tiles yellow, sides a faded murrey colour, the wall itself grey).

PLATE CCXCIX

Fig. 748. The skyline of Peking; looking north-west towards the yellow-tiled roofs of the halls of the Imperial Palace (orig. photo., 1952).

Fig. 749. The inner garden of a Taoist temple (Boerschmann (2), pl. 1, (3a), pl. 105); the votive temple of Chang Liang (Huang Shih Kung, cf. Vol. 2, p. 155) at Miao-thai-tzu, on the road through the Chhinling Mountains, in northern Shensi. See p. 75.

PLATE CCC

Fig. 750. A private garden in Suchow, now open to the public (orig. photo., 1964). The Liu Yuan, founded by Hsü Shih-Thai in +1522; a corner of the artificial lake.

PLATE CCCI

Fig. 751. Air view of the axial plan of the capital city of Peking (from Gutkind (1), pl. 60, photograph of *c.* 1925). We are looking south from a point just north of the Coal Hill gardens, with the old buildings of Peking University to the left. The Forbidden City (the Imperial Palace) is sharply demarcated by its broad moat; beyond it can be seen on the left the Imperial Ancestral Temple and on the right the Temple of the Land and Grain. Far in the distance, at the top of the picture, one can make out on the left the wooded *temenos* of the Temple of Heaven, and on the right that of the Temple of Agriculture. The outer and inner city walls, between which these lie, cross the picture almost horizontally. Cf. Fig. 752 on p. 78.

PLATE CCCII

Fig. 753. The five marble bridges across the Stream of Aureate Water (no. 9 in Fig. 752)
between the Meridian Gate (no. 8) and the Gate of Supreme Harmony (no. 10).

Fig. 754. The Meridian Gate, or Noon Gate, of the Purple Forbidden City
(photo. Vergassov, in Mirams (1), about 1936).

PLATE CCCIII

Fig. 755. Air view of the *temenos* of the Altar and Temple of Heaven, from the south (Anon. (*37*), pl. 104). Cf. Fig. 752 on p. 78. In the foreground the Orbed Concentric Platforms of the Altar, then the Hall of the Infinite Canopy of Heaven (the smaller round building), and at the northern end of the causeway, surmounting platforms round and square, the Hall of Prayer for the Year. To the right at the top, the complex of buildings for the fasting and preparation of the imperial celebrant.

Fig. 756. Image of Kungshu Phan (Lu Pan), patron saint of artisans and architects (cf. Vol. 4, pt. 2, p. 44 and Fig. 354), one of those in the temple at Mai-chi Shan in southern Kansu (orig. photo., 1958.)

PLATE CCCIV

Fig. 758. One of the Buddhist paradises depicted in the frescoes of the Tunhuang (Chhien-fo-tung) cave-temples (cf. the discussion of Waley (19), pp. 126 ff.). A Thang representation, painted about +700, of the Western Heaven of Amida Buddha (cave no. 172; from Anon. (10), pl. 37). Most of these pictures are full of architectural detail worth careful study, and generally contain more than one system of perspective, as in this scene, where one can find axonometric as well as optical drawing, and pick out at least five separate vanishing-points. See the discussion on perspective, pp. 112 ff.

PLATE CCCV

Fig. 761. One of the simplest forms of the corbel bracket system, part of the woodwork at the +8th-century temple of Nan-Chhan Ssu in the foothills of Wu-thai Shan, Shansi (orig. photo., 1964). Cf. p. 95 and Figs. 760, 762, 773, 774.

Fig. 764. Woodwork at Fo-Kuang Ssu, another Buddhist temple in the foothills of Wu-thai Shan, Shansi. Built in +857. Roof detail showing the lever arms or cantilever principal rafters (*ang*) piercing the corbel bracket clusters (orig. photo., 1964). Cf. pp. 95, 100, and Figs. 760, 765.

PLATE CCCVI

Fig. 766. The theatre-temple for New Year and other plays at the Ming fort of Chia-yü-kuan at the western end of the Great Wall (cf. Figs. 729, 730 and Fig. 14 in Vol. 1). The hipped gable roof curves gracefully up at all corners (orig. photo., 1943).

Fig. 767. Typical Chinese roof assembly under construction. Cross-beams (*liang*), posts (*cho*) and purlins (*hêng*) for a small gate-house, in a glade of the oasis at Chhien-fo-tung (orig. photo., 1958).

PLATE CCCVII

Fig. 768. A larger construction, the skeleton framework of a new commune headquarters at Ho-thang near Lo-phing in north-eastern Chiangsi, showing columns, tie-beams, cross-beams, and posts for the purlins (orig. photo., 1964).

PLATE CCCVIII

Fig. 769. Weight-bearing inverted V-braces (*jen tzu kung, chha shou*) at Fo-Kuang Ssu, a Thang building of +857 (photo. Liang Ssu-Chhêng, 1). See p. 100.

Fig. 770. Workers' houses under construction at Lanchow, Kansu, in 1943. Although at first sight they have a Western look, the walls are not weight-bearing, and the whole timber framework follows the traditional style (orig. photo.).

PLATE CCCIX

Fig. 771. Miniature buildings as interior decoration; the flying bridge pavilion of the Sūtra Repository at the Lower Hua-Yen Ssu temple at Ta-thung in Shansi (orig. photo., 1964). The great richness of the corbel bracket clusters in this model, made in +1038, is noteworthy, as also the presence of two rows of *ang* in its roof. See p. 106.

PLATE CCCX

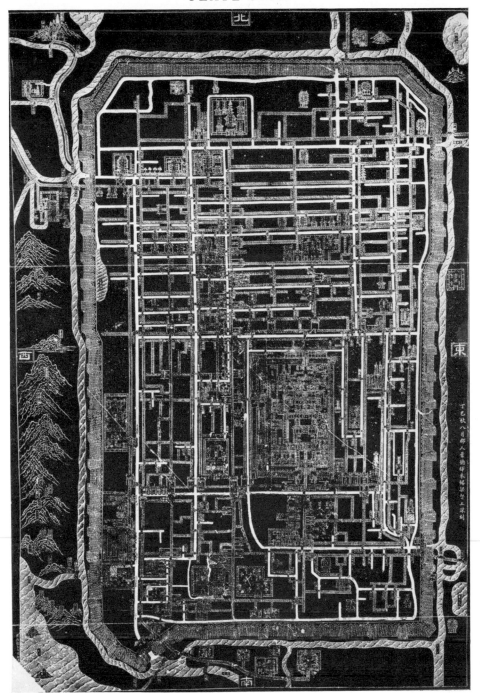

Fig. 772. The map of the city of Suchow carved on stone in +1229 by Lü Yen and two other local cartographers (from Liu Tun-Chên (5) and Chavannes (8), pl. 9). North at the top. This stele is still preserved in the old College (now a Middle School) attached to the Confucian Temple (Wên Miao) in the lower left, south-west, quarter of the city. The Grand Canal (Yün Ho) comes in at the top left hand corner as the third stream down, then it flows past the western city wall and out at the bottom just beside the large character Nan (south). Other waterways surround the city, providing canals within it almost as numerous as the streets, and crossed by 272 bridges. Marco Polo, who may well have seen the stele itself, referred to 6000 bridges, but his scribe probably exaggerated by two powers of ten. On the role of canals in East Asian urban planning see p. 309. Part of Lake Thai-hu is visible in the extreme left-hand bottom corner; above it there are the hills which protect the city from the lake, and at the top left hand-corner the famous hill of Hu-chhiu Shan is shown. The Buddhist temple of Pao-Ên Ssu can be made out in an enclosure at the extreme north of the city, and near the centre, above the inner walled complex of government buildings labelled Phing-chiang Fu (the old name of Suchow), is the great Taoist temple Hsüan-Miao Kuan. See p. 107.

PLATE CCCXI

Fig. 775. An example of mixed perspective principles in an +18th-century scroll-painting of the Madonna, half-European and half-Chinese in character; probably the work of the Jesuits, either Joseph Castiglione (Lang Shih-Ning) or Jean Denis Attiret (Wang Chih-Chhêng), painters at the court of Chhien-Lung; or of a Chinese artist influenced by them (Laufer (28), pl. 9). The colonnade has an obvious vanishing-point, but the interior of the house follows an axonmetric projection. Cf. Fig. 642 in Vol. 4, pt. 2. On Castiglione (+1688 to +1766) and Attiret (+1702 to +1768) see Pfister (1), pp. 635 ff., 787 ff. On perspective in general in China and the West see pp. 111 ff.

PLATE CCCXII

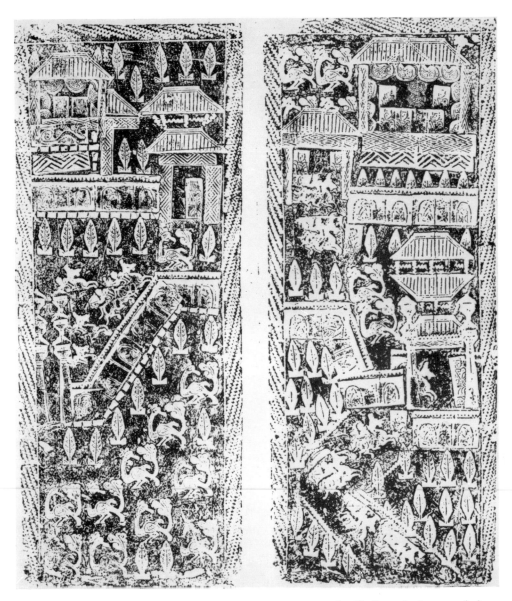

Fig. 777. One of the oldest Chinese landscape pictures extant; a pair of hollow-tile door-panels from a Han tomb of *c.* −60 excavated near Chêngchow. Continuous depth recession is conveyed to the on-looker by the zigzag lines of a road and a garden wall, so that he seems to be seeing the view from the top of a hill. The scene is constructed by the repeated impression of standard stamps on the clay while still soft. From Bulling (11).

PLATE CCCXIII

Fig. 780. Excavating a rectangular building, with internal structures and post-holes, storage spaces and ovens, in the large late Neolithic settlement of Pan-pho-tshun near Sian in Shensi (1954). Some of the buildings of this Middle Yang-shao culture, dating from about −2500 (cf. Vol. 1, pp. 81 ff. and Chêng Tê-Khun (9), vol. 1, pp. 75 ff.) are impressive in size. Photo. CPCRA and BCFA. See p. 121.

Fig. 781. A Shang royal tomb excavated at Hou-chia-chuang in the area of the capital of Anyang; c. −1350 (Chêng Tê-Khun (9) vol. 2 pl. 3a). Photo. Academia Sinica. See p. 121.

PLATE CCCXIV

Fig. 783. A two-storied hall depicted in the Hsiao-thang Shan bas-reliefs (*c.* +125); the rubbing published by the Fêng brothers in 1822, *Chin Shih So, Shih* sect. ch. 1, (pp. 90, 91). In the architecture of this Later Han reception hall the clusters of corbel brackets at the top of each column are noteworthy, as also the absence of any curvature of the roof lines. The same features are seen in Fig. 725.

PLATE CCCXV

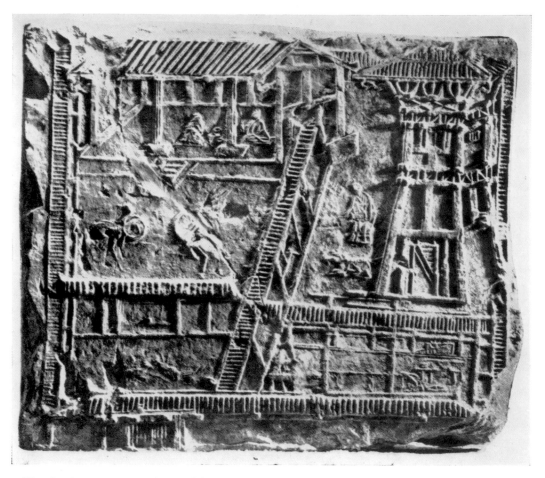

Fig. 784. A country manor-house of the Han period depicted on a Szechuanese moulded brick, probably intended for the decoration of a tomb-shrine (Anon. *22*). Entrance gate at lower left, to the right the kitchen with well and stove, behind, a look-out tower with a ladder-staircase and in the courtyard a watchdog and a servant with a broom; at the back on the left the master is entertaining a guest while a couple of cranes dance in the garden. The corbel brackets at the top of the tower are to be noted, and the transverse framework of the reception hall. The roofing of all the tamped earth walls is also characteristic.

PLATE CCCXVI

Fig. 785. One of the tomb gate steles with architectural tops found in Szechuan, the column at the tomb of Fêng Huan (+121). The corbel bracketing under the simulated wooden roof is very bold and clear (photo. Sickman & Soper, 1). Cf. the stele of Shen Fu-Chün shown in Fig. 507, Vol. 4, pt. 2. All these are modelled on gate watch-towers (*chhüeh*).

Fig. 786. Tomb model of a fortified manor-house of the Han period dated +76 found in Kuangtung (Anon. *43*). Towers at each corner and two pavilions on the central axis enclose two model buildings of two rooms each; these when taken out reveal figures engaged in various farming and domestic activities

PLATE CCCXVII

Fig. 787. Dragon columns of the main hall (Shêng Mu Tien, Hall of the Holy Mother) at the great Taoist temple of Chin Tzhu, south of Thaiyuan in Shansi (orig. photo., 1964). This magnificent hall is of great interest in many ways. Built originally about +1030, its present form dates from the restoration of +1102. The front porch pillars lean markedly inwards, and the whole front sags in the middle, perhaps intentionally. In the structural woodwork, elaborately painted, true *ang* and 'false' beaked *ang* alternate over the columns; three examples of each kind can clearly be distinguished in the picture. This would be one of the earliest appearances of the 'false' or horizontal beaked *ang* (see the discussion on pp. 95 ff. above). Inside the temple there is a remarkable set of some 30 wood and plaster statues, approximately life-size, representing the attendants of the goddess and dating from the Sung period. Many illustrations of these, with other temple buildings and their contents, can be found in Anon. (*66, 67*).

PLATE CCCXVIII

Fig. 788. Origins of the pagoda; Han pottery model of a tower (Anon. *37*). At each storey of the tower (*lou*) corbel bracket woodwork supporting balconies and roofs can be seen. From a +1st- or +2nd-century tomb at Wang-tu in Hopei. See pp. 128, 137 ff.

PLATE CCCXIX

Fig. 789. General view of the front of Fo-Kuang Ssu (Buddha's Aureole Temple), standing among the misty foothills of Wu-thai Shan, the second oldest wooden structure still extant in China, dating from +857 (orig. photo. 1964). The double and triple *ang* complexes are conspicuous among the plain *tou-kung* assemblies. Cf. pp. 95, 100, and Figs. 760, 764, 765.

Fig. 790. Thang woodwork at Tunhuang; the balcony outside cave no. 305 (T 196, P 63), probably of the same date as the frescoes and therefore, according to an inscription, +892 (orig. photo. 1943). Since then it has been preserved and restored.

PLATE CCCXX

北鎮縣 觀音閣內景 辽统和二年(公元984年)

Fig. 791. Tenth-century woodwork; the Kuan-Yin Hall of the Tu-Lo Ssu (Joy-in-Solitude Temple) at Chi-hsien in Hopei, built under the Liao rule in +984 (photo. Anon. 26). More than a dozen different kinds of bracket arms were used to suit the different positions in a lantern hall permitting the sixty-foot image to penetrate all three floor-levels. Several diagonal corbel brackets can be seen in the picture.

PLATE CCCXXI

Fig. 792. Octagonal wooden pagoda tower, some 200 ft. in height, at the Fo-Kung Ssu (Buddha Palace Temple) at Ying-hsien in Shansi, built in +1056 (photo. Anon. *26*). Just under sixty different kinds of bracket arms were used. See pp. 131, 137 ff.

PLATE CCCXXII

Fig. 793. Wood and brick domestic architecture possibly from the tenth century; the Chhen family mansion at Shantan in Kansu, showing the east end of the main block and a part of the east wing (photo. Alley). Perhaps because of the marked battering of the external columns of the transverse frames, this building has stood through many destructive earthquakes, but in date it may not be earlier than the +13th century.

PLATE CCCXXIII

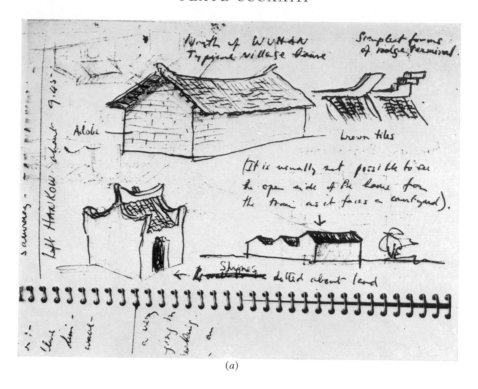

(a)

(b)

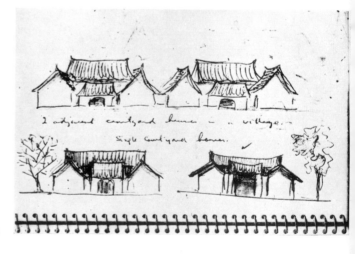

(c)

Fig. 794. Leaves from the sketch-book of an English architect in China; drawings of provincial domestic buildings by Francis Skinner (1955).

a Honan and Hupei; above left, a dwelling or barn with adobe brick walls and brown-tiled roof, above right, ridge terminals, below left, a shrine to a tutelary deity (*thu ti miao*), below right, a courtyard farmhouse.

b Hopei; top, adobe brick house with stone quoining, flat thatched roof and brick dentils along the eaves; bottom, house with brick walls and plastered roof with tile eaves, prominent lattice windows appearing at the back of a recessed portico.

c Honan and Hupei; courtyard farmhouses in a row forming a village façade, below, isolated courtyard houses often, as on the right, with noticeably lop-sided gables on the wings.

PLATE CCCXXIV

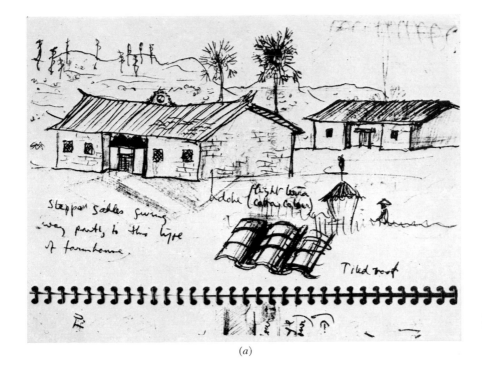

(a)

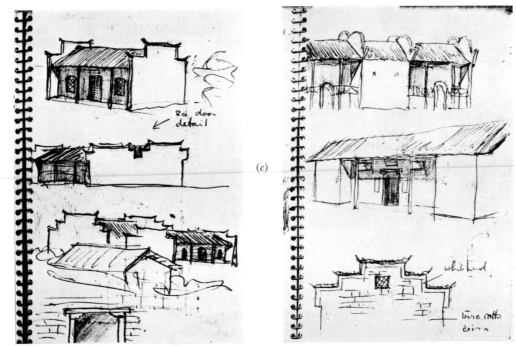

(b)

(c)

Fig. 795. Further leaves from Francis Skinner's sketch-book. See pp. 132 ff.

a   Kuangtung; free-standing farmhouses with light terra-cotta adobe brick walls, tiled roofs with central and terminal ridge ornaments, and recessed bays in the centre of the entrance side having decorated doors and decorative carving above them.

b   Hunan; top and centre, homes with portico on the entrance side and prominent stepped gables, bottom, typical house half-door entrance with lintel supported on shaped brackets.

c   Hunan; top, part of a row of houses with alternate blank and recessed bays, the latter with eaves supported on posts, and arched entrances flanked by openwork screens; the main transverse walls ending in shaped gables. Centre, house with recessed bay forming portico. Bottom, stepped and decorated gable outlined with a white band; this is very characteristic of Szechuan and Yunnan also.

PLATE CCCXXV

Fig. 796. A typical manor-house in Szechuan, with its tiled roof, ornamental finials, half-timbered and white plastered walls, standing on a solid foundation of large stone blocks. The building in this picture was for years during the second world war the home of the Institute of History and Philology of Academia Sinica (orig. photo., 1943). My collaborator in those years, Huang Hsing-Tsung, is standing in front of it.

(a)

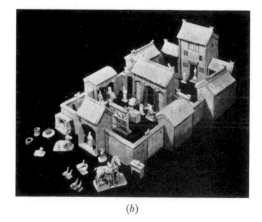

(b)

Fig. 797. Two Ming sets of tomb-models representing farmstead buildings, probably of the +15th century (photos. ILN).

a Set of seven pieces in baked clay, forming a courtyard. The smaller gate should be placed forward, with the larger one and the spirit-wall outside it. The reception hall is reminiscent of that in Fig. 784 but it has the fully developed roof-curve lacking in the Han picture. Eumorfopoulos Collection, British Museum.

b A more elaborate set, with walls and many models of farm servants and domestic animals, including the master's nag and groom in the foreground. Ontario Museum, Toronto.

PLATE CCCXXVI

Fig. 798. Interior view of a four-storey apartment house of the Hakka people in Fukien, one of the most remarkable types of Chinese rural dwelling. The need for security among an originally somewhat hostile indigenous population is thought to have led to the development of these large circular communal buildings with walls mainly blank on the outside, defensible entrances, and public service facilities in the centre of the ring. Near Yungting. See p. 133.

PLATE CCCXXVII

Fig. 799. A typical mounting for aeolian bells, the nine-storey façade of cave no. 96 at Chhien-fo-tung, Tunhuang (orig. photo. 1958). The Buddha image which it protects is 102 ft. high, set in a cave 130 ft. high, 29 ft. deep and 55 ft. broad.

PLATE CCCXXVIII

Fig. 800. The fifteen-storey brick pagoda at Sung-Yo Ssu (Temple of the Sacred Mountain of Sung) in Honan, built in +523 under the Northern Wei (photo. Anon. 26). Elegantly polygonal with twelve sides, it exemplifies the statement that for a pagoda one must pile Indian shrines on the top of a Chinese tower (p. 138).

PLATE CCCXXIX

Fig. 801. The six-storey Satmahal-prāsādaya at Polonnaruwa in Ceylon, an archaic structure built *c.* +1185 and now considered a particular type of *dagoba* (photo. Thomas). The simplest of early Chinese pagodas may have been something like this, and the pattern persisted into later times (see Fig. 802).

Fig. 802. The Ta-Yen Tha (Great Pagoda of the Wild Geese) at the Tzhu-Ên Ssu (Loving-Kindness Temple) at Sian, built in +652 after the return of the pilgrim-scholar Hsüan-Chuang to this abbey from India, and repaired in +704 (orig. photo. 1958).

PLATE CCCXXX

Fig. 804. Crumbling *stūpa* tombs on the desert across the dry river-bed from the Chhien-fo-tung cave-temples, some of which can be seen in the escarpment of the background hills. On the left the northern end of the oasis can be seen, and the tomb of Wang Tao-Shih, the discoverer of the famous Tunhuang library, is very near by. These *stūpa* tombs, commemorating monks of the Thang, Sung and Yuan periods, decay extremely slowly because of the dryness of the climate, and their coloured plaster-work is scoured and polished by the sandstorms of many ages. They extend over the desert for miles, up to and within the foothills of the San-wei Shan. Each one has a different design, and every one is beautiful (orig. photo., 1943, taken in the early morning).

Fig. 805. A number of model *stūpas* or *dagobas* fallen out from the heart of one of the large *stūpa* tombs on the desert near Chhien-fo-tung, as it decayed. Such a tomb may contain as many as a hundred, all formed from the ashes of the dead monk mixed with a suitable clay and pressed into a mould, then dried in the sun like adobe brick. The thumb-marks of the makers in the +13th or +14th centuries are still visible on the base of the models, and if broken open they may contain a *dhāraṇī* charm (*chen yen*) or magical prayer. Orig. photo. 1943.

PLATE CCCXXXI

Fig. 806. *Stūpa* tombs (*chhuang*) of Ming and Chhing date in the graveyard of a Buddhist abbey in Shantung (photo. Forman & Forman (1), p. 101). The superimposed sections, round, square, octagonal, lotus-petalled, etc., were interpreted as symbolising the five elements.

PLATE CCCXXXII

Fig. 807. A stone triumphal gateway (*phai-lou*), one of many in the streets of Chhü-fou in Shantung, site of the tomb and Temple of Confucius (photo. Forman & Forman (1), p. 103). In somewhat simpler forms this kind of stone gateway is widely found, imitating woodwork in its roofs and bracket arms. See p. 142.

Fig. 808. A wooden triumphal gateway (*phai-lou*), that at the main south entrance of the Confucian temple (Wên Miao) at Sian, Shensi, now the Provincial Museum, with its famous 'Forest of Steles' (Pei Lin). The inner side of the city wall can be seen in the background, and in front the marble balustrade of the semi-circular pool (*phan chhih*), canonical in Confucian temples, crossed by its arched bridge (*yuan chhiao*). The *phai-lou*, supported by struts on each side, blossoms in a wealth of seven rows of corbel brackets one above the other, holding up the massive roofs that cap the whole (orig. photo., 1958).

PLATE CCCXXXIII

Fig. 809. The tomb of the Thang empress Wu Tsê Thien (Wu Hou, *r.* +684 to +704), the Chhien Ling, at Chhien-hsien, north-west of Sian in Shensi. The tomb, within a steep-sided natural hill, not a tumulus, is approached by a horizontal adit some 200 ft. in length, but has not yet been opened for archaeological study. Below there is the long avenue of ministers, officials and Buddhist monks here seen, as well as about a hundred smaller figures representing envoys and tribute-bearers from various nations to the Thang. Orig. photo., 1964.

Fig. 810. The Ming imperial tombs north of Peking; a pair of camels in the long double row of figures flanking the avenue of approach, in the distance the great pavilion housing an inscribed stele (Pei Thing). Orig. photo., 1952. See pp. 143 ff.

PLATE CCCXXXIV

Fig. 811. Ming imperial tombs; the tomb-complex (cf. Arlington & Lewisohn (1), pp. 316 ff.) of the Yung-Lo emperor (*r.* + 1403 to + 1424) known as the Chhang Ling. A pavilion on the right-hand side of the courtyard inside the great entrance-gate. Orig. photo., 1952; since that date the temple and gardens have been completely recovered from the wilderness, and now form a public park much visited by the people of Peking.

Fig. 812. Ming imperial tombs; the inscribed stele within the pavilion shown in Fig. 811. This was erected by the first Chhing emperor to record an edict that the tombs of the rulers of the conquered dynasty should be well maintained in perpetuity (orig. photo., 1952).

PLATE CCCXXXV

Fig. 813. Ming imperial tombs; the tomb-temple of the Yung-Lo emperor at Chhang Ling. It stands behind the main Hall of Sacrifice and in front of the broad tumulus containing the burial chambers surrounded by a rampart; it is known as the Brilliant Tower (Ming Lou) and shelters a large stele mounted on a tortoise of stone. The figures on the wall give the size (orig. photo., 1952). The object in the middle foreground is an open-air altar.

Fig. 814. A coloured panorama set up at the Ming Tombs dam site in 1958 (cf. p. 144), showing the thirteen tomb-temples set in the valleys of the Thien-shou Shan range and the artificial lake which now covers the main valley floor (orig. photo.).

(e) BRIDGES

When the architect Frontinus, writing on the aqueducts of Rome in the + 1st century, completed his description, he added the following words: 'With such an array of indispensable structures, carrying so many waters, compare, if you will, the idle pyramids, or the useless, though famous, works of the Greeks.'[a] His Chinese counterparts would have had some conscious sympathy for the attitude of mind which lay behind this remark, but no small part of the genius of their civilisation lay in a subtle combination of the rational with the romantic, and this had its result in structural engineering. No Chinese bridge lacked beauty and many were remarkably beautiful.[b]

In describing the achievements of China in bridge-building it will be necessary to follow a logical classification.[c] Presumably the simplest type of bridge is a beam of wood or any rigid material laid transversely across the stream or other obstacle to be bridged. Here the limiting factors are soon reached, for before any considerable span has been attained, the material will cease to support the weights which its builders will want to send over it. We shall see shortly that the Chinese explored the possibilities of this simple method up to the maximum strength of the strongest natural material available, in a series of remarkable megalithic bridges. Release from the narrow confines of the capacities of single beams or blocks came only with the development of the truss, in which many component members, each taking tension or compression only, are jointed together in reticular geometrical systems. This was fully developed only at the Renaissance in Europe.[d] As Parsons suggests,[e] it was perhaps originally an

---

[a] *De Aquis Urbis Romae*, I, 16. Frontinus (+35 to +104) was a contemporary of Wang Chhung. For a modern treatment of his subject see Ashby (1). Cf. Vol. 4, pt. 2, p. 128.

[b] Today some of the most important are carefully restored and preserved as national monuments; cf. Mao I-Shêng (2).

[c] A few words on the literature available may be in order here. Books on bridges tend to be folkloristic (e.g. Robins, 2) or, if artistic, technically misleading (Brangwyn & Sparrow, 1). What is said to be the best history of bridge engineering (Tyrrell, 1) has not been accessible for us, but there is much of value in Straub (1) and Uccelli (1). Shirley-Smith (1) is authoritative. As usual, help is also to be obtained from old-fashioned books which have been banished from current shelves, e.g. Jenkin (1) and Fidler (1). Modern popular accounts of bridge-building are not very helpful, e.g. A. Black (1) or Steinman & Watson (1), but one could begin no better than by reading the *tour d'horizon* articles of Steinman (2). Florange (1) has an interesting analysis of medieval European bridges as depicted on coins—but space precludes mention of further items in the large European specialist literature. Most general books on Chinese architecture, e.g. Anon. (*37*); Sirén (1); Mirams (1), etc., have something on bridges, but there is a special monograph by Fugl-Meyer (1) which is much more informative than it appears at first sight. The counterpart to this in Chinese is the larger and very professional work of Lo Ying (*1*), published more than twenty years later. An excellent album of photographs of Chinese bridges of all periods, with good explanatory text, is due to Thang Huan-Chhêng (1). The earlier work of Liu Tun-Chên (1) also covers the field, but with particular attention to beam and cantilever bridges. The Department of Oriental Antiquities at the British Museum possesses the Eagle Collection of photographs of Chinese bridges, catalogued by Mr William Willetts, but it concerns mainly arched structures and has not yet been made generally available. Lo Ying (*2*) deals exclusively with this type.

[d] Especially by Andrea Palladio, *c.* +1570; cf. Davison (11); Uccelli (1), p. 682. But Leonardo had been familiar with the principle. As we have seen (pp. 101, 141 above), the truss had not been entirely lacking in Chinese constructional technique, for triangular and trapezoidal forms were used for many centuries in the support systems of the heavy overhanging roofs. Instead of giving rise to more complex forms, however, they died out as heavier woodwork came in. Was this, perhaps, yet another consequence of the Chinese lack of deductive geometry? On the triangular roof-trusses of the Greeks, see Hodge (1).

[e] (2), p. 486.

empirical discovery arising from the study of the timber false work or centering which had been used for centuries (in China also) in the construction of all arches. From geometry the Renaissance engineers knew that the triangle was the only figure which could not be deformed or distorted without changing the length of at least one of its parts; hence the elaborate combinations of triangles in trusses.[a] Drawbridges must come also under the heading of beam bridges, and something must be said too about the different kinds of piers employed when the bridge has more than one opening. These may be of wooden piles, of stone in various designs, or occasionally of wooden tripods, and last but not least, of boats, making those pontoon or floating bridges which appear very early in history.

The next class of bridges is that in which the cantilever beam is employed. Such a beam is one which is rigidly fixed at one end and free at the other,[b] so that it can move slightly according to its flexibility. In cantilever bridges, a series of such beams are thrown out from both sides of the gap and connected by a beam or truss in the centre. The home of these bridges seems to be the Himalayan region, and they were early known and used in China.

Arches have formed perhaps the most frequent and widespread bridge type. They were originally semicircular, and long remained so. In Europe there was a persistent belief that the Roman and Norman semicircular arch was indispensable because it directed the line of thrust vertically downward at the piers or abutments,[c] and this theory was not affected by the use of pointed arches in the Middle Ages, which have the lines of two much larger circles or other curves intersecting at the crown. The great departure from precedent occurred in the +14th century, when the base diameter of the semicircle was allowed to sink, as it were, far below the river, and the bridge itself became segmental, leaping forth from its abutments as if resembling the flying gallop motif in art.[d] In the present section it will appear that this fundamental advance had been anticipated by a Chinese engineer of genius some seven centuries before its appearance in Europe. Other shapes of curves, such as ellipses, could of course be used, and were.

The last important class of bridges is that of suspension bridges. Here the support comes from above, in the shape of a hanging catenary curve embodied in ropes or chains. In all the more primitive forms the passengers and animals follow the curve in crossing, but perhaps as a development from hand-rails connected at short intervals with the deck, there gradually arose the true flat-deck suspension bridge. The suspension bridge is native in many parts of the world, both Old and New, but only in one of them did eotechnic engineers make the transition from the use of ropes to that of iron chains. This occurred at an early date in the very mountainous country of south-west China bordering on Tibet, Assam and Burma, and the later suspension bridges of Europe derive from these remarkable iron-chain bridges.

[a] Cf. Uccelli (1), p. 295.
[b] It corresponds logically to the corbel in architecture.
[c] Parsons (2), p. 485. In fact, the semicircular arch has the great disadvantage that it needs maximum loading on the haunches and minimum at the centre. Hence the mortality of Norman towers.
[d] Cf. Vol. 1, pp. 166 ff. For such bridges the abutments had to be made stronger.

We may now summarise the classification, inserting some figures for the spans which traditional Chinese technique bridged by the various methods.

			Maximum span in ft.
BEAM	iron		10
	wood		20
	stone		70
CANTILEVER	wood		130
ARCH	stone	semicircular	90
		pointed or two-centered ('Gothic')	70
		segmental	200
SUSPENSION	catenary	single rope	
		V-section rope	
		tubular rope network	
		decked bamboo rope	450
		iron chain	
	flat deck	bamboo rope	
		iron chain	

It is of interest that the existence of different types of bridges seems to be betrayed in the structure of certain characters. Hopkins (14) considers that the earliest graph for *liang*,[1] which means a beam or a bridge, was a drawing of a plank crossing a stream (see inset). No early version of the present character itself has been found, but a cognate

| *liang* | K 1138*b* | K 1138*f* |

form contains the components of water and rice, with a third which seems to have been originally a drawing of a man doing something (K 738). Perhaps he was building an irrigation dam across which one could walk. In the *Shih Ching*[a] the character is generally used to mean a fixed palisade fish-trap, which could very naturally have formed a bridge.[b] The most usual term for bridge, *chhiao*,[2] adds the radical for wood to a form *chhiao*[3] which meant high and arched, as evidently appears from the graph of ancient times (K 1138*a, b, f, g*). A few words will come later concerning the origin of arches and vaults; although they are not at all prominent in Chou times, their abundant use in Chhin and Han suggests that the Chinese knew and used them, perhaps only for special purposes, during most of the −1st millennium. Further search among the ancient forms of written characters might unearth references to the cantilever and suspension types (see p. 186).

[a] Mao nos. 35, 63, 104 etc.; Legge (8), vol. 1, pp. 56, 106, 159; Karlgren (14), pp. 22, 43, 67; Waley (1), pp. 100, 46, 79.
[b] Such structures are still commonly used by the Chinese as they are among so many other peoples (cf. Thomazi (1), p. 120; Korrigan (1), p. 109). Permanent fish-traps of this kind are prominently visible from the air when one is flying over the lakes of north-western Manchuria.

[1] 梁   [2] 橋   [3] 喬

Foreign admirers of Chinese bridges could be adduced from nearly every century of the empire. Between $+838$ and $+847$ Ennin never found a bridge out of commission, and marvelled at the effective crossing of one of the branches of the Yellow River by a floating bridge 330 yards long, followed by a bridge of many arches, when on his way from Shantung to Chhang-an.[a] In the last decades of the $+13$th century Marco Polo reacted in a similar way, and speaks at length of the bridges in China, though he never mentions one in any other part of the world.[b] The 12,000 bridges of Hangchow, famous as an exaggeration of 'Marco Millione', probably arose from the omission of a line of manuscript and a confusion between city-gates and bridges; in fact there were in his time exactly 347. Of these 117 were within the city-walls, no less than 46 having been built during the previous century since $+1170$. A special edict issued in $+1268$ called for the increase and repair of the capital's bridges, and under the Governor Chhien Yüeh-Yu[1] more than half were demolished and rebuilt within a year, low ones being raised high to allow canal traffic to pass, and narrow ones broadened. Thus another Rialto awaited the traveller from the Far West.[c]

The first Renaissance visitors to China also conceived great admiration for the bridges which they found there. One of the earlier of them, Galeote Pereira, wrote about $+1577$ as follows concerning the Fukien megalithic type, which we shall shortly view more closely:

As you come in to either of these cities (near Chincheo in Fukien),[d] there standeth so great and mightie a bridge that the lyke thereof I have never seene in Portugall nor els where. I heard one of my felowes say, that he told in one bridge, 40 arches. The occasion wherfore these bridges are made so great, is for that the countrey is toward the sea very plaine and low, and overwhelmed ever as the sea water encreaseth. The breadth of the bridges, although it be well proportioned unto the length thereof, yet they are equally buylt, no higher in the middle than at eyther end, in such wyse that you may directly see from the one end to the other; and the sydes are wonderfully well engraved after the maner of Rome workes. But that wee did most marveyle at, was therwithall the hugenesse of the stones, the lyke whereof as we came into the citie, we dyd see many set up in places dishabited by the way, to no small charges of theyrs, howbeit to little purpose, whereas no body seeth them but such as doe come bye. The arches are not made after our fashion, vaulted with sundry stones set together, but paved, as it were, whole stones reaching out from one piller to another, in such wyse that they lye both for the arches heades, and galantly serve also for the high waye. I have been astunned to behold the hugenesse of these aforesaid stones, some of them are xii pases long and upwarde, the least a good xi pases long, and an halfe.[e]

[a] See Reischauer (2), pp. 280, 282, 283, (3), p. 120.
[b] Ch. 157 (Moule & Pelliot ed.). Unfortunately Marco Polo did not see or describe them with the eye and pen of an engineer.
[c] The discrepancies in Marco Polo's accounts, and the bridge-building activities of his Chinese contemporaries, have been discussed by Moule (9), (15), pp. 23, 27; cf. Gernet (2), p. 45.
[d] Chhüanchow or Changchow; see Boxer (1), p. 313.
[e] Eden ed. p. 238; Boxer (1), p. 7. It is not easy from his account to identify which two bridges Pereira had in mind, but very probably the Lo-Yang (Wan-An) Bridge near Chhüanchow was one of them. An equally colourful description of this written by Domingo de Navarrete, who passed over it in $+1659$, will be found in Cummins (1), vol. 1, pp. 143 ff. See Table 66.

[1] 潛說友

Parallel passages are frequent in Gonzales de Mendoza and most of the writers of the + 16th and + 17th centuries. In + 1773 Sir William Chambers gave an account of Chinese bridges in his most romantick vein, and as late as + 1796 van Braam Houckgeest showed admiration just as great, describing in particular one of 100 arches, and another of 15.[a] Indeed, when Peter the Great sent an embassy to China in May + 1675, one of the requests made to the Chinese by the envoy Nikolaie Milescu Spătarul was that expert bridge-builders should be sent westwards to teach their methods to the Russians.[b]

It is interesting that one of the things which the early Portuguese visitors to China in the + 16th century found most extraordinary about the bridges was the fact that they existed along the roads often far from any human habitations. 'What is to be wondered at in China', wrote Gaspar da Cruz, the Dominican who was there in + 1556, 'is that there are many bridges in uninhabited places throughout the country, and these are not less well built nor less costly than those which are nigh the cities, but rather they are all costly and very well wrought.'[c] In such wise did the works of an all-pervading imperial bureaucracy impress the visitors from an essentially city-state civilisation.

(1) Beam Bridges

The simple wooden beam bridge with trestle piers (*chia mu chhiao* [1]) is found in most parts of China.[d] Its chief interest is that it seems to have persisted unchanged from high antiquity.[e] The famous scene of the 'Battle on the Bridge' in the Wu Liang tombshrine carvings, dating from about + 150, shows clearly the graded approaches and the central span.[f] Probably it was this kind of bridge that Mencius was speaking of in the − 4th century, when he mentioned[g] the seasonal repairing of footbridges (*thu kang* [2])[h] and carriage bridges (*yü liang* [3]). Spans could not well exceed 15–20 ft. between the trestle piers,[i] but over shallow water spans might be numerous.

This was necessarily the case when rivers of considerable width had to be bridged, as they were very successfully from quite early times. After the rise of the Kuan-nei

---

[a] Pp. 138, 147. So also Macartney (Cranmer-Byng (2), pp. 92, 108, 175, 194).

[b] This remarkable fact was brought to our notice by H.E. Sardar K. M. Panikkar. Cf. Ysbrants Ides, pp. 63, 65; Panikkar (1), p. 235; Baddeley (2), vol. 2, pp. 351, 385. On the background of Milescu's embassy see Cahen (1); and, for his own account, Milescu (2). It is generally known as the Spathary embassy, from the Russian form of the official title of this Rumanian scholar in the service of the Tsar.

[c] Boxer (1), p. 105.

[d] Cf. Fugl-Meyer (1), fig. 19.

[e] The fact that the Chinese never passed beyond the use of simple beams to the compound reticular frameworks of wood with which very large truss bridges can be built, could be ascribed, as already suggested, to their lack of theoretical deductive geometry. But they used timber centering for arch construction just as Europeans did (Fig. 833*f*), and moreover their lattice design and other ornamentation was very geometrical (see Vol. 3, pp. 95, 112). As for other materials, there are a few rare examples of the use of iron bars as bridge beams (Hutson (1); Fugl-Meyer (1), p. 81), but this could not lead far, unless indeed it was what gave the idea of the *rod*-linked suspension bridges (cf. below, pp. 151, 196).

[f] *Chin Shih So*, Shih sect., ch. 3 (pp. 107 ff.). Cf. Bulling (13).

[g] *Mêng Tzu*, IV (2), ii (Legge (3), p. 193).   [h] Footbridges were later called *chio*. [4]

[i] But for light pedestrian and pack-animal use wooden beam bridges may span 35 ft. or so, as in the example with one log-crib pier given by Mock (1), p. 30.

[1] 架木橋   [2] 徒杠   [3] 輿梁   [4] 榷

region as the centre of Chinese culture the crossing of the Wei River became particularly important, and a beam structure of many spans, the Hêng C.[1],[a] was built by
Prince Chao Hsiang[2] of Chhin[b] soon after his accession in −305. Linking the capital
of Hsien-yang with the lands and passes south of the river, it retained all its importance
during the Han, when Chhang-an on the southern side became the capital.[c] Since its
length, some 2,000 ft., and the number of spans, 68, are known from the sources,[d]
they must have been approximately 29 ft. in length. While all its beams were of wood,
giving a deck width of 55 ft.,[e] its piers were of stone in its northern section, which was
called Shih-Chu C.[3,4] We can visualise this rude but noble bridge in two ways, first by
studying the representations of bridges on Han bricks still in Szechuanese museums
(Fig. 815, pl.),[f] as old as the Wu Liang relief, if not somewhat older; and secondly by
looking at structures of the same type still existing today. Three rivers fall into the Wei
near Chhang-an, the Pa[5] and the Chhan[6] to the east, and the Fêng[7] to the west, and
their old bridges of Han type with up to sixty-seven spans[g] remain still as Ennin saw
them when he walked across in +840 (Figs. 816, 817, pls.). A drawing by Thang Huan-
Chhêng explains their simple construction.[h] All such bridges remain close to water-level
and disappear from sight at flood seasons.

   Throughout Chinese history, this style persisted. We hear of a trestle structure built
by Chang Chung-Yen[8] in +1158 ten *li* in length. In pictures by Sung painters such as
Hsia Kuei[9] we find elegant wooden beam bridges with pavilions crowning their central

[a] Henceforward we shall abbreviate the word *Chhiao* (bridge) in this way.
[b] This same prince it was who in −257 effected the first bridge-crossing of the Yellow River,
certainly by pontoons (*Shih Chi*, ch. 5, p. 34*a*; Chavannes (*1*), vol. 2, p. 94).
[c] Two others, however, were then built, both to the east of the Hêng C.; see Adachi Kiroku (*1*); Lo
Ying (*1*), p. 56. All were described at some length in +1655 by Martini in his *Atlas Sinensis*, p. 45, but he
gave them high arches.
[d] *San Fu Huang Thu*, ch. 34, and *San Fu Chiu Shih*.
[e] Thus a nine-lane highway; cf. pp. 5 ff. above. Worthy of a royal, if not yet an imperial, capital. We
convert according to Chhin and Chhien Han measure.
[f] Liu Chih-Yuan (*1*), fig. 58; Anon. (*22*), pls. 34, 35; Thang Huan-Chhêng (*1*), fig. 6. The scene has
been reproduced on a postage-stamp in our time.
[g] Thang Huan-Chhêng (*1*), figs. 11, 12, cf. 14; Lo Ying (*1*), pp. 28 ff., 276 ff.; Geil (*3*), p. 172. The
memoir of Liu Tun-Chên (*1*) is primarily devoted to these. The Pa bridge was, we know, repaired many
times, e.g. in +22 and again in +582; I visited it briefly in 1958, when it was as Ennin saw it, but found
in 1964 that it had been entirely encased in concrete with a widened deck to take heavy motor traffic. It
was still possible, however, to study in detail a smaller bridge of 26 spans, identical in type, just north-
west of Sian, the Fêng C.[10] (Thang Huan-Chhêng (*1*), fig. 13; Lo Ying (*1*), p. 29; Fugl-Meyer (*1*), p. 70).
Each pier is constructed thus; a foundation of cypress-wood piling supports a transverse row of three
discoidal roller-mill base-plates (*nien phan*[11]); and each of these supports two columns of four superimposed stone cylinders (*lu chu*[12]) 2 ft. high and just under 2 ft. in diameter, the lowest cylinders but one
in each of the pairs being tied together by a wrought-iron band. Each pier thus has six columns. Measurements made in farmyards of the region showed, as suspected, that the cylinders were standard-size
threshing-rollers. Each pair of columns is capped by an oblong stone plate (*shih liang*[13]) bearing two
transverse wooden beams, upon which nine longitudinal beams are laid; above come transverse planks
and an earth covering. It is a great pleasure to be able to examine minutely a structure so simple, sturdy
and workmanlike as to have needed no essential alteration since the time of Christ.
[h] (*1*), p. 14.

[1] 橫	[2] 昭襄	[3] 石柱橋	[4] 石軸橋	[5] 灞	[6] 滻
[7] 灃	[8] 張中彥	[9] 夏珪	[10] 灃橋	[11] 碾盤	[12] 轆軸
[13] 石梁					

spans (Fig. 818, pl.).[a] And when in +1221 the Taoist Chhiu Chhang-Chhun was on his way to visit Chinghiz Khan at Samarqand, he and his party followed a road through a defile in the Thien Shan mountains east of Kuldja which had 'no less than 48 timber bridges of such width that two carts can drive over them side by side'. It had been built by Chang Jung[1] and the other engineers of Chagatai some years before.[b] The wooden trestles of Chinese bridges from the −3rd century onwards were no doubt similar to those supposed to have been employed in Caesar's bridge of −55 across the Rhine,[c] or drawn by Leonardo,[d] or found in use in Africa. But where in +13th-century Europe could a two-lane highway like Chang Jung's have been found?

Besides wooden trestles, wooden piles or stone piers, all kinds of other supports for beam bridges have been used in China at one time or another. We have just noted an example of the log-crib used as a pier,[e] and in western Szechuan, in the neighbourhood of Yachow (Ya-an), it has long been customary to rest wooden beam bridges crossing the Chhing-i[2] River and its tributaries on the gabions[f] so much used in works of river control.[g] One can even find instances of the use of iron columns as bridge piers. Some time during the Sung, a Chiangsi man, Tsang Hung,[3] built a bridge based upon twelve of these at Fou-liang Hsien[4] in that province,[h] but they were replaced by stone towards the end of the dynasty.[i] The record has come down to us in the local topographies, and as many as five more bridges of the same kind, each from a different province, have been discovered in this gazetteer literature.[j] One was built by a magistrate, Li Chên,[5] at O-mei Shan in Szechuan, c. +1470, and another in Yunnan had seven spans averaging 45 ft. set on 40 ft. columns, presumably composite. For four of the six cast iron is distinctly stated, and it was probably used in all.[k] We may even find references to such structures in the Old China Hand literature, for Hance (1) described in 1868 the ruins of two which he and a friend had found in Chekiang and Fukien. So far as we know, these facts have not been mentioned by those who have written on the history of cast iron in architecture and bridge-building.[l]

---

[a] Cf. Anon. (32), painting no. 38, pl. 9. Cf. Yang Jen-Khai & Tung Yen-Ming (1), vol. 1, pls. 59, 67.
[b] *Chhang-Chhun Chen Jen Hsi Yu Chi*, ch. 1, p. 18a (Waley (10), p. 85).
[c] *De Bello Gallico*, IV, 16–19; Neuburger (1), p. 463; Davison (11).
[d] Uccelli (1), p. 279.
[e] In some cases these may attain considerable heights, as in the bridge at Lhasa figured by Thang Huan-Chhêng (1), fig. 21.
[f] Open-work bamboo baskets filled with stones—see further, pp. 295, 339 below.
[g] See the photographs in Liu Tun-Chên (1), pl. 2a, and Thang Huan-Chhêng (1), fig. 3. Also Lo Ying (1), p. 23.
[h] This place is more gloriously known as Ching-tê-chen,[6] the greatest centre of the pottery and porcelain industry in China, active century after century. There may be significance in a connection between kilns and blast-furnaces.
[i] Liu Tun-Chên (1), p. 36.   [j] Lo Ying (1), p. 28.
[k] Needham (31, 32) may be consulted on this. Tsang Hung probably kept his iron piers well painted while they could be reached during the dry season—or perhaps his bridge did not last because he overlooked this. The differential susceptibility of the various kinds of iron to rusting is a complex question of some importance in the history of Chinese siderurgy, and we shall return to it in Sect. 30d. We have already noted an instance of the use of cast iron in the pillars of a palace hall (p. 103).
[l] E.g. Gloag & Bridgwater (1); Watterson (1); Skempton (1); Skempton & Johnson (1); Giedion (2), pp. 101 ff. The oldest cast-iron bridge in the West is that which spans the Severn between Shrewsbury and Bridgnorth in Shropshire. Built by the third Abraham Darby in +1779 with a span of 100 ft. it

[1] 張榮   [2] 青衣   [3] 臧洪   [4] 浮梁縣   [5] 李禎   [6] 景德鎮

Already for this simplest type of bridge effective pile-driving is needed, and until very recent times the Chinese continued to use an apparatus like the *fistuca* of Vitruvius,[a] in which the hammer or 'monkey' is raised to the top of its travel by a cable forking into many pulling-ropes so that the combined efforts of a number of men may be used.[b] Fig. 360 (in Vol. 4, pt. 2) shows one of these[c] at work at a washout on the Old Silk Road in 1943. For smaller jobs, four to eight men operated a punner or rammer (a cylindrical stone with bamboo handles),[d] while themselves standing on a small platform attached near the top of the pile, so that their weight added to the blows.

Stone beam bridges are familiar to English people because of the small 'clapper' bridges of the West Country.[e] But in China the principle was used on a much greater scale. Even the standard type of small stone beam bridge (Fig. 819, pl.)[f] is considerably larger than the clapper bridges. When the span is less than 16 ft. and the height above high water 6–10 ft. this is the type most commonly used. These bridges are always built in a dry enclosure formed by a cofferdam. This is constructed by erecting a double row of bamboo poles, tied together and covered with matting, the intervening space being filled with clay. Water is then pumped out by means of square-pallet chain-pumps (cf. Vol. 4, pt. 2, pp. 339 ff. above) until the dry bed of the stream can be worked on. The cofferdams are made convex against the main water-pressure. Three chief types of piers (*tun*[1]) are used with these bridges. In the first and simplest form, long flat stones like planks are set up vertically parallel to the stream-direction, and mortised into a long foundation-stone beneath. At the top they are assembled and mortised into another long horizontal stone, longer than the width of the bridge which it supports. This system offers very little resistance to the current, and can well sustain the impacts of boats. The abutments look massive, but are made with great economy, being no more than thin retaining-walls with a filling of clay and stone chips. A second form of pier is constructed of stone in pyramidal form, again presenting its narrowest axis to the current. The third method combines these two when it is desired to lift the bridge well above water level.

Occasionally, these bridges were used for many spans. Although unsuitable for single breadths exceeding 20 ft., they are to be found from time to time on the inland waterways, including the Grand Canal, but in such cases the abutments extend far into the stream and Fugl-Meyer thinks that this was done on purpose as a kind of crib for retarding the current. The first bridge to be built in China wholly of stone, spanning a

consists of a series of semi-circular ribs each cast in two pieces and tied by struts to larger concentric ribs which support the slightly humpback deck and add rigidity. It is still in use for pedestrians. Cf. Shirley-Smith (1), p. 64.

[a] III, 4; cf. Frémont (13), pp. 37 ff. for many details.

[b] Another application of the same method will later be seen in the hand-operated mangonels which were so popular in the artillery of medieval Chinese armies (Sect. 30).

[c] Medieval European parallels in Salzman (1), pp. 86, 328.

[d] Cf. pp. 38 ff. above.

[e] See Robins (2), pp. 63 ff.

[f] Fugl-Meyer (1), fig. 24; Mirams (1), p. 103; Thang Huan-Chhêng (1), figs. 5, 36; Boerschmann (3a), p. 115; Liu Tun-Chên (1), pl. 2b; Forman & Forman (1), pp. 131, 221.

[1] 墩

transport canal at Loyang in +135, was very probably of this kind, though it may have had one or more arches.[a] Among the many which remain in the cities of the eastern provinces there are two veritably of the Sung period at Shao-hsing, one of which, dating from +1256, the Pa Tzu C.,[1] has two approach ramps with balustrades on each side.[b]

All this has relatively little of engineering interest, but during the Sung period there was an astonishing development, the construction of a series of giant beam bridges, especially in Fukien province. Nothing like them is found in other parts of China, or anywhere outside China. These structures were (and are) very long, some of them more than 4,000 ft., and the spans extraordinarily large, up to 70 ft., a duty which necessitated the handling of masses of stone weighing up to 200 tons.[c] The art of building these bridges was afterwards forgotten, and no record was kept either of the quarries which had provided the stones or the techniques by which they had been hauled to the site and set in position.[d]

Much obscurity surrounds the names of the engineers responsible for these mega-lithic bridges. Inscriptions which some of them bear merely record the provincial officials under whose aegis they were constructed and repaired. One may sense the existence of some master bridge-builder and a school or tradition founded by him. According to local sources, the Wan-An (Lo-Yang) bridge at Chhüanchow was built under the superintendence of Tshai Hsiang[2] (+1012 to +1067), a scholar who was prefect in Fukien for some time,[e] and who is well known to us for his books on the lichi fruit and on tea.[f]

---

[a] See p. 172 below.

[b] Description in Chhen Tshung-Chou (1). Not far from Shao-hsing I found and photographed in 1964 another of this kind over the old Grand Canal extension, the Thai-Phing C.,[3] with a double stair-case on one side and a series of beam spans on the other (Fig. 820, pl.).

[c] These are certainly megalithic, but not more so than building components used in other civilisations. The west wall of the main temple at Baalbek in Syria contains three colossal blocks each weighing 850 tons, which must have been put in place about the +3rd century.

[d] Lo Ying (1), pp. 382 ff., suggests very plausibly that advantage was taken of the differences in tidal level, as also of the manipulation of buoyancy according to the technique of Huai-Ping (c. +1067, cf. Vol. 4, pt. 1, p. 40).

[e] The local legend about the origin of his connection with bridge-building has been given in English by Dukes (1), reproduced in Boxer (1), pp. 334 ff. See overleaf.

[f] His statue ornaments the south-western end of the bridge (photograph in Ecke, 5). The date often given for the construction, +1023, is wrong. Not till +1041 did Li Khuan[4] and Chhen Chhung[5] 'build dressed stone approaches and a floating bridge'. Then in +1053 Wang Shih,[6] Lu Hsi[7] and the monk Tsung-I[8] took the building of a megalithic beam bridge in hand but could not finish it, whereupon Tshai Hsiang brought greater resources to bear and the work was successfully accomplished by +1059. We have a record of it from his own brush, the *Wan-An Chhiao Chi*.[9] Perhaps it was he, or one of his assistants, Hsü Chung[10] and the monks I-Po[11] and Tsung-Shan,[12] who added an indispensable engineer-ing element of some interest previously missing. By +1078 we learn that a later prefect, Wang Tsu-Tao,[13] obtained an order forbidding people from carrying away stone from the 'oyster-shells' or 'oyster-clusters' (*li fang*,[14] cf. R 216 and Vol. 4, pt. 1, p. 88). These were nothing else than flat foundations of stone in the river-bed considerably larger than the piers which were built on them, and similar in prin-ciple to the concrete rafts of the present day. These rafts are not mentioned in the relevant passage of *TSFY*, ch. 99, p. 42b, but in the *Fu-chien Thung Chih* and other texts cited by Lo Ying (1), pp. 252 ff.

---

[1] 八字橋   [2] 蔡襄   [3] 太平橋   [4] 李寬   [5] 陳寵
[6] 王實   [7] 盧錫   [8] 宗已   [9] 萬安橋記   [10] 許忠
[11] 義波   [12] 宗善   [13] 王祖道   [14] 蠣房

There is a famous folk-tale about this which it would be impossible to omit.

Before the Loyang bridge was built, everyone had to cross over by boat. In the reign of the Emperor Shen Tsung (+998 to +1022) of the Sung dynasty, a pregnant woman from Fu-chhing was crossing the river to Chhüanchow; but just as the ferry was in the middle of the stream the bad tortoise and snake spirits sent suddenly a strong wind and high waves to upset and sink it. All at once a voice cried from the sky: 'Professor Tshai is in the boat. Spirits must behave decently with him.' And scarcely had the word been spoken when the wind and waves died down. All the passengers had heard exactly what was said, but there was no one of that family in the boat except the woman from Fu-chhing. They therefore congratulated her, and she said: 'If I really give birth to a son who becomes a professor, I will charge him to build a bridge over the Loyang river.'

Several months later she did bear a son, who was named Tshai Hsiang. Later on (about +1035) he did in fact become a professor. His mother related to him her experience on the river, and begged him to fulfil her vow. Being a dutiful son he immediately gave his consent, but at that time there was a law against anyone being appointed an official in his own province, and as Tshai Hsiang was a native of Fukien, he could not become Governor of Chhüanchow, which is in that province. Fortunately a friend of his, the Chief of the Eunuchs in the Imperial Palace, conceived a wonderful plan. One day, when it was announced that the Emperor would walk in the garden, he took some sugared water, and wrote on a banana leaf 'Tshai Hsiang must be appointed Governor of Chhüanchow'. Ants immediately smelt the honey and gathered on the characters in vast numbers, to the stupefaction of the Emperor, who happened to pass the banana tree and saw them drawn up in the form of eight characters. The Chief of the Eunuchs watched him reading them over, and drew up a decree which the Emperor signed.

And so eventually the bridge did get built, with the help of various immortals.[a]

But megalithic bridges, sea-walls, and other public works in this province are connected particularly closely with the names of Buddhist monks, for whom bridge-building[b] was a beatific work (*puṇyakṣetra*), and who were probably themselves the engineers. Fa-Chhao,[1] a contemporary of Tshai Hsiang's, built one of the bridges at Chhüanchow, and another contemporary, the Abbot Kho-Tsun,[2] was a great financial encourager of such projects. About +1178 another monk, Shou-Ching,[3] built a 1,000-ft. long beam bridge at Nan-an.[4] But the most famous of the group was Tao-Hsün[5] (*d.* +1278), who built more than 200 bridges of various magnitudes in the province, and who was said to have completed what Tshai Hsiang had begun. His best work was the Phan-Kuang C.[6] near Chhüanchow, but he also built dykes and sea-walls. So did Po-Fu[7] in the following century (*d.* +1330), and it is interesting to read of him that he always used to sleep, doubtless under very rough conditions, with the workers 'on the job'.[c] Intensive research in the local histories and topographies

---

[a] Tr. Eberhard (5), no. 28, mod. auct.

[b] *Hsiu chhiao pu lu*[8] is the classic phrase.

[c] The tradition continued, though in the north-west rather than Fukien, with the notable Chhan monk Fu-Têng[9] (+1540 to +1613) who specialised (apart from brickwork temple halls and cast-iron pagodas) in long multiple-arch bridges. One at least of these (+1599) was called 'reinforced', and much

[1] 法超   [2] 可遵   [3] 守靜   [4] 南安   [5] 道詢   [6] 盤光橋
[7] 伯福   [8] 修橋補路   [9] 福登

of the region may well bring to light many more facts about these monastic technologists.[a]

Some idea of the appearance of these bridges may be gained from Figs. 821 (pl.) and 822, taken from Ecke (3 and 5).[b] These illustrate the Chiang-Tung Bridge, on the post-road from Kuangtung to Chhüanchow, and the Wan-An (Lo-Yang) Bridge north-east of that city, details of which, together with some of the other bridges of the region, are given in the accompanying table.[c] Both Marco Polo and John of Monte Corvino must

Fig. 822. Megalithic beam bridges; the Wan-An or Lo-Yang bridge across the Lo-Yang R. north-east of Chhüanchow in Fukien. Built +1053 to +1059; length 3,600 ft. A Chinese drawing from the *Wu Chiang-chün Thu Shuo* (c. +1690) by Wu Ying, after Ecke (5). Wu Ying was a military and naval commander on the Chhing side in the fighting against the Ming forces under Koxinga (Chêng Chhêng-Kung) and his successors in the sixties and seventies of the +17th century. In these campaigns the bridges had much strategic importance. The mêlée depicted in the foreground is an episode in Wu Ying's exploits against the ablest of the Chêng-Ming generals, Liu Kuo-Hsüan, in +1678. Repairs to the bridge are proceeding in the background, with the aid of a shear-legs derrick.

metal was used in them, though we do not now know exactly how. Thirty years earlier Fu-Têng had been in the south-east and perhaps learnt some of his arts there. We have to thank Dr Else Glahn for advance knowledge of her biography of him.

[a] The whole subject of the economic and technological role of Fukienese Buddhism at this time has been studied by Chikusa Masa-aki (1). The bridges are freely ornamented with *stupas*, auspicious Tantric *bīja* syllables (cf. Vol. 4, pt. 2, p. 231), and other signs of religious devotion.

[b] Cf. also Sirén (1), vol. 4, pl. 99A, B, C.

[c] Our list is of course far from complete. Many more bridges can be seen in the +18th-century coastal maps of Chhen Lun-Chhiung (see Vol. 3, p. 517). One can count as many as thirty-three in the charts attached to the *Fu-chien Yün Ssu Chih*[1] (Records of the Transportation Bureau of Fukien province), written by Chiang Ta-Kun[2] in +1553. Moreover, there are several discrepancies in the various accounts, which could only be cleared up by thorough field-work in the province itself. Both places and

[1] 福建運司志   [2] 江大鯤

Table 66. Some megalithic beam bridges of Fukien

	Name of bridge	Location	River spanned	Date of construction	Total length (ft.)	Breadth (ft.)	No. of spans	Greatest span length (ft.)	Weight of largest beams (tons)	Builder	References
1	Wan-Shou C.[1] or Wên-Chhang C.[5]	Fuchow[2] (Min-hou[6])	Min chiang[3]	+1297 to +1322	2,050 (including shorter continuation south of Nanthai[7] Island) 2,620	14½ 14½	36 46	45 —	c. 80	Wang Fa-Chu[4]	P3; FM, p. 79; L, 72, 264 ff.; B, pp. 333, 338[a]
2	Hung-Shan C.[8]	A little further up river Fu-chhing	Min chiang[3]	+1476	—	—	28	—	—	—	P3; mentioned by de Mendoza, +1577
3	Fu-Chhing C.[9]		Local tidal estuary	Between late Sung and early Ming	c. 800	—	17	c. 27	—	—	E5, pl. 21, fig. 3[b]
4	Lo-Yang C.[10,c] or Wan-An C.[12]	N.E. of Chhüanchow[11]	Lo-yang chiang	+1023, or more prob. +1053 to +1059	3,600	15	47 (now 121)	c. 65	c. 150	Tshai Hsiang	FM, p. 77; P3; E5, pl. 21, fig. 1; THC, p. 18, fig. 17; D; B, pp. 333 ff.; L, 74, 249 ff.
5	Phan-Kuang C.[13] or Wu-Li C.[14]	N.E. of Chhüanchow[11]	Lo-yang chiang	c. +1255	> 4,000	16	—	—	—	Tao-Hsün	THC, p. 19, fig. 18
6	Shun-Chih C.[15]	S.W. of Chhüanchow[11]	Chin chiang[16]	+1190 to +1211 (repaired +1341, +1472 and c. +1650)	1,500	—	—	—	—	—	P3; B, pp. 253, 332; mentioned by de Rada, +1575
7	Fou C.[17]	S.W. of Chhüanchow[11]	Chin chiang[16]	c. +1050 as a floating bridge; in stone, +1160	800 (now strengthened to take a main road)	17	130	c. 27	—	Fa-Chhao	ED, p. 94; THC, p. 19, figs. 19, 20
8	An-Phing C.[18] or Wu-Li C.[14]	An-hai[19]	Local tidal estuary	+12th	5,000	—	360	15	—	—	E5, p. 272
9	Thung-An C.[20]	Thung-an	Local tidal estuary	+1094, repd. +1294	1,000	—	18	60	—	—	P3

10	Chiang-Tung C.[21] or Chhien-Tu C.[24,d] (= 'Po-Lam' br.)	20 miles up river from Amoy, E. of Changchow[25]	Chiu-lung chiang[22]	+1190 as a floating bridge; stone piers +1214 and beams +1237	1,100	18	19	>70	>200	Li Shao[23]	FM, p. 75; P3; E3; E5; ED, pl. 71b; L, 35, 75, 382
11	Thung-Chin C.[26] or Lao-Chhiao-thou[28]	S. of Chang-chow[25]	Lung chiang[27]	End +12th	900	—	28	40	—	—	P3; E5, pl. 21, fig. 2; ED, pl. 72a
12	Kuang-Chi C.[29] or Hsiang-Tzu C.[34]	Chhaochow[30] (in Kuang-tung)	Han chiang[31]	Sung +1170 to +1192	1,630	20	21	—	—	Ting Chiu-Yuan,[32] Shen Tsung-Yü[35] et al.	Yao Yu-Chih[33],e THC, fig. 16; L, 255

(This has a 270 ft. pontoon segment to permit the passage of large ships)

a As in private duty bound, I celebrate the Yuan builder of this most impressive bridge, the only one of the group that I myself have seen, remembering that sunny May morning of 1944 when Dr Huang Hsing-Tsung and I were borne in rickshaws over its granite baulks to spend the first of several enjoyable days among the bookshops and hot springs of the city of Fuchow. The greater part of our collection of the mathematical classics was purchased there.

b The early Portuguese travellers described a very great bridge north of Hsing-hua[36] (mod. Phu-thien[37]) between Fu-chhing and the Lo-Yang C. (cf. Boxer (1), p. 333), but it does not seem to exist now.

c Here the characters follow the reading on an +18th-century MS. coastal map described by Mills (7, 8), which was similar to that reproduced as Fig. 210 in Vol. 3; but the more usual modern form is Lo-Yang.[38]

d The name 'Tiger Ferry', Hu-Tu,[39] though sanctioned by Ecke and others, appears to be wrong. The classical numinous phrase, a 'devout ferrying over into the beyond', was much more appropriate for a bridge designed by a Buddhist monk (cf. Fêng Yu-Lan (2), p. 129).

e Essay in TSCC, Chih fang tien, ch. 1340, i wên I, p. 84; cf. ch. 1335, Chhaochow hui khao 3.

ABBREVIATIONS

B   Boxer (1)
D   Dukes (1)
E3   Ecke (3)
E5   Ecke (5)
ED   Ecke & Demiéville (1)
FM   Fugl-Meyer (1)
P3   Phillips (3)
L   Lo Ying (1)
THC Thang Huan-Chhêng (1)

1 萬壽橋	2 福州	3 閩江	4 王法助
5 女昌橋	6 閩侯	7 南台	8 洪山橋
9 福清	10 洛楊橋	11 泉州	12 萬安橋
13 盤光橋	14 五里橋	15 順治橋	16 晉江
17 浮橋	18 安平橋	19 安海	20 同安橋
21 江東橋	22 九龍江	23 李韶	24 慶渡橋
25 漳州	26 通津橋	27 韓江	28 老橋頭
29 廣濟橋	30 潮州	31 韓江	32 丁久元
33 姚友直	34 湘子橋	35 沈宗禹	36 興化
37 莆田	38 洛陽	39 虎渡	

have crossed these bridges.[a] The number of spans in relation to the total length of the Fukienese structures is very variable because as the huge stones of the deck failed, later generations were unable to replace them, and had to have recourse to the insertion of new piers between the old ones. In many cases the piers are corbelled or cantilevered out from their bases, to give additional support to the trusses, and the piers themselves are in general markedly boat-shaped. The largest beams (which Pereira so much admired), those exceeding 200 tons in weight, have cross-sections 5 ft. high and 6 ft. broad. With bridges of this type the greatest difficulty was the foundations, which in many cases proved insufficient, so that in different places in Fukien a number of ruined giant beam structures are found. It is clear that whoever was really responsible for them, the main period of their construction was the +11th to the +13th centuries.

An interesting piece of experimental archaeology was done on these megalithic beams by Fugl-Meyer. With the facilities of the Huangpo Conservancy Board's yards at his disposal, he tested the strength of bars of stone similar to those which the medieval builders had used. The maximum tensile stress calculated on the usual theory of bending proved to vary from 437 lb. per sq. in. for the red granite to 1,010 lb. per sq. in. for the grey.[b] Assuming then the weight of the stone to be 160 lb. per cu. ft., and the highest superimposed load to be 80 lb. per sq. ft. of the bridge deck, it was possible to calculate the upper limit of length for a single-span beam, using the best figure (1,010 lb. per sq. in.) obtained above. The resulting figure was 74 ft., exactly corresponding to the maximum lengths of spans met with in the bridges of the Fukien type.[c] The Sung builders had therefore reached the utmost limit of practicability, for baulks of any greater length will break under their own weight. An interesting historical problem remains as to how they found it; whether by bitter experience of failures in practice, or perhaps by the preliminary setting-up of actual empirical strength-of-materials tests at the quarries.

The Fukienese type of bridge-building gives one an impression of uncouth strength and determination, with a prodigious waste of material, as regardless of cost as the most massive Roman masonry. Yet it was a purely provincial style, for, as the sequel will show, the builders of arch bridges in Northern China had already attained six or seven centuries earlier an economy of materials and a grace of design which was not approached in Europe until the dawn of the Renaissance. Nevertheless the giant beam bridge is not to be despised, and holds its own for certain purposes at the present day. The Sone Bridge in Uttar Pradesh, with its twenty-two pre-stressed concrete spans of 150 ft. each, is a true descendant of the granite bridges of medieval China.

---

bridges have more than one name each, and there are the usual difficulties with authors who give only romanised forms of names. The general facts, however, which are all we need, are not in dispute. See further Lo Ying (1), pp. 74 ff.

[a] So also Domingo de Navarrete in +1659, whose graphic description, complete with legends about Tshai Hsiang, will be found in Cummins (1), vol. 1, pp. 143 ff. Sir Wm. Chambers (2) described the megalithic beam bridges of Fukien again in +1773.

[b] Experimental span lengths varied from 5 to 10 ft., and baulk dimensions from about 6 in. to 1 ft. 6 in. cross-section.

[c] For several of these it could be calculated that the maximum tensile stress with load is 900 lb. per sq. in. and without load 670 lb. per sq. in., assuming a cross-section 5 ft. high and 6 ft. broad.

Before leaving the realm of the beam, a few moments must be given to drawbridges and pontoon bridges. Drawbridges did not figure prominently in Chhing fortification practice, bridges over moats being just made easily dismountable at need. But in earlier times mechanically movable bridges had been used a good deal. Miguel de Loarca, in his 'Verdadera Relación' of the overland journey of the Spanish envoys from Amoy to Fuchow in +1575, described a great bridge of many spans outside Chhüanchow, perhaps the Shun-Chih C., which had a drawbridge at its end, and the Spaniards must have recognised a device of this kind when they saw one.[a] Nearly a thousand years earlier, the *Thai Pai Yin Ching*, that Taoist military encyclopaedia of +759, had an entry for the subject in its section on the defence of cities. Li Chhüan wrote:

Turning Portal Bridge (*chuan kuan chhiao*[1]). For this kind of bridge flat planks are used as the deck. At (one) end of the bridge there is a horizontal bolt (*hêng thien*[2]), and when this is removed the (whole) bridge turns away from the gate, so that soldiers and horses cannot pass over, and all of them fall into the moat. (Formerly, the King of) Chhin used such a bridge as this (in the attempt) to kill Tan, Prince of Yen.[b]

The story about Prince Tan occurs in the ancient fictional biography *Yen Tan Tzu*, written about the end of the +2nd century. The prince, escaping in −232 from the ruler of Chhin State,[c] where he had been ill-treated as a hostage, had to cross a bridge equipped with some release mechanism (*chi fa chih chhiao*[3]) whereby the king intended to kill him, but he passed over before it could operate (*Tan kuo chih chhiao wei pu fa*[4]).[d] Unfortunately the mechanism of these ancient 'drawbridges' is not quite clear from Li Chhüan's description. They were evidently not raised into the air like the draw-bridges of medieval Europe, nor were they swing-bridges like Leonardo's, nor escalator-bridges like those of Martini.[e] It seems clear that they either turned over sideways (which might have been rather effective), or else swung on hinges so as to drop into the moat probably at the end nearer the gate and wall. Later on (p. 190) we shall refer to a +5th-century submersible suspension bridge which acted as a drawbridge.

Floating pontoon bridges are of high antiquity in Europe since the Greek Mandrocles of Samos is credibly reported to have constructed one over the Bosphorus for an expedition of Darius I against the Scyths in −514.[f] But this date (*pace* Sarton) may well be anticipated by the first Chinese reference, which takes us back to a −8th-century text. Let us listen to Kao Chhêng discussing the matter in the Sung.[g]

---

[a] See Boxer (1), p. 332.

[b] Ch. 36, p. 3*b*, tr. auct. The passage is also found in *Hsü Shih Shih*, p. 43*b*, but somewhat garbled and indeed misleading.

[c] Chuang Chêng, later the first emperor of united China, Chhin Shih Huang Ti. Cf. Vol. 1, pp. 97 ff., 100 ff., and here, pp. 551 ff.

[d] P. 1*a*, tr. Chêng Lin (1); Franke (11).

[e] Cf. Uccelli (1), p. 280. A late Chinese example will appear on p. 347 below.

[f] Sarton (1), vol. 1, p. 76; Diels (1); Herodotus, IV, 87, 88. See also Herodotus on Xerxes I and his engineer Harpalus, *c.* −480, VII, 25, 33–6; and for the background of both works Huart & Delaporte (1), pp. 251 ff., 263 ff.

[g] *Shih Wu Chi Yuan*, ch. 7, p. 11*a*, tr. auct.

[1] 轉關橋   [2] 橫栓   [3] 機發之橋   [4] 丹過之橋爲不發

The *Chhun Chhiu Hou Chuan*[1] says[a] that in the 58th year of the Chou High King Nan[2] (−257), there was invented in the Chhin State the floating bridge (*fou chhiao*[3]) with which to cross rivers. But the Ta Ming[4] ode in the *Shih Ching* (Book of Odes)[b] says (of King Wên) that he 'joined boats and made of them a bridge' (*tsao chou wei liang*[5]) over the River Wei.[6] Sun Yen[c] comments that this shows that the boats were arranged in a row, like the beams (of a house), with boards laid (transversely) across them, which is just the same as the pontoon bridge of today. Tu Yü[d] also thought this...Chêng Khang-Chhêng[e] says that the Chou people invented it and used it whenever they had occasion to do so, but the Chhin people, to whom they handed it down, were the first to fasten it securely together (for permanent use).

For the Chhin State's bridge of boats across the Yellow River there is much better authority than Chhen Fu-Liang, for, as we have just seen (p. 150), Ssuma Chhien himself recorded its building by King Chao Hsiang of Chhin at the date here given.[f] With periodical renewals this Phu-Chin C.,[7] as it came to be called from its geographical position near the border of Shansi north of the great bend at Thungkuan, lasted for very many centuries, the senior and most famous of the three great floating bridges across the Yellow River.[g] The *Chhu Hsüeh Chi* encyclopaedia of +700 mentions it[h] and just now we found the Japanese monk Ennin admiring it in +840.[i] He estimated its length at about 1,000 ft. We have long been familiar with this bridge, for in Vol. 4, part 1, we read about the casting of iron anchor weights in the form of oxen, for fixing the cables, in +724, and how the monk Huai-Ping cleverly retrieved these in +1065 after they had been carried away in a severe flood.[j] From the +8th century also we have much information about the upkeep of these important bridges, notably in the MS. fragment of the Thang government ordinances of the Department of Waterways (+737).[k] Here we learn of the bridgekeepers (*shou chhiao*[8]), watermen (*shui shou*[9]) and maintenance artisans (*chu mu chiang*[10]) kept permanently on duty and exempt from military or other service, active in the defence against dangerous floating lumber in flood time and watchful to undo all fastenings when the ice set hard. Replacement

---

a By Chhen Fu-Liang[11] of the Sung.

b III (1), 2; Mao no. 236; Legge (8), vol. 2, p. 435; Karlgren (14), p. 188; Waley (1), p. 262.

c The +3rd-century philologist.

d Eminent engineer connected with water-mills (cf. Vol. 4, pt. 2, pp. 370, 393 above), astronomer, historical geographer, etc.

e Chêng Hsüan, the +2nd-century commentator.

f *Shih Chi*, ch. 5, p. 34a (cf. Chavannes (1), vol. 2, p. 94). *Chhu Hsüeh Chi*, ch. 7, p. 23b, gives the name of the actual builder as Chen Pên-Chin,[12] a son of the reigning prince of Chhin.

g The others were the Ta-Yang C.[13] at Shenchow[14] built by Chhiu Hsing-Kung[15] in +637 at a Northern Chou check-point, and the Ho-Yang C.[16] near Mêng-hsien[17] built by Tu Yü in +274. Together with the Hsiao-I C.[18] across the Lo River below Loyang they formed the four celebrated bridges of boats enumerated in Thang sources such as the *Thang Liu Tien* summarised in *Jih Chih Lu* (+1673), ch. 12, p. 19. See Lo Ying (1), pp. 280 ff., 301 ff.

h Ch. 7, p. 23a, b. Extensive repairs were carried out in +721.

i Reischauer (2), p. 280; and p. 148 above.   j Pp. 40 ff.

k Studied and translated by Twitchett (2). The articles on bridges are 20, 24, 26, 29, 30, 31, 32, 33.

---

1 春秋後傳	2 報	3 浮橋	4 大明	5 造舟爲梁
6 渭	7 蒲津橋	8 守橋	9 水手	10 竹木匠
11 陳傳瓦	12 鍼奔晉	13 大陽橋	14 陝州	15 丘行恭
16 河陽橋	17 孟縣	18 孝義橋		

pontoons (*fou chhiao chiao chhuan*[1]) were built in special shipyards and kept available up to half the total number, while provision was made for the manufacture, storage and periodical testing of the necessary bamboo hawsers.[a] The Phu-Chin C. was always an important strategic crossing. In +1049, for example, during one of the campaigns of the Chhi-tan Liao against the Hsi-Hsia State, it was successfully held as the main line of communication by the Liao general Hsiao Phu-Nu,[2] and his country's troops retreated across it.[b] As in all other civilisations, pontoon bridges often figured in military operations because so easily constructed and so readily removed—thus the Yangtze was bridged twice in this way, in +33 and +36.

The mention in the passage quoted above of Tu Yü,[3] the eminent engineer of the Chin period (+222 to +284), is interesting, because he himself constructed the second famous floating, or pontoon, bridge across the Yellow River, the Ho-Yang C. north-east of Loyang in +274. According to the story in the *Chin Shu* many of the imperial advisers urged that the plan was impossible, since the saints and sages had done nothing like it (*li shêng hsien erh pu tso chê, pi pu kho*[4]), but Tu Yü had the emperor's confidence, and successfully completed the work.[c] The dynastic history records the modest reply he made when the emperor came to inspect the bridge and drank a toast to him in the presence of all the court.[d]

References to pontoon bridges in the Thang and other periods are not infrequent in the encyclopaedias.[e] For example, there was Chang Chung-Yen's[5] Jurchen Chin pontoon bridge of +1158, Thang Chung-Yu's[6] use of iron chains for securing one in +1180, and another made with inflated skin rafts by the Chhi-tan engineer Shihmo An-Chê[7] in the Yuan period. Many floating bridges are mentioned in the account of the travels of the Taoist Chhiu Chhang-Chhun about +1221.[f] A famous one over the Amu Darya was constructed in a single month by Chang Jung,[8] the chief engineer of Chagatai, Chingiz Khan's second son.[g] In the +15th century there was another at Lanchow which greatly impressed an embassy from the Timurid Shāh Rukh,[h] and a

---

[a] Cf. Vol. 4, pt. 2, p. 64; and p. 191 below.

[b] *Liao Shih*, ch. 87, pp. 4*b*, 5*a*; tr. Wittfogel & Fêng (1), p. 166.

[c] Ch. 34, p. 9*b*.

[d] One event in connection with this bridge illustrates a general principle of action adopted by the bureaucratic administration in many dynasties. About +512 one of the princes of the Northern Wei, Yuan Chhang,[9] acting as governor of Honei, ordered that all empty carts leaving the capital at Loyang should transport stone for the abutments or causeways at each end of the bridge. Commandeering of transport, especially when done so as not to interfere with normal movements, was an important adjunct to the *corvée* itself. The account is in *Wei Shu*, ch. 14, p. 8*a*.

[e] E.g. *TSCC, Khao kung tien*, ch. 31. Ch. 32, p. 4*a*, reproduces a literary-philosophical essay by the Thang scholar Chang Chung-Su[10] (*fl.* +813) on the bamboo cables for floating bridges—the power of small things individually weak, such as bamboo strips, when plaited together in numbers. For similar ideas in relation to technology, cf. Vol. 4, pt. 2, p. 313, and below, p. 641.

[f] Tr. Waley (10), pp. 90, 112, 120.

[g] *Yuan Shih*, ch. 151, p. 19*b*. So close was the connection in the Mongolian mind between inflated animal skins and bridges, Prof. O. Lattimore tells us, that the same word, *khur*, came to be used for both.

[h] Yule (2), vol. 1, p. 278. This was the Chen-Yuan C.,[11] composed of 24 pontoons secured by iron chains. A model is in the Lanchow Museum.

---

[1] 浮橋脚船	[2] 蕭蒲奴	[3] 杜預	[4] 歷聖賢而不作者必不可
[5] 張中彥	[6] 唐中友	[7] 石抹按只	[8] 張榮 [9] 元萇
[10] 張仲素	[11] 鎮遠橋		

century later the Portuguese traveller, Galeote Pereira, coming into the country from the opposite end, had something to say of the bridge of boats, 'linked all together with two mighty chains' at Ganhsien in Chiangsi.[a] Huge cast-iron mooring-posts were also used.[b]

While travelling in China during the Second World War, I came across several floating bridges, mostly in Chiangsi and Kuangtung.[c] A special type occurs on the western border of Szechuan, near Yachow, where at certain times of the year a raft of bamboo is moored afloat from shore to shore, and across it a gangway, good at least for pedestrians, is laid.[d] Bridge pontoons are shown in Ming technical books.[e]

All this, however, has thrown no light on the oldest use of the device. Although the ode in the *Shih Ching* certainly refers to the founder of the Chou dynasty Wên Wang in the −11th century, to insist on dating it at that time would be most unwise, and the −8th or −7th centuries would be quite enough to do it justice. Its phrase about the bridge echoes down through the Han period, as in the *Tung Tu Fu* (Ode on the Eastern Capital), written by Pan Ku about +87, and much later still.[f] There can thus be no doubt that the pontoon bridge is a very old institution in China, and since the *Shih Ching* ode must date from the first half of the −1st millennium, it looks as if the 'artisans of King Wên', whoever they were, took precedence of Mandrocles of Samos. But it would not be at all surprising to find that Babylonian skill outdated both.

(2) CANTILEVER BRIDGES

In all southern and western parts of China, and especially near the Himalayas and Tibet, the people responded to the challenge of gorges and turbulent streams up to 150 ft. wide by the construction of cantilever bridges (*kung mu chhiao*[1]). The two cantilever arms were built out with superimposed timbers from the sides in various ways (see Fig. 823), and then connected in the middle of the span by long wooden beams. Additional struts (strainer-beams) may or may not be present. The abutments were, and often are, of timber filled and weighted with stones,[g] but heavy masonry works in which the cantilever arms are embedded also occur.[h] The principle is applied, moreover, to long bridges with many piers, the cantilever arms then springing from the

---

[a] Boxer (1), p. 33, who gives other references to it. On the bridge at Seoul in Korea see Underwood (1), p. 64.

[b] A floating bridge near Kweilin built in +1507 had iron chains 1,000 ft. long, iron anchors, and iron mooring-posts each 18 ft. in length; see the essay of Pao Yü[2] in *TSCC, Khao kung tien*, ch. 32, p. 15b. So also R. D. Thomas (1), p. 43, for Wuchow in Kuangtung.

[c] Photographs of examples still in use will be found in Thang Huan-Chhêng (1), figs. 110, 112, 113, 114, 115 and 116. I crossed one in 1964 near Ching-tê-chen.

[d] Descriptions in Hosie (4), p. 93, and Upcraft (1); photograph in Thang Huan-Chhêng (1), fig. 111.

[e] E.g. *Lung Chiang Chhuan Chhang Chih* (+1553), ch. 2, p. 22a.

[f] *Wên Hsüan*, ch. 1, p. 19a; tr. Hughes (9), p. 57.

[g] See, e.g., Thang Huan-Chhêng (1), figs. 21, 23, 24.

[h] See, e.g., Thang Huan-Chhêng (1), figs. 22, 26, 27, 35; Lo Ying (1), pp. 35, 58 ff., 62 ff. Lord Macartney met with one of each of these types on his travels: cf. Cranmer-Byng (2), pp. 108, 194. For another photograph, the bridge at Tien-tang across the Ta-tung Ho near Sining, see Farrer (2), opp. p. 112.

[1] 肱木橋　　　[2] 包裕

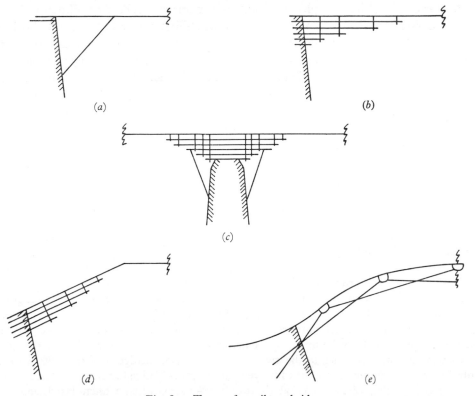

Fig. 823. Types of cantilever bridges.

(a) Simplest type of strainer-beam support (*pa tzu chhêng chia*[1]), used in small bridges, e.g. at Lungyen in Fukien (Thang Huan-Chhêng (1), fig. 30).

(b) Horizontal cantilever, used over river gorges, crossing e.g. the Brahmaputra at Phomi in Tibet (Thang Huan-Chhêng (1), fig. 24). Another, on the Kansu–Chhinghai road, is figured by Liu Tun-Chên (1), pl. 6b. A very fine example of a bridge of this kind with several spans is the covered bridge at Hsin-ning in Hunan (Anon. (37), fig. 149; Thang Huan-Chhêng (1), figs. 41, 42); and an even more magnificent one (Fig. 825, pl.) is the Chhêng-Yang C. at the place of that name north of San-chiang in northern Kuangsi (see overleaf). This bridge, though little known, was rightly included in the postage-stamp series issued by the Chinese government in 1962.

(c) Combination of horizontal cantilever and strainer-beam supports, often found in the many-spanned bridges of Hunan, e.g. at Li-ling over the Lu-shui (Fig. 824, pl.), south-east of Chhangsha (Boerschmann (3a), fig. 198; Parsons (1), p. 211; Thang Huan-Chhêng (1), fig. 22, cf. also figs. 34, 35); but also over the Mo-chiang and elsewhere in Yunnan to the south-west (Thang Huan-Chhêng (1), fig. 25; Mock (1), p. 33; Fugl-Meyer (1), fig. 53); and in Chiangsi to the south-east, as at Ganhsien (Beaton (1), p. 14), a bridge well known to me.

(d) Soaring cantilever, used across gorges or routes of water-traffic, e.g. the famous covered bridge in the city of Lanchow, Kansu (Geil (2), p. 319; Thang Huan-Chhêng (1), figs. 26 and 27, and the 'Extended Arm' bridge (Shen-Pei C.[2]) in the Muli District of south-west Szechuan (Thang Huan-Chhêng (1), fig. 23; Fugl-Meyer (1), fig. 54). This form occurs also in Tibet, e.g. the Lhasa bridge built on tall log-crib piers (Thang Huan-Chhêng (1), fig. 21).

(e) Multi-angular soaring cantilever, in which the main beams spring forth at two or more angles with the abutments, known mainly from the painting by Chang Tsê-Tuan (c. +1120) of a great bridge at the Sung capital Khaifêng (Chêng Chen-To (3); Anon. (37), fig. 58; Thang Huan-Chhêng (1), figs. 28, 29). See Fig. 826 (pl.).

[1] 八字撐架   [2] 伸臂橋

heads of each pier (Fig. 824, pl.). Bold single-span structures occur commonly all over western China from Kansu to Yunnan as well as in Tibet, the arms sometimes horizontal but often 'soaring' upwards from the abutments at an angle of about 25° to join the level connecting deck beams. Horizontal-arm cantilever bridges of many spans occur particularly in Hunan. These are impressive enough in themselves, but when combined with covered housings and pavilions over each pier, as in the south-west of the province and over the border in Kuangsi, they produce some of the most superb structures of traditional Chinese bridge-building (Fig. 825, pl.).[a]

In spite of the fact that boarding is usually placed in position to protect the cantilever arms from the weather, there is bound to be decay in the woodwork, and for this reason none of these bridges is very old. However, in the late +11th-century *Shêng Shui Yen Than Lu*[1] (Fleeting Gossip by the River Shêng) due to Wang Phi-Chih,[2] there is a clear reference to the building of a cantilever bridge.[b]

In the south-west of the city of Chhingchow[3,c] the land is very hilly, and for long the town was cut through by the Yang River (Yang Shui[4]) into two parts. Originally wooden piers were set up in the stream, and a bridge supported upon them, but when the water rose in its autumn floods, these piers were frequently damaged, and the bridge became unsafe. The magistrates were always worried about this. In the Ming-Tao reign-period (+1032 to +1033) the governor of Chhingchow, Hsia Ying-Kung,[5] was greatly desirous of overcoming the difficulty (and gave encouragement to) a certain retired prison guard, who was known for his ingenuity. He piled up large stones to make firm abutments, and then by connecting several dozen great beams together, he threw across the river a kind of 'flying bridge' (*fei chhiao*[6])[d] with no central pier. Though now more than fifty years in use, this bridge has never suffered injury. Then in the Chhing-Li reign-period (+1041 to +1048), when Chhen Hsi-Liang[7] was governor of Su (-chou)[8,e] he noted that the bridges on the Pien[9] Canal[f] often fell into disrepair, so that they damaged or destroyed official shipping, hurting and even killing travellers. He therefore ordered that (they should be rebuilt) following the pattern of the Flying Bridge of Chhingchow. Now all the bridges between the Fên[10] River[g] and the Pien[9] Canal are of this type, which is a great benefit to communications. The common people call them 'rainbow bridges' (*hung chhiao*[11]).

The cantilever principle was thus most welcome since it obviated the necessity for central piers, obstacles particularly liable to flood damage and prone to get in the way

---

[a] This is the Chhêng-Yang C.[12] across the Lin-chhi Ho in the autonomous district of the Thung people in Kuangsi. Each of the piers between its 75 ft. spans is crowned by a pavilion of four or five storeys, and the decks are roofed by a high and spacious gallery. See Thang Huan-Chhêng (1), p. 35, fig. 39; Lo Ying (1), p. 27; Anon. (37), fig. 173.

[b] Ch. 9, p. 6a, tr. auct. Cf. *CCL*, no. 9, p. 171 and *Sung Shih*, ch. 298, p. 22a.

[c] Presumably modern I-tu,[13] in Shantung (like the Shêng River), on the north-eastern slopes of the Thai-Shan massif.

[d] This term is loose and not technical, for it is found applied also to what were certainly tall humpbacked arch bridges across canals, and segmental arch bridges springing across gorges.

[e] In Anhui province near the Shantung border.

[f] See pp. 307 ff. below.

[g] In Shansi province, running past Thaiyuan. Thus what Wang Phi-Chih is saying is that from Shansi through Honan and Anhui to Chiangsu everybody began to build these large-span timber bridges.

[1] 澠水燕談錄   [2] 王闢之   [3] 青州   [4] 洋水   [5] 夏英公
[6] 飛橋   [7] 陳希亮   [8] 宿州   [9] 汴   [10] 汾
[11] 虹橋   [12] 程陽橋   [13] 益都

of navigation. Forming a veritable illustration to this passage is the depiction of a great bridge outside the city of Khaifêng, the capital of the Northern Sung, painted in a scroll by Chang Tsê-Tuan about +1120 on the very eve of its capture by the Chin Tartars. With its wonderful detail of daily life, this Chhing-Ming Shang Ho Thu[1] (Going up the River at the Spring Festival) has helped us before and will do so again.[a] The bridge which it shows (Fig. 826, pl.) is of the multi-angular soaring cantilever type borne not only upon about ten great beams rising out of the abutments at some 40° and supporting a series of central horizontal members, but also upon another set interdigitating with them and rising at some 55° to sustain corresponding pairs of sloping members which meet at the crown of the structure; the whole being trussed together with bars and collars similar to those used with the more ordinary bundles of parallel cantilever beams.[b] The deck of the bridge is thronged with busy life, and large river ships are being worked to and fro underneath.[c]

The further historical elucidation of these beautiful forms of bridge-building is not a very easy matter. A clear description of a timber cantilever bridge with stone abutments is given in the *Sha-Chou Chi*[2] written by Tuan Kuo[3] in the +4th or +5th century; it refers to one then built by the Thu-yü-hun[4] people (a Hsien-pi tribe) across the river of the Tunhuang oasis. This seems to be the oldest textual record. According to local tradition at Lanchow, however, the cantilever bridge in the western suburbs (Hsi-Chin C.[5]) dated from the Thang period.[d] By the time of the Ming (+16th century) the technique which had produced the Khaifêng bridge seems to have been lost (or was at any rate unknown to metropolitan scholars), for the Ming copies of Chang Tsê-Tuan's painting replace the cantilever structure by a single great arch.[e] But possibly in the meantime its simple lattice-like geometry had travelled to Europe, for an almost identical design was proposed by Leonardo da Vinci (*c.* +1480) for a military bridge.[f] Searching back in the Chinese literature, much will depend on the extent to which the term 'rainbow bridge', mentioned at the end of the quotation from Wang Phi-Chih, is to be taken as a technical rather than a popular descriptive term. For example, in the late +3rd century Chou Chhu[6] wrote in his *Fêng Thu Chi*[7] (Record of People and Places): 'In front of Yang-hsien[8,g] there is a great bridge 720 ft. long from north to south, and very high in the middle so that it looks like a rainbow. It was built there by the Lord Yuan[9].'[h] It is not difficult to imagine this as a series of

[a] Cf. Vol. 4, pt. 2, pp. 273, 318, and below, Figs. 923, 976, pls. Cf. pp. 115, 359, 463, 648.

[b] An interesting modern model is figured by Thang Huan-Chhêng (*1*), fig. 29.

[c] The painting of the bridge is now often reproduced, e.g. in Anon. (*37*), fig. 58. Lo Ying (*1*), pp. 45 ff., 67 ff., treats this type of bridge as a wooden arch, and says that some small examples still exist in Fukien and Chekiang. It would join with his hemi-decagonal stone cantilever arches as precursor of the 'Gothic' arch. See p. 171 below.

[d] Yang Chhun-Ho (*1*), quoted by Lo Ying (*1*), p. 59. I knew and admired this covered bridge (cf. Thang Huan-Chhêng (*1*), figs. 26, 27) in 1943, but afterwards it was taken down. A model is in the Provincial Museum. Mao I-Shêng (*1*) mentions another cantilever bridge in Kansu, but gives no reference.

[e] As is seen, for example, in the Ming set reproduced in full by Priest (*1*).

[f] See Uccelli (*1*), p. 281, from Cod. Atl.

[g] Modern I-hsing,[10] a town up-country behind the Thai Hu Lake in Chiangsu.

[h] Cit. *Chhu Hsüeh Chi*, ch. 7, p. 25*a*, tr. auct.

[1] 清明上河圖　　[2] 沙州記　　[3] 段國　　[4] 吐谷渾　　[5] 西津橋
[6] 周處　　[7] 風土記　　[8] 陽羨　　[9] 袁君　　[10] 宜興

short spans of parallel cantilever pattern like those of the Hunan type, with a single soaring, perhaps multi-angular, cantilever span in the centre to allow the passage of traffic under sail. But it may equally well have been a series of arches with a tall central arch.

One fact possibly relevant to the origin of the cantilever bridge in China is that in Han times pavilions were sometimes built out over lakes on corbelled brackets resembling cantilevers. Such structures were called *thi chhiao*[1] (ladder bridges). Chavannes reproduced a relief showing this many years ago,[a] and in 1958 I photographed an extremely similar one which is preserved at the Temple of Confucius at Chhüfou in Shantung (Fig. 827, pl.). A pavilion somewhat of this kind still exists at the Yen-Yu Temple at Hanoi in Vietnam and has been described by Yü Ming-Chhien (1). Dating from +1049 (repaired in +1105), it is borne on a cylindrical stone column rising out of a pool, and upheld by wooden cantilever corbels and strainer-beams. But the roof-supporting system of corbel brackets in Chinese architecture as a whole (cf. pp. 92 ff. above), and not only such corbels as these, presents itself as the background of the Chinese cantilever bridge.

The comparative approach does not get us very far either, at least in the analytical sense. The area of this type of bridge extends well outside China throughout the Himalayan region, as we have already seen. Robins illustrates a good example from Srinagar in Kashmir,[b] and Uccelli gives one from Bhutan.[c] Besides Leonardo, Villard de Honnecourt drew several types of cantilever bridge;[d] and if the structure which Trajan threw across the Danube in +104, and which is depicted on his famous Column of +113, was (as it seems to have been) a cantilever bridge of many spans on stone piers, the computed distance of 170 ft. for each span may not have been impossible.[e] It would thus have been similar to the Hunan type, but it seems to have been an isolated effort in occidental antiquity. The enormous wooden bridges erected in Germany by the Grubenmann brothers between +1755 and +1758 were a combination of the cantilever and arch reticulate truss principles; the longest span is said to have attained 390 ft.[f] But simple cantilever bridges had been built throughout the Middle Ages in the Savoy Alps. It would be interesting to know the original home of the basic design, and the stages of its spread, but the invention seems to have taken place so early that it is difficult now to trace its course through history. Perhaps if the Himalayan massif was the zone of origin of both cantilever and suspension bridges we

[a] (9), no. 1246; reproduced in Bulling (2), p. 28 and fig. 98/104. The stone was from Liang-chhêng Shan.[2] Other examples will be found in Pao Ting et al. (1). It was Mr Courtney Archer who first drew our attention to this point.

[b] (2), p. 94. Cf. Gill (1), pp. 116, 273; Kingdon Ward (4), (16), p. 192, for Tibet; Tyrrell (1), p. 71; H. S. Smith (1), fig. 2. I am grateful to Sir Evelyn Howell for a photograph of another.

[c] (1), p. 278. Cf. Mock (1), p. 31 (an old print).

[d] Hahnloser (1); Davison (11); Uccelli (1), p. 278.

[e] Davison (11); Jenkin (1); Neuburger (1), p. 468.

[f] Jenkin (1); Uccelli (1), p. 683; Brangwyn & Sparrow (1), p. 141 (who confuse them with suspension bridges). But there is doubt whether any spans actually built exceeded 200 ft. (cf. Mock (1), p. 37).

[1] 梯橋   [2] 雨城山

might expect them to occur combined, and this does in fact occasionally happen.[a] It is also possible, though in China unusual, to build a cantilever bridge in stone.[b] This brings us to the arch.

(3) ARCH BRIDGES

The generally accepted view used to be that the arch was an Etruscan invention of perhaps the −5th century, and that its Italian origin accounted for the fact that the Romans made such great use of it while the Greeks used it hardly at all. It is true that Greek architects did not employ it before about −300 and never much even then.[c] Argument has also centred round the use of the arch in India, for some, such as Fergusson, believed that it was very rare in pre-Muslim times, being associated only with Buddhism, while later writers (e.g. Havell) held this to be an over-statement.[d] All such discussions[e] have now entered a different phase with the establishment of the fact that the arch, the vault, and the dome were alike familiar to Sumerian Mesopotamia.[f] This makes it easier to understand the appearance of the arch and the barrel-vault in full flower in Chhin and Han tombs,[g] and to suppose that the use of the arch for city-gates and bridges was well understood by the Chou people, if not indeed by the Shang.[h]

[a] For a picture of a vertiginous unrailed packhorse bridge with at least one point of suspension at the end of a cantilever arm, see Bonatz & Leonhardt (1), p. 9; it crosses the Yarkhun in Chitral.

[b] A Chekiang example is given by Thang Huan-Chhêng (1), fig. 47; Liu Tun-Chên (1), pl. 7b. See further Lo Ying (1), pp. 38 ff.

[c] Cf. Robertson (1).

[d] The question is in fact complicated. The earliest surviving specimens of structural vaults in brick or masonry date from between the +4th and the +6th centuries (Ter, Chezarla, Bhitargaon, Bodh Gaya). But there are many representations of timber 'arches' in rock-cut caves (Karle, Bhaja, Barabar) going back to the −3rd century, and in reliefs (Bharhut, Mathura, Amaravati), which go back to the −1st. This is the 'caitya-arch'. Perhaps the most satisfactory recent reference is Coomaraswamy (6), p. 73. We are indebted to the kindness of Dr F. R. Allchin for advice on this subject.

[e] Till recently the true arch was not suspected in the Amerindian cultures, which avoided it by the extensive use of corbel vaulting. But Befu & Ekholm have now reported a true barrel-vault in a Maya structure of the late classic period at Campeche described by Ruppert & Denison.

[f] Cf. Woolley (2).

[g] True arches in the form of brick barrel-vaulting with a high semi-ovoid or beehive-shaped cross-section (but undeveloped keystone arrangements) were first found at Ying-chhêng-tzu near Ta-lien (Port Arthur) in Manchuria (Liaoning), by Mori & Naito (1) in a Han tomb of the ±1st century with fresco-paintings. Many more perfect examples are now known, notably that at Wang-tu in Hopei, in the tomb of +182 which has similar Han paintings on the walls (cf. Ao Chhêng-Lung et al. (1); Watson (1), pl. 98 and pp. 28 ff.). A barrel-vault covering a tomb near Peking was reported and figured by Chêng Chen-To (4); and many such structures at Loyang have been systematically reported on in Anon. (62). In another tomb, also of the Later Han, a model of which is to be seen in the Kuangchow Museum (Canton), the commoner corbel-vaulting is surmounted by a flat true dome (cf. p. 140 above). Barrel arches continued in use for tombs until a very late date; at Ta-ying-tzu in Jehol, for instance, there is a fine vault of this kind in the tomb of a Liao prince who died in +959 (Watson & Willetts (1), site 37). And some Sung tombs (e.g. one of +1056) have beehive-shaped domes (An Chin-Huai, 1).

[h] In a special paper Iida Sugashi (1) has considered the origin and development of the arch in Chinese building technology. His discussion centres round the words kuei[1] or khuei[1] and kuei tou[2] or kuei tou[3] which occur quite often in pre-Han classical writings. Kuei[1] is defined as a small door rounded at the top like a kuei[3] 'sceptre' or symbol of office, and tou[2,3] is an opening, so they may be taken to mean arched orifices. The relative unimportance of the arch in Chinese building technology is of course connected with the preference for using wood, brick, plaster and tile, rather than stone, in architecture

[1] 閨   [2] 閨竇   [3] 圭竇

In antiquity, both West and East, arches were invariably semicircular.[a] This type will not conveniently bridge gaps of more than about 150 ft. at the most,[b] and the average spans of Roman semicircular bridges (such as the Pons Milvius of −109 and the Pons Fabricius of −62) range between 60 and 80 ft., while aqueduct arches are generally much less, about 20 ft.[c] The longest surviving Roman arch is that of the Pont St Martin near Aosta which spans 117 ft.[d]

The essentials of the arch itself, with its ring of shaped voussoir stones culminating in the key-stone at the crown, were naturally identical in Roman and Chinese work. But Fugl-Meyer, himself a bridge engineer, found when he examined Chinese methods that the whole construction was so unlike in the two kinds of structure that there had never been, he was convinced, any contact between Roman and Chinese bridge-builders. The Roman bridge arch, he said, is of very massive masonry, so bulky that in some cases the chalk in the middle of the structure has remained plastic to the present day and will still harden when exposed to air. It was also over-dimensioned, unnecessarily wasteful of material. On the other hand the characteristic Chinese bridge arch is a thin stone shell loaded with loose filling, and this is kept in place by side-walls of stone, and topped with stone slabs to form the deck and approaches.[e] While such a method was most economical of materials it was also much more liable to deformation caused by rise or fall of the foundations, and the Chinese were therefore driven to invent a number of subsidiary devices, bonding-stones to hold the casing together, and built-in shear-walls running through the structure to counteract deforming forces. 'Since the Chinese bridge was constructed of a minimum of material', wrote Fugl-Meyer, 'it was an ideal engineering product, fulfilling both technical and engineering requirements.'

It will be worth while to look more closely at these structural details before considering a few of the great arch bridges (kung chhiao[1]) of China.

For the arches themselves several different types of bonding (chhüan[2]) were used as the stones were laid down upon the wooden centering (Fig. 828). In the transverse method (ping lieh,[3] 'abreast') a number of essentially separate arches were built up side by side in a series till the required width of the bridge was reached (I). The stones were often elongated like arcs of the circle, and the adjacent arches always bonded by alternate jointing. This was the method used for the great segmental arch bridges

(cf. p. 90 above); except, of course, for tombs, and major gates in the walls of cities, palaces and temples. We may find a reflection of this in the remark of a modern Chinese engineer, Tan Pei-Ying (1), p. 124, that the semicircular arch was never used in domestic architecture, because it was not considered lucky.

[a] With a qualification concerning arches not free-standing (see p. 179 below).

[b] Except with the use of modern theories and methods. One of the largest masonry spans in the world, the Cabin John Aqueduct Bridge at Washington, D.C., stretches 218 ft. Structures of reinforced concrete are also here excluded.

[c] See Robertson (1); Jenkin (1); Neuburger (1); Gauthey (1). The oldest surviving Roman bridge, on an old Etruscan road, the Via Amerina, antedates the Pons Milvius by about twenty years, but its span is very small (Perkins, 1).

[d] So Robertson (1); later authorities prefer the Ponte d'Augusto at Narni with an arch of 105 ft. span (Goodchild & Forbes), or 142 ft. (Tyrrell). It depends what one counts as extant. However, the range is clear.

[e] We have seen a very similar contrast with regard to roads, cf. pp. 2, 7 above.

[1] 拱橋　　　　[2] 券　　　　[3] 並列

which we shall shortly discuss; it had the advantages that the collapse of one or two particular arches would not put the bridge out of action, and as the building of each arch required only a narrow centering, moved on as each arch was completed, there was a great saving in timber. The longitudinal method (*tsung lien*,[1] 'lengthwise linked') was different (II); the shaped stones were built up row by row from both end-supports

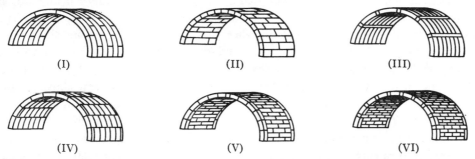

Fig. 828. Types of stone bonding in bridges (after Mao I-Shêng and Lo Ying). Explanation in text.

of the arch, alternating from side to side like headers and stretchers in brickwork over the whole width of the bridge. A third method (III) was in a way a combination of these two (*lien so*,[2] 'chain-linked'); in this system long curved arc-stones were prepared according to the radius intended, and made to alternate with tie-stones forming part of the arch lining but extending fully across its width. This was more suitable for small bridges.[a] The arch rings of old Chinese bridges can bear a greater load than the modern stress analysis of the arch alone would seem to justify, because the walls and fillings of spandrels (*kung fu*[3]) and abutments (*thu*[4]) exert a 'passive pressure' on the arch, increasing its strength and rigidity. Moreover, the wooden centering was traditionally given a little 'camber' so that when it was removed the stones sank under their own weight and pressed the arch more closely together.[b]

Mortar was seldom used in the arch or side-walls, and never in the foundations, as the composition of hydraulic mortar was not known.[c] The stones were, and are, mortised into each other with crossets, or held together by iron cramps in double dovetail shape, sunk into indentations of the same shape cut in the stones.[d] In the side-

---

[a] An example of it has already appeared in Fig. 820 (pl.), and another is in Fig. 830 (pl.).

[b] Cf. Thang Huan-Chhêng (*1*), p. 62; Mao I-Shêng (*1*). Three further methods are described by Lo Ying (*1*), pp. 42 ff. *Fên chieh ping lieh*[5] (IV) combines I with II above in that the stones of the arch are elongated as in the first but imbricate laterally as in the second. *Wai ping lieh nei tsung lien*[6] (V) combines I and II in a different way, using elongated arc-stones at the edges but transverse stonework internally over the soffits. *Khuang shih tsung lien*[7] (VI) combines II and III, using arc- and tie-stones at the edges but transverse stonework internally.

[c] The Romans could make this as they had supplies of pozzolana, a kind of volcanic clay; cf. Briggs (*3*), p. 407. In the absence of such resources the Chinese did well to grind their stones smooth and lay them dry, for thus they avoided the disastrous effects of expansion and contraction of mortar joints in periodical contact with water. On cement and concrete cf. Vol. 4, pt. 2, p. 219, and p. 38 above.

[d] This was an old device also in the West. Cf. p. 174 below.

---

[1] 縱聯   [2] 聯鎖   [3] 拱腹   [4] 撓   [5] 分節並列
[6] 外並列內縱聯   [7] 框式縱聯

walls between the arches, square ends of stone can often be seen; these are the through-binders (*chhang shih liang*[1]), or long stone ties running from wall to wall to prevent the walls from bulging outwards.[a] Since in many parts of China, such as the Yangtze delta, the subsoil was so bad that it was impossible for the Chinese engineers to make the foundations unyieldingly firm, the built-in shear-wall (*chien pi*[2]) was devised—namely a vertical stone wall running buried through the fill at right angles to the bridge axis and on each side of each arch. This can be well seen in Fig. 829 (pl.), a bridge at Khunshan in Chiangsu, taken from Mirams.[b] The shear-walls consist of long stone slabs, placed vertically side by side, mortised into the foundation and gathered at the top into a long horizontal stone which ties once again with the side-walls.[c] They are therefore geneti-cally identical with the simple rows of vertical slabs which, as we saw above (p. 152), constitute one of the forms of piers for the smaller kinds of stone beam bridges; but their combination with the arch was a brilliant idea. For while the arch itself is a loose chain which cannot resist any bending moment, the forces tending to deform the arch are transferred through the fill of clay and chips to the two shear-walls. Occasionally the side-walls and the fill of the arch section finish at the shear-walls, so that only the centre part of the bridge is built, connection being made with the banks by means of two beam spans. Fig. 830 (pl.), from Sirén,[d] shows another bridge at Suchow, to illustrate the degree of deformation which these structures can stand without collaps-ing.[e] The arch has become almost elliptical, but it continues to do duty. The rings of these arches, Fugl-Meyer found, are almost daringly thin; in an ordinary small bridge 1/30th the span, but on the larger ones as little as 1/60th. The bond of the stones used in the retaining side-walls varies somewhat, but usually rows of horizontal and up-ended slabs alternate, and in any case a number of the stones are placed with their long axes inwards, like so many spikes pointing into the fill, so as to act as additional ties between fill and wall.[f]

Another remarkable invention of Chinese bridge engineers was that of the complete circle structure, the arch of the bridge above being mirrored by a corresponding inverted arch (*fan kung*[3]) springing from the same abutments deep under the water. Great stability, especially valuable when rock or clay foundations are not to be had, can be obtained by such rings of masonry.[g] The first bridge of this kind in China was built

[a] Somewhat analogous with the lacing courses of flat bricks with which the Romans strengthened their city-walls filled with rubble and mortar. The Roman wall of London had these, and so do the walls of Byzantium (Istanbul) today, as I have myself seen.

[b] (1), fig. opp. p. 104; cf. Boerschmann (3a), fig. 282; Thang Huan-Chhêng (1), fig. 54; Little (1), p. 362; Schulthess (1), pls. 10, 13.

[c] Typical elevation, plan, and longitudinal section in Fugl-Meyer (1), fig. 44.

[d] (1), vol. 4, pl. 98B.

[e] Another view in Fugl-Meyer (1), fig. 45. Other examples, quite extreme, in Fugl-Meyer (1), fig. 46; Liu Tun-Chên (1), pl. 9A.

[f] Drawings in Fugl-Meyer (1), fig. 27. There is an important paper by Wang Pi-Wên (1) on the government regulation methods of bridge-building during the Chhing dynasty.

[g] The practice is quite common today for segmental arches of reinforced concrete. We have an example in Cambridge, the road-bridge where Fen Causeway crosses the Cam.

[1] 長石梁   [2] 間壁   [3] 反拱

between +1465 and +1487 in Chiangsu at a place called Chio-chih-chen near Wuhsien.[a] Known as the Tung-Mei C.,[1] it is still in active use.

To give an idea of some of these Chinese bridges, a few illustrations are offered. Most people are familiar with the exquisite bridges of the I Ho Yuan (Summer Palace) at Peking, the Jade Girdle Bridge (Yü-Tai C.[2]),[b] a single-span edifice resembling a breaking wave, and the Seventeen Arches Bridge (Shih-chhi Khung C.[3]),[c] a gentle and graceful curve formed by a series of arches of varying sizes, crescendo and diminuendo. Omitting these here, we will mention humbler, more workaday and more typical examples, e.g. the bridge at Tzu-thung[4] on the Szechuan–Shensi road,[d] or the Chia-Hsi C.[5] at Kweilin in Kuangsi,[e] which dates from about +1453. Fig. 831 (pl.), shows a well-known arch bridge at Wan-hsien, the San-Hsing C.[6] spanning a tributary of the Yangtze.[f] While most Chinese bridge arches are semicircular and often a little stilted (i.e. springing from a point slightly higher than the top of the pier), the use of the pointed 'Gothic' arch, possibly derived from the Graeco-Indian (Gandhāra) 'lotus' niche, was not at all unknown.[g] I photographed one bridge of pointed arches spanning a mountain stream beside the road north of the Chien-Mên Kuan[7] (Sword-gate Pass) between Szechuan and Shensi in 1943 (Fig. 832, pl.).[h] Travellers in China have always remarked on the frequency of very humpbacked decks and approaches, but this was a natural feature in a civilisation which depended a good deal on porters and pack-animals, and appears noticeably also in medieval European bridge-building.[i] One obvious reason was the facilitation of the passage of ships without lowering their masts and sails; a point specifically mentioned by Marco Polo.

Very few arched bridges remain from the Han and San Kuo periods, if any; the Wan-

---

[a] Lo Ying (1), p. 45.

[b] Mao I-Shêng (1); Thang Huan-Chhêng (1), fig. 69; Anon. (37), fig. 130.

[c] Thang Huan-Chhêng (1), fig. 82; Anon. (37), figs. 123, 128.

[d] The Road of the Golden Oxen, cf. p. 24 above. Cf. Wiens (2).

[e] Phan Ên-Lin (1), p. 34; Thang Huan-Chhêng (1), fig. 84; Forman & Forman (1), pp. 161, 163.

[f] Fugl-Meyer (1), fig. 8; Thang Huan-Chhêng (1), fig. 50; both give the span as about 94 ft. Edkins & Gregory (1) gave as dimensions span 78 ft., height 40 ft., breadth 22 ft. Cf. Carey (2); Little (1), p. 255.

[g] See Vol. 1, Fig. 20 for a Northern Wei (c. +5th-century) example of ogival arch decoration at the Tunhuang cave-temples. Baltrušaitis (1, 2) has enlarged upon the Asian origins of the ogival form of the pointed arch (the arc en accolade) in European Gothic architecture. He finds its roots in India at Nasik (−3rd century) and at Ajanta (+5th), then traces its spread eastwards to Tunhuang and Lungmên, following it also westwards through Islam (+10th) to reach Western Europe by the end of the +13th century, just at the time of transition between the 'Early English' and the 'Decorated' styles. In the 'Flamboyant' it became particularly popular. Lo Ying (1), however, pp. 40, 42, 395 ff., is not at all disposed to look for a foreign origin of the Chinese 'Gothic' bridge arch, and suggests that it arose indigenously from the stone cantilever style when the sides met in the middle to form an inverted V (cf. p. 165 above). There was a ready analogy for this jen tzu[8] system in the inverted V-braces of architecture (cf. p. 99 above). In seeing all round-headed arches as a further evolutionary stage Lo Ying is less convincing, since these, on the contrary, would seem to be the most primitive and ancient type (cf. p. 167 above).

[h] Cf. p. 24. Other examples, from Hopei, Shensi and Yunnan, will be found in Thang Huan-Chhêng (1), figs. 71, 72 and 87, 88.

[i] Cf. Uccelli (1), pp. 291 ff.

---

[1] 東美橋   [2] 玉帶橋   [3] 十七孔橋   [4] 梓潼   [5] 嘉熙橋

[6] 三星橋   [7] 劍門關   [8] 人字

Li C.[1] outside the south gate of Chhêngtu is claimed as one of them.[a] The Yang-Chhü C.[2] at Loyang, built by Ma Hsien[3] (*fl.* +135), has always been considered the oldest Chinese bridge entirely of stone, but one cannot be sure whether it had beams or arches; from the inscription recorded in the *Shui Ching Chu* (Commentary on the Waterways Classic)[b] the former interpretation seems the more likely. It must certainly be true of the bridge built a few years afterwards for the 'king-maker' Liang Chi[4] (*d.* +159) across a water-way in his vast park at Loyang.[c] The *Shui Ching Chu* however has preserved an account of a great stone bridge at Loyang which distinctly mentions a lofty arch; this was built towards the end of the +3rd century. Li Tao-Yuan says:[d]

As for the many bridges (of Loyang) they are all made of (dressed) stones piled up into high and gallant structures. Although with the passage of time there is some decay, they never fail to do their office. When Chu Chhao-Shih[5] was travelling (in his military campaigns) he wrote to his elder brother saying, 'Outside the palaces of Loyang some six or seven *li* away there is a bridge built all of great blocks of stone, and underneath it is rounded so that not only does the water pass, but also large ships can go through (*hsia yuan i thung shui, kho shou ta fang kuo yeh*[6]). An inscription on it says that it was built in the 3rd year of the Thai-Khang reign-period (+282). In the construction over 75,000 men were employed each day, and after five months it was finished.' In the course of years this bridge fell into disrepair and was then completely restored, but the inscription is no longer to be seen.

Such was the Lü-Jen C.[7] (Bridge of Wayfaring Men). Chinese historians of engineering have found no textual evidence for arched bridges earlier than this, but if the one here described had a span of some 80 ft. like the Pons Fabricius, as would seem to follow from what is said of the river traffic, it cannot possibly have been the first of its kind, and we could reasonably conclude that small arched bridges must have been constructed first in the Later Han period if not before.

The Thien-Chin C.[8] at Loyang was reconstructed with streamlined piers towards the end of the +7th century by Li Chao-Tê,[9] a renowned engineer who built many other bridges at the same capital.[e] This century is indeed something of a turning-point, for while hardly any extant structures can reliably be considered earlier in date, some of the noblest and most interesting that China possesses derive indubitably from its initial decades.[f] We can therefore entertain no doubt that between Ma Hsien and Li

[a] Not perhaps very convincingly. During the Second World War I often passed over this bridge, and dined in the well-known restaurant beside it, but I failed to realise its interest and never carefully examined it. It is referred to in a famous poem by Tu Fu (tr. Alley (6) p. 99).

[b] By Li Tao-Yuan, *c.* +500; ch. 16, p. 11*b*. Cf. *TSFY*, ch. 48, p. 45*a*.

[c] 'A flying bridge', says the *Hou Han Shu* (ch. 64, p. 13*b*), 'with stone steps (or piers) bestriding a gap across a watercourse (*fei liang shih têng ling khua shui tao*[10]).' Cantilever construction might be thought of here, but surely it was a beam bridge at some height above the water.

[d] Ch. 16, pp. 20*b*, 21*a*, tr. auct. Chu Chhao-Shih was a general who served the first ruler of the Liu Sung dynasty and was killed by Holien Pho-Pho (cf. p. 42 above) about +423. His letter is also to be found in *CSHK* (Chin sect.), ch. 141, p. 8*a*.

[e] *CTS*, ch. 87, p. 7*b*; *TSFY*, ch. 48, p. 47*a*.   [f] See immediately below, p. 175.

[1] 萬理橋   [2] 陽渠橋   [3] 馬憲   [4] 梁冀   [5] 朱超石
[6] 下圓以通水可受大舫過也   [7] 旅人橋   [8] 天津橋   [9] 李昭德
[10] 飛梁石磴陵跨水道

Chhun there was a powerful and vigorous tradition of arched bridge-building though few or no identifiable examples have happened to survive. Different parts of the country have had their traditionally famous bridge-builders, e.g. Phêng Jung,[1] Mao Fêng-Tshai[2] and the monk Tsu-Yin[3] in Szechuan, or Ko Ching[4] and Huang Phan-Lung[5] in Kweichow—but most of the names which have come down to us are of the Sung or later.

The major works of arched bridge-building in China are of course those in which many spans had to be built. Beautiful bridges of three or four arches occur in almost every city,[a] but pre-eminent among those with long arcades is the Wan-Nien C.[6] at Nanchhêng[7] in Chiangsi,[b] partly because it has had a special book devoted to it.[c] The *Wan-Nien Chhiao Chih*[8] (Record of the Bridge of Ten Thousand Years), written by Hsieh Kan-Thang[9] in 1896, is a full story of this bridge, as detailed as any of the local histories and topographies (cf. Vol. 3, p. 517 above).[d] It is a work of twenty-four arches, 1,803 ft. in length, which crosses the Ju Shui[10] River, six *li* east of the city where Hsieh Kan-Thang lived. It was always an important communications link between Chiangsi and Fukien, and from literary references we know that between +1271 and +1633 the crossing was made by a pontoon bridge. But the drowning of some thirty persons in this latter year induced the judicial commissioner of the province, Wu Lin-Jui,[11] to undertake the building of a long bridge of stone arches, and this was completed in +1647. Floods severely damaged it in +1724 but it was repaired by the then prefect, Li Chhao-Chu.[12] Finally in 1887 three arches were again swept away and others damaged, and this was the occasion on which Hsieh Kan-Thang took in hand the repairs which gave rise to his book. Of its great sociological interest this is not the place to speak, how Hsieh and some friends worked as supervisors without pay, how everything was financed by subscription without help from the local or provincial government, how Buddhist piety motivated many of those who volunteered their work, how believers in *fêng-shui*[e] objected to the quarrying of stone and felling of trees, etc., etc. For us the interest lies in the fact that although Hsieh's book was written at such a late date, everything seems to have been done by traditional eotechnic methods showing little or

[a] Cf. the photographs and drawings of bridges at Khun-shan and elsewhere in Chiangsu reproduced by Bonatz & Leonhardt (1), figs. 55, 56; H. S. Smith (1), fig. 8; Fugl-Meyer (1), figs. 5, 28, and Thang Huan-Chhêng (1), figs. 52, 53. Some of these clearly show the use of the third method of making arch linings (p. 169 above). Further illustrations of noteworthy bridges with semicircular arches will be found in Forman & Forman (1), pp. 115, 211.

[b] Thang Huan-Chhêng (1), fig. 92 and p. 70. It closely resembles the Tzu-thung bridge mentioned on p. 171. I had the pleasure of seeing this bridge myself in 1944, but did not then know how interesting it was. It is not to be confused with another of the same name at Fên-i in the same province, built with 11 arches by Hsü Tshung-Lung[13] about +1556 to replace a floating bridge. On this see Chhen Po-Chhüan (1).

[c] Further down river there is another important bridge at Fu-chou,[14] i.e. modern Lin-chhuan, called Wên-Chhang C., on which we have a special paper by Liu Tun-Chên (2). This is not to be confused with the other of the same name at Fuchow[15] in Fukien (see Table 66). Cf. Lo Ying (1), pp. 72, 385 ff.

[d] We are indebted to Hummel (16) for an analysis of it, as it is not easily available. Cf. Lo Ying (1), pp. 79 ff.

[e] Cf. Vol. 2, p. 359 and Vol. 4, pt. 1, p. 239 above, for an account of this system of geomancy.

[1] 鵬榮	[2] 毛鳳彩	[3] 祖印	[4] 葛鏡	[5] 黃攀龍
[6] 萬年橋	[7] 南城	[8] 萬年橋誌	[9] 謝甘棠	[10] 汝水
[11] 吳麟瑞	[12] 李朝柱	[13] 許從龍	[14] 撫州	[15] 福州

no trace of modern influences. The techniques used have been discussed in a special paper by Liu Tun-Chên (3), who reproduces many of the engineering diagrams of the book itself. Thus for example, Fig. 833a shows the picture of the dredger (*pha sha*[1]), Fig. 833b the gravel-filled bamboo gabions (*sha nang*[2]),[a] Fig. 833c the cofferdam for working on the foundations (*shui kuei*[3]),[b] Fig. 833d the dowelling and cramping of the stones of the piers (*hsiang shih*[4]), Fig. 833e the piers themselves (*chhi tun*[5]) with their rostral shearwater ends (*fên shui*[6]) and their upper parts forming the arch spandrels (*shan hua tun*[7]); finally Fig. 833f the wooden half-centering (*phien wêng*[8]).[c]

This may be the place for a word about the use of metal dowels and cramps in Chinese stonework.[d] 'Mutual inlaying' was common practice from the Thang and Sung onwards, as for instance in the stone causeway which impressed the Korean traveller, Chhoe Pu, in +1487;[e] and in the Thien-Chin C. in Honan, which was strengthened in this manner during the +10th-century repairs made by Hsiang Kung.[9] As we shall see in a moment, iron cramps were used in an arch bridge of particular interest erected early in the +7th century, but before that point it is more difficult to trace the technique back. One would hardly expect to find it in the stone tomb-shrines of the Han, which were not subject to any great stresses, and no stone bridges of that period have survived.

Another bridge, of sixty-eight arches and 1,000 ft. long, crossed the Wei River near Chhang-an (Sian), but it was destroyed before the Sui. Barrow, on his way south in +1793, was very impressed by a bridge of ninety-one semicircular arches which seemed to run parallel with the Grand Canal on which his party was travelling, and crossed an arm of the Thai Hu Lake south of Suchow at a point where it was about half a mile broad.[f] This was probably the Pao-Tai C.,[10] originally built in +806, which has central arches higher than the rest just as Barrow described[g] (Fig. 834, pl.). And indeed Chiangsu is rich in such many-spanned structures, for not far away, at Wuchiang,

---

[a] For river-bank control; see p. 339 below.

[b] Each cofferdam was drained by eight square-pallet chain-pumps (cf. Vol. 4, pt. 2, p. 339 above). For other uses of the term *shui kuei*, cf. p. 315 below.

[c] It is noteworthy that this is all made of right-angle joints, like the posts and beams of a Chinese house-roof, and does not involve slanting joints.

[d] On this general subject see Briggs (2), pp. 78 ff. Dowels and cramps were much used in the temple building of classical antiquity, and I remember being struck by the strange effects produced in the walls of the temples at Palmyra due to the 'quarrying' of later metal-thirsty generations. Cf. p. 178.

[e] Meskill tr. p. 104.

[f] (1), p. 520. Cf. Macartney himself, in Cranmer-Byng (2), pp. 175, 373. The arm of the lake is really the outflow of the Wusung River, which crosses the Grand Canal here.

[g] He gave them a span of some 40 ft. But the Pao-Tai C. has only fifty-three arches, and its largest span does not exceed 22½ ft. Here Chhoe Pu gave the figure of fifty-five arches, nearly right. A photograph will be found in Thang Huan-Chhêng (1), fig. 81; cf. Fugl-Meyer (1), pp. 44 ff. Though Chinese bridges often rose to a maximum height over their central spans, the decks were always smoothly graded; unlike some Turkish structures figured by Bonatz & Leonhardt (1), figs. 44 and 45, in which the deck follows a veritable switchback course depending on the height of the arches. The Pao-Tai C. appeared on a Chinese postage-stamp in 1962; I had the pleasure of paying it a visit two years later. It gets its name from the jewelled belt which the Thang governor, Wang Chung-Shu,[11] donated towards the expenses of its building. Fullest description is that of Lo Ying (1), pp. 242 ff.

| [1] 爬沙 | [2] 沙蘿 | [3] 水櫃 | [4] 鑲石 | [5] 砌墩 | [6] 分水 |
| [7] 山花墩 | [8] 駢甃 | [9] 向拱 | [10] 寶帶橋 | [11] 王仲舒 | |

there is another of the same kind, the Chhui-Hung C.,[1] also crossing an arm of the lake.[a] This, we know,[b] was at first a wooden bridge from +1047 onwards, rebuilt afterwards in stone by Chang Hsien-Tsu[2] under the Mongols about +1324.[c] It then had sixty-two arches on pile foundations, each secured by a steel tie-bar 13 ft. long.[d]

The greatest invention in arched bridge-building, both for engineering merit in economy of materials and for aesthetic quality, came when builders dared to abandon the idea that security demanded semicircular arches in which the ring approached the pier-line tangentially and so 'conducted the weight vertically downwards'. When it was realised that the curve of the arch could be made much flatter if it was a segment of a much larger semi-circle, and that the bridge could thus be made to fly forth from its abutments as if tending to the horizontal, the segmental bridge was born.[e] This discovery, which in Europe may have had something to do with the parallel development of the flying buttress, seems to have been made by Westerners late in the +13th century, for there are several structures of that period, such as the long Pont St Esprit across the Rhône and the small Abbot's Bridge at Bury St Edmunds in East Anglia, which show it. But it was not applied daringly and widely until the early part of the +14th century, and its growth, which can be followed in the illustrated pages of Uccelli, was then rapid. The Ponte Vecchio of Florence (+1345) was quickly followed by the covered bridge at Pavia (+1351), the Castelvecchio bridge at Verona (+1354), and the largest though most short-lived of all, the bridge at Trezzo (+1375).[f] This reached the maximum span of the group, 243 ft. Sixteenth-century bridges such as the Rialto at Venice, and the Santa Trinità at Florence, simply carried on the tradition, adding elegancies. The height above the chord-line was reduced from a full radius to less than one half (see Table 67).

Since all these bridges are rightly regarded as great achievements it must remain an extraordinary fact that a comparable bridge of even more advanced character, together with a number of smaller ones, was built about +610 by a Chinese engineer of outstanding quality, Li Chhun.[3] His activity, and that of his pupils—for it is clear that he founded a school and style which lasted for centuries—left its mark mostly in the

---

[a] In fact another mouth of the Wusung River.

[b] From the essay of a contemporary, Chhien Kung-Fu,[4] entitled *Chhui-Hung Chhiao Chi*[5] and quoted by Lo Ying (*1*), p. 246.

[c] From another essay with the same title written in the Yuan period by Yuan Chio[6] we know that the name of the engineer really responsible was Yao Hsing-Man.[7]

[d] Today it has seventy-two arches, one at the centre larger than the rest, to aid navigation. This is probably due to the rebuilding of +1799. The *Suchow Fu Chih*, ch. 34, pp. 16a ff., mentions 85; and the Korean traveller Chhoe Pu (cf. p. 360), who passed by in +1487, gave it in his excitement 'at least 400' (Meskill tr. p. 91). Perhaps he was counting in other bridges as well.

[e] Pippard & Baker (*1*), p. 243.

[f] The beautiful Charles Bridge at Prague, begun in +1357 and finished about +1370, is sometimes classed among these segmental arch structures, but in fact its appearance is somewhat illusory for the haunches of the almost semicircular arches are hidden by the shear-water ends of the protruding piers. See Gauthey (*1*), vol. 1, p. 32 and pl. III, fig. 40; Tyrrell (*1*), p. 50; Uccelli (*1*), p. 290, fig. 69. The special monograph by Novotný, Poche & Ehm is concerned with the artistic rather than the engineering aspects of the bridge.

[1] 垂虹橋　　　[2] 張顯祖　　　[3] 李春　　　[4] 錢公輔　　　[5] 垂虹橋記
[6] 袁桷　　　[7] 姚行滿

爬沙圖

架以順水
為穗麻繩
約九丈餘
每晚蒸過
以免燃爛

(a)

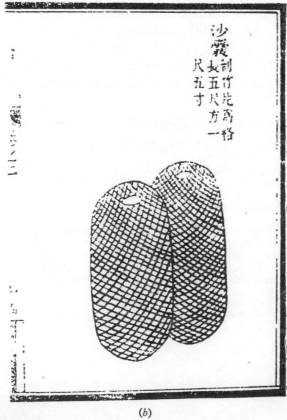

沙礨剉竹片為將
長五尺方一
尺五寸

(b)

水櫃後萬不可逾

此次安櫃不遵舊式未用篾折以致漏孔難塞以

先折決櫃棕
中後插板在
櫃折不舡釘罷
折不加版安蓋水篾
輕便且必宜嚴
短不舡在水長
中方可定位在水長

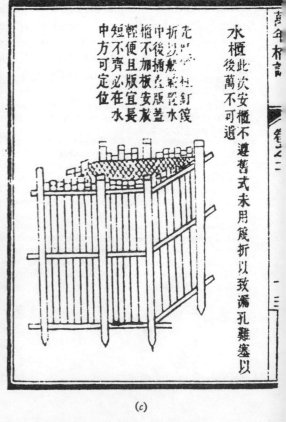

(c)

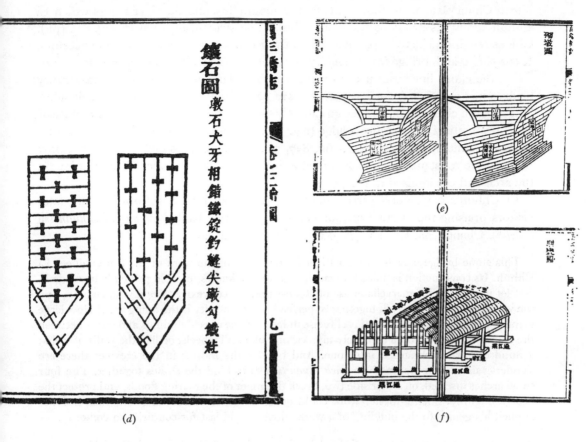

Fig. 833. Illustrations from the *Wan-Nien Chhiao Chih* (Record of the Bridge of Ten Thousand Years) by Hsieh Kan-Thang.

a The dredger (*pha sha*). 'The winch mounting should be set up in conformity with the current. The hemp cable more than 90 ft. long, should be heat-dried every night to avoid damp rot.'

b The gabions (*sha nang*). 'Bamboo strips are woven into loose nets 5 ft. long and 1·5 ft. across.'

c The cofferdam (*shui kuei*). 'This is a divergence from old methods; we did not use bamboo matting (alone), because it is extremely difficult to stop leakage in that way—that is a point not to be forgotten. First the main framework is set up, and then the bamboo matting nailed on; in this state it is light and convenient to handle, and a boat takes it out to its position. Later (after draining) wooden boards of unequal length are inserted within the framework; it is only when it is in position that the lengths of the boards can be fixed (to fit the inequalities of the river-bed).'

d Dowelling and cramping the stones of the piers (*hsiang shih*). 'The pier stones are inlaid with metal alternately, like dog teeth, and iron anchor pieces (dowels) hook the seams together. At the pointed (upstream) ends, the stones are fastened with hooks (cramps).'

e The piers (*chhi tun*). See text.

f The half-centering (*phien wêng*). 'Each side has eight "rafters". The curvature changes in eight steps.' Notable here is the use of pillars, cross-beams, tie-beams and king- and queen-posts, exactly as in the transverse framework of a Chinese building, but with the cross-section of the array of purlins arranged to be convex rather than concave.

The two last illustrations are full double-page openings like *a*, but reduced to half-scale.

provinces of Hopei and Shansi, centring on his finest work, the Great Stone Bridge (Ta-Shih C.[1]) named An-Chi C.[2] near Chao-hsien.[3] The country here is the edge of the North China plain at the foot of the Shansi mountains, and the river across which Li Chhun threw his bridge is the Chiao Shui.[4] The structure is seen in Fig. 835 (pl.). Chinese engineers justly regard it as one of the greatest achievements of their ancestors. It is a single-arch bridge (*tan khung chhüan chhiao*[5]) 123 ft. in span, and rising 23·7 ft. above the chord line.[a] But it even surpasses the galaxy of European +14th-century bridges just mentioned, for its spandrels are perforated by two arches on each side.[b] The modern character of this design can only be appreciated by comparing a picture of a +14th-century Renaissance bridge (e.g. Fig. 836, pl., Castelvecchio) with one of a twentieth-century structure (e.g. Fig. 837, pl., Salcano). This construction not only reduced the resistance to the flow of water in floods, but also lessened the burden on the arch and economised in materials.

Li Chhun's bridge still carries inscriptions written by Thang officials and later visitors praising the 'system of four holes near the two banks' (*liang yai chhuan ssu hsüeh*[6]). Chang Chia-Chên's[7] (*fl.* +675), for example, says:[c]

This stone bridge over the Chiao River is the result of the work of the Sui engineer Li Chhun. Its construction is indeed unusual, and no one knows on what principle he made it. But let us observe his marvellous use of stone-work. Its convexity (*lêng*[8]) is so smooth, and the voussoir-stones (*chen*[9]) fit together so perfectly...How lofty is the flying arch! How large is the opening, yet without piers!...Precise indeed are the cross-bondings and joints between the stones, masonry blocks delicately interlocking like mill wheels, or like the walls of wells; a hundred forms (organised into) one. And besides the mortar in the crevices there are slender-waisted iron cramps (*yao thieh shuan tshu*[10]) to bind the stones together. The four small arches inserted, on either side two, break the anger of the roaring floods, and protect the bridge mightily. Such a master-work could never have been achieved if this man had not applied his genius to the building of a work which would last for centuries to come....

And in the Ming the author of the *Piao I Lu* said that the bridge looked like a new moon rising above the clouds, or a long rainbow hanging on a mountain waterfall.[d] Ambassadors and other important travellers used to go out of their way in order to see it.[e] Indeed they still do. After the time of Li Chhun spandrelled arches were used in many +12th-century Chinese bridges,[f] but they did not appear in Europe until the

---

[a] The arch lining is composed of twenty-five individual arches built according to the first or transverse method (p. 169 above).

[b] And be it noted that the larger outer arch in each case is itself segmental.

[c] *TSCC, Chih fang tien*, ch. 95, *hui khao* 3, p. 2*b, Khao kung tien*, ch. 32, *i wên* 1, p. 1*b*; copied also in Thang Huan-Chhêng (*1*), p. 45; tr. auct. The readings in the different versions vary a little.

[d] Ch. 2, p. 2*a*. Legends accreted around it, cf. Anon. (73), vol. 5, p. 125.

[e] See for example the *Pei Hsing Jih Lu*[11] (Diary of a Journey to the North), written by Lou Yo[12] in +1169, ch. 1, p. 26*b*.

[f] Nos. 14, 16, 19, 20 in Table 67, for example. In no. 18, a small bridge, the spandrel arches are imperforate and had become an ornamental feature.

[1] 大石橋	[2] 安濟橋	[3] 趙縣	[4] 淡水	[5] 單孔券橋
[6] 兩涯穿四穴	[7] 張嘉貞	[8] 楞	[9] 磌	[10] 腰鐵栓礤
[11] 北行日錄	[12] 樓鑰			

+14th century, when we find them in the Pont de Céret over the Tech near Perpignan.[a] This is a large semicircular arch of 148 ft. span having one small semicircular arch in each spandrel, and three others of similar small size in the approaches, two on one side and one on the other.[b] Li Chhun's spandrel-arch construction was thus the ancestor of those many modern bridges of reinforced concrete which dispense with all filling between the arch ring and the deck, connecting them only by vertical pillars or a reticulate construction of concrete members. In our own time it has undergone a thorough restoration fitting it to stand for centuries more.[c] One cannot help comparing Li Chhun with Anthemius of Tralles, whose 100 ft. diameter dome of the Cathedral of the Holy Wisdom at Constantinople had been finished rather less than a century earlier, in +537. Great though his achievement was,[d] it had been a natural development from the older domed basilica, and what Li Chhun did was really much more original.

In order to appreciate the full originality of the work of Li Chhun we must take a closer look at the statement just made that the invention of segmental arches appeared in Europe no earlier than the +14th century. This is true only if we refer to marked segmentality in free-standing bridges, for segmental arches embodied in buildings go back to the Hellenistic period. At the temple of Deir al-Medineh in Egypt, built in Ptolemaic times (early −2nd century), stone segmental arches with brick walls above are used in the roofing of chambers, probably chapels;[e] but they are almost certainly Coptic, for the temple afterwards became a Christian church, and therefore not older than the +4th or +5th centuries.[f] However, at the Roman port of Ostia segmental arches made of large thin bricks appear to occur in a number of buildings dating from the −1st and −2nd centuries, notably the ancient *trattoria* known as the Thermopolium, though some at least of them may be true semicircular arches the haunches of which are hidden in the brickwork piers from which they seem to spring.[g] In any case, many real segmental arches are to be seen at Ostia embedded in brickwork or serving as the lintels of doors, and all this is characteristic of Roman architecture in other places also. Surprisingly, it may turn out to be characteristic of ancient Chinese brickwork too, for traces of two built-in segmental arches or vaults have come to light in the

---

[a] Two spandrels of the Pont d'Avignon have interior vaulted chambers, but imperforate. The practice of lightening the structure by inserting a small arch in the upper part of a pier between two main arches had begun much earlier; it occurs in the Pons Fabricius and in two of the medieval European bridges listed in Table 67. It is also to be seen among Chinese bridges in no. 15 therein.

[b] Cf. Gauthey (1), vol. 1, p. 57 and pl. IV, fig. 64. There is much difference among the authorities about the date of this bridge; Gauthey and Uccelli give +1336, Tyrrell +1494 and Mao +1321.

[c] See Liang Ssu-Chhêng (5); Yü Chê-Tê (1); Li Chhüan-Chhing (1); Mao I-Shêng (1); Lo Ying (1), pp. 224 ff. Remains recovered from the river-bed permitted the reconstruction of the carved balustrades. An excellent model is to be seen in the Imperial Palace Museum at Peking. The bridge was honoured with a Chinese postage-stamp in 1962.

[d] For its background see Diehl (1), pp. 166 ff.

[e] See Choisy (2), pp. 46, 47 and pl. XIII a. Egyptian architecture also has a false segmental arch formed by one or more corbel vault stones cut to its similitude, as at Abydos (cf. p. 67 and pl. XIII b) but with this we are not concerned.

[f] We are much indebted to Professor J. M. Plumley for this expert opinion, based *inter alia* on a recent personal visit to the site.

[g] See Calza (1), p. 24 and figs. 8, 10, 24, 31, 38, 40, 49. I had the pleasure of studying Ostia at leisure with Dr Dorothy Needham in the summer of 1957.

tomb of Liu Yen, the Chien Prince of Chungshan[1] (*d.* +88 or +90),[a] as well as in the larger vaulted tomb of similar or earlier date at Ying-chhêng-tzu.[b] But what could be done in buildings was not so easy to apply to bridges which had to launch forth into the void.[c]

The attempt has sometimes been made[d] to show that many medieval European bridge arches were segmental and that the departure of the +14th century was not so very revolutionary. Fortunately, this question is susceptible of quantitative answer. A simple measure of the 'flatness' of an arch form can be obtained by calculating the ratio of the rise to the half-span, $s/\frac{1}{2}l$, where s is the sagitta (or, in the limit case of the semicircular arch, the radius), and l the chord (in the limit case, the diameter). A further property is given by what is called Spangenberg's 'Audacity Factor', i.e. the square of the chord or span length (l) divided by the sagitta (s), i.e. the rise from the chord-line to the crown, as the soffit or under-surface of the key-stone is termed.[e] This is a measure of the horizontal thrust on the abutments and in the arch ring for a given load per unit length.[f] Data and derived characteristics for a number of bridges of known dimensions in East and West at different periods are assembled in Table 67 and Fig. 838.

It thus appears that in the West very slight approaches to segmentality occurred from Roman times onward, though often only to be seen by close inspection and measurement. The rising base-line in Fig. 838 shows the increase of the factor for semicircular arches of increasing size, and the degree of segmentality can be gauged from the height of the point above this line. There was thus only one important bridge with any substantial segmentality in Europe before the +14th century, no. 4; and its flatness ratio was only 0·51, contrasting with the Florentine bridges of after +1340 with their ratios of 0·38 and 0·36. Meanwhile all had been outdone long before not only by the Sui bridges of the +7th century, including the masterpiece of Li Chhun (no. 13) with its flatness ratio of 0·38, but also by the later Sung bridges of the +12th century, one of which (no. 14) gives the lowest ratio of any in the table. It is thus demonstrated that the segmental bridges of +7th-century China can be placed without hesitation among the best constructions of the kind in +14th-century Europe; indeed when we allow for the subsidiary segmental spandrel arches Li Chhun's school has a priority of more than a millennium, for not until the railway age (the seventies of the nineteenth century) did comparable, if larger, Western bridges arise in the work of

[a] See Ao Chhêng-Lung *et al.* (2), p. 130.

[b] Mori Osamu & Naito Hiroshi (*1*), frontispiece and figs. 19, 22, 38, pl. 22 (ii). Cf. Yeh Hsiao-Yen (*1*) on Han tombs at Shen-hsien in Honan.

[c] Small segmental arch bridges, supposed to be of Roman date, have been described and figured by Rittatore (*1*); they once carried paved roads over the upper Arno and its tributaries at Ponticino and Ponte Romito near Arezzo. But their date needs confirmation.

[d] As for instance by Prof. A. W. Skempton in his comments in the discussion on Chatley (36), and in subsequent correspondence.

[e] Straub (1), p. 11.

[f] It is also in a sense a measure of the economy of materials, since the flatter the arch the more stone is saved in the spandrels, but if the sides of the gap do not give good support, this saving will be compensated by the much stronger abutments necessary.

[1] 劉焉中山簡王

No.	Date of construction	Province	Place	Bridge	l (ft.)	s (ft.)	$s/\tfrac{1}{2}l$	l^2/s	References
	—	—	—	Semicircular arches	20	10	1·0	40	—
	—	—	—	Semicircular arches	50	25	1·0	100	—
	—	—	—	Semicircular arches	100	50	1·0	200	—
1	-62	—	Rome	Pons Fabricius	81·5	33·8	0·83	204	G, I, fig. 2; B & L, fig. 34; S, fig. 5
2	+1187	—	Avignon	Pont d'Avignon	110·5	46·1	0·83	264	G, v, fig. 90; M, p. 14; B & L, fig. 50; U, p. 285, fig. 53
3	+14th (main part, +1245)	—	Lyon	Pont de la Guillotière	109·0	38·0	0·70	314	G, v, fig. 72; cf. Vol. 4, pt. 2, Fig. 628
4	+1285 (finished +1305)	—	Nr. Bollène	Pont St Esprit	111·5	28·4	0·51	438	G, v, fig. 71; U, p. 286, fig. 55
5	+1345	—	Florence	Ponte Vecchio	98·0	18·4	0·38	526	G, I, fig. 1; U, p. 288, fig. 64
6	+1351	—	Pavia	Ponte Coperto	c. 100	c. 20	0·40	500	U, p. 261, fig. I
7	+1354	—	Verona	Castelvecchio	159	45	0·57	580	G, I, fig. 20; B & L, fig. 39; U, p. 289, fig. 65
8	+1375 (destroyed +1417)	—	Trezzo	Visconti Bridge	243	68	0·56	875	U, pp. 288 ff., figs. 67, 68
9	+1404	—	Sisteron	Pont de Castellane	92	23·4	0·51	363	G, VIII, fig. 149
10	c. +1525	—	Florence	Michael Angelo's single-span bridge	139	28·2	0·41	688	G, I, fig. 17
11	+1569	—	Florence	Santa Trinità	106	19·0	0·36	600	G, I, fig. 5; M, p. 18; B & L, fig. 49; U, pp. 688 ff., fig. 78
12	+1591	—	Venice	Rialto	94	22·6	0·48	397	G, I, fig. 25; U, p. 687, fig. 74
13	Sui, c. +610	Hopei	Chao-hsien[1]	An-Chi C.[2]	123	23·7	0·38	640	LSC (2, 3, 5); Li; Yü; Mao; Mir, pp. 101, 107; E; THC, figs. 55, 56; Anon. (26), p. 458, (37), fig. 26; M, p. 26
14	+1130	Hopei	Chao-hsien[1]	Yung-Thung C.[3]	c. 85	9·85	0·23	734	LSC (2); THC, fig. 61; Mir, p. 102
15	Probably +12th	Hopei	Chao-hsien[1]	Chi-Mei C.[4] (2 arches)	c. 27	5·5	0·41	135	LSC (2); THC, fig. 64
16	Sui, c. +615	Hopei	Ching-hsing[5]	Lou-Tien C.[6,a]	c. 60	c. 11	0·37	327	THC, fig. 67
17	J/Chin, c. +1175	Hopei	Luan-chhêng[7]	Ling-Khung C.[8]	c. 65	c. 10	0·31	422	THC, fig. 66
18	Probably +12th	Hopei	Luan-chhêng[7]	Ku-Ting C.[9]	c. 22	c. 6·5	0·59	75	THC, fig. 65
19	Probably +12th	Shansi	Kuo-hsien[10]	Phu-Chi C.[11]	31	12	0·77	80	THC, fig. 58
20	Probably +12th	Shansi	Chin-chhêng[12]	Ching-Tê C.[13]	c. 60	c. 13·8	0·46	261	THC, fig. 59
21	Probably +12th	Shansi	Chin-chhêng[12]	Chou-Tshun C.[14]	c. 28	c. 13	0·93	60	THC, fig. 60
22	Chhing, +18th	Kweichow	Hsing-i[15]	Mu-Chhia C.[16]	c. 67	c. 18·7	0·56	240	THC, fig. 68
23	+1191	Hopei	Nr. Peking	Lu-Kou C.[17] (11 arches)	36	c. 12·5	0·69	104	Mao; THC, fig. 91

ABBREVIATIONS

B & L Bonatz & Leonhardt (1)
E Ecke (4)
G Gauthey (1), vol. I, plate nos. and fig. nos.
LSC Liang Ssu-Chhêng (2, 3, 5)
Li Li Chhüan-Chhing (1)
M Mock (1)
Mao Mao I-Shêng (1)
Mir Mirams (1)
S H. S. Smith (1)
THC Thang Huan-Chhêng (1)
U Uccelli (1)
Yü Yü Chê-Tê (1)

NOTES

(1) s, sagitta (rise, radius); l, length of span (diameter or chord).

(2) No high degree of accuracy can be claimed for these figures for many have to be taken from published scale drawings. Small differences between the figures given by responsible authorities have been averaged. But the broad lines of the general picture cannot be wrong.

[a] The dimensions here are probably underestimated, for two spandrel arches, partly hidden by the cliff sides, can be made out in photographs.

1 趙縣　2 安濟橋　3 永通橋　4 濟美橋
5 井陘　6 樓殿橋　7 欒城　8 凌空橋
9 古丁橋　10 崞縣　11 普濟橋　12 晉城
13 景德橋　14 周村橋　15 興義　16 木卡橋
17 蘆溝橋

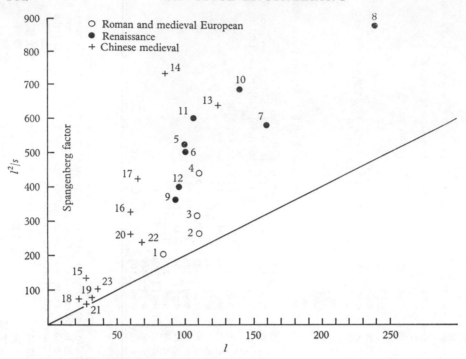

Fig. 838. Graph expressing comparatively the form of Chinese and European medieval and Renaissance segmental arch bridges, the Spangenberg factor plotted against the span length in feet. The rising solid line indicates the change in the factor for semicircular arches of increasing size, all points for segmental ones falling necessarily above it. The numbers attached to the individual points refer to the entries in Table 67. For further explanation and analysis see text.

engineers such as Paul Séjourné and Robert Maillart.[a] It is not often that we are able to make a graph of achievement with a chronological coordinate, but the segmental arch begs for illustration in these terms, and Fig. 839 portrays the results of plotting the flatness ratio of bridges against their dates of construction. The brilliance of the Chinese anticipation is here made plainly manifest.

Li Chhun's great bridge was not at all an isolated phenomenon, for nearly twenty others exist in various parts of China, mostly in the northern provinces but not exclusively.[b] Those included in Table 67 vary widely, some, such as no. 21, being almost semicircular, but others (e.g. nos. 14 and 17), the Yung-Thung C.[1] (Ever Communicating; Fig. 840, pl.) and the Ling-Khung C.[2] (Soaring across the Void), exceedingly segmental. Both of these date from the +12th century and both were

[a] Cf. Uccelli (1), pp. 693 ff.; Giedion (2), pp. 371 ff. The many elegant segmental arch bridges of the time of Thomas Telford (cf. Wilson, 1) do not have spandrel arches; and by the same token Leonardo da Vinci concedes the honour to Li Chhun (cf. Ucelli di Nemi (3), no. 35).

[b] Worthy of special mention is no. 16, a bridge of the early +7th century which spans a chasm at Ching-hsing in Hopei carrying a hall of two storeys. It is accompanied by another segmental bridge without spandrel arches and of uncertain date. At some time seemingly during the Ming or Chhing the segmental arch style spread to Kweichow and Kuangsi (cf. no. 22). An example not listed in Table 67 in one or other of these provinces has been illustrated, without details, by Groff & Lau (1).

[1] 永通橋   [2] 凌空橋

built under the Jurchen Chin dynasty, the former by an engineer whose name has come down to us, Phou Chhien-Erh.[1] Some are large, with spans of up to 70 ft. or so, others hardly larger than culverts, but all in the same segmental style. Particularly beautiful is no. 15, the Chi-Mei C.,[2] a low bridge of four spans, two very flat and two semicircular, the central pier being perforated by an arch of its own.[a] This brings us to

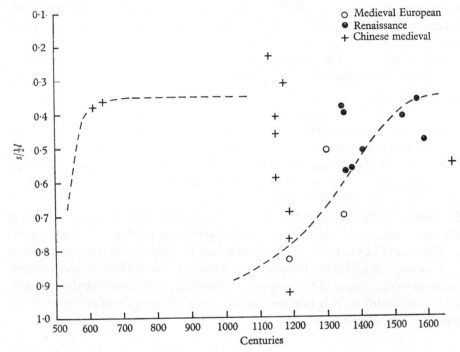

Fig. 839. Another graph expressing comparatively the form of Chinese and European segmental arch bridges, the flatness ratio plotted against a chronological scale. The degree of flatness of the 'flying' bridges of Renaissance Europe (+14th to +16th centuries) can thus be seen fully anticipated in those of early and late medieval China (+7th and +12th centuries). For further details see text and Table 67.

the longest many-spanned bridge of segmental arches in China. The celebrated Lu-Kou C.[3] which crosses the Yungting R.,[4] and gives its name to a small town near Peking, dates from +1189, and must have been standing by about +1280 in exactly its present form since Marco Polo describes it in detail.[b] In modern times foreigners have given his own name to it. Always strategically placed, it added to its renown by being the scene of incidents which started the Sino-Japanese war in 1937. It was thought by Marco Polo to be probably unequalled by any other bridge in the world (Fig. 841, pl.). Some 700 ft. long, it has eleven segmental arches averaging 62 ft. span each,[c] abutting on piers pointed both up and down stream; and the Venetian visitor was greatly

[a] Actually a cylindrical tunnel.
[b] Ch. 105 (Moule & Pelliot ed.). Cf. Fugl-Meyer (1), p. 24; Lo Ying (1), pp. 260 ff.
[c] Range from 52·5 ft. near the banks to 71 ft. in the middle.

[1] 裹錢而　　[2] 濟美橋　　[3] 蘆溝橋　　[4] 永定河

impressed by the fact that it was possible for ten mounted men to ride abreast upon the deck without inconvenience. The parapets of carved marble with their 283 stone lions, all different, were also a delight to him.[a] The name of the original builder has not been handed down, and we know only the name of Yang Chhi[1] (*fl.* +1510 to +1525), an engineer who repaired the bridge. It still serves a main road carrying a considerable amount of heavy truck and bus traffic.[b]

On several occasions elsewhere in this book we have instanced the segmental arch bridge as one among the inventions transmitted from China to Europe. Though almost nothing can be said about the details of the transmission we are not disposed to doubt the reality of the influence. The technical upsurge of the European +14th century in this particular respect points very clearly to travellers of Marco Polo's own time, though possibly the only message they brought was that the flying arch had been made by Asian men, and that it stood safe. Beyond this, what emerges from the analysis here outlined is that conceivably a similar message had come through in the +12th century (the time of the magnetic compass and the stern-post rudder) so that perhaps the Pont St Esprit as well as the Ponte Vecchio was really fathered at Chao-hsien. We must hope that future researches will enable us to ascertain exactly to which cluster of transmissions the segmental arch bridge belongs.

Could Galeote Pereira have seen the segmental arch bridges of Hopei as well as the megalithic beam bridges of Fukien, he would have been doubly confirmed in his opinion: 'This causeth us to think, that in all the world there be no better workmen for buildings, than the inhabitants of China.'[c] One wonders what he would have thought of the iron suspension bridges of the western provinces, logically in a way the converse of the eastern arch bridges. It is to these, passing from the domain of compression to that of tension, that we must now turn.

(4) SUSPENSION BRIDGES

The idea of spanning a mountain river-gorge by a suspended rope instead of a solid bridge must be very old in the history of human techniques; certainly it has been widely implemented.[d] It was put into practice in the New World by many Amerindian peoples inhabiting the southern part of the continent (e.g. the Peruvians)[e] where cables of

---

[a] Elsewhere Marco Polo drew attention to the roofed bridges common in Szechuan and other parts of Western China.

[b] I had the pleasure of visiting this fine structure with Mr Rewi Alley, Miss Ma Hsiao-Mi and other friends in the summer of 1952. Photographs in Thang Huan-Chhêng (1), fig. 91; Mao I-Shêng (2), pl. 2, (1); Sirén (1), vol. 4, pl. 92a.

[c] Boxer (1), p. 8.

[d] As an accompaniment to the ensuing sub-section the reader may like to make use of an introduction to the theory of suspension bridges such as the recent treatment by a master of the subject (Pugsley, 1). The only contribution specifically devoted to the history of the suspension bridge in China is that of Goodrich (16); it appeared long after the present sub-section had been written, but we were glad to find that little change was necessary.

[e] Cf. Mason (2); Robins (2); and especially von Hagen (3), pp. 106 ff., 113 ff., 131, 141 ff., 157 ff. The oldest Inca suspension bridge site he studied was of +1290. The classical description is that of Garcilasso de la Vega el Inca, written about +1610, (1), pp. 573 ff.

[1] 楊麒

lianas were used. One would naturally be inclined to assume independent invention, but facts later to be brought forward necessitate caution on this.[a] Among the high cultures of the ancient Old World such rope-bridges were particularly frequent in the eastern Himalayan region, where they have been described by many writers.[b] Those which Rock found still used in the old Nakhi kingdom in western Yunnan are of very primitive character; they consist of two bamboo cables, each securely attached to high points at opposite sides of the gorge, and declining in catenary curves so that their arrival points are much lower than their points of departure (Fig. 842, pl.). Men and animals pass across hanging in various kinds of cradles suspended from bamboo tubes greased with yak butter, and assistance from the further shore is only necessary sometimes in the concluding stages of the transit. In certain cases this method closely approached the invention of the cable railway.[c] Of a bridge north of Tung-chhuan[1] over the Niu-lan Chiang[2] in northern Yunnan we are told that it was customary to have wooden tubes (*mu thung*,[3] probably oak) as the runners supporting the cradles on the cable.[d] Gill saw one of these about 1880 near Lifan.[e] Similar methods spread to Japan, and were immortalised by the great artist Hiroshige, figuring in one of the prints of his set 'Sixty Famous Places'.[f] We are able to illustrate this technique by a couple of photographs taken on a recent Chinese scientific expedition to the Tibetan border; one of which (Fig. 843, pl.) shows a member crossing by the aid of one of these tubular slides (*liu thung*[4]).[g] Gill saw at Ta-chien-lu a form of double-rope bridge, one component of which had a series of 'strap-hangers' hanging from it so that the rider could pull himself forward towards the end of the span.[h] The eastern part of the Tibetan massif must surely be the focus of origin of Old World suspension bridges, not only on account of their total frequency, but because so many forms, from the simplest to the most advanced pre-modern types, are found there.

[a] See pp. 542 ff. below.

[b] E.g. Hutson (3); Kingdon Ward (2, 3, 4, 7); Little (1); Gill (1); and especially Rock (1), vol. 2, pp. 314 ff., 325, pls. 162, 164, 166.

[c] See Feldhaus (5), who attributes it to nineteenth-century Europe, and Kingdon Ward (12), opp. p. 145. If we take motive power at each end as the criterion for this, eighteenth-century China will qualify. In +1774 the acting magistrate of Chaohua[5] in Szechuan, Hsieh Thai-Fêng,[6] wishing to ensure the quick movement of despatches, spanned a ravine (impassable in summer flood seasons) by a 160 ft.-long windlass-tautened cable carrying an iron ring from which hung a wooden box in an iron cage. Hempen cords as long as the cable itself connected this with the bank on each side, so that the letters could be conveyed rapidly from one side of the gorge to the other by winding in one or other of the cords. An audio-visual signal system and post-horse riders in readiness completed the arrangements. The device was recorded in the *Chao-hua Hsien Chih* and Lo Ying (1), p. 85, quotes the passage. Then true funicular railways or cable-ways were used along the Yangtze gorges in the nineteenth century for bringing coal down to the waterside, sometimes in two stages. The descending full container brought up the empty one on a parallel cable, braking devices being applied to the wheel at the top. See the description of Blakiston (1), p. 265 (with engraving), quoted in Williams (1), vol. 1, p. 305.

[d] *TSCC, Khao kung tien*, ch. 31, *hui khao*, p. 12a.

[e] (1), p. 120.

[f] Horwitz (12). The suspended cable could also be used to carry the runner of a ferry-boat back and forth; for a Japanese illustration of this see Toyoda Toshitada (1), ch. 4, pp. 15b, 16a (unpag.).

[g] Thang Huan-Chhêng (1), pp. 73 ff., discusses the single-rope 'cable railways' in some detail, with quotations from local topographies and a photograph (fig. 96). Kingdon Ward (4, 7) illustrates a similar bridge used by the Nung people of Upper Burma.

[h] (1), p. 169.

[1] 東川　　　[2] 牛欄江　　　[3] 木筒　　　[4] 溜筒　　　[5] 昭化　　　[6] 謝泰奉

The next step in the development of the suspension bridge was the fixing of the rope to points at more or less equal height on each side of the river, and the adding of arrangements which would permit travellers to cross without hanging in a cradle or acquiring the skill of a tightrope walker. One of the simplest ways in which this was done was to suspend additional ropes as hand-rails so that the set of three formed a V-section, the rails being attached to the tread-rope at short intervals; this type is attested for China[a] as well as India, Burma, Gilgit, the Celebes, Borneo and Sumatra.[b] An improvement consisted in adding to the hand-rails an overhead rope and plaiting the whole together so as to form a continuous tubular structure, 3 to 5 ft. in diameter. Such bridges are made by the Abor tribespeople on the Assam–Tibet border some hundred miles up from Dibrugarh, and reliable descriptions speak of spans of as much as 800 ft. with a swing of 50 ft. from side to side.[c] The Nagas also construct impressive bridges with various combinations of ropes.

Fig. 845. Shooting birds with arrows to which long strings and weights are attached so that the arrows can afterwards be recovered. This technique, implicit in bronze inscriptions as far back as the Shang period, may have been important in solving the problem of getting suspension bridge cables in position across almost impassable gorges. The present scene, drawn, from a photograph, by Yetts (15), is taken from an inlaid bronze bowl of the Chou period in the Freer Gallery at Washington. Another scene of the same kind has already been given incidentally in Fig. 300.

When we consider the techniques which could have led to the establishment of such bridges, that which immediately comes to mind is the use of arrows with cords attached to them, so that the arrow could be recovered with the prey. Hopkins (5, 25) has suggested that the character *i*,[1] which came to mean a south-western barbarian, was originally a picture of an arrow with a string attached to it (K 551). This method of fowling (Fig. 845), which was called *i shê*,[2] is referred to not only in the *Chou Li*[d] but in much more ancient documents such as the *Shih Ching*, *Lun Yü* and *Mêng Tzu*, even, it may be, in late Shang bronze inscriptions of the −11th century.[e] It is quite often

[a] They are common on the Yunnan–Burma border (Fig. 844, pl.), where they are called 'grapevine bridges' (*thêng chhiao*[3]).

[b] Horwitz (12); Howell (private communication) with photograph; von Plessen (1); Kingdon Ward (11), opp. p. 80, (12), opp. p. 172; van Hasselt (1).

[c] Anon. (7); Bower (1), pp. 153, 205; Thang Huan-Chhêng (1), fig. 103; Kingdon Ward (4), (6), opp. p. 28, (15), opp. p. 177.

[d] Ch. 8, p. 14*a* (ch. 32), tr. Biot (1), vol. 2, p. 242.

[e] See Yetts (15).

[1] 夷   [2] 弋射   [3] 藤橋

depicted on Chou and Han bronzes[a] and tiles.[b] There was a special technical term for the arrows used, *tsêng shih*,[1] and another, *cho*,[2] for the string attached to them (K 1258*e*). Hsü Chung-Shu (*4*) has devoted a special paper to this subject, and we shall return to it in Sect. 30*d* on military technology. We came upon it once before because of a suggestion that it had something to do with the origin of the kite,[c] but surely its true technological interest lies here, namely as the ancestral technique which permitted a pilot-cord to be shot across the river, after which successively heavier ropes could be pulled over and made fast. Chao Ju-Kua tells us in the +13th century that the natives of Southern Formosa attached cords more than 100 ft. long to their javelins.[d] There may be significance, of course, in the fact that the I were south-western barbarians, people who lived in the very home of Old World suspension bridges, and it is agreeable to find that one of these famous bamboo-rope bridges is still called the I-Hsing S.C.[3];[e] it is on the Tibetan border west of Chhêngtu in Szechuan.

From the *Lu I Chi*[4] (Strange Matters) of the early Sung writer Tu Kuang-Thing[5] we learn of the special term applied to the bridge-ropes, *tso*.[6] He tells of a Taoist, Mao I-Huan,[7] who could cross ravines on the loose cables (*huan tso*[8]) of suspension bridges (*tso shêng chhiao*,[9] *chu so chhiao*[10]) when no one else was able to do so owing to their derelict state.[f] But the kind of bridges here referred to (also called *thêng chhiao*[11]) were certainly no longer V-shaped or tubular; long before the Sung the practice had grown up of setting as many as half-a-dozen bamboo cables horizontally a short distance apart, and then laying a proper deck of transverse planks, or lengths of bamboo, upon them. Ropes stretched alongside on either hand to form a rail were added, and the bridge was then fit for pack-animals as well as pedestrians, if too many did not come on at one time, and the bridge was not swaying too much in the wind.

We might seek the earliest Chinese reference to bridges of this kind in a text written about +90, the passage where the *Chhien Han Shu* is speaking of the Hindu Kush mountains (*Hsien-tu chê shih shan yeh*;[12] full of cliffs and rock-faces).[g] 'There the gorges and ravines allow of no connecting road, but ropes and cables are stretched across from side to side and by means of these a passage is effected (*chhi ku pu thung, i shêng so hsiang yin erh tu yün*[13]).' This occurs in the chapter about foreign countries in Central Asia and refers to Wu-Chha[14] which Chavannes (6) identified with Tashkurgan.[h] The

---

[a] Cf. Yang Tsung-Jung (*1*), fig. 20; the Yen-Yo Yü-Lieh Thu Hu (*c.* −4th cent.) reproduced in Fig. 300 (Vol. 4, pt. 1, p. 144).

[b] Rudolph & Wên (*1*), fig. 8 and pl. 76. [c] Vol. 4, pt. 2, p. 576 above.

[d] *Chu Fan Chih*, ch. 1, p. 39*b* (tr. Hirth & Rockhill, p. 165).

[e] S.C. will be used hereinafter as the abbreviation for *So Chhiao*[15] (rope bridge).

[f] We suspect that this association of a Taoist with suspension bridges was not quite so miraculous as it was made to sound; they were always involved in anything technical (cf. Vol. 2, pp. 34 ff. above).

[g] Ch. 96A, p. 9*a*; tr. Wylie (10), p. 31. Wylie translated 'chains', but the words used do not authorise this meaning.

[h] Herrmann (8) confirms this view, somewhere in Sarikol south of Yarkand, he says. These places command the eastern approaches of the passes which lead through the Pamirs to Gilgit, the Indus Valley and Afghanistan.

[1] 矰矢	[2] 繳	[3] 夷星索橋	[4] 錄異記	[5] 杜光庭
[6] 笮	[7] 毛意歡	[8] 絙笮	[9] 笮繩橋	[10] 竹索橋
[11] 藤橋	[12] 縣度者石山也	[13] 谿谷不通以繩索相引而度云		
[14] 烏秅	[15] 索橋			

very name of the Hindu Kush (Hsien-tu = Hsüan-tu[1]) means 'suspended crossings or passages', a testimony to the antiquity of the invention.

The passage just quoted is not, however, the earliest that we can find, for a few pages later on the same chapter of the *Chhien Han Shu* quotes[a] the speech of Tu Chhin[2] made in −25 and directed against the sending of Chinese diplomatic missions to Chi-Pin (Gandhāra, mod. Afghanistan),[b] on account of the extreme difficulties of the trans-Himalayan journey. The reader may remember, perhaps because of its description of mountain-sickness, the translation of this speech given in Vol. 1, p. 194, but we shall repeat a small portion of it here in an alternative and almost certainly better form.

Then comes the road through the San-chhih-phan gorge, 30 *li* long, where the path is only 16 or 17 inches wide, on the edge of unfathomable precipices. Travellers go step by step here, clasping each other (for safety), and rope suspension (bridges) are stretched across (the chasms) from side to side (*shêng so hsiang yin*[3]). After 20 *li* one reaches the Hsien-tu (mountain pass)...Verily the difficulties and dangers of the road are indescribable.[c]

Most geographers agree that this road (if road it could be called) was essentially the route between Yarkand and Gilgit. A great and lovable man traversed it in +399, Fa-Hsien,[4] first of the line of Chinese Buddhist pilgrims to India, and he did not fail to describe the engineering aspects of his journey.[d]

Keeping on through the (valleys and passes of the) Tshung-ling[5] mountain range (the Pamirs, the eastern parts of the Hindu Kush), we travelled south-westwards for 15 days. The road is difficult and broken, with steep crags and precipices in the way. The mountain-sides are simply stone walls standing straight up 8,000 ft. high. To look down makes one dizzy, and when one wants to move forward one is not sure of one's foothold. Below flows the Hsin-thou Ho[6] (the Sindhu, the Indus). Men of former times bored through the rocks here to make a way, and fixed ladders at the sides of the cliffs, seven hundred of which one has to negotiate.[e] Then one passes fearfully across (a bridge of) suspended cables to cross the river (*nieh hsüan kêng kuo ho*[7]), the sides of which are here rather less than 80 paces (*c.* 480 ft.) apart. (Access to this place) is precluded by the 'ninefold interpreters (necessary)' (i.e. by the enormous distance), which is why neither Chang Chhien nor Kan Ying in the Han period[f] ever reached this spot.

Further information was collected by Kuo I-Kung[8] early in the +4th century in his *Kuang Chih*[9] (Extensive Records of Remarkable Things),[g] and all the foregoing sources

---

[a] P. 12*b*.

[b] Chi-Pin included Kashmir only later on; see the interesting excursus of Petech (1), pp. 63 ff., on this region and its name.

[c] Tr. auct.; we diverge here from Wieger's interpretation of roped mountaineers (*TH*, p. 556), which we followed before, but also from Petech (1), p. 16, who did not notice the phrase used on p. 9*a*. The passage is also found in *TCKM*, ch. 6, p. 115*b*, and (in condensed form) in *Shui Ching Chu*, ch. 1, p. 4*b*.

[d] *Fo Kuo Chi*, ch. 7; tr. auct. adjuv. Beal (1), p. 21, (2), pp. xxix ff.; H. A. Giles (3), p. 10; Legge (4), p. 26; Rémusat (1), p. 35; Petech (1), pp. 15 ff. On Fa-Hsien and his travels cf. Vol. 1, pp. 207 ff.

[e] Lofty *chan tao* (pp. 20 ff.), called *ovrìng* in the Pamirs (cf. Polovtsov (1), pp. 135 ff.).

[f] On these two famous travellers see Vol. 1, pp. 173, 196.

[g] See *YHSF*, ch. 74, p. 35*a*, *b*. He agrees that the mountains got their Chinese name from the bridges.

[1] 懸度　　　[2] 杜欽　　　[3] 繩索相引　　　[4] 法顯　　　[5] 葱嶺
[6] 新頭河　　　[7] 躡懸絚過河　　　[8] 郭義恭　　　[9] 廣志

were made use of by Li Tao-Yuan about +500 in the opening chapter of his *Shui Ching Chu*.[a] Within twenty years thereafter more intrepid monks were daring the suspension bridges of the Himalayas, but for a reason which will presently appear we shall reserve their statements for a few pages.[b] Two and a half centuries later such bridges figured prominently in the campaigns of the great Korean general of the Thang, Kao Hsien-Chih[1], in western Sinkiang and the neighbouring regions.[c]

Well has it been said that the suspension bridge was almost a condition *sine qua non* for intercourse in historical times between the people of China and those of Tibet, Afghanistan, Kashmir, Nepal, India, Assam, Burma and Thailand.[d]

A seventeenth-century description occurs in the *Chin Chhuan So Chi*[2] (Fragmentary Notes on the Chin-chhuan Valley) by Li Hsin-Hêng,[3] who says:[e]

In this region (Chang-ku, now Tan-pa,[4] on the Szechuan–Sikang border) there are three suspension bridges. Hundreds upon hundreds of stakes and piles are driven in on the two banks of the river, and stones heaped over them. Long bamboo cables are suspended between them, with wooden boards laid down, and large ropes at the sides to help the traveller to support himself. Passengers walking over these bridges feel their feet declining and sinking as if they were on soft mud. But such bridges can be built where no stone structure is possible.

About the same time Athanasius Kircher, the Jesuit scientist, spoke in his *China Monumenta Illustrata* (+1667) of a 'flying bridge in Shensi having but one span 400 cubits long',[f] but he illustrated it by a copper-plate of a gigantic semicircular arch. His source was his colleague Martin Martini, who in the *Novus Atlas Sinensis* of +1655 had told of a bridge across the Yellow River '40 Chinese perches from end to end' (400 ft.) in the neighbourhood of Ninghsia.[g] And indeed there is the site of a cable suspension bridge (*so chhiao*[5]) marked just at the point where the line of the Great Wall crosses the Yellow River south-west of Ninghsia and turns north-west to cross the Gobi and protect the Old Silk Road (cf. p. 49 above). Bamboo cable suspension bridges are still extremely common in the mountainous parts of Yunnan, Szechuan, Sikang, Sikkim, Tibet[h] and Nepal.[i] But their distribution covers also Burma and Indonesia.[j]

---

[a] Ch. 1, p. 4*a*, *b*, tr. Franke (12), p. 58; tr. & comm. Petech (1), pp. 15 ff.
[b] Cf. p. 197 below.   [c] Cf. Chavannes (14); Stein (7).
[d] Goodrich (16).   [e] Tr. auct.

[f] (1), p. 215, 'Pons volans ex monte ad montem unico arcu exstructus 400 cubit. altitud. 500 cubit... in Xensi prope Chogan ad ripam Fi.' Cf. Nieuhoff (1), p. 247.

[g] (2), p. 52, by the Hiaikeu hills (possibly Hei-shan Hsia[6]?) near Ninghsia; 'at quod magis mirare, prope Chegan ad ripam, Fi pons est de monte ad montem unico exstructus arcu, cujus longitudo 40 est sinensium perticarum, hoc est 40 cubitorum, altitudo vero seu perpendiculum, sub quo fluit Croceus, 50 perticarum est, triennium in ejus fabricam impensum referunt, Sinae pontem volantem haud incongruo dixere nomine'. Surely *Fi pons* stands for *fei chhiao* here, and Kircher misunderstood it; he was also misled by the use of the words *unico arcu*. We cannot locate Chogan. As there is no mention of iron chains we insert this European echo here.

[h] Naturally they occur also in Tibetan religious iconography; cf. Highet (1), reproducing a Paradise of Padmasambhava.

[i] Horwitz (12); Sarat Chandra Das (1); Hooker (1); Kingdon Ward (7), opp. p. 257; Daniell & Daniell (1); Thang Huan-Chhêng (1), figs. 97, 100, 102. I am also indebted to Mr M. C. Gillett of the Foreign Service for first-hand accounts. Mr Gillett's numerous photographic negatives are deposited at the Royal Geographical Society.

[j] Robins (2).

[1] 高仙芝   [2] 金川瑣記   [3] 李心衡   [4] 丹巴   [5] 索橋   [6] 黑山峽

Most remarkable is the fact that identical developments took place, as we have seen, in South America, and Robins (2) reproduces a photograph of an Inca suspension bridge in the Huancayo Valley, Peru, made with cables of maguey fibre and hide, and spanning 150 ft., which, with its porches and abutments on each side of the river, resembles quite closely the type of Chinese bridge seen in Fig. 852 (pl.).[a]

Already at this stage we have to note the distinction between what we are calling the 'catenary' bridge and the suspension bridge as we know it, with its flat deck hanging level from the ropes or chains. In the first case the passengers travel along the curve, however tightened the longitudinal lines may be; in the second they move horizontally. It seems probable that the second form arose out of the first by a hypertrophy of the hand-rails. Even in the bamboo-cable bridges we begin to see a tendency for the hand-rails to take their origins from points higher up the banks in relation to the dipping deck than that which they occupy at midstream. This can be seen in a photograph of a bridge over the Dhauli River in Nepal on the Tibetan border,[b] or in another of a bridge built by the Nung tribespeople of Upper Burma.[c]

Cable suspension bridges were sometimes put to military uses in China, quite apart from the strategic importance which always tended to make them centres of combat operations.[d] The *Wei Shu* contains an account[e] of an ingenious submersible suspension bridge which could also be used as a boom. It was constructed by a Wei general, Tshui Yen-Po,[1] who was one of the commanders fighting Hsiao Yen[2] (later Liang Wu Ti) about +494. Tshui was guarding certain places on the Huai River,[f] his object being to deny the use of its waters and its banks to the enemy. So

he took the wheels of some carts, removed their rims, and cut short the spokes (to make cog-wheels) so that they would engage with one another. Strips of bamboo were twisted together to make ropes, and more than ten of these cables were strung together in parallel so as to form a bridge (with cross-planks). At both ends large windlasses were set up so that the bridge was submersible at will.[g] It could therefore neither be burnt nor cut. In this way the line of retreat of (Hsiao Yen's generals), Tsu Yüeh[3] and the rest, was blocked, and moreover ships and boats could not get past. Thus the forces of (Hsiao) Yen could not go to the rescue, and finally Tsu Yüeh's brigade was all captured.[h]

[a] Cf. Brigham (1) for Guatemala, and Shippee (1) for the Colca Valley in Peru. The relation between this kind of suspension bridge and that characteristically Amerindian device, the hammock, should not be overlooked.

[b] Smythe (1).   [c] Anon. (7).

[d] For example, the cable bridge built across the Imjin River by the Korean and Chinese allies in the campaign of +1593 against the Japanese invaders; cf. Hulbert (1), vol. 2, pp. 8, 9. This was not quite, as Hulbert thought, 'the first suspension bridge that history records'. Cf. pp. 188, 199.

[e] Ch. 73, p. 11a. We are indebted to Dr Arthur Waley for directing our attention to this passage.

[f] This region was quite outside the usual area of distribution of suspension bridges. But one does occasionally come across mentions of them in East China—for instance the south gate of Kao-yu on the Grand Canal (cf. p. 314) had a hanging-bridge lock (*tiao chhiao cha*[4]) outside it in +1632 (cf. Gandar (1), p. 30).

[g] Weighting could of course have been used, but there are some Chinese hardwoods, such as the Yunnanese chestnut, which sink like a stone, as the bridge engineers of the Burma Road found out during enemy bombings in the Second World War (Tan Pei-Ying (1), p. 114).

[h] Tr. auct.

[1] 崔延伯   [2] 蕭衍   [3] 祖悅   [4] 弔橋閘

For this Tshui Yen-Po received a well-deserved title.[a] Later on, in connection with iron chains, there will be something more to say about the military aspects of suspension bridges and booms.[b]

The most careful study of the catenary suspension bridges of West China is that of Fugl-Meyer, and we cannot do better than follow it here. On the remarkable properties of plaited bamboo strips something has already been said (Vol. 4, pt. 2, p. 64), and the subject will necessarily recur in connection with marine cordage (pp. 597, 664 below). The bamboo cables of the bridges are made in the same way as those used for towing ships against the current of rivers, but of larger dimensions. Bamboo strips taken from the inner part of the culm form a core in the centre of the rope, and round them is woven a thick plaiting of bamboo strips taken from the outer silica-containing layers. The plaiting is so done that the outer portion grips the core the more tightly the higher the tension. Such ropes (so[1]) are generally about 2 in. thick, and three or more twisted together form one of the bridge cables (tso[2]).[c] When placed in a testing machine, the straight inner strands break first, while the plaited material shows very great strength, not rupturing until a stress of 26,000 lb. per sq. in. is reached,[d] though an ordinary 2-in. hemp rope can carry a stress of only about 8,000 lb. per sq. in. Moreover, the silica-containing outer surface is very resistant to wear, e.g. against rock surfaces, which is naturally important both in towing and bridge cables.

The majority of West China suspension bridges[e] consist of a single span. At each end of the bridge a substantial bridge house is built on a stone foundation (cf. Fig. 846, pl.), its roof being supported either by wooden poles or stone walls. Inside this house or porch are placed two rows of stout vertical wooden columns, one for each of the side-cables of the bridge (Fig. 847, pl.), and these columns act as rotating capstans for the tightening of the bridge ropes. The columns are socketed into the foundations below and into beams of wood above, the whole structure being kept in place by the

---

[a] It is not quite clear why the bridge could neither be burnt nor cut if one of the great windlasses was on the hostile bank. Perhaps the bridge was secured just under water on that side, while the windlasses were both on the other side. It may be that our assumption about cog-wheels, which could imply mechanical advantage with the use of certain smaller wheels, may read too much into the text. Its words might also be compatible with the use of the cut-down cart-wheels as hand-spike wheels for manual power, assisted by ratchets. But in fact both interpretations may be right, the man-power being exerted upon long-spoked cart-wheels used as capstans, and these engaging with the windlass by means of short-spoked wheels set at right angles. In any case, a notable piece of field engineering.

[b] Pp. 202, 687.

[c] Torrance (2) has pointed out that this technical term and character gave a name to one of the border peoples living west of Szechuan, as we find already in the *Hua Yang Kuo Chih* (+4th century). Cf. *Shih Chi*, ch. 116, p. 2b (tr. Watson (1), vol. 2, p. 291).

[d] The figures given by Straub (1), p. 196, show that the bamboo cables of the eotechnic Chinese had attained about half the tensile strength of cables made of mild steel, i.e. 56,000 lb. per sq. in. We know now that the incorporation of other elements in the metal, such as nickel and chromium, and the use of modern cold-drawing processes, can allow of cables able to take up to 256,000 lb. per sq. in., as in the George Washington (Hudson) and Golden Gate suspension bridges.

[e] Owing to their profusion we make no attempt to list them here, but brief descriptions will be found in all the historical geographies. For instance, the *TSFY* speaks of three cable bridges over the Lung-chhuan River in Yunnan alone (ch. 118, p. 47b). A mass of information, not only about suspension bridges but about bridges of all kinds, is contained in *TSCC*, *Khao kung tien*, chs. 31–34 inclusive, much of which deserves translation.

---

[1] 索   [2] 笮

weight of a formidable crib of stones fitted underneath the roof.[a] Tightening is done by hand-spikes, and the whole process resembles the adjustment of a violin string by the turning of the key. The main cables under the deck are fastened to the capstan-columns nearest the span, or to horizontal winch-columns under the floor-boards, and in front of the bridge-house they pass through hardwood leads to keep them in their proper positions. When the deck cables begin to deteriorate, they are replaced by new ones, and then serve for a time as hand-rail cables. The number of cables varies from 6 to 12, and the spans crossed range from 130 to 250 ft.

The most famous bamboo suspension bridge is that at Kuanhsien,[1] the site of the remarkable — 3rd-century irrigation works which render fertile the plain of Chhêngtu in Szechuan, and which will be described in the following Section. The An-Lan S.C.[2] (or Chu-Pu S.C.,[3] as it used to be called)[b] consists of no less than eight major spans, the greatest of which is 200 ft., and its total length is somewhat over 1,050 ft. Good photographs of it are not plentiful owing perhaps to the mists and clouds of the region, but Figs. 848–851 (pls.) give a fairly good idea of this splendid bridge.[c] The bridge is 9 ft. wide, and supported on ten bamboo cables each $6\frac{1}{2}$ in. in diameter. The hand-rails are unusually elaborate, consisting of five of the same ropes each. One of the piers is built of granite masonry and crowned by a decorative gate and wooden roof, but the others are all hardwood trestles, roofed, and surrounded by additional piles to prevent scouring. Fortunately there are human figures visible in Fig. 850 to give some scale, and in fact the height of the bridge at the piers is about 50 ft. above the level of water or dry gravel bank of the Min River. One should remember, as Fugl-Meyer says, that not a single piece of metal is used in the whole structure. Further details are available in the papers of Liu Tun-Chen (1), Liang Ssu-Chhêng (3) and others;[d] all of whom show an admiration for the Kuanhsien bridge that may easily be understood.[e]

A document of +1177 records the crossing of this bridge by the great scholar and traveller Fan Chhêng-Ta. In the diary of his journey from Szechuan to the south-east, he wrote:[f]

Here I passed over the great suspension bridge (*shêng chhiao*[4]). This has five spans each 120 ft. long and 12 ft. wide. All the planks of the passage-way are held together by ropes. Bamboo fences are arranged at each side, and the bridge is borne upon several tens of huge wooden posts erected in the river and strengthened by piles of rocks. The bridge thus hangs between them in mid-air. When there is a strong wind, it sways up and down. It is like the nets strung out by fishermen for drying, or the loops of coloured silk suspended by the dyers.

---

[a] Cf. Fugl-Meyer (1), fig. 56, location unfortunately not given.

[b] In the Sung, yet another name, Phing-Shih S.C.[5]

[c] I use the adjective advisedly, from personal acquaintance, in 1943 and 1958. Other pictures in Phan Ên-Lin (1), pp. 156, 157.

[d] E.g. Lo Ying (1), pp. 273 ff.; Lane (1); Stevenson (1); Thang Huan-Chhêng (1), pp. 75 ff.; Mock (1), p. 32. Accounts differ as to the lengths of spans because in recent years additional trestle piers have been introduced, and if two small approach sections at each end are counted, the number of spans is now 10.

[e] It was naturally included in the series of pictorial stamps issued by the Chinese Post Office in 1962.

[f] *Wu Chhuan Lu*, ch. 1, p. 2b, tr. auct.

[1] 灌縣   [2] 安瀾索橋   [3] 珠浦索橋   [4] 繩橋   [5] 評事索橋

I had to give up my litter, and walked across swiftly in a manner which might have seemed graceful, but really I was trembling so much that I could hardly stand upright. All my company turned pale....

Elsewhere in the same book, other bamboo bridges of smaller size are mentioned. The original date of construction of the Kuanhsien bridge is not known, but in view of the folk-character of the principle, and the engineering capacity of Li Ping and his −3rd-century engineers,[a] there seems little reason why it should not go back as far as that time. It is certainly pre-Sung.

The renovation of the Kuanhsien bridge puts it out of action for more than two months each year, and since this must necessarily be the case also for all smaller bridges, even though for a shorter time, if they receive proper maintenance,[b] it was not unnatural that some more durable material should have been sought. The decisive step in the perfecting of suspension bridges was undoubtedly the use of wrought-iron chains, and as we shall shortly see, it looks as if this invention was made in south-west China not later than the end of the +6th century and quite probably in the +1st. The essential pre-conditions were there—traditional catenary cable bridges and an advanced metallurgy of iron.[c] To accompany the ensuing discussion we add one more provisional table which gives data on some of the more interesting iron-chain suspension bridges in the provinces of China (Table 68).[d] We call it provisional because the material is hard to assemble from literary sources alone; extensive travels would be required to do justice to the facts, though further detailed study of documents would undoubtedly bring more information to light. Here no attempt has been made, for instance, to work through the relevant chapters of the encyclopaedias, in which so much has been collected; this must be left to others.[e] To introduce the subject visually, and to suggest something of the great beauty of these iron-chain suspension bridges, we show in Fig. 852 (pl.) the Chi-Hung T.C. across the Mekong River on the old road to Burma, seeming to match with its tension the lowering weight of the surrounding mountains. Of similar nobility is the ancient bridge across the Hua Chiang which carries the road from the north in Kweichow to join the modern Kweiyang-Kunming highway at Phan-hsien. And to show this type of bridge in a calmer rural setting Fig. 853 (pl.) depicts the 'Fishermen's Bridge' at Pin-chhuan in Yunnan between Tali and the Yellow River.[f]

The iron-chain suspension bridges generally have no tightening arrangements,[g] the

---

[a] See below, pp. 288 ff.

[b] There were special guilds of builders and repairers of suspension bridges in West China (private communication from Mr M. C. Gillett).

[c] See Sect. 30d below, and meanwhile Needham (31, 32).

[d] Henceforward we abbreviate *Thieh Chhiao*[1] ('iron bridge') as T.C.

[e] For example, the *Phei Wên Yun Fu* and the *Thu Shu Chi Chhêng* (*Khao kung tien*). Lo Ying (1), pp. 82 ff., mentions over 30, and Mr Tsang Chi-Mou (private communication, April 1949) has details of 48 iron-chain suspension bridges. Goodrich (16) counts 118 suspension bridges of all kinds. No list approaching completeness has ever been attempted.

[f] Cf. Kingdon Ward (14), opp. p. 66; Farrer (1), opp. p. 140.

[g] But the chains may be hauled into position by powerful windlasses, as we read in the account of the building of the Ta-chou bridge (no. 10 in Table 68), given by the *Sui-ting Fu Chih* (Thang Huan-Chhêng (1), p. 78).

[1] 鐵橋

Table 68. Some iron-chain suspension bridges

No.	Province	Location	Name of bridge	River	Width (ft.)	Longest span (ft.)	No. of chains	Date	References
1	Yunnan	100 li S.W. of Ching-tung[1]	Lan-Chin T.C.[2]	Lan-tshang ch.[3] (Mekong)[a]	—	c. 250	20	Traditionally +65, perhaps Sui, repd. +1410	TSCC; Sa, 208; Go; L, 57
2	Yunnan	Yuan-chiang[4]	Yuan-chiang T.C.[4]	Yuan-chiang[4] (Hung ho[5])	—	c. 300	—	Repd. in Ming	THC, 80, fig. 104
3	Yunnan	350 li N.W. of Lichiang[6]	Tha-chhêng Kuan T.C.[7]	Chin-sha ch.[8] (Yangtze)	—	—	—	Sui, c. +595, temp. destr. in +794, captured +1252	TSCC; Ro; Sa, 193; Go
4	Yunnan	85 li W. of Lichiang[6]	Shih-mên Kuan T.C.[9]	Chin-sha ch.[8] (Yangtze)	—	—	—	Sui or early Thang, temp. destr. in +794, captured +1382	Ro; Go
5	Yunnan	E. of Lichiang[6]	Ching-li T.C.[10] = Tzu-li T.C.[11] = Chin-Lung T.C.[12]	Chin-sha ch.[8] (Yangtze)	c. 10	328	18	Late Chhing	Ro, 246, pls. 110, 111; Gu
6	Yunnan	N. of Tung-chhuan[13]	Chün-Min (T).C.[14]	Niu-lan ch.[15]	—	—[b]	12 (+2 rails) ¾ in. links, 1 ft. long	—	TSCC
7	Yunnan	Between Yung-phing[16] and Paoshan[17]	Chi-Hung T.C.[18] = Kung-Kuo T.C.[19] (?)[c]	Lan-tshang ch.[3] (Mekong)	—	225	14 (+2 rails) ¾ in. links, 1 ft. long	Ascr. San Kuo;[d] iron chains by +1470	Hor; Gi, 265; Go; Da, 75
8	Yunnan	Between Paoshan[17] and Thêng-yüeh[20]	Hui-Jen T.C.[21]	Nu ch.[22] (Salween)	—	219 (+156)[e]	9	—	Hor; Gi, 276; Ha, 334; Da, 55
9	Yunnan	Near Hsiakuan[23]	—	Yang-pi ch.[24]	—	120	2–4	—	Gi, 259; Ke
9a	Yunnan	Pin-chhuan[25]	Tiao T.C.[26]	—	c. 6	c. 65	—	—	THC, fig. 101
10	Szechuan	Near Tachow[27] (Suiting[28])	—	Thung ch.[29]	—	180	6 (+4 rails)[f]	—	THC, 78
11	Szechuan	Around Omei shan[30]	Several relatively small bridges	—	—	—	Some of linked rods	Before +1360	Hos; Lit; Phelps
12	Szechuan	San-hsia[31] (N. of Chia-ting[32])	San-Hsia T.C.[31]	Min ch.[33] (?)	—	—	—	—	TSCC
13	Szechuan	Jung-ching[34] (S. of Ya-an[35])	—	Jung-ching ho[34]	—	—	4 (rods)	—	Hor; Ri, 3/71
14	Szechuan	Lu-shan[36] (N. of Ya-an[35])	Lu-Shan T.C.[36]	Chhing-i ch.[37]	—	c. 430 (+?)[e]	—	—	THC, figs. 108, 109
15	Szechuan	Hsiao-ho-chhang[38]	—	Fou ch.[39]	—	75 (+3 × 75)[l]	7 (flat deck) (Rods)	—	Gi, 136
16	Szechuan	'Kiai-tsu-chang'[g]	—	A tributary of Min ch.[33] towards Sungphan[40],[h]	—	—	—	—	FM, 122; THC, 78, 81, fig. 107
17	Szechuan	Huai-yuan-chen[41] (nr. Chungking[J])	Ku T.C.[42]	—	—	—	—	Thang or pre-Thang	Go; THC, 81
18	Sikang[k]	Lu-ting[43]	Lu-ting T.C.[43]	Ta-tu ho[44]	9½	328 (formerly 361)	9 (+4 rails) ⅞ in. links, 10 in. long	+1701 to +1706	Up; Gi, 165; Lit; Hor; Go; THC, fig. 106; LTC, pl. 3a; Ha, 76; L, 269
19	Sikang[k]	Between Ya-an[35] and Ta-chien-lu[45] (Khang-ting[46])[l]	—	—	—	—	(Rods)	—	Hor; Ha, 57

20	Sikang[k]	Chhang-tu[47]		10		4		Tiao T.C.[48]	THC, 80, fig. 105
21	Kweichow	Between Anshun[50] and An-nan[51]	132 165	36	+1629[m]	30 or 36	Kuan-ling T.C.[52]	Hu; THC, 78; L, 20
21a	Kweichow	Between Shui-chhêng[54] and Phan-hsien[55]	200	—	—	4 (+6 rails)	Hua ch.[56]	GL
22	Kweichow	Chung-an[57] (E. of Kweiyang)	—	—	—	—	Chung-an T.C.[57]	PEL, fig. 20
22a	Kweichow	Between Tsunyi and Kweiyang	c. 200	—	—	—	Wu ch.[59]	Bo
23	Shensi	Ma-tao-i[50] (N. of Pao-chhêng)	50	6	—	6	Ma-tao-i T.C.[60]	Ri, 2/574; Hor; L, 57; W, 53
24	Shansi	Phu-chi[62] (nr. Tatung)	—	8	+1541	—	Phu-chi T.C.[62]	Gr

ABBREVIATIONS

Bo Bourne (1)
Da Davies (1)
FM Fugl-Meyer (1)
Gi Gill (1)
GL Groff & Lau (1)
Go Goodrich (16)
Gr Grootaers (1)
Gu Goullart (1)
Ha Hackmann (4)
Hor Horwitz (12)
Hos Hosie (4)
Hu Hummel (17)

Ke Kemp (1)
L Lo Ying (1)
Lit Little (1)
LTC Liu Tun-Chen (1)
PEL Phan Ên-Lin (1)
Ri von Richthofen (2)
Ro Rock (1)
Sa Sainson (1)
THC Thang Huan-Chhêng (1)
TSCC *Thu Shu Chi Chhêng*
Up Upcraft (1)
W Wiens (2)

The numbers refer to pages, preceded in the case of von Richthofen (2) by a volume number; and figure-numbers are indicated as such.

[a] Chiang is abbreviated to ch.

[b] This may have been only a bamboo cable bridge.

[c] The name Kung-Kuo T.C. on modern maps may refer only to the modern suspension bridge on this road. Older maps have a Fei-Lung (T.)C.[63] further up river, probably too far for it to be a synonym of the Chi-Hung T.C.; if so, this was probably a bamboo-cable bridge.

[d] This is the traditional association with Chuko Liang.

[e] Two spans.

[f] The account of this bridge in the *Sui-ting Fu Chih*, quoted by Thang Huan-Chhêng (1), p. 78, describes the way in which the deck planks were made to interlock by means of solid male and female joints secured with 'iron buttons', and also the provision of solid wooden side railings. It also mentions the windlasses used for bringing the chains to the right tension, and their fastening in position with 'stone columns'. This bridge was intended, and used, for vehicular traffic.

[g] Place-name unidentifiable. T'ang Huan-Chhêng (1), fig. 107, has Lao-chün-chhi,[64] perhaps another bridge of the same kind.

[h] The bridges in this region were regarded by the *Sui-ting Fu Chih* as the oldest of the kind.

[i] Four spans.

[j] Characters uncertain; there is a place of this name in Kuangsi. Goodrich (16) gave only the romanised form of the Szechuanese place-name.

[k] This former southern province of the Tibetan marches has become Chhangtu Territory (Chhangtu Ti-Chü), with loss of its eastern third to Szechuan.

[l] Another iron-chain suspension bridge in this region is described by Edgar (2).

[m] This bridge is said to have been preceded by a floating bridge made by a Taoist from Lo-fou Shan. See p. 203.

1 景東
2 蘭津鐵橋
3 瀾滄江
4 元江鐵橋
5 紅河
6 戛江
7 塔城關鐵橋
8 金沙江
9 石門關鐵橋
10 井里鐵橋
11 牛欄江鐵橋
12 金龍橋
13 東川
14 軍民(鐵)橋
15 功果鐵橋
16 永平
17 保山
18 靈虹鐵橋
19 怒江
20 騰越
21 惠人鐵橋
22 下關
23 達州
24 漾濞江
25 賓川
26 釣鐵橋
27 三峽鐵橋
28 綏定
29 通江
30 峨眉山
31 嘉定鐵橋
32 嘉定
33 帆江
34 樂經河
35 雅安
36 嘉山鐵橋
37 青衣江
38 小河場
39 浩江
40 松潘
41 潰溢鎮
42 古鐵橋
43 瀘定鐵橋
44 大渡河
45 打箭鑪
46 康定
47 昌都
48 吊鐵橋
49 吉曲河
50 安順
51 安寧
52 關鎖鐵橋
53 老盤江
54 水城
55 盤縣
56 花江
57 重安鐵橋
58 重安江
59 烏江
60 馬道驛鐵橋
61 白河
62 普濟鐵橋
63 飛龍(鐵)橋
64 老君溪

problem involved being one of anchorage only. For this purpose massive stone abutments are built to contain the chain ends, as for example in the photographs of bridges
across the Mekong and the Yangtze (Figs. 852 and 854, pls.). In the Lu-ting bridge, the
chains are embedded 40 ft. into the pillars on both sides. This problem still remains in
the most modern suspension bridges (cf. Steinman). No Chinese examples of chains
covering more than one span are known, and when a pier occurs, as at one of the
Mekong crossings, it is built on a natural island, and the two iron-chain bridges do not
form a continuous way right across. The chains are always hand-forged, with welded
links made of bar iron 2–3 in. in diameter. Owing to the constant lateral movement
caused by the winds of the gorges, the links close to the abutments tend to wear out, and
in the old days replacement was not easy, since the country is such that communications were very difficult even if funds sufficed. Chains were not, however, the only
types of catenary used—at least three bridges are known (and there were probably
many more, some still existing) which are constructed of linked iron bars. Deep in the
mountains of the Szechuan–Sikang border, such a bridge 300 ft. in length, described
by Fugl-Meyer,[a] has round bars 2·25 in. in diameter in lengths of 18 ft. joined by pin
connections. It crosses one of the tributaries which enter the Min Chiang[1] before it
issues from the mountains at Kuanhsien, and in this case three intermediate stone
piers have been built, the bars passing smoothly in channels over their rounded convex
tops, so that the bridge presents a gently undulating line.[b]

Of all these bridges, perhaps the most famous, if not the most interesting technically, is the Lan-Chin T.C.[2] near Ching-tung in Yunnan. As the local history and
topography says:[c]

> On the west bank of the (Lan-)tshang river there are perpendicular cloud-piercing cliffs
> reflecting the waters, and near by a waterfall plunges down over beautiful but dangerous
> precipices. The prospect is wonderful. Here iron chains are fixed north and south to form a
> bridge. Local tradition says that this was built in the time of the emperor Ming of the Han
> (c. +65). It was repaired in the Yung-Lo reign-period (c. +1410).

This raises the important question of the age of the earliest iron-chain bridges. Later
generations certainly believed that this one dated from the Han, for example Chang
Chia-Yin,[3] who wrote a poem about the songs of the Han soldiers resounding in the
gorge (c. +1545).[d] Modern historians, however, have shown some disinclination to
believe in the possibility of so early a date. They have pointed out that pre-Thang
texts mention only the crossing of the Mekong River at this point and do not distinctly

---

[a] (1), p. 122; cf. Hutson (3). We cannot identify the exact place, as Fugl-Meyer gives its name only in
aberrant romanised form, another example of the lamentable practice of omitting characters for Chinese
place-names. It may or may not be identical with the Lao-chün-chhi[4] of Thang Huan-Chhêng (1),
p. 81.

[b] Cf. Thang Huan-Chhêng (1), fig. 107.

[c] *TSCC, Chih fang tien,* ch. 1490, p. 3a, tr. auct. From *Nan Chao Yeh Shih* (+1550), ch. 26, tr.
Sainson (1), p. 208.

[d] *TSCC, Khao kung tien,* ch. 33, *i wên,* p. 6, which writes Yin.[5]

[1] 岷江　　　[2] 蘭津鐵橋　　　[3] 張佳胤　　　[4] 老君溪　　　[5] 引

describe any bridge.[a] Some have even suggested that the attribution to Ming Ti sounds like a story emanating from the local Buddhist abbey. For this was the emperor who traditionally patronised the first beginnings of Buddhism in China, and since the building of bridges was a religious duty, his name might well have become associated with an important one.[b] In any case there is no ground whatever for doubting its repair in the early +15th century, and that event alone antedates any iron-chain suspension bridge in Europe. Moreover, as we shall see in a moment, there is some evidence for the building of other famous iron-chain bridges in the Sui, and that might possibly be the date of the Ching-tung bridge also. But the best claim to renown which this bridge has is that it was described by several of the early travellers in China, and knowledge of it reaching the West through Jesuit channels, it became the accepted precursor of all successful eighteenth- and nineteenth-century iron-chain suspension bridges. The story of this, however, will come better at the end of our account.

Perhaps we have been too sceptical about the ability of the Han engineers to throw an iron-chain suspension bridge across the Mekong River.[c] Since the criticism of the traditional attribution much new knowledge has come to light about the iron and steel technology of Han times,[d] and if it was possible then to make cast iron more than a millennium before Europe could do so, it was surely not impossible to manufacture 250-ft. lengths of chain formed of substantial wrought-iron links. Since China was the focus of advanced siderurgical skill, not Tibet or Gandhāra, we must view in a new light the indubitable existence of long-established iron-chain suspension bridges on the route to India by the beginning of the +6th century. Fa-Hsien had not found them, but Sung Yün[1] and Hui-Sêng,[2] who travelled that way in +519, did. The narrative of their journey says:[e]

From the country of Po-Lu-Lo[3] (Bolor, modern Gilgit) to the kingdom of Wu-Chhang[4] (Udyāna, modern Swat, Chitral, etc.), they use iron chains for bridges. These are suspended in the void, in order that one may cross (over the mountain chasms). If one looks downward no bottom can be seen, and there is nothing to grasp at in case of a slip, so that in an instant a man may be hurled down ten thousand fathoms. On this account travellers will not cross over if a wind is blowing.

---

[a] Goodrich (16) instances *Hou Han Shu*, ch. 116, p. 18*b*; *Hua Yang Kuo Chih*, ch. 4, and *Shui Ching Chu*, ch. 36, p. 6*a*. According to the first of these, the bridge was in the country of the Ai-Lao tribes (cf. Vol. 4, pt. 1, p. 100).

[b] This suggestion was made by the late Professor Haloun. The attribution to Han Ming Ti entered the European literature, and will be found, for example, in the pages of Feldhaus (1), col. 152.

[c] The traditional connection between the Lan-Chin crossing and the time of Han Ming Ti is remarkably strong. On the Po-nan Shan[5] between Yung-phing and the Lan-Chin bridge there is an old temple called Yung-Kuo Ssu.[6] Stone tablets preserved in this temple, authenticating the *Li chiang Fu Chih*, ch. 2, p. 69*a*, say that a road over these mountains was first opened in Han Ming Ti's reign. From that time was handed down a song about the crossing of the Lan-tshang River which went somewhat as follows: 'So far-reaching is the virtue of the Han that a way has been made even into the wild regions beyond Lan-Chin where one can cross the Mekong.' See further in Rock (1), vol. 1, pp. 167 ff.

[d] Cf. Needham (31, 32) and more fully in Sects. 30*d* and 36*c* below.

[e] *Lo-yang Chhieh-Lan Chi*, ch. 5 (Chiao Shih ed. p. 102); cf. Vol. 1, p. 207; date +530. Tr. auct. adjuv. Beal (1), p. 187, (2), p. xciii.

[1] 宋雲   [2] 惠生   [3] 鉢盧勒   [4] 烏場   [5] 博南山   [6] 永國寺

After this it is natural to find that these bridges were still in use in Hsüan-Chuang's[1] time. In his account of Udyāna, written about +646, he says:[a]

> Re-ascending the Hsin-tu[2] River (Sindhu, the Indus) the roads are craggy and steep; the mountains and valleys dark and gloomy. Sometimes one has to cross the gorges by (bridges of) cables, sometimes on chains of iron stretched (from side to side). There are galleries (*chan tao*[3])[b] along the edges of abysses, vertiginous flying bridges, and wooden ladders or stone steps up which one has to climb....[c]

It seems most reasonable to suppose that the making of iron-chain suspension bridges radiated from the regions of most advanced iron technology, and it may well be therefore that the Ching-tung bridge was actually the predecessor of these bridges on the upper Indus between Gandhāra and Sinkiang.[d]

Concrete, indeed decisive, evidence for the early building of iron-chain suspension bridges comes from a time intermediate between Sung Yün and Hsüan-Chuang. It concerns an interesting cluster of these bridges in north-western Yunnan in the region of Lichiang,[4] a city which stands at the base of a tongue of mountainous country some sixty miles long formed by a northern detour of the Yangtze (here the River of Golden Sand, the Chin-sha Chiang[5]) as far as the Szechuanese border.[e] Above the sharp bend where the detour begins, the river was crossed by two famous iron-chain suspension bridges. The first of these was between Chhi-tsung[6] and Chü-tien[7] near the Tha-chhêng Kuan[8] pass, formerly the border between the Nakhi tribesfolk of the Nan Chao kingdom and the Tibetans; here there was a Thieh-chhiao Chhêng[9] (Iron-bridge Town), the present village of Tha-chhêng.[f] This was without doubt on an important line of communications through Chung-tien[10] into Tibet,[g] and apparently a special official, the Iron Bridge Commissioner (Thieh Chhiao Chieh-Tu[11]), was at one time in charge of it. Now encyclopaedias preserve a statement by Mu Kung[12] (the Honourable Mr Mu), of the Yuan, himself a Nakhi from this part of the country, which reads as follows:[h]

> *Hua-Ma Kuo.*[13] This (tribal region) was formerly called *Chü-chin-chou.*[14] The Yuan (emperor), Shih Tsu (Khubilai Khan), came here on a tour of inspection, and (afterwards) enfeoffed (Tuan Hsing-Chih[15] as hereditary Governor-General of Nan Chao, Yunnan, in

---

[a] *Ta Thang Hsi Yü Chi*, ch. 3, p. 6*b*; cf. Vol. 1, pp. 207 ff. Tr. auct. adjuv. Beal (2), p. 133; a further reference to flying bridges and bracketed galleries from the same page of Chinese text follows on p. 134.

[b] Cf. pp. 20 ff. above.

[c] We note from this that bamboo-cable bridges persisted along with the stronger iron-chain ones, probably for shorter spans.

[d] An excellent parallel is the teaching of iron-casting to the people of Ferghana and Parthia by artisan 'deserters' from a Chinese embassy in the −2nd century (*Shih Chi*, ch. 123, p. 15*b*; *Chhien Han Shu*, ch. 96A, p. 18*b*, tr. in Vol. 1, pp. 234 ff.).

[e] Besides the exhaustive account of this region by Rock (1) with its magnificent illustrations, there is a very living description by Goullart (1) who was Depot Master of the Chinese Industrial Cooperatives at Lichiang during the Second World War.

[f] Rock (1), vol. 2, pp. 292 ff., pls. 155, 156.　　　　[g] Cf. Liebenthal (8).

[h] *TSCC, Chih fang tien*, ch. 1505, *i wên* 2, p. 2*a*, tr. auct.

[1] 玄奘	[2] 信度	[3] 棧道	[4] 麗江	[5] 金沙江
[6] 其宗	[7] 巨甸	[8] 塔城關	[9] 鐵橋城	[10] 中甸
[11] 鐵橋節度	[12] 木公	[13] 華馬國	[14] 巨津州	[15] 段興智

+1253).[a] Continuing his tour to the west he visited Hua-Ma Kuo... (and passed) over the iron bridge south to Stonegate Pass (Shih-mên Kuan[1]). This bridge, which spans the River of Golden Sand, was built, according to the *Sui Shih*[2],[b] by (Shih) Wan-Sui[3] and Su Jung.[4] Going north (from the river) one comes to the Hei Shui[5] (Black River) which connects with Pa and Shu (Szechuan) flowing out to the east, and eventually to the San Wei (Shan)[6] (Mountains of the Three Dangers, near Tunhuang)—in all, ten thousand *li* of mountains.

Khubilai Khan was really in these parts, but it was not exactly on a serene tour of inspection, it was as a military commander of the Yunnan expedition of the Mongols under his reigning elder brother Mangu Khan in +1252, and he was leading one of three columns in the attack on Tali.[c] This was probably the occasion when the bridge was captured by A-Tsung A-Liang,[7] a Nakhi chieftain who was fighting on the Mongol side.[d] But by the time of Khubilai it was already old, and must have been long since repaired, for we know that it had been destroyed in +794 by the Nan Chao king I-Mou-Hsün[8] during the course of a campaign which he undertook in alliance with the Thang dynasty against the Tibetans.[e] In this way he cut off the retreat of their forces, and gained a considerable victory. As for the original builders of the bridge, there seems no reason for doubting the attribution to a general and a military engineer of the Sui dynasty. Other sources date the construction in the Khai-Huang reign-period (+581 to +600) under Sui Wên Ti.[f] And we know that from +594 to +597 Shih Wan-Sui was in command of an expeditionary force in the south-west sent to conquer the Man tribes of Yunnan, so that there was every reason for him to improve communications, like General Wade in another mountainous country.[g] Indeed the dynastic histories specifically mention his river crossings in the campaign. Su Jung was probably his military engineer, and the bridge would have been built in the last years of the century. Holes in the rocks for securing the chains can still be seen, but the bridge itself has long since disappeared.

Further down river between Chü-tien and the sudden northward bend at Shih-ku[9] there was another great iron-chain suspension bridge across the Chin-sha Chiang near

---

[a] See *Nan Chao Yeh Shih*, ch. 22, tr. Sainson (1), pp. 110 ff., 113 ff. He had been the last Nan Chao king of the Hou Li dynasty.

[b] This is not the dynastic history, the *Sui Shu*, but another work, by Wu Ching[10] (+713 to +755), apparently no longer extant.

[c] *Nan Chao Yeh Shih*, ch. 21, tr. Sainson (1), pp. 108 ff.; cf. Rocher (1), vol. 1, pp. 170 ff. Yang Shen mentions goatskin rafts in connection with the passage across the river, but naturally they would have supplemented the bridge in the case of an army.

[d] Rock (1), vol. 2, p. 95.

[e] *Wên Hsien Thung Khao*, ch. 329, pp. 14a, 17b (p. 2585.1); tr. Hervey de St Denis (1), vol. 2, pp. 190, 206; also *Li-chiang Fu Chih Lüeh* (+1743), pt. 1, ch. 4, p. 27a, cf. Rock (1), vol. 2, pp. 292 ff.; Rocher (1), vol. 1, p. 162; Bushell (3), pp. 507, 533. I-Mou-Hsün was the sixth king of Nan Chao (r. +778 to +808), cf. *Nan Chao Yeh Shih*, ch. 15, tr. Sainson (1), pp. 48ff., 53, 55, also p. 272. He organised many institutions in his country on the Chinese model, and employed Chinese cartographers. A faithful ally of the Thang.

[f] *Nan Chao Yeh Shih*, ch. 26, tr. Sainson (1), p. 193; the names of the builders are given in *Li-chiang Fu Chih Lüeh*, pt. 1, ch. 4, p. 27a.

[g] *Sui Shu*, ch. 2, p. 11a, ch. 53, pp. 5b, 6b; *Pei Shih*, ch. 73, pp. 10b, 12a; *Nan Chao Yeh Shih*, ch. 14, tr. Sainson (1), pp. 30 ff.

[1] 石門關	[2] 隋史	[3] 史萬歲	[4] 蘇榮	[5] 黑水
[6] 三危山	[7] 阿琮阿良	[8] 異牟尋	[9] 石鼓	[10] 吳兢

Shih-mên Kuan. If it was not also built by Shih and Su it must have been made not much later,[a] for in the same Tibetan campaign of +794 it too was destroyed, by the Thang general Wei Kao[1] guided by the local Nakhi prince Phu-Mêng Phu-Wang,[2] allies of I-Mou-Hsün.[b] But it was rebuilt in due course, for in +1382 it was captured by A-Chia A-Tê,[3] another Nakhi chief, who had taken up the cause of the Ming against the Mongols. For this he was rewarded by Thai Tsu with the surname of Mu and the hereditary magistracy of Lichiang, so that as Mu Tê[4] he is none other than the Honourable Mr Mu whose words we read just now.[c] But this bridge also no longer exists. Finally, another iron-chain suspension bridge, the Chin-Lung T.C.,[5] still very much in use (Fig. 854, pl.), links Lichiang to the east across the now south-flowing river with Yungpei,[6] a mountain town under the shadow of 13,000 ft. Kuang-mao Shan.[d]

Those who, like the writer, were familiar with the 'Burma Road' during the Second World War,[e] knew the modern suspension bridges which crossed the Salween and the Mekong rivers on each side of Paoshan[7] at Hui-thung-chhiao[8] and Kung-kuo-chhiao[9] respectively.[f] Not all were aware, however, that the latter gorge at least had been bridged since ancient times, and with iron chains since about +1470, hence the name Chi-Hung T.C.[10] ('a rainbow in a clear sky').[g] The provincial history avers that the boring of the holes in the rocks for the attachment of the cables or chains was done under Chuko Liang, the great Captain-General of Shu in the Three Kingdoms period, during his conquest of Yunnan[h] in +225 to +227. Though confirmatory evidence is

[a] An alternative tradition, attributed to the *Thang Shu* in *Ta Chhing I Thung Chih*, ch. 382, p. 3*b*, attributes the building of one or other of these two bridges to a king of Nan Chao who wanted to improve communications with Tibet in +751; cf. Chavannes (20), pp. 603 ff.

[b] Rock (1), vol. 1, pp. 25, 57, 62, 89, 275, 285 and pl. 153.

[c] Rock (1), vol. 1, pp. 99 ff., 101, 154 ff. On the Ming conquest of Yunnan see *Nan Chao Yeh Shih*, ch. 22, tr. Sainson (1), pp. 146 ff. We assume this identification because *TSCC*, at the place quoted, dubs Mu Kung a man of the Yuan period. Otherwise it would be much more plausible to suppose that the writer was a later Mu whose personal name was Kung, i.e. A-Chhiu A-Kung,[11] the fourteenth hereditary Nakhi chief (+1494 to +1553), the most literary and cultured of his line, the main compiler of the MS. 'Chronicle of the Nakhi Chieftains', a notable poet, and the intimate personal friend of Yang Shen. On him see Rock (1), vol. 1, pp. 74, 115 ff., 156. But although the date of the statement would thus be later, its authority would be hardly less good, an account of A-Chhiu A-Kung's scholarship and profound local knowledge.

[d] This is the bridge at Ching-li or Tzu-li mentioned by Rock (1), vol. 1, pp. 171, 174, 177, 246, and figured in pls. 110, 111. Goullart (1), pp. 199, 202, gives a vivid account of its temporary dismantling in 1949 when a rebellion on the Yungpei side in the interregnum between the Kuomintang and the Communists threatened the Nakhi city of Lichiang. An interruption of the same kind had happened twenty years before. Another late Chhing iron-chain suspension bridge is that across the Wu Chiang[12] between Kweiyang and Tsunyi in Kweichow (cf. Bourne (1), p. 77). Well do I remember that gloomy and romantic gorge from my travels during the second world war.

[e] See Anon. (81).

[f] See Tan Pei-Ying (1), pp. 105 ff., 144 ff. and opp. pp. 73, 168 and 185. The longest single-span suspension bridge on the road was 410 ft.; all were built under the greatest difficulties by Hsü Yi-Fang and Chhien Chang-Kan between 1937 and 1940.

[g] *Yunnan Thung Chih*, ch. 50, pp. 52*a* ff. We are not sure of the exact location of the old bridge or bridges; a Fei-Lung C.[13] is marked on some maps higher up than the point where the road now crosses.

[h] Cf. Rocher (1), vol. 1, pp. 156 ff.

[1] 韋臯	[2] 普蒙普王	[3] 阿甲阿得	[4] 木得	[5] 金龍鐵橋
[6] 永北	[7] 保山	[8] 惠通橋	[9] 功果橋	[10] 霽虹鐵橋
[11] 阿秋阿公	[12] 烏江	[13] 飛龍橋		

lacking, there is nothing at all impossible in this tradition.[a] About +1540 the Chi-Hung T.C. inspired a poem by Yang Shen:[1]

> Fearful of step, on the flying ladder one advances,
> Woven of iron, a lonely thread running straight through the sky,
> In malarious mists above one the cloud-dragons wander,
> In the abyss below the peacocks drink of the river's spray;
> South from the Lan-Chin Bridge went the road to the Ai-Lao Country,[b]
> West from the borders of Phu[2] ran the chain of camps of Chuko Liang—
> Far, far, a myriad *li* from here, is China's heartland,
> Thinking of ancient deeds, how could one's own heart not be full?[c]

The name of the engineer who effected the conversion to iron chains has fortunately been preserved. In his account of this bridge, Ku Tsu-Yü[3] says: 'Wang Huai[4] it was who began to link iron chains together (here), adding cross-planks over which people could walk as if on level ground.'[d] Presumably during the alterations a cable-bound pontoon bridge was provided, for he goes on to say,

According to the *Chih* (the local history), the Chi-Hung bridge over the Lan-tshang River was formerly of bamboo cables, but afterwards at the beginning of the Ming, when (Yunnan was) pacified, Hua Yo[5] cast iron (mooring-) posts and set them up on each bank so that boats could be connected together (to form a bridge).[e]

This period was one which saw many such conversions, for we know of yet another engineer, Chao Chhiung,[6] who replaced at least one of the cable bridges over the Lung-

---

[a] To commemorate his exploits Chuko Liang set up a stele with an inscription. The story goes that when Shih Wan-Sui went to see it he found a writing on the back which said, 'Hereafter Wan-Sui will conquer Yunnan, but not so gloriously as I did.' He therefore ordered his men to overturn it, but underneath there was another sentence saying, 'Shih Wan-Sui is not to uproot my stele'; so he had it replaced, offered a sacrifice and left hurriedly. More relevant to us here is that Chuko Liang set up also a commemorative column of cast iron. We find many references in subsequent Yunnanese records to such columns, for instance one erected by an early Nan Chao king, Chang Lo-Chin-Chhiu,[7] in +649, later re-cast in larger size by Mêng-Shih-Lung.[8] Another was set up in +632 at the San-Tha Ssu[9] temple at Tali to commemorate its building by the royal architect Chhih Ching-Tê.[10] Yet another was cast in +872 by the eleventh Nan Chao king Shih-Lung[11] to replace the column of Chuko Liang which had disappeared. These facts may not be without significance for the state of the iron and steel industry with relation to bridge-building in medieval Yunnan (cf. Sainson (1), pp. 30 ff., 62, 74, 210; Rock (1), vol. 1, p. 54). Rocher, who gives a precious account of the traditional metallurgical technology of the province, (1), vol. 2, pp. 195 ff., says, 'Iron is so common in Yunnan that we hardly know of any district without deposits of it.'

[b] Cf. Vol. 4, pt. 1, p. 100. Rock-crystal was one of its products.

[c] Yang Shen, the author of the *Nan Chao Yeh Shih*, was himself an enforced exile in Yunnan. His poem appears twice in *TSCC*, in *Chih fang tien*, ch. 1505, *i wên* 2, p. 2b (where it is misleadingly given in the Lichiang section), and in *Khao kung tien*, ch. 33, *i wên* 2, p. 6b. His biography, and an interesting account of his votive temple near Kunming, may be read in Rock (1), vol. 1, pp. 162 ff., and pls. 47, 48, 49. Yang Shen exerted great cultural influence in the province, where he founded a college, and as he was interested in many subjects such as botany, zoology, pharmacy and alchemy, we often have occasion to mention him elsewhere. Tr. auct.

[d] *TSFY*, ch. 118, p. 4b (+1667). He places the event, however, a little later than the *Yunnan Thung Chih*, in the Hung-Chih reign-period, +1488 to +1505.

[e] P. 5b; cf. p. 16b. Tr. auct.

[1] 楊慎　　　[2] 蒲蠻　　　[3] 顧祖禹　　　[4] 王槐　　　[5] 華岳
[6] 趙烱　　　[7] 張樂進求　　[8] 蒙世隆　　　[9] 三塔寺　　　[10] 遲敬德
[11] 世隆

chhuan River in northern Yunnan by an iron-chain bridge just about the same time as Wang Huai.[a]

It would be tedious to descant much further on the iron-chain suspension bridges of China, and a few more words must suffice. The Ku T.C.[1] at Huai-yuan-chen[b] near Chungking was celebrated in a poem by a Buddhist cleric of the Thang period, Chih-Mêng,[2] but as he called it a *shêng chhiao*[3] it was then presumably, though not necessarily, of bamboo cables. The San-Hsia T.C.,[4] also in Szechuan, must be at least Yuan, for Chieh Hsi-Ssu[5] wrote a poem on it before the +13th century was out.[c] Those in Shensi[d] and Shansi are interesting (Table 66) because they show the extreme geographical limits of the suspension bridge zone. The Lu-ting T.C.[6] across the Ta-tu Ho[7] in the mountains of western Szechuan[e] is known to have been put into its definitive form by Hsiung Thai[8] and the monk I-Fan[9] in +1705, but it is likely that there were the usual earlier versions at that place. After various +18th-century literary references,[f] this bridge acquired great fame in modern times as the scene of a heroic military exploit. In 1935 on the circuitous route of the Long March the Red Army was swinging north up this valley towards Yenan in northern Shensi, and succeeded in storming the bridge after the planking had been largely removed and in the face of heavy enemy fire. The way to the north was thus opened.[g]

Accounts of military operations reveal a close connection between suspension bridges, pontoon bridge cables and defensive harbour booms; all form a single technological complex. Interesting examples of the use of iron chains may now be added to the case of the submersible bamboo-cable bridge-boom of the late +5th century already described (p. 190). There were many ways of using catenaries besides capturing existing bridges or denying their use to hostile forces. The following account, from the *Wu Tai Shih Chi*,[h] tells of a battle in +928 when the second king of the Nan Han State, Liu Yen,[10] gained a victory over his enemies.

In the 4th year of the Pai-Lung reign-period the army of Chhu attacked Fêngchow[11] (on the West River near the western border of modern Kuangtung) with numerous ships, and

---

[a] P. 10a, b.

[b] No Huai-yuan-chen[12] is identifiable on maps at my disposal except a village in Kuangsi, but the *Chhung-chhing Fu Chih*, ch. 1, p. 27b, is not to be gainsaid, and I only regret that during my long period of residence at Chungking I failed to see this bridge.

[c] *TSCC, Khao kung tien*, ch. 33, *i wên* 2, p. 5a.

[d] Ma-tao-i had no suspension bridge when I passed through it in 1943 and 1944 so presumably the one which von Richthofen saw had been replaced by a steel bridge or abandoned.

[e] Cf. Thang Huan-Chhêng (1), fig. 106; Liu Tun-Chên (1), pl. 3a. Many Western travellers have given descriptions of this bridge; references are given by Goodrich (16).

[f] For example, the story of an unfortunate young courtesan, Chao San-Shou,[13] in the late +18th-century book *Chhin Yün Hsieh Ying Hsiao Phu*[14] (Minor Records of Heroines of the Western Provinces), prefaced by Wang Chhang.[15] In an often recurring motif, the poor scholar Tshao Jen-Hu,[16] though deeply in love, could not afford to buy her from her owners. The Lu-ting bridge, identifiable from the geographical description, is the setting for an incident important in her life.

[g] The story has been told, with some inaccuracies, by Snow (1), pp. 185 ff.; and by Yang Chêng-Wu (1). The scene was commemorated recently on a postage-stamp issued by the Chinese Government.

[h] Ch. 65, p. 5a, tr. auct. adjuv. Schafer (4).

[1] 古鐵橋   [2] 智猛   [3] 繩橋   [4] 三峽鐵橋   [5] 揭傒斯
[6] 瀘定鐵橋   [7] 大渡河   [8] 熊泰   [9] 一番   [10] 劉龑   [11] 封州
[12] 懷遠鎮   [13] 趙三壽   [14] 秦雲擷英譜   [15] 王昶   [16] 曹仁虎

defeated the defending forces on the Ho River.[1] (Liu) Yen was alarmed. Making rhabdomantic divination[a] about it with the aid of the *Book of Changes* he encountered the *kua Ta Yu* ('Greater Abundance') so he proclaimed an amnesty throughout his realm and changed the reign-period name to Ta-Yu. Then he sent his general Su Chang[2] with a 'Magic Crossbow Division' (*shen nu chün*[3])[b] of 3,000 men to the relief of Fêngchow. (Su) Chang took a pair of iron chains and sank them deeply in the Ho River, with very large winches (or capstans, lit. wheels, *lun*[4]) on each bank (to tighten them), and tamped earth redoubts to conceal (the winches and their crews). Then he invited battle with light boats which, pretending defeat, fled, hotly pursued by the men of Chhu. At the right moment (Su) Chang set in motion the great wheels, which hauled up the booms and cut off the Chhu ships, exposing them to the cross-fire of powerful arcuballistae set up on each bank, so that hardly a man of the Chhu forces escaped to tell the tale.

Chinese military engineers must have had recourse many times through the centuries to these techniques. For again in +1371, when the armies of the Ming were invading the west to re-establish imperial authority over Szechuan, the technicians of Shu flung boldly across one of the greatest of the Yangtze gorges three cable suspension bridges commanding as many iron-chain booms, and equipped with bomb-throwing trebuchets and all kinds of firearms. Later on we shall tell this story fully in another context (p. 687).[c]

Just as the arch bridge found its Boswell in Hsieh Kan-Thang so the iron-chain suspension bridge also rose, in one case, to the dignity of a book. This concerned the bridge at Kuan-ling[5] over the Northern Phan Chiang,[6] in south-western Kweichow, and Hummel (17) has given us an account of it. The *Thieh Chhiao Chih Shu*[7] (Record of the Iron Suspension Bridge) was written by Chu Hsieh-Yuan[8] and printed in +1665, though it concerned the erection of the bridge which had taken place already in +1629. Chu Hsieh-Yuan's father, Chu Chia-Min,[9] had been the prefect under whose auspices the bridge had been built, so that he himself had access to all the official documents. His book was illustrated by a panoramic drawing of the structure and its approaches, part of which is shown in Fig. 855. This may be compared with the description given by the famous explorer and traveller, Hsü Hsia-Kho (cf. above, Vol. 3, p. 524), who visited the bridge in +1638, not long after it was finished. His diary has the following:[d]

The Phan Chiang bridge is held up by iron chains which connect the cliffs on the eastern and western sides of the river, a distance of 150 (Chinese) feet.[e] The warp so made has a weft of planks. The cliffs themselves are about 300 ft. high and between them a swift raging stream

---

[a] Cf. Vol. 2, pp. 347 ff.

[b] This term most probably refers to what we might call an artillery division, for the adjective was applied to powerful arcuballistae before becoming common in the Ming for gunpowder weapons. Cf. Vol. 4, pt. 2, p. 425.

[c] And see Sect. 30*i*. For another (+13th-century) example of an iron-chain boom in Chinese literature see the *Chu Fan Chih*, ch. 1, p. 7*a* (tr. Hirth & Rockhill, p. 62).

[d] *Hsü Hsia-Kho Yu Chi*, ch. 8, p. 32*b*, tr. Hummel (17), mod. auct.

[e] At this time almost equivalent to ours.

[1] 賀江　　　[2] 蘇章　　　[3] 神弩軍　　　[4] 輪　　　[5] 關嶺
[6] 北盤江　　[7] 鐵橋志書　[8] 朱燮元　　　[9] 朱家民

Fig. 855. An illustration from the *Thieh Chhiao Chih Shu* (Record of the Iron Suspension Bridge) by Chu Hsieh-Yuan. It depicts the Kuan-Ling bridge in the gorge of the Northern Phan Chiang, between An-shun and An-nan in south-western Kweichow. Thirty to thirty-six chains bridged the span of some 165 ft. The structure, erected in +1629, was replaced in 1943 by a steel-chain suspension bridge at a broader but easier place about half a mile downstream. Chu's drawing has many interesting features. Under the bridge we read: 'The water is so deep here that it has no bottom.' To the right in the background there are a number of temples, including a Kuan-Yin pagoda and a library for the sūtras. To the left there is a masonry embankment against which break the 'hundred-foot waves'. In the foreground on the right there is a Buddhist statue, and on the left the 'stone of weeping', a monument to those who perished in the crossing before the bridge was built.

of water, of unfathomed depth,[a] rushes along. In earlier years ferry boats were often in grave danger of capsizing, whereupon people tried to span it by a stone structure, but they failed. Then in the 4th year of the Chhung-Chên reign-period the present Governor, then a judge, Chu Chia-Min, asked Major Li Fang-Hsien[1] to build (a suspension bridge). So now several tens of great iron chains are suspended (from towers) on each bank, and on them two layers of boards, about eight inches thick and more than eight feet long, are laid. The bridge looks flimsy and unsubstantial, but when people tread on it, it is as immovable as a mountain-peak; daily hundreds of oxen and horses with heavy loads pass over it. Each side of the

---

[a] The local people must have taken a pride in telling people this, for the legend in the picture says the same thing.

[1] 李芳先

bridge is protected by a high iron railing woven with smaller chains. On each bank there crouch two stone lions, three or four feet high, which clench these railing-chains tightly in their mouths.[a]

Hsü noted a point of technical interest here, namely that the railings were also made of chains, thereby taking a portion of the weight and inviting a transition to a flat deck suspended entirely on the catenary. From other sources we know that Li Fang-Hsien's bridge was partly destroyed in the turmoil at the end of the Ming in + 1644, but repaired in + 1660 and many times later. In 1939 a modern steel structure replaced the old one, but this was destroyed by the Japanese a year later. In 1943 a steel suspension bridge, 390 ft. long, was erected about half a mile downstream, but when I passed back and forth over it in the following years I had no idea how interesting the spot was, and did not try to find the old abutments.[b]

It is possible that the work of the Kweichow bridge-builders was brought to the notice of Europeans only a few decades after the description of Hsü Hsia-Kho. For in the map of this province in the sixth volume of Blaeu's great *Atlas* (entitled *Novus Atlas Sinensis*) Martin Martini in + 1655 marked an iron-chain bridge over a River 'Puon' to the west of a place, so far unidentifiable but probably somewhere near Kuan-ling, called 'Picie'. Then in his description of the monuments of Kweichow, he wrote:

Ad Picie occidentalem partem supra profundissimam vallem, per quam torrens ingenti aquarum in praeceps ruentium lapsu, atque impetu volvitur, ut viam sternerent Sinae, crassissimis ferreis catenis annulos aliquos ita hamis, uncisque ex utraque montium parte firmarunt, ut superimpositis asseribus pontem efformarint.[c]

Though Martini (Wei Khuang-Kuo[1]) may not have seen this himself in the course of his travels in China, the information would easily have reached him through the intelligence service of his Jesuit colleagues. In a moment we must take up the influence which such accounts had on the engineers of Europe.

Fig. 855 is by no means the oldest Chinese representation of an iron-chain suspension

---

[a] These are also visible in the picture. Hsü then goes on to discuss the inscriptions set up at each end of the bridge. Hsü Hsia-Kho took a great interest in bridges, and what he says about them would be worth careful study. Like us, he worried about the dates of suspension bridges and their conversions from bamboo to iron; see, for instance, his remarks on the Lung-chhuan River bridges in northern Yunnan, ch. 16, pp. 33*a*, 35*a*.

[b] Mr Tsang Chi-Mou informs me that Chu Chhao-Yuan's book attributes the idea of iron-chain suspension bridges to Chuko Liang. And indeed Hsü Hsia-Kho noted that one of the inscriptions of the Phan Chiang T.C., on a stele in large characters, was Hsiao Ko Chhiao,[2] 'A Little Bridge like Chuko Liang's'. Hsü remarked: 'This is because Chuko Liang is said to have built iron-chain bridges over the Mekong River a thousand and more years ago...But the Mekong did not (always) have iron-chain bridges. The oldest are in the neighbourhood of Lichiang, and it was not Chuko (Liang) who built them.' Chu Chhao-Yuan's book also states explicitly that to get the chains across, a pilot cord was first shot over by bow and arrow. Modern line-throwing still does the same. In Brangwyn & Sparrow (1), Phan-hsien has been corrupted to 'Auhsien'.

[c] Latin ed. p. 154, French ed. p. 146. Englished, it runs: 'West of Picie across a deep valley through which rolls a great torrent of water with impetuous violence, the Chinese, to make a way, have fixed certain several rings and hooks to each side of the ravine and attached very large iron chains; and so laying planks and treads upon them they have made a bridge.'

[1] 衛匡國   [2] 小葛橋

bridge. Wang Chen-Phêng[1] depicts one in a painting which gives his imaginative reconstruction of a Thang palace called the Ta-Ming Kung, done between +1312 and +1320. Here an iron-chain bridge is slung across the mouth of a huge cavern, over which a dragon head spouts a great waterfall. Access to it is gained by a *chan tao* (cf. pp. 20ff.), and together with an imposing fountain pavilion, it affords cool promenades for the court in summer.[a]

The question of the approach of these traditional catenary suspension bridges to the flat-deck type needs further investigation. Gill found a bridge of this kind in northern Szechuan at Hsiao-ho-chhang[2] above Phing-wu[3] on the upper waters of the Fou Chiang.[b] Fugl-Meyer maintained that in the Tibetan and Himalayan bridges the load was generally carried by two slack main cables with a flat deck slung below,[c] but he was not personally acquainted with the region and gave no illustrations. His only reference was to the travels of Sarat Chandra Das (ed. Rockhill), but while this book mentions several bamboo-cable bridges,[d] and at least three iron-chain ones,[e] nothing is said about the flat-deck type. However, one such bridge has been described by Waddell.[f] It crosses the Brahmaputra with a span of 450 ft. at Chak-sam-ch'ö-ri, and is known from a sketch to have had a flat deck in 1878, though this had been removed by 1903. The bridge had a tower at each end, but the deck was only wide enough for pedestrians. The date of construction is given as in the close neighbourhood of +1420, and the engineer is supposed to have been Thaṅ-ston-rgyal-po, who lived (according to tradition) from +1361 to +1485.[g] He was associated with the Tantric and ascetic forms of Lamaism, and his pontifical importance (in the constructive sense) is shown by his Tibetan title Lcags Zam Pa, the Builder of Iron Bridges.[h] Numerous other Tibetan and Bhutanese iron-chain suspension bridges have been reported by travellers,[i] but unfortunately as yet there has been no systematic study of them. Such a work would be well worth undertaking by a Tibetanist, even if only on account of the social interest of so advanced a technique in so theocratic a culture. It would enlighten us too on the first origin of the iron-chain bridge in Bod-Yul. At present we have nothing except the mention in the Tibetan chronicle found at Tunhuang which relates to the year +762. It runs as follows:

---

[a] This painting, in the Crawford Collection, was exhibited in London in 1965; see Sickman *et al.* (1), p. 36, no. 44. On fountains, see Vol. 4, pt. 2, pp. 132ff.

[b] (1), p. 136.   [c] (1), pp. 7, 8, 115.

[d] E.g. one at Rungit reminiscent of the Kuanhsien bridge.

[e] Pp. 143, 204, 228.

[f] (1), p. 312. See Fig. 856 (pl.).

[g] For these dates we are indebted to Dr Li An-Chê. A more probable range is given by Tucci (3), vol. 1, p. 163; i.e. +1385 to +1464.

[h] See Tucci (3), pp. 163, 550. A biography of this great lama was written in +1588 by Ṅag-gi-dbaṅ-phyug, entitled *mTshuṅs-med grub-pai Dbaṅ-phyug Lcags-zam-pair Nam-thar*, but it consists mainly of legends and miracles; it replaced an older biography which may well have been better. Thaṅ-ston-rgyal-po was a restless wanderer over Tibet, but had the favour of the ruling families, who collected iron for his bridges and organised *corveé* labour to build them.

[i] E.g. Hooker (1); Teichman (2), p. 112; Holdich (1), p. 116, a flat-deck bridge of +1450; Bogle, in Markham (1), pp. 20 ff., at Chuka in Bhutan; the Pundit of 1865, in Markham (1), p. cxi; Huc (1), vol. 2, p. 301, etc.

[1] 王振鵬   [2] 小河場   [3] 平武

In the late winter the Chinese emperor died (i.e. Su Tsung), and a new emperor (i.e. Tai Tsung) was installed. As the Chinese government had collapsed, it was not the right time for presenting tax-silk and maps of the country. (On the contrary) Zaṅ-rgyal-zigs and Zan-stoṅ-rtsaṅ, having crossed the iron bridge at Bum-liṅ, invested with their forces many Chinese cities ...which fell into their hands...[a]

This is but a gleam of light in the darkness, however, for we do not know how old the bridge was at this time,[b] nor whether it had been built by Tibetans or Chinese, nor

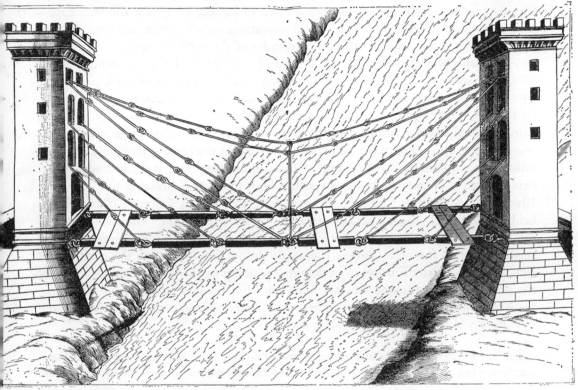

Fig. 857. Transmission of the idea of iron suspension bridges to Europe; the design of Faustus Verantius for a bridge of linked rods in +1595. No bridge of this general type was built in Europe, however, either in that or the following century.

even its exact location. From the context it was probably somewhere on the upper Yellow River south of Sining in the Thang-hsiang[1] or Tangut country, and doubtless commanded the approaches to Kansu.

We come at last to the history of suspension bridges in Europe. Not the +6th century, but the +16th, brought the first Western specification for a suspension bridge. Faustus Verantius in +1595 proposed two towers (cf. Fig. 857), a flat deck and a system of linked rods or inverted brackets, formed of eye-bar chains.[c] The fact that

[a] Tr. Bacot, Thomas & Toussaint (1), p. 65, mod. auct.; cf. Goodrich (16).
[b] From the text it does not sound a new thing; cf. the estimate in Vol. 1, p. 124.
[c] His pl. xxxiv; cf. Beck (1), p. 524; Parsons (2), p. 506, fig. 171; Davison (11). Also a cable one in Verantius' pl. xxxv.

[1] 党項

bridges of rods had already been used in south-west China is rather remarkable in this connection. But we have no proof that Verantius gathered anything from the cargo of tales, experiences and wonders which the Portuguese travellers of the early part of the century brought back with them. Any suspicion we may entertain on this score must for the present arise only from the contiguity of dates. In any case, Verantius did not actually construct such a bridge.[a] Meanwhile Martin Martini, in the sixth part of the Blaeu *Atlas* already mentioned, published at Amsterdam in +1655, marked also the Ching-tung iron-chain suspension bridge, and described it,[b] giving the Han date— 'hunc pontem Mingus Hanae familiae Imperator condidit, circa annum a Christo nato quintum supra sexagesimum'. The wonder of the work was greatly emphasised by Athanasius Kircher in his *China Illustrata* of +1667.[c] In the +18th century it was illustrated several times, notably by J. B. Fischer v. Erlach in his *Historia Architectur* (+1725), here reproduced in Fig. 858 (pl.); and more fantastically by Schramm (+1735).[d] According to Robert Stevenson, the earliest iron-chain suspension bridge in Europe was the Winch bridge over the Tees (+1741), and perhaps, significantly, this was of the catenary, not of the flat-deck, type.[e] About the same time the Saxon army built temporary military ones (+1734).[f] But it is astonishing that the first suspension bridge capable of carrying vehicles was not built until 1809, crossing the Merrimac River in Massachusetts by a span of 244 ft.[g] Next there was Telford's Menai Straits bridge (580 ft.), of 1819–26, and after that they became common.[h] One feels driven to the conclusion that there must have been, in this whole succession of events, a real series of stimuli from the Chinese iron-chain suspension bridges to the engineers of Renaissance and later Europe, even though we cannot as yet elucidate all the stages of the process.[i] Indeed it was almost acknowledged in the generous, though pleasantly realistic, words of Athanasius Kircher, who, after speaking of the structure and dimensions of the Ching-tung bridge, goes on:

Quem cum plures simul transeunt, pons titubat ac hinc inde movetur, non absque trans- euntium metu ruinae perculsorum, horrore et vertigine; ut proinde satis mirari non possim

[a] Darmstädter says that the first suspension bridge was constructed by Andrea Palladio in +1550 over the River Cismone, but this is a mistake; it was a compound truss bridge (Uccelli (1), p. 681).

[b] 'Catenae ejusmodi viginti sunt, duodecim perticarum longitudine singulae' (there are twenty of the same chains, each twelve perches (*c.* 200 ft.) long).

[c] P. 215. Cf. Nieuhoff (1), p. 283.

[d] P. 59, fig. 13. In his 'Dissertation on Oriental Gardening' of +1773 Sir Wm. Chambers gave it an ecstatic description. Later, Thomas Telford studied the Chinese bridges (Hague, 1).

[e] Cf. Uccelli (1), p. 709, fig. 139; Pugsley (1), p. 2. Roy (1) reports similar catenary bridges about the same time in the Appalachian Mountains of North America. A few are still in use in Europe, e.g. at Carrick-a-rede in Ulster (photo. in Deane, 1). Stevenson, writing in 1821, knew that there had long been iron-chain bridges in China, but had no details of them. The Winch bridge collapsed in 1802.

[f] Feldhaus (1), col. 152.

[g] The builder was James Finley, and his work still stands, in reconstructed form.

[h] Cf. Pugsley (1); Straub (1), pp. 170, 191.

[i] It has been suggested that suspension bridges were known in that strange Turkic-Jewish kingdom of the Khazars which, until the middle of the +10th century, occupied the lands north of the Caucasus between the Don and the Volga (cf. Vol. 3, p. 681). Arabic sources seem to say that the royal tombs of the Khazars were hung upon chains across flowing water (pers. comm. from Dr A. N. Poliak) but it is more likely that they were dug out under it (Dunlop (1), pp. 111, 115).

Key to Fig. 859

RESERVOIRS AND DAMS

▽1 Shao Pei; Peony Dam (Sunshu Ao)

▽2 Shê Kung Pei (Shen Chu-Liang)

▽3 Chhien-lu Pei (Shao Hsin-Chhen)

▽4 Hsin-fêng Hu (Chang Khai)

▽5 Lien Hu (Chhen Min)

▽6 Mu-lan Pei (Chhien Ssu-Niang)

⨿⨿⨿⨿ Chhien-thang Sea-Wall (Hua Hsin, Chhien Liu and Chang Hsia)

WEIRS AND DIVERSION PROJECTS

◇1 Chang River Irrigation System (Hsimên Pao and Shih Chhi)

◇2 Fên River Irrigation System (Phan Hsi) [unsuccessful]

◇3 Shouhsien System (Chhen Têng) [remains still extant]

◇4 Meihsien System (Khung Thien-Chien)

◇5 Ninghsia System (Mêng Thien)

◇6 Chêngkuo Canal and Wei-Pei Irrigation System (Chêng Kuo)

◇7 Kuanhsien System (Li Ping & Li Erh-Lang)

◇8 Kunming System (Shan-Ssu-Ting)

◇9 Shantan System

CANALS

[1] Hung Kou = Pien (= Pan) [Chou to end of Nan Pei Chhao]

[2] Lang-Tang Chhü

[3] Pien Ho = Thung Chi Chhü (Yüwên Khai) [Sui to Yuan]

[4] Kuang-Chi Chhü (Chhi Huan)

[5] Han Kou (Fu Chhai)

[6] Thao Chhü (Fu Chhai)

[7] Tshao Chhü = Kuang Thung Chhü (Chêng Tang-Shih & Hsü Po)

[8] Chhu Chhü

[9] [9a] Hu-Tho R. and Fên R. canals [unsuccessful, replaced by cart-road portages]

[10] Yang Chhü (Wang Liang & Chang Shun)

[11] Khai-Yuan Hsin Ho (Li Chhi-Wu) [not long successful]

[12] Pai Kou etc. (Tshao Tshao)

[13] Shan-yang Yün-Tao = Shan-yang Tu = Li Yün Ho (Chhen Min) [incorp. in Gd Canal]

[14] Tung-a Chhü = Ching Chi Tu (Hsün Hsien) [incorp. in Gd Canal]

[15] Ling Chhü (Shih Lu)

[16] Chan Chhü

[17] Chiang Nan Ho (Yüwên Khai)

[18] Yung Chi Chhü (Yüwên Khai)

[19] Thung Hui Ho (Kuo Shou-Ching) [pt of Gd Canal]

[20] Pai Ho [pt of Yung Chi Chhü, then of Gd Canal]

[21] Yü Ho [pt of Yung Chi Chhü, then of Gd Canal]

[22] Hui Thung Ho (Chang Khung-sun & Loqsi) [pt of Gd Canal]

[23] Chi Chou Ho (Kuo Shou-ching & Oqruqči) [summit section of Gd Canal]

[24] Huan Kung Kou (Huan Wên) [incorp. in Gd Canal with extensions]

[25] Yün Yen Ho [a network]

COURSES OF THE YELLOW RIVER
(for full details see Table 69, p. 242 below)

⓪ antiquity to −602

① −602 to +11

② +11 to +1048

②a +11 to +70

②b +70 to +1048

③ +893 to +1099

③a +1060 to +1099

④ +1048 to +1194

⑤ +1194 to +1288 (some flow till +1495)

⑥ +1288 to +1324

⑦ +1324 to 1855 (entirely after +1495)

⑧ 1855 to present date

⑧a 1887 to 1889 and 1938 to 1947

N.B. Numbers indicating towns and cities are the same as those in Table 60 above, and reference there will identify them.

Sinensium architectorum dexteritatem, qua ad itinerantium commoditatem tot ac tam ardua opera attentare sint ausi.[a]

Kircher admired, and Kircher was writing three quarters of a century before Europeans had constructed even the smallest viable suspension bridge, with its modest span of 70–80 ft. With all his optimism as a man of the scientific renaissance he could hardly have dreamed that the application of scientific principles in years to come would permit of the 4,200 ft. span of the Golden Gate Bridge. But if he had been able to do so, he would not have been averse, as a Jesuit, from saluting its first origins in the kingdom of Nan Chao.

(5) GEOGRAPHICAL DISTRIBUTION OF TYPES

If we glance at the map (Fig. 859) on which the locations of many of the bridges here discussed have been marked, we can see that Fugl-Meyer was broadly right in dividing China into three regions from the point of view of bridge engineering.[b] In the northern zone, extending down to the northern parts of Chekiang, Chiangsi and Hunan, arch building was predominant, beam bridges being reserved for minor or decorative structures.[c] This was the zone of the segmental arches[d] and the long multiple-span arch bridges.[e] In the western zone, on the other hand, which includes besides Yunnan, Szechuan and Kweichow, the two former provinces of the Tibetan marches (Sikang and Chhinghai), and also Kansu and part of Shensi, all bridges of importance were cantilever structures[f] or suspension bridges. Beam bridges are rarely found there. In the southern zone, centring on Fukien but including the two Kuangs, beam bridges are the commonest type, culminating in the megalithic giants of the Fukienese coast. Here arch bridges are only for minor or decorative purposes, and cantilevers and suspension bridges never seen. While obvious reasons of a topographical nature may be adduced to account for this distribution, it does not seem impossible that it might have some genetic connection with those local cultures to the fusion of which Chinese society owes its existence (cf. Vol. 1, p. 89 above). But the nature of the terrain and the materials available must have played a part at least equal to invention and its subsequent stylisation within particular ethnic or social groups.

[a] 'When several people cross the bridge at one time it moves and sways and oscillates up and down in such a way as to evoke in them no small dizziness and fear of the danger of falling off; yet I find it impossible sufficiently to admire the skill of the Chinese engineers, who have executed so many and such arduous works for the greater convenience of wayfaring men.'

[b] In +1488 Chhoe Pu had remarked in his *Phyohae Rok* (cf. p. 360) on a distribution of bridge-types (Meskill tr., p. 152). He associated stone bridges with the regions south of the Huai River, and floating or cantilever wooden bridges with the north. But he did not diverge much from his route home to Korea from Hangchow along the Grand Canal.

[c] Except for the long many-spanned structures of ancient Shensi, some still extant. One must remember also the pontoon bridges which were for so many centuries the standard crossings of the Yellow River.

[d] Especially in Hopei and Shansi, but reaching down later into Kweichow.

[e] Especially in the lower Yangtze valley.

[f] But there are subsidiary areas of cantilever bridges in the north-east, historically in Shantung and Honan, now also in the centre and south-west, notably in Hunan and Chiangsi, with many spans, but also in Yunnan. Strainer-beam structures penetrate into Fukien.

(f) HYDRAULIC ENGINEERING (II), CONTROL, CONSTRUCTION AND MAINTENANCE OF WATERWAYS

If there was one feature of China which impressed the early modern European travellers there more than any other, it was the great abundance of waterworks and canals. In +1696 Louis Lecomte wrote:

Tho' China were not of it self so fruitful a Country as I have represented it, the Canals which are cut thro' it, were alone sufficient to make it so. But besides their great usefulness in that, and the way of Trade, they add also much Beauty to it. They are generally of a clear, deep, and running Water, that glides so softly, that it can scarce be perceived. There is one usually in every Province, which is to it instead of a Road, and runs between two Banks, built up with flat course Marble Stones, bound together by others which are let into them, in the same manner as we use to fasten our strong wooden Boxes at the Corners.

So little Care was taken, during the Wars,[a] to preserve Works of Public Use, that this, tho' one of the Noblest of the Empire, was spoiled in several places, which is a great pity; for they are of no little use, both to keep in the Waters of these Canals, and for those to walk on who drag the Boats along. Besides these Cawseys they have the conveniency of a great many Bridges for the Communication of the opposite Shoars; some are of three, some five, and some seven, Arches, the middlemost being always extraordinary high, that the Boats may go through without putting down their Masts. These Arches are built with large pieces of Stone or Marble, and very well framed, the Supporters well fitted, and the Piles so small that one would think them at a distance to hang in the Air. These are frequently met with, not being far asunder, and the Canal being strait, as they usually are, it makes a Prospective at once stately and agreeable.

This great Canal runs out into smaller ones on either side, which are again subdivided into small Rivulets, that end at some great Town or Village. Sometimes they discharge themselves into some Lake or great Pond, out of which all the adjacent Country is watered. So that these clear and plentiful Streams, embellished by so many fine Bridges, bounded by such neat and convenient Banks, equally distributed into such vast Plains, covered with a numberless multitude of Boats and Barges, and crowned (if I may use the expression) with a prodigious number of Towns and Cities, whose Ditches it fills, and whose Streets it forms, does at once make that Country the most Fruitful and the most Beautiful in the World.

Surprised and as it were astonished at so Noble a Sight, I have sometimes bore a secret envy to China in Europe's behalf; which must own that it can boast nothing in that kind to be compared to the former. What would it be, then, if that Art which in the wildest and most unlikely Places has raised magnificent Palaces, Gardens, and Groves, had been employed in that rich Land, to which Nature has been lavish of her most precious Gifts.

Lecomte was thus full of admiration for the hydraulic engineers of China. He realised that their work went back even into the legends, for he went on:

The Chinese say their Country was formerly totally overflowed, and that by main Labour they drained the Water by cutting it a Way through these useful Canals. If this be true I cannot enough admire at once the Boldness and Industry of their Workmen who have thus made great Artificial Rivers, and a kind of Sea, and as it were created the most Fertile Plains in the World.

[a] I.e. the Manchu conquest half a century previously.

Lecomte was also quite clear about the dual function of the waterways, for transportation and for irrigation. Thus he says that the 'Great Canal' was

necessary for the Transportation of Grain and Stuffs, which they fetch from the Southern Provinces to Pekin. There are, if we may give credit to the Chineses, a Thousand Barks, from Eighty to a Hundred Tun, that make the Voyage once a year, all of them Freighted for the Emperor, without counting those of particular Persons, whose number is infinite. When these prodigious Fleets set out, one would think they carry the Tribute of all the Kingdoms of the East, and that one of these Voyages alone was capable of supplying all Tartary wherewithal to subsist for several years; yet for all that Pekin alone hath the benefit of it; and it would be as good as nothing, did not the Province contribute besides to the Maintenance of the Inhabitants of that vast City.

The Chineses are not only content to make Channels for the Convenience of Travellers, but they do also dig many others to catch the Rain-Water, wherewith they water the Fields in time of Drought, more especially in the Northern Provinces. During the whole Summer, you may see your Country People busied in raising this Water into abundance of small Ditches, which they contrive across the Fields. In other places they contrive great Reservatories of Turf, whose Bottom is raised above the Level of the Ground about it, to serve them in case of Necessity. Besides that, they have everywhere in Chensi and Chansi, for want of Rain, certain Pits from Twenty to an Hundred feet deep, from which they draw Water by an incredible Toil. Now if by chance they meet with a Spring of Water, it is worth observing how cunningly they husband it; they Sustain it by Banks in the highest places; they turn it here and there an Hundred different ways, that all the Country may reap the benefit of it; they divide it, by drawing it by degrees, according as every one hath occasion for it, insomuch that a small Rivulet, well managed, does sometimes produce the Fertility of a whole Province.[a]

Even allowing for a little enthusiasm, Lecomte was perfectly right that the Chinese people have been outstanding among the nations of the world in their control and use of water.[b] The purpose of this Section will be to examine more closely the nature of their achievements and the engineering techniques which they developed in the process. To begin with we must sketch a broad view of the problems they had to face and the solutions they adopted. There was the climate, with its special features of rainfall.[c] There was the topography, and the peculiarities of the great river-systems which formed the given framework for human enterprise, involving the first of the great requirements, protection from floods. The second great requirement, that of irrigation-systems, was dictated partly by the nature of the loess soil of the upper Yellow River basin, and partly by the widespread adoption of wet rice agriculture (Fig. 860). The Chinese may constitute a fifth of the world's population, but their irrigated land is a third of the world total, approximately 100 million acres out of 300 million acres.[d]

[a] Pp. 104, 108; 2nd ed. pp. 101 ff.
[b] As modern general accounts of what they did the well-illustrated articles of Anderson (1) or King (1) may serve as introduction, and there is a short historical paper by Hsüeh Pei-Yuan (1). Boerschmann (10) made a survey in terms of human geography. We shall discuss the literature further presently (p. 216).
[c] Cf. Vol. 3, Sect. 21, pp. 462 ff. above, and p. 219 below.
[d] This 100 million acres accounts for 30% of China's arable land and some 50% of her gross agricultural output. 62 million acres lie south of the 32nd parallel of latitude, 12 million in the north, north-west and north-east, and 24 million in the west. Rice cultivation occupies 64 million acres, and cotton-growing 1·5 million. About 5·5 million acres consist of irrigation-systems covering individually

Fig. 860. A late Chhing representation of Yü the Great exhorting irrigation workers, from *SCTS*, ch. 5, I Chi. The caption quotes the words: 'I caused the channels and canals to be dug and deepened...' (cf. Medhurst (1), p. 66; Karlgren (12), p. 9).

Thirdly, the more centralised the feudal-bureaucratic State became, the more essential was the construction of waterways along which tax-grain could be transported;[a] and this led naturally to the fourth factor, namely the military aspects of defence. Centralised granaries and arsenals could furnish army supplies at need, and canals were themselves an important obstacle to the penetration of the Chinese agrarian civilisation area by cultures of the nomad type. All this is comprehended in the classical Chinese term for hydraulic engineering, still used today, *shui li*,[1] 'benefit of water'. While its general and social aspects have been much enlarged upon,[b] historians have been perhaps too content to leave the technological principles and practices to relatively restricted professional engineering circles—canal planning and river training, silt and scour, dredging and dykes, gabions, stanches and sluices. All this of course is our legitimate material for the history of science and technology, but as in other fields, economic and political history itself must remain in a measure superficial without some real understanding of it.[c] Indeed, if there had been less subjective speculation about the 'hydraulic foundations of oriental despotism', and more objective study of the development of hydraulic engineering itself, we might now be better informed than we are about the real origins of feudal-bureaucratic society.[d]

To complete this introduction a few words may be added in clarification of what hydraulic engineers could hope to do with water throughout the centuries. The fundamental physiographic unit is of course the river-valley, whether it descends slowly or rapidly, shooting in cascades or broadening out into miles of lake and marsh. The harnessing of a river can take place in one or more of the following ways:

(*a*) Construction of a *dam* across a valley, forming a *reservoir* or *tank*, with one or more *spillways* to take care of any excess, and derivative irrigation canals to lead the water away for

more than 2,000 acres each. Since the institution of the People's Republic 40 million further acres have been irrigated, and it is planned to increase the total to 220 million acres. Already about 1 million acres are electrically pumped. These figures are derived from Kovda (1), writing in 1959.

[a] Thus from the point of view of origins, irrigation canals were a contribution of the northern States (Chhi, Chêng and Chhin), while canals for water-transport were rather a southern contribution (Wu and Chhu). This is discussed by Hsü Chung-Shu (3).

[b] See, for example, the literary treatments of Brittain (1) and Payne (1), easy reading but no help in depth.

[c] Those who have not previously given much attention to the subject-matter of this Section may like to know of a few books which might be consulted by way of background to the reading of it. A history of hydraulic engineering, which concentrates on the basic principles of hydrodynamics, has been written by Rouse & Ince (1), but it does not devalue the useful articles of Skempton (4), Pilkington (1) and Hadfield (1). On particular areas such as Lombardy or Ceylon there are a number of monographs which we shall refer to in their place. On the most ancient works see the review of Biswas (1).

In the technical literature old textbooks such as that of Vernon-Harcourt (1) are almost more relevant to the earlier history of hydraulic civil engineering than more up-to-date books such as those of Barrows (1) or Linsley, Kohler & Paulhus (1). But the writings of Leliavsky (1, 2) furnish an ideal contemporary scientific orientation on the works here to be described.

Of books on the great dams there is no end; that of de Roos (1) may be mentioned as an example, serving to describe the most spectacular developments of what was started so long ago in the Middle East, China, India and Ceylon. On the history of dams see Schnitter (1); on that of arch dams Goblot (2).

[d] A glimpse of the present state of play in this fundamental sociological field may be gained from Adams (2) and Crevenna (1). Cf. the interesting critical study of Leach (1) on hydraulic engineering in Ceylon and its relation to the sociological history of that culture. We shall make a few comparisons between the Chinese and Sinhalese achievements later on (pp. 368 ff.).

[1] 水利

use. This procedure is very ancient all over Asia, being especially suited to rivers with marked seasonal differences of flow. Spring or autumn flooding is thus retained for gradual and efficient disposal instead of being lost by rapid clearance from the area. The method, converting as it does inundatory to perennial waters, keeps all its importance today, for the water-supply of large modern cities, the unified control of large river-systems subject to disastrous floods in their lower reaches, and of course the production of hydro-electric power.

(b) Arrangement of *retention basins* whereby the upper waters of an inundatory river in the flood season are made to submerge agricultural or other land for a limited time, restoring fertility by silt deposition. This was characteristic of ancient Egypt. Here the main stream itself is not obstructed by a dam, but side-areas banked by low dykes, often with sluice gates, prevent the onward flow of the water until it can be discharged by the lowered river. Retention basins may also be used, however, for quite another purpose, namely to take the strain off dykes along lower reaches of great rivers in peak flood periods. All forms of barrage have classically aided irrigation and conservancy rather than navigation, but in some circumstances traffic may be aided, either because boats may utilise irrigation canals or because the inconvenience of extreme fluctuations of water-level in the river is removed by the dam.

(c) *Canalisation* of the main stream. Here the stretches are first levelled by the building of *weirs* (submerged embankments, often aslant to the river's main axis); these facilitate the abduction of lateral irrigation canals on either side. We shall see how skilled the ancient Ceylonese were in this art. As soon as it was desired, however, to use the river for navigation at the same time, it was necessary to couple each weir with a double *slipway* or a single gate[a] (the *stanch* or *flash-lock*).[b] These boat portages and passages were not always associated necessarily with weirs, but might be set alone at distances of a mile or so along the river's course. Lastly, the Chinese invention of the *pound-lock*[c] assured quiet and efficient transfer of shipping from one level to another, two gates being set close together to open and close alternately.

(d) In other situations, as when the bed or banks of a river were unsuitable for navigation, a *lateral transport canal* was cut, accompanying the main stream at the same or a somewhat higher level. Though reaching the same terminus in the end, with the same average gradient, the waterway was split up into many horizontal reaches by flash-locks, and later pound-locks. Here advantage could often be taken of tributary streams to supply water to the canal.[d]

(e) Derivation of a *lateral irrigation canal* high up the valley of a perennial river, which then descends more gradually than the main stream, following higher contours and branching repeatedly. Here long winding stretches may be almost horizontal, and as in (d), tributary streams of the river may usefully be captured for the canal. The flow of water in it, and the distribution of water by branch canals from it, is controlled by suitably adjustable gates (*sluices*). This method is also very ancient in Asia, where it was carried out on a grand scale. In medieval times, both East and West, the same principle was used to feed the races of

---

[a] The oldest form is probably the stop-log gate (cf. p. 347 below) in which a number of horizontal baulks are let down into, or drawn up from, two vertical grooves made in the abutments. Later this system was combined with the principle of the swinging gate, as by de Bélidor in the +18th century.

[b] Flash = flush = scour. Boats going upstream were hauled through against the current by winches, while those coming down 'shot the rapids' with the 'flash' of the water.

[c] So called because water is impounded between the gates in the short basins or the long reaches. See pp. 350 ff. below.

[d] This is perhaps the largest class of transport canals; cf. Skempton (4), figs. 280, 286, 288, 289; Pilkington (1), p. 351; Hadfield (1), fig. 309. Many of them were built in Europe from the +14th to the +18th centuries, as they had been in China during the previous dozen centuries. Significantly, one thinks of the Duke of Bridgewater's Canal, paralleling the Mersey and finished by James Brindley in +1767 (see Smiles (1), vol. 1, pp. 153 ff.) as the type specimen of the kind of England—but in China it was the Chhang-an Canal (cf. p. 273 below), paralleling the Wei and finished by Hsü Po in −130.

water-mills, and today it appears with great effect in the layout of some kinds of hydro-electric generating stations.

(*f*) When a lateral canal has its terminus in a water-course other than that from which it came it is termed a *contour canal*. Deriving from the upper waters of a river, it can wind gently round high contours and over a saddle among the hills into a second river-valley, effecting a junction there. If both rivers are, or have been rendered, navigable, such a canal will give traffic communication between two whole river-systems. This was first accomplished in China before the beginning of the present era (cf. pp. 299 ff.). In the same way a contour canal can serve for the irrigation of otherwise un-utilisable land outside the river-valley in which it originated. It may also go to feed a reservoir in another valley, and may thus combine the waters of perennial and inundatory river-systems, as in Assyria and ancient Ceylon.

(*g*) Connection between two river-systems is finally afforded by the *summit-level canal*, which scales the contours directly on each side of a range. If the water-shed was a very low one double slipways might suffice for handling the traffic, but steeper gradients could not be attempted until it was possible to build stanches or flash-lock gates with some success; and such routes did not become fully practicable until the invention of the pound-lock, which could raise or lower one or two vessels at a time some ten or twenty feet, avoiding haulage over slopes or against rushing discharges. There was of course always the difficulty, sometimes overcome by very ingenious expedients, of ensuring an adequate water-supply for the summit levels. Again this was first accomplished in China, as we shall see (pp. 314, 359).

Half a millennium before the beginning of our era, Chinese rulers and engineers were conscious of the high mechanical efficiency of water transport. Not until the coming of the steam locomotive and the internal combustion engine was there any other satisfactory means of carrying heavy loads from place to place. This may at once be seen from comparative figures of the loads carried or drawn by a single horse:[a]

	Ton
Pack horse	0·125
Horse harnessed to wagon—'soft' road	0·625
—macadamised road	2·0
—iron rails	8·0
Horse harnessed to barge—river	30·0
—canal	50·0

In the following pages we shall trace the epic development of the transport canal in China, sometimes evoked by the need for the movement of military supplies, but more often inspired by the nature of the fiscal system, which sought to concentrate the product of the people's labour at the bureaucratic nerve centre of a vast terrestrial empire.

It only remains to say something about the books and reviews which are available on the subject of this Section. The outstanding work in a European language on the history of Chinese hydraulic engineering is that of Chi Chhao-Ting (1), while its Chinese counterpart is due to Chêng Chao-Ching (*1*), but the former work is much more aware of the social and economic aspects of the various engineering achievements, indeed it was a classic contribution to social as well as technological history.[b]

---

[a] These figures were compiled by Skempton (4, 5) from authorities such as Smeaton and Telford.
[b] Cf. the opinion of Bielenstein (2), p. 93.

Without these two books as guides the present Section could not have been written.[a] A shorter account in German by Li Hsieh (Li I-Chih) exists (*1*), valuable on the engineering side but sinologically uncritical. Li I-Chih was one of the greatest Chinese engineers of modern times, and some idea of the state of hydraulic science in China in the earlier part of this century can be gained from the collective works which were published in commemoration of him.[b] In two Russian papers, much more recently, Nesteruk (*1, 2*) has traced the history of the Chinese works from the −12th century to the present era of hydro-electric power. In Chinese there is also now an excellent brief account, with maps, of the major works, by Fang Chi (*2*), and contributions by eminent experts such as Chang Han-Ying (*2*) at various levels of popularisation. There are also a growing number of studies of hydraulic works at particular historical periods; thus Yang Khuan (*10*) has made a contribution on the Warring States time, Lo Jung-Pang (*6*) has surveyed the Han, while Chhuan Han-Shêng (*1*) and Aoyama Sadao (*7*) have discussed the water conservancies of the Thang and the Sung. Other books, on the hydraulic engineering problems of China in general, may have interesting historical introductions, as does that of Li Shu-Thien *et al.* (*1*). Since the revolution there has been a flood of publications on the new works of civil engineering undertaken and in most cases successfully completed during the years of preparation of this book—we shall make reference to them in specific contexts as they arise and mention here only the general surveys of Fu Tso-Yi (*1*); Têng Tsê-Hui (*1*); Chang Han-Ying (*1*); and Sun Li *et al.* (*1*).[c]

(1) PROBLEMS AND SOLUTIONS

The most fundamental factor with which those who controlled the waterways had to deal was the Chinese climate. This has already been briefly sketched in the Section on meteorology, and towards the end of this work we shall have to speak of it again in comparing China with Europe, linking it with geographical factors such as the contrast of continents with archipelagoes, the 'isolation' of the Chinese culture-area, and so on. Here we are concerned with it only so far as precipitation of rain determined the size and character of the natural rivers.

The rainfall (Fig. 861) has a highly seasonal distribution, some 80% of it occurring in the three summer months. At the same time the prevailing wind direction changes. This is the phenomenon of the monsoon. In winter the air masses over inner Asia are cooled and tend to sink, expelling the moist oceanic air from China, and producing dry, cold, north-westerly winds. In summer the reverse takes place; the central air masses are warmed and rise, so that convectional circulation takes place, bringing in the humid air of the south-eastern seas with south-easterly winds. The regularity of this process can be seen in Fig. 862 which shows percentage wind-frequency according to season in

[a] The recent monograph of Sung Hsi-Shang (*2*) forms an excellent companion volume but it appeared too late to help us much.

[b] Anon. (*49, 50*). His nephew, Li Fu-Tu (*1*), has given us an interesting account of the current multiple-purpose Yellow River Control Scheme.

[c] On irrigation projects see especially Anon. (*53*). Anon. (*72*) is an album of photographs published by the Ministry of Water Conservancy in 1956.

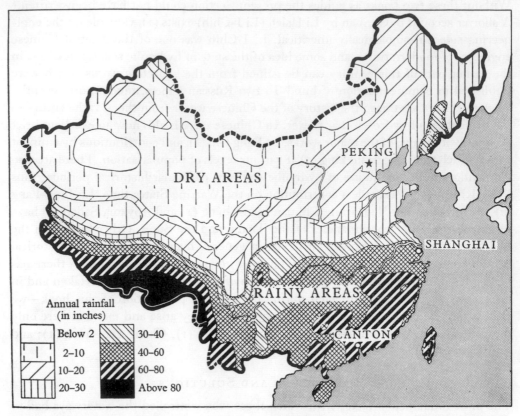

Fig. 861. Rainfall map of China (after Lo Kai-Fu (1), modified by Cedric Dover). The stippled border
separating the two great areas indicates very nearly the effective northern limit of wet rice cultivation,
though this is to some extent a customary division since rice will mature in all parts of China. The
Tibetan region to the west is too high for such agriculture. Extremes of rainfall here shown range from
50 to 2,000 mm. per annum.

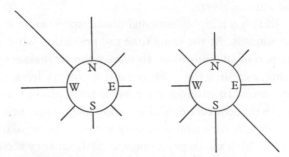

Fig. 862. Wind-roses showing percentage wind frequencies according to season in North China.
Data from Kendrew (1) and Chi Chhao-Ting (1).

Left, winter; right, summer.

North China. An example of the mean monthly precipitation is given in Fig. 863, taken from the records of the Kuangsi Agricultural Experiment Station at Shatang; here the four summer months are much wetter than the rest. This implied that water-courses would often be nearly dry for most of the year, and that flooding would be sudden; it was necessary, therefore, to build works which would withstand torrents much greater than the winter flow. And although the total rainfall was on the whole

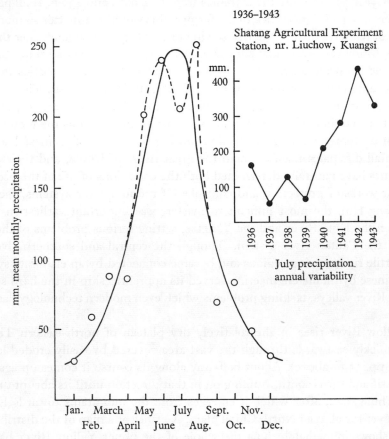

Fig. 863. Mean monthly precipitation at Shatang in Kuangsi (data from the Kuangsi Agricultural Experiment Station, 1944). The summer peak may be in reality a double one (cf. Sion (1), p. 13). Inset: annual variability of the July precipitation over an eight-year period at Shatang.

greater in China than in Europe,[a] its intensely seasonal character set a tremendous problem—the building of enough reservoirs to prevent the water from running uselessly away.

Still greater difficulties were caused by the fact that the monsoon climate shows an annual fluctuation of rainfall much greater than in other parts of the world.[b] Works

[a] It varies from less than 300 mm. per annum in Kansu and Ninghsia north of the Yellow River to 1,800 mm. per annum in Kuangtung (cf. Cressey (1), p. 61; Hu Huan-Yung, Hou Tê-Fêng & Chang Han-Ying (1), vol. 1, p. 14).

[b] Cf. Sion (1).

were therefore necessary on a scale sufficient to take care of even the most exceptional years, and naturally it was a long time before this ideal was achieved. In Europe the ratio of the wettest to the driest years hardly ever exceeds 2 for any given place; at Shanghai the ratio for a fifty-year period was 2·24. For individual months it might be much higher; for example, in 1886 there were only 3 mm. in July as against 306 in 1903. Fig. 863 shows the fluctuation at Shatang in the July precipitation during a recent eight-year period; here the extremes were 51·4 mm. and 430·6. If all periods of the year were of equal significance to the farmer, this would not matter so much, but a dry period of ten days during the rice sowing or re-planting may endanger the whole year's crop. Moreover, dykes and similar works had to be ready at all times to receive heavy and sudden pressure, although with *corvée* labour in most dynasties the necessary maintenance could not always be carried out in time. For most regions of China the wettest/driest ratio does not exceed 3 (though in India it may reach 9), but owing to the fact that the time of arrival of rain is almost as important as the total quantity, this has not prevented many grave famines during the course of Chinese history.

The rainfall dissipates itself through the great rivers of China, and their physical characteristics have naturally determined both the conditions of life of the people and the greatest works of protection and control.[a] Of the four river-systems, the Yellow River (Huang Ho), the most difficult to master, was important earliest in Chinese history, then the Huai River and the Yangtze, setting serious problems of their own, and finally the Canton river-system. Though the central and southern rivers flow through fertile rice-growing regions and became connected by an elaborate system of canals, Chinese hydraulic engineering served its apprenticeship in the hard school of the Yellow River valleys, tackling problems which even modern technology has not yet solved.[b]

The Yellow River rises in the relatively dry plateau of north-eastern Tibet and descends quickly eastwards through the vast area covered by easily eroded loess soil (cf. Vol. 1, pp. 55 ff. above). About half-way along its course it comes up against the Shansi massif and turns south, running on in that direction until its abrupt turn eastwards at Thungkuan, after which it emerges from the mountains towards Khaifêng and pours over the alluvial North China plain. A glance at a map of the distribution of the loess[c] shows immediately how the whole of the upper Yellow River basin, the cradle of Chinese culture, is loess-covered, with the exception of the Ordos Desert through which it runs at the top of its Great Bend. Similarly, the low-lying plain east of the mountains is all alluvial loess. By a curious chance, the debouchment of the river from the mountains brings it out just opposite another massif, that of the Shantung peninsula, so that it must reach the sea either north or south of that region. It has in fact taken both of these courses at different times, as we shall see.

From its source to the ocean, the Yellow River is about 2,890 miles in length,[d] with a

---

[a] The best recent monograph on the fluvial geography of China is that of Sung Hsi-Shang (*1*).

[b] One of the best short accounts of the hydrography of this river is that of Chatley (*25*), and I follow it here. The most recent researches have been reviewed by Jen Mei-O (*1*).

[c] For example, Cressey (*1*), p. 186 (Vol. 1, Fig. 6 above).

[d] But only the lower 500 miles is navigable.

drainage area of some 297,000 sq. miles. But this drainage area does not now include the alluvial plain through which it runs for 435 miles after it emerges from the mountains; and some 60,000 sq. miles of this land cannot be included in the water-shed figure[a] since the river has risen above the surrounding plain, from which it is cut off by dykes.[b] Indeed, the bed of this outstanding example of an 'elevated river' now constitutes a ridge separating two drainage-basins. In early times, however, the river must have wandered about freely over this plain, and accepted drainage from it. Because of the great fluctuations of rainfall, there are tremendous variations in the discharge; in the dry season the river is no more considerable than the lower Thames, but at its greatest, the discharge may be of the order of 1,000,000 cu. ft. per second.[c] This figure is only equal to the average discharge of the Yangtze, but the problem it sets is much more serious, for the Yangtze (though also liable to devastating floods)[d] is surrounded throughout its length by higher ground forming a narrower channel. What makes the Yellow River almost unique in the world is the prodigious quantity of silt which it carries down, perhaps as much as 1,000 million tons per annum.[e] This has probably been growing throughout historic times, for reasons we shall discuss shortly. In the light of these figures one can begin to understand the magnitude of the questions which faced the civil engineers when from the Chhin and Han onwards State centralisation began to give them the possibility of concentrating large masses of human manpower on the construction of flood protection works. The difficulty was to know exactly what best to do.

The Yangtze is a longer river than the Huang Ho, attaining a length of some 3,450 miles, and drains a much larger area, about 1,220,000 sq. miles.[f] Though the water is generally coffee-coloured, both at Chungking and down at the estuary, it carries far less silt than the Yellow River,[g] in fact not more than 0·2%. This contrasts with the figure of 10–11% which is common for the Yellow River's main stream,[h] while in the tributaries of the loess area it may exceed 30%. After issuing from the Tibetan mountains and passing through the Red Basin of Szechuan, the Yangtze plunges through its famous gorges with a mean gradient of about 1 ft. per mile. Rapids due to syenite dykes make it dangerous for navigation at low water. At one point a difference of as much as 200 ft. exists between high and low water levels, and at Chungking the average difference exceeds 100 ft.; in the gorges some of the pools have a

---

[a] The last 435 miles of the river drain only 1,140 sq. miles, plus the 4,600 sq. miles cleared by the Ta-Wên Ho tributary from the Shantung massif, which is the only tributary received during this course. Water is thus lost by evaporation and seepage, so that the discharge in volume/sec. at Chungmou is only 44% of what it is at Shenhsien (Jen Mei-O, 1).

[b] The northern dyke runs the full 435 miles; the southern one for 355 miles. The average width at the top of the dykes is 50 ft.

[c] About 28,300 cumecs (cu. m. per sec.). This is difficult to estimate since the maximum discharge usually coincides with a break in the dykes. The total flow is considered one twentieth of that of the Yangtze.   [d] Cf. Chatley (33).

[e] Jen Mei-O reports (1) 1,890 million metric tons. Cf. Kovda (1), p. 469.

[f] Cf. Chatley (24). Also Carles (1); it is the third longest in the world, and the fourth greatest in flow. It is navigable at least as far as Suifu, about 1,580 miles from the sea.

[g] See Chatley (29), (31).

[h] Hu Huan-Yung, Hou Tê-Fêng & Chang Han-Ying (1), vol. 3, p. 26, give a figure of 27–52 tons per second for the silt carriage of the Yellow River in Shensi.

depth of 200 ft., with currents of 10 knots or more.[a] After leaving the gorges at I-chhang, the Yangtze receives several great tributaries:

(a) the Hsiang from Hunan in the south-west, at Yoyang on the Tung-thing Lake,
(b) the Han from the north-west at Hankow,
(c) the Gan from Chiangsi in the south, at Chiuchiang on the Poyang Lake, and
(d) the Huai from the north, near Chenchiang below Nanking.

Between I-chhang and Nanking there are wide areas on each side of the river liable to flooding, and after Chenchiang there is no more safe high ground before the sea is reached, but the total area which the river can inundate is smaller than that which is at the mercy of the Yellow River. This is good, for the Yangtze valley is the most populous in the world, containing 250 million people, more than a third of China's total, and producing 70% of its rice.

The drainage area of the Huai River (about 106,000 sq. miles) has been included in the figure for the Yangtze, since for many centuries, after the blocking of its own exit to the sea on the coast south of Shantung, it has delivered its waters into the Yangtze. The north–south line of communication so formed became, as we shall see, one of the component parts of the Grand Canal.[b] In comparison with the two great rivers, the Huai, and the West River of Kuangtung (1,220 miles long and draining 269,000 sq. miles) are not considerable, yet the former has set many intractable problems to the experts in water-conservancy.[c] Only in our own time has the Huai valley question been firmly taken in hand, and the construction of a vast system of dams in the headwater valleys is one of the best-known achievements of the present Government.[d]

The paramount social importance of floods and their prevention is seen especially well in ch. 57 of the *Kuan Tzu* book, a text not one of its older portions, but dating probably from the late − 3rd or early − 2nd century. Duke Huan is interrogating Kuan Chung on the best location for a State capital, and the minister mentions the 'five harmful influences' (*wu hai*[1]), flood, drought, unseasonable wind, fog, hail and frost, finally pestilences and insect plagues—but of these floods are by far the worst.

---

[a] This helps us to understand the Thang poem of Li Pai entitled Tsao Fa Pai Ti Chhêng[2] (Leaving Pai-ti City at Dawn), in which he says that in a single day one can return to Chiang-ling[3] (in Hupei), a thousand *li* downstream. Pai-ti is in the gorges, in Szechuan, about 1,200 *li* or some 400 miles above Chiang-ling, so that if he meant a full twenty-four hour day at the most favourable state of the current his average speed would have been about 16½ knots. But he probably meant a day from just before dawn to midnight, perhaps 18 hours, in which case his average would have been over 20 knots, the current being assisted by sail and oar.

The same statement occurs also in a poem by Tu Fu (tr. Alley (6), p. 140), and often elsewhere, e.g. *Thang Kuo Shih Pu*, ch. 3, p. 21 *b*, another +8th-century work. It was probably proverbial long before.

[b] For comparison it may be noted here that the total length of the Grand Canal in its completed form was 1,290 miles.

[c] The Mekong and the Salween Rivers, each with some 1,250 miles within China's boundaries, stand somewhat outside the picture, since neither is navigable there, and occupying as they do very deep-cut gorges, could only be used on a very limited scale for irrigation purposes.

[d] On the general plan and the Jun-ho-chi control installations see Fu-Tso-Yi (1); Kao Fan (2); Sun Li *et al.* (1); Chang Han-Ying (1); and Hu Huan-Yung (1). On the Fu-tzu-ling dam and reservoir see Hsimên Lu-Sha (1). On the Sanho regulator-sluice dam near the Hungtsê Lake, and the North Chiangsu canal from the lake direct to the ocean, see Anon. (68). On the transverse contour canals and interlocking drainage system of the North Huai valley in Anhui, see Chhen Han-Sêng (1).

[1] 五害   [2] 早發白帝城   [3] 江陵

'I should like to hear', said Duke Huan, 'about the harmful influences of water'...

Kuan Chung replied...[a] 'It is the nature of water to flow, but when it reaches a bend (in its channel) it is retarded, and when the bend is full (the water) behind pushes forward that which is ahead. Where the land slopes downward it flows along smoothly, but where it rises (the water) is impeded. (In some places) where the bank curves (the water) pounds against it and crumbles it away, (in other such places) (the water) becomes agitated and leaps up. When it leaps up it runs to one side. On running to one side it forms whirlpools. After forming whirlpools it returns to its central course. On returning to its central course (and slowing) it deposits its silt, and when this has occurred (the channel) becomes obstructed. Obstruction leads to a change of course. Change of course brings fresh stoppage. Thus impeded, (the water) runs wild. Running wild, it injures men. When it injures men, there arises great distress among them. In great distress they treat the laws lightly. Laws being treated lightly, it is difficult to maintain good order. Good order lapsing, filial piety disappears. And when people have lost filial piety, they are no longer submissive...'[b]

Thus a characteristic sorites[c] passes smoothly from the hydraulic to the sociological and ends with a twist which gives us the viewpoint of the ancient feudal lords. But there is a way out; men must act in disciplined unity. Kuan Chung goes on to say:[d]

I request that you establish Water Conservancy Offices (Shui Kuan [1]) (in each district) and staff them with men who are experienced in the ways of water. There should be one high official (Ta Fu [2]) and one Deputy (Ta Fu Tso [3]), with just enough labour corps brigadiers (Shuai Pu [4]), section commanders (Hsiao Chang [5]) and administrative assistants (Kuan Tso [6]) (to meet the need). Then for the area on both sides of each river select one man as chief hydraulic engineer (Tu Chiang Shui Kung [7]). Order all these to inspect the waterways, the walls of cities and their suburbs, the dykes and rivers, canals and pools, government buildings and cottages; and to supply those who are to carry out the repair work in the districts with just enough men.

Presently there will be a little more to say about the titles and positions of the hydraulic engineers at various times (p. 267); here the military nuance of the nomenclature is worth noting. And indeed the rest of the chapter relates how these officials, working together with the village elders (San Lao [8]), must not only muster the *corvée* of able-bodied men and women, but also summon suitable strengths of armoured weapon-bearing soldiers (*chia shih* [9]) for the repair and maintenance of the dykes.[e] The number of spades, baskets, earth-tampers, planks and carts for each detachment is specified. The summer and autumn, times of hoeing and harvest, are never to be used for public works, but the winter is suitable for inspection and the accumulation of stores, such as brushwood for fascines. Most of the work must be done during the spring agricultural lull when water-levels are low. So much for the first great necessity, protection from

---

[a] At this point there intervenes the discussion on the principle of the inverted siphon for piped water-supplies, already mentioned, Vol. 4, pt. 2, p. 128.

[b] Pp. 7b, 8a, tr. Rickett (1), mod. auct.   [c] Cf. Vol. 4, pt. 1, p. 205, and Sect. 49.

[d] P. 8b, tr. Rickett (1), mod. auct.

[e] The Chinese Army's contemporary practice of rendering powerful assistance in the construction of great public works is thus a very ancient tradition indeed. A picture remains in my mind of the army engineers I met at the Ming Tombs dam north of Peking in 1958.

[1] 水官   [2] 大夫   [3] 大夫佐   [4] 率部   [5] 校長
[6] 官佐   [7] 都匠水工   [8] 三老   [9] 甲士

floods. But there were two other pressing incentives for hydraulic engineering in ancient China.

Chinese agrarian society was from the beginning based upon intensive agriculture, and for this to succeed, irrigation works both small and large were as necessary in its eotechnic culture as coal-mining and iron-metallurgy were for the palaeotechnic age (Fig. 864, pl.). In the north-western loess region, says Chi Chhao-Ting,[a] the problem was primarily one of leading off contour canals from natural water-courses, making them follow a slower gradient of fall, and distributing the water to the fields.[b] In the Yangtze valley and the valleys of Kuangtung, the problem was the drainage of the fertile but swampy alluvial land, and the maintenance of a system of canals for the purpose.[c] In the basins of the Huai and the Yellow River, the main necessity was the construction of works stout enough to withstand or delay the greatest flood volumes, and dams and reservoirs to retain the water of the rainy season and release it gradually.[d]

Modern studies of the loess (e.g. Barbour, 1, 2, 3) show that this spongy soil, with its high porosity and great capillary capacity, facilitates the rise of mineral elements from the subsoil to the roots of the plants. Rich in potassium, phosphorus and lime, it only needs abundant water, with the addition of organic manure, to show great fertility. And what is true of the loess soil *in situ* is true also of the loess being carried down towards the sea in the form of silt in the great rivers. Chinese peasants and officials appreciated long ago, in the − 1st millennium, the importance of this silt as a fertiliser, and faced very consciously the complex of problems relating to its control, e.g. how to prevent its obstructing the course of the river and necessitating ever higher dykes, how to distribute or retain it by means of sluice-gates, etc. At an early date, too, there was a recognition of the relation of this silt flow to the deforestation and denudation of the mountains. But all such questions were inextricably bound up with social difficulties. In the north, peasants could not easily be dissuaded from settling on the rich land within the great dykes and only occasionally flooded. In the south, landowners and rich farmers tended to encroach upon the lands recovered from drained swamps and lake bottoms (i.e. land nominally belonging to the State but unclaimed by any individuals), with the result that reservoir space available in flood times was greatly diminished.

Nevertheless, irrigation and the fertilising properties of silt give the clue to the achievement of China in maintaining so-called 'permanent agriculture' over the centuries. Though the periodical floods were tragic, they left the soils, says Wolfanger,

[a] (1), p. 12.

[b] This can be seen well in projects currently accomplished, such as the Thao River canal (Anon. 60) and the Nieh River canal (Kan Chi-Chai, 1), both in Kansu; as also the numerous contour canals derived from the Tutsao River in North-western Hupei (Yu Chêng, 1).

[c] This too can be seen in current examples, notably the drainage of the low-lying country in Eastern Hopei near Tientsin (Li Hsi-Fan, 1), and the interesting canals of Northern Anhui (Chhen Han-Sêng, 1). Cf. p. 284 below.

[d] Again this is clear in the new integrated Huai River Project, see Fu Tso-Yi (1); Kao Fan (2); Hsimên Lu-Sha (1); Anon. (68); as well as in the Yellow River Projects, see Kao Fan (1); Têng Tsê-Hui (1); Anon. (67); Shang Kai (1). Two works in Northern Hopei, the Kuanthing dam (Anon. 69) and the Miyün dam (Huangfu Wên, 1), exemplify the same principles. Their part in the Hai-Ho Basin control system as a whole is discussed by Hsiang Wên-Hua (1).

young, fresh, tillable and renewed. Even on the terraces among the hills the residual erosion of upper levels restored immaturity to their soils. For Simkhovitch, contrasting Chinese with European agriculture, the history of the former showed that owing to the rejuvenation of its productive and unleached soils, intensive crops could be obtained indefinitely without the necessity of recourse to mineral fertilisers.[a] But these advantages could not be made use of without an efficient system of irrigation, and in any given region this naturally depended on all kinds of purely social, dynastic and strategic factors; from time to time one or another region was favoured by the rulers, and therefore developed and maintained—elsewhere, or subsequently, there was neglect and abandonment.

Hydraulic engineering was doubly demanded in China. While in the north irrigation projects were suggested by the nature of the loess soil, though the main crops were dry, wheat, millet, etc., in the centre and south abundant water was indispensable for wet rice cultivation. All the agricultural treatises (cf. Vol. 4, pt. 2, pp. 166 ff.) emphasise the interest which the farmer should take in the supply of water to his rice-fields. In +1313 Wang Chên wrote, in the *Nung Shu*:[b]

Cultivators of rice build surface tanks and reservoirs to store water, and dykes and sluices to stop its flow (when necessary)...The land is divided into small patches, and after ploughing and harrowing, water is let into the fields and the seeds sown. When the plants grow five or six inches tall, they are planted out. All farmers south of the River now use this method. When the plants attain a height of seven or eight inches, the ground is hoed, and after hoeing the water is let go from the fields, so as to dry them.[c] Then when the plants begin to flower and seed, water is again let in...

The regulation of water is indeed the *sine qua non* of wet rice farming. Many Chinese drawings show work being done on the dykes and channels, a constant occupation of countrymen in off seasons.[d]

Throughout Chinese history the taxes essential for unified State power were collected in kind, and most of this was grain. Such grain tribute was the fundamental source of supply for the imperial clan, the central bureaucracy, and the army with its headquarters at the capital. Indeed, throughout the period of bureaucratic feudalism the government generally considered the interests of grain-transport as overriding those of irrigation and flood-control. One can see this particularly well in the case of the Grand Canal, which crossed the Huai valley at right angles, and often interfered with the rational conservation of Huai waters. As Chi Chhao-Ting puts it, the tax transport was primarily an act of appropriation, connected with the immediate enjoyment of the fruit of rulership and the obvious necessity for the maintenance of power by armed force. Irrigation and flood-control were more a question of peasant welfare, and

---

[a] Of course at the cost of (*a*) periodical devastating floods, (*b*) a very large labour force, (*c*) the use of human manure with all its dangers (in the uncomposted state) for human health, and (*d*) delay of the stimulus to improve crop plant varieties.

[b] Ch. 7, p. 5*a*, *b*, tr. Chi Chhao-Ting (1), p. 27, mod. auct.

[c] As it might often have to be pumped out or in, one can see the importance of the various forms of water-raising machinery described in Sect. 27*g* above, Vol. 4, pt. 2.

[d] Cf. *SCTS*, ch. 31, Tzu Tshai, p. 6*a* (Karlgren (12), p. 48).

therefore connected more remotely, though still vitally, with appropriation and State power. It is significant that Ssuma Chhien, after describing the early achievements of canal building, adds:[a]

All the canals were used for the passage of boats, and if there was sufficient water, they were employed for irrigation. The country people enjoyed the advantage of this. Wherever the canals passed, the peasants made use of the water, leading and guiding it to irrigate the fields —of these ditches there were tens and hundreds of thousands, nay, an incalculable number of them.

The importance of water-transport was by no means limited to times of peace; even more in times of disturbance and war, whether civil or external, the command of the waterways as supply routes was of inestimable value. After the fall of the Chhin dynasty, Liu Pang, founder of the house of Han, owed much of his success in defeating his powerful rival, Hsiang Yü, to his control of the region 'within the passes' (Kuan-Chung[1]), i.e. the Wei[2] valley. From that secure base Hsiao Ho[3] was able to send continual supplies of grain to the Han armies confronting the forces of Chhu in Honan. When, at the beginning of the Han dynasty, Hsiao Ho received the highest reward, O Chhien-Chhiu[4] spoke of him thus:[b]

When the armies of Han and Chhu held their ground against each other at Jungyang (in Honan) for several years, the army (of Han) had no ready supply of food. Hsiao Ho despatched grain from Kuanchung by water transport, and supplied food (for the army), thereby preventing shortages. Your Majesty lost Shantung several times, but Hsiao Ho continuously held Kuanchung at your disposal. This is an achievement which will endure for ten thousand generations.

The pattern was repeated some two centuries later, when Liu Hsiu established the Later Han dynasty. But there was a difference in that the base from which supplies were sent forth by water transport in different directions was no longer Kuanchung with its capital at Chhang-an west of the mountains, but rather Honei, i.e. the lower Yellow River valley, with its capital at Loyang east of the mountains.[c] Here Khou Hsün[5] was quite consciously entrusted with the same part to play as Hsiao Ho formerly.[d]

At an earlier stage (Vol. 1, pp. 114 ff.) something was said of the conception of 'key economic areas' by means of which Chi Chhao-Ting has thrown so much light on Chinese history. At different times State power in China has been centred on various different regions. These have constituted economic areas the agricultural productivity and strategic communication facilities of which were so superior to all other areas at

---

[a] *Shih Chi*, ch. 29, p. 2b, tr. Chavannes (1), vol. 3, p. 523; Chi Chhao-Ting (1), p. 66; Watson (1), vol. 2, p. 71.
[b] *Chhien Han Shu*, ch. 39, p. 4a, b, tr. Chi Chhao-Ting (1), p. 79. Cf. ch. 40, p. 8a.
[c] Hence Chinese historians speak of the Former Han dynasty as the Western Han, and the Later Han as the Eastern Han.
[d] *Hou Han Shu*, ch. 46, p. 19a.

[1] 關中   [2] 渭   [3] 蕭何   [4] 鄂千秋   [5] 寇恂

the time that whoever controlled the key area controlled all China. The power of the Chhin State in the −3rd century was largely built on extensive irrigation works (cf. pp. 285 ff. below) in the Wei valley, and the Chhin and Former Han empires were also based on this upper Yellow River region, Kuanchung. The Later Han, however, established its power on the lower Yellow River valley and that of the Huai, shifting the centre of gravity towards Shantung in the east; but meanwhile there occurred a great development both of Szechuan and of the lower Yangtze and Huai valleys. The result was that in the +3rd century three territories were evenly matched (West, North, and East-Central), hence the Three Kingdoms period. In spite of temporary unifications (e.g. the Chin), the centrifugal tendency continued for another three centuries, during which the north and west were often under the control of semi-Chinese Hunnish or Turkic 'nomadic' dynasties. The great unification of the Sui and the Thang at the beginning of the +7th century was connected with the fact that the productivity and transport system of the lower Yangtze valley had outstripped all other regions so much that a key economic area on a new level was produced. The whole story of the Grand Canal, which took definitive form first in the Sui, was essentially the building of a main artery to bring tax grain from the economic to the political centre of gravity of the country. The pattern is particularly clear because after centuries of further division between east-central and northern areas when the Sung contended with the Liao and the Chin Tartars, the Yuan (Mongol) unification once again brought the Grand Canal to a high level of efficiency, altering its course however so that it served Peking instead of Loyang. No subsequent disruption occurred.

The military significance of canal-building was not confined to the problems of supply. Networks of waterways and ditches formed a kind of defence in depth which gave almost insuperable difficulties to nomadic armies chiefly composed of cavalry. This comes out particularly well in the wars between the Liao (Chhi-tan Tartars) and the Sung.[a] The former were often under the necessity of losing time while ditches were filled in or bridged, or while barges were built for ferrying, and the walled cities of China with their moats, connected by good canals but inferior roads, constituted an almost ideal series of strong points scattered over the country which it was difficult to reduce. This was very different from the steppe country for which armies of horsemen were well suited.

(2) SILT AND SCOUR

Let us now turn to more detailed questions. The fertilising effects of silt on saline and alkali land[b] were appreciated as early as −246, when the Chêngkuo[1] Canal was completed, the first of a long series of projects for the irrigation of the Wei valley (cf. the quotation below, p. 285).[c] After the successful inauguration of a later one in the series,

---

[a] Cf. Wittfogel & Fêng (1), pp. 532, 535.

[b] Every ton of erosion loess silt, according to Kovda (1), p. 469, provides 3·3 lb. of nitrogen, the same amount of phosphorus, and 44 lb. of potassium.

[c] *Shih Chi*, ch. 29, p. 3a, b, *Chhien Han Shu*, ch. 29, p. 3a.

[1] 鄭國

the Pai[1] Canal (−95), the country people made a song which has come down to us in the *Chhien Han Shu*:[a]

> Where is the land? At the mouth of the Chhih-yang gorge.
> Chêng Kuo set the example and Pai Kung has followed suit.
> Picks and shovels are as good as clouds.
> And opening a canal is like bringing rain.
> A *tan*[2] has several *tou*[3] of silt (*ni*[4]).[b]
> How it irrigates (*kai*[5]) and fertilises (*fên*[6])!
> It makes our crops grow high
> And feeds millions of mouths at the capital.

In these early days silt was generally regarded as wholly advantageous, but as time went on it was perceived that too much might be a disaster. If it originates from topsoil and is not deposited too thickly it is good, but it may damage land severely when it comes from eroded subsoil (Lowdermilk & Wickes, 1). The dangers of silt can be seen recognised in many official reports, as for example in the late +10th century.[c]

Already in Wang Mang's time (−1 to +22) some very clear statements were made on silt problems. Chang Jung[7] was the first to give a quantitative estimate of the silt content of loess-bearing water, and to point out that its rate of sediment-deposition was proportional to the speed of flow. The following account[d] relates to about −1.

One of the officers of the Ta Ssuma (Grand Marshal), Chang Jung, of Chhang-an, said: 'The nature of water is to flow downwards; if it moves fast it will itself scour its bed (*tzu kua chhu*[8]), and will scoop out hollows and rather deep places. Now the water of the river is heavy with sediment (*cho*[9]); one *tan*[2] of water will even contain 6 *tou*[3] of silt (*ni*[4]). In every district both west and east of the capital, the people all conduct away the water coming from the mountains which flows into the Yellow River and the Wei River, to irrigate their fields. In spring and summer the streams dry up, and this is a time of little water, therefore the water flows slowly and (deposits more silt which) blocks up (*chu yü*[10]) all openings, and it is shallower. But when the rains come the water rushes violently down so that there are floods and breaks in the dykes. And then the government and the people go on building dykes (*ti sai*[11]) until the level (of the rivers) becomes slightly higher than the surrounding country. It is like building (ever higher) dams (*chu yuan*[12]) to store up water. It would be better to follow the nature of water, and not (permit so much) irrigation; then the hundred rivers would flow freely and the water-ways would keep themselves in order (*shui tao tzu li*[13]), and there would be much less danger from floods breaking through, with all the harm that they bring about.' The Yü Shih (Imperial Secretary) Han Mu[14] of Lin-Huai, considered that it might be

---

[a] Ch. 29, p. 8*a*, tr. Chi Chhao-Ting (1), p. 88, mod. auct.
[b] One Han picul (*tan* or *shih*), about 60 modern lbs., contained from 10 to 20 pecks (*tou*), more appropriately translated quarts, according to the grain or liquid concerned. Here we assume the former figure.
[c] E.g. *Sung Shih*, ch. 94.
[d] *Chhien Han Shu*, ch. 29, p. 16*b*, tr. auct. adjuv. Pokora (9), p. 49.

[1] 白   [2] 石   [3] 斗   [4] 泥   [5] 溉
[6] 糞   [7] 張戎   [8] 自刮除   [9] 濁   [10] 貯淤
[11] 隄塞   [12] 築垣   [13] 水道自利   [14] 韓牧

done. Although one might not be able to make as many deep channels (for letting away the water) as the 'Nine Rivers' of Yü the Great, still, even only four or five would do much good.[a]

This passage tells us a good deal. Chang Jung's actual estimate of silt-content (60%) seems at first sight impossibly high, and has often been regarded as a literary exaggeration. But he has been confirmed by recent measurements such as those recorded by Jen Mei-O (1); the highest figure for the main stream is 46·14% at Shen-hsien, but for tributaries 67·5%.[b] Besides, concrete figures of any kind show clearly that efforts were already being made in the Han to gain some quantitative measure of the silt-content of the water in which everyone was interested. He goes on to make his statement about the speed of flow, and connects it with the problem of dredging, that is to say, of keeping the stream beds as low as possible instead of building up ever higher dykes and embankments to contain the flood flow. As our topic develops we shall see what anxiety this decision caused in the earliest stages of Chinese hydraulic engineering, the choice between high dykes or deep beds. Here Chang Jung is arguing that advantage ought to be taken of the swift flow of the flood to scour out the channel, and one can see that the interests of flood-protection could be at variance with those of irrigation. Seasonal compromises had to be sought. The idea of self-scour is also a key one—the *Chou Li*, for example, says:[c]

Every canal should be made so as to take account of the characteristic forces of water; every embankment should take account of the characteristic strengths of earth (*Fan kou pi yin shui i; fang pi yin ti i*[1]). A good canal is scoured by its own water; a good embankment is consolidated by the sediment brought against it (*Shan kou chê, shui sou chih; shan fang chê, shui yin chih*[2]).

These views, and those of Chang Jung, inspired many great hydraulic engineers in later Chinese history, such as Wang Ching[3] in the +1st century,[d] Chia Lu[4] in the Yuan, Phan Chi-Hsün[5] in Ming and Chin Fu[6] in Chhing.

Here there is no space for more than a glance at the controversies which surrounded the deposition of silt and the flow of water in canals and rivers.[e] One finds echoes of them everywhere in the dynastic histories, though the full story has never been told even in the Chinese technical literature. Mêng Chien,[7] for example, when governor of

[a] The passage continues with the views of another notable engineer Wang Hêng.[8] When Huan Than[9] became Departmental Director of the Ministry of Works about +2 he made preparations for dredging and dyke-building, but under Wang Mang nothing could succeed. On Huan Than cf. Vol. 2, p. 367, Vol. 4, pt. 2, p. 392; and Pokora (9, 13).

[b] Modern measurements on the Ching River; cf. p. 287 below. For general comparisons see p. 366 below.

[c] Khao Kung Chi, ch. 12, p. 20b (ch. 43), tr. auct., adjuv. Biot (1), vol. 2, p. 570.

[d] His biography (*Hou Han Shu*, ch. 106, p. 6a) speaks of the success which he achieved by his 'method of stop and let flow' (*yen liu fa*[10]), which involved the building of sluice-gates.

[e] For treatments of these questions by modern engineers, one may consult papers by Griffith (1); Chatley (27, 28, 30, 32, 34, 35).

[1] 凡溝必因水埶防必因地埶 [2] 善溝者水漱之善防者水淫之 [3] 王景
[4] 賈魯 [5] 潘季馴 [6] 靳輔 [7] 孟簡 [8] 王橫 [9] 桓譚
[10] 塥流法

Chhangchow in the Thang,[a] succeeded in clearing the blocked bed of the Mêng River, and at the same time arranged for the irrigation of thousands of acres of fields. The question was of course intimately bound up with the invention and construction of sluice-gates and weirs which could be adjusted to the seasonal levels of the main stream, but these developments also served two other purposes, the derivation of irrigation canals and, with flash-locks, the facilitation of barge traffic. We shall reserve to a later stage (pp. 344 ff.) the detailed development of these inventions, and a comparison with what happened in other parts of the world; for the present it may be taken for granted that adjustable sluices (*shui mên*[1]) had been introduced at least as early as the Chhin and Former Han if not before. Let us listen to two old friends discussing the problems of the silt fertilisation method (*yu thien fa*[2]) in the latter half of the +11th century. In his *Tung-Pho Chih Lin*, the poet and official Su Tung-Pho wrote,[b] about +1060:

> Several years ago the government built sluices (*tou mên*;[3] lit. dipping gates) for the silt fertilisation method, though many people disagreed with the plan. In spite of all opposition it was carried through, yet it had little success. When the torrents on Fan Shan were abundant, the gates were kept closed, and this caused damage (by flooding) of fields, tombs and houses. When the torrents subsided in the late autumn the sluices were opened, and thus the fields were irrigated with silt-bearing water, but the deposit was not as thick as what the peasants call 'steamed cake silt' (*chêng ping yu*[4]) (so they were not satisfied). Finally the government got tired of it and stopped. In this connection I remember reading the *Chia I Phan* of Pai Chü-I (the poet) in which he says that he once had a position as Traffic Commissioner. As the Pien River was getting so shallow that it hindered the passage of boats he suggested that the sluice-gates along the river and canal should be closed, but the Military Governor pointed out that the river was bordered on both sides by fields which supplied army grain, and if these were denied irrigation (water and silt) because of the closing of sluice-gates, it would lead to shortages in army grain supplies. From this I learnt that in the Thang period there were government fields and sluice-gates on both sides of the river, and that irrigation was carried on (continuously) even when the water was high. If this could be done (successfully) in old times, why can it not be done now? I should like to enquire further about the matter from experts.

Here the poet's words show clearly the conflict between agricultural and transportation interests, though Su Tung-Pho himself evidently did not quite realise the significance of Pai Chü-I's contention. As a further illustration one may take a passage from the *Mêng Chhi Pi Than* of Shen Kua, written some twenty years later.[c]

> In the Hsi-Ning reign-period (+1068 to +1077) much emphasis was placed on the silt fertilisation method. Scholars discussed the statement in the *Shih Chi*[d] that one *hu*[5] of water of the Ching River[e] contains several *tou* of silt, which will fertilise the crops and make them

---

[a] Biographies in *Chiu Thang Shu*, ch. 163, *Hsin Thang Shu*, ch. 160.
[b] Ch. 7, p. 4*b*, tr. auct.
[c] Ch. 24, para. 10, tr. auct. Cf. Hu Tao-Ching (*1*), vol. 2, p. 755.
[d] A mistake for the *Chhien Han Shu*, the song already quoted, p. 228.
[e] This was the river harnessed in the Chêngkuo Canal system, on which see p. 285 below. The *hu* (bushel) was in Sung times a capacity measure equivalent to half the Sung picul (*tan*), i.e. 5 quarts (*tou*), or about 79 modern lbs.

1 水門   2 淤田法   3 斗門   4 蒸餅淤   5 斛

grow tall. I remember that when I was at Suchow[1] on official business I saw a monument on which there was an inscription recording how certain people of the Thang dynasty had built six sluice-gates (*tou mên*[2,3])[a] which distributed silt-bearing water from the Pien River to benefit the inhabitants near the lower reaches. Such were the origins of the silt fertilisation method.

There are contemporary records, too, of farmers such as Hsing Yen[4] coming to the authorities and demanding that silt-bearing water should be diverted to their fields.[b]

When we look at modern studies of fluvial hydraulics and silt-transportation[c] we are likely to feel some surprise at the degree of awareness of the old Chinese engineers (many further examples of which await the reader of the following pages), coupled with a sobering realisation of the prominent part still played by empirical observation and formulation in spite of three centuries of post-Renaissance mathematical hydrodynamics.[d] Some of Fargue's laws, e.g. that the thalweg (the line of deepest soundings in a river) hugs the concave bank, opposite which shoals form; and that the depth increases with the degree of curvature, they already knew—they knew also that in general the deposition of silt occurred in a manner inversely proportional to the velocity of flow. It was not for them to analyse the forces acting upon individual silt particles (impact, turbulence, drag, friction and hydraulic lift), as we do now, but they consciously sought for conditions which would give the 'silt-stable régime' so that a waterway would neither choke itself up nor erode in a migratory manner. It was not for them to establish the principles of helicoidal flow and centrifugal spin which account for the wave-form of the meander plan, nor to demonstrate its natural appearance in model experiments, but they knew (as we shall see, p. 249 below) that dykes were not as good as dredging, and that training works should encourage the natural scour of shoals. Indeed, even if they had distinguished between bed-load (the traction of the larger particles along the bottom) and suspension-carriage (the flotation of the water-borne smaller ones), they would not have been able to prevent the elevation of the bed of the Yellow River, an almost geological process which can hardly be reversed today. Yet in the debates which we shall mention later between the Chinese proponents and antagonists of the channel contraction method (p. 235), we may find advance echoes of Kennedy's principle today accepted, that shallower sections will take the greatest silt-loads, probably because of the greater vorticity, turbulence and cross-currents engendered in them.

[a] 'Head-gate' or 'sudden gate'; the adjective originally referred to a precipitous drop in space, such perhaps as would be represented by a ten-foot head of water, but later it acquired the meaning of 'suddenly' in time. Hence one cannot be quite sure of the meaning of its application to a sluice-gate, nor does it give us much clue as to the gate's actual construction. Cf. p. 349 below.

[b] It may be held that every social phenomenon can be found in all civilisations and that what is hypertrophied in one organism occurs in rudimentary form in others. France in +1559 saw on a small scale something reminiscent of Chinese styles, when the engineer Adam de Craponne built a 40-mile irrigation canal taking the silt-laden water of the Durance River from Cadenet to Arles. De Craponne understood the value of the silt, and his Provençal project, which still exists, was fully successful.

[c] E.g. Leliavsky (1); Brown (1).

[d] The founding fathers were Castelli (+1628), Mersenne (+1644), Torricelli (+1644) and Mariotte (+1686). See further Rouse & Ince (1); Leliavsky (3).

[1] 宿州　　[2] 陡門　　[3] 陡門　　[4] 邢晏

(3) THE RIVER AND THE FORESTS

The sorest point, where all controversies reached their acutest phase, was the handling of the Yellow River on the North China plain. The oldest attempts to control it known to us in historical times were the dykes (*yen*[1]) built along the lower reaches under the superintendence of Duke Huan of the State of Chhi (Chhi Huan Kung[2]), the ruler who figures so often in the books of the philosophers such as *Chuang Tzu*, *Lieh Tzu*, and *Kuan Tzu*.[a] This was in the first half of the −7th century.[b] Though it rests upon local tradition current in the Han[c] rather than upon documentary evidence, the attribution is quite acceptable. The dykes of Duke Huan had the effect of uniting the nine streams of the previous delta into one, and their remains still existed in Han times.[d]

From the Chhin and Han onwards the greatest possible efforts were made to prevent the river from overrunning the plain during high-discharge periods. It must have been noticed that in the uncontrolled state the river tended to build up a kind of low bank on each side of its winter channel, and so for two thousand years the Yellow River has been enclosed with dykes, constantly increasing in dimensions (Fig. 865). Every few years the river would rise to levels which threatened to overflow them, or else the meanders of the low-water channel would carry the river against an embankment and cause its collapse. Some fifty times or more the river has escaped from all control and formed a new channel on the plain, destroying in the process vast tracts of cultivation and settlement, and burying some land under excessive silt deposits. The earliest accounts we have of these troubles are perhaps in the 29th chapter of the *Shih Chi*, where Ssuma Chhien speaks as follows:[e]

The Han had been in power for thirty-nine years when, in the time of the emperor Hsiao Wên Ti, the River overflowed at Suan-tsao and broke through the 'Metal Dyke' (*chin thi*[3]) (in −168).[f] Great levies of soldiers were raised in the Eastern Commandery to close the breach.

Rather more than forty years later, in the present reign, in the Yuan-Kuang reign-period (−132), the River again overflowed at Hu-tzu, pouring off to the south-east in the Chü-yeh marshes, and communicating with the Huai and Ssu rivers. The Son of Heaven therefore commissioned Chi Yen[4] and Chêng Tang-Shih[5] to recruit men to close the breach, but it suddenly opened again.

Ssuma Chhien goes on to say that Thien Fên,[6] the marquis of Wu-An, whose fiefs (he is careful to add) lay on the north of the river, urged that such great floods were direct acts of Heaven with which it would be unwise to interfere, so nothing was done for

---

[a] For an example already familiar to us see Vol. 2, p. 122.

[b] It is worth noting that this type of public work seems thus to have begun before the earliest large reservoir dams (cf. p. 271 below, on Sunshu Ao).

[c] Reported in the *Shui Ching Chu*, ch. 5, p. 15 b.

[d] Cf. Maspero (28), p. 342.

[e] P. 3 b, tr. Chavannes (1), vol. 3, p. 525, eng. auct.; cf. Watson (1), vol. 2, p. 72.

[f] The lost +7th-century geography, *Kua Ti Chih*,[7] says that this was another name for the 'Thousand-Li Dyke'.

[1] 堰   [2] 齊桓公   [3] 金隄   [4] 汲黯   [5] 鄭當時   [6] 田蚡
[7] 括地志

Fig. 865. A late Chhing representation of river conservancy work. A dyke is being strengthened and sandbanks removed; baskets and punners are seen in use. The picture (from *SCTS*, ch. 6, Yü Kung) is intended to illustrate the words 'He (Yü the Great) led out the Hei Shui (Blackwater River) through the San Wei (Shan) (Mountains of the Three Dangers) and so into the Southern Sea' (cf. Medhurst (1), p. 110), but it is now thought that the meaning of these passages should be 'He travelled along...' (cf. Karlgren (12), p. 17). The editors of *SCTS* interpreted the Hei Shui as the Salween R., but more probably it was one of the rivers in southern Kansu.

twenty years.[a] Eventually, however, the condition of the provinces became so bad that the emperor, Wu Ti, made a tour of inspection himself; and Ssuma Chhien gives a graphic description[b] of the raising of myriads of men by Chi Jen[1] and Kuo Chhang,[2] the sacrificing of a white horse and a jade ring, and the carrying of bundles of wicker-work and faggots by high officials and commoners alike in order to fill up the breach. Finally the fill at Hsüan-fang was successfully accomplished (−109), and a triumphal pavilion erected at the place. It was a triumph destined to need perpetual renewal.

Apparently, in spite of the centralisation of the Han bureaucracy, the level of planning and control for such vast projects still did not suffice. About a hundred and fifty years later, a great Han engineer, Chia Jang,[3] revealed in a famous memorial to the throne (−6) something like chaos in the location of the Yellow River dykes.[c]

At the present time, the nearer embankments stand at a distance of several hundred paces only from the water, and even the furthest are only several *li* from it. South of Liyang the old 'Great Metal Dyke' stretched north-westwards from the west bank of the Yellow River to the southern foot of the western mountains. It also ran eastwards to meet the eastern mountains. People built their cottages on the eastern side of the dyke. After they had been living there a little over ten years, another dyke was thrown out from the eastern mountains southward to connect with the Great Dyke. Again, in the prefecture of Neihuang, a swamp with a circumference of several tens of *li* was drained by building a dyke round it, and the governor of the district then gave the land within the dyke to the people after they had lived there for more than ten years. Now people build cottages in it. These things I have myself seen. In the prefectures of Tungchün (Eastern Commandery) and Paima, the old 'Great Embankment' is paralleled by several other embankments (outside it), and people live in between them. From the north of Liyang to the border of (the former State of) Wei, the old 'Great Embankment' lies several tens of *li* from the river, but inside it there are also several rows of dykes which were built in earlier generations. Thus when the Yellow River flows from Honei north to Leiyang there is a stone embankment (*shih thi*[4]) forcing it eastwards. When it reaches Tung-chün and Phingkang, there is another stone embankment to force it north-west. When it arrives at Leiyang and Kuanhsia it meets a third, changing its flow north-eastwards again. At Tungchün and Chinpei it is diverted north-west, and at Weichün and Chaoyang north-east again—all by stone embankments. Thus in a distance of only a little over a hundred *li*,[d] it is turned westward twice and eastward three times...

Chia Jang was in fact an advocate of the channel expansion theory, a Taoist in hydraulics, who believed that the great river should be given plenty of room to take

---

[a] The historian adds that Thien Fên was supported by the meteorognostics and numerological diviners (*wang chhi yung shu chê*,[5]), Taoist adepts whose motto was evidently *wu wei*.

[b] P. 6*b*, tr. Chavannes (1), vol. 3, pp. 532 ff.; Watson (1), vol. 2, pp. 75 ff. Cf. p. 378 below.

[c] *Chhien Han Shu*, ch. 29, p. 13*b* ff., tr. auct. adjuv. Chi Chhao-Ting (1), p. 91. Cf. *TH*, pp. 584 ff., with its unjustified comment.

[d] The exact length of the *li* at different times in Chinese history is a somewhat involved question; it depended on the length of the foot as by one dynasty after another established. The most helpful study is no doubt that of Wu Chhêng-Lo (2), of which there is now a revised edition. Between the Chhin and the Chin the length of the *li* varied from 0·415 to 0·498 km.; after the confusion of the Nan Pei Chhao period, i.e. from Sui to Ming, it oscillated between 0·531 and 0·560 km. For rough estimates, therefore, half a kilometre, or 1,645 ft., is a tolerable figure; but in the Chhing dynasty it was longer than at any previous time, 0·575 km., or 1,894 ft. Cf. Vol. 4, pt. 1, p. 52.

[1] 汲仁   [2] 郭昌   [3] 賈讓   [4] 石隄   [5] 望氣用數者

whatever course it wanted.[a] Rivers, he said, were like the mouths of infants—if one tried to stop them up they only yelled the louder or else were suffocated. *Wu wei* was the best watchword:[b] 'those who are good at controlling water give it the best opportunities to flow away, those who are good at controlling the people give them plenty of chance to talk'.[c] In the Warring States period, when the oldest dykes had been built, people were allowed to cultivate the silt-laden land within them but not to settle there, for after villages had grown up there was always a natural urge to make new dykes ever nearer and nearer the low-water channel. Chia Jang recommended the wholesale resettlement of the populations of prefectures bordering the river. If the emperor was not prepared to do this, his second plan proposed the making of a great network of irrigation canals to relieve the flood pressure, but these would have to be controlled by sluice-gates (*shui mên*[1]) with stone revetments, much more stoutly built than the affairs of wood and tamped earth which already existed in the irrigation district of Jung-yang.[d] Last and worst of the three possibilities would be to go on repairing the protective dykes and embankments. Chia Jang's memorial implies a deep distrust of the policy of raising such works instead of trying to lower the bed, but if this was to be done there was urgent need for better co-ordination and effective central direction.[e] Whether the resources available to the government of the young emperor Ai Ti (Liu Hsin[2]), hardly more than a boy, could have tackled the radical re-alignment and strengthening that was necessary is very doubtful; in any case a certain amount of resettlement in no way solved the problem and the stage was set for the historic breakout of +11.

If there were hydraulic engineers of Taoist tendency there were also Confucian ones. Those who believed primarily in low dykes set far apart were opposed by those who believed in the main strength of high and mighty dykes,[f] set nearer together. Equally those who believed in giving a river's lower reaches the maximum degree of freedom were opposed by those who believed in contracting the channel so as to make the river dig its own bed. The former argued that with widely separated dykes there would be ample storage space between them for the summer flow.[g] The latter held that with a constricted channel the water, flowing more rapidly, would itself scour out the deep thalweg desired by the former.[h] The contractionists generally had the advantage of the expansionists because their plans, though more expensive, raised no difficult

[a] Tshen Chung-Mien (2), pp. 262 ff., has a full discussion of his views.

[b] Cf. Vol. 2, *sub voce*.

[c] *Shan wei chhuan chê, chüeh chih shih tao; shan wei min chê, hsüan chih shih yen.*[3]

[d] Cf. the saying of a distinguished engineer of the Ming long afterwards, Hsü Chên-Ming,[4] 'The northerners are not familiar with water benefits but are harassed by water damage; they do not know that failure to eliminate the water damage is due precisely to the non-development of water benefits. Accumulation of water is what brings damage; its distribution is what brings benefits' (*Ming Shih*, ch. 223, p. 19b).

[e] His memorial became classical for centuries afterwards; cf. *Chih Ho Fang Lüeh*, ch. 2, pp. 33b ff. and ch. 8, pp. 1b ff.

[f] The sociological parallel was explicitly made in Confucian writing (see Vol. 2, p. 544, cf. p. 256).

[g] This was often associated with the name of Chia Jang (*fl.* −50 to +10).

[h] This was often associated with the name of Wang Ching (*d.* +83), cf. p. 346.

[1] 水門   [2] 劉欣   [3] 善爲川者決之使道善爲民者宣之使言   [4] 徐貞明

social problems of resettlement of population. But in so far as the control of the Yellow River today has necessitated, as we shall see, the construction and evacuation of vast retention-basins alongside the main stream, the expansionists have not been entirely unjustified by modern technology. During twenty centuries the two schools contended,

Fig. 866. Bend erosion (from *Nung Chêng Chhüan Shu*). The picture is entitled *yin kou* because it illustrates a section concerned with underground afferent or efferent tunnels (cf. p. 333) for irrigation.

and neither proved wholly successful. The deep channel is liable to approach and undermine the dykes at bends, and rises of water-level occur with inconvenient speed. The Chinese were extremely conscious of bend-erosion; in the Section on geology we noted a technical term for the scooped-out bend (*khan*;[1] 'niche'),[a] and Fig. 866 shows

---

[a] Vol. 3, p. 604.

[1] 龕

a picture suggesting the effect.[a] On the other hand, the wide separation of dykes[b] allowed so great a deposition of sediment that the storage capacity was reduced very quickly; possession was then taken of the tempting new land, and smaller parallel dykes built to keep out ordinary floods, thereby annulling the whole object originally sought. Modern arguments about the best course, then not finally decided,[c] may be followed in the technical papers of engineers such as Freeman (1) and Todd & Eliassen (1), or Chatley (33),[d] supported by more popular accounts such as those of Clapp (2, 3); or preferably in Chinese books (e.g. Chang Han-Ying (1) and the collective work edited by Hu Huan-Yung, Hou Tê-Fêng & Chang Han-Ying). The theory of channel contraction, the greatest advocate of which was perhaps Phan Chi-Hsün[1] in the Ming (+1521 to +1595),[e] has not been entirely confirmed by modern laboratory experiments on model scale, and while dams and retention-basins are being built research is still proceeding.

What is generally conceded, however, is that the bed of the Yellow River has risen by about 3 ft. per century. In some places it has been necessary to build dykes up 6 ft. per annum for stretches of 20 or 30 miles to combat the huge silt deposits.[f] At the present time, the bed is only a few feet below the level of the plain, and in some places level with it or even as much as 12 ft. above it,[g] but in cross-section this central channel is very small compared with the miles of deposit on each side. At the Chêngchow railway bridge, for example, the bottom of the bed is level with the plain, and the average level of the deposit 24 ft. higher, while the top of the dykes on each side is 30 ft. with an abrupt descent to the plain on their outer escarpments.[h] Consequently when the river in high flood escapes through the dykes, the residual flow in the old channel fills that up with silt, and plugs it against future free flow in that course. It is therefore very difficult to get the river to resume its old course after a break.

To illustrate this, Fig. 869 shows a cross-section at a point just above the railway bridge between Chêngchow and Hsinhsiang. The bed bottom is about level with the

[a] NCCS, ch. 17, p. 32a, reproduced in TSCC, I shu tien, ch. 5, hui khao 3.

[b] At maximum 15·6 miles (Jen Mei-O, 1).

[c] An interesting echo of the age-old debate is to be found in the novel Lao-Tshan Yu Chi[2] (The Travels of Lao-Tshan), written in 1904 by Liu Ê[3] and now translated into English by Shadick (1). Liu Ê was himself an engineer and promoter of industrialisation projects, as well as a great humanitarian and a brilliant archaeologist (one of the first to recognise the significance of the oracle-bones). The problems of hydraulic engineering form part of the plot of the novel, which takes a permanent place among the literature strongly critical of the old officialdom.

[d] There is a curious compilation of opinions, mostly of Western advisers, edited by Cross & Freeman.

[e] Though he expected breakthroughs, and provided for defence in depth by outer parallel reserve dykes.

[f] Many charts of the Yellow River dykes at different periods are extant, e.g. those reproduced by Chi Chhao-Ting (1), pls. 5, 6. Here we reproduce (Figs. 867, 868, pls.) two original scrolls of late Ming or early Chhing date, at one time in the collection of Mr Rewi Alley, showing respectively the lower Yellow River in the neighbourhood of Chi-nan, and its earlier course between the San-mên Gorge and Hsia-i city, past Khaifêng. We are much indebted to Mr Alley for these photographs and for the permission to reproduce them. Rather older are the charts engraved on stone steles in the Pei Lin at Sian, which I had the pleasure of studying in 1958 and 1964. One of these Huang Ho Thu,[4] dating from +1535, commemorates the work of the eminent engineer Liu Thien-Ho.[5]

[g] Jen Mei-O (1).

[h] Hu Huan-Yung, Hou Tê-Fêng & Chang Han-Ying (1), vol. 3, opp. p. 32. At this point the banked deposit stretches away about five miles on one side and one mile on the other.

[1] 潘季馴   [2] 老殘遊記   [3] 劉鶚   [4] 黃河圖   [5] 劉天和

surrounding plain, and on each side huge silt deposits are heaped up, topped at their edges by dykes. For such a situation neither of the ancient methods is really applicable, since the dykes cannot be indefinitely heightened, and the work of removing a 25 ft. deep mass of silt 1,000 miles long and 5 miles broad would not be very feasible or economic even for the most modern excavating machinery. Already when Freeman

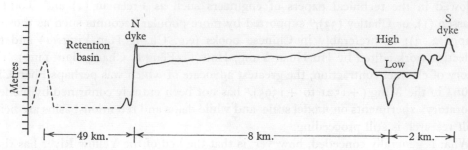

Fig. 869. Cross-section of the bed of the Yellow River just above the railway bridge linking modern Chêngchow with Hsinhsiang; i.e. just west of the mouth of the old Pien Canal.

was writing, the best solution seemed to be the building of retention-basins, i.e. long strips of plain-level 'wash-land' parallel to the river and protected on their outer sides by dykes massive enough and high enough to cope with any probable transient contents. This work is now well advanced, according to recent information (Fig. 870).[a] An outer embankment (the Pei-Chin Dyke) and the northern embankment of the river (the Ling-Huang Dyke) enclose an area of low-lying land 91·65 miles (150 km.) long and averaging 30·5 miles (49 km.)[b] in breadth. This runs across the whole of a new province called Phingyuan[c] in a north-easterly direction and ends at the Tung-phing Lake near the old crossing of the Grand Canal. Formerly, flood crests passed this distance in three days, but now the time taken is as much as eight; thus the speed of flow is greatly reduced, and the danger of dyke-failure almost entirely obviated.[d] Another smaller retention-basin has been built nearer to the mouth of the river.[e] The principle involved, that of controlled silt-deposition, is hardly a new one (cf. the rows of parallel dykes in Figs. 867, 868, pls.), but probably not since the time of Chhin Shih Huang Ti has there been a government sufficiently centralised, and never before sufficiently popular, to organise without excessive social strain the removal of towns and villages which must have been necessary.[f] Modern sluice-gates also matter.

[a] Cf. Kao Fan (1); Li Fu-Tu (1); Têng Tsê-Hui (1); Nesteruk (1), fig. 30. The same principle has been applied to the Yangtze near Hankow, cf. Anon. (67); Su Ming (1).

[b] This figure seems to mean the total distance from the dykes on the southern side of the river to the new dyke northwards, but the greatest width of the actual basin cannot be much less than 25 km.

[c] Located between Shantung, Shansi, Hopei and Honan. It has since been discontinued.

[d] In a normal year there are four flood seasons, 1,500 cumecs (cf. p. 221) discharging in April from melting snows, peaks of about 7,000 cumecs each in July and October from the rains, and a smaller flood crest of some 800 cumecs in January, again from snow. Low water flow is 60 to 80 cumecs. Figures from Kovda (1), p. 172.

[e] Now (1965) the total retention volume available is estimated at 50 billion cubic metres of overflow, enough to take as bad a flood as that of 1954. In addition, electric power-station reservoirs are being built on the upper waters of many tributaries. The potential amounts to 230 million kilowatts.

[f] Apparently more than twenty-five large villages and at least one city.

What was the origin of these mountains of silt? The blanket of loess extends for about 150,000 square miles over the drainage area of the Yellow River, with an average depth of 100 ft., varying from extremes of 1,000 ft. to a few inches. As long as it was covered by forests of trees, bush and wild grasses, the soft soil was protected from the cutting effects of the heavy monsoon rains. But with increasing population pressure

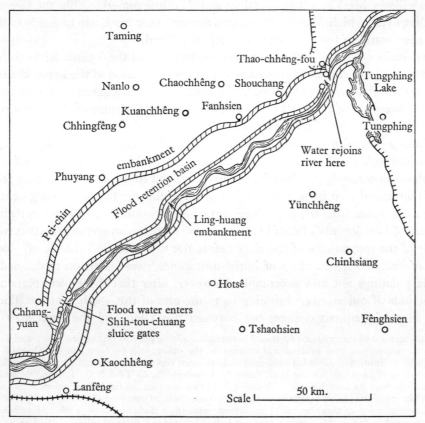

Fig. 870. The modern strategy of controlled silt-deposition in retention basins parallel with the Yellow River, which retard and lower flood crests. The map (after Kao Fan) shows the basin which has been completed along the left (north-west) bank of the river between its sudden turn east of Khaifêng and the point where it crosses the north–south line of the present Grand Canal north of Tung-phing. The retention basin in Fig. 869 is a projected one.

there has occurred progressive deforestation, denudation and erosion, until today there are few trees to be seen on the uplands of Shensi and Kansu. In north Shensi erosion gullies occupy no less than 50% of the total ground space, and the tracks often make their way along ridges of eroded loess as narrow as 3 or 4 ft. across. I myself have had the experience many times in Kansu of seeing literally torrents of liquid brown mud, the consistency of thin porridge, pouring down gullies and across roads after a cloudburst higher up in the hills. Fig. 871 (pl.) gives some idea of what the eroded countryside looks like from the air.[a] The average annual silt loss for the whole basin is

[a] From Fisher (1); cf. also the photographs in Koester (1).

estimated at 1·78 mm. if considered as a uniform layer. In a number of papers, Lowdermilk has urged that this erosion has been the dominant factor in the difficulties of control of the Yellow River,[a] and it follows that the vast programmes of re-afforestation of the western provinces now on foot will greatly decrease them. This re-vegetation is being started in the talus slopes of the canyons, using suitable indigenous trees such as willows, black locust and catalpa; and if in time something like the forests, long since destroyed, which Marco Polo passed through near Sian, can be made to flourish anew, the control of the river will be much simplified.[b]

Lowdermilk (3) believes that some agency has caused the Yellow River to become progressively more restless in its bed, and that the excavation of the upper channel has not been enough to account for this. Deforestation and consequent erosion must at any rate have been one of the most important factors. The measure of the river's restlessness is the wanderings of its lower reaches, sometimes flowing into the sea north, and sometimes south, of the Shantung peninsular massif. These unmanageable changes are summarised in Table 69 which should be compared with Fig. 859.[c] A brief study of these shows how slowly the river found out its present route after debouching from the hills; it advanced step by step through a series of four or five small stages each preceding a deflection from the final alignment which it would come to occupy. The chief weakness of Lowdermilk's belief in a progressive restlessness seems to be that we know so little of the movements of the river before the first recorded change of −602; his figure of just over 1,600 years of initial quiescence[d] was based on traditional semi-legendary datings not now acceptable. However, after that there were certainly two long periods of quiescence,[e] bringing us to the end of the +9th century. The Sung was a time of frequent change, but between the Yuan and the latter part of the

---

[a] A full account of the relations surmised between the silting process on the one hand, and deforestation, slope-cultivation, over-grazing and erosion on the other, will be found in Lowdermilk (1) and Lowdermilk & Smith (1). Detailed field studies have been reported for Shansi by Lowdermilk (4) and Lowdermilk & Wickes (2); for the Huai valley by Lowdermilk, Li Tê-I & C. T. Ren (1); and for the Wu-thai Shan area by Lowdermilk & Wickes (3). In this last case the field observations were compared with the history of the region drawn in much detail from the local gazetteers (cf. Vol. 3, p. 517). The first wave of deforestation of these beautiful mountains, with their Buddhist abbeys where Ennin had wandered, occurred in the +16th century and was halted by two enlightened officials, Hu Lai-Kung[1] and Kao Wên-Chien[2] in +1580. But in the late +17th and +18th centuries the destruction of the forests was resumed, and by the end of the nineteenth they were completely barren and unproductive. Specific aspects of erosion relating to forestry have been discussed by Lowdermilk & Li Tê-I (1), and interesting measurements of rainfall and run-off in forest areas have been reported by Lowdermilk (5, 6). It is often only around famous temples that forests have been allowed to survive in China; I remember particularly the magnificent surroundings of the Taoist abbey of Lou-kuan-thai (cf. Vol. 2, p. 164). But extensive re-planting is now radically altering this picture. In 1945 I had the pleasure of visiting, with Dr Dorothy Needham and Dr Tshao Thien-Chhin, the Soil Conservation Experiment Station at Thienshui, Kansu, established by Chinese experts in collaboration with Dr W. G. Lowdermilk.

[b] Current progress in the soil conservation of the loess highlands can be followed in papers such as those of Fang Hua-Yung (1) and Chhen Hsüeh-Nung (1).

[c] These are compiled from data given by Tshen Chung-Mien (2); Chêng Chao-Ching (1); Li Hsieh (1), and the maps of Kuo Hui-Ling and Nesteruk (1). Cf. also Lowdermilk (3). The pioneer work of Biot (21) is still well worth reading.

[d] Followed by Nesteruk (1).

[e] From the Chhun Chhiu period to the end of Early Han ①, and from the Later Han to the end of the Thang ②. Courses of the river are indicated in this way.

[1] 胡來貢   [2] 高文薦

Chhing more than 500 years (or at least 300) passed without serious alteration ⑦. During this period there was strongly centralised government and considerable improvements in civil engineering technique.[a] Yet after the change of 1852-5 there was a further catastrophic overflow to the south-east in 1887-9, and this pattern was repeated by the deliberate breaking of the dykes for military purposes in 1938. In general, therefore, the restlessness of the river seems to have occurred in a series of waves rather than as a continually increasing quantity.

Bielenstein (2) has presented a reasoned and rather convincing case for the view that the change of +11 was the fundamental cause of the failure of Wang Mang to found a long-enduring Hsin dynasty. The Yellow River disaster produced population migrations, widespread famines and all kinds of disturbances, including the rebellion of the 'Red Eyebrows',[b] which permitted the gentry supporters of the Han to regain control.[c] This conclusion has important implications for the social and economic history of the period. But if space permitted, one could illustrate each one of the river's great changes by echoes from Chinese political and literary sources. Those, for instance, who have enjoyed the Ming opera entitled *Pai Shê Chuan*[1] (The Story of the White Snake Pagoda) will remember the Abbot of Chin-shan, Fa-Hai,[2] withstanding the waves conjured up by the goddesses or serpent-spirits to overwhelm his temple—as indeed in +1194 the full force of the waters did break upon Liangshan before dividing to the north and to the south.[d]

To what extent people were conscious of the dangers of deforestation (Fig. 872) in ancient China is a question hard to answer, for one must beware of reading too much of our own ideas into their words.[e] Still, it is hard to doubt the conservation significance which has traditionally been ascribed to the Mencian adages:[f]

If the seasons of husbandry be not interfered with, the grain will be more than can be eaten. If close-meshed nets are prohibited in the pools and lakes, the fishes and turtles will be more than can be consumed. If axes and hatchets are used in the mountain forests only at suitable times, there will be more wood than people know what to do with (*Fu chin i shih ju shan lin, tshai mu pu kho shêng yung yeh*[3]).

---

[a] The history of the dyke breakages during the Chhing dynasty has been studied from the engineering point of view by Su Tsung-Sung, Hsüeh Li-Than & Lo Thêng (1). The Yellow River Administration during the same period has been the subject of an interesting paper by Hu Chhang-Tu (1), who believes it possible to trace the catastrophe of 1855 directly to preceding bureaucratic inefficiency. Hou Jen-Chih (1, 2) has written on the outstanding engineers of the earlier period, Chin Fu (d. +1692) and Chhen Huang (d. +1688).

[b] Cf. Vol. 1, p. 109, Vol. 2, p. 138.

[c] See especially his pp. 145 ff., 153 ff., 162, 165. Wang Mang was killed in +23 after half the population had been affected by floods and famine.

[d] Chang Hsin-Chêng (2) tells us that there are in fact historical grounds for the mention of Chin-shan, the island in the Yangtze two miles north-west of Chinchiang. Both in +713 and in +1539 violent windstorms (and probably seismic disturbances) shifted the channel of the great river for a few days only. This was recorded by Lang Ying in his *Chhi Hsiu Lei Kao*, ch. 2, (p. 44). Hence perhaps further food for legend.

[e] There is now a useful review of the history of soil and water conservation in China by Hsin Shu-Chhih (2). See also Schafer (12).

[f] *Mêng Tzu*, I, (1), iii, 3; tr. Legge (3), p. 6, mod. auct.

[1] 白蛇傳   [2] 法海   [3] 斧斤以時入山林材木不可勝用也

Table 69. *Changes of course of the Yellow River* (see Fig. 859)

Stability periods, with dates of sudden changes	No. of years in each period	Line of flow	Identification no. on Map	References to the sudden changes which opened or closed periods
Antiquity to −602	?	Much to the north of the present bed ⑧, joining the later line of the Grand Canal about Chhing-hsien[1] some distance south of mod. Tientsin and then flowing out through the Hai-ho estuary. One branch deviated eastwards, crossed the line of the Grand Canal between Tungkuang[2] and Têchow,[3] and emptied into the sea near the subsequent mouth of ③ₐ, i.e. probably along the line of the present Ssu-nü-ssu Ho[4]	⓪	Ch, p. 4; Ts, pp. 127 ff.; *TH*, p. 129; Maspero (28), p. 342 discussing *Shui Ching Chu*, ch. 5, p. 16a
−602 to +11*	613	Also to the north of the present bed ⑧, but more easterly than ⓪, deviating from ⓪ near Hua-hsien,[5] crossing the later line of the Grand Canal south of its sharp bend near Lin-chhing,[6] and returning to that line again about the point where the branch of ⓪ had crossed it, i.e. south of Tungkuang[2]	①	Ch, p. 9; Ts, p. 256; *TH*, p. 620; Bielenstein (2)
		This period includes the great break-outs of −168 and −132 mentioned on p. 232, when there was overflow into the Huai valley (see courses ⑤, ⑥ and ⑦ below)		Ts, pp. 244 ff.
+11 to +1048	1,037	Almost along the present course ⑧, but running parallel some distance north of it, and deriving from ① near Ta-ming[7] and Phuyang.[8] Its vestige is the present Hsi-hsieh Ho[9]	②	Ts, p. 344
+11 to +70	59	A downstream derivation from ①, from the neighbourhood of Ta-ming[7]	②ₐ	
+70 to +1048	978	A derivation further upstream from ①, passing near Phuyang[8]	②ᵦ	Ch, p. 25; Ts, p. 321; *TH*, p. 1607
+893 to +1099	206	A new deviation to the present course ⑧ east of Chêngchow, followed by an entirely new channel straight to the sea diverging from the present course ⑧ east of Khaifêng and running approximately parallel with ② but some distance north of it. Perhaps the present Ma-chia Ho[10] is its vestige	③	Ch, p. 25; Ts, p. 353; *TH*, p. 1607
+1060 to +1099	39	A long additional deltaic channel of ③ formed, running for some distance west of the old bed of ①, and emptying into the sea north of ③. Perhaps the present Ssu-nü-ssu Ho[4] is its vestige. A new feeder also tapped the residual flow in ① upstream near Ta-ming[7]	③ₐ	
+1048 to +1194	146	Reversion north-westwards (entirely so after +1099) into the later line of the Grand Canal, new beds forming south-west of Phuyang[8] and between Ta-ming[7] and Lin-chhing[6]	④	Ch, pp. 19, 28; Ts, p. 396; *TH*, p. 1607
+1194 to +1288	94	A new deviation to the present course ⑧ from the points of origin of ③ and ④, but after a short distance the river ran north of the present course, crossed over it and directed itself straight at the Shantung mountain massif. Upon meeting the hill of Liangshan[11] (south of Shouchang[12] and Chang-chhiu[13]) it divided into two almost equal flows and ran out both north and south of Shantung. In the north it	⑤	Ch, p. 30; Ts, p. 426; *TH*, p. 1607

Date	No.	Description		Reference
+1288 to +1324	36	ran west of the later line of the Grand Canal, then joined the present course ⑧. In the south it discharged into the sea north-east of Huaiyin¹⁴ The river had already adopted this double vent in a calamity in +1077, but had been induced to return to courses ③ and ④ by artificial blockage of the southern arm	⑥	HTCKM, ch. 7, pp. 50b ff.; TH, p. 1603
		Again a new deviation to the present course ⑧ from the point of origin of ⑤, but then a great overflow south-eastwards upstream of the Khaifêng bend, taking most of the water far to the south to join the Huai River and flow out through the Hung-tsê Lake into the Yangtze. This was partly the line of the new Pien Canal (cf. p. 307)		Ch, p. 30; Ts, p. 428; TH, p. 1728
		Some flow continued in ⑤ until +1495, however	⑦	Ch, pp. 42, 91; Ts, p. 478
+1324 to 1855	531	South-eastwards from the Khaifêng bend, entirely so after +1495; cutting off the great circuit with Liangshan¹¹ at its apex, joining the former southern bed ⑤ near Hsüchow (thus following approximately the line of the Lunghai Railway), and discharging to the sea north-east of Huaiyin¹⁴ as before. This was close to the route of the ancient Hung Kou Canal (cf. p. 269)		Ch, pp. 42, 91; Ts, p. 574; TH, pp. 1607, 1728, 1818
1855 to the present day	110	Entirely northwards, in the present course, both above and below Khaifêng, with the exception of two periods of devastating flow into the Huai valley	⑧	Ch, p. 91; Ts, p. 574; TH, p. 1607, 1818
1887 to 1889	2	South-eastward discharge along most of the tributaries of the Huai River, through the Hung-tsê Lake and into the Yangtze	⑧a	Ch, p. 95; Ts, p. 584
1938 to 1947	9	The same, but brought about for military reasons by the artificial breaking of the dykes at Chung-mou,¹⁵ just west of Khaifêng. Northern flow was restored after the Second World War and the Huai valley set free for extensive control works (cf. pp. 222 and 224)	⑧a	Anon. (9); Rivière (1); Peck (1); Belden (1); Hsüeh Tu-Pi (1), p. 543; Tung Hsien-Kuang (1), p. 5

KEY TO ABBREVIATIONS

Ch   Chao-Ching (1)
Ts   Tshen Chung-Mien (2)
TH   See p. xli
HTCKM   See p. xxxix

1 靑縣	2 東光	3 德州
4 四女寺河	5 淸縣	6 臨淸
7 大名	8 濮陽	9 徒駭河
10 馬頰河	11 梁山	12 壽張
13 張秋	14 淮陰	15 中牟

* There is reason to think that the change of course ascribed to +11 had actually begun in +2 to +6, and that it involved a temporary discharge into the Huai valley as in −168 and −132; cf. Bielenstein (2), p. 150; Chêng Chao-Ching (1), pp. 10, 193; Tshen Chung-Mien (2), p. 257, commenting on the edict of +70 in *Hou Han Shu*, ch. 2, pp. 15a ff., which Bielenstein translates.

Fig. 872. Deforestation in the mountains, a picture from the *Wang Kung Chung Chhin Lu*, printed about
+1590 (from the album of Chêng Chen-To, 5). The artist, Li Wên, was one of the foremost wood-block
illustrators of his time. On the procuration of timber see Yang Lien-Sheng (11), pp. 38 ff.

Elsewhere Mencius has an interesting passage comparing deforestation to the forcible
debauchery of man's natural goodness.[a]

Mêng Tzu said: 'The trees of Niu Shan were once beautiful, but as it was situated near the
borders of a large State, they were hewn down with axes and hatchets. How could the forests
retain their beauty? Still, through the ceaseless activity (of vegetal forces) day and night, and
the fertilising influences of the rain and the dew, they were not without buds and sprouts
springing forth, but then came cattle and goats to browse upon them. To these things is
owing the bare and stripped appearance (of the mountain), and people seeing it, suppose that
it was never finely wooded. But is this the nature of the mountain? (So also) of what properly
belongs to man—shall it be said that the mind (of any man) was without benevolence and
righteousness? The way in which a man loses his proper goodness of mind is like the way in
which the mountain is denuded of trees by axes and hatchets. Hewn down day after day, how
can (the mind) retain its beauty?'

This shows that Mencius regarded the denuded state as artificial and ominous, and the
passage has the added interest that the factor of over-grazing is clearly recognised. The

---

[a] *Mêng Tzu*, VI, (1), viii, 1, 2; tr. Legge (3), p. 283, mod. auct.

destruction of forests must have been severe in ancient times on account of the practice of burning clearings before planting ('*milpa* agriculture', cf. Sect. 41 below). To this the *Huai Nan Tzu* book, in one of its diatribes against the luxurious decadence of feudalism as opposed to the communalism of the golden age, adds deforestation for the needs of metallurgical fuel.[a]

Whole forests were burned for the chase, great tree-trunks being scorched and charred. Bellows were violently worked to send the blast through the tuyères in order to melt the bronze and the iron; metals flowed forth wastefully for hardening and forging—the work did not cease for a single day.[b] No tall trees were left on the mountains, and the silkworm-oaks (*chê*[1])[c] and lindera trees (*tzu*[2])[d] disappeared from the groves. (Untold amounts of) wood were burnt to make charcoal, and (great quantities of) plants turned to white ash in bonfires (for potash), so that the anise (*mang tshao*[3])[e] and the jasmine (*pai su*[4])[f] could never reach their perfection. Above (the smoke) obscured the very light of heaven, and below the riches of the earth were utterly exhausted. All this (devastation) was due to (the extravagant) use of fire.

By the sixteenth century, however, the direct relationship of denudation, erosion and flood problems was well recognised. The Ming scholar, Yen Shêng-Fang,[5] wrote:[g]

Before the Chêng-Tê reign-period (+1506 to +1521) flourishing woods covered the south-eastern slopes of the Shang-chih and Hsia-chih mountains (in the Chhi[6] district of Shansi). They were not stripped because the people gathered little fuel. Springs flowed into the Pan-to stream, and passing in long waves and powerful sweeps through the villages of Lu-chi and Fên-cha, entered the Fên River at Shangtuan-to as the Changyuan River...It was never seen dry at any time of year. Hence villages from afar and in the north of the district all cut branch canals and ditches which irrigated several thousand *chhing* of land. Thus Chhi became prosperous.

But at the beginning of the Chia-Ching reign-period (+1522 to +1566) people vied with each other in building houses, and wood was cut from the southern mountains without a year's rest. Presently people took advantage of the barren mountain surface and converted it into farms. Small bushes and seedlings in every square foot of ground were uprooted. The result was that if the heavens send down torrential rain, there is nothing to obstruct the flow of the water. In the morning it falls on the southern mountains; in the evening, when it reaches the plains, its angry waves swell in volume and break through the embankments, frequently changing the course of the river...Hence the district of Chhi was deprived of seven-tenths of its wealth.

This region of Shansi lies south of Thaiyuan and a good deal south of the Wu-thai Shan mountains to which reference was made a little earlier.

[a] Ch. 8, p. 10*a*, tr. auct. adjuv. Morgan (1), pp. 95 ff.; cf. Vol. 2, pp. 100 ff., 127 ff., Vol. 3, p. 609.
[b] Cf. Vol. 4, pt. 2, pp. 135 ff.; and Sect. 30*d* in Vol. 5.
[c] *Cudrania* spp., R 599.   [d] *Catalpa* spp., R 98. Cf. pp. 414, 645.
[e] *Illicium religiosum*, R 505. Well known in antiquity as a valuable insecticide, cf. Needham & Lu Gwei-Djen (1), also Sect. 38 below.
[f] Perhaps *Jasmimum* spp., R 178.
[g] *Shansi Thung Chih*,[7] ch. 66, p. 31*a*, tr. Chi Chhao-Ting (1), p. 22. Here we can go no further into forestry matters, and I will only mention the reviews of Têng Shu-Chun (1) and F. Y. Chang (1) on the history of forestry in China. Cf. Sect. 41 in Vol. 6.

[1] 柘   [2] 梓   [3] 莽草   [4] 白素   [5] 闇繩芳   [6] 祁
[7] 山西通志

However the balance swung between frugality and extravagant use of natural resources during the successive centuries, the Chinese farmers at any rate had always been very conscious of the need for dexterous utilisation of soil moisture. Here we must not anticipate what we shall have to say in Section 41 on agriculture, but a few words on pit-cultivation and terracing are indispensable, for everything that retained water in the upland soils helped the water-conservancy problem of the lower river-valleys. The climate of north China consists of a dry and windy spring, a hot dry summer with showers at long intervals, a very wet late autumn bringing two-thirds of the whole year's precipitation and causing great erosion if not serious floods, then finally a severe dry winter with wind-borne snow.[a] Farming in these conditions could only be successful if everything possible was done to conserve water in the soil. And indeed the oldest agricultural books, such as the *Fan Shêng Chih Shu*[1] ('The Book of Fan Shêng-Chih on Agriculture),[b] written about −10, and the *Chhi Min Yao Shu*[2] (Important Arts for the People's Welfare), written by Chia Ssu-Hsieh[3] about +540, are full of recommendations of means of collecting and conserving soil moisture. The right times for ploughing, hoeing and harrowing, the ways of retaining snow and even dew, 'dust mulch' and leafage canopy, all are intelligible in terms of reducing evaporation and retaining water.

One of the oldest systems for doing this on sloping ground was pit-cultivation (*ou thien*[4]). Fan Shêng-Chih says:[c]

During the reign of the emperor Thang[d] there was a long and severe drought. (One of his ministers), I-Yin,[5] (therefore) developed the *ou thien* system (i.e. cultivation of crops in shallow pits and ditches), and taught the people to treat the seeds and carry water for irrigating the crop. Cultivation in shallow pits depends mainly on the fertilising power of the soil (*fên chhi*[6]), so good land is not at all necessary. Mountain-sides, the edges of cliffs, steep places near villages, and even the inside slopes of city ramparts, all can be used for making shallow pits...Pit-cultivation can start directly on waste land, without need for any (other) preparatory work.

And the text goes on to give in elaborate detail the dimensions of the pits, usually some six inches deep, and the ditches running along the contours, adding figures for the number of them which could conveniently be made on different kinds of land. The pits were to be well manured as well as watered, and for some crops, such as melons, an earthen jar to be kept full of water was buried in the centre of each pit. Often mixed cultivation was practised, scallions or beans being planted in the pits along with the

[a] This description, with much else in the paragraph, comes from Shih Shêng-Han (1), pp. 40 ff.

[b] Fan Shêng-Chih's *floruit* was −32 to −7; he did outstanding service as one of the Instructors on Agricultural Affairs (an officially established position) and ended as an Imperial Censor.

[c] *YHSF*, ch. 69, pp. 54b ff., tr. Shih Shêng-Han (2), pp. 29 ff.; cf. pp. 63 ff. Fan's description of the *ou thien* system was extensively quoted in the *Chhi Min Yao Shu*, ch. 3, pp. 10b ff.; see Shih Shêng-Han (1), pp. 21, 52 ff.

[d] The semi-legendary first emperor (or high king) of the Shang dynasty, to be placed, in so far as historical, in the −16th century.

[1] 氾勝之書   [2] 齊民要術   [3] 賈思勰   [4] 區田   [5] 伊尹
[6] 糞氣

gourds.[a] The value of such devices for the retention of water and plant nutrients is obvious.

Clearly the *ou thien* system partook of the nature of very small terraces, and must have contributed greatly to the reduction of rapid run-off. Terracing itself (*thi thien*,[1] 'laddered fields'), with stone or earth walls, was simply a great extension of the same principle (Fig. 873, pl.). We suspect that this system was very ancient, perhaps prehistoric, in the south and south-west, and that it reached the north and north-west during the Middle Ages. The term itself first appears, according to Hsin Shu-Chhih (2), in Fan Chhêng-Ta's[2] *Tshan Luan Lu*[3] (Guiding the Reins—a narrative of his three month's journey from the capital to Kueilin in +1172),[b] but the system is described in Chhen Fu's[4] *Nung Shu*[5] (Treatise on Agriculture)[c] in +1149. It may also be inferred from landscape descriptions in Thang poems such as those of Tu Fu[6] and Chang Chiu-Ling[7] on Szechuan and Chiangsi respectively. This must mean that it was already widespread in the +8th century, and probably then extending northwards.

(4) ENGINEERING AND ITS SOCIAL ASPECTS IN THE CORPUS OF LEGEND

Ta Yü chih shui[8]—Yü the Great controlled the waters—such is the phrase which has become proverbial, and which one encounters all over China.[d] Probably no other people in the world have preserved a mass of legendary material into which it is so clearly possible to trace back the engineering problems of remote times. The essentials of the story are that two culture-heroes of civil engineering were successively put in charge of regulating and controlling the floods and rivers by the legendary emperors; the first, Kun, failed, but the second, Yü, succeeded. Many of the incidental features of the accounts which have come down to us are so revealing that they are well worth examining in detail.[e]

In the time of the legendary emperor Yao[9] there were great floods,[f] and Kun,[10,11] recommended by Ssu Yo[12] (Four-Mountains[g]), was put in charge of all works of

---

[a] Fan Shêng-Chih gives surprisingly high production figures for this kind of horticultural crop-raising, but they have been confirmed experimentally in our own time. See Wan Kuo-Ting (3).

[b] P. 14*b*.   [c] Ch. 2. Cf. Rudofsky (1), figs. 28, 30.

[d] I had an opportunity of meditating on it in the beautiful temple dedicated to Yü the Great at Lichuang in Szechuan, which directly overlooks the Yangtze, and housed at the time the headquarters of Tongchi University (1943).

[e] Granet (1), p. 483, serves well as the focal point of our investigations. But it is indispensable to be acquainted with the revolutionary modern studies of Ku Chieh-Kang, Ting Wên-Chiang and other scholars in *KSP*, vol. 1. A more recent summary of Chinese views will be found in the book of Hsü Ping-Chhang (1).

[f] *Shih Chi*, ch. 1 (Chavannes (1), vol. 1, p. 51); based on *Shu Ching* (ch. 1, Yao Tien), tr. Karlgren (12), p. 3, and *Shu Ching* (ch. 5, I Chi), tr. Karlgren (12), p. 9, cf. Medhurst (1), p. 65. Maspero (8), p. 47, sounds a caution against analogising in any way these legends with the Hebrew flood-stories derived from Babylonia. However, the masterly comparative analysis of flood-legends by Frazer (2) has attracted Chinese interest, and was translated as an appendix to Hsü Ping-Chhang's book.

[g] Commentators debate whether this means one or four persons or spirits, or an official title or titles. No one really knows.

[1] 梯田	[2] 范成大	[3] 驂鸞錄	[4] 陳旉	[5] 農書
[6] 杜甫	[7] 張九齡	[8] 大禹治水	[9] 堯	[10] 鯀
[11] 鮌	[12] 四岳			

protection and control.[a] But after nine years his efforts were unsuccessful;[b] the water rose as fast as he could build his dykes. On account of this complete failure, Kun's work was not considered praiseworthy,[c] he was exiled,[d] killed by Yao[e] and his body cut into pieces. Very significantly, Kun was later regarded as the patron and inventor of dykes, embankments and walls.[f] He was supposed to have received his instruction from a kite and a tortoise.[g]

Afterwards, Yü,[1] the son of Kun (born from a grain or a stone), was appointed by Shun[2] (successor of Yao),[h] also on the recommendation of the Ssu Yo, and in the space of thirteen years by titanic labours he opened the courses of the nine rivers, conducting them to the four seas and deepening the canals.[i] The motive of dredging the beds of waterways is always connected with the work of Yü. That he passed the door of his house many times during this period but never once went in to rest, is a favourite embellishment of his saga of devotion, classical formulations of which are found in Mencius.[j] Yü was supposed to have been fully successful in his operations, and came to occupy the permanent position of irrigation culture-hero which Kun had not attained. Other technical patronages accreted around him; thus he was a map-maker,[k] a maker of bronze weapons,[l] a fighter against pestilence,[m] and so on.[n]

Besides these two chief characters there was a subsidiary figure connected with hydraulic engineering, namely Kung-Kung[3] (literally 'communal labour'), whose other name was Chhui[4] (a word related to *chhui*,[5] the tuyère of a metallurgical furnace or forge). He was proposed, by Ssu Yo as usual, to Yao for the office of controller of the waters,[o] and also recommended by Huan-Tou[6],[p] but rejected by Shun. Finally he was banished and killed.[q]

There is method in all this, if rightly interpreted. Much insight was shown by Granet when he suggested that the contrast between Kun and Yü betrays to us the

---

[a] Chavannes (1), vol. 1, pp. 51, 67.

[b] *Shih Chi*, ch. 1 (Chavannes (1), vol. 1, pp. 51, 67, 98); based on *Shu Ching* (ch. 1, Yao Tien), tr. Karlgren (12), p. 3, cf. Medhurst (1), p. 10, and *Shu Ching* (ch. 24, Hung Fan), tr. Karlgren (12), p. 29; Legge (1), p. 139.

[c] *Shih Chi*, ch. 1 (Chavannes (1), vol. 1, p. 99); *Shu Ching* (ch. 2, Shun Tien).

[d] *Chu Shu Chi Nien*, ch. 1.

[e] *Shu Ching* (ch. 2, Shun Tien), tr. Karlgren (12), p. 5, cf. Medhurst (1), p. 27.

[f] *Lü Shih Chhun Chhiu*, ch. 94 (vol. 2, p. 45), also *Huai Nan Tzu*, ch. 1, p. 4b (tr. Morgan (1), p. 7).

[g] According to the *Thien Wên* ode of Chhü Yuan (Maspero (8), p. 48, cf. Hawkes (1), p. 48).

[h] *Shih Chi*, ch. 2 (Chavannes (1), vol. 1, p. 99); based on *Shu Ching* (ch. 2, Shun Tien), tr. Karlgren (12), p. 5.

[i] *Shih Chi*, ch. 2 (Chavannes (1), vol. 1, pp. 100, 154); based on *Shu Ching* (ch. 5, I Chi), tr. Karlgren (12), p. 9.

[j] *Mêng Tzu*, III, (2), ix, 3, 4 (Legge (3), p. 155); III, (1), iv, 7 (Legge (3), p. 126). Cf. *SCTS*, ch. 5, p. 14b.

[k] Granet (1), p. 482.   [l] Granet (1), p. 503.

[m] Granet (1), p. 486.

[n] Yü is the earliest of the legendary figures for whom there is any historical documentary evidence, as he is referred to in the *Shih Ching* (Mao nos. 210, 261; cf. Waley (1), pp. 212, 146). All those traditionally supposed to be earlier were in fact invented much later. Quite probably Yü was originally a god of the soil who piled up the mountains and arranged the courses of the rivers.

[o] *Shih Chi*, ch. 1 (Chavannes (1), vol. 1, p. 50); cf. Granet (1), p. 520.

[p] *Shih Chi*, ch. 1 (Chavannes (1), vol. 1, p. 67). Cf. Vol. 2, p. 117.

[q] *Shu Ching* (ch. 2, Shun Tien), tr. Karlgren (12), p. 5; Medhurst (1), p. 27.

[1] 禹   [2] 舜   [3] 共工   [4] 倕   [5] 錘   [6] 驩兜

existence of two rival schools of hydraulic engineering thought in ancient China.[a]
A conflict has existed throughout Chinese history, as he said, between the partisans of
high dykes and the partisans of deep channels.[b] Moreover, it took the form of a
conflict between two systems of morality, one in favour of confining and repressing
Nature, the other in favour of letting Nature take her course, or even assisting her to
return to it if necessary. This has only to be stated to show itself immediately in
consonance with much that we have already come across. Confucian jurists used the
analogy of dykes and embankments when discussing law.[c] The forceful repression of
Nature by the erection of convex 'masculine' ridges along the rivers was a case of what
the Taoists called *wei*[1] as opposed to *wu wei*[2] ('no action contrary to Nature'; cf.
Vol. 2, pp. 68 ff. above).[d] The deepening of river beds by excavation of 'feminine'
concavities was, on the contrary, a going along with Nature, an epiphany of some of the
most numinous Taoist archetypes, such as the 'Valley Spirit' and the female receptive-
ness of water (Vol. 2, pp. 57 ff. above). This appears rather clearly in a long discussion
in the *Kuo Yü* (Discourses concerning the Warring States);[e] Kun and Kung-Kung
applied force to Nature, hence their punishment, Yü the Great adopted the right
methods. That these methods were essentially Taoist throws further light on the role
played by that great sect in all early Chinese scientific and technological advances.

Not every school of engineering thought has had its favourite maxims or discovered
laws engraved on stone and venerated for centuries. Yet such is the case for the followers of Yü the Great. At the town of Kuanhsien,[3] some distance to the north-west of
the capital of Szechuan, Chhêngtu, there exists what must be one of the most remarkable irrigation works in the world.[f] A great river, the Min Chiang,[4] comes tumbling
out of the mountainous country of the Tibetan borderland, and is diverted at certain
seasons of the year by movable dams and spillways into an enormous cutting through a
hillside, forming thereby an artificial river which fertilises half a million acres of first-
class agricultural land by means of some 735 miles of irrigation channels. A project of
such magnitude, visible even on very large-scale maps of all China, will demand
presently a page or two to itself (p. 288); here the point to be made is that it dates from
the middle of the −3rd century, having been started by the then governor of the
province, Li Ping,[5] and completed by his son Li Erh-Lang.[6] In the temple dedicated to
the latter, which overlooks the famous suspension bridge (cf. p. 192 above), there are a
series of stone-cut inscriptions (Fig. 874, pl.) probably of no great age in themselves,

[a] (1), p. 484.
[b] Similar perplexities are quite easy to find in all ages. At the time of the draining of the East Anglian
fens, around +1630, two views came into sharp conflict. Some thought that the existing water-courses
should be deepened and embanked; others, including Sir Cornelius Vermuyden, that new and artificial
straight cuts, like the New Bedford Level, should be made. See L. E. Harris (1); Darby (1); Steers (1).
[c] Cf. Vol. 2, pp. 544, 256.
[d] This was the gravamen of the semi-legendary self-reproach of Mêng Thien, the general of Chhin
Shih Huang Ti; in organising the building of the Great Wall, he must have 'cut through the veins of the
earth'. See p. 53 above.
[e] *Chou Yü*, ch. 3, pp. 8*b* ff.; Prince Ling, 22nd year.
[f] There are many descriptions, e.g. Hutson (1); Lowdermilk (2); cf. below, p. 288.

[1] 爲   [2] 無爲   [3] 灌縣   [4] 岷江   [5] 李冰   [6] 李二郎

but perpetuating, as we know, certain key phrases some of which are attributable to the Chhin engineer himself. Of these the oldest is the following: *Shen Thao Than; Ti Tso Yen*,[1] i.e. 'Dig the channel deep, and keep the dykes (and spillways) low'.[a] The second may be as old, or nearly so: *Fêng Chih Chhou Hsin; Yu Wan Chieh Chio*,[2] i.e. 'Where the channel runs straight, dredge the centre of it; where it curves, cut off the corners'.[b] Again the teaching of Yü, and measures against erosion at bends. The third is probably a later addition: *Khuan Chhi Ti; Hsieh Chieh Mien*,[3] i.e. 'Make the (canal) beds broad, with gradually sloping profiles'.[c]

But now, returning to the legends, we come upon a very curious thing. The unsuccessful, or at any rate disapproved, irrigation engineers are identical with the corps of legendary rebels which we discussed at some length in the Section on Taoism (above, Vol. 2, pp. 115 ff.). Kung-Kung, always one of them, is recommended by Huan-Tou, another; and Kun himself is frequently identified[d] with Thao-Wu.[4] The suggestion made there was that all these (and related) figures, who fight with the legendary emperors and are overcome by them, constitute the fabulous residuum of those leaders of the people in primitive collectivist society who most strongly opposed the first institution of bronze-age proto-feudalism. It was therefore quite natural that some of them should have become heroes for the Taoists in later ages, since the Taoists opposed feudalism root and branch, urging a return to the collectivist golden age. Consequently it is of great interest that Chhü Yuan, in the *Li Sao*[e] and especially the *Thien Wên* odes[f] (*c.* −300), strongly takes the part of Kun, commiserating with him and saying that his death was unjust, having met with failure through no fault of his own.[g] Other examples could be adduced. We are thus led to the suggestion that perhaps the legend of the failure and death of Kun conceals the real failure of primitive collectivist society to cope with the greater problems of river-control and irrigation. Presumably that form of social order was unable to organise the human labour-force, and bring it to bear upon the urgent engineering tasks in view, as effectively as the *corvée* system under feudal society. If this were so, we might expect to find in the legends distinct traces of the connection of Yü with the origins of feudalism, and this is in fact the case. In the I Chi chapter of the *Shu Ching*,[h] Yü is made to say, for example, that as the result of his work, 'the ten thousand states have become well-governed (*wan pang tso i*[5])'; and that 'wherever I went, I established the five (classes of) chiefs (feudal lords) (*hsien chien wu chhang*[6])'. This point has already been appreciated by

---

[a] This is known as the 'Six-Character Teaching' (*Liu Tzu Chüeh*[7]). It was re-engraved in stone by imperial order in +972. Another photograph of it is in Phan Ên-Lin (1), p. 158.

[b] This is known as the 'Eight-Character Rule' (*Pa Tzu Ko*[8]). Some of the other inscriptions date from the time of the engineer Lu I[9] (*c.* +1510).

[c] Cf. above, p. 231, on the optimum cross-section for silt-carriage.

[d] Granet (1), p. 240.

[e] Lin Wên-Chhing tr., p. 86; Hawkes (1), p. 26.

[f] Hawkes (1), pp. 48, 50.   [g] Granet (1), p. 266.

[h] Karlgren tr. (12), pp. 9, 12.

¹ 深淘灘低作堰   ² 逢直抽心遇灣截角   ³ 寬砌底斜結面   ⁴ 檮杌
⁵ 萬邦作乂   ⁶ 咸建五長   ⁷ 六字訣   ⁸ 八字格
⁹ 盧翊

H. Wilhelm in so far as he wrote that the Yü legend incorporates two basic features of Chinese agrarian society, the regulation of waterways and the use of organised *corvée* labour for accomplishing it.[a] If such interpretations are correct, the conclusion might well follow that the transition from primitive collectivism to proto-feudalism took place under the same environmental compulsions as the transition later on from proto-feudalism to feudal bureaucratism. For the tasks of hydraulic engineering are set by geography itself, and the interconnected aims of relief from flooding, maintenance of water-supplies for irrigation, and attainment of convenient means of bulk transport, always tended towards strong centralised government as the only effective instrument. What the primitive 'sand-heap' (to appropriate Sun Yat-Sen's phrase) could not do, the forced levies of proto-feudal 'high kings' and their enfeoffed lords could; and what they in turn were incapable of accomplishing could be done by the high officials of the united empire from Chhin times onward, with their maps and surveys and developed techniques—in so far as it could be done at all before the advent of post-Renaissance technology. It is indeed tempting to read this significance into the character *jun*[1] which means to irrigate, where we see the water radical attached to the combination of 'gate' and 'king' as phonetic. This alone, also pronounced *jun*, means an intercalary month, i.e. something which was certainly given out by royal decree. There is no doubt that all three components are derived from ancient graphs with these meanings, but before accepting the general semantic significance of 'water flowing through gates regulated by kings' one would have to be sure that the word was not a corruption of something quite different.[b]

A few minor points of interest may be added here. The great cut at Kuanhsien reminds us that large cuttings occur also in the legends.[c] Yü is supposed to have made them or superintended their making[d]—the Lung Mên[2] gorge of the Yellow River, and the gorges of Mêng Mên[3] and Lü Liang.[4] Many place-names are associated in this way not only with Yü but with Kun.[e] A particularly striking legend of a great cut is that of the separation of the mountains Thai-Hsing[5] and Wang-Wu[6] by the giant sons of Khua O[7] (an otherwise unknown character, perhaps identical with the 'Boaster', Khua Fu[8]).[f] The story as it is given in the *Lieh Tzu* book is worth reproducing if only to indicate something of the indomitable spirit with which the Chinese of old approached their great engineering projects.[g] For men in community nothing was impossible.

[a] (1), p. 23.

[b] K12510, *p*, is no help, but the point is interesting in connection with the first invention and use of sluice-gates (see pp. 263, 349 below).

[c] Granet (1), p. 484.

[d] Cf. *Shih Tzu*, ch. 1, p. 19*a*; *Huai Nan Tzu*, ch. 8, p. 17*a*; *Lü Shih Chhun Chhiu*, ch. 25 (vol. 1, p. 51); discussed by Maspero (8), p. 51.

[e] Maspero (8), p. 71.

[f] *Shan Hai Ching*, ch. 8. Khua O would mean something like 'Ant King', a name which might connect not unreasonably with the message of the following quotation.

[g] Ch. 5, p. 9*a*, tr. L. Giles (4), p. 86; R. Wilhelm (4), p. 51; mod. auct. It was quoted with great effect by Mao Tsê-Tung in one of his speeches in 1945; see (2), vol. 4, p. 316.

[1] 潤   [2] 龍門   [3] 孟門   [4] 呂梁   [5] 太形   [6] 王屋

[7] 夸蛾   [8] 夸父

The two mountains Thai-Hsing and Wang-Wu, covering an area of 700 square *li* and rising to an enormous height, originally stood to the south of Chichow and north of Hoyang. The Simpleton of the North Mountain, an old man of ninety, dwelt opposite these hills, and was vexed in spirit because their northern flanks blocked the way to travellers, who had to go all the way round. So he called his family together and put forward a plan. 'Let us', he said, 'put forth our utmost strength to cut through this obstacle, and make a passage through the mountains to Hanyin. What say you?' They all assented except his wife, who made objections, saying: 'My good-man has not the strength to sweep away a dunghill, let alone two such mountains as Thai-Hsing and Wang-Wu. Besides, where will you put all the earth and stones that you dig up?' The others replied that they would throw them on the Pho-Hai promontory to the north of the Dark Lands.

So the old man, followed by his son and grandson, sallied forth with their pickaxes, and the three of them began hewing away at the rocks, cutting up the soil, and carting it away in baskets to the Pho-Hai promontory. A widow living near by had a little boy, who came skipping along to give them what help he could, though he was only just shedding his milk teeth. Engrossed in their toil, they never went home except once at the turn of the season.

The Wise Old Man of the River-bend burst out laughing and urged them to stop. 'How absurd is your behaviour!' he said. 'With the poor remaining strength of your declining years you will not succeed in removing a hair's breadth of the mountain, much less the whole vast mass of rock and soil.' The Simpleton sighed and said: 'Surely it is you who are hard of heart and narrow-minded. You are not worthy to be compared with this widow's son, despite his puny strength. Though I myself must die, I leave a son behind me and through him a grandson. That grandson will beget sons in his turn, and those sons will also have sons and grandsons. With all this posterity, my line will not die out, while on the other hand the mountain will receive no increment or addition. Why then should I despair of levelling it to the ground at last?' In reply to this, the Wise Old Man of the River-bend had nothing to say.

One of the serpent-brandishing deities[a] heard of the undertaking, and fearing that it might never be finished, went and told the emperor (of Heaven) who, touched by the old man's sincerity, commanded the two sons of Khua O to transport the mountains, one to the extreme north-east, the other to the southern corner, of Yung. Ever since then, the region lying between Chi in the north and Han in the south has been an unbroken plain.

There were other local culture-heroes besides this Pei Shan Yü Kung.[1] There was Thai-Thai[2] of Shansi, who made the Fên[3] and Thao[4] Rivers run properly, and became their tutelary god.[b] To this day there is a Thai-Thai Tsê[5] (marsh, now reclaimed) near Thaiyuan. The female pseudo-creator spirit, Nü-Kua, [6,7] also comes into

---

[a] Commentators refer to the *Shan Hai Ching*.

[b] The *locus classicus* for him is the *Tso Chuan*, Duke Chao, 1st year (Couvreur (1), vol. 3, pp. 30 ff.), i.e. −540. At that time he was supposed to be a disease-causing spirit, responsible for an illness of the Prince of Chin. Kungsun Chhiao[8] (cf. Vol. 2, pp. 365, 522) explained his story, saying that he had been the son of the hydraulic engineer Mei,[9] one of the descendants of the legendary emperor Shao Hao. Thai-Thai in his turn had constructed great public works, as a reward for which he was ennobled by the legendary emperor Chuan Hsü as Lord of Fên-Chhuan. Thus in time he became the spirit of the Fên River. Characteristically Kungsun Chhiao rejected any intervention of spirits in the prince's illness, attributing it to faulty hygiene and regimen. The story of Mei and Thai-Thai has obvious parallelism with that of Kun and Yü, though the son rather than the parent shares the sinister colour of Kun and Kung-Kung. Today Thai-Thai has a shrine in the great Taoist temple of Chin Tzhu south-west of Thaiyuan, and there I encountered him in 1964. See further Maspero (8), pp. 51, 73.

[1] 北山愚公　　[2] 臺駘　　[3] 汾　　[4] 洮　　[5] 臺駘澤　　[6] 女媧
[7] 女娃　　[8] 公孫僑　　[9] 昧

the picture, since she used the ashes of reeds (*lu*[1]) for making dykes; this is perhaps an echo of some clay-binding technique.[a]

Relations with afforestation in the legends are also curious. Ancient ideas of the value of vegetation-covered hills have already been noticed (pp. 241 ff. above). But the traditions make Yü the 'attacker' of forests. The *Shan Hai Ching* has him 'attacking' the Clouds and Rain Mountains (Yün-Yü chih Shan[2])[b] and the Chhêngchow[3] Mountains;[c] all commentators suppose that this means cutting down trees, or at least marking them for felling. The Yü Kung chapter of the *Shu Ching* opens with the statement that he went along the mountains and felled trees (*sui shan li mu*[4]).[d] Conceivably forest-removal may have been thought to reduce rainfall and so to make the task of the engineers easier lower down, though it is hard to believe that Brückner's principle could have been in the minds of the ancients.[e] The fearful erosion produced was far more important than any slight changes in rainfall which could result from deforestation. One senses a conflict of opinion in the *Tso Chuan*,[f] when in −525 envoys sent to make a rain-sacrifice in time of drought on Sang Shan[5] cut down many trees; the rationalist statesman Tzu-Chhan (Kungsun Chhiao[6]) said that this was a great evil and had them punished. Yet later, Chhin Shih Huang Ti, who had conceived a grudge against Mount Hsiang, had it completely deforested in −219.[g]

Another peculiar feature of the legends is the relation of Kun and Yü to some kind of special earth, 'living earth' (*hsi thu*[7] or *hsi jang*[8]), which they used for their dykes. Kun was supposed to have stolen it, or not to have known how to use it; Yü was able to use it successfully. Granet thought that it referred to some sort of inexhaustible clay-pits;[h] H. Wilhelm aligned it with the Prometheus legend, earth taking the place of fire.[i] Maspero adduces late authors who suppose that it was a kind of earth which swelled of itself, but this may be only a piece of symbolism comprehensible in view of what has already been said.[j] The subject merits further investigation.[k]

All the above is not incompatible with Maspero's conviction that the legends of floods and their control have elements of creation myths, such as he himself transcribed

[a] Cf. the use of these particular ashes in connection with the pitch-pipes, Vol. 4, pt. 1, pp, 188 ff. above. Maspero (8), pp. 52, 74.

[b] Ch. 15, p. 4*b*. [c] Ch. 17, p. 2*b*.

[d] Karlgren (12), p. 12; repeated in *Shih Chi*, ch. 2 (Chavannes (1), vol. 1, pp. 100, 154).

[e] Chatley (25). When evaporation on the windward side is the chief cause of rain, deforestation will to some extent reduce rainfall.

[f] Duke Chao, 16th year (Couvreur (1), vol. 3, p. 272).

[g] *Shih Chi*, ch. 6 (Chavannes (1), vol. 1, p. 156).

[h] (1), p. 485. [i] (1), p. 21.

[j] (8), p. 49. Cf. Bodde (21), pp. 399 ff. See also Sect. 38 below.

[k] One wonders whether there could be any connection between these ancient traditions and some of the engineering studies now being made which show that loess soil is an excellent dam material if used in the right way. At each level a network of small dykes is built, and into these 'ponds', containing about 2 ft. depth of water, the wetted loess soil is dumped. Wetting softens the bonds between the particles and causes collapse of the porous soil structure, then as the water evaporates compaction soon follows. If the clay content is low, and sand drains are left in the dam body, quite high dams can be built in this way. For further information see papers such as that of Huang Wên-Hsi & Chiang Phêng-Nien (1).

[1] 蘆 [2] 雲雨之山 [3] 程州之山 [4] 隨山利木 [5] 桑山
[6] 公孫僑 [7] 息土 [8] 息壤

from the oral traditions of the Thai peoples of Annam.[a] They also spoke of one person failing to control the waters, and a second succeeding. It is certainly interesting that both the Mencius passages begin suddenly with Yao and Yü, omitting the earlier stages invented later, and of course the ancient scholars tried as usual to make history out of myth, suppressing as much as possible of the folk-tale improbabilities. But it would be difficult to accept Maspero's view that the practical control of Yellow River floods never had anything to do with the corpus of Chinese legend.

(5) THE FORMATIVE PHASES OF ENGINEERING ART

At an earlier stage, in connection with roads (p. 4 above), we made the acquaintance of the classification of irrigation canals in the *Chou Li*. In his commentary on the entry concerning the Ssu Hsien (Director of Communications) Chêng Hsüan enumerated the five standard categories of canals,[b] and the text itself, when dealing with the

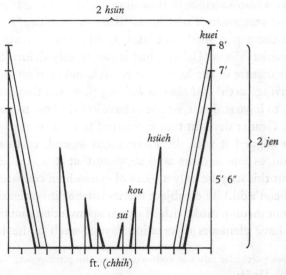

Fig. 875. Dimensions of irrigation canals according to the *Chou Li*. The scale of this diagram in feet is the same vertically as horizontally, and no difficulty of interpretation arises about the intentions regarding the three smaller channels (*sui*, *kou* and *hsüeh*). The largest one, however, the *kuei*, is defined as being two fathoms (*hsün*) broad, and two statures (*jen*) deep; since the latter measure may have been either 5 ft. 6 in., 7 ft. or 8 ft., the depth of the *kuei* may have been defined as 11 ft., 14 ft. or 16 ft. and this is represented on the diagram for these varying estimates of the *jen*.

functions of the Sui Jen (Grand Extensioner, or Minister of Agriculture) reported the same series in parallel with a set of roadways of standard widths.[c] In its Khao Kung Chi, or 'Artificers' Record' chapter, the *Chou Li* enlarges further on hydraulic engineering designs, always keeping to the same range of terms and giving dimensional specifications for each, figures which we find again in Chêng Hsüan's commentary on the second passage just mentioned. From Fig. 875, in which these are diagrammatically

---

[a] (8), p. 60.   [b] Ch. 7, p. 26a (ch. 30), cf. Biot (1), vol. 2, p. 199.
[c] Ch. 4, p. 25a (ch. 15), cf. Biot (1), vol. 1, p. 342.

shown, it seems clear that the literary tradition was rather stylised and schematic, figures being doubled at each step so as to produce cross-sections impracticably deep before the days of concrete.[a] Nevertheless, the text in the 'Artificers' Record', which occurs under the heading of 'Builders' (*chiang jen*[1]), gives interesting details of irrigation technology.[b] These are presumably also to be taken as idealised standard concepts and terms familiar to the Chhi or Han compilers of the book.[c]

It is the Builders who make conduits and canals (*kou hsüeh*[2]). The breadth of the (standard) spade is five tenths of a foot;[d] two of them together make a double cut one foot broad and one foot deep, and this is called a furrow (*chüan*[3]). At the head of a field (*thien*[4]) they make a double one of these, two feet broad and two feet deep, and this is called a ditch (*sui*[5]).

Nine lots of cultivated land are considered a Ching[6] (lit. well). The channels of water between the Ching are four feet broad and four feet deep, and they are known as conduits (*kou*[7]). A square piece of cultivated land with sides ten *li* long is known as a Chhêng[8] (lit. achievement), and its exterior artificial watercourse, eight feet broad and eight feet deep, is known as a small canal (*hsüeh*[9]). A square piece of cultivated land with sides one hundred *li* long is known as a Thung[10] (lit. together), and its exterior artificial watercourse, two fathoms (*hsün*[11]) broad and two statures (*jen*[12]) deep,[e] is known as a medium canal (*kuei*[13]). Only these connect with the great canals (*chhuan*[14]), each of which is given a name of its own.[f]

According to the general constitution of the earth, there will be a (natural) watercourse wherever there is a valley between two hills, and along all such watercourses there will be a way for a road (*thu*[15]). When an artificial canal encounters a rise of ground, we term it a 'point of no flow' (*pu hsing*[16]), and when (planned) watercourses are not in agreement with the principles (of hydraulic engineering) we use the same expression. Canals lying in a straight line, with no (derivatory) branches, should be doubled in breadth every thirty *li*. To reduce the speed of a stream, one arranges that it should flow round bends, and these should be like the two limbs of a ringing-stone (musical instrument, the stone-chimes)[g] (i.e. at an obtuse angle), the relation of the two limbs being as 3 to 5. If one wishes to make a pool (for navigation or as a reservoir) (*yuan*[17]), one gives a circular form to the bed (so that the current will keep it scoured out).

[a] These are by no means the only places in the *Chou Li* which make mention of hydraulic engineering works. For example, the Yung Shih[18] and Phing Shih[19] were both Superintendents of Waterways Police (ch. 10 pp. 3*b*, 4*a* (ch. 37), Biot (1), vol. 2, pp. 379 ff.). The former kept watch on the canals, organised the *corvée* for construction works, guarded hills from deforestation, and prevented the pollution of waterways. The latter organised the *corvée* for urgent flood protection works, kept watch for dangerous places, and protected fisheries out of season. We shall return to these texts in Sect. 44. There was also the Yeh Lu Shih,[20] or Superintendent of Traffic Police (ch. 10, p. 2*a* (ch. 37), Biot (1), vol. 2, pp. 376 ff.), whose men were in control of traffic on the canals as well as the roads. More will be said presently (p. 267) of the administration of the hydraulic engineering works, from the *Chou Li* onwards.

[b] Ch. 12, pp. 18*a* ff. (ch. 43), tr. Biot (1), vol. 2, pp. 565 ff., eng. auct. mod.

[c] Needless to say, they gave rise to a great deal of commentary, much of which is available in translation by Biot (1), vol. 2, pp. 565 ff.

[d] Note once again a standardised unit or module (cf. p. 67 above).

[e] Interpretation here depends on the values of these units. The *hsün* can safely be taken as 8 ft.; but the *jen*, essentially the height of a man (hence the *ad hoc* term introduced above), is variously defined in ancient texts as 8 ft., 7 ft. and 5 ft. 6 in.

[f] Cities had also encircling drainage canals internally and moats externally (*Kuan Tzu*, ch. 57, cf. Rickett (1) for a translation).   [g] Cf. Vol. 4, pt. 1, Figs. 304, 305, *et passim*.

[1] 匠人	[2] 溝洫	[3] 畎	[4] 田	[5] 遂	[6] 井
[7] 溝	[8] 成	[9] 洫	[10] 同	[11] 尋	[12] 仞
[13] 澮	[14] 川	[15] 涂	[16] 不行	[17] 淵	[18] 雍氏
[19] 萍氏	[20] 野盧氏				

Every canal should be made so as to take account of the characteristic forces of water; every dyke should take account of the characteristic strengths of earth. A good canal is scoured by its own water; a good dyke is consolidated by the sediment brought against it.[a]

(The general principle applying to) the making of dykes (*fang*[1]) is that the breadth and the height are equal. The reduction (*shan*[2]) of breadth at the top is one third, but for dykes of great size, it is more than this.[b]

In constructing canals or dykes one should first take the depth of one day's work as a test. Then one measures (the distance) in *li*, and after (the calculations are made) one can apply whatever combined labour force may be necessary.[c]

(As for the ramming down of the earth) the boards which contain the earth are bound together with ropes, but these should not be stretched too tight, as the wood will bend and the boards will not bear the weight...[d]

If this portion of the *Chou Li* had been written by a Chinese counterpart of Vitruvius a century or two before him, it would perhaps have been more informatory, since there were certainly very able experts in dyke and canal building in China at that time; yet even as seen through the redaction of a literary scholar, the passage has its value.

One point which it raises is the question of the Ching, or 'well-field' system (*ching thien*[3]). This belongs more properly to the development of land tenure and therefore to the realm of social and economic history,[e] but it has a certain connection with the origins of irrigation and therefore demands notice here. In the −6th century, Confucius makes no mention of large irrigation works, but twice refers to 'ditches and conduits' (*kou tu*[4] and *kou hsüeh*[5]).[f] The classical description of the well-field system was given by Mencius about −300. The Duke of Thêng[6] had sent Pi Chan[7] to ask him about it, and this was his reply:[g]

Since your prince, wishing to put in practice a benevolent government, has made choice of you and put you into this employment, you must exert yourself to the utmost. Benevolent government must begin with land boundaries. If the land boundaries be not defined correctly, the division of the land into the Well System will not be equal, and the produce available for salaries will not be equitably distributed. On this account, oppressive rulers and ministers with foul designs are sure to neglect the definition of boundaries. But if once the boundaries have been defined correctly, then the division of the fields and the regulation of allowances can be done while you sit comfortably in your office.

Though the territory of Thêng is rather small, its soil is rich. There must be both lords (*chün tzu*[8]) and peasants (*yeh jen*[9]) in it. If there were no lords there would be no one to rule the peasants; if there were no peasants there would be no one to support the lords.

I would propose to you that in the remoter districts, let the nine-lot division be observed,

---

[a] One should not fail to note the extremely Taoist character of these maxims, already mentioned above, p. 229.

[b] A few sentences further on, the *Khao Kung Chi* says that the battening or backward slope of city-walls should be such that the reduction at the top is one sixth of the total breadth. Cf. Vol. 3, Fig. 56.

[c] Cf. below, pp. 329 ff.

[d] Dyke construction thus also used the pisé method.

[e] See Sect. 48 below.

[f] *Lun Yü*, VIII, 21 (with reference to Yü the Great) and XIV, 18, iii.

[g] *Mêng Tzu*, III, (1), iii, 13–20, tr. Chi Chhao-Ting (1), p. 52; Legge (3), p. 119; mod. auct.

[1] 防   [2] 綯   [3] 井田   [4] 溝瀆   [5] 溝洫   [6] 滕
[7] 畢戰   [8] 君子   [9] 野人

and let one lot be set apart as a 'helping' lot (*chu*[1]) (i.e. a lot which the eight peasant families help each other to cultivate as land for the lord). In the city and suburb areas, let the people pay a tax of a tenth part of their produce, and render military service.

Then from the highest officials down to the lowest, each must have his glebe (*kuei thien*[2]), consisting of 50 *mou* (*c*. 8½ acres). Let supernumerary officials have each 25 *mou*.

On occasions of death, or removal from one place to another, let there be no quitting the district. The fields of a district are all within the Well unit. The peasants attached to this unit of land should render friendly offices to one another in their going out and their coming in, help one another in keeping watch and ward, and sustain one another in sickness. Thus the people are brought to live in affection and harmony.

A square *li* covers nine square lots of land, the area of the whole being 900 *mou*.[a] The central square lot is the 'public' field (*kung thien*[3]), and eight families, each having its private hundred *mou*, cultivate in common the 'public' field. And not till the 'public' work is finished, may they presume to attend to their private affairs (*ssu shih*[4]). This is how peasants are distinguished (from lords).

These are the general outlines of the system. Any special modifications, such as irrigation and fertilisation (*jun tsê*[5]), must be made by the prince and yourself.[b]

This passage has given rise to many debates and will probably long continue to do so.[c] The translation 'well-field' for *ching thien*, which has long been customary in English, is rather misleading, for quite apart from any connection with water-supply, the Chinese phrase simply signifies a pattern, i.e. the lines of division between nine square plots, lines which intersect just in the shape of the character *ching*.[6] According to the lexicographers the word has two distinct meanings—its usual one of a well of water, and the technical one of a nine-plot land division. The form of the character, strikingly evocative of such a plan, is very ancient, for the graph is found already in Shang bone inscriptions. It is evidently related to *thien*,[7] the fields,[d] equally ancient, a word which some believe originally designated hunting territories before it was applied to agricultural plots. Yet water-supply is never far from the semantic centres of these words; not only does it provide the primary meaning of *ching*,[6] but some — 9th-century bronze forms of the character actually indicate the well or water-source by a dot in the

[a] Note the combination of a triple and decimal system in this and the preceding passage. Each of the nine lots is divided into 100 squares, and above the level of the Ching the progression is decimal. Hsü Chung-Shu (6) and Hsü Hsü-Sêng (1) suggest that such differences may be vestiges of different metrological systems of Shang and Chou.

[b] It is interesting that this classical exposition of proto-feudalism is immediately followed by the episode of the 'Diggers', on whom see Vol. 2, pp. 120 ff. above.

[c] See, conveniently, the discussion of Yang Lien-Shêng (7). Classical debates are recorded in Anon. (51); Chhi Ssu-Ho (3); Hsü Chung-Shu (6); Li Chien-Nung (1, 2) and Wu Chhi-Chhang (4). Cf. the comments of Eberhard (21), pp. 6 ff.; J. Gray (1), p. 208; and the bibliography of H. D. Fêng (1). Among the most recent contributions are the paper of Hsü Hsü-Sêng (1) and the monograph of Vassiliev (1). Amano Motonosuke (2) sees in the *ching thien* system primarily an ancient method of land survey. During the past forty years it has played a central role in the discussions of Chinese scholars on ancient social and economic history, most of them declining to follow Hu Shih in the view that it was a pure figment of Mencius' imagination, and believing that it was an indication of something that had really existed. The question was, what? Did *kung thien* mean communal land, implying a residue of primitive communalism (and hence an anticipation of socialism), or did it mean demesne land, implying a contemporary feudalism or proto-feudalism? As will be seen, I incline to the latter view. The controversy has been described in detail by Levenson (4), acute, witty, and patronising, as usual.

[d] K362. In the most ancient forms the periphery of the graph is rounded, not square.

[1] 助   [2] 圭田   [3] 公田   [4] 私事   [5] 潤澤   [6] 井   [7] 田

centre.[a] Such dots occur also within the spaces of *thien*.[b] It may well be, too, that the 'ordinate and abscissa' lines in both these characters preserve traces of the layout of the irrigation ditches that ran alongside the plots, as in the *Chou Li* description.[c] In any case, the topography of north and north-west China is such that it would be quite natural for wells, small reservoirs, dew-ponds or shallow pits supplied by springs of underground water, to become the centres of patches of cultivated land before any extensive efforts had been made to capture the lower waters of natural streams.[d] Although co-operative work by the peasants on the lord's land is not in doubt, most scholars now regard the details of Mencius' nine-lot system as a schematic 'Utopian' policy rather than a description of a form of land allocation which ever had any real existence. It was of course Utopian only from the point of view of the feudal lords, since one ninth of the produce was due instead of a tithe. Although no doubt rectangular plots were used when possible for convenience of mensuration, in general the division of the land must have followed the contours of the terrain. There is also no reason for thinking that the lord's land was generally central to the village land as a whole. Why Mencius wanted it there may perhaps be explained by a concern, probably very ancient, that the lord's land should get the best water-supply. And as he was not at all the only ancient person to apply the term *ching* to land settlement, it must have been an ideal conception current in the Warring States period if not before.

The *Shih Ching* songs contain references (perhaps as early as the −9th or −8th century)[e] to 'public' and private fields (*kung thien*[1] and *ssu thien*[2] respectively). The former would have been the land, favourably placed in relation to the source of water, which the peasants came together under *corvée* to work for the feudal lord; the latter that which they worked for themselves. This conception of the usages of old certainly existed in the Han, as may be seen from an interesting fragment, the *Chhun Chhiu Ching Thien Chi*,[3] written by some unknown Confucian in the +1st century.[f] It involves certain important technical terms in ancient Chinese land utilisation; *chhê*,[4] cultivation in common; *chu*[5] or *chieh*,[6] mutual aid in tilling the lord's land; and *kung*,[7] feudal land-dues not necessarily derived from well-field layout or proportions. Mencius associated the first of these with the Chou, the second with the Shang, and the third with the Hsia,[g] but Li Chien-Nung has shown that this was really an inversion of

[a] K819*e*, interpreted so by Karlgren in the first edition. See also Jung Kêng (*3*), p. 279.

[b] Jung Kêng (*3*), p. 57.

[c] More obviously they represent the balks and headlands, as we should say, of unploughed turf, between the fields. There are in Chinese two special terms for these, *chhien*[8] and *mo*,[9] usually north–south and east–west respectively, which we shall soon meet with in various contexts. Cf. p. 72.

[d] It may be significant that there was an ancient State called Ching[10] in the upper Wei valley near modern Paochi.

[e] Pt. II, bk. vi, no. 8 in Legge (8); Mao no. 218; Waley (1), p. 171; Karlgren (14), p. 166. Even the same characters are used as in Mencius. The song speaks of rain coming down first on the lord's fields and then on those of the peasants. This ode was quoted by Mencius (*Mêng Tzu*, III, (1), iii, 9; Legge (3), p. 118), in a passage immediately preceding that just given.

[f] *YHSF*, ch. 39, p. 9*a*.

[g] *Mêng Tzu*, III, (1), iii, 6 (Legge (3), p. 116).

[1] 公田   [2] 私田   [3] 春秋井田記   [4] 徹   [5] 助   [6] 藉
[7] 貢   [8] 阡   [9] 陌   [10] 井

the true sequence.[a] The first dates probably from the age of tribal collectivism,[b] the second is certainly associated with the system of joint work on the lord's land which originated in the early Chou time, and the third belongs to the well-developed feudalism of the Warring States period. Another important reference to *ching-thien* organisation occurs in the account of the great land survey of Chhu in −547, where it is closely connected with irrigation works, still no doubt relatively small in scale.[c]

In Chhu, Wei Yen,[1] being Minister of War, was charged by the Premier to regulate the dues (*phi fu*[2]) and to count the number of cuirasses and weapons (which each fief had to provide). On a *chia-wu* day, he (began to) make a register of arable land. He assessed (the products of) mountains and forests, collected together the marshes and lakes (for fowling, under the princely authority), distinguished among the hills and valleys (as sites for tombs), marked the damp and briny places (for the extraction of salt), computed (the extent of) land subject to floods on the frontiers of the State (*shu chiang lao*[3]), measured the area of the dyked reservoirs (*kuei yen chu*[4]),[d] divided the flat lands between embankments into fields separated by balks or headlands (*thing yuan fang*[5]), set apart dry places beside the water for pasturage (*mu hsi kao*[6]), and laid out fertile flourishing country in Ching units (*ching yen wu*[7]). He then determined the contributions due, fixed the number of chariots and horses to be raised, (and assessed for each place) its levy in terms of chariot-riders, foot-soldiers, and armoured men with shields. When all this was done, Wei Yen handed in his report to the Premier, and it was approved.

This shows that in the −6th century a *ching-thien* (well-field) system of some kind, not necessarily identical with Mencius' specification, was in practical use for land settlement. Military service and supplies had long been apportioned on a territorial basis according to *ching*. For in −589 the prince of Lu ordered that each *chhiu*[8] (16 or 64 *ching*, according to different authorities) should furnish one *chia*[9] (a military unit of men with cuirasses).[e]

It is indeed probable that those are right who, with Kuo Mo-Jo[f] and Chi Chhao-Ting,[g] see in the well-field system something essentially similar to that of the manor, demesne or glebe. The word translated above 'allowances' (*lu*[10]) assuredly meant, in Mencian times and earlier, the income of the feudal lord derived from the land which the peasants worked together for him, i.e. the land referred to later in the same passage as 'public' land. So long as the productivity of labour underwent no change, the feudal incomes would have been determined purely by the number of peasant-serf families under the control of the lord in question. But the introduction of iron about the −6th

---

[a] (*1*); (*2*), pp. 122 ff.

[b] See the discussions in Sect. 10*f*, *g* above.

[c] *Tso Chuan*, Duke Hsiang, 25th year, tr. Couvreur (*1*), vol. 2, p. 439; eng. auct. mod.

[d] Note the mention of this type of public work, here about half a century after the earliest recorded example of a large reservoir dam (p. 271 below, on Sunshu Ao). The embankments along rivers were a still earlier type (cf. p. 232 above, on Chhi Huan Kung).

[e] Cf. *Tso Chuan*, Duke Chhêng, 2nd year; Couvreur (*1*), vol. 2, p. 2.

[f] (*1*), pp. 37 ff. This was a change of view, for in earlier writings, (*2*) for instance, Kuo Mo-Jo had doubted the real existence of the *ching-thien* system at any time.

[g] (*1*), pp. 54 ff. Cf. Chhi Ssu-Ho (*3*); Li Chien-Nung (*1*, *2*); and Hsü Chung-Shu (*6*).

---

[1] 蔿掩　　　[2] 庀賦　　　[3] 敤疆潦　　　[4] 規堰豬　　　[5] 町原防
[6] 牧隰皋　　　[7] 井衍沃　　　[8] 丘　　　[9] 甲　　　[10] 祿

century and its application as cast iron to ploughs soon afterwards, together with other changes such as the greater use of manure, increased agricultural productivity towards the beginning of the Warring States period, and this must have tempted the lords to reduce the size of the 'private' fields and increase that of the 'public' ones. This may have been what Mencius had in mind when he spoke of oppressive rulers and wicked ministers. Restoration of the original well-field system, with clear definition of land boundaries, would lighten and equalise the burden on the peasant-serfs. Of course, the 'private' fields were not private property in the modern sense. From Mencius it is clear that the peasants could not migrate away from their land, which they held only in 'return' for labour on the lord's land and military service. Thus the well-field unit was the lowest economic and administrative cell in the feudal body politic.[a]

It is now possible to understand more clearly what the *Chou Li* writer was talking about when he built up his series of technical irrigation terms from the nine-lot Ching or well-field. By his time the well itself had long evaporated to a mere symbol, and the fields were supplied with water according to a system more central or southern than northern in character. Han legend of course attributed even the refinements of this to the teaching of Yü the Great himself.

The basic social prerequisite for large-scale water-control projects was the possibility of extending the ancient *corvée* system, whereby work was done on the lord's field, to a vast mobilisation of labour for public works. So long as all peasants were closely attached to well-field units, there could be no great number of surplus and unattached labourers. But the system was already breaking down at the beginning of the −6th century, for it was in −593, according to the *Tso Chuan*,[b] that the State of Lu began to levy a tax according to the *mou*[c] of land occupied (*chhu shui mou*[1]), irrespective of the identity of the holder.[d] Such a tax would supersede a direct contribution of labour on the lord's land, and as the productivity of labour increased, there gradually grew up a large labouring population which, at any rate at certain times of the year, could be withdrawn from agriculture for the construction of great engineering works. When Mencius was speaking, about −300, it was quite natural for him to suggest that in the central regions of the State of Thêng the peasants should pay a tithe of their produce as land-tax. This system of tax, says Chi Chhao-Ting, was the essential device which cut the string tying the feudal lords to the agrarian routine, and freed them from concern over the harmful effects on agricultural production of large-

---

[a] Elsewhere in Mencius (v, (2), 2; Legge (3), p. 249) there is an account of the numbers of land units appropriated to the different ranks of the feudal hierarchy, though he says that in his time the old records were no longer available, since the nobles, considering them injurious to their interests, had destroyed them.

[b] Duke Hsüan, 15th year; Couvreur (1), vol. 1, p. 659.

[c] One *mou* is now equivalent to 0·164 acre, but in the Chou period it was much smaller, only 0·047 acre. At this rate the nine *ching-thien* squares amounted to 42·7 acres.

[d] Again in the State of Chêng in −537, on the order of Kungsun Chhiao (*Tso Chuan*, Duke Chao, 4th year; Couvreur (1), vol. 3, p. 87). In both cases there was, as one would expect, great complaint. The *Tso Chuan* adds regarding the change in Lu that it was not according to good custom, for previously State grain had come only from the 'public' fields cultivated by the peasants jointly.

[1] 初 稅 畝

scale long-term *corvée* labour.[a] Great engineering works would therefore hardly be expected much before the latter part of the − 5th century, and that is just the time to which later tradition referred the beginnings of large-scale irrigation canals,[b] associating them with the name of Li Khuei[1] (*fl.* − 440 to − 380) of the State of Wei,[2] a figure already known to us.[c]

In other words it was the coming of private land ownership, together with the increase of agricultural productivity in the new iron age, which 'liberated' the labour force required for the greater works. The first State to recognise the decay of the well-field system and to hasten its disappearance was Chhin. In the twelfth year of Duke Hsiao (− 350), when Kungsun Yang (Shang Yang)[d] was Premier, the feudal well-field system was abolished, the transfer of land by sale and purchase was legalised, especially for the benefit of those commoners who had performed meritorious military services, and the State divided into 31 *hsien*[3] districts, each under the control of an official, replacing former fiefs.[e] The unploughed balks or headlands between the fields were 'opened' (*khai chhien mo*[4]), i.e. the ancient regular divisions were all thrown together so that lots of any size might be held by larger or smaller landowning families. In − 348, the *fu*,[5] another kind of land-tax, was added. In these ways the State of Chhin, a century before the first unification of China, began to crystallise into the perennial pattern of bureaucratic feudalism, which all other regions were to follow. It can hardly be a coincidence that this change was accompanied by an exceptional development of successful irrigation and transportation projects (shortly to be described), such as the Chêngkuo Canal in the northern Wei valley, the Kuanhsien system in Szechuan, and later the Ling Chhü Canal uniting north and south.[f]

One cannot too much emphasise the role of man-power in the ancient and medieval achievements of Chinese civil engineering. Work in an eotechnic age had to be done somewhat on the basis of 'a million men with tea-spoons'.[g] I am able to illustrate this

---

[a] (1), p. 63.

[b] E.g. *Shih Wu Chi Yuan*, ch. 1, p. 38a.

[c] Cf. Vol. 2, pp. 210, 523 above. He was the first codifier of positive law (cf. Pokora, 1), and a remarkable economist, the first to plan a family budget and the first to suggest ever-normal granaries. Some of the *Kuan Tzu* book, in which there is a good deal on ditches and canals, may date from this time (cf. Forke (13), p. 79; Rickett (1), pp. 6 ff., 12 ff.). Meanwhile, cf. Vol. 2, pp. 42 ff., with Than Po-Fu *et al.* (1), pp. 86 ff. and Rickett (1), pp. 72 ff. See also Swann (1), pp. 136 ff.

[d] Cf. Vol. 2, pp. 205, 215 above.

[e] The sources are *Shih Chi*, ch. 5 (the Chhin annals) and ch. 68 (the biography of Shang Yang), as also *Chhien Han Shu*, ch. 24A (the chapter on Food and Commodities). There are various obscurities in the accounts, on which see Swann (1), p. 144, and Duyvendak (3), pp. 44 ff. The institution of the right to buy and sell land at this time seems to be first distinctly mentioned by Tung Chung-Shu (*Chhien Han Shu*, ch. 24A, p. 14b; Swann (1), p. 179) in a memorial to Han Wu Ti about − 100. On the whole subject see the recent study of Felber (1), and Moriya Mitsuo (1).

[f] Under the new régime of unrestricted alienability of land, and the growth of a class of large land-owners, whether aristocratic or mercantile in type, the construction of great irrigation-projects did not of course mean that the water was likely to be distributed equitably to everyone. The 'powerful gentry' (*thu hao*[6]), now developing, were likely to get more than their share of it. This may be the significance of the remark of Mencius (I, (2), v, 3; Legge (3), p. 38) that 'under true royal government there were no prohibitions respecting ponds and weirs—(*tsê liang wu chin*[7])'.

[g] The phrase is due to my friend Ritchie Calder.

¹ 李悝   ² 魏   ³ 縣   ⁴ 開阡陌   ⁵ 賦   ⁶ 土豪
⁷ 澤梁無禁

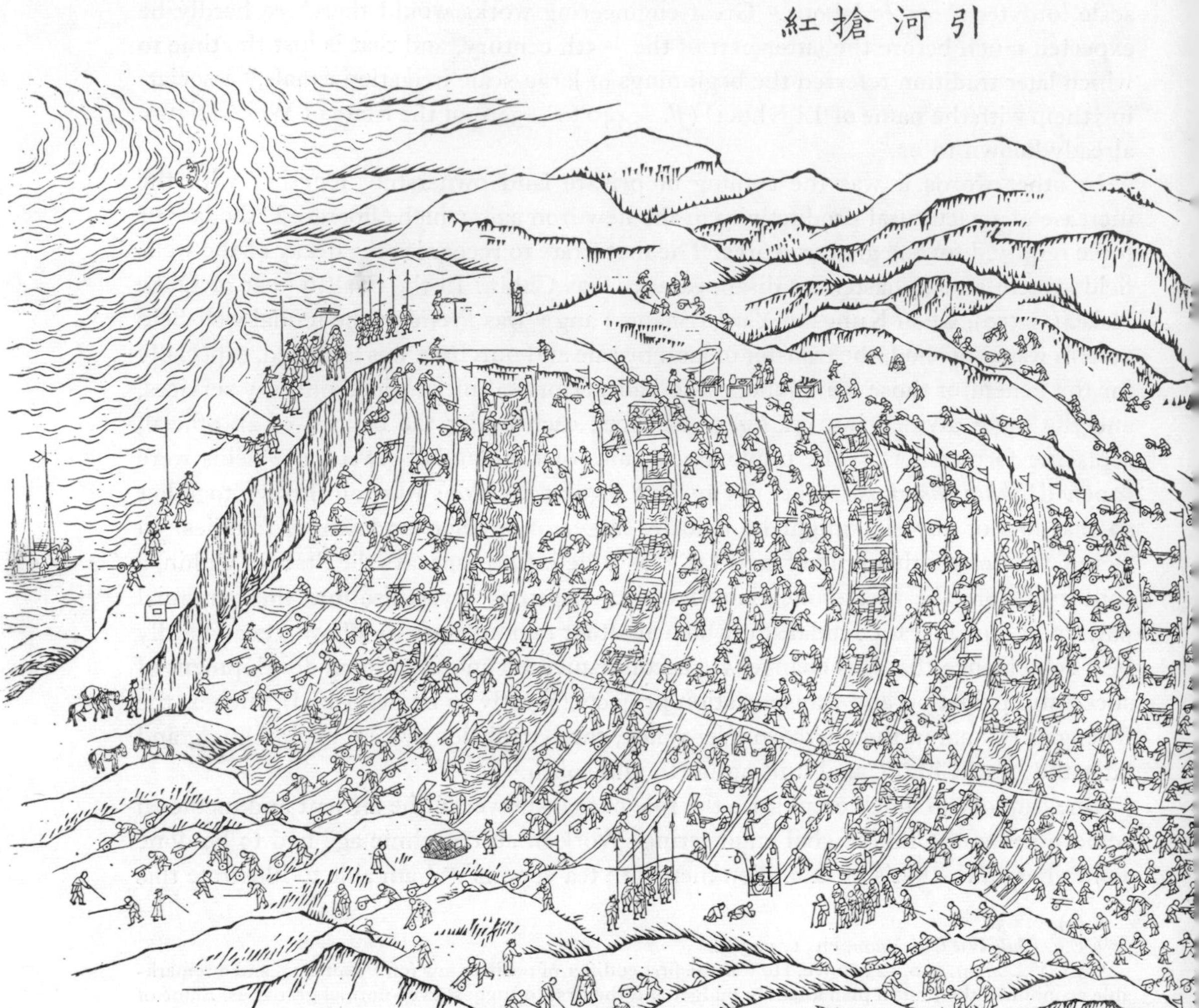

紅搶河引

Fig. 876. The Chinese genius for organising very large numbers of workers in civil engineering operations illustrated by a unique drawing from the autobiography of the high official and hydraulic engineer Lin-Chhing (*Hung Hsüeh Yin Yuan Thu Chi*, ch. 3, pp. 81 *a*, *b*, 82 *a*). The drawing is probably by his friend Chhen Chien. The somewhat enigmatic caption '"Wrestling for the Red" when Cutting a Canal' (Yin Ho Chhiang Hung) is explained in the accompanying text. When the job is rather more than half done, says Lin-Chhing, the superintendents begin to 'hang up the red', i.e. to organise competitions of work speed for prizes, of meat and wine, of boots and hats. When it is nine-tenths done they all set up a large umbrella-like lantern of red silk as a thank-offering to the local gods for the absence of mishaps, and upon this lantern the names of the winners are inscribed. This is an old custom, says Lin-Chhing, but far better than the threats and favouritism of some officials, which lead only to strikes.

The actual work shown in progress was the cutting of a canal at Chung-mou, a place south of the Yellow River between Khaifêng and Chêngchow in Honan, doubtless between 1833 and 1842 while Lin-Chhing was Director-General of River Conservancy centred at Huai-an in Chiangsu. Apart from many wheelbarrows, one can see more than a dozen square-pallet chain-pumps in the picture, these being apparently hand-worked and not by the feet (*pa chhê* therefore rather than *tha chhê*, cf. Figs. 579 and 580 in Vol. 4, pt. 2, as also p. 339). Other men are using hand-swung buckets for draining the excavation (cf. Vol. 4, pt. 2, p. 331, f.n. *c*). At the top on the left one can see a 'theodolite', levelling staff and chain-measure (cf. pp. 329 ff.). Further left, the engineer-in-charge, on a visit of inspection, is riding a horse on the neck of land which will be washed away when the canal is opened. In the foreground there is an altar to the local deity, with some guards on one side and a group of old people on the other.

point by a superb illustration in traditional style (Fig. 876), included by the secretaries of Lin-Chhing[1] in his *Hung Hsüeh Yin Yuan Thu Chi*[2] (Illustrated Record of Memoirs of the Events which had to happen in my Life) of 1849. Lin-Chhing was himself an able director of engineering works,[a] and wrote or edited the *Ho Kung Chhi Chü Thu Shuo*[3] (Illustrations and Explanations of the Techniques of Water Conservancy and Civil Engineering) in 1836. In the picture the excavation of earth from the bed of the projected canal can be seen in full activity, mostly with the use of wheelbarrows. Drainage is going on, in some places with the two-handled swung bailing buckets, elsewhere with batteries of hand-operated square-pallet chain-pumps one above the other. Flags mark boundaries. At the top may be seen surveyors with theodolite[b] and chains, as also a paymaster's table. To the left is an inspecting official, with his guards. Below one can see packhorses, a mat-shed, and an altar to the *genius loci*, or perhaps to Yü the Great. The bold vision of the Simpleton of the Northern Mountain could hardly be better illustrated.[c]

The account so far given of the great social and economic changes which accompanied the transition from bronze-age and early iron-age proto-feudalism (the *fêng chien*[4] system) to feudal bureaucratism (the *kuan liao*[5] system) has been very summary, and later on its complexities will have to be shaded in. Here we are only concerned with the social setting of the great waterworks, but we must be alive to the existence of great differences of opinion about both forms of society, and to the danger of identifying either of them with any feudalism which Europe ever knew. Wu Ta-Khun (1), for example, urges that the characteristic of the 'feudal' States of Chou (and presumably also Shang) times was that they were tribal, led by members of clans and sub-clans or septs related to that of the 'imperial' or 'High King' house; hence perhaps the term *kuei tsu*[6] (clan nobles) for the feudal lords. But in so far as all the land (and the slaves) remained the property of the tribal State, the lords were not enfeoffed so firmly as in Europe. Such questions cannot be pursued here.[d]

The right moment has now come to take up again a thread which we had in hand when dealing with the legends. I recur to the word *jun*[7]—specifically the *government's* channels and canals. The geo-climatic conditions of China exerted an irresistible influence upon Chinese society in the direction of strengthening centralised govern-

---

[a] A descendant of the royal house of the Jurchen Chin dynasty, his full name was Wanyen Lin-Chhing.[8] Van Hecken & Grootaers (1) have written a charming account of his Peking home and family history. Baylin (1) has translated a small part of his *Memoirs*.

[b] Strictly speaking, this is an anachronism; they are probably using some kind of circumferentor. But see p. 332 below.

[c] The Chinese have always had a genius for the efficient organisation of mass man-power. I saw it during the Second World War in the building of airfields (cf. Koester, 1); and again at the Ming Tombs Reservoir in 1958 (Fig. 877, pl.). On this latter work see Yen Yao-Ching *et al.* (1). 'I am of the opinion', wrote John Bell of Antermony in 1715, 'that no nation in the world was able for that undertaking [the building of the Great Wall] except the Chinese. For though some other kingdom might have furnished a sufficient number of workmen... none but the ingenious, sober and parsimonious Chinese could have preserved order amidst such labour.'

[d] See Sect. 48 below.

[1] 麟慶　　[2] 鴻雪因緣圖記　　[3] 河工器具圖説　　[4] 封建　　[5] 官僚
[6] 貴族　　[7] 濬　　[8] 完顏麟慶

ment.[a] This was so for the very simple reason that any effective treatment of the engineering problems set by the rivers, and the desired intercommunicating watercourses, tended, at every stage, to transcend the boundaries of the smaller feudal units.[b] Take the case of one of the earliest works, successfully executed before the first unification of the empire, the Chêngkuo Canal; it could not have achieved anything unless its planners were prepared to think in terms of a main course not less than a hundred miles long, with a service area of perhaps fifty miles in breadth. Fortunately, we have certain statements from the Former Han period which make the point quite consciously. Thus in the *Yen Thieh Lun* (Discourses on Salt and Iron) of about −80, we find:[c]

The Lord Grand Secretary said: 'The feudal lord, whose fief can be considered as forming but one household, has his concern limited to what is within it. But the emperor, whose domain is bounded only by the Eight Extreme Limits (of the world), has concerns extending beyond such small areas, indeed far and wide. Thus under the small (manorial) roof, expenses are trifling, in comparison with the great expenditure necessitated by the immense undertaking (of ruling the empire). Herein lies the reason for the Government's opening up of cultivatable land and reservoirs (*yuan chhih*[1]), and its concentrating under one hand the mountains and the lakes (*shan hai*[2]) to secure income that can be used to supplement tribute and taxes. Thus we improve canals and conduits (*kou chhü*[3]), promote various kinds of agriculture, extend farm and pasture lands, and develop nature reservations and walled hunting-parks (*yuan yu*[4]). The offices of the Thai-Phu,[5] the Shui-Hêng,[6] the Shao-Fu[7] and the Ta-Nung[8] compute annually the revenue derived from farm and pasture, and the rentals from leases of lakes and fishponds (*chhih yü chih chia*[9]). Now up to the limits of the empire in the north, supervisors of fields have been appointed, yet with all these efforts, there is still a deficit...'

So also, about thirty years before, there had been an imperial edict which used the following words:[d]

Agriculture is the basis of the whole world. Springs and rivers, irrigation ditches and reservoirs (*chhüan liu kuan chhin*[10]) make possible the cultivation of the five grains. In the empire there are innumerable mountains and rivers, but the small (ordinary) people do not understand their proper use (*wei chih chhi li*[11]). Hence (the Government) must cut canals and ditches (*kou tu*[12]) and build dykes and reservoirs (*pei tsê*[13]) to prevent drought...

This was in −111. The document was an order to carry out certain improvements to the Chêngkuo Canal which had been proposed by Erh Khuan.[14]

  [a] Of course only up to the point which was possible before the neotechnic age.
  [b] And also, in later ages, the boundaries of landholdings of any private individuals or groups of individuals. There is an important contrast between Chinese and European society here, for in Holland and England most of the fen drainage (e.g. Sir Cornelius Vermuyden in East Anglia) was done by private enterprise. The only analogy to this in China was the quite exceptional province of Shansi, where after the +14th century there were a number of private water-conservancy works (Chi Chhao-Ting (1), p. 44). While not denying a certain role of initiative to non-official local worthies in medieval China, we find the attempt of Eberhard (21), pp. 35 ff., to magnify it rather unconvincing, though we sympathise with his motives.
  [c] Ch. 13, p. 1*a*, tr. Gale (1), p. 81, mod. auct.
  [d] *Chhien Han Shu*, ch. 29, p. 7*b*, tr. Chi Chhao-Ting (1), p. 83, mod. auct.

[1] 園池	[2] 山海	[3] 溝渠	[4] 苑囿	[5] 太僕
[6] 水衡	[7] 少府	[8] 大農	[9] 池籞之假	[10] 泉流灌浸
[11] 未知其利	[12] 溝瀆	[13] 陂澤	[14] 兒寬	

But the transcendence of the boundaries of land belonging to 'small ordinary' landowners, or formerly to feudal lords, was not the only factor which tended inevitably to centralisation. At an early stage in the Warring States period, it was realised that water might be a weapon.[a] The sight of rooftops swirling round in flood-waters, with people clinging to them, in the midst of floating trees and bodies of dead animals, must have suggested to the 'defence' authorities of the feudal States that a strategic construction or destruction of dykes and watercourses might bring about gratifying results.[b] We realise this from another passage in the *Chhien Han Shu*, an earlier part of the speech of Chia Jang already quoted in connection with the Yellow River.[c]

The building of dykes and embankments began in recent ages during the period of the Warring States, when the various States blocked up the hundred rivers for their own benefit. (For example) Chhi, Chao and Wei all bordered the Yellow River. The frontiers of Chao and Wei rested on the foot of the mountains, while that of Chhi was on the low plain. (The State of) Chhi, therefore, constructed an embankment 25 *li* from the river, so that when rising waters approached this dyke in the east, they would be held back and forced to flood Chao and Wei in the west. Whereupon Chao and Wei also constructed an embankment 25 *li* from the river (to counteract the effect).

The feudal states also competed violently with one another for irrigation water, and in appropriating drained areas for growing crops.[d]

As Li Hsieh has pointed out, there is an unmistakable reference to this in Mencius. One of the characters who appears in the conversations recorded in the *Mêng Tzu*, which would have taken place a decade or so before −300, is Pai Kuei,[1] a man from Chou State. The deliciously Johnsonian flavour of Legge's translation demands reproduction intact.[e]

Pai Kuei said: 'My management of the waters is (I consider) superior to that of Yü (the Great).'

Mencius replied: 'You are wrong, Sir. Yü's regulation of the waters was according to the Tao of water. He therefore made the Four Seas their receptacle. But you make (the territories of) neighbouring States their receptacle. Water flowing in a contrary direction (to that in which it ought to flow) may be called an inundation. Inundations are a vast waste of water, and every benevolent man detests them. You are wrong, my good Sir.'

Such confusion could not but dispose people to welcome as an essential safeguard the imperial unification of the Chhin.

All subsequent governments controlled the rivers and canals as parts of a unified system, success achieved varying both with political power, the extent of the realm, and

---

[a] Here was another trait in Chinese culture which had been characteristic of ancient Mesopotamia also; cf. Drower (1), p. 554.

[b] We came upon an example of this at a much earlier stage (Vol. 1, p. 234) when in −101 Chinese hydraulic engineering experts (*shui kung*[2]) attached to the army besieging the capital of Ferghana were called upon to deprive the city of water by diverting a river, or to sap its walls by directing a greater stream against them. They successfully employed the former technique.

[c] *Chhien Han Shu*, ch. 29, p. 13b, tr. Chi Chhao-Ting (1), p. 64, mod. auct.

[d] Cf. the Ming historian Chêng Hsiao[3] (+1499 to +1566), in *Hsing Shui Chin Chien*, ch. 3, p. 5a (p. 37).

[e] *Mêng Tzu*, VI, (2), xi; tr. Legge (3), p. 319, mod. auct.

[1] 白圭   [2] 水工   [3] 鄭曉

the arbitrary conditions set by nature. But in times of war, both civil and foreign, there were often severe temptations to undo what had so laboriously been done. In +923 the Later Liang general in the field against the Later Thang, Tuan Ning,[1] broke the Yellow River dykes deliberately.[a] In +1020 Li Chhui[2] suggested a similar use of the flood weapon on the part of the Sung against the Chin Tartars,[b] and just a century later (+1128) this was actually done by Tu Chhung,[3] ruining most of the Grand (Pien) Canal south of the Huang Ho for several years.[c] In our own time the pattern repeated itself. The breaking of the Yellow River dykes in 1938 was undoubtedly a strategic move in the Sino-Japanese war, and a great deal of the friction between the Kuomintang and Kungchhantang after the Second World War was concerned with their rebuilding.[d] The former government, aided by the United Nations Relief and Reconstruction Administration (UNRRA), exerted itself to the utmost to return the Yellow River to its old bed before the latter government could make preparations to receive it (including the resettlement of hundreds of thousands of people who were farming the drained land), or to control it when it had reverted.[e]

All this of course is to say nothing of the military supplies motivation which led so often to the construction of great transport canals—plenty of examples will come to light in the following pages. Certain canals were expressly built too for the passage of warships; besides the Ling Chhü (cf. p. 299 below) there were several cases of this in the following centuries.[f]

Of course, individual quarrelling about water-rights was bound to be a permanent factor in Chinese rural life. The very word 'rival' in our own language derives from riparian settlement along rivers, where different people want to use the same water. An early anecdote concerning this occurs in the *Chan Kuo Tshê*.[g] At some time in the −4th century the people of the eastern part of the imperial Chou domain wanted to sow rice, but for one reason or another those of the western part refused to let the waters come down. Su Chhin[4] or one of his brothers managed to settle the matter, and was rewarded by both sides.[h] We have of course met with Su Chhin before (Vol. 2, p. 206) as one of the two political philosophers responsible for the alliances between the feudal States known as the Vertical and Horizontal Axes (*tsung hêng*[5]),[i] but the historicity of all the activities attributed to him is none too sure. The first certain system of regulation of water-rights recorded in Chinese history is associated with the

---

a Chêng Chao-Ching (*1*), p. 13.
b Chêng, *loc. cit.* p. 17; *Sung Shih*, ch. 91, pp. 8*b* ff.
c Chêng, *loc. cit.* p. 26.   d Cf. Anon. (9).
e Further information will be found in Peck (1); Rivière (1); Belden (1).
f E.g. a twenty-mile waterway dug by a J/Chin general in +1130 conducting naval operations against Han Shih-Chung, the celebrated Sung commander; cf. *WHTK*, ch. 158 (p. 1381.3).
g Ch. 2, p. 4*b*.   h Cf. Maspero (28), p. 349, (14), pp. 55 ff.
i Hence the ancient term often applied to the school of political philosophers, Tsung-Hêng Chia. Su Chhin is supposed to have died in −317, but Maspero (30) has revealed the confusions in the *Shih Chi* and *Chan Kuo Tshê* material about him, and believes that it derived from a romance with the title *Su Tzu*,[6] based on the life of a doubtless real person and written some time in the second half of the −3rd century. Still later the romance was continued maladroitly about his two younger brothers.

¹ 段凝   ² 李垂   ³ 杜充   ⁴ 蘇秦   ⁵ 縱橫   ⁶ 蘇子

name of Shao Hsin-Chhen,[1] who was governor of Ling-ling and of Nanyang and held other offices in the decades before his death in +4. His biography in the *Chhien Han Shu* depicts him in an attractive light;[a] he was fond of discussing agriculture with the farmers, he himself ploughed to encourage them, he liked living in cottages rather than in his official residence, and he was always wandering through the farm lands (*chhu ju chhien mo*[2]). He directed the construction of sluices (*shui mên*[3]) for fertilisation by silt (*yu*[4]), and he 'arranged the equitable distribution of water for the farmers by appointed mutual restraints (*wei min tso chün shui yo shu*[5])'. Then he set up inscribed stones in the fields, to perpetuate these customs and to avoid disputes.[b] In later times various special devices were used for distributing the water. For instance, during repairs to a 'Nine-Dragon Canal' (*Chiu lung chhü*[6]) in the Thang period (+9th century), a number of bronze dragon-heads were dug out at one point, with an inscription showing that they must have been first set up in +271.[c] It would be interesting to have a comparative study of the water-distribution regulations in different cultures.[d]

The 'Shui-Hêng', or 'Water Balancer', the title of one of the imperial accountants, just met with in a quotation from the *Yen Thieh Lun*, must originally have been concerned with regulation of water-rights. Although this title does not occur in the *Chou Li*, there are others which may be mentioned. The Chhuan-Hêng[7] were inspectors of all rivers and canals;[e] they seem to have been in charge of the river police and canal guards, Phing Shih[8] and Yung Shih,[9] who also (as we have seen, p. 255) patrolled embankments.[f] There was a parallel service of inspectors called Tsê-Yü[10] who administered lakes and their fisheries,[g] and lastly, certain Chhuan-Shih[11] officials who concerned themselves with fixing geographical names for all watercourses, and with listing their effects and products.[h] More reliable historically is the title Tu-Shui[12]—Director of Water Conservancy[i]—which seems to date from the beginning of the Han;[j] his office was then called Tu-Shui Thai.[13] In the San Kuo period (Wei) he was Tu-Shui Lang[14] and had only a department in the general secretariat.[k] But as the importance of water-conservancy increased, the controllers and planners grew into an

---

[a] Ch. 89, p. 14*a*.

[b] On the use of steles for recording traditional water-rights, see also Chhen Ching (*1*).

[c] *Thang Yü Lin*, ch. 8, p. 22*b*. There is also the question of what time-measuring instruments were used for controlling the switching of the channels. We have touched upon this subject in Vol. 3, p. 315, in connection with sinking-bowl clepsydras; according to Abercrombie (*1*) these are still used today for this purpose in the Yemen.

[d] For ancient Ceylon there is an interesting article by Paranavitana (*2*). In 1960 celebrations were held to mark the thousandth anniversary of the Tribunal de las Aguas in Valencia, where a large irrigation system for fruit-growing orchards goes back to the time of the Cordoban Caliphs.

[e] *Chou Li*, ch. 4, p. 36*a* (ch. 16; Biot (*1*), vol. 1, p. 374).

[f] Ch. 9, p. 5*a*; ch. 10, pp. 3*b*, 4*a* (chs. 34, 37; Biot (*1*), vol. 2, pp. 297, 379).

[g] Ch. 4, p. 36*b* (ch. 16; Biot (*1*), vol. 1, p. 374).

[h] Ch. 8, p. 31*a* (ch. 33; Biot (*1*), vol. 2, p. 283).

[i] Although the word *tu* alone (K45*e–g'*) can mean 'all', its principal meaning was from earliest times the capital of a State. This is yet another example of the inevitable association between waterworks and political centralisation. Cf. p. 264 above.

[j] *Shih Wu Chi Yuan*, ch. 6, p. 43*a*.   [k] *Shih Wu Chi Yuan*, ch. 5, p. 11*a*.

[1] 邵信臣	[2] 出入阡陌	[3] 水門	[4] 淤	[5] 爲民作均水約束
[6] 九龍渠	[7] 川衡	[8] 萍氏	[9] 雍氏	[10] 澤虞
[11] 川師	[12] 都水	[13] 都水臺	[14] 都水郎	

independent ministry. In the Thang this was at first called Tu-Shui Chien,[1] but changed its name in +685 to Shui-Hêng Chien,[2] 'Directorate of Hydraulic Control'.[a] Among its departments were the Ho Chhü Shu[3] (Bureau of Rivers and Canals), the Chu Chin Chien[4] (Sub-Directorate of Fords, Ferries and Bridges), and, till +738, the Chou Chi Shu[5] (Bureau of Boats and Oars, i.e. Canal Transport). Among its greater perambulating officials were the Ho-Thi Shih-Chê[6] (Comptrollers of River and Dyke Works), the strengthening of whose position was one of the first acts of the Sung dynasty (in +967);[b] and among its lesser ones were the Tou-Mên Chhang,[7] the officials in charge of sluice gates. All this personnel was theoretically in the province of the Ssu-Khung[8] or Minister of Works, one of the four great ancient officers of State.

These pages may conclude with a suggestion that as the years went by, the whole of Chinese theoretical thought came to be permeated by certain ideas proper to the control of water-courses, so important a feature of the civilisation, for example 'flow' and 'blockage'. In medicine, there was a doctrine of stasis as the cause of disease,[c] and a special term for it, yü,[9] which Chuang Chou frequently uses. Injurious effects spring from obstruction (yung[10]) of passages or pores, and this leads to stoppage (sai[11]) and choking (kêng[12]).[d] If the Tao be repressed (yü[9]), development may be inhibited,[e] or perhaps the chhi of earth may be blocked up (yü[9]),[f] producing earthquakes. Moreover, the connection of these concepts with actual observation of hydraulic works cannot be concealed; for example, Chuang Tzu says:[g]

It is the nature of water, when free from admixture, to be clear, and when not agitated to be level; while if obstructed and not allowed to flow, it cannot preserve its clearness (yü pi erh pu liu, i pu nêng chhing[13]).

So also the Lü Shih Chhun Chhiu says:[h]

The establishment of diseases and the arising of evil are due to a blocking up of the seminal essence (ching[14]) and the chhi. If water is stagnant it becomes slimy. If the sap in trees is dammed up, worms find their way in; if grass is similarly affected, it decays.

And earlier it was pointed out (Vol. 2, pp. 145 ff.) that the chief purpose of Taoist medical gymnastics was to unblock the pores of the body. Finally, this general cast of thought appears also in literature, as may be seen from the celebrated Wên Fu[15]

---

[a] Des Rotours (1), pp. 490 ff.
[b] *Shih Wu Chi Yuan*, ch. 6, p. 23b. This title goes back to the Chin (+3rd century).
[c] Later on (Sect. 44) the question will arise of similarities here with the medical theories of the Greeks.
[d] *Chuang Tzu*, ch. 26 (Legge (5), vol. 2, p. 139).
[e] Ch. 33 (Legge (5), vol. 2, p. 217).   [f] Ch. 11 (Legge (5), vol. 1, p. 301).
[g] Ch. 15 (Legge (5), vol. 1, p. 366).
[h] Ch. 121 (vol. 2, p. 106), tr. R. Wilhelm (3), p. 358, eng. auct. Cf. *Lieh Tzu*, ch. 4, p. 12b (L. Giles (4), p. 79) where the heart of someone who had nearly attained sagehood had only one orifice still occluded pu ta[16]).

[1] 都水監   [2] 水衡監   [3] 河渠署   [4] 諸津監   [5] 舟楫署
[6] 河隄使者   [7] 斗門長   [8] 司空   [9] 鬱   [10] 壅
[11] 塞   [12] 哽   [13] 鬱閉而不流亦不能清   [14] 精
[15] 文賦   [16] 不達

(Ode on the Art of Letters) written by Lu Chi[1] in +302. There the terms 'flow' and 'blockage' (*thung sai*[2]) are strikingly applied to literary composition and inspiration.[a]

(6) SKETCH OF A GENERAL HISTORY OF OPERATIONS

One of the songs in the *Shih Ching* contains the words:[b] 'How the water from the Piao pool flows away to the north! Flooding the rice-fields... (*Piao chhih pei liu, chhin pi tao thien*[3]).'[c] Since this may indeed be of the −8th century, it is perhaps one of the earliest mentions of irrigation in Chinese history, but it can only refer to a reservoir on some rather small scale. A couple of centuries later, impressive works were well under way.[d]

One of the larger, or at least longer, works of Chinese hydraulic engineering, however, had the reputation of being so ancient that its first origins were lost in the depths of time. This was the Hung Kou[4] (Canal of the Wild Geese, or Far-Flung Conduit), which connected the Yellow River near Khaifêng with the Pien[5] and Ssu[6] Rivers, and ultimately formed the model, as we shall see, for part of the Sui Grand Canal.[e]

Although the conduit may have been first made to bring irrigation water to the upper Huai basin rather than for transport, it linked the Huai River with the Yellow River at an early date and permitted the navigation of barges between the east-central and the northern key economic areas (see Fig. 36 (map) and Vol. 1, pp. 114 ff.). Ssuma Chhien said of it[f] that it connected the feudal States of Sung, Chêng, Chhen, Tshai, Tshao and Wei.[7] If indeed it began as an irrigation system, its availability as a long-distance transportation channel was ominous for their survival as independent States. Strictly speaking it was a summit canal, but the term would be inappropriate, for the watershed between the Yellow River and Huai basins is extremely low and flat, and the difference in levels was lessened almost to non-existence by the building-up of the Yellow River's bed (cf. p. 237 above), already well advanced.[g] Strictly speaking also the Hung Kou was a complex of artificial waterways rather than a single canal.[h] It took off eastwards from the Yellow River at a point downstream from the entry of the Lo

---

[a] See Hughes (7), pp. 107, 177, 179. Text in *CSHK* (Chin), ch. 97, p. 3a.

[b] Mao no. 229; Legge (8), II, 8, v; Waley (1), no. 110.

[c] The pool is perhaps to be identified with a lake which existed for many centuries west of Sian in Shensi.

[d] On the hydraulic engineering works of the Chou and Chan Kuo periods Yang Khuan (10) is a useful companion, but his work came out too late to help us in writing.

[e] In spite of Herrmann (1), we do not feel sure that Pien, the name later applied to the canal itself, was ever the name of a river. Ssuma Chhien does not mention it; he says that the canal joined the Yellow River to the Chi,[8] Ju,[9] Ssu[6] and Huai[10] Rivers.

[f] *Shih Chi*, ch. 29, p. 2a (Chavannes (1), vol. 3, p. 522; Watson (1), vol. 2, p. 71).

[g] This does not mean that no double slipways (*diolkoi* or haul-overs) were necessary, together with stop-log gates or flash-locks, but it does mean that the routes could be worked for many centuries without pound-locks.

[h] Lo Jung-Pang (6), pp. 50 ff., has made a gallant attempt to work out the details, but the historical geography is very difficult. Cf. Twitchett (4), p. 186.

---

[1] 陸機   [2] 通塞   [3] 瀌池北流浸彼稻田   [4] 鴻溝   [5] 汴
[6] 泗   [7] 宋鄭陳蔡曹衛   [8] 濟   [9] 汝   [10] 淮

River from Loyang, just beside the city of Jungyang,[1] where there was a great granary[a] in Han times, the hub of the tax-grain transport system. It then swung south in an arc of some 260 miles so as to connect with the head-waters of that gridiron of parallel south-eastward-flowing rivers entering the Huai from the north—the Ho,[2] the Tzu,[3] the Sui,[4] the Kuo[5] and the Ying.[6] This canal, called the Lang-Tang Chhü,[7] following almost imperceptible contours, joined together the upper reaches of the last three of these rivers, not without successive improvements in the form of a number of branches;[b] but it was never the part most used for traffic. That function devolved upon its most northerly branch, the Pien[8] (or Pan[9]) Canal or River, some 500 miles long, which joined the head-waters of the Ho[2] and so passing near Hsüchow[10] gave access through the Ssu[11] to the Huai. From +70 onwards its mouth on the Yellow River was protected by substantial dykes[c] built under the supervision of Wang Ching.[12] This then was the Hung Kou proper, or the Pien Canal of the centuries before +600.[d]

There remains the question of the date of its construction, still very uncertain. So old was it supposed to be that Ssuma Chhien gave it pride of place after the works of the Great Yü himself, assigning no clear origin to it. The diplomatist Su Chhin[13] mentioned it about −330 in discussing State boundaries.[e] According to some scholars it was first built between −361 and −353, but if that is so it seems curious that Ssuma Chhien should have known nothing about such an important work carried out under Prince Hui of Liang (i.e. of Wei[14]), often referred to by Mencius and Chuang Tzu.[f] Alternatively the late −6th century or the early part of the −5th century might be a reasonable estimate, i.e. the time of Confucius himself. It may be significant that Ssuma Chhien does not mention the State of Wei,[14] which was not founded until −403, in connection with the canal, but only smaller and more ancient States. Chhen and Tshai were absorbed by Chhu in −479 and −447 respectively, Tshao fell to Sung as

[a] The Ao Tshang.[15] It will be seen that this position was very well chosen, for tax-grain could meet there coming not only from the more southerly east-central economic area, then imperfectly developed, but also from the richer northern economic area, collected and despatched from the city of Thao[16] on the Chi[17] River in Shantung.

[b] Cf. the historical account of Chang Chi in +995, preserved in *Sung Shih*, ch. 93, pp. 17a ff.

[c] Cf. p. 308 below, as also Chêng Chao-Ching (1), p. 193. At an earlier time these protection works had been the scene of an operation well exemplifying the taste for hydraulic engineering which characterised ancient Chinese generals and strategists (cf. p. 265 above). In −225 Wang Pên,[18] one of Chhin Shih Huang Ti's commanders, broke down the dykes at a flood season so as to inundate the State of Wei[14] and sap the city-wall of its capital Ta-liang[19] which lay near modern Khaifêng beside the Hung Kou Canal. The scheme was successful and one of the last remaining great competitor States of Chhin was liquidated (cf. Vol. 1, p. 94; the story is told in *Shih Chi*, ch. 6, p. 9a, ch. 44, pp. 21b, 22a; tr. Chavannes (1), vol. 2, p. 121, vol. 5, pp. 195, 196; cf. Chêng, *loc. cit.* p. 5).

[d] At this distance of time its efficiency is naturally hard to estimate but according to Kêng Shou-Chhang[20] (as reported in *Chhien Han Shu*, ch. 24A, p. 17a, cf. *Sung Shih*, ch. 93, p. 18b) it brought to the capital in his time, *c.* −60, some 122,000 tons of grain annually. This was only about a quarter of what the Pien Canal of the Northern Sung could do; cf. p. 311 and pp. 352, 360. On Kêng Shou-Chhang see further Vol. 3, p. 24, etc.

[e] *Shih Chi*, ch. 69, p. 10b. Cf. p. 266 above.   [f] See Vol. 2, *sub voce*.

[1] 滎陽	[2] 獲	[3] 苢	[4] 濉	[5] 渦	[6] 潁
[7] 滾瀓渠	[8] 汴	[9] 汳	[10] 徐州	[11] 泗	[12] 王景
[13] 蘇秦	[14] 魏	[15] 敖倉	[16] 陶	[17] 濟	[18] 王賁
[19] 大梁	[20] 耿壽昌				

early as −487, and Chêng to Han by −375.[a] He may of course have meant only the country that had belonged to these ancient states, but it is clear that a Confucian dating is by no means excluded.

The date of the earliest known irrigation reservoir, though prior to this, is one which can be accepted with little reserve. In northern Anhui, south of the city of Shou-hsien,[1] there still exists a great tank, some 62 miles in circumference, known today as the Anfêng Thang,[2] but anciently as the Ssu-Ssu Pei[3] or Shao Pei[4] (Peony Dam). Both Ssuma Chhien[b] and the author of the *Huai Nan Tzu* book[c] make clear that the dam had first been constructed under the superintendence of Sunshu Ao,[5] minister to Duke Chuang of the Chhu State when Ting was the reigning Chou High King (−606 to −586). The dam simply flooded a rather flat valley, catching the north-flowing water from a considerable part of the mountains north of the Yangtze, and supplied eventually no less than six million acres. It was repeatedly repaired during the Han and Thang dynasties. Sunshu Ao therefore ranks as the most ancient historical figure among all the hydraulic engineers of China.

In the −5th century, besides the works of Li Khuei already mentioned, of which we know little,[d] there were two especially important projects. Hsimên Pao,[6] the rationalist statesman who abolished human sacrifices to river-gods,[e] organised a diversion of the Chang[7] River, which formerly flowed into the Huang Ho near An-yang, so that it met the great stream, then in course ① lower down, i.e. nearer the bend at modern Tientsin.[f] Thus the Chang, which rises in mountainous Shansi and flows south-eastwards, was led away north-eastwards as a lateral contour canal to irrigate a large region of Honei instead of wastefully adding to the burden of the Yellow River. This work must have been begun between −403 and −387, when the State of Wei was under Duke Wên.[g] But on account of resistance to *corvée* service or for some other reason, it was not fully completed until the time of the grandson of that ruler (−318 to −296), when Shih Chhi[8] was put in charge.[h] The people made a song in honour of this which Pan Ku recorded in his history.

The other work of this century was the Han Kou[9] (Han-Country Conduit) which connected the Huai River with the Yangtze. It had been initiated earlier by Fu Chhai,[10] King of the State of Wu[11] at the time, in −486, as a military measure to enable

---

[a] See Vol. 1, Fig. 12 and Table 6.

[b] *Shih Chi*, ch. 119, p. 1*b*, cf. Watson (1), vol. 2, p. 414.

[c] Ch. 18, pp. 19*b*, 20*a*, cf. Chêng Chao-Ching (1), p. 251; Chi Chhao-Ting (1), pp. 66 ff.

[d] And a reservoir made by Shen Chu-Liang,[12] the Lord of Shê[13] in the State of Chhu[14] about −500. Shen was a rather sympathetic noble who had conversations with Confucius (*Lun Yü*, VII, xviii; XIII, xvi, xviii; Legge (2), pp. 65, 133 ff.; cf. Vol. 2, pp. 9, 545; also Creel (4), pp. 54, 136).

[e] Cf. Vol. 2, p. 137.

[f] It now joins the Wei[15] River, which runs approximately in the old bed of course ④ of the Yellow River.

[g] *Shih Chi*, ch. 29, p. 2*b*; cf. Hsimên Pao's biography, ch. 126, p. 15*b* (Watson (1), vol. 2, p. 71; Yetts (14), p. 32).

[h] *Chhien Han Shu*, ch. 29, p. 2*b*.

---

[1] 壽縣	[2] 安豐塘	[3] 斯思陂	[4] 芍陂	[5] 孫叔敖
[6] 西門豹	[7] 漳	[8] 史起	[9] 邗溝	[10] 夫差
[11] 吳	[12] 沈諸梁	[13] 葉	[14] 楚	[15] 衛

supplies to reach Wu troops attacking more northern States such as Sung and Lu.[a] This line of communication thus became the second oldest segment of the Grand Canal, but its original route ran to the east of the present one, winding deviously through the Shê-Yang Hu[1] and other lakes. Ssuma Chhien does not refer to it by name, but after mentioning a canal between the Yangtze and the Huai, adds that the Wu State also initiated many waterways around the Thaihu lake, thus starting the immensely complicated system in modern Chiangsu province.[b]

Not many years after the work of Shih Chhi in Honan came the famous system of Li Ping in Szechuan (about −270) and the Chêngkuo Canal (about −246), but these were so important that we shall reserve them for fuller description. A third project of the Chhin State and empire, much less well known, deserves a few words. The Yellow River, after passing Lanchow, flows north with the Ordos Desert on its right bank for some 350 miles. Near Ninghsia (mod. Yin-chhuan) it traverses a broad valley with low hills to the west, having much land on both sides which is fertile if irrigated. Here was a splendid outpost against the Huns. Accordingly Chhin Shih Huang Ti sent his general Mêng Thien[2] in −215 to take possession of the hither ground,[c] and in the following year to pass across, build forts, and make new prefectures with a population of transported prisoners.[d] All this was done, but more also, for derivatory lateral irrigation canals were built, starting from the town of Chung-wei[3] and running more or less parallel with the river on both sides (Fig. 878, pl.).[e] With the extensions which were made to them in the Han and Thang, they now cover an area about 100 miles long and 20 miles or so broad.[f] Like the Kuanhsien system, this is clearly visible on small-scale maps, and like it also, is still today in full use.[g]

[a] *Kuo Yü (Wu Yü)*, ch. 19, pp. 5a, 16a; *Tso Chuan*, Duke Ai, 9th year (Couvreur (1), vol. 3, p. 657); *Wên Hsien Thung Khao*, ch. 158 (p. 1379.2); cf. Chi Chhao-Ting (1), p. 117. Whether or not this was before the construction of the Hung Kou giving access to the Yellow River westwards, Fu Chhai also constructed another low-gradient summit canal placing the upper reaches of the Ssu[4] River in communication with the Chi[5] near the city of Thao,[6] so that his ships could sail up to join those of the Prince of Chin. This was quite near the later line of the Grand Canal, but the work had no long life. Its date of completion was −483 (cf. *Kuo Yü (Wu Yü)*, ch. 19, pp. 8b, 9a, 16a).

[b] *Shih Chi*, ch. 29, p. 2a (Chavannes (1), vol. 3, p. 522; Watson (1), vol. 2, p. 71). Tradition ascribed these works to the direction of Wu Tzu-Hsü[7] (cf. Vol. 3, pp. 485 ff.). They were doubtless a model for the later Grand Canal in this region, if not exactly on the same alignment. Cf. Liu Tshai-Yü (1).

[c] *Shih Chi*, ch. 6, p. 21b, tr. Chavannes (1), vol. 2, p. 167.

[d] *Shih Chi*, ch. 6, p. 22a, tr. Chavannes (1), vol. 2, pp. 168 ff.

[e] This early Chhing chart was at one time in the collection of Mr Rewi Alley, to whom we are much indebted for the photograph and for permission to reproduce it here.

[f] A fuller description will be found in Fang Chi (2). Good photographs of the remains of the ancient works as they are today may be seen in the Kansu Provincial Museum at Lanchow. About a hundred miles further down the Yellow River, where it slowly turns from its northward to its eastward course above the Ordos Desert, similar systems of canalisation have been constructed at various times since the Han, also on the river's left bank. Large areas around Têng-khou[8] and Shenpa[9] are thus watered. Missionaries of the Latin Church were long active in recent times in the formation of villages and the promotion of canal-building in these districts, hence the detailed studies of one of their clergy, van Hecken (1, 2). His work is particularly instructive in connection with the conflict between Mongolian pastoral life and Chinese irrigated agriculture (cf. Vol. 1, pp. 67, 100 ff., 224).

[g] For an account of the great improvements which have been made during the past twenty years, see Shen Su-Ju (1). On the pedology, Wang Chi-Chih (1).

| [1] 射陽湖 | [2] 蒙恬 | [3] 中衛 | [4] 泗 | [5] 濟 | [6] 陶 |
| [7] 伍子胥 | [8] 磴口 | [9] 陝壩 | | | |

Coming to the Han period, the time of Han Wu Ti was a period of much civil engineering activity; a description of the great struggles against the Yellow River breaches has already been given (p. 234). By now the grain tribute coming from east of the passes had enormously increased.[a] In −133, therefore, Chêng Tang-Shih[1] proposed that a canal of about 100 miles in length should be cut between Chhang-an the capital and the Yellow River, so as to shorten the distance normally traversed along the Wei River by two-thirds and reduce the time by a half. He said:[b]

'If we take water from the Wei River to fill a canal dug from Chhang-an to the Yellow River, running along the foot of the southern mountains for a little more than 300 *li* (in a straight line), transportation will be greatly eased. Passages would take only three months (instead of six as hitherto). Besides, the people living below the canal would be able to irrigate more than 10,000 *chhing* (about 166,000 acres). Thus in one sweep we shall shorten the time taken by the (grain) transports, reduce the number of men required, augment the fertility of lands within the passes, and obtain fine harvests.' The emperor approved the project. He commissioned the hydraulic engineer (*shui kung*[2]) Hsü Po,[3] a man from Chhi, to survey the course of the proposed canal, and to recruit all available men, to the number of several tens of thousands, to excavate it. In three years the work was completed, and great advantage accrued to transportation, which gradually increased in volume, while the people living below the canal found its water most beneficial for their fields.

Seven centuries later, the Sui engineers reconstructed this canal and used it as part of their great system whereby Chhang-an was in direct water communication with Hangchow, but we do not know whether they used exactly the same course as that of Hsü Po.[c] In any case, the work, which received the name of Tshao Chhü[4] (Grain Traffic Waterway, *par excellence*), remains the classical Chinese example of the lateral transport canal.

Not all works of this period were equally successful. Phan Hsi,[5] governor of Ho-tung (Shansi, across the Huang Ho), urged that owing to the famous difficulties of navigation at Ti-Chu[6] on the Yellow River, it would be advisable to increase cultivated land nearer the capital. He therefore proposed in −129 that the Fên[7] River, coming down south-westwards out of Shansi, should be made to give rise to canals irrigating both the north and south sides of the valley where it flows into the Yellow River—furthermore, that water from the Huang Ho itself should be led off to irrigate several *hsien* of poor land within the sharp bend where it goes east at Thung-kuan.[d] Several tens of thousands of men were accordingly set to work but unfortunately a few years later the

---

[a] *Shih Chi*, ch. 30, pp. 2*b*, 3*a* (Watson (1), vol. 2, p. 81).
[b] *Shih Chi*, ch. 29, p. 4*a*, tr. auct. adjuv. Chavannes (1), vol. 3, p. 527; Watson (1), vol. 2, p. 73.
[c] It had also been repaired about +480 in the time of Hsüeh Chhin (cf. p. 278 below). Later, in the Thang, it retained great importance, and was thoroughly put in order again by Wei Chien[8] in +741, who organised a remarkable exhibition of the produce of all the provinces on ships in the docks at Chhang-an which he had constructed or enlarged. Cf. Twitchett (4), pp. 90, 308 ff.; Pulleyblank (1), pp. 36 ff., 207.
[d] *Shih Chi*, ch. 29, p. 4*b* (Chavannes (1), vol. 3, p. 528; Watson (1), vol. 2, p. 73).

---

[1] 鄭當時　　[2] 水工　　[3] 徐伯　　[4] 漕渠　　[5] 番係　　[6] 砥柱
[7] 汾　　[8] 韋堅

Yellow River changed its course and spoiled the whole scheme. All the land reclaimed by the Fên River Project was thus laid waste.[a]

This is the place to say a few words more about the navigation of the Yellow River upstream from the Hung Kou junction to the great bend at Thung-kuan and then on by the Wei or its parallel canal to the capital at Chhang-an (Sian). Perhaps the most dreadful and perennial headache for the transport authorities century after century was the rocky defile at a gorge called San-mên Hsia[1] (Three Gates). Here the river

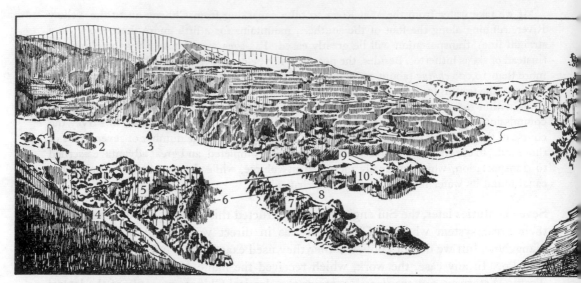

Fig. 879. The redoubtable gorge of the Three Gates (San-mên Hsia) on the Yellow River, for centuries a grave obstacle to safe navigation and now the site of a great modern dam (redrawn from Anon. *33*).

1 Dressing-table Island (Shu-chuang Thai)
2 Lord Chang's Island (Chang Kung Shih)
3 The Grindstone, rock (Ti-chu Shih)
4 The New Canal of the Khai-Yuan reign-period (Khai-Yuan Hsin Ho); a black arrow marks its upper entrance
5 Gate-of-Man Promontory (Jen Mên Tao)
6 Gate of Man (Jen Mên)
7 Gate-of-Spirits Island (Shen Mên Tao)
8 Gate of Spirits (Shen Mên)
9 Gate of Devils (Kuei Mên)
10 Gate-of-Devils Island (Kuei Mên Tao)

The river is flowing from left to right; the vista on the right is downstream. The thin suspension bridges were part of the work for the dam.

swings in a broad curve past promontories of porphyritic diorite where two rocky islands divide the stream into three roaring rapids, dedicated respectively to spirits, devils and men. Whenever the presence of the capital graced Chhang-an, as notably in the Early Han and the Thang periods, the grain traffic had to be worked upstream

[a] Another unsuccessful project concerned the Lo River, which joins the Huang Ho between the Fên and the Wei, but mention of this will come more appropriately in the sub-section on techniques (p. 334).

[1] 三門陝

through this pitiless place, resulting in untold losses of ships, men and produce.[a] The name Ti-chu, which became a byword, was that of a pinnacle of rock which stuck out of the water, with others, upstream of the two large islands, as if placed purposely by Nature to trap the ascending barque which had successfully negotiated the rapids, or, in worse case, carried away from the hauling cables of the trackers (cf. Fig. 879).[b] The place would be fit subject for an epic narrative, not least among the parts of which would be the praise of the enormous multi-purpose dam now being completed at the spot.[c]

As we have just seen, Phan Hsi wanted to reduce the traffic past this danger-point by increasing productivity west of it. Other suggestions, many times repeated, envisaged vast detours in order to avoid it. This explains the significance of the proposal we already studied (p. 21 above) for the Pao-Yeh Road. The bold idea of Chang Thang[1] was to bring grain from the south-east (the east-central key economic area) right over the Chhin-ling Mountains, using a road portage between the head-waters of the Pao[2] and the Yeh[3] Rivers. The latter flowed northward quickly into the Wei[4] and so communicated with the capital; the former descended southward to join the Mien[5] and hence the Han,[6] the Yangtze's greatest tributary. It is possible that Chang Thang had in mind to avoid the complicated region of lakes near modern Hankow and the swift current of the great river by making use of a small 'back-door'. Traffic proceeding up the Huai River into modern Honan could turn right into the Ju[7] and then left into the Chen.[8] One had then only to cross a low and narrow watershed to find the Pi[9] (now the Thang[10]) flowing westwards to join the Han River at Hsiangyang, and there is some reason to think that the barrier was crossed by another of the low-gradient summit canals, in this case built as early as −603, when Chhu State invaded Chêng.[d] It would be interesting to know if any traces of it are left today; at any rate, this gap in the hills along the Honan–Hupei border would have provided a much easier route than the Yangtze itself, even if road portage had always been necessary. Unfortunately the Pao–Yeh route failed in its main object because the canalisation of the Pao and Yeh streams themselves proved too tough a job in the absence of explosives and pound-locks or even sufficiently strong and numerous flash-locks.[e]

This detour was a vast southern arc, but a similar conception was attempted in the north. One usually thinks of the province of Shansi as bisected lengthwise by the Fên[11]

[a] During the Thang period, for example, losses ranged between 20 and 50%.

[b] A splendid account of the whole place, with its rock-cut trackers' galleries and the inscriptions in them, has been issued recently, Anon. (33). Intensive archaeological work was done because much of the remains will be submerged by the new dam.

[c] Cf. Têng Tsê-Hui (1); Li Fu-Tu (1). With a length of 2,600 ft. and a height of nearly 500 ft., its storage capacity will exceed that of the Boulder and Grand Coulee dams combined. Besides mitigating the risk of floods it will irrigate an area equal to a third of all the arable land in England, make the river navigable up to Lanchow, and produce 1·1 million kilowatts of power.

[d] *Tso Chuan*, Duke Hsüan, 5th year (Couvreur (1), vol. 1, p. 589).

[e] The Han valley route was used occasionally in later centuries, e.g. around +756, when normal communications were interrupted in the rebellions of An Lu-Shan and others. Cf. Twitchett (4), pp. 91, 309.

[1] 張湯 [2] 襃 [3] 斜 [4] 渭 [5] 沔 [6] 漢 [7] 汝
[8] 溱 [9] 泌 [10] 唐 [11] 汾

River from north to south, but in fact the Fên arises not far north of Thai-yuan, and the uplands above are drained by another river, the Hu-Tho,[1] which skirts the Fên valley in a sickle-shaped valley of its own and falls down on to the Hopei plain to empty into the old bed of the Yellow River (the present line of the Grand Canal) just before the Tientsin bend. Grain having reached the hub point at Jungyang it was therefore possible to take it not directly westwards up the Yellow River, but north-eastwards to the point of entry of the Hu-Tho, after which it could go slowly past Chêng-ting[2] up into the hill-country of Shansi. In the Early Han (−1st century), when the Yellow River was still in course ①, an attempt was made to join the head-waters of one of the Hu-Tho's tributaries, the Yeh,[3] with those of the Tung-Kuo[4] Ho, a tributary of the Fên. A summit canal is marked across this pass near Shouyang[5] on contemporary maps, but in antiquity it was not a practical proposition and was soon abandoned, except for portage traffic. Between +69 and +78 a more determined attempt was made to connect the Hu-Tho and the Fên. Further up the Hu-Tho itself, a tributary, the Mu-Ma[6] Ho, led up past Hsin-hsien[7] to the Shih-ling Kuan[8] pass, on the other side of which the Lo-Yin[9] Ho led downwards to the Fên. By this time the Yellow River had shifted to course ②, but enough water must have been still flowing in course ① to make the detour possible, if not very attractive. After nearly a decade of effort, the attempt was given up, and a portage cart-road substituted. Most probably these audacious schemes were frustrated not only by the inadequacy of the gates and the absence of pound-locks but by the great difficulty of supplying water to the summit levels—nevertheless when we see that both these passes are traversed by railways today we admire the perspicacity of the Han surveyors.

Of course the remedy for all this was not to have the capital at Chhang-an at all. It was certainly a material reason why the Later Han dynasty fixed its capital at Loyang, and why in the Thang the imperial seat oscillated continually between the two.[a] To improve transport to Loyang a lateral navigation canal was cut along the Lo[10] River in +48, the Yang Chhü[11] already noted in the sub-section on bridges (p. 172 above). Its builder, Chang Shun,[12] the Minister of Works at the time, made it terminate at the west of the city,[b] passing south of the walls somewhere near the place where Pi Lan's[13] water-raising installations were to arise a century or so later.[c] The need had been felt for some time, but Chang Shun was successful where Wang Liang[14] had failed. For in +29 Wang Liang had opened a canal to lead the Ku Shui[15] stream north of the capital to join the Lo River lower down towards the Huang Ho and so to form a navigable waterway, but when the work was completed it was found that the levels had not been well taken and the water refused to flow.[d] Learning the hard way, perseverance, and man-power were the features of ancient Chinese civil engineering.

[a] Cf. Chhüan Han-Shêng (1), pp. 20 ff.; Twitchett (4), pp. 86 ff.; Pulleyblank (1), p. 33.
[b] He had an intake of water from the Lo River itself.
[c] Vol. 4, pt. 2, p. 345.
[d] *Hou Han Shu*, ch. 52, p. 5 b; *Thang Yü Lin*, ch. 8, p. 23 a.

[1] 滹沱	[2] 正定	[3] 冶	[4] 洞渦	[5] 壽陽	[6] 牧馬
[7] 忻縣	[8] 石嶺關	[9] 洛陰	[10] 洛	[11] 陽渠	[12] 張純
[13] 畢嵐	[14] 王梁	[15] 穀水			

While we are on the subject of the San-mên gorge, we had better follow it through succeeding centuries. In −17 Yang Yen[1] tried to remove the obstacles by main force. The *Chhien Han Shu* says:[a]

In the 4th year of the Hung-Chia reign-period Yang Yen submitted that 'in sailing up and down the River there are continual calamities in the narrows of Ti-chu, but it would be possible to widen and deepen it there'. The emperor followed his advice and sent him to carry out the work. (His men) cut and hacked away at the Ti-chu rock until it was almost submerged by the waters, but they could not get rid of it altogether and the river boiled in fury worse than before, so the danger was increased instead of lessened.

The question has been raised whether the technique of steel-making was at this time sufficiently developed to give adequate tools for such an operation; in our opinion it certainly was,[b] but for work at low-water periods time pressed, and the job was surely impossible without explosives. After this the San-mên problem was traditionally handled in three different ways, by cutting trackers' galleries in the cliffs, by building detour roads to avoid the passage, and by short-circuiting the rapids with the aid of a rock-cut canal.

At an earlier point (p. 22) we said a good deal about the gallery roads on brackets (*chan tao*[2]) built anciently through mountain defiles in various parts of China, and these galleries at San-mên were a special case. Their remains can still be seen, and the multifarious inscriptions they bear have been carefully collected.[c] Wooden planks floored the decks of half-tunnel ways excavated in the perpendicular cliff-faces, widened by wooden baulks supported on projecting beams (Fig. 880, pl.), but naturally without balustrades because of the hauling cables. These galleries are generally from 15 to 30 ft. above the average river level. At some places there are sunk bollards cut out of the living rock. The oldest inscription, dated +150, bears the name of Li Erh.[3] From the +3rd century (e.g. +221, +240, +260, +281) they are numerous, and the artisans and engineers are given their titles—the Shih Kung[4] (Mason) Liu Fang;[5] the Shih Chiang[6] (Master Mason) Chang Ling-Hsien;[7] the Tu Chiang[8] (Director of Waterways Engineering) Yao Shih;[9] the Chih Ho Tu Chiang[10] (Director of Waterways and River Conservancy Engineering) Tso Kung;[11] all these men of the San Kuo and Chin periods. The galleries were continually being repaired in later times, in the Sui (+595), in the Thang particularly between +684 and +707 by the military engineer Yang Wu-Lien[12] (already met with in other connections, Vol. 4, pt. 2, p. 163),[d] again in the Sung (+1066) and even as late as 1809, when chains were added along some sections for the pullers.

As early as +195 Li Yo[13] wanted a road and carts to circumvent the gorge, and Liu Ai[14] supported him, saying that no good junk captains were left, but nothing was done

---

[a] Ch. 29, p. 11*b*, tr. auct.
[b] Sect. 30*d* below; meanwhile Needham (*32*).   [c] Anon. (*33*).
[d] *Chhao Yeh Chhien Tsai*, ch. 2, pp. 19*a* ff.; cf. Pulleyblank (*1*), p. 128; Twitchett (*4*), pp. 86, 302.

[1] 楊焉	[2] 棧道	[3] 李兒	[4] 石工	[5] 劉方
[6] 石匠	[7] 張令仙	[8] 都匠	[9] 藥世	[10] 治河都匠
[11] 左貫	[12] 楊務廉	[13] 李樂	[14] 劉艾	

at that time. The old Han road from Thung-kuan to Loyang (map, Fig. 711) of course remained available, but transport along it was very expensive.[a] About +480, under the Northern Wei, Hsüeh Chhin[1] therefore strongly urged a return to the water route along the Yellow River and through the gorge, together with adequate repair of the Chhang-an Canal; this found favour and the necessary work was put in hand though not completed. After fifty years, however, the galleries again fell into disrepair and traffic returned to the road. Not until +656 was a specific short-circuit road around the gorge attempted, when Chhu Lang[2] made one to the south, but the approaches to the quays were badly planned and swept away in floods.[b] The problem was solved successfully so far as it could be in +733, when Phei Yao-Chhing[3] took over as Imperial Transport Commissioner and built not only the six-mile road on the northern side through rock cuttings some of which can still be seen today,[c] but also large granaries one above and one below the gorge as warehouses for the trans-shipment.[d] After earthquake damage the road was restored and double-tracked in +785 by Li Pi[4] and at some periods greatly used, but the land portage remained a nuisance.

It was in order to avoid it that a few years later, in +741, Li Chhi-Wu,[5] Prefect of Shenchow, constructed the remarkable rock-cut canal west of Man-Gate 'Island' (Jen Mên Tao[6]) still to be seen (Fig. 881), but soon to be submerged behind the San-mên dam.[e] This is a regular steep-sided cutting some 840 ft. long with an average width of 22 ft. and an average depth of 30 ft.; it has a trackers' gallery in its eastern wall and was crossed by a bridge in Thang or Sung times. Unfortunately it was never deep enough to carry a great deal of traffic, perhaps because Li Chhi-Wu was starved of funds, perhaps because of rapid silting or the excavation of the Yellow River's own bed, so that it was used mainly during the summer flood season. Here again it is known that ample use was made of fire-setting with vinegar to split the rocks by steam.[f] On account of its date it was known as the Khai-Yuan Hsin Ho[7] or, by dedication to a Taoist goddess,[g] the Niang-Niang Ho.[8]

In course of time the navigation of the San-mên gorge ceased to occupy its position of central conundrum in the communications network. Significantly the capital of the Chinese empire at the time of the next unification, that of the Sung, was fixed immedi-

---

[a] The difficulties and dangers of this road in Thang times have been studied by Chhüan Han-Shêng (1) and Twitchett (4), pp. 84, 90 ff., 301, 308.

[b] *Hsin Thang Shu*, ch. 53, p. 1*a*; *Thang Hui Yao*, ch. 87 (p. 1595); cf. Twitchett (4), pp. 86, 302; Pulleyblank (1), p. 128.   [c] Anon. (*33*), pl. 40*a*.

[d] Cf. the account of his work and its background by Pulleyblank (1), pp. 34 ff., 129, 183 ff., 186, 201, and Twitchett (4), pp. 87 ff., 303, 306 ff.

[e] *Hsin Thang Shu*, ch. 53, p. 2*a*; cf. Twitchett (4), pp. 89, 307; Pulleyblank (1), pp. 131, 206.

[f] Cf. p. 26 above. Fire-setting is also recorded for Yang Wu-Lien's work on the galleries. One of the oldest references to the technique must be the passage in ch. 3 of the *Hua Yang Kuo Chih* referring to the widening of a river gorge in Warring States times (Shu). The use of vinegar instead of water seems to appear first in the Thang; cf. *Liu Pin-Kho Wên Chi*, ch. 8 (p. 67). For other references of Han and Thang date see Anon. (*33*), p. 69; for a Sung one, *Kung Khuei Chi*, ch. 91, p. 5*b*. Cf. Sect. 36 below and meanwhile Sandström (1), p. 29.

[g] Perhaps Shui Mu Niang-Niang,[9] originally a spirit of the Ssu River and the old Pien Canal (Doré (1), vol. 10, p. 796).

¹ 薛欽   ² 褚朗   ³ 裴耀卿   ⁴ 李泌   ⁵ 李齊物
⁶ 人門島   ⁷ 開元新河   ⁸ 娘娘河   ⁹ 水母娘娘

ately at Khaifêng, just beside the hub of the transport system, and not far away to the west. When it had to be withdrawn to Hangchow it was still at the centre of a great key economic area, even though the northern one fell to the dynasties of J/Chin and Liao. In the unifications of the Yuan and Ming the eastern plains continued to be predominant, and the ancient stronghold of Kuan-chung declined steadily to provincial status. The

Fig. 881. Another view of the gorge of the Three Gates (San-mên Hsia) on the Yellow River, showing to the left the rock-cut by-pass canal engineered by Li Chhi-Wu in +741 (redrawn from Anon. 33). The numbers are the same as those in Fig. 879, and the black arrow again marks its upstream entrance.

San-mên gorge thus sinks to minor importance, carrying no more traffic than its galleries or circuit roads can bear, and after the coming of the railway deserted, until in our own time its walls are not of rock only but of concrete, and the roar of its rapids is replaced by the hum of dynamos.

Let us now return to the Han and Chin periods. In +274 there was a proposal to link the Yellow River with the upper valley of the Lo in order to facilitate water transport between Chhang-an and Loyang, avoiding the unpopular road.[a] But such a summit canal would have been quite impossible with the techniques available at the time, and nothing came of it. Nevertheless bold ideas had long been in the air. About three centuries earlier, in −95, there had been a proposal for the most grandiose (or imaginative) single project of all ancient and medieval Chinese civil engineering, nothing less than the diversion of the entire Yellow River so as to cut off its vast northern excursion into the Gobi Desert area. The idea was due to a man named

[a] Anon. (33), p. 65.

Yen Nien,[1] of whom otherwise nothing is known.[a] In the *Chhien Han Shu* we read:[b]

At that time (the 2nd year of the Thai-Shih reign-period) there was concern regarding the Huns (Hsiung-Nu). Those who were eager for achievements and profits, and who discoursed of (possible) advantages, were very numerous. A man from Chhi,[c] Yen Nien, offered a written proposal which declared: 'The (Yellow) River emerges from the Khun-Lun (Mountains) and passes through China to flow into the Pho Hai (Eastern Sea). This is its geographical setting, sloping from the north-west southwards and eastwards; one may observe the nature of the territory according to the maps (*thu shu*[2]). If now Your Majesty should order the hydraulic engineers (*shui kung*[3]) to survey the high and low places, and to open a great river which would come forth from the mountainous plateau (of Tibet), traverse the middle of the Hu (Huns') country, and flow eastwards to the sea—in this way the land east of the passes would be perpetually freed from flood disasters, and the northern frontier would not suffer from the Huns. Much labour would be saved by not having to make dykes and embankments, and by not needing so many men to guard the frontiers. The Hun robbers are a calamity, invading and plundering us, overthrowing our armies and slaughtering our commanders, so that their bones lie exposed in the wildernesses. The empire is always warding off the Huns, but it does not suffer from the hundred Yüeh (peoples, in the south-east) because streams separate them and cultivated lands divide them. This project will confer the greatest benefit upon ten thousand generations.'

When the proposal had been presented the emperor praised it, replying as follows: 'Yen Nien's suggestion seems to be a thoroughly well-considered one. But the (Yellow) River was directed into its present course by the Great Yü. When sages carry out their enterprises, they act for the benefit of ten thousand generations. Whatever has been done with divine perspicacity is, we fear, difficult to alter.'

If we understand this aright, Yen Nien's project would have been an immense work deriving from the Yellow River at some point between Lanchow and Ninghsia, and rejoining its lower course above or below the Lungmên gorges, at any rate before its great eastern bend at Thung-kuan. It would thus have followed roughly part of the course of the Great Wall, and would have abandoned only the Ordos Desert to the Huns. In a straight line alone the distance would have been 300 miles or more, and the emperor's advisers doubtless felt (quite rightly) that it was far beyond the powers of the empire at the time. What is particularly noteworthy is that Yen Nien saw that a shortening of the river's course through the loess country would lessen its silt-content and make its control much easier, quite apart from the military advantages of having a great river as a frontier.

Towards the end of the Early Han dynasty, a fine work was carried out (−38 to −34) by Shao Hsin-Chhen in southern Honan, namely the Chhien-Lu Pei[4] (reser-

[a] It is possible that Yen Nien was the same person as Chhêngma Yen-Nien[5] who came into prominence some sixty years later, for Yen as a family name would be unusual, but the time span makes it rather improbable (see p. 330 below).

[b] Ch. 29, p. 8a, tr. Bates (1), mod. auct. Cf. Tshen Chung-Mien (2), pp. 259 ff., who judges Yen Nien's plan impracticable in any age.

[c] The prominence of engineers from Chhi, Hsü Po and Yen Nien, as well as so many other scientific and technical experts from that ancient State, should not be overlooked (cf. Vol. 2 above, *passim*).

[1] 延年　　　[2] 圖書　　　[3] 水工　　　[4] 鉗盧陂　　　[5] 乘馬延年

voir). This dammed up one of the larger northern tributaries of the Han River, and irrigated 30,000 *chhing* (about half a million acres). It interests us for several reasons: first, Shao, who was prefect of Nanyang, introduced six sluice-gates set in stone which greatly aided in distributing the water. His work on water-rights we have already noted (p. 267), and so popular did he become that he was known as 'Father Shao'.[a] Secondly, one of his immediate successors in office was Tu Shih, equally beloved,[b] and none other than the introducer in +31 of water-wheels for metallurgical blowing-engines (cf. Vol. 4, pt. 2, pp. 370 ff. above). Hence the name of the dam, the 'Tongs-and-Furnaces Dam', can hardly be a coincidence. Nanyang remained for a long time the centre of an iron industry which employed water-power.

Though the Later Han dynasty was less active in hydraulic projects, valuable things were done. Wang Ching[1] rebuilt the Shao Pei (Peony Dam)[c] between +78 and +83.[d] His greatest work, however, carried out with Wang Wu[2] in the previous decade, was the thorough reconstruction of the Pien Canal, with the introduction of numerous flash-lock gates.[e] When the Yellow River was out of control, as it had been in +11 (cf. p. 241), there was always a tendency for it to escape eastwards and southwards, flooding the Chi[3] River and spoiling the works of the canal. Reinforcement of its banks with stone revetments in the years +1 to +5 had not sufficed to prevent this, but now Wang Ching's work was well done and endured for a considerable time. A century later (+189), Chhen Têng[4] built a series of weirs from the city of Shou-hsien[5] westwards, collecting the water from 36 streams over an area about 20 miles in diameter, and irrigating 10,000 *chhing*.[f] The remains of some of these structures came to light in 1959 and have since been excavated. It thus appears that they were made of alternate layers of rice straw and clayey soil based upon a gravel foundation, the stalks being laid parallel to the current flow, and the whole supported by wooden piling and coffering denser at the centre than at the ends.[g] Military canals were also built at this time. When Tshao Tshao was campaigning against Yuan Shao in +204 he built a whole series of small waterways running along the foot of the Shansi Mountains northward in Hopei, and making use of old Yellow River arms near Ta-ming.[6] Rivers such as the Chang[7] and the Hu-Tho[8] were thus connected by contour canals (with names such as the Li Tshao Chhü[9] and the Phing Lo Chhü[10]). After the army of the north had been subdued, these supply routes seem to have played little role in communications and some were overrun when the Yellow River assumed course ④, but part of the

---

[a] *Chhien Han Shu*, ch. 89, p. 15a.

[b] Someone made up a saying, 'Shao was a father to us, and Tu a mother', the historians record.

[c] *Hou Han Shu*, ch. 106, p. 8a.

[d] Wang Ching came of a family that had long been settled in Korea; later, p. 562, we shall come upon one of his ancestors, also a Taoist technician. His biography (*Hou Han Shu*, ch. 106, p. 6b) says that he was fond of techniques, studied astronomy and mathematics, and understood the principles of water (*nêng li shui chê*[11]).

[e] Cf. p. 346 below.

[f] Chi Chhao-Ting (1), p. 94.

[g] *NCNA*, 1961, item 06723. Cf. *TKKW*, ch. 1, p. 15a (+1637).

---

[1] 王景   [2] 王吳   [3] 濟   [4] 陳登   [5] 壽縣   [6] 大名
[7] 漳河   [8] 滹沱河   [9] 利漕渠   [10] 平虜渠   [11] 能理水者

system still exists in the form of the Pai Kou[1] between Peking and Paoting.[a] In its prime it reached even as far north as Ku-pei-khou[2] on the Great Wall. In the +3rd and +4th centuries much attention was concentrated on Chiangsu. Chang Khai[3] built several important reservoirs south of the Yangtze, notably the Hsin-fêng[4] Tank,[b] and Chhen Min[5] excavated the Lien[6] Lake, both in +321. About +350, Chhen Min greatly improved the communication between the Huai and the Yangtze, by cutting a new canal, the Shan-yang Yün-Tao,[7] so that the ancient Han Kou fell completely into disuse.[c] Another Chin official, Hsün Hsien,[8] built a short canal in Shantung in +352 near Tung-a,[9] using the water of the Wên[10] River.[d] The importance of these works was that they were both destined to become portions of the Grand Canal.

The Grand Canal as a unit was the creation first of the Sui, when it led to Loyang, and then of the Yuan, when it led to Peking. We shall soon give it the attention it deserves. Before proceeding to the more detailed description of such greater works, however, let us take a glance at the historical picture as a whole. It would obviously be impossible in the space available to treat of the works of all succeeding dynasties in detail such as that which has been devoted to those of the Chhin and Han, all the more so since they continually grew in number and extent. However, an interesting attempt has been made to treat the matter statistically. Chi Chhao-Ting worked through the topographical histories (cf. Vol. 3, pp. 517 ff. above) of all the provinces in a systematic way, noting down all references to hydraulic engineering undertakings of whatever kind.[e] The results are summarised from his data in Fig. 882 (an arithlog plot)[f] and the accompanying table.

	w/y ratio[g]
Chou and Chhin	0·0175
Han (both)	0·131
San Kuo	0·545
Chin	0·110
Nan Pei Chhao	0·118
Sui	0·932
Thang	0·88
Wu Tai	0·245
Sung	3·48
Chin (Jurchen)	0·166
Yuan	3·50
Ming	8·2
Chhing	12·0

[a] On this subject see Lo Jung-Pang (6), pp. 48 ff. and Chêng Chao-Ching (1), p. 194.
[b] *Chin Shu*, ch. 76, p. 11*b*.
[c] Chêng Chao-Ching (1), p. 196; Chi Chhao-Ting (1), p. 112.
[d] The object was military supply, to get Mujung Lan[11] out of Khaifêng.
[e] (1), pp. 36 ff.
[f] The breadth of the columns give the number of years of the dynasty, and their height the number of waterworks undertaken.
[g] Here w is the number of works undertaken and y the number of years.

[1] 白溝	[2] 古北口	[3] 張闓	[4] 新豐	[5] 陳敏	[6] 練
[7] 山陽運道	[8] 荀羨	[9] 東阿	[10] 汶	[11] 慕容蘭	

From this one or two fairly obvious conclusions can be drawn. For the earliest periods records are sparse and all one can say is that the beginnings are there; it was the formative period of Chinese society. The real start came in the Han, especially the Former Han, and the important economic areas of the time are clearly revealed by the

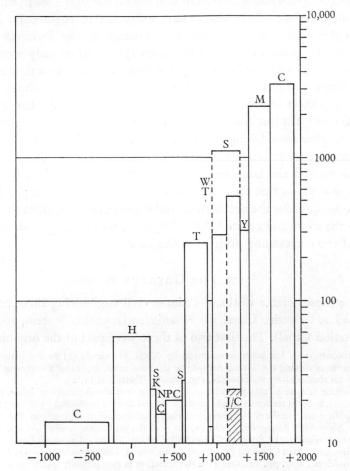

Fig. 882. Arithlog plot of the number of hydraulic engineering works undertaken in successive dynasties (data from Chi Chhao-Ting). For discussion, see text. C, Chou; H, Han; SK, San Kuo; C, Chin; NPC, Nan Pei Chhao; S, Sui; T, Thang; WT, Wu Tai; S, Sung; J/C, Jurchen Chin; Y, Yuan; M, Ming; C, Chhing. Chronological scale horizontal. The total Sung figure is higher than its two component columns for the Northern and Southern Sung periods respectively, because there were a certain number of projects which could not positively be allotted to either.

fact that Shensi and Honan have a big lead over all other provinces. The Three Kingdoms period, in spite of its short duration, shows a distinct volume of activity, attributable no doubt to strategic motives. Continued unsettlement from the +4th to the +7th centuries explains the somewhat poor performance of the lesser dynasties then regnant, but the unification of the Sui led immediately to a great development of civil engineering, especially with regard to the Grand Canal, and this was sustained throughout the Thang.[a] By now provinces such as Hupei and Fukien have appeared in the records, and during the Thang for the first time a southern province, Chekiang, outstripped every northern province. The Sung dynasty continued this progress, and works in certain provinces now first reached three figures. The breakdown shows that there was more activity after the removal of the capital to the south than before, so that the nomadic invasions may be said to have stimulated more intensive cultivation of the lower Yangtze valley[b] and the valleys of Kuangtung and Fukien.[c] While northern 'nomadic' empires such as that of the Chin (Jurchen) dynasty were not absolutely barren of hydraulic engineering works,[d] the figures indicate that they never understood their importance as did the truly Chinese dynasties. For example, no large water-conservancy project of any kind was put in hand during the two centuries of Liao (Chhi-tan) rule.[e] Yet this does not apply to the Yuan (Mongol) dynasty, which undertook the complete remodelling of the Grand Canal, involving works of great scale, and began irrigation in Yunnan. Heightened technical ability from the +15th century onwards is seen in the large figures for the Ming and Chhing, but the analysis by provinces also shows that these dynasties tried hard to increase production in Chihli (Hopei) so as to render the northern capital region as independent as possible of transport from the south and centre.[f] With this, then, we may begin our tour of inspection of some of the outstanding projects of the past.

(7) THE GREATER WORKS

Among the most notable works of Chinese civil engineering three are of Chhin date (the Chêngkuo Irrigation Canal, the Kuanhsien Irrigation System, and the Ling Chhü Transportation Canal). The first two of these were part of the organisation of the key

[a] An elaborate work has been consecrated by Naba Toshisada (2) to the history of irrigation and water-conservancy during the Thang period; he uses materials from the Tunhuang MSS, and includes information on sluice-gates, water-mills, etc. So also Twitchett (2, 4).

[b] A particular feature of this period and this region was the drainage of lakes, swamps and bayous (old river-beds), to form reclaimed land (yü thien[1] or wei thien[2]) protected by dykes; cf. Chi Chhao-Ting (1), pp. 134 ff., who translates the interesting memorial of Wei Ching[3] on the subject (c. +1200), recorded in Shou Shih Thung Khao, ch. 12, pp. 12b ff.; and Sudō Yoshiyuki (1), who shows that the process had begun under the Southern Thang. Governor Chhien (of sea-wall fame, cf. pp. 320 ff. below) was very active in it (Miu Chhi-Yü (1) expounds his work). Chi reproduces two old Chhing charts of the reclaimed 'polders', opp. pp. 136 and 148. Reclamation is still actively going on today (cf. Hung Hsia-Tien, 1), no longer causing the severe social stresses that it did in Sung times.

[c] In Sect. 28b (p. 32 above), we met with a woman road engineer, a Fukienese Taoist of the +14th century. She had had a predecessor in hydraulic constructions—a girl of the same province, named Chhien Ssu-Niang,[4] who in +1064 began the Mu-lan[5] dam near Phu-thien.[6] In bringing the project to completion she was assisted by a local scholar and a military officer. The story has been told by Kuo Khêng-Jo (1). On the role of local initiative see Twitchett (6).

[d] For example, Khung Thien-Chien[7] did good work in this dynasty (+1196) on the irrigation system at Meihsien,[8] south of the Wei River in Shensi (see Fu Chien, 1). It is also from this region and this century that we have one of the few examples of an aqueduct bridge in China, the Hui-Yuan C.[9] at Hungtung[10] in the Fên River valley in southern Shansi, built in +1136 and carrying an irrigation channel high above a tributary (see Lo Ying (1), p. 92). Cf. Vol. 4, pt. 2, p. 128.

[e] Cf. Wittfogel & Fêng (1), pp. 122, 365, 371, 373, 374.

[f] Some allowance has to be made in the Chhing figures for the fact that many works were repairs, improvements or enlargements, and for the fact that the statistics do not come down as far as 1911.

[1] 圩田   [2] 圍田   [3] 衛涇   [4] 錢四娘   [5] 木蘭陂
[6] 莆田   [7] 孔天監   [8] 郿縣   [9] 惠遠橋   [10] 洪洞

economic areas (north-western and western) of the −3rd century, and the third was a brilliant achievement connecting the north and centre with the extreme southern regions. These compare in greatness of conception, if not in length, with the Grand Canal itself, a work mainly of Sui and Yuan times (the +7th and the +13th centuries). For this reason we have reserved them for more detailed consideration now.[a] In connection with the Grand Canal a few words will be added about a type of project rather different from any of the others, the Chhien-Thang Sea-wall, which protected its southern terminus, Hangchow.

(i) *The Chêngkuo irrigation canal* (*Chhin*)

The story of the Chêngkuo Canal is carefully told both in the *Shih Chi* and in the *Chhien Han Shu*[b].

(The prince of) Han,[1] hearing that the State of Chhin was eager to adventure profitable enterprises, desired to exhaust it (with heavy activities), so that it should not start expanding to the east (and making attacks on Han). He therefore sent the hydraulic engineer (*shui kung*[2]) Chêng Kuo[3] (to Chhin) to persuade deceitfully (the king of) Chhin to open a canal from the Ching[4] River, from Chung-shan[5,c] and Hu-khou[6] in the west, all along the foot of the northern mountains, carrying water to fall into the river Lo[7] in the east. The proposed canal was to be more than 300 *li* long, and was to be used for irrigating agricultural land.

Before the construction work was more than half finished, however, the Chhin authorities became aware of the trick. (The king of Chhin) wanted to kill Chêng Kuo, but he addressed him as follows: 'It is true that at the beginning I deceived you, but nevertheless this canal, when it is completed, will be of great benefit to Chhin. [I have, by this ruse, prolonged the life of the State of Han for a few years, but I am accomplishing a work which will sustain the State of Chhin for ten thousand generations.]'[d] The (king of) Chhin agreed with him, approved his words, and gave firm orders that the canal was to be completed. When it was finished, rich silt-bearing water (*thien o chih shui*[8]) was led through it to irrigate more than 40,000 *chhing* (667,000 acres) of alkali land (*lu chih ti*[9]). The harvests from these fields attained the level of one *chung*[10] (= 64 pecks, *tou*[11]) per *mou* (i.e. they became very abundant). Thus Kuanchung[12] (the land within the passes) became a fertile country without bad years. (It was for this reason that) Chhin became so rich and powerful, and in the end was able to conquer all the other feudal States. And ever afterwards the canal (bore the name of the engineer and) was called the Chêngkuo Canal.[e]

---

[a] The selection of works for description in this sub-section is of course quite arbitrary. Lack of space has necessitated the omission, for instance, of the elaborate irrigation system of the upper Han valley, near Hanchung; see Chhen Tsê-Jung (*1*); Yang Ping-Khun (*1*); Chhen Ching (*1*). This was started by Hsiao Ho[13] in the −3rd century, and repaired and enlarged in many subsequent ages, e.g. by Chou Mi[14] in the Sung, Phu Yung[15] in the Yuan, Chhiao Chhi-Fêng[16] in the Ming, and Chang Kung-I[17] in the Chhing.

[b] Ch. 29, p. 3*a*, in both works, tr. Chavannes (*1*), vol. 3, p. 524; Watson (*1*), vol. 2, p. 71; mod. auct.

[c] Modern Ching-yang.[18]

[d] These words appear only in the *Chhien Han Shu* text.

[e] Ssuma Chhien's estimate of the acreage irrigated is considered impossibly high by Eliassen & Todd (*1*), but they grant 400,000 acres as the Han optimum. Today some 85,000 acres are irrigated by 87 miles of main canals. The minimum, in the worst circumstances at the end of the Chhing, was less than 2,000 acres, and this was largely from springs.

[1] 韓	[2] 水工	[3] 鄭國	[4] 涇	[5] 中山	[6] 瓠口
[7] 洛	[8] 填閼之水	[9] 鹵之地	[10] 鍾	[11] 斗	[12] 關中
[13] 蕭何	[14] 周宻	[15] 蒲庸	[16] 喬起鳳	[17] 張拱翼	[18] 涇陽

One could hardly have a more interesting record from the beginnings of such a typically Chinese technique as irrigation engineering. The fact that the Han people thought that those of Chhin could be deceived indicates, as Chi Chhao-Ting says, that the pros and cons of large-scale public works were not yet always clear to the feudal rulers. The willingness of the Chhin State to adopt Legalist innovations (cf. Vol. 2, pp. 204 ff. above) perhaps suggested that it would easily swallow a new and grandiose irrigation project which might be unsuccessful. But there is no reason to think that Chêng Kuo sabotaged it; he seems to have been professionally honest, and perhaps only realised himself during the course of the work what it would mean for Chhin once it had been completed. Ssuma Chhien understood perfectly the fundamental importance of the increase of productivity, the supply potential, for the ultimate political success of Chhin, and must be regarded as one of the first historians to appreciate that the supply potential is at least as important as military power in great international conflicts.[a] Moreover, he was fairly near the events in question, for he laid down his brush in −90, and Chêng Kuo's work had been brought to completion in −246, the year which witnessed the crowning of the future First Emperor as King of Chhin.

Only a few years before Ssuma Chhien was writing, extensive additions to the Chêngkuo Canal had been proposed and carried out by Erh Khuan (−111),[b] supplementary lateral contour branches irrigating land higher than the main canal. Again, in −95, another high official, Pai Kung,[1] pointed out[c] that as the Chêngkuo Canal had become so silted up as to lose much of its value, he therefore proposed to tap the Ching River a good deal higher up, and to carry a new canal for some 62 miles following a contour above that of its predecessor. This was successfully accomplished,[d] and we have read at an earlier point the popular song that the peasants made to commemorate Pai's achievement,[e] which they named Pai Chhü[2] in his honour.

Re-cutting the canals and tapping the Ching River higher up was a process which had to be done continuously for twenty centuries.[f] The Wei Pei[3] irrigation area, as it is called (Fig. 883),[g] is unusual in that although one of the first really large projects in China it has existed in use to the present day, and has been thoroughly studied by modern engineers.[h] The re-cutting of the canals has been due to the constant battle

---

[a] His father, Ssuma Than, was doubtless of the same opinion, for there is mention of the Ching and Wei irrigation systems in the biography of the 'unsuccessful Brutus' of China, Ching Kho (*Shih Chi*, ch. 86, p. 11 b), which, from internal evidence, must have been written by Ssuma Than. On this see Bodde (15); and on the general subject, Walker (1), pp. 41 ff.; Perelomov (1).

[b] See above, p. 264.

[c] *Chhien Han Shu*, ch. 29, p. 8 a. Naturally Ssuma Chhien does not mention this.

[d] The work involved at least one rather deep cutting. This still exists, and good photographs of it are to be seen in the Shensi Provincial Museum at Sian.

[e] P. 228 above. Brief general accounts of the history of the system will be found in Fang Chi (2), pp. 17 ff.; Chi Chhao-Ting (1), pp. 75 ff., 87 ff.; Chêng Chao-Ching (1), pp. 270 ff.; Nesteruk (1), pp. 52 ff.; Lowdermilk (7); Todd (1).

[f] Works of importance were carried out in, e.g., +377, +823, +958 and +1074.

[g] Description and map in Hu Huan-Yung, Hou Tê-Fêng & Chang Han-Ying (1), vol. 3, pp. 139 ff.

[h] Lowdermilk & Wickes (1); Eliassen & Todd (1); Li I-Chih (1); Chhen Tsê-Jung (1).

[1] 白公   [2] 白渠   [3] 渭北

with silt, which is never won,[a] and the reason for moving the intake higher and higher up the Ching is that the river has been continually eroding its bed. The original Chhin or Han intake is still identifiable, but this point is now no less than 50 ft. above the present level of the river.[b] As time went on there were many modifications; thus in

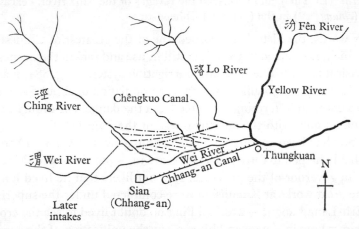

Fig. 883. Sketch-map of the Chêngkuo Canal irrigation system, first completed in −246 and still now in use under the name Wei Pei, or Ching-Hui Chhü, Irrigation Area. The small south-flowing river within the area is the Shih-chhuan Ho. The Lo R. now falls into the Wei R. and not directly into the Yellow River. East of the Lo, within its sharp bend, there is another ancient irrigation system, dating from *c.* −110 and still in use; this is now called the Lo-Hui Chhü after its main canal. Though not shown here, it is discussed in some detail on p. 333 below. Scale approx. 1:4,200,000.

+995 Liang Ting[1] and Chhen Yao-Sou[2] introduced 176 new sluice-gates,[c] while in +1310 we hear of a new rock cut being made by the use of fire-setting,[d] under Wang Chhêng-Tê.[3] Today the intake is far up the Ching gorges where the river has a rocky bed, and includes a 1,300 ft. tunnel above a massive dam; lower down the canal is carried under torrent streams from the northern mountains by eleven bridges.[e]

  [a] The burden rises to as much as 51 % during summer rains. Flood flow is of dramatic onset, rising 50 ft. in 10 minutes.
  [b] A photograph of what remains of the Chhin or Han head-works is given by Nesteruk (1), fig. 24. A considerable amount of well-cut stone masonry is still to be seen.
  [c] *Sung Shih*, ch. 94, p. 20b.
  [d] *Yuan Shih*, ch. 65, p. 13b. Two 'fire-experts' (*huo chiang*[4]) were employed; just possibly this may be a reference to the use of gunpowder in excavation, but fire-setting and steam rock-splitting is more probable. On this cf. pp. 26 and 278 above and p. 343 below, as also Sect. 36 below.
  [e] Until his death in 1940 Li Hsieh (Li I-Chih), one of the most outstanding of Chinese hydraulic engineers (cf. p. 217), was chief engineer of the Wei Pei system. Sigurd Eliassen from Norway was resident engineer during the construction of the new intake between 1930 and 1933; he afterwards wrote a vivid and readable (if possibly somewhat embroidered) account (1) of the life and the times, as seen by an engineer working under the auspices of the International Famine Relief Commission.

  [1] 梁鼎   [2] 陳堯叟   [3] 王承德   [4] 火匠

(ii) *The Kuanhsien division-head and cut* (*Chhin*)

In the country of Shu[1] (Szechuan), wrote Ssuma Chhien:[a]

(Li) Ping,[2] the governor, cut through the (shoulder of a mountain, so as to make the) 'Separated Hill' (Li-Tui[3]), and abolished the ravages of the Mo[4] river,[b] excavating the two great canals (*chiang*[10, 11]) in the (plain of) Chhêng-tu.[5]

In these few words, the historian recorded one of the greatest of Chinese engineering operations which, now 2,200 years old, is still in use and makes the deepest impression on all who visit it today.[c] The Kuanhsien irrigation system (Fig. 884) made it possible for an area of some 40 by 50 miles to support a population of about five million people, most of them engaged in farming, and free from the dangers of drought and floods. It can be compared only with the ancient works of the Nile.[d]

In −316 Shu had been conquered by the Chhin generals[e] Chang I[6] and Chang Jo.[7] Li Ping probably helped to fortify its chief cities in −309. In −250 Prince Hsiao Wên appointed him governor of the province. It is not likely that he lived long after about −240, so the great works at Kuanhsien were completed under the superintendence of his son Li Erh-Lang,[8] about −230.[f] Li Ping no doubt lived to see the crowning of the ruler of Chhin as king in −246, and his son saw the unification of the empire in −221, but in any case there can be no doubt that the project they carried through was (like the Chêngkuo Canal) one of the great sources of strength of the Chhin State and Empire.[g]

At Kuanhsien the Min Chiang[9] River flows into the basin of Szechuan from its source in the hills of the extreme north of the province surrounding Sungphan. Li Ping decided to divide it into two great Feeder Canals, the Inner (Nei Chiang[10]) and the Outer (Wai Chiang[11]), by means of a division-head of piled stones, known as the Fish Snout (Yü Tsui,[12] Fig. 887, pl.), about the point where the famous suspension bridge (see p. 192 above) crosses the river. The general layout of these head-works can be appreciated from the map in Fig. 884, the model[h] in Fig. 888 (pl.) and the views in Figs. 885 and

[a] *Shih Chi*, ch. 29, p. 2*a*, tr. Chavannes (*1*), vol. 3, p. 523; Watson (*1*), vol. 2, p. 71, mod. auct.

[b] Now a name for a tributary of the Min[9] R., then perhaps used for the main stream.

[c] I have had the pleasure and privilege of visiting Kuanhsien twice, in 1943, after which I recorded my impressions (*4, 21*), and again in 1958. My first visit was made in the company of Prof. Fêng Yu-Lan, Prof. Ho Wên-Chün, Dr Phêng Jung-Hua and Commissioner Kuo Yu-Shou; and I learnt much from Dr Chang Yu-Ling, then resident engineer. On the second occasion I studied the works more fully with my collaborator Dr Lu Gwei-Djen, and we are glad to record our indebtedness to the kind help of Cde. Li Chün-Chu, the resident engineer, and Cde. Yang Chhun.

[d] Much has been written on Kuanhsien, even in European languages, e.g. Esterer (*1*); Hutson (*1, 2*); Little (*2*); Lowdermilk (*2, 7*); Richardson (*1*); Worcester (*1*), pp. 86 ff. But the most authoritative and detailed treatments are of course in Chinese, e.g. Pao Chio-Min (*1*); Ho Pei-Hêng (*1*); Fang Chi (*2*); and perhaps best, Anon. (*52*). A brief history of the works will be found in *Sung Shih*, ch. 95, pp. 24*b* ff.

[e] On the colonisation of Shu by Chhin, see Hisamura Yukari (*2*).

[f] The names of one or two of their semi-legendary assistants, such as Wang Cho,[13] Comptroller of Labourers, have come down to us.

[g] In these datings we follow the +4th-century *Hua Yang Kuo Chih*, ch. 3, though modern authors often place Li Ping half a century earlier.

[h] An earlier model, photographed in 1943, can be seen in Needham (*4*), fig. 34.

[1] 蜀	[2] 李冰	[3] 離堆	[4] 沫	[5] 成都	[6] 張儀
[7] 張若	[8] 李二郎	[9] 岷江	[10] 內江	[11] 外江	[12] 魚嘴
[13] 王墾					

886 (pls.). Thirty years ago the effluents of the two feeder canals consisted of subsidiary anastomosing conduits with a total length of some 730 miles, irrigating 500,000 acres of excellent agricultural land, i.e. about 72% of the area of 14 *hsien* cities.[a] By 1958 so many new distribution canals had been built that some 930,000 acres were being supplied with water, and the Authority's estimate was that when all the possible extensions of Li Ping's system are completed, no less than 4,400,000 acres will be irrigated.

The works, dams and spillways are known in general as the Tu-Chiang Yen,[1] or the Dam on the Capital's River. Its primary feature is the division-head of piled stones[b] which separates the two feeder canals. The inner one is used wholly for irrigation; while the outer one, which follows the old course of the river, and is also known (because it flows due south while the other deviates eastwards) as the Chêng-Nan Chiang,[2] acts as a flood channel as well, and also carries some boat traffic. In order to construct the inner canal, following a slightly higher contour, Li Ping had to make a great rock cut through the end of a ridge of hills[c] on part of which the city of Kuan-hsien is built. This is known as the 'Cornucopia Channel' (Pao-Phing Khou[3]).[d] The height of the 'Separated Hill'[e] on which the temple dedicated to him now stands (Fig. 890, pl.) is about 90 ft. from the canal bed, so that the total height of the 90-ft. wide cut would have been 130 ft. or more.[f] Between the primary division-head and the rock cutting the two channels are separated by the Chin-Kang Thi[4] (Diamond Dyke) and the Fei-Sha Yen[5] (Flying Sands Spillway). The top of the Diamond Dyke is made higher than the flood level of the Min River, to aid in the division of the water, while

[a] Fig. 889 (pl.) shows a map of the Kuanhsien System painted on a wall in the Temple of Li Erh-Lang; a stone-inscribed plan exists there also (see Hutson, 2). Not all the minor canals date from Li Ping's time; many were added in the +3rd century under Tshui Yuan.[6] The Kuanhsien works were not by any means the only hydraulic undertakings of Li Ping and his son; all are described in the *Hua Yang Kuo Chih*, ch. 3, in a passage which Torrance (2) has translated.

[b] A large protective piece for this in the shape of a tortoise, made of iron and weighing about a ton, was cast and placed in position by Chi Tang-Phu[7] between +1277 and +1294, but before long it was washed away by a flood and now lies buried somewhere in the river-bed. A similar attempt was apparently made between +1522 and +1566 when another engineer-in-charge, Shih Chhien-Hsiang,[8] shod the left secondary division-head (the Thai-Phing Yü Tsui[9]) with some 41 tons of cast iron in the shape of two oxen with heads conjoined and tails separated; but this also was washed away.

[c] Yü-lei Shan[10] (Jade Rampart Mountain). Approximate dimensions of the cross-section are given in Fig. 891.

[d] This name is rather evocative. In Vol. 2, p. 142, we read the account of the Taoist paradise in *Lieh Tzu*, ch. 5, p. 12a, centred upon Amphora Mountain (Hu Ling Shan[11]). This was shaped like a vase (*tan chui*[12]), and at the top there was an opening (*khou*[13]) in the form of a round ring called Hydraulica (lit, Aspergent Flood, *tzu jung*[14]) because streams of living water came out of it continually. Li Ssu-Shun (1) now tells us of the irrigation deity worshipped in Szechuan, Kuan Khou Erh Lang,[15] the Younger Son (cf. Li Erh-Lang himself) of the Irrigation Channel Mouth. This personage is tentatively to be considered the deification of Yang Nan-Tang,[16] a king of the Ti Chhiang[17] tribal people of Chhiu-chhih,[18] who died in +464.

[e] Hence the name Li-Tui mentioned by Ssuma Chhien. Kuanhsien (the Irrigation City) must equally have got its name at the same period.

[f] The rock is a kind of conglomerate which though not extremely hard seems to have weathered very little since Li Ping's time. Richardson (1) called it a 'natural concrete'.

[1] 都江堰	[2] 正南江	[3] 寶瓶口	[4] 金剛堤	[5] 飛沙堰
[6] 崔瑗	[7] 吉當普	[8] 施千祥	[9] 太平魚嘴	[10] 玉壘山
[11] 壺領山	[12] 甗甀	[13] 口	[14] 滋冗	[15] 灌口二郎
[16] 楊難當	[17] 氐羌	[18] 仇池		

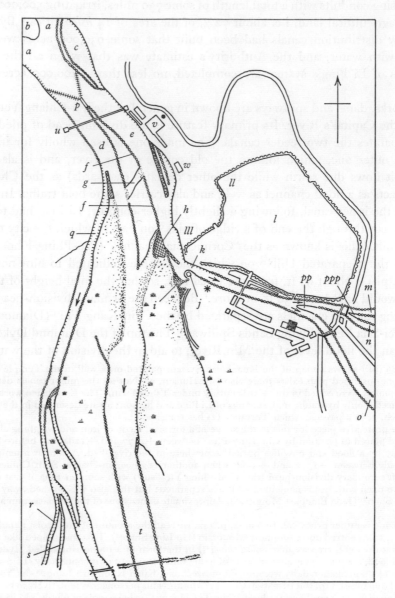

Fig. 884. Plan of the Kuanhsien irrigation system headworks, Tu-Chiang Yen (after Anon. 52). The intake works here shown beside the city and temples of Kuanhsien distribute water in a myriad canals all over the plain of Chhêngtu in Szechuan.

Plan of the Kuanhsien Irrigation System Headworks (Tu-Chiang Yen)

a, a Min Chiang[1] (Min River)
b Han-Chia Chü[2] (Han-family island)
c Pai-Chang Thi[3] (Thousand-foot Dyke)
d Wai Chiang[4] (Outer Feeder Canal; old course of river)
e Nei Chiang[5] (Inner Feeder Canal)
f, f Chin-Kang Thi[6] (Diamond Dyke)
g Phing-Shui Tshao[7] (Water-level By-pass or Adjusting-flume)
h Fei-Sha Yen[8] (Flying Sands Spillway)
i Jen-Tzu Thi[9] (V-shaped spillway)
   Li-Tui[10] (Separated Hill) and Fu Lung Kuan[11] (Tamed-Dragon Temple; the votive temple of Li Ping)
k Pao-Phing Khou[12] (Cornucopia Channel; the rock cut)
l Kuanhsien[13] City
ll Yü-lei Shan[14] (Jade Rampart Mountain)
lll Fêng-Lou Wo[15] (Phoenix Nest Cliff)
m Phu-Yang Ho[16] (derivatory canal)
n Po-Thiao Ho[17] (derivatory canal)
o Tsou-Ma Ho[18] (derivatory canal). New sluice-gates were installed across this, as shown, in 1952.
p Tu-Chiang Yü Tsui[19] (Fish Snout; primary division-head of piled stones)
pp Thai-Phing Yü Tsui[20] (left secondary division-head)
ppp Ting Kung Yü Tsui[21] (left tertiary division-head)
q Sha-Hei Tsung Ho[22] (right main derivatory canal)
r Sha-Kou Ho[23] (derivatory canal)
s Hei-Shih Ho[24] (derivatory canal). The feed into both of these is assisted by a spillway higher up, overflow rejoining the Chêng-Nan Chiang
t Chêng-Nan Chiang[25] (old course of river, flood course, etc.)
u An-Lan So Chhiao[26] (suspension bridge, cf. p. 192 above)
v Erh Wang Miao[27] (Temple of the Second Prince; the votive temple of Li Erh-Lang)
w Yü Wang Kung[28] (Temple of Yü the Great)

Note (i). The two lines, drawn in the convention usual for railways, which connect the Thousand-foot Dyke (*c*) and the right bank of the Min R. respectively with the Fish-Snout primary division-head (*p*), represent the positions of the temporary *ma chha* dams set up when the water is low for clearing the beds of the Nei Chiang (*e*) and the Wai Chiang (*d*), the former in January, the latter in November.

Note (ii). A steel cable, anchored at the point marked with an asterisk, guides rafts, floating tree-trunks, etc., into the Phu-Yang Ho (*m*), avoiding the Tsou-Ma Ho (*o*).

Note (iii). On the hill above the Temple of the Second Prince (*v*) there is a small but beautiful Taoist temple to Lao Tzu. An inscription says:

> 'The highest excellence does not lie in the highest place;
> In changes and transformations let nothing be contrary to Nature.'

[1] 岷江	[2] 韓家埧	[3] 百丈堤	[4] 外江	[5] 內江
[6] 金剛堤	[7] 平水槽	[8] 飛沙堰	[9] 人字堤	[10] 離堆
[11] 伏龍觀	[12] 寶瓶口	[13] 灌縣	[14] 玉壘山	[15] 鳳樓窩
[16] 蒲陽河	[17] 柏條河	[18] 走馬河	[19] 都江魚嘴	
[20] 太平魚嘴	[21] 丁公魚嘴	[22] 沙黑總河	[23] 沙溝河	[24] 黑石河
[25] 正南江	[26] 安瀾索橋	[27] 二王廟	[28] 禹王宮	

that of the Flying Sands Spillway is adjusted to the elevation required for the optimum supply of irrigation-water to the inner canal. When flood waters rise above this level, they overflow and automatically regulate the flow in the Nei Chiang.[a] Immediately after passing the city of Kuanhsien, the inner canal begins to give off its laterals and

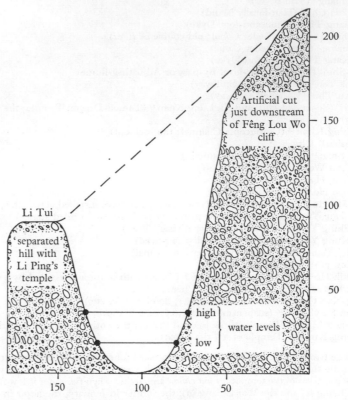

Fig. 891. Schematic cross-section of the Cornucopia Channel (data collected in 1958). Scales in ft.

sub-laterals, of which there are in all 526 and 2,200 respectively.[b] Some of them pass through and beside the city of Chhêngtu, and all ultimately find their way with the Min River into the Yangtze (past Chiating).

[a] An additional spillway, the Jen-Tzu Thi,[1] takes off excess water just before the Li-Tui hill is reached, and another (the Phing-Shui Tshao[2]) has been long incorporated in the middle of the Chin-Kang Thi (Diamond Dyke). This is the last to discharge as flood-level rises.

[b] Their total length is now well over 750 miles. At the Thai-Phing Yü Tsui[3] the Nei Chiang divides into the Phu-Yang Ho[4] on the left and the Tsou-Ma Ho[5] on the right; almost immediately, at the Ting Kung Yü Tsui[6] (named after a nineteenth-century engineer), the former gives off to the right the Po-Thiao Ho.[7] A similar tertiary division-head was introduced in the Tsou-Ma Ho as recently as 1957, generating a new feeder stream for the Chiang-An Ho.[8] Formerly the Wai Chiang made no branch until south of the Li-Tui hill (and this was still so when I first saw it), but now it gives off the Sha-Hei Tsung Ho[9] opposite the Erh Wang Miao temple, and this in due course feeds the Sha-Kou Ho[10] and the Hei-Shih Ho.[11]

[1] 人字堤	[2] 平水槽	[3] 太平魚嘴	[4] 蒲陽河	[5] 走馬河
[6] 丁公魚嘴	[7] 柏條河	[8] 江安河	[9] 沙黑總河	[10] 沙溝河
[11] 黑石河				

Each year there is a cycle of operations corresponding to the flow of water. From December to March the river is at the low water stage, with an average flow of 200 cumecs (cf. pp. 221, 238), falling sometimes to 130 cumecs. From April onwards, when planting starts, the flow increases, till at 585 cumecs the full requirements of the works of both the outer canal (280 cumecs) and the inner one (305 cumecs)[a] are satisfied. In summer, i.e. June and July, high water is reached, with a maximum flow of 7,500 cumecs in the river as a whole, after which there is a slow decline till November, faster thereafter. Throughout the centuries, the advice of Li Ping to clear out the beds and keep the dykes and spillways low has been faithfully followed,[b] and if it has been possible to preserve the system so closely as he left it, this is partly because the river is not extremely silt-laden,[c] and partly because its annual fluctuations have permitted incessant and effective maintenance. Every year, about mid-October, the annual repairs begin. A long row of weighted wooden tripods (*ma chha*[1]) is placed across the outer feeder canal at its inlet (Fig. 892, pl.) and covered with bamboo matting plastered with mud to form a cofferdam, thus diverting all the flow into the inner canal.[d] The bed of the outer canal is then excavated very actively to a predetermined depth, and any necessary repairs to the division-heads are carried out with the aid of gabions (*chu lung*[2]).[e] About mid-February, the stockade-dam is removed and re-erected at the intake of the inner canal, so that all water flows to the right, and similar maintenance of the inner system is effected.[f] On the 5th of April, the ceremonial removal of the coffer-dam marks the opening of the irrigation season and gives opportunity for a general celebration, even in these days of slide-rules and plans for power-stations.[g] What it was like towards the end of the last century, under the Chhing dynasty, we may glimpse from the following account:[h]

Generally about the 1st April or, a little earlier, when the repairs are nearing completion, the Water Inspector (Shui Li Fu[3]) selects a lucky day and invites the Intendant of Circuit (Lung Mao Tao[4]) to come and open the river. If the latter reckons the day chosen to be an unlucky one for himself, or for the people, he will choose a more auspicious one. The Water Inspector then notifies the people by proclamation as to the day and time when the dam will be opened.

---

[a] This is its minimum effective discharge, i.e. its 'duty', but it can and does carry 1,000 cumecs in July.

[b] Cf. p. 250 above, and immediately below.

[c] What it brings down is chiefly bed-load—current-rolled boulders and gravel. Cf. p. 231 above.

[d] A detailed colour photograph of the temporary dam in position will be found in Lowdermilk (7), pl. xv.

[e] See immediately below, p. 295.

[f] A set of three iron bars, known as the *wo thieh*,[5] each 10–12 ft. long and weighing about two-thirds of a ton, is fixed in the bed of the Nei Chiang just below the Fêng-Lou Wo cliff as a guide for the excavation depth; cf. Fig. 893 (pl.), from Lowdermilk (7). These bars are said to be those recovered by Shih Chhien-Hsiang in the middle of the +16th century, and believed by him to be the originals laid down by Li Ping or Li Erh-Lang.

[g] The opening now takes place rather earlier, in March. Pictures taken at the moment of breaching the dam will be found in Phan Ên-Lin (1), p. 157, and Lowdermilk (7).

[h] Hutson (2). Photographs in Lowdermilk (7), pp. 644, 645, taken in 1943, add life to it.

---

[1] 榪槎   [2] 竹籠   [3] 水利府   [4] 龍茂道   [5] 臥鐵

The Intendant of Circuit is met on the day of his arrival by all the city officials, who conduct him over the works, which he officially inspects, eventually escorting him to the residence which has been prepared for him for the night. While there he often receives complaints from the people regarding the supply of water for their fields, and other cases of dispute. The Water Inspector then presents his superior officer with a feast ready prepared, which is generally declined with thanks. The local official then presents his offering, a feast ready prepared, which is received and eaten. All the expenses of receiving, entertaining, escorting, and so on, are borne by the local officials, amounting to about 600 taels in all, the Water Inspector giving only a few presents amounting to about 100 taels.

The morning after the arrival His Excellency the Intendant of Circuit rises early before dawn and proceeds to the Temple erected for the memory and worship of Li Erh-Lang, and called Erh Wang Miao. Here he burns incense and makes obeisance to the Gods. Afterwards he proceeds to the river-bank, just below where the barrier is erected, and where everything is in readiness for the opening ceremony. A long bamboo tracking-rope has been attached to a few of the tripods, and a band of strong coolies stands at the end of it. An altar has been erected where incense and candles are already burning, and a sacrificial pig and goat are already offered—then, just as the sun begins to show its golden tints over the horizon, the great man kneels down and worships the god of the river, the coolies give one long shout and a longer pull, down fall the barriers, and the waters of the Min rush into the artificial course with such violence that almost every year a man or two is carried away and drowned. His Excellency gives a reward of about 50,000 cash to be divided among the workmen and coolies, but be it noticed that he is liberal at someone else's expense, for the local official has to pay all the money...By the time the waters have reached Kuanhsien, the Intendant of Circuit has passed on his return to the capital, but if there is a fairly good supply, the waters reach the capital before him.

Such were the customs which must have continued with little change for many centuries.

The only technical detail of which mention has not yet been made is that at various times 'nilometers' or gauges for the water levels[a] were incorporated at suitable positions.[b] One of these was a figure inscribed *Chieh pu chih tsu, shêng pu mei yao*[1]— 'In the dry season let the feet be covered, in flood let the level not pass the waist'. This referred to the proper height for the spillways. The ancient iron bars in the bed of the Nei Chiang we have just noted, and at various points there are old measuring scales from which the water-levels in the various canals can daily be observed.

These gauges are referred to, among other things, in an inscription carved in stone in the Erh Wang Miao and still to be seen today (Fig. 894, pl.).[c] Entitled the Chih Ho San Tzu Ching[2] (The Trimetrical Classic of River Control), it makes reference to the 'Six-Character Teaching' and the 'Eight-Character Rule', those other Kuanhsien inscriptions of engineering interest which we have already discussed (p. 250). The date of these 'trimetrical' or three-character verses is not known, but it will hardly be

---

[a] For ancient Egyptian, Islamic, and European parallels, see Feldhaus (1), col. 1303; Willcocks (4), p. 150.
[b] The earliest was due to Li Erh-Lang himself (*Hua Yang Kuo Chih*, ch. 3).
[c] Another photograph is given by Phan Ên-Lin (1), p. 158.

[1] 竭不至足盛不沒腰   [2] 治河三字經

earlier than the +13th century, when Wang Ying-Lin produced the pattern, his celebrated *San Tzu Ching*.[a] The inscription runs:[b]

> 'Dig the channel deep,
> And keep the spillways low';
> This Six-Character Teaching
> Holds good for a thousand autumns.
> Dredge out the river's stones
> And pile them on the embankments,
> Cut masonry to form 'fish snouts' (*yü tsui*[1]),[c]
> Place in position the 'sheep-folds' (*yang chüan*[2]),[d]
> Arrange rightly the spillways (*phai chhüeh*[3]),[e]
> Maintain the overflow pipes in the small dams (*lou kuan*[4]).[f]
> Let the (bamboo) baskets (*chu lung*[5])[g] be tightly woven,
> Let the stones be packed firmly within them.
> Divide (the waters) in the four-to-six proportion,[h]
> Standardise the levels of high and low water
> By the marks made on the measuring-scales (*shui hua fu*[6]);[i]
> And to obviate floods and all disasters
> Year by year dredge out the bottom
> Till the iron bars clearly appear.
> Respect the ancient system
> And do not lightly modify it.

[a] 'Trimetrical Primer', the classical Confucian catechetical instruction for schoolboys, in rhyming couplets for memorisation, used down to our own time. There is a translation by H. A. Giles (4). Wang Ying-Lin's gift to the young may be dated *c.* +1270. Cf. Vol. 2, p. 21.

[b] Tr. auct. Anon. (*52*) reproduces it, leaving out the last two lines.

[c] The division-heads, main, secondary, tertiary, etc.; see p. 291. Stone and iron have proved less satisfactory than gabions, but concrete is now used as foundation.

[d] These are cylindrical gabions formed of parallel wooden slats some 10–20 ft. long, and filled with stones the size of hen's eggs.

[e] Lit. gaps to let out the roaring billows. The modern term is *i hung tao*.[7]

[f] Such pipes were also sometimes placed at the bottom of small irrigation dams in order to prevent silting-up of the floor of the reservoir.

[g] These are the famous gabions of bamboo basketwork so much used in Chinese technology. Some are in sausage form about 10–20 ft. long (Fig. 914), others may assume bulkier shapes (Fig. 895, pl.). Stone revetments around the division-heads were built at Kuanhsien over and over again; as by Chi Tang-Phu in the Sung (late +13th century), Hu Kuang[8] in the Ming between +1488 and +1505, and Ting Pao-Chên[9] & Lu Pao-Tê[10] as late as 1877, but in comparison with gabions they always failed. These strong 'sausages' of woven bamboo slats packed with stones and built up on a stone or concrete foundation have two great advantages, (*a*) the water can drain in and out of them slowly, thus reducing sudden bursting pressures on the walls of canals, and (*b*) they are not too heavy for the light alluvial soil. They have always been used by those engineers at Kuanhsien whose work was most successful, e.g. Chao Pu-Yu[11] in the Sung, Lu I[12] in the Ming, and Hang Ai[13] in the Chhing (+1681). Permanent masonry revetments are of course quite satisfactory at many parts of the works. Cf. pp. 321 ff., 339 ff. below.

[h] According to the ancient prescription, at low water the Nei Chiang should carry 60% of the flow and the Wai Chiang 40%, while at flood time these proportions are reversed. The adjustment must be made by *ma chha* dams, and of course during dredging periods the entire flow passes down one or the other channel completely.

[i] The most important of these is in the Cornucopia Channel and can be seen from Li Ping's temple.

[1] 魚嘴　　　　　[2] 羊圈　　　　　[3] 湃缺　　　　　[4] 漏罐　　　　　[5] 竹籠
[6] 水劃符　　　[7] 溢洪道　　　[8] 胡光　　　　[9] 丁寶楨　　　[10] 陸葆德
[11] 趙不憂　　　[12] 盧翊　　　　[13] 杭愛

Truly an epitome of the hydraulic art. Perhaps one need only add that the benefit which the Szechuanese received from the Kuanhsien works was not limited to irrigation and flood protection. A stone tablet of the Yuan period records that 'water-wheels for hulling and grinding rice, and for spinning and weaving machinery, to the number of tens of thousands, were established along the canals (in the plain of Chhêngtu), and operated throughout the four seasons'.[a] Thus the economic life of a whole Chinese province at a time contemporary with Villard de Honnecourt and Roger Bacon depended on those noble works of civil engineering among the misty mountains of western Szechuan.

We cannot leave Kuanhsien without alluding to a subject which in some sense oversteps the boundaries of engineering as such. The Chinese were never content to regard notable works of great benefit to the people from a purely utilitarian point of view. With their characteristic ability to raise the highest secular to the level of the numinous, they built upon the top of the 'Separated Hill' a magnificent temple, the Fu Lung Kuan,[1] to commemorate Li Ping's heroic virtue;[b] and further back, in a scarcely less beautiful situation, the wooded hillside downstream from the suspension bridge, another one to that of his son Li Erh-Lang, the Erh Wang Miao[2] (Fig. 897, pl.).[c] When I was at Kuanhsien in 1943 the contrasts were deeply impressive, for incense sticks were glowing in Li Ping's temple before his impassive statue (Fig. 896, pl.), tended by an amiable Taoist, but in its second court were numerous models (cf. Fig. 888, pl.) of the improvements planned in the system—a control dam with large gates instead of the fish snout, power-stations and so on.[d] Li Erh-Lang's temple provided living quarters for the engineers, its beauty in no way diminished, while a smaller temple dedicated to Yü the Great (the Yü Wang Kung[3]) housed the Water Conservancy Bureau. Overlaid though it may formerly have been by some superstition and ignorance among the masses of the people, the cult at Kuanhsien seemed to demonstrate one of the most attractive aspects of Chinese culture, namely that synthesis of Confucianism and Taoism in which, whatever might be thought of gods and spirits, divine honours were certainly owing, and paid, to the great benefactors of human-kind.

[a] The text is given in *Szechuan Thung Chih*, ch. 13, pp. 27*a* ff., tr. Chi Chhao-Ting (1), p. 97. On water-power in general see Vol. 4, pt. 2, pp. 362 ff.; and on the use of water-power in China for working textile machinery in Marco Polo's time, see Vol. 4, pt. 2, p. 404.

[b] This is not in doubt, but a certain amount of indigenous Szechuanese semi-legendary material exists which suggests that there had been attempts earlier than Li Ping's to harness the Min River, notably by Khai Ming,[4] who lived at some time during the Warring States period. Torrance (2) has tried to elucidate this question, but it needs further study. Could it be another case of Kun and Yü? A fine photograph of Li Ping's temple, taken from the Phoenix Nest Cliff above the Cornucopia Channel, will be found in Phan Ên-Lin (1), p. 160.

[c] Other photographs are in Phan Ên-Lin (1), pp. 158, 159. This is first recorded in +494, but in its present form dates from +1078 to +1085 (*Shih Wu Chi Yuan*, ch. 7, p. 23*b*). How late the admirable system of votive shrines and temples continued may be seen from the fact that Erh Wang Miao contains a chapel to Ting Pao-Chên, the great and good Governor of Szechuan from 1876 to 1886. On my second visit in 1958 all the temples had been excellently repaired and repainted by the government. Material incense, it is true, was lacking, but it will be found necessary, even under socialism, if only to keep down the smell of bats.

[d] Since 1950 modern steel sluice-gates have been installed at the openings of all the main derivative canals.

[1] 伏龍觀   [2] 二王廟   [3] 禹王宮   [4] 開明

(iii) *The Kunming reservoirs (Yuan) and the Shantan system (Ming)*

As a postscript to the story of Kuanhsien, we shall mention two other projects which, though not so spectacular, greatly benefited smaller circumscribed pieces of territory. One is a reservoir system; the other was a derived irrigation canal issuing from its valley of origin through a saddle in the mountains. First let us consider the irrigation works of the Kunming plain in Yunnan.[a] This plain or basin surrounded by rolling uplands centres upon the provincial capital and the Kunming Lake, with its western hills crowned with woods and temples.[b] The essential problems here were first the assurance of free passage for the waters of the lake, which were otherwise liable to flood large areas, and secondly the formation of reservoirs and artificial canals so as to distribute as widely as possible the waters of the six small rivers which flow down into the lake. All this was achieved by the Yuan governor of the province, a Persian or Arab by origin, Sai-Tien-Chhih Shan-Ssu-Ting[1] (Sa'īd Ajall Shams-al-Dīn) also called Wu-Ma-La[2] ('Umar), in collaboration with a local engineer Chang Li-Tao,[3] and building upon minor works previously carried out by the former indigenous and independent Thai dynasty of Nan Chao.[c]

Sa'īd Ajall attached himself to Chinghiz Khan during the Mongol expeditions in the west, and in due course attained to many positions of importance under the Yuan régime,[d] as may be seen in the interesting monograph of Vissière (2). He was sent to Yunnan as governor in +1274, where he greatly exerted himself to raise the cultural level of the backward population, erecting impartially Confucian temples and Muslim mosques.[e] A stele with an inscription by one of his assistants, Chao Tzu-Yuan,[4] commemorates his benevolent administration.[f] A dozen reservoirs with dams in the hills, more than forty sluices, and a network of dyked canals lined with beautiful trees[g] still remain in function testifying to the enlightened rule of Hsien-Yang Wang[5].[h]

The second system of works selected for mention here is the Pai-Shih Yai[6] (White Rock Cliff) irrigation scheme, a contour canal which formerly watered a great space of fertile land (enough for more than a thousand farms) between the mountains and the desert near Shantan[7] in Kansu province. As one travels up the Old Silk Road from

---

[a] As in private duty bound, for these works were among my first impressions of China when I arrived there towards the end of 1942.

[b] Geomorphological description by Chhen Shu-Phêng (*1*).

[c] Cf. pp. 198 ff. above. It had been conquered by the Mongols in +1253.

[d] His biography is in *Yuan Shih*, ch. 125. He is not to be confused with another Shan-Ssu (Shams al-Dīn), the mathematician and geographer, who flourished in the following century, and whose biography is in *Yuan Shih*, ch. 190.

[e] The Confucian temple at Chhêng-kung,[8] south-east of Kunming, a place of great beauty which housed the Institute of Statistics during the second world war, was built by him. Cf. Vol. 1, Fig. 37.

[f] Chao indicates princely descent in Thai.

[g] Cf. Beaton (*1*), p. 3, for a photograph. Prof. Thang Phei-Sung and I have cycled and motored for many miles along these avenues, and bathed in such reservoirs as that near the village of Tapuchi. The system of canals attained its present form substantially by +1279.

[h] His tomb also still exists in open country south-east of Kunming, and the temple dedicated to him is now in the grounds of the University of Yunnan.

---

[1] 賽典赤瞻思丁   [2] 烏瑪喇   [3] 張立道   [4] 趙子元   [5] 咸陽王
[6] 白石崖   [7] 山丹   [8] 呈貢

Lanchow, going north-westwards, one has always upon one's right the desert (and sometimes what remains of the Wall), while to the left is the glittering snow-capped chain of the Chhilien Shan (or Nan Shan) mountains. For 200 miles or more in each direction from Shantan, the road passes through steppe country or desert scrub, crossing a great number of watercourses which take their origin in the mountains and run down to lose themselves in the sands of the Gobi. Water and its retention must have been the great problem here for many centuries. Most of what we know about the Pai-Shih Yai Chhü[1] is summarised upon a commemorative stele[a] erected at Shantan in +1503 by a (Taoist) hermit, Wang Chhin-Thieh.[2] The project was a bold one, for it tapped the Ta-Thung Ho[3] River[b] at a precipitous place, the White Rock Cliff, more than 200 *li* to the south-west of the Shantan plain. To understand this, one must realise that the Ta-Thung Ho flows in the deep valley behind the first range of the Chhilien Shan, running south-eastwards parallel with the Old Silk Road, to join the Yellow River above Lanchow. The division-head must thus have been constructed at an altitude of some 12,500 ft., for a pass of this height had to be crossed by the canal before it could descend to the Shantan plain (itself mostly about 6,000 ft.), and the canal must have followed the high contour for a long way before beginning its descent.[c] When the works were first constructed we do not know, but by the end of the +15th century they had silted up and ceased to function. An engineer named Liu Chen[4] was then asked by one of the provincial officials, Li Kho,[5] to take charge of radical repairs; workers were collected from an area of 1,200 *li* around, and began operations encouraged by drumming and dancing. In three years, not without some difficulties, everything was perfectly restored. Mr Wang the hermit ended his inscription with a poem:

> The White Cliff towers a thousand feet high,
> For a hundred years the desert lay waste,
> But then came a lover of the people
> Calling for a mighty engineer,
> And (Li) Ping and Yü (the Great) lived again,
> Building new dams, new dykes.
> The rolling waters, how everyone longingly wished for them!
> Drawn forth in curving course, like the Han river, they came.
> And the men of the three Armies could settle on the plain, and plough.

But in after years, during the disturbances of the last century, the works again fell into disrepair, and now there is little trace of them in the neighbourhood of Shantan. The upper valley of the Ta-Thung Ho is only wild mountain pasture, rarely visited save by shepherds and perhaps an occasional geologist, but it would indeed be interesting to seek there for the traces, which must still remain, of the work of Liu Chen and his predecessors.

[a] For a copy of this inscription we are much indebted to Mr Rewi Alley. The story is all the more worth telling in that the stele itself was taken by the military only a few years ago for use in some construction work, and even the rubbing has probably not survived the Shantan earthquake of 1953.
[b] Or one of its upper tributaries.
[c] Intervening streams and gullies were crossed by timber aqueducts.

[1] 白石崖渠   [2] 王欽鉄   [3] 大通河   [4] 劉振   [5] 李克

(iv) *The 'Magic Transport Canal'* (*Chhin and Thang*)

The Ling Chhü[1] (Magic Canal)[a] was a work of quite different character, meant to serve, not primarily irrigation, but the need for freight communication across one of the principal ranges of high mountainous country which separate the north and centre from the south. It connected two rivers in Kuangsi, one flowing northwards and one southwards, so that through transport was made possible between the Yangtze, the Tungthing Lake, and the West River flowing down to Canton.[b]

In the *Shih Chi* there is no mention of the Ling Chhü by name, either in the chapter on waterways or elsewhere. But there are accounts of its building, which show that its primary purpose was to keep up a flow of water-borne supplies to the armies which in −219 were sent south for the conquest of the people of Yüeh. It may also have served for the transport of a fleet of war-boats.[c] The *Shih Chi* says:[d]

> (Chhin Shih Huang Ti) sent the Commanders (Chao) Tho[2] and Thu Chü[3] to lead forces of fighting-men on boats with deck-castles (*lou chhuan*[4])[e] to the south to conquer the countries of the hundred tribes of Yüeh. He also ordered the Superintendent (Shih) Lu[5] to cut a canal so that supplies of grain could be sent forward far into the region of Yüeh.

This may not be the oldest reference to the canal, for the *Huai Nan Tzu* book has a page or two on the First Emperor's campaigns, and gives this statement about the engineer Shih Lu in almost the same words.[f] They go back, then, to −120, just a century after the work itself. We also have a passage in the biography of Yen Chu,[6] an official concerned with grain transport who flourished about −135. In one of his memorials he reported that 'elderly gentlemen said' that the Magic Canal had been dug by Shih Lu in the first emperor's time.[g] Thus the date of the original project may be considered as firmly established.

There are also passages in the annals of Chhin Shih Huang Ti which may refer to it indirectly.[h] In the 32nd year of the first emperor, i.e. −215:

> He had an inscription carved on the 'gates' of Chieh-Shih[7] (Stone Pillar Mountain), recording how he had broken through (all) walls and defences (of feudal fortresses), and had pushed lines of communication through (all) obstacles and barriers (*thung thi fang*[8]).

---

[a] Such was the significance of its name that came down through the ages, and quite rightly so (cf. p. 375), but according to Fang Chi (2), the original name of the Li River was the Ling, and this was the reason for the appellation in the first place.

[b] And also, of course, via the Han Kou and the Hung Kou (cf. pp. 271, 269), with Kuanchung and the Yellow River valley.

[c] This it certainly did during the later campaigns of the −1st and +1st centuries, see immediately below, p. 303.

[d] Ch. 112, p. 10*b*, tr. Aurousseau (2); Watson (1), vol. 2, p. 232, mod. auct. This text is part of a memorial by Yen An[9] directed against Han Wu Ti's aggressive foreign policy. The statement is repeated in Yen An's biography (*Chhien Han Shu*, ch. 64B, p. 3*a*).

[e] On these, see pp. 441, 445 below.

[f] Ch. 18, p. 16*a*.   [g] *Chhien Han Shu*, ch. 64A, p. 6*b*.

[h] *Shih Chi*, ch. 6, p. 20*b* (Chavannes (1), vol. 2, pp. 165, 166).

[1] 靈渠   [2] 趙佗   [3] 屠睢   [4] 樓船   [5] 史祿   [6] 嚴助
[7] 碣石   [8] 通隄防   [9] 嚴安

Part of the text, as recorded by Ssuma Chhien, ran thus:

The Sovereign Emperor (Huang Ti)...has overturned and destroyed outer walls and castle keeps, he has opened passages for waterways through obstacles (*chüeh thung chhuan fang*[1]), he has beaten down and done away with all difficulties and barriers (*i chhü hsien tsu*[2]), and the face of the earth being well regulated (*ti shih chi ting*[3]), the black-haired people are no longer overwhelmed with *corvée* labour (*li shu wu yao*[4])...

Most of our information about the Magic Canal is to be found collected in the eighteenth-century commentaries of Chhüan Tsu-Wang and Chao I-Chhing on the *Shui Ching Chu* (Commentary on the Waterways Classic).[a] The text of the *Shui Ching* itself (perhaps +3rd century) does not mention it, but Li Tao-Yuan's +5th-century commentary says that at the headwaters of the Li[5] River (flowing south) there is a pass (*yu kuan*[6]), and speaks of a Ling Chhi Shui Khou[7] (Magic Stream Water Opening) which must be the canal.

To understand what kind of a work it was (and still is) we must cast a glance at Figs. 898 and 899.[b] The northward-flowing river is the Hsiang,[8] which takes its rise in the high ground of Hai-yang Shan[9] and finds its way down to the plains of Hunan, passing Hêngyang and Chhangsha[c] on its way to the Tungthing Lake and the Yangtze. The southward-flowing one is the Li,[5] which rises in the hills called Yüeh-chhêng Ling[10] and joins another stream to form the Kuei Chiang,[11] that particularly beautiful river known to all who have visited Kweilin[12] city with its landscape of karst pinnacles,[d] a tributary of the West River quickly leading down to Canton. Between the oppositely-facing streams of Hsiang and Li there is a saddle in the hills which gave Shih Lu the opportunity for the first of all contour transport canals. The part of the Ling Chhü which justifies this designation was called the Nan Chhü[13] and branched off from the Hsiang River to run along a suitable level or slightly falling contour for about 3 miles till it met the upper waters of the Li River. These themselves had to be canalised for another $17\frac{1}{2}$ miles downstream as far as the junction with the Kuei Chiang. Meanwhile in the other valley a $1\frac{1}{2}$-mile lateral transport canal (the Pei Chhü[14]) was dug at a more even gradient than that of the untrained Hsiang. How the works were constructed may be appreciated from Fig. 899. A division-head called a 'spade-snout' (*hua tsui*[15]), reminiscent of those used at Kuanhsien, was built in the middle of the Hsiang and backed by two spillways (Ta Thien-Phing[16] and Hsiao Thien-Phing[17])[e] discharging into

---

[a] Ch. 38, p. 16a. Apart, that is, from the local topographical histories. A brief account occurs in *Sung Shih*, ch. 97, pp. 25b, 26a.

[b] The only photographs at all adequate are those in Wu Lien-Tê (*1, 1*), pp. 118 ff.

[c] Places of vivid memory to me, for in 1944 during the second world war I crossed the Hsiang to the west by the last train over the Hêngyang railway bridge before it was blown up to slow the Japanese advance. For an eye-witness account of the canal in 1911 see Lapicque (1).

[d] Cf. Vol. 1, Fig. 4.

[e] The greater and lesser 'balancers' or 'level-equalisers'; cf. Vol. 4, pt. 1, p. 24. Length in all 1,443 ft., height of retaining wall of dam *c.* 8–10 ft. Masonry of 'fish-scale' character.

[1] 決通川防	[2] 夷去險阻	[3] 地勢既定	[4] 黎庶無繇	[5] 灘
[6] 有關	[7] 靈溪水口	[8] 湘	[9] 海陽山	
[10] 越城嶺	[11] 桂江	[12] 桂林	[13] 南渠	[14] 北渠
[15] 鏵嘴	[16] 大天平	[17] 小天平		

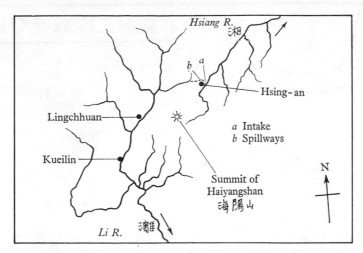

a Scale 1:2,000,000. The canal begins (a) near the little town of Hsing-an, and runs 3 miles westwards
   (b) to join the uppermost waters of the Li R., which had to be trained and canalised for a further
   17½ miles.

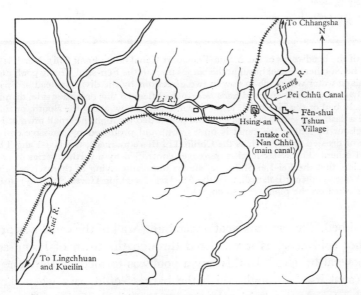

b Scale c. 1:450,000 (after Fang Chi). This shows the lateral canal dug to give a more even gradient
   without rapids on the east (right) bank of the Hsiang R. The division-head in the pool above it gave
   its name to the village of Fên-shui Tshun. The modern railway crosses the ancient canal and then
   runs along to the south of the canalised Li R.

Fig. 898. Sketch-maps of the geography of the Ling Chhü (Magic Canal) in north-eastern Kuangsi, first
built in −215. This, the most ancient of all contour transport canals, joins a northward-flowing river,
the Hsiang, with a southward-flowing one, the Li (called lower down the Kuei), by means of a canal cut
and constructed for some twenty miles through a saddle in the high hills.

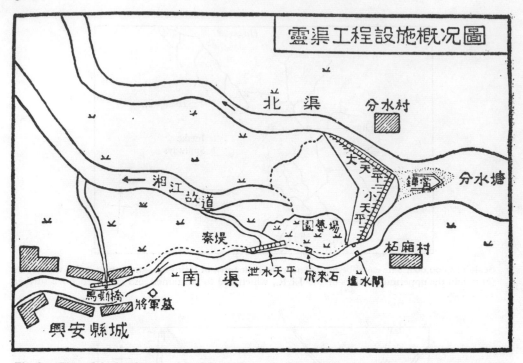

靈渠工程設施概況圖

Fig. 899. Plan of the headworks of the Magic Transport Canal (after Fang Chi). North is at the top left-hand corner. The Northern Canal brought the barges up into the Fên-shui Thang pool, just below which the Hsiang Chiang water was made to flow in two directions by the division-head or Spade Snout (*hua tsui*). Two spillways (*ta thien-phing* and *hsiao thien-phing*) flanked this on either side, dismissing overflow water into the old bed of the Hsiang R. lower down. Passing now along the South, or main, Canal, the vessels entered a flash-lock gate (*chin shui cha*) and proceeded through the small town of Hsing-an westwards. Throughout its length the canal is built upon, and protected by, massive embankments and retaining walls of dressed stone, such as the Chhin Thi (here marked), the Hai-Yang Thi, the Huang-Lung Thi, and others. The water-level is moreover controlled by a further series of spillways on the north-eastern side, first the Fei-Lai Shih, the 'stones that came flying of themselves', then the I-Shui Thien-Phing (Water-emitting Balance), then the Ma-Hsi (Neighing Horse Spillway) in Hsing-an itself (all these are marked on the plan), and so on.

the river's old bed. The embankment or retaining-wall of the canal was provided with several further spillways[a] as it wandered through the town of Hsing-an,[1] and was crossed by several bridges.[b] In this way a pool was formed on the same level as the connecting canal itself through which the barges could pass when they had been worked up that far. As the division-head was placed in the Hsiang River, considerably the larger of the two, most of the water in the canal (some 3 ft. deep and 15 ft. wide) was derived from it, and the local saying was that three-tenths of the Hsiang water went into the southern or connecting canal (Nan Chhü) while seven-tenths flowed

[a] Fei-Lai Shih,[2] I Shui Thien-Phing,[3] and another in Hsing-an town itself.
[b] Including the modern railway bridge on the line between Hêngyang and Kweilin. I passed over this several times during the war, but unfortunately at that time had no knowledge of the interest of the site and did not look out for the crossing. The railway also follows the canal between the hills.

[1] 興安   [2] 飛來石   [3] 泄水天平

down the northern lateral one (Pei Chhü).[a] All this was in working order, and much used, during the Early Han period, especially for naval purposes between −140 and −87 (with a peak traffic about −111) when Han Wu Ti was campaigning in the south for the final reduction of Yüeh. Another heavy duty period occurred about +40 in connection with the important expedition against Annam,[b] and it is recorded that on this occasion the commander-in-chief Ma Yuan[1] extended the canalisation of the Hsiang; this must mean that he improved its navigation for some distance downstream of the lower entrance of the Pei Chhü.[c]

The most extensive and classical account that we have found is given by Chou Chhü-Fei,[2] in his *Ling Wai Tai Ta*[3] (Information on What is Beyond the Passes) of +1178. He speaks as follows:[d]

The upper reaches of the Hsiang[4] River flow out into Hunan in the north, and the Jung[5] River (now called the Kuei Chiang[6]) comes south into Kuangsi (and Kuangtung). At the watershed (of these rivers) the highest part lies in the Hsing-an[7] *hsien* district of the prefecture of Chingchiang.[8]

In olden times when (Chhin) Shih Huang Ti was annexing the frontiers of the Five Ranges in the South, Shih Lu[9] (was sent to) cut a canal to connect a (small) tributary of the Li[10] River with the upper reaches of the Hsiang[4] River. (This canal) traverses Hsing-an district and falls into the Jung[5] River to the south. It was intended to facilitate the transport of army rations.

The situation was that water from the north was flowing southwards (in the Li), but the boats from the north came up against the barrier of the mountain. (Shih) Lu's method of cutting the canal was as follows. In the upper reaches (of the Hsiang River), in the midst of gravel banks and sunken rocks, he caused stones to be piled up layer upon layer making a division-head (*hua tsui*,[11] lit. spade-like snout). Thus the waters of the Hsiang River were divided into two (branches) by means of this sharp-pointed current-cleaving structure. (Then), following round the mountain (contours), he built an embankment (*thi*[12]) so as to make a smooth-gliding canal, along which the water was skilfully induced to run for 10 *li*, after which it reached more level ground. From there onwards, the canal was so dug that it continued to wind round the curves of the mountain-side, the whole distance amounting to 60 *li* (about 20 miles). Thus it came to the Jung[5] River, and so to all the south. This is what is now known as the Kuei Shui.[13]

The reason why the Li[10] River is so called is because it is a stream separate (*li*[14]) from the Hsiang[4] (and yet connected with it)—there is the Hsiang and there is the Li.

---

[a] By this very ingenious expedient, Shih Lu avoided the great difficulty (which at that time would have been insuperable) of supplying water to the uppermost levels of a true summit canal.

[b] On all these campaigns, which led to the incorporation of Kuangsi, Kuangtung and Annam within the empire, see pp. 441 ff. below. The story begins in *Shih Chi*, ch. 6, p. 21*b*, tr. Chavannes (1), vol. 2, p. 168. Cf. Cordier (1), vol. 1, pp. 209 ff., 235 ff., 259 ff.

[c] *Shui Ching Chu*, ch. 38, p. 16*b*.

[d] Ch. 1, pp. 10*a* ff.; tr. auct., adjuv. J. V. Mills, unpub. tr. Chhüan and Chao (see p. 300 above) quote this passage from the *Kuei Hai Yü Hêng Chih* of Fan Chhêng-Ta, about +1175, but it is not to be found in the editions of that book which I have seen. Fan Chhêng-Ta, however, refers to the Ling Chhü in his *Tshan Luan Lu*[15] (Guiding the Reins), a narrative of his journey from the capital to Kweilin (+1172), p. 23*a*, saying that Shih Lu built it and that the little town of Ling-chhuan (Lingchow) was named after it.

---

[1] 馬援	[2] 周去非	[3] 嶺外代答	[4] 湘	[5] 融
[6] 桂江	[7] 興安	[8] 靜江	[9] 史祿	[10] 灘水
[11] 鏵嘴	[12] 堤	[13] 桂水	[14] 離	[15] 驂鸞錄

Passengers travelling along at certain points are sometimes scared out of their wits, for about 2 *li* from the intake where the 'spade-snout' divides the waters and makes one branch enter the embanked canal, there is another spillway (*hsieh shui than* [1]) (which lets off excess water). Without this spillway, the raging force of the spring freshets could damage the retaining-wall and the water would never reach the south. But by its aid, the violence of the water is abated, the embankment is unbroken, and the water in the canal flows smoothly on. Thus the extra water drawn from the Hsiang penetrates into the Jung. This may really be called an ingenious device.

The canal waters meander through Hsing-an district, and (accordingly) the people rely on them for irrigating their fields.

The depth (of the canal) is not (more than) a few feet, and the breadth may be about 20 ft.; it is adequate to float a vessel of 1,000 *hu* (bushels of cargo).[a]

In the canal there are 36 lock gates (*tou mên* [2]). As each vessel enters one of these lock gates, (the people) immediately restore it (*tsê fu cha chih* [3]) to its closed position and wait while water accumulates (within the lock), so that by this means the ship gradually progresses. In such a way they are able to follow the mountainside and move upwards.

On the descent, it is like water flowing down the stepped groove of a roof, and thus there is communication for the boats between north and south. I myself have seen (I am happy to say) the historic traces of the work of (Shih) Lu.

The cruelty and suspicion of (Chhin) Shih Huang (Ti) I venture to deplore. But his despotic authority had power to trap the waters permanently, so that vessels could travel (overland). For ten thousand generations (people) have relied upon (his canal). But (the merit) belongs not only to (Chhin) Shih Huang (Ti)—(Shih) Lu was also a hero—and (on account of all these things) it is called the 'Magic Canal'.

This eye-witness account, with its clear description of pound-locks in the late +12th century,[b] is of great interest, and necessarily raises the question of the time of introduction of gates into the Ling Chhü system. Chou Chhü-Fei speaks as though locks of one kind or another had been there since the Chhin, but this can hardly be taken for granted. Although we shall have to return later on to the history of the development of sluice and lock gates in China, it will be best to consider the matter first here while the Ling Chhü's topography is fresh in the mind. One can have no difficulty in attributing the spillways to an engineer so ingenious as Shih Lu must have been, a younger contemporary, perhaps, of Li Ping, and certainly of his son, who had used them with so much effect at Kuanhsien; there is moreover the close similarity of the division-head 'snout' in the two cases. But whether Shih Lu installed any gates remains uncertain. Evidence will be presented later[c] which shows that sluice-gates were a familiar technique in the −1st century, and references of that time make clear that the Hung Kou of Han times[d] had flash-lock gates, especially at the junction with the Yellow River near Jungyang. If so, the Hung Kou of Warring States and Chhin times may have had them too, perhaps indeed it could hardly have worked without them, and in that case

---

[a] I.e. just over 35 tons; cf. pp. 230 and 645.

[b] Contrast the date of the earliest pound-locks in a European transport canal which surmounted a gradient, now established as those of the Stecknitz waterway, built *c.* +1395 (Skempton, 4). Cf. pp. 358 ff. below.

[c] Cf. pp. 344 ff. below.   [d] Cf. pp. 270, 281 above.

[1] 洩水灘   [2] 斗門   [3] 則復闑之

they may have been well known to Shih Lu. Nevertheless there is no positive textual or other evidence that he installed them in the Ling Chhü or its approaches.

If he did not, then one has to picture southbound barges being towed up the trained Hsiang River by gangs of trackers,[a] into the Pei Chhü section, and so into the pool by the great spillway. Having passed the division-head they would reverse into the canal where there was no rapid current, for the level course had been made as winding as possible precisely in order to slow it. No reverse was needed at the other end, where the boats would glide fairly fast down the trained Li. The first information that we have of gates comes from a period still relatively early, the Thang, and concerns the important restoration of the canal carried out in +825 by Li Pho.[1] In the *Shui Ching Chu* commentary we read:[b]

During the Thang dynasty, at the beginning of the Pao-Li reign-period (the banks of) the canal were collapsed and broken so that boats could not get through. The Inspector-General Li Pho therefore caused stones to be piled up in courses to make a dyke like a 'spade snout' to split and divide (the Hsiang River) into two streams. In each stream there were set up stone (abutments for) flash-lock gates (lit. dipping gates, *tou mên*[2]), with one keeper in charge (of each), who let people freely open and shut (as required). When the Li River (gate, or gates, was, or were) open, (the water) all flowed into the Kuei Chiang (to the south); when the (gate or gates, was, or were) closed on the Kuei Chiang side, then all (the water) returned to the Hsiang River (northwards). And moreover at the Hsiang River he cut a 'dividing-water canal' (*fên shui chhü*[3]) 35 paces long (175 ft.), in order to facilitate the movement of the ships.

Taking this last improvement first, the most obvious suggestion is that Li Pho isolated the division-head in the form of an island quay, making a canal behind it along the top of the dam so that it was no longer necessary for the boats to reverse. But the flash-locks were no less important. From other sources we know that there were 18 of them, first set up in rough construction of wood,[c] but then greatly strengthened by Yü Mêng-Wei[4] in the further improvements which he carried out[d] in +868. Their exact positions we do not know, but from the text given it seems sure that at least one of them was at the Li end of the canal, and at least one other at the Hsiang end. There was more need for them, however, on the graded approaches rather than on the level canal, so it is probable that the majority were erected actually in the course of the Li, the Pei Chhü section parallel with the Hsiang, and in the Hsiang's course lower down as well. From the +9th century at latest onward, then, barges ascending on either side were hauled through the flash-locks, probably with winches, and towed by a much reduced company of trackers on the relatively level intermediate sections.

Then comes the information that the number of gates was no longer 18 but 36. This is still the approximate number today, and after the account of Chou Chhü-Fei in

[a] As the *Chou Li* said, 'Wherever there is a stream in a valley there will be a place for a towpath' (p. 255).
[b] Ch. 38, p. 16*b*, tr. auct. The passage comes from *Thai-Phing Huan Yü Chi* (c. +980), ch. 162, p. 8*b*.
[c] Fang Chi (2), p. 44.
[d] Yü Mêng-Wei also rebuilt the division-head in better masonry, and left an account of his work with the title *Hsiu Chhü Chi*.[5] It is preserved in *Chhüan Thang Wên*, ch. 104, pp. 10*a* ff. This monograph may be compared with others on particular engineering operations mentioned on pp. 173 and 203.

[1] 李渤   [2] 斗門   [3] 分水渠   [4] 魚孟威   [5] 修渠記

+1178 the same number is repeated in mentions of +1396 and +1485 when again the stone abutments were well rebuilt.[a] The doubling of numbers in this way is always strongly indicative of the change from flash-locks to pound-locks, for there was everywhere a natural tendency to make use of facilities already installed, and use each existing stanch as one of the pair of pound-lock gates. The account of a later traveller, Kuang Lu,[1] is also worthy of notice. Writing about +1585, he said:[b]

On the Ling Chhü from north to south there are 32 lock-gates (*tou* (*mên*[2])), i.e. from the Li to the Thung-Ku Shui.[3] From east to west, entering Yung-fu,[4] there are 6. In the winter (the canal) dries up and one cannot pass through. But when I made the passage through these lock-gates there was plenty of water, and under the moonlight they looked like steps leading up to some high platform, or like tiers of walls and terraces coming down one behind another from the sky.

One could hardly have a better description of the pound-locks of a summit canal—for into such a waterway had the contour Ling Chhü been converted when pound-locks were installed in its approaches. On the evidence as a whole it seems clear that the +10th or +11th centuries are the most probable periods at which we should place the introduction of pound-locks in the Ling Chhü system;[c] and as we shall see later on, this dating fits the rest of the Chinese evidence extremely well.

The significance of the Ling Chhü Canal as a link in a chain of communications altogether extraordinary for the −3rd century in any part of the world should not be overlooked. By the navigation of the lower Yellow River, the Hung Kou and Han Kou Canals, the Yangtze, the Hsiang River leading south from the Tungthing Lake, the Ling Chhü Canal and the West River, the first Han emperor in −200 found himself in possession of a single trunk waterway extending from the 40th to the 22nd parallel of latitude, that is to say a distance of some 1,250 miles in a direct line, and doubtless more than double that as the vessels sailed. Few if any other ancient civilisations could show the like. And lastly the Ling Chhü resembles Kuanhsien in that though a work of the −3rd century it has been repaired and set in order once again in our own time, and continues to carry out heavy duties. Few if any other civilisations could demonstrate a work of hydraulic art in continuous use for 2,185 years.

(v) *The Grand Canal* (*Sui and Yuan*)

Our notes on the Grand Canal as a whole (the 'Traffic River', Yün Ho[5],[d] Fig. 859) can be all the more brief because several of its precursor or component sections have already been mentioned. The work of combining these together was done by the Sui

---

[a] Fang Chi (2), pp. 44 ff.

[b] *Chhih Ya*, ch. 2, p. 21 *b*, tr. auct. What Kuang Lu says about the numbers of the gates agrees with the general conclusion that there were many more locks on the approaches than in the level canal itself, in which he seems to have found three.

[c] The exact date may well be +1059, for we know from *Sung Shih*, ch. 97, pp. 25 *b*, 26 *a*, that in that year general repairs were carried out by Li Shih-Chung,[6] who, among other things, *chhung phi*,[7] which might mean 'doubled the openings'.

[d] Or Yün Liang Ho[8] (Tribute Traffic River).

[1] 鄺露   [2] 陡門   [3] 銅鼓水   [4] 永府   [5] 運河   [6] 李師中
[7] 重闢   [8] 運糧河

dynasty (+581 to +618) when the need to link the capital at Loyang with the key economic area of the lower Yangtze valley became imperative. In the last decades of the +13th century, under the Yuan emperors, the same need continued, but as the capital was now Peking, a vast remodelling of the canal was carried out, so that it finally formed a continuous waterway following the 118th meridian in an S-shaped course from Hangchow in the south to the furthest northern parts of the North China plain. In order to visualise this major work in its final stages, it is only necessary to remember that it covered 10° of latitude, and would be comparable with a broad canal extending from New York to Florida. Its total length attained nearly 1,100 miles. Its summit, reached when skirting the mountains of Shantung, was some 138 ft. above sea level.[a]

The oldest pioneer section was the Hung Kou,[1] later known as the Pien Chhü[2] Canal, which connected the Yellow River with the Huai valley. As we have seen,[b] this was at least as old as the −4th century. Besides civilian transport it saw naval activities from time to time, as in +280 when a famous fleet and army passed along it in the campaign of Wang Chün[3] against the State of Wu,[c] and again in +417 when Liu Yü[d] conquered Yao Hsing[e] of the Later Chhin.[f] But in the course of ages it had silted up so much that about +600 the chief engineer of the Sui,[g] Yüwên Khai,[4] decided upon an entirely different alignment[h] for the new canal (the Thung Chi Chhü[5]).[i] Broadly speaking, this ran parallel with the Hung Kou but south-west of it after diverging at Chhen-liu (9);[j] passing not Hsüchow (10) but Suchow (120), and joining the Huai River directly, west of the Hung-tsê Lake, without making use of the Ssu[6] River.[k] Its length was some 630 miles. In the north-west, between Chhen-liu, near Khaifêng, and the junction with the Yellow River at Ssu-shui[7] just downstream from the entry of the Lo[8] River from Loyang, the course of the Sui Canal was much the same as that of the

---

[a] Much has of course been written on the Grand Canal; for Chinese books, the bibliographies in papers mentioned below should be consulted. The best survey is no doubt that of Chu Hsieh (1), but it reached us too late to be of much assistance. The chief work in a European language is the historical monograph of Gandar (1), but though copious it is confusedly written, and it concentrates almost exclusively on the part between the Yangtze and the Huai R. Among minor contributions, Middleton Smith (2) and Price (1) may be mentioned. On the Grand Canal as it is today, improved and in full use for much of its length, see Yang Min (1). For the Ming period there is a fine monograph by Huang Jen-Yü (1).

[b] Cf. above, p. 269. [c] Cf. pp. 694 ff. below, and *TH*, p. 836.
[d] Cf. Vol. 4, pt. 2, p. 432. [e] Cf. Vol. 4, pt. 2, p. 287.
[f] Liu Yü found the canal in very poor repair, and had much of it dredged out, but the good effects of this did not last long (*Sung Shih*, ch. 93, pp. 17a ff.; a speech by Chang Chi[9] giving a brief history of the Pien Canal).

[g] He will be familiar to readers of Vol. 4, pt. 2, cf. pp. 32, 253.

[h] It has been elucidated, not without much difficulty, by Aoyama Sadao (6); Chhüan Han-Shêng (1); Tshen Chung-Mien (2); Chi Chhao-Ting (1); Twitchett (4) and others. There still remains some uncertainty about the exact course.

[1] At the same time, the old Pien Canal must have remained repairable long afterwards, because Tu Yu,[10] the great financial official and historian, suggested its use as an alternative route as late as +781.

[j] The present Hui-chi Ho[11] River preserves a trace of its course in this region.

[k] Two neighbouring Huai tributaries qualify as candidates for the line of the Sui Canal, the evocatively named Thang[12] River passing north of Suchow, and the river with the name that itself means a medium-sized canal (Kuei,[13] cf. p. 255 above), passing south of it.

[1] 鴻溝	[2] 汴渠	[3] 王濬	[4] 宇文愷	[5] 通濟渠	[6] 泗
[7] 氾水	[8] 洛	[9] 張洎	[10] 杜佑	[11] 惠濟河	[12] 唐
[13] 澮					

Hung Kou (the old Pien Chhü), if not identical with it, but special works were now erected to protect its opening.[a] In +587 another eminent engineer of the Sui, Liang Jui,[1] built on the south bank of the Yellow River a massive westward continuation of the 'Metal Dyke' (Chin Thi,[2] cf. p. 234 above), which was named in his honour the Liang Kung Yen;[3] it contained lock-gates for regulating the water levels, and double slipways for hauling boats over when the differences were too great for flash-lock operation.[b] The main project, the new Pien Chhü Canal, was completed in +605, more than five million men and women having been mobilised under Ma Shu-Mou[4] to carry out the excavations.[c] A detailed account of the work has come down to us in the anonymous *Khai Ho Chi*[5] (Record of the Opening of the Canal).[d] An imperial road was constructed along its banks, which were planted everywhere with trees.[e]

Throughout the Thang and the Sung (+7th to +13th centuries) the New Pien, the 'Grand Canal' of the Sui, continued in active, indeed heavy, use. One can penetrate behind the dry statistics of the dynastic histories to the realities of canal life in these times by reading some of the diaries and memoirs, notably those of foreigners such as the Japanese monk Ennin.[6] In +838 he journeyed from the coast to Yangchow on one of the lateral feeder canals from the numerous salt works, which had been built or repaired from the San Kuo to the Sui.[f] On a canal which was straight as far as the eye could see (like the Bedford Level in the East Anglian fen country), a train of forty boats, many lashed two or three abeam, was pulled slowly but efficiently by two water buffaloes. Once a bank had caved in but Ennin's party got through by digging, and when they came near the Grand Canal, the salt boats, three to five lashed abeam, followed one after the other continuously mile after mile.[g] In the following year he passed up the Grand Canal to the Huai River, and then towards the end of his stay, in +845, journeyed down the Pien Chhü (the New Pien) from Khaifêng (Chhen-liu) to Yangchow.[h] This took him nine days.

---

[a] Cf. Twitchett (4), pp. 182 ff.; Pulleyblank (1), pp. 128 ff.

[b] These works were rebuilt in +714 by Li Chieh.[7] Another opening was attempted by Liu Tsung-Chhi[8] in +726, but it was quite unsuccessful, and the old junction was put in order again by Fan An-Chi.[9] Difficulties about the entrance continued all through the Sung, and yet another opening was made in +1071 by Ying Shun-Chhen.[10] Though it quickly silted up, he made it work, at least for a time, by inserting many sluice-gates to take off flood-water, and digging subsidiary feeder canals to keep up its level in the low-water season (cf. Chêng Chao-Ching (1), pp. 208 ff.; *Sung Shih*, ch. 93, p. 22b).

[c] Cf. Vol. 1, p. 123.

[d] Preserved in *Hsing Shui Chin Chien*, ch. 92. Cf. p. 326 below. But most of the *Khai Ho Chi* is fiction rather than history (cf. Wright, 7).

[e] As a whole, this great achievement of the Sui dynasty merits in no way the characteristic (and here indeed ludicrous) sneers of Wieger (*TH*, p. 1275).

[f] Cf. Fig. 903 (pl.).

[g] Tr. Reischauer (2), pp. 16, 18, 20; cf. (3), pp. 72 ff., 153.

[h] Tr. Reischauer (2), pp. 81 ff., 371 ff.; cf. (3), p. 262. Though Ennin was now on his way home his journey was sorrowful, partly because in the persecution of Buddhism then going on he had been forced to return to lay life. Reischauer, followed by Twitchett (4), p. 188, thinks that Ennin's vessel followed the Kuo[11] River, a canalised waterway parallel to the New Pien but further to the south-west; we have not been able to see justification for this view.

[1] 梁睿	[2] 金隄	[3] 梁公堰	[4] 麻叔謀	[5] 開河記
[6] 圓仁	[7] 李傑	[8] 劉宗器	[9] 范安及	[10] 應舜臣
[11] 渦				

The engineers of the Thang introduced some modifications of the system but no fundamental change. In +689 a branch (the Chan Chhü[1]) was made, striking off north-eastwards from Chhen-liu to connect the trunk waterway with Yenchow[2] in Shantung.[a] Then between +734 and +737 Chhi Huan[3] constructed another new canal (the Kuang Chi Chhü[4]), which skirted the north shore of the Hung-tsê Lake and brought the traffic from the New Pien Canal directly to Huai-yin, thus short-circuiting the dangerous rapids of the Huai.[b] At the same time he also made a short cut near Yangchow (the I Lou Ho[5]) which saved some 40 li.[c] But this was on the more southern section between the Huai River and the Yangtze.

As we saw at an earlier stage, the ancient Han Kou, first built in the early −5th century, ran between these points. But its original route was circuitous, connecting several lakes, and an important straightening had been made about +350 by Chhen Min. His Shan-yang Yün-Tao (Shanyang Transport Canal) was restored by the first Sui emperor in +587 as the Shanyang Tu,[6] and then incorporated in the whole system in +605. This is known to have been 40 paces (200 ft.) wide, and bordered with trees like the more north-western sections.[d] Its length was about 120 miles.

The portion of the Grand Canal south of the Yangtze was not an entirely new enterprise of the Sui, because there had been earlier artificial waterways in the region (p. 272), but it took a new course, completed in +610. Eight hundred li in length,[e] it skirted the eastern side of the Thai-hu[7] Lake, and put Hangchow in direct communication with the north, thus enabling the supplies and products of the south-eastern coastal regions to flow to the capital. In order to assure traffic between the Yellow River and Chhang-an, Yuwên Khai restored the old canal in Shensi which Chêng Tang-Shih and Hsü Po had dug in −133, and renamed it the Kuang Thung Chhü.[8]

Here we may pause for a moment to take notice of the rather intensive development of urban waterways in the Venetian style in these cities, so useful as the capillaries of the traffic system.[f] Many of these may still be seen in Suchow (Chiangsu) today, encircled as it is by the Grand Canal (cf. Fig. 772, pl.), and one can find the remains of

---

[a] Chêng Chao-Ching (1), pp. 201 ff.

[b] Chiu Thang Shu, ch. 190B, p. 17a; Hsin Thang Shu, ch. 128, pp. 7a ff. The former account gives us a good instance of the use of ox-teams for hauling the barges with bamboo cables up these rapids.

[c] It brought the canal to Kuachow,[9] nearly opposite the entrance at Chen-chiang[10] on the south side, and thus saved much danger and difficulty in the shoals of the river itself as well as shortening the route.

[d] Eastwards a number of feeder canals were built, as we have just seen, to link it with the salt works of the Chiangsu coast, while at the junctions with Huai and Yangtze there were flash-lock gates and double shipways. For Chhen Min see p. 282.

[e] I.e. c. 250 miles, considerably longer than the route existing today (cf. Table 70 below). This stretch was commonly known as the Chiang Nan Ho.[11]

[f] Sumet Jumsai na Ayutya, who is working on this subject in a comparative way, has published (1, 2, 3) several interesting accounts of the waterway systems of Angkor Thom and the city of Ayut'ia, in Cambodia and Thailand respectively. There are many similarities with China here, as he shows.

---

[1] 湛渠   [2] 兗州   [3] 齊澣   [4] 廣濟渠   [5] 伊婁河
[6] 山陽瀆   [7] 太湖   [8] 廣通渠   [9] 瓜洲   [10] 鎮江
[11] 江南河

an elaborate network in Hangchow.[a] Of the docks at Chhang-an (Sian) and the Peking terminal we make mention elsewhere (pp. 273, 313, 355).

The portion of the Canal north of the Huang Ho, rather more than 620 miles in length, was truly a new enterprise of the Sui. It took advantage of a short river descending southwards from the Shansi mountains and falling into the left of the Huang Ho a little east of Ssu-shui, namely the Chhin;[1] and rendering the lower reaches of this navigable, it struck across country by a short branch to join the head waters of the canalised Wei[2] River which flowed north-eastwards in a long course to the neighbourhood of Tientsin.[b] Water from the Ta Tan[3] and Chhi[4] Rivers also helped. This route, finished in +608, was called the Yung Chi Chhü.[5] At Ta-ming (123) it joined the line of the later course ④ of the Yellow River, and at Lin-chhing (124) the line of the present Grand Canal. Thus the new waterway connected the region of Peking, no less than that of Hangchow, with the capital at Chhang-an (Sian). The immense Y-shaped system so formed extended for approximately 1,560 miles—a distance equivalent to $22\frac{1}{2}$ degrees of latitude, and resembling that from Stockholm to Syracuse, or Greece to Greenwich.[c] In Isidore of Seville's Europe one does not find the like.

During the Thang and Sung dynasties the highest officers of state occupied themselves with the engineering problems involved in conserving the Pien Canal as a great artery of tax-grain transport from south to north-west,[d] e.g. the erection and management of lock-gates at certain points, especially the junctions of the canal with the two great rivers which it crossed. Phei Yao-Chhing[6] (+681 to +743)[e] and Liu Yen[7] (+715 to +780)[f] were two among such men; during the administration of the former, about +735, the canal carried annually no less than 165,000 tons of grain.[g] Granaries were established at many places along the route, so that the grain could be stored in good conditions if flood or unduly low water hindered its onward carriage. In +733, as we have seen (p. 278), an arrangement was also made whereby the grain was portaged 6 miles overland to avoid the rocks and islands in the Huang Ho at the San-mên[8] gorge, which made navigation so dangerous, and granaries were built at each

[a] Charts of 1800 and 1900 are in the British Museum, catalogued as Maps Tab. 1, d, 3 & 4. Cf. Moule (15), fig. 1. Unique here was a tidal basin, connecting with the Chhien-thang estuary and closed by lock-gates (hun shui cha[9]); this filled at high tide, and after a few hours wait to let the silt settle the water was flushed through part of the canal system back to the sea in the north-west (cf. Moule (15), p. 21, earlier version, p. 116). On the origins and development of the famous West Lake at Hangchow, see Chang Hung-Chao (4, 5).

[b] Today (1953) a large scheme has just been completed for bringing the waters of the Yellow River from higher upstream to enter the upper reaches of the Wei; this will irrigate some 70,000 acres around Hsinhsiang,[10] and greatly increase the navigability of the Wei River; cf. Kao Fan (1).

[c] And all completed within a couple of decades; though for five hundred years it was to be the main trunk line of China's communication network.

[d] For general accounts of the fiscal and economic aspects of its history in Thang times the monographs of Pulleyblank (1) and Twitchett (4) are valuable.

[e] Cf. Pulleyblank (1), pp. 34 ff., 129, 183 ff. (where he translates in full two important memorials by Phei), 201 (a biographical notice); Twitchett (4), pp. 87 ff., 303, 306 ff.

[f] Cf. Twitchett (4), pp. 92 ff., 310 ff.

[g] Hsin Thang Shu, ch. 53, pp. 2b ff.; Chiu Thang Shu, ch. 49, p. 2b. Cf. a poem of Tu Fu's (tr. Alley (6), p. 131).

[1] 沁   [2] 衛   [3] 大丹   [4] 淇水   [5] 永濟渠
[6] 裴耀卿   [7] 劉晏   [8] 三門   [9] 混水閘   [10] 新鄉

transhipment point. With Liu Yen as Transport Commissioner (+763 to +779) the highest pitch of efficiency was perhaps reached, for special boats were constructed to suit each different section of the canal, and every part of it was kept in good order.

In the Five Dynasties period, which followed the collapse of the Thang,[a] the economic areas served by the Canal were split up into different political units, but there was little or no destruction of the works themselves. The Sung people after +960 intensified the care of the waterway as part of their expanded water-conservancy programme, which outdistanced that of any previous dynasty (cf. Figs. 882 and 900, pl.).[b] The capital was moved further east to be fixed logically at Khaifêng, just about the central point of the Y-system,[c] and we have many records of the complicated web of ship canals which ramified through and around the capital from the Pien or Grand Canal itself.[d] But by the +12th century the hold on the capital was becoming precarious. After the collapse of its defence against the Chin Tartars in +1126,[e] the Sung general staff deliberately broke the dykes south of the Yellow River, delaying indeed the Jurchen advance to the Yangtze but also destroying most of the works of the Pien Canal. After the establishment of the Sung capital at Hangchow in +1135, the lands between the Huai valley and the Yellow River became for centuries a battleground, so that the Pien Canal of the Sui must have been in a very dilapidated condition when the Yuan (Mongol) people overwhelmed Chin, Liao and Sung alike, and obliged historians to count the years of their dynasty from +1280.

Since the Yuan empire included so much northern territory which was outside the bounds of China proper, Peking was the natural choice for the capital. But since the long-established pattern of Chinese social and economic life persisted unchanged, simply with the imposition of Mongols (and other foreigners)[f] at the top, Peking could not become the central administrative ganglion without a 'stream-lining' of the tax-

[a] Already by about +800, after the rebellion of An Lu-Shan, the annual grain consignment had fallen as low as 28,200 tons (*Sung Shih*, ch. 93, p. 19a).

[b] The reader will remember the interesting passage from the *Mêng Chhi Pi Than* (ch. 25, para. 8), translated in Vol. 3, p. 577, where Shen Kua described his work as hydraulic engineer and surveyor in the restoration of the Pien Canal between Khaifêng and the Huai about +1070. Just after this, it is remarkable to read of ice-breakers on the canal. In his *Tung Hsien Pi Lu*, written towards the end of the century, Wei Thai says: 'The Pien Chhü always closed by the 10th month of each year, so that no traffic could use it. But when Wang Ching-Kung[1] (i.e. Wang An-Shih) was prime minister he wished to have transportation carried on in the winter as well, and kept it open. Apart from the low water levels, which hindered movement, much damage was caused to the vessels by floating ice. So he had several dozen 'foot-boats' equipped with trip-hammers (*tui*[2]) at the bows to break up the ice, but the men on duty suffered greatly from the cold and many died of it. Thus a saying went round in the capital: "We've heard of mill-stones being used for making gruel, but now we see trip-hammers used for pounding winter ice"' (ch. 7, p. 2b). On tilt-hammers and trip-hammers, and their various uses, see Vol. 4, pt. 2, pp. 51, 390 ff. As for the 'foot-boats' (*chiao chhuan*[3]) they may well have been paddle-wheel boats worked by treadmills. Similar terminology has already been encountered; see Vol. 4, pt. 2, pp. 417 ff. When the friar Domingo de Navarrete travelled down the Grand Canal in the winter of +1665 ice-breaking operations were being carried on, as he noted in his autobiography (Cummins tr., vol. 2, p. 227).

[c] According to an estimate of Wang Chi-Ong,[4] a Sung official who went over to the Mongols, 424,000 tons of grain reached Khaifêng annually during the Northern Sung period. Compare this figure for the record in the Thang of 165,000 tons just noted, and with the best figures for the sea route in later times (p. 478 below). Wang's figure is confirmed in *Sung Shih*, ch. 93, p. 17a.

[d] E.g. *Fêng Chhuang Hsiao Tu*, ch. 1, p. 22b.   [e] Cf. Vol. 4, pt. 2, p. 497.

[f] Cf. Vol. 1, p. 141.

[1] 王荊公   [2] 碓   [3] 脚船   [4] 王積翁

grain vascular system. For this purpose the essential thing was to make a short cut between the open arms of the Y, in such a way as to bring the east-central economic area more directly in touch with the north. From Hangchow to the Huai no great change was needed, but north of the Huai it was necessary to abandon the westward trend towards the age-old hub[a] of Jungyang near Khaifêng and to plan a more direct route. The line of the old Pien Canal was thus entirely replaced by a more easterly one which made its way over the shoulder of the Shantung mountains and crossed the Yellow River far to the east of the former junction points. This new alignment was truly a summit canal in a way which the Pien Chhü had never been, and doubtless its planning drew some inspiration from the known success of the Ling Chhü.[b] Though sea-transport now began to make serious competition,[c] the Yuan dynasty still relied largely on inland waterway traffic,[d] and the remodelling of the Grand Canal which it brought about lasted through the Ming and Chhing periods down to the present day. Who could fail to appreciate the political and economic consequences of so bold a conception[e]—an artificial river running north and south in a country where most of the natural rivers run from west to east (Fig. 901, pl.)?

The Great Khan [wrote Marco Polo] has made very great channels, both broad and deep, from the one river to the other and from the one lake to the other; and makes the water go through the channels so that they seem a great river; and quite large ships go there with the said grain loaded from this city of Caigiu up to the city of Cambaluc in Cathay.[f]

The northernmost part was comparatively short, the Thung Hui Ho[1] (Channel of Communicating Grace), running 164 li[g] from the capital to Thungchow,[2] and completed in +1293 under the direction of an old friend of ours,[h] the astronomer Kuo

[a] This very expression occurs in the *Sung Shih* (ch. 93, p. 19b), *Thien-Hsia chih shu*.[3]
[b] Cf. p. 304 above and p. 355 below. It will be remembered that pound-locks had been installed there in the +10th or +11th century.
[c] Throughout the Yuan there were extended controversies as to the respective merits of the sea route and the canal. It would be no exaggeration to say that they formed ample material for party politics. The essential point was that while the sea route was certainly capable of delivering a greater tonnage it involved much more risk. Cf. pp. 478 ff. below, and for further detail, Lo Jung-Pang (4); Schurmann (1), pp. 109 ff. See also Hummel (18); Wu Chhi-Hua (2).
[d] The government organised in +1283 a Directorate of Grain Transport (Tshao Yün Ssu[4]), and much information about its work has come down to us in the *Ta Yuan Tshang Khu Chi*[5] (Record of the Granaries (and the Grain Transport System) of the Yuan Dynasty).
[e] If one looks for the real spiritual fathers of the scheme, one finds them in two local officials in Shantung, Yao Yen[6] and Han Chung-Hui,[7] together with the former Sung governor, Wang Chi-Ong,[8] who was adviser to the Yuan government. In +1284 Wang died at sea when sailing as ambassador to Japan, probably murdered by the 'sea-route party', who detested canal schemes.
[f] Ch. 148, Moule & Pelliot ed. Caigiu is plausibly identified by Pelliot (47), vol. 1, p. 129, with Kuachow,[9] the town on the north bank of the Yangtze at the point where the Grand Canal opens into it. It is near the site of the Battle of Tshai-shih (cf. p. 416 below).
[g] I.e. some 51 miles. The reason for the discrepancy between this figure and Table 70 is that the distance here given was measured from the sources in the hills west of Peking which supplied the water for the canal and not from the actual terminus itself. The Jurchen Chin people had tried a project of this kind earlier (the Kao Liang Ho[10]) but had not been able to find enough water for it, especially as their lock-gates were poorly constructed.
[h] Cf. above, Vol. 3, *passim*.

[1] 通惠河　　[2] 通州　　[3] 天下之樞　　[4] 漕運司　　[5] 大元倉庫記
[6] 姚演　　[7] 韓仲暉　　[8] 王積翁　　[9] 瓜洲　　[10] 高梁河

Shou-Ching,[1] who was a brilliant civil engineer as well. This section needed about twenty lock-gates and Kuo Shou-Ching provided them.[a] Two hundred and fifty years later (+1558) a special monograph, the *Thung Hui Ho Chih*,[2] was consecrated to this noble waterway by Wu Chung.[3] It was being written about in Persian within a dozen years of its completion. In +1307, Rashīd al-Dīn al-Hamdānī, after giving a good account of Peking with its lakes and streams in his *Jāmi al-Tawārīkh*, goes on to say:[b]

We hear that since the river was so narrow, the capital could not be reached by ships, and that people were obliged to use beasts of burden for the transport of merchandise to and from the city of Khanbaliq. However, the geometers and philosophers of Cathay assured (the Khan) that it would be possible to bring to the capital the vessels of all the provinces of Cathay, and those also from the capital of the kingdom of Mātchin (the former Sung dominions), and even those from the cities of Khingsai (Hangchow) and Zeitun (Chhüan-chow) and other places.

Rashīd al-Dīn then goes on to describe the completion of the Thung Hui Ho, and of the whole Grand Canal system, mentioning especially its flash-lock or pound-lock gates and capstan slipways, and the planting and protection of trees alongside. Thus Saracens and Franks alike were impressed by what Chinese engineering could do.

The second part, (*b*) in Table 70, running from Thungchow to a point near modern Tientsin, had originally belonged to the Sui system, but was greatly improved in Kuo's time, and called the Pai Ho[4] (White River). The third section (*c*) also followed the Sui route, being no more than the Wei River again made navigable (though it was renamed the Yü Ho,[5] Imperial River); but about half the distance along it, at Lin-chhing[6] (124),[c] an entirely new canal (*d*, *f*), the Hui Thung Ho[7] (Union Link Channel), was thrown off southwards to cross the northern course of the Yellow River at right angles (cf. Fig. 902, pl.). The idea of this had arisen first in +1275, when Khubilai's general Bayan had enquired into its possibility; he was assured by Ma Chih-Chên[8] that such a canal was feasible if water from the Wên[9] River was used, and this was confirmed in a preliminary survey made by Kuo Shou-Ching himself. The project then hung fire, partly because of the use of the sea route, until +1289, when it was revived by the magistrate of Shou-chang, Han Chung-Hui,[10] and another astronomer, Shih Pien-Yuan.[11] The latter was authorised, with Ma Chih-Chên, to make a definitive survey, and this being quickly accomplished, charts and plans were presented to the emperor. Chang Khung-Sun[12] and a Mongol Loqsi[13] were then placed in charge of the work, which was completed within the year; it had 31 locks (*mu chha*[14]) in a distance of some 250 li (about 80 miles).[d] It therefore got the popular name of Chha Ho.[15]

[a] This is a subject to which we shall return below (p. 355).
[b] Tr. Klaproth (3), p. 341.
[c] A romantic engraving of this place and its pagoda will be found in Staunton (1), pl. 33.
[d] Chêng Chao-Ching (1), p. 216; Lo Jung-Pang (4). It is very probable that some of these locks were pound-locks, but the texts are not precise on the point. Eight of them were rebuilt with stone abutments between +1296 and +1302, and all by +1327.

[1] 郭守敬	[2] 通惠河志	[3] 吳仲	[4] 白河	[5] 御河
[6] 臨清	[7] 會通河	[8] 馬之貞	[9] 汶	[10] 韓仲暉
[11] 史邊源	[12] 張孔孫	[13] 樂師	[14] 木牐	[15] 牐河

One reason perhaps for this success was that in the same year a change in the course of the Yellow River occurred (from ⑤ to ⑥, cf. the map in Fig. 859), its northern arm being greatly reduced in volume, and most of the water running into the Huai and the Yangtze. The canal crossed the diminished northern arm[a] at a point south of the city of Tung-a[1] in western Shantung, and reached as far as An-Shan[2] near Tung-phing,[3] where it met the summit section (f, g) completed six years before. In doing this it was able to incorporate the remains of the short canal called Chhing Chi Tu[4] which (as already noted, p. 282) had been built by Hsün Hsien in Chin times (+352).[b]

The important summit section itself was the work of a Mongol military engineer, Oqruqči,[5] in +1283 and the following years, in accordance with plans drawn up by Kuo Shou-Ching; known as the Chi Chou Ho,[6] it connected Chi-ning[7] with the Chhing Chi Tu and so with the sea by means of the river north-eastwards, a stretch of more than 140 li (c. 44 miles). On the other side, to the south, it linked up with another work of the Chin period, the Huan Kung Kou,[8] which had been first constructed in +369, when Marshal Huan Wên[9] was campaigning northward against Mujung Chhui,[10] afterwards first king of the Later Yen dynasty. When its original object of carrying military supplies from Huai-yin[11] as far north as Chi-ning[7] had been fulfilled it served little purpose in later centuries until in +1283 it was restored and extended as part of the new north–south system (sections h, i).[c] This portion is seen in the Ming or Chhing chart of Fig. 903 (pl.).[d] Before the Hui Thung Ho was completed, therefore, the Yuan government could bring up grain supplies from the east-central economic area by canal boat as far as the northern mouth of the Yellow River, and then tranship them to sea-going vessels for the rest of the way to Tientsin. The Huan Kung Kou used for its water supply a number of lakes with which it connected on its way up a markedly rising slope (cf. Fig. 906), and covered (excluding them) some 300 li (c. 94 miles).[e] Early in the Yuan its southernmost portion ran near the southern arm of the Yellow River (course ⑤), but the change of +1289 brought those waters to the Huai at a more westerly point, and most of the discharge, after passing through the Hung-tsê Lake,[f] accompanied the Shan-yang Yün-Tao to Yangchow (course ⑥). In +1324 the

[a] At this time called Ta-Chhing Ho.[12]   [b] Chêng Chao-Ching (1), p. 214.

[c] Preliminary restoration work between Huai-yin and Chi-ning had been carried out from +1072 onwards in the Northern Sung period (Sung Shih, ch. 96, pp. 3b ff.; cf. Lo Jung-Pang, 4). After alternative plans proposed by Lo Chêng-Fu[13] and Chiang Chih-Chhi,[14] the survey and then the works were carried out by Chhen Yu-Fu.[15] On the complex Huai-yin region today see Wang Chun-Kao (1).

[d] At one time in the collection of Mr Rewi Alley, to whom we are much indebted for the photograph and permission to reproduce it.

[e] There is some obscurity about the route of Huan Wên's canal, which may have depended on the navigation of the Ssu River, and in any case was certainly not identical with sections (h) and (i) of the Grand Canal of today. Cf. Fang Chi (2), p. 47; Chêng Chao-Ching (1), p. 197.

[f] Later, in +1694, this was surrounded by a 62-mile dyke built by Yü Chhêng-Lung,[16] still to be seen. The general tendency was to separate off the canal from the lakes by dykes as time went by, so as to avoid danger or inconvenience in bad weather. The Kao-yu Lake was thus canalised by Li Phu[17] in +1007, but later on the dyke broke in many places, so in +1488 a detour canal section entirely separate from the

[1] 東阿	[2] 安山	[3] 東平	[4] 清濟瀆	[5] 奧魯齊
[6] 濟州河	[7] 濟寧	[8] 桓公溝	[9] 桓溫	[10] 慕容垂
[11] 淮陰	[12] 大清河	[13] 羅拯復	[14] 蔣之奇	[15] 陳祐甫
[16] 于成龍	[17] 李溥			

river reverted north again to course ⑦, so that once more the canal traffic had to cross it at Huai-yin[1].[a]

There is more of interest to be said about the summit section (the Chi Chou Ho) where the water surface attains a height of 138 ft. above the mean level of the Yangtze at the junction point. As Oqruqči left it, the water supply for the highest levels (always the greatest problem of summit canals) was unsatisfactory—so much so indeed that throughout the Yuan period the canal could not always compete with the alternative advantages of sea transport. After it had long been evident that something would have to be done, the necessary remodelling of sections (d) to (i), which brought the most difficult part of the Grand Canal to a high level of efficiency, was carried out by the Ming engineer Sung Li[2] in +1411, at the proposal of the Assistant Administrator of Chi-ning, Phan Chêng-Shu.[3] It is recorded that Sung Li, a former student of the Imperial University,[b] was helped to solve his problems by the advice of an old country-man (probably an irrigation-worker) of Wên-shang, Pai Ying,[4] who showed how the waters of the Wên[5] and Kuang[6] Rivers could be used more effectively (cf. Fig. 905, pl.).[c] Pai Ying suggested that a new mile-long bund or dam should be constructed on the latter north of Ning-yang to form a reservoir which would always keep the canal full, with the aid of a forking lateral canal from the former, and these major works were successfully completed in 200 days by a force of 165,000 men.[d] Besides the large reservoir, Sung Li also installed four smaller ones near the canal itself under the name of 'water boxes' (shui kuei[7]).[e] The Ming Shih has a rather significant passage on this.[f]

lake was built, the lake still being used when conditions were favourable. See Fig. 904 (pl.), from Nieuhoff (1), p. 148; cf. van Braam Houckgeest (1); Fr. ed. vol. 1, p. 312, Eng. ed. vol. 2, p. 126.

[a] Or, more correctly, leave it there, for traffic followed the river from Hsüchow (10) to the Huai. The route of the canal shown in Fig. 859 was not opened till +1609.

[b] Biography in Ming Shih, ch. 153, pp. 1a ff. At this time it was a general practice to use students and former students of the Hanlin Academy and the Imperial University in public works projects design and management (cf. Yang Lien-Shêng (11), p. 12). Sung Li, however, had already been Minister of Works.

[c] Ming Shih, ch. 85, p. 4b; ch. 153, p. 2a.

[d] The new status of the Grand Canal as a thoroughly practical proposition had a stimulating effect on sailing-barge construction (cf. p. 410 below) and a correspondingly inhibiting one on maritime ship-building and sea-power in general (cf. pp. 478, 484, 524 below).

[e] We thought for a long time that these were pound-locks, but they must have been feeder reservoirs. The oldest use of the term shui kuei so far found occurs in a statement concerning the second year of the Sung dynasty (+961). In his Hsü Tzu Chih Thung Chien Chhang Pien[8] of c. +1180, Li Tao[9] says: 'On a chia-hsü day in the second month of the 2nd year of the Chien-Lung reign-period, the emperor went to inspect the repairs being carried out to the shui kuei outside the southern watch-tower (of the capital)' (ch. 2, p. 2b). As this was Khaifêng, far from any hilly country, the purpose of the 'water box' here is not obvious—possibly it was some kind of dock or lock-basin. One suspects a change in the meaning of this technical phrase between Li Tao and Sung Li. This text was kindly signalised to us by Professor Twitchett. Then in +1087 Su Chhê[10] demanded that all districts east of the Pien Canal mouth should build shui kuei reservoirs for their water supply, some of the water being available for irrigation from time to time upon payment of a fee (Sung Shih, ch. 94, p. 4a, b). Of course shui kuei often means just tanks, as in the nautical literature. And conversely, as we have noticed already (Fig. 833 above), it was in later times at any rate applied to caissons that kept water out rather than in.

[f] Ch. 85, p. 5a, tr. auct. The 'water box' reservoirs with their proper names are mentioned again on pp. 11b, 24a, b, and 26b. They were repaired and amplified by Wang I-Chhi[11] in +1540, and again in +1616.

[1] 淮陰	[2] 宋禮	[3] 潘正叔	[4] 白英	[5] 汶	[6] 洸
[7] 水櫃	[8] 續資治通鑑長編		[9] 李燾	[10] 蘇轍	[11] 王以旂

He [Sung Li] also made near Wên-shang, Tung-phing, Chi-ning and Phei-hsien artificial reservoirs (lit. lakes), setting up 'water boxes' and sluice gates (*tou mên* [1]). On the west side of the Canal (Tshao Ho [2]) there are what are known as 'water boxes', on the east side there are what are called 'sudden declivity' sluice gates; the boxes are to store the water supplies, and the gates to let go any excess.

Although the exact arrangement could only be clarified by examination of the works themselves, it would seem that the 'water boxes' were something like the side ponds of modern pound-locks (cf. p. 345), perhaps the earliest of the kind, while the sluices delivered straight out of the eastern side of the canal; at any rate it is clear that Sung Li secured for the high-altitude levels a permanently adequate water-supply. But in order not to waste it the provision of better lock-gate equipment on each gradient was essential. Oqruqči had put in 14 gates on the northern side of the water-shed summit (the 'water spine', *shui chi* [3]), but Sung Li now rebuilt them and increased the number.[a] On the northern side, between An-shan Chen [4] and Lin-chhing, with a fall which he reckoned as 90 ft., he built 17 *cha* [5];[b] and on the southern side, between Nan-wang Chen [6] and somewhere about Su-chhien [7] (near Hsüchow), with a fall which he reckoned as 116 ft., he built 21.[c] The sluice gates of the feeder channels were arranged in pairs so as to deliver above or below the lock-gates along the summit levels as desired.[d] It was estimated that of the summit water supply 60% flowed northwards towards Lin-chhing while 40% went to the south past Chi-ning.[e] In the light of all this it is interesting to read the account of Staunton's passage on the Grand Canal in +1793:[f]

On the 25th October (two days after reaching Tong-whang-foo)[g] the yachts arrived at the highest part of the canal, being about two-fifths of its entire length. Here the river Luen,[h] the largest by which the canal is fed, falls into it with a rapid stream in a line which is perpendicular to the course of the canal. A strong bulwark of stone supports the opposite western bank, and the waters of the Luen striking with force against it, part of them flow to the northern and part to the southern course of the canal. A circumstance which not being generally explained or understood gave the appearance of wonder to the assertion, that if a bundle of

---

[a] Oqruqči himself had been warned by Ma Chih-Chên that his complement of lock-gates would not be sufficient. Cf. Lo Jung-Pang (4).

[b] This included the whole of the Hui Thung Ho, and crossed the present or northern line of the Yellow River (courses ⑤ and ⑧ in Fig. 859).

[c] *Ming Shih*, ch. 153, p. 2a. Whether these were flash-lock gates or pound locks we do not know, but from evidence shortly to be given (p. 360 below), it is probable that at least some of them were the latter. This southern slope took the canal down about as far as the point where the Yellow River, running in course ⑦, approached it, to run alongside as far as Huai-yin. It will be remembered that since +1324 the Yellow River had adopted this southern bed (see Fig. 859), discharging to the sea by the old mouth of the Huai R.

[d] Chêng Chao-Ching (1), p. 219. See the interesting picture in *Hsing Shui Chin Chien*, illus. p. 144.

[e] After Sung Li's death in +1422, a votive temple was built for him on the Nan-wang Lake, about +1496, and in +1572 the high posthumous honour of Grand Protector of the Crown Prince was bestowed upon him.

[f] (1), vol. 1, p. 387.

[g] Presumably Tung-chhang [8] (mod. Liao-chhêng [9]).

[h] Presumably a dialect form of Wên.[10]

[1] 陡門	[2] 漕河	[3] 水脊	[4] 安山鎮	[5] 閘	[6] 南旺鎮
[7] 宿遷	[8] 東昌	[9] 聊城	[10] 汶		

sticks be thrown into that part of the river, they would soon be separated and take opposite directions.[a]

It is, no doubt, from this elevated surface, that the author of this canal saw...the possibility of forming this important communication between the different parts of the Chinese Empire by measuring from thence the inclination of the ground to the north and south, and uniting the devious streams which descended from the heights on every side, into one great and useful channel; preventing by flood-gates occasionally dispersed upon it, any sudden and useless deperdition of its waters; and supplying the loss necessarily sustained by opening such flood-gates for the passage of vessels through them, from the plentiful source of the Luen, situated higher than the highest part of it, and falling by proportionate divisions into its opposite branches.[b]

The Shan-yang Yün-Tao (section *j*), or Shan-yang Tu, which Staunton and Macartney travelled down a few days later, was of course the improved alignment of the very ancient Han Kou, and incorporated subsequent ameliorations such as the I Lou Ho (p. 309); it was not essentially changed in the Yuan re-organisation.[c] A Ming or Chhing chart of it is reproduced in Fig. 903 (pl.). The Sui dynasty's Chiang Nan Ho on the south side of the Yangtze (sections *l* and *m*) also remained unaffected. Nevertheless constant repairs to all sections, and unceasing maintenance, were necessary during the Yuan, Ming and early Chhing periods.

By +1327, then, the Grand Canal had attained definitive form, extending in all to approximately 1,060 miles,[d] distributed as shown in Table 70. Here we see the distances of the various sections, the altitudes reached at each stage, and the average depths of bed. A diagrammatic longitudinal profile is added in Fig. 906. Since the canal crosses or touches five great rivers it would be reasonable to imagine that it crossed four watershed summits, but in fact it only crosses one because the Huang Ho and the Huai River at the junction-points are at considerably higher altitudes than the Hai Ho, the Yangtze or the Chhien-thang River.[e] Nevertheless the summit between sections (*f*) and (*g*) is at quite a notable height, the water-level being 138 ft. above the Yangtze mean, and the ground level (pierced by a cutting) 170 ft.[f] Moreover, the northern terminus, Peking, was not at sea-level but elevated 118 ft. above it, so that section (*a*) had to make a considerable ascent. A quite appreciable gradient also existed on the oldest of all the stretches, section (*j*), and an important cutting was necessary even in the relatively flat country of sections (*l*) and (*m*). What all this implied in terms of lock-gate and similar equipment we propose to reserve for the sub-

[a] A rather muddled account of this same place was given by the Korean traveller Chhoe Pu (cf. p. 360), who passed it in +1487 (Meskill tr. pp. 109, 151). He was invited to sacrifice to the Dragon King at the Watershed Shrine, but his strict Confucianism prevented him from doing so.

[b] Cf. Macartney (1), pp. 170 ff., (2), pp. 267 ff.

[c] The last name which this section acquired, a name which is still used today, is that of the Inner Grand Canal (Li Yün Ho[1]). At right angles to its course (as already noted, p. 308) there had long been a system of feeder canals connecting it with the coastal areas where sea-salt was made; these Salt Canals (Yün Yen Ho[2]) now reached their fullest development. See Fig. 903 (pl.).

[d] Fang Chi (2) gives a figure of 1,782 km.

[e] And because the land between the two last-named rivers is a plain lying very little above sea-level.

[f] Though the cutting was a little more than 30 ft. in depth, this still left a rather rapid fall of over 20 ft. at each end of it.

[1] 裏運河   [2] 運鹽河

Table 70. *Details of the Grand Canal in its final form (late Chhing)* [a]

| Section | Miles (cumulative) | Intervals (in miles) | Elevations above Yangtze (mean water level) | | Depth of Canal (ft.) |
			Surrounding ground (ft.)	Water level (ft.)	
Peking	0		118	112	
a (Thung Hui Ho)		18			10
Thungchow	18		92	85	
b (Pai Ho)		77			10–26
Tientsin	95		26	23	
c (Yü Ho)		240			10–33
Lin-chhing	335		118	115	
d (Hui Thung Ho)		70			10
Yellow River, north side	405		125[b]	115	
e		—			c. 30[c]
Yellow River, south side	405		125[b]	138[d]	
f (Hui Thung Ho, Chhing Chi Tu and Chi Chou Ho)		43			13–14
Nan-wang Chen[1] summit	448		170	138[e]	
g (Chi Chou Ho)		13			13–22
Chi-ning[2]	461		130	115	
h (Huan Kung Kou)		89			10
Lin-chia Pa[3] (south end of lakes today)	550		118	115	
i (Huan Kung Kou)		155			10–33
Huai-yin	705		60	54	
j (Shan-yang Yün-Tao)		116			13–26
Yangtze River, north side	821		16	0	
k		—			40–50
Yangtze River, south side	846		16	0	
l (Chiang Nan Ho)		29			13
Tanyang[4]	875		56	6[e]	
m (Chiang Nan Ho)		185			13
Hangchow	1,060		12	0	
		1,035			

[a] Prepared in conjunction with A. W. Skempton, adjuv. Chêng Chao-Ching (*1*).
[b] The ground level, not that of the top of the dyke.
[c] Very variable.
[d] Canal running on an embankment. This is not shown in Fig. 906.
[e] Canal in cutting.

[1] 南旺鎮   [2] 濟寧   [3] 藺家壩   [4] 丹陽

section devoted to that problem (p. 344 below). It may truly be said that the work of Kuo Shou-Ching and Oqruqči in +1283 constituted the oldest successful fully artificial summit canal in any civilisation; but we may remember that the installation of pound-locks in the Ling Chhü approaches some time between +900 and +1170 had, in a sense, converted that system into one of the summit kind.

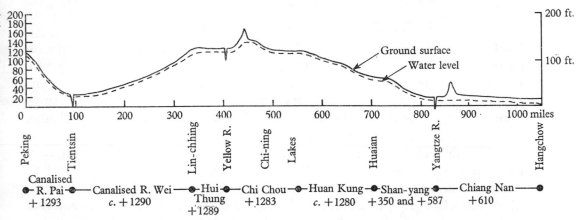

Fig. 906. Diagrammatic longitudinal profile of the Grand Canal (with A. W. Skempton, after Chêng Chao-Ching and others).

Here was a great work of engineering indeed, all the more remarkable when one remembers that in its course it had to connect with two of the greatest rivers in the world and one of the most changeable. Nevertheless, it was primarily a one-way channel, serving essentially the collection and concentration of grain-tax rather than the exchange of products among the masses of the people. The considered judgment of Chinese historians is that the governments in all dynasties invariably considered the interests of tax transport above those of irrigation or flood-control. Fiscal appropriation always came first in their minds. This had consequences particularly disastrous in the Huai valley region, where the ever-higher dykes of the Grand Canal blocked up the outlet of Huai waters, and caused severe floods when periodically swept away.[a] Eventually the canal fell upon very evil times.[b] Today it has come into its own again and is beginning to fulfil under socialism a national and public function not contemplated by its feudal-bureaucratic planners.[c] A 105-mile branch canal from Huai-an to the sea,[d] a 44-mile spur to the north Chiangsu coal mines,[e] and a 94-mile

[a] At the same time much work was often done in building drainage canals, sluices and flood-gates, as in 1489 after the Ming had come into power, and again in the early Chhing.

[b] Carles (2) gives us a picture of the Canal in the days of its decline (1896). Some parts of it were dry, or held water only a few inches deep; in other parts the current reached 10 knots. Many parts were flooded only to let the grain ships pass, then drained and excavated in readiness for the following year. In spite of all gates, great difficulties were met with when entering and leaving the Yellow River and the Yangtze, owing to differences in water-level and to currents. Haul-over capstans at various places were manned by as many as 400 men. After 1911 only the sections south of Huai-yin were utilisable.

[c] I regret that in my time in China I have seen so little of the Grand Canal. But even as one views it from the railway in Chiangsu it is much more impressive than anything similar in Europe.

[d] Anon. (68).   [e] NCNA, 5 Nov. 1961.

continuation to Ningpo,[a] have recently been built, while the main line is being straightened and shortened to some 985 miles. Equipped with new dock and lock facilities,[b] widened uniformly to at least 150 ft., lit in certain stretches with electric light, and affording passage for 2,000-ton motor barges, it serves the twentieth century better than it did the thirteenth.

(vi) *The Chhien-thang sea-wall* (*Han, Wu Tai and Sung*)

Lastly, a few words about a work of rather different character to any so far mentioned, namely the sea-wall which gives its name to the Chhien-thang[1] estuary (Fig. 907, pl.). It is upon this fifty-mile long inlet that Hangchow, the southern terminus of the Grand Canal, is situated, and through it the waters of the Fu-Chhun Chiang[2] (Happy Spring River) flow to the sea. The sea-wall was in a way an embankment like all other embankments, but it had to face particularly violent attacks, namely the serious storms for which the bay is notorious, and the well-known bore which has already been described in the Section on meteorology and sea tides.[c]

The first beginnings of a dyke, which probably only protected the settlement from which Hangchow grew, are associated with the name of Hua Hsin,[3] who was governor of the region between +84 and +87. History records his work as follows.[d]

Long ago, Hua Hsin, governor of the commandery, planned to build a dyke to keep out the sea. He first invited all who could do so to bring earth and stones, offering a thousand pieces of money for every load. Within ten days men flocked together (with materials) in clouds. But before the embankment was completed, he told them artfully that he would receive no more, so they threw down their earth and stones and went away. There was then enough to finish the dyke.[e]

As the centuries went by, the sea-wall gradually grew,[f] and about +436, Liu Tao-Chen,[4] who was governor at the time, wrote a book on it, the *Chhien-Thang Chi*.[5] His successor in +822 was none other than the poet Pai Chü-I,[6] and he also added to the wall, extending it to protect the irrigation systems of the region. Without going into too much detail, it may be of interest to give an account of some engineering controversies which centred round the wall, as recorded by an author of the early +13th century. The *Fêng Chhuang Hsiao Tu* (Maple-Tree Window Memories) says:[g]

In Hangchow the estuary sea-wall (*thi*[7]) was really first built by Chhien Shih-Po[8] in the fourth year of the Khai-Phing reign-period (of the Hou Liang dynasty; +910)...[h] Bamboo

---

[a] Fang Chi (2).   [b] Cf. Chi Yu-Ching (1).

[c] Cf. Vol. 3, pp. 483 ff.

[d] *Hou Han Shu*, ch. 101, p. 10*b*; *Thung Tien*, ch. 182 (p. 966.2), tr. Moule (3), p. 173, cf. (15), p. 29.

[e] Hence the name 'Money-Dyke Estuary'.

[f] There is a detailed map in Moule (3), and a sketch-map in Fang Chi (2).

[g] Ch. 1, p. 15*b*, tr. auct.

[h] This Chhien Liu[9] (+851 to +932) was a curious and interesting personality. Originally a salt smuggler, his military merits in the suppression of the rebellion of Huang Chhao led to rapid promotion, and in +907 he became Prince of Wu and Yüeh. In the time of the Later Liang and Later Thang dynas-

[1] 錢塘   [2] 富春江   [3] 華信   [4] 劉道眞   [5] 錢塘記
[6] 白居易   [7] 堤   [8] 錢始伯   [9] 錢鏐

gabions filled with stones were used, and large wood baulks set in among them. But after various periods of years it was always severely damaged.

In the seventh year of the Hsiang-Fu reign-period (+1014) Chhi Lun[1] and Chhen Yao-Tso[2] had a discussion as to whether it was better to use brushwood and earth mixed with straw[a] for the protective embankment, (or earth with a stone facing); some thought one way was more convenient and others preferred the other.[b] Li Phu,[3] when consulted, advised that it would be best to follow the old system of Chhien Shih-Po, and drive piles of wood between gabions containing stones. This plan was followed, but the sea-wall could not be finished, so that finally in some places earth bonded with straw was used.[c]

In the fourth year of the Thien-Shêng reign-period (+1026), Fang Chin[4] suggested that the two sluice gates (tou mên[5]) should be repaired.[d] In the sixth year of the Chhing-Li reign-period (+1046), Tu Chhi[6] extended the Chhien-thang Sea-Wall from Kuan-phu to Sha-ching, and it withstood tides and storms...Thus (we may say that) the sea-wall was first begun by Chhien (Shih-Po).

In the Ching-Yu reign-period (+1034 to +1037) Chang Hsia,[7] an official of the Ministry of Works, used soldiers to pile stones for the facing of the sea-wall.[e]...(Lastly), in the seventh year of the Chhien Tao reign-period (+1169), Shen Chhiung[8] lengthened the sea-wall by 9,400 ft....[f]

The interpretation of this seems to be that there was much reluctance to undertake the very large task of making a cut stone facing throughout; the sources indicate that the reinforced earth type of dam was what Chhen Yao-Tso wanted, as he had found it suitable in other places.[g] Ting Chin-Kung,[9] however, preferred Li Phu's opinion, and stuck to bamboo gabions. Some decades later Chang Hsia introduced more durable gabions of wood, and this was so successful for a time that it earned him a votive temple and the grateful title of the 'River-Pacifying Duke'.

It is clear, therefore, that there were three alternatives. There can be no doubt that at the beginning of the +10th century the great Legate Chhien Liu piled up stone-

ties he ruled from Hangchow nominally as governor but in practice as an almost independent sovereign. Chhien Liu was a remarkable example of the good administrators that some of the former Thang 'Legates Regulant and Mensurant' turned out to be when circumstances obliged them to carry on in isolation. Besides his great work on the sea-wall, he gave to the city of Hangchow in the years following +893 new walls and roads which formed the setting for the later Sung capital, and all its descriptions including that of Marco Polo.

[a] Cf. p. 281 above.

[b] We hear more of this argument in the *Chhing Po Tsa Chih*[10] (Green Waves Memoirs) by Chou Hui[11] (+1193), ch. 2, p. 23b.

[c] Perhaps the supply of stone was a difficulty.

[d] These were for letting out surplus water except at high tides. They were again repaired and enlarged by the great writer Su Tung-Pho in +1090, when he was Governor of Hangchow. He also did much for the canal system of the city, and greatly improved the navigation of the Chhien-thang River (cf. *Tung-Pho Chhüan Chi* (*Tsou i chi*), chs. 7 and 9; summarised by Lin Yü-Thang (5), pp. 266 ff., 270 ff.). Maps and plans are mentioned in connection with these projects. The most impressive tide-gates, however, are on the south side of the estuary, near Shao-hsing, with 28 arches; these date from +1537. Cf. p. 310.

[e] There is more about the work of this engineer in *Ssu Chhao Wên Chien Lu* (Things Seen and Heard at Four Imperial Courts), ch. 1, p. 35a; and also in *Chhing Po Tsa Chih*, ch. 2, p. 23b. His operations continued to about +1050.

[f] Rather over 1½ miles. The total length of the sea-walls of Chekiang province is now c. 200 miles.

[g] He built successful dykes elsewhere.

| [1] 戚綸 | [2] 陳堯佐 | [3] 李溥 | [4] 方謹 | [5] 斗門 | [6] 杜杞 |
| [7] 張夏 | [8] 沈肩 | [9] 丁晉公 | [10] 清波雜志 | [11] 周輝 | |

filled bamboo gabions (*chu lung*[1]) after the manner of Kuanhsien,[a] anchoring them
with wooden piles (*mu chhun*[2]) and binding them with iron chains. A second method,
later on proposed, was to build the embankment of tamped and reinforced earth such
as was used for so many dams and weirs inland; but this though doubtless less expensive, was less long-lasting. Finally, in the +14th century (+1368) a wall of stone rubble
was built,[b] but this had the failing that for a long time it was not backed by an earth-fill
embankment of sufficient dimensions, so that the facing was often left with nothing
behind it—the fault, however, was remedied by the +15th century. At this time too
(in +1448), Yang Hsüan[3] introduced the practice of building the masonry in steplike
fashion (*tieh shih chhi fa*[4]), the better to break the force of the waves.[c] Rather later in
the Ming period (+1542), Huang Kuang-Shêng[5] made a special study of bonding
methods (especially the 'fish-scale bond', *yü lin thang*[6]), which he described in his *Hai
Thang Shuo*[7] (Discourse on Sea-Walls), and these proved effective down to the present
time. Many auxiliary techniques, such as the use of iron clamps (*thieh sun*[8]) for the
blocks of stone, and the building of groynes (*thiao shui pa*[9]) and drainage canals (*pei
thang ho*[10]), have been practised since the time of Yang and Huang. Finally, late in the
Chhing, the experience and theory of so many centuries was crystallised in the book of
Chhen Hsü-Tsung[11] entitled *Hsiu Hai Thang I*[12] (Discussions on Sea-Wall Repair and
Maintenance).

Some further light is thrown on the sea-wall in the Sung period by Shen Kua, who
knew this part of the country well. His passage about it is worth reproducing.[d]

The Chhien-thang estuary sea-wall was originally, in the time of Governor Chhien himself, a stone (gabion) dyke. Along the whole length of the dyke, at its outer edge, were placed
more than ten rows of piles, heavy wooden baulks, which were known as 'ocean screen
pillars' (*huang chu*[13]).

In the Pao-Yuan and Khang-Ting reign-periods (+1038 to +1040), someone suggested
that the wooden piles should all be taken out, so that several hundreds of thousands of pieces
of good timber would be available. The commanding officer in Hangchow agreed, so this was
done, but the timber, once out of the water, proved to be decayed and useless.

After the wooden piles had been removed, the stone dyke began to break and burst every
year on account of the force of tides and waves. Men of former times had set these piles in
place to deflect and break the force of the water, instead of simply opposing it flatly, and thus
the waves did no harm.[e]

---

[a] This is evidenced not only by the quotation above, but in other texts also, e.g. the *Wu Yüeh Pei
Shih*,[14] by Chhien Yen[15] (*d. c.* +1000); cf. Moule (15), p. 15.

[b] The 'artificial hill' method (*pho tho fa*[16]).

[c] The dressed stone wall now rested on wooden piling some way back from its face, with pisé foundations for the mass of the embankment. A large part of this was constituted by the rubble wall, with earth
packed up behind it.

[d] *MCPT*, ch. 11, para. 22, tr. auct., adjuv. Yang Lien-Shêng (11), p. 44. Cf. Hu Tao-Ching (1), vol. 1,
pp. 429 ff.

[e] Cf. the remarks made above on the resilient, porous, shock-absorbing nature of gabion defences.

[1] 竹籠   [2] 木樁   [3] 楊瑄   [4] 叠石砌法   [5] 黃光昇
[6] 魚鱗塘   [7] 海塘說   [8] 鐵筍   [9] 挑水壩   [10] 備塘河
[11] 陳訏總   [12] 修海塘議   [13] 滉柱   [14] 吳越備史   [15] 錢儼
[16] 坡陀法

When Tu Wei-Chhang[1] (Tu Chhi) was Intendant of Circuit, someone suggested to him that a crescent-shaped breakwater or artificial headland (*yüeh thi*[2]) should be built several miles away to the east of the customs station (i.e. further out in the estuary), so as to deflect the rushing tide. All the water-conservancy foremen thought that it would be a good idea except one old man, who disagreed, telling his colleagues secretly that if such a work were erected there would be no more floods and that they would thus lose their livelihoods. They were all convinced by this and supported his opinion. Tu Wei-Chhang never discovered their dishonesty, and spent much money, but the trouble with the old sea-wall still happened every year.

In recent years the benefits of the crescent-shaped outworks (breakwaters, groynes and piers) have been carefully studied, and the harm done by the sea has certainly been reduced. But their efficiency is not as great as that of the old wooden pile (and gabion) system. Unfortunately it would now cost too much to replace all the piles.

According to this, therefore, the original dyke was ruined when in some moment of timber-shortage the wooden piles which anchored the stone-filled gabions in place were all removed. This of course was about the time (+1035 to +1040) when Chang Hsia was replacing the gabions of bamboo by gabions of wood. Tu Chhi then accepted the idea of protective breakwaters jutting out from the shore,[a] and though the opposition of the water-conservancy workers staved this plan off for a time,[b] it was certainly put into execution during the next two decades, before Shen Kua came upon the scene about +1070.

The sea-wall was visited by Major Edwards in 1865 after it had suffered from neglect for four years owing to the Thaiphing revolution and civil war. His criticisms (1) of the traditional engineering represented by the structure were that the lowest stone course was not sufficiently below low water level to prevent serious undermining in several places, that the piles were not stout enough, and that there were no counterforts to break the force of the waves. In the Chhing dynasty important histories of the sea-wall were written, notably the well-illustrated *Liang Chê Hai Thang Thung Chih*[3] by Fang Kuan-Chhêng,[4] and the *Hai Thang Lu*[5] by Tsê Chün-Lien.[6] In our own time there has been a memorable report on it by von Heidenstam (1). Today it serpentines along both sides of the estuary for just under 200 miles, its masonry rising to an average of 26 ft. above low water level, and efficiently performs its ancient task.[c]

(8) THE LITERATURE ON CIVIL ENGINEERING AND WATER CONSERVANCY

From all the foregoing material it will have been noticed that the chief sources for information on hydraulic engineering in ancient and early medieval times are the

[a] For a glimpse of current work on coastal processes see Silvester (1).

[b] Cf. Vol. 4, pt. 2, p. 28, where the remarks on the refusal of technical innovations because of fear of unemployment might need modifying in the light of the present example. Nevertheless, this social situation seems to have been exceedingly rare in traditional China.

[c] In 1964 I had the pleasure of visiting it in the neighbourhood of Haining in the company of one of the engineers of the conservancy, Mr Huang Jung-Tao, to whom I record my grateful thanks.

[1] 杜偉長　　[2] 月堤　　[3] 兩浙海塘通志　　[4] 方觀承　　[5] 海塘錄
[6] 瞿均廉

dynastic histories. The *Shih Chi* and the *Chhien Han Shu* contain long chapters on canals and waterworks, while some of the later ones, such as the *Sung Shih*,[a] have whole sections incorporating a great bulk of data which has never been fully analysed. Few fragments of Han date, for example, on water conservancy survive, and our knowledge of Han ideas and practice has to come from the speeches and memorials recorded in the histories. Most of the technical literature dates from the +14th century and later, as may be seen from modern descriptive bibliographies such as those of Mao Nai-Wên (*1*);[b] Chêng Hao-Shêng (*2*) and Chu Chhi-Chhien (*1*). An idea of its extent can be gained from the fact that the first of these lists about 200 books and the second 400. The *Thung Chih Lüeh* bibliography of *c.* +1150 has a section of 31 books under Chhuan Tu,[1] many of which deal with 'benefit of water', *shui li*.[2]

It is hard to make any sharp line of distinction between what we have called 'hydrographic' treatises,[c] and books specifically devoted to the technique of hydraulic engineering, for they overlap, though the latter are commoner in the latest periods. The oldest hydrographic survey which has come down to us is the *Shui Ching*[3] (Waterways Classic), attributed to Sang Chhin[4] of the Former Han, but more probably compiled in the San Kuo time (+3rd century). At the end of the +5th or the beginning of the +6th it was very greatly enlarged by the geographer Li Tao-Yuan,[5] who entitled it the *Shui Ching Chu*[6] (Waterways Classic Commented). But it is primarily geographical, and only incidentally informs us about canals. Three great Chhing scholars, Chhüan Tsu-Wang, Chao I-Chhing, and Tai Chen, devoted much labour to commenting upon it.[d]

From the Sung dynasty (+10th to +13th centuries) quite a number of relevant books survive. In +1059 the *Wu Chung Shui Li Shu*[7] (Water-Conservancy of the Wu District), by Shan O,[8] appeared; it was the result of many years' study of the canals of Chiangsu.[e] In +1242 Wei Hsien[9] produced the *Ssu Ming Tho-Shan Shui Li Pei Lan*[10] (Irrigation Canals of the Mount Tho District), a historical account of their development in the neighbourhood of Ningpo.[f] Also from the Sung, though not all easy to date exactly, are the following (among others):

Shui Li Shu[11] (Treatise on Water Conservancy) by Fan Chung-Yen[12] (about +1030).[g]

Chhing-Li Ho Fang Thung I[13] (General Discussion of the Flood-Protection Works in the Chhing-Li reign-period) by Shen Li[14] (between +1041 and +1048).

Shui Li Thu Ching[15] (Illustrated Manual of Civil Engineering) by Chhêng Shih-Mêng[16] (+1060).[h]

---

[a] Chs. 91–7.   [b] Much fuller detail will be found in Mao Nai-Wên (*2*).
[c] See Vol. 3, pp. 514 ff. above.
[d] Cf. Hu Shih (*5*).
[e] Cf. *Tung-Pho Chhüan Chi* (*Tsou i chi*), ch. 9, for the postscript by Su Tung-Pho.
[f] Wei Hsien had himself been active in hydraulic work, having organised the repair by Tung Hsing[17] of the canals built as far back as +833 by the governor Wang Yuan-Chang.[18]
[g] Already met, cf. Vol. 4, pt. 2, pp. 347, 468.
[h] The circumstances in which this work was written are detailed in *Sung Shih*, ch. 95, p. 22*b*.

[1] 川瀆	[2] 水利	[3] 水經	[4] 桑欽	[5] 酈道元
[6] 水經注	[7] 吳中水利書		[8] 單鍔	[9] 魏峴
[10] 四明它山水利備覽		[11] 水利書	[12] 范仲淹	[13] 慶曆河防通議
[14] 沈立	[15] 水利圖經	[16] 程師孟	[17] 董興	[18] 王元暐

Ho Shih Chi[1] (Collected Materials on River Control Works) by Chou Chün[2] (+1128).
Chih Ho Tshê[3] (The Planning of River Control) by Li Wei.[4]
Tshao Yün Fu Khu Tshang Yü[5] (Tax-Grain Water Transport and Granaries) by Wang
   Ying-Lin[6] (about +1270).[a]

Such are some of the works which remain to mark a period exceptionally rich in civil
engineers. From the Yuan may be mentioned an important book, the *Ho Fang Thung
I*[7] (A General Discussion of the Protection Works along the Yellow River), dated
+1321, and contributed by a Persian or Arab scholar in the Mongolian service, Sha-
Kho-Shih,[8] whose Chinese name was Shan Ssu.[9] Shan Ssu was an excellent mathematician and geographer, who used the Thien Yuan algebraic methods in his engineering calculations.[b] His book was really a revision and enlargement of two of the older
texts which we have just mentioned, those of Shen Li and Chou Chün. The +14th
century gave, as further examples:[c]

Chih-Chêng Ho Fang Chi[10] (Memoir on the Repair of the Yellow River Dykes in the Chih-
   Chêng reign-period) by Ouyang Hsüan[11] (about +1350).
Chih Ho Thu Lüeh[12] (Illustrated Account of Yellow River Floods and Measures against them)
   by Wang Hsi.[13]

The first great compendium dates from the Ming. Phan Chi-Hsün[14] served four
terms of office between +1522 and +1620 as Director-General of the Yellow River
Works and Grand Canal, becoming the greatest authority of his age on hydraulic
engineering. His book, the *Ho Fang Chhüan Shu*[15] (or *Ho Fang I Lan Chio*;[16] General
View of Water Control), includes many maps, copies of memorials, edicts and other
official documents, and many interesting discussions by the author, written in the
form of dialogues with imaginary opponents. Phan's own preface is dated +1590.[d]
The *Ssu Khu Chhüan Shu* catalogue of the late +18th century says that 'although
changes in methods were afterwards necessary to fit changing circumstances, yet
experts in river control always take this book as a standard guide'.[e] In the following
century the literature continued to grow. The *Shui Li Wu Lun*[17] (Five Essays on Water
Conservancy) of +1655 by Ku Shih-Lien[18] may be taken as a worthy example. But the
greatest hydraulic engineer of the +17th century was Chin Fu[19] (+1633 to +1692).

From +1677 until his death, Chin Fu was actively engaged in conservancy work,
especially on the Yellow River and the Grand Canal. He superintended extensive
dredging operations, and improved the form and strength of embankments, was

---

   [a] Often met with already, e.g. Vol. 2, p. 21; Vol. 3, p. 208; Vol. 4, pt. 2, p. 166.
   [b] See Vol. 3, pp. 129 ff.; there is a special study of Shan Ssu's applied mathematics by Yabuuchi
Kiyoshi (23). He wrote many other books, including a *Hsi Yü Thu Ching*[20] (Illustrated Treatise on the
Western Regions).
   [c] Cf. the study of these books by Yang Lien-Shêng (11), pp. 45 ff.
   [d] For information on other civil engineering works of about this date see Hummel (19).
   [e] Ch. 69, p. 54a.

[1] 河事集	[2] 周俊	[3] 治河策	[4] 李渭	[5] 漕運府庫倉庚
[6] 王應麟	[7] 河防通議	[8] 沙克什	[9] 贍思	[10] 至正河防記
[11] 歐陽玄	[12] 治河圖略	[13] 王喜	[14] 潘季馴	[15] 河防全書
[16] 河防一覽榷		[17] 水利五論	[18] 顧士璉	[19] 靳輔
[20] 西域圖經				

involved in a great dispute concerning the drainage of the lower Huai area, and constructed a successful 95-mile canal, the Chung Ho,[1] which short-circuited a bad patch of the Yellow River. Chin Fu wrote the *Chih Ho Fang Lüeh*[2] (Methods of River Control) in +1689, but though presented to the throne at that time, it was not printed till +1767. The book is regarded as comparable with that of Phan Chi-Hsün, which it much resembles in contents, and long exerted great authority. Appended to it as ch. 10 is an essay on the techniques of water control (*Ho Fang Tsê Yao*[3]) by Chhen Huang,[4] the eminent engineer who acted as Chin Fu's secretary and assistant throughout his career.[a] In ch. 9 the two men engage in a discussion of fundamental principles in the dialogue style of the *Huang Ti Nei Ching*.

The eighteenth century was very rich in books on this subject. Just before it opened, there were the *Chih Ho Shu*[5] and the *Huang Yün Liang Ho Thu*[6] (Maps and Diagrams of the Yellow River and the Grand Canal) from the pen of Chhêng Chao-Piao[7] (+1690). In +1705 there came the great historical hydrographic analysis of Hu Wei,[8] the *Yü Kung Chui Chih*,[9] already mentioned (Vol. 3, p. 540). Twenty years later Fu Tsê-Hung[10] produced his monumental *Hsing Shui Chin Chien*[11] (Golden Mirror of the Flowing Waters), the most comprehensive treatment of all Chinese waterways, natural and artificial. Fig. 908 shows one of his illustrations of a section of the Grand Canal. The book consists mostly of extracts from the chief primary sources, and was more than doubled in size by supplementary volumes compiled by Lei Shih-Hsü,[12] Yü Chêng-Hsieh[13] and Phan Hsi-Ên[14] in 1832. About the same time as Fu Tsê-Hung's book came the *Fang Ho Tsou I*[15] of Chin Fu's successor, Hsi Tsêng-Yün (+1733), a work on the contents of water-conservancy memorials to the throne. Fu Tsê-Hung's descriptive work was continued by Chhi Shao-Nan[16] with his *Shui Tao Thi Kang*,[17] printed in +1776. And indeed it has never ceased, since from the following century one might mention the *Chiang Pei Yün Chhêng*[18] (Handbook on the Course of the Grand Canal North of the Yangtze) by Tung Hsün[19] in 1867; while the work in several volumes on the Yellow River by our own contemporaries, Hu Huan-Yung, Hou Tê-Fêng & Chang Han-Ying, has already been mentioned and used.

Of works devoted rather to engineering techniques than historical description, mention should be made of Chang Phêng-Ho's[20] *Ho Fang Chih*[21] (River Protection Works) of +1725. Towards the end of the century, the mathematician Chhêng Yao-Thien,[22] a friend of Tai Chen, produced a short theoretical study of canal construction, the *Kou Hsüeh Chiang Li Hsiao Chi*.[23] Then there was Khang Chi-Thien's[24] *Ho Chhü*

---

[a] Though during his life he experienced many vicissitudes and trials of favour and disfavour, Chin Fu was posthumously canonised, and shared a temple with three other civil engineers, including Hsi Tsêng-Yün[25] (+1671 to +1739) and Kao Pin[26] (+1683 to +1755). Hou Jen-Chih (*1, 2*) has written technical biographies of Chin Fu and Chhen Huang respectively.

[1] 中河	[2] 治河防略	[3] 河防摘要	[4] 陳潢	[5] 治河書
[6] 黃運兩河圖	[7] 程兆彪	[8] 胡渭	[9] 禹貢錐指	[14] 潘錫恩
[10] 傅澤洪	[11] 行水金鑑	[12] 黎世序	[13] 俞正燮	
[15] 防河奏議	[16] 齊召南	[17] 水道提綱	[18] 江北運程	
[19] 董恂	[20] 張鵬翩	[21] 河防志	[22] 程瑤田	[25] 嵇曾筠
[23] 溝洫疆理小記	[24] 康基田			[26] 高斌

Fig. 908. A drawing from the *Hsing Shui Chin Chien* (illus. p. 141) of Fu Tsê-Hung (+1725). The Grand Canal, seen from the south-west, is snaking horizontally across the middle of the picture, and the city in the foreground is Yü-thai, on the border of Shantung and Chiangsu. The waters of the Chao-Yang Hu (lake) are seen on the right and those of Tu-Shan Hu (lake) above; the canal winds between them to this day. In the background loom the Shantung mountains, above the town of Tsou-hsien near the Temple of Confucius at Chhü-fou. The positions of named stop-log flash-lock gates are marked by a characteristic symbol based on the vertical grooves in which the baulks slide up and down; one such guards an opening in the centre of the picture leading to two watercourses looping to left and right, and each of these again is similarly guarded. The right-hand one is an old disused alignment of the Grand Canal south-west of the lakes, the left-hand one is a parallel stream, the Chung-thou Ho, coming down from the Nan-Wang Hu (lake) near the summit level and the 'point where the waters divide' (see p. 316). Along the stretch of the Canal shown in the picture there are three flash-lock traffic gates and three further side sluice-gates, all termed *cha* and named. A study of the adjoining pages, on which the chart is continued, shows that Fu depicted, between the Yellow River crossing to the north and the southern end of the lakes at Han-chuang on the Shantung-Chiangsu border to the south, a total of thirty-one flash-lock traffic gates and fourteen side sluice-gates. All but two of these last are called 'single gates' (*tan cha*), which may suggest that some of the traffic ones were double or pound-locks; an impression enhanced by the fact that one pair of traffic gates, those at Nan-wang, are called 'upper and lower' (*shang hsia*). On this subject see pp. 355, 360.

Chi Wên[1] (Notes on Rivers and Canals) of 1804, the author of which was a man of great experience in the field as well as in the administrative office. It is considered one of the best books on the subject in the whole Chinese literature. Finally, the *Ho Kung Chhi Chü Thu Shuo*[2] (Illustrations and Explanations of the Techniques of Water Conservancy and Civil Engineering), compiled by Lin-Chhing[3] and his assistants in 1836, must be commended.[a] Like some of the fore-mentioned books, but even more fully, it

[a] Cf. Fig. 876 and p. 263 above.

[1] 河渠紀聞   [2] 河工器具圖說   [3] 麟慶

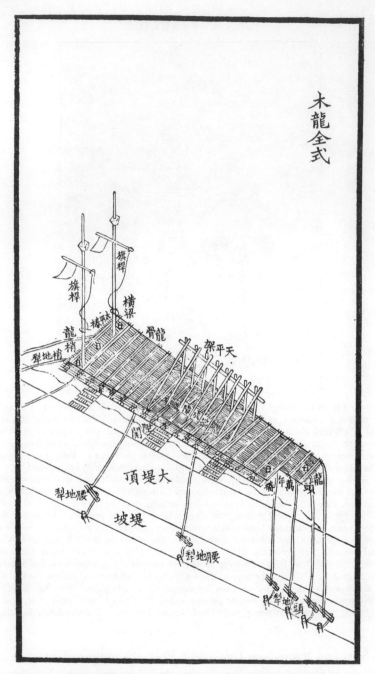

Fig. 909. An illustration from the *Ho Kung Chhi Chü Thu Shuo* of Lin-Chhing (ch. 3, p. 30a). The *mu lung* (Wooden Dragon) is a multiple-layer raft (*fa*) moored by numerous cables alongside a dyke (*thi*) endangered by erosion or internal fissures. By means of a derrick arrangement (*thien phing chia*) a series of bamboo-matting (*shui cha*) frames (*ti chhêng chang*) is driven into the bed of the waterway through a slot in the raft. The device is thus an ancestral form of sheet-piling. When the frames are in place, a fill (*chien*) of fascines, earth, loose stones, etc., is sunk between the dyke and the floating platform; the pile-frames, acting as a temporary cantilever wall, retain this in place until it settles into a compact mass. On the remarkable tensile strength of bamboo laths and cables, cf. Vol. 4, pt. 2, pp. 63 ff., and also pp. 191, 339, 415, 597, 664 of the present volume. The Wooden Dragon could also be used as an ordinary pile-driver.

Lin-Chhing tells us that the method was first employed in the Sung, in +1021, by Chhen Yao-Tso, and then in the Yuan by Chia Lu, *c.* +1350; but later it fell into disuse and had to be re-designed at the beginning of the Chhien-Lung reign (*c.* +1745). This was done by Li Ping, and Kao Wên-Ting made very effective use of it, so much so that when the emperor was on a tour of inspection in the south he himself wrote a poem on the Wooden Dragon. Li Ping produced a monograph on the technique, entitled *Mu Lung Chhêng Kuei*, and specialist artisans called Kou shou were trained in its management.

is supplied with pictures of coffer-dams, dykes, sluice-gates, gabions, ships, tools, and so on (cf. Fig. 909). Although of late date, the book is very little indebted to Western influences since so much of the technique was highly traditional.

(9) TECHNIQUES OF HYDRAULIC ENGINEERING

(i) *Planning, calculation and survey*

It is obvious that works on the scale which has been described could never have been undertaken without a considerable staff of ingenious planners.[a] The engineers whom I knew at Kuanhsien, with their slide-rules and desk computers, were only continuing with modern methods what Li Ping had started more than two thousand years before, relying perforce on crude counting-boards and a liberal measure of intuitional genius.[b] Unfortunately we have few indications of exactly how the planning problems presented themselves to the men of old, and how far they were able to get in solving them without having to fall back on guesswork.[c]

There is one passage in the *Chhien Han Shu*, however, which reveals to us something of the situation behind the scenes.[d] Kuo Chhang[1] and Fêng Chün,[2] about −30, had been trying to relieve the pressure on the lower Yellow River dykes by building canals, and the latter had pointed out that one important cause of trouble was the silting up of the Thun-shih[3] River, one of the channels of the delta.

These affairs were brought to the attention of the ministers of state, who referred them to the learned doctor Hsü Shang,[4] a man expert on the Shang Shu (chapters of the *Shu Ching*, the Historical Classic)[e] and skilful at mathematics (*shan wei suan*[5]) so that he was able to compute labour force problems (and the like).

Hsü Shang was sent down to inspect the situation. He agreed that the Thun-shih River was the cause of the floods. But because of the expense, which the financial situation of the government would (it was thought) not permit, the deepening and dredging were postponed.

After three years, the Yellow River again caused floods at Kuan-thao and the Eastern

---

[a] Here we cannot discuss the economic and financial aspects of the great public works of China. The reader is referred to an interesting series of lectures on this subject by Yang Lien-Shêng (11).

[b] While in Ceylon in 1958, where I had the opportunity of studying the ancient hydraulic engineering works on the spot (cf. pp. 368 ff. below), I discussed the problem of what survey instruments might have been used with my friend Mr William Delay, of the Ceylon Irrigation Department. Though confident that elementary levelling devices at least must have been employed from the beginning of our era onwards, he told me of experiences that he had had with an old Sinhalese irrigation worker who had an appreciation of contour so strong as to be like a sixth sense. This man could walk a way through low hilly country which would turn out to be a perfect trace as good as if made with levelling equipment. He also had an extraordinary flair for rock which could be used as foundations. When seeking such sites for bridges, embankments, etc., Mr Delay and his foreman many times came upon traces of the works of the ancient Sinhalese, who had found and used just the same locations. Cf. what Golab (1) has to say of the *düzlüq bashi* (the Master of the Horizontal) in Sinkiang.

[c] We have noted above several examples of projects which failed—for example the first attempt to build a lateral transport canal between Loyang and the Yellow River (p. 276), and the efforts to surmount the Shensi ranges in order to short-circuit (though it was a very long circuit) the San-mên Gorge (p. 275). Chassigneux (1) has given us a study of another failure, the Cu'u-Yên Canal in Annam.

[d] Ch. 29, pp. 9b ff., tr. auct. Cf. *TH*, p. 576.

[e] This would mean that he had deeply studied the chapters on the works of Yü the Great.

[1] 郭昌   [2] 馮逡   [3] 屯氏   [4] 許商   [5] 善爲算

Commandery, breaking the 'Metal Dyke' and overwhelming four commanderies and thirty-two *hsien* districts. More than 150,000 *chhing* (about 2·5 million acres) were under water, thirty foot deep at the worst parts. Of houses, official and private, some 40,000 were destroyed. This (proved how) mistaken had been the opinion of Yin Chung,[1] the Imperial Censor [who had presumably been responsible for the financial stringency], and the emperor reproached him so much that he committed suicide. Fei Tiao,[2] the Minister of Agriculture, was ordered to see to the problem of fair taxation to help the flooded districts, and two Inspectors-General were despatched to arrange for 500 grain-transport ships from Honan and the east to move more than 97,000 inhabitants to the hills.

Wang Yen-Shih,[3] Comptroller of Water-Conservancy Works (Ho-Thi Shih-Chê[4]), was then asked to close the breaches in the river dykes. He used for the fill bamboo 'sausages' containing stones (*chu lo*;[5] lit: bamboo plugs, or falling bamboo crates); these were 4 *chang* (40 ft. approximately) long, and 9 spans (arm-stretches, *wei*[6]) in circumference (of cross-section, i.e. *c.* 17·2 ft. diam.) all packed with small stones. They were deposited in place by being suspended between two barges (before dropping). After 36 days the dyke was fully repaired. The emperor thereupon said: 'Although the bursting of the Yellow River dykes in the Eastern Commandery caused floods over two provinces, now our Master-Comptroller (Wang) Yen-Shih has blocked the breaches in (hardly more than) thirty days. Let us change the 5th year of the present reign-period into the 1st year of a new reign-period, to be called Ho-Phing[7] (The River Pacified) [−28]. All soldiers who have taken part in the water-control operations shall receive exemption from six months' frontier service. Due to (Wang) Yen-Shih's excellence in planning, the expense has been minimal, and the time short. Desiring to encourage him I bestow upon him the title of Kuang-Lu Ta-Fu, with an appanage of 2,000 *tan* as a Kuan-Nei Marquis, and 100 catties of gold.'

But two years later the river broke out again [−26], at Phing-yuan, with floods which reached to Chinan and Chhien-chhêng, where half the buildings were destroyed. So Wang Yen-Shih was again sent to control it.

However, Tu Chhin,[8] speaking to the Commanding General Wang Fêng,[9] said: 'Formerly, when the Yellow River broke out, the Vice-Minister Yang Yen[10] told me that Wang Yen-Shih really learnt the technique of blocking dyke breaches from him, but nevertheless Yang Yen remained unknown (as a hydraulics expert). Now you are giving the responsibility solely to Wang Yen-Shih, but as he closed the dykes before with such ease, I fear he will not give very serious consideration to the harm (which the River can do). If these things are so, the skill (*chhiao*[11]) of Wang Yen-Shih is not as great as that of Yang Yen. Now the effects of water are various, and if their advantages and disadvantages are not the subject of wide discussion, if you give to one man only responsibility for the task, and if he himself is not quite up to the standard called for; then come winter come spring, when the early freshets arrive (*thao hua shui*[12]), there will (inevitably) be damage due to excess of silting. Then the spring sowings will not be possible in several commanderies, the people will flee away, and banditry will arise. Passing sentences of death on Wang Yen-Shih then will be no use at all. (My opinion is, therefore, that) you should appoint Yang Yen, the Director of Engineering Works (Chiang-Tso Ta-Chiang[13]) Hsü Shang, and the Imperial Counsellor Chhêngma Yen-Nien,[14] to be associated with Wang Yen-Shih as colleagues. He and Yang Yen will certainly have violent disputes; there will be deep discussions and mutual criticism. Hsü Shang and Chhêngma Yen-Nien are both excellent mathematicians, able to calculate labours and results, to dis-

[1] 尹忠	[2] 非調	[3] 王延世	[4] 河隄使者	[5] 竹落
[6] 圍	[7] 河平	[8] 杜欽	[9] 王鳳	[10] 楊焉
[11] 巧	[12] 桃華水	[13] 將作大匠	[14] 乘馬延年	

tinguish truth from error (*fên pieh shih fei*[1]), and to select the best plan to follow. In this way the works will certainly be successful.'

Wang Fêng did as Tu Chhin suggested, and sent Yang Yen and others to help start the work. In six months it was completed, and again Wang Yen-Shih received 100 catties of gold, and his soldiers, if they had not hired paid substitutes, were granted exemptions as before.

The upshot of this long but entertaining description would seem to be that in the Han there was some realisation of the necessity of associating good mathematicians and engineers with administrators in water-conservancy and control works. The latter alone must have been capable of making gross miscalculations and costly mistakes. Wang Yen-Shih seems to have been a successful career bureaucrat, skilled at retaining the limelight in all circumstances; the others more competent technologists. To some extent a conflict is revealed between the 'boys in the back room' and the regular officials (cf. Vol. 4, pt. 2, pp. 39 ff.). The passage is also of importance in that it is the oldest description of the bamboo gabions filled with stones which played so important a part throughout Chinese hydraulic engineering, and to which we shall return in a moment. That they were unfamiliar about −28, two men seeming to claim their invention, may mean that they were really introduced about this time.

One would naturally expect that specimen computations about canals and dykes would be given in the mathematical literature.[a] This is indeed the case; the Han *Chiu Chang Suan Shu* (Nine Chapters on the Mathematical Art) contains many problems concerning dyke-building, giving results in material, labour, time, and so on.[b] Fig. 56 (in Vol. 3, p. 43) accompanies a discussion[c] of dyke-building in the celebrated *Shu Shu Chiu Chang* (Mathematical Treatise in Nine Sections) of Chhin Chiu-Shao (+1247), while Figs. 248 and 249 (Vol. 3, p. 578), from the same work, state a problem in the distribution of irrigation water.[d]

Something has already been said of Chinese survey methods in the geographical Section with relation to the flourishing of the quantitative grid system in cartography.[e] Plumb-lines and the *groma*, chains, cords, graduated poles and water-levels with floating sights (cf. Fig. 245, in Vol. 3, p. 570) were early in use.[f] A passage from Shen

---

[a] Cf. Vol. 3, pp. 25 ff., 43, and other details there given, pp. 97 ff., on standard figures in solid geometry, for the volumes of which there were empirical formulae. A special article has been devoted by Wang Hu-Chen (*1*) to earthwork estimates in ancient times.

[b] Cf. Vol. 3, p. 26. Hsü Shang himself is known to have written a mathematical book, but it did not survive (cf. Vol. 3, p. 28).

[c] Ch. 13 (vol. 4, p. 334).   [d] Ch. 6 (vol. 2, p. 155).

[e] Sect. 22*e*; Vol. 3, pp. 569 ff. We now have also the excellent paper of Shen Khang-Shen (*1*).

[f] Cf. *Wu Ching Tsung Yao* (*Chhien chi*), ch. 11, p. 2*a*, *b*; *Wu Pei Chih* omits the pole. The traditional floating-sights water-level continued in use very late; cf. *Ho Kung Chien Yao*, ch. 4, p. 25*a*; *Ho Kung Chhi Chü Thu Shuo*, ch. 1, pp. 2*a*, *b*, 11*a*, *b*; *Hsiu Fang So Chih*, etc. How much can be done with the simplest apparatus is seen in Yu Chêng's description (*1*) of the contemporary irrigation works (lateral canals feeding multiple small reservoirs with perennial water, as well as bringing inundatory supply) built by the local people in the Tu-tsao River valley near Kuanghua in N.W. Hupei. A countryman, Li Ta-Kuei, who took the lead in this (1957) used successfully as theodolite a split bamboo with septa sights floating on the convex meniscus of water in a rice-bowl (Fig. 910). True, his army service had acquainted him with rifle sights, but the method has an age-old look about it. We have already sung the praises of the bamboo as an invitation to technical advances (Vol. 3, pp. 333, 352, Vol. 4, pt. 2, pp. 61 ff.). Cf. p. 415 below.

[1] 分別是非

Fig. 910. A floating-sights water-level proto-theodolite used by a Hupei countryman, Li Ta-Kuei, in 1957, for surveying irrigation-works undertaken by agricultural cooperatives (redrawn from Yu Chêng). The bamboo tube with septa bored as sights floats on the convex meniscus of water in a rice-bowl. This method is probably very old (cf. Fig. 245 in Vol. 3), as well as very practical.

Kua about his own work as a surveyor along the Pien Canal about +1070 was quoted, showing the use of graduated poles, and altazimuth sighting-tubes mounted in various ways.[a] Calculations involving similar right-angled triangles were in common use by the time of Liu Hui, whose book, the *Hai Tao Suan Ching* (Sea Island Mathematical Classic) of +263, gives many different examples of them.[b] The *baculum* or cross-staff (Jacob's Staff), generally ascribed to early +14th-century Europe, was shown by another citation from Shen Kua to have been in use in +11th-century China.[c] Yet a third excerpt from the *Mêng Chhi Pi Than* (Dream Pool Essays) indicated that compass bearings were also in use for mapping at that time.[d] Whether the hodometer, available from Han times onwards, was actually used in practice for measuring distances, remains uncertain.[e]

[a] Vol. 3, p. 577, from *MCPT*, ch. 25, para. 8; cf. Hu Tao-Ching (*1*), vol. 2, pp. 795 ff.; Chu Kho-Chen (*4*). Cf. the discussion of the sighting-tube in Vol. 3, pp. 332 ff.

[b] Cf. Vol. 3, pp. 30 ff.

[c] Vol. 3, p. 574, from *MCPT*, ch. 19, para. 13; cf. Hu Tao-Ching (*1*), vol. 2, p. 635.

[d] Vol. 3, p. 576, from *MCPT*, *Pu* appendix, ch. 3, para. 6; cf. Hu Tao-Ching (*1*), vol. 2, pp. 991 ff. In Vol. 3, pp. 514 ff., 517 ff., we had a good deal to say of the special geographical literature on water-ways and local topography. From the Sung onwards, starting in Chiangsu, official documents known as Yü Lin Thu[1] ('Fish-Scale Maps', i.e. dissected cadastral survey charts) were regularly compiled in every district, and in these particular attention was paid to irrigation systems, river control works, and natural products, classes of feature being marked by cartouche inscriptions in different colours. These mapping practices, described in Sung texts such as the anonymous *Chou Hsien Thi Kang*[2] (Complete Account of City and County (Government)), and Lü Pên-Chung's[3] *Kuan Han*[4] (Handbook for Magistrates), both of the +12th century, have been studied by Niida Noboru (*2*). Prof. D. C. Twitchett has also given much attention to them, and we have to thank him for the information contained in this note.

[e] See Vol. 4, pt. 2, pp. 281 ff. In its paddle-wheel form as proposed for ships (Vitruvius x, ix, 5; cf. Diels (*1*), p. 67; Torr (*1*), p. 101), it could have been useful perhaps as a stationary current-meter. On the history of such hydrometrical devices see Lanser (*1*), a subject which we have touched upon in Vol. 3, pp. 632, 635. But though the Chinese were the earliest of peoples to make practical use of the ad-aqueous paddle-wheel (see Vol. 4, pt. 2, pp. 413 ff.), we have not come upon any such ex-aqueous applications of it among them. According to Leliavsky Bey (*3*), p. 467, the hydrometrical paddle-wheel was first mentioned

[1] 魚鱗圖   [2] 州縣提綱   [3] 呂本中   [4] 官箴

After the route of a canal had been provisionally decided upon, trial borings were made along it to ascertain the nature of the ground. In his *Pei Hsing Jih Lu* (Diary of a Journey to the North) Lou Yo says,[a] about +1169, speaking of the Sung Canal constructed a century previously (see p. 314 above) between the Huai River and the Shantung lakes:[b]

> From Hung-Tsê[1] Lake to Kuei-shan,[2] a well was dug every one or two *li* to investigate the nature of the earth and bedrock. When the results had proved satisfactory, the emperor was memorialised, and eventually the canal was opened. This demonstrates the careful planning of our ancestors, whose decisions should not be altered lightly.

Further, in order to secure adherence to the original alignments, fiducial marks in the shape of stone or iron plates or statues were fixed in the sides and beds of canals as a guide for periodical deepening and silt-removal. We have already seen one case of this at Kuanhsien,[c] and Wang Kung[3] records others for the Pien Canal in his *Wên Chien Chin Lu*[4] (New Records of Things Heard and Seen).[d] A 'nilometer' in the stricter sense is the 'Two Fishes' gauge rock which has measured the Yangtze water-levels since +763 at Fou-ling in Szechuan.[e]

(ii) *Drainage and tunnelling*

From time to time the engineers were faced by difficulties due to springs,[f] underground water-courses, and loose or shaly soil which was prone to landslides. We have an example of efforts made to cope with these things from quite an early time, namely that of Han Wu Ti. Ssuma Chhien writes:[g]

> After that (i.e. about −120) Chuang Hsiung-Phi[5] declared that the people of Lin-chin wanted to open a canal which would leave the river Lo[6] and irrigate 10,000 *chhing* of land lying east of Chung-chhüan.[h] This land had previously been salty, but if one could really succeed in getting water for it, harvests of ten *tan* for every *mou* could be obtained. Accordingly more than ten thousand labourers were recruited for the work, and they cut a canal

in Leupold's *Theatrum Machinarum Generale* of +1724, and it was used by Muratori, Gennete and Michelotti for measuring surface velocities; its defect was that it could give no information about centre and bed flow. So far as we are aware, it was left for du Buat (+1734 to 1809) to make the first correlation of sediment motion with velocity of flow, using a submerged current-meter about +1779 (cf. Rouse & Ince (1), p. 129).

[a] Ch. 2, p. 12*a*, tr. auct.

[b] This afterwards became section *i* (cf. Table 70) of the Grand Canal in its final form.

[c] Pp. 293, 294 above.

[d] P. 10*b*. This book relates to events occurring between +954 and +1085. Cf. *Hsing Shui Chin Chien*, ch. 97 (p. 1428).

[e] It bears some 160 inscriptions beginning at +988; see Kung Thing-Wan (1).

[f] A brief note on springs appeared in Vol. 3, p. 606 above, and Soymié (2) has given us a monograph on the folklore of water-finding in ancient and medieval China.

[g] *Shih Chi*, ch. 29, pp. 5*b*, 6*a*, tr. Chavannes (1), vol. 3, p. 531, eng. mod. auct. adjuv. Watson (1), vol. 2, p. 75. Parallel passage in *Chhien Han Shu*, ch. 29, p. 5*a*, *b*. Watson's interpretation differs a little, as he takes the line of vertical shafts and the underground channel to have been only around Mount Shang-yen.

[h] This is the Lo River in Shensi, which flowed into the Yellow River from the north-west between the Fên and the Wei junction points; not the river of Loyang. Cf. Fig. 883.

[1] 洪澤   [2] 龜山   [3] 王鞏   [4] 聞見近錄   [5] 莊熊羆   [6] 洛

which, starting from (the city of) Chêng [1],[a] brought the waters of the Lo to the foot of Shang-yen mountain.

As the banks were liable to slide and crumble easily, (a series of) wells was dug, the deepest of which was as much as 400 ft.; and there were wells all along at regular intervals. At the bottom they communicated with each other (by a tunnel) through which the water flowed. The water came down until it met, and flowed round, the Shang-yen mountain, east of which the canal continued more than 10 *li* until it reached the hills. This was the first time that a subterranean canal with well-openings (*ching chhü* [2]) was built.

While the canal was being excavated, the bones of a dragon were found, hence it was called the 'Dragon-Head Canal'. More than ten years after its completion, the water was coming through all right, but very little benefit had accrued from it for agriculture.

Although this work of Chuang Hsiung-Phi's (if indeed he was the engineer)[b] seems to have had only qualified success, it must be admitted to be of considerable technical interest.[c] Though the whole length of the canal is not clearly stated, it seems at first sight to have been an expedient to avoid the collapse of the sides of a cut, perhaps through some narrow valleys. But one is strongly reminded of the traditional *qanāts* of Persia, so characteristic a feature from the air when one is flying over the Teheran region,[d] but also practised in other parts of the Muslim world.[e] The *qanāt* or (Tk.) *kārīz* [f] was (and is) a device to utilise mountain water sources which normally sink out of sight when they reach the foot-hills, losing themselves in the porous valley soil of confluent alluvial fans. The flow is then tapped near the base of the hills and led along a subterranean channel above the impervious clay strata and under a succession of vertical ventilation and excavation shafts, until it debouches into a reservoir from which fields and settlements can be continuously supplied.

The early history of the Persian *qanāt* seems obscure, but Marco Polo refers to streams of sweet water running underground in the province of Kerman, and these were very probably *qanāts*.[g] Underground pipe-lines, perhaps derived from *qanāts*, were also a feature of Iraq.[h] Indeed a work rather like that of Chuang Hsiung-Phi was carried out during the Sassanian period (+3rd to +7th centuries) at the north-eastern edge of the Tigris Valley. From the upper waters of the River Diyala (cf. p. 366) a derivate canal was led southwards by a long tunnel through the Jebel

[a] Modern Chhêng-chhêng.[3]
[b] The account in the *Chhien Han Shu* gives his name as Yen Hsiung.[4]
[c] For a general history of tunnel engineering see Sandström (1).
[d] Cf. Stein (8); B. Fisher (1); A. Smith (1); Beckett (2); Goblot (1); Wulff (1, 4); Drower (1), pp. 532 ff., fig. 348. Beckett (1) also gives some good photographs, including one of the fireclay rings which are baked on the spot and used for the shaft and tunnel linings. Colour photographs will be found in Eller (1), p. 510.
[e] E.g. in Chinese Turkestan, significantly enough, on which see Golab (1), and in certain Saharan oases, on which see Anon. (10).
[f] The Greeks knew these systems as *hyponomoi* (ὑπόνομοι); Polybius x, 28, 3 ff., also a −2nd-century reference, like that in the *Shih Chi*.
[g] Ch. 38, Moule & Pelliot ed.
[h] In the time of the caliph al-Mutawakkil (+9th century) no less than 300 miles of such subterranean conduits were constructed; cf. Ahmed Sousa (1); Krenkow (1). A description of +1000 still exists in the *Inbāṭ al-Miyāh al-Khafīya* (The Bringing of the Hidden Waters to the Surface) by Muḥammad ibn al-Ḥasan al-Ḥāsib.

[1] 徵   [2] 井渠   [3] 澄城   [4] 嚴熊

Hamrīn range of hills so as to bring irrigation water to the land at the edge of the plain below.[a] A Chinese reference of +1259 to *qanāts* occurs in the *Hsi Shih Chi*[1] (Notes on an Embassy to the West) by Chhang Tê,[2] who tells us that in the country of the Malāḥida, or Ismailite 'Assassins', otherwise the Elburz Mountains in Kuhistan, 'the land is destitute of water, (so) the local people dig wells at the edge of the mountains, and conduct the water down for several tens of *li* (to the plain), for the purpose of irrigating their fields'.[b]

We do not have sufficient information on the topography of Chuang Hsiung-Phi's −2nd-century project to see it clearly, but the similarity between Chinese and Persian techniques is notable and needs further study. At a much earlier stage the attention of the reader was drawn to possible influences and transmissions in this field across Central Asia.[c] Which way, one wonders, did they travel? In this case probably from west to east, if Lassøe is right in recognising *qanāts* in −8th-century Urartu (Armenia).

Another mention in China (also Shensi) occurs at a much later date. Lu Jung[3] in his *Shu Yuan Tsa Chi*[4] of +1475 (The Bean-Garden Miscellany)[d] tells us that:

> In the capital of Shensi there had formerly been very little water within the city, and the wells were so few that the inhabitants generally fetched their water from outside the west gate. When Yü Tzu-Chün[5] became governor of Sian, he reflected that Kuanchung (the Wei Valley) was a strategic region, and that if the city were besieged for several days, the inhabitants would hardly be able to live. So he bored an underground canal leading the waters of the rivers Pa[6] and Chhan[7] into the city from the east, and letting them flow out to the west. The water was obtained by means of (a series of) shafts with a masonry lining (*huan chou*[8]), the water flowing (in the tunnel) below and the ground being quite level above...

But the *qanāt* system never became widespread in China proper, presumably because the physiographic and geological circumstances did not call for it.[e] Yet when the history of the technique is fully known, China may be found to have made a contribution.[f]

(iii) *Dredging*

In considering problems of scour (cf. pp. 229 ff. above) it was altogether natural that from time to time the Chinese medieval engineers should have attempted mechanical as well as hydrodynamical means. A remarkable instance occurred in the Sung, when in +1073 determined efforts were made to clear the accumulation of silt and bed load in

[a] See Adams (1), p. 75, and maps. Cf. p. 372 below.

[b] In *Yü Thang Chia Hua*, ch. 2, p. 6a, tr. auct. adjuv. Bretschneider (2), vol. 1, p. 133.

[c] Vol. 1, pp. 235 ff.   [d] Ch. 1, p. 6b, tr. auct.

[e] Cf. Fang Chi (2), pp. 26 ff., who discusses mainly Sinkiang, where there are some 1,500 *qanāts* irrigating over 48,000 acres (Kovda, 1). But recently the system has been spreading in Kansu and Inner Mongolia; see Wang Wei-Hsin (1).

[f] One suspects some connection here with the magnetic compass and its travels. Calder (1), pp. 59 ff., has drawn attention to the prominent use of the floating iron needle by the Persian *qanāt* miners. As this could hardly have reached them from China before the +6th century it cannot have been the limiting factor for the technique, but the association is curious, and may have something to do with an overland passage of the compass to the West, which has been surmised for various other reasons (cf. Vol. 4, pt. 1, pp. 330 ff.).

[1] 西使記   [2] 常德   [3] 陸容   [4] 菽園雜記   [5] 余子俊
[6] 灞   [7] 滻   [8] 環甃

the Yellow River near Ta-ming. After devastating floods in Hopei and energetic promotion by the prime minister Wang An-Shih [1] a Yellow River Dredging Commission (Chün Huang-Ho Ssu [2]) was set up with Fan Tzu-Yuan [3] at its head.[a] The candidate-official Li Kung-I [4] then came forward with a suggestion for an 'iron dragon-claw silt-dispersing machine (*thieh lung chao yang ni chhê fa* [5])', i.e. a weighted and toothed rake trawled by two boats up and down to keep the loose bottom material on the move. A leading eunuch, Huang Huai-Hsin,[6] despatched to report on the proposal, approved of it but thought the specification rather too light, so he and the inventor were charged to produce something more serviceable. The resulting 'river-deepening harrow' (*chün chhuan pa* [7, 8]) was a beam 8 ft. long fitted with iron spikes each 1 ft. long, and sunk to its work by windlasses on two ships. It was confidently affirmed by the local magistrates that either the depth would be too great for the cables or else in shallow places the scarifier would get stuck in the mud, but in spite of adverse reports and some diplomacy on the part of Fan Tzu-Yuan at the capital, several thousands of the scarifiers were made and put into service, Li Kung-I being appointed his assistant. Unfortunately the historians recorded no informed judgment on the success of the operation.

A second example may be taken from a time five centuries later. In +1595, towards the end of the Ming, an Imperial Censor, Chhen Pang-Kho,[9] gave his views on the best ways of keeping rivers clear, and especially the Grand Canal.[b] If, he said, one does nothing but strengthen dykes and allows the bed to build itself up more and more, there will be no benefit of self-scour and every invitation to the flood water to break out. Three methods should therefore be used—first the traditional excavation of the bed as far as possible during low-water seasons (just as at Kuanhsien, cf. p. 293 above). Secondly,

let all official and private boats coming and going tow 'bed-harrowing ploughs' (*pa li* [10]), and sail with the wind, scraping the bottom as they go, so that the sand has no peace to sink and settle. Thirdly, imitating the hydraulic mill and the hydraulic trip-hammer, let wooden machines be made which use the current of the water to roll and vibrate, so that the sand is constantly stirred up and cannot accumulate.

The second method was just that of Li Kung-I; the third, more interesting perhaps yet more difficult to visualise, suggests moored paddle-wheel vessels like ship-mills, with bottom-agitating rakes worked from eccentrics, with or without connecting-rods, on each side. Unfortunately again the historian did not record the use, or the efficacy, of Chhen Pang-Kho's third method. In general, however, this late Ming discussion illustrates again certain perennial engineering convictions which we have already discussed (pp. 234 ff. above). As for the second method, it continued in use down to

---

[a] *Hsü Thung Chien Kang Mu*, ch. 7, pp. 13 b ff., 19 a, quite well summarised in Williamson (1), vol. 1, pp. 292 ff. Cf. Todd & Eliassen (1), p. 353. A considerable amount of straightening of the river's meanders was carried out at the same time.

[b] *Ming Shih*, ch. 84, p. 12 a, b, tr. auct.

[1] 王安石　　　　[2] 濬黃河司　　　　[3] 范子淵　　　　[4] 李公義
[5] 鐵龍爪揚泥車法　　　　[6] 黃懷信　　　　[7] 濬川杷　　　　[8] 濬川耙
[9] 陳邦科　　　　[10] 耙犁

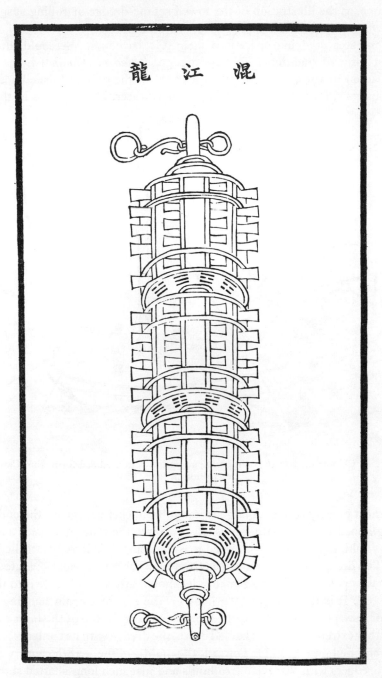

龍　江　混

Fig. 911. Towed scrape-dredge or rolling suspensifier, the *hun chiang lung* illustrated in the *Ho Kung Chhi Chü Thu Shuo* (ch. 2, p. 30*a*). Drawn along the bottom by a vessel proceeding upstream, it raised clouds of silt from the bed (hence the name), and so increased suspension clearance.

our own time, as the illustration of the towed scrape-dredge or rolling suspensifier in the *Ho Kung Chhi Chü Thu Shuo* (Fig. 911) testifies.[a]

Today one sees dredging operations mainly in harbours. We could illustrate the commonest form of traditional Chinese dredger (*chiao ni chhuan*[1]) by photographs taken at Canton in 1958, but a sketch (Fig. 912) in the only published account by a Western observer, that of Carmona,[b] is perhaps clearer. Descriptions in the Chinese

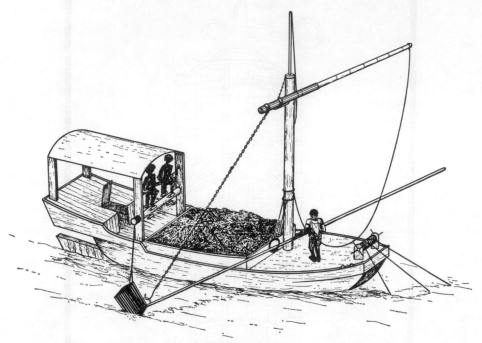

Fig. 912. A Chinese dredger of traditional type (redrawn in modified form from Carmona).
Description in text.

literature must be very scarce. A large rectangular bucket dredge on the end of a long spar strengthened and shod with iron is let down into the water alongside a barge with a capacious hold, to the forward-stepped mast of which it is loosely attached by a line and loop. The dredge is connected at the front by a cable to a pedal-operated windlass situated at the opening of the steersman's deckhouse aft, which brings it to the surface when it is full. It is then caught by a hook on the end of a chain hanging from one extremity of a swape-like lever[c] attached to a point near the top of the mast-gantry, and when the other extremity of this is hauled down, the dredge with its contents is swung inboard and emptied into the hold. Towards the middle of the + 18th century de Bélidor illustrates dredgers with Vitruvian treadmills and just such long-shafted shovels in his

[a] A double-hulled dredger is also shown in ch. 4, p. 5 *a, b.*
[b] (1), p. 36. He figures also a good model now in the Maritime Museum at Lisbon.
[c] Cf. Vol. 4, pt. 2, pp. 331 ff., and Forbes Taylor (1).

[1] 絞泥船

Architecture Hydraulique,[a] but whether they stem from a Chinese or a European original remains uncertain.

(iv) *Reinforcement and repair*

The need for internal bonding in structures which ultimately led to reinforced and pre-stressed concrete was of course felt by ancient and medieval Chinese engineers, and we have already seen examples of the methods they employed. Wooden piling was incorporated in walls such as the Great Wall (p. 51) and the Chhien-thang sea-wall (p. 322). But rods and weights of iron were used also when the supply of the metal permitted. In a Thang repertory[b] we are told that

in the Liang period great difficulties were encountered with a dam called the 'Floating Mountain Dam' (Fou-shan Yen[1]), which was always being rebuilt, and as often collapsing.[c] Finally, several thousands and ten thousands of catties of iron were placed underneath, and it was successfully completed.

The text goes on to say that in +762 there was similar trouble with a dam south of Haichow, which was not overcome until Li Chih-Yuan[2] acted upon suggestions that iron weights and bars should be incorporated in the foundations.

Without exception, however, the most important material used in China for bonding the interior of solid structures was bamboo. The remarkable tensile strength of plaited strips of this plant stem has already been referred to[d] in connection with tow-ropes, driving-belts, suspension bridges, and so on. Since it was available in such unlimited quantities, the earliest technique was probably to leave in position the baskets or skips in which stones or earth had been carried to the spot, instead of taking them away to fetch more. Then as time went on, the elongated gabion, or sausage-shaped open-work crate of bamboo packed with stones, was developed; and this stage must have been reached by the time of Yang Yen and Wang Yen-Shih (−28) whose breach-blocking achievements have just been quoted.[e] The great advantages of this invention have already been emphasised in the description of the Kuanhsien works; the relative lightness of the gabions permitted their use on alluvial subsoils without deep foundations, and their porosity gave them a most valuable shock-absorbing function, so that surges and sudden pressures did no damage to the defences.[f] It is interesting to find that the same device was employed in Europe, no doubt independently, at least from the +14th century onwards, especially in the sea-dykes of the Netherlands.[g] Here bales of

---

[a] (1), vol. 4, pls. 20, 21, 25. Models worth studying are to be seen in the Tekniska Museum at Stockholm, especially in the Christopher Polhem collections. Cdr. Waters tells us that the R. Thames was dredged for ballast near London in the +16th century by a 'spoon-and-pole' method.

[b] *Thang Yü Lin*, ch. 5, p. 31*b*, tr. auct.

[c] This dam, in the upper Huai valley, had been built by Wang Tsu[3] in +514.

[d] Vol. 4, pt. 2, pp. 63ff., 126, 129ff., 144, 153; here also pp. 191, 328, 415, 466, 597, 664.

[e] The dimensions of the gabions used by Wang Yen-Shih seem to have been 40 ft. in length and 17 ft. in diameter. Today 60 ft. lengths are commonly used.

[f] On modern Chinese practice regarding gabions see Sung Hsi-Shang (1), vol. 1, p. 42, fig. 4.

[g] See Forbes (17), figs. 622, 623. A classical description was given by Andries Vierlingh in +1579. It would be interesting to know how much of hydraulic engineering technique Europe eventually absorbed from China. De Bélidor, for instance, (1), vol. 4, pl. 35, gives excellent drawings of river dykes with fascines, etc.; and as his pp. 354 ff. show, he knew a good deal about the Chinese water-ways and their embankments.

[1] 浮山堰   [2] 李知遠   [3] 王足

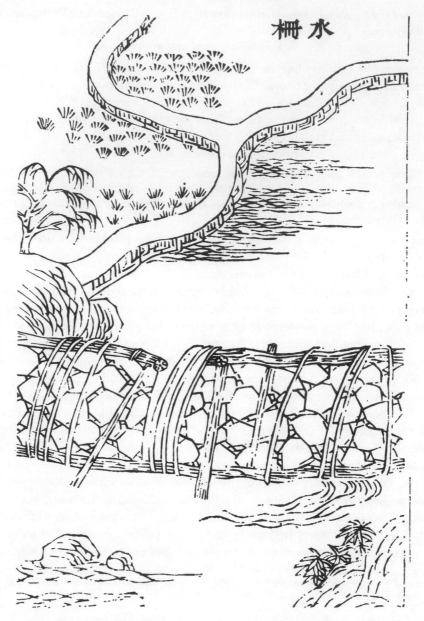

Fig. 914. A drawing of bamboo gabions in the *Nung Shu* of +1313. The *shui cha*, or 'water palisades', are acting as a small weir.

compressed seaweed, or a screen of compressed seaweed within piling, or bundles of reeds fixed down with their roots pointing seaward, were used outside the Dutch clay polder dykes. Such shock-absorbers, less resistant to decay than bamboo basket-work, had to be renewed every five years.

Variants of the sausage-shaped stone-filled bamboo crate bear different names; the commonest is *chu lung*,[1] but another, perhaps older, term is *shui cha*.[2] Great use has always been made of them at Kuanhsien (cf. Figs. 894, 895, pls.), and they are very noticeable when one travels anywhere in China. Fig. 913 (pl.) shows one of my own photographs of a weir near Chhêngtu, in which layers of them can be seen. Fig. 914 reproduces the picture from the *Nung Shu*[a] of +1313, which is superior to those of the later *Nung Chêng Chhüan Shu*[b] and the *Thu Shu Chi Chhêng*.[c] Elongated gabions formed of wooden slats (*yang chüan*[3]) are also used (cf. p. 295 above).[d] Gabions are prominent in Fig. 915 (pl.), taken from a painting of +1417 by Ma Wan entitled 'Yü of the Hsia Controlling the Waters'.[e]

Besides the gabions, huge fascine bundles of kaoliang stalks fastened with bamboo rope were also developed. These were very convenient when the water was heavily silt-laden, for solid material would quickly be deposited in the interstices of the mass as the water filtered through, and in time it would become very compact. Such brushwood fillers (a Sung invention) were termed *sao*,[4] and Fig. 916 taken from a work of *c.* +1775, the *Hsiu Fang So Chih*[5] (Brief Memoir on Dyke Repairs) by Li Shih-Lu[6], shows a drawing in Chinese style of the method of handling them.[f] Esterer was himself present at the closing of a gap in the Yellow River dykes in 1904, and wrote an impressive account of it.[g] The dyke was about 30 ft. broad at the bottom and 11 ft. at the top; its height was 33 ft. The gap to be filled was 36 ft. wide at the bottom and 54 ft. at the top, with water pouring through it. Gabions and fascines were used, handled by 20,000 men hauling on cables 100 ft. long—eotechnic but on the heroic scale. Fig. 917 (pl.), from Todd (1), shows the use of giant bales of kaoliang stalks (cross-section 20 × 50 ft.) for stopping a breach.[h] The technique may be studied in more detail in the report of Todd & Eliassen, who give also some archaeological data on what breakthroughs and washouts have revealed of the dyke construction of the +10th and +11th centuries. It seems that there was little or no stone facing used at that time (cf. the story of the sea-wall), but kaoliang fascines, bamboo and stone gabions, hemp rope, willow stakes, and bags of clay, were employed on an almost incredible scale to supplement and strengthen the earthen dykes themselves.[i]

[a] Ch. 18, p. 2*a*.　　　　　[b] Ch. 17, p. 30*a*.

[c] *I shu tien* (ill. no. 26), ch. 5, *hui khao* 3, p. 30*b*.

[d] These are perhaps identical with what *Sung Shih*, ch. 91, p. 10*b*, calls *mu lung*.[7] But in *Ho Kung Chhi Chü Thu Shuo*, *mu lung* is a pile-driver raft used in dyke repairs (ch. 3, pp. 30*a*, *b* ff.); see Fig. 909.

[e] In the Philadelphia Museum; see Bishop (8).

[f] The origin of the term *sao* is interesting. The word occurs in the *Shih Ching* as the name of a kind of plant which grew on walls and fixed itself so tightly that it was impossible to pull it out. Hence its adoption by the engineers as a technical term. On the various kinds of *sao* see *Ho Kung Chien Yao*, ch. 2, pp. 11*a* ff., ch. 3, pp. 1*a* ff.; *Chih Ho Fang Lüeh*, ch. 1, pp. 16*a* ff.; *Ho Kung Chhi Chü Thu Shuo*, ch. 3, pp. 1*a*, *b*.

[g] (1), p. 141.

[h] Cf. Anon. (75), no. 10, a contemporary Chinese painting, and Nesteruk (1), p. 19, fig. 7.

[i] In the *Sung Shih* we have an account (ch. 91, pp. 8*b* ff., 11*b* ff.) of the extraordinary preparations of materials—beams, bamboo ropes, piles of brushwood, etc.—accumulated at various Conservancy yards along the dykes (*sao*[4]). When the country people were not busy with farming they were drafted to do this kind of work. The description applies to the period between +1017 and +1021. Different categories of

[1] 竹籠　　　[2] 水柵　　　[3] 羊圈　　　[4] 埽　　　[5] 修防瑣志　　　[6] 李世祿
[7] 木龍

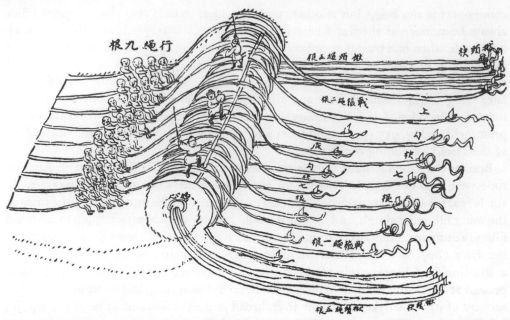

根九繩行

根五繩頭戟

槍頭戟

根二繩繩戟

上

勾

校

七

勾

根

七

根

根一繩繩戟

根五繩繩戟

根民戟

Fig. 916. Manhandling a giant fascine bundle into position; a sketch in traditional style, entitled *chüan sao thu*, from the *Hsiu Fang So Chih* of Li Shih-Lu (c. +1775). The nine hauling ropes (*hsing shêng*), with the hauliers, are seen on the left, the side towards the water. The five 'end-head bundle cables' (*chhou thou shêng*) pass through the centre of the fascine, and being slowly paid out, act as brakes. The seven 'overhook cables' (*shang kou chüeh*), continuous with the seven 'underhook cables' (*ti kou chüeh*), form a safety cage in which the fascine bundle (*sao*) can roll; they would have to be re-pegged from time to time.

A graphic description of a closure of breaches in the Yellow River dykes occurs in the *Mêng Chhi Pi Than* of Shen Kua, and is worth reproducing in full.[a]

In the Chhing-Li reign-period (+1041 to +1048) the (Yellow) River burst through its dykes at Shang-hu near the northern capital,[b] and for a long time no closure could be made. Kuo Shen-Hsi,[1] Vice-Minister of Finance at the time, went there to take charge of operations himself.

Now in closing gaps made by a river in embankments, success chiefly depends on the filling of the final gap when the last gabion or fascine bundle (*sao*[2]) (is placed in position). This is called 'Closing the Dragon Gate' (Ho Lung Mên[3]).[c] Sometimes many unsuccessful efforts have to be made.

(In this case) the *sao* (prepared for closing) the 'Dragon Gate' was (no less than) 60 paces (about 300 ft.) in length. Kao Chhao,[4] one of the assistant engineers (*shui kung*[5]), offered a suggestion. He said that the *sao* was too long, so long that it was beyond the power of human

material mentioned are given special technical names, such as *chhun liao*,[6] 'spring supplies', etc., and the different kinds of timber needed are distinguished. A mass of information on the technical terms of traditional Chinese hydraulic engineering, never yet properly studied in the West, though of comparatively recent date (1887), is contained in the interesting book of Chhiu Pu-Chou (1) entitled *Ho Kung Chien Yao*,[7] already several times quoted.

[a] Ch. 11, para. 19, tr. auct.; cf. Hu Tao-Ching (1), vol. 1, pp. 420 ff.
[b] Cf. Table 69 above.
[c] Cf. *Chih Ho Fang Lüeh*, ch. 10, pp. 34a ff.

[1] 郭申錫   [2] 埽   [3] 合龍門   [4] 高超   [5] 水工   [6] 春料
[7] 河工簡要

strength to press it down (ya[1]) into position at the bottom of the gap (against the force of the current). The water was rushing through incessantly, so that the cables frequently broke. 'What we ought to do', he said, 'is to separate the sao into three sections each 20 paces in length, connected together by ropes. After the first sao has been put into place on the bottom, the deposition of the second sao can be started, and finally the third section (will close the gap).' Some of the old-fashioned workers disagreed with him, and insisted that the suggestion should not be carried out, saying that a sao of only 20 paces length would not stop the water's flow, and having three sections would only double the expense to no benefit. (Kao) Chhao replied: 'The first sao will of course not stop the flow of water entirely, but it will abate its impetuosity by half, so that the second sao will need only half the force to be pushed into position (and held there). After the second sao has been fixed, the flow will again be less, so that the third (and last) sao can be pushed into place from level ground, and the full force of all available man-power can be applied to it. The two upper sao will soon be silted up with sand and mud from the turbid river, so that little further work will be required.'[a]

(Kuo) Shen-Hsi did not accept (Kao) Chhao's views, and adhered to previous advice. At this time Chia Wei-Kung[2] was military commander of the northern districts. He alone thought that (Kao) Chhao was right, and sent several thousand people with confidential instructions to points lower down to collect stones and brushwood (to demonstrate that Kuo Shen-Hsi's filling material was simply being carried away as fast as it was put down). And after the sao was put in position the sao itself (began to break up, and fragments of it were) carried away. And the (Yellow) River continued to flow through the gap worse than before. So (Kuo) Shen-Hsi was recalled and exiled, and eventually the suggestions of (Kao) Chhao were put into practice, so that Shang-hu had peace (from floods).

This gives us a revealing insight into the discussions which must have gone on for many centuries among those concerned with water control. Kuo Shen-Hsi's advisers seem not to have been practical men, while Kao Chhao was evidently a man of the people,[b] whose achievement reminds us of the famous phrase in the *Shen Tzu* book already quoted:[c] 'As for those who protect and manage the dykes and channels of the nine rivers and the four lakes, they are the same in all ages; they did not learn their business from Yü the Great, they learnt it from the waters.'

From the + 10th century onwards there was a succession of great engineers who achieved remarkable results in controlling the Yellow River, and brought the utmost ingenuity to bear on the closing of dyke breaches. Gabions and fascines were delivered from boats, boats were filled with stones and sunk where necessary, and mobile scrape-dredgers were (as we have seen) constructed.[d] The first of the series was Li Ku,[3] who

[a] Unfortunately it is impossible to tell from the account whether the three sao were to be placed in the gap one on top of another, or side by side. The word 'upper' in the last sentence might perhaps mean 'earlier placed in position'.

[b] In this connection it is interesting to note the remarkable upsurge of invention and technical innovation among the masses of the Chinese people. Country folk working on irrigation and conservancy projects in recent years have developed many useful devices such as semi-mechanised earth-pounders, self-dumping conveyors and the like (cf. Yang Min, 2). We have touched upon this subject before (Vol. 4, pt. 2, pp. 173 ff.).

[c] Vol. 2, p. 73.

[d] According to H. Li (1), p. 72, gunpowder was first used for blasting rock obstructions in the Yellow River by Chhen Mu in + 1541, but we have so far not been able to locate the textual authority for this statement.

[1] 歷   [2] 賈魏公   [3] 李毅

worked for the Northern Chou (*c.* +954), while after the Sung had come into power, Tu Yen-Chün[1] (*fl.* +994) was also successful in river control. The most eminent man about the time Shen Kua was describing, in the passage above, was Li Chung-Chhang,[2] but the greatest reputation was won by Chia Lu[3] who repaired dykes and made relief canals during the Yuan period (+1330 to +1360), and whose methods were described in a work already mentioned, the *Chih Chêng Ho Fang Chi*,[4] by one of his assistants, Ouyang Hsüan.[5] In the Ming, Liu Ta-Hsia[6] (+15th century), and Liu Thien-Ho,[7] Ong Ta-Li,[8] Li Hua-Lung[9] and Wan Kung[10] (+16th) were all famous, but their acknowledged past-master was Phan Chi-Hsün[11] about whom something has already been said.[a]

How far what would now be called river-training was included in their art is a question in need of deeper study than we have been able to devote to it. Revealing accounts are to be found in the sources however, and we would guess that the empirical knowledge was very considerable. For example, in +1015 an imperial edict laid down that in certain sections the depth of the Pien Canal was to be not less than 7 ft. 5 in., but Ma Yuan-Fang,[12] then Vice-Director of the Court of Imperial Sacrifices, urged that 5 ft. would be enough as long as an average breadth of 50 ft. was adhered to.[b] Besides towpaths (*tao tao*[13]) he made 'horse-heads' (*ma-thou*[14]), i.e. jetties, groynes or spurs, and 'towards the secluded parts of the banks where the water was shallow, he made saw-teeth (*chü-ya*[15]) in order to restrain (or control) the force of the water (*i shu shui shih*[16])'. Elsewhere[c] we encounter the same objects, together with wooden piling along the banks (*mu an*[17]), 'with the purpose of restraining the current and protecting the embankments (*i tsu shui shih hu thi*[18])'. The simplest interpretation of these works would suggest river-training projections or spurs, submerged or clear-standing, placed suitably to direct the force of the current away from the concave eroding bend so as to scour the shoals on the convex side, with the result of saving the banks and improving the fairway for vessels.[d] An abundance of such deflecting groynes, semi-diagrammatically drawn, can be seen in Figs. 867, 902 (pls.). All this was certainly standard practice during the Ming and Chhing.[e]

(v) *Sluice-gates, locks and double slipways*

After it was realised that water, even on a large scale, could profitably be made to run along channels between permanent ridges, the need must soon have been felt for some

[a] P. 325. Many of their most famous discourses and addresses to the throne are collected in ch. 8 of *Chih Ho Fang Lüeh*.

[b] *Sung Shih*, ch. 93, p. 21*a*. Cf. p. 231.

[c] *Sung Shih*, ch. 91, p. 12*a*, referring to the Yellow River.

[d] A modern example of the technique is figured and explained in Leliavsky (1), p. 104. Though the terminology differs one can recognise the process in *Ho Kung Chien Yao*, ch. 3, pp. 8*a*, 19*a* ff.; and *Chih Ho Fang Lüeh*, ch. 10, pp. 23*a* ff., 25*a*, *b*, 26*b* ff., and on *ma-thou* 41*a* ff.

[e] See the lucid account, with diagrams, in Sung Hsi-Shang (1), vol. 1, pp. 37 ff. and fig. 3.

[1] 杜彥鈞	[2] 李仲昌	[3] 賈魯	[4] 至正河防記	[5] 歐陽玄
[6] 劉大夏	[7] 劉田和	[8] 翁大立	[9] 李化龍	[10] 萬恭
[11] 潘季馴	[12] 馬元方	[13] 踏道	[14] 馬頭	[15] 鋸牙
[16] 以束水勢	[17] 木岸	[18] 以蹙水勢護隄		

kind of obstruction which was not permanent, but readily movable at will. Thus did the water-gate come into being. From the irrigation and flood-control function of hydraulic works derived the sluice-gate, while their transportation function gave rise to the lock-gate. The only essential difference between the two is that the latter must be so built as to allow the passage of shipping. In elucidating the invention of the lock[a] several stages have to be considered, first the placing of widely spaced 'flash-lock' gates along canals or canalised rivers, then the appearance of the pound-lock not much longer than the barge or boat which it is intended to raise and lower (an arrangement which obviously greatly reduces the time which the boat must wait while the water-level changes before it can proceed); and finally, improvements to pound-locks, such as the mitre type of gate construction, closing so as to oppose maximal strength to the current,[b] or ground sluices,[c] or side ponds.[d]

It may be significant that no mention of gates occurs in the *Shih Chi*, either in the chapter on rivers and canals, or elsewhere. But in the *Chhien Han Shu* there are references, some of which we have already quoted. In the last few decades of the −1st century (c. −36) Shao Hsin-Chhen incorporated a number of sluice-gates (*shui mên*[1]) in his Chhien-Lu dam (p. 280 above) and canals near Nanyang.[e] Moreover, the following speech of Chia Jang in −6 shows that the use of gates cannot have been a new idea at that time.[f] As we have already seen (p. 235) the emperor had issued an edict inviting useful suggestions about the control of the Yellow River and Chia Jang submitted a written memorial containing three plans in preferential order. His second alternative was a network of irrigation canals.

Now we can make [he said] a dyke of stone from Chhi-khou eastwards and build many sluice-gates (*shui mên*[1])...I fear that critics may think that the (Yellow) River is too large to be controlled. However, we can predict (our chances) by (our experience of) the Pien Canal (Tshao Chhü[2]) at Jung-yang. There the lock or sluice gates were built only of wood set into the earth (dyke; and yet they lasted for a long time without being destroyed). Now if we build stone dykes in this place, where there are firm foundations, we shall certainly be safe. The (derivation) head (gates) of the canal at Chichow should be modelled on those (at Jung-yang). The proper way of controlling the flow in canals involves more than merely digging out earth...During the dry season, the lower or eastern sluice-gates should be opened for irrigating the land of Chi chow, and during flood time, the great western sluice-gates should be opened to divert and drain away some of the river...

For our purpose it is not necessary to examine the details of the system which Chia Jang proposed; the point is that he envisaged sluice-gates large and small, set in stone piers,

---

[a] This subject was touched upon in Chatley (36) and especially in the discussion which followed his paper at the Newcomen Society. The exchange of views which then took place is superseded by the material now brought forward. An even fuller treatment of the evidence will be found in Needham (57).

[b] Cf. p. 358 below.

[c] Passages contrived in the masonry of the lock to admit or remove water without the necessity for the movement of the gates themselves, or of wicket-doors in them.

[d] Reservoirs installed on both sides of the lock to conserve half the water at each operation.

[e] Ch. 89, p. 14*b*. 'Several tens' of them are mentioned.

[f] Ch. 29, p. 15*b*, tr. auct. Cf. *TH*, p. 586. The speech is also to be found in *CSHK*, Chhien Han sect., ch. 56, p. 6*a*.

[1] 水門　　[2] 漕渠

not as a new suggestion, but as an improvement on wood and earth ones which had long been used. How far back the practice dated from remains uncertain. Nor is the silence of the *Shih Chi* conclusive, for it says nothing directly about the 'Magic Canal' of Shih Lu (see pp. 299 ff. above). It is true that the first definite evidence of flash-lock gates which we have for this canal is from the +9th century, and pound-locks were certainly there when Chou Chhü-Fei and Fan Chhêng-Ta travelled over it late in the +12th, but if there had been gates at the entrances of the Pien Canal in the −2nd or −1st centuries, and perhaps also at various points along its course, there remains always a certain possibility that Shih Lu introduced them on the approaches to the Ling Chhü as early as the −3rd century. In any case it is indubitable that sluice- and flash-lock gates were well known in Chinese hydraulic engineering by the last half of the −1st.

There is no need to labour this, but a few more touches may be brushed in. When the emperor Ming Ti inspected the repaired Pien Canal in +70, the work of Wang Ching,[1] he issued an edict reading in part as follows:[a]

Since the (dykes at the opening of) the Pien Canal burst, more than sixty years have elapsed. During some of this time the rains have not confined themselves to their proper seasons. The stream of the Pien encroached eastwards, conditions worsening day by day and month by month. The old emplacements of the flash-lock gates (*shui mên*[2]) were all forlorn in the midst of the river. Vast sheets of water overflowed so that one could not recognise where the banks had originally been. The whole pattern of the countryside was whelmed in chaos...

But now (the workers) have rebuilt the dykes, repaired the canal, cut off the waters, and established flash-lock gates (*li mên*[3]).[b] The (Yellow) River and the Pien (Canal) flow separated and are again in their old beds... Therefore (We) have sacrificed excellent jade and pure animals to the Spirit of the River.

This makes clear the fact that at least by the end of the −1st century there had been flash-lock gates along the Pien Canal, and that Wang Ching restored them and increased their number.[c] Then seven centuries later it is interesting to find in the MS. Ordinances of the Thang Department of Waterways many articles dealing with sluices and flash-lock gates.[d] One reads: 'To the east of Lan-thien[4] where there are water-mills, the mill-owners should be made to construct gates to regulate the flow of water, and allow free passage along the waterway (for traffic).' This belongs to +737, but it was just the arrangement in +16th- and +17th-century Europe, where mill weirs were

---

[a] *Hou Han Shu*, ch. 2, pp. 15a ff., tr. Bielenstein (2), p. 147, mod. auct.

[b] From *Hou Han Shu*, ch. 106, p. 7b, we know that Wang Ching established one every 10 *li*. From p. 270 above we can see that he must have built at least 202 gates, quite a considerable operation. His interest in this technique is shown from the fact (p. 6b) that he had a *yen liu fa*[5] (method of measuring out the flow), which Wang Wu,[6] the Inspector-General of Works (Chiang-Tso Yeh-Chê[7]), also used. These flash-lock gates were further strengthened with stone abutments in +171 (*Sung Shih*, ch. 93, pp. 17a ff.).

[c] In +72 he was made Inspector-General of Rivers and Dykes (Ho-Thi Yeh-Chê[8]). Besides rich presents of silk and money, chariots and horses, the emperor presented him with a number of books, including the *Shan Hai Ching* (cf. Vol. 3, pp. 503 ff.), the *Ho Chhü Shu*, i.e. Ssuma Chhien's work on the rivers and canals now incorporated in the *Shih Chi*, and a set of maps illustrating the 'Tribute of Yü' chapter of the *Shu Ching* entitled *Yü Kung Thu*.[9]

[d] Nos. 1, 2, 4, 6, 7, 8, 10; see Twitchett (2).

[1] 王景   [2] 水門   [3] 立門   [4] 藍田   [5] 揭流法
[6] 王吳   [7] 將作謁者   [8] 河堤謁者   [9] 禹貢圖

accompanied by stanches or flash-lock gates to permit vessels to pass up or down stream.[a] The 'flash' of water released by opening the gate was often essential for taking a barge over the shallows lower down, but after opening there might be a wait of a couple of hours for the 'abatement of the fall' before a barge could go up.[b] The Lan-thien Canal was built in +623 by Yen Hsü[1] in connection with a road across the Chhin-ling Mountains still more easterly than the Tzu-wu Tao described above (p. 22). It was in effect the canalisation of an upper tributary of the Chhan[2] River (cf. pp. 150, 335), which brought boats down to the close outskirts of Chhang-an.

There are plenty of poetical references to flash-lock gates in later literature. Thus about +1200 Chang Tzu[3] has a poem written beside a river or canal near a temple at Ling-yin Shan;[4] when the gate was opened to let a sampan through, the water came pouring down with a thunderous noise.[c]

In the seventeenth century foreign travellers began to take notice of the flash-lock gates in use on the Grand Canal and other Chinese waterways,[d] and there are many graphic nineteenth-century descriptions.[e] The practice was to haul vessels through upstream, generally with manned capstans and tow-ropes, against currents of as much as nine or ten knots, and in the reverse direction to let them 'shoot the rapids'. In Fig. 918 (pl.) we reproduce the drawing appended to Staunton's account of the Macartney embassy of +1793.

There is no doubt that throughout Chinese history the most typical form of sluice- and lock-gate was what is called the stop-log gate. Fig. 919 shows a small field example of it from the *Thien Kung Khai Wu*,[f] and Fig. 920 (pl.) the larger version used for flash-locks in canals and canalised rivers.[g] Two vertical grooves fashioned in wood or stone face each other across the waterway, and in them slide a series of logs or baulks let down or withdrawn as desired by ropes attached to each end. Windlasses or pulleys in wood or stone mountings like cranes on each bank helped to fit or remove the gate planks. This system was sometimes improved by fastening all the baulks together to form a continuous surface and then raising or lowering it in the grooves by means of

[a] Hence perennial conflicts between millers and transport officials, often echoing in the dynastic histories (cf. *Sung Shih*, ch. 94, pp. 3b, 4a, for +1086; ch. 96, p. 5a, for +1097). We have already touched upon this matter (cf. Vol. 4, pt. 2, p. 400).

[b] Cf. Skempton (5). An example of a similar kind, but with what was probably a pound-lock, will be mentioned on p. 354 below.

[c] *Nan Hu Chi*,[5] ch. 2, p. 4a—*khai cha fang san-pan*.[6]

[d] Cf. Lecomte (1), p. 104; de Navarrete (1) in Cummins ed. vol. 2, pp. 225 ff.

[e] Cf. Macartney (1), pp. 169 ff., (2), p. 268; Staunton (1), figs. 34, 35; Davis (1), vol. 1, pp. 141, 143. Dr Herbert Chatley told me that he had often waited at the gates on Chinese canals, commonly placed about a kilometre apart.

[f] Ch. 1, p. 15b. The oldest illustration of this kind that we have found is in the *Nung Shu*, ch. 18, p. 4b, the date of which (+1313) deprives Jacopo Mariano Taccola of the honour of having been the first to illustrate a dam with a sluice-gate (MS. of +1438) given him by Sarton (1), vol. 3, p. 1552. Cf. also *Nung Chêng Chhüan Shu*, ch. 56, p. 6a.

[g] Attention may be drawn in passing to the gangway which was rolled across the channel on grooved rails; this seems not to have been noticed so far by writers on the history of railways, e.g. Lee (1). Further description in Gandar (1), p. 59. Cf. p. 159 above.

[1] 顏旭   [2] 滻   [3] 張鎡   [4] 靈隱山   [5] 南湖集
[6] 開闔放三板

counterweights on the ends of the cables. Pulleys mounted on stone crane arms, obliquely set, can still be seen at many places, e.g. at Kao-pei-tien near Peking (Figs. 921, 922, pls.).[a] That grooved gates were nearly always used[b] is suggested by many of

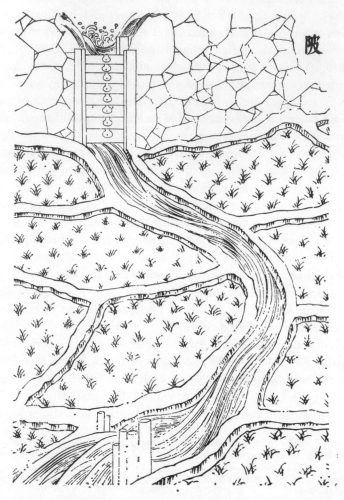

Fig. 919. The stop-log gate. A small example in a dyke, hence the caption *pei*, forming part of a minor irrigation system. From the *Thien Kung Khai Wu* (+1637), in versions after +1726.

the illustrations in the *Hsing Shui Chin Chien* (e.g. Fig. 908),[b] where grooves are shown in plan to mark symbolically the location of locks and sluices. Apparently gates swinging laterally, such as those familiar to us because usual in European practice, were not employed in China until modern times. Today, however, counterweighted steel shutter gates rising and falling after the Chinese style are used on waterways all over the world.

[a] See the study of Wang Pi-Wên (2). This place is on or near the line of Kuo Shou-Ching's Thung Hui Ho Canal. Its lock-gates were examined by Mr N. W. Pirie, F.R.S., in 1952.
[b] Vol. 1, p. 141. Cf. p. 326 above.

The simplest name for water-gate in Chinese literature is just that, *shui mên*.[1] As time went on, many other terms were introduced, e.g. *tou mên*[2] (dipping gates), and then *tou mên*[3] (sudden declivity, or 'head of water' gates), the latter arising, it might be permissible to infer, at some time when a difference of levels between 10 and 20 ft. made the structure particularly impressive. We also find the term *cha*[4] for a lock (etymologically, armour in a gate), and *pan cha*,[5] a lock-gate made of boards, i.e. a stop-log gate. Then comes *chha*[6] (or *cha*[6]), an 'inserting' gate, the phonetic being an old word for a grain-husking pestle and mortar (K631), with its combinations, *thi chha*[7] (a gate in a dyke), *shui chha*[8] (a gate for water), and *pa chha*[9] (a gate in a dam). The term *hsüan mên*,[10] or 'hanging gate', first found (so far as we know) in +984, is significant as indicating the permanent installation of windlasses. Indeed it is interesting to study the whole course of chronological distribution of these technical terms. There can be no doubt that *shui mên*[1] was the most ancient phrase, current in the Han and San Kuo periods but not much used thereafter.[a] The expression 'dipping gate' (*tou mên*[2]), which followed, became obsolete in its turn by the +14th century.[b] All these could be used indifferently for sluices and flash-lock gates. In the Sung period, about the beginning of the +11th century, *cha*[c] and *chha*[d] both make their first appearance.[e] They do so just about the time of the invention of the pound-lock, but the correspondence seems to be coincidental. The term *tou mên*[3] is tardy in general use, if not in first

---

[a] See, e.g., *Chhien Han Shu*, ch. 29, p. 15b, ch. 89, p. 14b. Cf. *Hou Han Shu*, ch. 2, p. 15a, ch. 106, pp. 6a ff. The term is used for the sluices in Têng Ai's[11] dams, c. +235.

[b] See, e.g., the MS. Ordinances of the Waterways Department of +737 (tr. Twitchett, 2). The term is quite common in Thang texts. The *Shu Hsü Chih Nan* (+1126), ch. 14, p. 6b, defines the use of sluices (*tou mên*[2]) as 'gathering the waters and releasing them according to the seasons' (*shih chhi chung hsieh*[12]), a phrase which occurs in the biographies of Kao Yü[13] (*Chiu Thang Shu*, ch. 162, p. 7b, *Hsin Thang Shu*, ch. 171, p. 6b), who in +768 built 180 *li* of irrigation reservoir embankments. Cf. *Thang Yü Lin*, ch. 3, p. 29b, on canal sluices at the capital. The early +13th-century *Ssu Chhao Wên Chien Lu*, ch. 3, p. 46b, tells of a city canal with sluices or flash-locks in +988 at Chhang-tê. About +1190 Lu Yu says that there were more than 170 *tou mên*[2] sluices in the Chêngkuo canal system (*Lao Hsüeh An Pi Chi*, ch. 5, p. 15a). In the *Chi Shan Chi* (ch. 4, p. 1a) there is a description of eight flash-lock gates built by Wang Ling-Chün[14] in a canal in +1203, and rebuilt as *pan cha*[5] with 24 stop-log baulks each by Phi Hou-Yüan[15] in +1305. The term *tou mên*[2] is used for sluice-gates set up by Hsieh An[16] in connection with double slip-ways on the Shan-yang Yün-Tao section of the Pien or Grand Canal as early as +385.

[c] According to Chu Chün-Shêng (1), the word was originally a verb meaning to open a closed door.

[d] In ancient texts the word means a wooden partition.

[e] *Cha*[4] is found in *MCPT*, ch. 12, para. 1, the passage translated immediately below, i.e. +1086 referring to +1025, on pound-lock gates. Flash-lock gates on the Pien Canal are mentioned repeatedly in travel diaries, e.g. in +1169 the *Pei Hsing Jih Lu*, ch. 2, p. 13a, b, and in +1170 the *Ju Shu Chi*, ch. 1, pp. 3a, 7b, 8a, 10a. Some of these may well have been pound-locks. But the term is also clearly applied to sluice-gates, as in an irrigation system in Chiangsu c. +1200, by Kung Ming-Chih[17] in his *Chung Wu Chi Wên*[18] (ch. 1, p. 14a), and for the gates controlling the lake levels at Hangchow around +1270 (*Wu Lin Chiu Shih*, ch. 5, p. 22a, and *Mêng Liang Lu*, ch. 11, p. 14a). A similar usage occurs in connection with watercourses near Kweilin c. +1585 (*Chhih Ya*, ch. 2, p. 18b). *Chha*[6] is rather less common, but lock-gates in *Sung Shih*, chs. 91 ff., are often so termed (+1345). An early +13th-century occurrence is in *Fêng Chhuang Hsiao Tu*, ch. 1, p. 12a, which tells how the egregious Chu Mien[19] (cf. Vol. 4, pt. 2, p. 501) broke down lock-gates and bridges in transporting huge pieces of stone to the capital in +1123 for Hui Tsung's gardens and collections. Cf. the painting reproduced in Parias (1), vol. 1, pl. 59.

---

[1] 水門	[2] 斗門	[3] 陡門	[4] 閘	[5] 板閘
[6] 牐	[7] 隄牐	[8] 水牐	[9] 壩牐	[10] 懸門
[11] 鄧艾	[12] 時其鍾洩	[13] 高瑀	[14] 汪令君	[15] 皮侯元
[16] 謝安	[17] 襲明之	[18] 中吳記聞	[19] 朱勔	

occurrence,[a] rather rare before the Ming. Amidst much terminological fluctuation it is hard to ascertain whether precise technical differences were at any time implied by the different words, and if so what they were. Possibly *cha*, or at least *chha*, meant stop-log gates with permanently fitted windlasses, as opposed to the simpler forms in which each bar had to be manhandled in and out of its grooves.[b] In the absence of any qualification of the terms, flash-lock gates may generally be assumed (unless the context indicates pound-locks clearly), for in the description of the latter further adjectives enter in, as we shall see.

The invention of the pound-lock is a question of substantial importance in the history of civil engineering. Industry and commerce in palaeotechnic Europe were greatly affected by the simple and convenient device of arranging gates so close together as to admit only one or two barges, changes of water-level taking thus a minimal time to achieve, and losses of water from the upper levels being reduced to the smallest proportions.[c] This key invention, so simple and yet so vital for intensified waterborne traffic in hilly country, invites a comparative study of its appearance in different civilisations. The fact that foreign travellers on the Grand Canal from the seventeenth to the nineteenth centuries spoke only of flash-lock gates or double slipways[d] has seemed to some sufficient evidence that pound-locks were never developed in China. But this would be to fall into the trap of supposing that inventions once made were necessarily utilised in Chinese culture whether or not the need for them continued. In fact it is possible to show that the pound-lock actually originated in China earlier than anywhere else, but was little used in later times because the need for the device ceased as conditions changed. We may be able to suggest how it was that this happened.

The oldest example of a pound-lock or locks in China dates from the beginning of the Sung dynasty, and is connected with the name of Chhiao Wei-Yo,[1] Assistant

---

[a] One finds it in the description of Sung Li's water-supply gates (cf. p. 316 above), *Ming Shih*, ch. 85, p. 5*a*; as also in the account of the pound-locks on the approaches to the Ling Chhü (cf. p. 306 above) written *c.* +1585 (in *Chhih Ya*, ch. 2, p. 21*b*). But it also occurs in *MCPT*, ch. 24, para. 10, with reference to irrigation sluice-gates from the Pien Canal, and apparently even in a Thang inscription there (cf. p. 231 above).

[b] Wang Pi-Wên (*2*) has given us an interesting account of the regulation methods of the Chhing for designing locks and culverts. All the former were stop-log types.

[c] Shortage of water in the upper levels was a chronic malady of the medieval Chinese canals. Sometimes it was necessary to have recourse to batteries of square-pallet chain-pumps (cf. Vol. 4, pt. 2, p. 339), 'pedalling up the water' from the neighbouring land to let the boats pass, hence the expression *chhê shui*[2] (see *Sung Shih*, ch. 96, p. 5*b*, for a case in +1098, and pp. 10*b*, 11*a*, for another in +1120; cf. Chêng Chao-Ching (*1*), p. 210). It was said that 'water was like gold' (*loc. cit.* p. 11*b*), and gates were allowed to be opened only once in three days. The expression *kuei shui ao*,[3] here also found, may mean that at some pound-locks chain-pumps mounted on pontoons were permanently stationed, and water returned to the upper level at each passage.

[d] A graphic account of +1698 will be found in Lecomte (*1*), pp. 104 ff.; he disliked the flash-lock gates but rather admired the double slipways. So did Dr Dinwiddie, in a note which Ambassador Macartney (+1793) appended to his *Observations* (*2*), p. 269. There are many references to the flash-lock gates in his *Journal*, (*1*), pp. 169, 171, 173, etc. He speaks of 'sluices and bridges of singular workmanship and beauty'. Occurring, he says, at 'the distance of a few miles from one another', the sluices 'properly form locks of that distance. The boats collect in great numbers at the sluice, the valves open, and in a few minutes the whole fleet passes through; the flood-gates are then let down and the canal soon recovers its former level.' For Dinwiddie on flash-lock gates see Proudfoot (*1*), pp. 59 ff.

[1] 喬維嶽   [2] 車水   [3] 歸水澳

Commissioner of Transport for Huainan in +983, a man who deserves to be remembered. He was concerned with the barge traffic problem at the northern end of the Shan-yang Yün-Tao section of the Pien or Grand Canal between the Yangtze and Huai-yin, and it is interesting to find that his invention arose from a social cause, his exasperation with the thefts of tax-grain made possible by the high casualty-rate of ships crossing the double slipways. The *Sung Shih* says[a] that in +984:

Chhiao Wei-Yo also built five double slipways (lit. dams, *yen*[1]) between An-pei and Huai-shih (or, the quays on the Huai waterfront). Each of these had ten lanes for the barges to go up and down. Their cargoes of imperial tax-grain were heavy, and as they were passing over they often came to grief and were damaged or wrecked, with loss of the grain and peculation by a cabal of the workers in league with local bandits hidden near by.

Chhiao Wei-Yo therefore first ordered the construction of two gates (*tou mên*[2]) at the third dam along the West River (near Huai-yin). The distance between the two gates was rather more than 50 paces (250 ft.), and the whole space was covered over with a great roof like a shed. The gates were 'hanging gates' (*hsüan mên*[3]); (when they were closed) the water accumulated like a tide until the required level was reached, and then when the time came it was allowed to flow out.

He also built a horizontal bridge between the banks, and added dykes of earth with stone revetments to protect their (or its) foundations. After this was done (to all the double slipways) the previous corruption was completely eliminated, and the passage of the boats went on without the slightest impediment.

Such were the first pound-locks in the history of any culture.[b] Large enough to take several vessels at a time, they must have been somewhat similar to those depicted in a well-known illustration by Zonca in +1607 showing a lock basin on the canal which connected Padua with the River Brenta.[c] With hanging or 'portcullis' gates, they must have resembled more or less those drawn as flash-lock gates in another familiar picture, the canalised river of the Laurenziano Codex produced about +1475.[d] The arrangement implies pulleys, windlasses and perhaps counter-weights.

Contrary to an impression that has sometimes prevailed, Chhiao Wei-Yo's pioneering led to a great wave of interest in the new technique. This we know partly from an informative passage in Shen Kua's *Mêng Chhi Pi Than*, finished in +1086. In this book, so often quoted by us,[e] he says:[f]

On the Grand Canal (the Pien Canal)[g] in Huai-nan, double slipways (*tai*[4]) were built to prevent wastage of water. No one knows when this method was first invented. According to

---

[a] Ch. 307, pp. 1*b* ff., tr. auct.; cf. ch. 96, p. 1*b*. See also Chêng Chao-Ching (*1*), p. 207.

[b] So acknowledged by Skempton (4), p. 439, following valuable discussions between Professor Skempton and ourselves in 1955.

[c] Cf. Beck (1), p. 316; Parsons (2), p. 396; Forbes (17), fig. 625. See also Skempton (4), p. 451, and +16th-century locks in figs. 284, 285.

[d] Cf. Parsons (2), fig. 132; Forbes (17), fig. 626; Skempton (4), fig. 281. This codex (Ashburnham no. 361), entitled 'Trattato dei Pondi, Levi e Tirari', has MS. notes in Leonardo's own hand. It is now considered the work of Francesco di Giorgio (*fl.* +1464 to +1497) whom we meet with elsewhere in other connections; cf. Papini (1), vol. 2, pl. 292.

[e] Cf. Vol. 1, p. 135.

[f] Ch. 12, para. 1, tr. auct.; cf. Hu Tao-Ching (*1*), vol. 1, pp. 432 ff.; Fang Chi (*2*), p. 53. See also *Hsü Tzu Chih Thung Chien Chhang Pien*, ch. 104, p. 23*a, b*.

[g] At the southern end and intermediate points on the Shan-yang Yün-Tao section.

¹ 堰　　² 斗門　　³ 懸門　　⁴ 埭

tradition the Shao-po [1] (now Shao-po [2]) double slipway was built by Hsieh Kung [3] (i.e. Hsieh An, [4] premier of Chin, in +385).[a] But from the account of Li Ao [5] (who described his journey over this part of the canal in +809 in his) *Lai Nan Lu* [6] (Record of a Journey to the South)[b] it was still a plain waterway in Thang times without any haul-overs, so it seems impossible that this double slipway was built in the time of Hsieh Kung.[c]

In the Thien-Shêng reign-period (+1023 to +1031), the Transport Commissioner and Palace Intendant stationed at Chen-chou [7],[d] Thao Chien, [8] suggested that 'double gates' (*fu cha* [9])[e] should be built both to prevent waste of water and to save the labour of hauling the barges over. At that time the Director of the Ministry of Works, Fang Chung-Hsün, [10] and the Fine Craftsmanship Bureau Commissioner, Chang Lun, [11] were appointed as Chief and Deputy Industrial Transport Commissioners respectively, and were authorised to proceed with the construction (of double gates). They began with the locks (*cha* [9]) at Chen-chou.[f] (It was found that) the work of five hundred labourers was saved each year, and miscellaneous expenditure amounting to 1,250,000 (cash) as well. With the old method of hauling the boats over, burdens of not more than 300 *tan* of rice per vessel (21 tons) could be transported, but after the (double) gates were completed, boats carrying 400 *tan* were brought into use (28 tons), and later on the cargo weights increased more and more. (Nowadays) Government boats carry up to 700 *tan* (49½ tons), and private boats as much as 800 bags each weighing 2 *tan* (i.e. 113 tons).

From that time onwards, at Pai-shen, Shao-po, Lung-chou and Chu-yü, the double slip-ways were all disused, and one after another replaced (by double, i.e. pound-lock gates). The advantages of this have continued down to the present day.

Once during the Yuan-Fêng reign-period (+1078 to +1085) I myself passed through Chen-chou, and saw an overturned monument lying among dungheaps at the back of the River Pavilion. This stele bore an inscription by Hu Wu-Phing [12] about the (first) building of the Chen-chou (double) lock gates, entitled *Shui Cha Chi*.[13] It was not very detailed but it did record the affair.

Hu's inscription,[g] more poetical than precise, has in fact been preserved,[h] and we can share to some extent Shen Kua's disappointment at its lack of technical information. He starts with a concealed reference to Chhiao Wei-Yo, saying that in the first decades of the dynasty those concerned with canal traffic were becoming extremely dissatisfied with the double slipways worked by ox-whim capstans (*niu tai* [14]),[i] and the water wastage of the flash-lock gates which led in most years to the drying-out of the canal so that it looked like a thousand-*li* wall. But after Thao Chien insisted that double gates

[a] Cf. Chêng Chao-Ching (1), p. 197.　　　　[b] Cf. Vol. 2, pp. 452, 494.

[c] In spite of Shen Kua's scepticism, there is good ground for believing that Hsieh An's nephew, the general and philosopher Hsieh Hsüan, [15] built 7 double slipways on the Ssu River section of the Pien Canal in +384. Cf. p. 363.

[d] Mod. I-chêng [16] on the Yangtze upstream of Yangchow and Kuachow (cf. p. 309).

[e] Etymologically, *fu* is the lining, the *doublure*, of a garment.

[f] Presumably those at Kuachow connecting the canal with the Yangtze, which varied considerably in level.

[g] His primary name was Hu Su.[17] An official meritorious for building schools and conservancy works, he was also a naturalist with views on earthquakes, and took an interest in alchemy, which he studied with a monk.

[h] In ch. 35 of his collected works, reproduced by Hu Tao-Ching, *loc. cit.* p. 435.

[i] Witness also an eloquent speech by Chia Tsung [18] in +1018 (*Sung Shih*, ch. 96, p. 1b).

[1] 召伯	[2] 邵伯	[3] 謝公	[4] 謝安	[5] 李翺
[6] 來南錄	[7] 眞州	[8] 陶鑑	[9] 覆閘	[10] 方仲荀
[11] 張綸	[12] 胡武平	[13] 水閘記	[14] 牛埭	[15] 謝玄
[16] 儀徵	[7] 胡宿	[18] 賈宗		

PLATE CCCXXXVI

Fig. 815. Chariot and horseman crossing a pile-and-beam bridge with balustrade; scene on a moulded brick of Han date from Chhêngtu, Szechuan (Anon. *22*). Cf. pp. 150 ff.

PLATE CCCXXXVII

Fig. 816. The Fêng R. bridge near Sian to the north-west, a pier-and-beam structure of Han type (orig. photo., 1964). Each pier consists of three pairs of pillars built of stone cylinders like threshing-rollers, and based on three discoidal roller-mill base-plates (hidden in the photograph by later added concrete), these in turn supported on cypress piling.

Fig. 817. Detail of the Fêng R. bridge (orig. photo., 1964). Each pair of columns is capped by an oblong stone plate, bearing two transverse wooden beams upon which the nine longitudinal beams of the deck are laid.

PLATE CCCXXXVIII

Fig. 818. A wooden beam bridge the centre span of which is topped with a pavilion. This type of structure, humped slightly to facilitate the passage of craft under the central span, was perennial through the ages in China; the present example is from a painting by the eminent Southern Sung artist Hsia Kuei (+1180 to +1230). Anon. (32).

PLATE CCCXXXIX

Fig. 819. A stone beam bridge near Hangchow (photo. Mirams).

PLATE CCCXL

Fig. 820. Combination of nine stone beam spans and an arch; the Thai-Phing bridge over the Grand Canal extension near Shao-hsing in Chekiang (orig. photo., 1964). The arch has a particularly elegant T-shaped double staircase giving access from the towpath.

(a)   (b)

Fig. 821. Megalithic beam bridges of south-east China; the Chiang-Tung or Chhien-Tu bridge across the Chiu-Lung R. upstream from Amoy in Fukien (photos. Ecke, 3). Built +1214 to +1237; length 1100 ft. Cf. pp. 153 ff. and Table 66.
a   Stone beam span at the sixth pier, from the north-east.
b   Another pier, from upstream (north-west).

PLATE CCCXLI

Fig. 824. Horizontal-cantilever and strainer-beam bridge of six spans over the Lu-Shui R. at Li-ling in Hunan (photo. Boerschmann, 3a). Cf. Fig. 823 on p. 163.

PLATE CCCXLII

Fig. 825. Horizontal-cantilever bridge of four spans over the Lin-chhi R. north of San-chiang in northern Kuangsi (Anon. *37*). This structure, the Chhêng-Yang bridge, is notable for the elaboration of features also used elsewhere, the crowning of the piers by pavilions with many-tiered roofs, and the provision of a spacious roofed gallery over the deck.

PLATE CCCXLIII

Fig. 826. The great bridge at Khaifêng in the scroll of Chang Tsê Tuan, Chhing-Ming Shang Ho Thu (Going up the River at the Spring Festival), painted about +1125. Photo. Chêng Chen-To (3). So far as is known, there are now no examples of this multi-angular soaring cantilever construction extant in China, but in pre-Ming times there seem to have been many. A towpath gallery can be seen under the bridge along the further abutment, and the stern-sweep of a great barge which has just passed under the bridge appears on the near side of the river. Cf. Fig. 823 and p. 165.

PLATE CCCXLIV

Fig. 827. Han stone relief of a pavilion built out over a lake and supported by corbel bracket-arms on the cantilever principle (*thi chhiao*). Preserved at the Temple of Confucius at Chhüfou in Shantung (orig. photo., 1958). See p. 166.

PLATE CCCXLV

Fig. 829. An arch bridge at Khunshan (Quinsan) in Chiangsu, showing the ends of the shear-walls in the piers (photo. Mirams). See pp. 167 ff.

Fig. 830. A small arch bridge at Suchow in Chiangsu, illustrating the degree of deformation which structures with shear-walls can stand without collapsing (photo. Sirén).

PLATE CCCXLVI

Fig. 831. The San-Hsing bridge at Wan-hsien in eastern Szechuan, a tall arch spanning a seasonal torrent and capped with a covered gallery (photo. F. T. Smith, in Carey, 2). A trace of pointedness is detectable.

Fig. 832. Bridge with three pointed arches north of the Sword-gate Pass (Chien-mên Kuan) between Szechuan and Shensi (orig. photo., 1943).

PLATE CCCXLVII

Fig. 834. The Pao-Tai bridge (+806) alongside the Grand Canal south of Suchow in Chiangsu (orig. photo., 1964). It crosses an arm of the Thai-Hu Lake, and aroused the admiration of many foreign travellers from Marco Polo's time onwards.

PLATE CCCXLVIII

Fig. 835. The oldest segmental arch bridge in the world; the An-Chi bridge across the Chiao Shui R. near Chao-hsien in southern Hopei, built by Li Chhun about +610 (photo. Mao I-Shêng, 1). Span, 123 ft. See pp. 175 ff. and Table 67.

PLATE CCCXLIX

Fig. 836. One of the Renaissance segmental arch bridges of Europe; the Castelvecchio bridge at Verona, built +1354 to +1357. Longest span, 160 ft. From Uccelli (1), p. 288, fig. 65.

Fig. 837. A modern railway bridge with arches in the spandrels, like Li Chhun's. The bridge at Salcano in Italy (now Slovenia, Yugoslavia), near Gorizia north of Trieste, built c. 1900; span just under 280 ft. From Uccelli (1), p. 694, fig. 92. This graceful segmental arch was destroyed in the first world war but afterwards reconstructed. During the twenties and thirties many beautiful variations of the same design were built all over the world in ferro-concrete.

PLATE CCCL

Fig. 840. A Chinese segmental arch bridge of the +12th century, the Yung-Thung bridge near Chao-hsien in Hopei, built by Phou Chhien-Erh in +1130 under the J/Chin dynasty. Span 85 ft. From Mirams (1).

Fig. 841. The Lu-Kou bridge across the Yung-ting R. a short distance west of Peking, at the small town of Lu-kou-chhiao, to which it gives its name. About 700 ft. in length, it is carried on 11 segmental arches with an average span of 62 ft. each (photo. Sirén, 1). This was the bridge which Marco Polo thought the finest in the world, partly because of its breadth and the elaborate carving of its stone balustrade; in modern times foreigners in China have called it after him. In 1937 it was the scene of incidents which started the Sino-Japanese hostilities preluding the second world war. The bridge is still in heavy use for general traffic, though built in +1189 under the J/Chin dynasty as part of their improvements of the metropolitan area.

PLATE CCCLI

Fig. 842. The earliest forms of suspension bridge; a double rope bridge at Lo-ta (Lo-ndu) crossing the Mekong in Nakhi country (northern Yunnan), from Rock (1), pl. 162. Here the separate cables are arranged so that their arrival points are much lower than their points of departure, crossing being effected by the attachment of men or animals to tubular runners of bamboo or wood greased with butter or oil. The runners are then returned by means of a separate cord. See pp. 184 ff.

PLATE CCCLII

Fig. 843. The suspension bridge approximating to a cable railway; a crossing on the Tibetan border. The two ends of the bamboo cable are at approximately equal heights on each side of the river, and the tubular runner carries a kind of rope cradle to support the passenger, as well as cords on each side to pull it to and fro. The photograph was taken in the course of a Chinese scientific expedition to the border country about 1947 led by Dr Tsêng Chao-Lun. The botanist, Dr Phei Li-Chhün, is beginning the crossing, advised by one of the local tribesmen.

Fig. 844. A vine or creeper bridge characteristic of Yunnan and the Burma border. Two bamboo cables form the handrails and a third the deck or tread-rope, then creepers are thickly plaited so as to form a walking way of V- or U-shaped cross-section.

PLATE CCCLIII

Fig. 846. The bamboo-cable suspension bridge at Kuanhsien in Szechuan, crossing the Min Chiang (cf. Fig. 884). One of the bridge-head buildings (west side), from which the hand-rail cables can be seen issuing on the right (photo. Boerschmann, 3a). Cf. pp. 192 ff. and Fig. 884 on p. 290.

Fig. 847. Four of the capstans for the rail cables within the bridge-head building on the east side (orig. photo., 1958). The deck cables themselves are tightened by winches underneath the floor-boards

PLATE CCCLIV

Fig. 848. General view of the An-Lan suspension bridge at Kuanhsien, with its eight major spans, total-ling rather over 1050 ft. in length (longest span, 200 ft.), taken from the hill overlooking it to the east (orig. photo., 1958).

PLATE CCCLV

Fig. 849. A closer view from nearer water level, showing the nature of the decks and the capped trestle piers (orig. photo., 1958).

Fig. 850. View on the Kuanhsien An-Lan suspension bridge looking eastwards along the eastern-most span (orig. photo., 1958). The fastenings of the hand-rail cables and deck planks are clearly seen.

PLATE CCCLVI

Fig. 851. Under surface of the An-Lan suspension bridge; the fourth span from the east, photographed from the island (orig. photo., 1958).

Fig. 852. The Chi-Hung iron-chain suspension bridge in one of the gorges of the Mekong R. (photo. Popper, FZ 287). With its bridge-head anchor-houses towering out of the swirling waters, the taut catenary of its deck, and the temples to the tutelary deities of the place visible on the left, it again affords an example of the beauty of traditional Chinese bridge-building. Cf. pp. 193 ff. and Table 68.

PLATE CCCLVII

Fig. 853. A small single-span iron-chain suspension bridge over a tributary of the Chin-Sha Chiang, at Chi-tsu Shan near Pin-chhuan in Yunnan. Here the hand-railing is all of woodwork.

Fig. 854. A traditional iron-chain suspension bridge of the Chhing period, the Chin-Lung or Tzu-Li bridge (Dsi-Li Shu-ĕr Ndso in Nakhi) linking Lichiang with Yungpei across the Chin-Sha Chiang (the Yangtze) in Yunnan (from Rock (1), pl. 111). Eighteen chains carry the road across a single span of 328 ft. The great river is here running at an altitude of 4600 ft. and the masonry of the bridge-heads is built up, as can be seen, to withstand an annual rise and fall of some 60 to 70 ft. or more.

PLATE CCCLVIII

(*a*)

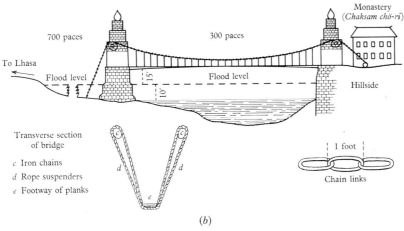

(*b*)

Fig. 856. The invention of the flat deck suspended from the catenary, the suspension bridge across the Brahmaputra at the Chak-sam-chö-ri lamasery, Tibet, first built in about +1420.

a   Photograph from the south, *c.* 1900. The river having flooded the flat ground beyond the island, the bridge was out of use and the deck had been removed.

b   Drawing made in 1878 when the bridge was in full use.

Both taken from Waddell (1).

PLATE CCCLIX

Fig. 858. A more or less imaginary drawing of the Lan-Chin bridge near Ching-tung in Yunnan, over the Mekong R. (actual span *c.* 250 ft.), set among alpine scenery, in an architectural work of the European eighteenth century, J. B. Fischer von Erlach' *Historia Architektur* (+1725), pl. 15. The oldest suspension bridges in Europe date from the two subsequent decades.

PLATE CCCLX

Fig. 864. Irrigated fields depicted in a Han tomb model of black pottery from Phêngshan (Chekiang Provincial Museum at Hangchow, orig. photo., 1964). From the reservoir on the left, identifiable because of two fishes carved in the clay (not well visible here), a channel runs to the right between four fields in which are piles of rice-straw. It is led under the low embankments by two culverts, perhaps sluice gates.

PLATE CCCLXI

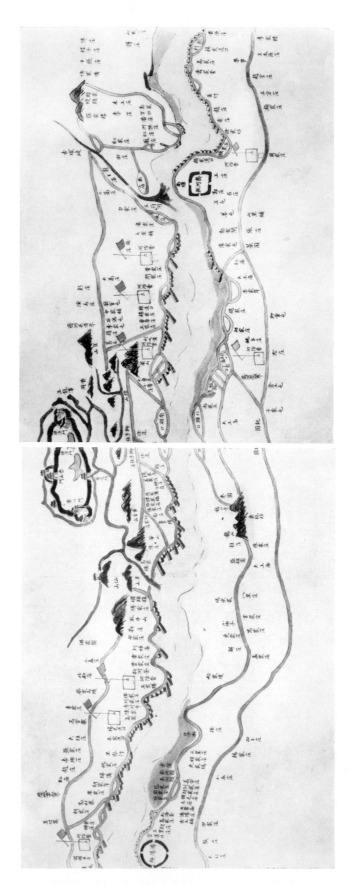

Fig. 867. Part of the lower Yellow River conservancy shown in a scroll map of late Chhing date formerly in the collection of Mr Rewi Alley. We are looking down on the northern course of the great river, course ⑧, in Shantung, from the north-west, with the large city of Chinan on the further side and two smaller cities on the nearer side, to the right Chhi-ho and downstream to the left Chiyang. Dykes and embankments, often as many as three in parallel, are shown as dark bands, and especially along the further (south-eastern) shore they generate an abundance of masonry groynes, some long, some short, designed to prevent bank erosion. Opposite Chhi-ho a tributary, the Wang-fu Ho (now called the Loshui) falls into the river. The depots (Ho Fang Ying) of the Conservancy Authority are shown as square enclosures with a symbolical *phai-lou* in front and a flag on a flagpole; no less than eight of them can be seen along this stretch, mostly on the further shore. Photos. Alley, kindly communicated. See pp. 220, 232 ff.

PLATE CCCLXII

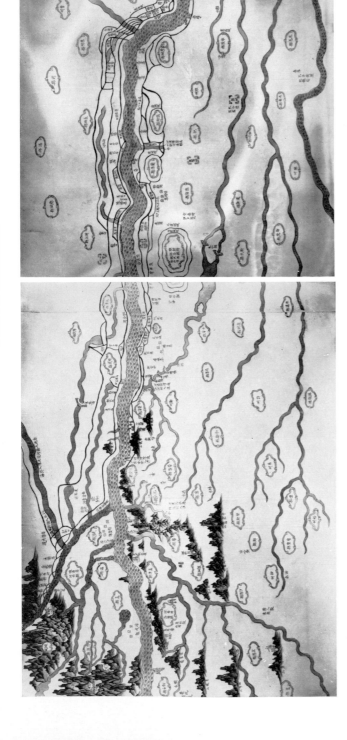

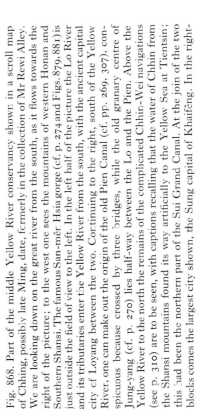

Fig. 868. Part of the middle Yellow River conservancy shown in a scroll map of Chhing, possibly late Ming, date, formerly in the collection of Mr Rewi Alley. We are looking down on the great river from the south, as it flows towards the right of the picture; to the west one sees the mountains of western Honan and Southern Shansi. The famous San-mên Hsia gorge (cf. p. 274 and Figs. 879, 881) is just outside the field of view to the left. In the left half of the picture the Lo River and its tributaries enter the Yellow River from the south, with the ancient capital city of Loyang between the two. Continuing to the right, south of the Yellow River, one can make out the origin of the old Pien Canal (cf. pp. 269, 307), conspicuous because crossed by three bridges, while the old granary centre of Jung-yang (cf. p. 270) lies half-way between the Lo and the Pien. Above the Yellow River to the north the remains of the complicated Chhin-Wei navigations (see p. 310) are to be seen, with captions recalling that the water of Chhin from the Shansi mountains found its way artificially to the Yellow Sea at Tientsin; this had been the northern part of the Sui Grand Canal. At the join of the two blocks comes the largest city shown, the Sung capital of Khaifêng. In the right-

hand half of the picture we see in the south the many almost parallel rivers that flow through Honan and Anhui to the Huai R.; and noticeable is the fact that the Yellow River accompanies them, that is to say, it is flowing in its great southern stream on course ⑦, passing such towns as Yü-chhêng, Hsüa-i, etc. Directed north-eastwards diagonally is the northern bed, course ⑧, smaller or almost dry when the map was made, and with an opening barred, or at least controlled, by six successive embankments; yet now again the main bed (see Table 69). As in Fig. 867 the Yellow River is seen almost throughout its length ribbed, cabined and confined within massive dykes, in some places as many as eight in parallel. These are denoted by thin black lines, joining occasionally with thin ladder-like thickenings along the river's edge; this difference in convention seems to indicate earthworks as opposed to masonry, judging by a caption just to the right of the junction with the north-eastern course. All cities are indicated by walls and gates, and a number of forts, passes, and places of historical (especially Confucian) interest, are shown. Photos. Alley, kindly communicated. Cf. pp. 220, 232 ff.

PLATE CCCLXIII

Fig. 871. Erosion in the loess country of North-west China. An air photograph of the north bank of the Yellow River east of Lanchow (photo. Fisher, 1).

PLATE CCCLXIV

Fig. 873. Typical rice-field terracing in Szechuan (Anon. *26*).

PLATE CCCLXV

Fig. 874. The oldest of three 'six-character teachings' at the temple of Li Erh-Lang at Kuanhsien (orig. photo., 1958). Placed in a commanding position on the monumental stairway, it reads: *Shen thao than, ti tso yen*—'Dig the channel deep, and keep the dykes and spillways low'. See pp. 250,294.

PLATE CCCLXVI

Fig. 877. The genius for man-power organisation illustrated by a photograph of work at the Ming Tombs Reservoir Dam in June 1958 (orig. photo.). Although two belt conveyors can be seen in the middle of the picture, and a few bulldozers on and near the crest of the barrage, with cranes, and inclined planes for light railways, in the background, and although several lines of larger light railways were bringing up material hauled by steam locomotives outside the picture to the right; a large proportion of the total work was being done, in fully traditional style, by manual labour. The successful use of this necessarily involves a perfection of assembly control and spontaneous discipline. Cf. p. 144.

PLATE CCCLXVII

Fig. 878. The Ninghsia irrigation system, first begun in the Chhin period, shown in a scroll map of Chhing, possibly late Ming, date, formerly in the possession of Mr Rewi Alley. The view is taken from the east, and the Yellow River is flowing northwards, i.e. from left to right. The whole irrigated area, some 100 miles in length, is circled to the west and north by the Alashan Mountains and the Gobi Desert, to the east, this side of the river, by the Ordos Desert. Bounding it on the south, the Great Wall approaches from the east, as seen in the bottom of the picture, enclosing several small walled towns such as Hêng-chhêng; then on the right it continues beyond the river across the end of the area, guarding Phing-lo city and resting on the Alashan (off the picture at the top left it continues down to the Lanchow salient). Outside the wall to the right (north) of Hêng-chhêng encampments of Mongol yurts are seen. The river enters from the left, flowing past the towns of Chung-wei and Kuang-wu into the defile called Chhing-thung Hsia (Green Bronze Gorge), here ornamented by many Buddhist *stūpas*. Immediately on emerging, it is made to give rise on the left bank to three main lateral derivate canals, the Thang-lai Chhü, the Han-yen Chhü and the Hui-nung Chhü, all of which rejoin the Yellow River about 100 miles downstream. Three much smaller and shorter ones take off from the right bank. Four important bridges

PLATE CCCLXVII

cross the left canals soon after their origin, and about forty other named bridges are marked. Roads and paths are indicated by thin dotted lines. The most westerly canal, the Thang-lai, waters the capital city of Ninghsia itself. Many subsidiary channels are taken off from all of them, and six aqueducts convey some of the water of the Han-yen across the main course of the Hui-nung. Some half-a-dozen important sluice-gate sites are marked, and the four chief bridges already mentioned may well have been sluices also. Between and among the canals a communicating system of 'lakes' is shown in darker colour; these presumably indicate low ground liable to flooding in seasons when the river is very high. One of the largest of these bears the name 'Thanks to the Officials Lake' (Hsieh Kuan Hu). A large number of forts, villages, temples and pagodas are also marked. Each of the villages (*pao*) has a character inside it indicating to which city's jurisdiction it belongs. Near the derivations one can make out four Dragon-King Temples (Lung Wang Miao) dedicated to the god of the waters, so important for hydraulic projects; and beyond the wall, on the far right, where the river passes out between the Alashan mountains, there is appropriately a temple of the God of War (Kuan Ti Miao). Photos. Alley, kindly communicated. Cf. p. 272.

PLATE CCCLXVIII

Fig. 880. A stretch of one of the trackers' paths or half-tunnel towpaths cut in the rock-faces of the Yangtze gorges, here at the defile commonly known as Wind-box Gorge, above I-chhang (photo. Popper, RO/109/13). Those at San-mên Hsia on the Yellow River are closely similar. Cf. pp. 23, 277.

PLATE CCCLXIX

Fig. 885. The Min River and Kuanhsien irrigation system head-works, a view looking upstream taken from the Jade Rampart hill (from Boerschmann (2), pl. 12, fig. 2 (3a), fig. 119). In the background, the heights of the Pa-lang Shan mountains. In the middle distance the Han-family island, and on the right the Thousand-foot Dyke; then the main division-head (the Tu-Chiang Fish Snout) and the suspension bridge. In the foreground the Diamond Dyke (Chin-Kang Thi) separating the Nei Chiang on the right from the Wai Chiang on the left, cut through by the 'water-level adjusting spillway' (Phing-Shui Tshao). On the right the roofs of the temples of Li Erh-Lang and Lao Tzu can be seen among the trees. See pp. 288 ff.

Fig. 886. The Min River and Kuanhsien irrigation system head-works, a view looking downstream taken from the hillside above the temple of Li Erh-Lang, the roofs of which can be seen nestling among the trees (orig. photo., 1958). The Nei Chiang is disappearing to the left round the Phoenix Nest Cliff of the Jade Rampart Hill into the Cornucopia Channel. Just beyond it there are the trees of the Li-Tui Hill hiding Li Ping's own temple. Below them the Jen-Tzu Thi spillway is functioning, and nearer the spectator the broader expanse of the Fei-Sha Yen spillway, also in function, can be made out. Below him the upper part of the Nei Chiang is invisible because of the woods, but the Wai Chiang can be seen for a long stretch from right to left. See Fig. 884 on p. 290.

PLATE CCCLXX

Fig. 887. The main division-head (Yü Tsui, or Tu-Chiang Yü Tsui) seen from the point where the sus-pension bridge crosses the artificial peninsula or island of the Diamond Dyke (orig. photo., 1958). Across the river, the Thousand-foot Dyke. See Fig. 884 on p. 290.

Fig. 888. Model of the Kuanhsien headworks in the exhibition-room of the Authority at the back of Li Ping's temple on the Li-Tui hill (orig. photo., 1958). The Min river can be seen dividing near the suspen-sion bridge into the Nei Chiang on the right and the Wai Chiang on the left. The Cornucopia Channel cutting is well shown between the old city walls to the right and the temple of Li Ping on the Li-Tui hill to the left. Li Erh-Lang's temple appears to the right of the suspension bridge. On the extreme left the new intake of the Sha-Kou Ho and Hei-Shih Ho derivatory canals is indicated, while at the extreme right at the bottom we see the new sluice gates of the Tsou-Ma Ho derivatory canal. Cf. Fig. 884.

PLATE CCCLXXI

Fig. 889. Diagrammatic map of the Kuanhsien irrigation system painted on a wall of one of the buildings of the temple of Li Erh-Lang (photo. Richardson, 1942). The division of the Min River into the Nei Chiang and the Wai Chiang is shown at the top right-hand corner; the capital city of Chhêngtu itself is seen within a U-shaped confluence of two canals about two-thirds of the way along the map to the left and about halfway up. The title reads: 'Bird's-eye view of the Tu-Chiang Yen Irrigation Area in Szechuan.'

Fig. 890. Cornucopia Channel, looking upstream from the terrace of Li Ping's temple on the Li-Tui hill (orig. photo., 1958). The Phoenix Nest Cliff towers on the right; on the left the Flying Sands spillway is strongly overflowing.

PLATE CCCLXXII

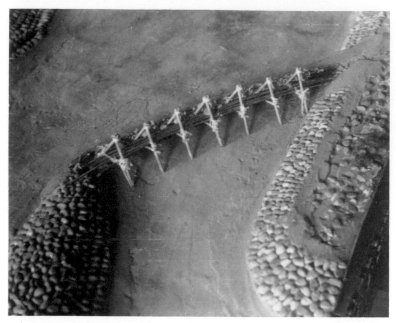

Fig. 892. One of the temporary *ma chha* cofferdams used seasonally at Kuanhsien for clearing and dredging the beds of the Nei Chiang and the Wai Chiang; a model in the exhibition-room of the Authority (orig. photo., 1958).

Fig. 893. The set of three iron bars (*wo thieh*) fixed in the bed of the Nei Chiang just opposite the Flying Sands spillway as a guide for the standard depth of excavation required (photo. Lowdermilk, 1943).

PLATE CCCLXXIII

Fig. 894. The 'Trimetrical Classic of River Control', an inscription in the temple of Li Erh-Lang at Kuanhsien (orig. photo., 1958). The calligraphy is that of Wên Shen, Governor of Chhêngtu in 1906, but the text may be as old as the +13th century. A translation is given on p. 295.

Fig. 895. An empty gabion of spheroidal shape beside one of the piers of the suspension bridge at Kuanhsien (orig. photo., 1958). Mr Li Chün-Chu, one of the engineers-in-charge, stands beside it. Cf. pp. 295, 321 ff., 339 ff. and Figs. 913, 914.

PLATE CCCLXXIV

Fig. 896. The image of Li Ping in his votive temple, the Fu-Lung Kuan, on the Li-Tui hill beside the Cornucopia Channel (photo. Boerschmann). The tablet reads: *Kung Chao Shu Tao* 'His achievement is the glory of the province of Szechuan'. See p. 296.

Fig. 897. The main courtyard of the temple of Li Erh-Lang at Kuanhsien (orig. photo. 1958). A picture of this place crowded with people on the occasion of the ceremonial opening of the Nei Chiang coffer-dam in 1943 is given in Lowdermilk (7).

PLATE CCCLXXV

Fig. 900. 'The Imperial Equipage on a Visit of Inspection to the Pien Canal', a work by an unknown Sung painter (from Anon. *32*). The incident occurred in the time of the emperor Chen Tsung, in +1006, on the occasion of a flood when the canal nearly broke its banks. The imperial entourage is seen on the left, with punner gangs and a chantyman on the right, while other men are bringing fill materials from various directions. The people who worked all night to save the dykes, adding five inches to their height, were personally rewarded with presents of money by the emperor, and some who had been drowned were buried at the public expense. This picture is one of the same set as that in Fig. 4 of Needham, Wang & Price (1).

PLATE CCCLXXVI

Fig. 901. General map of the Grand Canal in its final form (from Chêng Chao-Ching, *1*). From north to south one may trace its course as follows: Peking, Tientsin, Tê-chou, Lin-chhing, Crossing of the Yellow River near Tung-a, Chi-ning, Han-chuang (at the southern tip of the long Chao-yang Lake, and the crossing of the N–S Chin-phu Rly.), then Thai-erh-chuang, followed by the crossing of the E–W Lung-hai Rly. near Phei-hsien, then Huai-yin and the crossing of the Huai R., then Yangchow, Crossing of the Yangtze at Kuachow, Chen-chiang, the crossing of the E–W Hu-ning Rly., Suchow and finally Hangchow. The less important extension to Shao-hsing and Ningpo, though certainly in part as old as the +17th century, is not shown. See pp. 306 ff.

PLATE CCCLXXVII

Fig. 902. The crossing-point of the Grand Canal and the Yellow River, shown in a scroll map of late
Chhing date formerly in the collection of Mr Rewi Alley. This chart adjoins the right-hand edge of the
right-hand section given in Fig. 867. Again we are looking down on the river from the north-west so that
it flows across the page from right to left. The square city at the bottom on the right is Shou-chang,
the round one at the top on the left is Old Tung-a (today the name Tung-a is given also to a previously
smaller place, Thung-chhêng-i, on the hither bank). The Grand Canal enters from the bottom, splitting
into three approaches; the right one, silted up, is marked 'old mouth', the left ones are stated to have
lock-gates. Across the great river the canal starts again, also guarded with lock-gates, and passes off at the
top into the Tung-phing Lake (not shown). To its right there is a depot of the Conservancy Authority
and a few groyne works. The canal is drawn tapering off at both ends, but this was the convention of the
draughtsman, whose mind was set on the River. Opposite Old Tung-a there is a mariner's beacon drawn
on the hill of Chiang-chuang. The dark bands, as in Fig. 867, indicate the course of the protective dykes.
Photo. Alley, kindly communicated.

PLATE CCCLXXVIII

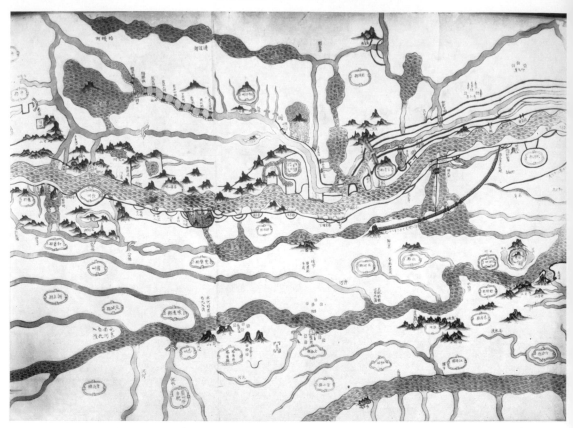

Fig. 903. The Grand Canal, the lower Yellow River and the lower Yangtze shown in a scroll map of Chhing, possibly late Ming, date, formerly in the collection of Mr Rewi Alley. It gives a schematic and horizontally stretched view of the provinces of Anhui and Chiangsu from the south, and the left edge of its left half adjoins the right-hand edge of the right half of the chart in Fig. 868. From top to bottom the three great streams are, (a) the Yellow River, on its southerly course (7) taken from +1354 to 1855, discharging through the Huai River estuary past Huaiyin (marked on the map by another of its names, Chhing-ho), (b) the Huai River itself, spreading out into the stomach-shaped Hung-Tsê Hu (lake) before meeting the Yellow River opposite Huaiyin, and (c) the Yangtze, flowing out to the sea north of Shanghai. Dykes are shown as continuous black lines of varying thickness, and many cities, temples, pagodas, beacons, etc. are marked, but much detail is lost in the reproduction.

The course of the Grand Canal can, however, be followed clearly. It comes in at the top left-hand corner, looping down towards the Yellow River past the Hui-Shan Hu (lake), through many flash-lock gates (cha) of which nine are marked and named, from Han-chuang to Thai-erh-chuang. After passing Phei-hsien it has on its left (east) bank seven named 'flying dykes' (pa, see p. 362) some of which open into the sumps still marked on modern maps, Lo-Ma Hu and Huang-Tun Hu. After this there follows a complex anastomosis with the waters of the Yellow River which captured the area around Su-chhien (the island city) when it embarked on its southern course here shown; just above Su-chhien a notable beam bridge is depicted in semi-perspective. The Grand Canal then recovers its identity and proceeds across the division of the picture paralleling the Yellow River amidst five long dykes until it suddenly takes a turn to the right and falls into the Yellow River just before Chhing-ho (Huaiyin). We have thus traced it for about 200 miles. The whole of the section so far described, h, i, in Table 70, followed, when rebuilt in +1283, the alignment of the Huan Kung Kou, a smaller canal dating from as far back as +369, and from this it gained its name.

Next the Grand Canal turns slowly due south, making straight for the Yangtze along the ancient way known as the Shan-yang Yün-Tao, j in Table 70, based again on an alignment of about +350, which had replaced the more winding −5th-century Han Kou canal. The distance now covered is about 115 miles, passing Huai-an, Pao-ying, Kao-yu and Yangchow. For some 35 miles the Canal has the Kao-Yu Hu (lake) on its right, but from this it was separated by a dyke built in +1007 (cf. Fig. 904). Five important named lock-gates are placed near the junction with the Huai and Yellow Rivers; and six more named ones regulate the double debouchment into the Yangtze beside the city of Kuachow. A prominent pagoda, the Chang Fêng Tha, doubtless a mariner's light or marker, is shown at this S-shaped bend. In addition to the lock-gates, this stretch of the Canal is equipped with eighteen named sluice-gates (cha) in its banks to relieve high water-levels, nineteen flying dykes (pa) for sudden release of heavy flood water, and several bridged outlets (chhiao) normally open and flowing, probably with weirs.

PLATE CCCLXXVIII

Finally the Grand Canal starts again on the Yangtze's southern bank directly opposite the main N–S line of the Shan-yang Yün-Tao, beside the city of Chen-chiang, where a lock-gate is marked. To reach this, traffic sailed past Chinshan Island (cf. p. 241) in the Yangtze. But of the southern portion of the Canal, the Chiang Nan Ho, *l* in Table 70, we see no more than the beginning.

The wealth of further detail must pass without adequate comment. In the west, however, the tributaries of the Huai R. are well drawn, and the cities near them, some of which, such as Shouchow (Shou-hsien), are referred to in the text (p. 281). In the east the bewildering maze of canals between the Shanyang Yün-Tao and the sea is striking, and it should be noted that most of these were artificially constructed, often as early as the San Kuo period, partly for easing the ever-intractable problem of the drainage of the Huai valley, but principally for transporting inland the large quantities of salt which were always made along the coast. Thus the main E–W canal just north of the Yangtze was called the Shang Yün-Yen Ho; it met another, the Yün-Yen Ho itself (shown here as passing under three great arch bridges), beside the city of Thaichow. Another such canal, the Yün-Yen Hsiao Ho, is marked on the map as running parallel with, and just south of, the Yellow River, all the way from the sea to the Grand Canal. More prominent than any of these, however, is the N–S salt artery which runs parallel with the Shan-yang Yün Tao, nearer the sea and inside an important dyke or sea-wall about 45 miles long (afterwards extended to just over 180 miles), the Fan Kung Thi, built by Fan Chung-Yen in +1027. This is the Yün-Yen Chhuan-Chhang Ho (Saltern-Connecting Transport Canal), dominated at its northern end by Salt City (Yen-chhêng), as the map shows, and able to discharge excess water into the sea through eighteen sluice-gates.

The city in the furthest south-east corner of the chart is Thungchow (now called Nan-thung), the birthplace, I add, as in private duty bound, of my collaborator Wang Ling. Just above and to the left of it some derrick-like structures are drawn in, and marked 'the Eight Beacons', while just below and to the right of it there is a mariner's mark pagoda on the hill of Chao-shan. Across the Yangtze, also by the mouth, near modern Kao-chhiao, a lighthouse or guard-post called Ju-shan Tun is shown. Further upstream, beyond Chenchiang and the Grand Canal crossing, a hill on the south bank named Yen-Tzu Chi (Swallow's-Nest Bluff) is to be seen; this is a well-known markpoint in navigation just north of the birthplace of my collaborator Lu Gwei-Djen, the capital city of Nanking, the position of which on the chart is thus fixed.

A map or maps very like this one, though not identical with it, must have been at the disposition of Gandar for his monograph (1) on the Grand Canal; cf. his pls. 12, 13, 14, 15 and especially 16. There are differences in all such maps concerning the lakes in northern Chiangsu province, but this is not necessarily due to inaccurate geography; it is rather because the land is so low-lying that many lakes have been impermanent, depending on the hydrographic conditions, and the engineering works undertaken, which changed decade after decade. Photos. Alley, kindly communicated.

PLATE CCCLXXIX

Fig. 904. An engraving of the Grand Canal running along the edge of the Kao-Yu Lake but separated from it by an embankment to facilitate traffic in bad weather. From Nieuhoff (+ 1665), as his embassy saw it in + 1656. 'In former times', he wrote (p. 148), 'all vessels coming from Nanking and the Yangtze, and bound for Peking, were obliged to wait below the walls of this city (Kao-yu) in stormy or foggy weather. But as these delays were very vexatious for trade, it was thought good, to avoid the perils of the lake, to build along its eastern side a canal 60 stadia long; and this is done with square white dressed stone blocks of such a size that one cannot imagine where they were got from, seeing that in the neighbouring provinces there are no great stone hills or quarries.' Note the cross-sections of the dykes indicated by the artist in the foreground. Cf. p. 314.

PLATE CCCLXXX

Fig. 905. A modern analogue of the story of Pai Ying (+1411); engineering students of Chhinghua University learning from the experience of an old countryman skilled in irrigation, c. 1961 (photo. Anon. 68). The rapids of a mountain river can be seen on the left, and behind the group is a great noria of traditional type (cf. Vol. 4, pt. 2, pp. 356 ff.). The perennial Chinese way of paying attention to sagacious empirical experience independent of book-learning has been greatly extended and emphasised in contemporary life. Cf. p. 315.

PLATE CCCLXXXI

Fig. 907. The Chhien-thang sea-wall near Hai-ning (orig. photo., 1964). Near this spot stands a hexagonal six-storeyed pagoda dating from the Chhien-Lung reign-period, and a 'platform for watching the tidal bore' (cf. Vol. 3, pp. 483 ff.). In the foreground can be seen one of the remaining cast-iron geomantic bulls placed along the wall in +1730.

Fig. 913. A weir of gabions near Chhêngtu in Szechuan (orig. photo., 1943). The plaited bamboo cylinders filled with stones can be built into any desired formation; in structures such as this the successive layers are covered with bamboo matting. Cf. pp. 295, 321 ff., 339 ff. and Figs. 895, 914.

PLATE CCCLXXXII

Fig. 915. Hsia Yü Chih Shui Thu; a Ming painting by Ma Wan (Ma Wên-Pi) dated +1417, on the
semi-legendary theme of Yü the Great organising the control of the waters (cf. p. 247). The part shown
gives a vivid impression of the Chinese people struggling with might and main to protect their country
from flood and yet irrigate their crops. Yü of the Hsia appears with his retinue, giving instructions, off
the scene to the right, but one of his foremen conveying them is just visible at the right-hand edge. On
the right a gabion is being rushed to a danger-spot, and in the foreground on the left another is being
sunk within the dyke. Carrying-poles, earth-baskets and mattocks are all busily in use. The upper part
of the picture (not shown) depicts, besides mountains and water, Yü's home (which according to the
story, he never had time to go and stay in), some scenes of agriculture, and two very curious boats,
perhaps dredgers.

PLATE CCCLXXXIII

Fig. 917. A great bale (*sao*) of kaoliang stalks being lowered into a dyke breach on the Yellow River near Liao-chhêng and Tung-a (photo. Todd, 1935). The cordage and pegging of these veritably mobile hay-stacks gives life to the sketch in Fig. 916. When such a bale is set in place a foot or so of earth is deposited on top of it so that it settles down more firmly into the mud and silt of the bottom.

PLATE CCCLXXXIV

Fig. 918. Flash-lock gates on the Grand Canal in +1793 as seen by the Macartney Embassy
(Staunton (1), pl. 35).

PLATE CCCLXXXV

PLAN AND SECTION OF A SLUICE OR FLOOD GATE ON THE GRAND CANAL OF CHINA.

and of an INCLINED PLANE by which VESSELS are made to pass between CANALS of different levels.

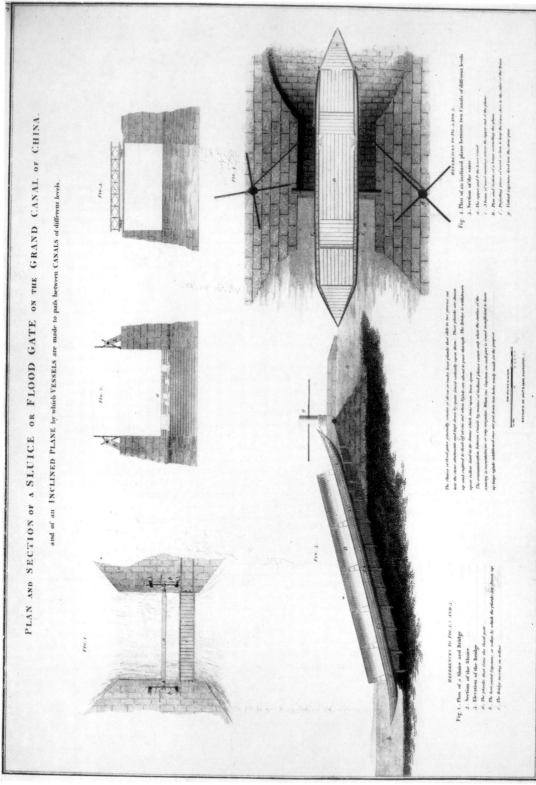

Fig. 920. Plans, sections and elevations of Chinese flash-lock gates and double slipways (cf. p. 344) as encountered by the Macartney Embassy in +1793 (Staunton (1), pl. 34).

fig. 1 Plan of flash-lock (stop-log) gate, looking down on the baulks and roller winch hoists, with rolling gangway in position.
fig. 2 Section of flash-lock, showing stop-logs and hoists.   fig. 3 Transverse elevation, with gangway in position.
fig. 4 Plan of double slipway, showing stone revetments of the inclines, and two capstans.

PLATE CCCLXXXVI

Fig. 921. The Chhing-fêng flash-lock gate at Kao-pei-tien near Peking (from Wang Pi-Wên, 2),
a photograph showing the grooves in the stonework for the stop-log baulks.

Fig. 922. The Chhing-fêng flash-lock gate; one of a pair of stone crane arms with roller hoists for the
lifting and lowering of the stop-log baulks (Wang Pi-Wên, 2). The canal bed in the background is almost
dry.

PLATE CCCLXXXVII

Fig. 923. River ships from the scroll painting Chhing-Ming Shang Ho Thu, by Chang Tsê-Tuan (*c.* +1125). This gives some idea of the size and approximate tonnage of the craft for which pound-locks developed in the +10th and +11th centuries under the Sung dynasty. See p. 359.

PLATE CCCLXXXVIII

Fig. 924. A *diolkos* (double slipway) on a canal in China *c.* 1800 (Davis, 1). A boat is just coming over the top. There are capstans on each side, and a watchman's booth on the left. Cf. pp. 363 ff.

Fig. 925. A *diolkos* still at work *c.* 1926, (photo. Fitch 1). The usual pair of capstans is pulling over a small boat laden with produce near Hangchow.

were the answer Fang and Chang made their computations and collected financial resources. Then, says Hu Su:[a]

for the outer gate they have piled up good masonry for the foundations, and made a strong dyke to take the force of the water, setting horizontal baulks (across the entrance), with two pillars (the crane arms) rising (one on each side). The lock basin is deep as the home of a sleeping black dragon, and like a dragon the water rises in the pool, so that the ships come and go continually, borne on waves like the tide flowing and ebbing. When the great gates are closed the water forms a whirlpool as the lock fills, and the white foam washes sides that never dry. Vessels pass through without any friction, and with very little expenditure of energy great advantage is received. At the north end there is the inner gate so that a basin is formed with well-built walls... When (either of the) wooden gates is opened, the boats go lightly through with oars propelled—not like the trouble of the slipways of old.

As Hu's inscription was dated +1027 the first part of the work, at Chen-chou, was probably accomplished in +1025. Perhaps the most significant part of Shen Kua's account is what he says about the increase of tonnage, an economic factor as important, no doubt, as the water-wastage question or the damage to the ships at slipway crossings. We shall come back to this point.

After such clear descriptions we are certainly predisposed to recognise pound-locks in Chinese statements during the following three or four centuries even when less clear. The next evidence, however, is equally demonstrative. It comes from the travel diary of a Japanese monk, Jōjin,[1] who journeyed northward on the Pien Canal in +1072 and back south again in the following year.[b] The following entries occur in his book entitled *San Tendai Gotaisan-Ki*[2] after the place of his pilgrimage. He is northbound on the Chiang Nan Ho section from Hangchow.

Hsi-Ning reign-period, 5th year, 25 Aug.

Weather fine. At the *mao* double-hour (about 5 a.m.) our boat cast off. By the *wu* double-hour (11 a.m.) we got to Yen-kuan hsien, arriving at the Chhang-an[3] double slipway (*yen*[4]).[c] About the *wei* double-hour (1 p.m.) the magistrate came, and we took tea at the Chhang-an rest-house. About the *shen* double-hour (3 p.m.) two of the lock-gates (*shui mên*[5]) were opened (in succession), in order to let the boat through. When it had passed through, the stop-logs (*kuan mu*[6]) were dragged back so as to close (the middle gate), and then the stop-logs of the third lock-gate (were lifted out) to open it, and the boat was let through. The surface of the succeeding part of the canal was a little more than five feet lower (than the upper part). After (each) gate was opened, the (water from the) upper section fell and the water level became equal, whereupon the boat proceeded through.

Here then we have a description of a two-stage pound-lock with three confining gates,[d] the gates being set so close together as to take Jōjin's barque alone, or with a few

---

[a] Tr. auct.

[b] We are much indebted to Professor E. Pulleyblank for communicating to us knowledge of Jōjin's descriptions, and for elucidating the monk's mixture of Japanese with the colloquial Chinese of his time.

[c] This small place still exists with the same name. Cf. *Sung Shih*, ch. 97, pp. 9b ff.

[d] Either there was fear of too great a head of water at the gates, or some auxiliary supply fed one or both of the lock-basins. A similar triple-gate pound-lock was built at Gouda in +1413 by Jan van Rhijnsburch (Skempton (4), p. 442), and a staircase of eight gates with a total rise of 70 ft., built by P. P. Riquet c. +1675, can still be seen on the Aude-Garonne Canal (p. 377 below) at Béziers (*ibid.* p. 467).

[1] 成尋   [2] 參天台五臺山記   [3] 長安   [4] 堰   [5] 水門   [6] 關木

others. In the entries for the following month he describes three pound-locks of the more normal two-gate type, for example:

14 Sep.

Weather fine. At the *mao* double-hour (about 5 a.m.) we cast off, and by the *chhen* double-hour (7 a.m.) our boat stopped at Shao-po village [a]... Towards the *wei* double-hour (1 p.m.) the two lock-gates were opened, and when the second was opened the boat passed right through.

It is noteworthy that throughout his writing Jōjin reserves the term *shui mên* [1] for what are clearly pound-locks, and speaks of *cha thou* [2] when he means gates at the end of very large lock-basins, or just *cha* [3] when he means flash-lock gates. Here is an instance of the second case:

17 Sep.

About the *ssu* double-hour (9 a.m.) we started, the boat returning to Chhu-chou [4] (city), [b] and reached a lock-gate (*cha thou* [2]). This is situated 9 *li* 300 paces from the city... After ten *li* we got to the (next) lock-gate, but water was lacking so the gate was not opened... By the *hsü* double-hour (7 p.m.) there was enough water, so the lock-gate was opened. A hundred boats were first let through, which took one double-hour, so that our boat did not get through until the *hai* double-hour (9 p.m.). They did not open the second lock-gate (at this place) that same evening, so we spent the night within (the lock-basin).

And for the third we may take an entry from his return journey:

6th year, 25 Apr. (+ 1073).

A letter came from the Transport Bureau to say that the gate-keepers must wait for a hundred boats to collect, but if in three days' time that number had not been reached, then they might open the gates (*cha* [3]). For the needs of irrigation were not to be neglected. As this was already the third day (we thought that) in the evening they must open the gate, but they did not.

A few years after Jōjin's memoirs we can find another pound-lock in an official Chinese source. The *Sung Shih* says: [c]

In the 4th year of the Yuan-Yu reign-period (+ 1089) the Transport Bureau officials of the Eastern Region said that the Chhing [5] River (in the Huai valley) communicated (most usefully) with many districts in Chiangsu, Chekiang and Huainan, but after two floods at Hsüchow, Lü-liang and Pai-pu it had become shallow and infested with dangerous rapids so that many ships had been wrecked. Consequently the sailors, the towing-oxen and ass drivers, and the longshoremen, were doing everything in their power to prevent merchants going that way. Now the government had already appointed the Prefectural Vice-Administrator of Chhichow, Shêng Hsi-Ching, [6] and the Magistrate of Chhangchow, Chao Sung [7] of Chin-ling, to survey the place with a view to engineering operations. If therefore it should prove possible to repair the stone-faced dam across the Yüeh [8] River, and to set up lock-gates (*chha* [9]) alongside it above and below, to be opened and closed at certain times so that ships should be able to go

---

[a] Cf. p. 352 above.   [b] Mod. Huai-an.
[c] Ch. 96, pp. 4*b*, 5*a*, tr. auct. Cf. Chêng Chao-Ching (*1*), p. 209.

[1] 水門   [2] 閘頭   [3] 閘   [4] 楚州   [5] 清河   [6] 滕希靖
[7] 趙竦   [8] 月河   [9] 牐

through; that would be a long-term plan which would bring great benefits. They (the officials) therefore begged that superintendents should be sent to make a start. The emperor agreed and the suggestion was carried out.

Here then was a clear case of a pound-lock on a canalised river just after William the Conqueror's time.[a] By the end of the century we are back at the pound-locks on the Ling Chhü (cf. p. 306 above). It is obviously unnecessary to prove the point further, and we shall only add one more example, the building of the Thung Hui Ho section of the Grand Canal between Peking and Thungchow by Kuo Shou-Ching in +1293. It will be remembered from Table 70 that the rise over this stretch of country was some 30 ft. Describing Kuo's specification for the locks on the canal the *Yuan Shih* says:[b]

(From Peking to) Thungchow let there be established a flash-lock gate (*chha*[1]) every ten *li*.[c] There will thus be in all seven *chha*.[1] At a distance of one *li* or more[d] from each *chha*[1] let there be set up double lock-gates (*chhung tou mên*[2]), so arranged as to open (lit. lift, *thi*[3]) and shut (*o*[4]) interconnectedly in order that the boats may pass but the water be held back.[e]

Thus the head of water at each pound-lock would have been of the order of 4 or 5 ft. Unfortunately the canal silted up long ago, and no modern study of its remains appears to have been published.[f] But we are now within a century of the datings of the first evidence for pound-locks in Europe, so it is time to turn to what happened at the other end of the Old World.

In the light of the facts so far described, it is clear that the conclusions of those who have confined their enquiries to Europe will have to be much modified. A general view, expressed in useful papers such as that of Wreden (1), is that while sluices and

[a] Further information on locks at this time will be found in the article of Nagase Mamoru (1). It is interesting to see again the social strains at work, as in the case of the original invention of Chhiao Wei-Yo.

[b] Ch. 164, p. 12b, tr. auct. with E. Pulleyblank. A parallel passage occurs in *Yuan Wên Lei*, ch. 50 (Kuo Shou-Ching's necrology).

[c] It will be remembered (p. 312 above) that the distance between Peking and Thungchow was much shorter than the length of the canal usually given, because that was reckoned from the water-sources in the hills west of the capital. *Hsin Yuan Shih*, ch. 53, reads 'ten *chha*[1] emplacements and a total of twenty lock sites (*tso*[5])'.

[d] Upwards of 600 yards.

[e] Another passage in the *Yuan Shih*, ch. 64, p. 3a, b, recounts the whole list of lock-gates with their names, starting from the west side of the capital and reaching some distance beyond Thungchow to the junction with the next section of the Grand Canal. It mentions 8 double gates and implies two more. In place of *chha*[1] and *tou mên*[6] it speaks of *cha*[7] and *chha*[1] respectively, internal evidence showing that the first in each case applies to flash-lock gates and the second to double or pound-lock ones. Thus although the terminology is confusing the meaning can be made out on careful study. Indeed, a kind of binomial phrasing continually recurs in medieval texts on hydraulic engineering, though the components differ. For instance, in +1118 Liu Thing-Chün[8] memorialised that on the Shan-yang Yün-Tao section of the Pien Canal, between Huaiyin and Yangchow, there had been 79 *tou mên*[6] and *shui chha*[9] 'and the like', but most were out of repair, so he was authorised to restore them (*Sung Shih*, ch. 96, p. 9b, cf. Chêng Chao-Ching (1), p. 210). Without further information one cannot identify which of these were the sluices and flash-lock gates, and which the pound-locks, but at any rate the historian, one feels, did not mention two things for nothing. We are indebted to Prof. E. Pulleyblank for friendly and helpful discussions on the material concerning the Thung Hui Ho Canal and its locks.

[f] In 1955 Dr Chhüan Han-Shêng told us that he once made a special visit to Thungchow to see the junction docks, much of which still exist.

[1] 牐   [2] 重斗門   [3] 提   [4] 閼   [5] 座   [6] 斗門
[7] 閘   [8] 柳廷俊   [9] 水牐

simple gates may go back to the +13th century at least, the idea of the pound-lock is to be ascribed to Leonardo da Vinci. In the elaborate (but circumscribed) investigation of Parsons the problem resolved itself into deciding just who, among Leonardo's immediate predecessors, was responsible—'the lock', he says, 'the greatest single contribution to hydraulic construction ever made, is unquestionably of Italian origin'.[a] He overlooked the achievements of the Dutch, but now it is clear that neither can compete with those of the Chinese.

It is possible that the sluice-gate may have originated in the Fertile Crescent before the −1st millennium. A passage in the 'Epic of Gilgamesh' (c. −2000)[b] has been thought to mention the device,[c] but the authorities do not agree about the meaning of the words.[d] Still, Mesopotamia might well be its home. There seems little reason to doubt the existence of sluice-gates in the impressive sweet-water canal system that Sennacherib gave to Nineveh in −690, but the case is not quite proved.[e]

Evidence somewhat stronger for the −2nd millennium is provided by the Phoenician harbour-works still extant at Sidon, where rock-cut grooves (one only now remaining) indicate the former presence of four sluice-gates.[f] Guarding swell- and spray-filled basins cut in the rock of the reef, a natural enclosing pier, these would have been used, it is thought, for flushing out the haven and preventing silt-deposition. They could well go back to the −12th to −8th centuries, and later examples of the use of streams and sluice-gates for flushing harbours can probably be adduced,[g] but on this tideless shore the surmised system falls perhaps just short of conviction. Nevertheless the gate slot is there, amidst other Phoenician rock-cut work, even if its sluice retained a bath or fish-pond only.[h]

Before going further, a word is necessary about the claim that there were ship locks on the ancient Egyptian canal which connected the Nile with the Red Sea.[i] This had

[a] (2), p. 372.
[b] It concerns something that the hero did before diving, like a pearler (cf. p. 668), to collect a submarine plant of immortality.
[c] Sandars (1), p. 113, cit. Frost (1), p. 195. The reference is to Tablet XI, ll. 265–76.
[d] The interpretation seems to be based on emendations of a neighbouring passage (l. 298) suggested by R. Campbell Thompson; see his (3), pp. 41, 55, 56; (4), pp. 83, 88. But all other translations either leave a blank or take the reference to be to other things, a channel, a girdle, sandals, etc. (Ungnad & Gressmann; Ranke; Contenau; Heidel; Lucas). Dr J. V. Kinnier-Wilson (in private correspondence) believes 'sluices' almost certainly a mistranslation of the Akkadian.
[e] Near modern Mosul; the water came from a tributary of the Great Zab R. (cf. p. 366). Jacobsen & Lloyd (1); Luckenbill (1); Thompson & Hutchinson (1, 2); cf. Drower (1), p. 531; Forbes (10), vol. 1, pp. 155 ff.; Garbrecht, in Biswas (1).
[f] See Frost (1), pp. 88 ff. and figs. 18, 19, 21, 23; Poidebard, Lauffray & Mouterde (1), pp. 43, 70 ff., 88, pl. XIV; Renan (1), pls. LXVI, LXVII, LXVIII. In 1966 I was so fortunate as to be able to spend a day studying the Sidon harbour-works under the expert guidance of Miss Honor Frost, and here record my warmest thanks.
[g] Bridport (West Bay) is apposite but late. Marseille (Vieux Port) had streams but no sluices (Bouchayer). Seleuceia-in-Pieria, near Antioch and the mouth of the Orontes, is claimed for both, but the remains are extremely complex and their reconstruction difficult (see Chesney; Chapot; Lehmann-Hartleben).
[h] Another report (Rao, 1) has mentioned sluice-gates in the overflow channel of a dock of the Harappa civilisation at Lothal north of Bombay (c. −1500), but the photographs so far published do not show the grooves claimed. As for the regulator sluices on the Baḥr Yūsuf in Egypt, where it spills into the Faiyūm Depression, often attributed to the −19th century, see p. 365 below.
[i] Herodotus, II, 158; IV, 39; cf. Kees (1), pp. 113 ff.; Toussoun (1, 2).

been begun by Necho (-610 to -595) and continued by Darius (-521 to -486) but was not completed till the time of Ptolemy Philadelphus ($c.$ -280). Although it is often said[a] that a true pound-lock with double gates was then installed, at least at the Suez end, the best authorities (e.g. de la Blanchère) can find no evidence for anything more than a single flood-gate (the 'ingenious barrier', *philotechnon diaphragma*, φιλότεχνον διάφραγμα, of Diodorus Siculus).[b] This would have been sufficient to handle tidal differences at Arsinoe (Suez) or for adjustment to flood or dry-season conditions at the Nile end.[c]

After this, there is complete silence in the Western world concerning sluices or locks until the Dutch archives begin to deliver their information from the $+11$th century onwards.[d] Naturally stanches or flash-lock gates come first. Stanches are attested for the R. Rotte in Holland in $+1065$ and for the R. Scarpe in Flanders in $+1116$, but it is not certain that these were used for aiding navigation. Quite probably they did so, like the flash-lock gates in Italy on the R. Mincio near Mantua, by $+1198$ (Lecchi, 1). Flood-gates or tide sluices, permitting vessels to pass between canals and tidal rivers, developed early in the Low Countries, as at Damme near Bruges in $+1168$,[e] at Spaarndam on the Zuyder Zee soon after $+1285$, and the well-known 'magnum slusam'[f] at Nieuwpoort, active in $+1184$. In general it may be said that flash-lock gates were common on European canals and canalised rivers by the end of the $+13$th century.[g]

Next it is possible now to pin-point the development of pound-locks with some accuracy. They first arose in places where there were differences of water-level only, i.e. on the highly tidal shores and estuaries of the North Sea. The first certain date is $+1373$, when a pound-lock was built at Vreeswijk in Holland at a point where a canal from Utrecht joined the R. Lek. A similar basin at Spaarndam (replacing the tide sluice) may possibly have existed from $+1315$ but was definitely there in $+1375$, the better dating.[h] Both these are known to have been large basins, accommodating twenty or thirty ships, and therefore something like the locks of Chhiao Wei-Yo and Zonca, but larger. The earliest European small basin, and therefore pound-lock of truest type, dates from $+1396$ at Damme near Bruges, where again it replaced tide sluices; its

[a] E.g. by Sarton (1), vol. 3, p. 1849. Sarton seems also to have erred in the interpretation of the term *euripos* (εὐρῖπος), which probably means the masonry-sided channel in which the gate was fixed, a 'strait' indeed.

[b] 1, 33; cf. Strabo, XVII, 25, 26; Pliny, *Nat. Hist.* VI, xxxiii; Ptolemy, *Geogr.* IV, 5.

[c] Diodorus says only that it was placed at the most suitable spot. It will be noted that such single gates would have been little, if anything, earlier than the flash-lock gates of the Chhin or Han periods (cf. p. 346 above).

[d] Excellent accounts by Skempton (4) and Forbes (11), pp. 55 ff., (25), are now available. Cf. Doorman (1).

[e] Correct Gille's (3) date of $+1180$.

[f] The word 'sclusa' had been used in Europe at least from the $+6$th century (as in the Life of St Gregory of Tours, $+580$), but it meant then a weir and not a sluice. On the Nieuwpoort gates see Doorman (1), pp. 81 ff.

[g] There is thus less interest attaching to the Italian examples on the Milan canal system ($+1438$, $+1445$, etc.; Parsons (2), p. 373), or late $+14$th-century French ones (Gille, 3).

[h] Correct Feldhaus (1), col. 962, who gives $+1253$ for Spaarndam. About $+1350$ may turn out to be acceptable.

length was just 100 ft. So far variations of water-level were alone concerned, but it was just at this time that the first successful attempt was made in Europe to overcome variations of the ground-level, that is to say, by the building of a true summit canal; and again this was executed not in Italy but in Germany, possibly because of Hanseatic connections with the Low Countries. The Stecknitz Canal, completed in +1398, bestrode a watershed elevation of some 56 ft., with the aid of two considerable pound-locks (*Kammerschleuse*).[a] Only now came the upsurge of Italian civil engineering, destined to lead to such substantial improvements. A great builder of pound-locks was Bertola da Novate (*c.* +1410 to +1475); he constructed 18 on the Bereguardo Canal (part of the Milan system) between +1452 and +1458, and 5 more near Parma between +1456 and +1459. Possibly it was these which L. B. Alberti[b] described in his famous *De Re Aedificatoria*, finished about +1460 but not printed till +1485. He suggested gates moving horizontally, not vertically, and this was taken up by Leonardo da Vinci, who certainly invented the mitre gate,[c] and built several examples of it before +1497, when he had been Ducal Engineer of Milan for fifteen years. Leonardo also included a wicket door for admitting water with a valve balanced eccentrically like a Chinese rudder (cf. p. 655 below).[d]

The installation of a series of locks for carrying a traffic canal over a watershed summit was proposed by Leonardo between +1497 and +1503 when he was working for the City of Florence on schemes for navigability and flood-control of the Arno.[e] One proposal involved a cutting through hills which would have been 225 ft. deep,[f] and a second suggested the series of locks on both sides of the summit. This, says Parsons,[g] 'was the first suggestion for a double or reversed action by locks, something which was not consummated in practice until the Canal de Briare'.[h] Yet we know that the Stecknitz Canal, built by artisans now unknown, had been a successful construction a hundred years earlier, as also that the Grand Canal of the Yuan had preceded Leo-

[a] Cf. Skempton (4), fig. 279.

[b] +1404 to +1472; cf. Parsons (2), p. 375.

[c] Leonardo's drawings of this have often been reproduced, as by Feldhaus (18); Parsons (2); Skempton (4); Forbes (11). A model is illustrated in Ucelli di Nemi (3), no. 25.

[d] Parsons (2), p. 390.

[e] It is interesting that the early European segmental bridges (cf. p. 175 above), such as the Ponte Vecchio of Taddeo Gaddi, were designed so as to cause minimum obstruction to the Arno. On this bridge, cf. p. 181.

[f] Cf. the work of Li Ping so long before (p. 291 above).

[g] (2), p. 330. Doubtless the residence of Leonardo at the court of Francis I at Amboise was not without influence on the later canal enterprises of France.

[h] This connected the Seine (through the Loing) with the Loire, was started in +1604, largely accomplished by +1611, and finally finished in +1642. Cf. Skempton (4), fig. 291, and the account of G. Espinas (1), who supposed it was the first summit canal in history. Of considerable interest is the fact that the Briare Canal was inspected in +1764 by two Chinese Jesuits, Kao Lei-Ssu[1] and Yang Tê-Wang,[2] in the course of their study of French industrial technology. They reported to the minister Bertin that the locks were very inferior in size and construction to those which they knew on the Grand Canal in China. In general their judgment was quite objective; they much admired many European arts. On Chinese bridge-building they furnished information to the celebrated bridge engineer J. R. Perronet (+1708 to +1794). For further details see Bernard-Maître (9); cf. Huard & Huang Kuang-Ming (5). British river navigations were little developed before the +17th century (see Skempton, 5), and canals not before the +18th.

[1] 高類思   [2] 楊德望

nardo's suggestion by two hundred years, and the conversion of the Ling Chhü to a summit system by more than three hundred years.

The successive end-points of our titration of cultures in this particular field now begin to appear in some clarity. After the small sluices which were probably common in all the ancient civilisations of the Middle East come the traffic-carrying tide- or flood-gates of the Nile/Red Sea Canal in the −3rd century; these bear comparison well enough with the fluvial entrance gates of the Pien Canal in Chou, Chhin and Han China, though there the partition principle is much more rapidly generalised by the erection of very numerous flash-lock gates. More than a millennium passes before these begin to multiply upon the canalised rivers of Europe, probably because of the fact that the Chinese built so great a mileage of purely artificial waterways before the end of the Han (c. +200). Entry and exit gates were also fitted, quite probably, on the Ling Chhü Canal of the −3rd century, by far the oldest transport contour canal in any civilisation, and comparable only with the irrigation contour canals of Ceylon, not with any work in Europe. Pound-lock gates incorporated into this system somewhere about +1060 converted it into the earliest example of a summit canal. The two inventions are of course intimately related, but the times of appearance in East and West are each different, and so of much interest. The first recognisable pound-lock built in China dates from the decade following +980, the first in Europe from that following +1370; the former for a summit canal albeit the rise was fairly small (cf. Fig. 906), the latter for a tidal difference of water-level only. Leaving aside the conversion of the Ling Chhü, the first summit canal in China was (*sensu stricto*) the Pien Canal of antiquity and the Sui, but since the watershed was there so flat it is better to take the Grand Canal of the Yuan dating from the years following +1280; this must contrast with the first successful summit canal in Europe, the Stecknitz, of the decade following +1390. Priority falls to China in all these achievements.[a]

How paradoxical it is then that (as we have seen, p. 347) post-Renaissance European travellers on the great Chinese canals should have found only flash-lock gates and double slipways. The explanation is, I believe, already implicit in the quotation from Shen Kua on p. 352 above concerning the tonnages of tax-grain boats using the Pien Canal, and how they increased so rapidly when pound-locks were substituted for slip-ways.[b] In the *Sung Hui Yao Kao* (Codex of Administrative Law of the Sung Dynasty)

---

[a] Even the best admirers of Chinese hydraulic engineering, such as Middleton Smith (2), have failed to realise its merit in the invention of the true or pound-lock basin. Some observers have found it strange that the invention took so long in coming, but one must remember that the gradient of the land along the course of the North China plain was very gentle. Stop-log gates every three or four miles met the need, therefore, for many generations. Eventually, however, the Shan-yang Yün-Tao gradient, and its junctions with Huai and Yangtze, needed something more as tonnages rose; so also the Ling Chhü approaches, and above all the direct route to Peking across the Shantung foothills.

[b] Quite large ships are indicated by the expressions *ta po*[1] (in a speech of +1089, *Sung Shih*, ch. 94, p. 5b), and *hsien*[2] (speeches of +1018 and +1139, loc. cit., ch. 96, p. 1b, ch. 97, p. 2b). Cf. pp. 462, 487 below. To get a clear idea of the sizes of the cargo and passenger ships used on the inland waterways of the +11th and +12th centuries, nothing is better than the Chhing-Ming Shang Ho Thu, a scroll painted by Chang Tsê-Tuan just before +1125, and often elsewhere referred to; see Fig. 923 (pl.) and pp. 115, 165, 463, 648.

[1] 大舶　　[2] 艦

there is a passage which speaks of 'pairs of gates' (*cha i chhung*[1]),[a] saying that formerly only official convoys and heavy vessels were sent through these pound-locks, the others being hauled over the slipways, but now (i.e. in +1167) the local authorities, being eager to increase the revenue from taxation, were sending all craft through the locks whatever their flag or burden, and had abolished the use of the slipways.[b] Pound-locks were in fact essentially the response to vessels of heavy burden on the canals, and if a time came when this stimulus should be withdrawn they might well not be renewed, junction points reverting to flash-lock gates and slipways. This is exactly what happened. When in the late +13th century the Yuan government fixed its capital at Peking, the canal system could not at first, and indeed throughout that dynasty never wholly, carry the weight of the northward traffic, so that a large part of it was shipped by sea.[c] Though annual sea shipments never exceeded what the canals could carry when at their highest pitch of efficiency, they often equalled the inland totals and quite often exceeded them. After +1450 Ming rule let drop the naval might that the Southern Sung and Yuan houses had laboriously built up,[d] but during some three centuries the need for really heavy craft on the Grand Canal, like those which the Northern Sung had employed with such striking results, had been in abeyance. Hence it had become customary to use a multitude of smaller vessels, and as this tradition established itself, the pound-locks fell into decay one by one and were not replaced.[e]

[a] *Shih Huo* section, ch. 8, p. 43a (*Tshê* no. 125). This enactment was noticed, and kindly communicated to us, by Professor E. Pulleyblank.

[b] At this time of course they had no control over the more northerly sections of the Pien Canal. As Jōjin (cf. p. 353) negotiated slipways when southbound in +1073 and not when going north in the previous year, we might guess that his return was made in a smaller vessel than that of his previous journey.

[c] Cf. pp. 312, 313, 478.   [d] Cf. pp. 524 ff. below.

[e] The question of whether there were pound-locks on the Grand Canal (or other Chinese waterways), and if so how many, between +1300 and +1700, is rather a difficult one. For this period foreign travellers bear more witness than indigenous texts. Thus in +1487 a Korean official Chhoe Pu,[2] travelling in a small coasting vessel, was blown by a tempest to the Chinese coast near Ningpo, and so travelled home with his attendants from Hangchow via Peking. In his *Phyohae Rok*[3] (Maritime Odyssey), the fascinating travel account which he wrote in the following year (tr. Meskill (1); cf. Makita Tairyō), he gives a number of details about the Grand Canal which are helpful in the present context. He speaks, for example, of Ku-thou Lower Lock (-gate, which implies an upper one, p. 106); he says of the Kuan-Yin temple at Lin-chhing that it stood on a promontory at the junction of two 'rivers' where 'to east and west four lock (-gates) had been built to hold water' (p. 111); and in a general summary he refers to 'the water in the locks' (p. 150) as he would hardly have done if they were miles apart. Besides this, Chhoe Pu mentions various sluices and at least 16 flash-lock gates by name, gives a lucid account of the procedure at them (p. 153), and quotes in full an inscription of c. +1470 set up on a stele at Huang-chia Lock several stations south of the summit. From this we know that the distance between flash-lock gates varied from 5 to about 11 *li*.

For the +16th century further evidence may be found in an itinerary from Hangchow to Peking printed in +1535, the *Yen Thu Shui I*,[4] known only from the unique copy which the monk-envoy Sakugen[5] took back with him to Japan (cf. Moule, 16). Here one reads of slipways, sluices, 'opening and shutting passages' and 'level-water locks'. Such a term would clearly exclude the flash-lock gate with its rush of water from one level to the next. Of these 'level-water locks' the Grand Canal had at this time eleven from south to north. Further information is contined in Sakugen's travel account, the *Nyū-Min Ki*[6] (Diary of Travels in the Ming Empire), reproduced and studied by Makita Tairyō (1).

In the +17th century we may note two pieces of evidence. The *Chih Ho Fang Lüeh* of +1689 has a design which may represent a pound-lock (ch. 3, p. 32a, b). The lock is 84 ft. long and 24 ft. wide, the underwater approach at each end consisting of 25 levels of stepped stonework; depths of water are not

[1] 閘一重   [2] 崔溥   [3] 漂海錄   [4] 沿途水驛   [5] 策彥
[6] 入明記

The decline and fall of the pound-lock in China notwithstanding, what influences could Chinese hydraulic engineering have exerted on Europe, given the datings just established? Ought the pound-lock and the summit canal to be inscribed among what we have already termed the ' + 14th-century cluster' of technical transmissions westwards, and if so, how could the notions have come? It is always conceivable that the + 12th-century Netherlanders and Italians knew by Crusader gossip of the Chinese flash-lock gates of Wang Ching, Li Pho and their innumerable colleagues going back at least to the + 1st century, but the device was of such natural simplicity that they could surely have originated it themselves. Perhaps however the pound-lock is another matter. It may not be a mere coincidence that the beginnings of pound-locks and summit canals in the West occur in the + 14th century, just after the time when the Pax Mongolica permitted the free travel and intercourse of merchants typified by Marco Polo. At a much earlier stage in this work[a] we noted the statement of Yule[b] that towards the latter part of the + 13th century 'Chinese engineers were employed on the banks of the Tigris'. This has often been copied by sinologists;[c] unfortunately, in spite of the grounds for suspecting westward transmission of hydraulic techniques just at this time, no one has yet been able to substantiate the statement from the sources which Yule quoted.[d] That many Chinese were in the first wave of the Mongolian conquest of Iran and Iraq is indeed well known—a Chinese general, Kuo Khan[1] (Kuka-ilka), was (with a Mongol brother officer Kiti-buka) first governor of Baghdad after its capture in + 1258 by Hūlāgu Khan.[e] That Chinese expert trebuchet artillerists and Chinese naphtha grenadiers formed part of Hūlāgu's armies is also well established.[f] As the Mongols had a habit of destroying irrigation and water-conservancy works of all kinds to annoy their more agricultural enemies, it would have been most natural of them to turn to their Chinese technical colleagues for the rehabilitation of Mesopotamia as soon as government exploitation was ready to replace military operations. In another place[g] we have mentioned a statement said to have been made by an Arabic writer of earlier date, al-Jāḥiẓ (d. + 869), that Chinese hydraulic engineers were brought to Iraq in his

given. That there were two gates is suggested by the words *chhien hou so khou*,[2] but there may have been only one. The lock is termed a *cha*,[3] which may be significant (cf. p. 355 above). Additional evidence that pound-locks were still in use in the + 17th century may be found in the travel narrative of Nieuhoff (1). Between Huai-an and Chi-ning, he says (p. 156): 'J'y ay conté un grand nombre d'escluses basties de pierre carrée; chacune d'icelles a une porte par laquelle entrent les navires; on la ferme avec des aix fort grands et épais; puis les ayant levés par le moyen d'une roue et d'une machine avec beaucoup de facilité, on donne passage à l'eau et aux navires, jusques à ce qu'on les ait fait passer par la seconde avec le mesme ordre, et la mesme methode, et ainsi en suite par toutes les autres...' There is still something slightly ambiguous about this description, but it would be rather hard to maintain that what he saw at Ney-nemiao (Niang-niang-miao?), also in the neighbourhood of Huai-an, was nothing but a flash-lock, for his words are (p. 152): '...après avoir franchi une forte escluse, qui estoit defendue de deux rangs de portes...'. This was in + 1656, some thirty years before the writing of Chin Fu's book.

[a] Vol. 1, p. 217.　　　　　　　　　[b] (2), vol. 1, p. 167.
[c] E.g. Carter (1), 2nd ed., p. 169.
[d] Notably d'Ohsson (1), vol. 2, p. 611, but he offers no contemporary evidence.
[e] Cf. *Yuan Shih*, ch. 149, pp. 14a ff., and the *Hsi Shih Chi* (Notes of an Embassy to the West) already described in Vol. 3, p. 523 above; both translated in part, and discussed, by Bretschneider (2), vol. 1, pp. 109, 111, 120, 122 ff.
[f] Cf. Spuler (1), pp. 411 ff.　　　　　[g] Vol. 4, pt. 2, p. 243.

[1] 郭侃　　[2] 前後鎖口　　[3] 閘

time. This is in fact a mere typographical error; the text itself says that they were Byzantine. Nevertheless there really were Chinese artisans in +8th- and +9th-century Baghdad, notably the paper-makers, textile workers and others who settled there after the Battle of the Talas River in +751.[a] There may well have been water engineers among them. In general therefore, though little specific can be adduced, there is good reason for thinking that the Middle East may have been a way-station for ideas on how to make ships go up and down hill, between the time of Li Pho and Bertola da Novate. In view of the dates now firmly established it is highly unlikely that these ideas were travelling eastwards.

In foregoing pages spillways have several times been encountered—the stabilising device which kept water-levels at the heights desired, whether in the Kuanhsien irrigation system (p. 291) or the Ling Chhü transport canal (p. 302).[b] Apart from the technical terms used in those cases, others are more commonly found. One is implied, for instance, in the Shih Tha Yen [1] (lit. stone gate dam) which was built in the time of Han Wu Ti (-120) to store water in the artificial Khun-ming Chhih [2] south-west of Chhang-an, a lake intended for naval combat exercises.[c] The same term occurs again later on, as at Haichow in +1020, when Wang Kuan-Chih [3] wanted to take water from such a spillway dam for the Pien Canal.[d] The commonest phrases, perhaps, are *shih to* [4] or *shui to*,[5] mentioned many times in Sung texts, often in connection with the Pien, which in certain stretches had rows of ten or thirteen spillways one after another along it.[e] On a much smaller scale culvert spillways (*tung* [6]), of stone, three or four feet square, were often contrived at suitable heights in the embankments enclosing a canal; and of course, as we have seen (Fig. 908), side sluice stop-log gates (*cha* [7]) were commonly in use for rapid evacuation of flood water.[f]

For emergency clearance an unusual technique was developed. On several occasions already[g] we have met with the term *pa*,[8, 9] defined in dictionaries simply as a dam, dyke or embankment,[h] and no doubt sometimes loosely so used. More precisely, however, it meant a 'flying dyke', i.e. a very long shallow U-shaped spillway with a stone revetment and masonry cheeks, running along the bund of a canal or lake. Normally this was kept filled with reed bundles, fascines, earth, etc., forming part of the embankment; but in time of emergency a small breach could be artificially made in the centre of the fill, whereupon the force of the water would quickly wash it all away, flood and débris discharging together through previously prepared channels. Eleven flying dykes

---

[a] Cf. Vol. 1, p. 236 above.

[b] We have already considered the role of the overflow constant-level device, doubtless the simplest and oldest of all homoeostatic devices, in connection with the time-keeping clepsydra; cf. Vol. 3, Fig. 138 and p. 324. It seems clear that the first application was in hydraulic engineering, and then in horological.

[c] Cf. Dubs (2), vol. 2, p. 63.   [d] *Sung Shih*, ch. 96, p. 2a.

[e] There are references for +1020 (Chang Lun,[10] cf. p. 352), +1058, +1069, +1137, +1139 and +1194, among others; cf. *Sung Shih*, ch. 96, p. 2b, ch. 97, p. 2b, and Chêng Chao-Ching (1), pp. 208, 211, 212, etc.

[f] Gandar (1), p. 17, describes both.

[g] Pp. 272, 318, 322, 349.

[h] Its shorter form is not to be confused with *chü*,[11] which also means an embankment.

[1] 石闥堰   [2] 昆明池   [3] 王貫之   [4] 石䃩   [5] 水䃩   [6] 洞
[7] 閘   [8] 壩   [9] 坝   [10] 張綸   [11] 坝

of this kind, built between +1680 and +1757 in the defences of the Hung-Tsê Hu (lake) and the Grand Canal, measured an average length of 400 ft. and an average height of 9 ft.[a]

At some quite early time it must have been realised that if the ramp of a spillway were made to slope at a reasonably gentle gradient it would be possible to drag canal-boats up and over it to the higher level, wastage of water at the same time being prevented. In this manner there arose the double slipway, a pair of inclined stonework aprons over which boats were hauled, generally in China with the use of capstans, from a waterway at one level to a waterway at another. The thing was partly also inspired, no doubt, by the primitive rough portages which ancient peoples had undertaken to avoid rapids on rivers, or to connect two arms of the sea across an isthmus. The Greeks called this a *diolkos* (διόλκος). The commonest Chinese term for the double slipway was *tai*,[1] but unfortunately it seems also to have been sometimes used (like *pa*) as a general word for dams or dykes, so that its interpretation may give rise to difficulties. However, when in +384 the philosophical Chin general Hsieh Hsüan[2] built seven *tai* on the Pien Canal at Lü-liang[3] (near modern Thungshan), he did it at the suggestion of Wênjen Shih[4] 'to facilitate transportation', so that slipways are certainly meant.[b] A contemporary writer refers to the use of capstans turned by oxen for drawing boats over these slipways,[c] which were therefore called *chhien tai*,[5] a term often subsequently found, e.g. in the poems of Wang An-Shih. Sung people named the slipways *tai chhêng*[6].[d] They were so characteristic of Chinese communications that Rashīd al-Dīn mentioned them in his description of the Grand Canal in +1307 (p. 313 above), and Chhoe Pu gave a detailed account of them in +1488.[e]

Such slipways were still in widespread use when the European travellers began to write their narratives of Chinese journeys. Thus Lecomte, for example, says in +1696:[f]

I have observed in some Places in China, where the waters of two Canals or Channels have no Communication together; yet for all that, they make the Boats to pass from the one to the other, notwithstanding the Level may be different above fifteen Foot: And this is the way they go to work. At the end of the Canal they have built a double Glacis, or sloping Bank of Free-stone, which uniting at the point, extends itself on both sides down to the Surface of the Water. When the Bark is in the lower Channel they hoist it up by the help of several Capstanes to the plane of the first Glacis, so far, till being raised to the Point, it falls back again by its own Weight along the second Glacis, into the Water of the upper Channel, where it skuds

---

[a] Gandar (1), pp. 17, 34, 39, 49. He studied them in 1893.
[b] *Chin Shu*, ch. 79, p. 8*b*, cf. Chêng Chao-Ching (1), p. 197.
[c] In the *Chin Chung Hsing Shu*[7] by Hsi Shao[8] or Ho Fa-Shêng.[9]
[d] As in the *Ho Pi Shih Lei*[10] by Hsieh Wei-Hsin,[11] or in the writings of Tsêng Kung[12] around +1070. It was just at this time that Jōjin passed over the Chhang-an *diolkos* (*yen*[13]) drawn by winches (*lu-lu*[14]) on his return journey (*San Tendai Gotaisan-Ki*, entry for 19 May +1073). This was near Hangchow.
[e] *Phyohae Rok*, Meskill tr., pp. 67, 153.
[f] (1), p. 107, 2nd ed. pp. 104 ff.

[1] 壩	[2] 謝玄	[3] 呂梁	[4] 聞人奭	[5] 牽壩
[6] 壩程	[7] 晉中興書	[8] 郤紹	[9] 何法盛	[10] 合璧事類
[11] 謝維新	[12] 曾鞏	[13] 堰	[14] 轆轤	

away during a pretty while, like an Arrow out of a Bow; and they make it descend after the same manner proportionably. I cannot imagine how these Barks, being commonly very long and heavy laden, escape being split in the middle, when they are poised in the Air on this Acute Angle; for, considering that length, the Lever must needs make a strange effect upon it; yet I do not hear of any ill Accident happen thereupon. I've pass't a pretty many times that way, and all the Caution they take, when they have a mind to go ashoare is, to tye ones self fast to some Cable for fear of being toss't from Prow to Poup.

These slipways have often been described and illustrated, e.g. by Davis;[a] Allom & Pellé;[b] Staunton;[c] Barrow;[d] Dinwiddie[e] and others (cf. Figs. 924, 925, pls.).[f]

Similar arrangements go far back in European history. The classical instance is the slipway across the isthmus of Corinth, which, though perhaps not completed by the first designer of the work, the tyrant Periander (-625 to -585), dates undoubtedly from the early -6th century.[g] This was originally, and long remained, a masonry road across a large part of the isthmus terminating in a true slipway at each end. Recent excavations reported by Verdelis (1, 2) show that the road of stone blocks, 12 to 16 ft. wide and some 4 miles long, which passes at 130 ft. through a ridge some 260 ft. above sea-level, had two parallel grooves some 5 ft. 6 in. apart all along its length, so that the ships must have been carried on wheeled cradles running in veritable tramway tracks.[h] There is even apparently a passing place on a curve, where there are traces of double tracking. This famous *diolkos*, connecting the Corinthian and the Saronic Gulfs, remained in use at least until the $+9$th century.[i] Soon afterwards the double slipway, now named 'overtoom', appears in Holland, where in $+1148$ there were two examples in use on the Nieuwe Rhijn Canal near Utrecht; probably another at Spaarndam in

[a] (1), vol. 1, p. 138. We have already considered the alleged application of wind-power to these capstans with some scepticism (Vol. 4, pt. 2, p. 559), but further evidence may come to light.

[b] (1), vol. 4, nr. p. 20. Id. Allom & Wright.

[c] (1), fig. 34.

[d] (1), p. 512.

[e] In Proudfoot (1)., p. 72 and in Macartney (2), Cranmer-Byng ed. p. 269. He considered the system 'preferable to English locks in every situation where the canal is nearly level, and constructed at a quarter of the expense... besides being cheaper they are much more expeditious'. Passages observed by him took $2\frac{1}{2}$–3 min. The device is still to be seen in England *in parvulo*, as those who pass back and forth across Coe Fen in Cambridge every day are well aware.

[f] Audemard (4), pp. 46 ff., is to be avoided; he confused the stop-log flash-lock gates with the double slipways. It is true that the fulcrum of the latter was often equipped with a rounded wooden baulk, but not with stop-logs.

[g] Strabo VIII, 2, i; 6, iv; 6, xxii; Thucydides II, 93, i, ii; comments in de la Blanchère (1). Cf. Neuburger (1), p. 500; Broneer (1).

[h] The figures given above derive partly from my own measurements in the autumn of 1966; at that time, alas, Dr Verdelis was dying, and I could not meet him. Nor could I see the double-track section, which was inside the grounds of a military engineering establishment. Verdelis' gauge figure was 4 ft. 11 in., but this must have been measured from the inner edges of the grooves, which are very broad and shallow. The curious similarity of this gauge to the modern standard railway gauge of 4 ft. $8\frac{1}{2}$ in. will not escape notice (cf. Vol. 4, pt. 2, pp. 250, 253 and pp. 5 ff. above). It is known to few that the remains of a stone railway of 4 ft. 3 in. gauge, first begun in $+1776$, still exist in this country—the ten-mile line from the granite quarries at Haytor on Dartmoor down to the R. Teign (cf. Amery Adams, 1). I had the opportunity of visiting this with the Newcomen Society in May 1963. Junction points as well as rails are all of cut granite blocks, and when in 1905 it was actually proposed to electrify the line using Bovey Tracey lignite as the power source, the most ancient shook hands with the most new. See also Lee (1, 2).

[i] A canal was attempted by the emperor Nero at this spot in $+67$, but not successfully accomplished till 1893, when exactly the same alignment was used.

+ 1220 and certainly at Ypres in + 1298.[a] A particularly well-known one was that on the Brenta Canal at Fusina near Venice erected in + 1437, a remodelling of which was illustrated by Zonca in + 1607.[b] Here European continuity seems to exclude the possibility of Chinese influence, but whether Periander's work could have echoed in China might admit of a wide solution. The most grandiose successor of these designs was the ship railway proposed by J. B. Eads in 1884 for the isthmus of Tehuantepec in Mexico, where ocean-going ships were to be drawn across on cradles hauled by three double-ended 2-14-0 × 2 Mallet locomotives on parallel tracks.[c] Owing to the Panama Canal this project came to nothing, but if it had ever been built it would have been indeed the apotheosis of the *diolkos* and the *tai*, delighting the hearts of Periander and Hsieh Hsüan.

(10) COMPARISONS AND CONCLUSIONS

The comparisons of the preceding paragraphs invite a more extended survey or balance-sheet of the achievements of Chinese hydraulic engineering in relation to that of other times and places, ancient Mesopotamia and Egypt, Greece and Rome, the Renaissance, and so on, for which this book is hardly the place. The works of Willcocks; Merckel; de la Blanchère, and others, suggest that when allowance is made for the available technical means, only the systems of Babylonia and ancient Egypt, and in later times Ceylon and India, can be compared with what was accomplished in China. To confront the size of the works excavated in length, breadth and tonnage, with Chinese parallels, would require an elaborate research in itself.

Irrigation engineers distinguish between perennial and inundatory canals, the former drawing water for the fields at all times of the year, the latter filled only during the flood season.[d] This distinction, as Willcocks (3) pointed out, differentiates also between the irrigation systems of ancient Mesopotamia and ancient Egypt, for the former watered the plain of the Tigris and Euphrates at all seasons, while the latter involved a vast chain of retention basins for silt deposition, and these were (and are) filled for only 45 days in the year.[e] The original plan of the − 2nd millennium, which has always been retained, was to build a dyke along one side of the Nile and to constitute basins by tying this dyke by a series of transverse dykes to the hills at the edge of the valley. The average area of each basin is some 7,000 acres, and the average depth of water in them when the river is at its maximum of 30 ft. above its bed is about 3 ft.; after the water has departed the land is sown and yields good crops.[f] The Nile, says

---

[a] Cf. Feldhaus (1), col. 944; van Houten (1), p. 138; Forbes (11), (25); Skempton (4) and private communication to us, May 1955. Some are still in use in Holland.

[b] Cf. Parsons (2), pp. 383, 396; Beck (1), p. 316. Designs were still being produced in the + 18th century, using water-power; cf. de Bélidor (1), vol. 4, pl. 42.

[c] See Corthell (1); Vernon-Harcourt (1), p. 397; and especially Covarrubias (1), p. 170 and pl. 42.

[d] Vernon-Harcourt (1), p. 425. Cf. pp. 214 ff. above.

[e] (4), esp. pp. 299 ff., (5). Cf. also now conveniently Drower (1).

[f] It is sometimes said that the greatest retention basin of all time was formed by the artificial conversion of the Faiyūm Depression into Lake Moeris (cf. Payne (1), pp. 14 ff.). Whoever thinks so has the authority of Herodotus (II, 149, 150), but Herodotus was wrong; cf. Kees (1), pp. 219 ff. and R. H. Brown (1). On its northward course the Nile throws off a natural deltaic channel, 'Joseph's arm' (Baḥr Yūsuf), formerly from Asyūṭ, now from Dairūt; this runs parallel with the river on its left or western side

Willcocks (1), is 'the most gentlemanly of all rivers', for it gives ample warning of its rise and fall, it makes no abrupt changes, it has enough silt to enrich the land annually but not enough to choke the canals, it is free from salt, and it flows between sandstone and limestone hills which furnish unlimited building-stone. The Tigris and the Euphrates possess few of these qualities, and the Yellow River almost none. With its silt content of about 10%, for example, may be compared the 0·75% of the Meso-potamian rivers and the 0·15% of the Nile.[a] Now the Mesopotamian irrigation pro-jects differed completely from the Egyptian, for apparently from the most ancient times they consisted of canals radiating from the main rivers and conveying away the water from derivations or division-heads. They were thus perennial, and in a sense aimed at converting the entire river valley into a delta.[b] Archaeologists have succeeded in tracing the elaborate works of the 250-mile Nahrawān Canal, which sprang from the left bank of the Tigris some 125 miles above Baghdad and rejoined it some 125 miles below.[c] This canal, of ancient origin, was perfected and adjusted to capture incoming left-hand tributaries under the Sassanians (+226 to +637); it has now long been for the most part in ruins. Running with a breadth of 400 ft. much of the way, it was the greatest work of the kind in Middle Eastern antiquity.[d] Canals of similar type, however,

near the edge of the desert for some 150 miles (cf. Budge (1), p. 6). Towards Illahun, near ancient Arsinoe (Medīnet al-Faiyūm), it leaves the Nile valley, however, and flows into the Depression north-westwards past two famous pyramids, splitting up into a great number of irrigation canals, which all go to form Lake Moeris (now called Birket Qārūn). The surface of this is no less than 147 ft. below sea-level, and though much smaller than of old, it still attains 90 sq. miles. It is certain that from the time of Sesostris II (−1906 to −1888) and Amenemḥāt III (−1850 to −1800) in the XIIth Dynasty, flood water was used to irrigate the Faiyūm (cf. Budge (1), p. 217), and it is equally certain that none of it could ever have been got back again into the Nile; but by the time of Diodorus Siculus at least (−30; see I, 52) and Strabo (+20; see XVII, i, 37) there was some sort of arrangement at Illahun (92 ft. above sea-level) by means of which the water could be routed either into the Depression or northward through a 50-mile lateral canal to rejoin the river near Cairo. Since Strabo speaks of 'artificial barriers' (*kleithra*, κλεῖθρα), and Diodorus of a 'skilful and costly device' (*kataskeuasma*, κατασκεύασμα), translators have not hesitated to use the term 'lock-gates', and Egyptologists such as Drioton & Vandier (1), p. 254, have freely attributed them to Sesostris himself; but Diodorus adds the significant information that it took the large sum of fifty talents (£10,000) to open or close the work. We should surely think therefore (with Hayes (1), p. 50) of some system of temporary dams (cf. p. 293), and not admit sluice-gates before Assyrian–Phoenician times if then. Probably what Sesostris and Amenemḥāt did was to cut the side-branch into the Depression, the northern lateral canal being the original course of the Baḥr Yūsuf. In any case, the Faiyūm has always been very fertile; rich olive-groves and vineyards decorate its slopes, and agriculture shares the bottom with fowling and marsh products. Thus Lake Moeris was a special case, a natural sump and never an artificial retention basin—yet highly useful to man for more than four millennia. We have to thank Prof. J. M. Plumley and Dr I. E. S. Edwards for help in elucidating this question.

[a] Willcocks (4), p. 546. Cf. pp. 228 ff. above.

[b] Perennial irrigation was not introduced to Egypt till the 19th century (Willcocks (4), p. 368).

[c] Jones (1); Willcocks (2), and now most lucidly Adams (1, 2). Three main intake derivate canals (*kātūl*) and at least three lesser ones took off from the great river in and near the city of Sāmārra, forming the Nahr al-Rāsāsi Canal. This captured most of the water of the R. 'Adheim, and fifty miles further on joined the course of the R. Diyala canalised to make the Nahr Tāmarrā; then all these waters, upon reaching the town of Nahrawān (mod. Sifwah) some 20 miles from the Tigris, bore off to the left to follow a long parallel course, irrigating widely the land on the way. Towards the end of this section, the Nahrawān Canal proper, an important division-head, the al-Qantara Weir, radiated its waters in many different directions. All drainage found its way back to the left bank of the Tigris.

[d] A general treatise on the history of the works of Iraq is available (in modern Arabic) by Ahmed Sousa (1). Excerpts from +11th-century monographs on them have been translated by Krenkow (1) and Cahen (3). On the ancient Assyrian achievements see Jacobsen & Lloyd (1); Luckenbill (1) and Thompson & Hutchinson (1, 2). One very important consideration in Iraq has been salinisation due to over-irrigation of unsuitable soils (cf. Jacobsen & Adams). This has not been unknown in China.

proliferated all over the arid lands later to constitute the Islamic culture-area—
Chorasmia, for example, south of the Aral Sea,[a] or Tunisia[b] or the Yemen.[c] In some of
these regions, but especially in Persia, annual systems were also necessary. The
Iranian rivers did not carry abundant water all the year round, but they were small
enough to be harnessed, unlike the Nile, with dams (*band-e āb*) and weirs feeding lateral
canals (*jūy*). Many of these works, dating at least from Sassanian times onwards, are
impressive in size and beautiful in design.[d]

Broadly speaking, the hydraulic works of the great civilisations of South and East
Asia combined in various proportions the Egyptian and the Babylonian patterns to
form more mixed and flexible systems. In India the fate of water-conservancy projects
was far less happy than in China owing partly to the absence of continuing political
and linguistic unity and centralisation, and partly to the recurring disruptions of the
country by foreign invasions.[e] There can be no doubt that canal-building is ancient in
India, and the waterway made by Nandivardhana in the − 5th century may well not
have been the first. Numerous references in the *Arthaśāstra* suggest that waterworks
were important and substantial in the time of the Mauryas (− 3rd century), and certainly imply this for their successors the Sunga and Āndhra dynasties before the + 3rd.
From Maurya times may date the beginnings of the canals of the Ganges delta in
Bengal.[f] And the Grand Anicut or weir on the Cauvery River near Tanjore in the south-
east, 360 yards long, 12 ft. high and 50 ft. wide, forming the headwork of a series of
irrigation canals, dates from *c.* + 130. Over the greater part of India, however, the
bund or dam forming a seasonal inundatory tank or reservoir in a valley was the most
characteristic type of waterwork.[g] Structures of this kind were built by the Pāla
dynasty in Bengal in the + 9th century, by the Chandēl dynasty (*c.* + 850 to + 1150) in
Bundēlkhund, and in the south by the Pallava dynasty under Mahendravarman I, a
Tamil contemporary of the Sui emperor in China (*c.* + 610). One of the Pallava dams in
Madras, $4\frac{1}{2}$ miles long and 40 ft. high, commands 6,300 acres. The + 11th century saw
the construction of the 250-sq. mile artificial Bhojpur Lake by the Bhoja dynasty in the
north,[h] and the 16-mile dam near Jayamkonda-cholapuram by the Chola dynasty in
the south. Under Islam there was at first a decline in hydraulic engineering operations, but later they revived, fostered by Fīrūz Shāh Tughluq (*r.* + 1351 to + 1388),
one of whose canals is still used, and the Bahmanī sultans of the Deccan. Not to be
outdone, the Hindu Vijayanagar dynasty in the south built under Bukka II in + 1406

[a] See the accounts of the elaborate Soviet expeditions to this region by Tolstov (1, 2, 3), summarised
in part by Ghirshman (2); Frumkin (1) and Mongait (1), pp. 235 ff.
[b] Solignac (1).   [c] Bowen (12).
[d] See, e.g., E. F. Schmidt (1); Houtom-Schindler (1); le Strange (4); Lambton (1); Ghirshman (4),
fig. 174; summary in Wulff (1).
[e] Background information for this paragraph will be found conveniently in V. A. Smith (1).
[f] Described by Willcocks (3). These canals were strangely misunderstood during the British occupation. Their irrigation function was forgotten, and breaches in the river dykes forbidden, though owing to
the absence of the technique of sluice-gates, the breaching of the dykes in certain places at specified times
had been the traditional method of filling the network of canals with water.
[g] Cf. K. L. Rao (1); Shrava (1); Biswas (2).
[h] This was the house of Rājā Bhōja (*r. c.* + 1018 to + 1060), that ingenious prince already met with in
connection with the mechanical sciences (Vol. 4, pt. 2, p. 156), a not unworthy peer of Hui Tsung,
Frederick II, Ulugh Beg and Alfonso el Sabio.

the Tungabhadra Dam, with its 15-mile aqueduct cut in solid rock most of the way. It will be evident even from this roughest of sketches that the achievements of Indian civil engineers in ancient and medieval times are quite worthy to be compared with those of their Chinese colleagues, though not to win the palm. Yet it was never in India that the fusion of the Egyptian and Babylonian patterns achieved its most complete and subtlest form.

This took place in Ceylon, the work of both cultures, Sinhalese and Tamil, but especially the former.[a] Its interest is such that it merits description a little more at large. The invitation was set forth in the first place by the meteorological and geographical nature of that incomparable island. The central mountains of Ruhunu and Maya Ratta are surrounded almost on three sides by 'dry jungle', watered only seasonally by the south-west and north-east monsoons respectively,[b] but the mountains themselves, together with the tract of country constituting the south-west quarter of the island, receive a rich precipitation and some perennial rainfall as well.[c] Hence a challenge to devise irrigation works of such a nature as to take advantage of both these sources of water. The process of evolution which is thought to have occurred[d] may be described as follows (Fig. 926); first the farmers made numerous small tanks in the hills and foothills near their fields or terraces to catch the run-off water, which they baled out at leisure. Then numbers of small dams (bunds, *bemma*)[e] forming small reservoirs (tanks, *wewa*, (Tam.) *kulam*) were built, often in series, on the upper reaches of tributaries of the greater rivers, thus retaining the annual or inundatory flow, and discharging it as desired by small canals (*ela*) along the valley sides. As time went on, larger dams were built, submerging or rendering unnecessary the smaller ones.[f] The next step was revolutionary: a weir (*anicut*, (Tam.) *tekkam*) was built much higher up the main river (*ganga*, *oya*, (Tam.) *aru*) to form the headwork for a long lateral trunk derivation-canal (*yodi-ela*), which thus brought perennial water to join the annual monsoon supplies in the great reservoir. This method, ambitious as well as scientific, had numerous advantages: (*a*) it harnessed a greater volume of water than any local

[a] During my stay in Ceylon in 1958 I had the valued privilege of personal discussions with the leading historian of Ceylonese civil engineering, Mr R. L. Brohier, and other members of the Survey and Irrigation Departments such as Mr W. Delay. The principal features of the story may be unravelled from the extensive technical publications which exist, notably Brohier (1, 2); Brohier & Abeywardena (1); Brohier & Paulusz (1), etc., but nothing can substitute for the opportunity of inspecting the ancient works themselves, and of listening to those who have devoted to them a lifetime of study. In viewing the great achievements of their ancestors it was a great pleasure to be accompanied by my colleague Mr Mahinda Silva and the doyen of Ceylonese archaeologists Prof. Senarat Paranavitana.

[b] For most of the year the term 'dry' is fully deserved, but a precipitation of 30 in. (or occasionally double that) comes and goes within two or three weeks (cf. Sion (1), fig. 5).

[c] See the rainfall maps in Brohier (1), vol. 1, opp. p. 2; vol. 2, opp. p. iv. In the footnotes to the following paragraphs we shall dispense with this reference and give only volume and page numbers.

[d] I follow the views of Brohier here, lucidly expressed in his semi-centennial presidential address of 1956 to the Engineering Association of Ceylon (2). Cf. vol. 1, p. 2; vol. 2, p. 5.

[e] When not identified as Tamil, or obviously English, the technical terms inserted in these paragraphs are Sinhalese.

[f] This was the point at which the Ceylonese went beyond what the terrain had generally rendered possible in South India. But it must be remembered that the Cauvery valley had the largest reservoir in the world for nearly a millennium, the Virānam-kulam near Jayamkonda-cholapuram already mentioned, built by the Cholas, a command of 22,000 acres.

catchment area could yield, (*b*) it put both monsoons as well as other rainfall to full use, (*c*) it secured a resource in drought periods as well as an even supply in normal years, and (*d*) it lessened the silt accumulation problem because the feeder canals could be cleared periodically much more easily than the tanks. Such *yodi-ela* canals, dropping

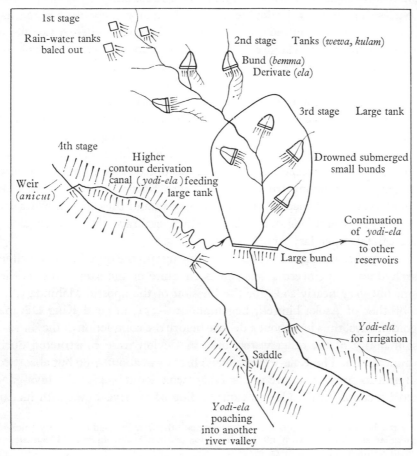

Fig. 926. Sketch of the evolution of hydraulic engineering works in Ceylon (constructed from R. L. Brohier's conversation, 1958).

very slowly along the contours, often passed across one or more watersheds en route.[a] They were generally dyked only on one side (*kandiya*),[b] sometimes spreading out into small lakes as they went, but in certain places a double embankment was needed (*depā-ela*). Smaller tributaries and gullies were crossed by means of spillway dams (*galwāna*) with wing-walls, sufficiently ample to take care of the greatest freshets,[c] but

[a] This had first been done in Assyria under Sennacherib for Nineveh's water-supply (−703 to −690). A long aqueduct led the canal to a low cutting across the watershed; see Jacobsen & Lloyd (1); Garbrecht (in Biswas, 1).

[b] These embankments could be as much as 90 ft. high, as at places on the Elahera-ela (vol. 1, p. 28).

[c] There is reason to think that flume (aqueduct) bridges of wood may sometimes have been used, e.g. on the Yodiye-bendi-ela before its junction with the Amban Ganga.

also so arranged as to deliver a constant supply from the canal in dry periods, thus converting the fitful tributary into a perennial stream, and saving labour by avoiding the construction of purely artificial distribution canals.[a] Elsewhere the *yodi-elas* ran (and run) for many miles over apparently flat country. For all these works can be traced today after many centuries, and many still function.

The type-specimen of this double tank-and-canal arrangement is perhaps the Walawe Ganga catchment in the south above Hambantota,[b] where a 14-mile *yodi-ela* from the river helps to feed a large reservoir with over 400 small tanks in its hinterland, the Pandik-kulam.[c] The greatest is no doubt the 25-mile Elahera-ela,[d] which draws from the Amban Ganga to discharge into the Minneriya-wewa, a famous 4,600-acre reservoir[e] not far from one of the two classical capitals in the northern plain, Polonnaruwa. The next step was the extension of the feeder canal so that it linked a chain of great reservoirs successively. The Elahera-ela was itself lengthened in this way to a total distance of 54½ miles, replenishing two other tanks, Kaudulla-wewa and Kantalai-wewa. But before it had reached this length it was outstripped by the artificial 54-mile long Jaya-ganga, originating from the 4,400-acre Kala-wewa on the Kala Oya and running northward to connect with the Tissa-wewa and the Abhaya-wewa one after the other, those elegant man-made lakes which embrace the western side of the alternative capital, legendary Anurādhapura.[f]

As the Abhaya-wewa[g] dates from *c.* − 300 it brings up the question of our time-scale. Sinhalese hydraulic engineering was not born quite at the same time as Sinhalese Buddhism, but very nearly so.[h] For the mission of the apostle Mahinda (*d.* − 204), younger brother of Asoka himself, began about − 251, in good King Dēvānampiya Tissa's golden days, thus just about a decade before the completion of the Tissa-wewa.[i] The vast majority of the other waterworks in Ceylon were constructed during the ensuing millennium. The Kala-wewa took its first form about + 80 but absorbed other tanks to reach its present stature under Dhātusena about + 479; the Jaya-ganga had been started a couple of decades previously. The Minneriya-wewa with its canal the

---

[a] Brohier & Abeywardena (1), pp. 6, 23. In some cases the farmers used temporary 'pick-up' stick dams on these streams lower down to fill their irrigation ditches, analogous to the Kuanhsien practice, but on a smaller scale (cf. *loc. cit.* diagr. 6).

[b] This is the coastal town where Leonard Woolf was once Assistant Government Agent (cf. Woolf, 1).

[c] See vol. 3, pp. 15 ff., maps opp. pp. 16, 18. For another excellent example of this type pattern, Mahātabbowa-wewa, see vol. 2, pp. 31 ff. and map opp. p. 2.

[d] See vol. 1, pp. 14 ff. Its average width is 100 ft.

[e] Its earth dam is 50 ft. high. On the whole region, which is known as Tamankaduwa, see vol. 1, pp. 19 ff.; Brohier & Abeywardena (1), diagr. 2, and many photographs.

[f] See vol. 2, pp. 7 ff. The gradient of the Jaya-ganga is only 6 in. per mile, but it crosses two watershed saddles in deep cuttings. Its average width is 40 ft. Later the Jaya-ganga was tapped on the right to help feed a second series of great tanks lying east of Anurādhapura, Nachchaduwa-wewa (late + 9th century), Nuwara-wewa (− 1st century) and Mahāgalkadawela-wewa. It was also tapped on the left to help feed the two Wilachchiya-wewas down the Talawe Oya valley (cf. vol. 2, pp. 28 ff.).

[g] Alt. (Tam.) Basavak-kulam.

[h] A doubtful tradition (vol. 1, p. 4, vol. 3, p. 8) takes the Panda-wewa (near Chilaw on the west coast) back to the − 6th century, approaching the time of Sunshu Ao, but a − 4th-century dating is quite acceptable.

[i] Note the striking similarity in date with the projects of the State and Empire of Chhin—the Chêngkuo and Kuanhsien irrigation systems and the Ling Chhü Canal.

Elahera-ela date from rather earlier, in the time of Mahāsēna (r. +277 to +304), but the latter did not reach its full length till the end of the +6th century.[a] The Walawe valley system may go back earlier than Mahāsēna, to about the +2nd century. Among the latest of the works is the Parākrama Samudra (Sea of Parākrama) reservoir, built by Parākrama Bāhu I, adjoining Polonnaruwa, between +1153 and +1186; it submerged the old Topa-wewa of Upatissa II (+368) and other earlier tanks. Parākrama's most grandiose conception, however, the culmination of Ceylonese hydraulics, was that of the Giants' Tank near Mannar in the north-west, an artificial lake with a 6½-mile embankment on three sides of a square, because sited on a sloping plain and not in a river-valley at all. A very fine *anicut*[b] diverted the waters of the Malwatu Oya, but the 17-mile feeder canal, falling only 12 ft. in 11½ miles, was apparently never completed,[c] doubtless owing to the king's death. Though little new was built after the +12th century, the dams and tanks of the ancients were so excellently planned that many, if not most, of them have been restored in our own time, and are again in use today.[d]

The designs so far mentioned were far from exhausting the inventiveness of the Sinhalese engineers of old. Some trunk canals (*yodi-ela*) do not end in reservoirs directly, but convey water from a high division-head along a lateral course above a river, providing irrigation water on the way. It is believed that some of these eventually discharged into a large tributary at a point above an *anicut* on it serving a great tank still lower down.[e] The outstanding example of this is the +5th-century Minipe-ela,[f] which parallels the Mahaveli Ganga (Ceylon's greatest river) for some 50 miles and then almost certainly ran into the Amban Ganga above the take-off for the +4th-century Tōpa-wewa (and the later Parākrama Samudra), thus greatly augmenting its perennial supplies.[g] Another system was to lead off two canals from the same weir; this was done at the Kalinga *anicut* on the Mahaveli Ganga, a work of the +5th century, probably due to Dhātusena. All kinds of combinations can be found. Thus the Wahalkada-wewa in the Yan Oya basin far out in the northern plain was fed partly by a *yodi-ela* poaching across the watershed from the neighbouring Ma Oya valley, but as

[a] The Kaudulla-wewa, later incorporated in the system, was an earlier work, dating from about +100, while the Kantalai-wewa was built by Aggabodhi II c. +590, who at the same time restored the Elahera-ela.

[b] There is a striking photograph in vol. 2, opp. p. 20. The downstream profile is stepped and sloping almost in flume meter style, as if the builders had had an inkling of the principle of the 'hydraulic jump' (cf. Brohier & Abeywardena (1), opp. p. 20). Such structures are easier and cheaper to maintain than those with almost perpendicular falls. On the first analysis of the hydraulic jump by Giorgio Bidone (+1731 to 1839) see Rouse & Ince (1), p. 143. A study of the structure of the ancient Chinese weirs and spillways would be of great interest, for many of the old works still remain, but we have not come across anything throwing light on this point.

[c] See vol. 2, pp. 18 ff., 23 ff. The +17th-century Dutch thought of reconditioning the Giants' Tank, but preferred to make transport-canals in the flat country near Colombo, and it was eventually done by the British in 1897. Though holding only 4,400 acres of water instead of Parākrama's planned 6,400, it was still at that time the third largest reservoir in the world. The Dutch also tried, not very successfully, to reclaim the swamp lands north of Colombo (cf. Abeyasekara, 1).

[d] One tank alone remained unbreached through the ages whether by flood or war, Minneriya-wewa.

[e] See vol. 1, pp. 9, 13.

[f] For diagrams of the headworks and a chart see vol. 1, opp. p. 10.

[g] Similarly the Yodiye-bendi-ela almost certainly ran into the Amban Ganga just above the Elahera take-off.

the lay of the land did not favour direct irrigation of rice-fields from the tank it mainly discharged into the Yan Oya, which was then made to generate lateral canals at four *anicuts*, three on the right bank and one on the left.[a] The Ma Oya is notable for its Padawiya-wewa, built by Duttha Gāmani in the −2nd century and enlarged by Parākrama Bāhu in the +12th, an exceptionally large tank, some 10,000 acres in extent.[b] Lastly, resort was had on occasion to tunnelling. The Buhu-ela at Pattipola in the mountains originates from an *anicut* on the Kotmale Oya, traverses a couple of tributaries with the usual spillway crossings, and then plunges through a ridge by means of a 220-yard tunnel (*buhu-kottu*, dug through compacted quartz gravel, with five shafts) to irrigate terraced rice-fields in a quite different valley, that of the Uma Oya.[c]

It remains only to mention some of the more interesting special devices of the Sinhalese engineers.[d] Already in the +1st century they understood the principle of the oblique weir, and had their *anicuts* traversing the stream at angles never more than 45° to the line of current flow, thus guarding against shocks that might dislodge the masonry.[e] Only later, when better stonework with hydraulic cement on rock foundations could be used, did they transgress this rule. The outer layers of *anicut* stones also had raised lips or mouldings so that each course was retained in position not only by its own weight but by the difficulty of forcing it forward by pressure from behind. Wootz steel tools were used in the dressing. The heights of dam spillways (*gal-pennuma*) were adjusted by removable pillars (*kalingula*) which would hold up with boarding an extra depth of water if larger retention was desired.[f] Sluices were well understood, as grooved stone abutments remain to testify. The inside surfaces of reservoir embankments were faced with ripple-bands, i.e. stone revetments (*relāpana*) of pitched work to prevent wave-erosion, and some of the greater tanks had submerged bunds which acted as wave-breaker groynes.[g] But perhaps the most striking invention was the intake-towers or valve-towers (*bisi-kottuwa*) which were fitted in the reservoirs, perhaps from the −2nd century onwards, certainly from the +2nd.[h] These were built of close-fitting cyclopean masonry half in and half out of the bund, so that the water spilled down within their walls and left the tank by double horizontal sluice-tunnels or culverts (*horowwa*, *sorowwa*) contrived at the base of the embankment. In this way silt- and scum-free water could be obtained, and at the same time the pressure-head was so

[a] See vol. 1, pp. 25 ff. Perhaps these works were as much for flood control as for irrigation.

[b] See vol. 1, p. 23, vol. 2, p. 25.

[c] See vol. 3, p. 33. Cf. p. 334 above.

[d] Of their surveying instruments we know practically nothing (cf. p. 331 above, and vol. 2, p. 21). Brohier writes (vol. 1, p. 3) that 'most of the irrigation schemes in Ceylon are on tracts of land which when estimated by the eye appear quite flat. Yet we know that channels were traced mile upon mile along gradients which call into use for their establishment the most precise instruments of the modern age. The same baffling ingenuity which no available present-day means can surpass traced out the bunds and contours of the larger tanks.'

[e] See vol. 3, p. 24. Though also an ancient Indian practice, this was not adopted in Europe until much later times.

[f] See vol. 2, p. 13, vol. 3, p. 21. Cf. the Chinese stop-log gate.

[g] Brohier & Abeywardena (1), p. 7.

[h] One of the best of these can still be seen at Vendarasan-kulam in the northern plain (private communication from Mr Delay). This tank conserves overflow from Kantalai-wewa. Cf. vol. 1, p. 19.

reduced as to render the out-flow controllable.[a] How the heights of the sluice-towers were anciently adjusted is not exactly known, but the larger reservoirs seem to have had three or four of different heights, while other traces suggest that the tops could be raised or lowered by detachable woodwork or fireclay rings.[b] Finally, the Sinhalese engineers were not without their charts, though hardly any have survived. We possess however a rare map of the Elahera *anicut* and canal leaving the Amban Ganga, with a contribution from the Kalu Ganga by way of the Yodiye-bendi-ela (one of those canals planned to arrive at *anicuts*), and making its way across a number of tributaries in the usual manner towards the great Minneriya-wewa.[c]

Such were the highest developments of South Asian hydraulic engineering in ancient and medieval times. When we turn to the East Asian theatre, and look back upon all the Chinese achievements which have been described in this Section, we are struck at once by a marked difference of emphasis. Irrigation, though important, is no longer supreme, and has to share its prominence with an unceasing struggle for river-control and a constant preoccupation with inland water transport.[d] The tank simple, as we might call it, begins, analogously, with Ceylon and India, though earlier,[e] in the −8th century or so, as the early Chou *Shih Ching* bears witness (p. 269),[f] and reaches substantial development by the early −6th in the Anfêng dam of Sunshu Ao (p. 271).[g] Though similar projects continued to be built down to the +13th century and beyond (p. 297),[h] the great reservoir was not destined to be the most characteristic form of Chinese hydraulic technique. The urgent need for a conservancy of rivers greater than almost any in the Indian culture-area made itself felt equally early, if we may accept the −7th-century dating for the first dykes of the Yellow River (p. 232).[i] From that

[a] In modern practice the water of irrigation tanks is generally drawn off from the bottom, because powerful mechanically operated sluices are now available. In one medieval Ceylonese case (vol. 1, p. 17) the openings of the sluices under the *bisi-kottuwa* were found to be seven times smaller than their openings on the delivery side of the bund—again a device for moderating the pressure.

[b] Brohier & Abeywardena (1), p. 56, describe the socket holes and slots for timberwork at one such tower, dating from *c.* +600. It is probable that ancient Chinese dams were also provided with devices equivalent to *bisi-kottuwas*. The Nanking Museum possesses an interesting Han pottery model of a small dam from a tomb at Phêngshan in Szechuan, the sluice-tunnel of which is backed by an internal wall forming a triangle with the dam itself but not built quite so high. Water would spill over this, and so issue through the sluice-tunnel at low pressure. The height of the intake wall would no doubt have been adjustable by some kind of shuttering. A filter-grid of cast iron for such a culvert, about 2 ft. 6 in. square, is in the Shensi Provincial Museum at Sian; it is considered Thang in date.

[c] See Brohier & Paulusz (1), pp. 192 ff. and pl. 54; Brohier & Abeywardene (1), opp. p. 4.

[d] The Elahera canal system in Ceylon was certainly used for boat traffic, according to ancient tradition (vol. 1, pp. 13, 16), and probably some of the others; but it seems clear that the craft used were very small, and none of the canals in Ceylon was planned with any obvious transportation purpose in mind.

[e] The date of the most ancient hydraulic works in the Old World as a whole is still obscure; we need to know more, for instance, about the nature of the irrigation projects of Achaemenian Persia (cf. von der Osten, 1).

[f] How far did the people of the Shang kingdom practise irrigation? Apparently very little (Chêng Tê-Khun (9), vol. 2, p. 197), though they understood something of town drainage (*loc. cit.* p. 48). The question is distinctly important for the comparative study of the genesis of civilisations from neolithic cultures, and needs further investigation. The loess soil which the Shang cultivated is highly amenable to irrigation (cf. p. 225 above), and we know that they had rice as well as other crops.

[g] Mem. also the reservoirs enumerated in the Domesday Book of Chhu in −547 (cf. p. 259 above).

[h] Indeed until the present time, but now in integrated groups of T.V.A. style (Fu Tso-Yi, 1).

[i] One would like to know how far at that time the Yellow River had already built up its bed across the plain, but I have not seen any convincing computation.

time onwards incredible efforts had to be devoted to the embanking and training of the Ho and the Chiang and their tributaries—a work the consummation of which is not yet even now complete.[a] In the course of it there arose quite naturally (if indeed it was not an importation from the Fertile Crescent)[b] the conception of lateral derivate irrigation canals with many branches,[c] and these indeed we find from the − 5th century onwards; associated with the names of Hsimên Pao (d. − 387) who diverted the Chang River north-eastwards to irrigate the left lower Huang Ho basin (p. 271), and Li Khuei (d. − 380), both of the State of Wei.[d] By the − 3rd century such works had attained, as we have seen, an astonishing level of sophistication, as in the Chêngkuo project (p. 285), the Kuanhsien system (p. 288) and the Ninghsia desert reclamation (p. 272). In the + 2nd century we have noticed a system of *anicuts* and canals very similar to those of Ceylon (p. 281). Thus was the Babylonian pattern transplanted to Chinese soil.[e]

But now a part of the orchestra unknown in South Asia, and almost unknown in the Middle East, took up the theme, for surprisingly soon the Chinese were building their wonderful transport canals.[f] Here we need not enlarge upon the social motivations of this, already sufficiently evident (p. 225), but seemingly far stronger than in any other ancient and medieval civilisation.[g] For nearly twenty centuries the Chinese alone appreciated the great mechanical advantage which artificial navigable waterways could offer for the systematic transportation of heavy goods,[h] anticipating long in this the industrial revolution of the + 18th century and dazzling all foreign observers who visited their works prior to that time. The Hung Kou (p. 269), even if of the − 4th rather than the − 6th century (and there are arguments for the earlier dating), has good claims to be the first important practical inland artificial waterway in human history,[i] and the

[a] To gain some idea of what floods have meant in Chinese life, even in our own time, with all its resources, see Wan Nung (1); Yang Wei-Chun (1) and Alley (7).

[b] Cf. p. 334 above, on hints of close connections with Iranian culture in ancient times. But where the technological 'elements' of water, earth and wood are concerned, people in very different parts of the world may well have developed their own traditions fairly independently—for metal I would not like to say as much.

[c] We have seen the connection clearly enough in the famous speech of Chia Jang on his three alternative projects for Yellow River control (p. 235 above).

[d] By this time Ta Yü, Yü the Great, originally a god who had made the earth rise above the surface of the waters, was turning into a human culture-hero of hydraulic engineering (cf. Vol. 1, pp. 87 ff.), and soon after became honoured as the founder of the Hsia kingdom. Perhaps he was part of the southern 'proto-thai' component of Chinese culture (Vol. 1, p. 89).

[e] The Egyptian one found no home there exactly (except in the form of the river-valley dams) until the retention-basins of the great rivers which our own contemporaries have built (p. 238).

[f] Note how Ssuma Chhien places the emphasis in his words on canals quoted on p. 226 above.

[g] The primary drive was of course the transport of taxes in kind (grain, textiles, etc.) from the periphery to the centre. At the same time water communications were important for the movements of officials and personnel, especially after the decay of the Chhin and Han road system (p. 30). But also one must not forget a considerable military element (cf. pp. 265 and 299 above). This took at least four forms, the diversion of natural waterways to sap city-walls, the weapon of floods caused by deliberate dyke-breaking, the defensive use of artificial canal networks against cavalry, and most significantly the construction of important canals chiefly for the transport of army supplies but also for the passage of warships. The first two of these had Babylonian antecedents, but the other two, not at all developed in South Asia, are rather characteristically Chinese.

[h] Cf. p. 216 above.

[i] The Nile/Red Sea Canal has been taken into consideration on p. 356 above and we shall hear more of it below (pp. 465, 609). Although said to have been begun in the time of Sunshu Ao (c. − 600) and con-

Han Kou's early − 5th-century date is in no doubt at all (p. 271). These works, helped by lakes and flat country, were the ancestors of the Pien Canal (p. 307) and ultimately of the more daring Grand Canal, oldest summit canal in any civilisation (+ 13th century, p. 312). In the − 3rd century Shih Lu had scored another triumph with the oldest contour transport canal in any culture, the Ling Chhü (p. 299), taking the First Emperor's barges and warships across a mountain range, just as some centuries earlier the Assyrians had led their irrigation water poaching from valley to valley. In the − 2nd century, again, Chêng Tang-Shih and Hsü Po of the Han set a pattern cardinal for later ages in the West when they designed and built a transport canal parallel to a navigable river but more suitable for heavy traffic (p. 273).[a] Thus this history affords plenty of laurels—if the Chinese ever felt inclined to rest upon them.

A couple of basic questions remain. If one asks where the works first arose one has to visualise initially, in early Chou times, a wide area from the lower Yangtze basin northward to the lower Yellow River, i.e. the North-east and the East-central key economic areas (cf. p. 226 above, and Fig. 36); then in the Warring States time the addition of the North-west (Kuan-chung, the Wei valley) and Western (Szechuan) areas. Already in the Chhin dynasty (− 3rd century) all parts of China, except perhaps Fukien and Yunnan, were open to operations. If one asks what the earliest works were, it would seem that the impounding of run-off water from hill valleys in tanks by dams, with small derivatory canals, came first, quickly followed by the river-controlling dykes; then after no long time the cross-country navigation canal, perhaps even preceding the river-derived lateral irrigation canal system. It is quite clear therefore that the Chinese achievements, though conditioned of course by physiographical and social features, were deeply original, a symphony on benefit of water by a different composer.[b]

If lastly one asks how the Chinese picture fits in to the framework of the annual–perennial concept, the answer is not very easy. Nowhere in China were two highly contrasting hydrographic zones in that proximity which pertained in Ceylon. With a climate essentially monsoon in character[c] all China's rivers had a marked rise and fall, e.g. the 100 ft. excursion of the Yangtze at Chungking. They were therefore annual in this sense, but perennial in that the low-water levels were themselves formidable—the

tinued in the Confucian age (c. − 500), it was not completed till the hegemony of the Chhin (c. − 280). How far it was ever used for heavy traffic is difficult to say, but as late as + 760 it permitted direct sailing without transhipment from India to the Atlantic, and may well have been important in the spread of nautic techniques. According to Bourdon (1) and Toussoun (1, 2) it ran in its prime for 60 miles with a depth of 17 ft. and a width of some 120 ft., carrying traffic till its final destruction before + 775. It is just conceivable that Chinese ships occasionally used it (cf. p. 465 below). If so, their masters may have found it less striking than the Pien Canal at home, some 750 miles in length with numerous lock-gates. The predecessor canals, the Hung Kou and Han Kou, had certainly been no shorter.

[a] There were many subsequent examples of this in Chinese history, including notable short stretches to circumvent bad river passages, e.g. the + 8th-century San-mên gorge by-pass of Li Chhi-Wu (p. 278).

[b] The thought lies near that if that intimate relation suspected by so many between feudal-bureaucratism and hydraulic works should prove to be well-founded, the precise character of the hydraulic works may have great sociological importance. If Ceylon seems not to have generated that form of society very markedly, while China undoubtedly did, the differences considered in this Section may well be meaningful relative to this diversity.

[c] Cf. Vol. 3, pp. 462 ff.

Tibetan snow-fields saw to that. In the great river-valleys there was always the tendency to lead off water for irrigation when the annual flood crest was passing, though this was opposed by many engineers, who wanted the strong current to scour out the bed. On the other hand there were works like the Chêngkuo and Ninghsia systems which could draw upon a reliable perennial flow.[a] As for the most interesting of Chinese systems, that of Kuanhsien, it partook of both types, for while not purely inundatory (since a minimal flow continued in the Neichiang even during the dry season, except for the period when the canal was deliberately closed for the dredging of its bed), it was not strictly perennial either (since the difference between flood and dry season conditions is so large). In a word, all possible variations and combinations could be found in China, and it was perhaps for this very reason that the invention and widespread use of sluices and lock-gates took place there so early (cf. pp. 344 ff. above).[b] It was the merit of the Ceylonese to fuse topographically separated annual and perennial waters by a hydraulic *tour de force*; the Chinese faced a fusion of the two implicit in all their problems from the beginning, and they too found ways to harness them for the good of mankind.[c]

Since European climate and agriculture rendered irrigation almost unnecessary, and since European rivers presented few serious control problems, European interest, when eventually it arose, centred mainly on transport canals. With one exception already mentioned (p. 357) the Hellenistic world accomplished relatively little in any of these fields.[d] Medieval society was always too incoherent and decentralised to permit thought of such projects,[e] and the Renaissance was rather slow off the mark in starting hydraulic works of any considerable size. In view of all that has been here set forth, we shall now be quite unable to accept the view of Parsons that until Leonardo's proposals for the Arno canal 'no engineer had planned a piece of excavation on such a scale'.[f] Nothing at all comparable with the works of Cathay was accomplished until the four

[a] Perhaps this was the reality behind the tradition in Mencius III, (1), iv, 7, where he says that Yü the Great 'cut a vent' for the Ju and the Han Rivers, and 'arranged' the Huai and the Ssu, so that their waters all fell into the Yangtze (cf. Legge (3), p. 127). See the comment of Fu Yin on this, c. +1160, in his *Yü Kung Shuo Tuan*, ch. 2, p. 16a. The Chhien-Lung editors, in their evaluation from *SKCS/TMTY*, ch. 11, p. 4a, prefaced to his book, say that this was the oldest of all methods of irrigation.

[b] Ancient Egyptian projects, it has been noted, suffered severely from the lack of that control which is given by well-built and well-placed gates (de la Blanchère, 1).

[c] It would be good to know more of the hydraulic history of the peripheral cultures of the Chinese culture-area. Literature on this subject appears to be scarce, and what there is suggests that irrigation was their chief interest. See the monograph on Korean works from the +14th century onwards by Yi Kwangrin (1, 1).

[d] A convenient conspectus of what it did will be found in Forbes (17). It must be granted noble aqueducts (especially Roman), good piping of bronze and lead (Phoenician, Pergamene and Roman), a few striking conduit tunnels (notably the 1,100-yard Samian one, but also Roman), some singularly unsuccessful swamp drainage (Boeotian, Pontine and East Anglian), and a short-lived canalisation of one of the mouths of the Rhône. This about completes the balance-sheet. The near absence of aqueducts in China must certainly be related to the abundance of waterways surrounding and penetrating most of the principal cities—fortunately the Chinese early learnt to boil their drinking-water, hence the apotheosis of tea. Bamboo pipe-lines were prominent, however (cf. Vol. 4, pt. 2, p. 129).

[e] With the exception of the coastal defence works and marshland drainage of the Low Countries, which began as early as the +7th century; see Forbes (17) again.

[f] (2), p. 330. Cf. Ucelli di Nemi (3), no. 62. Leonardo's inventions for the mechanising of excavation were, of course, brilliant; cf. Uccelli (1), pp. 341 ff.; Gille (3). He also had several projects for dredgers (Ucelli di Nemi (3), nos. 7, 8).

great seventeenth-century canals of France,[a] the last of which was not finished till as late as +1775. None of these was longer than 150 miles.[b] There were only 630 miles of canals in all France by the end of the eighteenth century, and even by 1893 the total mileage in that country was only three times that of the Grand Canal in +1300.[c] The canals of all Europe probably still fall short of the Chinese artificial navigable water-ways in mileage.[d] Similarly the dimensions of the early nineteenth-century canals in England (about 5 ft. deep and 45 ft. broad)[e] were less than that of the Yuan Grand Canal (from 10 to 30 ft. deep and often 100 ft. broad). Again, the locks on the Cale-donian Canal in the same period were of just about the same dimensions as those on the Ling Chhü and Pien Canals in the +11th, i.e. 170 ft. long and 40 ft. wide.[f]

It is tempting to make a comparison between the canals of Lombardy, among the first artificial navigable routes in Europe,[g] and the waterways of the Chhêngtu plain (p. 289 above), since both cover an area about 50 miles square. The Naviglio Grande originated in the irrigation needs of the Duchy of Milan. Water was brought to the city from a point on the Ticino River 31 miles away by a canal with a 110-ft. fall which was finished in +1209. Sixty years later its cross-section was enlarged and flash-locks built to allow of the traffic with Lake Maggiore from which the canal then took its name. In +1359 it was extended, for irrigation purposes only, down the Ticino valley half-way to Pavia. After the cathedral of Milan was begun in +1387 marble was brought south by way of the Naviglio Grande, and at an early date in the +15th century this was put into communication with the city's old moat, which thus acquired the name of the Naviglio Interno. But as the moat was fed from a different source it stood normally at a different level, so that there were long waits at the single stanch while the levels equalised. In +1438 the first pound-lock in Italy was installed there as the solution of this problem. After +1451 much expansion took place under the house of Sforza, when Bertola da Novate (cf. p. 358 above) was Ducal Engineer. A new extension of the Naviglio Grande to another point halfway to Pavia, for traffic this time, was built with the use of 18 pound-locks (+1458). A decade later Bertola built the Martesana Canal, connecting Milan with the Adda River; for crossing two tributaries at right angles he adopted not the Sinhalese and Chinese spillway technique, but a 3-arch aqueduct in one case (the first of its kind) and a large culvert in the other. The Martesana was intended to link Milan with Lake Como, but the Adda needed a by-pass at Paderno, and this though well planned was never completed. All these water-ways were supplemented by a considerable number of subsidiary irrigation canals and

[a] These were (a) the Aude–Garonne, planned +1516 by Francis I and the lord of the manor of Clos Lucé (Leonardo da Vinci), surveyed +1539 and +1559, completed +1681; (b) the Loire–Saône–Rhône, planned +1516, completed +1765; (c) the Seine–Yonne–Saône–Rhône, planned +1574, com-pleted +1775; (d) the Seine–Yonne–Loing–Loire, planned +1603, completed +1642. Details in Morel (1); Parsons (2); Skempton (4); Espinas (1); Pinsseau (1); Andreossy (1). The Canal de Briare (d) was not, as others besides Espinas have claimed, the first summit canal in the world.

[b] (a) 144 miles, (b) 72 miles, (c) c. 100 miles, (d) 37 miles.   [c] Vernon-Harcourt (1), p. 481.

[d] Unfortunately we have not been able to find even an approximate figure for this.

[e] Vernon-Harcourt (1), p. 351. Cf. Rolt (3); Hadfield (1, 2, 3); de Maré (1).

[f] Vernon-Harcourt (1), p. 377.

[g] Detailed accounts of these will be found in R. B. Smith (1); Parsons (2), pp. 367, 399; Skempton (4), pp. 444 ff. and fig. 280; see also Calvert (1); Hadfield (4).

conduits. Nothing could have been more different than the problems, aims and operations in the Szechuanese and Lombard areas, yet the same basic devices were needed, division-heads, weirs, dams, excavations, bridges, and the like. We may thus accept some parallelism between them—the only thing one can hardly avoid noticing is that the European works were carried out with various deficiencies, false starts and failures, more than a millennium and a half after the successful inauguration of the former.

All our comparisons can be as yet only preliminary and tentative. But it is evident that a comparative world survey on quantitative as well as qualitative lines would set the Chinese achievements in a highly favourable light. Thus we face once again a paradox. The lack of developed Euclidean deductive geometry did not prevent the Chinese engineers from executing works successfully on a scale hardly to be called anything but colossal. Nor did the absence of mathematical hydrodynamics inhibit them from developing effective training and control techniques. Astonishing were the successes of a cumulative empirical tradition from which the intuitive and the rational had never been excluded.

In effect, the story of the hydraulic works of China is nothing short of an epic. If the nature of the climate and the soil dictated them, if the form which Chinese society inevitably took necessitated and fostered them, they were nevertheless not achieved without the toil, more often willing than otherwise, of unnumbered millions of men and women, nor yet without the devotion, skill and ingenuity of successive generations of civil engineers worthy to be compared with those of any other people. The only fitting words with which this Section can conclude are those used by Ssuma Chhien twenty centuries ago when he wrote the last lines of his chapter on the River and the Canals:[a]

In the south I have climbed upon Mount Lu[b] and seen the Nine Rivers with the courses which Yü the Great gave them; I have visited the Kuei-chi mountain[c] and been to Thaihuang; I have sat on the Ku-su (terrace)[d] and contemplated the Five Lakes. In the east I have viewed from Mount Ta-phei[e] the place where the Lo River joins the Huang Ho. I have sailed upon the Yellow River itself, and travelled on the canals which join the rivers Huai, Ssu, Chi, Tha and Lo.[f] In the west I have looked towards the mountains whence the Min River flows forth, and seen the 'Separated Hill' (at Kuanhsien) in the country of Shu (Szechuan). In the north I have been beyond the Lungmên gorges (of the Yellow River) even as far as Shuofang.[g]

And I say again, inconceivably great are the benefits and the destruction which water can produce.

And in my time I myself, when in the imperial suite, carried brushwood bundles to help to close the breach at Hsüanfang; I shared in the sorrow expressed by the emperor's ode that he made at Hu-tzu[h] and now at last I have written this treatise on the (Great) River and the Canals.

---

[a] *Shih Chi*, ch. 29, p. 8*a*, tr. auct. adjuv. Chavannes (1), vol. 3, p. 537; Watson (1), vol. 2, p. 78.
[b] In Chiangsi beside the Poyang Lake, just south of Chiuchiang on the Yangtze.
[c] In Chekiang, south of Shao-hsing.   [d] South-west of Wu-hsien in Chiangsu.
[e] South-east of Chün-hsien in Honan.   [f] The Pien Canal (Hung Kou).
[g] A Mongol region north-west of the Yellow River's northernmost bend, and the name of a city on that river below Ninghsia.
[h] Near Pu-yang in Honan (cf. p. 232 above).

29. NAUTICAL TECHNOLOGY

(a) INTRODUCTION

'Navigation', said Louis Lecomte, writing in the last years of the seventeenth century,[a]

is another Point that shews the address of the Chinese; we have not always seen in Europe such able and adventurous Sailors as we are at present; the Ancients were not so forward to venture themselves upon the Seas, where one must lose sight of Land for a long time together. The danger of being mistaken in their Calculation (for they had not then the use of the Compass) made all Pilots circumspect and wary.

There are some who pretend that the Chinese, a long time before the Birth of our Saviour Christ, had sail'd all the Seas of India, and discover'd the Cape of Good Hope; However that be, it is most certain, that from all Antiquity they had always stout Ships; and albeit they have not perfected the Art of Navigation, no more than they have done the Sciences, yet did they understand much more of it than the Greeks and Romans; and at this Day they sail as securely as the Portuguese.

The remarkable justness of this appraisal will, we hope, be apparent towards the end of the present account.[b] In the preceding part of this book, we have stood, in thought at least, upon the bridges of Quinsay and Zayton; and shared (if only in imagination) the anxiety of those who through the centuries watched the onslaughts of the seas and rivers upon their dykes and sluice-gates. It is now an appropriate moment to launch forth upon these floods, and taking our station aboard some Chinese craft, to examine what part was played by Chinese sailors in the development of the techniques of the mariner and the shipbuilder.

Ships and boats have indeed often been mentioned on earlier pages. In the Warring States period, Chuang Tzu spoke about his boat[c] which one man could carry. In the Liang, trading junks brought lenses of glass,[d] and later the writer of the *Kuan Yin Tzu* book admired the empiricism of their crews.[e] In the Sung we found evidence of the use of the magnetic compass in navigation,[f] and this was later studied in detail.[g] On more than one occasion, the maritime exploits of the Ming dynasty had to be described.[h] More than once we had a presentiment that something was wrong with the idea quite often entertained that the Chinese were never a sea-going people.

---

[a] (1), p. 230.

[b] It was expressed differently, but just as forcefully, by Rudyard Kipling, a poet not inclined to underrate occidental pre-eminence, in his perceptive verses on nautical invention, 'The Junk and the Dhow' (definitive edition, p. 738).

[c] Above, Vol. 2, p. 66.

[d] Above, Vol. 4, pt. 1, p. 114.

[e] Above, Vol. 2, p. 73.

[f] Above, Vol. 2, pp. 361, 494; Vol. 3, pp. 541, 559, 576.

[g] Above, Vol. 4, pt. 1, pp. 279 ff.

[h] In the historical introduction, Vol. 1, p. 143, and the geographical Section, Vol. 3, pp. 556 ff.

Our aim must now be to place the junks[a] and sampans[b] of China in their proper relationship, so far as is possible, with the ships and boats developed in other parts of the world. The history of Chinese nautical technology can mean little if it is not set forth in a comparative way, so that the distinctive contributions of the civilisation can be seen. Distinctive characteristics there certainly were; indeed Chinese shipping has generally been thought to stand very much apart from all other water-borne creations. Whether priorities in time also imply transmissions, even if only of ideas, to other peoples, is a problem which will again arise acutely, and a final answer may well be impossible. Ethnologists and historians of technology studying diffusion and convergence should pay more attention to ships and ship gear, since the material, though highly complex,[c] is often strangely precise, and its very complexity may help to lessen the plausibility of double or simultaneous invention.[d]

The Chinese texts which give us information about shipping will speak for themselves as we go on, but we have been able to choose only a few from a wide literature which has never yet been fully explored. The sources from which they have to be collected are in fact many and various, for unlike some other technical subjects such as agriculture or pharmacy, systematic nautical treatises did not arise in Chinese culture, or at least did not get into print.[e] Let us glance briefly then at the varieties of Chinese

---

[a] The word 'junk' applied to the Chinese ship is, according to Yule & Burnell, one of the oldest in the Eurasian vocabulary. It occurs in the travels of Odoric of Pordenone (*c.* +1330) and of Ibn Baṭṭūṭah (*c.* +1348), and appears as 'Iūchi' on the Catalan Atlas (+1375, see below, p. 471). It undoubtedly derives either from Chinese *chhuan*,[1] a ship (Cantonese *shuēn*, *suēn*), or from the cognate Javanese and Malay words *jong* and *ajong*. It cannot come from *tsung*,[2] which means a fleet or squadron (Pao Tsun-Phêng, 1), cf. p. 491.

[b] This word is generally said to derive from the expression *san-pan*,[3] meaning 'three boards'. It made its way into western languages as a loan-word from certain local usages; the terms much more general in China for a small boat are *hua tzu*[4] or *thing tzu*.[5] Peri (2) thought that the expression *san-pan* was first mentioned in Chinese books in the late +17th century, applied to a kind of landing dinghy used in the Pescadores Islands. Some Chinese writers took the word to be a Malay one, and modern authorities on South-East Asia have been equally sure that it was Chinese. Peri believed that it originated from *chamban*, a word for a small boat in the language of the Colombian Indians of South America, which would have travelled westwards across the Pacific with the Spaniards by way of the Philippines. But the transmission may conceivably have been the other way round, or the similarity fortuitous, for Aurousseau (3) succeeded in finding a couple of examples of the term *san-pan* in Chinese poems of the +8th and +13th centuries. To this we can add a clear use of *san-pan chhuan* for fishing-boats in Wu Tzu-Mu's description of Hangchow in +1274 (*Mêng Liang Lu*, ch. 12, p. 15a). Old Hirth (16) noted this long ago, we find. Cf. too the poem of *c.* +1210 quoted on p. 347 above from the *Nan Hu Chi* (ch. 2, p. 4a). Perhaps then it was really part of sailor's language, which only very occasionally got into print. On this general situation we shall have more to say later (p. 403 below).

[c] And it must be admitted, somewhat forbidding to those without sea experience, on account of its specialised terminology. But there are plenty of glossaries available, e.g. Jal (2); Gruss (1); Ansted (1); Course (1); Adm. Smyth (1) and Layton (1).

[d] Sir Walter Raleigh used to think about these matters. '...But whosoever devised the Canoa among the Danubians, or among the Gaules, sure I am, that the Indians of America never had trade with either of these Nations, and yet from Frobishers straits to the straits of Magalaine, those Boats are found, and in some parts, of that length, As I have seene them rowed with 20 Oars of a side. The truth is, that all Nations how remote soever, being all reasonable creatures, and enjoy(ing) one and the same Imagination and Fantasie, have devised according to their means and materials the same things.'

[e] The *Chung-Kuo Tshung-Shu Tsung Lu* of 1961, for example, finds only three or four late minor works to list in its relevant entry (p. 799).

---

¹ 船   ² 艐   ³ 三板   ⁴ 划子   ⁵ 艇子

literature which have to be considered, thinking first of nomenclature, then successively of shipping in general, navigation, and ship-building.[a]

Ancient dictionaries and encyclopaedias[b] are of course primary stores of sea terms. But lexicographers of all times and tendencies may also be useful, as witness for instance the Buddhist monks Hsüan-Ying and Hui-Lin with their +8th-century *I Chhieh Ching Yin I* (Sounds and Meanings in the *Vinaya*). A thousand years later the study of seafaring men and their works was still continuing when an eminent scholar Hung Liang-Chi produced at the end of the eighteenth century a discussion of nautical nomenclature, the *Shih Chou*. Besides linguistics there developed also an iconographic tradition, strongly oriented towards warships rather than civilian traders, doubtless because the former were the care of the official bureaucracy while the others could look after themselves. Oldest in this line is Tsêng Kung-Liang's service encyclopaedia *Wu Ching Tsung Yao* (Collection of Military Techniques) of +1044, but as we shall see, its descriptions of ships go back much earlier to the *Thai Pai Yin Ching* (Manual of the Martial Planet, i.e. Venus) compiled in +759. Tsêng's work was truly a focal point, for from it stemmed the pictures not only of a series of later naval handbooks, e.g. the *Wu Pei Chih* (Treatise on Armament Technology) of +1628, but also those of a long succession of naval sections of encyclopaedias on the grand scale.[c] These in their turn were gathered in by eighteenth-century Japanese books[d] which combined them with very different illustrations of ships of the island culture, ships which embodied an altogether distinct tradition of naval architecture.

Passages about shipping of one kind or another are liable to turn up almost anywhere in Chinese historical writing, whether in the dynastic histories themselves and the great compilations based on them, or in unofficial collections of memorabilia like the *Thang Yü Lin*, or in the private memoirs of individual scholars.[e] An outstanding example of the latter is the *Phing-Chou Kho Than* (Phingchow Table-Talk) written in +1119 by a man whose father had been Port Superintendent and later Governor of Canton, and dealing with the maritime life of the coast during the last decades of the +11th century. Naturally enough, there is much about shipping in the literature which dealt with foreign countries and took it in hand to describe the way thither. Thus the +3rd-century books *Nan Chou I Wu Chih* (Strange Things of the South) and *Wu Shih Wai Kuo Chuan* (Records of Foreign Countries in the time of the State of Wu) both preserve important information. So also does the record of the embassy to Korea in +1124 by Hsü Ching.[f] Perhaps the best of all classical depictions of a

---

[a] Chinese characters for the names of books and authors mentioned in the following paragraphs will be found on later pages where they have more detailed consideration.

[b] Such as *Erh Ya, Fang Yen, Shuo Wên, Kuang Ya*, etc.

[c] E.g. *San Tshai Thu Hui* (+1609), *Thu Shu Chi Chhêng* (+1726), etc. Cf. p. 427.

[d] For instance Nishikawa Joken's *Ka-i Tsūshō-kō* (Studies on the Intercourse and Trade with Chinese and Barbarians) of +1708, or better Kanazawa Kanemitsu's *Wakan Senyōshū* (Collected Studies on the Ships used by the Japanese and Chinese) of +1766. By 1808 Motoki Masahide is writing his *Gunkan Zusetsu* (Illustrated Account of (European) Warships), and the tide of modern technology is sweeping in.

[e] Of course the difficulty is to disentangle the technical information about ships from the much more copious material on naval strategy and tactics, naval engagements, and commercial and economic affairs.

[f] *Hsüan-Ho Fêng Shih Kao-Li Thu Ching* (Illustrated Record of an Embassy to Korea in the Hsüan-Ho reign-period), finished in +1167.

Chinese sea-going ship is contained, not in any of the encyclopaedias already mentioned, but in the *Liu-Chhiu Kuo Chih Lüeh*, Chou Huang's eighteenth-century ethnological account of a visit to those islands (Fig. 939).

For navigation and shipbuilding we are unusually dependent on manuscript material. At an earlier stage[a] we saw how rutters and sailing-directions were beginning to be preserved in late Sung, Yuan and Ming, after the use of the mariner's compass had become universal on Chinese ships. By the +17th century substantial books on navigation were printed both in China and Japan—the same year, +1618, saw the appearance of the *Tung Hsi Yang Khao* (Studies on the Oceans East and West) by Chang Hsieh, and the *Genna Kōkaisho* (Navigation Manual of the Genna reign-period) by Ikeda Kōun.[b] Of course the books of earlier date are still more interesting; one such is preserved in MS. at Oxford, the anonymous *Shun Fêng Hsiang Sung* (Fair Winds for Escort) which dates probably from the first half of the +15th century, the time of the great voyages commanded by Chêng Ho.[c] We shall give an analysis of it later on.

For ship-building we are again beholden to a remarkable manuscript (now at Marburg), but it is a very late one, the anonymous *Min Shêng...Chan Shao Chhuan... Thu Shuo*, or Fukien Shipbuilding Manual, which belongs to the end of the +18th century. Other useful material is contained in the *Thien Kung Khai Wu* (+1637) by China's Diderot, Sung Ying-Hsing; and there is a +16th-century record of the yards near Nanking which had built many of the ships of Chêng Ho's fleets a hundred or more years earlier,[d] but it has not yet been properly studied. How regrettable it is that Chinese naval architecture never found its Li Chieh, its systematising scholar! At any rate one would not be far wrong in believing that the shipwrights of the Ming were probably the most accomplished artisans of any age in any civilisation who were at the same time illiterate and unable to record all their skill.

So much for the texts. As we shall find, archaeological evidence comes often to our aid, in the form both of models and pictures. This we shall discuss as it arises.

Though Chinese literature itself is thus not rich in works dealing specifically with the construction of ships, there is abundant material in the works of sailors and scholars of an even more maritime civilisation. General accounts and histories of shipping, such as those of Charnock (1), Jal (1), Moll (1) or la Roërie & Vivielle; of navigation, such as that of Marguet (1) or Hewson (1); and of shipbuilding, such as those of Abel (1) and van Konijnenburg (1), can all be brought under contribution.[e] About a hundred years ago a French admiral, F. E. Paris (1, 2), laid the foundations of our knowledge of the craft of Asia, and such work has been continued by a gifted

[a] Vol. 4, pt. 1, pp. 284 ff.

[b] It would be interesting to compare these in detail and to contrast them with the similar works of Arabs and Europeans. The studies of Mr J. V. Mills in this field are awaited with eagerness. We may add mention of Sakabe Kōhan's *Kairo Anshinroku* (Safe Journeys on the High Seas), 1816 but traditional.

[c] Cf. Vol. 1, p. 143; Vol. 3, pp. 556 ff. and below, p. 581.

[d] This is the *Lung Chiang Chhuan Chhang Chih* (Record of the Shipbuilding Yards on the Dragon River) by Li Chao-Hsiang (+1553). See pp. 404, 482 below.

[e] Albums of illustrations, such as that of Toudouze *et al.* (1), have also their uses.

namesake who was our contemporary.[a] Minute analysis of traditional Chinese ships and boats has been undertaken in many papers by Cdr. D. W. Waters, I. A. Donnelly, H. Lovegrove, and above all by G. R. G. Worcester (1–3), who has given us a superb series of scale drawings and descriptions of nearly 250 types of craft. With this must be mentioned the work of Audemard (3–7), notable also for the clarity and beauty of its diagrams and sketches.[b] Firm foundations having thus been laid upon the Chinese side, other writers, such as the biologist H. H. Brindley and the fisheries inspector J. Hornell,[c] have made much progress towards a general classification of the build of the boats and ships of all cultures; while similar comparative studies on sails and rigs have been made by Sir Alan Moore, H. W. Smyth and R. le Baron Bowen. The work of J. Poujade on the ships of the sea-routes of the Indies, though suffering from a somewhat inchoate presentation and too little use of texts and datings, constitutes one of the most original and stimulating books ever written on such a subject. Most fortunately, besides all the results of ethnological work on spatial distribution, we possess also certain historical documents of vital importance, both epigraphic and textual; these we shall examine in their place. But first it will be advisable to acquaint ourselves with the general evolution of the forms of boats and ships.

(b) COMPARATIVE MORPHOLOGY AND EVOLUTION OF SAILING CRAFT

What is the fundamental relationship of the Chinese junk to all other types of craft which men have used? This can best be understood by means of a summary such as that embodied in the accompanying chart (Table 71), which is based upon the survey of Hornell (1).[d] In the first place, following this scheme, we must be concerned with hull structure, reserving methods of propulsion for a later sub-section.

At the outset it is necessary to dispose briefly of a number of primitive types of craft which had no great future before them, but some of which played their modest role in history. As will be evident, the chart is arranged according to the various natural or artificial objects which ancient people must have seen floating on the water, and upon which they conceived the idea of launching forth themselves. Thus floating baskets gave rise to a number of small boats some of which are still used at the present day.[e] They could be simply caulked, either with bitumen if this was available, as in the *quffa*

---

[a] On F. E. Paris see Sigaut (2). A biography of Pierre Paris comes in the second edition of Paris (3).

[b] This and other monographs of the same author have the great merit, unusual in these subjects, of giving Chinese characters. Unfortunately editorial efforts to bring the sinology of the late Capt. Audemard's papers into ship-shape form have not been too successful. But this matters much less here in the account of his own observations, than it does in his attempt (2) to sketch a history of Chinese shipping solely from the illustrations of ships in late encyclopaedias (cf. p. 427 below). The latter work is prefaced by a biography of this excellent man.

[c] See the brief biography by Burkill (2).

[d] The numbers in the chart refer to pages in this book, for the assistance of readers who wish to know more of the data upon which the classification is based. The only other chart of the kind seems to be that in Brindley (1), much less satisfactory for our present purpose. Both had a predecessor in Pitt-Rivers (2), still worth reading.

[e] Review by Hornell (9).

Table 71. *Chart of the development of boat construction* (based on Hornell, 1)

Floating objects

elongated objects

reeds — reed rafts, 38, 69 — ancient Egyptian boats, 48 — modern tankwa, etc. (Upper Nile), 53 — Titicaca balsa, 41

bamboos — bamboo rafts, 69, 76, 78, 79 (with centre-board, 80, 82) — ⯑

double canoe

(CARVEL-BUILT) JUNK — ⯑ only, 85 ff. — sf/n

log
— several — ⯑ log rafts, 61 ff. — sailing catamaran, 61 ff.
— single — dugout canoe, 189
— added strakes abutted — CARVEL-BUILT BOAT or SHIP, 189
— added strakes overlapping — CLINKER-BUILT BOAT or SHIP, 195

CARVEL-BUILT BOAT or SHIP, 189:
Indian, 192, 236, 248, 249 — if/s if/n
Pers./Arab., 234, 235 — if/s
S. Eu. (Med.), 194 — pf/n if/n

CLINKER-BUILT BOAT or SHIP, 195:
Oceanian, 207 — if/s if/n
N. Eu. 196, 199 — if/s if/n

balance-boards, 260
outriggers, single and double, 253, 254
outboard poling gangways, 265

cleats on strakes, bifid stem and stern, special bailers, etc.

basket

caulked with mud or bitumen (quffa, 102; hisbiya, 57; ⯑ Tongking boat, 109)

covered with bark — bark canoes, 182

covered with skin or hide — coracle, 111; curragh, 133; ⯑ Tibetan type, 99 — elongated — umiak, 155; kayak, 163–

half-fruit

⯑ tub or half-barrel, 108
pottery bowl, 98

fruit, inflated gourd, skin, or closed earthenware pot

⯑ swimming, 13
float raft, 20 — pots, 37 ⯑ pontoons — skins, 25 ⯑

and *hisbiya* of Iraq,[a] or otherwise with a kind of mud, as in Tongking in Indo-China.[b] No evidence of the use of this simple form in China proper remains, but it is probably

[a] Hornell (1), pp. 57, 102; Nishimura (1).

[b] Hornell (1), p. 109; Nishimura (1); Dumoutier (3), p. 138; Poujade (1), p. 183. The strange *ghe-song*, *ghe-gia* and *ghe-nang* sailing-boats of Annam have been described by several writers (P. Paris (3), pp. 27 ff.; Poujade (1), pp. 184, 188). These boats have stemposts, sternposts, strakes held apart by thwart timbers, and sometimes even bulkheads, but the bottoms are nothing but basketwork caulked with a special mixture of resinous substances and cow-dung. The name of an inventor of these craft has survived; he was Trân Ung-Long[1] whose *floruit* was in the neighbourhood of +968 (cf. Dumoutier (2), pp. 97 ff.; Huard & Durand (1), pp. 61, 228). Such vessels are much more seaworthy than might be supposed, and the shark-fishermen of certain Indo-Chinese beaches make use of sailing-boats the hulls of which are wholly of caulked basketwork.

[1] 陳應龍

Key and notes for Table 71

if/s frames inserted; strakes sewn together previously
if/n frames inserted; strakes nailed, pinned or morticed together previously
pf/n frames preconstructed; strakes nailed or pinned
sf/n transom bulkhead 'frames' preconstructed; strakes nailed and clamped
中 indicates that the craft occurs in the Chinese culture-area

The numbers indicate pages in Hornell (1)

It should be appreciated that the fundamental characteristic which differentiates Chinese junks and sampans from all other traditional craft throughout the world is the system of transverse water-tight bulkheads, derived as the chart suggests from a natural model ubiquitous in East Asia, the longitudinally split bamboo.

The table is explained in the accompanying pages, and the significance of most of the dotted lines will be clear from the discussion. A few words are necessary, however, on topics peripheral to it.

Thus the balance-board consists of a plank laid athwartships and projecting a considerable distance out to either side. It still exists today off the Somali coast and between India and Ceylon (Hornell (1), p. 260, (17), pp. 225 ff.) and as far east as Annam (P. Paris (3), p. 46). When loaded on the weather side by one or more of the crew, it can give an effective counterpoise to the wind. So far as we know, it was never a Chinese practice. But it was connected with the complex problem of outriggers, i.e. floats of one kind or another, fixed outboard to thwart poles (like balance-boards), and giving stability to craft too narrow or unstable in themselves to attempt long voyages otherwise; cf. Hornell (1), pp. 253 ff.; (2), (24). The outriggers may be single or double. Their focus of origin, whence they spread by sea as far west as Africa and east to Polynesia, was undoubtedly Indonesia, but the idea itself, according to Hornell, derived probably from some fresh-water technique, and he believed (1, p. 265) that this was the outboard punting or poling gangway which is an ancient feature of the river ships of South China (Fig. 966, pl.). It is simply a light and narrow platform resting on the projecting ends of a number of booms laid at gunwale level athwart the hull. All degrees of intermediate stages are found between this and the double outrigger, among the more interesting being those in which each outrigger carries a number of men paddling. On this view, therefore, the double outrigger derived from the poling gangway of the Indo-Chinese cultural contact zone, as well as from the balance-board of early sea-going Indian Ocean boats; and the single outrigger was a secondary modification which developed at both the oceanic ends of its distribution zone (Madagascar and Polynesia) on account of its greater ability to endure bad weather.

Some craft are known (on Indo-Chinese and South American rivers) which have what might be called 'aerial non-floating out-riggers', i.e. lengths of buoyant material (bamboo, balsa-wood, etc.) lashed alongside the boat but not reaching the water unless the craft heels over when rolling (Hornell (1), pp. 267 ff.). It is interesting that the *Kao-Li Thu Ching* of +1124 describes (ch. 34, p. 5a) cylinders (*tho*[1]) of plaited bamboo which were attached to the top of the hull on each side, and which 'gave protection against the waves'. These were on the ships which took the embassy to Korea (cf. Vol. 4, pt. 1, p. 280).

The sailing-raft of logs, or catamaran (from the Tamil *kattu-maram*, tied logs), is connected with the Indian carvel-built ship by a dotted line because there is some reason for thinking that while the oldest carvel-built ships of Europe derived from the ancient Egyptian raft of reeds, some of the boats of India may have derived from wooden rafts. On the coasts of the peninsula (Malabar, Coromandel) there is a tendency to build rafts of odd numbers of logs so fixed that the central one is the lowest, thus approximating to a keel (Hornell (1), pp. 62, 71, 194, 198). Cf. de Zylva (1), fig. 8.

[1] 橐

25

the *manashi-katama*[1] of Japanese legend. When the basket was covered with skin or hide, the familiar coracle or curragh came into existence.[a] The distribution of this craft, both in time and space, is much wider than has often been realised. It appears on a number of well-known Assyrian bas-reliefs,[b] from which a Mesopotamian origin seems indisputable, and it is the characteristic vessel of the rapidly flowing head-waters of the great rivers which take their rise in Tibet. Skin-covered coracles are found near Batang, on the Yalungchiang, the upper Yangtze and the upper Mekong.[c] Travellers both Western and Chinese have often described them, e.g. Yao Ying[2] in 1845 and Kingdon Ward[d] more recently. In China they are known as 'skin boats' (*phi chhuan*[3]),[e] but apart from the Tibetan borders now used only in Manchuria (*cha-ha*[4]) and Korea. Nishimura supposes that they were the *kagami-no-fune*[5] (berry-boats) of Japanese legend, and there is a reference in *Pao Phu Tzu* (+4th century) to a man paddling himself across a broad river in a 'wickerwork boat' (*lan chou*[6]);[f] much earlier Chuang Chou (−4th century) spoke of a boat which a man could come and carry away, thus probably a coracle.[g] Not long after Ko Hung's time cowhide coracles were used in considerable numbers by the first king of the Later Yen dynasty, Mujung Chhui,[7] for feint attacks during military operations along the Yellow River[h] about +386. Three coracles are depicted on the walls of a Sui cave[i] at Chhien-fo-tung. And there is much evidence that boats of this type were abundantly used by the Mongols in their +13th-century conquests.[j] But in the Chinese culture-area such craft never developed into the elongated skin-covered and decked boats of the Eskimos and northern Siberian peoples, with all the curious arts pertaining thereunto.[k] Bark canoes[l] are also remote from our sphere, yet interesting in that Hornell found reason[m] for regarding them as possible ancestors of dugout canoes since in some localities the latter show vestigial rib-like internal ridges running in relief transversely across the bottom and up the sides.

Other basket-shaped objects may also float. The pottery bowl, if large enough, may be used to carry a man, as in Bengal,[n] but its place is taken in China by the wooden

---

[a] Hornell (1), pp. 111, 133, (11).   [b] Des Noëttes (2), figs. 20, 21.

[c] Hornell (1), p. 99; Rockhill (2); Teichman (2); Rin-Chen Lha-Mo (1); Donnelly (6).

[d] (3), p. 129 and pl. 25. On Yao Ying see Hummel (2), p. 239.

[e] Chinese encyclopaedias illustrate them as means used by soldiers for crossing rivers, e.g. *TSCC*, *Jung chêng tien*, ch. 98, *shui chan pu*, *hui khao* 2, p. 4*b*; and also in use by fishermen handling traps for fish and crayfish (*I shu tien*, ch. 14, *yü pu*, *hui khao*, pp. 15*a*, 16*a*). A number of references concerning them are collected in *KCCY*, ch. 28, p. 12*b*. The oldest picture of the military coracle is doubtless in *Wu Ching Tsung Yao* (*Chhien chi*), ch. 11, pp. 16*b*, 17*a*.

[f] *PWYF*, ch. 26*a* (p. 1336.1).   [g] Ch. 6 (Legge (5), vol. 1, p. 243).

[h] *Chin Shu*, ch. 123, p. 8*a*.   [i] No. 303.

[j] Cf. Sinor (6).

[k] Hornell (1), pp. 155, 163; Brindley (8). Nor did this kind of boat-building affect the Japanese, though old pictures show that they knew of them (Nishimura). In Chinese too there is a reference of +812 about the boats of a northern tribe which may be so interpreted (*Thung Tien*, ch. 200 (p. 1084.1); cf. Sinor (6), p. 161).

[l] Hornell (1), p. 182.   [m] (1), p. 187.

[n] Hornell (1), p. 98.

---

[1] 無目籠   [2] 姚瑩   [3] 皮船   [4] 札哈   [5] 蘿藦船   [6] 籃舟

[7] 慕容垂

tub, which many writers have described.[a] It is often called the *hu chhuan*[1] or kettle-boat, and is much used for collecting aquatic food plants; Japanese legend knew it as the basin-boat (*tarai-bune*[2]).

Buoyant objects, more or less spherical, formed the beginning of another line of descent. Swimmers supporting themselves upon gourds or inflated skins are frequently seen on Assyrian reliefs of the −9th century,[b] and the method survived for centuries in many parts of the world,[c] especially for campaigning purposes, as in the Mongol armies, whence it entered Chinese military books from the +11th-century *Wu Ching Tsung Yao* onwards,[d] and was perpetuated in later encyclopaedias[e] (*fou nang fa*[3]). People are shown using them in the reliefs at Sanchi in India,[f] and four centuries earlier Chuang Chou had referred to them. The logician Hui Tzu,[g] who had received a present of large gourds (*ta hu*[4]) from the Prince of Wei, did not know quite what to do with them, and the more practical Chuang Tzu suggested to him that he should use them for crossing rivers.[h] Gourds (*hisago*[5]), inflated deer-skins (*kako-no-kawa*[6]) and closed clay vessels (*hani-bune*[7]) all occur in Japanese legend (Nishimura, 1). In +12th-century China such aids were known as waist-boats (*yao chou*[8]).[i] They are used to this day in Japan by women fishers and divers.

The transition to what might be called craft in the strict sense came when a number of buoyant objects were attached to a framework of wood to form a buoyed raft. Lagercrantz (1), who has studied the present distribution of these rafts, concludes that their original focus was the region of fast-flowing rivers rising in Central Asia. The *phi fa tzu*[9] (skin raft), composed of thirteen goatskins, is indeed common in North-west China, on the Yellow River and all its tributaries;[j] I myself have journeyed often on them in the province of Kansu (Fig. 927, pl.). Still today one sees them being carried on men's backs beneath the walls of Lanchow. I have not, however, viewed the greater rafts of which Nishimura speaks, borne upon as many as 700 sheepskins, each of which is (or was) stuffed with a cargo of camel-hair or wool.[k] Closed pottery vessels as floats seem not now to be used in China, but they certainly were in antiquity, for the Han general Han Hsin[10] achieved a famous crossing of the Yellow River with his army by the aid of such rafts (*mu ying*[11]).[l] The device is described[m] in the *Wu Ching Tsung Yao* of +1044, and late encyclopaedias also illustrate it.[n]

[a] Worcester (3), vol. 2, pp. 290, 370; Hornell (1, 9).
[b] Des Noëttes (2), fig. 21; Hornell (1), pl. 1B. [c] Hornell (1), pp. 6 ff.
[d] *Chhien Chi*, ch. 11, pp. 15 b, 16 a.
[e] E.g. *TSCC*, *Jung chêng tien*, ch. 98, *shui chan pu*, *hui khao* 2, p. 3 b. This is the origin of the figure in Hornell (1), p. 13.
[f] Mukerji (1), p. 32.   [g] Cf. Vol. 2, p. 189 above.
[h] Ch. 1 (Legge (5), vol. 1, p. 172).   [i] In Chhêng Ta-Chhang's *Yen Fan Lu*, ch. 15.
[j] Teichman (3), pp. 169, 177; Donnelly (6); Worcester (12). Cf. Sinor (6). On the conventional shipping of the Yellow River around the Great Bend see Imabori Seiiji (1).
[k] Hornell (1), p. 25. The *Huai Nan Wan Pi Shu*, probably not much later than the −2nd century, says that one can swim across rivers on skins stuffed with wild goose feathers (*I Lin*, quoted in *Shuo Fu*, ch. 11, p. 32 b; *TPYL*, ch. 704, p. 3 b, ch. 916, p. 9 a).
[l] *Shih Chi*, ch. 92, p. 5 a.   [m] *Chhien chi*, ch. 11, pp. 17 b, 18 a.
[n] *STTH*, *Chhi yung* section, ch. 4; *TSCC*, *Jung chêng tien*, ch. 98, *shui chan pu*, *hui khao* 2, p. 5 b.

| [1] 壺船 | [2] 盤舟 | [3] 浮囊法 | [4] 大瓠 | [5] 匏 | [6] 鹿皮 |
| [7] 埴土舟 | [8] 腰舟 | [9] 皮筏子 | [10] 韓信 | [11] 木罌 | |

The real key-point in the origin of most ships is thought by many to have been the observation of the floating single log, and its conversion to a convenient vehicle by hollowing out to form the dugout canoe.[a] It would then be a natural notion to increase its freeboard by adding first a wash-strake, and later gradually building upwards a succession of strakes which became the sides of the ship.[b] Thus in most kinds of ships the ghost of the dugout canoe still lives on in the shape of the keel, not indeed vestigially however, since the keel gives necessary longitudinal strength, and if projecting much below the hull, has importance for the sailing properties of the craft. An early invention was the flaring of the sides of the dugout underbody by steaming the wood, after which they were retained in place by the insertion of U-shaped frames. Finally the keel became purely a beam. From the forward end there grew out of it the stempost, and aft the sternpost by a like extension. Boat-building then diverged into two recognisably different traditions, in one of which the strakes were laid overlapping each other, while in the second they were abutted edge to edge so as to form comparatively smooth external and internal surfaces. For these two methods the terms clinker-built and carvel-built are used respectively.

This particular classification is one into which all vessels in the world will fit, but it is far from being the whole story. Broadly speaking, the clinker-build is characteristic of northern Europe alone, while the ships of all other regions (the Mediterranean, the Persian–Arab culture area, India, Oceania, and China) are carvel-built.[c] Further distinctions then arise; the strakes may be sewn together with vegetable fibres[d] or they may be nailed or secured by wooden pins (trenails). Frames and ribs, thwarts and longitudinal stretchers, may be constructed first and the strakes attached to them, or on the other hand the planking may be fitted together first and the frames inserted afterwards. This second method was that of the ancient peoples of northern Europe throughout the Viking period. From the Als boat of the −4th century to the Oseberg and Gokstad ships of the +9th, the overlapping strakes were lashed to transverse framing

[a] As the *Huai Nan Tzu* book says (ch. 16, p. 9a): 'He who (first) looked at a hollow wooden log floating on the water, knew how to make boats (*Chien khuan mu fou shui, chih wei chou*[1]).' Actual specimens of dugout canoes from Liu An's time and much earlier are now coming to light in China. At the Nanking Museum in 1958 I saw one of the Warring States period from Chhangchow, about 35 ft. long. Single tree-trunks fashioned into canoes were also used for the boat-burials in Pa and Shu (see opposite, p. 389). It is interesting that the concavities in these craft are not pointed at each end but squared off fore-and-aft. Neolithic dugout canoes excavated in Japan have been described by Nishimura (1) and Matsumoto *et al.* (1). Cf. p. 392.

[b] Most authorities agree about this process, e.g. Hornell (1), pp. 189 ff.; Brindley (2); Poujade (1), pp. 187, 214 ff. It can actually be seen in operation in various parts of the world at the present day, cf. Haddon & Hornell (1).

[c] Hornell (8); Brindley (1). But, as everywhere in this subject, there are exceptions. Sometimes, as in this case, they are few. Certain boats on the Ganges (the *patela*, *melni* and *ulakh*) are clinker-built (Hornell (9, 10), (1), p. 250). These are also anomalous in that they have balanced rudders of Chinese type. Since the clinker-build was never adequate for ships of really large size, the south European system spread to the north fairly early. Excavations of wrecked ships from the Zuyder Zee (cf. van der Heide, 1) are expected to throw light on the process.

[d] We shall see (below, p. 459) that though the Chinese early knew of this method, they never used it. When the European travellers of the +13th century first met with it (p. 465 below) they formed a poor opinion of it, not knowing that their own ancestors had also used it. For colour photographs of sewn boats see Eller (1), p. 519. Further information in Hourani (1), p. 92.

[1] 見竅木浮水知爲舟

subsequently inserted, by means of cleats left upon their inner sides during their shaping.[a] The square-sailed longships of the Bayeux tapestry (late +11th century) were still of approximately Viking form,[b] but by that time nails had doubtless taken the place of the original sewn structure, as in modern clinker-built boats. Inserted framing co-existed, however, with carvel-build in India,[c] Arabia,[d] and Oceania;[e] as also among the Egyptians, Phoenicians, Greeks and Romans. It was only later, in the medieval Mediterranean, that the rib framework was raised first, and the planking nailed on to it afterwards.[f] Although it was not the custom in late European carvel-building to join the side strakes together directly, this was frequently done in Asia, either by sewing, or by clamps and nails of many different kinds.[g]

Compared with the distinction we have now to make, all these details are relatively unimportant. For it is clear that the ships of East Asia cannot be genetically explained on the theory of the simple floating hollow log. Bamboo is their ancestral material, not wood at all, for as we shall see, the Chinese hull (however its sides are built) is an elongated structure as full of transverse bulkheads as the stem of the bamboo is of those partitions which botanists call septa. These, they would say, are the transverse solid joints at the nodes of the culms or haulms of the arborescent grasses which constitute the sub-family Bambuseae. And this construction it is which sets the Chinese ship apart from the ships of the rest of the world.

[a] Brøgger & Shetelig (1); Anderson & Anderson (1), pp. 55, 66 ff.; Hornell (1), pp. 200 ff.; P. Gille (1); Marcus (3).

[b] Anderson & Anderson (1), p. 80; des Noëttes (2), fig. 65; Panels 4, 5, 24, 35 at Bayeux.

[c] Hornell (1), pp. 236, 248, 249.

[d] Hornell (1), pp. 193, 235.

[e] Hornell (1), p. 207. One of the most remarkable discoveries in this field was that of Hornell (6) which revealed far-reaching parallels between Scandinavian and Oceanian boat construction. Certain boats of that region still in use not only look like Viking longships, but have the identical method of lashing the strakes by cleats to transverse ribs. Bifid stems and sterns, a special kind of boat-bailer, and the custom of boat-burial, are also common to both areas. It seems impossible that such a complex of techniques could have evolved independently, but there is no very satisfactory explanation of transmission from either area to the other, unless we fall back on the prehistoric 'Pontic Migration' from north-west to south-east (see von Heine-Geldern, 1, 4, 6), which might seem too ancient. Extraordinary, too, is the fact that one boat with bifid stem and stern exists in China, namely, a certain kind of sampan used near Hangchow (G. R. G. Worcester (14), p. 98; unpublished material, no. 21). Its 'rams', which form a true keel, are employed by the Chinese boatmen for carrying it on their shoulders. What foreign influence could this be due to, and at what time did it act? On the problem of bifid stems and sterns in general see Noteboom (1); he thinks that they could have arisen very naturally during the process of adding wash-strakes to the dugout canoe, but that the opportunities which the form presented for development into the heads and mouths of mythical animals were too attractive to be overlooked by the symbolists of indigenous cultures. Boat-burial is also reported from China (Fêng Han-Chi, 1) especially from the Szechuanese countries of Pa and Shu in and before the −4th century; Dr Lu and I had opportunity of first-hand study of these remains at Chungking in 1958.

[f] Hence the dual origin of British boats, some elements deriving from the Norse, and some from the Mediterranean influence; cf. Hornell (10, 12); Davies & Robinson (1). When did the change arise and spread in Europe? Apparently at some period between the +7th and the +15th centuries. We know that European nautical technology was greatly influenced from the East in other ways (cf. p. 698). Would it be a wild surmise to think this change too was an echo, a stimulus from Chinese preconstructed bulkhead shipbuilding?

[g] Cf. Hornell (13) on the tongue-and-groove seams of Gujerati shipwrights; and Worcester (2), Audemard (3) on the clamps and oblique head-recessed nails of Chinese junks. 'European' here means 'post-classical'. For in recent years a wealth of evidence has been recovered from the depths of the sea showing that the shipwrights of Roman and Hellenistic times relied much on 'skin strength'; they fixed

It must have occurred very early to primitive men that instead of hollowing out a single log to make a canoe, they could bind a number of logs together and obtain a 'ship' of considerable size. Rafts made of bundles of reeds or rushes need not much concern us; they were important for the development of sailing-craft in Egypt and the New World,[a] but not in China, though not unknown there (*phu fa*[1]).[b] Rafts of logs joined together in various ways achieved wide distribution and were of great use to many peoples, especially in the form of the sailing catamaran, so common on the coasts of India and the East Indies.[c] In China these are seldom used at sea, but very great rafts of wood still descend the Yangtze and many of its tributaries.[d] We have exact descriptions of the *shan-mu fa tzu*[2] of pine wood which comes down to Chungking,[e] the raft of the Min River which has to negotiate the works at Kuanhsien,[f] the huge *mu phai*[3] of the Lower Yangtze,[g] and the small rafts of the Miao tribesmen in Kweichow.[h] Their prototypes must have been in use in the Chou period; at any rate we hear of them in the Han, for in +47 the prince of the Ai-lao[4] barbarians, Hsien Li,[5] ordered his forces to sail downstream upon rafts (*phai chhuan*[6]) to make an attack on the Lu-to[7] tribe.[i] Nishimura (1) plausibly finds mentions of wood rafts in Japanese legend, under the terms *ame-no-ukihashi*[8] and *uki-takara*.[9] But Chinese wooden rafts were not of genetic importance; bamboo rafts assuredly were, and to them we shall shortly return.

The moment has now come to describe the basic characteristics of the Chinese junk and sampan.[j] In doing this it is necessary to put aside, as it were, much that we have learnt in the last few pages, for everything which was done in shipbuilding in the rest of the world failed to exhaust the ways in which the men of old found it possible to make ships. In Europe and southern Asia the basal beam, the keel, was scarfed at each end to another stout beam which turned upwards to form the stempost and sternpost respectively. The strakes of the hull, which connected them, were held apart in the

their hull strakes together side by side with numerous mortice and tenon joints—as if in a 'monocoque fuselage' construction. Such hulls were attached to the ribs with copper nails driven through the centres of wooden pegs. All this has been revealed by the systematic study of submerged Mediterranean wrecks; see Frost (1), pp. 225 ff.; Benoit (4, 5); Casson (3), p. 195, (5, 6, 7); Bass (1); Hasslöf (1, 2, 3).

[a] Hornell (1), pp. 41, 48, 53; Poujade (1), p. 199; Reisner (1); Boreux (1).
[b] Cf. *Wu Ching Tsung Yao (Chhien chi)*, ch. 11, pp. 13b, 14a.
[c] Hornell (1), pp. 61 ff.
[d] In +1656 Nieuhoff sketched a 'floating village', i.e. a great bamboo raft, on the Yellow River near Huai-an, and afterwards gave a copper-plate of it, (1), pp. 154 ff. Reprod. Rudofsky (1), fig. 3.
[e] Worcester (1), p. 70.   [f] Worcester (1), p. 86; cf. p. 290 above.
[g] Worcester (3), vol. 2, p. 388.
[h] Worcester (3), vol. 2, p. 470.
[i] *Hou Han Shu*, ch. 1B, p. 17b; ch. 116, p. 17b. Another example, a military crossing in a campaign against the Chhiang people (+88), occurs in *Hou Han Shu*, ch. 46, p. 10b. In this case the rafts seem to have had bulwarks made of leather. A text of about the same time, the *Yüeh Chüeh Shu*, tells (ch. 8, p. 4b) of the building of a great fleet of rafts by 2,800 sailors of Kou Chien,[10] King of Yüeh (cf. Vol. 2, pp. 275, 555 above) in −472. These may well have been sailing-rafts (cf. p. 393).
[j] The precise statement of the principles of junk construction owes much to Hornell's classical paper (7). Cf. (8) and (1), pp. 86 ff. Of course, many Chinese statements exist, and some of them we shall note hereafter (p. 462), but Chinese writers never knew enough about the ship-building of the rest of the world to realise the true originality of their own people.

¹ 蒲筏	² 杉木筏子	³ 木箄	⁴ 哀牢	⁵ 賢栗
⁶ 箄船	⁷ 鹿茤	⁸ 天浮橋	⁹ 浮寶	¹⁰ 句踐

desired profile by an internal skeleton of bent timbers. But junk design, exemplified in the oldest and least modified types,[a] has a carvel-built hull wanting in all the three components which elsewhere were regarded as essential—keel, stempost and sternpost. The bottom may be flat or slightly rounded, and the planking does not close in towards the stern, but ends abruptly, giving a space which would remain open if it were not filled by a solid transom of straight planks. In the most classical types there is no stem either, but a rectangular transom bow.  The hull may be compared to the half of a hollow cylinder or parallelepiped, bent upwards towards each end, and there terminated by final partitions—like nothing so much as a longitudinally split bamboo. Moreover, frames or ribs are replaced by solid transverse bulkheads (analogous to the nodal septa) of which the stem and stern transoms may be regarded as the outward units.[b] This is clearly a much firmer method of construction than that found in other civilisations.[c] Fewer bulkheads were required than frames or ribs to give the same degree of strength and rigidity. It was obviously also possible for these bulkheads to be made watertight, and so to give compartments which would preserve most of the buoyancy of a vessel if a leak should occur, or damage below the waterline. In other ways, also, the bulkhead structure involved corollaries of great importance, for example in providing the essential vertical components necessary for the appearance of the hinged axial rudder.[d] These, together with cognate inventions in the domain of propulsion by sail (of surprisingly early date), will be examined in due course. Here we need only stop to remark the striking similarity between the bulkhead structure of the Chinese ship and the prominence of the transverse partitions or frameworks so fundamental in Chinese architecture (cf. p. 102 above). If the latter prevented a longitudinal vista and permitted the classical curve of the roof, the former provided distinct holds, rendered the vessel extremely strong,[e] and gave it the typical bluff bow and stern of large Chinese craft. One cannot but feel that both systems were inspired by the bamboo, that plant so familiar to every Chinese from a thousand uses,[f] with its transverse nodal septa. The sampan (shan[1]) is reminiscent of the bamboo stem just as much as the junk. It is an open punt-like skiff, wedge-shaped in plan, shallow, keelless, and very broad in the beam aft, where the gunwale rail and side strakes are often continued beyond the stern as an upwardly curved projection, endowing the craft with cheeks or wings facing

---

[a] It is necessary to make this reservation, since centuries of culture-contact have influenced the design of Chinese ships a good deal, especially those which are built in the southern parts of the culture-area; a few words about such hybrid types will be given further on, p. 433.

[b] Pure bulkhead construction is seen mainly in the ships of inland waterways; sea-going ships are often further strengthened by frames and half-bulkheads.

[c] In modern times it was first fully appreciated by Admiral Paris (1).

[d] See p. 653 below.

[e] The search for hull strength led in Korean ship-building to the ingenious idea of making the transverse bulkhead timbers 'sprung', i.e. bowed upwards, each one pressing like a spring upon its strakes at each end (see Underwood (1), p. 26, and figs. 27c and 30).

[f] Cf. Vol. 4, pt. 2, p. 64.

[1] 舢

astern.[a] It was the roofing of the space between these projections that led to the over-hanging stern-gallery of the junk.

There have been several theories about the origin of the junk and sampan. One sug-gests that the design derives from a double canoe system,[b] in which twin hulls were placed parallel a short distance apart, and connected by planking to form a new bottom with square ends. Craft of this kind, however, have not been found anywhere. Nor is there any evidence of the longitudinal bulkheads to which it would have given rise. On the other hand, a double-canoe build, as such, has existed, and still does, in various parts of the world.[c] It is curious, too, that the Chinese language contains a number of old words (*huang*,[1] *fang*,[2,3] *pang*,[4] *hung*,[5] and perhaps originally *hang*,[6,7] which came to mean sailing in general), all indicating two boats lashed or secured together with cross-timbers side by side.[d] Moreover, such devices are in current use in China, notably for transporting reed-stacks downstream, and for fishing (the Ichang 'Watershoes').[e] Nevertheless, this line of approach is not convincing.

A modified suggestion is that a process may have taken place like that by which the seining-boats of Ceylon are still made. A dugout hull of the required length is sawn lengthwise into two halves, and these are then connected at the required beam-breadth by frames, to which intermediate bottom-planks are then nailed.[f] There is, however, no evidence for the use of such a method in China at any time. Moreover, while it is not possible to say that dugout canoes occur nowhere in the Chinese culture-area, they are in occurrence and distribution exceedingly sparse.[g] Generally speaking, too, they seem to have disappeared during or before the Han.[h]

[a] Examples in Worcester (3), vol. 2, pp. 316, 373. That this is an ancient feature is proved by many old paintings—for the Thang, the boats on the Tunhuang frescoes (see on, p. 455), and a painting by Wang Wei reproduced by Sirén (6), vol. 1, pl. 58; for the Sung, the painting by Hsia Kuei reproduced by Waley (19), pl. 43; for the Yuan, the paintings of Ma Yuan (Waley (19), pl. 42; Sirén (6), vol. 2, pl. 59; Binyon (1), pl. 17) and Wu Chen (Sirén (6), vol. 2, pl. 123).

[b] This is favoured, e.g., by Gibson (2), pp. 16, 32.

[c] Hornell (1), pp. 44 (Peru), 78 (Fiji), 191, 248 (India), 263 (Polynesia). Double-hulled vessels have been used for various purposes all through naval history (cf. the invention of Sir William Petty in +1662), and they are now employed for fast-sailing pleasure-boats (Brown, 1). *The Times* has published a photograph of such a craft (2 Jan. 1958), terming it, however, a catamaran. This is a 'common and deplorable error' (Hornell), for the word is properly applied not to double-canoes or outrigger-canoes, but only to log rafts.

[d] Definitions in *Shuo Wên, Erh Ya, Fang Yen*, etc. *Locus classicus* in *Huai Nan Tzu*, ch. 9, p. 12a.

[e] Worcester (3), vol. 2, pp. 488, 491; Donnelly (6). That they were common in the Thang appears from the eye-witness evidence of the Japanese monk Ennin, who noted in his diary for +838 that the transport of his party on the Grand Canal consisted of two or three boats 'joined to form a single craft' (Reischauer (2), p. 16). Later on we shall see military examples of the practice on a much larger scale, veritable floating batteries (p. 694). Cf. the double-hulled fighting-ship in *TSCC, Jung chêng tien*, ch. 97, p. 28a (text tr. Audemard (2), p. 77).

[f] Hornell (1), p. 89.

[g] Worcester knew of none, until he visited Thaiwan (cf. (14), p. 79). A report by Donnelly (6) of dug-out canoes on the Chien-Yu River, a tributary of the Han, requires further investigation. F. H. Wells (1) reported dugout canoes on the Yellow River, but gave few particulars. Tisdale (1) saw them on the Yalu River, dividing Korea from Manchuria. For Mongolia see Sinor (6). Cf. pp. 388, 389.

[h] We have come across very few references to them in Chinese literature. However, certain personages, notably the Thang poet Chhang Chien[8] (*fl.* +727), are said to have had a predilection for dugout canoes, doubtless a symbol of Taoist simplicity and primitivity. Chhang is often seen paddling such craft in the paintings and carvings of traditional art (cf. Jenyns (2), and Vol. 2, pp. 99 ff.).

[1] 艎　　[2] 方　　[3] 舫　　[4] 舽　　[5] 艭　　[6] 航　　[7] 航　　[8] 常建

The conclusion to which Hornell came (7) was quite different. He was convinced that we should look to the bamboo raft as the origin of the junk and sampan.[a] In Thaiwan (Formosa) the Old World sea-going sailing-raft,[b] he said, attains its highest development (Fig. 928, pl.).[c] Here the 'hull' is formed of nine or eleven curved bamboo poles about 18 ft. in length, strongly sheered up forward, less so aft. The sides also curve upward so that there is a concavity athwartships as well as fore and aft.[d] The fore end is narrower than the stern as the thin ends of the bamboos point forward. The bamboos are lashed to eight curved wooden bars which cross the bamboo platform, giving at each point the desired transverse profile, while along each side, beginning from about half-way between the mast and the bow, run a pair of bamboo 'bulwark' rails, projecting a little beyond the stern like the cheeks of a sampan. The mast (some 17 ft. high), which carries a typical Chinese battened lug-sail,[c] is stepped in a solid wooden block. Two pairs of paddles are carried, with one or two steering-oars at quarter or stern, and there are no less than six centre-boards[f] to prevent leeward drift and act as auxiliary rudders.

The Formosan sailing-rafts are mentioned from time to time in old Chinese litera-ture, especially because of the raids of Thaiwan aborigines on Chinese coastal villages in the + 12th century. About + 1225 Chao Ju-Kua, speaking of the people of Southern Formosa (Phi-Shê-Yeh [1]), wrote that 'they do not sail in junks or rowed (boats) but lash bamboo into rafts, which can be folded up like screens, so that when hard pressed, a number of them can lift up (the component parts) and escape by swimming off with them'.[g] Another reference to such events is in the *Sung Shih*[h] which describes the sailing-rafts used by the Liu-Chhiu islanders in piratical raids on the Chinese coast between + 1174 and + 1189. But our oldest picture of these vessels[i] was drawn by a Japanese sailor, Hata Sadanori, as late as 1803.

If an evolution of the junk from the bamboo raft form is envisaged, it is only necessary to suppose a conversion of the wooden thwart-beams into bulkheads, the

---

[a] Poujade (1), p. 246 concurs.

[b] *Tek-pai*, i.e. *chu phai*[2] or *chu fa*.[3] Also called *fan fa*.[4] There have been many descriptions of it (e.g. Nishimura (1); Worcester (9), etc.), but the most recent and complete account is that of Ling Shun-Shêng (1, *1*). A well-known model is depicted by Hornell (1) in pl. XIIIA. The craft is used primarily for deep-sea fishing.

[c] But the type is equally characteristic of Northern Annam (Vietnam), where it may carry as many as three masts, curved in a special way and mountable in three or four sockets in a thwartwise timber, according to the weather. Descriptions in Claëys (1, 2); Paris (3), pp. 59 ff., figs. 45 to 52 and 233; Huard & Durand (1), figs. 106, 107. We are much indebted to the late Mr Paris for information in correspon-dence and for a photograph of the three-masted sailing-raft (Piétri), which recalls in its proportions the Lake Thaihu five-master (see Fig. 1017 below). The Indo-Chinese sailing-rafts are sometimes equipped with a small rudder.

[d] This tendency is also markedly seen in certain of the wood rafts of Japan, especially those called *nabe-buta* (Nishimura, 1).

[e] See below, p. 595. One Formosan type has a sprit-sail, see below, p. 589.

[f] Generally only three of them are in use at one time. The centre-board is, we believe, another invention of the Chinese culture-area; see below, p. 618.

[g] *Chu Fan Chih*, ch. 1, p. 39*b*; tr. auct. adjuv. Hirth & Rockhill (1), p. 165.

[h] Ch. 491, p. 1*b*.

[i] It is reproduced in Ling Shun-Shêng (1).

[1] 毗舍耶   [2] 竹桴   [3] 竹筏   [4] 帆筏

substitution of wood planks for bamboo in bottom and sides, and the addition of decking. Such a process can actually be seen at work in the catamaran log rafts of Madras, some of which have plank strakes pegged on along each side.[a] In many Chinese vessels, notably the Liu-phêng chhuan [1] of Kuangtung, the lines of the sailing-raft are preserved or exaggerated.[b] The conception of transverse bulkheads would have grown naturally out of the septa of the bamboo stem itself. Indeed, a length of bamboo cut in half longitudinally and floated on water gives a striking model of the constructional principle of all Chinese craft (see inset on p. 391).

It is not necessary to insist upon the sea-going bamboo sailing-raft of Thaiwan[c] as the only ancestor of all junks, for many other forms of bamboo raft are regularly using Chinese rivers at this day.[d] One of the most interesting is the Ya [2] River raft of Szechuan, which moves both up and down 100 miles of intractable waterway between Yachow and Chiating, carrying Tibetan trade.[e] This Chu-fa chhuan [3] must be one of the lightest-draught general cargo-carriers in the world, for its depth below the water-line when loaded (with a cargo of 7 tons) is often as little as 3 inches and never exceeds 6, owing to the buoyancy of the bamboos. In length the rafts, which are quite unsink-able, vary between 20 and 110 ft., and are built throughout of the culms of the giant bamboo (*nan chu*;[4] *Dendrocalamus giganteus*) which grows as high as 80 ft. with a diameter of as much as a foot. The bow is narrowed, and bent upwards in a curve by heating, so that the raft can slide over rocks which may be almost level with the water surface (Fig. 930, pl.). In other provinces there are also interesting bamboo rafts (*chu phai*[5]),[f] some of which have the upturned bow (Fig. 931, pl.).[g] Moreover, certain boats, such as the 'fan-tail' (Shen-po-tzu[6]) of western Szechuan,[h] seem to have transferred this ancient device to the stern as a protection against shipping water when descending rapids.

[a] Hornell (8). 　　　　　　　　　　　　　　　[b] Worcester (8).

[c] Extremely similar craft are in use, as we have seen, on the coast of Indo-China. All these closely parallel the famous balsa-wood or log sailing-rafts of the Amerindian culture-area, especially the Inca coast of Peru and Ecuador, and the northern coast of Brazil (cf. Lothrop (1); Hornell (1), pp. 81 ff., (22); Heyerdahl (2), pp. 513 ff.; Clissold (1), etc.). In their traditional form, however, the Amerindian sailing-rafts were manœuvred by the use of the centre-boards only, and knew no steering-oar or rudder. Ling Shun-Shêng (1) attacks the thorny problem of the relations between East Asian, Polynesian and Amer-indian sailing-rafts, pointing out the remarkable similarity between the Formosan and Ecuadorian–Peruvian forms. He goes so far as to suggest that even the names of such craft in the southern and western Pacific areas derive from Chinese roots, but this is linguistically unconvincing. Moreover, part of his argument for a basically Chinese origin rests on an apparent acceptance of the traditional −3rd and −4th millennium datings for the legendary emperors and culture-heroes of China. Nevertheless it seems to us much more probable for many reasons that the sailing-raft with centre-boards did traverse the Pacific from Asia to America rather than in the opposite direction. This opinion was also held by P. Paris, cf. (3), pp. 34, 64, 67, and by Hornell (22). Bowen (2), p. 108, supports it too.

[d] And long have done so. Cf. Fig. 929 from *TSCC, Khao kung tien*, ch. 178, p. 12b, originally from *STTH*.

[e] Worcester (1), pp. 91 ff., (3), vol. 1, p. 222, (14), pp. 179 ff.; Donnelly (6). A graphic description of a journey on one of these rafts has been given by Llewellyn (1).

[f] Worcester (3), vol. 2, pp. 304, 440; (1), p. 42; Audemard (7), pp. 74 ff., for the Yangtze; Donnelly (6) for Ninghsia; W. E. Fisher (1). Cf. R. D. Thomas (1), p. 47.

[g] A very similar illustration of rafts on a river in Kuangtung is given by Eigner, Alley et al. (1).

[h] Worcester (1), p. 47.

[1] 六篷船　　　　[2] 雅　　　　[3] 竹筏船　　　　[4] 南竹　　　　[5] 竹簰　　　　[6] 神駁子

Fig. 929. A bamboo raft from the *Thu Shu Chi Chhêng* (+1726), redrawn from *San Tshai Thu Hui*
(+1609). Note the bamboo side-rails, which could be anti-hogging trusses (see p. 437).

What Hornell never knew was that there are indigenous Chinese traditions that the
junk was developed from the raft. The +3rd- or +4th-century book *Shih I Chi* says:[a]
'Hsien-Yuan[1] changed the custom of floating on rafts (*fu*[2]), for he invented boats
(*chou*[3]) and oars (*chi*[4]).' To which a lexicographer commented:[b] 'So that before there

[a] Ch. 1, p. 2b. Attention was drawn to this passage by Ferguson (6).
[b] *STTH, Chhi yung* section, ch. 4, pp. 19a ff. So also *Pai Pien*, s.v.

[1] 軒轅   [2] 桴   [3] 舟   [4] 楫

were boats, people crossed rivers by means of rafts. Since the term *fu*[1] means the same thing as *fa*,[2] these rafts must have been known before Huang Ti's[a] time. Nowadays people call any raft (*phai*[3]) consisting of bamboo or timber a *fa*.[2]' Bamboo rafts are probably meant by the classical word *wei*;[4] the *Shih Ching* has a verse 'I can cross the river on a (bundle of) reeds' (*i wei hang chih*[5]),[b] and it continued through Sung times for poetic language, as in Su Tung-Pho. Confucius and the rhapsodists of Chhu were both concerned with rafts. Sighing at the recalcitrance of contemporaries to accept his ethical and social teaching, the Master said that he would embark upon a (sailing-) raft (*fu*[1]) and visit the Nine Barbarian Peoples in the hope of finding a better audience.[c] A couple of centuries later, one of the writers of the Chiu Chang series of odes speaks of 'riding downstream on a floating raft (*chhêng fan fu*[6])'.[d] The view that the junk developed from the bamboo raft is not contradicted by the famous passage in the *I Ching* where it is said[e] of the sages that they 'hollowed out logs to make boats, and hardened wood in the fire for oars (*khu mu wei chou; yen mu wei chi*[7])'. This usual translation lays too much weight on the meaning of the first word, which can also signify to rip, to cut off a slice, to cut up into several parts, to cut up an animal, and so on. Here it could refer equally to the cutting of a log into planks. And as we shall see before long, the ancient pictogram for a boat shows ends which are square and not pointed.[f] Moreover, it may be significant that several ancient books refer to the use of large bamboos for making boats.[g]

(c) CONSTRUCTIONAL FEATURES OF THE JUNK AND SAMPAN

The best way to proceed from the point we have now reached will be to examine more closely a few typical examples of marine architecture. At the same time we shall be able to gain an idea of some of the most important technical terms which were, and are, used by the shipwrights and sailors.

[a] Hsien-Yuan was one of the names of the legendary emperor Huang Ti.

[b] Mao no. 61, tr. Karlgren (14), p. 42; Legge (8), p. 104.

[c] *Lun Yü*, v, vi. The part about the barbarians was added by later scholiasts, as e.g. in *Shuo Wên*, s.v. The great Legge, in his translation (2), p. 38, made out that Confucius intended to get on a raft and drift aimlessly at sea. Doubtless he did not know of the existence of excellent sailing-rafts, but it was a pity to generate yet one more unnecessarily fatuous occidental conception of China. In fact, the picture of the sage's tall lug-sail breasting the waves of a stormy sea to bring the message of rational social order to men still slaves of superstition has a real sublimity. Well might such a vessel have merited the epithet of 'Starry Raft' (*hsing chha*[8]) applied long afterwards in Chinese usage to the ships of ambassadors. And well might it have voyaged to the Mexic shore.

[d] Hsi Wang Jih ode (*Chhu Tzhu Pu Chu*, ch. 4, p. 27a; tr. Hawkes (1), p. 76).

[e] Great Appendix, pt. II, ch. 2; tr. R. Wilhelm (2), vol. 2, p. 254; Baynes tr. vol. 1, p. 357.

[f] Cf. p. 439 below. It even looks as though the 'septum' bulkhead idea reacted back on some makers of dugout 'canoes'. On the Yalu and Tuman Rivers in North Korea large hollowed-out tree-trunks are used as ferry-boats, and in order to give extra strength, septa of wood are left between the hollowed-out sections (Underwood (1), p. 6 and fig. 1).

[g] *Shan Hai Ching*, ch. 17, p. 1a; *Shen I Ching*, ch. 3, cit. in *Sun Phu*, p. 11b; *Chu Phu*, pp. 1b, 3b; cf. *Yuan Chien Lei Han*, ch. 417, p. 3b.

[1] 桴　　　[2] 筏　　　[3] 棑　　　[4] 葦　　　[5] 一葦杭之　　　[6] 乘氾泭
[7] 刳木爲舟剡木爲楫　　　[8] 星槎

(1) Type-specimens

Let us begin with a cargo-boat of the Upper Yangtze,[a] the Ma-yang-tzu[1] (Figs. 932, 933, pl.). Like all river-junks, it is quite variable in size, from bow (*i*,[2,3] *shou*[4]) to stern (*chu*,[5] *shao*[6]) measuring between 35 and 110 ft. Formerly they were built as large as 150 ft. As will be seen from the elevation, there are no less than 14 bulkheads (*liang thou*[7]), forming so many separate holds (*tshang*[8]). In the oldest and most characteristic build, there is no basic longitudinal strengthening member (i.e. keel) at all, the structure depending for lengthwise rigidity only on the planking nailed to the bulkheads, and on very solid wales (*chhuan pien chia ta chin*,[9] lit. 'grasper sinews') upon the sides. These are still fitted, taking their place among the strakes (*chia chin*[10]), but modernisation has now sometimes induced a 'keel' (*lung ku*,[11] see p. 429) even in remote river types of build. Or there may be a kelson and two side kelsons at the turn of the bilges,[b] inside their planking (*wan chio pan*[12]).

Between the bulkheads, there may be some frames, half-frames, or ribs (*ya yü*[13]) though it is doubtful if these are of ancient origin. Floor-boards (*ti yü*[14]) in the holds can lie on these above the structural bottom planking (*chhuan ti pan*[15]). The bulkheads themselves nearly always include vertical members or stiffening bars (*liang thou chia pan*[16]). Many types of Chinese craft have a considerable tumblehome like European ships of the +18th century, in other words they may be said to be turret-built, and this is the case with the Ma-yang-tzu. The deck (*chü mien pan*[17]) does not, therefore, occupy by any means the whole beam breadth. As is almost invariably the case on Chinese ships, superstructures are aft of the main mast (*chhiang*[18]), their topside planking (*han phi*[19]) continuing into rails or bulwarks (*han phi mien chih chin*[20]), though these are seen more on sea-going ships, and indeed generally absent forward of the mast on river junks, thus giving uninterrupted space for handling oars, yulohs,[c] tracking gear, and so on. The deck is carried on transverse beams (*chü liang*[21]) placed at intervals along the top of the hull; some of these project outboard and serve for the fulcra of different kinds of oars, so that they may clear the guard deck or whaleback (i.e. the sloping upper boarding of the hull; *chhuan wai chü mien pan*[22]). The square bow ends in a massive projecting cross-beam (*lun*[23]) useful for all kinds of purposes. Hatch coaming (*tshang khou pien pan*[24]) is fitted from the bow to the deckhouses, and the deck planks run athwartships above it.[d]

[a] This is convenient, but I do it also as in private duty bound, since in former days I made a number of journeys on this type of vessel. What understanding I have of Chinese river seamanship I owe to Captain Wu[25] of Szechuan. Description in Worcester (1), p. 21. Claudel & Hoppenot have published (1) pl. 31, a striking photograph of models of Ma-yang-tzu and their makers.

[b] Worcester (1), p. 37.

[c] See p. 622 below.

[d] On Chinese hull construction in general see Audemard (3), pp. 10 ff. He has also an excellent general statement on masts and tabernacles, pp. 31 ff.

1 麻秧子 2 舰 3 艋 4 艏 5 舳 6 艄
7 樑頭 8 艙 9 船邊夾大筋 10 夾筋 11 龍骨
12 彎角板 13 軋玉 14 底玉 15 船底板 16 樑頭夾板
17 柜面板 18 艢 19 旱舦 20 旱舦面直筋 21 柜樑
22 船外柜面板 23 艌 24 艙口邊板 25 巫

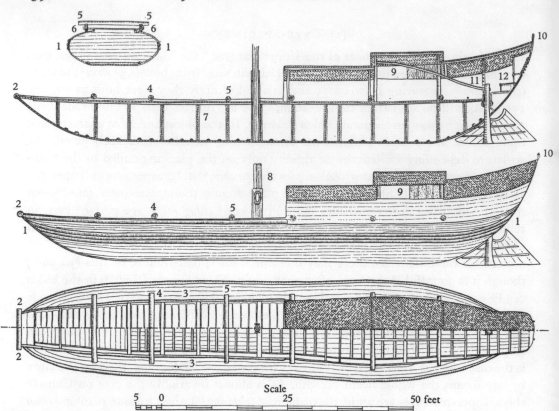

Fig. 932. The Ma-yang-tzu, a river-junk of the Upper Yangtze, here taken as a prototype of all Chinese shipping (from Worcester, 1).

1   long and heavy wales running the length of the vessel
2   projecting cross-beam at transom bow
3   low coaming from bow to deckhouse
4, 5   projecting cross-beams holding the thole-pins for the yulohs (see p. 622 and Fig. 933, pl.)
6   transoms
7   fifth bulkhead supporting hardwood capstan
8   80-ft. *shan-mu* mast and its tabernacle, with carved halyards cleat
9   tiller-room with forward view (but in some forms the tiller may rise above this deck-house and be worked from a thwartwise gangway above it)
10   two tall pins on which spare bamboo rope is coiled
11   25-ft. tiller of balanced rudder (cf. p. 655 and Fig. 1043, pl.)
12   after-house (skipper's cabin and home)

The pinewood mast, some 80 ft. in height, is stepped in tall tabernacles which rise as high as 6 ft. above deck level. A tenon on the heel of the mast fits into a socket in a movable timber of considerable strength, bearing on half-frames below and fitting snugly against the sides of the bulkheads, thus distributing the thrust (Fig. 934). On modern versions of the build, coffer-dams or small compartments kept free of cargo are introduced between bulkheads, into which bilge-water can drain for removal by bamboo pumps. It will be seen that the rudder (*tho* [1,2,3,4]) is a balanced one (that is to

¹ 舵   ² 舺   ³ 柂   ⁴ 柂

say, part of its blade is forward of its axis), and carries a tiller 25 ft. long, which may require as many as three men to handle in a difficult rapid. There may also be a bow-sweep. The mast will carry a single lofty lug-sail. While the permanent crew of a large river-junk of this kind may be only eight, fifty or sixty more men will be engaged from time to time; and for towing upstream in certain places, as many as 400 trackers may be necessary.

Of this kind of vessel there are many variations, such as the Tho-lung-tzu[1] and the Nan-ho chhuan[2] which ply west of Chungking.[a] Some of them, such as the latter, supplement their rudders with a very massive bow-sweep.

As our type-specimen of the sea-going junk we may take the Chiangsu freighter[b] or Sha-chhuan[3] (sand-ship).[c] Formerly these reached in size a length of as much as 170 ft. The pinewood hull (*chhuan kho*[4]) is flat-bottomed (see Fig. 935), the central longitudinal timber being somewhat larger than the others and substituting for a keel. As many bulkheads are present as in the up-river boat just described, and the sides of the hull are strengthened by wales. Since the curves of the turret-built hull con-

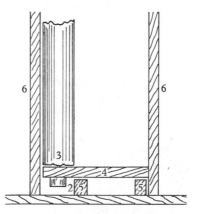

Fig. 934. Tabernacling of typical Chinese mast (from Worcester, 1). A tenon (2) on the heel (3) of the mast fits into a socket in a horizontal timber (4) bearing on the ribs (5) and fitting snugly against the sides of the bulkheads (6). This prevents the heel of the mast moving forward, and helps to distribute the thrust. Cf. Fig. 935, and for a +17th-century description, p. 413.

verge at bow and stern, the foremost and after compartments are masterpieces of construction,[d] and the curved deck beams are rabbeted with great ingenuity into the curved frames of the hull.[e] Certain of the stem and stern ribs (longitudinal members) are actually grown to shape.[f] Both bow and stern are bluff and capable of withstanding the worst weather, but abaft the stern a kind of 'false stern' (*tho-lou*[5]) is built on by extending the sides of the hull in a rising curve beyond the final transom, to terminate in a shorter false transom about 7 ft. above the water-line. The decked surface of this structure prolongs the deckhouses and carries a windlass for hoisting or lowering the rudder, which is slung within this enclosed space. Such an arrangement has been for centuries particularly characteristic of Chinese ships. The rudder-post works in three open-jawed wooden gudgeons, and the tiller is handled either on the roof of the deckhouse or from within it. Still aft of the false stern is a long stern-gallery.

[a] Worcester (1), pp. 61 and 78.   [b] Sometimes called the Pechili freighter.
[c] Worcester (3), vol. 1, p. 114; Waters (2).
[d] See, for example, Fig. 936 (pl.), which shows the bows of a Fuchow pole-junk; cf. Worcester (3), vol. 1, p. 139; Donnelly (5). And Fig. 937 (pl.), those of a Hangchow Bay freighter; cf. Worcester (3), vol. 1, p. 137.
[e] What Smyth said of Dutch vessels is applicable here: 'As with the case of the Chinese junk, however bluff or unwieldy the upper works appear, the underwater lines are generally very sweet, and Neptune, to his credit be it said, has ever a soft heart for a full sweet curve' (p. 82).
[f] A remarkably Taoist practice. But it was also used by European shipwrights.

[1] 舵籠子   [2] 南河船   [3] 沙船   [4] 船殼   [5] 舵樓

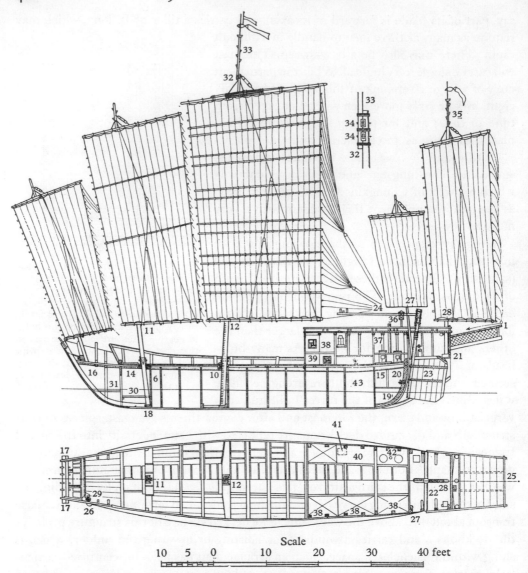

Scale

10   5   0   10   20   30   40 feet

Fig. 935. The Sha-chhuan or Chiangsu freighter, a sea-going junk which was probably the parent type for many kinds of Chinese ships (from Worcester (3), vol. 1). It was often nearly 200 ft. in length, approaching the size of the great wooden ships of the Ming navy (cf. pp. 479 ff. below), but the example here drawn is of 85 ft.

  1   block and multiple sheets of aft mizen-sail (cf. p. 597 below)
  6   first internal bulkhead
 10   one of numerous strong deck beams
 11   46-ft. midship foremast (forward rake)
 12   70-ft. mainmast (slight aft rake)
 14   fore compartment, usable for living quarters and cordage
 15   after compartment, abaft of the twelfth and last internal bulkhead
 16   three fore-and-aft stem ribs grown to shape
 17   bow cross-beam
 18   heel of foremast

Attention should be given to the masts, of which there are five, for (as we shall see) this system was surprising for the Europeans of the +13th century, and seems to have exerted a great effect upon their subsequent ship design. Generally speaking, all sea-going junks of any size were, and are, provided with multiple masts, river craft rarely having more than two. A feature, however, which did not spread outside the Chinese culture-area was the system of staggering the masts in port and starboard positions.[a] Thus in the present case, the foremast (*thou wei*[1]) is placed off centre to port, while the second foremast (*erh wei*[2]) is amidships; both are raked forward. The main-mast (*chung ta wei*[3]) is also amidships but raked slightly aft; then come the mizen mast (*ssu wei*[4]), on the port side with a marked forward rake, and lastly the bonaventure mast (*wei wei*[5]), considerably taller than the mizen and stepped with no rake at all. The raking is not the same on all ships, but the general tendency is to have the masts radiating like the sticks of a fan.[b] There is also individual variation in the construction of the tabernacles. As in nearly all traditional Chinese sea-going ships, the masts are completely devoid of stays,[c] but in some types the heavy primary masts are provided with single or Y-shaped struts about deck level which transfer part of the thrust of the sail to the junctions of hull and

[a] This may have originated in the desire to obtain more room for the handling of gear and stowage of cargo, but it also has the effect of preventing the smaller sails, used mainly for steering, being becalmed by the others. The principle of staggering seems to be attractive to some designers of modern yachts (private communication from Mr Wells Coates).

[b] In Korean ships this becomes very exaggerated (Underwood (1), p. 18 and figs. 18, 19). It is well shown in the picture of a Fukien trader in *Wakan Senyōshū*, ch. 3, pp. 19b ff.

[c] This must have been a great help in the development of fore-and-aft sailing techniques (cf. pp. 591, 608). But the masts of river-ships used for tracking have them (Figs. 933, 957, 971, pls., 1047), and sometimes many (Figs. 976, 1024, 1032, pls.). Shrouds are very rare (Fig. 972, pl.).

[1] 頭桅　　　[2] 二桅　　　[3] 中大桅　　　[4] 四桅　　　[5] 尾桅

Key to Fig. 935 (*continued*)

19 fore-and-aft stern timbers grown to shape
20 bluff and rounded transom stern
21 false stern, consisting of an 8-ft. extension of the sides of the hull beyond the transom, ending in a shorter false transom
22 windlass for hoisting the rudder (cf. p. 632)
23 rudder, iron-bound and non-balanced
24 16-ft. tiller
25 10-ft. stern gallery
26 31-ft. port foremast (forward rake)
27 28-ft. port mizen mast (forward rake)
28 48-ft. aft mizen mast (vertical), slightly to port of amidships
29 port foremast tabernacle (inside bulwarks)
30, 31 fore-and-aft baulks securing midship foremast heel
32 hounds of mainmast
33 light topmast of mainmast
34 sheave pins passing through both masts and securing double halyard sheaves
35 light topmast of aft mizen mast
36 navigation light
37 galley
38, 39 cabins with bunks and sliding doors
40 rice bin and stores
41 shrine to Kuan Yin
42 cooking-stoves
43 below-deck living quarters

bulkheads forward (Fig. 938, pl.).[a] All the masts carry balanced lug-sails,[b] and some of them have top-masts on which, today as in the Middle Ages, topsails are set under suitable sailing conditions. Such junks as these, which today carry a crew of about twenty, must resemble fairly closely the prototypes which journeyed to the Indian Ocean in the Sung period.[c] Their beauty when in full sail has fascinated many observers.[d]

(2) TECHNICAL TERMS

As we have already pointed out, the elucidation of Chinese shipbuilding and nautical nomenclature has its difficulties. So far as we know there is nothing in Chinese literature which does for this subject what the great *Ying Tsao Fa Shih*[e] did in the field of building technology. Again the trouble is that the practical men never committed anything to writing, and the literary men had little or no knowledge of the building and handling of ships; they could only make commentaries on technical terms which even their predecessors had perhaps only half understood.[f] So although Chinese encyclopaedias, from the *Erh Ya* onwards, generally contain sections devoted to shipping terms,[g] it is noticeable that the majority of these concern types of boats and ships long obsolete (or not very easily identifiable), such as *mêng*,[1] *tiao*,[2] *thung*,[3] *tang*,[4] and so on. The number of technical terms for distinct parts of the ship and its gear is smaller. Even then, much space is devoted to the identification of dialect phrases or local usages, so that the task of selecting out items of information which really prove the existence of any given technique at a given time in history will require prolonged research. Although the field is so inviting, we have naturally not been able to attempt this here. For example, in +1126 Jen Kuang produced his *Shu Hsü Chih Nan* (Literary South-Pointer), the 15th chapter of which is partly devoted to explanations of terms concerning ships, but this has never been studied. An investigation of the sort we need would

---

[a] Sometimes the heavy masts are compound, i.e. built up of several separate longitudinal spars bound together with iron straps. In 1842 British naval officers were astonished at the size of the main-masts of Shanghai junks. The circumference of one, taken a little above the deck, was 11 ft. 6 in., its height 141 ft., and its main yard 111 ft. long. Very strong spars were necessary for the enormous sail, and there were no shrouds or stays. See Bernard (1), vol. 2, p. 365.

[b] Cf. below, pp. 595 ff. Those of the main- and fore-masts are always stiffened with battens; the others may be.

[c] A word of caution should be spoken here regarding the reconstruction of a sea-going junk in Clowes & Trew (1), p. 81, just because it occurs in a collection of drawings otherwise so excellent. I suspect that it was taken from Adm. Paris (2), pt. IV. But the rudder 'slit' is too narrow, the side-galleries too long and prominent, the multiple sheets wrongly drawn, and the hull also out of drawing.

[d] Cf. Ommanney (2), p. 111.

[e] Cf. pp. 84 ff. above.

[f] There appear to be important shipbuilding manuals and books on shipping in other cultures, but none seem to have been translated into Western languages, or even described at all fully. For example, there is the *Yuktikalpataru* compiled by Bhōja Narapati and often quoting from Rājā Bhōja of Dhara (c. +1050); this is still only in MS. (Mukerji (1), p. 19). And in Japanese there is the *Wakan Senyōshū*[5] by Kanazawa Kanemitsu[6] (+1766), well illustrated and containing much of interest. Other versions of some of these pictures are in the *Ka-i Tsūshō-kō*[7] of Nishikawa Joken[8] (+1708). Just a century later Motoki Masahide[9] wrote his illustrated account of Western warships, the *Gunkan Zusetsu*.[10]

[g] E.g. *STTH, Chhi yung* sect., ch. 4, pp. 9a ff.

| [1] 艋 | [2] 舠 | [3] 舸 | [4] 艚 | [5] 和漢船用集 | [6] 金澤兼光 |
| [7] 華夷通商考 | | [8] 西川如見 | | [9] 本木正榮 | [10] 軍艦圖說 |

also have to make use of the best + 18th-century discussion, the *Shih Chou*[1] (Nautical Nomenclature)[a] of Hung Liang-Chi[2] (+1746 to 1809). This mentions many unusual words, such as *ling*[3] for hold floors or bulkheads, and *i*[4] for strakes, though today the latter means long oars, and for Jen Kuang distinctly steering-oars.

Western sinology is not much more helpful. Something may be learnt from Edkins (12) and Doolittle (2); more deduced from the plain perusal of dictionaries. Had sinologists applied one tenth of the effort devoted during the past hundred years to translations of belles-lettres, to investigating the rise and progress of the useful arts in China, we should now be better off. Conversely, it seems at first sight regrettable that the copious and careful works of Worcester do not give us the Chinese equivalents for every technical term used.[b] Here, however, an unexpected difficulty presented itself. Even the finest shipwrights and sea-captains, with whom he worked, generally could not write, nor could any members of their crews or families. There are certainly many spoken craft terms for which no written form exists at all.[c] We shall see in a moment that an official eighteenth-century scribe probably had to invent characters for certain words used by his sea informants. Moreover, the Chinese seamen seem not to have troubled to elaborate that infinity of technical terms, covering the smallest parts of gear, in which Europeans have delighted.[d] And lastly the terms vary from port to port.[e]

It is therefore essential to use the 'ethnological' method side by side with the historical one.[f] This of course would be equally necessary for any traditional art or industry of Europe.

Although it is not specifically a treatise on the shipbuilding art, a great mass of information (not yet digested by historians) is to be found in a book already mentioned

[a] Contained in *Chüan Shih Ko Wên Chia I Chi*[5] (part of his collected works), ch. 3.

[b] The same deficiency was noted (Vol. 4, pt. 2, p. 50 above) regarding the otherwise excellent book of Hommel on Chinese tools and technical processes. The same explanations apply, but Hommel is more open to criticism in that he gave no Chinese characters at all. It may be added here that reliance is not always to be placed on the sinological–historical discussions in Worcester's monographs; lacking, unfortunately, the collaboration of a Chinese historian, he allowed the incorporation of a certain amount of legendary and semi-legendary material. But it was hard to interest academically trained scholars in anything so low and tarry as ships. Dr Chhen Chen-Han told me in 1944 that he intended to write a history of Chinese shipping, and we must hope that he will. On English shipwrights' tools see Salaman (1).

[c] I can confirm this from personal experience, especially among the dialects of the north-west. Now that interest in, and respect for, indigenous crafts has so greatly increased in China, we may expect that a multitude of new characters will be coined in technical dictionaries so that old craft terms can be written down.

[d] Such, at least, was the thirty-year experience of Worcester himself (personal communication). Why did this difference exist between Chinese and European sailors? Perhaps the European attention to detail in nomenclature was a direct result of the dominance of the scientific view of the world during the past three centuries. Elsewhere, Needham (2), p. 71, (27), instances were given of the way in which failure to develop adequate scientific terminology was characteristic of medieval European science, and this was one of the limiting factors which the upsurge of the Renaissance swept to one side. Above (Vol. 2, pp. 43, 260) we saw the same thing with regard to the Taoists. If this is true, then the 'main spencer outhaul' is as much a sign of analysed complexity as the 'inguinal aponeurotic falx'.

[e] Publication of the first-hand work of Miss Barbara Ward upon the shipping of Hongkong, embodying much traditional and living technical information, will be awaited with interest. In the meantime, see Ward (1). Similar studies of Chinese sailors' ways from the life are contained in Worcester (14).

[f] But one should not confuse the two. A history of Chinese shipping cannot be written from the models of junks in the Science Museum, as the title of Worcester (15) seems to imply.

[1] 釋舟   [2] 洪亮吉   [3] 笒   [4] 杝   [5] 卷施閣文甲乙集

on the Yangtze shipyards, the *Lung Chiang Chhuan Chhang Chih*,[1] written by Li Chao-Hsiang[2] in +1553. Discussion of this will be postponed to a more convenient place (p. 482 below). The work is illustrated by a number of drawings of ships and boats, but only one or two of them help to explain technical terms and the rest are rather roughly sketched.[a] The best picture of a Chinese ship which we have been able to find in a Chinese work is that contained in the *Liu-Chhiu Kuo Chih Lüeh*[3] (Account of the Liu-Chhiu Islands) written by Chou Huang[4] in +1757.[b] This is pictured in Fig. 939.[c] The drawing is particularly valuable because the artist added a number of technical terms (see the accompanying key). The transom stem and stern are clearly shown, as also the longitudinal strengthening member (*lung ku*, 'dragon spine', cf. p. 429) of the hull. Four masts are stepped, and the characteristic mat-and-batten sails are well drawn, with the topping lifts on both fore- and main-sail, and the multiple sheets on the fore-sail.[d] Additional sails, and masts or spars for setting them, as in Marco Polo's time (cf. p. 467), are carried—bowsprit-sail,[e] spinnaker, topsail,[f] and a notably bellying mizen-sail, all of these being cotton, reinforced with vertical roping. One cannot help recalling the combination, in modern racing-yacht practice, of a taut battened fore-and-aft mainsail with a bellying spinnaker; and of observing too the contrast with the 'full-rigged ship' of Renaissance Europe, where the square-sails were dominant and the fore-and-aft sail only on the mizen (cf. p. 609). Our junk is obviously running merrily before the wind. The use of the deck winches (*liao*) for hoisting sail will be appreciated from Fig. 940 (pl.). Then the slung rudder, partially raised to reduce water-resistance, should be noted, and the bousing-to tackle which runs from the foot of the rudder under the bottom of the vessel to a windlass in the forecastle and holds the rudder against the transom, where it rotates in wooden jaws (cf. p. 632). This was mentioned by Lecomte in the +17th century (p. 635). Two grapnel anchors are shown at the bow, and another, without a stock (i.e. a cross-bar), on the port side forward. Portholes are present.[g]

While there exists no great published work on Chinese shipbuilding,[h] some MS. material is available, and there must be a good deal more still dormant in Chinese provincial archives. In Europe, the Library at Marburg now possesses a most interesting manuscript[i] which appears to have been a manual for Fukienese officials concerned

[a] The best study of them so far is that of Pao Tsun-Phêng (1, 1).
[b] With a colleague, he had been an official envoy to the islands in the previous year.
[c] *Thu hui* sect. (preceding ch. 1), pp. 33*b*, 34*a*.
[d] Cf. pp. 595 ff. below.   [e] A small square-sail bent to a yard under the bowsprit.
[f] See below, pp. 591, 602.
[g] Interesting, for this also had been one of the surprises for Europeans in the +13th century.
[h] There are of course occasional diagrams of ships in Chinese books. For example, rather rough ones exist in the *Chung-Shan Chhuan Hsin Lu*[5] (Travel Diary of an Embassy to the Liu-Chhiu Islands), written by Hsü Pao-Kuang[6] in +1721. These seem indeed to be earlier specimens of an iconographic tradition which culminated in the picture of Fig. 939. A *Shui Shih Chi Yao* (Essentials of Sea Affairs) was partly translated into Russian a century ago by K. A. Skachkov (1), but we have not been able to find any information on it in Chinese sources, and the paper itself has been inaccessible to us. Cf. p. 424.
[i] No. 5 in the Hirth Collection, formerly at the Royal Library in Berlin. We are much indebted to Dr W. Seuberlich for providing a microfilm for our use.

[1] 龍江船廠志   [2] 李昭祥   [3] 琉球國志略   [4] 周煌
[5] 中山傳信錄   [6] 徐葆光

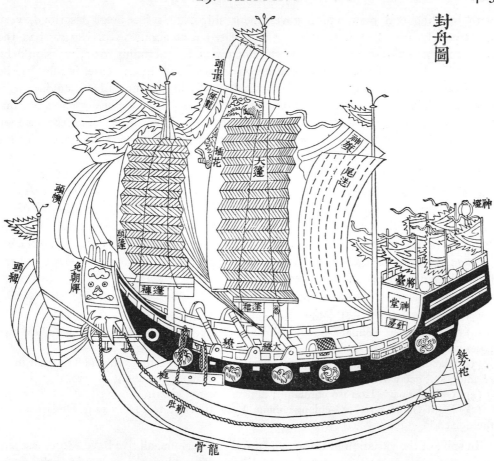

封舟圖

Fig. 939. The ocean-going junk in the *Liu-Chhiu Kuo Chih Lüeh* of +1757, one of the best pictures of a Chinese ship in the Chinese style to be found in literature.

fêng chou   government ship
thou po   'kerchief' or 'headcloth', spinnaker (sail)
thou chhi   'pursuer', water-sail or bowsprit-sail. In the West the bowsprit-sail is a descendant of the Roman *artemon* sail (cf. Torr (1); Chatterton (1), p. 112). The *Santa Maria* of Columbus (+1492) had it, but it did not become general on occidental ships until the end of the +16th century
mien chhao phai   'avoidance of courtesy' notice, i.e. 'bound on important government business'
thou phêng   foremast mat-and-batten sail
phêng khu   'trousers', foot of the foresail, or perhaps foresail boom
ting   grapnel anchor
tu lê   bousing-to tackle securing the rudder
lung ku   'dragon spine', central longitudinal strengthening member of hull
erh liao
ta liao }   winches ('winders') for halyards of foremast and mainmast
phêng chhün   'skirt', foot of the mainsail, or perhaps mainsail boom
ta phêng   mainsail (of bamboo matting and battens)
chha hua   inserted ensign
thou chin ting   topsail (of cloth or canvas)
i thiao lung   dragon ensign   *shen thang*   chapel
shen chhi   'spirit flag'   *chiang thai*   poop
wei sung   mizen-sail   *shen têng*   'spirit light'
chen fang   compass cabin   *thieh li tho* ironwood rudder

with building and maintaining government ships. This has been described, very inadequately, by Moll & Laughton,[a] who dated it at about 1850. Having had the opportunity of a re-examination of the document in microfilm form, we would be inclined rather to place it more than fifty years earlier;[b] in any case, it fully merits publication by editors competent both sinologically and technically. Including as it does about sixty drawings of the component timbers of five classes of ships, it constitutes the nearest approach in the domain of shipbuilding to the standard set long before in architecture by Li Chieh. The dry descriptions, too, are reminiscent of his style.[c]

The MS. is entitled *Min Shêng Shui-Shih, Ko Piao Chen Hsieh Ying, Chan Shao Chhuan Chih Thu Shuo*[1] (Illustrated Explanation of the (Construction of the Vessels of the) Coastal Defence Fleet (Units) of the Province of Fukien stationed at each of the Headquarters of the several Grades). It opens with diagrams of the five classes of small corvettes (as they might be called), then proceeding in turn to details of their mustering (*chhuan chih hao shu*[2]), the itemisation of their parts (*khuan hsiang ming mu*[3]), and the methods and dimensions in construction (*tso fa chhih tshun*[4]).

The five classes referred to are the following:

(1) 'Arrow Pursuit Vessel' (Kan-tsêng chhuan[5]) 40 ft. × 12 ft. to 83 ft. × 21·2 ft.

(2) 'Two-Master' (Shuang-phêng chhuan[6]) 34 ft. × 9 ft. to 61·6 ft. × 16·6 ft. These two types were first fixed in +1688.

(3) 'Flat-Bottomed Ship' (Phing-ti chhuan[7]) 42 ft. × 11 ft. to 48 ft. × 14·8 ft. Said to be steadier, and good with yulohs, but not sea-going. Design fixed about +1730.

(4) 'Official Boat' (Hua-tso chhuan[8]); dimensions not given.

(5) 'Eight-Oared Boat' (Pa-chiang chhuan[9]) 32·9 ft. × 9 ft. to 40 ft. × 12 ft. Design of both these fixed in +1728.

In spite of the variation of these names and dimensions, all five have a foremast and mainmast, with a small mizen lacking a sail and stepped on the port side of the poop. All have *lung ku*,[10] or stout longitudinal hull bottom members, except the third, here reproduced as Fig. 941. Since the component parts are labelled rather clearly, it is possible to confirm a number of important words, e.g. sheets (*phêng so*[11]), yard and boom (*shang* and *hsia phêng tan*[12]), halyards winches (*ta hsiao lao niu*[13]), tabernacle (*lu erh*[14]), crowsnest (*wei li*[15]). The bottom of the transom bow is called the 'wave-lifting board' (*tho lang pan*[16]). But some of the characters are peculiar, not to be found even in the Khang-Hsi dictionary, for example[d] *chhao*,[17] apparently the bow-sweep; and

---

[a] The drawings of the ships and their parts are ill reproduced, no Chinese characters are given, and in their stead only unintelligible romanisations, many of the pieces of woodwork and their functions are not identified, and what appears to be a translation is only an abbreviated paraphrase, often erroneous.

[b] The latest date mentioned in the MS. is +1730, and decisions taken in +1688 are referred to as if they were still authoritative.

[c] At the same time, we do not feel sure that the official who wrote the text always fully understood what the shipwrights were trying to explain to him.

[d] Pronunciation here is surmised.

[1] 閩省水師各標鎮恊營戰唪船隻圖說		[2] 船隻號數	[3] 欵項名目
[4] 做法尺寸	[5] 赶繒船	[6] 雙篷船	[7] 平底船	[8] 花座船
[9] 八槳船	[10] 龍骨	[11] 篷繂	[12] 上下篷檐	[13] 大小繚牛
[14] 鹿耳	[15] 梡笠	[16] 托浪板	[17] 𣏾	

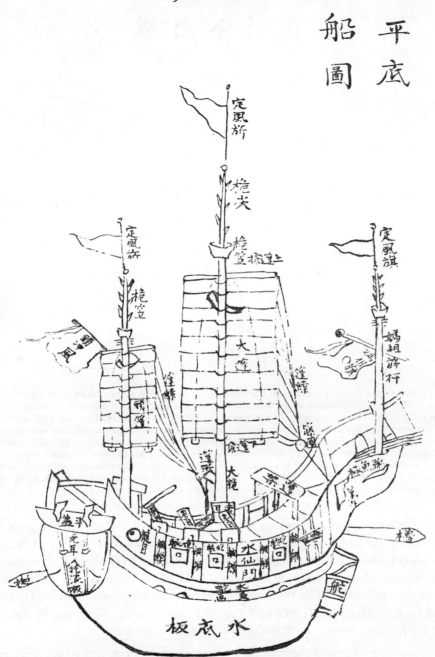

Fig. 941. Drawing of a ship from the most important manuscript shipbuilding manual, the *Min Shêng Shui-Shih. . . Chan Shao Chhuan Chih Thu Shuo*, preserved in the Prussian Staatsbibliothek at Marburg. Though the MS. may be of the middle nineteenth century, the contents belongs to the first half of the eighteenth. This 'flat-bottomed ship' (Phing-ti chhuan) bears considerable resemblance to that in Fig. 939. For explanations of some of the Chinese technical terms, see text; but not all those in the MS. are either legible or identifiable in the best dictionaries. However, the many diagrams of component parts often tell their own story, and permit the identification of sea-terms which the lexicographers ignored. Also noteworthy in this picture are the 'weather-vane pennants' (ting fêng chhi), the 'standard of (the goddess) Ma-Tsu' (cf. p. 523) at the stern, the rudder (tho) and sculling-oar (lu, cf. p. 620), the oculus (lung mu), and the bulwark gate (shui hsien mên).

圖形分面艙

Fig. 942. One of the detailed component drawings from the Fukien shipbuilding manuscript; a plan at deck or waterline. The bows are to the left, the stern to the right, and the greater beam astern is very evident (cf. p. 417). Not counting the transom stem and stern, there seem to be seven bulkheads, four of which are marked (*khan niu*). From left to right one sees, first the emplacement and tabernacle of the foremast (*shu thou wei*), then that of the mainmast (*shu ta wei chhu*), a succession of holds and hatches amidships, and two water-tanks (*shui kuei*) to port and starboard. Aft of these there is a section marked 'combat deckhouse' (*chan phêng*), presumably a place of comparative safety, and still further aft (showing that this is not the same ship as that in Fig. 941) the halyard winches (*lao niu*). At the stern an entrance leads into the stateroom and chapel (*pai phêng*), where the compass was kept.

po-mo,[1] vertical supporting timbers for the bulwarks which sheltered the guns on the deck,[a] with their loopholes (*phao yen*[2]). All of these vessels were built with bulkheads (*khan niu*[3]) forming fifteen compartments in each (Fig. 942). Among the component pieces mentioned are those (*lu tu lê*[4]) which hold fast at the bows the tackle retaining in place the rudder aft. The transverse gantry on the afterdeck in Fig. 941 is one of several frames (*phêng chia*[5]) on which the sails were stowed. The value of this manuscript for further research on these questions will be obvious.

[a] We have in our collection a photograph of a traditional war-junk's topsides, taken in 1929 on one of the Cantonese rivers. It might however be a launch-towed passenger-boat, for these were also armed (G. R. G. Worcester, unpublished material, no. 109). Probably the best photographs ever taken aboard such war-junks are to be found in the book of Lilius (1), a Finnish journalist who investigated the people of the Bias Bay region in the twenties of the present century. There is a vivid account of a woman bandit sea-captain, but unfortunately Lilius was not interested in the nautical technology which she and her colleagues understood so well.

[1] 皴模   [2] 炮眼   [3] 艦牛   [4] 鹿肚勒   [5] 篷架

Chinese literature contains several interesting accounts of the building of model ships for instructional purposes.[a] About +1158 Chang Chung-Yen[1] was in the service of the Jurchen Chin. Of him we read:[b]

When they began to build ships, the artisans did not know how. So (Chang) Chung-Yen made with his own hands a small boat several (tens of) inches long. Without the use of glue or lacquer it all fitted together perfectly from bow to stern. He called it his 'demonstration model'. Then the astonished artisans showed him the greatest respect. Such was his intelligence and skill.

After the large ships had been built and were ready to be launched, people were to come from all the surrounding districts to drag them to the water, but Chung-Yen ordered several tens of his craftsmen to build sloping runways leading down to the river. Fresh millet stalks were collected and spread thickly on these slipways, which were supported on each side by large beams. Early in the morning, when there was a frost, he led the men to the launching, and because of the slipperiness the work was accomplished with very little effort.

A contemporary in the service of the Sung down south was Chang Hsüeh.[2]

When he was prefect of Chhuchow he wished to construct a large ship, but his advisers were not able to estimate the cost. (Chang) Hsüeh therefore showed them how to make a small model vessel, and then when its dimensions were multiplied by ten (the cost of the full-size ship) was successfully estimated.[c]

The text goes on to tell how his artisans estimated a cost of 80,000 strings of cash for the walls of a temple park, but he had them build an experimental 10 ft. length, which proved it could be done for 20,000. Evidently a man not to be trifled with.

One of the few serious literary descriptions of nautical technology is to be found in the relevant chapter of the *Thien Kung Khai Wu* (Exploitation of the Works of Nature), written by Sung Ying-Hsing in +1637. It is so informative that sample quotations could hardly be omitted.[d] Sung Ying-Hsing describes first a typical grain-carrying ship[e] (Tshao fang[3]) of the Grand Canal towards the end of the Ming (Fig. 943), and then goes on to refer more briefly to sea-going junks similar to that which we have already examined in Worcester's modern drawings (Fig. 935).[f]

[a] Of recreational models we have heard a little already; Vol. 4, pt. 2, p. 162.
[b] *Chin Shih*, ch. 79, p. 9*a*, tr. auct. with Lo Jung-Pang.
[c] *Sung Shih*, ch. 379, p. 13*a*, *b*, tr. Lo Jung-Pang.
[d] It occurs in ch. 9, pp. 1*b* ff.
[e] On this subject an old paper of Playfair (1) remains well worth reading. From official documents he translated much information on the grain-transport service of the Grand Canal, the number and nature of the personnel, the details of loading the tax-grain, and the official regulations regarding the build and maintenance of the junks employed.
[f] A previous translation was given by Ting & Donnelly (almost certainly this was Ting Wên-Chiang, the eminent geologist, 1887 to 1936, who wrote the biography of Sung Ying-Hsing appended to the best edition of the book). But Ting Wên-Chiang was not very familiar with seamanship and Donnelly was apparently not capable of checking the Chinese text. Accordingly there were many uncertainties of meaning, and one long and interesting passage concerning tacking was completely omitted (without warning). Very recently the text was all translated again by Sun Jen I-Tu & Sun Hsüeh-Chuan (1); theirs is the most literal rendering, but it makes the feeblest attempts to use the correct English technical terms. For a deeper understanding see the translation into Japanese in Yabuuchi Kiyoshi (*11*), together with his annotative essay, pp. 168 ff. (Chinese trs., Peking ed., pp. 190 ff., Thaiwan-Hongkong ed., pp. 193 ff.).

1 張中彥 2 張觷 3 漕舫

Fig. 943. A grain freighter of the Grand Canal at the beginning of the Ming; the Tshao fang as depicted in the *Thien Kung Khai Wu* of +1637 (Chhing illustration).

Sung Ying-Hsing says that the design of his standard inland navigation ship or sailing barge dated back to the beginning of the +15th century, when in the Yung-Lo reign-period, on account of the grain-transport losses on the sea route, it was decided to revert once more to the use of the Grand Canal.[a] 'Accordingly, the present shallow-draught canal boats were introduced by a certain Mr Chhen,[1] lord of Phing-Chiang.[2]'[b]

[a] For the technological background of this, see p. 312 above, and pp. 478, 526 below.

[b] This was Chhen Hsüan,[3] who made his name as a military commander against the southern tribal peoples, but later worked mainly as a hydraulic engineer, especially in the Huai region, where he built nearly fifty locks as well as protective dykes for the Grand Canal. In +1403 or soon afterwards he was ennobled, taking his title from Suchow, and in +1415 received order to build 3,000 sailing barges for the inland grain transport (*Ta Ming Hui Tien*, ch. 200). As can be seen, the designs produced under his auspices remained authoritative for more than two centuries. We now realise that in fact it went back to his predecessor Sung Li (cf. p. 315, and *Ming Shih*, ch. 79, p. 2a, also ch. 153); and Yabuuchi (*11*) must be right in his view that it was much older than that in essence, indeed deeply traditional. However, Chhen Hsüan's fame was noted by the Korean traveller Chhoe Pu (cf. p. 360) in +1487 (Meskill tr. p. 106).

¹ 陳   ² 平江伯   ³ 陳瑄

The general construction of a canal ship [Sung continues][a] is as follows: a bottom (*ti*[1]) (of stout planking) serves as the foundation, there are (thwart and fore-and-aft) timbers (*fang*[2]) like the walls (*chhiang*[3]) of a building,[b] and there is bamboo tiling (*yin yang chu*[4])[c] (to cover the hold) as if it were a roof. (The compartments) forward of the mast framework (*fu shih*[5],[d] i.e. its tabernacle and associated structures) are like the main gates, and (the compartments) aft of it are like the sleeping quarters. The mast (*wei*[6]) is like (the stock of) a crossbow, and the halyards (*hsien*[7])[e] and sails (*phêng*[8]) are like wings. Oars (*lu*[9]) (may also be motive power) as the horse is to the cart; hauling cables (*than chhien*[10])[f] are as the shoe is to the walker. The cordage (*yü so*[11]) adds strength like the bones and sinews of a hawk. The bow-sweep (*chao*[12])[g] goes before like a spearhead, the rudder (*tho*[13]) (at the stern guides the direction of the vessel) like a commander, and the anchors (*mao*[14]) call a halt like an army encamping for the night.

On the original specification, the grain-ship is built 52 ft. long with planks 2 in. thick. The choicest timber for it is large baulks of *nan-mu*[15],[h] but chestnut (*li*[16])[i] is also used as second best. The bow and stern are each 9·5 ft. long,[j] while the breadth of the bottom is 9·5 ft. at the middle, 6 ft. at the bow, and 5 ft. at the stern.[k] The breadth of beam at the fore-mast tabernacle is 8 ft., and that at the main-mast tabernacle is 7 ft.[l] Fourteen bulkheads (*liang thou*[17]) are built across the ship;[m] (that forward of the main hatch) the 'dragon's mouth' bulkhead (*lung khou liang*[18]) is 10 × 4 ft.; (that beside the main-mast) the 'wind-using' bulkhead (*shih fêng liang*[19]) is 14 × 3·8 ft.; and that towards the stern, the 'cut-water' bulkhead (*tuan shui liang*[20]) is 9·5 × 4·5 ft.[n] The width of two granary holds (*ao*[21]) is 7·6 ft.

[a] Tr. auct., adjuv. Ting & Donnelly (1); Sun & Sun (1).

[b] Sung Ying-Hsing explains in landsman's terms. If we had not studied building technology (cf. p. 92 above) we should not know that *fang* means tie-beams, both along a structure and from side to side of it. Most Western dictionaries fail to give this most important meaning of the word.

[c] So called because made of imbricated half-bamboo sections.

[d] Lit. 'lion-tamer', a good name for a weather-withstander.

[e] This word means of course a bowstring or the string of a musical instrument, hence the chord of an arc, but also (pronounced *hsüan*) the hypotenuse of a right-angled triangle (cf. Vol. 3, pp. 22, 96, 104, 109 and *passim*). Here it fits the halyards very well, the other two sides of the triangle being the mast and the distance along the deck between the mast and the halyard winch. Sung Ying-Hsing himself makes just this analogy elsewhere, see below, p. 604.

[f] For tracking and towing; cf. pp. 662 ff. below.

[g] An unusual term—but the thing is clearly depicted in the Ming illustration though not in the Chhing one.

[h] *Persea nanmu* (B11, 512), a tree-laurel, often miscalled cedar, oak, etc.; syn. *Machilus nanmu* (R502, cf. Wang Kung-Wu (1), p. 106).

[i] *Castanea vulgaris*, as in Europe (B11, 494). The classical passage in Europe about timber for ship-building is Theophrastus, v, vii, 1–3.

[j] Presumably from the furthest forward point to the bow water-line or to what corresponded to the fore-foot gripe, and the same reckoning *mutatis mutandis* at the stern.

[k] There may be room for a suspicion that the text has inverted the beam dimensions as between bow and stern.

[l] As we shall see later (p. 415) it was obligatory to have more than one mast on ships over 100 ft. long, but these of half that length also had two, as the illustrations show. They are better differentiated as to height in the Ming than in the Chhing one.

[m] Note that in the Ming the number of bulkheads was about the same as on the modern-built ships described above (p. 398).

[n] These dimensions may imply some degree of turret-build. Donnelly however (in Ting & Donnelly) made a reconstruction according to them, without this; the resulting lines look quite like a modern river-junk, perhaps rather broader in the beam, and lower, amidships.

[1] 底	[2] 枋	[3] 牆	[4] 陰陽竹	[5] 伏獅	[6] 桅
[7] 弦	[8] 篷	[9] 櫓	[10] 菆縴	[11] 絆索	[12] 招
[13] 舵	[14] 錨	[15] 楠木	[16] 栗	[17] 梁頭	[18] 龍口梁
[19] 使風梁		[20] 斷水梁		[21] 厫	

The general plan is thus uncompromisingly flat-bottomed and keel-less, with numerous transverse bulkheads each specifically named.

Such a ship could take a cargo of nearly 2,000 piculs of rice[a] [though in practice only 500 piculs were delivered to each boat],[b] but another type designed independently later on by the army transport service was 20 ft. longer and more than 2 ft. wider both at bow and stern, so that it could carry 3,000 piculs. As the flash-lock gates (*cha khou*[1]) in the (Grand) Canal are only 12 ft. wide, these craft are just slim enough to negotiate them.[c] The boats used nowadays for travelling officials (Kuan-tso chhuan[2]) are of just the same type (as the freighters), but their windows, doors and passageways are made somewhat wider, and besides they are more elegantly painted and finished; that is all.[d]

It is interesting to compare these figures with others for the nineteenth century.[e] A burden of 2,000 piculs (*tan*[3]) is about 140 tons. According to Playfair (1), 670 grain junks in 1874 transported to the capital 1·36 million piculs of grain (96,000 tons),[f] so that the lading of each was about 143 tons. The average size of the ships thus remained at that time just about what it had been at the beginning of the +17th century, and indeed the beginning of the +15th. From evidence given above (p. 352) in connection with the invention of the pound-lock, it can be seen that this tonnage had already been attained very nearly by the middle of the +11th century. But it had probably dropped greatly during the three centuries when the sea route was predominant.[g] Indeed, just after mentioning his figure of 2,000 piculs, Sung Ying-Hsing adds, in one of his commentating 'footnotes', that upon comparing his specification with actual practice, he found that the burden was generally not more than 500 *tan* (*c.* 35 tons).

Sung Ying-Hsing next tell us what he saw on his visits to the shipyards.[h]

The construction of a boat begins with the bottom.[i] The strakes of the hull[j] (*chhiang*[4])[k] are built up on both sides from the bottom (planking) to a height (equivalent to that of the future) deck (*chan*[5]). Bulkheads (*liang*[6]) are set at intervals to divide the vessel (into separate compartments), and (the holds have) sheer vertical sides which are (also) called *chhiang*.[4] The hull is covered at the top (or, surmounted) by great longitudinal members (*chêng fang*[7]).[l] The (winches for the) halyards (*hsien*[8])[m] are fixed above these. The position of the mast just for-

---

a Or of course other tax goods such as silk.   b Sung Ying-Hsing's own 'footnote'.

c On locks see pp. 347 ff. above.

d Ch. 9, p. 2a, tr. auct. adjuv. Ting & Donnelly (1); Sun & Sun (1).

e The question of ship tonnages often arises in this Section, cf. pp. 304, 441, 452, 466, 467, 481, 509, 600, 641, 645. It is one of the most intractable of historical problems.

f This was hardly more than half what the Old Grand Canal of the Thang period had carried in +735 (cf. p. 310).

g See the discussion on p. 360 above.

h Ch. 9, p. 2a, b, tr. auct. adjuv. Ting & Donnelly (1); Sun & Sun (1).

i There is information concerning the tools of the shipwrights in Audemard (3), p. 18; Hommel (1) and Mercer (1). Worcester (1, 3) and Waters (1, 2) give details also about the different kinds of iron clamps and nails (*ting*[9]) used. The date when iron fastenings were first used in Chinese shipbuilding is rather a difficult question which I have often discussed with Mr J. V. Mills, who thinks it was in the Thang.

j Most probably held together by vertical (though curved) stiffening bars or bilge frames (see Worcester (3), pl. 39).

k Sung uses this word as interchangeable with *chhiang*,[10] but a sailor would probably not have done so, for its real meaning is mast, yard or boom.

l Certainly wales or top-timbers; see Worcester (1), pl. 7, (3), pl. 39.

m See fn. e on p. 411 just above.

1 開口   2 官坐船   3 石   4 檣   5 棧   6 梁
7 正枋   8 弦   9 釘   10 牆

ward of one of the bulkheads is called the 'anchor altar' (*mao than*[1]). The horizontal bars (*hêng mu*[2]) which grasp the mast's foot below this are called 'ground dragons' (*ti lung*[3]), and these are connected by components called 'lion-tamers' (*fu shih*[4]), while underneath them lies another called the 'lion-grasper' (*na shih*[5]).[a] Under the 'lion-tamers' are the 'closure pieces' (*fêng thou mu*[6]) otherwise known as the 'triple tie-bars' (*lien san fang*[7]).[b] On the deck towards the bow there is a square hatchway (*shui ching*[8]) [in which ropes, cables and miscellaneous gear are kept].[c] At the forward quarter (lit. the eyebrows)[d] on each side, two (strong) posts are placed symmetrically to serve as bollards (for making cables fast, etc.); these are called the 'two generals' (*chiang-chün chu*[9]). The part where the stern slopes upwards is called the 'grass-sandal bottom' (*tshao hsieh ti*[10]); it is composed of the short transverse timbers (*tuan fang*[11]) which close the top of the stern, under which is the transom stern itself, or 'sandal-strap bulkhead' (*wan chiao liang*[12]). The deck at the stern is the place where the helmsman stands to manage the rudder, having above him a bamboo platform (*yeh chi phêng*[13])[e] [on which, when sail is hoisted high, someone sits to manipulate the sheets (*phêng so*[14])[f] according to the wind].[c]

We have a number of eye-witness accounts of Chinese shipbuilding in the present century,[g] all confirming and elaborating what Sung Ying-Hsing had to say. Many observers have been struck by the fact that the Chinese traditional shipwrights used no templates or blueprints, depending rather upon the skill and sureness of eye of the oldest and most experienced craftsmen. Although, as we have seen, some technical manuals have existed,[h] the greater part of the industry must always have been based upon the personally transmitted 'know-how' of the masters.[i] Elsewhere in the yards Sung Ying-Hsing sees the finishers at work:[j]

The joints (*fêng*[15]) between the planks are caulked (*nien*[16]) by first forcing in ravelled floss jute[k] fibre (*pai ma*[17]) with a blunt chisel. Then a (putty-like) composition of fine sifted lime

---

[a] Probably the piece of timber which grasps the tenon at the base of the mast in its socket or mortice.

[b] These are all parts of the structure in which the mast is stepped, and which distributes its thrust. A diagram of Worcester's, Fig. 934, assists the translation to some extent, but the various components are not defined clearly enough in the text to permit exact identification.

[c] Commentary as before, presumably Sung Ying-Hsing's own 'footnotes'.

[d] The expression has additional significance because of the oculus so often painted at the bows of Chinese vessels (cf. p. 436).

[e] The name for this 'pheasant's roost' or poop deck-house, often only a framework, is confusing, but arises because *phêng* means bamboo matting as well as a sail.

[f] See fn. n on p. 414. They are labelled clearly in Fig. 941.

[g] For example Lovegrove (1), for Kuangtung; Worcester (1), p. 4 and (3), vol. 1, pp. 33 ff. for Upper and Lower Yangtze respectively; Donnelly (in Ting & Donnelly); Audemard (3), etc. The desired curvature of the bottom planks was often produced by weighting with stones (G. R. G. Worcester, unpublished material, no. 98).

[h] Cf. also p. 480.

[i] Of course this was also the practice in Europe down to the industrial revolution. The earliest known English shipwright's draughts are those of Matthew Baker in the last quarter of the +16th century (cf. p. 418).

[j] Ch. 9, p. 4*b*, the last sentence from p. 5*b*, tr. auct. adjuv. Ting & Donnelly (1); Sun & Sun (1).

[k] This is 'Chinese jute' or 'Tientsin jute' (*chhing*[18]), i.e. *Abutilon Avicennae* (Malvaceae), a very ancient Chinese fibre-plant (see Sect. 38 in Vol. 6). R274.

[1] 錨壇	[2] 橫木	[3] 地龍	[4] 伏獅	[5] 挐獅
[6] 封頭木	[7] 連三枋	[8] 水井	[9] 將軍柱	[10] 草鞋底
[11] 短枋	[12] 挽脚梁	[13] 野雞篷	[14] 篷索	[15] 縫
[16] 撚	[17] 白麻	[18] 苘		

and tung oil[a] (is applied to complete the job).[b] In Wênchow and Thaichow, and in Fukien and Kuangtung, the ash of oyster-shells is used in place of lime...For sea-going ships (*hai chou*[1]) the caulking is done with a mixture of fish oil and tung oil, why, I do not know.

Then he has a look at the stores:[c]

The timber for the mast is usually fir (*shan-mu*[2]),[d] which must be straight and sound. If the natural size of the spar is not long enough for the mast, two pieces can be coupled together by means of a series of iron bands placed around the joint a few inches apart. An open space is left in the deck for the mast. For the stepping of the main-mast, its upper part is laid across several large boats brought alongside, and the top hoisted up (into position) with a long rope.

Hull timbers and bulkheads are made of *nan-mu*[3],[e] *chu-mu*[4],[f] camphor wood (*chang-mu*[5]),[g] elm (*yü-mu*[6]), or sophora wood (*huai-mu*[7]).[h] [Camphor wood, if taken from a tree felled in spring or summer, is liable to be attacked by boring insects or worms.][i] Deck planks can be made of any wood. The rudder-post is made of elm, or else of *lang-mu*[8],[j] or of *chu-mu*.[4] The tiller should be of *chou-mu*[9],[k] or of *lang-mu*.[8] The oars should be of fir (*shan-mu*[2]) or juniper (*kuei-mu*[10])[l] or catalpa wood (*chhiu-mu*[11]).[m] These are the main points.

Sheets and halyards (*phêng so*[12])[n] carried are made of retted hemp fibres (*huo ma*[13]) [in other words *ta ma*[14],[o]][i] twisted roughly together until they reach a diameter of more than an inch;

[a] From *Aleurites Fordii* (Euphorbiaceae), R321.

[b] This is the classical Chinese caulking material, employed not only for ships, but in the Szechuanese brine industry (cf. below, Sect. 37). 'As for Okam to caulk withal', said Lecomte in +1698, 'they do not use melted Pitch and Tar, but a Composition of Lime and Oil, or rather of a particular Gum with Flax of rasp'd Bambou—this Matter is not subject to the Accidents of Fire, and the Okam is so good, that the Vessel seldom or never leaks...' ((1), p. 232). He made a good point, but Marco Polo had praised the stuff long before. Every early European traveller was impressed by it, e.g. Mendoza, who wrote (p. 150): 'The pitch wherewith they do trimme their shippes...is found in that kingdome in great aboundance; it is called in their language Iapez, and is made of lyme, oyle of fish, and a paste which they call Uname. It is verie strong and suffereth no wormes, which is the occasion that one of their shippes doth twice last out one of ours...' The fish-oil was originally Indo-Chinese or Indonesian rather than Chinese (cf. Ling Shun-Shêng, *1*, 1). The mixture mainly referred to, however, is what is known in China today as 'chunam' (*yu shih hui*[15]). Its qualities are indeed excellent; as Lovegrove says, it dries stone hard, and has been proved to last thirty years. Cf. Audemard (3), p. 20.

[c] Ch. 9, pp. 4b, 5a, tr. auct. adjuv. Ting & Donnelly (1); Sun & Sun (1).

[d] *Cunninghamia sinensis*, sometimes also *Cryptomeria japonica* (R786b; B II, 228).

[e] *Persea* or *Machilus nanmu* (R502; B II, 512). Now *Phoebe nanmu* (Chhen Jung (*1*), p. 345).

[f] A kind of oak, probably *Quercus sclerophylla* (R616; B II, 539).

[g] *Laurus* now *Cinnamomum Camphora* (R492; B II, 513). Its perfume fills the streets of the wood-workers in southern cities. I always remember it in Kweilin in 1944.

[h] *Sophora japonica* (B II, 288, 546). The yellow-berry or pagoda tree (R410). References to it are very frequent in Chinese literature, cf. Vol. 4, pt. 1, p. 73.

[i] Sung Ying-Hsing's own commentary.

[j] A kind of elm, *Ulmus parvifolia* (R608; B II, 304).

[k] A kind of oak, *Quercus glauca* (R614; B II, 538).

[l] *Juniperus sinensis* (R787; B II, 506).

[m] *Catalpa Kaempferi* (R99; B II, 508). This choice goes back to the +3rd century at least, see below, p. 645.

[n] Sung Ying-Hsing means here sail cordage in general. He is not quite consistent in his technical terminology for while halyards have been *hsien* and sheets *phêng-so* (pp. 411, 413 above), the latter phrase is used here for all, and elsewhere (p. 604 below) means indubitably halyards. Topping lifts he seems not to mention.

[o] Common *Cannabis sativa* of course; see Sect. 38 in Vol. 6, pt. 1.

1 海舟	2 杉木	3 楠木	4 櫧木	5 樟木	6 榆木
7 槐木	8 榔木	9 桐木	10 檜木	11 楸木	12 篷索
13 火蔴	14 大蔴	15 油石灰			

then the rope can sustain a weight of 10,000 *chün*.[a] Anchor cable is made of thin strips of the outer parts of the stems of green bamboo, which after being boiled in water are twisted into rope. The tracking (towing) cables (*than*[1]) are made in the same way. Cables more than 100 ft. long come in sections with loops at both ends to join them together,[b] so that they can be uncoupled immediately if obstacles are encountered. It is in the nature of bamboo to be 'straight' (i.e. to have a high tensile strength), so that one such skin strip can sustain a weight of 1,000 *chün*.[a] When ships are going up the Yangtze gorges to Szechuan they do not use the twisted cables, but rather bamboo laths cut to the width of about an inch and joined to form a long flexible spar (or chain of rods); this is necessary as twisted cable can easily be cut or broken by the sharp rocks.[c]

Finally he makes some remarks about the general rig of the vessels of the inland waterways.[d]

Ships more than 100 ft. long must have two masts. The main-mast (*chung wei*[2]) is stepped two bulkheads forward of the mid-point, and the fore-mast (*thou wei*[3]) some 10 ft. or more further forward. On the grain-ships the main-mast is about 80 ft. high, but spars shorter by one or two tenths of this are also used. The heel is stepped into the hull by a depth of some 10 ft. The position (of the halyard-blocks) of the sail is at a height of 50 to 60 ft. The fore-mast is less than half the height of the main-mast, and the dimensions of the sail which it carries are not more than about one third (of the main-sail). Rice transport boats in the six prefectures of Huchow and Suchow[e] have to pass under stone arch bridges, along waterways without the dangers of the great rivers like the Yangtze and the Han, so the dimensions of their masts and sails are much reduced. Ships which navigate in the provinces of Hunan, Hupei and Chiangsi, crossing the great lakes and rivers, amidst incalculable winds and waves, have to have their anchors, cables, sails and masts well found according to the standard regulations exactly—then there need be no anxiety.[f]

About sea-going ships Sung Ying-Hsing had little to say (they were evidently peripheral to his experience), but he recorded some interesting names.[g]

Sea-going junks (*hai chou*[4]) were used for grain transport during the Yuan dynasty and at the beginning of the Ming. One type was called the 'shallow-draught ocean ship' (Chê-yang-

[a] These two estimates are interesting. A *chün* is a weight of 30 catties (*chin*), and in Ming times therefore was equivalent to about 40 lb. Consequently the first figure given would represent just under 180 tons and the second just under 18. As Sung Ying-Hsing forgot to specify the cross-section in either case, assessment of what he says is tricky, but we know from Vol. 4, pt. 2, p. 64, that modern measurements give an average of about 3·3 tons per sq. in. (the same order as for steel wires). Most probably therefore he was thinking quite approximately of floating tonnages retained against adverse currents; certainly not in terms of modern strength of materials tests. Sun & Sun (1) translate *chün* dangerously as catties, believing that the weights were liable to be confused; if so, the first means 5·9 tons and the second 0·59. Although these are temptingly nearer to the modern test figure, it is probably better not to interpret the text in this way. Cf. pp. 191, 664.
[b] Probably with toggle connections. Cf. Audemard (3), pp. 48 ff.; also here p. 662.
[c] No doubt this property of toughness is connected with the silica naturally present in the untreated stems of bamboos.
[d] Ch. 9, pp. 2b, 3a, tr. auct. adjuv. Ting & Donnelly (1); Sun & Sun (1).
[e] In Chiangsu province.
[f] Still today, according to Donnelly, the boats of these provinces have loftier spars than others, but he thinks that this may also be due to the necessity for catching wind coming over high river banks at low water seasons.
[g] Ch. 9, pp. 5a, b, 6a, tr. auct. adjuv. Ting & Donnelly (1), Sun & Sun (1); sentence order modified.

[1] 䉶　　[2] 中桅　　[3] 頭桅　　[4] 海舟

chhien chhuan [1]), another the 'boring-into-the-wind ship' (Tsuan-fêng chhuan[2])[a] [or the 'sea-eel' or 'sea-serpent' (Hai-chhiu[3])[b] ship].[c] Their voyages did not exceed 10,000 *li* along the coasts, across the Dark Sea[d] and past the Sha-mên[4] islands.[e] No great danger was encountered in the voyage.[f] As compared with the junks which sailed to Japan, the Liu-Chhiu islands, Java and Borneo for trade, these (coastal transports) were not one-tenth as large or expensive.

The build of the first type of sea-going junk is similar to that of the canal junk, save that it is some 16 ft. longer, and 2·5 ft. broader in the beam. All else is the same except that the rudder-post must be made of iron-wood (*thieh-li-mu*[5])[g]...[h]

(All) the ships that sail to foreign countries (*wai-kuo hai po*[6]) are rather similar in specification to the (great) sea-going junks (just mentioned).[i] Those which hail from Fukien and Kuangtung [Hai-chhêng in Fukien and Macao in Kuangtung][c] have bulwarks of half-bamboos for protection against the waves. The ships from Têngchow and Laichow (in Shantung) are of a different type again. In Japanese ships the rowers are completely under cover, in Korean vessels not so. All these types, however, have in common two mariners' compasses, one at the bow and the other at the stern, to indicate the direction of the course. They also have in common 'waist-rudders' (i.e. leeboards)...[j] They carry several piculs of fresh water, enough for the whole ship's company for two days, in bamboo barrels, and when they touch at islands they get more.

The bearings of all the different islands can be found by the magnetic compass, which is indeed a wonderful invention almost beyond human power. (To undertake such voyages) helmsman, crew and master must be of good judgment as well as steadfast in the highest degree. Blind bravery is no use at all.

Everything that has been said in the preceding pages about the construction of the junk is applicable, having regard to size, to the larger and smaller sampans whether decked or otherwise. These have been studied by several observers, whose interesting

[a] No doubt so termed because of its capacity for sailing into the wind close-hauled.

[b] This name goes back certainly as far as the Southern Sung, for Hai-chhiu were used (with the larger Mêng-chhung) as warships in the Battle of Tshai-shih (+1161) against the Jurchen Chin. See *WHTK*, ch. 158 (p. 1382.1), quoting Yang Wan-Li's[7] *Hai Chhiu Fu Hou Hsü*[8] (in *Chhêng-Chai Chi*,[9] ch. 44, pp. 6b ff.). Both types were there equipped with treadmill-operated paddle-wheels; cf. Vol. 4, pt. 2, pp. 421, 724. *Hai-chhiu* commonly means 'whale'.

[c] Sung Ying-Hsing's own commentary.

[d] I.e. the Yellow Sea. One recalls the *mare tenebroso* feared by the early Portuguese.

[e] Off Shantung.

[f] A euphemism. Cf. Schurmann (1), pp. 111, 122, etc.

[g] Perhaps a palm growing in Kuangtung and Annam; *Sagus Rumphii, Arenga Engleri*, etc. (R 715, 716, Stuart, p. 389). According to the *Chhou Hai Thu Pien* of +1562, Cantonese junks were built entirely of this very hard wood, and if a collision with a Fukienese or a Japanese vessel occurred, the latter came off very much the worse (ch. 13, pp. 2b ff., tr. Mills (6); text abridged in *TSCC, Jung chêng tien*, ch. 97, p. 11b, tr. Audemard (2), p. 49). See also the entry in *KCCY*, ch. 66, p. 5a; and p. 646 below.

This wood is difficult to identify. If it was not a palm, it may have been *Mesua ferrea* (Guttiferae); see Chhen Jung (1), p. 849. This tree, which grows to 100 ft., gives a very hard and durable wood, used for buildings and furniture; it is native to Kuangsi. Again, *thieh li mu* appears to be a local name for the hemlock-spruce *Tsuga sinensis* (Pinaceae), *loc. cit.* p. 42.

[h] There follows the remark on caulking, already given on p. 414 above.

[i] In this connection it is interesting that a European ship with three masts (and a lateen sail on the mizen, cf. below, p. 512) is depicted in *TSCC, Khao kung tien*, ch. 178, *hui khao*, p. 24b. But we agree with Sun & Sun (1) that Sung Ying-Hsing did not mean foreign ships here. The analogy is with 'China clippers', which were not Chinese.

[j] There follows the passage on leeboards translated on p. 619 below.

[1] 遮洋淺船   [2] 鑽風船   [3] 海鰍   [4] 沙門   [5] 鐵力木
[6] 外國海舶   [7] 楊萬里   [8] 海鰌賦後序   [9] 誠齋記

memoirs will satisfy the reader desiring full technical detail.[a] They have won the admiration of the sailors of other cultures for their excellent adaptation to their fishing, ferrying or freight-carrying duties, no less than the largest and most majestic traditional Chinese vessels.

(3) HULL SHAPE AND ITS SIGNIFICANCE

The ship of approximately rectangular cross-section, with rounded corners, had indeed the future before it. The nearest illustrated encyclopaedia will be enough to demonstrate to the reader that this is the form of the iron and steel steamships of our own time. It was already apparent to Charnock's contemporaries a century and a half ago that such a section had very great advantages, particularly in stability under lading.[b] But Chinese junks had another characteristic which surprised Europeans at the first encounter, yet which has also been adopted by them (if in forms less extreme), namely a build in which the broadest beam at water level exists aft of the midship line.

The ways of moulding the hull of a ship are limitless, but a simple distinction can be made between forms which are broader forward and those which are broader aft. Most symmetrical are those in which the master-couple (i.e. the ribs enclosing the greatest area) coincides with the midship frame, from which there extends in both directions a gradual taper. Fullness aft or forward may be obtained by rounding the taper more gradually in one direction than in the other. But extremer forms exist in which the master-couple occurs well forward or aft of the middle of the ship. Broadly speaking, the European tendency has always been to set the greater fullness of the ship towards the bow, while the Chinese tendency was to set it towards the stern.[c] So natural was thought the former practice that John Dryden could write:

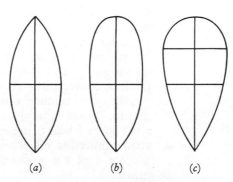

Fig. 944. Horizontal longitudinal sections of a ship at water-level.

a   symmetrical
b   fuller at bow (or stern)
c   master-couple towards bow (or stern)

> By viewing nature Nature's handmaid, Art,
>   Makes mighty things from small beginnings grow;
> Thus fishes first to shipping did impart
>   Their tail the rudder and their head the prow.[d]

Arguments about the best position of the master-couple go back to Isaac Vossius in the +17th century, who wrote a commentary on the marine architecture of the

[a] E.g. Worcester (1, 3, 7, 8, 9, 13); Carmona (1); Waters (9, 10).
[b] Cf. Charnock (1), vol. 3, p. 340.   [c] Poujade (1), p. 210.
[d] 'On Shipping, Navigation and the Royal Society', from *Annus Mirabilis* (+1667).

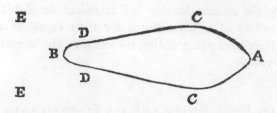

A fit prora, B puppis. Aquæ itaque motu navis congeftæ
& coacervatæ ad proram A, primo quidem fegnius defluunt
ad utrunque navis latus CC, defcenfu vero ipfo augent ce-
leritatem, donec perveniant ad D D, ubi intenfiffimus eft
aquarum lapfus. Inde repulfæ occurfu aquarum pone
relabentium declinant ad E E, ubi demum fiftuntur & lo-
cum dant aquis à tergo & utroque latere venientibus. Sed
vero conftat gubernacula non ab infequentibus, fed ab
affluentibus à prora regi undis, unde planum fit aptius collo-
cari ad D D, quam ad B, quo vix attingunt fubterlabentes
undæ, cum ad D D maxima fiat percuffio & fufficiens ad
regendam totam navem, quæ quanto longior eft tanto facilius regitur, & quanto velocius procedit tanto magis obfe-
quitur gubernaculis, & hinc eft quod longæ naves facilius
& citius convertantur & circumagantur, quam minus longæ, & quod minutæ etiam cymbæ, majori egeant guber-
naculo, quam quævis maximæ naves, cum enim non
alte aquam fubeant, nec multas propellant undas, utique
etiam imbecillis eft aquarum relabentium affluxus, qua-
propter regi vix poffunt, nifi enorme habeant guberna-
culum.

Fig. 945. A diagram and paragraph of text from the *Variarum Observationum Liber* of Isaac Vossius
(+1685). In the section entitled *De Triremium et Liburnicarum Constructione* he gave this drawing of a
hull with its master-couple well forwards. The text sets forth an argument in favour of quarter-rudders.

ancients.[a] Charnock gives a diagram to illustrate his ideas.[b] It suffices to look over the
boats and small ships in any European harbour, or the models in any great naval
museum, to see how widespread was the conviction that the greatest fullness should be
towards the bow. One of the drawings in the Baker MS. of +1586 compares the under-
water body of a ship with the form of a fish, according to the famous maxim 'a cod's
head and a mackerel tail' (Fig. 946, pl.).[c] Following this parallel, much used in older
Western writings, Poujade says[d] that while the Europeans felt they ought to construct
their boats with the outline of a fish,[e] the Chinese felt that they ought to follow the
outline of a swimming bird at water-level. Admiral Paris in 1840 was perhaps the
first observer to note this great difference. He wrote:[f]

a *De Triremium et Liburnicarum Constructione*, in *Variarum Observationum Liber*, pp. 59 ff.; diagram
of hull shape showing greatest master-couple forward, on p. 135 (Fig. 945).
  b (1), vol. 1, p. 103.   c Clowes (2), p. 69.
  d (1), pp. 210, 248; cf. Charnock (1), vol. 3, p. 402.
  e Typical example in Charnock (1), vol. 3, opp. p. 364.   f (1), p. 3.

Some peoples have made the keel follow a convex curve, others have made it concave; some make the hull float lower forward than aft, or they place the broadest beam forward, amidships, or aft. For our best hulls we have taken the fishes as models, always larger at the cephalic end, but the Chinese, who also copied Nature, imitated the palmipeds, which float with the greatest breadth behind, for somewhat obscure reasons. In this they were acute, for aquatic birds, like boats, float between the two media of air and water, while fish swim only in the latter. These strange people seem to have done everything in the opposite way to which it is done at the other end of the continent, and they copied Nature still further in seeking to exert the greatest possible propulsion on the stern, instead of applying, as it were, a tractive force to the prow. This led them to the use of those strong paddles (the yulohs)[a] which imitate in position the web-feet of the palmipeds—a position which must have been very important for swimming, since it deprived such birds of the facility of walking easily on land, and even, in the case of the best swimmers, prevented this altogether. These very simple observations (which the Chinese have utilised) will find one day, perhaps, a happy application to the steam-boat, which, set in motion as it is by an internal force not coming like the wind from without, finds itself in exactly the same situation as the swimming bird, and might gain from a closer approximation to the latter's form.

These prophetic words of Paris were justified within a couple of decades by the coming of the screw propeller; indeed in the year before they were written, F. P. Smith's 45 h.p. screw steamer 'Archimedes' had made her trial run.[b] As for the hull shape, the pioneer water-tank experiments of Gore in +1797 had seemed to prove the superiority of shapes with the fullness forward, but his trial objects were totally immersed.[c] That precisely the opposite is true we now know; the largest master-couple of a good sailing-ship should be located between 3 and 8% aft of the middle of the flotation line.[d] The racing yacht 'America' demonstrated this in the last years of the nineteenth century.

That the lines of Chinese vessels do show their greatest fullness aft has been confirmed by Worcester in a range of accurate drawings far beyond the data which Admiral Paris had at his disposal. A cursory examination reveals about thirty-five types which are built in this way. Some merely continue the master-couple for varying distances aft of the mid-line,[e] but others show distinctly the master-couple itself well aft.[f] All the Ma-yang-tzu[1] varieties are so built, and the name itself means 'hemp sprout', something more pointed at one end than the other. Several of the familiar names for the river junk types allude to the form bulging aft—the 'hempseed oil bottle' (Ma-yu

[a] See below, p. 622.

[b] On the history of the screw see de Loture & Haffner (1), p. 203; Feldhaus (1); McGregor (1); Seaton (1); Bourne (1); G. S. Graham (1). There are claims as early as the seventies and eighties of the previous century; see Gibson (2), p. 148. But they were not practical successes. Cf. Vol. 4, pt. 2, p. 125.

[c] Described in Charnock (1), vol. 3, pp. 377 ff.

[d] The question is of course complicated, for the best hull shape depends upon the speed at which it is expected to move through the water (see the discussion of Birt, 1). The superiority of the Chinese type of hull is seen at slow speeds; for higher speeds an almost symmetrical outline is more effective.

[e] Worcester (1), pp. 38, 51, 63, 81; (3); vol. 1, pp. 126, 134, 137, 139, 200, 203; vol. 2, pp. 276, 279, 284, 288, 321, 322, 336, 366, 421, 436, 463, 466, 489.

[f] Worcester (1), pp. 21, 65, 68, 80; (3), vol. 2, pp. 295, 393, 408, 422, 429, 456, 494. While a good number of forms are approximately symmetrical, there is only one which has the greatest fullness forward (Worcester (3), vol. 2, p. 457).

[1] 麻秧子

hu-tzu[1]), the 'ingot' (Chin-yin ting[2]), the 'red slipper' (Hung-hsiu hsieh[3]), and, as Admiral Paris would have been delighted to learn, the 'duck's tail' (Ya-shao[4]).

Chinese shipwrights were in fact quite well aware of what they were doing. In his *Hai Yün Hsin Khao*[5] (New Investigation of Sea Transport) of +1579, Liang Mêng-Lung[6] alludes to a famous shipbuilder of the Yuan period,[a] Lo Pi,[7] who constructed junks with the significant name of Hai-phêng.[8] The reference was to the gigantic bird which could take off from the water and fly to the south, so famous in Chuang Tzu.[b] Lo Pi's motto was *kuei shen, shê shou*[9], i.e. 'with the body like a tortoise and the head like a snake'. No more definite statement could be imagined.

(4) WATER-TIGHT COMPARTMENTS

It is possible to discern another valuable shipbuilding technique which Europeans adopted from China, namely the water-tight compartment.[c] From all that has gone before it will have become evident that bulkhead construction was the most fundamental principle of Chinese naval architecture (cf. Fig. 947, pl.), and water-tight compartments were its corollary. This is one of those cases where it is also possible to be fairly precise as to the date when the device was adopted.[d] It was 'in the air' during the closing decades of the +18th century, and nearly every reference of the period recalls its use in Chinese ships. In +1787 Benjamin Franklin wrote a letter concerning mail packets projected between the United States and France, in which he said:

> As these vessels are not to be laden with goods, their holds may without inconvenience be divided into separate apartments after the Chinese manner, and each of these apartments caulked tight so as to keep out water. In which case if a leak should happen in one apartment ...This being known would be a great encouragement to passengers.[e]

In 1845 Guppy read a paper to the Institution of Civil Engineers in London describing the trial voyages of the iron steamship 'Great Britain', and in the ensuing discussion a communication was read from Lady Bentham about the improvements in naval architecture which had been introduced by Sir Samuel Bentham (+1757 to 1831).[f] The merit of 'the first introduction of that great improvement of fixed water-tight bulkheads' was, it appeared, his.[g] In the year +1795 he had been entrusted by the Lords Commissioners of the Admiralty to design and construct six vessels of entirely new type, and he

---

[a] Ch. 1, p. 32a. Of Lo Pi's activities as a sea-captain we shall have more to say below, p. 478.

[b] Ch. 1, tr. Legge (5), vol. 1, p. 164; cf. Vol. 2, p. 81 above.

[c] This has often been pointed out, as by Gilfillan (1) and Audemard (3).

[d] It is true that Leonardo da Vinci had proposed a ship construction with double sides to minimise the danger of holing (cf. Ucelli di Nemi (3), no. 82), but the idea, probably never implemented, was nearer to the double-hulled craft already discussed (p. 392) than to transverse water-tight compartments.

[e] Quoted by Playfair (2). Another letter of Franklin's on the same subject is quoted by Charnock (1), vol. 3, p. 361. See also Franklin's *Maritime Observations*, published in +1786, p. 301.

[f] Brother of Jeremy.

[g] Actually many naval improvers had been suggesting it, e.g. Captain Schanck (Charnock (1), vol. 3, p. 344).

---

[1] 麻油壼子   [2] 金銀錠   [3] 紅繡鞋   [4] 鴨艄   [5] 海運新考
[6] 梁夢龍   [7] 羅璧   [8] 海鵬   [9] 龜身蛇首

built them with 'partitions contributing to strength, and securing the ship against foundering, as practised by the Chinese of the present day'.

The circumstances of Bentham's introduction are so interesting with regard to the diffusion of techniques that they are worth a closer look.[a] From Lady Bentham's biography of him (1862), which drew upon his early journals, we find that though in the latter part of his life he was for long chief engineer and architect of the British Navy, his earlier years had been spent in the Russian service, which he left with the rank of Brigadier-General. In the course of extensive Siberian travels[b] undertaken about + 1782, he had reached the frontiers of China at Kiachta where he had met the local Chinese governor and observed large boats on a river which must have been the Shilka.[c] Some of these, if not all, would have had the usual transverse bulkheads.

This was no invention of General Bentham's [wrote his wife long afterwards]; he has said himself officially that it was 'practised by the Chinese of the present day, as well as by the ancients', yet to him is due the merit of having appreciated the advantage of water-tight compartments, and of having introduced the use of them. Shipwrights, perhaps, may not be familiar with classic lore, but could they all have been ignorant of the expedient so common in Chinese vessels?[d]

Still more interesting is the fact that Bentham copied another Chinese technique. About + 1789, when serving as colonel of a regiment in the Crimea, he constructed an articulated boat, the 'Vermicular', for negotiating the shallow and winding reaches of certain rivers down which it was desired to bring natural products.[e] As we shall shortly see,[f] this was a characteristic Chinese practice.[g] The case is analogous to one which we have already studied.[h]

A writer (Anon. 20) in the *Mechanics' Magazine* for 1824 said: 'There is a method of making it almost impossible to sink ships which was known to the ancients[i] and is now employed by the Chinese. The hold is divided into a number of compartments, so that if the ship spring a leak, or her sides be stove in...' she will still remain afloat. The only remarkable thing about this transmission is that it should have taken so long to be adopted in the West. Marco Polo (as we shall see below, p. 467) distinctly described the water-tight compartments of Chinese ships about + 1295, and his information was repeated by Nicolò de Conti in + 1444—'These Shippes are made with Chambers,

[a] We are indebted to Mr D. R. Bentham for calling our attention to Samuel Bentham's Chinese contacts.

[b] Samuel Bentham was a most ingenious man; at Perm he built amphibious vehicles and a hodometer, and improved salt mines. Later in life he was one of the pioneers of vacuum distillation.

[c] Bentham (1), pp. 49, 51. The place was 10 versts (about 10 km.) from Nerchinsk. Today the frontier between Manchuria and Siberia follows the River Argun, a good deal more to the south-east.

[d] Bentham (1), p. 107.

[e] This craft had six component sections coupled together with a total length of 252 ft., a breadth of 16 ft. 9 in., and a draught of 4 in.

[f] P. 432 below.

[g] It would seem quite probable that Samuel Bentham saw something similar on the river between Siberia and Manchuria.

[h] That of Simon Stevin at the beginning of the + 17th century (Vol. 4, pt. 1, p. 227 above), whose double borrowing, if such it was, concerned the sailing-carriage idea and the mathematical formula for equal temperament in music.

[i] This was certainly not the case.

after such a sorte, that if one of them should breake, the others may goe and finish the Voyage.'[a] Cressey (1) is quite right, therefore, in remarking that this practical safety device was not adopted until five hundred years after knowledge of it first reached Europe.

Less well known is the interesting fact that in some types of Chinese craft the foremost (and less frequently also the aftermost) compartment is made free-flooding. Holes are purposely contrived in the planking.[b] This is the case with the salt-boats which shoot the rapids down from Tzuliuching in Szechuan,[c] the gondola-shaped boats of the Poyang Lake, and many sea-going junks.[d] The Szechuanese boatmen say that this reduces resistance to the water to a minimum, and the device must certainly cushion the shocks of pounding when the boat pitches heavily in the rapids, for she acquires and discharges water ballast rapidly just at the time when it is most desirable to counteract buffeting at stem and stern. The sailors say that it stops junks flying up into the wind. It may be the reality at the bottom of the following story, related by Liu Ching-Shu[1] of the +5th century, in his book *I Yuan*[2] (Garden of Strange Things):[e]

In Fu-Nan (Cambodia) gold is always used in transactions. Once there were (some people who) having hired a boat to go from east to west, near and far, had not reached their destination when the time came for the payment of the pound (of gold) which had been agreed upon. They therefore wished to reduce the quantity (to be paid). The master of the ship then played a trick on them. He made (as it were) a way for the water to enter the bottom of the boat, which seemed to be about to sink, and remained stationary, moving neither forward nor backward. All the passengers were very frightened and came to make offerings. The boat (afterwards) returned to its original state.

This, however, would seem to have involved openings which could be controlled, and the water pumped out afterwards.[f]

One reason for using water-tight compartments with a free-flooding portion is to enable fishing-boats to bring their catch to port and market in live condition. This was easily effected in China,[g] but the practice was known also in England, where the compartment was called a 'wet-well', and the boat in which it was built, a 'well-smack'.[h] If the tradition is right that such boats date in Europe from +1712 then it may well be that the Chinese bulkhead principle was introduced twice, first for small coastal fishing-boats at the end of the seventeenth century, and then for large ships a century later.

[a] (1), p. 140.
[b] Sometimes they permit also the passage of the 'stick-in-the-mud' anchor; see below, p. 660.
[c] Cf. p. 430 below.
[d] Worcester (2), p. 25, (3), vol. 1, p. 41; vol. 2, p. 414; Waters (5).
[e] Tr. Pelliot (29), eng. auct.
[f] Water ballast under control was, of course, one of the cardinal inventions accompanying the European development of the submarine (Bushnell, +1775; Fulton, 1800); cf. de Loture & Haffner, pp. 261 ff.
[g] Fishing-boats so equipped are common there especially in the estuaries of Kuangtung (G. R. G. Worcester, unpublished material, no. 100), and at Hongkong (B. Ward, private communication).
[h] E. W. White (1), p. 23.

[1] 劉敬叔   [2] 異苑

(d) A NATURAL HISTORY OF CHINESE SHIPS

'There are those who affirm', said Domingo de Navarrete in +1669, 'that there are more Vessels in China than in all the rest of the known World. This will seem incredible to many Europeans, but I, who have not seen the eighth part of the Vessels in China, and have travel'd a great part of the World, do look upon it as most certain.'[a] All the early travellers to China remarked upon the abundance of shipping, from Marco Polo and Ibn Baṭṭuṭah onwards (cf. Fig. 948, pl.). It was natural, therefore, that a great variety of types of ships should have developed, and modern research has done a good deal towards their analysis and classification. Our frame of reference will not allow us to embark upon an extended review of its achievements, but a few pointers to the literature may be given, and some of the more remarkable vessels mentioned.

Systematic lists of shipping types occur on several occasions in Chinese literature throughout the ages, and we mention some of them elsewhere in this Section.[b] One of the oldest descriptions of classes of warships which has survived dates from +759, the work of a Taoist military and naval encyclopaedist in the Thang; we shall translate it fully later on in the context of the history of ship armour.[c] Sung descriptions of types also exist (cf. pp. 381, 602). For the Ming period, there is the *Shui Chan I Hsiang Lun*[1] (Advisory Discourse on Naval Warfare) of Wang Ho-Ming[2] written late in the +16th century, before +1586, which describes nine warship types in detail.[d] At the same time Wang Chhi[3] wrote a classification of the suitability of the different types for the defence of the different parts of the coast.[e]

Sung Ying-Hsing, in +1637, at the conclusion of the passage above quoted (pp. 411 ff.), went on to describe nine other types of ship; in the course of which he mentioned the yuloh (self-feathering propulsion oar),[f] the cloth sails of the Chhien-thang River,[g] and the bipod masts (*shuang chu*[4]) of the river-boats of Kuangtung province.[h] These occur only in the south. Two Chinese MSS in the British Museum,[i] dating from about +1700, give, according to Donnelly (2), drawings and paintings of some eighty-four vessels; these documents have not yet received scholarly description. Mayers (5) reported the names and very brief descriptions of thirty-five different sorts

[a] *Tratados...de la Monarchia de China*, Cummins ed., vol. 2, p. 227. Worcester (3) quotes this from the *New History of China* (+1688) of Gabriel de Magalhaens, but we could not find it there. The statement is akin to one of Ricci's, in Trigault (Gallagher tr., pp. 12, 13) and in d'Elia (1), vol. 1, pp. 20, 23. Cf. Ysbrants Ides, p. 163. Other quotations will be found in Audemard (2), pp. 19 ff. Perhaps the first statement of a modern European is that of Cristovão Vieira, writing from Canton in +1524; 'ships and boats without number' he said he had seen (D. Ferguson (1), p. 19). Barrow in 1804 was impressed just as deeply, (1), p. 399. William Bourne tells us, in his *Regiment for the Sea* (+1580), how Master John Dee encouraged him to plan navigations to Cathay by showing him passages from Marco Polo describing the vast number of ships which flew the flag of the Great Khan (see Taylor (13), p. 313).

[b] Pp. 406, 424 ff., 465, 482, 685 ff.   [c] Pp. 685 ff. below.

[d] This is quoted at length in *Hsü Wên Hsien Thung Khao*, ch. 132 (pp. 3972.2 ff.). By this time occidental influence was perhaps manifesting itself, as two of the types described had 'keels'. Some were also armed with culverins in Portuguese style. Cf. pp. 481 ff.

[e] *Hsü Wên Hsien Thung Khao*, ch. 132 (pp. 3974.1 ff.).

[f] Cf. pp. 622 ff. below.   [g] Cf. p. 457 below.

[h] Cf. p. 435 below.   [i] Lansdown no. 1242 and Egerton no. 1095.

¹ 水戰議詳論   ² 王鶴鳴   ³ 王圻   ⁴ 雙柱

of craft nearly a century ago. He had been preceded by K. A. Skachkov (1) who in a little-known paper discussed a MS. or printed book in the Rumiantzov Museum entitled *Shui Shih Chi Yao*[1] (Essentials of Sea Affairs), seemingly on warships of a traditional kind but sinologically so far unidentifiable. In recent years a number of types have been given detailed descriptions,[a] the greatest collection of which, that of Worcester (1–3), has analysed no less than 243 vessels, ranging from the smallest sampans to the largest sea-going junks. The catalogue of the Maze Collection of Chinese ship models in the Science Museum at South Kensington (Anon. 17), together with the fine photographs issued by the Museum, affords further aid.[b]

Paintings of Chinese vessels by Westerners began only in the early years of the last century, when in 1801 William Alexander's excellent coloured pictures, made upon the occasion of Lord Macartney's embassy (+1793), appeared.[c] Afterwards came the beautiful drawings and diagrams of Admiral Paris (1, 2), and the remarkable early photographs of J. Thomson (1) which include at least two good views of sea-going junks.[d] Long before modern times, indeed, there had been sketches of Chinese ships incorporated in late medieval Western world maps, but these we must reserve for mention in a more historical context shortly.[e]

Too little study has been devoted to the Chinese traditions of nautical iconography. Here the centrepiece is, as we have said, the *Wu Ching Tsung Yao*[2] (Collection of the most important Military Techniques)[f] of +1044, in which Tsêng Kung-Liang[3] carefully describes six warship types. But his text is closely based on a much earlier set of descriptions, those of the *Thai Pai Yin Ching*[4] (Manual of the Martial Planet),[g] compiled by Li Chhüan[5] in +759, and copied, with the inevitable variations, into Tu Yu's[6] *Thung Tien*[7] (Comprehensive Institutes)[h] of +812. These six types of ships, which constantly recur throughout ten following centuries, are (in order of decreasing size):[i]

(1) 'Tower-Ships' (battleships with fortified upper-works),[j] Lou chhuan.[8]

(2) 'Combat-Junks' (less protected), Chan hsien[9] or Tou hsien.[10]

(3) 'Sea-Hawk Ships' (converted merchantmen), Hai-hu (or -ku).[11]

[a] E.g. Audemard (1, 5, 6, 7); Farrère & Fouqueray (1); Carmona (1) on craft typical of the regions near Macao; Waters (1–5) especially on the Antung, Pechili, and Hangchow traders of the north; Sigaut (1) on the Tsingtao freighter; Donnelly (1, 3–5) on ships of the Yangtze and of Fukien; Lovegrove (1) on those of the Canton and West River systems in the south. See also the catalogue of Nooteboom (2).

[b] Cf. Worcester (10) on the late Sir Frederick Maze, and (16) on the junks of Chiangsi.

[c] These have been reproduced in accessible form (Anon. 19). Those of Capt. Drummond may date from about the same time. On George Chinnery (+1774 to 1852) see Berry-Hill & Berry-Hill (1).

[d] In vol. 1, fig. 14 and vol. 2, fig. 46.   [e] See pp. 471, 473 below.

[f] *Chhien Chi*, ch. 11, pp. 4b ff.   [g] Ch. 40, pp. 10a ff., tr. p. 685 below.

[h] Ch. 160, p. 16b (pp. 848.3, 849.1).

[i] The order of enumeration varies in the different sources. In the *WCTY* the description of the patrol boat (Yu thing[12]) was conflated with that of quite another type of ship, the 'five-flagged cruiser' (Wu-ya hsien[13]), equipped with 'grappling-irons' (cf. p. 689 below), which Yang Su[14] had built for Sui Kao Tsu about +595. This came in from a source other than *TPYC*. The picture in the Ssu Khu Chhüan Shu edition of *WCTY* shows the five-flagged cruiser wrongly labelled as the patrol-boat. *STTH* put this confusion right, but *TSCC*, in its usual garbling way, brought it back again.

[j] Presently we shall see that this term is much older still.

[1] 水事輯要	[2] 武經總要	[3] 曾公亮	[4] 太白陰經	[5] 李筌
[6] 杜佑	[7] 通典	[8] 樓船	[9] 戰艦	[10] 鬪艦
[11] 海鶻	[12] 遊艇	[13] 五牙艦	[14] 楊素	

(4) 'Covered Swoopers' (fast 'destroyers'), Mêng-chhung.[1]

(5) 'Flying Barques' (smaller fast ships), Tsou ko.[2]

(6) 'Patrol Boats', Yu thing.[3]

Unfortunately no corresponding Thang pictures seem to have survived, and the *Wu Ching Tsung Yao* set is illustrated by two series obviously drawn at much later periods. Those of the +1510 edition, recently reproduced, may well be contemporary, i.e. Ming, and therefore of considerable value, but those of the Ssu Khu Chhüan Shu edition, which was taken from a MS. (old but of uncertain date) in the Imperial Library, degenerated a good deal in the redrawing, presumably about +1780. We represent in Fig. 949 the Ming version of the 'tower-ship' (Lou chhuan[4]) or battleship with fortified and crenellated upper-works on several levels.[a] The artist was so enamoured of the warlike aspect of his task that he omitted the masts and sails altogether, and drew the balanced rudder very poorly, but did not fail to show a counterweighted trebuchet on the top deck.[b]

From the beginning of the +17th century encyclopaedias both military and civil continued to copy all this ancient material while at the same time enlarging it by many new descriptions and accounts of ship types. Wang Chhi's[5] *San Tshai Thu Hui*[6] of +1609 gave some thirty of these,[c] including nine civilian vessels and one with a bipod or sheerleg mast (the 'ship of the immortals', Hsien chhuan[7]). In spite of its romantic name, this craft[d] was said to ply upon the southern rivers and lakes as a transport for high officials. About twenty of Wang Chhi's items were incorporated by Mao Yuan-I[8] into the thirty-four of his *Wu Pei Chih*[9] (Treatise on Armament Technology)[e] of +1628. This was concerned entirely with war vessels, but in most editions the drawings are somewhat crude, and the artists do not seem to have been sufficiently interested in sea matters to differentiate much between the ships which they were supposed to be depicting. Some however bear reproduction, such as the 'sand-boat' (Sha chhuan[10]) of the Chiangsu coast, with its lug-sails set (Fig. 950).[f] Other pictures show craft with what seem to be rudders at the bow as well as at the stern (the 'eagle-ship', Ying chhuan[11] and the 'double-headed ship', Liang-thou chhuan[12]).[g] There may have been

[a] It is interesting that in Chinese pictures the super-structures are almost always placed amidships, in contrast to the 'fore-castles' and 'after-castles' of medieval Europe.

[b] The +18th-century draughtsman quite lost the thread here, and made it look like a flag at the end of a leaning pole supported in a crutch. The late *WCTY* picture and that in *TSCC* are reproduced in Audemard (2), pp. 35 ff., with translation of the *TSCC* text by Shih Chun-Shêng *et al.*

[c] *Chhi yung* sect., ch. 4, pp. 19 a ff.

[d] Reproduced, with description, in *TSCC*, *Khao kung tien*, ch. 178, pp. 18 b ff.

[e] Chs. 116–118.

[f] The text concerning this in *TSCC*, *Jung chêng tien*, ch. 97, p. 25 b, is translated by Shih Chun-Shêng *et al.* in Audemard (2), p. 73. 'Sand-folk' was the name given to the fishermen and sailors of this part of the coast.

[g] The text concerning these in *TSCC*, *Jung chêng tien*, ch. 97, pp. 21 b and 24 b, is translated by Shih Chun-Shêng *et al.* in Audemard (2), pp. 68, 70. The description of the 'double-headed ship' was based on a passage in Chhiu Chün's[13] *Ta Hsüeh Yen I Pu*,[14] written about +1480. This is a book which we shall encounter again in connection with the history of gunpowder. Cf. Vol. 4, pt. 2, p. 430, Fig. 638.

[1] 蒙衝	[2] 走舸	[3] 遊艇	[4] 樓船	[5] 王圻
[6] 三才圖會	[7] 仙船	[8] 茅元儀	[9] 武備志	[10] 沙船
[11] 鷹船	[12] 兩頭船	[13] 丘濬	[14] 大學衍義補	

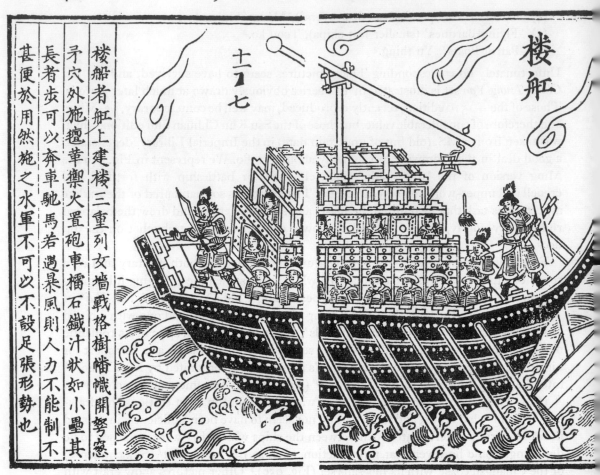

甚便於用然施之水軍不可以不設足張形勢也

長者步可以奔車馳馬若遇暴風則人力不能制不

矛穴外施氈革禦火置砲車檑石鐵汁狀如小壘其

樓船者舟上建樓三重列女牆戰格樹幡幟開弩窻

十一七

楼舡

Fig. 949. The battleship (Lou chhuan) in the +1510 edition of the *Wu Ching Tsung Yao* (+1044).
Comments in text. For a translation of the passage on the left, see p. 685.

a confusion here with the bow-sweep, which has certainly existed on river ships for
many centuries; but it seems rather more likely that the reference was to types of craft
(now mainly Indo-Chinese) which have a rudder at the stern and a centre-board very
far forward or even sliding on the stempost.[a] Another illustration shows two barges
made fast alongside supporting a turret or turrets above (the 'mandarin-duck rowed
assault boat', Yuan-yang-chiang chhuan[1]),[b] and there are of course also rafts with
explosives (*pho chhuan fa*[2]), fire-ships (*huo lung chhuan*[3]) and the like. Ten of Wang
Chhi's drawings and two of Mao Yuan-I's were incorporated into the *Khao kung tien*

    [a] See P. Paris (3), pp. 50 ff., 69, and figs. 102, 155. On bow-sweeps see p. 630 below.
    [b] The text concerning this in *TSCC, Jung chêng tien*, ch. 97, p. 28b, is translated by Shih Chun-
Shêng *et al.* in Audemard (2), p. 77. The crews were supposed to cast off on approaching the enemy ship
and board her on both sides. On twin hulls see p. 392 above.

    ¹ 鴛鴦槳船   ² 破船筏   ³ 火龍船

section of the 'Great Encyclopaedia' of +1726, the *Thu Shu Chi Chhêng*,[a] but this included also two much better drawings from the *editio princeps* of the *Thien Kung Khai Wu* (Exploitation of the Works of Nature)[b] of +1637. It added a picture of a three-masted European ship taken from the *Khun Yü Thu Shuo*[1] (Explanation of the World Map),[c] by Nan Huai-Jen[2] (Ferdinand Verbiest, S.J.) in +1672. Then in its military section (*Jung chêng tien*) the *Thu Shu Chi Chhêng* gave a further series of thirty-one illustrations of shipping, largely but not exclusively naval. But this contained nothing that had not already been given in the books of the early +17th century.[d]

The analysis of these valuable materials (and they are far from exhausting the available Chinese literature) has not been particularly happy. Long ago Krause (1) translated the brief descriptions in the *Thung Tien*. Our translation of the basic source, the *Thai Pai Yin Ching*, will be found on p. 685 below, reserved thereto because it fits in better with the story of ship armour, and projectile (as opposed to close-quarter) tactics, in Chinese naval history. The work of Krause seems to have been unknown to Worcester (4), who essayed a study of the Chinese war-junk,[e] and still more regrettably to Audemard, in whose posthumously published work (2) we have a full translation of the relevant passages of the *Thu Shu Chi Chhêng*[f] made by Shih Chun-Shêng. Audemard was wisely cautious about attributing any date to the texts and drawings which he and Shih studied, but the futility of attempting a 'history of the junk' on the basis of texts copied in *c*. +1780, without knowing that some of the most important go back to just before +760, will readily be apparent. Such a proceeding (especially in view of the inability or unwillingness of Chinese book-illustrators in olden days to depict machinery accurately)[g] could hardly fail to give Western readers yet one more crass misconception of Chinese technology; for what may be laudable in the +8th century becomes scandalous in the +18th—in 'unchanging' China quite as much as everywhere else. Moreover, the use of the word 'encyclopaedia' for translating the title of the *Thu Shu Chi Chhêng*, though traditional, is very unsatisfactory, since it invites comparison with modern works such as the *Encyclopaedia Britannica*. The compilers of the *Thu Shu Chi Chhêng* did not set out to produce a work of reference giving the most up-to-date information on all subjects, but rather to compile a vast anthology or

---

[a] Ch. 178, *hui khao*, pp. 5*a* ff. Boats are also shown in connection with fishing procedures in *I shu tien*, ch. 14, *yü pu, hui khao*, pp. 4*a* ff.

[b] Ch. 9, pp. 10*b* ff.

[c] P. 215. See Fig. 988.

[d] The pictures which stemmed from the +11th century were still being reproduced in 1842 in the valuable work on coastal defence, *Fang Hai Chi Yao*[3] by Yü Chhang-Hui,[4] ch. 15, alongside others of a more diagrammatic character. Cf. Sect. 27*i* above.

[e] He worked only with the pictures of *STTH* and *TSCC*, apparently unaware of their long historical background. See preferably Worcester (3), vol. 2, pp. 350 ff.

[f] *Jung chêng tien* series only. Audemard & Shih used no other encyclopaedia, but their anonymous editor added mention of *STTH* and supplied the +18th-century pictures of *WCTY*. He also seems to have checked the translation in a different edition of *TSCC*. But unfortunately the result is as a whole unreliable, and sometimes gravely at fault.

[g] On this cf. Vol. 4, pt. 2, pp. 1, 373.

[1] 坤輿圖說　　　[2] 南懷仁　　　[3] 防海輯要　　　[4] 俞昌會

沙船

Fig. 950. A smaller warship of Chiangsu style (Sha chhuan) depicted in the *Wu Pei Chih* of +1628. The batten lug-sails with their multiple sheets are graphically, if roughly, drawn, and the vessel, probably a coastal defence patrol-boat, carries a small bombard at the bows. Cf. Fig. 979 (pl.).

florilegium of all the classical historical and literary remains. One must remember the proper meaning of the Chinese title: 'Imperially Commissioned Compendium of Literature and Illustrations, Ancient and Modern'. To suggest that the ship designs of the *Thu Shu Chi Chhêng* represented either the nautical conceptions or the actual practice of the Chinese of +1726 (or even indeed of +1426) was only to give

Westerners still further historically unwarranted grounds for their customary self-satisfaction. In fact, an ocean of further research stretches out before us.

Geographical factors have had considerable influence in differentiating the craft found along the coasts of China. This was clearly seen already in the +17th and +18th centuries by acute observers of local customs.

A scholar of that time, Hsieh Chan-Jen,[1] commenting on a passage in the *Jih Chih Lu* (Daily Additions to Knowledge) of Ku Yen-Wu, itself finished in +1673, wrote as follows:[a]

> The sea-going vessels of Chiang-nan are named 'sand-ships' (Sha chhuan[2]) for as their bottoms are flat and broad they can sail over shoals and moor near sandbanks, frequenting sandy (or muddy) creeks and havens without getting stuck...Chekiang ships...(are built in the same way) and can also sail among sandbanks, but they avoid shallow waters as they are heavier than the sand-ships. But the sea-going vessels of Fukien and Kuangtung have round bottoms and high decks. At the base of their hulls there are large beams of wood in three sections called 'dragon-spines' (*lung ku*[3]). If (these ships) should encounter shallow sandy (water) the dragon-spine may get stuck in the sand, and if wind and tide are not favourable there may be danger in pulling it out. But in sailing to the South Seas (Nan-Yang) where there are many islands and rocks in the water, ships with dragon-spines can turn more easily to avoid them.

Here the reference to the better sailing qualities of ships with deep hulls and centre-boards is rather clear. With this passage in mind we may look again at Fig. 939, where the *lung ku* is the central strengthening member of the hull of a Fukienese or Cantonese sea-going junk, with rounded bottom and high decks. Such a timber is still called *lung ku* by Chinese shipwrights, but it should not be regarded as a keel in the European sense (to which they sometimes apply the same term), for it is not the main longitudinal component of the vessel, this function devolving rather on three or more enormous hardwood wales which are built into the hull at or below the water-line.[b] The real value of the passage is that it points up the differences in hull shape induced historically by the geographical differences between the northern and southern parts of the Chinese culture-area. North of Hangchow Bay (Lat. 30° N., cf. Fig. 711) the coastal and sea-going craft are flat-bottomed and hard-chined with relatively large, heavy and square rudders which can be lowered well below the ship's bottom or raised up high. They are thus fitted for frequent beaching in the shallow harbours or muddy estuaries of the north, where the tidal effects are most noticeable, while at sea the rudder acts as an efficient drop-keel. South of Hangchow Bay the coastal waters are deeper, the inlets fjord-like, and the islands more numerous. Here the under-water lines of the vessels become progressively more curvaceous, with a sharper entry, softer chine and rounder stern; at the same time the rudders, often supplemented by centre-boards, become sometimes narrower and deeper, sometimes fenestrated and rhomboidal. All this is implicit in the words of Hsieh Chan-Jen.

[a] Ch. 29 (p. 89), tr. auct., adjuv. Lo Jung-Pang, p.c.
[b] Cf. Worcester (3), vol. 1, pp. 140, 141, pl. 50.

[1] 謝占壬   [2] 沙船   [3] 龍骨

As for the flat bottom and the rectangular cross-section, it is certainly interesting that this build became generally adopted throughout the world for iron steamships in modern times. Among medieval shipbuilders in wood, the Chinese alone had had it. But Chinese ships, as we have said, were not always flat-bottomed; though lacking any true keel, their sides sometimes rose up in a quite rounded way from the lowest main longitudinal timbers. This we see from texts of a much earlier date, e.g. the *Kao-Li Thu Ching* (Illustrated Record of an Embassy to Korea)[a] of +1124. Speaking of the Fukien-built Kho chou[1] or Retainer Ships, which carried the ambassador's staff, and were somewhat smaller than his own 'Sacred Ship' (Shen chou[2]), Hsü Ching[3] says:[b]

The upper parts of the ship (the deck) are level and horizontal, while the lower parts sheer obliquely like the blade of a knife; this is valued because it can break through the waves in sailing (*kuei chhi kho i pho lang erh hsing yeh*[4])[c]...When the ship is at sea the sailors are not afraid of the great depth of water, but rather of shoals, for since the bottom of the vessel is not flat, she would heel over if she went aground on an ebb tide. For this reason they always use a lead weight on a long rope to take soundings.

Such a shape of hull could be seen in modern times in certain types of Chinese construction, e.g. the fishing-boats of Chusan called 'pairers' (Tui chhuan[5])[d] and the smaller naval junks used towards the end of the Chhing dynasty (Khuai tu[6]).[e] But all the sea-going junks of the south exemplify it.

Among the most extraordinary of all Chinese craft are the crooked-bow and crooked-stern boats of the upper Yangtze region, to which a special monograph has been devoted (Worcester, 2).[f] The crooked-stern junks (Wai-phi-ku[7]) centre upon Fouchow,[8] east of Chungking, at the mouth of the Kungthan River; but the crooked-bow boats are used further west, for bringing the salt down along an almost un-navigable river from Tzu-liu-ching.[9] In both these cases the 'rectangular' bows or sterns are slewed round in such a way as to bring one of their corners more or less in line with the main axis of the vessel. This is done at Fouchow by bending the bottom planks under 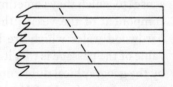 heat and steam[g] along a line making an angle of some 60° with their long axis, not at right angles. The terminal bulkheads, moreover, are not quite vertical. The final result is as shown in Figs. 951 and 952. In the Lu chhuan[10] of Tzuliuching, the bow only is slewed; in the Fouchow Hou-pan[11] or Huang-shan[12] the process affects only the

---

a See below, pp. 602, 641.

b Ch. 34, pp. 4*a*, 5*b*, tr. auct.

c Paik (1) translated this as 'their ability to sail against the wind'. Although they certainly could and did do this (see p. 603 below), it cannot be the meaning of this particular sentence.

d See Worcester (3), vol. 1, p. 134.　　　　　　e See Worcester (3), vol. 2, p. 353.

f Also (1), pp. 51 ff., (14), pp. 202 ff., 230 ff. Cf. Carey (2).

g Similar procedures are described for Korean shipyards by Underwood (1), p. 24. Slightly twisted or crooked bows and sterns are also known in that country (ibid. pp. 7, 11). The bamboos of the Formosan sailing-raft are also steam-bent (Ling Shun-Shêng, 1).

¹ 客舟　　　² 神舟　　　³ 徐兢　　　⁴ 貴其可以破浪而行也　　　⁵ 對船
⁶ 快渡　　　⁷ 歪屁股　　　⁸ 涪州　　　⁹ 自流井　　　¹⁰ 橹船　　　¹¹ 厚板
¹² 黃鱔

stern. In all these types of boat, very large stern-sweeps (*hou shao* [1]) are carried,[a] and in the latter region two, which owing to the construction do not interfere with one another (Fig. 953, pl.). Whether the balance of the whole vessel is so affected by the unsymmetrical lines as to confer real advantages in the negotiation of rapids has not been scientifically investigated,[b] but the boatmen maintain this, and there is at any rate

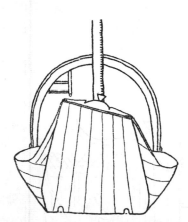

Fig. 951. The twisted bow of the Lu chhuan river-junk of Tzu-liu-ching (Tzu-kung, cf. Fig. 948, pl.) in central Szechuan. Explanations in text. Drawing from Worcester (2).

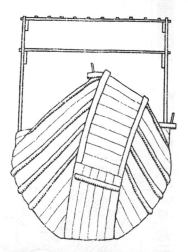

Fig. 952. The twisted stern of the Wai-phi-ku river-junk of Fouchow (Fou-ling) in eastern Szechuan. Explanations in text. Drawing from Worcester (2).

no reason for denying antiquity to this strange build. The most obvious explanation would be that at some time unknown, it was desired, in these parts of China, to obtain a fulcrum for the stern-sweep firmer and more intimately linked with the whole structure of the boat, than could be obtained by mounting it centrally on the usual transom stern. Such a view would not be in contradiction with an invention of the axial 'stern-post' rudder even earlier, since this would not necessarily have penetrated to these western provinces. In order to appreciate the rapids which these boats have to ascend and descend, it is necessary to read the graphic descriptions of Worcester.

Another remarkable vessel is the Liang-chieh-thou [2] or articulated junk, which works on the Grand Canal.[c] This is a very long and narrow barge of shallow draught, built in two separate component sections which are detachable. Coupling and un-coupling of the bow and stern portions is simply done, by the aid of a wire becket and handspikes. A collapsible mast and leeboards are fitted. The invention of the articulation doubtless took place as the Canal became silted up, for the two halves can separately negotiate shallow channels where larger boats would be forced to await a rising water level. The two halves also bank in alongside each other. Although this

[a] As long as the boats themselves.
[b] Experiments with models indicate this (personal communication from Mr G. R. G. Worcester).
[c] Worcester (3), vol. 2, p. 333.

[1] 後梢   [2] 兩節頭

聯環舟

刱勢卷二卄四　水　戰船二

Fig. 954. An articulated barge of the late +16th century in use for military purposes; from the *Wu Pei Chih Shêng* (MS. in the Cambridge University Library, copied in 1843). Such vessels (Liang-chieh-thou or Lien-huan chou) were used as minelayers, the forward portion being left beside the target with a time fuse while the aft portion silently withdrew.

invention may not be very old, it was perhaps the first instance of the use of the articulation principle which in the railway age became so important.[a] In any case, it certainly goes back beyond the early +18th century, for we see it depicted in the *Thu*

---

[a] All 'trains of carriages' exemplify it, but especially the articulated locomotive, first introduced by the Swiss engineer Anatole Mallet in 1888.

Shu Chi Chhêng encyclopaedia[a] adapted for war purposes; the forward section is filled with bombs and other explosives, while the crew and marines occupy the aft section. Presumably the idea was to work the vessel (here called Lien-huan chou[1]) up some creek on a dark night to a point beside some city-wall or under a bridge, then uncouple and silently withdraw, having set a time fuse. Since the boat is described and figured under the same name in the *Wu Pei Chih*,[b] the late +16th century would seem to be a likely time for the origin of this device.[c] Fig. 954 is taken from a MS. of a related work by the same author, *Wu Pei Chih Shêng*,[2] in the Cambridge University Library.

(1) AFFINITIES AND HYBRIDS

It may be that those who have denied any connection between Chinese and ancient Egyptian shipping have been rather too hasty.[d] The most typical and characteristic Egyptian ship was that familiar type which has an extremely long bow and stern sloping and tapering in each direction above the water-line. The flotation length may be not much more than half the total length. But this was not the only kind of hull known in ancient Egypt. Especially during the VIth dynasty (*c.* −2450), a quite different type was prevalent, and it presents some extraordinary similarities with certain river-junks still in use in China today. Egyptologists (e.g. Boreux; Reisner and Klebs) know these two types as the Naqadian and the Horian respectively, for the former can be traced back to the pre-dynastic pottery designs of Naqad,[e] and the latter are associated with a conquering people who came from further east and worshipped the god Horus (Fig. 955). Both types of ship are sometimes seen in one and the same carving, for example at the Deir al-Gebrawi necropolis, where two funerary reed-bundle Naqadian boats are being towed by a Horian ship with square-sail set.[f] One has only to compare the latter type, with its high stern, stern gallery, low bow, relatively truncated ends, and mast set forward of the mid-line, with existing Chinese vessels, to see the

[a] *Jung chêng tien*, ch. 97, *shui chan pu*, *hui khao* 1, p. 33*a*; cf. Audemard (2), pp. 84 ff.

[b] Ch. 117.

[c] Not long after writing this paragraph, we were interested to see an articulated set of pontoon-barges in three sections used by the contractors engaged in the repair of the river-wall of Magdalene College. Such barges were much used by army engineers during the second world war. We have already referred (p. 421 above) to the introduction of the device to Europe by Samuel Bentham at the end of the +18th century, and we have shown that he had had prior Chinese contacts. According to Toudouze *et al.* (1), p. 323, the same suggestion was made by Robert Fulton in 1803, and they illustrate his proposal. So also James White in +1795 patented a scheme for a 'serpentine' canal boat, i.e. a string of barges articulated in such a way as to reduce tractive effort, and apparently he carried out some successful experiments (see Dickinson, 5). Thus the idea was very much in the air at that time.

[d] F. H. Wells (1), who devoted a paper to alleged similarities between Chinese and ancient Egyptian nautical techniques, concluded that there was very little in common, but his level of analysis was somewhat superficial.

[e] Boreux (1), p. 17. They have some resemblance to the boats on the Indo-Chinese bronze drums (cf p. 446 below), and with the bows of extant Siamese dugout canoes. Occupying an important intermediary position, and much older than these, are the rare remaining representations of Mohenjodaro ships, one of which is clearly 'Naqadian' and the other not unlike it (see Bowen, 8).

[f] Boreux (1), p. 153. More Horian ships escorting, p. 158; discussion on this, p. 491. Another Horian ship is seen in Klebs (1), fig. 86, p. 105; and again in Klebs (2), fig. 102, p. 138. The latter is early Middle Kingdom, *c.* −2200. Cf. also Winlock (2).

[1] 聯環舟   [2] 武備制勝

similarity.[a] All the Ma-yang-tzu varieties answer to it, especially the Nan-ho chhuan,[1] but a number of others are also relevant.

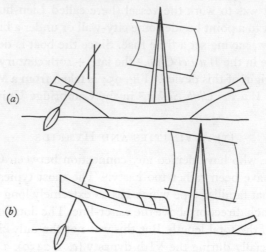

Fig. 955. The two principal types of ancient Egyptian ship: (a) Naqadian; (b) Horian.

Fortunately, a number of tomb-models of both kinds of boats have survived,[b] and have been examined from the point of view of naval archaeology by Reisner (1) and Poujade (3). The latter immediately emphasised one point of fundamental importance, namely that some of the Horian boats are square-ended (Fig. 956, pl.), like all true Chinese ships.[c] However, so far as can be ascertained (for many of the models are solid blocks), the use of the bulkhead was unknown; at any rate when the model is hollow

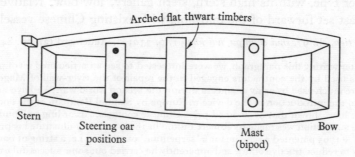

there are only a number of arched flat thwart timbers.[d] The foremost of these is likely to carry the bipod mast, and the aftermost has also two sockets for the two posts supporting the steering-oars (see cut). The stern gallery is as characteristically Chinese

[a] Worcester (1), opp. p. 45, also pp. 21, 78; (3), vol. 2, pp. 350, 428, 430, 463, 486, 494.
[b] The Horian models are, e.g. nos. 4802, 4882, 4886, 4887, 4888, 4910, 4918, 4955, in the Cairo Museum (Reisner (1), pls. XIII, XIV).
[c] So also is the ritual ship of masonry constructed by the Vth-dynasty kings at Memphis (Boreux (1), p. 104).
[d] Reisner (1), p. 53, concerning no. 4882.

[1] 南河船

as the transom stem and stern. Boreux himself did not see these striking similarities between the Horian and Chinese build, but believed that the Horian people had come from Mesopotamia;[a] in this case certain elements of Chinese shipbuilding might join with other proto-scientific material which elsewhere we have seen reason to believe were diffused from Mesopotamia to the beginnings of Chinese civilisation.[b] According to Boreux and Reisner, the Horian type of ship died out before −2000, but perhaps it would be safer to say that after that time it appeared no more in carvings or tomb-models.

Poujade,[c] inspired by a beautiful terra-cotta lamp in the form of an Athenian galley,[d] with bifid rostrum, *stolos*, and forward-curving stern, found examples all over south-eastern Asia of builds which might have derived from these Mediterranean conceptions.[e] The only Chinese vessel affected is the so-called fan-tail junk of the Ta-Ning[1] River in Szechuan, the 'lighter of the gods' (Shen-po-tzu[2]), which has a forward-curving stern.[f] But this might equally well be a relic of Egypt.[g]

The bipod masts of southern Chinese shipbuilding (see Fig. 957, pl.) have already been mentioned (p. 423).[h] Hornell (18) drew attention to the fact that such masts were quite characteristic of ancient Egyptian ships;[i] a resemblance which had not escaped the eagle eye of the prince of diffusionists.[j] There really seems no reason for rejecting this trait as a possible transmission eastwards in ancient times, but if so, it is curious that it did not spread or survive further north in the Chinese culture-area. The bipod mast is, of course, far from being a mere historical curiosity; from the engineering point of view it was excellent, and it has come back in the days of metal tubing. Moreover, a bipod mast is recommended by modern yacht designers as avoiding interference with the

[a] (1), p. 517.

[b] E.g. astrological doctrines (Vol. 2, pp. 353 ff.), equatorial astronomy (Vol. 3, pp. 254 ff.), and the fundamentals of acoustics (Vol. 4, pt. 1, pp. 176 ff.), not to mention such things as bronze technology, the wheel and the chariot. One wonders whether Hornell's theory of an origin from the bamboo raft would apply also to the Horian ships? Bamboo of an inferior kind was known and used in Africa and Assyria (cf. R. C. Thompson, 2).

[c] (1), pp. 275, 277 ff.

[d] Des Noëttes (2), fig. 50. It is in the Museum at Athens. The *stolos* was the upper part of the ship's 'beak', at the bows.

[e] E.g. Burmese royal barges, the doubly bifid canoes of Madura Island, those of Bawean Island; and certainly remarkably Greek-looking, those of Talaud Island; all these are in the Javanese area (pp. 282, 284). Cf. p. 389 above.

[f] Worcester (1), p. 47; Poujade (1), pp. 285 ff.

[g] The local shipwrights preserve the memory of a priest, presumably Taoist, Wang Yeh[3] (Grandfather Wang), who taught them how to build these boats, in the Thang period or before.

[h] Here we illustrate (Fig. 957, pl.) from the Li River near Kweilin (Groff & Lau, 1). Another good photograph from Yangshuo further down will be found in Forman & Forman (1), pl. 179. Cf. Schulthess (1), pl. 165.

[i] See Boreux (1); Reisner (1). There is a hint of the bipod mast in the Mohenjodaro culture (c. −1800); see Bowen (8).

[j] Elliott-Smith (2). But he missed the resemblance of the Horian ships to Chinese river-junks. Many of his views are of course quite unacceptable today, e.g. that Red Sea shipping reached the Chinese coast in the −7th century. He also maintained that East Asian craft in ancient times succeeded in crossing the Pacific to the American continent—this too was unacceptable in the pre-Heyerdahl era, but whether it still is we shall discuss briefly later (pp. 540 ff.).

[1] 大寧河   [2] 神䰈子   [3] 王爺

airflow over the leading edge of the sail.[a] And two lesser spars could substitute for a more massive one if that was not available. Of course it may be said that the greater the practical value of a simple invention the more likely it was to have been made independently—but just how simple are any inventions? Diffusion may perhaps be thought more probable when a practice is magical rather than practical, as in the case of the oculus, i.e. the painting of an eye on each bow quarter of the hull;[b] this must certainly have spread outwards in all directions from ancient Egypt or Mesopotamia. When it reached China we do not know, but the fact that it is confined there to the southern and central regions suggests that it came relatively late, perhaps not before the Han.

Several general questions may now be asked. Were there any exceptions in the Chinese culture-area to the basic principle of boats without keel, stem or stern posts? There is one, and it is of great interest. The dragon-boats (Lung chhuan[1]),[c] used for those races (Fig. 958, pl.) which form such an important feature of the Fifth Month festival (*tuan wu chieh*[2]),[d] supposed to commemorate the poet Chhü Yuan[3] (*d.* −288),[e] are (in many cases at least) built with a true keel or a kelson.[f] This takes the form of a *shan-mu*[4] (fir) pole the length of the boat, which may be as much as 115 ft. and resembles an English 'eight', though longer and narrower.[g] There may be thirty-six or more paddlers. Although bulkheads slotted to the kelson are built in, we are here clearly in presence of an archaic element of one of the constituent cultures which fused to give Chinese civilisation. This is well seen in the ethnological study of Bishop (7) on

[a] Curry (1), p. 81.

[b] The oculus is considered in Hornell (1), pp. 285 ff., cf. also (23), and its diffusion was of course noted by Elliott-Smith (2). It has recently been the subject of an epic controversy between Bowen (7) and Quigley (1); the former maintaining that it passed from Mesopotamia or Egypt to India by way of the + 1st-century Roman (actually Graeco-Syrian and Graeco-Egyptian) trade, the latter that it had spread east long before, either through the maritime Mesopotamian-Mohenjodaro contacts in the − 3rd millennium, or through Phoenician or pre-Islamic Arab influence in the − 2nd. Cf. Audemard (4). For recent reviews of ancient Middle Eastern shipping cf. Barnett (1); Šmíd (1).

[c] Descriptions will be found in Worcester (1), pp. 31 ff., tr. Poujade (1), pp. 288 ff.; Worcester (3), vol. 1, pp. 220, vol. 2, p. 490; Audemard (4), pp. 23 ff. They are strictly ritual craft. The term *lung chhuan*, however, cannot always be taken to mean these very elongated racing 'canoes', for ornamental vessels of broader beam were often built (usually in connection with imperial entertainments) having large dragon figure-heads at the bows, pavilions amidships, and a tapering tail at the stern (cf. Audemard (4), pp. 29 ff.; Lin Yü-Thang (7), fig. 72). These were also called *lung chhuan*. The oldest representation of them known to me occurs in a painting by Wang Chen-Phêng (done between + 1312 and + 1320), which shows three on a lake beside a Thang imperial palace (see Sickman *et al.* (1), p. 36, no. 44). It was probably craft of this kind that Nieuhoff saw in + 1656, and afterwards illustrated, (1), p. 147. They also give the form for the large paper *lung chhuan* that are floated down river on rafts with lights at festivals in some provinces. Cf. *Huai Nan Tzu*, ch. 8, p. 9b (tr. Morgan (1), p. 94). The 'dragon-ship' of Sui Yang Ti (*r.* +605 to +616) was particularly renowned; cf. *Ta Yeh Tsa Chi*, in e.g. *Shuo Fu*, ch. 57, p. 30b, *Lei Shuo*, ch. 4, p. 18a.

[d] The writer will never forget the races which he saw at Liuchow in Kuangsi in 1944. Full details of the folkloristic aspect of the festival are given by Chao Wei-Pang (3) and Mulder (2).

[e] The real background of homage to the water-gods of the ancient south-eastern 'proto-thai' component of Chinese civilisation (cf. Vol. 1, p. 89) has been elucidated by Wên Chhung-I (1).

[f] An internal keel, i.e. one which does not come in contact with the water.

[g] Average beam about 4 ft. In one case studied, the bottom of the boat was markedly concave below, like that of a modern racing motor-boat, but apparently this feature was not general (G. R. G. Worcester, unpublished material, no. 131).

¹ 龍船   ² 端午節   ³ 屈原   ⁴ 杉木

long-houses and dragon-boats. The culture in question was in fact the south-eastern Malayo-Indonesian one. But a further point of particular interest is that in order to prevent hogging[a] in so long and narrow a vessel, a long bamboo cable is slung from the projecting bow end of the kelson to the stern, exactly as was the common practice in ancient Egyptian ships.[b] This 'anti-hogging truss' is again perhaps an early transmission eastwards, and one is hardly surprised to find it in so archaic a form.[c] The dragon-boat might indeed be described as a canoe-derivate which has survived alone in a world of raft-derivates.[d]

Another possible exception is constituted by the boats of the Erh Hai[1] Lake in south-western Yunnan, described by Fitzgerald (6). These seem to have frames without bulkheads, but the presence of keels needs further investigation. The area is a very isolated one, and could well have been influenced from India.

To what extent did the rectangular construction spread from China to other lands? Information here is scanty, yet so far as may be ascertained all Japanese hulls before the seventeenth century followed approximately the build of the Chinese shipwrights (cf. Fig. 1038, pl.).[e] Clowes & Trew give a reconstruction of a medieval Japanese ship, based upon the drawings of Admiral Paris. An English shipwright, Will Adams, worked in Japan during the first two decades of the century, however, and later vessels were often copies of occidental models, notably Russian schooners.

A somewhat more interesting, though much more modest, problem concerns the origin of the punt, as we know it today in Western countries. This familiar rectangular craft is extremely similar to the Chinese sampan, and it is curious that no one has investigated its origin. The word itself is, of course, related to the Latin root signifying a bridge. There were 'pontones' in Roman times, but apparently they were merchant-ships,[f] if we may judge from the remarkable +2nd-century mosaic at Althiburus in

[a] Drooping of bow and stern portions of a boat.

[b] See, e.g., Boreux (1), p. 475; Neuburger (1), p. 479; Klebs (1), fig. 85, p. 103; Winlock (2); Chatterton (1), p. 20.

[c] Poujade (1), pp. 277 ff., thought he could recognise traces of the anti-hogging truss in the ornamentation of the stems and sterns of a number of Asian boats, and in the stern-embracing mast-stay used on certain Chinese river-junks when tracking. Dr J. S. Morrison tells me that a truss was part of the normal gear of the standard −4th-century trireme in ancient Greece. What is perhaps more astonishing is to find that a form of it survived in the construction of Mississipi stern-wheeler steamboats where the boilers were right forward and the engines right aft (King-Webster, 1). And we may see a form of it every time we go to a railway station if we look at the inverted trapezoidal trusses underneath the frames of bogie goods trucks and wagons.

[d] Here we must remember the occasional dugout canoes already mentioned (pp. 388 ff. above). And there is also the unique 'snake-boat' (Shê chhuan[2]) which survives as a carrier of freight and passengers on the Kungthan River (Worcester (1), p. 56). Its stem and stern are high and narrow, and though it has no keel, it is unique in having no bulkheads either.

[e] Such at any rate is the general view of western students such as Purvis (1); Peri (1) and Elgar (1). Yet very many of the pictures of boats and ships in the *Wakan Senyōshū* (+1766) clearly show stemposts, especially the river craft and fishing vessels in chs. 5 and 6, as also the shipbuilding yard drawings (ch. 3, pp. 6b, 7a; ch. 4, pp. 2b, 3a), and a good cross-section with true keel (ch. 10, p. 12b). The same is true of the vessels depicted in Hokusai's *Fugaku Hyaku-kei* (1834), e.g. ii, 10, 19; iii, 27. They have stem-posts combined with flat bottoms, transom sterns and prominent rudders. Cf. Dickins (1).

[f] Torr (1), p. 121. The use of the term 'punt' (especially as 'quay-punt') for a keeled and sharp-ended boat continues to this day in local usage (see White (1), pp. 19, 20, 30, 32, 35, 36, 48).

[1] 洱海   [2] 蛇船

Tunis, which incorporates the names, as well as the pictures, of a number of different craft.[a] Caesar mentions a Gaulish transport boat under the name,[b] Aelfric about +1000 had a 'pontonium', and 'pontebots' begin to appear in East Anglian records about +1500. A punt is mentioned in a London law case report of +1371.[c] The oldest illustrations of punts which we have found are in the pictures of the MS. *Fischereibuch* of the Emperor Maximilian I (+1459 to +1519) conserved at Vienna. Yet Isidore of Seville (c. +630) described the 'ponto' as a simple rowed boat of rectangular shape,[d] with sloping sides and flat bottom; and the *Pandects* (+533) refer to it as being used for ferry services.[e]

The problems of affinity and transmission are thus rather obscure. Clowes, commenting in the discussion on Brindley (1), alluded to the widespread use of punts in Kashmir and the Upper Indus valley. If the Chinese build found its way overland already in Bactrian times, punts in our sense may well have existed in Roman Europe. On the other hand if Roman boats with the same name were not what we should call punts, those of St Isidore may have been introduced through Byzantine contacts. Or perhaps Chinese technicians following the Mongol armies constructed rectangular keel-less boats for military purposes in +13th-century Eastern Europe[f]—indeed the punt, like the crossbow, may have been introduced several times.[g] Or lastly the European punt may always have existed as a descendant of the square-ended boats of ancient Egypt just described.[h]

Did hybrid types develop in south-eastern Asia in the regions where Chinese mingled with Indian and indigenous influences? Abundantly, as we know from the exhaustive account given by P. Paris (3) of the ships of Indo-China. Individual types have been carefully described too, by Poujade (2); Piétri (1); and others. A keel was perhaps the

---

[a] Bonnard (1), pp. 151, 154.   [b] *De Bello Civili*, III, 209.
[c] Callendar (1).

[d] *Etymologiarum sive Originum*, XIX, 1, 24.

[e] VIII, 3, 38.

[f] The propulsion of flat-bottomed boats with the aid of punt-poles on the West Lake at Hangchow is described by Marco Polo; cf. Gernet (2), p. 59.

[g] Scattered about in small local areas in the Western part of the Old World there are interesting hybrid craft which distinctly recall the characteristic Chinese build. Two of these, encountered by chance in the course of ordinary travels, deserve mention. In 1957 I had the opportunity of studying, with Dr Lu Gwei-Djen, the strange *fuhr* gondola-bowed punts traditional on the waters of the Hallstättersee in Austria. They have a transom stem and stern, without keel, but the bow is drawn out very fine (pointing slightly upwards) so that the forward transom is very small. These boats are propelled by four spade-shaped paddles lashed to holes in high rowlock-boards made in one piece with the strakes. Then in 1960 I had a chance of examining, with Cdr. D. W. Waters, the flat-bottomed *chata* or *xavega* fishing-boats of the beach of Nazaré in Portugal. These are markedly concave fore-and-aft, but while the stern is a transom one the bow points almost vertically with a 'pseudo-stempost' originating at the junction of the boat's sides, a junction which also bears a hook by which the boat can be wound up the beach. We could not observe a keel. Excellent models of these boats can be seen in the Ethnological and Folk Art Museums at Lisbon. Boats of similar general pattern to the *fuhr* and *xavega* occur in old Japanese paintings, notably one by Tosa Yukihide (+14th century) which shows cormorant fishing, but the pointed bows may conceal a transom build which the artist did not bother to indicate clearly. Beaudouin (1) has studied the origin of the Nazaré *netinha* and *barco do candil* build, but without adding anything to the solution of our problem. Cf. Marsden & Bonino (1).

[h] Such descendants may be much more widespread than is generally realised, for except in China they never grew out of the small boat class. Ceylon has the punt form with sewn hull construction (Science Museum photographs SM 2720, 2721, 2722). Cf. de Zylva (1), figs. 6, 10; Greenhill (1).

first feature to be adopted,[a] then bulkheads were abandoned for rib frames (*wan chhai*[1]), as in the *twaqo* of Singapore (Waters, 4). Cases are known, such as the Malayan *tongkung* (Waters, 3), in which the ships are designed, built, owned and worked by Chinese, yet the structure and rig is entirely European. The Chinese sails, however, on account of their great efficiency, are among the last components to disappear, and indeed, as in the famous Portuguese 'lorcha' (*hua thing*[2]) of Macao and Hongkong,[b] have co-existed since the + 16th century with a slender hull of normal European type.

(e) THE CHINESE SHIP IN PHILOLOGY AND ARCHAEOLOGY

If it is not putting too much weight on what might have been only an arbitrary convention of scribes, the Shang and Chou forms of words for boats indicate that already in remote antiquity the characteristic build of the Chinese vessel had developed. They seem to give a remarkable pictorial statement of the basically rectangular shape, though a design with pointed ends would hardly have been too difficult to draw.[c] The earliest form of the word *chou*[3] (Rad. no. 137; K 1084), as seen on the oracle-bones, shows the transom and bulkhead construction, without any sharp stem or stern—or, at the least, a rectangular slightly concave raft with thwartwise members. The original significance of the other component in *chhuan*,[4] a ship (K 229 *e, f*), is not known, but the mouth probably represented the crew, and the two lines river-banks. The word *pan*[5] (K 182 *a, b, c*), which eventually acquired many meanings (to transport, change

K 1084   K 229 *e, f*   K 182 *a, b, c*   K 893 *f–i*

place, turn round in general), distinctly shows an oar and a hand beside a boat; it is, indeed, as Edkins (12) knew, an old steersman's word, now often written with the hand radical, *pan*.[6] *Pan shao*[7] means to pull the stern-sweep or steering-oar to oneself (i.e. to port), *thui shao*[8] to push it away (i.e. to starboard).[d] The word *chen*[9] (K 893 *f–i*),

[a] As, apparently, in the three-masted fishing-boats of Amoy (Worcester, 7), and in the sea-going junks of Hainan island and many other parts of Kuangtung (Worcester, 8). The Maze Collection embodies a fine model of a 106-ft. Kuangtung freighter of this kind (Anon. (17), no. 3) which we illustrate in another connection at Fig. 1040 (pl.). Cf. p. 429 above.

[b] Carmona (1); Worcester (3), vol. 2, pp. 375 ff. References to this attractive type of vessel are found as early as + 1605 (Peri (1), p. 107). It is a remarkable illustration of the rule enunciated by Poujade (1), pp. 170, 177, that hulls tend to change more easily than rigs. He gives good sociological reasons for this— the shipwright and the sail-maker are quite different people. Hulls, he says, are sensitive to commercial contact; rigs, more closely associated with the traditions and life of the sea-faring folk, change only under political dominance. But there is much contrary evidence, and all 'rules' in this field are probably premature.

[c] Attention has been drawn to the interest of the ancient forms by Gibson (4) and Worcester (3), vol. 1, p. 6, but strangely they failed to see the significance of the shape.

[d] How old these expressions are we do not know. While they seem to appear but rarely in written records, they may well belong to ancient oral tradition.

[1] 彎柴　　[2] 划艇　　[3] 舟　　[4] 船　　[5] 般　　[6] 搬　　[7] 般梢
[8] 推梢　　[9] 朕

formerly written *chen*[1] and now meaning a seam, originally meant to caulk the seams of a boat, since we see indeed a boat accompanied by two hands holding something, presumably the caulking chisel.[a] So common a word as *shou*,[2] to receive, was also once written in such a way as to depict a boat-like object and two hands (K 1085), but this was more probably a weaving shuttle; if so, the pictogram does not indicate, as some have thought, the loading and unloading of boats. People of the Chinese culture-area were always surprised when they saw the sharp-ended boats and ships of other civilisations. For example, the Chin Tartar Wukusun Chung-Tuan,[3] who had been on diplomatic missions for his country, wrote in his travel notes about +1220 of the Islamic lands of the West (Yin-Tu Hui-Ho[4]): 'Their boats resemble a shuttle.'[b] And in +1259, when Chhang Tê[5] crossed the Syr Daria River on his embassy from Mangu Khan to Hūlāgu Khan, he was much surprised to find that the boat 'resembled a Chinese woman's pointed and crescent-shaped shoe'.[c]

K 1085

(i) FROM ANTIQUITY TO THE THANG

There seems to be nothing very revealing among the mentions of boats in the *Shih Ching* or other classics, and the *Tso Chuan*'s accounts of naval battles do not help much. Nevertheless, there is no reason to doubt the historicity of the fleet which was sent northwards by the State of Wu,[d] under the admiral Hsü Chhêng,[6] to attack the State of Chhi in −486. What he probably commanded was a number of large paddled canoes, some perhaps large enough to carry deck-castles for archers, and they certainly kept close in-shore.[e] We have already had occasion to mention the great fleet of sailing-rafts

[a] Later the word came to be written with the moon radical in place of the boat; this was a corruption. Eventually it was used as a personal pronoun appropriated to the emperor (the 'imperial We'), and its origins were forgotten. In the *Chou Li* it occurs in connection with sewing the seams of armour, an interesting point in view of the prevalence of sewn boats in some parts of the world (Hornell (1), many refs.).

[b] Cf. Bretschneider (8), p. 105, translating his *Pei Shih Chi*.[7] Cf. Vol. 3, p. 522.

[c] Cf. Bretschneider (2), vol. 1, p. 130, translating his *Hsi Shih Chi*.[8] Cf. Vol. 3, p. 523.

[d] *Tso Chuan*, Duke Ai, 10th year (Couvreur (1), vol. 3, p. 659). Hsü Chhêng's amphibious operation met with failure, however; comments in Hsü Chung-Shu (3). There had been a long background to this naval activity of the State of Wu. For example: Duke Hsiang, 24th year (−548), Chhu fought afloat against Wu, unsuccessfully (Couvreur (1), vol. 2, p. 412); Duke Chao, 17th year (−524), the indecisive battle of Chhang-an,[9] between Wu and Chhu, in which the Wu forces lost and later recovered their 'flagship', the 'Yü-Huang',[10] in a famous feat of derring-do (Couvreur (1), vol. 3, pp. 282 ff.); Duke Ting, 6th year (−503), Wu destroyed the Chhu fleet at last (Couvreur (1), vol. 3, p. 530). According to tradition (see p. 678 below), the naval importance of Wu was due to the organising powers of the minister Wu Tzu-Hsü[11] (*fl.* −530, d. −484, already often met with, cf. e.g. Vol. 3, pp. 485 ff., Vol. 4, pt. 1, p. 269). He would of course have built upon the age-old expertise with boats of the 'proto-thai' fraction of the Chinese people.

The Chhu and Wu fleets were the precursors of the Yüeh navy, which by the time of king Kou Chien (r. −496 to −470) consisted of 300 fighting-ships (lit. halberd boats, *ko chhuan*[12]) manned by 8,000 men and a fleet of decked (or deck-castle) ships (*lou chhuan*) manned by more than 3,000 men (*Wu Yüeh Chhun Chhiu*, ch. 10).

[e] I say this because later texts, even some purporting to be of the Later Han, such as *Yüeh Chüeh Shu* (cf. p. 679 below), attribute 120-ft. long sailing-ships to the Warring States period, and this, in my view, is unlikely on general developmental grounds. Sailing-ships of substantial size may have been occasion-

[1] 舟片	[2] 受	[3] 烏古孫仲端	[4] 印都回紇	[5] 常德
[6] 徐承	[7] 北使記	[8] 西使記	[9] 長岸	[10] 艅艎
[11] 伍子胥	[12] 戈船			

(p. 390) built by the King of Yüeh, another southern State, in −472. The vessels of the Warring States period, however, were not all naval, and we can be sure that there were trading expeditions at least along the coasts of Siberia, Korea and Indo-China.[a] There were also some explorations of the Pacific itself.[b] And of course, as ever, inland water transport.[c]

For the Han time there is a good deal more information, and only a few points can be touched upon.[d] In −219 the emperor Chhin Shih Huang Ti had sent a great military expedition to conquer the southern peoples of Yüeh,[e] under the command of Chao Tho[1] and Thu Chü;[2] its main strength consisted of 'marines' based upon war-boats with deck-castles (*lou chhuan*[3]).[f] A century later, when the province founded by Chao Tho showed signs of becoming a permanently independent kingdom,[g] Han Wu Ti, in −112, had to send another expeditionary force, and this again employed a fleet of 'the ships of the south which have deck-castles (or, more than one deck) (*nan fang lou hsiang*[4])'.[h] It was commanded by Yang Phu[5] and Lu Po-Tê,[6] whose titles as admirals indicate perhaps the growing importance of naval techniques.[i] One such title gives us

ally built, however, in those times, and perhaps the 'Yu-Hüang' was one of them. I suspect that the nearest parallel for the Warring States naval ships is to be found in the Viking longships, which never exceeded 80 ft. in length before the +10th century. For the *ko chhuan* 50 ft. and for the *lou chhuan* 70 ft. might be good guesses.

[a] Literary references have been assembled by Wei Chü-Hsien (4), pp. 5 ff., but his conclusions are subject to caution and reserve.   [b] Cf. pp. 551 ff. below.

[c] Lo Jung-Pang (6), p. 29, has drawn attention to the invasion of Chhu State by Chhin in −312 and −311 under the general Ssuma Tsho,[7] one of Ssuma Chhien's ancestors (cf. Chavannes (1), vol. 2, p. 74). The chief sources are *Shih Chi*, ch. 70, pp. 10b, 11a, and *Hua Yang Kuo Chih*, ch. 3. Water-transport was organised by Chang I,[8] a minister famed as one of the School of Politicians (cf. Vol. 2, p. 206). Double-hulled boats (*fang chhuan*[9]), allegedly 100,000, bore fifty warriors each down 3,000-*li* river routes, and 10,000 small freighters (*sao*[10]) carried 6,000,000 bushels of army grain supplies. If inference from these round figures is justifiable, the average burthen would have been 16·35 tons, which may give an idea of the size of the craft. The power of Chhu was greatly diminished by this campaign.

[d] Krause (1) did a service in collecting, translating and commenting on some of the texts of the Han and San Kuo periods describing naval warfare.

[e] *Shih Chi*, ch. 112, p. 10b. This was the occasion when the great engineer Shih Lu[11] built the Ling Chhü Canal (see p. 299 above) to allow of the through passage of waterborne supplies from the north. Cf. Aurousseau (2).

[f] To this type of vessel a special monograph has been devoted by Pao Tsun-Phêng (2, 2), largely based upon the tomb-model ships described below (pp. 447 ff.).

[g] Cf. Cordier (1), vol. 1, pp. 235 ff.; Fitzgerald (1), p. 181.

[h] *Shih Chi*, ch. 30, p. 17a (tr. Chavannes (1), vol. 3, p. 592); ch. 113, pp. 7b ff.; *Chhien Han Shu*, ch. 6, pp. 19a ff. (tr. Dubs (2), vol. 2, pp. 79 ff.). Also ch. 24B, p. 16b (tr. Swann (1), p. 306). A few pages earlier (p. 15a; Swann, p. 298) the exercises of these boats, ornamented with flags and standards, on the Khunming Lake in −115, are described, but chiefly as yet another example of imperial extravagance, since the chapter in question is that on economics. The expedition against Nan Yüeh for which these preparations were made was the occasion when the wealthy merchant of Chhi, Pu Shih,[12] volunteered to take charge of the fleet, with a number of technical experts from that province (*Shih Chi*, ch. 30, p. 17b; Chavannes (1), vol. 3, p. 594). But the offer was not accepted. We shall meet again with Pu Shih below. The expedition was completely successful and the admiral of Nan Yüeh, Lü Chia,[13] was captured after fleeing to the West with what remained of his ships.

[i] For example, Lou Chhuan Chiang-Chün[14] (Commander of the Embattled Ships) for Yang Phu, Fu Po Chiang-Chün[15] (the Wave-Subduing Commander) for Lu Po-Tê, and Ko Chhuan Chiang-Chün[16] (the Fighting-Ship Commander) for a former marquis of Yüeh named Yen[17] who had changed allegiance and returned to the service of the Han. On Fu-Po see Kaltenmark (3).

[1] 趙佗	[2] 屠睢	[3] 樓船	[4] 南方樓舡	[5] 楊僕	[6] 路博德
[7] 司馬錯	[8] 張儀	[9] 舫船	[10] 艘	[11] 史祿	[12] 卜式
[13] 呂嘉	[14] 樓船將軍		[15] 伏波將軍	[16] 戈船將軍	[17] 嚴

a useful glimpse of a particular kind of seamanship important at the time.[a] In the following year another substantial river and coastal fleet was organised, and sent under Han Yüeh[1] and others to suppress a rebellion in eastern Yüeh.[b] Then in −108 Yang Phu led a sea force against Korea.[c] Thus naval operations were conducted on a considerable scale in the reign of Han Wu Ti.[d]

At the beginning of the Later Han period, Kungsun Shu[2] tried to set up an independent kingdom in Szechuan. Three of his military engineers built a remarkable fortified floating bridge and boom in Hupei, but this was destroyed by a Han fleet of thousands of vessels including many 'castled ships'.[e] We shall describe this action more closely on p. 679 below, on account of the interest of the types of craft involved. This was in +33. Just ten years later there was the great expedition of Ma Yuan[3] to Chiao-Chih[4] (Tongking) which involved a fleet of 2,000 'castled ships' (*lou chhuan*[5]).[f] Subsequently many naval fights between the Chinese and the Champa people (of the Lin-I[6] kingdom; modern Annam) are recorded.[g] The term *lou chhuan*[5] for warships of large size persisted down all through the centuries. By the +8th, the best Thang source says[h] that they had three decks, with bulwarks, arms, flags, and catapults,[i] but were not very handy in rough weather. When one remembers that Han Yüeh's chief expedition took place just about fifty years before the Battle of Actium, one would like to know much more about these Han warships, and how far (for example) their tactics compared with the boarding technique used by the 'marines' of Rome during the First Punic War (*c.* −260 to −240).[j] There is some evidence that they engaged in ramming, like Greek triremes.[k]

Sea communications with Kuangtung, Indo-China and Malaya began to acquire importance from the beginning of the present era. In +2 Wang Mang got tribute of a live rhinoceros from those parts, and it certainly came some of the way by boat.[l] This

---

[a] A certain Tsu Kuang-Ming[7] was appointed Hsia Lai Chiang-Chün[8] (the Torrent-Descending Commander). He was no doubt an expert in the management of ships in those rapids so abundant on the Chinese rivers. See *Chhien Han Chi*, ch. 14, p. 1b.

[b] His title was Hêng Hai Chiang-Chun[9] (the Ocean-Traversing Commander); *Chhien Han Shu*, ch. 6, p. 21a (tr. Dubs (2), vol. 2, p. 82).

[c] *Chhien Han Shu*, ch. 6, p. 24b (tr. Dubs (2), vol. 2, pp. 90 ff.). Although the expedition was successful and Korea was divided into four Han commanderies Yang Phu's losses were so severe that he himself fell into disgrace and was dismissed.

[d] Ch. 95 of the *Chhien Han Shu*, which gives the fullest account of all the campaigns and their political background, is available to Western readers only in the antiquated translation of Pfizmaier (51).

[e] *Hou Han Shu*, ch. 47, p. 17b. Cf. Kungsun Shu's biography in ch. 43, pp. 22a ff.

[f] *Hou Han Shu*, ch. 54, p. 10a. Cf. *Shui Ching Chu*, ch. 37, p. 9a, and Maspero (18).

[g] Details in Ferrand (3). Among the dates are +248, +359, +407. In +431 more than a hundred Cham ships, with superstructures, ravaged the province of Tongking, but were finally beaten off.

[h] *TPYC*, cit. e.g. in *Thung Tien*, ch. 160 (p. 848.3), cf. p. 685 below. There are many references to these war boats or ships in earlier books; as, for instance, the +4th-century *Hai Nei Shih Chou Chi* (end of introduction). This passage concerns the expedition of young men and girls led by Hsü Fu[10] which set forth in search of the Magical Islands in the Eastern Sea, and the herbs of immortality which were supposed to grow there, at the command of the First Emperor. Cf. p. 552 and Sect. 33a below.

[i] Trebuchets, perhaps also large crossbows fixed to stands. Cf. Sect. 30i below.

[j] Cf. p. 693 below.   [k] Cf. p. 679 below.

[l] *Chhien Han Shu*, ch. 12, p. 2a; ch. 28B, pp. 39a, 40a. Cf. Duyvendak (8).

---

[1] 韓說	[2] 公孫述	[3] 馬援	[4] 交趾	[5] 樓船
[6] 林邑	[7] 祖廣明	[8] 下瀨將軍	[9] 橫海將軍	[10] 徐福

tribute was repeated in +84 and +94, and continued intermittently as late as the Thang.[a] An interesting passage in the *Chhien Han Shu* describes Han trade with the south seas:[b]

From the barriers of Jih-Nan[1] (Annam), or from Hsü-wên[2] and Ho-phu[3] (in Kuang-tung),[c] going by boat for five months, there is the Tu-Yuan[4] kingdom...[and four other kingdoms are then mentioned, all of which had offered tribute from Han Wu Ti's time onwards].

There are superintendent interpreters (*i chhang*[5]) belonging to the civil service personnel (*huang mên*[6]),[d] who recruit crews and go to sea to trade for brilliant pearls, glass,[e] strange gems and other exotic products, giving in exchange gold and various silks. In the countries where they come the officials and their followers are provided with food and handmaidens. Merchant-ships (*ku chhuan*[7]) of the barbarians (may) transport them (part of the way) home again. But (these barbarians) also, to get more profit (sometimes) rob people and kill them. Moreover (the travellers) may encounter storms and so drown. Even if nothing (of this kind happens, they are) away for several years.

As for the great pearls, they may measure as much as two inches in circumference...

This probably refers to the two centuries preceding the time when Pan Ku was writing (i.e. about +90), so we may take it as well applicable to the −1st century, indeed back to the time of Han Wu Ti. Since the furthest country is said in the text to require a sea voyage of just over twelve months, and since the whole account is quite devoid of any legendary quality, Pelliot felt that one should visualise Chinese missions penetrating already at this time as far as the western extremity of the Indian Ocean.[f] Further evidence for these extensive contacts has come from archaeological investigations in South-east Asia; thus Chinese coins of the first quarter of the +1st century have been found in the tombs at Dôngsón (northern Annam).[g] Chinese pottery of the Former Han, one piece bearing an inscription dated −45, occurs in Sumatra, Java and Borneo.[h] And certain stone sculptures of Sumatra bear a close similarity to those of the Han.[i]

Indeed it is more than likely that the foundations of this maritime trade had already been built (as suggested above, p. 441) by the people of Yüeh in the Warring States period. A passage in *Chuang Tzu*, seemingly often misunderstood,[j] may be brought

[a] Cf. Pelliot (30); Laufer (15), p. 80.
[b] Ch. 28B, p. 39b, tr. auct. adjuv. Pelliot (30); Ferrand (3); Duyvendak (8); Wang Kung-Wu (1).
[c] Cf. p. 669 below.
[d] On this interesting term cf. Vol. 3, p. 358.
[e] *Pi-liu-li*,[8] see Vol. 4, pt. 1, p. 105 above. Pelliot considered this term equivalent to the Skr. *vaiḍūrya*, and that it meant glass here. On glass in Indo-China see the remarks of Janse (5), vol. 1, pp. 51 ff.
[f] Note that this estimate would place Chinese long-distance navigation two or three centuries earlier than the date normally accepted (see Vol. 1, p. 179). But the text need not mean that Chinese ships themselves went all the way. Still, it is arresting enough to think that Chinese merchant-officials walked with Roman citizens from Greece, Syria and Egypt on the quays of Arikamedu (Vīrapatnam; cf. Vol. 1, p. 178 and Wheeler (4), pp. 137 ff.).
[g] Goloubev (1).   [h] De Flines (1).
[i] Van der Hoop (1).
[j] Ch. 24 (*Pu Chu*, ch. 8B, p. 3a), tr. auct. adjuv. Legge (5), vol. 2, p. 93. The parallel passage in *Lü Shih Chhun Chhiu*, ch. 65 (vol. 1, p. 126), makes it quite clear that seafarers are intended.

[1] 日南   [2] 徐聞   [3] 合浦   [4] 都元   [5] 譯長   [6] 黃門
[7] 賈船   [8] 璧流離

forward in witness of this. The Taoist recluse Hsü Wu-Kuei,[1] having had an interview with Duke Wu of Wei, is discussing his good reception with the Duke's minister.

Have you not heard [he says], of the wanderers of Yüeh? When they have been gone from their country several days, they are glad when they see anyone whom they knew there. When they have been absent for weeks or months, they are happy if they meet anyone whom they had formerly seen at home. But by the time they have been away for a whole year they are delighted if they meet with anyone who even looks like a compatriot. The longer they are gone, the more affectionately they think of their own people—is this not so?

Thus the Duke has wandered far from his true native land of the Tao—no wonder he welcomes a messenger from there. But for us the interest lies in the sails of the merchantmen of Yüeh flitting about the isles of the Indies.[a]

It may have been not only the Indies. The possibility is still open that Han trading envoys got as far as the Axumite kingdom of Ethiopia. Reading further in the text of the *Chhien Han Shu* just given, we find:[b]

In the Yuan-Shih reign-period (+1 to +6) under the Emperor Phing Ti, (the minister) Wang Mang, assisting the government, desired to glorify his majestic virtue. He (therefore) addressed rich presents to the King of Huang-Chih[2,c] enjoining him to send an embassy with a rhinoceros as tribute. From the kingdom of Huang-Chih, going by ship about eight months, one reaches Phi-Tsung.[3] Going on further by ship about two months, one gets to the frontier of Hsiang-Lin[4] in Jih-Nan.[5] It is said that south of Huang-Chih there is the country of Ssu-Chhêng-Pu.[6] It was from there that the envoy-interpreters (*i shih*[7]) of the Han returned.

We still have no definitive identification of any of these countries, but Huang-Chih is generally believed to be Kāncīpura (mod. Conjeveram in Madras, then capital of the Pallava State).[d] This would fit in with the itinerary through four kingdoms omitted in the previous quotation, for it includes a ten-day land journey sandwiched between months of sailing, and this could very reasonably be interpreted as a traverse of the Kra isthmus in southern Siam. Judging by the timings given, however, Herrmann (4) suggested that Huang-Chih could have been the port of Adulis (mod. Massawa in the Red Sea), in which case Ssu-Chhêng-Pu would be the oldest Chinese mention of East Africa. Most sinological geographers have frowned upon this view, though not all, and it is still on the agenda.

Representations of boats and ships of the Warring States and Han times were until lately scarce. Some quite small boats are shown on the tomb-shrine reliefs of Hsiao-thang Shan and Wu Liang Tzhu (carved between +125 and +150); they are all sampans carrying two or three people each.[e] It is hard to be sure about their nature;

---

[a] Wang Kung-Wu (1) has recently given us a useful monograph on the Chinese South Seas trade between −220 and +960 which brings up to date the older treatment of Fêng Chhêng-Chün (1).

[b] Ch. 28B, p. 40a, tr. auct. adjuv. Pelliot (30); Duyvendak (8). The mention of Annam in the midst of Indian Ocean places is puzzling, and no one has proposed a convincing explanation of the whole route.

[c] One of the kingdoms mentioned in the previous passage.

[d] From many papers we mention only Wang Kung-Wu (1), pp. 16 ff.; Lo Jung-Pang (7); Duyvendak (8); Yü Ying-Shih (1), pp. 172 ff.

[e] See *Chin Shih So* (Shih sect. ch. 1, pp. 110 ff., 114 ff., ch. 3, pp. 108 ff., ch. 4, pp. 6 ff.); also Chavannes (9, 11); Chhang Jen-Chieh (1), pl. 17, etc.

[1] 徐無鬼   [2] 黃支   [3] 皮宗   [4] 象林   [5] 日南   [6] 巳程不
[7] 譯使

some might be dugout canoes, others look more like reed-bundle craft, and P. Paris (1) surely goes too far in seeing an assured stem-post and stern-post in some of them. More interesting are the Warring States and Early Han bronzes depicting war-boats which show distinctly above the rowers an upper deck carrying spearmen, halberdiers and archers. This is the first appearance of the 'castled ships' or *lou chhuan*[1] which have already been mentioned. In Vol. 4, pt. 1, for a different purpose, we reproduced in Fig. 300 a bronze vase of the −4th century, the Yen-Yo Yü-Lieh Thu Hu preserved in the Imperial Palace Museum, Peking.[a] Below on the left a naval engagement is proceeding; the two ships are meeting bow to bow, their rowers in the characteristically Chinese forward stance, their pennants flying. The 'marines' at the bow fight with

Fig. 959. Representation of a decked ship (*lou chhuan*) of the Former Han period (−2nd or −1st century) on a flat bronze bowl (*chan chien*) described and figured by Sun Hai-Po (2), pl. 14*b* (a), whence here redrawn. The four sailors within the ship, all facing forward to row, as usual in China, are armed with short swords in scabbards at their waists. The marines on the upper deck are armed not only with the typical dagger-axe 'halberds' (*ko*), but alternately with bows. As in the rather older (−4th-century) bronze depicted in Fig. 300, to which the present engraving bears great similarity, there is a drummer at the stern, here beating on two drums, a large and a small. Aft of the ship there is a figure (not shown), seemingly in the water among the fishes, which appears to be pushing the warship along—perhaps a guardian spirit favourable to the warriors.

short swords, supported by 'dagger-axe' halberdiers at longer reach,[b] while at the stern of the right-hand ship a small figure beats a drum. The curved and over-arching sterns of the boats are noteworthy,[c] and below there are swimmers among fishes. A very similar design occurs on a bronze described by Sun Hai-Po (2) and attributed to the Early Han period (Fig. 959), but archers are more prominent in it.[d] There would have been nothing particularly progressive about such ships in −3rd-century China,

[a] Yang Tsung-Jung (1), pl. 20.
[b] See further in Sect. 30*c* below.
[c] Cf. pp. 394, 435.
[d] This has also been described and figured by Bulling (1), fig. 338*a*.

[1] 樓船

since the Phoenician or Greek ships sculptured on Sennacherib's palace at Nineveh about −700 have an analogous structure.[a]

A sidelight on the 'castled ships' of the Han may be obtained from the inscribed bronze drums of contemporary date which have been found in Indo-China. Fig. 960 shows some of their pictures of war-boats, done in the peculiar and characteristic style of the Dôngsón culture.[b] Here again there are superstructures, forerunners of the 'fore-

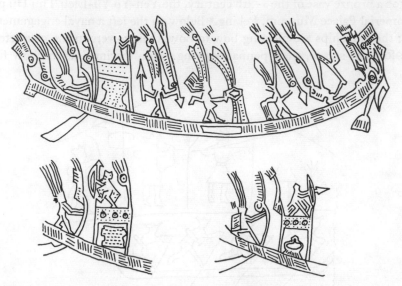

Fig. 960. War-boats on the Dôngsón bronze drums of Indo-China; tracings from the photographs of Parmentier (2) and Goloubev (1). Through the peculiar and characteristic style of these −1st-century engravings one can easily make out the helmsman at the stern with his steering-oar, the ritual bronze drum within the after-castle and the archers on its roof, warriors further forward with spears and javelins, and perhaps a stone anchor hanging from the bows. In the complex bow structure an oculus is distinctly visible, and just aft a figure which could be an enchanter casting a spell.

castles' and 'after-castles' such as appeared so much later in European shipbuilding when the Viking long-ship was turning into the medieval ship.[c] The bronze-drum boats seem to have had steering-oars at both ends (unless the object at the bow is a stone anchor), and warriors are generally much more in evidence than the rowers, who perhaps were hidden by the vessel's sides. The 'castles' are manned especially by archers, and each seems to contain a bronze drum, carried perhaps as a kind of magical Palladium or Ark of the Covenant. No masts or sails occur.[d] In date these drums are

---

[a] G. Holmes (1), p. 26; Anderson & Anderson (1), p. 35; des Noëttes (2), figs. 23, 24; Šmíd (1). And on Etruscan vases nearly as old. A good Greek vase-painting showing warriors on an upper deck is reproduced by la Roërie & Vivielle (1), vol. 1, p. 45; cf. des Noëttes (2), fig. 32.

[b] See Parmentier (2); Goloubev (1). Some think them ritual 'spirit-boats' (Christie, 2), either conveying the souls of the dead, or serving as a means of communication with them, but they may represent real craft none the less.

[c] As late as the +12th and +13th centuries; cf. the seals of Dover and of Sandwich (+1238), des Noëttes (2), fig. 67, G. Holmes (1), p. 67.

[d] Though in Fig. 960 there is something suspiciously like a tabernacle just forward of the centre of the vessel.

ascribed to the −1st century, so that the earliest of them could be contemporary with the Chinese conquest of −111 while the later ones would have been saved from the invaders of +43. Thus we have in these stylised representations a rather precious record of the kind of war-boats with which the 'Wave-Subduing Admirals' Lu Po-Tê and Ma Yuan successively had to contend. The Dôngsón drums are more closely associated with Chinese culture than was previously thought, for very similar bronzes, but even more brilliantly designed and cast, have been obtained in profusion from royal tombs of the State of Tien¹ excavated at Shih-chai Shan² near Chin-ning beside the Kunming Lake in Yunnan province.[a] These finds date from the −1st and −2nd centuries and are therefore contemporary with those of the Dôngsón culture and with the Chinese conquests of it.[b] In the rubbings so far published, however,[c] the boats are distinctly less large and elaborate than those depicted on the Dôngsón drums, products of so maritime a region.

Until recently it was thought that no model boat had survived from any Han tomb, but a magnificent specimen has now been recovered from a princely tomb of the Former Han period (−1st century) at Chhangsha (Fig. 961, pl.).[d] This river ship, which is 4 ft. 3 in. long and delicately made of wood, was found complete with sixteen oars and a great stern-sweep twice their length. The build is typically Chinese, flat-bottomed, with strictly rectangular bow and stern, and that part of the bow projecting above the water is very elongated. It is a pity that the model was not fitted with a mast and sails, for this would have given us very important information. Published photographs show considerable differences in the assembly of the parts; in one reconstruction a couple of deckhouses rise one above the other, in a second they follow on fore and aft. The former shows a large stern gallery, the ancestor of those elaborate structures on sea-going junks which cover the rudder housing; the latter, less plausibly, omits it and leaves no room for the steersman. One reconstruction has included the bulwarks pieces, but apparently upside down, for the oar-ports should be lower and the sides would then give protection to the rowers.[e] It is not easy to pick out any existing Chinese craft which resemble this boat in detail, but there is something archaic (indeed truly Egyptian) about the long projecting bow and stern, as in certain Indian boats, such as the *malar panshi* of Bengal.[f] But they are perhaps more 'Naqadian', while the Han boat is more 'Horian' (cf. p. 434 above), and seems almost to prefigure the Ma-yang-tzu. From accompanying inscribed objects, the Chhangsha boat model seems to be datable about −49.

[a] Archaeological report, Anon. (*28*); short account in English, Wang Chiung-Ming (*1*). These discoveries are of the first importance, not only as revealing a comparatively high culture the very existence of which was previously unsuspected, but because the Tien style has close affinities with the 'animal style' of the Eurasian steppes (cf. Vol. 1, pp. 159, 167) and even with the art of the Mayas and Aztecs, as well as with that of Nan Yüeh beyond the mountains to the south-east.

[b] Even the gold seal of the King of Tien, presented to him by Han Wu Ti in −109 (*Shih Chi*, ch. 116, p. 5b), was recovered.

[c] Anon. (*28*), pl. 126.

[d] See Hsia Nai (*1*) and Anon. (*11*), pp. 154 ff. and pl. 103.

[e] As in the case, perhaps, of Tshen Phêng's crews (p. 621 below).

[f] Smyth illustrates one of these, (*1*), p. 365; also Hornell (*1*), p. 249. See Fig. 969 below.

¹ 滇   ² 石寨山

By great good fortune, excavations in the city of Canton since 1954 have given us remarkable models of +1st-century craft which complement the river-ship, since they evidently belong to estuaries and to the sea.[a] Some of these are of red pottery rather roughly finished (Fig. 962, pl.),[b] but showing most clearly the typical bluff square-ended bow and stern and the flat bottom of the hull. The thwart timbers overhang the sides and on the arched deck there are the figures of four rowers, but no mast. Both fore and aft the uppermost strake timbers are prolonged so as to form galleries, the former no doubt for handling the anchor, the latter for fixing the steering apparatus, though what this was these models do not show.

To this particular question the greater ships of grey pottery (also in the Canton Museum) have a dazzling answer to make, but we shall say no more of it here since it belongs to the debate on the rudder and its origins.[c] The general appearance of one of these ships, recovered from a Later Han tomb in the Tung-Chiao district of Canton, is shown in Fig. 963 (pl.), a photograph taken from the starboard side.[d] The hull is of the standard Chinese type, as can also well be seen from Fig. 964 (pl.), which shows the bows, and the anchor[e] hanging from a bollard on them just forward of an ornamental structure which may well be the ancestor of the 'prow yokes' now mostly found on Indo-Chinese vessels.[f] Passing aft we find a sort of cowl, doubtless made of bamboo matting, to provide protection for the forward well-deck;[g] and then three bollards or rowlocks on each side, between which stand two figures amidships. For two-thirds of its length the vessel is covered over by a barrel-vaulted roof (perhaps of matting), out of which stand three deckhouses, the aftermost one being clearly the steersman's cabin (*tho lou*[1]).[h] The covered part is flanked by three tall vertical pointed timbers of uncertain function,[i] and by well-marked outboard poling galleries[j] supported on projections of the thwart timbers, with another member of the crew standing on the starboard side. Where the mast was stepped remains a mystery, but the most likely place would seem to be the gap just forward of the covered deck where the poling galleries end (see Fig. 965, pl., a photograph taken from above). It is sad that all

[a] Canton tombs of the +1st century have also now yielded further river-ship models like the Chhangsha boat, but smaller. A photograph of one of these has been published by Hsia Nai (2). The full description is in Mai Ying-Hao (*1*), pp. 26 ff., pl. 8, fig. 5. Length *c*. 2 ft. 7 in.

[b] With Dr Lu Gwei-Djen I had the opportunity of studying one of these in 1958 at the Canton Museum. This is housed in the Wang Hai Lou, a remarkable Ming building of five storeys built about +1380 and standing on the hill called Yüeh-hsiu Shan in the north of the city. We are greatly indebted for much kind assistance and the provision of valuable photographs to Dr Wang Tsai-Hsin, the Director, and to his colleagues Mr Tsêng Hai-Shêng and Mr Liao Yen-Yu.

[c] Pp. 649 ff. below.

[d] Total length 1 ft. 9¼ in. This beautiful model has been described but briefly by its Chinese discoverers (Anon. *31*), and figured without comment by Lanciotti (2). The excellent photograph of it on a glass base in Anon. (*26*), pl. 444*a*, is ruined because the rudder was assembled the wrong way round. No proper nautical study of it has yet appeared.

[e] Again typically Chinese in form (cf. pp. 656 ff. below).

[f] See P. Paris (3), pp. 56 ff. and especially figs. 67, 169 and 172.

[g] This is frequently seen in later medieval Chinese pictures of ships, and is still in use today.

[h] This has a galley on its starboard side and a latrine aft.

[i] Perhaps they are supports for the gantry framework on which sails, oars and punt-poles are stowed.

[j] Cf. p. 385 above.

[1] 柁樓

evidence of mast and rig has gone, but the general resemblance with a traditional Cantonese ship of a few years ago being worked upstream can be glimpsed from Fig. 966 (pl.).[a]

The +3rd century, the San Kuo period, is rich in naval history, but unfortunately the annalists were sparing of detail about the ships themselves. Doubtless the *lou chhuan* were larger, and we begin to hear of fast fighting ships called Mêng-chhung,[1] Tsou ko[2] and Tou hsien[3].[b] Bulwarks were covered with wetted hides to prevent the effect of incendiaries, but sometimes these weapons were very effective, as in the famous case of the Battle of the Red Cliff (+207), when Tshao Tshao's fleet was destroyed.[c] Sea-going ships as well as river craft must have been developing, for in +233 a fleet of the State of Wu was lost in a storm in the Yellow Sea.[d] From a number of descriptions it is clear that detachments of crossbowmen (unknown until much later in medieval Europe) firing from the ships often played a more important part than boarding or ramming.[e]

In the San Kuo period the word *po*[4] for sea-going ships or junks, appears. Khang Thai,[5] who travelled to the kingdoms of south-east Asia about +260 on behalf of the ruler of the Wu State, left a book called *Wu Shih Wai Kuo Chuan*[6] (Account of Foreign Countries contemporary with Wu), fragments of which have been preserved in encyclopaedias. Thus we find[f] that the Malayan princes of that time used to get horses from the Indo-Scythians (Yüeh-chih);[g] and we hear about the ambassador Su Wu[7] sent by Fan Chan,[8] King of Cambodia, about +250 to a kingdom in India (Thien-Chu[9]), whence he returned escorted by an Indian envoy named Chhen Sung[10] and bringing four fine Yüeh-chih horses.[h] We also learn about the earlier arrival in Cambodia (Fu-Nan[11]) of the Hindu cultural missionary Kaundinya (Hun-Thien[12]) who married a *nāga* princess and founded a dynasty.[i] Another lost book of Khang Thai's, his *Fu-Nan Chuan*,[13] told of a Chinese merchant Chia Hsiang-Li[14] who traded all the way to and from India, and once on his way back recounted to Fan Chan at length the

---

[a] Other good photographs of ships with poling galleries will be found in Forman & Forman (1), pl. 179; Schulthess (1), pls. 166, 167.

[b] Among the earliest appearances of these types was the fleet built by Liu Piao,[15] governor of Ching-chow (d. +208), for a campaign against Tshao Tshao, the founder of the Wei State, cf. *WHTK*, ch. 158 (p. 1379.3). Liu Piao was a patron of astronomy and astrology, cf. Vol. 3, p. 201.

[c] *San Kuo Chih*, ch. 32, p. 8b; the story is told also in ch. 54, pp. 2b ff. Another instance of fire-ships (*yu chhuan*[16]), used by Wei generals this time, in +222 (Chuko Chhien[17] and Wang Shuang[18]), is in *San Kuo Chih*, ch. 56, p. 12a; but they failed.

[d] *San Kuo Chih*, ch. 26, p. 12a. It was on active service against the Wei kingdom.

[e] E.g. Sun Chhüan[19] fighting Huang Tsu,[20] in *San Kuo Chih*, ch. 55, p. 10a (+208). We shall mention this captain again, p. 657 below. Cf. what is said below upon the 'shock cycle' and the 'projectile cycle' in tactics (Sect. 30 in Vol. 5).

[f] *TPYL*, ch. 359, p. 1b.

[g] Cf. Vol. 1, pp. 173 ff.

[h] *Liang Shu*, ch. 54, pp. 22b ff., tr. Pelliot (16), p. 271.

[i] *Liang Shu*, ch. 54, pp. 8a ff., tr. Pelliot (16), p. 265. Also *TPYL*, ch. 347, p. 7b. Cf. Grousset (1), pp. 557 ff.

---

[1] 蒙衝	[2] 走舸	[3] 鬥艦	[4] 舶	[5] 康泰
[6] 吳時外國傳	[7] 蘇物	[8] 范旃	[9] 天竺	[10] 陳宋
[11] 扶南	[12] 混填	[13] 扶南傳	[14] 家翔梨	[15] 劉表
[16] 油船	[17] 諸葛虔	[18] 王雙	[19] 孫權	[20] 黃祖

customs of that populous country and the success of Buddhism there.[a] All these people travelled in *po*.[1] To the Chinese, for whom the building of sharp-ended boats with stem and stern posts was only a commemorative ritual act, Khang Thai brought knowledge of such craft in full use.[b]

In the kingdom of Fu-Nan they cut down trees for the making of boats. The long ones measure 12 fathoms in length (about 70 ft.) and their breadth is 6 ft. The stem and the stern resemble (the head and tail of) a fish, and they are decorated all over with ornaments of iron.[c] The large (boats) can carry a hundred men. Each man has a long oar, a short oar (i.e. a paddle), and a pole for quanting. From stem to stern there are 50 men, or more than 40, according to the boat's size. In full motion they use the long oars; when they sit down (to row) they use the paddles; and when the water is shallow they quant with the poles. They all raise (their oars) and respond to the shouts in perfect unison.

Such boats have obvious relation with Chinese dragon-boats (p. 436). They continued in use in south-east Asia for centuries. The bas-reliefs of the Banteai-Chmar, and the Bayon of Angkor Thom itself (probably commemorating the victories of the Khmers over the Chams from +1177 onwards), which have been described by Parmentier (1) and P. Paris (2), show similar large canoes with about 20 rowers and steered by quarter-paddles.[d] They have some resemblance to the broader-beamed Viking longships, and with their warriors on an upper deck might be larger versions of the Warring States and Early Han war-boats.[e] Their relation to the Tien and Dôngsón war-boats is also fairly clear. All this throws into bright relief the enormous progress in shipbuilding made by the Chinese during the +1st millennium.

Chou Ta-Kuan gave a further description of Khmer shipbuilding in +1297. Speaking of Cambodia he said:[f]

The great boats are made of hard wood. The carpenters have no saws and work with hatchets, so that it takes much wood and much trouble to produce a single plank. Knives are also used, even in house-building. For the boats they use iron nails, and cover them (i.e. roof them) with *chiao*[2] leaves,[g] held in place by strips from the *pin-lang*[3] trees.[h] Such a boat is called a *hsin-na*[4] and is rowed. For caulking they use fish oil and lime. Small boats are made of great trees hollowed out in the form of a scoop (*tshao*[5]) softened by fire, (and water), and enlarged by ribs, so that they are broad in the centre and pointed at the ends.[i] They have no sail (*phêng*[6]), but can carry several persons, and they row them. These are called *phi-lan*[7] boats.

[a] *Shui Ching Chu*, ch. 1, p. 9a, tr. Petech (1), p. 40; Pelliot (16), p. 277.

[b] *TPYL*, ch. 769, p. 5a; parallel passage, condensed, in *Nan Chhi Shu*, ch. 58, p. 15a. Tr. Pelliot (29), eng. auct.

[c] This probably refers to the iron clamps joining the strakes together.

[d] Attention was drawn to the similarity by Groslier (1), p. 109. Though Khang Thai does not explicitly mention 'deck-castles' they are implied by what he says, and the Khmer and Cham boats are certainly *lou chhuan* in the Early Han sense. I had the chance of studying them *in situ* in 1958.

[e] Of course we do not know the shape of the Warring States hulls, whether pointed to stem- and stern-posts or square-ended in classical Chinese style.

[f] *Chen-La Fêng Thu Chi*, cit. in *Shuo Fu*, ch. 39, p. 25a; tr. Pelliot (9), p. 172, (33), p. 32, eng. auct.

[g] Doubtless *chiao chang*,[8] i.e. kajang, Malaysian matting made from *Pandanus* spp. (the screw-pines, cf. Burkill (1), vol. 2, pp. 1644 ff.).

[h] *Areca Catechu*, the betel-nut palm (R719). Cf. Fig. 990.   [i] Cf. p. 388.

[1] 舶   [2] 茭   [3] 檳榔   [4] 新拿   [5] 槽   [6] 篷   [7] 皮蘭
[8] 茭葦

His description of the dugout and ribs indicates that these were things unfamiliar to the Chinese of his time.

There is one other text of the +3rd century which must be quoted. It is of capital importance in that it establishes a date in the earliest development of fore-and-aft sailing, but it will be more conveniently reserved till the discussion of the history of propulsion by sails (p. 600 below).

For the Sui and Thang periods (+6th to +10th centuries) information is hard to assemble. It remains easier to study epigraphic evidence outside China and to read what the Chinese say about the shipping of other countries rather than about their own. As P. Paris (1) and Kuwabara (1) have pointed out, it must be significant that few of the Indian missionaries or Buddhist pilgrims[a] (+5th to the +7th centuries) speak of taking passages on Chinese ships, though Indian, Persian and Malay (Khun-Lun[1]) vessels are frequently mentioned. It was not, apparently, until the close of the period that Chinese ships grew to their full stature, and indeed Paris is probably right in seeking the most rapid development of the great sea-going junk during the +9th to the +12th centuries. Marco Polo, Odoric of Pordenone and Ibn Baṭṭūṭah all took passages on Chinese ships. Nevertheless, Yang Su,[2] about +587, was constructing vessels with as many as five decks, more than a hundred feet in height.[b]

No one seems to have noticed a rather important passage in the *Thang Yü Lin* (Miscellanea of the Thang Dynasty), compiled in the Sung from Thang documents by Wang Tang, which is worth giving in full.[c] It refers to the +8th century.

In the south-eastern districts there are no places lacking means of communication by water. Therefore merchandise is mostly transported by boat. Every year the Transport Commissioners move two million piculs of rice (about 141,000 tons) to Kuan-chung (i.e. the capital in Shensi), through the Thung Chi Canal[d] to the Yellow River. The boatmen of Huai-nan, however, cannot navigate upon (lit. enter) the Yellow River. The most dangerous places (of the different provinces) are the Three (Yangtze) Gorges in Szechuan, the San-mên (rocks) in Shensi,[e] the O-Chhi river in Fukien and Kuangtung, and the Kan-shih rapids at Nan-khang. At all these places the local people do the work (as pilots)...

On the Yangtze and the Chhien-thang Rivers they set sail according to the two tides, and Chiangsi is the place where shipping flourishes most.

Sails are plaited from rushes (*phu*[3]), and the largest of them exceeds 80 sections (*fu*[4]). From Pai-sha (White Sands) the ships go upstream when they have a north-east wind; this is called the *hsin-fêng*[5] (reliable seasonal wind).[f] In the seventh and eighth months there is the

---

[a] Cf. Vol. 1, pp. 207 ff. above.

[b] *Sui Shu*, ch. 48, p. 3 a, abridged in *WCTY* (*Chhien chi*), ch. 11, p. 6 a. These were the Wu-ya hsien battleships with 'grappling-irons' (cf. p. 424 above and p. 689 below). They carried 800 marines besides the seamen.

[c] Ch. 8, pp. 23 b ff.; tr. auct. But we find that a parallel passage was translated by Moule (3), p. 183, from the +8th-century *Thang Kuo Shih Pu*,[6] ch. 3, pp. 21 a ff. The general testimony is the same. Cf. Hirth & Rockhill (1), p. 9.

[d] See p. 307 above.   [e] See pp. 274 ff. above.

[f] The reference may be to sailing out of the Poyang Lake and up the Yangtze River in winter, the N.E. monsoon season. Cf. pp. 462, 511, 571 below, and Vol. 3, p. 463.

---

[1] 崑崙   [2] 楊素   [3] 蒲   [4] 幅   [5] 信風   [6] 唐國史補

'upper reliable (seasonal wind)'; in the third they rely on (migrating) birds, and in the fifth they look for (the wind on) the wheat. 'Catapult clouds' presage storm.[a]

The sailors worship Pho-Kuan [1] (Goddess of Wind and Wave) with sacrifices, and ask the monks to pray for them.

The sailors have a saying: 'Water won't carry 10,000'—this means that the largest ship cannot exceed a load of 8,000 to 9,000 piculs (i.e. 562 to 635 tons).[b]

In the Ta-Li and Chen-Yuan reign-periods (+766 to +779 and +785 to +804) there were the (large) ships of Yü Ta-Niang [2].[c] The crews of these ships lived on board, they were born, married and died there. The ships had, as it were, lanes (between the dwellings), and even gardens (on board). Each one had several hundred sailors. South to Chiangsi and north to

[a] Cf. Vol. 3, p. 470 above (in meteorology).

[b] This statement is important. As will be seen below (p. 466), Marco Polo, in the +13th century, described river junks ranging in cargo burthens from 224 to 672 tons. Those engaged in the +17th-century grain-transport on the Grand Canal, according to Sung Ying-Hsing (p. 412 above), were smaller, having average ladings of between 140 and 210 tons. The figure for the Thang ships preserved by Wang Tang compels attention, for if we are to accept it they were remarkably large for their time.

As is well known, the interpretation of all ancient and medieval tonnage figures bristles with difficulties (cf. Gibson (2); Clowes (2); Lyman (1); P. Gille (2); Braudel (1), pp. 249 ff. etc.). 'Tuns burthen' and 'tuns and tunnage' were terms which arose from the Bordeaux wine trade in the +12th century, the former having reference to the number of tuns or barrels which the vessel could carry, and the latter including also the empty spaces between the barrels. For some Chinese parallels of these packing problems cf. Vol. 3, pp. 142 ff. above, in the mathematics Section. One tun was reckoned as being equivalent to 60 cu.ft. or 2,000 lb. We must presumably regard the figures of Wang Tang, Marco Polo, and Sung Ying-Hsing as analogous to 'tuns and tunnage'. The trouble is that in old statements the way of calculating the tonnage given is not always specified. Confusion with modern definitions of tonnage may arise if it is not understood that the practice of giving the weights of ships in terms of the weight of water displaced ('displacement tonnage' if without cargo; 'deadweight tonnage' if with it) grew up only after the middle of the nineteenth century. The most suitable unit for comparison with medieval figures is perhaps the 'gross tonnage', i.e. the weight calculated from the cubic capacity of all enclosed spaces on the ship at the rate of 100 cu.ft. to the ton. But such comparisons are difficult because of the space occupied by engines, allowed for in 'net, or registered, tonnage' by a percentage reduction.

A ship of 200 'tuns and tunnage' burthen was exceptionally large in +14th-century Europe (cf. Clowes (2), pp. 56 ff.; Gibson (2), p. 110). In the following century Prince Henry's famous caravels are reckoned at c. 50 tons, and only the caravela redonda reached the larger figure (da Fonseca, 1). None of Vasco da Gama's ships exceeded 300 tons and some were much less (Prestage, 1); and for the Santa Maria of Columbus 280 tons is an acceptable figure (Braudel, 1). Yet in the middle of the +8th century Chinese ships had reached nearly 600 'tuns and tunnage', and by the middle of the +13th, 700. These sizes were about the same as those of the largest Venetian carracks of the +15th century (Lane (1), pp. 47, 102, 246 ff.). The average size of the ships of the Spanish Armada of +1588 was still only 528 tons (presumably 'tuns and tunnage'), as against an average of about 177 in the English fleet (Charnock (1), vol. 2, pp. 11, 66). Only seven out of 132 Spanish ships were over 1,000 tuns burthen. In +1602 the largest ship in the English navy was of 995 tons. It is probable that junks of about this size or rather larger were common between the +13th century (Marco Polo's time) and the +18th. Casson (1) claims a figure of 1,200 tons or thereabouts for the great Roman grain ships of Lucian's time (+2nd century), but this is still subject to discussion. In Europe about +1500 only a few large Venetian warships attained or surpassed this figure (Lane (1), pp. 47 ff.). Though exceptional dimensions of 414 ft. and 8,000 tons have existed, the practical upper limit for wooden-hulled sailing-ships reached in the West in the mid-nineteenth century proved to be about 3,100 tons (cf. Gibson (2), pp. 110 ff., 121 ff., 129). But if the arguments of Mills (9) and others are sound (and we believe that they are), this limit had already been approached by the larger vessels of the Ming navy (the Treasure-ships, cf. pp. 480 ff.) under Chêng Ho in the first half of the +15th century. For reasonable assumptions lead to the conclusion that these junks had a burthen of about 2,500 tons and a displacement of about 3,100 tons. Nicolò de Conti, an eye-witness about +1438, reckoned the larger five-masted Chinese junks at about 2,000 tons (Penzer ed. p. 140). Our general impression is that throughout the medieval period the tonnages of Chinese shipping were consistently higher than those of Europe.

[c] Was she a shipowner, a captain, a shipwright or a goddess? Or just someone whom these large ships were named after?

[1] 婆官   [2] 俞大娘

Huainan they made one journey in each direction every year, with great profit. This almost amounted to 'carrying 10,000'.[a]

In Hupei many people live entirely on the water, in boats nearly half as big as houses. The large boats are always owned by merchants, who have bands of musicians on them, as well as their slave-girls, and all these people live under the poop (*tho-lou*[1]).[b]

The sea-going junks (*hai po*[2]) are foreign ships. Every year they come to Canton and An-i. Those from Ceylon are the largest, the companion-ways alone being several tens of feet high. Everywhere the various kinds of merchandise are stacked up. Whenever these ships arrive, crowds come forth into the streets, and the whole city is full of noise. There is a foreign Headman (Fan-Chhang[3]) in charge. The Trade Commissioner (Shih Po Shih[4])[c] has to make a record of his name, and the merchandise carried, to receive customs duty, and to forbid (the sale of) pearls and strange (things).[d] Among the merchants some have been sent to prison because of their deceit. When these ships go to sea, they take with them white (homing) pigeons, so that in case of shipwreck the birds can return with messages.[e]

The passage attests the use of mat-and-batten sails and the existence of remarkably large river-junks working back and forth between Chiangsi and Anhui; it also strongly indicates that the axial rudder was in use. It seems to suggest, too, that these vessels had not been entirely successful, and had been replaced in the writer's own time by rather smaller ships. Evidence collected by Reischauer (1) shows that about this period the Arab, Persian, and Sinhalese merchants came no further north than Yangchow, while the Shantung ports and the old estuary of the Huai River were the resort of Japanese and Koreans. The name of one Thang sea-captain trading to Japan, who was known also as a shipbuilder, has come down to us, Chang Chih-Hsin.[5] From what Wang Tang says one must not infer that all the coastal and sea-going ships were foreign, since from the beginning of the +8th century it became the practice to carry large amounts of grain and other commodities from the south to Hopei, the northern province menaced by Chhi-tan and Koreans.[f]

It was this period, indeed, which saw the apogee of maritime intercourse between China, Japan and Korea.[g] Many Japanese monks and scholars lived for years at the centres of Chinese religion and learning (Ennin,[6] for instance, was there between

[a] This may explain the statement of Lecomte (1), p. 233. Perhaps he misunderstood what had originally been an apotropaic proverb. The water-gods might well have drawn the line at 10,000 piculs.

[b] The great importance of this remark will be explained later (p. 644 below).

[c] This office had been first established about +700. It was held afterwards by many distinguished men, e.g. Chu Fu, the father of the author of the *Phing-Chou Kho Than*, important in the history of the magnetic compass, towards the end of the +11th century (cf. Vol. 4, pt. 1, p. 279); and by Phu Shou-Kêng towards the end of the +13th (cf. p. 465 below, and the classical paper of Kuwabara, 1).

[d] Presumably because these were destined for the imperial court.

[e] So our author. But as Hornell (16) has stated, in an interesting paper on the role of birds in early navigation, the Indians had carried birds with them at sea since the −5th century (there are many references in Sanskrit and Pali), and used them for shore-sighting when doubtful of their position, rather than as message-carriers. Cf. Pliny, *Nat. Hist.* VI, 22. See also p. 555 below.

[f] Some of the evidence is summarised by Pulleyblank (1), pp. 80, 159.

[g] We have now an excellent history of the cultural relations between Japan and China in the book of Kimiya Yasuhiko (1), while the embassies of the Japanese to the Thang court have been carefully reviewed by Mori Katsumi (1). As early as +300 the King of Silla had sent skilled shipbuilders to Japan (*Nihongi*, Aston tr., vol. 1, p. 269; Tamura Sennosuke (1), p. 117).

¹ 舵樓　　² 海舶　　³ 番長　　⁴ 市舶使　　⁵ 張支信　　⁶ 圓仁

+838 and +847),[a] there were numerous Japanese embassies, and a great deal of commercial activity went on. One of the most colourful figures of the time was the Korean Chang Pogo,[1] a great shipowner and merchant prince. Having laid the basis of his fortune as an official in China, he returned to Korea in +828 and settled on Wando Island, at the south-western tip of the peninsula, a strategic control point for the Sino-Japanese sea traffic. But later he became involved in the politics of the Silla Kingdom, first as Commissioner of Chhŏnghaejin to suppress a slave trade in Korean peasants,

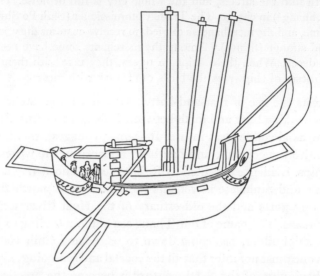

Fig. 967. The Ajanta ship, a tracing from the reproduction of Yazdani & Binyon (1).
The fresco painting is dated in the +6th or +7th century.

and eventually as king-maker victorious over an usurping pretender. He died by political assassination in +841, after which his maritime commercial network collapsed.[b] His career typifies the importance of the Koreans in the sea trade between the three countries in the +9th century. Unfortunately we have not found any pictorial evidence which would allow us to form an idea of the nature of Chang Pogo's ships, but there are valuable representations from other Asian regions which have a bearing on the vessels of the Thang time.

Much interest attaches to the ship represented upon the painted walls of one of the Ajanta caves (Fig. 967). It is reproduced in colour by Yazdani & Binyon.[c] If Hadi Hasan (1) is right, it may be identified with a ship bringing Persian ambassadors to the court of Pulakēṣin II of the Deccan, just before +628, and the painting would not be

[a] Ennin tells us a good deal about nautical matters. We have noted one of his pieces of information already (p. 308), and shall find more important ones below (pp. 555, 619, 643).
[b] An account of Chang Pogo's life and work has been given by Reischauer (2), p. 100 and (3), pp. 287 ff. The primary source for it is the +12th-century Korean chronicle, *Samguk Sagi*, chs. 10, 11, and 44. On Korean naval history in general see Anon. (69).
[c] Cave no. 2. Vol. 2, pl. 42. Moll (1), A, II, *a*, 12, reproduces a version which wrongly shows four masts.

[1] 張寶高

long posterior. Others make it derive from the early years of the previous century.[a] No one, however, has ventured to be very positive as to what country or nautical tradition it represents.[b] The sail at the bow is hoisted on a spar intermediate between a bowsprit and a fore-mast, i.e. on something very like the Roman *artemon*.[c] The hull has a Sinhalese air about its build.[d] The quarter steering-paddles, one on each side, and seemingly connected by some kind of mechanism, resemble those of many ancient Egyptian ships.[e] Behind this there is a space, roofed but open-sided, containing sacks or jars. The galleries at bow and stern (if that is what they are) are rather Chinese. Most puzzling are the three sails, narrow and lofty, but not drawn sufficiently clear to inform us without doubt whether they are balanced lug-sails.[f] The significance of this enquiry will be fully apparent a few pages further on; we shall have to bear the Ajanta ship in mind and think of it again. If the straight set of the sails does betray the Chinese mat-and-batten system, there is a further hint in the stepping of the masts, apparently in tabernacles and radiating like the spokes of a fan. The probability is that the artist was unthinkingly combining characteristics from a number of different sea-going vessels which he had seen, but not looked at, and it is not surprising to find certain Chinese elements among them.[g]

Less confusing, but still apparently not quite characteristic of their culture, are the beautiful paintings of small sailing-boats preserved in the complementary Chinese frescoes of the cave-temples at Tunhuang in Kansu province (Mo-kao-khu or Chhien-fo-tung). There are a good many of these little ships and they occur in all kinds of contexts. For example, the vessel in Fig. 968 (pl.) is obviously the Buddhist Ship of Faith; it is being seen off from the shores of this world by rampaging devils, and its destination is certainly the Western paradise of Amida Buddha which is visible in the background.[h] The square-ended bow and stern of this early Thang (+7th-century) ship are indeed very Chinese, with a characteristic aft prolongation of the uppermost longitudinal timbers. But here the Chinese elements end—except for the beehive hut of straw[i] which someone thought would be the right thing for a deck-house.[j] The bellying square-sail is exceedingly un-Chinese, though suitable for a boat of the Ganges, and the latter would also have arrangements for working the quarter steering-paddles

---

[a] Such as Mukerji (1), p. 39.

[b] Perhaps the literary parallel for this uncertainty is the passage in Cosmas Indicopleustes (XI, 337) about Ceylon (Taprobane), written c. +547. 'From the remotest countries, I mean Tzinista (China) and other trading places, it receives silk, aloes, cloves, sandalwood, and other products...' (tr. McCrindle (7), pp. 365 ff.). Was the silk brought by Chinese merchants in their own ships, picking up the other goods on the way? Unfortunately we do not know.

[c] As Bowen (7) has also pointed out.

[d] Cf. Bowen (2), p. 194.

[e] E.g. des Noëttes (2), figs. 12, 16.

[f] Bowen (7) strongly favours the latter. But for a Chinese square-sail of just these proportions see Fig. 971 (pl.).

[g] Representations of vessels, perhaps related to the Ajanta ship but equally difficult to interpret, are found carved at Aurangabad and Ellora; for references and discussion see Paris (1).

[h] Cave no. 55. Pelliot (25), pl. 237; photograph by Pelliot in the Musée Guimet collection (Paris), no. 45153/45, B 237 △ A/84. The mast and sail are very inadequate in spite of a massive tabernacle framework, as no doubt the three laymen, awkwardly rowing at the bow, fully agree.

[i] Cf. above, p. 121, Fig. 779, in the building technology Section.

[j] And a thwartwise bar by the two steering-oars which recalls the typical halyard winches.

closely similar to those which the painter drew here at Tunhuang. The resemblance can be seen by comparing this picture with Smyth's sketch (1) of a *malar panshi* (Fig. 969). Perhaps the monastic artist, who had quite possibly journeyed overland from India and never seen either the sea or the great rivers of China, was remembering the ships of Bengal rather than depicting those of Chinasthan.[a]

Fig. 969. The *malar panshi* of Bengal, a drawing by Smyth (1), for comparison with the ships in Figs. 961 and 968 (pls.). The sheets of the square-sails are echoed clearly in the Tunhuang ship fresco.

All the representations of ships at Tunhuang follow this same pattern. Another large sailing-boat is found in cave no. 45, painted in the +8th or +9th century but rather less well preserved; it closely resembles the one just described save that the mast and sail, stepped almost equally far forward, and bellying almost equally markedly, are in better proportion.[b] The box-built hull with its projecting upper strakes also appears well in another painting of the early Thang period[c] which tells the legend of the arrival of a ferry-boat carrying Buddhist images belonging to the great Indian king Aśoka.[1] The bellying elongated square-sails of his country are seen even on the small sampans which the Tunhuang artists of the Thang and Sung included in many of their

[a] Something like this was also the view of Hornell (17), p. 188, both for Ajanta and Tunhuang.
[b] Studied personally in 1958; photograph by Mrs Anil da Silva available. Pelliot (25), pl. 248; photograph by Pelliot in the Musée Guimet collection (Paris), no. 45153/56, B 248.
[c] Cave no. 323, also seen in 1958; reproduced by Yeh Chhien-Yü (1), fig. 9, and Sirén (10), pl. 63. The story refers most probably to an event of +313, when two statues (of Vipaśyin and Kāśyapa), thought to be of Aśoka's time, were found by a fisherman (cf. Zürcher (1), p. 278). A few years later, a statue of Aśoka himself turned up in similar circumstances and embarrassed an old friend of ours, the military governor Thao Khan[2] (cf. Vol. 4, pt. 1, p. 62, and Zürcher (1), pp. 243, 279, 405). A similar ferry-boat, but with a small sail, and carrying a Buddhist pavilion of relics, occurs in cave no. 468 dating from the +8th century (mid-Thang).

[1] 阿育王   [2] 陶侃

frescocs.[a] But the most seamanlike of the Buddhist ships is not at Chhien-fo-tung at all, it is sculptured on a relief of the style of the Liang period (+6th century),[b] reproduced in Fig. 970 (pl.). The build of the hull is excellently shown, the mast rakes forward, and the nymphs of the breezes are gallantly filling a well-proportioned square-sail. One is left with a lingering doubt. If all these craft are to be taken as truly Chinese they might seem to contradict other evidence which points to a respectable antiquity for the taut mat-and-batten sail.[c] In all probability both types coexisted for centuries; loose square-sails are not unknown on traditional Chinese craft free from external influences, and if the boats of the Chhien-thang River had cloth sails in the Thang as they do now, they might well have served as the model for these representations (cf. Fig. 971, pl.). Sung paintings still depict such sails occasionally, but as time goes on they give way more and more to the characteristic flat lugs.[d]

Of great interest also are the numerous sculptures of ships on the walls of the great monument of Borobodur in Java, the most probable date for which would be about +800 or just before.[e] There are seven of these ships in all, and five of them are clearly and recognisably of the same type (see Fig. 973, pl.).[f] It is not our direct concern, since

[a] E.g. cave no. 323 (early Thang), reproduced in Anon. (*10*), pl 50; cave no. 126 (middle Thang); cave no. 98 (Wu Tai), and caves nos. 61 and 146 (early Sung). As time goes on, the forward-pointing 'cheeks' of the sampans become more and more exaggerated. The yards of the sails are always clear but sometimes booms can be made out as well.

[b] Dr Lu and I had the pleasure of studying a good rubbing of it at the Historical Museum of Szechuan University, Chhêngtu, in 1958; we thank Dr Fêng Kuo-Ting and Mr Chou Lo-Chhin for their kind welcome and for providing us with a photograph. The relief had been published long ago by Ti Phing-Tzu (*1*), ch. 1, p. 34*a*, but its present location is not known to us. Some authorities make it Liu Sung (+5th century). A very similar ship, but not so distinctly carved, can be seen in another Buddhist relief of the Liu Chhao period found at Wan-Fo Ssu and now in the Chhêngtu Municipal Museum.

[c] See pp. 599 ff. below.

[d] On my travels in 1964 I came upon two beautiful bronze mirrors depicting ships, which from the close similarity of style would seem likely to be close in date. One is in the Shensi Provincial Museum at Sian (Fig. 972, pl.), the other in the Sumitomo Collection at Kyoto. For the latter see Sumitomo Tomozumi, Taki Seiichi & Naitō Torajirō (*1*), vol. 10, no. 137, and the comments in Hamada Kōsaku, Harada Yoshito & Umehara Sueji (*1*), no. 137; Taki Seiichi, Naitō Torajirō & Hamada Kōsaku (*1*), p. 217, entry 132. A broad-beamed ship with a single mast, secured (in a very un-Chinese way) with prominent standing rigging, is scudding through the deep. Three of the crew are at the stern with the steering-oar, and three more people, perhaps pilgrims, at the bow, while each of the two deckhouse windows has a person looking out. In the Sian mirror the bellying narrow square-sail carries an inscription '(name unreadable), Inspector of Thien-hsing city in Fênghsiang prefecture'. In the Kyoto mirror the foot of the sail has come adrift and it is streaming out horizontally in the wind. One gains the impression that these mirrors rehearse the dangers safely overpassed by the Inspector either as a Buddhist pilgrim or an imperial envoy. As for the date, the Sian museum assigns it (no doubt for sound art-historical reasons) to the neighbourhood of +1100 in the Sung, but if the history of technology can have anything to say, a date in Wu Tai or late Thang, *c.* +900, would be much preferable. Perhaps a Sung craftsman was being consciously archaic. The Kyoto scholars, on the other hand, regard their mirror as Korean, and place it in the Kōrai period, i.e. before +668; this seems rather too early.

[e] According to Krom & van Erp (*1*), who have published the collection of photographs of the bas-reliefs.

[f] The best is Krom (in Krom & van Erp), Ser. 1*b*, pl. XLIII, fig. 86; used also by Mukerji (*1*) as frontispiece, and redrawn by him, p. 48, no. 3. Less clear in detail is Krom, Ser. 1*b*, pl. XLIV, fig. 88; I think this is Mukerji's p. 46, no. 1. Once the ship is shown next to a small boat, with the great sails being set; Krom, Ser. 1*b*, pl. LIV, fig. 108; Mukerji (redrawn), p. 48, no. 5. A valuable version with the sails furled is Krom, Ser. II, pl. XXI, fig. 41; Mukerji (re-drawn), p. 48, no. 6. The fifth, heading in the opposite direction, is Krom, Ser. 1*b*, pl. XXVII, fig. 53; also in Hornell (*1*), pl. XXXIIIB. There is also a much smaller craft of much the same kind, re-drawn by Mukerji, p. 46, no. 2. Reconstruction of the general type in Hornell (*14*), fig. 5.

purely Indonesian in nature,[a] recalling the vessels with which the colonisation of Madagascar must have been completed.[b] Nevertheless it is worth close examination. The hull is probably sewn, not nailed, as is suggested by its panelled appearance; the stem- and stern-posts are much emphasised. To the hull are attached large and complex outriggers upon which the crew could climb, to aid stability, if required.[c] The masts, two of which are fitted in each ship, are drawn (at any rate in some cases) as bipod or tripod in form.[d] They carry very elongated canted square-sails of the characteristic Indonesian type—important because canting was in all probability, as we shall see, the earliest device in the development of fore-and-aft sailing.[e] From the ship with the furled sails we can deduce that some method of roller-reefing was used. Yet the sails belly considerably when set, so P. Paris (1) can hardly be right in believing that they were of matting. In every case a small *artemon* in Roman style is mounted.

But these are not the only craft pictured at Borobodur. A radically different type intrudes.[f] It is shown but once, with a rather numerous crew, and it has no similarity at all to the others. There is no outrigger and no panelling, the stern-post is not emphasised but thwart timbers are prominent, there is only one mast consisting of a very stout spar; and above one can just make out a rigid Chinese lug-sail with the matting texture clearly indicated (Fig. 974, pl.). The artist failed to show the luff of the sail, slightly forward of the mast, but this only emphasises its fore-and-aft character,[g] and the difference from the sails shown on the other ships is extraordinary. Modest though it is, we may have in this picture the oldest representation of a Chinese sea-going ship in history.[h]

It looks as if there was some mutual influence between Chinese and South-east Asian shipbuilding methods during the +8th century. Half a century before the Borobodur carvings, we have an interesting statement by a Chinese monk who wrote about southern ships in a commentary on the Vinaya Canon, the *I Chhieh Ching Yin I*[1]

---

[a] Hornell (17), p. 221, concurs, as also Bowen (2, 9).

[b] Mukerji (1) and others have seen in the Borobodur ships Indian vessels of the time. But it does not follow, even if the artists intended to illustrate 'Indian adventurers sailing out to colonise Java' (which they almost certainly did not; as Vogel (1) has shown), that they would have done anything but copy the designs of the ships most closely known to them. And the Indonesian character of these vessels is overwhelmingly strong.

[c] Cf. p. 385 above, on outriggers. It is to be noted that on the Borobodur reliefs, whichever way the ship is pointing, the outrigger is shown. Can we take this to indicate that the gear was fitted on both sides? Or did these ships go about by reversing (cf. p. 612 below)? Vessels rather similar to those of Borobodur still exist, or did so until recently, e.g. the Sinhalese *yathra dhoni*, *yathra oruwa*, or coasting boat, figured by Hornell (1), p. 257, fig. 60, (14), pp. 247 and fig. 4.

[d] Here, then, is an intermediate link between the ancient Egyptian vessels and the Kuangtung riverboats of today, as Hornell (18), p. 39, well saw.

[e] Cf. p. 608.

[f] Krom (in Krom & van Erp), Ser. 1 b, pl. XII, fig. 23; Mukerji (re-drawn), p. 48, no. 4.

[g] Krom agrees, vol. 2, p. 238. Van Erp also, see the summary of his views in vol. 2, p. 235. The carving appears to be unfinished.

[h] Java had been well known to the Chinese since the +5th century at latest (cf. Hirth & Rockhill (1), p. 78; Fêng Chhêng-Chün (1), pp. 132 ff.). But apart from trade relations it is pleasant to recall the astronomical expedition under Ta-Hsiang and Yuan-Thai which was sent to the south seas in +724 to observe and chart all the constellations of the southern hemisphere (cf. Vol. 3, pp. 274, 293). This may well have visited Java as well as Sumatra, and the time was not long before the building of Borobodur.

[1] 一切經音義

(Dictionary of Sounds and Meanings of Words in the *Vinaya*).[a] Here Hui-Lin[1] said:

[a gloss on a text]

Pho-Po.[2] The second word is pronounced *po*. Ssuma Piao,[3] in his commentary on *Chuang Tzu*, says 'The great ships of the sea are called *po*'.[b]

According to the *Kuang Ya* (dictionary), the *po* is a sea-going ship.[c]

(These ships) lie 6 or 7 ft. deep in the water. They are fast and can transport more than 1,000 men, apart from cargo.[d] They are also called 'Khun-Lun'[4] ships.[e] Many of those who form the crews and technicians of these ships are Ku-Lun[5] people.

With the fibrous bark of the coconut (tree) (*yeh tzu*[6]),[f] they make cords which bind the parts of the ship together. And they caulk them with a paste made of *ko-lan*,[g] stopping up the openings and preventing the water from coming in. Nails and clamps are not used,[h] for fear that the heating of the iron would give rise to fires.

(The ships) are constructed by assembling (several) thicknesses of side-planks,[i] for the boards are thin and they fear that they would break. (The ships) are several...[j] long, and divided fore and aft into three sections. Sails are hoisted to make use of the wind, and (indeed, these ships) cannot be propelled by the strength of men (alone).[k]

The reference is certainly to southern (Indo-Chinese, Malay or Indonesian) ships, and though there is no mention of outriggers, the description applies not badly to the

---

[a] TW2178; N1605; Japanese Tripiṭaka, Wei sect., IX, p. 155a. This was Hui-Lin's enlargement of a work originally due to Hsüan-Ying[7] (*fl. c.* +649).

[b] This scholar lived from +240 to +305. The commentary is lost. No one has been able to guess what passage he could have been commenting on, especially as the word *po* does not occur in *Chuang Tzu*.

[c] Compiled about +230 by Chang I.[8] The word is there, in the Shih Shui[9] section.

[d] Certainly an exaggeration. In the previous line also the unamended text says 60–70 ft.

[e] This word meant Malaysian, as also Ku-Lun in the next line. See Hirth & Rockhill (1), pp. 32, 84; Stein (1), pp. 65 ff.; Link (1), pp. 9 ff. The binome *khun-lun* was originally connected with the idea of primeval chaos (cf. *hun-tun*, Vol. 2, pp. 107, 114, 119 ff.) and thus came to mean the central mountain of ancient cosmology, the Chinese Mt Meru (Vol. 3, pp. 565 ff.). Then either because it was used also as the name for Poulo Condor Island transliterating Pulo Kohnaong (Hirth & Rockhill (1), p. 50), or because it came to stand for the name of a Thai People Komr or Krom, it eventually signified Malaya and Malays in general (cf. Ferrand, 3). One of the earliest uses of it in this sense occurs in the biography of Tao-An (cf. Vol. 3, p. 566) in the *Kao Sêng Chuan* written by Hui-Chiao about +530, in which Tao-An is called by his enemies Khun-Lun Tzu, i.e. 'Little Darkie', no doubt on account of his complexion. Thang statuettes which may represent the dusky Khun-Lun southerners still exist (Lips, 1). Another theory of the transference of Khun-lun from Central to South-east Asia suggests that it first became associated with Indian merchants who reached China by the overland route round the Tibetan massif, and then when commerce shifted to the maritime route the term retained its connection with Indian or Indic trade and traders (Maspero (14), developed by Christie).

[f] *Cocos nucifera* (R720).

[g] This *ko-lan*[10] is probably a mistake for *kan-lan*[11] (*Canarium album*, the Chinese olive, R337), or *kan-lan*[12] (*Brassica oleracea*, kohlrabi, related to colza, R475); whatever it was, it was some plant from which a drying oil could be prepared. Neither of these plants appears in the pharmacopoeias until the Thang, however.

[h] Parallel statement in *Ling Piao Lu I* (*c.* +895). The point is important because it shows that iron fastenings were normal in +8th-century Chinese ships (cf. p. 412 above).

[i] Cf. on this Marco Polo, below, p. 468.

[j] The word in the text is *li* (half-kilometres), but either this crept in afterwards or the monk was spinning a long yarn. Wang Kung-Wu (1), following Kuwabara, interprets 200 ft.

[k] Tr. Pelliot (29), eng. auct.

[1] 慧琳　　[2] 破舶　　[3] 司馬彪　　[4] 崑崙　　[5] 骨論　　[6] 椰子
[7] 玄應　　[8] 張揖　　[9] 釋水　　[10] 葛覽　　[11] 橄欖　　[12] 甘藍

main Borobodur type (Fig. 973, pl.). But a south-east Asian sewn hull seems to have co-existed here with Chinese bulkheads. As for the sails, we reserve for the appropriate place (p. 600 below) a most striking piece of +3rd-century evidence concerning them.

Thus we may assume that these multiple-masted ships of the East Indies and the South China seas had been developing for the better part of a millennium before +800. According to Christie (1) they may be recognised in an expression in the *Periplus of the Erythraean Sea* (written *c.* +110)[a] which until now baffled all commentators.[b] For the word κολανδιοφῶντα (*kolandiophonta*), applied to the large ocean-going ships of South-east Asia, turns out to be nothing else than a corrupted Greek form of *Khun-lun po*.

(2) FROM THE THANG TO THE YUAN

Specific research would certainly discover much of interest for nautical technology between the +8th and the +12th centuries, but as yet only a few notes can be given. Our oldest elaborate descriptions of warships come from just before +760 but we shall postpone the translation of this text until p. 685 below in connection with ship armour. There was great activity in canal and river boat construction about +770, associated with the name of Liu Yen,[1] who established ten shipwrights' yards and offered competitive rewards.[c] Many-decked naval vessels were used, as by Chu Ling-Yün,[2] in the fighting of the Wu Tai period (+934).[d] The first Sung emperor attached importance to shipbuilding and often visited the yards (*tsao chhuan chhang*[3]).[e] The name of one of his chief shipwrights, Fan Chih-Ku,[4] has come down to us. In +1048 the Liao State, conscious also of sea- (or rather river-) power,[f] commissioned Yehlü To-Chen[5] to build 130 warships which would carry horses below decks and soldiers above; these worked effectively as landing-craft in operations along the Yellow River. In +1124 two very large vessels were built for the embassy to Korea, and we read of the excitement of the people of that country when they arrived in port.[g] In +1170 a traveller on the Yangtze watched naval manœuvres carried out by 700 ships each about 100 ft. long, with castles, towers, flags flying and drums beating. They sailed rapidly even against the stream, and his description was enthusiastic.[h]

For the +12th century there happens to be a cluster of important documents, both pictorial and literary. Let us start with one far away from China proper, the carving in relief on the Bayon temple built by Jayavarman VII at Angkor Thom in Cambodia (Fig. 975, pl.).[i] This wonderful representation, dating from *c.* +1185, is entirely dif-

---

[a] Cf. Vol. 1, p. 178 etc., and p. 518 below.　　　[b] Cf. also Stein (1), pp. 65 ff.

[c] *Thang Yü Lin* (Miscellanea of the Thang Dynasty), ch. 1, p. 23b.

[d] *Tiao Chi Li Than* (Talks at Fisherman's Rock), p. 30b.

[e] *Shih Wu Chi Yuan* (Record of the Origins of Affairs and Things), ch. 10, p. 5a.

[f] *Liao Shih* (History of the Liao Dynasty), ch. 93, p. 6a; Wittfogel & Fêng (1), p. 166.

[g] *Kao-Li Thu Ching* (Illustrated Record of an Embassy to Korea in the Hsüan-Ho reign-period), ch. 34, p. 4a.

[h] *Ju Shu Chi* (Journey into Szechuan), ch. 4, p. 12b.

[i] The carving has often been reproduced, as by Claëys & Huet (1), from whom this photograph is taken; Carpeaux (1), pl. xxv, fig. 31, pl. xxvi, fig. 32; and Marchal (2), p. 84. Cf. pl. XLVIII in Paris (2); Ec. Fr. d'Extr. Or. photos 5063, 5064, and enlargement 5627.

[1] 劉晏　　　[2] 朱令贇　　　[3] 造船場　　　[4] 樊知古　　　[5] 耶律鐸軫

ferent from all the other craft depicted on the monument, which are paddled canoe-like boats similar to those already described (p. 450 above). There is general agreement that it shows a Chinese junk from Kuangtung or Tongking.[a] The strakes of the carvel-built hull are clearly seen, and the reduced transom bow typical of the South China junk. The stern-gallery overhangs, in true Chinese fashion. There are two masts, each carrying Chinese matting sails (*li*,[1] *shuang*,[2] each with six to eight battens), the texture of which is beautifully reproduced, nor has the artist omitted to show the equally characteristic multiple sheets by which the set of the sails is controlled.[b] At the mast-heads there are two square objects which might either be topsails or basketwork crows-nests. At bow and poop there are flagstaffs bearing flags of typical Chinese design, with serrated edges.[c] From the bow there hangs an anchor which uses an elongated stone as a stock, and it is being hoisted by means of a windlass.[d] Most interesting of all is the deep and narrow axial 'stern-post' rudder lowered well below the bottom of the ship. We must scrutinise it further when dealing systematically with the history of this great invention (p. 648 below).[e]

Poujade (2) has made a detailed comparison between the Bayon junk and a certain type of craft still sailing today, built and worked by the Chinese in Siam. Most of the particulars are closely similar, but the hull is hybrid, bulkheads coexisting with a keel and true stem- and stern-posts.[f] He believes that the hull of the Bayon ship was already hybrid in this sense. Even the deckhouses are in the same position. But the Sino-Siamese junks of today carry sails very much larger than those shown on the monument, and with rounded leeches.[g] On the whole it seems far more likely that the Bayon ship was a true Chinese vessel in all respects, and not a hybrid at that early date.[h]

Some sixty years before the Bayon junk was being carved, Chu Yü[3] was writing (+1119) his *Phing-Chou Kho Than*[4] (Table-Talk at Phingchow). His father had been Superintendent of Merchant Shipping at Canton and his notes refer to a period from

[a] P. Paris (1, 2); Groslier (2); Poujade (2).

[b] The same kind of sails, together with multiple sheets, but no battens, are seen in another carving of three smaller boats in a regatta (Ec. Fr. d'Extr. Or. photos 5377, 5378). Note that all these sails are closely similar to the Chinese sail at Borobodur, and like it, fail to represent the smaller portion of the lug-sail forward of the mast, again emphasising its fore-and-aft character. The sails of Fuchow sailing sampans today look from a distance exactly like those on the carvings (cf. G. R. G. Worcester, unpublished material, no. 64).

[c] For comparison, see the picture of Chao Po-Chü[5] (+1127 to +1162), 'The First Han Emperor entering the Chhin Capital'. The flags are closely similar. Sirén (6), vol. 2, pls. 40, 41, 42. He also reproduces some of the flags on the Tunhuang frescoes, as in vol. 1, pl. 31; cf. Yeh Chhien-Yü (1), fig. 33.

[d] On anchors, see below, p. 656.

[e] Together with Dr Lu, I was fortunate enough to have the opportunity of studying this great carving *in situ* in the summer of 1958.

[f] As in the Singapore junk described by Smyth (1), p. 474 (with Maxwell Blake's drawings). Cf. Waters (3); P. Paris (3), and p. 438 above.

[g] These boats are of some 25 tons. A hundred years after the carving of the Bayon ship, Chou Ta-Kuan said in the *Chen-La Fêng Thu Chi* that each of the boats which he described brought back to China some 50-60 tons of beeswax (Pelliot (9), p. 167, (33), p. 26).

[h] An interesting sidelight on the sort of place these merchantmen were sailing to is afforded by recent archaeological studies in Sarawak. Mr Tom Harrison and Dr Chêng Tê-Khun have told us of their discovery of a Chinese ironworks site and port at Shan-tu-wang, clearly occupied and functioning throughout the Thang and Sung periods.

[1] 悝   [2] 雙   [3] 朱彧   [4] 萍州可談   [5] 趙伯駒

about +1086 onwards. We have already quoted his words (part of the same passage) on the use of the magnetic compass in navigation (Vol. 4, pt. 1, p. 279 above). This is what he says:[a]

The Pavilion of the Inspector of Foreign Trade (Shih-Po[1]) is by the waterside near the Hai-Shan Tower, facing the Five Islands. Below this, the river is called the 'Little Sea'. In midstream for some ten feet or so the merchant-ships (po chhuan[2]) take on water for use on their voyages; this water does not spoil, but water from outside this limit, and all ordinary well-water, cannot be stored (on board ship), for after a time it breeds worms. What the principle is underlying this I do not know.[b]

Ships sail in the eleventh or twelfth months to avail themselves of the north wind (the north-east monsoon), and return in the fifth or sixth months using the south wind (the southwest monsoon).

The ships are built squarely like rectangular wooden grain-measures (hu[3]).[c]

If there is no wind, they cannot move. Their masts (chhiang[4]) are firmly stepped, and the sails are hoisted beside them. One side of the sail is close to the mast, (around which it moves) like a door on its hinges.[d] The sails (fan[5]) are made of matting (hsi[6]).

These ships are called chia-thu[7]—a local expression.[e]

At sea they can use not only a wind from abaft, but winds from onshore or offshore can also be used. It is only a wind (directly) contrary (ni[8]) which can not be used. This is called 'using the winds of the three directions' (shih san mien fêng[9]). When the wind is dead ahead they cast anchor and stop...[f]

According to the government regulations for sea-going ships, the larger ones can carry several hundred men, and the smaller ones may have more than a hundred on board...[g]

Sea-going ships are several tens of fathoms in breadth and depth.

The greater part of the cargo consists of pottery, the small pieces packed in the larger, till there is not a crevice left.

At sea (the mariners) are not afraid of wind and waves, but of running aground, for if this happens there is no way of getting off again. If the ship suddenly springs a leak, they cannot mend it from the inside, but they order their foreign blackamoor slaves (kuei nu[10])[h] to take chisels and oakum (hsü[11]) and mend it from outside, for these men are expert swimmers and do not close their eyes when under water...[i]

---

[a] Ch. 2, pp. 1b ff.; tr. Hirth & Rockhill; mod. auct.

[b] Probably the sailors chose brackish water from bottom springs, not too salt to drink, but salt enough to prevent the growth of algae and animal life.

[c] This is an admirable statement of the characteristic rectangular build. 'Like a baking-trough', as Osbeck was to say later (+1751, p. 190).

[d] Again, an admirable statement of fore-and-aft rig; presumably here lug-sails. Cf. pp. 603, 611.

[e] This term has not been explained. It could well have been of foreign origin, since the Cantonese had intercourse with foreigners for so long, but we cannot assume this (cf. pp. 678 ff.).

[f] Cf. p. 602. There follows a mention of the provincial governor regularly offering prayers for wind.

[g] There follows a description of the organisation of the merchants and crew, given already, Vol. 4, pt. 1, p. 279 above. It continues with an account of the difficulties of trading in the south sea kingdoms.

[h] Khun-Lun men, no doubt.

[i] There follows the passage about the use of the compass, already quoted. The whole text concludes with details about catching fish, seeing 'dragon monsters' (whales?), and making offerings to Buddhist priests upon safe return.

---

¹ 市舶   ² 舶船   ³ 斛   ⁴ 檣   ⁵ 帆   ⁶ 席

⁷ 加突   ⁸ 逆   ⁹ 使三面風   ¹⁰ 鬼奴   ¹¹ 絮

This passage, the literary counterpart of the Bayon junk, is of much interest. For the end of the +11th and the beginning of the +12th century it attests the bulkhead build, the fore-and-aft lug (not in Western Europe until *c.* +1500), the use of taut mat-sails and beating to windward. So also do other books of the same period, such as the *Kao-Li Thu Ching* of +1124 (Illustrated Record of an Embassy to Korea) already mentioned.[a]

It is fortunate, too, that we have a document of high importance relating to the river-ships of China, almost contemporary with the description of Chu Yü. This is the painting entitled Chhing-Ming Shang Ho Thu,[1] 'Going up the River to the Capital at the Spring Festival', by Chang Tsê-Tuan,[2] made a little before +1126, when Khaifêng, the capital in question, fell to the Chin Tartars. Some relevant parts of it, depicting the vessels in the river, are shown in Fig. 976 (pl.). The detail of the painting is drawn with meticulous care, almost as if the artist had thought kindly of future historians of technology.[b] One ship is shown lowering its bipod mast before passing under the great bridge;[c] others are loading and unloading along the banks or being tracked upstream. The junks are broadly speaking of two different types, freighters with narrow sterns, and passenger-boats and smaller craft with broad ones, but both are provided with large and prominent slung rudders. These are all balanced—a remarkably advanced piece of technique which will be discussed later.[d] Large stern-sweeps and bow-sweeps are in use on two or three of the boats, some worked by as many as eight men.[e]

The history of Chang Tsê-Tuan's great work has been much discussed by scholars;[f] so famous was it that at least one Yuan emperor wrote a poem on his copy. The oldest existing copy of the painting (that reproduced, Figs. 826, 923, 976, 1034, pls.) has been published by Chêng Chen-To (3). Conserved in the Imperial Palace Museum, its silk is recognisably Sung, and among its MS. colophons it bears an inscription by a scholar of the Jurchen Chin which is dated +1186, i.e. just sixty years after the completion of the picture-scroll.[g] Scholars consider therefore that this is almost certainly the original

---

[a] Vol. 4, pt. 1, p. 280. We shall quote this later on (pp. 602 and 641) with regard to sails and to the stern-post rudder. Some of the pages of this interesting book which concern shipping have been translated by Paik (1), but he went astray on several technical points. He also (p. 92) interpreted 'the Court' as meaning the Korean Court, implying that the large ships of the Chinese ambassador (cf. pp. 430, 460 above) were Korean ships, though the text says that the governors of Fukien and the two Chekiang provinces had been responsible for their building.

[b] No less than twenty-three junks and smaller craft are to be seen in panels 5 to 13.

[c] On this bridge cf. p. 165 above.

[d] See below, pp. 655 ff.

[e] An interesting parallel to this scene from our own time is the group of tea-junks at Ganhsien in Chiangsi (type description in Worcester (3), vol. 2, p. 411) photographed by Beaton (1) during the Second World War.

[f] E.g. Waley (20) and now exhaustively by Whitfield (1). There are numerous later copies (mostly Ming), some of which may be studied in Binyon (2), pl. XLII, and (partially) in Moll (1), A, v, 8. The most sumptuous presentation is that by Priest (1) of a version attributed to Chhou Ying[3] (*fl.* +1510 to +1560). All these show the much more lofty sterns of the late Sung and Yuan periods; one of the ships is in full sail upstream, with mat-and-batten lug-sails and multiple sheets on all three masts, fore, main and mizen. Others are shown coming alongside the bank or moored, one with the sail just being lowered and falling into its accordion-like pleats. Far away, two can be seen being tracked up the river.

[g] The date of *c.* +1125 has usually been adopted because the capital fell in +1126 and the picture is not mentioned in the imperial catalogue (*Hsüan-Ho Hua Phu*) of +1120. But as Whitfield points out, the

[1] 清明上河圖   [2] 張擇端   [3] 仇英

painting. In any case we can accept it with perfect confidence as a testimony of the techniques of +12th-century shipping. The second oldest colophon, by another Chin scholar *c.* +1190, took the form of a poem, worth reproducing here:

> Through the streets carts and horses are rumbling and thronging—
> We are back in a year of the Hsüan-Ho reign-period.
> One day a Han-Lin scholar presented this painting,
> Worthy of handing down the ways and works of a peaceful time.
> Going east from the Water-gate one comes to the Canal of the Sui,
> The streets and the fields are alike incomparable
> (But Lao Tzu formerly warned against prosperity
> And today we know it has all become waste-land).
> Yet the vessels that sail ten thousand *li* on their voyages
> With rudders of timber from Chhu and their masts from Wu,
> Fine scenery north of the bridge and south of the bridge,
> Recall for a time the dream of halcyon days,
> One can hear the flutes and drums; the towers seem close at hand.[a]

Some decades later (+1178) we have Chou Chhü-Fei[1] writing on the sea-going ships of the south again.[b]

The ships which sail the southern sea and south of it are like houses. When their sails are spread they are like great clouds in the sky. Their rudders (*tho*[2]) are several tens of feet long.[c] A single ship carries several hundred men, and has in the stores a year's supply of grain. Pigs are fed and wine fermented on board. There is no account of dead or living, no going back to the mainland when once the people have set forth upon the caerulean sea. At daybreak, when the gong sounds aboard the ship, the animals can drink their fill, and crew and passengers alike forget all dangers. To those on board everything is hidden and lost in space, mountains, landmarks, and the countries of the foreigners. The shipmaster may say 'To make such and such a country, with a favourable wind, in so many days, we should sight such and such a mountain, (then) the ship must steer in such and such a direction'. But suddenly the wind may fall, and may not be strong enough to allow of the sighting of the mountain on the given day; in such a case, bearings may have to be changed. And the ship (on the other hand) may be carried far beyond (the landmark) and may lose its bearings. A gale may spring up, the ship may be blown hither and thither, it may meet with shoals or be driven upon hidden rocks, then it may be broken to the very roofs (of its deckhouses). A great ship with heavy cargo has nothing to fear from high seas, but rather in shallow water it will come to grief.[d]

This gives us relatively little technical information, and most other passages have the same defect. In all his discussion of foreign countries and their exports to China, Chao Ju-Kua,[3] for example, about +1225, has little to say concerning the ships them-

prosperity of the scene would make the previous reign-period, *c.* +1110 to +1115, a better guess, and there may be other explanations, such as the painter's youth, or his political sympathies, for the exclusion of the painting from the catalogue.

  [a] Tr. Whitfield (1), mod. auct.

  [b] *Ling Wai Tai Ta*[4] (Information on what is Beyond the Passes), ch. 6, p. 7*b*; tr. Hirth & Rockhill mod. auct.   [c] Cf. Fig. 980 on p. 481.

  [d] Excellent statement of a truth about early sailing too often unappreciated by landsmen obsessed with the illusory safety of 'coasting voyages'.

  [1] 周去非   [2] 柂   [3] 趙汝适   [4] 嶺外代答

selves,[a] though he mentions the use of whale oil on the Somali coast (Chung-Li[1]), with lime, for caulking,[b] and the names and sizes of junks at Hainan.[c] The period was one of great maritime commercial activity, and our best light on it is due to the elaborate monograph of Kuwabara (1, 1)[d] on Phu Shou-Kêng,[2] a Chinese of Arab or Persian descent,[e] who was Commissioner of Merchant Shipping[f] at Chhüanchow between +1250 and +1275, at the end of which time he transferred allegiance to the Mongols and died a great local worthy. It was just after this that Marco Polo was in China.

One of the difficulties of the argument of this Section is that lacking systematic treatments of the subject in Chinese it is necessary to fall back upon the words of a number of persons who gave, at different dates, general descriptions of Chinese nautical technology. If these are divided up in strict accordance with their content, their 'special colour', as the Chinese say, evaporates. So far we have listened to Wang Tang and Hui-Lin, the oldest, to Chu Yü, Chou Chhü-Fei and Sung Ying-Hsing.[g] We shall have to allow ourselves to be buttonholed by two more Ancient Mariners, more ancient at least than Sung Ying-Hsing—I mean Marco Polo and Ibn Baṭṭūṭah.

The former was in China during the years +1275 to +1292 and the following accounts were set down[h] after his return to Italy about +1295. On the way out, he saw the sewn ships of the Persian Gulf, 'ships of the worst kind' he called them, at Hormuz.[i] Yarn or twine from the coconut palm, or its nuts, was used, and fish-oil

[a] *Chu Fan Chih*[3] (Records of Foreign Peoples).
[b] Ch. 1, p. 27a; Hirth & Rockhill, pp. 131, 132.   [c] Ch. 2, p. 17a; Hirth & Rockhill, p. 178.
[d] There is now a Chinese translation by Chhen Yü-Ching, and further study by Lo Hsiang-Lin (1). On the commodities themselves see the monograph of Wheatley (1).
[e] Phu is probably from Abū.
[f] He had to collect customs, issue licences (cf. Vol. 4, pt. 1, p. 279), look after foreign merchants, enforce the prohibition on the export of currency, etc. The office had been founded about +700, one of the first holders being Chou Chhing-Li,[4] who was a friend of the Persian monk Chi-Lieh referred to in Vol. 1, p. 188 above. Cf. p. 453 above.
[g] One could also quote the long account by Wu Tzu-Mu in his *Mêng Liang Lu* of the shipping working out of Hangchow in +1274 just before Marco Polo arrived on the scene. A full translation of this will be found in Hirth (16), and we have quoted part in Vol. 4, pt. 1, p. 284.
[h] Ch. 15, 16; Penzer ed. p. 265. Ch. 19 in Yule (1), p. 108, whose notes (p. 117) are worth reading. He made a gallant effort to sketch the history of the rudder, but was handicapped by his inability to distinguish between rudders and steering-oars.
[i] Cf. the colour photographs by Eller (1), p. 519, of sewn boats still in use in India. Their actual ocean-going capacities are still disputed today (cf. the discussion between van Beek (1) and Hourani).
In this connection Dr R. Mauny of Dakar has brought two interesting texts to our notice. In his *Kitāb al-Buldān* (Book of the Countries) written in +889 al-Yaʿqūbī says: 'Near the mosque of Bahlūl at Massa is the anchorage of the sewn ships which are built at Ubulla and which sail as far as China' (tr. Wiet (1), p. 226). Now Massa is on the Atlantic coast of Morocco between Agadir and Cape Nun. It gives pause for thought that Arabic shipping, five centuries and a half before the Portuguese 'explorations' of this coast (cf. p. 505 below), should have connected China with the region of the Canaries. Another remarkable statement occurs in the *Murūj al-Dhahab* (Meadows of Gold) of al-Masʿūdī (+947), who says that 'teak-wood planks, pierced with holes and sewn together with coconut fibre', had been found on the coasts of Crete from wrecked ships which must have been built in the Indian Ocean. He concluded that there had been a connection with the Mediterranean, and indeed we know that the canal from the Red Sea to the Nile was re-opened in the +7th century. Cf. Casson (7).
The ancient history of this has already concerned us (pp. 356, 374 above) in relation to the origins of lockgates. The second orthodox caliph, 'Umar I (*r.* +634 to +644), prevailed upon 'Amr ibn al-Āṣ, after his conquest of Egypt in +641, to re-open the canal, which, with a re-alignment in the +2nd century under

[1] 中理   [2] 蒲壽庚   [3] 諸番志   [4] 周慶立

mixed with oakum, instead of pitch, for caulking. These ships had but one mast, one deck, and one rudder. By contrast, his admiration for the Chinese ships, of all kinds, was unbounded. He described[a] the wealth of the cities of Yangiu and Siugiu (Yang-chow and Chhuchow) with the 'marvellous great shipping' that frequented them. Of the Yangtze he said that 'more dear things, and of greater value, go and come by this river, than go by all the rivers of the Christians together, nor by all their seas'. He estimated that the lower reaches of this river were navigated by 15,000 ships, not counting the many rafts of good timber.

And all the large ships of this river are made as I shall tell you. They are covered with only one deck and have only one mast with one sail, but they are of great tonnage, for I tell you that they carry cargo for the most part from 4,000 quintals up to 12,000 (which some of them carry) in weight, by the count of our country of Venese, varying between the said numbers according to the size of the ship...[b]

Now you may know that all the ships have not all the tackle of ropes and hemp, except indeed that they have the masts and the sails rigged with them.[c] But I tell you that they have the hawsers, or to speak plainly, tow-lines, of nothing else but of canes, with which the ships are towed upstream by this river. You may know that these ships which go on this river, those which go against the current of the water, are towed because the current of water is strong, otherwise they could not go. And you understand that these canes are the thick and long canes of which I have told you, which are quite 15 paces long.[d] They take these canes and split them from one end to the other into many thin strips, and bind them the one end with the other, and make them ropes as long as they wish, twists quite 300 ells, that is paces, long, and it is much stronger than hempen rope would be,[e] with so great care are they made.[f]

Marco Polo was chiefly impressed, therefore, by the high freight capacity of the river-junks, and by the cables made of bamboo, as they still are today.

His account of the sea-going junks written in connection with his description of Zayton (Chhüanchow in Fukien) is of great interest, and we must give it in full.[g]

Trajan, had been carrying traffic till the end of the +6th. The object was the relief of a famine in the Hedjaz, and this was successfully accomplished. After the time of 'Umar II (r. +717 to +720) the canal silted up again with blown sand, and about +761 the caliph al-Manṣūr ordered it to be closed and filled in for an exactly opposite reason, the denial of supplies to the 'Alid (Shī'ah) leader Muḥammad at Medina. Though after this it never regained any importance, the existence of such a waterway long before the building of the Suez Canal must always be borne in mind in considering east–west exchanges in maritime matters (cf. p. 609 below). On the Arab reopening see Hitti (1), pp. 165, 290 and especially Toussoun (1), pp. 171 ff., pl. v, (2), pp. 230 ff., pl. xvi. The canal ran from the Pelusian arm of the Nile in the delta a little above Bubastis (mod. Zagazig), or later from the Trajanic canal parallel to it, along the Wadi Tumilat, just as road, railway and fresh-water canal do now, to Lake Timsah, after which its course southwards to the Red Sea passed through the Bitter Lakes like that of the present Suez Canal. Some remains are still visible.

[a] Ch. 147, Moule & Pelliot ed.; Penzer ed. p. 303.

[b] If the quintal be taken as about equivalent to the hundredweight, the two figures given would amount to 224 tons and 672 tons respectively.

[c] He means that many cables were made of bamboo, which he calls 'cane'.

[d] The Italian pace was very variable, but ranged around a value of about 5 ft., like the Chinese double-pace.

[e] As to this question, see above, pp. 191, 415, and below, p. 664.   f Ch. 147.

[g] It is ch. 158, Moule & Pelliot ed.; Penzer ed. p. 314. Cf. Beazley (1), vol. 3, pp. 126 ff. Some writers, such as Mukerji, have believed that the description applied to contemporary Indian ships, but they were misled by Marco Polo's loose use of the word 'Indies'. In his time, China was part of the 'Further Indies'. The description, moreover, is full of characteristics which, upon other evidence, we know to be distinctively Chinese. This mistake was disposed of by Hornell (17), p. 203.

We shall begin first of all to tell about the great ships in which the merchants go and come into Indie through the Indian sea. Now you may know that those ships are made in such a way as I shall describe unto you.

I tell you that they are mostly built of the wood which is called fir or pine.

They have one floor, which with us is called a deck, one for each, and on this deck there are commonly in all the greater number quite 60 little rooms or cabins, and in some, more, and in some, fewer, according as the ships are larger and smaller, where, in each, a merchant can stay comfortably.

They have one good sweep or helm, which in the vulgar tongue is called a rudder.

And four masts and four sails, and they often add to them two masts more, which are raised and put away every time they wish, with two sails, according to the state of the weather.

Some ships, namely those which are larger, have besides quite 13 holds, that is, divisions, on the inside, made with strong planks fitted together, so that if by accident that the ship is staved in any place, namely that either it strikes on a rock, or a whale-fish striking against it in search of food staves it in...And then the water entering through the hole runs to the bilge, which never remains occupied with any things. And then the sailors find out where the ship is staved, and then the hold which answers to the break is emptied into others, for the water cannot pass from one hold to another, so strongly are they shut in; and then they repair the ship there, and put back there the goods which had been taken out.

They are indeed nailed in such a way; for they are all lined, that is, that they have two boards above the other.

And the boards of the ship, inside and outside, are thus fitted together, that is, they are, in the common speech of our sailors, caulked both outside and inside, and they are well nailed inside and outside with iron pins. They are not pitched with pitch, because they have none of it in those regions, but they oil them in such a way as I shall tell you, because they have another thing which seems to them to be better than pitch. For I tell you that they take lime, and hemp chopped small, and they pound it all together, mixed with an oil from a tree. And after they have pounded them well, these three things together, I tell you that it becomes sticky and holds like birdlime. And with this thing they smear their ships, and this is worth quite as much as pitch.

Moreover I tell you that these ships want some 300 sailors, some 200, some 150, some more, some fewer, according as the ships are larger and smaller.

They also carry a much greater burden than ours.

And formerly in time past the ships were larger than they are now at present, because the violence of the sea has so broken away the islands in several places that in many places water was not found enough for those ships so great, and so they are now made smaller, but they are so large that they carry quite 5,000 baskets of pepper, and some 6,000.[a]

Moreover I tell you that they often go with sweeps, that is, with great oars, and four sailors row at each oar.

And these larger ships have such large tenders that they carry quite 1,000 baskets of pepper. But I tell you that they take 40, 50, some 60, some 80, some 100 sailors, and these go with oars and with sails when there is opportunity. And often again they help to tow the great ship with ropes, that is, hawsers, when they are moved with oars, and also when they are moved with sails, if the wind prevails rather from the beam, because the smaller go in front of the larger, and tow it tied with ropes, but not if the wind blows straight (abaft), for the sails of the larger

---

[a] One would very much like to know the weight and dimensions of one of these units. Elsewhere (Latham ed. p. 224), Marco Polo says that the Chinese ocean-going ships had a draught of about 4 paces—equivalent to some 20 ft.

ship would prevent the wind from catching the sails of the smaller, and so the larger would overtake the smaller. They take 2 or 3 of these large tenders, but the one is larger than the other. And of small ships which we call boats, also they take quite 10, to anchor and to catch fish and to wait upon the large ship in many other ways. And the ship carries all these boats through the water lashed to her sides outside, and when necessary they put them in the water, but they tow the two large ones astern, which each have their mariners and their sails and all that is needed for themselves and for them. And again I tell you that the said two large tenders also carry small boats.

Moreover I tell you again that when the great ships wish to be decorated, that is, to be repaired, and it has made a great voyage or has sailed a whole year or more and needs repair, they repair it in such a way. For they nail yet another board over the aforesaid original two all round the ship, without removing the former at all, and then there are three of them over the whole ship everywhere, the one nailed above the other, and then, when it is nailed, they also caulk and oil it with the foresaid mixture, and this is the repair which they do. And at the end of the second year, at the second repair, they nail yet another board, leaving the other boards, so that there are four. And in this way they go each year from repair to repair up to the number of six boards, the one nailed upon the other. And when they have six boards the one upon the other nailed, then the ship is condemned and they sail no more in her on too high seas but in near journeys and good weather, and they do not overload them; until it seems to them that they are no more of any value, and that one can make no more use of them. Then they dismantle and break them up.

For +13th-century junks, then, Marco Polo attests cabins (naturally the first thing a travelling merchant would notice), rudder (this had already been in use for eighty years or so in Europe), multiple masts (not yet used in Europe), and bulkhead-built hull. He makes a particular point of the repair of the ship by the continual overlaying with new layers of caulked strakes.[a] This system of superimposed timbers ('doubling') was afterwards employed in European warships of the +18th and early +19th centuries.[b] The traveller duly mentions the use of tung-oil and lime (*chunam*),[c] and significantly exclaims at the size and tonnage of the ships, which he thought remarkable.[d] The oars rowed by four men each may well have been yulohs, though the description leaves it uncertain. As for sailing, it would appear that for some reason or other the smaller ships could sail better to windward than the larger ones, so that under such circumstances towing by the smaller tenders was resorted to. Marco Polo seems unduly astonished at the accompanying of the great ship by so many smaller boats, pinnaces and dinghies as we should say, and one can only suppose that such arrangements were uncommon in the Mediterranean of his day.

When Marco Polo left China in +1292 (it must have been a hard parting for him, after seventeen years, even though he was going as imperial envoy-extraordinary in

---

[a] This was mentioned also by several other later travellers and writers, e.g. Jordanus Catalanus (+1322); de Conti (about +1438); Duarte Barbosa (*c.* +1520); de Castanheda (+1554); the number of layers ranging up to seven. The practice goes back to the +9th century at least, as we know from Ennin (Reischauer (2), p. 8).

[b] G. Holmes (1), p. 133; Adm. Smyth (1), p. 258.

[c] This alone would be enough to establish the cultural area from which Marco Polo's ships came. Nobody else had tung-oil.

[d] So also Nicolò de Conti: 'They doe make bigger Shippes than wee do, that is to say, of 2000 tons, with five sayles, and so many mastes' (Penzer ed. p. 140). Cf. Cordier (5).

charge of a princess), 'The Great Khan caused to be armed and set forth 14 great ships, and every one of them had four masts... In every ship he put 600 men, and provision for two years...'[a] It was almost certainly a much braver flcct than any European country, including the England of Edward I, and the France of St Louis, could have launched for the occasion.

While Marco Polo's words are fresh in our minds, let us listen to the description of the great Arab traveller, Ibn Baṭṭūṭah, who was in China just over half a century later (+1347).[b] His opening statement is significant.

People sail on the China seas only in Chinese ships, so let us mention the order observed upon them.

There are three kinds: the greatest is called 'jonouq', or, in the singular, 'jonq' (certainly *chhuan*[1]); the middling sized is a 'zaw' (probably *tshao*[2] or *sao*[3]); and the least a 'kakam'.[c]

A single one of the greater ships carries 12 sails, and the smaller ones only three. The sails of these vessels are made of strips of bamboo, woven into the form of matting. The sailors never lower them (while sailing, but simply)[d] change the direction of them according to whether the wind is blowing from one side or the other. When the ships cast anchor, the sails are left standing in the wind.[e]

Each of these ships is worked by 1,000 men, 600 sailors and 400 marines, among whom there are archers and crossbowmen furnished with shields, and men who throw (pots of) naphtha.

Each great vessel is followed by three others, a 'nisfi', a 'thoulthi' and a 'roubi'.[f]

These vessels are nowhere made except in the city of Zayton[g] (Chhüanchow) in China, or at Sin-Kilan, which is the same as Sīn al-Ṣīn (Canton).

This is the manner after which they are made; two (parallel) walls of very thick wooden (planking) are raised, and across the space between them are placed very thick planks (the bulkheads) secured longitudinally and transversely by means of large nails, each three ells in length.[h] When these walls have thus been built, the lower deck is fitted in, and the ship is launched before the upper works are finished.

The pieces of wood, and those parts of the hull, near the water (-line) serve for the crew to wash and to accomplish their natural necessities.[i]

On the sides of these pieces of wood[j] also the oars are found; they are as big as masts, and are worked by 10 to 15 men (each), who row standing up.[k]

[a] Ch. 3, Penzer ed. p. 24; Moule & Pelliot ed. vol. 1, p. 90.
[b] Defrémery & Sanguinetti (1), vol. 4, pp. 91 ff; Lee (1), p. 172.
[c] This word may have come from *ko-kang*,[4] or might indeed be the Italian *cocca* or French *coque*; cf. cog as a ship-name.
[d] This excellent interpretation is due to P. Paris (1).
[e] This probably means simply that in a light breeze the ship was brought up into the wind before casting anchor; if the breeze then died down, the sailors might not trouble to lower sail. It cannot possibly mean that the sails were always left hoisted in harbours.
[f] Suggestions for these tenders might be *i-fu*,[5] a pinnace for going ashore; *tho-tieh*,[6] a small boat fitted with a rudder; and *jao-pu*,[7] a rowing-boat.
[g] Elsewhere, he says 'Zayton is one of the largest in the world—no, I am wrong, it *is* the largest of all ports. I saw about 100 large junks, with small vessels innumerable'. Defrémery & Sanguinetti (1), vol. 4, p. 269; Lee (1), p. 212. Cf. D. H. Smith (1).
[h] This is not so exaggerated as it seems at first sight if Ibn Baṭṭūṭah was referring to long clamps.
[i] This is obscure, unless perhaps the stern galleries are meant.
[j] Here gunwales must be meant.
[k] Probably yulohs (self-feathering propulsion oars). Cf. pp. 622 ff. below.

[1] 船   [2] 䑓   [3] 艘   [4] 舸舩   [5] 艤艑   [6] 柂艓   [7] 橈艀

The vessels have four decks, upon which there are cabins and saloons for merchants. Several of these 'misriya' contain cupboards and other conveniences; they have doors which can be locked, and keys for their occupiers. (The merchants) take with them their wives and concubines. It often happens that a man can be in his cabin without others on board realising it, and they do not see him until the vessel has arrived in some port.[a]

The sailors also have their children in such cabins;[b] and (in some parts of the ship) they sow garden herbs, vegetables, and ginger in wooden tubs.

The Commander of such a vessel is a great Emir; when he lands, the archers and the Ethiops[c] march before him bearing javelins and swords, with drums beating and trumpets blowing. When he arrives at the guesthouse where he is to stay, they set up their lances on each side of the gate, and mount guard throughout his visit.

Among the inhabitants of China there are those who own numerous ships, on which they send their agents to foreign places. For nowhere in the world are there to be found people richer than the Chinese.

Ibn Baṭṭūṭah had had his experiences with these ships. At an Indian port the unfortunate man embarked, with a number of his concubines, on a junk, but all the suitable cabins had been reserved by Chinese merchants, so the party transferred to a *kakam*; then before he himself went on board, the junk with the presents for the emperor sailed out into a storm and was lost with all hands. The captain of the *kakam* then also left without him, and he never recovered any of the girls or his valuable merchandise. And on the way home he experienced a tempest and escaped a 'rukh'.

The Moor confirms the Venetian in a number of particulars, such as the cabins, the multiple masts, and the bulkheads, concerning which he speaks as if he had himself visited the Cantonese shipwrights' yards.[d] He was also impressed by the number of subsidiary boats. But he complements Marco Polo most usefully. He attests the great mat-and-batten lug-sails, much greater in number than were carried by any European or Arab craft of the time, and their capacity to make use of wind coming from almost any quarter. And he tells us about the huge oars worked by several men, which must have been yulohs,[e] as we see by his development of the subject:[f]

This sea (on the way to China) has no wind, waves, or motion, in spite of its great extent; hence each Chinese junk is accompanied by three boats as we have already said. They serve to make it proceed by rowing and towing. Besides, there are in the junk about 20 very great oars, like masts, each of which have about 30 men to work them, standing in two rows facing each other. The oar, like a club, is provided with two strong cords or cables, and one of the two rows of men pulls on the cable and then lets it go, while the others pull on the second cable. And as they work, these rowers raise good voices in a chanty, generally saying ''la, 'la, 'la 'la'.

The rope connections make the identification fairly certain.[g]

---

[a] Evidently the pursers were not as efficient as the navigation department.

[b] This is still the case on thousands of Chinese ships; the captain's cabin is the family home. So it was on Captain Wu's Upper Yangtze junk (p. 397 above).

[c] Cf. the 'black slaves' of pp. 459, 462 above. Probably these men-at-arms were Malays in Ibn Baṭṭūṭah's time.

[d] Elsewhere he describes the ship-caulkers' quarter at Hangchow (Lee (1), p. 219).

[e] McRobert (1) drew attention to this passage, and though the identification has been questioned (Waters, 6), he seems to be right.

[f] Defrémery & Sanguinetti (1), vol. 4, p. 247; Lee (1), p. 205.

[g] See below, p. 622.

In the meantime, Marco Polo's information (and that of others less well known) had been spreading in Europe. It is agreed that the famous Catalan world-map[a] of +1375 and the world-map of Fra Mauro Camaldolese[b] of +1459 were based upon this information. Here we are not concerned with the geographical material, but with the small drawings of ships which appear, most fortunately, on both these documents.[c] Tracings of them are given here taken from the edition of Santarem (2).

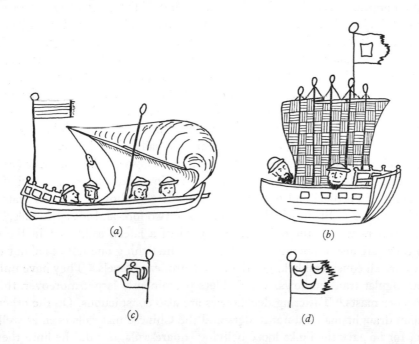

(a)   (b)

(c)   (d)

Fig. 977. Ships, Chinese and Western, on the Catalan world-map of +1375; tracings from Santarem (2).
a   the European ship near the Canary Is.
b   one of three large Chinese junks in the eastern seas, all five-masted with taut sails
c   flag flown by the junk-like ship in the Caspian Sea, and over all Russian-Mongolian cities
d   flag of three crescents marking all Chinese cities from Zayton to Lop

The eastern portion of the Catalan map (see Fig. 977) shows in the seas three large vessels which differ from one another only in slight details. They are clearly and

a The Carta Catalana is in the Mazarin Gallery at the Bibliothèque Nationale in Paris (MS. Espagnol 30; cf. Yule (2), vol. 1, pp. 299 ff.; Anon. (47), p. 14). It was drawn for King Charles V of France by Abraham Cresques, the great Jewish cartographer and instrument-maker of Majorca. His son Jafuda (Mestre Jacome de Malhorca) entered the service of Portugal about +1425 and must have worked for Prince Henry the Navigator (cf. p. 503 below). A good copy of the map may be seen in the Maritime Museum at Barcelona. Besides Marco Polo, Odoric of Pordenone (cf. Vol. 1, pp. 189 ff.) was another probable source of the Cresques' information. On the East Asian names in this world-map see Cordier (6) and the special glossary of Hallberg (1).

b See Zurla (1). Besides Marco Polo, Nicolò de Conti (cf. Yule (2), vol. 1, pp. 174 ff.) was an important source for Fra Mauro. East Asian names on this map are also in Hallberg's dictionary (1).

c A third map, dating from +1445, in the Estense Library at Modena, has been studied by Kretschmer (3). It is like Fra Mauro's, but the drawings of the ships are less clear. However, they agree with what Fra Mauro said: 'Le Nave over zonchi che navegant questo Mar portano quatro albori...'

recognisably junks.[a] They have the transom bow and stern, the rails of a stern-gallery, portholes,[b] and notably as many as five masts, with unmistakable mat-and-batten sails, though their cut and rig was evidently not understood by the draughtsman.[c] He was also vague as to where they ought to be placed in the seas of 'the Indies'. One of them is, reasonably, off Java, a second is off the Rann of Cutch (Gujerat), but the third is in the Caspian. The two former are flying an ensign with a square device, and this is the flag which appears over all the Persian cities; that of the cities of China bears three crescents. The junk in the Caspian flies a flag with an indecipherable sign looking like an attempt at Arabic writing, and this is common to all the Russian (Mongol?) cities. Too much importance should probably not be attached to these confusions, for the build of the junks is not to be mistaken. Fortunately, also, the draughtsman provided posterity with a sketch of a European ship near the Canaries, off the African coast. It resembles the Norman ships of the Bayeux Tapestry, and is evidently descended from the Viking longships; its stem- and stern-posts are well shown, and it has a great bellying square-sail oddly in contrast with the stiff flat sails of the junks.

In the following century the map of Fra Mauro shows similar contrasts (Fig. 978). The western section has several sharp-stemmed ships with square-sails, including one off the coast of Portugal and one in the Baltic. It also shows a curious craft by the north-west corner of Spain which seems to have two fore-and-aft sprit-sails,[d] and in Egyptian waters we see the obliquely set yard of a lateen sail.[e] But in the eastern section there are also numerous ships,[f] and the first thing one notices about them is that they are all considerably larger than the European vessels.[g] They have unmistakable rectangular transom bows, and rudders prominently large; moreover, they have four or more masts.[h] Towering deckhouses are also conspicuous. On the other hand, the Italian draughtsman did not understand the Chinese mat sails even as well as the Catalan, for he gave the junks loose bellying square-sails, nor did he limn them very justly. These are in the Indian Ocean, but far to the north, in the Yellow Sea, we find a smaller ship of the same type, towing behind her one of those tenders or pinnaces

[a] As has often been realised (Waters (2); Worcester (3), vol. 1, p. 16). Jal (1), vol. 1, pp. 39 ff., came near to doing so more than a hundred years since, but he demanded too much of the +14th-century draughtsman and refused to recognise either a European ship or a Chinese junk in the two types of drawing on the map.

[b] At least a century earlier than European ships had them, according to Clowes (1).

[c] Probably he could not imagine any sails other than square-sails.

[d] This would not be surprising, as the earliest generally accepted date for them in Europe (but see p. 615) is +1416 or a little before. For Turkish craft of very similar appearance, see Moore (1), figs. 152, 154.

[e] Further on this, see pp. 609 ff. below.

[f] They have long been recognised as junks, e.g. by Clowes (1).

[g] 'The Vessels which they navigate in Cathay', wrote Jordanus Catalanus, about +1325, 'be very big, and have upon the ship's hull more than 100 cabins, and with a fair wind they carry ten sails; and they are very bulky, being made of three thicknesses of plank, so that the first thickness is as in our great ships, the second crosswise, and the third again longwise; in sooth, 'tis a very strong affair' (Yule (3), pp. 54 ff.).

[h] Actually the use of more than one mast in Asia was a very old tradition, if we may judge from the coins of the Āndhra dynasty (c. −230 to +225), contemporary with the Han (Mukerji (1), p. 50; Schoff (3), p. 244; Chakravarti (1), p. 59). Some of those of King Pulumāyi's reign (c. +100) clearly show vessels with two masts of about equal height. In the +3rd century Wan Chen and Khang Thai talked of vessels carrying up to seven sails on at least four masts (see below, pp. 600ff.).

Fig. 978. Ships, Chinese and Western, on the world-map of Fra Mauro, +1459; tracings from Santarem (2).

a–d   vessels shown in European seas
e–g   vessels shown in the eastern seas

a   a ship off the west coast of Portugal showing marked stem-post, and square-sails on main and (apparently) mizen masts
b   a ship with evident stem-post and curious stern off the north-west cape of Spain, apparently an attempt to represent two sprit-sails
c   bellying square-sails in the Baltic
d   obliquely set yard of a lateen sail off Egypt
e   a ship in the Indian Ocean west of Ceylon with four masts, transom bow and prominent rudder
f   another of the same kind in a more northerly position, also showing the towering poop superstructures
g   a very square-ended junk in the Yellow Sea north of Shantung, towing a boat just as Marco Polo described

which had so much impressed Marco Polo. And indeed very near the two best junks a small scroll inscription in the middle of the Indian Ocean bears words which obviously derive from him.[a]

The importance of the discovery by Europeans that really large ships with multiple masts had been built and could do useful work has been emphasised by Clowes (1).[b] Of course there had been two- and three-masted ships in Hellenistic times (+1st to +3rd centuries), for the raking bowsprit called the *artemon* gradually turned into a foremast, and a small mizen was added aft,[c] but these did not survive the collapse of the Roman Empire. The earliest precisely datable medieval three-masted ship in Europe is found about the same time as the map of Fra Mauro, i.e. in +1466, on the seal of Louis de Bourbon.[d]

It was the introduction of the three-masted ship [says Clowes] with its improved ability to contend with adverse winds,[e] which made possible the great voyages of discovery of the end of the +15th century, of Columbus to the West Indies, of Vasco da Gama to India, and of the Cabots to Newfoundland; and it is a curious thought that this great development may really have been due to the introduction into Europe of accounts of the multiple-masted Chinese junks which traded so effectively in the Indian Ocean... at any rate, it is a fact that no one has yet been able to explain satisfactorily the cause of the extraordinary and rapid development— in the course of only a hundred and fifty years—of the one-masted vessel of +1350, able only to run before the wind, into the three- or four-masted ship of +1500, similar in all essential principles of rig to the three-masted ship of the +17th century.[f]

In connection with this we must recall that Chinese sea-going ships reached their apogee in the voyages of discovery under Chêng Ho in the first half of the +15th century, just in the lifetime of Prince Henry. To this remarkable fact we shall shortly return (p. 487). But who could the intermediaries have been? Presumably Nicolò de Conti (who probably visited South China in +1438) and other travellers of his time.[g]

---

[a] Quoted below, p. 572.

[b] The wording in the following quotation from him has been amended so as to conform with modern knowledge.

[c] One can see them on the Ostia mosaics, and in an Etruscan tomb-painting (Moretti, 1). Bowen (9), pp. 274 ff., believes, in view of the dates of Roman–Indian commercial contacts (cf. Vol. 1, pp. 176 ff.), that in this case also the idea of multi-masted rigs was brought west from the Indian Ocean. Conversely, the *artemon* itself appears both at Ajanta and at Borobodur (Figs. 967, 973, pl.).

[d] But there is some evidence that the larger Portuguese caravels had carried three masts as early as +1436 (see da Fonseca, 1). Until the time of the death of Prince Henry, lateen sails were always used on them; after that, lateen sails and square-sails were combined in mixed rigs. Two-masted ships go back in Europe to the +13th century (see Lethbridge (1), fig. 533, from the *Lapidario* of Alfonso X), and they may have derived their foremast from the old Roman *artemon* (see Lethbridge (1), fig. 530a, from the Pisa campanile, datable at +1174).

[e] This was because the mizen mast gave opportunity of using a lateen sail, which permitted the ship to sail nearer the wind than was possible with square-sails alone. Or such at least is the opinion of some experts, but others (e.g. Cdr. George Naish, in private communication) believe that the mizen lateen made the bigger ships handier in steering, but did not enable them to sail closer to the wind. Adam & Denoix, for their part, emphasise the importance of the stern-post rudder in this evolution, (1), p. 103, and regard it as one of the cardinal factors which permitted the adoption of multiple masts by European ships. Cf. p. 637 below.

[f] Cf. another quotation from Clowes on p. 610; and Clowes (2), p. 71.

[g] Between +1450 and +1550 there seems also to have been a great increase in the tonnage of European ships, accompanying the increase in the number of masts. It would be interesting to investigate this trend statistically. Cf. Gibson (2), pp. 110 ff., 121 ff.; and Baratier & Reynaud (1) and Mollat (1), both reviewed in Anon. (24).

It is a curious fact that just as the Europeans were struck by the larger size of the Chinese ships, so also the Chinese were (or had been) under the impression that the ships of the Far West were also larger than their own. In +1178 Chou Chhü-Fei had written:[a]

Beyond the Ocean west of the Arab countries there are countless countries more, but Mu-Lan-Phi[1] is the only one which is visited by the great ships (*chü hsien*[2]) of the Arabs. Its ships (*chou*[3]) are the biggest of all. Putting to sea from Tho-phan-ti[4] in the Arabic country one sails due west for full an hundred days and so arrives. A single ship carries a thousand men; on board there are stores of wine and provisions, looms and shuttles (for weaving), and a market-place. If it does not encounter favourable winds it does not get back to port for years. No ship but a very big one could make such voyages. So nowadays the term 'Mu-Lan' ship is used (in China) for designating the largest kind of junk. If one speaks of big ships, there are none as large as those of Mu-Lan-Phi.

And all this was repeated by Chao Ju-Kua in +1225.[b] But there was a seemingly legendary element here, for the accounts go on to mention grains of wheat three inches long, melons six feet around, and sheep the fat of which could be harvested from time to time by surgical operation. It seems that people at both ends of the Old World thought that the other end had the largest ships, but objectively the Europeans were right in this opinion while the Chinese seem to have been wrong.[c]

Or were they? There may be more than meets the eye in the story of Mu-Lan-Phi. The usual identification of the place accepted by sinologists has been Spain, the name deriving from the Almoravid dynasty (+1061 to +1147) of the al-Murābiṭūn. But the botanist Li Hui-Lin (1), perceiving that the time of 100 days seems impossibly long for an east–west Mediterranean transit,[d] suggests that in fact the journey was a transatlantic crossing, and that the strangeness of the plants and animals described may conceal species typical of the Americas.[e] If one follows him in taking the descriptions seriously, the huge cereal grains which keep so long in storage must be maize, the melons could be *Cucurbita pepo*, which can weigh as much as 240 lb., the unheard-of fruits could be the pineapple and the avocado, while the tall 'sheep' might perhaps be llamas and alpacas. Li Hui-Lin associates the idea of Arab transatlantic navigations with an old story reported by al-Idrīsī that in the +10th century some Spanish Muslim sailors set out westwards from Lisbon but were never seen again.[f] The greatest difficulties in accepting Li's suggestion arise, however, on the nautical side, for everything that we know of the sewn ships of Arabic culture (cf. pp. 388, 465) precludes us from believing that they could ever have been built stoutly enough to withstand a return transatlantic

a *Ling Wai Tai Ta*, ch. 3, p. 4a, and ch. 6, p. 8a; tr. Hirth & Rockhill (1), pp. 34, 142; mod. auct.
b *Chu Fan Chih*, ch. 1, p. 30a. With embroideries, e.g. 'several thousand men'.
c Cf. Parkinson (1), p. 321: 'Europeans first arrived in the +16th century in crank and leaky ships, vastly inferior to the Chinese junks, and a great deal smaller.'
d He identifies Tho-phan-ti as Dimyāṭ (Damietta, on the shore of the Nile delta). But, on his theory, the name Mu-Lan-Phi is unidentifiable.
e Presently (pp. 540 ff. below) we shall have more to say about pre-Columbian contacts with the American continent.
f Dozy & de Goeje (1). Cf. pp. 503, 511.

¹ 木蘭皮   ² 巨艦   ³ 舟   ⁴ 陁盤地

crossing. Moreover, in order to get back to Europe the Muslim sailors would have had to discover the régime of winds and currents in the Atlantic which was laid bare five centuries later by the Portuguese, and we have no evidence whatever that they did so. For the present, then, we may retain the view that Chou Chhü-Fei and Chao Ju-Kua were talking about the Mediterranean, not the Atlantic, and the ships, large perhaps but slow, which sailed therein.

Probably none of the travellers in Sung and Yuan China had sufficient historical perspective to realise that in the Southern Sung a great event had occurred, the creation of the Chinese navy.[a] The development of the maritime south had been a sociological consequence of the wars, invasions and political unrest, even of the climatic changes, in the north, which drove down masses of the population to the Fu and Kuang coastal provinces with their innumerable rivers, fjords, creeks and havens. Since agriculture supported the people less readily here, commercial cities, actively backed by the State, began to flourish, and this encouraged shipbuilding, navigation and all else that follows when men, for trade or defence, go down to the sea in ships. Thus it came about naturally that in the first half of the +12th century, after the fall of Khaifêng and the removal of the capital to Hangchow, when government was centred in the southeastern quarter of China, the rise of a permanent navy first took place. China must now, wrote Chang I[1] in +1131, regard the Sea and the River as her Great Wall, and substitute warships for watch-towers.[b]

The first Admiralty was set up in the following year at Ting-hai under the name Yen Hai Chih-Chih Shih-Ssu[2] (Imperial Commissariat for the Control and Organisation of Coastal Areas).[c] From a total of 11 squadrons and 3,000 men it rose in one century to 20 squadrons totalling 52,000 men, with its main base near Shanghai. The regular striking force could be supported at need by substantial merchantmen; thus in the campaign of +1161 some 340 ships of this kind participated in the battles on the Yangtze. The age was one of continual innovation; in +1129 trebuchets throwing gunpowder bombs were decreed standard equipment on all warships,[d] between +1132 and +1183 a great number of treadmill-operated paddle-wheel craft, large and small, were built, including stern-wheelers and ships with as many as 11 paddle-wheels a side (the invention of a remarkable engineer Kao Hsüan[3]),[e] and in +1203 some of these were armoured with iron plates (to the design of another outstanding shipwright Chhin Shih-Fu[4]).[f] In Kao Tsung's time, about +1150, China was sea-minded as never before, unless indeed one might say that the ancient spirit of Wu and Yüeh had now come into its own,[g] so that a typical Nan Sung scholar such as Mo Chi,[5] the

---

[a] We owe to Lo Jung-Pang (1) a noteworthy pioneer study of this process. Cf. also Din Ta-San & Munido (1), which has not been available to us.

[b] Li Tai Ming Chhen Tsou I, ch. 334, p. 5.

[c] Sung Shih, ch. 167, p. 3b.

[d] Sung Hui Yao Kao (Ping sect.), ch. 29, pp. 31, 32. See further on this whole subject Sect. 30k below.

[e] See Lo Jung-Pang (3). The story has already been told in Vol. 4, pt. 2, pp. 418 ff. above.

[f] For the development of metal armament in Chinese and Korean naval history see pp. 682 ff. below.

[g] An interesting collection of ancient texts illustrating the sea-faring qualities of the Yüeh peoples will be found in Wei Chü-Hsien (4), pp. 47 ff.

[1] 章宜   [2] 沿海制置使司   [3] 高宣   [4] 秦世輔   [5] 莫汲

Director of the Imperial University, used to go out sailing when free from office, and compelled his crews to follow him far to the north.[a] In sum, the navy of the Southern Sung held off the Chin Tartars and then the Mongols for nearly two centuries, gaining complete control of the East China Sea. Its successor, the navy of the Yuan, was to control the South China Sea also—and that of the Ming the Indian Ocean itself.

(3) FROM THE YUAN TO THE CHHING

Under the Mongolian rule in the Yuan dynasty naval operations were particularly prominent.[b] First, owing to the nature of the country, the reduction of the resistance of Sung forces in the south had necessarily to take the form of coastal and river fighting. The campaign of +1277 involved large fleets on both sides, and in the final naval battle two years later near Canton, which had been the last temporary Sung capital, no less than 800 warships were captured by the Mongols. The death of the nine-year-old emperor and of his ministers with their families on this occasion occurred because their junk was too large and heavily-laden to escape in the fog with the rest of their squadron. But all this was but the beginning of the naval activities of the unexpectedly sea-minded Mongol government. At the same time as the war against the Sung in South China, Khubilai Khan's urge for world dominion was impelling him to engage in a series of formidable expeditions against Japan. In that of +1274 the fleet was composed of 900 warships, which transported a quarter of a million soldiers across the sea. In +1281 a larger armada of 4,400 ships set sail, but each time the Japanese, aided by typhoons and bad weather, succeeded in repulsing the invaders and inflicting on them very great losses.[c] The emperor intended to mount a third attack in +1283 but had to desist from it on account of the strength of popular disapproval. Undeterred in other directions, however, he despatched a fruitless expedition to Champa in +1282 and another (of 1,000 ships) to Java in +1292; though these were large operations they were too far from home to effect anything permanent. Lastly in +1291 there took place an abortive attempt to appropriate the Liu-Chhiu Islands. Unfortunately, the historical sources which tell us of the extensive naval activities of the Yuan period have never been investigated from the point of view of nautical technology, and much may be expected when this attempt is made. We can at any rate be sure of our ground in viewing the Yuan navy as the continued development of what had been started in the Sung, and the predecessor of the glories of the early Ming.

[a] The account of Mo Chi is in *Chhu Tung Yeh Yü*, ch. 17, pp. 22a ff.

[b] A convenient summary is given by Cordier (1), vol. 2, pp. 296 ff. As early as +1270 the minister Liu Chêng[1] had advocated a powerful navy (*Yuan Shih*, ch. 161, pp. 12a ff.). In +1283 alone no less than 4,000 warships were built.

[c] Interesting illustrations of the ships of the Yuan navy occur in a scroll painting attributed to a contemporary Japanese artist Takezaki Suenaga[2] and entitled Mōko Shūrai Ekotoba[3] (Poetical Portrayal of the Mongol Invasion). Some of these have been reproduced in Ikeuchi Hiroshi (1). This is the scroll which is so important in the history of gunpowder weapons (cf. Sect. 30k). A small excerpt is reproduced in Purvis (1), fig. 8.

[1] 劉整　　[2] 竹崎季長　　[3] 蒙古襲來繪詞

When the Sung empire was finally conquered the sailors in the Mongol service were called upon to perform a new task, the shipment in guarded convoy of grain supplies from the southern provinces to the northern capital.[a] As far back as the Sui the tribute grain had been carried by a fairly adequate system of waterways from the region of Nanking north-westwards, but now, in +1264, Khubilai Khan fixed his capital far in the north at or near modern Peking, and until the Grand Canal could be remodelled to cope with the new transport requirements,[b] the stability of the new dynasty itself depended upon the success of an alternative route. So successful in fact did the sea route become that an acrimonious controversy arose between the protagonists of the 'blue water' and the canal routes, which lasted for fifty years, much longer indeed than the reign of the Great Khan. We know about it largely because of the survival of an official collection entitled *Ta Yuan Hai Yün Chi*[1] (Records of Maritime Transportation of the Yuan Dynasty) which originally formed part of *Yuan Ching Shih Ta Tien*, and a smaller work, Wei Su's[2] *Yuan Hai Yün Chih*[3] (A Sketch of Sea-Transport during the Yuan).[c]

The first success of the naval service occurred in +1282. A fleet of a hundred and forty-six vessels was gathered by two former privateer commanders who had joined the Mongol forces in the coastal campaign against the Sung, Chu Chhing[4] and Chang Hsüan,[5] and another naval officer Lo Pi[6].[d] After wintering in a Shantung port they unloaded some 3,230 tons of grain at the mouth of the Wei River near modern Tientsin. Very soon the grain transported came to equal that brought up by the waterways, some 19,800 tons, but party politics intensified, especially after the loss of a great grain fleet in a typhoon in +1286, and Chu and Chang were removed from their commands while work on the canals was pushed on more energetically. Nevertheless throughout the dynasty the sea route remained the more effective, and in +1291 the two old pirates, now admirals, regained control of it.[e] Although they did not long survive the death of the Great Khan, their successors the Muslims Qobis and Muḥammad (Ho-Pi-Ssu[7] and Ma-Ha-Mo-Tê[8]) carried the service to still greater efficiency, reaching a record annual shipment of some 247,000 tons in +1329.[f] After that date the sea transport gradually declined, partly because of increased use of the canals, partly because of foreign pirates, and with the coming of the Ming dynasty the capital shifted again to

a This resumed an +8th-century practice; cf. p. 453 above. An account of the whole history of the sea route is given by Wu Chhi-Hua (*1*).

b This story has already been sketched on pp. 311 ff. above.

c These and other sources have been analysed and drawn upon in the excellent study of Lo Jung-Pang (*4*). See also Schurmann (*1*), pp. 108 ff., who gives a translation of the cardinal source, *Yuan Shih*, ch. 93, pp. 14*b* ff.

d We have met with him already (p. 420) as an outstanding shipbuilder.

e From +1293 onwards the fleet stood well out to sea east of the Shantung peninsula, reducing the time between the Yangtze and Tientsin to ten days. Hence probably the significance of the sailing-directions already mentioned (Vol. 4, pt. 1, p. 285). About this time the large sea-going junks were carrying loads of some 640 tons each. Cf. pp. 412, 452, 466.

f It is interesting to compare this figure with that for the record period centering upon +735 (cf. p. 310 above), 165,000 tons, all canal-borne.

¹ 大元海運記   ² 危素   ³ 元海運志   ⁴ 朱清   ⁵ 張瑄
⁶ 羅璧   ⁷ 和必斯   ⁸ 瑪哈默德

Nanking.[a] Even in later times, however, after its return to Peking in +1409, the sea route never regained the predominance which it had had in the days of the Yuan navy.[b]

The archaeological centre-piece of this period was quite unknown to us when this Section was first drafted in 1953, for it still reposed deep in the mud of a swampy dead tributary of the Yellow River at Liang-shan Hsien,[1] about 200 miles from Chinan. It was found three years later by countryfolk planting lotus roots. The village school-master recognised its value as the complete hull of a +14th-century ship, and by the time the provincial archaeologists arrived the peasants were enthusiastically digging it out. The vessel is now preserved in a special building in the Shantung Provincial Museum at Chinan (Fig. 979, pl.).[c] There is no doubt about the dating, for the anchor bears an inscription of +1372, and a bronze cannon another of +1377. The hull is typically Chinese in character, transom-ended with thirteen bulkheads, and very long and narrow, about 66 ft. from stem to stern, and about 10 ft. in the beam. One can make out the emplacement for the slung rudder and the remains of two masts. The vessel seems built for speed,[d] and as the remains of helmets and other accoutrements were found in her, it is believed that she was a government patrol boat of the naval police on the Grand Canal and associated waters. Although not one of the greater craft of the time, this relic is of deep interest because contemporary with the Catalan world-map and only a few decades later than the time when Ibn Baṭṭūṭah was in China.

Of what was accomplished by the peaceful maritime expeditions of the Ming in the early +15th century much has already been said,[e] and we must shortly take another view of them in the perspective of the Portuguese voyages of exploration which were proceeding exactly at the same time. But a few details may be given here about the shipbuilding aspect of that remarkable navy which Chêng Ho[2] commanded, and which may have influenced Europe much more than has generally been supposed. Chêng Hao-Shêng (1) tells us that the yards, which were mostly on the Yangtze near Nanking, were at the height of their activity between +1403 and +1423. They were called the Pao Chhuan Chhang.[3] Their first order was for 250 vessels (hsien po[4]), many larger than ever previously built.[f] Authority varied, sometimes it was military (the

---

[a] We shall see later on (p. 526) that the preference for canal transport, after the capital had been removed to the north, was one of the factors which stopped oceanic shipbuilding and played a part in the discontinuance of the expeditions of Chêng Ho. On the sea route in comparison with the canal route in the Ming period see the monograph of Wu Chhi-Hua (2).

[b] Neither in Ming nor Chhing did the annual shipments exceed 106,000 tons. The record for steamer transport was in 1909 with 212,000 tons. Such figures give a striking impression of the scale of the Yuan operations. After the +14th century the Liaoning region of Manchuria supplied much grain.

[c] Where I had the pleasure of studying it with Dr Lu Gwei-Djen in June 1958. We are much indebted to the Vice-Director, Dr Chhin Kang-Chhing, for his kind assistance here. The Liang-shan ship has been described by Liu Kuei-Fang (1), and some less satisfactory photographs of it have appeared in Li-Shih Chiao-Hsüeh for May 1958 (no. 5).

[d] The breadth of the ship's bottom narrows to only 3 ft. or so.

[e] Notably in Vol. 1, pp. 143 ff. and Vol. 3, pp. 556 ff.

[f] So at least says the Shu Yü Chou Tzu Lu[5] by Yen Tshung-Chien,[6] ch. 8, p. 25b. But there is better authority in the Ming Shih Lu, chs. 20–116. From these official records we learn that between +1403 and +1419 the Chinese government shipyards produced all told 2,149 sea-going vessels, including 94 first-rate Treasure-ships. In addition to these, 381 freighters were withdrawn from the grain-transport

[1] 梁山縣   [2] 鄭和   [3] 寶船廠   [4] 艦舶   [5] 殊域周咨錄
[6] 嚴從簡

Chün Wei Yu Ssu;[1] Defence Production Board), sometimes civil (Kung Pu;[2] Ministry of Works). Other yards were also hard at work, and between +1405 and +1407 those in Fukien, Chekiang, and Kuangtung produced no less than 1,365 ships of various rates. In +1420 large-scale naval architecture attained the status of a Board of its own, the Ta Thung Kuan Thi-Chü Ssu,[3] and we know the name of the chief designer and builder, Chin Pi-Fêng,[4] who made many working drawings (*thu yang*[5]). The 'Treasure-ship Yards' became so famous that they afterwards figured in a novel,[a] written in +1597, by Lo Mou-Têng.[6] Taoists were appointed to select fortunate days (*hao jih tzu*[7]) for laying down ships,[b] and there were offices for the various kinds of work, organising the carpenters, metal-workers, etc. The best artisans, selected by examinations, were transferred from other work, such as the building and repair of palaces and temples. Thirteen provinces contributed in special taxation. As for the size of the junks, the biography of Chêng Ho in the *Ming Shih*,[c] which is likely to be reliable, tells us that the 62 largest ships were 440 ft. long, and at broadest beam 180 ft.[d] Each one carried a crew of 450–500 men.[e] The poop had three superimposed decks, and there were several decks below the main one. From other sources, no less than nine masts were stepped in the largest Treasure-ships.[f]

The true size of these great vessels constitutes a cardinal problem of naval archaeology and has aroused much discussion. In the general anxiety to reduce the dimensions, some have suspected that the beam figure may have included the overhang of the sails when set furthest out, but this is quite unlikely. What is more to the point is the fact that in the typical Chinese build the upper decks and poop can over-ride the bottom timbers by some 30%, so that for the dimensions given a bottom length of about 310 ft. could be assumed, and individual timbers up to 80 ft. long. The shape is very

service (cf. p. 478) and re-fitted for naval work in Indian and African waters. The first order of +1403 went to Fukienese shipyards for 137 ocean-going vessels, and the first to the Nanking base was in the following year for 50 such ships. This was in +1411, and soon afterwards, in +1415, the transportation of the order given to Brigadier Wang Hao[8] in +1407 to convert 249 grain-transports for service on the high seas. In any case, the right figure for the ships laid down in +1403 was 361 rather than 250. We are greatly indebted to Dr Lo Jung-Pang for placing at our disposal the results of his searches in these official records.

After +1419 there was an almost complete cessation of maritime shipbuilding orders. This must surely have had something to do with the success of the engineer Sung Li, described in detail on p. 315 above, in perfecting the Yuan Grand Canal summit section by rendering it a fully practical proposition at all seasons. This was in +1411, and soon afterwards, in +1415, the transportation of grain by sea was suspended altogether (*Ming Shih Chi Shih Pên Mo*, ch. 24, p. 26). At the same time my Lord Chhen of Phing-chiang (already met with on p. 410 above) was authorised to build 3,000 shallow-draught sailing barges for the canal. Thus the shipbuilding energies of the State were diverted to other objectives, leaving the maritime yards idle just when that nursery of deep-water sailors, the grain-transport service, was also temporarily disbanded. All this weakened the sea party in the trial of strength that was to come (cf. pp. 524 and 527 below).

[a] *San-Pao Thai-Chien Hsia Hsi-Yang Chi Thung Su Yen I*[9] (Voyages and Traffics...in the Western Oceans). See p. 494.
[b] As still at Hongkong.
[c] Ch. 304, p. 2*b*.
[d] Ming units (1·02 ft.); 449 ft. and 184 ft. our measure respectively.
[e] It is not clear whether this figure includes marines and other passengers, or whether they were as many again.
[f] Pao Tsun-Phêng (1, 1).

[1] 軍衛有司   [2] 工部   [3] 大通關提舉司   [4] 金碧峯   [5] 圖樣
[6] 羅懋登   [7] 好日子   [8] 汪浩   [9] 三寶太監下西洋記通俗演義

broad-beamed (a length–beam ratio of 1 to 2·45), but this is confirmed by data in another source,[a] which give the dimensions of the second-rate 8-masted ships (Ma chhuan[1]) as 370 ft. long and 150 ft. abeam. None of the sources specifies the draught, but several less official texts confirm the dimensions in the *Ming Shih*. Collecting them all together, Pao Tsun-Phêng (1, *1*) has drawn up a list of all the ship sizes probably standard in Chêng Ho's fleets; they come in 23 rates, ranging from the largest 9-masters down to small 1-masted vessels of one tenth the length, and whenever a beam figure is given, the length–beam ratio remains about the same.

The credibility of the figures for the great Treasure-ships has been discussed by Mills (9) and Lo Jung-Pang (1, 5) as well as by Pao, and what they say should be considered in the context of our note on tonnage (p. 452 above). The *Ming Shih* figure suggests, according to Mills, a burden of some 2,500 tons and a displacement of about 3,100 tons. Other sources, however,[b] may be taken to imply that the highest tonnage used in the expeditions had a value of 2,000 *liao*.[2] If Lo Jung-Pang is right in interpreting the *liao* as a shipwright's cargo unit equivalent to roughly 500 lb., this means a burthen of 500 tons (i.e. a displacement tonnage of only about 800)—still much greater, however, than that of contemporary Portuguese ships. Lo supports this conclusion by arguments from the numbers of crew and marines carried on the individual ships of Chêng Ho's fleets.[c] At the same time he inclines to believe that some of the Sung ships had been larger, notably vessels of 5,000 *liao* (1,250 tons burthen) mentioned, e.g. in the *Mêng Liang Lu*[d] of +1275. This would certainly agree with Marco Polo's evidence of almost the same time (p. 467 above), hard though that is to interpret exactly.

A startling new development occurred in 1962 when an actual rudder-post of one of Chêng Ho's Treasure-ships was discovered at the site of one of the Ming shipyards near Nanking. As described by Chou Shih-Tê (*1*) this great timber, 36·2 ft. long and of 1·25 ft. diameter, shows a rudder attachment length of 19·7 ft. Assuming the usual Chinese 7/6 length–breadth proportions for the rudder blade, this means an area of no less than 452 sq.ft. Chou could therefore calculate, using accepted formulae, the

Fig. 980. A reconstruction of the rudder of one of the great ships of Chêng Ho (c. +1420) based on the dimensions of an actual rudder-post recovered in 1962 near Nanking. The size may be judged from the figure.

[a] *Kho Tso Chui Yü*, ch. 1, p. 29*b*. Cf. Duyvendak (9), p. 357.

[b] Especially the *Lung Chiang Chhuan Chhang Chih*, and a stele of Chêng Ho described by Kuan Ching-Chhêng (*1*) and Hsü Yü-Hu (*1*), though now lost.

[c] More on this subject will be found in Pao Tsun-Phêng (1, *1*) but the conclusions are necessarily speculative, for different items of essential data are lacking in the accounts of each voyage. We are grateful to Dr Lo Jung-Pang for joining in correspondence on this difficult subject.

[d] Ch. 12, p. 15*a*.

[1] 馬船   [2] 料

approximate length of the vessel on which it had been used, and obtained lengths of 480 ft. and 536 ft.[a] depending on different assumptions about draught. The discovery of the rudder-post shows that the Ming texts are not 'spinning a yarn' when they give dimensions at first sight hard to believe for the flagships of Chêng Ho's fleets.[b]

Then, before +1450, as we shall see, came a fundamental change in policy. The anti-maritime party at court, for reasons still somewhat obscure,[c] got the upper hand, and the long-distance navigations were brought to an end. That it never completely destroyed the traditions of the sea, however, is indicated by the fact that in +1553 a full-dress history of Chêng Ho's shipyards was written. This was the *Lung Chiang Chhuan Chhang Chih*[1] already mentioned (Record of the Shipbuilding Yards on the Dragon River), near Nanking,[d] by Li Chao-Hsiang,[2] whose work must be regarded as one of the treasures of Chinese technological literature.[e] It opens by a brief history of shipbuilding during the Ming dynasty, with an account of the officials who were entrusted with the organisation of it. Then come a number of illustrations and descriptions of ships,[f] but these indicate a decline in the size of vessel built, for only one is a four-masted ship (Hai chhuan[3]),[g] and most of them are two-masters (Ta huang chhuan,[4] Ssu pei liao chan (or hsün) tso chhuan[5,6]);[h] there are also a number of single-masted river boats and smaller craft.[i] The fourth chapter describes the yards, giving plans (Fig. 981), and then there follow specifications and dimensions of materials, tabulated with costs in ounces of silver, and details of the number of shipwrights and workmen required for each particular job. Finally, the eighth and last chapter gives one of the best collections of literary and historical references to ships and shipping in all Chinese literature.[j]

A related book which has come down to us from about the same time is the *Tshao Chhuan Chih*[7] (Records of River and Canal Shipping) compiled by Hsi Shu[8] in +1501 and enlarged by Chu Chia-Hsiang[9] in +1544. This deals with shipbuilding yards in various parts of the country, and gives lists of types of junks, but there are no illustrations, and in general the text is more administrative than technological.[k]

[a] Huai units (1·12 ft.), not the usual Ming units; 538 ft. and 600 ft. our measure respectively.

[b] To gain some idea of what these ships must have looked like, see Figs. 986, 987 (pl.) below.

[c] We shall discuss this further below, pp. 524 ff.

[d] Traces of these docks remain still to this day, and in 1964 I had the pleasure of visiting them in company with Mr Sung Po-Yin and Dr Yao Chhien, to whom my best thanks are due. One sees six large ponds, no longer connected with the Yangtze, each some 600 yards long and 100 yards broad. Here were found the great rudder-post (see p. 481), anchors and other iron objects, stone pestles and mortars for caulking material, etc.

[e] Attention was first called to this book by W. Franke (3), no. 256.

[f] Ch. 2, pp. 12b ff.   [g] Ch. 2, p. 36a.

[h] This measure of 400 may of course refer to amount of timber necessary rather than to cargo tonnage, *pace* the views of Lo Jung-Pang. Ch. 2, pp. 17b, 23a.

[i] In most cases the text quotes at some length a *Nan Chhuan Chi*[10] (Record of Southern Ships) by Shen Tai,[11] probably now lost, which requires further investigation.

[j] Much information about this work will be found in Pao Tsun-Phêng (1, 1).

[k] Cf. also another Ming book, *Chin-Ling Ku Chin Thu Khao*[12] by Chhen I,[13] where the shipyards figure in Nanking's local topography.

[1] 龍江船廠志   [2] 李昭祥   [3] 海船   [4] 大黃船
[5] 四百料戰座船   [6] 四百料巡座船   [7] 漕船志   [8] 席書
[9] 朱家相   [10] 南船記   [11] 沈岱
[12] 金陵古今圖考   [13] 陳沂

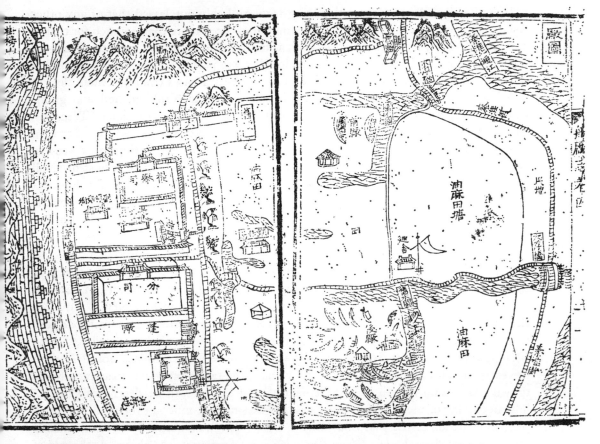

Fig. 981. One of the plans of the shipyards near Nanking where the great ships of Chêng Ho's fleets were built and fitted out, from the *Lung Chiang Chhuan Chhang Chih*, ch. 4, pp. 1*b*, 2*a*.

This view of part of the yards is taken looking approximately south at the strip of land between the walls of the city of Nanking on the left and the Chhin-huai Ho (R.) debouching into the Yangtze at the bottom on the right (as the legend at the top on the right says). The Chhin-huai R., coming from Chiangning, flows all round the south of Nanking outside the city walls, sending a loop through the southern quarter; it got its name from the belief that it had first been canalised in Chhin times, and if not salubrious it was famous for the painted boats of sing-song girls which were moored along the loop end to end. But at its mouth it was big enough to float the great hulls of the ocean-going Treasure-ships of the +15th century.

At the top of the plan is Ma-an Shan, a hill now within the city-walls, and to its left, inside them, Kuapang Shan is labelled, the Hill of Hanging up the Pass-Lists of Successful Candidates. In the left-hand half of the picture from top to bottom we can make out first the Main Gate (Ta Mên), then the Intendant's Headquarters (Thi-Chü Ssu), the Foremen's Offices (Tso Fang), various administrative sections (Fên Ssu), the Sail Loft (Phêng Chhang), and the Naval Liaison Command (Chih-Hui Chü) marked by a flag. All around are wide fields (*yu ma thien*) in which hemp was grown to yield oakum for caulking. In the right-hand half of the picture two shipyards are seen with their slipways and docks, the Chhien Chhang above and the Hou Chhang below; between them there is a Guard Post (Hsün Shê) marked by another flag. The entrances of the channels are crossed by two floating bridges, the smaller (Hsiao Fou-Chhiao) above, and the larger (Ta Fou-Chhiao) below; these carry the road along the bank of the Chhin-huai R.

The remains of the docks and shipbuilding yards are still visible today near the suburb of Chung-paotshun as basins no longer in connection with the Yangtze, from which they are separated by a high dyke. Excavations at the place have yielded valuable results, cf. Fig. 980.

In its heyday, about +1420, the Ming navy probably outclassed that of any other Asian nation at any time in history, and would have been more than a match for that of any contemporary European State or even a combination of them. Under the Yung-Lo emperor it consisted of some 3,800 ships in all, 1,350 patrol vessels and 1,350 combat ships attached to guard stations (*wei*[1] and *so*[2]) or island bases (*chai*[3]), a main fleet of 400 large warships stationed at Hsin-chiang-khou[4] near Nanking, and 400 grain-transport freighters. In addition there were more than 250 long-distance 'Treasure-ships' or galleons (Pao chhuan[5]), the average complement of which rose from 450 men in +1403 to over 690 in +1431 and certainly overstepped 1,000 in the largest vessels. A further 3,000 merchantmen were always ready as auxiliaries, and a host of small craft did duty as despatch-boats and police launches.[a] But the peak of the development which had started in +1130 came in +1433, and after the great reversal of policy the navy declined much more rapidly than it had grown, so that by the middle of the +16th century almost nothing was left of its former grandeur.[b]

With the +17th and +18th centuries we reach the period of intensified intercourse with the West and the end of our archaeological tour. The Chinese literature of this time has a good deal to say on the building and handling of ships, but it can be mentioned more conveniently in specific connections elsewhere in this survey. Towards the end of the +16th century much was written on the problem of grain transport and the relative advantages of the sea route, a perennial debate in the bureaucracy.[c] Liang Mêng-Lung's[6] *Hai Yün Hsin Khao*[7] of +1579 could be taken as an example.[d] Then in the early decades of the +17th century we have the technical encyclopaedias already drawn upon,[e] and certain important navigational manuals to be described shortly.[f] The beginning of the +18th century saw the assembly of much of the medieval shipping information in the *Thu Shu Chi Chhêng* (+1726), and later there were valuable travel books such as the *Liu-Chhiu Kuo Chih Lüeh* used on p. 404 above. At the end of it there were the philological researches of Hung Liang-Chi, and the Fukien ship-building manual.[g] Throughout these centuries Chinese shipping continued to play a highly important part in the maritime trade of South-east Asia; the study of this has been begun in the useful, though small, monographs of Thien Ju-Khang (*1, 2*).

In the +18th century Europeans began to see a good deal of Chinese ships in practical use. Sometimes also they made use of the services of Chinese shipwrights.

---

[a] Sources for much of the information in this and the preceding paragraphs have been assembled by Chêng Hao-Shêng (*1*) and Lo Jung-Pang (*1, 2*).

[b] Cf. pp. 524 ff. below. Much information on the navy between +1368 and +1575 is contained in *Hsü Wên Hsien Thung Khao*, ch. 132. Cf. von Wiethof (*1, 2*).

[c] For the Yuan background of this see the account of Schurmann (*1*), ch. 6.

[d] This work gives two maps of the coast between Hangchow and Tientsin, and includes sections on the distances sailed (ch. 1, pp. 23*b* ff.), trials of newly built vessels (p. 32*a*), and enlistment of ship-wrights (p. 32*b*).

[e] *San Tshai Thu Hui* in +1609, *Wu Pei Chih* in +1628, *Thien Kung Khai Wu* in +1637, and so on.

[f] Especially the *Tung Hsi Yang Khao* of +1618.

[g] See pp. 403, 406 above.

---

[1] 衛   [2] 所   [3] 寨   [4] 新江口   [5] 寶船   [6] 梁夢龍

[7] 海運新考

In his voyage from Macao to explore the north-west coast of America in +1788, John Meares took a company of them along with him. Of course

the Chinese artificers in this branch had not the least idea of our mode of naval architecture. The vessels of their nation, which navigate the China and adjacent seas, are of a construction peculiar to them. In vessels of a thousand tons burthen, not a particle of iron is used;[a] their very anchors are formed of wood, and their enormous sails made of matting. Yet these floating bodies of timber are able to encounter any tempestuous weather, hold a remarkable good wind, sail well, and are worked with such facility and care, as to cause the astonishment of European sailors.[b]

Such was the opinion of a great English navigator at the end of the century.

One final scene may complete the story. In 1848 readers of the *Illustrated London News* learnt that on 28 March a Chinese junk had arrived in the Thames, after having been navigated successfully across the Pacific and Atlantic Oceans. This was the 'Keying', named after the Manchu official and diplomat, favourable to intercourse with foreigners, Chhi-Ying,[1] who had been Commissioner at Canton.[c] She was a junk of 750 tons, 160 ft. long and of 33 ft. beam, teak-built with 15 water-tight compartments. The main-mast was 90 ft. in height, the fore 75 and the mizen 50; the mainsail yard was 67 ft. long and the lug-sails had battens at 3 ft. intervals. The mainsail weighed 9 tons and took some time to hoist, but could be reefed almost instantaneously. The rudder was slung in the classical way. Her master, Capt. C. A. Kellett, described her as 'a good sea boat and remarkably dry'.[d] But she was never to leave England, and was eventually broken up. Other long voyages were subsequently made with Chinese junks, such as that of the 'Whangho' from Hongkong to Sydney in 1908, and that of the 'Ningpo' from Shanghai to San Pedro, California, in 1912–13;[e] all proved, to the satisfaction of modern sailors,[f] the seaworthiness of Chinese ships.[g] Such voyages continue still. Not long before the completion of this Section the members of the International Congress of the History of Science assembled at Barcelona were

---

[a] Here he was certainly exaggerating.

[b] Meares (1), p. 88. Cf. the impressions of Macartney (+1794), Cranmer-Byng ed., pp. 81, 179, 200, 274 ff.

[c] On the voyage of the 'Keying' see Brindley (9); Chhen Chhi-Thien (2); Donnelly (5); Orange (1), p. 440, fig. 13; Audemard (4), pp. 32 ff.; Anon. (22).

[d] 'A most beautiful and easy sea-boat, never having shipped a drop of water since leaving China, or leaking' (Anon. (21), p. 282).

[e] Full details in Pritchard (1). She was of the Fuchow pole type and dated from early in the +19th century.

[f] One of the more recent of such experiments was that of de Bisschop & Tatibouët in 1934 (see des Noëttes (2), p. 141), but they had trouble with their slung rudder. Another is that described by E. A. Petersen (1) who sailed a junk across the Pacific in 1953.

[g] One cannot omit mention of the opinion of one of the greatest masters of seamanship, Capt. Joshua Slocum. He was the first man to sail a small sloop (the 'Spray') round the world single-handed (1895–8). But before that he had built, at Guarakasava in Brazil in 1888, after the loss of his barque 'Aquidneck', a 35-ft. 3-masted boat with bamboo-batten lug-sails (including parrels and multiple sheets) and a European-style hull, but hoisting a fenestrated rudder. 'Her rig', he wrote afterwards, 'was the Chinese sampan style, which is, I consider, the most convenient boat rig in the whole world' (p. 330). In this craft, the 'Liberdade', he safely made Washington, with all his family, before the end of the year. In his earlier life he had often visited the Chinese coast.

[1] 耆英

privileged to see the arrival at that port in September 1959 of the 'Rubio', a 60-ton South Chinese lorcha with typical round-leeched sails, navigated from Hongkong with all success by Capt. José-Maria Tey Planas.[a]

(4) THE SEAS THEY SAILED

'Westerners', wrote Cordier,[b] 'have singularly narrowed the history of the world in grouping the little that they knew about the expansion of the human race around the peoples of Israel, Greece and Rome. Thus have they ignored all those travellers and explorers who in their ships ploughed the China Sea and the Indian Ocean, or rode across the immensities of Central Asia to the Persian Gulf. In truth the larger part of the globe, containing cultures different from those of the ancient Greeks and Romans but no less civilised, has remained unknown to those who wrote the history of their little world under the impression that they were writing world history.' Here then within the compass of this sea-chapter we must do something to redress the balance and to mark how far from home the Chinese captains were prepared to go, voyaging on their own wine-dark waters. On a previous occasion Gibbon's words were recalled:[c] 'if the Chinese', he said, 'had possessed the genius of the Greeks and Phoenicians, they might have spread their discoveries over the southern hemisphere.' Allowing for the fact that this hemisphere is very largely ocean, that was just what they did (cf. the map in Fig. 989a, opp. p. 560).

It has been suggested that Asian sailors never rounded the Cape of Good Hope because of want of courage rather than of technical equipment.[d] Assuming for the moment that they did not, it is extremely doubtful whether either of these propositions is true in any sense. The sewn ships of Arabia and India were doubtless too uncertain for really long voyages, yet the Indonesians accomplished the colonisation of Madagascar by the ways of the sea.[e] There was much less reason, indeed none, why the great ships of China should not have discovered the west coast of Africa, and the Australian continent too, but social or political circumstances were assuredly the inhibiting factors rather than nautical technology. From Basra to Borneo, and from Zanzibar to Kamchatka, was not an insignificant span for the furthest ranges of the Chinese flag. And the less said about courage the better, as any modern sailor would feel if he found himself invited to undertake a voyage with the same equipment and the same facilities as the Buddhist pilgrims or the Emirs of + 14th-century Zayton.

Whoever has had in his life the good fortune both to wander on those Fukienese and Cantonese shores which saw the passage of the great barques of Chêng Ho, and to stand also on that hill which overlooks the Tower of Belem and the Praia de Restelo on

---

[a] His account of the journey was published soon afterwards (Tey, 1).

[b] (1), vol. 1, p. 237, eng. auct.

[c] In Sect. 26i (Vol. 4, pt. 1, p. 231). *Decline and Fall*, vol. 7, p. 95. Gibbon added, characteristically, I am not qualified to examine, and I am not disposed to believe, their distant voyages to the Persian Gulf or the Cape of Good Hope'. His opinion will be weighed in the ensuing pages, and found wanting.

[d] Parkinson (1), p. 6. Smyth took a very different view: 'Next to the Chinese', he said, 'the Arabs have been the most skilful and daring sailors of the ancient East...' (1), 1st ed. p. 301, 2nd ed. p. 346.

[e] Cf. Moreland (1).

the banks of the Tagus, cannot but be powerfully impressed by the strange contemporaneity of the great Portuguese and Chinese voyages of discovery. It is indeed an extraordinary historical coincidence that Chinese long-distance navigation from the Far East reached its high-water mark just as the tide of Portuguese exploration from the Far West was beginning its spectacular flow.[a] These two great currents almost met, but not quite, and in a single region, the coasts of the African continent. Their wind-angels, their inspirers, were two equally extraordinary men active in maritime affairs, on the one side a royal patron of navigators, on the other an imperial eunuch, ambassador and admiral. The contrast is inescapable, for this was the apogee of Chinese maritime enterprise.

(i) *The Admiral of the Triple Treasure*

To the assiduous reader of this book Chêng Ho[1] and his assistant commanders will be no strangers, for there was need to mention their exploits in the discussion of inter-cultural contacts, and more fully in the Section on geography and cartography.[b] Let us quote once again from the *Li-Tai Thung Chien Chi Lan*[2] (Essentials of the Comprehensive Mirror of History),[c] compiled by a group of scholars under imperial order in +1767.

In the third year of the Yung-Lo reign-period (+1405), the Imperial Palace Eunuch (Chung-Kuan[3]) Chêng Ho, [gloss: commonly known as the 'Three-Jewel Eunuch' (San-Pao Thai-Chien[4]),[d] a native of the province of Yunnan], was sent on a mission to the (countries of the) Western Oceans.

[Comm.] The emperor Chhêng Tsu, under the suspicion that (his nephew) the (previous) Chien-Wên emperor (Hui Ti) might have fled beyond the seas, commissioned Chêng Ho, Wang Ching-Hung[5] and others,[e] to pursue his traces. Bearing vast amounts of gold and other treasures, and with a force of more than 37,000 officers and men under their command, they built great ships (*ta po*[6]) [gloss: sixty-two in number], and set sail from Liu-chia Kang[7,f] in the prefecture of Suchow, whence they proceeded by way of Fukien to Chan-Chhêng[8] (Champa, Indo-China), and thence on voyages throughout the western seas.[g]

---

[a] This was noticed long ago by Mayers (3), who said that the great Chinese voyages of the early +15th century were 'singularly coincident in time with the heroic undertakings urged on from the opposite side of the globe by Henry the Navigator, Prince of Portugal...' (1875). So also Lévi (2) spoke of these analogous movements, 'provoked by the rhythm of history', sixty years ago.

[b] Vol. 1, pp. 143 ff., 180; Vol. 3, pp. 556 ff., etc.

[c] Ch. 102, pp. 4a ff., tr. Mayers (3), mod. auct. The present translation amplifies that in Vol. 3, p. 557, and corrects an error in it.

[d] Chêng Ho's famous title (borne also by a few other high palace officials) has a markedly Buddhist background, for San-Pao means the Three Jewels (*triratna*) of pious ejaculation, analogous to 'the strong name of the Trinity', i.e. the Buddha (Fo[9]), the Dharma (Fa[10]), and the Saṅgha (Sêng[11]). Yet it is quite certain that Chêng Ho was a Muslim by birth. Such was the syncretising tendency of Chinese folk religion; perhaps, too, jewels spiritual were confused with jewels temporal.

[e] See on, p. 491. The text writes erroneously Ho[12] for Hung.

[f] Near present Shanghai.

[g] Down to this point the commentary has followed the official biography of Chêng Ho in *Ming Shih*, ch. 304, pp. 2b ff., henceforward it is a paraphrase in shortened form. The biography is translated in Pelliot (2a), pp. 273 ff., 277 ff., 290 ff., 294 ff., 299, 300 ff. and 302. The passage in the dynastic history

---

[1] 鄭和	[2] 歷代通鑑輯覽	[3] 中官	[4] 三保太監	[5] 王景弘
[6] 大舶	[7] 劉家港	[8] 占城	[9] 佛	[10] 法
[11] 僧	[12] 和			

Here they made known the proclamations of the Son of Heaven, and spread abroad the knowledge of his majesty and virtue (*wei tê*[1]). They bestowed gifts (*tzhu*[2]) upon the kings and rulers, and those who refused submission they over-awed by the show of armed might (*pu fu tsê i wu shê chih*[3]). Every country became obedient to the imperial commands, and when Chêng Ho turned homewards, sent envoys in his train to offer tribute. The emperor was highly gladdened, and after no long time commanded Chêng Ho to go overseas once more and scatter largesse among the different States (*pien lai chu pang*[4]). On this, the number of those who presented themselves before the throne grew ever greater.

Chêng Ho was commissioned on no less than seven diplomatic expeditions, and thrice he made prisoners of foreign chiefs [gloss telling who, see p. 515 below]. His exploits were such as no eunuch before him, from the days of old, had equalled. At the same time, the different peoples, attracted by the profit of Chinese merchandise, enlarged their mutual intercourse for purposes of trade, and there was uninterrupted going to and fro. Thus it came to pass that in those days 'the Three-Jewel Eunuch who went down into the West' (*San-Pao Thai-Chien hsia hsi-yang*[5]) became a proverbial expression; and all who, in after times, were sent as bearers of commissions to the countries by sea, were wont to impress the outer nations with the name of Chêng Ho.

In this interesting summary we see at the outset some of the primary motives of the voyages. There was the search for the deposed emperor, but overshadowing it was the clear desire to impress upon foreign countries even beyond the limits of the known world the idea of China as the leading political and cultural power. There was also the encouragement of overseas trade. Had not one of the greatest of Sung emperors, Kao Tsung, the founder of the Chinese navy, said—[a] 'The profits from maritime commerce are very great. If properly managed they can amount to millions (of strings of cash). Is this not better than taxing the people?' That had been about +1145, when in falling back to Hangchow the government had first become fully conscious of the importance of sea power,[b] but it was no less applicable now, when Tīmūr Lang (Tamerlane) had just completed his general devastation of Western Asia, and all the lands and routes of Turkestan were closed again to Chinese commerce.[c]

has 27,800 men instead of 37,000, and gives the dimensions of the largest ships. To this point we shall return (p. 509 below). After the mention of Hui Ti it also adds: 'and being desirous of demonstrating to foreign countries the power and glory of China...'

[a] *Sung Hui Yao Kao*, ch. 44, pp. 20, 24; tr. Lo Jung-Pang (1).

[b] There had, of course, been semi-governmental trading intercourse with the south seas for a thousand years already (cf. Vol. 1, Sect. 7, and the monograph of Wang Kung-Wu (1) which gives a list of return tribute missions from +2 to +960). In +987 the Sung court had sent eight palace officials in four fleets to the countries of the south seas (South-east Asia and probably India) to purchase drugs, ivory, rhinoceros horns, pearls, etc.; and they carried blank patents to be conferred on the heads of States who were invited to send trade missions to China. The Yuan government had also sent many missions abroad beginning in +1278, and the fleet of fourteen great ships which carried Marco Polo homeward, reaching Ormuz in +1294, will, of course, be remembered. Another such fleet, of 25 ships, arrived at Ormuz in +1301. Facts such as these make it clear that there was nothing essentially new in principle in the voyages of Chêng Ho. The scale and scope of the operations were what was unprecedented.

[c] Born in +1335, this famous conqueror seized the sultanate of Transoxiana in +1365, and fifteen years later embarked upon a series of campaigns which subdued Persia, Afghanistan, Iraq, Syria and Turkey with great destruction and immense loss of life. In +1398 he invaded India and established what became later on the Mogol Empire with its centre at Delhi. He died, to everyone's relief, in the very year of Chêng Ho's first commission. Cf. Grousset (1), vol. 2, p. 487; Hitti (1), pp. 699 ff.; Smith (1), p. 252; and more copiously, Hookham (1).

[1] 威德   [2] 賜   [3] 不服則以武懾之   [4] 徧賚諸邦   [5] 三保太監下西洋

The long-distance navigations involved at least three specialised activities. On the naval side there was the conduct of large fleets of junks, the greatest vessels then afloat, over many thousands of sea miles to regions where no organised Chinese fleets had been before, working them safely in and out of little-known ports and havens, with a great deal of handling in the narrow waters of the South-east Asian archipelago as well as direct passages on the high seas from Malaya to Africa.[a] On the military side there was the organisation of marines and gunners at sea and ashore, with commanders who proved efficient and successful in certain unexpected actions, as we shall presently see, though the duties of the troops were primarily ceremonial. As for the diplomatic or prestige function, what it involved in practice was the giving of rich presents to rulers into whose domains the envoys came, at the same time inducing them to acknowledge the nominal suzerainty or overlordship of the Chinese emperor, and to despatch, if possible, tribute-bearing missions to the Chinese court. Under the rubric of tribute a great deal of State trading was carried on, and besides this there may have been some desire to foster the activities of private traders and merchants. Lastly there was a proto-scientific function. An increase in knowledge of the coasts and islands of the Chinese culture-area was looked for, and the routes to the Far West were to be surveyed. Furthermore, the search for rarities of all kinds was to be actively prosecuted and gems, minerals, plants, animals, drugs and the like were to be collected for the imperial cabinets. All these functions were the expression of motives, and we shall come back to them when we compare them with those of the Portuguese pioneers. One gets the impression that the more the voyages of exploration developed and the further they reached out, the more important became the collection of natural curiosities and the less important the securing of acknowledgements of tributary status on the part of local princes, while the search for the missing emperor faded into the background altogether.

The seven expeditions from China extended progressively westwards. The first (+1405 to +1407) visited Champa (Indo-China), Java and Sumatra, going the other way as far as Ceylon, and Calicut on the west coast of India. The second (+1407 to +1409), under another commander, Chêng Ho being absent, visited Siam and added Cochin to its Indian ports of call.[b] On the third the fleet went to all the usual places in the East Indies using Malacca as a base,[c] added Quilon in south-west India, and became involved in affairs both distressing and comic on the island of Ceylon (+1409 to +1411). At this time a third able leader, the eunuch Hou Hsien,[1] was joined to Chêng

[a] It is not at all easy to tell from the sources what were the approximate overall losses on the expeditions. They were certainly exaggerated by the propaganda of the anti-maritime party (cf. Vol. 3, p. 557).

[b] Chêng Ho's absence on this expedition results from the argumentation of Duyvendak (9), pp. 363 ff., but it is very uncertain. Modern Chinese scholars such as Hsü Yü-Hu (1), Chêng Hao-Shêng (4) and Lo Jung-Pang (private communication) find good evidence to show that Chêng Ho led this expedition like all the other principal ones (cf. the date of the Ceylon stele, p. 523 below).

[c] Here the ground had been prepared by the successful diplomatic mission of the eunuch Yin Chhing[2] in +1403 (Ming Shih, ch. 325, p. 6a, cf. Purcell (3), p. 17). The friendship of the prince Parameśvara and his people assured the use of the port facilities by the Chinese throughout the +15th century. In al-Idrīsī's time, three centuries earlier, Almaid Island had been the place 'where gather and stay the Chinese ships' (Jaubert tr., vol. 1, pp. 89 ff.), but the identification of this seems still uncertain.

[1] 侯顯　　[2] 尹慶

Ho and Wang Ching-Hung. India was left behind by the fourth expedition in +1413 to +1415 which took different directions. While some squadrons visited all the East Indies again, others (based on Ceylon) explored Bengal, the Maldive Islands, and reached the Persian sultanate of Ormuz. Such interest was aroused in the Arabic culture-area at this time, including the Arab city-States on the East African coast,[a] that in +1416 a remarkable flow of ambassadors converged on Nanking, with the result that a fleet greater than ever was organised in the following year to conduct them home again. Between then and +1419 while its Pacific squadrons went as far as Java, Ryukyu and Brunei, its Indian ones ploughed the seas from Ormuz to Aden, visiting Mogadishiu in Somaliland, Malindi north of Mombasa,[b] and other places on the Zanguebar coast. This was the time when giraffes went back to Peking to favour the court with their auspicious presence and to be the joy of Chinese naturalists. The sixth expedition covered the same ground as before (+1421 to +1422) including calls at al-Aḥsā'[c] and Ẓafār in south Arabia, and Brawa on the East African coast. To visit thus thirty-six States from Borneo in the east to Zanguebar in the west within two years suggests that the fleet must have separated into a number of different groups (with all the difficulties of rendezvous which that would have raised), and that for the prosecution of its scientific function as well as the encouragement of trade a single diplomatic mission did not suffice.[d] Then came in +1424 the first warning of the Ming navy's future calamities. The Yung-Lo emperor died and his successor Jen Tsung, strongly inclined to the anti-maritime party, countermanded the voyage which had been ordered for that year. But almost at once he too expired, and the following monarch, Hsüan Tsung, was left to preside over the last and perhaps most dazzling of the Grand Treasure-Ship Fleets. It left in +1431 and before it returned in +1433 its commanders, with their 27,550 officers and men, had established relations with more than twenty realms and sultanates from Java through the Nicobar Islands to Mecca in the north and the coast of al-Zanj (East Africa) in the south. Exactly how far down this coast the Chinese penetrated is not quite sure; we shall return to it in a moment. But regarding their visits to the Persian Gulf and the Red Sea one should remember that these were not new in the +15th century,[e] indeed that Chinese ships had already been frequenting those waters for a thousand years—what was quite new was the appearance of organised naval forces with junks of great size, not simply small and isolated merchant-ships. In view of this the fundamentally pacific character of the Ming voyages seems all the more remarkable. Two years after the return of Chêng Ho

[a] Glimpses of how things looked from the Arabic side may be found in Darrag (1), pp. 196 ff., 217 ff.

[b] Malindi is a fateful name, for here it was that some few decades later Vasco da Gama would find his Arab pilot to take his ships across to Calicut, thus opening Asian seas to European penetration.

[c] Some have identified this place with modern al-Hufūf, near Bahrein Island in the Persian Gulf, others thought Muscat; cf. the discussion in Duyvendak (19). Mr J. V. Mills believes that we should locate it on the shore of the Hadhramaut just west of Mukalla. Al-Idrīsī showed a place called Lasa'a and Lis'a at this spot in his world-maps of +1154 and +1161 (cf. Fig. 239 in Vol. 3).

[d] The chief rendezvous port was probably Malacca on all occasions; indeed the *Ying Yai Shêng Lan* distinctly says so (pp. 36, 37). Many Chinese, of course, live there now, but the weight of the evidence indicates that there was little settlement, if any, before the period of Portuguese dominion which began in +1511. See Purcell (1), pp. 282 ff., (3).

[e] Cf. Vol. 1, pp. 179 ff.

and his ships the Hsüan-Tê emperor died and now the Chinese Admiralty was doomed at last. Ying Tsung and his successors listened to the Confucian 'agriculturists', the scholar-landlords, so that official maritime activities were reduced to the minimum needed to protect coasts and grain-ships from the predations of Japanese pirates (and often below it). This was a decision which had far-reaching results not only for Chinese but also for world history.[a]

It is clear that the Grand Fleets (*Ta tsung*[1]) separated into a number of squadrons (*Fên tsung*[2]) with particular missions, and that they used a variety of overseas ports as bases, never seeking, however, to establish themselves there in forts and dockyards obtained by military means. Malacca, Ceylon (probably Beruwala rather than Galle harbour), Calicut and Aden were evidently among these. Moreover, their activities were only the culmination of naval diplomatic missions which had been increasing since the end of the Yuan, and which paralleled missions to the western countries by land. Hou Hsien, who was considered the next most important diplomat (Chêng-Shih Thai-Chien[3]) after Chêng Ho himself,[b] was in Tibet in +1407 and Nepal in +1413, while in +1415 he headed a special embassy to Bengal. In +1403 Ma Pin[4] had presented compliments (and rich gifts) in a similar way to the Chola King of Tamil Coromandel (modern Madras).[c] Bengal received Yang Min[5] in +1412 and Yang Chhing[6] in +1421,[d] while another eunuch, Hung Pao,[7] who had been in Siam in +1412, organised an important mission to Mecca[e] in +1432. Chhang Kho-Ching[8] and Wu Pin[9] specialised in relations with the ruler of Quilon, while Chou Man[10] dealt with the ruler of Aden.[f] Most of these men ranked as Assistant Ambassadors and Grand Eunuchs (Fu-Shih Thai-Chien[11]). By exclusion one may surmise that Wang Ching-Hung was primarily a naval commander, and that this was also the role of men such as Li Hsing,[12] Chu Liang,[13] Yang Chen,[14] Chang Ta[15] and Wu Chung.[16] A few names of Brigadiers of Marines (Tu-Chih-Hui[17]) have come down to us, e.g. Chu Chen,[18] and Wang Hêng.[19] Geomancers, astronomers[g] and physicians[h] were certainly carried,

---

[a] For further details cf. p. 524 below.

[b] His biography follows that of Chêng Ho in *Ming Shih*, ch. 304, pp. 4b ff. See on him Pelliot (2a), pp. 314, 320, (2b), p. 286.

[c] Pelliot (2a), p. 328.

[d] Pelliot (2a), pp. 240, 272, 319, 321, 342, (2b), p. 311, (2c), p. 214; Duyvendak (9), p. 380.

[e] Pelliot (2a), pp. 342 ff.   [f] Duyvendak (9), p. 386.

[g] Liu Ming-Shu (2) has identified the principal 'Yin-Yang expert' as Lin Kuei-Ho,[20] who sailed on five voyages with the great admiral. He was certainly concerned with weather forecasting and other meteorological and calendrical matters, no doubt also with divinations of success or failure in various special enterprises, and quite probably with astronomical navigation too. It would be quite surprising if such a man did not take lively interest in all kinds of natural phenomena observable during the travels of the fleets. Cf. p. 562 below.

[h] Remarkably enough, quite a number of biographies have come down to us (cf. *TSCC, I shu tien*, chs. 531, 532, 534). One of the chief medical officers of many voyages must have been the Imperial Physician Chhen I-Chhêng,[21] known also as a poet and calligrapher; another who gained high promotion because of his services afloat was Yü Chen.[22] Many of these naval physicians, such as Phêng Chêng[23] and

---

[1] 大椶	[2] 分椶	[3] 正使太監	[4] 馬彬	[5] 楊敏
[6] 楊慶	[7] 洪保	[8] 常克敬	[9] 吳賓	[10] 周滿
[11] 副使太監	[12] 李興	[13] 朱良	[14] 楊眞	[15] 張達
[16] 吳忠	[17] 都指揮	[18] 朱眞	[19] 王衡	[20] 林貴和
[21] 陳以誠	[22] 郁震	[23] 彭正		

though most of the best-known figures now are those interpreters and secretaries whose written accounts are among the most important of the surviving sources.

Ma Huan,[1] an Arabist from Kuei-Chi in Chekiang and a Muslim like Chêng Ho himself,[a] was the first to start writing. His account of the expeditions was begun in +1416 and completed about +1435, though not printed till +1451 when it appeared with the title *Ying Yai Shêng Lan*[2] (Triumphant Visions of the Ocean Shores).[b] The following century found this rough text lacking by the standards of polite literature, so it was rewritten in condensed classical style by a scholar named Chang Shêng[3] and in +1522 came out again with a slightly different title.[c] Meanwhile, towards the end of the great voyages, two other sea-faring secretaries had written valuable books, Kung Chen[4] with his *Hsi-Yang Fan Kuo Chih*[5] (Record of the Foreign Countries in the Western Oceans)[d] of +1434, and Fei Hsin[6] with his romantically entitled *Hsing Chha Shêng Lan*[7] (Triumphant Visions of the Starry Raft)[e] of +1436. This was the literary term for ships carrying ambassadors. If other books were written by actual participants in the expeditions they have not yet come to light.[f] But certain scholars living a little later devoted themselves to collecting information about Chêng Ho's achievements, notably Huang Shêng-Tsêng,[8] whose *Hsi-Yang Chhao Kung Tien Lu*[9] (Records of the Tributary Countries of the Western Oceans) of +1520 is a valuable contribution.[g] Another book, the *Chhien Wên Chi*[10] (Traditions of Past Affairs) by Chu Yün-Ming,[11] written about +1525, preserves precious details about the composition of the ships' companies, the names and classes of ships, and the exact times taken in the various sailings from port to port.[h] Besides all these works there are of course the standard

Chang Shih-Hua,[12] hailed from the neighbourhood of Suchow, the home of Fei Hsin and one of the fleet's shore bases. Chhen Chhang,[13] who went overseas three times between +1425 and +1435, was also interested in navigation, on which he afterwards used to discourse. The presence of many learned naturalists in Chêng Ho's fleets is a point of much importance to which we must presently return (p. 530). On one voyage there were 180 medical officers (Hsü Yü-Hu (1), p. 26).

In connection with this it has been little noticed that there were books on naval medicine in the medieval Chinese literature. The Sung bibliography contains at least two—a *Hai Shang Ming Fang*[14] (Famous Shipboard Prescriptions) by one Chhien Yü,[15] and a *Hai Shang Chi Yen Fang*[16] (Collected Well-Tried Shipboard Prescriptions) by Tshui Hsüan-Liang[17] (see *Sung Shih*, ch. 207, pp. 25 a, 27 b). I should not be at all surprised if these works, now long lost, told sailors how to escape scurvy.

[a] Chêng Ho's original family name was also Ma; it had been changed as a mark of imperial favour. When we find that his father had made the pilgrimage to Mecca we can appreciate one at least of the qualities which made him so suitable for the management of contacts with Western Asia on the grand scale.

[b] Partial translations of this have been made by Groeneveldt (1) and Phillips (1, 2). Cf. Pelliot (2 a), pp. 241 ff. The full study by Mr J. V. Mills is eagerly awaited.

[c] *Ying Yai Shêng Lan Chi*.[18] This was the text used by Rockhill (1) for his translations. Copious excerpts are found in *TSCC, Pien i tien*, chs. 58 to 106.

[d] Till recently known only by quotations (cf. Pelliot (2 a), p. 340), but now edited from a complete MS. by Hsiang Ta.   [e] Cf. Pelliot (2 a), pp. 264 ff.

[f] There is mention in the postface to Ma Huan's book of a Muslim collaborator, Kuo Chhung-Li.[19] Perhaps this was the Kuo Wên[20] who was in Siam in +1416 (Pelliot (2 a), p. 263).

[g] Tr. Mayers (3); cf. Pelliot (2 a), pp. 344 ff.

[h] See Pelliot (2 a), pp. 305 ff., amplifying Mayers (3).

[1] 馬歡	[2] 瀛涯勝覽	[3] 張昇	[4] 鞏珍	[5] 西洋番國志
[6] 費信	[7] 星槎勝覽	[8] 黃省曾	[9] 西洋朝貢典錄	
[10] 前聞記	[11] 祝允明	[12] 張世華	[13] 陳常	[14] 海上名方
[15] 錢竽	[16] 海上集驗方	[17] 崔玄亮	[18] 瀛涯勝覽集	
[19] 郭崇禮	[20] 郭文			

official publications (*Ming Shih*, *Ming Shih Lu*, etc.). The whole of the available literature has been elaborately studied by sinologists in East and West,[a] but further discoveries may yet be looked for. Our documentation would have been even more
abundant if the archives of the Ming navy had not been purposely destroyed at a subsequent time (cf. p. 525 below). But what remains is amply sufficient to give us as good
a knowledge of the Chinese as of the Portuguese ventures, for which, curiously enough,
there are also serious documentary gaps (cf. p. 528).

The expeditions of Chêng Ho's time had considerable influence on Chinese literature, parallel in a smaller way to the spread of knowledge of the Portuguese discoveries
through Europe. Some MS. books of sailing-directions have come down to us (cf.
p. 583 below). Though none of the works just mentioned were furnished with maps, a
number of valuable 'portolans' or route-maps in a distinctively Chinese style were preserved in the *Wu Pei Chih*[1] (Treatise on Armament Technology)[b] compiled by Mao
Yuan-I[2] and presented to the throne in +1628. These undoubtedly derive from the
cartographers attached to Chêng Ho, and we accordingly discussed them in an earlier
Section.[c] The illustrated geographical encyclopaedias of the age were greatly affected
by the new knowledge brought back by Chêng Ho, for instance the *I Yü Thu Chih*[3]
(Illustrated Record of Strange Countries) compiled between +1420 and +1430 prob-
ably under the supervision of the learned Ming prince Chu Chhüan[4] (Ning Hsien
Wang[5]).[d] This too we have already discussed.[e] Another book with a very similar title,
I Yü Chih,[6] written before the end of the +14th century, had inspired Ku Pho,[7] who
(at the request of a mutual friend Lu Thing-Yung[8]) wrote the postface to Ma Huan's
'Triumphant Visions of the Ocean Shores', and whose words are worth quoting in
illustration of the enlightened Chinese attitude of the time towards foreign parts.[f]

In my youth [he said], by reading books such as the *I Yü Chih* (Record of Strange Countries), I learnt the vastness of the surface of the earth, the differences in customs, the diversity
of human beings, and the variety of natural products—which are all truly astounding, lovable,
admirable and impressive. However I had some suspicion that these books were perhaps the

[a] See Pelliot (*2a, b, c*); Duyvendak (8, 9, 10, 11); Mills (3), etc., who built on the earlier investigations
of Mayers (3) and Rockhill (1). Cf. Hsiang Ta (1). The best Chinese biography of Chêng Ho is that of
Chêng Hao-Shêng (*1*), cf. (*4*). Maps and sea-routes have been intensively studied by Fan Wên-Thao (*1*)
and Chou Yü-Sên (*1*) in special monographs; cf. also the paper of Liu Ming-Shu (*2*). Fêng Chhêng-
Chün (*1*) places the voyages in the general context of Chinese knowledge of the south seas. Filesi (1, 2)
assays them in the language of Marco Polo.

[b] And in similar later books such as the *Wu Pei Pi Shu*[9] (Confidential Treatise on Armament Technology) by Shih Yung-Thu.[10]

[c] Vol. 3, pp. 559 ff. It is good to see the value of these maps now widely recognised, as in the popular
treatise of Debenham (1), p. 122. New ed. Hsiang Ta (*4*).

[d] Alchemist, botanist, mineralogist, pharmacist and an expert in acoustics and music, cf. Vol. 1,
p. 147 and Sections 33 and 38.

[e] Vol. 3, pp. 512 ff.

[f] Tr. Duyvendak (10), p. 11; mod. auct. Cf. Pelliot (*2a*), p. 260; *SKCS/TMTY*, ch. 78, p. 3*a*. It
seems that the original name of the *I Yü Chih* was *Lo Chhung Lu*[11] (Record of the Naked Creatures, i.e.
the Barbarian Peoples) and that it was probably written by Chou Chih-Chung[12] in the neighbourhood of
+1366. The name was apparently changed just before +1400 by the elder brother of the official Khai
Chi,[13] who perhaps re-wrote or expanded the book.

[1] 武備志	[2] 茅元儀	[3] 異域圖志	[4] 朱權	[5] 寧獻王
[6] 異域誌	[7] 古朴	[8] 陸廷用	[9] 武備秘書	[10] 施永圖
[11] 臝蟲錄	[12] 周致中	[13] 開濟		

work of busybodies with too much imagination, and I doubted whether there really existed such things. But now I have read the notes which Mr Ma Tsung-Tao[1] (Ma Huan) and Mr Kuo Chhung-Li have made of their experiences among the foreign countries, and I realise that what the *I Yü Chih* reported deserved confidence and was no fable...

Lastly the great voyages provided material for one of the famous Ming novels, the *Hsi-Yang Chi*[2] written by Lo Mou-Têng[3] and published in +1597. Though containing much fabulous material, it is also, as Duyvendak (19) has shown, a source of reliable information concerning the organisation of tribute missions and their gifts, together with interesting technical details about spectacles, gunnery and the like.[a]

So great was the fame of Chêng Ho and his companions in South-east Asia that they entered at last, like Kuan Yü, into the realms of heroic hagiology. For the admiral was adopted as a tutelary deity by the Chinese communities of the Malayan diaspora, and incense burns to this day in the temple of Sam-Po-Tai-Shan[4] at Malacca.[b]

(ii) *China and Africa*

But Chinese relations with East Africa were far older than the day of Chêng Ho. From ancient Egyptian times onwards there had been trade down the coast, and Ptolemy's Promontorium Prassum was probably Cape Delgado. Permanent non-indigenous settlement began in the +8th century with the foundation of Arab trading centres such as Mogadishiu in Somaliland about +720 and Sofala south of the Zambezi River about +780. Gradually these developed into mercantile city-states[c] and from them Arab exploration spread out, Madagascar and the Comoro Islands in the Mozambique Channel being known in the +9th. What is much more unexpected is that descriptions of this part of the world (Azania, al-Zanj) can be found in Chinese literature[d] as early as about +860. When Tuan Chhêng-Shih[5] was compiling his *Yu-Yang Tsa Tsu*[6] (Miscellany of the Yu-Yang Mountain Cave) at this time, he included in his accounts of foreign countries an interesting passage on Po-Pa-Li[7]—none other than Berbera, the south coast of the Gulf of Aden.[e] With varying orthography and steadily increasing detail this country is described in the subsequent books on overseas lands, such as the *Chu Fan Chih*[8] (Records of Foreign Peoples) by Chao Ju-Kua[9] in +1225, where it is called[f] Pi-Pha-Lo.[10] This book has also an elaborate description of the Somali coast

---

[a] Its full title is *San-Pao Thai-Chien Hsia Hsi-Yang Chi Thung Su Yen I*[11] (Popular Instructive Story of the Voyages and Traffics of the Three-Jewel Eunuch (Admiral, Chêng Ho), in the Western Oceans). Cf. Vol. 4, pt. 1, p. 119 and in Sect. 30*k*. It is probable that these voyages really were the means of introducing eye-glasses (spectacles) to China.

[b] Purcell (3), pp. 17, 123.

[c] Cf. Wainwright (2); Mathew (2); Freeman-Grenville (4). Another example: Gedi just south of Malindi was founded about +1100, and its first great mosque built *c*. +1450 (Kirkman, 3).

[d] Hirth (13), a pioneer paper; Rockhill (1); elaborated by Duyvendak (8). Cf. the curious but stimulating article of E. H. L. Schwarz (1). See also Hirth & Rockhill (1); Wheatley (1); Filesi (1).

[e] Ch. 4, p. 3*b*, abridged in *Hsin Thang Shu*, ch. 221B, p. 11*b*; tr. Duyvendak (8).

[f] Ch. 1, p. 25*b*; tr. Hirth & Rockhill (1), p. 128; Duyvendak (8).

[1] 馬宗道   [2] 西洋記   [3] 羅懋登   [4] 三寶太神   [5] 段成式
[6] 酉陽雜組   [7] 撥拔力   [8] 諸蕃志   [9] 趙汝适   [10] 弼琶囉
[11] 三寶太監下西洋記通俗演義

(Chung-Li[1]).[a] At a date which must have been very soon after the first foundation of Kenyan Malindi, the *Hsin Thang Shu* (+1060) has an account[b] of a city and country of Malindi (Mo-Lin[2]), perhaps those on the coast of Kenya.[c] Still more surprising, Chao Ju-Kua has an article on Tshêng-Pa,[3] i.e. the whole Zanguebar coast between the Juba River in Somalia and the Mozambique Channel;[d] while in his *Ling Wai Tai Ta*[4] (Information on what is beyond the Passes), written in +1178, Chou Chhü-Fei[5] describes Madagascar (Khun-Lun Tshêng-Chhi[6]) at some length.[e] When one comes to the +14th century these regions are all well known. Wang Ta-Yuan,[7] who was travelling widely between +1330 and +1349, deals with most of them in his *Tao I Chih Lüeh*[8] (Records of the Barbarian Islands), including not only Berbera, and the Zanguebar coast (al-Zanj) as Tshêng-Yao-Lo,[9] but also the Comoro Islands in the Mozambique Channel.[f]

How many Chinese merchants and sailors actually themselves visited these waters between the +8th and the +14th centuries we have no means of telling.[g] Apart from texts such as those just mentioned there is only the mute evidence of Chinese objects scattered up and down the coast. And these are many indeed, so many that it is rather hard to believe they all came there through the hands of intermediate traffickers. Before we briefly consider them, let us look at one positive testimony, from an Arabic source,

---

[a] Ch. 1, p. 26 a; tr. Duyvendak (8), p. 20; Hirth & Rockhill (1), pp. 130 ff.

[b] Ch. 221 B, pp. 11 a, b, translated by Duyvendak (8), first translated by Hirth (1), p. 61.

[c] It was reached by crossing a desert and lay 2,000 *li* south-west of the Arab and Byzantine realms, together with a neighbouring unidentified country called Lao-Pho-Sa.[10] Both were inhabited by fierce black people. In spite of Duyvendak's opinion, Mr J. S. Kirkman assures us that the arid climate and desert features of this Mo-Lin cannot fit Kenyan Malindi; perhaps there was some other place with a similar name on the Somali or Berbera coast further north. In the Ming, Kenyan Malindi was written differently, Ma-Lin.[11] Duyvendak also overlooked an interesting point to which Lo Jung-Pang (7) has drawn attention. The account in the *Hsin Thang Shu* must derive from the *Ching Hsing Chi*[12] (Record of My Travels), written by the Chinese officer Tu Huan[13] about +763 when he returned from his eleven years of captivity among the Arabs after the Battle of the Talas River (cf. Vol. 1, pp. 125, 236); for parallel passages occur in excerpts from this (*Thung Tien* (c. +812), ch. 193 (p. 1041.3), and *Wên Hsien Thung Khao*, ch. 339 (p. 2659.2), tr. Hirth (1), p. 84). If then one could locate more clearly the native heath of these xenophobic Africans, Tu Huan's account would be the earliest description of the Somali littoral, antedating the *Yu-Yang Tsa Tsu* by about a century. Perhaps in any case it ought to be considered so.

[d] Ch. 1, p. 25 a, tr. Hirth & Rockhill (1), p. 126. The *Sung Shih*, ch. 490, pp. 21 b, 22 a, describes the same region, and records the arrival of ambassadors from it in +1071 and +1083; they were sent home with rich gifts. About a century before Chao Ju-Kua, it had also been described by an anonymous writer in a book called *Tao I Tsa Chih*[14] (Miscellaneous Records of the Barbarian Islands), preserved because quoted at length in the *Shih Lin Kuang Chi* encyclopaedia. Wada Kyutoku (1) reproduces the text.

[e] Ch. 3, p. 6 a, repeated (with variations) in *Chu Fan Chih*, ch. 1, pp. 32 b, 33 a, tr. Duyvendak (8); Hirth & Rockhill (1), p. 149. On Khun-lun as a term for dark people (who could be enslaved) cf. p. 459 above.

[f] All the relevant passages are translated in Rockhill (1), who compares them with the statements in the books of the Ming naval secretaries.

[g] An occasional spot of light is cast by a particular name. Very likely Wang Ta-Yuan met another far-ranging scholar, Li Nu,[15] who was at Ormuz in +1337, where he married an Arab or Persian girl and adopted Islam. Li Nu was an ancestor of the great Ming reformer Li Chih[16] (cf. Vol. 1, p. 145, and further Hucker (1), p. 144; Needham (43), p. 293; Pokora (6) and Franke (4), still the standard study). Other clansmen of Li Nu participated in the voyages of Chêng Ho's time.

[1] 中理　　　　[2] 磨鄰　　　　[3] 層拔　　　　[4] 嶺外代答　　　[5] 周去非
[6] 崑崙層期　　[7] 汪大淵　　　[8] 島夷志略　　[9] 層搖羅　　　　[10] 老勃薩
[11] 麻林　　　　[12] 經行記　　　[13] 杜環　　　　[14] 島夷雜志　　　[15] 李駑
[16] 李贄

of the presence of Chinese merchants on the shores of +12th-century East Africa. The great Sicilian geographer Abū 'Abdallāh al-Idrīsī, writing about +1154, says:[a]

Opposite the coasts of Zanj are the Zalej (or Zanej) Islands, large and numerous; their inhabitants are very dark, and everything that they grow is dark—dhorra, sugar-cane, camphor, etc. One of these isles is called Sherbua...Another is al-Anjebi, where the chief town is called in the Zanguebar language al-Anfuja, its inhabitants being mostly Muslims though of mixed descent...This island is very populous, with many villages and domestic animals; rice is grown there. There is much commerce, and markets to which all kinds of things for sale and use are brought. It is said that once when the Chinese affairs were troubled by rebellions, and when tyranny and confusion became intolerable in India, the Chinese moved their commercial centre to Zalej and the other islands which belong to it, entering into familiar relations with the inhabitants because of their equity, uprightness, amenity of customs and aptitude for business. This is why the island is so populous and so frequented by strangers.

Here we have only a glimpse, for it is not quite clear what al-Idrīsī had in mind. The Chinese rebellion to which he refers sounds like that of Huang Chhao[1] (+875 to +884) during which the Arab quarter of Canton was destroyed,[b] but trouble on the East African mainland would have been a much more likely cause of the removal of Chinese trading stations there to an island. Nor is al-Idrīsī's reference to India easily understandable. Nevertheless what he says about the story of the Chinese 'factory' is itself quite precise, and we may accept it as a picture of such activities about +1000. If there was one such Chinese station on the coast in Sung times there were probably several, and merchant-junks too, to connect them with home. As for the identity of the Zalej or Zanej Islands, they are believed to be the Mafias, off the Tanzanian coast about 150 miles south of Zanzibar.[c]

Among the things which the Chinese wanted from Africa were elephant tusks, rhinoceros horns, strings of pearls, aromatic substances, incense gums and the like.[d] Statistics preserved in the *Sung Shih* show that these imports increased ten times between +1050 and +1150. Al-Idrīsī, on the other hand, tells us what Aden (and hence the Coast) received from China and India—iron, damascened sabres, musk and porcelain (typical Chinese exports), saddles, 'velvety and rich textiles' (probably silk), cotton goods, aloes, pepper and South Sea spices.[e] Fortunately some of this was hardware and has survived until today. 'I have never seen', wrote Wheeler (6) in 1955, 'so much broken china as in the past fortnight between Dar-es-Salaam and the Kilwa Islands; literally fragments of Chinese porcelain by the shovelful...I think it is fair to say that as far as the middle ages is concerned, from the +10th century onwards, the buried history of Tanganyika is written in Chinese porcelain.'

---

[a] *Nuzhat al-Mushtāq fī Ikhtirāq al-Āfāq* (Recreation of those who long to know what is beyond the Horizons), tr. Jaubert (1), vol. 1, pp. 59 ff. The better translation of Dozy & de Goeje (1) does not deal with this part of Africa.

[b] Cf. Shih Yu-Chung (1).   [c] Revington (1).

[d] See Duyvendak (8), p. 16, and the elaborate study of Wheatley (1).

[e] Jaubert tr., vol. 1, p. 51. A very preliminary sketch of the Chinese–African trade from the African end has been assayed by Fripp (1). Cf. Wainwright (2). David Livingstone (1), p. 50, came across a trade in furs from Botswana to China. On the porcelain finds at Aden see Lane & Serjeant (1); Doe (1).

[1] 黃巢

Archaeological research in East Africa is now in full swing, and general conclusions can only be provisional. But the positive acquisitions are already extraordinary.[a] Along the entire Swahili coast from Somalia to Cape Delgado 'an unexpected and improbably large quantity of Chinese porcelain' has been found and is under study.[b] A single Tanganyikan collector found 400 shards from thirty sites between the Kenya border and the Rufiji River near the Mafias.[c] On these islands themselves, and in the neighbourhood of Kilwa, Wheeler himself saw great quantities of porcelain fragments. But the porcelain is not always broken, for whole pieces are found inset in the plastered walls of houses and mosques, where also there are niches designed to contain them. A pillar tomb near Bagamoyo (opposite Zanzibar Island) was decorated with sea-green bowls of the Yuan period,[d] exactly contemporary with the descriptions of Wang Ta-Yuan. Broadly speaking (and perhaps as might be anticipated) the oldest periods are represented most strongly in the north, where Sung celadon finds have been plentiful.[e] Further south the evidence points to a great upsurge of the importation of Chinese wares from the middle of the +14th century, after which no reign-period during the Ming and Chhing is unrepresented.[f] Possibly this may be attributed to the decline of the Middle Eastern kilns after the collapse of the Abbasid caliphate in the Mongol invasions.[g] Nor are the finds restricted only to the coastal areas, for many pieces have appeared far inland.[h] Exactly how far south this influence went is as yet hard to determine since few investigations have been reported from Mozambique, but it must at least have reached Sofala.[i] In any case the products of Chinese culture are celebrated in Swahili literature. The late +18th-century poet al-Inkishāfī, describing the wealth of the city of Paté before its fall, says:

> Wapambaye Sini ya kutuewa
> Na kula kikombe kinakishiwa
> Kati watiziye kazi ya kowa
> Katika mapambo yanawiriye.[j]

[a] Cf. Kirkman (7, 8).

[b] Mathew (3). See also Kirkman (1, 2, 3, 6, 9, 10); Mathew (1, 2); Freeman-Grenville (2, 6).

[c] Freeman-Grenville (2).

[d] Hunter (1); cf. Kirkman (5).

[e] The largest amount of good +13th-century celadon has been found at twelve deserted towns on the borders of Somaliland and Ethiopia which belonged to the medieval sultanate of Adal and were destroyed in the +16th century. Here it occurs as far as 200 miles inland (Mathew, 3). At Kilepwa, just south of Malindi, such porcelain occurs at contemporary levels (Kirkman, 2), as also at Mogadishiu.

[f] Freeman-Grenville (2). In a tomb with an inscription dated +1399 at Gedi, Kirkman (4) found sherds of Chinese porcelain of all dates from the +13th century onwards, becoming abundant after c. +1325 (Fig. 982, pl.).

[g] Hūlāgu Khan sacked Baghdad in +1258; cf. Vol. 1, p. 224.

[h] Chinese porcelain from the +13th century onwards has been found at the famous site of Zimbabwe in the south of Rhodesia; McIver (1); Caton-Thompson (1); Stokes (1). This is quite far south, about the same latitude as Sofala on the Mozambique coast. On inland finds see also Davidson (1), p. 239.

[i] Chinese +14th-century celadon and later porcelain has been reported from Madagascar; cf. Deschamps (1); Grandidier & Grandidier (1), pl. 4.

[j] Tr. W. Hichens, mod., in Freeman-Grenville (2). This is as much as to say:

> 'Their feasts were decked with Seric porcelain bright
> Each bowl with finest graving overlaid
> And in the midst the crystal pitchers made
> Glitter and glow above the napery white.'

The other kind of Chinese hardware on the East African coast is monetary—coins and coin-hoards, always so fascinating and yet so difficult to interpret. Out of a total of 506 foreign coins found on the coasts of Kenya and Tanzania and dating from before 1800, no less than 294 are Chinese,[a] and the great majority of them, curiously enough, are of the Sung period.[b] This may not mean that trade was more intense then than at other times, but only that African merchandise was for a period bought with money rather than bartered. The earliest coins date from about +620. An important hoard of Chinese coins was found at Kajengwa in Zanzibar; was it the savings of a settler, or of a Zanzibari who had visited India or China? Such Chinese settlers, generally fishermen now speaking only the local languages, have been reported in the Bajun Islands off the Somali and Kenya coasts.[c] In the north, Thang coins, and many of the Sung (+11th century), have been found at Mogadishiu and other places in Somalia.[d] Further discoveries will be of great interest. Meanwhile many other evidences of Chinese contacts with East Africa are being explored.[e]

It is thus clear that before the appearance of European ships in the Indian Ocean Chinese trading influence extended down the eastern coast of Africa almost as far as Natal, certainly to the mouth of the Zambezi, and that the Mozambique Channel was ploughed by Chinese hulls. How far south the Ming fleets carried their planned investigations, however, is uncertain. The naval secretaries refer in much detail to Mogadishiu, Brawa and Malindi; there is also mention of al-Jubb (Chu-Pu,[1] Jubaland).[f] The *Wu Pei Chih* maps[g] end just south of a port called Ma-Lin-Ti[2] but mark Mombasa (Man-Pa-Sa[3]) north of it. Since the whole coast was known to the Portuguese later on as Melinde, it is probable that Ma-Lin-Ti in this chart means not modern Malindi in Kenya but rather Mozambique (lat. 15° S.).[h] The *Ming Shih* knows of a place called Chhi-Erh-Ma,[4] and it has been suggested that this was Kilwa (nearly 10° S.) south of Zanzibar.[i] Moreover, the official history also names two places as being at the most extreme distance from China, saying that the Admiral (or some of his lieutenants) went there, but that no tribute missions were ever sent by their rulers.

---

[a] See the surveys of Freeman-Grenville (1, 3, 5), admittedly only a beginning, and (Fig. 983, pl.) from Hulsewé (3). Cf. Mathew (1).

[b] Both of the coins found at Gedi were Southern Sung (Kirkman, 3).

[c] Grottanelli (1), cf. Elliott (1). The old city of Paté was on one of these islands.

[d] De Villard (1).

[e] For instance the spread of cultivated plants. A Chinese character has been recognised doing duty as a wall decoration motif in traditional African buildings in Rhodesia (Dart, 1). But this would be more convincing if it had not been the simple geometrical form *thien*,[5] which anyone could have thought of.

[f] *Ming Shih*, ch. 326, p. 11*a*. Cf. Rockhill (1).

[g] A special contribution to Yusuf Kamal's atlas was made by Duyvendak (11) on this subject. Little attention has been paid, however, to the interesting names of mountain-ranges marked on these maps inland from the coast. Could Chhi-ta-erh[6] (perhaps for Chhi-ha-erh[7]) be the Chyula hills, and could Chê-lang-la-ha-lang-la[8] be Mt Kenya or Mt Kilimanjaro?

[h] Mr J. V. Mills identifies the Ko-Ta-Kan[9] of the maps as Quitangonia Island, in Conducia Bay, 10 miles north of Mozambique.

[i] Goodrich (14). The reference is to ch. 332, p. 29*a*. Cf. Bretschneider (2), vol. 2, p. 315. Fripp's 'Quiloa' (1), for Kilwa, is an error; the place in question is Indian Quilon.

---

[1] 竹步   [2] 麻林地   [3] 慢八撒   [4] 乞兒麻   [5] 田
[6] 起苔兒   [7] 起哈兒   [8] 者耶剌哈耶剌   [9] 葛苔幹

These were Pi-La[1] and Sun-La.[2] Though they were almost certainly on the African coast, and if so, farthest south, they cannot as yet be identified, unless the latter should be taken to mean Sofala.[a] Since this was an Arab trading-centre it is quite likely that a squadron of the Treasure-Ships went down that far, and if it did, the position reached was 20° S.[b]

Early Chinese knowledge of Africa is evinced by considerations of quite another kind. Nearly a century ago Rockhill (1) was impressed by the fact that in the *Yü Ti Tsung Thu*[3] (General World Atlas) of Shih Ho-Chi,[4] printed in +1564, South Africa was shown in its right shape, i.e. with its tip pointing to the south.[c] As he well knew, the European cartographical tradition before the Portuguese discoveries was to have it pointing to the east.[d] In fact, Shih's atlas was a relatively late derivative of the *Kuang Yü Thu*[5] (Enlarged Terrestrial Atlas), also printed in the +16th century but stemming from the fundamental work of the great cartographer Chu Ssu-Pên,[6] in progress about +1312 and finished a few years later.[e] Fuchs (1) has established that already at that time Chu Ssu-Pên drew it correctly—an achievement on which no small light is thrown by the facts just summarised concerning Chinese intercourse with Swahili Africa from the Thang onwards. But Chu was not the only Yuan geographer to draw the continent as it really is, there were also Li Tsê-Min[7] and the Buddhist monk Chhing-Chün,[8] one working about +1325 and the other about +1370. When their world-maps reached Korea at the end of the century they were combined into a magnificent planisphere entitled *Hon-il Kangni Yŏktae Kukto chi To*[9] (Map of the Territories of the One World and the Capitals of the Countries in Successive Ages)[f] by the cartographer Yi Hoe[10] and the astronomer Kwŏn Kŭn.[11] Here in +1402, before the first Portuguese caravel had sighted Cape Nun, Africa was made to point south, with a roughly correct triangular shape, and some 35 place-names including Alexandria were marked upon it.[g] This world-map (Fig. 985, pl.) is greatly superior to the Catalan Atlas of +1375 and even to the map of Fra Mauro (+1459), presumably because the knowledge of Europe and Africa which the Chinese scholars obtained from their Arab informants was better and more abundant than all that Marco Polo and the other Western

[a] *Ming Shih*, ch. 326, p. 14b; see Pelliot (2a), pp. 326 ff., (2b), p. 285. For Pi-La, Lo Jung-Pang (7) suggests Zembere at the mouth of the Zambezi, or a town named Belugaras in Delagoa Bay somewhere near Lourenço Marques (c. 26° S.).

[b] The *Wu Pei Chih* maps give polar altitudes (and Hua kai altitudes) for all the chief places they mark on the East African coast; see below, p. 567.

[c] Cf. Vol. 3, p. 560.

[d] Cf. Skelton, Marston & Painter (1).

[e] The map containing South Africa is in ch. 2, p. 87a (Fig. 984). Cf. Vol. 3, pp. 551 ff. Identifications in Fuchs (1), p. 14.

[f] Cf. Vol. 3, pp. 554 ff. and Figs. 234, 235. Important studies on it have been made by Ogawa (1); Aoyama (1, 2, 3); Unno (1, 4) and Miyazaki (1). Kwŏn Kŭn's astronomical work has been discussed in Vol. 3, p. 279 and Fig. 107.

[g] A pagoda-like object represents the Pharos (cf. p. 661). On South Africa in this map see Fuchs (6).

[1] 比剌　　[2] 孫剌　　[3] 輿地總圖　　[4] 史霍冀　　[5] 廣輿圖
[6] 朱思本　　[7] 李澤民　　[8] 清濬　　[9] 混一疆理歷代國都之圖
[10] 李薈　　[11] 權近

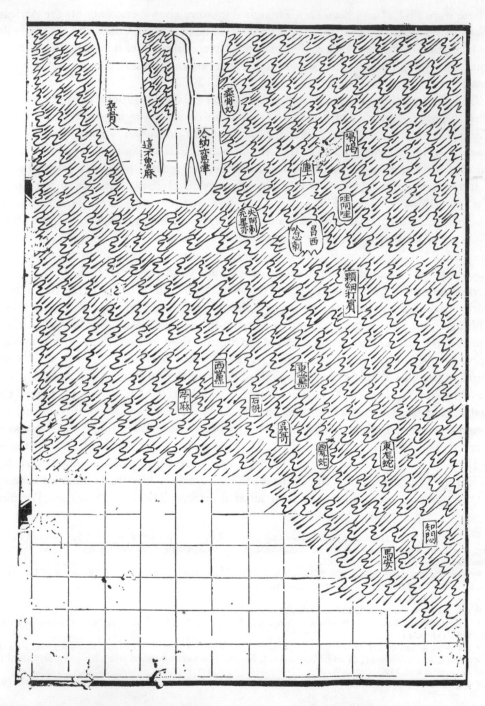

Fig. 984. The representation of the southern part of Africa in the world atlas *Kuang Yü Thu* by Chu Ssu-Pên, *c.* +1315, first printed *c.* +1555 (ed. of +1799). Already in Chu's time the Chinese knew that the continent pointed due south and not eastwards. It is interesting to find that the centre of the land-mass is shown as a vast lake, probably because the existence of one or more of the great East-Central African

travellers could bring home about East Asia. The Chinese were in fact a good century ahead.[a] In Europe, one may perhaps see conceptions of Africa in the act of changing, if Kimble (2) is right in his interpretation of the Laurentian world-map.[b] Here two drawings are superimposed, a medieval ink outline with Africa's tip pointing east, and over it a coloured L-shaped continent pointing south.[c] He supposes that the first dates from +1351, while the latter was painted on top of it after +1450, perhaps after +1500.

At the outset of this digression (if such it is) we were content to acquiesce in the conventional view that the Portuguese were the first to double the Cape of Good Hope.[d] But the atlas of Fra Mauro carries two among its many inscriptions which in this context are very curious. In a scroll on the East African coast near Diab (the Cape), the first of these says:

About the year +1420 a ship or junk of the Indies[e] passed directly across the Indian Ocean in the direction of the Men-and-Women Islands beyond Cape Diab, and past the Green Islands and the Dark (Sea), sailing (thereafter) west and south-west for forty days and finding nothing but air and water. According to the estimate of her (company) she travelled 2,000 miles. Then, conditions worsening, she returned in 70 days to the aforesaid Cape Diab. When the sailors went on shore to satisfy their needs, they saw an egg of the bird called roc, the which was the size of the belly of an amphora; and so great was the size of the bird that the span of its wings was sixty paces. This bird, which can carry off an elephant with ease, as well

[a] Cf. Fuchs (6), but they did not show the continent's north-west bulge.

[b] This is the Portolano Laurenziano-Gaddiano, part of the Medicean Atlas in the Laurentian Library at Florence.

[c] The first printed map of Africa (+1508) by Fracanzano di Montalboddo has just the same shape (see Anon. (47), no. 41, and pl. III b), with the southern projection also inadequately extended. It is interesting, but quite natural in view of the directions from which the continent was approached, that the Europeans should have thought of it primarily as an east–west land-mass, while the Chinese thought of it primarily as stretching north–south.

[d] Unless we accept the *periplus* of the Phoenicians under the Pharaoh Necho II (−609 to −594) recorded in Herodotus, IV, 42, and still maintained by modern geographers (e.g. Debenham (1), p. 30), but which, like Gibbon, I have never been 'disposed to believe'. Cf. Germain (1).

[e] For him, of course, China was part of the Indies, 'extra Gangem'. The word junk is clearly in the original.

Legend to Fig. 984 (*continued*)

lakes (Nyasa, Tanganyika, Victoria, etc.) was known. The words Chê-pu-lu-ma (Chê-pu-lu-ha-ma in some versions) undoubtedly stand for Ar. Jebel al-Qamar, the Montes Lunae (Mountains of the Moon) of Ptolemy, as Takahashi (1) and Unno (5) have seen; probably mod. Mt Ruwenzori, on the Uganda–Congo frontier, possibly Mt Kilimanjaro in N. Tanzania, hardly the Drakensberg range in Basutoland (Chang Kuei-Shêng, 2), in spite of its size. The large island off the east coast marked Sang-ku-nu, lit. 'the slaves of al-Zanj', is clearly Zanzibar or Madagascar; but the similar phrase on the west of the continent (Sang-ku-pa) is puzzling, unless Chang's identification of it as the Congo (after all Songo in Portuguese) may be accepted. The most difficult is the name along the great north-flowing river Ha na-i-ssu-chin (or -chih), for which Fuchs (1) offered 'ford of Hanais' (a mere construct); Chang the more plausible Ar. al-Nīl al-Azraq, the Blue Nile; and Takahashi and Unno surprisingly Ar. khaṭṭ al-istiwā', i.e. the equator. This, it is true, passes along the north shore of Lake Victoria. Fifteen other islands are scattered in the Indian Ocean, possibly Réunion, Mauritius, the Seychelles, the Maldives, etc., but their Chinese names are not easily interpretable. Few will follow Chang Kuei-Shêng in making Chhang-hsi-ha-pi-la Kerguelen Is. near the Antarctic, still fewer his suggestion that Chêng Ho's men (or any Arabs either) went there. In this far southern region the sea gives over, and the characteristic 400-*li* grid resumes for a *terra incognita*.

as all other large beasts, does a great deal of harm to the inhabitants of those parts and is extremely rapid in flight.[a]

And the cartographer continued in another, more southerly, scroll, as part of a passage maintaining the continuity of the Indian and Atlantic Oceans:

Moreover I have had speech with a person worthy of belief who affirmed that he had passed in a ship of the Indies through a raging storm 40 days out of the Indian Ocean beyond the Cape of Sofala and the Green Islands more or less south-west and west. And according to the calculations of her astronomers, his guides, this person sailed 2,000 miles. Whence assuredly we may take him as sincere as those who say that they have sailed 4,000 miles (down the west coast of Africa and back) [i.e. the Portuguese explorers whose charts Fra Mauro says earlier in the same inscription that he had had at his disposal].[b]

This is all that we know. A porthole opens to disclose some sea-going junk flying before wind and rain in the Agulhas Current round the Cape, then caught in the South-eastern Trade-winds till finding no land her master comes down again into the Westerlies and their current further south. And so he finds himself back again in the Indian Ocean where at some landfall his crew stumble upon the traces of giant birds. Then quickly the view is shut off. But we need have no doubt that the junks of Chêng Ho's time (and probably of a couple of centuries earlier) could have accomplished this; the opinion of a most weighty mariner supports us.[c] Curious, too, that the Venetian friar should have named the date of +1420 as that of one of these peregrinations. Yet even if Chêng Ho had nothing to do with them, may he not in some sense be called the Vasco da Gama of China? Exactly in what sense is a question which we shall try to answer a little further on.[d]

The Indian Ocean, Fra Mauro averred, was quite continuous with the Atlantic, no Austral or Antarctic continent standing in the way. In this he was but following the consensus of opinion among the great Arabic geographers from Ibn Khurdādhbih in the +9th century through al-Bīrūnī in the +11th, down to their disciple Marino Sanuto (+1306).[e] The Arabs, moreover, denied the common Western opinion that the south-

---

[a] Tr. auct. after Yusuf Kamal (1), vol. 4, pt. 4, pp. 1409 ff. Everyone was interested in the ostrich eggs of these parts. Says the *Ling Wai Tai Ta* (+1178) of Madagascar: 'There are great *phêng*[1] birds there, which so mask the sun in their flight that the sundial shadow shifts. If one of these finds a wild camel, it swallows it whole. And if a man should chance to find a *phêng*'s feather, he can cut a water-butt from its quill' (cf. 3, p. 6a); cf. *Chu Fan Chih*, ch. 1, p. 32b. Cf. also Vol. 2, p. 81.

[b] Tr. auct.

[c] Villiers (2), p. 102. Long after writing this paragraph I was glad to find that Chang Kuei-Shêng (1) had argued that Ming ships circumnavigated Madagascar and that Lo Jung-Pang (7) had independently proposed the same interpretation of Fra Mauro.

[d] Yuan Chia-Ku (1) called him 'of the line of Columbus and Magellan', cf. Pelliot (2b), p. 280. But the 'Vasco da Gama of China' was a title not given him by a Chinese writer, rather by one of the most eminent of Western geographers, Debenham (1), p. 121. And he had been preceded by one of the greatest of Western orientalists, Lévi (2), p. 440. The commonly received opinion, however, continues to flourish. In a leading article in *The Times* of 27 June 1964 we find: 'In spite of the long coast-line there is no record of any Chinese Columbus wondering what lay beyond those heaving waters.' See further pp. 551 ff. below.

[e] This matter has been studied by Issawi (1). Al-Idrīsī was one of the minority who did not believe in a southern passage; cf. his world-map reproduced in Fig. 239 of Vol. 3 (c. +1150).

[1] 鵬

ern hemisphere would be too hot to permit human life, even if one could get there. And now we enter upon the second half of the story which we are following. The Arab geographers knew well of a land in the west, bordering upon the occidental ocean, which they called al-Burṭuḳāl (from Portus Cale, modern Oporto) and which their Muslim brothers had taken from the Visigoths during the general conquest of the Iberian peninsula early in the +8th century.[a] But during the time of al-Idrīsī the Christian princes of the north succeeded in reconquering most of Portugal, and by +1185 were in possession of all except the southern province of Algarve. From this the Muslims were expelled in +1249, though the last battle which secured Portugal for Christendom was not fought till +1340.

Of the great epic of Portuguese maritime discovery and expansion in the +15th century we need not speak in detail here, since it is so widely known, and described in so many useful books.[b] But in order to make the comparison which is demanded by the contemporaneity of the Chinese navigations, we must give at any rate a summary of it. During the first half of the +15th century the Portuguese were creeping down the west coast of Africa while the Chinese were examining the east coast at least as far south as Mozambique; during the second half the Portuguese found their way round into the Indian Ocean to meet no one but Arabs and Africans, since a change of policy in China had withdrawn the treasure-fleets for good.

(iii) *The Sea-Prince of the Five Wounds*

When Chêng Ho was a boy of about seventeen, revelling in the sunny climate of upland sub-tropical Yunnan, there was born at the other extreme end of the Old World in an equally beautiful country a boy whom it pleases us now to think of as a key figure of parallel historical importance.[c] Henry of Avis, called by historians the Navigator, was of birth more exalted than Chêng Ho, since his father was King João I, a monarch whose reign achieved the definitive emancipation of Portugal from the recurring threat of Castilian sovereignty. At the Battle of Aljubarrota (+1385) which effected this, the Portuguese were assisted by English archers, and next year King João married an English girl, Philippa of Lancaster, the daughter of John of Gaunt. Dom Henrique 'the Navigator' (+1394 to +1460) was their third son,[d] and while he was brought up in all the usual chivalric mentality of the time, he developed into something altogether more original—a prince who realised the value of ships and could talk

[a] Dunlop (3). Cf. pp. 475, 511.

[b] One need only instance Prestage (1) or the more introductory Parry (1). But the best and most recent chronicle of the Great Discoveries is that of Peres (1), published both in Portuguese and English on the occasion of the Quincentenary of Prince Henry.

[c] How strange it is to think, not only that they never met, but that they never even knew of each other's existence. Probably neither of them knew anything of a third, equally great, contemporary, the astronomer prince Ulūgh Beg (+1393 to +1449) working away in Samarqand (cf. Sayili, 3).

[d] Prince Henry has inspired biographers all worth reading: Major (1); Beazley (2), and the romancée account of Sanceau (1). For the Quincentenary an excellent album of illustrations with a reliable text was produced by Marjay (1). The latest biography is by Nemésio (1). Henry's fostering of the university as a nursery of cosmographers is described in the monograph of de Sà (1). Cf. Anon. (47, 74); Brochado (4) in Anon. (53).

with the rough sea-farers and fishermen who handled them, a medieval noble who did not disdain to commune for many years with men of learning, astronomers and cosmographers, and in the end a visionary imbued with a world-shaking idea. Causes and motives will occupy us later, but there can be no doubt that the men around Prince Henry were convinced of the truth of the Arab belief that south of Africa there was a cape which could be rounded. They knew, moreover, all that +15th-century Westerners could know about the Indies and the Further Indies. They hoped to find help from Prester John in Abyssinia or else in Central Asia.[a] If then the Cape could be rounded, and one came to the Red Sea and the Persian Gulf from the south, establishing links with the Indian and East Indian producers of silks and spices, what would be the position of Islam? The flank of the Arab world would have been turned.[b] And thus the confabulations of the 'School of Sagres', the group of navigational planners gathered round Prince Henry during his long governorship of Algarve at Lagos and at the scientific citadel on the Sagres promontory,[c] with its aim of using all the powers of heaven and earth to find a way of taking the Saracens in the rear, constituted a veritable transition from the medieval mentality of the Crusades to the Renaissance out-reaching towards all possible knowledge. The search for the southern passage was to be the new secret weapon of Christendom in its age-long battle against the 'infidel', and Sagres was the Los Alamos of the +15th century. Most of the time the 'Commander of the Faithful' had headed what now seems to us a clearly superior civilisation. Perhaps too secure in that knowledge, the Arabic world did not appreciate that the pans of the balance were slowly shifting; and no Muslim ruler thought to act upon the teaching of his own geographers, sending down expeditions to strengthen the cities of al-Zanj and perhaps to build a giant fortress or a powerful fleet to deny to the Franks that fateful southern portal. Instead, the methodical exploration of the West African coasts proceeded unopposed. Even the definitive European discovery of the Americas was in a way its by-product.[d]

The opening rounds were fired at Ceuta. Perhaps the first idea of the House of Avis was for a conquest of Morocco; in any case the city, just across the Straits, was a centre of Moorish naval strength which would have to be reduced if Portuguese ships were to pass freely back and forth along the south-western coasts. The city was therefore taken and sacked in +1415, the year of Chêng Ho's return from his fourth voyage, but doubt-

[a] Cf. Vol. 1, p. 133; Vol. 4, pt. 1, p. 332. The book of Sanceau (3) considers him only in his Ethiopian manifestation. But Coimbra (1) shows that the aim of Prince Henry already was to get to the real Indies.
[b] On the ever-recurring theme of political travel for strategic flanking movements see Vol. 1, pp. 223 ff.
[c] The 'School of Sagres' has perhaps become something of a legend, but there can surely be no doubt that from +1435 onwards Prince Henry's court was settled at Raposeira between Lagos and the Sagres promontory, if not nearer the present fort itself, the first form of which was doubtless built by him; and that he did welcome and entertain for long periods astronomers, cartographers, cosmologists and sea-captains. Sagres is one of the most interesting sites in all Europe; I had the pleasure of visiting the place in 1960. Whether the stone compass-rose now to be seen there dates from Henry's own time is a subject still under discussion (cf. Madeira, 1). Madeira believes that it was intended to centre round a vertical gnomon as a kind of sundial, but the view of Fanning (1) that it was used to enable ships to check and adjust their compasses by the reciprocal bearings technique seems at least as plausible. De Zurara says in his chronicle that ships would be able to 'take their bearings' at Sagres (Beazley & Prestage tr. p. 21).
[d] Cf. p. 513 below.

less no word of the great event reached the Chinese flagship lying in the roads of Ormuz. After the fall of the city the Infantes Henry and his brothers wrote a remarkable letter. John, Duke of Bourbon, had challenged them to meet him in single combat, with sixteen knights and esquires 'de nom et darmes sans reprouches' on each side. But now they had no time for jousting. They answered that having with God's help won Ceuta they intended to continue and win for Him many other cities and towns in Africa.[a] Though still in a crusading context, their attitude symbolises the change from the departing Middle Ages to something radically new.[b] But perhaps this was not all gain. 'With the capture of Ceuta', wrote Parry, revealingly, 'the crusading movement passed from its medieval to its modern phase; from a war against Islam in the Mediterranean basin to a general struggle to carry the Christian faith and European commerce and arms around the world.'[c]

After Ceuta ships were sent out almost every year to sound the ocean, for Henry 'had a wish to know the land that lay beyond the Isles of Canary and a cape that is called Bojador, because until that time, neither in writing nor in men's memory had it been definitely known what was the nature of the land beyond that cape'.[d] For the first dozen years they were concerned mainly with Madeira and the Azores, but in +1426 Frei Gonçalo Velho surveyed the coast of the Anti-Atlas mountains, and in +1434 Gil Eanes doubled Cape Bojador (26° N.). In the former year the last and greatest of the Chinese voyages to Africa was being prepared, in the latter it had just gone home. In +1444 Nuno Tristão reached the mouth of the Senegal River (16° N.), and two years later Álvaro Fernandes was on the Guinea coast (12° N.). The year +1453 was marked by two events, one colossal blow and one seemingly minor affair: Byzantium fell to the Turks, as if to show the Portuguese that they were none too soon in their endeavours,[e] while along the African coast sailed Cid de Sousa, leading the first expedition to Guinea with the primary object of trade.[f] Then came the death of Prince Henry, just after Pedro de Sintra had reached Sierra Leone (8° N.). A lull of some ten years followed, but in +1471 João de Santarém pushed on to Ashanti (5° N.)[g] and three

---

[a] Brit. Mus. Add. MS. 18,840, cf. Anon. (47), no. 8.

[b] To understand the crusader spirit of Portugal at this time one should remember that in the Iberian peninsula itself there were Muslim states, including the highly cultured Emirate of Granada, which did not fall till +1492. In a sense, the discoveries of West and South Africa were only a continuation of the *reconquista*. If Fuchs (7) is right in his interpretation of evidence which we shall consider in Sect. 33, there had been an embassy from Granada to China about +1317.

[c] (1), p. 11.

[d] From the chronicle of de Zurara; Beazley & Prestage tr. pp. 27, 32. Cf. L. Bourdon (1).

[e] This was also a mortal blow for the Venetians, who had always tried to keep on good terms with the Turks, and had hoped that a way for Asian commerce could be kept open for Christendom through the Middle East.

[f] A persistent claim has stated that French sea-captains from Dieppe established trading posts on the West African shores as early as +1364 (see e.g. de Loture & Haffner (1), pp. 65 ff.), but the evidence for it is extremely weak. De Santarém (3) criticised it devastatingly more than a century ago, but no one reads him (cf. Godinho, 1). De Sousa's trading expedition was soon followed by others, notably those of the Venetian Alvise Ca' da Mosto and the Genoese Antonio Usodimare, merchant-captains in Prince Henry's service (cf. Prestage (1), pp. 94 ff.). On the historiography of the Portuguese discoveries see Brochado (3).

[g] Here the Portuguese completed, by +1482, their most important West African fortress, El-Mina (Peres (1), p. 58). Its present remains are described in A. W. Lawrence (1).

years afterwards the equator was crossed when Lopo Gonçalves came first to Cape Lopez (now in the French Congo, and 2° S.). This was a latitude equivalent to that of al-Jubb, where the Chinese had passed on their parallel south-eastern journeys.

After a period of internecine struggle between Portuguese and Spaniards (who wanted to participate in the profits of West African gold and slaves), the explorations started again on a distinctly larger scale. Between +1482 and +1486 Diogo Cão made two remarkable journeys, marking the places he visited by stone crosses (*padrões*) which he took out with him from Lisbon, fixing first Cape Sta. Maria in Angola (14° S.), then Cape Cross in Damaraland (22½° S.). Now the Portuguese were almost beyond the limit of the Chinese explorations on the other side. At last came the historic voyage of Bartolomeu Dias in +1488, rounding the Cape of Good Hope (about 35° S.) and naming the truly most southerly point, Cape Agulhas. The way was now open for the culminating cruise, that of Vasco da Gama, who left Lisbon in +1497, was off the mouth of the Zambezi early in the following year, and then entering the 'Chinese' area, touched at Malindi in April, just about fifty years after the Ming navy had ceased to frequent those shores. At Malindi he was fortunate enough to obtain the services of one of the leading Arab pilots of the age, Aḥmad ibn Mājid, who brought the Portuguese admiral to Calicut in India the following month.[a] The die was now cast, the Europeans were in the Indian Ocean for good or evil—and of the latter much.

After this, exploration and the extension of European geographical knowledge went rapidly forward. The second Portuguese fleet which sailed to India, under the command of Pedro Alvares Cabral in +1500, touched at Brazil on the way, and took a closer look at the Arab city-States of the East African coast, especially Sofala and Kilwa. A few years later Vicente Sodré landed on Socotra Island off the tip of Somalia.[b] By +1507 the title 'Governor of India' was arrogated to Alfonso de Albuquerque, and by +1510 he had captured the Arab port at Goa on the west coast of Bijapur, one of the Deccan sultanates into which the Bahmani kingdom had split.[c] Within five years, all that remained of the Governor's own life, it had become already a large city, and after +1530 the chief base of the Portuguese empire in Asia.[d] European historians credit him as the first sea commander to appreciate the complex relationships between

[a] An anonymous *Roteiro* of this voyage has been preserved, and translated with commentary by Ravenstein (2). On Ibn Mājid see Szumowski (1, 2) and Brochado (1).

[b] It is almost invariably said that Sodré 'discovered' Socotra. It is high time that this tiresome convention of European historians was discontinued. A Portuguese ship could not 'discover' an island which had been colonised by Arabs and visited by Chinese for many centuries previously. The junks of the Ming navy certainly went there, and as far back as +1225 there had been an account of the island (*Chu Fan Chih*, ch. 1, p. 27a, cf. Hirth & Rockhill, pp. 131 ff.), in which Chao Ju-Kua took care to mention its most characteristic export, the red resin called 'dragon's-blood'. This comes from the fruits of palms (*Daemonorhops* spp.) and of trees and shrubs (*Dracaena* spp.); see Burkill (1), vol. 1, pp. 747 ff., 857 ff. It is used as a dye. Let us say: 'After Sodré's visit, Socotra became known in Europe.'

[c] From this time (+1509) dates the oldest extant Portuguese chart of the Indian Ocean, on which see Uhden (2). On Portuguese cartography in general see Cortesão (3).

[d] On Goa see Penrose (1) and the more popular books of Collis, especially (2). On the history of Portuguese relations with the Malabar coast see Panikkar (2). One of the better points of Albuquerque's policy was not only the general absence of colour bar in Portuguese settlements in Asia but the positive encouragement of marriages between Portuguese and Arabic, Indian and Indonesian girls. Couples were given dowries of boats and other bread-winning gear. Cf. Baiao (1), pp. 10, 62, 74, 139; and, in general, Freyre (1). This was in striking contrast with the attitudes of other European nations in such matters.

fleets and ports,[a] at any rate he certainly saw Portugal's need for a permanent fleet and naval bases in the Indian Ocean. One suspects that most of these ideas would have been equally obvious to Chêng Ho and Wang Ching-Hung—but there was a slight difference: they were not engaged in a war of conquest in the regions between Malaya and Natal, and they could count on the co-operation of local rulers when they needed to careen, re-caulk, or even build, when far from home.[b] The year before Goa was taken, Malacca, that old rallying-point, had first been visited by Portuguese, who also 'discovered' the Nicobar Islands on the way, and in the year after, it was stormed by the Governor and appropriated to the empire. Two final features will suffice to complete this summary: in +1512 Francisco Serrão attained at last to the 'spice islands', exploring the Moluccas beyond the Celebes and annexing Timor, and significantly he used a junk to do so; secondly Jorge Álvares came at last to China with a trading ship in +1513. Thence he and his men could look across in the direction of the Philippines which from +1521 onwards were to be Spanish.[c] And so the whole world was 'bounden in a bond', and if four hundred rather than 'four thousand winter' would put it right again, many of the consequences were permanent and irreversible.[d]

It is of much interest that the former Chinese presence on Indian Ocean coasts was soon known, but much misunderstood, in Europe. In the *Lusiad* Camoens represents certain civilised Africans living between Sofala and Mozambique as telling the Portuguese of da Gama's ships that they were not the only light-skinned navigators of those seas.[e]

> In the Arabick-Tongue (which they speak ill,
> But Fernand Martyn understandeth though)
> They say, in Ships as great as these we fill,
> That sea of theirs is travers't to and fro,
> Even from the rising of the Sun, untill
> The Land makes Southward a Full Point, and so
>   Back from the South to East, conveying, thus,
>   Folks of the colour of the Day, like Us.

[a] E.g. Parry (1), p. 41. The Portuguese policy of setting up overseas bases was no doubt forced upon them by the hostility of the Muslim traders who were *a priori* antagonistic to any European activities in Indian waters. They naturally felt that their position as middlemen of the Eastern trade with Italy and other parts of Europe was threatened. In this they were quite right, but the whole situation was a direct legacy of the Crusades. The irreconcilable antagonisms of the Levant were thus transferred to poison the whole of the Indies.

[b] Certain Chinese sources such as the *Chhien Wên Chi* give us a picture of what seems like an exceptionally large proportion of carpenters, iron-workers and all kinds of artisans in the crews of the Treasure-ships of the expeditions under Chêng Ho; cf. Pelliot (2a), p. 306.

[c] This is of course the date of the first 'discovery' under Magellan; the islands were not colonised in force until the 'eighties; cf. Masiá (1), p. 583. The Portuguese had first encountered Chinese junks at Malacca in +1509, and came first to Canton in +1517. Macao was founded (by Portuguese 'squatters', the Chinese always used to say) much later, in about +1555, but its importance as a cultural entrepôt proved very great, especially because of the Jesuits (cf. Vol. 3, pp. 437 ff., Vol. 4, pt. 2, p. 436). See the entertaining account of life there by Boxer (4).

[d] For instance the fusion of medieval European and Asian sciences into one universal modern science (cf. Vol. 3, pp. 448 ff., and Needham, 59).

[e] v, 77, Fanshawe tr. p. 166; cf. Aubertin tr. vol. 1, p. 279; Atkinson tr., p. 135. The epic was begun in Macao about +1556 but not printed till +1572.

Actually the information had come from India. After the return to Lisbon of Nicolau Coelho, one of Vasco da Gama's captains, but before the arrival of the Admiral himself, a Florentine merchant residing there, Girolamo Sernigi, wrote a report of the great voyage to a colleague in Italy. In this he mentioned that 'certain vessels of white Christians' had visited the Malabar ports regularly during the preceding century. Writing in July +1499 he said:

It is now about 80 years since there arrived in this city of Chalicut certain vessels of white Christians, who wore their hair long like Germans and had no beards, except round the mouth, such as are worn at Constantinople by cavaliers and courtiers. They landed wearing a cuirass, helmet and visor, and carrying a certain weapon like a sword on the end of a spear.[a] Their vessels are armed with bombards shorter than those in use with us. Every other year they come back with 20 or 25 vessels. They (the Indian informants) are unable to tell what people they are, nor what merchandise they bring to this city, save that it includes very fine linen-cloth (probably silk) and brass-wire. They load spices. Their vessels have four masts like those of Spain. If they were Germans it seems to me that we should have had some notice about them; possibly they may be Russians if they have a port there. On the arrival of the Captain (da Gama) we may learn who these people are...[b]

This letter was first published by di Montalboddo (2) in +1507, but whether Sernigi ever gained any further enlightenment the available documents do not say. It is quite evident now, however, that the 'white Christians' were the Chinese, if only from the description of their characteristic hand weapon. It may well be, as some have thought,[c] that the Arabs of al-Zanj welcomed the Portuguese at first precisely because they thought they were Chinese, and only became hostile when they found that they were Christians from Frankistan. A comment in itself, this, sad and paradoxical, on that culture which bore upon its banners 'On earth peace, and goodwill toward men'.

(iv) *Contrasts and comparisons*

Now comes the moment for comparisons and contrasts between the sea-farers from East and West. Unanswered questions present themselves in some such order as this—nautical aspects, war, trade and religion. Let us try to elucidate them with all justice to both sides, captains and crews that never met on earth, but only now in the court of history.

From the maritime point of view, central to the present Section, we can see at once that the Chinese achievement of the +15th century involved no revolutionary techni-cal break with the past, while that of the Portuguese was more original. The Chinese had had their fore-and-aft lug-sails since the +3rd century at least,[d] and already in the time of Marco Polo and Chao Ju-Kua their ships were many-masted.[e] If they used

---

[a] Obviously the crescentic bill or gisarme (*chi*[1]). Cf. Sect. 30c.
[b] From Ravenstein (2), p. 131.
[c] Osorius Silvensis, (1), p. 29b, quoted by Mickle (1), p. 26. For these interesting references we are indebted to Prof. Donald Lach and Dr Lo Jung-Pang.
[d] See pp. 600 ff. below.
[e] Cf. p. 467 above.

[1] 戟

their mariner's compasses in the Mozambique Channel, they were only doing what their predecessors had done in the Straits of Thaiwan right back to the foundation of the Sung navy at the beginning of the +12th century.[a] Though their stern-post rudders were attached in weaker fashion to the hulls than those of the Westerners with pintle and gudgeon, they were highly efficient in more ways than one, and descended from patterns as early as the +1st century.[b] The most obvious difference which would have struck everyone if the vessels of da Gama had met those of Chêng Ho lay in the much greater size of the Treasure-ships of the Grand Fleet (Ta-tsung Pao chhuan[1]) for many of these were of 1,500 tons if not considerably more,[c] while none of Vasco's were over 300 tons and some were much less.[d] In shipbuilding China was far ahead of Europe.[e] But while the Chinese vessels were the culmination of a long evolutionary development, those of the Portuguese were relatively new in type.[f] At the end of the +14th century European ships were equipped only with square-sail rig; the *barca* might be some 30 tons with one mast, or up to 100 if with two. Of this kind, doubtless, were the early vessels sent out by Prince Henry. But finding the North-east Trade-winds quite contrary to their return from the Guinea coast, the Portuguese threw overboard the square-sail rig, and for their famous caravels[g] adopted a fore-and-aft one, in the form of the lateen sail, from their enemies the Arabs. This permitted sailing much closer into the eye of the wind. For the past half-century, since the fundamental work of Lopes de Mendonça,[h] it has been clear that the caravel was a capital invention.[i] By +1436 (i.e. by the end of the Chinese period) these ships carried triangular lateen sails on as many as three masts and averaged between 50 and 100 tons. Then, as the century went on, the superior advantages of the square-sail for running before the wind reasserted themselves, and ships began to be built which combined both rigs. Thus about +1500 the *caravela redonda* (up to 200 tons) carried two rather small square-sails on the foremast, and lateens on the three after-masts, while the *nau redonda* (of about the same burthen) had multiple square-sails on the fore- and main-masts with a lateen only on the mizen. Columbus' *Santa Maria* was so rigged. But the originality of the Portuguese seems somewhat qualified when we remember that of the basic inventions they used, the mariner's compass and the stern-post rudder were

<hr />

[a] The Chinese contribution to the development of the mariner's compass has been related in Vol. 4, pt. 1, pp. 249 ff. A summary will be found in Needham (39).

[b] See pp. 637 ff. below, summarised in Needham (40).

[c] Mills (9) estimates as high as 2,500 tons burthen for the largest. Cf. the discussions on pp. 452 and 480 ff. The average complement rose during the first half of the +15th century from under 450 to over 700 men (Lo Jung-Pang, 2). Cf. Kuan Chin-Chhêng (1).

[d] Prestage (1), p. 250. The Chinese ships probably looked something like the junks shown in Figs. 986, 987 (pl.), but much larger.

[e] I am glad that Thien Ju-Khang (2) also emphasises this.

[f] An admirable account of the Portuguese contribution to ship design and rig in the +15th century is given by da Fonseca (1).

[g] The word itself is said to be of Arabic origin (da Fonseca (1), p. 43). Oriental influence is also claimed for the hull (Amsler, in Parias (1), vol. 2, pp. 25 ff.). On the lateen sail, see pp. 609 ff. below.

[h] Many students confirmed him—Navarrete, la Roërie, Guillén y Tato, da Fonseca.

[i] At the same time it is clear from comparative study that everything the Portuguese did with their caravels could have been accomplished equally effectively with small sea-going junks. Voyages might have taken a little longer—but the crews would have kept much drier.

[1] 大䑸寶船

Fig. 986. Drawing of a five-masted Pei-chi-li freighter (Landström) to give some indication of the probable type of build of the much larger Treasure-ships of the Grand Fleet of the +15th century. Cf. pp. 480 ff. See also Figs. 935, 939, and 936, 938, 1010, 1027, 1042 (pls.).

transmissions from much earlier Chinese practice,[a] the principle of multiple masts was characteristically Asian,[b] and the lateen sail was taken directly from the Arabs. Of the comments of the first Chinese sailors who examined European vessels little or nothing has remained, but from the middle of the +16th century there was a slight admixture of types, as we have seen,[c] though Chinese shipbuilding remained for the most part unchanged. Since Chinese pictures of European ships are rather uncommon, we reproduce in Fig. 988 that of the *hai po*[1] which Fr. Verbiest included in his *Khun Yü Thu Shuo*[2] (+1672).[d]

[a] The importance of these two discoveries for the great geographical explorations has been repeatedly stressed by historians; cf. a typical statement in Trend (1), cit. p. 652 below.
[b] Cf. p. 474 above and p. 602 below.   [c] P. 438 above.
[d] Ch. 2 (p. 215); cf. p. 428 above. The picture was reproduced in *TSCC, Khao kung tien*, where it brings up the rear of the shipping illustrations in ch. 178. On the comparative superiority of European ocean-going ships over those of South Asia after +1500 see our comments on Cipolla (1), pp. 513, 514 below.

[1] 海舶   [2] 坤輿國說

There was another matter in which the Portuguese showed seemingly more originality than the Chinese, namely the understanding and use of the régime of winds and currents. It would be truer to say that the problems set for them by Nature were much more difficult, and that they rose gallantly to the occasion. For the Atlantic Ocean had never been explored, and these were the seas of which it could most truly be said: 'por mares nunca de antes navegados'.[a] The general situation can be at once appreciated from any good world-maps of winds and currents (cf. the companion Fig. 989 b).[b] Almost as far south as Madagascar the Chinese were within the realm of the monsoon winds,[c] the 'junk-driving winds' with which they had been familiar in their own home waters for more than a millennium. One sailed south in the winter and north in the summer (broadly speaking).[d] Once free of the narrow waters of the East Indies, one had the North Equatorial Current to help one across from Sumatra to Zanzibar (if one was not calling at Calicut); and the more southerly Equatorial Counter-current might help one back.[e] But the inhospitable Atlantic had never encouraged sailors in the same way, and though there had been a number of attempts to sail westwards,[f] that ocean had never been systematically explored. First the Guinea coast proved to be a trap, for while the North-east Trade-winds helped the ships down, and the Canaries Current and the Guinea Current lent their aid, the journey back meant endless tacking. This (as well as disease) might be the origin of the English nautical adage:

> Beware, beware the Bight of Benin,
> Whence few come out, though many go in.

But the Portuguese knew well that above the 'Horse Latitudes' under the Azores, there were strong westerly winds (accompanying indeed the Gulf Stream) which would blow them home, so in returning from El-Mina, the fortress-factory they had built on the Volta River, they bore out far westwards into the Atlantic with the Trades on their starboard beam, and then came north into the Westerlies to run for the Tagus. This

---

[a] Camoens, *Lusiad*, I, 1. Of course sailors the world over had been working according to prevailing winds and currents since time immemorial. Casson (1) has shown that the Roman grain-ships returning from Alexandria used to sail east of Cyprus and touch at North Syria on the way home instead of taking the direct route to Ostia in western Italy by which they had come. Similarly the Peruvian sailing-rafts were found to be using the inshore current when southbound, but going out to sea to take advantage of the Humboldt Current when northbound (cf. Heyerdahl (2), p. 615, and Col. B. Kennon in Leland (1), p. 71, for the Mexican–Californian coast, as also p. 547 below). But the scale and boldness of the Portuguese hydrographic work in the Atlantic astonishes.

[b] Opp. p. 560. See A. A. Miller (1); Sverdrup, Johnson & Fleming (1). Cf. G. R. Williams (1).

[c] The full strength of the monsoon is said to end about the latitude of Malindi (Fripp, 1).

[d] Cf. Duyvendak (9), p. 358. Round Singapore the S.W. monsoon blows from April to October, and the N.E. monsoon from November to March. Summer brings the 'line squalls' or 'sumatras', winter day after day of continuous rain and low cloud, as in the graphic description of Ommaney (1), p. 119, well worth reading. The Pao chhuan sailed from Malacca homeward bound in the fifth month (*Ying Yai Shêng Lan*, pp. 36, 37). Cf. Marco Polo's statement in Moule & Pelliot (1), vol. 1, p. 161, vol. 2, p. lxii; and Chhoe Pu's remarks in Meskill (1), p. 47. See also pp. 451, 462, 571 here.

[e] It seems unlikely that Chinese ships ever made much use of the South Equatorial Current running westwards, and its accompanying easterly Trade-winds, though perhaps the junks of which Fra Mauro heard (p. 501) may have done so.

[f] We hear of at least one such attempt by Moors from Arabic Lisbon (cf. pp. 475, 503) and there were the Genoese Vivaldi brothers in +1291. Usually no one returned. Cf. Dunlop (4).

海
舶
圖

Fig. 988. The oldest Chinese picture of a European ship, that in the *Khun Yü Thu Shuo* (Explanation of the World Map *Khun Yü Chhüan Thu*), by the Jesuit Ferdinand Verbiest (Nan Huai-Jen) printed in +1672. The vessel is recognisably a full-rigged ship, with the yards on the mizen very canted, by which the artist must have intended to represent lateen sails. These were, however, by this time out of date; characteristic of the +16th century but now being replaced by the gaff- or yacht-sail, so he must have depended on older sketches.

course was known as the Volta da Guiné or the Sargasso Arc (Fig. 989 a).[a] A chronicler of King João II (r. +1481 to +1495) relates an interesting event at his court.[b] Talk turned at the royal table to the sea-ways to Guinea and back, and a notable navigator, Pero d'Alenquer (*muito grande piloto de Guiné*),[c] boasted that he would bring back a square-sailed *nau* safely, however large she was. But the king insisted that this might no wise be, and that only lateen-sailed caravels could make the return journey. In fact the Arc was a State secret, all the more important because the Portuguese had acquired particular skill in the building and handling of caravels. Their sale abroad was forbidden, and other countries constructed and sailed them only with much difficulty.

As the Portuguese found their way down the African coast they encountered the opposite hydrographic situation. The northward Benguela Current opposed their progress, and the strong South-east Trade-winds were equally unfriendly. But once beyond lat. 35° S. the Westerly Polar Winds, very strong in summer, would help them on their way. Thus towards the end of the +15th century another great Arc was ventured, the Volta do Bresil, i.e. the Brazil, or Cape St Roque, Arc. Leaving the African coast about Sierra Leone, the ships bore out far into the Atlantic with the wind on their port beam, made landfall if necessary south of that Brazilian cape, and so, sailing on south, circumvented the South-eastern Trade-winds until the Roaring Forties shot them through into the Indian Ocean. Such essentially was the navigation of da Gama and Cabral. The fact that they sailed with barques rather than caravels is highly significant, for it can only mean that previous Portuguese explorers had plotted out the route.[d] How far the +15th-century Portuguese advanced beyond

[a] An excellent account of the Portuguese oceanographic discoveries, hitherto insufficiently appreciated by historians of science, has been written by Costa Brochado (2).

[b] J. Cortesão (1), p. 32; Prestage (1), p. 201.

[c] He was the pilot of Vasco da Gama's flagship.

[d] This is what we had in mind when we said a few pages above that the discovery of the Americas was perhaps a by-product of the 'Crusade'-route efforts. The evidence suggests that the discovery of Brazil must have been made between +1486 and +1497 (Brochado (2), pp. 55 ff.). Columbus' voyage in +1492 seems to have been only one item in a series of which most of the details have been lost, probably because they were secret. Writing in +1505 Duarte Pacheco already knew that the three Americas from

---

Legend to Fig. 988 (*continued*)

In the book the accompanying text says: '(A Western) ocean-going ship (*hai po*) looks both broad and large; it can carry more than 1,000 men; it has more than ten sails needing in all about 24,000 ft. of (canvas) cloth; the masts are 200 ft. in height; the iron anchor weighs over 6,350 catties (c. 3.7 tons); the cordage weighs more than 14,300 catties (c. 8.4 tons). Further details can be found in the last section of the *Hai Po Shuo* (a chapter earlier in the book).'

It must be remembered that this was a far more advanced type of ship than those in which the Portuguese had first entered the Indian Ocean. Already at that time however they were probably superior to those of any South Asian nation, and by the time, thirty years later, when European–Chinese maritime contacts began in earnest, two things had happened: the Ming navy had been completely run down both in individual size and numbers (cf. p. 524), and the ships of the West, built with the new technology and above all armed with the improved ordnance of Renaissance capitalist Europe, could outclass in turn their East Asian rivals. Since the fleets of Chinese bureaucratic society never regained their +15th-century apogee, by a hundred and sixty years later, the time of this picture, the superiority of Western warships was unquestionable, and indeed the period of the Opium Wars was already adumbrated, a period only now in our own time ending.

the level of the Arabs and the Chinese in astronomical navigation during this period is a difficult question, and we may return to it later (pp. 557, 567).

Now we can turn to consider briefly war and trade. Here the contrast is an extraordinary one, for while the entire Chinese operations were those of a navy paying friendly visits to foreign ports, the Portuguese east of Suez engaged themselves in total war.[a] Already in +1444 the first casualties of their campaign occurred, when Gonçalo de Sintra and six others were killed at the Gulf of Arguim in Mauretania while trying to capture some of the inhabitants.[b] But so long as the Portuguese were working down the West African coast their aggressive activities were (apart from slaving) relatively restrained, and it was only after +1500, when they were in a position to carry on terrorist warfare against the East African Arabs, and then against the Indians and other Asians, that European naval power showed what it could do in earnest. 'In the bloodstained history of mankind upon this earth, that portion of the East African shore successively described as Azania, the land of Zanj and the Swahili Coast, certainly had its full share of strife.'[c] Before the coming of the Portuguese the Arab city-States had no defence-works;[d] these only arose when it became clear that it was the settled policy of the Westerners to destroy the Arab African–Indian trade root and branch.[e] It would be tedious to recount all the savage attacks which took place, the sacking of Mombasa in +1505, the devastation of Oja, Brawa and Socotra in the following year, the burning of Mombasa again in +1528, and so on.[f] In the 'eighties two attempts were made by the Turks under Mir Ali Bey to recapture the coast, but they failed, and in +1587 Faza, Mombasa and Manda were again burnt to the ground by the Portuguese. The records repeatedly say that in many of these sieges no living thing was spared. In India and on the way there the Portuguese behaved in the same way. They battered and sank Arab pilgrim-ships, fired the limbs of executed Muslims over Indian cities, treacherously killed their ally the head of the Javanese colony at Malacca, and blinded with red-hot bowls the relations of the Sultan of Ormuz.[g] 'Cruelties', wrote Whiteway, 'were not confined to the baser sort, but were deliberately adopted as a line of terrorising policy by Vasco da Gama, Almeida and Albuquerque, to take no mean examples. Da Gama [and Cabral] tortured helpless fishermen; Almeida tore out the eyes of a Nair who had come in with a safe-conduct because he suspected a design

Labrador to Uruguay were a single continent, and thus reveals important sources of non-Columbian information which have perished (J. Cortesão (1), pp. 165 ff.). There are suggestive hints about Brazil as early as +1448, on which see Prestage (1), pp. 227 ff.

[a] This has two aspects, the psychology and the technique. In an interesting monograph, Cipolla (1) urges that the rapid rise of gun-founding in Europe and the mounting of such cannon upon ships soon more advanced in type than those of any South Asian people gave the Europeans an overwhelming advantage from the first irrespective of their level of culture and civilisation. This is surely true, and perhaps best understandable as yet another aspect of the fact that Europe developed capitalism and that India and China did not.

[b] Peres (1), p. 39; de Zurara, tr. Beazley & Prestage, p. 91.

[c] Boxer & de Azevedo (1), p. 13.   [d] Kirkman (1, 2).

[e] Mathew (3).

[f] Details will be found in works such as those of Strandes (1); Coupland (1); Whiteway (1); Boxer & de Azevedo (1). No wonder that Aḥmad ibn Mājid bitterly regretted āt the end of his life the service which he had rendered to the Franks.

[g] Cf. Elgood (1), p. 384.

on his own life; Albuquerque cut off the noses of the women and the hands of the men who fell into his power on the Arabian coast. To follow the example of Almeida and sail into an Indian harbour with the corpses of unfortunates, often not fighting-men, dangling from the yards, was to proclaim oneself a determined fellow.'a I mention these facts with reluctance, but they certainly do something to correct the stereotyped image (still met with in Europe) that Asians have been more cruel and barbarous than Europeans. Some of the greatest occidental scholars of the time approved these activities. João de Barros wrote:b

It is true that there does exist a common right to all to navigate the seas, and in Europe we acknowledge the rights which others hold against us, but this right does not extend beyond Europe, and therefore the Portuguese as lords of the sea by the strength of their fleets are justified in compelling all Moors and Gentiles to take out safe-conducts under pain of confiscation and death. The Moors and Gentiles are outside the law of Jesus Christ, which is the true law that all must keep under pain of damnation to eternal fire. If then the soul be so condemned, what right has the body to the privileges of our laws?... It is true that the Gentiles are reasoning beings, and might if they lived be converted to the true faith, but inasmuch as they have not shown any desire as yet to accept this, we Christians have no duties towards them.

What could the Chinese show in comparison with all this? 'They bestowed gifts upon the kings and rulers', we read above (p. 488), 'and those who refused submission they over-awed by the show of armed might.' This statement demands examination. But on all their expeditions there were only three occasions when they got into difficulties and had to fight.c The first was in +1406 when Chhen Tsu-I,[1] a tribal chief of Palembang who had been pillaging merchants, made a surprise attack upon their camp; but he was defeated, captured and subsequently executed in Nanking.d The third happened seven or eight years afterwards, when Su-Kan-La,[2] one of the pretenders to the throne of north-western Sumatra, quarrelled over the distribution of the Chinese gifts. He too led forces against those of Chêng Ho, but he was beaten at Lambri and captured with his family.e The second occasion was much the most serious. In +1410 the King of Ceylon, Ya-Lieh-Khu-Nai-Erh,[3] i.e. Alagakkonāra (probably, but

---

a Evidences are in Whiteway (1), pp. 87, 91 ff., 93, 119, 125, 144, 155, 165. The quotation is from p. 22. The fate of the Nair is given by de Castanheda, História, 11, 28; cf. Panikkar (2), p. 51.

b Décadas de Asia, 1, i, 6.

c Two other examples of conflict between Chinese and Indian Ocean peoples we exclude, since there is no mention of them in any oriental sources. In his Novus Orbis of +1555 (p. 208) Joseph of Cranganore reported that there had been serious fighting formerly at Calicut on account of a trade dispute, after which the Chinese came only to Mailapetam. In the Décadas da Asia (+1552), 11, ii, 9 and IV, v, 3, João de Barros stated that Diu had been built by a Gujerati king in memory of a sea victory over the Chinese. On both these texts see the comments of Yule (1), vol. 2, pp. 391 ff. Their authenticity is hard to estimate, and they may refer to earlier periods than that of Chêng Ho, with whose activities there is really nothing positive to connect them.

d Cf. Pelliot (2a), p. 274, (2b), p. 281.

e Cf. Pelliot (2a), p. 290, (2c), p. 214; Duyvendak (9), pp. 376 ff. Sekandar was taken to Nanking like Chhen Tsu-I, and also ultimately executed.

---

[1] 陳祖義   [2] 蘇幹剌   [3] 亞烈苦奈兒   33-2

not certainly, Bhuvaneka Bāhu V),[a] enticed Chêng Ho's expeditionary guard into the interior and then demanded excessive presents of gold and silk, meanwhile sending troops to burn and sink his ships, which lay, in all probability, in Galle harbour or Beruwala roads. But Chêng Ho pushed on to the capital (undoubtedly Kotte, as Kandy had not then been founded), took the king and his weakly defended court by surprise, and then fought his way back to the coast with the captives, routing the Sinhalese army on the journey. The prisoners were taken to Nanking, where they were kindly treated and sent home again after an arrangement had been made to choose a relative of the king as his successor.[b] Naval armed might thus meant something very different in the Chinese and the Portuguese interpretations.[c]

[a] This corrects Vol. 3, p. 558, but the question is a complicated one as the Chinese names do not quite agree with those in the Sinhalese histories, nor do the personalities and their reputations coincide. See Pelliot (2a), p. 278; (2b), p. 284; Hsü Yü-Hu (1), pp. 103 ff.; Lévi (2), pp. 429 ff. There are several versions of what happened in Ceylon, the fullest and not the least interesting being a Buddhist commentary on Hsüan-Chuang's *Hsi Yu Chi* preserved in *TSCC, Pien i tien*, ch. 66, pp. 9b, 10a, tr. Lévi (2). Here the Sinhalese king is represented as a Hinduiser, and reproved by Chêng Ho for paying inadequate honour to the Tooth relic. Miracles attend the embarkation and the safe journey home. Whatever exactly happened, there was nothing to sully Chêng Ho's reputation as a humane emissary and commander-in-chief.

[b] The literature contains much more combat material than this, but it stems from a fictional, not a historical, source, the novel *Hsi-Yang Chi* of Lo Mou-Têng, and is therefore discounted by purists. One narrative in this concerns the siege and bombardment of al-Aḥsā' somewhere on the Hadhramaut coast. Here Duyvendak (19) went astray in taking the mention of *Hsiang-yang phao* to imply large cannon, for as we shall show in Sect. 30i, this was a technical term for large counter-weighted trebuchets. That these were often mounted on battleships of Sung and Ming times we have already seen (p. 425 above). None the less his main contention that the fleets of Chêng Ho were armed with bombards and other gunpowder weapons is undoubtedly correct. Many dated Chinese cannon of the later half of the +14th century are known (cf. Sect. 30k, and meanwhile Sarton & Goodrich (1); Goodrich (15); Goodrich & Fêng Chia-Shêng (1) and Wang Ling (1), perhaps the best general account of the development of explosives in Chinese warfare). A bronze bombard 1 ft. 2 in. long, now in the Peking Historical Museum, with an inscription of +1332, is certainly the oldest dated metal cannon in existence, East or West. For an account of it see Wang Jung (1); and correct Vol. 1, p. 142.

Another piece now also in Peking, at the Military Museum, is of particular interest here because it was intended for naval use. Dr Lu Gwei-Djen and I examined it at the Nanking Museum in 1958. In length just over 1 ft. 5 in., weighing 34·8 lb. and with a muzzle calibre of 4·34 in., it bears on its bronze barrel the following inscription: 'Left Naval Station (Shui Chün Tso Wei[1]). Chin[2] Series, no. 42. Large bowl (size) muzzle tube. Cast on an auspicious day in the twelfth month of the 5th year of the Hung-Wu reign-period (+1372) by the Pao-Yuan Chü[3] (Bureau).' This Naval Station near Nanking was still active in Chêng Ho's time and provided many of his escort warships and the marines who manned them. There is evidence that these bombards were mounted on a stool-like carriage so that they could be made to swivel for aiming. Lastly, a bombard with a Chinese inscription bearing the date +1421 was found in Java and is now in the Museum f. Völkerkunde in Berlin (cf. Partington (5), pp. 275 ff.; Feldhaus (1), col. 424, (2), p. 59). It is remarkable as having a covered touch-hole, yet unknown on contemporary European specimens. One moral of the story is that what the old Chinese novelists said is by no means always to be discounted. Archaeology, which always has the last word, is liable to justify them. But the existence of gunpowder weapons in Chêng Ho's fleets is one thing, and the extent to which they were used is another.

[c] It seems almost inconceivable that Toussaint (1), in an otherwise meritorious book on the history of the Indian Ocean, can say, in the context of Chinese activity there: 'The Chinese, for their part, proceeded by conquest and annexation; their soldiers occupied the country, and every place so conquered was forced to adopt the institutions, customs, religions, language and script of the Chinese.' These words, a blunted echo of Coedès (5), pp. 64, 66, were by him applied to a contrast between the peaceful, if partial, penetration of Hindu culture into South-east Asia from the +4th to the +12th century, and the supposed relations of China with Vietnam, Korea and Japan. They are not justified there, and wholly unjustifiable for the Indian Ocean in the +15th century.

[1] 水軍左衛   [2] 進   [3] 寶源局

As for trade, our knowledge of the inner workings of that matter is as usual still very deficient, but it was only natural that what was done both by the Chinese and the Portuguese was done under the aegis of their respective economic systems, and these were very different. While much further research is needed, it seems at least clear that the Portuguese activities were from the start much more concerned with private enterprise.[a] The search for the 'El-Dorado' which would make one's personal fortune was an integral part of the *conquistador* mentality. Well before the death of Prince Henry expeditions primarily commercial were wending their way round the West African coasts,[b] and the more Africa was explored, the further the expeditions had to go and the more necessary it became to make them financially at least self-supporting. Trading ventures were thus certainly encouraged and licensed by the Portuguese court, but behind it stood the international finance, indeed the developing capitalism of all Europe; and the role of such support is now being actively investigated.[c] By contrast the Chinese expeditions were the well-disciplined naval operations of an enormous feudal-bureaucratic State the like of which was not known in Europe, their impetus was primarily governmental, their trade (though large) was incidental,[d] and the 'irregular' merchant-mariners whose trafficking was to be encouraged were mostly humble men of small means. The bureaucracy in China generally saw to that. Only their numbers made them important. And what was true of trade in general was true also of the slave-trade in particular. The Chinese and other Asian nations had been using negro slaves for many centuries,[e] but the fact that their slavery was basically domestic kept the practice within bounds. Not so the use of Africans in agricultural plantation labour, especially in the New World, which brought it about that between +1486 and +1641 no less than 1,389,000 slaves were taken by the Portuguese from Angola alone.[f] One has only to read the celebrated chronicle of de Zurara himself to see that the Portuguese expeditions down the West Coast of Africa involved kidnapping and slave-raiding forays from the very first.[g] This had been the mutual custom of Moors and Christians all round the Mediterranean throughout the Middle Ages, but now in mournful augury it was extended to those who had never had any part in that quarrel. But the results of this we cannot follow further here.

Thus the paradox appears that while the feudal State of Portugal, hardly emerged

[a] Prince Henry seems to have planned a merchant port town near Sagres but it never developed (de Zurara, tr. Beazley & Prestage, p. 21).

[b] The first 'factory' was established at Arguim in +1448, and the famous fortress-factory of El-Mina de Ouro on the Volta River in Ashanti in +1482. These were the predecessors of Fort Jesus at Mombasa and all the fortified trading-posts of the East Indies.

[c] Verlinden & Heers (1); Rau (1); da Silva (1); Kellenbentz (1); Otte (1).

[d] Cf. Thung Shu-Yeh (1); Lo Jung-Pang (2, 5).

[e] A good summary of this trade will be found in Duyvendak (8), p. 23. Much further material is scattered through Hirth & Rockhill (1).

[f] Davidson (1), pp. 119 ff.

[g] I.e. +1433 (Beazley & Prestage tr., p. 33), cf. the account of the captures made by Antão Gonçalves in +1441 (tr. p. 43). It is true that Prince Henry instructed his captains to take prisoners for the express purpose of learning everything possible about the country and the language (tr. p. 35). De Zurara has a harrowing account of the distribution of Moroccan and Mauretanian families as chattel slaves (chs. 25, 26) in +1444, at which Prince Henry himself was present, with all its separations of husbands and wives, parents and children. De Zurara is delighted with their readiness to become Christians—on the whole they seem to have been rather well treated, and in time even came to like bread and cheese.

from the Middle Ages, founded an empire of mercantile capital, bureaucratic feudalism, though certainly not the economy of the future, gave to China the lineaments of an empire without imperialism. But we must beware of doing any injustice to the first Portuguese merchants battling in the Indian Ocean; perhaps they were caught in a mesh of intractable economic necessity? During da Gama's first visit to Calicut in + 1498 a highly significant event occurred. When the Portuguese presented the goods which they had brought, consisting of striped cloth, scarlet hoods, hats, strings of coral, hand wash-basins, sugar, oil and honey, the king laughed at them, and advised the admiral rather to offer gold.[a] At the same time the Muslim merchants already on the spot affirmed to the Indians that the Portuguese were essentially pirates, possessed of nothing that the Indians could ever want, and prepared to take what the Indians had by force if they could not get it otherwise.[b]

There is something very familiar about this scene. In fact it symbolised perfectly a fundamental pattern of trade imbalance which had been characteristic of relations between Europe and East Asia from the beginning, and which was destined to continue so until the industrial age of the late nineteenth century.[c] Broadly speaking, Europeans always wanted Asian products far more than the Easterners wanted Western ones, and the only means of paying for them was in precious metals.[d] This process occurred at many places along the east–west trade-routes, but primarily of course in medieval times at the Levantine borders between Christendom and Islam.[e] The Chinese on the other hand probably never had to face an adverse balance of trade, for silk and lacquer were everywhere esteemed, and good in exchange for anything the Chinese ever wanted to buy.[f] As Domingo de Navarrete said in his book of + 1676:[g]

[a] Prestage (1), pp. 262 ff., based on the anonymous *Roteiro* ed. Ravenstein (2).

[b] The Moorish merchants also said that if the Portuguese were entertained 'other vessels would stay away', meaning perhaps those of the Chinese as well as their own, and the country would be ruined.

[c] Cf. Gibbon, *Decline*, vol. 1, pp. 88 ff. There are two outstanding treatments of this subject, the book of Warmington (1), now old, but good, esp. pp. 180 ff., and that of Braudel (1), a luminous exposition far wider than its title implies, esp. pp. 361 ff. The full panorama has been well appreciated in an unpublished work by Purcell (2), cf. esp. pp. 159 ff.

[d] To show the continuity of this historical pattern it is instructive to compare the lists of desirable Asian goods given in the *Periplus* of +70 or +110 (Schoff (3), pp. 284 ff., cf. Vol. 1, pp. 178 ff.) and in the letters of Albuquerque, *c.* +1513 (de Almeida (1), vol. 3, pp. 558 ff.). J. Pirenne (1) now proposes the later date of +246 for the *Periplus*, long ago suggested by Reinaud, but that does not affect our argument.

[e] In the +16th century there was a sharp separation at this frontier, for letters of credit, so common and useful within Europe, did not run in the Islamic or Asian countries (Braudel (1), p. 363). The latter had of course their own systems but, as it were, 'outside the sterling area'.

[f] On the chain of east–west exchanges see Lopez (4), p. 309, but I very much doubt whether, as he thinks, the Chinese had an unfavourable balance in their trade with the Arab and South-east Asian countries in the +11th and +12th centuries. He seems to have misunderstood a passage in Duyvendak (8), p. 16 (taken from Hirth & Rockhill (1), p. 19 and derived from *Sung Shih*, ch. 186, pp. 20*a, b,* 27*b*) where 'units of count' refers to the merchandise, and does not mean coins or weights of precious metal paid by the Chinese.

It is of much interest that the idea of enriching the country by foreign trade is found in Chinese literature as early as −80. The *Yen Thieh Lun* (ch. 2, p. 6*b*; cf. Vol. 2, p. 251) says: 'Good rulers exchange the non-essential for the fundamental, and secure what is substantial with their own emptiness... Insignificant articles are a means of inveigling foreign countries and snaring the treasures of the Chhiang and the Hu. Thus a piece of Chinese plain silk can be exchanged with the Huns for articles worth several pieces of money... Foreign products (thus) keep flowing in while our wealth is not dissipated...'; tr. Gale (1), p. 14, mod.

[g] Cummins ed. vol. 1, p. 137.

My Design is only to give some hints of what is most remarkable, which will suffice to make known how bountifully God has dealt with those People...giving them all they can desire, without being necessitated to seek for any thing abroad; we that have been there can testify this Truth.

In the present work we have already met with the European–Asian imbalance twice. During the Roman empire, from about −50 to +300, gold and silver were drained away from Europe to pay for the silk of China and the spices of India,[a] a process which may be followed not only in the Roman coins scattered throughout Asia[b] but in the gold stocks of the Han.[c] Nearly two thousand years later the opium trade with South China (and hence the Opium Wars) arose because the East India Company,[d] alarmed at the drain of silver from Europe to pay for its silk, tea and lacquer, sought for some substitute commodity.[e] Indeed, the Mediterranean region acted for two millennia as a kind of monstrous centrifugal pump continually piping off towards the East all the gold and silver which entered into it.[f] Alexandria might pay partly with glass,[g] medieval Western Europe partly with slaves, Venice with mirrors and England with tin, but the Arabs, the Chinese and the Indians took little interest at any time in the most typical European products such as woollen cloth[h] or wine, and when all the barter was over

[a] See Vol. 1, p. 109. The most recent treatment of this subject is that of Schwartz (1), which supplements Warmington (1), Charlesworth (2) and Schoff (1–5, esp. 3), etc., with a large amount of fresh archaeological evidence, some of which is also to be found in Wheeler (1, 4). All that we now know gives much support to two key passages, long noted, in Pliny (*Nat. Hist.* VI, xxvi, 101 and XII, xli, 84), which say (about +75) that as much as a hundred million sesterces went annually to Arabia, India and China, to pay for imports such as silk, spices and perfumes. And indeed almost every recipe in the cookery book *Artis Magiricae*, attributed to Apicius but actually compiled between +35 and +435, includes pepper (cf. the critical translation of Flower & Rosenbaum). Thus the spice trade was already in full swing in Roman times. As Persius wrote (*Sat.* v, about +54):

'The greedy merchants, led by lucre, run
To the parch'd *Indies* and the rising sun;
From thence hot Pepper and rich Drugs they bear,
Bart'ring for Spices their *Italian* ware...' (tr. Dryden).

And making up the difference with good metal too.
[b] From a large literature we need only quote Piganiol (2), p. 389 or H. W. Codrington (1), p. 32 for the background, with Wheeler (4) and Schwartz (1) for the present state of the question. Of course, some European products were welcomed in Asia, as witness the Arretine pottery of Arikamedu (Vīrapatnam) and the glass of Begram (cf. Vol. 1, pp. 179, 182; and e.g. Hackin *et al.*).
[c] This matter has been studied by Dubs (4) and Lo Jung-Pang (8). Sources of high reliability (signally *Chhien Han Shu*, ch. 99c, p. 29b, tr. Dubs (2), vol. 3, p. 458) disclose that the gold reserve of the Hsin emperor in +23 amounted to no less than 4,706,880 oz. troy. The State as a whole may well have had a further stock. Han Wu Ti in −123 must have had at least 1,568,000 oz. (*Shih Chi*, ch. 30, p. 4b, tr. Chavannes (1), vol. 3, p. 553), probably two or three times as much, since this was given away in rewards to his troops. Yet the total stock of gold in medieval Western Europe is estimated at about 3,700,000 oz., and all the silver brought to Spain from the Americas between +1503 and +1660 was no more than 5,829,996 oz. (Hamilton (1), p. 42). The Han gold reserves derived partly from Siberia, partly from lode and placer sources in China itself, but largely from trade with Europe especially in silk. It will be remembered that only a few years after the death of Wang Mang, Tiberius in Rome (r. +14 to +37) prohibited the wearing of silk by men, in order to minimise the eastward 'haemorrhage' of specie (Tacitus, *Ann.* II, 33).
[d] A fact here relevant, but not perhaps as well known as it might be, is that the Hon. East India Company, founded in +1600, developed directly out of the Levant Company, founded in +1581.
[e] See Vol. 4, pt. 2, p. 600; the facts have been elucidated with great success by Greenberg (1).
[f] Cf. Braudel (1), p. 362. And it was never adequately supplied with either metal.
[g] Cf. Hackin & Hackin (1); Hackin, Hackin *et al.* (1), on the Begram depot.
[h] Cf. Carus-Wilson (2).

there remained a large perennial insoluble deficit.[a] And even after the discovery of the Americas and the exploitation of Potosi silver, the Asian suction still asserted itself and much of the metal found its way westward through the Philippines instead of continuing to fill the coffers of Spain.[b] Thus in the presence of the somewhat discomfited admiral at the term of one of the greatest sea-journeys of history, the imbalance of the Levantine trade reappeared exactly in its classical form.[c]

What then were the Portuguese sea-captains to do at the end of the +15th century? First, crusading apart, spices they must have. The European demand for pepper was sure and certain, it was the expression of a very real need, it was not (as is so often said) a 'luxury trade'.[d] Until the full development of winter fodder a couple of centuries later, European animal husbandry could keep each year only the animals needed for labour and reproduction; the others had to be killed and their meat preserved by salting.[e] This was the process which needed pepper by the shipload,[f] the pepper which

---

[a] In a word, Asia was indifferent to European staples. As the Chhien-Lung emperor wrote to George III in often-quoted words (+1793): 'Strange and costly objects do not interest us. As your Ambassador [Lord Macartney] can see for himself, we possess all things (that we need). We set no value on rare and ingenious objects [i.e. 'sing-songs', cf. Vol. 4, pt. 2, pp. 522 ff.], and have no use for your country's manufactures...There is, therefore, no need to import the goods of foreign barbarians in exchange for our produce. But as the tea, silk and porcelain of the Chinese Empire are necessities for European nations, we have permitted as a mark of special favour that foreign warehouses should be established at Canton so that your wants may be supplied and your country thus participate in our benevolence' (tr. Backhouse & Bland (1), pp. 322 ff., mod.).

[b] This was one of a number of factors in the drying-up of the Peruvian–Bolivian consignments of silver to Spain in the middle of the +17th century. Cf. Hamilton (1), pp. 36 ff., Braudel (1), p. 415.

[c] Braudel's trenchant phrase, (1), p. 371.

[d] Even by Pirenne (1), p. 143; Runciman (2), pp. 88, 89; Postan (1), p. 169; Lopez (4), p. 261.

[e] This predicament is acknowledged by all students of medieval Western agriculture (e.g. Gras (1), p. 15; Franklin (1), pp. 48, 121; Curwen (6), p. 84; Parain (1), pp. 123, 127, 132, 153 ff.). Turnips, native to Europe, could have been used from antiquity onwards, but the open-field system imposed too great a uniformity on crops, and implied general autumn stubble grazing incompatible with a planned winter-feed policy. Clover and lucerne (alfalfa, cf. Vol. 1, p. 175), the 'artificial grasses' from the Middle East, came into use slowly during the +17th century, and rapeseed also, though it was not known until later that cattle would eat the oil-cake after pressing. Maize and potatoes were of course introductions from the New World. On the origin of all these cultivated plants see Vavilov (2), and on the development of winter fodder Forbes (18).

[f] The usual idea (cf. e.g. Prestage (1), p. 267; Jensen (1), pp. 85 ff.; Drummond & Wilbraham (1), p. 34) is that pepper and other spices were simply for table condiments or sauces designed to disguise the taste of tainted meat. But this could never have accounted for the vast imports of the Western Middle Ages. A couple of examples—in +1504 five Portuguese ships arrived at Falmouth with 380 tons of pepper from Calicut (Braudel (1), p. 422). About the same time, each fleet of Venetian galleys returning from Alexandria brought 1,250 tons (Lane (1), p. 26), i.e. some 3,100 tons of total mixed spices annually (Lane, 2). So we are bound to suppose that as in traditional China and the Islamic lands the pepper was actually mixed with the salt for the salting of the meat to be preserved. All the directions for making jerked and salted meat (fu la[1]) in the +6th-century treatise on rural economy, Chhi Min Yao Shu, ch. 75, include powdered pepper. Or take a piece of contemporary evidence—our collaborator, Dr Lu Gwei-Djen, grew up at Nanking in a family of some 40 persons, where pork, beef and duck meat was preserved in large amounts at the beginning of winter; for this purpose pepper, cinnamon, etc., was bought not by the ounce but by the pound. The mixed powder of salt with about a third of its weight of pepper (chiao yen[2]) is familiar to all who have frequented restaurants in China (cf. the recipe for 'aromatic salt' in Apicius; Flower & Rosenbaum (1), p. 55), but for salting, less pepper was used. Salting, one could say, was part of the body of knowledge, empirical but profound, handed down in the medieval food industry all over the Old World. If too much salt was used before the air-drying process the meat would eventually have to be soaked in water before consumption, with consequent tastelessness and loss of nutrients; if too

[1] 脯臘   [2] 椒鹽

now lay just within the grasp of the Portuguese—if they could pay for it. Hence, secondly, the importance of the gold of Guinea. Portugal could not disburse, like Venice, the gold and silver from Central Europe,[a] and the wealth of the New World was not yet available. Whether or not Prince Henry had foreseen it, the Portuguese began to tap the gold of the Sudan[b] successfully about +1445, when they reached Arguim; and then after the opening of the Guinea coast most of it started to flow west towards their oceanic ships.[c] Thus were they provided with precious metal necessary for the Eastern trade.[d] It does not seem that the Portuguese defrauded the Africans greatly in this exchange, for, unlike Asians, they were pleased to get horses, wheat, wine and cheese, copper ware and other metals, blankets and strong cloth.[e] Unfortunately the gold of Africa was never anything like enough for the ambitions of the Portuguese in the eastern seas, and out of the need for bases there arose naturally the temptation to accumulate further treasure by appropriating at sword-point the wealth of coastal port cities such as Malacca.[f] The real criticism of their operations is that they were not content with a reasonable share of Asian trade, what they wanted before long was the complete domination of the trade and the traders;[g] this it was given only to others to achieve,[h] and then but for a time.

The last difference between these sailors of East and West was religion, and fortunately, if unexpectedly, it provides some comic relief. On the one side all is grim indeed. Missionary activity, well-intended, accompanied the Portuguese explorations

little, then it would become slimy and even less palatable. In any case, rancidity would develop. The addition of spices in the correct amount permitted the use of these lesser quantities of salt without any evil effects, probably because of inhibition of the autolytic enzymes as well as bacteriostatic action and an anti-oxidant effect on fats. Modern biochemical work supports this view; cf. Jensen (2), pp. 378 ff.; Deans (1); Webb & Tanner (1); Chipault *et al.* (1). Of course the Chinese needed pepper just as the Europeans did, but they had it, so to say, on their doorstep.

[a] See on this P. Grierson (1). The sources were in Saxony, Bohemia, Hungary, Serbia and Transylvania.

[b] The 'Sudan' means properly not, as for many people, the 'Anglo-Egyptian Sudan' of the upper Nile, now independent, but the whole vast savannah territory between the Sahara in the north and the Congolese equatorial rain-forests in the south (cf. Davidson (1), pp. 61 ff.).

[c] Cf. Braudel (1), pp. 369 ff. Around +1510 they were getting some 15,000 oz. a year. This diversion of an age-old north-flowing gold current which had nourished the Maghrib and the Muslim kingdoms of Spain (cf. Bovill, 1, 2) led to a great decay of culture in North Africa. It was also a crushing blow to French and Italian commerce with the Maghrib and the Sudan which had been carried on by overland routes. There followed a marked gold shortage in non-Atlantic Europe, hence greater prosperity for the German mining industry, with all that that implied for the science of the great Agricola (cf. Vol. 3, p. 649 and *passim*). However, a good deal of gold continued to reach Turkish Islam in the +16th century by way of Egypt (cf. Braudel (1), pp. 364 ff.). And El-Mina seems to have tapped rich sources of Ashanti gold which had formerly not supplied North Africa and Europe at all.

[d] Not always directly, however, for since Chinese and South-east Asians preferred silver to gold, the Portuguese, it is thought, exchanged the metals at Antwerp (Magalhães Godinho, in Braudel (1), p. 371).

[e] Braudel (1), p. 370.

[f] Whiteway (1), pp. 141 ff.

[g] In this the Portuguese were never entirely successful, and by +1550 the Muslim Red Sea route had largely recovered. Thereafter the importance of the two routes oscillated in uneasy balance until the end of the spice age. Cf. Braudel (1), pp. 421 ff.

[h] The Portuguese empire collapsed as quickly as it had sprung up. The first Hollander in the Indian Ocean, Cornelius Houtman, was under sail in +1596; by +1625 the Dutch were its masters. Their victory was immediately more fatal to the Levantine spice trade than all the Portuguese endeavours. Cf. Braudel (1), pp. 441 ff.; Collis (2), pp. 250 ff., and p. 524 below.

of course from an early time;[a] the first mass in West Africa was celebrated by a Fr. Polono at Arguim in +1445. But before the end of the century the war against all Muslims was being extended to all Hindus and Buddhists too, save those with whom the Portuguese might find it expedient to arrange a temporary alliance. In +1560 the Holy Inquisition was established at Goa, where it soon acquired a reputation even more unsavoury than that which it had in Europe. It subjected the non-Christian as well as the Christian subjects of the empire to all those forms of secret-police terror which have disfigured our own century, yet more abominable here perhaps because enlisted in the interests of high religion.[b] On board the Chinese ships what a contrast. Without forsaking the basic teachings of the sages Khung and Lao, Chêng Ho and his commanders were 'all things to all men'; in Arabia they conversed in the tongue of the Prophet and recalled the mosques of Yunnan, in India they presented offerings to Hindu temples, and venerated the traces of the Buddha in Ceylon.

Ceylon provided the scene for a particularly interesting example of this almost excessive urbanity. In 1911 a stele with inscriptions in three languages (Chinese, Tamil and Persian) was unearthed by road engineers within the town of Galle.[c] This had commemorated, as was soon clear, one of the visits of the Ming navy under Chêng Ho, and took the form of an address accompanying religious gifts. Owing to differential weathering (or perhaps the greater legibility of Chinese in such circumstances) the Chinese version was the first to be deciphered, and it is worth reproducing. It reads like this:

His Imperial Majesty, Emperor of the Great Ming, has despatched the Grand Eunuchs Chêng Ho, Wang Chhing-Lien[1] and others,[d] to set forth his utterances before the (Lord) Buddha, the World-Honoured One, as herein follows.

Deeply do we reverence Thee, Merciful and Honoured One, of bright perfection wide-embracing, whose Way and virtue passes all understanding, whose Law pervades all human relations, and the years of whose great *kalpa* rival the river-sands in number; Thou whose controlling influence ennobles and converts, inspiring acts of love and giving intelligent insight (into the nature of this vale of tears); Thou whose mysterious response is limitless! The temples and monasteries of Ceylon's mountainous isle, lying in the southern ocean far, are imbued and enlightened by Thy miraculously responsive power.

Of late we have despatched missions to announce our Mandate to foreign nations (*pi chê chhien shih chao yü chu fan*[2]), and during their journeys over the oceans they have been favoured with the blessing of Thy beneficent protection. They have escaped disaster or misfortune, journeying in safety to and fro, ever guided by Thy great virtue.

Wherefore according to the Rites we bestow offerings in recompense, and do now reverently present before the (Lord) Buddha, the World-Honoured One, oblations of gold and silver,

---

[a] See the monograph of Brasio (1). The Canaries were a bishopric as early as +1479; São Thomé had one by +1534; cf. Anon. (47), pp. 117 ff. on the 'Apostolate of the Indies'.

[b] It is best to read the accounts of those who suffered under it, notably the French doctor, Charles Dellon, (1), whose story has been retold by Collis (2).

[c] The stone is now in the Colombo Museum, where I had the pleasure of seeing it in 1958. See the paper of E. W. Perera (1).

[d] This name is apparently a mis-reading for Wang Kuei-Thung,[3] who was also envoy to Champa in +1407 (*Ming Shih*, ch. 324, p. 4a). The rectification is due to Yamamoto Tatsuro (1) and Naito Torajiro (2).

[1] 王清濂   [2] 比者遣使詔諭諸番   [3] 王貴通

gold-embroidered jewelled banners of variegated silk, incense-burners and flower-vases, silks of many colours in lining and exterior, lamps and candles, with other gifts, in order to manifest the high honour of the (Lord) Buddha. May His light shine upon the donors.[a]

And the inscription concludes with a list of the presents, including 1,000 pieces of gold, 5,000 pieces of silver, 100 rolls of silk, 2,500 catties of perfumed oil, and all kinds of bronze ecclesiastical ornaments gilded and lacquered. It is clearly dated in the 7th year of the Yung-Lo reign-period, i.e. +1409.

A particular devotion to Buddha or to the Tooth Relic on the part of Chêng Ho (though a Muslim by birth) might be surmised on account of the votive stele which he and his commanders erected at Chhang-lo in Fukien in +1432 to the Buddhist-Taoist goddess of the sea, Thien-Fei[1] or Ma-Tsu,[2] giving thanks for all their preservations.[b] Clearly this was not quite Islam as it was understood in the lands of the Caliph. But a surprise was in store. The Galle slab proved to be no Rosetta Stone repeating the same text in different languages. When the Tamil version was deciphered and translated twenty years later,[c] it turned out to say that the Chinese emperor, having heard of the fame of the god Tenavarai-nāyaṇār (equivalent to Sinhalese Devundara Deviyo, an incarnation of Vishnu), caused the stone to be set up in his praise. Still more remarkable, the Persian version, though the most damaged of the three, is clear enough that the presentation was to the glory of Allah and some of his (Muslim) saints. But while the texts are thus different, they all agree in one thing—the lists of the presents are almost exactly identical. There is thus hardly any escape from the conclusion that three identical sets of gifts were brought out by sea and handed over to the representatives of the three most important religions practised on the island. Moreover, the Galle inscriptions are not likely to have been the result of any local trickery, for one historical analysis at least indicates that[d] the Chinese one was carved in China on the date named (+1409) but that the stele was not erected in Ceylon until +1411. Such humanistic catholicity contrasts indeed with the *autos-da-fé* of Goa later on.

In one of history's entertaining footnotes, the stone was seen by the Portuguese before it became buried. The Jesuit Fernão de Queiroz[e] tells us that his countrymen found 'stone pillars (*padrões*) which the Kings of China ordered to be set up there, with Letters of that Nation, as a token, it seems, of their devotion to those idols'.[f]

[a] Tr. auct. adjuv. Sir Edw. Backhouse, in E. W. Perera (1).

[b] We gave a translation of part of their inscription in Vol. 3, pp. 557, 558, and the whole will be found in Duyvendak (8, 9). The cult of Thien-Fei, the Heavenly Consort, patroness of sailors and sea-farers briefly described in Doré (1), vol. 11, pp. 914 ff., has recently been the subject of illuminating studies by Li Hsien-Chang (1, 3–7). She started her career in the +11th century, and until recent years was still worshipped on the coast and islands of south-east China. Chêng Ho was unquestionably attracted by Buddhism and Taoism.

[c] Paranavitana (3).   [d] Duyvendak (9), pp. 369 ff.

[e] Himself a humane and generous man as well as a notable historian. In his *Conquista Temporal e Espiritual de Ceylão*, finished in +1687, he shows how horrified he was by the conduct of the Portuguese in that country. He believed that the 'heretical' Dutch were permitted by God to possess it because the Roman Catholic Portuguese had proved utterly unworthy to hold it.

[f] Tr. S. G. Perera (1), vol. 1, p. 35. For a contrast with the behaviour of the Portuguese themselves in Ceylon see vol. 3, pp. 1005 ff. As Dr Dinwiddie remarked, at the end of the eighteenth century, 'However superstitious the Chinese may be deemed, they possess one convenient quality—they never interfere with strangers about religious matters' (Proudfoot (1), p. 82).

[1] 天妃   [2] 媽祖

(v) *The captains and the kings depart*

How did it all end? Greater robbers came to prey upon the lesser. During her temporary dynastic union with Spain the empire of Portugal remained intact, but its foundations were insecure. When the Portuguese shook off this unwanted association in + 1640 it was too late to save her Asian dominion, essentially a long trade route dotted with fortresses—from Mombasa to Muscat and Goa, to Cochin and on to Colombo, thence to Malacca, north to Macao or east to Timor. They could no longer be supplied sufficiently with men and munitions; Portuguese long-distance shipping could no longer bear the burden. The great eastern centres fell to the Dutch one by one; Malacca in + 1641, Muscat in + 1648, nodes of the network, then Colombo in + 1656, Quilon in + 1661 and Cochin in + 1662. Goa, the 'Rainha do Oriente', declined into a grass-grown poverty. The Dutch had no high-flown ideas about the conversion of Asia to Christianity, what they minded was business; but others too could play at that game, perhaps with more subtlety, perhaps with better resources, and the Netherlands empire in turn passed first to the French and then the British. With the decline of colonialism in our own time the wheel has come full circle and Asia resurgent takes a rightful place in the counsels of the world.

The decline of Chinese long-distance shipping had set in even faster. In Portugal there had at first been some critics of the West African explorations but they were soon silenced by the evident profits arising from gold, slaves and other commodities. In China the critics had always been far more numerous and determined. The Confucian bureaucracy, with its country landlord basis, was always liable to look askance at any intercourse with foreign countries. These countries were of no interest in themselves and could offer nothing but unnecessary luxuries. But according to the classical Confucian motif of scholarly austerity, to which in the national ethos the imperial court itself was theoretically bound, unnecessary luxuries were deeply wrong.[a] And since all real needs of food and clothing, including even the magnificent products of Chinese craftsmanship, were available in abundance at home, what good could it possibly do to spend money on seeking strange jewels or other things with dubious properties abroad? The Grand Fleet of Treasure-ships swallowed up funds which, in the view of all right-thinking bureaucrats, would be much better spent on water-conservancy projects for the farmers' needs, or in agrarian financing, 'ever-normal granaries' and the like. Indeed, the Confucians were not in favour of too much aggrandisement of the Court, for in practice it meant of course the aggrandisement of the Grand Eunuchs thereof. It was thus no accident that the admirals of the fleet (strange though it may seem in Western eyes) were mostly eunuchs; in fact the whole episode of the great Chinese navigations was only one engagement in that administrative battle between Confucian bureaucrats and Imperial eunuchs which had been going on since the Han

---

[a] We have already seen a signal instance of this ideology in the refusal by the first Ming emperor about + 1368 to accept a splendid rock-crystal clock with a water-wheel or sand-wheel escapement of Chinese type, presented by the Astronomer-Royal of the time (Vol. 4, pt. 2, p. 510; Needham, Wang & Price (1), p. 156). 'Thai Tsu considered it to be a useless (extravagance) and had it broken up.'

at least and would still for many years go on.[a] The sympathy of modern students is generally on the side of the former, but it must be recognised that in this case at least (and it is probably not the only one) the eunuchs were the architects of an outstanding period of greatness in China's history.

The champion of retrenchment and economy was Hsia Yuan-Chi,[1] who had financed the expeditions at the beginning. Imprisoned under the Yung-Lo emperor, he was immediately released by Jen Tsung upon his accession in the summer of +1424, and urged that the fleet movements should be stopped immediately. 'If there are any ships already anchored in Fukien or Thai-tshang they must at once return to Nanking, and all building of sea-going ships for intercourse with barbarian countries is to cease forthwith.'[b] But as we have seen, before a year was out, Jen Tsung died, and his successor, the Hsüan-Tê emperor, was of a different opinion, so that the last great voyage of Chêng Ho was able to set forth and to return in peace. Not long afterwards the new emperor died, and once again the policy was reversed, this time *en permanence*.

Chêng Ho and his associates must certainly have presented the fullest records of their voyages to their imperial master. But before the century ended these were burnt and destroyed by administrative thugs in the service of the Confucian anti-maritime party. In his *Kho Tso Chui Yü*[2] (Memorabilia of Nanking), compiled in +1628, Ku Chhi-Yuan[3] tells us that in the Chhêng-Hua reign-period (+1465 to +1487) an order was given to search in the State archives for the documents concerning Chêng Ho's expeditions to the western world. But Liu Ta-Hsia,[4] then Vice-President of the War Office, took them and burnt them, considering their contents 'deceitful exaggerations of bizarre things far removed from the testimony of people's eyes and ears'. Other sources, giving further details,[c] say that Liu's activities were covered up and protected by his successive chiefs Hsiang Chung[5] and Yü Tzu-Chün,[6] so that the affair must have happened in the neighbourhood of +1477. At this time the Confucian bureaucrats were strongly opposing the plans of the eunuch Wang Chih,[7] who had been appointed Inspector of the Frontiers, and who called for the records of Chêng Ho's time in connection with his policy of restoring Chinese power in South-east Asia.[d]

All this does not of course exhaust the causes of the decline of the Ming navy, to which Lo Jung-Pang (2) has devoted a brilliant study. Economic factors were at least

---

[a] As yet there is no full sociological study of the eunuchs in Chinese history, but a good account of another engagement, also in the Ming, but later, is given by Hucker (1).

[b] See Duyvendak (9), pp. 388 ff.

[c] References and a full account will be found in Duyvendak (9), pp. 395 ff. Some important passages are contained in the *Shu Yu Chou Tzu Lu*[8] (Record of Despatches concerning the different Foreign Countries) compiled from official documents by Yen Tshung-Chien.[9] Minor officials were beaten for not being able to find dossiers which had been made away with by their own superiors.

[d] It is fair to say that not everyone accepts the account of these events which Duyvendak worked out. According to the *Ming Shih*, ch. 182, p. 14b, the documents which Liu Ta-Hsia hid (not burnt) were concerned only with the planning of the invasion of Annam in the Yung-Lo reign-period, an invasion which Wang Chih wanted to repeat, and not with the naval expeditions of Chêng Ho. *Ming Shih Chi Shih Pên Mo*, ch. 22 (p. 15) and ch. 37 (p. 71) dates the matter precisely in +1480. We are indebted to Dr Lo Jung-Pang for this *caveat*. But Duyvendak's case is still a strong one.

[1] 夏原吉	[2] 客座贅語	[3] 顧起元	[4] 劉大夏	[5] 項忠
[6] 余子俊	[7] 汪直	[8] 殊域周咨錄	[9] 嚴從簡	

as important. Very great profits accrued to the Chinese State from the tribute-trade system of Chêng Ho's time, but by the middle of the century a severe currency depreciation had set in, the value of the paper notes falling to 0·1 % of their face value. Had the long-distance voyages been continued, China would have had to export precious metals. At the same time there was an increase in private trade beyond what had been contemplated at the beginning of the century, porcelain and new cotton products being exchanged directly at southern and western ports for the goods needed at home. There was also an unexpected technological revolution. For centuries canal carriage and sea transport had competed in the essential function of grain shipment from south to north, and now a heavy oscillation took place in favour of the former. It was in +1411 (as we saw above, p. 315) that the engineer Sung Li perfected the water-supply of the summit section of the Grand Canal, thus converting it into a full-capacity all-seasons proposition at last, and in +1415 the maritime grain-transport service was abolished while thousands of Chhen Hsüan's new sailing-barges were put in hand. Thus at one and the same time a great nursery of deep-water sailors was lost, and orders flowed no longer into the maritime ship-building yards, which were run down to maintenance level.[a] Military events also intervened. The serious deterioration of the north-western frontiers diverted all attention from the sea, and in +1449 at the disastrous Battle of Thu-mu, that Chinese emperor who had suppressed the Treasure-ship fleets was himself taken into captivity by the Mongol and Tartar armies.[b] At the same time there was a significant shift of population from the south-eastern seaboard provinces, reversing the trend which had been so strong at the beginning of the Southern Sung. Finally one should not overlook the development during the +15th century of a sterile conventionalised version of Neo-Confucianism, markedly idealist in metaphysics and Buddhist in religion,[c] which led to a loss of interest in geographical science and maritime techniques, replacing the energetic valour of the early Ming by an introspective culture and a political lethargy. This was, indeed, but one aspect of a general decline which reflected itself severely in many branches of science and technology.[d]

The navy simply fell to pieces.[e] By +1474 only 140 warships of the main fleet of 400 were left. By +1503 the Têngchow squadron had dropped from 100 vessels to 10. Desertions occurred wholesale and the corps of shipwrights disintegrated. In the +16th century the anti-maritime party grew ever more powerful. Perhaps the government feared the disturbing social consequences of large ships in the hands of private merchants,[f] new horizons, change, sedition, progress, and so on. Perhaps they thought

---

[a] The internal rise in prices was also an important factor, crippling all shipyards, while the worsened foreign exchange position hindered the use of imported timber (Lo Jung-Pang, 5).

[b] Cordier (1), vol. 3, p. 44. Even under the Yung-Lo emperor, the absence of the court at campaign headquarters in the north permitted the rise of grave corruption in the naval administration, especially in the timber supply agencies (Lo Jung-Pang, 5).

[c] Cf. Vol. 2, pp. 509 ff.

[d] Cf. Vol. 3, pp. 437, 442, 457. But not, as we shall see, in botany and medicine.

[e] Details in this paragraph are derived from Chêng Hao-Shêng (1) and Lo Jung-Pang (2).

[f] Actually during the second half of the +15th century Chinese private merchant shipowners had a temporary flourishing period. Former government shipwrights entered their service, and built a con-

that coastal provinces might make league with foreign powers. Perhaps they felt that losses of tax-grain on the sea-route could not be risked. In any case, the shipyards were more and more closed down, the workers diverted to other employments, and maritime activity discouraged as much as possible. By +1500 regulations were in force which made it a capital offence to build a sea-going junk with more than two masts. An edict of +1525 authorised coastal officials to destroy all ships of this kind, and to arrest mariners who continued to sail them. Another of +1551 declared that whoever went to sea in multiple-masted ships, even to trade, was committing a crime analogous to espionage by communicating with foreigners.[a] This was the agoraphobic mentality which closed Japan for two centuries from outside intercourse, and though it never prevailed in China to the same extent, the great naval possibilities had been done to death.[b] Gradually Chinese shipping recovered and most of the traditional types of build were preserved. But what a difference it might have made to Lin Tsê-Hsü[1] and his friends such as Phan Shih-Chhêng,[2] feverishly studying Western shipbuilding

siderable mercantile fleet, so that by the beginning of the +16th century some substantial venturers, such as Lin Yü,[3] owned as many as fifty large sea-going ships; cf. Lo Jung-Pang (5); Fang Chi (1). Passing Hangchow in +1487, a Korean eye-witness, the official Chhoe Pu (cf. p. 360), wrote that 'foreign ships stand as thick as the teeth of a comb', and that the masts of the river shipping were 'crowded thickly as a forest' (Meskill tr. pp. 88, 89). See further von Wiethof (1, 2).

[a] All these enactments are preserved in the *Ta Ming Hui Tien*. They did not prevent a curious system of semi-governmental trading, on which again see von Wiethof (1, 2).

[b] It is interesting that +16th-century Europe knew dimly that there had been a great Chinese withdrawal from sea power. Most of de Mendoza's ch. 7 is devoted to this (Staunton ed. pp. 92 ff.). In +1585 he wrote as follows: '...They (the Chinos) have found by experience that to go forth of their owne kingdome to conquer others, is the spoile and loss of much people, and expences of great treasures, besides the travaile and care which continually they have to sustaine that which is got, with feare to be lost againe; so that in the meane time whilest they were occupied in strange conquests [i.e. those of Chêng Ho], their enimies the Tartarians and other kings borderers unto them, did trouble and invade them, doing great damage and harm...(So) they found it requisit for their quietnes and profite...to leave all they had got and gained, out of their own kingdome, but especially such countries as were farre off. And from that day forwards not to make wars in any place; for that from thence did proceed a known damage and a doubtfull profite...(Thus the Chinese emperor) commanded upon great penalties, that al his subjects and vassals naturall that were in any strange countries, should, in a time limited, return home...(Likewise he commanded that his governors) should in his name abandon and leave the dominion and possession (of foreign countries) excepting such as would of their owne good will acknowledge vassalage, and give him tribute and remaine friends, as unto this day the Lichios (the Liu-Chhiu islanders) and other nations do. All the which that is said, seemeth to be true, for that it is cleerely found in their histories and books of navigations of old antiquitie; whereas it is plainely seene that they did come with shipping into the Indies, having conquered al that is from China, unto the farthest part thereof...So that at this day there is a great memory of them in the Ilands Philippinas and on the coast of Coromande...and the like notice and memory there is in the kingdom of Calicut, where be many trees and fruits, that the naturals of that countrie do say, were brought thither by the Chinos when that they were lords and governours of that countrie.' De Mendoza adds that now (in his time) Chinese merchant-captains can, and do, trade overseas under confidential licences from their own government. Three Chinese merchants had been in Mexico, he said, as he wrote, and even went on to Spain and other parts of Europe. Professor Donald Lach of Chicago, to whom we owe a reminder of this interesting passage, suggests that de Mendoza emphasised what he knew of the Chinese *répli* so strongly because he wanted to condemn by indirect comparison the current idea of the feasibility and desirability of a conquest of China by the Spanish, Portuguese or other European nations. Details of this are given on p. 534 below. If the example of an empire without imperialism was thus held up before the eyes of Europe, Europe took very little notice of it.

[1] 林則徐　　　[2] 潘仕成　　　[3] 林昱

methods at the time of the Opium Wars, in the 'forties of the last century,[a] if the short-sighted landsmen of the Ming court had not won the day. Indeed before the +16th century was out, their policy laid the coast open to ferocious attacks by Japanese pirates which were overcome only with great difficulty.[b] The resulting increase of naval strength proved valuable at the end of the century, when between +1592 and +1598 squadrons from Shantung, Fukien and Kuangtung fought side by side with those of the gallant Korean admiral Yi Sunsin,[c] successfully repelling the fleets of Japan. Then in the +17th century the last remnants of the Ming navy fought under Chêng Chhêng-Kung[1] (Koxinga) against the Manchus and their allies the Dutch, whom he had expelled from Formosa in +1661. The Chhing emperors were not at all interested in the sea, and under them the navy languished.

We have seen how Chinese officials in the +15th century destroyed documents of incalculable historical value. At the other end of the Old World an earthquake did the same job. But the destruction of a large part of Lisbon in +1755 would not have been able to effect this if there had not been a persistent practice of secrecy on the part of the Portuguese rulers of the +15th century. The disappearance of documentary proof of many alleged Portuguese discoveries has caused great heartburnings among modern historians, but it seems that those who have maintained the existence of a policy of secrecy are being justified. For example, in +1485, three years before the voyage of Bartolomeu Dias, a 'speech of obedience' was made at Rome by the representative of the King of Portugal, Vasco Fernandes de Lucena, in which he said that his country-men had attained to the Gates of India, 'almost to the Promontorium Prassum, where the Arabian Gulf begins'.[d] Nothing further was divulged. These words have always been difficult to explain unless some caravel captains had been exploring the East African coast a few years before the success of Dias. Similarly, Vasco da Gama has always been regarded as the first European to sail up the coast to Sofala. Yet when the manuscript rutter of Aḥmad ibn Mājid came to light a few years ago in Leningrad, he was found to say quite distinctly that a Frankish (i.e. Portuguese) expedition had been shipwrecked near Sofala as early as +1495, three years before the coming of da Gama.[e] Thus there is another interesting parallel between China and Europe in the destruction of records of the greatest naval age of both.

  [a] The monograph of Chhen Chhi-Thien (1) gives many particulars about the revival of shipbuilding at this time, and another one (3) on Tso Tsung-Thang[2] describes the foundations of the Fuchow Dockyard. See also Rawlinson (1); Anon. (72).

  [b] Cf. Vol. 3, p. 517. These incursions became severe from +1515 onwards. On the navy during this period see *Hsü Wên Hsien Thung Khao*, ch. 132.

  [c] Cf. pp. 683 ff. below.

  [d] See J. Cortesão (1), p. 92; Brochado (1), pp. 98 ff.; for another interpretation see Peres (1), pp. 63 ff.

  [e] Brochado (1), pp. 79, 102.

¹ 鄭成功   ² 左宗棠

(vi) *Motives, medicines and masteries*

It is reasonable to put the question whether, if the debacle of the Chinese Admiralty had not taken place, their expeditions would have continued. Would they have gone on round the Cape of Good Hope? Perhaps. But the Chinese motive was never primarily geographical exploration[a]— what they sought rather was cultural contacts with foreign peoples, even if quite uncivilised. The Chinese voyages were essentially an urbane but systematic tour of inspection of the known world. The Portuguese motive was not primarily geographical exploration either. Their discoveries of the coasts of South Africa (and indeed also of Brazil) were really secondary achievements in the great attempt of a nation which believed itself the champion of Christendom in an unceasing war, to find a way round to the Indies and so to take the Islamic world in the rear. The chronicler de Zurara puts this fourth in his famous list of Prince Henry's motives, beginning with the State-supported quest for cosmographical knowledge, and mentioning thereafter the profit of traffic and commerce.[b] Economic historians such as Magalhães Godinho are now revealing many other reasons which impelled the Portuguese to their great efforts.[c] A lack of coin impeding mercantile activities induced a thirst for gold. A monetary devaluation caused the nobility to hanker for new domains, and drove the sons of the petty aristocracy to 'make their fortunes' (by brigandage if not otherwise). A deficit in cereals meant importing, for which it was difficult to pay. Plantation culture led to a growing need for slaves, and other industries were interested in expansion; the textile trade, for instance, needed gums and dyes. On the Chinese side, it is hard to say whether the search for the dethroned emperor was ever more than a pretext;[d] the main motive was surely the demonstration of China's prestige by obtaining the nominal allegiance (and rich exchange) of far-away princes,[e] and this was exactly what the Confucian bureaucrats thought so unnecessary. But when one compares the Chinese and the Portuguese ventures over the whole range of our knowledge of them, it does seem as if the proto-scientific function of collecting natural rarities, strange gems

[a] Their conscious attitude to Africa would probably have been to let the Arabs do the exploring of it, if it ever interested them to do so. The Far West had been known to the Chinese well enough at second hand since the days of the Roman Empire, but they knew of nothing vital which would have added to their own civilisation so much that it was worth going there to get. Debenham's comment (1), p. 123, that 'Asians seemed to have little need to explore', is only part of a half-truth. In the person of Chang Chhien, China discovered Europe before Europe discovered her. In the person of Dom Henrique, Europe discovered America while trying to do something else. Of course, after the Renaissance, man's outlook on the unknown parts of the world changed rapidly.

[b] See Beazley & Prestage (1), p. xiii, tr. pp. 27 ff.: 'It seemed to the Lord Infante that if he or some other lord did not endeavour to gain that knowledge, no mariners nor merchants would ever dare to attempt it, for it is clear that none of them ever trouble themselves to sail to a place where there is not a sure and certain hope of profit.' To the search for the longed-for ally, Prester John, de Zurara added a reconnaissance motive, the wish to find out how far south in Africa the Moorish power extended. His fifth motive was missionary zeal, and his sixth was more a cause than a motive, Prince Henry's noble horoscope.

[c] See Godinho (1), pp. 41 ff.; cf. (2).

[d] Considered opinion is (Moule & Yetts (1), pp. 106 ff.) that he died, if not in +1402, then in some unknown place before +1423. Cf. Pelliot (2b), pp. 303 ff.; Duyvendak (8), p. 27.

[e] This was distinctly characteristic of Chinese civilisation. One must not forget, however, the subsidiary motives of encouraging the South Seas trade, and prospecting the coasts and islands with a view to defence.

and animals, was more marked in the former. Before long, of course, the humanistic Renaissance curiosity about all exotic things asserted itself strongly in the West,[a] but this was an old Chinese tradition too which had been very powerful, for example, during the Thang period.[b] Said Huang Shêng-Tsêng:[c]

Amid the thundering billows and surges rearing mountain-high, helped by their flying masts and labouring oars, now with their cordage tightly strained and now under loosened sails, the envoys journeyed many myriads of *li*, and in their voyaging to and fro spent well-nigh thirty years...

Then, their vessels filled with pearls and precious stones, with eagle-wood and ambergris, with marvellous beasts and birds—unicorns and lions, halcyons and peacocks—with rarities like camphor and gums and essences distilled from roses, together with ornaments such as coral and divers kinds of gems, the envoys returned.

Some of the specimens had a particular symbolic value for the Chinese court; the giraffe, for example, was identified with the mythical animal *chhi-lin*[1] which according to age-old legend was one of the greatest auspicious signs appearing in Nature to signalise an imperial ruler of perfect virtue.[d] Even the morose Hsia Yuan-Chi joined in the praises of the giraffe. Collecting went on everywhere—

When the Chinese ships arrived [says Ma Huan, speaking of Zafār in Arabia], after the reading of the Imperial Rescript and the offering of presents, the King despatched his chiefs all over the country to command the people to bring incense, dragon's-blood (resin), aloes, myrrh, benzoin, liquid storax, *Momordica* seeds (*mu pieh tzu*[2]), etc. in exchange for hempen cloth, silks and porcelain ware.[e]

Our attention is arrested here by the mention of a drug-plant, a cucurbitaceous vine much appreciated and used by Chinese physicians and pharmacists.[f] Could Chêng Ho's men have been looking for materia medica? There were many able physicians among them.[g] We suspect that it would be worth while to investigate whether the search for new drugs was not a more important motive than has usually been thought both for the Portuguese and for the Chinese ventures. Both in Europe and China the +14th century had been a time of extremely serious epidemics. The notorious Black Death had ravaged Europe,[h] and similar outbreaks including bubonic and pneumonic plague had occurred in China eleven times[i] between +1300 and +1400. It was only reasonable to think that for new and dreadful diseases new and powerful drugs must be

---

[a] One remembers the lion cages at Queluz, and the rhinoceros which Albrecht Dürer drew.

[b] The beautiful treatise of Schafer (13) on this subject has met with much interest.

[c] *Hsi-Yang Chhao Kung Tien Lu*, preface, tr. Mayers (3), p. 223.

[d] Full details of this episode are given in Duyvendak (9), pp. 399 ff., and summarised in (8), pp. 32 ff.

[e] *Ying Yai Shêng Lan Chi*, entry for Tsu-fa-erh, tr. Duyvendak (10), p. 59. See also in (10), pp. 11, 60, 74, etc. Panegyrics on ostriches are discussed in (9), p. 382, and the favourable views of the Yung-Lo emperor about foreign trade are quoted on p. 357. On the search for rarities cf. Pelliot (2a), p. 445.

[f] *Momordica cochinchinensis*, see Stuart (1), p. 265; Burkill (1), vol. 2, p. 1486. The seeds contain two large oily cotyledons, green on the outside but yellow internally; these are prescribed in the form of a paste for abscesses, ulcers and wounds, as well as in other ways for other affections.

[g] Cf. p. 491 above.

[h] Cf. Hecker (1); Garrison (1), p. 188. The social effects of the devastations of the plague in Portugal in the period before Prince Henry are described by Nemésio (1), p. 28.

[i] *Hsü Wên Hsien Thung Khao*, ch. 228 (pp. 4646, 4647).

[1] 麒麟   [2] 木鼈子

found.[a] That the intrepid exploration of previously unknown lands did in fact produce these is a matter of common knowledge—we speak not of syrup of Tolu, Peruvian balsam or sarsaparilla, but of the coca leaves from which the Andean Indians chewed their cocaine, and above all the 'fever-bark tree', *Cinchona* spp., specific for the greatest evil, perhaps, to which man in historical times was victim.[b] A noble literature resulted in which two books stand out, Garcia da Orta's 'Colloquies on the Simples and Drugs of India'[c] printed in Goa in +1563, and Nicholas Monardes' 'Joyfull Newes out of the New-Found Worlde' (i.e. the Americas),[d] first printed, with a less exciting title, at Seville in +1565. The Chinese parallels are of course much less well known. Just after Chêng Ho's time there appeared the last great edition of the *Chêng Lei Pên Tshao*[1] (+1468),[e] but it seems to have been a faithful revision of the +1249 edition with little or no innovation. However, the great contemporary interest in Arabic drugs and therapeutic methods is shown by the publication, some time during the early decades of the +15th century, of a work entitled *Hui-Hui Yao Fang*[2] (Pharmaceutical Prescriptions of the Muslims), all the more interesting in that some parts of it are printed in Persian. Sung Ta-Jen (1), who has described the unique copy in the Peking National Library, unfortunately incomplete, regards it as the work of an Arabic or Persian physician of the Yuan period, translated into Chinese about +1360. It is quite probable, too, that some of the inorganic products of the voyages were discussed in the *Kêng Hsin Yü Tshê*[3] (Precious Secrets of the Realm of Kêng and Hsin, i.e. all things concerned with metals and minerals, symbolised by these two cyclical characters, including alchemy and pharmaceutics), produced in +1421 by the Ming prince Ning Hsien Wang[4] (Chu Chhüan[5]).[f] We have noted already that the *I Yü Thu*

---

[a] Actually the Chinese had been perennially interested in drugs from foreign countries. As we shall see in Sect. 44, the incidence of disease was always changing through the centuries, and new remedies were always being sought for. Already in Vol. 1, p. 188, we mentioned Li Hsün,[6] the early +10th-century scholar of Persian extraction, whose *Hai Yao Pên Tshao*[7] (Materia Medica of the Countries Beyond the Seas) was still available, at least in part, to Li Shih-Chen. The *Nan Hai Yao Phu*[8] (A Treatise on the Materia Medica of the South Seas), recorded in the Sung bibliography (*Sung Shih*, ch. 207, p. 23a), may or may not have been the same book under an alternative title. In Vol. 1 the authorship was attributed to Li Hsün's younger brother Li Hsien; this is to be corrected.

[b] Duran-Reynals (1) has provided us with a history, both witty and moving, of the discovery and utilisation of quinine. Unfortunately, following the lamentable example of some of the best medical historians, such as Zinsser, it is undocumented. It is interesting to recall that in +1693 the Jesuit Jean de Fontaney (Hung Jo-Han[9]) presented some quinine to the Khang-Hsi emperor and cured his intermittent fever (Pfister (1), p. 428). This does not mean, however, that there were no genuine anti-malarial drugs in the Chinese pharmacopoeias (cf. Sect. 45 below).

[c] Garcia da Orta was a Portuguese from Elvas; his book, cast in the form of dialogues, is most readable still today. An English translation was made by Sir Clements Markham. Among the drugs which da Orta described was Indian hemp, *Cannabis*; the thorn-apple with its atropine and hyoscyamine, *Datura*; a variety of *Rauwolfia* found in Ceylon; ginger, etc. Cf. Mieli (2), vol. 5, pp. 136 ff.

[d] Nicholas Monardes was a Spanish physician of Seville; he was among the first to ascertain the true nature of bezoar. Among the American drugs which he described was Peruvian balsam from *Myroxylon*, favoured for dressing ulcers until very recent times; but the really important natural drugs from the New World were not introduced until the following century, ipecacuanha from Brazil and quinine from Ecuador and Peru. Cf. Mieli (2), vol. 5, pp. 142 ff.; Sarton (9), p. 129.

[e] We are very fortunate to have a copy of this rare Ming edition at our disposal in Cambridge. See Fig. 990.

[f] We have been trying to get a sight of this book for many years, but so far without success.

[1] 證類本草　　[2] 回回藥方　　[3] 庚辛玉册　　[4] 寧獻王　　[5] 朱權
[6] 李珣　　　　[7] 海藥本草　　　南海藥譜　　[9] 洪若翰

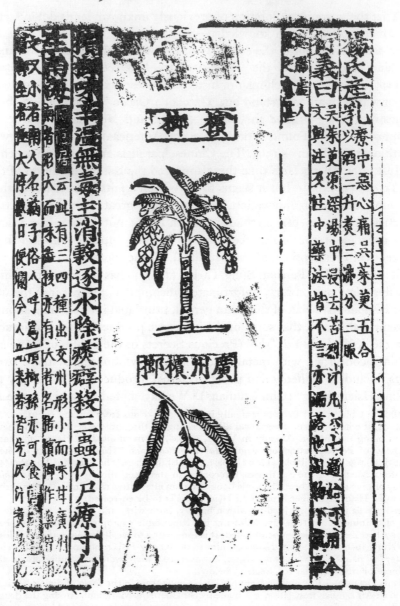

Fig. 990. To illustrate the role of natural history in the Chinese discoveries and navigations of the early +15th century—a page from the rare +1468 edition of the *Chêng Lei Pên Tshao* (Classified Pharmaceutical Natural History). The intense interest of the Chinese in possibly useful plants and animals of foreign countries did not begin in the time of Chêng Ho, but it was certainly one of the motives which sent out the fleets of Treasure-ships. Here is illustrated the betel-nut (*pin-lang* or *ping-lang*) tree, *Areca Catechu*, on which Bretschneider (6), pp. 27 ff., Stuart (1), p. 46, and Watt (1), pp. 83 ff., may with advantage be consulted. Above is the whole palm, and below a drawing of the panicle with its hundred egg-sized white fruits, each containing nut and pulp within its coriaceous rind. The description, beginning in the first column to the left of the drawings, states as usual the basic characteristics of the product, that it is bitter, harsh and warming, that it contains no powerful active principle, that it cures or helps to cure such and such diseases, and that it grows in the lands of the South Seas. Two quotations from earlier writers follow. In the edition figured ch. 13, p. 13*b*, in the +1249 edition, p. 11*b*.

Chih (Illustrated Record of Strange Countries), which almost certainly profited from knowledge gained by the voyages, was also connected with him.[a] For materia medica in general the place to look will be the *Pên Tshao Phin Hui Ching Yao*[1] (Essentials of the Pharmacopoeia Ranked according to Nature and Efficacity),[b] imperially commissioned by the Hung-Chih emperor, and finished by Liu Wên-Thai[2] and others in +1505. Between +1485 and +1565, the date of Monardes' book, there were at least three substantial Chinese pharmaceutical works, and then in +1596 came the 'Great Pharmacopoeia' (*Pên Tshao Kang Mu*[3])[c] of Li Shih-Chen.[4] In view of the very rapid spread of plants such as tobacco and maize in China after the discovery of the New World,[d] it would even be surprising if the great voyages had brought no new drugs home.

The time has come to draw all these threads together. Sofala, by the sea-routes, was just about half-way between Lisbon and Nanking. Might it not perhaps have changed the course of history if the first Portuguese vessels coming past Sofala to Malindi had met much greater fleets of bigger ships with company more numerous than their own[e]—and people, too, with very different ideas about the proper relations between civilised men and barbarians? 'Coming into contact with barbarian peoples', wrote Chang Hsieh in +1618 in a passage we shall shortly read,[f] 'you have nothing more to fear than touching the left horn of a snail. The only things one should really be anxious about are the means of mastery of the waves of the seas—and, worst of all dangers, the minds of those avid for profit and greedy of gain.' And indeed it was in accordance with this enlightened conception of inter-cultural contact that the Chinese set up no factories, demanded no forts, made no slave-raids,[g] accomplished no conquests.[h] Their total lack of any proselytising religion precluded friction from that source. The governmental character of their enterprises helped to restrain individual avarice and the crimes to which it could give rise. On the other hand, it is clear enough that the Portuguese behaviour originated as a development of the Crusader mentality.[i] They were at war. But if the naval struggle against the Muslim mercantile States on the shores of the Indian Ocean was a continuation of the 'Holy War' for the holy places, it

[a] See p. 493 above. [b] Cf. Bertuccioli (1).
[c] It may not be without significance in the present context that Li's friend Wang Shih-Chên,[5] who wrote the preface for his book, was also interested in the voyages of Chêng Ho, and wrote several notices about them. For Li's biography see Lu Gwei-Djen (1).
[d] Cf. Sect. 41 below.
[e] One must remember here the good order and discipline of the Ming. The dynasty was not decadent in the +15th century, not comparable with the late Chhing and its position of rank inferiority *vis-à-vis* the Western powers.
[f] P. 584 below.
[g] It is regrettably evident that the doctrine of the curse of Cain, which made all Africans slaves by nature, was already familiar in de Zurara's time (Beazley & Prestage tr., p. 54).
[h] Filesi, too, well emphasises this, (1), p. 42.
[i] That the Chinese did not have a legacy of century-long religious wars is a feather in the cap of Chinese humanism. It would not be unfair, however, to say that the millennial struggles of the Chinese against the north-western nomads, from the Han and the Huns onwards, paralleled in a way the Crusades. But these were never wars of religious fanaticism, and (more important in the present context) they in no way affected Chinese attitudes to the peoples of southern Asia.

[1] 本草品彙精要   [2] 劉文泰   [3] 本草綱目   [4] 李時珍   [5] 王世貞

turned insensibly into something quite different, an insatiable thirst for gold, not only Muslim gold, and an obsessive desire for power over all African and Asian peoples, whether or not they had anything to do with Islam. And now too there began to come into play that decisive superiority of armaments which the scientific Renaissance was to give to Europeans, enabling them to dominate the Old World and the New for three centuries. 'From the Cape of Good Hope onwards', wrote Capt. João Ribeiro in + 1685, 'we were unwilling to leave anything outside of our control; we were anxious to lay hands on everything in that huge stretch of over 5,000 leagues from Sofala to Japan...There was not a corner which we did not occupy or desire to have subject to ourselves.'[a] And if China had been any weaker than she was, this vaulting ambition would not have spared her.[b]

[a] Quoted by Boxer (3) in his refutation of an eminent white-washer, from the *Fatalidade Historica da Ilha da Ceilão*, bk. 3, ch. 1. Cf. Anon. (47), no. 129.

[b] It is not generally known that even the earliest Portuguese relations of China contained estimates of her 'conquerability'. Approaching as they did by sea, the first European visitors noticed particularly the low state of defence of the coastal provinces which the decline of the Ming navy had brought about. The first ambassador, Tomé Pires, wrote in + 1515, before he himself had been to China: 'Such is their fear of Malays and Javanese that it is quite certain that one (of our) ship(s) of 400 tons could depopulate Canton, and this would bring great loss to China...It certainly seems...that the Governor of Malacca would not need as much force as they say in order to bring it under our rule, because the people are very weak and easy to overcome. And the principal travellers who have been there affirm that with ten ships the Governor of India who took Malacca (i.e. Albuquerque, see Whiteway (1), pp. 141 ff.) could take the whole of the Chinese coast' (A. Cortesão (2), vol. 1, p. 123). Yet Tomé Pires was a most sympathetic character, an apothecary and 'Factor of the Drugs of India'. His estimate was grotesquely falsified by his own fate; landed at Canton in + 1517 his embassy failed because of the news of the fall of Malacca and the lawless actions of Portuguese sea-captains upon the coasts (especially Simão de Andrade, who was sent out to collect him), he himself was for a time imprisoned, and finally died in exile in a small Chinese city consoled by a Chinese spouse (A. Cortesão (2), vol. 1, pp. xviii ff., lxi ff.). Among these actions were the building of forts and the kidnapping of Chinese children (A. Cortesão (2), p. xxxvi). Portuguese prisoners in Canton continued to plot, and in + 1524 one of them, Cristovão Vieira, got a letter out urging that the capture of the whole province would need only some 2,500 men and a fleet of 10 to 15 sail (D. Ferguson (1), pp. 29 ff.). Half a century later, in + 1576, Francisco de Sande, Governor of the Philippines, formally proposed to the Spanish king a plan for the conquest of China, but Philip II rejected it (Blair & Robertson (1), vol. 4, pp. 21 ff.; Boxer (1), p. l). On 13 September 1584 the great Jesuit Matteo Ricci wrote, on behalf of himself and Ruggieri, a letter to Juan-Baptista Román, the Royal Factor or Treasurer of the Philippines, who was staying at Macao. This gave a brief but brilliant account of China (it is printed in Venturi (1), vol. 2, pp. 36 ff.). Ricci alluded in passing to the apparent military weakness of the Chinese *vis-à-vis* for example the Japanese pirates, which he found surprising in view of all their other qualities. Román sent this back to Europe with a covering letter, in the course of which he said: 'With 5,000 Spaniards, at the most, the conquest of this country might be made, or at least of the maritime provinces, which are the most important in all parts of the world. With half-a-dozen galleons and as many galleys one would be master of all the coast of China, as well as of all that sea and archipelago which extends thence to the Moluccas' (Colin & Pastells (1), vol. 3, pp. 448 ff.; englished by Sir George T. Staunton (3), p. lxxx). These words were written at a time when Spain was contemplating a full-scale embassy to China (see d'Elia (2), vol. 1, p. 216; Trigault (Gallagher tr.), pp. 170 ff.; Colin & Pastells (1), vol. 2, pp. 520 ff.). As neither the Chinese nor the Portuguese wanted this, it fell through. Sentiments similar to those of Román are found, if in disguised form, in the great book of Juan Gonzales de Mendoza published in Rome the year after Ricci wrote his letter. De Mendoza wrote: 'I do not here declare the industrie that might (with the favour of God) be used to win and overcome this people, for hat the place serveth not for it; and I have given large notice thereof, unto (those to) whom I am bound. And again, my profession is more to bee a meanes unto peace, than to procure any warres; and if that which is my desire might be done, it is, that with the word of God, which is the sworde that cutteth the hearts of men, wherewith I hope in the Lorde to see it' (Staunton (3), p. 89). Thus de Mendoza had passed on information about China's military weakness relative to European arms at this time, derived no doubt in the main from the Augustinians, Franciscans and Dominicans whom he had met in the Philippines. Just over a century later, the Jesuit Louis Lecomte, addressing Cardinal Furstenberg, wrote: 'I confess, my Lord, that in viewing all those Cities which their Inhabitants esteem the strongest

As we have seen, the 'Portuguese Century' was also the 'Chinese Century'. Our feelings of ambivalence towards the Portuguese explorers and *conquistadores* cannot be overcome. Their great and courageous actions compel all admiration. Their behaviour and policy towards Arabs and Asians is often excused by the roughness and violence of the time.[a] But the Chinese crews and captains were exact contemporaries of the Portuguese empire-builders, and their proceedings were not under the sign of Mars.[b] Let us cherish the memories of those Lusitanians who were truly great, not so much Albuquerque and Almeida, but the navigators and cartographers, the astronomers and the naturalists. The stature of Dom Henrique is nothing dimmed; he remains for ever an inspiring and lovable figure. And from lesser men we have plenty to choose: João Fernandes who lived friendly with the Arabs and negroes of Mauretania,[c] Fernão Queiroz who wept for the doings of his countrymen in Ceylon,[d] Tomé Pires the amiable and unfortunate apothecary-ambassador,[e] Sebastião Manrique the Augustinian who did not hesitate to take the yellow robe with a bhikku his companion to visit some Portuguese sailors exiled in the mountains of Arakan.[f] Let us celebrate too the insight of a man not always approved of, the writer of the first autobiographical novel, Fernão Mendes Pinto.[g] His famous *Peregrinaçam* of +1614 was not a plain unvarnished account like those of the other early Portuguese travellers in and near China, but rather a great work of art, a dramatic judgment on the exploits of his nation. A veiled criticism of the Western attitude to Asians runs all through it, a conviction that imperialism rested on impiety. These are the voices of the just men who suffice to save Lisboa for our perpetual affection.

So there we leave them—voyagers from the East, the Chinese, calm and pacific, unencumbered by a heritage of enmities; generous (up to a point), menacing no man's livelihood; tolerant, if more than a shade patronising; in panoply of arms, yet conquering no colonies and setting up no strongholds—voyagers from the West, the Portuguese, crusader-traders out to take hereditary enemies in the rear and wrest a mercantile foothold from unsympathetic soil; hostile to other faiths yet relatively free from racial prejudice; hot in the pursuit of economic power, and heralds of the Renaissance. In all the maritime contacts between Europe and Asia in that dramatic age our forefathers were quite sure who the 'heathen' were. Today we suspect that these were not the less civilised of the two. And here we shall leave also that great arena the Indian Ocean, and turn our eyes to the seas of other continents and those who may have sailed them.

in the World, I have often with no little pleasure reflected how easily Lewis the Great would subdue those Provinces, if Nature had made us a little nearer Neighbours to China; he whom the stoutest Places in Europe can at best withstand but during a few days' (1, p. 73). Finally even Macartney in +1794 had his ideas about invasion projects (Cranmer-Byng ed., pp. 203, 211). How fortunate it was for the future cultural synthesis of the world that Chinese civilisation was never overcome by European arms.

[a] It was certainly sharpened by the fact that they were so unwelcome in the Indian Ocean. But they expected enmity, and were prepared to use all the force at their command to overcome it.

[b] We cannot forbear from referring to a former page which spoke of a strange astrological difference between the Romans and the Seres (Vol. 1, p. 157).

[c] Prestage (1), pp. 76 ff.   [d] S. G. Perera (1).

[e] A. Cortesão (2).   [f] Collis (2).

[g] See the brilliant interpretation and paraphrase of Collis (1).

(vii) *China and Australia*

Who first navigated the waters of Australia? We know that New South Wales was named by Captain Cook in +1770 and that Captain Dampier had explored the north-west and western coasts of the continent in the years +1684 to +1690. Before that, a long series of Dutch surveys is equally established—first contact in +1606, study of the northern, western and southern shores by Zeachen, Edels, Nuyts and Tasman between +1618 and +1627, the naming of Western Australia as New Holland in +1665. Sixteenth-century recognitions are more controversial, but it seems quite probable that either Cristóvão de Mendonça in +1522 or Gomes de Sequeira in +1525 trod upon Australian land and met its aboriginal people.[a] A French claim for +1503 is still more shadowy.[b] In recent times, however, the question of a possible pre-European discovery of the great island continent by Chinese sailors has been raised in serious form.

The subject is interesting partly because of the wide area of the Southern Seas over which Chinese discovery and traffic did certainly extend. The Chinese had maritime and commercial relations with the Philippines,[c] Java,[d] Bali,[e] Borneo and Sarawak,[f] and the Moluccas and Timor,[g] not only in the time of the great Ming expeditions, but also at least as far back as the Sung, when Chao Ju-Kua wrote his classic description of sea trade (*Chu Fan Chih*) about +1225. Of the traffic with Nan Hai,[1] the South Sea Islands or East Indies, the most complete account is probably that of Wang Ta-Yuan, who wrote his book, the *Tao I Chih Lüeh*, about +1350, from notes gathered during his own travels in those parts from +1330 to +1349. Chinese influence in these far-flung Indonesian island countries is shown today (as in East Africa) by the omni-

---

[a] See Peres (1), pp. 120 ff.

[b] Cf. Stefánsson & Wilcox (1), p. 626, referring to Binot de Gonneville. It is also claimed that Guillaume le Testu sighted Australia in +1531. But the Lusitano-French maps of the +16th century (+1536 to +1550) may be interpreted as implying early Chinese knowledge of Australia rather than early French. These maps depict consistently (as no earlier ones do) a large continent (Greater Java) south of Java. Collingridge (1), p. 306, after exhaustive study, came to the conclusion that the early Portuguese navigators in the East Indies must have got this information from Chinese or Malay maritime sources (cf. his pp. 166 ff., 180 ff., 192, 220), and passed it on to the French. The tradition of two island Javas, a 'Great' and a 'Lytil', goes back much further, for it is found in Marco Polo (Yule (1), vol. 2, pp. 272, 284) as well as the later travellers Odoric, de Conti and Jordanus Catalanus (Yule (3), pp. 30 ff.). Before +1536 these appear on European maps just as two large islands (cf. Collingridge (1), pp. 26 ff., 44, 106, 120). It is true, as Lo Jung-Pang (7) has pointed out, that there was a similar confusion between two Javas (Shê-pho[2] and Chao-wa[3]) in the Chinese literature, but Schlegel (9) adduced much evidence to show that while one was modern Java the other was some place on the Malayan coast; moreover they were not distinguished as the greater and the lesser. Though the question is complex, there is much interest in Lo's suggestion that one of the Chinese Javas might have been Australia.

[c] *CFC*, ch. 1, pp. 36b ff., tr. Hirth & Rockhill (1), pp. 159 ff. See also Laufer (29); Wada (1); Rockhill (1), pp. 267 ff.

[d] *CFC*, ch. 1, pp. 10b ff., tr. Hirth & Rockhill (1), pp. 75 ff.

[e] *CFC*, ch. 1, p. 13b, tr. Hirth & Rockhill (1), p. 84.

[f] *CFC*, ch. 1, pp. 34b ff., tr. Hirth & Rockhill (1), pp. 155 ff. Cf. p. 461 above.

[g] *Tao I Chih Lüeh* (+1350), pp. 62b ff., tr. Rockhill (1), pp. 257 ff., 259 ff.; cf. Fêng Chhêng-Chün (1), p. 87. Wang Ta-Yuan tells the story of the early +14th-century merchant captain Wu Chai,[4] whose unfortunate voyage to Timor ended with the loss of most of his crew.

[1] 南海　　　[2] 闍婆　　　[3] 爪哇　　　[4] 吳宅

presence of ceramic pieces, many of high quality and great beauty.[a] A rich Chinese traffic with Borneo, for example, especially active in the Thang,[b] traded ceramics, beads and metal tools for the edible birds' nests of the Niah caves, for hornbill ivory,[c] and for rhinoceros horn.[d] The abundance of fine Thang ware, such as jars, in various parts of Borneo, demonstrates that the trade was already old in the time of Chao Ju-Kua.[e] But Sarawak affords dated pieces of Chêng Ho's century also.[f] And much further evidence assuredly remains to be discovered.

Since Timor is only just over 400 miles from Port Darwin, there seems no inherent improbability in a visit of Chinese ships to that part of the Australian coast at any time from the +7th century onwards.[g] Hence the interest of the new study of Fitzgerald (7a, b), who, after disposing of several baseless claims, drew attention to the finding, undisputedly authentic, of a Chinese Taoist statuette, about 4 in. high (cf. Fig. 991, pl.), near the shore at Port Darwin. It represents Shou Lao [1] (the 'spirit of longevity'), mounted on his vehicle the deer, and carrying the peaches of immortality in his hand.[h] The discoverers in 1879 found this 4 ft. below the surface of the ground among the roots of a banyan tree at least 200 years old which had to be removed in the course of road-making.[i] Black with age when unearthed, the statuette is in style Ming or early Chhing, quite reasonably contemporary with Chêng Ho.[j] Its deposition may thus well have antedated the earliest European discoveries of Australia. That the image is Chinese is certain, but it would be hard to prove that it was left there by the crew of a Chinese junk rather than by Malay or Sunda fishermen who, like all South-east Asians, have treasured cult-objects of Chinese origin. The Macassarese and Buginese used to make annual visits to the Australian coast, following the monsoons in going and returning, and written records of their periodical residences are plentiful from the +18th century onwards. In exchange for natural products such as turtle-shell, fish and pearls, they traded food, cloth, tools, tobacco and similar goods. These visits were broken off by the Australian Government in 1907, but the aboriginal inhabitants still look back (not, it seems, altogether justifiably) at their contacts with the Malays as if to a Golden Age. That the Chinese themselves were not far out of the picture, however, is shown by the fact that of all the things which the Northerners came for, trepang was

[a] For the Philippines see Cole & Laufer (1).

[b] Details in Harrisson (1), from whom these words are taken. Domingo de Navarrete noted much china-ware in Borneo in +1657; cf. Cummins ed. vol. 1, p. 111.

[c] This comes from the helmeted hornbill, *Rhinoplax vigil*. An interesting account of Chinese carvings in hornbill ivory has been given by Cammann (6).

[d] Harrisson (6) describes the trade, and Jenyns (2) reviews Chinese carvings in rhinoceros horn as well as the old beliefs about its magico-medical efficacy. The classical treatment of this still remains, however, Laufer (15).

[e] See Harrisson (4, 5). On Sung ware see Noakes (1); Harrisson (2, 3, 7); Sullivan (5, 6, 7, 8).

[f] See Thien (1); Pope (1).

[g] Cf. the evidence given in Vol. 3, p. 274, that an astronomical expedition from China in +724 went as far south from Sumatra as about 15° S. See also p. 567 below.

[h] See Doré (1), vol. 11, pp. 966 ff.

[i] The site, fixed by chain measurements and therefore still precisely identifiable, is near one of the only two fresh-water springs around the land-locked harbour of Darwin, and in a gully leading down to a small cove with a sandy beach.

[j] Wei Chü-Hsien (4), pp. 99 ff., concurs, with art-historical arguments.

[1] 壽老

perhaps the most important.[a] This fishery produces the dried and smoked body-walls of sea-slugs (*Holothuria edulis* and many other genera and species), known in Chinese as *hai shen*[1] and made into glutinous soups, a delicacy, be it noted, which has always been exclusively Chinese. Moreover, the Chinese alone have been successful in the preparing-trade, which involves drying and smoking by means of mangrove-wood fires. There is thus much significance in the report of Worsley (1) that according to aboriginal tradition the Macassarese were preceded by a people they call the Baijini, much lighter in colour and possessing an advanced technology.[b] If these were indeed from China,[c] then perhaps the Shou Lao statuette is a true record of their visits, and the latter half of the +15th century remains a possible time.

So far little light has been thrown by Chinese cartography on early contacts with Australia, but that is because no serious investigation has yet been made.[d] Contrary to certain published statements, the lacquer David Globe (see p. 586 below), made in

---

[a] See Burkill (1), vol. 1, pp. 1181 ff. He says that by the end of the Malayan trade era the tropical coasts of Australia were all overworked by the trepang fishery. *Hai shen thang* remains popular in Chinese cuisine today, and I have myself often enjoyed it.

[b] The Baijini brought looms and carried on agricultural operations during their stays, which would be quite in character for Chinese. If they were getting what they wanted (water-supplies, trepang and other products) the inhospitability of the coast may well have discouraged further exploration, especially as the ships were probably rather small.

[c] Is it possible that 'Baijini' could have been derived from *pai jen*,[2] 'white folk', as opposed to the black aboriginals; or *pei jen*,[3] 'northerners'? Even *Pei-ching jen*[4] suggests itself, if perhaps the first trepang-hunters came south with semi-official authority from Peking—and after all, in traditional China everything was semi-official. The interpretation in terms of skin colour gains perhaps a little plausibility in view of the misunderstanding mentioned above (p. 507) on the 'vessels of white Christians' on the Malabar coast.

[d] This desideratum is certainly not fulfilled in the interesting though curious book of Wei Chü-Hsien (4), whose concern it is to show that the Chinese were in contact with Australia from very early times. Towards this end he has assembled a large collection of quotations from ancient and medieval Chinese texts, which do sometimes suggest, but never conclusively prove, a knowledge of the inhabitants and the fauna and flora of the Australasian region. Thus for example a number of passages from the Liu Chhao period onwards concerning the throwing of crooked knives (*fei tao*[5]) are interpreted as allusions to weapons of the boomerang type. Similarly the *chhiung-chhiung*[6] of the *Shan Hai Ching* or the *chüeh*[7] of the *Erh Ya* and *Lü Shih Chhun Chhiu*, an animal like a rat in front and like a rabbit or a hare behind, which carries its young and leaps about, is identified with the kangaroo. More convincingly, perhaps, Wei Chü-Hsien draws attention to a country called Chiao-Yao Kuo,[8] mentioned in many texts from the *Shan Hai Ching* onwards, where the people are but three feet tall; these may well be references to the negrito pygmies of New Guinea, though they cannot refer to the Australian aborigines. For Chiao-Yao means essentially 'dusky dwarfs', having as synonyms Chu-Ju Kuo[9] and Hsiao-Jen Kuo.[10] The Aëta pygmies of the Philippines were clearly described by Chao Ju-Kua in +1225 (cf. Hirth & Rockhill (1), p. 161). Associated with these races are lands in which the seasons are reversed, summer there corresponding to winter in China. This statement first occurs in a lost Chin (+4th-century) book, the *Wai Kuo Thu*[11] (Illustrated Account of Foreign Countries), quoted by Tao-Shih[12] in +668 in his *Fa Yuan Chu Lin*.[13] If it was not a pure cosmographical deduction (which seems unlikely), it must refer to experiences in regions about 30° lat. south, and if South Australia was not the place the only other possibility is South Africa—also a land approached through regions where pygmies dwelt. It is thus of particular interest that the *Kua Ti Chih*[14] (Comprehensive Geography; cf. Vol. 3, p. 520) of +638 says that the Chiao-Yao pygmies live south of the Roman Empire (Ta-Chhin, cf. Vol. 1, p. 186 *et passim*), i.e. in Africa, adding a reference to the famous Western fable of their battles with the cranes (Vol. 3, p. 505). Cf. *TPYL*, ch. 796, p. 7*a*, and on the fable itself Laufer (9) and de Mély (2). An illustration can be found in *TSCC*, *Pien shang tien*, ch. 42, p. 6*b* (Fig. 992). It is remarkable that knowledge of this kind could travel so far in those ancient times.

---

[1] 海參	[2] 白人	[3] 北人	[4] 北京人	[5] 飛刀
[6] 邛邛	[7] 蟨	[8] 焦僥國	[9] 侏儒國	[10] 小人國
[11] 外國圖	[12] 道世	[13] 法苑珠林	[14] 括地志	

Fig. 992. 'The Country of the Pygmies', a drawing of dwarf tropical people from the *Thu Shu Chi Chhêng* encyclopaedia of +1726, but illustrating reports embodied in the Chinese literature from Chhin and Han times onwards.

+1623, shows only New Guinea and Antarctica;[a] Australia is indeed present on the enamel Rosthorn Globe (though joined to New Guinea and to Tasmania) but the date of this may be as late as +1770. The date of the first appearance of Australia in Chinese maps, and its first mention in books concerned with geography and navigation,[b] presents an intriguing problem which further research will no doubt solve.

(viii) *China and pre-Columbian America*

The alleged discovery of the American continent by Buddhist monks from China in the +5th century is one of those youthful indiscretions at which modern sinology is accustomed to blush. As usual, Joseph de Guignes was the *enfant terrible*; having proved in +1758 that the Chinese were a colony of the ancient Egyptians (1),[c] he announced three years later (4) that he had evidence of Chinese navigations on the west coasts of America in pre-Columbian times. This, he thought, would account for the 'politesse' (a rather curious term in this connection) which distinguished the indigenous Mexicans, the Aztecs in fact, from the other barbarians of the continent. De Guignes astonished his contemporaries by appending elaborate and beautiful copper-plate maps showing the journeys of the Chinese to Alaska and California in +458, countries which they knew, he said, under the name of Fu-Sang.[1]

The texts from which de Guignes drew his conclusions are perfectly sound as far as they go. The basic description in the *Liang Shu*[d] dates from the neighbourhood of +629; what it says is that in the year +499 a monk named Hui-Shen[2] appeared in the capital and gave a circumstantial account of what he had seen in Fu-Sang, a country east of China and lying 20,000 *li* east of Ta-Han[3] (the Buriat region of Siberia). He described the curious trees from which the land took its name, saying that they afforded food and that bark-cloth and a kind of writing-paper was made from them. The people lived in unfortified wooden houses and were un-warlike, they had oxen and horses, and drank the milk of deer. Gold and silver were not esteemed among them, and they had copper but no iron. Hui-Shen further described their mating and funeral customs, the absence of taxes, and the cyclical changes of the colour of the ruler's vestments. He

---

[a] Matteo Ricci's Chinese world-maps of +1584 and +1600 are also like this; the former has a confused jumble of continents in the south seemingly labelled Magellanica, the latter knows New Guinea but makes it a promontory of Antarctica. See d'Elia (2), vol. 2, pp. 58, 60; Wei Chü-Hsien (4), p. 179.

[b] One thinks of the *Hsien Pin Lu* of +1590 and the *Tung Hsi Yang Khao* of +1618, both discussed in Sect. 22 above. Mr J. V. Mills informs me that he could find nothing relevant in the former (private communication, April 1956), nor is there any clue in the informative introduction of Wada (1) to the latter, which needs detailed study. News of an Indonesian tradition of a south-eastern antipodean continent probably reached the Jesuits in China from Malacca by about +1610, but whether this influenced any Chinese cartographers remains uncertain. Chhen Lun-Chiung's *Hai Kuo Wên Chien Lu* of +1744, however, seems to contain a reference (p. 31a) to the French claim for the first sight of Australia. A certain gulf between the practical sailors and the literary scholars can always be assumed in Chinese culture.

[c] Cf. Vol. 1, p. 38.

[d] Ch. 54, pp. 35a ff., fully translated by Schlegel (7a). The passage was copied in *Nan Shih*, ch. 79, pp. 7a ff. and in *Wên Hsien Thung Khao*, ch. 327 (p. 2569.1), whence de Guignes no doubt first got it. It is also to be found in abridged form in various encyclopaedias. An elaborated and fictionalised version occurs in the *Liang Ssu Kung Chi* (c. +695).

[1] 扶桑   [2] 慧深   [3] 大漢

added that formerly the Fu-Sang people did not know the law of Buddha, but that in the year +458 five Kashmiri monks had gone there, and that since that time their way of life had much improved. He also appended to his story an account of a country of Amazons (Nü-Kuo[1]), even more fabulous, which lay beyond Fu-Sang to the east yet further.[a] And the texts conclude with a statement that in +507 a Fukienese ship was blown by a tempest far east in the Pacific to an island where the men had faces like dogs and the people subsisted mainly on small beans.[b]

Now Fu-Sang had a long background in Chinese literature which de Guignes did not quite know. In the *Shan Hai Ching* (Classic of the Mountains and Rivers), that ancient magico-geographical text of the late Chou and early Han periods, the Fu-Sang tree growing far in the east had branches upon which the ten suns perched before taking off on their journeys of the ten-day week.[c] Other Han texts such as the *Shang Shu Ta Chuan*[d] and the *Hai Nei Shih Chou Chi*[e] recount similar fables. But this does not mean that Fu-Sang was not thought of in later times as a real place. We have already met with a story of envoys bringing glass or rock-crystal from a country of that name in the Liang period,[f] and Yang Chiung[2] in +676 in his *Hun Thien Fu*[3] (Ode on the Celestial Sphere) mentions it as being somewhere on the shores of the Eastern Ocean (the Pacific).[g] The *Yu-Yang Tsa Tsu* in +863 again says[h] that in +581 a Korean was carried east to Fu-Sang in a storm, though the story is facetious and seems to refer to the Ainu. The +7th-century astronomer Li Shun-Fêng[4] is on record as saying that Fu-Sang is somewhere to the east of Japan just as Japan is east of China.[i] In short, no one knew exactly where Fu-Sang was. Schlegel (7a) in 1892 satisfied himself that its most probable location was the long island of Karafuto or Sakhalin, north-east of Japan,[j] and if Kamchatka and the Kuriles may also be considered there is no better means of identifying it at the present day.

It would be easy enough, but it is not worth while, to expatiate on the literature of controversy to which de Guignes' famous memoir gave rise;[k] we may notice only the book of Leland (1) who took up the cudgels on his behalf in 1875 and went to all kinds of lengths to show that features of the Amerindian civilisations corresponded to items

---

[a] *Nan Shih*, ch. 79, p. 1a, says also that a Taoist came to China from Fu-Sang between +520 and +526, but perhaps this refers to Hui-Shen.

[b] It is interesting that these stories of Amazons and Cynocephali were very much in the mind of Columbus when he came to Cuba, Venezuela and Honduras between +1492 and +1502 (cf. Laufer, 38). They were part of the common stock of Hellenistic, Indian and Chinese legend (cf. Vol. 3, pp. 505 ff.). It is even said that 'cannibal' is a corruption of Carib because of *canis*. Cf. the River Amazon.

[c] Ch. 9, p. 3a, ch. 14, p. 5a, etc. Cf. the 'Arbores Solis et Lunae' in Vol. 3, Figs. 212, 213, 228, 242.

[d] Cf. Schlegel (7a), p. 109. The orthography varies, e.g. Fu-Sang.

[e] Cf. Schlegel's translation (7a), pp. 118 ff. See also *Shih I Chi*, ch. 3, p. 2a, b.

[f] Vol. 4, pt. 1, p. 114; the source is the *Liang Ssu Kung Chi* of +695 referring to about +520, in *TSCC, Pien i tien*, ch. 41, pp. 3b, 4a.

[g] The passage is not in the *Yü Hai* version, but occurs in *TSCC, Pien i tien*, ch. 41, p. 5b.

[h] Ch. 14, p. 8b.

[i] *Wên Hsien Thung Khao*, ch. 324 (p. 2547.1), comm.

[j] In late Chinese literary usage Fu-Sang was often used as a loose poetical term for Japan itself.

[k] Already in 1831, Klaproth (4) had dismissed de Guignes' theory as incredible. Neumann ten years later, and Paravey, seem to have defended it, but we have not been able to see their papers.

[1] 女國　　[2] 楊烱　　[3] 渾天賦　　[4] 李淳風　　[5] 榑桑

in Hui-Shen's account.[a] Thus the *fu-sang* tree was the maguey plant, the lack of iron pointed clearly to the Mayas and Aztecs, the milking of hinds was noted by the early travellers in Central America,[b] and the 'small beans' were certainly frijoles. Yet although Leland and his comrades-in-arms could be discredited on every particular item, there remained an insoluble residuum, the implicit belief that connections of some kind between the Asian and Amerindian cultures had existed; and this is still with us. The sinologists however,[c] undeterred, exposed without mercy all the nonsense in de Guignes and Leland, and by the time of the First World War, with the criticisms of Laufer,[d] Cordier[e] and others, the Fu-Sang thesis was stone dead.[f]

In November 1947 I found myself in Mexico City as one of the secretariat of a United Nations specialised organisation, and for a couple of months had golden opportunities of tasting a culture not only Spanish but also still profoundly Amerindian. I sat at the feet of great Amerindianists such as Alfonso Caso and Sylvanus Morley, studied the wonderful collections of the National Museum in the capital with Miguel Covarrubias and Julian Huxley; and visited the imposing remains of Aztec and Mayan culture from Teotihuacán and Xochicalco to Chichén-Itzá where my friend Alberto Ruz-Lhuillier was then Resident Archaeologist. When I consult my shelf of books on Amerindian studies I vividly recall the sense of keen intellectual adventure which a personal acquaintance with these cultures can give to one who comes to them with an Asian background. This adventure, indeed, had some of the quality of the *déjà vu*, and I was deeply impressed during my stay with the palpable similarities between many features of the high Central American civilisations and those of East and South-east Asia.[g] Was it not striking, to begin with, that the former all arose on the

[a] Leland was a general writer and populariser on many subjects; an American himself, he was keenly interested in the archaeology of his own continent. His book was afterwards more ridiculed than read—the copy in the Cambridge University Library sat quietly on its shelf for 85 years with uncut pages. In fact it is not without many meritorious insights.

[b] Leland (1), p. 154. The authority was d'Eichthal (1), p. 199, relaying de Bourbourg (1), p. xl, in the introduction to his translation of the *Popul Vuh*, the sacred book of the Quiché Mayas; cf. the new translation by Recinos, Goetz & Morley (1). Cf. de Landa (1), vol. 1, p. 99.

[c] Notably Bretschneider (11) and Sampson (1).

[d] (14), p. 198; (38).

[e] (1), vol. 1, pp. 558 ff.

[f] Except, perhaps significantly, in the literature of exploration, for Stefánsson & Wilcox took it quite seriously as late as 1947, see (1), pp. 107 ff. It may perhaps have been, as they think, one of the reasons why Peter the Great sent out the expedition of Vitus Behring to search for land east of Kamchatka in +1725 (pp. 443 ff.).

About 1958 there appeared a privately printed book by Henriette Mertz (1), devoted not only (in apparent ignorance of the work of Leland) to proving that Hui-Shen's Fu-Sang was indeed America, but also that the account in the *Shan Hai Ching* (of −2250!) of what lay 'across the great eastern sea' (chs. 4, 9, 14) was a description of mountain ranges and identifiable places in the southern U.S.A. Mertz knew of the Chhin explorations (see p. 551 below) and the 'Kamchatka Current', she spoke justifiably of the 'uncanny resemblance' of Amerindian artifacts and ideas to those of East and South Asia; but she gave no consideration at all to the nautical means or possibilities of the implied trans-oceanic travel. One may retain as a paradigm of hardihood the equation of Quetzalcoatl with some Chinese person, doubtless Hui-Shen himself, landing near Los Angeles (p. 34), but the proposed identities in general require a heroic suspension of disbelief.

In 1961 the monk appeared again, in Russian and Chinese popular literature, even attaining the dignity of a *Times* leading article on 29 January 1962 (cf. *CKHW Bulletin*, 7 June 1962).

[g] Already in 1933 Chinese–Maya resemblances had deeply impressed the pioneer socialist scholar Chiang Khang-Hu, who had written of them in a little-known article (see (1), p. 380).

western side of the continent, as if fertilised or induced or stimulated from across the Pacific (see Fig. 993)?[a] Then the predominance of the horizontal line in the terraces and monumental stairways of Central Amerindian temple and town patterns, the pyramidal *teocalli* notwithstanding,[b] the omnipresent sky-dragon motifs,[c] amphisbaenas,[d] the split-face designs[e] resembling *thao-thieh*,[1] the *teponatzli* drums[f] like *mu-yü*,[2] the tripod pottery reminiscent of *li*[3] forms,[g] the terra-cotta figures and groups,[h] and even paintings,[i] so similar to those of Chhu and Han, the dresses made of feathers,[j] the double permutation system of

Fig. 993. Distribution of the high cultures of Central America (after Krickeberg, 1).

[a] From Krickeberg (1), after Heyerdahl (2). De Guignes himself remarked on this long ago, (4), p. 518. Cf. Covarrubias (2), p. 71.

[b] See Anon. (48), figs. 19, 148, 204; Castillo (1), pp. 13, 14, 15, 20 and all good albums of relevant photographs. The importance of the horizontal in Chinese architectonics has been underlined on p 61 above. The parallelism between the *teocalli* and the sacred enclosures and stepped pyramidal platforms of the altars of Heaven and Earth, etc., in Chinese culture has been explored by Ling Shun-Shêng (2, 2, 3, 3).

[c] Paralleling *nāgas* and *lung*,[4] cf. Vol. 3, p. 252 and many other references. On the Amerindian dragons cf. Anon. (48), fig. 22; Morley (1), p. 215; Vaillant (2), pp. 52, 57, 175 ff., 182 ff., pls. 23, 53; Spinden (1), pp. 89 ff., 206; Covarrubias (1), p. 130; (2, 3); Soustelle (1), p. 47; Noguera (1), p. 32; finally Combaz (7), esp. pp. 262 ff., who wrote 'les rapprochements...sont très troublants'. Correspondences in the worship of dragon rain-gods and in the complicated rain-making ceremonies, down to minute detail, between the ancient Chinese and Mayan cultures, have been revealed by the thorough study of Liu Tun-Li (1, 1). The famous *cenote* sacrifice at Chichén-Itzá was essentially identical with that of the brides of the river-gods in −5th-century China (cf. Vol. 2, p. 137). Had Central Amerindian civilisation been allowed to develop in freedom, would not a Hsimên Pao have arisen in due time to despatch the cruelty of religion with the sword of rationalism?

[d] Spinden (1), p. 98; Combaz (7). Such a double-headed serpent, symbolic in ancient China of the rainbow (Vol. 3, p. 473), has been seen in Vol. 1, Fig. 19, sculptured in a +5th-century Buddhist cave-temple at Chhien-fo-tung. Cf. Covarrubias (2), pp. 45, 169, (3), pp. 176 ff.

[e] See Anon. (48), fig. 52; Spinden (1), pp. 167 ff., 223; Adam (1). References to the Chinese iconography will be found in Vol. 2, p. 117. Cf. Hentze (3); Covarrubias (2), pp. 31, 35, (3), pp. 235, 238.

[f] See Anon. (48), figs. 201, 202, 241, 249, 250; Schaeffner (1), pp. 72 ff. The fish-shaped wooden drums of Chinese Buddhist temples have been noted in Vol. 4, pt. 1, p. 149.

[g] See Anon. (48), figs. 155, 167, 182, 195, 212, 253. The ancient Chinese counterparts have been discussed in Vol. 1, p. 82. Many Amerindian examples in Covarrubias (2, 3); Ekholm (4).

[h] See Anon. (48), fig. 216, some dancers. The National Museum in Mexico City has a complete set of figures playing the ceremonial or divinatory ball-game (p. 546 below). In style these figurines recall the recently discovered Yunnan bronze set-pieces (Anon. 28) as much as any other Han tomb-figure material. Abundant description of Amerindian figurines in Covarrubias (2, 3).

[i] See Pijoán (1), vol. 10, pl. XII; Vaillant (1), p. 60, to be compared with fresco and lacquer paintings in Mizuno Seiichi (1); Yang Tsung-Jung (1), figs. 1, 15; Chhang Jen-Chieh (1), figs. 1, 4, 5, 6; Shang Chhêng-Tsu (1); O. Fischer (2). Further on frescoes see Covarrubias (2), pp. 56, 86, 116, (3), pp. 253 ff., 287 ff., 303; and on the use of coloured lac (2), pp. 56, 117, (3), p. 95.

[j] Cf. Vaillant (1), pp. 66, 72 ff. and compare with Vol. 1, p. 202, Vol. 4, pt. 1, p. 149. See also Covarrubias (2), pp. 55, 100.

[1] 饕餮   [2] 木魚   [3] 鬲   [4] 龍

the Maya and Aztec calendars,[a] ideographic script,[b] and the far-reaching parallelisms in symbolic correlations (colours, animals, compass-points, etc.)[c] and cosmological legend[d]—all combined to give an overwhelming impression of cultural influences exerted upon the Amerindian cultures by those of Asia.[e] I was aware also of long-known ethnological facts, as in the sphere of games,[f] divination (even extending to scapulimancy),[g] computing devices,[h] art forms,[i] etc., which pointed unmistakably to Asian influences. It was strange enough that jade should have been so treasured by the Aztecs and Mayas, as dearly as by the Chinese,[j] but even stranger that on both sides of the Pacific, jade beads or cicadas should have been placed in the mouth of

---

[a] Morley (1), pp. 269 ff.; Caso (1), pp. 39 ff.; Spinden (1), pp. 111 ff.; Soustelle (1); Rock (1). The parallel between the enmeshed cogwheels of the various cyclical series in the Chinese and Amerindian culture-areas has already been drawn in some detail (Vol. 3, pp. 397, 407). On the Chinese sexagesimal cycle and its characters cf. also Vol. 1, p. 79; Vol. 2, p. 357; Vol. 3, p. 82. On parallelisms concerning the animal cycle (Vol. 3, pp. 405 ff.), the lunar mansions (Vol. 3, pp. 242 ff.), etc., research is actively proceeding; cf. J. E. S. Thompson (1); Kelley (1, 2); Soustelle (1), pp. 79 ff.; Kirchhoff (1).

[b] Morley (1), pp. 260 ff.; Thompson (1). Cf. Vol. 1, pp. 27 ff.

[c] Spinden (1), pp. 126, 231, 234; Soustelle (1), pp. 30, 56 ff., 75, where the association of the colours with the spatial directions is particularly impressive (cf. Vol. 2, pp. 261 ff.). The reader will not need to be reminded of the fundamental importance of the symbolic correlations in all ancient and medieval Chinese thinking. For a 'Yin-Yang' design cf. Spinden (1), p. 243, and further, Léon-Portilla (1).

[d] Of this we have already seen one example (Vol. 3, p. 215), the opening and closing of the gap in the rims of sky and earth; cf. Hatt (1); Erkes (17). But the Aztecs also had the rabbit in the moon (Soustelle (1), p. 19; Liu Tun-Li (1), p. 67); cf. Vol. 3, p. 228 and other references. Close relatives of the brothers Hsi and Ho (see Vol. 3, p. 188) appear again among the Mayas (de Landa (1), vol. 2, p. 15). Cf. the history of de Sahagún (1) and the translation of the *Popol Vuh* by Recinos, Goetz & Morley (1).

[e] Nothing in these statements should be taken as intending to deny the profound originality of the high cultures of Central America. Whether in their religious systems, such as the personifications of rain and maize, or in their social organisation, so highly developed in the Inca State, or in the techniques peculiar to themselves, e.g. the working of obsidian (volcanic glass), the metallurgy of platinum, or the technology of rubber, their personality was entirely their own. I speak only of significant influences from Asia which may have helped them to construct it. Some Amerindianists (such as my friend Prof. Gene Weltfish, to whom I am greatly indebted for discussions of these problems) take the view that one must envisage external influence on a particular culture only when its technical development shows some palpable and otherwise inexplicable discontinuity. I feel that this is too severe, and that it is fair to draw attention to striking similarities between different cultures if they make their appearance. Independent invention can perhaps never be disproved (as in the case of printing, which arose in Europe later than in China), but when a technique has a long history in one culture prior to its appearance in another, I feel that the onus of proof lies on the adherents of independence—provided always that it can be shown with fair probability that cultural transmission was physically and geographically possible. In this connection the contrasting theses of Heine-Geldern (16) and Caso (2) should be pondered.

[f] See Tylor's classical paper (1) on the backgammon-like game of *patolli*, the Aztec form of which so closely resembles prior Asian forms.

[g] On scapulimancy see Vol. 2, pp. 347 ff.; Cooper's finding of it (1, 2) among the Algonquin Indians was unexpected.

[h] Computing and recording by means of knotted cords is best known by the Inca *quipu* (cf. Locke, 1), but there are many evidences of the same device in ancient China (cf. Vol. 1, p. 164; Vol. 2, pp. 100, 327, 556; Vol. 3, pp. 69, 95).

[i] For Khmer art, Marchal (1); for Indonesian and Indo-Chinese art, d'Eichthal (1); Kreichgauer (1). The prominent appearance of the squared spiral motif in Chou and Amerindian decoration is most striking; cf. Covarrubias (1), pp. 110 ff.; (2, 3) *passim*. Indonesian parallels have also been adduced by Weltfish (2), who believes however that the similarities may be independent derivations from the techniques of weaving and basket-making, notably twill-plaiting with flat strips.

[j] Cf. Morley (1), pp. 425 ff., pls. 91 to 93; Vaillant (1), pp. 75 ff., (2), p. 128, pls. 3, 16; Anon. (48), figs. 199, 200; Spinden (1), pp. 89, 160, 162, 243; Covarrubias (1), pp. 107 ff.; (2, 3) *passim*. The Maori love of jade is also to be remembered; cf. Chapman (1); Ruff (1); Duff (1).

the dead;[a] and astonishment turned to conviction when one learnt that in all these civilisations the jade corpse-amulets were sometimes painted with the life-giving colour of red cinnabar or haematite.[b]

If this conviction was a half-reluctant one it was due to the fact that Amerindianists cherished for many years a kind of Monroe Doctrine, denying any external influences on the development of the high indigenous cultures of Central America.[c] But this orthodoxy is now, twenty years later, fast disappearing, and periodical Asian influences are becoming every day more and more accepted. A mountain of evidence is accumulating[d] that between the -7th century and the $+16$th, i.e. throughout the pre-Columbian ages, occasional visits of Asian people to the Americas took place, bringing with them a multitude of culture traits, art motifs and material objects (especially plants), as well as ideas and knowledge of different kinds. Current researches in haematology,[e] ethno-botany,[f] and ethno-helminthology[g] are throwing much light on

[a] Morley (1), p. 205; Caso (1), p. 38; Covarrubias (1), p. 108. The Amerindian peoples mostly placed jade beads in the mouth, but they also carved jade cicadas to go alongside, cf. Anon. (48), fig 244; Pijoán (1), vol. 10, pl. XI and fig. 250; Noguera (1), p. 39. For the parallel Chinese practices cf. Laufer (8), pp. 294 ff.; Biot (1), vol. 1, pp. 40, 389; Wieger (2), p. 90, etc.

[b] Covarrubias (1), p. 108, (2), pp. 48, 79, 104, (3), p. 55. A Chinese parallel is in Laufer (8), p. 301. Other uses of red paints in the disposal of the Aztec dead are mentioned by Vaillant (2), p. 37. Another remarkable parallel between Amerindian and Chinese culture in the field of jade is the belief in 'exhalations' which help the prospector to find it (cf. Vol. 3, p. 677). This is in de Sahagún (1), vol. 11, pp. 277 ff.; cf. Covarrubias (1), p. 108, (2), p. 105.

[c] It was not of course denied that America had been peopled from Asia across the Behring Straits in the palaeolithic, for the 'Mongolian' character of the indigenous peoples had long been an assured finding of physical anthropology; the debate concerned only cultural influences during the past three millennia.

[d] For reviews see von Heine-Geldern (3, 7, 10, 11); Covarrubias (2); M. W. Smith (1); Ekholm (3); as also Gladwin (1) and Raglan (1), pp. 154 ff. The characteristics of the Amerindian cultures can be studied in de Landa (1); de Sahagún (1); Spinden (1); Morley (1); Recinos, Goetz & Morley (1); Ruz-Lhuillier (1); Armillas (1); Vaillant (2).

[e] See Layrisse & Arends (1); Lewis, Ayukawa et al. (1) on the Diego blood group antigen.

[f] This subject has recently been reviewed by von Heine-Geldern (9) and G. F. Carter (1). Here the pièce de résistance is maize, classically regarded as native to the American continent (de Candolle (1), pp. 387 ff.; Vavilov (2), p. 40). The usual view, established in a classical paper by Laufer (36), is that it was unknown in East Asia until the $+16$th century when it was introduced to China from America (presumably by Europeans) via India and Burma, perhaps also directly to coastal areas. This is strongly upheld by Mangelsdorf & Reeves (1, 2, 3); Reeves & Mangelsdorf (1, 2); Ames (1), pp. 92 ff.; Ho Ping-Ti (1) and others. But a primitive race of maize has been found deeply acclimatised in Assam by Stonor & Anderson (1) (cf. Hatt (3), pp. 902 ff.), who concluded that this must either have originated in Asia, or have been taken there in pre-Columbian times. As Heyerdahl said, it could have 'spread with man in drifting craft to...Oregon or California by a southern branch of the Kuroshio or North Pacific Current' (2, p. 494). Against this, Mangelsdorf (1, 2) has found fossil maize pollen some 80,000 years old in Mexico, while Mangelsdorf & Oliver (1) could match the Assamese material with South American counterparts. On which side of the Pacific maize was first cultivated thus still remains undecided; the balance of evidence seems in favour of an American priority, but pre-Columbian culture in Asia has not been disproved. The other outstanding example is that of cotton. According to the work of Hutchinson, Silow & Stephens (1); Hutchinson (1) and Sauer (1), which has been criticised but not overthrown, Old World cotton (Gossypium arboreum) with 13 large chromosomes must have been brought somehow to cross with New World cotton (G. Raimondii) with 13 small chromosomes, so as to give the cultivated Peruvian Inca cotton which is a polyploid of 26 chromosomes, half of them large and half small. Again this points to pre-Columbian voyages across the Pacific. Stebbins (1) suggests China as the way-station, but transmission must have occurred before the -1st millennium if the earliest Peruvian textile datings are correct. Many other plants useful to man also figure in this still continuing argument. At present the evidence for American–Polynesian contact seems very strong; see the review of Heyerdahl (7).

[g] The distribution of patterns of hookworm infestation has ethnological significance; cf. Darling (1, 2); Soper (1). Mixed infestations of Ancylostoma duodenale and Necator americanus exist both in the

these events, while comparative studies in metallurgy,[a] paper-making,[b] religious art[c] and architecture,[d] 'Great Walls' and roads,[e] musicology,[f] folklore,[g] divinatory games,[h] agricultural practices,[i] social organisation,[j] and perhaps even dress[k] are elaborating the general picture.[l] We have to visualise the arrival from time to

Chinese culture-area and among the Amerindians. It is argued therefore that the latter must have originated in Asia between lats. 20° and 35° N., or have been in contact with people from there. Such infestation patterns would not persist through a slow migration along the cold latitudes of the Behring Straits, but indicate more rapid transmissions in sea-going craft across the Pacific (cf. Heyerdahl (2), p. 508).

[a] See the elaborate study of von Heine-Geldern (4), whose comparative juxtapositions are in many cases quite as striking as those of bronze-age objects from the western and eastern ends of the Old World which we gave in Vol. 1, pp. 160, 162. The *cire-perdue* process was used in the New World as much as the Old; cf. Covarrubias (2), pp. 86, 125, 220 (3), pp. 99, 310. On pre-Columbian metallurgy in general the reader is referred to Rivet & Arsendaux (1).

[b] See von Hagen (1).

[c] See von Heine-Geldern (7); v. Heine-Geldern & Ekholm (1); Ekholm (2); Covarrubias (2, 3). Cf. Hentze (1, 5). In this connection it is interesting that in the cabinet of antiquities of H.M. the King of Sweden there is a Chou bronze 'totem pole' some 7 in. high, extremely evocative of those of the N.W. Canadian Indians, in miniature.

[d] See von Heine-Geldern (12); W. Müller (1); Ekholm (1); Covarrubias (2, 3); Ling Shun-Shêng (2, 2, 3, 3, 4, 4, 5, 5). It now appears that the Mayas even knew the true arch, though using it sparingly (Befu & Ekholm; Ruppert & Davison). Cf. p. 167 above.

[e] See Shippee (2); von Hagen (2, 3). Cable suspension bridges are common to the Andean and the Tibetan massifs.

[f] Cf. Schaeffner (1), pp. 72 ff., 249 ff., 265, 284, 288 ff., 387; Sachs (2), pp. 192 ff., 202. 'It would be hard to deny', says Sachs, 'a connection between the Chinese and South American notched flutes and the Chinese and South American pan-pipes...The notched flute, outside America, is confined to the Far East including Mongolia...Some other common features are also the lack of stringed instruments, the occurrence of nailed drums and the use of sonorous stones...Except for a few universal instruments, all relatives of American instruments are exclusively found in a territory comprising China, the arc between China and India, the Malay archipelago and the Pacific islands. More than 50 % can be located in the Burmese hinterland and the adjacent countries—that is, among peoples supposed to have been the aboriginals of what is called China today, who were driven southward by the invading Chinese [i.e. Hsia Shang and Chou] in neolithic [and later] times.' In other words, as a sinologist might say, the Ti and the Jung. With regard to the pan-pipes a remarkable discovery was made by von Hornbostel (3) who found that the absolute pitches and scales of American, Melanesian and East Asian examples were in striking conformity. The Chinese pan-pipes were of course tuned to the 12 *lü* (cf. Sect. 26*h*), six being Yang and six Yin; and the American forms are also often in two sets, male and female, sometimes consisting of two frames connected together by a long cord (Mead, 1). Lastly, in view of the exceptional antiquity and importance of the metal bell in Chinese culture, its occurrence in pre-Columbian America is very remarkable (cf. Covarrubias (3), pp. 99 ff.; Hurtado & Littlehales).

[g] See Soustelle (1); Hatt (1, 2, 3); Liu Tun-Li (1).

[h] See for example Löffler (1); Schroeder (1) and Krickeberg (2). One of the most remarkable features of Central Amerindian culture was the ball game played in elaborate courts forming striking architectural patterns, as at Chichén-Itzá. Here the two sides had religious and divinatory significance, the struggle symbolising that between day and night, between light and darkness—indeed between Yin and Yang. The connection between these and the sexual jousts or mating festivals in ancient China so much studied by Granet (2); cf. Vol. 2, p. 277 above; with their ball-throwing between teams of young men and girls, has been fully pointed out. But I suspect also a connection with that divinatory 'star-chess' developed in +6th-century China, about which much was said in Sect. 26*i* (Vol. 4, pt. 1, pp. 318 ff.).

[i] E.g. the use of night soil as fertiliser (Vaillant (2), p. 135), and the elaborate terracing of hillsides (Covarrubias (2), p. 65).

[j] Has not the 'socialist empire of the Incas' always recalled the feudal-bureaucratic society of China? One may even find the ceremonial ploughing custom in it. See Baudin (1), esp. p. 83; Toscano (1).

[k] Compare Vaillant (1), pp. 8 ff., (2), pp. 136 ff., pl. 39; Morley (1), p. 191, etc., with the pictures in Anon. (28). The remarkable circum-Pacific distribution of slat, rod, and plate armour in Asia and America is also not to be forgotten; see Laufer (15), pp. 258 ff.; Covarrubias (2), pp. 150, 158 ff., 165.

[l] Cultural complexes in which a number of traits of probable trans-pacific origin are found together now win recognition. Ecuador seems to be a focal region, for already in the Formative Period (−2nd

time of small groups of men (and doubtless of women also) with a background of high culture, never any massive invasion like that of the Europeans in the +16th century.[a]

When the sinologists gave the quietus to the Fu-Sang story they were reckoning without the sailing-raft.[b] Ignorance of practical techniques and their development proved once again the Achilles heel of literary history. Moreover they were writing in the pre-Heyerdahl era. For in earlier months of the same year that brought our secretariat to Mexico, Heyerdahl and a number of companions, desirous of demonstrating that the isles of Polynesia had been peopled from South America, made a courageous journey themselves from the Peruvian coast to Raroia, one of the most northerly islands of the Tuamotu archipelago, not far from Tahiti. Their vessel was a sailing-raft of balsa logs built as nearly as possible like those of the ancient Peruvians.[c] The successful accomplishment of this voyage depended on the north-westward flowing Humboldt Current and the westward South Equatorial Current, as well as on the South-eastern Trade winds blowing in the same direction.[d] With the main theory of Heyerdahl (2, 4, 5) regarding the origin of the Polynesians we are not here concerned; it is still impeached by severe difficulties, notably on the linguistic side,[e] but even its strongest opponents, such as von Heine-Geldern (8, 14, 15), are at one with Heyerdahl in admitting the navigability of the Pacific for sailing-rafts, to say nothing of more conventional

millennium) the neolithic pottery closely resembles that of the Jōmon period in contemporary Japan; see Estrada & Evans (1) and Kidder (1). Then in the Bahía culture (c. −500 to +500) there occur pottery house models (as in Han tombs), neck-rest pillows, seated statuettes in the Buddhist *vajrapariyanka* position, rectanguloid four-hole pottery net-weights (as in Han Indo-China), Jōmon-style ear-plugs, pan-pipe assemblies with the shortest tube-lengths central, and evidence of carrying-poles after the Chinese manner. Sailing-rafts are to be presumed in addition. Descriptions will be found in Estrada & Meggers (1); Estrada, Meggers & Evans (1), who accept the idea of small groups of culture-transmitters periodically arriving. The effects of such a group, in this historical case negroes just before the Spanish conquest, are described by de Balboa (1), p. 133.

[a] Cf. our remarks in Vol. 1, p. 248. The chronology of the high Amerindian cultures is of course still far from settled—a review of the present state of the question will be found in the two large volumes of Covarrubias (2, 3). According to Spinden (1), p. 136, the 7th *baktun* of the Mayan day count started from −613, and the month count from −580, but the shorter time-scale accepted by Morley (the Thompson–Goodman–Martinez–Hernandez correlation) preferred the date −353. A comparative table is given in Covarrubias (3), p. 219. The oldest dated Mayan object belongs to +60 on the Spinden system, or to +320 if we adopt the shorter scale (Morley (1), pp. 47, 284). But there are a number of other self-dated objects found outside the Maya area which carry the Maya datings back as far as −287 on the longer scale (Covarrubias (3), pp. 51, 241). After a period of dominance of the shorter time correlation, radio-carbon and other measurements at first favoured the longer one (cf. Bennett (1); G. R. Willey (1); Covarrubias (2), p. 23, (3), p. 218); but now support again with some decisiveness the shorter; cf. Coe (1); Satterthwaite & Ralph (1); Bushnell (1). At the same time they place the early city culture of the Toltecs between −800 and −400. In any case there can be no doubt that the high Amerindian cultures were competent to accept cultural influences from the late, if not the early, Chou period. The above dates are of much interest in comparison with those of the voyages of the Chhin and Han periods discussed below on pp. 551 ff.

[b] This has been discussed at length on p. 393 above, but the reader may again be referred to the excellent paper of Ling Shun-Shêng (1).

[c] See Heyerdahl (1). Other demonstrations include Heyerdahl & Skjölsvold (1) on the Galápagos Islands, and Heyerdahl (3) on Easter Island.

[d] In what follows we rely upon the current map in Sverdrup, Johnson & Fleming (1), chart VII. See also an earlier study of Hambruch (1) on currents in relation to ethnic migrations, and Sittig (1). For winds we follow the maps of A. A. Miller (1), figs. 9 and 10.

[e] Cf. A. S. C. Ross (1).

non-European vessels.[a] And it is quite clear that what could be done by balsa rafts, with sails and centre-boards, from east to west along lats. 0° to 25° S., could also be done by sailing-rafts of the South Chinese and Annamese pattern navigating (or drifting) from west to east along lats. 25° to 45° N.[b] For they would take advantage of the strong Kuroshio Current and the North Pacific Current flowing eastwards, as well as the westerly winds, particularly powerful in the winter and early spring (Fig. 989 b).[c] They would be helped, too, by the climate of the North Pacific at these times, exceptionally warm for its latitudes.[d]

It is interesting to ask how far back a knowledge of the eastward currents can be traced in Chinese and Japanese literature. The Japanese name of the Kuroshio [1] Current (the 'black tide'), which diverges into the Sub-arctic and North Pacific Currents, has got into world geographical literature as such, but the Chinese name for it was different, Wei-Lü.[2, 3] Already in the Warring States period this great gulf stream

[a] During the nineteenth century East Asian junks were driven to American shores as often as about once every five years (v. Heine-Geldern, 10). Reliable eighteenth-century evidence also exists (Sittig, 1). Similar facts were reported by Col. B. Kennon (in Leland (1), p. 77; cf. pp. 43 ff.) and there is a wealth of data in the reports of C. W. Brooks (1) and H. C. Davis (1). So numerous were these wrecks that they formed the main source of iron and copper for the Indians of British Columbia (Rickard, 1). It is not generally known that an occurrence of this kind played a part in the emergence of Japan as a modern nation. In the early 1850s Nagahama Manjūrō,[4] a fisherman, was blasted by current and storm to the waters of Hawaii, where he was picked up by an American ship. Befriended by a clergyman, he was taken to San Francisco, where he learnt English, and later on making his way back to Japan he became a counsellor to the Shogunate and adviser on foreign affairs in the first years after the opening of the country following Cdr. Perry's famous visit. For knowledge of this story we are indebted to Dr Charles D. Sheldon. Finally on 12 August 1962 Heri Kenichi, a Japanese 23 years of age, arrived in San Francisco Bay having crossed the Pacific single-handed in a 19-ft. sailing-boat (Birrell, 1). His advent could not have been more appropriately timed, as the International Congress of Americanists was on the point of discussing trans-Pacific contacts.

[b] Cf. Heyerdahl (2), pp. 77, 81, 494, 509, (3), p. 356, all envisaging west–east transmissons across the Pacific. Such sailing-raft voyages are now commonly accepted, as for instance by Liu Tun-Li (1). Those who knew the Pacific well saw their feasibility long ago, e.g. Col. B. Kennon, in 1874 (in Leland (1), p. 74). The archaeological 'emptiness' of the Aleutians (cf. Covarrubias (2), pp. 157, 163) also emphasises the importance of possible maritime transmissions, which would have by-passed them.

[c] For further discussion of the physical oceanography see the review of Heyerdahl (6) on all the possible routes to and from the Americas, strongly supporting the view here expressed. The fundamental fact about the hydrographic and meteorological situation in the Pacific is that it favours west-to-east transmissions north of the equator and east-to-west transmissions south of it. This seems an almost insuperable objection to all theories such as that of Rivet (1) which postulate migrations from Asia to America by way of Oceania, or even casual cultural influences taking that route. On the other hand few will wish to follow C. W. Brooks (2, 3) in his idea of 1875 that the Chinese were a colony of the ancient Mayas (cf. Vol. 1, p. 38).

Bowen tries to save the southern hemisphere route, (2), p. 104, by postulating the use of the great westerly current and wind-belt below lat. 40° S., but the temperatures there, even in summer, would surely have been too cold for the sailing-rafts. Birrell's transmission 'from China to Peru' is dogged by this difficulty.

[d] See charts of isanomalous lines, as in Miller (1), figs. 1, 2. There is no need of course to assume that the journeys envisaged would always have been pleasurable, and one can count on great individual fortitude among the people of the Chinese culture-area; cf. the epic of Phêng Lin, a sailor who survived on a raft for 133 days in the Atlantic in 1942 and 1943, recounted by Lo Hsiao-Chien (1) and Harby (1). As for the Equatorial Counter-current, which sets eastward between the North and South Equatorial Currents, it seems to be rather a hydrographer's formality than an aid to navigators (cf. de Bisschop, 1), so we will not consider it further here. Its most important part, the newly discovered Cromwell Current, is in fact entirely submarine (Knauss, 1).

[1] 黑潮   [2] 尾閭   [3] 泥潤   [4] 長濱萬次郎

perpetually flowing north-eastwards seems to have been known, for in the *Chuang Tzu* book we read:[a]

Of all the waters under Heaven there is none so great as the ocean. A myriad rivers flow into it without ceasing, yet it is never full, and the Wei-Lü (current) carries it continually away, yet it is never empty. Springs and autumns cause no change in it; it takes no notice of floods and droughts. Its pre-eminence over the Yangtze and the Yellow River no measures nor numbers can express.

Wei-Lü might be translated 'the ultimate drain' or 'cosmic cloaca',[b] and another term for it was Wu-Chiao,[1] 'converging and pouring away'. People sometimes said that it was a great rock, with an abyss or whirlpool into which the sea evermore discharged itself.[c] In +1067 Ssuma Kuang was quite sure that the Fu-Sang country was to the west of the Wei-Lü Current, i.e. on its hither side, a fact which had much influence on later European sinologists.[d] By +1744 Chhen Lun-Chiung spoke with the voice of centuries-long tradition when he said that the Wei-Lü was the ancient name of the current now known as the Kuroshio.[e] Possibly the fable of the abyss discouraged now and then some sailors of the Chinese coasts from launching forth on to the main—but very similar stories plagued the mariners of Europe.[f]

A search in the maritime literature of old China would certainly fish up much about currents as well as the belief in the maelstrom, the 'old notion', as Hirth & Rockhill put it,[g] 'of a hole in the Pacific into which the waters of the Ocean emptied'. In his *Ling Wai Tai Ta*, speaking of Java (Shê-Pho[2]), Chou Chhü-Fei says:[h] 'East of Shê-Pho is the Great Eastern Ocean Sea, where the waters begin gradually to slope downwards. The Kingdom of Women (Nü-Jen Kuo[3]) lies there. Still further east is the place where the Wei-Lü drains into the world from which men do not return.' The statement about the point of origin of the Kuroshio current was right enough, though we should say the Philippines instead of Java; and perhaps the 'bourne from whence no traveller returns' was the American continent rather than the abyss. In the legends

---

[a] Ch. 17, tr. Legge (5), vol. 1, p. 375, mod. auct.

[b] This second appellation could lead us off upon another tack, for *wei-lü* was adopted as a technical term in Taoist mystical (microcosmic) anatomy for the proctodaeal orifice (rectum and anus), and later by extension the coccyx. The former use is prominent in the Nei Ching Thu,[4] a remarkable diagram of uncertain date inscribed on stone at the Pai Yün Kuan[5] Taoist temple near Peking, and reproduced in scroll paintings as well as rubbings. A copy of the latter, white on red, I owe to the great kindness of Mr Rewi Alley in the capital in 1952. The terminology for the microcosmic cloaca has been mentioned by R. A. Stein (4), but the only treatment of the Nei Ching Thu as a whole remains that of Rousselle (4, 5). We intend to deal with it in Sect. 43 below. A reference to *wei-lü* as the coccyx will be found in *Shêng Chi Tsung Lu*,[6] ch. 191, p. 2a; this was the great medical encyclopaedia issued by imperial authority in +1111.

[c] E.g. Chhêng Hsüan-Ying, in his Thang commentary on the above passage; *Chuang Tzu Pu Chêng*, ch. 6B, p. 3a.

[d] *Wu Yin Chi Yün*,[7] quoted in *Khang-Hsi Tzu Tien* (p. 1493), sub *lü*.

[e] *Hai Kuo Wên Chien Lu*, p. 12b.

[f] Cf. Lloyd Brown (1), pp. 42, 95; Brochado (2), p. 14; Anon. (53), pp. 15, 57.

[g] (1), p. 26.

[h] Ch. 2, p. 9b, tr. auct. adjuv. Hirth & Rockhill (1), p. 26.

[1] 沃焦　　　　[2] 闍婆　　　[3] 女人國　　　[4] 內經圖　　　[5] 白雲觀
[6] 聖濟總錄　　　[7] 五音集韻

about the Country of Women there are many features which point to Japan.[a] Chou's statement of +1178 was copied in identical words by Chao Ju-Kua in +1225,[b] who added, in his entry for Nü-(Jen) Kuo, '(Here) the water constantly flows to the east, and once in several years it overflows and drains away'.[c] But the best account of the ideas of the sailors of the Sung is in the *Ling Wai Tai Ta*. Chou Chhü-Fei wrote:[d]

> South-west of the four commanderies of Hai-nan is the Great Sea called the Sea of Chiao-Chih (Tongking). In it there are three currents (*san ho liu*[1]) carrying away the scudding waves and spume in three directions. A southward one communicates with the seas where the various barbarian countries lie, and a northward one flows past the provinces of Kuangtung, Fukien and Chekiang. The third makes eastward into that boundless deep the Great Eastern Ocean. Southern ships in their voyages have to breast these three currents; if they have favouring breezes they are safe, but if they meet with danger, when there is no wind, they cannot avoid it, and must drift to destruction with the currents. I have heard say that in the Great Eastern Ocean there is a bank of sand and rocks some myriads of *li* in length, and near by is the Wei-Lü, the place where the water pours down into the Nine Underworlds (*chiu yu*[2]). In times gone by a certain ocean-going junk was driven by a great westerly wind to within hearing distance of the roar and thunder of the waters (falling into) the Wei-Lü of the Great Eastern Ocean. No land was to be seen. Suddenly there arose a strong easterly wind, and so (the junk) was saved.

The distinction between three currents was quite reasonable, for though inshore the current sets southwards from Hongkong to Singapore, further out in the South China Sea it runs northwards towards Formosa,[e] while east of Luzon and Thaiwan the Kuroshio Current begins. The thousand-*li* sand-bank is at first strange, but not if we remember the concentric ring-continents beloved of Buddhist cosmography, seen already in Fig. 242 of Vol. 3. As for the erring junk, when Edgar Allen Poe wrote his tale 'A Descent into the Maelstrom' in 1841 he can hardly have imagined how widespread, as well as how old, was the superstition of the sea which he pressed into his service.

'The ambassador of the Han, Chang Chhien,' wrote Chang Hua about +285, 'won through across the Western Seas to reach Ta-Chhin (the Roman Empire)...but the Eastern Ocean is yet more vast, and we know of no one who has crossed it.'[f] Perhaps this was because no one ever came back. A corollary of the hydrographic situation just described is that once any group of people had sailed from Asia to America on primitive craft there was very little chance of their ever returning, since before relatively modern times no general understanding of the régime of winds and currents could develop. Of those who made the journey, within the span of the −1st and +1st millennia, many were probably fishermen or traders, involuntary carriers of culture to the Americas, but assuredly sometimes the great voyage was undertaken purposively for

---

a Cf. Schlegel (7 *t*).

b *Chu Fan Chih*, ch. 1, p. 10 *b*; Hirth & Rockhill (1), pp. 75, 79.

c Ch. 1, p. 33 *a*, tr. auct. adjuv. Hirth & Rockhill (1), p. 151.

d Ch. 1, pp. 13 *b*, 14 *a*, tr. auct. adjuv. in part. Hirth & Rockhill (1), p. 185.

e See Sverdrup, Johnson & Fleming (1), chart VII.

f *Po Wu Chih*, ch. 1, p. 5 *a*. On Chang Chhien, cf. Vol. 1, pp. 173 ff.

¹ 三合流   ² 九幽

one reason or another, though not with knowledge of the landfall. Surprise is often expressed that the Chinese did not 'explore' the Pacific, but only by those who lack acquaintance with Chinese literature.[a] In a moment we shall return to this, meanwhile the problem of the origin of the sailing-raft remains.[b] In view of the much greater anti-quity of high culture along the Chinese and Indo-Chinese coasts than of anything which America can show, it seems almost quixotic to believe that the craft originated first in Central America. It is surely preferable to suppose that the sailing-raft was not only the oldest form of transport between the two culture-areas, but one of the first of the gifts of the Old World to the Amerindian peoples.

All these considerations cast rather a new light on the statements in ancient Chinese literature about voyages into the Pacific. To be brief, we must concentrate attention on the Chhin period, when the rulers of China were convinced that drug-plants giving longevity or immortality were to be found on islands in the Eastern Ocean. In the late −3rd century many sea-captains were sent out in search of these, generally without success, but although the only name which has come down to us is that of Hsü Fu,[1,2] the whole story of their activities is of such interest for the early maritime history of China that it is worth examining in some detail.[c] Ssuma Chhien recurs to it four times in his great *Shih Chi*, the oldest of the dynastic histories, finished in −90, and his texts have much to tell us. First, in his chapter on the national Fêng and Shan sacrifices, he says:[d]

From the time of the Kings of Chhi (State), Wei (r. −378 to −343) and Hsüan (r. −342 to −324), and King Chao of Yen (State) (r. −311 to −279), people were sent out into the ocean to search for (the islands of) Phêng-Lai,[3] Fang-Chang,[4] and Ying-Chou.[5] These three holy mountain (isles) were reported to be in the midst of Po-Hai[6] (the Gulf of Pei-chih-li within the Yellow Sea), not so distant from human (habitations), but the difficulty was that when they were almost reached, boats were blown away from them by the wind. Perhaps some succeeded in reaching (these islands). (At any rate, according to report) many immortals (*hsien*[7]) live there, and the drug which will prevent death (*pu ssu chih yao*[8]) is found there. Their living creatures, both birds and beasts, are perfectly white, and their palaces and gate-towers are made of gold and silver. Before you have reached them, from a distance they look like clouds, but (it is said that) when you approach them, these three holy mountain (isles) sink down below the water—or else a wind suddenly drives the ship away from them. So no one can really reach them. Yet none of the lords of this age would not be delighted to go there.

When Chhin Shih Huang Ti, having united the empire, came to the shores of the (Eastern) Sea, the magicians and technicians reported all kinds of (extraordinary) things about (lands in the ocean). As the emperor considered that if he himself went to sea he would probably not be successful (in finding the magic islands), he ordered a certain person to embark with a crew of

---

[a] It is true that by the time when Chinese shipping developed to a technical level which made it fully capable of crossing the Pacific back and forth, the predominantly agrarian tone of the civilisation had set hard. The isles of the far north were unattractive, and the Pacific seemed quite empty.

[b] We have already discussed it on p. 394 above.

[c] We have touched upon the subject at an earlier stage, cf. Vol. 2, pp. 83 ff., 133, 240 ff.

[d] Ch. 28, pp. 11 b ff., tr. auct. adjuv. Chavannes (1), vol. 3, pp. 436 ff.; Dubs (5), p. 66. The passage is repeated in *Chhien Han Shu*, ch. 25 A, p. 11 a.

[1] 徐市　　　[2] 徐福　　　[3] 蓬萊　　　[4] 方丈　　　[5] 瀛洲　　　[6] 孛海
[7] 僊　　　[8] 不死之藥

young men and girls, and to search for them. Their ships cruised in the ocean, but after some time they returned, saying in excuse that although they had seen (the isles) from afar, they could never get near them on account of (contrary) winds.

Here the atmosphere is all legendary, Taoist, magical, alchemical. Hsü Fu is not mentioned by name. But more substantial facts follow. The next passage refers precisely to the year −219, when Chhin Shih Huang Ti was on one of his periodical excursions along the eastern coasts.[a]

After this affair (the setting-up of commemorative steles on the shore at Lang-ya and elsewhere), Hsü Fu of Chhi and others made request (to the emperor) saying: 'In the midst of the (Eastern) Sea there are three magic mountain islands, Phêng-Lai, Fang-Chang and Ying-Chou, inhabited by immortals. We beg to be authorised to put to sea, after due purification, and accompanied by (a suitable number of) young men and girls, go forth in search of these islands.' (The emperor approved this petition and) despatched Hsü Fu with several thousand young men and maidens to go and look for (the abodes of) the immortals (hidden) in the Eastern Ocean.

One would give a good deal to know what kind of craft they sailed in, and one would not be surprised to discover that whole fleets of sailing-rafts were employed.[b] These voyages have been well compared with those of the Far West which sought for the 'Fortunate Isles', both having good foundation in fact; Japan, the Liu-Chhius, Micronesia, Hawaii and America on the one side, and Madeira, the Canaries, the Azores and America on the other.[c] They continued to fascinate Chhin Shih Huang Ti until his death. Whatever the vessels of Hsü Fu were, they cost an enormous sum, and in −212 the emperor bitterly complained about the expenses, always fruitless, which were being incurred.[d] In the following year he was at Lang-ya again, and after mention of the searches of 'the magician (*fang shih*[1])' Hsü Fu and others for the drugs of immortality, the historian continues:[e]

Several years having passed without any success, and great outlay having been made, (the sea-captains) feared that they would be blamed, so they spun a yarn and said: 'To obtain the drugs of Phêng-Lai is quite possible, but we have always had difficulties with great sharks (*chiao*[2]), and this is why we have never succeeded. We beg that excellent archers should be sent out with us so that when these fishes appear they can be killed by the use of multiple-bolt arcuballistas (*lien nu*[3]).'[f]

In the bizarre and dramatic sequel, Chhin Shih Huang Ti dreamt that he was fighting a sea-god with a human face, after which he ordered that the ships of those who went to sea should be suitably armed to kill the objectionable fishes. Meanwhile he himself patrolled the coast with multiple-bolt arcuballistas, and in the end he killed a great sea-animal at Mt Chih-fou. Soon afterwards he fell ill and died in −210 at Sha-chhiu on the way back to the capital.

---

[a] *Shih Chi*, ch. 6, pp. 18*a* ff., tr. auct. adjuv. Chavannes (1), vol. 2, pp. 151 ff.
[b] Cf. pp. 390, 441 above, on the fleet of rafts built in Yüeh in −472.
[c] The comparison was first made by Yetts (4).
[d] *Shih Chi*, ch. 6, p. 26*a*, cf. Chavannes (1), vol. 2, p. 180.
[e] *Shih Chi*, ch. 6, p. 29*b*, tr. auct. adjuv. Chavannes (1), vol. 2, pp. 184, 190 ff.
[f] On these see Sect. 30*i*.

[1] 方士   [2] 鮫   [3] 連弩

But the best account of Hsü Fu's proceedings is given by Ssuma Chhien in his biography of the Prince of Huai-Nan. There incidentally he says:[a]

Chhin Shih Huang Ti also sent out Hsü Fu by sea to search for magical beings and strange things. When he returned he invented excuses, saying: 'In the midst of the ocean I met (on an island) a great Mage who said to me "Are you the envoy of the Emperor of the West?", to which I replied that I was. "What have you come for?" said he, and I answered that I sought for those drugs which lengthen life and promote longevity (*yen nien i shou yao*[1]). "The offerings of your Chhin King", he said, "are but poor; you may see these drugs but you may not take them away." Then going south-east we came to Phêng-Lai, and I saw the gates of the Chih-Chhêng Palace, in the front whereof there was a guardian of brazen hue and dragon form lighting the skies with his radiance. In this place I did obeisance to the Sea Mage twice, and asked him what offerings we should present to him. "Bring me young men", he said, "of good birth and breeding, together with apt virgins, and workmen of all the trades; then you will get your drugs."' Chhin Shih Huang Ti, very pleased, set three thousand young men and girls at Hsü Fu's disposal, gave him (ample supplies of) the seeds of the five grains, and artisans of every sort, after which (his fleet again) set sail. Hsü Fu (must have) found some calm and fertile plain, with broad forests and rich marshes, where he made himself king—at any rate, he never came back to China.

Ssuma Chhien thus suggests that though Hsü Fu humoured the emperor's Taoist beliefs, he really knew that there were good and vacant lands away in the east, and planned to make off there. Later generations sometimes believed that he had settled in Japan, and a tomb shrine of Jofuku (as he is there called) exists to this day at Shingū in Wakayama prefecture.[b] But this has not the value of an independent tradition since Japanese scholars all through the ages were familiar with the *Shih Chi*. Archaeological evidence is more compelling, for Chinese influence reveals itself strongly in the artifacts of the middle Yayoi period (−1st and +1st centuries).[c] Nevertheless it may be almost equally likely that the story of Hsü Fu's disappearance conceals one voyage at least to the American continent.[d] Where he and his people went we shall probably never know.[e] But what sails the settlers had, or what means they took to steer their vessels over the broad waters, are matters not beyond all conjecture.

[a] *Shih Chi*, ch. 118, p. 11*a*, tr. auct. adjuv. Chavannes (1), vol. 2, pp. 152 ff.; Yetts (4).

[b] This is on the coast south of Kyoto and Osaka. See Davis & Nakaseko (1, 2). A book by Wei Thing-Sêng on the subject has not been available to us.

[c] See Kidder (1); Yü Ying-Shih (1), pp. 185 ff.

[d] A few particulars about the Taoist adepts who were sent to sea by the first Chhin emperor to search for the immortals and the drug of everlasting life have come down to us. We have already encountered Master Lu[2] of Yen (Vol. 3, p. 56) who set sail in −215. Later in the same year three further flotillas seem to have gone forth, carrying Han Chung,[3] the Venerable Hou,[4] and Master Shih[5] respectively. Upon his return after some months Lu sêng presented his magical book to the emperor, and continued to advise him in speeches till −212, but then the adepts began to despair of Chhin Shih Huang Ti's autocratic and managing ways, and soon afterwards disappeared into impenetrable hermitages. One of them, Hou Kung, returned to the stage briefly as an envoy from Liu Pang to Hsiang Yü in −203, but the others, who found court life as empty as the Pacific, were seen by men no more. All that we know of them was recorded by Ssuma Chhien in *Shih Chi*, ch. 6 (see Chavannes (1), vol. 2, pp. 164, 167, 176, 178, 180 ff., 312 ff.).

[e] Exploratory voyages into the Pacific continued through most of the Former Han dynasty, but especially under Han Wu Ti. Thus the alchemist Li Shao-Chün[6] (cf. Sect. 33 below), favoured at court in −133, had spent much time at sea communing with the immortals, whose islands he claimed to have

[1] 延年益壽藥   [2] 盧生   [3] 韓終   [4] 侯公   [5] 石生   [6] 李少君

(f) CONTROL (I), NAVIGATION

(1) THE THREE AGES OF PILOTRY

Some short account of the techniques of Chinese navigation can now no longer be postponed. We have come close to the subject several times, notably in the first volume where the maritime trade-routes were described,[a] then in the third with regard to maps and sailing-charts,[b] again in Vol. 4, pt. 1, in connection with the history of the magnetic compass,[c] and now with great force in the contrast between the Portuguese and the Chinese long-distance navigations. It might perhaps have been more logical to defer it until after the description of the development of the forms of the steering-gear itself,[d] but the topic is so intimately concerned with the history of Chinese shipping and with all that was involved in men, routes and destinations, that it seems good to place it here. We shall give detailed documentation only for new material, and for the rest refer the reader to the relevant foregoing Sections.

To recapitulate or even condense the rich literature on the history of navigational methods in the western part of the Old World would of course be out of the question, yet we must try to summarise in a few paragraphs what is known of it so that we may not lack standards of comparison.[e] The conventional periodisation of this history has caused some discomfort to its students, and we shall therefore prefer to distinguish three periods, those of (a) primitive navigation, (b) quantitative navigation, and (c) mathematical navigation. I propose to date the beginning of the second period at about + 1200 in the Mediterranean (as we shall see, it was nearer + 900 in East Asia), and the beginning of the third period at, or a very little before, + 1500. Let us survey their characteristics in turn.

It cannot be said that the mariners of the primitive period were without astronomical guidance; from very early times they steered (i.e. they oriented themselves) by the stars and the sun.[f] By night they could tell the time by the circumpolars and the cul-

---

visited. In the same year Khuan Shu,[1] a level-headed official who was afterwards Minister of Sacrifices, was despatched to sea in charge of an expedition to search for Phêng-Lai. In −113 experts in the interpretation of clouds or emanations (chhi) were sent out with the ships—conceivably they were probing the foggy Aleutians, a disappointing area for magic palaces. At the same time another thaumaturgical technician, Luan Ta[2] (important for the history of magnetism, cf. Vol. 4, pt. 1, p. 315), left with a flourish of trumpets for the coast to head another expedition, but in −112 it was found that he had not dared to embark, preferring to offer sacrifices on Thai-shan; and as his magical arts seemed to be failing him this soon led to his downfall. By −98 it could be concluded that all the sea-voyages were yielding nothing of interest to the Han court and bureaucracy. What proportion of sailing-rafts or ships never returned would be of much interest to us, but that the records do not say.

[a] Pp. 176 ff. [b] Pp. 532 ff., 556 ff.
[c] Sect. 26i. [d] See pp. 627 ff. below.

[e] Among the books which have been most useful in this study are those of Taylor (7, 8); Waters (15); Ferrand (7); Marguet (1) and Hewson (1). Those of Iberian authors, which occupy a specially important place, will be mentioned shortly. An unusual and interesting study on the Baltic is that of Drapella (2).

[f] This has been well emphasised recently by Adam (1). It is sufficiently demonstrated by the famous passage in Aratus, Phaenomena, ll. 31–44 (based on Eudoxus of Cnidus and echoed by Strabo), which says, c. −275, that the Greeks steered by Helice (the Great Bear) while the Phoenicians, more accurately, by Cynosura (the Little Bear and its Guards). We have already noted this, Vol. 3, p. 230. Cf. Taylor (8), p. 43. Cf. Odyssey, v, Rieu tr. p. 94.

[1] 寬舒   [2] 欒大

minations or risings of decan-stars,[a] gaining an idea of their latitude by rough assessment of the height of the pole seen against masts and rigging;[b] by day the varying relations of ecliptic and horizon helped them to construct their wind-rose.[c] Time and distance estimation was still crude, no more than a count of day and night watches with a guess at the way made; but these ancient pilots were observant men, men who took soundings,[d] noted sea-bottom samples,[e] marked the prevailing winds and currents, and recorded in their early rutters (such as the *periploi*)[f] depths, anchorages, landmarks and tides—nor did they forget to use the services of shore-sighting birds.[g] If they made any charts not a single one has survived. The art of such a man was summed up in a +4th-century Indian text which says of a famous pilot Supāraga:

He knew the courses of the stars and could always readily orient himself; he also had a deep knowledge of the value of signs, whether regular, accidental or abnormal, of good and bad weather. He distinguished the regions of the ocean by the fish, by the colour of the water, by the nature of the bottom, by the birds, the mountains (landmarks), and other indications.[h]

And we ourselves voyaged in this way when we followed in imagination[i] the adventures of the Chinese Buddhist pilgrims to India between c. +380 and +780. Primitive though the arts of their pilots were, we must not miss the significant point that this navigation was already oceanic.

Measurement was the keynote of the second or quantitative period of navigation, when pilots, more and more adopting new inventions and practices, would no longer sail by guesswork and the help of the gods. After about +1185 in the Mediterranean the emphasis rapidly shifted from celestial to a terrestrial observation, for the discovery of the magnetic compass now became known and used there, causing a veritable revolution in the sea-faring art.[j] Not only could the way forward be known through days and nights of overcast cloud or storm, but also the quantitative accuracy of azimuth dial readings brought many important developments in its train. Naturally the wind-rose became more complex, but in addition it transferred itself to parchment in the form of those portolan charts with their interlaced loxodromes or rhumb-lines radiating from a series of centres, which we have already described.[k] The earliest dated example of such a chart (illustrated in Vol. 3, Fig. 219) is of +1311, but historians of cartography customarily date its elaborate counterpart, the Carta Pisana,[l] at about +1275, and there is some textual evidence for sea-charts probably of this type from the

[a] Thirty-six constellations selected in Egyptian astronomy, one for each 10° of celestial longitude; cf. Vol. 2, p. 356 and Vol. 3, s.v.; also Boll (1); Bouché-Leclercq (1).

[b] Lucan, *Pharsalia*, VIII, 172 ff., c. +64; cf. Taylor (8), p. 47.

[c] Cf. Taylor (8), p. 6, also Vol. 3, p. 305 above, and p. 583 below.

[d] Cf. Ennin's first landfall on the Chinese coast in +838 (Reischauer (2), p. 5).

[e] Cf. Vol. 4, pt. 1, pp. 279, 284, and Hirth (16), translating the *Mêng Liang Lu*.

[f] Cf. Vol. 3, p. 532, on Timosthenes of Rhodes, c. −266, a contemporary of Aratus.

[g] For an ancient Indian account of this practice see Rhys-Davids (4), quoted in Taylor (8), p. 72. A general review of it was written by Hornell (16). Cf. p. 453 above.

[h] Supāraga was none other than the Buddha in a previous incarnation. The source is the *Jātakamālā* by Āryasūra, translated into Chinese before +434 (Lévi (7), p. 86, quoted by Ferrand (7), p. 177).

[i] Vol. 1, pp. 207 ff.

[j] See our discussion in Vol. 4, pt. 1, Sect. 26 i. Lane (3) analyses the economic effects of this revolution.

[k] Vol. 3, pp. 532 ff.

[l] Cf. Motzo (1); Taylor (8), pp. 110 ff., (10). Bagrow (2), p. 49, pl. 27, places it nearer +1300.

year +1270 when St Louis of France was on his way from Aigues-Mortes to the Crusades.[a] We may freely date their beginning during the second half of the +13th century. In the new precision of these bearing-and-distance plots further practices were implicit, notably time-and-distance measurement, so that the more accurate determination of time spent on a particular course became vitally important. Hence we are not surprised to find much talk of 'orologes de mer' from about +1310 onwards, 'dyolls' from +1411, and 'running-glasses' from about +1490, all being one and the same thing, namely the hour-glass or sand-glass, regularly turned (with a chant) by the petty officer of the watch.[b] Moreover, the pilot might well be obliged by the winds, as of old, to follow a course rather different from what he intended, but from about +1300 onwards the new discipline (new to Europe) of trigonometry supplied him with sets of traverse tables from which he could calculate very easily how much of his intended course he had made good after a certain time, and how far he would have to sail to get back on to it.[c] Of this 'Toleta de Marteloio' we have no example before +1428, but we know from passages in the writings of that great though eccentric Catalan philosopher-alchemist Raymond Lull, c. +1290, that it was developing in his time—significantly contemporary with the rise of trigonometry in the work of men such as Richard of Wallingford. Doubtless an absolute limiting factor for the appearance of these computing pilots of the late +13th and early +14th centuries was the popularisation of the Hindu–Arabic numerals, which came to completion just after the first use of the magnetic compass by Mediterranean sailors (late +12th century), though the new figures had first become known to the West before the end of the +10th.[d] As for the compilation of rutters, our oldest extant example of this period is the Italian *Compasso da Navigare* of +1253, which describes the whole Mediterranean in terms of the bearing-and-distance system which the portolan charts were shortly to embody, together with much other valuable pilotage information.[e]

So far there is no dispute, but real difficulty arises when we come to the quantisation of that astronomical guidance which even the most ancient navigators had relied upon, the extension, in fact, of the new standards of measurement, in some practical form, to

---

[a] See Taylor (8), p. 109.

[b] This was the oldest marine clock since neither sundial nor clepsydra could conveniently be taken to sea. The best discussion of the sand-glass as a navigating instrument, with full references, is by Waters (11), who believes that it must have been in use in the +13th century because of its obvious association with the portolan charts. The meaning of the term 'dial' was for some time in doubt because of the presumed existence of Norse 'bearing-dials' or wooden wind-roses anterior to the magnetic compass, but an object found by Sølver (3) can be interpreted in other ways, and the suggestion of Naish (1) that the +15th-century 'dyoll' was the sand-glass is now generally accepted. The crucial evidence is perhaps the mention in a late +15th-century English rutter of 'smale diale sonde' on the sea-bottom in 65 fathoms off Belle-isle (Anon. (51), p. 21). It is natural enough to find the first use of the term 'dial' in +1411 (Moore, 2), because the new mechanical clock had acquired its dial in the just preceding century (cf. Vol. 4, pt. 2, p. 511). It is also natural that its name should have changed again at the beginning of the following century for then it became necessary to distinguish it from the new portable watches and combined sundial-compasses. We shall return presently to the question of the origin of the sand-glass.

From the +17th century, as Drover *et al.* (1) have shown in an interesting paper, very finely powdered calcite from egg-shells was used in sand-glasses rather than sand.

[c] See Taylor (8), pp. 117 ff., (10); Waters (15), pp. 37 ff.; Beaujouan & Poulle (1), p. 106, etc. The portolan, the magnetic compass, the sand-glass and the marteloio form a rather closely knit group of complementary techniques.

[d] Cf. Vol. 3, pp. 15, 146; Taylor (10).   [e] See Motzo (1); Taylor (8, 9, 10).

the celestial bodies. What happened in East and South Asia we shall recount, as far as we can, in due course; here the problem concerns chiefly the point in the +15th century at which the Portuguese navigators of the ocean sea began to measure polar altitudes accurately by the simplified seaman's astrolabe or the simple quadrant, and the matter is one of active controversy.[a]

After +1321, when it was first described in the West by Levi ben Gerson,[b] the use of the cross-staff (and Jacob's Staff, Fig. 247) for this purpose has been commonly assumed, but wrongly.[c] The Portuguese cross-staff or *balestilha* was probably not a direct development from this astronomical and survey instrument, but rather from the *kamāl* of the Arabs encountered in the Indian Ocean (cf. pp. 573 ff. below). Much evidence has been collected showing that the pilots of the Mediterranean, who were primarily concerned with east west transits, never took altitudes at all until a very late date.[d] Our oldest extant seaman's astrolabe dates from +1555[e] and our oldest dated drawing pushes this back only to +1525,[f] but we may accept its use from +1480 onwards.[g] The use of the sea-quadrant after +1480 is also acceptable enough,[h] but that it goes back to Prince Henry's prime is a pure assumption, and the generally accepted date of *c*. +1460 is uncertain.[i] The great services of the Portuguese in determining the régime of winds and currents in the Atlantic are, however, questioned by none,[j] and it is agreed on all hands that by about +1480 the Portuguese were effecting fairly precise measurements of polar altitude, especially when during the following years they made their latitude determinations along the whole length of the African coast.[k] This involved the refinement of the old observations of the circumpolars into the 'Rule of the Guards'.[l] It also chimed in with the recovery of Ptolemy's cartography of parallels and 'climates' which, as we know, dated from the beginning of the +15th century.[m]

[a] Splendid work has been done during the past half-century by the Iberian school of historians of navigation, including, e.g., A. Barbosa (1); Coutinho (1); Bensaude (1, 2); da Costa (1, 2); and Franco (1), but some of its members have gone too far in their claims. Discussion has been particularly difficult because of the theory of the secrecy policy of the Portuguese crown (cf. p. 528 above). However, the most judicious of the Portuguese scholars, such as Leite (1, 2) and Teixeira da Mota (1), are not far removed from the most judicious of the critics, such as Beaujouan (1); cf. Beaujouan & Poulle (1). For the reasons already given, we cannot accept the conventional statement (as in Taylor (8), p. 167) that the 'foundations of astronomical navigation' were laid by the +15th-century Portuguese—or by anyone else at that time. The foundations were both ancient and world-wide, the quantisation came about both in East and West.

[b] Cf. Vol. 3, p. 573.

[c] The oldest known example, a purely astronomical instrument, dates from +1571; cf. Price (12). As Beaujouan & Poulle (1), p. 112, point out, no text authorises belief in the use of the cross-staff for astronomical navigation in the West before the first years of the +16th century.

[d] This has been conclusively shown by A. Barbosa (1) and Cdr. T. da Mota (1).

[e] But down to the Second World War the Museum at Palermo possessed one dated +1540. On the whole subject see Waters (18).

[f] Price (12); Waters (14).   [g] As da Costa maintains, (1), p. 13.

[h] See Taylor (12); Bensaude (1), vol. 7.   [i] See Beaujouan (1).

[j] The elucidation of these was mainly the work of Adm. Gago Coutinho (1); conveniently summarised by Costa Brochado (2).

[k] Cf. Peres (1); Beaujouan (1). Observations with the early instruments could be made reasonably accurately on land, but hardly at all at sea.

[l] Or the 'Regiment of the North Star'; cf. Taylor (8), pp. 47, 130, 146, 163, (12); Waters (15), pp. 43 ff. and *passim*.

[m] Cf. Vol. 3, p. 533.

Most probably what came first were those empirical sailing arcs in the broad Atlantic which we have already described (pp. 511 ff.), and it was these long voyages far from sight of land which impressed upon the Portuguese that something more was necessary than the rhumb bearings of the portolan charts, namely a way of determining the latitude at sea.[a] The fact that the Guinea arc was already in full use by +1440 pleads in favour of the use of the mariner's astrolabe or quadrant already at that time, but no evidence, textual or archaeological, sustains it.[b] The use of such devices on the Brazil arc, which came about 47 years later, strains our credulity the less.[c] Besides all this, the question of Arabic influence on the Portuguese navigators of the +15th century has been raised, though not yet answered—with all the contacts which they had at the eastern end of the Mediterranean, Ibn Mājid was surely not the first Arab pilot ever encountered by them.[d] The hard problem is to know just what the Catalan influence in Portugal at the beginning of the century,[e] and the deliberations of the 'School of Sagres' towards its middle, really amounted to.[f] Certain it is that in the second half of the +15th century many navigators and teachers of navigation were following methods which went back to the beginning of the period—notably Alvise Ca' da Mosto on his Gambia voyage[g] in +1455; while in the north Pierre Garcie in +1483 still lacked even the portolan chart, though he made history by his sketches of prominent landmarks like those still used in Admiralty Sailing Directions, and his deep knowledge and tabulation of the tides.[h]

This is the time, however, the threshold of the Renaissance, which marks the transition from the second to the third period, from quantitative to mathematical navigation.

[a] As da Mota (1), p. 131, has emphasised. The intermediate stage was altitude navigation; comparing the polar or noon sun height at the place with that known for Lisbon.

[b] Criticising the conventional views (e.g. da Costa (1), p. 12; Taylor (8), p. 159; Waters (15), p. 46), Beaujouan (1) has analysed destructively the only three pieces of textual evidence for its use before the last two decades of the century. One of these is interesting because it shows how navigators then felt about the old portolans. Either Diogo Gomes or Martin Behaim is speaking, about +1483 (and as there is this doubt the passage cannot be dated earlier). 'And I myself had a quadrant when I went down to those places (Guinea), and I marked the altitude of the north pole upon its face (or table). I found it much better than the map, for while on the map you can certainly see the route to be followed (i.e. the bearing), once you go wrong you can never get back to your original objective.' In other words, 'finding and running down' the latitude is much better than trusting to compass and marteloio alone. This was the beginning in the West of a method that lasted for centuries, but it was what the sailors of the East had been doing for some time already (p. 567 below).

Another passage studied by Beaujouan (1) also unites Western and Eastern methods. When Alvise Ca' da Mosto was at the mouth of the Gambia in +1455 he found the Southern Cross standing 'at the height of a lance'. Evidently he had no quadrant. The only other known use of this expression occurs in a conversation of another Venetian, Marco Polo, about +1300, recorded by Pietro d'Abano; the great traveller said that a celestial body which must have been the main Magellanic Cloud in Doradus could be seen from various parts of the southern hemisphere, near the antarctic pole and 'at the height of a lance'. What this suggests to us is that Chinese navigators, knowing the great use of the gnomon by their own astronomers, not only at home but in many places abroad (cf. Vol. 3, pp. 274, 292 ff.), set up such styles when they touched at different places, and expressed star altitudes roughly by their aid. It was, after all, the simplest possible kind of cross-staff method.

[c] Cf. Peres (1), p. 46.

[d] Cf. T. da Mota (1), pp. 133, 135 ff., 140, (2); Beaujouan (1); Beaujouan & Poulle (1), p. 109; Brochado (1), pp. 111 ff. Cf. pp. 567, 572 below.

[e] De Reparaz is the protagonist of this, cf. his (3), (4).

[f] Cf. p. 504 above.

[g] Ed. Crone (1).

[h] Taylor (8), pp. 168 ff.; Waters (15), pp. 12 ff., (17).

From +1500 onwards new aids for the sea pilot came tumbling out of the cornucopia of the 'new, or experimental, philosophy' in a wealth almost as bewildering as that of the aerofoil or transistor age.[a] Beginning again with the astronomical field, first the tables multiplied—noon solar declination tables for computing polar altitude (+1485),[b] quadrennial tables (+1497), Southern Cross tables (+1505), star culmination tables (+1514),[c] sun amplitude tables (+1595),[d] all eventually leading to publications of the nautical almanac type (+1678 onwards). Then there were continual improvements in instruments, the first vernier-type graduations (+1542),[e] the supplementation of the cross-staff by the Davis back-staff (+1594),[f] and the coming of the reflecting sextant and octant (+1731). With the development at long last of accurate mechanical time-keeping machines, marine chronometers, suggested in +1530 but not accomplished till c. +1760, the problem of the longitude was solved;[g] and with the production of the marine barometer c. +1700, forecasting of weather became possible at sea. Meanwhile knowledge of magnetic phenomena steadily increased. Declination having become known to Europeans in the last decades of the +15th century,[h] its variation in different localities, first plotted by João de Castro and others from about +1535, was mapped on a world scale in Halley's voyage of +1699—important knowledge for mariners, though the hopes that it might solve the problem of the longitude were to prove illusory. From +1500 onwards the Cardan suspension[i] was taken to sea, and the compass swung in gimbals. At the same time great advances were made in the measurement of the ship's speed, and the old rough rules for estimating it were abandoned in favour of the log and knotted log-line (+1574),[j] or the accurate timing by minute-glass of the passage of an object in the water past two marks on the ship.[k] The continuously working screw log came in the eighteenth century.[l] Finally, there were advances in the written records of word, chart and terrestrial globe. Rutters

[a] See for broad surveys Taylor (7); Waters (15).
[b] The 'Regiment of the Sun'; see Bensaude (1), vols. 1, 2, 3, 5; da Costa (1), pp. 18 ff.; Taylor (8), p. 165; Mollat (2); Laguardia Trias (1). Sun observations constituted a great technical advance on shipboard for it was much easier to take a shadow position than to observe a star through alidades. Even then the errors were great, and in +1538 de Castro had as many of his crew as possible taking observations so as to minimise them (see da Mota (1), pp. 141 ff.).
[c] First in João de Lisboa's *Livro de Marinharia*; da Costa (1), pp. 19, 24.
[d] I.e. azimuth rising and setting points, in Thomas Hariot's MSS, especially useful for checking magnetic declination; see Waters (15), pp. 588, 590; cf. Taylor & Sadler (1).
[e] See Vol. 3, p. 296; cf. Taylor (8), pp. 175, 236; Waters (15), p. 304 and pl. LXXII. The concentric circles of Nunes were first, then the zigzag lines of Hommel and Chancellor.
[f] Waters (15), pp. 302 ff. and pl. LXXI; Taylor (8), pp. 220, 255.
[g] Gould (1); Taylor (8), pp. 204, 260 ff.
[h] Cf. Vol. 4, pt. 1, p. 308; cf. de Castro (1, 2, 3).
[i] Cf. Vol. 4, pt. 2, pp. 228 ff. and Fig. 474. The date of this application may have been a little earlier.
[j] See especially Waters (12); the date is that of the first mention, in William Bourne's *Regiment for the Sea*. On 'knots per hour' consult Waters (16).
[k] This method could have been already in use much earlier, by the beginning of the +15th century, since it is mentioned, as Feldhaus says, (1), col. 934, by Nicholas of Cusa. At this time the Chinese also employed it (cf. p. 583 below).
[l] The so-called perpetual or patent log, due to Foxon and Russell (+1773); see Hewson (1), p. 166. Beaufoy (2) attributed the idea to 'the celebrated Dr Hooke'. A small paddle-wheel had been suggested for this purpose by Vitruvius, x, ix, 5–7, and this 'sea-hodometer' was often illustrated in the Renaissance (cf. Vol. 4, pt. 2, p. 413). It was even tried experimentally a number of times, as by Saumarez in +1720 (cf. Spencer, 1), and Smeaton in +1754, but it never proved practical.

became ever more detailed from + 1500 onwards, gaining stimulus from the experiences of Europeans in the Indian Ocean, while latitude charts with graduated meridians[a] led directly to the projections of Mercator (+1569) and others, corrected by Hariot and Wright (+1599) so that the nautical triangle drawn on them would show latitude and longitude, bearing and course, approximately correctly.[b] Great circle sailing was explained by Pedro Nunes and many others from + 1537 onwards. Thus we are brought to the world of the nineteenth century and all our modern techniques of echo-sounding, ship's wireless, radar and the like. In what follows we shall see clearly enough that although Chinese pilots never by themselves entered the third of the three phases, mathematical navigation, they had sailed into the second phase, that of quantitative navigation, some two or three centuries ahead of Europeans. They too, then, deserve their praise.

> If Pilots painful toil be lifted then aloft
> For using of his art according to his kind,
> What fame is due to them that first this art outsought,
> And first instructions gave to them that were but blind.[c]

(2) STAR, COMPASS AND RUTTER IN THE EASTERN SEAS

It is hardly to be questioned that from the earliest times when Chinese shipmasters sailed their vessels out of sight of landmarks they steered by the stars and the sun. Chang Hêng was probably referring to their starcraft when he wrote in his *Ling Hsien* (+118): 'There are in all 2,500 (greater) stars, not including those which the sea people observe.'[d] Here *hai jen*[1] could equally well have been translated 'sailors', and so indeed we made it when we cited the passage in full.[e] This raises the ghost of a literature long lost, and now hard to interpret, but perhaps highly relevant. At that same place we mentioned in a footnote the fact that the *Khai-Yuan Chan Ching*,[2] the great treatise of the Thang on astronomy and astrology, often quotes an ancient *Hai Chung Chan*,[3] i.e. 'Astrology (and Astronomy) of (the People in) the Midst of the Sea (or, of the Sailors)'. The antiquity of a text of this name is certain because it appears (with ampler title) in the bibliography of the *Chhien Han Shu*, completed by the last decades of the + 1st century,[f] and is quoted several times in the *Hou Han Shu* commentary[g] written about

---

[a] See da Costa (1), p. 30; Taylor (8), p. 176, etc. There was also a steady improvement in the value adopted for the terrestrial degree and the nautical mile and league; see Moody (1); da Costa (1), p. 31.

[b] See Clos-Arceduc (1); George (1); Taylor (8), pp. 222 ff., (11); Taylor & Sadler (1); Waters (13), (15), pp. 223 ff. and pls. LIX, LX. By this correction, embodied in the 'Table of Meridional Parts', the loxodromes or rhumb lines, which Nunes had shown to be spirals on the globe, appeared as straight lines.

[c] From Robert Norman's 'Safegarde of Saylers' (+1590); see Waters (15), pp. 167 ff.

[d] In *YHSF*, ch. 76, p. 64a. Full translation of the passage in Vol. 3, p. 265. Cf. Vol. 2, p. 354 and Vol. 3, pp. 271 ff. On ancient Chinese fixed star observations see Po Shu-Jen (1).

[e] Vol. 3, p. 265.

[f] Ch. 30, p. 42b.

[g] Ch. 21, pp. 6b, 8b, 11a. This is the second astronomical chapter, written before the end of the +2nd century by Ssuma Piao[4] and incorporated in the *Hou Han Shu* at the beginning of the +6th. The quotations all concern planetary astrology.

¹ 海人   ² 開元占經   ³ 海中占   ⁴ 司馬彪

+502 by Liu Chao.[1] The books that we find in the Han bibliography are named as follows:

Hai Chung Hsing Chan Yen[2] (Verified Stellar Prognostications of the Sea People).
Hai Chung Wu-Hsing Ching Tsa-Shih[3] (Manual of the Five Planets for Divers Occasions according to the Usage of the Sea People).
Hai Chung Wu-Hsing Shun Ni[4] (The Forward Motions and Retrogradations of the Planets, a Sea People's Manual).
Hai Chung Erh-shih-pa Hsiu Kuo Fên[5] (Geographical Astrology of the 28 Lunar Mansions, a Sea People's Manual).
Hai Chung Erh-shih-pa Hsiu Chhen Fên[6] (Bureaucratic Personnel Astrology of the 28 Lunar Mansions, a Sea People's Manual).
Hai Chung Jih Yüeh Hui Hung Tsa Chan[7] (Miscellaneous Prognostications by Sun, Moon, Comets and Rainbows, according to the Usage of the Sea People).

There are three interpretations of the expression *hai chung*: (*a*) that it meant the people of some foreign countries or islands overseas, (*b*) the Chinese as opposed to overseas (*hai wai*[8]) people, (*c*) the sea-faring men of China's coastal provinces. Moreover the material of the books could have been primarily astrological or primarily concerned with navigation. The first alternative of the three has been favoured by Western sinologists (so often prone to doubt Chinese originality);[a] the second was argued by Ku Yen-Wu in the +17th century,[b] though he did not explain why Hai-Chung should have been used when Chung-Kuo was the normal term, and *hai nei* the obvious antonym. We agree with Lao Kan (2) that the third is the most reasonable.[c] So also thought the great scholar Wang Ying-Lin[9] about +1280, who in his *Han I Wên Chih Khao Chêng*[10] (Textual Criticism of the Former Han Bibliography) wrote:[d]

In the astronomical chapters of the *Hou Han Shu*, the commentary quotes a *Hai Chung Chan*, and this is cited again in the section on astronomy in the *Sui Shu*, together with a *Hai Chung Hsing Chan Hsing Thu*[11] (Star Charts and Stellar Prognostications of the Sea People), each composed of one chapter. These are what Chang Hêng referred to as the 'mariners' observations'.[e] In the astronomical monograph of the *Thang Shu* it says that in the 12th year of the Khai-Yuan reign-period (+724) the Astronomer-Royal was instructed by decree to proceed to Chiao-chou (mod. Hanoi) to measure sun-shadow lengths.[f] There while at sea (*hai chung*) in the eighth month, looking southwards, they observed the remarkably high altitude of Canopus (Lao jen).[g] Below it there were numerous stars brightly shining, including many large ones, but these had not then been recorded on the celestial maps, and their names were not known.

---

[a] Cf. Vol. 2, p. 354 above.
[b] *Jih Chih Lu*, ch. 30 (ch. 10, p. 7).
[c] Though it is rather strange that sailors should have occupied themselves with geographical astrology, the influences of stars on particular places (cf. Vol. 3, pp. 545 ff.); and on the officials of the bureaucracy. So Chinese a trait as this last pleads for a home rather than a foreign origin.
[d] In *Erh-shih-wu Shih Pu Pien*, p. 1425.2. Tr. auct.
[e] Or 'reckonings', or 'prognostications'.
[f] We are very familiar with this story from Vol. 4, pt. 1, pp. 44 ff.
[g] α Carinae.

[1] 劉昭   [2] 海中星占驗   [3] 海中五星經雜事   [4] 海中五星順逆
[5] 海中二十八宿國分   [6] 海中二十八宿臣分   [7] 海中日月彗虹雜占
[8] 海外   [9] 王應麟   [10] 漢藝文志考證   [11] 海中星占星圖

The words *hai chung* are in the *Chiu Thang Shu* text,[a] so that the phrase certainly meant star observations at sea in the +8th-century documents from which the +10th-century writer drew. Probably we shall not go far wrong if, with Lao Kan (*2*), we identify the Hai Chung ('sea-going') corpus as the work of those 'magicians' of the Warring States period and Early Han who lived along the coasts of Chhi and Yen,[b] the 'mathematical practitioners' of the earliest stages of Chinese navigation. Their skills were doubtless undifferentiated, and it would be impossible to disentangle in them the components which today we should call astrology, astronomy, stellar navigation, weather-prediction, and the lore of winds, currents and landfalls;[c] all the more so since (as in the work of Dee, Hartgill, Goad, Gadbury, and many others)[d] these elements were still wholly confused down to the end of the +17th century in Europe. At all events we can now form some idea what kind of men those 'sea-going magician-technicians' (*ju hai fang shih*[1]) were whom Chhin Shih Huang Ti interrogated in −215, nameless abstractions though they remain in the text of the Grand Historian.[e] A concrete figure presents itself in the person of Wang Chung,[2] the eighth-generation ancestor of the hydraulic engineer Wang Ching[3] already mentioned (p. 281). The *Hou Han Shu* tells us[f] that Wang Chung, who lived in Shantung, 'delighted in Taoist techniques and was well versed in astronomy',[g] so when trouble came during the rebellion of the Lü family about −180 he put out to sea with all his people and sailed eastward to Lo-lang in Korea, where he settled in the mountains.

Chinese pilots in the period of primitive navigation certainly made use also of all those other ancient aids which we have already mentioned.[h] But it was they who brought this period to an end by being the first to employ the magnetic compass at sea. This great revolution in the sailor's art, which ushered in the era of quantitative navigation, is solidly attested for Chinese ships by +1090, just about a century before its initial appearance in the West.[i] Our first text which shows this also mentions astro-

[a] Ch. 35, p. 6*a*. Slightly abbreviated in *Hsin Thang Shu*, ch. 31, p. 6*a*.

[b] These men we have often encountered already, cf. Vol. 1, pp. 91, 93; Vol. 2, pp. 240 ff.; Vol. 3, p. 197, etc. Cf. Vol. 4, pt. 2, p. 11. We have also just seen (p. 551 above) the part they played in the first Chinese explorations of the Pacific in search of the isles of the immortals.

[c] After all, no ancient mariner would have found it easy to distinguish between prognostication of a safe voyage based on planets and stars from that based on seasonal winds and other meteorological factors associated with the recurring positions of stars. Here we must not forget the *Hai Tao Suan Ching* (Sea Island Mathematical Manual) by Liu Hui, wholly devoted to survey geometry, which appeared in +263; cf. Vol. 3, pp. 30 ff.

[d] On John Dee see Taylor (8), pp. 195 ff.; on George Hartgill see Taylor (7); on John Goad, a particularly demonstrative case, see Thorndike (1), vol. 8, pp. 347 ff.; on John Gadbury see Thorndike (1), vol. 8, pp. 331 ff.; on William Bourne see Taylor (13), pp. xxiii, 325. In general cf. Thorndike (1), vol. 7, pp. 105, 473, 645, vol. 8, pp. 459, 483.

[e] *Shih Chi*, ch. 28, p. 12*a*, tr. Chavannes (1), vol. 3, p. 438. Watson (1), vol. 2, p. 26, imports an unjustified nuance.

[f] Ch. 106, p. 6*a*, tr. Sun & de Francis (1), p. 95, from Lao Kan (*4*).

[g] *Hao Tao shu ming thien wên*.[4]

[h] We lack studies of the opening phases of navigational practice in Chinese culture. There is nothing of any great value now in Joseph de Guignes (3), and the old papers of Hirth (14) and Hennig (8) deal rather with trade-routes and the like.

[i] The whole story of the development of Chinese knowledge of magnetic phenomena has been given in Vol. 4, pt. 1, Sect. 26*i*. Concise summary in Needham (39).

[1] 入海方士   [2] 王仲   [3] 王景   [4] 好道術明天文

nomical navigation and soundings, together with the study of sea-bottom samples.[a]
Two further accounts in the +12th century follow before the first European mention;[b]
each emphasising the value of the compass on nights of cloud and storm. The exact
date at which the magnetic compass first became the mariner's compass, after a long
career ashore with the geomancers, is not known, but some time in the +9th or +10th
century would be a very probable guess.[c] Before the end of the +13th (Marco Polo's
time) we have compass bearings recorded in print,[d] and in the following century,
before the end of the Yuan dynasty, compilations of these began to be produced.[e]

In all probability from the beginning of its use at sea, the Chinese compass was a
magnetised needle floating on water in a small cup. A thousand years earlier, the first
and oldest compass had been a spoon-shaped object of lodestone rotating on a bronze
plate. At some intervening period the frictional drag of the spoon on the plate had
been overcome by inserting the lodestone in a piece of wood with pointed ends, which
could be floated, or balanced upon an upward-projecting pin.[f] The dry-pivot compass
had thus been invented, but although these primitive arrangements seem still to have
been used as late as the +13th century, Chinese sailors did not (so far as we know)
employ them. For at some time between the +1st and +6th centuries the discovery
had been made that the directive property of the lodestone could be transferred by induction to the small pieces of iron or steel[g] which the lodestone attracted, and that
these also could be made to float upon the surface of water by suitable devices. The
earliest extant description of a floating compass of this kind dates from just before
+1044 and involves a thin leaf of magnetised iron with upturned edges, cut into the
shape of a fish.[h] To floating compasses of one kind or another Chinese navigators
remained faithful for nearly a millennium. We have detailed accounts of their use from
the +15th century.[i] But in the +16th there came Dutch influence, mediated in part
through the Japanese,[j] as a result of which the dry-pivoted needle and then the

---

[a] *Phing-Chou Kho Than*, ch. 2, p. 2a, tr. Vol. 4, pt. 1, p. 279. The oft-repeated statement that the
ships were of some other culture rests on a mistranslation and must be rejected.

[b] It may be significant that one of these (*Hsüan-Ho Fêng Shih Kao-Li Thu Ching*, ch. 34, pp. 9b, 10a)
specifically says that at night the pilots steer by the stars and the Great Bear (cf. Vol. 4, pt. 1, p. 280).
This may mean no more than was intended by Aratus in the −3rd century (p. 554 above, and Vol. 3,
p. 230), but on the other hand it might imply the beginnings of altitude measurements by +1124. See
further on p. 575 below. The complement of Aratus is doubtless the passage in the *Huai Nan Tzu* book
(ch. 11, p. 4b), c. −120, where we read: 'Those at sea who become confused and cannot distinguish east
from west, orient themselves as soon as they see the pole-star.'

[c] We say this because these centuries saw rather refined measurements of magnetic declination (Vol. 4,
pt. 1, pp. 293 ff.), which could only have been made by the use of the needle.

[d] *Chen-La Fêng Thu Chi*, ch. 1, p. 1a. Cf. Vol. 4, pt. 1, p. 284.

[e] E.g. *Hai Tao Chen Ching*,[1] *Chen Wei Pien*,[2] and *Yüeh Yang Chen Lu Chi*.[3] Cf. Vol. 3, p. 559, Vol. 4,
pt. 1, p. 285.

[f] *Shih Lin Kuang Chi*, ch. 10; see Vol. 4, pt. 1, pp. 255 ff.

[g] Good steel was available in China at least as early as the −1st century, and wootz steel was also
being imported from India by the +5th. See Sect. 30d below, and in the meantime Needham (32).

[h] *Wu Ching Tsung Yao (Chhien Chi)*, ch. 15, p. 15b.

[i] E.g. *Shun Fêng Hsiang Sung*; see Vol. 4, pt. 1, p. 286 and p. 582 below.

[j] The *Hai Kuo Wên Chien Lu*[4] (+1744) has a short passage comparing Chinese and Dutch methods
of navigation.

---

[1] 海道針經   [2] 鍼位編   [3] 粵洋針路記   [4] 海國聞見錄

compass-card[a] (doubtless an Italian invention) were adopted on Chinese vessels. The Chinese compass-makers, however, employed a very delicate form of suspension which automatically compensated for variations of dip, and still impressed western observers as late as the beginning of the nineteenth century.

The remarkable series of maritime expeditions led by the admiral Chêng Ho between +1400 and +1433 must always remain a focal point in the history of Chinese navigational technique. By great good fortune, certain maps of portolan character[b] dating from about this period, which trace the routes followed by these and other Chinese ships and convoys, have been preserved intact. Early in the +17th century they were printed as the last chapter of an important treatise on military and naval technology, the *Wu Pei Chih*,[c] and part of a map showing the Indian Ocean with the openings of the Persian Gulf and the Red Sea has been reproduced in Sect. 22*d* above.[d] These charts are extremely distorted but schematic, and ships' courses are drawn across their oceans like the tracks in the maps issued by modern steamship companies. The lines of travel are accompanied by legends giving detailed compass-bearings, with distances in numbers of watches (*kêng*[1] or *ching*[1]),[e] and notes of most of the coastal features which could be important in navigation. The bearings are always given in the form '*hsing ting wei chen*'[2] (sail with the needle between Ting and Wei azimuth points, i.e. due S.S.W.),[f] or '*yung kêng shen chen*'[3] (use the needle pointing between Kêng and Shen, i.e. due W.S.W.), while '*tan khun*'[4] (bear on red, or single, Khun) meant sailing with the needle pointing directly to Khun, i.e. within $3\frac{3}{4}°$ on each side of S.W.[g] Thus the general formula is: 'sail on $x°$ for y watches.' The notes include indications of half-tide rocks and shoals as well as all ports and havens. Routes are given for inner and outer

[a] The wind-rose actually attached to the magnet. Cf. Wang Chen-To (5), p. 133. For T. da Mota (2), pp. 16, 18, this alone is the 'boussole authentique' or 'véritable boussole'. Have we not here another example of the grudging attitude so often adopted by Western historians of science towards Asian inventions and discoveries? One recognises a standard exercise (cf. Vol. 4, pt. 2, p. 545). Obliged at last to acknowledge a non-European achievement, they retire in good order by re-defining (to their own advantage) what the achievement really was.

[b] The term is used here only loosely and non-technically, for the Chinese sailing-charts had no intersecting rhumb-lines and no grid. Cf. da Mota (2).

[c] Ch. 240. The work was completed by Mao Yuan-I in +1621 and presented to the emperor in +1628.

[d] The reader is referred to Fig. 236 in Vol. 3, p. 560.

[e] N.B. the same term as that used for the unequal night-watches ashore (cf. Needham, Wang & Price (1), p. 199), but these were not unequal. The nautical watches were defined in terms both of time and distance run. A *kêng* was generally 1/10 of the 12 double-hour day-and-night period, i.e. 2·4 of our hours (but apparently sometimes 1/12, i.e. the same as the double-hour itself, Liu Ming-Shu (4), p. 59). Here again we have an instance of the classical Chinese preference for decimal metrology (cf. Vol. 3, pp. 82 ff.). The *kêng* was also generally considered to be the equivalent of 60 *li* (*Hsi-Yang Chhao Kung Tien Lu, Liu-Chhiu Kuo Chih Lüeh* and many other authorities, cf. *TSCC, I shu tien*, ch. 531, p. 9*b*), but apparently sometimes 42 *li* (*Min Tsa Chi*,[5] a Chhing work quoted by Liu Ming-Shu (4), p. 59). The Ming *li* being 0·348 mile, the way made in an hour was 8·73 miles (or 10·45 miles if the *kêng* was assimilated to the double-hour). If the Later Han and Chin *li* of 0·258 mile persisted as a sea measure, the figures would be 6·5 miles and 7·7 miles respectively. If the Chhing *li* of 0·357 mile is used with a Chhing *kêng* of 42 *li*, the results are very similar, 6·25 miles and 7·5 miles respectively. A speed of 6 to 10 knots would be quite reasonable for Chêng Ho's ships (cf. Vol. 3, p. 561).

[f] For a description of the Chinese system of azimuth mensuration see Sect. 26*i* above (Vol. 4, pt. 1, p. 298, Table 51). It has 48 sections of $7\frac{1}{2}°$ each.

[g] On these matters see Mulder (1), with some reservations.

[1] 更   [2] 行丁未針   [3] 用庚申針   [4] 丹坤   [5] 閩雜記

passages of islands, sometimes with preferences if outward- or homeward-bound. Much attention has been given to the accuracy of these diagrams and descriptions by modern scholars,[a] and to the identification of the place-names in them,[b] with the result that a high opinion has been formed of the knowledge and precision of these Chinese navigators' records.[c] Some idea of the skill of the pilots may be gained by the fact that in circumnavigating Malaya they laid their course through the present Singapore Main Strait, which was not discovered (or at least not used) by the Portuguese until they had been in those waters for more than a hundred years.[d]

The interest of the last chapter of the *Wu Pei Chih*, however, is not exhausted by these schematic charts. Four instructive navigational diagrams are given, summarising the star positions to be maintained during as many regular voyages. Here we reproduce (Fig. 994) the diagram of pilot's directions for that between Ceylon (Hsi-lan Shan [1]) and Sumatra (Su-mên-ta-la,[2] Kuala Pasé, modern Samudra).[e] A reading of the notes concerning the 'Guiding Stars' (*chhien hsing* [3])[f] which are distributed round the central picture will lead us into the heart of the matter.[g]

[Above] The Pole Star (Pei chhen) to be 1 *chih* above the horizon, and the 'Imperial baldachin' (Hua kai) 8 *chih*.

[To the left] In the north-west the Pu ssu stars[h] to be 4 *chih* above the horizon; and the same in the south-west.

[Below] The 'Frame' or 'Bone' of the 'Lantern' (orthographically Têng lung ku[4]) (i.e. the Southern Cross)[i] to be $14\frac{1}{2}$ *chih* above the horizon. The twin stars of the 'Southern gate' (Nan mên)[j] to be level at 15 *chih*.

[To the right] In the north-east the 'Weaving girl' (Chih nü)[k] to be 11 *chih* above the horizon.

[a] Cf. Mills (1); Blagden (1).

[b] Here the chief work was done by G. Phillips (1).

[c] Best, in Mills (1), has explained the 'double bearings' by which some have been puzzled in these texts. They meant only the usual procedure of setting off on the first course, and then changing to the second upon raising the island or point which acted as the intermediate landmark. He also plotted and 'swung' the bearings given so as to fit a modern map; the excellent result indicated that the Chinese navigator of the +15th century was using a compass with a declination about 5° W. of true N. Cf. Table 52 in Vol. 4, pt. 1, p. 310; and Smith & Needham (1). 'Treble bearings' are also known.

[d] Cf. Duyvendak (1); Mills (1).

[e] Some five miles up the Krueng Pasai River on the north coast. On this place-name, see Gerini (1), pp. 642 ff.; Pelliot (2c), p. 214.

[f] A close parallel, this expression, with the 'Leiðarstjarna' of the Vikings (cf. Sølver, 1).

[g] The directions for this route have been minutely studied by Mills (10).

[h] This asterism cannot be found in lists such as those of Chhen Tsun-Kuei (3) or Schlegel (5); its name here used was probably current only among sailors. Yen Tun-Chieh (19) suggests α Aurigae.

[1] It is a common assumption (e.g. in Schlegel (5), pp. 553, 554) that the Southern Cross was unknown to the Chinese before the arrival of the Jesuits. No doubt the name Shih tzu chia[5] derives from them, and by this it has since been known in the astronomical repertories, but in Têng Lung we must have a sailors' name which was never included in such official lists. I have not found it on any of the old Chinese star-charts, nor in those of encyclopaedias (*TSCC*, *STTH*, etc.) compiled long after the +15th century. But it is obvious that the pilots could never have done the work they did without being familiar with it. Here may be another example of the gap between the knowledge of the semi-literate artisans and the all too literate scholars.

[j] α[n], ε or β Centauri and stars in Circinus.

[k] α, ε, ζ Lyrae, i.e. a small group including Vega. Cf. Schlegel (5), p. 196.

[1] 錫闌山　　　[2] 蘇門荅剌　　　[3] 牽星　　　[4] 燈籠骨　　　[5] 十字架

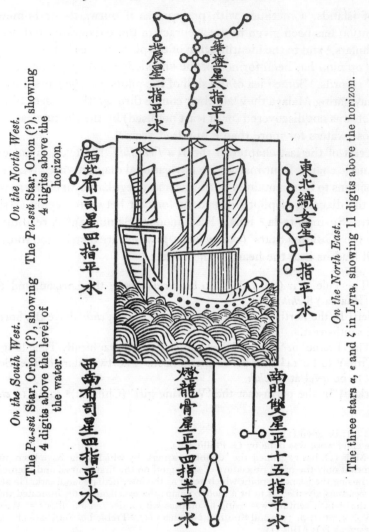

On the North West.
The *Pu-ssŭ* Star, Orion (?), showing 4 digits above the horizon.

華蓋星六指平水

晨星一指平水

東北織女星十一指平水

On the North East.
The three stars α, ε and ζ in Lyra, showing 11 digits above the horizon.

西北司星四指平水

西南布司星四指平水

On the South West.
The *Pu-ssŭ* Star, Orion (?), showing 4 digits above the level of the water.

燈籠骨星正十四指半平水

南門雙星平十五指平水

On the South.

The Southern Cross showing 14½ digits above the horizon, and the *Nan-mên-shuang-hsing* (Centaurus, α and β) showing 15 digits above the horizon.

Fig. 994. One of the navigational diagrams in the last chapter of the *Wu Pei Chih*, that for the Ceylon–Sumatra run, reproduced from Phillips (1), with his interpretative translations.

The explanation of all this lies in the fact that for measuring the altitudes of the pole-star[a] and other stars the pilots did not use the degrees of the astronomers but rather

---

[a] At this time the pole star was without doubt α Ursae Minoris. Synonyms in Schlegel (5), pp. 523 ff., who sets it, following the majority of old Chinese star-charts, actually in no constellation, but sur-rounded by the sickle-shaped portion of Kou chhen[1] (the Hooked array, or Angular arranger). Chhen Tsun-Kuei (3), however, lists it as Kou chhen 1. See Vol. 3, p. 261 and Fig. 97.

[1] 勾陣

another graduation in finger-breadths (*chih*[1]), each of which was divided, possibly into 8, more probably into 4, parts (*chio*[2]).[a] Moreover, for this voyage the pole-star was very low on the horizon or invisible, and it was therefore necessary to substitute for it a circumpolar markpoint, the Hua kai constellation.[b] Altitude on this would be measured each night when it culminated, and the altitudes of all the other guiding stars taken presumably at the same time. The *Wu Pei Chih* charts give Hua kai altitudes for a number of places.[c]

Once we realise that the navigators of the China Seas and the Indian Ocean depended quite as much on polar altitudes as the Portuguese came to do towards the end of the +15th century, a host of fascinating questions arises. Unfortunately we know as yet neither exactly how far back this quantitative oceanic navigation went in Eastern waters, nor how far the Europeans of the Atlantic border were influenced by it during the explorations of the West African coast. Certain it is that when the Portuguese showed him their astrolabes and quadrants in the summer of +1498 Ibn Mājid was not in the least surprised, saying that the Arabs had similar instruments, but the Portuguese were very astonished that he was not surprised.[d] Moreover, there are a number of points at which we may suspect East Asian influence on Europe, or where at least we have to grant considerable East Asian priority.

First, it is clear that the Chinese navigators of Chêng Ho's time, besides their compass-bearings, knew the method of finding and running down the latitude. In the *Hsi-Yang Chhao Kung Tien Lu*, for example, there is talk of a voyage from Bengal (probably Chittagong) to Malé in the Maldive Islands by way of Ceylon, and the polar elevations are given for every stage of the journey.[e] Thus a certain Ceylonese mountain will be sighted when the altitude has sunk to 1 *chih* 3 *chio*. We are still uncertain, however, as to the instruments which the Chinese used. By +1400 quadrants would have been quite possible—the armillary sphere had had a long and elaborate history in China,[f] and some such apparatus had been used overseas as far back as the beginning of the +8th century, when I-Hsing's meridian arc survey teams took altitude measurements from Indo-China to Mongolia.[g] That was the time, too, when a southern hemisphere astrographic expedition had been sent to map the constellations to about 20° from the

---

[a] Since such expressions as '1 *chih*, 3½ *chio*' occur, the *chio* was most likely a quarter of the *chih*. For it might be rather hard to believe that this measure could have been divided sufficiently accurately and practically into sixteenths. Besides, values above 3½ *chio* are never found.

[b] Sixteen stars in Cassiopeia and Camelopardus. Cf. Vol. 3, Figs. 99, 106, 107, 108, 109; and Schlegel (5), p. 533. Mills (priv. comm.) suggests that one of the two stars used was probably 50 Cassiopeiae, of N. declination 72° 12′ 18″ and R.A. 1 hr. 59 min., recognised by Chhen Tsun-Kuei (3) as no. 5 of the *kang*[3] portion of Hua kai.

[c] For instance, the Maldive Islands and the east coast of Africa. But though correct for the former, they seem in error for the latter (Mills, priv. comm.).

[d] Ferrand (7), p. 193, following João de Barros, *Dec.* 1, iv, 6; so also da Mota (2), p. 19.

[e] Ch. 2, p. 7*a*, *b*; best thanks are due to Mr Mills for drawing our attention to this. Cf. p. 558.

[f] See Vol. 3, pp. 339 ff. and Fig. 146.

[g] See Vol. 4, pt. 1, pp. 45 ff., briefly discussed in Vol. 3, pp. 292 ff. There are reservations about the northernmost station, but the arc comprised eleven stations over a total length of rather more than 2,500 km. Regrettably, no descriptions of the equipment used survived.

[1] 指   [2] 角   [3] 杠

antarctic pole.[a] Astrolabes as known in the West would not be in the picture, for reasons already given,[b] but a simplified armillary ring with swinging alidades or the characteristically Chinese sighting-tube[c] may well have been used ashore. Even more probable, seemingly, would be the simpler types of cross-staff, for elsewhere[d] evidence has been given that Jacob's Staff was known in China and used by surveyors three centuries before the description of Levi ben Gerson, i.e. by $+1086$ rather than $+1321$. This would also be more in line with the practices of the Arab and Indian pilots, as we shall see.

The problem of the maritime charts is also very obscure. That they existed is implicit in many Chinese texts,[e] but the only ones which have survived are the schematic diagrams, almost like the 'Peutinger Tables', preserved in the *Wu Pei Chih*. Nevertheless, the tradition of quantitative cartography was much stronger in China than it was in Europe,[f] so that already by $+1137$ a superb map on a scale of 100 *li* to the division could be produced,[g] and there is little reason to think that the much larger map of Chia Tan on a similar scale in $+801$ was any less good.[h] Indeed, the principle of the rectangular grid went back to Phei Hsiu in the $+$3rd century, and never gave place, as quantitative cartography did in Europe, to the discoidal fancies of the religious cosmographers.[i] It is thus of particular interest that in the work of Shen Kua late in the $+11$th century we do have a hint that the grid was combined with compass-bearing rhumb-lines, just as occurred two or three centuries later in the Mediterranean, but his work was terrestrial, not nautical, and it did not survive.[j] Lastly, the projection of

a See Vol. 3, p. 274. This expedition probably worked on the south coast of Java but it may have sailed due south some distance to raise the more southerly asterisms. Cf. p. 537 above.

b Vol. 3, pp. 375 ff.

c Vol. 3, pp. 332, 352 and Fig. 146. Cf. p. 576 below.

d Vol. 3, pp. 574 ff.   e See, e.g., pp. 576, 582.

f This has been clearly shown in Sect. 22. How it is possible for Western scholars to continue to say, as for instance Taylor does (10), that the Mediterranean portolan chart of about $+1300$ was 'the first map known to carry a scale', when decimal grids had been standard practice on Chinese maps at least from $+800$, and came to brilliant manifestation on extant examples from $+1100$, I do not understand.

g Vol. 3, Fig. 226 and p. 547.

h Vol. 3, p. 543.

i Vol. 3, pp. 528 ff., 538 ff. We have even offered evidence (Vol. 3, pp. 564 ff.) that European cartographers were influenced by the Chinese rectangular grid tradition, especially in its 'Mongolian' form, where place-names were inserted on a grid without any symbols for physical features. This influence appears to have been exerted both on Muslim and Christian geographers in the early decades of the $+14$th century, significantly just after the time of the Mongols and Marco Polo, and would have paved the way for the later portolans which combined rhumb-lines with Ptolemaic coordinates. Even Mercator may have been affected. As for the level of knowledge of the East Asian cartographers, it is a striking fact that in $+1402$, when Prince Henry was just eight years old, the Korean makers of the world-map described in Vol. 3, pp. 554 ff., were worrying about the shape of the Iberian peninsula and the names of European cities, when nobody at the Portuguese, or any other court of Christendom, had ever even heard of the existence of Chao-Hsien. Cf. p. 499 above, and Fig. 985 (pl.).

j Vol. 3, pp. 576 ff. This raises the much-discussed question of the existence of Arabic portolans, and their possible influence on both China and Europe (cf. Vol. 3, pp. 533, 561, 564 ff., 587). Apart from a very late copy of a Maldivian chart with rhumb-lines and rectangular grid described and figured by Brohier & Paulusz (1), pl. LI, p. 158, which may or may not be evidence, no certain Arabic portolan has ever come to light. Teixeira da Mota (2) has now developed an argument of some weight from textual sources to prove that the Arab navigators never used such charts. At the same time he demonstrates, however, that they did have rectangular-grid charts, often graduated in *iṣba'* (see p. 570), and sometimes possibly of 'Mongolian style' (Vol. 3, p. 564) with names of ports and places but little geographical detail.

Gerard Mercator in +1569 was a great advance, but he never knew that he had been preceded by Su Sung five centuries earlier in a celestial atlas,[a] in which the hour-circles between the *hsiu* (lunar mansions) formed the meridians, with the stars marked in quasi-orthomorphic cylindrical projection on each side of the equator according to their north polar distances. With such a brilliant background, we must hope that archaeological discoveries will yet reveal what charts were used by the master-mariners of the Sung, Yuan and Ming.[b]

As we saw above, the magnetic compass, the portolan chart, the sand-glass and the marteloio formed a closely connected knot of complementary techniques. Little can be said of traverse tables, which have not so far been recognised in Chinese rutters, but the use of sand for time-measuring opens up curious perspectives. Wang Chen-To (5), who went into the matter in his work on the history of the mariner's compass, concluded that the sand-glass was not known or used on Chinese ships until the end of the +16th century when they acquired it from the Dutch or the Portuguese.[c] But since the time of Wang's memoir much information has come to light about an important develop-ment in the history of Chinese mechanical clockwork which occurred about +1370, namely the substitution of sand for water in clocks of the classic scoop-wheel type.[d] Whether or not these continued the link-work escapement or adopted reduction gearing remains unclear, but they certainly involved something new for Chinese clock-work as well as for the more recent clocks of the West, which were acquiring it too, namely a stationary dial with a moving pointer.[e] This new look is associated with the name of Chan Hsi-Yuan, and there is no reason why one or more of his clocks should not have been carried (as the older water-wheel ones could not have been) on each of the great ships of Chêng Ho's fleet. In any case, it is clear that time-keeping by sand-

[a] Vol. 3, Fig. 104 and p. 278.

[b] Hope must not be set too high. As mentioned elsewhere (pp. 403, 413) the Chinese shipbuilders seem to have been particularly skilled in doing without any blue-prints or diagrams. Then there was family property in technical ideas and records, very noticeable in the medical world. And we have seen (p. 525) how government documents could be purposely destroyed. All the same, it passes belief that the pilots of the great Treasure-ships could have used no charts. Mr J. V. Mills indeed believes that those in the *Wu Pei Chih* approximate very closely to what they carried.

[c] Cf. Vol. 4, pt. 2, p. 509. We reproduce here the only Chinese illustration of a ship's sand-glass which we have come across, in Fig. 995 (pl.); it occurs in the *Liu-Chhiu Kuo Chih Lüeh* (+1757), *Thu hui* sect., p. 34b. Cf. p. 556 above.

[d] A full account of the history of clockwork in China, centring on the first invention of an escape-ment early in the +8th century, has been given in Sect. 27j (Vol. 4, pt. 2, pp. 435 ff.). A complementary and in some respects more detailed treatment will be found in Needham, Wang & Price (1). For sand-clocks herein see pp. 154 ff., as also Vol. 4, pt. 2, pp. 509 ff. The main reason for the introduction of sand was that it was a cheaper method of avoiding freezing than the use of mercury, the motive power of some instruments from the +10th century onwards.

[e] See Vol. 4, pt. 2, pp. 510 ff. We have just seen (p. 556) how the term 'dial' came to be applied to time-measurement by sand-flow at sea in +15th-century Europe, but the association of the things them-selves was closer in the +14th-century sand-clocks of China. Tracing back the dial in history leads us to the anaphoric clock of Hellenistic times. This was essentially a clepsydra with a float-and-pulley arrange-ment which, as the water-level fell, rotated a dial constituting a planisphere and seen through a network of wires representing the horizon, meridian, equator, etc. It has been shown that the anaphoric clock was the ancestor of the astrolabe with its *rete*, and it is generally assumed that the clock-dial, though station-ary, was also derived from it. There is some evidence for the use of anaphoric clocks in China before the time of Chan Hsi-Yuan. On the whole question see Vol. 3, pp. 376 ff., Vol. 4, pt. 2, pp. 466 ff., 503 ff. and Needham, Wang & Price (1), pp. 64 ff.

flow was very much in the minds of the Chinese at that time. It is necessary therefore to re-examine the Western traditions which make the sand-glass begin with Liutprand of Cremona in the +10th century,[a] and to reconsider the suspicion of Speckhart,[b] long ignored, that the hour-glass came to Europe from the East.[c] Liu Ming-Shu (4) mounts an argument of some weight to the effect that since nautical watches (*kêng*) are mentioned (or implicit) in many descriptions of Chinese navigation from the beginning of the +12th century onwards (cf. Vol. 4, pt. 1, pp. 279 ff.), the measurement of such units must have necessitated the sand-glass, since no form of water clepsydra would be imaginable at sea.[d] If Waters (11) is right in tracing the Western nautical sand-glass back to the Venetian glass industry of the late +12th century, the possibility presents itself that together with the magnetic compass itself and the stern-post rudder it might have formed part of one of those clusters of transmissions from Asia which we find in so many fields of applied science. But against this there is a serious argument and another way out. The sand-glass implies blown glass, and as we have earlier found, the glass-blowing art appears to be wholly European and Western, though by no means glass-making itself.[e] Is not the time-keeping 'joss-stick' the real answer to Liu Ming-Shu's paradox? Burning incense in stick-like form is a practice which goes far back into China's Middle Ages, and it would have been very easy to measure time approximately enough with the 'joss-sticks' that were kept alight in the ship's shrine where the compass lay also. In this case the use of combustion clocks at sea for watch-keeping gave a very practical and reliable 'proto-chronometer',[f] and its exact forms merit even more examination than that which Bedini (5, 6) has already devoted to it in a fascinating monograph. Yet the incense-stick was so characteristic of Chinese religion and culture that it may have been difficult for it to spread to mariners of other cultures even though they might have found it very useful.

Let us now return to the altitude measurements in *chih* and *chio*. The remarkable feature of this system is that it was practically identical with that in use among the Arabic shipmasters of the Indian Ocean, who expressed altitudes in *isba'* (the finger-breadth or inch), equalling 1° 36' 25",[g] and its eighth part, the *zām*.[h] The system was

---

[a] Lombard bishop and ambassador, c. +922 to +972. Thus, at least, I interpret the reference in Feldhaus (1), col. 1222, to 'Luitprand of Chartres, *fl.* +760', which in itself seems improbable to a degree.

[b] In his German translation of Saunier (1), p. 177. The reference is to an earlier author, unspecified, whose evidence Speckhart would like to have checked.

[c] There is no lack of Alexandrian antecedents for the sand-flow method—for example Heron's automobile and automatic puppet theatre, in which the motive power was derived from the outflow of sand or cereal grains from a large tank, with the consequent fall of a weighted float. See Needham (38).

[d] He makes a dubious exception for the sinking-bowl clepsydra (cf. Vol. 3, p. 315). It appears that coconut shells have been used in South-east Asia for this purpose, but whether at sea we know not. In the Chhing period mariner's hour-glasses were of pottery or porcelain rather than glass, which may plead for the antiquity of the device in Chinese sea-faring culture. There is a description in the *Min Tsa Chi*. Hsü I-Thang (1), p. 273, quotes Marco Polo as saying that there was a sand-glass in the watch-house on every bridge in Hangchow, but the text says only *un horiuolo* (Moule & Pelliot ed. p. 332 (ch. 152), cf. Moule (15), p. 23), and we cannot assume that this was anything other than a clepsydra.

[e] See Vol. 4, pt. 1, pp. 103, 104.

[f] See Vol. 3, p. 330; Vol. 4, pt. 2, pp. 127, 462, 526.

[g] Von Hammer-Purgstall (3), p. 770.

[h] See Ferrand (7), who printed de Saussure (36). There were 224 *isba'* in the complete circle.

long known to Europeans mainly from the *Muḥīṭ* (The Ocean), a compendium of nautical instructions[a] put together by the scholarly Turkish admiral Sīdī 'Alī Re'is ibn Ḥusain[b] when staying at Ahmedabad in India in +1553 on his epic journey home after the destruction of his fleet.[c] Later his chief sources became known, the treatise of Sulaimān al-Mahrī (+1511), and especially the *Kitāb al-Fawā'id* written about +1475 by Shihāb al-Dīn Aḥmad ibn Mājid,[d] the Arab pilot who joined Vasco da Gama at Malindi in +1498. We know now that the Portuguese navigators made use of the system for some time afterwards.[e] It will be seen that this tradition must have been in full employment at the time of Chêng Ho's voyages. Moreover, when the measurements in the *Muḥīṭ* and its sources are compared with those in the *Wu Pei Chih*, they are found, generally speaking, in good agreement.[f] The chief difference between the Arab and the Chinese systems seems to be that when a 'substitute' polar mark-point was desired in equatorial latitudes, the Arabs chose the classical 'Guards' (β and γ Ursae Minoris), which they called al-Farḳadain (the 'Calves'),[g] while the Chinese chose Hua kai; the declinations being very similar, but the right ascensions almost exactly 180° (12 hr.) apart.[h] Both Arabs and Chinese took a pole-star elevation of 1 finger-breadth as the point at which it was no longer safe to trust to pole-star measurements; they then changed from pole-star to circumpolar mark-point, the Arabs taking 8 finger-breadths of al-Farḳadain, and the Chinese 8 finger-breadths of Hua kai as equivalent to 1 finger-breadth of pole-star elevation.[i]

The first Europeans who visited the southern hemisphere found it very strange that the northern pole-star disappeared from sight. Marco Polo lost it in Sumatra on his

[a] Translated in part by von Hammer-Purgstall (3) with notes by Prinsep (3); parts also in Bittner (1) and Ferrand (1), vol. 2, pp. 484 ff.

[b] *D.* +1562; brief biography in Adnan Adivar (2), pp. 67 ff. His predecessor, the great admiral Piri Re'is (*d.* +1554, biogr. pp. 59 ff.), possessed one of the earliest maps of Columbus (+1498), as Kahle (4, 5) showed in a remarkable discovery.

[c] His account of this has been translated by von Diez (1) and Vambéry (1).

[d] Editions by Ferrand (6), vols. 1 and 2. Biographies in (7), pp. 176 ff. Since the work of Ferrand, hitherto unknown MSS. of Ibn Mājid of great importance have been published and reported by Szumowski (1, 2) and made more widely known by Costa Brochado (1). See also Hourani (1), pp. 107 ff., who takes the Arabic rutter tradition back to the +9th century.

[e] See da Mota (2), pp. 21 ff., 29 ff. The *iṣba'* was translated *polegada*.

[f] Private communication from Mr J. V. Mills on unpublished researches.

[g] Full details of the Arabic form of the 'Rule' in de Saussure (36).

[h] Surely the only likely explanation for this is that the Arabs and the Chinese were at some time or other accustomed to sail in these southerly latitudes at different times of the year. The practice which suited each group would then have become a convention. Perhaps the Chinese were avoiding the typhoon season and the Arabs the monsoon. Seen from the southern tropics Hua kai culminates about midnight towards the beginning of November, and would be usable between August and February; Ursa Minor culminates about midnight in early May, and would be usable between February and August. The typhoons draw northward as summer goes on (Cressey (1), p. 67) so that sailing south in late autumn or winter would have suited the Chinese, while in the Indian Ocean the S.W. monsoon gales (more violent than those of the China coast) do not begin until June or thereabouts, so that sailing east in the early spring would have suited Arab and Indian pilots. Cf. pp. 451, 462, 511 above. For discussion on thes matters we are indebted to the late Prof. F. J. M. Stratton, F.R.S.

[i] Both Arabic and Chinese texts record the pole-star altitude at many places. But owing to the change in the polar distance of the pole-star since the texts were written, it is necessary to add a certain factor to the recorded altitudes in order to make them coincide with the latitudes determined by modern observations. Prinsep (3), p. 444, computed this to be about 5° 31' (but see also his p. 780); Mills from Chinese data in the *Wu Pei Chih* maps gets about 4° 54'.

voyage home in $+1292$ and recovered it at Cape Comorin (lat. 8° N.);[a] Odoric of Pordenone remarked on the same phenomenon some twenty years later.[b]

Ye shall understand [wrote Mandeville about $+1360$] that in this land (Sumatra) and in many other thereabout, men may not see the star that is called Polus Arcticus, which stands even (i.e. due) north and stirs never, by which shipmen are led, for it is not seen in the south. But there is another star which is called Antarctic and that is even against (i.e. diametrically opposite to) the other star, and by that star are shipmen led there, as here by Polus Arcticus.[c]

Although none of these Western writers recorded the southern stars which the Asian pilots really sighted, they were greatly impressed by this astronomical navigation, so much so indeed that they gave the impression that the magnetic compass was not used in those waters. Nicolò de Conti, who was in the China seas just before $+1440$, said:

Commonly the Indians sayle by the guiding of the starres of the Pole Antartique, for seldome times do they see oure North Starre. They use not the Loademans-stone as wee do; they do measure their waye, and distance of places, according as their Pole riseth and falleth, and so they do knowe by this meanes what place they are in. They doe make bigger shippes than we doe...[d]

And twenty years afterwards Fra Mauro put the same information in his map. Very near the two clearest sea-going junks[e] a scroll inscription in the middle of the Indian Ocean says:[f]

The ships or junks which navigate these seas carry four masts or more, some of which can be raised or lowered, and they have 40 to 60 cabins for the merchants. They have but one single rudder, and navigate without a compass because they carry an astronomer who stands alone on the high (poop) and commands the navigation with an astrolabe.

And Mandeville, though he says nothing on this subject, found posthumously an illustrator ($c.$ $+1385$) who depicted the instrument on the poop of a ship in the south seas.[g] The words of Fra Mauro have not hitherto been much noticed, but their date gives them a special relevance to the contemporary development of nautical astronomy among the Portuguese (cf. pp. 558, 567). Presumably the *kamāl* was what he really meant (cf. p. 574).

The impression given about the compass in these passages, however, was undoubtedly wrong, and Taylor has plausibly explained how it arose.[h] The Mediterranean pilot of the $+14$th century never took his eyes off the needle, and gave orders to the helmsman

---

[a] Penzer ed. pp. 103, 112; Moule & Pelliot ed., vol. 1, pp. 373, 416; cf. pp. 417, 419 for other altitude readings, now not interpretable. Cf. also p. 558.

[b] Yule (2), 2nd ed., vol. 2, p. 146. Cf. Jordanus Catalanus (Yule (3), p. 34).

[c] Letts ed., vol. 1, p. 128.

[d] Penzer ed. p. 140. In view of de Conti's contemporaneity with Chêng Ho, his remark about the size of the ships has a special interest.

[e] Cf. p. 474.

[f] See Yusuf Kamal (1), vol. 4, pt. 4, p. 1409 or Almagià (2).

[g] Anon. (50), vol. 2, no. 158; the *Livre des Merveilles*, Bib. Nat., French MS. 2810. Beaujouan & Poulle (1), who reproduce it, urge that the drawing represents a mariner's compass rather than an astrolabe, but it seems to have a suspension cord attached to it.

[h] (8), p. 128, amplified in private correspondence, here gratefully acknowledged. Similar views had previously been adumbrated by de Saussure (35), p. 67, in Ferrand (7), p. 118. Da Mota (2), pp. 17 ff., supports Taylor's view with further evidence.

accordingly while working out his course by bearing and distance. For the Asian pilot the needle was only one of his instruments, and the determination of position by star (and possibly even sun)[a] sights was at least equally important. This was no doubt because the region sailed by the seamen of the Arabic tradition was one of relatively scanty, or at least very seasonal, rainfall, and frequent clear skies, so that orientation by the stars was more inviting and capable of more precision. At the same time their oceanic domain included both northern and southern hemispheres, with all that that implied of astronomical challenge. And as the interruption of overcast skies was less, there was not the same reason to wax enthusiastic about the leading of the lodestone. It is true that its only begetters, the more northerly Chinese, had done so, but their words were enclosed in the ideographic language, not to be understood or appreciated by Westerners until comparatively modern times.

The question may be raised[b] as to the mutual influence of the Arabic and Chinese navigators, but at present we hardly know enough to answer it. They had certainly been in contact for many centuries before +1400. Measurements of altitude were particularly prominent in all Arabic astronomy,[c] but on the other hand circumpolar mark-points for invisible stars were rather characteristically Chinese.[d] Again, *chih* and *chio* measurements are not common in early printed Chinese texts,[e] but that does not mean that they were not in widespread use by pilots, whose rutters were generally hand-written. Moreover, we can find possible Chinese mentions of surveyors' measurements in 'finger-breadths' (*chih*) a good deal earlier than anything similar in Arabic culture. For example, as has already been noted,[f] one of the Wei generals, Têng Ai,[1] was well known about +260 for his interest in military topography— 'whenever he saw a high mountain or a wide moor, he always estimated (heights and distances), measuring by finger-breadths, so as to sketch and plan the best positions for an army camp or fort'. His contemporaries, easily amused, thought this rather pedantic. But of course the system of finger-breadth units for altitudes could easily have arisen independently in the Arabic and Chinese culture-areas.

What instruments were used by the Chinese pilots of the Yuan and Ming for taking star altitudes was long a puzzle, but of those employed among the Arab sailors a good deal was known; they were all forms of the cross-staff or Jacob's Staff, including one,

[a] See de Saussure (35), p. 52, in Ferrand (7), p. 97. But da Mota (2), pp. 10, 20, in the light of further texts and studies, thinks it less and less likely that the Arabic navigators took solar altitudes.

[b] As recently by Goodrich (10).

[c] See Sect. 20*f* (Vol. 3, p. 267).

[d] On this see Sect. 20*e* above (Vol. 3, pp. 232 ff.).

[e] Pelliot (33), p. 79, noted them, apart from the *Wu Pei Chih* maps themselves, only rarely, another instance being Huang Shêng-Tsêng's[2] *Hsi Yang Chhao Kung Tien Lu*[3] (Record of the Tributary Countries of the Western Oceans), written just before +1520. Current investigations, such as those of Mr J. V. Mills, are now finding more, especially in MS. sources, such as the *Shun Fêng Hsiang Sung*. An interesting echo of these altitude measurements occurs in a Western source of the late +16th century. As we saw in Vol. 4, pt. 1, p. 225, many books were brought away from Fukien in +1575 by the Augustinian friar Martín de Rada; and among them was one 'for the making of ships of all sorts, and the order of navigation, with the altitudes of every port, and the qualitie of every one in particular'.

[f] Vol. 3, pp. 571, 572, referring to *TPYL*, ch. 335, p. 2*a*.

[1] 鄧艾　　[2] 黃省曾　　[3] 西洋朝貢典錄

the tablet or *kamāl* (Fig. 996), in which the stock was represented by a knotted string.[a]
Supposedly earlier was the set of nine square boards or plates extended on a string or
rod of standard length.[b] These devices
measured the angle between the star and the
horizon, not that between star and zenith,
which was much more difficult on shipboard.[c]
At the end of the +15th century the Portuguese pilots used the *kamāl* for some time,
calling it the *tavoleta*, or the *balistinha do
mouro*.[d] The fact that Arab pilots later on
called the cross-staff *al-bilistī*[e] must no doubt
mean that some of them had received it
from the West,[f] but this does not necessarily mean that its origin had been there, or
even that their forefathers had not transmitted it in the opposite direction. For as we
have already seen, there is evidence of the
existence of the cross-staff in China in the
+11th century, three hundred years before
its traditional invention in Provence.[g] It
remains extremely probable, therefore, that
the Chinese pilots of the +15th century
used some form of cross-staff.

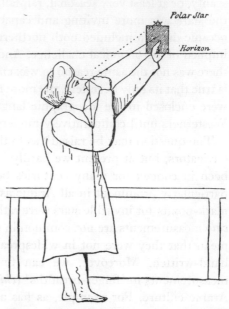

Fig. 996. The *kamāl* in use for taking star altitudes, a drawing made by Congreve in 1850.

   That they used a version of the *kamāl* as
one sort of cross-staff has now been proved
by Yen Tun-Chieh's brilliant interpretation (*19*) of a passage which we already gave at
an earlier stage without being able to explain it.[h] Here it demands re-translation. Li
Hsü[1] (+1505 to +1592), in his *Chieh An Lao Jen Man Pi* (An Abundance of Jottings
by Old Mr (Li) Chieh-An; printed in 1606),[i] has this to say:[j]

   The set of 'guiding star stretch-boards' (*chhien hsing pan*[2]) of Ma Huai-Tê[3] of Suchow has
twelve plates in all, made of ebony, ranging gradually from small to large. The largest is more

---

   [a] See Prinsep (2) and Congreve (1), both of whom were in personal contact with Arab pilots more
than a century ago. Both reprinted in Ferrand (7). De Conti (*c.* +1440) was the first Westerner to mention the *kamāl*. Cf. Kiely (1).
   [b] Normally the stretch of a man's arm. The first description of this so far found is in the *Muḥīṭ*
(+1554). See Prinsep (2, 3); Kahle (6); Kiely (1).
   [c] A point well made by de Saussure (36) in Ferrand (7), pp. 160 ff.; yet the use of astrolabes at sea is
often illustrated, and still more often believed to have been common.
   [d] Da Mota (2), pp. 21 ff.
   [e] According to Prinsep (2).
   [f] Port. *balhestilha* and Fr. *arbalestrille*, from the Low Latin Roman–Greek hybrid form *arcuballista*.
   [g] Vol. 3, p. 574. And later on, in Vol. 5, we shall find that the cross-bow itself was more ancient in
China than in Europe, two successive westward transmissions occurring.
   [h] In connection with counting-rods, Vol. 3, p. 74.
   [i] Correct the reference to *Chieh An Chi* at loc. cit.
   [j] Tr. auct.

   [1] 李詡   [2] 捽星板   [3] 馬懷德

than seven inches square (lit. long). They are labelled 'one *chih*', 'two *chih*' etc., up to 'twelve *chih*', all marked in fine script upon them; and they differ regularly just as a foot is divided into inches. There is also one ivory piece, two inches square (lit. long), and cut off at the corners so that it indicates half a *chih* (i.e. 2 *chio*), half a *chio*, one *chio* and three *chio*. This may be turned on one side or another facing you (in conjunction with one of the larger plates), and these lengths must be the measurements (required for right-angle triangle calculations according to the methods of the) *Chou Pei* (*Suan Ching*) (Arithmetical Classic of the Gnomon and the Circular Paths of Heaven).[a]

Evidently we have here a set of standard ebony tablets held at a fixed distance from the eye, not the single one with its knotted string that constituted the typical *kamāl*;[b] plus the interesting addition of a 'fine adjustment' in the shape of an ivory tablet with corners truncated to small standard edge lengths, held up at the same time to allow the measurement of fractions of a *chih*. Yen Tun-Chieh's calculations showed that the series of tablets described corresponds to a range of from 1° 36′ to 18° 56′ of altitude, with an average difference of 1° 34′ 30″ representing the *chih* (cf. p. 570). It is equally clear that the Chinese pilots at this time at any rate had 4 *chio* to a *chih*, not 8 (cf. pp. 567, 570), though the half-*chio* was marked on the ivory fine adjustment plate. How long before Li Hsü's time this system had been in use the text does not say, but the mention of Ma Huai-Tê is intriguing, for a commander of that name was active in the Sung, *c.* +1064. Since he was a Khaifêng man, however, and our 'mathematical practitioner' came from Suchow, it is more probable that he lived at a somewhat later date, but whether it was in the Sung, Yuan or Ming we do not as yet know.[c] We can be sure, at any rate, that the Chinese pilots were using his method in the +15th century, and they may well have been doing so in the +14th or even the +13th.

Evidence indeed seems to be growing that they were taking altitudes by the beginning of the +12th. A text of +1124 has already suggested this (Vol. 4, pt. 1, p. 280, and p. 563 above), and strange confirmation comes from a passage noticed by Lo Jung-Pang in the *Sung Hui Yao Kao* (Drafts for the History of the Administrative Statutes of the Sung Dynasty).[d] There we read:

In the 3rd year of the Chien-Yen reign-period (+1129) the Supervising Censor Lin Chih-Phing[1] was appointed to take charge of the defences of the (Yangtze) River and the sea, with authorisation to appoint his staff in the region under his command, i.e. from Hangchow to Thai-phing...[e] (Lin) Chih-Phing spoke of the need for sea-going ships, and requested that they be chartered from the coastal ports of Fukien and Kuangtung (and re-fitted)... These ships should each be equipped with a 'Dipper-observer' (*wang tou*[2]), bulwarks for protection against arrows (*chien ko*[3]), iron(-shod) striking-arms (*thieh chuang*[4]), stores of projectiles (*ying*

---

[a] See Vol. 3, pp. 19 ff., for an account of this, the oldest, Chinese mathematical classic.

[b] Yen doubts whether a cord was used with the Chinese set.

[c] Alas, he is listed neither in the *Chhou Jen Chuan* nor the *Chê Chiang Lu*. The *Su-Chou Fu Chih* would be the only hope, but he is not in the edition of +1691.

[d] *Ping* sect., ch. 29, pp. 31*b*, 32*a*, tr. Lo Jung-Pang. We are much indebted to Dr Lo for communicating to us a knowledge of this interesting passage.

[e] Presumably Thai-phing in Anhui, in which case Lin's bailiwick covered most of the marches between Sung and Chin as well as the open sea.

[1] 林之平   [2] 望斗   [3] 箭隔   [4] 鐵撞

tan[1]), trebuchets (*shih phao*,[2] for hurling them), gunpowder bombs (*huo phao*,[3] to be hurled likewise), incendiary arrows (*huo chien*,[4] at this date very probably rockets), and also other weapons, together with fire-fighting equipment (*fang huo*[5]).[a]

Though we have not elsewhere encountered the expression *wang tou*, its obvious meaning is a sighting-tube for determining the positions and altitudes of the stars of the Great Bear.[b] We shall remember in this connection the sighting-tube (*wang thung*[6]) and quadrant already illustrated in Vol. 3, Fig. 146, drawings taken from a book very close in date to the above document—the *Ying Tsao Fa Shih* (Treatise on Architectural Methods)[c] of +1103. But the 'Dipper-observer' might equally well have been a cross-staff or *kamāl*. Perhaps therefore the quantisation of stellar altitudes followed closely upon the quantisation of azimuth directions by the Chinese pilots.[d]

Summing up the present state of our knowledge about the development of quantitative navigation in the eastern seas, we have to start with the introduction of the mariner's compass on Chinese ships some time before +1050, possibly as early as +850. How soon this spread to the Indian Ocean we still do not know. Before +1300 there is hardly any evidence for the taking of star altitudes at sea by instrument, whether among Arabic or Indian pilots, and only very little for the Chinese navigators.[e] But the *Shun Fêng Hsiang Sung* tells us[f] that from +1403 'the drawings of the guiding stars were compared and corrected', which suggests a considerable previous development during the +14th century. Broadly speaking, therefore, we may not be far off the truth if we say that when Ibn Mājid met Vasco da Gama at Malindi, fully quantitative navigation was some two or three centuries old 'East of Suez' but hardly one century old in the West.

The diagrammatic illustration in Fig. 997 shows how star altitude measurements continued to interest Chinese groups concerned with the navigation of traditional craft down to our own times.[g] It comes from the *Ting-Hai Thing Chih*[7] (Local Gazet-

[a] We shall refer again to various items in this list below as occasion offers; cf. pp. 687, 690, 693.

[b] It will be remembered that the Little Bear was not seen as a constellation by the Chinese (Vol. 3, p. 261).

[c] Ch. 29, p. 2b. Cf. ch. 3, pp. 1a ff. Description, pp. 84ff. above.

[d] Lo Jung-Pang (7) recalls the surprise expressed by Hirth & Rockhill (1), p. 29, at the absence of any reference to the use of the seaman's astrolabe by Chinese navigators. Their assumption that it was normally used by Arab navigators was of course ill-founded, and they did not realise, perhaps, that although an astrolabe reached Peking with the scientific mission of +1267, it was not congruent with Chinese astronomy and aroused little interest (cf. Vol. 3, pp. 374 ff.). The sea astrolabe was in any case a far simpler thing—the difficulty was, as we have seen, to use it at sea.

[e] Here the impressions of Marco Polo, though negative, carry conviction to many. But cf. p. 558 above on the possible use of gnomons by landing parties. Negative evidence must always cede to positive when the latter exists. It would easily be possible to compile an anthology of unfavourable opinions about Chinese navigation (cf. de Navarrete (+1657), Cummins ed., p. 111, or Macartney (+1794), Cranmer-Byng ed., pp. 81, 275), but the failure of any particular observer to notice something cannot outweigh positive evidence that it was in fact there.

[f] P. 5a; cf. Duyvendak (1), p. 232.

[g] This picture was brought to our attention in a curious way. We are indebted for it to Mr P. H. Daniels, technician in the Radiotherapeutics Department at Cambridge. Coming of a family of printers, he had been interested, with his father Mr H. G. F. Daniels, of Harleston, to take some pulls of Chinese wood blocks which had been brought to England early in the present century by the father of Mr A. E. Lambden. These were the diagrams of the *Ting-Hai Thing Chih*. We should also like to thank Dr B. E. Holmes for directing Mr Daniels to us.

¹ 硬彈   ² 石炮   ³ 火炮   ⁴ 火箭   ⁵ 防火   ⁶ 望筒   ⁷ 定海廳志

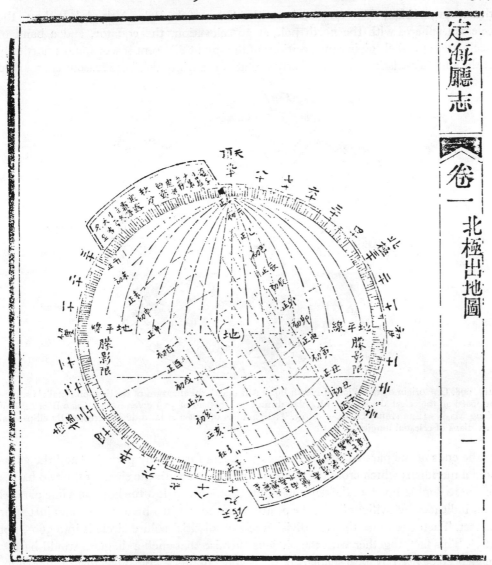

Fig. 997. A Chinese diagram of navigational astronomy, the 'Plan of North Polar Altitudes' (Pei Chi chhu Ti Thu) in the *Ting-Hai Thing Chih*. Explanation in text.

teer of the Sub-Prefecture of Ting-hai). This place is the chief town of a large island, Chou-shan (Chusan), situated off the coast of Chekiang province just on lat. 30° N., and protecting the straits and estuary leading to Ningpo.[a] The only edition of its local history and geography with this title appeared[b] in 1902, but it was doubtless based on

[a] It figured rather prominently, as will be remembered, in our story of the paddle-wheel boat in China (Vol. 4, pt. 2, pp. 428 ff.).

[b] Under the editorship of Shih Chih-Hsün[1] and Wang Hsün.[2]

[1] 史致馴   [2] 汪洵

the +1715 edition.[a] The diagram is entitled 'Plan of the Polar Altitude'. It shows the celestial sphere, with the north pole at 30° elevation, the equator, and a band of declination parallels giving the positions of the sun at different seasons, most northerly at the summer solstice, most southerly at that of the winter.[b] These seasons are marked

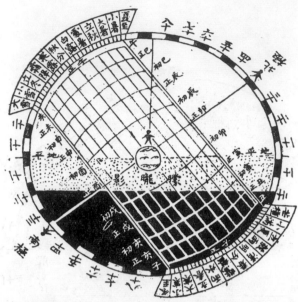

Fig. 998. One origin of the 'Plan of North Polar Altitudes', the diagram of declination parallels super-imposed on the celestial sphere (with polar altitude 35° instead of 30°) given in the *Thien Wên Lüeh* of Yang Ma-No (Emmanuel Diaz), + 1615. The parallels of celestial latitude are crossed by ellipsoidal meridians of celestial longitude.

at the ends of the parallels by the names of the 24 fortnightly periods;[c] and the ellipsoidal meridians which cross the parallels are labelled with the signs of the two halves of the 12 double-hours of the day and night.[d] Some 16° below the horizon a line parallel to it indicates the twilight limit; the positions of the sun just before dawn and just after sunset. Then one notes that the division of the solstitial colure circle is into 360°, not 365¼. This fact, together with the obvious surmise that such a diagram would hardly have been devised by local scholars of the island,[e] points suspiciously to a Jesuit origin. And indeed one prototype of the diagram can easily be found in the *Thu Shu Chi Chhêng* encyclopaedia,[f] there reproduced from the *Thien Wên Lüeh*[1] (Explicatio Sphaeris Caelestis)[g] of Yang Ma-No,[2] i.e. Emmanuel Diaz, published in + 1615 (Fig. 998). But an important difference is that the Ting-hai diagram adds a whole

---

[a] Which had been compiled by Miu Sui[3] and Chhen Yü-Wei[4] as a *Hsien Chih*. The oldest recorded edition is that of +1563.

[b] This was a graphical representation of the data contained in the 'Regiment of the Sun', cf. p. 559.

[c] See Vol. 3, p. 405, Table 35.   [d] See Needham, Wang & Price (1), pp. 200, 202.

[e] It is a pity that they did not add an altitude scale graduated in *chih* and *chio*.

[f] *Chhien hsiang tien*, ch. 2, p. 18b.

[g] Pp. 48, 50, 52, 57.

[1] 天問略   [2] 陽瑪諾   [3] 繆燧   [4] 陳于渭

family of 16 quarter-ellipses, equally spaced, in that part which represents the visible sky. Similar, but much simpler, diagrams were used to demonstrate navigational astronomy in +16th-century England.[a] The complications of this one raise some difficult questions.

There is no problem about the band of declination parallels, for it forms part of a particular astrolabic projection. A few years before Diaz' exposition it had appeared in a Chinese tractate on a planisphere[b] written by the Jesuit Sabbatino de Ursis (Hsiung San-Pa[1]) in +1611, entitled *Chien Phing I Shuo*.[2] This band of parallels between the tropics is the same as that of the orthogonal astrolabe projection described in +1550 by Juan de Rojas Sarmiento.[c] Now the 'ordinary' astrolabe[d] plate (or *tympanum*) uses a stereographic projection of the celestial sphere from one of the poles on to the plane of the equator;[e] a practice that stems in theory from Ptolemy's *Planisphaerium* of the +2nd century,[f] though no instrument earlier than the late +9th century has survived. The disadvantage of this is that for every latitude one must use a different plate under the rotating fretwork star-map (or *rete*); so that naturally the demand grew up for a 'universal' astrolabe usable without substantial change in any latitude. One of the answers to that was Rojas' projection, not stereographic but orthogonal (or orthographic). In this the celestial sphere is projected from the vernal point on to the plane of the solstitial colures, with the result that the parallels become straight lines like the equator, and the meridians semi-ellipses. Naturally the intervals between both parallels and meridians become less the further they are away from the centre of the whole. These unequal intervals can clearly be seen in the Jesuit diagrams (cf. Fig. 998) and must have been intended in the Ting-hai one (Fig. 997) though in this respect its meridian horary spacings within the band are not as well drawn as those of its declination parallels. What any meridian quarter-ellipses with their unequal intervals should look like is seen in the simple diagram of Fig. 999, taken from the +1551 edition of the *Cosmographia* of Petrus Apianus.

Actually Rojas never claimed the invention of the 'Rojas projection'. Simpler forms of it had been sketched in antiquity, notably a sundial analemma described by Vitruvius.[g] Textual evidence suggests that Abū al-Raiḥān al-Bīrūnī thought of the same thing about +998, though his explanation has never been properly studied.[h] Further-

---

[a] See Waters (15), p. 134; Taylor (13), pp. 215 ff. A cognate diagram entitled 'Differences in Rising and Setting Positions of the Sun at the Two Solstices' will be found (strangely for a dynastic history) in *Ming Shih*, ch. 33, p. 28*b* (begun +1646, finished +1736, published +1739).

[b] Wylie (1), p. 87; Pfister (1), p. 105; it is illustrated by two diagrams. Correct the translation of the title in Vol. 3, pp. 446, 694, 814, to 'Description of a Simple Planisphere'. An example of +1680 is in the Paris Observatory Museum. Cf. pl. 52 of Verbiest's *Chu I Hsiang Pien Yen* (+1674); Pfister (1), p. 359.

[c] In his book *Commentariorum in Astrolabium libri sex*, discussed and elucidated in great detail by Maddison (2). Rojas was, it seems, much aided by a Dutch assistant, Hugo Helt. Cf. Waters (15), pl. XXVII, p. 165.

[d] Cf. Vol. 3, pp. 375 ff. One must remember that the astrolabe was not used at all in traditional Chinese astronomy.

[e] A glance at Fig. 85 in Vol. 3 will help in the understanding of this and what follows.

[f] Ed. Heiberg (2), vol. 2.

[g] IV, 7, Granger ed., pl. 50.

[h] See Sachau (2), pp. 357 ff.; Maddison (2), p. 21.

[1] 熊三拔   [2] 簡平儀說

more several instruments have come down to us antedating Rojas but inscribed with the same orthogonal projection, e.g. a splendid astrolabe of +1462 in the National Maritime Museum in Greenwich,[a] one of +1480 in the Collegium Maius at Cracow,[b] and a third of +1483 at Florence.

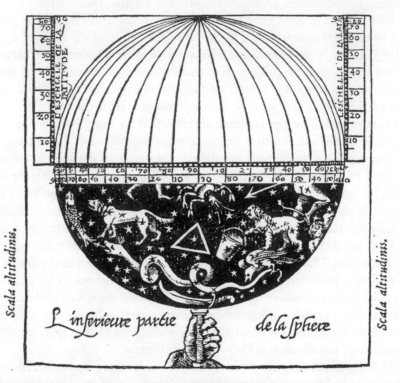

Fig. 999. Another origin of the 'Plan of North Polar Altitudes', the diagram of altitude quarter-ellipses in the *Cosmographia* of Petrus Apianus (Paris, +1551).

Of course the Rojas projection was not the only universal one. Best known and most commonly found is that of 'Alī ibn Khalaf, a Toledan astronomer[c] who was working about +1040. This 'lamina universal' is based on a stereographic projection of the celestial sphere from the vernal point on to the plane of the solstitial colures, with the result that both parallels and meridians become arcs of circles, the intervals between them (contrary to the Rojas projection) becoming larger the further away they are from the centre of the whole.[d] Later in the same century (c. +1070) the greater astronomer Abū Isḥāq al-Zarqālī (Azarquiel, Azarchel, called al-Naqqāsh, the engraver),[e]

[a] Description by Price (15). [b] See Zakrzewska (1).
[c] See Mieli (1), p. 186.
[d] Description in the *Libros del Saber*, ed. Rico y Sinobas (1), vol. 3, pp. 1–237, fig. opp. p. 10. See also Michel (3), pp. 18 ff.
[e] See Mieli (1), p. 184; Suter (1), no. 255.

author of the influential Toledan Tables,[a] improved the design by making a double grid for ecliptic as well as equatorial coordinates; and this is the ṣafīḥa or plate (açafeha, saphaea) that we find illustrated and minutely described in the famous *Libros del Saber de Astronomia* of Alfonso X, King of Castile, produced about +1276.[b]

If the *saphaea* had needed mention only as one of the probable influences upon Rojas (and hence upon de Ursis and Diaz) we could have passed it over in silence, but it has a strange connection with the 16 quarter-ellipses of the Ting-hai diagram. The front of the second instrument described in the relevant part of the *Libros del Saber* is indeed engraved with the *saphaea*, but the back bears a diagram which has not yet been adequately explained. While one quadrant is ruled sexagesimally with lines giving the sines of the angles of the scale of degrees,[c] the other three contain a series of semi- or quarter-ellipses, at first sight like those of the orthogonal projection but spaced at equal intervals apart, so that they cannot represent the meridians of that projection.[d] The *Libros del Saber* is far from being alone in giving this construction, for we find it also on earlier astrolabes, one made by Muḥammad ibn Futūḥ al-Khamā'irī in +1212 (with 20 equal divisions),[e] and another by Muḥammad ibn Hudhail in +1252 (with 24 equal divisions).[f] Hence the question may be raised whether the equally spaced quarter-ellipses of the Ting-hai diagram do not derive from earlier direct contacts between Chinese and Arabic astronomical navigators rather than from later Jesuit intermediation. Since the spacing of the parallels in Fig. 997 has so clearly been made unequal (as it has to be on the Rojas projection), the equal spacings of the altitude ellipses seem designedly inconsistent with this, and point perhaps to older contacts with the astrolabists and sea-captains of Islam.

Before taking leave of the Chinese pilots it may be of interest to glance at the contents of two or three typical rutters or navigational compendia. The first of these is the *Shun Fêng Hsiang Sung*[1] (Fair Winds for Escort), composed by an anonymous mariner some time about +1430 or at the close of the period of Chêng Ho's expeditions.[g] The second is the *Tung Hsi Yang Khao*,[2] compiled by Chang Hsieh[3] in +1618, a few years after Emmanuel Diaz had produced his explanation of the celestial sphere, but showing no evidence of any occidental influences. The writer of these 'Studies on the Oceans, East and West' was much more scholarly as a historian and geographer than the +15th-century sea-captain, but seems also to have had personal acquaintance with

---

[a] Cf. Vol. 4, pt. 2, p. 544.

[b] See Rico y Sinobas (1), vol. 3, pp. 135 ff., fig. opp. p. 148. One of the Latin MSS of al-Zarqālī's book on the *saphaea* is in the College Library at Caius. On the *Libros del Saber* the reader will recall Vol. 4, pt. 2, p. 443.

[c] Cf. Michel (3), p. 40.

[d] See Rico y Sinobas (1), vol. 3, p. 143 and opp. p. 149, with the discussion in Maddison (2), pp. 25 ff.

[e] Sauvaire & de Rey Pailhade (1).   [f] Millás Vallicrosa (1).

[g] Bodleian Library, Oxford, Laud Orient. MS. no. 145; cf. Duyvendak (1). We have been privileged to consult a draft translation prepared by Mr J. V. Mills (5). The +15th-century dating stems also from Dr E. R. Hughes, though Hsiang Ta & Hughes (1) placed the MS. itself between +1567 and +1619. Some statements in it may be after +1571 for on p. 65 a it is said that there are Fo-Lang (Frankish) foreigners established at Nagasaki (priv. comm. from Mr Mills). It is not claimed that the MS. itself is earlier than the last half of the +16th century. The text is now available in Hsiang Ta (5).

[1] 順風相送   [2] 東西洋考   [3] 張燮

the sea. His ninth chapter[a] is entitled Chou Shih Khao,[1] i.e. on the Ship's Master and what he should know.[b]

Part of the introduction of the *Shun Fêng Hsiang Sung* reads as follows:[c]

In bygone days, the Duke of Chou discovered and worked out the principles of the south-pointing needle. Throughout the centuries from ancient times until to-day, these principles have circulated far and wide. Yet if you ignore the increase or decrease in the number of watches, or their divisions, you will be at fault. Thus it was that charts were drawn, and all details of voyages recorded.

Now these old documents get worse worn every year, and it is difficult to judge from them what is the truth of the matter. If later people make copies from these originals, they will, I fear, fall into error. (So) availing myself of leisure, I have made a comparison of the calculated (number of) watches for every day, and have investigated the respective (number of) days for (each) through voyage. And I have collected and written down the number of the watches, the directions of the compass-needle, the appearances of mountains and the conditions of the water, whether there are bays and islands, shoals and deeps; with regard to all the places from the directly governed (district) of the Southern Capital (Nanking) to Thai-Tshang[2] and Wu-Li-Yang[3] (the Gulf of Siam, the Sumatran Seas and the Indian Ocean) (where are) the barbarian countries and other such places—in order to hand down to later generations the way and manner of making good voyages.

Looking first at what the two texts have in common, they are found to give abundant information on landmarks and general sailing directions (Yang Chen Lu[4]) with compass-bearings, and soundings in fathoms (*tho*[5]). Chang Hsieh's compendium includes as destinations Indo-China, Malaya, Siam, Java and Sumatra, Borneo, Timor, the Moluccas and the Philippines. The Anonymus goes even further afield—to Aden, Ormuz, India, Ceylon and Japan. Both give tables of monthly and seasonal winds (Chu Yüeh Fêng[6])[d] with copious advice on weather-signs (Chan Yen[7]), observing the shapes of clouds,[e] the behaviour of wind and rain, together with other meteorological phenomena such as solar haloes.[f] Both give a kind of tide table (Chhao Hsi[8]), adding other signs such as the colour of the water, and any objects likely to be floating on it. Both supply the master with liturgical instructions (Chi Ssu[9]), but the emphasis differs somewhat since the Anonymus is more concerned with the patron saints and tutelary deities of the compass, while Chang Hsieh likes to speak of Thien-Fei,[10] the Mistress of the Heavens, a sailor's goddess who had, as we know, the devotion of Chêng Ho and his fleets.[g]

---

[a] We have again had the benefit of a draft translation (Mills, 4).

[b] Other quotations from these texts will be found in Sect. 26*i* above. Among later books may be mentioned Yü Chhang-Hui's *Fang Hai Chi Yao*, ch. 13 of which is devoted to meteorological forecasting and tide-craft.

[c] P. 4*a*; tr. auct. adjuv. Mills (5), Duyvendak (1).

[d] The sea was supposed to be particularly 'choppy' (*shui hsing*[11]) on certain days of the month.

[e] Cf. Sect. 21 above, Vol. 3, p. 470.

[f] Note the section-heading in the light of p. 560 above. The non-astrological quality is comparable to what we find in Wm. Bourne (+1581), cf. Taylor (13), pp. 317, 399 ff.

[g] Cf. p. 523 above.

[1] 舟師考	[2] 太倉	[3] 巫里洋	[4] 洋針路	[5] 托	[6] 逐月風
[7] 占驗	[8] 潮汐	[9] 祭祀	[10] 天妃	[11] 水醒	

The material which is only in the +15th-century text has some features of special interest. We hear of the method of selection of water for the floating compass, and of the proper way to make the needle float on it. Besides three tables of the 24 azimuth points, there is one listing only 14 of them as the 'palaces of the heavens', and another associating them with winds. Most interesting with regard to possible Arab influence is a small table entitled 'Principles of Star Observations', which gives the azimuth rising and setting points of four constellations, the Great Bear, Hua kai (significantly), the Southern Cross (Têng lung ku), and Shui phing hsing [1] (possibly Canopus). Now such points of rising and setting were the elements of which the Arab sailor's azimuth circle graduation (lacking of course the abstract Chinese cyclical characters) was wholly constructed.[a] Another table lists the azimuth rising points of sun and moon throughout successive months, with corresponding clepsydra divisions and lengths of the night and day.[b] The fact that mnemonic verses on this, and on the moon's times of rising and setting, follow, suggests that the pilots were widely accustomed to pay attention to these data. The Anonymus also gives mnemonic verses about lightning as a weather-sign. Lastly he adds something about the determination of currents and tides, and the calculation of watches. It is here that we find mention of the floating wood method of logging the speed of the ship.[c]

Another anonymous rutter entitled *Chih Nan Chêng Fa*[2] (General Compass-Bearing Sailing Directions) exists in MS. form[d] appended to a military encyclopaedia, the *Ping Chhien*[3] (Key of Martial Art), by Lü Phan[4] and Lu Chhêng-Ên,[5] the preface of which is dated +1669. Besides much weather-lore and rhumb bearings for various voyages not yet analysed,[e] it has a sub-section on star sights (Kuan Hsing Fa[6]) with diagrams of constellations (Fig. 1000, pl.). On the page reproduced, Hua kai, Altair (the Herd-boy) and Vega (the Weaving girl),[f] the Southern Cross and Canopus (or else Achernar) can all be seen, with rising and setting azimuth points, though not altitudes, recorded at the foot of some of the columns. As we saw above (p. 566) Chinese pilots were accustomed to take altitude sights of many stars apart from the circumpolars, and the stars of the equatorial lunar mansions were doubtless among them. Indeed in Chinese astronomy, as we have seen, the correlation of circumpolar stars with equatorial mark-point constellations was particularly prominent.[g] Observations of the *hsiu* would give not only the time of night but also the latitude by simple calculation. It is very probable

[a] On this see de Saussure (35), pp. 49 ff., who tabulates the complete series. Reprinted, with additional illustration, in Ferrand (7), pp. 91 ff. De Saussure believed that this system must have originated about the +8th century (before the knowledge of the magnetic compass) and for use about 10° lat. N.

[b] This might also seem at first sight to indicate Arabic influence. But as was shown by Maspero (4), pp. 283 ff., series of azimuth determinations of this kind had been important for the regulation of water-clocks already in Han and Sui times, whence indeed tables survive. Cf. Vol. 3, p. 306.

[c] P. 6b; cf. p. 559 above. Also in *Min Tsa Chi*, and *Chhou Hai Thu Pien*, ch. 2, p. 6b.

[d] Bodleian Library, Backhouse Orient. MS. no. 578, *pên* 7. We are much indebted to Mr J. V. Mills for telling us of this. The text has since been printed by Hsiang Ta (5).

[e] Further investigations may well reveal the Chinese equivalent of the Arabic *tirfāt* tables, i.e. the distance required on various compass-bearings to raise or lay an *iṣba'* (cf. de Saussure (36) in Ferrand (7), pp. 171 ff.). For Western equivalents using the degree see Waters (15), e.g. p. 137.

[f] Cf. Vol. 3, p. 251 *et passim*.   [g] See Vol. 3, pp. 232 ff.

[1] 水平星   [2] 指南正法   [3] 兵鈐   [4] 呂磻   [5] 盧承恩
[6] 觀星法

therefore that further researches in the literature will bring to light nautical tables correlating *hsiu* culminations with the positions of unseen circumpolars, similar to the Arabic lists of the *manāzil* described and analysed by de Saussure.[a]

Lastly, a word on tide-tables. Since several of the extant Chinese rutters include forms of these, it is worth recalling that the phenomena of sea-tides were carefully studied in China earlier than in Europe.[b] Authoritative histories still inform us[c] that the oldest tide-table for a particular port is the early +13th-century 'fflod at london brigge', but in an earlier Section of this book it was shown that Yen Su's[1] *Hai Chhao Thu Lun*[2] of +1026 contained a detailed tide-table for Ningpo. Not much later, in +1056, Lü Chhang-Ming[3] drew up a tide-table for Hangchow which was inscribed on the walls of a pavilion on the banks of the Chhien-thang River. The Chinese pilots from the Yuan to the Chhing had thus a great tradition behind them.[d]

The spirit of these navigators, bent primarily, even under the orders of so great an admiral as Chêng Ho, on peaceful intercourse with the other inhabitants of Asia and Africa, is well seen in the concluding words of Chang Hsieh's chapter on navigation.

According to the writer's opinion [he says], those who make carriages build them in workshops, but when they come forth to the open road, they are already adjusted to the ruts. So it is with good sea-captains. The wings of cicadas make no distinction between one place and another, while even the small scale of a beetle will measure the vast empty spaces. If you treat the barbarian kings like harmless seagulls (i.e. without any evil intentions),[e] then the trough-princes and the crest-sirens will let you pass everywhere riding on the wings (of the wind). Verily the Atlas-tortoise with mountain-islands for its hat is no different from (an ant) carrying a grain of corn.[f] Coming into contact with barbarian peoples you have nothing more to fear than touching the left horn of a snail. The only things one should really be anxious about are the means of mastery of the waves of the seas—and, worst of all dangers, the minds of those avid for profit and greedy of gain.[g]

(3) TERRESTRIAL GLOBES

If we wanted to find a terrestrial globe at sea on board a modern liner there might be an ornamental one in the reading-room but we should certainly not expect to find one on the bridge. Yet there was a time, in the late +16th and early +17th centuries, when

---

[a] (36), in Ferrand (7), pp. 138 ff.   [b] See Vol. 3, pp. 483 ff., esp. p. 492.

[c] E.g. Taylor (8), p. 136.

[d] To complete the history of East Asian navigation the traditions of Japan and Korea will have to be investigated. Exactly contemporary with the book of Chang Hsieh is the *Genna Kōkaisho*[4] of Ikeda Kōun,[5] which includes illustrations of instruments such as a double quadrant. Later on, navigators were taught spherical trigonometry in the *Kairo Anshinroku*[6] (Safe Journeys on the High Seas) by Sakabe Kōhan.[7] Unfortunately neither time nor space permit further study of these works by us. Nor have we been able to see the *Sŭngsŏn Chipchirok*[8] (Guide for Shipmasters) printed in Korea in +1416 (cf. Tamura Sennosuke (*1*), p. 92).

[e] The allusion is to the story in *Lieh Tzu* (ch. 2, p. 16*b*) about the sailor for whom the seagulls were quite tame when he swam in their midst, until one day after he had promised his father to catch one—they knew it, and would not even alight on the water. Trs. Wieger (7), p. 93; L. Giles (4), p. 49.

[f] Here is the Buddhist visualisation of untold universes in a single drop of water, galaxy beyond galaxy in which the vast immensities of sky and sea form but an infinitesimal speck.

[g] Ch. 9, p. 16*a*; tr. auct.

[1] 燕肅   [2] 海潮圖論   [3] 呂昌明   [4] 元和航海書   [5] 池田好運
[6] 海路安心錄   [7] 坂部廣胖   [8] 乘船直指錄

such objects figured prominently among navigational instruments—as may be seen from the group of treatises of which Robert Hues' *Tractatus de Globis* of +1594 may be taken as representative.[a] Although we have touched upon terrestrial globes before, this then is the place to conclude the matter.

It is very improbable that such a globe could have been found on a Chinese vessel, even on Chêng Ho's flagship, for representations of this kind were not in the Chinese tradition; or not exactly, for the statement is a half-truth and needs explaining. First let us recall what happened in the West. The usual belief that the first maker of a terrestrial globe was the Stoic Crates of Mallos (*c.* −160) is authorised by the words of Strabo.[b] Here the *oikoumene* was shown as but one of four great continents separated by oceans—yet another example of those strange parallelisms between the thought of the Far West and the Far East, for Tsou Yen about −290 had been saying that there were nine of such continents sundered in just the same way.[c] After Strabo little or nothing is heard of terrestrial globes in the West, but the tradition must have been transmitted to the Arabs for in +903 the Persian geographer Aḥmad ibn Rustah gave a good description of the terrestrial as well as the celestial sphere.[d] A few centuries later the Latins could do the same, as is shown by the Englishman Sacrobosco's *Tractatus de Sphaera* (*c.* +1233), popular till the Renaissance.[e] Now it was later in this same century that Jamāl al-Dīn, heading a scientific co-operation mission from the Ilkhan of Persia to the court of China in +1267, took to Peking a terrestrial globe (or at least the design of one); 'a globe to be made of wood', says the *Yuan Shih*,[f] 'upon which seven parts of water are represented in green, three parts of land in white, with rivers, lakes, etc. Small squares are marked out so as to make it possible to reckon the sizes of the regions and the distances along roads.' But the idea did not catch on.

Why it did not is very hard to say, for there was much in the Chinese tradition to welcome it. The great cosmologists of the Han repeatedly said that the earth floated in the heavens like the yolk in a hen's egg, or that the earth was 'as round as a crossbow bullet' suspended in space.[g] When Hsiung Ming-Yü fifteen hundred years later illustrated his treatise on astronomy and geography with a pleasant picture of Chinese junks circumnavigating an upside-down ocean[h] he significantly made use of exactly the same phraseology. Evidence is accumulating, moreover, that the belief in the sphericity of the earth was much more widespread in medieval Chinese culture than has often been thought.[i] And in fact the astronomers of China had actually been making terrestrial globes for centuries, but not on the same scale of size as their celestial ones, rather as quite small earth models held on a pin within their demonstrational

ᵃ Ed. & tr. Markham (2). On the general practice see Waters (15), pp. 145, 189 ff., 207 ff., etc.; Hewson (1), pp. 88 ff.; Stevenson (1), vol. 1, pp. 190 ff.

ᵇ II, v, 10. On Crates see Sarton (1), vol. 1, p. 185; Stevenson (1), vol. 1, pp. 7 ff.

ᶜ See Vol. 2, p. 236.

ᵈ Sarton (1), vol. 1, p. 635; Hitti (1), p. 385.

ᵉ Sarton (1), vol. 2, p. 617; Stevenson (1), vol. 1, p. 43.

ᶠ Ch. 48, p. 12a. On this and the other instruments and plans presented from Persia, see Vol. 3, p. 374, and among other references there given Hartner (3).

ᵍ E.g. Chang Hêng and Yü Sung about +120. See Vol. 3, p. 217, etc.

ʰ Reproduced in Vol. 3, Fig. 203, p. 499. Parallel Japanese example in Oberhummer (1), p. 108.

ⁱ See, e.g., Wang Yung (2), pp. 72 ff.; Wei Chü-Hsien (4).

armillary spheres.[a] What arrangement exactly the pioneer instruments of Chang Hêng (+125) and Lu Chi (+225) had is uncertain,[b] but it is quite clear that Ko Hêng in +260 placed a model earth inside his armillary,[c] and it is also certain that about the same time Wang Fan preferred to represent the earth horizon by sinking the sphere to the right extent within a flat-topped box. Ko Hêng's plan however was followed by many other instrument-makers[d] down to Kêng Hsün in +590, after which came the epoch of the escapement-controlled orreries which also depicted the earth in a variety of ways.[e] At least one of these, a Korean example from the eighteenth century, still probably exists,[f] and its small inner earth model is marked with all the earth-masses known to modern geography. Thus paradoxically although the terrestrial globe in its fully enlarged form was unfamiliar to Chinese culture, there had been earth models within armillary spheres from the +3rd century onwards, a practice which did not begin in Europe until the end of the +15th century.

It only remains to describe the two most important extant Chinese terrestrial globes, both of Renaissance type, and to be counted among the successors of the famous globe of Martin Behaim, now the senior of those still preserved,[g] since it dates from +1492. The first is a product of the Jesuit times (Fig. 1001, pl.), a painted lacquer globe constructed in +1623 under the guidance of Yang Ma-No (Emmanuel Diaz) and Lung Hua-Min (Nicholas Longobardi) whose names appear in the cartouche visible at the bottom. It is known as the David Globe.[h] The photograph shows China at the centre, with the Malayan peninsula, Sumatra, Java and Borneo below; Japan is rather distorted, New Guinea much too large, and there is of course no sign of Australia.[i] Two elaborate drawings of European ships with lateen sails on the mizen-masts occupy much of the Indian Ocean and the Pacific, besides abundant inscriptions. The David Globe was not the first terrestrial globe to be made in connection with the Jesuit Mission, for Lingozuon[j] had already constructed one in +1603 with such success that Matteo Ricci himself described it as 'very fine'.[k] But it appears not to have survived.

[a] This is discussed in considerable detail in Vol. 3, pp. 343, 345 ff., 350, 383 ff., 386, etc. The presence of a model earth within a demonstrational armillary sphere was often associated with arrangements for rotating it automatically, on which see Vol. 4, pt. 2, pp. 481 ff.

[b] On Lu Chi see also Needham, Wang & Price (1), pp. 23, 61.

[c] We do not have positive proof that these ancient earth models were not flat plates, but in view of the clear statements of the classical cosmologists this seems unlikely. Cf. Needham, Wang & Price (1), p. 96.

[d] E.g. Liu Chih (+274), Chhien Lo-Chih (+436), and Thao Hung-Ching (+520).

[e] I-Hsing and his collaborators (+721) adopted the box of Wang Fan, but models of the earth seem to have been present in the astronomical clocks of Chang Ssu-Hsün (+979) and (still more certainly) Wang Fu (+1124).

[f] Description in Vol. 3, p. 389 and Fig. 179; Vol. 4, pt. 2, p. 519; and more fully in a forthcoming monograph by Combridge, Lu, Maddison & Needham.

[g] See Stevenson (1), vol. 1, pp. 47 ff.; Ravenstein (1). In 1957 Dr Lu and I had the pleasure of seeing it in the National Museum at Nürnberg.

[h] For it was in the private collection of Sir Percival David, to whom our warmest thanks are due for permission to reproduce the accompanying illustration. It is now exhibited at the British Museum. Description by Wallis & Grinstead (1); Wallis (1).

[i] But it was good to show the Torres Strait, discovered in +1607 but ignored by nearly all European cartographers till as late as +1770.

[j] I.e. Li Wo-Tshun,[1] better known by his other name Li Chih-Tsao,[2] the very gifted scientific collaborator of the Jesuits.　　　　[k] Venturi (1), vol. 1, p. 396; d'Elia (2), vol. 2, p. 178.

[1] 李我存　　　[2] 李之藻

The Rosthorn Globe (Fig. 1002, pl.) is a very different object,[a] considerably smaller (just under 1 ft. in diameter) and made of silver sheet metal on which the map and the inscriptions were incised before being covered all over with translucent cloisonné enamel in bright blue, green, violet and other colours. It carries no statement of origin, and internal evidence can only date it as of some time probably between Tasman and Cook, i.e. between +1650 and +1770; but a considerable similarity in place-names shows that it shares a common source with the world-map of Chuang Thing-Fu,[1] one edition at least of which was printed as late as 1800.[b] The place-names are meant to be read with the south pole uppermost. Instead of political boundaries, spaces between the meridians[c] and parallels on a single land-mass are picked out rather fancifully in different colours, as for instance in the case of Australia (bottom right in Fig. 1002), which is wrongly joined both to New Guinea and to Tasmania (Fig. 1003, pl.). Here the legend reads: 'The new Western records say that this is New Holland (Hsin Wo-a-lang-ti-ya[2]), a great continent all desert with nobody living there.' The East Indies are also badly drawn, for Borneo is placed between Malaya's tip and Java, a feature which suggests the absence of any Jesuit influence on this globe, and makes it more probably a product of some northern Chinese cartographer unfamiliar with the Nan Hai. A large Antarctic continent (wrongly attached to New Zealand) is shown however. The drawing of California as an island might indicate a date around +1700, for Halley's magnetic chart is one of the last to represent it in this way.

A closer look at the map of China (Fig. 1004, pl., with south at the top) shows in the colouring a strange perpetuation of the old division between Cathay in the north (light) and Manzi in the south (dark). The windings of the Yellow River are marked well enough, those of the Yangtze less so, and the Gobi Desert prominently separates China from Mongolia. On the continent many names such as Shantung and Kansu can easily be made out, while in the China Seas Tsushima (Tui-ma tao[3]) and the Ryukyus (Liu-Chhiu Kuo[4]) with 'the eight mountain isles' (Pa-chhung shan[5]) are the most obvious. Lastly an interesting feature of the Rosthorn Globe is that it bears twelve 30° segments of longitude each stating the time difference between local and Chinese time in double-hours. Thus although some have been inclined to consider it an *objet d'art* rather than an instrument of precision, it cannot be denied a considerable measure of scientific interest.

[a] Bought in Peking by Prof. Rosthorn in 1902, it is now in the Österr. Museum f. angew. Kunst at Vienna (H.I. 28769/Go. 1827). The first description was given by Oberhummer (1). For the photographs here reproduced we are greatly indebted to Director V. Griessmaier, Mr J. V. Mills and Dr N. Mikoletzky. Furthermore Mr Mills most kindly placed at our disposal the materials for his forthcoming study of Chinese globes with Miss Wallis, which is eagerly awaited, and visited Cambridge to discuss them with us.

[b] Royal Geogr. Soc., World/251, title *Ta Chhing Thung-Shu Chih Kung Wan-Kuo Ching-Wei Ti-Chhiu Shih*,[6] hand coloured.

[c] The prime meridian runs through the Canary Islands, about 20° W. from Greenwich. On the history of the prime meridian see Hewson (1), pp. 9 ff.; Greenwich longitude, though now so familiar, was internationally adopted only very late in the last century.

[1] 莊廷尃   [2] 新窩阿即地亞   [3] 對馬島   [4] 琉球國   [5] 八重山
[6] 大清統屬職貢萬國經緯地球式

(g) PROPULSION

(1) Sails; the Position of China in the Development of the Fore-and-Aft Rig

(i) *Introduction*

Sails may be defined as pieces of textile fabric held outstretched upon ships in various ways so that the pressure and flow of the wind can be utilised to drive the vessel upon its course. Sails, and combinations of sails, which exist and have existed at diverse times and places, are of manifold shapes and rigs, almost bewildering in their complexity. If this had been the result of caprice or mere local custom, the subject could hardly interest anyone other than the nautical dilettanti; but in fact there runs throughout the series one single creative thread, the desire of man to release himself from natural servitude by sailing, not only with favourable following breezes, but directly into the eye of the wind. Though sails alone were never to permit him to accomplish this feat, there was a maximum efficiency point which could be reached in sailing to windward, and that he ultimately attained.[a] The history of this branch of nautical technology might thus be epigrammatically described as the advance from S.W. to N.W. by N. in the face of a due north wind—a difference of nine points of the compass, yet one which took three millennia to achieve.

For the understanding of what follows it is necessary that the reader should bear in mind the chief varieties of what have been called the 'primary' types of sail.[b] These are sketched in Fig. 1005. In these diagrams the masts and spars are shown by a heavy line, while the free edges of the sail are left light. First, there is the square-sail, oldest and simplest, symmetrically hoisted, necessitating always a yard, but at different times and places with or without a boom (*A, A'*). The square-sail is the only principal sail which always receives the wind upon the same surface. With a wind from behind and to the right of the ship (i.e. aft, from the starboard quarter), the right-hand side of the sail (the starboard side) would be braced forward while the left-hand side (the port side) would be aft of the mast. Then when the wind changed to the other, port, quarter, or when the ship changed course, the position of the sail would be reversed, in other words, the forward or weather yard-arm became the after or lee yard-arm as the sail was trimmed to face the wind. But broadly speaking the limitations of these manœuvres were soon reached, and sailors of all seas and cultures sought perennially for some escape from the essentially transverse character of the square-sail. Only by devising arrangements which would permit their canvas to be mounted more in line with the long axis of the vessel, i.e. fore-and-aft, could they hope to take advantage of beam or contrary winds.[c]

---

[a] Or very nearly. Prof. Bryan Thwaites, to whom we are much indebted for advice and drafting in the present sub-section, thinks that we may hope to improve even now upon the capacity of yachts to sail into the wind.

[b] For fuller information: Moore (1); Smyth (1); Anderson & Anderson (1); and many other books, including the nautical dictionaries such as that of Gruss (1); Ansted (1); Adm. Smyth (1).

[c] The square-sail runs through the whole of nautical history from ancient Egypt to the clippers. It is, as Casson (2) has said, 'without a peer for voyages, especially long ones, made with a following wind. It

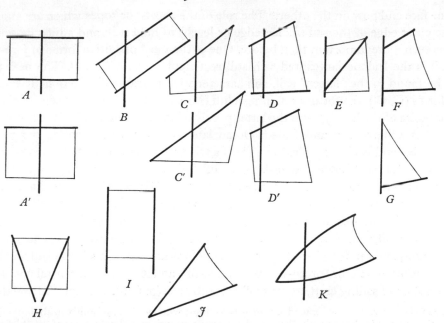

Fig. 1005. Principal sail types. Masts, yards, booms, sprits, gaffs, etc., in thick lines; sail edges in thin. No attempt is made to indicate relative sizes of sails.

A	square-sail with boom
A′	square-sail without boom (loose-footed)
B	Indonesian canted 'square' (rectangular) sail, with boom
C	lateen sail with short luff edge
C′	lateen sail without luff edge
D	lug-sail with considerable luff area forward of the mast
D′	lug-sail with reduced luff area forward of the mast
E	sprit-sail
F	gaff- or yacht-sail
G	leg-of-mutton sail
H	Indian Ocean bifid-mast sprit-sail
I	Melanesian double-mast sprit-sail
J	Oceanic (Polynesian) sprit-sail
K	Pacific boom-lateen sail

For detailed explanations see text.

All these other types of sail, which constituted successive approximations to the ideal fore-and-aft rig, differ fundamentally from the square-sail in being placed assymetrically with respect to the mast, so that the surface area differs on the two sides. Swivelling thus round the mast as an axis,[a] they receive the wind, as opportunity serves, now

offers every inch of its surface to the wind's thrust, the vessel rides comfortably and safely, and the canvas needs a minimum of handling.' Moreover, a square-rigged ship can hoist twice the sail area of a fore-and-aft rigged one, which is very valuable in light airs, and chafing does not occur since yards and sails strain away from the masts. These points have often been emphasised to us by Cdr. George Naish in conversation and in correspondence.

[a] In the less developed types this may necessitate lowering the sail and hoisting it again each time the vessel goes about, as in the case of dipping lugs. Sometimes there may be no mast at all in the strict sense, but the principle remains the same.

on one face and now on the other.[a] The role of the sheets, or ropes which are attached to the outer edge of the sail (its lee-edge or leech) to hold it in and adjust its set, becomes even more important than before. One of the most primitive forms of fore-and-aft sail is the Indonesian canted sail, still rectangular,[b] shown in B. The next stage may be found in the 'lateen' sail, so characteristic of Arabic civilisation; it exists in two forms (C, C′), the former retaining a luff (i.e. a short fore or inner edge of the sail, the opposite of the leech),[c] and the latter purely triangular, the head of the sail joining its foot. Mediterranean and Indian Ocean lateens never have a boom, but various South Asian and Pacific peoples use sails of a triangular shape bent to booms as well as yards (K). These 'Pacific boom-lateens' of Indonesia, Micronesia, Fiji, etc., are believed to derive from a kind of sprit-sail (J) termed 'Oceanic' because characteristic of Polynesia, in which the upper sprit performs the office of a more or less aft-raking mast.[d] This in turn originated, it would seem, from a still more ancient form of rig (H) in which the sail, though approximately square, is held aloft by two sprits equivalent to a bifid mast, and can thus be set approximately fore-and-aft. This is the Indian Ocean 'bifid-mast sprit-sail' or 'proto-Oceanic sprit-sail', and it occupies a central position in the evolution of sailing craft, as we shall presently see (p. 606). It is certainly related to

[a] The following remarks are intended primarily for those who are already familiar with the subject of this section. Bowen (2) has recently challenged accepted conceptions and definitions of the chief classes of sails. He wishes to adopt the narrow definition of Webster's Dictionary that a fore-and-aft sail is 'any sail not supported by a yard or yards, usually carried on a gaff, or stay, with or without a boom'. For him, therefore, those only are true fore-and-aft sails in which the luff edge is attached to the mast, and which can be tacked head to wind without shifting anything but the sheet. Bowen admits that lateen sails (and of course lug-sails) can be, and often are, set fore and aft to the ship, but he emphasises the fact that lateens rarely tack, performing what is for them the safer operation of wearing about. He further says that lateens and lugs are more like square-sails in that they can run before the wind with the sail across the boat without danger of gybing, and also that they will not sail to windward when the sail is backed against the mast. He is brought into difficulty, however, by having to admit that there are certain sails which may be tacked head to wind by shifting only the sheet, yet according to his definition are not fore-and-aft sails because they are set on yards; for example, many lugs.

We are not able to adopt his standpoint here, or to follow him in his criticisms of Anderson, Chatterton, and others. It is surely much better to adhere to the definition of fore-and-aft sails as those which, because of their longitudinal relation to the hull, allow of working to windward. We know this involves classifying the sail of a 'Humber keel', which is rigorously a square-sail, as a fore-and-aft sail when braced almost along the ship. But that does not affect the historical treatment, since the example is altogether exceptional, the adaptation of a survival which was made possible by wire stays and mechanical devices unknown in antiquity or the Middle Ages. Bowen's definitions are all insufficient for the Chinese material; Chinese lug-sails do accomplish tacking by shifting only the sheets, as Ibn Baṭṭūṭah said already in the +14th century, and they do draw well in windward sailing when backed against the mast, because of their stiffness. The Webster definition was derived purely from familiarity with the late occidental full-rigged ship. As Anderson said in his reply to Bowen, it is preferable to retain the older definition according to which fore-and-aft sails are those which receive the wind on either surface and always keep the same edge to windward. 'The Chinese lug', writes Hasler (1), 'is a pure fore-and-aft sail.'

In a later contribution, Bowen (9) maintained his terminology. But his two memoirs contain a mass of information on the evolutionary connections of sails and rigs which compels not only a large measure of assent to his conclusions but also admiration for his services to nautical scholarship.

[b] For Bowen (2), pp. 199 ff., (9), pp. 163, 192, 197, this is a balance lug.

[c] Clearly, the luff of a fore-and-aft sail is always the luff, but the luff of a square-sail changes each time the sail is re-set. When running before the wind it would not be appropriate to call either edge of a square-sail the luff. In the terminology of Bowen (2), pp. 186 ff., (9), pp. 184 ff., form C is a jib-headed dipping lug and C′ the only true lateen.

[d] One of the first illustrations of this rig in a Western book must surely be that in de Herrera's Novus Orbis of +1622.

the strange Melanesian 'double-mast sprit-sail' (*I*), where the sail is held up by both edges on spars which one may call indifferently masts or sprits. These South Asian and Pacific rigs were formerly neglected by those who wrote on the development of sail, but it is now clear that no general scheme can omit them.[a]

More advanced, and suitable for far larger vessels, is the lug-sail or ear-shaped sail ('voile aurique'), beloved of Chinese shipmen. This is a development of both the ancient square-sail and the canted sail, for it could retain the almost horizontal boom of the former but cants the yard like the latter, while forward of the mast its luff area may be quite small (*D*, *D'*). The true sprit-sail (*E*) is one in which the upper angle of the leech (the peak) is held out by a spar rising from near the foot of the mast; and finally the gaff-sail (*F*), or yacht-sail with which we are so familiar, is sustained by a half-sprit or gaff near the top of the mast, and a boom below like that of the lug-sail. Both of these types are fore-and-aft sails in the fullest sense, since they pivot upon the mast and leave no structures forward of it. So also is the 'leg-of-mutton' sail (*G*), a triangular piece of canvas of obscure origins, secured to mast and boom alone.

All else is elaboration. Jib-sails, set on a bowsprit or jib boom, are developments of stay-sails, which took advantage of the stays of masts.[b] Topsails are smaller sails hoisted above the regular sails.[c] The Romans had a triangular topsail above their square mainsail, and European ships from the +16th century onwards hoisted topsails in ever-increasing numbers and size, crowding on canvas but wisely reducing the height of each sail. So also the Hellenistic sailors of the −1st century introduced the bowsprit-sail in the form of the *artemon*, predecessor of their small forward-raking foremast. Lastly, as we have often remarked, the full-rigged ship of Renaissance Europe and later times combined massive square-sails on the forward masts with a lateen (and afterwards a gaff-sail) on the mizen; the former more efficient for running, and the latter for beating to windward.

It will be worth while to consider for a moment the effects of wind on a fore-and-aft sail.[d] One might easily think that only that part of the wind which strikes the windward, hollow, side of the sail has any effect; but in truth the wind which flows round the lee-ward, convex, side also contributes its share of the driving force. For what matters is that there should be a difference of pressure between corresponding points on the two surfaces of a sail, so that a total force results. This difference of pressure is a natural consequence of the way in which wind flows past a curved obstacle such as a sail;[e] but

[a] See particularly the studies of Brindley (1); Haddon & Hornell (1); and Bowen (2, 9).

[b] Jib-sails, though not unknown in China, are rare, being found only on the southern sea-going junks of Kotak and Pakhoi (G. R. G. Worcester, unpublished material, no. 160). Although fitted with battens (see on, p. 597), they are assuredly a late introduction from foreign contact. Stay-sails have also been seen on Hongkong junk-trawlers (B. Ward, private communication), but the same applies, for classical Chinese masts used neither stays nor shrouds.

[c] Cf. Fig. 939 on p. 405 above. Antung traders hoist topsails (Cdr. Waters and Mr Stunson), cf. pp. 404, 602.

[d] What follows here has been formulated in collaboration with Professor Bryan Thwaites, then of Southampton University. The book of Marchaj (1) may be consulted, and for experiments in towing tanks and wind tunnels Herreshoff (1); Herreshoff & Newman (1); Herreshoff & Kerwin (1).

[e] There is an analogy here with the partial vacuum on the upper surface of an aeroplane's wings, essential for lift and flight (cf. Vol. 4, pt. 2, pp. 580 ff., 590 ff.), but the quantitative relations are very different.

the curvature of the sail is less important than the fact that the wind is blowing at an angle to the boom. Indeed it is not yet generally agreed whether it is better in practice for a fore-and-aft sail to be very tightly bent to the boom so that the whole sail is as flat as possible, or for it to be allowed to set in a distinct curve.[a] The sail is essentially an aerofoil,[b] and many believe that it gives its best results when relatively taut; loose bellying sails lose energy by air turbulence, and perfectly flat ones would not give the differential flow effect.[c]

The action of the wind's force on a sail may be simply described (see Fig. 1006). The wind W strikes the sail—or (for simplicity) the boom AB—at a certain angle of incidence. The difference in pressure between the two sides of the sail to which we have already referred produces a lift force L perpendicular to AB and a drag force D along AB. It is the conscious purpose of aerodynamic design, and it has been the unconscious aim of ship-builders throughout the centuries, that L should be much larger than D. When a boat is making good to windward, L and D may be resolved into two other components, T and S, T being the force which drives the boat forward on its course into the eye of the wind, and S the force which tends to drive the boat sideways—to make leeway—a force which it is the function of the hull, keel and leeboards to counteract.[d] But if the boat is steered too close to the wind, then the drag D will be so

a Some theoretical studies of two-dimensional sails, that is, sails with luff and leech parallel and vertical to an infinite extent, indicate that an advantage lies with the sail which can belly to a certain extent even when close-hauled. This information comes from Professor Thwaites. Taut main-sails and balloon-bellying spinnakers are now of course habitually combined on many modern racing yachts. A striking picture of 'Sovereign' with both set appeared in *The Times* as this sub-section was undergoing revision (22 July 1963). Yet in practice it is a commonplace that the chief use of the one is in beating to windward and that of the other in running before the wind.

b It is now understood that one of the most important functions of a jib-sail is that it provides a space like the slot in the aeroplane wing (cf. p. 656 below) and guides the flow of air over the main-sail (cf. Curry (1), pp. 29, 60).

c Relative tautness has long been regarded as more efficient aerodynamically (Curry (1), pp. 38 ff.). It was not until the middle of the 19th century that the American Gloucester schooners started the move for sails as flat as possible (Chatterton (1), p. 207). But, as we shall see, Chinese mat-and-batten sails had been approximating to this aerodynamic virtue during the preceding fourteen or fifteen centuries. Dr Dinwiddie, with his usual perspicacity, sensed the value of tautness at the end of the +18th century; 'the (Chinese) sails on horizontal bamboos', he wrote, 'have one advantage over the European—they do not belly' (Proudfoot (1), p. 65).

Some of the most modern yacht sails based on scientific principles have adopted Chinese devices such as the battens, and sometimes the multiple sheets (cf. Curry (1), pp. 210, 213, 311; (2), pp. 85, 89; (3), pp. 70, 77, 83, 95, 100, 110; Budker (1); Wells Coates, etc.). Among the first to have battens were the racing 'canoe yawls' which Linton Hope was designing in the eighteen-nineties (private communication from Cdr. George Naish). An ingenious adaptation was to make the battens of inflatable tyre tubing so that their rigidity could be adjusted by the pressure according to weather conditions (private communication from the late Mr H. E. Tunnicliffe). But they are now generally short lengths holding out the leech rather than continuing across the full breadth of the sail. They can be seen in the picture of 'Sovereign' just mentioned.

The climax of the Chinese battened lug-sail in modern yacht practice was perhaps attained when in the summer of 1960 Lt-Col. H. G. Hasler finished second after a magnificent voyage in the single-handed transatlantic race which he had himself originated. His 'Jester' was equipped with a single unstayed mast and carried a single tall five-battened terylene lug of pure Chinese design, modified in certain ways but retaining the classical multiple sheets, topping lifts, etc. (see Hasler, 1).

d The total aerodynamic force on the sail is in a direction which makes an angle only very slightly more by 1 or 2 degrees, certainly not more than 5°, than a right angle with the wind direction. This force then has a large sideways component which the hull, keel, etc., balance out, and a small forward component which drives the boat.

large and the lift L so small that the force T will exert itself backwards (as T') rather than forwards, and the boat will make no headway to windward. One can now see why there was a historical tendency towards ever loftier sails, for the ratio L/D steadily increases as the height of the sail rises in relation to the length of the boom.[a] One can also see why the loose-footed square-sail, even when braced very much forward, could not easily work to windward.[b]

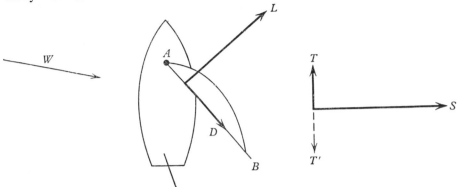

Fig. 1006. Diagram to explain the action of the wind on a vessel under sail. Explanation in text; the case considered is that of an almost beam wind and a fore-and-aft sail.

From the earliest days of sailing it was found that since one could not proceed directly into the wind, one must approach it by a series of passages as near the wind as possible, and as everyone knows, this zigzag movement is termed in general tacking. But the movement of turn has varied with the type of sail. Square-rigged ships could not often tack; they then had to 'wear about' by turning the stern to the wind (see Fig. 1007), and the same is true for lateen-rigged ships, though in these the manœuvre of tacking is occasionally done.[c] With all the more developed types of fore-and-aft sail, however, the helm has simply to be put over, bringing the bows up into the wind, and the sail 'luffs', hanging loose, until it catches the wind on the other tack. How near these courses could be to the wind is shown on the accompanying diagram (Fig. 1008). Ancient (e.g. Egyptian) square rig could not even make full use of a beam wind,[d] and post-Renaissance square-sailed ships could come no nearer to the wind than six or seven points ($79°$), while the fore-and-aft rigs (including the lateen) can sail as near as four points, and a well-designed modern yacht[e] may make good four points ($45°$). The

---

[a] Cf. Curry (1), pp. 34 ff. Here again, Chinese empiricism found long ago that lofty sails were advantageous, and they are still characteristic of many types of Chinese craft (cf. Fig. 987, pl.).

[b] Since it has no rigid luff it cannot withstand the high aerodynamic loading near that edge. There have been square-sailed vessels, it is true, renowned for their capacity of sailing into the wind, notably the 'Humber keel' (Moore (1); White (1), p. 18). Others, such as the 'Deal pilot-galley' (White (1), p. 29), have had lug-sails so symmetrical to the mast as to be almost square-sails. But these are boats which were in their prime during the past two centuries, when comparatively modern techniques could be used for bracing their sails fore and aft. Thus securely held, the leading edge, or luff, could correspond to the mast of a fully fore-and-aft sail. This was never possible for ancient and medieval sailors, even if they could have rigged their stays and shrouds to allow of it. Finally the form of the hull cannot be left out of account. Even if ancient square-sails had been braceable in this way, ancient hulls were unfitted for opposing leeward drift.

[c] See the excellent description by Bowen (1).

[d] According to the experimental work of Bowen (10).   [e] Such as a Bermudan cutter.

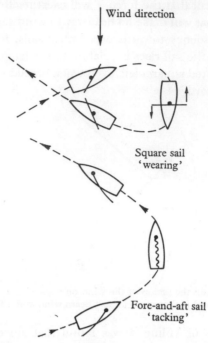

Fig. 1007. Diagram to show the principles of tacking and wearing. Fore-and-aft rigs usually tack; a ship with a square-sail more often works to windward by wearing.

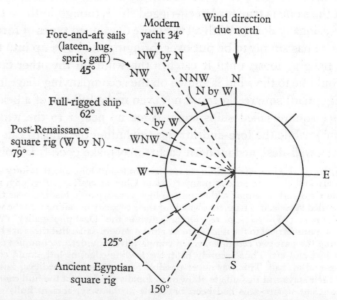

Fig. 1008. The capacity of different craft for sailing near the wind; angles projected upon a rose. The aerodynamic limit, analogous to stalling in aircraft (cf. Vol. 4, pt. 2, p. 592), would be of the order of 30°. Leeway is the difference between 'sailing' and 'making good a course' (see p. 618).

full-rigged ships, using their lateen mizens, could make good six.[a] With this information in our minds we are in a position to study the historical evidence and the Chinese
contribution.

(ii) *The mat-and-batten sail; its aerodynamic qualities*

The most characteristic Chinese sail is the balanced stiffened lug-sail. Fig. 1009*a*, from
Worcester,[b] gives a clear diagram of it. It shows a more or less northern type, for in the

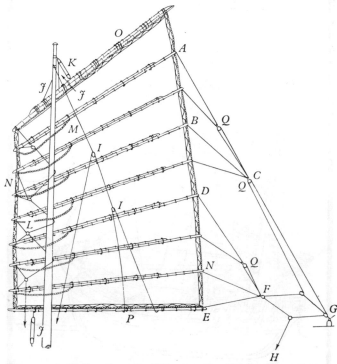

Fig. 1009*a*. Delineation of a Chinese mainsail (after Worcester, 3).

ABDE	Some of the points of attachment of the multiple sheets to the battens
ABC	Upper flexible section of sheets
DEF	Lower flexible section of sheets
CG, FG	Main sheets
G	Ring-bolt on deck
H	Feeding part of main sheets
I, I	Topping lifts and their euphroes
J, J	Mainsail halyards
K	Secondary mainsail halyards (for setting the peak of the yard)
L, L	Hauling parrels
M, M	Parrels
N, N	Bolt-ropes along the free edges of the sail
O	Yard
P	Boom
Q	Euphroes of the sheets

[a] These are generally accepted estimates.

[b] (3), vol. 1, pp. 65, 81 ff., vol. 2, pp. 256, 501; also (1), p. 11 ff.; Audemard (3), pp. 36 ff.; Poujade
(1), pp. 159 ff. Smyth, (1), p. 461, gives an excellent description from the point of view of a practical
sailor. The drawings of Adm. F. E. Paris (1), pls. 49 to 68, should also be studied.

船　條

Fig. 1009b. A standard transport junk (Thiao chhuan) in Lin-Chhing's *Ho Kung Chhi Chü Thu Shuo* (ch. 4, p. 3a); the best traditional Chinese drawing which shows the system of battens, the multiple sheets with their euphroes, the halyards and the topping lifts. As usual, the stern (to the right) is higher than the bow (where the capstan and grapnel may be noted), and though a long tiller is shown the object astern is probably a sampan rather than the top of a long rudder. A transport of this kind could do coastal as well as river work.

south the leeches of the sails are rounded.[a] As we know from the historical material already discussed, the Chinese sail, in its most typical form, is stiffened by transverse

[a] See especially Lovegrove (1); Waters (5), and below, p. 598.

battens or laths (*thiao*[1]) of bamboo, the ends of which are secured to bolt-ropes suspended from the yard so as to take the weight of what might be called the 'sail-frame', a kind of skeletal ladder. The fabric of the sail is laced to the perimeter of this frame and to each batten (Figs. 1010, 1011, pls.), so that it is kept very flat. The widespread use of bamboo matting sails (*li*[2])[a] necessitated this form of frame and led naturally to the balance lug shape. The aerodynamic importance of tautness is, as we have seen, considerable, yet such a design, which doubtless arose because of the easy availability of a material so light and at the same time so strong, never arose in any other culture.[b] We shall come back to this point.[c] The battens have at least five other uses: they permit of precise and stepwise reefing, they allow immediate furling of sail, which falls into pleats; their setting system obviates the need for cloth or canvas as strong as on other sails, and they act as ratlines giving access for the crew to any desired part. Above all, they are a complete protection against tearing and carrying away; a Chinese sail may have half its surface full of holes, and still draw well.

The multiple sheets (*liao ssu*[3]) represent a device of great interest. Each batten is connected with all the others by a system of bights and leads, which, gathered up by means of pulley-blocks (*kuan-lieh*[4]) and euphroes,[d] terminate finally in one main sheet on deck, which originates from a fixed point, *G* in Fig. 1009*a*. Thus the sail is divided into sections (e.g. *ABC, DEF*). Naturally the variations in the sheeting are legion, but Worcester & Sigaut have given good accounts of many of them.[e] The greater the number of battens, the greater the number of multiple sheets, the flatter the sail, and the finer the adjustment of its leech (Figs. 1011, 1014, pls.). The halyards (*li lan*[5]), *J, K*, pass through blocks at masthead (*wei thou*[6]) and yard (*fan kang*[7]). The sail is held to the mast by a parrel, *M*, for each batten, and there is a hauling parrel, *L*, arranged as shown, which assists in this and helps to peak the sail when reefed. In a squall only the halyard need be touched, for the sail falls neatly into the lazy lines formed by the double topping-lifts (*fan kang shêng*[8]), the lowest batten-contained sections collapse, the sheets automatically slacken, and can easily be re-set.[f] The sail never jams. As Audemard thought fit to emphasise, this system 'avoided the sending of men aloft to take in reefs, always a dangerous operation in bad weather'.[g]

'I can speak', wrote Smyth,[h] 'from some experience in handling this form of sail, and

[a] Cf. Figs. 939, 943, 975 (pl.), 977, 1009*b*.
[b] As a corollary of the sail's tautness, the Chinese never needed, or used, bowlines.
[c] P. 599 below.
[d] A euphroe is a pulley-block which is so extended in length as to include a number of holes through which separate bights of rope may pass, not themselves on pulleys. On the interesting term *kuan-lieh* and cognate expressions in mechanical engineering with varying orthography see Vol. 4, pt. 2, p. 485, as also Needham, Wang & Price (1), pp. 103 ff. I suspect that a better reading here would be *kuan-li*.[9]
[e] In Worcester (3), vol. 1, pp. 81 ff. Cf. Fig. 986 and Figs. 987, 1010 (pls.).
[f] The *Kao-Li Thu Ching* (+1124) talks of 'furling the sail into serpentine (folds), more, or less, according to the strength of the wind' (ch. 34, p. 7*a*). Cf. Fig. 937 (pl.).
[g] (4), p. 33. It is true that one of Sung Ying-Hsing's 'footnotes' (*TKKW*, ch. 9, p. 2*b*) says that in high winds a sailor was stationed in a kind of crowsnest to watch over the tackle, but we have found nothing to confirm this, either textual, iconographic or experiential.
[h] (1), p. 465, 1st ed. p. 406.

| [1] 條 | [2] 䉓 | [3] 繚絲 | [4] 關捩 | [5] 䉓纜 | [6] 桅頭 |
| [7] 帆杠 | [8] 帆杠繩 | [9] 關棙 | | | |

I may say that once having learned the set and balance of the sails for various points of sailing, nothing can surpass the handiness of the rig.' Another expert, Capt. Fitzgerald, called Chinese ships 'the handiest vessels in the world'. Elsewhere Smyth added, in an often quoted statement:[a] 'As an engine for carrying man and his commerce upon the high and stormy seas as well as on vast inland waterways, it is doubtful if any class of vessel is more suited or better adapted to its purpose (than the Chinese junk), and it is certain that for flatness of sail and for handiness, the Chinese rig is unsurpassed.' And Admiral Paris spoke of the rig as 'one of the most ingenious of Chinese inventions'. The only criticism which he and others have made of it is that it is liable to be rather heavy. Much detail can be made out in the snaps of a Pao-chhing Chhiu-tzu[1] reproduced in Fig. 1012a, b (pls.).[b]

As has just been mentioned, the sail-peaks and leeches of all Chinese vessels hailing from south of the Yangtze are very rounded, so that the gallant quadrilateral shape (D, D' in Fig. 1005) tends to turn into the softer contour of the quarter disc, giving a quadrantal form. So far does this tendency go that in the fishers and freighters of Yeungkong (Yangchiang), a port halfway between the Pearl River and the Leichow peninsula, the battens are gathered at the base of the luff edge so that the sails give the elegant impression of a fan (Fig. 1013, pl.). Some noteworthy details of a Yangchiang foresail appear in Fig. 1014 (pl.).

The Chinese balance lug ranks indeed among the foremost achievements in man's use of wind power. Of the circumstances which gave rise to it we know very little, but the suggestion has been made that it arose from plaiting together successive palm branches so that the central stem or mid-rib provided a natural built-in batten.[c] As we shall see in a moment, our oldest Chinese text concerning fore-and-aft sailing mentions just such plaiting. The lug-sail is not the only type used by Chinese ships, however. On the upper Yangtze the square-sail has persisted; its stiffening (accomplished by light ropes sewn in) is vertical, not horizontal, and a roller-reefing technique is used.[d] Square-sails of cloth, tall and narrow, are used on the Chhien-thang River (Fig. 971, pl.). Equally interesting is the fact that the true sprit-sail exists also in China (Fig. 1015, pl.).[e] It differs from its European analogues in being stiffened horizontally, though without battens, and since the usual system of multiple sheets applies, vangs are unnecessary.[f] Nor do the Chinese have brails, for the sail is set or taken in by hoisting or lowering the halyards or tending the sprit purchase.[g] The existence of the true sprit-sail in China is certainly of great interest, though evidence is still lacking whereby

---

[a] (1), p. 455, 1st ed. p. 397.

[b] Cf. Worcester's photographs (3), vol. 1, pl. 14, vol. 2, pl. 199.

[c] Bowen (9), pp. 194 ff.

[d] Worcester (1), p. 12. This suggests a genetic connection with the Indonesian canted square-sail, see on, pp. 608, 612.

[e] Worcester (3), vol. 1, pp. 74, 84, 162; Fitch (1).

[f] Vangs are guys or ropes leading from the end of a sprit or gaff to the gunwales; their function is to hold the peak of the sail in the desired position.

[g] Brails are ropes leading from the leech or the foot of a sail through blocks on the mast or yard, so that it can be trussed or gathered for furling.

[1] 寶慶邸子

its historical significance may be assessed. It will be understood that this sail is one of those which embody the fullest development of fore-and-aft sailing, since nothing is forward of the mast.

Outside the Chinese culture-area the mat-and-batten sail as such never widely spread.[a] It was of course used from fairly early times in Japan,[b] where copies of +12th-century pictures show it particularly clearly (Fig. 1033a below). It spread, too, to the Maldive Islands, where it is habitually used.[c] The Portuguese in the +16th century appreciated it,[d] and rigged their *lorchas* with it, as we have seen, but there was some obstacle to its diffusion to Europe and other regions, probably the lack of bamboo or other suitable material for the battens. In 1829, however, at least one British steamship working out of Calcutta, the 'Forbes', was rigged with large Chinese batten lug-sails.[e] An entirely different question concerns the invention and diffusion of the lug-sail considered as a cut, i.e. a shape, and this we shall try to answer shortly.[f]

(iii) *Chinese sails in history*

What information can we now assemble concerning the history of Chinese sails? Among the ancient characters on the oracle-bones, *fan*,[1] later meaning 'all, every', and used as an initial particle, 'generally speaking', signified, in its original form (K625), a sail. It is interesting that this graph seems clearly to depict the 'double-mast sprit-sail' now known only in Melanesia (cf. Fig. 1005, *I*, and Fig. 1016).[g] Probably this was one of the contributions of the

K625

south-eastern or oceanic component of ancient Chinese culture,[h] during the second half of the −2nd millennium. It seems a less likely ancestor of Chinese lug-sails than the canted Indonesian square-sail, but the Chinese true sprit-sail could be its direct descendant. The fact is that we have no clear evidence on the origin of the Chinese balance lug, but one cannot help recognising that it was the ideal answer to the problem of making headway against the monsoon winds which blow so regularly up and down the coasts of China, using a material so abundant and cheap as bamboo matting for the sail's surface (cf. Figs. 989b and 1009a, b).

Fig. 1016. Sketch of a Melanesian double-mast sprit-sail made at Port Moresby, Papua (after Bowen (2), drawing from Haddon & Hornell (1), fig. 132).

---

[a] As such, that is to say, for we have just mentioned (p. 592) the adoption of battens and half-battens in the sails of modern racing yachts.

[b] Purvis (1); Elgar (1). The Japanese remained curiously fond of tall square-sails, however, cf. Noteboom (1).

[c] Hornell (17), p. 181; Bowen (2), p. 195.

[d] Cf. the remark of de Rada (+1575) in Boxer (1), p. 294.

[e] Worcester (3), vol. 1, p. 69.

[f] Pp. 612ff. below.

[g] Cf. Bowen (2), pp. 86ff., 101, 110, (9), p. 167.

[h] Cf. Vol. 1, pp. 89 ff.

[1] 凡

The earliest Han mentions seem to be generally to matting, as in the expression *kua hsi chhien li*,[1] 'blown along by mat(-sails) a thousand *li*', while the term *fan*,[2] later usual, does not often occur with *pu* as *pu fan*[3] (cloth sails) until the middle of the Later Han.[a] Archaic ways of writing the word brought out the connection with the wind, e.g. *fan*[4] in the +2nd-century *Shuo Wên*, and *fan*[5] in the +6th-century *Yü Phien* dictionary.[b] The *Shih Ming*[6] dictionary, compiled by Liu Hsi[7] about +100, says that a sail is 'like a curtain, held up to the wind, so that the boat goes lightly and swiftly'.[c] It is at any rate clear that in the last decade of the +4th century cloth sails were reserved for the boats of officials; this we know from a story in the life of the painter Ku Khai-Chih,[8] who was aide-de-camp to a governor at the time.[d] 'Cloth sails' became a stock phrase in later poets to indicate an atmosphere of luxury, pomp or buoyancy. Presumably mat-sails were regarded as too rough for the official junks—irrespective of the performance of the two types.[e] Nothing here throws much light on the shape of sail or kind of tackle used.

A +3rd-century text of capital importance does so, however. It occurs in the *Nan Chou I Wu Chih*[9] (Strange Things of the South), written by Wan Chen,[10] and runs as follows:[f]

> The people of foreign parts (*wai yü jen*[11]) call *chhuan*[12] (ships) *po*.[13] The large ones are more than 20 *chang* in length (up to 150 ft.), and stand out of the water 2 or 3 *chang* (about 15 to 23 ft.). At a distance they look like 'flying galleries' (*ko tao*[14])[g] and they carry from 600 to 700 persons, with 10,000 bushels (*hu*[15]) of cargo.[h]
>
> The people beyond the barriers (*wai chiao jen*[16]),[i] according to the sizes of their ships, sometimes rig (as many as) four sails, which they carry in a row from bow to stern. From the leaves

---

[a] Such at least is the impression gained by Dr Chhen Shih-Hsiang; further research on the point would be profitable. Cf. *CSHK* (Hou Han Sect.), ch. 18, p. 12*a*; *TPYL*, ch. 771, p. 6*a*.

[b] Ch. 18, p. 56*a*, ch. 20, p. 66*b*.

[c] Ch. 25, *Shih Ming Su Chêng Pu* (p. 379).

[d] *Chin Shu*, ch. 92, p. 21*a*; tr. Chhen Shih-Hsiang (2), pp. 13, 25. Cf. *Shih Shuo Hsin Yü*, ch. 3B, p. 13*b*.

[e] In later centuries there is mention of sails of oiled silk. According to Hsü Mêng-Hsin, the ships of the Chin Tartars cruising off Shantung in +1161 were equipped with these, but they were easily set on fire by the rockets and incendiary arrows of the Sung vessels (*San Chhao Pei Mêng Hui Pien*, ch. 237, pp. 1*a* ff.; noted by Lo Jung-Pang, 5).

[f] Cit. *TPYL*, ch. 769, p. 6*a*, ch. 771, p. 5*b*. Tr. Pelliot (29), eng. et mod. auct. adjuv. Wang Kung-Wu (1), p. 38. It is remarkable that Pelliot himself seems not to have appreciated the technological importance of the passage which he had found. Others, however, did so later, e.g. P. Paris (1).

[g] This is a reference to the bridge-corridors or open galleries which appear to have existed in palaces such as that of Chhin Shih Huang Ti.

[h] Assuming the bushel here is equivalent to the picul, this would mean about 260 tons.

[i] This phrase is important. If Wan Chen had wished to refer to foreigners from far away, he would have used the phrase with which the first paragraph begins. But the region which is now Kuangtung was rather loosely attached to the State of Wu in the San Kuo period, and the regions of Annam and Tongking still less so. The probability is, therefore, that the reference is to sailors of these coasts rather than those of Indonesia. This view is strongly supported by the phraseology used in the account of Lü Tai's[17] conquests in the Indo-Chinese region (c. +230) in *San Kuo Chih* (*Wu Chih*), ch. 15, p. 9*a*.

---

[1] 颸蓆千里	[2] 帆	[3] 布帆	[4] 颿	[5] 𩗗
[6] 釋名	[7] 劉熙	[8] 顧愷之	[9] 南州異物志	[10] 萬震
[11] 外域人	[12] 船	[13] 舶	[14] 閣道	[15] 斛
[16] 外徼人	[17] 呂岱			

of the *lu-thou*[1] tree,[a] which have the shape of '*yung*'[2],[b] and are more than 1 *chang* (about
7½ ft.) long, they weave the sails.

'The four sails do not face directly forwards, but are set obliquely, and so arranged that they
can all be fixed in the same direction,[c] to receive the wind and to spill it[d] (*Chhi ssu fan pu
chêng chhien hsiang, chieh shih hsieh i hsiang chü, i chhü fêng chhui fêng*[3]). Those (sails which are)
behind (the most windward one)[e] (receiving the) pressure (of the wind), throw it from one to
the other, so that they all profit from its force (*Hou chê chi erh hsiang shê, i ping tê fêng li*[4]). If
it is violent, they (the sailors) diminish or augment (the surface of the sails) according to the
conditions. This oblique (rig), which permits (the sails) to receive from one another the
breath of the wind, obviates the anxiety attendant upon having high masts. Thus (these ships)
sail without avoiding strong winds and dashing waves, by the aid of which they can make
great speed.

This is indeed a striking passage. It establishes without any doubt that in the +3rd
century southerners, whether Cantonese or Annamese, were using four-masted ships
with matting sails in a fore-and-aft rig of some kind. The Indonesian canted square-sail
is not absolutely excluded, but it would be unwieldy on a vessel with several masts,[f]
and some kind of tall balanced lug-sail seems much more probable.[g] The writer was

Fig. 1017. Sketch of a Lake Thai-Hu trawler (after Audemard, 3). This remarkable craft,
though only some 25 ft. long, carries five small masts with lug-sails.

[a] Not easily to be identified now, but resembling reeds or rushes used for basket-work.

[b] A word not even in the Khang-Hsi dictionary, and of uncertain significance, but possibly a kind of
vase. *Yu*,[5] lattice of windows, is more likely.

[c] I.e. approximately parallel to each other. Lit. 'moved so that they are mutually adjusted'.

[d] The writer probably had in mind a ship sailing with a wind from the aft quarter, and observed from
that direction, so that most of the sail area was visible; and he imagined the air stream being diverted
from the aftermost sail to the other sails successively, doing work upon each.

[e] I.e. further forward in the ship. It will be seen that we prefer a punctuation different from that accepted by Pelliot (29), Fêng Chhêng-Chün (1), p. 19, and others, at this point.

[f] Moreover, no such examples are known.

[g] Mr R. le B. Bowen concurs (private communication). A miniature parallel, still existing, is of much
interest (Fig. 1017), the Lake Thai-Hu trawler (cf. Audemard (3), pp. 45, 47). Though not exceeding
25 ft. in length, it steps five masts with lug-sails, all controlled by one reefing-line.

[1] 盧頭   [2] 矚   [3] 其四帆不正前向皆使邪移相棨以取風吹風
[4] 後者激而相射亦並得風力   [5] 驕

perhaps a little confused about the purpose of the rig, but his emphasis on the oblique-
ness of the sails necessitates a cut whereby they would not have got in each other's way.

Additional contemporary testimony for ships with multiple masts may be derived
from the lost book *Wu Shih Wai Kuo Chuan*[1] (Record of Foreign Countries in the
time of the State of Wu) by Khang Thai[2] (*fl.* +260), passages from which are preserved in encyclopaedias. Thus we hear that in the seas off Chia-Na-Tiao-Chou,[3]
some country which at present cannot be identified,[a] there were great junks (*ta po*[4])
with no less than seven sails.[b] These were used by people travelling to and from Syria
(Ta-Chhin).

Evidence concerning sails from subsequent centuries has been noted already in
passing. There is certainly no reason why the Ajanta artist (+628) should not have
been absent-mindedly drawing a Chinese lug-sail, lofty and taut.[c] At Borobodur
(+800)[d] and at Angkor Thom (+1185) the absence of a luff forward of the mast might
suggest sprit-sails,[e] yet lugs were almost certainly intended. The multiple masts of
Wan Chen and Khang Thai occur again, of course, in Marco Polo and Ibn Baṭṭūṭah, as
also in the derivative world-maps. The statement of Chu Yü for +1090 that Chinese
ships could use wind from almost any quarter (*shih san mien fêng*[5])[f] is highly significant.

The *Kao-Li Thu Ching* of approximately +1124 is quite informative on sails and
sailing. Speaking of the 'retainer ships' (*kho chou*[6]) which accompanied the larger ships
of the diplomatic envoy and carried his staff, Hsü Ching says:[g]

When the ship makes land and enters a harbour (*khai shan ju kang*[7]) she usually does so on
the flood tide, and then all the sailors row, singing to keep the time. Those using poles also
jump and shout and exert themselves to the utmost, but the ship cannot move nearly so fast as
when sailing with a good wind. The main-mast is 100 ft. high, and the foremast 80 ft. When
the wind blows favourably, they hoist the cloth sails (for running fast) (*pu fan*[8]) made of fifty
strips (*fu*[9]).[h] But when the wind blows from the side they use the advantageous mat sails (*li
phêng*[10]), set to the left or to the right like wings according to the direction of the wind. At the
top of the main-mast they may add a small topsail (*hsiao fan*[11]), made of ten strips of cloth.
This is called the 'wild fox sail' (*yeh hu fan*[12]), and is used in light airs, when there is almost no
wind. Of all the eight quarters whence the wind may blow, there is only one, the dead ahead

---

[a] The first chapter of the +6th-century *Shui Ching Chu* refers to it as an island. In his translation of
the passages concerned, Petech (1), pp. 15, 53, interprets the name as Gaṇadvīpa, and places it somewhere in Malaya or Indonesia.

[b] *TPYL*, ch. 771, p. 5b. Also quoted in the commentary of *Shih Ming Su Chêng Pu*, ch. 25 (p. 380).

[c] As Bowen always urges, cf. (2), p. 194, (9), p. 192. Alternatively, a square-sail three times as high as
its width, like those still used on the Chhien-thang River (cf. Fig. 971, pl.). But would not this have been
unsuitable for a three-master like the Ajanta ship?

[d] The second type, cf. p. 458 above.

[e] As Poujade (1), p. 256, proposed. But Bowen (9), p. 193, prefers our view.

[f] Cf. p. 462. This phrase, we believe, distinctly implies sailing to some four points near the wind
(cf. Fig. 1008). And Hsü Ching's words explicitly confirm it.

[g] Ch. 34, p. 5a; tr. auct. adjuv. Paik (1).

[h] Taking *fu* as classifier, Lo Jung-Pang (5) interprets 'fifty sails', which seems too many.

[1] 吳時外國傳	[2] 康泰	[3] 加那調州	[4] 大舶	[5] 使三面風
[6] 客舟	[7] 開山入港	[8] 布颿	[9] 幅	[10] 利篷
[11] 小颿	[12] 野狐颿			

quarter, which cannot be used to make the ship sail.[a] The sailors also attach some bird feathers to the top of an upright pole as a weathercock; this is called the 'Five Ounces' (*wu liang*[1]).[b] To get a favourable wind is not easy, so that the great cloth sails are not as useful as the mat sails, which, when skilfully employed, will carry men wheresoever they may wish to go.

This important passage clearly shows that the aerodynamic properties of taut mat-and-batten sails were empirically appreciated by a scholar early in the +12th century, and that they were used for beating to windward, while sails of cloth, or of silk,[c] were hoisted additionally when running before the wind. This combination we have seen already in Fig. 939 and its description (p. 405). The use of topsails (also in Fig. 939) is interesting at this time, but Hsü Ching further describes what may have been something like detachable bonnets on the sails of the Korean 'official boats' which came out to welcome the Chinese ambassador.[d]

Our series of texts may be completed by a passage from the +16th-century historian of the Portuguese conquest of the Indian Seas, de Castanheda.[e] He spoke of the Chinese junks which had come to Malacca during the previous hundred years and earlier, bringing gold, silver, rhubarb, all kinds of silks, satins and damasks, porcelains, gilded boxes and fine furniture. They took back with them pepper, Indian cottons, saffron, coral, cinnabar and mercury, drugs and cereals.[f]

These junks, as the ships of this region are called, are very great and very different from those of all the other countries in the world, for the bow and the stern have the same shape, with one rudder forward and a rudder at the stern.[g] They have but one mast,[h] and a sail of 'Bengal matting' (made from small reeds) which turns around the mast as if pivoted on a spindle (*dobadeira*).[i] For this reason the junks never wear around as our ships do. When it is desired to reef (the sails), it is not necessary to fold them, for they fall all in a single piece. Hence the junks sail well; they take much greater burden than our ships; they are much stouter, and they have such great internal beams that a camel could hardly carry one...

Here there can be no doubt that lug-sails were meant, as in the later description of Sung Ying-Hsing, which runs as follows:[j]

[a] *Jan fêng yu pa mien, wei tang thou pu kho hsing.*[2] Cf. p. 462 and the diagram on p. 594 above.
[b] Cf. Vol. 3, p. 478 above.
[c] This refers to the sails (embroidered with designs) of the ambassador's own vessel, the 'sacred ship' (*shen chou*[3]), ch. 34, p. 4a. The actual term is *chin fan*,[4] which perhaps could also mean embroidered cloth, but less probably.
[d] Ch. 33, p. 2a. The passage is curious; it is said that their sails were of more than 20 strips of cloth but that the pieces in the lower fifth of the sail were not sewn together. Perhaps he was trying to describe some arrangement like that still seen in China (p. 613 below) where a small canted square-sail is hoisted like a spinnaker in addition to the main lug-sail.
[e] (1), ch. 112; cf. Ferrand (4); Paris (1).
[f] This Chinese commerce had been notified to the Portuguese King Dom Manuel by Albuquerque in +1512 in an interesting covering letter for a Javanese map which showed all the regular routes and ports of the trading junks (Baião ed. pp. 76 ff.).
[g] On this, see p. 619 below.
[h] De Castanheda or his informants may have seen the junks only in port, when their fore and mizen masts were unstepped. For as we know, multiple masts have always been typical of sea-going junks.
[i] Cf. p. 462 above, and p. 611 below.
[j] *TKKW*, ch. 9, p. 3a, b, tr. auct. adjuv. Ting & Donnelly (1); Sun & Sun (1).

¹ 五兩   ² 然風有八面唯當頭不可行   ³ 神舟   ⁴ 錦帆

The size of the sail (*fêng phêng*[1]) depends upon the beam of the vessel.[a] If the sail is too large the ship will be endangered; if it is too small it will be ineffective. The sail is made by weaving together thin and narrow strips of the outer parts of the stems of bamboo, and (this matting is) divided into sections grasped by (parallel) bamboo battens (*thiao*[2]). Thus the sail folds in tiers (*tieh*[3]), ready to be (bent to yard and boom and) hoisted. A large main-sail (*chung wei phêng*[4]) in a grain-ship needs ten men to hoist it, but for the fore-sail (*thou phêng*[5]) two suffice. In order to get ready to sail, the halyards (*phêng so*[6])[b] (have to be passed through) a pulley-block (*kuan-lieh*[7])[c] with a one-inch gauge pulley fixed to the top of the mast, and brought down to the halyard winches (on the deck) amidships (*yao chien yuan mu*[8]). The adjustment of the height of the sail (according to the wind force, etc.) is thus like the varia-tion of the three sides of a triangle (*san ku chiao tsho erh tu chih*[9]). For equal sections (*yeh*[10])[d] the upper part of the sail is thrice as effective as the lower.[e] What matters is the adjustment to the conditions. When the wind is favourable the sail is hoisted to its full height and the boat moves at a good speed like a racing horse, but if the wind freshens the sail is reefed (coming down by its own weight) in due order (section by section one after another). [In a squall, the sail may have to be pulled down by long hooks.][f] In a gale only one or two sections of the sail are hoisted.

Next follows the passage on tacking which Ting & Donnelly omitted. Sung Ying-Hsing explains it clearly enough, having evidently watched the process, but he seems to have confused it a little with the handling of the ship in a current.[g]

A beam wind is called a 'tacking wind' (*chhiang fêng*[11]).[h] When a boat is sailing with the current, the sail is hoisted and the vessel wanders about (i.e. takes a zigzag course). Some-times a difference of an inch in setting the sail when tacking to the east will make all the dif-ference between a safe passage and a setback of several hundred feet. Before she reaches the bank, the rudder is turned (*lieh tho*[12])[i] (hard over), and the sail is reset (*chuan phêng*[13]). The

---

a Cf. the modules and modulor proportions which we saw laid down in the books on architecture (pp. 67 ff., 82, 89 above).

b See fn. n on p. 414 above.

c This technical term is now an old friend; cf. the orthographic note on p. 597 above. For a discussion of its many related usages and their origins the reader is referred to Vol. 4, pt. 2, p. 485, as also Needham, Wang & Price (1), pp. 103 ff. On Chinese ships' pulleys and pulleyblocks see Audemard (3), pp. 50 ff.; on the employment of these devices ashore, rarely illustrated, see Vol. 4, pt. 2, p. 96, fn. (h).

d I.e. parts of the sail between two battens.

e A significant remark in view of the physical theory of sails, cf. p. 593 above.

f Sung Ying-Hsing's commentary.

g One remembers that he was writing in +1637 or a little before, and that there were no fore-and-aft rigged yachts in England till the end of the century, when some were presented to Charles II by the Dutch (Chatterton, 1, 4).

h Though Sung Ying-Hsing uses an expression implying right-angle incidence, it is obvious that his nautical informants were referring to winds from port and starboard forward quarters, and that their technical term is a precious reference to fore-and-aft sailing. Obviously in tacking, a ship moves to right and left seemingly broadside to the wind.

The term *chhiang fêng*, which could be englished as 'stealing the wind', is in fact an ancient one, considered by lexicographers to mean simply 'contrary wind'. The earliest example they give of it occurs in Yü Shan's[14] +4th-century ode *Yang Tu Fu*[15] (cf. p. 86). In view of the evidence presented on p. 600 about the first origin of fore-and-aft sailing in the Chinese culture-area this date might be quite signifi-cant, but it must be admitted that the boatmen in the poem lost heart at the contrary wind and let their vessel drift.

i This phrase we shall note presently in the +11th century (p. 640 below).

1 風蓬	2 篠	3 疊	4 中桅蓬	5 頭蓬	6 蓬索
7 關捩	8 腰間緣木	9 三股交錯而度之		10 葉	11 搶風
12 捩舵	13 轉蓬	14 庾闡	15 揚都賦		

vessel is now tacking to the west, making use both of the strength of the current and the force of the wind. Forging ahead, creaking and seething, in a short while the boat will have sailed more than ten *li*. When sailing on lakes with no current one can make headway slowly with the same methods. But if the current is strongly adverse to the ship one may not be able to make headway at all.

Later in the same century the Jesuit Louis Lecomte added his description, which shall be our last:[a]

The low Sails are of very thick Matt, trimm'd up with Laths[b] and long Poles, to strengthen 'em, from two foot to two foot, fasten'd to the whole length of the Masts by several little Loops;[c] they are not fasten'd in the Middle, but have three-quarters of their breadth loose,[d] that they may be accommodated to the Wind, and readily tack about as occasion serves. A great many small Cords hanging at the side of the Sail, where they are plac'd at several distances from the Sail-yard to the bottom, are gather'd up, and keep tight the whole length of the Mat, and further the Motion when the Ship's Course is to be changed.[e]

Lastly, a few niceties of terminology. A few pages back, at the opening of the description of Chinese sails, the word *li*[1] was mentioned in passing as the technical term for the mat-and-batten type. One may suspect, however, that this properly applies only to canvas sails strengthened with battens, for the character contains the 'cloth', not the 'bamboo', radical. The really correct term would seem to be *shuang*,[2] in which this radical appears; as its oldest authority the Khang-Hsi dictionary gives the *Nan Yüeh Chih*[3] (Records of the South), written by Shen Huai-Yuan[4] in the +4th century. This book also said that the mat-sails were woven from the leaves of the *lu-thou* cane. Another word, meaning blinds or awnings but constantly used for sails, e.g. in the *Kao-Li Thu Ching*,[f] is of course *phêng*,[5] and this also has the bamboo radical. Again, an older expression for halyards or sheets, instead of those already noted, was *fan chhien*;[6] thus the *Than Yuan Thi Hu*[7] (A Delicious Dish of Talk) by Yang Shen[8] (c. +1510) says that these were made of green silk on the imperial ship of one of the Chhi emperors.[g]

(iv) *The place of Chinese sails in world nautical development*

How do these Chinese inventions compare with parallel progress made elsewhere in solving the problem of windward sailing? We shall try to make our answer as brief as possible, with the aid of the chart in Table 72. Sail types are arranged on it as if against an invisible map of the Old World, and the lines represent the diffusion of forms or the voyaging of stimuli.[h]

[a] (1), p. 231. Only Donnelly (2) has so far seen the value of Lecomte's nautical descriptions.
[b] Cf. Ysbrants Ides (1), p. 66.   [c] The parrels.
[d] A clear statement that the sails were lug-sails, not square-sails.
[e] The 'small cords' are of course the multiple sheets.
[f] E.g. ch. 34, pp. 6b, 9a, b; ch. 39, pp. 2a, 4b.   [g] Quoted in *KCCY*, ch. 28, p. 15a.
[h] A great deal of patient work has been done since Elliott-Smith (2) first attacked the problem with his heroic obsession, and a great deal more remains to be done, but in this field at least his Egyptophilia finds no small justification.

[1] 帾   [2] 籗   [3] 南越志   [4] 沈懷遠   [5] 篷   [6] 帆綷
[7] 譚苑醍醐   [8] 楊慎

Table. 72. Chart showing the distribution and possible genetic derivation of types of sails

Masts, yards, booms, sprits, gaffs, etc., in thick lines, sail edges in thin. No attempt is made to indicate relative sizes of sails. Explanation in text. The chart is to be thought of as superimposed upon a map of the Old World.

There is general agreement that the oldest sailing-ships, those of ancient Egypt, were invariably rigged with square sails.[a] This can be seen on what is perhaps the oldest known record of a sail, a painting of a boat with high stem and stern preserved on a piece of 1st-dynasty pottery and dating from about −3000.[b] Many accept the view that since the prevailing wind in the Nile valley is from north to south, the boats could travel downstream relying on the current, while on the return journey they could run before the wind all the way.[c] Hence the problem of windward sailing did not arise for a long time.

From this focus the square-sail radiated in all directions—to the Mediterranean, where it was universal throughout Greek, Roman and Hellenistic times;[d] to the north, where it was the only automotive device of the ships of the Vikings and Normans,[e] and to all Asia, including India and China. In many parts of the world it still survives, as in the 'Humber keels' of our own country,[f] the *barcos rabêlos* of the Douro,[g] and the freighters of Lake Como,[h] certain Norwegian boats,[i] ships of considerable size on Indian[j] and Burmese[k] rivers, and (as we saw, pp. 457, 598, 602) the junks of the Upper Yangtze and the Chhien-thang.[l]

There has been some debate concerning the extent to which square-sailed ships could use quarter or beam winds, or even make headway to windward, and the means whereby they achieved what they did.[m] Chatterton speaks of the 'utter incapacity' of

[a] Jal (1), vol. 1, pp. 47 ff.; Wilkinson (1), vol. 1, pp. 412 ff., vol. 2, pp. 120 ff.; G. C. V. Holmes, (1), p. 23; Chatterton (1), pp. 13 ff.; Koester (3); Boreux (1); Bowen (11). Reisner (1), pl. VII, no. 4,841, and p. 28, describes a beautiful model in which the sail is still preserved with much of its gear. Pictures in des Noëttes (2), figs. 9, 12, 13a; Moll (1); reconstruction in Clowes & Trew (1), p. 51.

[b] Brit. Mus. vase no. 35,324; Boreux (1), p. 66; Frankfort (1), pl. 13, reproduced in Bowen (1, 11). A pottery model ship from Eridu in Mesopotamia carries the invention of mast and sail some four centuries further back, for it is equipped with a socket for a mast and two holes in the gunwales which would have secured the shrouds (Lloyd & Safar (1); cf. Casson (3), p. 2). But here there is no clue to the shape of sail used.

[c] Holmes (1), p. 23; Chatterton (1), p. 13. Koester (2) has been able to explain in this connection the puzzling passage in Herodotus, II, 96. A flat plate of boards was used as a kind of sea-anchor to keep the current working on the boat, at times when high northern winds were having the effect of negating its northward drift with the Nile current. Such a table or board appears again in connection with one of the upstream-working paddle-boats of the Renaissance engineers (cf. Vol. 4, pt. 2, p. 412 above); Feldhaus (1), fig. 611. A stone was also dragged astern to keep the bow facing the waves.

[d] Torr (1); Cook (2); Tarn (3); des Noëttes (2); reconstructions in Clowes & Trew (1), pp. 55, 59; cf. la Roërie & Vivielle (1), vol. 1, pp. 48, 52, 54, 77; Casson (3); Bowen (9).

[e] Anderson & Anderson (1), p. 80; Jal (1), vol. 1, pp. 121 ff.; reconstruction in Clowes & Trew (1), p. 65.

[f] Moore (1), pp. 28 ff.; Smyth (1), p. 139; Clowes & Trew (1), p. 67.

[g] Filgueiras (1). Almost alone of European craft, these freighters have tall steering galleries athwartships like many Chinese river-junks, for handling the long stern-sweeps, almost as long as the vessels themselves.

[h] Private communication from Professor Luciano Petech.

[i] Smyth (1), p. 50.

[j] Smyth (1), pp. 365 ff.; Chatterton (1), p. 13; Clowes & Trew (1), p. 53. The boats of the Upper Indus have square-sails of dimensions and proportions almost identical with those of ancient Egypt (Smyth (1), p. 336).

[k] Smyth (1), pp. 371 ff.; Chatterton (1), p. 13.

[l] Of course, in combination with other sails, it has also survived, though now moribund, in the large barques and brigantines still occasionally used. And the Maldive Island trader preserves the exact rig and size of the late +16th-century *nau redonda* (Smyth (1), p. 369, 1st ed. p. 320) except that the mizen lateen is replaced by a gaff-sail.

[m] See Bowen (1); Gibson (2), etc.

Viking ships to do this,[a] and gives some remarkable examples of the dependence of highly developed sailing-ships on favourable winds in recent times. As late as 1800 they sometimes had to wait as long as three months at Hamoaze in order to get into Plymouth Sound, and this was long after the introduction of a lateen sail on the mizen-mast. T. R. Holmes maintained that Roman ships could work to windward, but his argument depended largely on complex indirect evidence. An interesting passage in Aristotle's *Mechanica* has been interpreted[b] to mean that Greek and Roman ships managed to get a little nearer to the wind by furling, brailing, or reefing in one way or another, a half of the square-sail, thus giving an approximately triangular character to the other half.[c] What was perhaps another way of doing the same thing may be seen in a Norse carving[d] which shows a number of ropes attached to the foot of a Viking square-sail, which in this way may have been compressed like the drawing of a curtain to one side. Such practices might conceivably be among the origins of the lateen sail. But that arose comparatively late, and in earlier times one just had to be content to wait for a fair wind.[e] All this constitutes a historical problem on which much light could be thrown by actual experiment, and here a beginning has at last been made by Bowen (10)—with results very unfavourable to the powers of ancient Egyptian square-sails. Further experiments, however, will have to take more account of the possible hull shapes.

It can hardly be doubted that the earliest development from the square-sail was the canted square-sail, which consists simply in so adjusting the yardarm and foot as to get the greatest possible area of the sail to one side of the mast (Fig. 1018, pl.).[f] This is the typical Indonesian arrangement which we saw in the main Borobodur type (Fig. 973, pl.). Although it is thus attested for the end of the +8th century, we are inclined to think that the invention was very much older there. But it occurs also in the celebrated frameless boats of the Middle Nile, the *naggar* and the *markab*.[g] In all these cases the proportions of the sail (long and narrow) are significantly the same as those of the square-sails of ancient Egyptian ships. The *naggar* may be their direct descendant,[h] or alternatively it may derive from the Javanese invention mediated through the Indonesian

---

[a] (1), p. 40.

[b] By Verwey (2), though the reconstruction of the passage so that it makes sense requires a larger number of words in brackets than almost any other, even from the laconic Chinese, which we have come across. Cf. Torr (1), p. 96. The *Mechanica* is not contemporary with Aristotle, though it may contain some of his work; it is best placed about −250 at the earliest (cf. Sarton (1), vol. 1, p. 132).

[c] The Romans developed a complicated system of furling lines passing through rings sewn on the sails (Poujade (1), pp. 125, 130 ff.). These brail rings are most clearly seen in carvings of Roman ships on stone in the Musée Lapidaire at Narbonne (Esperandieu (1), pp. 37, 40). Cf. Table 72.

[d] Gibson (2), pl. 9 and p. 87.

[e] If this never came and if oars alone could not make the passage, there was nothing to be done at all. According to Carpenter (1), this was just what oars were incapable of in early times for the Bosphorus, where the current runs at 4–6 knots, and the wind was nearly always contrary. The invention of 50-oared longships (penteconters) by Ameinocles in the −7th century solved the problem, and may have given rise to the legend of the Argonauts.

[f] Cf. Smyth (1), 1st ed. pp. 325, 328, 342; Chatterton (1), p. 21; drawings in Adm. Paris (1), pls. 46, 86; Clowes & Trew (1), p. 63. Such a sail may also be called a rectangular balance lug, as by Bowen (2), pp. 199 ff., (9), pp. 163, 193, 197. Besides Indonesia it is found in Indo-China (Poujade (1), pp. 149, 150; Piétri); and much in Malaya (Hawkins & Gibson-Hill).

[g] Hornell (1), p. 214, (4); Smyth (1), p. 341; Chatterton (4), p. 29.

[h] Hornell (18).

cultural influence on East Africa.[a] Such influence is surely also shown by the canted square-sails of the fishing-fleet of Mukalla in the Hadhramaut.[b]

The most prominent Western fore-and-aft sail was the lateen, so characteristic of the Islamic culture-area.[c] This sail exists (Fig. 1005) in two forms, the purely triangular, and the 'quasi-lug' type which retains a small luff edge. The first of these is found only in the Mediterranean, the second throughout the Indian Ocean.[d] Much information exists about the historical appearance of the lateen sail, which is first clearly depicted in a Byzantine manuscript dating approximately from +880.[e] There is thus no doubt that the triangular form was coming into general use in the Mediterranean in the +9th century, but it may be a few centuries older.[f] A determined attempt has recently been made to prove the existence of the lateen sail in Hellenistic times, from an Eleusinian tomb relief of the +2nd or +3rd century, but it has not carried conviction.[g] On general grounds, the 'quasi-lug' form would seem to be the more primitive or transitional of the two, and if we are justified in deriving it from the canted square-sail of South-east Asia we may envisage it as generating westwards in its turn the purely triangular form.[h]

An important turning-point occurred when the square-sail of Northern Europe was associated in a combined rig, on vessels of more than one mast, with the lateen. The lateen was hoisted on the mizen. Some think that the process began when ships from Bayonne entered the Mediterranean about +1304,[i] but progress in the adaptation was

---

[a] Hornell (4), p. 137, (19).　　　　[b] Bowen (4).

[c] There are elaborate descriptions of its use, by Vence (1); Bowen (1, 3); P. Paris (4); and Villiers (1, 4). Drawings in Adm. Paris, pl. 5; Smyth (1), pp. 275 ff.; Clowes & Trew (1), p. 111. Discussion in Jal (1), vol. 2, pp. 1 ff.; Moore (1), pp. 86 ff.; Poujade (1), p. 141; Hourani (1), pp. 101 ff. The origin of the term is most obscure. Höver (1) tries to derive it from *velum laterale*, not very convincingly.

[d] Smyth (1), pp. 352, 360 ff.; Bowen (2, 9). But this need not mean that the 'quasi-lug' type was never present in Mediterranean waters. If it had never been known there it would be hard to explain why the sails of the peculiar Iberian windmills (see Vol. 4, pt. 2, p. 556) are in fact precisely of this form.

[e] From the Sermons of St Gregory Nazianzen (Bib. Nat. Gk. MS., no. 510) discovered by Brindley (6). Good reproductions are given by Hourani (1), pls. 5, 6. There is an earlier picture, a rude drawing on the outer stone wall of a ruined church of the pre-Muslim period at El-Auja in South Palestine, which might be of the +6th or +7th century (Palmer), but this shows only an oblique yard with the sail indistinct, yet apparently a lateen. Sottas (1) and Kaeyl (1) have drawn attention to a text of c. +533 in Procopius, *De Bello Vandalico*, I, 13, 3, which says that the ships carrying commanding officers were to have a third of the upper angle of their sails painted red; but critics point out that this may well have meant the small triangular topsails of Roman times, likely still to be in use. These continued even into the eighteenth century, as in the *tartana* rig (Poujade (1), p. 126). Dolley (1) deduced the existence of lateen sails in the Arab fleet at the siege of Thessalonica in +904 from the description of the floating siege towers which were there improvised; this also was subjected to criticism, but the case for it seems to be a better one, and it is intrinsically much more likely.

[f] One wonders whether its westward spread could have been connected with the ship-canal between the Nile and the Red Sea. That ancient waterway was working again between +643 and +760, restored by the second orthodox caliph (cf. pp. 356, 374, 465 above).

[g] Casson (2), with a misleading photograph corrected in Casson (3), pl. 15c, after amelioration by Bowen. The yard of this stationary ship is certainly canted very high, but it is attached to the mast centrally, and as the foot of the sail is not seen, part of it could be lying in folds. A square-sail may therefore have been intended. In any case a triangular lateen can never have been meant, since the luff edge is clearly visible on the relief. I remain unconvinced, even by Casson (4).

[h] An interesting example of the association of these types of sails occurred when, towards the end of the +18th century, a captain of the Honourable East India Company, Thomas Forrest, made two famous journeys of exploration in a 'galley' which carried two lateens and a canted Indonesian sail, with a tripod mast (1, 2).

[i] Anderson & Anderson (1), p. 110. It would have been natural for the Basque and Portuguese sailors, owing to their geographical location, to try to combine what was best in both Atlantic and Mediterranean

slow, and it was not generally adopted until the latter half of the +15th century, when the impression created by the multiple-masted Chinese junks had borne full fruit.[a] The ships of Columbus (+1492), and all the three-masters of the +16th and +17th centuries, carried a lateen aft,[b] so that they could either run with their mainsails or beat with their mizens.[c] Such a ship, of about +1525, can be seen in the stained-glass windows of King's College Chapel in Cambridge.[d] Clowes has written:[e]

In +1400 northern ships were entirely dependent on a fair wind, and were quite unable, indeed never attempted, to make headway against an adverse one. Before +1500 (European) ships had been able to make the long ocean voyages which had resulted in Columbus' discovery of America, Dias' doubling of the Cape of Good Hope, and the opening of the Indian trade route by Vasco da Gama. Other scientific advances, such as the introduction of the mariner's compass from China,[f] bore their part in making such voyages possible, but without the far-reaching improvements in masts and sails the great discoverers could never have accomplished their work.

The typically Chinese quadrilateral lug-sail exists in Europe, almost the same in shape, and it became so common there that some have described it as the universal sail of the Atlantic coasts of France.[g] It is also well known in English, Italian, Greek and Turkish waters.[h] But any evidence for it before the late +16th century is difficult to adduce; the earliest representation found by Brindley (2) was +1586, and the name not till a century later.[i] We shall return to its origins in a moment (p. 613).

Strangely enough, a fore-and-aft sail type in the strictest sense, with nothing forward of the mast, is now known to have been in Europe much earlier. This was the sprit-sail.[j] The first illustration of it was long supposed to be datable at +1416,[k] and

practice. We have already seen how the Portuguese *barca* of about +1430, with its two square-sails, proved unsuitable for beating up and down the west coasts of Africa, so that the caravel of two or three lateens, quite large by +1460, was substituted for it. Well before +1500 the synthesis had been achieved in the equilibrium form of the *nau redonda*, which had square-sails on fore- and main-masts, keeping a lateen only on the mizen. An alternative, less popular and widespread, was the *caravela redonda*, with square-sails only on the foremast and great lateens on two or three others. For further details on the Portuguese development see da Fonseca (1).

[a] Jal (1), vol. 2, pp. 134 ff.; Brindley (2); Chatterton (1), pp. 56, 66, (2); Smyth (1), p. 279; Gibson (2), p. 108. Cf. the models of Chatterton (3); pl. 6 of a +12th-century ship with square-sail alone, pls. 8, 9, 14 of ships after +1450 with lateen mizens.

[b] Model of the 'Santa Maria' in Chatterton (3), pls. 11, 12; full-scale reconstruction in la Roërie & Vivielle (1), vol. 1, pp. 238 ff., 248 ff. Clowes & Trew illustrate similarly Drake's 'Golden Hind' of +1577.

[c] For the reservations to which this interpretation is subject, see pp. 589, 593 above. Perhaps the mizen lateen was really only a 'handling' sail, primarily for use in inshore waters. Probably the square-sail rig survived until steam because owing to the prohibitive amount of work involved in continually trimming sail, ocean voyagers have always looked for, and waited for, fair winds. Even trading junks work monsoons. The tea-clippers could have been rigged as multiple-master fore-and-aft schooners, but they were not, because for running before the wind without change of sail for many days nothing superior to the square-sail was ever developed. Yet this is not to deny that there were solid advantages in the mizen lateen, which greatly increased the control of the ship at low speeds. For this note we are indebted to Cdr. George Naish.

[d] Harrison & Nance (1).   [e] (2), p. 54.   [f] Cf. pp. 562 ff. above.

[g] Smyth (1), pp. 246 ff., 261, 267, etc., 306 ff.; Chatterton (4), pp. 38 ff.; Moore (1), pp. 206 ff. Model in Chatterton (3), pl. 137.

[h] Smyth (1), pp. 98, 134, 188, 196, 207, 319 ff., 329; Chatterton (4), pp. 227 ff., 310 ff.

[i] Moore (1), p. 206.   [j] Best discussion in Moore (1), pp. 147 ff.

[k] Brindley (4); Chatterton (1), p. 165. The picture occurs in the *Très Belles Heures de Notre Dame*, by H. van Eyck. See also Brindley (3); the MS. was lost in a fire at the Turin Library.

certainly the following decades show several examples,[a] but we know now of a kind of sprit-sail in northern Europe from the previous century, and tomb reliefs take the evidence back to the Hellenistic period. For reasons which will explain themselves as we go on, discussion of this is postponed for a few paragraphs.

There followed eventually the gaff-sail, so familiar in our own time as the typical sail of yachts.[b] Since the gaff was spoken of as a half-sprit, its origin from the sprit-sail is generally considered likely;[c] this took place about the beginning of the +16th century in Holland, and spread to England at the Restoration.[d] Gradually the more efficient gaff-sail replaced the lateen on the mizen masts of full-rigged ships,[e] a process which, carried further, led to the graceful schooner.[f] Gaff-sails may have been a purely European development, advantageous because easier to handle than sprit-sails; but one cannot be sure, since certain Indo-Chinese[g] and Melanesian boats have something very like them.[h]

It now remains only to add a few keystones into the arches of speculation which we daringly throw across the abysses of our ignorance. Let us look again at the chart (Table 72) which shows the sail forms superimposed upon an imaginary map of the Old World. We see the ancient Egyptian square-sail radiating in all directions, north, north-east and east. We may assume that the canted square-sail really is Indonesian (perhaps of the −1st century) and that the Middle Nile *naggars* are part of a cultural backwash to the African continent. If we suppose that the canted sail gave off genetically a kind of lateen sail in each direction, east and west, we can explain the appearance of sails of triangular lateen form, but fitted with booms as well as yards, in the Pacific.[i] These were first found among the Ladrone Islands by Magellan[j] in +1521, and as the eastern offspring of the canted sail, they would correspond to the 'quasi-lug' lateen of the

---

[a] Nance (1) adduced another example from the late +15th century (MS. Eg. no. 1065, BM). Massa (1) shows one about +1475 in a picture of the St Elizabeth legend at Amsterdam. We have noticed that in the world-map of Fra Mauro, +1459 (p. 473 above).

[b] Moore (1), pp. 167 ff.; Anderson & Anderson (1), pp. 164 ff.; Chatterton (1), pp. 165 ff.

[c] Moore (1), p. 168; Moore & Laughton (1); Bowen (9), pp. 161 ff. But others, e.g. Gibson (2), p. 123, point out that the jib-sail came with the gaff-sail, which suggests that both were, in a sense, the component parts of a lateen sail divided vertically into two. See further Clowes (2), p. 80.

[d] The oldest illustration concerns a siege of Stockholm in +1523 (Nance, 2). Drawings and models of the Stuart Royal Yachts in Chatterton (3), pl. 137; Clowes & Trew (1), p. 137.

[e] Some warships continued for a long time, however, to carry the long lateen yard as a useful spar, with a gaff-sail bent to the aft portion of it; so Nelson's flagship at the Battle of the Nile (+1798).

[f] How little the principles of fore-and-aft sailing were understood by some Europeans in the era of the full-rigged ship may be gauged by the astonishment with which Charnock (about +1798) speaks of the rig of Chinese junks: 'the yard is brought in contact with the mast', he says (vol. 3, pp. 290 ff.), 'not at the centre but near the end'. 'Three-quarters of the sail is on the sheet side, so that they turn round the mast as on a pivot.' Of course, Charnock was a naval architect, so it would perhaps be too much to expect him to be very intelligent on sails. Cf. pp. 462, 603 above.

[g] Poujade (1), pp. 149, 150; the boats of Phan-Tiet. See pp. 614 ff. below.

[h] Bowen plausibly suggests, (2), pp. 205, 208, that the forked boom of the gaff-sail may have been copied by the Dutch from prototypes existing in the Admiralty Islands of Melanesia when discovered in +1616.

[i] Haddon & Hornell (1); Brindley (1); Malinowski (1); Bowen (2, 9). This is also the rig of the *prahu* or *prao* of the Java seas; Smyth (1), p. 414.

[j] Hourani (1), p. 105. Late eighteenth-century Europeans much admired the 'flying *proas*' with their boomed lateens, as witness Charnock (1), vol. 3, pp. 314 ff. For Englishmen, however, the classical description is no doubt that of Anson (1), p. 339. There were many tales of their remarkable speed.

Erythraean Arabs, which would have been its western offspring.[a] There are, of course, difficulties in thinking of the genesis of the Arab lateen in this way, for it has almost invariably no boom, and it does not seem to occur as a primary cultural trait in those eastern Indian waters which form the intermediate zone between 'Erythraea' and Indonesia. Still, contact could have occurred at least as early as the +2nd century between Roman Syrian sailors, pre-Muslim Arabs, and Persians, with sailors from regions east of India (Chryse; the Golden Chersonese, etc.).[b]

Support for this belief may be derived from a study of P. Paris (5), who made an interesting analysis of two obscure passages in Strabo and Pliny. About +75 Pliny knew of boats from Taprobane (or points further east) 'with a prow at each end'.[c] These can only have been craft like the *oruwa*, with Indonesian canted square-sails, Indian Ocean sprit-sails or Pacific lateens, which tack, not by going about, but by reversing.[d] He also refers to boats which must have been similar to the *yathra dhoni* of Ceylon.[e] Strabo's words (about +23) imply a knowledge of boats with double out-riggers,[f] and these could only have come from the Indonesian culture-area. The same passage of Pliny contains the description of an ambassador, one Rachias, from Tapro-bane (here probably Sumatra), who journeyed to Rome about +45, and spoke of the commerce which his people had with the Seres.[g] There seems really little historical difficulty in believing that all lateens originated from the canted square-sail.[h]

Meanwhile, we must presumably suppose that the Chinese balanced lug-sail was another development from the canted rig.[i] This process was apparently occurring at

[a] Poujade (1), pp. 145 ff., 157 ff., seems to agree more or less with this theory of the origin of the lateen. Bowen, on the other hand, believes that the Arab lateen must have descended, through some kind of lug-sail, from some form of square-sail which had no boom, (2), p. 188. The difficulty here is to find the intermediate forms. Bowen thinks that we may see them in the jib-headed dipping lugs of the western Indian Ocean, (2), p. 187. Bowen also thinks, (2), pp. 101, 110, that the Pacific boom-lateen developed not so much from the canted and boomed Indonesian square-sail as from the 'proto-Oceanic bifid-mast sprit-sail' of the Indian Ocean, (2), pp. 84 ff., which we shall consider shortly in connection with the origin of all sprit-sails; transmission occurring through the Polynesian–Micronesian triangular sprit-sail, (2), pp. 87 ff., perhaps also through the Melanesian 'double-mast sprit-sail', (2), pp. 86 ff., 101. But this strange type—one of the putative ancestors, as we have seen, of the Chinese lug-sail—might itself be an 'Indonesian' rectangular sail canted so far as to stand bolt upright. A peculiar standing lug of the Melanesian Bismarck Archipelago, still retaining a short mast as well as the almost vertical boom and yard, has almost reached the 'mastless' (or 'double-mast') stage (cf. (2), p. 205). It is interesting to see how the traditional terminology tends to break down in this field of bewildering but fascinating variety.

[b] This, after all, was about the beginning of the Indonesian migration to Madagascar (Ferrand, 3).

[c] *Nat. Hist.* VI, xxiv, 82, 83.

[d] Charnock (1), vol. 3, p. 314; Hornell (20). Description of the process in Bowen (2), pp. 115, 190, 209.

[e] Hornell (1), p. 257, (14).

[f] *Geogr.* XV, 1, XV.

[g] It will be remembered that Pliny confused them here (VI, xxiv, 85 ff.) with the Yüeh-chih, describing them as fair-haired blue-eyed people. As we saw above (p. 449) there were commercial contacts later on between the Yüeh-chih and the Malays, under the Kushān empire in India. Cf. Vol. 1, p. 206, on Chinese–Kushān diplomatic relations just after Pliny's time.

[h] If one prefers the view that the *naggar* sail was an indigenous Egyptian development, and that all Arab lateens originated from it, one is left with no explanation for the Pacific lateens except independent invention and convergence. Of course this cannot be excluded.

[i] This was also the view of P. Paris (3), p. 44. Here Bowen (9), pp. 192 ff., too, concurs, tending however to emphasise the direct descent of the Chinese balance lug from ancient tall rectangular square-sails. Elsewhere (p. 197) he traces the descent of Siamese and Indo-Chinese standing lugs from the Indonesian canted square-sail or ancient balance lug. On lug-sail terminology cf. Ansted (1), p. 168.

least as early as the +3rd century, as we have seen.[a] Here the mechanics of the evolution are a little easier to understand. Besides, several traces of it have survived. For example, some of the junks of the Chhien-thang River carry square-sails, but when wishing to sail to windward, hoist them canted.[b] Roller-reefing,[c] a feature characteristic of the canted sail both in Indonesia[d] and on the Middle Nile,[e] has persisted on some of the square-sailed junks of the Upper Yangtze.[f] And certain Chinese boats carry a vestigial canted sail as a kind of spinnaker additional to their lug (Fig. 1019, pl.).[g]

If one cannot spend much time rambling in the Southern Seas oneself, one has only to look at good photographs of the craft of Indonesia and Malaya to see the Indonesian canted rig turning almost visibly into the lug. Thus Hawkins & Gibson-Hill illustrate a boat of Kuala Trengganu which has its square-sail canted so that the lowest corner (the tack) comes in line with the mast (Fig. 1020, pl.). Another, the *Prahu Buatan barat* of Kelantan, has a sail similarly set, but the boom seems no longer quite parallel to the yard, in other words the lengthening process of the leech foot is taking place.[h] Exactly the same thing can be seen happening on Poujade's chart of the sails of Indo-Chinese craft.[i]

A major problem is presented by the European lugger. If it really originated indigenously as late as the end of the +16th century, its spread around all the coasts of western Europe was astonishingly rapid. The suggestion is at any rate worth entertaining that it came directly from the Chinese junks, losing its battens and multiple sheets on the way.[j] A suspicious circumstance lies in the fact that the Adriatic, Marco Polo's home waters,[k] seems to be the geographical centre of the distribution of the rig in Europe. The *trabaccolo* (Fig. 1021) and the *braggozzi*, both luggers,[l] still predominate in Venetian ports like Chioggia, carrying two standing lugs[m] and a jib. Still more remarkable, they have other Chinese features; they are very flat-bottomed, and they have an enormous rudder which descends some distance below the keel, and which can be triced up when shallow water is approached. 'These boats', says Smyth,[n] 'are of beautiful lines and great power, constituting one of the finest forms of sea-going lugger

---

[a] Above, p. 600.

[b] Worcester (3), vol. 1, p. 67.

[c] It is of interest that this practice is so old, since in the middle of the nineteenth century Cunningham patented 'self-reefing topsails' using a roller mechanism (Gillespie, 1). Cf. Wailes (3), p. 94.

[d] Hornell (4), p. 136; Hawkins & Gibson-Hill (1); Poujade (1), p. 126. At least one of the Borobodur ship carvings indicates the process in operation.

[e] Hornell (1), p. 214, (4), p. 131, (18).

[f] Worcester (1), p. 12.

[g] Spencer (2) also gives a photograph of this practice.

[h] The perfect Chinese analogue to these pictures is seen in the photographs of lug-sail ships on the Hsiang River published by Forman & Forman (1), pls. 145, 210.

[i] (1), pp. 149, 157. Cf. Noteboom (1).

[j] Note Smyth (1), p. 310, 'The typical Italian lug-sail seen in the Adriatic, and thence carried to the far corners of the Mediterranean by its enterprising seamen, is what we term a balance lug—a Chinese lug without the battens, laced to boom as well as yard, and when hoisted "set up" by the tack purchase.' This last is a rope or tackle attached to the fore lower corner of a fore-and-aft sail to tauten the luff edge.

[k] Smyth (1), pp. 306 ff.; Chatterton (4), pp. 38 ff.; Moore (1), p. 230; Gillmer (1). He is sometimes said to have been born at Korčula.

[l] They were immortalised in the paintings of Turner. Cf. la Roërie & Vivielle (1), vol. 2, p. 196.

[m] Always with booms, like Chinese lugs.

[n] (1), p. 313.

in the world.' And of the Morbihan luggers, he says,[a] 'At a distance their sails (lofty and rectangular) are in outline strangely like those of some of the two-masted Amoy fishing-junks.' Further hints may be gained by the fact that the only instance in Europe of multiple sheeting occurs in Turkish luggers, and there not as sheets at all,

Fig. 1021. The *trabaccolo* of Venetian waters, one of the best examples of traditional lug-sail rig in Europe (sketch after Smyth, 1). Description in text.

but as bowlines,[b] which with batten-sails in China had never been necessary. And on certain Turkish boats, a pair of side-wings or cheeks abaft the stern, just as on sampans, could be seen.[c] Such indications permit us, therefore, to suggest that evidence may later be forthcoming that at some time after the return of Marco Polo to his native city, the Chinese lug-sail was copied in Europe.[d]

Another major problem concerns the origin of the first sprit-sail rigs in Europe. Those who have discussed the matter have omitted to notice that rigs of almost exactly the same kind occur in China (Fig. 1015, pl.).[e] That they occur widely there prevents one from lightly assuming them to have been a transmission from Europe in the +17th century. The assumption often made that in Europe they must have derived from the Mediterranean lateen is also unattractive, for on general evolutionary principles the reduction of a square-sail to a triangular one would not be readily reversible. More interesting is the possibility that all sprit-sails derive from the 'bifid-mast' sprit rigs of

[a] (1), p. 267.

[b] A rope leading forward from the weather edge of certain sails, to tauten it. On the Turkish multiple bowlines, see Moore (1), p. 44.

[c] Smyth (1), p. 327, who gives a drawing of +1671.

[d] After this was written similar conclusions were reached by Bowen (2), pp. 192, 205, 208, (9), pp. 192, 198. There is not much difficulty in thinking of intermediaries; they may well have been Turkish. We have seen that the cartographer of +1375 knew that there was something very special about Chinese sails, and there could well have been many seamen, less articulate, but perspicacious practitioners, contemporary with him or even earlier. Another possibility is the close relation between France and Siam in the +18th century, as Bowen suggests.

[e] Worcester (3), vol. 1, pp. 74, 84, 162, vol. 2, p. 442, etc.; Fitch (1).

the Indian Ocean which carry a square-sail.[a] Many examples are known—from Ceylon, from Madagascar, from the Pacific also.[b] In this case, the Chinese sprit-sails (information about the earliest date of which is urgently required) would be one derivative, while the European would be another.[c] Here the discovery of Brindley (5) is of exceptional interest; the earliest sprit-sail which he found in Europe, that on the seal of Kiel (+1365), seems in fact to be an elongated square-sail carried on a bifid spar, though a normal mast was stepped as well (Fig. 1022). There was apparently neither yard nor boom, just as in the Indian Ocean form.[d] Furthermore, Brindley (6) found that in Tahiti and Hawaii also a conversion of the bifid-mast square-sail to the sprit-sail had occurred, perhaps independently. Again there may be significance in the fact that one of the chief foci of the sprit-sail in Europe is Turkish.[e] The exact means of transmission, however, remain unknown.

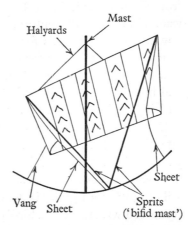

Fig. 1022. The sail on the seal of the city of Kiel (+1365); explanation in text.

If Casson (2) is right in his identifications, as he seems to be, Indian influence on Mediterranean sails may have been exerted a thousand years earlier than the ship in Brindley's seal. For in an important discovery Casson has figured four Hellenistic tomb reliefs of about the +2nd century which do seem to represent sprit-sails of some kind or other, though two of them resemble the seal's arrangement more closely than the true sprit-sail of later times. Still, this seems acceptable in two cases.[f] The lack of any evidence for the long intervening period remains however somewhat mysterious, and the hypothesis of two successive introductions cannot as yet be excluded.[g] That

[a] I.e. the 'proto-Oceanic sprit-sail' in Bowen's terminology; a strange geometrical contrast with the Egyptian and Cantonese bipod masts. It must certainly be one of the most ancient fore-and-aft rigs.

[b] And on the Malabar Coast and the Yemen (Adm. Paris (1), pl. 16). See Hornell (2); Brindley (6); Clowes & Trew (1), p. 93; Bowen (2, 4, 9).

[c] Independent origin has of course its adherents, e.g. Moore (1), p. 146. Another suggestion has been that the European sprit-sail had something to do with a spar called the vargord or beitiáss, which was used by Norse and Cornish sailors as an alternative to bowlines to keep taut the leading edge of a lug-sail or a primitive square-sail (Marcus (2); Moore (1), p. 255). Here again, however, this has South-east Asian parallels (Smyth (1), 1st ed., p. 338; Bowen (2), p. 204).

[d] Clowes (3), it is true, has doubted Brindley's interpretation; as none of the lines show any sag, he thought that the drawing might have been intended to show a yard, two braces and two sheets.

[e] Smyth (1), pp. 325 ff.

[f] All the more so as a Hellenistic textual reference to tacking has been discovered by de Camp (1). In Achilles Tatius' late +3rd-century novel *Kleitophon and Leukippe*, III, I, there is 'a landlubber's account of an effort by the crew of a ship to keep from being blown ashore by tacking against the wind'. The attempt failed, but the description is fairly clear, and we are glad of it. Conviction grew during the discussion of Casson's paper when it was pointed out that in all four of his examples the mast is stepped very far forwards, quite unlike the position in the usual Roman square-sailed ships, and furthermore his sprit-sails show none of the brails and rings so prominent on the latter, (3), p. 219. In one relief, that from Ostia, these differences appear strikingly as the two types of rig are seen side by side. This is reproduced also by van der Heyden & Scullard (1), fig. 308.

[g] We shall encounter other examples of this in Sect. 30e, notably in connection with the crossbow. And we have already seen in discussing equine harness (Vol. 4, pt. 2, pp. 315 ff.) how experimentation seems to have gone on in Hellenistic times without leading to any widespread usage of improved devices.

adequate Roman–Indian contacts for the exchange of techniques existed is indubitable; of the trading stations on the Indian coasts much is now known.[a] Bowen (2) already saw that the Indian Ocean bifid-mast sprit-sail could not possibly have been derived from Europe, for its luff is loose, and by the time that Europeans came again to Indian waters in the + 16th century, their sprit-sails had long been laced to the mast. He has now abandoned (9) his former reluctance to believe that these themselves derived from the Indian Ocean form, which is coming to occupy a central position in evolutionary theory. It may well be also, as he thinks, the ancestor of the triangular sprit-sails and boom-lateens of Oceania,[b] as well as of the Melanesian 'double-mast' sprit-sail which philological evidence connects with the earliest sails of China (p. 599).

One last kind of sail has so far not been mentioned: the triangular peakless fore-and-aft sail (G in the diagram on p. 589). This has a modern air because of its use on racing yachts, where it is called the Bermuda rig or 'leg-of-mutton' sail. Its wide application in modern times has come about because of the realisation that most of the work of a sail of optimum rigidity is done by the leading edge, and its aerodynamics have been compared with those of a bird's wing.[c] But sails of this triangular shape are very much older, for it is found in the Chinese culture-area, notably on some of the Indo-Chinese craft described by Adm. Paris and others.[d] Here the yard is lashed to the mast and stands upright like a backward-curving antenna, while the boom converges to it with or without a very small luff. It might be regarded as an 'Arab' lateen sail shifted horizontally leechwards so far that its yard could become an almost vertical continuation of the mast; and since indeed it was almost certainly derived from standing lug-sails of more normal shape,[e] this is really what it is. Of course the presence of the boom betrays its origin from the canted Indonesian square-sail. This was at first sight not the only originator of the triangular sail with vertical leading edge, however, for it has at least two other foci, the Lake of Geneva and the West Indies. On the Lake of Geneva the lateen sail reaches the northern limit of its distribution,[f] and its conversion there to the leg-of-mutton sail is thought to have been an adaptation to catch light airs in a mountain region.[g] The New World (the West Indies and Bermuda) offers more of a problem. The lateen may well have been concerned, for it seems that this was taken to

[a] Cf. Vol. 1, pp. 177 ff., and more recently Wheeler (4, 7). The Roman–Syrian factory of Arikamedu (Vīrapatnam), for example, was in full function between −50 and +200. Here Pliny's knowledge of reversing craft in +77 is highly significant.

[b] If these were derived, as Brindley and others used to urge, from the Indonesian canted square-sail, the forward foot of the Pacific boom-lateen would not be fastened to the hull as it is, but rather held by a tack-line; Bowen (2), p. 100.

[c] Curry (1), p. 74, etc.

[d] F. E. Paris (2), pt. 1; Piétri (1); Poujade (1), pp. 149, 150, 157; P. Paris (3), pp. 43 ff. and figs. 130, 134, 145, 153, 220, etc.

[e] See Bowen (9), p. 197, who expounds also the interesting intermediate forms, photographs of which can be seen in P. Paris (3), figs. 201, 208. In order to give these standing lugs and even gaff-sails of a kind, as well as the great Chinese balance lugs, and eventually the extreme form of the leg-of-mutton sail, there had to be a progressive enlargement of the leech area of the sail.

[f] Cf. the paintings of François Boçion (1828–90), as also Curry (3), p. 106; Poujade (1), p. 155. It is known that Italian shipwrights took service with the city-States of Geneva and Berne in their struggles against the House of Savoy. On the Rhône, lateen sails come no further north than Arles (Benoit, 2).

[g] Cf. the point made by Sung Ying-Hsing (p. 604 above).

the Antilles and to Canada by the French.[a] On the other hand a Pacific origin may be invoked, for the earliest drawings of the sea-going balsa sailing-rafts of Peru (cf. pp. 394, 547 above), due to van Spilbergen in +1619, show sails like triangular lateens without masts. Bowen thought at first[b] that the artist here inadvertently omitted the booms, in which case the rig would have been essentially the triangular sprit-sail of Polynesia and Micronesia, with more leech divergence of the sprits; but subsequently[c] admitted that the Peruvians could indeed have abandoned the boom.[d] Since the oldest evidence for leg-of-mutton sails in Europe is Dutch and dates from +1623 Bowen is prepared to envisage a direct influence from Peru.[e] Thus ultimately the leg-of-mutton sails of Europe and of Indo-China would both have been derived from the canted Indonesian square-sail through the luff convergence of the lug's yard and boom.[f]

It would be hard to summarise all the complex facts which we have here been surveying. Yet it seems sure enough that after the first development of canted square-sails, wherever and whenever that may have taken place, the earliest fore-and-aft rigs appeared in the Indian Ocean about the time of Aśoka (−3rd century), spreading in different forms both to the Mediterranean area and to the South China seas by the +2nd.[g] Seemingly in the West the sprit-sail was forgotten, but in East Asia the tall balanced lug-sail went from strength to strength. Next following were the lateen sails of Arabic culture from about the +7th century onwards. European sailors of medieval times were slow to adopt these more advanced techniques, but the +15th-century Portuguese pioneered nobly in windward sailing. Whether or not the European lug-sail was derived directly from China or South-east Asia remains to be proved, though it seems exceedingly likely. At all events the credit for the oldest decisive inventions which permitted more and more efficient sailing near the wind must go to the peoples of Indian, Indonesian and especially Sino-Thai culture.

[a] Poujade (1), p. 159. Pell (1) has published a MS. drawing of lateen sails on galleys on Lake Champlain in +1776. Morris (1), pp. 20 ff., maintains that sails of this kind were common enough in North America before that time, and could adduce pictorial evidence for triangular sails with booms in American waters as early as +1629, or at least +1671. Some pictures, such as those of Pell, seem to show in progress the transformation effected by lashing the long lateen yard to a short mast so that it projected vertically far above it. Cf. Laughton (1).

[b] (2), p. 90.

[c] (9), p. 167.

[d] In either case he envisages influence exerted from Asia or Oceania, assuming it to have come by way of the South Pacific, though we would think the North Pacific much more likely (cf. the discussion on pp. 548 ff. above).

[e] Possibly reaching Holland rather earlier by way of Spain after Pizarro.

[f] For Bowen of course the primary New World forms would have descended rather from the Indian Ocean bifid-mast sprit-sail.

[g] The situation, so far as we can now see it, resembles that already encountered for rotary milling and for water-wheels, inventions appearing and spreading almost contemporaneously at the two ends of the Old World (cf. Sect. 27 in Vol. 4, pt. 2). In these cases, however, no intermediate focus has yet been discovered. But for the sailor's art it looks as if the coasts of India played a central disseminating role not unlike that of Babylonia where mathematics, astronomy and acoustics were concerned (cf. Sects. 14 in Vol. 2; 19 and 20 in Vol. 3; and 26h in Vol. 4, pt. 1).

(v) *Leeboards and centre-boards*

In all fore-and-aft sailing there is likely to be a marked tendency for the ship to drift to leeward. The reason for this was explained at an earlier stage (p. 593) when we discussed the forces acting on a ship so rigged that it can sail near the wind. The tendency can only be counteracted by making the hull of such a shape as to minimise this drift. Here the occidental keeled construction had great potential advantage over the Chinese flat-bottomed box build. But other means of opposing leeward loss may take the form of movable pivoted flat boards, either attached to the sides of the vessel or let down through a trunk along the midship line—these are known as leeboards and centre-boards.[a]

Many types of Chinese river craft are equipped with these devices,[b] for example the Yao-wang chhuan[1] described by Worcester.[c] Such leeboards tend to be about one sixth of the length of the ship, and occur in various fan-like shapes. The Cantonese sea-going trawlers have conspicuous centre-boards.[d] Japanese drawings of the +18th century show leeboards on ships from Nanking.[e] Indeed there is good reason for thinking that the Chinese culture-area is their original home, for as was noted above (p. 393) the Formosan bamboo sailing-raft, perhaps the most ancient existing proto-type of all Chinese vessels, has both leeboards and centre-boards.[f] These are essential for its work (it can actually beat to windward with its lug-sail), and though now assisted by steering-oars they are capable of carrying all the burden of the helm, as they still do in its Peruvian counterpart (cf. pp. 394, 548).[g] Then as Chinese shipbuilding developed it would have been most natural to continue the use of leeboards when the flat-bottomed junk construction came to be powered by efficient fore-and-aft rigs.

While medieval Chinese literary references to them are rare, a rather clear one can be seen in the description of one of the warships (Hai-hu[2]) in the *Thai Pai Yin Ching* of +759. A passage which we shall quote in full presently[h] says that there are attached to the hull on both sides of the ship 'floating-boards' (*fou pan*[3]) shaped like the wings of the bird (a hawk or grebe) after which this class of vessels was named. These attach-ments 'help the ships, so that even when wind and wave arise in fury, they are neither (driven) sideways, nor overturn'.[i] The writer, Li Chhüan, may not himself perhaps

---

[a] Cf. Gilfillan (1), pp. 39 ff.　　　　　　　　　[b] Cf. Worcester (3), vol. 2, p. 257; Donnelly (4).
[c] (3), vol. 1, p. 124.　　　　　　　　　[d] Private communication from Miss B. Ward.
[e] *Wakan Senyōshū*, ch. 4, pp. 37*b*, 38*a* (cf. ch. 7, p. 12*a*); *Ka-i Tsūshō-kō*, pp. 242, 244.

[f] The centre-boards are the largest, inserted just fore and aft of the mast; the twin leeboards at bow and stern differ of course from leeboards in the normal sense because they project downwards between the bamboos near, but not at, the sides of the vessel.

[g] Cf. Adam & Denoix (1), pp. 96 ff.　　　　　　[h] Ch. 40, p. 10*b*; cf. pp. 642, 686 below.

[i] The Chinese phrase is given in its place. Hai-hu chhuan are also referred to in *Ko Chih Ching Yuan*, ch. 28, p. 10*a*, quoting a lost book with the title *Hai Wu I Ming Chi*[4] (Strange Sea Names and Strange Sea Things). We do not know the author or date of this, but it must be older than +1136 because it is excerpted in the *Lei Shuo*.[5] What it says is that the Yüeh people have a warship of this name, which flies rapidly over the waves and cannot sink. Here a reference to the people of ancient Yüeh (cf. p. 441 above) has usually been assumed, but one wonders whether perhaps the Wu-Yüeh kingdom of the first half of the +10th century may not have been meant. The iconographic tradition helps very little. The *Wu Ching Tsung Yao* of +1044, quoting of course TPYC, gives an illustration (*Chhien chi*, ch. 11, p. 11*a*,

[1] 搖網船　　　[2] 海鶻　　　[3] 浮板　　　[4] 海物異名記　　　[5] 類説

have been perfectly clear about the function of these devices,[a] but what he says can hardly refer to anything but leeboards.[b] There is no evidence of their existence in Europe before about +1570 at the earliest, and this has led Chatterton (4) to suggest that the device was brought to Europe from Chinese waters in the +16th century.[c] In favour of this view, perhaps, would be the fact that they then appeared first upon Portuguese and Dutch ships.[d]

For writers of the late Ming leeboards were a commonplace. Speaking of river and canal ships Sung Ying-Hsing says:[e]

If the ship is rather long (in relation to the depth of the rudder with which she is fitted), then in a strong cross-wind the power of the rudder will be insufficient (to prevent leeward drift). Then broad boards (*phien phei shui pan*[1]) are lowered into the water as quickly as possible, and this counteracts the influence (of the wind).

And of sea-going ships:[f]

Amidships there are large horizontal beams set thwartwise and projecting outboard several feet; these are for letting down the 'waist rudders' (*yao tho*[2]). All sorts of ships have them. These 'waist rudders' (i.e. leeboards) do not resemble the proper stern rudders in form, but are fashioned out of broad boards into the shape of knives, and when lowered into the water they do not turn, but help to keep the boat steady. At the top there are handles fitted on to the cross-beams. When the ship sails into shallow waters, these 'waist rudders' (leeboards) are raised, just as the rudders themselves are, hence the name.

Curiously relevant here is the quotation given above (p. 603) in another connection, from de Castanheda. Referring to the period +1528 to +1538, he spoke of two rudders on junks, one at the stern and another at the bow. P. Paris has suggested that

reproduced in Audemard (2), p. 48) showing a bulwarked mastless ship towing nine spars roughly in the position of oars and each attached by a cable to the edge of the deck. In *TSCC*, *Jung chêng tien*, ch. 97, p. 8*b* (his p. 47), the spars, six in number, are apparently pinned to the edge. Evidently the nature of the 'floating-boards' was already misunderstood when the drawings were made.

[a] Another possible mention about this time occurs in the diary of the Japanese monk Ennin. In the early part of the storm which shipwrecked him at his first coming to China in +838, he notes that the 'flat irons' (*phing thieh*[3]) were carried away (tr. Reischauer (2), p. 4). The translator conjectures iron plates used as reinforcements on the sides of the ship, or, less plausibly, the gongs used as ship's bells, but if leeboards of that age were strengthened with iron bands they might well have earned that name, and they would fit the description of the monk. On his adventures see further, p. 643 below.

[b] The alternative is, of course, outriggers of some kind, as Dr Lo Jung-Pang (private communication) is inclined to believe. This deserves serious consideration since the few texts we have dwell so much upon the fact that these ships would not capsize. That quite large vessels in medieval South-east Asia had outriggers we know from the Borobodur carvings (already discussed, p. 458 above). The difficulty about such an interpretation is that nothing at all like this survived even vestigially among the multifarious Chinese ship designs of modern times. Is it possible that in the Thang some sea-going ships had both leeboards and outriggers?

[c] Pp. 71, 73. Cf. Poujade (1), p. 243 n. More recently the subject has been intensively investigated by Doran (1) whose paper, written in the first place without knowledge of the present text, proves the transmission with ethnographical rigour.

[d] For example, the famous *muleta* (Clowes & Trew (1), p. 57), and the *barinho* (F. E. Paris, 2). The first representation in Europe is said to be in the picture of Amsterdam harbour by Joh. Saenredam (+1600). Chatterton (3) figures a model of a +17th-century Dutch yacht with gaffsail and leeboards (pl. 58). But they were slow to spread. In +1790 the latest thing was the trial cutter which Capt. Schanck fitted with three centre-boards for the Admiralty (diagrams in Charnock (1), vol. 3, pp. 352 ff.). Earlier experiments had taken place in +1771.

[e] *TKKW*, ch. 9, p. 4*a*, tr. auct. adjuv. Ting & Donnelly (1); Sun & Sun (1). Cf. p. 634 below.

[f] *TKKW*, ch. 9, p. 5*b*, tr. auct. adjuv. Ting & Donnelly (1); Sun & Sun (1). Cf. p. 416 above.

[1] 偏披水板   [2] 腰舵   [3] 平鐵

what he or his informants had actually seen were certain southern vessels, such as the *tang-way*[a] of Kuangtung, and the *ghe-nang* of Annam, which have a movable board sliding up and down at the bows.[b] This bow centre-board[c] does give, from a distance, the impression of a rudder.[d] Paris also sees significance in the fact that the lee- or centre-board, in any form, is practically unknown in traditional Mediterranean craft.

An alternative method of accomplishing the same end consists in having the rudder relatively large, and lowering it well below the bottom of a flat-bottomed craft. As Poujade[e] has rightly recognised, this was certainly one of the reasons for the widespread Chinese medieval use (which still continues) of rudders deeply slung, and capable of being raised when navigating shallow waters.[f] An extreme example from Korea is given in Fig. 1031. But that is no reason for setting East Asian rudders on one side and excluding them from the history of rudders in general.[g]

(2) OARS

(i) *Rowing and the handled oar*

Since paddles and oars used by rowers go back to prehistoric times, it is quite natural to find that the Chinese language contains a number of terms for them. The exact meaning of these is subject to local variation, and it is hard to identify the significance which they had two thousand years ago. Perhaps the central word is *chi*[1] or *chieh*;[2,3] this goes far back into the − 1st millennium, being found in the *Shih Ching* and the *Shu Ching*.[h] Related words, which fluctuate orthographically between the boat and wood radicals, are *chiang*,[4,5] *cho*,[6] and *cho* (or *chao*).[7,8] More clearly defined is the steering-oar or stern-sweep,[i] *shao*,[9] a word which is differentiated from that for the stern of a boat, *shao*.[10] The term *jao* (or *nao*)[11,12] originally meant a curved piece of wood, and may therefore have been an early appellation of the curved 'propeller' or angle-oar which will shortly be discussed in some detail. The name for this has long been *lu*, a

---

[a] The characters for these words were not given by Paris, and have not been identified.

[b] (1), (3), pp. 50 ff., figs. 3, 95, 98, 102, 104, 155, etc. In the *ghe-nang* the bow transom has become so narrow that it forms with the forward ends of the strakes a slot of pear-shaped cross-section in which the sabre-like centre-board can slide up and down.

[c] It is the equivalent of the extension into a plate-like ridge of the fore-foot or gripe of the stem of European boats. In the absence of stem-posts this process of hypertrophy could not occur in China.

[d] G. W. Long (1).

[e] (1), p. 258.

[f] Vessels with lateen sails traditionally have rudders deeper than the boat's keel (Vence (1), pp. 31, 97). And the flat-bottomed *rascona* of the Venetian lagoon (P. Paris (3), p. 49, figs. 232, 234; Smyth (1), 1st ed. pp. 274 ff.; Adam & Denoix (1), p. 99) has a deep-set rudder as its only 'centre-board'; this necessitates a mast stepped unusually far aft, and results in a somewhat crabwise motion. Such a pattern impels the assumption of Chinese influence in Marco Polo's home waters, and takes its place beside the European focus of lug-sails, which is precisely in this Adriatic region also (cf. p. 613 above).

[g] As Adam & Denoix (1) attempt to do.

[h] Cf. *TPYL*, ch. 771, p. 2*b*.

[i] These are not the same thing; we regard a steering-oar as equivalent to a quarter-paddle, while the stern-sweep is mounted at the true stern of the boat; see on, pp. 627 ff.

¹ 楫	² 檝	³ 艥	⁴ 槳	⁵ 艪	⁶ 櫂	⁷ 棹
⁸ 艣	⁹ 艄	¹⁰ 艄	¹¹ 橈	¹² 蟯		

word which can be written in an unusually large number of different ways.[1,2,3,4,5] Lastly, the punt or quant pole, *kao*,[6] has always been clearly distinguished.

Since in Chinese vessels the deckhouses were always placed towards the stern, while the rowers took their places on the deck forwards, the word *lu* written in a slightly different manner[7] came to mean the bows. In the *Chhien Han Shu*, for example;[a] with reference to −106: 'Han Wu Ti had his ships (lit. sterns and bows, *chu lu*[8]) floating on the River for 1,000 *li*.' Specific study of Han literature, indeed, would throw much light on nautical practices. Thus in the accounts of the fighting which accompanied the restoration of the Han dynasty, about +33, the biography of a general Tshen Phêng[9] tells us[b] that 'There were on the transport ships more than 60,000 rowers working the oars (*cho*[10])... Several thousands of boats with *lu-jao*[11] rushed to the attack.' The Thang commentator, Li Hsien, explains the latter term as meaning that the oars (*chieh*[12]) were outside the ship while the men were inside. This seeming platitude probably implies that the rowers were invisible from outside because of built-up gunwales for protection against arrows, the oars issuing through ports (cf. the tomb-model in Fig. 961, pl.).[c]

Probably in the Han time, as still today, Chinese sailors generally rowed standing and facing forwards. This was a fairly common ancient Egyptian practice,[d] and in China would have arisen naturally from the paddling of the long South-east Asian canoes which gave rise to the dragon boats.[e] Oars are sunk much more deeply in the water at each stroke than is customary in Europe, and often carry a transverse handle at the end of the loom, convenient for this kind of rowing. These T-handled oars go back at least to the Sung, as we know from paintings.[f] Facing rowing has a wide distribution, and though not characteristic of Europe, occurs in connection with Venetian gondolas and the boats of Austrian and Hungarian lakes; a detailed investigation of it might well prove interesting.[g]

During the Sung, Yuan and Ming periods quite large ships were often classified according to the number of oars carried. Thus speaking of the grain-transports called 'wind-borers' (Tsuan-fêng[13]), surely in allusion to their quality in beating to windward, the *Mêng Liang Lu* (+1275) divides them into the greater and smaller 'eight-oared' types (Ta Hsiao pa lu[14]), besides some with six oars only.[h] These were almost certainly not ordinary oars but large sweeps like the 'yulohs' next to be described. The *Hsüan-*

[a] Ch. 6, p. 25*b*. This was the famous occasion on which the emperor went hunting for alligators or crocodiles.

[b] *Hou Han Shu*, ch. 47, p. 17*b*; quoted in *Piao I Lu*, ch. 7, p. 7*a*; tr. auct. Cf. p. 680 below.

[c] This arrangement is seen in many of the illustrations of warships given in the *Wu Pei Chih*, *San Tshai Thu Hui*, and *Thu Shu Chi Chhêng*. Oar-ports were used both in ancient Mediterranean and Viking ships; their persistence in English traditional types has been discussed by Hornell (10).

[d] Cf. des Noëttes (2), fig. 15; Boreux (1), pp. 312, 313.

[e] The crews of the ritual dragon-boats always use paddles (*pa*[15]), not oars (*cho*[10]). Near relatives of these boats can be found all over South-east Asia, e.g. Sarawak and Siam, often connected with rituals or royalty.

[f] E.g. one in the Strehlneck Collection (Musée Guimet photo no. 644213/5) by Li Chhêng.

[g] Cf. Poujade (1), p. 296.

[h] Ch. 12, p. 15*a*. All these, the text says, carry more than 100 men. Cf. p. 416 above.

1 櫓 2 艪 3 樐 4 艣 5 樚 6 篙
7 艫 8 舳艫 9 岑彭 10 棹 11 露橈 12 檝
13 鑽風 14 大小八櫓 15 扒

Ho Fêng Shih Kao-Li Thu Ching (+1124) reports that each ship of the embassy fleet carried ten oars, continuing:

When making a landfall or entering port, or taking advantage of the tide to negotiate a channel, the ship proceeds with the sound of creaking oars. The sailors throw themselves back and forth, shouting and straining, but the vessel moves much more slowly than when sailing before the wind.[a]

Sometimes the oars were useful for short bursts of speed under other conditions, as in a flat calm on open waters. There is an account of a sea-fight with pirates 'ten thousand *li* from China' on one of the Ming expeditions, when a squadron of Chinese ships succeeded in getting away by rowing. Probably this testifies to the efficiency of the 'yuloh'.

(ii) *Sculling and the self-featuring 'propeller'*

Besides the propulsive effect produced by oars worked in the ordinary way, motion forward can also be obtained by mounting an oar approximately in the line of the main axis of the boat, most conveniently at the stern,[b] but also elsewhere, and moving it from side to side of that axis. This technique is used only for small dinghies in the West, but the Chinese elaborated it into an extremely ingenious heavy duty system. Known as the 'yuloh'[c] (actually a verb *yao lu*,[1] 'to shake the oar'), it has often been described;[d] we have already had occasion to refer to it several times. Its great effect[e] is due to three special modifications: (*a*) the loom or handle is brought more nearly parallel to the deck by means of a curve or angle inboard of the fulcrum, (*b*) the fulcrum consists of a thole-pin[f] or bumkin fitting into a hole in a block attached to the oar, and (*c*) the loom or handle is attached to a fixed point on the deck by a short length of rope. The cup-and-pin system constitutes an approach to the universal joint.[g] It is obvious from a consideration of the parallelogram of forces of such a sculling-oar that the effective work desired is accomplished only during a certain part of the stroke, and for the remainder the oar must be feathered so as to oppose no resistance to the water. The 'yuloh' accomplishes this without any need for work with the wrist; it is simply pushed and pulled to and fro, the handle necessarily describing an arc of a circle, while at the moment required for the

---

a Ch. 34, p. 5*a*, tr. auct. adjuv. Lo Jung-Pang, p.c. Cf. p. 602 above.

b The conjecture is obvious that 'sculling', as it is called, arose from the steering-oar.

c The Cantonese form of the words.

d Scarth (1); Maze (1); P. Paris (1); Worcester (1), p. 6; (3), vol. 1, pp. 57 ff.; Dimmock (1); Ward (2); Waters (8); Audemard (3), p. 64. Cf. the model in Fig. 933 (pl.).

e Sampans with four yulohs have competed against ships' boats with four oars and won.

f Hornell (10) found oars with thole-pins and blocks used in Ireland; elsewhere in Europe he noted them only in the fishing-boats of Portugal and Madeira. Since they are apparently unknown in the Mediterranean, it is possible that they may have been a +16th-century Portuguese introduction from China.

g It is one among several Chinese empirical tentatives towards the contrivance; others have already been encountered in connection with the Cardan suspension (Vol. 4, pt. 2, pp. 231 ff., 235) and the spinning wheel (*op. cit.* pp. 103, 115). Cf. the method of pivoting the needle in Chinese mariners' compasses (Vol. 4, pt. 1, p. 290).

1 搖櫓

feathering, the rope is given a jerk (Fig. 1023, pl.).[a] The whole process is difficult to describe, but can be quickly appreciated when it is seen in action.[b]

How old the yuloh system is cannot be said with certainty but it seems to go back as far as the Han. The word *lu* itself is Han. The *Shih Ming* dictionary of Liu Hsi (about +100) says:[c] '(That which is) at the side (of the boat) is called *lu*.[1] Now *lu* is connected with *lü*[2] (the backbone). (Thus) when the strength of (men's) spines is used, then the boat moves (forward).' This does not exclude the self-feathering propulsion oar because it has long been a practice in China to mount two or more of them near the stern or even near the bows. This is seen in the diagram which Worcester gives of a Chekiang Khuai-pan[3] boat,[d] and clearly in the famous picture of +1125 which we discuss in several connections (cf. Figs. 826, 976, 1034, pls.). The *lu*[1] occurs again, for example, in the *Tung Ming Chi* (Light on Mysterious Things)[e] which would be evidence for, perhaps, the +5th century.

One of the first references to the *yao-lu*, as a phrase, is that in the *Fang Yen*[4] (Dictionary of Local Expressions) of Yang Hsiung,[5] but as the book may contain some later interpolations we cannot be quite sure that this was in the original text of about −15. However, it would certainly indicate a time not subsequent to the Later Han. The passage says:[f] 'That which sustains an oar (*cho*[6]) is called a *chiang*.[7] *Chiang*[7]—a small wooden peg (*chüeh*[8]) for the *yao-lu*.[9] The people of Chiangtung also call it the 'foreigner' (*hu jen*[10]). That which ties down the oar (*cho*[6]) is called the *chhi*.[11] *Chhi*[11]—the rope attached to the head of the oar.' A later term for the thole-pin or bumkin was *shuai*,[12] and today the boatmen speak of it as the *lu-ni-thou*.[13] The binome *yao-lu*[14] appears again in the accounts of the battles between the States of the Three Kingdoms period. When Lü Mêng,[15] a general of Wu, was campaigning against Kuan Yü of Shu he employed at one point (+219) a ruse in which he dressed his best soldiers in white like merchants and mounted them on merchant-ships (*kou-lu*[16]) which they sculled back and forth.[g] Since the text dates from before the end of the same century, when many contemporary documents must still have been available, it is rather firm evidence for the dating of the technical phrase.

In the diary of his journey to Szechuan in +1170 Lu Yu said:[h] 'On the twentieth day they unshipped the mast and set up oar-stands (*lu chhuang*[17]), because when

[a] So at least in the north, but in the south the movement is performed entirely by handling the loom of the sweep, without any pulling on the rope (B. Ward, private communication).

[b] It looks easier to perform than it really is, however; there is a knack in it. There is nothing to keep the oar on the pivot, which often has but a shallow penetration into the block. As Worcester says, the motion of the yuloh amounts to that of a reversible screw-propeller. Cf. the discussion above (Vol. 4, pt. 2, pp. 55 ff., 102 ff.) on the development of continuous as opposed to alternating rotary motion, and (pp. 119 ff.) on the screw (and even the screw-propeller, p. 125) in Chinese culture.

[c] Ch. 25, (p. 378), tr. auct.

[d] (3), vol. 1, p. 162.

[e] Ch. 1.

[f] Ch. 9, p. 8*a*, tr. auct. The echoing sentences are those of Kuo Pho's early +4th-century commentary.

[g] *San Kuo Chih*, ch. 54, p. 21*b*; *WHTK*, ch. 158 (p. 1380.1).

[h] *Ju Shu Chi*, ch. 5, p. 9*b*, tr. auct.

1 櫓	2 膂	3 快板船	4 方言	5 揚雄	6 櫂
7 槳	8 橛	9 搖樐	10 胡人	11 緝	12 [illegible]
13 櫓泥頭	14 搖櫓	15 呂蒙	16 䑦艫	17 艣牀	

ascending the gorges they use only oars, and the "thousand-footer" (i.e. the trackers' rope), not sails.' The difficulty is to distinguish between the yuloh fulcrum and rowlocks in general,[a] but probably the former was meant. From the +14th century we have a painting of a fishing-boat by Yen Hui,[1] which clearly depicts the yuloh and its rope.[b] Just about the same time there is a classical reference to the working of massive yulohs by many men, in the account of Ibn Baṭṭūṭah (already quoted, p. 469 above).[c] His description of a dozen or more men working each one has been questioned,[d] but even today on comparatively small river craft it is quite common to see yulohs worked by a team of eight, with two men giving the jerk on the rope, one from each side, which brings them in turn almost flat on their backs. Under yulohs, a speed of $3\frac{1}{2}$–4 knots is easily maintained.

At the end of the +17th century, the yuloh much impressed Lecomte, whose words[e] may end this brief account:

Notwithstanding these Barks be extraordinary big, and tho' they always be either under Sail, or tugg'd along by Ropes, yet they do by times make use of Oars, when they are upon great Rivers, or cross Lakes. As for ordinary Barks, they do not row them after the European manner; but they fasten a kind of a long Oar to the Poup, nearer one side of the Bark than to the other, and sometimes another like it to the Prow, that they make use of as the Fish does of its Tail, thrusting it out, and pulling it to them again, without ever lifting it above Water. This work produces a continual rolling in the Bark; but it hath this advantage, that the Motion is never interrupted, whereas the Time and Effort that we employ to lift up our Oars is lost, and signifies nothing.

The yuloh also impressed the British Navy, for we learn from Admiralty papers that in +1742 experiments were made in which a sloop was fitted with 'a set of Chinese sculs'.[f] Something rather similar to the yuloh figured among the nautical inventions of the third Earl Stanhope[g] about +1790; it was called the 'Ambi-Navigator or Vibrator'. The intention was to apply steam-power to the device, but it never proved successful.[h] Then in 1800 Edward Shorter arranged 'a two-bladed screw, at the end of a revolving shaft, set at an angle like an oar in sculling, and having a universal joint connection with a horizontal shaft on the deck'.[i] Two years later, a deeply laden naval transport ship, worked by eight men turning a capstan, made $1\frac{1}{2}$ knots with this device. More research is needed regarding the relation of the earliest Western tentatives at screw-

[a] In the Section on optics above we saw how the motion of the oar on the fulcrum helped Shen Kua in his thinking on light-beams (Vol. 4, pt. 1, pp. 97 ff.).

[b] Described by Chêng Tê-Khun (5), p. 11, in whose collection it is.

[c] Apparently Marco Polo did not describe yulohs, only ordinary oars (p. 467 above).

[d] The Kao-Li Thu Ching of +1124 says that the 'retainer ships' (cf. p. 602 above), which were 100 ft. long, had 10 sculls (lu[2]), presumably five a side. In our opinion, the many references elsewhere to 'eight-oared ships' also refer to the number of yulohs carried.

[e] P. 234. Cf. Osbeck (1), p. 191. Staunton (2), +1797, vol. 2, p. 46, has a nice description of yulohs being worked on moonlight nights to the sound of chanteys echoing over the water.

[f] MSS Collection of the National Maritime Museum, Greenwich, letter A/2316 of 17 January 1742, and B/145 of 15 January 1752. We are much indebted to Cdr. George Naish for the discovery and communication of copies of these documents.

[g] Life by Stanhope & Gooch (1).   [h] Description in Cuff (1).

[i] Seaton (1), ch. 1.

[1] 顏暉   [2] 艣

propulsion with knowledge of the Chinese yuloh. Mention has already been made[a] of the curious story reported by McGregor, that 'a Chinese screw-propeller was brought to Europe and seen by a Col. Beaufoy in 1780', some thirty years after Bernoulli had first suggested the possible superiority of the screw over the paddle-wheel. It may well turn out that the yuloh played a certain part in the early inspirations of screw-propulsion.

(iii) *The human motor in East and West*

Students of the writings of naval archaeologists soon become familiar with that rather tedious field of controversy which centres upon the construction of the Greek and Roman galley, the interpretation of the technical terms for the banks of oars, and what is to be made of the apparent existence of galleys with forty banks superimposed.[b] The only relevance which this has for us is that it prompts us to ask why such problems never arose in the Chinese culture-area. Apart from the dragon-boats which were used only for racing in folk-festivals, the many-oared galley (though obvious prototypes in the long canoes of the south-east, with their numerous paddlers, were near at hand) remained throughout recorded history absolutely foreign to Chinese civilisation.

The conclusion that this indicates a clear technical superiority of Chinese seamanship seems almost unavoidable.[c] Good reasons have already been given (p. 600 above) for believing that South-east Asian waters were the scene of the earliest really successful attempts at sailing into the eye of the wind. If the first lug-sails and sprit-sails were of the +2nd or +3rd century, they antedated the lateens by four or five hundred years,[d] and gaff-sails by seven hundred more. After all, the object of using the human motor on a mass scale was not only to overcome flat calms, but to make headway against strongly adverse winds. The reader has only to imagine what it must have been like for sailors in ancient times when they tried to escape from the proximity of a lee shore in heavy weather with a ship which could make almost no headway to windward. The greater aerodynamic efficiency of the rigid Chinese mat-and-batten sails must at least have reduced the number of occasions when recourse to other motive power was essential. The naval requirements were not so widely different;[e] there is plenty of

---

[a] Vol. 4, pt. 2, p. 125 above.

[b] E.g. Charnock (1), vol. 1, pp. 44 ff.; Jal (1); A. B. Cook (1); Torr (1); des Noëttes (2), pp. 35 ff.; Poujade (1), p. 233; la Roërie & Vivielle (1); de Loture & Haffner (1); Moll (1); Tarn (4), pp. 128 ff., etc. Satisfactory solutions are now emerging in the studies of Morrison (1, 2); Morrison & Williams (1).

[c] It may indeed be argued that the harbours of Greece and Ionia were mostly land-locked and difficult of access, so that oars were as necessary as auxiliary engines would be for modern sailing-ships in those waters, while on the other hand Chinese harbours were mostly estuarine. Such a contrast might be worth investigating more closely, but I doubt if it would hold good.

[d] If the bifid-mast sprit-sails are as old in the Indian Ocean as they seem to be (cf. pp. 606, 614), this figure increases to eight or nine. The priority of the canted Indonesian square-sail will also probably be of about the same order.

[e] But one great difference there was, namely that Greek ships had keels which were easily extended forward into rams, and therefore Greek naval vessels for centuries attacked by ramming, an action which needed maximum motive power exerted for short periods of time. Warships in China on the contrary had no keels, and the ramming tactic seems to have been used there only at certain periods, e.g. the −5th to the +2nd centuries (pp. 678 ff. below), and again with the large paddle-wheel ships in the +12th century (cf. Vol. 4, pt. 2, p. 420). Moreover, as we shall see later on (pp. 682 ff.), Chinese naval commanders always had a preference for projectile weapons rather than close combat.

combat afloat recorded in the Chinese official histories. And when the Chinese did turn their attention to other motive power, they produced, with that mechanical inventiveness often only grudgingly accorded them by other peoples in modern times, the subtler yuloh (from about the +2nd century), and then the treadmill-operated paddle-wheel boat (at least from the +8th, if not before).[a]

There is nothing here in contradiction with the fact that human motive power was used on a tremendous scale in ancient and traditional China. All sailors rowed, every seaman could handle the yuloh, but their work was not done under the conditions of organised inhumanity which often prevailed in Europe. There were also the trackers, whose work along the banks of the great rivers was sometimes gruelling, but again in China they were not slaves.[b] Elsewhere in this Section we have frequently come across ship types named from the number of their 'oars' (the term includes yulohs).[c] When these were quite large, as some of them certainly were, the gear was carried primarily for manœuvring in landlocked waters, or when way was urgently required in a flat calm.[d] Moreover, there were many types of fast official-carrying boats (like admirals' gigs) and patrol boats, which were thus powered.[e] But put it all together and it adds up to nothing approaching the galley-slave pattern characteristic of Mediterranean Europe for some two thousand years.

Des Noëttes, completing his studies of the history of the axial rudder (2, 4), shortly to be referred to, boldly entitled them, in analogy with his work on the history of the efficient animal harness (cf. Vol. 4, pt. 2, pp. 304, 329), 'Contributions to the Study of the History of Slavery'.[f] In the nautical context the relation was doubtless not so obvious, and his critics, especially la Roërie (1), pointed out that while the Athenian galleys, which were equipped only with steering-oars, had been rowed by free citizens,[g] the galleys of the +17th century, which all had axial rudders, were propelled by slaves under some of the worst conditions ever recorded in the annals of slavery.[h] Neverthe-

[a] Cf. Vol. 4, pt. 2, pp. 416 ff. above.
[b] See further on this, p. 662 below, and mem. p. 415 above.
[c] Pp. 406, 447, 621.
[d] Cf. the use of steering-oars on large modern sailing-ships, pp. 629, 636 below.
[e] For example the 'centipede-boat' (Wu-kung chhuan[1]), with several tens of oars on each side, very fast, mentioned in *Ming Shih*, ch. 92, p. 16b and many other sources. In his *Chi Hsiao Hsin Shu*[2] (New Treatise on Military and Naval Efficiency), c. +1575, Chhi Chi-Kuang[3] describes the Pa chiang chhuan[4] as having 16 oars a side, but the *Wu Pei Chih* gives it only four a side (ch. 18, p. 21a and 117, p. 5a, b respectively). We suspect that oars proper were meant in the former text and yulohs in the latter. For the same reason the *Wu Pei Chih* allots the 'centipede-boat' nine 'oars' a side, not twenty or thirty (ch. 117, pp. 12b, 13a, b). Good sketches from the life will be found in Audemard (4), pls. 65, 66. He reproduces, (2), p. 66, the drawing of a Pa chiang chhuan in *TSCC, Jung chêng tien*, ch. 97, p. 20a.
[f] Esp. (2), p. 110.
[g] This was in any case only a half-truth. Cook (2) and Mayor (2) have shown that on the Greek and Hellenistic galleys slave labour was frequently employed, and the more so as time went on. According to Tarn (3) the Roman galleys first used allied peoples, then more and more frequently slaves. The 'slave's lack of interest in improving the tools he handles' (as Gibson (2), p. 60, put it) can hardly be unconnected with the failure of Graeco-Roman civilisation to adopt effective means of sailing into the wind. But this raises questions too large for treatment here.
[h] French naval historians have studied this in detail; there are dreadful descriptions in Kaltenbach (1); Garnier (1); la Roërie & Vivielle (1), vol. 1, pp. 150 ff. and in de Loture & Haffner (1), pp. 108 ff. The Renaissance rowed galley rose and fell like the witch-mania or the Inquisition; its heyday was from

[1] 蜈蚣船   [2] 紀效新   [3] 戚繼光   [4] 八槳船

less, there can be little doubt that the coming of the axial rudder, which permitted great increases in the size of sailing-ships, was one of the factors which did eventually lead to the final disappearance of the rowed galley and the merciless exploitation of the human motor. There were, of course, other factors too, probably at least as important, such as the advances in shipbuilding which multiplied the number and height of masts; and especially the use of gunpowder, since cannon could devastate an attacking fleet of galleys, and they themselves, being long and narrow, could not support the relatively stable platform which gunners required.[a] It is a remarkable fact that each of these technical developments, if traced back to its origin, can be shown, with probability or certainty, to have emanated from Asian, usually Chinese, civilisation. Some of these demonstrations have already been given; others will follow in due course.[b]

(h) CONTROL (II), STEERING

(1) INTRODUCTION

As every Cambridge undergraduate well knows, who has paddled his canoe beyond Byron's Pool, the simplest form of steering-gear is an oar or paddle held steady at a desired angle on either aft quarter (most conveniently the right or starboard quarter) of the boat. In this way, the streamline flow of the water is deflected so as to impart a turning moment to the hull. On the other hand, the most highly developed form of steering-gear is a great vane pivoting upon the stern-post of a ship, and controlled from its bridge by means of chains to which a source of power is applied and which substitute more effectively for the leverage exerted in small ships by the bar known as the tiller.[c] In the West, the terminology of the successive stages in direction-control presents little difficulty. First there were steering-oars or quarter-paddles, or occasionally stern-sweeps centrally fixed,[d] then came rudder-shaped paddles permanently

---

the +15th to the end of the +17th centuries; by +1748 it was dead. It was essentially a Mediterranean phenomenon, unsuited to the choppy seas and cold weather of the North, where it consequently played little part. But weather can hardly be the reason why such ships were not generated by Chinese civilisation.

[a] La Roërie & Vivielle (1), vol. 1, p. 159. The Battle of Lepanto (+1571) was the last major sea combat between rowed ships. The turning-point came in +1684, when a French vessel, the 'Le Bon', held off 36 Spanish galleys throughout a long windless day, till with the evening breeze she made her escape (de Loture & Haffner (1), p. 118). Cf. Gibson (1).

[b] As to masts, see p. 474 above; as to rudders, the immediately following sub-section. For gunpowder, see Sect. 30 below.

[c] It will be evident that while the steering-oar has the mechanical status of a simple lever, the stern-post rudder is essentially a crank, so held that it rotates about the axis of its centre piece rather than that of one of its end pieces. In terms of the kinematics of machinery, 'point closure' has been replaced by 'line closure' (cf. Vol. 4, pt. 2, p. 68). What this means is that with the greater security of the elongated hinge attachment a much larger movable plane surface can be set up to guide the streamline flow than could ever be possible with a single point fulcrum attachment.

[d] As the undergraduate knows, the steering effect is intensified if the paddle is moved away from the boat. Its action then resembles that of an ordinary oar working at right angles to the boat's axis, but only at the stern. This is the principle of the stern-sweep, which enters the water as far from the hull as possible (cf. Clowes (2), p. 36). Moreover, as Adam & Denoix (1) emphasise, the steering-oar was a very versatile tool, capable of acting (a) as a sweep when rotated about its fulcrum, (b) as a primitive and inefficient rudder when rotated on its own axis, and (c) as a leeboard when held vertically alongside the boat.

attached to one of the stern quarters, and lastly the stern-post rudder itself, hung on pintle and gudgeon.[a] When a Lincoln MS. of +1263 differentiates tolls between 'navi cum handerother' and 'navi cum helmerother'[b] we can guess pretty well what was meant. But the Chinese terminology is a more difficult matter, since the thing changed while the words did not, as we shall soon see.

A classical monograph was devoted to the invention of the stern-post rudder by des Noëttes (2). He claimed that because of the weakness of the steering-oar a cardinal limiting factor to nautical development existed before the beginning of the +13th century.[c] Until that turning-point the capacity of ships was restricted to about 50 tons.[d] Lack of manœuvrability also kept them slow, and the fact that in heavy weather any kind of steering-oar would inevitably take charge, interfering with the handling of the sails, meant that ships were constrained to keep within reach of shelter and could not venture to any extent on ocean passages. The chief critic of des Noëttes, la Roërie (1, 2), maintained that the stern-post rudder had little or no advantage over the steering-oar, but the consensus of qualified nautical opinion crystallised almost unanimously against him,[e] though des Noëttes, who was admittedly a landsman, often failed to receive the credit which he deserved.[f]

The steering-oar, however, has always remained of value in rapid rivers and narrow landlocked waters, hence its continued use in China today. To respond to the rudder, a boat must have way on her, must, in other words, be moving relatively to the surrounding water, for otherwise there is no streamline flow to be diverted. But when

---

[a] These are the traditional names for the components of the rudder's hinge; cf. p. 632 below.

[b] F. B. Brooks, cit. in la Roërie (2).

[c] Cf. Febvre (3).

[d] See especially des Noëttes (2), pp. 48, 58, 69, (4). This remains as a staked claim, though the figure must certainly be too small (cf. p. 452 above), and there are some demands extreme in the other direction, notably the thousand-ton Roman grain-ships of the +2nd century visualised by Casson (1). Perhaps the most urgent need of naval archaeology today is a systematic, sober and definitive study of estimated tonnages in all historical periods and cultures. Obviously this work cannot be done here.

[e] E.g. Poujade (1); Smyth (1); Gilfillan (1); Anderson & Anderson (1); de Loture & Haffner (1), pp. 11, 17, 49. Among la Roërie's supporters Carlini (1) was the most prominent. As Adam & Denoix point out (1), it was rather quixotic to defend the theoretical value of the steering-oar, since in historical fact it was supplanted by the stern-post rudder quite quickly, and this could not have been without good cause. They show by simple calculations, however, that the force required to manage the steering-gear (other things being equal) is considerably less for a steering-oar, with its smaller surface, than for a stern-post rudder. But though more flexibly attached, the former was much more fragile, needed much greater skill on the part of the helmsman, and was limited to relatively small vessels.

[f] This may be a convenient point to warn readers that both des Noëttes (2) and de Loture & Haffner (with less excuse) are quite untrustworthy guides in matters of East Asian nautical technology. Des Noëttes was wrong on almost all points; he believed (a) that there were no rudders in China before the coming of the Portuguese, taking at face value an imaginary sketch in van Linschoten, (b) that Chinese iron-working was rudimentary, (c) that Chinese ships made only coasting voyages, and (d) that junks could sail only before the wind. De Loture & Haffner are even worse (though twenty years later); they say (a) that the magnetic compass is 'proved' to have been used by the Chinese in −2698, (b) that in −1398 the eunuch 'Chien-Ho' made voyages and touched at California, (c) that the Chinese junk can only run before the wind, (d) that when the Portuguese arrived the Chinese had only steering-oars, often double, (e) that junks had bottoms so flat that they drifted to leeward, making little headway, (f) that their sails have multiple bowlines (instead of sheets!), (g) that northern junks have triangular sails and southern ones rectangular; finally (h) that the Chinese 're-invented', and used, the astrolabe. A more remarkable collection of howlers could hardly be imagined. Yet their book is sound and usable where Europe is concerned.

descending rapids, a boat may be moving at almost the same speed as the water, and in such a case it is highly advantageous to have a long stern-sweep, so long that its effect depends not on streamline flow but on reaction to water resistance, just as in the case of an ordinary oar. The lever, in such a stern-sweep, is much longer on each side of the fulcrum than it is in the rudder.[a] Imparting to the boat's stern a strong transverse movement, it can equally well be used for turning the vessel about when stationary in a lake or harbour.[b] We have already seen several examples (Figs. 826, 933, 953, 961, pls.) of the massive stern-sweeps of Chinese river ships.

Of all the wonders of China which so much impressed the Jesuit Louis Lecomte at the end of the +17th century, none was greater than the seamanship of the river junk-men, and this he recounted in a vivid and eloquent passage.[c]

The knack the Chinese have to sail upon Torrents is somewhat wonderful and incredible; They in a manner force Nature, and make a Voyage without any dread, which other People dare not so much as look upon without being seiz'd with some Apprehension. I speak not of those Cataracts they ascend by meer strength of Arm, to pass from one Canal to another, which in some Relations are call'd Sluices; but of certain Rivers that flow, or rather run head-long quite cross abundance of Rocks, for the space of three or fourscore Leagues. Had I not been upon these perilous Torrents my self, I should have much ado to believe, upon another's report, what I my self have seen. It is a Rashness for Travellers to expose themselves, if they have been but never so little informed of it; and a kind of Madness in Sailors to pass their life in a Trade, wherein they are every moment in danger of being destroy'd.

These Torrents whereof I speak, which the people of the Country call Chan, are met with in several places of the Empire; many of them may be seen when one travels from Namtcham (Nan-chhang) the Capital City of Kiamsi (Chiangsi) to Canton. The first time I went that way with Father Fontaney, we were hurry'd away with that Rapidity, that all the Endeavours of our Mariners could not withstand it—our Bark, abandon'd to the Torrent, was turn'd round like a Whirlegigg for a long time, amongst the sinuosities and windings that the Course of the Water form'd; and at last dasht upon a Rock even with the Water, with that violence, that the Rudder, of the thickness of a good Beam, broke like a piece of Glass, and the whole Body of it was carry'd by the force of the Currant upon the Rock, where it remain'd immoveable: If instead of touching at the Stern, it had hit sideways, we had been infallibly lost—nor yet are these the most dangerous places.

In the Province of Fokien, whether one comes from Canton, or Hamcheu (Hangchow), one is, during Eight or Ten days, in continual danger of perishing. The Cataracts are continual, always broken by a Thousand Points of Rocks, that scarce leave breadth enough for the passage of the Bark; there are nothing but Turnings and Windings, nothing but Cascades and contrary Currants, that dash one against another, and hurry the Boat along like an Arrow out of a Bow; you are always within Two foot of Shelves, which you avoid only to fall foul upon

---

[a] This is what vitiated the argument of des Noëttes (2), p. 43; he assumed that with a steering-oar the shorter arm of the lever was necessarily inboard. But Chinese stern-sweeps are often as long inboard as they are outboard, or nearly so; many examples are illustrated in Worcester (1, 2, 3), some as long overall as 100 ft. Nevertheless des Noëttes was basically right in his dictum that the steering-oar has always been essentially on the human scale, while the axial rudder can be on the scale of the ship, however large.

[b] A recent invention accomplishes this by mounting a small auxiliary screw on the blade of a steel balanced rudder (cf. p. 655 below); it is called the 'active rudder'.

[c] (1), p. 234. There is a Chinese parallel in *Kuangtung Hsin Yü*, ch. 6, p. 12a, tr. Kaltenmark (3), p. 10.

another, and from that to a third; if the Pilot, by an address not sufficiently to be admired, does not escape from Shipwrack that threatens him every moment.

There are none in all the World besides the Chinese capable of undertaking such like Voyages, or so much engag'd therein, as not to be discourag'd, maugre all the Accidents that befal them, for there passes not a day that is not memorable for Shipwracks; nay, and I wonder all Barks do not perish. Sometimes a Man is so fortunate as to split in a place not far distant from the shoar, as I chanc'd twice to do, then indeed one escapes by swimming, provided one has strength enough to struggle out of the Torrent, which is usually very strait. Other times the Bark runs adrift, and in a moment is upon the Rocks, where it remains aground with the Passengers; but sometimes it happens, especially in some more rapid Vortices, that the Vessel is in pieces, and the Crew bury'd, before one has time to know where they are. Sometimes also when one descends the Cascades form'd by the River, that altogether runs headlong, the Boats by falling all on a sudden, plunge into the Water at the Prow, without being able to rise again, and disappear in a trice. In a word, these Voyages are so dangerous, that in more than Twelve thousand Leagues that I have sail'd upon the most tempestuous Seas in the World, I don't believe I ever ran through so many Dangers for Ten years, as I have done in ten days upon these Torrents.

The Barks they make use of are built of a very thin light Timber, which makes it more fit to follow all the impressions one had a mind to give them. They divide them into five or six Apartments separated by good Partitions, so that when they touch at any place, upon any point of a Rock, only one part of the Boat is full, whilst the other remains dry, and affords time to stop the hole the Water has made.[a]

For to moderate the Rapidity of the Motion, in places where the Water is not too deep, six Seamen, three on each side, hold a long Spret or Pole thrust to the bottom, wherewith they resist the Currant, yet slackning by little and little, by the help of a small Rope made fast at one end to the Boat, and twin'd at the other round the Pole, that slips but very hardly, and by a continual rubbing, slackens the motion of the Bark, which, without this Caution, would be driven with too much Rapidity; insomuch that when the Torrent is even, and uniform, how precipitous soever its Course be, you float with the same slowness, as one does upon the calmest Canal; but when it winds in and out, this Caution is to no purpose; then indeed they have recourse to a double Rudder, made in fashion of an Oar, of forty or fifty Foot long, one whereof is at the Prow,[b] and the other at the Poop. In the plying of these two great Oars consists all the Skill of the Sailors, and safety of the Bark; the reciprocal Jerks and cunning Shakes they give it, to drive it on, or to turn it right as they would have it, to fall just into the Stream of the Water, to shun one Rock, without dashing on another, to cut a Currant, or pursue the fall of Water, without running headlong with it, whirles it about a thousand different ways—It is not a Navigation, it is a Manège; for there is never a manag'd Horse that labours with more fury under the hand of a Master of an Academy, than these Boats do in the hands of these Chinese Mariners: So that when they chance to be cast away, it is not so much for want of Skill as Strength; and whereas they carry not above eight Men, if they would take fifteen, all the

<hr/>

[a] Watertight compartments; cf. p. 420 above.

[b] The bow-sweep is not uncommon on certain types of river-junk. Examples: Worcester (1), pp. 44, 54, (2), 50, (3), vol. 2, p. 472; Audemard (3), p. 67. Its use is mentioned in a poem by Wang Chou[1] about +1510 and one sees it in the Chhing-Ming Shang Ho Thu. Moreover the principle was extended in traditional Chinese shipbuilding to rudders—a ship with one at bow and one at stern is figured in *LCCCC*, ch. 2, p. 38a. Possibly this was a paddle-wheel ship, for we have already seen an early nineteenth-century Chinese double-ruddered design of just this kind in Fig. 638 (Vol. 4, pt. 2). Nor is the principle obsolete, for modern special-purpose steamships are not infrequently equipped with bow-rudders conforming to their lines. Cf. Fig. 933 (pl.).

[1] 王周

Violence of the Torrents would not be capable to carry them away. But it is a thing common enough in the World, and especially in China, rather to hazard a Man's Life, and run the risk to lose all he has, than to be at indifferent Charges one thinks not absolutely necessary.

Lecomte's description, so graphic as to be almost incoherent, is not at all exaggerated, as may be seen from the accounts of our own contemporaries,[a] and from a picture such as that in Fig. 1024 (pl.) which shows a boat negotiating rapids in the Yangtze upstream. It illustrates very well the kind of conditions which must have sharpened men's minds to seek ever more effective vessel control, some specific needs leading to the development of the steering-oar into the greater stern-sweep, and others leading to the invention of the axial rudder. We owe to Li Sung[1] (fl. +1185 to +1215) a dramatic painting (Fig. 1025, pl.) showing boats descending a river out of Szechuan and entitled Pa Chhuan Hsia Hsia Thu.[2] The date will be of interest to us in the light of evidence shortly to be adduced, for one of the boats shows a powerful rudder, seemingly unbalanced (cf. p. 655 below). It is also interesting that we have a literary passage of almost exactly the same period which may serve as a companion piece to this painting, though it does not quite differentiate decisively between stern-sweep and rudder. The passage occurs in the *Yün Lu Man Chhao*[3] (Random Jottings at Yün-lu), written by Chao Yen-Wei[4] about +1206, but referring to events of about +1170 onwards. Chao Yen-Wei, like Louis Lecomte, gives a long and graphic account of the descent of river rapids by the boats of Chekiang and Fukien.[b] He says that the water rushes furiously past the rocks and shoals like a cataract, and therefore that the sailors use methods very different from those of palace pleasure-boats. He then goes on to describe the various kinds of oars and punt-poles, and the different commands or methods of using them. One of these is called *chhiang-kao*[5] ('in with the punt-poles!').

At the tail (stern) of the boat there is a hole (*chou wei yu hsüeh*[6]), and when the captain shouts 'Chhiang-kao!' all the punt-poles are immediately (taken) out of the water, while one man hurriedly manipulates a large projecting cudgel-like timber (*thing*[7]) at the stern. Otherwise they are afraid that the boat will sink (if it strikes a rock)...Between Chhing-thien and Wên-hsiang the water rushes around the rocks, so the boat has to be manœuvred in a curving snakelike course; otherwise it will be dashed to pieces and everybody drowned. So the local people have a proverb, saying: 'A boat may be made of paper as long as it has an iron helmsman (*thieh shao kung*[8])'...[c]

What exactly Chao Yen-Wei was trying to describe here turns largely on his mention of the 'hole' at the stern. Although it is true that numerous pictures of Roman ships show steering-oars coming out through ports in the hull,[d] such an arrangement occurs on no extant type of Chinese boat or ship, nor have we found references to it either

---

[a] E.g. Worcester (2), p. 30. And I am not without my own experiences.

[b] Ch. 3, pp. 2a ff.

[c] Tr. auct. The proverb may contain an implicit reference to the strengthening of rudders with iron bands; see immediately below, p. 633.

[d] See the illustrations in Daremberg & Saglio referred to below, pp. 635 ff.; as also Carlini (1), p. 31. La Roërie was very sceptical about these ports, however—artist's mistakes, he thought them, for the aft opening of the *parodos*, (2), pp. 39 ff.

[1] 李嵩   [2] 巴船下峽圖   [3] 雲麓漫抄   [4] 趙彥衛   [5] 搶篙

[6] 舟尾有穴   [7] 椗   [8] 鐵梢工

pictorial or literary. Openings for oars or punt-poles would be very improbable in this context. Thus the most likely interpretation would be that the river-boats on which Chao Yen-Wei travelled had axial rudders, as in Li Sung's painting, and that his 'projecting cudgel-like timber' was either the tiller or the rudder-post itself.

The limitations of the steering-oar or stern-sweep became particularly severe at sea, or upon great lakes where rough weather was likely to be met with. A ship of any size required a very considerable spar for this duty, and all the worse the consequences would be when it broke under the impact of heavy seas. Other disadvantages attended upon the attachment of a short but heavy paddle to the aft starboard quarter; it made an inconvenient projection liable to foul other ships or come into collision with quays, and that this was felt is shown by the fact that the Roman vessels were characteristically built with a kind of streamlined shield, an after extension of the *parodos*, to protect the quarter-paddle. The chief value of the quarter-paddle lay in its balanced character, a flat blade existing on both sides of the axis and not on one side only (cf. p. 655 below), but the universal adoption of the stern-post rudder in the late Middle Ages shows how the weight of advantage lay, and in China, as we shall see, the developed axial rudder retained a balanced form.

Before going further we must pause for a moment to consider what is known of the way in which rudders were attached to Chinese and European ships, and how this is still done in shipbuilding of the traditional kind. In Western antiquity steering-oars and quarter-paddles were slung in various forms of tackle,[a] probably in China also, for China remains the realm of the slung sliding rudder *par excellence*.[b] Indeed there is no evidence that any Chinese rudders were ever attached by eyes or gudgeons[c] so as to hinge with pintles[d] on the hull. In Western ships and boats the pintle was always erect, standing parallel with the stern-post and pointing upwards if attached thereto, downwards if attached to the rudder itself. Such hooks and eyes were foreign to Chinese usage. Throughout the ages their rudders have been held to the hull primarily in wooden jaws or sockets, and, if large, suspended from above by a tackle pulling on the shoulder so that they can be raised or lowered in the water (cf. Figs. 1026, 1027, pl.). Sometimes the foot of the largest type of rudder is even connected with the fore-part of the ship by bousing-to tackle which holds it in place.[e] The gudgeon-like fittings

---

[a] Cf. Carlini (1); la Roërie (1, 2).

[b] Cf. Worcester (1, 3); Lovegrove (1); Waters (5); Underwood (1).

[c] In the sense of the bearings of shafting the first use of this word is c. +1408, but by +1496 'the ring or eye in the heel of a gate, which turns on the hook in the gate-post'. The first recorded nautical use is +1558. As Gr. γόμφος (gomphos) meant a bolt or pin, there may have been a transference of meaning from the male to the female element in the hinge, perhaps because of the introduction of a stronger male term.

[d] The word has the same anatomical derivation as pencil and penicillin. The first recorded use of 'pyntle and gogeon' is +1486.

[e] This ingenious arrangement (seen in Fig. 939, cf. p. 404) does not seem to have been recorded in print by Worcester, Donnelly, Waters, or other modern observers, but it has certainly not died out. Charnock noted it, (1), vol. 3, pp. 290 ff., and it appears in two of the Maze Collection models (nos. 1 and 11). The bousing-to cables or chains are traditionally made fast at the bows by a windlass, and when therefore we find a very prominent one pictured there (e.g. in Fig. 1033a) we should bear in mind that it may be for this purpose rather than for weighing the anchor, as we naturally tend to assume. If the bousing-to tackle does not extend as far forward as the bows it may be taken inside the ship's stern and tightened by a windlass fixed to a bulkhead there (Fig. 1026).

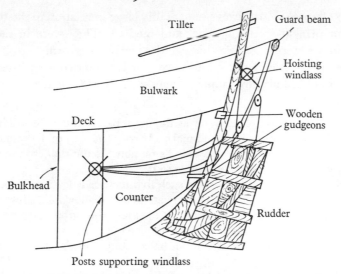

Fig. 1026. The rudder and tiller of a Hangchow Bay freighter (sketch after Waters, 5). See text.

(bearings, as it were, for the main rudder post) can be open, half-open, or occasionally altogether closed by outer pieces of shaped timber (cf. Fig. 1028). Thus they have cor-responded to the braces (eyes, gudgeons) of occidental ships, though what turned in them was not the pintle but the rudder-post itself, suitable apertures being present in the blade, when necessary, to allow of the helm being put hard over without inter-ference from the socket.[a] The system of open jaws was what permitted the rudder to be raised and lowered.

Though cable and wood thus took the place of iron hinges, it should not be thought that iron was absent from rudders in tra-ditional China, for in fact the larger ones,

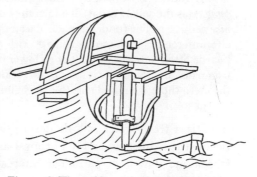

Fig. 1028. The rudder and tiller of a Cantonese Kwailam junk (sketch after Lovegrove, 1). See text.

weighing many tons, were, and are, heavily bound with iron straps and other strength-enings.[b] The Chinese rudder was by no means necessarily located at the aftermost point of the hull or upperworks, but sometimes considerably forward of this, its post indeed frequently descending through a rudder-trunk built into the hull. Such a con-struction was facilitated by the transom-and-bulkhead anatomy so typical of the Chinese junk, and we shall see in the dénouement how closely the whole conception of the vertical steering mechanism was connected with this.[c] The essential point is that

[a] See Worcester (3), vol. 1, pp. 106 ff.
[b] The reader will remember the recent recovery of one of the Ming Treasure-ships' rudder-posts, over 36 ft. in length, by Chou Shih-Tê (1); cf. p. 481 above.
[c] The archetype can be grasped immediately by a glance at Fig. 1028, a drawing by Lovegrove (1) of the manner of attachment of the rudder on a Cantonese Kwailam junk. A similar arrangement is seen in

the rudders of Chinese ships, always remarkably large in relation to the total size of the vessel, were in principle vertical, axial and median. They were in fact 'stern-post rudders' without a stern-post. To this paradox we shall return.

To end this introduction let us quote a few words from two +17th-century witnesses, both of whom are already familiar friends. In his *Thien Kung Khai Wu*, Sung Ying-Hsing wrote:[a]

The nature of a ship is to follow water as the grass bends under the wind.[b] Therefore a rudder (*tho*[1]) is constructed to divide and make a barrier (*chang*[2]) to the water, so that it will not itself determine the direction of the vessel's motion. As the rudder is turned, the water turbulently presses on it, and the boat reacts to it.

The dimensions of the rudder should be such that its base is level with the bottom of the (inland transport) ship. If it is deeper, even by an inch, a shallow may allow the hull to pass but the stern with its rudder may stick firmly in the mud (thus grounding the vessel); then if the wind is at gale strength that inch of wood will give rise to indescribable difficulties. If the rudder is shorter, even by an inch, it will not have enough turning force to bring the bows round.

The water divided and obstructed (*chang*[2]) by the rudder's strength, is echoed (*hsiang ying*[3])[c] as far as the bows; it is as if there were underneath the hull a swift current carrying the vessel in the very direction desired. So nothing needs to be done at the bows. All this is marvellous beyond words.

The rudder is worked by a tiller attached to the top of its post, a 'door-bar' (*kuan mên pang*[4]) (as the sailors call it). To turn the boat to the north the tiller is thrust to the south, and *vice versa*...[d] The rudder is made of a straight post of wood [more than 10 ft. long and 3 ft. in circumference for the grain-ships][e] with the tiller at the top, and an axe-shaped blade of boards fitted into a groove cut at its lower end.[f] This blade is firmly fastened to the post with iron nails, and the whole is fixed (with tackle) to the ship to perform its function. At the end of the stern there is a raised part (for the helmsman) which is also called the 'rudder-house' (*tho lou*[5]).[g]

This is a fresh-water sailor's description, and Sung himself must often have stood beside the helmsman on the Poyang Lake or the Grand Canal; his attempt to describe streamline flow is of particular interest. Half a century later Louis Lecomte wrote likewise of sea-going junks:[h]

Their Vessels are like ours of all Rates, but the Model is not so fine; they are all flat-bottom'd; the Fore-castle is cut short without a Stem, and the Stern open in the middle, to the

Fig. 1026, a longitudinal section, drawn by Waters (5), of the stern of a Hangchow Bay freighter. See further in Worcester (1), pp. 5 ff., (3), vol. 1, pp. 144 ff.; Audemard (3), pp. 22 ff.

  [a] Ch. 9, pp. 3*b*, 4*a*, tr. auct. adjuv. Ting & Donnelly (1); Sun & Sun (1).|

  [b] An age-old Confucian analogy, with the people as the grass and the prince with his officials as the wind.

  [c] On this term cf. Vol. 2, pp. 89, 282, 304, Vol. 4, pt. 1, pp. 130, 159, 234. Here we see again the wide significance of action and reaction in Chinese conceptions of natural processes.

  [d] Here follows the passage on leeboards already given, p. 619 above.

  [e] Sung Ying-Hsing's own commentary.

  [f] He was clearly describing the balanced rudder—see below, p. 655 and Fig. 1043 (pl.).

  [g] On the 'wheelhouse' and the poop cf. p. 453 above, p. 644 below. 'Also', because there were other terms, cf. p. 405 above.

  [h] (1), pp. 230 ff.

  [1] 舵   [2] 障   [3] 相應   [4] 關門棒   [5] 舵樓

end that the Rudder, which they shut up as in a Chamber, may be defended on the Sides from the Waves: this Rudder, much longer than ours, is strongly ty'd to the Stern-post[a] by two Cables that pass under the whole length of the Vessel to the fore-part, and two other suchlike Cables hold it up, and facilitate the hoisting or lowering it, as occasion serves. The Bar (i.e. the tiller) is as long as is necessary for the guiding it; the Seamen at the Helm are also assisted by Ropes fastened to the Larboard and Starboard, and roll'd upon the extremity of the Bar they hold in their Hand, which they fasten or slacken as they see occasion, to thrust or stop the Helm.

Lecomte thus aids us by showing that some mechanical assistance or advantage was applied to the tiller in large junks in his time.[b]

(2) FROM STEERING-OAR TO STERN-POST RUDDER IN THE WEST

Jal (1) was the first to notice that the stern-post or axial rudder began to appear in Europe early in the +13th century. All subsequent investigations have established that before that time there was no trace of it in the West.[c] Ancient Egyptian boats are generally shown with steering-paddles at the stern, sometimes as many as five a side, each worked by one man.[d] Or there may be two quarter-paddles connected together by a framework and bar (like that on the Ajanta ship later).[e] The stern-sweep was also known in ancient Egypt; it was fixed to the end of the high stern, supported by a post at its forward end and furnished with a 'tiller' by which it could be rotated by the helmsman standing on the deck.[f] Steering-oars on each quarter appear also on the double-decked Phoenician or Greek ships of the −7th century shown in the famous Sennacherib reliefs.[g] Single steering-oars are found in Etruscan tomb-paintings[h] and Greek vases,[i] though Greek ships generally had one on each side.[j] Steering-oars were also universal on Roman and Hellenistic ships, sometimes single,[k] sometimes double

[a] Here of course he went astray—there was none.
[b] The point is of some importance, cf. p. 651.   [c] See on, p. 637.
[d] Boreux (1), pp. 21, 34, 160, 162, 260, opp. p. 272, 395; des Noëttes (2), pp. 10 ff., figs. 1 to 10; Wilkinson (1), vol. 1, pp. 412, 414; la Roërie (2).
[e] Moll (1); des Noëttes (2), figs. 13, 16. This last is the reconstruction by von Bissing & Borchardt of the 60 ft. masonry ship of the sun-god Ra, built by 5th-dynasty Memphis kings (Boreux (1), p. 104).
[f] Des Noëttes (2), figs. 13 a, b; Wilkinson (1), vol. 2, p. 124, figs. 400, 401, p. 128, fig. 402; Hornell (1), p. 219; Reisner (1), no. 4,951, pl. XXIII; Boreux (1), pp. 273, 400. Model paddles exist. Steering-gear of exactly this type is still to be found on some Indian boats, such as those of the Ganges depicted by Crealock (Fig. 1029, pl.). It may also be seen in a Gallo-Roman tomb-carving (Bonnard (1), fig. 6, opp. p. 146). Adam & Denoix (1), p. 96, suggest that all these steering-paddles could be disengaged from the frameworks and used as free sweeps when the ship was almost without way on her.
[g] Des Noëttes (2), figs. 23, 24; Daremberg & Saglio, fig. 5263; Adam & Denoix (1), p. 101.
[h] Des Noëttes (2), fig. 29; Bartoccini (1), pl. v; Moretti (1); Bloch (1), pl. 48; Lawrence (1), opp. p. 68. A very curious oblong object with two long projections, depicted in coloured relief on one of the columns in the −3rd-century Etruscan 'Tomb of the Reliefs' at Cerveteri (see Pallettino (1), p. 35, (2), pl. 43; Giglioli (1), pls. 342, 343, p. 64) has been taken for a rudder with its projecting eyes or gudgeons. But the shape makes this impossible, and we can hardly go beyond old Dennis, who described it, (1), vol. 1, p. 254, as 'a nondescript piece of furniture more like a double lamp-bracket than anything else'.
[i] Des Noëttes (2), figs. 35, 36; Daremberg & Saglio, figs. 3664, 3665.
[j] Des Noëttes (2), figs. 30, 31, 33, 34, 37, 43; la Roërie & Vivielle (1); Daremberg & Saglio, figs. 5265, 5282, 5288.
[k] Des Noëttes (2), figs. 53, 54, 55; Poujade (1), p. 133; Daremberg & Saglio, figs. 884, 885, 5271, 5272, 5273, 5274, 5290, 5294.

and united by a bar.[a] Evidence exists that they now began to be slung permanently in position on the quarter by tackle, and the streamlined shield was introduced.[b] Nothing new was developed by the Byzantine culture, but the Viking long-ships, beginning with the steering-oar, went on to attach it to a pivot,[c] and ended by converting the paddle to a rudder-shaped form and hingeing it on the side of the boat.[d] Some of these quarter-rudders were also slung.[e] Norman ships continued the same methods, but the steering-oars on the Bayeux tapestry of *c.* +1080 are still very primitive in type.[f] Indeed, this remains the case until the end of the +12th century,[g] and even long afterwards artists and sculptors continued to represent steering-oars instead of rudders,[h] although the latter were becoming universal.[i] During this period steering-oars and rudders sometimes co-existed on the same vessel, each being found useful for different purposes.[j]

The oldest European MS. illustration of a stern-post rudder, with tiller, is in a Latin commentary on the Apocalypse preserved at Breslau, of +1242, as des Noëttes (2) recognised.[k] But Brindley (5) had already pushed the date of introduction somewhat further back by finding a notable iron-bound rudder on the ship depicted in the seal of Ipswich, which came into use in the close neighbourhood of +1200.[l] A number of other seals of the +13th century (e.g. Elbing, +1242; Wismar, +1256; Stubbkjoep-

---

[a] Des Noëttes (2), fig. 58.

[b] Des Noëttes (2), fig. 56; Moll (1); la Roërie (1, 2); Daremberg & Saglio, figs. 5289, 5291, 5293, 5295. Seyrig (1) figures an ex-voto bronze of +121 which seems almost a scale model.

[c] As on the Oseberg ship of about +900, cf. Mercer (1), p. 251; des Noëttes (2), fig. 64; Gille (1).

[d] As on the Rebaek ship; Sølver (2).

[e] On the Nydam and Gokstad ships' steering arrangements see la Roërie (2); la Roërie & Vivielle (1), vol. 1, p. 177; Gille (8).

[f] Des Noëttes (2), fig. 65.

[g] For example, a +9th-century French MS. (des Noëttes (2), fig. 63), and a +12th-century Venetian one (fig. 66).

[h] Seal of Dover, late +13th (des Noëttes (2), fig. 67), and Lat. MS., Bib. Nat. no. 8,846 (fig. 68); for the +14th, figs. 69 and 71; for the +15th, fig. 71 a.

[i] At sea, that is to say. Stern-sweeps have persisted on some European river-ships until the present day, notably the interesting *barco rabêlo* of the Douro (cf. Filgueiras, 1), with its high and very 'Chinese' steersman's bridge. Another curious survival of steering-oars to our own day exists in those of the boats used for the traditional water-jousts at Lyons. Cf. Boreux (1), p. 206.

[j] The steering-oar or stern-sweep can help not only in harbour and estuary manœuvres when the ship has hardly any way on her, but also at sea if all way is lost on tacking (cf. la Roërie (1), p. 579; Carlini (1), p. 7). Adam & Denoix mention three +15th-century illustrations of ships clearly showing steering-oars as well as rudders; two of these are reproduced in la Roërie (2), figs. 18, 19. The combination, though not very common, has long been known in Chinese practice; la Roërie (2), fig. 20, figured a Thaiwan fishing-boat using two steering-oars (though equipped with rudder) while attending to nets. Adam & Denoix maintain that the disappearance of the steering-oar was brought about not by the rudder itself, but by the whip-staff and the wheel, which applied mechanical advantage to the control of it. In spite of the illustrations just mentioned, this thesis is at variance with the facts. By about +1500 steering-oars had quite gone out, but the whip-staff did not come in until about +1600, and the wheel not till the neighbourhood of +1700.

[k] His fig. 75; also reproduced in Alwin Schultz (1), vol. 2, p. 335 (fig. 149), who did not point out its importance. The MS. is *Alexandri Minoritae Apocalypsis Explicata*; I have not been able to trace further details about it in Potthast or Chevalier. About the same date is the rudder of the ship in one of the stained-glass window medallions in the Lady Chapel of the cathedral at Le Mans.

[l] See Jewitt & Hope (1), vol. 2, p. 331. This is just about the time at which were carved, according to Lynn White (14), p. 161, the font reliefs mentioned on the opposite page. The hardness of the stone is thought by some to have imposed a ruder, more archaic, style than was by then usual.

ing, Harderwyk, +1280; Damme, +1309)[a] also show ships with stern-post rudders.[b] But even the Ipswich seal is still not the first representation, for the ships which appear carved on certain fonts in Belgium and England show them, and these sculptures, made by a school of artisans from Tournai, are datable at about +1180. The two best examples still exist at Zedelghem (Fig. 1030, pl.) and Winchester.[c]

Subsequent developments related rather to the control devices than to the rudder itself.[d] In the +17th century it was common to have a second, vertical, lever (known as the whip-staff) attached to the forward end of the tiller.[e] This system may have derived originally from the crank at the end of the steering-paddle which we see in the sculptured ship of the shrine of St Peter Martyr in the basilica of St Eustorgio at Milan.[f] It disappeared about +1710, with the general introduction of the wheel and tiller ropes of hide (later chains).[g]

In sum, therefore, we may take +1180 as the dateline for the first introduction of the stern-post rudder in Europe.[h] Evidence will now be presented which indicates that at this period the axial rudder had already been known for a very long time in China. This has been sensed (on inadequate grounds) by a number of occidental scholars,[i] and van Nouhuys even cited a 'Far Eastern document of +1124' which appeared to concern Korean craft, but for which he gave no reference.[j] Of the importance of the question in the history of technology there can be little doubt. The rapidity of the effect may perhaps be seen in the fact that the crusades of the +13th century, unlike the earlier ones, were maritime expeditions. And the +15th century saw the beginning of the great explorations round the coast of Africa, leading to the Western domination of the Indian Ocean and the discovery of America. It is quite noteworthy that the stern-post rudder is attested first for European ships within a very few years, perhaps less than a decade, of the first occidental mention of the mariner's compass.[k] This fact

[a] The date +1309 is also that of the earliest mention of the stern-post rudder in the Mediterranean (the *Chronicle of Villani*; la Roërie, 2). The first use of the word rudder in English is +1303.

[b] Some of these seals seem to show a double rudder, i.e. two rudders placed very far aft on each quarter, or so Brindley thought; but others (e.g. Anderson and la Roërie) do not agree with this view, attributing the effect to imperfections in the dies. Cf. des Noëttes (2), figs. 73, 74, 76, 78. Quarter rudders have been re-invented for some Rhône barges (Benoit, 2).

[c] Descriptions by Swann (1); Eden (1). The story which the Winchester font illustrates is that in the *Legenda Aurea*, vol. 2, p. 120. Cf. Brindley (3); Anderson & Anderson (1); la Roërie & Vivielle (1), vol. 1, p. 193. Some experts still decline to accept the Winchester carving as showing anything more than a quarter-rudder (e.g. Clowes (2), p. 48).

[d] On terminology see the curious studies of Drapella (1).

[e] La Roërie (1); Clowes (2), p. 79; Halldin & Webe (1).

[f] Des Noëttes (2), figs. 72, 86. Cf. Anderson & Anderson (1). This was slung. Variable depth rudders lasted on very late on the Rhône at Arles (Benoit, 2).

[g] Chatterton (1); Anderson & Anderson (1), p. 164; Gilfillan (1), p. 70.

[h] All attempts to find evidence of it earlier have proved unconvincing (cf. Febvre, 4). Nordmann (1) and Laurand (1) discussed a passage in Lucian which turned out to refer to the double steering-oar connected by a bar. Nothing else emerged from the studies of de St Denis (1) on the vocabulary of ancient Mediterranean sailors. Verwey (1) described earthenware models of boats dug out of a peat bog amongst +10th-century Saxon pottery; these had an unusual vertical stern-post with a hole in it. But naval archaeologists who discussed the find did not agree that it provided evidence of a stern-post rudder.

[i] E.g. Smyth (1), p. 373; Gilfillan (2); la Roërie (2); Elliott-Smith (2); Landström (1), pp. 218 ff.

[j] Commenting on Verwey (1). The explanation is on p. 642 below.

[k] See Sect. 26i in Vol. 4, pt. 1, and on 'transmission clusters', Vol. 4, pt. 2, pp. 544, 584.

alone might arouse one's suspicion that the stern-post rudder was not an auto-chthonous development either, but made its appearance as the result of long travel from somewhere else.

(3) CHINA AND THE AXIAL RUDDER

The problem of the history of the rudder in China presents us with a classical case of the difficulty which arises when there is reason to believe that one single word has done duty through the centuries for two or more devices technologically quite distinct. The word *tho* (or *to*)[1, 2, 3, 4] certainly meant 'steering-oar' or 'steering-paddle' in the −3rd century; equally certainly it meant the axial 'stern-post' rudder in the +13th. Since, as we have just seen, the first appearance of the latter in Europe antedates +1200 by very little, any investigation of a possible Chinese contribution cannot rely upon names and words alone. It is necessary to see what everyone who used the word actually said about the thing. If any reader should find such an enquiry tedious, he has only to turn to p. 649 where the result is stated, but the method is unavoidable.

(i) *Textual evidence*

The simplest procedure is to group the relevant texts into separate classes. Let us deal first, (*a*) with early mentions, probably mostly referring to the steering-oar, then (*b*) consider what verb is used in connection with the *tho*, (*c*) what its shape and length, (*d*) how it is said to be fixed, and (*e*) of what material it is made.

One of the oldest references must be that in the *Huai Nan Tzu*, about −120, and here the archaic word *to*[5] is employed. The writer says:[a] 'If the will is there, people are capable of destroying a boat to make a *to*; or melting down a big bell to make a little one.' After that, the word *tho* takes its place,[b] as in the following examples:

Chhien Han Shu,[c] referring to −107:
> 'Han Wu Ti had his boats (*chu-lu*) floating on the water for a thousand *li*.'
>> Commentator Li Phei[6] of the San Kuo period, +3rd century, says: 'The *chu*[7] is the after part of the boat where the *tho* is held (*chhih*[8]); the *lu*[9] is the forward part of the boat where the oars (*cho*[10]) are shipped (*tzhu*[11]).'

Yen Thieh Lun (Discourses on Salt and Iron),[d] c. −80, by Huan Khuan:
> '(The promotion of mediocrities to positions of great responsibility) is like setting out to cross rivers or seas without oars or steering-gear (*ju wu chieh chu chi chiang hai*[12]), only to be carried away by the first storm encountered, and sunk in the abysses full hundred fathom deep, or blown eastward into the shoreless Ocean.'[e]

---

[a] Ch. 17, p. 11*b*, tr. auct. The reference to the bell is perhaps to be explained by an incident already related (Vol. 4, pt. 1, pp. 170, 204).

[b] Or, still more vaguely, *chu*.

[c] Ch. 6, p. 25*b*; cit. *TPYL*, ch. 768, p. 3*a*; *Tzu Shih Ching Hua*, ch. 158, p. 12*a*; cf. Dubs (2), vol. 2, p. 95.

[d] Ch. 21, p. 7*a*, tr. Gale, Boodberg & Lin (1).

[e] This last remark is of particular interest in view of the early Chinese explorations of the Pacific discussed elsewhere, pp. 551 ff.

---

[1] 柂   [2] 舵   [3] 杝   [4] 舳   [5] 杕   [6] 李裴   [7] 舳
[8] 持   [9] 艫   [10] 櫂   [11] 刺   [12] 如無檝舳濟江海

Fang Yen (Dictionary of Local Expressions),[a] −15, by Yang Hsiung:

'The stern of a boat is called *chu*.[1] The *chu* controls (*chih*[2]) the water.'

The commentary, ascribed to Kuo Pho (early +4th century) says 'Nowadays Chiang-tung people pronounce it like *chu*[3] (axle)'. Yen Shih-Ku (early +7th century), commenting on the above *Chhien Han Shu* passage, repeats this. The idea of something pivoting was evidently not far from their minds.

Shih Ming (Dictionary of Terms),[b] *c.* +100, by Liu Hsi:

'The tail of a ship is called *tho*. Now *tho* is like *tho*,[4] to pull. At the stern (indeed) you see the (man) pulling; this is to direct the ship. By the aid (of the *tho*) you can make the boat go in the desired direction, and not let it drift obliquely.'

Tzhu Shih Chi Hsieh Fu[5] (Ode to Cure quickly the Evils of the Age)[c] by Chao I[6] (*fl.* +178):

'It is like a ship travelling in the sea and losing its *tho*, or like somebody sitting on a pile of fuel waiting for it to burn.'

Hou Han Shu,[d] quoting a poem by Chungchhang Thung,[7] *c.* +210:

'Let the vital force (in meditation) be my boat,
And the light breezes (of renunciation) my *tho*.'

Yu Hsien Shih[8] (Poem of the Wandering Hsien),[e] by Chang Hua,[9] *c.* +285:

'The wandering immortal comes (at last) to the furthest West,
Where the "Weak Water" lies beyond the Shifting Sands,[f]
Then with the clouds for rowers, and the dew for his *tho*,
His boat begins to glide over the flying waves.'

Chiang Fu[10] (The River Ode),[g] by Kuo Pho, early +4th century:

'...riding over the waves and using the *tho*...'

Sun Chho Tzu[11] (The Book of Master Sun Chho),[h] early +4th century:

'Action without control and without principle is like a boat without a *tho*.'

Sun Fang Pieh Chuan[12] (Unofficial Biography of Sun Fang),[i] late +4th century:

'The Venerable Mr Yü[13] established a school. Sun Fang was the youngest and always took the last place when the boys lined up. When Yü Kung asked him why, he said: "Have you not seen the *tho* on a boat? It comes last but it controls the course".'

Evidently some of these passages give no technical help at all; they certainly show that poets now and then had a clear appreciation of the importance of steering-gear, but no one thought of describing it in detail. The best clue is the statement of Kuo Pho that the word *chu* for stern was in some districts pronounced exactly the same as the cognate word which means axle—and indeed both characters share the same phonetic,

[a] Ch. 9, p. 8*a*; cit. *TPYL*, ch. 771, p. 4*b*.
[b] Ch. 25 (p. 378); cit. *TPYL*, ch. 771, p. 4*b*.
[c] *CSHK*, Hou Han section, ch. 82, p. 8*b*.
[d] Ch. 79, p. 14*b*; tr. Balazs (1) who, however, used the word 'gouvernail'.
[e] Cit. *TPYL*, ch. 771, p. 4*b*.　　　　　[f] Cf. Vol. 3, pp. 607 ff.
[g] *CSHK*, Chin section, ch. 120, p. 3*b*. We are indebted to Prof. H. H. Dubs for this passage.
[h] *YHSF*, ch. 71, p. 11*a*.
[1] In the *Chin Shu* bibliography, without author's name. The official life of Sun Fang is in *Chin Shu*, ch. 82. Our passage is cited in *TPYL*, ch. 771, p. 4*b*, and by Yü Shih-Nan in his +7th-century *Pei Thang Shu Chhao*.

[1] 軸　　　　[2] 制　　　　[3] 軸　　　　[4] 拖　　　　[5] 刺世疾邪賦
[6] 趙壹　　　[7] 仲長統　　　[8] 遊仙詩　　　[9] 張華　　　[10] 江賦
[11] 孫綽子　　　[12] 孫放別傳　　　[13] 庾公

differing only in radical.[a] The scent grows warm, for here is something 'turning' like an axle at the stern. But the force of this is still insufficient to distinguish between the steering-oar or stern-sweep pivoted on bumkin or rowlock and the rudder held in bearings like those of a rotating shaft or axle. In other words (the language of machinery) we still do not know whether 'point-closure' or 'line-closure' was meant.[b]

We can perhaps make headway a little further by studying the verbs used to describe the pivoting motion. The word *chuan*,[1] for instance, has the nuance of something swinging round an axis rather than of something pivoting on a single point. And we meet with this already in the +3rd century, in a story about Sun Chhüan,[2] afterwards emperor of the Wu State.[c]

Sun Chhüan, when at Wu-chhang, took out a large newly built ship called the 'Chhang-An' on trials at Tiao-thai. There came a strong wind, and Ku Li[3] asked the helmsman to steer (the ship) to Fan-khou. But Sun Chhüan said that they ought to go to Lochow. (Ku) Li then drew his dagger and faced the helmsman, saying that if he did not make Fan-khou he would be beheaded. So he immediately turned the helm (*chuan tho*;[4] whatever it was), and sailed into Fan-khou.

At any rate, this verb came naturally to a late +13th-century poet, Yü Po-Sêng,[5] who used exactly the same expression for what was by then (+1297) undoubtedly an axial rudder.[d] A more important piece of evidence occurs in a poem (Phêng Li Shih[6])[e] of the great radical minister Wang An-Shih[7] (+1021 to +1086) where he says: *Tung hsi lieh tho wan chou hui*[8]—'turning their rudders from east to west, the ten thousand ships returned'. The axial force of the word *lieh* can be appreciated from the fact that we have already come across it in the +17th-century expression for a pulley-block (*kuan-lieh*[9]),[f] and in multifarious combinations of engineering technical terms, including pivots, much earlier than that.[g]

Next comes the consideration of shape and length. In a lost +5th-century book by Shan Chhien-Chih[10] called the *Hsün-Yang Chi*[11] (Memoirs of the Chiu-chiang District), we find the following passage:[h]

Among the western ranges of Lu Shan[12] there are springs of sweet water.[i] A *tho* was once seen floating down the stream (from the lake fed by these springs), so people called it the

---

[a] The semantic range of this phonetic (cf. K 1079) is wide, and we do not know the meaning of its archaic graph, but unless some transference took place, it appears in *chou*[13] as the drawing of a spiked hat or helmet. Is it possible that this word-component could have been borrowed by the artisans for the bearings or journals at the ends of shafts, often taking the form of spikes rotating in sockets?

[b] Cf. Vol. 4, pt. 2, p. 68.

[c] From the *Chiang Piao Chuan*[14] of Yü Phu,[15] quoted in the commentary of the *San Kuo Chih*, ch. 47 (*Wu Shu*, ch. 2), p. 18*b*, tr. auct.

[d] Cit. *Lu Thang Shih Hua*, p. 2*a*.   [e] Cit. *Khang-Hsi Tzu Tien*, art. *lieh*.

[f] Cf. above, p. 604.

[g] Cf. Vol. 3, p. 314 and Vol. 4, pt. 2, pp. 235, 292, 485. One of these technical terms, it will be remembered, had to be translated as 'projecting lug'. This would be well suited to the tiller.

[h] Cit. *TPYL*, ch. 771, p. 4*b*, tr. auct.

[i] This mountain overlooks the Poyang Lake, bearing the modern resort of Kuling. A fine +13th-century picture of it is found in Anon. (*32*), no. 30.

[1] 轉	[2] 孫權	[3] 谷利	[4] 轉柁	[5] 虞伯生	[6] 彭蠡詩
[7] 王安石	[8] 東西捩柂萬舟回		[9] 關捩	[10] 山謙之	[11] 尋陽記
[12] 廬山	[13] 胄	[14] 江表傳	[15] 虞溥		

Tho-hsia Chhi[1] (Stream of the Downcoming *Tho*). Men sent by (Dukes) Hsüan and Mu found the remains of a flat-bottomed skiff (*phien*[2]) by the lake, so it was seen that the story was true.

We need not take this report as evidence for the times of the Chou rulers mentioned, but it does say something about the steering devices used in Shan's own period. For it implies that it was now quite possible to distinguish a rudder from an oar by its shape. This would not have been the case if the steering-oar had still been the only device in use, for it would not readily have been distinguished from any other oar, and one of the ordinary words for oar would have been employed. So by about +450 at any rate there was a clear difference in shape.

More important, perhaps, is the passage in the *Kuan shih Ti Li Chih Mêng*[3] (Master Kuan's Geomantic Instructor), an obscure book which has already been mentioned in connection with the history of the magnetic compass.[a] Though attributed to Kuan Lo[4] of the +3rd century, it cannot be earlier than the Thang, and does not seem to be later than the end of the +9th. In speaking of the proper depth for tombs, which should be neither too deep nor too shallow, the writer says:[b]

If the hairpin is shorter than it ought to be, its ornamented end will not show. If the key is shorter than it ought to be, the lock of the box cannot be secured. If the *tho* goes deeper than it ought to, then the end of the boat will not carry its cargo (because it will go aground or strike a rock).

This gives practical certainty, for the typical Chinese rudder has always been slung adjustably so that it can hang down well below the level of the ship's bottom, and aid in preventing leeward drift.[c] Moreover, in the following century, we have a remark by Than Chhiao[5] in the *Hua Shu*[6] (about +940) which also concerns the length of a rudder. There he says:[d] 'The control of a ship carrying ten thousand bushels of freight is assured by means of a piece of wood no longer than one fathom (*hsün*[7]).' This length (8 ft. in Chinese measurement, here just under 8 ft. 2 in. in ours) is clearly much too short for a steering-oar or stern-sweep (often over 50 ft. on comparatively small river boats), but agrees well with an axial rudder.

The most decisive passage, however, occurs in the *Hsüan-Ho Fêng Shih Kao-Li Thu Ching*[8] (Illustrated Record of an Embassy to Korea in the Hsüan-Ho reign-period), written by Hsü Ching[9] about the mission of +1124. This must be the text which van Nouhuys had heard of somewhere. It is full of references to rudders, often to mishaps when they broke,[e] or to changing them,[f] but the main sentences are these:[g]

---

[a] See Vol. 4, pt. 1, pp. 276, 302, 310.
[b] Cit. *TSCC*, *I shu tien*, ch. 657, *hui khao* 7, p. 1*a*, tr. auct.
[c] Besides, the Chinese steering-oar or stern-sweep offers only a trailing resistance which would be raised by any submerged object and could not catch on it.
[d] P. 9*b*, tr. auct. The weight mentioned corresponds to 700 tons; cf. pp. 230, 304, 441, 452, 600, 645.
[e] Ch. 34, p. 10*b*; ch. 39, pp. 3*a*, 4*b*; ch. 40, p. 2*a*.
[f] Ch. 39, p. 3*b*.
[g] Ch. 34, p. 5*a*, tr. auct.

---

[1] 柂下溪   [2] 艑   [3] 管氏地理指蒙   [4] 管輅   [5] 譚峭
[6] 化書   [7] 尋   [8] 宣和奉使高麗圖經   [9] 徐兢

At the stern, there is the rudder (*chêng tho*[1]), of which there are two kinds, the larger and the smaller. According to the differences in the depth of the water, the larger is exchanged for the smaller, or vice versa. Abaft the deckhouse (*chhiao*[2]) two oars (*cho*[3]) are stuck down into the water from above, and these are called 'Third-Assistant Rudders' (*san fu tho*[4]). They are only used when the ship begins to sail in the ocean.[a]

These passages leave no doubt in the mind that by the beginning of the +12th century, and on Chinese ships,[b] several sizes of axial rudders[c] were carried and used under different conditions, while at the same time steering-oars may have been retained for special purposes. This was sometimes done long afterwards on European vessels,[d] and there are Persian drawings of the +15th century which show ships with two steering-oars as well as a quarter-rudder.[e]

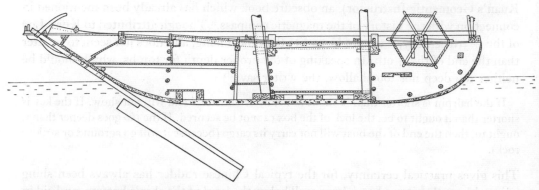

Fig. 1031. Longitudinal section of a medium-sized (*c.* 57 ft.) fishing vessel of traditional type as constructed at the small havens on Kanghwa Island in Kyonggi-do (the metropolitan) province on the West coast of Korea (after drawings in Underwood (1), figs. 14, 32). The extremely elongated rudder, which acts as a centre-board when fully lowered, slides up and down in jaws, and its post has a series of sockets into any one of which the tiller may be placed to suit the helmsman's convenience. There are only two bulkheads but the transom bow and stern are extremely bluff. The two masts carry very tall Chinese lug-sails trimmed only slightly assymetric. The size of the anchor winch is also notable.

Korean craft of the early +12th century also had axial rudders, for Hsü Ching, describing the coastguard patrol boats (*hsün chhuan*[5]) which came out to welcome the ambassador's flotilla, says that they were one-masted, with no deckhouse, and that 'they have only a (sculling) oar (*lu*[6]) and a rudder (*tho*[7]) (at the stern)'.[f] If the latter had not been something different from an oar, he would hardly have mentioned the

---

[a] This suggests that these 'oars' were really leeboards (cf. p. 618). It is unlikely that additional steering-oars (though sometimes carried as well as rudders, cf. p. 636) would have been used 'in the ocean'. The 'assistance' was in this case against leeward drift.

[b] The ships of the ambassadors were certainly Chinese, and we are told that the Koreans were somewhat surprised at their size. Cf. p. 463 above.

[c] There is nothing in the text which proves that the rudders were axial ones rather than quarter-rudders, but to assume the latter would be gratuitous since no trace of them in China exists, either in literature or surviving traditional practice. Poujade (1), p. 259, agrees.

[d] Two excellent +15th-century illustrations showing this will be found in la Roërie (2), figs. 18, 19.

[e] Moll (1); des Noëttes (2), fig. 93.

[f] Ch. 33, p. 1*b*.

[1] 正柁　　[2] 嵩　　[3] 棹　　[4] 三副柁　　[5] 巡船　　[6] 艣
[7] 柁

two together in this way.[a] The full meaning of the change described from one rudder to another according to the depth of the water can be appreciated by a glance at the extremely deep rudders, acting also as centre-boards, which Korean ships have retained until the present day (Fig. 1031).[b]

A Japanese record, as graphic and personal as that of Hsü Ching, permits us to take back his technical statements fully as far as the time of the book of Master Kuan the geomancer. This is the *Nittō Guhō Junrei Gyōki* (Record of a Pilgrimage to China in search of the Buddhist Law) by the Japanese monk Ennin. What he says of the sea-going ships in which he travelled (and in which he was sometimes shipwrecked) shows that early in the +9th century axial rudders co-existed with steering-oars or stern-sweeps.[c] Accompanying a Japanese embassy to the Chinese court in +838, his landfall on the coast of the mainland was highly inauspicious.[d] A fierce storm drove the ship on to a shoal on the shores of Huainan (Chiangsu, north of the Yangtze estuary), where 'the corners of the rudder (*tho chio*[1]) snapped in two places'. Soon afterwards the blade of the rudder stuck fast in the muddy sand, so the crew abandoned it and cut down the mast, whereupon the ship bumped shorewards and was left, partly broken up, by the withdrawing tide when the tempest slackened in the morning. Here was no steering-oar. But there is a probable mention of one in the diary a year later, when Ennin was returning home[e] in another Japanese ship and sailing east or north-east along the southern coast of Shantung. There, in +839, he says that he saw the sun setting 'in the middle of the great oar (*ta cho*[2])'. In the next breath, however, he tells us that the moon set 'behind the stern rudder house',[f] so it would seem that an auxiliary steering-oar was carried as well as the rudder with its tiller worked in the *tho-lou*.[g] This is just what we have seen could sometimes happen with medieval ships both European and Chinese. Finally, a couple of months later, when Ennin was staying ashore with the Korean monks of Chhih Shan[3] in Shantung, a storm hit this same ship so that it was 'blown on to rough rocks, and the rudder board (*tho pan*[4]) broken off'.[h] Thus all the eye-witness testimony of Ennin vividly confirms the conclusions that have been drawn from the words of Master Kuan and the Taoist Than Chhiao.

After the time of the Chinese embassy to Korea there is the description of Marco

[a] Another mention of this period occurs in the *Wu Lin Chiu Shih* (Institutions and Customs at Hangchow), ch. 3, p. 4a, where we read of a small boat with a flat bottom and a simply constructed *tho*. Here the *tho* was probably a rudder, for a steering-oar would hardly have been mentioned in such close connection with the shape of the hull. This book deals with life at Hangchow from +1165 onwards.

[b] Underwood (1), p. 15 and figs. 13, 14, 32.

[c] Most of the details which he describes refer to Japanese vessels, but similar arrangements on Chinese and Korean ships may undoubtedly be inferred.

[d] Translation of the passage in Reischauer (2), p. 6; cf. (3), p. 69.

[e] He did not in fact return to Japan at this time, but spent eight more years on pilgrimage in China, having managed to get permission to stay, and so leaving the embassy which went home without him. The following passage is translated in Reischauer (2), p. 115, but we cannot quite accept all his interpretations. Presumably Ennin's reference to the 'great oar' must mean 'in line with it'.

[f] *Lu tho tshang*,[5] presumably a slip for *chu tho tshang*.[6]

[g] See immediately below.

[h] Tr. Reischauer (2), p. 132; cf. (3), p. 93. A similar disaster had occurred on the occasion of the abortive Japanese embassy of +836; Reischauer (3), p. 61.

[1] 柂角   [2] 大櫂   [3] 赤山   [4] 柂板   [5] 艫柂倉   [6] 舳柂倉

Polo, who is quite specific in what he says (see p. 467 above), but as he wrote a century after the European dateline, it is not relevant to our argument here.[a] We must turn to the remaining types of evidence, how the steering device was fixed and of what material it was made.

The *Yü Phien*[1] (Jade Page Dictionary) of Ku Yeh-Wang,[2] compiled in the middle of the +6th century (about +543), says[b] of the *tho* that it is 'a piece of timber for regulating (the direction of) a ship. It is set (*shê*[3]) at the stern'. While many of the earlier references mentioned above, as well as others, refer to the stern, the verb used here implies something very permanently established there, more so, perhaps, than would be warranted by the lashings of a steering-oar. However, this is only a hint. The *Thang Yü Lin* (Miscellanea of the Thang Dynasty) has to be taken more seriously, for referring to about +780 it speaks of the *tho-lou*[4] ('rudder-tower'), i.e. the stern gallery or extension of the poop.[c] This is the term always later used for the projecting after-castle in or on which the helmsman stood (and still stands) to work the tiller (*tho-kang*[5]), and which also contains the winches or other arrangements by which the rudder is raised and lowered (cf. Fig. 1026 above). So far as we know, this is the first appearance of the expression. Its significance cannot well be overlooked, for with steering-oars there is no necessity for a *tho-lou*, indeed it would be in the way, and among all the hundreds of existing types of Chinese boats and ships there is no instance of a *tho-lou* co-existing with stern-sweep or steering-oars. These elongated instruments, where present, run some distance forward over the after part of the deck to a kind of light bridge on which the helmsman stands. This, therefore, constitutes strong evidence for axial rudders as early as the +8th century.

It is tolerably sure that Chinese rudders were never attached to the stern with pintles and gudgeons of iron. But that they were early strengthened with that metal is certain enough. It may be worth quoting a couple of passages from Chou Mi's *Kuei-Hsin Tsa Chih* (Miscellaneous Information from Kuei-Hsin Street), written about the last decade of the +13th century. In one place he relates that

Li Shêng-Po[6] used to say: 'I always followed (the seamanship of) old Chang Wan-Hu[7] when sailing in the sea. Between Chang-chia-ping and Yen-chhêng there are eighteen sandbanks. If a sea-going junk should go aground, the cargo of grain must be thrown overboard to lighten her. If the ship still cannot be moved, rafts of timber should be prepared to save the lives (of the crew), for she will go to pieces and give no protection. The (best) wood at the bottom of the rudder (*tho shao*[8]) is called the "iron corner" (*thieh lêng*[9]). Sometimes the wood called *wu-lan-mu* from Chhinchow is used. One measure of this costs 500 (ounces of) silver...!'[d]

---

[a] Still, it is interesting to hear him on the raising and lowering of the rudders. His Chinese fleet being somewhere in the East Indies, he says: 'And in the middes of this Iland about forty miles, there is but four passes of water, therefore the great Shippes do take off (take up) their Rudders...' (Penzer ed. p. 103).

[b] Ch. 12, p. 17*a*, cit. in *Khang-Hsi Tzu Tien*, art. *tho*.[10]

[c] Ch. 8, p. 24*b*, cf. p. 453 above. Master-helmsmen (*tho shih*[11]) are mentioned in the Thang Waterways Department Ordinances of +737 (cf. Twitchett (2), p. 55).

[d] *Hsü Chi*, ch. 1, p. 36*b*, tr. auct. On the grain transport service by the sea-route, cf. p. 478 above.

[1] 玉篇　　　[2] 顧野王　　　[3] 設　　　[4] 舵樓　　　[5] 柁杠　　　[6] 李聲伯
[7] 張萬戶　　　[8] 柂梢　　　[9] 鐵稜　　　[10] 柁　　　[11] 拖師

Elsewhere, referring to +1291, he says:[a]

In the *hsin-mao* year Mr Commissary-General Chu[1] was transporting grain to the capital …and met with a furious storm on the way. An anchor (*ting*[2]) hurriedly lowered was lost, and three or four iron grapnels (*mao*[3]) were broken one after the other. The rudder-post (*tho kan*[4]) and the 'iron corner' (*thieh lêng*[5]) made a terrible noise, 'ya-ya, ya-ya', and seemed to be on the point of breaking…

The ironmasters of this time would have been perfectly capable of arranging pintles and gudgeons if that had been required, but Chinese seamen always preferred to have the rudder movable up and down in guides, no doubt because at the lowest position it greatly improved the windward sailing quality of their ships.[b] Moreover, when sailing in heavy monsoon weather it was good to be able to keep the rudder down so as to protect it from breaking seas. And of course it was essential to be able to raise it when beaching or sailing in the shallow waters of estuaries and the Yellow Sea.

There remains the question of the material of which the *tho* was made. The use of special wood seems to go far back in history. In the *San Fu Huang Thu* (Description of the Three Cities of the Metropolitan Area), which is probably of the late +3rd century, there is the following story:[c]

Someone presented a small sampan (*tou tshao*[6]) to the emperor, but he said, 'Cinnamon wood for oars, pine for hulls, make a boat heavy enough; how could one sail in this?' And he ordered that catalpa wood should be used for the boat's hull, and magnolia wood for the *tho*.

At the other end of history, we may remember the details given by Sung Ying-Hsing fourteen centuries later about iron-wood for rudders (p. 416 above). Intermediate in date is an interesting description in the *Ling Wai Tai Ta* of Chou Chhü-Fei (+1178) of special South Chinese woods much sought after for rudders. He says:[d]

The Chhinchow coastal mountains[e] have strange woods, of which there are two remarkable kinds. One is the *tzu-ching-mu*[7] (purple thorn tree)[f] as hard as iron and stone, in colour red as cosmetic paint, and straight-grained; as large in girth as two men's reach, and when used for roof beams will last for centuries. The other kind is the *wu-lan-mu*[8]; it is used for the rudders (*tho*[9])[g] of large ships, for which it is the finest thing in the world. Foreign ships (*fan po*[10]) are as big as a large house. They sail the southern seas for several tens of thousands of *li*, and thousands or hundreds of lives depend upon one rudder. Other varieties of (timber for) rudders are not more than 30 ft. in length, and are good enough for junks with a capacity of 10,000 bushels, but these foreign ships carry several times this amount, and might break in two if they encountered storms on the deep sea.[h] But this Chhinchow timber is dense and tough, comes in 50 ft. lengths[i] and is not affected by the anger of the winds and waves. It is as

---

[a] Ch. 1, p. 41a, tr. auct. The story is concerned with the efficacy of certain Taoist prayers and talismans.

[b] Cf. Poujade (1), p. 258. Cf. pp. 429, 620 above.   [c] Ch. 4, tr. auct. Cf. p. 414 above.

[d] Ch. 6, p. 9b, tr. auct. adjuv. Hirth & Rockhill (1), p. 34.

[e] In the farthest west corner of Kuangtung.

[f] Apparently *Cercis sinensis*, R380; Stuart (1), p. 101; Li Shun-Chhing (1), p. 628.

[g] Hirth & Rockhill confused the passage by translating *tho* throughout as 'logs' or 'timbers'. One thing at least cannot be doubted, namely that we have to do here with special timber for steering-gear.

[h] Just over 350 tons; would not 'several times' imply some 1,500 tons? Cf. pp. 230 and 304. The ambassador's ship in +1124 could carry about 70 tons of grain, but she was primarily a passenger vessel.

[i] Probably the lengths in which the timber came to the shipwrights, not those of the finished *tho*.

[1] 朱	[2] 釘	[3] 錨	[4] 柁幹	[5] 鐵稜	[6] 豆槽
[7] 紫荆木	[8] 烏婪木	[9] 柂	[10] 番舶		

if one could use a single silk thread to hoist a thousand *chün*[a] or sustain the weight of a mountain landslide—this (wood) is truly a treasure for those who ride the stormy seas. A couple of these (logs for) rudders at Chhinchow are worth only a few hundred strings of cash. But at Canton and Chhüanchow they are worth ten times as much, for only one or two tenths of the supply of this timber is sent there, its length making transport by sea difficult.

This '*wu-lan-mu*' wood[b] was probably the same as the iron-wood used in later times, but the passage is rather ambiguous about the kind of *tho* for which it was destined. Chou Chhü-Fei has just told us, in a passage already quoted,[c] that the ships which sail south of the Southern Sea, 'like great clouds in the sky when their sails are spread', have *tho* several tens of feet long. It was unfortunate that he expressed himself just in this way, for if he meant 20 or 30 ft. the length would not be at all excessive for a true axial rudder of substantial size with the rudder-post included (cf. p. 481 above), but if he meant 70 or 80 ft., longer even than the lengths of wood from Chhinchow and elsewhere, he could only have been referring to a great stern-sweep.

(ii) *Iconographic and archaeological evidence*

The foregoing argument, based solely on the words of ancient and medieval writers, was first sketched out in 1948. It has to be completed by the evidence from pictorial representations which have survived, and finally by the archaeological evidence. The former, though of great interest, proved incapable of taking us much further than the point to which the texts had already led us, but the latter, just a decade later, turned out to be quite decisive, settling the matter in a way more radical than anyone had anticipated. It has shown that the other lines of argument were fully justified, and has proved what they could only surmise.

Let us proceed in the same way as before, starting with the earliest times and coming forwards. We may then also work backwards from the most reliable Chinese pictures of ships which date from times later than the European appearance of the stern-post rudder. In this way we shall be focusing, as it were in a microscope, both from below and from above. The epigraphic counterparts of the Han and San Kuo passages which we have been reading are of course the familiar relief carvings of the tomb-shrines, in which small boats with steering-paddles are frequently seen (cf. Fig. 394).[d] There are also the boats, large and small, of the Indo-Chinese bronze drums (cf. Fig. 960), invariably directed with steering-oars.[e] This takes us down to the +3rd century. From then until the end of the Thang we are mainly dependent upon the paintings and carvings of Buddhist iconography, such as the frescoes of the Tunhuang cave-temples (Fig. 968, pl.) and the steles of the Liu Chhao period (Figs. 970, 972, pls.). In these again the steering-oar uniformly persists, even when the craft are seemingly

---

[a] A *chün* weighs 30 catties (lbs.), cf. p. 415 above.

[b] Not identifiable. But there are a number of tropical woods known as iron-wood: e.g. *Casuarina equisetifolia*, *Fagraea gigantea*, *Intsia bakeri*, *Maba buxifolia*, and *Mesua ferrea* (see Burkill (1), pp. 491, 995, 1243, 1380, 1458 respectively). See above, p. 416.

[c] P. 464 above. One thinks of his words as referring to Malayan vessels of the Borobodur type (cf. Fig. 973, pl.), but he must also have meant Annamese and South Chinese sea-going ships.

[d] In Vol. 4, pt. 2, opp. p. 94. Cf. Chavannes (9); Fairbank (1); des Noëttes (2), figs. 104, 105.

[e] Cf. des Noëttes (2), fig. 103.

quite large. One might be inclined to suspect Indian influence here, since bellying square-sails are shown and never the Chinese mat-and-batten sails, yet perhaps we need not seek to deny them Chinese nationality since the stern-sweep persisted so long and so powerfully in China (at least in river craft) until today. In any case no rudders appear in these waters.

What of the other end? Here we are troubled at times by the question of authenticity, for Chinese and Japanese artists did not always faithfully reproduce the technical detail in the paintings which they copied, and the number of authentic examples from the Sung is now relatively small. Nevertheless Yuan paintings of the +13th and +14th centuries always show rudders below the high curving sterns of the ships, as for instance a famous one by Wang Chen-Phêng[1] reproduced by Japanese art historians.[a] The Suzuki collection in Japan has a Sung painting dating from before +1180 of two junks with fine balanced rudders (Fig. 1032, pl.);[b] and another in Chinese possession, entitled Chiang Fan Shan Shih[2] (Sailing Ships making for Market by a Mountain Shore), the work of an anonymous painter, is placed at or before c. +1200.[c] Japanese

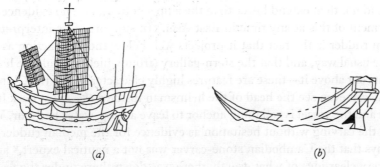

(a)   (b)

Fig. 1033. (a) The great warship in the scroll-painting of the Battle of Dan-no-Ura (+1185) preserved in the Akama-no-Miya shrine at Shimonoseki, and copied from an earlier original by Mitsunobu Ukone-no-Shōgen (+1434 to +1525) of the Tosa-ryū school of painters, in the first years of the +16th century. The simple (unbalanced) axial rudder is clearly drawn, and the parrels and topping lifts of the large lug-sails inserted with loving care. Note also the prominent anchor windlass at the bows.

(b) Rough sketch of one of the ships depicted in the scroll-painting entitled Mōko Shūrai Ekotoba (Illustrations and Narrative of the Mongol Invasion of Japan, +1281), done in +1292 and preserved in the Imperial Household Museum. The picture occurs in pên 3, p. 10a, b of the 1915–18 coloured reproduction edited by Kubota Beisan. The mastless boats are stuffed with Mongol fighting-men, and rowed by two or three sailors on each side standing on outboard projections. In the original the rope or chain (drawn in fact like a rod) of the slung (unbalanced) rudder tackle is prominent and noteworthy, though in the small sketch not clearly separated from adjoining parts of the hull. The sketch should also make the bows square and transom-ended.

[a] Harada & Komai (1), vol. 2, pl. III, no. 1 (from Kokka, no. 270, pl. III). Examples as late as this are not difficult to find; thus a famous scroll illustrating the Mongol invasion of +1281 shows a rudder under the tho-lou at the stern (Fig. 1033 b). Purvis (1), who reproduced this, described also another painting of ships with rudders depicting one of the voyages of the wandering Buddhist monk and preacher Ippen Shōnin[3] about +1280, but it is a +18th-century copy by Awataguchi Mimbu Hōgen Takamitsu.

[b] Harada & Komai (1), vol. 2, pl. IV, no. 1 (from Kokka, no. 537, pl. I). This is attributed to Ma Ho-Chih.[4] Compare the very similar painting by his contemporary Ma Lin[5] in the Liaoning Museum collection (Yang Jen-Khai & Tung Yen-Ming (1), vol. 1, pl. 60).

[c] Anon. (32), no. 43. Here the rudders are of the simple or unbalanced type.

[1] 王振鵬   [2] 江帆山市   [3] 一遍上人   [4] 馬和之   [5] 馬麟

scroll-paintings copied from medieval originals depict the Battle of Dan-no-Ura[1] in +1185 where the Minamoto clan defeated the Taira.[a] One of the great war-junks (Fig. 1033a) has rectangular lug-sails very clearly drawn, with parrels and topping lifts, as well as an axial rudder. But the authenticity of the details is not very sure.[b]

With the Chinese ship carved so grandly by Cambodian artists on the walls of the royal city of Angkor Thom[c] we are in a better position, for the date, almost exactly the same as that of the Battle of Dan-no-Ura, is not in doubt. Unfortunately this time the details of the relief have been regarded as difficult to interpret. That it shows a Chinese merchant-ship is certain from many features (mat-and-batten sails with multiple sheets, grapnel anchor, etc.), but at the after part of the hull there is seen (Fig. 975, pl.) what appears at first sight to be an axial 'stern-post' rudder turned round so that it is facing forward along the quarter. As such it has been accepted by many,[d] but others have been uncertain; P. Paris (2) for instance pointed out that the post seems to have been carved with nodes in it as if it were a piece of bamboo, suggesting rather that it was meant to be the mast of the small fishing-sampan alongside, yet with this it has no connection and the stone shows no sign of damage.[e] It could not be a quarter rudder since it would not then extend lower than the ship's bottom, and no evidence exists for the employment of this at any time in East Asia.[f] In support of its interpretation as a true median rudder is the fact that it projects well below the hull, acting as a centre-board in the usual way, and that the stern-gallery (from which it would be hoisted and lowered) juts out above it—these are features highly characteristic of the Chinese slung rudder. One can even see the head of the helmsman inside a *tho-lou* at deck level. The sails are set and the vessel is weighing anchor to leave a deep-water harbour. We therefore accept the carving without hesitation as evidence for the median rudder, remembering always that the Cambodian stone-carver was not a nautical expert,[g] and lacked perhaps a very clear idea of what exactly there was down there under the water at the stern of a Chinese merchantman.[h]

Then as our next document we have the celebrated scroll-painting by Chang Tsê-Tuan finished just before +1125 and entitled Chhing-Ming Shang Ho Thu (Coming up the River to the capital, Khaifêng, after the Spring Festival),[i] with all its wonderful

[a] Cf. Purvis (1).
[b] A parallel case is the painting of Chao Po-Chü[2] which shows a junk with a clearly drawn stern-post rudder, but it is a Yuan, perhaps even a Ming, copy (Hájek & Forman (1), pl. 176–7).
[c] Cf. p. 460 above.
[d] E.g. Groslier (2); Poujade (2).
[e] Personal observation in 1958.
[f] Except certain vessels of Sumatra (Poujade (1), p. 267), which probably derive from Arab influence. At one time Poujade believed (p. 256) that a quarter rudder was the best interpretation of the Bayon ship's gear.
[g] The beading at the right edge of the rudder should probably not be taken for the post but rather as a carver's convention. The post was hidden within the hull.
[h] Des Noëttes (4) supposed that the date of the Bayon ship corresponded closely with that of the Norman longships shown on the Bayeux tapestry. Their square-sails and steering-oars compare very unfavourably with the fore-and-aft sails and rudder of the Bayon ship. It was in fact a century later.
[i] Fully described above, p. 463. Cf. the study of Waley (20). Reproductions of later copies in Binyon (2), pl. XLII. We have already discussed the question of authenticity.

[1] 檀浦   [2] 趙伯駒

wealth of detail on the daily life of the people and their techniques. Here many slung and balanced rudders are depicted with the utmost clarity (Fig. 1034, pl.). Since the date corresponds almost exactly with the literary evidence of the embassy to Korea we have once again the mutual confirmation of painting and text.[a]

Lastly, if the existing Sung copies are to be trusted, we can find balanced rudders in the paintings of the famous Wu Tai artist Kuo Chung-Shu.[1] In his Hsüeh Chi Chiang Hsing Thu[2] (Sailing on the River while the Sky is clearing after Snow),[b] painted in +951, there are two large junks with well-drawn balanced rudders.

Between +950 and +100 only one other piece of iconographic evidence presents itself, a painting of a ship by Ku Khai-Chih,[3] the famous artist of the second half of the +4th century. It illustrates the *Lo Shen Fu*[4] (Rhapsodic Ode on the Nymph of the Lo River), written by Tshao Chih[5] a hundred and fifty years earlier.[c] While this painting (Fig. 1035, pl.) seems to embody certain archaic ideas of perspective characteristic of the draughtsmanship of Ku Khai-Chih's time,[d] the earliest extant version of it is probably a Sung copy of the +11th or +12th century.[e] It is therefore impossible to be sure that the rudder-like object shown was really in the original, for it might have been introduced in one of the Thang copies through which no doubt the famous picture was handed down. Besides, what is it? At the stern of the ship we see a curious trapeziform structure which looks quite like an axial rudder hoisted inboard to a position high out of the water, and beside it there is a backward and downward pointing spar which might be a 'yuloh' (a sculling-oar), or perhaps a stern-sweep quite appropriate for a river ship.[f] But one remains rather at a loss, and few were willing to take Ku Khai-Chih's picture as evidence for the 'stern-post' rudder in the +4th century, the time of Kuo Pho and Sun Fang.[g]

Since 1958 all such doubts could be set at rest. For in the previous years excavations undertaken by the Kuangtung Provincial Museum and Academia Sinica in the city of Canton, in connection with rebuilding operations, brought to light from a tomb of

[a] From about this time also, the first half of the +12th century, at any rate, we have a painting by Chhiao Chung-Chhang,[6] done to illustrate Su-Tung-Pho's second Ode on the Red Cliff, which shows the poet's friends drifting in a boat with a very clear triangular balanced rudder. This work, in the Crawford Collection, was exhibited in London in 1965; see Sickman *et al.* (1), p. 32, no. 15.

[b] Reproduced in Wu Lien-Tê (1), p. 501, from the Thaiwan section of the Imperial Palace Museum; another copy is in the Nelson Art Gallery at Kansas City.

[c] *CSHK*, San Kuo sect., ch. 13, p. 2a.

[d] See pp. 114, 116 above.

[e] See Yang Jen-Khai & Tung Yen-Ming (1), vol. 1, pl. 10, and Waley (19), pp. 59 ff. We reproduce the copy in the Freer Gallery of Art at Washington, by the kindness of Dr A. G. Wenley. Cf. Sullivan (3), p. 428.

[f] If the small upright piece of timber at the stern is not the bumkin of the yuloh or stern-sweep, it might be the post of the rudder. If the spar is not a yuloh or a stern-sweep, it might conceivably have been inspired by a slanting strut such as those which still exist on Hakka boats in the neighbourhood of Meihsien. This gives added strength by connecting directly the after part of the rudder with the middle of the tiller (G. R. G. Worcester, unpublished material, nos. 86, 88, 179). Some Dutch boats also have spars of this kind.

[g] See p. 639 above. Worcester (3), vol. 1, pp. 104 ff., was prepared to accept the picture as evidence for the axial rudder, doubting only its historical authenticity.

[1] 郭忠恕   [2] 雪霽江行圖   [3] 顧愷之   [4] 洛神賦   [5] 曹植
[6] 喬仲常

Hou Han date (+1st and +2nd centuries) a magnificent pottery ship model which demonstrated the existence of the axial rudder already at that time. Before these discoveries the tomb ship models recovered in modern archaeological research had all been of the Warring States or Early Han period (−4th to −1st centuries), and all had evidence of steering oars (cf. p. 447 and Fig. 961, pl.). Now however the pottery model, nearly two feet long, was found to be equipped in very modern style. We need not recapitulate the description already given (p. 448), suffice it to say that, as Fig. 965 (pl.) shows, deckhouses cover most of the beam, which is surrounded on both sides by a poling gallery.[a] The stern extends aft a considerable distance beyond the last transom-bulkhead in the form of an after-gallery (in fact a *tho-lou*), the floor of which is formed by a criss-cross of timbers through which the rudder-post descends into the water. This is seen particularly well in Fig. 1036 (pl.), a photograph taken from astern. The true rudder is indeed present, trapeziform as would be expected, and having no resemblance to a steering-oar,[b] but most clearly exemplifying the remark about 'eight foot of timber' which Than Chhiao was to make nearly a thousand years later.[c] Most gratifyingly, its shoulder is pierced by a hole exactly where the suspending tackle should be attached to it.[d] Possibly the original state of the model, made no doubt for some wealthy merchant-venturer and ship-owner of Han Canton, incorporated all the tackle by which the rudder was secured, but the little cables long ago rotted away and we can only guess now how it was done. A second hole is present at the top of the rudder-post.

---

[a] It has not as yet, we think, been pointed out that a curious similarity exists between the lines of this ship (in particular the position and shape of the rudder) and those to be seen in a representation of the ship of the famous Japanese minister Sugawara no Michizane[1] (+845 to +903), Ennin's younger but more exalted contemporary (Fig. 1038, pl.). According to Purvis (1) this many times copied scroll still exists at the Shinto temple of the minister at Kyoto, the Kitano Tenmangū, where he is deified as the god of learning. The same applies to another fine Japanese ship painting, that in the Kōzanji at Kyoto dated *c.* +1210 and classed as a national treasure (see Anon. (66), pl. 37). This work, perhaps from the brush of Enichibo Jōnin (a disciple of the famous abbot Myōe,[2] +1173 to +1232), depicts the taking of the Kegon School doctrines to Silla by Gishō and Gengyō. The general shape of hull and sails recalls the Bayon ship (cf. p. 461) and there is a Tunhuang-like pagodesque deckhouse at the poop (cf. p. 455), but the rudder is singularly like that of the Canton model. The vessel is guided by a dragon—the spirit of a Chinese girl of high birth who fell in love with Gishō unavailingly and threw herself into the sea. So also we may be led to find prototypes of these hulls and rudders in paintings of the Thang—I think particularly of Li Chao-Tao's[3] boats on the Chhü River (Sirén (10), pl. 81). Since Li's *floruit* was +670 to +730, a time of maximum Chinese influence upon Japan, may it not be permissible to suggest that some Han shipbuilding traditions lasted on there at least till the early Kamakura period?

[b] Pao Tsun-Phêng (2, 2) takes Watson (3) rightly to task for having missed the significance of the axial rudder, but he himself accepts Watson's statement that a steering-oar was mounted on the port quarter. This strange idea is certainly due to misinterpretation of a long-gowned human figure which stands on the poling gallery at the port forward corner of the helmsman's deckhouse. Of course neither Pao nor Watson had the opportunity of studying the model at first hand, or adequate photographic documentation either, but the point is not unimportant as we know that the combination of rudder and steering-oar did, and does, sometimes exist (pp. 636, 642), and the record needs keeping straight.

[c] The exact shape of the rudder, and the manner of its attachment within the after-gallery, are shown in my drawings made in the Museum at Canton in 1958 (Fig. 1037*a*, *b*, pl.). The landsmen in Chinese museums have since then frequently assembled the model wrongly. Thus the otherwise beautiful photograph in Anon. (26), pl. 444A, has the rudder reversed in the aft-forward direction; and in the National Historical Museum at Peking in 1964 the rudder was placed pointing out of the open after end of the after-gallery instead of through its floor.

[d] Cf. Chang Tsê-Tuan's drawing in Fig. 1034 (pl.) above.

[1] 菅原道眞   [2] 明惠   [3] 李昭道

It is very noteworthy also that the rudder is distinctly balanced, about one third of the breadth of the blade being forward of the axis of the post.[a]

Thus on the main issue guesswork is ended, and we have positive demonstration that by the +1st century the true median rudder had come into being. How strange it is that this was just the time, too, to which one may trace back the first beginnings of the magnetic compass.[b] How strange also that though the latter was much slower in its development than the steering mechanism, it appeared in Europe at the same time as the rudder, just a millennium later. The only difference between them is that the compass in the West is recorded first from the Mediterranean, the axial rudder first from Northern European waters.

(iii) *Transmissions and origins*

About the transmission of the technique (and surely such there was)[c] very little can be said. It seems overwhelmingly likely *a priori* that an invention of this kind would have come round by way of mariners' contacts in the South Asian seas, though it is not impossible that a Chinese artisan who had built ships for the Liao dynasty handed on certain ideas to Russian merchant-shipwrights trading to Sinkiang in the realm of the Western Liao (Qarā-Khiṭāi) between +1120 and +1160. This might explain the region in which the rudder first manifests itself in European culture, but support from Russian sources is so far lacking.[d] At the same time the Islamic world offers more light (though not very much) for the travels of the rudder than it does for the mariner's compass.

A famous illustration in a Baghdad MS. of +1237 shows an axial rudder on a sewn ship (Fig. 1039, pl.);[e] it comes from the *Maqāmāt* (Historical Anecdotes in Rhymed Prose) of Abū Muḥammad al-Qāsim al-Ḥarīrī (+1054 to +1122).[f] The tiller arrangement is not very clear, but medieval Asian seamen certainly used various kinds of relieving-tackle to secure it. A description of a related device has already been noticed in the quotation from Lecomte (p. 635 above), and a contemporary Chinese example of a tiller held firmly by adjustable ropes is well seen in a port bow view of the Maze

[a] We repeat mention of our warmly felt indebtedness to Dr Wang Tsai-Hsin and his colleagues of the Canton Museum for their kind help in these studies, pursued in 1958 with the collaboration of Dr Lu Gwei-Djen.

[b] The story has been told in Vol. 4, pt. 1, Sect. 26*i*.

[c] La Roërie (2), p. 31, was prepared to admit that the axial rudder was used in the Far East at least a century before it appeared in Europe—'mais tout ceçi n'a d'ailleurs qu'un mince intérêt de curiosité'. As Western naval historians and archaeologists become more aware of Chinese priorities, an increased disposition to deny their relevance may be expected. Thus already Adam & Denoix (1), while acknowledging fully the antiquity of the Chinese axial rudder, urge that it should be regarded as something *sui generis*, a 'rudder-centreboard', and implicitly deny that it had any influence on the stern-post rudders of the West. Needless to say, we have no sympathy with this view. It belongs, we feel, to the Department of Face-Saving Re-definitions, on which see Vol. 4, pt. 2, p. 545, and also p. 564 above.

[d] We have drawn attention to the appositeness of the Western Liao State in +12th-century East–West contacts also in connection with the magnetic compass (Vol. 4, pt. 1, p. 332).

[e] Bib. Nat. MS. Arabe no. 5,847. Cf. Blochet (1); Hourani (1), p. 98; des Noëttes (2), fig. 90.

[f] Partial tr. Preston (1). A literary parallel exists in John of Montecorvino (see Yule (2), vol. 3, p. 67). On al-Ḥarīrī see Mieli (1), p. 209. A very similar picture illustrates the book of the sea-captain Buzurj ibn Shahriyār al-Rāmhurmuzī on the marvels of India ('*Ajā'ib al-Hind*)written in +953; see the tr. of van der Lith & Devic (1), opp. p. 91.

Collection model of a South Chinese freighter (cf. Fig. 1040, pl.). Now for the past century and a half European observers have described elaborate tackle-controlled rudders on many types of Arabic sailing-ship.[a] Hence considerable interest attaches to a passage[b] in the *Aḥsan al-Taqāsīm fī Maʿrifat al-Aqālīm* (The best Divisions for the Knowledge of the Climates) written by Abū Bakr al-Bannāʾ al-Bashārī al-Muqaddasī[c] in +985. Describing a difficult passage in the Red Sea, he says:

> From al-Qulzum to as far down as al-Jār, the bottom is overspread with huge rocks which render navigation in this part of the sea most difficult. On this account, the passage is made only by day. The shipmaster takes his stand on the top[d] and steadily looks into the sea. Two boys are likewise posted on his right hand and on his left. On espying a rock he at once calls to either of the boys to give notice of it to the helmsman by a loud cry. The latter, on hearing the call, pulls one or other of two ropes which he holds in his hand to the right or the left, according to the directions. If these precautions are not taken, the ship stands in danger of being wrecked against the rocks.[e]

It seems almost impossible that this description could refer to lanyards attached to steering-oars, but on the contrary it would closely agree with the tackle-controlled axial rudders which have lasted in use in Arab waters to the present day;[f] in this case we have to conclude that the Chinese invention had already been introduced in the Arabic culture-area before the end of the +10th century. From all that we know of Arab trade in the eastern seas, this would not be at all extraordinary. But the transition from the Muslims to northern Europe remains at first sight more difficult to understand. Perhaps some sea-captain from northern Europe was more observant and alert during the Second Crusade (+1145 to +1149) than any of his colleagues from the Mediterranean.[g]

In spite of all controversies, the stern-post rudder, no less than the mariner's compass, was an essential pre-requisite for the oceanic navigation of large ships. Without it, the developments of the second and third periods of quantitative and mathematical pilotry (see pp. 554 ff. above) would have been long delayed if not completely inhibited. The historical implications of the stern-post rudder in the West are only now beginning to be understood, but some will be obvious from our account of the +15th-century voyages (pp. 511 ff. above).

The Portuguese success in these maritime undertakings [wrote Trend] was due to science; and the science of the day, however rudimentary, had led to a series of technical improvements

---

[a] E.g. Moore (1), p. 137; Hornell (1), p. 239, (3); Moll (2); Bowen (13).

[b] Attention was drawn to it by Hadi Hasan (1), p. 111, but he did not appreciate its significance. Cf. a parallel in al-Idrīsī (+1154); Jaubert (1), p. 135.

[c] Mieli (1), p. 115. Not to be confused with the medical writer al-Tamīmī al-Muqaddasī, a contemporary.

[d] Perhaps a crows-nest, on account of the great spread of the lateen sail.

[e] Tr. Ranking & Azoo (1), p. 16.

[f] One should not forget that relieving-tackle was used on European ships in the +17th century and earlier (Clowes (2), p. 79). And the lug-rigged boats of the Rio de Aveiro north of Coimbra have tiller lines running from the rudder to the bows. A good model is in the Museu de Arte Popular at Lisbon.

[g] It will be remembered (Vol. 4, pt. 2, p. 555) that the first recorded windmill in Europe dates from +1180. There is a long tradition that this introduction was due to the crusades. Yet it was in Northern, not Southern, Europe.

in ships and how to sail them. Most important were the axled, hinged, rudder, and the mariner's compass...Without these it is safe to say that the Portuguese discoveries would have been impossible, but their nautical science was up-to-date because they were aware of what was being done in other countries, and willing to invite foreign specialists to come and help them.[a]

And Magalhães Godinho, in a striking passage where he rejects the major criticisms directed against des Noëttes, writes:

Of course the stern-post rudder did not solve all the problems of steering ships—it would be astonishing if it had...(Western) ship tonnages did not rise suddenly, but between the +15th and the middle of the +16th centuries Portuguese average burthens doubled at the least. The decisive role of the stern-post rudder was really rather different; owing to its position on the median line and pivoting there, it became possible to stand a course in heavy weather, maintaining a constant angle between the median line of the ship and the wind direction. Hence it became possible to outflank the regions of the trade-winds far from the sight of any coasts. What was at stake was nothing less than the mastery of ocean navigation itself.[b]

Elaborating the words of Godinho (like Trend, a scholar rather than a practical sailor) one might add that by standing a course in heavy weather he meant beating to windward or making efficient use of a beam wind for days at a time without exhausting the crew; this was what the stern-post rudder permitted, the complete circuit of the trade-wind areas (cf. Fig. 989 a, map), neither desperately fighting them nor merely running before them. Thus an invention of the China Seas found its supreme application in the roaring Atlantic.

Now comes what we have called the dénouement. The invention of the stern-post rudder involves a remarkable constructional paradox—it was developed by a people whose ships had characteristically no stern-posts. If we look again at pictures of the ships of ancient Egypt, of the Greeks, or of the Norsemen, we see invariably that the stern sloped gradually upwards in a curve from the water-line. The slanting stern-post was in fact, to use anatomical terminology, a 'posterior sternum' corresponding to the 'anterior sternum' of the stem, and a direct prolongation, like the latter, of the keel. But the junk had never any keel. Its bottom, relatively flat, was joined to the sides, as we have seen (p. 391), by a series of bulkheads forming a set of watertight compartments, and instead of stem and stern posts there were transom ends. Now the bulkhead build provided the Chinese shipwrights with the essential vertical members to which the post of the axial rudder could conveniently be attached, not necessarily the aftermost transom but perhaps one or two bulkheads forward of it. This principle held good from the smallest to the largest sailing-ships.[c] It might be called that of the 'invisible stern-post'. Of course, in later times, rudders were fashioned in curving shape so as to fit various kinds of curving stern-posts, but our argument suggests that

---

[a] (1), p. 134.

[b] (1), pp. 19 ff., eng. auct. Godinho does justice also to the invention and spread of the mariner's compass.

[c] For instance, the great rudders of Fuchow sea-going junks (Figs. 1041, 1042, pl.), cf. Worcester (3), vol. 1, p. 144, and opp. p. 139.

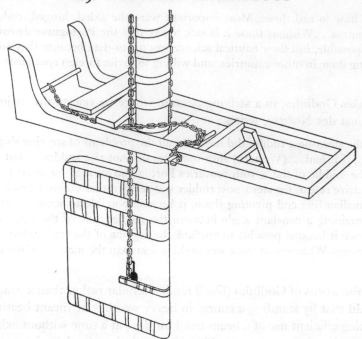

Fig. 1041. Drawing of the rudder of a Fuchow timber freighter or Hua-phi-ku (Worcester, 3). Cf. Figs. 936 and 1042 (pls.). The rudder, weighing some 4–8 tons, measures 32½ ft. in height and 11½ ft. in width; it is hoisted and lowered by a bight of chain passing through a sheave in the blade, both parts being wound round the barrel of the windlass above. The rudder-post, 15 ft. long, is iron-bound at intervals of a foot throughout.

the difficulty of doing this was one of the chief factors which inhibited any earlier development of the invention in the West.[a] The bulkhead-attached rudder posts can be seen clearly in many Sung pictures (Figs. 1025, 1034, pls.), as well as in drawings of contemporary Chinese craft. The Cantonese ship model of the Han does not show the vertical nature of the attachment so well, but this may be partly because we do not know exactly how it slung its rudder—in any case the shape of the latter speaks for itself. Alternatively we may be seeing the axial rudder here *in statu nascendi*, after it had acquired its very particular shape and just before it had found its home on the vertical bulkhead timbers. For it is noteworthy that although the Canton ship model has bulkheads its lines are rather like those of a punt, sloping gradually to the water-line at bow and stern;[b] and only when more upright and blunt-ended sea-going forms developed from this would the invitation to verticality really have asserted itself—with the additional advantage of the centre-board function in the lowered rudder position. To sum up the matter, we can not only feel sure that the 'stern-post' rudder originated in the Chinese culture-area at the beginning of our era, but we can form a pretty good idea of just why it did so.

a Even la Roërie saw this, (2), p. 35; in the sense that he commented on the tendency of the stern-posts of European ships to become straighter as soon as the axial rudder was adopted. 'The convex stern-post', he said, 'lent itself but poorly to the fixing of a line of hinges.'
b This is true also of the Japanese ship shown in Fig. 1038 (pl.).

(4) BALANCED AND FENESTRATED RUDDERS

The civilisation (so often miscalled 'static') which initiated axial rudders also gave them a far-reaching development. From time to time we have had occasion to mention the 'balancing' of rudders. People generally think of the rudder as an object in which the whole of the blade or flat part is abaft the post itself. But many large modern ships, on the contrary, have rudders in which there is a flat portion forward of the post as well, and this construction is termed 'balanced'.[a] Such an arrangement not only balances the weight on the bearings but also facilitates the work of the helmsman and the steering-power which may assist him, since the water exerts pressure in his favour on the forward portion. The value of this balanced structure was a leading point in the reasoning of Carlini and la Roërie, who admired the steering-oars of antiquity for exhibiting it, and reproached the medieval Western stern-post rudders for failing to conserve it. These writers, who were not very interested in what they called 'exotic pirogues', did not know that balanced rudders are common on many types of Chinese river-junk (Fig. 1043, pl.),[b] though in their simple forms they are unsuitable for sea-going vessels.[c] Although we have not been able to find any specific literary references to them there is now no room for doubt that they go back to the earliest stages of the invention in China. Indeed it seems quite likely that the balanced axial rudder was the first to evolve, for the placing of a steering-paddle in a median upright position against or near the aftermost bulkhead would lead directly to it.

Europeans were very slow, generally speaking, to adopt the principle, perhaps because they were mainly interested in sea-going ships, and until iron construction afforded ways of securing balanced rudders thoroughly (e.g. by pivoting the base of the post) they were not very feasible.[d] An 'Equipollent Rudder', however, was among the inventions of Lord Stanhope about +1790,[e] and Shuldham pressed the matter forward in 1819. One of the earliest ships with a modern balanced rudder was the 'Great Britain' of 1843. The strangest aspect of the situation is that one of the two oldest representations of rudders of any kind in Europe, that on the Winchester font, shows what looks like a balanced rudder. Could this conceivably have any significance with regard to the transmission?

The balanced rudder is also traditionally at home in India, especially on the Ganges, where certain classes of boats, the *ulakh*, the *patela*, etc., are equipped with striking triangular forms.[f] These seem more primitive than the Chinese ones however, because

---

[a] Cf. Attwood (1), p. 103.　　　　[b] Cf. Beaton (1), p. 14 and Worcester (3), vol. 1, pl. 35.

[c] Waters, reviewing Worcester (3), stated that some of the Chinese balanced rudders come within 10% of the theoretically correct proportions. In modern practice, about one third of the rudder blade is forward of the pivot. This is exactly what we find in the Han rudder of the Canton ship model.

[d] A balanced rudder is also liable to stop the way of a sailing-ship unless fenestrated (see immediately below).

[e] Cuff (1); Perrin (1).

[f] See Hornell (10), fig. 14b; (1), p. 249, pl. XXIX, fig. B; (9), p. 138; Crealock (1); Solvyns (1), vol. 3, in +1799; Mukerji (1), pp. 235 ff.; des Noëttes (2), fig. 101; F. E. Paris (1), pl. 35. Some of these ships present also the extraordinary peculiarity of being the only clinker-built vessels in Asia. A +17th-century sketch of one of them can be seen in Bowrey (1). The full history of these strange craft would be well worth investigation by Indian historians of technology.

entirely symmetrical fore and aft, and therefore less efficient. One hesitates whether to look to China or to ancient Egypt as the main source of influence on the rudders of these vessels. But perhaps they are best regarded as very exaggerated ancient Egyptian steering-paddles, all the more so because, though vertical, they are generally fixed to the counter as quarter-rudders and not to the stern-post at all.[a]

Perhaps the most remarkable of all these inventions was the fenestrated rudder. When European sailors first frequented Chinese waters, they were surprised to see some junk rudders which were riddled with holes. No doubt they found it difficult to believe that this had been designedly done. Such fenestrations, generally diamond-shaped and cut out at the edges of the planks, ease the steering by reducing the pressure against which the tiller has to act, and minimise the drag on the ship caused by turbulence in the hydrodynamic flow past the rudder.[b] But as the water is a viscous medium the efficiency of the rudder is very little impaired. The practice was remarked by Admiral Paris, though he did not fully appreciate its value. Fig. 1044 (pl.) shows the stern of a Hongkong fishing-junk in dry dock with a fenestrated rudder.[c] The device was probably quite empirical in origin, based upon knotty wood or damaged gear, but it is not at all too fanciful to suppose that some medieval Taoist sailor, finding that his work was eased and that his ship sailed better, was fully content to follow the principle of *wu wei*,[d] and letting well alone, recommended the arrangement to his friends.[e] The fenestrated rudder has been widely adopted in modern iron ships during the present century, having been brought to the attention of European marine engineers by Winterbotham[f] in 1901. Indeed, it may even have helped to stimulate the important invention of anti-stalling slots in the wings of aircraft.[g]

(i) TECHNIQUES OF PEACE AND WAR AFLOAT

(1) ANCHORS, MOORINGS, DOCKS AND LIGHTS

Much has been written on the history of the anchor, an essential device which goes back to prehistoric times. The ancient Egyptians used heavy stones combined with hooked branches to form grapnels,[h] but metal hooks were coming into use already in

---

[a] Exact replicas of the ancient Egyptian quarter steering-paddles with tillers occur on other Ganges boats (cf. p. 635 above; Hornell (1), figs. 55, 56). Yet a Chinese flavour asserts itself, for though the *patela* rudder is vertical it is not at all firmly fixed and it does not look at home on the craft that carry it. One senses the application of an extraneous idea to a vessel not really adapted for the purpose.

[b] Cf. Poujade (1), p. 258.

[c] Cf. Fitch (1); des Noëttes (2), fig. 111; Waters (4); Anon. (17), no. 3.

[d] Cf. Vol. 2, pp. 68 ff.

[e] Ultimately the principle was extended to gripes and keels; cf. the Hainan junk model in the Maze Collection, Anon. (17), no. 9.

[f] The holes, he said, 'reduce the labour of "putting hard over" to a minimum while not much affecting efficiency in steering; the stream lines, owing to the viscosity of the water, being deflected almost as well as though the small holes were non-existent'. Mr Hubert Scott described to me in 1959 the trial run in 1901 of one of the first coal-fired Parsons turbine torpedo-boat destroyers. So powerful was the streamline flow that when the helm was put hard over at 30 knots, the balanced rudder could not be reversed, the vessel continuing to speed round in circles. A fenestrated rudder was the answer.

[g] We have discussed this already (Vol. 4, pt. 2, p. 592).

[h] Boreux (1), p. 416; van Nouhuys (2), an article much preferable to those of Moll (2); Reisner (1), is the best authority. On the simplest and oldest anchors, holed stones, see Frost (1, 2).

PLATE CCCLXXXIX

Fig. 927. Crossing the Yellow River on a goatskin raft near Lanchow (photo. Gordon Sanders, 1944).

PLATE CCCXC

Fig. 928. The sea-going sailing-raft of Thaiwan and the south-eastern Chinese culture-area (photo. Ling Shun-Shêng, 1). Note the curved wooden bar at the bow, one of those which gives the bamboo platform its concave profile; the centre-boards, some of which are sticking up in position; the bamboo bulwark rail on each side of the craft; and the characteristically Chinese lug-sail with battens.

Fig. 930. Two of the large rafts of the Ya River in Szechuan (photo. Spencer, 2).

PLATE CCCXCI

Fig. 931. A flotilla of bamboo rafts used in cormorant fishing, on the Hsin-chiang R. near Huang-chin-pu, south-east of the Poyang Lake in Northern Chiangsi (orig. photo., 1964).

Fig. 933. Model of a Ma-yang-tzu (Maze Collection, Science Museum, Kensington). Anon. (17), no. 5. The powerful bow-sweep, generally carried, is shown. See p. 630.

PLATE CCCXCII

Fig. 936. Bow view of a Fuchow pole-junk or timber freighter (Hua-phi-ku), from the Waters Collection (National Maritime Museum, Greenwich), to show the complex construction of the forward part of the hull. The horns of the two wings at the bow rise some 10 ft. above the deck. Though rarely with more than three masts, these ships also attain lengths of nearly 200 ft. (note the two crew figures visible at the bows). Scale drawings of such a ship will be found in Worcester (3), vol. 1, pl. 50.

PLATE CCCXCIII

Fig. 937. Bow view of a Hangchow Bay freighter (Shao-hsing chhuan) in port, from the Waters Collection (National Maritime Museum, Greenwich), showing the build of the forward part of the hull. The blunt transom bow looms forward, and its thwartship planking rests on and slightly behind the rounded extremity of the ascending fore-and-aft bottom planking. The bow is usually decorated with a face in bright colours, and the counters bear paintings of the *pa-kua* and Yin-Yang symbols (cf. Vol. 2, pp. 273, 312) instead of oculi. Note the grapnel anchor and the furled batten foresail. These ships never exceed 90 ft. in length and are usually three-masted; they also carry leeboards (cf. p. 618), not here seen. Scale drawings in Worcester (3), vol. 1, pl. 48.

PLATE CCCXCIV

Fig. 938. Deck of a Swatow freighter (from the Waters Collection, National Maritime Museum, Green-wich). The mainmast strut, which transfers part of the thrust of the wind on the sail to the hull and bulkheads forward, is prominent. One of the usual iron bands and wedges on the mast can also be seen, and the halyard winches (one dismounted) on each side to port and starboard.

PLATE CCCXCV

Fig. 940. A small three-masted freighter on the Chhien-thang River near Hangchow (orig. photo., 1964). The crew are hoisting the mainsail by means of the transverse halyard winch so characteristic of Chinese ships (cf. Fig. 939). Grapnel anchor, oculus and steersman's shelter are noticeable. The identification-plate reads Chê Hang fan 23, i.e. 'Sailing-ship no. 23 of Hangchow in Chekiang Province'.

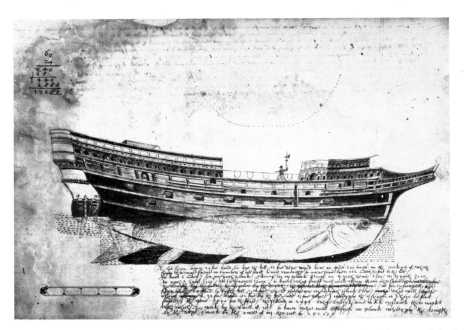

Fig. 946. A page from the Matthew Baker manuscript of +1586 in the Pepysian Library at Magdalene College, Cambridge. The shape of a fish is superimposed upon the drawing of a hull, to illustrate the famous shipwright's maxim of those times: 'a cod's head and a mackerel's tail.'

PLATE CCCXCVI

Fig. 947. A revealing photograph of the port quarter of a South Chinese freighter or 'trawler' under repair in the shipyards of Hongkong (from the Waters Collection, National Maritime Museum, Greenwich). Four or five bulkheads are seen, the aftermost one especially clearly, because of the removal of the strakes of the hull. Within the upcurving stern itself, an additional five ribs or frames are seen, and these also occur in various combinations between pairs of bulkheads further forwards. A Fuchow freighter such as that in Fig. 936 (pl.) would have some fifteen bulkheads and about 37 rib frames. In the present picture the slot for the slung rudder can just be made out under the overhang of the stern gallery. Figures give the scale.

PLATE CCCXCVII

Fig. 948. A picture to illustrate Marco Polo's encomium upon the extraordinary abundance of Chinese shipping; some of the salt transport boats awaiting cargo at Tzu-liu-ching in Szechuan (cf. Vol. 4, pt. 2, p. 129). The temple on the ridge to the right, and that on the projection of the city-wall, with the 'half-timbered' houses (cf. Fig. 796), are very characteristically Szechuanese (photo. Jukes Hughes, from the Salt Commissioner's house, 1920).

Fig. 953. Model of a Wai-phi-ku (Maze Collection, Science Museum, Kensington). See p. 430.

PLATE CCCXCVIII

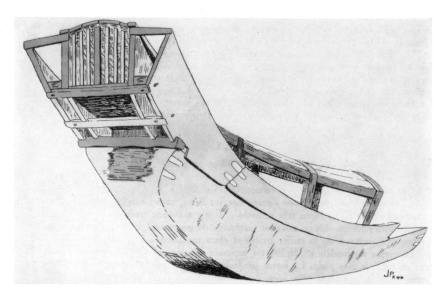

Fig. 956. Ancient Egyptian tomb-model of a Horian ship from the VIth Dynasty (Poujade, 3). The resemblance to characteristic Chinese builds is striking.

Fig. 957. River-junks starting downstream from Kweilin on the Li-chiang (R.) in the early morning mists (photo. Groff & Lau, 1). The bipod masts so typical of this region recall those of ancient Egypt.

PLATE CCCXCIX

Fig. 958. Dragon-boat (Lung chhuan) races at Hongkong.

Fig. 961. Wooden tomb-model of a Former Han river-ship excavated from a princely burial of the − 1st century at Chhangsha (Anon. (*11*), pl. 103). Length, 4 ft. 3 in. There is some uncertainty as to how the component pieces should be put together. The arrangement shown here is about the same as that adopted at the National Historical Museum in Peking (1964), but to the nautical eye it cannot be right, as it leaves no room either for helmsman or rowers (the black object amidships is the smallest of three deckhouses). The arrangement of Hsia Nai (1) is preferable. The U-shaped piece, here enclosing the after deckhouse, ought to project astern as a gallery, and a central notch in this indicates that the stern-sweep or steering-oar was intended to turn on it, though one would have expected it rather to work through it on the terminal hull transom bar. The two larger deckhouses ought perhaps to be superimposed, and the smallest one should be moved either aft or forward. Lastly the bulwarks are upside down, for the oar-ports must be along their lower not their upper edges; and this is proved because those along the line of ports are flat while the other edges are slightly convex. Other components of the tomb-model have not been incorporated in any reassembly we have seen—notably part of a fourth cabin or deckhouse, oval objects like shields, L-shaped objects, miscellaneous unidentified boards, and an extraordinary piece of carved and fretted woodwork that might have been a figurehead (cf. Fig. 960), or a 'prow-yoke' (cf. Fig. 964, pl.). There are no obvious bulkheads.

Fig. 962. Tomb-model ship in red pottery of the Later Han period (+1st century), in the Canton Museum (orig. photo., 1964). Length, 1 ft. 4 in. Four human figures (*yung*) are shown as crew. Description in text, p. 448.

Fig. 963. Tomb-model ship in grey pottery of the Later Han period (+1st century), excavated, like the preceding, from a burial under the city of Kuangchow (photo. Canton Museum). Length just under 1 ft. 10 in. This is a document of singular importance for the history of shipbuilding, cf. Fig. 1036. View from the starboard side, showing slung rudder under the overhanging poop or 'false stern' (cf. p. 399), steersman's cabin, several roofed or matting-covered deckhouses, a long poling gallery (cf. Fig. 966), bollards or yuloh thole-pins, and a projecting bow with an anchor (cf. p. 657) hanging from it. The mast was probably stepped just forward of the deckhouses.

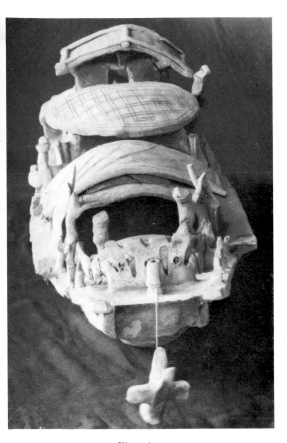

Fig. 964

Fig. 965

Fig. 964. Bow view of the model in the preceding illustration (photo. Canton Museum). The anchor is seen attached to its bollard, and just behind this an ornamental screen which recalls the 'prow-yokes' of Indo-Chinese craft. Lateral projections support narrow outboard decking and bitts or bumkins. The transom stem is well seen.

Fig. 965. View of the model in the preceding illustrations from above, bow at bottom (photo. Canton Museum). The removable roofing shows the cabins or holds, but no bulkheads are visible. The wide poling galleries break off only in one place, the probable position of the mast, but no tabernacling is provided.

Fig. 966. The use of the poling gallery, a picture of a Cantonese river-ship at Kukong (Chhü-chiang) on the Pei-chiang (R.), taken from a passing sampan (orig. photo., 1944). Gruelling labour upstream on windless days.

Fig. 968. The largest of the ships in the Tunhuang cave-temple frescoes; from cave no. 55 at Chhien-fo-tung (photo. Pelliot Collection, Musée Guimet, Paris). Description in text (p. 455). The Buddhist Ship of Faith sails from the shores of illusion in the foreground (the upright oblongs are inscription-bearing cartouches) to the Paradise of Amida.

Fig. 970. Carving of a ship on a Buddhist stone stele of the Liu Sung or Liang dynasties (+5th or +6th century) from the Wan-Fu Ssu temple at Chhêngtu (photo. Historical Museum, Szechuan University).

Fig. 971. A timber freighter of the Chhien-thang River, often used for transporting flood-protection fascines (cf. p. 341); one of the relatively few Chinese craft which make use of square-sails. Note also the sprit-sail on the small foremast (photo. Fitch (1), 1927).

Fig. 972. A ship in stormy seas depicted on the back of a bronze mirror of Thang, Wu Tai or Sung date, i.e. +9th to +12th century (photo. of rubbing, Shensi Provincial Museum, Sian). The inscription on the bellying square-sail reads: 'The Inspector of Thien-hsing city in Fêng-hsiang prefecture...(name illegible).' Note the very uncharacteristic shrouds. Description in text, p. 457.

Fig. 973. Typical Indonesian ship from the reliefs of the great temple of Borobodur in Java, *c.* +800 (photo. Krom & van Erp). Sewn hull, prominent stem- and stern-posts, large outrigger, bipod masts, and an *artemon* sail as well as the characteristic Indonesian canted square-sails. Further discussion in text, p. 458.

Fig. 974. A craft of a different type in the reliefs of Borobodur, *c.* +800 (photo. Krom & van Erp). Reasons are given in the text for regarding it as Chinese. The sail, for instance, appears to be a mat-and-batten lug-sail, and the hull has a square-ended look.

Fig. 975. The Chinese merchant-ship carved on the Bayon at Angkor Thom in Cambodia about +1185 under Jayavarman VII (photo. Claëys & Huet). A document of importance in the history of naval architecture, this is discussed in the text at several points, notably pp. 460 ff. and p. 648. The mat-and-batten sails, with their multiple sheets, the axial rudder slung below the level of the ship's bottom, the anchor with its winch, and the characteristic 'oriflamme' flag, are all to be noted. Many other vessels are depicted on the monument, but they are invariably of the paddled canoe type, even if substantial in size.

Fig. 976. One of the passenger-carrying river-junks in the painting Chhing-Ming Shang Ho Thu, a work by Chang Tsê-Tuan of about +1125 (from the reproduction of Chêng Chen-To, 3). The scene is one of the waterways near Khaifêng, perhaps the Pien Canal (cf. p. 311) itself. Judging by the figures on board, the length of the vessel from stem to stern would be about 65 ft. It is being tracked upstream by five men (out of the frame on the left), the bipod mast being supported by numerous stays. The large rudder, slung and balanced, is especially noteworthy. The poling gallery is in use at the port bow and to starboard while the master and his mates, interrupted at their lunch on the upper deck, are shouting instructions and warnings to the crew of a large boat (out of the frame on the left) which seems likely to collide with the junk negotiating, with lowered mast, the great bridge shown in Fig. 826.

Fig. 979. The government patrol-ship of +1377, recovered from the mud near Liang-shan Hsien and now preserved in the Shantung Provincial Museum at Chinan (photo. Liu Kuei-Fang, *1*). Description in text (p. 479).

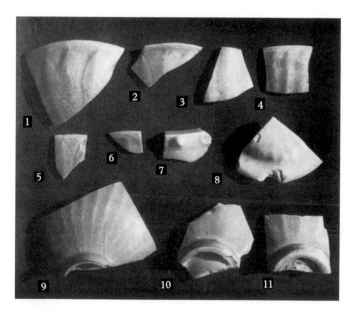

Fig. 982. A few examples of the massive quantity of sherds of Chinese porcelain along the coast of East Africa (Kirkman (4), pl. 6).

Celadon pieces of the +14th century or earlier from a Muslim tomb of +1399 at Gedi on the Kenya coast.

1, 2	brown	3	light grey
4, 9	dark green, crackled, lotus bowl		
5	sage grey	6	dark grey	7	blue-green
8, 10	basic tint			11	dull sea-green

White and blue-and-white porcelain was also found in the later strata.

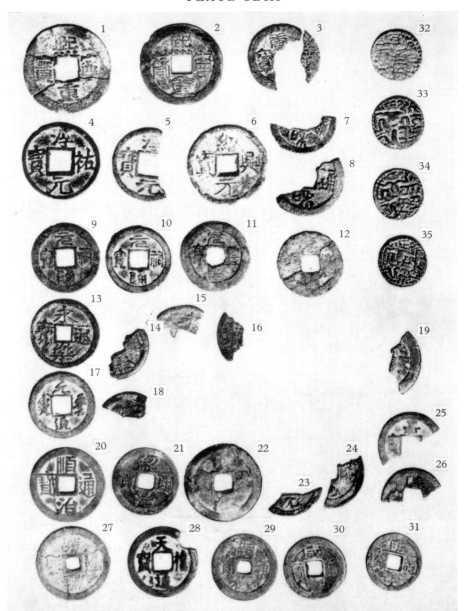

Fig. 983. Some Chinese coins from the East African coasts, a collection identified by Hulsewé (3). Nos. 5, 20, 21, 29 and 31 from the neighbourhood of Brawa, no. 16 from near Merca, and the rest in or near Mogadishiu.

1, 2	Sung	Hsi-Ning reign-period	+1068 to +1077 (cast +1071)
3	Nan Thang	(Pao-Ta, Chung-Hsing or Chiao-Thai r.p.)	+943 to +961
4, 5	Sung	Shun-Yu r.p.	+1241 to +1252 (cast +1249)
6	Sung	Shao-Hsing r.p.	+1131 to +1162 (cast +1145)
7, 8	Sung	Chêng-Ho r.p.	+1111 to +1117
9, 10	Sung	Yuan-Yu r.p.	+1086 to +1093
11, 17	Sung	Yuan-Fêng r.p.	+1078 to +1085
13, 14	Ming	Yung-Lo r.p.	+1403 to +1424 (cast after +1408)
20	Chhing	Shun-Chih r.p.	+1644 to +1661
22, 26	Sung	(Chien-Chung-Ching-Kuo or Chhung-Ning r.p.)	+1101 to +1106 (cast in +1101 or +1102)
28	Sung	Thien-Hsi r.p.	+1017 to +1021
30	Chhing	Hsien-Fêng r.p.	1851 to 1861

21, 27, 29, 31   Annamese nineteenth-century coins of four reign-periods

32, 33, 34, 35   Ceylonese +12th- and +13th-century coins (+1153 to +1296)

12, 15, 16, 18, 19, 23, 24, 25   Illegible or undatable.

Fig. 985. The European and north African portions of the Korean world-map of Yi Hoe & Kwǒn Kǔn, *Hon-il Kangni Yǒktae Kukto chi To* (Map of the Territories of the One World and the Capitals of the Countries in Successive Ages), first prepared in | 1402. The last three characters of this title (cf. p. 499) appear at the top. The map exemplifies the comparatively more advanced level of geographical knowledge in the Chinese culture-area at that time, for it shows much greater detailed knowledge of Europe and the Middle East (gained no doubt by close attention to Arab informants) than that of East and South Asia displayed in the Catalan map of +1375 or the map of Fra Mauro made in +1459. It has been briefly described already in Vol. 3, pp. 554 ff. and other parts of it reproduced in Figs. 234 and 235. Like all else, it had its own antecedents, particularly world-maps by Li Tsê-Min *c.* +1330 and the monk Chhing-Chün *c.* +1375, but the work of the two Koreans who combined and added to them in Chêng Ho's own prime carried the East Asian *mappamundi* to an entirely new level, unapproached elsewhere.

The version here reproduced is a copy dated +1502 preserved in the Ryūkoku University Library at Kyoto (from a photostat kindly provided by Dr Funakoshi Akio). Two other versions are also preserved in Japan, one (undated, but probably done in Korea soon after +1568) at the Tenri University Library; and the other, also Korean and traditionally a gift to Hideyoshi in +1592, in the Hommyōji Temple at Kumamoto in Kyushu. On these see Unno (1). In 1964 we had the opportunity of studying closely the Tenri University copy.

Besides these major codices there are many derivative maps of later date springing from the same tradition, including printed versions. One of these, dated +1663, and entitled *Thien-Hsia Chiu-Pien Wan-Kuo Jen-Chi Lu-Chhêng Chhüan Thu*, has been discussed and illustrated by Unno (4). Another, *Yü Chui Chhüan Thu*, published in Korea at least as late as the end of the +18th century, has been described and reproduced by Lowell (1); by this time of course it incorporates ideas and nomenclature of Jesuit origin, but it is quite clearly based on the map of Yi Hoe & Kwǒn Kǔn. Moreover, their work was so much admired that its name got attached mistakenly to maps of related traditions, e.g. to a Korean copy of Yang Tzu-Chhi's +1526 map of China and Korea (properly called *Ta Ming Yü Ti Thu*) now kept in the Myōshinji Temple at Kyoto (see Miyazaki, 1).

At the first glance one is astonished to see a very recognisable outline of the Mediterranean, with the Italian and Greek peninsulas, Sicily, Sardinia, the coasts of Palestine and Spain, and a pagoda marking the position of the Pharos of Alexandria (A-la-sai-i) in the land of al-Miṣr (Mi-ssu, Egypt). The copyist, however, was not quite sure that the Mediterranean was a sea, so he drew its outlines as if they were rivers, even though this meant giving some of the 'rivers' a circular moat-like course. The same is true of the Tenri version, but in the late Lowell version the Mediterranean is labelled Ti Chung Hai (the Sea in the Middle of the Land), and drawn accordingly, though with an outline more distorted than here. Yi Hoe & Kwǒn Kǔn must have had it right in +1402, for no one could have invented that outline for rivers, but copyists during the ensuing centuries had, we think, misgivings, and so went astray.

For the rest, the northern part of the continent of Africa is drawn here so that it fits conveniently on to the southern part shown in Fig. 984, for in this respect there are no great differences between the *Hon-il Kangni*...and the *Kuang Yü Thu*; but our cartographer was at fault in making the long river flow into the top of the Red Sea and not into the Mediterranean. This at least is assuming that it was intended for the Nile. If on the other hand the Nile is the shorter one flowing towards Alexandria out of the great central lake, it would be tempting to see the parallel river accompanying it as Joseph's arm (cf. p. 365), and indeed this does send off a side-branch to a lake on the left—thus the Arabs may have told the Koreans and Chinese about the Faiyūm and Lake Moeris. The long peninsula pointing to the south beside Africa is of course Arabia, and east of it the rivers of Mesopotamia are seen flowing into a very truncated Persian Gulf, while a large round island in the middle of the Indian Ocean is marked Hai Tao (Sea Isle).

A wealth of place-names, mostly in cartouches, can be seen, but no adequate study has yet interpreted them. One can make out transliterated forms of Hispania (I-ssu-pan-ti-na), Barcelona (Pai-la-hsi-na) and Tarragona (Ti-li-khu-na) on the Spanish peninsula; while in the Tenri version France is labelled Fa-li-hsi-na (= al-Afransīyah) and Germany A-lei-man-i-a (= al-Lamānīyah). In the Ryūkoku version, here reproduced, one finds satisfyingly Ma-li-hsi-li-na where Marseille should be. There is nothing easily identifiable in Italy, unless Ma-lu is Milan. England does not figure in the Ryūkoku version, but the Tenri one has a large island off France and Germany labelled Khun-lun Tao, no doubt the British Isles. It retains the same position in the late Lowell version (which does some justice to the Baltic), now becoming, many may be glad to learn, two islands, not only Ying-chi-li Kuo but also I-erh-lang-ta. What the Ryūkoku version does show, however, is even more extraordinary, namely the Azores Is. (Chi Shan) off the northwest coast of Spain, unknown to al-Idrīsī (cf. Fig. 239) and Ibn Khaldūn, and not re-discovered by the Portuguese till after +1394. It is interesting to find that the late Lowell version, absorbing old Western legends, Jesuit-transmitted, adds Fu Tao, the Fortunate Isles, off a modest northwest bulge given to Africa.

The chief object of interest in the northern part of the European zone is a mysterious great city shown as a dark crenellated disc, equivalent in importance (according to the symbols used) to Seoul, and marked fairly clearly Hsi-kho-na; this has been thought to be Budapest, but from its position suggests Moscow or, more probably for that date, Novgorod. The Caspian (Chiu-li-wan) is shown but the Black Sea omitted—unless it is represented by the blank space with the wiggly upper border north of Italy and Greece. Only two cities of the second class (dark discs without crenellations) are indicated along the tract of the Old Silk Road, Pieh-shih-pa-li (Bishbaliq) in Sinkiang, and Tu-a-pu-ni (Derbend) on the west coast of the Caspian. Both can be seen in the portion of the map reproduced. Between them come Pu-ha-la (Bokhara), not even in a cartouche, and Pu-lu-erh (Balkh), shown as an island in a lake.

For the interpretation of Hsi-kho-na it is necessary to remember that the map was made just at the end of the career of Timur Lang (+1336 to +1405), who with other Mongol leaders used to assemble armies in the seventies at a town called Sighnaq, near the Jaxartes R. (Syr Darya) in Turkestan or Kazakhstan, and on one of the western loops of the Old Silk Road (cf. Hookham (1), pp. 99, 125 ff.). If this were the explanation, Hsi-kho-na should have been placed between Bishbaliq and Derbend rather than west of them, but of course none of the cartographers concerned had ever been, so far as we know, within two thousand miles of the Aral Sea. The question remains open.

Strangely, there is no obvious mention of Byzantium. Other places such as Damascus (Tu-mi-shih), Hama (Ho-mi) on the Orontes in Syria, Mosul (Ma-shih-li), Afāq (A-fa) in Mesopotamia, Mecca (Ma-ho), and Medina (Mo-tê Kuo) in another version, are recognisable enough, and usually in just about their right positions.

We have no means of knowing exactly what the place-names on the original map of +1402 were, because changes were introduced as one version succeeded another. This was especially the case with the East Asian names since dynastic alterations were so common, but later on growth of knowledge about Europe also led to revision. It would be very desirable to have a systematic catalogue of identifications of all those on the versions that still exist, but this work has not yet been done. A tenacious tradition will certainly be revealed. The general conclusion is that largely as the result of friendly intercourse between the sea-going East and West Asian peoples, China and Korea knew far more about Europe in Chêng Ho's time than Europe knew about them.

Fig. 987. A moderate-sized five-masted Ta-pu-thou freighter (from Chiaochow Bay near Tsingtao) at Wei-hai-wei, under sail in light breezes (photo. Waters Collection, National Maritime Museum, Greenwich). This again may give some idea of the probable type of build of the much larger Treasure-ships of the Grand Fleet of the +15th century. See also Figs. 935, 936, 938, 939, 986, 1010, 1027, 1042. In this ship the forcmast is stepped to starboard so that from this angle only the luff edge of the foresail can be seen.

Fig. 991. The Chinese Taoist statuette, *c.* 4 in. high, found in 1879 at Port Darwin in Australia among the roots of a banyan tree at least 200 years old (photo. Fitzgerald). See p. 537.

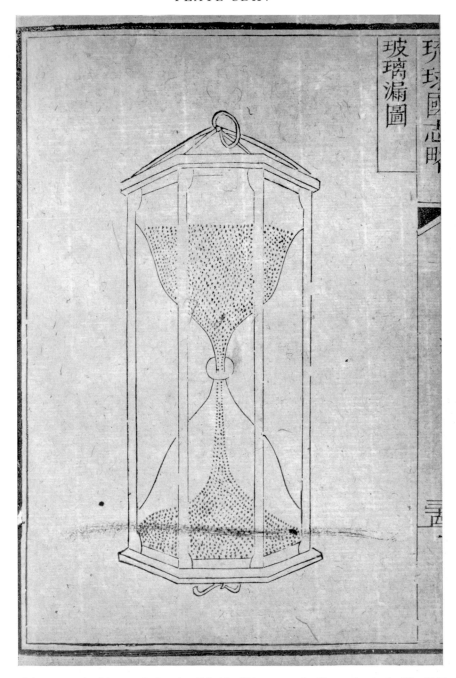

玻璃漏圖

Fig. 995. A 'seaman's dyoll' or sand-glass (*po-li lou*) in Chinese use, the illustration in the *Liu-Chhiu Kuo Chih Lüeh* of +1757. Whether or not Chinese mariners had the sand-glass before the middle or end of the +16th century remains doubtful, and it is surely more probable that what they used for their watch-keeping in Thang, Sung, Yuan and Ming times was the pyro-chronometer or combustion clock con-stituted by the carefully made incense-stick (cf. Bedini, 5, 6). This was very common in temples, yamens and homes ashore, and would have been highly convenient at sea. Cf. Vol. 3, p. 330, Vol. 4, pt. 2, pp. 127, 462, 526.

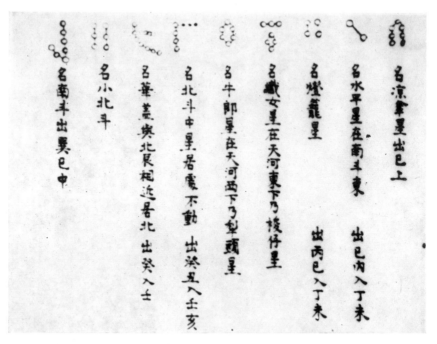

Fig. 1000. Part of a register of navigational constellations contained in an anonymous MS. rutter entitled *Chih Nan Chêng Fa* (General Compass-Bearing Sailing Directions) and appended to a military and naval encyclopaedia, the *Ping Chhien* (Key of Martial Art), by Lü Phan & Lu Chhêng-Ên (+1669).

From left to right:

(1) Nan tou (Southern dipper) = Tou *hsiu*, six stars in Sagittarius.
(2) Hsiao pei tou (Little northern dipper). According to Yen Tun-Chieh (*19*) β-Cassiopeiae, Ar. Naqeh.
(3) Hua kai (Imperial baldachin), sixteen stars in Cassiopeia and Camelopardus.
(4) Pei tou chung hsing (Star in the midst of the Northern dipper). A sailor's name; possibly Thai yang shou (Bright guardian) = χ Ursae Majoris.
(5) Niu lang hsing (Herd-boy), Altair in Aquila, and two neighbouring stars.
(6) Chih nü (Weaving girl), Vega in Lyra, and two neighbouring stars.
(7) Têng lung (Lantern), the Southern Cross, four stars in Crux Australis.
(8) Shui phing hsing (Level-with-the-water star). Possibly Canopus (α Carinae). According to Yen Tun-Chieh (*19*) α Eridani (Shui wei), Ar. Achernar; cf. Ideler (1), p. 233.
(9) Liang san hsing (Cool umbrella stars). According to Yen Tun-Chieh (*19*), α β Gruis, Ar. Hamārein, or possibly α, β Centauri (Nan mên).

All but three of these entries (nos. 2, 5 and 6) include rising and setting azimuth points, though not altitudes. Photo. Bodleian Library.

Fig. 1001. Chinese terrestrial globes; the David Globe (now in the British Museum), made in +1623 under the guidance of the Jesuits Emmanuel Diaz (Yang Ma-No) and Nicholas Longobardi (Lung Hua-Min), whose names appear as signatories of the inscription in the cartouche on the globe's under surface. A partial translation of this is given in Wallis & Grinstead (1). Looking at China as the centre we can see the outlines of Korea and Japan, Indo-China, Malaya, Sumatra, Java, Borneo and New Guinea, but not Australia. Far on the left the outlines of the Red Sea, Arabia, the Persian Gulf and the Caspian can also be made out. Far on the right at the bottom can be seen a group of islands which represents the Solomons, discovered by Mendaña in +1568, and the New Hebrides, Quiros' 'Austrialia del Espiritu Santo' of +1606.

The geography of the globe embodies notable improvements upon the world maps made in China by Matteo Ricci (Li Ma-Tou) between +1584 and +1603. Among other things it registers a very early record of the discovery of the Torres Strait, which as late as +1770 remained unknown to all but a few European cartographers, and it recognises the archipelagic nature of the land east of New Guinea. Now that this too was known to be an island, Diaz and Longobardi and their Chinese colleagues were probably very uncertain about the existence of a southern continent, the presumed 'Magellanica', so they found it convenient to use the space for their inscription. Dutch exploration of the Australian coasts had in fact been going on in +1606 and from +1616 onwards, but of this they did not know. Yet they placed the antarctic continent well south of South America, so they may have heard of the Dutch discovery of Cape Horn in +1616.

The globe, 1 ft. 11 in. in diameter, is made of painted lacquer on wood, to a scale of 1:21,000,000. The mounting shown is Copernican (inclined) but the globe has now been remounted in its original Ptolemaic (vertical) manner.

Fig. 1002. Chinese terrestrial globes; the Rosthorn Globe (in the Österreichisches Museum f. ange-wandte Kunst at Vienna), undated but attributable by internal evidence to the period + 1650 to + 1770. Just under 1 ft. in diameter, it is made of silver sheet metal on which the outlines and inscriptions were incised before being covered all over with translucent cloisonné enamel in bright colours. Around the South Pole the times in different parts of the world are given in Chinese double-hour signs (cf. Vol. 4, pt. 2, pp. 439, 461; Needham, Wang & Price (1), p. 200). Instead of political boundaries, the 'rectangles' formed by the meridians and parallels are depicted in different colours, e.g. on Australia, and Russia (O-lo-ssu Kuo).

The place-names are meant to be read with the south pole uppermost, but here and in the following two illustrations we present the photographs in the usual orientation. China, central in this picture, is portrayed in two colours, light for 'Cathay' north of the Yangtze, and dark for 'Manzi' south of it. Indo-China lies well enough below, but the East Indies are badly drawn, Sumatra being placed too far west and Borneo intervening between Malaya and Java, with New Guinea joined to Australia. The Indian peninsula is clear, the words Pang-ko-la Hai (Bay of Bengal) prominently appearing east of it, and on the far left one can make out the mouth of the Indus (Yin-tu Ho), the Persian Gulf, Arabia and the Red Sea.

Fig. 1003. A closer look at Australasia on the Rosthorn Globe. On the continent itself there are three chief inscriptions; in the north-west Wei-tê-ssu (de Witt's Land), and in the north-east Tieh-mên-ssu-an (van Diemen's Land) surrounded by conventional drawings of rocks or mountains—and in fact Australia is joined to Tasmania. In the south we read 'The new Western records say that this is New Holland (Hsin Wo-a-lang-ti-ya); a great continent all desert with nobody living there'. To the north Timor and the Celebes are roughly inserted, with a bloated Groote Eylandt in the Gulf of Carpentaria. Off Exmouth Gulf and the North-west Cape a large sausage-shaped island is shown, conceivably intended for the Cocos-Keeling Islands, discovered in +1609, or even for Christmas Island, first marked in +1666. Off Perth there is another, which might be meant for Rottnest Island, enormously magnified. North of the equator, which runs correctly through the middle of Borneo, the Philippines are shown as a too scattered conglomeration, while at the bottom on the right are the Solomons, etc., all individually much too large.

Fig. 1004. A closer look at China on the Rosthorn Globe. As mentioned already, the provinces south of the Yangtze are done in dark colour, those north of the River in light, and since north is here at the top, the Chinese place-names are all upside down. The Gobi desert is represented by the broad scaly-patterned band running north-east–south-west and labelled Sha-mo. To the left, at the edge of the light-coloured square, is marked Thu-lu-fan (Turfan in Sinkiang). In Manchuria we can see, among other names, Heilungchiang, Kirin, and Tu-erh-po-thê, i.e. the land of the Durbet Mongols, one of the four tribes or groups of banners of Inner Mongolia (cf. Gilbert (1), p. 907). The Korean peninsula, in light colour, is adequate, but Japan is badly drawn, having Kyushu hooked to the north; moreover Sakhalin, instead of being an island, appears as a fat black peninsula. Between Japan and Korea, Tsushima Island (Tui-ma Tao) is prominently marked, and Quelpart (Cheju) Island also appears among several that are unmarked. Further south the Liu-Chhiu island chain appears as a light-coloured lump just to the left of the 130° meridian, with a landmark, Pa-chhung Shan (the eight mountain isles) clearly incised on the billowing sea.

In China itself the Yellow River is following course ⑦ and discharging through the mouth of the Huai, though the old northern course is partly shown. A row of scale-like dots similar to those used for the Gobi desert traces the course of the Great Wall, with Peking (Ching Shih) just south of it. The light-coloured country north of the Yangtze bears provincial names such as Shantung, Shansi, Shensi and Kansu, clearly visible, but the southern provinces are so dark that no names can be read; the country between the Huai and the Yangtze is of an intermediate colour but has no inscription. South of Kuang-tung, however, some places are marked in the sea, notably Ao-mên (Macao), the Wan-li Sha (Thousand-mile Shoal, cf. Vol. 4, pt. 1, p. 284), and an island called Phêng which should stand for the Phêng-Hu or Pescadores Is. between Thaiwan and southern Fukien, but seems here to include Thaiwan itself. Hainan I. is also shown. All these are just outside the frame of this picture, but faintly visible in Figs. 1002 and 1003.

Fig. 1010. The mainsail of a Ta-pu-thou freighter (cf. Fig. 987) drawing nicely on a wet and windy day off the coast of Shantung (from the Waters Collection, National Maritime Museum, Greenwich). The details of the rig come clearly out. In the foreground there are the multiple sheets, with two of their euphroes visible, all attached to the leech ends of the battens, where the bolt-ropes can be seen along the sail edge. Behind them to port and starboard we see three ropes of the topping lift system (I, I in Fig. 1009a), and further away to port the feed of this, which includes a twofold purchase. The halyards (J, J in Fig. 1009a) are on the other side of the mainsail, but the four main sheets of the foresail (and even the crowfoot convergence to the lower euphroe) can be seen descending to their belaying-cleats on the port side of the deck (cf. Fig. 987). Three figures near by give the scale. The halyards of one of the mizen masts run up on the extreme left of the picture. On the starboard side an old-fashioned lantern appears.

Fig. 1011. Details of the sail leech edge of an Antung freighter visiting Wei-hai-wei (from the Waters Collection, National Maritime Museum, Greenwich). The attachments of the multiple sheets to the batten ends, the auxiliary bamboo battens, the bolt-ropes (*N, N* in Fig. 1009*a*), and the patched canvas can all clearly be seen.

(a)

(b)

Fig. 1012a, b. A Pao-chhing Chhiu-tzu freighter under sail, seen from the great bridge over the Hsiang Chiang (R.) at Hsiang-than south of Chhangsha in Hunan (orig. photo., 1964). A description and scale drawing of this smart and well-appointed type of river-ship, c. 75 ft. long, and named after a city (now Shao-yang) on a tributary, the Tzu R., is given by Worcester (3), vol. 2, p. 431, pl. 156. Here one can see many things, the prominent balanced rudder, the multiple sheets, the parrels securing the sail to the mast, halyards and secondary halyards, the ropes of the topping lift, some of the hauling parrels (cf. Fig. 1014), the lumber irons that bridge the mat-roofed cargo deckhouse, and the old-style capstan amidships forward, worked with rough handspikes and used both for hoisting sail and weighing the anchor. The large empty pottery jars which form part of the freight are similar to those universally used for the complicated preparation of 'soya-bean sauce' (chiang yu), on which see Sect. 40.

Fig. 1013. A Yangchiang fishing-junk becalmed, showing the extreme rounding of the sail leeches which is characteristic of the southern Chinese coast (from the Waters Collection, National Maritime Museum, Greenwich). The yard, battens and boom are all gathered towards the base of the luff edge, so that the lug-sails look at first sight like radial fans. The multiple sheets, which in the case of the mizen sail are attached some way short of the leech edge, are clearly seen, and topping lifts, parrels and hauling parrels can also be made out. An unusual feature of these ships is the prominent mast stays, perhaps adopted from European practice (cf. p. 401), in any case useful on the edge of a typhoon. In build, the stern recalls that of Hongkong junks (cf. Fig. 1044), the strakes being gathered by scarf joints so that the aftermost transom is relatively small, but this is hidden by a false aft transom under the projecting stern gallery. A great slot is left for the slung rudder, which here is deep in the water.

Fig. 1014. The foresail of a Yangchiang fishing-junk, just set, to show more clearly the nature of the rig (from the Waters Collection, National Maritime Museum, Greenwich). The sail shape and mast stays are as expected from Fig. 1013, but one can now see better that the battens are connected by a number of radial ropes which help to keep them just the right distance apart. The topping lift system is well seen, but also the four heavy wire parrels, preventing the natural tendency of fore-and-aft sails to move forwards, as does the heavy strop from the boom which is taken several turns round the mast. Further to be made out is the complicated reeving of the hauling parrels (cf. Worcester (3), p. 71), which assist in holding back the sail and preserve its balance; the system seems close to variant C in Worcester's pl. 15. The multiple sheets are of course hidden, but the purchase block of their feed can be seen behind the figure in shadow. A second is sitting at the bows beside the two anchors, a one-sided adze anchor and a Chinese non-fouling anchor with the stock at the crown (cf. p. 657). Behind, the mainsail is just being hoisted, coming up grandly out of its pleats.

Fig. 1015. A Chhangsha sprit-sail sampan on the Hsiang River (orig. photo. 1964). Description and scale drawing in Worcester (3), vol. 2, p. 442, pl. 167. Even so small a sail as this has multiple sheets, as can be seen.

Fig. 1018. An Indonesian *prao-mayang*, showing the characteristic canted square-sail rig (drawing of Adm. Paris). The steering-oar is also to be noted. In this picture the vessel is seen sailing as near the wind as possible, but by the direction of the flag the angle can hardly be much better than 70°. Nevertheless the canted square-sail is to be regarded as the first step in the development towards fore-and-aft sails.

Fig. 1019. A fishing-junk from Tolo Harbour (Thu-lu-wan, the great eastern inlet in the Hongkong New Territories) using a canted square-sail as a kind of spinnaker in addition to the two usual lug-sails (from the Waters Collection, National Maritime Museum, Greenwich).

Fig. 1020. A small boat of the Kuala Trengganu fishing fleet (North-eastern Malaya) running for port before the afternoon wind from the sea (photo. Hawkins & Gibson-Hill). The foot of the canted square-sail is here placed so far to starboard, or at least amidships, that it almost approximates to a lug.

Fig. 1023. Working a small lighter (probably at Canton) by the use of the self-feathering scull oar (photo. Fitch, 1927). The motion of this yuloh, which pivots on a fulcrum and is attached to the deck by a short length of rope, amounts to that of a reversible screw propeller. See pp. 622 ff.

Fig. 1024. A boat being tracked upstream in one of the rapids in the Yangtze gorges near Chungking (photo. Potts, *c.* 1938).

Fig. 1025. Pa Chhuan Hsia Hsia Thu (Boats descending the Gorges out of Szechuan), a painting by Li Sung, *c.* +1200 (Ming copy). The slung rudder and the outboard thole-pins for the yulohs are noteworthy.

Fig. 1027. The deck of a Swatow freighter, looking forward (from the Waters Collection, National Maritime Museum, Greenwich). In the foreground the massive iron-bound slung rudder can be seen entirely out of the water and stowed on the poop, since the vessel is at rest in harbour.

Fig. 1029. A stern-sweep with tiller, of ancient Egyptian type, still in use on the Ganges in India today (photo. Crealock, 1).

Fig. 1030. One of the two oldest depictions of an axial or stern-post rudder in Europe, the carving on the font at Zedelghem in Belgium, dating from c. +1180. It illustrates some legend of the saints, just as do the story-pictures in the Tunhuang frescoes.

Fig. 1032. Two passenger-carrying river-junks with stayed bipod masts moored side by side in the mists of evening outside a city wall; a fan painting attributed to Ma Ho-Chih, *c.* +1170 (in the Suzuki Collection, reproduction from Harada & Komai). Both ships show balanced rudders, but that on the left more clearly, that on the right being obscured to some extent by a bundle of poles, perhaps a fender, hanging over the side.

Fig. 1034. Details of the slung and balanced rudder system of one of the cargo-boats in the Chhing-Ming Shang Ho Thu (Coming up the River to the Capital after the Spring Festival), painted by Chang Tsê-Tuan c. +1125.

Fig. 1035. Ku Khai-Chih's painting of a ship, done about +380 to illustrate the poem of Tshao Chih 'Rhapsodic Ode on the Nymph of the Lo River', written *c.* +230. None of the early copies, nor the original, survive, so the illustration is of a Sung (+12th-century) copy preserved in the Freer Gallery of Art at Washington. There has been much speculation about the structures at the stern, which may or may not be evidence for an axial rudder as Ku Khai-Chih saw it, or as his copyists misinterpreted it. Archaeo-logical evidence now strongly indicates that he was trying to depict one (see text, p. 649).

Fig. 1036

Fig. 1037 (*a*)

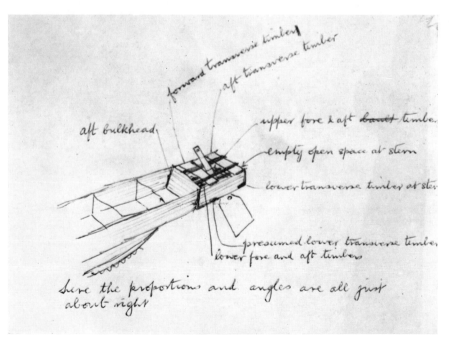

Fig. 1037 (*b*)

Fig. 1036. Stern view of the grey pottery tomb-model ship of the +1st century excavated at Kuangchow (photo. Canton Museum, cf. Figs. 963, 964, 965). The attachment of the axial rudder, with its eye for the slinging tackle, between the timbers of the floor of the after-gallery, is well seen.

Fig. 1037. Pages from my notebook with sketches of the shape and mode of attachment of the rudder of the +1st-century tomb-model ship, made in the Canton Museum, 1958.
a Shape and relative dimensions taken from photographs.
b Manner in which the slung rudder was secured within the after-gallery.

Fig. 1038. The ship of the great minister Sugawara no Michizane taking him into exile c. +900; a scroll-painting preserved in the Ueno Museum and said to be a copy of an older painting by Fujiwara no Nobuzane (d. +1264 or +1265) still in the Tenmangū at Kyoto (the Shinto temple dedicated to Sugawara as god of learning). In general build, and particularly in the shape and position of the rudder, this vessel distinctly recalls the Canton model ship of the +1st century. As in the ship sketched in Fig. 1033 (b) the sailors are rowing backwards on outboard projections, not forwards as in China. Reproduction from Purvis (1).

Fig. 1039. Ship in a +1237 MS. of the *Maqāmāt* of al-Ḥarīrī (Bibliothèque Nat., Paris, Ar. 5847). The axial, presumably stern-post, rudder seems to be provided with some kind of lateral control. It is natural to think of the Arabs as having been the transmitters of the invention of the axial or median rudder to the Europeans, but no illustrative material of the required time, the +12th century, has so far come to light.

Fig. 1040. A freighter or grain-carrier of Hongkong and other ports of South China (model in the Maze Collection, Science Museum, Kensington). Anon. (17), no. 3. The relieving tackle for the tiller is well seen at the stern. Ships of this kind are intermediate between the Chinese and European traditions, for they have hulls with keels and stempost, as here shown, while rig and rudder follow Chinese practice. The stern view of a ship such as this would be similar to those seen in Figs. 1013 and 1044. On hybrid types cf. pp. 433 ff.

Fig. 1042. The massive slung rudder of a Fuchow timber freighter stowed in the after-castle when in port (from the Waters Collection, National Maritime Museum, Greenwich). Cf. Fig. 936 and the drawing in Fig. 1041.

Fig. 1043. The rudder of a Ma-yang-tzu (cf. Figs. 932, 933 and Worcester (1), pl. 1) seen in a shipyard on the Upper Yangtze (photo. Spencer, 2). This beautiful shape, fitted on the right to the curve of the hull, is the Chinese balanced rudder *par excellence*.

Fig. 1044. The stern of a Hongkong fishing vessel in dry dock, showing the fenestrated rudder characteristic of many types of Chinese ship (from the Waters Collection, National Maritime Museum, Greenwich). The figures of workmen give the scale, and another fenestrated rudder being re-bladed is seen on
the right.

Fig. 1050. A reconstruction of one of the armoured 'Turtle ships' used by the Korean naval forces under the admiral Yi Sunsin against the Japanese in the last decade of the +16th century. The model is in the National Historical Museum at Peking (orig. photo. 1964). Cf. Underwood (1), figs. 46–9. It is believed that at least two masts were used, but lowered before battle through a central fore-and-aft slot in the armoured roof (not shown in this model) by some method of striking such as that described in Worcester (3), vol. 1, p. 79.

the Tène age.[a] Brindley (10) has studied a bronze-age anchor which would be about contemporary with the Homeric mentions,[b] and from about − 500 onwards the anchor of the Mediterranean peoples had attained approximately the familiar form, as is known from many coins on which it appears. The stock, however, was absent in the earlier periods. That a parallel evolution of forms occurred in China can be deduced from the names employed. The earliest expression for casting anchor[c] was *hsia shih*,[1] and the stone-weighted grapnel was called *ting*,[2] sometimes written

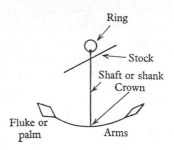

ting.[3] It was simply made by binding from one to four forked branches together with a piece of stone,[d] and this lasted long in use; we saw it above on the Bayon junk of + 1185 (Fig. 975, pl.).[e] When the use of metal hooks was introduced, these words were replaced by *mao* (or *miao*),[4] combining the idea of the shoots of plants or the claw of a cat with the metal radical, but the sound was perpetuated in *ting*,[5] though the original and still more general meaning of this is an ingot. The *Yü Phien* dictionary of + 543 seems to be the first in which the word *mao* appears, and if this may be taken as evidence that metal anchors were not used much earlier, their introduction in China would have occurred later than in the West.

Nevertheless the Chinese made a contribution of some importance to the development of the anchor. The most characteristic of their forms is the adze anchor, i.e. an anchor made in such a way that the arms form an acute angle with the shaft (about 40° or less) at the crown instead of diverging from it at a right angle or in an arc. Such forms were not unknown in Europe (the Roman temple-ships or barges on the Lake of Nemi had them),[f] but the Chinese passed the stock across the shaft not at the ring end but near the crown end (also of course at right angles to the plane of the arms). This serves the purpose of canting the anchor and ensuring that the arms bite, but it has also the great advantage of being almost non-fouling. The efficiency of this device was often

---

[a] Feldhaus (1), col. 930. A stone and metal anchor from the Swiss lake-dwelling culture is in the museum at Biel.

[b] But *Odyss.* xiii, 77, refers to a stone with a hole in it. On this kind see Frost (1), pp. 29 ff.

[c] This still occurs in the travel account of Fa-Hsien, *c.* +414; cf. the discussion between Brindley (11) and van Nouhuys (1), to which the intervention of a distinguished sinologist, Giles (10), brought little enlightenment; indeed he quite misled Moll.

[d] An early mention of the stone-weighted grapnel or killick is associated with the name of Huang Tsu,[6] a commander of the San Kuo period (early + 3rd century), but Worcester (3), vol. 1, p. 97, failed to give the exact reference, and we have not found it.

[e] It is mentioned also in the *Kao-Li Thu Ching* (ch. 34, p. 4b) earlier in the same century. The contemporary Norman ships depicted on the Bayeux tapestry are said to show the earliest flukes of modern type on their iron anchors (des Noëttes (2), fig. 65).

[f] Van Nouhuys (2), p. 37; Moretti (1). The dating of these ships to the reign of the emperor Caligula (+37 to +41) has been confirmed by radiocarbon analyses (Godwin & Willis; Godwin, Walker & Willis).

---

[1] 下石   [2] 矴   [3] 碇   [4] 錨   [5] 錠   [6] 黃祖

praised by European nautical writers,[a] and during the nineteenth century it was several times 're-invented', with addition of hinged stocks (Hawkins, 1821; Piper, 1822; Porter, 1838), so that some of the most modern 'stockless' types, such as the Danforth anchor, derive originally from the Chinese rather than the Graeco-Roman form.[b] We can trace the Chinese adze anchor back through the Fukienese Shipbuilding MS. (Fig. 1045) not only to the +17th-century *Wu Pei Chih*[c] (Fig. 1046) but to the +1st century in the Cantonese tomb-model ship of Fig. 964 (pl.).

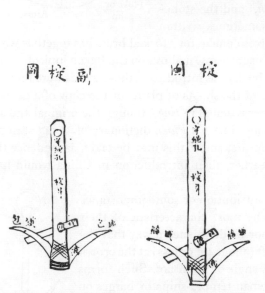

Fig. 1045. Two adze anchors from the Fukien Ship-building MS., a larger (*ting*) and a smaller (*fu ting*). The tips of the arms are shod with iron. Cf. p. 406 above.

Fig. 1046. An adze anchor from the *Wu Pei Chih* (printed +1628). The wooden label (*phai tzu*) says: 'Anchor belonging to Official Ship No. so-and-so.'

Anchor windlasses (i.e. vertically mounted drums)[d] are mentioned several times in the *Kao-Li Thu Ching* of +1124; those on the Korean 'official boats' Hsü Ching refers to[e] as *ting lun*,[1] while those on the larger Chinese 'retainer ships' which carried the personnel of the embassy he speaks of[f] as *chhê lun*.[2] The rope which was wound on these was, he says, as thick as the rafters of a house, and was made of twisted *thêng*[3] (some kind of vine or liana) 500 ft. long. A slightly later depiction of an anchor wind-lass we have noted already in the Bayon junk (Fig. 975, pl.).[g] Among modern Chinese

---

[a] E.g. Charnock (1), vol. 3, p. 297 (fig. opp. p. 292); F. E. Paris (1), p. 74.

[b] For many details concerning existing types of traditional anchors in China, see Worcester (3), vol. 1, pp. 99 ff. and Audemard (3), pp. 52 ff. Some have a ring on one of the palms for tripping.

[c] Ch. 117, p. 23*a*.

[d] Emphasis has already been placed on the distinction between vertical and horizontal mountings; cf. Vol. 4, pt. 2, s.v. Here we retain our definition; though a sailor would of course refer to a windlass as a horizontal drum.

[e] Ch. 33, p. 2*a*.   [f] Ch. 34, p. 4*b*.   [g] See p. 461 above.

¹ 矴輪   ² 車輪   ³ 藤

sailors' names for the anchor windlass *chu lung chiao*[1] ('the pig-basket reel') may be mentioned; this must derive from the custom[a] of making pulleys and drums without rims, the spoke ends being connected together with strong cords, and the whole looking therefore somewhat like a round basket. Such windlass drums may be seen in pictures of Korean ships of the present day,[b] where they are exceptionally large and prominent. But the largest of the type are found in the winch, or the capstan (horizontally turning drum), of the *mu phai*[2] (timber rafts) of the Lower Yangtze. This attains a height of 14 ft. and in construction is similar to the great winding-drums of the brine borehole workings at Tzu-liu-ching in Szechuan. Its function is to heave the large drogues or sea-anchors which keep the raft heading in the desired direction.[c] Sea-anchors are also referred to in +1124 by Hsü Ching,[d] under the name of *yu ting*,[3] but he erred in saying that they were just like ordinary anchors, though he knew they were used in stormy weather at sea. In fact, they must then have been what they are now, large bamboo baskets, in the use of which no sailors are more skilled than the Chinese.[e]

Sung Ying-Hsing has this to say of the anchors of the inland grain-transport freighters:[f]

Anchors of iron are dropped into the water to moor the ships; a grain-boat usually has five or six. The heaviest weighs about 500 catties and is called the 'watchdog' (*khan chia mao*[4]). In addition, two small ones are slung both at bow and stern. When a ship in midstream meets with too strong an adverse breeze, so that she cannot go forward, and also cannot tie up anywhere, [or where the river-bed near shore is rocky instead of sandy, and one cannot approach the shore, then one must anchor in deep water][g] letting go the hook so that it sinks (quickly) to the bottom. The hawser (*hsi yü*[5]) of the anchor is wound round the bollards (*chiang chün chu*[6]) on the deck (and made fast to them). When the flukes (*chao*[7]) of the anchor touch the mud and sand of the bottom they dig in and hold securely. The 'watchdog' is resorted to only when danger is imminent; its hawser is called 'ship's-self' (*pên shen*[8]) in order to indicate its importance. Or again, when the boat is under way in company and seems likely to collide with another vessel ahead which has had to slow down, the stern anchors are smartly lowered into the water to check the speed. As soon as the wind abates, the anchors are hoisted by means of a winch (*yün chhê*[9]).

A very different form of mooring is commonly found in use by the smaller river-junks and sampans; it consists of a trunk or tube built into one or more of the compartments of the vessel.[h] Through this a weighted pole is driven down into the mud of the lake or river bottom. Known as the 'water-eye' (*shui yen*[10]), this device has the advantage of continuous adaptation to fluctuating water-levels. In China this pole-setting or

---

[a] Prevalent not only in nautical technology, but also in many other fields, e.g. textile machinery, as we have already seen (Vol. 4, pt. 2, Fig. 404 and p. 103).

[b] Underwood (1), p. 16 and figs. 14, 19, 20, 32. See Fig. 1031.

[c] Worcester (3), vol. 2, p. 391.   [d] Ch. 34, p. 4*b*

[e] Worcester (3), vol. 1, p. 103.

[f] *TKKW*, ch. 9, pp. 4*a*, *b*, tr. auct. adjuv. Ting & Donnelly (1); Sun & Sun (1).

[g] Sung Ying-Hsing's own commentary.

[h] See Waters (5); Worcester (3), vol. 1, p. 98. Conceivably this practice helped in the invention of the rudder-trunk.

[1] 猪籠絞   [2] 木簰   [3] 遊矴   [4] 看家錨   [5] 繫縴
[6] 將軍柱   [7] 爪   [8] 本身   [9] 雲車   [10] 水眼

'stick-in-the-mud anchor' is at least as old as the Sung, for paintings of that period show it, but it is ancient in many parts of the world from New Guinea to +6th-century Holland. The principle is still used in modern dredgers.

To study the construction and layout of harbours and docks in Chinese history would require a whole chapter to itself, all the more difficult to write in that we have come across no studies of the subject either by Chinese or Western scholars.[a] One question of marked technical interest may however be raised, namely the development of dry docks for building and repair. The European aspect is a little obscure. For Darmstädter (1) the first dry dock in England, and probably, he thought, in the Western world, was that made at Portsmouth for Henry VII in +1495; it had no gates but was closed by piling filled in as required. Straub, on the other hand, puts the first dry docks in de Bélidor's time (c. +1710).[b] Neuburger[c] and Forbes,[d] as well as others, have claimed Alexandrian credit, even to the −3rd century, but the more recent survey of Goodchild & Forbes brings nothing to substantiate this. In any case, we have excellent evidence for the invention in the Sung period, with a circumstantial account from the pen of Shen Kua:[e]

At the beginning of the dynasty (c. +965) the two Chê provinces (now Chekiang and southern Chiangsu) presented (to the throne) two dragon ships[f] each more than 200 ft. in length. The upper works included several decks with palatial cabins and saloons, containing thrones and couches all ready for imperial tours of inspection. After many years, their hulls decayed and needed repairs, but the work was impossible as long as they were afloat. So in the Hsi-Ning reign-period (+1068 to +1077) a palace official Huang Huai-Hsin[1] suggested a plan.[g] A large basin was excavated at the north end of the Chin-ming Lake capable of containing the dragon ships, and in it heavy crosswise beams were laid down upon a foundation of pillars. Then (a breach was made) so that the basin quickly filled with water, after which the ships were towed in above the beams. The (breach now being closed) the water was pumped out by wheels[h] so that the ships rested quite in the air. When the repairs were complete, the water was let in again, so that the ships were afloat once more (and could leave the dock). Finally the beams and pillars were taken away, and the whole basin covered over with a great roof so as to form a hangar in which the ships could be protected from the elements and avoid the damage caused by undue exposure.[i]

Apparently Shen Tsung did not have swinging gates any more than Henry VII but he did have four centuries priority.

While speaking of harbours, havens and anchorages, a word may be said about the lighthouses[j] (têng tha,[2] though this term is probably not old) which help one to get

---

[a] Some records of Sung harbour-works have been collected, however, by Liu Ming-Shu (4).

[b] (1), p. 144.   [c] (1), p. 482.

[d] (2), p. 68.

[e] MCPT, Pu app. ch. 2, para. 19, cf. Hu Tao-Ching (1), vol. 2, p. 954, (2), p. 313; tr. auct. with Lo Jung-Pang. The importance of the passage has been recognised by Ku Chün-Chêng (1).

[f] Cf. p. 436 above.

[g] Already met with, p. 336 above, in connection with dredging techniques.

[h] Powered norias or square-pallet chain-pumps.

[i] Cf. the lock basins of Chhiao Wei-Yo (p. 351 above). In view of what we have seen concerning lock gates in the Sung, this dry dock could very probably have had stop-log gates.

[j] See de Loture & Haffner (1), pp. 25, 55, 276; Allard (1); Hennig (9).

[1] 黃懷信   [2] 燈塔

there. They are perhaps less prominent in Chinese than in ancient European literature, as might be expected from the relatively greater importance of maritime navigation in the West. References to beacons (*fêng sui*[1]) collected in encyclopaedias almost always refer to lights on hills or forts for military or other governmental signalling.[a] There was nothing at all similar to the Pharos of Alexandria, built in −270 by Sostratus of Cnidus for Ptolemy Philadelphus, which probably reached a height of 150 ft. and much of which was still standing in the +13th century.[b] Though coastwise and lakeside lights must have been used on a small scale in China, literary references generally speak of them in connection with foreign parts. Thus Chia Tan[2] the geographer, writing between +785 and +805, says, in his description of the sea route between Canton and the Persian Gulf, referring to some place near the latter's mouth, that 'the people of the Lo-Ho-I[3] country have set up ornamental pillars (*hua piao*[4]) in the sea, on which at night they place torches (*chü*[5]) so that people travelling on board ships shall not go astray'.[c] Independent confirmation of lighthouses in the Persian Gulf for a century later is available in Arabic authors such as al-Mas'ūdī,[d] and al-Muqaddasī.[e] It is perhaps of interest to read what a Chinese writer said of the Alexandrian Pharos in +1225:[f]

The country of O-Kên-Tho[6] (Alexandria) belongs to Egypt (Wu-Ssu-Li[7]). According to tradition, in olden times a stranger (*i jen*[8]), Chhu-Ko-Ni[9] by name,[g] built on the shore of the sea a great pagoda, underneath which the earth was excavated to make two rooms, well connected and thoroughly hidden. In one vault was stored grain, and in the other arms. The tower was 200 ft. high.[h] Four horses abreast could ascend (by a winding ramp) to two-thirds of its height. Below the tower, in the middle, there was a well of great size connected by a tunnel with the great river. To protect this pagoda from foreign soldiers, the whole country guarded it against all enemies. In the upper and lower parts of it twenty thousand men could readily be stationed as a guard or to make sorties. At the summit there was an immense mirror, and if

[a] E.g. *TPYL*, ch. 335, pp. 5 *a* ff. But when the forts were on a coast or waterside there was a gain to shipping because the universal practice was to keep one light burning perpetually in time of peace. The *Thai Pai Yin Ching* (+759) says: 'Every night when all is well one light is lit. If there is an alert the watchmen light two, when smoke or dust (etc.) give warning of an enemy's approach they light three, and when the enemy comes in sight the basket of fuel is set on fire. If at dawn and by night the peace light is not seen, the watchmen must have been captured by the enemy' (ch. 46, p. 2*b*). Similar beacons announced the arrival of the Chinese embassy to Korea in +1124 (*Hsüan-Ho Fêng Shih Kao-Li Thu Ching*, ch. 35, p. 2*b*). By +1562 there were no less than 711 beacon stations along the coasts from western Kuangtung to northern Chiangsu (*Chhou Hai Thu Pien*, chs. 3–6). Gandar (1), pp. 18 ff., describes the watch-towers on artificial mounds (*tun*[10]) still used in northern Chiangsu in the late nineteenth century for giving warning of tidal waves as well as pirates. For Chinese material we have unfortunately nothing comparable with Rosani's monograph (1) on maritime lights and signals. But information on naval formations, signals and combat positions is contained in Yü Chhang-Hui's *Fang Hai Chi Yao*, ch. 14 (1822).

[b] See Forster (1); Neuburger (1), p. 245; Feldhaus (1), col. 624; de Camp (2), etc.

[c] *Hsin Thang Shu*, ch. 43 ᴮ, p. 18*b*, tr. Hirth & Rockhill (1), p. 13. The location seems to be the coast of Baluchistan (Mekran).

[d] Tr. de Meynard & de Courteille (1), vol. 1, p. 230.

[e] Tr. Ranking & Azoo (1), p. 17.

[f] Chao Ju-Kua in *Chu Fan Chih*, p. 31*b*, tr. Hirth & Rockhill (1), p. 146, mod. auct. Cf. Fig. 985.

[g] Certainly Dhū al-Qarnayn ('he of the horns'), i.e. Alexander the Great himself.

[h] Amending *chang* (10 ft.) to *chhih* (1 ft.).

[1] 烽燧	[2] 賈耽	[3] 羅和異	[4] 華表	[5] 炬
[6] 遏根陀	[7] 勿斯里	[8] 異人	[9] 徂葛尼	[10] 墩

warships of other countries tried to make an attack, the mirror detected them beforehand, and the troops were ready to repel it. But in recent years there came (to Alexandria) a foreigner, who asked to be given work in the guardhouse below the tower, and he was employed to sprinkle and to sweep. For years no one entertained any suspicion of him, but suddenly one day he found an opportunity to steal the mirror and throw it into the sea, after which he made off.

With this background it is rather interesting that one of the most famous Chinese lighthouses was the minaret of a mosque in Canton. This was the Kuang Tha¹ of the Huai Shêng Ssu,² and we have already come across it, with its golden cock at the top of the tower, in connection with parachutes.ᵃ A detailed account of the buildings was given by Chhou Chhih-Shih at the beginning of the nineteenth century,ᵇ in words based partly on a much older description, that of Fang Hsin-Ju³ in his *Nan Hai Pai Yung*⁴ (A Hundred Chants of the Southern Seas), written about +1200. Rising to a height of 165 ft., it was called the 'Light Tower' apparently because a light was kept burning at the top to guide shipping. It was first built in the Thang by foreigners, our authorities say,ᶜ and had a spiral staircase inside it. Each year in the fifth and sixth months the Arab foreigners used to assemble to scan the estuary for their sea-going barques, and then at the fifth drum they ascended the tower to shout prayers for favourable winds. In +1468 the Imperial Censor Han Yung⁵ caused the minaret to be repaired, and arranged it for sending official messages, presumably by lights. However, Buddhist pagodas also served occasionally as lighthouses. The *Hang-chou Fu Chih*⁶ says that the Liu-Ho Tha⁷ on the Chhien-thang River was equipped with a permanent light from the early Sung onwards to guide ships seeking their anchorages at night.ᵈ Thus two at least of China's religions contributed something as the equivalent of the Brethren of Trinity House.

(2) TOWING AND TRACKING

Mention has often been made above of the tracking of junks up rivers in China.ᵉ For anyone who has lived near one of the great Szechuanese rivers, the Chialing at Chung-king for example, the cries of the trackers, and the sound of the drums which give them the time, remain among unforgettable memories. No natural difficulties defeated the Chinese boatmen, as can be seen from Fig. 880 (pl.) which shows one of the towing galleries along the gorges of the Yangtze.ᶠ Teams of as many as a hundred men may be

---

ᵃ See Vol. 4, pt. 2, p. 594.

ᵇ *Yang-chhêng Ku Chhao*, ch. 3, pp. 36b, 37a, ch. 7, p. 16a.

ᶜ Probably the Sung is a more likely date.

ᵈ Ch. 35, p. 20a. Cf. Liang Ssu-Chhêng (10). This local tradition is still very living, as I found on a visit in 1964. The pagoda was first built in +970 with 9 storeys, but rebuilt in +1136 with its present pattern of 7, made by intermediate wooden balconies to look like 13.

ᵉ Pp. 354, 415, 466, 626. Photograph in Fessler (1), p. 51.

ᶠ An elaborate study of the trackers' galleries in the San-mên gorge of the Yellow River has now been published (Anon. 33). On this see p. 277 above.

¹ 光塔   ² 懷聖寺   ³ 方信孺   ⁴ 南海百詠   ⁵ 韓雍
⁶ 杭州府志   ⁷ 六和塔

Fig. 1047. Salt boats being tracked upstream, an illustration from the *Szechuan Yen Fa Chih* (ch. 6, pp. 15*b*, 16*a*). The whirlpool adds to the drama of the drawing, but the rivers of Szechuan have them, as I know from sailing down from Chiating (Lo-shan) to Suifu (I-pin) in a sampan in 1943.

employed in the most difficult places. Fig. 1047 shows salt boats being tracked upstream in Szechuan. The paintings of Hsia Kuei[1] (+1180 to +1230) are often cited as evidence of tracking before the time of Marco Polo's description (p. 466 above), but we have much earlier representations, such as those among the Tunhuang frescoes (Fig. 547, opp. p. 311 in Vol. 4, pt. 2). A possible relation between the trackers' cloth-band harness and the development of the efficient harness for draught animals has already been suggested at the same place (p. 312).[a] In tracking (*la chhien*[2]), the cable (*so*[3]) is secured to a cross-beam aft, whence it passes to the mast and runs through a cast-iron snatch-block. This can be raised or lowered by halyards; normally it hangs at about one third of the height of the mast, but when overtaking another boat it is hoisted

[a] Worcester (2), pp. 59 ff., has devoted an interesting chapter to Chinese sailors' knots, among which those used by the trackers figure. Cf. the tome of Ashley (1).

[1] 夏圭   [2] 拉縴   [3] 索

to the masthead.[a] The bamboo cable used for tracking has already been mentioned in other connections;[b] one of its remarkable qualities is that while hempen ropes lose some 25% of their strength when wet, this plaited bamboo undergoes on saturation with water an increase of tensile strength of about 20%. Tests by Fugl-Meyer showed that a cable of $1\frac{1}{2}$ in. diam. would take a load of nearly 5 tons when dry and about 6 tons when wet.[c]

(3) CAULKING, HULL-SHEATHING AND PUMPS

The means used by traditional Chinese shipwrights and sailors to render hulls water-tight have been referred to from time to time[d] in foregoing sub-sections—generally speaking the classical mixture for caulking (*nien chhuan*[1]) was tow mixed with tung oil and lime, while additional security was obtained in old ships by nailing on season after season fresh layers of strakes to increase the hull's thickness. The marine drying putty was essentially the same as that used by the brine-works engineers of Szechuan for their piping and other containers, but the best varieties were of more complex composition, especially notable being the addition of a proportion of soya-bean oil.[e]

Protection against enemies other than water was a rather different matter.

The sheathing of hulls against the attacks of *Teredo* and other pests,[f] or the growth of sessile marine organisms, and the protection of upper works against the attacks of the enemy in battle, go naturally together. But metal plates were used for the former purpose long before their employment for the latter. Moll (3) has attempted to write the history of the conservation of wood in ships and buildings; from this and other sources[g] we know that the Romans tried sheathing ship bottoms with lead, as in the galley of Trajan on Lake Riccio (described by Alberti), and in the temple-ships of Lake Nemi.[h] This, however, was only exceptional,[i] and no examples of sheathing are known from the Middle Ages. Lead was tried again in Europe about +1525, but soon abandoned in favour of sheathing with a layer of boards (cf. p. 468) often with horse-hair packing; then from +1758 copper plates began to be widely employed.[j] Attention

---

[a] On all details, see Worcester (1), pp. 13 ff., (3), vol. 1, pp. 42, 62 ff., vol. 2, p. 296. Tracking by men along the river-sides lasted in France till 1830, so it is not surprising to find Gallo-Roman bas-reliefs which show it, but one does feel some surprise in finding that they also attached their cables to the mast (Benoit (2); Bonnard (1), fig. 18, opp. p. 240; Pobé & Roubier (1), pl. 210).

[b] Pp. 191, 328, 597.   [c] In Worcester (1), p. 15.

[d] Pp. 413, 414, 462, 467. Cf. Audemard (3), p. 20.

[e] Described in the book of techniques written by the monk Lu Tsan-Ning[2] about +980, the *Wu Lei Hsiang Kan Chih*[3] (On the Mutual Responses of Things according to their Categories), (p. 31). Quoted recently by Li Chhang-Nien (1), p. 76, no. 94.

[f] See Lane (1).

[g] Neuburger (1), p. 482, referring to Athenaeus, *Deipnosophists*, v, 40.

[h] Moretti (1); Ucelli di Nemi (1, 2).

[i] Or so it used to be thought, but the −3rd-century Greek merchant-ship sunk off Marseille which has been investigated by divers had its hull (and apparently part of its deck) completely covered with 20 tons of lead plating secured with lead-coated copper nails (Cousteau, 1).

[j] Charnock (1), vol. 1, p. 101; vol. 3, p. 201; de Loture & Haffner (1), p. 103; Clowes (2), pp. 85, 104. As also frequently on Chinese sea-going junks of the +18th century, if not rather earlier (see Chao Chhüan-Chhêng, 1). There was always much trouble with electrolysis where the copper came in contact with the iron of the nails.

[1] 艌船   [2] 錄贊寧   [3] 物類相感志

was drawn by Julien (4) to the fact that Chinese writings of the early +4th century refer to the covering of junk bottoms with copper. Thus the *Shih I Chi*, by Wang Chia, referring to an embassy from the Jan-Chhiu[1] kingdom in the legendary reign of Chhêng Wang,[2] says:[a] 'Floating on the seething seas, the ambassadors came on a boat which had copper (or bronze) (plates)[b] attached to its bottom, so that the crocodiles and dragons could not come near it.' Here the defence against organic life is clearly mentioned, and the passage would seem to prove that the idea, at least, existed in Wang Chia's time. Another Chin book, the *Chiao-chou Chi*[3] by Liu Hsin-Chhi,[4] says that at Anting a copper or bronze boat which had been built for the King of Yüeh lay for a long time buried in the sand, where it could be seen at low tide.[c]

It has now been shown that stories of metal boats occur abundantly in the early Chinese literature of folklore and legend.[d] They are particularly common in South China and Annam, where they often form part of the epic exploits of the Han general, Ma Yuan,[5] who restored the far south to Chinese allegiance in the campaign of +42 to +44. The bronze or copper boats of which people see the vestiges are thus associated with the setting up of bronze columns to mark the southern limits of the empire, the casting of bronze oxen as landmarks, and the building of canals to shorten sea voyages or make them more safe.[e] The evidential texts date from all periods between the +3rd and the +9th centuries, but the only one which specifically mentions the bottom of a ship is the early +4th-century *Shih I Chi*. Although it is quite possible, as sinologists tend to think, that the idea of using metal in the construction of boats was purely magical and imaginary in origin, it is at any rate equally possible that some southern group of shipwrights in those ages had the services of smiths who beat metal into plates fit for nailing to the hulls of their craft to protect the timbers. If so, the copper-bottomed junks of the +18th century derived from an indigenous tradition, and not from the Lake of Nemi in the Far West. We even hear of iron boats. In or before the Sung, a book of unknown authorship, the *Hua Shan Chi*,[6] speaks of a derelict iron boat beside a mountain lake.[f] This was doubtless a further echo of the same legend, or the same technique. But iron armour for warships was no legend, as we shall see (p. 682).

[a] Ch. 2, p. 6*a*, tr. auct. Cit. *KCCY*, ch. 28, p. 12*a*.

[b] The word is *po*,[7] a screen, or *po*,[8] a hoe, giving in either case the nuance of metal plates. A variant *po*,[9] beam or column, cedes to this.

[c] Cit. *Hou Han Shu*, ch. 33, p. 16*b*; also in the late +7th-century encyclopaedia *Chhu Hsüeh Chi*, ch. 7, p. 3*a*; *TPYL*, ch. 769, p. 6*a*; *KCCY*, ch. 28, p. 12*a*; and *Thai-Phing Huan Yü Chi*, ch. 171, p. 4*b*.

[d] See R. A. Stein (1), pp. 147 ff.; Kaltenmark (3), pp. 20 ff., 22 ff., 30, 32 ff.; Schafer (16), pp. 97 ff.

[e] Among the texts which mention boats of bronze or copper are the *Lin-I Chi*,[10] cit. *Shui Ching Chu*, ch. 37, pp. 6*b*, 7*a*; the *Nan Yüeh Chih*,[11] cit. *Thai-Phing Huan Yü Chi*, ch. 170, p. 6*b*; the *Fang Yü Chi*,[12] cit. *TPYL*, ch. 66, p. 7*a*, *Thai-Phing Huan Yü Chi*, ch. 169, p. 4*a*, *b*; and the *Yuan-Ho Chün Hsien Thu Chih* (+814), ch. 38, p. 5*a*.

[f] Cit. *KCCY*, ch. 28, p. 12*a*. In one of its tall stories about the south, the early +6th-century *Shu I Chi* describes a river in Tshang-chou[13] the water of which is so dense that metal and stone will not sink in it—the opposite of the 'weak water' (cf. Vol. 3, p. 608), and conceivably an echo of the Dead Sea (ch. 2, p. 12*a*; cit. also *KCCY*, ch. 28, p. 12*a*). So the people make boats of stoneware and iron when they want to cross it.

[1] 燃丘　　　　[2] 成王　　　　[3] 交州記　　　　[4] 劉欣期　　　　[5] 馬援
[6] 華山記　　　　[7] 薄　　　　[8] 鐯　　　　[9] 欂　　　　[10] 林邑記
[11] 南越志　　　　[12] 方輿記　　　　[13] 滄州

The machines used to keep the hulls of ships dry when afloat have been little studied. The large ship built about −225 by Hieron of Syracuse was said to have been fitted with an Archimedean screw or *cochlea*, worked by a single man, for pumping out the bilges, but the account of Athenaeus seems somewhat fabulous.[a]

After the end of the +16th century the Chinese used piston pumps (*chhou shui chhi*[1]), as Europeans did earlier.[b] But under eotechnic conditions such machines were probably much less effective than chain pumps, and indeed we find that Westerners who came in contact with Chinese shipping at this time greatly admired the methods employed. Our earliest statement is that of Gaspar da Cruz, a Portuguese Dominican who was in China for a few months in +1556. After explaining that the Chinese 'do use in all things more sleight than force', he goes on to say:[c]

A ship be it never so big, and have it never so great a leak, the pumps are made by such sleight that one man sitting alone moving his feet as one that goeth up a staircase, in very little space he pumps it out. These pumps are of many pieces made in the manner of water-wheels, laid alongside the side of the ship, between rib and rib, every piece having a piece of wood of half a yard (a) little more or less, one quarter well wrought; in the middest of this piece of wood is a square little board, almost of a hand's breadth, and they join one piece to another in such a manner as it may double well. The joints, which are all very close, whereby this manner of pump doth run, are within of the breadth of the little boards of every one of the pieces, for they are all equal; and this manner of pump bringeth so much water as may be contained between the two little boards.

The same appreciation was manifested by de Mendoza[d] in +1585:

The pumpes which they have in their shippes are much differing from ours, and are farre better; they make them of many peeces, with a wheele to draw water, which wheele is set along the shippes side within, wherewith they doe easily clense their shippes—for that one man alone going in the wheele, doth in a quarter of an hour cleanse a great shippe, although she leake verie much...

And later it was popularised by Isaac Vossius,[e] and still thought noteworthy at the beginning of the nineteenth century.[f] As we lack clear Chinese descriptions, it is a little difficult to be sure about the type of chain-pump used. At first sight a vertical *sāqīyah* might seem the most suitable for shipboard, but the description of da Cruz, though (as his editor says) rather incoherent, points unmistakably to an inclined square-pallet chain-pump of true Chinese type (see Vol. 4, pt. 2, p. 339). Before the middle of the +16th century, then, in this connection at any rate, the Chinese had not been much inhibited by the absence of the piston-pump.

---

[a] *Deipnosophists*, v, 43; cf. Torr (1), p. 61.

[b] It will be remembered from Sect. 27 that, with certain interesting exceptions, piston-pumps for liquids were not in the Chinese engineering tradition.

[c] Boxer (1), p. 121.   [d] (1), vol. 1, p. 150 (tr. Parke).

[e] (1), p. 139: 'Illud quoque (Lusitani) observandum in navibus Sinicis quod quamvis ruinosae fiant et multas admiserint aquas, non tamen mergantur; cum ab uno homine sedente, et tympanum costis navium appositum calcante, spatio unius horae plus aquae extruditur, quam in nostris navibus etiam complures integro exhauriant die' (+1685).

[f] Davis (1), vol. 3, p. 82.

[1] 抽水器

The two machines, in fact, travelled in opposite directions at this time, chain-pumps being adopted on Western ships, and piston-pumps attracting attention in China. Chain-pumps were mentioned by Sir Walter Raleigh, about +1600, in a list of improvements introduced in his own time into the ships of the British Navy (together with bonnets, studsails, and anchor capstans).[a] Ewbank, from later documents, confirms that chain-pumps came into general use on British naval ships towards the end of the seventeenth century, and were not ousted by piston-pumps until the eighteenth.[b]

There was another purpose for which pumps were valued at sea, namely for extinguishing conflagrations caused by incendiary enemy attack. This subject has already been mentioned from time to time in passing,[c] but it needs a word more here. In the discussion of navigation above (p. 576) we quoted a passage from the *Sung Hui Yao Kao* of date +1129 which referred *inter alia* to the equipping of warships with fire-prevention devices (*fang huo*[1]). Elsewhere there has been talk of wet leather curtains for protection against fire-arrows.[d] But in one account at least we seem to see some system of syringes or pumps for spraying water. In his *Kuo Chhao Wên Lei*[2] (Classified Documents of the Present Dynasty) of about +1360, Su Thien-Chio[3] speaks as follows concerning the Battle of Yai-shan[4] between the Sung and Yuan fleets which occurred early in +1279:[e]

(The Mongolian commander) (Chang) Hung-Fan[5] ordered Intendant Yüeh[6] to attack the Sung warships with his trebuchets from shore batteries, but they were so strongly built that they suffered no damage...Later (Chang) Hung-Fan, having captured a number of boats of the Tan-chia[7] river-people,[f] caused straw to be piled on them and soaked with oil; then, when the wind was favourable, he had them set adrift as fire-ships to burn the Sung fleet. But the Sung ships had previously been plastered all over with mud, and moreover countless numbers of 'water-tubes' (*shui thung*[8]) were suspended over (their sides). So when the piles of blazing straw came near, they were pulled to pieces with (long) hooks, and extinguished with water. Thus none of the Sung ships was harmed.

Although, then, the exact nature of the system remains unclear, there must have been tanks on the upper deck kept supplied by pumps and capable of providing water for sprays and hoses.

[a] Charnock (1), vol. 2, p. 68. 'It is not long since', wrote Raleigh, 'the striking of the Top-mast (a wonderful great ease to great ships both at Sea and Harbour) hath been devised, together with the Chain pumpe, which takes up twice as much water as the ordinary did; we have lately added the Bonnett, and the Drabler...The weighing of Anchors by the Capstone is also new.'

[b] (1), pp. 154 ff. Nevertheless the full-rigged Swedish warship 'Vasa', which sank in Stockholm harbour in +1628, had two large piston suction-pumps, as we know from her recovery almost intact in a great epic of salvage only recently completed (see Howander & Åkerblad, 1). The authorities say that one of these was a 'double-action' one (Cederlund *et al.* (1), p. 2; Ohrelius (1), p. 111); if so, the point would be of real importance in engineering history (cf. Needham, 48). But in fact the terminology is misleading, for the apparatus in question was a double-cylinder single-acting suction-pump. I am indebted to Lt. Bengt Hallvards and Mr E. Hamilton of the 'Vasa' Museum staff for clearing up this matter.

[c] Pp. 432, 576, 685.

[d] P. 449.

[e] Ch. 41, p. 19*a*, *b*, tr. auct. with Lo Jung-Pang.

[f] Cf. p. 672.

[1] 防火　　　[2] 國朝文類　　　[3] 蘇天爵　　　[4] 崖山　　　[5] 張弘範
[6] 樂　　　[7] 蛋家　　　[8] 水筒

(4) DIVING AND PEARLING

At one or two points already there has been mention of submersible craft, of course a late development in any civilisation. Here it may not be out of place to refer to the efforts of earlier times to enable human divers to remain under water at considerable depths and for as long as possible. China comes into the story in connection with pearling.[a] In one of the later chapters of the *Thien Kung Khai Wu* (+1637),[b] Sung Ying-Hsing describes the pearl fisheries, which in his time (apart from foreign countries of the south seas) were concentrated near Leichow and Lienchow in southern Kuangtung, north and north-west of Hainan Island. He tells us that the divers (*mu jen*[1]),[c] who belonged to the Tan,[2] an ancient southern people,[d] worked over special pearl-oyster beds, using broad-beamed boats peculiar to themselves. Sacrifices to the sea-gods, it was thought, gave them unusual powers of seeing under water,[e] and enabled them to avoid sharks and dragons (i.e. other dangerous fish).[f] They descended, he says, as much as 400 to 500 (Chinese) feet,[g] with a long rope from a winch secured to their waists, and breathed through a curving pipe strengthened by rings of tin (*i hsi tsao wan huan khung kuan*[3]), which was fastened over the face with a leather mask (Fig. 1048). If anything went wrong, they pulled on the rope as a signal and were quickly hauled up, but many were unfortunate and 'found a tomb in the bellies of fishes', while others died of cold after leaving the water. Apparently to relieve these hardships, a Sung inventor, Li Chao-Thao,[4] devised a kind of weighted dragnet with iron prongs like a plough, and a hempen bag the mouth of which was held open to receive the oysters, so that the pearl-fishermen could tow their dredge while under sail (Fig. 1049).[h] In Sung Ying-Hsing's time, both methods were used. He adds that

---

[a] But there has also been much traditional diving, mostly by women of the Ama clan, in the northern parts of the Chinese culture-area, i.e. the islands of Japan and Korea, not for pearls but for sea-food. Their work has attracted much interest in recent times; see Maraini (2) and Hong Sukki & Rahn (1).

[b] Ch. 18, pp. 1*a* ff., 8*a* ff.; tr. Sun & Sun (1), pp. 295 ff.

[c] More often written *mu jen*[5] or *mo jen*.[5]

[d] Probably the same as the modern Tanka (Tan-chia[6]) boat people, so numerous in and around Canton. Cf. Kaltenmark (3), p. 93.

[e] Cf. the account given earlier (p. 462) of the dark 'slaves', probably Malays or Tamils, who were sent overboard to repair Chinese ships under water in the +11th century because they could see so well when diving.

[f] The danger from sharks must have been very real indeed. It is emphasised in many of the Chinese literary references, as e.g. most vividly in *Ling Wai Tai Ta* (+1178), ch. 7, pp. 6*b* ff., tr. Schafer (10), p. 164. But certain jellyfishes, sea-urchins and clams are also greatly feared by Asian pearl-divers; cf. Bowen (5, 6).

[g] This depth would have been quite impossible without the use of a modern diving-suit and helmet, which was not on the cards. It seems that 70 ft. is about the limit at which divers without respiratory aids actually work, though they may go down as far as 120 ft. for very short periods; cf. Hornell (21); Thomazi (1); Diolé (2); Frost (1). Free divers with oxygen can work down to 300 ft. but below 100 ft. deep-water narcosis is to be feared, leading to dangerous errors of judgement. Attached divers with pumping machinery have rarely worked below 500 ft. A single air-pipe would not be much use except at shallow depths, but possibly Chinese divers in the early Chhing may have used paired ones with inlet and outlet valves, and a double-acting piston-bellows working in the boat.

[h] Further details are known of subsequent improvers, e.g. Lü Hung-Yin[7] about +1410; see *TSCC*, *Shih huo tien*, ch. 324, *tsa lu*, p. 4*b*.

---

¹ 沒人   ² 蛋   ³ 以錫造彎環空管   ⁴ 李招討   ⁵ 及人
⁶ 蛋家   ⁷ 呂洪因

没水採珠船

Fig. 1048. Pearl divers in the *Thien Kung Khai Wu*, ch. 18, pp. 8*b*, 9*a*. The caption says: 'The ship which carries the divers who go down into the sea to collect the pearls.' Breathing-tubes and some kind of masks are in use, but in the corresponding picture of the Ming edition, only the latter are seen.

sometimes, for tens of years at a time, there was a sort of close season, to allow the pearls to grow.

The history of pearling on the coasts of Kuangtung has been sketched in an interesting paper by Schafer (10), from which we may glean some indications about the techniques employed at different periods. The centre of the industry lay in the Lienchow region, the old name of which was Ho-phu[1] commandery,[a] and here on the coast among islands there were 'pearl lagoons' or indentations of the sea (*chu chhih*[2]) so famous that at one time the whole district was known by that name. This wealth had been exploited at least as far back as −111, when the armies of Han Wu Ti annexed the

[a] This was the geographical clue which led Schafer to so much information. But the very large literature in Chinese on pearling has hardly yet been examined, especially from the technological point of view. A beginning might be made with the quotations assembled in *KCCY*, ch. 32, esp. pp. 11*a* ff.

[1] 合浦   [2] 珠池

底沉芭竹

珠採帆揚

Fig. 1049. Li Chao-Thao's dredge or drag-net for collecting pearl-oysters (*Thien Kung Khai Wu*, ch. 18, pp. 9*b*, 10*a*). The caption on the right says: 'Hoisting sail to collect pearls', and on the left: 'The bamboo drag that goes down to the bottom.'

old kingdom of Yüeh (cf. p. 441), and the abundant production of pearls is recorded already in the *Chhien Han Shu*.[a] Towards the end of the −1st century people from other parts of China made fortunes by organising the work of the pearl-divers.[b] So also did successive governors, with the result that over-fishing produced scarcity, and it took the action of a wise and good man about +150 to retrieve the situation. Speaking of Ho-phu, the *Hou Han Shu* (History of the Later Han Dynasty) says:[c]

[a] Ch. 28 B, pp. 12*b*, 39*a*. Cf. *Yen Thieh Lun*, ch. 2, p. 7*b*.
[b] About −30 the wife and children of an official, Wang Chang,[1] migrated to Ho-phu after his death, where, helped by a local assistant governor, Wang Shang,[2] they amassed great quantities of pearls in a short time (*Chhien Han Shu*, ch. 76, p. 31*a*; cit. also *TPYL*, ch. 802, pp. 8*b*, 9*a*, tr. Pfizmaier (94), p. 622).
[c] Ch. 106, p. 13*b*; parallel passage from another Later Han history in *TPYL*, ch. 802, p. 9*b*. Tr. Schafer (10), mod. auct.

[1] 王章   [2] 王商

The province did not produce grain or fruit, but the sea gave forth treasure of pearls. As it bordered on Chiao-chih, there was constant coming and going of merchants and dealers who brought in cereals. Formerly, all the governors had been in general avaricious or corrupt, requiring the people to gather and search (for pearls) without regard for any limit. Consequently the pearl(-oysters) gradually migrated to the confines of Chiao-chih province. The travellers therefore ceased to come, people were without resources, and the destitute died of starvation in the streets. But when (Mêng) Chhang[1] took up his office he radically altered the former evil practices, and sought out what would restore the well-being of the community. Before a year was out, the departed pearl(-oysters) returned again, the people resumed their normal work, and commerce circulated. This was regarded as quite a miracle.

Thus Mêng Chhang (perhaps appreciating the true cause of the disappearance of the pearls rather better than the historian) decreed a temporary cessation of pearling, and stands out as a successful exponent of nature-protection and fisheries conservancy.[a] In characteristic Chinese fashion he became later on the tutelary deity of the industry, and long afterwards Thao Pi[2] (+1017 to +1080) wrote an inscription for his temple:[b]

> In bygone times good Governor Mêng,
> Loyal and honest, walked by this distant shore.
> He did not rob the wombs of the oysters,
> And the waters' depths abounded with returning pearls.

In the +3rd century, after the end of the Han, the pearling districts became part of Wu State, and from this time dates what may be the earliest account of the divers' work. In his *Nan Chou I Wu Chih*,[3] Wan Chen[4] wrote:[c]

There are people in Ho-phu who excel in swimming to search for pearls. When a boy is ten or more years of age he is instructed in pearl-diving. The officials forbid the folk to gather pearls (except for the government). But certain skilful robbers, crouching on the sea-bottom, split open the oysters and get fine pearls, whereupon they swallow them and so come forth.

Smuggling was thus keeping pace with government control. And indeed from +228 the whole district had been renamed for a while Chu-kuan,[5] i.e. (the domain of the) Director of Pearling. This official title, persisting centuries afterwards, struck the +9th-century poet Lu Kuei-Mêng[6] as a peculiar piece of local colour, and he described the far south as a place

> Where most of the men are herbalists, and practise mad sorcery,
> And the bureaucracy includes a Director of Pearling, who disburses the salary cash.[d]

Over the ages the industry was troubled from time to time by waves of Confucian austerity at court which injured all such luxury trades, and during the Thang pearling

---

[a] Exactly the same policy was advocated by another high official, Thao Huang,[7] early in the Chin period, c. +280; see *Chin Shu*, ch. 57, p. 6b, tr. Schafer (10), p. 159; Pfizmaier (94), p. 627. The passage also occurs in *TPYL*, ch. 802, p. 11a.

[b] Preserved in *Yü Ti Chi Shêng*,[8] ch. 120, pp. 6b ff., tr. Schafer (10), mod. auct.

[c] Text preserved in *TPYL*, ch. 803, p. 10b; tr. Schafer (10), mod. auct.; cf. Pfizmaier (94), p. 653.

[d] From *Thang Fu-Li Hsien-sêng Wên Chi*,[9] ch. 9, p. 27b, tr. Schafer (10), cf. (16), pp. 160ff.

[1] 孟嘗   [2] 陶弼   [3] 南州異物志   [4] 萬震   [5] 珠官
[6] 陸龜蒙   [7] 陶璜   [8] 輿地紀勝   [9] 唐甫里先生文集

was stopped several times. No such inhibitions weighed upon Liu Chhang,[1] however, the last of the emperors of the Nan Han dynasty in the Wu Tai period, who stationed a whole division of soldiers near Lienchow and had them instructed in pearl-diving. The texts concerning these men [a] say that they weighted themselves with stones and dived below 500 ft. (which must be an exaggeration), so that one after another died of drowning or sharks. But as soon as the Sung armies took Canton in +971, this use of troops was abolished.

One of the earliest texts which attributes to the Tan people the greatest role in the pearling industry is the *Thieh Wei Shan Tshung Than*[2] (Collected Conversations at Iron-Fence Mountain) written about +1115 by Tshai Thao.[3] In a long and interesting passage[b] he tells us that the fishermen arrange ten or more of their boats (*hai thing*[4]) over the pearl beds in a ring, and let down on both sides mooring-cables attached to rocks which lie as anchors on the bottom. Then the Tan diver, having attached a small rope to his waist,

takes a deep breath and plunges straight down from 10 to 100 ft., after which he leaves the mooring-cable and feels his way to collect the pearl-oysters (lit. pearl-mothers, *chu mu*[5]) After what seems only a few moments he urgently needs air, so he gives a big jerk to the waist-rope, and the sailors on the boat, seeing the signal, wind this rope in, while at the same time the diver climbs up along the mooring-cable (as fast as he can).[c]

From this it would seem that windlasses were employed, and that the waist-rope probably remained attached to the main cable by a smooth loose ring, so that the diver was rapidly brought back to his way of escape when the winding-in began. Tshai Thao continues with a graphic account of the agonies of divers who overstepped by accident the narrow limits of safety, and the means taken to revive them,[d] saying that among those who see and admire pearls in ordinary society, very few have any conception of what it costs to get them. The same emphasis, especially concerning the dangers from sharks and other evil beasts of the sea, is found in the long account of Chou Chhü-Fei in his *Ling Wai Tai Ta* sixty years later.[e] He adds little to the technicalities however, except to say that baskets are also let down on long cords with the divers themselves, a further

---

[a] E.g. *Sung Shih*, ch. 481, p. 2*b*; *Wên Hsien Thung Khao*, ch. 18 (p. 179.2), ch. 22 (p. 220.2); *Ling Wai Tai Ta*, ch. 7, p. 7*b*, etc.; full details in Schafer (10).

[b] Ch. 5, pp. 22*a* ff.; much abridged tr., from a later quotation, in Schafer (10), p. 164.

[c] Tr. auct.

[d] Since 40 ft. is the maximum depth for working long periods (as is now possible using oxygen equipment) without decompression halts on the way up, the unassisted pearl-divers of old, going as much as three times as deep though for shorter times, may sometimes have suffered from 'the bends', i.e. caisson-disease, and diver's palsy. At high environing pressures the blood dissolves much more of the inert gases (especially nitrogen) than normally, and when the pressure is reduced these separate as actual bubbles in the blood-stream and the nervous system, causing dire effects. Chinese powers of observation and inference being what they were, one would hardly be surprised to find in some early Chhing text a recognition of the desirability of a decompression halt, the need for which might have become apparent after the introduction of air-pipes.

[e] Ch. 7, pp. 6*b* ff., tr. Schafer (10). The divers 'often meet with some marvellous creature which gapes its mouth, exhales and inhales...'—giant clams? The worst fish is the 'tiger-fish' or spiny elasmobranch (*tzhu sha*[6])—not now identifiable.

---

[1] 劉鋹   [2] 鐵圍山叢談   [3] 蔡絛   [4] 海艇   [5] 珠母   [6] 刺鯊

precaution since these could be wound in at leisure. One begins to get a picture of slow but continuous improvements in diving technique through the centuries, leading to the inventions of the Ming with which we started.

There is nothing improbable in the techniques described in the *Thien Kung Khai Wu*. Indeed, they may be quite ancient. A passage from *Pao Phu Tzu* (*c.* +320) includes, among magical recipes, the following:[a] 'Take a real rhinoceros horn more than 1 ft. long and carve on it the shape of a fish, then put one end in the mouth and enter the water—the water will open out 3 ft. on all sides, and you will be able to breathe in the water.' Perhaps this is a concealed reference, in the alchemical manner, to a diver's tube. In any case both breathing-tubes and diving-bells of a kind are alluded to by Aristotle[b] and other ancient writers such as Vegetius. A German ballad of +1190 (i.e. about Li Chao-Thao's time) mentions the breathing-tube of a diver,[c] and the first European illustration occurs in the work of the anonymous Hussite engineer about +1430. It is of much interest that between this time and that of Sung Ying-Hsing, Leonardo sketched, in the *Codex Atlanticus*, a breathing-tube such as was used by Indian Ocean pearl-divers;[d] and this, besides having spikes to keep off fishes, is strengthened against collapse under pressure by just such metal rings as those referred to in the *Thien Kung Khai Wu*. But the pressure of the water on the diver's lungs must always have been the great limiting factor for the attempt to use atmospheric air. It is therefore interesting that bellows for pumping it down the tube are mentioned in an Arabic work on hydraulic engineering of about +1000, and it would be interesting to know whether the Chinese of the Sung period also used them.[e] The first European mention of a double pipe for breathing in and out occurs in the works of Borelli (+1679),[f] and Halley in +1716 combined such pipes with a diving-bell. We have not

---

[a] *TPYL*, ch. 890, p. 2*b*, tr. auct. This text was attributed to the *Huai Nan Wan Pi Shu* by an oversight in Yeh Tê-Hui's reconstruction of that book.

[b] *Problemata Physica*, XXXII, 5 (960 *b* 21 ff.). The *Problemata* as we have it now is not the book that Aristotle wrote, but it certainly stems from the Peripatetic school, and belongs to the −3rd and −2nd centuries (just as much of the *Mo Tzu* book was written by the Mohists and not by Mo Ti, cf. Vol. 2, p. 166). The passage on diving techniques could have been a still later interpolation but there is really no adequate stylistic or other philological reason for thinking so. Moreover there is a reference to these techniques in an Aristotelian text of undoubted genuineness, the *De Partibus Animalium*, II, 16 (659 *a* 9 ff.). Talking of elephants, Aristotle says: 'Just, then, as divers are sometimes provided with instruments for respiration, through which they can draw air from above the water, and thus remain for a long time under the sea, so also have elephants been furnished by Nature with their lengthened nostril...' Thus one kind of artificial aid at least was known in −4th-century Greece. It was not the 'diving-bell' or 'diving-helmet' of the *Problemata*, but all these devices are so simple that there seems no need for hesitation in ascribing them to the −2nd century. We owe thanks to Mr Geoffrey Lloyd for helping to clear up this question.

[c] See Feldhaus (1), col. 1119. It is that of Salman and Morolf, and the passage is: 'Eyn rore in daz schiffelin ging, da mit Morolf den atem ving.'

[d] Folios 7Ra, 333Va and 386Rb; an illustration is reproduced by Feldhaus (1), col. 1120, (18), pp. 136 ff. See also McCurdy (1), vol. 2, pp. 162, 215 ff.; Ucelli di Nemi (3), no. 78. Leonardo himself makes (in another MS.) the reference to the Indian Ocean. Many other designs date from this time, e.g. that of Francesco di Giorgio, reproduced in Brinton (1), fig. 27.

[e] See Krenkow (1).

[f] That other component of the modern deep diver's outfit, the dress, also has Chinese antecedents. Domingo de Navarrete, when journeying down the Grand Canal in the winter of +1665, was much impressed by the fur-lined hide overalls which permitted fishermen to stay casting their nets for a long time while immersed up to the neck in ice-cold water. Such 'diving-suits', with gloves attached to the oars, were also used by rowers. See Cummins (1), vol. 2, p. 227.

so far found any references to the latter in Chinese literature,[a] but the idea was current in Europe much earlier than is generally supposed, and remarkable pictures of 'diving-bells',[b] some with breathing-tubes, are to be found in +14th- and +15th-century MSS of the Alexander-Romance.[c] The adventurous king was supposed to have descended into the depths of the sea in a sphere or barrel of glass, emerging safe and sound in spite of the treachery of his queen Roxana, who let go the cables. The pearling fields of Asia, that continent which Alexander aspired to rule, especially those of India and China, may well have been the original home of all diving techniques, and their early history may have to be discovered from there.[d] Indeed, it has been plausibly suggested that the breathing exercises so prominently associated with Hindu, Buddhist and Taoist meditational, mystical and thaumaturgical disciplines, had a close connection with the practices of the divers who earned a hard living, from antiquity onwards, by ravishing the treasures of an element not natural to man.[e] At any rate, the Buddhist *matsya-dharma* (law of the fishes) caused pathetic dismay to Alexander, according to one of the legends of his descents:[f]

Seigneurs, Barons, fait-il, bien je me suis aperçu
Que le monde entier est damné et perdu,
Les violents grands poissons devorent les menus.[g]

So far we have been thinking only of those natural pearls which were sought at such risk by the Cantonese divers century after century. But there are also cultivated pearls and artificial pearls, the former made by the bivalves under adventitious stimulus, the latter made entirely by man. The invention of cultured or induced pearls, formed by the implantation of a small foreign body which the oyster then coats with nacre,[h]

---

[a] With the exception of the curious story in the *Shih I Chi* (Memoirs on Neglected Matters), a book of wonders written by Wang Chia about +370. Here we find (ch. 4, p. 6a) the following: 'Chhin Shih Huang Ti was fond of matters concerning spirits and immortals. (In his time) people from Yuan-Chhü[1] arrived in "conch-boats" (*lo chou*[2]). These had the form of spiral conch-shells and made their way on the sea-bottom, no water being able to enter them although submerged (*chhen hsing hai ti, erh shui pu chin ju*[3]). Another name for these craft was "under-the-waves boats" (*lun po chou*[4]).' The passage also occurs in collections, e.g. *Lei Shuo*, ch. 5, p. 19a. It is hard to know now whether we should treat this kind of report as pure legend and 'wish-fulfilment' thinking, or whether it had any basis in diving-bell experiments. Perhaps further evidence will emerge from the texts. Attention was drawn to this passage long ago by items in the *Journal des Débats* and the *Antiquitäten Rundschau*; we owe our knowledge of it to the notes of the late Prof. Fritz Jäger, kindly placed at our disposal by Prof. W. Franke.

[b] The history of the diving-bell has been sketched by D. W. Thomson (1).

[c] Reproductions are given by Feldhaus (2), pl. IX and figs. 295, 296, 297, and by Cary (1), pl. VII. On the Alexander Romance see Cary (1); Thorndike (1), vol. 1, pp. 551 ff.; Tarn (1), p. 429. The legendary corpus concerning Alexander the Great is of course quite post-classical; it seems to have started in Alexandria in the +3rd or +4th century, and the early versions are known as pseudo-Callisthenes, since some of them were attributed to the real historian of that name (d. −328), who accompanied Alexander on his Asian campaigns. The corpus passed, with continual accretions, into many Levantine and European languages. On the descent into the sea in particular, see Cary (1), pp. 237, 341. Dr D. J. A. Ross has promised us a literary and iconographic study of Alexander's bathyscaphe. Cf. p. 56 above.

[d] On the Persian Gulf industry see Bowen (5, 6); Mokri (1).

[e] Diolé (1), p. 264. See also Vol. 2, pp. 143 ff.

[f] P. Meyer (1); the +12th-century *Alexandriade*, cit. Frost (1).

[g] Cf. Vol. 2, p. 102.

[h] Nacre or mother-of-pearl is a deposit of crystalline calcium carbonate laid down upon a network of protein (conchiolin); cf. Grégoire (1).

[1] 宛渠   [2] 螺舟   [3] 沉行海底而水不浸入   [4] 淪波舟

seems to be essentially Chinese.[a] In 1825 J. E. Gray, while studying mollusc shells in the British Museum, noticed that some good pearls were still attached to a shell of *Barbala plicata*, and that they had clearly been artificially induced by the introduction of small pieces of nacre as nuclei. They had come from China. Later Gray reported that other examples had been formed round minute pieces of silver wire. Thirty years afterwards Hague (1) gave an eye-witness account of the industry at Huchow, where foreign bodies of all kinds were used in fresh-water mussels including very small Buddhist images.[b] The local people attributed the invention to a +13th-century inhabitant named Yeh Jen-Yang.[c]

It is indeed possible to find a clear account of the technique in Chinese literature at least a couple of centuries before the time ascribed to this worthy. It occurs in the *Wên Chhang Tsa Lu*[1] (Things Seen and Heard by an Official at Court), written by Phang Yuan-Ying[2] in +1086. He says:[d]

Hsieh Kung-Yen,[3] an Executive Official of the Ministry of Rites, found out a way of cultivating pearls (*yang chu fa*[4]). The way this is done now is to make (first) 'false pearls' (from pieces of nacre, etc.). The smoothest, roundest and most lustrous of these are then selected, and inoculated into fairly large oysters kept in clean sea-water, as soon as they open their valves. The clean sea-water is repeatedly renewed, and at night the oysters take up the best influences of the moon.[e] Then after two years real pearls are fully formed.

Thus the irritating chip or granule of silica or calcium carbonate was made to do duty in the service of man. We cannot be sure that Hsieh Kung-Yen was really the first to accomplish this, for certain passages from still earlier books form a kind of background to his achievement. Liu Hsün,[5] in his *Ling Piao Lu I*[6] (Strange Southern Ways of Men and Things), finished about +895, gave a brief account of the Lienchow pearl fisheries, in the course of which, after saying that tribute pearls are cut out from the older and larger oysters, he went on:[f]

Moreover, (the fishermen) take the flesh of the small oysters, pierce them with bamboo splinters, and dry them in the sun; these are called 'pearl matrices' (*chu mu*[7]). People of Jung(-hsien) and Kuei(-lin)[g] (over the passes in Kuangsi) like to take them and roast them, for offering (to guests) at wine parties. Within the flesh there are minute pearls like grains, wherefore it is known that the oysters of the Pearl Lagoons have pearls in their wombs in accordance with their size, whether large or small.

[a] A preliminary statement of the material here presented was published by Sarton & Needham (1).
[b] A photograph of some of these will be found in *EB*, p. 422.
[c] We have not been able to ascertain the characters of this name. Cf. McGowan (6).
[d] Text in *Shuo Fu*, ch. 31, p. 12*b*, tr. auct. Also in *KCCY*, ch. 32, p. 12*b*; *TSCC*, *Shih huo tien*, ch. 324, p. 4*b*.
[e] On the influence of the moon on marine animals cf. Vol. 1, p. 150; Vol. 4, pt. 1, pp. 31 ff., 90; and in Sect. 39 below.
[f] Text in *Shuo Fu*, ch. 34, pp. 23*a* ff., *KCCY*, ch. 32, p. 11*b*, tr. Schafer (10), mod auct. Schafer's translation of the whole passage is abridged somewhat, as he did not use all the extant versions of the text.
[g] We are not quite sure of this interpretation, as the versions vary. It is possible that Jung Kuei-Shuai[8] was a person, the first to make use of a by-product of the pearling.

[1] 文昌雜錄   [2] 龐元英   [3] 謝公言   [4] 養珠法   [5] 劉恂
[6] 嶺表錄異   [7] 珠母   [8] 容桂率

This makes it clear that in the +9th century the growth of the pearl within the lamellibranch was well enough understood to suggest the implantation of an inductive nucleus. Indeed, this may already have been practised in some restricted circles, for still earlier we read of presentations of pearls of special shapes. For example, the *Nan Chhi Shu* records[a] for +489: 'The Yüeh Prefecture presented (to the throne) a white pearl shaped naturally like the image of a "meditating Buddha", three inches in height.' This was handed over to the Chhan-Ling Ssu temple for safe-keeping. Even if the size is exaggerated in the record, the nature of the object recalls the practices of the Huchow people in modern times. Finally, as far back as the −2nd century, we have a remarkably perspicacious statement in the *Huai Nan Tzu* book (c. −120):[b] 'Though luminous pearls are an advantage to us, they are a disease for the oyster (*Ming yüeh chih chu, hui chih ping erh wo chih li*[1]).' Once this was realised, the inoculation of an irritant to cause the disease need not have been a far step in thought, though many long years may have passed before anyone managed to do it in practice.

Information about the art spread somehow to Europe before the middle of the +18th century, for the great botanist Linnaeus took advantage of it, and acknowledged the origin of it.[c] When a young man he saw fresh-water mussel pearling going on at Purkijaur on Lake Luleå in Lappland. Twenty years later (+1751) he wrote that he had read of a Chinese method of producing cultured or induced pearls, and in another ten years he had been able to demonstrate the feasibility of the method on the fresh-water mussels of Sweden, using small pieces of silver wire together with tiny balls of plaster or limestone. Eventually he sold the process for a substantial sum.[d] The invention was further re-discovered and perfected in Japan where it has become a very great industry, employing some 100,000 people and producing no less than 36 tons of cultured pearls every year, worth some £6,500,000 to the country's export trade.[e] The father of this industry, Mikimoto Kokichi, who died in 1954, had succeeded, not without great trouble, in placing the art on an industrial basis in 1905.

The real secret of the technique had probably always been to implant small pieces of the nacre-secreting epithelium of the mantle along with the inorganic nucleus so that a closed cyst was formed within the parenchymatous tissue. Naturally this was not revealed to Phang Yuan-Ying. Otherwise only 'blister-pearls' are formed as the outer epithelium covers the intruded object inside the shell with mother-of-pearl,[f] and most of the Buddhist images are of this kind. The present procedure is to form a sac with epithelium from another oyster around the nucleus and to implant this into the

---

[a] Ch. 18, p. 19b, tr. Schafer (10).

[b] Ch. 17, p. 12a. The word here used for the oyster came afterwards to signify the tapeworm, but the meaning is unmistakable. *Khang-Hsi Tzu Tien*, s.v., says that pearls were once found in the stomachs of 'dragon fishes' long and thin, which had presumably eaten the oysters.

[c] One communication at least is known in the letter of the Jesuit missionary d'Entrecolles (1), written from Peking in +1734.

[d] The details are related by Gourlie (1), pp. 87, 200, 243, who gives the references to the original correspondence.

[e] The species used is *Pinctada Martensii*.

[f] The resulting flat surface is mentioned in the +3rd-century *Nan Fang Tshao Mu Chuang*; see *TPYL*, ch. 803, p. 10a (tr. Pfizmaier (94), p. 652).

[1] 明月之珠蚘之病而我之利

sub-epidermal or other tissue, after which the host is cultured for as long as seven years.[a]

Lastly, to pursue this digression to its furthest end, we ought to consider for a moment the production of purely artificial or imitation 'pearls', not made within the bivalve at all. There is considerable reason for thinking that this technique, like that of the cultivation or induction of true pearls, goes back a long way in China. It depended on the isolation of minute pearly crystals from natural sources and their deposition in the form of a stable film upon a spherical base of glass or other material. In Europe it has been current since + 1680 when Jacquin, a Paris rosary-maker, made a preparation called (curiously enough) 'essence d'orient' from the silvery scales of the bleak-fish, *Alburnus lucidus*, a cyprinid teleost.[b] Having caused a film of the thick mass of suspended crystals to adhere to the inner walls of small glass drops, he filled the cavity with white wax and produced the artificial pearls. Now for a century past it has been known that the crystals are those of the purine base guanine,[c] and to this day they are used for the same purpose. But Jacquin's performance strikes a note of memory. Did we not read something of the same kind in connection with ancient Chinese glass? From Wang Chhung's *Lun Hêng* of +83 we did indeed quote the following passage:[d]

Similarly, pearls from fishy oysters are like the bluish jade of the Tribute of Yü; all true and genuine (natural products). But by following proper timing (i.e. when to begin heating and how long to go on) pearls can be made from chemicals (*yao*[1]), just as brilliant as genuine ones. This is the climax of Taoist learning and a triumph of their skill...[e]

The whole passage, it will be remembered, concerns the making of mirrors or lenses of glass, imitating not only jade but those highly polished mirrors of bronze which were used to ignite tinder with 'fire from the sun'. It may well be, then, that in ancient times the Taoists also found a way of extracting and suspending the guanine crystals from fish skin, depositing them on glass to make 'false pearls'.[f]

Although the modern sources are primarily the teleostean or bony fishes, the existence of a long-enduring corpus of Chinese legend associating pearls with sharks is just worth noticing. Thus the *Chiao-chou Chi*,[2] a Chin work, says that the shark (*chiao yü*[3]) has on its back armour 'patterned with pearls'.[g] The *Phi Ya*[4] (+1096) says that

[a] Cf. Jameson (1); Kawakami (1), and the classical review of Biedermann (1), pp. 720 ff. A great deal of scientific work has been done in Japan on the physiology of pearl-production in lamellibranchiate molluscs, and special Institutes exist devoted to the subject.

[b] The iridocytes, especially numerous in the silvery part of the skin known as the argenteum, are so stuffed with the crystals that the cell-nucleus can hardly be seen. See the review of Fuchs (1), pp. 1410 ff.

[c] The original discovery was due to Barreswil (1), followed by Voit (1) and Bethe (1). Oya (1) has given us an account of the chemistry of the lustrous particles in the epidermis of the hair-tail fish *Trichiurus haumela*, which serves as the principal source of 'pearl-essence' in the Japanese industry.

[d] Vol. 4, pt. 1, p. 112.

[e] Ch. 8, tr. auct. adjuv. Forke (4), vol. 1, pp. 377 ff. For a discussion of the words translated 'proper timing', the reader is referred to the page where the passage was given in full. The crucial sentence about making pearls from chemicals is often found quoted, as e.g. in *TPYL*, ch. 803, p. 5a; cf. p. 7b.

[f] The glass part is referred to in the +4th-century *Kuang Chih*[5] (*YHSF*, ch. 74, p. 38b; *TPYL*, ch. 803, p. 9a, ch. 809, p. 3b): '"Mineral pearls" are made by the melting of minerals; some call them "court beads" (*Shih chu chu shih wei chih, i ming chhao chu*[6]).'

[g] Ch. 1, (p. 3).

[1] 藥   [2] 交州記   [3] 鮫魚   [4] 埤雅   [5] 廣志   [6] 石珠鑄石為之一名朝珠

while oyster pearls are in the belly, shark pearls are in the skin.[a] Abundance of texts describe the shark people (*chiao jen*[1]), who live at the bottom of the sea, give lodging to pearl-divers, and sometimes come ashore to wander about and sell their pongee silk.[b] On departing they pay their bills by weeping tears which turn to pearls (*lei chu*,[2] *chhi chu*[3]). Perhaps this was just what certain fishes, in the hands of the Taoist alchemists, metaphorically did.

(5) THE RAM

There remains still something to be said about naval techniques as opposed to the peaceful practices of the sea. On various occasions above (pp. 442, 625) we have touched upon the question of whether or not Chinese ships of war in ancient and medieval times engaged in the typically Graeco-Roman technique of ramming. *A priori* this would seem unlikely on account of the basic blunt-ended and flat-bottomed build where no keel invited elongation into a sharp under-water weapon of offence.[c] Nevertheless the attachment of one or two such pointed protuberances to hole the enemy under the water-line was perfectly possible, and indeed there is a certain amount of evidence that these were in fact used.[d] It does not appear, however, that this technique ever held the dominating position that it had in the ancient Mediterranean.

Among the ancient classes of warships there may have been one called 'Stomach-strikers' (*Thu-wei*[4]). It appears for instance in a passage from the *Yüeh Chüeh Shu* (Lost Records of the State of Yüeh)[e] now preserved only in collections such as the *Yuan Chien Lei Han* encyclopaedia.[f]

Ho Lu (king of Wu, *r.* −514 to −496) had an interview with (Wu) Tzu-Hsü[g] and asked him about naval preparedness. (Tzu-Hsü) answered: 'The (classes of) ships are named Ta-i[5] (Great wing), Hsiao-i[6] (Little wing), Thu-wei[4] (Stomach-striker),[h] Lou chhuan[7] (Castled ship) and Chhiao chhuan[8] (Bridge ship). Nowadays in training naval forces we use the tactics of land forces for the best effect. Thus Great-wing ships correspond to the army's heavy chariots, Little-wing ships to light chariots, Stomach-strikers to battering-rams (*chhung chhê*[9]), Castled ships to mobile assault towers (*hsing lou chhê*[10]) and Bridge ships to the light cavalry.'

---

[a] Ch. 1, (p. 17).

[b] *Po Wu Chih* (*c.* +290), ch. 2, p. 3*b*; *Shu I Chi* (early +6th), ch. 2, p. 20*b*; *Tung Ming Chi* (+5th or +6th) in an account of the Fei-Lê[11] country; *TPYL*, ch. 790, p. 10*b*, ch. 803, pp. 7*a*, 8*a*, quoting from *Po Wu Chih* and *Sou Shen Chi* (+348).

[c] As elsewhere it may well have done. Cf. the ethnographic study of Noteboom (1) on the bifid prow.

[d] We are much indebted to Dr Lo Jung-Pang for discussing this with us *in extenso* by correspondence.

[e] Attributed to Yuan Khang and datable *c.* +52, but the present text contains much that cannot have been in the original book, and encyclopaedias preserve fragments of uncertain date that are not in all versions of the present text.

[f] Ch. 386, p. 1*a*, *b*. The passage is also quoted in *TPYL*, ch. 770, p. 1*b*, and in Hung Mai's[12] *Jung Chai Sui Pi*[13] of *c.* +1200, ch. 11, p. 6*a*. Tr. auct. with Lo Jung-Pang.

[g] Often encountered already, cf. Vol. 4, pt. 1, p. 269, and better, Vol. 3, pp. 485 ff.

[h] Many versions write Thu-mao[14] here (Colliding swooper), cf. immediately below.

[1] 鮫人	[2] 淚珠	[3] 泣珠	[4] 突胃	[5] 大翼	[6] 小翼
[7] 樓船	[8] 橋船	[9] 衝車	[10] 行樓車	[11] 吠勒	[12] 洪邁
[13] 容齋隨筆		[14] 突冒			

With one of these types we are already familiar, the castled ships full of marines, hard though it is to say whether oars or sails were their principal means of propulsion. The two 'wing' classes must obviously be sailing-ships, and the 'bridge' ships may well have been small rowed boats pressed into service from normal use in pontoon bridges. As for the 'stomach-strikers' or 'colliding swoopers' it is difficult to interpret them as anything else than ships fitted with rams. But the date of the passage may not be quite what it seems. Another excerpt purporting to be from the same book occurs elsewhere in the *Thai-Phing Yü Lan*;[a] it describes the 'Great wing' man-o'-war, following what may have been a lost book called *Wu Tzu-Hsü Shui Chan Fa*[1] (Wu Tzu-Hsü's Manual of Naval Tactics). Each ship, 120 ft. long and 16 ft. in the beam, carried 26 marines and 50 rowers, with 3 sailors and helmsmen each at bow and stern, and an equipment of 4 long 'grappling-hooks' (*chhang kou*[2]), 4 spears and 4 long-handled axes, under the command of four petty officers, together with five officers and master-archers, making in all a crew of 91.[b] Similar details occur in a quotation from another lost book, *Shui Chan Ping Fa Nei Ching*[3] (Esoteric Manual of Naval Tactics), due to Hung Mai,[c] who towards the end of the +12th century remarked that after all the ships of Chou times must have been quite large. We may feel some hesitation in accepting such a conclusion for the −6th century, and prefer to regard these accounts as revealing the practice of the Han San Kuo or even Chin periods, but in any case the presence of rams on certain craft seems fairly sure. At least equally important to note here, however, is the mention of 'grappling-hooks', a subject shortly to lead us in unexpected directions.

No doubt there was an ancient tradition that the fleets of Wu and Yüeh had engaged in ramming. From a book of which little is now left, *Wan Chi Lun*[4] (The Myriad Stratagems), written by Chiang Chi[5] about +220, the following words have come down to us:[d]

When Wu and Yüeh were fighting on the Five Lakes (mod. Thai-hu), they used ships with oars, which butted into each other as if with horns (*hsiang chhu*[6]). Whether handled bravely or timidly all were overturned, whether blunt or sharp all capsized (and sank).

Rams again, perhaps, yet the description does not quite suggest holing below the water-line. More reliably matter-of-fact is the description in the *Hou Han Shu* of the battles on the Yangtze River in +33 when Kungsun Shu[7] was trying to set up an independent kingdom in Szechuan at the beginning of the Later Han period.[e] He ordered his generals Jen Man,[8] Thien Jung[9] and Chhêng Fan[10] to lead twenty or thirty

---

[a] Ch. 315, p. 2a. A parallel passage, rather corrupt now, occurs in *Mo Tzu*, ch. 58, p. 17a, but we need not assume that it was there originally.

[b] Further stores: 32 crossbows, 3,300 quarrels, and 32 helmets and suits of armour.

[c] *Loc. cit.*

[d] *TPYL*, ch. 769, p. 3a. The book was reconstituted as far as possible by Ma Kuo-Han in *YHSF*, ch. 73, pp. 57a ff., but he overlooked this passage apparently. For the battles on the Five Lakes cf. *Kuo Yü*, ch. 21, p. 2a.

[e] Ch. 47, pp. 17b, 18a. Cf. p. 621 above.

[1] 伍子胥水戰法　　　　[2] 長鉤　　　　[3] 水戰兵法內經　　　　[4] 萬機論
[5] 蔣濟　　　[6] 相觸　　　[7] 公孫述　　　[8] 任滿　　　[9] 田戎　　　[10] 程汎

thousand troops down river on rafts to attack the Han army under the command of Tshen Phêng.[1] Having defeated three of his captains they took up position at a suitable spot where they constructed a floating bridge with fortified posts on it; this was protected by a boom which blocked the river and by forts on the surrounding slopes. When Tshen Phêng came up he attacked several times but without avail, so he waited and equipped himself with castled ships (Lou chhuan[2]), rowed assault boats (lu-jao[3]) and 'colliding swoopers' (Mao-thu[4]), a fleet of several thousand vessels, after which he successfully broke through and routed the Szechuanese. Li Hsien's commentary of +676 explains that the advantage of the Mao-thu was that they could butt (chhu[5]) in violent collision, i.e. that they could ram.

At this point we come to a curious gradual shift in technical terminology. It is clear that the earlier Mao-thu or Thu-wei ships were the predecessors of the attack ships later called Mêng-chhung.[6] Already in +100 the Shih Ming says:[a] '(Vessels that are) long and narrow in appearance are called Mêng-chhung;[7] they dash (like battering-rams) against the ships of the enemy (i chhung thu ti chhuan yeh[8]).' Towards the end of the +18th century Wang Nien-Sun, commenting at length on a list of ship names in the Kuang Ya of +230, remarked that 'among ships there is the Mêng-chhung[6] just as among vehicles there is the battering-ram (chhung chhê[9]), for mêng[6] is the same thing as mao[10] and chhung[9] has the same sense as thu[11]'.[b] But as we shall shortly see (p. 686) by +759 the term mêng[6] in this combination has come to mean primarily 'armoured' protection for the crew (whether of wet leather, wooden planks or iron plates) and not an impetuous course leading to collision; we shall there translate the binome as 'Covered swooper'. The explanation is that the old word mao[10] had two quite distinct meanings, not only 'rushing and colliding' but also 'hat or covering'. By +208, when Tung Hsi[12] acquired glory in naval fights between the Wu and Wei forces, Mêng-chhung[6] ships had become common,[c] but it is permissible to think that while the emphasis changed in these centuries from the ram function to the armour build there was a continuity of nomenclature because shipwrights were under the impression that mao[10] had always been used in the sense of 'hat' rather than the sense of 'rush'. Presently we shall see how this would fit in with a general trend from close-quarters to projectile warfare discernible all through the history of Chinese sea tactics.

The question has been raised whether another argument for rams in early times would not arise from the term ko chhuan[13] (dagger-axe ships, or halberd ships).[d] We have met with the name in Han times quite often (cf. p. 440 above), taking it in its most obvious meaning as ships manned with marines using dagger-axe halberds (as in the expedition against Yüeh in −112). But the +3rd-century commentator Chang Yen understood that the halberds had been fixed to the hulls of the boats to keep off Yüeh swimmers and dangerous marine animals, and his early +7th-century successor Yen

---

[a] Ch. 25, (p. 381), (Shih Ming Su Chêng Pu).   [b] Kuang Ya Su Chêng, ch. 9B, p. 17a.
[c] San Kuo Chih, ch. 55, p. 10a; cf. p. 449 above, and Schafer (16), p. 242.
[d] On this weapon see Sect. 30c below.

[1] 岑彭   [2] 樓船   [3] 露橈   [4] 冒突   [5] 觸   [6] 蒙衝
[7] 艨衝   [8] 以衝突敵船也   [9] 衝車   [10] 冒   [11] 突
[12] 董襲   [13] 戈船

Shih-Ku agreed with him. Chhen Tsan (*c.* +300), on the other hand, followed some authority derived from Wu Tzu-Hsü in holding that each ship carried men-at-arms equipped with shields and halberds, and the Sung scholars Sung Chhi and Liu Pin took his side—'Yen Shih-Ku', said the latter, 'was a northern landlubber, and knew nothing about how ships sail.'[a] Far be it from us to adjudicate between the literati, yet may not some memory of ancient ramming tactics be the background of the argument?

That the Chinese build of hull did not lend itself to the sticking on of 'halberds' or any other sharp-pointed part under the water-line is, as has been said, obvious,[b] but in this connection we may recall one most unusual type of boat still extant in modern China which possesses a bifid stem and stern. This is a sampan used near Hangchow, and its origin is quite unknown.[c] With its ram-like projection it resembles the boats of ancient Scandinavia and traditional Indonesia-Polynesia, the close connection of which was one of Hornell's strangest discoveries.[d] Perhaps this sampan is a solitary relic of the Indonesian component in southern Chinese culture, but perhaps also it may be relevant to what evidence there is for the use of the ram in Chinese antiquity.

The passage which has generally been thought to give the clearest proof of rams in the Warring States period seems to us rather to prove something else, and to lead off in a different direction. It occurs in the *Mo Tzu* book, and tells how the famous engineer Kungshu Phan[e] came south about −445 and reorganised the navy of Chhu.[f]

Formerly the people of Chhu and the people of Yüeh had a battle on the River.[g] The Chhu forces advanced with the current in their favour, but it was against them when they wanted to retreat. With success in sight they came on, but when defeat threatened them they found withdrawal quite difficult. Conversely the Yüeh forces had to advance upstream but could retreat downstream. In favourable conditions they could push slowly forwards but if the luck of battle turned they could get away quickly. With this advantage the Yüeh people greatly defeated those of Chhu.

Master Kungshu then came south from Lu to Chhu and began making naval warfare implements called 'hook-fenders' (*kou chhiang*[1]).[h] When an (enemy ship) was about to retreat, one used the hook (part); when an (enemy ship) came on, one used the fender (part). The length

---

[a] The discussion will be found in *Chhien Han Shu*, ch. 6, p. 19*b*.

[b] Nevertheless it is possible to find instances of ramming in Sung naval warfare. Indeed we have encountered them already. In Vol. 4, pt. 2, p. 420, we saw how in +1134 some of the big paddle-wheel warships used by the revolutionary forces of Yang Yao were equipped with rams (*chuang kan*[2]) which sank many government vessels; see *Sung Shih*, ch. 365, pp. 8*a* ff., esp. p. 10*b*. So also on p. 422 we read of the shipwright Chhin Shih-Fu, who in +1203 built four-wheeled warships with iron-armoured sides and 'spade-beak' (*hua tsui*[3]) rams. His designs will be discussed more fully on p. 688 below. Since the speed of these ships was probably of the order of 4 knots, the impetus of their considerable weight would have made broadside-on impact rather effective. As late as the end of the +16th century, Wang Ho-Ming tells us, in his account of the Fu chhuan[4] (Fukienese warships), that they could sink smaller enemy vessels with their 'plough-share' (*li*[5]), presumably some sharp offensive weapon fixed to the bows (*Hsü Wên Hsien Thung Khao*, ch. 132, p. 3972.3).

[c] Cf. p. 389 above, and Worcester (14), p. 98.

[d] (1), pp. 202 ff.; cf. p. 389 above. [e] Cf. Vol. 4, pt. 2 *passim*.

[f] Ch. 49, p. 9*a*, *b* (pp. 301 ff.), tr. auct., adjuv. Mei Yi-Pao (1), pp. 254 ff.

[g] Cf. p. 440 above.

[h] *TPYL*, ch. 334, p. 3*b*, quoting these two sentences, has *kou chü*[6] instead of *kou chhiang*, but the meaning is the same. It will be seen that we take the term as a binome, not two separate things, the hook and the ram, as do Mei Yi-Pao and others.

[1] 鈎强　　　[2] 撞竿　　　[3] 鏵嘴　　　[4] 福船　　　[5] 犁　　　[6] 鈎拒

of this weapon was adopted as a standard for the ships, so that the vessels of Chhu were all standardised while those of Yüeh were not. With this advantage the Chhu people greatly defeated those of Yüeh.

Master Kungshu was proud of his ingenuity and asked Master Mo, saying: 'My warships have the "hook-fender" device (one part to pull and one part to push). Do you have anything like this in your (philosophy of) righteousness?' Master Mo replied: 'The "grappling-and-ramming" device in my (philosophy of) righteousness is much better than your war-boat gear...'a

And Mo Tzu goes off into a sermon (excellent in its way) on 'pulling with love and pushing with respect', for which the whole story was no doubt a pretext. That of course does not mean that we should reject it as a valid account of something that happened in −4th-century naval warfare. But the 'hook-fender' device was not exactly a ram; we believe it was a heavy T-shaped iron piece (like a dagger-axe in shape) fitted at the end of a long spar pivoted in derrick fashion at the base of the mast, and capable of being either dropped heavily on the retreating enemy's deck to pin it at a desired distance,b or lowered into position to fend off his closing ship at about the same distance away. In both situations the enemy was held at the best crossbow range. This leads directly to our next subject.

(6) ARMOUR-PLATING AND 'GRAPPLING-IRONS'; PROJECTILE TACTICS *VERSUS* CLOSE COMBAT

If, as we believe, the application of thin sheets of metal to the under-water parts of hulls with a preservative purpose was at least as old in China as it was in the Mediterranean region, and if, as here (and in Sect. 30) we shall hope to show, the military and naval tactics most highly favoured by the Chinese in all periods tended to be those of projectile exchange and accurate fire rather than those of close hand-to-hand combat,c it would be expected that a development of 'armour-plating' above the water-line would occur among them at quite an early date, extending the concept of city-walls to the bulwarks of fighting-ships.d And indeed this is just what we find. In these late medieval contexts, of course, ship-armour does not mean anything like what it came to mean in the world of the nineteenth century, but its perfectly legitimate ancestor may be observed in the thin plates of forged wrought iron which were sometimes fixed to the upper works of Chinese war-junks from the Sung onwards.e Since projectile weapons generally evoke more important works of defence than those required against

---

a On Mohism see Vol. 2, pp. 165 ff.

b From this it was not a long step to lengthening and sharpening the weapon so that it would go right through the hull and hole it from above, cf. p. 690 below.

c Our conclusions here differ widely from those of Cipolla (1), but he relied mainly on the impressions of late foreign visitors, Chinese literature and archaeology not being available to him.

d This very analogy was used by a +12th-century writer, as we shall remember from p. 476 above.

e It would be a nice question whether any of the iron-armoured ships of the Sung or later employed cast-iron plates. As Sects. 30 and 36 will show, the siderurgical industry of the time would have been perfectly capable of providing them, and plates of this kind were used on a considerable scale in the iron pagodas of the +11th century, some of which are still extant. In the meantime see Needham (32), p. 20 and figs. 34, 35.

shock weapons, this development was entirely natural.[a] It was accompanied by another, even stranger for Western preconceptions, namely the use of 'grappling-irons' not for the purpose of laying gangways for boarding-parties, but rather as great pick-axes for holing the enemy ship, or alternatively as clamps for the opposite purpose of keeping her crew at a distance where they could be shot down by point-blank fire.

At several points earlier in this Section mention has already been made of the wooden bulwarks which concealed the rowers and sailors on Chinese war-junks, and discouraged boarders.[b] The next logical step was to roof over the whole deck, leaving a slot through which the mast or masts could be raised or lowered, and then to give further protection by armouring the roof and sides with iron or copper plates. This was seen fully developed in the fleets of the great Korean admiral Yi Sunsin [1] at the time of the Japanese expeditions of Hideyoshi to Korea (+1592 to +1598).[c] For Yi Sunsin then built a number of 'Turtle ships' (Kuei chhuan [2]),[d] which proved very effective in the sea battles against the invaders off Chemulpo and in Fusan Sound.[e] A whole flotilla of them is seen, together with other Korean war-junks, in a screen-painting of the +17th century.[f]

It is rather difficult to get a clear idea of these Turtle ships, but though the sources are not very informative,[g] Underwood gave much attention to them, assisted by Korean scholars such as Cheung Inpo.[h] His conclusions were that a typical warship of this kind was about 110 ft. in length with a beam of 28 ft. and one main deck $7\frac{1}{2}$ ft. above the ship's bottom (Fig. 1050, pl.). The rowers' positions were within the hull, on each side of a central line of cabins for stores and bunks, leaving the main deck free for the gunners and musketeers to maintain fire through 12 gunports and 22 loopholes. They in turn were protected by the sloping roof, the slot in which was probably closed by a sliding hatch before entering battle. The roof was certainly studded with spikes and knives, and though no contemporary text has been found which proves that it was always covered with metal plates, strong local traditions, dating back to the early

[a] One may take a side-glance here at the practice of arming the water-gates of Chinese city-walls with plates of iron. In +1487 Chhoe Pu, the Korean traveller, saw eight or more of these at Ningpo and was much impressed by them (cf. Meskill tr., pp. 68, 69).

[b] Pp. 407, 447, 621 etc. Another instance may be added here. At the Battle of Huang-thien-tang in +1130, some Fukienese sailors advised the Chin Tartars to build protective bulwarks with oar-ports on their ships. Then at a time when the Sung warships were becalmed these galleys rowed towards them discharging incendiary arrows. The tactic was successful, but it presaged exactly what the Sung paddle-wheel ships were afterwards to do even more successfully. See *Sung Chih*, ch. 364, pp. 7b, 8a, cf. *WHTK*, ch. 158 (p. 1381.3). Cf. Vol. 4, pt. 2, p. 416.

[c] See Underwood (1), pp. 71 ff.; Hulbert (1), vol. 1, pp. 349 ff., 375 ff., 399 ff.; vol. 2, pp. 15, 29 ff., 33 ff., 39 ff., 48 ff.; Osgood (1), pp. 198 ff. Yi was born in the same year as Drake, whom he might have met if Drake had visited Korea instead of the Philippines in +1579. And like Nelson, he died aboard his flagship in the last battle of a successful campaign.

[d] Strictly speaking, 'Tortoise ships'.

[e] Cf. Purvis (1), and the discussion on his paper; also Underwood (1).

[f] From the Prince Yi Household collection. Opp. preface in Underwood (1).

[g] Even the (*Yi*) *Chhungmu Kong Chŏnsŏ* [3] (Complete Writings of the Loyal and Martial Duke), i.e. Yi Sunsin, collected in 1795.

[h] See his figs. 46–9. The curious picture of von Pawlikowski-Cholewa (1), p. 32, is given, tantalisingly, with no references. Cf. Yi Unsang (1).

[1] 李舜臣　　　[2] 龜船　　　[3] 忠武公全書

+17th century, affirm this. It would at any rate tally with the fact that none of these ships could be set on fire by the incendiary weapons of the Japanese, and that it was difficult, if not impossible, to pierce their walls with the projectiles of the time. On the other hand, it is certain that each Turtle ship carried an animal figure-head with a tube through which dense toxic smoke could be emitted, the result of the activities of chemical technicians hidden in the bows.[a] Such compositions, popularly supposed to contain sulphur and saltpetre, had long been known in China (as in the *Wu Ching Tsung Yao* of +1044) and we shall examine them in a later place.[b] Apparently there were 10 oars on each side, each worked by two men,[c] but lug-sails were generally used as motive power, except when entering battle or making port.

Perhaps the most interesting feature of the whole episode is that Yi Sunsin was carrying to its ultimate conclusion the projectile tradition of Chinese sea warfare as opposed to shock tactics.[d] The men of his Turtle ships were fully protected against arrows, musket-shots, and incendiary weapons, but also above all against the boarding-parties beloved of the Japanese.[e] Keeping the beam fairly narrow, he built for speed so as to command the range, and adding smoke-screen equipment, he gained valuable superiority in surprise manœuvre. Finally it seems that his armament outweighed the Japanese guns and hand-guns by a factor of forty to one. The tactics were therefore no longer to close and board, but to stand off and pound with projectiles. A line-ahead formation was used to deliver successive broadsides, and ramming might follow only if the enemy was disabled.

About the same time similar factors were leading in Europe to similar developments.[f] According to Rudlov (1) the Dutch, during the siege of Antwerp in +1585, gave partial protection with iron plates to a man-of-war, the 'Finis Bellis', but it was much less successful than the Turtle ships of Yi Sunsin, for it went aground at once and was captured by the Spaniards, who made no use of it. Rudlov also reports a tradition that when Tunis was being besieged by Charles V in +1530 an attempt was made to protect a carrack, the 'Santa Anna', against incendiary weapons by attaching sheets of lead, but this seems inherently less likely. In spite of gunpowder, ship-armour was curiously slow in acquiring importance in Europe, and it would take us too far to follow here its use in the floating batteries of the +18th century until the famous duel between the 'Merrimac' and the 'Monitor' ushered in the modern ironclad period.[g]

As late as +1796 there was still at least one Korean Turtle ship, with two smoke-emitting heads, at the port of Yohsu in Chŭlla province. But it is possible to trace this type of vessel, and the tactics which it implies, much further back than Yi Sunsin's

[a] The conception was strikingly similar to the 'siphons' of Byzantine ships carrying Greek Fire. Another Byzantine analogy is near at hand, too, cf. p. 693.

[b] Sect. 30*k*.

[c] Note that this was double the number of those used in the ships of equivalent size in +1124. One would like to know whether they were yulohs in both cases.

[d] Whether by Greek ramming or Roman boarding. See Sect. 30*c*, *e* for further explanations.

[e] According to Sansom (2), vol. 2, p. 309, Oda Nobunaga, the ruthless precursor of the Tokugawa shogunate, experimented about +1580 with iron-armoured ships, but without the success of Yi Sunsin.

[f] Leonardo designed some partially armoured galleys, at least with protection for the rowers (Ucelli di Nemi (3), no. 5).

[g] Cf. Daly (1).

time, and hence much further back than the parallel developments in Europe. To begin with, the Korean ruler in +1414 (Thaejong of the Yi dynasty) inspected a type of warship, then thought new, called a Turtle-boat.[a] The usual +17th- and +18th-century Chinese sources (*Wu Pei Chih*,[b] *Thu Shu Chi Chhêng*,[c] etc.) all contain collections of illustrations of war-junks with varying degrees of protection for crews and marines, but so also does the *Wu Ching Tsung Yao*[d] of +1044. In fact, the preference for projectile tactics against close combat in naval fighting (and the associated appetite for defensive armour) probably goes back as far as one needs to look in China, as far back indeed as the archers on the Lou chhuan of the Chhin and Han.[e]

The oldest text we have found which seems to have been intended to accompany such a collection of illustrations occurs in the *Thai Pai Yin Ching*[1] (Canon of the White and Gloomy Planet of War),[f] compiled by Li Chhüan[2] in +759. It seems worth while to give translations of his brief descriptions, for they show rather clearly that even as early as the +8th century there was a tendency in Chinese naval construction to roof over completely the upper decks of certain types of combat craft, thereby giving protection from boarders while enabling all projectile weapons to have full play. Here then is what he says.[g]

Tower-Ships[h] (Lou chhuan[3]);[i] these ships have three decks (*lou san chung*[4]) equipped with bulwarks (*nü chhiang*[5]) for the fighting-lines, and flags and pennants flying from the masts. There are ports and openings for crossbows and lances, [and at the sides there is provided felt and leather (*chan ko*[6]) to protect against fire],[j] while (on the topmost deck) there are trebuchets[k] for hurling stones, set up (in appropriate places) (*chih phao chhê lei shih*[7]). And there are also (arrangements for making) molten iron (for throwing in containers from these catapults). (The whole broadside) gives the appearance of a city wall (*chhêng lei*[8]). In the Chin period the Prancing-Dragon Admiral, Wang Chün,[9] invading Wu, built a great ship 200 paces (1,000 ft.) in length,[l] and on it set flying rafters and hanging galleries[m] on which chariots and horses could go.[n] But if [all of a sudden][o] a violent wind is encountered, (such ships are likely to) get out of human control, so they were judged inconvenient in practice [for warlike action].[p] But the fleet cannot fail to be furnished with such ships, in order that its overawing might may be perfected (*i chhêng hsing shih*[10]).[q]

---

[a] Underwood (1), p. 74.
[b] Chs. 116–118. Note, for instance, the 'Eagle ship' (Ying chhuan[11]), ch. 117, p. 6a, b.
[c] *Jung chêng tien*, ch. 97.   [d] *Chhien chi*, ch. 11.
[e] Cf. pp. 441 ff. above.   [f] On this see further in Sect. 30 below.
[g] Ch. 40, pp. 10a ff., tr. auct.   [h] Or 'Castled battleships'.
[i] For this first paragraph the text has been conflated with parallel passages in the *Thung Tien* (+812), ch. 160, p. 16b (pp. 848.3, 849.1; tr. also by Krause (1), p. 69), and in *TSCC*, *Jung chêng tien*, ch. 97, p. 6a, of nine centuries later, in order to give an idea of the extent of variation. The *Wu Ching Tsung Yao* text is also added.
[j] *WCTY* and *TSCC* only.   [k] See Sect. 30i below.
[l] The explanation of this seemingly incredible statement will be given shortly (p. 694). *WCTY* omits the reference to the historical events of +280 and makes a figure of 100 paces seem to refer to the ships under description.
[m] *TPYC* only.   [n] In *TSCC*, abridgement here unintelligible.
[o] *TPYC* and *TTN* only.   [p] *TTN* and *TPYC* only.
[q] *TTN* only; wording slightly different in *TPYC*, *WCTY* and *TSCC*.

[1] 太白陰經   [2] 李筌   [3] 樓船   [4] 樓三重   [5] 女牆
[6] 氈革   [7] 置拋車播石   [8] 城壘   [9] 王濬   [10] 以成形勢
[11] 鷹船

Covered Swoopers[a] (Mêng-chhung[1]); these are ships which have their backs roofed over and (armoured with) a covering of rhinoceros hide[b] (*i hsi ko mêng fou*[c] *chhi pei*[2]). Both sides of the ship have oar-ports; and also both fore and aft, as well as to port and starboard, there are openings for crossbows and holes for spears. Enemy parties cannot board (these ships) (*ti pu tê chin*[3]), nor can arrows or stones injure them. This arrangement is not adopted for large vessels because higher speed and mobility are preferable, in order to be able to swoop suddenly on the unprepared enemy. Thus these (Covered swoopers) are not fighting-ships (in the ordinary sense).

Combat-Junks (Chan hsien[4]);[d] combat-junks have ramparts and half-ramparts[e] above the side of the hull, with the oar-ports below. Five feet from the edge of the deck (to port and starboard) there is set a deckhouse with ramparts, having ramparts above it as well. This doubles the space available for fighting. There is no cover or roof over the top (of the ship). Serrated pennants are flown from staffs fixed at many places on board, and there are gongs and drums; thus these (Combat-junks) are (real) fighting-ships (in the ordinary sense).

Flying Barques (Tsou ko[5]); another kind of fighting-ship. They have a double row of ramparts on the deck, and they carry more sailors (lit. rowers) and fewer soldiers, but the latter are selected from the best and bravest. These ships rush back and forth (over the waves) as if flying, and can attack the enemy unawares. They are most useful for emergencies and urgent duty.[f]

Patrol Boats (Yu thing[6]) are small vessels used for collecting intelligence. They have no ramparts above the hull, but to port and starboard there is one rowlock every four feet, varying in total number according to the size of the boat. Whether going forward, stopping, or returning, or making evolutions in formation, the speed (of these boats) is like flying. But they are for reconnaissance, they are not fighting ships.[f]

Sea-Hawks (or *Sea-Grebes*)[g] (Hai-hu[7]); these ships have low bows and high sterns, the forward parts (of the hull) being small and the after parts large,[h] like the shape of the *hu* bird (when floating on the water).[i] Below deck level, both to port and starboard, there are 'floating-boards' (*fou pan*[8]) shaped like the wings of the *hu* bird. These help the (Sea-hawk) ships, so that even when wind and wave arise in fury, they are neither (driven) sideways, nor overturn.[j] Covering over and protecting the upper parts on both sides of the ship are stretched

---

[a] Perhaps this word requires the reader's indulgence. It is necessary to translate the impression of rushing violent motion. The word 'destroyer' seems natural only because we are accustomed to it in a naval context, just as we are used to the idea that a submarine is not an apprentice sea-soldier.

[b] A material of famed use in ancient Chinese armour, see Sect. 30*e* below.

[c] Note that this word, so pronounced, had meanings such as a hen sitting on eggs, or soldiers in ambush.

[d] *WCTY/CC*, ch. 11, p. 10*a*, and *TSCC*, *Jung chêng tien*, ch. 97, p. 8*a*, have Tou hsien[9] here.

[e] It is to be presumed that some kind of crenellated bulwarks are meant.

[f] Although nothing is said in either of these descriptions of mast and sails we are reluctant, in view of all else that is known of Chinese shipping, to assume that anything like galleys purely rowed was intended here.

[g] For the birds in question see R258 and 314.

[h] The text must be inverted here; we translate in corrected form.

[i] The significance of this remark, in connection with what has already been said about hull shape of Chinese craft (p. 417 above), will not be missed.

[j] This is a most important passage concerning leeboards (see p. 618 above). Conflating *TPYC* with *WCTY* and *TSCC*, it runs: '*chu chhi chhuan sui fêng thao nu chang erh wu yu tshê chhing*'.[10] The existence of leeboards in the +8th century gives China a long priority in their invention.

---

¹ 蒙衝　　　² 以犀革蒙覆其背　　　³ 敵不得近　　　⁴ 戰艦　　　⁵ 走舸
⁶ 遊艇　　　⁷ 海鶻　　　⁸ 浮板　　　⁹ 鬬艦
¹⁰ 助其船雖風濤怒漲而無有側傾

raw ox-hides, as if on a city wall.[a] There are serrated pennants, and gongs and drums, just as on the fighting-ships.

This text seems to show us that the general principle of projectile warfare at sea from 'armoured' ships which could approach their targets rapidly, deliver a broadside, and make away again, can be traced back to the +8th century at least. Although the Covered swoopers (Mêng-chhung), surely the lineal ancestors of the Turtle ships of Yi Sunsin, are said to be hardly fighting-ships in the strict sense, while the Combat-junks are specifically so termed, this may have been an explanation intended for military readers used to close combat on *terra firma*, and it would be unwise to conclude that grappling and boarding had always been typical of Chinese naval practice in earlier times. Indeed the term Mêng-chhung itself as a designation of naval vessels goes back at least to the +2nd century.[b] Then the tendency to cover over the upper deck appears also in the sixth type, the Sea-hawks (Hai-hu), which from the description suggest some kind of converted cargo-boat like the Ma-yang-tzu or the river junks in Sung pictures.[c] And the great majority of the crew and soldiers, except perhaps the artillerists on the topmost deck, were clearly protected in the 'battleships' known as Lou chhuan.

So much for the general principle of ship-armour. But there is more than this, for we have several Chinese records of plating with iron considerably earlier than the time of Yi Sunsin. One scene was the confusion at the end of the Yuan period. In +1366 Ming Shêng[1] had succeeded his father as ruler of an independent State of Shu (Szechuan) which was rising out of the ruins, but Chu Yuan-Chang mounted a Western Expedition against him which began to make its way up the Yangtze valley in +1370. The *Ming Shih* says:[d]

Next year Liao Yung-Chung[2] was deputy commander of the Western Expedition, and followed Thang Ho[3] as admiral of the (river) fleet against Shu. Thang Ho had his headquarters at Ta-chhi-khou.[e] (Liao) Yung-Chung set off in the vanguard, arrived at Old Khuei-fu[f] and routed its defenders under the (Szechuanese) general Tsou Hsing[4] and others; then going on he reached the Chhü-thang[5] Gorge. Here, where the cliffs are very precipitous and the water most dangerous, the Szechuanese had set up iron chains (*thieh so*[6]) (as booms), and bridges (*chhiao*[7]),[g] to block the gorge horizontally so that no ships could get through. (Liao)

---

[a] Moistened ox-hide was a well-known protection against incendiary projectiles. Cf. p. 449.

[b] In Sect. 27g above (Vol. 4, pt. 2, p. 416), we gave a passage from the *Sung Shu*, ch. 45, p. 7a, which uses the expression for fast assault craft apparently propelled by treadle-operated paddle-wheels. The date of the action was +418. Cf. p. 680 above.

[c] Cf. Figs. 933, 976, 1032 (pls.) above. But perhaps more probably a ship like the Swatow three-master in Anon. (17), pl. 10. See also Figs. 939, 950, 1013 (pl.), 1028.

[d] Ch. 129, p. 12a; tr auct.

[e] There is now a place of this name about 200 miles below Chungking, but the locality here referred to must have been somewhere between I-chhang and Pa-tung, i.e. below the gorges.

[f] Between present-day Wu-shan and Fêng-chieh.

[g] The text reads as if these were iron-chain suspension bridges (which they could well have been), but the account in the *Ming Shih Lu*, Hung-Wu (Thai Tsu) sect., ch. 63, p. 4a, shows that the chains were booms, and that three cable suspension bridges were set up to command them. It says that 'these were fixed to the cliffs on both sides, and provided with flat decks of wooden boards bearing the upright wooden

---

¹ 明昇   ² 廖永忠   ³ 湯和   ⁴ 鄒興   ⁵ 瞿塘   ⁶ 鐵鎖
⁷ 橋

Yung-Chung therefore secretly sent several hundred men with supplies of food and water to make a portage with small boats, so that they appeared up river beyond these defences.[a] Now the mountains of Szechuan are so well wooded that he had ordered the soldiers to wear green garments and sleeveless raincloaks (*so i*[1]) made of leaves; and thus they descended through the forests and rocks. When agreed stations had been reached, the best troops were ordered to attack at Mo-yeh Ferry. At the fifth night watch the general assault began both by water and land. The bows of the naval ships were sheathed with iron (*thieh kuo chhuan thou*[2])[b] and all kinds of firearms (*huo chhi*[3]) were made ready upon them. Only as dawn was breaking was the army's presence discovered by the Szechuanese, who threw in all their best troops for the defence, but unavailingly. When (Liao) Yung-Chung had captured all six of their positions, he assembled his commanders, including those who had led the boat-portage parties, and pushed on the fight, some attacking above (the defences) and some below, so that the Szechuanese were completely defeated, and Tsou Hsing lost his life.[c] The three bridges were burnt, and the iron chains all cut.

And thus Liao Yung-Chung was able to enter Khueichow in triumph. This engagement was certainly a remarkable one, and much credit must go to the defeated Szechuanese for the ingenuity and engineering skill of their defences, chiefly organised, it would seem,[d] by the Shu officials Mo Jen-Shou[4] and Tai Shou.[5] But the use of iron ship-armour by Liao Yung-Chung is also of interest, since it occurred a couple of centuries before the sea-going Turtle ships of Yi Sunsin.

It was however far from new in +1370. For at the time of most rapid development of the Southern Sung navy, in +1203, a remarkable shipwright, Chhin Shih-Fu,[6] built at the Chhihchow yards two prototype 'Sea-hawk' (Hai-hu) paddle-wheel warships the sides of which (and perhaps the flat roofs also) were armoured with iron plates. The decks were given complete protection by the overhead cover, and besides the usual arrangements for crossbows, arcuballistae, fire-lances, bomb-throwing trebuchets, etc., each ship was provided with a spade-shaped ram.[e] The smaller vessel, of 100 tons burden and with two treadmill-operated paddle-wheels, needed a propulsion crew of 28 men; the larger, of some 250 tons though not very much greater in

posts of trebuchets (*phao shih mu kan*[7]) and also iron barrel guns (*thieh chhung*[8]). At the ends of the bridges on the two banks were more trebuchets to resist our troops.' Another example of the use of iron-chain booms had been recorded by Chao Ju-Kua in the previous century (+1225). He described such a device in his chapter on Palembang, though it may perhaps have been used across the strait of Johore (*Chu Fan Chih*, ch. 1, p. 7*a*; cf. Hirth & Rockhill (1), p. 62; Steiger *et al.* (1), pp. 112 ff.). In Chinese design and practice suspension bridges and booms were closely related (cf. pp. 202 ff. in Sect. 28*e* above).

[a] From a third account, the *Phing Hsia Lu*[9] (written by Huang Piao[10] about +1544), p. 8*a*, it appears that these small-boat parties attacked the defences from the rear, using grenades and bombs (*huo phao*[11]) and fire-lances or hand-guns (*huo thung*[12]).

[b] This is confirmed in the same words by Huang Piao.

[c] Hit by a rocket or incendiary arrow, according to Huang Piao.

[d] According to the account in the *Ming Shih Lu*.

[e] Cf. p. 300 above, and Underwood (1), pp. 80 ff. The essential words of description of the two cruisers are: *Hsin yang thieh pi hua tsui phing mien Hai-hu chan chhuan*.[13] They come from the *Sung Hui Yao Kao*, tsê 146 (*Shih huo*, ch. 50), pp. 32*b* ff., where very full specifications are given. See Lo Jung-Pang (3), p. 199, who discovered the passage.

[1] 蓑衣	[2] 鐵裹船頭	[3] 火器	[4] 莫仁壽	[5] 戴壽
[6] 秦世輔	[7] 砲石木竿	[8] 鐵銃	[9] 平夏錄	[10] 黃標
[11] 火砲	[12] 火筒	[13] 新樣鐵壁鏵嘴平面海鶻戰船		

length, and apparently with four paddle-wheels, needed 42 men to work them and carried 108 marines.[a] The paddle-wheels were protected by housing above the water-line.

To defend thus the mechanism of progression as well as the projectile-firing combat elements themselves was a very natural and logical step in the development of armour, but (failing the invention of the screw-propeller) paddle-wheels alone permitted it. It could not be done either for masts and sails or for projecting oars. There is thus more connection than might at first sight appear between the relatively early flourishing both of paddle-wheel propulsion and of ship-armour in Chinese culture. The history of the paddle-wheel boat we have already sketched in Sect. 27*i* above,[b] and some of the texts there given show very clearly indeed how conveniently this motive device suited the tactics of rapid fire from a fast armoured ship. When paddle-wheel steam warships first approached the coasts of China in the early nineteenth century they brought nothing tactically novel, indeed they constituted the realisation of a dream which projectile-minded Chinese admirals had entertained for well over a millennium. Although the cruisers of Chhin Shih-Fu were said to be of 'new design', there is no special reason for regarding them as the first of all Chinese iron-armoured ships, and earlier examples may well come to light, though—at a guess—perhaps not before the beginning of the +12th century, when the foundation of the permanent navy occurred.[c]

We can follow out the contrast between projectile tactics and close-combat boarding-party tactics much further still in the context of 'grappling-irons'. If we trace down the warship descriptions of the +8th-century *Thai Pai Yin Ching* in later compilations of naval knowledge, we find that the *Wu Ching Tsung Yao* of +1044 inserted some very interesting material from a quite independent, and still older, source. Under its entry for the Patrol boats (Yu thing[1]) it included a description of the Wu-ya hsien[2] (Five-Ensign battleships) which should more properly have been put with the Lou chhuan[3] (Castled battleships). After a slightly abridged version of the Thang text on the Patrol boats, the passage continues by quoting from the *Sui Shu* (History of the Sui Dynasty), the biographical parts of which were finished by +636. It tells the story of the building of fleets for the emperor Kao Tsu by a celebrated engineer[d] Yang Su,[4] and of the naval battle in which they sealed the fate of the Chhen dynasty.[e]

---

[a] It is perhaps worth remembering that Chhin Shih-Fu's smaller cruiser was twice the size of one of Prince Henry's caravels, and that his larger one was about the same size as Vasco da Gama's flagship. Of course the Chinese paddle-wheel warships, so prominent in the Sung, worked primarily upon the lakes and rivers, where they were very effective. This means of propulsion was not suited to the sea until the days of iron hulls and steam engines.

[b] Vol. 4, pt. 2, p. 413. [c] Cf. Lo Jung-Pang (1), and p. 476 above.

[d] We have met with him already in Vol. 4, pt. 2, p. 400. Cf. Balazs (8), p. 88.

[e] In what follows we translate the parallel texts *Wu Ching Tsung Yao* (*Chhien chi*), ch. 11, p. 6*a* (Ming edition of +1510, ch. 11, p. 5*b*); *Sui Shu*, ch. 48, pp. 2*b*, 3*a*; *Yü Hai* (apparently reproducing another version of Yang Su's biography), ch. 147, pp. 11*b*, 12*a*. The *Sui Shu* text is much the fullest of these, but it does not contain all the statements which are relevant. For a full account of the battle, all three texts (and also other extant descriptions) would have to be conflated for translation, but we do not need it here. We follow mainly the *Wu Ching Tsung Yao* version, additions from *Sui Shu* being added in square brackets. Tr. auct. Cf. *WHTK*, ch. 158 (p. 1380.3).

¹ 遊艇   ² 五牙艦   ³ 樓船   ⁴ 楊素

[In the 4th year of the Khai-Huang reign-period (+584)] Sui Kao Tsu commissioned Yang Su (as commander-in-chief) to destroy the Chhen. So coming down the gorges to Hsinchow he built [at Yung-an] great war-junks (*ta hsien*[1]) called Wu-ya hsien (Five-Ensign battleships).[a] Above they had five decks, and their height was more than 100 ft.[b] To left and right, and fore and aft, he set up six *pho-kan*[2] (lit. striking- or patting-arms), each 50 ft. long. And of marines (each ship carried) 800 men, and many flags and banners fluttered aloft.

...When the ship comes alongside an enemy ship, they let go the striking-arms (*fa pho-kan*[3]) on top of her, and whether barque or barge she is all broken into fragments.[c]

Then there were the Yellow Dragon (Huang lung[4]) ships, each with 500 marines,[d] while others embarked on smaller boats (*chê hsien*[5]) of many different kinds. [(Yang) Su then led the fleet down through the gorges.] When they got down to Ching-mên the Chhen general Lü Chung-Su[6] was there with [more than 100] ships [...and a boom] to resist (Yang) Su, but he ordered the Man[7] (highlanders) of Pa[8] (in Szechuan)[e] to man four of the Five-Ensign battleships and fight them with the striking-arms (*pho-kan*[2]).[f] They thus destroyed more than ten of the great war-junks (*hsien*[1]) of the Chhen fleet, breaking them to pieces, and so the river route was forced open.

What were these 'striking-arms'? It seems fairly clear that they were long heavy pointed spikes, probably iron-shod, and fixed at right angles at the end of elongated arms like guy-derrick jibs, then suddenly released from an approximately vertical position to crash down through the deck and hull-timbers of the enemy vessel. They were certainly not intended to 'grapple', i.e. to provide gang-planks for boarders, and the best name for them would be 'holing-irons'.[g] We can actually see them, rather badly drawn as usual, in Fig. 1051,[h] looking like thin hammers, and in the unarmed position.[i]

a *Ya* here doubtless implies *ya chhi*,[9] 'dented' or 'indented' banners or standards, so called either because of the tooth or claw symbol at the top of the flagstaff, or because of their serrated edges. And indeed five can be seen at the bows in Fig. 1051.

b Whether this means from bottom to deck or from water-line to masthead we do not know.

c This sentence is part of an explanation, only found in *WCTY*, which precedes the point at which we begin, and duplicates the description of Yang Su's big ships. It is reproduced, in a most mystifying way, in *TSCC*, *Jung chêng tien*, ch. 97, p. 4*a*, where it stands alone.

d The *Sui Shu* text says 100.

e *Sui Shu*, interestingly, says the Tan[10] tribesmen, so that presumably skilled sailors were intended. We have heard much of the Tan people a few pages above in connection with pearl-diving; perhaps it was a mistake here to make them come from Pa (Szechuan).

f *Sui Shu* has *po-chhiang*[11] here, a striking spar or boom, but it is clearly equivalent, an inadvertent change of radical having occurred in the first word.

g Audemard (2), pp. 34 ff., puzzled by the isolated sentence in the *TSCC*, went astray, we think, in calling them 'flails'; though there are expressions for the instrument in which the first word occurs, other technical terms would have been more obvious. He was no nearer the mark in comparing them to the battering-rams used against castle gates in the West.

h The two editions of *WCTY* vary as to illustration. The Ming edition of +1510, recently reproduced, and said to derive directly from the Sung edition of +1231, gives a picture of what is evidently a patrol-boat, with a crew of 14; but the *SKCS* edition, derived from a manuscript in the imperial Wên Yuan Ko library copied in +1782, illustrates the Five-Ensign battleship, as shown, though still entitling it Yu Thing (Patrol boat). This confusion was recognised by Audemard (2), who knew only the Chhing *WCTY* and the *TSCC*, and never saw the Ming *WCTY*.

i We have already met with the striking-arms under the name *thieh chuang*[12] in the text of +1129 given above in the sub-section on navigation (p. 575).

Since, as we know from Vol. 4, pt. 2, p. 416, there were *pho chhuan*[13] (ships with striking-arms) in a Liang fleet in +552, the device must go back at least to the earlier part of the +6th century.

[1] 大艦	[2] 拍竿	[3] 發拍竿	[4] 黃龍	[5] 舴艋
[6] 呂仲肅	[7] 蠻	[8] 巴	[9] 牙旗	[10] 蜑
[11] 柏檣	[12] 鐵撞	[13] 拍船		

游 艇

Fig. 1051. An illustration of the Wu-ya hsien (Five-Ensign battleship) of the Sui dynasty, described in the *Sui Shu* (+636), and copied in the *Wu Ching Tsung Yao* (+1044) along with a corpus of traditional material on warships from the *Thai Pai Yin Ching* (+759). The description was mistakenly appended to the entry on Yu thing (Patrol boats) instead of that on Lou chhuan (Battleships, lit. Tower-ships), so that in the Chhing edition of *WCTY* the illustration (here given) was wrongly captioned 'Yu thing'. Although the highly fanciful creation of an architectural painter (cf. p. 106), it is valuable for one thing— it is almost the only traditional drawing we have which shows the 'striking-arms' or 'holing-irons' or 'fending-irons' (*pho-kan*), three on the port side in the lowered position. Though very badly drawn, we know that these were heavy pointed spikes on the end of long spars, which could be let go suddenly either to crash through the enemy's timbers and sink him, or at least to hold his ship off so that its deck could be raked by intense crossbow fire. For further discussion on this device and its relation to the projectile-mindedness of Chinese medieval naval commanders, see text. Cf. also Fig. 949.

44-2

If the shafts of the holing-irons were long enough, and if the enemy crews were not provided with gangways of the right length known beforehand, then even if they wanted to attack by boarding, they would be held, as it were, at arm's length, where the missiles of the battleship could play upon them and decimate them. Such 'grappling-irons for keeping people off', or what we may call 'fending-irons', are mentioned by Lu Yu[1] about +1190 in his *Lao Hsüeh An Pi Chi*,[2] where he describes the campaign which government forces had had to wage some sixty years earlier against an important equalitarian peasant revolt led by Chung Hsiang[3] and Yang Yao.[4] We have followed the fortunes of this before in connection with multiple paddle-wheel warships, which played so conspicuous a part in it;[a] here we need only glance at another of their techniques. Lu Yu says:[b]

The rebels in the Ting-li region, such as Chung Hsiang and Yang Yao [the way they pronounce Huan[5] in those parts] had fighting-ships (on the Tung-thing Lake), such as paddle-wheel ships (*chhê chhuan*[6]), oared (yuloh?) vessels (*chiang chhuan*[7]), and sailing junks (lit. Sea-eels, Hai chhiu[8]). Among their offensive weapons were the *na tzu*[9] [commonly pronounced *nao tzu*[10]], the 'fish forks' (*yü chha*[11]), and the *mu lao ya*.[1] The *nao tzu* and the *yü chha* were on bamboo poles 20 to 30 ft. long like handles, and prevented the (government) marines with hand-weapons from boarding and attacking at close quarters. Chhêng Chhang-Yü's[13] men, however, though from Tshaichow, were also skilled at the use of these, so they gained the day repeatedly. The *mu lao ya*, which was also called the 'no *corvée* log' (*pu chieh mu*[14]), was a piece of wood strong and heavy just a little over 3 ft. long and sharpened at both ends; these were used by the war-ships (on both sides?) and proved very effective...[c]

Lu Yu then goes on to talk about explosive bombs scattering lime, and other means of making poisonous smokes,[d] but he has said enough to show that the *na tzu*, 'kneaders' or 'pounders', were probably smaller versions of the holing-irons, and succeeded in keeping the enemy at an inconvenient distance. Another writer of the period tells us that the rebel ships mounted what might be called 'holing-derricks'. In his *Chi Yang Yao Pên Mo*,[15] written about +1140, Li Kuei-Nien[16] says:[e]

The rebel ships had two or three decks, and some could carry more than 1,000 men. They were equipped with 'holing derricks' (*pho-kan*[17]), which were like great masts over a hundred feet high. Large rocks were hoisted up to the top of these by means of pulleys, and when a government ship came close, they were suddenly let go to smash her...

Here then, though the same technical term is used as for the holing-irons of the Sui, the projectile mentality had again overcome the conception of a shock weapon, and

---

[a] Vol. 4, pt. 2, pp. 419 ff.

[b] Ch. 1, p. 2*a*, *b*, tr. auct. Square brackets indicate Lu Yu's own commentary interjections. On the Sea-eels cf. p. 416 above.

[c] We have not been able to solve the use of the 'no *corvée* logs', otherwise known as 'old crows'. But we can appreciate the name.

[d] This continuation of the passage has been translated in Vol. 4, pt. 2, p. 421. For further information on gunpowder warfare at this time see Sect. 30*k*.

[e] The passage is preserved in Hsiung Kho's[18] *Chung Hsing Hsiao Chi*,[19] ch. 13, p. 15*a*; tr. Lo Jung-Pang (3), mod. auct. We quoted it more fully in Vol. 4, pt. 2, p. 420.

[1] 陸游	[2] 老學庵筆記	[3] 鍾祥	[4] 楊么	[5] 幻
[6] 車船	[7] 槳船	[8] 海鰌	[9] 拏子	[10] 鐃子
[11] 魚叉	[12] 木老鴉	[13] 程昌寓	[14] 不藉木	[15] 記楊么本末
[16] 李龜年	[17] 拍竿	[18] 熊克	[19] 中興小紀	

heavy weights were dropped from jib arms above in the attempt to hole and sink the opposing craft.[a]

It is interesting that an exactly similar device is recorded as having been in common use at the same time, and perhaps somewhat earlier, in the Byzantine navy.[b] It had Hellenic precedents,[c] and Leonardo da Vinci was later to design one.[d] But what a contrast all these Chinese 'grappling-irons' were with the celebrated *corvus* of the Romans, on which a notable monograph has been written by Wallinga (1). In the wars with the Carthaginians, *c.* −260, the Roman tacticians decided to accept the necessity of head-on collisions, but to avoid the usual sequel in which the more agile ship would disengage and fatally ram the less agile. This they did by a device which would fix the ships together and at the same time give scope for their own superiority in hand-to-hand fighting by pouring their marines on to the Carthaginian decks.[e] It consisted of a gangway in the bows, 36 ft. long and 4 ft. wide, armed underneath at its outboard end with a massive iron spike, and held in readiness by suspension over a pulley at the top of a 24-ft. post. When the plank had fastened itself to the enemy ship, the Roman legionaries could board it two abreast. The comparison of their short swords with the crossbows and catapults of the Chinese gives the measure between two utterly different conceptions of naval warfare.

A word may be added here about the mounting of trebuchet or mangonel artillery on Chinese warships.[f] This is clearly attested for the +8th century in words which we read a few pages above (*chih phao chhê lei shih*[1]) in the *Thai Pai Yin Ching* (p. 685), and all the subsequent compendia confirm it. At that time, and also in the +11th century, when the *Wu Ching Tsung Yao* was compiled, the trebuchets must have been of the classical manned type, but the only illustrations we have are later, and show the counterweighted type (Fig. 949).[g] There can be no doubt that this was a constant practice through the ages.

---

[a] A variant of the 'striking-arm' technique was used at the Battle of Huang-thien-tang in +1130, when the Sung admiral Han Shih-Chung (cf. Vol. 4, pt. 2, pp. 418, 421, 432) equipped his warships with long 'plaited iron cables' or chains with long links (*thieh kêng*[2]) each with one or more hooks of iron on the end. As they passed the Chin ships, these were thrown on to them, by some trebuchet technique or from the masthead, whereupon they towed the enemy away playing upon him with crossbows as they went (*Sung Shih*, ch. 364, p. 7*b*, cf. *WHTK*, ch. 158 (p. 1381.3), etc.). Possibly some clue as to the nature of these stiff chains may be gained from the 'iron whip' (*thieh pien*[3]) described among the close-combat weapons in *WCTY* (*Chhien chi*), ch. 13, p. 14*a*, *b*.

[b] Casson (3), p. 244.

[c] Thucydides, VII, xli, the account of the 'dolphin-bearing cranes' used against Syracusan ships by the Athenians during the great siege of −413.

[d] See Ucelli di Nemi (3), no. 5.

[e] The classical description is in Polybius, I, 22, 3–11.

[f] We have met with it already in Vol. 4, pt. 2, p. 335 and Fig. 634. The Chinese did not use the torsion catapults of Hellenistic antiquity, but a swape-like lever pulled down sharply at one end either by manned ropes or by a counterweight, and lobbing over the projectile in a sling from the other (see Sect. 30*i*).

[g] From the Ming (+1510) edition of *WCTY*. The Chhing palace edition drawing makes the trebuchet unrecognisable, looking like a flag on the end of a long staff caught in a crotch. The Lou chhuan in *TSCC*, *Jung chêng tien*, ch. 97, p. 5*b*, has three counterweighted trebuchets on the upper deck, but the artist was half inclined to make them look like culverins.

---

[1] 置砲車擂石   [2] 鐵緪   [3] 鐵鞭

From the foregoing pages only one character still needs pinning down—the Prancing-Dragon Admiral.[a] What Wang Chün[1] did in +280 prefigured very closely the great naval victory of Yang Su in +585 about which we have just been reading, for in both cases a powerful fleet descending the Yangtze from the west overbore all the defences both of land and water which had been set up against it, and brought about a change of dynasty. Yang Su as the admiral of the Sui undid the house of Chhen, Wang Chün as the admiral of the Chin liquidated the former Three Kingdoms State of Wu. Needless to say, those three hundred years that intervened saw many changes in shipbuilding, and though in the +3rd century we hear nothing of holing-arms or paddle-wheel boats, we do find other techniques equally interesting, notably the building of floating fortresses borne on multiple hulls. This has perplexed many students because the seemingly extraordinary dimensions found their way into later texts describing the Castled battleships (Lou chhuan), as we have seen.[b] But the matter is soon unravelled when we look at the *Chin Shu*'s biography of Wang Chün, completed by +635. It runs as follows:[c]

The emperor Wu Ti (of Chin), meditating the overthrow of the Wu State, ordered (Wang) Chün to construct (a fleet of) war-ships. (Wang) Chün also built a great square composite floating fortress with multiple hulls (*ta chhuan lien fang*[2]), 120 paces (600 ft.) along each side, holding more than 2,000 men, with high towers (*lou tsêng*[3]) and four sally-port gates, set in wooden walls round the ramparts of which horses could be ridden back and forth. At the bows there were decorations of strange birds and animals to overawe the river spirits. For abundance of oars and skill of ship-craft, nothing like it had ever been seen before.

(Wang) Chün built his ships and the floating fortress in Szechuan from the wood of persimmon trees (*shih*[4]),[d] and some of the (smaller) timbers floated down the (Yangtze) river. The Wu governor of Chien-phing, Wu Yen,[5] noticed this, had the pieces collected from the water, and showed them to Sun Hao[6] (the last emperor of Wu), saying that the Chin ruler was evidently bent on war and advising an urgent increase in the army of defence. But his advice was not taken.

The rest of the story is told in the following pages of the *Chin Shu*.[e] The Chin army and navy did indeed descend the Yangtze in force. The Wu defenders used iron-chain booms, and iron stakes more than ten feet long stuck in the shallows to hole the attacking ships.[f] But Wang Chün also built several dozen very large rafts, each more than 100 paces (500 ft.) square, with dummy marines wearing what looked like armour and carrying fake weapons; these rafts, piloted by skilled sailors, crashed over the iron

---

[a] See p. 685 above.
[b] *STTH*, *Chhi yung* sect., ch. 4, p. 38b, is another case in point, and it was a query from Mr J. V. Mills on this text which led to the solution of the problem.
[c] Ch. 42, pp. 5a ff., tr. auct. Cf. *WHTK*, ch. 158 (p. 1380.1).
[d] R188; Li Shun-Chhing (1), pp. 886 ff.; Chhen Jung (1), pp. 975 ff. Some *Diospyros* spp. are noted for their hard and durable wood.
[e] Pending full translation, cf. *TH*, pp. 869 ff.
[f] This latter method was commonly used in Chinese littoral defence tactics. We have another account of it for +937 in a sea battle between the Nan Han king, Liu Yen,[7] and a rebellious commander Wu Chhüan[8] (*Wu Tai Shih Chi*, ch. 65, p. 6a, tr. Schafer (4), p. 357).

[1] 王濬   [2] 大船連舫   [3] 樓橧   [4] 柿   [5] 吳彥   [6] 孫晧
[7] 劉龑   [8] 吳櫂

spikes and bent or broke them. Then with the aid of quantities of hemp-seed oil poured on the water at several points, and ignited by torches more than 100 ft. long, the boats supporting the chain booms were set on fire, so that the chains themselves melted in the heat and no further obstacle remained to hold back the ships and the floating fortress. No doubt this huge craft might also be termed a floating battery, for trebuchet catapults were almost certainly mounted on it.[a] It is a pity that no details were given of the joint hulls by which it was supported, but we may surmise some sixteen vessels each 150 ft. long. The idea could have arisen very naturally from the floating-bridge principle, which had been much used in Chinese antiquity.[b] Although these facts have no bearing on the largest size ever reached by Chinese ships, as once was thought, they do supply yet another indication of the prevalent projectile-mindedness—if we have no strong-points or artillery-emplacements on the frontiers of the Wu territory, said the Chin strategists, we shall build a great one and float it right down into their midst.

(j) CONCLUSIONS

The most suitable ending for this Section on nautical technology will perhaps be a brief tabulation of the characteristics of Chinese craft, and the influence which they may have exerted upon Western practice through the centuries.[c] Let us begin with the view, so highly probable, that the basic principle of Chinese ship construction was derived from the example of the bamboo stem with its septa, and indeed that in actual fact the earliest vessels of East Asia were rafts of bamboo. This led directly to (A) the rectangular horizontal plan. The following corollaries resulted:

(i) The absence of stem-post, stern-post, and keel,

(ii) The presence of bulkheads, giving a hull very resistant to deformation, and leading naturally to

(iii) The system of water-tight compartments, with its many advantages. These were almost surely in use by the +2nd century, but were not adopted in the West until the end of the +18th. Provenance was then recognised.

(iv) The possibility of free-flooding compartments, found useful both on river rapids and at sea. This was not adopted to any extent in Europe.

(v) The existence of a vertical member to which the axial rudder could be attached, in 'line closure' rather than 'point closure'. See D (i).

Additional elements not essential to the design were:

(vi) Approximation to a flat bottom. This goes back many centuries in China, but was not adopted in Europe for ships of any size until the nineteenth century.

[a] For the evidence pointing to this, see Sect. 30*i*.
[b] Cf. pp. 160 ff. above.
[c] A similar attempt has been made by R. L. Dickinson (1). Navigational techniques and their diffusion we have considered on pp. 560 ff. above. Alan Villiers has written (3): 'I think the Chinese were the greatest of all Asian seamen, and their junk the most wonderful ship. Hundreds of years ago, the sea-going junk embodied improvements only relatively recently thought out in European ships—water-tight bulkheads to isolate hull damage and so keep the ship afloat, the balanced rudder which makes steering easier, and sails extended with battens.'

(vii) Approximation to a rectangular cross-section. This was also old in China, but again not adopted in Europe until the development of ships of iron and steel.

(viii) The placing of the largest master-couple well aft. This is still prevalent in traditional Chinese ships, and must be old, though how old has not yet been established; it was certainly current in the Thang (+8th century) and may well be earlier. Its interest for sailing vessels was not understood in the West until the end of the nineteenth century.

With regard to propulsion, Chinese seamanship had the lead of Europe for more than a millennium. First, concerning the use of (B) oars and paddles, we note:

(i) The invention, not later than the +1st century, of the self-feathering 'propeller' or sculling-oar (the yuloh). Though universally used in China, this was never adopted in the West.

(ii) The invention of the treadmill-operated paddle-wheel boat in the +8th if not the +5th century, and its great development in the Sung (+12th century) for warships with multiple paddle-wheels and trebuchet artillery. Though proposed in +4th-century Byzantium, and discussed in Western Europe in the +14th and +15th centuries, no practical use of the principle was made there until the +16th century in Spain.

(iii) The complete absence of the multi-oared galley from Chinese civilisation, whether powered by slaves or free-men (apart from small patrol craft and the paddled dragon-boats used only for ritual races). This must partly be regarded as the result of a relatively advanced development of

(C) sails and rig. Here several points are important.

(i) From at least the +3rd century onwards, ships of the Chinese culture-area were fitted with multiple masts. This may well have been a corollary of A (ii) above, since the bulkheads invited the placing of several tabernacles along the fore-and-aft mid-ship line. Europeans of the +13th century and later were greatly impressed by the large size and many masts of the sea-going junks, and in the +15th they adopted a system of three, which led in due course to the development of the full-rigged ship.

(ii) The Chinese also staggered their masts thwartwise in order to avoid the becalming of one sail by another. This is approved by modern sailing-ship designers, but was not adopted by Europeans during the period of importance of the sailing-ship. Nor did the Chinese practice of radiating the rakes of the masts like the spines of a fan win acceptance in other parts of the world.

(iii) One of the earliest solutions of the problem of sailing to windward in large vessels was due to the Chinese of the +2nd and +3rd centuries, or to their immediate Malayan and Indonesian neighbours at the zone of Sino-Indian culture-contact. This involved the development of fore-and-aft sails. The Chinese lug-sail arose, in all probability, from the Indonesian canted square-sail, and hence indirectly from the square-sail of ancient Egypt; perhaps also, as philological evidence suggests, it had something to do with the 'double-mast sprit-sail' (now known only in Melanesia), which in its turn had developed from the Indian Ocean 'bifid-mast sprit-sail'. The Chinese sprit-sail would have had similar origins. Roman–Indian contacts at the same period (+2nd and +3rd centuries) generated the sprit-sail in Mediterranean waters, but it seems to have fallen out of use there, and was introduced a second time from Asia at the beginning of the +15th century. Meanwhile, in the West, the lateen sail, characteristic of the Arabic culture-area, dominated in the Mediterranean from about the end of the +8th century, and spread to the ocean-going full-rigged ship in the latter part of the +15th. The later lug-sails of Europe derived in all probability from Chinese balance lugs.

(iv) The earliest tightly bent taut aerofoil sails were the mat-and-batten sails developed in China from the Han period onwards. The system involved many ingenious auxiliary techniques, such as multiple sheeting. Such sails were never used in the West during the period of importance of the sailing-ship, but modern research has demonstrated their value, and present-day racing yachts have adopted important elements of Chinese rig, including battens for tautening the sails, and the system of multiple sheets.

(v) As a corollary from A (vi) above, and having regard to C (iii), the use of leeboards and centre-boards developed in China, probably from the sailing-rafts of ancient times, and certainly by the beginning of the Thang (+7th century). They were adopted in Europe a thousand years later, in the late +16th century. Sailors of the Chinese culture-area raised and lowered them by tackle (especially when the large rudder served the purpose), by pivoting, or by sliding in grooves.

In the domain of vessel control, there was the great development of steering-gear which centred on the invention of (D) the rudder. Remembering A (v) above, we note:

(i) The fundamental invention of the axial or median rudder. This device was fully developed in China by the end of the +2nd century (probably the +1st), and the attachment to the transom stern, if not already achieved, followed soon afterwards, certainly by the end of the +4th; but its first appearance in Europe did not occur until the end of the +12th century. Steering-oars and stern-sweeps, however, though thus relegated to a secondary position so early in Chinese nautical technology, were never completely abandoned, partly because of their value in negotiating river rapids, partly for manœuvring when the ship was nearly stationary. Bow-sweeps lived on for the same reasons.

(ii) The invention of the balanced rudder, hydrodynamically more efficient than the unbalanced type. This was current in China at least as early as the +11th century, but was still regarded as a new and important device at the end of the +18th in Europe.

(iii) The further invention of the fenestrated rudder, also hydrodynamically advantageous. This was not adopted in Europe until the era of iron and steel ships.

Among miscellaneous ancillary techniques (E), the following are worthy of remark:

(i) Hull sheathing. Already in the +11th century the superimposition of layers of fresh strakes was usual in China; in Europe it became general in the +16th and later. Sheathing with copper plates was discussed, if not actually performed, in +4th-century China; lead had been used in Hellenistic Europe. The practical use of copper did not become general, both in East and West, until the +18th century.

(ii) Armour plating. The strong predilection of Chinese naval commanders throughout the ages for projectile tactics as opposed to reliance on the close combat of boarding-parties led to the armouring of hulls and upper-works above the water-line with iron plates by the end of the +12th century, and this continued in later times with notable Korean contributions in the +16th. By that time similar developments were occurring, though rather half-heartedly, in Europe.

(iii) The development of non-fouling 'stockless' anchors.

(iv) The invention of articulated or coupled trains of vessels, probably in the +16th century. This was not often employed in Europe for shipping, but greatly used in other fields of transportation.

(v) Ingenious dredging practices.

(vi) Advanced techniques of diving generated by the pearling industry.

(vii) Advanced techniques of bilge clearance by chain-pumps, admired by +16th-century Europeans.

So much for the genius of Chinese shipwrights and sailors.[a] Of the most ancient influences acting upon their ship construction, we have been able to detect certain affinities with the naval architecture of ancient Egypt. These include (a) square-ended hulls, (b) bipod masts, (c) the anti-hogging truss in dragon-boats, (d) stern galleries, (e) some forward-curving sterns, and (f) the practice of facing rowing. If we knew more about the shipping of ancient Mesopotamia we might find that some of these radiated outward from there in both directions,[b] yet on the whole the people of ancient Egypt seem to have been water-farers cleverer and more assiduous than any nations of the Fertile Crescent—save the Phoenicians, and they were much younger. Of Chinese connections with South-east Asian practices, the most important were (a) the interchanges in sails and rig, (b) the multi-paddled dragon-boats, and (c) the poling gangways, which may be connected closely with outriggers.

During the past two thousand years there seems to have been scarcely any century which did not witness the transmission of one or another element of nautical technology from Asia to Europe. It does no harm for us to realise this, children as we are of that occidental archipelago where maritime commerce arose and flourished so exceedingly, and whence the *conquistadores* set forth on their explorations of every strait and ocean. The succession may be sketched as follows:

+2nd century	The sprit-sail (from India to the Roman Mediterranean).
+8th	The lateen sail (from the Arabic culture-area to Byzantium).
Late +12th	The mariner's compass and the axial (stern-post) rudder (either by Arabic–Crusader contact or overland through the West Liao State in Sinkiang).
+7th to +15th	Preconstructed rib frames as opposed to strake-morticed hulls with inserted frames, possibly derived from bulkhead construction.
+15th	Multiple masts (from Chinese junks), the sprit-sail again (perhaps from Sinhalese craft), and the adoption of the lateen, first on all masts, then on the mizen with square-sails on the other masts.
+16th	The protection of the hull by additional strake layers.
Late +16th	Leeboards.

[a] It will be seen that we do not feel able to accept the contention of Poujade (1), p. 296, that little or nothing was ever given to, or borrowed from, the Chinese culture-area in nautics. We are also very uneasy about his principle of conservatism (pp. 170, 175), namely that sailors never change the rig of their ships except under foreign domination, though hulls (p. 176) change more easily. The Chinese sails of the Maldives disprove this. Nor are we much attracted by the converse of this 'law' enunciated by Bowen (2), p. 82, (7), pp. 269, 287, who maintains that sails are very easily influenced by culture-contact while hulls are most resistant to change. The *lorcha* suffices as witness to the contrary. Such generalisations seem premature; we are still constrained to accumulate examples and suspend our judgment.

[b] Coracles would be a good example.

Late +18th	Water-tight compartments, centre-boards, perhaps also copper hull-sheathing.
19th and 20th	Flat bottoms, hulls of approximately rectangular cross-section, balanced rudders, fenestrated rudders, non-fouling stockless anchors, aerodynamically efficient sails, thwartwise staggered masts, multiple sheets, and the placing of the master-couple aft of the midship line.

Some of these developments may of course have been partly independent. Even when we have good reason to believe in a transmission, we know very little of the means by which it took place. But as in all the other fields of science and technology the onus of proof lies upon those who wish to maintain independent invention, and the longer the period elapsing between the successive appearances of a discovery or invention in the two or more cultures concerned, the heavier that onus generally is. The techniques here discussed were, to be sure, often improved by the Europeans when they later adopted them. All that our analysis indicates is that European seamanship probably owes far more than has generally been supposed to the contributions of the sea-going peoples of East and South-east Asia. One would be ill-advised to undervalue the Chinese sea-captain and his crew. Those who know them best today are very willing to apply to them the words of an English sea-poet when he compared past and present on the waters:

> The thranite now and thalamite are pressures low and high,
> And where three hundred blades bit white the twin-propellers ply.
> The God that hailed, the keel that sailed, are changed beyond recall,
> But the robust and Brassbound Man, he is not changed at all!

Perhaps, *mutatis mutandis*, the same could be said of gunners and spagyrical adepts, of miners, smelters and iatrochemists, of the Georgic mages of the countryside and the leeches with their wisdom and wort-cunning? If so, the following volumes as they unfold will bear testimony of it.

BIBLIOGRAPHIES

A CHINESE AND JAPANESE BOOKS BEFORE +1800

B CHINESE AND JAPANESE BOOKS AND JOURNAL ARTICLES SINCE +1800

C BOOKS AND JOURNAL ARTICLES IN WESTERN LANGUAGES

In Bibliographies A and B there are two modifications of the Roman alphabetical sequence: transliterated *Chh-* comes after all other entries under *Ch-*, and transliterated *Hs-* comes after all other entries under *H-*. Thus *Chhen* comes after *Chung* and *Hsi* comes after *Huai*. This system applies only to the first words of the titles. Moreover, where *Chh-* and *Hs-* occur in words used in Bibliography C, i.e. in a Western language context, the normal sequence of the Roman alphabet is observed.

When obsolete or unusual romanisations of Chinese words occur in entries in Bibliography C, they are followed, wherever possible, by the romanisations adopted as standard in the present work. If inserted in the title, these are enclosed in square brackets; if they follow it, in round brackets. When Chinese words or phrases occur romanised according to the Wade–Giles system or related systems, they are assimilated to the system here adopted without indication of any change. Additional notes are added in round brackets. The reference numbers do not necessarily begin with (1), nor are they necessarily consecutive, because only those references required for this volume of the series are given.

Korean and Vietnamese books and papers are included in Bibliographies A and B.

ABBREVIATIONS

See also p. xxxix

AA	Artibus Asiae
AAA	Archaeologia
AAAA	Archaeology
AAL/RSM	Atti d.r. Accad. dei Lincei (Rendiconti, Ser. Morali)
AAN	American Anthropologist
AANTH	Archiv. f. Anthropologie
AAPSS	Annals of the American Academy of Political and Social Sciences
AART	Archives of Asian Art
AAS	Arts Asiatiques (continuation of Revue des Arts Asiatiques)
ABRN	Abr-Nahrain (Annual of Semitic Studies, Universities of Melbourne and Sydney)
ABSA	Annual of the British School at Athens
ACASA	Archives of the Chinese Art Soc. of America
ACLS	American Council of Learned Societies
AD	Architectural Design
ADAB	Aden Dept. of Antiquities Bulletin
ADVS	Advancement of Science
AEST	Annales de l'Est (Fac. des Lettres, Univ. Nancy)
AFFL	American Forests and Forest Life
AFLB	Annales de la Faculté de Lettres de Bordeaux
AFS	Africa South
AGNT	Archiv f. d. Gesch. d. Naturwiss. u. d. Technik (cont. as AGMNT)
AH	Asian Horizon
AHAW/PH	Abhandlungen d. Heidelberger Akad. Wiss. (Phil.-Hist. Klasse)
AHES	Annales d'Hist. Econ. et Sociale
AHES/AESC	Annales; Economies, sociétés, civilisations
AHES/AHS	Annales d'Hist. Sociale
AHES/MHS	Mélanges d'Hist. Sociale
AHOR	Antiquarian Horology
AHR	American Historical Review
AI/AO	Ars Orientalis (formerly Ars Islamica)
AIRS	Acta Inst. Rom. Regni Sueciae
AJA	American Journ. Archaeology
AJH	American Journ. Hygiene
AJP	American Journ. Philology
AJSC	American Journ. Science and Arts (Silliman's)
AJTM	American Journ. Tropical Medicine
AKML	Abhandlungen f. d. Kunde des Morgenlandes
AM	Asia Major
AMA	American Antiquity
AMBG	Annals of the Missouri Botanic Garden
AMNH/AP	Anthropological Papers of the American Museum of Natural History (New York)
AMSC	American Scientist
AMSR	American Sociol. Review
AN	Anthropos
ANEPT	American Neptune
ANI	Ancient India
ANP	Annalen d. Physik
ANTJ	Antiquaries Journal
AOAW/PH	Anzeiger d. Österr. Akad. Wiss. (Wien), (Phil.-Hist. Klasse)
APAW/PH	Abhandlungen d. preuss. Akad. Wiss. Berlin (Phil.-Hist. Klasse)
AQ	Antiquity
AQSU	Antiquity and Survival (Internat. Rev. of Trad. Art and Culture)
AREV	Architectural Review
ARK	Arkitektur (Stockholm)
ARLC/DO	Annual Reports of the Librarian of Congress (Division of Orientalia)
ARO	Archiv Orientalní (Prague)
ARSI	Annual Reports of the Smithsonian Institution
ARUSNM	Annual Reports of the U.S. National Museum
AS/BIE	Bulletin of the Institute of Ethnology, Academia Sinica (Thaiwan)
ASEA	Asiatische Studien; Études Asiatiques
ASIA	Asia
ASIC	Arts and Sciences in China (London)
ASPN	Archives des Sciences Physiques et Naturelles (Geneva)
ASR	Asiatic Review
ASRAB	Annales de la Soc. (Roy.) d'Archéol. (Brussels)
ASURG	Annals of Surgery
AT	Atlantis
AX	Ambix
BAFAO	Bulletin de l'Association Française des Amis de l'Orient
BAMM	Bulletin des Amis du Musée de la Marine (Paris)
BAVH	Bulletin des Amis du Vieux Hué (Indo-China)
BBMMAG	Bull. Belfast Municipal Museum and Art Gallery
BBSHS	Bulletin of the British Society for the History of Science

BCGS	Bulletin of the Chinese Geological Society	CHI	Cambridge History of India
BCHQ	British Columbia Historical Quarterly	CHJ/T	Chhing-Hua (T'sing-Hua) Journal of Chinese Studies (New Series, publ. Thaiwan)
BCIC	Bollettino Civico Instituto Colombiano (Genoa)	CINA	Cina (Ist. Ital. per il Medio ed Estremo Oriente, Rome)
BE/AMG	Bibliographie d'Études (Annales du Musée Guimet)	CJ	China Journal of Science and Arts
		CLR	Classical Review
BEFEO	Bulletin de l'École Française de l'Extrême Orient (Hanoi)	CMB	Canterbury Museum Bulletin (New Zealand)
BEP	Bulletin des Études Portugaises	CMIS	Chinese Miscellany
BGHD	Bulletin de Géographie Histor. et Descr.	COMP	Comprendre (Soc. Eu. de Culture, Venice)
BGTI	Beiträge z. Gesch. d. Technik u. Industrie (continued as Technik Geschichte—see BGTI/TG)	CQ	Classical Quarterly
		CR	China Review (Hongkong and Shanghai)
BGTI/TG	Technik Geschichte	CR/BUAC	China Review (British United Aid to China)
BH	Bulletin Hispanique		
BIBLOS	Biblos (Coimbra)	CR/MSU	Centennial Review of Arts and Science (Michigan State University)
BIHM	Bulletin of the (Johns Hopkins) Institute of the History of Medicine (cont. as Bulletin of the History of Medicine)	CRAS	Comptes Rendus hebdomadaires de l'Acad. des Sciences (Paris)
BIIEH	Bulletin de l'Inst. Indochinois pour l'Étude de l'Homme	CREC	China Reconstructs
		CTE	China Trade and Engineering
BIRSN	Bulletin de l'Institut Royal des Sciences Naturelles de Belgique	CUOIP	Chicago Univ. Oriental Institute Pubs.
BJPC	British Journal of Psychology	CUP	Cambridge University Press
BJSSF	Bulletin of the Japanese Society of Scientific Fisheries	CURRA	Current Anthropology
BLSOAS	Bulletin of the London School of Oriental and African Studies	D	Discovery
		DHT	Documents pour l'Histoire des Techniques (Paris)
BM	Bibliotheca Mathematica		
BMFEA	Bulletin of the Museum of Far Eastern Antiquities (Stockholm)	DSS	Der Schweizer Soldat
		DVN	Dan Viet Nam
BMFJ	Bulletin de la Maison Franco-Japonaise (Tokyo)		
BMQ	British Museum Quarterly	EA	Eastern Art (Philadelphia)
BN	Bulletyn Nautologyczny (Gdynia)	EAM	East of Asia Magazine
BNYAM	Bulletin of the New York Academy of Medicine	EB	Encyclopaedia Britannica
		EHOR	Eastern Horizon (Hongkong)
BQR	Bodleian (Library) Quarterly Record (Oxford)	EHR	Economic History Review
		EM	Ecological [Oecological] Monographs
BSG	Bulletin de la Société de Géographie (continued as La Géographie)	EN	Engineer
		END	Endeavour
BSKIY	British Ski Yearbook	EPJ	Edinburgh Philosophical Journal (continued as ENPJ)
BTG	Blätter f. Technikgeschichte (Vienna)		
		ESC	Engineering Society of China, Papers
BUA	Bulletin de l'Université de l'Aurore (Shanghai)	ETH	Ethnos
BUM	Burlington Magazine	EZ	Epigraphica Zeylanica
CAMR	Cambridge Review	FEQ	Far Eastern Quarterly (continued as Journal of Asian Studies)
CAS/PC	Cambridge Antiquarian Society, Proceedings and Communications	FLF	Folk Life
CE	Civil Engineering (U.S.A.)	FLS	Folklore Studies (Peiping)
CENAR	Central Asian Review (London)	FLV	Folk-Liv
CET	Ciel et Terre	FMNHP/AS	Field Museum of Natural History (Chicago) Publications; Anthropological Series
CEYHJ	Ceylon Historical Journal		
CEYJHS	Ceylon Journ. Histor. and Social Studies	FOODR	Food Research
CFC	Cahiers Franco-Chinois (Paris)	FOODT	Food Technology

G	Geography	JICE	Journ. Instit. Civil Engineers (U.K.) (continued from PICE)
GAL	Gallia		
GB	Globus	JIN	Journal of the Institute of Navigation (U.K.)
GE	The Guilds Engineer (London)		
GGM	Geographical Magazine	JJIE	Journ. Junior Instit. Engineers (U.K.)
GJ	Geographical Journal		
GM	Geological Magazine	JMEOS	Journ. Manchester Egyptian and Oriental Soc.
GR	Geographical Review		
GRSCI	Graphic Science	JMGG	Jahresbericht d. Münchener Geogr. Gesellsch.
GZ	Geographische Zeitschrift		
		JOSHK	Journal of Oriental Studies (Hongkong Univ.)
HBML	Harvard (University) Botanical Museum Leaflets		
		JRAI	Journal of the Royal Anthropological Institute
HCHTC	Hsin Chung-Hua Tsa Chih		
HH	Han Hiue (Han Hsüeh); Bulletin du Centre d'Études Sinologiques de Pékin	JRAS	Journal of the Royal Asiatic Society
		JRAS/B	Journal of the (Royal) Asiatic Society of Bengal
HJAS	Harvard Journal of Asiatic Studies	JRAS/KB	Journal (or Transactions) of the Korea Branch of the Royal Asiatic Society
HMSO	Her Majesty's Stationery Office		
HP	Hespéris (Archives Berbères et Bulletin de l'Institut des Hautes Études Marocaines)	JRAS/M	Journal of the Malayan Branch of the Royal Asiatic Society
		JRAS/NCB	Journal (or Transactions) of the North China Branch of the Royal Asiatic Society
HZ	Horizon (New York)		
IAQ	Indian Antiquary	JRCAS	Journal of the Royal Central Asian Society
ICE/MP	Institution of Civil Engineers; Minutes of Proceedings		
		JRGS	Journal of the Royal Geographical Society (London)
ILN	Illustrated London News		
IM	Imago Mundi; Yearbook of Early Cartography	JRIBA	Journ. Royal Institute of British Architects
IQ	Islamic Quarterly	JRS	Journal of Roman Studies
ISIS	Isis	JRSA	Journal of the Royal Society of Arts
ISRM	Indian State Railways Magazine	JSA	Journal de la Société des Americanistes
JA	Journal Asiatique		
JAFRS	Journ. African Society	JSPC	Journ. Social Psychol.
JAH	Journ. African History	JWCBRS	Journal of the West China Border Research Society
JAHIST	Journ. Asian History		
JAN	Janus	JWH	Journal of World History (UNESCO)
JAOS	Journal of the American Oriental Society		
		K	Keystone (Association of Building Technicians Journal)
JAS	Journal of Asian Studies (continuation of Far Eastern Quarterly, FEQ)		
		KBGJ	Jaarboek v. d. Koninklijke Bataviaasch Genootschap van Kunsten en Wetenschappen
JASA	Journ. Assoc. Siamese Architects		
JBASA	Journal of the British Astronomical Association	KDVS/AKM	Kgl. Danske Videnskabernes Selskab (Archaeol.-Kunsthist. Medd.)
JCUS	Journ. Cuneiform Studies		
JDAI/AA	Jahrb. d. deutsch. Archäologische Institut (Archäologische Anzeiger)	KDVS/HFM	Kgl. Danske Videnskabernes Selskab (Hist.-Filol. Medd.)
JEA	Journal of Egyptian Archaeology	KU/ARB	Asiatic Research Bulletin (Asiatic Research Centre, Korea Univ., Seoul)
JEGP	Journal of English and Germanic Philology		
JEH	Journal of Economic History		
JF	Journ. Forestry (U.S.A.)	LEC	Lettres Édifiantes et Curieuses écrites des Missions Étrangères (Paris, 1702 to 1776)
JGE	Journ. Gen. Education		
JGIS	Journal of the Greater India Society		
JGSC	Journal of the Geogr. Soc. China	LI	Listener (B.B.C.)
JHI	Journal of the History of Ideas	LM	Larousse Mensuel
JHMAS	Journal of the History of Medicine and Allied Sciences	LN	La Nature
		LP	La Pensée

MA	*Man*
MAAA	*Memoirs American Anthropological Association*
MAI/LTR	*Mémoires de Litt. tirés des Registres de l'Acad. des Inscr. et Belles-Lettres* (Paris)
MAPS	*Memoirs of the American Philosophical Society*
MAS/B	*Memoirs of the Asiatic Society of Bengal*
MC/TC	*Techniques et Civilisations* (formerly *Métaux et Civilisations*)
MCB	*Mélanges Chinois et Bouddhiques*
MCMG	*Mechanics Magazine*
MD	*MD (Doctor of Medicine), cultural Journal for physicians* (New York)
MDAI/ATH	*Mitteilungen d. deutschen Archäol. Instituts* (Athenische Abt.)
MDGNVO	*Mitteilungen d. deutsch. Gesellsch. f. Natur. u. Volkskunde Ostasiens*
MEJ	*Middle East Journal*
MFSKU/E	*Memoirs of the Faculty of Science, Kyushu University, Ser. E (Biology)*
MGGMU	*Mitteilungen d. geographische Gesellschaft München*
MGGW	*Mitteilungen d. geographische Gesellschaft Wien*
MGSC	*Memoirs of the Chinese Geological Survey*
MIE	*Mémoires de l'Institut d'Egypte* (Cairo)
MIFAN	*Mémoires de l'Institut Français d'Afrique Noire* (Dakar)
MIT	Massachusetts Institute of Technology
MJBK	*Münchner Jahrb. f. bildenden Kunst*
MJLS	*Madras Journ. of Lit. and Sci.*
MJPGA	*Mitteilungen aus Justus Perthes Geogr. Anstalt* (Petermann's)
MK	*Meereskunde* (Berlin)
MMI	*Mariner's Mirror*
MQ	*Modern Quarterly*
MRDTB	*Memoirs of the Research Dept. of Tōyō Bunko* (Tokyo)
MRMVK	*Mededelingen van het Rijksmuseum Voor Volkenkunde*
MS	*Monumenta Serica*
MS/M	*Monumenta Serica Monogr.*
MSAA	*Mem. Soc. American Archaeology* (supplements to *AMA*)
MSAF	*Mémoires de la Société (Nat.) des Antiquaires de France*
MSB	*Morskoe Sbornik*
MSOS	*Mitteilungen d. Seminar f. orientalischen Sprachen* (Berlin)
MSRGE	*Mém. Soc. Roy. Geogr. d'Égypte*
MUJ	*Museum Journal* (Philadelphia)
N	*Nature*
NADA	*Annual of the Native Affairs Department, Southern Rhodesia*
NAVC	*Naval Chronicle*
NAVSG	*Nouvelles Annales des Voyages et des Sciences Géographiques*
NC	*Numismatic Chronicle (and Journ. Roy. Numismatic Soc.)*
NCR	*New China Review*
NEPT	*Neptunia*
NGM	*National Geographic Magazine*
NH	*Natural History*
NION	*Nederlandsch Indië Oud en Nieuw*
NMM	*National Maritime Museum* (Greenwich)
NO	*New Orient* (Prague)
NQCJ	*Notes and Queries on China and Japan*
NS	*New Scientist*
NSEQ	*Nankai [University] Social and Economic Quarterly* (Tientsin)
NSN	*New Statesman and Nation* (London)
NU	*The Nucleus*
NV	*The Navy*
NVO	*Novy Orient* (Prague)
NZMW	*Neue Zeitschrift f. Missionswissenschaft (Nouvelle Revue de Science Missionnaire)*
OAV	*Orientalistisches Archiv* (Leipzig)
OAZ	*Ostasiatische Zeitschrift*
OE	*Oriens Extremus* (Hamburg)
OL	*Old Lore; Miscellany of Orkney, Shetland, Caithness and Sutherland*
OLL	*Ostasiatischer Lloyd*
ORA	*Oriental Art*
ORE	*Oriens Extremus*
OSIS	*Osiris*
OUP	Oxford University Press
PA	*Pacific Affairs*
PAAQS	*Proceedings of the American Antiquarian Society*
PAI	*Paideuma*
PARA	*Parasitology*
PASCE/JHD	*Proc. American Soc. Civil Engineers; Journ. Hydraulics Division*
PASCE/JID	*Proc. American Soc. Civil Engineers; Journ. Irrigation and Drainage Division*
PBA	*Proceedings of the British Academy*
PC	*People's China*
PCAS	*Proc. California Academy of Sciences*
PCC	*Proceedings of the Charaka Club*
PCPS	*Proc. Cambridge Philological Society*
PEFQ	*Palestine Exploration Fund Quarterly*

PGA	*Proc. Geologists' Assoc. (U.K.)*	*RPARA*	*Rendiconti della Pontif. Accad. Rom. di Archeologia*
PHY	*Physis* (Florence)		
PICE	*Proc. Instit. Civil Engineers (U.K.)* (absorbed in *JICE*)	*RPLHA*	*Revue de Philol., Litt. et Hist. Ancienne*
PKR	*Peking Review*	*RSO*	*Rivista di Studi Orientali*
PLS	*Proc. Linnean Soc.* (London)	*RTDA*	*Report and Trans. of the Devon-*
PMASAL	*Papers of the Michigan Academy of Sci., Arts and Letters*		*shire Association for the Advancement of Science, Literature and*
PNHB	*Peking Natural History Bulletin*		*Art*
PP	*Past and Present*	*RUNA*	*Runa (Archivo para las Ciencias*
PR	*Princeton Review*		*del Hombre),* Buenos Aires
PRGS	*Proceedings of the Royal Geographical Society*	*S*	*Sinologica* (Basel)
PROG	*Progress (Unilever Journal)*	*SA*	*Sinica* (originally *Chinesische*
PRPSG	*Proceedings of the Royal Philosophical Society of Glasgow*		*Blätter f. Wissenschaft u. Kunst)*
		SAAB	*South African Archaeological Bulletin*
PRSA	*Proceedings of the Royal Society* (Series A)	*SACAJ*	*Journal of the Sino-Austrian Cultural Association*
PRSB	*Proceedings of the Royal Society* (Series B)	*SAE*	*Saeculum*
PRSG	*Publicaciones de la Real Sociedad Geográfica* (Spain)	*SAFJS*	*South African Journal of Science*
		SAM	*Scientific American*
PTRS	*Philosophical Transactions of the Royal Society*	*SBE*	*Sacred Books of the East* series
		SC	*Science*
		SCI	*Scientia*
QJCA	*Quarterly Journal of Current Acquisitions* (Library of Congress, Washington)	*SCISA*	*Scientia Sinica* (Peking)
		SGM	*Scottish Geographical Magazine*
		SIS	*Sino-Indian Studies* (Santiniketan)
QSGNM	*Quellen u. Studien z. Gesch. d. Naturwiss. u. d. Medizin* (continuation of *Archiv. f. Gesch. d. Math., d. Naturwiss. u. d. Technik, AGMNT,* formerly *Archiv. f. d. Gesch. d. Naturwiss. u. d. Technik, AGNT)*	*SM*	*Scientific Monthly* (formerly *Popular Science Monthly)*
		SMC	*Smithsonian (Institution) Miscellaneous Collections (Quarterly Issue)*
		SMJ	*Sarawak Museum Journal*
		SP	*Speculum*
RA	*Revue Archéologique*	*SPAW*	*Sitzungsberichte d. preuss. Akad. d. Wissenschaft*
RAA/AMG	*Revue des Arts Asiatiques (Annales du Musée Guimet)*		
		SPR	*Science Progress*
RAI/OP	*Occasional Papers of the Royal Anthropological Institute*	*SSE*	*Studia Serica (West China Union University Literary and Historical Journal)*
RBS	*Revue Bibliographique de Sinologie*		
RDI	*Rivista d'Ingegneria*	*STE*	*Studia Etruschi*
REL	*Revue des Études Latines*	*STU*	*Studia* (Lisbon)
RFCC	*Revista da Faculdade de Ciências, Universidade de Coimbra*	*SUM*	*Sumer*
		SWAW/PH	*Sitzungsberichte d. k. Akad. d. Wissenschaften Wien* (Vienna) (Phil.-Hist. Klasse)
RGHE	*Revue de Géographie Humaine et d'Ethnologie*		
RGI	*Rivista Geografica Italiana*	*SWJA*	*Southwestern Journal of Anthropology*
RHES	*Revue d'Histoire Écon. et Soc.* (continuation of *Revue d'Histoire des Doctrines Écon. et Soc.)*		
		SYR	*Syria*
		SZ	*Spolia Zeylanica*
RHS	*Revue d'Histoire des Sciences* (Paris)	*TAP*	*Annals of Philosophy* (Thomson's)
		TAPA	*Transactions (and Proceedings) of the American Philological Association*
RHSID	*Revue d'Histoire de la Sidérurgie* (Nancy)		
RI	*Revue Indochinoise*	*TAPS*	*Transactions of the American Philosophical Society* (cf. *MAPS)*
RIIA	*Royal Institute of International Affairs*		
RMA	*Revue Maritime*	*TAS/J*	*Transactions of the Asiatic Society of Japan*
RP	*Revue Philosophique*		

TASCE	*Transactions of the American Society of Civil Engineers*	UNESCO	United Nations Educational, Scientific and Cultural Organisation
TCULT	*Technology and Culture*		
TEAC	*Transactions of the Engineering Association of Ceylon*	*VA*	*Vistas in Astronomy*
TG/K	*Tōhō Gakuhō, Kyōto (Kyoto Journal of Oriental Studies)*	*VAG*	*Vierteljahrsschrift d. astronomischen Gesellschaft*
TGUOS	*Transactions of the Glasgow University Oriental Society*	*VGEB*	*Verhandl. d. Gesellsch. f. Erdkunde (Berlin)*
TH	*Thien Hsia Monthly (Shanghai)*	*VKAWA/L*	*Verhandelingen d. Koninklijke Akad. v. Wetenschappen te Amsterdam (Afd. Letterkunde)*
TIAU	*Transactions of the International Astronomical Union*		
TIMEN	*Transactions of the Institute of Marine Engineers*	*VMAWA*	*Verslagen en Meded. d. Koninklijke Akad. v. Wetenschappen te Amsterdam*
TINA	*Transactions of the Institution of Naval Architects*	*VS*	*Variétés Sinologiques*
TJSL	*Transactions (and Proceedings) of the Japan Society of London*	*WBKGA*	*Wiener Beiträge z- Kunst- und Kultur-Gesch. Asiens*
TMIE	*Travaux et Mémoires de l'Inst. d'Ethnologie (Paris)*	*WBKGL*	*Wiener Beiträge z. Kulturgeschichte und Linguistik*
TNR	*Tanganyika Notes and Records*		
TNS	*Transactions of the Newcomen Society*	*WJK*	*Wiener Jahrb f. Kunstgesch.*
		WP	*Water Power*
TNZI	*Transact. New Zealand Inst.*		
TOCS	*Transactions of the Oriental Ceramic Society*	*Y*	*Yachting*
		YJBM	*Yale Journal of Biology and Medicine*
TP	*T'oung Pao (Archives concernant l'Histoire, les Langues, la Géographie, l'Ethnographie et les Arts de l'Asie Orientale, Leiden)*	*YM*	*Yachting Monthly*
		YW	*Yachting World*
TRIBA	*Transactions, Royal Institute of British Architects*	*Z*	*Zalmoxis; Revue des Études Religieuses*
TSE	*Trans. Society of Engineers (London)*	*ZBW*	*Zeitschr. f. Bauwesen*
		ZFE	*Zeitschr. f. Ethnol.*
TSNAMEN	*Trans. Society of Naval Architects and Marine Engineers*	*ZGEB*	*Zeitschr. d. Gesellsch. f. Erdkunde (Berlin)*
TYG	*Tōyō Gakuhō (Reports of the Oriental Society of Tokyo)*	*ZHWK*	*Zeitschrift f. historische Wappenkunde (continued as Zeitschr. f. hist. Wappen- und Kostumkunde)*
UAJ	*Ural-Altaische Jahrbücher*		
UC	*Ulster Commentary (Govt. Information Service, Belfast)*	*ZPC*	*Zeitschr. f. physiologischen Chemie*
		ZWZ	*Zeitschr. f. wissenschaftlichen Zoologie*
UIB	*University of Illinois Bulletin*		

A. CHINESE BOOKS BEFORE +1800

Each entry gives particulars in the following order:

(a) title, alphabetically arranged, with characters;

(b) alternative title, if any;

(c) translation of title;

(d) cross-reference to closely related book, if any;

(e) dynasty;

(f) date as accurate as possible;

(g) name of author or editor, with characters;

(h) title of other book, if the text of the work now exists only incorporated therein; or, in special cases, references to sinological studies of it;

(i) references to translations, if any, given by the name of the translator in Bibliography C;

(j) notice of any index or concordance to the book if such a work exists;

(k) reference to the number of the book in the *Tao Tsang* catalogue of Wieger (6), if applicable;

(l) reference to the number of the book in the *San Tsang* (Tripitaka) catalogues of Nanjio (1) and Takakusu & Watanabe, if applicable.

Words which assist in the translation of titles are added in round brackets.

Alternative titles or explanatory additions to the titles are added in square brackets.

It will be remembered (p. 700 above) that in Chinese indexes words beginning *Chh-* are all listed together after *Ch-*, and *Hs-* after *H-*, but that this applies to initial words of titles only.

Where there are any differences between the entries in these bibliographies and those in Vol. 1–3, the information here given is to be taken as more correct.

References to the editions used in the present work, and to the *tshung-shu* collections in which books are available, will be found at the end of this volume.

ABBREVIATIONS

C/Han	Former Han.
E/Wci	Eastern Wei.
H/Chhin	Later Chhin.
H/Han	Later Han.
H/Shu	Later Shu (Wu Tai).
H/Thang	Later Thang (Wu Tai).
J/Chin	Jurchen Chin.
L/Sung	Liu Sung.
N/Chou	Northern Chou.
N/Chhi	Northern Chhi.
N/Sung	Northern Sung (before the removal of the capital to Hangchow).
N/Wei	Northern Wei.
S/Chhi	Southern Chhi.
S/Sung	Southern Sung (after the removal of the capital to Hangchow).
S/Thang	Southern Thang.
W/Wei	Western Wei.

Chan Kuo Tshê 戰國策.
    Records of the Warring States.
    Chhin.
    Writer unknown.

Chao-Hua Hsien Chih 昭化縣志.
    Gazetteer of Chao-hua (in Szechuan).
    Chhing.
    Chang Shao-ling (ed.) 張紹齡.
    Revised 1845, 1864.

Chen-Chou Shui Cha Chi 眞州水閘記.
    See *Shui Cha Chi*.

Chen-La Fêng Thu Chi 眞臘風土記.
    Description of Cambodia.
    Yuan, +1297.
    Chou Ta-Kuan 周達觀.

Chen Wei Pien 鍼位編.
    Compass-Bearing Sailing Directions.
    Ming, late +15th or early +16th century.
    Writer unknown.
    Uncertain whether extant; see Vol. 3, p. 559.

Chêng Lei Pên Tshao 證類本草.
    [*Chhung Hsiu Chêng-Ho Ching Shih Chêng Lei Pei Yung Pên Tshao*.]
    Reorganised Pharmacopoeia.
    N/Sung, +1108, enlarged +1116; re-edited in J/Chin, +1204, and definitively re-published in Yuan, +1249; re-printed many times afterwards, e.g. in Ming, +1468.
    Original compiler: Thang Shen-Wei 唐愼微.
    Cf. Hummel (13); Lung Po-Chien (1).

Chi Hsiao Hsin Shu 紀效新書.
    A New Treatise on Military and Naval Efficiency.
    Ming, c. +1575.
    Chhi Chi-Kuang 戚繼光.

Chi Shan Chi 霽山集.
    Poetical Remains of the Old Gentleman of Chi Mountain.
    Sung, end +13th century.
    Lin Ching-Hsi 林景熙.

Chi Yang Yao Pên Mo 記楊么本末.
    The History of (the Rebellion of) Yang Yao from Beginning to End.
    Sung, c. +1140.
    Li Kuei-Nien (b) 李龜年.
    Preserved only in fragments.

Chi Yün 集韻.
    Complete Dictionary of the Sounds of Characters [cf. *Chhieh Yün* and *Kuang Yün*].
    Sung, +1037.
    Compiled by Ting Tu 丁度 *et al.*
    Possibly completed in +1067 by Ssuma Kuang 司馬光.

Chiang Fu 江賦.
    The River Ode.
    Chin, c. +310.
    Kuo Pho 郭璞.

Chiang Pei Yün Chhêng 江北運程.
    See *Tung Hsün* (1).

Chiang Piao Chuan 江表傳.
    The Story of Chiang Piao.

Chiang Piao Chuan (*cont.*)
Thang or pre-Thang.
Yü Phu 虞溥.
Chiang Tzu Wan Chi Lun 蔣子萬機論.
See *Wan Chi Lun*.
Chiao-Chou Chi 交州記.
Record of Chiaochow (District).
Chin.
Liu Hsin-Chhi 劉欣期.
Chieh An Lao Jen Man Pi 戒菴老人漫筆.
An Abundance of Jottings by Old Mr (Li) Chieh-An.
Ming, *c.* +1585, pr. +1606.
Li Hsü 李詡.
Chieh Tzu Yuan Hua Chuan 芥子園畫傳.
The Mustard-Seed Garden Guide to Painting.
Chhing, 1679, with later continuations.
Li Li-Ong 李笠翁 (preface) & Wang Kai 王概 (text and ills.).
Chien Chieh Lu 鑒誡錄.
Cautionary Stories [of events of the Thang and Wu Tai periods].
Sung, +10th century.
Ho Kuang-Yuan 何光遠.
Chien Phing I Shuo 簡平儀說.
Description of a Simple Planisphere.
Ming, +1611.
Hsiung San-Pa (Sabbatino de Ursis) 熊三拔.
Chih-Chêng Ho Fang Chi 至正河防記.
Memoir on the Repair of the Yellow River Dykes in the Chih-Chêng reign-period [+1341 to +1367].
Yuan, *c.* +1350.
Ouyang Hsüan 歐陽玄.
Chih Ho Fang Lüeh 治河防略.
Methods of River Control.
Chhing, +1689, pr. +1767.
Chin Fu 靳輔.
Chih Ho Shu 治河書.
A Treatise on River Control.
Chhing, *c.* +1690.
Chhêng Chao-Piao 程兆彪.
Chih Ho Thu Lüeh 治河圖略.
Illustrated Account of Yellow River Floods and Measures against them.
Yuan or Ming, +14th century.
Wang Hsi 王喜.
Chih Ho Tshê 治河策.
The Planning of River Control.
Sung.
Li Wei 李渭.
Chih Nan Chêng Fa 指南正法
(incl. *Kuan Hsing Fa* 觀星法).
General Compass-Bearings (and Star-Sights) Rutter and Sailing Directions.
Ming or Chhing (before +1675).
Compiler unknown.
Attached as the 7th *pên* to the *Ping Chhien* of Lü Fan & Lu Chhêng-Ên, q.v.

Bodl. Libr. Or. MS., Backhouse 578. Text in Hsiang Ta (5).
Chih Yen 至言.
Words to the Point [a hortatory essay addressed to Han Wên Ti].
C/Han, *c.* −178.
Chia Shan 賈山.
Reconstructed in *YHSF*.
Chin Chhuan So Chi 金川瑣記.
Fragmentary Notes on the Chin-chhuan Valley (on the Szechuan–Sikang border).
Chhing.
Li Hsin-Hêng 李心衡.
Chin Chung Hsing Shu 晉中興書.
The Renewal of the Chin Dynasty.
Chin.
Hsi Shao 郗紹 or Ho Fa-Shêng 何法盛.
Chin Hou Lüeh 晉後略.
Brief Records set down after the (Western) Chin (dynasty).
Chin, after +317.
Hsün Cho 荀綽.
Chin-Ling Ku Chin Thu Khao 金陵古今圖考.
Illustrated Study of the Historical Topography of Nanking.
Ming.
Chhen I 陳沂.
Chin Lou Tzu 金樓子.
Book of the Golden Hall Master.
Liang, *c.* +550.
Hsiao I 蕭繹.
(Liang Yuan Ti 梁元帝.)
Chin Shih 金史.
History of the Chin (Jurchen) Dynasty [+1115 to +1234].
Yuan, *c.* +1345.
Tho-Tho (Toktaga) 脫脫 & Ouyang Hsüan 歐陽玄.
Yin-Tê Index, no. 35.
Chin Shih So
See Fêng Yün-Phêng & Fêng Yün-Yuan (1).
Chin Shih Tshui Pien 金石萃編.
See Wang Chhang (1).
Chin Shu 晉書.
History of the Chin Dynasty [+265 to +419].
Thang, +635.
Fang Hsüan-Ling 房玄齡.
A few chs. tr. Pfizmaier (54–57); the astronomical chs. tr. Ho Ping-Yü (1). For translations of passages see the index of Frankel (1).
Ching Chhuan Pai Pien 荊川稗編.
See *Pai Pien*.
Ching Fu Tien Fu 景福殿賦.
Ode on the Ching-Fu Palace (at Hsü-chhang).
San Kuo (Wei), *c.* +240.

Ching Fu Tien Fu (*cont.*)
Ho Yen 何晏.
Tr. von Zaoh (6).

Ching Hsing Chi 經行記.
Record of My Travels.
Thang, *c.* +763.
Tu Huan 杜環.
Now preserved only in fragments in *Thung Tien*, ch. 193; *Thai-Phing Huan Yü Chi*, ch. 186; *Hsin Thang Shu*, ch. 221B; *Wên Hsien Thung Khao*, ch. 339, etc.

Chiu Chang Suan Shu 九章算術.
Nine Chapters on the Mathematical Art.
H/Han, +1st century (containing much material from C/Han and perhaps Chhin).
Writer unknown.

Chiu Chang Suan-Shu Yin I 九章算術音義.
Explanations of Meanings and Sounds of Words occurring in the *Nine Chapters on the Mathematical Art*.
Sung.
Li Chi 李藉.

Chiu Thang Shu 舊唐書.
Old History of the Thang Dynasty [+618 to +906].
Wu Tai, +945.
Liu Hsü 劉昫.
Cf. des Rotours (2), p. 64.
For translations of passages see the index of Frankel (1).

Cho Chêng Yuan Thu 拙政園圖.
Pictures [and Description] of the Garden of an Unsuccessful Official (the Cho Chêng Yuan at Suchow founded by Wang Huai-Yü 王槐雨).
Ming, +1533.
Wên Chêng-Ming 文徵明.

Chou Hsien Thi Kang 州縣提綱.
Complete Description of City and County (Government).
Sung, late +12th century.
Chhen Hsiang 陳襄.

Chou Li 周禮.
Record of the Institutions (lit. Rites) of (the) Chou (Dynasty) [descriptions of all government official posts and their duties].
C/Han, perhaps containing some material from late Chou.
Compilers unknown.
Tr. E. Biot (1).

Chu Fan Chih 諸蕃志.
Records of Foreign Peoples.
Sung, *c.* +1225. (This is Pelliot's dating; Hirth & Rockhill favoured between +1242 and +1258.)
Chao Ju-Kua 趙汝适.
Tr. Hirth & Rockhill (1).

Chu I Hsiang Pien Yen 諸儀象弁言.
Introductory Notes to Pictures of Scientific Instruments.
Chhing, +1674.

Nan Huai-Jen (Ferdinand Verbiest) 南懷仁.
Cf. *Hsin Chih Ling-Thai I Hsiung Chih*.

Chu Phu 竹譜.
A Treatise on Bamboos (and their Economic Uses; in verse and prose) [probably the first monograph on a specific class of plants].
L/Sung, *c.* +460.
Tai Khai-Chih 戴凱之.
Tr. Hagerty (2).

Chu Shu Chi Nien 竹書紀年.
The Bamboo Books [annals, fragments of a chronicle of the State of Wei, from high antiquity to −298].
Chou, −295 and before, such parts as are genuine. (Found in the tomb of An Li Wang, a prince of the Wei State *r.* −276 to −245; in +281.)
Writers unknown.
See van der Loon (1).
Tr. E. Biot (3).
Reconstruction of the genuine parts by Chu Yu-Tsêng and Wang Kuo-Wei; see Fan Hsiang-Yung (1).

Chüan Shih Ko Wên Chia I Chi 卷施閣文甲乙集.
First and Second Literary Collections of the Chüan-Shih Studio.
Chhing, prior to +1800 but not printed till *c.* +1889.
Hung Liang-Chi 洪亮吉.

Chuang Tzu 莊子.
[= *Nan Hua Chen Ching*.]
The Book of Master Chuang.
Chou, *c.* −290.
Chuang Chou 莊周.
Tr. Legge (5); Fêng Yu-Lan (5); Lin Yü-Thang (1).
Yin-Tê Index no. (suppl.) 20.

Chuang Tzu Pu Chêng
The Text of *Chuang Tzu*, Annotated and Corrected.
See Liu Wên-Tien (1).

Chung Chih Shu 種植書.
The Book of Crop-Raising.
See *Fan Shêng-Chih Shu*.

Chung-Hsing Hsiao Chi 中興小紀.
Brief Records of Chung-hsing (mod. Chiang-ling on the Yangtze in Hupei).
Sung, *c.* +1150.
Hsiung Kho 熊克.

Chung Shan Chhuan Hsin Lu 中山傳信錄.
Travel Diary of an Embassy to the Liu-Chhiu Islands.
Chhing, +1721.
Hsu Pao-Kuang 徐葆光.

Chung Thien-Chu Shê-Wei Kuo Chih-Yuan Ssu Thu Ching 中天竺舍衛國祇洹寺圖經.
Illustrated Description of the Jetavana

Chung Thien-Chu Shê-Wei Kuo Chih-Yuan Ssu Thu Ching (cont.)
　　Monastery in Śrāvastī in Central India.
　　Thang, +667.
　　Tao-Hsüan　道宣.
　　TW/1899.

Chung Wu Chi Wên　中吳記聞.
　　Record of Things heard in Chiangsu.
　　Sung, c. +1200.
　　Kung Ming-chih　龔明之.

Chhang-An Chih　長安志.
　　History of the City of Eternal Peace [Chhang-an (Sian), ancient capital of China].
　　Sung, c. +1075.
　　Sung Min-Chhiu　宋敏求.

Chhang-An Chih Thu　長安志圖.
　　Maps to illustrate the History of the City of Eternal Peace [Chhang-an (Sian), ancient capital of China].
　　Yuan, c. +1330.
　　Li Hao-Wên　李好文.

Chhang-Chhun Chen Jen Hsi Yu Chi
　　長春眞人西遊記
　　The Western Journey of the Taoist (Chhiu) Chhang-Chhun.
　　Yuan, +1228.
　　Li Chih-Chhang　李志常.

Chhang Wu Chih　長物志.
　　Notes on Life's Staples.
　　Ming, c. +1595.
　　Wên Chen-Hêng　文震亨.

Chhao Yeh Chhien Tsai　朝野僉載.
　　Stories of Court Life and Rustic Life [or, Anecdotes from Court and Countryside].
　　Thang, +8th century, but much remodelled in Sung.
　　Chang Tso　張鷟.

Chhêng Chai Chi　誠齋集.
　　Collected Writings of (Yang) Chhêng-Chai (Yang Wan-Li).
　　Sung, c. +1200.
　　Yang Wan-Li　楊萬里.

Chhi Kuo Khao　七國考.
　　Investigations of the Seven (Warring) States.
　　Ming.
　　Tung Yüeh　董說.

Chhi Min Yao Shu　齊民要術.
　　Important Arts for the People's Welfare [lit. Equality].
　　N/Wei (and E/Wei or W/Wei), between +533 and +544.
　　Chia Ssu-Hsieh　賈思勰.
　　See des Rotours (1), p. c; Shih Shêng-Han (1).

Chhi Tung Yeh Yü　齊東野語
　　Rustic Talks in Eastern Chhi.
　　Sung, c. +1290.
　　Chou Mi　周密.

Chhieh Yün　切韻.
　　Dictionary of the Sounds of Characters [rhyming dictionary].

　　Sui, +601.
　　Lu Fa-Yen　陸法言.
　　See *Kuang Yün*.

Chhien Han Chi　前漢紀.
　　Records of the Former Han Dynasty.
　　H/Han, c. +200.
　　Hsün Yüeh　荀悅.

Chhien Han Shu　前漢書.
　　History of the Former Han Dynasty [−206 to +24].
　　H/Han (begun about +65), c. +100.
　　Pan Ku　班固, and (after his death in +92) his sister Pan Chao　班昭.
　　Partial trs. Dubs (2), Pfizmaier (32–34, 37–51), Wylie (2, 3, 10), Swann (1).
　　Yin-Tê Index, no. 36.

Chhien Wên Chi　前聞記.
　　Traditions of Past Affairs.
　　Ming, c. +1525.
　　Chu-Yün-Ming　祝允明.
　　Partial tr. Pelliot (2a), p. 305.

Chhih Ya　赤雅.
　　Information about the Naked Ones [Miao and other tribespeople].
　　Ming, +16th or +17th century.
　　Kuang Lu　鄺露.

Chhin-Ting Ku Chin Thu Shu Chi Chhêng
　　欽定古今圖書集成.
　　See *Thu Shu Chi Chhêng*.

Chhin-Ting Shou Shih Thung Khao
　　欽定授時通考.
　　See *Shou Shih Thung Khao*.

Chhin-Ting Shu Ching Thu Shuo
　　欽定書經圖說.
　　See *Shu Ching Thu Shuo*.

Chhin-Ting Ssu Khu Chhüan Shu, etc.
　　See *Ssu Khu Chhüan Shu*.

Chhin Yün Hsieh Ying Hsiao Phu　秦雲擷英小譜.
　　A Bundle of Records of Heroines of the Western Provinces.
　　Chhing, +1778.
　　Ed. & pref. Wang Chhang　王昶.

Chhing I Lu　清異錄.
　　Records of the Unworldly and the Strange.
　　Wu Tai, c. +950.
　　Thao Ku　陶穀.

Chhing-Li Ho Fang Thung I　慶曆河防通議.
　　General Discussion of the Flood-Protection Works in the Chhing-Li reign-period.
　　Sung, between +1041 and +1048.
　　Shen Li　沈立.

Chhing Po Tsa Chih　清波雜志.
　　Green-Waves Memories.
　　Sung, +1193.
　　Chou Hui　周煇.

Chhou Hai Thu Pien　籌海圖編.
　　Illustrated Seaboard Strategy.
　　Ming, +1562.
　　Chêng Jo-Tsêng　鄭若曾.
　　Also attributed to Hu Tsung-Hsien　胡宗憲.

Chhu Hsüeh Chi 初學記.
    Entry into Learning [encyclopaedia].
    Thang, +700.
    Hsü Chien 徐堅.
Chhu Tzhu 楚辭.
    Elegies of Chhu (State) [or, Songs of the
        South].
    Chou, *c.* −300 (with Han additions).
    Chhü Yuan 屈原 (& Chia I 賈誼, Yen
        Chi 嚴忌, Sung Yü 宋玉, Huainan
        Hsiao-Shan 淮南小山 *et al.*).
    Partial tr. Waley (23); tr. Hawkes (1).
Chhu Tzhu Pu Chu 楚辭補註.
    Supplementary Annotations to the *Elegies of
        Chhu.*
    Sung, *c.* +1140.
    Ed. Hung Hsing-Tsu 洪興祖.
Chhüan Thang Wên.
    See Tung Kao (1).
Chhui Hung Chhiao Chi 垂虹橋記.
    A Record of the Chhui-Hung Bridge (at
        Wuchiang).
    Sung, *c.* +1060.
    Chhien Kung-Fu 錢公輔.
Chhui Hung Chhiao Chi 垂虹橋記.
    A Record of the Chhui-Hung Bridge (at
        Wuchiang).
    Yuan, *c.* +1330.
    Yuan Chüeh 袁桷.
Chhun Chhiu 春秋.
    Spring and Autumn Annals [i.e. Records of
        Springs and Autumns].
    Chou; a chronicle of the State of Lu kept
        between −722 and −481.
    Writers unknown.
    Cf. *Tso Chuan; Kungyang Chuan; Kuliang
        Chuan.*
    See Wu Khang (1); Wu Shih-Chhang (1);
        van der Loon (1).
    Tr. Couvreur (1); Legge (11).
Chhun Chhiu Ching Thien Chi 春秋井田記.
    Record of the 'Well-Field' Land System of
        the Spring and Autumn Period.
    H/Han.
    Writer unknown.
Chhun Chhiu Hou Chuan 春秋後傳.
    A History of the Ages since the Time of the
        Spring-and-Autumn Annals.
    Sung.
    Chhen Fu-Liang 陳傅良.
Chhün Ching Kung Shih Thu 羣經宮室圖.
    Illustrated Treatise on the Plans, Technical
        Terms, and Uses of the Houses, Palaces,
        Temples and other Buildings described in
        the Classics.
    Chhing, late +18th century but not printed
        till after 1800.
    See Chiao Hsün (1).
Chhung-Chhing Fu Chih 重慶府志.
    History and Topography of Chungking (in
        Szechuan).

Chhing, 1843 but based on +18th-century
    materials.
Repr. 1926.
Yu-Chhing 有慶.
Chhungmu Kong Chŏnsŏ 忠武公全書.
    Complete Writings of the Loyal and Martial
        Duke [the Korean Admiral Yi Sunsin,
        +1545 to +1598].
    Korea, +1795.
    Yi Sunsin 李舜臣.
    Compiled by royal order by Yun Haengim
        尹行恁.

Erh Ya 爾雅.
    Literary Expositor [dictionary].
    Chou material, stabilised in Chhin or
        C/Han.
    Compiler unknown.
    Enlarged and commented on *c.* +300 by
        Kuo Pho 郭璞.
    Yin-Tê Index no. (suppl.) 18.

Fa Yuan Chu Lin 法苑珠林.
    Forest of Pearls in the Garden of the Law
        [Buddhist encyclopaedia].
    Thang, +668.
    Tao-Shih (monk) 道世.
Fan Shêng Chih Shu 氾勝之書.
    [= *Chung Chih Shu.*]
    The Book of Fan Shêng-Chih on
        Agriculture.
    C/Han, late −1st century (*c.* −10).
    Fan Shêng-Chih 氾勝之.
    In *YHSF*, ch. 69, pp. 50*a* ff.
    Tr. Shih Shêng-Han (2).
Fang Hai Chi Yao.
    See Yü Chhang-Hui (1).
Fang Ho Tsou I 防河奏議.
    Water Conservancy Memorials to the
        Throne.
    Chhing, +1733.
    Hsi Tsêng-Yün 嵇曾筠.
Fang Yen 方言.
    Dictionary of Local Expressions.
    C/Han, *c.* −15 (but much interpolated
        later).
    Yang Hsiung 揚雄.
Fang Yü Chi 方輿記.
    General Geography.
    Chin, or at least pre-Sung.
    Hsü Chiai 徐鍇.
Fêng Chhuang Hsiao Tu 楓牕小牘.
    Maple-Tree Window Memories.
    Sung, late +12th century.
    Yuan Chhiung 袁褧.
    Completed by a later writer soon after
        +1202.
Fêng Thu Chi 風土記.
    Record of Airs and Places [local customs].
    Chin, +3rd century.
    Chou Chhu 周處.

Fo Kuo Chi 佛國記 .
　　[=*Fa-Hsien Chuan* or *Fa-Hsien Hsing Chuan*.]
　　Records of Buddhist Countries [also called
　　　　Travels of Fa-Hsien].
　　Chin, +416.
　　Fa-Hsien (monk) 法顯 .
　　Tr. Rémusat (1); Beal (1); Legge (4); H. A.
　　　　Giles (3); Li Yung-Hsi (1).
Fugaku Hyaku-Kei 富嶽百景 .
　　See Katsushika Hokusai (1).
Fukien Thung Chih 福建通志 .
　　History and Topography of Fukien Province.
　　Chhing, +1684.
　　Chin Hung (ed.) 金鋐 .
Fukien Yün Ssu Chih 福建運司志 .
　　Records of the Transportation Bureau of
　　　　Fukien Province.
　　Ming, +1553.
　　Chiang Ta-Kun 江大鯤 .

Genna Kōkaisho 元和航海書 .
　　Manual of Navigation of the Genna year-
　　　　period.
　　Japan, +1618.
　　Ikeda Kōun 池田好運 .
　　(Repr. *NKKZ*, vol. 12.)

Hai Chhao Thu Lun 海潮圖論 .
　　Illustrated Discourse on the Tides.
　　Sung, +1026.
　　Yen Su 燕肅 .
Hai-Chhiu Fu Hou Hsü 海鰌賦後序 .
　　Postface to the Ode on the 'Sea-Eel'
　　　　(Warships) [and their role at the Battle of
　　　　Tshai-shih, +1161].
　　Sung, *c.* +1170.
　　Yang Wan-Li 楊萬里 .
　　In *Chhêng-Chai Chi*, ch. 44, pp. 6*b* ff.
Hai Chung Chan 海中占 .
　　Astrology (and Astronomy) of (the People in)
　　　　the Midst of the Sea (or, of the Sailors).
　　Han.
　　Writer unknown.
　　Now extant only in the form of numerous
　　　　quotations in the *Khai-Yuan Chan Ching*.
Hai Kuo Thu Chih.
　　See Wei Yuan & Lin Tsê-Hsü (1).
Hai Kuo Wên Chien Lu 海國聞見錄 .
　　Record of Things Seen and Heard about the
　　　　Coastal Regions.
　　Chhing, +1744.
　　Chhen Lun-Chiung 陳倫炯 .
Hai Nei Shih Chou Chi 海內十洲記 .
　　Record of the Ten Sea Islands [or, of the
　　　　Ten Continents in the World Ocean].
　　Ascr. Han; prob. +4th or +5th century.
　　Attrib. Tungfang Shuo 東方朔 .
Hai Tao Chen Ching 海道針經 .
　　Seaways Compass Manual.
　　Yuan or Ming, +14th century.
　　Writer unknown.

Hai Tao Ching 海道經 .
　　Manual of Sailing Directions.
　　Yuan, +14th century.
　　Compilers unknown.
Hai Tao I Chih Chai Lüeh 海島逸誌摘略 .
　　Brief Selection of Lost Records of the Isles
　　　　of the Sea [or, a Desultory Account of the
　　　　Malayan Archipelago].
　　Chhing, between +1783 and +1790,
　　　　preface +1791.
　　Wang Ta-Hai 王大海 .
　　Tr. Anon. (37).
Hai Tao Suan Ching 海島算經 .
　　[Originally a supplement to the *Chiu Chang
　　　　Suan Shu* and known before the Thang
　　　　as *Chhung Chha*. The Sui bibliography
　　　　also has a *Chiu Chang Chhung Chha Thu*
　　　　九章重差圖 .]
　　Sea Island Mathematical Manual.
　　S/Kuo, +263.
　　Liu Hui 劉徽 .
Hai Thang Lu 海塘錄 .
　　History of the (Chhien-Thang) Sea Wall
　　　　(near Hangchow).
　　Chhing.
　　Tsê Chün-Lien 翟均廉 .
Hai Thang Shuo 海塘說 .
　　Discourse on Sea-Walls.
　　Ming.
　　Huang Kuang-Shêng 黃光昇 .
Hai Yao Pên Tshao 海藥本草 .
　　Drugs of the Southern Countries beyond the
　　　　Seas [or Pharmaceutical Codex of Marine
　　　　Products].
　　Thang, *c.* +775 (or early +10th century).
　　Li Hsün (acc. to Li Shih-Chen) 李珣 .
　　Li Hsien (acc. to Huang Hsiu-Fu) 李玹 .
　　Preserved in *Pên Tshao Kang Mu*, etc.
Hai Yün Hsin Khao 海運新考 .
　　New Investigation of Sea Transport.
　　Ming, +1579.
　　Liang Mêng-Lung 梁夢龍 .
Han I Wên Chih Khao Chêng 漢藝文志考證 .
　　Textual Criticism of the Bibliography in the
　　　　History of the Former Han Dynasty.
　　Sung, *c.* +1280.
　　Wang Ying-Lin 王應麟 .
Hang-Chou Fu Chih 杭州府志 .
　　Gazetteer [Historical Topography] of
　　　　Hangchow.
　　Chhing, +1686.
　　Ma Ju-Lung 馬如龍 .
　　Ma I *et al.* (ed.) 馬益 .
Hang Hai Chen Ching 航海針經 .
　　Sailors' Compass Manual.
　　Yuan or Ming, +14th century.
　　Writer unknown.
Ho Chhü Shu 河渠書 .
　　On the Rivers and Canals.
　　Ssuma Chhien's monograph incorporated
　　　　into the *Shih Chi*, q.v.

Ho Fang Chhüan Shu 河防全書
    (or *Ho Fang I Lan Chio* 河防一覽権).
    General View of Water Control.
    Ming, +1590.
    Phan Chi-Hsün 潘季馴.

Ho Fang Chih 河防志.
    River Protection Works.
    Chhing, +1725.
    Chang Phêng-Ko 張鵬翮.

Ho Fang Thung I 河防通議.
    A General Discussion of the Protection
    Works along the Yellow River [a revision
    and enlargement of two older texts, the
    Chhing-li Ho Fang Thung I of Shen Li (*c.*
    +1045) and the *Ho Shih Chi* of Chou
    Chün (+1128)].
    Yuan, +1321.
    Shan Ssu (Sha-Kho-Shih (a Persian or
    Arab) 瞻思 (沙克什).

Ho Fang Tsê Yao 河防摘要.
    Select Principles of Water Control.
    Chhing, *c.* +1680, pr. +1767.
    Chhen Huang 陳潢.

Ho Kung Chhi Chü Thu Shuo 河工器具圖說.
    See Lin Chhing (2).

Ho Kung Chien Yao 河工簡要.
    See Chhiu Pu-Chou (1).

Ho Pi Shih Lei 合璧事類.
    The 'Borrowed Jade Returned'
    Encyclopaedia.
    Sung, +1257.
    Hsieh Wei-Hsin 謝維新.

Ho Shih Chi 河事集.
    Collected Materials on River Control
    Works.
    Sung, +1128.
    Chou Chün 周俊.

Hou Han Shu 後漢書.
    History of the Later Han Dynasty [+25 to
    +220].
    L/Sung, +450.
    Fan Yeh 范曄.
    The monograph chapters by Ssuma Piao
    司馬彪 (d. +305), with commentary by
    Liu Chao 劉昭 (*c.* +510), who first
    incorporated them in the work.
    A few chs. tr. Chavannes (6, 16); Pfizmaier
    (52, 53).
    Yin-Tê Index, no. 41.

Hou Shan Than Tshung 後山談叢.
    Hou-Shan Table Talk.
    Sung, *c.* +1090.
    Chhen Shih-Tao 陳師道.

Hua Chhüan 畫筌.
    The Painting Basket.
    Chhing.
    Ta Chung-Kuang 笪重光.

Hua Shan Chi 華山記.
    Record of Mount Hua.
    Sung or pre-Sung.
    Writer unknown.

Hua Shu 化書.
    Book of the Transformations (in Nature).
    H/Thang, *c.* +940.
    Attrib. Than Chhiao 譚峭.
    TT/1032.

Hua Yang Kuo Chih 華陽國志.
    Records of the Country South of Mount
    Hua [historical geography of Szechuan
    down to +138].
    Chin, +347.
    Chhang Chhü 常璩.

Huai Nan Hung Lieh Chieh 淮南鴻烈解.
    See *Huai Nan Tzu.*

Huai Nan Tzu 淮南子.
    [= *Huai Nan Hung Lieh Chieh.*]
    The Book of (the prince of) Huai-Nan
    [compendium of natural philosophy].
    C/Han, *c.* −120.
    Written by the group of scholars gathered
    by Liu An (prince of Huai-Nan) 劉安.
    Partial trs. Morgan (1); Erkes (1); Hughes
    (1); Chatley (1); Wieger (2).
    Chung-Fa Index, no. 5.
    TT/1170.

Huai-Nan (Wang) Wan Pi Shu 淮南(王)萬
    畢術.
    [Prob. = *Chen-Chung Hung-Pao Yuan-Pi
    Shu* and variants.]
    The Ten Thousand Infallible Arts of (the
    Prince of) Huai-Nan [Taoist magical
    and technical recipes].
    C/Han, −2nd century.
    No longer a separate book but fragments
    contained in *TPYL*, ch. 736 and else-
    where.
    Reconstituted texts by Yeh Tê-Hui in
    Kuan Ku Thang So Chu Shu, and Sun
    Fêng-I in *Wên Ching Thang Tshung-Shu.*
    Attrib. Liu An 劉安.
    See Kaltenmark (2), p. 32.
    It is probable that the terms *Chen-Chung*
    枕中 Confidential Pillow-Book; *Hung-
    Pao* 鴻寶 Infinite Treasure; *Wan-Pi*
    萬畢 Ten Thousand Infallible; and
    Yuan-Pi 苑祕 Garden of Secrets; were
    originally titles of parts of a *Huai-Nan
    Wang Shu* 淮南王書 (Writings of the
    Prince of Huai-Nan) forming the Chung
    Phien 中篇 (and perhaps also the Wai
    Shu 外書) of which the present *Huai
    Nan Tzu* book (q.v.) was the Nei Shu
    內書.

Huang Ti Nei Ching, Ling Shu 黃帝內經靈
    樞.
    The Yellow Emperor's Manual of Corporeal
    Medicine; The Spiritual Pivot (or Gate,
    or Driving-Shaft, or Motive Power)
    [medical physiology and anatomy].
    Probably C/Han, *c.* −1st century.
    Writers unknown.
    Edited, +762 by Wang Ping 王冰.

Huang Ti Nei Ching, Ling Shu (*cont.*)
Analysis by Huang Wên (*1*).
Tr. Chamfrault & Ung Kang-Sam (*1*).
Huang Ti Nei Ching, Su Wên 黃帝內經素問 .
The Yellow Emperor's Manual of Corporeal
Medicine; The Plain Questions (and
Answers) [clinical medicine]. (Cf. *Pu Chu
Huang Ti Nei Ching, Su Wên.*)
Chou, remodelled in Chhin and Han,
reaching final form −2nd or −1st
century.
Writers unknown.
Partial trs. Hübotter (*1*), chs. 4, 5, 10, 11,
21; Veith (*1*); complete, Chamfrault &
Ung Kang-Sam (*1*).
See Wang & Wu (*1*), pp. 28 ff.; Huang
Wên (*1*).
Huang Ti Nei Ching Su Wên Chi Chu
黃帝內經素問集註 .
The *Yellow Emperor's Manual of Corporeal
Medicine; The Plain Questions* (*and
Answers*); with Commentaries [by Wang
Ping (Thang) and Ma Shih 馬蒔 (Ming)
as well as the editor's own].
Chhing, +1670.
Chang Chih-Tshung 張志聰 .
(Repr. in *TSCC, I Shu tien*, chs. 21–66.)
Huang Ti Nei Ching, Thai Su 黃帝內經太素 .
The Original Recension of the *Yellow
Emperor's Manual of Corporeal Medicine.*
Chou, Chhin & Han, essentially in present
form by −1st century, commented upon
in Sui, +605 to +618.
Ed. and comm. Yang Shang-Shan
楊上善 .
Identifications of chapters and component
passages with the corresponding texts in
the *Plain Questions* (*and Answers*) and the
Spiritual Pivot in Wang Ping's recension,
and in the *Chen Chiu Chia I Ching*
(Treatise on Acupuncture and Moxibus-
tion) of Huangfu Mi (by Hsiao Yen-
Phing 蕭延平 1924).
Huang Ti Su Wên Ling Shu Ching 黃帝素問
靈樞經 .
See *Huang Ti Nei Ching, Ling Shu.*
Huang Ti Su Wên Nei Ching 黃帝素問內經 .
See *Huang Ti Nei Ching, Su Wên.*
Huang Yün Liang Ho Thu 黃運兩河圖 .
Maps and Diagrams of the Yellow River
and the Grand Canal.
Chhing, *c.* +1690.
Chhêng Chao-Piao 程兆彪 .
Hui-Hui Yao Fang 回回藥方 .
Pharmaceutical Prescriptions of the Muslims
[contains some text in Persian, as well as
the chief part in Chinese].
Late Sung or Yuan, translated late Yuan
and printed early Ming.
Writer unknown, but probably an Arabic or
Persian physician.

Now only incompletely preserved; unique
copy in Peking Nat. Library.
Cf. Sung Ta-Jen (*1*), p. 85.
Hun Thien Fu 渾天賦 .
Ode on the Celestial Sphere.
Thang, +676.
Yang Chiung 楊烱 .
In *Yü Hai*, ch. 4, p. 27*a, b.*
Hunan Thung Chih 湖南通志 .
Historical Geography of Hunan (province).
See Ma Hui-Yü (*1*).
Hung Hsüeh Yin-Yuan Thu Chi.
See Lin Chhing (*1*).
Hwasŏng Sŏngyŏk Ŭigwe 華城城役儀軌 .
Records and Machines of the Hwasŏng
Construction Service [for the Emergency
capital at Suwŏn].
Korea, +1792, presented +1796, pr. 1801.
Chŏng Yagyong 丁若鏞 .
See Chevalier (*1*); Henderson (*1*).
Hsi Ching Fu 西京賦 .
Ode on the Western Capital (Chhang-an).
H/Han, +107.
Chang Hêng 張衡 .
Tr. von Zach (6); Hughes (9).
Hsi Shih Chi 西使記 .
Notes on an Embassy to the West.
Yuan, +1263.
Chhang Tê 常德 .
Writer, Liu Yü 劉郁 .
Preserved in *Yü Thang Chia Hua*, ch. 94
(ch. 2, pp. 4*a* ff.); cf. H. Franke (14).
Tr. Bretschneider (2), vol. 1, pp. 122 ff.
Hsi Tu Fu 西都賦 .
Ode on the Western Capital (Chhang-an).
H/Han, *c.* +87.
Pan Ku 班固 .
Tr. Hughes (9).
Hsi-Yang Chhao Kung Tien Lu 西洋朝貢典
錄 .
Record of the Tributary Countries of the
Western Oceans [relative to the voyages
of Chêng Ho].
Ming, +1520.
Huang Shêng-Tsêng 黃省曾 .
Ed. by Sun Yün-Chia 孫允伽 & Chao
Khai-Mei 趙開美 .
Tr. Mayers (3).
Hsi-Yang Chi 西洋記 .
See *San-Pao Thai-Chien Hsia Hsi-Yang
Chi Thung Su Yen I.*
Hsi-Yang Fan Kuo Chih 西洋番國志 .
Record of the Foreign Countries in the
Western Oceans [relative to the voyages
of Chêng Ho].
Ming, +1434.
Kung Chen 鞏珍 .
Hsi Yu Chi 西遊記 .
Story of a Journey to the West (or,
Pilgrimage to the West) [novel =
Monkey].

Hsi Yu Chi (*cont.*)
Ming, c. +1560.
Wu Chhêng-Ên 吳承恩.
Tr. Waley (17).
Hsi Yü Fan Kuo Chih 西域番國志.
Records of the Strange Countries of the
West.
Ming, c. +1417.
Chhen Chhêng 陳誠.
Hsi Yü Hsing Chhêng Chi 西域行程記.
Diary of a Diplomatic Mission to the
Western Countries (Samarqand, Herat,
etc.).
Ming, +1414.
Chhen Chhêng 陳誠 & Li Ta 李暹.
Hsien Pin Lu 咸賓錄.
Record of All the Guests.
Ming, +1590.
Lo Jih-Chhung 羅日褧.
Hsien-Shun Lin-An Chih 咸淳臨安志.
Hsien-Shun reign-period Topographical
Records of the Hangchow District.
Sung, +1274.
Chhien Yüeh-Yu 潛說友.
Hsin Chih Ling-Thai I Hsiang Chih 新製靈臺
儀象志.
Memoir on the (Theory, Use and Construc-
tion of the) Instruments of the New
Imperial Observatory.
Chhing, +1673.
Nan Huai-Jen (Ferdinand Verbiest)
南懷仁.
Hsin Thang Shu 新唐書.
New History of the Thang Dynasty
[+618 to +906].
Sung, +1061.
Ouyang Hsiu 歐陽修 & Sung Chhi 宋祁.
Cf. des Rotours (2), p. 56.
Partial trs. des Rotours (1, 2); Pfizmaier
(66–74). For translations of passages see
the index of Frankel (1).
Yin-Tê Index, no. 16.
Hsin Wu Tai Shih 新五代史.
New History of the Five Dynasties (+907
to +959].
Sung, c. +1070.
Ouyang Hsiu 歐陽修.
For translations of passages see the index of
Frankel (1).
Hsin Yuan Shih 新元史.
See Kho Shao-Min (1).
Hsing Chha Shêng Lan 星槎勝覽.
Triumphant Visions of the Starry Raft
[account of the voyages of Chêng Ho,
whose ship, as carrying an ambassador, is
thus styled].
Ming, +1436.
Fei Hsin 費信.
Hsing Shui Chin Chien 行水金鑑.
Golden Mirror of the Flowing Waters (cf.
Hsü Hsing Shui Chin Chien).

Chhing, +1725.
Fu Tsê-Hung 傅澤洪.
Hsiu Chhü Chi 修渠記.
Record of the Repairs to the (Ling) Chhü
[Magic Canal].
Thang, +868.
Yü Mêng-Wei 魚孟威.
Hsiu Fang So Chih 修防瑣志.
Brief Memoir on Dyke Repairs.
Chhing, before +1778.
Li Shih-Lu 李世祿.
Hsiu Hai Thang I 修海塘議.
Discussions on Sea-Wall Repair and
Maintenance.
Chhing.
Chhen Hsü-Tsung 陳訏總.
Hsü Han Shu 續漢書.
Supplement to the [Former] Han History.
San Kuo, +3rd century.
Hsieh Chhêng 謝承.
Hsü Hou Han Shu 續後漢書.
Supplement to the History of the Later
Han.
Sung, c. +1265.
Hao Ching 郝經.
Hsü Hsia-Kho Yu Chi 徐霞客遊記.
Diary of the Travels of Hsü Hsia-Kho.
Chhing, +1776 (written +1641).
Hsü Hsia-Kho 徐霞客.
Hsü Hua Phin Lu 續畫品錄.
Continued Records of the Classification of
Old Painters.
Liang, c. +550.
Yao Tsui 姚最.
Cf. Hirth (12); Wang Po-Min (1).
Hsü Shih Shih 續事始.
Supplement to the *Beginnings of All Affairs*
(cf. *Shih Shih*).
H/Shu, c. +960.
Ma Chien 馬鑑.
Hsü Thung Chien Kang Mu.
See *Thung Chien Kang Mu Hsü Pien* and
Thung Chien Kang Mu San Pien.
Hsü Tzu Chih Thung Chhien Chhang Phien
續資治通鑑長編.
Continuation of the *Comprehensive Mirror
(of History) for Aid in Government* [+960
to +1126].
Sung, +1180.
Li Tao 李燾.
Hsü Wên Hsien Thung Khao 續文獻通考.
Continuation of the *Comprehensive Study of
(the History of) Civilisation* (cf. *Wên
Hsien Thung Khao* and *Chhin-Ting Hsü
Wên Hsien Thung Khao*).
Ming, finished +1586, pr. +1603.
Ed. Wang Chhi 王圻.
This covers the Liao, J/Chin, Yuan and
Ming dynasties, adding some new material
for the end of S/Sung from +1224
onwards.

Hsüan-Ho Fêng Shih Kao-Li Thu Ching
宣和奉使高麗圖經 .
Illustrated Record of an Embassy to Korea
in the Hsüan-Ho reign-period.
Sung, +1124 (+1167).
Hsü Ching 徐兢 .

Hsüan-Ho Hua Phu 宣和畫譜 .
Hsüan-Ho (reign-period) Catalogue of the
Paintings (in the Imperial Collection) [of
the emperor Hui Tsung].
Sung, c. +1120.
Writer unknown.

Hsüan-Ho Po Ku Thu Lu 宣和博古圖錄 .
[= *Po Ku Thu Lu.*]
Hsüan-Ho reign-period Illustrated Record
of Ancient Objects. [Catalogue of the
archaeological museum of the emperor
Hui Tsung.]
Sung, +1111 to +1125.
Wang Fu 王黼 or 韍 , *et al.*

I Chhieh Ching Yin I 一切經音義 .
Dictionary of Sounds and Meanings of
Words in the *Vinaya* [part of the Buddhist
Tripiṭaka].
Thang, c. +649, enlarged c. +730.
Hsüan-Ying 玄應 .
Enlarged by Hui-Lin 慧琳 .
N/1605; TW/2178.

I Ching 藝經 .
Treatise on Arts and Games.
San Kuo (Wei), +3rd century.
Hantan Shun 邯鄲淳 .

I Li Shih Kung 儀禮釋宮 .
Explanations Concerning the Buildings in
the Personal Conduct Ritual.
Sung, +1193.
Li Jo-Kuei 李如圭 .

I Lin 易林 .
Forest of Symbols of the (*Book of*) *Changes*
[for divination].
C/Han, c. −40.
Chiao Kan 焦贛 .

I Tsung Chin Chien 醫宗金鑑 .
Golden Mirror of Medicine.
Chhing (imperially ordered), +1743.
Ed. O-Erh-Thai 鄂爾泰 .

I Yü Chih 異域誌 .
(Previously entitled *Lo Chhung Lu.*)
Record of Strange Countries.
Yuan & Ming, orig. written c. +1366, re-
titled just before +1400.
Attrib. Chou Chih-Chung 周致中 .
Re-titled and perhaps expanded probably
by the elder brother of Khai Chi 開濟 .

I Yü Thu Chih 異域圖志 .
Illustrated Record of Strange Countries.
Ming, c. +1420 (written between +1392
and +1430); pr. +1489.
Compiler unknown, perhaps Chu Chhüan
朱橚 .

Cf. Moule (4); Sarton (1), vol. 3, p. 1627.
(A copy is in the Cambridge University
Library.)

I Yuan 異苑 .
Garden of Strange Things.
L/Sung, c. +460.
Liu Ching-Shu 劉敬叔 .

Jih Chih Lu 日知錄 .
Daily Additions to Knowledge.
Chhing, +1673.
Ku Yen-Wu 顧炎武 .

Ju Shu Chi 入蜀記 .
Journey into Szechuan.
Sung, +1170.
Lu Yu 陸游 .

Jung Chai Sui Pi 容齋隨筆 .
Miscellanies of Mr [Hung] Jung-Chai
[collection of extracts from literature, with
editorial commentaries].
Sung, 1st part pr. c. +1185, 2nd part
+1192, 3rd part +1196, 4th part after
+1202.
Hung Mai 洪邁 .

Ka-i Tsūshō-Kō 華夷通商考 .
Studies on the Intercourse and Trade
between Chinese and Barbarians.
Japan, +1708.
Nishikawa Joken 西川如見 .
Repr. *NKS*, vol. 5.

Kan Chhüan Fu 甘泉賦 .
Rhapsodic Ode on the (Palace-Temple at
the) Sweetwater Springs.
C/Han, c. −10.
Yang Hsiung 揚雄 .
Tr. von Zach (6).

Kao-Li Thu Ching.
See *Hsüan-Ho Fêng Shih Kao-Li Thu Ching.*

Kao Sêng Chuan 高僧傳 .
Biographies of Outstanding (Buddhist)
Monks [especially those noted for learn-
ing and philosophical eminence].
Liang, between +519 and +554.
Hui-Chiao 慧皎 .
TW/2059.

Kêng Chih Thu 耕織圖 .
Pictures of Tilling and Weaving.
Sung, presented in MS., +1145, and
perhaps first printed from wood blocks at
that time; engraved on stone, +1210, and
probably then printed from wood blocks.
Lou Shou 樓璹 .
The illustrations published by Franke (11)
are those of +1462 and +1739; Pelliot
(24) published a set based on an edition
of +1237. The original illustrations are
lost, but cannot have differed much from
these last, which include the poems of
Lou Shou. The first Chhing edition was
in +1696.

Kêng Hsin Yü Tshê 庚辛玉册.
Precious Secrets of the Realm of Kêng and Hsin (i.e. all things connected with metals and minerals, symbolised by these two cyclical characters) [on alchemy and pharmaceutics. Kêng-Hsin is also an alchemical synonym for gold].
Ming, +1421.
Ning Hsien Wang (prince of the Ming) 寧獻王.

Khai Ho Chi 開河記.
Record of the Opening of the (Grand) Canal.
Sui.
Han Wu 韓偓.

Khai-Yuan Chan Ching 開元占經.
The Khai-Yuan reign-period Treatise on Astrology (and Astronomy).
Thang, +729.
(Some parts, such as the *Chin Chih* (*Navagrāha*) calendar, had been written as early as +718.)
Chhüthan Hsi-Ta 瞿曇悉達.

Khang-Hsi Tzu Tien 康熙字典.
Imperial Dictionary of the Khang-Hsi reign-period.
Chhing, +1716.
Ed. Chang Yü-Shu 張玉書.

Khao Kung Chi 考工記.
The Artificers' Record [a section of the *Chou Li*, q.v.].
Chou and Han, perhaps originally an official document of Chhi State, incorporated *c.* −140.
Compiler unknown.
Tr. E. Biot (1).
Cf. Kuo Mo-Jo (1); Yang Lien-Shêng (7).

Kho Tso Chui Yü 客座贅語.
My Boring Discourses to my Guests [memorabilia of Nanking].
Ming, *c.* +1628.
Ku Chhi-Yuan 顧起元.

Khun Yü Thu Shuo 坤輿圖說.
Explanation of the World Map.
Chhing, +1672.
Nan Huai-Jen (Ferdinand Verbiest) 南懷仁.

Ko Chih Ching Yuan 格致鏡原.
Mirror of Scientific and Technological Origins.
Chhing, +1735.
Chhen Yuan-Lung 陳元龍.

Kou Hsüeh Chiang Li Hsiao Chi 溝洫疆理小記.
A Short Theoretical Study of Canal Construction.
Chhing, late +18th century.
Chhêng Yao-Thien 程瑤田.

Ku Chin Chu 古今註.
Commentary on Things Old and New.
Chin, *c.* +300.

Tshui Pao 崔豹.
See des Rotours (1), p. xcviii.

Ku Hua Phin Lu 古畫品錄.
Records of the Classification of Old Painters.
S/Chhi, *c.* +500.
Hsieh Ho 謝赫.
Cf. Hirth (12); Wang Po-Min (1).

Ku Kung I Lu 故宮遺錄.
Description of the Palaces (of the Yuan Emperors).
Ming, +1368.
Hsiao Hsün 蕭洵.

Ku Wei Shu 古微書.
Old Mysterious Books [a collection of the apocryphal Chhan-Wei treatises].
Date uncertain, in part C/Han.
Ed. Sun Chio 孫瑴 (Ming).

Ku Wên Hsi I 古文析義.
Collection of Essays in the Old Clear Style, Classified and Elucidated.
Chhing, before +1697, pr. +1716.
Lin Yün-Ming (ed.) 林雲銘.

Kua Ti Chih 括地志.
Comprehensive Geography.
Thang, +7th century.
Wei Wang-Thai 魏王泰.
(Fragments reconstituted by Sun Hsing-Yen in +1797.)

Kuan-Chung Chhuang-Li Chieh-Than Thu Ching 關中創立戒壇圖經.
Illustrated Treatise on the Method of Setting up (Buddhist) Ordination Altars used in Kuan-Chung.
Thang, +667.
Tao-Hsüan 道宣.
TW/1892.

Kuan Han 官箴.
Handbook for Magistrates.
Sung, *c.* +1119.
Lü Pên-Chung 呂本中.

Kuan shih Ti Li Chih Mêng 管氏地理指蒙.
Master Kuan's Geomantic Instructor.
Ascr. San Kuo, +3rd century; prob. Thang, +8th century.
Attrib. Kuan Lo 管輅.

Kuan Tzu 管子.
The Book of Master Kuan.
Chou and C/Han. Perhaps mainly compiled in the Chi-Hsia Academy (late −4th century) in part from older materials.
Attrib. Kuan Chung 管仲.
Partial trs. Haloun (2, 5); Than Po-Fu *et al.* (1).

Kuan Yin Tzu 關尹子.
[= *Wên Shih Chen Ching.*]
The Book of Master Kuan Yin.
Thang, +742 (may be later Thang or Wu Tai). A work with this title existed in the Han, but the text is lost.
Prob. Thien Thung-Hsiu 田同秀.

Kuang Chih 廣志.
Extensive Records of Remarkable Things.
Chin, +4th century.
Kuo I-Kung 郭義恭.
YHSF, ch. 74.
Kuang Ya 廣雅.
Enlargement of the *Erh Ya*; *Literary Expositor* [dictionary].
San Kuo (Wei), +230.
Chang I 張揖.
Kuang Ya Su Chêng 廣雅疏證.
Correct Text of the *Enlargement of the Erh Ya*, with Annotations and Amplifications.
Chhing, +1796.
Wang Nien-Sun 王念孫.
Kuang Yü Thu 廣輿圖.
Enlarged Terrestrial Atlas.
Yuan, +1320.
Chu Ssu-Pên 朱思本.
First printed, and the word *Kuang* added, by Lo Hung-Hsien (羅洪先), Ming, c. +1555.
Kuang Yün 廣韻.
Enlargement of the *Chhieh Yün*; *Dictionary of the Sounds of Characters*.
Sung.
(A completion by later Thang and Sung scholars given its present name in +1011.)
Lu Fa-Yen 陸法言 *et al.*
Kuangtung Hsin Yü 廣東新語.
New Talks about Kuangtung Province.
Chhing, c. +1690.
Chhü Ta-Chün 屈大均.
Kuei Hai Yü Hêng Chih 桂海虞衡志.
Topography and Products of the Southern Provinces.
Sung, +1175.
Fan Chhêng-Ta 范成大.
Kuei-Hsin Tsa Chih 癸辛雜識.
Miscellaneous Information from Kuei-Hsin Street (in Hangchow).
Sung, late +13th century, perhaps not finished before +1308.
Chou Mi 周密.
See des Rotours (1), p. cxii; H. Franke (14).
Kuei-Hsin Tsa Chih Hsü Chi 癸辛雜識續集.
Miscellaneous Information from Kuei-Hsin Street (in Hangchow) (First Addendum).
Sung or Yuan, c. +1298.
Chou Mi 周密.
See des Rotours (1), p. cxii.
Kuei Thien Lu 歸田錄.
On Returning Home.
Sung, +1067.
Ouyang Hsiu 歐陽修.
Kung Khuei Chi 攻媿集.
Bashfulness Overcome; Recollections of My Life and Times.
Sung, c. +1210.
Lou Yo 樓鑰.

Kung Kuo Ko 功過格.
Gradation of Merits and Faults.
Ascr. Thang, +8th century, in fact Ming, +17th.
Attrib. Lü Tung-Pin 呂洞賓, actual writer Yuan Piao 袁表.
Kung Shih Khao 宮室考.
A Study of Halls and Buildings.
Chhing, early +18th century.
Jen Chhi-Yün 任啓運.
Kung Shih Tiao Cho Chêng Shih Lu Pan Mu Ching Chiang Chia Ching 工師雕斲正式魯班木經匠家鏡.
The Timberwork Manual and Artisans' Mirror of Lu Pan, Patron of all Carvers, Joiners and Wood-workers.
Date unknown, contents traditional and certainly partly medieval.
Repr. 1870 and many other dates.
Apparent author Ssuchêng Wu-Jung 司正午榮.
Editors Chang Yen 韋嚴 & Chou Yen 周言.
Kuo Chhao Wên Lei 國朝文類.
Classified Documents of the Present Dynasty (Yuan).
Yuan, c. +1360.
Su Thien-Chio 蘇天爵.
Cf. H. Franke (14).
Kuo Yü 國語.
Discourses on the (ancient feudal) States.
Late Chou, Chhin and C/Han, containing early material from ancient written records.
Writers unknown.

Lai Nan Lu 來南錄.
Record of a Journey to the South.
Thang, +809.
Li Ao 李翱.
Lan Phei Lu 攬轡錄.
Grasping the Reins [narrative of his embassy to the Chin Tartars].
Sung, +1170.
Fan Chhêng-Ta 范成大.
Cf. *Tshan Luan Lu*.
Lao Hsüeh An Pi Chi 老學庵筆記.
Notes from the Hall of Learned Old Age.
Sung, c. +1190.
Lu Yu 陸游.
Lei Shuo 類說.
A Classified Commonplace-Book [a great florilegium of excerpts from Sung and pre-Sung books, many of which are otherwise lost].
Sung, +1136.
Ed. Tsêng Tsao 曾慥.
Li Chi 禮記.
[= *Hsiao Tai Li Chi*.]
Record of Rites [compiled by Tai the Younger].

Li Chi (cont.)
(Cf. *Ta Tai Li Chi*.)
Ascr. C/Han, *c.* −70 to −50, but really
H/Han, between +80 and +105, though
the earliest pieces included may date
from the time of the *Analects* (*c.* −465 to
−450).
Attrib. ed. Tai Shêng　戴聖 .
Actual ed. Tshao Pao　曹褒 .
Trs. Legge (7); Couvreur (3); R. Wilhelm
(6).
Yin-Tê Index, no. 27.

Li-Chiang Fu Chih Lüeh　麗江府志略 .
Classified History and Topography of
Li-chiang (Yunnan).
Chhing, +1743.

Li Sao　離騷 .
Elegy on Encountering Sorrow [ode].
Chou (Chhu), *c.* −295.
Chhü Yuan　屈原 .
Tr. Hawkes (1).

Li Tai Ming Chhen Tsou I　歷代名臣奏議 .
Memorials to the Throne by Eminent
Ministers in all Ages.
Ming, +1416, for the Palace Libraries only,
printed for general circulation in +1635.
(Orig. eds.) Huang Huai　黃淮　& Yang
Ssu-Chhi　楊士奇,　(later ed.) Chang
Phu　張溥 .

Li Tai Thung Chien Chi Lan　歷代通鑑輯覽 .
Essentials of the *Comprehensive Mirror of
History*.
Chhing, +1767.
Ed. Fu Hêng　傅恆 .
Lu Hsi-Hsiung　陸錫熊 *et al.*

Liang Chê Hai Thang Thung Chih　兩浙海塘
通志 .
Comprehensive History and Geography of
the Sea-Walls of the Two Chekiangs
[two parts of Chekiang province].
Chhing, +1750.
Fang Kuan-Chhêng　方觀承 .

Liang Chhi Man Chih　梁溪漫志 .
Bridge Pool Essays.
Sung, +1192.
Fei Kun　費袞 .

Liang Ching Hsin Chi　兩京新記 .
New Records of the Two Capitals.
Thang, early +8th century.
Wei Shu　韋述 .
Only one chapter now extant.

Liang Kung Ting Chien Chi　兩宮鼎建記 .
Alternative title of *Tung Kuan Chi Shih*, q.v.

Liang Shu　梁書 .
History of the Liang Dynasty [+502 to
+556].
Thang, +629.
Yao Chha 姚察　and his son Yao Ssu-Lien
姚思廉 .
For translations of passages see the index of
Frankel (1).

Liang Ssu Kung Chi　梁四公記 .
Tales of the Four Lords of Liang.
Thang, *c.* +605
Chang Yüeh　張說 .

Liao Shih　遼史 .
History of the Liao (Chhi-tan) Dynasty
[+916 to +1125].
Yuan, *c.* +1350.
Tho-Tho (Toktaga) 脫脫　& Ouyang
Hsüan　歐陽玄 .
Partial tr. Wittfogel, Fêng Chia-Shêng *et al.*
Yin-Tê Index, no. 35.

Lieh Hsien Chuan　列仙傳 .
Lives of Famous Hsien (cf. *Shen Hsien
Chuan*).
Chin, +3rd or +4th century, though
certain parts date from about −35 and
shortly after +167.
Attrib. Liu Hsiang　劉向 .
Tr. Kaltenmark (2).

Lieh Tzu　列子 .
[= *Chhung Hsü Chen Ching*.]
The Book of Master Lieh.
Chou and C/Han, −5th to −1st centuries.
Ancient fragments of miscellaneous origin
finally cemented together with much new
material about +380.
Attrib. Lieh Yü-Khou　列禦寇 .
Tr. R. Wilhelm (4); L. Giles (4); Wieger
(7); Graham (6).
TT/663.

Lin-I (*Kuo*) *Chi*　林邑(國)記 .
Records of Lin-I (State) and Province.
Chin or at least pre-Sui.
Writer unknown.

Ling Piao Lu I　嶺表錄異 .
Southern Ways of Men and Things [on the
special characteristics and natural history
of Kuangtung].
Thang, *c.* +895.
Liu Hsün　劉恂 .

Ling Shu Ching.
See *Huang Ti Nei Ching, Ling Shu*.

Ling Wai Tai Ta　嶺外代答 .
Information on What is Beyond the Passes
(lit. a book in lieu of individual replies to
questions from friends).
Sung, +1178.
Chou Chhü-Fei　周去非 .

Liu-Chhiu Kuo Chih Lüeh　琉球國志略 .
Account of the Liu-Chhiu Islands.
Chhing, +1757.
Chou Huang　周煌 .

Liu Pin-Kho Wên Chi　劉賓客文集 .
Literary Collections of the Imperial Tutor
Liu.
Thang, after +842.
Liu Yü-Hsi　劉禹錫 .

Lo Chhung Lu　臝蟲錄 .
Record of the Naked Creatures (i.e. the
Barbarian Peoples).

Lo Chhung Lu (*cont.*)
Yuan, *c.* +1366.
Attrib. Chou Chih-Chung 周致中.
Re-titled *I Yü Chih* (q.v.) just before
+1400.
Lo Shen Fu 洛神賦.
Rhapsodic Ode on the Nymph of the Lo
River.
San Kuo, late +3rd century.
Tshao Chih 曹植.
Lo-Yang Chhieh Lan Chi 洛陽伽藍記
(or *Loyang Ka-Lan Chi*; *sêng ka-lan*
transliterating *sanghārāma*).
Description of the Buddhist Temples and
Monasteries at Loyang.
N/Wei, *c.* +547.
Yang Hsüan-Chih 楊衒之.
Lo-Yang Ming Yuan Chi 洛陽名園記.
Record of the Celebrated Gardens of
Loyang.
Sung, *c.* +1080.
Li Ko-Fei 李格非.
Lü Hsiang Kan Thung Chuan 律相感通傳.
Miscellaneous Temple Traditions according
with the Vinaya Regulations.
Thang, +667.
Tao-Hsüan 道宣.
TW/1898.
Lu I Chi 錄異記.
Strange Matters.
Sung, +10th century.
Tu Kuang-Thing 杜光庭.
Lu Ling-Kuang Tien Fu 魯靈光殿賦.
Ode on the Ling-Kuang Palace in the land
of Lu (near Chhü-fou in Shantung).
H/Han, *c.* +140.
Wang Yen-Shou 王延壽.
Tr. von Zach (6).
Lu Pan Ching 魯班經.
The Carpenter's Classic, or Manual of Lu
Pan (Kungshu Phan).
Date unknown.
Writer unknown.
Lu Pan Ching 魯班經.
See *Kung Shih Tiao Cho Chêng Shih Lu
Pan Mu Ching Chiang Chia Ching.*
Lü Shih Chhun Chhiu 呂氏春秋.
Master Lü's Spring and Autumn Annals
[compendium of natural philosophy].
Chou (Chhin), −239.
Written by the group of scholars gathered
by Lü Pu-Wei 呂不韋.
Tr. R. Wilhelm (3).
Chung-Fa Index, no. 2.
Lu Thang Shih Hua 麓堂詩話.
Foothill Hall Essays [literary criticism].
Ming, +1513.
Li Tung-Yang 李東陽.
Lun Hêng 論衡.
Discourses Weighed in the Balance.
H/Han, +82 or +83.

Wang Chhung 王充.
Tr. Forke (4); cf. Leslie (3).
Chung-Fa Index, no. 1.
Lun Yü 論語.
Conversations and Discourses (of Confucius)
[perhaps Discussed Sayings, Normative
Sayings, or Selected Sayings]; Analects.
Chou (Lu), *c.* −465 to −450.
Compiled by disciples of Confucius (chs. 16,
17, 18 and 20 are later interpolations).
Tr. Legge (2); Lyall (2); Waley (5); Ku
Hung-Ming (1).
Yin-Tê Index, no. (suppl.) 16.
Lung Chiang Chhuan Chhang Chih 龍江船廠
志.
Records of the Shipbuilding Yards on the
Dragon River (near Nanking).
Ming, +1553.
Li Chao-Hsiang 李昭祥.
Cf. W. Franke (3), no. 256; Pao Tsun-
Phêng (1).

Mêng Chhi Pi Than 夢溪筆談.
Dream Pool Essays.
Sung, +1086; last supplement dated
+1091.
Shen Kua 沈括.
Ed. Hu Tao-Ching (1); cf. Holzman (1).
Mêng Liang Lu 夢梁錄.
Dreaming of the Capital while the Rice is
Cooking [description of Hangchow
towards the end of the Sung].
Sung, +1275.
Wu Tzu-Mu 吳自牧.
Mêng Tzu 孟子.
The Book of Master Mêng (Mencius).
Chou, *c.* −290.
Mêng Kho 孟軻.
Tr. Legge (3); Lyall (1).
Yin-Tê Index, no. (suppl.) 17.
Miao Chih Thu Khao 廟制圖考.
Illustrated Study of (Imperial Ancestral)
Temple Planning.
Chhing, *c.* +1685.
Wan Ssu-Thung 萬斯同.
*Min Shêng Shui-Shih, Ko Piao Chen Hsieh
Ying, Chan Shao Chhuan Chih Thu Shuo*
閩省水師各標鎮協營戰哨船隻圖說.
Illustrated Explanation of the (Construction
of the Vessels of the) Coastal Defence
Fleet (Units) of the Province of Fukien
stationed at each of the Headquarters of
the several Grades.
Chhing, late +18th century.
MS. in the Marburg Library.
Min Shui Yen Than Lu.
See *Shêng Shui Yen Than Lu.*
Min Tsa Chi 閩雜記.
Miscellaneous Records of Fukien.
Chhing.
Shih Hung-Pao 施鴻保.

Ming Kung Shih 明宮史.
　An Account of the Palaces and Public
　　Buildings of the Ming Dynasty [at Peking].
　Ming, *c.* +1621.
　Liu Jo-Yü 劉若愚.
Ming Shih 明史.
　History of the Ming Dynasty [+1368 to
　　+1643].
　Chhing, begun +1646, completed +1736,
　　first pr. +1739.
　Chang Thing-Yü 張廷玉 *et al.*
Ming Shih Chi Shih Pên Mo 明史紀事本末.
　Narrative of Events throughout the Ming
　　Dynasty.
　Chhing, *c.* +1680.
　Ku Ying-Thai 谷應泰.
Ming Shih Lu 明實錄.
　Veritable Records of the Ming Dynasty.
　Ming, collected early +17th century.
　Official compilation.
Mo Chi 默記.
　Things Silently Recorded [affairs of the
　　capital city].
　Sung, +11th century.
　Wang Chih 王銍.
Mo Ching 墨經.
　See *Mo Tzu*.
Mo Tzu (incl. *Mo Ching*) 墨子.
　The Book of Master Mo.
　Chou, −4th century.
　Mo Ti (and disciples) 墨翟.
　Tr. Mei Yi-Pao (1); Forke (3).
　Yin-Tê Index, no. (suppl.) 21.
　TT/1162.
Mōko Shūrai Ekotoba 蒙古襲來繪詞.
　Illustrated Narrative of the Mongol
　　Invasion (of Japan).
　Japan, +1293, facsim. ed. ed. Kubota
　　Beisan, 1916.
　Orig. painter and writer unknown.
Mu An Chi 牧菴集.
　Literary Collections of (Yao) Mu-An.
　Yuan, *c.* +1310.
　Yao Sui 姚燧.
Mu Ching 木經.
　See *Kung Shih Tiao Cho Chêng Shih Lu
　　Pan Mu Ching Chiang Chia Ching*.

Nan Chao Yeh Shih 南詔野史.
　History of the Nan Chao Dynasty (Yünnan).
　Ming, +1550 (enlarged in +1775 by Hu
　　Wei 胡蔚).
　Yang Shen 楊愼.
　Tr. Sainson (1).
Nan Chhi Shu 南齊書.
　History of the Southern Chhi Dynasty
　　[+479 to +501].
　Liang, +520.
　Hsiao Tzu-Hsien 蕭子顯.
　For translations of passages see the index of
　　Frankel (1).

Nan Chhuan Chi 南船記.
　Record of Southern Ships.
　Ming, +15th or early +16th century.
　Shen Tai 沈岱.
Nan Chou I Wu Chih 南州異物志.
　Strange Things of the South.
　San Kuo or Chin, +3rd or early +4th
　　century.
　Wan Chen 萬震.
Nan Fang Tshao Mu Chuang 南方草木狀.
　An Account of the Plants and Trees of the
　　Southern Regions.
　Chin, +3rd century.
　Chi Han 稽含.
Nan Hai Pai Yung 南海百詠.
　A Hundred Chants of the Southern Seas.
　Sung, *c.* +1200.
　Fang Hsin-Ju 方信孺.
Nan Hu Chi 南湖集.
　Southern Lake Collection of Poems.
　Sung, +1210.
　Chang Tzu 張鎡.
Nan Hua Chen Ching 南華眞經.
　See *Chuang Tzu*.
Nan Shih 南史.
　History of the Southern Dynasties [Nan
　　Pei Chhao period, +420 to +589].
　Thang, *c.* +670.
　Li Yen-Shou 李延壽.
　For translations of passages see the index of
　　Frankel (1).
Nan Tshun Cho Kêng Lu 南村輟耕錄.
　See *Cho Kêng Lu*.
Nan Tu Fu 南都賦.
　Ode on the Southern City (Wan, Nanyang).
　H/Han, +110.
　Chang Hêng 張衡.
　Tr. von Zach (6).
Nan Yüeh Chih 南越志.
　Records of the South.
　Chin.
　Shen Huai-Yuan 沈懷遠.
Nan Yüeh Pi Chi 南越筆記.
　Memoirs of the South.
　Chhing, +1780.
　Li Thiao-Yuan 李調元.
Nittō Guhō Junrei Gyōki 入唐求法巡禮行記.
　Record of a Pilgrimage to China in Search
　　of the (Buddhist) Law.
　Thang, +838 to +847.
　Ennin 圓仁.
Nung Chêng Chhüan Shu 農政全書.
　Complete Treatise on Agriculture.
　Ming. Composed +1625 to +1628; printed
　　+1639.
　Hsü Kuang-Chhi 徐光啓.
　Ed. Chhen Tzu-Lung 陳子龍.
Nung Shu 農書.
　Treatise on Agriculture.
　Yuan, +1313.
　Wang Chên 王禎.

Nung Shu 農書.
Treatise on Agriculture.
Sung, +1149; printed +1154.
Chhen Fu (Taoist) 陳旉.

Nyū-Min Ki 入明記
(alternatively *Nyū-Tō Ki* 入唐記).
Diary of Travels in the Ming Empire.
Japan, *c.* +1555.
Sakugen 策彥.

Pai Pien 稗編.
Leaves of Grass [encyclopaedia].
Ming, +1581.
Ed. Thang Shun-Chih 唐順之.

Pao-Chhêng Hsien Chih 襃城縣志.
Gazetteer of Pao-chhêng (in Shensi; local topography and history).
Chhing, 1832 (collected from earlier records).
Ed. Kuang Chhao-Khuei 光朝魁.

Pao Phu Tzu 抱樸(朴)子.
Book of the Preservation-of-Solidarity Master.
Chin, early +4th century, prob. *c.* +320.
Tr. Ware (5). Partial trs. Feifel (1, 2); Wu & Davis (2); etc.
Ko Hung 葛洪.
TT/1171–1173.

Pei Chhi Shu 北齊書.
History of the Northern Chhi Dynasty [+550 to +577].
Thang, +640.
Li Tê-Lin 李德林, and his son Li Pai-Yao 李百藥.
A few chs. tr. Pfizmaier (60).
For translations of passages see the index of Frankel (1).

Pei Chou Shu 北周書.
See *Chou Shu*.

Pei Hsing Jih Lu 北行日錄.
Diary of a Journey to the North.
Sung, +1169.
Lou Yo 樓鑰.

Pei Shih 北史.
History of the Northern Dynasties [Nan Pei Chhao period, +386 to +581].
Thang, *c.* +670.
Li Yen-Shou 李延壽.
For translations of passages see the index of Frankel (1).

Pei Shih Chi 北使記.
Notes on an Embassy to the North.
J/Chin and Yuan, +1223.
Wukusun Chung-Tuan 烏古孫仲端.

Pei Thang Shu Chhao 北堂書鈔.
Book Records of the Northern Hall [encyclopaedia].
Thang, *c.* +630.
Yü Shih-Nan 虞世南.

Pên Tshao Kang Mu 本草綱目.
The Great Pharmacopoeia.
Ming, +1596.

Li Shih-Chen 李時珍.
Paraphrased and abridged tr. Read & collaborators (1–7) and Read & Pak (1) with indexes.

Pên Tshao Phin Hui Ching Yao 本草品彙精要.
Systematic Compendium of Materia Medica (Imperially Commissioned).
Ming, +1505.
Ed. Liu Wên-Thai 劉文泰, Wang Phan 王槃 & Kao Thing-Ho 高廷和.

Phei Wên Yün Fu 佩文韻府.
Encyclopaedia of Phrases and Allusions arranged according to Rhyme.
Chhing, +1711.
Ed. Chang Yü-Shu 張玉書 *et al.*

Phi Ya 埤雅.
New Edifications on (i.e. Additions to) the *Literary Expositor*.
Sung, +1096.
Lu Tien 陸佃.

Phing-Chou Kho Than 萍州可談.
Phingchow Table-Talk.
Sung, +1119 (referring to +1086 onwards).
Chu Yü 朱彧.

Phing Hsia Lu 平夏錄.
Record of the Pacification of Hsia (the conquest of the empire by Chu Yuan-Chang).
Ming, *c.* +1544.
Huang Piao 黃標.

Phyohae-Rok 漂海錄.
A Record of Drifting across the Sea; or, A Maritime Odyssey.
Korea, +1488.
Chhoe Pu 崔溥.
Tr. Meskill (1).

Piao I Lu 表異錄.
Notices of Strange Things.
Ming.
Wang Chih-Chien 王志堅.

Ping Chhien 兵鈐.
Key of Martial Art [military encyclopaedia].
Chhing, +1675.
Lü Phan 呂磻 & Lu Chhêng-Ên 盧承恩.
Bodl. Libr. Or. MS., Backhouse 578.

Po Ku Thu Lu 博古圖錄.
See *Hsüan-Ho Po Ku Thu Lu*.

Po Wu Chi 博物記.
Notes on the Investigation of Things.
H/Han, *c.* +190.
Thang Mêng (*b*) 唐蒙.

Po Wu Chih 博物志.
Record of the Investigation of Things (cf. *Hsü Po Wu Chih*).
Chin, *c.* +290 (begun about +270).
Chang Hua 張華.

Pukhagŭi 北學議.
Discussion on the Northern Learning [i.e. Chinese science and technology].
Korea, *c.* +1780.
Pak Chega 朴齊家.

Samguk Sagi 三國史記.
 History of the Three Kingdoms (of Korea)
  [Silla (Hsin-Lo), Kokuryo (Kao-Chü-Li)
  and Pakche (Pai-Chhi), −57 to +936].
 Korea, +1145 (imperially commissioned by
  King Injong), repr. +1394, +1512.
 Kim Pusik 金富軾.

San Chhao Pei Mêng Hui Pien 三朝北盟會
 編.
 Collected Records of the Northern Alliance
  during Three Reigns.
 Sung, +1196.
 Hsü Mêng-Hsin 徐夢莘.

San Fu Chiu Shih 三輔舊事.
 Tales of the Three Cities of the
  Metropolitan Area (Chhang-an (mod.
  Sian), Fêng-i and Fu-fêng).
 Between Chin and Thang, +4th to +8th
  centuries.
 Writer unknown.

San Fu Huang Thu 三輔皇圖.
 Illustrated Description of the Three Cities
  of the Metropolitan Area (Chhang-an
  (mod. Sian), Fêng-i and Fu-fêng).
 Chin, late +3rd century, or perhaps
  H/Han.
 Attrib. Miao Chhang-Yen 苗昌言.

San Kuo Chih 三國志.
 History of the Three Kingdoms [+220 to
  +280].
 Chin, *c.* +290.
 Chhen Shou 陳壽.
 Yin-Tê Index, no. 33.
 For translations of passages see the index of
  Frankel (1).

San Li Thu 三禮圖.
 Illustrations (Diagrams) of the Three
  Rituals [i.e. *Li Chi, Chou Li* and *I Li*].
 Originally H/Han.
 Chêng Hsüan 鄭玄 & Juan Shen
  阮諶.
 Edited in Liu Chhao period by Liang
  Chêng 梁正.
 Recension of +956 by Nieh Chhung-I
  聶崇義.
 Edited +1676 by Nalan Chhêng-Tê
  納蘭成德.

*San-Pao Thai-Chien Hsia Hsi-Yang Chi Thung
 Su Yen I* 三寶太監下西洋記通俗
 演義.
 Popular Instructive Story of the Voyages
  and Traffics of the Three-Jewel Eunuch
  (Admiral, Chêng Ho), in the Western
  Oceans [novel].
 Ming, +1597.
 Lo Mou-Têng 羅懋登.

San Tendai Gotaisan-Ki 參天台五臺山記.
 Record of a Pilgrimage to the Thien-Thai
  Temples on Wu-thai-shan (+1072 and
  +1073).
 Sung, +1074.

Jōjin (Japanese monk) 成尋.
 In *Dai Nihon Bukkyō Zensho*, vol. 115
  (Travel section, vol. 3, p. 46).

San Tshai Thu Hui 三才圖會.
 Universal Encyclopaedia.
 Ming, +1609.
 Wang Chhi 王圻.

San Tzu Ching 三字經.
 Trimetrical Primer.
 Sung, *c.* +1270.
 Wang Ying-Lin 王應麟.

Shan Hai Ching 山海經.
 Classic of the Mountains and Rivers.
 Chou and C/Han.
 Writers unknown.
 Partial tr. de Rosny (1).
 Chung-Fa Index, no. 9.

Shang Shu Ta Chuan 尚書大傳.
 Great Commentary on the *Shang Shu*
  chapters of the *Historical Classic*.
 C/Han, −2nd century.
 Fu Shêng 伏勝.

Shansi Thung Chih 山西通志.
 Provincial Historical Geography of Shansi.
 Chhing, +1733.
 Ed. Lo Shih-Lin 羅石麟 *et al.*

Shen I Ching 神異經.
 Book of the Spiritual and the Strange.
 Ascr. Han, but prob. +4th or +5th
  century.
 Attrib. Tungfang Shuo 東方朔.

Shen Tzu 愼子.
 The Book of Master Shen.
 Date unknown, probably between +2nd
  and +8th centuries.
 Attrib. Shen Tao (Chou philosopher)
  愼到.

Shêng Chi Chi 聖迹記.
 Records of Holy Places (Buddhist Temples).
 Sui, *c.* +585.
 Ling-Yü 靈裕.

Shêng Chi Tsung Lu 聖濟總錄.
 Imperial Medical Encyclopaedia [issued by
  authority].
 Sung, *c.* +1111.
 Ed. twelve Physicians.

Shêng Hsien Tao Thung Thu Tsan 聖賢道統
 圖贊.
 Comments on Pictures of the Saints and
  Sages, Transmitters of the Tao.
 Ming, +1629.
 Huang Thung-Fan 黃同樊.

Shêng Shui Yen Than Lu 澠水燕談錄.
 Fleeting Gossip by the River Shêng [in
  Shantung].
 Sung, late +11th century.
 Wang Phi-Chih 王闢之.

Shih Chi 史記.
 Historical Records [or perhaps better:
  Memoirs of the Historiographer(-Royal);
  down to −99].

Shih Chi (cont.)
C/Han, c. −90 [first pr. c. +1000].
Ssuma Chhien 司馬遷 and his father
Ssuma Than 司馬談.
Cf. Burton Watson (2).
Partial trs. Chavannes (1); Burton Watson
(1); Pfizmaier (13–36); Hirth (2);
Wu Khang (1); Swann (1), etc.
Yin-Tê Index, no. 40.
Shih Ching 詩經.
Book of Odes [ancient folksongs].
Chou, −9th to −5th centuries.
Writers and compilers unknown.
Tr. Legge (8); Waley (1); Karlgren (14).
Shih Chou 釋舟.
Nautical Nomenclature.
Chhing, c. +1790.
Hung Liang-Chi 洪亮吉.
In *Chüan Shih Ko Wên Chia I Chi*.
Shih Hsüeh 視學.
The Science of Seeing [on perspective].
Chhing, +1729, enl. ed. +1735.
Nien Hsi-Yao 年希堯.
Shih I Chi 拾遺記.
Memoirs on Neglected Matters.
Chin, c. +370.
Wang Chia 王嘉.
Cf. Eichhorn (5).
Shih Kung Hsiao Chi 釋宮小記.
Brief Record of Buildings and Palace
Halls.
Chhing, late +18th century, but not printed
till after 1800.
Chhêng Chêng-Chün 程徵君.
See Chhêng Yao-Thien (4).
Shih Lin Kuang Chi 事林廣記.
Guide through the Forest of Affairs
[encyclopaedia].
Sung, between +1100 and +1250; first pr.
+1325.
Chhen Yuan-Ching 陳元靚.
(A Ming edition of +1478 is in the
Cambridge University Library.)
Shih Ming 釋名.
Explanation of Names [dictionary].
H/Han, c. +100.
Liu Hsi 劉熙.
Shih Ming Su Chêng Pu 釋名疏證補.
See Wang Hsien-Chhien (3).
Shih Shuo Hsin Yü 世說新語.
New Discourses on the Talk of the Times
[notes of minor incidents from Han to
Chin]. Cf. *Hsü Shih Shuo*.
L/Sung, +5th century.
Liu I-Chhing 劉義慶.
Commentary by Liu Hsün 劉峻 (Liang).
Shih Wu Chi Yuan 事物紀原.
Records of the Origins of Affairs and
Things.
Sung, c. +1085.
Kao Chhêng 高承.

Shou Shih Thung Khao 授時通考.
Complete Investigation of the Works and
Days [Imperially Commissioned; a
treatise on agriculture, horticulture an
all related technologies].
Chhing, +1742.
Ed. O-Erh-Thai (Ortai) 鄂爾泰 with
Chang Thing-Yü 張廷玉, Chiang Phu
蔣溥 et al.
Shu Ching 書經.
Historical Classic [or, Book of Documents].
The 29 'Chin Wên' chapters mainly Chou (a
few pieces possibly Shang); the 21 'Ku
Wên' chapters a 'forgery' by Mei Tsê
梅賾, c. +320, using fragments of
genuine antiquity. Of the former, 13 are
considered to go back to the −10th
century, 10 to the −8th, and 6 not
before the −5th. Some scholars accept
only 16 or 17 as pre-Confucian.
Writers unknown.
See Wu Shih-Chhang (1); Creel (4).
Tr. Medhurst (1); Legge (1, 10); Karlgren
(12).
Shu Ching Thu Shuo 書經圖說.
The *Historical Classic* with Illustrations
[published by imperial order].
Chhing, 1905.
Ed. Sun Chia-Nai 孫家鼐 et al.
Shu Hsü Chih Nan 書敘指南.
The Literary South-Pointer [guide to style
in letter-writing, and technical terms].
Sung, +1126.
Jen Kuang 任廣.
Shu I Chi 述異記.
Records of Strange Things.
Liang, early +6th century.
Jen Fang 任昉.
See des Rotours (1), p. ci.
Shu Shu 蜀書.
See *San Kuo Chih*.
Shu Shu Chiu Chang 數書九章.
Mathematical Treatise in Nine Sections.
Sung, +1247.
Chhin Chiu-Shao 秦九韶.
Shu Tao I Chhêng Chi 蜀道驛程記.
Record of the Post Stages on the Szechuan
Circuit.
Chhing, +1672.
Wang Shih-Chên 王士禎.
Shu Tu Fu 蜀都賦.
Ode on the Capital of Shu State (Szechuan)
(Chhêngtu).
Chin, c. +270.
Tso Ssu 左思.
Tr. von Zach (6).
Shu Yü Chou Tzu Lu 殊域周咨錄.
Record of Despatches concerning the
Different Countries.
Ming, +1520.
Ed. Yen Tshung-Chien 嚴從簡.

Shu Yuan Tsa Chi 菽園雜記.
The Bean-Garden Miscellany.
Ming, +1475.
Lu Yung 陸容.

Shui Cha Chi 水閘記.
[= *Chen-Chou Shui Cha Chi*.]
Record of the Building of the Pound-Locks
in the Chenchow District (mod. I-chêng)
[on the Shan-yang Yün-Tao section of the
Grand Canal, south of Huai-yin,
c. +1023].
Sung, +1027.
Hu Su 胡宿.

Shui Chan I Hsiang Lun 水戰議詳論.
Advisory Discourse on Naval Warfare.
Ming, late +16th century, before +1586.
Wang Ho-Ming 王鶴鳴.

Shui Ching 水經.
The Waterways Classic [geographical
account of rivers and canals].
Ascr. C/Han, prob. San Kuo.
Attrib. Sang Chhin 桑欽.

Shui Ching Chu 水經注.
Commentary on the *Waterways Classic*
[geographical account greatly extended].
N/Wei, late +5th or early +6th century.
Li Tao-Yuan 酈道元.

Shui Hua Erh Ta 睡畫二答.
Replies to Questions on Sleep and on Painting.
Ming, +1629.
Pi Fang-Chi 畢方濟
(Francesco Sambiasi.)

Shui Li Shu 水利書.
Treatise on Water Conservancy.
Sung, *c.* +1030.
Fan Chung-Yen 范仲淹.

Shui Li Thu Ching 水利圖經.
Illustrated Manual of Civil Engineering.
Sung.
Chhêng Shih-Mêng 程師孟.

Shui Li Wu Lun 水利五論.
Five Essays on Water Conservancy.
Chhing, +1655.
Ku Shih-Lien 顧士璉.

Shui Pu Shih 水部式.
Ordinances of the Department of Waterways.
Thang, +737.
Writer unknown.
Tunhuang MS., P/2507, Bib. Nat. Paris.
Tr. Twitchett (2).

Shui Tao Thi Kang 水道提綱.
Complete Description of Waterways.
Chhing, +1776.
Chhi Shao-Nan 齊召南.

Shun Fêng Hsiang Sung 順風相送.
Fair Winds for Escort [pilot's handbook].
Ming, *c.* +1430.
Author unknown.
MS., Bodleian Library, Laud Or. no. 145.
Text in Hsiang Ta (5).

Shuo Fu 說郛.
Florilegium of (Unofficial) Literature.
Yuan, *c.* +1368.
Ed. Thao Tsung-I 陶宗儀.
See Ching Phei-Yuan (1).

Shuo Wên.
See *Shuo Wên Chieh Tzu*.

Shuo Wên Chieh Tzu 說文解字.
Analytical Dictionary of Characters.
H/Han, +121.
Hsü Shen 許愼.

Sou Shen Chi 搜神記.
Reports on Spiritual Manifestations.
Chin, *c.* +348.
Kan Pao 干寶.
Partial tr. Bodde (9).

Ssu Chhao Wên Chien Lu 四朝聞見錄.
Record of Things Seen and Heard at Four
Imperial Courts.
Sung, early +13th century.
Yeh Shao-Ong 葉紹翁.

Ssu Chou Chih.
See Liu Tsê-Hsü (1).

Ssu Khu Chhüan Shu 四庫全書.
Complete Library of the Four Categories
(of Literature).
Chhing Imperial MS. Collection.
A vast MS. collection commissioned by the
Chhien-Lung emperor in +1772. For ten
years some 360 scholars, headed by Chi
Yün 紀昀, were employed in collating the
texts of the 3,461 books regarded as the
most noteworthy and valuable. 6,793 books
of lesser interest were described in the
analytical catalogue but not embodied in
the collection. Each of the finished sets
comprised more than 36,000 *pên*. Of the 7
MS. sets 3 still exist in China, and a
selection has been printed as a *tshung-shu*.
See Mayers (1); Têng & Biggerstaff (1),
pp. 27 ff.

Ssu Khu Chhüan Shu Chien Ming Mu Lu
四庫全書簡明目錄.
Abridged Analytical Catalogue of the
Complete Library of the Four Categories
(*of Literature*) (made by imperial order).
Chhing, +1782.
There are two versions of this: (*a*) ed. Chi
Yün 紀昀, which contains mention of
nearly all the books in the *Thi Yao*; (*b*)
ed. Yü Min-Chung 于敏中, which
contains entries only for the books which
were copied into the imperial MS. sets.

Ssu Khu Chhüan Shu Tsung Mu Thi Yao
四庫全書總目提要.
Analytical Catalogue of the *Complete
Library of the Four Categories* (*of
Literature*) (made by imperial order).
Chhing, +1782.
Ed. Chi Yün 紀昀.
Indexes by Yang Chia-Lo; Yü & Gillis.
Yin-Tê index, no. 7.

Ssu Ming Tho Shan Shui Li Pei Lan 四明它
山水利備覽.
Irrigation Canals of the Mount Tho
District (near Ningpo).
Sung, +1242.
Wei Hsien 魏峴.

Su Tzu 蘇子.
The Book of Master Su.
Chhin or Han, late −3rd century.
Attrib. Su Chhin (d. −317) 蘇秦.
Actually a semi-fictional biography of him
by some other writer.
Only fragmentary now in *YHSF*.

Sui Shu 隋書.
History of the Sui Dynasty [+581 to +617].
Thang, +636 (annals and biographies);
+656 (monographs and bibliography).
Wei Chêng 魏徵 *et al.*
Partial trs. Pfizmaier (61–65); Balazs (7, 8);
Ware (1).
For translations of passages see the index of
Frankel (1).

Sui-Ting Fu Chih 綏定府志.
See *Ta-Chou Chih*.

Sun Chho Tzu 孫綽子.
The Book of Master Sun Chho.
Chin, c. +320.
Sun Chho 孫綽.

Sun Fang Pieh Chuan 孫放別傳.
Unofficial Biography of Sun Fang.
Chin, late +4th century.
Writer unknown.

Sun Phu 筍譜.
Treatise on Bamboo Shoots.
Sung, c. +970.
Tsan-Ning (monk) 贊寧.

Sung Hui Yao Kao 宋會要稿.
Drafts for the *History of the Administrative
Statutes of the Sung Dynasty*.
Sung.
Collected by Hsü Sung (1809) 徐松.
From the *Yung-Lo Ta Tien*.

Sung Shih 宋史.
History of the Syng Dynasty [+960 to
+1279].
Yuan, c. +1345.
Tho-Tho (Toktaga) 脫脫 & Ouyang
Hsuan 歐陽玄.
Yin-Tê index, no. 34.

Sung Shu 宋書.
History of the (Liu) Sung Dynasty [+420
to +478].
S/Chhi, +500.
Shen Yo 沈約.
A few chs. tr. Pfizmaier (58).
For translations of passages see the index of
Frankel (1).

Sŭngsŏn Chikchirok 乘船直指錄.
Guide for Shipmasters.
Korea, +1416.
Compiler unknown.

Szechuan Thung Chih 四川通志.
General History and Topography of
Szechuan Province.
Chhing, +18th century (pr. 1816).
Ed. Chhang Ming 常明, Yang Fang-
Tshan 楊芳燦 *et al.*

Szechuan Yen Fa Chih.
See Lo Wên-Pin *et al.* (1).

Ta Chhing I Thung Chih 大清一統志.
Comprehensive Geography of the (Chinese)
Empire (under the Chhing dynasty).
Chhing, c. +1730.
Ed. Hsü Chhien-Hsüeh 徐乾學.

Ta-Chou Chih 達州志.
Local History and Topography of Tachow
(Sui-ting, Szechuan).
Chhing, +1747.
Chhen Chhing-Mên 陳慶門.

Ta Hsüeh Yen I Pu 大學衍義補.
Restoration and Extension of the Ideas of
the *Great Learning* [contains many
chapters of interest for the history of
technology].
Ming, c. +1480.
Chhiu Chün (d. +1495) 丘濬.

Ta Ming Hui Tien 大明會典.
History of the Administrative Statutes of
the Ming Dynasty.
Ming, 1st edn. +1509, 2nd edn. +1587.
Ed. Shen Shih-Hsing 申時行 *et al.*

Ta Sung Chu Shan Thu 大宋諸山圖.
Drawings of the Various (Halls in the)
Mountain (Abbeys) of the Great Sung
Dynasty.
Alternative name for *Wu Shan Shih Chha
Thu*, q.v.

Ta Thang Hsi Yü Chi 大唐西域記.
Record of (a Pilgrimage to) the Western
Countries in the time of the Thang.
Thang, +646.
Hsüan-Chuang 玄奘.
Text by Pien-Chi 辯機.
Tr. Julien (1); Beal (2).

Ta Thang Wu Shan Chu Thang Thu
大唐五山諸堂圖.
Drawings of the Various Halls in the Five
Mountain (Abbeys) dating from the
Great Thang Dynasty.
Alternative name for *Wu Shan Shih Chha
Thu*, q.v.

Ta-Yeh Tsa Chi 大業雜記.
Records of the Reign of Sui Yang Ti [the
Ta-Yeh reign-period, +605 to +616].
Sui.
Tu Pao 杜寶.

Ta Yuan Hai Yün Chi 大元海運記.
Records of Maritime Transportation of the
Yuan Dynasty [originally part of the *Yuan
Ching Shih Ta Tien*].
Yuan, before +1331.

Ta Yuan Hai Yün Chi (*cont.*)
Compiler unknown.
Ed. Hu Ching (Chhing) 胡敬 .

Ta Yuan Tshang Khu Chi 大元倉庫記 .
Records of the Granaries (and Grain
Transport System) of the Yuan Dynasty
[originally part of the *Yuan Ching Shih
Ta Tien*].
Yuan, before +1331.
Compiler unknown.
Ed. Wên Thing-Shih (Chhing) 文廷式 .

Tao I Chih Lüeh 島夷志略 .
Records of the Barbarian Islands (in the
Pacific and Indian Oceans, including the
coasts of East Africa).
Yuan, +1350, based on notes made during
his travels from +1330 to +1349.
Wang Ta-Yuan 汪大淵 .

Tao I Tsa Chih 島夷雜志 .
Miscellaneous Records of the Barbarian
Islands.
Sung, early +12th century.
Writer unknown.
Text reconstructed in Wada Kyutoku (*1*).

Tao Tê Ching 道德經 .
Canon of the Tao and its Virtue.
Chou, before −300.
Attrib. Li Erh (Lao Tzu) 李耳(老子).
Tr. Waley (4); Chhu Ta-Kao (2); Lin Yü-
Thang (1); Wieger (7); Duyvendak (18);
and very many others.

Tao Tsang 道藏 .
The Taoist Patrology [containing 1,464
Taoist works].
All periods, but first collected in the Thang
about +730, then again about +870 and
definitively in +1019. First printed in the
Sung (+1111 to +1117). Also printed in
J/Chin (+1186 to +1191), Yuan
(+1244), and Ming (+1445, +1598 and
+1607).
Writers numerous.
Indexes by Wieger (6), on which see
Pelliot's review; and Ong Tu-Chien (Yin-
Tê Index, no. 25).

Tao Yuan Hsüeh Ku Lu 道園學古錄 .
Historical Essays of (Yü) Tao-Yuan.
Yuan, *c.* +1340.
Yü Chi 虞集 .

Thai Pai Yin Ching 太白陰經 .
Manual of the White (and Gloomy) Planet
(of War; Venus) [treatise on military
affairs].
Thang, +759.
Li Chhüan 李筌 .

Thai-Phing Huan Yü Chi 太平寰宇記 .
Thai-Phing reign-period General Descrip-
tion of the World [geographical record].
Sung, +976 to +983.
Yüeh Shih 樂史 .

Thai-Phing Yü Lan 太平御覽 .
Thai-Phing reign-period Imperial Encyclo-
paedia (lit. the Emperor's Daily Readings).
Sung, +983.
Ed. Li Fang 李昉 .
Some chs. tr. Pfizmaier (84–106).
Yin-Tê Index, no. 23.

Than Yuan Thi Hu 譚苑醍醐 .
A Delicious Dish of Talk.
Ming, *c.* +1510.
Yang Shen 楊愼 .

Thang Fu-Li Hsien-sêng Wên Chi 唐甫里先
生文集 .
Collected Literary Remains of Mr Fu-Li of
the Thang (i.e. Lu Kuei-Mêng).
Thang, before +881.

Thang Hui Yao 唐會要 .
History of the Administrative Statutes of
the Thang Dynasty.
Sung, +961.
Wang Phu 王溥 .
Cf. des Rotours (2), p. 92.

Thang Kuo Shih Pu 唐國史補
(orig. title: *Kuo Shih Pu*).
Supplementary Information for the History
of the present [i.e. Thang] Dynasty.
Thang, +8th century.
Li Chao 李肇 .

Thang Liu Tien 唐六典 .
Institutes of the Thang Dynasty (lit.
Administrative Regulations of the Six
Ministries of the Thang).
Thang, +738 or +739.
Ed. Li Lin-Fu 李林甫 .
Cf. des Rotours (2), p. 99.

Thang Shu.
See *Chiu Thang Shu* and *Hsin Thang Shu.*

Thang Yü Lin 唐語林 .
Miscellanea of the Thang Dynasty.
Sung, collected *c.* +1107.
Wang Tang 王讜 .
Cf. des Rotours (2), p. 109.

Thieh Chhiao Chih Shu 鐵橋志書 .
Record of the (Building of the) Iron(-Chain)
Suspension-Bridge [at Kuanling
nr. Phanhsien in Kueichow].
Cf. *Wan Nien Chhiao Chih.*
Ming, +1629 (first pr. +1665).
Chu Hsieh-Yuan 朱燮元 .

Thieh Chi Phu 蝶几譜 .
Discourse on Butterfly Tables [furniture].
Ming, +1617.
Ko Shan 戈汕 .

Thieh Wei Shan Tshung Than 鐵圍山叢談 .
Collected Conversations at Iron-Fence
Mountain.
Sung, *c.* +1115.
Tshai Thao 蔡條 .

Thien Kung Khai Wu 天工開物 .
The Exploitation of the Works of Nature.
Ming, +1637.
Sung Ying-Hsing 宋應星 .

Thien Wên 天問 .
   Questions about Heaven ['ode', perhaps a
     ritual catechism].
   Chou, generally ascr. late −4th, but perhaps
     −5th century.
   Attrib. Chhü Yuan 屈原, but probably
     earlier.
   Tr. Erkes (8); Hawkes (1).
Thien Wên Lüeh 天問畧 .
   Explicatio Sphaeris Coelestis.
   Ming, +1615.
   Yang Ma-No (Emanuel Diaz) 陽瑪諾 .
Thu Hua Chien Wên Chih 圖畫見聞志 .
   Observations on Drawing and Painting
     (from +841 to +1074).
   Sung, after +1074.
   Kuo Jo-Hsü 郭若虛 .
   Hirth (12), p. 109.
Thu Shu Chi Chhêng 圖書集成 .
   Imperial Encyclopaedia.
   Chhing, +1726.
   Ed. Chhen Mêng-Lei 陳夢雷 *et al.*
   Index by L. Giles (2).
Thung Chien Kang Mu 通鑑綱目 .
   Short View of the *Comprehensive Mirror (of
     History, for Aid in Government)* [the *Tzu
     Chih Thung Chien* condensed, with
     headings and sub-headings].
   Sung (begun +1172), +1189.
   Chu Hsi 朱熹 (and his school).
   With later continuations, *Thung Chien Kang
     Mu Hsü Pien* and *Thung Chien Kang Mu
     San Pien.*
   Definitive edition; with all commentaries,
     etc., *c.* +1630.
   Partial tr. Wieger (1).
   Ed. Chhen Jen-Hsi 陳仁錫 .
Thung Chien Kang Mu Hsü Pien 通鑑綱目續
   編 .
   Continuation of the *Short View of the
     Comprehensive Mirror (of History, for Aid
     in Government)* [covering the Sung and
     Yuan periods].
   Ming, +1476, pr. after +1500.
   Ed. Shang Lu 商輅 .
Thung Chien Kang Mu San Pien 通鑑綱目三編 .
   Continuation of the *Short View of the
     Comprehensive Mirror (of History, for Aid
     in Government)* [covering the Ming
     period].
   Chhing, +1746.
   Ed. Shen Tê-Chhien 沈德潛 & Chhi
     Shao-Nan 齊召南 .
Thung Chih 通志 .
   Historical Collections.
   Sung, *c.* +1150.
   Chêng Chhiao 鄭樵 .
   Cf. des Rotours (2), p. 85.
Thung Chih Lüeh 通志畧 .
   Compendium of Information [part of *Thung
     Chih*, q.v.].

Thung Hui Ho Chih 通惠河志 .
   Record of the Canal of Communicating
     Grace (part of the Grand Canal, from
     Peking to Thungchow, built by Kuo
     Shou-Ching, +1293).
   Ming, +1558.
   Wu Chung 吳仲 .
Thung Thien Hsiao 通天曉 .
   Book of General Information [including
     techniques, etc.].
   Chhing, end +18th century, pr. 1816, 1837,
     1856 at Canton.
   Wang Hsiang-Thang 王纕堂 .
Thung Tien 通典 .
   Comprehensive Institutes [reservoir of
     source material on political and social
     history].
   Thang, *c.* +812 (completed by +801).
   Tu Yu 杜佑 .
Tiao Chi Li Than 釣磯立談 .
   Talks at Fisherman's Rock.
   Wu Tai (S/Thang) & Sung, begun *c.* +935.
   Shih Hsü-Pai 史虛白 .
Ting-Hai Thing Chih 定海廳志 .
   Local Gazetteer of the Sub-Prefecture of
     Ting-hai (on Choushan Island).
   Chhing +1715, revised 1884 and 1902.
   Ed. Miu Sui 繆燧 & Chhen Yü-Wei
     陳于渭 .
   Revised Shih Chih-Hsün 史致馴 & Wang
     Hsün 汪洵 .
Ting-Li I Min 鼎澧逸民 .
   Recollections of Tingchow.
   Sung, *c.* +1150.
   Writer unknown.
   Cf. Chu Hsi-Tsu (2).
Tōdaiji Zōritsu Kuyōki 東大寺造立供養記 .
   A Record of the Rebuilding of the Tōdaiji
     (temple, at Nara) [+1168 to +1185].
   Japan, +1452.
   Writer unknown.
Tōdo Kōteiki 唐土行程記 .
   Japanese title of the translation (+1769) of
     Phyohae-Rok, q.v.
Tsao Chuan Thu Shuo 造甎圖說 .
   Illustrated Account of Brick-and-Tile-
     Making.
   Ming, between +1525 and +1565.
   Chang Wên-Chih 張問之 .
Tshan Luan Lu 驂鸞錄 .
   Guiding the Reins [narrative of a three
     months' journey from the capital to
     Kueilin].
   Sung, +1172.
   Fan Chhêng-Ta 范成大 .
   Cf. *Lan Phei Lu.*
Tshao Chhuan Chih 漕船志 .
   Records of Canal and River Shipping.
   Ming, +1501, enlarged +1544.
   Hsi Shu 席書 & Chu Chia-Hsiang
     朱家相 .

Tshao Yün Fu Khu Tshang Yü 漕運府庫倉度.
Tax-Grain Water Transport and Granaries.
Sung, about +1270.
Wang Ying-Lin 王應麟.

Tso Chuan 左傳.
Master Tsochhiu's Tradition (or Enlargement) of the *Chhun Chhiu* (*Spring and Autumn Annals*) [dealing with the period −722 to −453].
Late Chou, compiled from ancient written and oral traditions of several States between −430 and −250, but with additions and changes by Confucian scholars of the Chhin and Han, especially Liu Hsin. Greatest of the three commentaries on the *Chhun Chhiu*, the others being the *Kungyang Chuan* and the *Kuliang Chuan*, but unlike them, probably originally itself an independent book of history.
Attrib. Tsochhiu Ming 左邱明.
See Karlgren (8); Maspero (1); Chhi Ssu-Ho (1); Wu Khang (1); Wu Shih-Chhang (1); van der Loon (1); Eberhard, Müller & Henseling.
Tr. Couvreur (1); Legge (11); Pfizmaier (1–12).
Index by Fraser & Lockhart (1).

Tso Chuan Pu Chu 左傳補注.
Commentary on *Master Tsochhiu's Enlargement of the Chhun Chhiu*.
Chhing, +1718.
Hui Tung 惠棟.

Tu Chhêng Chi Shêng 都城紀勝.
The Wonder of the Capital (Hangchow).
Sung, +1235.
Mr Chao 趙氏 [Kuan Pu Nai Tê Ong 灌圃耐得翁; The Old Gentleman of the Water-Garden who achieved Success through Forbearance].

Tu Shih Fang Yü Chi Yao 讀史方輿紀要.
Essentials of Historical Geography.
Chhing, +1667.
Ku Tsu-Yü 顧祖禹.

Tung Ching Fu 東京賦.
Ode on the Eastern Capital (Loyang).
H/Han, +107.
Chang Hêng 張衡.
Tr. von Zach (6); Hughes (9).

Tung Ching Mêng Hua Lu 東京夢華錄.
Dreams of the Glories of the Eastern Capital (Khaifêng).
S/Sung, +1148 (referring to the two decades which ended with the fall of the capital of N/Sung in +1126 and the completion of the move to Hangchow in +1135), first pr. +1187.
Mêng Yuan-Lao 孟元老.

Tung Hsi Yang Khao 東西洋考

Studies on the Oceans East and West.
Ming, +1618.
Chang Hsieh 張燮.

Tung Hsien Pi Lu 東軒筆錄.
Jottings from the Eastern Side-Hall.
Sung, end +11th century.
Wei Thai 魏泰.

Tung Kuan Chi Shih 冬官紀事.
A Relation concerning the Ministry of Works [especially the upright career of his father, Ho Shêng-Jui, therein].
Ming, c. +1610.
Ho Chung-Shih 賀仲軾.

Tung Ming Chi 洞冥記.
Light on Mysterious Things.
Ascr. Han; prob. +5th or +6th century.
Attrib. Kuo Hsien 郭憲.

Tung-Pho Chhüan Chi or *Chhi Chi* 東坡全集 (七集).
The Complete (or Seven) Collections of (Su) Tung-Pho [i.e. Collected Works].
Sung, down to +1101, but put together later.
Su Tung-Pho 蘇東坡.

Tung-Pho Chih Lin 東坡志林.
Journal and Miscellany of (Su) Tung-Pho [compiled while in exile in Hainan].
Sung, +1097 to +1101.
Su Tung-Pho 蘇東坡.

Tung Tu Fu 東都賦.
Ode on the Eastern Capital (Loyang).
H/Han, c. +87.
Pan Ku 班固.
Tr. Hughes (9).

Tzhu Shih Chi Hsieh Fu 刺世疾邪賦.
Ode to Cure Quickly the Evils of the Age.
H/Han, c. +178.
Chao I 趙壹.

Tzu Chih Thung Chien 資治通鑑.
Comprehensive Mirror (of History) for Aid in Government [−403 to +959].
Sung, begun +1065, completed +1084.
Ssuma Kuang 司馬光.
Cf. des Rotours (2), p. 74; Pulleyblank (7).
A few chs. tr. Fang Chih-Thung (1).

Tzu Shih Ching Hua 子史精華.
Essence of the Philosophers and Historians [dictionary of quotations].
Chhing, +1727.
Yün Lu 允祿 *et al.*

Wakan Senyōshū 和漢船用集.
Collected Studies on the Ships used by the Japanese and Chinese.
Japan, +1766 (author's preface, +1761).
Kanazawa Kanemitsu 金澤兼光.
(Repr. *NKKZ*, vol. 12.)

Wan-An Chhiao Chi 萬安橋記.
A Record of the Wan-An (megalithic beam) bridge (near Chhüanchow, Fukien).
Sung, c. +1060.
Tshai Hsiang 蔡襄.

Wan Chi Lun 萬機論.
　　The Myriad Stratagems [naval and military].
　　San Kuo (Wei), *c.* +225.
　　Chiang Chi 蔣濟.
Wan Nien Chhiao Chih 萬年橋誌.
　　See Hsieh Kan-Thang (*1*).
　　Cf. *Thieh Chhiao Chih Shu.*
Wan Pi Shu 萬畢書.
　　See *Huai-Nan (Wang) Wan Pi Shu.*
Wei Lüeh 魏略.
　　Memorable Things of the Wei Kingdom
　　　　(San Kuo).
　　San Kuo (Wei) or Chin, +3rd or +4th
　　　　century.
　　Yü Huan 魚豢.
Wei Lüeh 緯畧.
　　Compendium of Non-Classical Matters.
　　Sung, +12th century (end).
　　Kao Ssu-Sun 高似孫.
Wei Shu 魏書.
　　See *San Kuo Chih.*
Wei Shu 魏書.
　　History of the (Northern) Wei Dynasty
　　　　[+386 to +550, including the Eastern
　　　　Wei successor State].
　　N/Chhi, +554, revised +572.
　　Wei Shou 魏收.
　　See Ware (*3*).
　　One ch. tr. Ware (*1, 4*).
　　For translations of passages, see the index of
　　　　Frankel (*1*).
Wei Tu Fu 魏都賦.
　　Ode on the Capital of Wei State (Yeh,
　　　　Hsiangchow).
　　Chin, *c.* +270.
　　Tso Ssu 左思.
　　Tr. von Zach (6).
Wên Chhang Tsa Lu 文昌雜錄.
　　Things Seen and Heard by an Official at
　　　　Court (during service in the Department
　　　　of Ministries).
　　Sung, +1086.
　　Phang Yuan-Ying 龐元英.
Wên Chien Chin Lu 聞見近錄.
　　New Records of Things Heard and Seen.
　　Sung, wr. +1085 to +1104, pr. +1163.
　　　　Deals with events from +954 to +1085.
　　[First of the three parts of the *Chhing Hsü
　　　　Tsa Chu,* q.v.]
　　Wang Kung 王鞏.
Wên Hsien Thung Khao 文獻通考.
　　Comprehensive Study of (the History of)
　　　　Civilisation (lit. Complete Study of the
　　　　Documentary Evidence of Cultural
　　　　Achievements (in Chinese Civilisation)).
　　Sung & Yuan, begun perhaps as early as
　　　　+1270 and finished before +1317,
　　　　printed +1322.
　　Ma Tuan-Lin 馬端臨.
　　Cf. des Rotours (2), p. 87.
　　A few chs. tr. Julien (2); St Denys (1).

Wên Hsüan 文選.
　　General Anthology of Prose and Verse.
　　Liang, +530.
　　Ed. Hsiao Thung (prince of the Liang)
　　　　蕭統.
　　Tr. von Zach (6).
Wên Shih Chen Ching 文始眞經.
　　True Classic of the Original Word (of Lao
　　　　Chün, third person of the Taoist Trinity)
　　　　(= *Kuan Yin Tzu,* q.v.).
Wu-Chê Wang Chhêng-Fu Chuan 圬者王承
　　福傳.
　　What I learnt from the Mason Wang
　　　　Chhêng-Fu [essay].
　　Thang, *c.* +810.
　　Han Yü 韓愈.
Wu Chhuan Lu 吳船錄.
　　Account of a Journey by boat to Wu [from
　　　　Chhêngtu in Szechuan to Chiangsu].
　　Sung, +1177.
　　Fan Chhêng-Ta 范成大.
Wu Ching Tsung Yao 武經總要.
　　Collection of the most important Military
　　　　Techniques [compiled by Imperial
　　　　Order].
　　Sung, +1040 (+1044).
　　Ed. Tsêng Kung-Liang 曾公亮.
Wu Chung Shui Li Shu 吳中水利書.
　　The Water-Conservancy of the Wu District.
　　Sung, +1059.
　　Shan O 單鍔.
Wu Lei Hsiang Kan Chih 物類相感志.
　　On the Mutual Responses of Things
　　　　according to their Categories.
　　Sung, *c.* +980.
　　Attrib. Su Tung-Pho 蘇東坡.
　　Actual writer, (Lu) Tsan-Ning (monk)
　　　　錄贊寧.
Wu Li Hsiao Shih 物理小識.
　　Small Encyclopaedia of the Principles of
　　　　Things.
　　Chhing, +1664.
　　Fang I-Chih 方以智.
　　Cf. Hou Wai-Lu (3, 4).
Wu Lin Chiu Shih 武林舊事.
　　Institutions and Customs of the Old
　　　　Capital (Hangchow).
　　Sung, *c.* +1270 (but referring to events
　　　　from about +1165 onwards).
　　Chou Mi 周密.
Wu Mu Ching 五木經.
　　Manual of the Five (Throws of the)
　　　　Wooden (Dice).
　　Thang, *c.* +810.
　　Li Ao 李翱.
Wu Pei Chih 武備志.
　　Treatise on Armament Technology.
　　Ming, +1628.
　　Mao Yuan-I 茅元儀.
Wu Pei Chih Shêng Chih 武備制勝志.
　　The Best Designs in Armament Technology.

Wu Pei Chih Shêng Chih (cont.)
    Ming, *c.* +1628.
    Mao Yuan-I 茅元儀.
    MS. of 1843 in the Cambridge University
        Library.
Wu Pei Pi Shu 武備秘書.
    Confidential Treatise on Armament
        Technology [a compilation of selections
        from earlier works on the same subject].
    Chhing, late +17th century (repr. 1800).
    Shih Yung-Thu 施永圖.
Wu Shan Shih Chha Thu 五山十刹圖.
    Drawings of the Five Mountain (Abbeys)
        and the Ten Priories [MS. preserved in
        Japan].
    Sung, *c.* +1259.
    Gikai (monk) 義介.
    Ed. Yokoyama Hidetoshi 橫山秀哉 *c.* 1940.
    See Tanabe Yasutake (1).
Wu Shih Wai Kuo Chuan 吳時外國傳.
    Records of the Foreign Countries in the
        Time of the State of Wu.
    San Kuo, *c.* +260.
    Khang Thai 康泰.
    Only in fragments in *TPYL*.
Wu Shu 吳書.
    See *San Kuo Chih*.
Wu Tai Shih Chi.
    See *Hsin Wu Tai Shih*.
Wu Tu Fu 吳都賦.
    Ode on the Capital of Wu State (Suchow).
    Chin, *c.* +270.
    Tso Ssu 左思.
    Tr. von Zach (6).
Wu Yin Chi Yün 五音集韻.
    The Five Notes Complete Dictionary of the
        Sounds of the Characters [a compilation
        of the *Kuang Yün* and the *Chi Yün*, q.v.].
    J/Chin, *c.* +1200.
    Han Tao-Chao 韓道照.
Wu Yüeh Chhun Chhiu 吳越春秋.
    Spring and Autumn Annals of the States of
        Wu and Yüeh.
    H/Han.
    Chao Yeh 趙曄.
Wu Yüeh Pei Shih 吳越備史.
    Materials for the History of Wu and Yüeh
        (in the Five Dynasties Period).
    Sung, *c.* +995.
    Chhien Yen 錢儼.

Yang-Chhêng Ku Chhao.
    See Chhou Chhih-Shih (1).
Yen Chi Thu 燕几圖.
    Diagrams of Peking Tables [furniture].
    Sung, *c.* +1090.
    Huang Po-Ssu 黃伯思.
Yen Fan Lu 演繁露.
    Extension of the *String of Pearls on the
        Spring and Autumn Annals* [on the mean-
        ing of many Thang and Sung expressions].

Sung, +1180.
    Chhêng Ta-Chhang 程大昌.
    See des Rotours (1), p. cix.
Yen Tan Tzu 燕丹子.
    (Life of) Prince Tan of Yen (*d.* −226) [an
        embroidered version of the biography of
        Ching Kho (q.v.) in *Shih Chi*, ch. 86, but
        perhaps containing some authentic details
        not therein].
    Probably H/Han, end of +2nd century.
    Writer unknown.
    Tr. Chêng Lin (1); H. Franke (11).
Yen Thieh Lun 鹽鐵論.
    Discourses on Salt and Iron [record of the
        debate of −81 on State control of com-
        merce and industry].
    C/Han, *c.* −80 to −60.
    Huan Khuan 桓寬.
    Partial tr. Gale (1); Gale, Boodberg & Lin.
Yen Thu Shui I 沿途水驛.
    Post Stations along the Roads and Streams
        [itinerary from Hangchow to Peking].
    Ming, +1535.
    Writer unknown.
Yi Chhungmu Kong Chŏnsŏ 李忠武公全書.
    See *Chhungmu Kong Chŏnsŏ*.
Ying Tsao Chêng Shih 營造正式.
    Right Standards of Building Construction.
    Ming, but may contain older material.
    Attrib. Lu Pan 魯般.
    See Liu Tun-Chên (6, 7).
Ying Tsao Fa Shih 營造法式.
    Treatise on Architectural Methods.
    Sung, +1097; pr. +1103; repr. +1145.
    Li Chieh 李誡.
Ying Yai Shêng Lan 瀛涯勝覽.
    Triumphant Visions of the Ocean Shores
        [relative to the voyages of Chêng Ho].
    Ming, +1451. (Begun +1416 and com-
        pleted about +1435.)
    Ma Huan 馬歡.
    Tr. Groeneveldt (1); Phillips (1);
        Duyvendak (10).
Ying Yai Shêng Lan Chi 瀛涯勝覽集.
    Abstract of the *Triumphant Visions of the
        Ocean Shores* [a refacimento of Ma Huan's
        book].
    Ming, +1522.
    Chang Shêng (*b*) 張昇.
    Passages cit. in *TSCC, Pien i tien*, chs. 58,
        73, 78, 85, 86, 96, 97, 98, 99, 101, 103, 106.
    Tr. Rockhill (1).
Yü Chien 寓簡.
    Allegorical Essays.
    Sung.
    Shen Tso-Chê 沈作喆.
Yü Hai 玉海.
    Ocean of Jade [encyclopaedia].
    Sung, +1267 (first pr. Yuan, +1351).
    Wang Ying-Lin 王應麟.
    Cf. des Rotours (2), p. 96.

Yü Kung Chui Chih 禹貢錐指.
A Few Points in the Vast Subject of the
Tribute of Yü [the geographical chapter
in the *Shu Ching*] (lit. 'pointing at the
Earth with an Awl') (including the set of
maps, *Yü Kung Thu*).
Chhing, +1697 and +1705.
Hu Wei 胡渭.

Yü Kung Shuo Tuan 禹貢說斷.
Discussions and Conclusions regarding the
Geography of the *Tribute of Yü*.
Sung, *c.* +1160.
Fu Yin 傅寅.

Yü-Phi Li Tai Thung Chien Chi Lan 御批歷
代通鑑輯覽.
Imperially Commissioned Essentials of the
Comprehensive Mirror of History.
See *Li Tai Thung Chien Chi Lan*.

Yü Phien 玉篇.
Jade Page Dictionary.
Liang, +543.
Ku Yeh-Wang 顧野王.
Extended and edited in the Thang (+674)
by Sun Chhiang 孫強.

Yü Thang Chia Hua 玉堂嘉話.
Refined Conversations in the Academy.
Yuan, +1288.
Wang Yün 王惲.
Cf. H. Franke (14).

Yü Ti Chi Shêng 輿地紀勝.
The Wonder of the World [geography].
Sung, +1221.
Wang Hsiang-Chih 王象之.

Yü Ti Tsung Thu 輿地總圖.
General World Atlas.
Ming, +1564.
Shih Ho-Chi 史臞冀.

Yü-Tsuan Pên Tshao Phin Hui Ching Yao
御纂本草品彙精要.
See *Pên Tshao Phin Hui Ching Yao*.

Yu-Yang Tsa Tsu 酉陽雜俎.
Miscellany of the Yu-yang Mountain (Cave)
[in S.E. Szechuan].
Thang, +863.
Tuan Chhêng-Shih 段成式.
See des Rotours (1), p. civ.

Yuan Chien Lei Han 淵鑑類函.
Mirror of the Infinite; a Classified Treasure-
Chest [great encyclopaedia; the conflation
of 4 Thang and 17 other encyclopaedias].
Chhing, presented, +1701, pr. +1710.
Ed. Chang Ying 張英 *et al.*

Yuan Ching Shih Ta Tien 元經世大典.
Institutions of the Yuan Dynasty.
Yuan, +1329 to +1331.
Partly reconstructed and ed. Wên Thing-
Shih (1916) 文廷式.
Cf. Hummel (2), p. 855.

Yuan Hai Yün Chih 元海運志.
A Sketch of Maritime Transportation
during the Yuan Period.

Yuan or Ming, late +14th century.
Wei Su 危素.

Yuan-Ho Chün Hsien Thu Chih 元和郡縣圖
志.
Yuan-Ho reign-period General Geography.
Thang, +814.
Li Chi-Fu 李吉甫.
Cf. des Rotours (2), p. 102.

Yuan Shih 元史.
History of the Yuan (Mongol) Dynasty
[+1206 to +1367].
Ming, *c.* +1370.
Sung Lien 宋濂 *et al.*
Yin-Tê Index, no. 35.

Yuan Shu Tsa Chi 宛署雜記.
Records of the Seat of Government at
Yuan(-phing) (Peking).
Ming, +1593.
Shen Pang 沈榜.

Yuan Wên Lei 元文類.
Classified Collections of Yuan (Dynasty)
Literature.
Yuan, *c.* +1350.
Su Thien-Chio 蘇天爵.

Yuan Yeh 園冶.
On the Making of Gardens.
Ming, +1634.
Chi Wu-Fou 計無否.

Yüeh Chüeh Shu 越絕書.
Lost Records of the State of Yüeh.
H/Han, *c.* +52.
Attrib. Yuan Khang 袁康.

Yüeh Yang Chen Lu Chi 粵洋針路記.
Compass-Bearing Rutter for the Cantonese
Seas.
Ming or Chhing.
Writer unknown.
See Pelliot (3b), p. 308.

Yün Lu Man Chhao 雲麓漫抄.
Random Jottings at Yün-Lu.
Sung, +1206 (referring to events of about
+1170 onwards).
Chao Yen-Wei 趙彥衛.

Yün Shih Chai Pi Than 韻石齋筆談.
Jottings from the Sounding-Stone Studio.
Chhing, early +17th century.
Chiang Shao-Shu 姜紹書.

Yung-Lo Ta Tien 永樂大典.
Great Encyclopaedia of the Yung-Lo reign-
period [only in manuscript].
Amounting to 22,877 chapters in 11,095
pên, only about 370 being still extant.
Ming, +1407.
Ed. Hsieh Chin 解縉.
See Yuan Thung-Li (1).

Yünnan Thung Chih 雲南通志.
General History and Topography of Yunnan
Province.
Chhing, +1691.
Enl. repr. 1894.
Ed. Fan Chhêng-Hsün 范承勳.

B. CHINESE AND JAPANESE BOOKS AND JOURNAL ARTICLES SINCE +1800

Adachi Kiroku (1) 足立喜六.
Chōan Shiseki no Kenkyū 長安史蹟の研究.
Researches on the History and Archaeology of Chhang-an (the Chhin and Han capital of China).
2 vols.
Tōyō Bunko Ronsō, Tokyo, 1933.

Amano Motonosuke (2) 天野元之助.
Shū no Hōkensei to Seidensei 周の封建制と井田制.
On the Feudal System of the Chou and the 'Well-Field' System.
JK, 1956, 7, 836.
Abstr. RBS, 1956, 2, no. 76.

Amano Motonosuke (3) 天野元之助.
Chūgoku ni okeru Suiri Kankō 中國における水利慣行.
Customs of Irrigation (Water-Rights, etc.) in China.
SN, 1955 (no. 6), 123 (559).

Amanuma Shun-ichi (1) 天沼俊一.
Nihon Kenchikushi Zuroku 日本建築史圖錄.
An Illustrated History of Japanese Architecture.
6 vols.
Tokyo, 1938.

An Chin-Huai (1) et al. 安金槐.
Chêngchow Nan-Kuan Wai Pei Sung Chuan Shih Mu 鄭州南關外北宋磚室墓.
A Brick-built Tomb of the Northern Sung period outside the South Gate of Chêngchow.
WWTK, 1958 (no. 5), 52.

Andō Kōsei (1) 安藤更生.
Kanshin 鑑眞.
Life of Chien-Chen (+688 to +763) [outstanding Buddhist missionary to Japan, skilled also in medicine and architecture].
Bijutsu Shuppansha, Tokyo, 1958, repr. 1963.
Abstr. RBS, 1964, 4, no. 889.

Anon. (3).
Ming Chhang Ling Hsiu Shan Kung Chhêng Chi Yao 明長陵修繕工程紀要.
Memoir on the Restoration of the Chhang Ling (Tomb of the Yung-Lo Emperor), [one of the (Thirteen)] Ming [Imperial Tombs (North of Peking)].
Ministry of the Interior Restoration Committee, Peiping, 1935.

Anon. (9).
Chōan to Raku-yō 長安と洛陽.

[(Thang) Maps (City-Plans) of] Chhang-an and Loyang.
In Tōdai Kenkyū no Shiori 唐代研究のしおり.
Aids to the Study of Thang History.
Jimbun Kagaku Kenkyūjō, vols. 5, 6 and 7.
Kyoto, 1957.

Anon. (10).
Tunhuang Pi Hua Chi 敦煌壁畫集.
Album of Coloured Reproductions of the fresco-paintings at the Tunhuang cave-temples.
Peking, 1957.

Anon. (11).
Chhangsha Fa Chüeh Pao-Kao 長沙發掘報告
Report on the Excavations (of Tombs of the Chhu State, of the Warring States period, and of the Han Dynasties) at Chhangsha.
Acad. Sinica Archaeol. Inst., Kho-Hsüeh, Peking, 1957.

Anon. (12).
Ti-Hsia Kung Tien—Ting Ling 地下宮殿—定陵.
A Palace Underground—the Ting Ling (Tomb of the Wan-Li emperor of the Ming).
Wên-Wu, Peking, 1958.

Anon. (22).
Szechuan Han Hua Hsiang Chuan Hsüan Chi 四川漢畫像磚選集.
A Selection of Bricks with Stamped Reliefs from Szechuan.
Wên-Wu, Peking, 1957.

Anon. (23).
Chêngchow Erh-Li-Kang 鄭州二里岡.
Report on the Erh-Li-Kang (Tombs) at Chêngchow.
Academia Sinica, Peking, 1959.

Anon. (24).
Honan Hsin-yang Chhu Mu Chhu Thu Wên Wu Thu Lu 河南信陽楚墓出土文物圖錄.
Illustrated Report on the Cultural Objects excavated from (Princely) Chhu State Tombs at Hsinyang in Honan.
Archaeological Institute, Honan Provincial Pub. House, Chêngchow, 1959.

Anon. (25).
Pan-pho I-chi Chieh Shao 半坡遺跡介紹.
The Relics of the [Neolithic Village] of Pan-pho.
Pan-pho Museum, Sian, 1958.

Anon. (26).
  Chung-Kuo 1959 中國 1959.
  China, a Pictorial Album (commemorating
    the Tenth Anniversary of the
    Foundation of the People's Republic).
  Peking, 1959.
Anon. (28).
  Yünnan Chin-Ning Shih-Chai Shan Ku Mu
    Chhün Fa-Chüeh Pao-Kao 雲南晉寧石
  寨山古墓羣發掘報告.
  Report on the Excavation of a Group of
    Tombs (of the Han period Tien Culture)
    at Shih-chai Shan near Chin-ning in
    Yunnan.
  2 vols.
  Yunnan Provincial Museum.
  Wên-Wu, Peking, 1959.
Anon. (31).
  Kuang-Chou Shih Tung-Chiao Tung Han
    Chuan Shih Mu Chhing Li Chi-Lüeh
  廣州市東郊東漢磚室墓清理紀略.
  Report on the Tomb Furniture of Brick
    Tombs of the Later Han period excavated
    in the Tung-Chiao district of Canton.
  WWTK, 1955 (no. 6), 61; *RBS*, 1957, **1**,
    no. 209.
Anon. (32).
  Chung Hua Mei Shu Thu Chi 中華美
  術圖集.
  Album of Chinese Painting and Calligraphy.
  2 vols.
  Chung-Hua Tshung-Shu Committee.
  Thaipei, 1955.
Anon. (33).
  San Mên Hsia Tshao Yün I Chi 三門峽
  漕運遺跡.
  The Remains of the Canal [and the
    Trackers' Galleries] in the San Mên
    Gorge (of the Yellow River).
  Kho-Hsüeh, Peking, 1959.
  (Academia Sinica, Archaeological Field
    Studies, no. 8.)
Anon. (36).
  Chhüan Kuo Kung-Lu Chan-Lan Hui Thê-
    Khan 全國公路展覽會特刊.
  The Highroads of China; A Companion to
    the Highways Exhibition.
  Chungking, 1944.
Anon. (37).
  Chung-Kuo Chien Chu 中國建築.
  Chinese Architecture (and Bridge-building)
    [album].
  Dept. of the History of Architecture,
    Academia Sinica, and the Architectural
    School of Chhinghua University.
  Wên-wu, Peking, 1957.
Anon. (38).
  Chung-Kuo Chien-Chu Tsai-Hua Thu An
  中國建築彩畫圖案.
  Painting and Decoration in Chinese
    Architecture [album].

Peking Cultural Restoration Committee.
  Jen-Min I-Shu, Peking, 1955.
Anon. (39).
  Lan-Kan Kung-Chhüan Chu-Chhu Chhiang
    Mien Chuang Shih 欄杆拱券柱礎牆
  面裝飾.
  The Ornamentation of Balustrades, Arches,
    Columns, Plinths and Walls (in Chinese
    Architecture).
  Ministry of Architecture and Engineering,
    Peking Bureau of Industrial Planning.
  Architectural and Engineering Press,
    Peking, 1955.
Anon. (40).
  Mei Shu Khao-Ku Hsüeh Yung Yü Chi,
    Chien-Chu Phien 美術考古學用語集
  建築篇.
  (Korean) Vocabularies in the Fields of Arts
    and Archaeology, Pt. 1 Architecture.
  Eul-Yu Pub. Co., Seoul, 1955 (for the
    Research Dept. Nat. Mus. of Korea).
  Pubs. Nat. Mus. Korea, Ser. A, no. 2.
Anon. (41).
  Nei Mêng-Ku Ku Chien-Chu 內蒙古古
  建築.
  The Traditional Architecture of Inner
    Mongolia.
  Wên-Wu, Peking, 1959.
Anon. (42).
  Kuang-Chou Chhu-Thu Han-Tai Thao Wu
  廣州出土漢代陶屋.
  Pottery Models of Dwellings excavated
    from Cantonese Tombs (including
    Granaries, Wellheads and Stoves).
  Wên-Wu, Peking, 1958.
Anon. (43).
  Hsin Chung-Kuo-ti Khao-Ku Shou-Huo
  新中國的考古收獲.
  Successes of Archaeology in New China.
  Wên-Wu, Peking, 1961.
  Names of the 22 writers in this collective
    work are given on p. 135.
  Crit. rev. Chêng Tê-Khun, *AQ*, 1964, **38**,
    179.
Anon. (47).
  Chiangsu chih Tha 江蘇之塔.
  The Pagodas of Chiangsu Province.
  Shanghai, 1960.
Anon. (49) (ed.).
  Li I-Chih hsien-sêng Chi Nien Khan
  李儀祉先生紀念刊.
  Volume Commemorating the Death of Li
    I-Chih [hydraulic engineer].
  Ministry of Agriculture and Forestry,
    Northwestern Scientific Branch.
  Wukung, Shensi, 1938.
Anon. (50) (ed.).
  Li I-Chih hsien-sêng Shih Shih Chou Nien
    Chi Nien Khan 李儀祉先生逝世週年
  紀念刊.
  Volume to commemorate the First Anni-

Anon. (*50*) (ed.) (*cont.*)
  versary of the Death of Li I-Chih
  [hydraulic engineer].
  Dept. of Hydraulic Engineering,
  Nat. NW Agricultural College.
  Wukung, Shensi, 1939 [mimeographed].

Anon. (*51*) (ed.).
  Ching-Thien Chih-Tu Yu Wu? 井田制度
  有無.
  Was there a 'Well-field' System [in
  Antiquity] or Not? (collective work).
  Hua-thung, Shanghai, 1930.

Anon. (*52*).
  Tu Chiang Yen Chieh Shao 都江堰介紹.
  The Traditions of the Kuanhsien Irrigation
  System (lit. of the Dam on the Capital
  River).
  Tu-chiang-yen Authority, Kuanhsien and
  Chhêngtu, 1954.

Anon. (*53*).
  Kuan-Kai Kuan-Li Kung-Tso Ching-Yen
  灌溉管理工作經驗.
  Experience in the Administration of
  Irrigation Projects.
  Shui-Li, Peking, 1958.
  (*Nung-Thien Shui-Li Tshung-Shu*, no. 1.)

Anon. (*54*).
  Chung-Kuo Tshung-Shu Tsung Lu 中國叢
  書綜錄.
  Register of Books in extant Tshung-Shu
  Collections.
  3 vols.
  Chunghua, Shanghai and Peking,
  1961.

Anon. (*55*).
  Chiangsu Hsü-Chou Han Hua Hsiang Shih
  江蘇徐州漢畫象石.
  Stone Reliefs and Engravings from the
  (Later) Han Period found at Hsüchow in
  Chiangsu.
  Kho-Hsüeh, Peking, 1959.
  (Academia Sinica, Archaeological
  Monographs, Ser. B, no. 10.)

Anon. (*62*).
  Lo-Yang Shao-Kou Han Mu 洛陽燒溝
  漢墓.
  The Han Tombs of Shao-Kou [in the
  northeast suburbs of] Loyang.
  Kho-Hsüeh, Peking, 1959.
  (Khao-Ku-Hsüeh Chuan-Khan, Ser. D,
  no. 6.)
  Abstr. in *RBS*, 1965, **5**, no. 379.

Anon. (*63*).
  Thang Chhang-An Ta-Ming Kung 唐長安
  大明宮.
  The Palace of Grand Resplendence at Sian
  [excavation].
  Kho-Hsüeh, Peking, 1959.
  (Khao-Ku-Hsüeh Chuan-Khan, Ser. D,
  no. 11.)
  Abstr. in *RBS*, 1965, **5**, no. 396.

Anon. (*66*).
  Chin Tzhu Fêng Kuang 晉祠風光.
  Scenes and Objects at Chin Tzhu (Taoist
  Temple south of Thaiyuan in Shansi)
  [album of colour photographs].
  Chin Tzhu Cultural Preservation Bureau,
  Thaiyuan n.d. (*c.* 1960).

Anon. (*67*).
  Chin Tzhu Su Hsiang 晉祠塑像.
  The (Sung dynasty) Wood and Plaster
  Images at Chin Tzhu (Taoist Temple
  south of Thaiyuan in Shansi) [album of
  colour photographs].
  Shansi Jen-Min, Thaiyuan, 1959.

Anon. (*68*).
  Chhing-Hua Ta-Hsüeh, 1911–1961 清華
  大學.
  Chhing-Hua University, Peking [semi-
  centennial album of photographs].
  Hsin Chhing-Hua, Peking, 1961.

Anon. (*69*).
  Hanguk Haeyang Sa 韓國海洋史.
  A History of Korean Sea Power.
  Haekun Benbu (Korean Naval Ministry),
  Seoul, 1955.

Anon. (*72*).
  *Ya-Phien Chan-Chêng Mo-Chhi Ying Chün
  tsai Chhang-Chiang Hsia-yu-ti Chin Lüeh
  Tsui Hsing* 鴉片戰爭末期英軍在長江
  下游的侵略罪行.
  Naval Engagements on the Yangtze in the
  Later Phases of the British Aggression
  in the Opium War (1842).
  Jen-Min, Shanghai, 1964.

Ao Chhêng-Lung *et al.* (*1*) 敖承隆.
  Wang-tu Erh Hao Han Mu 望都二号漢
  墓.
  The Han Tomb No. 2 at Wang-tu (+182).
  Wên-Wu, Peking, 1959.

Ao Chhêng-Lung *et al.* (*2*) 敖承隆.
  *Hopei Ting-hsien Pei-chuang Han Mu Fa-
  Chüeh Pao-Kao* 河北定縣北庄漢墓
  發掘報告.
  Report on the Excavation of a Han Tomb
  at Pei-chuang near Ting-hsien in Hopei
  (believed to be that of Liu Yen, Prince of
  Chungshan, *d.* +88 or +90) [with a
  collection of 174 inscriptions from Han
  tombs].
  AS/CJA, 1964, **34** (no. 2), 127.

Aoyama Sadao (*1*) 青山定雄.
  Kochishi Chizu Nado no Chōsa 古地誌地
  圖等の調査.
  In Search of Old Geographical Works and
  Maps.
  TG/T, 1935, **5** (Suppl. Vol.), 123.

Aoyama Sadao (*2*) 青山定雄.
  Gendai no Chizu ni tsuite 元代の地圖に
  ついて.
  On Maps of the Yuan Dynasty.
  TG/T, 1938, **8**, 103.

Aoyama Sadao (3) 青山定雄.
Ri-chō ni okeru ni-san no Chōsen Zenzu ni tsuite 李朝に於ける二三の朝鮮全圖 について.
On Some General Maps of Korea made during the Yi Dynasty.
TG/T, 1939, **9**, 143.

Aoyama Sadao (6) 青山定雄.
Tōsō Henga Kō 唐宋汴河考.
A Study of the Pien Canal in the Thang and Sung periods.
TG/T, 1931, **1** (no. 2), 1.

Aoyama Sadao (7) 青山定雄.
Tōdai no Chisui suiri Kōji ni tsuite 唐代の 治水水利工事について.
Flood Control and Irrigation Engineering under the Thang dynasty.
TG/T, 1944, **15**, 1, 205.

Chang Chung-I 張仲一, Tshao Chien-寶 曹見寶, Fu Kao-Chieh 傳高傑 & Tu Hsiu-Chün (1) 杜修均.
Hui-Chou Ming Tai Chu Tsê 徽州明代住 宅.
The Ming Dwelling-Houses in Huichow.
Architectural and Engineering Press, Peking, 1957.

Chang F. Y. (1).
Historical Sketch of Forestry in China.
JAAC, 1930 (no. 77), 4.

Chang Han-Ying (1) 張含英.
Li Tai Chih Ho Fang Lüeh Shu Yao 歷代治河方略述要.
Outline History of the Engineering Works for the Control of the Yellow River.
Com. Press, Chungking, 1945.

Chang Han-Ying (2) 張含英.
Chung-Kuo Ku-Tai Shui-Li Shih Yeh-ti Chhêng-Chhiu 中國古代水利事業的 成就.
Achievements of Hydraulic Engineering in Ancient and Medieval China.
Kho-Hsüeh, Peking, 1954.

Chang Hsing-Lang (1) 張星烺.
Chung Hsi Chiao-Thung Shih-Liao Hui Pien 中西交通史料滙篇.
Materials for the Study of the Intercourse of China with Other Countries.
6 vols.
(1) Ancient Times, and Europeans;
(2) Europeans;
(3) Africans and Arabs;
(4) Armenians, Jews and Persians;
(5) Turks and Central Asians;
(6) Indians.
Fu-Jen Univ. Press, Peiping, 1930.

Chang Hung-Chao (5) 章鴻釗.
Hangchow Hsi-Hu Chhêng Yin I-Chieh 杭州西湖成因一解.
On the Origins and Development of the West Lake at Hangchow.

BCGS, 1922, **3** (no. 1), 21, Engl. summ. 26.

Chao Chhüan-Chhêng (1) 趙泉澄.
Shih-pa Shih-chi Lü-Sung I-lao-ko Hang Chhuan lai Hua Chi 十八世紀呂宋一 咾哥航船來華記.
A Ship's Voyage from the Ilocos District in Luzon (Philippines) to China in the Eighteenth Century.
YK, 1937, **6** (no. 11), 1.
Tr. Sun & de Francis (1), p. 353.

Chêng Chao-Ching (1) 鄭肇經.
Chung-Kuo Shui-Li Shih 中國水利史.
History of River Conservancy, Transport Canals and Irrigation Engineering in China.
Com. Press, Chhangsha, 1939.

Chêng Chen-To (3) (ed.) 鄭振鐸.
Sung Chang Tsê-Tuan 'Chhing-Ming Shang Ho Thu' Chüan 宋張擇端清明上河 圖卷.
[Album of] Reproductions of the Painting 'Going up the River to the Capital at the Spring Festival' finished by Chang Tsê-Tuan in +1126, with Introduction.
Wên-Wu Ching-Hua, 1959 (no. 1) and separately, Wên-Wu, Peking, 1959.

Chêng Chen-To (4) 鄭振鐸.
Pei-Ching Chin Chiao Wên-Wu-ti Fa-Chüeh yü Pao-Hu 北京近郊文物的發掘與 保護.
The Excavation and Conservation of Cultural Antiquities in the Neighbour-hood of Peking.
HHJP, 1955 (31 Dec.).

Chêng Chen-To (5) (ed.) 鄭振鐸.
Chung-Kuo Pan Chhui Hsüan 中國版垂 選.
A Collection of Wood-block Illustrations from Old Books.
Peking, 1956.

Chêng Hao-Shêng (1) 鄭鶴聲.
Chêng Ho 鄭和.
Biography of Chêng Ho [great eunuch admiral of the +15th century].
Victory Pub., Chungking, 1945.

Chêng Hao-Shêng (2) 鄭鶴聲.
Chung Kuo Li-Tai Chih Ho Wên Hsien Khao Lüeh 中國歷代治河文獻考略.
Bibliography of Works on River Control and Hydraulic Engineering.
CLTC, **1** (no. 4), 433.

Chêng Hao-Shêng (4) 鄭鶴聲.
Chêng Ho I Shih Hui Pien 鄭和遺事彙 編.
Relics of Chêng Ho's Naval Expeditions.
Shanghai, 1948.

Chiao Hsün (1) 焦循.
Chhün Ching Kung Shih Thu 群經宮室 圖.
Illustrated Treatise on the Plans, Technical

Chiao Hsün (1) (cont.)
Terms and Uses of Houses, Palaces,
Temples and other Buildings mentioned
in the Classics.
c. 1825, repr. 1876.
In *Chiao shih I Shu* (Mr Chiao's Remains)
焦氏遺書 and HCCC (*Hsü Pien*), chs.
359, 360.

Chikusa Masa-aki (1) 竺沙雅章.
Sōdai Fukken no Shakai to ji-in 宋代福
建の社會と寺院.
The Role of Buddhist Temples in the
Economic and Social Life of Fukien in
the Sung Period.
TK, 1956, **15**, 170.

Chou I-Liang (2) 周一良.
*Chien-Chen ti Tung Tu yü Chung Jih Wên
Hua Chiao-Liu* 鑑眞的東渡與中日
文化交流.
The Mission of Chien-Chen (Kanshin) to
Japan (+735 to +748) and Cultural
Exchanges between China and Japan.
WWTK, 1963 (no. 9), 1.

Chou Shih-Tê (1) 周世德.
*Tshung Pao Chhuan Chhang Tho-Kan-ti
Chien Ting Thui Lun Chêng Ho Pao
Chhuan* 從寶船廠舵杆的鑑定推論鄭
和寶船.
An Estimate of (the Size of) the Great
Treasure Ships of Chêng Ho based on
the Discovery of a Rudder-Post at the
Site of a Ming Shipyard [Chung-pao on
the San-chha River near Nanking].
WWTK, 1962 (no. 3), 35.
Cf. WWTK, 1957 (no. 12), 80.

Chou Yü-Sên (1) 周鈺森.
Chêng Ho Hang Lu Khao 鄭和航路考.
A Study of the Sea-Routes and Navigation
of Chêng Ho.
Hai-Yün, Thaipei, 1959.

Chu Chhi-Chhien (1) 朱啓鈐.
*Tshun Su Thang Ju Tshang Thu Shu Ho
Chhü chih Pu Mu Lu* 存素堂入藏圖
書河渠之部目錄.
Catalogue of Works on Rivers and Canals
collected by Mr Chu Chhi-Chhien.
BSRCA, 1934, **5** (no. 1), 98.

Chu Chhi-Chhien (2) 朱啓鈐.
Phai-Lou Suan Li 牌樓算例.
A Study of Triumphal Gates and their
Accepted Measurements
Peking, n.d.

Chu Chhi-Chhien (3) 朱啓鈐.
Kuei Chi Thu 匡几圖.
Diagrams of Boxable Tables [furniture;
pieces which can be dismounted and
fitted into one another].
Society for Research in the History of
Chinese Architecture, Peking, 1931.
(Bound with *Yen Chi Thu* and *Thieh Chi
Phu*, q.v.)

Chu Chhi-Chhien 朱啓鈐 & Liang Chhi-
Hsiung 梁啓雄 (1 to 6).
Chê Chiang Lu [parts 1 to 6] 哲匠錄.
Biographies of [Chinese] Engineers,
Architects, Technologists and Master-
Craftsmen.
BSRCA, 1932, **3** (no. 1), 123; 1932, **3**
(no. 2), 125; 1932, **3** (no. 3), 91; 1933, **4**
(no. 1), 82; 1933, **4** (no. 2), 60; 1934, **4**
(nos. 3 and 4), 219.

Chu Chhi-Chhien 朱啓鈐, Liang Chhi-
Hsiung 梁啓雄 & Liu Ju-Lin (1)
劉儒林.
Chê Chiang Lu [part 7] 哲匠錄.
Biographies of [Chinese] Engineers, Archi-
tects, Technologists and Master-Crafts-
men (continued).
BSRCA, 1934, **5** (no. 2), 74.

Chu Chhi-Chhien 朱啓鈐 & Liu Tun-Chên
劉敦楨 (1, 2).
Chê Chiang Lu [parts 8 and 9] 哲匠錄.
Biographies of [Chinese] Engineers, Architects, Technologists and Master-Craftsmen (continued).
BSRCA, 1935, **6** (no. 2), 114; 1936, **6**
(no. 3), 148.

Chu Chün-Shêng (1) 朱駿聲.
Shuo Ya 說雅.
Ancient Meanings of Words (an appendix
to the *Shuo Wên*).
c. 1840.

Chu Hsieh (1) 朱偰.
Chung-Kuo Yün-Ho Shih-Liao Hsüan-Chi
中國運河史料選輯.
Selected Materials for the History of the
Grand Canal.
Chung-Hua, Peking, 1962.

Chu Kho-Chen (4) 竺可楨.
*Pei Sung Shen Kua tui yü Ti-Hsüeh chih
Kung-Hsien yü Chi-Shu* 北宋沈括對於
地學之貢獻與紀述.
The Contributions to the Earth Sciences
made by Shen Kua in the Northern Sung
Period.
KHS, 1926.

Chhang Jen-Chieh (1) 常任俠.
Han Tai Hui Hua Hsüan Chi 漢代繪畫選
集.
Selection of Reproductions of Han Draw-
ings and Paintings (including Stone
Reliefs, Moulded Bricks, Lacquer,
etc.).
Chao-Hua I-Shu, Peking, 1955.

[Chhang Shu-Hung] (1) (ed.) 常書鴻.
Tunhuang Mo-kao-khu 敦煌莫高窟.
(The Cave-Temples at) Mo-kao-khu
[Chhien-fo-tung] near Tunhuang.
Kensu Jen-min, Lanchow, 1957.

[Chhang Shu-Hung] (2) (ed.) 常書鴻.
Chung-Kuo Tunhuang I Shu Chan 中國敦
煌藝術展.

Chhang Shu-Hung] (2) (ed.) (cont.)
Catalogue of the Tokyo Exhibition of
Tunhuang (Chhien-fo-tung Cave-
Temples) Art.
Tokyo, 1958.

Chhang Shu-Hung (3) (ed.) 常書鴻.
Tunhuang Pi Hua 敦煌壁畫.
The Wall-Paintings of the Tunhuang
(Cave-Temples) [album].
Wên-Wu, Peking, 1959.

Chhen Ching (1) 陳經.
Chhiu Ku Ching Shê Chin Shih Thu
求古精舍金石圖.
Illustrations of Antiques in Bronze and
Stone from the Spirit-of-Searching-Out-
Antiquity Cottage.
1818.

Chhen Ching (1) 陳靖.
Han Nan Shui-Li Than 漢南水利談.
On the Irrigation Systems of the South
bank of the Han River.
SL, 1934, 6, 262.

Chhen Chung-Chhih (1) 陳仲篪.
Sung Yung-Ssu Ling Phing Mien Chi Shih
Tshang Tzu chih Chhu Pu Yen Chiu
宋永思陵平面及石藏子之初步
研究.
Preliminary Researches on the General
Plan and the Stone Vault of the Yung-Ssu
Ling, Imperial Tomb of the Sung
Dynasty.
BSRCA, 1936, 6 (no. 3), 121.

Chhen Chung-Chhih (2) 陳仲篪.
'Ying Tsao Fa Shih' Chhu Than 「營造法
式」初探.
A Study of the Genesis of the Treatise on
Architectural Methods.
WWTK, 1962 (no. 136), 12.

Chhen Jung (1) 陳嶸.
Chung-Kuo Shu Mu Fên Lei Hsüeh
中國樹木分類學.
Illustrated Manual of the Systematic
Botany of Chinese Trees and Shrubs.
Agricultural Association of China Series.
Nanking, 1937.

Chhen Ming-Ta (1) 陳明達.
Kuan-yü Han-Tai Chien-Chu-ti Chi-ko
Chung-Yao Fa-Hsien 關於漢代建築的
幾個重要發現.
On Some Important Discoveries concerning
the Architecture of the Han Period.
WWTK, 1954 (no. 9), 91.

Chhen Po-Chhüan (1) 陳柏泉.
Chi Chiangsi Fên-i Wan-Nien Chhiao
記江西分宜萬年橋.
On the Wan-Nien Bridge at Fên-i in
Chiangsi province.
WWTK, 1961 (no. 2), 22.

Chhen Shu-Phêng (1) 陳述彭.
Yünnan Thang-Lang Chhuan Liu Yü chih
Ti-Wên 雲南螳螂川流域之地文.

Geomorphology of the Thang-Lang River
Valley, Yunnan (Kunming and its lake).
JGSC, 1948, 15 (nos. 2, 3, 4), 1.

Chhen Tsê-Jung (1) 陳澤榮.
Chung-Kuo Ku Tai chih Kuan Kai Chhêng
Chi 中國古代之灌漑成績.
The Development of Irrigation Works in
Ancient China.
SL, 1931, 1, 237.

Chhen Tshung-Chou (1) 陳從周.
Shao-hsing-ti Sung Chhiao 紹興的宋橋.
Sung Bridges at Shao-hsing.
WWTK, 1958 (no. 7), 59.

Chhen Tsun-Kuei (3) 陳遵嬀.
Hêng Hsing Thu Piao 恆星圖表.
Atlas of the Fixed Stars, with identifications
of Chinese and Western Names.
Com. Press, Shanghai, 1937, for Academia
Sinica.

Chhêng Yao-Thien (4) 程瑤田.
Shih Kung Hsiao Chi 釋宮小記.
Brief Record of Buildings and Palace
Halls.
Pr. c. 1825.
In HCCC (Hsü Pien), ch. 535.

Chhi Ssu-Ho (3) 齊思和.
Mêng Tzu Ching-Thien Shuo Pien 孟子井
田説辨.
Analysis of the Statements of Mencius about
the 'Well-Field' (Land) System.
YCHP, 1948, 35, 101.

Chhi Ying-Thao (1) 祁英濤.
Hopei Shêng Hsin-chhêng-hsien Khai-Shan
Ssu Ta Tien 河北省新城縣開善寺大
殿.
The Great Hall of the Khai-Shan Temple at
Hsin-chhêng Hsien in Hopei province
[built under the Liao Dynasty between
+1004 and +1123].
WWTK, 1957, no. 10 (no. 86), 23.

Chhien Yung (1) 錢鏞.
Phing-chiang Thu Pei 平江圖碑.
On the Stele of the City-map of Phing-
chiang (Suchow, Chiangsu).
WWTK, 1959 (no. 2), 49.

Chhiu Pu-Chou (1) 邱步洲.
Ho Kung Chien Yao 河工簡要.
Essentials of River Conservancy and
Hydraulic Engineering.
1887.

Chhou Chhih-Shih (1) 仇池石.
Yang-chhêng Ku Chhao 羊城古鈔.
A Documentary History of Canton.
1806.

Chhü Shou-Yo (1) 曲守約.
Chung-Kuo Ku-Tai-ti Tao Lu 中國古代
的道路.
The Roads of China in Ancient Times
(Chou, Chhin and Han Periods).
CHJ/T, 1960, 2 (no. 1), 143.
Abstr. RBS, 1967, 6, no. 82.

Chhüan Han-Shêng (1) 全漢昇.
  Thang Sung Ti Kuo yü Yün-Ho 唐宋帝
    國與運河.
  The Thang and Sung Empires and the
    Grand Canal.
  Com. Press, Chungking, 1944.

Fan Wên-Thao (1) 范文濤.
  Chêng Ho Hang Hai Thu Khao 鄭和航海
    圖考.
  Study of the Maps connected with Chêng
    Ho's Voyages.
  Com. Press, Chungking, 1943.

Fang Chi (1) 方楫.
  Ming Tai Hai Yün ho Tsao Chhuan Kung
    Yeh 明代海運和造船工業.
  Maritime Transportation and the Ship-
    building Industry during the Ming
    Period.
  WSC, 1957 (no. 5), 46.

Fang Chi (2) 方楫.
  Wo Kuo Ku-Tai-ti Shui-Li Kung-Chhêng
    我國古代的水利工程.
  Hydraulic Engineering in Ancient and
    Mediaeval China.
  Hsin-Chih-Shih Pub. Co., Shanghai,
    1955.
  Abstr. RBS, 1957, 1, no. 459.

Fêng Chhêng-Chün (1) 馮承鈞.
  Chung-Kuo Nan-Yang Chiao-Thung Shih
    中國南洋交通史.
  History of the Contacts of China with the
    South Sea Regions.
  Com. Press, Shanghai, 1937.

Fêng Han-Chi (1) 馮漢驥.
  Szechuan Ku-Tai-ti Chhuan-Kuan Tsang
    四川古代的船棺葬.
  Boat-Burial in Szechuan (in the States of
    Pa and Shu, in and before the −4th
    century).
  AS/CJA, 1958 (no. 2), 77.

Fêng Yün-Phêng 馮雲鵬 & Fêng Yün-
    Yuan 馮雲鵷.
  Chin Shih So 金石索.
  Collection of Carvings, Reliefs and
    Inscriptions.
  (This was the first modern publication of
    the Han tomb-shrine reliefs.)
  1821.

Fu Chien (1) 傅健.
  Shensi Mei-hsien Chhü Yen chih Tiao-Chha
    陝西郿縣渠堰之調查.
  An Enquiry into the Canals and Dams of
    Mei-hsien (south of the Wei River) in
    Shensi Province.
  SL, 1934, 7, 239.

Fujita Toyohachi (1) 藤田豐八.
  Koshō ni tsukite 胡床につきて.
  On (the History of) the 'Barbarian Bed' (the
    Chair) in China.
  TYG, 1922, 12, 429; 1924, 14, 131.

Hamada Kōsaku 濱田耕作, Harada
    Yoshito 原田淑人 & Umehara Sueji
    梅原末治 (1).
  Senoku Seishō Kaisetsu 泉屋清賞解説.
  (Explanatory Notes for the Albums illustrat-
    ing the Sumitomo Collection of Bronzes
    at Kyoto).
  2 vols. (vols. 6 & 7) of the series, one in
    Japanese, one in English; see Sumitomo,
    Taki & Naitō (1).
  Osaka, 1923.

Haneda Tōru (1) 羽田亨.
  Genchō Ekiden Zakko 元朝驛傳雜考.
  A Study of the Post-Station System in the
    Yuan Dynasty.
  Toyo Bunko, Kyoto, 1930.

Harada Yoshito 原田淑人 & Komai
    Kazuchika 駒井和愛 (1).
  Shina Koki Zukō 支那古器圖考.
  Chinese Antiquities (Pt. 1, Arms and
    Armour; Pt. 2, Vessels [Ships] and
    Vehicles).
  Tōhō Bunka Gakuin, Tokyo, 1937.

Hibino Takeo (1) 日比野丈夫.
  Kan no Seihō Hatten to Ryōkan-setsu no
    jiki ni tsuite 漢の西方發展と兩關設
    の時期について.
  On the Period of Westward Expansion and
    the Dates of the Establishment of the Han
    Gate-Fortresses [Yü-mên Kuan and Yang
    Kuan at the Western end of the Great
    Wall].
  TG/K, 1957, 27, 31.
  Abstr. RBS, 1962, 3, no. 160.

Hibino Takeo (2) 日比野丈夫.
  Tonkō no Godaisan ni tsuite 敦煌の五臺
    山につれこ.
  The Panorama of Wu-thai Shan at Tun-
    huang (cave no. 61).
  ABA, 1958, 34, 75.

Hisamura Yukari (1) 久村因.
  Shin Kan Jidai no Nyū-Shoku-ro ni tsuite
    秦漢時代の入蜀路について.
  On the Road to Szechuan in Chhin and Han
    Times.
  TYG, 1955, 38 (no. 2), 178, 324.
  Abstr. RBS, 1956, 2, no. 94.

Hisamura Yukari (2) 久村因.
  Kodai Shisen ni Dochaku Seru Kan Minzoku
    no raireki ni tsuite 古代四川に土着せ
    る漢民族の來歷について.
  On the Colonisation of Szechuan by the
    Han People (in the −4th and −3rd
    centuries).
  RK, 1957 (no. 204), 1.
  Abstr. RBS, 1962, 3, no. 131.

Ho Pei-Hêng (1) (ed.) 何北衡.
  Tu Chiang Yen; Shui-Li Kung-Chhêng Shu
    Yao, Chi chhi Kai Shan Chi Hua Ta
    Kang 都江堰水利工程述要及其改
    善計劃大綱.

Ho Pei-Hêng (1) (ed.) (cont.)
The Tu Chiang Yen (Capital River Dam)
[at Kuanhsien, Szechuan], with an
account of its most important engineering
features, and of the Plans for its
Improvement (Engl. title: A Note on the
Tukiangyen Irrigation System).
Bureau of Hydraulic Engineering (Szechuan
Water Conservancy Bureau), Chêngtu,
1943.
(Szechuan Hydraulic Publication Series,
Vol. 1. no. 6.)

Hou Jen-Chih (1) 侯仁之.
Chin Fu Chih Ho Shih Mo 靳輔治河
始末.
The Story of the Water Conservancy Work
of Chin Fu.
YAHS, 1936, 2, 43; CIB, 1938, 2, 115.

Hou Jen-Chih (2) 侯仁之.
Chhen Huang, Chhing-Tai Chieh Chhu-ti
Chih-Ho Chuan-Chia 陳潢清代傑出的
治河專家.
Chhen Huang, a Distinguished Hydraulic
Engineer of the Chhing Dynasty.
KHSC, 1958, 1 (no. 2), 73.

Hu Huan-Yung (1) 胡煥庸.
Huai Ho 淮河.
The Huai River [Conservancy Works under
Construction for the Huai Valley
Authority].
Khaiming, Peking, 1952.

Hu Huan-Yung 胡煥庸, Hou Tê-Fêng
侯德封 & Chang Han-Ying (ed.)
張煥英 (1).
Huang Ho Chih 黃河志.
History of the Yellow River, 3 vols.
Vol. i. Climatic Factors (Hu).
ii. Geological Factors (Hou).
iii. Engineering Works (Chang).
Com. Press, Shanghai, 1936.

Hu Tao-Ching (1) 胡道靜.
'Mêng Chhi Pi Than' Chiao Chêng 夢溪
筆談校證.
Complete Annotated and Collated Edition
of the Dream Pool Essays (of Shen Kua,
+1086).
2 vols.
Shanghai Pub. Co., Shanghai, 1956.
Analyt. rev. Nguyen Tran-Huan, RHS,
1957, 10, 182.

Hu Tao-Ching (2) 胡道靜.
Hsin Chiao Chêng 'Mêng Chhi Pi Than'
新校正夢溪筆談.
New Corrected Edition of the Dream Pool
Essays (with additional annotations).
Chung-Hua, Peking, 1958.

Huang Shêng-Chang (1) 黃盛璋.
Pao-Yeh Tao yü Shih Mên Shih Kho
褒斜道石門石刻.
Stone Inscriptions on the Pao-Yeh Road.
WWTK, 1963 (no. 2), 29.

Huang Shou-Fu 黃綬芙 & Than Chung-Yo
譚鍾嶽 (1).
O Shan Thu Shuo (or Chih) 峨山圖說
(志).
An Illustrated Guide to O-mei Shan.
Chhêngtu, 1891.
Repr. with English tr. by D. L. Phelps (2)
and redrawn illustrations, Chhêngtu, 1936.

Hung Huan-Chhun (1) 洪煥椿.
Shih chih Shih-san Shih-Chi Chung-Kuo
Kho-Hsüeh-ti Chu-Yao Chhêng-Chiu
十至十三世紀中國科學的主要成就.
The Principal Scientific (and Technological)
Achievements in China from the +10th
to the +13th centuries (inclusive) [the
Sung period].
LSYC, 1959, 5 (no. 3), 27.

Hsia Nai (2) 夏鼐.
Khao-Ku-Hsüeh Lun Wên Chi 考古學論
文集.
Collected Papers on Archaeological Subjects.
Academia Sinica, Peking, 1961.

Hsiang Ta (4) 向達.
Chêng Ho Hang Hai Thu 鄭和航海圖.
The Charts of Chêng Ho's Naval Expedi-
tions (reproduction of ch. 240 of the Wu
Pei Chih, with introduction, commentary,
and geographical dictionary).
Chung-Hua, Peking, 1961.

Hsiang Ta (5) 向達.
Liang Chung Hai Tao Chen Ching 兩種海
道針經.
An Edition of Two Rutters [the Shun Fêng
Hsiang Sung (Fair Winds for Escort),
perhaps c. +1430, from a MS. of c.
+1575; and the Chih Nan Chêng Fa
(General Compass-Bearing Sailing
Directions) of c. +1660].
Chung-Hua, Peking, 1961.

Hsieh Kan-Thang (1) 謝甘棠.
Wan Nien Chhiao Chih 萬年橋志.
Record of the Bridge of Ten Thousand
Years [at Nan-chhêng in Chiangsi; and
the Repairs to it which Hsieh Kan-
Thang supervised; with elaborate descrip-
tion of bridge engineering techniques].
Nan-chhêng, 1896 [cf. Liu Tun-Chên, 3].

Hsieh Kuo-Chên (1) 謝國楨.
'Ying Tsao Fa Shih' Pan-Pên Yuan-Liu
Khao 營造法式版本源流考.
Historical Study of the Different MSS.
Editions of the Treatise on Architecture
(Sung).
BSRCA, 1933, 4 (no. 1), 1.

Hsin Shu-Chhih (2) 辛樹幟.
Wo Kuo Shui Thu Pao-Chhih-ti Li-Shih
Yen-Chiu (Chhu Kao) 我國水土保持
歷史研究(初稿).
Preliminary Studies on the History of Soil
and Water Conservation in China.
KHSC, 1958, 1 (no. 2), 31.

Hsü Chü-Chhing (1).
Pei Pien Chhang Chhêng Khao 北邊長城考.
A Study of the Northern Frontier Great
Walls.
YAHS, 1929, 1.

Hsü Chung-Shu (3) 徐中舒.
*Ku Tai Kuan Kai Kung-Chhêng Yuan Chhi
Khao* 古代灌溉工程源起考.
On the Origin of Irrigation Engineering in
Ancient China.
AS/BIHP, 1935, 5, 255.

Hsü Chung-Shu (4) 徐中舒.
*I Shê yü Nu chih Su Yuan chi Kuan-yü
tzhu Lei Ming Wu chih Khao-Shih*
弓射與弩之溯原及關於此類名物之
考釋.
A Study of the Origin of Archery (I Shê)
and of the Crossbow (Nu), and the
Etymology of the Names of their Related
Objects.
AS/BIHP, 1934, 4, 417.

Hsü Chung-Shu (6) 徐中舒.
Ching-Thien Chih Tu Than Yuan 井田制
度探源.
A Study of the Origin of the 'Well-Field'
(Land) System.
BCS, 1944, 4 (no. 1), 121.
Abridged tr. Sun & de Francis (1), p. 3.

Hsü Hsü-Sêng (1) 徐旭生.
*Ching-Thien Hsin Chieh Ping-Lun Chou
Chhao Chhien-Chhi Shih Nung Pu-Fên-ti
Han I* 井田新解並論周朝前期士農
不分的含意.
A New Explanation of the 'Well-Field'
System in Pre-Chou Times before the
Differentiation of Soldiers and
Husbandmen.
LSYC, 1961, 7 (no. 4), 53.

Hsü I-Thang (1) 徐益棠.
*Nan Sung Hang-Chou Chih Tu Shih ti
Fa-Chan* 南宋杭州都市的發展.
The Development of Hangchow as a
Metropolis during the Southern Sung
Dynasty.
BCS, 1944, 4, 231.

Hsü Ping-Chhang (1) 徐炳昶.
*Chung-Kuo Ku Shih ti Chhuan-Shuo Shih-
Tai* 中國古史的傳説時代.
The Legendary Period in Ancient Chinese
History.
Chung-Kuo Wên-Hua Fu-Wu, Chungking,
1943.
Revised and enlarged edition under the
name of Hsü Chhang.
Kho-Hsüeh, Peking, 1962.

Hsü Sung (2) 徐松.
Thang Liang Ching Chhêng Fang Khao
唐兩京城坊考.
Studies on the Districts of the Two Thang
Capitals.
1810.

Hsü Yü-Hu (1) 徐玉虎.
Chêng Ho Phing Chuan 鄭和評傳.
Critical Biography of (the Admiral) Chêng Ho.
Thaipei, Thaiwan, 1958.

Iida Sugashi (1) 飯田須賀斯.
*Shina Kenchiku no Kyō (Serimochi) ni
Kansuru Ikkōsatsu* 支那建築の拱に關
する一考察.
Some Considerations on the (Origin and
Development of the) Arch in Chinese
Architecture.
TG/T, 1940, 11, 733.
Abstr. *MS.*, 1942, 7, 375.

Iida Sugashi (2) 飯田須賀斯.
*Zui-Tō Kenchiku no Nihon ni oyobaseru
Eikyō* 隋唐建築の日本に及はせる
影響.
Echoes of Sui and Thang Architecture (and
Planning) in Japan.
BK, 1955, 19, 24.
Abstr. *RBS*, 1955, 1, no. 136.

Ikeuchi Hiroshi (1) 池內宏.
Genkō no Shin Kenkyū 元寇の新研究.
New Studies on the Yuan Invasions (+1274
and +1281).
2 vols.
Tokyo, 1931.

Imabori Seiji (1) 今堀誠二.
Shindai-igo ni okeru Kōga no Suiun ni tsuite
清代以後にずける黄河の水運について.
A Study of the Shipping on [the middle
stretch of] the Yellow River [in Kansu
and Shansi] since Chhing Times.
SGKK, 1959 (no. 72), 23.
Abstr. *RBS*, 1965, 5, no. 272.

Ishibashi Ushio (1) 石橋丑雄.
Tendan 天壇.
On the Altar of Heaven.
Yamamoto Shoten, Tokyo, 1957.
Abstr. *RBS*, 1962, 3, no. 512.

Ishida Mosaku 石田茂作 & Wada Gun-ichi
和田軍一 (ed.) (1).
Shōsōin 正倉院.
The Shōsōin (an Eighth-Century Treasure-
House containing all kinds of valuable
objects imperially dedicated in +756 and
at several later dates; attached to the
Tōdaiji temple at Nara).
Mainichi; Tokyo, Osaka and Moji, 1954.

Itō Seizō (1) 伊藤清造.
Shina no Kenchiku 支那建築.
(A History of) Chinese Architecture.
Osaka Yago Shoten, Tokyo, 1929.

Jen Mei-O (1) 任美鍔.
*Wo-Kuo Tsui-Chin tui-yü Huang Ho Wên-
Thi chih Hsin Yen-Chin* 我國最近對
於黄河問題之新研究.
Yellow River Project Studies—a Review.
JGSC, 1948, 15 (no. 1), 31.

Jung Kêng (1) 容庚.
  Han Wu Liang Tzhu Hua Hsiang Khao Shih
  漢武梁祠畫像考釋.
  Investigations on the Carved Reliefs of the
  Wu Liang Tomb-shrines of the [Later]
  Han Dynasty.
  2 vols.
  Yenching Univ. Archaeol. Soc. Peking,
  1936.
Jung Kêng (3) 容庚.
  Chin Wên Pien 金文編.
  Bronze Forms of Characters.
  Peking, 1925, repr. 1959.

Katsushika Hokusai (1) 葛飾北齋.
  Fugaku Hyaku-Kei 富嶽百景.
  A Hundred Views of Mt. Fuji.
  Prefaces by Tanchiko Ryūtei *et al.*
  Tokyo, 1834–36.
  Cf. Dickins (1).
Khang Chi-Thien (1) 康基田.
  Ho Chhü Chi Wên 河渠紀聞.
  Notes on Rivers and Canals.
  1804.
Kho Shao-Min (1) 柯紹忞.
  Hsin Yuan Shih 新元史.
  New History of the Yuan Dynasty (issued
  as an official dynastic history by Presiden-
  tial Order).
  Peiping, 1922.
Kimiya Yasuhiko (1) 木宮泰彥.
  Nikka Bunka Kōryūshi 日華文化交
  流史.
  A History of Cultural Relations between
  Japan and China.
  Fuzambō, Tokyo, 1955.
  Abstr. *RBS*, 1959, **2**, no. 37.
Koiwai Hiromitsu (1) 小岩井弘光.
  Sōdai Kyūkyaku Teihohei ni tsuite 宋代
  急脚遞鋪兵について.
  On the Express Couriers of the Govern-
  ment in Sung Times.
  TYGK, 1959, **1**, 25.
  Abstr. *RBS*, 1965, **5**, no. 176.
Ku Chün-Chêng (1) 顧均正.
  *Chung-Kuo Tsai Shih-i Shih-Chi Chiu Chhu
  Hsien Chhuan Wu* 中國在十一世紀就
  出現船塢.
  A Dry Dock in the Chinese Eleventh
  Century.
  TCKH, 1962 (no. 1), 5.
Kuan Chin-Chhêng (1) 管勁承.
  Chêng Ho Hsia Hsi Yang ti Chuan 鄭和下
  西洋的船.
  The Ships used by Chêng Ho in his Voyages
  to the Western Oceans.
  TFTC, 1947, **43** (no. 1), 47.
Kung Thing-Wan (1) 龔廷萬.
  *Szechuan Fou-ling 'Shih Yü' Thi-Kho Wên-
  Tzu ti Thiao-Chha* 四川涪陵石魚題刻
  文字的調查.

A Study of the Inscriptions on the 'Stone
  Fishes' Nilometer at Fou-ling in
  Szechuan.
  WWTK, 1963 (no. 7), 39.
Kuo I-Fu (1) 郭義孚.
  Han-Yuan Tien Wai Kuan Fu-Yuan
  含元殿外觀復原.
  A Reconstruction of the External Appear-
  ance of the Han-Yuan Hall (of the Ta-
  Ming Palace) (in Thang Chhang-an).
  KKTH, 1963 (no. 10), 567.
Kuo Khêng-Jo (1) 郭鏗若.
  Fu-chien Phu-thien Mu-lan Pei 福建莆田
  木蘭陂.
  On the Mu-lan Dam at Phu-thien in
  Fukien Province.
  SL, 1936, **11**, 20.
Kuo Mo-Jo (1) 郭沫若.
  Shih Phi Phan Shu 十批判書.
  Ten Critical Essays.
  Chün-i, Chungking, 1945.
Kuo Mo-Jo (2) 郭沫若.
  Ku Tai Shih-Hui chih Yen-Chiu 古代社
  會之研究.
  Studies in Ancient Chinese Society.
  Shanghai, *c.* 1927.
Kuo Mo-Jo (7) 郭沫若.
  Loyang Han Mu Pi-Hua Shih-Than 洛陽
  漢墓壁畫試探.
  A Study of the Wall-Paintings in a Han
  Tomb (*c.* −50) at Loyang [excavated by
  Li Ching-Hua *et al.* (1)].
  AS/CJA, 1964, **34** (no. 2), 1.

Lao Kan (2) 勞榦.
  Lun Han-Tai chih Lu Yün yü Shui Yün
  論漢代之陸運與水運.
  Transportation by Land and Water during
  the Han period.
  AS/BIHP, 1947, **16**, 69.
Lao Kan (3) 勞榦.
  Pei Wei Loyang Chhêng Thu ti Fu-Yuan
  北魏洛陽城圖的復原.
  Reconstruction of the Plan of [the Capital]
  Loyang, as it was during the Northern
  Wei period [according to the *Lo-yang
  Chhieh Lan Chi*].
  AS/BIHP, 1948, **20**, 299.
Lao Kan (4) 勞榦.
  Liang Han Hu-Chi yü Ti-Li chih Kuan-Hsi
  兩漢戶籍與地理之關係.
  Population and Geography in the Two Han
  Dynasties.
  AS/BIHP, 1935, **5** (no. 2), 179.
  Tr. Sun & de Francis (1), p. 83.
Li Chhang-Nien (1) (ed.) 李長年.
  *Chung-Kuo Nung-Hsüeh I-Chhan Hsüan
  Chi; Tou Lei* 中國農學遺產選集豆
  類.
  Chinese Agriculture Source-Book; Beans.
  Chung-Hua Shu Chü, Shanghai, 1958.

Li Chhêng-Fan (1) 李承範.
Fên-Huang Ssu Mu Chuan Tha 芬皇寺模
磚塔.
The Pagoda of Moulded Bricks at the
Punhoang (Fragrant-Kingship) Temple
(in Korea) [built under Queen Sŏndŏk of
Silla in +634].
HCH, 1956 (no. 11), 39.
Li Chhüan-Chhing (1) 李全慶.
Chaochow Chhiao Hsiu-Fu Kung-Chhêng
Chün-Kung 趙州橋修復工程竣工.
The Completion of the Engineering Opera-
tion of Restoring the (Great Segmental
Arch) Bridge at Chaochow.
WWTK, 1959 (no. 2), 35.
Li Chi (1) (ed.) 李濟.
An-yang Fa Chüeh Pao-Kao 安陽發掘報
告.
Reports of the Excavations at An-yang [one
of the Shang capitals].
Academia Sinica, 4 vols. consecutively
paged.
Vols. 1 and 2, Peiping, 1929; vol. 3,
Peiping, 1931; vol. 4, Shanghai, 1931–33.
Li Chien-Nung (1) 李劍農.
Chhê Chu Kung 徹助貢.
On Cooperative Cultivation [Chhê], Joint
Corvée Cultivation of the Lord's Land
[Chu], and Feudal Dues or Land Tax
[Kung].
WUQJSS, 1948, 9, 25.
Li Chien-Nung (2) 李劍農.
Chung-Kuo Ching Chi Shih Kao 中國經
濟史稿.
Draft for an Economic History of China.
Hankow, c. 1948.
Li Ching-Hua et al. (1) 李京華.
Loyang Hsi Han Pi-Hua Mu Fa Chüeh Pao-
Kao 洛陽西漢壁畫墓發掘报告.
Report on the Excavation of a Western
Han Tomb with Wall-Paintings (c. −50)
at Loyang.
AS/CJA, 1964, 34 (no. 2), 107.
Cf. Kuo Mo-Jo (7).
Li Chu-Chün et al. (1) 李竹君
Lin-ju Pai-Yün Ssu 臨汝白雲寺.
(An Architectural Study of the Buildings of
the) Pai-Yün Temple at Lin-ju (in
Honan north of the Huai Valley)
[including a J/Chin hall of the +12th or
early +13th century].
WWTK, 1961 (no. 2), 17.
Li Hsien-Chang (1) 李獻璋.
Boso Densetsu no Gensho Keitai 媽祖傳説
の原初形態.
The Original Form of the Legend of the
Goddess Ma-Tsu [protectress of seafarers
and travellers, originating in the +11th
century].
TS, 1956 (no. 11), 61.
Abstr. RBS, 1959, 2, no. 519.

Li Hsien-Chang (3) 李獻璋.
'Sankyō Sōjin Daizen' to 'Tempi Jōboden' wo
Chūshin to Suru Boso Densetsu no Kōsatsu
三敎搜神大全と天妃娘媽傳を中心
とする媽祖傳説の考察.
A Study of the Legend of the Goddess Ma-
Tsu as it developed in the (Ming) books
San Chiao Sou Shen Ta Chhüan and Thien-
Fei Niang Ma Chuan [patroness of sailors].
TYG, 1956, 39, 76.
Abstr. RBS, 1959, 2, no. 519.
Li Hsien-Chang (4) 李獻璋.
Gen Min no Chihōshi ni arawareta Boso
Densetsu no Emben 元明の地方志に現
われた媽祖傳説の演變.
The Legends of the [Sailors'] Goddess Ma-
Tsu in the Gazetteers of Yuan and Ming
Times.
THG, 1957, 13, 29.
Abstr. RBS, 1962, 3, no. 788.
Li Hsien-Chang (5) 李獻璋.
Ryūkyū Sai Koba Densetsu Kōshō; Boso
Densetsu no Kaiten ni Kanren Shite
琉球蔡姑婆傳説考證; 媽祖傳説の
開展に關聯して.
On the [late Ming] Liu-Chhiu Island Cult
of Tshai Ku-Pho (Matron Tshai) [a
Sailor's Goddess]; and its Relation with
the Earlier Development of the Ma-Tsu
Cult.
TK, 1957, 16, 154.
Abstr. RBS, 1962, 3, no. 789.
Li Hsien-Chang (6) 李獻璋.
Sōtei no Hōshi Kara Mita Boso Shinkō no
Hattatsu 宋廷の封賜から見た媽祖
信仰の發達.
A Study of the Honorific Titles accorded to
[the Sailors' Goddess] Ma-Tsu during the
Sung Dynasty.
SKKK, 1959, 32, 416.
Abstr. RBS, 1965, 5, no. 786.
Li Hsien-Chang (7) 李獻璋.
Shindai no Boso densetsu ni taisuru Hihanteki
Kenkyū; Tenbi Kenseiroku to sono Ryūden
no sukashite 清代の媽祖傳説に對す
る批判的研究; 天妃顯聖錄とその流
傳を透して.
Critical Researches on the Legend and
Hagiography of [the Sailors' Goddess] Ma-
Tsu during the Chhing Period; especially
after the appearance of a fictional bio-
graphy in the Thien Fei Hsien Shêng Lu.
TS, 1958, 13–14, 65; 1959, 15, 53.
Abstrs. RBS, 1964, 4, no. 930; 1965, 5,
no. 787.
Li I-Chih (1) 李儀祉.
Shensi Ching Hui Chhü Kung-Chhêng Pao-
Kao 陝西涇惠渠工程報告.
Report on the Engineering Work of the
Ching-Hui Canal in Shensi Province.
SL, 1932, 3, 3.

Li Shu-Thien *et al.* (*1*) 李書田 .
Chungkuo Shui Li Wên-Thi 中國水利問
題 .
The Problem of Water-Conservancy in
    China.
2 vols.
Com. Press, Shanghai, 1937.

Li Ssu-Shun (*1*) 李思純 .
Chiang-tshun Shih Lun 江村十論 .
River Village Essays.
Jen-Min, Shanghai, 1957.
Abstr. *RBS*, 1962, **3**, no. 56.

Liang Ssu-Chhêng (*1*) 梁思成 .
*Chêng-Ting Tiao Chha Chi Lüeh (Lung-hsing
    Ssu)* 正定調查紀略 (隆興寺) .
The Ancient Architecture of Chêng-Ting
    [Hopei]; [On the Revolving Library at]
    the Lung-hsing Temple.
BSRCA, 1933, **4** (no. 2), 1 (14 ff.).

Liang Ssu-Chhêng (*2*) 梁思成 .
*Chao-hsien Ta-Shih Chhiao chi An-Chi
    Chhiao (fu Hsiao Shih Chhiao [Yung-
    Thung Chhiao], Chi Mei Chhiao)* 趙縣
    大石橋即安濟橋 (附小石橋 [永通橋]
    濟美橋) .
The Great Stone Bridge of Chao-hsien, also
    called the An-Chi Bridge; with an appen-
    dix on some of the smaller segmental arch
    bridges (of the same region), the (Yung-
    Thung Bridge and the) Chi-Mei Bridge.
BSRCA, 1934, **5** (no. 1), 1.

Liang Ssu-Chhêng (*3*) 梁思成 .
*Wo Kuo Wei Ta ti Chien Chu Chhuan
    Thung yü I Chhan* 我國偉大的建築
    傳統與遺產 .
The Great Achievements of Traditional
    Chinese Building Methods, and Some of
    the Buildings which have come down to
    us in *Wo-mên Wei Ta ti Tsu Kuo* 我們
    偉大的祖國 Our Great Country,
    Peking, 1951.
Repr. *HHYK*, 1951, **4** (no. 1), 190.

Liang Ssu-Chhêng (*4*) 梁思成 .
*Chi Wu-thai Shan Fo-Kuang Ssu Chien-
    Chu* 記五臺山佛光寺建築 .
On the Architecture of Fo-Kuang Ssu (the
    Buddha Light Temple) on Wu-thai Shan.
BSRCA, 1945, **7** (no. 1), art. 1, (no. 2), art. 4
    (mimeographed).

Liang Ssu-Chhêng (*5*) 梁思成 .
*I-chhien-san-pai To Nien Chhien ti Shih
    Chhiao* 一千三百多年前的石橋 .
A Stone Bridge built more than 1,300 years
    ago [the great segmental arch bridge at
    Chaochow].
JMHP, 1958 (no. 8), (no. 98), 30.

Liang Ssu-Chhêng (*6*) 梁思成 .
*Wo-mên so Chih-Tao-ti Thang Tai Fo Ssu
    yü Kung Tien* 我們所知道的唐代佛
    寺與宮殿 .
What we know of (the Architecture of)

Buddhist Temples and Palace Halls in the
    Thang Period.
BSRCA, 1932, **3** (no. 1), 48.

Liang Ssu-Chhêng (*7*) 梁思成 .
*Po-Hsi-Ho hsien-sêng Kuan-yü Tunhuang
    Chien-Chu-ti i fêng Hsin* 伯希和先生
    關於燉煌建築的一封信 .
A Letter from Professor Pelliot on an
    Architectural Matter at Tunhuang.
BSRCA, 1932, **3** (no. 4), 125; Eng. abstr. 10.

Liang Ssu-Chhêng (*8*) 梁思成 .
*Chêng-ting Tiao Chha Chi Lüeh (Hsien Wên-
    Miao Ta Chhêng Tien)* 正定調查紀略
    (縣文廟大成殿) .
The Ancient Architecture of Chêng-ting
    [Hopei]; the main hall of the Confucian
    temple [probably late Thang in date].
BSRCA, 1933, **4** (no. 2), 1 (39 ff.).

Liang Ssu-Chhêng (*9*) 梁思成 .
*Chi-Hsien Tu-Lo-Ssu Kuan-Yin-Ko Shan
    Mên Khao* 薊縣獨樂寺觀音閣山門
    考 .
The Kuan-Yin Pavilion and the Mountain
    Gate at the Tu-Lo Ssu Temple, Chi-
    hsien (Two Liao Dynasty Structures).
BSRCA, 1932, **3** (no. 2), 1–92.

Liang Ssu-Chhêng (*10*) 梁思成 .
*Hangchow Liu-ho Tha Fu-Yuan Chuang Chi
    Hua* 杭州六和塔復原狀計劃 .
Plans for the Restoration of the Liu-ho
    Pagoda at Hangchow.
BSRCA, 1935, **5** (no. 3), 1.

Liang Ssu-Chhêng (*11*) 梁思成 .
*Hsien-Hua Wên-Wu Chien-Chu ti Chhung-
    Hsiu yü Wei-Hu* 閒話文物建築的重修
    與維護 .
Rambling Comments on the Restoration and
    Protection of Architectural Monuments.
WWTK, 1963 (no. 7), 5.

Liang Ssu-Chhêng 梁思成 & Liu Chih-
Phing 劉致平 (*1*).
Chien Chu Shê Chi Tshan Khao Thu Chi
    建築設計參考圖集 .
Portfolios of Photographs illustrating typical
    features of Chinese Building: (*a*) Terraces
    and Pedestals, (*b*) Balustrades, (*c*) Shop-
    Fronts, (*d*) Wooden Corbelling.
Soc. for Res. in Ch. Archit., Peiping, 1936,
    1937.
Crit. rev. Hummel (*21*).

Lin Chhing (*1*) 麟慶 .
Hung Hsüeh Yin-Yuan Thu Chi 鴻雪因
    緣圖記 .
Illustrated Record of Memories of the
    Events which had to happen in My Life.
1849.
Cf. Hummel (*2*), p. 507.

Lin Chhing (*2*) 麟慶 .
Ho Kung Chhi Chü Thu Shuo 河工器具
    圖說 .
Illustrations and Explanations of the

Lin Chhing (2) (*cont.*)
  Techniques of Water Conservancy and
    Civil Engineering.
  1836.
  Cf. Hummel (2), p. 507
Lin Hui-Yin 林徽因 & Liang Ssu-Chhêng
    梁思成 (1).
  Chin Fên Ku Chien-Chu Yü-Chha Chi Lüeh
    晉汾古建築預查紀略 .
  Brief Report of a Preliminary Enquiry into
    the Ancient Architecture of the Upper
    Fên River Valley (in Shansi).
  BSRCA, 1935, **5** (no. 3), 12.
Lin Tsê-Hsü (1) 林則徐 .
  Ssu Chou Chih 四洲志 .
  Information on the Four Continents
    [especially on the West].
  c. 1840.
  One of the chief sources of Wei Yuan &
    Lin Tsê-Hsü (q.v.).
Ling Shun-Shêng (1) = (1) 凌純聲 .
  Thaiwan-ti Hang Hai Fan-Fa chi chhi Chhi-
    Yuan 臺灣的航海帆筏及其起源 .
  The Formosan Sea-going Sailing Raft and
    its Origin in Ancient China.
  AS/BIE, 1956 (no. 1), 1.
Ling Shun-Shêng (2) = (2) 凌純聲 .
  Pei-phing-ti Fêng Shan Wên-Hua 北平的
    封禪文化 .
  The Sacred Enclosures and Stepped Pyra-
    midal Platforms [Altar of Heaven and
    Earth, etc.] at Peking [and the History of
    Chinese Cosmic Religion and its Temples].
  AS/BIE, 1963 (no. 16), 1.
  Cf. Liu Tzu-Chien (1).
Ling Shun-Shêng (3) = (3) 凌純聲 .
  Chung-Kuo Ku-Tai Shê chih Yuan-Liu
    中國古代社之源流 .
  The Origin of the Shê [Holy Place] in
    Ancient China.
  AS/BIE, 1964 (no. 17), 1.
Ling Shun-Shêng (4) = (4) 凌純聲 .
  Chhin Han Shih-Tai chih Chih 秦漢時代
    之時 .
  The Sacred Places of the Chhin and Han
    Periods.
  AS/BIE, 1964 (no. 18), 113.
Ling Shun-Shêng (5) = (5) 凌純聲 .
  Chung-Kuo-ti Fêng Shan yü Liang-Ho Liu-
    Yü-ti Khun-Lun Wên-Hua 中國的封
    禪與兩河流域的昆侖文化 .
  A Comparative Study of the Ancient Chinese
    Fêng and Shan [Sacrifices] and the
    Ziggurats of Mesopotamia.
  AS/BIE, 1965 (no. 19), 1.
Liu Chih-Phing (1) 劉致平 .
  Chung-Kuo Chien-Chu Lei-Hsing Chi Chieh-
    Kou 中國建築類型及結構 .
  Systematic Treatise on Chinese Architecture.
  Architectural and Engineering Press,
    Peking, 1957.

Liu Chih-Phing 劉致平 & Fu Hsi-Nien (1)
    傅熹年 .
  Lin-Tê Tien Fu-Yuan ti Chhu-Pu Yen-Chiu
    麟德殿復原的初步研究 .
  Preliminary Researches on the Reconstruction
    of the Lin-Tê Hall (of the Ta Ming Kung
    Palace at the Thang capital, Chhang-an).
  KKTH, 1963 (no. 7), 385.
Liu Chih-Yuan (1) (ed.) 劉志遠
  Szechuan Shêng Po-Wu-Kuan Yen-Chiu
    Thu Lu 四川省博物館研究圖錄 .
  Illustrated Studies on the (Reliefs, Bricks,
    and other Objects in the) Szechuan
    Provincial Museum (Chungking and
    Chhêngtu).
  Ku-Tien I-Shu, Peking, 1958.
Liu Ê (1) 劉鶚 .
  Lao-Tshan Yu Chi 老殘遊記 .
  The Travels of Lao-Tshan (Mr Derelict)
    [novel].
  Shanghai, 1904-7.
  Tr. Shadick (1).
Liu Hai-Su (1) 劉海粟 .
  Ming Hua Ta Kuan 名畫大觀 .
  Album of Celebrated Paintings.
  4 vols.
  Chung-Hua, Shanghai, 1935.
Liu Hsien-Chou (9) 劉仙州 .
  Wo Kuo Tu-Lun-Chhê ti Chhuang Shih
    Shih-Chhi Ying Shang Tui Tao Hsi Han
    Shih-Tai 我國獨輪車的創始時期應
    上推到西漢晚年 .
  The Time of the First Appearance of the
    Wheelbarrow in China Traced to the
    latter part of the Former Han Dynasty.
  WWTK, 1964 (no. 6), 1.
Liu Kuei-Fang (1) 劉桂芳 .
  Shantung Liang-shan Hsien Fa-Hsien-ti
    Ming Chhu Ping Chhuan 山東梁山縣
    發現的明初兵船 .
  A Warship from the Beginning of the Ming
    Dynasty (+1377) Discovered and
    Excavated at Liang-shan Hsien in
    Shantung Province.
  WWTK, 1958 (no. 2), 51.
Liu Ming-Shu (2) 劉銘恕 .
  Chêng Ho Hang Hai Shih Chi Chih Tsai
    Than 鄭和航海事蹟之再探 .
  Further Investigations on the Sea Voyages
    of Chêng Ho.
  BCS, 1943, **3**, 131.
  Crit. ref. by A. Rygalov, *HH*, 1949, **2**, 425.
Liu Ming-Shu (4) 劉銘恕 .
  Sung-Tai Hai-Shang Thung Shang Shih Tsa
    Khao 宋代海上通商史雜考 .
  Miscellaneous Studies on the Seafaring and
    Commerce of the Sung Period.
  BCS, 1945, **5**, 49.
Liu Tshai-Yü (1) 劉彩玉 .
  Lun Fei-shui Yuan yü Chiang-Huai Yün-Ho
    論肥水源與江淮運河 .

Liu Tshai-Yü (*1*) (*cont.*)
On the Source of the Fei River in relation
to the Canal (of Warring States times)
linking the Yangtze and the R. Huai.
LSYC, 1960, **7** (no. 3), 69.

Liu Tun-Chên (*1*) 劉敦楨.
Shih Chou Chu Chhiao Shu Yao (*Sian Pa
Chhan Fêng San Chhiao*) 石軸柱橋述
要(西安灞滻豐三橋).
Notes on Stone Pier Bridges, with Special
reference to Three Bridges at Sian.
BSRCA, 1934, **5** (no. 1), 32.

Liu Tun-Chên (*2*) 劉敦楨.
*Fu-Chün Wên-Chhang Chhiao Chih chih
Chieh-Shao* 撫郡文昌橋志之介紹.
A Bibliographical Chronicle of the Wên-
chhang Bridge (at Fuchow (Lin-chhuan)
in Chiangsi).
BSRCA, 1934, **5** (no. 1), 93.

Liu Tun-Chên (*3*) 劉敦楨.
Wan-Nien Chhiao Chih Shu Lüeh 萬年橋
志述略.
Brief Account of the Wan-Nien Bridge [at
Nan-chhêng in Chiangsi].
BSRCA, 1933, **4** (no. 1), 22.

Liu Tun-Chên (*4*) 劉敦楨.
Chung-Kuo Chu Tsê Kai Shuo 中國住宅
概說.
A Short Study of Chinese Domestic
Architecture.
Based on material collected by the Chinese
Architectural Research Unit, formed
jointly by the Architectural Research
Institute and the Nanking College of
Engineering.
Architectural and Engineering Press,
Peking, 1957.
Abridged translation, without illustrations,
by Liao Hung-Ying & R. T. F. Skinner,
Collet, London, 1957.

Liu Tun-Chên (*5*) 劉敦楨.
Suchow Ku Chien Chu Tiao-Chha Chi
蘇州古建築調查記.
Record of an Investigation of the Ancient
Architecture of Suchow.
BSRCA, 1936, **6** (no. 3), 17.

Liu Tun-Chên (*6*) 劉敦楨.
*Ming Lu Pan 'Ying Tsao Chêng Shih'
Chhao-pên Chiao Tu Chi* 明魯般'營造
正式'鈔本校讀記.
Notes on a MS. copy of a Ming edition of
the 'Right Standards of Building
Construction' attributed to Lu Pan.
BSRCA, 1937, **6** (no. 4), 162.

Liu Tun-Chên (*7*) 劉敦楨.
Lu Pan 'Ying Tsao Chêng Shih' 魯班'營
造正式'.
On the 'Right Standards of Building
Construction' attributed to Lu Pan.
WWTK, 1962 (no. 136), 9.

Liu Tun-Li (*1*) 劉敦勵.

*Ku-Tai Chung-Kuo yü Chung Mei Ma-Yeh
Jen ti Chhi Yü yü Yü Shen Chhung Pai*
古代中國與中美馬耶人的祈雨與雨
神崇拜.
On Rain-Making Ceremonies and the
Worship of Rain Gods among the Ancient
Chinese and the Mayas of Central America.
AS/BIE, 1957 (no. 4), 31.

Liu Tzu-Chien (*1*) 劉子健.
*Fêng Shan Wên-Hua yü Sung-Tai Ming
Thang Chi Thien* 封禪文化與宋代明
堂祭天.
Two Forms of the Worship of Heaven in
the Sung Dynasty (the open-air Fêng
Shan Sacrifice and the Rites of the Hall of
Enlightenment or Ming Thang).
AS/BIE, 1964 (no. 18), 45.
Cf. Ling Shun-Shêng (*2*).

Liu Wên-Chhi (*1*) 劉文淇.
Yangchow Shui Tao Chi 揚州水道記.
A History of the Waterways in the
Yangchow Region.
1845.

Liu Wên-Tien (*1*) 劉文典.
Chuang Tzu Pu Chêng 莊子補正.
Emended Text of *The Book of Master
Chuang*.
Com. Press, Shanghai, 1947.

Lo Chê-Wên (*1*) = (*1*) 羅哲文.
Yün-kang Shih Khu 雲岡石窟.
The Cave-Temples of Yünkang.
Wên-Wu, Peking, 1957.

Lo Chê-Wên (*2*) 羅哲文.
*Lin-thao Chhin Chhang Chhêng, Tunhuang
Yümên Kuan, Chiu-chhüan Chiayükuan
Khan-Chha Chien Chi* 臨洮秦長城, 燉
煌玉門關, 酒泉嘉關, 勘查簡記.
A Brief Report on Investigations of the
Chhin Great Wall at Lin-thao (south of
Lanchow), of the Yümên Gate near
Tunhuang, and of the Chia-yü-kuan Gate
near Chiu-chhüan (Suchow).
WWTK, 1964 (no. 6), 46.

Lo Hsiang-Lin (*1*) 羅香林.
Phu Shou-Kêng Yen-Chiu 蒲壽庚研究.
Researches on Phu Shou-Kêng
[Superintendent of Overseas Shipping at
Chhüanchow in the late Sung period].
Chung-Kuo Hsüeh-Shê, Hongkong, 1959.
Abstr. *RBS*, 1965, **5**, no. 183.

Lo Jung-Pang (*1*) 羅榮邦.
Chung-Kuo chih Chhê Lun Chhuan
中國之車輪船.
(The History of) the Paddle-Wheel Boat in
China.
CHJ/T, 1960 (n.s.), **2** (no. 1), 213.
Eng. tr. Lo Jung-Pang (*3*).

Lo Wên-Pin *et al.* (*1*) 羅文彬.
Szechuan Yen Fa Chih 四川鹽法志.
Memorials of the Salt Industry of
Szechuan and its control.

Lo Wên-Pin *et al.* (*1*) (*cont.*)
Compiled officially at the request of Ting
Pao-Chên, Governor-General of the
province, 1882.

Lo Ying (*1*) 羅英 .
Chung-Kuo Chhiao Liang Shih Liao (*Chhu
Kao*) 中國橋梁史料 (初稿).
A History of Bridge-Building in China
(Preliminary Draft).
[Peking, 1964.]

Lo Ying (*2*) 羅英 .
Chung-Kuo Shih Chhiao 中國石橋 .
Chinese Bridge-Construction in Masonry
[and its History].
Jen-Min, Peking, 1959.
Rev. *FET*, 1962, **17** (no. 2), 181.
Abstr. *RBS*, 1965, **5**, no. 831.

Lou Tsu-I (*1*) 樓祖詒 .
Chung-Kuo Yu I Shih Liao 中國郵驛史料 .
Materials on the History of the Chinese
Post-Station System.
Jen-Min Yu-Tien, Peking, 1958.

Lu Kung (*1*) (ed.) 路工 .
Mêng Chiang Nü Wan Li Hsün Fu Chi
孟姜女萬里尋夫記 .
Collection of Material on the Ballad of
Mêng Chiang Nü going to seek for her
Husband at the Great Wall.
Shanghai Chhu-Pan, Shanghai, 1955.
Abstr. *RBS*, 1957, **1**, no. 308.

Lung Fei-Liao (*1*) 龍非了 .
Hsüeh Chü Tsa Khao 穴居雜考 .
A Study of Cave-Dwellings (in China).
BSRCA, 1934, **5** (no. 1), 55.

Lung Fei-Liao (*2*) 龍非了 .
Khai-fêng chih Thieh Tha 開封之鐵塔 .
The Iron(-Coloured) Pagoda at Khaifêng.
BSRCA, 1932, **3** (no. 4), 53.

Ma Hui-Yü (*1*) (ed.) 馬慧裕 .
Hunan Thung Chih 湖南通志 .
Historical Geography of Hunan (province).
1820.
Revised and enlarged edition by Li Han-
Chang 李瀚章 .
1885.

Ma Kuo-Han (*1*) (ed.) 馬國翰 .
Yü Han Shan Fang Chi I Shu 玉函山房
輯佚書 .
The Jade-Box Mountain Studio Collection
of (Reconstituted) Lost Books.
1853.

Mai Ying-Hao (*1*) *et al.* 麥英豪 .
*Kuang-chou Huang Ti Kang Hsi Han Mu
Kuo Mu Fa-Chüeh Chien-Pao* 廣州皇
帝岡西漢木槨墓發掘簡報 .
Preliminary Report on the Excavation of a
Wooden-Coffin Tomb of the Early Han
Period at Huang-Ti Ridge in Canton.
KKTH, 1957 (no. 4), 22 (26 ff.).
Abstr. *RBS*, 1962, **3**, no. 408.

Makita Tairyō (*1*) 牧田諦亮 .
Sakugen 'Nyū-Min Ki' no Kenkyū 策彥
入明記の研究
Researches on the Travel Diaries of
Sakugen [Japanese monk in China in the
+16th century; with comparative
material on Chhoe Pu's *Phyohae Rok*].
2 vols.
Hōzōkan, Kyoto, 1955, 1959.
Abstrs. *RBS*, 1962, **3**, no. 899; 1965, **5**,
no. 792.

Mao I-Shêng (*2*) 茅以昇 .
*Chung-Tien Wên-Wu Pao-Hu Tan-Wei
chung ti Chhiao; Lu-ting Chhiao, Lu-kou
Chhiao, An-phing Chhiao, An-chi Chhiao,
Yung-thung Chhiao* 重點文物保護單
位中的橋；瀘定橋，蘆溝橋，安平橋；
永通橋 .
(Ancient) Bridges Scheduled for Preservation
as Cultural Monuments; the Lu-ting
(suspension) bridge, the Lu-kou (multiple arch) bridge, the An-phing (megalithic beam) bridge, and the An-chi and
Yung-thung (segmental arch) bridges.
WWTK, 1963 (no. 9), 33.

Mao Nai-Wên (*1*) 茅乃文 .
*Chung-Kuo Ho Chhü Shui Li Kung Chhêng
Shu Mu* 中國河渠水利工程書目 .
Bibliography of Works on River Control,
Hydraulic Engineering and Irrigation
(titles, authors and dates only).
Nat. Library, Peiping, 1935.

Mao Nai-Wên (*2*) 茅乃文 .
Chung-Kuo Ho Chhü Shu Thi Yao 中國
河渠書提要 .
Descriptive Bibliography of Works on River
Control, Hydraulic Engineering and
Irrigation.
SL, 1936, **11**, 52, 108, 172, 218, 292; 1937,
12, two instalments at pages not known,
then 226, 329, 395, 475; 1938, **13**, 94,
162.

Miu Chhi-Yü (*1*) 繆啟愉 .
*Wu Yüeh Chhien Shih tsai Thai-Hu Ti-
Chhü-ti Yü-Thien Chih-Tu ho Shui-Li
Hsi-Thung* 吳越錢氏在太湖地區的
圩田制度和水利係統 .
On the Methods used by Governor Chhien
[in the Wu Tai Period] for the Draining
of Polder Land in the Thai-Hu [Lake]
District of Wu and Yüeh, and their
Significance for the [History of] Irrigation
Engineering.
AHRA, 1960, **2**, 139.

Miyazaki Ichisada (*1*) 宮崎市定 .
*Myōshinji Rinshōin zō 'Konitsu Rekidai
Kokuto Kyōri Chizu' ni tsuite* 妙心寺麟
祥院藏「混一歷代國都疆理地圖」につ
いて .
On a world-map in the Rinshōin (Library)
of the Myōshinji (Temple) entitled 'Map

Miyazaki Ichisada (*1*) (*cont.*)
of the Capitals and Territories of the One
World in Successive Ages'.
Art. in *Kanda Hakase Kanreki Kinen
Shoshigaku Ronshū* 神田博士還曆記
念書誌學論集 (Essays in Honour of
Kanda Kiichirō; Kanda Memorial
Volume), p. 577, Kyoto, 1957.
Abstr. *RBS*, 1962, **3**, no. 284.
Miyazaki Ichisada (*2*) 宮崎市定.
*Chūgoku ni okeru Sonsei no Seiritsu; Kodai
Teikoku Hōkai no Ichimen* 中國にお
ける村制の成立；古代帝國崩壞の
一面.
The Development of the Village System in
Ancient China; an Aspect of the
Breakdown of Imperial Power.
TK, 1960, **18**, 569.
Abstr. *RBS*, 1967, **6**, no. 39.
Mizuno Seiichi (*1*) 水野清一.
Kandai no Kaiga 漢代の繪畫.
Painting of the Han Dynasty.
Kyoto, 1957.
Mizuno Seiichi 水野清一 & Nagahiro
Toshio (*1*) = (*1*) 長廣敏雄.
Unkō Sekkutsu 雲岡石窟.
The Cave-Temples of Yünkang.
16 vols. in 31 parts.
Jimbun Kagaku Kenkyūjō, Kyoto, 1950–56.
Mori Katsumi (*1*) 森克巳.
Kentō-shi 遣唐史.
On (Japanese) Embassies to Thang China.
Shibundō, Tokyo, 1955.
Abstr. *RBS*, 1957, **1**, no. 133.
Mori Osamu 森修 & Naitō Hiroshi 內藤寬
(*1*).
*Ying Chhêng Tzu; Chhien-Mu-Chhêng-I
fukin no Kandai hekiga sembo* 營城子；
前牧城驛附近の漢代壁畫甎墓.
Ying Chhêng Tzu; (Two) Han Brick
Tombs with Fresco Paintings near
Chhien-mu-chhêng-i (in South Man-
churia).
With preface by Kusaka Tatsuta and appen-
dices by Hamada Kōsaku & Mizuno
Seiichi.
Tōa Kōkogaku Kwai, Tokyo and Kyoto,
1934 (Archaeologia Orientalis, no. 4).
Moriya Mitsuo (*1*) 守屋美都雄.
Art. in *Chūgoku Kodai no Shakai to Bunka;
Sono Chi-ikibetsu Kenkyū* 中國古代の
社會と文化；その地域別研究.
Ancient Chinese Society and Culture;
Regional Differences among the States.
Daigaku Shuppankai, Tokyo, 1957.
Abstr. *RBS*, 1962, **3**, no. 117.
Motoki Masahide (*1*) 本木正榮.
Gunkan Zusetsu 軍艦圖說.
Illustrated Account of (European) Warships.
Japan, 1808.
(Repr. *NKKZ*, vol. 12.)

Mu Shou Chhi (*1*) 慕壽祺.
Chhang Chhêng Khao 長城考.
A Study on the [Origins of] the Great Wall.
SWYK, 1944, **4**, 631.
Murakami Yoshimi (*1*) 村上嘉實.
Rikuchō no Teien 六朝の庭園.
(Houses and) Gardens in the Six Dynasties
Period.
KDK, 1955, **4**, 41.
Abstr. *RBS*, 1957, **1**, no. 255.
Murakami Yoshimi (*2*) 村上嘉實.
Tōdai Kizoku no Teien 唐代貴族の庭園.
(Houses and) Gardens of the Upper Classes
in the Thang Period.
TG/T, 1955, **11**, 71.
Abstr. *RBS*, 1957, **1**, no. 256.
Murata Jirō (*1*) 村田治郎.
Dai Tōa Kenchiku Rombun Sakuin 大東亞
建築論文索引.
A Bibliography of East Asian Architecture.
Tokyo, 1944.
Murata Jirō 村田治郎, Fujieda Akira
藤枝晃 *et al.* (*1*).
Kyoyōkan 居庸關.
On Chü-yung-kuan [the gate of the Great
Wall near Nan-khou, ornamented with
Buddhist carvings and inscriptions,
+1343].
2 vols. Kyōto Daigaku Kōgakubu.
(Faculty of Engineering, Kyoto Univ.),
1957.
Abstr. *RBS*, 1962, **3**, no. 513.

Naba Toshisada (*2*) 那波利貞.
*Tōdai no Nōden-suiri ni Kansuru Kitei ni
tsukite* 唐代の農田水利に關する規
定に就きて.
On the History of Irrigation and Water-
Conservancy during the Thang dynasty
[including materials from the Tunhuang
documents].
SGZ, 1943, **54**, 18, 150, 249.
Nagahiro Toshio (*2*) 長廣敏雄.
Yen Li-Tê to Yen Li-Pên ni tsuite 閻立德
と閻立本について.
On Yen Li-Tê and Yen Li-Pên [great
architects and artists of the Thang].
TG/K, 1959, **29**, 1.
Nagase Mamoru (*1*) 長瀨守.
*Hokusō-matsu ni okeru Chō Rin no suiri
Seisaku ni tsuite* 北宋末における趙
霖の水利政策について.
On Chao Lin's Policy for the Utilisation of
Water at the end of the Northern Sung
Dynasty.
In *Chūgoku no Shakai to shūkyō* 中國の
社會と宗教.
Chinese Society and Religion (ed.
Yamazaki Hiroshi).
Tokyo Educational College (Fumaidō
Shoten), 1954, p. 121.

Naitō Torajirō (2) 內藤虎次郎.
'*Hsing Chha Shêng Lan*' *Chhien Chi Chiao Chu* 「星槎勝覽」前集校注.
Commentary on the First Part of [Fei Hsin's] *Triumphant Visions of the Starry Raft* [+1436].

Niida Noboru (2) 仁井田陞.
Shina no Tochi Daichō '*Gorinzusetsu*' *no Shiteki Kenkyū* 支那の土地台帳「魚鱗圖」冊の史的研究.
An Investigation of the History of the 'Fish-Scale Maps' (dissected cadastral survey charts) in China.
TG/T, 1936, **6**, 157.

Ogawa Takuji (1) 小川琢治.
Kinsei Seiyō Kōtsū Izen no Shina Chizu ni tsuite 近世西洋交通以前の支那地圖に就て.
(A Historical Sketch of) Cartography in China before the modern Intercourse with the Occident.
CZ, 1910, **22**, 407, 512, 599.
Reprinted in Ogawa (2).

Okuyama Tsunegorō 奧山恆五郎, Itō Chūta 伊東忠大, Tsuchiya Jun-ichi 土屋純一 & Ogawa Kazumasa (1) 小川一眞.
Shinkoku Pekin Shikin-jō-denmon no Kenchiku (and other reports) 清國北京紫禁城殿門の建築.
Architecture of the Buildings and Gates of the Purple Forbidden City in Peking.
SRSETU, 1903, no. 4; 1906, no. 7.

Ōshima Toshikazu (1) 大島利一.
Chūgoku Kodai no Shiro ni tsuite 中國古代の城について.
On the Walled Cities of Ancient China.
TG/K, 1959, **30**, 39.
Abstr. *RBS*, 1965, **5**, no. 67.

Pai Shou-I (1) 白壽彝.
Chung-kuo Chiao Thung Shih 中國交通史.
History of Communications in China.
Com. Press, Shanghai, 1937.

Pao Chio-Min (1) 鮑覺民.
Chhêng-tu Phing-Yuan chih Shui-Li 成都平原之水利.
The Irrigation System of the Chhêngtu Plain [Szechuan].
NQJEP, 1937, **5**, 1.

Pao Ting 鮑鼎, Liu Tun-Chên 劉敦楨 & Liang Ssu-Chhêng (1) 梁思成.
Han-Tai-ti Chien-Chu Shih-Yang yü Chhuang-Shih 漢代的建築式樣與裝飾.
The Architectural Style of the Han Dynasty.
BSRCA, 1934, **5** (no. 2), 1.

Pao Tsun-Phêng (1) = (1) 包遵彭.
Chêng Ho Hsia Hsi-Yang chih Pao Chhuan Khao 鄭和下西洋之寶船考.
A Study of the 'Great Treasure Ships' used in Chêng Ho's Voyages in the Western Oceans.
Chung-Hua Tshung-Shu, Thaipei and Hongkong, 1961.
(Nat. Historical Museum Collected Papers on the History and Art of China, 1st ser., no. 6.)

Pao Tsun-Phêng (2) = (2) 包遵彭.
Han-Tai Lou-Chhuan Khao 漢代樓船考.
A Study of the 'Castled Ships' of the Han Period.
Nat. Historical Museum, Thaipei, Thaiwan, 1967 (Coll. Papers on the History and Art of China, no. 2).

Phan Ên-Lin (1) (ed.) = (1) 潘恩霖.
Hsi-Nan Lan Shêng 西南攬勝.
Scenic Beauties of Southwest China.
China Travel Service, Shanghai, 1939.

Phan Nien-Tzhu (1) 潘念慈.
Kuan-yü Yuan-Tai-ti I Chhuan 關于元代的驛傳.
On the Post-Station System of the Yuan Dynasty.
LSYC, 1959, **5** (no. 2), 59.
Abstr. *RBS*, 1965, **5**, no. 190.
Minor criticisms by Chhen Tê-Chih & Shih I-Khuei, *LSYC*, 1959 (no. 7), 57.

Po Shu-Jen (1) 薄樹人.
Chung-Kuo Ku-Tai Hêng Hsing Kuan Tshê 中國古代恆星觀測.
Ancient Chinese Observations of Fixed Stars.
KHSC, 1960, **1** (no. 3), 35.

Sakabe Kōhan (1) 坂部廣胖.
Kairo Anshinroku 海路安心錄.
Safe Journeys on the High Seas.
Japan, 1816.
(Repr. *NKKZ*, vol. 12.)

Sekino Tadashi 關野貞 & Takeshima Takuichi 竹島卓一 (1).
Ryō-Kin Jidai no Kenchiku to sono Butsuzō 遼金時代の建築と其佛像.
The Architecture and Buddhist Images of the Liao and J/Chin Periods.
3 vols. (1 of text, 2 of plates).
Tōhō Bunka Gakuin, Tokyo, 1934, 1935, 1944.

Shang Chhêng-Tsu (1) 商承祚.
Chhang-sha Chhu Thu Chhu Chhi Chhi Thu Lu 長沙出土楚漆器圖錄.
Album of Plates and Description of Lacquer Objects from the State of Chhu excavated at Chhangsha (Hunan).
Shanghai, 1955.

Shen Khang-Shen (1) 沈康身.
Wo Kuo Ku-Tai Tshê-Liang Chi-Shu-ti Chhêng-Chiu 我國古代測量技術的成就.
Ancient and Mediaeval Chinese Achievements

Shen Khang-Shen (1) (cont.)
　　in Surveying and Survey Instrumenta-
　　tion.
　　KHSC, 1965 (no. 8), 28.
Shen Yao (1) 沈垚.
　　Lo Fan Lou Wên Chi (or Kao) 落帆樓文
　　集(or稿).
　　Harbour View Essays.
　　c. 1830.
Shih Shu-Chhing (3) 史樹青.
　　*Yu Kuan Han-Tai Tu-Lun-Chhê ti Chi-Ko
　　Wên-Thi* 有關漢代獨輪車的幾个問
　　題.
　　Some Questions concerning the Wheel-
　　barrow in the Han Period.
　　WWTK, 1964 (no. 6), 6.
Shih Yen (1) 史岩.
　　*Kuan-yü Kuang-Yuan Chhien-Fo-Yai Tsao
　　Hsiang ti Chhuang-Shih Shih-Tai Wên-
　　Thi* 關于廣元千佛崖造象的創始時
　　代問題.
　　On the Dating of the Buddhist images of
　　the Thousand-Buddha Cliff Cave-
　　Shrines at Kuangyuan (in northern
　　Szechuan) [+6th to +8th century].
　　WWTK, 1961 (no. 2), 24.
Shou Phêng-Fei (1) 壽鵬飛.
　　Li-Tai Chhang Chhêng Khao 歷代長城考.
　　A Study of the History of the Great Wall.
　　No place or date (the author is of Shao-
　　hsing), with large folding map or maps at
　　the back, 1961.
Su Tsung-Sung 粟宗嵩, Hsüeh Li-Than
　　薛履坦 & Lo Thêng (1) 駱騰.
　　*Chhing Shun-Khang-Yung San Chhao,
　　Chhien-Lung, Chia-Tao Liang Chhao,
　　Huang Ho Chüeh Khou Khao* 清順康
　　雍三朝乾隆嘉道兩朝黃河決口考.
　　Historical Researches on the Dyke Break-
　　ages of the Yellow River, 1644 to 1850.
　　SL, 1936, **10**, 335, 348, 378.
Sudō Yoshiyuki (1) 周藤吉之.
　　*Sōdai no Uden to Shōensei; Toku ni Kōnan
　　Tōro ni tsuite* 宋代の圩田と莊園制;
　　特に江南東路について.
　　A Study of the Reclamation of Land from
　　Lakes and Old River-Beds in the lower
　　Yangtze valley during the (Southern
　　Thang and) Sung.
　　TBK, 1956, **10**, 229.
　　Abstr. *RBS*, 1956, **2**, no. 185.
Sugimura Yūzō (1) 杉村勇造.
　　Chūgoku no Niwa 中國の庭.
　　Chinese Gardens (and Dwellings).
　　Kyuryudo, Tokyo, 1966.
Sumitomo Tomozumi 住友友純, Taki
　　Seiichi 瀧精一 & Naitō Torajirō (1)
　　内藤虎次郎.
　　Senoku Seishō 泉屋清賞.
　　(Albums illustrating the Sumitomo Collection
　　of Bronzes at Kyoto.)

7 vols. (vols. 1–5 and Supplementary 9 and
　　10).
　　Osaka, 1918–26.
　　See Hamada, Harada & Umehara (1).
Sun Hai-Po (2) 孫海波.
　　Honan Chi Chin Thu Chih Shêng Kao
　　河南吉金圖志賸稿.
　　Further Notes, with Illustrations, on
　　Bronzes found in Honan.
　　Khao Ku Hsüeh Shê Chuan Khan
　　考古學社專刊 no. 19.
　　Peiping, 1939.
Sung Hsi-Shang (1) 宋希尚.
　　Chung-Kuo Ho Chhuan Chih 中國河川
　　誌.
　　China's River-Systems.
　　Thaipei, 1954.
Sung Hsi-Shang (2) 宋希尚.
　　Li-Tai Chih Shui Wên Hsien 歷代治水
　　文獻.
　　Hydraulic Engineering [in China] and its
　　History.
　　Thaipei, 1954.
Sung Ta-Jen (1) 宋大仁.
　　*Chung-Kuo ho A-La-Po ti I-Yao Chiao
　　Liu* 中國和阿拉伯的醫藥交流.
　　Mutual Influences in Medical and Pharma-
　　ceutical Science between China and the
　　Arabic Nations.
　　LSYC, 1959 (no. 1), 79.

Takahashi Tadashi (1) 高橋正.
　　*Tōzen seru Chūsei Isurāmu Sekaizu; Shu to
　　shite 'Kon-itsu Kyōri Rekidai Kokuto no
　　Zu' ni tsuite* 東漸せる中世イスラー
　　ム世界圖;主として混一疆理歷代國
　　都之圖について.
　　The World of Islam; the Religion expanded
　　to the East; a Commentary on the *Hon-il
　　Kangni Yŏktae Kukto chi To* [the Korean
　　world-map of +1402].
　　RDR, 1963 (no. 374), 77.
Taki Seiichi 瀧精一, Naitō Torajirō
　　内藤虎次郎 & Hamada Kōsaku (1)
　　濱田青陵.
　　Santei Senoku Seishō 删訂泉屋清賞.
　　(New and Revised Explanatory Notes
　　for the Albums illustrating the Sumi-
　　tomo Collection of Bronzes at
　　Kyoto.)
　　Kyoto, 1934.
Tamura Sennosuke (1) 田村專之助.
　　Tōyōjin no Kagaku to Gijutsu 東洋人の
　　科學與技術.
　　(Essays on the History of) Science and
　　Technology among East Asian Peoples
　　[mainly astronomical and meteorological,
　　with much on Korea as well as China and
　　Japan].
　　Awaji Shobō Shinsha, Tokyo, 1958.
　　Abstr. *RBS*, 1964, **4**, no. 936.

Tanabe Yasuaki (1) 田邊泰著.
'Ta Thang Wu Shan Chu Thang Thu' Khao
「大唐五山諸堂圖」考.
A Study of the 'Drawings of the Various
Halls in the Five Mountain (Abbeys)
dating from the Great Thang Dynasty'
(MS. of +1259 preserved in Japan).
Tr. Liang Ssu-Chhêng 梁思成.
BSRCA, 1932, 3 (no. 3), 71.

Thang Chin-Yü (1) 唐金裕.
Hsi-an Hsi-chiao Han-Tai Chien-Chu
I-Chih Fa-Chüeh Pao-Kao 西安西郊
漢代建築遺址發掘報告.
Report on the Excavation of Architectural
Remains [a pi yung, College and Hall for
Venerating Elders] in the Western
Suburb of Sian [from the Chhien Han
Period].
AS/CJA, 1959 (no. 2), 45.
Abstr. RBS, 1965, 5, no. 376.

Thang Huan-Chhêng (1) 唐寰澄.
Chung-Kuo Ku Tai Chhiao-Liang
中國古代橋梁.
Ancient (and Medieval) Chinese Bridges.
Wên-wu, Peking, 1957.
Abstr. RBS, 1962, 3, no. 514 (with serious
misunderstandings).

Thien Hsiu (1) 田秀.
I Fu Sung-Tai Hui Hua; Fang-Chhê Thu
一幅宋代繪畫紡車圖.
A Sung Scroll-Painting of a Spinning-
Wheel.
WWTK, 1961 (no. 2), 44, with plate.

Thien Ju-Khang (1) 田汝康.
Shih-chhi Shih-Chi chih Shih-chiu Shih-Chi
chung-yeh Chung-Kuo Fan-Chhuan tsai
Tung-Nan Ya-Chou Hang-Yün ho Shang-
Yeh shang ti Ti-Wei 十七世紀至十九
世紀中葉中國帆船在東南亞洲航運和
商業上的地位.
The Place of Chinese Sailing-Ships in the
Maritime Trade of Southeast Asia from
the 17th to the 19th Centuries.
LSYC, 1956, 2 (no. 8), 1.
Sep. published in book form, Jen-min,
Shanghai, 1957.

Thien Ju-Khang (2) 田汝康.
Tsai Lun-Shih-chhi chih Shih-chiu Shih-chi
chung-yeh Chung-Kuo Fan-Chhuan Yeh ti
Fa-Chan 再論十七至十九世紀中葉
中國帆船業的發展.
Further Studies on Chinese Sailing-Ships
and the Development of Maritime Trade
from the 17th to the 19th Centuries.
LSYC, 1957, 3 (no. 12), 1.
Abstr. RBS, 1962, 3, no. 330.

Thung Shih-Hêng (1) 童世亨.
Li Tai Chiang Yü Hsing Shih I Lan Thu
歷代疆域形勢一覽圖.
Historical Atlas of China.
Com. Press, Shanghai, 1922.

Thung Shu-Yeh (1) 童書業.
Chhung Lun Chêng Ho Hsia Hsi Yang Shih
Chien chih Mao-I Hsing-Chih 重論鄭
和下西洋事件之貿易性質.
A Further Discussion of the Commercial
Nature of Chêng Ho's Voyages to the
Western Oceans.
YK, 1937, 7 (no. 1), 239.

Ti Phing-Tzu (1) 狄平子.
Han Hua (1st series) 漢畫 (It is
doubtful if any more was ever published.)
Collection of Drawings [Inscribed Bronzes
and Stones, Moulded Bricks, etc.] of the
Han period.
Shanghai (probably issued by a curio
dealer), n.d. (c. 1928?).

Tokiwa Daijō 常葉大定 & Sekino Tadashi
關野貞(1).
Shina Bukkyō Shiseki 支那佛教史蹟.
Monuments of Buddhism in China.
Tokyo, 1926–29.

Toyoda Toshitada (1) 豐田利忠.
Zenkōji-Michi Meisho Zue 善光寺道名所
圖會.
Pictorial Description of Noted Places along
the Route to the Zenkōji Temple (in
Shinano Province).
Minoya Iroku, Nagoya, 1849.

Tsêng Chao-Yü 曾昭燏, Chiang Pao-
Kêng 蔣寶庚 & Li Chung-I (1)
黎忠義.
I-nan Ku Hua Hsiang Shih Mu Fa-
Chüeh Pao-Kao 沂南古畫像石墓發掘
報告.
Report on the Excavation of an Ancient
[Han] Tomb with Sculptured Reliefs at
I-nan [in Shantung] (c. +193).
Nanking Museum, Shantung Provincial
Dept. of Antiquities, and Ministry of
Culture, Shanghai, 1956.

Tshen Chung-Mien (2) 岑仲勉.
Huang-Ho Pien Chhien Shih 黃河變遷
史.
History of the Changes of Course of the
Yellow River.
Jen-Min, Peking, 1957.

Tu Hsien-Chou (1) 杜仙州.
I-Hsien Fêng-Kuo Ssu Ta Hsiung Tien
Thiao-Chha Pao-Kao 義縣奉國寺大
雄殿調查報告.
Report of a Study of the Ta Hsiung Hall at
the Fêng-Kuo Temple at I-hsien (in
Western Liaoning) [a J/Chin building of
the early +12th century].
WWTK, 1961 (no. 2), 5.

Tung Hsün (1) 董恂.
Chiang Pei Yün Chhêng 江北運程.
Handbook on the Course of the Grand
Canal North of the Yangtze.
1867.

Tung Kao (*1*) (ed.) 董誥 *et al.*
    Chhüan Thang Wên 全唐文.
    Collected Literature of the Thang Dynasty.
    1814.
    Cf. des Rotours (2), p. 97.

Uchida Gimpū (*2*) 內田吟風.
    *Kodai Yūboku Minzoku ni okeru Doboku
    Kenzō Gijutsu* 古代遊牧民族に於ける
    土木建造技術.
    Civil Engineering [Fortification] Techniques
    of the Ancient [Asian] Nomads.
    TK, 1951, **11** (no. 2), 111.
Unno Kazutaka (*1*) 海野一隆.
    *Tenri Toshokan Shozō Daiminkoku Zu ni
    tsuite* 天理圖書館所藏大明國圖につ
    いて.
    On an Anonymous Map of Ming China
    [Korea and Japan] preserved in the Tenri
    Central Library [actually a copy dating
    from about +1550 of the Korean world
    map of +1402].
    MOULA, 1958 (no. 6), 60.
Unno Kazutaka (*4*) 海野一隆.
    Tōyō Chirigaku Shi 東洋地理學史.
    History of Geography in East Asia.
    Pt. 1, Sect. 1, chs. 1–3 of 'History and
    Methods of Geography' *Chirigaku no
    Rekishi to Hōhō* 地理學の歷史と方法
    by S. Noma, M. Matsuda & K. Unno.
    Taimeidō, Tōkyō, 1959.
Unno Kazutaka (*5*) 海野一隆.
    'Kuang Yü Thu' no Shiryō to natta Chizurui
    「廣輿圖」の資料となつた地圖類.
    On the Original Sources of the *Kuang Yü
    Thu* World Atlas.
    SHSS, 1967, **15**, 21.

Wada Kyntoku (*1*) 和田久德.
    Sōdai Nankai Shiryo to shite no 'Toi Sasshi'
    宋代南海史料としての「島夷雜誌」.
    The *Tao I Tsa Chih*; a New Chinese Source
    for the History of the Eastern Archipelago
    and the Coasts of the Indian Ocean during
    the Sung Dynasty [part of the *Shih Lin
    Kuang Chi*, q.v.].
    OUSS, 1954, **5**, 27.
Wan Kuo-Ting (*3*) 萬國鼎.
    Ou Thien Fa ti Yen-Chui 區田法的研究.
    Researches on the Pit- or Basin-Method of
    Cultivation [and strip-planting, in
    alternate depressed water-collecting
    squares or troughs, to increase crop
    yields].
    AHRA, 1958, **1**, 5.
    Abstr. *RBS*, 1964, **4**, no. 951.
Wang Chen-To (*5*) 王振鐸.
    *Ssu-Nan Chih-Nan Chen yü Lo-Ching Phan
    (Hsia)* 司南指南針與羅經盤(下).
    Discovery and Application of Magnetic
    Phenomena in China, III (Origin and

Development of the Chinese Compass
    Dial).
    AS/CJA, 1951, **5** (n.s., **1**), 101.
Wang Chhang (*1*) 王昶.
    Chin Shih Tshui Pien 金石萃編.
    Collection of Inscriptions on Bronze and
    Stone (from the earliest times down to
    +1279).
    1805.
Wang Chi-Chih (*1*) 王吉智.
    *Yin-chhuan Phing-Yuan Chu-Yao Thu-
    Jang ti Hsing-Chhêng Thê-Chêng, chi chhi
    Kai-liang Li-Yung Thu-Ching* 銀川平
    原主要土壤的形成特征及其改良利
    用途徑.
    Genesis and Properties of the Main Soil
    Types of the Plain of Yin-chhuan [Ning-
    hsia in the Upper Yellow River valley]
    and Means for their Reclamation and
    Utilisation.
    APS, 1964, **12**, 23.
Wang Hsien-Chhien (*3*) 王先謙.
    Shih Ming Su Chêng Pu 釋名疏證補.
    Revised and Annotated Edition of the
    [Han] *Explanation of Names* [dictionary].
    Peking, 1895.
Wang Hu-Chên (*1*) 汪胡楨.
    Ku Tai Thu Kung Chi Chia Fa 古代土
    工計價法.
    On Earthwork Valuation in Ancient Times.
    SL, 1936, **11**, 439.
Wang Jung (*1*) 王榮.
    *Yuan Ming Huo-Thung-ti Chuang Chih Fu-
    Yuan* 元明火銃的裝置復原.
    On the Restoration of the Carriage Mount-
    ings of the Yuan and Ming Bombards.
    WWTK, 1962 (no. 3), 41.
Wang Kuo-Liang (*1*) 王國瓦.
    Chung-kuo Chhang Chhêng Yen Ko Khao
    中國長城沿革考.
    A Study of the Development of the Great
    Wall.
    Com. Press, Shanghai, 1931.
Wang Pi-Wên (*1*) 王璧文.
    Chhing Kuan Shih Shih Chhiao Tso Fa
    清官式石橋做法.
    Regulation Methods of Stone Bridge-
    Building in the Chhing Dynasty.
    BSRCA, 1935, **5** (no. 4), 58.
Wang Pi-Wên (*2*) 王璧文.
    *Chhing Kuan Shih Shih Cha chi Shih Han
    Tung Tso Fa* 清官式石閘及石涵洞做
    法.
    Official Regulations of the Chhing Dynasty
    for the Designing of Locks and Culverts.
    BSRCA, 1935, **6** (no. 2), 49.
Wang Po-Min (*1*) 王伯敏.
    'Ku Hua Phin Lu'; 'Hsü Hua Phin Lu'
    古畫品錄; 續畫品錄.
    The *Records of the Classification of Old
    Painters* [by Hsieh Ho, *c.* +500] and the

Wang Po-Min (1) (cont.)
   Continued Records of the Classification of
   Old Painters [by Yao Tsui, c. +550]
   [edited, annotated and translated into
   modern colloquial].
   Jen-Min, Peking, 1959.
   Abstr. RBS, 1965, 5, no. 433.
Wang Shih-Jen (1) 王世仁.
   Chi Hou Thu Tzhu Miao Mao Pei 記后土
   祠廟貌碑.
   On the Stone Stele [of +1011] inscribed
   with a Perspective Plan of the Hou-Thu
   Tzhu Temple [at Wan-jung (Jung-ho) in
   Shansi, where the R. Fên joins the Yellow
   River].
   KKTH, 1963 (no. 5), 273.
Wang Shih-Jen (2) 王世仁.
   Han Chhang-an Chhêng-Nan Chiao Li Chih
   Chien-Chu (Ta-thu-mên Tshun I-Chih)
   Yuan-Chuang-ti Thui-Tshê 漢長安城
   南郊禮制建築 (大土門村遺址) 原狀的
   推測.
   Reconstruction of the Ceremonial Buildings
   [Ming Thang, etc.] of the Han Dynasty
   in the Southern Suburb of the Han City
   of Chhang-an (on the basis of the
   remains of the foundations still extant at
   Ta-thu-mên village).
   KKTH, 1963 (no. 9), 501.
Wang Yü-Kang (1) (ed.) 王與剛.
   Chêng-chou Nan-Kuan I-pei-wu-shih-chiu
   hao Han Mu ti Fa-chhüeh 鄭州南關
   159 號漢墓的發掘.
   Excavation of a Han Tomb (No. 159) at
   Nan-Kuan, Chêngchow.
   WWTK, 1960 (nos. 8–9), 19.
Wang Yung (2) 王庸.
   Chung-Kuo Ti-Thu Shih Kang 中國地圖
   史綱.
   Brief History of Chinese Cartography.
   Peking, 1958.
Wei Chü-Hsien (1) 衞聚賢.
   Chung-Kuo Khao-Ku-Hsüeh Shih 中國考
   古學史.
   History of Archaeology in China.
   Com. Press, Shanghai, 1937.
Wei Chü-Hsien (3) 衞聚賢.
   Ku Shih Yen-Chiu 古史研究.
   Studies in Ancient History [researches and
   discussions on the interpenetration of
   Indian and Chinese civilisations before
   the Chhin Dynasty].
   Shanghai, 1934.
Wei Chü-Hsien (4) 衞聚賢.
   Chung-Kuo Jen Fa-Hsien Ao-Chou 中國
   人發現澳洲.
   The Chinese Discovery of Australia.
   Wei-Hsing, Hongkong, 1960.
Wei Yuan & Lin Tsê-Hsü (1)
   Hai Kuo Thu Chih 海國圖志.
   Illustrated Record of the Maritime

[Occidental] Nations. 1844, enlarged ed.,
   1847; further enlarged edition, 1852. For
   the problem of authorship see Chhen
   Chhi-Thien (1).
Wên Chhung-I (1) 文崇一.
   'Chiu Ko' chung-ti Shui Shen yü Hua-Nan-
   ti Lung Chou Sai Shen 九歌中的水神
   與華南的龍舟賽神.
   The Water-Gods of the Nine Songs and the
   Spirits connected with the Dragon-Boat
   Races of South China.
   AS/BIE, 1961 (no. 11), 51, 121.
Wu Chhêng-Lo (2) 吳承洛.
   Chung-Kuo Tu Liang Hêng Shih 中國度
   量衡史.
   History of Chinese Metrology [weights and
   measures].
   Com. Press, Shanghai, 1937; 2nd ed.
   Shanghai, 1957.
Wu Chhi-Chhang (4) 吳其昌.
   Chhin i-chhien Chung-Kuo Thien-Chih Shih
   秦以前中國田制史.
   On the History of the Chinese Land System
   before the Chhin Dynasty.
   WUQJSS, 1935, 5, 543 and 833.
   Abridged tr. Sun & de Francis (1), p. 55.
Wu Chhi-Hua (1) 吳緝華.
   Yuan Chhao yü Ming Chhu ti Hai-Yün
   元朝與明初的海運.
   Maritime Grain Transport in the Yuan and
   early Ming Periods.
   AS/BIHP (Thaiwan ser.), 1956, 28, 363.
   Abstr. RBS, 1962, 3, no. 264.
Wu Chhi-Hua (2) 吳緝華.
   Ming Tai Hai-Yün Chi Yün-Ho ti Yen-Chiu
   明代海運及運河的研究.
   A Study of Transportation by Sea and by
   the Grand Canal in the Ming Dynasty.
   Acad. Sinica History and Linguistics
   Institute, Thaipei, 1961.
   (Special Publications no. 43.)
Wu Lien-Tê (1) (ed.) (= 1) 伍聯德.
   Chin Hsiu Chung-Hua 錦繡中華.
   Magnificent China (album of photographs,
   many in colour, with accompanying text).
   Liang-Yu Book Co., Hongkong, 1966.

Yabuuchi Kiyoshi (11) (ed.) 藪內清.
   'Tenkō Kaibutsu' no Kenkyū 天工開物
   の研究.
   A Study of the Thien Kung Khai Wu
   (Exploitation of the Works of Nature,
   +1637). A Japanese translation of the
   text, with annotative essays by several
   hands.
   Tokyo, 1953.
   Chinese translations of the eleven essays:
   'Thien Kung Khai Wu' chih Yen-Chiu
   by Su Hsiang-Yü 蘇薌雨 et al.
   Tshung-Shu Wei Yuan Hui, Chi-Shêng,
   Thaiwan and Hongkong, 1956; 'Thien

Yabuuchi Kiyoshi (*11*) (ed.) (*cont.*)
   Kung Khai Wu' Yen-Chiu Wên Chi by
   Chang Hsiung 章熊 & Wu Chieh
   吳傑. Com. Press, Peking, 1961.
Yabuuchi Kiyoshi (*23*) 藪內清.
   '*Kahō Tsūgi*' *ni tsuite* 河防通議について.
   On the *Ho Fang Thung I* [and its applied
   mathematics].
   SBK, 1965, **13** (Suppl.), 297.
Yamamoto Tatsuro (*1*) 山本達郎.
   Tei Wa no Seisei 鄭和の西征.
   The Western Expeditions of Chêng Ho.
   TYG, 1934, **21** (no. 3), 90.
Yang Chhun-Ho (*1*) 楊春和.
   Hsi Chin Chhiao Thi-An Chi Lüeh 西津
   橋堤岸記略.
   An Account of the Renewals of the Abut-
   ments of the Hsi-chin Bridge (in the
   Western suburbs of Lanchow).
   c. 1810.
Yang Hung-Hsün 楊鴻勛 & Wang Shih-
   Jen (*1*) 王世仁.
   Chhü-fou '*Yen Shêng Kung Fu*' *Chhu Chha*
   Pao-Kao 曲阜「衍聖公府」初查報告.
   Preliminary Report on the (Mansion and
   Gardens called) Yen Shêng Kung Fu at
   Chhü-fou (in Shantung) (dating from the
   early +16th century).
   WWTK, 1957, no. 10 (no. 86), 14.
Yang Jen-Khai 楊仁愷 & Tung Yen-Ming
   董彥明 (*1*) (ed.).
   Liaoning Shêng Po-Wu-Kuan Tshang Hua
   Chi 遼寧省博物館藏畫集.
   Album of Pictures illustrating the Collection
   of Paintings in the Liaoning Provincial
   Museum.
   Wên-Wu, Peking, 1962.
Yang Khuan (*10*) 楊寬.
   Chan-Kuo Shih-Tai Shui-Li Kung-Chhêng
   ti Chhêng Chiu 戰國時代水利工程的
   成就.
   Achievements of Hydraulic Engineering in
   the Warring States Period.
   Essay in Li Kuang-Pi & Chhien Chün-Yeh
   (*1*), p. 99.
   Peking, 1955.
Yang Mo-Kung 楊陌公 & Hsieh Hsi-Kung
   (*1*) (ed.) 解希恭.
   Shansi Phing-lu Tsao-yuan-tshun Pi-Hua
   Han Mu 山西平陸棗園村壁畫漢墓.
   A Han Dynasty Tomb with Painted Walls
   at Tsao-yuan-tshun, near Phing-lu in
   Shansi.
   KKTH, 1959 (no. 9), 462 and pl. 1.
Yang Ping-Khun (*1*) 楊炳堃.
   Han-chung Chhü Nan Pao-chhêng Yang têng
   Hsien Shui-Li Thiao-Chha Pao-Kao Shu
   漢中區南褒城洋等縣水利調查報告
   書.
   Report of an Investigation of the Hydraulic

Engineering Works at Pao-chhêng,
   Yang[-chou] and other hsien in the south
   of Han-chung district.
   SL, 1931, **I**, 459.
Yang Tsung-Jung (*1*) 楊宗榮.
   Chan Kuo Hui Hua Tzu-Liao 戰國繪畫
   資料.
   Materials for the Study of the Graphic Art
   of the Warring States Period.
   Ku-tien I-shu, Peking, 1957.
Yang Yu-Jun (*1*) 楊有潤.
   Chhêngtu Yang-Tzu Shan Thu Thai I-Chih
   Chhing-Li Pao-Kao 成都羊子山土臺
   遺址清理報告.
   Report on the Remains of Yang-tzu Shan
   near Chhêngtu (an artificial mound once a
   thai or *ziggurat* of retaining walls in three
   tiers, probably dating from the early
   years of the State of Shu, *c.* −660).
   AS/CJA, 1957, no. 4 (no. 18), 17.
   Abstr. *RBS*, 1962, **3**, no. 437.
Yao Chhêng-Tsu 姚承祖, Chang Chih-
   Kang 張至剛 & Liu Tun-Chên (*1*)
   劉敦楨.
   Ying Tsao Fa Yuan 營造法原.
   The Characteristics of Chinese Architecture
   and their Development.
   Architectural and Engineering Press, Peking,
   1959.
   Contains a valuable glossary of technical
   terms.
   Abstr. *RBS*, 1965, **5**, no. 830.
Yao Ying (*1*) 姚瑩.
   Khang Yu Chi Hsing 康輶紀行.
   Travel Diary of the Tibetan Border.
   1845.
Yeh Chhien-Yü (*1*) 葉淺予.
   Tunhuang Pi Hua 敦煌壁畫.
   The Wall-Paintings at Tunhuang (the
   Chhien-fo-tung Cave-Temples).
   Chao-Hua I-Shu, Peking, 1957.
Yeh Hsiao-Yen (*1*) *et al.* 葉小燕.
   Honan Shen-hsien Liu-chia chhü Han Mu
   河南陝縣劉家渠漢墓.
   An Excavation of [46] Han Tombs at Shen-
   hsien in Honan.
   AS/CJA, 1965, (no. 1), 107.
Yen Kho-Chün (*1*) (ed.) 嚴可均.
   Chhüan Shang-Ku San-Tai Chhin Han San-
   Kuo Liu Chhao Wên 全上古三代秦
   漢三國六朝文.
   Complete Collection of Prose Literature
   (including Fragments) from Remote
   Antiquity through the Chhin and Han
   Dynasties, the Three Kingdoms and the
   Six Dynasties.
   Finished 1836; published 1887-93.
Yen Tun-Chieh (*19*) 嚴敦傑.
   Chhien Hsing Shu; Wo-Kuo Ming Tai Hang-
   Hai Thien-Wên Chih-Shih i Phieh
   牽星術;我國明代航海天文知識一瞥.

Yen Tun-Chieh (*19*) (*cont.*)

    The 'Stretching-Out Art'; a Glance at the Knowledge of Astronomical Navigation in the Ming Period [on the set of tablets used for measuring stellar altitudes].

    KHSC, 1966 (no. 9), 77.

Yi Kwangnin (*1*) 李光麟.

    Yijo Suri Sa Yŏn-gu 李朝水利史研究.

    History of Irrigation during the (Korean) Yi Dynasty.

    Korean Research Centre, Seoul, 1961 (Korean Studies series, no. 8).

Yi Unsang (*1*) 李思相.

    Yi Chhungmu Kong Ildae-ki 李忠武公一代記.

    Records of the Time of the Loyal and Martial Duke (i.e. Yi Sunsin).

    Seoul, 1946.

Yü Chê-Tê (*1*) 余哲德.

    Chaochow Ta Shih Chhiao Shih Lan ti Fa-Hsien chi Hsiu-Fu ti Chhu Pu I Chien 趙州大石橋石闌的發現及修復的初步意見.

    Discoveries concerning the Balustrades of the Great Stone (Segmental Arch) Bridge at Chaochow, and the Progress of its Restoration Works.

    WWTK, 1956 (no. 3), 17, with 10 pls.

Yü Chhang-Hui (*1*) 俞昌會.

    Fang Hai Chi Yao 防海輯要.

    Essentials of Coast Defence.

    1822.

Yü Ming-Chhien (*1*) 余鳴謙.

    Yüeh-Nan Ku Chi Chi Yu 越南古蹟記游.

    An Archaeological Study Tour in Vietnam.

    WWTK, 1959, no. 2 (no. 102), 53, 59.

Yuan Chia-Ku (*1*) 袁嘉穀.

    Tien I 滇繹.

    The Story of Yunnan.

    Kunming, 1923

C. BOOKS AND JOURNAL ARTICLES IN
WESTERN LANGUAGES

ABEL, SIR WESTCOTT (1). *The Shipwright's Trade.* Cambridge, 1948.

ABERCROMBIE, T. J. (1). 'Behind the Veil of Troubled Yemen.' *NGM*, 1964, **125**, 403.

ABEYASEKERA, H. P. (1). *Muturajavela [Wela]* (the swamp land along the coast north of Colombo). Ceylon Govt. Press, Colombo, 1954.

ADAM, L. (1). 'Das Problem der asiatisch-altamerikanischen Kulturbeziehungen mit besonderer Berücksichtigung der Kunst.' *WBKGA*, 1931, **5**, 40.

ADAM, P. (1). 'Navigation Primitive et Navigation Astronomique.' Art. in Proc. 5th International Colloquium of Maritime History, Lisbon, 1960. Abstract distributed in mimeographed form.

ADAM, P. & DENOIX, L. (1). 'Essai sur les Raisons de l'Apparition du Gouvernail d'Etambot.' *RHES*, 1962, **40**, 90.

ADAMS, E. AMERY (1). 'The Old Heytor Granite Railway.' *RTDA*, 1946, **78**, 153.

ADAMS, R. McC. (1). *Land Behind Baghdad; a History of Settlement on the Diyala Plains.* Univ. Chicago Press, Chicago and London, 1965. Rev. S. N. Kramer, *TCULT*, 1966, **7**, 74.

ADAMS, R. McC. (2). 'Early Civilisations, Subsistence and Environment.' Art. in *City Invincible; a Symposium on Urbanisation and Cultural Development in the Ancient Near East.* Univ. of Chicago, Chicago, 1960.

ADAMS, R. McC. See also Jacobsen, T. & Adams, R. McC.

ADAMS, WILL. (1) (shipwright). *Memorials of the Empire of Japan in the +16th and +17th Centuries (The Kingdome of Japonia); Letters of W.A. +1611 to +1617*, with commentary by T. Rundall. Hakluyt Society Pubs. Series 1, no. 8, 1850, repr. 1965.

ADNAN ADIVAR (1). On the *Tanksuq-nāmah-ī Īlkhān dar Funūn-i 'Ulūm-i Khiṭāi.' ISIS*, 1940 (appeared 1947), **32**, 44.

ALBERTI, L. B. (1). *De Re Aedificatoria.* Rembolt & Hornken, Paris, 1512. Ital. tr. *I Dieci Libri di Architettura.* Venice, 1546. Fr. tr. Paris, 1553. Eng. tr. London, 1726. See Olschki (5).

ALDRED, C. (1). 'Furniture, to the End of the Roman Empire.' Art. in *A History of Technology*, ed. C. Singer *et al.* Oxford, 1956, vol. 2, p. 220.

ALEX, W. (1). *Japanese Architecture.* Prentice-Hall, London; Braziller, New York, 1963.

ALLARD, E. (1). *Les Phares; Histoire; Construction; Éclairage.* Rothschild, Paris, 1889.

ALLEY, REWI (1). 'Notes on some Highways in China.' *CJ*, 1937, **26**, 240.

ALLEY, R. (3) (tr.). *Peace through the Ages; translations from the Poets of China.* Pr. pub. Peking, 1954.

ALLEY, R. (5). 'Pagodas and Towers in China.' *EHOR*, 1962, **2** (no. 5), 20.

ALLEY, R. (6) (tr.). *Tu Fu, Selected Poems.* Foreign Languages Press, Peking, 1962.

ALLEY, REWI (7). *Man against Flood; a story of the 1954 Flood on the Yangtze, and of the Reconstruction that followed it.* New World Press, Peking, 1956.

ALLOM, T. & PELLÉ, C. (1). *China, its Scenery, Architecture, Social Habits, etc. described and illustrated.* Fisher, London & Paris, n.d. [about 1830].

ALLOM T. & WRIGHT, G. N. (1). *China, in a Series of Views, displaying the Scenery, Architecture and Social Habits of that Ancient Empire, drawn from original and authentic Sketches by T.A— Esq.*, with historical and descriptive Notices by Rev. G.N.W—. 4 vols. Fisher, London & Paris, 1843.

ALMAGIÀ, R. (2). *Il Mappemonde di Fra Mauro.* Ist. Poligrafico dello Stato & Libreria dello Stato, Rome, 1954.

DE ALMEIDA, FORTUNATO (1). *História de Portugal.* 3 vols. Coimbra, 1925.

AMES, OAKES (1). *Economic Annuals and Human Cultures.* Botanical Museum, Harvard Univ., Cambridge, Mass., 1939, repr. 1953.

ANDERS, LESLIE (1). *The Ledo Road; General Joseph W. Stilwell's Highway to China.* Univ. Oklahoma Press, Norman, Okla, 1965.

ANDERSON, A. R. (1). *Alexander's Gate, Gog and Magog, and the Enclosed Nations.* Cambridge, Mass., 1932. (Pubs. Medieval Acad. of America, no. 12; Monograph Ser. no. 5.)

ANDERSON, G. F. (1). 'The Wonderful Canals of China.' *NGM*, 1905, **16**, 68.

ANDERSON, R. & ANDERSON, R. C. (1). *The Sailing Ship; Six Thousand Years of History.* Harrap, London, 1926.

ANDERSON, R. C. (1). 'Correspondence with H. H. Brindley on Mediaeval Rudders.' *MMI*, 1926, **12**; 1927, **13**, 181.

ANDRADE, E. N. DA C. (1). 'Robert Hooke' (Wilkins Lecture). *PRSA*, 1950, **201**, 439. (The quotation concerning fossils is taken from the advance notice, Dec. 1949.) Also *N*, 1953, **171**, 365.

ANDREOSSY, F. (1). *Histoire du Canal du Midi, ou Canal de Languedoc; considéré sous les Rapports d'Invention, d'Art, d'Administration, d'Irrigation, et dans ses Relations avec les Étangs de l'Intérieur des Terres qui l'avoisinent . . .* 2 vols. Crapelet, Paris, 1804.

ANON. (7). 'Cane Bridges of Asia.' *NGM*, 1948, **94**, 243.

ANON. (9). 'UNRRA Relief for the Chinese People; a Report by CLARA.' Information Dept. CLARA, Shanghai, 1947 [n.b. UNRRA = United Nations Relief and Rehabilitation Administration; CNRRA = Chinese National Relief and Rehabilitation Administration (Kuomintang); CLARA = Chinese Liberated Areas Relief Association (Kungchangtang)].

ANON. (10). 'The Persian "Qanāts" (lines of boreholes connected by a tunnel below) in the Saharan oasis of Adrar.' *NGM*, 1949, **95**, 226.

ANON. (16). *La Charpente Chinoise (Toitures, Collonnes, Poteaux, Fermes, Balustrades, Plafonds) et la Menuiserie Chinoise (suite à la Charpente, comprenant de plus, Portes, Lambries, Caissons, Ornements, Sculptures, Balustrades); documents extraits de cahiers des Maitres Charpentiers de la Dynastie du Sung (10ᵉ et 11ᵉ siècles).* 2 vols. 260 plates. Peiping, 1931–33.

ANON. (17). *Illustrated Catalogue of the Maze Collection of Chinese Junk Models in the Science Museum, London.* Pr. pr. Shanghai; pr. pub. London, 1938.

ANON. (19). 'George III's Embassy under Chinese Convoy.' *ASIA*, 1920, **20**, 877.

ANON. (20). 'On Watertight Compartments in Ships.' *MCMG*, 1824, **2**, 224.

ANON. (21). *London as it is Today; Where to go and What to see during the Great Exhibition.* Clarke, London, 1851.

ANON. (22) [perhaps Capt. Kellett]. *Description of the Junk 'Keying', printed for the Author, and Sold on Board the Junk.* Such, London, 1848.

ANON. (24). 'Deux Études Nouvelles sur les Techniques Maritimes aux 14ᵉ et 15ᵉ siècles.' *MC/TC*, 1952, **2**, 90.

ANON. (28). 'Report on the past twenty-five years work in Japan on the History of Chinese and Japanese Astronomy (to the Commission for the History of Astronomy of the International Astronomical Union).' *TIAU*, 1954, **8**, 626.

ANON. (37) (tr.). 'The Chinaman Abroad; or, a Desultory Account of the Malayan Archipelago, particularly of Java, by Ong-Tae-Hae' (Wang Ta-Hai's *Hai Tao I Chih Chai Lüeh* of +1791). *CMIS*, 1849 (no. 2), 1.

ANON. (41). *An Outline History of China.* Foreign Languages Press, Peking, 1958.

ANON. (47). *Prince Henry the Navigator and Portuguese Maritime Enterprise; Catalogue of an Exhibition at the British Museum, Sept. to Oct. 1960.* BM, London, 1960.

ANON. (48). *Pre-Hispanic Art of Mexico.* Instituto Nacional de Antropologia e Historia, Mexico City, 1946.

ANON. (50) (ed.). *Livre des Merveilles (Marco Polo, Odoric de Pordenone, Mandeville, Hayton, etc.); Reproduction des 265 Miniatures du MS français 2810 de la Bibliothèque Nationale.* 2 vols. Berthaud, Paris, 1908.

ANON. (51). *Sailing Directions for the Circumnavigation of England* (late +15th century). Ed. J. Gairdner & E. D. Morgan. Hakluyt Soc. London, 1889. (Hakluyt Society Pubs. 1st ser., no. 79.)

ANON. (53) (ed.). *Henri le Navigateur.* Comissão Executiva das Comemorações do Quinto Centenário da Morte do Infante Dom Henrique, Lisbon, 1960.

ANON. (54). 'The First Trains in Szechuan.' *CREC*, 1952, **1** (no. 2), 32.

ANON. (55). 'The Chungking–Chhêngtu Railway Completed.' *CREC*, 1952, **1** (no. 5), 19.

ANON (56). 'New Railway for the Nation's Birthday.' *CREC*, 1952, **1** (no. 6), 11.

ANON. (57). *Travellers' Guide to the Lung-Hai Railway.* Lung-Hai Rly Admin. Chêngchow, 1935.

ANON. (58). *China's Railways; a Story of Heroic Reconstruction.* For. Lang. Press, Peking, n.d. (1950).

ANON. (59). *Labour and Struggle; Glimpses of Chinese History.* Museum of Chinese History, Peking. (Supplement to *CREC*, Apr. 1960.)

ANON. (60). 'A Canal through the Mountains' (the utilisation of the water of the Thao River in Southern Kansu). *PKR*, 1958 (no. 24), 17.

ANON. (66). *L'Art Japonais à travers les Siècles.* Paris, 1958. (Art et Style ser., no. 46.)

ANON. (67). 'Miracle on the Yangtze' (an account of the Ching River retention-basin north of the Tung-thing Lake, with its three long regulator-sluice dams). *CREC*, 1952, **1** (no. 5), 6.

ANON. (68). 'A New Dam and Canal irrigating North Chiangsu' (account of the Sanho regulator-sluice dam near the Hungtsê Lake, and the North Chiangsu Irrigation Canal from the Lake direct to the sea). *CREC*, 1954, **3** (no. 4), 40.

ANON. (69). 'The Taming of the Yungting River' (account of the Kuanthing dam and reservoir on the Yungting River in Northern Hopei). *CREC*, 1953, **2** (no. 6), 10.

ANON. (70). '*Pukhagŭi*' (*Pei Hsüeh I*; a treatise by Pak Chega, +1750/1809, on Chinese technology, recommending its adoption in Korea). *KU/ARB*, 1962, **5** (no. 5), 20.

ANON. (72). *Water-Conservancy in New China* (album of photographs with bilingual captions and text). For the Ministry of Water-Conservancy; People's Art Pub. Ho., Shanghai, 1956.

Anon. (73) (ed.). *Folk-Tales from China.* 5 vols. For. Lang. Press, Peking, 1957– .

Anon. (74). *Exposição Henriquina* (Catalogue). Comissão Executiva das Comemorações do Quinto Centenario da Morte do Infante Dom Henrique. Lisbon, 1960.

Anon. (75) (ed.). *Modern Paintings in the Chinese Style.* For. Lang. Press, Peking, 1960. (Supplement to *CREC*, 1960, no. 9.)

Anon. (81). 'Hand-built Highway [the Burma and Ledo Roads].' *MD*, 1967, **11** (no. 1), 320.

Anon. (82). *History of Road Development in India.* Central Road Research Inst., New Delhi, 1964. Rev. S. K. Ghaswala, *TCULT*, 1965, **6**, 301.

Anson, G. A. Admiral (1). *A Voyage Round the World in the Years 1740–1744.* Ed. R. Walter. London, 1748.

Ansted, A. (1). *A Dictionary of Sea Terms.* Gill, London, 1898; Brown & Ferguson, Glasgow, 1933.

Anthiaume, A. (1). *Le Navire, sa Construction en France, et principalement chez les Normands.* Dumont, Paris, 1924.

Anthiaume, A. (2). *Le Navire, sa Propulsion en France, et principalement chez les Normands.* Dumont, Paris, 1924.

Apianus, Petrus [Peter Bienewitz] & Frisius, Gemma [van der Steen] (1). *Cosmographia...sive Descriptio Universi Orbis Petri Apiani et Gemmae Frisii, jam demum integritati suae restituta...* Birckmann, Köln, 1574; Arnold Bellerus, Antwerp, 1584. The 1st edition of Apianus' *Cosmographia* was pr. Weyssenburg, Landshut, 1524. From the 2nd edition onwards (Birckmann, Antwerp; and de Sabio, Venice, 1533), the *Libellus de Locorum Describendorum Ratione* was combined with it. E.g. Paris, 1551.

Apicius. See Flower & Rosenbaum (1).

Arlington, L. C. & Lewisohn, W. (1). *In Search of Old Peking.* Vetch, Peiping, 1935.

Armbruster, G. (1). 'Das *Shigisan Engi Emaki*; ein japanisches Rollbild aus dem 12 Jahrhundert.' *MDGNVO*, 1959, **11**, 1–290.

Armillas, P. (1). 'Teotihuacán, Tula y los Toltecas; las Culturas post-arcaicas y pre-Aztecas del Centro de Mexico — Excavaciones y Estudios 1922–1950.' *RUNA*, 1950, **3**, 37.

Arnaiz, G. (1). 'Construcción de los Edificios en las Prefecturas de Čoan-čiu [Chhuanchow?] y Čian-čiu, Fû-Kien sur, China.' *AN*, 1910, **5**, 907.

Ashby, T. (1). *The Aqueducts of Ancient Rome.* Ed. I. A. Richmond. Oxford, 1935. Rev. R. C. Carrington, *AQ*, 1936, **10**, 127.

Ashley, C. W. (1). *Book of Knots.* Faber & Faber, London, 1944.

Aston, W. G. (1) (tr.). '*Nihongi*', *Chronicles of Japan from the Earliest Times to +697.* Kegan Paul, London, 1896; repr. Allen & Unwin, London, 1956.

Atkinson, W. C. (1) (tr.). *Camoens' 'The Lusiads'.* Penguin, London, 1952.

Attwood, E. L. (1). *The Modern Warship.* CUP, Cambridge, 1913.

Aubertin, J. J. (1) (tr.). *The Lusiads of Camoens, translated...* 2 vols. (with Portuguese text on facing pages). Kegan Paul, London, 1878.

Auboyer, J. (1). 'L'Influence Chinoise sur le Paysage dans la Peinture de l'Orient et dans la Sculpture de l'Insulinde.' *RAA/AMG*, 1935, **9**, 228.

Audemard, L. (1). 'Quelques Notes sur les Jonques Chinoises.' *BAMM*, 1939, **9** (no. 1), no. 33, 205.

Audemard, L. (2) (with the assistance of Shih Chun-Shêng). *Les Jonques Chinoises; I. Histoire de la Jonque* (posthumously edited by C. Nooteboom). Museum voor Land- en Volken-Kunde & Maritiem Museum Prins Hendrik, Rotterdam, 1957.

Audemard, L. (3). *Les Jonques Chinoises; II. Construction de la Jonque.* Museum voor Land- en Volken-Kunde & Maritiem Museum Prins Hendrik, Rotterdam, 1959.

Audemard, L. (4). *Les Jonques Chinoises; III. Ornementation et Types.* Museum voor Land- en Volken-Kunde & Maritiem Museum Prins Hendrik, Rotterdam, 1960.

Audemard, L. (5). *Les Jonques Chinoises; IV. Description des Jonques.* Museum voor Land- en Volken-Kunde & Maritiem Museum Prins Hendrik, Rotterdam, 1962.

Audemard, L. (6). *Les Jonques Chinoises; V. Haut Yang-tse Kiang.* Museum voor Land- en Volken-Kunde & Maritiem Museum Prins Hendrik, Rotterdam, 1963.

Audemard, L. (7). *Les Jonques Chinoises; VI. Bas Yang-Tse Chiang.* Museum voor Land- en Volken-Kunde and Maritiem Museum Prins Hendrik, Rotterdam, 1965.

Aurousseau, L. (2). 'La Première Conquête Chinoise des Pays Annamites.' *BEFEO*, 1923, **23**, 137.

Aurousseau, L. (3). 'Le Mot Sampan est-il Chinois?' *BEFEO*, 1920, **20**.

Ayscough, F. (2). 'Notes on the Symbolism of the Purple Forbidden City.' *JRAS/NCB*, 1921, **52**, 51; repr. as ch. 4 of *A Chinese Mirror.* Boston, 1925.

Baber, E. C. (1). *Travels and Researches in the Interior of China.* London, 1886.

Bachhofer, L. (1). 'Die Raumdarstellung in der chinesischen Malerei der ersten Jahrtausends n. Chr.' *MJBK*, 1931, **8**, 197.

BACKHOUSE, E. & BLAND, J. O. P. (1). *Annals and Memoirs of the Court of Peking.* Heinemann, London, 1914.

BACOT, J., THOMAS, F. W. & TOUSSAINT, C. (1). *Documents de Touen-houang relatifs à l'Histoire du Tibet.* Geuthner, Paris, 1940–46. *BE/AMG*, no. 51.

BADDELEY, J. F. (2). *Russia, Mongolia, China; being some Record of the Relations between them from the Beginning of the +17th century to the Death of the Tsar Alexei Mikhailovitch (+1602 to +1676), rendered mainly in the form of Narratives dictated or written by the Envoys sent by the Russian Tsars or their Voevodas in Siberia to the Kalmuk and Mongol Khans and Princes, and to the Emperors of China; with Introductions Historical and Geographical, also a series of Maps showing the Progress of Geographical Knowledge in regard to Northern Asia during the +16th, +17th and early +18th Centuries; the Texts taken more especially from Manuscripts in the Moscow Foreign Office Archives...* 2 vols. Macmillan, London, 1919.

BAGROW, L. (2). *Die Geschichte der Kartographie.* Safari, Berlin, 1951.

BAIÃO, A. (1) (ed.). *Afonso de Albuquerque Cartas para el-Rei D. Manuel I.* Sá da Costa, Lisbon, 1942.

BAKER, MATTHEW (1). 'Fragments of Ancient English Shipwrightry.' MS. (draughts and plans) collected by Samuel Pepys and preserved in the Pepysian Library, Magdalene College, Cambridge, 1586.

BALAZS, E. (= S.) (1). 'La Crise Sociale et la Philosophie Politique à la Fin des Han.' *TP*, 1949, **39**, 83.

DE BALBOA, MIGUEL CABELLO (1). *Obras.* Ed. J. Jijón y Caamaño. Quito, 1945.

BALD, R. C. (1). 'Sir William Chambers and the Chinese Garden.' *JHI*, 1950, **11**, 287.

BALLARD, G. A. (1). 'Egyptian Square Sails.' *MMI*, 1919, **5**, 6.

BALTRUŠAITIS, J. (1). *Le Moyen Age Fantastique; Antiquités et Exotismes dans l'Art Gothique.* Colin, Paris, 1955.

BALTRUŠAITIS, J. (2). *Aberrations; quatre Essais sur la Légende des Formes.* Perrin, Paris, 1957.

BALTRUŠAITIS, J (3). *Anamorphoses ou Perspectives Curieuses.* Perrin, Paris, 1955.

BALTZER, (ADOLF W.) FRANZ (1). *Das japanische Haus; eine bautechnische Studie.* Ernst, Berlin, 1903. (Reprinted from *ZBW*.)

BALTZER, (ADOLF W.) FRANZ (2). *Die Architektur d. Kultbauten Japans.* Ernst, Berlin, 1907.

BANNISTER, T. C. (1). 'The First Iron-Framed Buildings.' *AREV*, 1950, **107**, 231.

BARATIER, E. & REYNAUD, F. (1). *Histoire du Commerce de Marseille (1291–1480).* 2 vols. Paris, 1951.

BARBOSA, A. (1). *Novos Subsidios para a Histórica da Ciencia Náutica Portuguesa da Epoca dos Descobrimentos.* Lisbon, 1938; Porto, 1948.

BARBOSA, DUARTE (1). *A Description of the Coasts of East Africa and Malabar in the Beginning of the +16th Century, by D.B., a Portuguese.* Eng. tr. from a Spanish MS., by E. H. J. Stanley (Lord Stanley of Alderley). Hakluyt Society, London, 1866 (Hakluyt Soc. Pubs., 1st ser., no. 35). Eng. tr. from the Portuguese, by Hakluyt Society, London, 1918 (Hakluyt Soc. Pubs., 2nd ser., no. 44).

BARBOUR, G. B. (1). (a) 'The Loess of China.' *CJ*, 1925, **3**, 454, 509; *ARSI*, 1926, 279. (b) 'The Loess Problem of China.' *GM*, 1930, **67**, 458.

BARBOUR, G. B. (2). 'Physiographic History of the Yangtze.' *GJ*, 1936, **87**, 17. (Based on a survey expedition with Teilhard de Chardin & C. C. Yang.)

BARBOUR, G. B. (3). 'Recent Observations on the Loess of North China.' *GJ*, 1935, **86**, 54.

BARNETT, R. D. (1). 'Early Shipping in the Near East.' *AQ*, 1958, **32**, 220.

BAROCELLI, P. (1). 'Appunti sugli scavi della Terramare parmense del Castellazzo di Fontanellato.' *RPARA*, 1943, **20**, 193.

BARRESWIL (1). 'Sur le Blanc d'Ablette qui sert à la Fabrication de Perles Fausses.' *CRAS*, 1861, **53**.

DE BARROS, JOAõ (1). *Décadas da Asia.* Década I (+1420 to +1505), Galharde, Lisbon, 1552; Década II (+1506 to +1515), Galharde, Lisbon, 1553; Década III (+1516 to +1525), Galharde, Lisbon, 1563; all republished together, Lisbon, 1628. Década IV (+1526 to +1538), Lisbon, 1615; Madrid, 1615.

BARROW, JOHN (1). *Travels in China.* London, 1804. German tr. 1804; French tr. 1805; Dutch tr. 1809.

BARROWS, H. K. (1). *Floods; their Hydrology and Control.* McGraw-Hill, New York, 1948.

BARTOCCINI, R. (1). *Le Pitture Etrusche di Tarquinia.* Martello, Milan, 1955.

BASS, G. F. (1). *Underwater Excavations at Yassı Ada; a Byzantine Shipwreck [dated by coins between +610 and +641].* *JDAI/AA*, 1962, 538. (The strake-morticing of the hull not reported here, but known by private communications to other scholars later.)

BATES, M. S. (1). 'Problems of Rivers and Canals under Han Wu Ti.' *JAOS*, 1935, **55**, 303.

BAUDIN, L. (1). 'L'Empire Socialiste des Inka [Incas].' *TMIE*, 1928, no. 5.

BAYLIN, J. (1) (tr.). *Extraits des Carnets de Lin K'ing [Wanyen Lin-Chhing]; Sites de Pékin et des Environs vus par un Lettré Chinois.* Lim. ed., Nachbaur, Peiping, 1929. Reproduction of 26 illustrations and descriptions from the *Hung Hsüeh Yin Yuan Thu Shuo.*

BEAL, S. (1) (tr.). *Travels of Fah-Hian [Fa-Hsien] and Sung-Yün, Buddhist Pilgrims from China to India (+400 and +518).* Trübner, London, 1869.

BEAL, S. (2) (tr.). ' *Si Yu Ki [Hsi Yü Chi]*', *Buddhist Records of the Western World, translated from the Chinese of Hiuen Tsiang [Hsüan-Chuang]*. 2 vols. Trübner, London, 1881, 1884; 2nd ed. 1906. Repr. in 4 vols. with new title, *Chinese Accounts of India....* Susil Gupta, Calcutta, 1957–8.

BEAL, S. (3) (tr.). *The Life of Hiuen Tsiang [Hsüan-Chuang] by the Shaman [Śramana] Hwui Li [Hui-Li]*, *with an Introduction containing an account of the Works of I-Tsing [I-Ching]*. Trübner, London, 1888; Kegan Paul, London, 1911.

BEASLEY, W. G. & PULLEYBLANK, E. G. (1) (ed.). *Historians of China and Japan.* Oxford Univ. Press, London, 1961. (Historical Writing on the Peoples of Asia, Far East Seminar; Study Conference of the London School of Oriental Studies, 1956.)

BEATON, C. (1). *Chinese Album* (photographs). Batsford, London, 1945.

BEAUDOUIN, F. (1). 'Recherches sur l'Origine de deux Embarcations Portugaises.' *AHES/AESC*, 1965, **20**, 564.

BEAUFOY, [MARK] COL. (1). *Nautical and Hydraulic Experiments, with numerous Scientific Miscellanies*, *Vol. 1* [all published]. Pr. pr. London, 1834.

BEAUFOY, [MARK] COL. (2). 'On the Spiral Oar; Observations on the Spiral as a Motive Power to impel Ships through the Water, with Remarks when applied to measure the Velocity of Water and Wind.' *TAP*, 1818, **12**, 246.

BEAUJOUAN, G. (1). 'Science Livresque et Art Nautique au 15e Siècle.' Art. in Proc. 5th International Colloquium of Maritime History, Lisbon, 1960. Distributed in mimeographed form.

BEAUJOUAN, G. & POULLE, E. (1). 'Les Origines de la Navigation Astronomique aux 14e et 15e Siècles.' Art. in Proc. 1st International Colloquium of Maritime History, Paris, 1956. Ed. M. Mollat & O. de Prat, p. 103.

BEAZLEY, C. R. (1). *The Dawn of Modern Geography.* 3 vols. (vol. 1, +300 to +900; vol. 2, +900 to +1260; vol. 3, +1260 to +1420). Vols. 1 and 2, Murray, London, 1897 and 1901. Vol. 3, Oxford, 1906.

BEAZLEY, C. R. (2). *Prince Henry the Navigator.* Putnam, New York, 1895; London, 1923.

BEAZLEY, C. R. & PRESTAGE, E. (1) (tr.). *The Chronicle of the Discovery and Conquest of Guinea [Gomes Eanes de Zurara's 'Crónica do Descobrimento e Conquista da Guiné']*. 2 vols. London, 1896, 1899. (Hakluyt Society Pubs. nos. 95 and 100.) The chronicle of de Zurara ends at +1448, twelve years before the death of Prince Henry.

BEBA, K. (1). 'Tibet Revisited' (on the China–Tibet roads). *CREC*, 1957, **6** (no. 6), 9.

BECK, T. (1). *Beiträge z. Geschichte d. Maschinenbaues.* Springer, Berlin, 1900.

BECKETT, P. H. T. (1). 'Waters of Persia.' *GGM*, 1951, **24**, 230.

BECKETT, P. H. T. (2). 'Qanats around Kerman.' *JRCAS*, 1953, **40**, 47; *JIS*, 1952 (Jan.).

BECKFORD, W. (1). *Vathek.* London, 1786.

BEDINI, S. A. (5). 'The Scent of Time; a Study of the Use of Fire and Incense for Time Measurement in Oriental Countries.' *TAPS*, 1963 (n.s.), **53**, pt. 5, 1–51. Rev. G. J. Whitrow, *A/AIHS*, 1964, **17**, 184.

BEDINI, S. A. (6). 'Holy Smoke; Oriental Fire Clocks.' *NS*, 1964, **21** (no. 380), 537.

VAN BEEK, G. W. (1). 'Ancient South Arabian Voyages to India.' *JAOS*, 1958, **78**, 147. Criticism by G. F. Hourani, 1960, **80**, 135, with rejoinder by G. W. van Beek, 1960, **80**, 136.

BEER, A., HO PING-YÜ, LU GWEI-DJEN, NEEDHAM, JOSEPH, PULLEYBLANK, E. G. & THOMPSON, G. I. (1). 'An 8th-century Meridian Line; I-Hsing's Chain of Gnomons and the Pre-History of the Metric System.' *VA*, 1961, **4**, 3.

BEFU, HARUMI & EKHOLM, G. F. (1). 'The True Arch in pre-Columbian America?' *CURRA*, 1964, **5**, 328.

BELDEN, J. (1). *China Shakes the World.* London, 1950.

DE BÉLIDOR, B. F. (1). *Architecture Hydraulique; ou l'Art de Conduire, d'Elever et de Menager les Eaux, pour les différens Besoins de la Vie.* 4 vols. Jombert, Paris, 1737–53.

BELL OF ANTERMONY, JOHN (1). *Travels from St Petersburg in Russia to Diverse Parts of Asia.*
  Vol. 1. *A Journey to Ispahan in Persia, 1715 to 1718; Part of a Journey to Pekin in China, through Siberia, 1719 to 1721.*
  Vol. 2. *Continuation of the Journey between Mosco and Pekin; to which is added, a translation of the Journal of Mr de Lange, Resident of Russia at the Court of Pekin, 1721 and 1722, etc., etc.* Foulis, Glasgow, 1763.
  Repr. as *A Journey from St Petersburg to Pekin*, ed. J. L. Stevenson. University Press, Edinburgh, 1965.

BELL, R. (1) (tr.). *The Holy Qu'rān.* Clark, Edinburgh, 1937.

BELOCH, J. (1). *Die Bevölkerung der griechisch-römischen Welt.* Leipzig, 1886.

BENGTSON, H., MILOJČIĆ, V. *et al.* (1). *Grosser Historischer Weltatlas.* Bayerischer Schulbuch Verlag, München, 1958.

BENNETT, W. C. (1) (ed.). 'A Reappraisal of Peruvian Archaeology' (symposium). *MSAA*, no. 4; suppl. to *AMA*, 1948, **13** (no. 4).

BENOIT, F. (2). 'Un Port Fluvial de Cabotage; Arles et l'ancienne Marine à Voile du Rhône.' *AHES/ AHS*, 1940, **2**, 199.

BENOIT, F. (4). Nouvelles Épaves de Provence. *GAL*, 1958, **16**, 5.

BENOIT, F. (5). 'Fouilles Sous-Marines; l'Épave du Grand Congloué à Marseille.' *GAL*, 1961, Suppl. **14**, 1–211.

BENSAÚDE, J. (1). *Histoire de la Science Nautique Portugaise à l'Époque des Grandes Découvertes; Collection de Documents*... Kuhn, later Obernetter, Munich, 1914–16. Facsimile reproductions, with introductions.

Vol. 1, *Regimento do Estrolábio e do Quadrante* and *Tractado da Spera do Mundo*. Munich copy, *c.* +1509.

Vol. 2, *Tratado da Spera do Mundo* and *Regimento da Declinaçam do Sol.* Evora copy, +1518.

Vol. 3, *Almanach Perpetuum Celestium Motuum Astronomi Zacuti cuius Radix est 1473 (Tabulae Astronomicae Raby Abraham Zacuti...in Latinum translatae per Magistrum Joseph Vizinum...).* Augsburg copy, Leiria, +1496.

Vol. 4, *Tratado del Esphera y del Arte del Marear*, by Francisco Faleiro, Seville, +1535.

Vol. 5, *Tratado da Sphera, com a Theorica do Sol a da Lua e ho Primeiro Livro da Geographia de Claudio Ptolemeo Alexandrino...*', by Pedro Nunes, Lisbon, +1537.

Vol. 7, *Reportorio dos Tempos*, by Valentim Fernandes, Lisbon, +1563.

BENSAÚDE, J. (2). *L'Astronomie Nautique au Portugal à l'Époque des Grandes Découvertes*. Drechsel, Berne, 1912.

BENTHAM, LADY M. S. (1). *Life of Sir Samuel Bentham, formerly Inspector-Genera lof Naval Works, lately a Commissioner...with the distinct duty of Civil Architect and Engineer of the Navy.* Longman Green, London, 1862.

BENTHAM, SIR SAMUEL (1). See Guppy.

BERGIER, N. (1). *Histoire des Grands Chemins de l'Empire Romain*... 2 vols. Paris, 1622; Rheims, 1637; Brussels, 1728, 1736. Eng. tr. *The History of the Highways in all Parts of the World, more particularly in Great Britain*... London, 1712.

BERNARD, W. D. (1). *Narrative of the Voyages and Services of the 'Nemesis' from 1840 to 1843, and of the combined Naval and Military Operations in China; comprising a complete account of the Colony of Hongkong, and Remarks on the Character of the Chinese, from the Notes of Cdr. W. H. Hall, R.N., with personal observations.* 2 vols. Colburn, London, 1844.

BERNARD-MAÎTRE, H. (9). 'Deux Chinois du 18e siècle à l'École des Physiocrates Français.' *BUA*, 1949 (3e sér.), **10**, 151.

BERRY-HILL, H. & BERRY-HILL, S. (1). *Artist of the China Coast [George Chinnery, 1774 to 1852].* S. Lewis, Leigh-on-Sea, 1963 (lim. ed.).

BERTUCCIOLI, G. (1). 'A Note on Two Ming Manuscripts of the *Pên-Tshao Phin-Hui Ching-Yao*.' *JOSHK*, 1956, **3**, 63.

BETHE, A. (1). 'Ü. d. Silbersubstanz in d. Haut von *Alburnus lucidus*.' *ZPC*, 1895, **20**, 472.

BEVERIDGE, W. M. (1). (*a*) 'Racial Differences in Phenomenal Regression.' *BJPC*, 1935, **26**, 59; (*b*) 'Some Racial Differences in Perception.' *BJPC*, 1940, **30**, 57.

BIEDERMANN, W. (1). 'Physiologie d. Stütz- und Skelett-substanzen.' Art. in *Handbuch d. vergl. Physiol.* Ed. H. Winterstein. Vol. 3, sect. 1, pt. 1, pp. 319–1185. Fischer, Jena, 1914.

BIELENSTEIN, H. (2). 'The Restoration of the Han Dynasty.' *BMFEA*, 1954, **26**, 1–209 and sep. Göteborg, 1953.

BINYON, L. (1). *Chinese Paintings in English Collections*. Van Oest, Paris & Brussels, 1927. (Eng. tr. of the French text in Ars Asiatica series, no. 9.)

BINYON, L. (2). *The George Eumorphopoulos Collection; Catalogue of the Chinese, Korean and Siamese Paintings.* Benn, London, 1928.

BIOT, E. (1) (tr.). *Le Tcheou-Li ou Rites des Tcheou [Chou].* 3 vols. Imp. Nat., Paris, 1851. (Photographically reproduced Wêntienko, Peiping, 1930.)

BIOT, E. (3) (tr.). *Chu Shu Chi Nien* (Bamboo Books). *JA*, 1841 (3e sér.), **12**, 537; 1842, **13**, 381.

BIOT, E. (21). 'Mémoire sur les Déplacements du Cours Inférieur du Fleuve Jaune.' *JA*, 1843 (4e sér.), **1**, 432; **2**, 84, 307.

BIRK, A. (1). *Die Strasse; ihre Verkehrs u. bautechnische Entwicklung im Rahmen der Menschheitsgeschichte.* Karlsbad-Drahowitz, 1934.

BIRK, A. (2). 'Die Strassen des Altertums.' *BGTI/TG*, 1934, **23**, 6.

BIRRELL, V. (1). *Transpacific Contacts and Peru.* Proc. 35th Internat. Congress of Americanists, Mexico City, 1962. Vol. 1, p. 31.

BIRT, D. H. C. (1). *Sailing Yacht Design.* Ross, Southampton, 1951.

BISHOP, C. W. (6). 'An Ancient Chinese Capital; Earthworks at Old Chhang-An.' *AQ*, 1938, **12**, 68.

BISHOP, C. W. (7). 'Long-houses and Dragon-boats.' *AQ*, 1938, **12**, 411.

BISHOP, C. W. (8). 'Two Chinese Bronze Vessels.' *MUJ*, 1918, **9**, 99.

DE BISSCHOP, E. (1). *Kaimiloa; d'Honolulu à Cannes par l'Australie et le Cap à bord d'une Double Pirogue Polynésienne*. Paris, 1939.

BISWAS, A. K. (1). 'Hydrological Engineering prior to −600.' *PASCE/JHD*, 1967, **93**, 115. Discussion by G. Garbrecht, G. J. Requardt & N. J. Schnitter, 1968, **94**, 612.

BISWAS, A. K. (2). 'Irrigation in India; Past and Present.' *PASCE/JID*, 1965, **91**, 179.

BITTNER, M. (1) (tr.). *Die Topographischen Kapitel d. Indischen Seespiegels 'Moḥīṭ' ['Muḥīṭ (The Ocean) of Sidi 'Alī Reïs]; mit einer Einleitung sowie mit 30 Tafeln versehen, von W. Tomaschek*. K. K. Geographischen Gesellschaft, Vienna, 1897. (Festschrift z. Erinnerung an die Eroffnung des Seeweges nach Ostindien durch Vasco da Gama, +1497.)

BLACK, A. (1). *The Story of Bridges*. McGraw-Hill, New York & London, 1936.

BLAGDEN, C. O. (1). 'Notes on Malay History.' *JRAS/M*, 1909, no. 53.

BLAIR, E. H. & ROBERTSON, J. A. (1). *The Philippine Islands, +1493 to +1898*. 4 vols. Cleveland, Ohio, 1903.

BLAKE, M. E. (1). *Ancient Roman Construction in Italy from the Prehistoric Period to Augustus*. Carnegie Institution, Washington, 1947 (Pub. No. 570).

BLAKISTON, T. W. (1). *Five Months on the Yang-tsze*. London, 1862.

DE LA BLANCHÈRE, M. R. (1). Art. 'Fossa' (Canals and Hydraulic Works). In Daremberg & Saglió, II, 1321.

BLOCH, R. (1). *Etruscan Art*. New York Graphic Society, New York, 1959.

BLOCHET, E. (1). *Mussulman Painting, +12th to +17th Century*. Tr. C. M. Binyon, introd. E. D. Ross. Methuen, London, 1929.

BLÜMNER, H. (1). *Technologie und Terminologie der Gewerbe und Künste bei Griechern und Römern*. 4 vols. Teubner, Leipzig & Berlin, 1912.

BODDE, D. (15). *Statesman, Patriot and General in Ancient China*. Amer. Or. Soc., New Haven, Conn. 1940. (Biographies of Lü Pu-Wei, Ching Kho and Mêng Thien.)

BODDE, D. (21). 'Myths of Ancient China.' Art. in *Mythologies of the Ancient World*, ed. S. N. Kramer. Doubleday, New York, 1961.

BOEHLING, H. B. H. (1). 'Chinesische Stampfbauten' (making of pisé-de-terre walls). *S*, 1951, **3**, 16.

BOERSCHMANN, E. (1). *Chinesische Architektur*. 2 vols. Wassmuth, Berlin, 1925.

BOERSCHMANN, E. (2). *Baukunst und Religiöse Kultur der Chinesen*. 2 vols; vol. 1, P'u T'o Shan (the famous island with its many Buddhist temples off the coast of Chiangsu); vol. 2, Gedächtnistempel (memorial temples both Taoist and Confucian, esp. those of Chang Liang at Miaot'ai-tzu in Shensi, of Li Ping and Li Erh-Lang at Kuanhsien in Szechuan, of Confucius in Shantung, etc. etc.). Reinier, Berlin, 1911.

BOERSCHMANN, E. (3). *Baukunst und Landschaft in China*. Wassmuth, Berlin, n.d. (about 1912, 1919, 1925). Fr. edn. *La Chine Pittoresque*. Calavas, Paris, n.d. (about 1920). (Photographs taken from 1906 to 1909.)

BOERSCHMANN, E. (3a). *China; Architecture and Landscape—a Journey through Twelve Provinces*. Studio, London, n.d. (1928–1929). Eng. ed. of Boerschmann (3).

BOERSCHMANN, E. (4). *[Chinesische] Pagoden, Pao-Tha* (vol. 1; only one vol. published). de Gruyter, Berlin & Leipzig, 1931. (Die Baukunst und religiöse Kultur der Chinesen, vol. 3.)

BOERSCHMANN, E. (5). *Chinesische Baukeramik*. Lüdtke, Berlin, 1927.

BOERSCHMANN, E. (6). 'K'uei-sing [Khuei Hsing]-türme und Fêngshui-Säulen.' *AM*, 1925, **2**, 503.

BOERSCHMANN, E. (7). 'Pagoden d. Sui- u. frühen Thang-Zeit.' *OAZ*, 1924, **11**, 195.

BOERSCHMANN, E. (8). 'Chinese Architecture and its Relation to Chinese Culture.' *ARSI*, 1911, 539 (tr. from *ZFE*, 1910, **42**, 390).

BOERSCHMANN, E. (10). 'Beobachtungen über Wassernutzung in China.' *ZGEB*, 1913, 516.

BOGLE, G. See Markham, C.-R. (1).

BOLL, F. (1). *Sphaera*. Teubner, Leipzig, 1904.

BON, A. M. & BON, A. (1). 'Les Timbres Amphoriques de Thasos.' In *Études Thasiennes*, vol. 4. École Fr. d'Athènes, Paris, 1957.

BONATZ, P. & LEONHARDT, F. (1). *Brücken*. Langewiesche, Königstein i/Taunus 1951.

BONNARD, L. (1). *La Navigation Intérieure de la Gaule à l'Époque Gallo-Romane*. Picard, Paris, 1913.

BOOKER, P. J. (1). *A History of Engineering Drawing*. Chatto & Windus, London, 1963. Rev. D. Chilton, *TCULT*, 1965, **6**, 128.

BOREUX, C. (1). *Études de Nautique Égyptienne*. Instit. Français d'Archéol. Orient. Cairo, 1924.

BOSWELL, J. (1). *The Life of Samuel Johnson Ll.D.* 6th ed. 4 vols. Cadell & Davies, London, 1811.

BOUCHAYER, A. (1). *Marseille; ou la Mer qui Monte*. Paris, 1931.

BOUCHÉ-LECLERCQ, A. (1). *L'Astrologie Grecque*. Leroux, Paris, 1899.

BOUILLARD, G. (1). *Les Tombeaux Impériaux*. Peking, 1931.

BOUILLARD, G. & VAUDESCAL, C. (1). 'Les Sépultures Imperiales des Ming (Che-san Ling) [Shih San Ling].' *BEFEO*, 1920, **20**, no. 3.

DE BOURBOURG, E. C. BRASSEUR (1). '*Popol Vuh*', ou Livre Sacré...des Quichés*. Durand, Paris, 1861.

BOURDON, C. (1). 'Anciens Canaux, Anciens Sites et Ports de Suez.' *MSRGE*, 1925, **7**, 1.

BOURDON, L. (1). 'Introduction à la Traduction du *Chronique de Guinée* (Gomes Eanes de Zurara) par L. Bourdon & R. Ricard.' *MIFAN*, 1960 (no. 60).

BOURNE, F. S. A. (1). *Report of a Journey in South-west China, presented to both Houses of Parliament.* HMSO, London, 1888.

BOURNE, WILLIAM (1). *A Regiment for the Sea, conteining very necessary Matters for all sorts of Seamen and Travellers, as Masters of Ships, Pilots, Mariners and Marchaunts, newly corrected and amended by the Author; whereunto is added a Hidrographicall Discourse to goe unto Cattay [Cathay], five severall Wayes...* East & Wight, London, 1580. Earlier editions, not including the Hydrographical Discourse, 1574, 1577. Reprinted 1584, 1587, revised 1592, repr. 1596, 1601, 1620, 1631. Dutch tr. 1594 repr. 1609. See Taylor (13).

BOVILL, E. W. (1). *Caravans of the Old Sahara.* Oxford, 1933.

BOVILL, E. W. (2). *The Golden Trade of the Moors.* Oxford, 1958.

BOWEN, R. LE B. (1). 'Arab Dhows of Eastern Arabia.' *ANEPT*, 1949, **9**, 87. Also separately, as pamphlet, enlarged, pr. pr. Rehoboth, Mass. U.S.A., 1949.

BOWEN, R. LE B. (2). 'Eastern Sail Affinities.' *ANEPT*, 1953, **13**, 81 and 185. Comment by R. C. Anderson, 1953, **13**, 213.

BOWEN, R. LE B. (3). 'The Dhow Sailor.' *ANEPT*, 1951, **11** (no. 3).

BOWEN, R. LE B. (4). 'Primitive Watercraft of Arabia.' *ANEPT*, 1952, **12** (no. 3).

BOWEN, R. LE B. (5). 'Pearl Fisheries of the Persian Gulf.' *MEJ*, 1951, April.

BOWEN, R. LE B. (6). 'Marine Industries of Eastern Arabia.' *GR*, 1951, July.

BOWEN, R. LE B. (7). 'Maritime Superstitions of the Arabs.' *ANEPT*, 1955, **15**, 5; 'Origin and Diffusion of Oculi.' *ANEPT*, 1957, **17**, 262; 'The Origin and Diffusion of Oculi.' *ANEPT*, 1958, **18**, 235.

BOWEN, R. LE B. (8). 'Boats of the Indus Civilisation.' *MMI*, 1956, **42**, 279.

BOWEN, R. LE B. (9). 'The Origins of Fore-and-Aft Rigs.' *ANEPT*, 1959, **19**, 155 and 274.

BOWEN, R. LE B. (10). 'Experimental Nautical Research; Third-Millennium B.C. Egyptian Sails.' *MMI*, 1959, **45**, 332. Crit. King-Webster (1).

BOWEN, R. LE B. (11). 'Egypt's Earliest Sailing-Ships.' *AQ*, 1960, **34**, 117.

BOWEN, R. LE B. (12) (with appendices by G. W. van Beek & A. Jamme). *Researches [on Ancient Irrigation] in South Arabia (the Yemen).* Reprint from R. le B. Bowen & F. P. A. Albright, *Archaeological Discoveries in South Arabia.* Johns Hopkins Press, Baltimore, 1958. Pr. pub. with grants from Mellon Trust and Scaife Foundation, 1958.

BOWEN, R. LE B. (13). 'Early Arab Ships and Rudders.' *MMI*, 1963, **49**, 303. A consideration of the al-Ḥarīrī picture.

BOWER, URSULA G. (1). *The Hidden Land.* Murray, London, 1953. (The Abor country near Karko in Northern Assam on the Tibetan border; suspension-bridges.)

BOWREY, T. (1). *A Geographical Account of the Countries round the Bay of Bengal, 1669–1675.* Hakluyt Society, London, 1905. (Hakluyt Society Pubs. 2nd series, no. 12.)

BOXER, C. R. (1) (ed.). *South China in the Sixteenth Century; being the Narratives of Galeote Pereira, Fr. Gaspar da Cruz, O.P., and Fr. Martin de Rada, O.E.S.A. (1550–1575).* Hakluyt Society, London, 1953. (Hakluyt Society Pubs. 2nd series, no. 106.)

BOXER, C. R. (3). 'S. R. Welch and the Portuguese in Africa.' *JAH*, 1960, **1**, 55.

BOXER, C. R. (4). *Fidalgos in the Far East.* Nijhoff, The Hague, 1948.

BOXER, C. R. & DE AZEVEDO, C. (1). *Fort Jesus and the Portuguese in Mombasa, +1593 to +1729.* Hollis & Carter, London, 1960.

[BOYD, ANDREW] (1). *Chinese Architecture* (Introduction to the Catalogue of the Exhibition prepared by the Architectural Society of China and shown at the Royal Institute of British Architects, 1959). R.I.B.A., London, 1959. Rev. R. Banham, *NSN*, 1959, 79.

BOYD, A. (2). *Chinese Architecture and Town Planning, −1500 to 1911.* Tiranti, London, 1962. For a biography of Andrew Boyd, see Hollamby (1).

VAN BRAAM HOUCKGEEST, A. E. (1). *An Authentic Account of the Embassy of the Dutch East-India Company to the Court of the Emperor of China in the years 1794 and 1795 (subsequent to that of the Earl of Macartney), containing a Description of Several Parts of the Chinese Empire unknown to Europeans; taken from the Journal of André Everard van Braam, Chief of the Direction of that Company, and Second in the Embassy.* Tr. L. E. Moreau de St Méry. 2 vols., map, but no index and no plates; Phillips, London, 1798. French ed. 2 vols, with map, index and several plates; Philadelphia, 1797. The two volumes of the English edition correspond to vol. 1 of the French edition only.

BRANGWYN, F. & SPARROW, W. S. (1). *A Book of Bridges.* Lane, London, 1915.

BRASIO, ANTONIO (1). *A Acção Missionária no Período Henriquino.* Comissão Executiva das Comemorações do Quinto Centenário da Morte do Infante Dom Henrique. Lisbon, 1958. (Colecção Henriquina, no. 9.)

BRAUDEL, F. (1). *La Mediterranée et le Monde mediterranéen à l'Epoque de Philippe II*. Colin, Paris, 1949.

BRAZIER, J. S. (1). 'Analysis of Brick from the Great Wall of China.' *JRAS/NCB*, 1886, **21**, 232.

BRETSCHNEIDER, E. (1). *Botanicon Sinicum; Notes on Chinese Botany from Native and Western Sources*. 3 vols. Trübner, London, 1882 (printed in Japan). (Repr. from *JRAS/NCB*, 1881, **16**.)

BRETSCHNEIDER, E. (2). *Mediaeval Researches from Eastern Asiatic Sources; Fragments towards the Knowledge of the Geography and History of Central and Western Asia from the +13th to the +17th Century*. 2 vols. Trübner, London, 1888.

BRETSCHNEIDER, E. (3). 'Chinese Intercourse with the Countries of Central and Western Asia during the 15th Century' [Introduction]. *CR*, 1875, **4**, 312. Reprinted in Bretschneider (2), vol. 2, p. 157.

BRETSCHNEIDER, E. (4). 'Chinese Intercourse with the Countries of Central and Western Asia during the 15th century. II. A Chinese Itinerary of the Ming Period from the Chinese Northwest Frontier to the Mediterranean Sea'. *CR*, 1876, **5**, 227. (Reprinted (abridged) in Bretschneider (2), vol. 2, p. 329.)

BRETSCHNEIDER, E. (5). *Recherches Archéologiques et Historiques sur Pékin et ses Environs*. Tr. V. Collin de Plancy, 1879.

BRETSCHNEIDER, E. (7). 'Chinese Intercourse with the Countries of Central and Western Asia during the 15th century: I, Accounts of Foreign Countries and especially those of Central and Western Asia, drawn from the *Ming Shih* and the *Ta Ming I Thung Chih*.' *CR*, 1875, **4**, 385; 1876, **5**, 13, 109, 165.

BRETSCHNEIDER, E. (8). *Notes on Chinese Mediaeval Travellers to the West*. American Presbyterian Mission Press, Shanghai, and Trübner, London, 1875 (with an appendix by A. Wylie). Reprinted from *CR*, 1874, **5**, 113, 173, 237, 305; 1875, **6**, 1, 81.

BRETSCHNEIDER, E. (11). 'Über das Land Fu-Sang.' *MDGNVO*, 1876, **2**, 1.

BREUSING, A. (1). *Die nautischen Instrumente bis zur Erfindung des Spiegelsextanten*. Bremen, 1890.

BRIGGS, M. S. (1). *Muhammadan Architecture in Egypt and Palestine*. Oxford, 1924.

BRIGGS, M. S. (2). *A Short History of the Building Crafts*. Oxford, 1925.

BRIGGS, M. S. (3). 'Building Construction [in the Mediterranean Civilisations and the Middle Ages].' Art. in *A History of Technology*, ed. C. Singer *et al.* Vol. 2, p. 397. Oxford, 1956.

BRIGHAM, W. T. (1). *Guatemala*. 1887.

BRINDLEY, H. H. (1). 'Primitive Craft; Evolution or Diffusion?' *MMI*, 1932, **18**, 303.

BRINDLEY, H. H. (2). 'The Evolution of the Sailing Ship.' *PRPSG*, 1926, **54**, 96.

BRINDLEY, H. H. (3). 'Some Notes on Mediaeval Ships.' *CAS/PC*, 1916, **21**, 83.

B[RINDLEY], H. H. (4). 'Early Sprit-Sails.' *MMI*, 1914, **4**, 221; 1920, **6**, 248.

BRINDLEY, H. H. (5). 'Mediaeval Rudders' [and the earliest sprit-sail rig]. *MMI*, 1926, **12**, 211, 232, 346; 1927, **13**, 85.

BRINDLEY, H. H. (6). 'Early Pictures of Lateen Sails.' *MMI*, 1926, **12**, 9.

BRINDLEY, H. H. (7). 'The Sailing Balsa of Lake Titicaca and other Reed-Bundle Craft.' *MMI*, 1931, **17**, 7.

BRINDLEY, H. H. (8). 'Notes on the Boats of Siberia.' *MMI*, 1920, **5**, 66, 101, 130, 184, **6**, 15.

BRINDLEY, H. H. (9). 'The "Keying"' [a Chinese junk which was sailed round the world in 1848]. *MMI*, 1922, **8**, 305.

BRINDLEY, H. H. (10). 'A Bronze-Age Anchor.' *MMI*, 1927, **13**, 5.

BRINDLEY, H. H. (11). 'Chinese Anchors.' *MMI*, 1924, **10**, 399.

BRINTON, S. (1). *Francisco di Giorgio Martini of Siena; painter, sculptor, engineer, civil and military architect*. 2 vols. Besant, London, 1934.

BRITTAIN, R. (1). *Rivers, Man and Myths*. Doubleday, New York, 1958; Longmans, London, 1958.

BROCHADO, COSTA (1). *O Piloto Árabe de Vasco da Gama*. Comissão Executiva das Comemorações do Quinto Centenário da Morte do Infante Dom Henrique. Lisbon, 1959.

BROCHADO, COSTA (2). *The Discovery of the Atlantic* (tr. of 'O Descobrimento do Atlántico'). Comissão Executiva das Comemorações do Quinto Centenário da Morte do Infante Dom Henrique. Lisbon, 1960. (Colecção Henriquina, no. 3.) French tr. in *Henri le Navigateur*, ed. Anon. (53), p. 57.

BROCHADO, COSTA (3). *Historiógrafos dos Descobrimentos*. Comissão Executiva das Comemorações do Quinto Centenário da Morte do Infante Dom Henrique. Lisbon, 1960. (Colecção Henriquina, no. 12.)

BROCHADO, COSTA (4). 'La Vie et l'Oeuvre du Prince Henri le Navigateur.' Art. in *Henri le Navigateur*, ed. Anon. (53), p. 9. Lisbon, 1960.

BRØGGER, A. W. & SCHETELIG, H. (1). *The Viking Ships*. Oslo, 1951.

BROHIER, R. L. (1). *Ancient Irrigation Works in Ceylon*. 3 vols. Ceylon Govt. Press, Colombo.
Vol. 1. *Tamankaduwa district, Polonnaruwa systems, Parākrama Samudra etc., Minipe-ela, Elahera-ela, Minneriya-wewa, Padawiya- and Wahalkada-wewas*. 1934. Repr. 1949.
Vol. 2. *Kala-wewa, Jaya-ganga and the Anurādhapura tanks, the Vanni and the Jaffna peninsula, Mannar, Pomparippu, the Giants' Tank and the Tabbowa-wewa projects*. 1935. Repr. 1950.
Vol. 3. *The Maya Ratta and Ruhunu Ratta, the Walawe Ganga catchment, and the works of the Eastern Seaboard*. 1935.

BROHIER, R. L. (2). 'Some Structural Features of the Ancient Works in Ceylon for Storing and Distributing Water.' *TEAC*, 1956, **50**, 29.

BROHIER, R. L. & ABEYWARDENA, D. F. (1). *The History of Irrigation and Agricultural Colonisation in Ceylon; the Tamankaduwa District and the Elahera-Minneriya Canal.* Ceylon Govt. Press, Colombo, 1941.

BROHIER, R. L. & PAULUSZ, J. H. O. (1). *Land Maps and Surveys; Descriptive Catalogue of Historical Maps in the Surveyor-General's Office, Colombo.* 2 vols. Ceylon Govt. Press, Colombo, 1950, 1951.

BROMEHEAD, C. E. N. (6). 'The Early History of Water Supply.' *GJ*, 1942, **99**, 142 and 183.

BRONEER, O. (1). 'The Corinthian Isthmus and the Isthmian Sanctuary.' *AQ*, 1958, **32**, 80.

BROOKS, C. W. (1). 'A Report on Japanese Vessels Wrecked in the North Pacific Ocean, from the Earliest Records to the Present Time.' *PCAS*, 1876 (1875), **6**, 50.

BROOKS, C. W. (2). 'Early Migrations—the Ancient Maritime Intercourse of Western Nations before the Christian Era, ethnologically considered and Chronologically arranged; illustrating Facilities for Migration among early types of the Human Race.' *PCAS*, 1876 (1875), **6**, 67.

BROOKS, C. W. (3). 'The Origin and Exclusive Development of the Chinese Race—an Inquiry into the Evidence of their American Origin, suggesting a great Antiquity of the Human Races on the American Continent.' *PCAS*, 1876 (1875), **6**, 95.

BROWN, A. C. (1). *Twin Ships.* Mariners' Museum, Newport News, 1939.

BROWN, C. B. (1). 'Sediment Transportation.' Art. in *Engineering Hydraulics* (Proc. 4th Hydraulics Conference, Iowa Inst. of Hydraulics Research, 1949). Ed. H. Rouse. P. 769. Wiley, New York; Chapman & Hall, London, 1950.

BROWN, LLOYD A. (1). *The Story of Maps.* Little Brown, Boston, 1949.

BROWN, R. H. (1). *The Fayum and Lake Moeris.* Stanford, London, 1892.

BRUHL, ODETTE & LÉVI, S. (1). *Indian Temples.* Oxford Univ. Press, Bombay, 1937; Calcutta, 1939.

BRUNET, P. & MIELI, A. (1). *L'Histoire des Sciences (Antiquité).* Payot, Paris, 1935.

DU BUAT, P. L. G. (1). *Principes d'Hydraulique.* Paris, 1779. Enlarged ed. 2 vols. 1786; enlarged posthumous ed. 3 vols. 1816.

BUCK, J. LOSSING (1). *Land Utilisation in China.*

BUCK, J. LOSSING (2). *Chinese Farm Economy.*

BUDGE, E. A. WALLIS (1). *Guide to the Egyptian Collections in the British Museum.* Brit. Mus. Trustees, London, 1909, and subsequent editions.

BUDKER, P. (1). 'Entretien avec Manfred Curry.' *NEPT*, 1949, no. 14, 8.

BUFFET, B. & EVRARD, R. (1). *L'Eau Potable à travers les Ages.* Solédi, Liége, 1950.

BULLING, A. (1). 'Descriptive Representations in the Art of the Chhin and Han Period.' Inaug. Diss., Cambridge, 1949.

BULLING, A. (2). 'Die Chinesische Architektur von der Han-Zeit bis zum Ende der Thang-Zeit.' Inaug. Diss., Berlin, 1936. Subsequently privately published (Imprimerie Franco-Suisse, Lyon), in two fascicules, the first of text, the second of illustrations, and for some time available at the International Chinese Library, Geneva.

BULLING, A. (3). 'Two Models of Chinese Homesteads.' *BUM*, 1937, **71**, 153.

BULLING, A. (5). 'Die Kunst der Totenspiele in der östlichen Han-zeit.' *ORE*, 1956, **3**, 28.

BULLING, A. (9). 'Buddhist Temples in the Thang Period.' *ORA*, 1955 (n.s.), **1**, 79 and 115.

BULLING, A. (11). 'A Landscape Representation of the Western Han Period (*c.* −60, in a tomb near Chêngchow).' *AA*, 1962, **25**, 293.

BULLING, A. (12). 'Hollow Tomb Tiles; Recent Excavations and their Dating.' *ORA*, 1965, **11**.

BULLING, A. (13). 'Three Popular Motifs in the Art of the Later Han Period; the Lifting of the Tripod, the Crossing of a Bridge, Divinities.' *AART*, 1967, **20**, 25.

BURFORD, A. M. (2). 'The Economics of Greek Temple Building.' *PCPS*, 1965 (no. 191), 21.

BURKILL, I. H. (1). *A Dictionary of the Economic Products of the Malay Peninsula* (with contributions by W. Birtwhistle, F. W. Foxworthy, J. B. Scrivenor & J. G. Watson). 2 vols. Crown Agents for the Colonies, London, 1935.

BURKILL, I. H. (2). 'James Hornell (1865 to 1949).' *PLS*, 1949, **161**, 244.

BURTON, SIR RICHARD F. (2) (tr.). *Os Lusiadas (The Lusiads), englished by R.F.B....* 2 vols. Ed. I. Burton. Quaritch, London, 1880.

BUSCHAN, G., BYHAN, A., VOLZ, W., HABERLANDT, A. & M., & HEINE-GELDERN, R. (1). *Illustrierte Völkerkunde.* 2 vols. in 3. Stuttgart, 1923.

BUSH, ROWENA E. & CULWICK, A. T. (1). *Illustration for Africans.* Technical Report to the Government of Tanganyika, undated (*c.* 1950), available for consultation in the United Africa Company Library, London.

BUSHELL, S. W. (3). 'The Early History of Tibet.' *JRAS*, 1880, n.s., **12**, 435, 538.

BUSHNELL, G. H. S. (1) 'Radio-carbon Dates and New World Chronology.' *AQ*, 1961, **35**, 286.

CAHEN, C. (3). 'Le Service de l'Irrigation en Iraq au Début du 11e Siècle.' *BEO/IFD*, 1950, **13**, 117.

CAHEN, G. (1). *Some Early Russo-Chinese Relations*. Tr. and ed. W. S. Ridge. National Review, Shang-hai, 1914. Repr. Peking, 1940. Orig. ed. *Histoire des Relations de la Russie avec la Chine sous Pierre le Grand* (+*1689 à* +*1730*). 1912.

CALDER, RITCHIE (1). *The Inheritors*. Heinemann, London, 1961.

CALDER, W. M. (1). 'The Royal Road in Herodotus.' *CLR*, 1925, **39**, 7.

CALLENDAR, G. (1). 'Punts and Shouts.' *MMI*, 1923, **9**, 117.

CALVERT, R. (1). *Inland Waterways of Europe*. Allen & Unwin, London, 1963.

CALZA, G. (1). *Ostia*. Libreria dello Stato, Rome, 1950. (Ministero della Pubblica Istruzione; Itinéraires des Musées et Monuments d'Italie.)

CAMMANN, S. VAN R. (6). 'Chinese Carvings in Hornbill Ivory.' *SMJ*, 1951, **5**, 293.

DE CAMOENS, LUIS (1). *Os Lusiados*. Lisbon, 1572 (facsimile Lisbon, 1943). Eng. tr. see Fanshawe, R. (1); Aubertin, J. J. (1) (with Portuguese); Burton, R. F. (2); Atkinson, W. C. (1) (prose); Mickle, W. J. (1) (Popian couplets, not recommended, and contains insertions and deletions, but with interesting notes).

ÇAMORANO, RODRIGO (1). *Compendio de la Arte de Navegar*. Seville, 1581; 2nd ed. 1588.

DE CAMP, L. SPRAGUE (1). 'Sailing Close-Hauled.' *ISIS*, 1959, **50**, 61.

DE CAMP, L. SPRAGUE (2). 'The "Darkhouse" of Alexandria.' *TCULT*, 1965, **6**, 423.

CAMPBELL, D. T. (1). 'Distinguishing Differences of Perception from Failures of Communication in Cross-Cultural Studies.' Art. in *Cross-Cultural Understanding; Epistemology in Anthropology* (a Wenner-Gren Foundation Symposium). Ed. F. S. C. Northrop & H. H. Livingston. Harper & Row, New York, 1964, p. 308.

DE CANDOLLE, ALPHONSE (1). *The Origin of Cultivated Plants*. Kegan Paul, London, 1884. Tr. from the French ed. Geneva, 1882.

CAPART, J. (1). *L'Art Égyptien*. Vromant, Brussels & Paris, 1922.

CAPOT-REY, R. (1). *Géographie de la Circulation sur les Continents*. Gallimard, Paris, 1946.

CAREY, H. F. (1). 'Transportation on the Yangtze Kiang.' *CJ*, 1929, **10**, 249.

CAREY, H. F. (2). 'Romance on the Great River' (the Yangtze). *CJ*, 1932, **17**, 276.

CARLES, W. R. (1). 'The Yangtze Kiang.' *GJ*, 1898, **12**, 225.

CARLES, W. R. (2). 'The Grand Canal of China.' *JRAS/NCB*, 1896, **31**, 102.

CARLINI, CAPT. (1). *Le Gouvernail dans l'Antiquité*. Communication to the Association Technique Maritime et Aéronautique, 1935. (Brochure without bibliographical identifications.)

CARMONA, A. L. B. ADM. (1). *Lorchas, Juncos e outros Barcos usados no Sul da China, a Pesca em Macau e Arredores*. Imprensa Nacional, Macao, 1954.

CARPEAUX, C. (1). *Le Bayon d'Angkor Thom; Bas-Reliefs publiés...d'après les documents receuillis par le Mission Henri Dufour*. Leroux, Paris, 1910.

CARPENTER, R. (1). 'On Greek ships.' *AJA*, 1948, **52**, 1.

CARTER, G. F. (1). 'Plants across the Pacific.' Art. in *MSAA*, no. 9; suppl. to *AMA*, 1953, **18** (no. 3), p. 62.

CARTER, T. F. (1). *The Invention of Printing in China and its Spread Westward*. Columbia Univ. Press, New York, 1925, revised ed. 1931. 2nd ed. revised by L. Carrington Goodrich. Ronald, New York, 1955.

CARTER, T. F. (2). 'The Westward Movement of the Art of Printing.' In *Yearbook of Oriental Art and Culture*, ed. A. Waley, vol. 1, p. 19. Benn, London, 1925. (Rev. B. Laufer, *JAOS*, 1927, **47**, 71; A. C. Moule, *JRAS*, 1926, 140.)

CARUS-WILSON, E. M. (2). 'The Woollen Industry [of Mediaeval Europe].' Art. in *Cambridge Economic History of Europe*, Ed. M. Postan & E. E. Rich, vol. 2, p. 355. Cambridge, 1952.

CARY, G. (1). *The Medieval Alexander*. Ed. D. J. A. Ross. Cambridge, 1956. (A study of the origins and versions of the Alexander-Romance; important for medieval ideas on flying-machine and diving-bell or bathyscaphe.)

CASO, ALFONSO (1). *The Religion of the Aztecs*. Editorial Fray B. de Sahagun, Mexico City, n.d. (1947) (Pasado e Presente ser.).

CASO, ALFONSO (2). *Relations between the Old and New World; a Note on Methodology*. Proc. 35th Internat. Congress of Americanists, Mexico City, 1962. Vol. 1, p. 55.

CASSON, L. (1). 'The "Isis" and her Voyage.' *TAPA*, 1950, **81**, 43. Crit. B. S. J. Isserlin, *TAPA*, 1955, **86**, 319; reply, 1956, **87**, 239.

CASSON, L. (2). 'Fore-and-Aft Sails in the Ancient World.' *MMI*, 1956, **42**, 3. With correction by R. le B. Bowen, p. 239, and discussion by G. la Roërie, p. 238, continued by J. Lyman, R. le B. Bowen, Sir Alan Moore and G. la Roërie, 1957, **43**, 63, 160, 241, 329. More popular presentation of the same material but with further comparative illustration, *AAAA*, 1954, **7**, 214.

CASSON, L. (3). *The Ancient Mariners; Sea-farers and Sea Fighters of the Mediterranean in Ancient Times*. Gollancz, London, 1959.

CASSON, L. (4). 'The Lateen Sail in the Ancient World.' *MMI*, 1966, **52**, 199. (Evidence from a graffito on a Hellenistic amphora handle from Thasos; see Bon & Bon (1), vol. 4, no. 2274; Thasos Museum, no. 1606.)

CASSON, L. (5). 'Ancient Shipbuilding; New Light on an Old Source [the tombstone of Longidienus, *c.* +200, showing him inserting a frame in a strake-morticed hull].' *TAPA*, 1963, **94**, 28.

CASSON, L. (6). 'Odysseus' Boat [a new interpretation of the Homeric description in terms of a strake-morticed hull].' *AJP*, 1964, **85**, 61.

CASSON, L. (7). 'New Light on Ancient Rigging and Boat-building.' *ANEPT*, 1964, **24**, 81. (Strake-morticing, sewn boats in the ancient Mediterranean (by literary evidence), the Etruscan two-master (cf. Moretti, 1), etc.)

DE CASTANHEDA, FERNÃO LOPES (1). *História do Descobrimento e Conquista da India pelos Portuguezes.* Lisbon, 1552, 1554, 1561; mod. edn. 1833.

VON CASTEL, GRAF (1). *Chinaflug* [unequalled air photographs]. Atlantis, Zürich.

CASTELLI, BENEDETTO (1). *Delli Misure dell'Acque correnti*... 1628; 2 vols. Parma, 1676.

CASTILLO, A. C. (1). *Archaeology in Mexico Today.* Petroleos Mexicanos (Pemex), Mexico City, n.d. (1947).

DE CASTRO, JOÃO (1). *Roteiro de Lisboã a Goa.* Annotated by João de Andrade Corro. Lisbon, 1882.

DE CASTRO, JOÃO (2). *Primo Roteiro da Costa da India desde Goa até Dio; narrando a viagem que fez o Vice-Rei D. Garcia de Noronha en socorro deste ultima Cidade, 1538–1539.* Köpke, Porto, 1843.

DE CASTRO, JOÃO (3). *Roteiro em que se contem a viagem que fizeran os Portuguezes no anno de 1541 partindo da nobre Cidade de Goa atee Soez que he no fim e stremidade do Mar Roxo....* Paris, 1833.

CATON-THOMPSON, G. (1). *The Zimbabwe Culture; Ruins and Reactions.* Oxford, 1931.

CEDERLUND, C. O., HAMILTON, E., LUNDSTRÖM, P. & SOOP, H. (1). *The Warship* Wasa (*Catalogue of the* Vasa *Museum in Stockholm*). Tr. J. Herbert. Stockholm, 1963.

CESCINSKY, H. (1). *Chinese Furniture; a series of Examples from the Collections in France.* London, 1922.

CHAKRAVARTI, P. C. (1). *The Art of War in Ancient India.* Univ. of Dacca Press, Ramma, Dacca, 1941. (Univ. of Dacca Bulletin, no. 21.)

CHAMBERS, SIR WM. (1). *Designs of Chinese Buildings, Furniture, Dresses, Machines and Utensils; to which is annexed, A Description of their Temples, Houses, Gardens, etc.* London, 1757.

CHAMBERS, SIR WM. (2). *A Dissertation on Oriental Gardening...; To which is annexed, An Explanatory Discourse by Tan Chet-Qua, of Quang-chew-fu, Gent.* 2nd ed., with additions, Griffin, Davies, Dodsley, Wilson, Nicoll, Walter & Emsley, London, 1773.

CHANG CHI-HSIEN (ed.) (1). *The Chinese Yearbook, 1943.* Council of International Affairs, Chungking, 1943; Thacker, Bombay, 1943.

CHANG CHING-CHIH (1). 'Nothing stops the Railway Builders' (on the Pao-chhêng Line). *CREC*, 1956, **5** (no. 7), 6.

CHANG FO-KUEI (1). *Chinese Architecture of the Chhing Dynasty.* Peking, 1935.

CHANG HAN-YING (1). 'New View of Water Conservancy' (account of the Huai River Control Project, the North-western Hupei contour canal irrigation systems, and reforestation). *CREC*, 1959, **8** (no. 8), 2.

CHANG HSIN-CHÊNG (2). *Chinese Popular Literature from the 13th to the 19th Century* (in the press).

CHANG KUANG-CHIH (1). *The Archaeology of Ancient China.* Yale Univ. Press, New Haven, 1963. Rev. Chêng Tê-Khun, *AQ*, 1964, **38**, 179.

CHANG KUEI-SHÊNG (1). *Chinese Great Explorers.* Inaug. Diss. Univ. of Michigan, 1955.

CHANG KUEI-SHÊNG (2). 'A Re-Examination of the Earliest Chinese Map of Africa.' *PMASAL*, 1957 (1956), **42**, 151.

CHANG PO-CHUN (1). 'Our Shipping and Highways.' *CREC*, 1952, **1** (no. 3), 15.

CHANG PO-CHUN (2). 'The First Highways to Tibet.' *CREC*, 1955, **4** (no. 5), 2.

CHAO WEI-PANG (3). 'The Dragon-Boat Race in Wu-ling, Hunan.' *FLS*, 1943, **2**, 1.

CHAO YUNG-SHEN (1). 'The Ming Tombs Reservoir—after Three Years.' *CREC*, 1962, **11** (no. 1), 24.

CHAPMAN, F. R. (1). 'On the Working of Greenstone or Nephrite by the Maoris.' *TNZI*, 1892, **24**, 479.

CHAPOT, V. (1). 'Seleucie de Piérie.' *MSAF*, 1907 (7e sér.), **6** (66), 149–226.

CHARLESWORTH, M. P. (2). *Trade-Routes and Commerce of the Roman Empire.* Cambridge, 1924; 2nd ed. 1926. Fr. tr. *Les Routes et le Trafic Commercial de l'Empire Romain.* Paris, 1938.

CHARNOCK, J. (1). *An History of Marine Architecture.* 3 vols. Faulder *et al.* London, 1800–2.

CHASSIGNEUX, E. (1). 'Le Canal Cu'u-Yên.' Art. in *Etudes Asiatiques, publiées à l'occasion du 25e Anniversaire de l'Ecole Française d'Extrême-Orient [à Hanoi].* 2 vols. van Oest, Paris & Brussels, 1925. (Pubs. Ec. Fr. d'Extr. Or. nos. 19, 20.) Vol. 1, p. 125.

CHATLEY, H. (24). 'The Hydrology of the Yangtze River.' *JICE*, 1939, 227 and 565 (Paper no. 5223).

CHATLEY, H. (25). 'The Yellow River as a Factor in the Development of China.' *ASR*, 1939, 1.

CHATLEY, H. (27). 'The Properties of Clay and Silt.' *ESC*, 1923, **22** (Paper no. 4).

CHATLEY, H. (28). 'The Stability of Dredged Cuts in Alluvium.' *JJIE*, 1927, **37**, 525.

CHATLEY, H. (29). 'Silt' [at the mouth of the Yangtze]. *PICE*, 1921, **212**, 400 (Paper no. 4380). 'Silt Equilibrium.' *PICE*, 1925 (Paper no. 4493).

CHATLEY, H. (30). 'The Physical Properties of Clay Mud.' *TSE*, 1922, 133.

CHATLEY, H. (31). 'Silt Subsidence and Saecular Change.' *CJ*, 1928, **8**, 150.

CHATLEY, H. (32). 'Some Problems on Silt.' *ESC*, 1919, **18** (Paper no. 5).

CHATLEY, H. (33). *Floods and Flood Prevention* (the Yangtze). Privately pr. Shanghai, n.d. (*c.* 1920).

CHATLEY, H. (34). 'Mud and similar Granular Mixtures.' *ESC*, 1929, **28** (Paper no. 4).

CHATLEY, H. (35). 'River Discharge Formulae in relation to the Dimensional Theory of Fluid Resistance.' *ESC*, 1930, **29** (Paper no. 3).

CHATLEY, H. (36). 'Far Eastern Engineering.' *TNS*, 1954, **29**, 151. With discussion by J. Needham, A. Stowers, A. W. Skempton, S. B. Hamilton *et al.*

CHATTERTON, E. K. (1). *The Ship under Sail*. Fisher Unwin, London, 1926.

CHATTERTON, E. K. (2). *Sailing Ships, the Story of their Development from the Earliest Times to the Present Day*. London, 1909. *Sailing Ships and their Story*. London, 1923.

CHATTERTON, E. K. (3). *Ship Models*. Studio, London, 1923.

CHATTERTON, E. K. (4). *Fore and Aft; the Story of the Fore-and-Aft Rig from the Earliest Times to the Present Day*. Seeley Service, London, 1912; 2nd ed. 1927.

CHAVANNES, E. (1). *Les Mémoires Historiques de Se-Ma Ts'ien [Ssuma Chhien]*. 5 vols. Leroux, Paris, 1895–1905. (Photographically reproduced, in China, without imprint and undated.)
  1895 vol. 1 tr. *Shih Chi*, chs. 1, 2, 3, 4.
  1897 vol. 2 tr. *Shih Chi*, chs. 5, 6, 7, 8, 9, 10, 11, 12.
  1898 vol. 3 (i) tr. *Shih Chi*, chs. 13, 14, 15, 16, 17, 18, 19, 20, 21, 22.
           vol. 3 (ii) tr. *Shih Chi*, chs. 23, 24, 25, 26, 27, 28, 29, 30.
  1901 vol. 4 tr. *Shih Chi*, chs. 31, 32, 33, 34, 35, 36, 37, 38, 39, 40, 41, 42.
  1905 vol. 5 tr. *Shih Chi*, chs. 43, 44, 45, 46, 47.

CHAVANNES, E. (6) (tr.). 'Les Pays d'Occident d'après le Heou Han Chou.' *TP*, 1907, **8**, 149. (Ch. 118, on the Western Countries, from *Hou Han Shu*.)

CHAVANNES, E. (8). 'L'Instruction d'un Futur Empereur de Chine en l'an 1193' [on the astronomical, geographical, and historical charts inscribed on stone steles in the Confucian temple at Suchow, Chiangsu]. In *Mémoires concernant l'Asie Orientale* (publ. Acad. des Inscriptions et Belles Lettres), Leroux, Paris, 1913, vol. 1, p. 19.

CHAVANNES, E. (9). *Mission Archéologique dans la Chine Septentrionale*. 2 vols. and portfolios of plates. Leroux, Paris, 1909–15. (Publ. de l'École Franç. d'Extr. Orient. no. 13.)

CHAVANNES, E. (11). *La Sculpture sur Pierre en Chine aux Temps des deux dynasties Han*. Leroux, Paris, 1893.

CHAVANNES, E. (14). *Documents sur les Tou-Kiue (Turcs) [Thu-Chüeh] Occidentaux, receuillis et commentés par E.C.* ... . Imp. Acad. Sci., St Petersburg, 1903.

CHAVANNES, E. (20). 'Documents historiques et géographiques relatifs à Lichiang.' *TP*, 1912, **13**, 565.

CHÊNG LIN (1) (tr.). *Prince Dan of Yann [Yen Tan Tzu]*. World Encyclopaedia Institute, Chungking, 1945.

CHÊNG TÊ-KHUN (5) (ed.). *Illustrated Catalogue of an Exhibition of Chinese Paintings from the Mu-Fei Collection* (held in connection with the 23rd International Congress of Orientalists). Fitzwilliam Museum, Cambridge, 1954.

CHÊNG TÊ-KHUN (9). *Archaeology in China*.
  Vol. 1, *Prehistoric China*. Heffer, Cambridge, 1959.
  Vol. 2, *Shang China*. Heffer, Cambridge, 1960.
  Vol. 3, *Chou China*. Heffer, Cambridge, and Univ. Press, Toronto, 1963.
  Vol. 4, *Han China* (in the press).

CHESNEY, LT. COL. (1). 'On the Bay of Antioch and the Ruins of Seleucia Pieria.' *JRGS*, 1838, **8**, 228. French tr. *NAVSG*, 1839 (4e sér.), **2**, 42.

CHEVALIER, H. (1). *Cérémonial de l'Achèvement des Travaux de Hoa-Syeng (Corée), 1800, traduction et résumé* (an illustrated Korean work on city fortifications described). *TP*, 1898, **9**, 394.

CHHEN CHHI-THIEN (1). *Lin Tsê-Hsü; Pioneer Promoter of the Adoption of Western Means of Maritime Defence in China*. Dept. of Economics, Yenching Univ., Vetch (French Bookstore), Peiping, 1934. ([Studies in] Modern Industrial Technique in China, no. 1.)

CHHEN CHHI-THIEN (2). *Tsêng Kuo-Fan; Pioneer Promoter of the Steamship in China*. Dept. of Economics, Yenching Univ., Vetch (French Bookstore), Peiping, 1935. ([Studies in] Modern Industrial Technique in China, no. 2.)

CHHEN CHHI-THIEN (3). *Tso Tsung-Thang; Pioneer Promoter of the Modern Dockyard and the Woollen Mill in China*. Dept. of Economics, Yenching Univ., Vetch (French Bookstore), Peiping, 1938. ([Studies in] Modern Industrial Technique in China, no. 3.)

CHHEN HAN-SÊNG (1). 'Chequerboard of Canals' (the drainage system of the North Huai valley catchment area in Anhui, part of the Huai River Control Project). *CREC*, 1959, **8** (no. 2), 22.

CHHEN HSÜEH-NUNG (1). 'Transforming a Poor Hill Village [check dams in loess ravines as well as terracing].' *PKR*, 1964 (no. 25), 28.

CHHEN SHIH-HSIANG (2) (tr.). 'Biography of Ku Khai-Chih' [*Chin Shu*, ch. 92]. Univ. Calif. Press, Berkeley, Calif. 1953. (Inst. East Asian Studies, Univ. of Calif. Chinese Dynastic History Translations, no. 2.)

CHHEN SHOU-YI (2). 'The Chinese Garden in 18th Century England.' *TH*, 1936, **2**, 321.

CHHEN TSU-LUNG (1). 'Table de Concordance des Numérotages des Grottes de Touen-Hoang [Tun-huang].' *JA*, 1962, **250**, 257.

CHHIU KHAI-MING [CHIU KAIMING] (1). 'Agriculture' (Chinese). In *China*, ed. H. F. McNair, p. 466. Univ. of Calif. Press, Berkeley, 1946.

CHI CHHAO-TING (1). *Key Economic Areas in Chinese History, as revealed in the Development of Public Works for Water-Control*. Allen & Unwin, London, 1936. See Lattimore, Boyd Orr *et al.* (1).

CHI YU-CHING (1). 'Rebuilding the Grand Canal.' *CREC*, 1963, **12** (no. 7), 5.

CHIANG KHANG-HU (1). *On Chinese Studies*. Com. Press, Shanghai, 1934.

CHIANG SHAO-YUAN (1). *Le Voyage dans la Chine Ancienne, considéré principalement sous son Aspect Magique et Religieux*. Commission Mixte des Oeuvres Franco-Chinoises (Office de Publications), Shanghai, 1937. Transl. from Chinese by Fan Jen.

CHIN SHOU-SHEN (1). 'The Great Wall of China.' *CREC*, 1962, **11** (no. 1), 20.

CHINA HANDBOOK. See Tong, Hollington, K.

CHINESE YEARBOOK. See Chang Chi-Hsien.

CHIPAULT, J. R. *et al.* (1). 'The Anti-oxidant Effects of Spices.' *FOODR*, 1952, **17**, 46; *FOODT*, 1956, **10**, 209.

CHIU. See Chhiu.

CHOISY, A. (1). *Histoire de l'Architecture*. 2 vols. Rouveyre, Paris, 1906.

CHOISY, A. (2). *L'Art de Bâtir chez les Egyptiens* (portfolio). Baranger, Paris, 1904.

CHOISY, A. (3). *L'Art de Bâtir chez les Romains*. Baranger, Paris, 1904 (?).

CHOISY, A. (4). *L'Art de Bâtir chez les Byzantins*. Baranger, Paris, 1904 (?).

CHRISTIAN, V. (1). 'Die Beziehungen der altmesopotamischen Kunst zum Osten.' *WBKGA*, 1926, **1**, 41.

CHRISTIE, A. (1). 'An Obscure Passage from the *Periplus*; κολανδιοφωντα τα μέγιδτα.' *BLSOAS*, 1957, **19**, 345.

CHRISTIE, A. (2). 'The Sea-Locked Lands; the Diverse Traditions of Southeast Asia.' Art. in *The Dawn of Civilisation; the First World Survey of Human Cultures in Early Times*, p. 277. Ed. S. Piggott. Thames & Hudson, London, 1961.

CHU CHHI-CHHIEN & YEH KUNG-CHAO (1). '[Chinese] Architecture; a brief Historical Account based on the Evolution of the City of Peiping.' In *Symposium on Chinese Culture*, p. 97. Ed. Sophia H. Chen Zen. Inst. Pacific Relations, Shanghai, 1931.

CHUMOVSKY, T. A. See Szumowski, T. A.

CIPOLLA, C. M. (1). *Guns and Sails in the Early Phase of European Expansion, + 1400 to + 1700*. Collins, London, 1965.

CLAËYS, J. Y. C. (1). 'L'Annamite et la Mer.' *BIIEH*, 1942; 'L'Annamite devant la Mer.' *INC*, 1943 (March).

CLAËYS, J. Y. C. (2), with CÔNG-VAN-TRUNG & PHAM-VAN-CHUNG. 'Les Radeaux de Pêche de Luong-nhiêm (Thanh-hoa) en Bambous flottants.' *BIIEH*, 1942.

CLAËYS, J. Y. C. & HUET, M. (1). *Angkor* (album of photographs). Hoa-Qui, Paris, n.d. (1948).

CLAPP, F. G. (1). 'Along and across the Great Wall of China.' *GR*, 1920, **9**, 221.

CLAPP, F. G. (2). 'The Huang Ho.' *GR*, 1922, **12**, 1.

CLARK, GRAHAME (1). 'Water in Antiquity.' *AQ*, 1944, **18**, 1.

CLAUDEL, P. & HOPPENOT, H. (1). *Chine* (Album of photographs with introduction). Ed. d'Art Albert Skira, Paris, 1946.

CLISSOLD, P. (1). 'Early Ocean-going Craft in the Eastern Pacific.' *MMI*, 1959, **45**, 234.

CLOS-ARCEDUC, A. (1). 'La Génèse de la Projection de Mercator.' Art. in *Proc. 3rd International Colloquium of Maritime History*, p. 143. Ed. M. Mollat, L. Denoix, O. de Prat, P. Adam & M. Perrichet. Paris, 1958.

CLOWES, G. S. LAIRD (1). 'Ships of Early Explorers.' *GJ*, 1927, **69**, 216.

CLOWES, G. S. LAIRD (2). *Sailing Ships; their History and Development as illustrated by the Collection of Ship Models in the Science Museum*. Pt. I, *Historical Notes*. Science Museum, London, 1932 (reprinted 1951). [Pt. II is the Catalogue of Exhibits.]

CLOWES, G. S. LAIRD (3). 'Comment on the Kiel spritsail.' *MMI*, 1927, **13**, 89.

CLOWES, G. S. LAIRD & TREW, C. G. (1). *The Story of Sail*. Eyre & Spottiswoode, London, 1936.

COATES, WELLS (1). Design of a 'wingsail catamaran' with rigid sail (adopting certain Chinese principles). In *Designers in Britain*, I, p. 241. Wingate, London, 1947. 'A Wingsail Catamaran.' *YM*, 1946, **81**, 329.

CODRINGTON, H. W. (1). *A Short History of Ceylon* (with a chapter by A. M. Hocart). Macmillan, London, 1939.

COE, M. D. (1). 'Cultural Development in Southeastern Mesoamerica.' Art. in *Aboriginal Cultural Development in Latin America; an Interpretative Review*, ed. B. J. Meggers & C. Evans, p. 27.

COEDÉS, G. (5). *Les États Hindouisés d'Indochine et d'Indonésie*. Boccard, Paris, 1948. (*Histoire du Monde*, ed. E. Cavaignac, vol. 8, pt. 2.)

COIMBRA, C. (1). *O Infante e o Objectivo Geográfico dos Descobrimentos*. Communication to the Congresso Internacional do História dos Descobrimentos. Lisbon, 1960.

COLE, F. C. & LAUFER, B. (1). 'Chinese Pottery in the Philippines.' *FMNHP/AS*, 1912, no. 162.

COLIN, F. & PASTELLS, P. (1). *Labor Evangélica de los Obreros de la Compañia de Jesus en las Islas Filipinas*. 3 vols. Barcelona, 1902.

COLLINGRIDGE, G. (1). *The Discovery of Australia; a Critical, Documentary and Historic Investigation concerning the Priority of Discovery in Australasia by Europeans before the Arrival of Lt. James Cook in the 'Endeavour' in the year 1770*. Hayes, Sydney, 1895.

COLLIS, M. (1). *The Grand Peregrination; the Life and Adventures of Fernão Mendes Pinto*. Faber & Faber, London, 1949.

COLLIS, M. (2). *The Land of the Great Image; Experiences of Friar Manrique in Arakan*. Faber & Faber, London, 1953; 1st ed. 1943.

COMBAZ, G. (1). *L'Inde et l'Orient classique*. 2 vols. Geuthner, Paris, 1937.

COMBAZ, G. (2). 'Les Sépultures Impériales de la Chine.' *ASRAB*, 1907, **21**, 381. Pub. sep. Vromant, Brussels, 1907.

COMBAZ, G. (3). 'Les Palais Impériaux de la Chine.' *ASRAB*, 1908, **21**, 425. Pub. sep. Vromant, Brussels, 1909.

COMBAZ, G. (4). 'Les Temples Impériaux de la Chine.' *ASRAB*, 1912, **26**, 223. Pub. sep. Vromant, Brussels, 1912.

COMBAZ, G. (5). 'L'Evolution du Stūpa en Asie.' *MCB*, 1932, **2**, 163 (Étude d'Architecture Bouddhique); 1935, **3**, 93 (Contributions Nouvelles et Vue d'Ensemble); 1937, **4**, 1 (Les Symbolismes du Stūpa).

COMBAZ, G. (6). 'La Peinture Chinoise, vue par un Peintre Occidental; Introduction à l'Histoire de la Peinture Chinoise.' *MCB*, 1939, **6**, 11.

COMBAZ, G. (7). 'Masques et Dragons en Asie' (includes as App. 11, 'Masques et Dragons dans l'Amérique pre-columbienne', pp. 262 ff.). *MCB*, 1945, **7**, 1.

COMBRIDGE, J. H., LU GWEI-DJEN, MADDISON, F. & NEEDHAM, J. (1). *The Hall of Heavenly Records; Stars, Clocks and Instruments in the Yi Dynasty of Chosŏn (Korea), +1400 to +1750*. In the Press.

CONGREVE, H. (1). 'A Brief Notice of some Contrivances practiced by the Native Mariners of the Coromandel Coast, in Navigating, Sailing and Repairing their Vessels.' *MJLS*, 1850, **16**, 101. Reprinted in Ferrand (7), pp. 25 ff.

CONTENAU, G. (1). *L'Épopée de Gilgamesh; Poème Babylonien*. l'Artisan du Louvre, Paris, 1929.

DE CONTI, NICOLÒ (1). In *The Most Noble and Famous Travels of Marco Polo, together with the Travels of Nicolò de Conti, edited from the Elizabethan translation of J. Frampton (1579), etc.*, by N. M. Pewzer. Argonaut, London, 1929. Bibliography by Cordier (5).

COOK, A. B. (1). *Zeus*. 3 vols. Cambridge, 1914, 1925, 1940.

COOK, A. B. (2). '[Ancient Greek] Ships.' In *A Companion to Greek Studies*, ed. L. Whibley, Cambridge, 1905.

COOMARASWAMY, A. K. (6). *History of Indian and Indonesian Art*. New York, 1927.

COOPER, J. M. (1). 'Northern Algonkian Scrying and Scapulimancy.' Art. in W. Schmidt, *Festschrift*. Ed. W. Koppers, p. 205. Vienna, 1928.

COOPER, J. M. (2). 'Scapulimancy.' Art. in *Essays in Anthropology* Kroeber Presentation Volume. Ed. R. H. Lowie, p. 29. Univ. Calif. Press, Berkeley, 1936.

LE CORBUSIER, . (1). *The Modulor, a Harmonious Measure to the Human Scale universally applicable to Architecture and Mechanics*. Tr. from the French by P. de Francia & A. Bostock. Faber & Faber London, 1954.

CORDIER, H. (1). *Histoire Générale de la Chine*. 4 vols. Geuthner, Paris, 1920.

CORDIER, H. (5). 'Deux Voyageurs dans l'Extrême-Orient au 15e et 16e Siècles.' *TP*, 1899, **10**, 380. (de Conti and Varthema; bibliographical only.)

CORDIER, H. (6). 'L'Extrême Orient dans l'Atlas Catalan de Charles V, Roi de France.' *BGHD*, 1895, 1.

CORTESÃO, A. (1). *Cartografia e Cartógrafos Portugueses dos Séculos...XVe e XVIe*. 2 vols. Seara Nova, Lisbon, 1935.

CORTESÃO, A. (2) (tr. and ed.). *The Suma Oriental of Tomé Pires, an Account of the East from the Red Sea to Japan...written in...1512 to 1515....* London, 1944. (Hakluyt Society Pubs., 2nd series, nos. 89, 90.)

CORTESÃO, A. (3). *Cartografia Portuguesa Antiga.* Comissão Executiva das Comemorações do Quinto Centenário da Morte do Infante Dom Henrique. Lisbon, 1960. (Colecção Henriquina, no. 8.)

CORTESÃO, A. & DA MOTA, A. TEIXEIRA (1). *Portugaliae Monumenta Cartographica.* 4 vols. Lisbon, 1960. *Tabularum Geographicarum Lusitanorum Specimen.* Lisbon, 1960. (Coloured plates excerpted from the main work.)

CORTESÃO, J. (1). *A Política de Sigilo nos Descobrimentos nos Tempos do Infante Dom Henrique e de Dom João II.* Comissão Executiva das Comemorações do Quinto Centenário da Morte do Infante Dom Henrique. Lisbon, 1960. (Colecção Henriquina, no. 7.)

CORTHELL, E. L. (1). *The Tehuantepec Ship Railway.* Franklin Institute, Philadelphia, 1884.

DA COSTA, FONTOURA (1). *La Science Nautique des Portugais à l'Époque des Decouvertes.* (Report to the International Congress of the History of Science, Coimbra, 1934.) Agência Geral das Colónias, Lisbon, 1941. Reprinted in *Henri le Navigateur*, Lisbon, 1960 (Anon. (53), p. 161). Portuguese tr., with illustrations, *A Ciencia Náutica dos Portugueses na Epoca dos Descobrimentos.* Comissão Executiva das Comemorações do Quinto Centenário da Morte do Infante Dom Henrique, Lisbon, 1958. (Colecção Henriquina, no. 4.)

DA COSTA, FONTOURA (2). *A Marinharia dos Descobrimentos.* Armada, Lisbon, 1933. Crit. C. R. B[oxer], *MMI*, 1935, **21**, 214.

COULING, S. (1). *Encyclopaedia Sinica.* Kelly & Walsh, Shanghai; Oxford and London, 1917.

COUPLAND, R. (1). *East Africa and its Invaders, from the Earliest Times to the Death of Seyyid Said in 1856.* Oxford, 1938, 1956.

COURSE, A. G. (1). *A Dictionary of Nautical Terms.* Arco, London, 1962.

COUSTEAU, J. Y. (1). 'Fish-Men discover a 2200-year-old Greek Ship.' *NGM*, 1954, **105**, 1.

COUTINHO, ADM. GAGO (1). *A Náutica dos Descobrimentos; os Descobrimentos Maritimos vistos por um Navegador.* 2 vols. Lisbon, 1951–2.

COUVREUR, F. S. (1) (tr.). '*Tch'ouen Ts'iou*' [*Chhun Chhin*] et '*Tso Tchouan*' [*Tso Chuan*]; *Texte Chinois avec Traduction Française.* 3 vols. Mission Press, Hochienfu, 1914.

COUVREUR, F. S. (3) (tr.). '*Li Ki*' [*Li Chi*], *ou Mémoires sur les Bienséances et les Cérémonies.* 2 vols. Hochienfu, 1913.

COVARRUBIAS, M. (1). *Mexico South; the Isthmus of Tehuantepec.* Knopf, New York, 1946, 1947.

COVARRUBIAS, M. (2). *The Eagle, the Jaguar, and the Serpent; Indian Art of the Americas—North America (Alaska, Canada, the United States).* Knopf, New York, 1954.

COVARRUBIAS, M. (3). *Indian Art of Mexico and Central America.* Knopf, New York, 1957.

CRANMER-BYNG, J. L. (2) (ed.). *An Embassy to China; being the Journal kept by Lord Macartney during his Embassy to the Emperor Chhien-Lung, +1793 and +1794.* Longmans, London, 1962. Includes Macartney's Voyage Notes and Journal (1), his Observations on China (2), and the Observations of Dr Gillan (1) on the state of Medicine, Surgery and Chemistry in China.

CREALOCK, W. E. W. (1). 'A Living Epitome of the Genesis of the Rudder.' *ISRM*, 1938, **11**, 515.

CREEL, H. G. (1). *Studies in Early Chinese Culture* (1st series). Waverly, Baltimore, 1937.

CREEL, H. G. (2). *The Birth of China.* Fr. tr. by M. C. Salles, Payot, Paris, 1937. (References are to page numbers of the French ed.)

CREEL, H. G. (4). *Confucius; the Man and the Myth.* Day, New York, 1949; Kegan Paul, London, 1951. Rev. D. Bodde, *JAOS*, 1950, **70**, 199.

CRESSEY, G. B. (1). *China's Geographic Foundations; A Survey of the Land and its People.* McGraw-Hill, New York, 1934.

CRESSEY, P. F. (1). 'Chinese Traits in European Civilisation: a Study in Diffusion.' *AMSR*, 1945, **10**, 595.

CRESSWELL, K. A. C. (2). *Early Muslim Architecture.* 2 vols. London, 1932, 1940.

CRESSWELL, K. A. C. (3). *A Short Account of Early Muslim Architecture.* Pelican, London, 1958.

CREVENNA, T. R. (1) (ed.). *Irrigation Civilisations; a Comparative Study. A Symposium on Method and Result in Cross-Cultural Regularities* [1953]. Pan-American Union, Washington, D.C., 1955, reprinted 1960 (Social Science Monographs, no. 1). Contributions on Mesopotamian, East Asian, and Amerindian civilisations by J. H. Steward, R. McC. Adams, D. Collier, A. Palerm, K. A. Wittfogel & R. L. Beals.

CRONE, G. R. (1). *The Voyages of Cadamosto* [*Alvise Ca' da Mosto*]. London, 1937.

CROSS, H. & FREEMAN, J. R. (1) (ed.). *River Control and the Yellow River of China; a Collection of the Opinions of China.* 2 vols. Brown Univ., Providence, R.I., 1918.

CUFF, E. (1). 'The Naval Inventions of Charles, Third Earl Stanhope (1753 to 1816).' *MMI*, 1947, **33**, 106.

CUMMINS, J. S. (1) (ed.). *The Travels and Controversies of Friar Domingo Navarrete, +1618 to +1686.* 2 vols., Cambridge, 1962. (Hakluyt Society Pubs., 2nd series, nos. 118, 119.)

CURRY, MANFRED (1). *Yacht Racing; the Aerodynamics of Sails, and Racing Tactics.* Tr. from the German. Bell, London, 1928 (later editions 1930 and 1948). Germ. title, *Die Aerodynamik des Segels und die Kunst des Regatta-Segelns.* Fr. tr. *l'Aerodynamique de la Voile et l'Art de Gagner les Régates*, tr. P. B[udker]. Chiron, Paris, 1930 (2nd ed. 1949).

CURRY, MANFRED (2). *Clouds, Wind and Water* [photographs]. Country Life, London, 1951.

CURRY, MANFRED (3). *Wind and Water* [photographs]. Country Life, London, 1930.

CURWEN, E. C. (6). *Plough and Pasture.* Cobbett, London, 1946. (Past and Present, Studies in the History of Civilisation, no. 4.) Re-issued in Curwen & Hatt (1).

DALY, R. W. (1). *How the 'Merrimac' Won.* Crowell, New York, 1958.

DANIELL, T. & DANIELL, W. (1). *Oriental Scenery.* London, 1814.

DARBY, H. C. (1). *The Draining of the Fens,* 2nd ed. London, 1955.

DAREMBERG, C. & SAGLIO, E. (1). *Dictionnaire des Antiquités Grecques et Romains.* Hachette, Paris, 1875.

DARLING, S. T. (1). 'Observations on the Geographical and Ethnological Distribution of Hookworms.' *PARA,* 1920, **12**, 217.

DARLING, S. T. (2). 'Comparative Helminthology as an aid in the solution of Ethnological Problems.' *AJTM,* 1925, **5**, 323.

DARMSTÄDTER, L. (1) (with the collaboration of R. du Bois-Reymond & C. Schäfer). *Handbuch der Geschichte d. Naturwissenschaften u. d. Technik.* Springer, Berlin, 1908.

DARRAG, AHMAD (1). *L'Egypte sous le Règne de Barsbay (+1422 à +1438).* Inst. Fr. de Damas, Damascus, 1961.

DART, R. A. (1). 'A Chinese Character as a Wall Motive in Rhodesia.' *SAFJS,* 1939, **36**, 474.

DAS, SARAT CHANDRA (1). *Journey to Lhasa and Central Tibet,* ed. W. W. Rockhill. Murray, London, 1902.

DAUMAS, M. (2) (ed.). *Histoire de la Science; des Origines au XXe Siècle.* Gallimard, Paris, 1957. (Encyclopédie de la Pléiade series.)

DAVEY, N. (1). *A History of Building Materials.* Phoenix, London, 1961.

DAVIDSON, BASIL (1). *Old Africa Rediscovered.* Gollancz, London, 1959.

DAVIDSON, D. S. (1). 'The Snowshoe in Japan and Korea.' *ETH,* 1953, **18** (no. 1/2), 45.

DAVIDSON, D. S. (2). 'Snowshoes.' *MAPS,* 1937, no. 6.

DAVIES, A. & ROBINSON, H. (1). 'The Evolution of the Ship in relation to its Geographical Background.' *G,* 1939, **24**, 95.

DAVIES, R. M. (1). *Yunnan, the Link between India and the Yangtze.* Cambridge, 1909.

DAVIS, H. C. (1). 'Records of Japanese Vessels driven upon the Northwest Coast of America.' *PAAQS,* 1872, **1**, 1.

DAVIS, J. F. (1). *The Chinese; a General Description of China and its Inhabitants.* 1st ed. 1836. 2 vols. Knight, London, 1844, 3 vols., 1847, 2 vols. French tr. by A. Pichard, Paris, 1837, 2 vols. Germ. trs. by M. Wesenfeld, Magdeburg, 1843, 2 vols. and M. Drugulin, Stuttgart, 1847, 4 vols.

DAVIS, TENNEY L. & NAKASEKO ROKURO (1). 'The Tomb of Jofuku [Hsü Fu] or Joshi [Hsü Shih]; the Earliest Alchemist of Historical Record.' *AX,* 1937, **1**, 109.

DAVIS, TENNEY L. & NAKASEKO ROKURO (2). 'The Jofuku [Hsü Fu] Shrine at Shingu, a Monument of Earliest Alchemy.' *NU,* 1937, **15** (no. 3).

DAVISON, C. St C. (11). 'Bridges of Historical Importance.' *EN,* 1961, **211**, 196.

DAWSON, H. CHRISTOPHER (1). *Progress and Religion; an Historical Enquiry.* Sheed & Ward, London, 1929.

DAWSON, R. (1) (ed.). *The Legacy of China.* Oxford, 1964.

DEANE, C. D. (1). 'An Open-Air Museum of the Future' (the Ulster coastal nature preservation project). *UC,* 1967 (no. 258), 10.

DEANS, R. (1). *The Preservative Action of Spices; a Review of the Literature.* Food Research Reports of the British Food Manufacturers Research Association, 1945, no. 53 (confidential, supplied in principle to members only).

DEBENHAM, F. (1). *Discovery and Exploration; an Atlas-History of Man's Journeys into the Unknown.* With introduction by E. Shackleton. Belser, Stuttgart, 1960; Hamlyn, London, 1960.

DEFRÉMERY, C. & SANGUINETTI, B. R. (1) (tr.). *Voyages d'ibn Batoutah.* 5 vols. Soc. Asiat., Paris, 1853–9. (Many reprints.)

DELLON, C. (1). *Relation de l'Inquisition de Goa.* Leiden, 1687; Amsterdam, 1719, 1737; Köln, 1759 etc. Eng. tr. H. Wharton, London, 1688; repr. 1815.

DEMBER, H. & UIBE, M. (1). (a) 'Versuch einer physikalischen Lösung des Problems der sichtbaren Grössenänderung von Sonne und Mond in verschiedenen Höhen über dem Horizont.' *ANP,* 1920 (4th ser.), **61**, 353. (b) 'Über die Gestalt des sichtbaren Himmelsgewälbes.' *ANP,* 1920 (4th ser.), **61**, 313.

DEMIÉVILLE, P. (4). General account of the Chinese architectural literature; in a review of the 1920 edition of the *Ying Tsao Fa Shih. BEFEO,* 1925, **25**, 213–64.

DENNIS, G. (1). *Cities and Cemeteries of Etruria.* 2 vols. 3rd ed. Murray, London, 1883.

DESCHAMPS, H. (1). *Histoire de Madagascar*. Berger-Levrault, Paris, 1960.

DICK, T. L. (1). 'On a Spiral Oar.' *TAP*, 1818, **11**, 438.

DICKINS, F. V. (1). *Fugaku Hiyaku-Kei; or, A Hundred Views of [Mt.] Fuji(-yama), by Hokusai; Introduction and Explanatory Prefaces, with Translations from the Japanese, and Descriptions of the Plates.* Batsford, London, 1880. Bound in Japanese style to accompany the three volumes of the original printed in 1875. 4 vols. in all.

DICKINSON, R. L. (1). 'Sketching Boats on the China Coast.' *PCC*, 1931, **7**, 97.

DIEHL, C. (1). 'Byzantine Art.' Art. in *Byzantium*. Ed. N. H. Baynes & H. St L. B. Moss, p. 166. Oxford, 1949.

DIELS, H. (1). *Antike Technik*. Teubner, Leipzig & Berlin, 1914; enlarged 2nd ed., 1920 (rev. B. Laufer, *AAN*, 1917, **19**, 71).

VON DIEZ, H. F. (1) (tr.). Translation of *The Mirror of the Countries* in *Denkwürdigkeiten von Asien*, vol. 2, pp. 733 ff. French tr. by M. Morris; *Miroir des Pays, ou Relation des Voyages de Sidi Aly fils d'Housaïn nommé ordinairement Katibi-Roumy, amiral de Soliman II [Mir'at al-Mamālik]*. *JA*, 1826 (1e sér.), **9**, 2. Eng. tr. by Vambéry, see Vambéry (1).

DIMMOCK, L. (1). 'The Chinese "Yuloh".' *MMI*, 1954, **40**, 79.

DINSMOOR, W. B. (1). *The Architecture of Ancient Greece*. Batsford, London, 1950.

DIN TA-SAN & MUNIDO F. OLESA (1). *El Poder Naval Chino desde su Origine hasta Caida de la Dinastia Ming*. Barcelona, 1965.

DIOLÉ, P. (1). *Promenades d'Archéologie Sous-Marine*. Michel, Paris, 1952. Eng. tr. by G. Hopkins, *Four Thousand Years under the Sea*. Sidgwick & Jackson, London, 1954.

DIOLÉ, P. (2). *Under-water Exploration*. Tr. from the French by H. M. Burton. Elek, London, 1954.

DOE, D. B. (1). 'Pottery Sites near Aden.' *JRAS*, 1963, 150. Repr., without pagination but with some additional illustrations, together with Lane & Serjeant (1), in *ADAB*, 1965, no. 5.

DOLLEY, R. H. (1). 'The Rig of Early Mediaeval Warships.' *MMI*, 1949, **35**, 51. Crit. R. le B. Bowen, 1950, **36**, 88; reply by Dolley, p. 158; rejoinder by Bowen to Dolley (1) and (2), 1953, **39**, 224; further reply by Dolley, 1954, **40**, 76 and comment by R. C. Anderson, p. 77; long justification by Bowen, 1954, **40**, 315, comment by R. C. Anderson, 1955, **41**, 67; continuation of discussion by D. W. Waters, 1956, **42**, 147.

DOLLEY, R. H. (2). 'The "Nef" Ships of the Ravenna Mosaics [c. +1204].' *MMI*, 1952, **38**, 315.

DONNELLY, I. A. (1). *Chinese Junks and other Native Craft*. Kelly & Walsh, Shanghai, 1924. Critique and rev. H. H. B[rindley], *MMI*, 1923, **9**, 158, 318; 1925, **11**, 331.

DONNELLY, I. A. (2). 'Early Chinese Ships and Trade.' *CJ*, 1925, **3**, 190; *MMI*, 1925, **11**, 344.

DONNELLY, I. A. (3). *Chinese Junks, a Book of Drawings in Black and White*. Kelly & Walsh, Shanghai, 1924.

DONNELLY, I. A. (4). 'River Craft of the Yangtzekiang.' *MMI*, 1924, **10**, 4.

DONNELLY, I. A. (5). 'Fuchow Pole Junks.' *MMI*, 1923, **9**, 226.

DONNELLY, I. A. (6). 'Strange Craft of Chinese Inland Waters.' *MMI*, 1936, **22**, 410.

DOOLITTLE, J. (2). '[Glossary of Chinese] Shipping and Nautical Terms.' In Doolittle, J. (1), II, p. 557.

DOORMAN, G. (1). *Techniek en Octrooiwezen in hun Aanvang*. Nijhoff, The Hague, 1953.

DORAN, E. (1). 'The Origin of Leeboards.' *MMI*, 1967, **53**, 39. Discussion: Lord Riverdale, W. A. King-Webster, J. A. Clare, pp. 142, 170, 209.

DORÉ, H. (1). *Recherches sur les Superstitions en Chine*. 15 vols. T'u-Se-Wei Press, Shanghai, 1914–29.
    Pt. I, vol. 1, pp. 1–146: 'Superstitious' practices, birth, marriage and death customs (*VS*, no. 32).
    Pt. I, vol. 2, pp. 147–216: talismans, exorcisms and charms (*VS*, no. 33).
    Pt. I, vol. 3, pp. 217–322: divination methods (*VS*, no. 34).
    Pt. I, vol. 4, pp. 323–488: seasonal festivals and miscellaneous magic (*VS*, no. 35).
    Pt. I, vol. 5, sep. pagination: analysis of Taoist talismans (*VS*, no. 36).
    Pt. II, vol. 6, pp. 1–196: Pantheon (*VS*, no. 39).
    Pt. II, vol. 7, pp. 197–298: Pantheon (*VS*, no. 41).
    Pt. II, vol. 8, pp. 299–462: Pantheon (*VS*, no. 42).
    Pt. II, vol. 9, pp. 463–680: Pantheon, Taoist (*VS*, no. 44).
    Pt. II, vol. 10, pp. 681–859: Taoist celestial bureaucracy (*VS*, no. 45).
    Pt. II, vol. 11, pp. 860–1052: city-gods, field-gods, trade-gods (*VS*, no. 46).
    Pt. II, vol. 12, pp. 1053–1286: miscellaneous spirits, stellar deities (*VS*, no. 48).
    Pt. III, vol. 13, pp. 1–263: popular Confucianism, sages of the Wên miao (*VS*, no. 49).
    Pt. III, vol. 14, pp. 264–606: popular Confucianism historical figures (*VS*, no. 51).
    Pt. III, vol. 15, sep. pagination: popular Buddhism, life of Gautama (*VS*, no. 57).

DOZY, R. & DE GOEJE, M. J. (1) (tr.). *Description de l'Afrique et de l'Espagne par Idrisi*. Leiden, 1866. Partial translation of Al-Idrīsī's *Nuzhat al-Mushtāq-fī Ikhtirāq al-Āfāq* (Recreation of those who long to know what is beyond the Horizons, +1154). Cf. Jaubert (1).

DRACHMANN, A. G. (7). 'Ancient Oil Mills and Presses.' *KDVS/AKM*, 1932, **1** (no. 1). Sep. publ. Levin & Munksgaard, Copenhagen, 1932.

DRACHMANN, A. G. (9). *The Mechanical Technology of Greek and Roman Antiquity; a Study of the Literary Sources*. Munksgaard, Copenhagen, 1963.

DRAPELLA, W. A. (1). *Ster; ze Studiów nad Kształtowaniem się Pojęć Morskich, Wiek* XV–XX. Zakład Imienia Ossolińskich we Wrocławiu, Gdańsk, 1955. (Towarzystwo Przyjaciół Nauki i Sztuki w Gdańsku — Komisja Morska — Podkomisja Językowa. Prace i Materiały z Zakresu Polskiego Słownictwa Morskiego, ed. L. Zabrocki, no. 3.)

DRAPELLA, W. A. (2). *Żegluga — Nawigacja — Nautika; ze Studiów nad Kształtowaniem się Pojęć Morskich, Wiek* XVI–XVIII. Zakład Imienia Ossolińskich we Wrocławiu, Gdańsk, 1955. (Towarzystwo Przyjaciół Nauki i Sztuki w Gdańsku — Komisja Morska — Podkomisja Językowa. Prace i Materiały z Zakresu Polskiego Słownictwa Morskiego, ed. L. Zabrocki, no. 5.)

DRESBECK, LEROY J. (1). 'The Ski; its History and Historiography.' *TCULT*, 1967, **8**, 467.

DROVER, C. B., SABINE, P. A., TYLER, C. & COOLE, P. G. (1). 'Sand-Glass "Sand"'; Historical, Analytical and Practical.' *AHOR*, 1960.

DROWER, M. S. (1). 'Water-Supply, Irrigation and Agriculture [from Early Times to the End of the Ancient Empires].' Art. in *A History of Technology*, ed. C. Singer *et al.* vol. 1, p. 520. Oxford, 1954.

DRUMMOND, CAPT. (1). 'Drawings of Chinese Junks brought home by Capt. Drummond, in the East India Company's Service.' MS. Album of about 1800 preserved at the National Maritime Museum, Greenwich.

DRUMMOND, SIR JACK C. & WILBRAHAM, A. (1). *The Englishman's Food; Five Centuries of English Diet*. Cape, London, 1939.

DUBS, H. H. (2) (tr., with assistance of Phan lo-Chi and Jên Thai). *History of the Former Han Dynasty, by Pan Ku, a Critical Translation with Annotations*. 2 vols. Waverly, Baltimore, 1938.

DUBS, H. H. (3). 'The Victory of Han Confucianism.' *JAOS*, 1938, **58**, 435. (Reprinted in Dubs (2), pp. 341 ff.)

DUBS, H. H. (4). 'An Ancient Chinese Stock of Gold [Wang Mang's Treasury].' *JEH*, 1942, **2**, 36.

DUBS, H. H. (5). 'The Beginnings of Alchemy.' *Isis*, 1947, **38**, 62.

DUBS, H. H. (7). *Hsün Tzu; the Moulder of Ancient Confucianism*. Probsthain, London, 1927.

DUFF, R. (1). 'The Moa-Hunter Period of Maori Culture.' *CMB*, 1950 (no. 1).

DUKES, E. J. (1). *Everyday Life in China; or Scenes along River and Road in Fukien*. London, 1885. (Gives an account of Tshai Hsiang and the building of the Loyang or Wan-an megalithic beam bridge at Chhüanchow, pp. 144–50.)

DUMONTIER, G. (2). 'Les Cultes Annamites.' *RI*, 1905, 690. Sep. pub. Schneider, Hanoi, 1905.

DUMONTIER, G. (3). 'Essai sur les Tonkinois.' *RI*, 1907, 454.

DUNLOP, D. M. (1). *The History of the Jewish Khazars*. Princeton Univ. Press, Princeton, N.J., 1954.

DUNLOP, D. M. (3). 'Burtuḳāl.' Art. in *Encyclopaedia of Islam*, p. 1338.

DUNLOP, D. M. (4). 'The British Isles according to Mediaeval Arabic Authors.' *IQ*, 1957, **4**, 11.

DUPONT, M. (1). *Les Meubles de la Chine*. 2nd series. Paris n.d. (1950). See Roche.

DURAN-REYNALS, M. L. (1). *The Fever-Bark Tree*. Allen, London, 1947.

DUYVENDAK, J. J. L. (1). 'Sailing Directions of Chinese Voyages' (a Bodleian Library MS.). *TP*, 1938, **34**, 230.

DUYVENDAK, J. J. L. (3) (tr.). *The Book of the Lord Shang; A Classic of the Chinese School of Law*. Probsthain, London, 1928.

DUYVENDAK, J. J. L. (8). *China's Discovery of Africa*. Probsthain, London, 1949. (Lectures given at London University, Jan. 1947; rev. P. Paris, *TP*, 1951, **40**, 366.)

DUYVENDAK, J. J. L. (9). 'The True Dates of the Chinese Maritime Expeditions in the Early Fifteenth Century.' *TP*, 1939, **34**, 341.

DUYVENDAK, J. J. L. (10). 'Ma Huan Re-examined.' *VKAWA/L*, 1933 (n.s.), **32**, no. 3.

DUYVENDAK, J. J. L. (11). 'Voyages de Tchêng Houo [Chêng Ho] à la Côte Orientale d'Afrique, +1416 à +1433.' In Yusuf Kamal, *Monumenta Cartographica*, 1939, vol. 4, pt. 4, pp. 1411 ff.

DUYVENDAK, J. J. L. (19). 'Desultory Notes on the *Hsi Yang Chi* [Lo Mou-Têng's novel of +1597 based on the Voyages of Chêng Ho]' (concerns spectacles and bombards). *TP*, 1953, **42**, 1.

DYE, D. S. (1). *A Grammar of Chinese Lattice*. 2 vols. Harvard-Yenching Institute, Cambridge, Mass., 1937. (Harvard-Yenching Monograph Series, nos. 5, 6.)

DYE, D. S. (2). 'Some Elements of Chinese Architecture, with Notes on Szechuan Specialities.' *JWCBRS*, 1926, **3**, 162.

EASTWOOD, T. (1). 'Roofing Materials through the Ages.' *PGA*, 1951, **62**, 6.

EBERHARD, W. (2). *Lokalkulturen im alten China*. Pt. 1, Northern and Western, *TP* (Suppl.), 1943, **37**, 1–447; Pt. 2, Southern and Eastern, *MS*, Monograph no. 3, 1942. (Crit. H. Wilhelm, *MS*, 1944, **9**, 209.)

EBERHARD, W. (3). 'Early Chinese Cultures and their Development, a Working Hypothesis.' *ARSI*, 1937, 513. (Pub. no. 3476.)

EBERHARD, W. (5) (coll. and tr.). *Chinese Fairy Tales and Folk Tales.* Kegan Paul, London, 1937.

EBERHARD, W. (9). *A History of China from the Earliest Times to the Present Day.* Routledge & Kegan Paul, London, 1950. Tr. from the Germ. ed. (Swiss pub.) of 1948 by E. W. Dickes. Turkish ed. *Çin Tarihi,* Istanbul, 1946. (Crit. K. Wittfogel, *AA,* 1950, **13**, 103; J. J. L. Duyvendak, *TP,* 1949, **39**, 369; A. F. Wright, *FEQ,* 1951, **10**, 380.)

EBERHARD, W. (20). 'Chinesische Bauzauber.' *ZFE,* 1940, **71**, 87.

EBERHARD, W. (21). *Conquerors and Rulers; Social Forces in Mediaeval China* (theory of gentry society). Brill, Leiden, 1952. Crit. E. Balazs, *ASEA,* 1953, **7**, 162; E. G. Pulleyblank, *BLSOAS,* 1953, **15**, 588.

EBERHARD, W. (24). 'Zweiter Bericht über die Ausgrabungen bei Anyang.' *OAZ,* 1933, **19**, 208.

ECKE, G. V. (2). 'Wandlungen des Faltstuhls; Bernerkungen Z. Geschichte d. Eurasischen Stuhlform.' *MS,* 1944, **9**, 34.

ECKE, G. V. (3). 'Chiang Tung Chhiao; eine Brücke in Sud-Fukien aus der Zeit d. Nan Sung.' *OAZ,* 1929, **15**, 110.

ECKE, G. V. (4). 'The Institute for Research in Chinese Architecture; Summary of Field Work 1932–1937.' *MS,* 1937, **2**, 468.

ECKE, G. V. (5). 'Zaytonische Granitbrücker; ihr Schmuck u. ihre Heiligtümer.' *SA,* 1931, **6**, 270, 296.

ECKE, G. V. (6). *Chinese Domestic Furniture.* Peking, 1944.

ECKE, G. V. (7). 'Contributions to the study of Sculpture and Architecture; Shen-Thung Ssu and Ling-Yen Ssu [both in Shantung] once more.' *MS,* 1942, **7**, 295.

ECKE, G. V. & DEMIÉVILLE, P. (1). *The Twin Pagodas of Zayton; a Study of Later Buddhist Sculpture in China.* Harvard-Yenching Institute, Peking; Harvard University Press, Cambridge, Mass., 1935. (Harvard-Yenching Monograph Series, no. 2.) Rev. J. B[uhot], *RAA/AMG,* 1935, **9**, 237.

EDEN, C. H. (1). *Black Tournai Fonts in England.* Elliot Stock, London, 1909.

EDGAR, J. H. (2). 'From Ta-tsien-lu to Mu-phing via Yü-thung.' *CJ,* 1932, **17**, 282.

EDKINS, J. (12). 'Chinese Names for Boats and Boat Gear; with Remarks on the Chinese Use of the Mariner's Compass.' *JRAS/NCB,* 1877, **11**, 123. (Rev. *CR,* 1877, **6**, 128.)

EDKINS, J. (15). 'Chinese Architecture.' *JRAS/NCB,* 1890, **24**, 253.

EDKINS, J. & GREGORY, W. (1). 'Bridges in China.' *CR,* 1896, **22**, 738.

EDWARDS, MAJOR (1). 'Extracts from a Report on the Present Condition of the [Chhien-Thang] Sea Wall.' *JRAS/NCB,* 1865, **1**, 136.

EGGERS, G. (1). 'Wasserversorgungstechnik im Altertum.' *BGTI/TG,* 1936, **25**, 1.

D'EICHTHAL, G. (1). 'Des Origines Asiatico-Bouddhiques de la Civilisation Américaine.' *RA,* 1864, **10**, 187, 370; 1865, **11**, 42, 273, 486.

EIGNER, J., ALLEY, R. *et al.* (1). 'China's Inland Waterways [Photographs].' *CJ,* 1937, **26**, 250.

EKHOLM, G. F. (2). 'A Possible Focus of Asiatic Influence in the late Classic Cultures of Meso-america.' Art. in *MSAA,* no. 9; suppl. to *AMA,* 1953, **18** (no. 3), p. 72.

EKHOLM, G. F. (3). 'Is American Culture Asiatic?' *NH,* 1950, **59**, 344.

EKHOLM, G. F. (4). 'The Possible Chinese Origin of Teotihuacan Cylindrical Tripod Pottery and Certain Related Traits.' Proc. 35th Internat. Congress of Americanists. Mexico City, 1962. Vol. 1, p. 39.

ELGAR, F. (1). 'Japanese Shipping.' *TJSL,* 1895, **3**, 59.

ELGOOD, C. (1). *A Medical History of Persia and the Eastern Caliphate from the Earliest Times....* Cambridge, 1951.

D'ELIA, PASQUALE (1). 'Echi delle Scoperte Galileiane in Cina vivente ancora Galileo (1612–1640).' *AAL/RSM,* 1946 (8ᵉ ser.), **1**, 125. Republished in enlarged form as 'Galileo in Cina. Relazioni attraverso il Collegio Romano tra Galileo e i gesuiti scienzati missionari in Cina (1610–1640).' *Analecta Gregoriana,* **37** (Series Facultatis Missiologicae A (N/I)) Rome, 1947. Reviews: G. Loria, *A/AIHS,* 1949, **2**, 513; J. J. L. Duyvendak, *TP,* 1948, **38**, 321; G. Sarton, *ISIS,* 1950, **41**, 220.

D'ELIA, PASQUALE (2) (ed.). *Fonti Ricciane; Storia dell'Introduzione del Cristianesimo in Cina.* 3 vols. Libreria dello Stato, Rome, 1942–9. Cf. Trigault (1); Ricci (1).

ELIASSEN, S. (1). *Dragon Wang's River.* Methuen, London, 1957. Tr. by K. John from the Norwegian: *Gamle Drage Wangs Elv.* Gyldendal, Oslo, 1955.

ELIASSEN, S. & TODD, O. J. (1). 'The Wei River Project.' *CJ,* 1932, **17**, 170.

ELISSÉEV, S. (1) (tr.). 'La Révélation des Secrets de la Peinture [by Wang Wei, +699/+759].' *RAA/AMG,* 1927, **4**, 212.

ELLER, E. M. (1). 'Troubled Waters East of Suez.' *NGM,* 1954, **105**, 483 (colour photographs of qanats).

ELLIOTT, J. A. G. (1). 'A Visit to the Bajun Islands.' *JAFRS,* 1926, **25**, 10, 147, 245 (261), 338.

ELLIOTT-SMITH, SIR GRAFTON (2). 'Ships as Evidence of the Migrations of Early Culture.' *JMEOS,* 1916, 63 and sep. pub. Manchester Univ. Press, 1917.

ENGEL, H. (1), with introduction by W. GROPIUS. *The Japanese House; a Tradition for Contemporary Architecture*. Tuttle, Rutland, Vt. and Tokyo, 1964.

D'ENTRECOLLES, F. X. (1). *Lettre au Père Duhalde* (on alchemy and various Chinese discoveries in the arts and sciences, porcelain, artificial pearls and magnetic phenomena) 4 Nov. 1734. *LEC*, vol. 22, pp. 91 ff.

ENTWISTLE, C. (1). 'How to use the Modulor.' *AD*, 1953, **23**, 72.

VON ERDBERG, E. (1). *Chinese Influence on European Garden Structure*. Cambridge, Mass., 1936.

ERKES, E. (17). 'Chinesische-Amerikanische Mythenparallelen.' *TP*, 1925, **24**, 32.

VON ERLACH, J. B. FISCHER (1). *Historia Architectur*. 2nd ed. Leipzig, 1725. 1st ed. von Erlachen, J. B. Fischer: *Entwürff einer historischen Architectur. In Abbildung unt erschiedener beruhmten Gebaüde des Alterthums und fremder Völcker...In dem Ersten Büche, die von der Zeit vergrabene Bau- arten der Alten, Jüden Egÿptier, Syrer, Perser, und Griechen. In dem Andren, Alte unbekante Römische. In dem Drütten, einige fremde, in- und aüser-Europaische, als die Araber, und Turcken etc. auche neüe Persian- ische, Siamitische, Sinesische, und Japonesische Gebaüde. In dem Vierten, einige Gebaüde von des Autoris Erfindung und Zeichnung*. Fischer, Vienna, 1721. (A fifth volume, not mentioned in the title, depicted *Divers Vases Antiques...et Modernes*.)

VAN ERP, T. (1). 'Voorstellingen van vaartuigen op de Reliefs van den Boroboedoer.' *NION*, 1923, **8**, 227. (English summary in Krom & van Erp, II, p. 235.)

ESCHER, M. C. (1). *The Graphic Work of M. C. Escher* (album). Oldbourne, London, 1961.

ESPÉRANDIEU, E. (1). *Souvenir du Musée Lapidaire de Narbonne*. Commission Archéologique, Narbonne, n.d.

ESPÉRANDIEU, E. (2). *Receuil Général des Bas-Reliefs, Statues et Bustes de la Gaule Romaine*. Imp. Nat., Paris, 1908, 1913.

ESPINAS, G. (1). 'Comment on Faisait un Canal au XVIIIe Siècle; le Canal de Briare.' *AHES/AESC*, 1946, **1**, 347.

ESTERER, M. (1). *Chinas natürliche Ordnung und die Maschine*. Cotta, Stuttgart and Berlin, 1929. (Wege d. Technik Series.)

ESTRADA, E. & EVANS, C. (1). 'Cultural Development in Ecuador.' Art. in *Aboriginal Cultural Develop- ment in Latin America; an Interpretative Review*, p. 77. Ed. B. J. Meggers & C. Evans.

ESTRADA, E. & MEGGERS, B. J. (1). 'A Complex of Traits of Probable Trans-Pacific Origin on the Coast of Ecuador.' *AAN*, 1961, **63**, 913.

ESTRADA, E., MEGGERS, B. J. & EVANS, C. (1). 'Possible Trans-Pacific Contact on the Coast of Ecuador.' *S*, 1962, **135**, 371.

EWBANK, T. (1). *A Descriptive and Historical Account of Hydraulic and other Machines for Raising Water, Ancient and Modern*.... Scribner, New York, 1842. (Best ed. the 16th, 1870.)

FABRE, M. (1). *Pékin, ses Palais, ses Temples, et ses Environs*. Librairie Française, Tientsin, 1937.

FABRI, C. (1). *An Introduction to Indian Architecture*. Asia, London, 1963.

FAIRBANK, WILMA (1). 'A Structural Key to Han Mural Art.' *HJAS*, 1942, **7**, 52.

FANG HUA-YUNG (1). 'Soil Conservation on Loess Highlands.' *CREC*, 1962, **11** (no. 7), 15.

FANNING, A. E. CDR. (1). Note on the use of the compass-rose at Sagres. *JBASA*, 1959, **69**, 272.

FANSHAWE, RICHARD (1) (tr.). *The Lusiad, or Portugalls Historicall Poem*.... Moseley, London, 1655. Ed. and repr. J. D. M. Ford. Harvard Univ. Press, Cambridge, Mass., 1940.

FARRER, R. (1). *On the Eaves of the World*. 2 vols. Arnold, London, 1917.

FARRER, R. (2). *The Rainbow Bridge*. Arnold, London, 1926.

FARRÈRE, C. & FONQUERAY, C. [pseudonym for F. C. BARGONE]. *Jonques et Sampans*. Horizons de France, Paris, 1945.

FAULDER, H. C. (1). 'Chinese Model Junks.' *CJ*, 1938, **28**, **29**.

FAVIER, A. (1). *Pékin; Histoire et Description*. Peking, 1897. For Soc. de St Augustin, Desclée & de Brouwer, Lille, 1900.

FEBVRE, L. (3). Editorial on the problem of the invention of the stern-post rudder (cf. des Noëttes (2); la Roerie (1)]. *AHES*, 1935, **7**, 536.

FEBVRE, L. (4). 'Toujours le Gouvernail.' *AHES/MHS*, 1941, **2**, 60.

FELBER, R. (1). *Die Reformen des Shang Yang und das Problem der Sklaverei in China*. Contribution to the 2nd Conference of Ancient Historians, Stralsund, 1962.

FELDHAUS, F. M. (1). *Die Technik der Vorzeit, der Geschichtlichen Zeit, und der Naturvölker* [encyclo- paedia]. Engelmann, Leipzig and Berlin, 1914. Photographic reprint; Liebing, Würzburg, 1965.

FELDHAUS, F. M. (2). *Die Technik d. Antike u. d. Mittelalter*. Athenaion, Potsdam, 1931. (Crit. H. T. Horwitz, *ZHWK*, 1933, **13** (N.F. 4), 170.)

FELDHAUS, F. M. (5). *Zur Geschichte d. Drahtseilschwebebahnen*. Zillessen, Berlin-Friedenau, 1911. (Monogr. z. Gesch. d. Technik, no. 1; ed. Quellenforschungen z. Gesch. d. Technik und Wissen- schaften; Schriftleitung F. M. Feldhaus & C. von Klinckowström.)

FELDHAUS, F. M. (18). *Leonardo der Techniker u. Erfinder*. Diederichs, Jena, 1913.

FELDHAUS, F. M. (24). *Geschichte des technischen Zeichnens*. 2nd ed., with E. Schruff. Kuhlmann, Wilhelmshafen, 1959. (Rev. R. S. Hartenberg, *TCULT*, 1961, **2**, 45; Eng. tr. in *GRSCI*, 1960.)

FÊNG, H. D. (1). *Bibliography on the Land Problems of China* (deals especially with the *ching thien* system). *NSEQ*, 1935, **8**, 325.

FÊNG YU-LAN (2). *The Spirit of Chinese Philosophy*, tr. E. R. Hughes. Kegan Paul, London, 1947.

FERGUSON, D. (1). 'Letters from Portuguese Captives in Canton.' *IAQ*, 1901, **30**, 421, 467; 1902, **31**, 10, 53.

FERGUSON, J. C. (6). 'Transportation in Early China.' *CJ*, 1929, **10**, 227.

FERGUSSON, J. (1). *History of Indian and Eastern Architecture*. 2 vols. Murray, London, 1910.

FERRAND, G. (1). *Relations de Voyages et Textes Géographiques Arabes, Persans et Turcs relatifs à l'Extrême Orient, du 8e au 18e Siècles, traduits, revus et annotés etc.* 2 vols. Leroux, Paris, 1913.

FERRAND, G. (2) (tr.). *Voyage du Marchand Sulaymān en Inde et en Chine redigé en +851; suivi de remarques par Abū Zayd Ḥaṣan (vers +916)*. Bossard, Paris, 1922.

FERRAND, G. (3). 'Le K'ouen-Louen [Khun-Lun] et les Anciennes Navigations Interocéaniques dans les Mers du Sud.' *JA*, 1919 (11e ser.), **13**, 239, 431; **14**, 5, 201.

FERRAND, G. (4). 'Malaka, le Malāyu et Malāyur.' *JA*, 1918 (11e ser.), **11**, 391; **12**, 148.

FERRAND, G. (5). 'Les Relations de la Chine avec le Golfe Persique avant l'Hégire.' Art. in *Mélanges offerts à Gaudefroy-Demombynes par ses amis et anciens élèves, 1935–1945*. Maisonneuve, Paris, 1945.

FERRAND, G. (6). *Instructions Nautiques et Routiers Arabes et Portugais des 15e et 16e Siècles*. 2 vols. Geuthner, Paris, 1921–5. Vols. 1 and 2 *Le Pilote des Mers de l'Inde, de la Chine et de l'Indonesie;* facsimile texts of MSS. of Shihāb al-Dīn Aḥmad al-Mājid (*c.* +1475) and of Sulaimān al-Mahrī (*c.* +1511).

FERRAND, G. (7). *Instructions Nautiques et Routiers Arabes et Portugais des 15e et 16e Siècles*. Geuthner, Paris, 1928. Vol. 3 *Introduction à l'Astronomie Nautique Arabe*. Consists of reprints of Prinsep (2, 3); Congreve (1); de Saussure (35, 36), and excerpts from Reinaud & Guyard (1), etc., with biographies of Ibn Mājid and al-Mahrī by Ferrand.

FESSLER, L. *et al.* (1). *China*. Time-Life International, Amsterdam, 1968 (pr. Verona, Italy).

FIDLER, T. C. (1). *A Practical Treatise on Bridge Construction*. Griffin, London, 1887.

FILESI, T. (1). *I Viaggi dei Cinesi in Africa nel Medioevo*. Ist. Ital. per l'Africa, Rome, 1961.

FILESI, T. (2). *Le Relazioni della Cina con l'Africa nel Medio-Evo*. Giuffré, Milan, 1962.

FILGUEIRAS, O. L. (1). *Rabões da Esquadra Negra* (the barcos rabelos of the Douro River). Porto, 1956.

FILLIOZAT, J. (2). 'Les Origines d'une Technique Mystique Indienne.' *RP*, 1946, **136**, 208.

FILLIOZAT, J. (3). 'Taoisme et Yoga.' *DVN*, 1949, **3**, 1.

FISCHER, E. S. (1). *The Sacred Wu-Thai Shan, in connection with Modern Travel from Thai-yuan Fu via Mount Wu-Thai to the Mongolian Border*. 1925.

FISCHER, J. (1). 'Fan Chung-Yen (+989/+1052); das Lebensbild eines chinesischen Staatsmannes.' *OE*, 1955, **2**, 39, 142.

FISCHER, OTTO (2). *Chinesische Malerei der Han Dynastie*. Neff, Berlin, 1931.

FISHER, B. (1). 'The *Qanāts* of Persia.' *GR*, 1928, **18**, 302.

FISHER, W. E. (1). 'Wings over China.' *CJ*, 1937, **26**, 250.

FITCH, R. F. (1). 'Life Afloat in China' [populations living on boats in rivers and ports]. *NGM*, 1927, **51**, 665.

FITZGERALD, C. C. P. (CAPTAIN, R.N.) (1). *Boat Sailing and Racing*. Griffin, Portsmouth, 1888.

FITZGERALD, C. P. (1). *China; a Short Cultural History*. Cresset Press, London, 1935.

FITZGERALD, C. P. (6). 'Boats of the Erh Hai Lake, Yunnan.' *MMI*, 1943, **29**, 135.

FITZGERALD, C. P. (7a). 'A Chinese Discovery of Australia?' Art. in *Australia Writes*, p. 76. Ed. T. Inglis Moore. Cheshire, Melbourne, 1953.

FITZGERALD, C. P. (7b). 'Evidence of a Chinese Discovery of Australia before European Settlement.' Paper delivered to the 23rd International Congress of Orientalists, Cambridge, 1954 (27th Aug.). Followed by a correspondence in *The Times* to which contributions were made by A. Christie, Sir Percival David, F. H. Dampier Atkinson, J. Bastin and others (1st to 7th Sept.). Printed as abstract in *Proceedings of the 23rd Congress*, p. 293.

FITZGERALD, C. P. (9). *Son of Heaven; a Biography of Li Shih-Min, Founder of the Thang Dynasty*. Cambridge, 1933.

FITZGERALD, C. P. (10). *Barbarian Beds; the Origin of the Chair in China*. Cresset, London, 1965.

FLETCHER, B. (1). *A History of Architecture on the Comparative Method*. Batsford, London, 1948.

DE FLINES, E. W. V. O. (1). 'De Keramische Verzameling [d. Koninklijke Bataviaasch Genootschap van Kunsten en Wetenschappen].' *KBGJ*, 1936, **3**, 206; 1937, **4**, 173; 1938, **5**, 159.

FLORANGE, C. (1). *Études sur les Messageries et les Postes, d'après les Documents Métalliques et Imprimés, précédée d'un Essai Numismatique sur les Ponts et Chaussées*. Florange, Paris, 1925.

FLOWER, B. & ROSENBAUM, E. (1) (tr.) (with contributions by V. Scholderer, J. Liversidge & K. Wilczynski). *The Roman Cookery Book; a critical translation of 'The Art of Cooking'* [Artis Magiricae] *by Apicius, for use in the Study and the Kitchen.* Harrap, London, 1958.

DA FONSECA, QUIRINO (1). *Os Navios do Infante Dom Henrique.* Comissão Executiva das Comemorações do Quinto Centenario da Morte do Infante Dom Henrique, Lisbon, 1958. (Colecção Henriquina, no. 5.)

FORBES, R. J. (2). *Man the Maker; a History of Technology and Engineering.* Schuman, New York, 1950. (Crit. rev. H. W. Dickinson & B. Gille, *A/AIHS*, 1951, **4**, 551.)

FORBES, R. J. (6). *Notes on the History of Ancient Roads and their Construction.* A. P. Stichting, Amsterdam, 1934; Brill, Leiden, 1934. [Archaeologische-Historische Bijdragen d. Allard Pierson Stichting, Amsterdam, no. 3.] Crit. revs. R. F. Jessup, *AQ*, 1935, **9**, 381; I. A. Richmond, *JRS*, 1935, **25**, 114.

FORBES, R. J. (10). *Studies in Ancient Technology.* Vol. 1, *Bitumen and Petroleum in Antiquity; The Origin of Alchemy; Water Supply.* Brill, Leiden, 1955. (Crit. Lynn White, *ISIS*, 1957, **48**, 77.)

FORBES, R. J. (11). *Studies in Ancient Technology.* Vol. 2, *Irrigation and Drainage; Power; Land Transport and Road-Building; The Coming of the Camel.* Brill, Leiden, 1955. (Crit. Lynn White, *ISIS*, 1957, **48**, 77.)

FORBES, R. J. (17). 'Hydraulic Engineering and Sanitation [in the Mediterranean Civilisations and the Middle Ages].' Art. in *A History of Technology*, ed. C. Singer *et al.*, vol. 2, p. 663. Oxford, 1956.

FORBES, R. J. (18). 'Food and Drink [from the Renaissance to the Industrial Revolution].' Art. in *A History of Technology*, ed. C. Singer *et al.*, vol. 3, p. 1. Oxford, 1957.

FORBES, R. J. (20). *Studies in Early Petroleum History.* Brill, Leiden, 1958.

FORBES, R. J. (22). 'Land Transport and Roadbuilding (+1000 to 1900).' *JAN*, 1957, **46**, 104. The first section of this paper is identical with Forbes (11), pp. 156–60.

FORBES, R. J. (25). Introductions to Chapters in Vol. 5 (Engineering) of the *Principal Works of Simon Stevin* including discussions of drainage-mills (wind-power), hydraulic engineering (sluices, locks, dredgers, double slipways), and Simon Stevin's patents.

FORKE, A. (3) (tr.). *Me Ti [Mo Ti] des Sozialethikers und seine Schüler philosophische Werke.* Berlin, 1922. (*MSOS*, Beibände, **23–25**.)

FORKE, A. (4) (tr.). '*Lung-Hêng*', *Philosophical Essays of Wang Chhung.* Vol. 1, 1907. Kelly & Walsh, Shanghai; Luzac, London; Harrassowitz, Leipzig. Vol. 2, 1911 (with the addition of Reimer, Berlin). Photolitho Re-issue, Paragon, New York, 1962. (*MSOS*, Beibände, **10** and **14**.) Crit. P. Pelliot, *JA*, 1912 (10e sér.), **20**, 156.

FORKE, A. (6). *The World-Conception of the Chinese; their astronomical cosmological and physico-philosophical Speculations* (Pt. 4 of this, on the Five Elements, is reprinted from Forke (4), vol. 2, App. I). Probsthain, London, 1925. German tr. *Gedankenwelt des Chinesischen Kulturkreis.* München, 1927. Chinese tr. *Chhi Na Tzu Jan Kho Hsueh Ssu Hsiang Shih.* Critique: B. Schindler, *AM*, 1925, **2**, 368.

FORKE, A. (13). *Geschichte d. alten chinesischen Philosophie* (i.e. from antiquity to the beginning of the Former Han). de Gruyter, Hamburg, 1927. (Hamburg. Univ. Abhdl. a. d. Geb. d. Auslandskunde, no. 25 (Ser. B, no. 14).)

FORMAN, W. & FORMAN, B. (1). *Das Drachenboot* (album of photographs of Chinese places, buildings, vessels, etc.). Artia, Prague, 1960.

FORREST, THOS. (CAPT. H. E. I. C.) (1). *A Voyage to New Guinea and the Moluccas, etc., performed in the 'Tartar' Galley belonging to the Honourable East India Company, during the Years 1774, 1775 and 1776.* Scott, London, 1780. Fr. tr. *Voyage aux Moluques et à la Nouvelle Guinée, fait sur la Galère 'La Tartare' en 1774, 1775 & 1776, par le Capitaine Forrest.* Thou, Paris, 1780.

FORREST, THOS. (CAPT. H. E. I. C.) (2). *A Voyage from Calcutta to the Mergui Archipelago.* Robson, London, 1792.

FORSTER, E. M. (1). *Alexandria; a History and a Guide.* Morris, Alexandria, 1922.

FOSTER, J. (1). 'Crosses from the Walls of Zaitun [Chhüanchow].' *JRAS*, 1954, 1.

FOX, C. (1). 'Sleds, Carts and Waggons.' *AQ*, 1931, **5**, 185.

FRANCK, H. A. (1). *Roving through Southern China.* Century, New York, 1925.

FRANCO, S. GARCIA (1). *Historia del Arte y Ciencia de Navegar.* 2 vols. Madrid, 1947.

FRANKE, O. (4). *Li Tschi [Li Chih], ein Beitrag z. Geschichte d. chinesisches Geisteskämpfe in 16-Jahrh.* *APAW/PH*, 1938, no. 10.

FRANKE, O. (11) (intr. & tr.). *Kêng Tschi T'u [Kêng Chih Thu]; Ackerbau und Seidegewinnung in China, ein Kaiserliches Lehr. u. Mahn-Buch.* Friederichsen, Hamburg, 1913. (Abhandl. d. Hamburgischen Kolonialinstituts, vol. 11; Ser. B, Völkerkunde, Kulturgesch. u. Sprachen, vol. 8.)

FRANKE, O. (12). *Beiträge z. Kenntnis der Türkvölker und Skythen Zentralasiens.* Berlin, 1904.

FRANKE, W. (3). *Preliminary Notes on Important Literary Sources for the History of the Ming Dynasty.* Chhêngtu, 1948. (SSE Monographs, Ser. A, no. 2.)

FRANKEL, H. H. (1). *Catalogue of Translations from the Chinese Dynastic Histories for the Period +220 to +960*. Univ. Calif. Press, Berkeley and Los Angeles, 1957. (Inst. Internat. Studies, Univ. of California, East Asia Studies, Chinese Dynastic Histories Translations, Suppl. no. 1.)

FRANKFORT, H. (1). 'Studies in Early Pottery of the Near East.' *RAI/OP*, 1924, no. 6.

FRANKFORT, H. (3). 'On Egyptian Art.' *JEA*, 1932, **18**, 33.

FRANKLIN, BENJAMIN (1). 'Maritime Observations' (a letter to Mr Alphonsus le Roy dated Aug. 1785). *TAPS*, 1786, **2**, 294 (p. 301); abstracted in *NAVC*, 1803, **9**, 32.

FRANKLIN, T. B. (1). *A History of Agriculture*. Bell, London, 1948.

FRASER, E. D. H. & LOCKHART, J. H. S. (1). *Index to the 'Tso Chuan'*. Oxford, 1930.

FRAZER, SIR J. G. (2). *Folk-lore in the Old Testament; Studies in Comparative Religion, Legend and Law*. Macmillan, London, 1923

FREEMAN, J. R. (1). 'Flood Problems in China.' *TASCE*, 1922, **85**, 1405.

FREEMAN-GRENVILLE, G. S. P. (1). 'East African Coin Finds and their Historical Significance.' *JAH*, 1960, **1**, 31.

FREEMAN-GRENVILLE, G. S. P. (2). 'Chinese Porcelain in Tanganyika.' *TNR*, 1955 (no. 41), 62.

FREEMAN-GRENVILLE, G. S. P. (3). 'Coinage in East Africa before Portuguese Times.' *NC*, 1957 (6th ser.), **17**, 151.

FREEMAN-GRENVILLE, G. S. P. (4). *The East African Coast; Select Documents from the +1st to the earlier 19th Centuries*. Oxford, 1962.

FREEMAN-GRENVILLE, G. S. P. (5). 'Coins from Mogadishiu; c. +1300 to c. +1700' [but including Chinese ones from the +10th to the 19th centuries]. *NC*, 1963 (7th ser.), **3**, 179.

FREEMAN-GRENVILLE, G. S. P. (6). 'Some Recent Archaeological Work on the East African Coast.' *MA*, 1958, **58**, 108.

FRÉMONT, C. (13). *Études Expérimentales de Technologie Industrielle, No. 64: le Marteau, le Choc, le Marteau Pneumatique*. Paris, 1923. (Hammers and vibrators.)

FREYRE, G. (1). *The Portuguese and the Tropics*. Tr. H. M. d'O. Matthew & F. de Mello Moser. Exec. Committee for the Commemoration of the Vth Centenary of the Death of Prince Henry the Navigator, Lisbon, 1961.

FRIEDLÄNDER, L. (1). *Roman Life and Manners under the Early Empire*. 4 vols. Routledge, London, n.d. (1909–13). Tr. from the German by L. A. Magnus & J. H. Freese.

VON FRIES, S. (1). 'The Tent-Theory of Chinese Architecture.' *JRAS/NCB*, 1890, **24**, 303.

FRIPP, C. E. (1). 'A Note on Mediaeval Chinese–African Trade.' *NADA*, 1940 (no. 17), 88.

FROST, HONOR (1). *Under the Mediterranean; Marine Antiquities*. Routledge & Kegan Paul, London, 1963.

FROST, HONOR (2). 'From Rope to Chain; on the Development of Anchors in the Mediterranean.' *MMI*, 1963, **49**, 1.

FRUMKIN, G. (2). 'Archaeology in Soviet Central Asia; v, The Deltas of the Oxus and Jaxartes; Khorezm and its Borderlands.' *CENAR*, 1965, **13**, 69.

FU TSO-YI (1). *Ending the Flood Menace* (an account of the Huai River Project, especially the Jun-ho-chi Control Installations). *CREC*, 1952 (no. 1), 4.

FUCHS, R. F. (1). 'Der Farbenwechsel u. d. chromatische Hautfunktion d. Tiere.' Art. in *Handbuch d. vergl. Physiol*. Ed. H. Winterstein. Vol. 3, sect. 1, pt. 2, pp. 1189–1652. Fischer, Jena, 1924.

FUCHS, W. (1). *The 'Mongol Atlas' of China by Chu Ssu-Pên and the Kuang Yü T'hu*. Fu-Jen Univ. Press, Peiping, 1946. (MS/M series, no. 8.) (Rev. J. J. L. Duyvendak, *TP*, 1949, **39**, 197.)

FUCHS, W. (4). 'Huei-Ch'ao's Pilgerreise durch Nordwest-Indien und Zentral-Asien um 726.' *SPAW/PH*, 1938, **30**, 426.

FUCHS, W. (6). 'Was South Africa already known in the +13th Century?' *IM*, 1953, **10**, 50.

FUCHS, W. (7). 'Ein Gesandschaft u. Fu-Lin in chinesischer Wiedergabe aus den Jahren +1314 bis +1320.' *OE*, 1959, **6**, 123.

FUGL-MEYER, H. (1). *Chinese Bridges*. Kelly & Walsh, Shanghai, 1937.

FULLER, M. L. & CLAPP, F. G. (1). 'Loess and Rock Dwellings of Shensi.' *GR*, 1924, **14**, 215.

GADBURY, JOHN (1). *Nauticum Astrologicum; or, the Astrological Seaman; directing Merchants, Mariners, Captains of Ships, Ensurers, etc. How (by God's Blessing) they may escape divers Dangers which commonly happen upon the Ocean; Unto which is added a Diary of the Weather for 21 Years together, Exactly Observed in London, with sundry Observations thereon*... Sawbridge, London, 1691, 1710.

GALE, E. M. (1) (tr.). *Discourses on Salt and Iron ('Yen Thieh Lun'), a Debate on State Control of Commerce and Industry in Ancient China, chapters 1–19*. Brill, Leiden, 1931. (Sinica Leidensia, no. 2.) Crit. P. Pelliot, *TP*, 1932, **29**, 127.

GALE, E. M., BOODBERG, P. A. & LIN, T. C. (1) (tr.). 'Discourses on Salt and Iron (*Yen Thieh Lun*), Chapters 20–28.' *JRAS/NCB*, 1934, **65**, 73.

GALLAGHER, L. J. (1) (tr.). *China in the 16th Century; the Journals of Matthew Ricci, 1583–1610.* Random House, New York, 1953. (A complete translation, preceded by inadequate bibliographical details, of Nicholas Trigault's *De Christiana Expeditione apud Sinas* (1615). Based on an earlier publication: *The China that Was; China as discovered by the Jesuits at the close of the 16th Century: from the Latin of Nicholas Trigault.* Milwaukee, 1942.) Identifications of Chinese names in Yang Lien-Shêng (4). (Crit. J. R. Ware, *ISIS*, 1954, **45**, 395.)

GANDAR, D. (1). *Le Canal Impérial; Étude Historique et Descriptive.* T'ou-Sé-Wé, Shanghai, 1894 (VS, no. 4).

GARNIER, M. l'ABBÉ (1). 'Galères et Galéasses à la Fin du Moyen Age.' Art. in *Proc. 2nd International Colloquium of Maritime History*, p. 37, ed. M. Mollat, L. Denoix & O. de Prat. Paris, 1957.

GARRISON, F. H. (1). 'History of Drainage, Irrigation, Sewage-Disposal, and Water-Supply.' *BNYAM*, 1929, **5**, 887.

GARRISON, F. H. (2). 'History of Heating, Ventilation and Lighting.' *BNYAM*, 1927, **3**, 57.

GAUBIL, A. (11). *Description de la Ville de Pékin.* Ed. de l'Isle and Pingré, Paris, 1763, 1765. Russ. tr. by Stritter; Germ. tr. by Pallas; Eng. tr. (abridged), *PTRS*, 1758, **50**, 704.

GAUTHEY, E. M. (1). *Traité de la Construction des Ponts.* 2 vols. Didot, Paris, 1809–13.

GEIL, W. E. (2). *The Eighteen Capitals of China.* Constable, London, 1911.

GEIL, W. E. (3). *The Great Wall of China.* Murray, London, 1909.

GEORGE, F. (1). 'Hariot's Meridional Parts.' *JIN*, 1956, **9**, 66.

GERINI, G. E. (1). *Researches on Ptolemy's Geography of Eastern Asia (Further India and Indo-Malay Peninsula).* Royal Asiatic Society and Royal Geographical Society, London, 1909. (Asiatic Society Monographs, no. 1.)

GERMAIN, G. (1). 'Qu'est-ce que le *Périple* d'Hannon; Document, Amplification ou Faux Intégral?' *HP*, 1957, **44**, 205.

GERNET, J. (1). *Les Aspects Economiques du Bouddhisme dans la Société Chinoise du 5ᵉ au 10ᵉ siècles.* Maisonneuve, Paris, 1956. (Publications de l'Ecole Française d'Extrême-Orient, Hanoi & Saigon.) (Revs. D. C. Twitchett, *BLSOAS*, 1957, **19**, 526; A. F. Wright, *JAS*, 1957, **16**, 408.)

GERNET, J. (2). *La Vie Quotidienne en Chine à la Veille de l'Invasion Mongole (+1250 à +1276).* Hachette, Paris, 1959.

GHIRSHMAN, R. (2). 'Essai de Recherche Historico-archéologique' [digest of the work of S. P. Tolstov]. *AA*, 1952, **16**, 209 and 292.

GHIRSHMAN, R. (4). *Iran; Parthes et Sassanides.* Gallimard, Paris, 1962.

GIBB, H. A. R. (3) (tr.). *The Travels of Ibn Baṭṭūṭah (+1325 to +1354); translated with Revisions and Notes from the Arabic Text edited by C. Defrémery & B. R. Sanguinetti.* 2 vols. Cambridge, 1958, 1962. (Hakluyt Society Pubs., 2nd Series, no. 110, 117.) Rev. G. F. Hourani, *JAOS*, 1960, **80**, 269.

GIBBON, EDWARD (1). *The History of the Decline and Fall of the Roman Empire.* 12 vols. Strahan, London, 1790 (1st ed. 1776–88).

GIBBS, C. D. I. (1). 'The River-Life of Canton.' *NV*, 1930, p. 73.

GIBSON, C. E. (1). 'The Ship and Society.' *MQ*, 1947 (n.s.), **2**, 163, with comments by G. Lee, p. 254.

GIBSON, C. E. (2). *The Story of the Ship.* Schuman, New York, 1948; Abelard-Schuman, New York, 1958.

GIBSON, H. E. (4). 'Communications in China during the Shang Period.' *CJ*, 1937, **26**, 228.

GIEDION, S. (2). *Space, Time and Architecture; the Growth of a New Tradition.* Oxford, 1954.

GIGLIOLI, G. Q. (1). *l'Arte Etrusca.* Treves, Milan, 1935.

GILES, H. A. (1). *A Chinese Biographical Dictionary.* 2 vols. Kelly & Walsh, Shanghai, 1898; Quaritch, London, 1898. Supplementary Index by J. V. Gillis & Yü Ping-Yüeh, Peiping, 1936. Account must by taken of the numerous emendations published by von Zach (4) and Pelliot (34), but many mistakes remain. Cf. Pelliot (35).

GILES, H. A. (3) (tr.). *The Travels of Fa-Hsien.* Cambridge, 1923.

GILES, H. A. (4) (tr.). *San Tzu Ching, translated and annotated.* Kelly & Walsh, Shanghai, 1900.

GILES, H. A. (10). 'Chinese Anchors.' *MMI*, 1924, **10**, 399; 1925, **11**, 328.

GILES, L. (4) (tr.). *Taoist Teachings from the Book of Lieh Tzu.* Murray, London, 1912; 2nd ed. 1947.

GILES, L. (5). *Six Centuries of Tunhuang.* China Society, London, 1944.

GILES, L. (13). *Descriptive Catalogue of the Chinese Manuscripts from Tunhuang in the British Museum.* British Museum, London, 1957. (Rev. J. Průsek, *ARO*, 1959, **27**, 483.)

GILFILLAN, S. C. (1). *Inventing the Ship.* Follett, Chicago, 1935.

GILFILLAN, S. C. (2). *The Sociology of Invention.* Follett, Chicago, 1935.

GILL, W. (1). *The River of Golden Sand, being the narrative of a Journey through China and Eastern Tibet to Burmah*, ed. E. C. Baber & H. Yule. Murray, London, 1883.

GILLE, B. (3). 'Léonard de Vinci et son Temps.' *MC/TC*, 1952, **2**, 69.

GILLE, P. (1). 'Les Navires des Vikings.' *MC/TC*, 1954, **3**, 91.

GILLE, P. (2). 'Jauge et Tonnage des Navires.' Art. in *Proc. 1st International Colloquium of Maritime History*, Paris, 1956. Ed. M. Mollat & O. de Prat, p. 85.

GILLESPIE, R. ST J. (1). 'Cunningham's Self-Reefing Topsails.' *MMI*, 1945, **31**, 7. Note by G. F. Howard, 1946, **32**, 120.

GILLMER, T. C. (1). 'Present-Day Craft and Rigs of the Mediterranean.' *ANEPT*, 1941, **1**, 352; 1942, **2**, 56.

GIQUEL, P. (1). 'Mechanical and Nautical Terms in French, Chinese and English.' In Doolittle, J. (1), vol. 2, p. 634.

GLADWIN, H. S. (1). *Men out of Asia*. McGraw-Hill, New York, 1947. (Crit. rev. B. Lasker, *PA*, 1948, **21**, 439. Summary of the theory in Covarrubias (2), pp. 25 ff.)

GLOAG, J. & BRIDGWATER, D. (1). *A History of Cast Iron in Architecture*. Allen & Unwin, London, 1948.

GOAD, JOHN (1). *Astro-Meteorologica; or, Aphorisms and Discourses of the Bodies Coelestial, their Natures and Influences, Discovered from the Variety of the Alterations of the Air, Temperate or Intemperate, as to Heat or Cold, Frost, Snow, Hail, Fog, Rain, Wind, Storm, Lightening, Thunder, Blasting, Hurricane, Tuffon, Whirlwind, Iris, Chasme, Parelij, Comets their Original and Duration, Earthquakes, Vulcano's, Inundations, Sickness epidemical, Maculae Solis, and other Secrets of Nature*. Rawlins & Blagrave, London, 1686.

GOBLOT, H. (1). 'Dans l'ancien Iran; les Techniques de l'Eau et la Grande Histoire.' *AHES/AESC*, 1963, 499.

GOBLOT, H. (2). 'Sur Quelques Barrages Anciens, et la Genèse des Barrages-Voûtes.' *DHT*, 1967 (no. 6), 109. (Special number) *RHS*, 1967, **20** (no. 2), 109.

GODINHO, V. MAGALHÃES (1). *Les Grandes Découvertes*. Coimbra, 1953. Reprinted from *BEP*.

GODINHO, V. MAGALHÃES (2). *A Expansão Quatrocentista Portuguesa; Problemas das Origens e da Linha de Evolução*. Testemunho Especial, Lisbon, 1945.

GODINHO, V. MAGALHÃES (3). *Documentos sobre a Expansão Portuguesa*. 3 vols. Vols. 1 and 2, Gleba, Lisbon; vol. 3, Cosmos, Lisbon, 1943 to 1956.

GODWIN, H., WALKER, D. & WILLIS, E. H. (1). 'Radiocarbon Dating and Post-Glacial Vegetational History; Scaleby Moss.' *PRSB*, 1957, **147**, 352.

GODWIN, H. & WILLIS, E. H. (1). 'Cambridge University Natural Radiocarbon Measurements, 1.' *AJSC* (Radiocarbon Supplement), 1959, **1**, 63. 'Radiocarbon Dating of the Late-Glacial Period in Britain.' *PRSB*, 1959, **150**, 199.

DE GOEJE, E. (1). 'De Muur van Gog en Magog.' *VMAWA*, 1883, **3** (no. 5), 87.

DE GOEJE, M. J. (1) (ed.). *Bibliotheca Geographorum Arabicorum* (texts). 8 vols. Brill, Leiden, 1870–94.

GOLAB, L. WAWRZYN (1). 'A Study of Irrigation in East Turkestan.' *AN*, 1951, **46**, 187.

GOLOUBER, V. (1). 'L'Age du Bronze au Tonkin et dans le Nord Annam.' *BEFEO*, 1929, **29**, 1.

GOMBRICH, E. H. J. (1). *Art and Illusion; a Study in the Psychology of Pictorial Representation* (Mellon Lectures, 1956). Phaidon, London, 1960; 2nd ed. 1962. (Nat. Art Gallery, Washington, Mellon Lectures, ser. no. 5.)

GOODCHILD, R. G. & FORBES, R. J. (1). 'Roads and Land Travel [including Bridges, Cuttings, Tunnels, Harbours, Docks and Lighthouses] (in the Mediterranean Civilisations and the Middle Ages).' Art. in *A History of Technology*, ed. C. Singer *et al.* vol. 2, p. 493. Oxford, 1956.

GOODRICH, L. CARRINGTON (1). *Short History of the Chinese People*. Harper, New York, 1943.

GOODRICH, L. CARRINGTON (10). 'Query on the Connection between the Nautical Charts of the Arabs and those of the Chinese before the days of the Portuguese Navigators.' *ISIS*, 1953, **44**, 99.

GOODRICH, L. CARRINGTON (14). 'A Note on Professor Duyvendak's Lectures on China's Discovery of Africa.' *BLSOAS*, 1952, **14** (no. 2).

GOODRICH, L. CARRINGTON (15). 'Firearms among the Chinese; a supplementary note.' *ISIS*, 1948, **39**, 63.

GOODRICH, L. CARRINGTON (16). 'Suspension-Bridges in China.' *SIS*, 1957, **5** (nos. 3–4), 1.

GOODRICH, L. CARRINGTON & FÊNG CHIA-SHÊNG (1). 'The Early Development of Firearms in China.' *ISIS*, 1946, **36**, 114. With important addendum, *ISIS*, 1946, **36**, 250.

GOTHEIN, M. L. (1). 'Die Stadtanlage von Peking.' *WJK*, 19, **7**.

GOULD, R. T. (1). (*a*) *The Marine Chronometer; its History and Development*. Potter, London, 1923. (Rev. F. D[ ], *MMI*, 1923, **9**, 191.) (*b*) *The Restoration of John Harrison's Third Timekeeper*. Lecture to the British Horological Institute, 1931. Reprint or pamphlet, n.d.

GOULLART, P. (1). *Forgotten Kingdom* (the Lichiang districts of Yunnan). Murray, London, 1955.

GOURLIE, NORAH (1). *The Prince of Botanists* [Linnaeus]. London, 1953.

GRAHAM, A. C. (6) (tr.). *The Book of Lieh Tzu*. Murray, London, 1960.

GRAHAM, DOROTHY (1). *Chinese Gardens*. New York, 1938.

GRAHAM, G. S. (1). 'The Transition from Paddle-Wheel to Screw Propeller.' *MMI*, 1958, **44**, 35.

GRANDIDIER, A. & GRANDIDIER, G. (1). *L'Ethnographie de Madagascar*. Paris, 1908.

GRANET, M. (1). *Danses et Légendes de la Chine Ancienne*. 2 vols. Alcan, Paris, 1926.

GRANET, M. (2). *Fêtes et Chansons Anciennes de la Chine*. Alcan, Paris, 1926 ; 2nd ed. Leroux, Paris, 1929.

GRANET, M. (5). *La Pensée Chinoise*. Albin Michel Paris, 1934. (Evol. de l'Hum. series, no. 25 *bis*.)

GRANGER, F. (1) (ed. & tr.). *Vitruvius On Architecture*. 2 vols. Heinemann, London, 1934. (Loeb Classics edn.)

GRANTHAM, A. E. (1). *The Ming Tombs* (Shih San Ling). Wu Lai-Hsi, Peiping, 1926.

GRAS, N. S. B. (1). *A History of Agriculture in Europe and America*. 2nd ed. Crofts, New York, 1946.

GRAY, B. & VINCENT, J. B. (1). *Buddhist Cave-Paintings at Tunhuang*. Faber & Faber, London, 1959.

GRAY, J. (1). 'Historical Writing in Twentieth-Century China; Notes on its Background and Development.' Art. in *Historians of China and Japan*, ed. W. G. Beasley & E. G. Pulleyblank, p. 186. Oxford Univ. Press, London, 1961.

GRAY, J. E. (1). (*a*) 'On the Structure of Pearls and on the Chinese Mode of producing them of a Large Size and Regular Form.' *TAP*, 1825 (2nd ser.), **9**, 27; (*b*) 'On the Chinese Manner of forming Artificial Pearls.' *TAP*, 1826 (2nd ser.), **10**, 389.

GREENBERG, M. (1). *British Trade and the Opening of China, 1800–1842*. Cambridge, 1951.

GREENHILL, B. (1). 'A Boat of the Indus [the *quantel battella*].' *MMI*, 1963, **49**, 273. Punt build, yet with vertical rudder at the stern, used for cargo-carrying as well as in small sizes.

GRÉGOIRE, C. (1). 'Further Studies on the Structure of the Organic Components in Mother-of-Pearl, especially in Pelecypods.' *BIRSN*, 1960, **36**, no. 23.

GREGORY, J. W. (2). *The Story of the Road, from the Beginning to the Present Day*. 2nd ed. revised & enlarged by C. J. Gregory. Black, London, 1938.

GREGORY, RICHARD (1). 'How the Eyes Deceive.' *LI*, 1962, **68** (no. 1736), 15.

GRIERSON, P. (1). 'La Moneta Veneziana nell'Economia Mediterranea del Trecento e Quattrocento.' Art. in *La Civittà Veneziana del Quattrocento*, p. 77. Sansoni, Florence, 1957.

GRIFFITH, W. M. (1). 'A Theory of Silt and Scour.' *PICE*, 1926, **223** (Paper no. 4545).

GROENEVELDT, W. P. (1). 'Notes on the Malay Archipelago and Malacca.' 1876. In *Miscellaneous Papers relating to Indo-China*. 2nd series, 1887, vol. 1, p. 126.

GROFF, G. W. & LAU, T. C. (1). 'Landscaped Kuangsi; China's Province of Pictorial Art.' *NGM*, 1937, **72**, 700.

GROOTAERS, W. A. (1). 'La Géographie Linguistique de la Chine, II.' *MS*, 1945, **10**, 389.

GROSLIER, G. (1). *Recherches sur les Cambodgiens*. Challamel, Paris, 1921.

GROSLIER, G. (2). 'La Batellerie Cambodgienne du 8e au 13e siècle de notre Ère.' *RA*, 1917 (5e ser.), **5**, 198.

GROTTANELLI, V. L. (1). *Pescatori dell'Oceano Indiano*. Cremonese, Rome, 1955.

GROUSSET, R. (1). *Histoire de l'Extrême-Orient*. 2 vols. Geuthner, Paris, 1929. (Also appeared in *BE/AMG*, nos. 39, 40.)

GRUSS, R. (1). *Petit Dictionnaire de Marine*. Challamel, Paris, 1945.

DE GUIGNES, [C. L. J.] (1). 'Idée générale du Commerce et des Liaisons que les Chinois ont eus avec les Nations Occidentales.' *MAI/LTR*, 1784 (1793), **46**, 534.

DE GUIGNES, [JOSEPH] (3). 'Réflexions générales sur les Liaisons et le Commerce des Romains avec les Tartares et les Chinois.' *MAI/LTR*, 1763 (1768), **32**, 355.

DE GUIGNES, [JOSEPH] (4). 'Recherches sur les Navigations des Chinois du Coté de l'Amérique, et sur quelques Peuples situés à l'extremité orientale de l'Asie.' *MAI/LTR*, 1761, **28**, 503.

GUILLAN, F. (1). 'Les Châteaux-Forts Japonais.' *BMFJ*, 1942, **13**, 1–216.

GUPPY, T. R. (1). 'Description of the "Great Britain" Iron Steamship, with an Account of the Trial Voyages.' *ICE/MP*, 1845, **4**, 178. (Includes communication by J. Field of information provided by Lady Bentham concerning the use of watertight compartments in ship construction by Sir Samuel Bentham.)

GUTKIND, E. A. (1). *Revolution of Environment*. Kegan Paul, London, 1946.

GUTSCHE, F. (1). 'Die Entwicklung d. Schiffsschraube.' *BGTI/TG*, 1937, **26**, 37.

HACKIN, J. & HACKIN, J. R. (1). *Recherches archéologiques à Begram, 1937*. Mémoires de la Délégation Archéologique Française en Afghanistan, vol. 9. Paris, 1939.

HACKIN, J., HACKIN, J. R., CARL, J. & HAMELIN, P. (with the collaboration of J. Auboyer, V. Elisséeff, O. Kurz & P. Stern) (1). *Nouvelles Recherches archéologiques à Begram (ancienne Kāpiśi), 1939–1940*. Mémoires de la Délégation Archéologique Française en Afghanistan, vol. 11. Paris, 1954. (Rev. P. S. Rawson, *JRAS*, 1957, 139.)

HACKMAN H. (4). *Von Omi bis Bhamo*. Galle a/d Saale, 1905.

HADDAD, SAMI I. & KHAIRALLAH, AMIN A. (1). 'A Forgotten Chapter in the History of the Circulation of the Blood.' *ASURG*, 1936, **104**, 1.

HADDON, A. C. & HORNELL, J. (1). *Canoes of Oceania*. Bernice P. Bishop Museum, Honolulu, Hawaii, 1936, 3 vols.
    vol. 1, Polynesia, Fiji and Micronesia.

vol. 2, Melanesia, Queensland and New Guinea.

vol. 3, Definition of Terms, General Survey and Conclusions.

(Bishop Museum Special Pubs. no. 27.)

HADFIELD, E. C. R. (1). 'Canals; Inland Waterways of the British Isles [in the Industrial Revolution].' Art. in *A History of Technology*, ed. C. Singer *et al.* vol. 4, p. 563. Oxford, 1958.

HADFIELD, E. C. R. (2). *Introducing Canals; a Guide to British Waterways Today*. Benn, London, 1955.

HADFIELD, E. C. R. (3). *British Canals; an Illustrated History*. Phoenix, London, 1950, 1959. 3rd ed. David and Charles, Newton Abbot, 1966.

HADFIELD, E. C. R. (4). *Canals of the World*. Blackwell, Oxford, 1964.

HADI HASAN (1). *A History of Persian Navigation*. Methuen, London, 1928.

VON HAGEN, V. W. (1). *The Aztec and Maya Paper-makers*, with an introduction by Dard Hunter and an appendix by Paul C. Standley. Augustin, New York, 1944. Enlarged Eng. tr. of *La Fabricacion del Papel entre los Aztecas y los Mayas*, with introductions by Alfonso Caso and Dard Hunter, Nuevo Mundo, Mexico City, 1935.

VON HAGEN, V. W. (2). 'America's Oldest Roads; the Highways of the Incas.' *SAM*, 1952, **187** (no. 1), 17.

VON HAGEN, V. W. (3). *Highway of the Sun*. Travel Book Club, London, n.d. (1953). Popular report of the Inca Highway Expedition. (American Geographical Society.) The full scientific report was apparently never published.

HAGERTY, M. J. (2) (tr. and annot.). 'Tai Khai-Chih's *Chu Phu*; a Fifth-Century Monograph on Bamboos written in Rhyme with a Commentary.' *HJAS*, 1948, **11**, 372.

HAGUE, D. B. (1). *The Conway Suspension Bridge*. Curwen Press, for the National Trust, London, n.d. (1968).

HAGUE, F. (1). 'On the Natural and Artificial Production of Pearls in China.' *JRAS*, 1856, **16**, 280.

HAHNLOSER, H. R. (1) (ed.). *The Album of Villard de Honnecourt*. Schroll, Vienna, 1935.

HÁJEK, L. & FORMAN, W. (1). *Chinese Art*. Artia, Prague; Spring Books, London, n.d. (1953).

HAKEWILL, GEO. (1). *An Apologie or Declaration of the Power and Providence of God in the Government of the World*. 3rd ed. Turner, Oxford, 1635.

DU HALDE, J. B. (1). *Description Géographique, Historique, Chronologique, Politique et Physique de l'Empire de la Chine et de la Tartane Chinoise*. 4 vols. Paris, 1735, 1739; The Hague, 1736. Eng. tr. R. Brookes, London, 1736, 1741. Germ. tr. Rostock, 1748.

HALLBERG, I. (1). *L'Extrême-Orient dans la Littérature et la Cartographie de l'Occident des 13e, 14e et 15e siècles; Etudes sur l'Histoire de la Géographie*. Inaug. Diss., Upsala, Zachrisson, Göteborg, 1907.

HALLDIN, G. & WEBE, G. (1). *Statens Sjöhistoriska Museum (Guide to the National Maritime Museum, Stockholm)*. N.M.M., Stockholm, 1952.

HAMBRUCH, P. (1). 'Das Meer in seiner Bedeutung für die Völkerverbreitung.' *AANTH*, 1909, **35**, 75.

HAMILTON, E. J. (1). *American Treasure and the Price Revolution in Spain, +1501 to +1550*. Harvard University Press, Cambridge, Mass., 1934. (Harvard Econ. Studs. Monogr. ser. no. 43.)

HAMILTON, S. B. (1). 'Building and Civil Engineering Construction' [1750 to 1850]. Art. in *A History of Technology*, ed. C. Singer *et al.* Vol. 4, p. 422. Oxford, 1958.

HAMILTON, S. B. (3). 'The Use of Cast Iron in Buildings.' *TNS*, 1941, **21**, 139.

HAMILTON, S. B. (4). 'The Structural Use of Iron in Antiquity.' *TNS*, 1958, **31**, 29.

HAMILTON, S. B. (5). 'Building Materials and Techniques' [1850 to 1900]. Art. in *A History of Technology*, ed. C. Singer *et al.* Vol. 5, p. 466. Oxford, 1958.

VON HAMMER-PURGSTALL, J. (3). 'Extracts from the *Mohit* [*Muḥīṭ*] (The Ocean), a Turkish Work on Navigation in the Indian Seas' [by Sidi 'Ali Reïs, *c.* +1553]. *JRAS/B*, 1834, **3**, 545; 1836, **5**, 441; 1837, **6**, 805; 1838, **7**, 767; 1839, **8**, 823; with notes by J. Prinsep.

H[ANCE], H. F. W. 'The Use of Iron Cylinders in Bridge-Building.' *NQCJ*, 1868, **2**, 180.

HARADA, J. (1). *The Gardens of Japan*. London, 1928.

HARADA, YOSHITO & KOMAI, KAZUCHIKA (1). *Chinese Antiquities*. Pt. 1, *Arms and Armour*; Pt. 2, *Vessels [Ships] and Vehicles*. Academy of Oriental Culture, Tokyo Institute, Tokyo, 1937.

HARBY, S. F. (1). 'They Survived at Sea.' *NGM*, 1945, **87**, 617.

HARRIS, L. E. (1). *Vermuyden and the Fens; a study of Sir Cornelius Vermuyden and the Great Level*. Cleaver-Hume, London, 1953. Rev. H. C. Darby, *N*, 1954, **173**, 913.

HARRISON, K. P. & NANCE, R. M. (1). 'The King's College Chapel Window Ship.' *MMI*, 1948, **34**, 12.

HARRISSON T. (1). 'New Archaeological and Ethnological Results from the Niah Caves, Sarawak.' *MA*, 1959, **59**, 1.

HARRISSON, T. (2). 'Some Ceramics excavated in Borneo; with some asides to Siam and London' (Sung ware). *TOCS*, 1954, **28**, 11.

HARRISSON, T. (3). 'Some Ceramic Objects recently acquired for the Sarawak Museum' (Sung pieces). *SMJ*, 1951, **5**, 541.

HARRISSON, T. (4). 'Some Borneo Ceramic Objects' (Thang jars). *SMJ*, 1950, **5**, 270.

HARRISSON, T. (5). 'Ceramics penetrating Borneo.' *SMJ*, 1955, 6, 549. (Thang and Sung pieces treasured among the Dusuns of North Borneo.)

HARRISSON, T. (6). 'Rhinoceros in Borneo, and traded to China.' *SMJ*, 1956, 7, 263.

HARRISSON, T. (7). 'Japan and Borneo; some Ceramic Parallels.' *SMJ*, 1957, 8, 100.

HARTGILL, G. (1). *Astronomical Tables showing the Declinations, Right Ascentions, and Aspects of 365 of the most principall fixed Stars, and the number of them in their Constellations after Aratus; as also, the true Oblique Ascentions and Descentions of all the said Stars, upon the Cusps of every one of the 12 Houses of Heaven according to their Latitude; first invented by George Hartgill, Minister of the Word of God, and now reduced to this our Age by John and Timothy Gadbury*. Company of Stationers, London, 1656.

HARTNER, W. (3). 'The Astronomical Instruments of Cha-Ma-Lu-Ting, their Identification, and their Relations to the Instruments of the Observatory of Maragha.' *ISIS*, 1950, 41, 184.

HASLER, H. G. (1). 'Technically Interesting [an account of the sailing-boat *Jester*, which finished second in the 1960 single-handed transatlantic yacht race, rigged with a Chinese lug-sail].' *YW*, 1961, 113 (no. 2624), 14. See also 'Unusual Rig', *YW*, 1958, 110 (no. 2589), 13.

VAN HASSELT, A. L. (1). *Atlas Ethnographique d'une Partie de l'Ile de Sumatra*. Pr. pr. Batavia, 1886.

HASSLÖF, O. (1). 'Wrecks, Archives and Living Tradition; Topical Problems in Marine Historical Research.' *MMI*, 1963, 49, 162.

HASSLÖF, O. (2). 'Sources of Maritime History and Methods of Research.' *MMI*, 1966, 52, 127.

HASSLÖF, O. (3). 'Carvel Construction Technique; its Nature and Origin.' *FLV*, 1957–8, 21–22, 49. (Strake-morticed hull construction, with inserted frames, persisting into the Middle Ages.)

HATT, G. (1). 'Asiatic Motifs in American [Amerindian] Folklore.' In Singer Presentation Volume, *Science, Medicine and History*, vol. 2, p. 389, ed. E. A. Underwood. Oxford, 1954.

HATT, G. (2). 'Asiatic Influences in American [Amerindian] Folklore.' *KDVS/HFM*, 1949, 31, no. 6.

HATT, G. (3). 'The Corn Mother in [Amerindian] America and in Indonesia.' *AN*, 1951, 46, 853.

HAVELL, E. B. (1). (*a*) *Indian Architecture*. . . . Murray, London, 1913. (*b*) *Ancient and Mediaeval Architecture of India*. Murray, London, 1915.

HAVERFIELD, F. (1). *Ancient Town Planning*. Oxford, 1913.

HAWKES, D. (1) (tr.). '*Chhu Tzhu*'; the Songs of the South—an Ancient Chinese Anthology. Oxford, 1959. (Rev. J. Needham, *NSN*, 18 July 1959.)

HAWKINS, G. & GIBSON-HILL, C. A. (1). *Malaya*. Govt. Printing Office, Singapore, 1952.

HAYES, L. N. (1). *The Great Wall of China*. Kelly & Walsh, Shanghai, 1929.

HAZARD, B. H., HOYT, J., KIM HA-TAI, SMITH, W. W. & MARCUS, R. (1). *Korean Studies Guide*. Univ. of Calif. Press, Berkeley and Los Angeles, 1954.

VAN HECKEN, J. L. (1). 'Les Réductions Catholiques du Pays des Alashan.' *NZMW*, 1958, 14, 29.

VAN HECKEN, J. L. (2). 'Les Réductions Catholiques du Pays des Ordos; une Méthode d'Apostolat des Missionaires de Scheut.' *NZMW*, Schöneck-Beckenried, Switzerland, 1957. (Schriftenreihe d. Neuen Zeitschr. f. Missionswissenschaft (Cahiers de la Nouvelle Revue de Science Missionnaire), no. 15.)

VAN HECKEN, J. L. & GROOTAERS, W. A. (1). 'The "Half-Acre Garden (Pan Mou Yuan)", a Manchu Residence in Peking.' *MS*, 1959, 18, 360.

HECKER, J. F. C. (1). *The Epidemics of the Middle Ages*, tr. from the German by B. G. Babington. Sydenham Society, London, 1844.

HEIBERG, J. L. (2) (ed.). *Claudii Ptolomaei Opera quae exstant Omnia*. 3 vols. Teubner, Leipzig, 1907.

VAN DER HEIDE, G. D. (1). 'Archaeological Investigations on New Land; II, The Excavation of Wrecked Ships in Zuyder Zee Territory.' *AQSU*, 1959, 31.

HEIDEL, A. (1). *The Gilgamesh Epic and Old Testament Parallels*. Chicago Univ. Press, Chicago, 1946.

VON HEIDENSTAM, H. (1). *Report on The Hydrology of the Hangchow Bay and the Chhien-Thang Estuary*. Whangpoo Conservancy Board, Shanghai Harbour Investigation, 1921 (series I, no. 5).

VON HEINE-GELDERN, R. (1). 'Prehistoric Research in the Netherlands East Indies' (cultural connections between Indonesia and S.E. Europe). In *Science and Scientists in the Netherlands Indies*. Ed. P. Honig & F. Verdoorn. Board for the Netherlands Indies, Surinam and Curaçao; New York, 1945. (*Natuurwetenschappelijk Tijdschrift voor Nederlandsch Indïe*, Suppl. to 102.)

VON HEINE-GELDERN, R. (3). 'L'Art pre-bouddhique de la Chine et de l'Asie du Sud-Est, et son influence en Océanie.' *RAA/AMG*, 1937, 11, 177.

VON HEINE-GELDERN, R. (4). 'Die asiatische Herkunft d. südamerikanische Metalltechnik.' *PAI*, 1954, 5, 347.

VON HEINE-GELDERN, R. (6). 'Das Tocharerproblem und die Pontische Wanderung.' *SAE*, 1951, 2, 225.

VON HEINE-GELDERN, R. (7). 'Cultural Connections between Asia and Pre-Columbian America.' *AN*, 1950, 45, 350. Account of a discussion at the International Congress of Americanists, New York, 1949.

VON HEINE-GELDERN, R. (8). 'Some Problems of Migration in the Pacific.' *WBKGL*, 1952, 9, 313.

VON HEINE-GELDERN, R. (9). 'Kulturpflanzengeographie und das Problem vorkolumbische Kulturbeziehungen zwischen alter und neuer Welt.' *AN*, 1958, 53, 361.

VON HEINE-GELDERN, R. (10). 'Theoretical Considerations concerning the Problem of pre-Columbian Contacts between the Old World and the New.' *Proc. (Selected Papers), Vth Internat. Congr. Anthropol. & Ethnol.*, p. 277. Philadelphia, 1956.

VON HEINE-GELDERN, R. (11). 'Das Problem vorkolumbischen Beziehungen zwischen alter und neuer Welt, und seine Bedeutung für die allgemeine Kulturgeschichte.' *AOAW/PH*, 1954, **91**, 343.

VON HEINE-GELDERN, R. (12). 'Weltbild und Bauform in Südostasien.' *WBKGA*, 1930, **4**, 28.

VON HEINE-GELDERN, R. (14). 'Heyerdahl's Hypothesis of Polynesian Origins; a Criticism.' *GJ*, 1950, **116**, 183.

VON HEINE-GELDERN, R. (15). 'Voyaging Distance and Voyaging Time in Pacific Migration.' *GJ*, 1951, **118**, 108.

VON HEINE-GELDERN, R. (16). 'Traces of Indian and Southeast Asian Hindu–Buddhist Influences in Meso-America.' *Proc. 35th Internat. Congress of Americanists, Mexico City, 1962.* Vol. 1, p. 47.

VON HEINE-GELDERN, R. & EKHOLM, G. F. (1). 'Significant Parallels in the Symbolic Arts of Southern Asia and Middle America.' *Proc. XXVIIth Internat. Congr. Americanists*, vol. 1, p. 299. New York, 1949 (1951).

HEJZLAR, J. (1). 'The Return of a Legendary Work of Art; the most famous Scroll in the Peking Palace Museum, "On the River during the Spring Festival" by Chang Tsê-Tuan (+1125).' *NO*, 1962, **3** (no. 1), 17.

HENNIG, R. (4). *Terrae Incognitae; eine Zusammenstellung und Kritische Bewertung der wichtigsten vorcolumbischen Entdeckungsreisen an Hand der darüber vorliegenden Originalberichte.* 2nd ed. 4 vols. Brill, Leiden, 1944. (Includes most of the Chinese voyages of exploration, Chang Chhien, Kan Ying, etc.)

HENNIG, R. (7). *Rätselhafte Lände.* Berlin, 1950.

HENNIG, R. (8). 'Zur Frühgeschichte des Seeverkehrs im indischen Ozean.' *MK*, 1919, no. 151.

HENNIG, R. (9). 'Beitrag z. ält. Gesch. d. Leuchttürme.' *BGTI*, 1915, **6**, 35.

HENTZE, C. (1). *Mythes et Symboles Lunaires (Chine Ancienne, Civilisations anciennes de l'Asie, Peuples limitrophes du Pacifique)*, with appendix by H. Kühn. de Sikkel, Antwerp, 1932. Crit. *OAZ*, 1933, **9** (19), 33.

HENTZE, C. (3). 'Le Culte de l'Ours ou du Tigre et le T'ao-T'ie.' *Z*, 1938, **1**, 50.

HENTZE, C. (5). *Objets Rituels, Croyances et Dieux de la Chine Antique et de l'Amérique.* Antwerp, 1936.

HENTZE, C. (6). *Das Haus als Weltort der Seele; ein Beitrag zur Seelensymbolik in China, Grossasien, und Altamerika.* Klett, Stuttgart, 1961.

DE HERRERA, ANTONIO (etc.) (1). *Novus Orbis, sive Descriptio Indiae Occidentalis...Metaphraste C. Barlaeo, accesserunt et aliorum Indiae Occidentalis Descriptiones et Navigationes nuperae Australis Jacobi Le Mire, uti et navigationum omnium per Fretum Magellanicum Succincta Narratio.* Colin, Amsterdam, 1622.

HERRESHOFF, H. C. (1). 'Hydrodynamics and Aerodynamics of the Sailing Yacht.' *TSNAMEN*, 1964, **72**, 445–92.

HERRESHOFF, H. C. & KERWIN, J. E. (1). 'Sailing Yacht Research.' *Y*, 1965, **118** (no. 1), 51.

HERRESHOFF, H. C. & NEWMAN, J. N. (1). 'The Study of Sailing Yachts.' *SAM*, 1966, **215** (no. 2), 60.

HERRMANN, A. (1). *Historical and Commercial Atlas of China.* Harvard-Yenching Institute, Cambridge, Mass., 1935.

HERRMANN, A. (4). 'Ein Alter Seeverkehr zw. Abessinien u. Süd-China bis zum Beginn unserer Zeitrechnung.' *ZGEB*, 1913, 553.

HERRMANN, A. (8). 'Die Westländer in d. chinesischen Kartographie.' In Sven Hedin's *Southern Tibet; Discoveries in Former Times compared with my own Researches in 1906–1908*, vol. 8, pp. 91–406. Swedish Army General Staff Lithographic Institute, Stockholm, 1922. (Add. P. Pelliot, *TP*, 1928, **25**, 98.)

HERRMANN, A. (12). 'Das geographische Bild Chinas im Altertum.' *SA* (Forke-Festschrift Sonderausgabe), 1937, p. 72.

HERTWIG, A. (1). 'Aus der Geschichte der Strassenbautechnik.' *BGTI/TG*, 1934, **23**, 1.

D'HERVEY ST DENYS, M. J. L. (1) (tr.). *Ethnographie des Peuples Étrangers à la Chine; ouvrage composé au 13e siècle de notre ère par Ma Touan-Lin...avec un commentaire perpétuel.* Georg & Mueller, Geneva, 1876–1883. 4 vols. [Translation of chs. 324–48 of the *Wên Hsien Thung Khao* of Ma Tuan-Lin.] Vol. 1. Eastern Peoples; Korea, Japan, Kamchatka, Thaiwan, Pacific Islands (chs. 324–7). Vol. 2. Southern Peoples; Hainan, Tongking, Siam, Cambodia, Burma, Sumatra, Borneo, Philippines, Moluccas, New Guinea (chs. 328–32). Vol. 3. Western Peoples (chs. 333–9). Vol. 4. Northern Peoples (chs. 340–8).

HERVOUET, Y. (1). *Un Poète de Cour sous les Han; Sseu-ma Siang-Jou [Ssuma Hsiang-Ju].* Presses Univ. de France, Paris, 1964. (Biblioth. de l'Inst. des Htes. Études Chinoises, no. 19.) Rev. T. Pokora, *ARO*, 1967, **35**, 334.

HEWSON, J. B. (1). *A History of the Practice of Navigation.* Brown & Ferguson, Glasgow, 1951.

VAN DER HEYDEN, A. A. M. & SCULLARD, H. H. (1). *Atlas of the Classical World*. Nelson, London, 1959.

HEYERDAHL, T. (1). *The Kon-Tiki Expedition; by Raft across the South Seas*. Allen & Unwin, London, 1950. Eng. tr. of *Kon-Tiki Ekspedisjonen*. Gyldendal Norsk, Oslo, 1948.

HEYERDAHL, T. (2). *American Indians in the Pacific; the Theory behind the Kon-Tiki Expedition*. Allen & Unwin, London, 1952; Rand McNally, New York, 1953.

HEYERDAHL, T. (3). *Aku-Aku; the Secret of Easter Island*. Allen & Unwin, London, 1958.

HEYERDAHL, T. (4). 'The Voyage of the Raft *Kon-Tiki*.' *GJ*, 1950, **115**, 20.

HEYERDAHL, T. (5). 'Voyaging Distance and Voyaging Time in Pacific Migration.' *GJ*, 1951, **117**, 69.

HEYERDAHL, T. (6). 'Feasible Ocean Routes to and from the Americas in Pre-Columbian Times.' *AMA*, 1963, **28**, 482. Orig. in *Proc. 35th Internat. Congress of Americanists*, Mexico City, 1962. Vol. 1, p. 133.

HEYERDAHL, T. (7). 'Plant Evidence for Contacts with America before Columbus.' *AQ*, 1964, **38**, 120.

HEYERDAHL, T. & SKJÖLSVOLD, A. (1). 'Archaeological Evidence of pre-Spanish visits to the Galápagos Islands.' *MSAA*, no. 12; suppl. to *AMA*, 1956, **22** (no. 2).

HEYMANN, R. E. (1). *An Approach to Early Art from the Psychology of Technical Drawing*. Communication to the Xth Internat. Congress of the History of Science, Ithaca, N.Y., 1962.

HIGHET, G. (1). 'An Iconography of Heavenly Beings.' *HZ*, 1960, **3** (no. 2), 39.

HILDEBRAND, H. (1). *Der Tempel Ta-Chüeh-Sy [Ta-Chio Ssu, near Peking]*. Asher, Berlin, 1897.

HIRTH, F. (1). *China and the Roman Orient*. Kelly & Walsh, Shanghai; G. Hirth, Leipzig and Munich, 1885. (Photographically reproduced in China with no imprint, 1939.)

HIRTH, F. (9). *Über fremde Einflüsse in der chinesischen Kunst*. G. Hirth, München and Leipzig, 1896.

HIRTH, F. (10). 'Biographisches nach eigenen Aufzeichnungen.' *AM*, 1922, **1**, 1. (Hirth Presentation Volume.) (With complete bibliography of the writings of F. Hirth appended.)

HIRTH, F. (12). 'Scraps from a Collector's Notebook.' *TP*, 1905, **6**, 373 (biographies of Chinese painters and archaeologists). Subsequently reprinted in book form, Stechert, New York, 1924.

HIRTH, F. (13). 'Early Chinese Notices of East African Territories.' *JAOS*, 1909, **30**, 46.

HIRTH, F. (14). 'Über den Seeverkehr Chinas im Altertum nach chinesischen Quellen.' *GZ*, 1896, **2**, 444.

HIRTH, F. (16). 'Über den Schiffsverkehr von Kinsay [Quinsay, Hangchow] zu Marco Polo's Zeit', also 'Der Ausdruck So-Fu' and 'Das Weisse Rhinoceros'. *TP*, 1894, **5**, 386.

HIRTH, F. (17). 'Bausteine zu eine Geschichte d. chinesischen Literatur, als Supplement zu Wylie's "Notes on Chinese Literature".' *TP*, 1895, **6**, 314, 416; 1896, **7**, 295, 481.

HIRTH, F. (18). 'The Word "Typhoon"; its History and Origin', *JRGS*, 1881, **50**, 260; 'Teifun', *OLL*, 1896, **10**, 1132.

HIRTH, F. (21). 'Contributions to the History of Oriental Trade during Antiquity', *CR*, 1889, **18**, 41; 'Zur Geschichte des antiken Orienthandels', *VGEB*, 1889, **16**, 46.

HIRTH, F. (22). 'Contributions to the History of Oriental Trade during the Middle Ages', *CR*, 1889, **18**, 307; 'Zur Geschichte des Orienthandels im Mittelalter', *GB*, 1890, **56** (nos. 14–15), 209, 236.

HIRTH, F. (23). 'Ü. den Seehandel Chinas im Altertum und Mittelalter.' *JMGG*, 1894–5, cxv.

HIRTH, F. & ROCKHILL, W. W. (1) (tr.). *Chau Ju-Kua; His work on the Chinese and Arab Trade in the 12th and 13th centuries, entitled 'Chu-Fan-Chi'*. Imp. Acad. Sci., St Petersburg, 1911. (Crit. G. Vacca, *RSO*, 1913, **6**, 209; P. Pelliot, *TP*, 1912, **13**, 446; E. Schaer, *AGNT*, 1913, **6**, 329; O. Franke, *OAZ*, 1913, **2**, 98; A. Vissière, *JA*, 1914 (11ᵉ sér.), **3**, 196.)

HITTI, P. K. (1). *History of the Arabs*. 4th ed. Macmillan, London, 1949; 6th ed. 1956.

HO PEI-HUNG (ed.) (1). *A Note on the Tukiangyien Irrigation System [the Tuchiangyen Works at Kuan-hsien, Szechuan]*. Szechuan Provincial Water Conservancy Bureau, Chêngtu, 1943. [Szechuan Hydraulic Publication Series, **1**, no. 6.]

HO PING-TI (1). 'The Introduction of American Food Plants into China.' *AAN*, 1955, **57**, 191.

HO PING-TI (3). 'Loyang (+495 to +534); a Study of Physical and Socio-Economic Planning of a Metropolitan Area.' *HJAS*, 1966, **26**, 52.

HODGE, A. TREVOR (1). *The Woodwork of Greek Roofs*. Cambridge, 1960.

HODGE, A. TREVOR (2). 'A Roof at Delphi.' *ABSA*, 19, **49**, 202.

HODOUS, L. (1). *Folkways in China*. Probsthain, London, 1929.

HOLDICH, T. (1). *Tibet the Mysterious*. Rivers, London, n.d.

HOLLAMBY, T. (1) (ed.). 'Andrew Boyd; his Life and Work.' *K*, 1962, **36** (no. 3), 1 (memorial issue).

HOLMES, G. C. V. (1). *Ancient and Modern Ships*. Chapman & Hall, London, 1906. (Victoria & Albert Museum Science Handbook.)

HOLMES, T. R. (1). 'Could Ancient Ships Sail to Windward?' *CQ*, 1909, **3**, 26.

HOLZMAN, D. (1). 'Shen Kua and his *Mêng Chhi Pi Than*.' *TP*, 1958, **46**, 260.

HOLZMAN, D. (3). 'The *Lo-Yang Chhieh Lan Chi* and its Author [Yang Hsüan-Chih].' *Proc. 14th Conference of Junior Sinologists, Breukelen, 1962*

HOMMEL, R. P. (1). *China at Work; an illustrated Record of the Primitive Industries of China's Masses, whose Life is Toil, and thus an Account of Chinese Civilisation*. Bucks County Historical Society, Doylestown, Pa., 1937; John Day, New York, 1937.

HONG SUKKI & RAHN, H. (1). 'The Diving Women of Korea and Japan.' *SAM*, 1967, **216** (no. 5), 34.

HOOKER, J. D. (1). *Himalayan Journals; Notes of a Naturalist in Bengal, the Sikkim and Nepal Himalayas, the Khasia Mountains, etc.* Colosseum, Glasgow, n.d. (1869?).

HOOKHAM, H. (1). *Tamburlaine the Conqueror.* Hodder & Stoughton, London, 1962.

VAN DER HOOP, A. N. J. TH. À TH. (1). *Megalithic Remains in South Sumatra.* Zutphen, 1932.

HOPKINS, L. C. (5). 'Pictographic Reconnaissances, I.' *JRAS*, 1917, 773.

HOPKINS, L. C. (25). 'Metamorphic Stylisation and the Sabotage of Significance; a Study in Ancient and Modern Chinese Writing.' *JRAS*, 1925, 451.

VON HORNBOSTEL, E. M. (3). 'Über ein akustische Kriterium für Kultur-zusammenhänge.' *ZFE*, 1911, **43**, 601.

HORNELL, J. (1). *Water Transport; Origins and Early Evolution.* Cambridge, 1946. Rev. M. J. B. Davy, *N*, 1947, **159**, 419; P. Paris, *MRMVK*, 1948 (no. 3), 39.

HORNELL, J. (2). 'Balancing Devices in Canoes and Sailing Craft.' *ETH*, 1945, **1**, 1.

HORNELL, J. (3). 'A Tentative Classification of Arab Sea-Craft.' *MMI*, 1942, **28**, 11.

HORNELL, J. (4). 'The Frameless Boats of the Middle Nile.' *MMI*, 1939, **25**, 417; 1940, **26**, 125.

HORNELL, J. (5). 'The Boats of Lake Menzala, Egypt.' *MMI*, 1947, **33**, 94.

HORNELL, J. (6). 'Constructional Parallels in Scandinavian and Oceanic Boat Construction.' *MMI*, 1935, **21**, 411.

HORNELL, J. (7). 'The Origin of the Junk and Sampan.' *MMI*, 1934, **20**, 331.

HORNELL, J. (8). 'Origins of Plank-built Boats.' *AQ*, 1939, **13**, 35.

HORNELL, J. (9). 'Primitive Types of Water Transport in Asia; Distribution and Origins.' *JRAS*, 1946, 124.

HORNELL, J. (10). 'The Significance of the Dual Element in British Fishing-Boat Construction.' *FLV*, 1946, **10**, 113.

HORNELL, J. (11). *British Coracles and Irish Curraghs; with a Note on the Quffah of Iraq.* Society for Nautical Research, London, 1938.

HORNELL, J. (12). 'Evolution of the Clinker-built Fishing Lugger.' *AQ*, 1936, **10**, 341.

HORNELL, J. (13). 'The Tongue and Groove Seam of Gujerati Boatbuilders.' *MMI*, 1930, **16**, 309.

HORNELL, J. (14). 'Sea Trade in Early Times.' *AQ*, 1941, **15**, 233.

HORNELL, J. (15). 'Naval Activity in the Days of Solomon and Rameses III.' *AQ*, 1947, **21**, 66.

HORNELL, J. (16). 'The Rôle of Birds in Early Navigation.' *AQ*, 1946, **20**, 142.

HORNELL, J. (17). 'The Origins and Ethnological Significance of Indian Boat Designs.' *MAS/B*, 1920, **7**, 139.

HORNELL, J. (18). 'The Sailing Ship in Ancient Egypt.' *AQ*, 1943, **17**, 27.

HORNELL, J. (19). 'Indonesian Influence on East African Culture.' *JRAI*, 1934, **64**, 305.

HORNELL, J. (20). 'The Fishing and Coastal Craft of Ceylon.' *MMI*, 1943. 'The Pearling Fleets of South India and Ceylon.' *MMI*, 1945.

HORNELL, J. (21). *Fishing in Many Waters.* Cambridge, 1950.

HORNELL, J. (22). 'South American *balsas*; the problem of their Origin.' *MMI*, 1931, **17**, 347.

HORNELL, J. (23). 'Survivals of the Use of Oculi in Modern Boats.' *JRAI*, 1923, **53**, 298. 'Boat Oculi Survivals; Additional Records.' *JRAI*, 1938, **68**, 347.

HORNELL, J. (24). 'The Outrigger-Nuggar of the Blue Nile.' *AQ*, 1938, **12**, 354.

HORWITZ, H. T. (6). 'Beiträge z. aussereuropäischen u. vorgeschichtlichen Technik.' *BGTI*, 1916, **7**, 169.

HORWITZ, H. T. (12). 'Über Urtümliche Seil-, Ketten- und Seilbahn-Brücken.' *BGTI/TG*, 1934, **23**, 94.

HOSIE, A. (4). *Three Years in Western Szechuan; a Narrative of Three Journeys in Szechuan, Kweichow and Yunnan.* Philip, London, 1890.

HOUCKGEEST, A. E. VAN BRAAM. See van Braam Houckgeest.

HOURANI, G. F. (1). *Arab Seafaring in the Indian Ocean in Ancient and Early Mediaeval Times.* Princeton Univ. Press, Princeton, N.J., 1951. (Princeton Oriental Studies, no. 13.)

HOURANI, G. F. (2). 'Direct Sailing between the Persian Gulf and China in pre-Islamic Times.' *JRAS*, 1947, 157.

VAN HOUTEN, J. H. (1). 'Protection contre la Mer et Assèchements en Hollande.' *MC/TC*, 1953, **2**, 133.

HOUTOM-SCHINDLER, A. (1). 'Note on the Kur River in Fārs; its Sources and Dams and the District it irrigates.' *PRGS*, 1891, **13**, 287.

HÖVER, O. (1). 'Das Lateinscgcl — *Velum Latinum* — *Velum Laterale*.' *AN*, 1957, **52**, 637.

HOWANDER, B. & ÅKELBLAD, H. (1). 'Wasavarvet i Stockholm (the *Vasa* Museum in Stockholm).' *ARK*, 1962, **62** (no. 9), 237.

HSIA NAI (1). 'New Archaeological Discoveries.' *CREC*, 1952, **1** (no. 4), 13.

HSIA NAI (2). 'Tracing the Thread of the Past.' *CREC*, 1959, **8** (no. 10), 45.

HSIA NAI (4). 'Opening an Imperial Tomb.' *CREC*, 1959, **8** (no. 3), 16.

HSIA NAI (5). 'Our Neolithic Ancestors.' *CREC*, 1956, **5** (no. 5), 24.

HSIANG TA (1). 'A Great Chinese Navigator.' *CREC*, 1956, **5** (no. 7), 11.

HSIANG TA & HUGHES, E. R. (1). 'Chinese Books in the Bodleian Library.' *BQR*, 1936, **8**, 227.

HSIANG WÊN-HUA (1). 'River Control Benefits Hopei Farmers' (the Hai River System near Tientsin). *CREC*, 1963, **12** (no. 2), 30.

HSIAO YÜ (1). 'Recherches sur Mong Kiang Niu [history of the *Mêng Chiang Nü* ballad].' *S*, 1948, **1**, 189.

HSIMÊN LU-SHA (1). 'Giant against Flood' (the Fu-tzu-ling dam on the Pi River, a tributary of the Huai River, part of the latter's Control Project). *CREC*, 1955, **4** (no. 2), 28.

HSÜ CHI-HÊNG (1). 'The Man who built the first Chinese Railway [Chan Thien-Yu].' *CREC*, 1955, **4** (no. 7), 26.

HSÜ CHING-CHIH (SU GIN-DJIH)(1). *Chinese Architecture; Past and Contemporary*. Sin Poh Amalgamated H.K. Ltd, Hongkong, 1964; Swindon Book Co., Kowloon, Hongkong, 1964.

HSÜ MING (1). 'The Excavation of the Thang capital (Chhang-an).' *EHOR*, 1963, **2** (no. 9), 12.

HSÜEH PEI-YUAN (1). 'Water Conservancy Two Thousand Years Ago' (the Kuanhsien Works, the Chêngkuo Canal, and the Chhin irrigation canal along the Yellow River in Ninghsia). *CREC*, 1957, **6** (no. 10), 9.

HSÜEH TU-PI (1). 'Water Conservancy.' Art. in *Chinese Yearbook, 1943*, ed. Chang Chi-Hsien, p. 530. Thacker, Bombay, 1943.

HU CHHANG-TU (1). 'The Yellow River Administration in the Chhing Dynasty.' *FEQ*, 1955, **14**, 505.

HU CHIA (1). *Peking Today and Yesterday*. Foreign Languages Press, Peking, 1956.

HU SHIH (5). 'A Note on Chhüan Tsu-Wang, Chao I-Chhing and Tai Chen; a Study of Independent Convergence in Research as illustrated in their works on the *Shui Ching Chu*.' In Hummel (2), p. 970.

HUA LO-KÊNG (1). 'Operational Research blossoms in China.' *CREC*, 1961, **10** (no. 8), 24.

HUANG, JEN-YÜ (1). 'The Grand Canal during the Ming Dynasty.' Inaug. Diss. Univ. of Michigan, Ann Arbor, Mich., 1964.

HUANG, RAY. See Huang Jen-Yü.

HUANG WÊN-HSI & CHIANG PHÊNG-NIEN (1). 'Research on Characteristics of Materials of Dams Constructed by Dumping Soils into Ponded Water.' *SCISA*, 1963, **12**, 1213.

HUANGFU WÊN (1). 'North China's Biggest Reservoir' (account of the Miyün dam and reservoir on the Chhao and Pai Rivers in Northern Hopei). *CREC*, 1960, **9** (no. 2), 6.

HUARD, P. & DURAND, M. (1). *Connaissance du Việt-Nam*. Ecole Française d'Extr. Orient, Hanoi, 1954; Imprimerie Nationale, Paris, 1954.

HUARD, P. & HUANG KUANG-MING (M. WONG) (5). 'Les Enquêtes Françaises sur la Science et la Technologie Chinoises au 18e Siècle.' *BEFEO*, 1966, **53**, 137–226.

HUART, C. & DELAPORTE, L. (1). *L'Iran Antique, Élam et Perse, et la Civilisation Iranienne*. Albin Michel, Paris, 1943. (Evol. de l'Hum. Series, Prehist. no. 24.)

HUC, R. E. (1). *Souvenirs d'un Voyage dans la Tartarie et le Thibet pendant les Années 1844, 1845 & 1846* [with J. Gabet], revised ed., 2 vols. Lazaristes, Peiping, 1924. Abridged ed., *Souvenirs d'un Voyage dans la Tartarie le Thibet et la Chine...*, ed. H. d'Ardenne de Tizac, 2 vols. Plon, Paris, 1925. Eng. tr., by W. Hazlitt, *Travels in Tartary, Thibet and China during the years 1844 to 1846*. Nat. Ill. Lib. London, n.d. (1851–2). Also ed. P. Pelliot, 2 vols. Kegan Paul, London, 1928.

HUC, R. E. (2). *The Chinese Empire; forming a Sequel to 'Recollections of a Journey through Tartary and Thibet'*. 2 vols. Longmans, London, 1855, 1859.

HUCKER, C. O. (1). 'The Tung-Lin Movement in the Late Ming Period.' Art. in *Chinese Thought and Institutions*, ed. J. K. Fairbank, p. 132. Univ. Chicago Press, Chicago, 1957.

HUDEMANN, E. E. (1). *Geschichte d. römischen Postwesens während die Kaiserzeit*. Berlin, 1878. (Calvary's Philol. & Archaeol. Bibliothek, nos. 32 and 43.)

HUDSON, G. F. (1). *Europe and China; A Survey of their Relations from the Earliest Times to 1800*. Arnold, London, 1931 (rev. E. H. Minns, *AQ*, 1933, **7**, 104).

HUES, ROBERT. See C. R. Markham (2).

HUGHES, E. R. (7) (tr.). *The Art of Letters, Lu Chi's 'Wên Fu', A.D. 302; a Translation and Comparative Study*. Pantheon, New York, 1951. (Bollingen Series, no. 29.)

HUGHES, E. R. (9). *Two Chinese Poets; Vignettes of Han Life and Thought*. Princeton Univ. Press, Princeton, N.J., 1960.

HULBERT, H. B. (1). *History of Korea*. Seoul, 1905. (Revised edition ed. C. N. Weems, 2 vols. Hilary House, New York, 1962.)

HULLS, L. G. (1). 'The Possible Influence of Early Eighteenth Century Scientific Literature on Jonathan Hulls, a Pioneer of Steam Navigation.' *BBSHS*, 1951, **1**, 105.

H[ULSEWÉ], A. F. P. (3). Chinese Coins found in Somaliland (East Africa). *TP*, 1959, **47**, 81.

HUMMEL, A. W. (2) (ed.). *Eminent Chinese of the Chhing Period.* 2 vols. Library of Congress, Washington, 1944.

HUMMEL, A. W. (14). 'Chinese and other Asiatic Books added to the Library of Congress, 1928–1929.' *ARLC/DO*, 1928/1929, 285.

HUMMEL, A. W. (16). 'The History of a Bridge' [the Wan Nien Chhiao at Nanchhêng]. *ARLC/DO*, 1940, 159.

HUMMEL, A. W. (17). History of the Kuangling Iron Suspension-Bridge [over the Northern Phan Chiang in Kweichow]. *QJCA*, 1948, **5**, 23.

HUMMEL, A. W. (18). 'Ocean Transport in the Sixteenth Century [in China].' *ARLC/DO*, 1938, 235. [On the *Hai Yün Hsin Khao* of Liang Mêng-Lung (+1579).]

HUMMEL, A. W. (19). 'River Control and Coast Defence.' *ARLC/DO*, 1937, 187. [On the *Huai Yin Shih Chi* and the *Hai Fang Thu I* of Chang Chao-Yuan (+1600).]

HUMMEL, A. W. (20). 'A Rare MS. on the Construction of Imperial Palaces.' *ARLC/DO*, 1927–1928, 279.

HUMMEL, A. W. (21). 'Chinese Architecture.' *ARLC/DO*, 1937, 177.

HUNG HSIA-TIEN (1). 'From Marsh to State Farm.' *CREC*, 1962, **11** (no. 10), 34.

HUNTER, G. (1). 'A Note on some Tombs at Kaole (near Bagamoyo, East Africa).' *TNR*, 1954 (no. 37), 134.

HURST, H. E. (1). *The Nile; a general account of the River and the Utilisation of its Waters.* Constable, London, 1952.

HURTADO, E. D. & LITTLEHALES, B. (1). 'Into the Well of Sacrifice; Return to the Sacred Cenote [at Chichén-Itzá]—a Treasure-Hunt in the Deep Past.' *NGM*, 1961, **120**, 540 and 550.

HUTCHINSON, J. B. (SIR JOSEPH) (1). 'The History and Relationships of the World's Cottons.' *END*, 1962, **21** (no. 81), 5.

HUTCHINSON, J. B., SILOW, R. A. & STEPHENS, S. G. (1). *The Evolution of Gossypium and the Differentiation of the Cultivated Cottons.* Oxford, 1947.

HUTSON, J. (1). 'The Shu Country [Szechuan].' *JRAS/NCB*, 1922, **53**, 37; 1923, **54**, 25.

HUTSON, J. (2). 'West Szechuan's Most Remarkable Work; the Artificial Irrigation of Kuanhsien.' *EAM*, 1905, **4**, 145.

HUTSON, J. (3). 'Bridges of West China.' *EAM*, 1905, **4**, 356.

ICHIDA, MIKINOSUKE (2). 'A Biographical Study of Giuseppe Castiglione (Lang Shih-Ning), a Jesuit Painter at the Court of Peking under the Chhing Dynasty.' *MRDTB*, 1960, **19**, 79.

IDELER, L. (2). *Untersuchungen ü. den Ursprung und die Bedeutung der Stern-namen; ein Beytrag z. Gesch. des gestirnten Himmels.* Weiss, Berlin, 1809.

IDES, E. YSBRANTS[ZOON] (1). *Three Years Travels from Moscow overland to China, thro' Great Ustiga, Siriania, Permia, Sibiria, Daour, Great Tartary, etc. to Peking, containing an exact and particular Description of the Extent and Limits of those Countries, and the Customs of their Barbarous Inhabitants; with reference to their Religion, Government, Marriages, daily Imployment, Habits, Habitations, Diet, Death, Funerals, etc., written by his Excellency E. Y . . . I . . ., Ambassador from the Czar of Muscovy to the Emperor of China; to which is annex'd an accurate Description of China, done Originally by a Chinese Author [Dionysius Kao, surgeon, who embraced the Christian faith and travelled thro' Siam and India (p. 210)]; With several Remarks by way of Commentary, alluding to what our European Authors have writ of that Country [By a Learned Pen].* Eng. tr. from the Dutch. Freeman, Walthoe, Newborough, Nicholson & Parker, London, 1706. (E.Y.I. set out 14 March 1692 and had audience of the emperor Khang-Hsi, 16 Nov. 1693.)

IMBERDIO, F. (1). 'Les Routes Médiévales; Mythes et Realités Historiques.' *AHES/AHS*, 1939, **1**, 411.

INN, HENRY & LEE SHAO-CHANG. See Juan Mien-Chhu & Li Shao-Chhang (1).

INNOCENT, C. F. (1). *Development of English Building Construction.* Cambridge, 1916.

IORGA, N. (1) (ed.). *Oeuvres Inédites de Nicolas Milescu.* Cult. Naṭ., Bucarest, 1929. (Acad. Roum. Etudes et Rech. no. 3.)

ISSAWI, C. (1). 'Arab Geographers and the Circumnavigation of Africa.' *OSIS*, 1952, **10**, 117.

ITO, CHUTA. *Architectural Decoration in China* (tr. by Jiro Harada from *Shina Kenchiku Shoshoku*; see Index). 5 vols. Tokyo, 1941–5.

JACKSON, T. G. (1). *Byzantine and Romanesque Architecture.* 2 vols. Cambridge, 1920.

JACKSON, T. G. (2). *Gothic Architecture in France, England and Italy.* 2 vols. Cambridge, 1915.

JACOBSEN, T. & ADAMS, R. McC. (1). 'Salt and Silt in Ancient Mesopotamian Agriculture; Progressive Changes in Soil Salinity and Sedimentation contributing to the Break-up of Past Civilisations.' *SC*, 1958, **128**, 1251.

JACOBSEN, T. & LLOYD, S. (1). 'Sennacherib's Aqueduct at Jerwan [−690].' *CUOIP*, 1935, **24**, 1–52. (The oldest instance of an irrigation contour canal.)

JACOT, A. (1). 'Perspective and Chinese Art.' *CJ*, 1927, **7**, 236.

JAL, A. (1). *Archéologie Navale.* 2 vols. Arthus Bertrand, Paris, 1840. (Crit. R. C. Anderson, *MMI*, 1920, **6**, 18; 1945, **31**, 160; A. B. Wood, 1919, **5**, 81.)

JAL, A. (2). *Glossaire Nautique; Repertoire Polyglotte de Termes de Marine Anciennes et Modernes.* Didot, Paris, 1848.

JAMESON, H. LYSTER (1). 'The Japanese Artificially Induced Pearl.' *N*, 1921, **107**, 396, 621; **108**, 528.

JANSE, O. R. T. (5). *Archaeological Research in Indo-China.* 2 vols. (Harvard-Yenching Monograph Series, nos. 7 and 10). Harvard Univ. Press, Cambridge, Mass., 1947 and 1951. (Also in *RAA/AMG*, 1935, **9**, 144, 209; 1936, **10**, 42.)

JAUBERT, P. A. (1) (tr.). *Géographie d'Edrisi, traduite de l'Arabe en Français.* 2 vols. Impr. Roy. Paris, 1836. (Receuil de Voyages et de Mémoires publiés par la Société de Géographie, nos. 5 and 6.) This, though defective, is the only complete translation of Al-Idrīsī's *Nuzhat al-Mushtāq-fi Ikhtirāq al-Āfāq* (+1154) (Recreation of those who long to know what is beyond the Horizons). Cf. Dozy & de Goeje (1).

JENKIN, F. (1). *Bridges; an Elementary Treatise on their Construction and History.* Black, Edinburgh, 1878 [reprint from *EB*].

JENSEN, L. B. (1). *Man's Foods.* Gerrard, Champaign, Ill., 1953.

JENSEN, L. B. (2). *The Microbiology of Meats.* 3rd ed. Gerrard, Champaign, Ill., 1954.

JENYNS, R. SOAME (1). *A Background to Chinese Painting.* Sidgwick & Jackson, London, 1935.

JENYNS, R. SOAME (2). 'The Chinese Rhinoceros and Chinese Carving in Rhinoceros Horn.' *TOCS*, 1954, **29**, 31.

JEWITT, L. & HOPE, W. H. ST J. (1). *The Corporation Plate and Insignia of Office of the Cities and Towns of England and Wales.* London, 1895.

DE JODE, G. (1). *Speculum Orbis Terrarum* [atlas]. +1578.

JOHNSON, H. R. & SKEMPTON, A. W. (1). 'William Strutt's Cotton Mills, 1793 to 1812.' *TNS*, 1956, **30**, 179.

JONES, J. F. CDR. (1). *Narrative of a Journey, undertaken in April 1848, by Cdr. J. F. Jones, I.N. for the purpose of determining the Tract of the Ancient Nahrawān Canal* [on the left or north bank of the Tigris]. Selections from the Records of the Bombay Government, 1857. N.S. no. 43, 33–134.

JORDANUS CATALANUS (1), Bp. of Quilon. See Yule (3), and additional notes in Yule (2), vol. 3.

JOUSSE, MATHURIN (1). *Le Théatre de l'Art de Carpentier.* Griveau, La Flèche, 1627.

JUAN MIEN-CHHU & LI SHAO-CHHANG (1), with contributions by Chhen Shou-Yi, Thung Chün, Chhen Jung-Chieh *et al. Chinese Houses and Gardens.* Honolulu, Hawaii, 1940; 2nd ed. Hastings House, New York, 1950.

JULIEN, STANISLAS (4). 'Notes sur l'Emploi Militaire des Cerfs-Volants, et sur les Bateaux et Vaisseaux en Fer et en Cuivre, tirées des Livres Chinois.' *CRAS*, 1847, **24**, 1070.

JUMSAI NA AYUTYA, SUMET (1). 'Some Comparative Aspects of [the Hydraulic Engineering of] Angkor Thom and Ayutya [Ayut'ia].' *JASA*, 1966 (no. 2), unpaged.

JUMSAI NA AYUTYA, SUMET (2). *Water Towns; Forms and Societies—A Comparative Study of Water Towns and Cities as Revealed by their Physical and Sociological Forms.* Inaug. Diss. Cambridge, 1966.

JUMSAI NA AYUTYA, SUMET (3). 'Ayutya—Venice of South-east Asia; the Restoration of the Capital of Ancient Siam.' *UNESC*, 1966, **19** (no. 10), 4.

KAEYL, G. P. 'The Lateen Sail.' *MMI*, 1956, **42**, 154. Discussion by G. la Roërie, 1957, **43**, 76.

KAHLE, P. (4). 'A Lost Map of Columbus.' *GR*, 1933, **23**, 621. Reprinted in Kahle (3), p. 247.

KAHLE, P. (5). *Die verschollene Columbus-Karte von +1498 in einer türkischen Weltkarte von +1513.* Berlin & Leipzig, 1933.

KAHLE, P. (6). 'Nautische Instrumente der Araber im indischen Ozean.' Art. in *Oriental Studies in Honour of Dasturji Sahib Cursetji Pavry*, p. 176. Oxford, 1934. Reprinted in Kahle (3), p. 266.

KALTENBACH, J. (1). *Les Protestants sur les Galères et dans les Cachots de Marseille de 1545 à 1750.'* Église Reformée, Marseille, n.d. (1950?).

KALTENMARK, M. (2) (tr.). *Le 'Lie Sien Tchouan' [Lieh Hsien Chuan]; Biographies Légendaires des Immortels Taoistes de l'Antiquité.* Centre d'Etudes Sinologiques Franco-Chinois (Univ. Paris), Peking, 1953. (Crit. P. Demiéville, *TP*, 1954, **43**, 104.)

KALTENMARK, M. (3). 'Le Dompteur des Flots' (on the Han title Fu-Po Chiang-Chün). *HH*, 1948, **3** (nos. 1–2), 1–113.

KAMAL, YUSSUF (PRINCE) (1) (ed.). *Monumenta Cartographica Africae et Aegypti.* 14 vols. Privately published, 1935–9.

KAN CHI-CHAI (1). 'The Water came over the Mountains' (the building of a long irrigation contour canal with aqueducts from the Nieh River in the Tungliang Mts. in Kansu). *CREC*, 1958, **7** (no. 5), 24.

KAO FAN (1). 'Taming the Yellow River.' *CMR*, 1952, **122**, 341.

KAO FAN (2). 'The Huai River Project in its Second Year.' *CMR*, 1952, **122**, 437.

KARLGREN, B. (1). *Grammata Serica; Script and Phonetics in Chinese and Sino-Japanese. BMFEA*, 1940, **12**, 1. (Photographically reproduced as separate volume, Peiping, 1941.) Revised edition, *Grammata Serica Recensa*, Stockholm, 1957.

KARLGREN, B. (12) (tr.). 'The Book of Documents' [*Shu Ching*]. *BMFEA*, 1950, **22**, 1.

KARLGREN, B. (14) (tr.). *The Book of Odes; Chinese Text, Transcription and Translation.* Museum of Far Eastern Antiquities, Stockholm, 1950. (A reprint of the text and translation only from his papers in *BMFEA*, **16** and **17**; the glosses will be found in **14**, **16** and **18**.)

KATES, G. N. (1). *Chinese Household Furniture.* New York and London, 1948.

KAUFMAN, L. & ROCK, I. (1). 'The Moon Illusion.' *SAM*, 1962, **207** (no. 1), 120.

KAUTILYA. See Shamasastry.

KAWAKAMI, ITSUE K. (1). 'Studies on Pearl-Sac Formation; I, On the Regeneration and Transplantation of the Mantle Piece in the Pearl Oyster.' *MFSKU/E*, 1952, **1**, 83.

KEES, H. (1). *Ancient Egypt; a Cultural Topography*, ed. T. G. H. James. Faber & Faber, London, 1961. Tr. from the German: *Das alte Aegypten*, Klotz, Berlin, 1960.

KELLENBENTZ, M. (1). 'La Participation des Capitaux de l'Allemagne Meridionale aux Entreprises Portugaises de Découverte aux Environs de 1500.' Communication to the Vᵉ Colloque International d'Histoire Maritime. Lisbon, 1960.

KELLEY, D. H. (1). *Parallelisms in astronomy and calendar science between Amerindian and Asian Civilisations.* Unpublished material personally discussed, 1956.

KELLEY, D. H. (2). 'Calendar Animals and Deities.' *SWJA*, 1960, **16**, 317.

KELLING, R. (1). *Das chinesische Wohnhaus; mit einem II Teil über das fruhchinesische Haus unter Verwendung von Ergebnissen aus Übungen von Conrady im Ostasiatischen Seminar der Universität Leipzig, von Rudolf Keller und Bruno Schindler.* Deutsche Gesellsch. für Nat. u. Völkerkunde Ostasiens, Tokyo, 1935. (*MDGNVO*, Supplementband XIII.) Crit. P. Pelliot, *TP*, 1936, **32**, 372.

KELLING, R. & SCHINDLER, B. (1). *Das fruhchinesische Haus....* See Kelling (1).

KELLY, M. N. (1). 'Russia before the Mongols.' *GGM*, 1952, **25**, 400. *Mirror to Russia.* Country Life, London, 1951.

KEMP, E. G. (1). *The Face of China.* Chatto & Windus, London, 1909.

KENDREW, W. G. (1). *Climate.* Oxford, 1930.

KENNEDY, ADMIRAL SIR WILLIAM (1). *Hurrah for the Life of a Sailor; Fifty Years in the Royal Navy.* Nash, London, 1910.

KERBY, K. & MO TSUNG-CHUNG (MO ZUNG CHUNG) (1). *An Old Chinese Garden; a Threefold Masterpiece of Poetry, Calligraphy and Painting* (description, reproductions and translations of the *Cho Chêng Yuan Thu* by Wên Chêng-Ming, +1533). Chung-Hua Book Co. Shanghai, n.d. (1922).

KEYNES, SIR GEOFFREY (2). *The Life of William Harvey.* Oxford Univ. Press, Oxford, 1966.

KIDDER, J. E. (1). *Japan before Buddhism.* Praeger, New York, 1959; Thames & Hudson, London, 1959.

KIMBLE, G. H. (2). 'The Laurentian World-Map (+1351) with special reference to its Portrayal of Africa.' *IM*, 1935, **1**, 29.

KING, F. H. (1). 'The Wonderful Canals of China.' *NGM*, 1912, **23**, 931.

KING, F. H. (2). *Irrigation in Humid Climates.* U.S. Dept. of Ag. Farmers' Bull. no. 46. Govt. Printg. Off. Washington, 1896.

KING, F. H. (3). *Farmers of Forty Centuries; or, Permanent Agriculture in China, Korea and Japan.* Cape, London, 1927.

KING, MRS LOUIS. See Rin-Chen Lha-Mo.

KING-WEBSTER, W. A. (1). 'Experimental Nautical Research; —3rd-Millennium Egyptian Sails.' *MMI*, 1960, **46**, 150.

KIPLING, RUDYARD (1). [*Collected*] *Verse* (definitive edition). Hodder & Stoughton, London, 1912.

KIRBY, J. F. (1). *From Castle to Teahouse; Japanese Architecture of the Momoyama* [late +16th-century] *Period.* Tuttle, Rutland, Vt. and Tokyo, 1962.

KIRCHER, ATHANASIUS (1). *China Monumentis qua Sacris qua Profanis Illustrata.* Amsterdam, 1667. (French tr. Amsterdam, 1670.)

KIRCHHOFF, P. (1). *The Diffusion of a Great Religious System from India to Mexico* [the calendrical animal cycle]. Proc. 35th Internat. Congress of Americanists Mexico City, 1962. Vol. 1, p. 73.

KIRKMAN, J. S. (1). 'Historical Archaeology in Kenya.' *ANTJ*, 1957, **37**, 16.

KIRKMAN, J. S. (2). 'The Excavations at Kilepwa; an Introduction to the Mediaeval Archaeology of the Kenya Coast.' *ANTJ*, 1952, **32**, 168.

KIRKMAN, J. S. (3). *The Arab City of Gedi* [*near Malindi*]; *Excavations at the Great Mosque.* Oxford, 1954.

KIRKMAN, J. S. (4). 'The Tomb of the Dated Inscription at Gedi.' *RAI/OP*, 1960, no. 14.

KIRKMAN, J. S. (5). 'The Great Pillars of Malindi and Mambrui.' *ORA*, 1958 (n.s.), **4**, 55.

KIRKMAN, J. S. (6). 'Excavations at Ras Mkumbuu on the Island of Pemba.' *TNR*, 1959, **51**, 161.

KIRKMAN, J. S. (7). 'The Culture of the Kenya Coast in the Later Middle Ages; some Conclusions from Excavations 1948 to 1956.' *SAAB*, 1956, **11**, 89.

KIRKMAN, J. S. (8). 'Azanici Centri.' Art. in *Enciclopedia Universale dell'Arte*. Ist per la Collab. Cult., Venice & Rome, 1958. Vol. 2, p. 286.

KIRKMAN, J. S. (9). 'Mnarani of Kilifi; the Mosques and Tombs.' *AI/AO*, 1959, **3**, 95.

KIRKMAN, J. S. (10). 'Kinuni, an Arab Manor on the Coast of Kenya.' *JRAS*, 1957, 145.

KLAPROTH, J. (3). 'Description de la Chine sous la Règne de la Dynastie Mongole, traduite du Persan et accompagnée de Notes.' *JA*, 1833 (2ᵉ sér.), **11**, 335, 447. (From the *Jāmiʿal-Tawārīkh* of Rashīd al-Dīn al-Hamdānī, +1307; with criticisms of a previous translation by Hammer, *BSG*, 1831, **15**, 265.)

KLAPROTH, J. (4). *Recherches sur le Pays de Fousang*. Paris, 1831.

KLEBS, L. (1). 'Die Reliefs des alten Reiches (2980–2475 v. Chr.); Material zur ägyptischen Kultur-geschichte.' *AHAW/PH*, 1915, no. 3.

KLEBS, L. (2). 'Die Reliefs und Malereien des mittleren Reiches (7–17 Dynastie, *c.* 2475–1580 v. Chr.); Material zur ägyptischen Kulturgeschichte.' *AHAW/PH*, 1922, no. 6.

KLEMM, F. (1). *Technik; eine Geschichte ihrer Probleme*. Alber, Freiburg and München, 1954. (Orbis Academicus series, ed. F. Wagner & R. Brodführer.) Engl. tr. by Dorothea W. Singer, *A History of Western Technology*. Allen & Unwin, London, 1959.

KNAUSS, J. A. (1). 'The Cromwell Current.' *SAM*, 1961, **204** (no. 4), 105.

KOESTER, A. (1). *Das antike Seewesen*. Schoetz & Parrhysius, Berlin, 1923.

KOESTER, A. (2). *Studien z. Geschichte d. antiken Seewesens*. Dieterich, Leipzig, 1934. (Klio Beiheft no. 32 (NF no. 19).)

KOESTER, A. (3). *Schiffahrt und Handelsverkehr des östlichen Mittelmeeres im 3 und 2 Jahrtausend v. Chr.* Hinrichs, Leipzig, 1924. (Der Alte Orient, Beiheft no. 1.)

KOESTER, H. (1). 'Four thousand hours over China.' *NGM*, 1938, **73**, 571.

VAN KONIJNEUBURG, E. (1). *Shipbuilding from its Beginnings*. 3 vols. Exec. Cttee. of the Permanent International Association of Congresses of Navigation, Brussels, 1913.

KORRIGAN, P. (1). *Causerie sur la Pêche Fluviale en Chine*. T'ou-Sé-Wé, Shanghai, 1909.

KOVDA, V. A. (1). *Ocherki Prirody i Bochv Kitaya*. Acad. Sci. Moscow, 1959. Eng. tr. *Soils and the Natural Environment of China*. U.S. Joint Pubs. Research Service, Washington D.C., 1960. (Photocopied typescript.)

KRAMRISCH, S. (1). *The Art of India; Traditions of Indian Sculpture, Painting and Architecture*. Phaidon, London, 1955.

KRAUSE, F. (1). 'Fluss- und Seegefechte nach Chinesischen Quellen aus der Zeit der Chou- und Han-Dynastie und der Drei Reiche.' *MSOS*, 1915, **18**, 61.

KREICHGAUER, D. (1). 'Neue Beziehungen zwischen Amerika und der alten Welt.' Art. in *W. Schmidt Festschrift*, ed. W. Koppers, p. 366. Vienna, 1928.

KRENKOW, F. (1). 'The Construction of Subterranean Water Supplies during the Abbasid Caliphate.' *TGUOS*, 1951, **13**, 23.

KRETSCHMER, K. (3). 'Die Katalanische Weltkarte d. Bibliotheca Estense zu Modena.' *ZGEB*, 1897, **32**, 65, 191.

KRICKEBERG, W. (1). 'Beiträge zur Frage der alten Kulturgeschichtlichen Beziehungen zwischen Nord- und Süd-Amerika.' *ZFE*, 1934, **66**, 287.

KRICKEBERG, W. (2). 'Das mittelamerikanische Ballspiel und seine religiöse Symbolik.' *PAI*, 1949, **3**, 118.

KROEBER, A. L. (5). 'Structure, Function and Pattern in Biology and Anthropology.' *SM*, 1943, **56**, 105.

KROEBER, A. L. (6). 'The Concept of Culture in Science.' *JGE*, 1949, **3**, 182.

KROM, N. J. & VAN ERP, T. (1). *Barabudur; Archaeological and Architectural Description*. 3 large port-folios and 3 vols. text. Nijhoff, The Hague, 1927 and 1931.

KU CHIEH-KANG (1). *Autobiography of a Chinese Historian* (preface to *Ku Shih Pien*, q.v.), tr. A. W. Hummel. Leiden, 1931. (Sinica Leidensia series, no. 1.)

KU LEI (1). 'Tsaidam (Basin)' (and the new roads there). *CREC*, 1957, **6** (no. 4), 2.

KUWABARA, JITSUZO (1). 'On Phu Shou-Kêng, a man of the Western Regions, who was the Superintendent of the Trading Ships' Office in Chhüan-Chou towards the end of the Sung Dynasty, together with a general sketch of the Trade of the Arabs in China during the Thang and Sung eras.' *MRDTB*, 1928, **2**, 1; 1935, **7**, 1 (revs. P. Pelliot, *TP*, 1929, **26**, 364; S. E[lisséev], *HJAS*, 1936, **1**, 265). Chinese translation by Chhen Yü-Ching, Chunghua, Peking, 1954.

LAGERCRANTZ, S. (1). 'Inflated Skins and their Distribution.' *ETH*, 1944, **2**, 49.

LAMBTON, A. K. S. (1). *Landlord and Peasant in Persia*. Oxford, 1953.

LAMPREY, J. (Surgeon, 67th Regiment) (1). 'On Chinese Architecture.' *TRIBA*, 1866–7, 157.

LAN TIEN (1). 'Seventy Years Young' (autobiography of one of the railway engineers constructing the Pao-chhêng Railway). *CREC*, 1958, **7** (no. 4), 11.

LANCIOTTI, L. (2). 'L'Archeologia Cinese, Oggi.' *CINA*, 1958, **4**, 3.

LANCIOTTI, L. (3). 'Un Palazzo Imperiale Thang recentemente Scoperto.' *CINA*, 1961, **6**, 3.

DE LANDA, DIEGO (1). *Relation des Choses de Yucatan*. Spanish text of *Relacion de las Cosas de Yucatan* and French tr. by J. Genet, with annotations. 2 vols. Genet, Paris, 1928. (Collection de Textes relatifs aux anciennes Civilisations du Mexique et de l'Amérique Centrale, no. 1.) Eng. tr. by W. Gates, Baltimore, 1937.

LANDSTRÖM, BJÖRN (1). *The Ship; a Survey of the History of the Ship from the Primitive Raft to the Nuclear-Powered Submarine, with Reconstructions in Words and Pictures*. Tr. from *Skeppet*, by M. Phillips. Allen & Unwin, London, 1961.

LANE, A. & SERJEANT, R. B. (1). 'Pottery and Glass Fragments from the Aden Littoral, with Historical Notes.' *JRAS*, 1948, 108. Repr. without pagination, together with Doe (1) and some additional illustrations, in *ADAB*, 1965, no. 5.

LANE, C. E. (1). 'The *Teredo*.' *SAM*, 1961, **204** (no. 2), 132.

LANE, E. WILLARD (1). Description of the Kuanhsien Suspension Bridge. *CE*, 1931 (no. 1), 399.

LANE, F. C. (1). *Venetian Ships and Shipbuilders of the Renaissance*. Johns Hopkins Univ. Press, Baltimore, Md., 1934.

LANE, F. C. (2). 'Venetian Shipping during the Commercial Revolution.' *AHR*, 1933, **38**, 228.

LANE, F. C. (3). 'The Economic Meaning of the Invention of the Compass.' *AHR*, 1963, **68**, 605.

LANE, R. H. (1). 'Waggons and their Ancestors.' *AQ*, 1935, **9**, 140.

LANSER, O. (1). 'Zur Geschichte d. hydrometrischen Messwesens.' *BTG*, 1953, **15**, 25.

LAPICQUE, P. A. (1). 'Note sur le Canal de Hing-ngan [Hsing-an] en Kouang-si [Kuangsi].' *BEFEO*, 1911, **11**, 425.

LASKE, F. (1). *Der Ostasiatische Einfluss auf die Baukunst des Abendlandes, vornehmlich Deutschlands in 18 Jahrhundert*. Berlin, 1909.

LASSØE, J. (1). 'The Irrigation System at Ulḥu [−714].' *JCUS*, 1951, **5**, 21.

LATHAM, R. E. (1) (ed.). *The Travels of Marco Polo*. Penguin, London, 1958.

LATTIMORE, O. (1). *Inner Asian Frontiers of China*. Oxford Univ. Press, London and New York, 1940. (Amer. Geogr. Soc. Research Monograph Series, no. 21.)

LATTIMORE, O. (2). 'Origins of the Great Wall of China; a Frontier Concept in Theory and Practice.' *GR*, 1937, **27**, 529.

LATTIMORE, O., BOYD ORR, LORD, ROBINSON, JOAN, NEEDHAM, JOSEPH & KESWICK, J. (1). 'Chi Chhao-Ting—Scholar Revolutionary.' *ASIC*, 1964, **2** (no. 1), 9.

LAUFER, B. (1). *Sino-Iranica; Chinese Contributions to the History of Civilisation in Ancient Iran*. *FMNHP/AS*, 1919, **15**, no. 3 (Pub. no. 201) (rev. and crit. Chang Hung-Chao, *MGSC*, 1925 (ser. B), no. 5).

LAUFER, B. (3). *Chinese Pottery of the Han Dynasty*. (Pub. of the East Asiatic Cttee. of the Amer. Mus. Nat. Hist.) Brill, Leiden, 1909. (Photolitho re-issue, Tientsin, 1940.)

LAUFER, B. (8). 'Jade; a Study in Chinese Archaeology and Religion.' *FMNHP/AS*, 1912. Repub. in book form, Perkins, Westwood & Hawley, South Pasadena, 1946. Rev. P. Pelliot, *TP*, 1912, **13**, 434.

LAUFER, B. (9). 'Ethnographische Sagen der Chinesen.' In *Aufsätze z. Kultur u. Sprachgeschichte vornehmlich des Orients Ernst Kuhn gewidmet* (Kuhn Festschrift). Marcus, Breslau (München), 1916, p. 199.

LAUFER, B. (14). 'Optical Lenses' (in China and India). *TP*, 1915, **16**, 169, 562.

LAUFER, B. (15). 'Chinese Clay Figures, Pt. I; Prolegomena on the History of Defensive Armor.' *FMNHP/AS*, 1914, **13**, no. 2 (Pub. no. 177).

LAUFER, B. (28). 'Christian Art in China.' *MSOS*, 1910, **13**, 100.

LAUFER, B. (29). 'The Relations of the Chinese to the Philippine Islands.' *SMC*, 1907, **50**, 248.

LAUFER, B. (36). 'The Introduction of Maize into Eastern Asia.' Proc. XVth Internat. Congr. Americanists, Quebec, 1906 (1907), vol. 2, p. 223.

LAUFER, B. (37). 'The Introduction of the Ground-Nut into China.' Proc. XVth Internat. Congr. Americanists, Quebec, 1906 (1907), vol. 2, p. 259.

LAUFER, B. (38). 'Columbus and Cathay; the Meaning of America to the Orientalist.' *JAOS*, 1931, **51**, 87.

LAUFER, B. (39). 'The Reindeer and its Domestication.' *MAAA*, 1917, no. 4 (2), p. 91.

LAUGHTON, L. G. C. (1). 'The Bermuda Rig.' *MMI*, 1956, **42**, 333.

LAURAND, L. (1). 'Note sur le Gouvernail Antique.' *RPLHA*, 1937, **63**, 131.

LAWRENCE, A. W. (1). *Trade Castles and Forts of West Africa*. Cape, London, 1964.

LAWRENCE, D. H. (1). *Etruscan Places*. Secker, London, 1932.

LAYRISSE, M. & ARENDS, T. (1). 'The *Diego* Blood Factor in Chinese and Japanese.' *N*, 1956, **177**, 1083.

LAYTON, C. W. T. (1). *A Dictionary of Nautical Words and Terms*. Brown & Ferguson, Glasgow, 1955.

LEA, F. M. (1). *The Chemistry of Cement and Concrete*. Arnold, London, 1956.

LEACH, E. R. (1). 'Hydraulic Society in Ceylon.' *PP*, 1959 (no. 15), 2.

LECCHI, A. (1). *Trattato de Canali Navigabili*. Milan, 1776.

LECOMTE, LOUIS (1). *Nouveaux Mémoires sur l'État présent de la Chine*. Anisson, Paris, 1696. (Eng. tr. *Memoirs and Observations Topographical, Physical, Mathematical, Mechanical, Natural, Civil and Ecclesiastical, made in a late journey through the Empire of China, and published in several letters, particularly upon the Chinese Pottery and Varnishing, the Silk and other Manufactures, the Pearl Fishing, the History of Plants and Animals, etc. etc. translated from the Paris edition, etc.* 2nd ed. London, 1698. Germ. tr. Frankfurt, 1699–1700. Dutch tr. 's Graavenhage, 1698.

LEE, See Li.

LEE, C. E. (1). 'Some Railway [History] Facts and Fallacies.' *TNS*, 1960, **33**, 1.

LEE, C. E. (2). 'The Haytor Granite Tramroad.' *TNS*, 1963, **35**, 237.

LEE, S. (1) (tr.). *The Travels of Ibn Baṭṭuṭah*. Oriental Translation Cttee, Royal Asiatic Soc., London, 1829.

LEE SHAO-CHANG. See Juan Mien-Chhu & Li Shao-Chhang (1).

LEGER, A. (1). *Les Travaux Publics, les Mines et la Metallurgie aux Temps des Romains, la Tradition Romaine jusqu'à nos Jours*. 2 vols. Paris, 1875.

LEGGE, J. (1) (tr.). *The Texts of Confucianism, translated*: Pt. 1. The '*Shu Ching*', *the religious portions of the '* Shih Ching', the '*Hsiao Ching*'. Oxford, 1879. (*SBE*, no. 3; reprinted in various eds. Com. Press, Shanghai.) For the full version of the *Shu Ching* see Legge (10).

LEGGE, J. (2) (tr.). *The Chinese Classics, etc.*: Vol. 1. *Confucian Analects, The Great Learning, and the Doctrine of the Mean*. Legge, Hongkong, 1861; Trübner, London, 1861. Photolitho re-issue, Hongkong Univ. Press, Hongkong, 1960 with supplementary volume of concordance tables, etc.

LEGGE, J. (3) (tr.). *The Chinese Classics, etc.*: Vol. 2. *The Works of Mencius*. Legge, Hongkong, 1861; Trübner, London, 1861. Photolitho re-issue, Hongkong Univ. Press, Hongkong, 1960 with supplementary volume of concordance Tables, and notes by A. Waley.

LEGGE, J. (4) (tr.). *A Record of Buddhistic Kingdoms; being an account by the Chinese monk Fa-Hsien of his Travels in India and Ceylon (+399 to +414) in search of the Buddhist Books of Discipline*. Oxford, 1886.

LEGGE, J. (5) (tr.). *The Texts of Taoism*. (Contains (a) *Tao Tê Ching*, (b) *Chuang Tzu*, (c) *Thai Shang Kan Ying Phien*, (d) *Chhing Ching Ching*, (e) *Yin Fu Ching*, (f) *Jih Yung Ching*.) 2 vols. Oxford, 1891; photolitho reprint, 1927. (*SBE*, nos. 39 and 40.)

LEGGE, J. (7) (tr.). *The Texts of Confucianism*: Pt. III. *The '* Li Chi'. 2 vols. Oxford, 1885; reprint, 1926. (*SBE* nos. 27 and 28.)

LEGGE, J. (8) (tr.). *The Chinese Classics, etc.*: Vol. 4, Pts. 1 and 2. '*Shih Ching*'; *The Book of Poetry*. 1. The First Part of the *Shih Ching*; or, the Lessons from the States; and the Prolegomena. 2. The Second, Third and Fourth Parts of the *Shih Ching*; or the Minor Odes of the Kingdom, the Greater Odes of the Kingdom, the Sacrificial Odes and Praise-Songs; and the Indexes. Lane Crawford, Hongkong, 1871; Trübner, London, 1871. Repr., without notes, Com. Press, Shanghai, n.d. Photolitho re-issue, Hongkong Univ. Press, Hongkong, 1960 with supplementary volume of concordance tables, etc.

LEGGE, J. (9) (tr.). *The Texts of Confucianism*. Pt. II. *The '* Yi King' [*I Ching*]. Oxford, 1882, 1899. (*SBE*, no. 16.)

LEGGE, J. (10) (tr.). *The Chinese Classics, etc.* Vol. 3, Pts. 1 and 2. *The '* Shoo King' (*Shu Ching*). Legge, Hongkong, 1865; Trübner, London, 1865. Photolitho re-issue, Hongkong Univ. Press, Hongkong, 1960 with supplementary volume of concordance tables, etc.

LEGGE, J. (11). *The Chinese Classics, etc.* Vol. 5, Pts. 1 and 2. *The '* Ch'un Ts'ew' with the '*Tso Chuen*' ('*Chhun Chhiu*' and '*Tso Chuan*'). Lane Crawford, Hongkong, 1872; Trübner, London, 1872. Photolitho re-issue, Hongkong Univ. Press, Hongkong, 1960 with supplementary volume of concordance tables, etc.

LEHMANN-HARTLEBEN, K. (1). *Die antiken Hafenanlagen des Mittelmeeres; Beiträge zur Geschichte d. Städtebaues im Altertum*. Dieterich, Leipzig, 1923. (Klio Beiheft no. 14.)

LEITE, DUARTE (1). 'Lendas na Historia da Navegação Astronomica em Portugal.' *BIBLOS*, 1950, **26**, 413.

LEITE, DUARTE (2). *Historia dos Descobrimentos*, ed. V. Magalhães Godinho. Lisbon, 1958.

LELAND, C. G. (1). *Fusang; or, the Discovery of America by Chinese Buddhist Priests in the +5th Century*. Trübner, London, 1875.

LELIAVSKY BEY, S. (1). *An Introduction to Fluvial Hydraulics*. Constable, London, 1955. Rev. G. H. Lean. *N*, 1956, **178**, 711.

LELIAVSKY BEY, S. (2). *Irrigation and Hydraulic Design.* 3 vols. Chapman & Hall, London, 1955–60.

LELIAVSKY BEY, S. (3). 'Historic Development of the Theory of the Flow of Water in Canals and Rivers.' *EN*, 1951, **191**, 466, 498, 533, 565, 601.

LÉON PORTILLA, M. (1). 'Philosophy in the Cultures of Ancient Mexico.' Art. in *Cross-Cultural Understanding; Epistemology in Anthropology* (a Wenner-Gren Foundation Symposium), ed. F. S. C. Northrop & H. H. Livingston. Harper & Row, New York, 1964, p. 35.

LETHABY, W. R. (1). *Architecture, Nature and Magic.* Duckworth, London, 1956.

LETHBRIDGE, T. C. (1). 'Shipbuilding [in the Mediterranean Civilisations and the Middle Ages].' Art. in *A History of Technology*, ed. C. Singer *et al.* Vol. 2, p. 563. Oxford, 1956.

LETTS, M. (1) (ed. & tr.). *Mandeville's 'Travels'; Texts and Translations.* 2 vols. Hakluyt Society, London, 1953. (Hakluyt Society Pubs., 2nd ser. nos. 101, 102.)

LEUPOLD, J. (1). *Theatrum Machinarum Generale.* Leipzig, 1724. *Theatrum Machinarum Molarium.* Deer, Leipzig, 1735.

LEVENSON, J. R. (4). 'Ill-Wind in the Well-Field; the Erosion of the Confucian Ground of Controversy.' Art. in *The Confucian Persuasion*, ed. A. F. Wright. Stanford Univ. Press, Palo Alto, Calif., 1960, p. 268.

LÉVI, S. (2). 'Ceylan et la Chine.' *JA*, 1900 (9e sér.), **15**, 411. Part of Lévi (1).

LÉVI, S. (5). 'Les Marchands de Mer et leur Rôle dans le Bouddhisme Primitif.' *BAFAO*, 1929, 19.

LÉVI, S. (6). *Le Nepal.* 3 vols. Paris, 1907.

LÉVI, S. (7). 'Pour l'Histoire du *Rāmāyaṇa*.' *JA*, 1918 (11e sér.), **11**, 5 (86).

LEWIS, M., AYUKAWA HIROKO, CHOWN, B. & LEVINE, P. (1). 'The Blood-Group Antigen *Diego* in North American Indians and in Japanese.' *N*, 1956, **177**, 1084.

LEWIS, NORMAN (1). *Dragon Apparent; Travels in Indo-China.* Cape, London, 1951.

LI CHI (2). *The Formation of the Chinese People; an Anthropological Enquiry.* Harvard Univ. Press, Cambridge, Mass., 1928.

LI FU-TU (1). 'The Yellow River will Run Clear' (brief account of the multiple-purpose project for control and utilisation). *CREC*, 1955, **4** (no. 11), 2.

LI HSI-FAN (1). 'Putting our Old Enemy to Work' (account of the drainage of the low-lying country in Eastern Hopei near Tientsin). *CREC*, 1958, **7** (no. 11), 8.

LI HSIEH (LI I-CHIH) (1). 'Die Geschichte des Wasserbanes in China.' *BGTI*, 1932, **21**, 59.

LI HUI (1). 'A Comparative Study of the "Jew's Harp" among the Aborigines of Formosa and East Asia.' *AS/BIE*, 1956, **1**, 137.

LI HUI-LIN (1). 'Mu-Lan-Phi; a Case for Pre-Columbian Transatlantic Travel by Arab Ships.' *HJAS*, 1961, **23**, 114.

LI I-CHIH. See Li Hsieh.

LI SHUN-CHHING (LEE SHUN-CHING) (1). *Forest Botany of China.* Com. Press, Shanghai, 1935.

LI, S. T. (1). *Shipping in China; its Early Days.* Mei-Hua (for China Merchants Steam Navigation Co.), Thaipei, 1962.

LI YUNG-HSI (1) (tr.). *A Record of the Buddhist Countries*, by Fa-Hsien. Chinese Buddhist Association, Peking, 1957.

LIANG SSU-CHHÊNG (1). 'China's Architectural Heritage and the Tasks of Today.' *PC*, 1952 (Nov.), 30.

LIANG SSU-CHHÊNG (2). 'China's Oldest Wooden Structure.' *ASIA*, 1941, **41**, 384.

LIEBENTHAL, W. (8). 'The Ancient Burma Road—a Legend?' *JGIS*, 1956, **16**, 1.

LILIUS, ALEKO E. (1). *I Sailed with Chinese Pirates.* Arrowsmith, London, 1930.

LIN CHAO (1). 'The Tsinling [Chhin-ling] and Tapashan [Ta-pa Shan] (Mountains) as a Barrier to Communications between Szechuan and the Northwestern Provinces.' *JGSC*, 1947, **14**, 5.

LIN YÜ-THANG (5). *The Gay Genius; Life and Times of Su Tung-Pho.* Heinemann, London, 1948.

LIN YÜ-THANG (7). *Imperial Peking; Seven Centuries of China* (with an essay on the Art of Peking, by P. C. Swann). Elek, London, 1961.

LING SHUN-SHÊNG (1) = (1). 'The Formosan Sea-going Raft and its Origin in Ancient China.' *AS/BIE*, 1956, **1**, 25.

LING SHUN-SHÊNG (2) = (2). 'The Sacred Enclosures and Stepped Pyramidal Platforms [Altars of Heaven and Earth, etc.] at Peking, [and the History of Chinese Cosmic Religion and its Temples].' *AS/BIE*, 1963, **16**, 83.

LING SHUN-SHÊNG (3) = (3). 'The Origin of the Shê [Holy Place] in Ancient China.' *AS/BIE*, 1964, **17**, 36.

LING SHUN-SHÊNG (4) = (4). 'The Sacred Places [Chih] of the Chhin and Han Periods.' *AS/BIE*, 1964, **18**, 136.

LING SHUN-SHÊNG (5) = (5). 'A Comparative Study of the Ancient Chinese Fêng and Shan [Sacrifices] and the Ziggurats of Mesopotamia.' *AS/BIE*, 1965, **19**, 39.

LINK, A. E. (1). 'Biography of Shih Tao-An (+312 to +385).' *TP*, 1958, **46**, 1.

VAN LINSCHOTEN, JAN HUYGHEN (1). *Itinerario, Voyage ofte Schipvaert van J. H. van L. naer oost ofte Portugaels Indien* (+1579 to +1592). Amsterdam, 1596, 1598; ed. H. Kern, 's Gravenhage, 1910. Repr. C. E. Warnsinck-Delprat, 5 vols. 's Gravenhage, 1955. Eng. tr. by W. Phillip, *John Huighen Van Linschoten his discours of voyages into ye Easte and West Indies*. Wolfe, London, 1598. (Hakluyt Society Pubs., 1st ser., nos. 70, 71. London, 1885.) Ed. A. C. Burnell & P. A. Tiele. (Information about the Chinese coast dating from *c.* +1550 to +1588 collected at Goa *c.* +1583 to +1589.)

LINSLEY, R. K., KOHLER, M. A. & PAULHUS, J. L. H. (1). *Applied Hydrology*. McGraw-Hill, New York, 1949.

LIPS, J. E. (1). 'Foreigners in Chinese Plastic Art.' *ASIA*, 1941, **41**, 377.

VAN DER LITH, P. A. & DEVIC, L. M. (1) (tr.). *Le Livre des Merveilles de l'Inde* (the *'Agā'ib al-Hind* by Buzurg ibn Shahriyār al-Rāmhurmuzī, +953). Brill, Leiden, 1883.

LITTLE, ALICIA BEWICKE (MRS ARCHIBALD LITTLE) (1). *The Land of the Blue Gown*. Fisher Unwin, London, 1902.

LITTLE, A. J. (1). *Mount Omi [Omei], and Beyond*. Heinemann, London, 1901.

LITTLE, A. J. (2). 'The Irrigation of the Chêngtu Plateau.' *EAM*, 1904, **3**, 189.

LIU, JAMES T. C. See Liu Tzu-Chien.

LIU TUN-LI (LOU WING-SOU) (1). 'Rain[-God] Worship [and Rain-Making ceremonies] among the Ancient Chinese and the Nahua and Maya Indians.' *AS/BIE*, 1957 (no. 4), 31–108.

LIU TZU-CHIEN (1). 'An Early Sung Reformer; Fan Chung-Yen.' Art. in *Chinese Thought and Institutions*, ed. J. K. Fairbank. Univ. Chicago Press, Chicago, 1957, p. 105.

LIVINGSTONE, DAVID (1). *Missionary Travels and Researches in South Africa*. London, 1857.

LLEWELLYN, B. (1). 'A Chinese Cyclops; Down the Rapids in Chinese Tibet.' *CR/BUAC*, 1949, **2** (no. 2), 8. Reprinted as ch. 14 of *I left my Roots in China*. Allen & Unwin, London, 1953.

LLOYD, S. (1). 'Building in Brick and Stone [from Early Times to the Fall of the Ancient Empires].' Art. in 'A History of Technology', ed. C. Singer *et al*. Oxford, 1954. Vol. 1, p. 456.

LLOYD, S. & SAFAR, F. (1). 'Eridu; a preliminary Communication on the Second Season's Excavations.' *SUM*, 1948, **4**, 115.

LO CHÊ-WÊN (1) = (1). *The Yünkang Caves*. English supplement (tr.) issued at Yünkang. Cultural Objects Press, Peking, 1957.

LO HSIAO-CHIEN (K. H. C. Lo) (1). 'The Poon Lim Epic.' *CR/BUAC*, 1949. Also in *Penguin New Writing*, 1945, no. 24, p. 63.

LO JUNG-PANG (1). 'The Emergence of China as a Sea-Power during the late Sung and early Yuan Periods.' *FEQ*, 1955, **14**, 489. Abstract, *RBS*, 1955, **1**, 66.

LO JUNG-PANG (2). 'The Decline of the Early Ming Navy.' *OE*, 1958, **5**, 149.

LO JUNG-PANG (3). 'China's Paddle-Wheel Boats; the Mechanised Craft used in the Opium War and their Historical Background.' *CHJ/T*, 1960 (n.s.), **2** (no. 1), 189. Abridged Chinese tr. Lo Jung-Pang (1).

LO JUNG-PANG (4). 'The Controversy over Grain Conveyance during the Reign of Khubilai Khan (+1260 to +1294).' *FEQ*, 1953, **13**, 262.

LO JUNG-PANG (5). *Ships and Shipbuilding in the Early Ming Period*. Unpub. MS.

LO JUNG-PANG (6). *Communications and Transport in the Chhin and Han Periods*. Unpub. MS. (In the press.)

LO JUNG-PANG (7). 'Chinese Explorations of the Indian Ocean before the Advent of the Portuguese.' Unpub. MS. (A paper read at the Pacific Coast Branch of the Amer. Histor. Assoc., Seattle, Sept. 1960.)

LO JUNG-PANG (8). 'The Han Stock of Gold and what happened to it; a Variation on a Theme by H. H. Dubs.' Unpub. MS. (A paper read at the Western Branch of the Amer. Oriental Soc., Seattle, Apr. 1959.)

LO JUNG-PANG (9). *The Sung Navy, +960 to +1279*. (In the press.)

LO JUNG-PANG (10). *The Art of War in the Chhin and Han Periods*. Unpub. MS. (In the press.)

LO JUNG-PANG (11). 'Chinese Shipping and East–West Trade from the +10th to the +14th Century.' Communication to the International Congress of Maritime History, Beirut, 1966.

LO KAI-FU (1). 'The Basic Geography of China.' *CREC*, 1956, **5** (no. 12), 18.

LO, KENNETH H. C. See Lo Hsiao-Chien.

LO WU-YI (1). 'The Art of Ming Dynasty Furniture.' *CREC*, 1962, **11** (no. 5), 39.

LOCKE, L. L. (1). *The Quipu*. Amer. Mus. Nat. Hist, New York, 1923.

LOEWE, M. (2). 'The Orders of Aristocratic Rank in Han China.' *TP*, 1961, **48**, 97.

LÖFFLER, L. G. (1). 'Das Zeremonielle Ballspiel im Raum Hinterindiens.' *PAI*, 1955, **6**, 86.

LONG, G. W. (1). 'Indochina faces the Dragon.' *NGM*, 1952, **102**, 287 (302).

VAN DER LOON, P. (1). 'The Ancient Chinese Chronicles and the Growth of Historical Ideals.' Art. in *Historians of China and Japan*, ed. W. G. Beasley & E. G. Pulleyblank, p. 24. Oxford Univ. Press, London, 1961.

LOPEZ, R. S. (2). 'L'Evoluzione dei Transporti Terrestri nel Medio Evo.' *BCIC*, 1953, **1**. Eng. tr. 'The Evolution of Land Transport in the Middle Ages.' *PP*, 1956 (no. 9), 17.

LOPEZ, R. S. (4). 'The Trade of Mediaeval Europe; the South.' Art. in *Cambridge Economic History of Europe*, ed. M. Postan & E. E. Rich, vol. 2, p. 257. Cambridge, 1952.

LOTHROP, S. K. (1). 'Aboriginal Navigation of the North-west Coast of South America.' *JRAI*, 1932, **62**, 237.

DE LOTURE, R. & HAFFNER, L. (1). *La Navigation à travers les Ages; Évolution de la Technique Nautique et de ses Applications*. Payot, Paris, 1952.

LOU WING-SOU, DENNIS. See Liu Tun-Li.

LOVEGROVE, H. (1). 'Junks of the Canton River and the West River System.' *MMI*, 1932, **18**, 241.

LOVEJOY, A. O. (3). 'The Chinese Origin of a Romanticism.' In *Essays in the History of Ideas*, p. 99. Johns Hopkins Univ. Press, 1948. Also *JEGP*, 1933, **32**, 1.

LOWDERMILK, W. G. (1). 'Relation of deforestation, slope-cultivation, erosion and overgrazing, to increased silting of Yellow River.' *AFFL*, 1925, **31** (July), no. 379.

LOWDERMILK, W. G. (2). 'The Kuanhsien Irrigation System.' *AFFL*, 1943 (Sept.).

LOWDERMILK, W. G. (3). 'Erosion and Floods in the Yellow River Watershed.' *JF*, 1924, **22** (no. 6), 11.

LOWDERMILK, W. G. (4). 'Forest Destruction and Slope Denudation in the Province of Shansi.' *CJ*, 1926, **4**, 127.

LOWDERMILK, W. G. (5). 'Measurements of Rainfall and Run-off in Temple Forests.' Proc. 3rd Pacific Science Congress. Tokyo, 1926, p. 2122.

LOWDERMILK, W. G. (6). 'The Changing Evaporation-Precipitation Cycle of North China.' *ESC*, 1926, **25**, Paper no. 5.

LOWDERMILK, W. G. (7). 'China Fights Erosion with U.S. Aid.' *NGM*, 1945, **87**, 641.

LOWDERMILK, W. G. & LI TÊ-I (1). 'Forestry in Denuded China.' *AAPSS*, 1930, **152**, 127.

LOWDERMILK, W. G., LI TÊ-I & REN, C. T. (1). *A Cover and Erosion Survey of the Huai River Catchment Area*. 1926, MS. deposited in the Office of the Soil Conservation Service, Washington DC, USA.

LOWDERMILK, W. G. & SMITH, J. R. (1). 'Notes on the Problem of Field Erosion.' *GR*, 1927, **17**, 227.

LOWDERMILK, W. G. & WICKES, D. R. (1). 'Ancient Irrigation in China brought up to date.' *SM*, 1942, **55**, 209.

LOWDERMILK, W. G. & WICKES, D. R. (2). 'China and America against Soil Erosion. I. The Fate of Conservation in Northern Shansi. II. Losses and Gains.' *SM*, 1943, **56**, 393 and 505.

LOWDERMILK, W. G. & WICKES, D. R. (3). 'History of Soil Use in the Wu-Thai Shan Area.' *JRAS/NCB*, Special Monograph, 1938.

LOWELL, P. (1). *Chosön, the Land of the Morning Calm; a Sketch of Korea*. Trübner, London, n.d. [1888].

LU GWEI-DJEN (1). 'China's Greatest Naturalist; a Brief Biography of Li Shih-Chen.' *PHY*, 1966, **8**, 383 Abridgement in *Proc. XIth. Internat. Congress of the History of Science*, Warsaw, 1965, vol. 5, p. 50.

LU KUEI-CHEN. See LU GWEI-DJEN.

LUCAS, F. L. (1). *Gilgamesh, King of Erech*. Golden Cockerel, London, 1948.

LUCKENBILL, D. D. (1) (tr.). 'The Annals of Sennacherib.' *CUOIP*, 1924, **2**, 1–196.

LUM, PETER (1). *The Purple Barrier; the Story of the Great Wall of China*. Hale, London, 1960.

LUTHER, C. J. (1). 'Hippopodes (Horse-footed Men), the World's Early Skiers; Prehistoric and Early Records of Ski-ing.' *BSKIY*, 1952, **15**, 57.

LYMAN, J. (1). 'Registered Tonnage and its Measurement.' *ANEPT*, 1945, **5**, 223, 311. 'Tonnage-Weight and Measurement.' *ANEPT*, 1948, **8**, 99.

MACARTNEY, GEORGE (LORD MACARTNEY) (1). *Journal kept during his Embassy to the Chhien-Lung Emperor (+1793 and +1794)*, ed. J. L. Cranmer-Byng (2). Longmans, London, 1962.

MACARTNEY, GEORGE (LORD MACARTNEY) (2). *Observations on China*, ed. J. L. Cranmer-Byng (2). Longmans, London, 1962.

McCRINDLE, J. W. (7) (tr.). *The Christian Topography of Cosmas [Indicopleustes], an Egyptian monk* (written c. +547). London, 1897. (Hakluyt Society Pubs., 1st ser., no. 98.)

McCURDY, G. G. (1). *Human Origins*. 2 vols. New York, 1924.

McGOWAN, D. J. (6). *Pearls and Pearl-making in China*. 1854.

McGREGOR, J. (1). 'On the Paddle-Wheel and Screw Propeller, from the Earliest Times.' *JRSA*, 1858, **6**, 335.

McIVER, D. R. (1). *Mediaeval Rhodesia*. Macmillan, London, 1906.

McKAY, E. (1). *Early Indus Civilisations*, ed. D. McKay. Luzac, London, 2nd ed. 1948.

McROBERT, I. (1). 'The Chinese Yuloh [self-feathering propulsion oar].' *MMI*, 1940, **26**, 313.

MADDISON, F. (2). 'Hugo Helt and the Rojas Astrolabe Projection.' *RFCC*, 1966, **39**, 5. (Junta de Investigações do Ultramar, Agrupamento de Estudos de Cartografia Antiga, Secção de Coimbra, no. 12.)

MADEIRA, J. A. (1). 'Estudo Histórico-Cientifico, sob o aspecto gnomónico, da Figura Radiada de pedra tosca suposta coeva do Infante Dom Henrique, existente na sua antiga "Vila de Sagres".' In Resumo das Comunicações do Congresso Internacional de História dos Descobrimentos. Lisbon, 1960, p. 37. Actas, vol. 2, p. 451.

MAGAILLANS. See De Magalhaens.

DE MAGALHAENS, GABRIEL (1). *A New History of China, containing a Description of the Most Considerable Particulars of that Vast Empire.* Newborough, London, 1688. Tr. from *Nouvelle Relation de la Chine.* Barbin, Paris, 1688. The work was written in 1668.

MAJOR, R. H. (1). *The Life of Prince Henry of Portugal, surnamed the Navigator.* Asher, London, 1868.

MALINOWSKI, B. (1). *Argonauts of the Western Pacific.* London, 1922.

MALONE, C. B. (1). 'Current Regulations for Building and Furnishing Chinese Imperial Palaces, 1727–1750.' *JAOS*, 1929, **49**, 234.

MALONE, C. B. (2). *History of the Peking Summer Palaces under the Chhing Dynasty.* Urbana, Ill., 1934. (Sep. from *UIB*, 1934, **31** (no. 41), 1–247.) Rev. J. H. Shryock, *JAOS*, 1934, **54**, 443; photolitho re-issue, Paragon, New York, 1966.

MANDEVILLE, SIR JOHN (+1362). See Letts, M.

MANGELSDORF, P. C. (1). 'Reconstructing the Ancestor of Corn [i.e. Maize].' *ARSI*, 1959, 495 (Pub. no. 4408).

MANGELSDORF, P. C. (2). 'The Mystery of Corn.' *SAM*, 1950, **183** (no. 1), 20.

MANGELSDORF, P. C. & OLIVER, D. L. (1). 'Whence came Maize to Asia?' *HBML*, 1951, **14**, 263.

MANGELSDORF, P. C. & REEVES, R. G. (1). 'The Origin of Corn [Maize]; I, Pod Corn, the Ancestral Form.' *HBML*, 1959, **18**, 329.

MANGELSDORF, P. C. & REEVES, R. G. (2). 'The Origin of Corn [Maize]; III, Modern Races, the Product of Teosinte Introgression.' *HBML*, 1959, **18**, 389.

MANGELSDORF, P. C. & REEVES, R. G. (3). 'The Origin of Corn [Maize]; IV, The Place and Time of Origin.' *HBML*, 1959, **18**, 413.

MANNING, THOMAS. See Markham, C. R. (1).

MAO I-SHÊNG (1). 'The Stone Arch—Symbol of Chinese Bridges.' *CREC*, 1961, **10** (no. 11), 18.

MAO TSÊ-TUNG (2). 'Selected Works.' 5 vols. Lawrence & Wishart, London, 1954– .

MARAINI, F. (1). *Meeting with Japan.* Hutchinson, London, 1959.

MARAINI, FOSCO (2). *Hekura; the Diving Girls' Island,* tr. from the Italian by E. Mosbacher. Hamilton, London, 1962.

MARCH, B. (1). 'A Note on Perspective in Chinese Painting.' *CJ*, 1927, **7**, 69.

MARCH, B. (2). 'Linear Perspective in Chinese Painting.' *EA*, 1931, **3**, 113.

MARCH, B. (3). *Some Technical Terms of Chinese Painting.* Amer. Council of Learned Societies, Waverly, Baltimore, 1935. (ACLS Studies in Chinese and Related Civilisations, no. 2.)

MARCHAJ, C. A. (1). *Sailing Theory and Practice.* Dodd Mead, New York, 1964.

MARCHAL, HENRI (1). 'Rapprochements entre l'Art Khmer et les Civilisations polynésiennes et pre-colombiennes.' *JSA*, 1934, **26**, 213.

MARCHAL, HENRI (2). *Les Temples d'Angkor.* Guillot, Paris, 1955.

M[ARCUS], G. J. (2). 'A Note on the Beitiáss.' *MMI*, 1952, **38**, 139.

M[ARCUS], G. J. (3). 'Mast and Sail in the North.' *MMI*, 1952, **38**, 140.

DE MARÉ, E. S. (1). *The Canals of England.* Architectural Press, London, 1960.

MARGOULIÉS, G. (1). *Le Kou Wen [Ku Wên] Chinois; Receuil de Textes avec Introduction et Notes.* Geuthner, Paris, 1925. [Inaug. Diss. Paris.] (Rev. H. Maspero, *JA*, 1928, **212**, 174.)

MARGOULIÉS, G. (2). *Le 'Fou' [Fu] dans le Wen-Siuan [Wên Hsüan]; Etude et Textes.* Geuthner, Paris, 1925. [Supplementary Inaug. Diss. Paris.]

MARGOULIÉS, G. (3). *Anthologie Raisonnée de la Littérature Chinoise.* Payot, Paris, 1948.

MARGOULIÉS, G. (4). *Histoire de la Littérature Chinoise.* 2 vols. i (Prose), 1949; ii (Poésie), 1951. Payot, Paris.

MARGUET, F. (1). *Histoire Générale de la Navigation du 15e au 20e siècles.* Soc. d'Ed. Geogr. Maritimes et Colon. Paris, 1931.

DE MARICOURT, PIERRE. See Peregrinus, Petrus.

MARIOTTE, EDME (1). *Traité du Mouvement des Eaux.* Paris, 1686.

MARJAY, F. (ed.) (1). *Dom Henrique the Navigator.* Executive Committee for the Quincentenary Com-memorations of the Death of the Infante Dom Henrique, Lisbon, 1960. Contributions by Costa Brochado, Vitorino Nemésio, Fr. Maurício, Joaquim Bensaúde, Damião Peres, Teixeira da Mota and F. Marjay.

MARKHAM, C. R. (1) (ed.). *Narratives of the Mission of George Bogle to Tibet, and of the Journey of Thomas Manning to Lhasa.* Trübner, London, 1876. Germ. tr. *Aus dem Lände der lebenden Buddhas; Erzählungen von der Mission George Bogle's nach Tibet und Manning's Reise nach Lhasa (1774 u. 1812),* by M. von Brandt, 1909. (Bibl. denkw. Reisen, no. 3.)

MARKHAM, C. R. (2) (ed.). *Tractatus de Globis et eorum Usu* [+1594]; *a Treatise descriptive of the Globes constructed by Emery Molyneux and published* [*i.e. issued, at the end of*], +1592; *by Robert Hues*. Hakluyt Soc. London, 1889. (Hakluyt Society Pubs., 1st ser., no. 79.) The first English edition was London, 1639, repr. 1659.

MARQUART, J. (1). *Osteuropäische und Ostasiatische Streifzüge*. Leipzig, 1903. (Das Itinerar des Mis'ar ben al-Muhalhil nach der chinesischen Hauptstadt, pp. 74 ff.)

MARSDEN, P. R. V. & BONINO, M. (1). 'Roman Transom Sterns.' *MMI*, 1963, **49**, 143, 302.

MARTINI, M. (2). *Novus Atlas Sinensis*. 1655. See Schrameier (1) and Szczesniak (4).

MASIÁ, ANGELES (1). *Introducción a la Historia de España*. Apolo, Barcelona, 1943.

MASON, O. T. (1). 'Primitive Travel and Transportation.' *ARUSNM*, 1894, 237.

MASON, O. T. (2). *The Origins of Invention; a Study of Industry among Primitive Peoples*. Scott, London, 1895.

MASPERO, H. (4). 'Les Instruments Astronomiques des Chinois au temps des Han.' *MCB*, 1939, **6**, 183.

MASPERO, H. (8). 'Légendes Mythologiques dans le Chou King [*Shu Ching*].' *JA*, 1924, **204**, 1.

MASPERO, H. (10). 'Le Serment dans la Procédure Judiciaire de la Chine Antique.' *MCB*, 1925, **3**, 257.

MASPERO, H. (14). *Études Historiques; Mélanges Posthumes sur les Religions et l'Histoire de la Chine*, vol. 3, ed. P. Demiéville. Civilisations du Sud, Paris, 1950. (Publ. du Mus. Guimet, Biblioth. de Diffusion, no. 59.) Rev. J. J. L. Duyvendak, *TP*, 1951, **40**, 366.

MASPERO, H. (17). 'La Vie Privée en Chine à l'Époque des Han.' *RAA/AMG*, 1931, **7**, 185.

MASPERO, H. (18). 'Études d'Histoire d'Annam.' *BEFEO*, 1916, **16**, 1; 1918, **18** (no. 3), 1.

MASPERO, H. (28) (posthumous). 'Contribution à l'Etude de la Société Chinoise à la Fin des Chang [Shang-Yin] et au Début des Tcheou [Chou].' *BEFEO*, 1954, **46**, 335.

MASPERO, H. (30). 'Le Roman de Sou Ts'in [Su Chhin].' Art. in *Etudes Asiatiques, publiées à l'occasion du 25e Anniversaire de l'Ecole Française d'Extrême-Orient* [*à Hanoi*]. 2 vols. van Oest, Paris & Brussels, 1925. Vol. 2, p. 127. (Pubs. Ec. Fr. d'Extr. Or. nos. 19, 20.)

MASSA, J. M. (1). 'La Brouette.' *MC/TC*, 1952, **2**, 93.

AL-MAS'ŪDĪ. See de Meynard & de Courteille.

MATHEW, G. (1). 'The Culture of the East African Coast in the Seventeenth and Eighteenth Centuries, in the Light of recent Archaeological Discoveries.' *MA*, 1956, **56**, 65.

MATHEW, G. (2). 'The East Coast Cultures [of Africa].' *AFS*, 1958, **2** (no. 2), 59.

MATHEW, G. (3). 'Chinese Porcelain in East Africa and on the Coast of South Arabia.' *ORA*, 1956 (n.s.), **2**, 50.

MATHYS, F. K. (1). 'Der Militärskilauf und seine historische Entwicklung.' *DSS*, 1955, **30** (no. 12), 287.

MATSUMOTO, N., FUJITA, R., SHIMIZU, J., ESAKA, T. *et al.* (1). *Kamo; a Study of the Neolithic Site and a Neolithic Dugout Canoe discovered in Kamo, Chiba Prefecture, Japan*. Mita Shigakukai, Tokyo, 1952. (Pub. Hist. Dept., Fac. of Lit., Keio University, Archaeol. & Ethnol. Ser. no. 3.)

MAUNY, R. (1). *Les Navigations Médiévales sur les Côtes Sahariennes*. Lisbon, 1960.

MAYERS, W. F. (1). *Chinese Reader's Manual*. Presbyterian Press, Shanghai, 1874; reprinted, 1924.

MAYERS, W. F. (3). 'Chinese Explorations of the Indian Ocean during the +15th century.' (Partly a translation of the *Hsi-Yang Chhao Kung Tien Lu* of Huang Shêng-Tsêng, +1520.) *CR*, 1875, **3**, 219, 331; 1875, **4**, 61, 173.

MAYERS, W. F. (5). 'Chinese Junk Building.' *NQCJ*, 1867, **1**, 170.

MAYOR, R. J. G. (2). 'Slaves and Slavery [in Ancient Greece].' In *A Companion to Greek Studies*, ed. L. Whibley, pp. 416, 420. Cambridge, 1905.

MAZE, SIR FREDERICK (1). 'Notes concerning Chinese Junks.' *BBMMAG*, 1949, **1**, 17. 'Note on the Chinese Yuloh [self-feathering propulsion oar].' *MMI*, 1950, **36**, 55.

MEAD, C. W. (1). 'The Musical Instruments of the Incas.' *AMNH/AP*, 1924, **15**, no. 3.

MEARES, JOHN (1). *Voyages made in the Years 1788 and 1789, from China to the North-west Coast of America...Narrative of a Voyage performed in 1786 from Bengal in the ship* Nootka...*Observations on the probable Existence of a North-west Passage; and some Account of the Trade between the North-west Coast of America and China, and the latter country and Great Britain*. Walter, London, 1790.

MEDHURST, W. H. (1) (tr.). *The 'Shoo King' [Shu Ching], or Historical Classic* (Ch. and Eng.). Mission Press, Shanghai, 1846.

MEGGERS, B. J. & EVANS, C. (1) (ed.). 'Aboriginal Cultural Development in Latin America; an Interpretative Review.' *SMC*, 1963, **146**, no. 1, 1–148.

MEI YI-PAO (1) (tr.). *The Ethical and Political Works of Mo Tzu*. Probsthain, London, 1929.

DE MÉLY, F. (2). 'Le "De Monstris" Chinois et les Bestiaires Occidentaux.' *RA*, 1897 (3ᵉ ser.), **31**, 353.

DE MENDONÇA, LOPES (1). *Estudos sobre Navios Portugueses nos Séculos XV e XVI*. Lisbon, 1892.

DE MENDOZA, JUAN GONZALES (1). *Historia de las Cosas mas notables, Ritos y Costumbres del Gran Reyno de la China, sabidas assi por los libros de los mesmos Chinas, como por relacion de religiosos y oltras personas que an estado en el dicho Reyno*. Rome, 1585 (in Spanish). Eng. tr. Robert Parke, *The Historie of the Great & Mightie Kingdome of China and the Situation thereof; Togither with the Great*

Riches, Huge Citties, Politike Gouvernement and Rare Inventions in the same [undertaken 'at the earnest request and encouragement of my worshipfull friend Master Richard Hakluyt, late of Oxforde']. London, 1588 (1589). Reprinted in Spanish, Medina del Campo, 1595; Antwerp, 1596 and 1655; Ital. tr. Venice (3 editions), 1586; Fr. tr. Paris, 1588, 1589 and 1600; Germ. and Latin tr. Frankfurt, 1589. New ed. G. T. Staunton, London, 1853 (Hakluyt Society Pubs., 1st ser., nos. 14, 15). Spanish text again ed. P. F. García, Madrid, 1944. (España Misionera, no. 2.)

MERCATOR, GERARD (1). *Atlas.* 1613.

MERCER, H. C. (1). *Ancient Carpenter's Tools illustrated and explained, together with the Implements of the Lumberman, Joiner, and Cabinet-Maker, in use in the Eighteenth Century.* Bucks County Historical Society, Doylestown, Pennsylvania, 1929.

MERCKEL, C. (1). *Die Ingenieur-Technik im Altertum.* Springer, Berlin, 1899.

MERDINGER, C. J. (1). *Civil Engineering through the Ages.* Soc. Amer. Milit. Engineers, New York, 1963. (Rev. J. K. Finch, *TCULT*, 1964, **5**, 435.)

MERSENNE, MARIN (2). *Phaenomena Hydraulica et Pneumatica.* Paris, 1644.

MERTZ, HENRIETTE (1). *Pale Ink; Two Ancient Records of Chinese Exploration in America.* Orig. pub. Ralph Fletcher Seymour, Chicago, but imprint cancelled by label giving pr. pr. address, Box 207, Old Post Office Station, Chicago 90, Ill., n.d. (c. 1958).

MESKILL, J. (1) (tr. & ed.). *Chhoe Pu's Diary; a Record of Drifting across the Sea* [the *Phyohae-Rok*, written in +1488]. Univ. Arizona Press, Tucson, Ariz., 1965. (Monographs and Papers of the Association for Asian Studies, no. 17.)

MEYER P. (1). *Alexandre le Grand dans la Littérature Française du Moyen-Age.* 2 vols. Paris, 1886.

MEYERHOF, M. (1). 'Ibn al-Nafīs und seine Theorie d. Lungenkreislaufs.' *QSGNM*, 1935, **4**, 37.

MEYERHOF, M. (2). 'Ibn al-Nafīs (+13th century) and his Theory of the Lesser Circulation.' *ISIS*, 1935, **23**, 100.

DE MEYNARD, C. BARBIER (2). 'Le Livre des Routes et des Provinces par ibn Khordadbih [Ibn Khurdādhbih's *Kitāb al-Masālik w'al-Mamālik*, +846].' *JA*, 1865 (6e sér.), **5**, 5, 227, 446.

DE MEYNARD, C. BARBIER & DE COURTEILLE, P. (1) (tr.). *Les Prairies d'Or* (the *Murūj al-Dhabab* of al-Mas'ūdī, +947). 9 vols. Paris, 1861–77.

MICHEL, H. (3). *Traité de l'Astrolabe.* Gauthier-Villars, Paris, 1947. (Rev. F. Sherwood Taylor, *N*, 1948, **162**, 46.)

MICHEL, H. (14). 'Les Tubes Optiques avant le Télescope.' *CET*, 1954, **70** (nos. 5/6), 3.

MICKLE, W. J. (1) (tr.). *The Lusiad, or the Discovery of India; an Epic Poem* [by Luis de Camoens], translated from the original Portuguese by W.J.M. Jackson & Lister, Oxford, 1776; repr. 1778; 5th ed. London, 1877.

MIDDETON, . (1). *The Engraved Gems of Classical Times.*

MIELI, ALDO (1). *La Science Arabe, et son Rôle dans l'Evolution Scientifique Mondiale.* Brill, Leiden, 1938. (Repr. Mouton, the Hague, 1966 with a bibliography and analytic index by A. Mazaheri.)

MIELI, A. (2). *Panorama General de Historia de la Ciencia.* Vol. 1, *El Mundo Antiguo; griegos y romanos;* vol. 2, *El Mundo Islámico e el Occidente Medieval Cristiano.* Espasa-Calpe, Buenos Aires, 1945 and 1946; 2nd ed. 1952. (Nos. 1 and 5 respectively of Colección Historia y Filosofia de la Ciencia, ed. J. Rey Pastor.)

MIELI, A. (2) (contd.). *Panorama General de Historia de la Ciencia.* Vol. 3, *La Eclosión del Renacimiento;* vol. 4, *Leonardo da Vinci, Sabio.* Espasa-Calpe, Buenos Aires and Madrid, 1951 and 1950.

MIELI, A. (2) (contd.). *Panorama General de Historia de la Ciencia.* Vol. 5, *La Ciencia del Renacimiento; Matemática y Ciencias Naturales.* Espasa-Calpe, Buenos Aires and Mexico City, 1952.

MIELI, A. (2) (with D. PAPP & JOSÉ BABINI). *Panorama General de Historia de la Ciencia.* Vol. 6, *La Ciencia del Renacimiento; Astronomía, Física y Biologia;* vol. 7, *La Ciencia del Renacimiento; Las Ciencias Exactas en el Siglo XVII.* Espasa-Calpe, Buenos Aires, 1952 and 1954.

MIELI, A. (2) (with D. PAPP & JOSÉ BABINI). *Panorama General de Historia de la Ciencia.* Vol. 8, *El Siglo del Illuminismo;* vol. 9, *Biologia y Medicina en los Siglos XVII y XVIII;* vol. 10, *Las Ciencias Exactas en el Siglo XIX.* Espasa-Calpe, Buenos Aires, 1955 and 1958.

MILESCU, NICOLAIE (SPĂTARUL) (1). *Descrierea Chinei* (in Rumanian, originally written in Russian, c. 1676), with preface by C. Bărbulescu. Ed. Stat pentru Lit. şi Artă, Bucarest, 1958. This work in 58 chs. is, with the exception of chs. 3, 4, 5, 10 and 20 essentially a Russian translation and adaptation of Martin Martini's text accompanying the maps in his *Atlas Sinensis* (Amsterdam, 1655). Milescu prepared it in the course of his diplomatic mission (1675 to 1677) as the Ambassador of the Tsar of Russia to the Emperor of China. See Baddeley (2).

MILESCU, NICOLAIE (SPĂTARUL) (2). *Jurnal de Călătorie în China* (in Rumanian, originally written in Russian, 1677, as the report to the Tsar from his Ambassador to the Chinese Emperor), with preface by C. Bărbulescu. Ed. Stat pentru Lit şi Artă, Bucarest, 1956; repr. with a new preface by C. Bărbulescu, Ed. pentru Lit. Bucarest, 1962. Eng. tr. Baddeley (2), vol. 2, pp. 242 ff.

MILLER, A. A. (1). *Climatology.* 8th ed. Methuen, London, 1953; Dutton, New York, 1953. 1st ed. 1931.

MILLS, J. V. (1). 'Malaya in the *Wu Pei Chih* Charts.' *JRAS/M*, 1937, **15** (no. 3), 1.

MILLS, J. V. (3). 'Notes on Early Chinese Voyages.' *JRAS*, 1951, 17.

MILLS, J. V. (4). Translation of ch. 9 of the *Tung Hsi Yang Khao*. (Studies on the Oceans East and West) Unpub. MS

MILLS, J. V. (5). Translation of *Shun Fêng Hsiang Sung* (Fair Winds for Escort). Bodleian Library, Land Orient. MS. no. 145. Unpub. MS.

MILLS, J. V. (6). Translation of part of ch. 13 of the *Chhou Hai Thu Pien* (on shipbuilding, etc.). Unpub. MS.

MILLS, J. V. (7). 'Three Chinese Maps. [Two Coastal Charts (*c.* 1840) and a copy of the Chhien-Lung map of China, +1775.]' *BMQ*, 1953, **18**, 65.

MILLS, J. V. (8). 'Chinese Coastal Maps.' *IM*, 1954, **11**, 151.

MILLS, J. V. (9). 'The Largest Chinese Junk and its Displacement.' *MMI*, 1960, **46**, 147.

MILLS, J. V. (10). 'The Voyage from Kuala Pasé [in Sumatra] to Beruwala [in Ceylon].' Unpub. MS.

MILNE, W. C. (1). 'Pagodas in China.' *JRAS (Trans.)/NCB*, 1855 (1st. ser.), **2** (no. 5), 17.

MINORSKY, V. F. (3) (tr.). *Ḥudūd al-'Ālam*, '*The Regions of the World*', a Persian geography [+982], with introduction by W. Barthold. Luzac, London, 1937. (E. J. W. Gibb Memorial Series, no. 11.)

MIRAMS, D. G. (1). *A Brief History of Chinese Architecture*. Kelly & Walsh, Shanghai, 1940.

MIRSKY, JEANETTE (1). *The Great Chinese Travellers*. Allen & Unwin, London, 1965.

MIYAZAKI ICHISADA (1). 'Les Villes en Chine à l'Époque des Han.' *TP*, 1960, **48**, 376. Abstr. *RBS*, 1967, **6**, no. 38.

MIZUNO SEIICHI & NAGAHIRO TOSHIO (1) = (1). *Unkō Sekkutsu; The Yünkang Cave-Temples*. 16 vols. in 31 parts. Jimbun Kagaku Kenkyūsō, Kyoto, 1950–6.

MOCK, E. B. (1). *The Architecture of Bridges*. Mus. of Modern Art, New York, 1949.

MOKRI, M. (1). 'La Pêche des Perles dans le Golfe Persique.' *JA*, 1960, **248**, 381.

MOLES, ANTOINE (1). *Histoire des Charpentiers*. Gründ, Paris, 1949.

MOLL, F. (1). *Das Schiff in der bildenden Kunst vom Altertum bis zum Ausgang des Mittelalters*. Schroeder, Bonn, 1929.

MOLL, F. (2). 'History of the Anchor.' *MMI*, 1927, **13**, 293. 'Die Entwicklung des Schiffsankers bis zum Jahre 1500 n. Chr.' *BGTI*, 1919, **9**, 41.

MOLL, F. (3). 'Holtzschütz, seine Entwicklung v. d. Urzeit bis zum Umwandlung des Handwerkes in Fabrikbetrieb.' *BGTI*, 1920, **10**, 66.

MOLL, F. & LAUGHTON, L. G. CARR (1). 'The Navy of the Province of Fukien.' *MMI*, 1923, **9**, 364.

MOLLAT, M. (1). *Le Commerce Maritime Normand à la Fin du Moyen Age*. Paris, 1952.

MOLLAT, M. (2). 'Soleil et Navigation au Temps des Découvertes.' Art. in *Le Soleil à la Renaissance; Sciences et Mythes*, p. 89. Presses Univ. de Bruxelles, Brussels, 1965; Presses Univ. de France, Paris, 1965. (Travaux de l'Institut pour l'Etude de la Renaissance et de l'Humanisme, no. 2.)

MOLLAT, M., DENOIX, L. & DE PRAT, O. (1) (ed.). *Le Navire et l'Économie Maritime du Moyen-Age au XVIIIe Siècle, principalement en Méditerranée*. Proc. 2nd International Colloquium of Maritime History, Paris, 1957. Sevpen, Paris, 1958. (Bib. Gén. de l'École Prat. des Hautes Études, VIe Section.)

MOLLAT, M., DENOIX, L., DE PRAT, O., ADAM, P. & PERRICHET, M. (1). *Le Navire et l'Économie Maritime du Nord de l'Europe du Moyen Age au XVIIIe Siècle*. Proc. 3rd International Colloquium of Maritime History, Paris, 1958. Sevpen, Paris, 1960. (Bib. Gén. de l'École Prat. des Hautes Études, VIe Section.)

MOLLAT, M. & DE PRAT, O. (1) (ed.). *Le Navire et l'Économie Maritime du XVe au XVIIIe Siècles*. Proc. 1st International Colloquium of Maritime History, Paris, 1956. Sevpen, Paris, 1957. (Bib. Gén. de l'École Prat. des Hautes Études, VIe Section.)

MONARDES, NICHOLAS (1). *Joyfull Newes out of the Newe-Founde Worlde...* Allde & Norton, London, 1577, 1596. Eng. tr. by J. Frampton of *Historia Medicinal de todas las Cosas que se traen de nuestras Indias Occidentales que sirven al Uso de Medicina*. Trugillo, Seville, 1565, 1569, 1571, 1574, 1580, etc. Latin ed. Plantin, Antwerp, 1574, repr. 1582, 1605. Ed. S. Gaselee, 2 vols. Constable, London, 1925; Knopf, New York, 1925. (Tudor Translations series, nos. 9 and 10.)

MONGAIT, A. L. (1). *Archaeology in the U.S.S.R.* Penguin (Pelican), London, 1961.

MONNIER, M. (1). *La Tour d'Asie*. Paris, 1895. Extracts in *CFC*, 1961 (no. 12), 40.

DI MONTALBODDO, FRACANZANO (1). *Itinerarium Portugallesium e Lusitania in India et inde in occidentem et denum ad aquilonem*. Minuziano (?), Milan, 1508.

DI MONTALBODDO, FRACANZANO (2). *Poesi Novamente Retrovati*. Vicenza, 1507.

MOODY, CDR. A. B. (1). 'Early Units of Measurement and the Nautical Mile.' *JIN*, 1952, **5**, 262.

MOOKERJI. See Mukerji.

MOORE, SIR ALAN (1). 'Last Days of Mast and Sail; an Essay in Nautical Comparative Anatomy.' Oxford, 1925.

MOORE, SIR ALAN (2). 'Accounts and Inventions of John Starlyng [+1411].' *MMI*, 1914, **4**, 20.

M[OORE, SIR ALAN] & L[AUGHTON, G. CARR] (1). Discussion on the origin of lug-sails in Europe, in answer to a query by F. K. I[ ]. *MMI*, 1923, **9**, 190 and 252.

MOREL, P. (1). *Petite Histoire du Languedoc*. Arthaud, Grenoble and Paris, 1941.

MORELAND, W. H. (1). 'Ships of the Arabian Sea about 1500 A.D.' *JRAS*, 1939, 63 and 173.

MORETTI, G. (1). *Il Museo delle Nave Romane di Nemi*. Libreria dello Stato, Rome, 1940.

MORETTI, M. (1). *Tarquinia; la Tomba delle Nave*. Lerici, Milan, 1961.

MORGAN, E. (1) (tr.). *Tao the Great Luminant; Essays from 'Huai Nan Tzu', with introductory articles, notes and analyses*. Kelly & Walsh, Shanghai, n.d. (1933?).

DE MORGAN, JACQUES (1). *Recherches sur les Origines de l'Egypte*. 2 vols. Leroux, Paris, 1896.

MORGAN, M. H. (1). *Vitruvius; the Ten Books on Architecture*. Harvard Univ. Press, Cambridge, Mass., 1914.

MORLEY, S. G. (1). *The Ancient Maya*. Stanford Univ. Press, Palo Alto, Calif., 1946; 2nd ed. Oxford Univ. Press, Oxford, 1947.

MORRIS, E. P. (1). *The Fore-and-Aft Rig in America*. Yale University Press, New Haven, Conn., 1927.

MORRISON, J. S. (1). 'The Greek Trireme.' *MMI*, 1941, **27**, 14.

MORRISON, J. S. (2). 'Notes on certain Greek Nautical Terms and on Three Passages in I.G. ii² 1632.' *CQ*, 1947, **41**, 121.

MORRISON, J. S. & WILLIAMS, R. T. (1). *Greek Oared Ships, −900 to −322*. Cambridge, 1968.

DA MOTA, A. TEIXEIRA (1). 'L'Art de Naviguer en Méditerranée du 13e au 17e Siècle et la Création de la Navigation Astronomique dans les Océans.' Art. in Proc. 2nd International Colloquium of Maritime History, Paris, 1957, p. 127. Ed. M. Mollat, L. Denoix & O. de Prat.

DA MOTA, A. TEIXEIRA (2). 'Méthodes de Navigation et Cartographie Nautique dans l'Océan Indien avant le 16e siècle.' *STU*, 1963 (no. 11), 49. Sep. pub. Junta de Investigações do Ultramar, Lisbon, 1963. (Agrupamento de Estudios de Cartografia Antiga, Secção de Lisboa, no. 5.)

MOTHERSOLE, J. (1). *Hadrian's Wall*. Lane, London, 1922.

MOTZO, B. R. (1) (ed.). *Il Compasso da Navigare; opera Italiana della Metà del Secolo XIII [+1253].'* Univ., Cagliari, 1947. (Annali d. Fac. di Lett. e Filosofia, Università di Cagliari, no. 8.)

MOULE, A. C. (3). 'The Bore on the Ch'ien-T'ang River in China.' *TP*, 1923, **22**, 135 (includes much material on tides and tidal theory).

MOULE, A. C. (5). 'The Wonder of the Capital' (the Sung books *Tu Chhêng Chi Shêng* and *Mêng Liang Lu* about Hangchow). *NCR*, 1921, **3**, 12, 356.

MOULE, A. C. (9). 'The Ten Thousand Bridges of Quinsay.' *NCR*, 1922, **4**, 32.

M[OULE], [A.] C. (11). 'The Fireproof Warehouses of Lin-An [+13th cent. Hangchow].' *NCR*, 1920, **2**, 207.

MOULE, A. C. (15). *Quinsai, with other Notes on Marco Polo*. Cambridge, 1957. An extension of a number of previous papers, notably 'Marco Polo's Description of Quinsai'. *TP*, 1937, **33**, 105.

MOULE, A. C. (16). 'Relics of the Monk Sakugen's Visits to China, +1539 to +1541 and +1547 to +1550.' *AM* (n.s.), **3**, 59.

MOULE, A. C. & PELLIOT, P. (1) (tr. and annot.). *Marco Polo (+1254 to +1325); The Description of the World*. 2 vols. Routledge, London, 1938. Further notes by P. Pelliot (posthumously pub.). 2 vols. Impr. Nat. Paris, 1960.

MOULE, A. C. & YETTS, W. P. (1). *The Rulers of China, −221 to 1949; Chronological Tables compiled by A. C. Moule, with an Introductory Section on the Earlier Rulers, c. −2100 to −249 by W. P. Yetts*. Routledge & Kegan Paul, London, 1957.

MO ZUNG-CHUNG. See Kerby & Mo Tsung-Chung (1).

MUKERJI, RADHAKAMUD (1). *Indian Shipping; a History of the Sea-Borne Trade and Maritime Activity of the Indians from the Earliest Times*. Longmans Green, Bombay and Calcutta, 1912.

MULDER, W. Z. (1). 'The *Wu Pei Chih* Charts.' *TP*, 1944, **37**, 1.

MULDER, W. Z. (2). 'Het Chineesche Drakenbootfest.' *CI*, 1944, **6**, 153.

MÜLLER, W. (1). 'Stufenpyramiden in Mexico und Kambodscha.' *PAI*, 1958, **6**, 473.

MULLIKIN, M. A. (1). 'Thai Shan, Sacred Mountain of the East.' *NGM*, 1945, **87**, 699.

AL-MUQADDASĪ. See Ranking & Azoo.

MURASAWA, FUMIO (1). *The Castle, the National Treasure*. Min. of Ed., Shokoku-sha, Tokyo, 1962.

MURATA, J. & FUJIEDA, A. (1) = (1). *Chü-yung-kuan, the Buddhist Arch of the +14th Century at the Pass of the Great Wall north of Peking*. 2 vols. Faculty of Engineering, Kyoto Univ., 1957.

MURPHY, H. K. (1). '[Chinese] Architecture.' In *China*, ed. H. F. McNair. Univ. Calif. Press, Berkeley & Los Angeles, 1946. Ch. 23, p. 363.

MYRDAL, J. & KESSLE, G. (1). *Report from a Chinese Village*. Heinemann, London, 1964.

NAISH, G. P. B. (1). 'The "dyoll" and the bearing dial.' *JIN*, 1954, **7**, 205. With comment by W. E. May.

N[ANCE], R. M. (1). 'Smack Sails in the Fifteenth Century.' *MMI*, 1920, **6**, 343.

NANCE, R. M. (2). 'Spritsails.' *MMI*, 1913, **3**, 155.

NANCE, R. M. (3). *Sailing-Ship Models; a Selection from European and American Collections.* Halton, London, 1924; 2nd ed. much enlarged. Halton & Truscott Smith, London, 1949.

DE NAVARRETE, DOMINGO (1). *Tratados Historicos, Politicos, Ethicos y Religiosos de la Monarchia de China.* Infançon, Madrid, 1676. See Cummins (1).

DE NAVARRETE, DOMINGO (2). *Controversias Antiguas y Modernas de la Mission de la Gran China.* Partially printed, Madrid, 1677. See Cummins (1).

DE NAVARRO, J. M. (1). 'The Amber Trade-Routes.' *GJ*, 1925, **66**, 481.

NEDULOHA, A. (1). 'Kulturgeschichte des technischen Zeichnens.' *BTG*, 1957, **19**; 1958, **20**; 1959, **21**. Sep. pub. Springer, Vienna, 1960. (Rev. L. R. Shelby, *TCULT*, 1963, **4**, 217.)

NEEDHAM, JOSEPH (2). *A History of Embryology.* Cambridge, 1934. Revised ed. Cambridge, 1959; Abelard-Schuman, New York, 1959.

NEEDHAM, JOSEPH (4). *Chinese Science.* Pilot Press, London, 1945.

NEEDHAM, JOSEPH (18). 'Science in Southwest China. I, the Physico-Chemical Sciences.' *N*, 1943, **152**, 9. Repr. in Needham & Needham (1).

NEEDHAM, JOSEPH (19). 'Science in Southwest China. II, the Biological and Social Sciences.' *N*, 1943, **152**, 36. Repr. in Needham & Needham (1).

NEEDHAM, JOSEPH (21). 'Science in Western Szechuan. I. Physico-Chemical Sciences and Technology.' *N*, 1943, **152**, 343. Repr. in Needham & Needham (1).

NEEDHAM, JOSEPH (22). 'Science in Western Szechuan. II. Biological and Social Sciences.' *N*, 1943, **152**, 372. Repr. in Needham & Needham (1).

NEEDHAM, JOSEPH (24). 'Science in Kweichow and Kuangsi.' *N*, 1945, **156**, 496. Repr. in Needham & Needham (1).

NEEDHAM, JOSEPH (27). 'Limiting Factors in the Advancement of Science as observed in the History of Embryology.' *YJBM*, 1935, **8**, 1. Carmalt Memorial Lecture of the Beaumont Medical Club of Yale University.

NEEDHAM, JOSEPH (31). 'Remarks on the History of Iron and Steel Technology in China' (with French translation; 'Remarques relatives à l'Histoire de la Sidérurgie Chinoise'). In *Actes du Colloque International 'Le Fer à travers les Ages'*, pp. 93, 103. Nancy, Oct. 1955. (*AEST*, 1956, Mémoire no. 16.)

NEEDHAM, JOSEPH (32). *The Development of Iron and Steel Technology in China.* Newcomen Soc., London, 1958. (Second Biennial Dickinson Memorial Lecture, Newcomen Society.) Précis in *TNS*, 1960, **30**, 141; rev. L. C. Goodrich, *ISIS*, 1960, **51**, 108. Repr. Heffer, Cambridge, 1964. French tr. (unrevised, with some illustrations omitted and others added by the editors). *RHSID*, 1961, **2**, 187, 235; 1962, **3**, 1, 62.

NEEDHAM, JOSEPH (33). 'The Peking Observatory in A.D. 1280 and the Development of the Equatorial Mounting.' Art. in *Vistas of Astronomy* (Stratton Presentation Volume), ed. A. Beer, vol. 1, p. 67. Pergamon, London, 1955.

NEEDHAM, JOSEPH (38). 'The Missing Link in Horological History; a Chinese Contribution.' *PRSA*, 1959, **250**, 147. (Wilkins Lecture, Royal Society.) Abstract, with illustrations, in *NS*, 1958, **4** (no. 108), 1481.

NEEDHAM, JOSEPH (39). 'The Chinese Contributions to the Development of the Mariner's Compass.' Abstract in *Resumo das Comunicações do Congresso Internacional de História dos Descobrimentos*, p. 273. Lisbon, 1960. *Actas*, Lisbon, 1961, vol. 2, p. 311. Also *SCI*, 1961, **96**, 225.

NEEDHAM, JOSEPH (40). 'The Chinese Contributions to Vessel Control.' Abstract in *Resumo das Comunicações do Congresso Internacional de História dos Descobrimentos*, p. 274. Lisbon, 1960. *Actas*, Lisbon, 1961, vol. 2, p. 325. Also *SCI*, 1961, **96**, 123, 163. Polish abridgement by W. A. Drapella, *BN*, 1963–4, **6–7**, 33. And (with illustrations), French tr. as art. in 'Les Aspects Internationaux de la Découverte Océanique aux 15e et 16e Siècles'. *Actes du Ve Colloque International d'Histoire Maritime*, Lisbon, 1960, Sevpen, Paris, 1966.

NEEDHAM, JOSEPH (43). 'The Past in China's Present.' *CR/MSU*, 1960, **4**, 145 and 281; repr. with some omissions, *PV*, 1963, **4**, 115. French tr.: *Du Passé Culturel, Social et Philosophique Chinois dans ses Rapports avec la Chine Contemporaine*, by G. M. Merkle-Hunziker. *COMP*, 1960, no. 21–2, 261; 1962, no. 23–4, 113; repr. in *CFC*, 1960, no. 8, 26; 1962, no. 15–16, 1.

NEEDHAM, JOSEPH (44). 'The Ways of Szechuan.' *AH*, 1948, **1** (no. 3), 62.

NEEDHAM, JOSEPH (45). 'Poverties and Triumphs of the Chinese Scientific Tradition.' Art. in *Scientific Change; Historical Studies in the Intellectual, Social and Technical Conditions for Scientific Discovery and Technical Invention from Antiquity to the Present*, ed. A. C. Crombie, p. 117. Heinemann, London, 1963. With discussion by W. Hartner, P. Huard, Huang Kuang-Ming, B. L. van der Waerden & S. E. Toulmin (Symposium on the History of Science, Oxford, 1961). Also, in modified form: 'Glories and Defects....' in 'Neue Beiträge z. Geschichte d. alten Welt.', vol. 1 'Alter Orient und Griechenland', ed. E. C. Welskopf, Akad. Verl. Berlin, 1964. French tr. (of paper only) by M. Charlot, 'Grandeurs et Faiblesses de la Tradition Scientifique Chinoise'. *LP*, 1963, no. 111. Abridged version, 'Science and Society in China and the West', *SPR*, 1964, **52**, 50.

NEEDHAM, JOSEPH (46). 'An Archaeological Study-Tour in China, 1958.' *AQ*, 1959, **33**, 113.

NEEDHAM, JOSEPH (49). 'The Snowshoe and the Ski in Chinese Literature.' *BSKIY*, 1962, **20**, 15.

NEEDHAM, JOSEPH (55). 'Time and Knowledge in China and the West.' Art. in 'The Voices of Time; a Cooperative Survey of Man's Views of Time as expressed by the Sciences and the Humanities' ed. J. T. Frazer. Braziller, New York, 1966, p. 92.

NEEDHAM, JOSEPH (56). 'Time and Eastern Man.' *RAI/OP*, 1964. (Henry Myers Lecture.)

NEEDHAM, JOSEPH (57). 'China and the Invention of the Pound-Lock.' *TNS*, 1964, **36**, 85.

NEEDHAM, JOSEPH (59). 'The Roles of Europe and China in the Evolution of Oecumenical Science.' *JAHIST*, 1966, **1**, 1. As Presidential Address to Section X, British Association, Leeds, 1967, in *ADVS*, 1967, **24**, 83.

NEEDHAM, JOSEPH & LIAO HUNG-YING (1) (tr.). 'The Ballad of Mêng Chiang Nü weeping at the Great Wall.' *S*, 1948, **1**, 194.

NEEDHAM, JOSEPH & LU GWEI-DJEN (1). 'Hygiene and Preventive Medicine in Ancient China.' *JHMAS*, 1962, **17**, 429. Abridgment in *HEJ*, 1959, **17**, 170.

NEEDHAM, JOSEPH & LU GWEI-DJEN (2). 'Efficient Equine Harness; the Chinese Inventions.' *PHY*, 1960, **2**, 121.

NEEDHAM, JOSEPH & LU GWEI-DJEN (4) 'A Further Note on Efficient Equine Harness; the Chinese Inventions.' *PHY*, 1965, **7**, 70.

NEEDHAM, JOSEPH & NEEDHAM, DOROTHY M. (1) (ed.). *Science Outpost*. Pilot Press, London, 1948.

NEEDHAM, JOSEPH, WANG LING & PRICE, D. J. DE S. (1). *Heavenly Clockwork; the Great Astronomical Clocks of Mediaeval China*. Cambridge, 1960. (Antiquarian Horological Society Monographs, no. 1.) Prelim. pub. *AHOR*, 1956, **1**, 153.

NEMÉSIO, V. (1). *Vida e Obra do Infante Dom Henrique*. Comissão Executiva das Comemorações do Quinto Centenario da Morte do Infante Dom Henrique, Lisbon, 1959. (Colecção Henriquina, no. 2.)

NESTERUK, F. Y. (1). 'Vodnoye Khozyaistvo Kitaia (The Waterways of China; Hydromechanical and Hydrotechnical Engineering in Chinese History)' (in Russian). Art. in *Iz Istorii Nauki i Tekhniki Kitaya* (Essays in the History of Science and Technology in China), pp. 3–109. Acad. Sci. Moscow, 1955.

NESTERUK, F. Y. (2). 'Razvitie Gydro-energetisheskogo Stroitelstva v Kitaiskoi Narodnoi Republike (The Growth of Hydro-electric Power in the Chinese People's Republic)' (in Russian). Art. in *Nauki i Tekhniki v Stranach Vostoka* (Science and Technology in the Lands of the East), vol. 2, p. 7. Acad. Sci. Moscow, 1961.

NEUBURGER, A. (1). *The Technical Arts and Sciences of the Ancients*. Methuen, London, 1930. Tr. by H. L. Brose from *Die Technik d. Altertums*. Voigtländer, Leipzig, 1919. (With a drastically abbreviated index and the total omission of the bibliographies appended to each chapter, the general bibliography, and the table of sources of the illustrations.)

NEWBERRY, P. E. (1). *Beni Hasan [Excavations]*. Archaeol. Survey, London, 1893, 1894.

NIEUHOFF, J. (1). *L'Ambassade [1655–1657] de la Compagnie Orientale des Provinces Unies vers l'Empereur de la Chine, ou Grand Cam de Tartarie, faite par les Sieurs Pierre de Goyer & Jacob de Keyser; Illustrée d'une tres-exacte Description des Villes, Bourgs, Villages, Ports de Mers, et autres Lieux plus considerables de la Chine; Enrichie d'un grand nombre de Tailles douces, le tout receuilli par Mr Jean Nieuhoff...* (title of Pt. II: *Description Generale de l'Empire de la Chine, ou il est traité succinctement du Gouvernement, de la Religion, des Mœurs, des Sciences et Arts des Chinois, comme aussi des Animaux, des Poissons, des Arbres et Plantes, qui ornent leurs Campagnes et leurs Rivieres; y joint un court Recit des dernieres Guerres qu'ils ont eu contre les Tartares*). de Meurs, Leiden, 1665.

NIEUWHOFF. See Nieuhoff.

NISHIMURA, SHINJI (1). *A Study of Ancient Ships of Japan*. Soc. of Naval Architects, Waseda University, Tokyo, 1917–30.

Vol.	Pt.	Sect.	
	i		Floats.
	iii	1 & 2	Ancient Rafts of Japan.
VII	iv	3	Kagami-no-fune or Wicker Boats.
v	iv	1	Manashi-Katama or Meshless Basket Boats.
II	i	1	Hisago-buné or Calabash Boats [and Float-supported Rafts].

(The numbering of the different parts of this work, which appeared in several forms and editions, is confusing, especially as no library in the U.K. seems to contain the full set.)

NOAKES, J. L. (1). 'Celadons of the Sarawak Coast.' *SMJ*, 1949, **5**, 25.

DES NOËTTES, R. J. E. C. LEFEBVRE (1). *L'Attelage et le Cheval de Selle à travers les Ages; Contribution à l'Histoire de l'Esclavage.* Picard, Paris, 1931. 2 vols. (1 vol. text, 1 vol. plates). (The definitive version of *La Force Animale à travers les Ages.* Berger-Levrault, Nancy, 1924.) Abstracts *LN*, 1927 (pt. 1).

DES NOËTTES, R. J. E. C. LEFEBVRE (2). *De la Marine Antique à la Marine Moderne; La Révolution du Gouvernail.* Masson, Paris, 1935. (Rev. A. Mieli, *A*, 1936, **18**, 270; H. de Saussure, *RA*, 1937 (6e ser.), **10**, 90.

DES NOËTTES, R. J. E. C. LEFEBVRE (4). 'Le Gouvernail; Contribution à l'Histoire de l'Esclavage.' *MSAF*, 1932 (8e ser.), **8 (78)**, 24.

DES NOËTTES, R. J. E. C. LEFEBVRE (8). Autour du Vaisseau de Borobodur; l'Invention du Gouvernail.' *LN*, 1932, i; 1934 (no. 2934, 1 Aug.), 97.

DES NOËTTES, R. J. E. C. LEFEBVRE (9). 'La Voie Romaine et la Route Moderne.' *RA*, 1925 (5e ser.), **22**, 105; *LM*, 1925, **6**, 771.

NOGUERA, E. (1). *Guide-Book to the National Museum of Archaeology, History and Ethnology* [*Mexico City*]. Central News, Mexico City, 1938. (Popular Library of Mexican Culture, no. 2.)

NOOTEBOOM, C. (1). *Trois Problèmes d'Ethnologie Maritime.* (1) *l'Origine des Proues Bifides;* (2) *la Signification de la Proue Bifide;* (3) *Quelques types de Voiles de l'Asie Orientale; une Etude de Diffusion.* Museum voor Land- en Volken-Kunde & Maritiem Museum Prins Hendrik, Rotterdam, 1952.

NOOTEBOOM, C. (2) (with illustrations by G. R. G. Worcester). *Tentoonstelling van Chinese Scheepvaart* (in Dutch). Museum voor Land- en Volken-Kunde, Rotterdam, 1950.

NORDMANN, P. (1). 'Note sur le Gouvernail Antique.' *RPLHA*, 1938, **64**, 330.

NORTHROP, F. S. C. & LIVINGSTON, H. H. (1) (ed.). *Cross-Cultural Understanding; Epistemology in Anthropology* (a Wenner-Gren Foundation Symposium). Harper & Row, New York, 1964.

VAN NOUHUYS, J. W. (1). 'Chinese Anchors.' *MMI*, 1925, **11**, 96.

VAN NOUHUYS, J. W. (2). 'The Anchor.' *MMI*, 1951, **37**, 17, 238.

NOVOTNÝ, K., POCHE, E. & EHM, J. (1). *The Charles Bridge of Prague.* Poláček, Prague, 1947.

OBERHUMMER, E. (1). 'Alte Globen in Wien.' *AOAW/PH*, 1922, **59** (nos. 19–27), 87.

OBERHUMMER, E. (2). 'Schanghai.' *MGGW*, 1932, **74**, 1.

OGAWA, KAZUMASA (1). *Photographs of the Palace Buildings of Peking, compiled by the Imperial Museum of Tokyo.* Tokyo, 1906.

OHRELIUS, BENGT, CDR. (1). '*Vasa*', *the King's Ship.* Cassell, London, 1962; Rabén & Sjögren, Stockholm, 1962.

D'OHSSON, MOURADJA (1). *Histoire des Mongols depuis Tchinguiz Khan jusqu'à Timour Bey ou Tamerlan.* 4 vols. van Cleef, The Hague and Amsterdam, 1834–52.

OLBRICHT, P. (1). *Das Postwesen in China unter der Mongolenherrschaft im 13 und 14 Jahrhundert.* Harrassowitz, Wiesbaden, 1954. (Göttinger Asiatische Forschungen, no. 1.) Rev. J. Průsek, *ARO*, 1959, **27**, 478; E. Balazs, *TP*, 1956, **44**, 449.

O'MALLEY, C. D. (1). 'A Latin Translation (+1547) of Ibn al-Nafīs, related to the Problem of the Circulation of the Blood.' *JHMAS*, 1957, **12**, 248. Abstract in Actes du VIIIe Congrès International d'Histoire des Sciences, Florence, 1956, p. 716.

OMMANNEY, F. D. (1). *Eastern Windows.* Longmans Green, London, 1960.

OMMANNEY, F. D. (2). *Fragrant Harbour; a Private View of Hongkong.* Hutchinson, London, 1962.

ORANGE, J. (1). *The Chater Collection; Pictures relating to China, Hongkong, Macao, 1655 to 1860, with Historical and Descriptive Letterpress. . . .* Butterworth, London, 1924.

DA ORTA, GARCIA (1). *Colloquies on the Simples and Drugs of India* (with the annotations of the Conde de Ficalho). Sotheran, London, 1913. Eng. tr. by Sir Clements Markham of *Coloquios dos Simples e Drogas he Cousas Mediçinais da India, compostos pello Doutor G. da O. . . .* de Endem, Goa, 1563. Latin epitome by Charles de l'Escluze, Plantin, Antwerp, 1567, repr. 1574, and later standard edition, ed. Conde de Ficalho, Lisbon, 1895.

ORTELIUS, ABRAHAM (1). *Theatrum Orbis Terrarum* [Atlas]. Several editions +1570 to +1601. The China sheet is dated +1584 and attributed to Ludovicus Georgius, apparently a pen-name for Ortelius himself (see Bagrow, 1).

OSGOOD, C. (1). *The Koreans and their Culture.* Ronald, New York, 1951.

OSÓRIO, JERÓNIMO, Bp. of Silves (+1506 to +1580) (1). *Epistolae de Rebus Emmanuelis Lusitaniae Regis.* Colon. Agripp. (Cologne), 1574. (Repr. Typ. Acad. Reg., Coimbra, 1791.)

OSORIUS SILVENSIS (OSORIUS HIERONYMUS SILVENSIS). See Osório, Jerónimo (1).

VON DER OSTEN, H. H. (1). *Die Welt der Perser.* Stuttgart, 1956.

OTTE, E. (1). *Une Source Inédite pour l'Histoire de la Première Navigation Américaine; le Registre des Changes de la Casa de la Contratación (+1508 à +1510).* Communication to the Ve Colloque International d'Histoire Maritime, Lisbon, 1960.

OYA, F. (1). 'The Chemistry of the "Pearl-Essence" from the Hair-tail Fish.' *BJSSF*, 1954, **19**, 1061, 1065, 1123, 1127, 1130.

PAIK, L. G. (1) (tr.). 'From Koryu to Kyung by Soh Keung, Imperial Chinese Envoy to Korea in 1124 A.D.' (excerpts concerning boats and ships from Hsü Ching's *Kao-Li Thu Ching*). Printed as appendix to Underwood (1), *JRAS/KB*, 1933, **23**, 90.

PAINE, R. T. & SOPER, A. (1). *Art and Architecture of Japan*. Penguin (Pelican), London, 1955.

PALLOTTINO, M. (1). *La Necropoli di Cerveteri*. Libreria dello Stato, Rome, 1939.

PALLOTTINO, M. (2). *Etruscologia*. Hoepli, Milan, 1947.

PALMER, E. H. (1). 'The Desert of the Tih and the Country of Moab.' *PEFQ*, 1871 (n.s.), **1**, 28.

PANIKKAR, K. M. (1). *Asia and Western Dominance*. Allen & Unwin, London, 1953.

PANIKKAR, K. M. (2). *Malabar and the Portuguese, being a History of the Relations of the Portuguese with Malabar from +1500 to +1663*. Taraporevala, Bombay, 1929.

PANNELL, J. P. M. (1). *An Illustrated History of Civil Engineering*. Thames & Hudson, London, 1964. Rev. C. J. Merdinger, *TCULT*, 1965, **6**, 447.

PAO TSUN-PHÊNG (1) = (1). *On the Ships of Chêng Ho*. Chung-Hua Tshung-Shu, Thaipei and Hongkong, 1961. (Nat. Historical Museum Collected Papers on the History and Art of China, 1st ser., no. 6.)

PAO TSUN-PHÊNG (2) = (2). *A Study of the 'Castled Ships' of the Han Period*. Nat. Hist. Mus., Thaipei, Thaiwan, 1967. (Coll. Papers on the History and Art of China, no. 2.)

PAPINI, R. (1). *Francesco di Giorgio, Architetto*. 3 vols. Electa, Florence, 1946.

PAPP, D. & BABINI, JOSÉ. *Panorama General de Historia de la Ciencia*. Vols. 6 to 10. See Mieli (2).

PARAIN, C. (1). 'The Evolution of Agricultural Techniques.' Art. in *Cambridge Economic History of Europe*, vol. 1, ed. J. H. Clapham & E. Power, p. 118. Cambridge, 1941.

PARANAVITANA, S. (2). 'Some Regulations concerning Village Irrigation Works in Ancient Ceylon.' *CEYJHS*, 1958, **1**, 1.

PARANAVITANA, S. (3). 'The Tamil Inscription on the Galle Trilingual Slab.' *EZ*, 1933, **3**, 331.

PARIAS, L. H. (1) (ed.). *Histoire Universelle des Explorations*. 4 vols.
    Vol. 1, *De la Préhistoire à la Fin du Moyen-Âge*, L. R. Nougier, J. Beaujeu & M. Mallat.
    Vol. 2, *La Renaissance, +1415 à +1600*, J. Amsler.
    Vol. 3, *Le Temps des Grands Voiliers*, P. J. Charliat.
    Vol. 4, *Époque Contemporaine*, J. Rouch, P. E. Victor & H. Tazieff. Sant'Andrea, Paris, 1955. (Nouvelle Librairie de France.)

PARIS, F. E., ADMIRAL (1). *Essai sur la Construction Navale des Peuples Extra-Européens*. Arthus Bertrand, Paris, n.d. (1841–3).

PARIS, F. E., ADMIRAL (2). *Souvenirs de Marine; Collection de Plans ou Dessins de Navires et de Bateaux Anciens ou Modernes, Existants ou Disparus*. Gauthier-Villars, Paris.
    I   1882   contains Japanese and Indo-Chinese material.
    II   1884   contains some Chinese material.
    III   1886
    IV   1889   contains some Chinese material.
    V   1892
    VI   1908   contains some Japanese material.

PARIS, P. (1). 'Quelques Dates pour une Histoire de la Jonque Chinoise.' Paper at Congrès International d'Ethnographie Brussels, 1948. MS. copy kindly placed at our disposition by the author in 1950; subsequently pub. *BEFEO*, 1952, **46**, 267.

PARIS, P. (2). 'Les Bateaux des Bas-Reliefs Khmers.' *BEFEO*, 1941, **41**, 335.

PARIS, P. (3). 'Esquisse d'une Ethnographie Navale des Pays Annamites.' *BAVH*, 1942 (no. 4, Oct. and Dec.), 351. (The stocks of this periodical were all lost during the troubles in Indo-China, but a copy is available in the libraries of the Royal Anthropological Institute and the Science Museum.) Reprinted, Museum voor Land- en Volken-Kunde & Maritiem Museum Prins Hendrik, Rotterdam, 1955.

PARIS, P. (4). 'Voile Latine? Voile Arabe? Voile Mystérieuse.' *HP*, 1949, **36**, 69.

PARIS, P. (5). 'Note sur Deux Passages de Strabon et de Pline dont l'intérèt n'est pas seulement Nautique.' *JA*, 1951, **239**, 13.

PARKINSON, C. N. (1). *Trade in the Eastern Seas, 1793–1813*. Cambridge, 1937.

PARMENTIER, H. (1). 'Les Bas-Reliefs de Banteai-Chmar.' *BEFEO*, 1910, **10**, 205.

PARMENTIER, H. (2). 'Anciens Tambours de Bronze.' *BEFEO*, 1918, **18** (no. 1), 1.

PARRY, J. H. (1). *Europe and a Wider World, +1415 to +1715*. Hutchinson, London, 1949.

PARSONS, W. B. (2). *Engineers and Engineering in the Renaissance*. Williams & Wilkins, Baltimore, 1939.

PARTINGTON, J. R. (5). *A History of Greek Fire and Gunpowder*. Heffer, Cambridge, 1960.

VON PAWLIKOWSKI-CHOLEWA, A. (1). *Die Heere des Morgenlandes*. de Gruyter, Berlin, 1940.

PAYNE, ROBERT (1). *The Canal Builders*. Macmillan, London & New York, 1959.

PECK, GRAHAM (1). *Two Kinds of Time*. Boston, 1950.

PEDLER, F. J. (1). 'Characteristics of African Populations.' *PROG*, 1961, **48**, 223, 258.

PELL, H. P. (1). 'Naval Action on Lake Champlain, 1776.' *ANEPT*, 1948, **8**, 255.

PELLIOT, P. (2*a*). 'Les Grands Voyages Maritimes Chinois au Début du 15ᵉ Siècle' (review of Duyvendak, 10). *TP*, 1933, **30**, 237. Chinese translation by Fêng Chhêng-Chün, Shanghai, 1935, entitled *Chêng Ho Hsia Hsi-Yang Khao*.

PELLIOT, P. (2*b*). 'Notes additionelles sur Tcheng Houo [Chêng Ho] et sur ses Voyages.' *TP*, 1934, **31**, 274.

PELLIOT, P. (2*c*). 'Encore à Propos des Voyages de Tcheng Houo [Chêng Ho].' *TP*, 1936, **32**, 210.

PELLIOT, P. (9) (tr.). 'Mémoire sur les Coutumes de Cambodge' (a translation of Chou Ta-Kuan's *Chen-La Fêng Thu Chi*). *BEFEO*, 1902, **2**, 123. Revised version: Paris, 1951, see Pelliot (33).

PELLIOT, P. (16). 'Le Fou-Nan' [Cambodia]. *BEFEO*, 1903, **3**, 57.

PELLIOT, P. (17). 'Deux Itinéraires de Chine à l'Inde à la Fin du 8ᵉ Siècle.' *BEFEO*, 1904, **4**, 131.

PELLIOT, P. (25). *Les Grottes de Touen-Hoang [Tunhuang]; Peintures et Sculptures Bouddhiques des Époques des Wei, des T'ang et des Song [Sung]*. Mission Pelliot en Asie Centrale, 6 portfolios of plates. Paris, 1920–4.

PELLIOT, P. (27). *Les Influences Européennes sur l'Art Chinois au 17e et au 18e siècle*. Imp. Nat., Paris, 1948. (Conférence faite au Musée Guimet, Feb. 1927.)

PELLIOT, P. (28). 'La Peinture et la Gravure Européennes en Chine au Temps de Matthieu Ricci.' *TP*, 1921, **20**, 1.

PELLIOT, P. (29). 'Quelques Textes Chinois concernant l'Indochine Hindouisée.' In *Études Asiatiques publiées à l'occasion du 25ᵉ Anniversaire de l'École Française d'Extrême-Orient*. van Oest, Paris, 1925, II, 243. (Pub. Éc. Fr. d'Extr. Or. nos. 19 and 20.)

PELLIOT, P. (30). Note on Han relations with South-East Asian countries, with tr. of a passage from *Chhien Han Shu*, ch. 28B, in review of Hirth & Rockhill. *TP*, 1912, **13**, 446 (457).

PELLIOT, P. (33) (tr.). *Mémoire sur les Coutumes de Cambodge de Tcheou Ta-Kouan [Chou Ta-Kuan]; Version Nouvelle, suivie d'un Commentaire inachevé*. Maisonneuve, Paris, 1951. (Oeuvres Posthumes, no. 3.)

PELLIOT, P. (47). *Notes on Marco Polo; Ouvrage Posthume*. 2 vols. Impr. Nat. Maisonneuve, Paris, 1959.

PENN, C. (1). 'Chinese Vernacular Architecture.' *JRIBA*, 1965, **72**, 502. A beautifully illustrated résumé of a paper by Wang Chi-Ming presented at the Peking International Scientific Symposium, 1964, entitled 'Dwelling-Houses of Chekiang Province'.

PENROSE, BOIES (1). *Goa—Rainha do Oriente; Goa—Queen of the East* (in Portuguese and English). Comissão Ultramarina, Lisbon, 1960. (Comemorações do Quinto Centenario da Morte do Infante Dom Henrique.)

PENZER, N. M. (1) (ed.). *The Most Noble and Famous Travels of Marco Polo, together with the Travels of Nicolò de Conti, edited from the Elizabethan translation of John Frampton (1579)*... Argonaut, London, 1929; 2nd ed. Black, London, 1937.

PEREGRINUS, PETRUS (PIERRE DE MARICOURT) (1). *Epistola de Magnete seu Rota Perpetua Motus*. 1269. First pr. by Achilles Gasser, Augsburg, 1558 (a MS. copy of this, with Engl. tr. by an unknown hand is in Gonv. and Caius Coll. MS. 174/95). Second pr. in Taisnier (1). See Thompson, S. P. (5); Hellmann, G. (6); Anon. (46); [Mertens, J. C.] (1); Chapman & Harradon (1).

PERELOMOV, L. S. (1). *Imperija Čin—pervoe Centralizovannoe Gosudarstvo v Kitae* (The Chhin Empire, the First Centralised State in China), in Russian. Izdatel'stvo Vostočnoj Lit., Moscow, 1962. Rev. T. Pokora, *ARO*, 1963, **31**, 165.

PERERA, E. W. (1). 'The Galle Trilingual Stone.' *SZ*, 1913, **8**, 122.

PERERA, S. G. (1) (tr.). *The Temporal and Spiritual Conquest of Ceylon* (tr. of Fernão de Queiroz' *Conquista*...). 3 vols. Richards (for the Govt. of Ceylon), Colombo, 1930.

PERES, DAMIÃO (1). *A History of the Portuguese Discoveries* (tr. of *Historia dos Descobrimentos Portugueses*, 1959). Comissão Executiva das Comemorações do Quinto Centenario da Morte do Infante Dom Henrique, Lisbon, 1960. (Colecção Henriquina, no. 1.)

PERI, N. (1). 'Essai sur les Relations du Japon et de l'Indochine au 16e et 17e siècles.' *BEFEO*, 1923, **23**, 1.

PERI, N. (2). 'A Propos du Mot "Sampan".' *BEFEO*, 1919, **19** (no. 5), 13.

PERKINS, J. B. WARD (1). 'Recording the Face of Ancient Etruria before Modern Agricultural Methods Destroy the Traces.' *ILN*, 1957, **230**, 774.

PEROWNE, STEWART (1). 'The Site of the Holy Sepulchre' (and the Madeba mosaic view of Hadrianic Jerusalem, a city lay-out of +135). *LI*, 1962, **68** (no. 1745), 351.

PERRAULT, CLAUDE (1) (tr.). *Les Dix Livres d'Architecture de Vitruve*. 2nd ed. Paris, 1684.

P[ERRIN], W. G. (1). 'The Balanced Rudder.' *MMI*, 1926, **12**, 232.

PERRONET, J. R. (1). *Description des Projets et de la Construction des Ponts de Neuilli, de Mantes, d'Orléans, etc.* Paris, 1788.

PERROT, G. & CHIPIEZ, C. (1). *Histoire de l'Art dans l'Antiquité*. Paris.

PETECH, L. (1). *Northern India according to the 'Shui Ching Chu'*. Ist. Ital. per il Medio ed Estremo Oriente, Rome, 1950. (Rome Oriental series, no. 2.)

PETERSEN, E. ALLEN (1). *In a Junk across the Pacific.* Elek, London, 1954.

PETRIE, W. M. FLINDERS (2). *Arts and Crafts of Ancient Egypt.* Edinburgh, 1910.

PFISTER, L. (1). *Notices Biographiques et Bibliographiques sur les Jésuites de l'Ancienne Mission de Chine* (*+1552 to +1773*). 2 vols. Mission Press, Shanghai, 1932 (*VS*, no. 59).

PFIZMAIER, A. (19) (tr.). 'Keu-Tsien, König von Yue, und dessen Haus' (Kou Chien of Yueh and Fan Li). *SWAW/PH*, 1863, **44**, 197. (Tr. ch. 41, *Shih Chi*; cf. Chavannes (1), vol. 4.)

PFIZMAIER, A. (31) (tr.). 'Das Ende Mung Tien's' (Mêng Thien). *SWAW/PH*, 1860, **32**, 134. (Tr. ch. 88, *Shih Chi*; not in Chavannes (1).)

PFIZMAIER, A. (51) (tr.). 'Die Eroberung der beiden Yue [Yüeh] und des Landes Tschao Sien [Chao-Hsien, Korea] durch Han.' *SWAW/PH*, 1864, **46**, 481. (Tr. ch. 95, *Chhien Han Shu.*)

PFIZMAIER, A. (94) (tr.). 'Beiträge z. Geschichte d. Perlen.' *SWAW/PH*, 1867, **57**, 617, 629. Tr. *Thai-Phing Yü Lan*, chs. 802 (in part), 803.

PHAN KUANG-CHHIUNG (1). 'Communications and Transportation [in modern China].' In *Chinese Yearbook*, p. 572. 1943.

PHELPS, D. L. (2) (tr.). *A New Edition of the Omei Illustrated Guide Book* ['*O Shan Thu Shuo*' or '*Chih*'] *by Huang Shou-Fu & Than Chung-Yo (1887 to 1891); with an English translation by D. L. P- - - - -, with pictures redrawn from the original plates by Yü Tzu-Tan.* Jih Hsin Yin-Shua Kung-Yeh Shê, Chhêngtu, 1936. (West China Union University, Harvard-Yenching Institute ser. no. 1.)

PHILLIPS, G. (1). 'The Seaports of India and Ceylon, described by Chinese Voyagers of the Fifteenth Century, together with an account of Chinese Navigation...' *JRAS/NCB*, 1885, **20**, 209; 1886, **21**, 30 (both with large folding maps).

PHILLIPS, G. (2). 'Précis translations of the *Ying Yai Shêng Lan.*' *JRAS*, 1895, 529; 1896, 341.

PHILLIPS, G. (3). 'Notable Fukien Bridges.' *TP*, 1894, **5**, 1.

PIÉTRI, J. B. (1). *Voiliers d'Indochine.* S.I.L.I. Saigon, 1943.

PIGANIOL, A. (1). 'Les Etrusques, Peuple d'Orient.' *JWH*, 1953, **1**, 328.

PIGANIOL, A. (2). *Histoire de Rome.* Presses Univ. de France, Paris, 1946. (Clio ser. no. 3.)

PIJOÁN, JOSÉ (1). *Summa Artis; Historia General del Arte.* 12 vols. Espasa-Calpe, Madrid, 1946, 1952.

PILKINGTON, R. (1). 'Canals; Inland Waterways outside Britain [in the Industrial Revolution].' Art. in *A History of Technology*, ed. C. Singer *et al.* Vol. 4, p. 548. Oxford, 1958.

PINSSEAU, M. (1). *Le Canal de Briare; 1604 à 1943.* Houzé, Orléans, 1943; Clavreuil, Paris, 1943. (Rev. Espinas, G. *AHES/AESC*, 1946, **1**, 347.)

PINTO, FERNAÕ MENDES (1). *Peregrinaçam de Fernam Mendez Pinto em que da conta de muytas e muyto estranhas cousas que vio e ouvio no reyno da China, no da Tartaria...* Crasbeec, Lisbon, 1614. Abridged Eng. tr. by H. Cogan: *The Voyages and Adventures of Ferdinand Mendez Pinto, a Portugal, During his Travels for the space of one and twenty years in the kingdoms of Ethiopia, China, Tartaria, etc.* Herringman, London, 1653, 1663, repr. 1692. Still further abridged edition, Unwin, London, 1891. Full French tr. by B. Figuier: *Les Voyages Advantureux de Fernand Mendez Pinto...* Cotinet & Roger, Paris, 1628, repr. 1645. Cf. M. Collis (1): *The Grand Peregrination* (paraphrase and interpretation), Faber & Faber, London, 1949.

PIPPARD, A. J. S. & BAKER, J. F. (1). *Analysis of Engineering Structures.* Arnold, London, 1936.

PIRENNE, H. (1). *Economic and Social History of Mediaeval Europe.* Kegan Paul, London, 1936.

PIRENNE, J. (1). 'Un Problème-Clef pour la Chronologie de l'Orient; la Date du *Périple de la Mer Érythrée*.' *JA*, 1961, **249**, 441.

PIRES, TOMÉ (1). *Suma Oriental.* See A. Cortesão (2).

PITT-RIVERS, A. H. LANE-FOX (2). 'On Early Modes of Navigation.' *JRAI*, 1874, **4**, 399. Reprinted in Pitt-Rivers (4).

PLAYFAIR, G. M. H. (1). 'The Grain Transport System of China; Notes and Statistics taken from the *Ta Chhing Hui Tien.*' *CR*, 1875, **3**, 354. [(1) The Personnel of the Transport Service; (2) The Itinerary of the Grand Canal; (3) Tribute; (4) White Rice Tribute; (5) The Building and Repairing of Junks; (6) Grain Fleets.]

PLAYFAIR, G. M. H. (2). 'Watertight Compartments in Chinese Vessels.' *JRAS/NCB*, 1886, **21**, 106. (Quotation of a letter of Benjamin Franklin.)

PLEDGE, H. T. (1). *Science since +1500.* HMSO, London, 1939.

v. PLESSEN, V. (1). 'The Dayaks of Central Borneo.' *GGM*, 1936, **4**, 17.

POBÉ, M. & ROUBIER, J. (1). *The Art of Roman Gaul; a Thousand Years of Celtic Art and Culture.* Galley, London, 1961.

POIDEBARD, A., LAUFFRAY, J. & MOUTERDE, R. (1). *Sidon.* Impr. Cath., Beirut, 1951–2.

POKORA, T. (1). 'The "Canon of Laws" by Li Khuei—a Double Falsification?' *ARO*, 1959, **27**, 96.

POKORA, T. (6). 'A Pioneer of New Trends of Thought at the End of the Ming Period; Marginalia on Chu Chhien-Chih's Book on Li Chih.' *ARO*, 1961, **29**, 469.

POKORA, T. (9). 'The Life of Huan Than.' *ARO*, 1963, **31**, 1.

POKORA, T. (13). 'La Vie du Philosophe Matérialiste Houan T'an [Huan Than].' Art. in *Mélanges de Sinologie offerts à Monsieur Paul Demiéville*. Presses Univ. de France, Paris, 1966. (Biblioth. de l'Inst. des Hautes Études Chinoises, no. 20.)

POLOVTSOV, A. (1). *The Land of Timur; Recollections of Russian Turkestan*. Methuen, London, 1932.

POPE, J. (1). 'Chinese Characters in Brunei and Sarawak Ceramics.' *SMJ*, 1958, **8**, 267.

POSTAN, M. (1). 'The Trade of Mediaeval Europe; the North.' Art. in *Cambridge Economic History of Europe*, ed. M. Postan & E. E. Rich, vol. 2, p. 119. Cambridge, 1952.

POTT, F. L. HAWKS (1). *A Sketch of Chinese History*. Kelly & Walsh, Shanghai, 1936.

POTTHAST, A. (1). *Bibliotheca Historica Medii Aevi; Wegweiser durch die Geschichtswerke des europäischen Mittelalters bis 1500*. 2 vols. Weber, Berlin, 1896.

POUDRA, N. G. (1). *Histoire de la Perspective ancienne et moderne*. Corréard, Paris, 1864.

POUJADE, J. (1). *La Route des Indes et Ses Navires*. Payot, Paris, 1946.

POUJADE, J. (2). *Les Jonques des Chinois du Siam* (in relation to the ship sculptured on the Bayon of the Angkor Vat). Publication du Centre de Recherche Culturelle de la Route des Indes, Gauthier-Villars, Paris, 1946. (Documents d'Ethnographie Navale, Fasc. 1.)

POUJADE, J. (3). *Trois Flotilles de la VIe Dynastie des Pharaons*. Publication du Centre de Recherche Culturelle de la Route des Indes, Gauthier-Villars, Paris, 1948; Centre I.F.A.N., Djibouti, 1948. (Documents d'Archéologie Navale, Fasc. 1.)

POWELL, FLORENCE L. (1). *In the Chinese Garden*. New York, 1943.

[POWELL, THOMAS] (1). *Humane Industry; or, a History of most Manual Arts, deducing the Original, Progress, and Improvement of them; furnished with variety of Instances and Examples, shewing forth the excellency of Humane Wit*. Herringman, London, 1661. Bibliography in John Ferguson (2).

PRESTAGE, E. (1). *The Portuguese Pioneers*. Black, London, 1933.

PRESTON, T. (1) (tr.). *Makamat [Maqāmāt], or Historical Anecdotes, of al-Ḥarīrī of Basra*. Madden & Parker, London, 1850; Deighton, Cambridge, 1850.

PRICE, D. J. DE S. (12). 'Two Mariner's Astrolabes [with check-list of specimens known].' *JIN*, 1956, **9**, 338.

PRICE, D. J. DE S. (15). 'The First Scientific Instrument of the Renaissance' (an astrolabe of +1462 in the National Maritime Museum, Greenwich, with a simplified 'Rojas' orthographic projection on the back). *PHY*, 1959, **1**, 26.

PRICE, WILLARD (1). 'Grand Canal Panorama.' *NGM*, 1937, **71**, 487.

PRIEST, A. (1). *Chhing Ming Shang Ho (Spring Festival on the River); a Scroll Painting (ex coll. A. W. Bahr) of the Ming Dynasty, after a Sung Dynasty subject, reproduced in its entirety and in its original size...with an Introduction and Notes*. Metropolitan Museum of Art, New York, 1948.

PRINSEP, J. (2). 'Note on the Nautical Instruments of the Arabs.' *JRAS/B*, 1836, **5**, 784. Reprinted in Ferrand (7), pp. 1 ff.

PRINSEP, J. (3). 'Notes [on von Hammer-Purgstall's translations from the *Mohit* (*Muḥīṭ*) (The Ocean) of Sidi 'Ali Reïs].' *JRAS/B*, 1836, **5**, 441; 1838, **7**, 774. Reprinted in Ferrand (7) (without very clear identification), pp. 12 ff.

PRIP-MØLLER, J. (1). *Chinese Buddhist Monasteries; their Plan and Function as a Setting for Buddhist Monastic Life*. Oxford, 1937. Repr. with biographical and bibliographical notes, Vetch, Hongkong, 1968.

PRITCHARD, L. A. (1). *The 'Ningpo' Junk* (voyage from Shanghai to San Pedro, Calif. 1912/1913). *MMI*, 1923, **9**, 89 and notes by H. Sz[ymanski], p. 312 and G. A. B[allard], p. 316.

PROUDFOOT, W. J. (1). *Biographical Memoir of James Dinwiddie, LL.D., Astronomer in the British Embassy to China (1792–4), afterwards Professor of Natural Philosophy in the College of Fort William, Bengal; embracing some account of his Travels in China and Residence in India, compiled from his Notes and Correspondence by his grandson....* Howell, Liverpool, 1868.

PRŮŠEK, J. (2). 'Some Chinese Studies.' *ARO*, 1959, **27**, 476.

PUGSLEY, SIR ALFRED (1). *The Theory of Suspension Bridges*. Arnold, London, 1957.

PUINI, C. (1). 'I Muraglione della Cina.' *RGI*, 1915, **22**, 481.

PULLEYBLANK, E. G. (1). *The Background of the Rebellion of An Lu-Shan*. Oxford, 1954. (London Oriental Series, no. 4.)

PULLEYBLANK, E. G. (7). 'Chinese Historical Criticism; Liu Chih-Chi and Ssuma Kuang.' Art. in *Historians of China and Japan*, ed. W. G. Beasley & E. G. Pulleyblank, p. 135. Oxford Univ. Press, London, 1961.

PURCELL, V. (1). *The Chinese in South-East Asia*. Oxford, 1951 (for the Royal Institute of International Affairs and the Institute of Pacific Relations).

PURCELL, V. (2). *Gibbon and the Far East*. Unpublished monograph.

PURCELL, V. (3). *The Chinese in Malaya*. Roy. Inst. Internat. Affairs & Inst. of Pacific Relations, London, 1948. Repr. Kuala Lumpur, 1967.

PURCHAS, S. (1). *Hakluytus Posthumus, or Purchas his Pilgrimes, contayning a History of the World in Sea Voyages and Lande Travells.* 4 vols. London, 1625. 2nd ed. *Purchas his pilgrimage, Or Relations of the world and the religions observed in all ages and places discovered.* London, 1626.
PURVIS, F. P. (1). 'Ship Construction in Japan.' *TAS/J*, 1919, **47**, 1; 'Japanese Ships of the Past and Present.' *TJSL*, 1925, **23**, 51.

DE QUEIROZ, FERNÃO (1). *Conquista Temporal e Espiritual de Ceylão* (+1687). Cottle (for the Govt. of Ceylon), Colombo, 1916.
QUIGLEY, C. (1). 'Certain Considerations on the Origin and Diffusion of Oculi.' *ANEPT*, 1955, **15**, 191; 'The Origin and Diffusion of Oculi; a Rejoinder.' *ANEPT*, 1958, **18**, 25; 'The Origin and Diffusion of Oculi.' *ANEPT*, 1958, **18**, 245.

DA RADA, MARTÍN (1). *Narrative of his Mission to Fukien (June–Oct. 1575). Relation of the things of China, which is properly called Taybin [Ta Ming].* Tr. and ed. Boxer (1).
RAGLAN, LORD (1). *How came Civilisation?* Methuen, London, 1939.
RALEIGH, SIR WALTER (1). *Judicious and Select Essays eand Observations by that Renowned and Learned Knight, Sir W. R. upon The First Invention of Shipping; The Misery of Invasive Warre; The Navy Royall and Sea-Service; with his Apologie for his Voyage to Guiana.* Humphrey Mosele, London, 1650. Full title of the first essay: *A Discourse of the Invention of Ships, Anchors, Compasse, etc.; the First Naturall Warre, the severall use, defects, and supplies of Shipping, etc.;* Full title of the third: *Excellent Observations and Notes concerning the Royall Navy and Sea-Service.*
RAMSAY, A. M. (1). 'The Speed of the Roman Imperial Post.' *JRS*, 1925, **15**, 73.
RANDALL-MACIVER. See McIver, D. R. (1).
RANKE, H. (1). *Das Gilgameschepos.* Lerchenfeld Press, for Friederichsen, Hamburg, 1924.
RANKINE, W. J. McQ. (1). *A Manual of the Steam Engine and other Prime Movers.* Griffin & Bohn, London, 1861.
RANKING, G. S. A. & AZOO, R. F. (1) (tr.). Eng. translation of al-Muqaddasī's *Aḥsan al-Taqāsīm fī Ma'arifat al-Aqālīm* (The Best Divisions for the Knowledge of the Climates). Asiatic Society, Calcutta, 1897–1910. (Bib. Ind. N.S. nos. 899, 952, 1001, 1258.)
RAO, K. L. (1). *Earth Dams, Ancient and Modern, in Madras State.* Proc. IVth Internat. Congress of Large Dams, 1951, vol. 1, p. 285.
RAO, S. R. (1). 'New Light on the Indus Valley Civilisation; Seals, Drains and a Dockyard in New Excavations at Lothal in India.' *ILN*, 1961, **238**, 302, 387.
RASMUSSEN, S. E. (1). *Towns and Buildings.* Univ. Press, Liverpool, 1951. (Translated from the Danish of 1949 by Eve Wendt.) Original edition different in many ways.
RAU, VIRGINIA (1). 'Les Marchands-Banquiers Étrangers sous le Régne de Dom João III (+1521 à +1557).' Communication to the Vᵉ Colloque International d'Histoire Maritime. Lisbon, 1960.
RAVENSTEIN, E. G. (1). *Martin Behaim; his Life and his [terrestrial] Globe.* London, 1908.
RAVENSTEIN, E. G. (2) (tr.). *A Journal of the First Voyage of Vasco da Gama, +1497 to +1499, [an anonymous Roteiro].* London, 1898. (Hakluyt Society Pub. no. 99.)
RAWLINSON, J. L. (1). *China's Struggle for Naval Development, 1839 to 1895.* Harvard Univ. Press, Cambridge, Mass., 1967. (Harvard East Asian ser. no. 25.)
READ, BERNARD E. (with LIU JU-CHHIANG) (1). *Chinese Medicinal Plants from the 'Pên Tshao Kang Mu'* A.D. *1596...a Botanical, Chemical and Pharmacological Reference List.* (Publication of the Peking Nat. Hist. Bull.). French Bookstore, Peiping, 1936 (chs. 12–37 of *Pên Tshao Kang Mu*) (rev. W. T. Swingle, *ARLC/DO*, 1937, 191).
READ, BERNARD E. (2) (with LI YÜ-THIEN). *Chinese Materia Medica; Animal Drugs.*

		Serial nos.	Corresp. with chaps. of *Pên Tshao Kang Mu*
Pt. I	Domestic Animals	322–349	50
II	Wild Animals	350–387	51 *A* and *B*
III	Rodentia	388–399	51 *B*
IV	Monkeys and Supernatural Beings	400–407	51 *B*
V	Man as a Medicine	408–444	52

PNHB, 1931, **5** (no. 4), 37–80; **6** (no. 1), 1–102. (Sep. issued, French Bookstore, Peiping, 1931.)
READ, BERNARD E. (3) (with LI YÜ-THIEN). *Chinese Materia Medica; Avian Drugs.*

		Serial nos.	Corresp. with chaps. of *Pên Tshao Kang Mu*
Pt. VI	Birds	245–321	47, 48, 49

	Serial nos.	Corresp. with chaps. of *Pên Tshao Kang Mu*

PNHB, 1932, **6** (no. 4), 1 101. (Sep. issued, French Bookstore, Peiping, 1932.)

READ, BERNARD E. (4) (with LI YÜ-THIEN). *Chinese Materia Medica; Dragon and Snake Drugs.*

Pt. VII Reptiles — 102–127 — 43

PNHB, 1934, **8** (no. 4), 297–357. (Sep. issued, French Bookstore, Peiping, 1934.)

READ, BERNARD E. (5) (with YU CHING-MEI). *Chinese Materia Medica; Turtle and Shellfish Drugs.*

Pt. VIII Reptiles and Invertebrates — 199–244 — 45, 46

PNHB (Suppl.), 1939, 1–136. (Sep. issued, French Bookstore, Peiping, 1937.)

READ, BERNARD E. (6) (with YU CHING-MEI). *Chinese Materia Medica; Fish Drugs.*

Pt. IX Fishes (incl. some amphibia, octopoda and crustacea) — 128–198 — 44

PNHB (Suppl.), 1939. (Sep. issued, French Bookstore, Peiping, n.d. prob. 1939.)

READ, BERNARD E. (7) (with YU CHING-MEI). *Chinese Materia Medica; Insect Drugs.*

Pt. X Insects (incl. arachnidae etc.) — 1–101 — 39, 40, 41, 42

PNHB (Suppl.), 1941. (Sep. issued, Lynn, Peiping, 1941.)

READ, BERNARD E. (8). *Famine Foods listed in the 'Chiu Huang Pên Tshao'.* Lester Institute, Shanghai, 1946.

READ, BERNARD E. & PAK, C. (PAK KYEBYŎNG) (1). *A Compendium of Minerals and Stones used in Chinese Medicine, from the 'Pên Tshao Kang Mu'.* PNHB, 1928, **3** (no. 2), i–vii, 1–120. (Revised and enlarged, issued separately, French Bookstore, Peiping, 1936 (2nd ed.).) Serial nos. 1–135, corresp. with chs. of *Pên Tshao Kang Mu*, 8, 9, 10, 11.

RECINOS, ADRIÁN, GOETZ, D. & MORLEY, SYLVANUS G. (1) (tr.). '*Popol Vuh*' (*Book of the People); the Sacred Book of the Ancient Quiché Maya.* Univ. of Oklahoma Press, Norman, Oklahoma, 1949; Hodge, London, 1951.

REEVES, R. G. & MANGELSDORF, P. C. (1). 'The Origin of Corn [Maize]; II, Teosinte, a Hybrid of Corn and *Tripsacum*.' HBML, 1959, **18**, 357.

REEVES, R. G. & MANGELSDORF, P. C. (2). 'The Origin of Corn [Maize]; V, A Critique of Current Theories.' HBML, 1959, **18**, 428.

REICHWEIN, A. (1). *China and Europe; Intellectual and Artistic Contacts in the Eighteenth Century.* Kegan Paul, London, 1925. Tr. from the German edn. Berlin, 1923.

REINAUD, J. T. & GUYARD, S. (1) (tr.). *Taqwīm al-Buldān of Abū'l-Fidā.* Paris, vol. 1, 1848 (Reinaud); vol. 2, 1883 (Guyard). Partially reprinted in the form of excerpts in Ferrand (7), pp. 18 ff.

REINAUD, J. T. & MAURY, A. (1). 'Introduction Générale à la Géographie des Orientaux.' In Reinaud & Guyard's *Géographie d'Aboulfeda*, vol. 1, pp. CDXXXIX ff. Partially reprinted in Ferrand (7), pp. 18 ff.

REINHARDT, H. (1) (with contributions by D. SCHWARZ, J. DUFT & H. BESSLER). *Der St Gallen Klosterplan* (with full-size eight-colour lithographic facsimile of the Carolingian abbey plan, *c.* +820). Fehr, St Gallen, 1952. (Neujahrsblatt d. histor. Vereins St Gallen, no. 92.) Rev. K. J. Conant, SP, 1955, **30**, 676.

REISCHAUER, E. O. (1). 'Notes on Thang Dynasty Sea-Routes.' HJAS, 1940, **5**, 142.

REISCHAUER, E. O. (2) (tr.). *Ennin's Diary; the Record of a Pilgrimage to China in Search of the Law* (the *Nittō Guhō Junrei Gyōki*). Ronald Press, New York, 1955.

REISCHAUER, E. O. (3). *Ennin's Travels in Thang China.* Ronald Press, New York, 1955.

REISNER, G. A. (1). *Models of [Ancient Egyptian] Ships and Boats.* Cat. Gen. des Antiq. Eg. du Mus. du Caire, Inst. Fr. d'Archéol. Orient. du Caire. Cairo, 1913.

RÉMUSAT, J. P. A. (1) (tr.). '*Foe Koue Ki [Fo Kuo Chi]*', on *Relation des Royaumes Bouddhiques; Voyage dans la Tartarie, dans l'Afghanistan et dans l'Inde, executé, à la Fin du 4e siècle, par Chy Fa-Hian [Shih Fa-Hsien].* Impr. Roy. Paris, 1836. Eng. tr. *The Pilgrimage of Fa-Hian [Fa-Hsien]; from the French edition of the 'Foe Koue Ki' of Rémusat, Klaproth and Landresse, with additional notes and illustrations.* Calcutta, 1848. (Fa-Hsien's *Fo Kuo Chi*.)

RENAN, E. (1). *Mission de Phénicie dirigée par Mons. E.R.* 1 vol. text, 1 vol. plates (ed. M. Thobois). Imp. Impér. Paris, 1864.

RENOU, L. & FILLIOZAT, J. (1). *L'Inde Classique; Manuel des Études Indiennes*. Vol. 1, with the collaboration of P. Meile, A. M. Esnoul & L. Silburn. Payot, Paris, 1947. Vol. 2, with the collaboration of P. Demiéville, O. Lacombe & P. Meile. École Française d'Extrême Orient, Hanoi, 1953; Impr. Nationale, Paris, 1953.

DE REPARAZ, G. (3). 'Les Sciences géographiques et astronomiques au 14e Siècle dans le Nord-Est de la Péninsule Ibérique et leur Origine.' *A/AIHS*, 1948, **3**, 434.

DE REPARAZ, G. (4). 'L'Activité maritime et commerciale du Royaume d'Aragon au 13ᵉ Siècle et son influence sur le développement de l'École Cartographique de Majorque.' *BH*, 1947, **49**, 422 (*AFLB*, 1947, **49**, 422).

REVINGTON, T. M. (1). 'Some Notes on the Mafia Island Group.' *TNR*, 1936, **1** (no. 1), 33.

RHYS-DAVIDS, T. W. (4). 'Early Commerce between India and Babylonia [and the use of shore-sighting birds].' *JRAS*, 1899, **31**, 432.

RIBEIRO, JOÃO (1). *Fatalidade Historica da Ilha da Ceilão*. 1685; ed. Lisbon, 1836.

RICCIOLI, J. B. (1). *Geographia et Hydrographia Reformata*. Bologna, 1661.

RICHARDS, J. M. (1). 'Off the Floor and into the West' (on traditional Japanese domestic architecture and current changes in it). *LI*, 1962, **68** (no. 1741), 205.

RICHARDSON, H. L. (1). 'Szechuan during the War' (World War II). *GJ*, 1945, **106**, 1.

RICHTER, G. M. A. (1). *Ancient Furniture; a History of Greek, Etruscan and Roman Furniture*. Oxford, 1926.

RICHTER, G. M. A. (2). *The Furniture of the Greeks, Etruscans and Romans*. Phaidon, London, 1966.

VON RICHTHOFEN, F. (2). *China: Ergebnisse eigener Reisen und darauf gegründeter Studien*. 5 vols. and Atlas. Reimer, Berlin, 1877–1912. (Teggart bibliogr. says 5 vols.+2 Atlas vols.)

VON RICHTHOFEN, F. (3). 'Über den Seeverkehr nach und von China in Altherthum und Mittelalter.' *VGEB*, 1876, **3**, 86.

RICKARD, T. A. (1). 'The Use of Copper and Iron by the Indians of British Columbia.' *BCHQ*, 1939, **3**, 25.

RICKETT, W. A. (1) (tr.). *The 'Kuan Tzu' Book*. Hongkong Univ. Press, Hongkong, 1965. Rev. T. Pokora, *ARO*, 1967, **35**, 169.

RICO Y SINOBAS, M. (1). *'Libros del Saber de Astronomia' del Rey D. Alfonso X de Castilla*. 4 vols. Aguado, Madrid, 1863–66.

RIN-CHEN LHA-MO (MRS LOUIS KING) (1). *We Tibetans*. London, 1926.

RITTATORE, F. (1). 'Resti Etrusco-Romani nell'Aretino.' *STE*, 1938, **12**, 257.

RIVET, P. (1). *Los Origenes del Hombre Americano*. Mexico City, 1943.

RIVET, P. & ARSENDAUX, H. (1). 'La Métallurgie en Amérique pre-Columbienne.' *TMIE*, 1946, no. 39.

RIVIÈRE, C. (1). 'Destruction et Reconstruction de la Digne de Hua Yuan Kou sur le Hoang-Ho (Fleuve Jaune).' *RGHE*, 1948, **1**, 76.

RIVIUS, G. H. (2). *Vitruvius Teutsch, Nemlichen des aller namhaftigsten und hocherfarnesten Römischen Architecti und Kunstreichen Werck oder Baumeisters Marci Vitruvii Pollionis Zehen Bucher von der Architektur und Künstlichem Bauen....* Petreius, Nuremberg, 1548.

ROBERTSON, D. S. (1). *Handbook of Greek and Roman Architecture*. Cambridge, 1929.

ROBINS, F. W. (1). *The Story of Water Supply*. Oxford, 1946.

ROBINS, F. W. (2). *The Story of the Bridge*. Cornish, Birmingham, n.d. (1948).

ROCHE, O. (1). *Les Meubles de la Chine*. 1st series, Paris, n.d. See Dupont.

ROCHER, E. (1). *La Province Chinoise du Yunnan*. 2 vols. (incl. special chapter on metallurgy). Leroux, Paris, 1879, 1880.

ROCK, F. (1). 'Kalendarkreise und Kalendarschichten in alten Mexico und Mittelamerika.' Art. in W. Schmidt Festschrift, ed. W. Koppers, p. 610. Vienna, 1928.

ROCK, J. F. (1). *The Ancient Na-Khi Kingdom of Southwest China*. 2 vols. (with magnificent collotype illustrations). Harvard University Press, Cambridge, Mass., 1947. (Harvard-Yenching Monograph Series, no. 9.)

ROCKHILL, W. W. (1). 'Notes on the Relations and Trade of China with the Eastern Archipelago and the Coast of the Indian Ocean during the +15th Century.' *TP*, 1914, **15**, 419; 1915, **16**, 61, 236, 374, 435, 604.

ROCKHILL, W. W. (2). 'Notes on the Ethnology of Tibet.' *ARUSNM*, 1893, 669.

LA ROËRIE, G. (1). 'Les Transformations du Gouvernail.' *AHES*, 1935, **7**, 564.

LA ROËRIE, G. (2). 'l'Histoire du Gouvernail.' *RMA*, 1938 (no. 219), 309; (no. 220), 481. Also sep. Soc. d'Editions Géographiques Maritimes et Coloniales, Paris, 1938.

LA ROËRIE, G. & VIVIELLE, J. (1). *Navires et Marins, de la Rame à l'Hélice*. 2 vols. Duchartre & van Buggenhoudt, Paris, 1930.

ROLT, L. T. C. (3). *The Inland Waterways of England*. Allen & Unwin, London, 1950.

LA RONCIÈRE, C. DE B. (1). *Histoire de la Marine Française*. 6 vols. Perrin (later Plon), Paris, 1899–1932.

DE ROOS, H. (1). *The Thirsty Land; the story of the Central Valley Project*. Stanford Univ. Press, Palo Alto, Calif., 1948.

ROSANI, S. (1). *La Segnalazione Marittima attraverso i Secoli*. Tipografio Stato Maggiore Marina, Rome, 1949.

ROSE, A. (1). *When all Roads led to Rome*. London, 1934.

ROSE, A. (2). *Public Roads of the Past*. Washington, D.C., 1952.

ROSETTI, C. (1). *Corea a Coreani impressioni e Ricerche sull'Impero del Gran Han*. Bergamo, 1905.

ROSS, A. S. C. (1). 'Comparative Philology and the "Kon-Tiki" Theory.' *N*, 1953, **172**, 365.

DES ROTOURS, R. (1) (tr.). *Traité des Fonctionnaires et Traité de l'Armée, traduits de la Nouvelle Histoire des T'ang* (chs. 46–50). 2 vols. Brill, Leiden, 1948. (Bibl. de l'Inst. des Hautes Études Chinoises, no. 6) rev. P. Demiéville, *JA*, 1950, **238**, 395.

ROUSE, H. & INCE, S. (1). *History of Hydraulics*. Iowa Inst. of Hydraulics Research, Univ. of Iowa, Iowa City, 1957. Repr. Dover Paperbacks, New York, 1964.

ROUSSELLE, E. (4). *Zur Seelischen Führung im Taoismus*. Wissenschaftl. Buchgesellsch., Darmstadt, 1962. (A collection of three reprinted articles, including footnotes and superscript references to Chinese characters, but omitting the characters themselves.)

ROUSSELLE, E. (5). 'Ne Ging Tu [Nei Ching Thu], "Die Tafel des inneren Gewebes"; ein Taoistisches Meditationsbild mit Beschriftung.' *SA*, 1933, **8**, 207.

ROY, CLAUDE (1). *La Chine dans un Miroir*. Clairefontaine, Lausanne, 1953.

RUDLOV, J. (1). 'Die Einführung d. Panzerung im Kriegsschiffbau und die Entwicklung d. erster Panzerflotten.' *BGTI*, 1910, **2**, 1.

RUDOFSKY, B. (1). *Architecture without Architects; an Introduction to Non-Pedigreed Architecture*. Museum of Modern Art, Doubleday, New York, 1964.

RUDOLPH, R. C. & WÊN YU (1). *Han Tomb Art of West China; a Collection of First and Second Century Reliefs*. Univ. of Calif. Press, Berkeley and Los Angeles, 1951. (Rev. W. P. Yetts, *JRAS*, 1953, 72.)

RUFF, E. (1). *Jade of the Maori*. London, 1950.

RUNCIMAN, S. (1). *Byzantine Civilisation*. Arnold, London, 1933.

RUNCIMAN, S. (2). 'Byzantine Trade and Industry.' Art. in *Cambridge Economic History of Europe*, ed. M. Postan and E. E. Rich, vol. 2, p. 86. Cambridge, 1952.

RUPPERT, K. & DENISON, J. H. (1). *Archaeological Reconnaissance in Campeche, Quintana Roo and Peten*. Carnegie Inst. Wash. Pub. no. 543. Washington, D.C., 1943.

RUZ-LHUILLIER, ALBERTO (1). *La Civilización de los antiguos Mayas*. Santiago de Cuba, 1957.

RYKWERT, J. (1). *The Idea of a Town*. de Boer, Hilversum, n.d. (1965) and St George's Gallery, London, n.d. (1965). Originally in *Forum*, pub. G. van Saane, Lectura Architectonica, Hilversum.

DE SÀ, A. MOREIRA (1). *O Infante Dom Henrique e a Universidade*. Comissão Executiva das Comemorações do Quinto Centenario da Morte do Infante Dom Henrique, Lisbon, 1960. (Colecção Henriquina, no. 11.)

SACHAU, E. (2) (tr.). *The Chronology of Ancient Nations; an English Version of the Arabic Text of the 'Athār-ul-Bākiya' of al-Bīrūnī (or 'Vestiges of the Past')*, collected and reduced to writing by the author in A.H. 390–1, i.e. A.D. 1000. London, 1879.

SACHS, C. (2). *The History of Musical Instruments*. New York, 1940; Dent, London, 1942.

SACKUR, E. (1). *Sibyllinische Texte und Forschungen*. Halle, 1898. (Pp. 1–96 contain the best text of the *Revelationes* of Pseudo-Methodius.)

SADLER, A. L. (1). *A Short History of Japanese Architecture*. Angus & Robertson, Sydney, 1941; 2nd ed. Tuttle, Rutland, Vt. and Tokyo, 1965.

SÄFLUND, G. (1). 'Le Terramare della provincie di Modena, Reggio Emilia, Parma, Piacenza.' *AIRS*, 1939, **7**, 1–265; 'Bemerkungen zur Vorgeschichte Etruriens.' *STE*, 1938, **12**, 17.

DE SAHAGÚN, BERNADINO (1). *Historia General de las Cosas de Nueva España*. Ed. C. M. de Bustamente, 3 vols, Mexico City, 1829–30; Eng. tr. F. R. Bandelier. Nashville, Tenn., 1932. Best Spanish ed. Robledo, Mexico City, 1938.

SAINSON, C. (1) (tr.). *Histoire particulière du Nan Tchao; 'Nan Tchao Ye Che' [Nan Chao Yeh Shih]; Traduction d'une Histoire de l'Ancien Yun-nan; accompagnée d'une Carte et d'un Lexique Géographique et Historique*. Imp. Nat. Leroux, Paris, 1904. (Pub. Ec. Lang. Or. Viv. (5ᵉ sér.) no. 4.)

DE ST DENIS, E. (1). 'Le Gouvernail Antique, Technique et Vocabulaire.' *REL*, 1934, **12**, 390; *Le Vocabulaire des Manœuvres Nautiques en Latin*. Protat, Mâcon, 1935.

DE ST DENYS. See d'Hervey St Denys.

SALAMAN, R. A. (1). 'Tools of the Shipwright, 1650 to 1925.' *FLF*, 1967, **5**, 19.

SALZMAN, L. F. (1). *Building in England down to 1540*. Oxford, 1952. Repr. 1966.

SALZMAN, L. F. (2). *English Industries of the Middle Ages*. Oxford, 1923.

SAMPSON, T. (1). 'Buddhist Priests in America.' *NQCJ*, 1869, **3**, 79. (Note in answer to queries by Y.J.A. and F.P.S. on p. 58.)

SANCEAU, E. (1). *Henry the Navigator*. Hutchinson, London, n.d. (1946); New York, 1947.

SANCEAU, E. (3). *Portugal in Quest of Prester John*. Hutchinson, London, n.d. (1943).

SANDARS, N. K. (1) (tr.). *The Epic of Gilgamesh.* Penguin, London, 1960.

VON SANDRART, JOACHIM (1). *Teutsche Akademie d. Bau-, Bild- und Mahlerey-Künste.* Amsterdam, 1675, 1679; *Deutsches Akademie d. edlen Bau-, Bild- und MahlereyKünste.* Nuremberg, 1675, 1679.

SANDSTRÖM, G. E. (1). *The History of Tunnelling; Underground Workings through the Ages.* Barrie & Rockliff, London, 1963.

SANSOM, SIR GEORGE (1). *Japan: a Short Cultural History.* Cresset Press, London, 1931; 2nd ed. 1946.

SANSOM, SIR GEORGE (2). *A History of Japan.* 3 vols. Vol. 1, to +1334; vol. 2, +1334 to +1615; vol. 3, +1615 to 1854. Cresset Press, London, 1958.

DE SANTARÉM, M. VISCONDE (1). *Essai sur l'Histoire de la Cosmographie et de la Cartographie pendant le Moyen Age, et sur les Progrès de la Geographie après les Grandes Découvertes du XVᵉ Siècle; pour servir d'introduction et d'explication à l'Atlas composé de Mappemondes et de Portulans et d'antres Monuments Geographiques depuis le VIᵉ siécle de notre Ére jusqu'au XVIIᵉ.* 3 vols. Maulde & Renou, Paris, 1849–52.

DE SANTARÉM, M. VISCONDE (2). *Atlas Composé de Mappemondes et de Cartes Hydrographiques et Historiques.* Maulde & Renou, Paris, 1845.

DE SANTARÉM, M. VISCONDE (3). *Memória sobre a Prioridade dos Descobrimentos Portugueses na Costa de África Ocidental.* 1841, repr. Comissão Executiva das Comemorações do Quinto Centenário da Morte do Infante Dom Henrique, Lisbon, 1958. (Colecção Henriquina, no. 6.)

SARTON, GEORGE (1). *Introduction to the History of Science.* Vol. 1, 1927; vol. 2, 1931 (2 parts); vol. 3, 1947 (2 parts). Williams & Wilkins, Baltimore. (Carnegie Institution Pub. no. 376.)

SARTON, GEORGE (9). *The Appreciation of Ancient and Mediaeval Science during the Renaissance (+1450 to +1600).* Univ. of Pennsylvania Press, Philadelphia, 1955. (Rosenbach Fellowship in Bibliography Pubs. no. 14.)

SARTON, GEORGE & GOODRICH, L. CARRINGTON (1). 'A Chinese Gun of +1378?', Query and Notes on the easliest dated Chinese cannon. *ISIS,* 1944, **35**, 177 and 211.

SARTON, GEORGE & NEEDHAM, JOSEPH (1). 'Who was the Inventor of Pearl Culture?' *ISIS,* 1955, **46**, 50.

SATCHELL, T. (1) (tr.). ['*A Tour on*] *Shanks' Mare', being a Translation of the Tōkaidō Volumes [i.e. Chapters] of the [Dōchū] Hizakurige; Japan's Great Comic Novel of Travel and Ribaldry [by Jippensha Ikku] [1802 to 1822].* Tuttle, Rutland, Vt. and Tokyo, 1965. With colour reproductions of a little-known version of the set of prints 'The Fifty-three stages of the Tōkaidō' by Ando Hiroshige.

SATTERTHWAITE, L. & RALPH, E. K. (1). 'New Radio-carbon Dates and the Maya Correlation Problem.' *AMA,* 1960, **26**, 165.

SAUER, C. (1). 'Cultivated Plants of South and Central America.' Art. in *Handbook of South American Indians,* vol. 6, p. 487. Washington, 1950.

SAUNIER, C. (1). *Die Geschichte d. Zeitmesskunst.* 2 vols. Hübner, Bautzen; Diebener, Leipzig, n.d. (1902–4). Tr. from the French by G. Speckhart.

DE SAUNIER, L. BAUDRY (1). *Histoire de la Locomotion Terrestre.* Paris, 1936. 2nd ed. de Saunier, L. Baudry, Dollfus, C. & Geoffroy, E. *Histoire de la Locomotion Terrestre; la Locomotion Naturelle, l'Attelage, la Voiture, le Cyclisme, la Locomotion Mécanique, l'Automobile.* Paris, 1942.

DE SAUSSURE, L. (35). 'L'Origine de la Rose des Vents et l'Invention de la Boussole.' *ASPN,* 1923 (5ᵉ sér.), **5** (nos. 3 and 4). Sep. pub. Luzac, London, 1923 and reprinted in Ferrand (7), pp. 31 ff. Emendations by P. Pelliot, *TP,* 1924, **23**, 51.

DE SAUSSURE, L. (36). 'Commentaire des Instructions Nautique de Ibn Mājid et Sulaimān al-Mahrī.' In Ferrand (7), pp. 128 ff.

SAUVAIRE, H. & DE REY PAILHADE, J. (1). 'Sur une "mère" d'Astrolabe Arabe du 13ᵉ Siècle (609 de l'Hégire) portant un Calendrier Perpétuel avec Correspondance Mussulmane et Chrétienne.' *JA,* 1893 (9ᵉ sér.), **1**, 5, 185.

SAVILLE, M. H. (1). 'The Ancient Maya Causeways of Yucatan.' *AQ,* 1935, **9**, 67.

SAYILI, AYDIN (3). *Uluğ Bey ve Semerkanddeki Ilim Faaliyeti Hakkinda Gryasüddin-i Kâşî' nin Mektubu (Ghiyāth al-Dīn al-Kāshī's Letter on Ulūgh Beg and the Scientific Activity at Samarqand),* in Turkish, English and Arabic. Türk Tarih Kurumu Basimevi, Ankara, 1960. (Türk Tarih Kurumu Yayinlarindan, 7th Ser. no. 39.) Abstract in Actes du IXᵉ Congrès International d'Histoire des Sciences, Barcelona, 1959, p. 586.

[SCARTH, JOHN, CANON] (1). *Twelve Years in China; the People, the Rebels and the Mandarins, by a British Resident.* Edmonston & Douglas, Edinburgh, 1860.

SCHAEFFNER, A. (1). *Origine des Instruments de Musique.* Payot, Paris, 1936.

SCHAFER, E. H. (4). 'The History of the Empire of Southern Han according to chapter 65 of the *Wu Tai Shih* of Ouyang Hsiu.' Art. in Silver Jubilee Volume of the Zinbun Kagaku Kenkyuso, Kyoto University, Kyoto, 1954, p. 338. (*TG/K,* 1954, **25**, pt. 1.)

SCHAFER, E. H. (10). 'The Pearl Fisheries of Ho-Phu.' *JAOS,* 1952, **72**, 155.

SCHAFER, E. H. (12). 'The Conservation of Natural Resources in Mediaeval China.' Contrib. to Xth Internat. Congr. of the History of Science, Ithaca, 1962. Abstract vol. p. 67.

SCHAFER, E. H. (13). *The Golden Peaches of Samarkand; a Study of Thang Exotics.* Univ. of Calif. Press. Berkeley and Los Angeles, 1963.

SCHAFER, E. H. (14). 'The Last Years of Chhang-an.' *OE*, 1963, **10**, 133–79.

SCHAFER, E. H. (16). 'The Vermilion Bird; Thang Images of the South.' Univ. of Calif. Press, Berkeley and Los Angeles, 1967.

SCHÄFER, H. (1). *Von Aegyptischer Kunst.* 2 vols. Leipzig, 1919. Repr. 1930.

SCHLEGEL, G. (5). *Uranographie Chinoise, etc.* 2 vols. with star-maps in separate folder. Brill, Leiden, 1875. (Crit. J. Bertrand, *JS*, 1875, 557; S. Günther, *VAG*, 1877, **12**, 28. Reply by G. Schlegel, *BNI*, 1880 (4ᵉ volg.), **4**, 350.)

SCHLEGEL, G. (7). 'Problèmes Géographiques; les Peuples Étrangers chez les Historiens Chinois.'
(*a*) Fu-Sang Kuo (ident. Sakhalin and the Ainu). *TP*, 1892, **3**, 101.
(*b*) Wên-Shen Kuo (ident. Kuriles). *Ibid.* p. 490.
(*c*) Nü Kuo (ident. Kuriles). *Ibid.* p. 495.
(*d*) Hsiao-Jen Kuo (ident. Kuriles and the Ainu). *TP*, 1893, **4**, 323.
(*e*) Ta-Han Kuo (ident. Kamchatka and the Chukchi) and Liu-Kuei Kuo. *Ibid.* p. 334.
(*f*) Ta-Jen Kuo (ident. islands between Korea and Japan) and Chhang-Jen Kuo. *Ibid.* p. 343.
(*g*) Chün-Tzu Kuo (ident. Korea, Silla). *Ibid.* p. 348.
(*h*) Pai-Min Kuo (ident. Korean Ainu). *Ibid.* p. 355.
(*i*) Chhing-Chhiu Kuo (ident. Korea). *Ibid.* p. 402.
(*j*) Hei-Chih Kuo (ident. Amur Tungus). *Ibid.* p. 405.
(*k*) Hsüan-Ku Kuo (ident. Siberian Giliak). *Ibid.* p. 410.
(*l*) Lo-Min Kuo and Chiao-Min Kuo (ident. Okhotsk coast peoples). *Ibid.* p. 413.
(*m*) Ni-Li Kuo (ident. Kamchatka and the Chukchi). *TP*, 1894, **5**, 179.
(*n*) Pei-Ming Kuo (ident. Behring Straits Islands). *Ibid.* p. 201.
(*o*) Yu-I Kuo (ident. Kamchatka tribes). *Ibid.* p. 213.
(*p*) Han-Ming Kuo (ident. Kuriles). *Ibid.* p. 218.
(*q*) Wu-Ming Kuo (ident. Okhotsk coast peoples). *Ibid.* p. 224.
(*r*) San Hsien Shan (the magical islands in the Eastern Sea, perhaps partly Japan). *TP*, 1895, **6**, 1.
(*s*) Liu-Chu Kuo (the Liu-Chhiu islands, partly confused with Thaiwan, Formosa). *Ibid.* p. 165.
(*t*) Nü-Jen Kuo (legendary, also in Japanese fable). *Ibid.* p. 247.
A volume of these reprints, collected, but lacking the original pagination, is in the Library of the Royal Geographical Society. Rev. F. de Mély, *JS*, 1904 (n.s.), **2**, 472. Chinese transl. under name Hsi Lo-Ko, by Fêng Chhêng-Chün, Shanghai, 1928.

SCHLEGEL, G. (9). 'Geographical Notes.'
(*a*) The Nicobar and Andaman Islands. *TP*, 1898, **9**, 177.
(*b*) Lang-ga-siu (Lang-ya-hsiu), Lang-ga-su (Lang-ya-hsü) and Sih-lan-shan (Hsi-lan-shan) (ident. Ceylon). *Ibid.* p. 191.
(*c*) Ho-ling (ident. Kaling). *Ibid.* p. 273.
(*d*) Maliur and Malayu. *Ibid.* p. 288.
(*e*) Ting-ki-gi (Ting-chi-i), (ident. Ting-gü). *Ibid.* p. 292.
(*f*) Ma-it, Mai-t-tung, Ma-iëp-ung. *Ibid.* p. 365.
(*g*) Tun-sun, or Tian-sun (Tien-sun), (ident. Tenasserim or Tānah-sāri). *TP*, 1899, **10**, 33.
(*h*) Pa-hoang (Pho-huang Kuo), Pang-khang (Phêng-khêng Kuo), Pang-hang (Phêng-hêng Kuo), (ident. Pahang or Panggang). *Ibid.* p. 38.
(*i*) Dziu-hut (Jou-fo Kuo), (ident. Djohor, Johore). *Ibid.* p. 47.
(*j*) To-ho-lo, or Tok-ho-lo (Tu-ho-lo), (ident. Takōla or Takkōla). *Ibid.* p. 155.
(*k*) Ho-lo-tan (Kho-lo-tan) or Ki-lan-tan, (Chi-lan-tan), (ident. Kelantan). *Ibid.* p. 159.
(*l*) Shay-po (Shê-pho), (ident. Djavā, Java). *Ibid.* p. 247.
(*m*) Tan-tan, or Dan-dan (ident. Dondin?). *Ibid.* p. 459.
(*n*) Ko-la (Ko-lo) or Ko-la-pu-sa-lo (Ko-lo-fu-sha-lo), (ident. Kora-bēsar). *Ibid.* p. 464.
(*o*) Moan-la-ka (Man-la-chia), (ident. Malacca). *Ibid.* p. 470.

SCHMIDT, E. F. (1). *Flights over the Ancient Cities of Iran.* Chicago, 1940.

SCHMITTHENNER, H. (2). *Chinesische Landschaften und Städte.* Strecker & Schröder, Stuttgart, 1925.

SCHNITTER, N. J. (1). 'A Short History of Dam Engineering.' *WP*, 1967, **19** (no. 4), 142.

SCHOFF, W. H. (1). *Parthian Stations by Isidore of Charax; an account of the overland Trade Routes between the Levant and India in the 1st century B.C.* Philadelphia, 1914.

SCHOFF, W. H. (2). *Early Communication between China and the Mediterranean.* Philadelphia, 1921.

SCHOFF, W. H. (3). '*The Periplus of the Erythraean Sea*'; *Travel and Trade in the Indian Ocean by a Merchant of the First Century, translated from the Greek and annotated, etc.* Longmans Green, New York, 1912.

SCHOFF, W. H. (4). 'Navigation to the Far East under the Roman Empire.' *JAOS*, 1917, **37**, 240.

SCHOFF, W. H. (5). 'Some Aspects of the Overland Oriental Trade at the (Beginning of the) Christian Era.' *JAOS*, 1915, **35**, 31.

SCHRAMM, C. C. (1). *Brücken*. Leipzig, 1735.

SCHREIBER, H. (1). *The History of Roads*. Barrie & Rockliff, London, 1962.

SCHROEDER, A. H. (1). 'Ball Courts and Ball Games of Middle America and Arizona.' *AAAA*, 1955, **8**, 156.

SCHULTHESS, E. (1). *China* (album of photographs, with commentary). Collins, London, 1966.

SCHULTZ, ALWIN (1). *Das höfische Leben zur Zeit der Minnesinger [12th & 13th cents.]*. 2 vols. 2nd edn. Hirzel, Leipzig, 1889.

SCHURHAMMER, G. (1). 'Fernão Mendez Pinto und seine *Peregrinaçam*.' *AM*, 1926, **3**, 71, 194.

SCHURMANN, H. F. (1) (tr.). *Economic Structure of the Yuan Dynasty; a translation of chs. 93 and 94 of the 'Yuan Shih'*. Harvard Univ. Press, Cambridge, Mass., 1956. (Harvard-Yenching Institute Studies, no. 16.) Rev. J. Průsek, *ARO*, 1959, **27**, 479; H. Franke, *RBS*, 1959, **2**, 84.

SCHWARTZ, J. (1). 'L'Empire Romain, l'Egypte, et le Commerce Oriental.' *AHES/AESC*, 1960, **15** (no. 1), 18.

SCHWARZ, E. H. L. (1). 'Chinese Connections with Africa.' *JRAS/B*, 1938 (3rd ser.), **4**, 175.

SEATON, A. E. (1). *The Screw Propeller, and other Competing Instruments for Marine Propulsion*. Griffin, London, 1909.

SEGALEN, V., DE VOISINS, G. & LARTIGUE, J. (1). *Mission Archéologique en Chine, 1914 à 1917*. 1 vol. with 2 portfolios plates. (The text volume is entitled *L'Art Funeraire à l'Époque des Han*.) Geuthner, Paris, 1923–5 (plates), 1935 (text).

SEYRIG, H. (1). 'Antiquités de Beth-Maré (ex-voto de bronze représentant un navire).' *SYR*, 1951, **28**, 101.

SHADICK, H. (1) (tr. & ed.). *'The Travels of Lao Tshan', by Liu Thieh-Yün (Liu Ê)*. Cornell Univ. Press, Ithaca, N.Y., 1952.

SHAMASASTRY, R. (1) (tr.). *Kautilya's 'Arthaśāstra'*. With introd. by J. F. Fleet. Wesleyan Mission Press, Mysore, 1929.

SHANG KAI (1). 'Taming the Serpent River' (the Mêng River in Northern Honan; by a dam; tanks, reservoirs and cisterns in the loess highlands; terracing and reforestation). *CREC*, 1958, **7** (no. 4), 6.

SHEN SU-JU (1). 'At the Great Bend of the Yellow River' (on the present state of the Ninghsia irrigation systems). *CREC*, 1964, **13** (no. 7), 31.

SHIH SHÊNG-HAN (1). *A Preliminary Survey of the book 'Chhi Min Yao-Shu'; an Agricultural Encyclopaedia of the +6th Century*. Science Press, Peking, 1958.

SHIH SHÊNG-HAN (2). *On the 'Fan Shêng-Chih Shu', a Chinese Agricultural Book written by Fan Shêng-Chih in the −1st Century*. Science Press, Peking, 1959.

SHIH, VINCENT. See Shih Yu-Chung.

SHIH YU-CHUNG (1). 'Some Chinese Rebel Ideologies.' *TP*, 1956, **44**, 150.

SHIPPEE, R. (1). 'A Forgotten Valley of Peru [the Colca Valley].' *NGM*, 1934, **65**, 111 (131).

SHIPPEE, R. (2). 'The "Great Wall" of Peru, and other Aerial Photographic Studies by the Shippee-Johnson Peruvian Expedition.' *GR*, 1932, **22**, 1.

SHIRLEY-SMITH, H. (1). *The World's Great Bridges*. Phoenix, London, 1953; revised ed. 1964. Rev. R. H. Macmillan, *N*, 1954, **174**, 667.

SHRAVA, S. S. (1). *Irrigation in India prior to the +17th Century*. Govt. of India (Central Board of Irrign.), New Delhi, 1951.

SHULDHAM, M. (1). 'On Balanced Rudders.' *TINA*, 1864, **5**, 123.

SHUMOVSKY, T. A. See Szumowski, T. A.

SICKMAN, L., LOEHR, M., YANG LIEN-SHÊNG & SULLIVAN, M. (1). *Chinese Painting and Calligraphy from the Collection of John M. Crawford Jr. [Catalogue of an Exhibition with Introductions]*. Arts Council of Gt. Britain, Victoria and Albert Museum, London, 1965.

SICKMAN, L. & SOPER, A. (1). *The Art and Architecture of China*. Penguin (Pelican), London, 1956. (Rev. A. Lippe, *JAS*, 1956, **11**, 137.) New ed. 1968.

SĪDĪ 'ALĪ REĪS. See Von Diez (1); Vambéry (1); von Hammer-Purgstall (3); Bittner (1).

SIGAUT, E. (1). 'A Northern Type of Chinese Junk [the low-decked Tsingtao freighter].' *MMI*, 1960, **46**, 161.

SIGAUT, E. (2). 'François Edmond Paris; French Admiral [1806 to 1893].' *MMI*, 1961, **47**, 255.

DA SILVA, J. G. (1). *L'Appel aux Capitaux Étrangers et le Processus de Formation du Capital Marchand au Portugal du 15e au 17e Siècle*. Communication to the Ve Colloque International d'Histoire Maritime. Lisbon, 1960.

SILVERBERG, R. (1). *The Long Rampart; the Story of the Great Wall of China*. Chilton, Philadelphia, 1966.

SILVESTER, R. (1). 'Coastal Processes.' *N*, 1960, **188**, 467; 1962, **196**, 819.

SIMKHOVITCH, V. G. (1). 'Rome's Fall Reconsidered.' In *Toward the Understanding of Jesus, and other Historical Studies*, p. 111. New York, 1921.

SINGER, C. (2). *A Short History of Science, to the Nineteenth Century.* Oxford, 1941. Cf. Singer (11).

SINGER, C. (11). *A Short History of Scientific Ideas to 1900.* Oxford, 1959. A complete re-writing of Singer (2).

SINGER, C., HOLMYARD, E. J., HALL, A. R. & WILLIAMS, T. I. (1) (ed.). *A History of Technology.* 5 vols. Oxford, 1954–8. (Revs. M. I. Finley, *EHR*, 1959, **12**, 120; J. Needham, *CAMR*, 1957, **78**, 299; 1959, **80**, 227; E. J. Bickerman & G. Mattingly, *AJP*, 1956, **77**, 96; 1958, **79**, 317.)

SINOR, D. (6). 'On Water-Transport in Central Eurasia.' *UAJ*, 1961, **33**, 156.

SION, J. (1). *Asie des Moussons.* Vol. 9 of *Geographie Universelle.* Colin, Paris, 1928.

SIRÉN, O. (1). (*a*) *Histoire des Arts Anciens de la Chine.* 3 vols. van Oest, Brussels, 1930. (*b*) *A History of Early Chinese Art.* 4 vols. Benn, London, 1929. Vol. 1, Prehistoric and Pre-Han; Vol. 2, Han; Vol. 3, Sculpture; Vol. 4, Architecture.

SIRÉN, O. (2). *Chinese Sculpture from the +5th to the +14th Century,* 1 vol. text, 3 vols. plates. Benn, London, 1925. French tr. *La Sculpture Chinoise du 5e au 14e Siècle.* van Oest, Paris & Brussels, 1926.

SIRÉN, O. (3). *The Imperial Palaces of Peking.* 3 vols. van Oest, Paris & Brussels, 1927.

SIRÉN, O. (4). *The Walls and Gates of Peking.* London, 1924.

SIRÉN, O. (5). 'Chinese Architecture.' *EB*, v, p. 556.

SIRÉN, O. (6). *History of Early Chinese Painting.* 2 vols. Medici Society, London, 1933.

SIRÉN, O. (7). *Les Peintures Chinoises dans les Collections Americaines.* van Oest, Paris & Brussels, 1927.

SIRÉN, O. (8). *Gardens of China.* Ronald Press, New York, 1949.

SIRÉN, O. (10). *Chinese Painting; Leading Masters and Principles.* Lund Humphries, London, 1956; Ronald, New York, 1956. 6 vols. Pt. I, The First Millennium, 3 vols. incl. one of plates; pt. II, The Later Centuries, 4 vols., incl. one of plates.

SIRÉN, O. (11). 'The Chinese Garden; a Work of Art in the Forms of Nature.' *ORA*, 1948, **1**, 24.

SITTIG, O. (1). 'Über unfreiwillige Wanderungen im Grossen Ozean.' *MJPGA*, 1890, **36**, 161, 184; Eng. tr. 'Compulsory Migrations in the Pacific Ocean.' *ARSI*, 1895 (1896), 519.

SKACHKOV, K. A. (1). 'O Voenno-Morskom Depe i Kitaiskev' [on the warships of China; a MS. or printed book entitled *Shui Shih Chi Yao* (Essentials of Sea Affairs), N 18/162 in the Library of the Rumiantzov Museum] (in Russian). *MSB*, 1858, **37** (no. 10), 289. (Skachkov Bibliography, no. 5337, Cordier (2), col. 1562.)

SKACHKOV, P. E. (1). *Bibliographia Kitaia* (in Russian). Gosudarstvennoe Socialvno-Economicheskoe Isdatelstvo, Moscow, 1932.

SKELTON, R. A., MARSTON, T. E. & PAINTER, G. D. (1), with a foreword by A. O. VIETOR. *The Vinland Map and the Tartar Relation.* Yale Univ. Press, New Haven and London, 1965.

SKEMPTON, A. W. (1). 'The Boat Store, Sheerness (1858–60), and its Place in Structural History.' *TNS*, 1959, **32**, 57.

SKEMPTON, A. W. (2). 'The Origin of Iron Beams.' *Actes du VIIIᵉ Congrès International d'Histoire des Sciences*, Florence, 1956, vol. 3, p. 1029.

SKEMPTON, A. W. (3). 'The Evolution of the Steel Frame Building.' *GE*, 1959, **10**, 37.

SKEMPTON, A. W. (4). 'Canals and River Navigations before +1750.' Art. in *A History of Technology*, ed. C. Singer *et al.*, vol. 3, p. 438. Oxford, 1957.

SKEMPTON, A. W. (5). 'The Engineers of the English River Navigations, +1620 to +1760.' *TNS*, 1954, **29**, 25.

SKEMPTON, A. W. (6). 'The History of Structural Iron, Steel and Concrete.' Three lectures given at Cambridge, May 1964.

SKEMPTON, A. W. & JOHNSON, H. R. (1). 'The First Iron Frames.' *AREV*, 1962.

SKINNER, R. T. F. (1). 'Peking, 1953.' *AREV*, 1953, Oct. 255.

SKINNER, R. T. F. (2). 'Chinese Domestic Architecture.' *JRIBA*, 1958 (3rd ser.), **65**, 430.

SLOCUM, CAPT. JOSHUA (1). *Sailing Alone Around the World [first pub. 1900]; and, the Voyage of the 'Liberdade' [first pub. 1894].* Hart Davis, London, 1948; Reprint Society, London, 1949.

ŠMÍD, MIRKO (1). 'Prvni Plavci na Širém Moři; Plavidla a Objevné Plavby Foiničanů' (The First Sailors on the High Seas, an Account of the Phoenician Navigators and their Vessels). *NVO*, 1960, **15** (no. 6), 126.

SMILES, S. (1). *Lives of the Engineers.* Murray, London, 1st ed. 1857. Vol. 1, *Early Engineering; Vermuyden, Middleton, Perry, James Brindley,* 1874; vol. 2, *Harbours, Lighthouses, Bridges; Smeaton and Rennie,* 1874; vol. 3, *History of Roads; Metcalfe, Telford,* 1874; vol. 4, *The Steam-Engine; Boulton and Watt,* 1874; vol. 5, *The Locomotive; George and Robert Stephenson,* 1877.

SMITH, ADAM (1). *An Inquiry into the Nature and Causes of the Wealth of Nations,* ed. J. S. Nicholson. Nelson, London, 1901.

SMITH, ANTHONY (1). *Blind White Fish in Persia.* Allen & Unwin, London, 1953.

SMITH, C. A. MIDDLETON (1). 'Chinese Creative Genius.' *CTE*, 1946, **1**, 920, 1007.

SMITH, C. A. MIDDLETON (2). 'The Age-Long Engineering Works of China.' *EN*, 1919, **127**, 72.

SMITH, D. H. (1). 'Zayton's Five Centuries of Sino-Foreign Trade.' *JRAS*, 1958, 165.

SMITH, G. ELLIOTT. See Elliott-Smith, Sir Grafton.

SMITH, H. S. See Shirley-Smith, H.

SMITH, M. W. (1) (ed.). 'Asia and North America; Trans-Pacific Contacts' [Symposium]. *AMA*, 1953, **18**, no. 3, pt. 2. (Memoirs of the Society for American Archaeology, no. 9.)

SMITH, P. J. & NEEDHAM, JOSEPH (1). 'Magnetic Declination in Mediaeval China.' *N*, 1967, **214**, 1213.

SMITH, R. BAIRD (1). *Italian Irrigation; a Report on the Agricultural Canals of Piedmont and Lombardy, addressed to the Honourable the Court of Directors of the East India Company.* 2 vols. Allen, London, 1852.

SMITH, V. A. (1). *Oxford History of India, from the earliest times to 1911.* 2nd ed. ed. S. M. Edwardes. Oxford, 1923.

SMYTH, H. WARINGTON (1). *Mast and Sail in Europe and Asia.* Blackwood, Edinburgh, 1906; 2nd ed. 1929.

SMYTH, W. H. ADM. (1). *The Sailor's Word-Book.* Blackie, London, 1867.

SMYTHE, F. S. (1). 'Suspension bridges on the Nepal–Tibet border.' *GGM*, 1938, **7**, 189.

SNOW, EDGAR (1). *Red Star over China.* London and New York, 1938.

SOLIGNAC, M. (1). *Recherches sur les Installations Hydrauliques de Kairouan et des Steppes Tunisiennes du 7e au 11e Siècle.* Carbonel, Algiers, 1953. (Publications de l'Instit. d'Études Orient. de la Faculté de Lettres de l'Univ. d'Alger, no. 13.)

SØLVER, C. V. (1). 'Leidarsteinn; the Compass of the Vikings.' *OL*, 1946, **10**, 293.

SØLVER, C. V. (2). 'The Rebaek Rudder.' *MMI*, 1946, **32**, 115.

SØLVER, C. V. (3). 'The Discovery of an Early Norse Bearing-Dial.' *JIN*, 1953, **6**, 294. Discussion by E. G. R. Taylor, W. E. May, R. B. Motzo & T. C. Lethbridge, 'A Norse Bearing-Dial?', *JIN*, 1954, **7**, 78.

SOLVYNS, F. B. (1). *Les Hindous* [a collection of coloured plates with descriptions]. Nicolle, Paris, 1808.

SOOTHILL, W. E. (5) (posthumous). *The Hall of Light; a Study of Early Chinese Kingship.* Lutterworth, London, 1951. (On the Ming Thang, and contains discussion of the *Pu Thien Ko.*)

SOPER, A. C. (1). 'Hsiang-Kuo Ssu; an Imperial Temple of the Northern Sung [at Khaifêng].' *JAOS*, 1948, **68**, 19.

SOPER, A. C. (2). *The Evolution of Buddhist Architecture in Japan.* Princeton Univ. Press, Princeton, N.J., 1942. (Princeton Monographs in Art and Archaeology, no. 22.)

SOPER, F. L. (1). 'The Report of a nearly pure *Ancylostoma duodenale* Infestation in native South American Indians, and a Discussion of its Ethnological Significance.' *AJH*, 1927, **7**, 174.

SOTTAS, J. (1). 'An Early Lateen Sail in the Mediterranean.' *MMI*, 1939, **25**, 229. Discussion by L. G. C. Laughton, p. 441. Cf. Bowen (1), p. 93.

SOUSA, AHMED (1). *The Irrigation System of Sāmarrah during the Abbasid Caliphate.* 2 vols. (in Arabic). Alma'arif Press, Baghdad, 1948.

SOUSTELLE, J. (1). *La Pensée Cosmologique des anciens Mexicains; Representation du Monde et de l'Espace.* Hermann, Paris, 1940. (Actualités Scientifiques et Industrielles, no. 881.)

SOWERBY, A. DE C. (4). 'Junks and Sampans of the Inland Waterways of China.' *CJ*, 1929, **10**, 243.

SOYMIÉ, M. (2). 'Sources et Sourciers en Chine.' *BMFJ*, 1961 (n.s.), **7**, no. 1.

SPECKHART, G. See Saunier, C. (1).

SPEISER, W. (1). *Oriental Architecture in Colour.* Thames & Hudson, London, 1965. Tr. C. W. E. Kessler from *Baukunst des Ostens.* Burkhard, Essen, 1964.

SPENCER, H. R. (1). 'Sir Isaac Newton on Saumarez' Patent Log.' *ANEPT*, 1954, **14**, 214.

SPENCER, J. E. (1). 'The Houses of the Chinese.' *GR*, 1947, **37**, 254.

SPENCER, J. E. (2). 'The Junks of the Yangtze.' *ASIA*, 1938, **38**, 466.

SPIELMANN, P. E. & ELFORD, E. J. (1). *Road Making and Administration.* Arnold, London, 1948.

VAN SPILBERGEN, J. (1). *Miroir Oost- & West-Indical.* Jansz, Amsterdam, 1619, 1621.

SPINDEN, H. J. (1). *Ancient Civilisations of Mexico and Central America,* 3rd ed. Amer. Mus. Nat. History, New York, 1946. (Handbook series, no. 3.)

SPRATT, H. P. (1). 'The Pre-natal History of the Steamboat.' *TNS*, 1960, **30**, 13 (paper read Oct. 1955).

SPRATT, H. P. (2). *The Birth of the Steamboat.* Griffin, London, 1959 (revs. H. O. Hill, *MMI*, 1960, **46**, 159; A. W. Jones, *N*, 1959, **183**, 1626).

SPULER, B. (1). *Die Mongolen in Iran; Politik, Verwaltung und Kultur der Ilchanzeit, +1220 to +1350.* Hinrich, Leipzig, 1939; 2nd ed. Akademie Verlag, Berlin, 1955.

STANHOPE, G. & GOOCH, G. P. (1). *Life of Charles, Third Lord Stanhope.* Longmans Green, London, 1914.

STAUNTON, SIR GEORGE LEONARD (1). *An Authentic Account of an Embassy from the King of Great Britain to the Emperor of China. . .taken chiefly from the Papers of H.E. the Earl of Macartney, K.B. etc.. . . .* 2 vols. Bulmer & Nicol, London, 1797; repr. 1798. Germ. tr., Berlin, 1798; French tr., Paris, 1804; Russian tr., St Petersburg, 1804. Abridged Eng. ed. 1 vol. Stockdale, London, 1797.

STAUNTON, SIR GEORGE THOMAS (2). *Notes on Proceedings and Occurrences during the British Embassy to Peking in 1816* [*Lord Amherst's*]. London, 1824.

STAUNTON, SIR GEORGE THOMAS (3) (tr. & ed.). J. G. de Mendoza's *History of the Great and Mightie Kingdome of China*. See de Mendoza (1).

STEBBINS, G. L. (1). 'Origin and Migrations of Cotton.' *EM*, 1947, **17**, 149.

STEERS, J. A. (1) (ed.). *The Cambridge Region, 1965*. Collective work prepared for the British Association Cambridge Meeting. Brit. Ass., London, 1965.

STEFÁNSSON, V. & WILCOX, O. R. (1). *Great Adventures and Explorations, from the Earliest Times to the Present, as told by the Explorers themselves....* Hale, London, 1947.

STEIGER, G. N., BEYER, H. O. & BENITEZ, C. (1). *A History of the Orient*. Ginn, New York & Boston, 1926.

STEIN, SIR AUREL (1). *Ruins of Desert Cathay; Personal Narrative of Explorations in Central Asia and Westernmost China*. 2 vols. Macmillan, London, 1912.

STEIN, SIR AUREL (2). *Innermost Asia; Detailed Report of Explorations carried out in Central Asia, Kansu and Eastern Iran....* 2 vols. text, 1 vol. plates, 1 box maps. Oxford, 1928.

STEIN, SIR AUREL (3). *On Ancient Central Asian Tracks; Brief Narrative of Three Expeditions in Innermost Asia and North-western China*. Macmillan, London, 1933.

STEIN, SIR AUREL (4). *Serindia; Detailed Report of Explorations in Central Asia and Westernmost China....* Oxford, 1921.

STEIN, SIR AUREL (7). 'A Chinese Expedition across the Pamirs and Hindukush, A.D. 747.' *NCR*, 1922, **4**, 161. (Reprinted, with Stein (6), Chavannes (12) and Wright, H. K., in brochure form, Peiping, 1940.)

STEIN, SIR AUREL (8). 'The Indo-Iranian Borderlands: their Prehistory in the Light of Geography and of Recent Explorations.' *JRAI*, 1934, **64**, 188 (196).

STEIN, SIR AUREL (9). 'Explorations in Central Asia, 1906–8.' *SGM*, 1910, **26**, 225, 281.

STEIN, SIR AUREL (10). 'Explorations in the Lop Desert.' *GR*, 1920, **9**, 1.

STEIN, R. A. (1). 'Le Lin-Yi; sa localisation, sa contribution à la formation du Champa, et ses liens avec la Chine.' *HH*, 1947, **2** (nos. 1–3), 1–300.

STEIN, R. A. (2). 'Jardins en Miniature d'Extrême-Orient; le Monde en Petit.' *BEFEO*, 1943, **42**, 1–104.

STEIN, R. A. (3). 'L'Habitat, le Monde et le Corps Humain, en Extrême-Orient et en Haute Asie.' *JA*, 1957, **245**, 37. For the illustrations desirable in following this paper see Stein (4).

STEIN, R. A. (4). 'Architecture et Pensée religieuse en Extrême-Orient.' *AAS*, 1957, **4**, 163.

STEINMAN, D. B. (1). *A Practical Treatise on Suspension Bridges*. Wiley, New York, 1929.

STEINMAN, D. B. (2). 'Bridges.' *SAM*, 1954, **191** (no. 5), 61. *AMSC*, 1954, **42**, 397 and 460.

STEINMAN, D. B. & WATSON, S. R. (1). *Bridges and their Builders*. Putnam, New York, 1941. Revised ed. Dover Paperbacks, New York, 1961.

STEVENSON, E. L. (1). *Terrestrial and Celestial Globes; their History and Construction....* 2 vols. Hispanic Soc. Amer. (Yale Univ. Press), New Haven, 1921.

STEVENSON, E. L. (2). *Portolan Charts; their Origin and Characteristics:....* Hispanic Soc. Amer., New York, 1911.

STEVENSON, P. H. (1). 'Description of the Kuanhsien Suspension Bridge.' *CJ*, 1927, **6**, 186.

STEVENSON, ROBERT (1). 'On the History and Construction of Suspension Bridges.' *EPJ*, 1821, **5**, 237.

STEWARD, JULIAN (1) (ed.). *Irrigation Systems*. Washington, D.C., 1956. (Pan-American Union, Social Science Monographs, no. 1.)

STOKES, F. M. C. (1). 'Zimbabwe.' *GGM*, 1935, **2**, 142.

STONE, L. H. (1). *The Chair in China*. Royal Ontario Museum of Archaeology, Toronto, 1952.

STONOR, C. R. & ANDERSON, E. (1). 'Maize among the Hill Peoples of Assam.' *AMBG*, 1949, **36** (no. 3), 355.

STRANDES, J. (1). *Die Portugiesenzeit von Deutsch- und Englisch-Ostafrika*. Berlin and Leipzig, 1899. Eng. tr. by J. Wallwork, annotated by J. S. Kirkman, *The Portuguese in East Africa*. Kenya History Society, Nairobi, 1961.

LE STRANGE, G. (4). *Description of the Province of Fārs (with a translation of al-Balkhī's account)*. London, 1912.

STRAUB, H. (1). *Die Geschichte d. Bauingenieurkunst; ein Überblick von der Antike bis in die Neuzeit*. Birkhäuser, Basel, 1949. 2nd ed. Liebing, Würzburg, 1964. Eng. tr. by E. Rockwell, *A History of Civil Engineering*. Leonard Hill, London, 1952.

STUART, G. A. See Wang Chi-Min (2). Biography no. 35.

STUART, G. A. (1). *Chinese Materia Medica; Vegetable Kingdom*, extensively revised from Dr F. Porter Smith's work. Amer. Presbyt. Mission Press, Shanghai, 1911. An expansion of *Contributions towards the Materia Medica and Natural History of China, for the use of Medical Missionaries and Native Medical Students*, by F. Porter Smith. Amer. Presbyt. Mission Press, Shanghai, 1871; Trübner, London, 1871.

Su GIN-DJIH. See Hsü Ching-Chih.

Su MING (1). 'A Victory on the Yangtze River' [in retention-basins near Hankow]. *PC*, 1952, no. 14, 28.

SULLIVAN, M. (1). *An Introduction to Chinese Art.* Faber & Faber, London, 1961.

SULLIVAN, M. (2). 'The Heritage of Chinese Art.' Art. in *The Legacy of China*, ed. R. Dawson, p. 165. Oxford, 1964.

SULLIVAN, M. (3). 'Notes on Early Chinese Landscape Painting.' *HJAS*, 1955, **18**, 422.

SULLIVAN, M. (4). 'Sandrart on Chinese Painting' (cf. Joachim von Sandrart, 1). *ORA*, 1949, **1**, 159.

SULLIVAN, M. (5). 'Archaeology in the Philippines.' *AQ*, 1956, **30**, 68.

SULLIVAN, M. (6). 'Notes on Chinese Export Wares in Southeast Asia.' *TOCS*, 1962.

SULLIVAN, M. (7). 'Chinese Export Porcelain in Singapore.' *ORA*, 1957, **3**, 145.

SULLIVAN, M. (8). 'Kendi' (drinking vessels, Skr. *kundika*, with neck and side-spout). *ACASA*, 1957, **11**, 40.

SUN, E-TU ZEN. See Sun Jen I-Tu.

SUN JEN I-TU & DE FRANCIS, J. (1). *Chinese Social History; Translations of Selected Studies.* Amer. Council of Learned Societies, Washington, D.C., 1956. (ACLS Studies in Chinese and Related Civilisations, no. 7.)

SUN JEN I-TU & SUN HSÜEH-CHUAN (1) (tr.). '*Thien Kung Khai Wu*', *Chinese Technology in the Seventeenth Century, by Sung Ying-Hsing.* Pennsylvania State Univ. Press; University Park and London, Penn., 1966.

SUN LI et al. (1). *China's Big Leap in Water Conservancy* (mostly on the relatively smaller projects). Foreign Language Press, Peking, 1958.

SVERDRUP, H. V., JOHNSON, M. W. & FLEMING, R. H. (1). *The Oceans; their Physics, Chemistry and General Biology.* Prentice-Hall, New York, 1942.

SWANN, NANCY L. (1) (tr.). *Food and Money in Ancient China; the Earliest Economic History of China to +25* (with tr. of [*Chhien*] *Han Shu*, ch. 24 and related texts [*Chhien*] *Han Shu*, ch. 91 and *Shih Chi*, ch. 129). Princeton Univ. Press, Princeton, N.J., 1950. (Rev. J. J. L. Duyvendak, *TP*, 1951, **40**, 210; C. M. Wilbur, *FEQ*, 1951, **10**, 320; Yang Lien-Shêng, *HJAS*, 1950, **13**, 524.)

SZUMOWSKI, T. A. (1). *Tres Roteiros Desconhecidos de Aḥmad ibn Mājid, o Piloto Árabe de Vasco da Gama.* Comissão Executiva das Comemorações do Quinto Centenário da Morte do Infante Dom Henrique, Lisbon, 1960. Portuguese translation by M. Malkiel-Jirmunsky of *Tri Neisvestnych Lotsı Aḥmada ibn Mājida Arabskogo Lotsmana Vasco da Gamvi* [Three Unpublished Nautical Rutters of A. ibn M., the Arab pilot of Vasco da Gama] (in Russian) facsimile and tr., Academy of Sciences, Moscow & Leningrad, 1957.

SZUMOWSKI, T. A. (2). 'An Arab Nautical Encyclopaedia of the +15th Century [*Book of Useful Chapters on the Basic Principles of Sea-faring, by Aḥmad ibn Mājid, c. +1475*].' In Resumo das Comunicações do Congresso Internacional de História dos Descobrimentos, Lisbon, 1960, p. 109. Actas, vol. 3, p. 43.

TAN AI-CHING (1). 'Socialist Labour Builds a Dam' (the Ming Tombs Reservoir). *CREC*, 1958, **7** (no. 8), 2.

TAN PEI-YING (1). *The Building of the Burma Road.* McGraw-Hill, New York and London, 1945.

TANGE KENZŌ & KAWAZOE NOBORU (1), with photographs by WATANABE YOSHIO and an introduction by J. BURCHARD. *Ise; Prototype of Japanese Architecture.* M.I.T. Press, Cambridge, Mass., 1965.

TARN, W. W. (1). *The Greeks in Bactria and India.* Cambridge, 1951.

TARN, W. W. (3). 'The Roman Navy.' In *A Companion to Latin Studies*, ed. J. E. Sandys, p. 489. Cambridge, 1913.

TARN, W. W. (4). *Hellenistic Military and Naval Developments.* Cambridge, 1930. (Lees-Knowles Lectures in Military History Cambridge.)

TAYLOR, E. G. R. (6). 'The South-Pointing Needle.' *IM*, 1951, **8**, 1.

TAYLOR, E. G. R. (7). *The Mathematical Practitioners of Tudor and Stuart England* (for Inst. of Navigation). Cambridge, 1954. (Rev. D. J. de S. Price, *JIN*, 1955, **8**, 12.)

TAYLOR, E. G. R. (8). *The Haven-Finding Art; a History of Navigation from Odysseus to Captain Cook.* Hollis & Carter, London, 1956. (Crit. rev. D. W. Waters, *MMI*, 1957, **43**, 256.)

TAYLOR, E. G. R. (9). 'The Oldest Mediterranean Pilot [*Il Compasso da Navigare, c. +1253*].' *JIN*, 1951, **4**, 81.

TAYLOR, E. G. R. (10). 'Mathematics and the Navigator in the +13th Century' (First Duke of Edinburgh Lecture). *JIN*, 1960, **13**, 1. Sep. Repr. for distribution at the International Congress in Commemoration of the Fifth Centenary of Prince Henry the Navigator, Lisbon, 1960, by the Royal Geographical Society, London.

TAYLOR, E. G. R. (11). 'John Dee and the Nautical Triangle, +1575.' *JIN*, 1955, **8**, 318.

TAYLOR, E. G. R. (12). 'The Navigating Manual of Columbus.' *JIN*, 1952, **5**, 42; *BCIC*, 1953, **1**, 32.

TAYLOR, E. G. R. (13) (ed.). 'A Regiment for the Sea', and other Writings on Navigation, by William Bourne of Gravesend, a Gunner (c. +1535 to +1582). Cambridge, 1963. (Hakluyt Society Pubs., 2nd ser., no. 121.)

TAYLOR, E. G. R. & RICHEY, M. W. (1). The Geometrical Seaman, a Book of Early Nautical Instruments. Hollis & Carter (for Inst. of Navigation), London, 1962.

TAYLOR, E. G. R. & SADLER, D. H. (1). 'The Doctrine of Nauticall Triangles Compendious [+1594].' JIN, 1953, 6, 131.

TAYLOR, F. R. FORBES (1). 'Heavy Goods Handling prior to the Nineteenth Century' [a history of cranes]. TNS, 1963, 35, 179.

TEICHMAN, SIR ERIC (2). Travels of a Consular Officer in Eastern Tibet. Cambridge, 1922.

TEICHMAN, SIR ERIC (3). Travels of a Consular Officer in Northwest China. London, 1921.

TEMKIN, O. (2). 'Was Servetus influenced by Ibn al-Nafīs?' BIHM, 1940, 8, 731.

TÊNG SHU-CHUN (1). 'The Early History of Forestry in China.' JF, 1927, 25, 564.

TÊNG SSU-YÜ (1). 'China's Examination System and the West.' In China, ed. H. F. McNair, p. 441. Univ. of Calif. Press, Berkeley, 1946.

TÊNG SSU-YÜ & BIGGERSTAFF, K. (1). An Annotated Bibliography of Selected Chinese Reference Works. Harvard-Yenching Instit., Peiping, 1936. (Yenching Journ. Chin. Studies, monograph no. 12.)

TÊNG TSÊ-HUI (1). Report on the Multiple-Purpose Plan for permanently controlling the Yellow River and exploiting its Water Resources. Foreign Language Press, Peking, 1955.

TEW, D. H. (1). 'Canal Lifts and Inclines, with particular reference to those in the British Isles.' TNS, 1951, 28, 35.

TEY, J. M. (1). Hongkong-Barcelona en el Junco 'Rubio'. Edit. Juventud, Barcelona, 1959.

THAN PO-FU, WÊN KUNG-WÊN, HSIAO KUNG-CHÜAN & MAVERICK, L. A. (tr.). Economic Dialogues in Ancient China; Selections from the 'Kuan Tzu' (Book)... Pr. pr. Carbondale, Illinois and Yale Univ. Hall of Graduate Studies, New Haven, Conn., 1954. (Rev. A. W. Burks, JAOS, 1956, 76, 198.)

THANG LAN (1). 'Palace of Emperors; now Palace of Art (the Imperial Palace in Peking).' CREC, 1956, 4 (no. 11), 24.

THÉRY, R. (1). 'Jouffroy d'Abbans et les Origines de la Navigation à Vapeur.' MC/TC, 1952, 2, 42.

THIEN, J. K. (1). 'Two Kuching Jars' (c. +1470, found in Sarawak). SMJ, 1949, 5, 23.

THOMAS, F. W. (2). 'Political and Social Organisation of the Maurya Empire.' CHI, i, ch. 19.

THOMAS, R. D. (1). A Trip on the West River [from Canton to Wuchow and Return]. Cover entitled Pastures New; in a Stern-Wheeler up the Si Kiang [Hsi Chiang]. China Baptist Publication Society, Canton, 1903.

THOMAZI, A. (1). Histoire de la Pêche. Payot, Paris, 1947.

THOMPSON, D'ARCY W. (2). Growth and Form. Cambridge, 1917; 2nd ed. 1942.

THOMPSON, E. A. & FLOWER, B. A Roman Reformer and Inventor; being a New Text of the Treatise 'De Rebus Bellicis', with a translation...introduction...and Latin index.... Oxford, 1952. This text is now generally conceded to have been written by a Latin of Illyria in the close neighbourhood of +370; see Schneider (3), Berthelot (7), Reinach (2), Neher (1) and Oliver (1).

THOMPSON, J. E. S. (1). Maya Hieroglyphic Writing; an Introduction. Washington, 1950. (Carnegie Institution Pubs. no. 589.)

THOMPSON, R. CAMPBELL (2). A Dictionary of Assyrian Botany. Brit. Acad. London, 1949.

THOMPSON, R. CAMPBELL (3). The Epic of Gilgamesh; a New Translation from a Collection of Cuneiform Tablets in the British Museum, rendered literally into English Hexameters. Luzac, London, 1928.

THOMPSON, R. CAMPBELL (4). The Epic of Gilgamesh; Text, Transliteration and Notes. Clarendon, Oxford, 1930.

THOMPSON, R. CAMPBELL & HUTCHINSON, R. W. (1). A Century of Explorations at Nineveh. Luzac, London, 1929.

THOMPSON, R. CAMPBELL & HUTCHINSON, R. W. (2). 'The Excavations on the Temple of Nabû at Nineveh.' AAA, 1929, 74, 103.

THOMPSON, SYLVANUS P. (1). 'The Rose of the Winds; the Origin and Development of the Compass-Card.' PBA, 1913, 6, 1.

THOMSON, D. W. (1). 'Two Thousand Years under the Sea; the Story of the Diving Bell.' ANEPT, 1947, 7, 261.

THOMSON, JOHN (1). Illustrations of China and its People; a Series of 200 Photographs with letterpress descriptive of the Places and the People represented. 4 vols. Sampson Low, London, 1873-4. French tr. by A. Talandier & H. Vattemare, 1 vol. Hachette, Paris, 1877.

THORNDIKE, LYNN (1). A History of Magic and Experimental Science. 8 vols. Columbia Univ. Press, New York: vols. 1 and 2, 1923, repr. 1947; 3 and 4, 1934; 5 and 6, 1941; 7 and 8, 1958. (Rev. W. Pagel, BIHM, 1959, 33, 84.)

THOULESS, R. H. (1). 'Phenomenal Regression to the Real Object.' BJPC, 1931, 21, 239; 1932, 22, 1.

THOULESS, R. H. (2). 'Individual Differences in Phenomenal Regression.' *BJPC*, 1932, **22**, 216. 'The Truth about Perspective.' *D*, 1930, **11**, 121.

THOULESS, R. H. (3). 'A Racial Difference in Perception.' *JSPC*, 1933, **4**, 330.

TIEN. See Thien.

TING WAN-CHÊNG (1). 'Building the Railway to Mongolia.' *CREC*, 1955, **4** (no. 6), 6.

TING WÊN-CHIANG & DONNELLY, I. A. (tr.). '"Things Produced by the Works of Nature", published 1639, translated by Dr V. K. Ting.' *MMI*, 1925, **11**, 234. (A translation, with notes, of that part of ch. 9 of *Thien Kung Khai Wu* (+1637) which deals with nautical technology.)

TISDALE, A. (1). 'Down the Yalu in a "Jumping Chicken" [boat].' *ASIA*, 1920, **20**, 902.

TODD, O. J. (1). 'Taming "Flood Dragons" along China's Huang Ho.' *NGM*, 1942, **81**, 205.

TODD, O. J. & ELIASSEN, S. (1). 'The Yellow River Problem.' *TASCE*, 1940, **105**, 346 (Paper no. 2064). (With discussion by many contributors, including H. Chatley.)

TOGAN, ZAKI VALIDI (2). *Ibn Fadlān's Reisebericht*. Brockhaus, Leipzig, 1939. (*AKML*, no. 24 (3).)

TOLSTOV, S. P. (1). *Drevniy Choresm* (Ancient Chorasmia) (in Russian). University Press, Moscow, 1948. Germ. tr. *Auf den Spuren d. altchoresmischen Kultur*, by O. Mehlitz. Kultur & Vorschritt, Berlin, 1953. (Rev. A. D. H. Bivar, *ORA*, 1955 (n.s.), **1**, 129.) See also Ghirshman (2). Mongait (1), pp. 235 ff.

TOLSTOV, S. P. (2). 'Les Résultats des Travaux de l'Expédition archéologique et ethnographique de l'Académie des Sciences de l'U.R.S.S. au Khorezm en 1951–1955.' *AAS*, 1957, **4**, 83 and 187.

TOLSTOV, S. P. (3). Lecture delivered at Cambridge 14 Mar. 1956 on the Soviet Expeditions in Chorasmia, dealing especially with irrigations and fortifications.

TONG, HOLLINGTON K. See Tung Hsien-Kuang.

TORR, C. (1). *Ancient Ships*. Cambridge, 1894. Repr. 1964, ed. and intr. A. J. Podlecki, with an appendix containing a series of articles on the Greek warship and the Greek trireme, by W. W. Tarn, A. B. Cook, C. Torr, W. Richardson & P. H. Newman; and many new illustrations (Argonaut, Chicago).

TORRANCE, T. (2). 'The Origin and History of the Irrigation Work of the Chêngtu Plain.' *JRAS/NCB*, 1924, **55**, 60. With addendum: 'The History of [the State of] Shu; a free translation of [part of] the *Shu Chih* [ch. 3 of *Hua Yang Kuo Chih*]'.

TORRICELLI, EVANGELISTA (1). *Del Moto di Gravi*. Florence, 1644.

TOSCANO, S. (1). *Derecho y Organización Social de los Aztecas*. Mexico City, 1937.

TOUDOUZE, G. G., DE LA RONCIÈRE, C., TRAMOND, J., RONDELEUX, C., DOLLFUS, C., LESTONNAT, R., SEBILLE, A. & LEFÉBURE, R. (1). *Histoire de la Marine*. l'Illustration, Paris, 1939.

TOUSSAINT, M. M. A. (1). *History of the Indian Ocean*, tr. from the French by J. Guicharnaud. Routledge & Kegan Paul, London, 1966.

TOUSSOUN, PRINCE OMAR (1). 'Mémoire sur les anciennes Branches du Nil.' *MIE*, 1922, no. 4.

TOUSSOUN, PRINCE OMAR (2). 'Mémoire sur l'Histoire du Nil.' *MIE*, 1925, nos. 8–10.

TOYNBEE, A. J. (1). *A Study of History*. RIIA, London, 1935–9. 6 vols.

TREND, J. B. (1). *Portugal*. Benn, London, 1957.

TRIAS, R. A. LAGUARDIA (1). *Comentarios sobre los Origenes de la Navegacion Astronomica*. Madrid, 1959.

TRIGAULT, NICHOLAS (1). *De Christiana Expeditione apud Sinas*. Vienna, 1615; Augsburg, 1615. Fr. tr.: *Histoire de l'Expédition Chrétienne au Royaume de la Chine, entrepris par les PP. de la Compagnie de Jésus, comprise en cinq livres... tirée des Commentaires du P. Matthieu Riccius, etc.* Lyon, 1616; Lille, 1617; Paris, 1618. Eng. tr. (partial): *A Discourse of the Kingdome of China, taken out of Ricius and Trigautius*. In *Purchas his Pilgrimes*. London, 1625, vol. 3, p. 380. Eng. tr. (full): see Gallagher (1). Trigault's book was based on Ricci's *I Commentarj della Cina* which it follows very closely, even verbally, by chapter and paragraph, introducing some changes and amplifications, however. Ricci's book remained unprinted until 1911, when it was edited by Venturi (1) with Ricci's letters; it has since been more elaborately and sumptuously edited alone by d'Elia (2).

TROUSDALE, W. (1). 'Architectural Landscapes attributed to Chao Po-Chü [fl. +1100/+1150].' *AI/AO*, 1961, **4**, 285.

TSUNG PAI-HUA (1). 'Space-Consciousness in Chinese Poetry and Painting' (a lecture given in Chinese and pub. in *HCHTC*, 1949, **12** (no. 10); Eng. tr. E. J. Schwarz). *SACAJ*, 1949, **1**, 25.

TUCCI, G. (3). *Tibetan Painted Scrolls*. 2 vols. and 1 vol. plates. Libreria dello Stato, Rome, 1949.

TUNG HSIEN-KUANG (1) (ed.). *China Handbook, 1937–1943*. Chinese Ministry of Information, New York, 1943; Macmillan, New York, 1943.

TWITCHETT, D. C. (2). 'The Fragment of the Thang "Ordinances of the Department of Waterways" [+737] discovered at Tunhuang.' *AM*, 1957, **6**, 23.

TWITCHETT, D. C. (4). *Financial Administration under the Thang Dynasty*. Cambridge, 1962. (Univ. of Cambridge Oriental Pubs. no. 7.) Rev. Yü Ying-Shih, *JAOS*, 1964, **84**, 71.

TWITCHETT, D. C. (6). 'Some Remarks on Irrigation under the Thang.' *TP*, 1960, **48**, 175.

TYLOR, E. B. (1). 'On the Game of *Patolli* in ancient Mexico and its probable Asiatic Origin.' *JRAI*, 1878, **8**, 116.

TYRRELL, H. G. (1). *History of Bridge Engineering*. Priv. pub., Chicago, 1911.

UCCELLI, A. (1) (ed.) (with the collaboration of G. SOMIGLI, G. STROBINO, E. CLAUSETTI, G. ALBENGA, I. GISMONDI, G. CANESTRINI, E. GIANNI & R. GIACOMELLI). *Storia della Tecnica dal Medio Evo ai nostri Giorni*. Hoeppli, Milan, 1945.

UCELLI DI NEMI, G. (1). 'Il Contributo Dato dalla Impresa di Nemi alla Conoscenza della Scienza e della Tecnica di Roma.' Art. in *Nuovi Orientamenti della Scienza*. XLI Reunione della Società Italiana per il Progresso delle Scienze, Rome, 1942.

UCELLI DI NEMI, G. (2). *Le Nave di Nemi*. Libreria dello Stato, Rome, 1940.

[UCELLI DI NEMI, G.] (3) (ed.). *Le Gallerie di Leonardo da Vinci nel Museo Nazionale della Scienza e della Tecnica [Milano]*. Museo Naz. d. Sci. e. d. Tecn., Milan, 1956.

UHDEN, R. (2). 'The Oldest Portuguese Original Chart of the Indian Ocean, +1509.' *IM*, 1939, **3**, 7.

UNDERWOOD, H. H. (1). 'Korean Boats and Ships.' *JRAS/KB*, 1933, **23**, 1–100.

UNGNAD, A. & GRESSMANN, H. (1). *Das Gilgamesch-Epos*. Vandenhoeck & Ruprecht, Göttingen, 1911. (Forschungen z. Rel. u. Lit. des Alt. & Neuen Test., no. 14.)

UPCRAFT, W. M. (1). 'Curious Bridges in Interior China.' *EAM*, 1904, **3**, 241.

USHER, A. P. (1). *A History of Mechanical Inventions*. McGraw-Hill, New York, 1929. 2nd ed. revised. Harvard Univ. Press, Cambridge, Mass., 1954. (Rev. Lynn White, *ISIS*, 1955, **46**, 290.)

VAILLANT, G. C. (1). *Artists and Craftsmen in Ancient Central America*. Amer. Mus. Nat. Hist. New York, 1935. (Guide leaflet series, no. 88.) *The History of the Valley of Mexico* (illustrated chart). Amer. Mus. Nat. Hist. New York, 1936. (Guide Leaflet series, no. 103; suppl. to 88.)

VAILLANT, G. C. (2). *Aztecs of Mexico; the Origin, Rise and Fall of the Aztec Nation*. Doubleday, New York, 1947.

VALLICROSA, J. M. MILLÁS (1). 'Un Ejemplar de "azafea" Árabe de Azarquiel.' *AAND*, 1944, **9**, 111.

VAMBÉRY, A. (1) (tr.). *The Travels and Adventures of the Turkish Admiral Sidi 'Ali Reïs in India, Afghanistan, Central Asia and Persia, +1553/+1556 [Mir'at al-Mamālik]*. Luzac, London, 1899.

VASSILIEV, L. S. (1). *Agrarie Otnoshenia i Obshina v Drevniem Kitai (Agrarian Relations in Ancient China from the −11th to the −7th Centuries and the Primitive Community)* (in Russian). Izdatelstvo Voistonoi Literaturi, Moscow, 1961. (Rev. T. Pokora, *ARO*, 1963, **31**, 171.)

VAVILOV, N. I. (1). 'The Problem of the Origin of the World's Agriculture in the Light of the Latest Investigations.' In *Science at the Cross-Roads*. Papers read to the 2nd International Congress of the History of Science and Technology. Kniga, London, 1931.

VAVILOV, N. I. (2). *The Origin, Variation, Immunity and Breeding of Cultivated Plants; Selected Writings*. Chronica Botanica, Waltham, Mass., 1950. (Chronica Botanica International Collection, vol. 13.)

DE LA VEGA, EL INCA, GARCILASSO (1). *The Florida of the Inca; a History of the Adelantado Hernando de Soto, Governor and Captain-General of the Kingdom of Florida, and of other heroic Spanish and Indian Cavaliers, written by the Inca Garcilasso de la Vega [+1539 to +1616], an officer of His Majesty and a Native of the Great City of Cuzco, Capital of the Realms and Provinces of Peru*. Tr. J. G. Varner & J. J. Varner. Univ. Texas Press, Houston, 1951.

VENCE, J. (1). *Construction et Manoeuvre des Bateaux et Embarcations à Voile Latine*. Challamel, Paris, 1897.

VENTURI, P. T. (1) (ed.). *Opera Storiche del P. Matteo Ricci*. 2 vols. Giorgetti, Macerata, 1911.

VERANTIUS. See Veranzio.

VERANZIO, FAUSTO (1). *Machinae Novae Fausti Verantii Siceni, cum Declaratione Latina, Italica, Hispanica, Gallica et Germanica* (written c. 1595). Florence, 1615; Venice, 1617. Account in Beck (1), ch. 22. Thorndike (1), vol. 7, p. 615, thinks that the assumed date of writing due, to Libri, is too early, but Veranzio did not enter the clergy till +1594 after the death of his wife.

VERBIEST, F. (1). *Astronomia Europaea sub Imperatore Tartaro-Sinico Cám-Hy [Khang-Hsi] appellato, ex Umbra in Lucem Revocata; à R. P. Ferdinando Verbiest Flandro-Belgica e Societate Jesu, Academiae Astronomicae in Regia Pe Kinensi Praefecto....* Bencard, Dillingen, 1687. This is a quarto volume of 126 pp., edited by P. Couplet. A folio volume with approximately the same title (Verbiest, 2) had appeared in 1668, consisting of some 15 pp. Latin text with title-pages all block-printed in Chinese style and on Chinese paper from Verbiest's own handwriting, together with 18 plates, only one of which, the general view of the re-organised Peking Observatory, was re-engraved in small format for the 1687 edition. (Cf. Houzean (1), p. 44; Bosmans (2).)

VERBIEST, F. (2). *Liber Organicus Astronomiae Europaeae apud Sinas Restitutae, sub Imperatore Sino-Tartarico Cám-Hy [Khang-Hsi] appellato....* Peking, 1668. This is the folio volume with Latin text and engraved plates. It is rare, and I know only the copy in the library of the London School of Oriental Studies. This copy is bound up with another production of Verbiest's, the *Chu I Hsiang Pien Yen* (Introductory Notes to Pictures of Scientific Instruments) the 117 engraved plates of which are numbered from the 'back', i.e. the Chinese beginning, of the volume. These plates follow after the two pages of introductory remarks in Chinese, dated 1674, and are numbered in Chinese erratically, some numbers being absent, and others being identical on two or three successive plates. A seventeenth-century hand has endeavoured to re-number them in red ink. The first 6 plates are the same as those of Verbiest (2), and all of these appear again in the *Huang Chhao Li Chhi Thu Shuo* (q.v.). The rest are concerned with all branches of the physical sciences as well as astronomy, and include many pictures of the making and positioning of the astronomical instruments. Pfister (1), p. 358, says that there should be 125 plates, but his own copy had only 117; cf. his p. 354. Perhaps the present *Chu I Hsiang Pien Yen* is only the illustration section or *thu phu* of Verbiest's *Hsin Chih Ling-Thai I Hsiang Chih* (Memoir on the (Theory, Use and Construction of the) Instruments of the New Imperial Observatory), 1673, all the text being absent. From what we know of the date of the re-fitting of the observatory at Peking, it would seem that the publication of 1668 was a plan of campaign while the later ones were detailed accounts of what had been accomplished.

VERDELIS, N. M. (1). 'The Corinthian Diolkos.' *ILN*, 1957, **231**, 649.

VERDELIS, N. M. (2). 'Der Diolkos am Isthmus von Korinth.' *MDAI/ATH*, 1956, **71**, 51.

VERLINDEN, C. & HEERS, J. (1). 'Le Rôle des Capitaux Internationaux dans les Voyages de Découverte au 15e et 16e Siècles.' Communication to the Ve Colloque International d'Histoire Maritime, Lisbon, 1960.

VERNET, J. (1). 'Influencias Musulmanas en el Origen de la Cartografía Náutica.' *PRSG*, 1953, ser. B, no. 289.

VERNON-HARCOURT, L. F. (1). *Rivers and Canals; the Flow, Control and Development of Rivers, and the Design, Construction and Development of Canals, both for Navigation and Irrigation*. 2 vols. Oxford, 1896.

VERWEY, D. (1). 'An Early Median Rudder and Sprit-Sail?' *MMI*, 1934, **20**, 230. Comments by R. Anderson, p. 373; J. W. van Nouhuys, 1936, **22**, 476.

VERWEY, D. (2). 'Could Ancient Ships Work to Windward?' *MMI*, 1936, **22**, 117.

VETTER, H. (1). 'Zur Geschichte d. Zentralheizungen bis zum Übergang in die Neuzeit.' *BGTI*, 1911, **3**, 276.

DE VILLARD, U. MONNERET (1). *Note sulle Influence Asiatische nell'Africa Orientale*. 1938.

VILLIERS, A. (1). *Sons of Sindbad; an Account of Sailing with the Arabs in their Dhows, in the Red Sea, round the Coasts of Arabia and to Zanzibar and Tanganyika; Pearling in the Persian Gulf; and the Life of the Shipmasters and Mariners of Kuwait*. Hodder & Stoughton, London, 1940.

VILLIERS, A. (2). *The Indian Ocean*. Museum, London, 1952. *Monsoon Seas; the Story of the Indian Ocean*. McGraw-Hill, New York, 1952.

VILLIERS, A. (3). 'Ships through the Ages; a Saga of the Sea.' *NGM*, 1963, **123**, 494.

VILLIERS, A. (4). 'Sailing with Sindbad's Sons.' *NGM*, 1948, **94**, 675.

VINCENT, I. V. (1). *The Sacred Oasis; the Caves of the Thousand Buddhas at Tunhuang*. Univ. of Chicago Press, 1953.

VIROLLEAUD, C. (1) (ed.). *Ourartou, Neapolis des Scythes, Kharezm*. Maisonneuve, Paris, 1954. (Review articles on Soviet archaeological discoveries by B. B. Piotrovsky, P. N. Schultz & V. A. Golovkina, and S. P. Tolstov.)

VISSIÈRE, A. (2). *Études Sino-Mahometanes*. 2 vols. (the second with G. Cordier, C. Huart & A. C. Moule Leroux, Paris, 1911, 1913, repr. from *Rev. du Monde Mussulman* 1909–11.

VITRUVIUS (MARCUS VITRUVIUS POLLIO). *De Architectura Libri Decem*. Ed. D. Barbari, Venice, 1567; ed. G. Philandri, Leiden, 1586, Amsterdam, 1649. For Eng. tr. see Morgan (1), Granger (1); French tr. Perrault (1); Germ. tr. Rivius (2).

VOGEL, J. P. (1). 'The Ship of Boro-Budur.' *JRAS*, 1917, 367.

VOGT, E. Z. & RUZ-LHUILLIER (1). *Desarrollo Cultural de los Mayas* (Wenner-Gren Foundation Symposium). Univ. Nac. Autonom. de Mexico, Mexico City, 1964.

DE VOISINS, G., LARTIGUE, J. & SEGALEN, V. (1). 'Account of the work of the French Archaeological Expedition in West China.' *JA*, 1915 (11th ser.), **5**, 467.

VOIT, C. (1). 'Ü. d. in den Schuppen und d. Schwimmblase von Fischen vorkommenden irisierenden Krystalle.' *ZWZ*, 1865, **15**, 515.

VOLPERT, P. A. (1). 'Die Ehrenpforten in China.' *OAV*, 1910, **1**, 140, 190.

VOLPICELLI, Z. (3). 'The Ancient Use of Wheels for the Propulsion of Vessels by the Chinese. *JRAS/NCB*, 1891, **26**, 127.

DE VORAGINE, JACOBUS (1). *The Golden Legend* (+1275, tr. from the French by W. Caxton *et al.* 1483) ed. F. S. Ellis. Dent, London (Temple Classics edition) 7 vols., 1900.

VOS, ISAAC (1). *Variarum Observationum Liber*. Scott, London, 1685. (Contains, *inter alia, De Artibus et Scientiis Sinarum*, p, 69; *De Origine et Progressu Pulveris Bellici apud Europaeos*, p. 86; *De Triremium et Liburnicarum Constructione*, p. 95.) Cf. Duyvendak (13).

VOSSIUS. See Vos, Isaac.

WADA S. [WADA, KIYOSHI] (1). 'The Philippine Islands as known to the Chinese before the Ming Dynasty.' *MRDTB*, 1929, **4**, 121. (Translated from *TYG*, 1922, **12**, 381.)

WADDELL, L. A. (1). *Lhasa and its Mysteries*; with a Record of the Expedition of 1903–1904. Murray, London, 1905.

WAILES, R. (3). *The English Windmill*. Routledge & Kegan Paul, London, 1954.

WAINWRIGHT, G. A. (2). 'Early Foreign Trade in East Africa.' *MA*, 1947, **47**, 143.

WALEY, A. (1) (tr.). *The Book of Songs*. Allen & Unwin, London, 1937.

WALEY, A. (10) (tr.). *The Travels of an Alchemist; the Journey of the Taoist [Chhiu] Chhang-Chhun from China to the Hindu-Kush at the summons of Chingiz Khan, recorded by his disciple Li Chih-Chhang*. Routledge, London, 1931. (Broadway Travellers Series.) Crit. P. Pelliot, *TP*, 1931, **28**, 413.

WALEY, A. (11). *The Temple, and other Poems*. Allen & Unwin, London, 1923.

WALEY, A. (13). *The Poetry and Career of Li Po (701 to 762 A.D.)*. Allen & Unwin, London, 1950.

WALEY, A. (18). *An Index of Chinese Artists, represented in the Sub-Department of Oriental Prints and Drawings in the British Museum*. BM, London, 1922.

WALEY, A. (19). *An Introduction to the Study of Chinese Painting*. Benn, London, 1923. Repr. 1958.

WALEY, A. (20). 'A Chinese Picture' (Chang Tsê-Tuan's 'Going up the River to Kaifêng at the Spring Festival' c. +1126). *BUM*, 1917, **30**, 3.

WALKER, R. L. (1). *The Multi-State System of Ancient China*. Shoestring Press, Hamden, Conn., U.S.A., 1953. (Yale Univ. Foreign Area Studies Monographs, no. 1.)

WALLACKER, B. E. (1) (tr.). *The 'Huai Nan Tzu' Book, [Ch.] 11; Behaviour, Culture and the Cosmos*. Amer. Oriental Soc., New Haven, Conn., 1962. (Amer. Oriental Series, no. 48.)

WALLINGA, H. T. (1). *The Boarding-Bridge of the Romans; its Construction and its Function in the Naval Tactics of the First Punic War*. Wolters, Groningen, 1956; Nijhoff, The Hague, 1956. Crit. J. S. Morrison, *AAAA*, 1958, **11** (no. 2), 142.

WALLIS, H. M. (1). 'The Influence of Father Ricci on Far Eastern Cartography.' *IM*, 1965, **19**, 38.

WALLIS, H. M. & GRINSTEAD, E. D. (1). 'A Chinese Terrestrial Globe, +1623.' *BMQ*, 1963, **25**, 83.

WAN NUNG (1). 'Fighting the Big Flood' (at Hankow on the Yellow River). *CREC*, 1954, **3** (no. 6), 9.

WANG CHI-MIN (2). *Lancet and Cross* (biographies of fifty Western physicians in 19th-century China). Council for Christian Medical Work, Shanghai, 1950.

WANG CHI-MING (1). 'The Style of Chekiang Houses.' *CREC*, 1963, **12** (no. 3), 12.

WANG CHIUNG-MING (1). 'The Bronze Culture of Ancient Yunnan.' *PKR*, 1960 (no. 2), 18. Reprinted in mimeographed form. Collet's Chinese Bookshop, London, 1960.

WANG CHUN-KAO (1). 'More Waterways for Huai-an.' *CREC*, 1962, **11** (no. 11), 3.

WANG GUNGWU. See Wang Kung-Wu.

WANG KUNG-WU (1). 'The Nanhai Trade; a Study of the Early History of Chinese Trade in the South China Sea [from later Han to Wu Tai +1st to +10th Centuries].' *JRAS/M*, 1958, **31** (pt. 2), 1–135.

WANG LING (1). 'On the Invention and Use of Gunpowder and Firearms in China.' *ISIS*, 1947, **37**, 160.

WANG WEI-HSIN (1). 'Tapping Sub-surface Water' (*qanāts* in Kansu and Inner Mongolia). *CREC*, 1962, **11** (no. 2), 17.

WANG YU-CHI (1). 'Railways Forge Ahead' (on the Pao-chhêng Line). *CREC*, 1953, **2** (no. 6), 28.

WARD, B. E. (1). 'A Hongkong Fishing Village.' *JOSHK*, 1954, **1**, 195.

WARD, B. E. (2). 'The Straight Chinese "Yuloh".' *MMI*, 1954, **40**, 321.

WARD, F. KINGDON (1). 'Tibet as a Grazing Land.' *GJ*, 1947, **110**, 60.

WARD, F. KINGDON (2). *From China to Hkamti Long*. Arnold, London, 1924.

WARD, F. KINGDON (3). *The Land of the Blue Poppy; Travels of a Naturalist in Eastern Tibet*. Cambridge, 1913.

WARD, F. KINGDON (4). 'Pflanzenbrücken im Himalaya.' *AT*, 1947, **19**, 83.

WARD, F. KINGDON (6). *Plant Hunting in the Wilds*. Figurehead, London, n.d. (1931).

WARD, F. KINGDON (7). *Burma's Icy Mountains*. Cape, London, 1949.

WARD, F. KINGDON (11). *Plant Hunting on the Edge of the World*. Gollancz, London, 1930.

WARD, F. KINGDON (12). *Return to the Irrawaddy*. Melrose, London, 1956.

WARD, F. KINGDON (13). *Plant Hunter's Paradise*. Cape, London, 1937.

WARD, F. KINGDON (14). *The Romance of Plant Hunting*. Arnold, London, 1924.

WARD, F. KINGDON (15). *Pilgrimage for Plants* (with introduction and bibliography by W. T. Stearn). Harrap, London, 1960.

WARD, F. KINGDON (16). *The Mystery Rivers of Tibet*.... Seeley Service, London, 1923.

WARMINGTON, E. H. (1). *The Commerce between the Roman Empire and India*. Cambridge, 1928.

WATERS, D. W. (1). 'Chinese Junks; the Antung Trader.' *MMI*, 1938, **24**, 49.

WATERS, D. W. (2). 'Chinese Junks; the Pechili Trader.' *MMI*, 1939, **25**, 62.

WATERS, D. W. (3). 'Chinese Junks, an Exception; the Tongkung.' *MMI*, 1940, **26**, 79.

WATERS, D. W. (4). 'Chinese Junks; the Twaqo.' *MMI*, 1946, **32**, 155.

WATERS, D. W. (5). 'Chinese Junks; the Hangchow Bay Trader and Fisher.' *MMI*, 1947, **33**, 28.

WATERS, D. W. (6). 'The Chinese Yuloh' [self-feathering propulsion oar]. *MMI*, 1946, **32**, 189.

WATERS, D. W. (8). 'The Straight, and other, Chinese "Yulohs".' *MMI*, 1955, **41**, 60.

WATERS, D. W. (9). 'Some Coastal Sampans of North China: I. The Sampan of the Antung Trader.' MS. Notes on Models in the National Maritime Museum, Greenwich, the Science Museum, South Kensington, and the Mystic Seaport Museum, Mystic, Conn., U.S.A.

WATERS, D. W. (10). 'Some Coastal Sampans of North China: II. The "Duck", "Chicken" and "Open Bow" Sampans of Wei-Hai-Wei.' MS. Notes on Models in the National Maritime Museum, Greenwich, the Science Museum, South Kensington, and the Mystic Seaport Museum, Mystic, Conn., U.S.A.

WATERS, D. W. (11). 'Early Time and Distance Measurement at Sea.' *JIN*, 1955, **8**, 153.

WATERS, D. W. (12). 'The Development of the English and the Dutchman's Log.' *JIN*, 1956, **9**, 70. Discussion by E. G. R. Taylor, A. H. W. Robinson & D. W. Waters. *JIN*, 1956, **9**, 357.

WATERS, D. W. (13). 'The Sea Chart and the English Colonisation of America.' *ANEPT*, 1957, **17**, 28.

WATERS, D. W. (14). 'A Tenth Mariner's Astrolabe [preserved in Japan].' *JIN*, 1957, **10**, 411.

WATERS, D. W. (15). *The Art of Navigation in England in Elizabethan and Early Stuart Times*. Hollis & Carter, London, 1958.

WATERS, D. W. (16). 'Knots per Hour.' *MMI*, 1956, **42**, 148.

WATERS, D. W. (17) (ed.). *The Rutters of the Sea; Sailing Directions of Pierre Garcie*. Yale Univ. Press, New Haven, Conn., 1967.

WATERS, D. W. (18). 'The Sea- or Mariner's Astrolabe.' *RFCC*, 1966, **39**, 5. (Junta de Investigações do Ultramar; Agrupamento de Estudos de Cartografia Antiga (Seccaõ de Coimbra), pub. no. 15.)

WATSON, BURTON (1) (tr.). '*Records of the Grand Historian of China*', translated from the '*Shih Chi*' of Ssuma Chhien. 2 vols. Columbia Univ. Press, New York, 1961.

WATSON, BURTON (2). *Ssuma Chhien, Grand Historian of China*. Columbia Univ. Press, New York, 1958.

WATSON, W. (1). *Archaeology in China*. Parrish, London, 1960. (An account of an exhibition of archaeological discoveries organised by the Chinese People's Association for Cultural Relations with Foreign Countries and the Britain–China Friendship Association, 1958.) Cf. Watson & Willetts (1).

WATSON, W. (2). *China before the Han Dynasty*. Thames & Hudson, London, 1961. (Ancient Peoples and Places, no. 23.)

WATSON, W. (3). 'A Cycle of Cathay; China, the Civilisation of a Single People.' Art. in *The Dawn of Civilisation; the First World Survey of Human Cultures in Early Times*, ed. S. Piggott, p. 253. Thames & Hudson, London, 1961.

WATTERS, T. (2). *A Guide to the Tablets in the Temple of Confucius*. Presbyt. Miss. Press, Shanghai, 1879.

WATTERSON, J. (1). *Architecture, Five Thousand Years of Building*. Norton, New York, 1950.

WEBB, A. H. & TANNER, F. W. (1). 'The Effect of Spices and Flavouring Materials on the Growth of Yeasts.' *FOODR*, 1945, **10**, 273.

WELLS, F. H. (1). 'How much did Ancient Egypt influence the Design of the Chinese Junk?' *CJ*, 1933, **19**, 300.

WELLS, W. H. (1). *Perspective in Early Chinese Painting*. Goldston, London, 1935; repr. 1945. (Crit. J. B[uhot], *RAA/AMG*, 1935, **9**, 238.)

WELLS, W. H. (2). 'Some Remarks on Perspective in Early Chinese Painting.' *OAZ*, 1933, **9**, 214.

WELTFISH, G. (2). *The Origins of Art [in Amerindian basket-making, weaving and pottery]*. Bobbs-Merrill, Indianapolis and New York, 1953.

WHEATLEY, P. (1). 'Geographical Notes on some Commodities involved in Sung Maritime Trade.' *JRAS/M*, 1959, **32** (pt. 2), 1–140.

WHEELER, SIR R. E. M. (1) (with GHOSH, A. & DEVA, K.). 'Arikamedu; an Indo-Roman Trading Station on the East Coast of India.' *ANI*, 1946 (no. 2), 17.

WHEELER, SIR R. E. M. (4). *Rome beyond the Imperial Frontiers*. Bell, London, 1954. (Crit. rev. J. E. van Lohuizen de Leeuw, *ORA*, 1955 (n.s.), **1**, 130.)

WHEELER, SIR R. E. M. (6). 'Archaeology in East Africa.' *TNR*, 1955 (no. 40), 43.

WHEELER, SIR R. E. M. (7). *Impact and Imprint; Greeks and Romans beyond the Himalayas*. King's College, Newcastle-on-Tyne, 1959. (39th Earl Grey Memorial Lecture.)

WHITE, E. W. (1). *British Fishing-Boats and Coastal Craft; Pt. I, Historical Survey*. Science Museum, London, 1950. (Pt. II is the Catalogue of Exhibits.)

WHITE, W. C. (2), Bp. of Honan. 'Chinese Home Life 1800 Years Ago; Model Houses [Farmsteads], in Pottery, from Chinese Tombs, ascribed to the +2nd century; possibly a pair from the graves of husband and wife, and said to be the first published examples.' *ILN*, 1934, **185**, 148. Subsequent investigation led to the conclusion that these models were more probably of Ming date, +15th century; see Bulling (3).

WHITE, W. C. (3), Bp. of Honan. *Bronze Culture of Ancient China; an archaeological study of Bronze Objects from Northern Honan dating from about −1400 to −771*. Univ. of Toronto Press, Toronto, 1956. (Royal Ontario Museum Studies, no. 5.)

WHITE, W. C. (4), Bp. of Honan. *Bone Culture of Ancient China; an archaeological study of Bone Material from Northern Honan dating from about the −12th century*. Univ. of Toronto Press, Toronto, 1945. (Royal Ontario Museum Studies, no. 4.)

WHITE, W. C. (5), Bp. of Honan. *Tomb-Tile Pictures of Ancient China; an Archaeological Study of Pottery Tiles from Tombs of Western Honan dating from about the −3rd century*. Univ. Toronto Press, Toronto, 1939. (Royal Ontario Museum Studies, no. 1.)

WHITEWAY, R. S. (1). *The Rise of Portuguese Power in India, +1497 to +1550*. Constable, London (Westminster), 1899.

WHITFIELD, RODERICK (1). *Chang Tsê-Tuan's 'Chhing Ming Shang Ho Thu'*. Inaug. Diss. Princeton, 1965.

WIEGER, L. (1). *Textes Historiques*. 2 vols. (Ch. and Fr.) Mission Press, Hsienhsien, 1929.

WIEGER, L. (2). *Textes Philosophiques*. (Ch. and Fr.) Mission Press, Hsienhsien, 1930.

WIEGER, L. (3). *La Chine à travers les Ages; Précis, Index Biographique et Index Bibliographique*. Mission Press, Hsienhsien, 1924. Eng. tr. E. T. C. Werner.

WIEGER, L. (4). *Historie des Croyances Religieuses et des Opinions Philosophiques en Chine depuis l'origine jusqu'à nos jours*. Mission Press, Hsienhsien, 1917.

WIEGER, L. (6). *Taoisme*. Vol. 1. *Bibliographie Générale*: (1) Le Canon (Patrologie); (2) Les Index Officiels et Privés. Mission Press, Hsienhsien, 1911. (Crit. by P. Pelliot, *JA*, 1912 (10 sér.), **20**, 141.)

WIEGER, L. (7). *Taoisme*. Vol. 2. *Les Pères du Système Taoiste* (tr. selections of Lao Tzu, Chuang Tzu, Lieh Tzu). Mission Press, Hsienhsien, 1913.

WIENS, H. J. (1). 'The "Shu Tao", or Road to Szechuan.' *GR*, 1949, **39**, 584.

WIENS, H. J. (2). *The 'Shu Tao' or, the Road to Szechuan; a Study of the Development and Significance of Shensi-Szechuan Road Communication in West China*. Ann Arbor, Mich., 1948. (Inaug. Diss. Univ. of Michigan.) Xerocopy Reprint, University Microfilms, Ann Arbor, Mich., 1966.

WIET, G. (1) (tr.). *Les Pays [Aḥmad ibn Wāḍiḥ al-Ya'qūbī's 'Kitāb al-Buldān' (Book of the Countries), +889]*. Cairo, 1937. (Pub. Inst. Fr. d'Archéol. Or.; Textes et Traductions d'Auteurs orientaux, no. 1.)

VON WIETHOF, B. (1). 'On the Structure of the [Chinese] Private Trade with Overseas [Countries] about +1550.' With chart. Proc. 14th Conference of Junior Sinologists, Breukelen, 1962 (unpaged mimeographed report).

VON WIETHOF, B. (2). *Die Chinesische Seeverbotspolitik und der private Überseehandel von 1368 bis 1567*. Wiesbaden, 1963.

WILHELM, HELLMUT (1). *Chinas Geschichte; zehn einführende Vorträge*. Vetch, Peiping, 1942.

WILHELM, RICHARD (2) (tr.). *'I Ging' [I Ching]; Das Buch der Wandlungen*. 2 vols. (3 books, pagination of 1 and 2 continuous in first volume). Diederichs, Jena, 1924. (Eng. tr. C. F. Baynes (2 vols). Bollingen-Pantheon, New York, 1950.) See Vol. 2, p. 308.

WILHELM, RICHARD (3) (tr.). *Frühling u. Herbst d. Lü Bu-We (the Lü Shih Chhun Chhin)*. Diederichs, Jena, 1928.

WILHELM, RICHARD (4) (tr.). *Liä Dsi; Das Wahre Buch vom Quellenden Urgrund; Tschung Hü Dschen Ging; Die Lehren der Philosophen Liä Yü Kou und Yang Sschu*. Diederichs, Jena, 1921.

WILHELM, RICHARD (6) (tr.). *Li Gi, das Buch der Sitte des älteren und jungeren Dai [i.e. both Li Chi and Ta Tai Li Chi]*. Diederichs, Jena, 1930.

WILKINSON, J. G. (1). *A Popular Account of the Ancient Egyptians*. 2 vols. Murray, London, 1854.

WILLCOCKS, SIR WILLIAM (1). *The Irrigation of Mesopotamia*. Spon, London, 1911; 2nd ed. 1917.

WILLCOCKS, SIR WILLIAM (2). *The Restoration of the Ancient Irrigation Works on the Tigris; or, the Re-Creation of Chaldea*. Nat. Printing Dept., Cairo, 1903.

WILLCOCKS, SIR WILLIAM (3). *Lectures on the Ancient System of Irrigation in Bengal and its Application to Modern Problems*. Univ. of Calcutta, Calcutta, 1930.

WILLCOCKS, SIR WILLIAM (4). *Egyptian Irrigation*. London, 1889; 3rd edn., 2 vols. Spon, London, 1913 (with J. I. Craig).

WILLCOCKS, SIR WILLIAM (5). *From the Garden of Eden to the Crossing of the Jordan*. Spon, London, 1919.

WILLETTS, W. Y. (1). *Chinese Art*. 2 vols. Penguin, London, 1958. New and revised edition in greatly enlarged format: *Foundations of Chinese Art, from Neolithic Pottery to Modern Architecture*. Thames & Hudson, London, 1965.

WILLEY, G. R. (1). 'Historical Patterns and Evolution in Native New World Cultures.' Art. in *Evolution after Darwin*, ed. S. Tax, vol. 2, p. 111. Chicago Univ. Press, Chicago, Ill., 1960.

WILLIAMS, G. R. (1). 'Hydrology [and Meteorology].' Art. in *Engineering Hydraulics*, ed. H. Rouse, p. 235. (Proc. 4th Hydraulics Conference, Iowa Inst. of Hydraulics Research, 1949.) Wiley, New York, 1950; Chapman & Hall, London, 1950.

WILLIAMS, S. WELLS (1). *The Middle Kingdom; a Survey of the Geography, Government, Literature [or Education], Social Life, Arts, [Religion] and History, [etc.] of the Chinese Empire and its Inhabitants*. 2 vols. Wiley, New York, 1848; later eds. 1861, 1900; London, 1883.

WILLIAMS-ELLIS, C., EASTWICK-FIELD, J. & EASTWICK-FIELD, E. (1). *Building in Cob, Pisé and Stabilised Earth.* Country Life, London, 1947 (1st edn, 1919).

WILLIAMSON, H. R. (1). *Wang An-Shih; Chinese Statesman and Educationalist of the Sung Dynasty.* 2 vols. Probsthain, London, 1935, 1937.

WILSON, C. (1). 'Thomas Telford (+1757 to 1834).' *PROG*, 1957, **46** (no. 256), 61.

WILSON, C. E. (1). 'The Wall of Alexander against Gog and Magog; and the Expedition sent out to find it by the Caliph al-Wāthiq in +842.' *AM*, Introductory Volume (Hirth Anniversary Volume), n.d. (1923), 575.

WILSON, E. H. (1). *China, Mother of Gardens.* 1929.

WINLOCK, H. E. (2). *Models of Daily Life in Ancient Egypt, from the Tomb of Meket-Rē at Thebes (c. −2000).* Harvard Univ. Press, 1955. (Pubs. Metropolitan Museum of Art, no. 18.)

WINSLOW, E. M. (1). *A Libation to the Gods; the Story of the Roman Aqueducts.* Hodder & Stoughton, London, 1963.

WINTERBOTHAM, W. G. (1). 'The Chinese Junk.' *TIMEN*, 1901, **13**, paper no. XCVII.

WITH, K. (1). *Japanische Baukunst.* Leipzig, 1921. (Bibliothek d. Kunstgeschichte, no. 10.)

WITTFOGEL, K. A. FÊNG, CHIA-SHÊNG *et al.* (1). 'History of Chinese Society (Liao), 907–1125.' *TAPS*, 1948, **36**, 1–650. (Rev. P. Demiéville, *TP*, 1950, **39**, 347; E. Balasz, *PA*, 1950, **23**, 318.)

WOLF, A. (1) (with the co-operation of F. DANNEMANN & A. ARMITAGE). *A History of Science, Technology and Philosophy in the 16th and 17th Centuries.* Allen & Unwin, London, 1935; 2nd ed., revised by D. McKie, London, 1950.

WOLF, A. (2). *A History of Science, Technology and Philosophy in the 18th Century.* Allen & Unwin, London, 1938; 2nd ed., revised by D. McKie, London, 1952.

WOLFANGER, L. A. (1). 'Major World Soil Groups and some of their Geographic Implications.' *GR*, 1929, **19**, 106.

WOLFF, G. (2). *History of Perspective down to the Year 1600*, vol. 1, p. 39. Actes du 8e Congrès International d'Histoire des Sciences, Florence, 1956.

WOLKENHAUER, A. (1). 'Beiträge z. Geschichte d. Kartographie und Nautik des 15 bis 17 Jahrhunderts.' *MGGMU*, 1906, **1**, 161.

WONG. See Wang.

WOOLF, LEONARD (1). 'Diaries in Ceylon, 1908–1911; being the official diaries maintained by L.W. while Assistant Government Agent of the Hambantota District.' *CEYHJ*, 1960, **9**, 1–250. Sep. pub., with appendix *Three Short Stories on Ceylon.* Ceylon Histor. Soc. Colombo, 1960.

WOOLLEY, L. (2). *The Development of Sumerian Art.* Faber & Faber, London, 1935.

WORCESTER, G. R. G. (1). *Junks and Sampans of the Upper Yangtze.* Inspectorate-General of Customs, Shanghai, 1940. (China Maritime Customs Pub., ser. III, Miscellaneous, no. 51.)

WORCESTER, G. R. G. (2). *Notes on the Crooked-Bow and Crooked-Stem Junks of Szechuan.* Inspectorate-General of Customs, Shanghai, 1941. (China Maritime Customs Pub., ser. III, Miscellaneous, no. 52.)

WORCESTER, G. R. G. (3). *The Junks and Sampans of the Yangtze; a study in Chinese Nautical Research.* Vol. 1, *Introduction, and Craft of the Estuary and Shanghai Area.* Vol. 2, *The Craft of the Lower and Middle Yangtze and Tributaries.* Inspectorate-General of Customs, Shanghai, 1947, 1948. (China Maritime Customs Pub., ser. III, Miscellaneous, nos. 53, 54.) (Rev. D. W. Waters, *MMI*, 1948, **34**, 134.)

WORCESTER, G. R. G. (4). 'The Chinese War-Junk.' *MMI*, 1948, **34**, 16.

WORCESTER, G. R. G. (5). 'The Origin and Observance of the Dragon-Boat Festival in China.' *MMI*, 1956, **42**, 127.

WORCESTER, G. R. G. (6). 'The Coming of the Chinese Steamer.' *MMI*, 1952, **38**, 132.

WORCESTER, G. R. G. (7). 'The Amoy Fishing Boat.' *MMI*, 1954, **40**, 304.

WORCESTER, G. R. G. (8). 'Six Craft of Kuangtung.' *MMI*, 1959, **45**, 130.

WORCESTER, G. R. G. (9). 'Four Small Craft of Thaiwan.' *MMI*, 1956, **42**, 302.

WORCESTER, G. R. G. (10). Appreciation of the late Sir Frederick Maze, K.B.E., K.C.M.G. *MMI*, 1959, **45**, 90.

WORCESTER, G. R. G. (11). 'The First Naval Expedition on the Yangtze River 1842.' *MMI*, 1950, **36**, 2.

WORCESTER, G. R. G. (12). 'The Inflated Skin Rafts of the Huang Ho.' *MMI*, 1957, **43**, 73.

WORCESTER, G. R. G. (13). 'Some Brief Notes on Fishing in China.' *MMI*, 1958, **44**, 49.

WORCESTER, G. R. G. (14). *The Junkman Smiles.* Chatto & Windus, London, 1959.

WORCESTER, G. R. G. (15). *Sail and Sweep in China; the History and Development of the Chinese Junk as illustrated by the Collections of Models in the Science Museum [London].* HMSO, London, 1966.

WORCESTER, G. R. G. (16). 'Four Junks of Chiangsi.' *MMI*, 1961, **47**, 187.

WORSLEY, P. M. (1). 'Early Asian Contacts with Australia.' *PP*, 1955 (no. 7), 1.

WREDEN, R. (1). 'Vorläufer u. Entstehen der Kammerschleuse; ihre Würdigung u. Weiterentwicklung.' *BGTI*, 1919, **9**, 130.

WRIGHT, A. F. (5). 'On Teleological Assumptions in the History of Science.' *AHR*, 1957, **62**, 918.

WRIGHT, A. F. (7). 'Sui Yang Ti; Personality and Stereotype.' Art. in *The Confucian Persuasion*, ed. A. F. Wright, p. 47. Stanford Univ. Press, Palo Alto, Calif., 1960.

WRIGHT, EDWARD (1). *Certaine Errors in Navigation*. London, 1610.

WU CHING-CHHAO (2). 'Economic Development' (of China). In *China*, ed. H. F. McNair, p. 455. Univ. of Calif. Press, Berkeley, 1946.

WU LIEN-TÊ (1) (ed.) (= *1*). *Magnificent China* (album of photographs, many in colour, with accompanying text). Liang Yu Book Co., Hongkong, 1966.

WU LUEN-TAK. See Wu Lien-Tê.

WU, NELSON I. See Wu No-Sun.

WU NO-SUN (1). *Chinese and Indian Architecture*. Prentice-Hall, London, 1963; Braziller, New York, 1963.

WU TA-KHUN (1). 'An Interpretation of Chinese Economic History.' *PP*, 1952 (no. 1), 1.

WULFF, H. E. (1). *The Industrial Arts [Crafts] of Persia; their Development, Technology and Influence on Eastern and Western Civilisation*. Inaug. Diss. Univ. of New South Wales, 1964. Pub. M.I.T. Press, Cambridge, Mass., 1966.

WULFF, H. E. (4). 'The *Qanāts* of Iran.' *SAM*, 1968, **218** (no. 4), 94.

WYLIE, A. (1). *Notes on Chinese Literature*. 1st ed. Shanghai, 1867. Ed. here used Vetch, Peiping, 1939 (photographed from the Shanghai 1922 ed.).

WYLIE, A. (5). *Chinese Researches*. Shanghai, 1897. (Photographically reproduced, Wêntienko, Peiping, 1936.)

WYLIE, A. (10) (tr.). 'Notes on the Western Regions, translated from the "Ts'een Han Shoo" [*Chhien Han Shu*] Bk. 96.' *JRAI*, 1881, **10**, 20; 1882, **11**, 83. (Chs. 96A and B, as also the biography of Chang Chhien in ch. 61, pp. 1–6, and the biography of Chhen Thang in ch. 70.)

WYMSATT, G. (1) (tr.). *The Lady of the Long Wall [Mêng Chiang Nü]*. Columbia Univ. Press, New York, n.d.

YANG CHÊNG-WU (1). 'The Fight for Lu-ting Bridge.' *CR*, 1957, **6**, 24.

YANG LIEN-SHÊNG (3). *Money and Credit in China; a Short History*. Harvard Univ. Press, Cambridge, Mass., 1952. (Harvard-Yenching Institute Monograph Series, no. 12.) (Rev. R. S. Lopez, *JAOS*, 1953, **73**, 177; L. Petech, *RSO*, 1954, **29**, 277.)

YANG LIEN-SHÊNG (7). 'Notes on N. L. Swann's "Food and Money in Ancient China".' *HJAS*, 1950, **13**, 524. Repr. in Yang Lien-Shêng (9), p. 85 with additions and corrections.

YANG LIEN-SHÊNG (11). *Les Aspects Économiques des Travaux Publics dans la Chine Impériale*. Collège de France, Paris, 1964.

YANG MIN (1). 'Reviving the Grand Canal.' *PR*, 1958 (no. 18), 14.

YANG MIN (2). 'Peasant Inventions for Irrigation.' *PKR*, 1958 (no. 15), 13.

YANG WEI-CHÊN (1). 'About the *Shih Shuo Hsin Yü*.' *JOSHK*, 1955, **2**, 309.

YANG WEI-CHUN (1). 'Kuangtung Fights the Floods.' *CREC*, 1959, **8** (no. 9), 16.

YAZDANI, G. & BINYON, L. (1). *Ajanta; Colour and Monochrome Reproductions of the Ajanta Frescoes based on photography, with explanatory text [by G. Y.] and introduction [by L. B.]*. 4 vols. Oxford, 1930–55.

YEN YAO-CHING et al. (1). *Builders of the Ming Tombs Reservoir*. For. Languages Press, Peking, 1958.

YETTS, W. P. (4). 'Taoist Tales, III; Chhin Shih Huang Ti's Expeditions to Japan.' *NCR*, 1920, **2**, 290.

YETTS, W. P. (6). 'Notes on Chinese Roof-Tiles.' *TOCS*, 1927, 13.

YETTS, W. P. (8). 'A Chinese Treatise on Architecture.' *BLSOAS*, 1927, **4**, 473.

YETTS, W. P. (9). 'A Note on the *Ying Tsao Fa Shih*.' *BLSOAS*, 1930, **5**, 855.

YETTS, W. P. (10). 'Writings on Chinese Architecture.' *BM*, 1927, **50**, 116.

YETTS, W. P. (11). 'Concerning Chinese Furniture.' *JRAS*, 1949, 125.

YETTS, W. P. (14). *An-yang; a Retrospect*. China Society, London, 1942. (Occasional Papers, no. 2.)

YETTS, W. P. (15). 'A Datable Shang-Yin Inscription' [the Yi Yu vessel, with an account of ancient fowling using arrows with cords attached to them]. *AM*, 1949 (n.s.), **1**, 75.

YETTS, W. P. (17). '*West*' and '*East*' and the *Chou Dynasty* (with a Memoir of the Author and a List of his Published Work relating to Chinese Studies by S. H. Hansford). China Society, London, 1958. (China Society Occasional Papers, no. 11.)

YI KWANGNIN (1). *History of Irrigation in the Yi Dynasty [Korea, +1392 to 1910]*. Extensive English summary of Yi Kwangnin (*1*). Korean Research Centre, Seoul, Korea. (Korean Studies Series, no. 8.)

YOSHIDA, T. (1). *Das japanische Wohnhaus*, 2nd ed. Tübingen, 1954.

YSBRANTS IDES. See Ides, E. Ysbrants.

YU CHÊNG (1). 'Irrigation by "Water-Melons"' (small tanks and reservoirs fed by lateral irrigation contour canals from dams and weirs on the Tutsao River, a tributary of the Han River near Hsianyang). *CREC*, 1959, **8** (no. 5), 17.

Yü Ying-Shih (1). *Trade and Expansion in Han China; a Study in the Structure of Sino-Barbarian Economic Relations*. Univ. Calif. Press, Berkeley and Los Angeles, 1967.

Yule, Sir Henry (1) (ed.). *The Book of Ser Marco Polo the Venetian, concerning the Kingdoms and Marvels of the East, translated and edited, with Notes, by H. Y...*, 1st ed. 1871, repr. 1875. 2 vols. ed. H. Cordier. Murray, London, 1903 (reprinted 1921). 3rd ed. also issued, Scribners, New York, 1929. With a third volume, *Notes and Addenda to Sir Henry Yule's Edition of Ser Marco Polo*, by H. Cordier. Murray, London, 1920.

Yule, Sir Henry (2). *Cathay and the Way Thither; being a Collection of Mediaeval Notices of China*. Hakluyt Society Pubs. (2nd ser.) London, 1913–15. (1st ed. 1866.) Revised by H. Cordier. 4 vols. Vol. 1 (no. 38), *Introduction; Preliminary Essay on the Intercourse between China and the Western Nations previous to the Discovery of the Cape Route*. Vol. 2 (no. 33), *Odoric of Pordenone*. Vol. 3 (no. 37), *John of Monte Corvino and others*. Vol. 4 (no. 41), *Ibn Baṭṭuṭah and Benedict of Goes*. (Photographic reprint, Peiping, 1942.)

Yule, Sir Henry (3) (tr.). '*Mirabilia Descriptio*'; *the Wonders of the East*, [*written c. +1330 by Jordanus Catalanus, O.P., Bp. of Columbum, i.e. Quilon in India*]. London, 1863. (Hakluyt Society Pubs., 1st ser., no. 31.)

Yule, H. & Burnell, A. C. (1). *Hobson-Jobson: being a Glossary of Anglo-Indian Colloquial Words and Phrases....* Murray, London, 1886.

Yule & Cordier. See Yule (1).

von Zach, E. (2) (tr.). 'Chang Hêng's poetische Beschreibung der westlichen Hauptstadt (*Hsi Ching Fu*) [Chheng-An].' In *Deutsche Wacht*. Batavia, 1953. (Cit. and partially reproduced in German; Bulling (2), p. 51; in English; Gutkind (1), p. 318.) (Repr. von Zach (6), vol. 1, p. 1.)

von Zach, E. (3). 'Das *Lu Ling Kuang Tien Fu* des Wang Wên-Ka'o' [Wang Yen-Shou]. *AM*, 1926, **3**, 467. (Repr. von Zach (6), vol. 1, p. 164.)

von Zach, E. (6). *Die Chinesische Anthologie; Übersetzungen aus dem 'Wên Hsüan'*. 2 vols. Ed. I. M. Fang. Harvard Univ. Press, Cambridge, Mass., 1958. (Harvard-Yenching Studies, no. 18.)

Zakrewska, M. (1). *Catalogue of Globes in the Jagellonian University Museum*. Inst. Hist. Sci. & Tech. (Polish Acad. Sci.), Cracow, 1965.

de Zurara, Gomes Eanes. See Beazley & Prestage.

Zürcher, E. (1). *The Buddhist Conquest of China; the Spread and Adaptation of Buddhism in Early Mediaeval China*. 2 vols. Brill, Leiden, 1959. (Sinica Leidensia, no. 11.)

Zurla, D. P. (1). *Il Mappamonde di Fra Mauro Camaldolese*. Venice, 1806.

de Zylva, E. R. A. (1). *The Mechanisation of Fishing Craft and the Use of Improved Fishing Gear*. Fisheries Research Station, Colombo, 1958. (Bulletin no. 7.)

GENERAL INDEX

by MURIEL MOYLE

NOTES

(1) Articles (such as 'the', 'al-', etc.) occurring at the beginning of an entry, and prefixes (such as 'de', 'van', etc.) are ignored in the alphabetical sequence. Saints appear among all letters of the alphabet according to their proper names. Styles such as Mr, Dr, if occurring in book titles or phrases, are ignored; if with proper names, printed following them.

(2) The various parts of hyphenated words are treated as separate words in the alphabetical sequence. It should be remembered that, in accordance with the conventions adopted, some Chinese proper names are written as separate syllables while others are written as one word.

(3) In the arrangement of Chinese words, Chh- and Hs- follow normal alphabetical sequence, and *ü* is treated as equivalent to *u*.

(4) References to footnotes are not given except for certain special subjects with which the text does not deal. They are indicated by brackets containing the superscript letter of the footnote.

(5) Explanatory words in brackets indicating fields of work are added for Chinese scientific and technological persons (and occasionally for some of other cultures), but not for political or military figures (except kings and princes).

CHINESE ABBREVIATIONS

ABA	*Ars Buddhica* (Tokyo)	*NKKZ*	*Nihon Kagaku Koten Zensho* (Collection of works concerning the History of Science and Technology in Japan)
AHRA	*Agric. History Research Annual* (*Nung Shih Yen-Chiu Chi-Khan,* formerly *Nung Yeh I-Chhan Yen-Chiu Chi-Khan*)		
APS	*Acta Pedologica Sinica*	*NQJEP*	*Nankai University Quarterly Journ. Econ. and Pol. Sci.* (Tientsin)
AS/BIHP	*Bulletin of the Institute of History and Philology* (*Academia Sinica*)	*OUSS*	*Ochanomizu University Studies*
AS/CJA	*Chinese Journal of Archaeology* (*Academia Sinica*)	*RDR*	*Ryūkoku Daikaku Ronshū* (*Journ. Ryūkoku Univ.,* Kyoto)
BCGS	*Bull. Chinese Geological Soc.*	*RK*	*Rekishigaku Kenkyū* (*Journ. Historical Studies*)
BCS	*Bulletin of Chinese Studies* (Chhêngtu)		
BK	*Bunka* (*Culture*), (Sendai)	*SBK*	*Seikatsu Bunka Kenkyū* (*Journ. Econ. Cult.*)
BSRCA	*Bulletin of the Society for Research in* [*the History of*] *Chinese Architecture*	*SGKK*	*Shigaku Kenkyū* (*Rev. Historical Studies*)
		SGZ	*Shigaku Zasshi* (*Historical Journ. of Japan*)
CHJ/T	*Chhing-Hua* (*T'sing-Hua*) *Journal of Chinese Studies* (New Series, publ. Thaiwan)	*SHSS*	*Studies in the Humanities and Social Sciences* (College of General Education, Osaka University)
CIB	*China Institute Bulletin* (New York)		
CLTC	*Chen Li Tsa Chih* (*Truth Miscellany*)	*SKKK*	*Shukyō Kenkyū* (*Research on Religion*), (Sendai)
CZ	*Chigaku Zasshi* (*Journ. Tokyo Geogr. Soc.*)	*SL*	*Shui Li* (*Hydraulic Engineering*)
		SN	*Shirin* (*Journal of History*), (Kyoto)
FET	*Far Eastern Trade* (London)	*SRSETU*	*Science Reports of the School of Engineering of the Imperial University of Tokyo*
HCH	*Hsin Chao-Hsien* (*New Korea*)		
HHJP	*Hsin Hua Jih Pao* (Peking)	*SWYK*	*Shuo Wên Yüeh Khan* (*Philological Monthly*)
HHYK	*Hsin Hua Yüeh Khan* (*New China Magazine*)		
		TBK	*Tōyō Bunka Kenkyūjo Kiyō* (*Memoirs of the Institute of Oriental Culture, Univ. of Tokyo*)
JAAC	*Journ. Agric. Assoc. China*		
JGSC	*Journal of the Geogr. Soc. China*		
JK	*Jimbun Kenkyū* (Osaka)	*TCKH*	*Ta Chung Kho-Hsüeh* (*Popular Science*), (Peking)
JMHP	*Jen Min Jih Pao* (*People's Daily*)		
		TFTC	*Tung Fang Tsa Chih* (*Eastern Miscellany*)
KDK	*Kodaigaku* (*Palaeologica*), (Osaka)		
KHS	*Kho Hsüeh* (*Science*)	*TG/K*	*Tōhō Gakuhō, Kyōto* (*Kyoto Journal of Oriental Studies*)
KHSC	*Kho-Hsüeh Shih Chi-Khan* (*Ch. Journ. Hist of Sci.*)		
KKTH	*Khao Ku Thung Hsün* (*Archaeological Correspondent*), (cont. as *Khao Ku*)	*TG/T*	*Tōhō Gakuhō, Tōkyō* (*Tokyo Journal of Oriental Studies*)
		THG	*Tōhōgaku* (*Eastern Studies*), (Tokyo)
		TK	*Tōyōshi Kenkyū* (*Researches in Oriental History*)
LSYC	*Li Shih Yen Chiu* (*Journal of Historical Research*), (Peking)	*TS*	*Tōhō Shūkyō* (*Journal of East Asian Religions*)
		TYG	*Tōyō Gakuhō* (*Reports of the Oriental Society of Tokyo*)
MOULA	*Memoirs of the Osaka University of Liberal Arts and Education*	*TYGK*	*Tōyōgaku* (Sendai)

WSC	*Wên Shih Chê (Literature, History and Philosophy)*, (Shantung University)	*YAHS*	*Yenching Shih Hsüeh Nien Pao (Yenching University Annual of Historical Studies)*
WUQJSS	*Wuhan University Quart. Journ. Social Science and Philosophy*	*YCHP*	*Yenching Hsüeh Pao (Yenching University Journal of Chinese Studies)*
WWTK	*Wên Wu Tshan Khao Tzu Liao (Reference Materials for History and Archaeology)*, (cont. as *Wên Wu*)	*YK*	*Yü Kung (Chinese Journal of Historical Geography)*

INTERIM LIST OF EDITIONS OF CHINESE TEXTS USED

By Léonie Callaghan *et al.*

INTERIM LIST OF EDITIONS OF CHINESE TEXTS USED

At the outset of the present work it was envisaged that many readers of these Volumes who understand Chinese would naturally wish to refer to the original sources. Especially where translations of passages have been given would they be likely to want to find the originals conveniently, and since editions of Chinese books are often multifarious, it would be desirable to know the edition used. The fulfilment of this promise (cf. Vol. 1, p. 20 on Bibliography A, entries (m) and (n); Vol. 4, Part 1, p. xxviii; and Vol. 4, Part 2, p. xlviii) could hardly be postponed until the end of Volume 7, so we present here an interim list, prepared with the assistance of Miss Léonie Callaghan. The following points of explanation are needed.

1. The book-titles in the main part of the list (A) are placed in alphabetical order just as they occurred in the bibliographies of the several Volumes, but conflated. The rules of Bibliography A continue to apply, such as all Chh- after Ch-, and all Hs- after H-.

2. The list continues (B) with the *tshung-shu* collections, each being given a standard abbreviation in roman letters, used in list (A) to indicate in which one a particular text is to be found. This list includes not only *tshung-shu* in the strict sense, but also a number of collections of separate books, even when these are the work of the same author.

3. Not all the books listed in the bibliographies of the several Volumes will be found here, for some could only be quoted by title and author, as their texts have never been available to us. The reader is reminded that books not now extant in any form do not appear in the bibliographies at all and references to them can be sought only through the general indexes.

4. Works in Chinese script published in Japan, Korea, Vietnam, etc., are placed here according to the Chinese pronunciation of the characters of the title and not, as in the previous Volumes themselves, according to the Japanese, Korean or Vietnamese pronunciations.

5. Sometimes more than one edition is given if these were all consulted by us.

6. Some texts have been seen only in microfilm form; when this is so the fact is noted.

7. Some editions at our disposal are almost impossible to identify, owing to loss of title-page or last page by rebinding or otherwise, but it is hoped that the information given will generally be sufficient.

8. The present list includes no entry in Bibliography B. There is of course a certain overlap between Bibliographies A and B, for the date of 1800 is an artificial one adopted for convenience, and some collections such as *HCCC* straddled it; however, in general the information in Bibliography B has always been more precise

as regards editions than that in Bibliography A, which gives the original date and author only.

9. With few exceptions all the book titles named are in the Cambridge University Library, or in our own working library at Caius College. J. N.

August 1970

A. BOOK TITLES

In alphabetical order according to the romanisations (not in stroke order);
including books in Japanese, Korean, etc. Abbreviations in capitals refer to
the list of *tshung-shu* collections following (List B).

瞀隅子歔歙琯微論 CPTC
安天論 YHSF

戰國策 HYHTS 校註
滙淵靜語 CPTC
張邱建算經 SCSS, CPTC, KCSHTS
張邱建算經細草 CPTC
掌故叢編
　　1928, 故宮博物院
張子全書 SPPY
張燕公集
　　武英殿聚珍版 TSHCC
肇論 SANT
折獄龜鑑 SSKTS
眞誥 SF, LS
眞臘風土記 TSCC, SF
證類本草 +1468 ed. facsimiles of +1249, and
　　SPPY
政論
　　(崔氏政論) YHSF
正蒙 in 張子全書
正蒙注 CSIS
正字通
　　江西南康府 1670
機巧圖彙
　　text in Yamaguchi Ryūji (1)
急就篇(章) YHAI, SPTK
汲冢周書
　　See 逸周書
幾何原本 KCSHTS
紀效新書
　　上海 1895, 醉經廔校印
集古今佛道論衡 SANT
集古錄 SF
集古錄跋尾
　　In 歐陽文忠公集 KHCP, and in 歐陽文
　　忠公全集, 澹雅書局 1893
緝古算經 SSSS, SCSS, CPTC
紀錄彙編 長沙 1938, 商務, and TSHCC
計倪子 (范子計然) YHSF
霽山集 CPTC
濟生方
　　北京 1956, 人民衛生 from 永樂大典
記楊么本末 in 中興小紀
集韻 LS
甲申雜記 CPTC
嘉祐雜志 (江鄰幾雜志) BH

江賦 WHS
江北運程
　　北京 1867
戒菴老人漫筆 TSH, CCH
芥子園畫傳 北京 1963
羯鼓錄 SSKTS, TTT
鑒誡錄 CPTC, SF
劍南詩稿 SPPY
簡平儀說 SSKTS, TSHCC
建炎以來繫年要錄 上海 1937, 商務, KYTS,
　　TSHCC
至正河防記 TSHCC, CHS
職方外紀 SSKTS
治河方略
　　1767, 聽泉齋藏板
治河圖略 TSHCC
治蝗全法 1857
志林新書 YHSF
芝田錄 SF, LS
志雅堂雜鈔 SF
至言 YHSF
金川瑣記 TSHCC
晉中興書 HHT, HSK
晉後略 HHT, HSK
金剛經 SANT
金光明最勝王經 SANT
金樓子 CPTC
錦囊啓蒙
　　In 永樂大典 Ms, Cambr. Univ. Lib., ch.
　　16, 343-4
金瓶梅
　　上海 repr. of 1695 ed. 1923 上海書局
金史 ESSS
金石索 WYWK
金師子章 SANT
晉書 ESSS
近思錄 KHCP, 集註
敬齋古今黈 TSHCC, WYTCC
荆楚歲時記 HWTS
景福殿賦 WHS
靖康緗素雜記 SSKTS
經書算學天文考 1797
經天該 THSCC
九章算術 SCSS
九章算術細草圖說
　　上海 1896, 文淵山房
九章算術音義 SCSS

九家晉書輯本　TSHCC
九宮行碁立成　Tunhuang Ms S/6164
救荒本草
　　In 農政全書, and facsimile of +1525 ed.
就日錄 (古杭夢遊錄)　SF
九穀考　HCCC
舊唐書　ESSS
九天玄女青囊海角經　TSCC
舊五代史　ESSS
輟耕錄
　　廣文堂藏板
州縣提綱　SKCS, TSHCC
周易集解　CTPS, HCTY, KCCHH, TSHCC
周易略例　HWTS
周易本義
　　In 易經, 浙江書局 1893, and in 易經,
　　揚郡二邮廟惜字局藏板
周易參同契分章註解
　　敦仁堂藏板 1876, and 上海, 錦章圖書局
　　ca. 1920
周易外傳　CSIS
周官義疏
　　紫陽書院藏版
周禮 上海, 涵芬樓 edition based on Ming
　　repr. by 宋岳氏, and 文祿堂 photocopy
　　1934
周禮正義　SPPY
周禮疑義舉要　SSKTS
周髀算經　KCSHTS, SCSS
周髀算經音義　KCSHTS
周書　ESSS
諸器圖說
　　1830, 來鹿堂藏板 and SSKTS
諸蕃志
　　1914, 東京, 民友社
橘錄
　　宋本, prob. orig. ed., with preface of +1178,
　　also BCSH and TSHCC
竹譜　BCSH, HWTS
諸病源侯總論 (巢氏)
　　湖北官書處 1886
　　北京 1955, 人民衛生出版社
竹書紀年　HWTS, and 上海 1956
朱子全書
　　Palace edition +1713/+1714
朱子學的　CITCS, TSHCC
諸子辨　KHCP and 北平 1928, 樸社
朱子文集　CITCS, TSHCC
卷施閣文甲乙集　HPC
莊氏算學　SKCS
莊子 (南華經解)
　　上海, ca. 1915, 廣益書局
莊子解　CSIS
郡齋讀書志　SPTK
郡齋讀書附志　SPTK
種植書 in 氾勝之書 YHSF
中西經星同異考　WYWK
中興小紀　KYTS, TSHCC
中華古今注　BCSH, HWTS
種藝必用
　　Mff. of 永樂大典 ch. 13, 194
中觀論疏　SANT

鐘律緯　YHSF
中論　HWTS
中山傳信錄　HSF
種樹郭橐馳傳　YIKW, TSCC
中吳記聞　CPTC, SSKTS
中原音韻
　　東洋文庫論叢, 東京 1925
中庸　text in Legge (2)
長安志圖　SKCS
長春眞人西遊記　SPPY
長物志　TSHCC
朝野僉載　TTT
陳書　CPTC
誠齋集　SKCS, SPTK
誠齋雜記　SF, CTPS
乘除通變算寶　TSHCC
成唯識論　SANT
七政推步　SKCS
奇器圖說 (遠西奇器圖說錄最)
　　來鹿堂藏板 1830, and SSKTS
菜經　SSKTS
七修類稿
　　上海 1961, 中華書局
七國考　SSKTS
齊民要術
　　北京 1956, 中華; SPPY and 1936, 商務
契丹國志
　　In 宋遼金元別史, 南沙席氏掃葉山房
　　刊本, 1795-7
齊東野語　BH
七緯
　　See 古微書
七曜星辰別行法　SANT
七曜攘災決 (訣)　SANT
樵香小記　SSKTS
切韻
　　see 廣韻
千金 (要) 方
　　北京 1955, 人民衛生
潛夫論　HWTS
前漢紀　SPTK, TSHCC
前漢書
　　四川 1871, repr. 1938
乾象曆術　HCCC
　　and in 五禮通考
潛虛　CPTC
前聞記　CLH
池北偶談 1701, repr.
赤雅　CPTC
欽定續文獻通考　ST
欽定書經圖說 1905
欽定四庫全書簡明目錄
　　see 四庫 etc.
欽定四庫全書總目提要
　　see 四庫 etc.
秦雲擷英小譜　SMCA
青箱雜記　BH
清虛雜著　CPTC
清虛雜著補闕　CPTC
清異錄　HYHTS, SF
青囊奧旨　TSCC
清波別志　CPTC

清波雜志　BH, CPTC
青烏緒言　HHLP
請雨止雨書　YHSF
穹天論　YHSF
籌海圖編　SKCS
疇人傳　KHCP
趨朝事類　SF
袪疑說纂　BH, BCSH
初學記
     古香齋鑒賞袖珍本
楚辭　HYHTS
楚辭補註　HYHTS
曲洧舊聞　CPTC
泉志　TSHCC
吹劍錄外集　CPTC
春秋(左傳)
     text in Couvreur (1) and 3 vols. 北京 1955
春秋井田記　YHSF
春秋繁露　SPPY
春秋後傳　SKCS
春秋穀梁傳　SPPY
春秋公羊傳(何氏解詁)　SPPY
春秋緯漢含孳　YHSF, SSKTS
春秋緯考異郵　YHSF
春秋緯說題辭　YHSF, SSKTS
春秋緯元命苞　YHSF, SSKTS
春渚紀聞
     涵芬樓本
羣經宮室圖　HCCC/HP
重修革象新書
     see 革象 etc.

二程粹言　TSHCC, SPPY
二十四山向訣　TSCC
爾雅
     申報館 1884, and 古逸叢書本 1884, and
     文莫室 1892

發蒙記　YHSF
發微論　CPTC
法言(揚子法言)
     1914 世德堂刊本
法苑珠林　TTT
梵天火羅九曜　SANT
范子計然
     see 計倪子
范文正公文集　TSHCC
方洲雜言　TSHCC
方廣大莊嚴經　SANT
方言　HWTS
方言疏證　SPPY
楓聰小牘　BH
封神演義上海,錦章圖書局
封氏聞見記　SF, LS
風俗通義　HWTS, SPTK
風土記　SF
佛國記　HWTS
佛說北斗七星延命經　SANT
佛祖統紀　SANT
伏侯古今注　YHSF
復性書　HLTCHT
傅子　HWTS

福建運司志　HSL
海潮輯說　TSHCC
海潮賦　TSCC
海涵萬象錄　PLHS
海國圖志 上海 1852
海國聞見錄　IHCC
海內十洲記　HWTS
海道經　TSHCC
海島逸誌摘略　CMIS
海島算經　SCSS
海島算經細草圖說
     上海 1896, 文淵山房
海塘錄　SKCS, SKC
海塘說　HSF
海運新考　HSL
韓非子 上海 1930
漢藝文志考證　YHAI
撼龍經 1892
韓詩外傳　HWTS
漢武故事　SF, LS
漢武帝內傳　HWTS, SSKTS
河防志　CKSL
河防通議　SSKTS, TSHCC
和漢三才圖會
     大阪 c. 1800, 東京 1906, and some text in
     de Mély (1)
鶡冠子　SPPY
河工器具圖說
     1836 南河節署藏板
何首烏傳　TSCC
鶴林玉露　TSHCC
和名類聚抄
     京都 c. 1617 and 東京 1954
河南程氏遺書　SPPY
河南程氏外書　SPPY
河朔訪古記　SSKTS
河圖緯稽耀鉤　SSKTS
河圖緯括地象　SSKTS
河源記　TSHCC
候鯖錄　BH
後漢書
     四川 1871, repr. 1938
後山談叢　TSHCC, SF
弧矢算術　LJIS, CPTC
弧矢算術細草　LJIS, CPTC
畫筌　TPTC
畫塍集　CPTC
華山記　SF
化書　SSKTS
華陽國志　HWTS
華嚴經　SANT
花營錦陣　text in van Gulik (3)
淮南天文訓補注
     湖北 1823,崇文書局
淮南子(淮南鴻烈解)　SPPY
淮南(王)萬畢術 問經堂叢書 1802
     觀古堂所箸書 1902 and many fragments
     in TPYL
皇朝禮器圖式 1766
皇極經世書　TT
黃書　CSIS

黃道總星圖　see 儀象考成
黃帝宅經　TSCC
黃帝九鼎神丹經訣　TT
黃帝內經靈樞　SPTK, ITCM,
　　and 浙江 1877
黃帝內經素問　ITCM,
　　and 浙江 1877,
　　and 1954, 商務
黃帝內經素問集註 1890
黃帝內經太素
　　北京 1955, 人民衛生出版社
皇祐新樂圖記　HCTY, TSHCC
晦菴先生朱文公集　SPTK
渾儀　YHSF
渾蓋通憲圖說　SSKTS, TSHCC
渾天賦　YHAI
渾天象說 (注)　CSHK
洪範五行傳
　　See 尙書緯五行傳
鴻雪因緣圖記
　　上海 1866, 同文書局
弘明集　SPTK, SPPY
火功挈要　1847 reprint of 1643 edition, and
　　TSHCC
火龍經 1644
西征記　SF
西征庚午元曆
　　In 元史
西溪叢語　BH
西清古鑑　LS
西京賦　WHS
西京雜記　BH, HWTS
歙州硯譜　BCSH, HCTY
西夏紀事本末　LCCS
西湖志
　　杭州 1734-5
西銘　in 張子全書 ch. 1 and HYHTS
西步天歌　TSCC
西使記　in 玉堂嘉話, ch. 2, SSKTS
西都賦　WHS
西洋朝貢典錄　CH, YYT
西洋番國志 北京 1961
西域番國志 (with 西域行程記)
　　據明鈔本影印, 國立北平圖書館
西域行程記 (with 西域番國志)
　　據明鈔本影印, 國立北平圖書館
西域聞見錄　Kučă 1777　京都 1801, MS of
　　early 19th cent., and 味經堂 1874
洗冤錄
　　童濂 ed. 1843
　　文晟 ed. 1847
夏小正　in 2nd 卷 of 大戴禮記 HWTS and
　　KHCP
夏侯陽算經　SCSS, KCSHTS
暇日記　SF
詳解九章算法纂類
　　In 永樂大典 Ms, Cambr. Univ. Lib., ch. 16,
　　344
湘中記　SF
詳明算法
　　In 永樂大典 Ms, Cambr. Univ. Lib., ch. 16,
　　343-4

象山全集　SPPY
湘山野錄　LS, SF
象緯新篇　HLTCHT
曉菴新法　SSKTS
孝經　BWCTS
小學紺珠　YHAI
小戴禮記　See 禮記
孝緯雌雄圖　YHSF
孝緯援神契　YHSF, SSKTS
斜川集　SPPY
蟹略　SF
蟹譜　BCHS
閑窗括異志　BH
賢奕編　TSHCC
咸賓錄　YCTS
咸淳臨安志　SKCS
心經　HSCC
新法表異　TSCC
新法算書 1669
新修本草　photographic facsimile
　　日本 1889, 嵏喜廬叢書
　　reduced re-issue 上海 1955, 羣聯
新儀象法要　SSKTS
新刻漏銘　WHS, YHAI
新論　HWTS
新書　HWTS
新唐書　ESSS
昕天論　YHSF
新五代史　ESSS
新語　HWTS
新元史　ESWS
星槎勝覽　HHLP
星經　HWTS, TSHCC
性理精義 1853
性理大全 [書]　Wieger (2)
星命溯源　SKCS
星命總括　SKCS
行水金鑑　KHCP
刑統　國務院法制局 1918
刑統賦　In 藉香零拾 1896
星宗　TSCC
修眞太極混元圖　TT
袖中記　SF
宿曜儀軌　SANT
修防瑣志　CKS
續漢書
　　In 七家後漢書 1882, 太平崔國榜等刊本
續後漢書　SKC, TSHCC
徐霞客遊記　上海　1928, 商務
續畫品錄　TTT
續高僧傳　SANT
續古摘奇算法　CPTC
續博物志　BH, LS
續神仙傳　IMKT, TSHCC, SF
續事始　SF
續世說　SSKTS
續資治通鑑長編　SKCS
續文獻通考　ST
續幽怪錄　TTT
懸解錄　YCCC
玄中記　YHSF
宣和奉使高麗圖經　CPTC, WYWK

58-3

宣和畫譜　SKCS, HSTY, THSCC
宣和博古圖錄
　　台北 1969, 亦正堂原刻本
宣和石譜　SF
玄女經　SMCA
　　text also in van Gulik (3)
玄圖　TPYL (fragments)
玄都律文　TT
學齋佔畢　BCSH
學古編　HCTY
詢芻錄　TSHCC
荀子　STTK

猗覺寮雜記　CPTC
一切經音義　SANT, TSHCC, HSHK
易經
　　1872, 山東書局, repr. 張宗昌 1925
藝經　YHSF
逸周書(汲冢周書)　SPPY
易傳(京房)　HWTS, TSCC
易傳(關朗)　HWTS, TSCC
夷夏論　YHSF
儀象考成　1744, enlarged 1757
醫心方
　　北京 1955, 入民衛生
易學啓蒙　HLTCHT
益古演段　SSSS, PFT, CPTC, KCSHTS
儀禮　SPTK, SPPY, YKT
儀禮釋宮　SSKTS
易林　HWTS, TSCC
意林　SPPY
易數鈎隱圖
　　通志堂經解, 康熙本
蛾術編　CP, 北京 1864
易圖明辨　SSKTS
易通書(周子通書)　HYHTS, SPPY
易洞林　YHSF
醫宗金鑑　上海 1892
易緯稽覽圖　SSKTS
易緯河圖數　SSKTS
易緯通卦驗　SSKTS
藝文類聚北京 1959, 中華書局
異物志　TSHCC
易音　YYH, HCCC
異域志　IMKT, TSHCC
異域圖志
　　orig. ed. +1489, Cambr. Univ. Lib.
異苑　CTPS, HCTY

入物志　HWTS, SSKTS
日知錄　KHCP
日本永代藏
　　岩波文庫, 東京 1958 and 1928
日聞錄　SSKTS, TSHCC
肉刑論　CSHK
儒林公議　BH
如實論　SANT
入蜀記　CPTC
入唐求法巡禮行記
　　東京 1926, 東洋文庫論叢
容齋隨筆　SKCS

甘泉賦　WHS
感應經　SF
感應類從志　SF
高厚蒙求
　　1809, 雲間徐氏藏板
高僧傳　SANT, HSHK
耕織圖
　　1696, 1742 (in 授時通考), and versions in
　　O. Franke (11), Pelliot (24), also 1883 ed.
耕織圖詩　CPTC
開方說　LJIS, PFT
開河記　TTT
開元占經
　　Mff. of 恒德堂藏板
堪輿漫興　TSCC
康熙字典
　　上海, 新鐫銅版印, 商務
亢倉子　SSKTS, SF
考工記
　　In 周禮
考工記圖　HCCC and 上海 1955
考工析疑　KHTS/LCCS
客座贅語　CLTS
晹車志　BH
愧郯錄　CPTC
困知記　HLTCHT, TSHCC
坤輿圖說　TSHCC
孔叢子　HWHTS, WYWK
孔子家語
　　冢田多門 1792, 東都書肆
格致鏡原 1735
革象新書　SKCS, HCH
格古要論　+1593, ed. and HYHTS
格物麤談　TSHCC
割圓連比例圖解　KCSHTS
割圓密率捷法　KCSHTS
溝洫疆理小記　HCCC
古泉滙 1864
古今註　BWCTS, HWTS
古今姓氏書辨證　SSKTS
古今律曆考　TSHCC
古今偽書考　CPTC
古今樂錄　YHSF
古刻叢鈔　CPTC
故宮遺錄　CPTC
穀梁傳
　　See 春秋穀梁傳
古算器考
　　In 藝海珠塵甲集, 聽彝堂原本
古微書　SSKTS
古文析義
　　1716, 牟璧堂藏板
古音表　YYH
古樂府　LS
怪石贊　MSTS
宦箴　BCSH, SF, SKC, HCTY, TSHCC
觀石錄　MSTS
管氏地理指蒙　TSCC
管子上海廣益書局, c. 1915, and 1876,
　　浙江書局
關尹子　SSKTS, SF
廣川書跋　TSHCC

律書 YLCS
麓堂詩話 CPTC
論衡
    1923, 掃葉山房
論天 CSHK
論語 In Legge (2)
龍江船廠志 HSLT
龍虎還丹訣 TT

麻姑山仙壇記
    In 古今遊記叢鈔, 上海 1936
蠻書 TSHCC
毛詩注疏 SPPY
毛詩古音考 HCTY, YYH
茅亭客話 LS
美人賦 CSHK
梅子新論 YHSF
蒙齋筆談 BH
夢占逸旨 IHCC, TSHCC
夢溪筆談論癡彭 刊 1885, and 胡道靜
    ed. 上海 1956, 1959 and 1963, 中華
    書局
夢梁錄(東京夢華錄) 上海 1958
孟子 In Legge (3)
孟子字義疏證 CH
廟制圖考
    1944, 約園刊本
妙法蓮華經 SANT
棉花圖(御題棉花圖) 1765
閩部疏 TSHCC
閩省水師各標鎮協營戰哨船隻圖說
    Ms in Marburg Library (see text)
閩雜記 HSF
明宮史 SKCS, HCTY
明皇雜錄 SSKTS
名醫別錄 CHY
明譯天文書 HFLPC
明儒學案
    紫筠齋藏板
明史 ESSS
明史紀事本末 SKCS, TSHCC
明實錄
    台北 1966, 中央研究院歷史語言研究所
明道雜志 SF
明堂大道錄 TSHCC, HCCC/HP
默記 CPTC
墨莊漫錄 BH
墨客揮犀 BH
摩登伽經 SANT
墨子
    靈巖山館本 1783, repr. 1876
牧菴集 YSHS
穆天子傳 HWTS

南齊書 ESSS
南州異物志 LSC
南方草木狀 BCSH, HWTS, SF and 上海
    1955, 商務
南海百詠 TSHCC
南濠詩話 CPTC
南湖集 CPTC
離儞計濕囀囉天說支輪經 SANT

南史 ESSS
南村輟耕錄 SPTK
南都賦 WHS
南越志 YHSF/P
南越筆記 TSHCC
能改齋漫錄 SSKTS
農政全書
    曙海樓藏板 1843 and
    北京 1942, 輔華齋南紙印刷局
農桑輯要 SPPY, TSHCC
農桑衣食撮要 SSKTS, TSHCC
農書 (宋; 陳旉)
    北京 1956, 中華
農書 (元; 王禎)
    武英殿聚珍版 1783, and CPTC
農書 (明; 沈氏) and TSHCC 北京 1956, 中華

噩夢 CSIS

白虎通德論
    平江 repr. of +1305 edition 何永宮刊, and
    HWTS
白石道人詩集歌曲 SPTK
般苦波羅蜜多經 SANT
抱樸 (or 朴) 子
    經綸元記校刊 1894, SPTK, TT
保生心鑑 Pref. of 1506
北齊書 ESSS
北溪字義 HYHTS
北戶錄 TTT
北行日錄 CPTC
北夢瑣言 BH
北史 ESSS
北堂書鈔
    南海孔氏三十有三萬卷堂 1888, SKCS
北斗七星念誦儀軌 SANT
本起經 Wieger (2)
本草和名
    Tokyo 1926, 長島豐太郎(日本古典全
    集)
本草綱目
    1885 edn., and 上海 1954, 商務
本草綱目拾遺
    1885 (with 本草綱目) and 上海 1954,
    商務
本草品彙精要
    上海 1956, 商務
本草衍義 1877 重刻元刻本,
    and 上海, 1936, 大東書局
佩文韻府
    上海 1937, 商務
蓬窗類紀 HFLPC
琵琶賦 CSHK
埤雅 LLSK, TSHCC
駢字類編 1728
    上海 1887, 同文書局
萍州可談 SSKTS
平夏錄 CLHP, SF, THSCC
普濟方 北京 1960
普曜經 SANT
蒲元別傳
    Fragments in CSHK, TPYL

秘傳花鏡 (=花鏡)
　　金閶文業堂本, and 皇都書林, 京都 1829,
　　and 北京 1956
避暑錄話　BH
表異錄　HYHTS
辨惑編　SSKTS
辨疑志　SF
泊宅編　BH
博物記　YHSF
博物志　pref. by 唐琳玉, and BH, HWTS
步里客談　SSKTS
伯牙琴　CPTC
卜筮正宗全書　TSCC
步天歌　TSCC

三朝北盟會編　SKCS
三秦記　EYT, SF
三輔舊事　EYT, TSHCC
三輔皇圖　HWTS
三國志　ESSS
三國志演義 1883
三國史記
　　京城 1928, 朝鮮史學會
三禮圖　YHSF
三柳軒雜識　SF
三命消息賦
　　珞琭子三命消息賦注　SSKTS
　　徐氏(珞琭子)
　　　三命消息賦注　SSKTS
三命通會　TSCC
三農紀
　　四川, 藜照書屋, from blocks of 1760
三寶太監下西洋記通俗演義　SPK
三才圖會 1609
三字經
　　Text in H. A. Giles (4); St Julien (9)
三字經訓詁
　　Comm. by 王晉升　蘇州 1666,
　　　綠蔭堂藏板
三餘贅筆　TSHCC
搔首問　CSIS
僧惠生使西域記
　　In 洛陽伽藍記
生神經　TT
山居新話　CPTC
山海經
　　四川　宏道堂藏版, 蜀北果城成或因繪
　　圖
　　and 立雪齋原本 1895
上清洞真九宮紫房圖　TT
上清握中訣　TT
商君書
　　子書二十三種 1897, 圖書集成局
上方大洞真元妙經圖　TT
上林賦　WHS
尚書釋天　HCCC
尚書大傳　SPTK
尚書緯璇璣鈐　YHSF, SSKTS
尚書緯考靈曜　YHSF, SSKTS
尚書緯五行傳　SSKTS
舍頭諫太子二十八宿經　SANT
申鑒　HWTS, SPPY

神相全編
　　李升培 pr. pr. 1927, and TSCC
神仙傳　HWTS
神仙通鑑
　　Pref. of 1712, 抔屋袋樓秘本
神異經(紀)　HWTS, SF, LS
神滅論　TSCC
神農本草經
　　日本, 森立之 ed. 1845 photo facsimile;
　　羣聯, 上海 1955; also 顧觀光 ed. 1883,
　　北京 1955

神農本草經疏
　　繆希雍 ed. 1625, 綠君亭
神通遊戲經　SANT
慎了　SSKTS
聖濟總錄
　　+1300 ed. repr. 上海, 文瑞樓 1919
聖賢道統圖贊 1629
聖門事業圖　BCSH
聖壽萬年曆　YLCS
澠水燕談錄　BH, CPTC
史記　ESSS, KHCP
十駕齋養新錄　上海 1935, 商務
詩經
　　Text in Legge (8)
十二杖法　TSCC
石湖詞　CPTC
拾遺記　BH, HWTS
食療本草　Tunhuang Ms ed. M. Nakao (1)
事林廣記　+1478 ed. in Cambr. Univ. Lib.
石林燕語　BH
釋名　HWTS
釋名疏證補　WYWK
世本　HWTS, TSHCC and 上海 1957, 商務
詩本音　HCCC
石品　HWTS
事始　SF, LS
十三經注疏
　　北京 1739/47
世說新語　HYHTS
尸子　in 子書二十三種
　　圖書集成局 1897, and
　　浙江書局據湖海樓本 1877
詩緯汜歷樞　YHSF, SSKTS
事物紀原　HYHTS
石藥爾雅　TT
時務論　YHSF
師友談記　SF
授時曆議經
　　In 元史
授時通考
　　1847, 四川蕭署藏板
授受五嶽圓法　Tunhuang Ms S/3750
庶齋老學叢談　CPTC
蜀錦譜　PYT, MHCH
書經
　　Texts in Medhurst (1); Karlgren (12) and
　　欽定書經圖說
書經圖說 1905
叔苴子(內篇外篇)　TSHCC
書敘指南　HYHTS, SSKTS

數學　SSKTS
述異記　BH, HWTS, SF, LS
數理精蘊　LLYY
鼠璞　BCSH, SF
數術記遺　KCSHTS, SCSS
數書九章　KHCP, TSHCC
書肆說鈴　SFH
蜀道驛程記　HSF
蜀都賦　WHS
庶物異名疏　Mff. Rare Books of the National
　　Library of Peking, no. 140 (pref. 1637)
殊域周咨錄
　　In 東藩輯畧　道光 Ms
水經　HWTS
水經注　SPPY
水滸傳
　　上海 1934, 商務
水利五論　LTT
水部式
　　In 鳴沙石室古佚書, 東京 1913
水道提綱
　　1898, 孟夏新化三昧書室
順風相送　in 兩種海道針經北京 1961
淳祐臨安志　WL
說文解字
　　北京 1963
說文通訓定聲
　　臨嘯閣藏版 Anhui 1870
說苑　HWTS, SPPY, SPTK, LS
搜神記　BH, HWTS, SF, LS
搜神後記　HWTS
搜采異聞錄　BH
四朝聞見錄　CPTC, SF
俟解　CSIS
思玄賦　WHS, CSHK
四庫全書簡明目錄　浙江 1795
四庫全書總目提要
　　存古　坐重印 1910 and 同文書局 1884
四明它山水利備覽　TSHCC
四民月令　BWCTS
思問錄　CSIS
四元玉鑑
　　Text in 四元玉鑑細草 with 四元玉鑑釋
　　例 ed. of 羅士琳 and 易之瀚 1838
素女經　SMCA
素女妙論
　　Text in von Gulik (3)
蘇沈良方　CPTC and 北京 1956
素書　HWTS, SF
蘇子　YHSF
蘇魏公文集　上海 1925
算法全能集
　　In 永樂大典, Ms, Cambr. Univ. Lib., ch.
　　16, 343–4
算法取用本末
　　In 楊輝算法
算法通變本末　THSCC
算法統宗
　　掃葉山房 1883
算學啓蒙
　　江南機器製造局影寫重刊 1871 repr. of
　　1839, and KCSHTS

算學新說　YLCS
遂初堂書目　SF
隨手雜錄　CPTC
隋書　ESSS
隨隱漫錄　BH
孫綽子上海 1916, 廣益書局 and YHSF
箚譜　BCSH, TSCC
孫子兵法　HWTS and 上海 1916, 廣益書局,
　　and 四川, 北溫泉本 1945 (中國辭典館),
　　and 北京 1962, 1964 (今譯)
孫子算經　SCSS, CPTC, KCSHTS
松窗百說　CPTC
宋會要輯稿
　　北京 1936, photolitho repr. of Ms fragments
宋遺民錄　CPTC
宋高僧傳　SANT
宋論　CSIS
宋史　ESSS
宋史紀事本末　LCCS
宋書　ESSS
宋司星子韋書　YHSF
宋四子抄釋　HYHTS
宋元學案　WYWK, HHKHTS (abridged)

大清會典
　　北京 1818
大清一統志　SPTK
大清歷朝實錄
　　新京 1937
大清律例
　　北京 1802
大方等大集經　SANT
大學
　　Text in Legge (2)
大學衍義補　SKCS
大孔雀咒王經　SANT
大明會典
　　北京 +1587
大明律
　　大阪 c. 1880
大寶積經　SANT
大戴禮記　HWTS
大唐西域記　SSKTS
大業雜記　SF, LS
大元氈罽工物記　HSTP
大元海運記　HHTK
大元倉庫記　KTHS
代醉編　SFH
彈茶經　TSCC
丹鉛總錄　KYTS, PYT
丹方鑑源　TT
島夷志略　SKCS
道德經
　　終南山古樓觀說經臺藏板 1877
　　上海 1931, 商務, BWCTS, HWTS, and
　　北平研究院 (考古專報, 1, 2), 1936
道園學古錄　SKCS, SPTK, SPPY
登真隱訣　TT
鄧析子
　　Text in H. Wilhelm (2) and SF
太清神鑑　SSKTS
太清石壁記　TT

太清導引養生經　TT
太極眞人雜丹藥方　TT
太極說　in 性理精義
太極圖解義　in 性理精義 and 宋四子抄釋
太極圖說　in 性理精義, 宋四子抄釋, 宋元
　　學案, 个思錄
太極圖說解(註)　in 性理精義, 今思錄
胎息經　HCTY
泰西水法
　　In 農政全書
太玄經　TSCC
太乙金鏡式經　SKCS
太一金華宗旨　TTHPCC
太白陰經　SSKTS
太平寰宇記　TSHCC
太平廣記　北京 1959
太平聖惠方
　　北京 1959, 入民衛生
太平御覽
　　1818, 鮑崇城本
太上黃庭外景玉經　TT
太上感應篇　TT
太上三天正法經　TT
坦齋通編　SSKTS
譚苑醍醐　SKCS
唐闕史　CPTC
唐甫里先生文集　SPTK
唐會要
　　北京 1955, 中華書局
唐國史補　CTPS, SKCS, HCTY
唐六典　SKCS
唐本草
　　See 新修本草
棠陰比事　SPTK
唐語林　SSKTS, HYHTS
唐韻正
　　In 音學五書
蝶几譜　SCHS
鐵山必要記事　NKKZ
鐵圍山叢談　CPTC
天鏡　YHSF
天竺靈籤　CKKT
田畝比類乘除捷法　TSHCC
天下郡國利病書　SPTK
天下山河兩戒考　HLC
天工開物　天津, 陶涉園本 1929, and text in
　　Yabuuchi (11), and 上海 1959, 中華書局
　　(photographic copy of 1st edn. of 1637), and
　　KHCP
天步眞原　SSKTS, TSHCC
天地陰陽大樂賦　SMCA
天對　TSCC
天問　HYHTS
天文忐　in 晉書
天文錄　in 開元占經
天問畧　TSCC, TSHCC
天文大成管窺輯要
　　三元堂藏板
天文大象賦
　　江陰六嚴校刊本 1856
天隱子　SF
天元曆理全書　Ms of +1425, Cambr. Univ. Lib.

天元玉曆祥異賦　Ms of +1425, Cambr. Univ.
　　Lib.
程史　BH
投壺新格　SF
投壺變　YHSF
透簾細草
　　In 永樂大典 Ms, Cambr. Univ. Lib., ch.
　　16, 343-4 and CPTC
圖經衍義本草　TT
圖畫見聞誌　CTPS, HSTY, SPTK/SP,
　　TSHCC
推篷寤語　SFH
推步法解　SSKTS, TSHCC
峒谿纖志　TSHCC
通鑑前編
　　Mff. Rare Books of the National Library of
　　Peking, no. 71
通鑑綱目
　　1864, 漁古山房珍藏
通鑑綱目續編
　　1864, 漁古山房珍藏
通鑑綱目三編　SKCS
通志　ST
通志略　SPPY
通惠河志　HSL
同話錄　SF
通藝錄 1803
通俗文　YHSF
通典　ST
同文算指　TSHCC
通雅 1800
通原算法
　　In 永樂大典, Ms, Cambr. Univ. Lib., ch.
　　16, 343-4
地鏡　YHSF
帝京景物畧　CKW
地鏡圖　YHSF
地理琢玉斧
　　青黎閣藏板
地理五訣
　　永順堂藏板
帝王世紀　TSHCC
釣磯立談　CPTC
丁巨算法　CPTC, TSHCC
葬書　TSCC
增補文獻備考
　　東國文化社, Seoul, 1958
聽鸎錄　CPTC, SF
蠶書　CPTC
參同契　(see also 周易參同契分章註解)
　　HCTY
參同契發揮　TT
參同契分章註解　TTCY
參同契考異　SSKTS
操縵古樂譜　YLCS
草木子　PLHS, SFH
冊府元龜
　　建陽, 福建, 1642 and
　　北京 1960, 中華書局
測量法義　TSHCC
測量異同　TSHCC
策算　SCSS

測圓海鏡　SSSS, CPTC, KCSHTS, PFT
左傳
　　　Text in Couvreur (1), and 3 vols, 北京 1955
左傳補注　SSKTS
都城紀勝
　　　Text in Moule (5), and 上海 1958
獨醒雜志　CPTC
獨異志　BH
度人經　TT
讀史方輿紀要
　　　二林齋藏板 1901
讀通鑑論　SCIS
杜陽雜編　BH, TTT
端溪硯譜　BCSH, HCTY
東京賦　WHS
東京夢華錄
　　　上海 1958
東西洋考　HYHTS
洞玄靈寶諸天世界造化經　TT
洞霄詩集　CPTC
洞霄圖志　TSHCC, CPTC
東軒筆錄　BH
洞玄子　SMCA
冬官紀事　TSHCC
東國輿地勝覽　Seoul 1930
洞冥記　HWTS
東坡全集(七集)　SPPY
東坡志林　BH, BCSH
洞天清錄(集)　SF
東都賦　WHS
剌世疾邪賦　CSHK
資治通鑑
　　　北京 1956, 古籍出版社
子華子　SSKTS
梓人遺制
　　　Mff. from 永樂大典
子史精華
　　　上海 1909, 朝記書莊

萬機論　YHSF
王忠文公集　CHTS, TSHCC
忘懷錄　SF
王文成公全書　SPTK
緯略　SSKTS
魏略　YHSF/P
唯識二十論　SANT
魏書　ESSS
魏都賦　WHS
聞見近錄　CPTC
文賦　WHS
文獻通考　ST
文心雕龍　HWTS
文始眞經(關尹子)　SSKTS, SF
文士傳　SF
文殊師利菩薩及諸仙所說吉凶時日善惡宿
　　　曜經　SANT
文子　SSKTS
勿菴曆算書目　CPTC
悟眞篇(紫陽眞人悟眞篇註疏)　TT
　　　and 上海 c. 1920, 錦章圖書局
悟眞篇直指祥說三乘祕要　TT
吳船錄　CPTC, TSHCC

五經算術　SCSS
武經總要　SKCS, and facsimile of 明本,
　　　上海 1959
吳郡志　SSKTS
吳中水利書　SSKTS
五星行度解　SSKTS
五行大義　ITTS, CPTC
物類相感志　TSHCC
物理小識　1664 ed.
物理論　TSHCC
吳禮部詩話　CPTC
五禮通考　1761 ed.
武林舊事　CPTC, and 上海 1958
吳錄　SF
五木經　TTT
武備制勝志
　　　1843 重刊
无能子　TT, TSHCC
武備志
　　　京都 1664 and
　　　廣東 1843
五雜俎
　　　國學珍本文庫 1935,
　　　北京 1959, 中華書局
五殘雜變星書　YHSF
五曹算經　SCSS, CPTC
吳都賦　WHS
五音集韻　SKCS
吳越春秋　HWTS

楊輝算法　ICT, TSHCC
鄴中記　HWTS
野客叢書　BH
演禽斗數三世相書
　　　facsimile ed. 東京 1933
燕几圖　TSHCC, SF
演繁露　SF
演連珠　WHS
鴈門公妙解錄　TT
燕北雜記　SF, LS
硯譜　BCSH
硯史　BCSH, HCTY
顏氏家訓　HWTS, CPTC
燕丹子
　　　四川北溫泉(重慶)本, 1945 (中國學辭
　　　典館) and text in H. Franke (11)
鹽鐵論　HWTS, SPTK, and 王利器校注,
　　　上海 1958, and 郭沫若讀本, 北京 1957
沿途水驛
　　　Facsimile of Ms
晏子春秋　SPPY
陰符經　HWTS, SSKTS, and 樓觀臺本
音學五書　1667, and YYH
音論　YYH, HCCC
尹文子　SSKTS
陰陽二宅全書
　　　1752 片山書樓藏板
應閒　CSHK
營造法式
　　　江蘇 1919, 朱氏石印本,
　　　上海 1920 石印本 large format,
　　　上海 1954, 4 vols repr.

瀛涯勝覽　TSHCC
瀛涯勝覽集　CLHP
樂記(戰國)　YIISF
樂記(前漢)　YHSF
樂經　YHSF
樂府雜錄　SSKTS, TTT
樂學軌範　facsimile reproduction of 1610 edition
樂律義　YHSF
樂書(北魏)　YHSF
樂書(宋)
　　Mff. Rare Books of the National Library of
　　Peking, no. A67-68
樂書註圖法
　　樂書要錄
樂書要錄　ITTS
玉照神應眞經　TSCC
玉照定眞經　SKCS
漁樵問對　BCSH
寓簡　CPTC
玉房指要
　　In 醫心方 and van Gulik (3)
玉房祕訣　SMCA
玉海
　　1806, 康基田本 but 80 冊
游宦紀聞　BH, CPTC, TSHCC
幽怪錄　TTT, SF, LS
禹貢
　　Texts in Medhurst (1); Karlgren (12) and
　　欽定書經圖説
禹貢錐指　HCCC
禹貢説斷　SSKTS
郁離子　HCTY, JYTS
羽獵賦　WHS
玉篇　SPPY
玉堂嘉話　SSKTS

與地紀勝
　　台北 1952, 文海出版社
玉音問答　CPTC
漁陽公石譜　SF
酉陽雜爼
　　小嫏嬛山館藏板, BH, TTT
元眞子　CPTC
淵鑑類函
　　上海 1887, 同文書局
元經　HWTS
遠鏡説　TSHCC
元海運志　HHLP, TSHCC
元和郡縣圖志　TSHCC
原人論　SANT
元史　ESSS
援鶉堂筆記 1835
宛署雜記 北京 1961
元代畫塑記　HSTP
元文類
　　上海 1899
園冶
　　北平 1932, 京城印書局
淵穎集　CHTS, TSHCC
粵劍編　HSLT
越絕書　HWTS, SPPY, 據明刻本
月令
　　See 小戴禮記 and 呂氏春秋, in which it is
　　incorporated
月令章句　YHSF
雲溪友議　TTT, BH
韻集　YHSF
雲笈七籤　SPTK
雲林石譜　CPTC
雲麓漫抄　BH
韻石齋筆談　CPTC
榕城詩話　CPTC

B. TSHUNG-SHU COLLECTIONS

In alphabetical order according to the romanisations (not in stroke order).

BCSH　百川學海
　　上海 1921, 博古齋刊本
BH　裨海
　　康熙本
BWCTS　北溫泉叢書
　　四川 1942

CCT　常州先哲遺書
　　武進 c. 1890, 盛宣懷刊本
CH　指海
　　上海 1935, 大東書局
CHTS　金華叢書 1869
　　胡氏退補齋刊本
CHY　陳修園先生醫書七十二種
　　1803, 錦章圖書局
CITCS　正誼堂全書
　　福州 1866, 正誼書院藏版

CKKT　中國古代版畫叢刊
　　上海 1958, 古典文學出版社
CKS　中國水利珍本叢書
　　南京 1937, 水利工程學會排印
CKSH　中國史學叢書
　　台灣 1964, 學生書局
CKSL　中國水利要籍叢編上海出版社
CKW　中國文學參考資料小叢書
　　上海 1957, 古典文學出版社
CKY　鄭開陽雜著
　　1932, 陶風樓
CLHP　紀錄彙編
　　長沙 1938, 商務
CLTS　金陵叢刻
　　1904, 傅晦齋刊本
CPLTS　枕碧樓叢書
　　1913, 沈氏刊本

CPTC 知不足齋叢書
　　1921, lithographic copy of 乾隆道光間本
　　(between 1787 and c. 1821)
CSHK 全上古三代秦漢三國六朝文
　　1836, repr. 1893
CSIS 船山遺書
　　上海 1933, 太平洋書店
CSYCC 金聲玉振集
　　嘉靖中 嘉趣堂刊本
CTPS 津逮祕書
　　上海 1922, 博古齋刊本

ESSS 二十四史
　　百衲本上海 民國 商務
ESWS 二十五史
　　上海 1935, 開明書店
EYT 二酉堂叢書
　　1821, 二酉堂藏板

HCCC 皇清經解
　　廣州 1829, 學海堂刊本
HCCC/HP 皇清經解續編
　　1888, 南菁書院刊本
HCH 續金華叢書
　　1924, 胡氏夢選廔刊本
HCTY 學津討原
　　1805, 曠照閣刊本 and
　　上海 1922, 商務
HFLPC 涵芬樓祕笈
　　上海 1916, 商務
HHKHTS 學生國學叢書
　　上海 1928, 商務
HHLP 學海類編
　　上海 1920, 涵芬樓
HHT 漢學堂叢書
　　Later 黃氏逸考書 1925
HLC 徐位山先生六種
　　1876, 本衙藏板
HLTCHT 性理大全會通
　　光裕聚錦堂藏板
HLTS 槐廬叢書
　　1887, 吳縣朱氏
HLW 海寧王靜安先生遺書
　　長沙 1940, 商務印書舘
HPC 洪北江全集
　　1877, 授經堂刊本
HSCC 西山全集 pr. 1737
HSF 小方壺輿地叢鈔
　　上海 1897, 著易堂
HSHK 海山仙館叢書
　　c. 1845, 潘氏刊本
HSL 玄覽堂叢書
　　上海 1947, 鄭振鐸景印本
HSLT 玄覽堂叢書續集
　　北平 1947, 國立中央圖書館
HSTP 學術叢編
　　上海 1916, 倉聖明智大學
HTTK 雪堂叢刻
　　上海 1915, 羅氏排印本
HWTS 漢魏叢書
　　1895, 黃元壽本 and
　　上海 1911, 大通書局, 張鶱本

HYHTS 惜陰軒叢書 1846

ICT 宜稼堂叢書
　　上海 1840–2
IHCC 藝海珠塵
　　嘉慶中 聽彝堂原本
IMKT 夷門廣牘
　　長沙 1940, 商務
ITCM 醫統正脈全書
　　北平 1923
ITTS 佚存叢書
　　日本 1788, 寬政至文化間刊本 repr. 1882

JYTS 榕園叢書
　　廣東 1874, 張氏刊

KCCHH 古經解彙函
　　1873, 粵東書局刊本
KCIS 古今逸史
　　上海 1937, 商務
KCSHTS 古今算學叢書
　　1898, 算學書局
KCW 古今文藝叢書
　　上海 1913, 廣益書局
KHCP 國學基本叢書
　　上海 1935
KHTS/LCCS 抗希堂十六種全書, 1750
　　康熙嘉慶間 抗希堂刊本
KSTTS 功順堂叢書
　　光緒中 潘氏刊
KTHS 廣倉學宭叢書
　　上海 1916, 倉聖明智大學排印本
KYTS 廣雅書局叢書
　　光緒中 廣雅書局刊

LCCS 歷朝紀事本末
　　上海 1899, 愼記書莊石印本
LCKTS 靈鶼閣叢書
　　1897, 湖南使院刊本
LJIS 李銳遺書(李氏算學遺書)
　　上海 1890, 醉六堂
LLSK 玲瓏山館叢書
　　1889, 文選樓刊本
LLYY 律曆淵源 1723
LS 類說
　　上海 1955, 文學古籍刊行社
LSC 麓山精舍叢書
　　1900, 陳氏刊本
LTT 婁東雜著
　　1833, 太倉東陵氏刊本

MHCH 墨海金壺
　　上海 1921
MSTS 美術叢書
　　上海 1911 and 1937, 神州國光社

NKKZ 1946, 日本科學古典全書 13 vols.
NLWH 中國內亂外禍歷史叢書
　　上海 1946, 神州國光社

OHLS 藕香零拾 1910

PFT 白芙堂算學叢書
    上海 1897, 文瀾書局石印本
PLHS 百陵學山
    上海, 涵芬樓
PYT 寶顏堂秘笈
    上海 1922, 文明書局石印本

SANT 三藏 Taishō Issaikyō ed.
SCHS 山居小玩
    1629, 汲古閣刊本
SCSS 算經十書
    微波榭本 and WYWK
SCTC 善成堂道書七種
    c. 1841, 善成堂梓
SF 說郛
    上海 1927, 据明鈔本, 涵芬樓藏板,
    pref. by 張宗祥 c. 1920
    See Ching Phei-Yuan (1)
SFH 說郛續 1646, 宛委山堂
    and repr. 台北 1964
SKCS 四庫全書珍本初集
    上海 1935, 商務
SMCA 雙梅景闇叢書
    長沙 1903, 葉氏郎園刊行
SPK 申報舘叢書餘集
    光緒中 申報舘排印本
SPPY 四部備要
    上海 1936, 中華書局
SPTK 四部叢刊
    上海 1919, 商務
    2nd ed. 1937
SPTK/SP 四部叢刊續編
    上海 1934, 商務
SSKTS 守山閣叢書
    上海 1889, 鴻文書局 repr. 上海 1922, 博古
    齋本
SSSS 古今十三算書
    1873-7, 荷沱精舍
ST 十通
    上海 1937, 商務
STTK 世德堂刊 1914
SY 適園叢書
    1913-17, 烏程張氏刊本

TCTS 檀几叢書
    惠州 1695, 霞舉堂刊本
TPTK 太平天國印書
    江蘇人民出版社
TSCC 圖書集成
    上海 1887, 中華書局影印

TSH 藏說小萃
    1606, 前書樓刊本
TSHCC 1935-7, 叢書集成初編
TSPC 遵生八牋
    1810, 金閶多父堂
TT 道藏
    上海 1923-6, 商務
TTCY 道藏輯要
    成都 1906
TTHPCC 道藏續編初集
    上海 1889, 醫學書局
TTT 唐代叢書
    上海 1806, 錦章圖書局

WHS 文選
    上海 1809, 會文堂
    facsim. ed. of 鄱陽胡氏本
WL 武林掌故叢編
    1883, 嘉惠堂
WYTCC 武英殿聚珍版書
    1773-83, repr. 1874, 江西書局刊本
    1936, 廣雅書局刊本
WYWK 萬有文庫
    上海 1930, 商務

YCCC 雲笈七籤 SKCS, SPTK, TT
YCTS 豫章叢書
    南昌 1917, 豫章叢書編刻局刊本
YHAI 玉海
    1687, repr. 1883
YHSF 玉函山房輯佚書 1883
YHSF/P 玉函山房輯佚書補編
    稿本
YIKW 英譯古文觀止
    重慶 1942, (extracts from H. A. Giles (12)
    without acknowledgement, but with Chinese
    texts added)
YKT 景刊堂開成石經
    1926, 二百忍堂
YLCS 樂律全書
    1596, 鄭藩本
YSHS 元詩選
    1694, 秀野草堂刊本
YYH 音韻學叢書
    成都 1937
YYT 粵雅堂叢書
    1855, 聯出版社
    and 台北 1965

夏	HSIA kingdom (legendary?)		*c.* −2000 to *c.* −1520
商	SHANG (YIN) kingdom		*c.* −1520 to *c.* −1030
周	CHOU dynasty (Feudal Age)	Early Chou period	*c.* −1030 to −722
		Chhun Chhiu period 春秋	−722 to −480
		Warring States (Chan Kuo) period 戰國	−480 to −221

First Unification 秦 CHHIN dynasty —221 to −207

漢	HAN dynasty	Chhien Han (Earlier or Western)	−202 to +9
		Hsin interregnum	+9 to +23
		Hou Han (Later or Eastern)	+25 to +220

三國 SAN KUO (Three Kingdoms period) +221 to +265

First		蜀	SHU (HAN)	+221 to +264
Partition		魏	WEI	+220 to +265
		吳	WU	+222 to +280

Second 晉 CHIN dynasty: Western +265 to +317
Unification Eastern +317 to +420

劉宋 (Liu) SUNG dynasty +420 to +479

Second Northern and Southern Dynasties (Nan Pei chhao)

Partition		齊	CHHI dynasty	+479 to +502
		梁	LIANG dynasty	+502 to +557
		陳	CHHEN dynasty	+557 to +589
	魏		Northern (Thopa) WEI dynasty	+386 to +535
			Western (Thopa) WEI dynasty	+535 to +556
			Eastern (Thopa) WEI dynasty	+534 to +550
	北齊		Northern CHHI dynasty	+550 to +577
	北周		Northern CHOU (Hsienpi) dynasty	+557 to +581

Third 隋 SUI dynasty +581 to +618
Unification 唐 THANG dynasty +618 to +906
Third 五代 WU TAI (Five Dynasty period) (Later Liang, +907 to +960
Partition Later Thang (Turkic), Later Chin (Turkic),
Later Han (Turkic) and Later Chou)

	遼	LIAO (Chhitan Tartar) dynasty	+907 to +1124
		West LIAO dynasty (Qarā-Khiṭāi)	+1124 to +1211
	西夏	Hsi Hsia (Tangut Tibetan) state	+986 to +1227
Fourth	宋	Northern SUNG dynasty	+960 to +1126
Unification	宋	Southern SUNG dynasty	+1127 to +1279
	金	CHIN (Jurchen Tartar) dynasty	+1115 to +1234
	元	YUAN (Mongol) dynasty	+1260 to +1368
	明	MING dynasty	+1368 to +1644
	清	CHHING (Manchu) dynasty	+1644 to +1911
	民國	Republic	+1912

N.B. When no modifying term in brackets is given, the dynasty was purely Chinese. Where the overlapping of dynasties and independent states becomes particularly confused, the tables of Wieger (1) will be found useful. For such periods, especially the Second and Third Partitions, the best guide is Eberhard (9). During the Eastern Chin period there were no less than eighteen independent States (Hunnish, Tibetan, Hsienpi, Turkic, etc.) in the north. The term 'Liu chhao' (Six Dynasties) is often used by historians of literature. It refers to the south and covers the period from the beginning of the +3rd to the end of the +6th centuries, including (San Kuo) Wu, Chin, (Liu) Sung, Chhi, Liang and Chhen. For all details of reigns and rulers see Moule & Yetts (1).

SUMMARY OF THE CONTENTS OF VOLUME 4

PHYSICS AND PHYSICAL TECHNOLOGY

Part 1, Physics

With the collaboration of Wang Ling and the special co-operation of Kenneth Robinson

Part 3, Civil Engineering and Nautics

With the collaboration of Wang Ling and Lu Gwei-Djen